THE UNIVERSITY OF ARIZONA SPACE SCIENCE SERIES
RICHARD P. BINZEL, GENERAL EDITOR

Comets III
K. J. Meech, M. R. Combi, D. Bockelée-Morvan, S. N. Raymond, and M. E. Zolensky, editors, 2024, 860 pages

The Pluto System After New Horizons
S. A. Stern, J. M. Moore, W. M. Grundy, L. A. Young, and R. P. Binzel, editors, 2021, 663 pages

Planetary Astrobiology
V. S. Meadows, G. N. Arney, B. E. Schmidt, and D. J. Des Marais, editors, 2020, 534 pages

Enceladus and the Icy Moons of Saturn
P. M. Schenk, R. N. Clark, C. J. A. Howett, A. J. Verbiscer, and J. H. Waite, editors, 2018, 475 pages

Asteroids IV
P. Michel, F. E. DeMeo, and W. F. Bottke, editors, 2015, 895 pages

Protostars and Planets VI
Henrik Beuther, Ralf S. Klessen, Cornelis P. Dullemond, and Thomas Henning, editors, 2014, 914 pages

Comparative Climatology of Terrestrial Planets
Stephen J. Mackwell, Amy A. Simon-Miller, Jerald W. Harder, and Mark A. Bullock, editors, 2013, 610 pages

Exoplanets
S. Seager, editor, 2010, 526 pages

Europa
Robert T. Pappalardo, William B. McKinnon, and Krishan K. Khurana, editors, 2009, 727 pages

The Solar System Beyond Neptune
M. Antonietta Barucci, Hermann Boehnhardt, Dale P. Cruikshank, and Alessandro Morbidelli, editors, 2008, 592 pages

Protostars and Planets V
Bo Reipurth, David Jewitt, and Klaus Keil, editors, 2007, 951 pages

Meteorites and the Early Solar System II
D. S. Lauretta and H. Y. McSween, editors, 2006, 943 pages

Comets II
M. C. Festou, H. U. Keller, and H. A. Weaver, editors, 2004, 745 pages

Asteroids III
William F. Bottke Jr., Alberto Cellino, Paolo Paolicchi,
and Richard P. Binzel, editors, 2002, 785 pages

Tom Gehrels, General Editor

Origin of the Earth and Moon
R. M. Canup and K. Righter, editors, 2000, 555 pages

Protostars and Planets IV
Vincent Mannings, Alan P. Boss,
and Sara S. Russell, editors, 2000, 1422 pages

Pluto and Charon
S. Alan Stern and David J. Tholen, editors, 1997, 728 pages

Venus II—Geology, Geophysics, Atmosphere, and Solar Wind Environment
S. W. Bougher, D. M. Hunten,
and R. J. Phillips, editors, 1997, 1376 pages

Cosmic Winds and the Heliosphere
J. R. Jokipii, C. P. Sonett,
and M. S. Giampapa, editors, 1997, 1013 pages

Neptune and Triton
Dale P. Cruikshank, editor, 1995, 1249 pages

Hazards Due to Comets and Asteroids
Tom Gehrels, editor, 1994, 1300 pages

Resources of Near-Earth Space
John S. Lewis, Mildred S. Matthews,
and Mary L. Guerrieri, editors, 1993, 977 pages

Protostars and Planets III
Eugene H. Levy and Jonathan I. Lunine, editors, 1993, 1596 pages

Mars
Hugh H. Kieffer, Bruce M. Jakosky, Conway W. Snyder,
and Mildred S. Matthews, editors, 1992, 1498 pages

Solar Interior and Atmosphere
A. N. Cox, W. C. Livingston,
and M. S. Matthews, editors, 1991, 1416 pages

The Sun in Time
C. P. Sonett, M. S. Giampapa,
and M. S. Matthews, editors, 1991, 990 pages

Uranus
Jay T. Bergstralh, Ellis D. Miner,
and Mildred S. Matthews, editors, 1991, 1076 pages

Asteroids II
Richard P. Binzel, Tom Gehrels,
and Mildred S. Matthews, editors, 1989, 1258 pages

Origin and Evolution of Planetary and Satellite Atmospheres
S. K. Atreya, J. B. Pollack,
and Mildred S. Matthews, editors, 1989, 1269 pages

Mercury
Faith Vilas, Clark R. Chapman,
and Mildred S. Matthews, editors, 1988, 794 pages

Meteorites and the Early Solar System
John F. Kerridge and Mildred S. Matthews, editors, 1988, 1269 pages

The Galaxy and the Solar System
Roman Smoluchowski, John N. Bahcall,
and Mildred S. Matthews, editors, 1986, 483 pages

Satellites
Joseph A. Burns and Mildred S. Matthews, editors, 1986, 1021 pages

Protostars and Planets II
David C. Black and Mildred S. Matthews, editors, 1985, 1293 pages

Planetary Rings
Richard Greenberg and André Brahic, editors, 1984, 784 pages

Saturn
Tom Gehrels and Mildred S. Matthews, editors, 1984, 968 pages

Venus
D. M. Hunten, L. Colin, T. M. Donahue,
and V. I. Moroz, editors, 1983, 1143 pages

Satellites of Jupiter
David Morrison, editor, 1982, 972 pages

Comets
Laurel L. Wilkening, editor, 1982, 766 pages

Asteroids
Tom Gehrels, editor, 1979, 1181 pages

Protostars and Planets
Tom Gehrels, editor, 1978, 756 pages

Planetary Satellites
Joseph A. Burns, editor, 1977, 598 pages

Jupiter
Tom Gehrels, editor, 1976, 1254 pages

Planets, Stars and Nebulae, Studied with Photopolarimetry
Tom Gehrels, editor, 1974, 1133 pages

Comets III

Comets III

Edited by

K. J. Meech, M. R. Combi, D. Bockelée-Morvan,
S. N. Raymond, and M. E. Zolensky

With the assistance of

Renée Dotson

With 86 collaborating authors

THE UNIVERSITY OF ARIZONA PRESS
Tucson

in collaboration with

LUNAR AND PLANETARY INSTITUTE
Houston

About the front cover:

A mosaic of four images taken by the Rosetta spacecraft navigation camera (NAVCAM) on January 31, 2015, when the spacecraft was 28 km from the nucleus of Comet 67P/Churyumov-Gerasimenko. The comet nucleus is roughly 5 km across the longest dimension from the lower left to just past the upper right of the image. A combination of rough and smooth terrain and boulders are seen as well as a dust jet appearing to come from the "neck" region of the "rubber-duck"-shaped nucleus between the lower larger lobe and the upper smaller lobe. This image shows the northern hemisphere, which had been sunward-facing since the comet's aphelion at 5.7 au, with the Sun beginning to rise overhead on the long equatorial region as the comet races toward perihelion at 1.2 au. Eight months later, after passing through perihelion, the Sun fully illuminated the comet's southern hemisphere, producing the maximum measured gas and dust production. The Rosetta mission ended on September 30, 2016, when the orbiter was allowed to fall onto the nucleus. Original release: https://sci.esa.int/web/rosetta/-/55409-comet-67p-on-31-january-2015-navcam-mosaic.

About the back cover:

Comet C/2021 A1 Leonard imaged from Farm Tivoli, Namibia, by Gerald Rhemann on December 27, 2021, when it was 0.635 au from the Sun. A 12-inch astrograph of 1058 mm focal length was used to create this composite image by combining separate images through four filters in visible light from the far red to blue. This long-period comet was discovered by G. J. Leonard at the University of Arizona Mount Lemmon Observatory on January 3, 2021. At discovery the comet had a heliocentric distance of 5.1 au and was one year away from perihelion at 0.615 au (January 3, 2022). The comet reached naked eye visibility in early December 2021 and developed spectacular dust and gas tails. The comet began to break up in mid-December 2021 after a rapid increase in water production and a series of outbursts near perihelion. By February 2022 the nucleus had disappeared and only a debris cloud remained. The ion tail began to show some structure during the second week of December, but didn't develop the spectacular structure seen in this image until about the third week in December. Typically, cometary ion tails are dominated by CO^+ and H_2O^+ emission, however, C/2021 A1 was depleted in CO. The development of the striking plasma tail in December may be related to the breakup and sudden increase in water production.

The University of Arizona Press
in collaboration with the Lunar and Planetary Institute

∞ This book is printed on acid-free, archival-quality paper.
Manufactured in the United States of America

29 28 27 26 25 24 6 5 4 3 2 1

Library of Congress Cataloging-in-Publication Data
Names: Meech, Karen J. (Karen Jean), editor. | Combi, Michael R., editor. | Bockelée-Morvan, Dominique, editor. | Raymond, Sean N., editor. | Zolensky, M. E. (Michael E.), editor.
Title: Comets III / edited by Karen J. Meech, Michael R. Combi, Dominique Bockelée-Morvan, Sean N. Raymond, and Michael E. Zolensky ; with the assistance of Renée Dotson ; with 86 collaborating authors.
Other titles: University of Arizona space science series.
Description: Tucson : University of Arizona Press ; Houston : Lunar and Planetary Institute, 2024. | Series: University of Arizona space science series | Includes bibliographical references and index.
Identifiers: LCCN 2024021638 (print) | LCCN 2024021639 (ebook) | ISBN 9780816553631 (hardcover) | ISBN 9780816553648 (ebook)
Subjects: LCSH: Comets.
Classification: LCC QB721 .C64845 2024 (print) | LCC QB721 (ebook) | DDC 523.6—dc23/eng/20240630
LC record available at https://lccn.loc.gov/2024021638
LC ebook record available at https://lccn.loc.gov/2024021639

Dedication

This work, Comets III, and the scientific advances documented within, are dedicated to Professor Michael F. A'Hearn (1940–2017), a fundamental pillar of our cometary science community. His influence and leadership inspired so many of us in our own journeys to better comprehend comets, their links to our origins in the molecular cloud, and their role in planetary formation. Mike's innate ability and passion for engaging colleagues across various career stages and scientific viewpoints played a pivotal role in fostering discussions that underpin many of the results presented here. He inspired young scientists, sharing his vast knowledge with genuine respect for their perspectives, and fostered an environment of growth and collaboration that has ensured the future of cometary science.

Contained in this book are nearly two decades of results that are informed by Mike's scientific vision. Many stem directly from his core research into the compositional variation and formation of comets and flow from his leadership on Deep Impact and its extended mission EPOXI, as well as his expert insights and guidance on the Stardust NExT and Rosetta missions. Mike was also a consistent force in advocating for population surveys. As he predicted, the first insights from such efforts, detailed here, have indeed revolutionized our perception of the physical and chemical makeup of comets.

With Mike's invaluable contributions and kindhearted, generous guidance, we have achieved remarkable strides in our understanding of the origin and evolution of comets. Yet, as he was fond of noting, our goal is not just to answer the scientific questions we set out with, but rather to learn, to synthesize, and to debate so that we can formulate the next set of inquiries. As Mike wrote in the overview to Comets II, he expected key insights on comets to come from study of related objects, such as Trojan asteroids, Centaurs, and Kuiper belt objects. With the advent of the James Webb Space Telescope, recent and upcoming missions, and the near-term explosion of telescopic surveys, we are poised to reach that vision.

Mike was with us for much of our journey thus far. Now, as we look to the future, we must continue to carry his legacy and enthusiasm with us as we take the advances and open issues in Comets III and chart the new questions we will pursue on our way to Comets IV.

With love and profound gratitude to Mike — a remarkable scientist, mentor, and friend to all.

— Jessica M. Sunshine

Contents

PART 5: GAS COMA

PART 6: PLASMA

PART 7: DUST

PART 8: INTERRELATIONS

List of Contributing Authors

Scientific Organizing Committee

Acknowledgment of Reviewers

The editors gratefully acknowledge the following individuals, as well as anonymous reviewers, for their time and effort in reviewing chapters in this volume:

Phil Armitage
Donia Baklouti
Sébastien Besse
Nicolas Biver
Boncho Bonev
Ramon Brasser
Cecilia Ceccarelli
Martin A. Cordiner
Tom Cravens
Bradley De Gregorio
Steve Desch
Tony L. Farnham
Lori Feaga
Julio A. Fernandez
Marc Fries
Perry A. Gerakines
Tamas Gombosi
Eberhard Grün
Olivier R. Hainaut
Pierre Henri
Zhenguang Huang
Wing-Huen Ip
Masateru Ishiguro
Anders Johansen
Toshi Kasuga
Horst Uwe Keller
Michael S. P. Kelley
Matthew M. Knight
Alexander N. Krot
Anny-Chantal Levasseur-Regourd
Javier Licandro
Wladimir Lyra
Matteo Massironi
Fernando Moreno
Amaya Moro-Martin
Johannes Markkanen
Michael J. Mumma
David Nesvorný
Joseph A. Nuth III
Richard Parker
Charles A. Schambeau
Jordan Steckloff
Jessica M. Sunshine
Chad Trujillo
Jeremie Vaubaillon
Ronald J. Vervack Jr.
Jean-Baptiste Vincent
Reika Yokochi
Vladimir Zakharov

Preface

Comets III is the third in the Space Science Series of books on comets and their connections to our solar system and the galaxy. The first book, *Comets*, was published in 1982 and written to document the state of knowledge about comets 30 years after the revolutionary works from around 1950 that provided the first modern and reasonably accurate picture of the nature of comets and their likely origin. From 1950 to 1980 modern astronomical techniques greatly expanded our detailed physical and chemical understanding of comets, and it was very timely to summarize this work before the 1986 apparition of Comet 1P/Halley and the first spacecraft to fly by Comets 21P/Giacobini-Zinner in 1985 and 1P/Halley in 1986. The next 20 years saw a considerable technical expansion with remote sensing observations extending from the X-ray through the ultraviolet, visible, infrared, and radio parts of the electromagnetic spectrum. Spacecraft exploration continued as well with the flybys of two Jupiter-family comets (JFCs): the Deep Space 1 mission to 19P/Borrelly in 2001 and the Stardust mission to 81P/Wild 2 in 2004, the latter returning the first grains from the coma of a known comet. As envisioned from the science goals for all these explorations, comets emerged as key tracers of processes in the early solar system, including the complex pathway between the interstellar medium and materials of prebiotic relevance to our solar system. This new level of detailed understanding, with considerable newly developed science questions, led the way to the publication of *Comets II* in 2004, culminating with the launch of the Rosetta mission to Comet 67P/Churyumov-Gerasimenko.

Two decades have passed since *Comets II* and the momentum of increasing remote sensing capabilities from the ground, in space, and *in situ* has continued unabated, along with new developments in laboratory experiments and sample analysis (see the chapter by Poch et al.). The emergence of all-sky surveys [Lowell Observatory Near-Earth-Object Search (LONEOS), Near-Earth Asteroid Tracking (NEAT), Lincoln Near Earth Asteroid Research (LINEAR)] in the 1990s increased the rate of comet discovery from fewer than 10 per year to around 30 per year [excluding those discovered by the Solar Terrestrial Relations Observatory (STEREO) and Solar and Heliospheric Observatory (SOHO) solar space observatories] at the time of *Comets II*. This new volume, *Comets III*, comes forth in an era of expanded discovery rates (about 80 per year) thanks to the Catalina, Asteroid Terrestrial-impact Last Alert System (ATLAS), and Panoramic Survey Telescope and Rapid Response System (Pan-STARRS) surveys (chapter by Bauer et al.). In addition, several space observatories — Spitzer (2003–2020) in the infrared, Herschel (2009–2013) in the far-infrared and submillimeter, Odin (2001–present) and Submillimeter Wave Astronomy Satellite (SWAS) (1998–2004) in the submillimeter, and SOHO's Solar Wind ANisotropy (SWAN) (1995–present) in the ultraviolet — as well as advances in instrumentation at groundbased observatories have enabled the study of JFC nuclei (chapter by Knight et al.), their dust comae (chapters by Engrand et al., Kolokolova et al., and Agarwal et al.), and chemistry for all comets (chapter by Biver et al.). The Atacama Large Millimeter Array (ALMA) began its science observations in 2011, providing key spatially resolved data on comet gas and dust and clues to their relationship to the protoplanetary disk (chapter by Aikawa et al.). Since *Comets II* the Deep Impact spacecraft conducted an active experiment on the nucleus of Comet 9P/Tempel 1 in 2005 and flew past Comet 103P/Hartley 2 during its extended mission in 2010, and the Stardust spacecraft was repurposed to fly past the impact site on Comet 9P/Tempel 1 in 2011 (chapter by Snodgrass et al.). Hugely impactful to our updated knowledge was the 2014 arrival of the Rosetta mission at Comet 67P/Churyumov-Gerasimenko, becoming the first spacecraft to rendezvous with a comet — landing on its surface, and following its activity and changes as it traveled from 3.6 au from the Sun before perihelion, through its perihelion at 1.24 au, and back out to 3.6 au after perihelion (chapter by Snodgrass et al.). Rosetta provided unprecedented detail about the processes of activity (chapters by Pajola et al., Filacchione et al., and Marschall et al.) and detected many new species and their isotopes in the coma, including the first measurements of noble gases, molecular oxygen, and a wealth of prebiotic organic compounds (chapter by Biver et al.). The mission showed that some of the diversity as seen with JFCs may be a consequence of specific orbits, nucleus shape, rotation, and seasonal effects. Results from the mission have significantly advanced our understanding of how comets form and work (chapters by Simon et al. and Marschall et al.). The advances in ground and space facilities, along with in situ observations, have provided a testbed for a deeper understanding of the physical processes in cometary comae and their relation to the interplanetary medium (chapters by Bodewits et al., Beth et al., and Götz et al.). The New Horizons mission is flying into the cometary realm, past Pluto and Arrokoth, providing previously unseen detail about icy outer solar system planetesimals. In 2017 the first interstellar comet, 1I/`Oumuamua, was discovered followed in 2019 by the second interstellar comet, giving us our first glimpse of planetary building blocks from other star systems (chapter by Fitzsimmons et al.). Answering bigger questions related to cometary origins, however, likely requires a sample return in combination with the results from the next generation of facilities on the ground and in space.

The current groundbased surveys along with observations from space have revolutionized our understanding of comets from the ultraviolet to the radio, and these will lay the groundwork for the next generation of surveys, the James Webb Space Telescope (JWST), and future missions. The characterization of the grains returned from the Stardust mission has added a new dimension to studies of the early solar system, particularly as these samples were not what was expected from studies of probable cometary grains captured in Earth's stratosphere. There is also now a new understanding of the importance of carbon dioxide in controlling the activity of comets (chapter by Biver et al.), and for the first time we are able to start to study the nuclei of long-period comets (LPCs) (chapter by Knight et al.). New dynamical models of the early solar system have drastically changed as compared to when *Comets II* came out (chapters by Kaib and Volk and Fraser et al.), and this is changing the interpretation of new discoveries. Likewise, laboratory studies of primitive cometary materials and new observations are creating a new understanding of the relationship of comets to the interstellar birth environment (chapters by Bergin et al., Aikawa et al., Poch et al., and Marty et al.). Advances in groundbased, spacebased, and *in situ* observations, combined with laboratory work, are providing a new understanding of the interiors of comets and how these bodies evolve (chapters by Fraser et al., Guilbert-Lepoutre et al., Jewitt and Hsieh, Ye and Jenniskens, and Prialnik and Jewitt). The Rubin Observatory begins the ten-year Legacy Survey of Space and Time (LSST) in 2024. During its lifetime, the population of JFCs is expected to increase by a factor of 10, and thousands of LPCs and interstellar objects will be discovered each year. This will enable population studies and early detection of LPCs so that we can understand how their activity evolves. We will begin to explore the diversity of interstellar objects and thus of planetary formation elsewhere (chapter by Fitzsimmons et al.). Early discovery of LPCs will be an enabling factor for a new style of fast, reactive missions that are being explored, such as the European Space Agency's (ESA's) Comet Interceptor mission being planned for launch in 2029 (chapter by Snodgrass et al.). Both the past and current Planetary Decadal surveys have listed a comet nucleus sample return as a high priority for the next New Frontiers class mission. The JWST successfully launched in December 2021, released its first stunning images in July 2022, and its spectra will revolutionize our understanding of comets and small icy bodies in the outer solar system. We are poised at the next increment in technological advances that will enable a new understanding of comets and what they can tell us about the formation of the solar system, so this now seems like the right time for a third retrospective look at comets with an eye to future studies, leading to the publication of *Comets III*.

While we pinpoint above some specific highlights in *Comets III*, we want to offer some detail on the enormous contributions and efforts that brought this entire volume into being. In response to interest from the comet community for the next comets volume in the University of Arizona Space Science Series, an invitation was sent to the cometary community in 2019 inviting expressions of interest to participate in the volume. One hundred and twenty members of the community responded, expressing interest as either an author, as a member of the scientific organizing committee (SOC) (54), or as an editor (41). Richard Binzel, the General Editor of the Space Science Series, approached two senior scientists from this list to take on the role of lead editor (K. Meech) and co-lead editor (M. Combi). Three additional editors (D. Bockelée-Morvan, S. Raymond, M. Zolensky) were invited from those who had expressed interest in this role to reflect diversity in subject matter expertise, research host institution, and gender. The five editors formed the core of the SOC and then they added eight additional scientists to the SOC to provide balance in all of these areas. The full SOC is listed elsewhere in the front matter of this volume.

In a series of virtual meetings, the SOC carefully considered all the community input that was proposed on the expression of interest forms to set up a chapter structure that both covered the major science areas from the previous books in the series but that also included some focus on interdisciplinary science. The SOC selected lead authors for each of the chapters from those who had expressed an interest in contributing, and the invitations went out to the lead authors in December 2020. Upon acceptance, the lead authors were given access to the list of scientists who had also expressed an interest in contributing, but it was up to each lead author to select a team of co-authors for each chapter. Through this process we assembled a set of world-renowned scientists to write 25 chapters covering 8 subsections: (1) From Interstellar Matter (ISM) to Solar Nebula (SN), (2) Orbital Dynamics of Comets, (3) Missions/Campaigns/Laboratory Experiments, (4) Nucleus, (5) Gas Coma, (6) Plasma, (7) Dust, and (8) Interrelations.

The editors would like to thank the authors of all the chapters and appendices in this book for their well-written contributions, the many chapter referees for their important work to improve the elements of this book, and the production staff at the Lunar and Planetary Institute for their efforts in the editing, formatting, and production of this volume. The editors would also like to thank and acknowledge Dr. Richard P. Binzel for his continued leadership, not only for the publication of this volume, but also for the entire Space Science Series.

Karen J. Meech, Michael R. Combi, Dominique Bockelée-Morvan,
Sean N. Raymond, and Michael Zolensky, editors

Part 1:

From Interstellar Medium to Solar Nebula

Bergin E., Alexander C., Drozdovskaya M., Gounelle M., and Pfalzner S. (2024) Interstellar heritage and the birth environment of the solar system. In *Comets III* (K. J. Meech, M. R. Combi, D. Bockelée-Morvan, S. N. Raymond, and M. E. Zolensky, eds.), pp. 3–31. Univ. of Arizona, Tucson, DOI: 10.2458/azu_uapress_9780816553631-ch001.

Interstellar Heritage and the Birth Environment of the Solar System

Edwin Bergin
University of Michigan

Conel Alexander
Carnegie Institution for Science

Maria Drozdovskaya
Universität Bern

Matthieu Gounelle
Muséum National d'Histoire Naturelle

Susanne Pfalzner
Jülich Supercomputing Center, Max-Planck-Institut für Radioastronomie

In this chapter, we explore the origins of cometary material and discuss the clues cometary composition provides in the context of the origin of our solar system. The review focuses on both cometary refractory and volatile materials, which jointly provide crucial information about the processes that shaped the solar system into what it is today. Both areas have significantly advanced over the past decade. We also view comets more broadly and discuss compositions considering laboratory studies of cometary materials, including interplanetary dust particles and meteoritic materials that are potential cometary samples, along with meteorites, and *in situ*/remote studies of cometary comae. In our review, we focus on key areas from elemental/molecular compositions, isotopic ratios, carbonaceous and silicate refractories, short-lived radionuclides, and solar system dynamics that can be used as probes of the solar birth environment. We synthesize this data that points toward the birth of our solar system in a clustered star-forming environment.

1. INTRODUCTION

Comets are generally posited as providing samples of the "most" primordial material in the solar system. In this context, it is important to understand what primordial means. Generically, the limit is when our solar system becomes isolated from the interstellar medium (ISM). Do the comets bear an imprint of the physicochemical processes active during planet formation or do they preserve a history of the stages that preceded the birth of the Sun? There is evidence in the cometary record of both aspects with processed solids, combined with potentially unaltered icy and refractory material. [In general, astronomical literature makes a distinction between refractory material, which is effectively the "rocky" solids, and volatile material, which is found as ice and gas. The distinction between these components is the condensation/sublimation temperature, which is substantially higher for refractories (e.g., silicate minerals and macromolecular organics) than for volatiles (e.g., H_2, H_2O, and CO).] This is of interest as the composition of primordial material might therefore relate to the birth environment. Our goal is to understand the full context of this perspective and leave the question of the chemistry that is active during the phase of planet formation to the chapter in this volume by Aikawa et al. For the remainder of this section, we will give a general description of galactic star formation and its stages, delineate some key cometary properties, discuss their meteoritic analogs from the inner (<5 au) solar system, and discuss explorations of cometary material in the laboratories on Earth. In section 2, we describe the general evolution from the diffuse ISM to the molecular cloud and explore which aspects of the cometary record reflect this stage, while section 3 discusses the protostar and protostellar disk stages and their relevance to cometary composition. In section 4, we summarize the current understanding of the birth environment of the solar system as traced by comets. In relevant subsections we directly summarize the available constraints.

1.1. Star, Planet Formation, and the Birth Environment

In Fig. 1 we provide a cartoon of the stages of star and planet formation and here provide a brief description of star and planet formation with specifics provided in subsequent sections. Stars are born in clouds of gas that are predominantly molecular in composition and small dust particles with an average size of 0.1 μm. The gas is H_2 rich with dilute but measurable quantities of other molecular species (e.g., CO, H_2O ice and vapor, CO_2, organics, etc.), while the dust is both silicate and carbonaceous in composition (*Draine,* 2003). Dense ($n_H > 10^4$ cm^{-3}) gas in molecular clouds is concentrated in filamentary structure. More centrally concentrated "prestellar cores" are embedded within the filaments in a phase that lasts for ~1 m.y. (*André et al.,* 2014). These cores condense and collapse to form protostars. Astronomers designate protostars as either Class 0 or Class I. Class 0 are younger protostars where a young *protostellar* disk forms surrounded by a dense collapsing envelope (*Andre et al.,* 2000). Young stars actively accrete from this envelope, perhaps episodically (*Audard et al.,* 2014). They also generate well-characterized bipolar outflows and outflows, along with disk winds. These collectively ablate the surrounding envelope (*Arce et al.,* 2007). This leads to a disk with a much-reduced natal envelope, thereby ensuring the Class I stage. The lifetime of the protostellar stage is short, lasting only on the order of 100,000 yr (*Kristensen and Dunham,* 2018). As time proceeds, the envelope dissipates and the disk-star system is exposed to the interstellar radiation field — the so-called protoplanetary disk or Class II stage, which has a half-life of ~2 m.y. Within the protostellar and protoplanetary disk dust grains coagulate to ~millimeter/centimeter sizes (called pebbles) and concurrently sink to the dust-rich disk midplane where, depending on their size, they are subjected to differential forces, which leads to inward pebble drift (*Andrews,* 2020). In the solar system, comets themselves formed either in the protostellar or protoplanetary phase in the outer reaches of the disk in proximity or beyond the location of gas and ice giant planets.

Statistics of star-forming regions demonstrate that most stars are born in clusters (*Lada and Lada,* 2003) and previous analyses of the birth environment of the solar system have suggested that the Sun was born in a cluster (e.g., *Adams,* 2010; *Pfalzner and Vincke,* 2020). A central facet of clustered star formation is that all massive (>1 $M_\odot$) stars are born in clusters and thus birth in a cluster could mean the nearby presence of one or more massive (>1 $M_\odot$) stars depending on the size of the cluster. Looking at local star formation, there is a question as to whether there are distinct clustered or distributed modes. *Bressert et al.* (2010) find a continuum of star-formation surface density with a log-normal function and a peak of young stellar object (YSO) surface densities at 22 YSO pc^{-2}. *Megeath et al.* (2022) explored the YSO surface density in eight nearby clouds. They drive a distinction between clouds with high >25 YSO pc^{-2} (e.g., Ophiuchus, located near massive stars including the Sco OB2 association) and clouds without clusters (e.g., Taurus with no nearby massive stars) with low values, <10 YSO pc^{-2}.

In this review, we explore a perspective for the overall chemistry in terms of the presence or absence of a nearby presence of massive star. This then would place limits on the birth environment in terms of a modestly to very rich cluster of stars. The alternate would be a small cluster of stars or more distributed star formation such as seen in some nearby star-forming clouds (e.g., Taurus). *For the purposes of this review we will refer to this difference as "clustered*

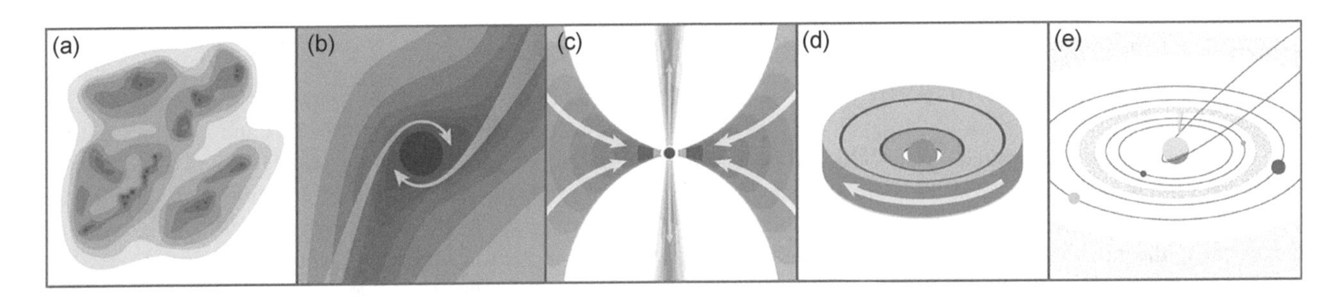

Fig. 1. Schematic of star and planet formation taken from Öberg and Bergin (2021). Stars and planets are born in **(a)** molecular clouds that have filamentary substructure and are comprised of gas and dust. **(b)** Rotating centrally concentrated dense cores are embedded within the filaments; these are labelled as prestellar cores. **(c)** The rotating prestellar core collapses leading to the formation of a protostar (Class 0 and I), surrounded by a forming *protostellar* disk that is still accreting material from the natal envelope. The presence of accretion onto the forming star produces a bipolar outflow. **(d)** Over time the stellar wind and outflow dissipates the surrounding envelope, leaving an exposed gas- and dust-rich *protoplanetary* disk (labeled as Class II). Giant planets and the seeds of terrestrial worlds form in this stage with comets forming in regions beyond the disk water ice sublimation front. **(e)** After gas disk dissipation and the final stages of terrestrial planet formation, a planetary system emerges. Credit: K. Peek.

vs. isolation," where the key dividing line is the presence or absence of a massive star. A nearby massive star leaves an imprint on the composition of cometary ices and refractories. If the Sun was born in a cluster with a nearby massive star the external energetic radiation field would indirectly heat the surrounding molecular cloud material, via reprocessed radiation, and directly expose surfaces to photodissociating/ionizing radiation. This can change the chemistry of cometary volatiles. Winds from the star, or the supernova explosion, could seed the forming cloud or core with active radionuclides and newly formed grains.

1.2. General Information About Comets

Comets are the most distant residents of our solar system. The two main reservoirs of cometary bodies are located in the outer sectors of the planetary system of the Sun. The first is the Kuiper belt, including the scattered disk, that extends at radii between ~30–50 au, beyond the orbit of Neptune, and encompasses objects of various sizes (from smaller than 1 km to several hundreds of kilometers in diameter), binaries, bodies with rings, and several dwarf planets such as Pluto. The second is the Oort cloud, located at distances of 2000–200,000 au and containing a much larger number of bodies (billions in comparison to several hundred thousands in the Kuiper belt), but of smaller dimensions (a few kilometers at most). The Kuiper belt is donut-shaped, dynamically stable, and generally gives rise to short-period comets upon small gravitational perturbations from Neptune. On the other hand, the Oort cloud is thought to be spherical and dynamically active [i.e., objects do not have well-defined orbits and will eventually leave (*Levison and Dones,* 2007)]. The Oort cloud likely formed as a result of migration of the giant planets, which caused surrounding small bodies to be scattered erratically and far out during early stages of arrangement of our planetary configuration (*Fernandez and Ip,* 1984; *Nesvorný,* 2018). The scattered disk is thought to result from scattering events of Kuiper belt objects with Neptune and other giant planets [after the formation of the Kuiper belt (e.g., *Pirani et al.,* 2021)]. Long-period comets stem from the Oort cloud. Models suggest that ~3% of cometary bodies in the Kuiper belt were originally trapped from the Oort cloud as well (*Nesvorný et al.,* 2017). Jupiter-family comets (JFCs) are short-period comets with periods of no more than 20 yr, whose current orbit is determined by Jupiter, but with origins in the Kuiper belt [not yet clear if dominated by the scattered disk (*Duncan et al.,* 2004; *Volk and Malhotra,* 2008)]. This relative framing matters as different cometary origins might trace different physical radii within the natal solar protoplanetary disk, thereby tracing the conditions of that location at cometary birth. However, it remains difficult to pinpoint exactly which disk radii the different cometary families sample, beyond the fact that Oort cloud comets are more pristine (less thermally processed) than JFCs.

Cometary comae are typically water-dominated and boast a chemically rich volatile inventory. This is amply discussed in the reviews by *Mumma and Charnley* (2011) and *Altwegg et al.* (2019). The relative ratio of volatiles and refractories is a fundamental parameter to characterize. It is important to realize that the dust-to-water mass ratio does not equal the dust-to-volatiles mass ratio nor the refractory-to-ice mass ratio. For the most well-studied comet, 67P/Churyumov-Gerasimenko (hereafter referred to as 67P), these three ratios span the 0.64–7.5 range (*Choukroun et al.,* 2020, and references therein), which makes it impossible to judge whether the comet is an "icy mudball" or a "dirty snowball." An accurate estimate can be considered the mission-integrated refractory-to-ice mass ratio of the nucleus obtained with the RSI instrument of 3–7, which implies that the nucleus of 67P is a highly porous, very dusty body with very little ice (*Pätzold et al.,* 2019).

Irrespective of the current dynamics of any particular comet, all comets are remnants of past planetesimal formation processes that took place in the protoplanetary disk that birthed our planetary system. An infant star-forming region does not contain any comet-sized objects prior to the onset of star and planet formation. The earliest phases contain dust grains predominantly smaller than 0.3 μm (*Weingartner and Draine,* 2001) that must assemble into objects with sizes of a few kilometers (comet-sized) to several thousands of kilometers (planet-sized). Although the exact sequence of steps in this assembly remains to be elucidated, in one way or another comets must either be stepping stones or byproducts of the physical processes that took place. Thanks to their current large distances from the Sun, and potentially early scattering into the outer sectors of our solar system, comets are the most pristine remnants of the protoplanetary disk (*A'Hearn,* 2011). Consequently, placing stringent constraints on the physical properties and chemical compositions of comets are of utmost relevance for understanding the original buildings blocks of our solar system.

1.3. Chondritic Meteorites

The chondritic meteorites are composed of three basic nebular components: chondrules, refractory inclusions [such as Ca-Al-rich inclusions (CAIs)], and fine-grained matrix (*Scott and Krot,* 2014). Chondrules and refractory inclusions are the products of relatively brief high-temperature (~1500–2100 K) processes. The fine-grained matrix is a complex mixture of materials, many of which were thermally processed in the solar nebula (e.g., crystalline silicates), as well as lesser amounts of materials like organic matter and presolar circumstellar grains that were not and were inherited from the protosolar molecular cloud. Based on their chemical, isotopic, and physical properties, the chondrites have been divided into four classes (ordinary, carbonaceous, enstatite, and Rumuruti) and subdivided into several groups (ordinary: H, L, LL; carbonaceous: CI, CM, CV, CO, CK, CR, CB, CH; enstatite: EH, EL). It is generally assumed that each group is derived from a separate parent body, but it cannot be ruled out that multiple parent bodies formed with similar compositions/properties.

The chondrites formed as unconsolidated "sediments" between ~2 Ma and ~4 Ma after CAIs (the oldest dated solar system objects). Various lithification processes (geologic mechanisms for generating a rock from a loose collection of grains) operating in their parent bodies after accretion produced "rocks" that were strong enough to survive impact excavation and atmospheric entry. Lithification was driven largely by internal heating due to the decay of ^{26}Al ($t_{1/2}$ ~ 0.7 Ma), although impacts may also have played a role. In asteroids that formed at ~2 Ma (ordinary, Rumuruti, enstatite, CK), only meteorites that were excavated from close to their parent-body surfaces will have escaped severe heating and preserved some primitive materials. Except for the enstatite chondrites, all these meteorite parent bodies accreted water-ice, but in most meteorites this water has either reacted by oxidizing Fe-metal or been driven off by the internal heating. Parent bodies that formed between 3 and 4 Ma (e.g., CI, CM, CR) had sufficient ^{26}Al to melt water-ice and drive reactions that generated hydrous minerals, carbonates, and Fe,Ni-oxides, but generally temperatures probably never exceeded ~150°C. Planetesimals that formed after ~4 Ma would not have had enough ^{26}Al to even melt ice and are unlikely to be the sources of meteorites, unless heated by impacts, but could be sources of some interplanetary dust.

Dynamical simulations indicate that in the early solar system the asteroid belt was a region of stability that accumulated planetesimals that formed over a wide range of orbital distances and had been scattered there, mostly by the giant planets. Based on small, systematic isotopic variations in multiple elements [e.g., Ca, Ti, Cr, Mo, Ru (*Warren,* 2011)] recording the diversity of stellar nucleosynthetic processes (r-, s-, p-processes and so forth), the chondrites and other objects are now often classified as either carbonaceous or noncarbonaceous (the latter including Earth, Moon, and Mars). The association of the noncarbonaceous group with Earth and Mars, along with the need to keep the carbonaceous and noncarbonaceous groups separate, has led researchers to propose that the noncarbonaceous group formed inside of Jupiter's orbit and the carbonaceous group beyond its orbit (*Budde et al.,* 2016; *Kruijer et al.,* 2017; *Warren,* 2011; *Desch et al.,* 2018).

The accretion of ices, organics, and presolar materials by chondrites — and in the case of the carbonaceous group formation in the outer solar system — all point to genetic links between chondrites and comets. The ability to analyze large samples in the laboratory means that chondrites can provide important information about the nature of the dust in the solar nebula that cannot be inferred from remote observations of comets. However, the likelihood that lithification processes in even the most primitive chondrites have modified to some degree the primordial materials that they accreted must always be borne in mind.

Two additional sources of extraterrestrial materials on Earth are interplanetary dust particles (IDPs) and micrometeorites. The definitions for these two types of particles are somewhat vague, but generally IDPs are taken to be <50 to 100 μm across, while micrometeorites are more in the 100–1000-μm range in diameter. IDPs with broadly chondritic compositions are divide into two categories, smooth (CS) and porous (CP) (*Bradley,* 2014). CS-IDPs are dominated by clay minerals but also contain carbonates and other minerals that are characteristic of aqueously altered meteorites, although whether they are genetically related to known meteorite groups is uncertain. CP-IDPs are anhydrous, very fine grained and heterogeneous, and are generally assumed to have cometary origins.

Most IDPs were heated to at least 500°C for a few seconds during atmospheric entry. The larger micrometeorites were generally heated more severely during atmospheric entry, exhibiting a range of textures from mildly heated to fully molten and partially evaporated. Micrometeorites have many mineralogical, textural, and chemical properties that resemble those of the major chondrite groups, especially the carbonaceous chondrites (*Genge et al.,* 2020). However, the ultracarbonaceous Antarctic micrometeorites (UCAMMS) are quite distinct and are believed to be cometary (*Duprat et al.,* 2010). As their name implies, they are dominated by carbonaceous material with variable D and ^{15}N enrichments, but also tend to contain other primitive materials such as presolar grains (*Dobrica et al.,* 2012; *Duprat et al.,* 2010). In any case, dynamical analyses of the zodiacal cloud dust suggests that a significant proportion of micrometeorites originate from comets (*Nesvorný et al.,* 2010).

1.4. Comets in the Laboratory

The Stardust mission brought back to Earth several micrograms of dust collected in the coma of the JFC Wild 2 (*Brownlee,* 2014). This material is the only *bona fide* cometary material that has been studied in the laboratory so far. To the surprise of the investigators, CAIs and chondrules that are mineralogically and chemically similar to those found in carbonaceous chondrites were both found in the Stardust samples (*Brownlee,* 2014). Furthermore, the O isotopic compositions of Wild 2 CAIs and chondrules are similar to those of their counterparts in carbonaceous chondrites (*Nakamura et al.,* 2008; *McKeegan et al.,* 2006; *Defouilloy et al.,* 2017). In addition, the mineralogy of the Wild 2 fine-grained fraction has been found to bear similarities with carbonaceous chondrites' matrixes and IDPs (*Zolensky et al.,* 2008; *Ishii et al.,* 2008; *Zhang et al.,* 2021). Finally, cubanite ($CuFe_2S_3$) and magnetite (Fe_3O_4) have been found among Stardust samples (*Berger et al.,* 2011; *Hicks et al.,* 2017). Because these secondary minerals are typical of CI chondrites that have endured extensive aqueous alteration (*King et al.,* 2020), the presence of these minerals among Stardust samples have been interpreted as hints of fluid circulation in the Wild 2 comet (*Berger et al.,* 2011; *Hicks et al.,* 2017) (although magnetite can also be a high-temperature product). Given the absence of phyllosilicates, the intensity of aqueous alteration would have to have been far more limited than in CM, CR, or CI chondrites (*Brownlee,* 2014).

Despite the violent capture process of the dust particles by the Stardust instrument, some organic particles were also found in the Stardust samples. The analysis of these particles proved to be very challenging, and it is not clear to what degree they were modified during their capture. Nevertheless, they were found to be heterogeneous, and compared to organics in chondrites and IDPs tended to be less aromatic, more O- and N-rich (*Cody et al.,* 2008; *Sandford et al.,* 2006), and have smaller D and ^{15}N enrichments (*McKeegan et al.,* 2006; *De Gregorio et al.,* 2010).

The fact that the cometary icy phase is so evidently visible to the public and astronomers alike, in the form of a spectacular coma, led to the concept that meteorites, which contain no ice and little water, could not come from comets. Meteorites have long been thought to originate from the ice-poor, inner solar system and therefore asteroids (*Kerridge and Matthews,* 1988). Another obstacle for the acknowledgment of cometary meteorites is that comets have long been perceived as primitive objects with components directly inherited from the ISM, while asteroids have been envisioned as processed solar system objects. This view has been challenged at the beginning of the twenty-first century by *Gounelle et al.* (2006), who established a cometary orbit for the highly processed Orgueil CI1 meteorite. Building on that observation and other arguments, these authors proposed that rather than a rigorous dichotomy, there existed a continuum between low-albedo asteroids and comets (*Gounelle et al.,* 2008). The asteroid-comet continuum view is now gaining acceptance (*Hsieh,* 2017; *Engrand et al.,* 2016), especially since small bodies such as comets and asteroids are known to have considerably moved between inner and outer solar system (*Levison et al.,* 2009) and because some asteroids are now considered to be volatile-rich (*Platz et al.,* 2016; *Nuth et al.,* 2020). If there is a continuum between dark asteroids and comets, it is possible that some carbonaceous chondrites might originate from cometary bodies as mentioned above. Most low-albedo asteroids might come from cometary regions (*Levison et al.,* 2009). Indeed, the results from the Stardust cometary missions support a strong link between carbonaceous chondrites and comets (*Brownlee,* 2014; *Berger et al.,* 2011). Moreover, it has been proposed that all the so-called carbonaceous meteorites (*Warren,* 2011; *Kleine et al.,* 2020; *Desch et al.,* 2018) formed beyond the orbit of Jupiter. This implies that at least some meteorites could come from zones of the nebula associated with ice-rich material (i.e., cometary formation zones). Perhaps the once hot debate about the possible existence of cometary meteorites has now abated, as it appears more and more evident that comets might not be as primitive as once thought. Also, getting closer to comets through the advent of space missions, it is realized that there is probably as much diversity among comets as among asteroids. So, it is very possible that not only the CI1 chondrites come from comets, but other meteorites groups originate from comets (*Van Kooten et al.,* 2016).

1.5. Summary of Observational Constraints on the Birth Environment

Throughout this review we will explore the perspective offered by laboratory analysis and observational results of cometary refractories and volatiles. In Fig. 2, we summarize these aspects with a brief pointer toward the potential connection to the birth environment of the solar system within a galactic context of star formation. We focus on several of these where substantive work has occurred and where connections to the birth environment can be made. These are discussed in detail in sections 2 and 3. This is not fully inclusive of the list in Fig. 2. To cover areas we do not discuss, here are some key references: For noble gases, the work of *Monga and Desch* (2015), *Marty et al.* (2017), and *Ciesla et al.* (2018) deserves mention, while for crystalline ices we refer the reader to *McClure et al.* (2015) and *Min et al.* (2016).

2. THE GALACTIC PERSPECTIVE OF THE ORIGINS OF STAR AND PLANETARY SYSTEMS

2.1. The Interstellar Medium

The initial stages of cometary material begin in the ISM through the cycle of stellar birth and death. Stellar winds and supernova explosions return material to the ISM, which is then available for the formation of the next generation of stars and planetary systems in the galaxy. We divide our discussion of the ISM into two components: dust and gas. In the following, we will distinguish between different components of the ISM based on density.

2.1.1. Interstellar gas. The majority of the mass of the ISM resides in hydrogen with dust comprising only 1% of the total mass. The galactic ISM is comprised of several well-characterized components (hot ionized medium, warm neutral medium, warm ionized medium, cold neutral medium). The isotopes of most elements heavier than H and He are produced in the interiors of stars that are more massive than our Sun. The nucleosynthetic pathways for the elements that are most relevant to this chapter are summarized in Table 1, together with the mechanisms that disperse them into space. The overall gradient of past star formation rates in the Milky Way has created a dependence of heavy elemental abundance that declines with increasing galactocentric distance (*Wilson and Rood,* 1994), but also must have local deviations.

There are known depletions of heavy elements from the gas (*Savage and Sembach,* 1996) when assuming an abundance standard associated with the Sun or nearby B stars (*Asplund et al.,* 2009; *Nieva and Przybilla,* 2012). This notably includes Fe, Si, Mg, and some O, and these elements are likely major constituents of interstellar dust. About 50%

Constraint	Method	Formation Mechanism or Origin	Clustered (with nearby massive star) vs. Isolated	Predictions
Molecular Ice Composition	Comparison of ice abundances in ISM with cometary volatiles	Ices form via adsorption of atoms and molecules onto grain surfaces in cold (T = 10–20 K) gas. Potentially altered depending on radiation exposure.	Nearby massive star would lead to ices forming in a warm environment with some fraction exposed to more UV radiation.	No clear prediction at present.
	Crystalline water ice	Amorphous water ice forms in the ISM. Detection of crystalline water ice requires processing via heating events and/or planetesimal collisions.	Crystalline ices have been detected in some systems and origin is suggested to be native to the disk system.	Does not conclusively point toward the birth environment.
Isotopic Ratios	D/H (in water, organics)	Low T (<30 K) ion-molecule chemistry parent body processes depending on ^{26}Al and size.	In a clustered environment, with massive stars, quiescent gas temperature is higher (e.g., T ~ 20–30 K as opposed to 10 K) and D fractionation reactions are slower. This results in lower deuterium enrichments in ices in warmer regions.	Higher water ice D/H ratio in regions of distributed or isolated star formation compared to those born in cluster with massive star.
	^{15}N/^{14}N	Low T ion-molecule chemistry and isotope selective photodissociation of N_2; both cases: followed by adsorption of atoms/ product molecules onto grain surfaces.	Self-shielding: active in surface of natal cloud or solar nebular disk surface. Depends on strength of UV field (external for cloud; stellar for disk).	No clear prediction at present.
	^{18}O/^{16}O, ^{17}O/^{16}O	Isotope selective photodissociation of CO followed by adsorption of oxygen atoms onto grain surfaces. Active on surface of natal cloud or the solar nebular disk surface.	Self-shielding: active in surface of natal cloud or solar nebular disk surface. Depends on strength of UV field (external for cloud; stellar for disk).	Self-shielding effects associated with cloud/ filament formation implants heavy O isotopes directly in ices during main phase of water ice formation. Elevated O isotopic ratios require enhanced UV radiation field (i.e., nearby presence of massive star).
Noble Gases	Ne, Ar, Kr, Xe abundance relative to water	Implanted as energetic particles or via adsorption during primary phase of ice formation. Potential probe of cometary contribution to Earth volatiles.	Implantation of noble gases requires low temperatures (<25 K) and water in the gas. Models suggest a link to photodesorption and hence the UV field.	No clear prediction at present.
Refractories	Composition of (unaltered) silicate minerals and carbonaceous organics	Partially formed in AGB stars (and/or supernovae) at high temperature and seeded to ISM with some fraction reforming in diffuse and/or dense ISM. Some large molecules (polycyclic aromatic hydrocarbons) form at low T.	If formed in dense ISM then there is a potential link to the birth environment via deuterium and ^{15}N enrichments.	No clear prediction at present.
	Pre-solar grains	Formed in AGB stars, J-type, and other C-stars, supernovae, and possibly novae at high T and seeded to ISM.	Can be used to glimpse range of stellar sources for refractory material including potential nearby massive star; but is not exclusive to that star.	If a significant overabundance of supernovae grains were present it could point to nearby supernova. However, this does not seem to be the case and is inconclusive.
	Crystalline silicates	Formed via annealing or condensation in solar nebula.	Thermal annealing is possible in the warm inner regions of the disk due to heating from parent star.	Does not conclusively point toward the birth environment.
Short-lived Radionuclides	^{26}Al, ^{60}Fe	Supplied to natal clouds or disk from galactic background or from nearby massive stars.	If traced to massive star, then provides direct information on birth environment.	Dispersed: low ^{26}Al and ^{60}Fe Modest size cluster: high ^{26}Al, low ^{60}Fe Massive cluster: high ^{26}Al, high ^{60}Fe
Dynamics	Outer edge of Kuiper belt, orbit of Sedna and outer solar system bodies	Gravitational interaction with nearby young stars in potential birth cluster.	Constrain properties of solar nebular birth cluster.	Edge of TNO population and distribution of orbits of Sedna-like objects point toward sculpting by stellar flyby in at least a modestly dense cluster.

Fig. 2. Probes of the solar birth environment.

of elemental C and <18% of N is missing from the gas (*Jensen et al.,* 2007; *Mishra and Li,* 2015; *Rice et al.,* 2018).

2.1.2. Interstellar dust. Late-stage stellar evolution is thought to play a key role in the formation of interstellar dust grains (*Henning,* 2010; *Cherchneff,* 2013; *Gobrecht et al.,* 2016; *Sarangi et al.,* 2018). There is substantial evidence for the presence of dust grains in the galactic ISM based on the wavelength dependence of extinction (*Mathis,* 1990; *Draine,* 2003) along with the aforementioned gas-phase depletion of heavy elements (*Savage and Sembach,* 1996). Interstellar grain models generally assume a steep size distribution following size $a^{-3.5}$ [known as "MRN" based on the last names of the three authors in *Mathis et al.* (1977)], and two compositions with contributions to extinction from silicates and carbonaceous materials (*Weingartner and Draine,* 2001). Models of carbonaceous grains have both aliphatic (carbon atoms in open chains) and aromatic (carbon atoms present in rings) components (*Jones et al.,* 2013, 2014; *Chiar et al.,* 2013). These are associated with broad spectral features seen in the diffuse ISM [aliphatic (*Günay et al.,* 2018)] and, for aromatics, in gas in close proximity to massive stars or other sources of UV emission [e.g., emission from polycyclic aromatic hydrocarbons (PAHs) (*Tielens,* 2008)].

The external sources of ISM silicate dust are evolved star envelopes, winds, and ejecta with elemental C/O < 1, where dust grains form by condensation from the hot (>1000 K) gas as it expands and cools. This is supported by the fact that a fraction (~20%) of silicates have crystalline structures and these are directly observed in circumstellar shells (*Molster et al.,* 2002; *Gielen et al.,* 2008). Carbonaceous grains form by analogous processes in stellar sources with elemental C/O > 1.

In the ISM there is processing of dust, as evidenced by the lack of crystalline silicates that is likely to be at

TABLE 1. The principal nucleosynthetic reactions and sources of the isotopes of H, C, N, O, Mg, Si, and S (*Clayton,* 2003).

	Nucleosynthesis	Sources
^{1}H	Big Bang	Big Bang
D	Big Bang	Big Bang
^{12}C	He-burning	Type II SN, AGB
^{13}C	CNO cycle	AGB
^{14}N	CNO cycle	AGB
^{15}N	CNO cycle	Novae, Type II SN, AGB
^{16}O	He-burning	Type II SN
^{17}O	CNO cycle	Novae, AGB
^{18}O	He-burning	Type II SN
^{24}Mg	C- and Ne-burning	Type II SN
^{25}Mg	He-burning	AGB
^{26}Mg	He-burning	AGB
^{28}Si	O-burning	Type II SN
^{29}Si	Ne-burning	Type II SN
^{30}Si	Ne-burning	Type II SN
^{32}S	O-burning	Type II SN
^{33}S	O-burning	Type II SN
^{34}S	O-burning	Type II SN
^{36}S	C-burning	Type II SN

SN = supernovae; AGB = asymptotic giant branch stars.

least in part due to amorphitization of crystalline silicates in shocks or via interactions with energetic particles (*Demyk et al.,* 2001; *Kemper et al.,* 2004). Processing can also be much more destructive as ISM dust grains are subject to a variety of erosive processes that reduce sizes or even destroy them. These include shocks (*Jones et al.,* 1994; *Micelotta et al.,* 2010), erosion in very hot (5000–6000 K) gas (*Bocchio et al.,* 2012), and cosmic-ray processing (*Micelotta et al.,* 2011). Estimates of the timescale of dust production in asymptotic giant branch (AGB) stars and overall destruction in supernova shocks find a general mismatch in that rates of destruction exceed production, requiring an additional means of production (*Dwek,* 1998; *Jones et al.,* 2014). Supernovae have been proposed as major contributors to dust production (*Dwek,* 1998; *Sugerman et al.,* 2006; *Zhukovska et al.,* 2016), which is supported by the detection of dusty galaxies at high ($z > 6$) redshift (*Bertoldi et al.,* 2003; *Watson et al.,* 2015) when AGB stars do not have time to evolve. Dust is clearly present in young supernova remnants, such as 1987A (e.g., *Matsuura et al.,* 2015); however, there are significant uncertainties in the yield (see extensive reviews by *Sarangi et al.,* 2018; *Micelotta et al.,* 2018, and references therein). There remains considerable uncertainty and some dust production in the denser ISM could be required (e.g., *Rouillé et al.,* 2014). For example, *Krasnokutski et al.* (2014) have shown experimentally that SiO clusters, etc., form at very low temperatures near absolute zero (it is barrierless), making grain formation in the ISM possible.

More recently, centimeter-wave observations have unambiguously detected emission from identified PAH molecules (1- and 2-cyanonaphthalene and others) forming *in situ* in dense ($n \sim 10^4$ cm^{-3}), cold ($T \sim 10$ K) dark molecular cloud cores (*McGuire et al.,* 2018, 2021; *Burkhardt et al.,* 2021a,b). Concurrently, *Cernicharo et al.* (2021b) discovered o-Bezene and a host of other building-block species (*Cernicharo et al.,* 2021a). This exciting discovery is shown in Fig. 3 and is enabled by a close coupling of laboratory spectroscopy and astronomical observations at centimeter wavelengths (*McCarthy and McGuire,* 2021). The formation route for PAHs in the dense cloud environment is currently uncertain (*Burkhardt et al.,* 2021b), but is unambiguously gas-phase in nature. Thus, some fraction of large molecules/small grains are created in the dense ISM. This is clearly important for cometary ices as these PAHs likely coat the surface of dust grains and are supplied to the planet-forming disk via collapse. At present, it is not clear whether there is feedback of these cloud-created PAHs into the overall ISM as a dust production term.

2.2 Molecular Cloud, Star, and Planetesimal Formation from the Diffuse Interstellar Medium

The journey of pre-cometary material, as shown schematically in Fig. 4, starts in the low-density warm neutral medium (density or $n \sim 1$ cm^{-3}, temperature or $T \gtrsim 1000$ K), which comprises a large fraction of the hydrogen mass of the galactic ISM (*Wolfire et al.,* 2003; *Heiles and Troland* 2003). ISM grains exist in this gas with an average size estimated via extinction of starlight to be on the order of 0.1 μm with $\langle n_{gr}\,\sigma_{gr}\rangle \sim 2 \times 10^{-21}$ (n/1 cm^{-3}) cm^{-1}. Here, n_{gr} is the space density of grains, n is the space density of hydrogen, and σ_{gr} is their cross-section. These grains remain bare due to the long collision timescales of atoms with grains ($t_{gas\text{–}gr} = 1/(n_{gr}\,\sigma_{gr}\,v) > 10^8$ yr, where v is the velocity of the gaseous atom or molecule) and the fact that abundant ultraviolet (UV) photons will desorb any atom that reaches the grain surface. In this medium most of the H is atomic, with other elements similarly as atoms that are either neutral (e.g., O, N) or singly ionized (e.g., C, Fe) depending on the first ionization potential relative to the Lyman limit of 13.6 eV. The galactic ISM is highly dynamic, and material becomes compressed behind colliding flows of gas or due to a shockwave induced by a nearby supernova (see *Ballesteros-Paredes et al.,* 2020, and references therein). The shock-heated gas cools down on a cooling timescale (on the order of one million years

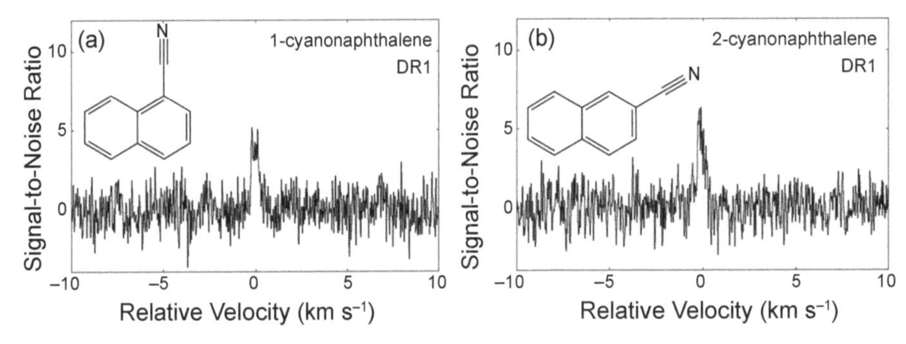

Fig. 3. Spectroscopic detection of 1- and 2-cyanonaphthalene toward the Taurus molecular cloud. From *McGuire et al.* (2021).

(1 m.y.), but depends on shock strength) leading to a higher density and colder medium (*Clark et al.,* 2012). Within this denser (n > 100 cm^{-3}) medium H_2 forms via catalytic chemistry on dust surfaces and CO subsequently forms as the UV background becomes diluted, due to strong UV photoabsorption by dust gains and self-shielding by CO molecules, within the dense parcel of gas (*Bergin et al.,* 2004; *Glover et al.,* 2010). The presence of CO denotes the observational presence of a molecular cloud.

This molecular cloud is dense enough such that atoms are colliding with grains as the formation of the H_2-dominated cloud requires grain-surface formation of H_2. Thus, a rudimentary grain mantle begins to form in this low-density slightly warmer (20–30 K) gas (*Hassel et al.,* 2010; *Furuya et al.,* 2015; *Ruaud et al.,* 2018). Simulations of this dynamic medium then show the formation of multiple filamentary systems with cores condensing within the filament (section 3). The increase in gas density associated within centrally concentrated dense core condensation leads to the rapid (<10^5 yr) formation of the ice mantles that are widely characterized in interstellar space via observation at infrared (IR) wavelengths (e.g., *Boogert et al.,* 2015) and the creation of D-enriched ices (*Ceccarelli et al.,* 2014). This is the main phase of interstellar ice formation where key volatiles are formed and implanted (e.g., H_2O, CO, and CO_2). Concurrently, there is evidence of modest grain growth in the centers of dense cores as suggested by *Steinacker et al.* (2015) (but see also *Ysard et al.,* 2016).

It is with the collapse of the core into the protostar and protostellar disk that significant dust evolution occurs. A central aspect is that the grains falling onto the disk from the surrounding envelope are covered in ices that appear

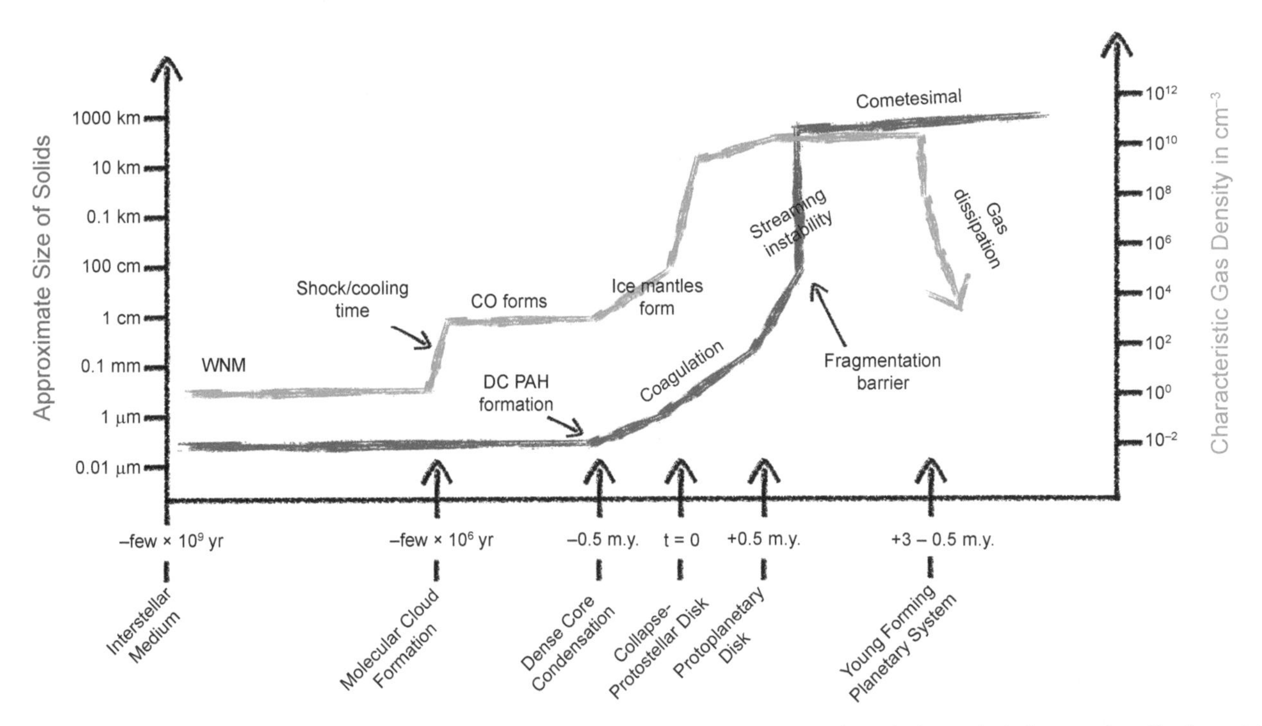

Fig. 4. Generalized schematic of solid growth (from micrometers to kilometers) and characteristic gas density (in cubic centimeters) tracing the formation of cometary bodies in a natal disk. WNM = warm neutral medium phase of galactic hydrogen; and DC PAH formation = formation of polycyclic aromatic hydrocarbons in quiescent dark clouds.

to arrive mostly unaltered (*Visser et al.,* 2009; *Cleeves et al.,* 2014; *Furuya et al.,* 2016). In the disk, grains grow to millimeter sizes and settle to a dust-rich midplane quite rapidly (*Weidenschilling and Cuzzi,* 1993; *Pinte et al.,* 2016). Observations by *van't Hoff et al.* (2020) are now demonstrating that protostellar disks appear to be warmer than the later-stage protoplanetary disk (as the result of accretion), which might lead to some chemical alteration for hypervolatile ices with the lowest sublimation temperatures [e.g., CO and N_2 (Öberg and Bergin, 2021)]. Within the disk these millimeter-sized pebbles are subject to differential forces that lead to inward drift (*Weidenschilling and Cuzzi,* 1993; *Birnstiel and Andrews,* 2014). Thus, the outer regions of the natal disk may be an important source of material to the comet-forming zones. Ultimately, our current understanding is that the grains grow up to the fragmentation barrier, i.e., ~centimeter-sized (*Blum and Wurm,* 2008; *Birnstiel et al.,* 2011), and cometesimals form in regions of high dust density (low gas/dust ratio) where the streaming instability (or some other physical mechanism) is activated (*Youdin and Goodman,* 2005; *Johansen et al.,* 2014).

2.3. Short-Lived Radionuclides

The presence of radioactive elements in meteorites has long been known (*Urey,* 1955; *Lugaro et al.,* 2018; *Desch et al.,* 2023). Some of them have short half-lives compared to the age of the Sun and have now decayed (*Davis et al.,* 2014). These so-called extinct short-lived radionuclides (SLRs) are identified through excesses of their decay products in meteorites (*Lee et al.,* 1976). Long- and short-lived radionuclides provided an internal heating source for early formed planetesimals. They can also be used for radiochronometric purposes, to probe the galactic history of the Sun's building blocks and the astrophysical context of the Sun's birth.

The initial solar system abundances of radioactive elements are usually assumed to be that measured in CAIs, which are the oldest dated solar system solids (*Connelly et al.,* 2012). However, many SLRs have not been detected in CAIs and their initial solar system values are only inferred (*Dauphas and Chaussidon,* 2011). Note that the very concept of *an* initial value implicitly assumes the homogeneous spatial distribution of the considered SLR. Given that abundance variations can be either interpreted in term of spatial or temporal variations, it is difficult to firmly establish spatial homogeneity. The existence of a significant portion of ^{26}Al-poor CAIs among a majority of ^{26}Al-rich CAIs seems to indicate that there was some degree of heterogeneity in the distribution of that important SLR (*Makide et al.,* 2013; *Krot,* 2019) whose half-life is 0.72 m.y.

Due to the difficulty of measurements and the scarcity of available material, no radioactive element has been detected in Wild 2 samples (*Matzel et al.,* 2010). However, if (some) carbonaceous chondrites come from comets, the presence of most radioactive elements in cometary regions is almost certain as most of the SLRs have been detected in carbonaceous chondrites.

2.4. What Aspects of Cometary Material Could Trace the General Interstellar Medium?

2.4.1. Initial ice mantles. The molecular cloud phase ensues once the gas density increases to >10^3 cm^{-3} and the temperature drops to <20 K (*Bergin and Tafalla,* 2007). In this phase, simple molecules (H_2O, NH_3, CO_2, CH_4) start to be efficiently formed via grain-surface chemistry (e.g., *Hiraoka et al.,* 1998; *Cuppen et al.,* 2017) and the initial ice mantle forms (e.g., Fig. 4). It has been experimentally demonstrated that under ISM-like physical conditions, H_2O is formed through hydrogenation of atomic O (*Dulieu et al.,* 2010), NH_3 through the hydrogenation of atomic N (*Fedoseev et al.,* 2015), CO_2 through the association of CO and OH (*Ioppolo et al.,* 2011), and CH_4 through hydrogenation of atomic C (*Qasim et al.,* 2020). Water is the dominant species forming the initial icy mantle on top of 0.1-µm dust grains in molecular clouds owing to its high desorption temperature (*van Dishoeck et al.,* 2014, 2021).

Deuterium formed alongside H in the Big Bang. In the diffuse ISM, it is locked up in HD. Cosmic-ray ionization of H_2 forms H_3^+ that in turn liberates D in the form of H_2D^+ for participation in chemical reactions.

$$H_2 \xrightarrow{\text{C.R.}} H_2^+ \xrightarrow{H_2} H_3^+(+H) \xrightarrow{\text{HD}} H_2D^+ + (H_2 + 230\ \text{K})$$

$$\xrightarrow{e^-} H_2 + D$$

H_2D^+ paves the way for gas-phase molecules to become deuterated in particular at temperatures <20 K, as its production is strongly favored due to the 230 K exothermicity of the reaction [the zero-point energy of the H-D bond is lower than that of H-H (*Watson,* 1974)]. Meanwhile, the final step of dissociative recombination with electrons increases the availability of atomic D in the gas. Upon adsorption onto grain surfaces, incorporation of D into solid-state molecules increases (*Roberts et al.,* 2003). Deuterium readily participates in addition reactions on grain surfaces thanks to its longer residence time on the grains in comparison to that of the lighter, more mobile H (*Tielens,* 1983). Deuteration of molecules is particularly enhanced once CO freezes out on grain surfaces, impeding destructive gas-phase reactions with H^+ and H_2D^+ (*Caselli and Ceccarelli* 2012). As the first H_2O ice layers form in molecular clouds, HDO is also synthesized via D addition reactions on grain surfaces. These initial ice layers will have reduced D/H ratios compared to later stages associated with colder (~10 K) prestellar cores (section 3.1).

Isotopic fractionation effects also occur for other major elemental pools: C, N, and O. Of the major elements, H has the largest zero-point energy difference as the mass change between H and D is larger than for other isotopes (e.g., ^{12}C vs. ^{13}C, ^{16}O vs. ^{18}O, ^{14}N vs. ^{15}N). Thus, the effects of fractionation via kinetic chemical reactions are reduced for the heavier elements. Kinetic effects for C and O are discussed by *Langer et al.* (1984), *Röllig and Ossenkopf* (2013), and *Loison et al.* (2019, 2020). Nitrogen is an interesting case as clear evidence for fractionation exists in star-forming dense cores within nitriles (C-N bonds), N hydrides (N-H bonds), and potentially within N_2 (*Hily-Blant et al.*, 2013; *Redaelli et al.*, 2018) along with ^{15}N enrichments in meteorites (*Alexander et al.*, 2012). However, the origin of these heavy isotope enrichments (and deficits) via gas-phase pathways in the dense ISM is uncertain (*Roueff et al.*, 2015; *Wirström and Charnley* 2018).

Much larger fractionation effects occur in the gas phase, induced by UV photons and molecular self-shielding. Most molecules have photodissociation cross-sections that are continuous across the UV spectrum (*Heays et al.*, 2017). Because formation rates are less than photodissociation rates, grains must provide the shielding from UV. H_2, CO, and N_2 are an exception in that their photodissociation cross-sections are through molecular bands or lines in discrete areas of the UV spectrum (*van Dishoeck and Black,* 1988). [H_2O can self-shield in some instances due to fast formation rates (*Bethell and Bergin,* 2009).] These spectral lines can be saturated with optical depths much greater than unity such that molecules downstream of the UV radiation could be self-shielded by the destructive effects of UV photons. Isotopologs have slighted shifted spectral lines from the corresponding main species and reduced abundances, hence the onset of self-shielding occurs in somewhat deeper layers (*van Dishoeck and Black,* 1988; *Heays et al.*, 2014). For example, $^{12}C^{18}O$ self-shields in deeper layers than $^{12}C^{16}O$. This leads to a layer with excess ^{18}O. If this were to be incorporated into water ice via surface reactions it would carry an isotopic enrichment. Self-shielding has been suggested as a mechanism to account O and N isotopic enrichments in meteorites (*Clayton,* 1993; *Marty,* 2012), which would be active on the surface of the molecular cloud to be provided by condensation first to the dense star-forming core and then by collapse to the young disk (*Yurimoto and Kuramoto,* 2004). *Lee et al.* (2008) explored the potential for cloud/core formation to contribute to O isotopic enrichments. They find that UV radiation fields need to be elevated above the average interstellar radiation field for significant enrichments to be provided via collapse; i.e., it requires close proximity of a massive star. But see the discussion in section 3.3 regarding protostellar disks as this is another potential area where self-shielding could be active (*Lyons and Young,* 2005).

Summary of potential links to birth environment: The cloud could be the source of heavy isotopic enrichments due to the enhanced effects of radiation on lower-density gas within molecular clouds (through self shielding). This would imply the presence of a massive star and formation in a cluster (as opposed to isolation). More work needs to be done to explore whether N might provide additional links and on the efficiency of similar processes with the young disk.

2.4.2. Presolar circumstellar grains. Presolar circumstellar grains are found in the fine-grained matrixes of the most primitive members of all the chondritic meteorite groups (*Floss and Haenecour,* 2016; *Nittler and Ciesla,* 2016; *Zinner,* 2014), as well as in micrometeorites and IDPs, some of which probably come from comets. The other major components of chondrites, chondrules and refractory inclusions, formed at high enough temperatures in the solar nebula to have destroyed any presolar materials. Circumstellar grains are recognized as such based on their isotopic compositions, which are so different from solar system materials that they cannot be explained by typical physical and chemical processes that operated in the early solar system. Instead, their isotopic anomalies are best explained by nucleosynthetic processes that operate in evolved stars toward the end of their lives.

Based on their isotopic compositions, the dominant stellar sources are AGB stars, supernovae, J-type C stars, born-again AGB stars, and possibly novae. The number of stellar sources represented in the grains remains uncertain, but must be at least in the many tens (*Alexander,* 1993, 1997). However, the actual number may be much larger, as dating of the residence times of individual grains in the ISM suggests that they had been accumulating in the ISM for ~1 Ga prior to formation of the solar system (*Heck et al.,* 2020).

The grains are typically micrometer to submicrometer in size, but grains up to 25 μm across have been found (*Gyngard et al.,* 2018). When found *in situ* in the meteorites, they are present as isolated grains that are mostly, but not always, composed of one phase. These grains almost certainly spent considerable time in the ISM but, contrary to the predictions of the Greenberg model for interstellar grains (e.g., *Li and Greenberg,* 1997), they do not have observable carbonaceous layers surrounding them. The main presolar circumstellar grain phases are silicates (both crystalline and amorphous), oxides [Al_2O_3, $Mg(Al,Cr)_2O_4$, $Ca(Al,Ti)_{12}O_{19}$, TiO_2], SiC, Si_3N_4, nanodiamonds, and graphite.

All the presolar grains are susceptible, to varying degrees, to destruction by the aqueous alteration and/or thermal metamorphism that lithified the chondrites. Nanodiamonds are arguably the most robust and, possibly, the most abundant type of presolar grain — their abundance is ~1400 ppm in chondrite matrixes. They are too small (<5 nm) to measure individually, but in bulk have Xe and Te isotopic compositions that suggest formation in supernovae. However, the abundances of Xe and Te are low and may only be carried by one in 10^5–10^6 grains. The bulk C and N isotopic compositions of the nanodiamonds are similar to those of the bulk solar values and, therefore, it is possible that the vast majority formed in the ISM and/or the early solar system. Nevertheless, to first order, the bulk abundances of the Xe-containing nanodiamonds are constant in the most primitive members of all chondrite groups, when corrected for their varying matrix contents.

The next most abundant and the most susceptible to destruction by parent body processes are the presolar circumstellar silicates. The highest silicate and oxide abundances (~175–250 ppm) have been found in the matrixes of the most primitive CM, CO, and CR carbonaceous chondrites.

The lower abundances reported for other members of these groups, as well as members of other chondrite groups, probably reflects the fact that they experienced more destructive parent body conditions. Presolar silicate and oxide abundances reported for CP-IDPs, the most likely IDPs to come from comets, range from 0 ppm to 15,000 ppm, with an average of 430 ppm (*Alexander et al.,* 2017b). This range must, to some extent, reflect the small areas analyzed for individual CP-IDPs, as well as variable atmospheric entry heating. However, it seems likely that CP-IDPs are composed of materials that experienced a range of reprocessing in the solar nebula and/or their parent bodies.

Based on their average presolar silicate and oxide abundances, IDPs contain roughly a factor of 2 more unprocessed molecular cloud material than chondrite matrixes. Presolar silicate/oxide abundances have been measured in some unambiguously cometary material, i.e., returned Wild 2 material. This material was severely affected by the high-velocity (~6 km s^{-1}) capture process. However, by comparing the survival efficiency of presolar grains in experiments designed to simulate the capture process with what is found in the returned samples, *Floss et al.* (2013) estimated presolar circumstellar silicate and oxide abundances in Wild 2 dust of 600–830 ppm (i.e., ~2.5–5× what is found in primitive chondrite matrixes). One other thing to bear in mind when discussing presolar circumstellar silicate/oxide abundances is that *Hoppe et al.* (2017) argue that typical *in situ* searches for silicate/oxide presolar grains underestimate their abundances by at least a factor of 2 because detection efficiencies decrease rapidly for grains with diameters below ~225 μm.

Roughly 40% of presolar silicates found in meteorites and IDPs are crystalline (*Alexander et al.,* 2017b). This compares to the <2% crystalline fraction in the diffuse ISM silicate dust (*Kemper et al.,* 2004, 2005). The higher abundances of crystalline silicates observed in meteorites, IDPs, and comets (e.g., Wild 2) than in diffuse ISM dust could have been produced by at least three processes: annealing, melting/crystallization, and condensation from a gas. Presumably this is also the case for extrasolar disks with crystalline silicates. All three processes for making crystalline silicates are likely to result in the O isotopic exchange of circumstellar grains with H_2O in the disk gas (e.g., *Yamamoto et al.,* 2018), and in many circumstances with the more abundant interstellar silicate dust, which would have destroyed any nucleosynthetic isotopic signatures. Hence the level of crystallinity of the presolar circumstellar silicates is almost certainly primary and is roughly on the same order as the 10–20% crystallinity estimated for silicates in stellar outflows (*Kemper et al.,* 2004). From the <2% crystallinity of ISM dust and the 40% crystallinity of circumstellar presolar grains, we can place an upper limit of <5.5% for unprocessed circumstellar silicates in the diffuse ISM, which would increase to <8–18% if we use the crystallinity estimates for stellar outflows.

Other estimates suggest that the fractions of circumstellar grains in the diffuse ISM may be ≤0.4–2 % (*Alexander et al.,* 2017b) and a few percent (*Hoppe et al.,* 2017). These values are similar to some model predictions for grain destruction and reformation in the ISM that suggest that circumstellar grains make up ~3% of ISM dust (*Zhukovska et al.,* 2016).

Assuming circumstellar grain abundances of 2–3% in the ISM means that 30–50× as much interstellar dust must have accompanied the presolar silicates found in extraterrestrial materials. It also suggests that the ≤250 ppm of presolar silicates in chondrite matrixes represent only ≤0.8–1.3% of the original ISM dust from which the chondrite matrixes ultimately formed, 1.4–2.2% for average CP-IDPs (430 ppm) and 2–4% for Wild 2 dust. These numbers would at least double if the *Hoppe et al.* (2017) detection efficiencies are applicable to most *in situ* presolar silicate/oxide searches. Nevertheless, the numbers give some indication of the level of reprocessing of the original protosolar molecular cloud material these primitive materials experienced. Although interstellar silicate grains should vastly outnumber the circumstellar ones, unambiguously identifying them is problematic because, if they formed in the ISM, they are expected to have similar isotopic compositions to the bulk solar values.

Summary of potential links to birth environment: The circumstellar grains in primitive extraterrestrial materials provide constraints on the range and number of stellar sources that contributed material to the protosolar molecular cloud in the last billion years or so. Most of the grains formed around AGB stars and other relatively low-mass, long-lived stars that are unlikely to have ended their lives in the protosolar molecular cloud. On the other hand, if a massive star within a clustered star-forming environment had ended its life as a type II supernova in the vicinity of the protosolar molecular cloud, one might expect an excess in supernova grains in the circumstellar grain population. At present, unlike with AGB grains, it is not possible to estimate how many supernovae are represented in the circumstellar grain population. In addition, estimates of supernova dust production rates are quite uncertain. Nevertheless, to date there is little indication that the solar system has a higher proportion of supernova-derived grains than expected for the solar neighborhood.

2.4.3. Refractory organics in chondrites and comets. There are roughly equal masses of silicate and carbonaceous dust in the diffuse ISM (*Compiègne et al.,* 2011; *Draine,* 2009; *Zubko et al.,* 2004). Using the abundances of circumstellar silicates in ISM dust estimated above, this means that there should be roughly 30–50× as much interstellar carbonaceous dust in the extraterrestrial materials as circumstellar silicates, or 0.75–1.3 wt.% carbonaceous dust in chondrite matrixes, 1.3–2.2 wt.% in average CP-IDPs, and 1.8–4.2 wt.% in Wild 2 dust. Again, these abundances may at least double if the *Hoppe et al.* (2017) circumstellar silicate detection efficiency estimates are typical for *in situ* searches. Primitive chondrite matrixes contain ~3–4 wt.%

organic C, which is similar to the estimates for interstellar carbonaceous dust that should have accompanied the circumstellar silicates. This raises the question of whether some or all of the organic C in chondrites, and presumably comets, could ultimately have an interstellar origin. The C contents of CP-IDPs tend to be significantly higher than would be predicted from their average circumstellar grain abundances (*Schramm et al.,* 1989; *Thomas et al.,* 1993), but there is a strong bias in favor of low densities (e.g., organic-rich) in particles that survive atmospheric entry relatively unheated. The abundance of carbonaceous material in returned Wild 2 dust could not be determined because much of it is likely to have been destroyed by the capture process.

While primitive chondrites contain complex suites of solvent extractable compounds, such as amino and carboxylic acids (*Glavin et al.,* 2018), the bulk of the organic C in chondrites is insoluble in typical solvents and so is thought to be macromolecular (*Alexander et al.,* 2017a; *Glavin et al.,* 2018). However, on average only about half of this C can typically be isolated from primitive meteorites for reasons that are not understood. The isolatable material is normally referred to as insoluble organic matter or IOM, and its properties and possible origins has recently been extensively reviewed by *Alexander et al.* (2017a) and are summarized below.

In the most primitive chondrites, IOM has an elemental composition of $C_{100}H_{75-79}O_{11-17}N_{3-4}S_{1-3}$ (relative to 100 Cs). Studies *in situ* and of IOM isolates show that the organic C is present in grains with a wide range of sizes and morphologies, and that there is no obvious spatial relationship between the organics and any inorganic materials. Most organic grains are submicrometer in size, with larger particles that are found *in situ* being aggregates of smaller grains that may result from fluid flow concentrating grains in veins. The most distinctive grain morphologies, although not the most abundant, are those of so-called nanoglobules that are typically roughly spherical and hollow. Nanoglobule-like objects have also been found in IDPs and Wild 2 samples.

Infrared and nuclear magnetic resonance (NMR) spectroscopy indicates that the IOM is predominantly composed of small, highly substituted polyaromatic units that are linked together by short, highly branched aliphatic material. IOM is also very rich in D and ^{15}N, with bulk compositions that range up to D/H $\approx$ 6.2–7.0 $\times$ 10^{-4} and $^{14}N/^{15}N \approx$ 190–240, and micrometer to submicrometer hotspots, probably associated with individual grains, that can range up to D/H $\approx$ 6.4 $\times$ 10^{-3} and $^{14}N/^{15}N \approx 70$. There is no straightforward correlation between the extent of D and ^{15}N enrichments, either in bulk IOM or in hotspots. The C, O, and S isotopic compositions of IOM do not seem to be very different from terrestrial values. The D and ^{15}N enrichments are generally interpreted as being due to the formation of the IOM or its precursors in cold and/or radiation-rich environments, although whether this was in the presolar molecular cloud or the outer solar system remains the subject of debate.

Comets 1P/Halley and 67P contain refractory (macromolecular?) carbonaceous material that is roughly as abundant by mass as the silicate dust (*Bardyn et al.,* 2017; *Jessberger et al.,* 1988). In 1P/Halley, the carbonaceous material has a bulk composition of roughly $C_{100}H_{80}O_{20}N_4S_2$ (*Kissel and Krueger,* 1987), while in 67P it has a composition of roughly $C_{100}H_{104}N_{3.5}$ (*Fray et al.,* 2017; *Isnard et al.,* 2019) (the O and S contents were not measured). The refractory carbonaceous material in 67P is also very D-rich with a D/H = 1.57 $\pm$ 0.54 $\times$ 10^{-3} (*Paquette et al.,* 2021). In other comets, the carbonaceous dust contents appear range from below those of 1P/Halley and 67P (*Lisse et al.,* 2006, 2007) to comparable to them (*Woodward et al.,* 2021), perhaps reflecting a range of primitiveness.

The similar elemental compositions of IOM and refractory carbonaceous dust in comets 1P/Halley and 67P suggest that there is a genetic link between these materials. The higher abundance of the carbonaceous material in the two comets, along with the higher H/C and D/H, all point to the dust in these comets having been less processed in the solar nebula or in their parent bodies. Indeed, the abundances of carbonaceous material in 1P/Halley and 67P is roughly consistent with their dust being predominantly composed of unprocessed molecular cloud material. If the carbonaceous material is not interstellar in origin, then mechanisms must be found that destroy the interstellar carbonaceous dust and then remake carbonaceous dust, at least in the outer solar system, very efficiently.

If there is a genetic relationship between IOM and the carbonaceous material in comets 1P/Halley and 67P, a possible interstellar connection can be explored by comparing the properties of IOM with what is known of carbonaceous dust in the ISM. Some of the early speculation that IOM might be related to interstellar carbonaceous dust came with the recognition that there is a striking resemblance between the 3–4-μm IR spectra (the aliphatic C-H stretch region) of IOM and diffuse ISM dust (*Ehrenfreund et al.,* 1991; *Pendleton et al.,* 1994; *Wdowiak et al.,* 1988). *Sandford et al.* (1991) interpreted the 3–4-μm IR spectra of diffuse ISM dust as being due to short aliphatic chains attached to electronegative groups, such as aromatic rings, O, and N. Subsequently, *Pendleton and Allamandola* (2002) concluded that the diffuse ISM dust contains few if any heteroatoms (e.g., O and N) and that the aromatic units must be large. They also pointed out that a range of materials give reasonably good fits to the diffuse ISM spectra, not just IOM.

The short aliphatic chains attached to aromatic units is certainly reminiscent of the structure of IOM, but a predominance of large polyaromatic units is inconsistent with what is seen in IOM. On the other hand, other models for the structure of diffuse ISM carbonaceous dust prefer small aromatic units (*Dartois et al.,* 2005; *Jones et al.,* 1990; *Kwok and Zhang,* 2011). The absence of heteroatoms in diffuse ISM carbonaceous dust is, potentially, more problematic given the abundance of O, in particular, in IOM and 1P/Halley carbonaceous dust. Isotopic compositions also pose a challenge to there being a link between diffuse

ISM dust and IOM as well as the refractory carbonaceous dust in 67P. Carbon stars are thought to be the major stellar sources of carbonaceous dust in the diffuse ISM. As a result of nucleosynthesis and dredge-up in these stars, their dust should be depleted in D and ^{15}N and exhibit a very wide range in C isotopic compositions (as is seen in presolar circumstellar SiC and graphite grains). This is the opposite of what is seen in IOM (and 67P dust), which is D and ^{15}N enriched and has normal C isotopic compositions. The C isotopes of IOM would be consistent with efficient destruction of circumstellar dust and reformation in the ISM, but the enrichments in D and ^{15}N are difficult to explain via isotopic fractionation in the diffuse ISM.

However, the solar system formed from molecular cloud material not diffuse ISM material. Is it possible that diffuse ISM dust is modified in molecular clouds to look more like IOM and the refractory carbonaceous material in comets 1P/Halley and 67P? The detection of PAHs forming in cold conditions (section 2.2) where D-enrichments would be active is suggestive of a potential contributor. Another possibility is that radiation damage by cosmic rays accumulates in very cold, ice-coated carbonaceous dust grains while resident in molecular clouds (*Alexander et al.,* 2008). These ices are likely to be enriched in both D and ^{15}N. Upon warming during or after formation of the solar system, radiation generated radicals in the ice will become mobile and able to react with radiation-damaged regions of the carbonaceous dust, thereby adding H, N, and O with their associated isotopic enrichments. Also, adding H to radiation-damaged interior regions of large aromatic units would generate both smaller aromatic units and short, branched aliphatic chains. This mechanism for providing a direct path from diffuse ISM dust to the carbonaceous material in comets and meteorites remains speculative at present, but at least in principle could be experimentally tested.

Other proposed mechanisms for making IOM are envisaged to have operated in the early solar system include Fischer-Tropsch-type synthesis, but the inefficiency of FTT synthesis under solar nebula conditions is problematic. Irradiation of ices is another potential source of condensed organic material (*Ciesla and Sandford,* 2012), but experiments on interstellar/disk ice analogs do not produce materials that resemble IOM or the refractory C in comets (*Nuevo et al.,* 2011). An alternative energy source would be irradiation by energetic particles and/or UV of ice-free carbonaceous particles (*Ciesla and Sandford,* 2012).

Summary of potential links to birth environment: While a solar system origin for the IOM cannot be ruled out, an interstellar origin seems more likely given the high abundance of IOM-like material in comets with volatile ice compositions that suggest a molecular cloud heritage. If the IOM did ultimately form in the protosolar molecular cloud, can it provide any constraints on the solar system's birth environment? The detection of PAHs forming in dark clouds offers a potential avenue for molecular cloud origin of the IOM. Although, the full import of this dark cloud PAH formation is currently uncertain, some aspects would be clear. PAHs are forming in dark clouds at cold (~10–20 K) temperatures (*McGuire et al.,* 2021; *Cernicharo et al.,* 2021b) where D-fractionation reactions are active. Thus, they will form with D-enrichments. This formation will occur via reactions linking to CH_2D^+ (*Millar et al.,* 1989). This pathway fractionates at higher temperatures than the reactions that form water, which are linked to H_2D^+ (*Roueff et al.,* 2015). At face value this would imply efficient D-fractionation routes for organics with less temperature dependence (i.e., harder to link to subtle changes in the gas physical state of the birth environment). Moreover, in the dark cloud environment, N self-shielding may be active at modest extinctions [~1.5^m (*Heays et al.,* 2014)]. Nitrogen-bearing PAHs are detected, and N is playing a part in the complex organic gas phase chemistry (*Burkhardt et al.,* 2021b). Thus, ^{15}N enrichments could be implanted as well. This would depend on the strength of the external UV field and, like O (*Yurimoto and Kuramoto* 2004; *Lee et al.,* 2008), the ^{15}N enrichment could relate to the presence of a nearby massive star. More work is needed to understand any potential predictions that might discriminate between various scenarios for the solar birth environment.

3. THE LOCAL BIRTH SITE

3.1. Star Cluster Dynamical Perspectives: Solar Birth in a Cluster?

Most stars form as part of a stellar group (*Lada and Lada,* 2003; *Porras et al.,* 2003); this is possibly true for the Sun as well. The following features have been considered as indicators of the Sun's cluster origin.

First, there is a steep drop in surface density of the disk at ≈30 au. Beyond, only the equivalent of 0.06× Earth's mass is contained in all Kuiper belt objects together (*Di Ruscio et al.,* 2020). It is likely that the solar system's disk was initially considerably more extended and then truncated (*Morbidelli and Levison,* 2004). Disk truncation could have happened by a close flyby of another star (*Ida et al.,* 2000; *Kenyon and Bromley,* 2004; *Pfalzner et al.,* 2018; *Batygin et al.,* 2020), external photoevaporation by nearby massive stars (*Adams et al.,* 2004; *Owen et al.,* 2010; *Mitchell and Stewart,* 2011; *Winter et al.,* 2018; *Concha-Ramírez et al.,* 2019, 2021), an earlier binary companion to the Sun (*Matese et al.,* 2005; *Siraj and Loeb,* 2020), and a nearby supernova explosion (*Chevalier,* 2000). All these processes require a strong interaction with neighboring stars, which is most easily realized in a cluster environment.

Second, most transneptunian objects (TNOs) move on highly inclined orbits, e.g., Sedna (*Brown et al.,* 2004; *Trujillo and Sheppard,* 2014; *Becker et al.,* 2018). Neptune could not have catapulted the Sedna-like objects to such distances, but an additional external dynamical mechanism is required. The close flyby of another star is one possibility. Sedna-like objects could either have been part of the primordial disk and flung out (e.g., *Ida et al.,* 2000; *Kenyon and*

Bromley, 2004; *Batygin et al.,* 2020) or they were captured from the disk of the perturber star (*Jílková et al.,* 2015). The periastron distances would have been 50–350 au (*Jílková et al.,* 2015; *Pfalzner et al.,* 2018; *Moore et al.,* 2020). Nowadays, such close flybys are infrequent for the solar system (*Bailer-Jones,* 2018), but were more common earlier on if the Sun formed within a young cluster.

Each of these features has been used to constrain the properties of the solar system birth cluster (*Mitchell and Stewart,* 2011; *Pfalzner,* 2013; *Li and Adams,* 2015; *Jílková et al.,* 2015; *Owen et al.,* 2010; *Pfalzner and Vincke,* 2020; *Moore et al.,* 2020). The general consensus is that clusters like Westerlund 2 or Trumpler 14 are excluded as solar birth environments due to their extremely high stellar density ($n_* > 10^5$ stars pc^{-3}) and very strong radiation fields ($G_0 \gg 1000$) (e.g., *Hester et al.,* 2004; *Williams and Gaidos,* 2007; *Adams,* 2010). $G_0 = 1$ is defined as the normal interstellar radiation field (*Habing,* 1968). Dynamical and radiation arguments alike (*Adams,* 2010; *Li and Adams,* 2015; *Pfalzner and Vincke,* 2020) agree that it is unlikely that the solar birth cluster contained $N_* > 10{,}000$ stars, because in such environments disks mostly are truncated to sizes well below 30 au. Similarly, there seems to be fair agreement on the lower limit independent of the constraining feature. Also, one cannot completely exclude that the Sun formed in a low-mass cluster (*Parker and Dale,* 2016), an environment similar to the one recently suggested by *Portegies Zwart* (2019) with $N = 2500 \pm 300$ stars and a radius of $r_{vir} = 0.75 \pm 0.25$ pc seems to be a more likely choice.

Recent studies find it more useful to define a typical local stellar density rather than limiting the number of stars. The existence of long- and short-lived clusters means that there exists no general one-to-one relation between the number of stars and stellar density (*Pfalzner,* 2009). Equally, substructure in the cluster means that local densities can considerably exceed average stellar densities (*Parker et al.,* 2014). Simulations show that in 90% of all cases solar system analogs are formed in areas where the local stellar density exceeds $\rho_{local} > 5 \times 10^4$ pc^{-3} (*Brasser et al.,* 2006; *Schwamb et al.,* 2010; *Pfalzner and Vincke,* 2020; *Moore et al.,* 2020).Clusters with an average cluster density of 100 pc^{-3} to a few 10^4 pc^{-3} usually contain areas within this density range. In recent years, the first solar sibling candidates have been identified (*Ramírez et al.,* 2014; *Bobylev and Bajkova,* 2014; *Liu et al.,* 2016; *Martínez-Barbosa et al.,* 2016; *Webb et al.,* 2020), strengthening the argument that the Sun was born as part of a cluster of stars.

3.2. Dense Filaments in Molecular Clouds

3.2.1. General structure and properties. Wide-field mapping of thermal continuum emission from dust by instruments such as the Submillimetre Common-User Bolometer Array (SCUBA) and Herschel Spectral and Photometric Imaging Receiver (SPIRE) and Photodetector Array Camera and Spectrometer (PACS) show molecular clouds are dominated by dense webs of filamentary substructure (*Motte et al.,* 1998; *André et al.,* 2010, 2014; *Arzoumanian et al.,* 2011; *Motte et al.,* 2018). *Hacar et al.* (2013) showed that filaments contain additional substructure (called fibers) in velocity when observed using the best tracer of dense molecular gas, N_2H^+ [for a discussion of N_2H^+, see *Bergin and Tafalla* (2007)]. A sample of this structure is shown in Fig. 5 where the thermal dust continuum image of the central regions of the Orion molecular cloud, the nearest site of massive star-cluster formation, is shown alongside an Atacama Large Millimeter/submillimeter Array (ALMA) image of N_2H^+ emission with a spatial resolution of 1800 au. For a larger-scale view of the filamentary structure in Orion, the reader is referred to *Bally et al.* (1987). The images in Fig. 5 illustrate the general sense of star formation as a series of nested filaments, associated with active starbirth, converging toward hubs where clusters are born (*Myers,* 2009). Another key perspective is that dense core collapse within filamentary structure could potentially lead to inhomogeneity within infall streams and different portions of the filament may contribute at different times to the disk (*Pineda et al.,* 2020).

In the denser ($n > 10^4$ cm^{-3}) filaments, e.g., Fig. 5, hundreds of ice layers are placed on top of the initial mantle formed with the cloud. As the thickness of ice increases, the deuteration fraction of H_2O and other carriers will increase with dropping temperatures and a larger availability of atomic D. Further, the freezeout of CO enhances the abundance of H_2D^+, resulting in this gas being the main formation phase of D-enrichments in precometary ices (*Taquet et al.,* 2014). The temperature of this gas matters. Even a slight 5–10-K difference in temperature has a chemical effect on the composition of volatile ices. For example, the D/H ratio of water ice would change by a factor of ~3 between 15 K and 20 K (see, e.g., *Lee and Bergin,* 2015).

A central distinction between clustered star-forming environments compared and those that are isolated is that prestellar cores in Orion appear warmer (e.g., >10 K) due to the presence of (passive) external heating sources [e.g., nearby massive stars (*Harju et al.,* 1993; *Li et al.,* 2003; *Kirk et al.,* 2017)] and perhaps embedded in filaments with higher average density (*Hacar et al.,* 2018). Herschel surveys of dust emission of nearby Gould belt star-forming clouds (*Andre et al.,* 2010) place this conclusion on sound statistical footing. For example, material in Ophiuchus has a higher overall dust temperature (*Ladjelate et al.,* 2020), than Taurus (*Palmeirim et al.,* 2013). [The Herschel studies are generally average temperatures along the line of sight. There does exist a line-of-sight temperature structure such that colder gas exists in the deep interior of centrally concentrated cores (*Crapsi et al.,* 2007; *Roy et al.,* 2014).] Clustered star-forming filaments also must have surfaces exposed to higher radiation fields, as, for example, the average UV field along the Orion filament exceeds 1000× the average interstellar radiation field (*Goicoechea et al.,* 2015, 2019). This is in stark contrast with more isolated regions (*Goldsmith et al.,* 2010), where the UV field is closer to

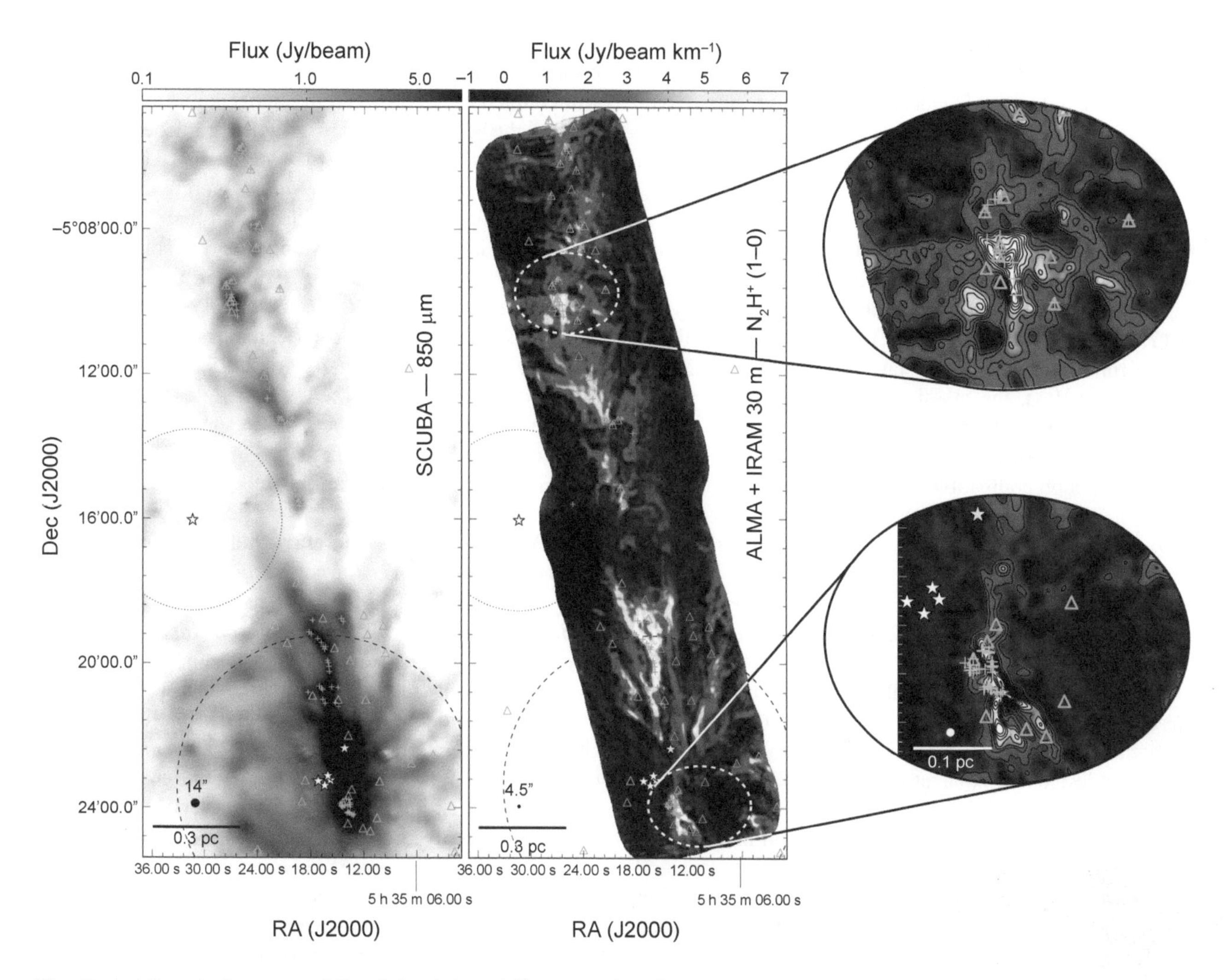

Fig. 5. N_2H^+ emission map of the Orion integral filament taken from *Hacar et al.* (2018) that includes 850-μm dust continuum emission from *Johnstone and Bally* (1999). White stars are the positions of the trapezium with the dashed line representing the 0.5-pc radius of the Orion nebular cluster, the open black star with circle shows the M43 nebula, the yellow star relates to the Orion BN source, and Spitzer protostars are shown as blue triangles. The beam sizes of the images are given on the bottom right.

the average interstellar radiation field (*Habing,* 1968). This difference, and the effects of self-shielding, would alter the O and N isotopic ratios on cloud or disk surfaces, and potentially implant heavy isotopes into forming ice mantles (*Yurimoto and Kuramoto,* 2004; *Lyons and Young,* 2005; *Lee et al.,* 2008).

3.2.2. Sources of short-lived radionuclides to molecular clouds. The longer-lived radioisotopes in the early solar system (T > 2.5 m.y.) result from the continuous production (and decay) in the galaxy (*Meyer and Clayton,* 2000) and will not be discussed here. This also the case for ^{60}Fe (T = 2.62 m.y.) whose abundance in the early solar system is lower than initially thought (*Trappitsch et al.,* 2018). Beryllium-10 (T = 1.39 m.y.) has been suggested to have been inherited from the molecular cloud (*Desch et al.,* 2004) or made by irradiation in the protoplanetary disk (*McKeegan et al.,* 2000) together with ^{41}Ca (T = 0.1 m.y.) (*Liu,* 2017) and ^{36}Cl (T = 0.3 m.y.) (*Tang et al.,* 2017). The observed heterogeneity of ^{10}Be is incompatible with an inherited origin (*Dunham et al.,* 2022; *Fukuda et al.,* 2019, 2021). Aluminum-26, by far the most documented SLR, is not made by irradiation (*Fitoussi et al.,* 2008). Given its short half-life compared to typical timescales of star formation, and its high abundance compared to the expectations of galactic nucleosynthesis (*Meyer and Clayton,* 2000), it likely requires a local last-minute stellar origin. If it had been inherited from the galaxy as suggested by *Young* (2016) and *Gaidos et al.* (2009), it would not exhibit any spatial heterogeneity (*Krot,* 2019; *Makide et al.,* 2013). AGB stars and supernovae have been considered as potential candidates. Supernovae can now be excluded, because they would yield a ^{60}Fe/^{26}Al ratio at least 1 order of magnitude larger to the one observed in the early solar system (*Gounelle and Meibom,* 2008).

The probability of associating an AGB star with a star-forming region is very small (*Kastner and Myers,* 1994). At present the most promising models are the ones involving massive star winds (*Arnould et al.,* 1997; *Gounelle and Meynet,* 2012). Iron-60 is indeed absent from winds, avoiding the supernova caveat. Because massive stars have short existence times and end their lives as supernovae, models that require a multiplicity of massive stars (*Gaidos et al.,* 2009) would also yield excesses of ^{60}Fe relative to ^{26}Al and what is observed in meteorites. A setting whereby a massive star injects ^{26}Al in a dense shell it has itself formed and within which a new star generation (*Gounelle and Meynet,* 2012; *Dwarkadas et al.,* 2017) is formed is more likely than a runaway Wolf-Rayet star (*Tatischeff et al.,* 2010). In such a context, the massive star source of ^{26}Al would belong to a cluster of a few thousand stars, in rough agreement with dynamical constraints (see section 3.1). Such a setting, although compatible with our present knowledge of star formation, would concern only a few percent of stars (*Gounelle,* 2015) and has not been directly observed at present (*Dale et al.,* 2015).

3.3. Protostellar Disks and Envelopes

Protostellar disks are a central phase of evolution as the early disk must have been hot enough to lead to the formation of CAIs and other refractory inclusions. CAIs are the oldest dated solar system materials (*Connelly et al.,* 2012) and, assuming that the Al-Mg system can be used as a chronometer, formed over a period on the order of 10^5 yr (*Liu et al.,* 2019; *Thrane et al.,* 2006). This and the high (>1400 K) formation temperatures suggest that the high-temperature phase in the disk must be short lived or associated with accretion bursts (e.g., FU Ori events) (*Boss et al.,* 2012; *Larsen et al.,* 2020; *MacPherson et al.,* 2022; *Kristensen and Dunham,* 2018). Current data do show that protostellar disks are warmer than their protoplanetary counterparts, but the temperatures are still moderate [tens to ~100 K (*van't Hoff et al.,* 2020)]. This does not imply that hotter material does not exist. Rather this material is difficult to isolate via astronomical observation due to the presence of high dust optical depth, both in the disk and the surrounding envelope, and the existence of hot shocked gas in close proximity to the forming star. A hot phase associated with solid state (ice or refractory) sublimation must be important and would operate to reset the chemistry. For example, the D-enrichments from ices would be erased (*Drouart et al.,* 1999; *Yang et al.,* 2013) and C-rich grains irreversibly destroyed (*Li et al.,* 2021). However, chemical reset cannot be absolute as infall, and the inward drift of pebbles would supply fresh primordial material during subsequent evolutionary stages. Much of the material exposed to the highest temperatures within a heating event will be accreted onto the star.

Observations have gradually revealed kinematically distinct protostellar disks within embedded young protostellar envelopes (*Tobin et al.,* 2012; *Murillo et al.,* 2013; *Sakai et al.,* 2014; *Maret et al.,* 2020). Dust grains coagulate and grow in the protostellar disk to at least millimeter-sized and these grains are highly settled to a geometrically thin midplane (*Pinte et al.,* 2016). During this stage the gas is still coupled to the dust and the budgets of highly volatile ices might be altered. Grain growth should enhance the penetration of UV photons, which are mediated by the small submicrometer grains in the system. Overall, the presence of species such as CH^+ provides hints that UV photons are present in protostellar envelopes and irradiate outflow cavity walls (*Kristensen et al.,* 2013; *Benz et al.,* 2016). In principle, this could increase the potential for self-shielding to produce O and N isotopic enrichments in the young disk as suggested by *Lyons and Young* (2005) and *Visser et al.* (2018). It is not clear that the UV penetrates to layers deep enough such that the products of UV photochemistry can be incorporated into forming ices to be carried down to the dust-rich protostellar disk midplane.

3.4. What Aspects of Cometary Material Trace Star and Disk Formation?

3.4.1. Calcium-aluminum-rich inclusions and crystalline silicates. Analysis of the Stardust samples established that a significant fraction of cometary solids is the result either of high-temperature processing in the disk (CAIs, chondrules) or low-temperature (370–420 K) aqueous alteration in the Wild 2 comet, demonstrating that cometary solids are not entirely composed of preserved interstellar dust as originally thought (*Brownlee,* 2014). They also showed that the differences between cometary solids and carbonaceous chondrites is comparable to the differences observed within the carbonaceous chondrites themselves, changing our perspective on possible cometary meteorites. The high-temperature materials must be products of processes operating in the early protostellar disk. Generally, they are assumed to have formed relatively close to the Sun and could have been carried outward by an initially hot (>1000 K) compact but expanding disk (*Ciesla,* 2015). This initially hot material will have had little D-enrichment and as it cooled would have made water ice with the cosmic D/H ratio (~10^{-5}). To account for elevated cometary D/H (≥10^{-4} in H_2O) ratios would have required mixing of this inner solar system water with pristine material provided by infall from the cold (T ~ 10–20 K) core.

Comets are known to have crystalline silicates (*Crovisier et al.,* 1997) intermixed with amorphous grains that must have formed at lower temperature (*Hanner,* 1999). Since most interstellar silicates are amorphous (*Kemper et al.,* 2004), this means that these grains must have crystallized in the protostellar or the protoplanetary disk. Thermal annealing of initially amorphous grains in the inner warmer regions of the disk combined with radial mixing is suggested as a potential process (*Henning,* 2010, and references therein), along with shocks (*Harker and Desch,* 2002). Annealing would have been more strongly active in the younger warmer protostellar disk, and astronomical observations of young disks suggest that crystalline silicate emission comes predominantly from the inner regions the disks (*van Boekel et al.,* 2004; *Bouw-*

man et al., 2008; *Watson et al.*, 2009).

Summary of potential links to birth environment: Both CAIs and crystalline silicates likely formed in the disk and thus are not linked to the birth environment.

3.4.2. Overall chemical inventory. There is an intriguing overlap between the volatile chemical composition of comets and young star-forming regions (*Greenberg and Li*, 1999; *Ehrenfreund and Charnley*, 2000; *Mumma and Charnley*, 2011). A young protostar heats up its surrounding environment (envelope and inner disk), resulting in a region where most volatiles are thermally desorbed into the gas, which is called the hot corino (or hot core in the case of high-mass protostars). The chemical inventories of hot corinos are a unique window on the complete volatile inventories in star-forming regions as volatile molecules are no longer hidden from observations in the ices. Groundbased observations of the Oort cloud comet C/1995 O1 (Hale-Bopp) showed a strong similarity between Hale-Bopp's abundances of CHO- and N-bearing molecules and those observed on large envelope- and cloud-scales [thousands of astronomical units (*Bockelée-Morvan et al.*, 2000)]. These similarities were confirmed based on *in situ* mass spectrometry measurements by the Rosetta spacecraft at the Jupiter-family comet 67P, and on small disk-scales (tens of astronomical units) probed with ALMA [Fig. 6 (*Drozdovskaya et al.*, 2019), and recently extended to S-bearing molecules].

Summary of potential links to birth environment: If other comets have chemical inventories that match those of C/1995 O1 (Hale-Bopp) and 67P, such similarities imply that the volatile chemical inventory of comets is to a degree set in the prestellar and protostellar stages, because the same molecules are found in both contexts with comparable relative ratios. More work needs to be done to explore the differences between ice abundances in clustered vs. isolated environments by, e.g., the James Webb Space Telescope (JWST).

3.4.3. Isotopic ratios. Isotopic ratios in volatiles also support the critical nature of early phases of star formation for cometary volatiles (*Bockelée-Morvan et al.*, 2015). Mass spectrometry measurement from the Rosetta mission to comet 67P on the ratio of D_2O/HDO to HDO/H_2O is much higher (17) than the statistically expected value of 0.25 (*Altwegg et al.*, 2017). Overall, monodeuterated water has a D/H ratio that is more than 2× higher than the Vienna standard mean ocean water (VSMOW) value and in closer agreement with the elevated deuteration seen in star-forming regions (*Altwegg et al.*, 2019). Observations carried out at high spatial resolution with ALMA suggest that the water D/H ratio can differ between clustered and isolated protostars (Fig. 7). Water in isolated regions has D/H ratios that are a factor of 2–4 higher than in clustered regions, which may be a result of either colder temperatures of the innate molecular cloud or longer collapse timescales in the isolated environments (*Jensen et al.*, 2019). The effect of temperature on the overall D chemistry is well documented via observations and theory (see, e.g., *Punanova et al.*, 2016). It should not be surprising that this could be reflected within ices that form as the result of this temperature-dependent process.

Recently, the O isotopic ratios of water (*Schroeder et al.*, 2019) and other species (*Altwegg et al.*, 2020) have been found to be enhanced in ^{17}O and ^{18}O, relative to ^{16}O and solar, in comet 67P. For water, this effect is modest: ^{18}O is enriched by 19 ± 9% ($\delta^{18}O_{VSMOW}$ = 121 ± 89%) and ^{17}O by 28 ± 10% ($\delta^{17}O_{VSMOW}$ = 206 ± 94%) (1σ errors). A similarly ^{16}O-poor isotopic signature has been inferred for the initial solar system water from meteoritic materials (*Sakamoto et al.*, 2007). This signature could have originated in the young protostellar disk or in the protoplanetary disk (*Lyons and Young*, 2005) or the protosolar molecular cloud (*Sakamoto et al.*, 2007; *Lee et al.*, 2008; *Alexander et al.*, 2017b).

Summary of potential links to birth environment: Given existing information, comets in our solar system show a closer agreement with clustered star-forming regions in the

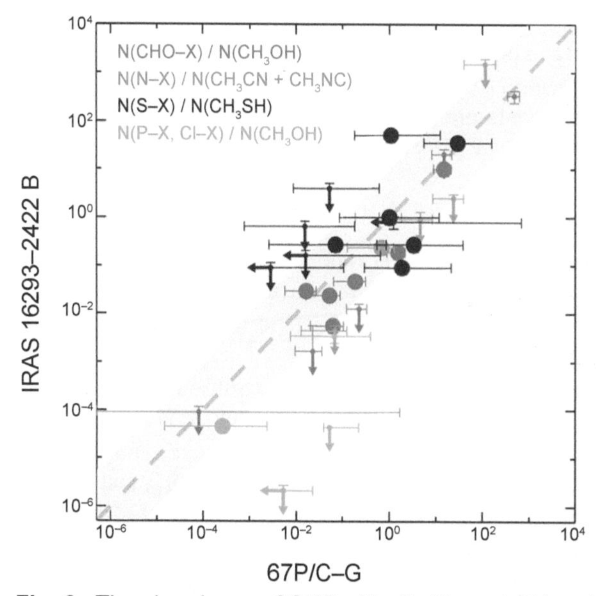

Fig. 6. The abundance of CHO-, N-, S-, P-, and Cl-bearing molecules relative to species specified in the top left corner (which include methanol, methyl cyanide, methyl isocyanide, and methyl mercaptan). The values along the ordinate were observed with ALMA toward an offset position near IRAS 16293-2422 B (0.5″ to the southwest from the source). These observations are probing volatiles on disk scales in this young low-mass protostar. The values along the abscissa were measured with Rosetta-ROSINA in the coma of comet 67P/Churyumov-Gerasimenko. Each chemical family has a unique color. The shaded region corresponds to an order of magnitude scatter about the one-to-one linear correlation of the two datasets. Smaller-sized data points pertain to an upper limit or an estimate of sorts either for the protostar, the comet, or both. A significant correlation is seen between cometary and protostellar volatile abundances, suggesting preservation of prestellar and protostellar volatiles into cometary bodies upon some degree of chemical alteration. Full details are available in the original source: *Drozdovskaya et al.* (2021).

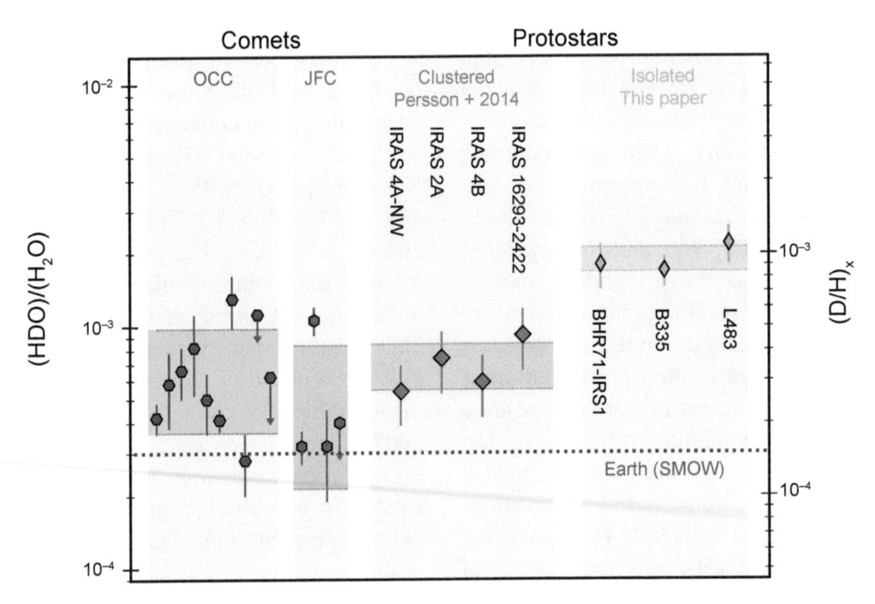

Fig. 7. The HDO/H_2O ratio along the left ordinate and the D/H ratio along the right ordinate for Oort cloud comets (OCC), Jupiter-family comets (JFC), and low-mass Class 0 protostars in clustered and isolated environments. Error bars are 1σ uncertainties. The colored regions are the standard deviations per category. In this publication, sources were classified as isolated if they are not associated with any known cloud complexes. The clustered regions pertain to star-forming regions with several low-mass protostars therein. Isolated sources display significantly higher levels of deuteration than clustered environments and solar system comets, suggesting that our Sun was born in a cluster. Full details are available in the original source: *Jensen et al.* (2019).

measured water ice D/H ratio. The O and N isotopic ratio also bear information on the birth environment; however, more work is needed to understand the potential contribution of the disk.

3.4.4. Sulfur as a unique tracer. Sulfur is found in volatile cometary species [H_2S, SO_2, SO, OCS, H_2CS, CH_3SH, C_2H_6S, CS_2, CS (*Biver et al.,* 2016; *Calmonte et al.,* 2016)] and in more refractory-like compounds (S_2, S_3), and even in atomic form (*Calmonte et al.,* 2016). This implies that S is a unique element that may allow the volatile and refractory phases to be understood in relation to one another. The partitioning of elemental S between the two phases in comets has not yet been established (*Drozdovskaya et al.,* 2018; *Altwegg et al.,* 2019). From the ISM perspective, S is even more of a mystery. In the diffuse ISM, the cosmic abundance of S is accounted for fully in the gas phase (*Lucas and Liszt,* 2002). However, in prestellar cores and protostellar regions, only a fraction of S is observed in volatiles species (*Ruffle et al.,* 1999; *Anderson et al.,* 2013). It is thought that a large fraction of S may be hidden as H_2S ice; however, it has thus far not been identified via IR observations (*Boogert et al.,* 1997, 2015). Alternatively, S may be predominantly found in more refractory phases [e.g., S_2, S_3, FeS (*Kama et al.,* 2019)]. S_2 can be produced from H_2S ice upon UV irradiation (*Grim and Greenberg,* 1987). Unfortunately, such species cannot be observed directly with remote facilities.

Summary of potential links to birth environment: Sulfur isotopic ratios could be a promising way forward toward understanding the chemistry of S and how it gets incorporated into comets. Mass spectrometry measurements made by Rosetta on comet 67P allowed the $^{34}S/^{32}S$ ratio to be determined in volatiles [H_2S, OCS, CS_2 (*Calmonte et al.,* 2017)] and in several dust particles (*Paquette et al.,* 2017). The ratio in the refractories is consistent with the terrestrial value, but that in the volatiles is lower (Fig. 8). The isotopic ratio could potentially be used to trace the amount of S harbored in the refractory phase and unlock the potential of this abundant element as a tracer of origins.

4. THE INCOMPLETE PICTURE OF OUR ORIGINS

4.1. A Broader Context of Cometary

For a long time, comets have been only looked at from a distance. Their differences with asteroids were obvious since, by construction, comets were defined as active bodies. In contrast, asteroids are usually defined as activity-free objects. The focus on cometary ices, which show strong links with the ISM, led to the idea that comets were primitive, unprocessed bodies. The identification of the highly processed Orgueil meteorites as a possible cometary meteorite, the discovery that Wild 2 cometary samples contained both high-temperature minerals and alteration phases, and the presence of active bodies in the outer belt (*Hsieh and Jewitt,* 2006), changed this view and lead to the idea that there might exist a continuum between asteroids and comets.

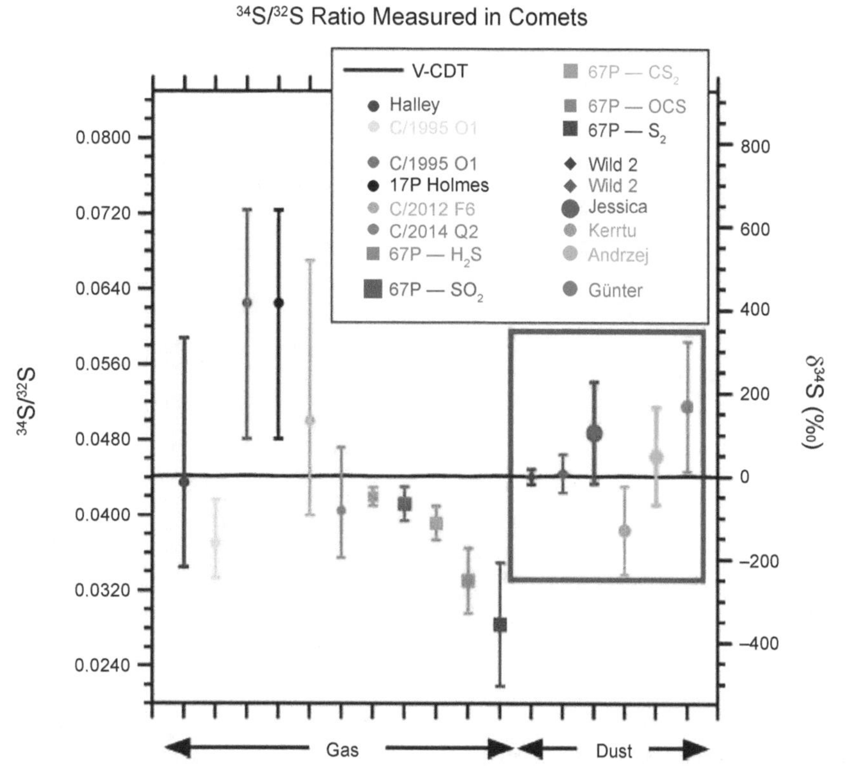

Fig. 8. The $^{34}S/^{32}S$ isotopic ratios measured in several different comets. The figure includes measurements obtained during the Rosetta mission to comet 67P/Churyumov-Gerasimenko with the ROSINA instrument measuring volatile gases (squares with molecules specified) and with the COSIMA instrument measuring refractory dust grains (circles corresponding to named particles). Vienna-Canyon Diablo Troilite (V-CDT) is shown as a reference. The COSIMA measurements agree within errors with the V-DCT value and STARDUST values, while volatiles show a more significant dispersion. Full details are available in the original source: *Paquette et al.* (2017).

The recent discovery that carbonaceous chondrites and related meteorites probably came from the outer solar system supported the asteroid-comet continuum and the idea that at least a significant fraction of cometary dust has been processed in the protoplanetary disk or in parent bodies. Since there is a large variability within the different carbonaceous chondrite groups, it should also lead to the realization that there might be as much variability in comets as in asteroids, and that speaking of comets in general might be misleading. Finally, one should keep in mind that the phenomenological definition of comets as active bodies might hinder the fact that some so-called asteroids are also ice-rich and formed in the outer solar system, leading to a reconsideration of the notion of primitivity.

4.2. Pinpointing t = 0 in a Solar Birth Cluster

The chemical composition of comets is diverse and complex. The similarity in composition between the chemical inventory of cometary volatiles and gases in protoplanetary disk-forming regions in the vicinity of protostars suggests partial inheritance of materials. These similarities have now been demonstrated for more than one comet and are strengthened by improvements in the quality of the available data. Deuteration of cometary molecules and the presence of highly volatile species at appreciable abundances in them strongly supports a pristine cometary nucleus, one that has never been fully heated to engender significant volatile loss. Moreover, the high levels of deuteration of cometary volatiles, and indeed in solar system water, can only be achieved in the darkest, coldest parts of the ISM: the prestellar cores. The alternative would be that these ratios are created in the cold outer parts of the disk, but the lack of a strong gradient in the water D/H ratios in the solar system, as would be expected given the temperature gradient, and the need for ionization (*Cleeves et al.,* 2014) argue against the outer disk (but see *Furuya et al.,* 2015). This *may* be pinpointing the t = 0 point of cometary volatile ice formation.

The D/H ratio is a key fingerprint. The origin of O and N isotopic anomalies in meteorites, which appear to be present in cometary volatiles (Schroeder et al., 2019; Altwegg et al., 2020), could also represent further important clues. Self-shielding will be active during molecular cloud and dense core formation as these effects are widely observed in the ISM. Models of the origin of heavy O isotope enrichment in interstellar ices within the molecular cloud require the birth of the Sun in a cluster (Lee et al., 2008), in agreement with dynamical evidence (section 3.1) and radionuclides (section 2.3). This is now in agreement with the water D/H ratio. Regions associated with star cluster birth have $(D/H)_{H_2O} \sim 10^{-4}$, comparable to the solar system, with higher ratios measured in more isolated systems (*Jensen et al.,* 2019). Thus, isotopic enrichment may be a key new piece in support of the Sun's birth in a clustered environment. However, at present, origin within the disk (e.g., *Lyons and Young,* 2005) cannot be ruled out. Clearly, more (and higher-precision) measurements of O, N, S, and H isotopic ratios in comets are needed. Further, more detailed chemical/dynamical models that encompass all isotopic systems need to be developed.

4.3. Looking Forward

There remain significant open questions. First, are the high-temperature components mixed outward — prior to the main phase of cometary assembly — to be combined with the grains with ice coatings from earlier phases, or are they somehow created in the comet-forming zone? *Bergner and Ciesla* (2021) points out that radial drift provides a strong supply term of primordial material to the inner tens of astronomical units. Thus, an inward supply of solids is clear, but supply via outward movement via, e.g., viscous spreading, is unsettled. Analysis of solar system samples provide information — with the support of models — of outward movement of material (*Yang et al.,* 2013; *Desch et al.,* 2018; *Nanne et al.,* 2019). Astronomically, evidence is less clear (*Najita and Bergin,* 2018; *Trapman et al.,* 2022), and which mechanism (spreading or drift) dominates needs to be understood. The carbonaceous refractory components are another important aspect with a new burst of information from Rosetta. This component carries the bulk of C and N within cometary nuclei. However, we know very little about its origin and need to develop remote sensing techniques to probe commonalities of this phase beyond the detailed *in situ* studies of 1P/Halley and 67P. The work of *Woodward et al.* (2021) provides a recent example of how to explore this and *Lisse et al.* (2020) summarizes the Spitzer mission legacy, which will be significantly expanded upon by the contributions of the JWST. There is some dispersion in the bulk cometary carbon content with potentially two Sun-grazing comets having bulk carbon consistent with chondrites (*Ciaravella et al.,* 2010; *McCauley et al.,* 2013). This is significantly below 1P/Halley and 67P (*Bergin et al.,* 2015; *Rubin et al.,* 2019) and hints at relatively unprobed diversity in the bulk content of cometary bodies.

More broadly, the D/H ratio of water vapor is known with high precision in a handful of comets, and determining whether there is a traceable link between D/H ratio (or other isotopic ratios) to, e.g., radius of formation appears to be difficult. It may be the case that radial drift wipes out any such signature, but more models, combined with a statistically significant sample observed with high precision, are needed.

Acknowledgments. E.A.B. acknowledges support from NSF Grant#1907653 and NASA Exoplanets Research Program, grant 80NSSC20K0259and Emerging Worlds Program, grant 80NSSC20K0333. M.N.D. acknowledges support of the Swiss National Science Foundation (SNSF) Ambizione grant no. 180079, the Center for Space and Habitability (CSH) Fellowship, and the IAU Gruber Foundation Fellowship. M.N.D. also acknowledges beneficial discussions held with the international team #461 "Provenances of Our Solar System's Relics" (team leaders Maria N. Drozdovskaya and Cyrielle Opitom) at the International Space Science Institute, Bern, Switzerland.

REFERENCES

Adams F. C. (2010) The birth environment of the solar system. *Annu. Rev. Astron. Astrophys., 48,* 47–85.

Adams F. C., Hollenbach D., Laughlin G., and Gorti U. (2004) Photoevaporation of circumstellar disks due to external far-ultraviolet radiation in stellar aggregates. *Astrophys. J., 611,* 360–379.

A'Hearn M. F. (2011) Comets as building blocks. *Annu. Rev. Astron. Astrophys., 49,* 281–299.

Alexander C. M. O. (1993) Presolar SiC in chondrites: How variable and how many sources? *Geochim. Cosmochim. Acta, 57,* 2869–2888.

Alexander C. M. O. (1997) Dust production in the galaxy: The meteorite perspective. In *Astrophysical Implications of the Laboratory Study of Presolar Materials* (T. J. Bernatowicz and E. Zinner, eds.), pp. 567–594. AIP Conf. Proc. 402, American Institute of Physics, Melville, New York.

Alexander C. M. O., Cody G. D., Fogel M., and Yabuta H. (2008) Organics in meteorites: Solar or interstellar? In *Organic Matter in Space* (S. Kwok and S. Sandford, eds.), pp. 293–298. IAU Symp. 251, Cambridge Univ., Cambridge.

Alexander C. M. O., Bowden R., Fogel M. L., Howard K. T., Herd C. D. K., and Nittler L. R. (2012) The provenances of asteroids, and their contributions to the volatile inventories of the terrestrial planets. *Science, 337,* 721.

Alexander C. M. O., Cody G. D., De Gregorio B. T., Nittler L. R., and Stroud R. M. (2017a) The nature, origin and modification of insoluble organic matter in chondrites, the major source of Earth's C and N. *Chem. Erde–Geochem., 77,* 227–256.

Alexander C. M. O., Nittler L. R., Davidson J., and Ciesla F. J. (2017b) Measuring the level of interstellar inheritance in the solar protoplanetary disk. *Meteoritics & Planet. Sci., 52,* 1797–1821.

Altwegg K., Balsiger H., Berthelier J. J., Bieler A., Calmonte U., De Keyser J., Fiethe B., Fuselier S. A., Gasc S., Gombosi T. I., Owen T., Le Roy L., Rubin M., Sémon T., and Tzou C.-Y. (2017) D_2O and HDS in the coma of 67P/Churyumov-Gerasimenko. *Philos. Trans. R. Soc. London Ser. A, 375,* 20160253.

Altwegg K., Balsiger H., and Fuselier S. A. (2019) Cometary chemistry and the origin of icy solar system bodies: The view after Rosetta. *Annu. Rev. Astron. Astrophys., 57,* 113–155.

Altwegg K., Balsiger H., Combi M., De Keyser J., Drozdovskaya M. N., Fuselier S. A., Gombosi T. I., Hänni N., Rubin M., Schuhmann M., Schroeder I., and Wampfler S. (2020) Molecule-dependent oxygen isotopic ratios in the coma of comet 67P/Churyumov-Gerasimenko. *Mon. Not. R. Astron. Soc., 498,* 5855–5862.

Anderson D. E., Bergin E. A., Maret S., and Wakelam V. (2013) New constraints on the sulfur reservoir in the dense interstellar medium provided by Spitzer observations of S I in shocked gas. *Astrophys. J., 779,* 141.

André P., Ward-Thompson D., and Barsony M. (2000) From prestellar cores to protostars: The initial conditions of star formation. In *Protostars and Planets IV* (V. Mannings et al., eds.), pp. 59–96. Univ. of Arizona, Tucson.

André P., Men'shchikov A., Bontemps S., Könyves V., Motte F., Schneider N., Didelon P., Minier V., Saraceno P., Ward-Thompson D., di Francesco J., White G., Molinari S., Testi L., Abergel A., Griffin M., Henning T., Royer P., Merín B., Vavrek R., Attard M., Arzoumanian D., Wilson C. D., Ade P., Aussel H., Baluteau J. P., Benedettini M., Bernard J. P., Blommaert J. A. D. L., Cambrésy L., Cox P., di Giorgio A., Hargrave P., Hennemann M., Huang M., Kirk J., Krause O., Launhardt R., Leeks S., Le Pennec J., Li J. Z., Martin P. G., Maury A., Olofsson G., Omont A., Peretto N., Pezzuto S.,

Prusti T., Roussel H., Russeil D., Sauvage M., Sibthorpe B., Sicilia-Aguilar A., Spinoglio L., Waelkens C., Woodcraft A., and Zavagno A. (2010) From filamentary clouds to prestellar cores to the stellar IMF: Initial highlights from the Herschel Gould Belt Survey. *Astron. Astrophys., 518,* L102.

André P., Di Francesco J., Ward-Thompson D., Inutsuka S. I., Pudritz R. E., and Pineda J. E. (2014) From filamentary networks to dense cores in molecular clouds: Toward a new paradigm for star formation. In *Protostars and Planets VI* (H. Beuther et al., eds.), pp. 27–52. Univ. of Arizona, Tucson.

Andrews S. M. (2020) Observations of protoplanetary disk structures. *Annu. Rev. Astron. Astrophys., 58,* 483–528.

Arce H. G., Shepherd D., Gueth F., Lee C. F., Bachiller R., Rosen A., and Beuther H. (2007) Molecular outflows in low- and high-mass star-forming regions. In *Protostars and Planets V* (B. Reipurth et al., eds.), pp. 245–260. Univ. of Arizona, Tucson.

Arnould M., Paulus G., and Meynet G. (1997) Short-lived radionuclide production by non-exploding Wolf-Rayet stars. *Astron. Astrophys., 321,* 452–464.

Arzoumanian D., André P., Didelon P., Könyves V., Schneider N., Men'shchikov A., Sousbie T., Zavagno A., Bontemps S., di Francesco J., Griffin M., Hennemann M., Hill T., Kirk J., Martin P., Minier V., Molinari S., Motte F., Peretto N., Pezzuto S., Spinoglio L., Ward-Thompson D., White G., and Wilson C. D. (2011) Characterizing interstellar filaments with Herschel in IC 5146. *Astron. Astrophys., 529,* L6.

Asplund M., Grevesse N., Sauval A. J., and Scott P. (2009) The chemical composition of the Sun. *Annu. Rev. Astron. Astrophys., 47,* 481–522.

Audard M., Ábrahám P., Dunham M. M., Green J. D., Grosso N., Hamaguchi K., Kastner J. H., Kóspál Á., Lodato G., Romanova M. M., Skinner S. L., Vorobyov E. I., and Zhu Z. (2014) Episodic accretion in young stars. In *Protostars and Planets VI* (H. Beuther et al., eds.), pp. 387–410. Univ. of Arizona, Tucson.

Bailer-Jones C. A. L. (2018) The completeness-corrected rate of stellar encounters with the Sun from the first Gaia data release. *Astron. Astrophys., 609,* A8.

Ballesteros-Paredes J., André P., Hennebelle P., Klessen R. S., Kruijssen J. M. D., Chevance M., Nakamura F., Adamo A., and Vázquez-Semadeni E. (2020) From diffuse gas to dense molecular cloud cores. *Space Sci. Rev., 216,* 76.

Bally J., Langer W. D., Stark A. A., and Wilson R. W. (1987) Filamentary structure in the Orion molecular cloud. *Astrophys. J. Lett., 312,* L45.

Bardyn A., Baklouti D., Cottin H., Fray N., Briois C., Paquette J., Stenzel O., Engrand C., Fischer H., Hornung K., Isnard R., Langevin Y., Lehto H., Le Roy L., Ligier N., Merouane S., Modica P., Orthous-Daunay F.-R., Rynö J., Schulz R., Silén J., Thirkell L., Varmuza K., Zaprudin B., Kissel J., and Hilchenbach M. (2017) Carbon-rich dust in comet 67P/Churyumov-Gerasimenko measured by COSIMA/Rosetta. *Mon. Not. Roy. Astron. Soc., 469,* S712–S722.

Batygin K., Adams F. C., Batygin Y. K., and Petigura E. A. (2020) Dynamics of planetary systems within star clusters: Aspects of the solar system's early evolution. *Astron. J., 159,* 101.

Becker J. C., Khain T., Hamilton S. J., Adams F. C., Gerdes D. W., Zullo L., Franson K., Millholland S., Bernstein G. M., Sako M., Bernardinelli P., Napier K., Markwardt L., Lin H. W., Wester W., Abdalla F. B., Allam S., Annis J., Avila S., Bertin E., Brooks D., Carnero Rosell A., Carrasco Kind M., Carretero J., Cunha C. E., D'Andrea C. B., da Costa L. N., Davis C., De Vicente J., Diehl H. T., Doel P., Eifler T. F., Flaugher B., Fosalba P., Frieman J., García-Bellido J., Gaztanaga E., Gruen D., Gruendl R. A., Gschwend J., Gutierrez G., Hartley W. G., Hollowood D. L., Honscheid K., James D. J., Kuehn K., Kuropatkin N., Maia M. A. G., March M., Marshall J. L., Menanteau F., Miquel R., Ogando R. L. C., Plazas A. A., Sanchez E., Scarpine V., Schindler R., Sevilla-Noarbe I., Smith M., Smith R. C., Soares-Santos M., Sobreira F., Suchyta E., Swanson M. E. C., Walker A. R., and the DES Collaboration (2018) Discovery and dynamical analysis of an extreme trans-neptunian object with a high orbital inclination. *Astron. J., 156,* 81.

Benz A. O., Bruderer S., van Dishoeck E. F., Melchior M., Wampfler S. F., van der Tak F., Goicoechea J. R., Indriolo N., Kristensen L. E., Lis D. C., Mottram J. C., Bergin E. A., Caselli P., Herpin F., Hogerheijde M. R., Johnstone D., Liseau R., Nisini B., Tafalla M., Visser R., and Wyrowski F. (2016) Water in star-forming regions with Herschel (WISH). VI. Constraints on UV and X-ray irradiation from a survey of hydrides in low- to high-mass young stellar objects. *Astron. Astrophys., 590,* A105.

Berger E. L., Zega T. J., Keller L. P., and Lauretta D. S. (2011) Evidence for aqueous activity on comet 81P/Wild 2 from sulfide mineral assemblages in Stardust samples and CI chondrites. *Geochim. Cosmochim. Acta, 75,* 3501–3513.

Bergin E. A. and Tafalla M. (2007) Cold dark clouds: The initial conditions for star formation. *Annu. Rev. Astron. Astrophys., 45,* 339–396.

Bergin E. A., Hartmann L. W., Raymond J. C., and Ballesteros-Paredes J. (2004) Molecular cloud formation behind shock waves. *Astrophys. J., 612,* 921–939.

Bergin E. A., Blake G. A., Ciesla F., Hirschmann M. M., and Li J. (2015) Tracing the ingredients for a habitable Earth from interstellar space through planet formation. *Proc. Natl. Acad. Sci. U.S.A., 112,* 8965–8970.

Bergner J. B. and Ciesla F. (2021) Ice inheritance in dynamical disk models. *Astrophys. J., 919,* 45.

Bertoldi F., Carilli C. L., Cox P., Fan X., Strauss M. A., Beelen A., Omont A., and Zylka R. (2003) Dust emission from the most distant quasars. *Astron. Astrophys., 406,* L55–L58.

Bethell T. and Bergin E. (2009) Formation and survival of water vapor in the terrestrial planet-forming region. *Science, 326,* 1675.

Birnstiel T. and Andrews S. M. (2014) On the outer edges of protoplanetary dust disks. *Astrophys. J., 780,* 153.

Birnstiel T., Ormel C. W., and Dullemond C. P. (2011) Dust size distributions in coagulation/fragmentation equilibrium: Numerical solutions and analytical fits. *Astron. Astrophys., 525,* A11.

Biver N., Moreno R., Bockelée-Morvan D., Sandqvist A., Colom P., Crovisier J., Lis D. C., Boissier J., Debout V., Paubert G., Milam S., Hjalmarson A., Lundin S., Karlsson T., Battelino M., Frisk U., Murtagh D., and the Odin Team (2016) Isotopic ratios of H, C, N, O, and S in comets C/2012 F6 (Lemmon) and C/2014 Q2 (Lovejoy). *Astron. Astrophys., 589,* A78.

Blum J. and Wurm G. (2008) The growth mechanisms of macroscopic bodies in protoplanetary disks. *Annu. Rev. Astron. Astrophys., 46,* 21–56.

Bobylev V. V. and Bajkova A. T. (2014) Search for kinematic siblings of the Sun based on data from the XHIP catalog. *Astron. Lett., 40,* 353–360.

Bocchio M., Micelotta E. R., Gautier A. L., and Jones A. P. (2012) Small hydrocarbon particle erosion in a hot gas: A comparative study. *Astron. Astrophys., 545,* A124.

Bockelée-Morvan D., Lis D. C., Wink J. E., Despois D., Crovisier J., Bachiller R., Benford D. J., Biver N., Colom P., Davies J. K., Gérard E., Germain B., Houde M., Mehringer D., Moreno R., Paubert G., Phillips T. G., and Rauer H. (2000) New molecules found in comet C/1995 O1 (Hale-Bopp): Investigating the link between cometary and interstellar material. *Astron. Astrophys., 353,* 1101–1114.

Bockelée-Morvan D., Calmonte U., Charnley S., Duprat J., Engrand C., Gicquel A., Hässig M., Jehin E., Kawakita H., Marty B., Milam S., Morse A., Rousselot P., Sheridan S., and Wirström E. (2015) Cometary isotopic measurements. *Space Sci. Rev., 197,* 47–83.

Boogert A. C. A., Schutte W. A., Helmich F. P., Tielens A. G. G. M., and Wooden D. H. (1997) Infrared observations and laboratory simulations of interstellar CH_4 and SO_2. *Astron. Astrophys., 317,* 929–941.

Boogert A. C. A., Gerakines P. A., and Whittet D. C. B. (2015) Observations of the icy universe. *Annu. Rev. Astron. Astrophys., 53,* 541–581.

Boss A. P., Alexander C. M. O., and Podolak M. (2012) Cosmochemical consequences of particle trajectories during FU Orionis outbursts by the early Sun. *Earth Planet. Sci. Lett., 345-348,* 18–26.

Bouwman J., Henning T., Hillenbrand L. A., Meyer M. R., Pascucci I., Carpenter J., Hines D., Kim J. S., Silverstone M. D., Hollenbach D., and Wolf S. (2008) The formation and evolution of planetary systems: Grain growth and chemical processing of dust in T Tauri systems. *Astrophys. J., 683,* 479–498.

Bradley J. P. (2014) Early solar nebula grains: Interplanetary dust particles. In *Treatise on Geochemistry, second edition* (A. M. Davis, ed.), pp. 287–308. Elsevier, Oxford.

Brasser R., Duncan M. J., and Levison H. F. (2006) Embedded star clusters and the formation of the Oort cloud. *Icarus, 184,* 59–82.

Bressert E., Bastian N., Gutermuth R., Megeath S. T., Allen L., Evans I., Neal J., Rebull L. M., Hatchell J., Johnstone D., Bourke T. L., Cieza L. A., Harvey P. M., Merin B., Ray T. P., and Tothill N. F. H. (2010)

The spatial distribution of star formation in the solar neighbourhood: Do all stars form in dense clusters? *Mon. Not. R. Astron. Soc., 409,* L54–L58.
Brown M. E., Trujillo C., and Rabinowitz D. (2004) Discovery of a candidate inner Oort cloud planetoid. *Astrophys. J., 617,* 645–649.
Brownlee D. (2014) The Stardust mission: Analyzing samples from the edge of the solar system. *Annu. Rev. Earth Planet. Sci., 42,* 179–205.
Budde G., Burkhardt C., Brennecka G. A., Fischer-Gödde M., Kruijer T. S., and Kleine T. (2016) Molybdenum isotopic evidence for the origin of chondrules and a distinct genetic heritage of carbonaceous and non-carbonaceous meteorites. *Earth Planet. Sci. Lett., 454,* 293–303.
Burkhardt A. M., Long K. L. K., Bryan C. P., Shingledecker C. N., Cooke I. R., Loomis R. A., Wei H., Charnley S. B., Herbst E., McCarthy M. C., and McGuire B. A. (2021a) Discovery of the pure polycyclic aromatic hydrocarbon indene (c-C9H8) with GOTHAM observations of TMC-1. *Astrophys. J. Lett., 913,* L18.
Burkhardt A. M., Loomis R. A., Shingledecker C. N., Lee K. L. K., Remijan A. J., McCarthy M. C., and McGuire B. A. (2021b) Ubiquitous aromatic carbon chemistry at the earliest stages of star formation. *Nature Astron., 5,* 181–187.
Calmonte U., Altwegg K., Balsiger H., Berthelier J. J., Bieler A., Cessateur G., Dhooghe F., van Dishoeck E. F., Fiethe B., Fuselier S. A., Gasc S., Gombosi T. I., Hässig M., Le Roy L., Rubin M., Sémon T., Tzou C.-Y., and Wampfler S. F. (2016) Sulphur-bearing species in the coma of comet 67P/Churyumov-Gerasimenko. *Mon. Not. R. Astron. Soc., 462,* S253–S273.
Calmonte U., Altwegg K., Balsiger H., Berthelier J. J., Bieler A., De Keyser J., Fiethe B., Fuselier S. A., Gasc S., Gombosi T. I., Le Roy L., Rubin M., Sémon T., Tzou C. Y., and Wampfler S. F. (2017) Sulphur isotope mass-independent fractionation observed in comet 67P/Churyumov-Gerasimenko by Rosetta/ROSINA. *Mon. Not. R. Astron. Soc., 469,* S787–S803.
Caselli P. and Ceccarelli C. (2012) Our astrochemical heritage. *Astron. Astrophys. Rev., 20,* 56.
Ceccarelli C., Caselli P., Bockelée-Morvan D., Mousis O., Pizzarello S., Robert F., and Semenov D. (2014) Deuterium fractionation: The Ariadne's thread from the precollapse phase to meteorites and comets today. In *Protostars and Planets VI* (H. Beuther et al., eds.), pp. 859–882. Univ. of Arizona, Tucson.
Cernicharo J., Agúndez M., Cabezas C., Tercero B., Marcelino N., Pardo J. R., and de Vicente P. (2021a) Pure hydrocarbon cycles in TMC-1: Discovery of ethynyl cyclopropenylidene, cyclopentadiene, and indene. *Astron. Astrophys., 649,* L15.
Cernicharo J., Agúndez M., Kaiser R. I., Cabezas C., Tercero B., Marcelino N., Pardo J. R., and de Vicente P. (2021b) Discovery of benzyne, o-C6H4, in TMC-1 with the QUIJOTE line survey. *Astron. Astrophys., 652,* L9.
Cherchneff I. (2013) The chemistry of dust formation in red supergiants. In *Betelguese Workshop 2012 – The Physics of Red Supergiants: Recent Advances and Open Questions* (P. Kervella et al., eds.), pp. 175–184. EAS Publ. Series, Vol. 60, Cambridge Univ., Cambridge.
Chevalier R. A. (2000) Young circumstellar disks near evolved massive stars and supernovae. *Astrophys. J. Lett., 538,* L151–L154.
Chiar J. E., Tielens A. G. G. M., Adamson A. J., and Ricca A. (2013) The structure, origin, and evolution of interstellar hydrocarbon grains. *Astrophys. J., 770,* 78.
Choukroun M., Altwegg K., Kührt E., Biver N., Bockelée-Morvan D., Drążkowska J., Hérique A., Hilchenbach M., Marschall R., Pätzold M., Taylor M. G. G. T., and Thomas N. (2020) Dust-to-gas and refractory-to-ice mass ratios of comet 67P/Churyumov-Gerasimenko from Rosetta observations. *Space Sci. Rev., 216,* 44.
Ciaravella A., Raymond J. C., and Giordano S. (2010) Ultraviolet spectra of the C-2003K7 comet: Evidence for dust sublimation in Si and C lines. *Astrophys. J. Lett., 713,* L69–L73.
Ciesla F. J. (2015) Sulfidization of iron in the dynamic solar nebula and implications for planetary compositions. *Astrophys. J. Lett., 800,* L6.
Ciesla F. J. and Sandford S. A. (2012) Organic synthesis via irradiation and warming of ice grains in the solar nebula. *Science, 336,* 452–454.
Ciesla F. J., Krijt S., Yokochi R., and Sandford S. (2018) The efficiency of noble gas trapping in astrophysical environments. *Astrophys. J., 867,* 146.
Clark P. C., Glover S. C. O., Klessen R. S., and Bonnell I. A. (2012) How long does it take to form a molecular cloud? *Mon. Not. R. Astron. Soc., 424,* 2599–2613.
Clayton D. (2003) *Handbook of Isotopes in the Cosmos: Hydrogen to Gallium.* Cambridge Univ., Cambridge. 328 pp.
Clayton R. N. (1993) Oxygen isotopes in meteorites. *Ann. Rev. Earth Planet. Sci., 21,* 115–149.
Cleeves L. I., Bergin E. A., Alexander C. M. O. D., Du F., Graninger D., Öberg K. I., and Harries T. J. (2014) The ancient heritage of water ice in the solar system. *Science, 345,* 1590–1593.
Cody G., Ade H., Alexander C. M. O., Araki T., Butterworth A., Fleckenstein H., Flynn G. J., Gilles M., Jacobsen C., Kilcoyne A. L. D., Messenger K., Sandford S. A., Tyliszczak T., Westphal A., Wirick S., and Yabuta H. (2008) Quantitative organic and light element analysis of Comet 81P/Wild 2 particles using C-, N-, and O-μ-XANES. *Meteoritics & Planet. Sci., 43,* 353–365.
Compiègne M., Verstraete L., Jones A., Bernard J.-P., Boulanger F., Flagey N., Le Bourlot J., Paradis D., and Ysard N. (2011) The global dust SED: Tracing the nature and evolution of dust with DustEM. *Astron. Astrophys., 525,* A103.
Concha-Ramírez F., Wilhelm M. J. C., Portegies Zwart S., and Haworth T. J. (2019) External photoevaporation of circumstellar discs constrains the time-scale for planet formation. *Mon. Not. R. Astron. Soc., 490,* 5678–5690.
Concha-Ramírez F., Wilhelm M. J. C., Portegies Zwart S., van Terwisga S. E., and Hacar A. (2021) Effects of stellar density on the photoevaporation of circumstellar discs. *Mon. Not. R. Astron. Soc., 501,* 1782–1790.
Connelly J. N., Bizzarro M., Krot A. N., Nordlund Å., Wielandt D., and Ivanova M. A. (2012) The absolute chronology and thermal processing of solids in the solar protoplanetary disk. *Science, 338,* 651.
Crapsi A., Caselli P., Walmsley M. C., and Tafalla M. (2007) Observing the gas temperature drop in the high-density nucleus of L 1544. *Astron. Astrophys., 470,* 221–230.
Crovisier J., Leech K., Bockelee-Morvan D., Brooke T. Y., Hanner M. S., Altieri B., Keller H. U., and Lellouch E. (1997) The spectrum of comet Hale-Bopp (C/1995 O1) observed with the infrared space observatory at 2.9 astronomical units from the Sun. *Science, 275,* 1904–1907.
Cuppen H. M., Walsh C., Lamberts T., Semenov D., Garrod R. T., Penteado E. M., and Ioppolo S. (2017) Grain surface models and data for astrochemistry. *Space Sci. Rev., 212,* 1–58.
Dale J. E., Haworth T. J., and Bressert E. (2015) The dangers of being trigger-happy. *Mon. Not. R. Astron. Soc., 450,* 1199–1211.
Dartois E., Muñoz Caro G. M., Deboffle D., Montagnac G., and D'Hendecourt L. (2005) Ultraviolet photoproduction of ISM dust: Laboratory characterisation and astrophysical relevance. *Astron. Astrophys., 432,* 895–908.
Dauphas N. and Chaussidon M. (2011) A perspective from extinct radionuclides on a young stellar object: The Sun and its accretion disk. *Annu. Rev. Earth Planet. Sci., 39,* 351–386.
Davis A. M., Alexander C. M. O. D., Ciesla F. J., Gounelle M., Krot A. N., Petaev M. I., and Stephan T. (2014) Samples of the solar system: Recent developments. In *Protostars and Planets VI* (H. Beuther et al., eds.), pp. 809–832. Univ. of Arizona, Tucson.
De Gregorio B. T., Stroud R. M., Nittler L. R., Alexander C. M. O., Kilcoyne A. L. D., and Zega T. J. (2010) Isotopic anomalies in organic nanoglobules from Comet 81P/Wild 2: Comparison to Murchison nanoglobules and isotopic anomalies induced in terrestrial organics by electron irradiation. *Geochim. Cosmochim. Acta, 74,* 4454–4470.
Defouilloy C., Nakashima D., Joswiak D. J., Brownlee D. E., Tenner T. J., and Kita N. T. (2017) Origin of crystalline silicates from Comet 81P/Wild 2: Combined study on their oxygen isotopes and mineral chemistry. *Earth Planet. Sci. Lett., 465,* 145–154.
Demyk K., Carrez P., Leroux H., Cordier P., Jones A. P., Borg J., Quirico E., Raynal P. I., and d'Hendecourt L. (2001) Structural and chemical alteration of crystalline olivine under low energy He^+ irradiation. *Astron. Astrophys., 368,* L38–L41.
Desch S. J., Connolly J., Harold C., and Srinivasan G. (2004) An interstellar origin for the beryllium 10 in calcium-rich, aluminum-rich inclusions. *Astrophys. J., 602,* 528–542.
Desch S. J., Kalyaan A., and Alexander C. M. O. (2018) The effect of

Jupiter's formation on the distribution of refractory elements and inclusions in meteorites. *Astrophys. J. Suppl., 238,* 11.

Desch S. J., Young E. D., Dunham E. T., Fujimoto Y., and Dunlap D. R. (2023) Short-lived radionuclides in meteorites and the Sun's birth environment. *Astron. Soc. Pacific Conf. Ser., 534,* 759.

Di Ruscio A., Fienga A., Durante D., Iess L., Laskar J., and Gastineau M. (2020) Analysis of Cassini radio tracking data for the construction of INPOP19a: A new estimate of the Kuiper belt mass. *Astron. Astrophys., 640,* A7.

Dobrica E., Engrand C., Leroux H., Rouzaud J. N., and Duprat J. (2012) Transmission electron microscopy of CONCORDIA ultracarbonaceous Antarctic micrometeorites (UCAMMs): Mineralogical properties. *Geochim. Cosmochim. Acta, 76,* 68–82.

Draine B. T. (2003) Interstellar dust grains. *Annu. Rev. Astron. Astrophys., 41,* 241–289.

Draine B. T. (2009) Interstellar dust models and evolutionary implications. In *Cosmic Dust: Near and Far* (T. Henning et al., eds.), pp. 453–473. ASP Conf. Ser. 414, Astronomical Society of the Pacific, San Francisco.

Drouart A., Dubrulle B., Gautier D., and Robert F. (1999) Structure and transport in the solar nebula from constraints on deuterium enrichment and giant planets formation. *Icarus, 140,* 129–155.

Drozdovskaya M. N., van Dishoeck E. F., Jørgensen J. K., Calmonte U., van der Wiel M. H. D., Coutens A., Calcutt H., Müller H. S. P., Bjerkeli P., Persson M. V., Wampfler S. F., and Altwegg K. (2018) The ALMA-PILS survey: The sulphur connection between protostars and comets: IRAS 16293-2422 B and 67P/Churyumov-Gerasimenko. *Mon. Not. R. Astron. Soc., 476,* 4949–4964.

Drozdovskaya M. N., van Dishoeck E. F., Rubin M., Jørgensen J. K., and Altwegg K. (2019) Ingredients for solar-like systems: Protostar IRAS 16293-2422 B versus comet 67P/Churyumov-Gerasimenko. *Mon. Not. R. Astron. Soc., 490,* 50–79.

Drozdovskaya M. N., Schroeder I. R. H. G., Rubin M., Altwegg K., van Dishoeck E. F., Kulterer B. M., De Keyser J., Fuselier S. A., and Combi M. (2021) Prestellar grain-surface origins of deuterated methanol in comet 67P/Churyumov-Gerasimenko. *Mon. Not. R. Astron. Soc., 500,* 4901–4920.

Dulieu F., Amiaud L., Congiu E., Fillion J. H., Matar E., Momeni A., Pirronello V., and Lemaire J. L. (2010) Experimental evidence for water formation on interstellar dust grains by hydrogen and oxygen atoms. *Astron. Astrophys., 512,* A30.

Duncan M., Levison H., and Dones L. (2004) Dynamical evolution of ecliptic comets. In *Comets II* (M. C. Festou et al., eds.), pp. 193–204. Univ. of Arizona, Tucson.

Dunham E., Wadhwa M., Desch S., Liu M., Fukuda K., Kita N., Hertwig A., Hervig R., Defouilloy C., Simon S., Davidson J., Schrader D., and Fujimoto Y. (2022) Uniform initial $^{10}Be/^{9}Be$ inferred from refractory inclusions in CV3, CO3, CR2, and CH/CB chondrites. *Geochim. Cosmochim. Acta, 324,* 194–220.

Duprat J., Dobrică E., Engrand C., Aléon J., Marrocchi Y., Mostefaoui S., Meibom A., Leroux H., Rouzaud J. N., Gounelle M., and Robert F. (2010) Extreme deuterium excesses in ultracarbonaceous micrometeorites from central Antarctic snow. *Science, 328,* 742–745.

Dwarkadas V. V., Dauphas N., Meyer B., Boyajian P., and Bojazi M. (2017) Triggered star formation inside the shell of a Wolf-Rayet bubble as the origin of the solar system. *Astrophys. J., 851,* 147.

Dwek E. (1998) The evolution of the elemental abundances in the gas and dust phases of the galaxy. *Astrophys. J., 501,* 643.

Ehrenfreund P. and Charnley S. B. (2000) Organic molecules in the interstellar medium, comets, and meteorites: A voyage from dark clouds to the early Earth. *Annu. Rev. Astron. Astrophys., 38,* 427–483.

Ehrenfreund P., Robert F., D'Hendencourt L., and Behar F. (1991) Comparison of interstellar and meteoritic organic matter at 3.4 microns. *Astron. Astrophys., 252,* 712–717.

Engrand C., Duprat J., Bardin N., Dartois E., Leroux H., Quirico E., Benzerara K., Remusat L., Dobrică E., Delauche L., Bradley J., Ishii H., and Hilchenbach M. (2016) The asteroid-comet continuum from laboratory and space analyses of comet samples and micrometeorites. *Proc. Intl. Astron. Union, 29A,* 253–256.

Fedoseev G., Ioppolo S., Zhao D., Lamberts T., and Linnartz H. (2015) Low-temperature surface formation of NH_3 and HNCO: Hydrogenation of nitrogen atoms in CO-rich interstellar ice analogues. *Mon. Not. R. Astron. Soc., 446,* 439–448.

Fernandez J. A. and Ip W. H. (1984) Some dynamical aspects of the accretion of Uranus and Neptune: The exchange of orbital angular momentum with planetesimals. *Icarus, 58,* 109–120.

Fitoussi C., Duprat J., Tatischeff V., Kiener J., Naulin F., Raisbeck G., Assunção M., Bourgeois C., Chabot M., Coc A., Engrand C., Gounelle M., Hammache F., Lefebvre A., Porquet M. G., Scarpaci J. A., de Séréville N., Thibaud J. P., and Yiou F. (2008) Measurement of the $^{24}Mg(^{3}He,p)^{26}Al$ cross section: Implication for ^{26}Al production in the early solar system. *Phys. Rev. C, 78,* 044613.

Floss C. and Haenecour P. (2016) Presolar silicate grains: Abundances, isotopic and elemental compositions, and the effects of secondary processing. *Geochem. J., 50,* 3–25.

Floss C., Stadermann F. J., Kearsley A. T., Burchell M. J., and Ong W. J. (2013) The abundance of presolar grains in comet 81P/Wild 2. *Astrophys. J., 763,* 140.

Fray N., Bardyn A., Cottin H., Baklouti D., Briois C., Engrand C., Fischer H., Hornung K., Isnard R., Langevin Y., Lehto H., Le Roy L., Mellado E. M., Merouane S., Modica P., Orthous-Daunay F.-R., Paquette J., Rynö J., Schulz R., Silén J., Siljeström S., Stenzel O., Thirkell L., Varmuza K., Zaprudin B., Kissel J., and Hilchenbach M. (2017) Nitrogen-to-carbon atomic ratio measured by COSIMA in the particles of comet 67P/Churyumov-Gerasimenko. *Mon. Not. Roy. Astron. Soc., 469,* S506–S516.

Fukuda K., Hiyagon H., Fujiya W., Takahata N., Kagoshima T., and Sano Y. (2019) Origin of the short-lived radionuclide ^{10}Be and its implications for the astronomical setting of CAI formation in the solar protoplanetary disk. *Astrophys. J., 886,* 34.

Fukuda K., Hiyagon H., Fujiya W., Kagoshima T., Itano K., Iizuka T., Kita N. T., and Sano Y. (2021) Irradiation origin of ^{10}Be in the solar nebula: Evidence from Li-Be-B and Al-Mg isotope systematics, and REE abundances of CAIs from Yamato-81020 CO3.05 chondrite. *Geochim. Cosmochim. Acta, 293,* 187–204.

Furuya K., Aikawa Y., Hincelin U., Hassel G. E., Bergin E. A., Vasyunin A. I., and Herbst E. (2015) Water deuteration and ortho-to-para nuclear spin ratio of H_2 in molecular clouds formed via the accumulation of HI gas. *Astron. Astrophys., 584,* A124.

Furuya K., van Dishoeck E. F., and Aikawa Y. (2016) Reconstructing the history of water ice formation from HDO/H_2O and D_2O/HDO ratios in protostellar cores. *Astron. Astrophys., 586,* A127.

Gaidos E., Krot A. N., Williams J. P., and Raymond S. N. (2009) ^{26}Al and the formation of the solar system from a molecular cloud contaminated by Wolf-Rayet winds. *Astrophys. J., 696,* 1854–1863.

Genge M. J., Van Ginneken M., and Suttle M. D. (2020) Micrometeorites: Insights into the flux, sources and atmospheric entry of extraterrestrial dust at Earth. *Planet. Space Sci., 187,* 104900.

Gielen C., van Winckel H., Min M., Waters L. B. F. M., and Lloyd Evans T. (2008) SPITZER survey of dust grain processing in stable discs around binary post-AGB stars. *Astron. Astrophys., 490,* 725–735.

Glavin D. P., Alexander C. M. O., Aponte J. C., Dworkin J. P., Elsila J. E., and Yabuta H. (2018) The origin and evolution of organic matter in carbonaceous chondrites and links to their parent bodies. In *Primitive Meteorites and Asteroids: Physical, Chemical, and Spectroscopic Observations Paving the Way to Exploration* (N. Abreu, ed.), pp. 205–271. Elsevier, Oxford.

Glover S. C. O., Federrath C., Mac Low M. M., and Klessen R. S. (2010) Modelling CO formation in the turbulent interstellar medium. *Mon. Not. R. Astron. Soc., 404,* 2–29.

Gobrecht D., Cherchneff I., Sarangi A., Plane J. M. C., and Bromley S. T. (2016) Dust formation in the oxygen-rich AGB star IK Tauri. *Astron. Astrophys., 585,* A6.

Goicoechea J. R., Teyssier D., Etxaluze M., Goldsmith P. F., Ossenkopf V., Gerin M., Bergin E. A., Black J. H., Cernicharo J., Cuadrado S., Encrenaz P., Falgarone E., Fuente A., Hacar A., Lis D. C., Marcelino N., Melnick G. J., Müller H. S. P., Persson C., Pety J., Röllig M., Schilke P., Simon R., Snell R. L., and Stutzki J. (2015) Velocity-resolved [CII] emission and [CII]/FIR mapping along Orion with Herschel. *Astrophys. J., 812,* 75.

Goicoechea J. R., Santa-Maria M. G., Bron E., Teyssier D., Marcelino N., Cernicharo J., and Cuadrado S. (2019) Molecular tracers of radiative feedback in Orion (OMC-1): Widespread CH^+ (J = 1–0), CO (10–9), HCN (6–5), and HCO^+ (6–5) emission. *Astron. Astrophys., 622,* A91.

Goldsmith P. F., Velusamy T., Li D., and Langer W. D. (2010) Molecular hydrogen emission from the boundaries of the Taurus molecular cloud. *Astrophys. J., 715,* 1370–1382.
Gounelle M. (2015) The abundance of ^{26}Al-rich planetary systems in the galaxy. *Astron. Astrophys., 582,* A26.
Gounelle M. and Meibom A. (2008) The origin of short-lived radionuclides and the astrophysical environment of solar system formation. *Astrophys. J., 680,* 781–792.
Gounelle M. and Meynet G. (2012) Solar system genealogy revealed by extinct short-lived radionuclides in meteorites. *Astron. Astrophys., 545,* A4.
Gounelle M., Shu F. H., Shang H., Glassgold A. E., Rehm K. E., and Lee T. (2006) The irradiation origin of beryllium radioisotopes and other short-lived radionuclides. *Astrophys. J., 640,* 1163–1170.
Gounelle M., Morbidelli A., Bland P. A., Spurný P., Young E. D., and Sephton M. (2008) Meteorites from the outer solar system? In *The Solar System Beyond Neptune* (M. A. Barucci et al., eds.), pp. 525–541. Univ. of Arizona, Tucson.
Greenberg J. M. and Li A. (1999) Morphological structure and chemical composition of cometary nuclei and dust. *Space Sci. Rev., 90,* 149–161.
Grim R. J. A. and Greenberg J. M. (1987) Photoprocessing of H_2S in interstellar grain mantles as an explanation for S_2 in comets. *Astron. Astrophys., 181,* 155–168.
Günay B., Schmidt T. W., Burton M. G., Afşar M., Krechkivska O., Nauta K., Kable S. H., and Rawal A. (2018) Aliphatic hydrocarbon content of interstellar dust. *Mon. Not. R. Astron. Soc., 479,* 4336–4344.
Gyngard F., Jadhav M., Nittler L. R., Stroud R. M., and Zinner E. (2018) Bonanza: An extremely large dust grain from a supernova. *Geochim. Cosmochim. Acta, 221,* 60.
Habing H. J. (1968) The interstellar radiation density between 912 Å and 2400 Å. *Bull. Astron. Inst. Netherlands, 19,* 421–431.
Hacar A., Tafalla M., Kauffmann J., and Kovács A. (2013) Cores, filaments, and bundles: Hierarchical core formation in the L1495/B213 Taurus region. *Astron. Astrophys., 554,* A55.
Hacar A., Tafalla M., Forbrich J., Alves J., Meingast S., Grossschedl J., and Teixeira P. S. (2018) An ALMA study of the Orion Integral Filament I. Evidence for narrow fibers in a massive cloud. *Astron. Astrophys., 610,* A77.
Hanner M. S. (1999) The silicate material in comets. *Space Sci. Rev., 90,* 99–108.
Harju J., Walmsley C. M., and Wouterloot J. G. A. (1993) Ammonia clumps in the Orion and Cepheus clouds. *Astronomy Astrophys. Suppl., 98,* 51–76.
Harker D. E. and Desch S. J. (2002) Annealing of silicate dust by nebular shocks at 10 au. *Astrophys. J. Lett., 565,* L109–L112.
Hassel G. E., Herbst E., and Bergin E. A. (2010) Beyond the pseudo-time-dependent approach: Chemical models of dense core precursors. *Astron. Astrophys., 515,* A66.
Heays A. N., Visser R., Gredel R., Ubachs W., Lewis B. R., Gibson S. T., and van Dishoeck E. F. (2014) Isotope selective photodissociation of N_2 by the interstellar radiation field and cosmic rays. *Astron. Astrophys., 562,* A61.
Heays A. N., Bosman A. D., and van Dishoeck E. F. (2017) Photodissociation and photoionisation of atoms and molecules of astrophysical interest. *Astron. Astrophys., 602,* A105.
Heck P. R., Greer J., Kööp L., Trappitsch R., Gyngard F., Busemann H., Maden C., Ávila J. N., Davis A. M., and Wieler R. (2020) Lifetimes of interstellar dust from cosmic ray exposure ages of presolar silicon carbide. *Proc. Natl. Acad. Sci. U.S.A., 117,* 1884–1889.
Heiles C. and Troland T. H. (2003) The millennium Arecibo 21 centimeter absorption-line survey. II. Properties of the warm and cold neutral media. *Astrophys. J., 586,* 1067–1093.
Henning T. (2010) Cosmic silicates. *Annu. Rev. Astron. Astrophys., 48,* 21–46.
Hester J. J., Desch S. J., Healy K. R., and Leshin L. A. (2004) The cradle of the solar system. *Science, 304,* 1116–1117.
Hicks L. J., MacArthur J. L., Bridges J. C., Price M. C., Wickham-Eade J. E., Burchell M. J., Hansford G. M., Butterworth A. L., Gurman S. J., and Baker S. H. (2017) Magnetite in comet Wild 2: Evidence for parent body aqueous alteration. *Meteoritics & Planet. Sci., 52,* 2075–2096.
Hily-Blant P., Bonal L., Faure A., and Quirico E. (2013) The ^{15}N-enrichment in dark clouds and solar system objects. *Icarus, 223,* 582–590.
Hiraoka K., Miyagoshi T., Takayama T., Yamamoto K., and Kihara Y. (1998) Gas-grain processes for the formation of CH_4 and H_2O: Reactions of H atoms with C, O, and CO in the solid phase at 12 K. *Astrophys. J., 498,* 710–715.
Hoppe P., Leitner J., and Kodolányi J. (2017) The stardust abundance in the local interstellar cloud at the birth of the solar system. *Nature Astron., 1,* 617–620.
Hsieh H. H. (2017) Asteroid-comet continuum objects in the solar system. *Philos. Trans. R. Soc. London Ser. A, 375,* 20160259.
Hsieh H. H. and Jewitt D. (2006) A population of comets in the main asteroid belt. *Science, 312,* 561–563.
Ida S., Larwood J., and Burkert A. (2000) Evidence for early stellar encounters in the orbital distribution of Edgeworth-Kuiper belt objects. *Astrophys. J., 528,* 351–356.
Ioppolo S., van Boheemen Y., Cuppen H. M., van Dishoeck E. F., and Linnartz H. (2011) Surface formation of CO_2 ice at low temperatures. *Mon. Not. R. Astron. Soc., 413,* 2281–2287.
Ishii H. A., Bradley J. P., Dai Z. R., Chi M., Kearsley A. T., Burchell M. J., Browning N. D., and Molster F. (2008) Comparison of comet 81P/Wild 2 dust with interplanetary dust from comets. *Science, 319,* 447.
Isnard R., Bardyn A., Fray N., Briois C., Cottin H., Paquette J., Stenzel O., Alexander C., Baklouti D., Engrand C., Orthous-Daunay F. R., Siljeström S., Varmuza K., and Hilchenbach M. (2019) H/C elemental ratio of the refractory organic matter in cometary particles of 67P/Churyumov-Gerasimenko. *Astron. Astrophys., 630,* A27.
Jensen A. G., Rachford B. L., and Snow T. P. (2007) Is there enhanced depletion of gas-phase nitrogen in moderately reddened lines of sight? *Astrophys. J., 654,* 955–970.
Jensen S. S., Jørgensen J. K., Kristensen L. E., Furuya K., Coutens A., van Dishoeck E. F., Harsono D., and Persson M. V. (2019) ALMA observations of water deuteration: A physical diagnostic of the formation of protostars. *Astron. Astrophys., 631,* A25.
Jessberger E. K., Christoforidis A., and Kissel J. (1988) Aspects of the major element composition of Halley's dust. *Nature, 332,* 691–695.
Jílková L., Portegies Zwart S., Pijloo T., and Hammer M. (2015) How Sedna and family were captured in a close encounter with a solar sibling. *Mon. Not. R. Astron. Soc., 453,* 3157–3162.
Johansen A., Blum J., Tanaka H., Ormel C., Bizzarro M., and Rickman H. (2014) The multifaceted planetesimal formation process. In *Protostars and Planets VI* (H. Beuther et al., eds.), pp. 547–570. Univ. of Arizona, Tucson.
Johnstone D. and Bally J. (1999) JCMT/SCUBA submillimeter wavelength imaging of the integral-shaped filament in Orion. *Astrophys. J. Lett., 510,* L49–L53.
Jones A. P., Duley W. W., and Williams D. A. (1990) The structure and evolution of hydrogenated amorphous carbon grains and mantles in the interstellar medium. *Q. J. R. Astron. Soc., 31,* 567–582.
Jones A. P., Tielens A. G. G. M., Hollenbach D. J., and McKee C. F. (1994) Grain destruction in shocks in the interstellar medium. *Astrophys. J., 433,* 797.
Jones A. P., Fanciullo L., Köhler M., Verstraete L., Guillet V., Bocchio M., and Ysard N. (2013) The evolution of amorphous hydrocarbons in the ISM: Dust modelling from a new vantage point. *Astron. Astrophys., 558,* A62.
Jones A. P., Ysard N., Köhler M., Fanciullo L., Bocchio M., Micelotta E., Verstraete L., and Guillet V. (2014) The cycling of carbon into and out of dust. *Faraday Discussions, 168,* 313.
Kama M., Shorttle O., Jermyn A. S., Folsom C. P., Furuya K., Bergin E. A., Walsh C., and Keller L. (2019) Abundant refractory sulfur in protoplanetary disks. *Astrophys. J., 885,* 114.
Kastner J. H. and Myers P. C. (1994) An observational estimate of the probability of encounters between mass-losing evolved stars and molecular clouds. *Astrophys. J., 421,* 605.
Kemper F., Vriend W. J., and Tielens A. G. G. M. (2004) The absence of crystalline silicates in the diffuse interstellar medium. *Astrophys. J., 609,* 826–837.
Kemper F., Vriend W. J., and Tielens A. G. G. M. (2005) Erratum: The absence of crystalline silicates in the diffuse interstellar medium (ApJ, 609, 826 [2004]). *Astrophys. J., 633,* 534–534.
Kenyon S. J. and Bromley B. C. (2004) Stellar encounters as the origin of distant solar system objects in highly eccentric orbits. *Nature,*

432, 598–602.
Kerridge J. F. and Matthews M. S., eds. (1988) *Meteorites and the Early Solar System.* Univ. of Arizona, Tucson. 1269 pp.
King A. J., Phillips K. J. H., Strekopytov S., Vita-Finzi C., and Russell S. S. (2020) Terrestrial modification of the Ivuna meteorite and a reassessment of the chemical composition of the CI type specimen. *Geochim. Cosmochim. Acta, 268*, 73–89.
Kirk H., Friesen R. K., Pineda J. E., Rosolowsky E., Offner S. S. R., Matzner C. D., Myers P. C., Di Francesco J., Caselli P., Alves F. O., Chacón-Tanarro A., Chen H.-H., Chun-Yuan Chen M., Keown J., Punanova A., Seo Y. M., Shirley Y., Ginsburg A., Hall C., Singh A., Arce H. G., Goodman A. A., Martin P., and Redaelli E. (2017) The Green Bank Ammonia Survey: Dense cores under pressure in Orion A. *Astrophys. J., 846*, 144.
Kissel J. and Krueger F. R. (1987) The organic component in dust from comet Halley as measured by the PUMA mass spectrometer on board Vega 1. *Nature, 326*, 755–760.
Kleine T., Budde G., Burkhardt C., Kruijer T. S., Worsham E. A., Morbidelli A., and Nimmo F. (2020) The non-carbonaceous carbonaceous meteorite dichotomy. *Space Sci. Rev., 216*, 55.
Krasnokutski S. A., Rouillé G., Jäger C., Huisken F., Zhukovska S., and Henning T. (2014) Formation of silicon oxide grains at low temperature. *Astrophys. J., 782*, 15.
Kristensen L. E. and Dunham M. M. (2018) Protostellar half-life: New methodology and estimates. *Astron. Astrophys., 618*, A158.
Kristensen L. E., van Dishoeck E. F., Benz A. O., Bruderer S., Visser R., and Wampfler S. F. (2013) Observational evidence for dissociative shocks in the inner 100 au of low-mass protostars using Herschel-HIFI. *Astron. Astrophys., 557*, A23.
Krot A. N. (2019) Refractory inclusions in carbonaceous chondrites: Records of early solar system processes. *Meteoritics & Planet. Sci., 54*, 1647–1691.
Kruijer T. S., Burkhardt C., Budde G., and Kleine T. (2017) Age of Jupiter inferred from the distinct genetics and formation times of meteorites. *Proc. Natl. Acad. Sci. U.S.A., 114*, 6712–6716.
Kwok S. and Zhang Y. (2011) Mixed aromatic-aliphatic organic nanoparticles as carriers of unidentified infrared emission features. *Nature, 479*, 80–83.
Lada C. J. and Lada E. A. (2003) Embedded clusters in molecular clouds. *Annu. Rev. Astron. Astrophys., 41*, 57–115.
Ladjelate B., André P., Könyves V., Ward-Thompson D., Men'shchikov A., Bracco A., Palmeirim P., Roy A., Shimajiri Y., Kirk J. M., Arzoumanian D., Benedettini M., Di Francesco J., Fiorellino E., Schneider N., Pezzuto S., Motte F., and the Herschel Gould Belt Survey Team (2020) The Herschel view of the dense core population in the Ophiuchus molecular cloud. *Astron. Astrophys., 638*, A74.
Langer W. D., Graedel T. E., Frerking M. A., and Armentrout P. B. (1984) Carbon and oxygen isotope fractionation in dense interstellar clouds. *Astrophys. J., 277*, 581–604.
Larsen K. K., Wielandt D., Schiller M., Krot A. N., and Bizzarro M. (2020) Episodic formation of refractory inclusions in the solar system and their presolar heritage. *Earth Planet. Sci. Lett., 535*, 116088.
Lee J.-E. and Bergin E. A. (2015) The D/H ratio of water ice at low temperatures. *Astrophys. J., 799*, 104.
Lee J.-E., Bergin E. A., and Lyons J. R. (2008) Oxygen isotope anomalies of the Sun and the original environment of the solar system. *Meteoritics & Planet. Sci., 43*, 1351–1362.
Lee T., Papanastassiou D. A., and Wasserburg G. J. (1976) Demonstration of ^{26}Mg excess in Allende and evidence for ^{26}Al. *Geophys. Res. Lett., 3*, 41–44.
Levison H. F. and Dones L. (2007) Comet populations and cometary dynamics. In *Encyclopedia of the Solar System, second edition* (L. McFadden et al., eds.), pp. 575–588. Elsevier, Oxford.
Levison H. F., Bottke W. F., Gounelle M., Morbidelli A., Nesvorný D., and Tsiganis K. (2009) Contamination of the asteroid belt by primordial trans-neptunian objects. *Nature, 460*, 364–366.
Li A. and Greenberg J. M. (1997) A unified model of interstellar dust. *Astron. Astrophys., 323*, 566–584.
Li D., Goldsmith P. F., and Menten K. (2003) Massive quiescent cores in Orion. I. Temperature structure. *Astrophys. J., 587*, 262–277.
Li G. and Adams F. C. (2015) Cross-sections for planetary systems interacting with passing stars and binaries. *Mon. Not. R. Astron. Soc., 448*, 344–363.
Li J., Bergin E. A., Blake G. A., Ciesla F. J., and Hirschmann M. M. (2021) Earth's carbon deficit caused by early loss through irreversible sublimation. *Sci. Adv., 7*, eabd3632.
Lisse C., Bauer J., Cruikshank D., Emery J., Fernández Y., Fernández-Valenzuela E., Kelley M., McKay A., Reach W., Pendleton Y., Pinilla-Alonso N., Stansberry J., Sykes M., Trilling D. E., Wooden D., Harker D., Gehrz R., and Woodward C. (2020) Spitzer's solar system studies of comets, centaurs and Kuiper belt objects. *Nature Astron., 4*, 930–939.
Lisse C. M., VanCleve J., Adams A. C., A'Hearn M. F., Fernández Y. R., Farnham T. L., Armus L., Grillmair C. J., Ingalls J., Belton M. J. S., Groussin O., McFadden L. A., Meech K. J., Schultz P. H., Clark B. C., Feaga L. M., and Sunshine J. M. (2006) Spitzer spectral observations of the Deep Impact ejecta. *Science, 313*, 635–640.
Lisse C. M., Kraemer K. E., Nuth J. A., Li A., and Joswiak D. (2007) Comparison of the composition of the Tempel 1 ejecta to the dust in Comet C/Hale Bopp 1995 O1 and YSO HD 100546. *Icarus, 191*, 223–240.
Liu F., Asplund M., Yong D., Meléndez J., Ramírez I., Karakas A. I., Carlos M., and Marino A. F. (2016) The chemical compositions of solar twins in the open cluster M67. *Mon. Not. R. Astron. Soc., 463*, 696–704.
Liu M.-C. (2017) The initial ^{41}Ca/^{40}Ca ratios in two type A Ca-Al-rich inclusions: Implications for the origin of short-lived ^{41}Ca. *Geochim. Cosmochim. Acta, 201*, 123–135.
Liu M.-C., Han J., Brearley A. J., and Hertwig A. T. (2019) Aluminum-26 chronology of dust coagulation and early solar system evolution. *Sci. Adv., 5*, eaaw3350.
Loison J.-C., Wakelam V., Gratier P., Hickson K. M., Bacmann A., Agúndez M., Marcelino N., Cernicharo J., Guzman V., Gerin M., Goicoechea J. R., Roueff E., Petit F. L., Pety J., Fuente A., and Riviere-Marichalar P. (2019) Oxygen fractionation in dense molecular clouds. *Mon. Not. R. Astron. Soc., 485*, 5777–5789.
Loison J.-C., Wakelam V., Gratier P., and Hickson K. M. (2020) Gas-grain model of carbon fractionation in dense molecular clouds. *Mon. Not. R. Astron. Soc., 498*, 4663–4679.
Lucas R. and Liszt H. S. (2002) Comparative chemistry of diffuse clouds III. Sulfur-bearing molecules. *Astron. Astrophys., 384*, 1054–1061.
Lugaro M., Ott U., and Kereszturi Á. (2018) Radioactive nuclei from cosmochronology to habitability. *Prog. Part. Nucl. Phys., 102*, 1–47.
Lyons J. R. and Young E. D. (2005) CO self-shielding as the origin of oxygen isotope anomalies in the early solar nebula. *Nature, 435*, 317–320.
MacPherson G. J., Krot A. N., Kita N. T., Bullock E. S., Nagashima K., Ushikubo T., and Ivanova M. A. (2022) The formation of type B CAIs: Evolution from type A CAIs. *Geochim. Cosmochim. Acta, 321*, 343–374.
Makide K., Nagashima K., Krot A. N., Huss G. R., Hutcheon I. D., Hellebrand E., and Petaev M. I. (2013) Heterogeneous distribution of ^{26}Al at the birth of the solar system: Evidence from corundum-bearing refractory inclusions in carbonaceous chondrites. *Geochim. Cosmochim. Acta, 110*, 190–215.
Maret S., Maury A. J., Belloche A., Gaudel M., André P., Cabrit S., Codella C., Lefévre C., Podio L., Anderl S., Gueth F., and Hennebelle P. (2020) Searching for kinematic evidence of Keplerian disks around Class 0 protostars with CALYPSO. *Astron. Astrophys., 635*, A15.
Martínez-Barbosa C. A., Brown A. G. A., Boekholt T., Portegies Zwart S., Antiche E., and Antoja T. (2016) The evolution of the Sun's birth cluster and the search for the solar siblings with Gaia. *Mon. Not. R. Astron. Soc., 457*, 1062–1075.
Marty B. (2012) The origins and concentrations of water, carbon, nitrogen and noble gases on Earth. *Earth Planet. Sci. Lett., 313*, 56–66.
Marty B., Altwegg K., Balsiger H., Bar-Nun A., Bekaert D. V., Berthelier J. J., Bieler A., Briois C., Calmonte U., Combi M., De Keyser J., Fiethe B., Fuselier S. A., Gasc S., Gombosi T. I., Hansen K. C., Hässig M., Jäckel A., Kopp E., Korth A., Le Roy L., Mall U., Mousis O., Owen T., Rème H., Rubin M., Sémon T., Tzou C. Y., Waite J. H., and Wurz P. (2017) Xenon isotopes in 67P/Churyumov-Gerasimenko show that comets contributed to Earth's atmosphere. *Science, 356*, 1069–1072.
Matese J. J., Whitmire D. P., and Lissauer J. J. (2005) A wide-binary

solar companion as a possible origin of Sedna-like objects. *Earth Moon Planets, 97,* 459–470.

Mathis J. S. (1990) Interstellar dust and extinction. *Annu. Rev. Astron. Astrophys., 28,* 37–70.

Mathis J. S., Rumpl W., and Nordsieck K. H. (1977) The size distribution of interstellar grains. *Astrophys. J., 217,* 425–433.

Matsuura M., Dwek E., Barlow M. J., Babler B., Baes M., Meixner M., Cernicharo J., Clayton G. C., Dunne L., Fransson C., Fritz J., Gear W., Gomez H. L., Groenewegen M. A. T., Indebetouw R., Ivison R. J., Jerkstrand A., Lebouteiller V., Lim T. L., Lundqvist P., Pearson C. P., Roman-Duval J., Royer P., Staveley-Smith L., Swinyard B. M., van Hoof P. A. M., van Loon J. T., Verstappen J., Wesson R., Zanardo G., Blommaert J. A. D. L., Decin L., Reach W. T., Sonneborn G., Van de Steene G. C., and Yates J. A. (2015) A stubbornly large mass of cold dust in the ejecta of supernova 1987A. *Astrophys. J., 800,* 50.

Matzel J. E. P., Ishii H. A., Joswiak D., Hutcheon I. D., Bradley J. P., Brownlee D., Weber P. K., Teslich N., Matrajt G., McKeegan K. D., and MacPherson G. J. (2010) Constraints on the formation age of cometary material from the NASA Stardust mission. *Science, 328,* 483–486.

McCarthy M. C. and McGuire B. A. (2021) Aromatics and cyclic molecules in molecular clouds: A new dimension of interstellar organic chemistry. *J. Phys. Chem. A, 125,* 3231–3243.

McCauley P. I., Saar S. H., Raymond J. C., Ko Y.-K., and Saint-Hilaire P. (2013) Extreme-ultraviolet and X-ray observations of comet Lovejoy (C/2011 W3) in the lower corona. *Astrophys. J., 768,* 161.

McClure M. K., Espaillat C., Calvet N., Bergin E., D'Alessio P., Watson D. M., Manoj P., Sargent B., and Cleeves L. I. (2015) Detections of trans-neptunian ice in protoplanetary disks. *Astrophys. J., 799,* 162.

McGuire B. A., Burkhardt A. M., Kalenskii S., Shingledecker C. N., Remijan A. J., Herbst E., and McCarthy M. C. (2018) Detection of the aromatic molecule benzonitrile (c-C_6H_5CN) in the interstellar medium. *Science, 359,* 202–205.

McGuire B. A., Loomis R. A., Burkhardt A. M., Lee K. L. K., Shingledecker C. N., Charnley S. B., Cooke I. R., Cordiner M. A., Herbst E., Kalenskii S., Siebert M. A., Willis E. R., Xue C., Remijan A. J., and McCarthy M. C. (2021) Detection of two interstellar polycyclic aromatic hydrocarbons via spectral matched filtering. *Science, 371,* 1265–1269.

McKeegan K. D., Chaussidon M., and Robert F. (2000) Incorporation of short-lived ^{10}Be in a calcium-aluminum-rich inclusion from the Allende meteorite. *Science, 289,* 1334–1337.

McKeegan K. D., Aléon J., Bradley J., Brownlee D., Busemann H., Butterworth A., Chaussidon M., Fallon S., Floss C., Gilmour J., Gounelle M., Graham G., Guan Y., Heck P. R., Hoppe P., Hutcheon I. D., Huth J., Ishii H., Ito M., Jacobsen S. B., Kearsley A., Leshin L. A., Liu M.-C., Lyon I., Marhas K., Marty B., Matrajt G., Meibom A., Messenger S., Mostefaoui S., Mukhopadhyay S., Nakamura-Messenger K., Nittler L., Palma R., Pepin R. O., Papanastassiou D. A., Robert F., Schlutter D., Snead C. J., Stadermann F. J., Stroud R., Tsou P., Westphal A., Young E. D., Ziegler K., Zimmermann L., and Zinner E. (2006) Isotopic compositions of cometary matter returned by Stardust. *Science, 314,* 1724.

Megeath S. T., Gutermuth R. A., and Kounkel M. A. (2022) Low mass stars as tracers of star and cluster formation. *Publ. Astron. Soc. Pacific, 134,* DOI: 10.1088/1538-3873/ac4c9c.

Meyer B. S. and Clayton D. D. (2000) Short-lived radioactivities and the birth of the Sun. *Space Sci. Rev., 92,* 133–152.

Micelotta E. R., Jones A. P., and Tielens A. G. G. M. (2010) Polycyclic aromatic hydrocarbon processing in interstellar shocks. *Astron. Astrophys., 510,* A36.

Micelotta E. R., Jones A. P., and Tielens A. G. G. M. (2011) Polycyclic aromatic hydrocarbon processing by cosmic rays. *Astron. Astrophys., 526,* A52.

Micelotta E. R., Matsuura M., and Sarangi A. (2018) Dust in supernovae and supernova remnants II: Processing and survival. *Space Sci. Rev., 214,* 53.

Millar T. J., Bennett A., and Herbst E. (1989) Deuterium fractionation in dense interstellar clouds. *Astrophys. J., 340,* 906.

Min M., Bouwman J., Dominik C., Waters L. B. F. M., Pontoppidan K. M., Hony S., Mulders G. D., Henning T., van Dishoeck E. F., Woitke P., Evans I., Neal J., and the DIGIT Team (2016) The abundance and thermal history of water ice in the disk surrounding HD 142527 from the DIGIT Herschel Key Program. *Astron. Astrophys., 593,* A11.

Mishra A. and Li A. (2015) Probing the role of carbon in the interstellar ultraviolet extinction. *Astrophys. J., 809,* 120.

Mitchell T. R. and Stewart G. R. (2011) Photoevaporation as a truncation mechanism for circumplanetary disks. *Astron. J., 142,* 168.

Molster F. J., Waters L. B. F. M., Tielens A. G. G. M., Koike C., and Chihara H. (2002) Crystalline silicate dust around evolved stars. III. A correlations study of crystalline silicate features. *Astron. Astrophys., 382,* 241–255.

Monga N. and Desch S. (2015) External photoevaporation of the solar nebula: Jupiter's noble gas enrichments. *Astrophys. J., 798,* 9.

Moore N. W. H., Li G., and Adams F. C. (2020) Inclination excitation of solar system debris disk due to stellar flybys. *Astrophys. J., 901,* 92.

Morbidelli A. and Levison H. F. (2004) Scenarios for the origin of the orbits of the trans-neptunian objects 2000 CR105 and 2003 VB12 (Sedna). *Astron. J., 128,* 2564–2576.

Motte F., Andre P., and Neri R. (1998) The initial conditions of star formation in the rho Ophiuchi main cloud: Wide-field millimeter continuum mapping. *Astron. Astrophys., 336,* 150–172.

Motte F., Bontemps S., and Louvet F. (2018) High-mass star and massive cluster formation in the Milky Way. *Annu. Rev. Astron. Astrophys., 56,* 41–82.

Mumma M. J. and Charnley S. B. (2011) The chemical composition of comets: Emerging taxonomies and natal heritage. *Annu. Rev. Astron. Astrophys., 49,* 471–524.

Murillo N. M., Lai S.-P., Bruderer S., Harsono D., and van Dishoeck E. F. (2013) A Keplerian disk around a Class 0 source: ALMA observations of VLA1623A. *Astron. Astrophys., 560,* A103.

Myers P. C. (2009) Filamentary structure of star-forming complexes. *Astrophys. J., 700,* 1609–1625.

Najita J. R. and Bergin E. A. (2018) Protoplanetary disk sizes and angular momentum transport. *Astrophys. J., 864,* 168.

Nakamura T., Noguchi T., Tsuchiyama A., Ushikubo T., Kita N. T., Valley J. W., Zolensky M. E., Kakazu Y., Sakamoto K., Mashio E., Uesugi K., and Nakano T. (2008) Chondrule-like objects in short-period comet 81P/Wild 2. *Science, 321,* 1664.

Nanne J. A. M., Nimmo F., Cuzzi J. N., and Kleine T. (2019) Origin of the non-carbonaceous-carbonaceous meteorite dichotomy. *Earth Planet. Sci. Lett., 511,* 44–54.

Nesvorný D. (2018) Dynamical evolution of the early solar system. *Annu. Rev. Astron. Astrophys., 56,* 137–174.

Nesvorný D., Jenniskens P., Levison H. F., Bottke W. F., Vokrouhlický D., and Gounelle M. (2010) Cometary origin of the zodiacal cloud and carbonaceous micrometeorites: Implications for hot debris disks. *Astrophys. J., 713,* 816–836.

Nesvorný D., Vokrouhlický D., Dones L., Levison H. F., Kaib N., and Morbidelli A. (2017) Origin and evolution of short-period comets. *Astrophys. J., 845,* 27.

Nieva M. F. and Przybilla N. (2012) Present-day cosmic abundances: A comprehensive study of nearby early B-type stars and implications for stellar and galactic evolution and interstellar dust models. *Astron. Astrophys., 539,* A143.

Nittler L. R. and Ciesla F. (2016) Astrophysics with extraterrestrial materials. *Annu. Rev. Astron. Astrophys., 54,* 53–93.

Nuevo M., Milam S. N., Sandford S. A., De Gregorio B. T., Cody G. D., and Kilcoyne A. L. D. (2011) XANES analysis of organic residues produced from the UV irradiation of astrophysical ice analogs. *Adv. Space Res., 48,* 1126–1135.

Nuth I., Joseph A., Abreu N., Ferguson F. T., Glavin D. P., Hergenrother C., Hill H. G. M., Johnson N. M., Pajola M., and Walsh K. (2020) Volatile-rich asteroids in the inner solar system. *Planet. Sci. J., 1,* 82–90.

Öberg K. I. and Bergin E. A. (2021) Astrochemistry and compositions of planetary systems. *Phys. Rept., 893,* 1–48.

Owen J. E., Ercolano B., Clarke C. J., and Alexander R. D. (2010) Radiation-hydrodynamic models of X-ray and EUV photoevaporating protoplanetary discs. *Mon. Not. R. Astron. Soc., 401,* 1415–1428.

Palmeirim P., André P., Kirk J., Ward-Thompson D., Arzoumanian D., Könyves V., Didelon P., Schneider N., Benedettini M., Bontemps S., Di Francesco J., Elia D., Griffin M., Hennemann M., Hill T., Martin P. G., Men'shchikov A., Molinari S., Motte F., Nguyen Luong Q.,

Nutter D., Peretto N., Pezzuto S., Roy A., Rygl K. L. J., Spinoglio L., and White G. L. (2013) Herschel view of the Taurus B211/3 filament and striations: Evidence of filamentary growth? *Astron. Astrophys., 550*, A38.

Paquette J. A., Hornung K., Stenzel O. J., Rynö J., Silen J., Kissel J., and Hilchenbach M. (2017) The $^{34}S/^{32}S$ isotopic ratio measured in the dust of comet 67P/Churyumov-Gerasimenko by Rosetta/COSIMA. *Mon. Not. R. Astron. Soc., 469*, S230–S237.

Paquette J. A., Fray N., Bardyn A., Engrand C., Alexander C. M. O., Siljeström S., Cottin H., Merouane S., Isnard R., Stenzel O. J., Fischer H., Rynö J., Kissel J., and Hilchenbach M. (2021) D/H in the refractory organics of comet 67P/Churyumov-Gerasimenko measured by Rosetta/COSIMA. *Mon. Not. R. Astron. Soc., 504*, 4940–4951.

Parker R. J. and Dale J. E. (2016) Did the solar system form in a sequential triggered star formation event? *Mon. Not. R. Astron. Soc., 456*, 1066–1072.

Parker R. J., Church R. P., Davies M. B., and Meyer M. R. (2014) Supernova enrichment and dynamical histories of solar-type stars in clusters. *Mon. Not. R. Astron. Soc., 437*, 946–958.

Pätzold M., Andert T. P., Hahn M., Barriot J.-P., Asmar S. W., Häusler B., Bird M. K., Tellmann S., Oschlisniok J., and Peter K. (2019) The nucleus of comet 67P/Churyumov-Gerasimenko — Part I: The global view — nucleus mass, mass-loss, porosity, and implications. *Mon. Not. R. Astron. Soc., 483*, 2337–2346.

Pendleton Y. J. and Allamandola L. J. (2002) The organic refractory material in the diffuse interstellar medium: Mid-infrared spectroscopic constraints. *Astrophys. J. Suppl., 138*, 75–98.

Pendleton Y. J., Sandford S. A., Allamandola L. J., Tielens A. G. G. M., and Sellgren K. (1994) Near-infrared absorption spectroscopy of interstellar hydrocarbon grains. *Astrophys. J., 437*, 683–696.

Pfalzner S. (2009) Universality of young cluster sequences. *Astron. Astrophys., 498*, L37–L40.

Pfalzner S. (2013) Early evolution of the birth cluster of the solar system. *Astron. Astrophys., 549*, A82.

Pfalzner S. and Vincke K. (2020) Cradle(s) of the Sun. *Astrophys. J., 897*, 60.

Pfalzner S., Bhandare A., Vincke K., and Lacerda P. (2018) Outer solar system possibly shaped by a stellar fly-by. *Astrophys. J., 863*, 45.

Pineda J. E., Segura-Cox D., Caselli P., Cunningham N., Zhao B., Schmiedeke A., Maureira M. J., and Neri R. (2020) A protostellar system fed by a streamer of 10,500 au length. *Nature Astron., 4*, 1158–1163.

Pinte C., Dent W. R. F., Ménard F., Hales A., Hill T., Cortes P., and de Gregorio-Monsalvo I. (2016) Dust and gas in the disk of HL Tauri: Surface density, dust settling, and dust-to-gas ratio. *Astrophys. J., 816*, 25.

Pirani S., Johansen A., and Mustill A. J. (2021) How the formation of Neptune shapes the Kuiper belt. *Astron. Astrophys., 650*, A161.

Platz T., Nathues A., Schorghofer N., Preusker F., Mazarico E., Schröder S. E., Byrne S., Kneissl T., Schmedemann N., Combe J. P., Schäfer M., Thangjam G. S., Hoffmann M., Gutierrez-Marques P., Landis M. E., Dietrich W., Ripken J., Matz K. D., and Russell C. T. (2016) Surface water-ice deposits in the northern shadowed regions of Ceres. *Nature Astron., 1*, 0007.

Porras A., Christopher M., Allen L., Di Francesco J., Megeath S. T., and Myers P. C. (2003) A catalog of young stellar groups and clusters within 1 kiloparsec of the Sun. *Astron. J., 126*, 1916–1924.

Portegies Zwart S. (2019) The formation of solar-system analogs in young star clusters. *Astron. Astrophys., 622*, A69.

Punanova A., Caselli P., Pon A., Belloche A., and André P. (2016) Deuterium fractionation in the Ophiuchus molecular cloud. *Astron. Astrophys., 587*, A118.

Qasim D., Fedoseev G., Chuang K. J., He J., Ioppolo S., van Dishoeck E. F., and Linnartz H. (2020) An experimental study of the surface formation of methane in interstellar molecular clouds. *Nature Astron., 4*, 781–785.

Ramírez I., Bajkova A. T., Bobylev V. V., Roederer I. U., Lambert D. L., Endl M., Cochran W. D., MacQueen P. J., and Wittenmyer R. A. (2014) Elemental abundances of solar sibling candidates. *Astrophys. J., 787*, 154.

Redaelli E., Bizzocchi L., Caselli P., Harju J., Chacón-Tanarro A., Leonardo E., and Dore L. (2018) $^{14}N/^{15}N$ ratio measurements in prestellar cores with N_2H^+: New evidence of ^{15}N-antifractionation. *Astron. Astrophys., 617*, A7.

Rice T. S., Bergin E. A., Jørgensen J. K., and Wampfler S. F. (2018) Exploring the origins of Earth's nitrogen: Astronomical observations of nitrogen-bearing organics in protostellar environments. *Astrophys. J., 866*, 156.

Roberts H., Herbst E., and Millar T. J. (2003) Enhanced deuterium fractionation in dense interstellar cores resulting from multiply deuterated H_3^+. *Astrophys. J. Lett., 591*, L41–L44.

Röllig M. and Ossenkopf V. (2013) Carbon fractionation in photodissociation regions. *Astron. Astrophys., 550*, A56.

Roueff E., Loison J. C., and Hickson K. M. (2015) Isotopic fractionation of carbon, deuterium, and nitrogen: A full chemical study. *Astron. Astrophys., 576*, A99.

Rouillé G., Jäger C., Krasnokutski S. A., Krebsz M., and Henning T. (2014) Cold condensation of dust in the ISM. *Faraday Discussions, 168*, 449.

Roy A., André P., Palmeirim P., Attard M., Könyves V., Schneider N., Peretto N., Men'shchikov A., Ward-Thompson D., Kirk J., Griffin M., Marsh K., Abergel A., Arzoumanian D., Benedettini M., Hill T., Motte F., Nguyen Luong Q., Pezzuto S., Rivera-Ingraham A., Roussel H., Rygl K. L. J., Spinoglio L., Stamatellos D., and White G. (2014) Reconstructing the density and temperature structure of prestellar cores from Herschel data: A case study for B68 and L1689B. *Astron. Astrophys., 562*, A138.

Ruaud M., Wakelam V., Gratier P., and Bonnell I. A. (2018) Influence of galactic arm scale dynamics on the molecular composition of the cold and dense ISM. I. Observed abundance gradients in dense clouds. *Astron. Astrophys., 611*, A96.

Rubin M., Altwegg K., Balsiger H., Berthelier J.-J., Combi M. R., De Keyser J., Drozdovskaya M., Fiethe B., Fuselier S. A., Gasc S., Gombosi T. I., Hänni N., Hansen K. C., Mall U., Rème H., Schroeder I. R. H. G., Schuhmann M., Sémon T., Waite J. H., Wampfler S. F., and Wurz P. (2019) Elemental and molecular abundances in comet 67P/Churyumov-Gerasimenko. *Mon. Not. R. Astron. Soc., 489*, 594–607.

Ruffle D. P., Hartquist T. W., Caselli P., and Williams D. A. (1999) The sulphur depletion problem. *Mon. Not. R. Astron. Soc., 306*, 691–695.

Sakai N., Sakai T., Hirota T., Watanabe Y., Ceccarelli C., Kahane C., Bottinelli S., Caux E., Demyk K., Vastel C., Coutens A., Taquet V., Ohashi N., Takakuwa S., Yen H.-W., Aikawa Y., and Yamamoto S. (2014) Change in the chemical composition of infalling gas forming a disk around a protostar. *Nature, 507*, 78–80.

Sakamoto N., Seto Y., Itoh S., Kuramoto K., Fujino K., Nagashima K., Krot A. N., and Yurimoto H. (2007) Remnants of the early solar system water enriched in heavy oxygen isotopes. *Science, 317*, 231–233.

Sandford S. A., Allamandola L. J., Tielens A. G. G. M., Sellgren K., Tapia M., and Pendleton Y. J. (1991) The interstellar C-H stretching band near 3.4 microns: Constraints on the composition of organic material in the diffuse interstellar medium. *Astrophys. J., 371*, 607–620.

Sandford S. A., Aléon J., Alexander C. M. O., Araki T., Bajt S., Baratta G. A., Borg J., Bradley J. P., Brownlee D. E., Brucato J. R., Burchell M. J., Busemann H., Butterworth A., Clemett S. J., Cody G., Colangeli L., Cooper G., d'Hendecourt L., Djouadi Z., Dworkin J. P., Ferrini G., Fleckenstein H., Flynn G. J., Franchi I. A., Fries M., Gilles M. K., Glavin D. P., Gounelle M., Grossemy F., Jacobsen C., Keller L. P., Kilcoyne A. L. D., Leitner J., Matrajt G., Meibom A., Mennella V., Mostefaoui S., Nittler L. R., Palumbo M. E., Papanastassiou D. A., Robert F., Rotundi A., Snead C. J., Spencer M. K., Stadermann F. J., Steele A., Stephan T., Tsou P., Tyliszczak T., Westphal A. J., Wirick S., Wopenka B., Yabuta H., Zare R. N., and Zolensky M. E. (2006) Organics captured from comet 81P/Wild 2 by the Stardust spacecraft. *Science, 314*, 1720–1724.

Sarangi A., Matsuura M., and Micelotta E. R. (2018) Dust in supernovae and supernova remnants I: Formation scenarios. *Space Sci. Rev., 214*, 63.

Savage B. D. and Sembach K. R. (1996) Interstellar abundances from absorption-line observations with the Hubble Space Telescope. *Annu. Rev. Astron. Astrophys., 34*, 279–330.

Schramm L. S., Brownlee D. E., and Wheelock M. M. (1989) Major element composition of stratospheric micrometeorites. *Meteoritics, 24*, 99–112.

Schroeder I. R. H. G., Altwegg K., Balsiger H., Berthelier J.-J., De Keyser J., Fiethe B., Fuselier S. A., Gasc S., Gombosi T. I., Rubin

M., Sémon T., Tzou C.-Y., Wampfler S. F., and Wurz P. (2019) $^{16}O/^{18}O$ ratio in water in the coma of comet 67P/Churyumov-Gerasimenko measured with the Rosetta/ROSINA double-focusing mass spectrometer. *Astron. Astrophys., 630,* A29.

Schwamb M. E., Brown M. E., Rabinowitz D. L., and Ragozzine D. (2010) Properties of the distant Kuiper belt: Results from the Palomar Distant Solar System Survey. *Astrophys. J., 720,* 1691–1707.

Scott E. R. D. and Krot A. N. (2014) Chondrites and their components. In *Treatise on Geochemistry, second edition* (A. M. Davis, ed.), pp. 65–137, Elsevier, Oxford.

Siraj A. and Loeb A. (2020) The case for an early solar binary companion. *Astrophys. J. Lett., 899,* L24.

Steinacker J., Andersen M., Thi W. F., Paladini R., Juvela M., Bacmann A., Pelkonen V. M., Pagani L., Lefèvre C., Henning T., and Noriega-Crespo A. (2015) Grain size limits derived from 3.6 μm and 4.5 μm coreshine. *Astron. Astrophys., 582,* A70.

Sugerman B. E. K., Ercolano B., Barlow M. J., Tielens A. G. G. M., Clayton G. C., Zijlstra A. A., Meixner M., Speck A., Gledhill T. M., Panagia N., Cohen M., Gordon K. D., Meyer M., Fabbri J., Bowey J. E., Welch D. L., Regan M. W., and Kennicutt R. C. (2006) Massive-star supernovae as major dust factories. *Science, 313,* 196–200.

Tang H., Liu M.-C., McKeegan K. D., Tissot F. L. H., and Dauphas N. (2017) In situ isotopic studies of the U-depleted Allende CAI Curious Marie: Pre-accretionary alteration and the co-existence of ^{26}Al and ^{36}Cl in the early solar nebula. *Geochim. Cosmochim. Acta, 207,* 1–18.

Taquet V., Charnley S. B., and Sipilä O. (2014) Multilayer formation and evaporation of deuterated ices in prestellar and protostellar cores. *Astrophys. J., 791,* 1.

Tatischeff V., Duprat J., and de Séréville N. (2010) A runaway Wolf-Rayet star as the origin of ^{26}Al in the early solar system. *Astrophys. J. Lett., 714,* L26–L30.

Thomas K. L., Blanford G. E., Keller L. P., Klöck W., and McKay D. S. (1993) Carbon abundance and silicate mineralogy of anhydrous interplanetary dust particles. *Geochim. Cosmochim. Acta, 57,* 1551–1566.

Thrane K., Bizzarro M., and Baker J. A. (2006) Extremely brief formation interval for refractory inclusions and uniform distribution of ^{26}Al in the early solar system. *Astrophys. J. Lett., 646,* L159–L162.

Tielens A. G. G. M. (1983) Surface chemistry of deuterated molecules. *Astron. Astrophys., 119,* 177–184.

Tielens A. G. G. M. (2008) Interstellar polycyclic aromatic hydrocarbon molecules. *Annu. Rev. Astron. Astrophys., 46,* 289–337.

Tobin J. J., Hartmann L., Chiang H.-F., Wilner D. J., Looney L. W., Loinard L., Calvet N., and D'Alessio P. (2012) A ~0.2-solar-mass protostar with a Keplerian disk in the very young L1527 IRS system. *Nature, 492,* 83–85.

Trapman L., Tabone B., Rosotti G., and Zhang K. (2022) Effect of MHD wind-driven disk evolution on the observed sizes of protoplanetary disks. *Astrophys. J., 926,* 61.

Trappitsch R., Boehnke P., Stephan T., Telus M., Savina M. R., Pardo O., Davis A. M., Dauphas N., Pellin M. J., and Huss G. R. (2018) New constraints on the abundance of ^{60}Fe in the early solar system. *Astrophys. J. Lett., 857,* L15.

Trujillo C. A. and Sheppard S. S. (2014) A Sedna-like body with a perihelion of 80 astronomical units. *Nature, 507,* 471–474.

Urey H. C. (1955) The cosmic abundances of potassium, uranium, and thorium and the heat balances of the Earth, the Moon, and Mars. *Proc. Natl. Acad. Sci. U.S.A., 41,* 127–144.

van Boekel R., Min M., Leinert C., Waters L. B. F. M., Richichi A., Chesneau O., Dominik C., Jaffe W., Dutrey A., Graser U., Henning T., de Jong J., Köhler R., de Koter A., Lopez B., Malbet F., Morel S., Paresce F., Perrin G., Preibisch T., Przygodda F., Schöller M., and Wittkowski M. (2004) The building blocks of planets within the 'terrestrial' region of protoplanetary disks. *Nature, 432,* 479–482.

van Dishoeck E. F. and Black J. H. (1988) The photodissociation and chemistry of interstellar CO. *Astrophys. J., 334,* 771.

van Dishoeck E. F., Bergin E. A., Lis D. C., and Lunine J. I. (2014) Water: From clouds to planets. In *Protostars and Planets VI* (H. Beuther et al., eds.), pp. 835–858. Univ. of Arizona, Tucson.

van Dishoeck E. F., Kristensen L. E., Mottram J. C., Benz A. O., Bergin E. A., Caselli P., Herpin F., Hogerheijde M. R., Johnstone D., Liseau R., Nisini B., Tafalla M., van der Tak F. F. S., Wyrowski F., Baudry A., Benedettini M., Bjerkeli P., Blake G. A., Braine J., Bruderer S., Cabrit S., Cernicharo J., Choi Y., Coutens A., de Graauw T., Dominik C., Fedele D., Fich M., Fuente A., Furuya K., Goicoechea J. R., Harsono D., Helmich F. P., Herczeg G. J., Jacq T., Karska A., Kaufman M., Keto E., Lamberts T., Larsson B., Leurini S., Lis D. C., Melnick G., Neufeld D., Pagani L., Persson M., Shipman R., Taquet V., van Kempen T. A., Walsh C., Wampfler S. F., Yıldız U., and the WISH Team (2021) Water in star-forming regions: Physics and chemistry from clouds to disks as probed by Herschel spectroscopy. *Astron. Astrophys., 648,* A24.

Van Kooten E. M. M. E., Wielandt D., Schiller M., Nagashima K., Thomen A., Larsen K. K., Olsen M. B., Nordlund Å., Krot A. N., and Bizzarro M. (2016) Isotopic evidence for primordial molecular cloud material in metal-rich carbonaceous chondrites. *Proc. Natl. Acad. Sci. U.S.A., 113,* 2011–2016.

van't Hoff M. L. R., Harsono D., Tobin J. J., Bosman A. D., van Dishoeck E. F., Jørgensen J. K., Miotello A., Murillo N. M., and Walsh C. (2020) Temperature structures of embedded disks: Young disks in Taurus are warm. *Astrophys. J., 901,* 166.

Visser R., van Dishoeck E. F., Doty S. D., and Dullemond C. P. (2009) The chemical history of molecules in circumstellar disks. I. Ices. *Astron. Astrophys., 495,* 881–897.

Visser R., Bruderer S., Cazzoletti P., Facchini S., Heays A. N., and van Dishoeck E. F. (2018) Nitrogen isotope fractionation in protoplanetary disks. *Astron. Astrophys., 615,* A75.

Volk K. and Malhotra R. (2008) The scattered disk as the source of the Jupiter family comets. *Astrophys. J., 687,* 714–725.

Warren P. (2011) Stable-isotopic anomalies and the accretionary assemblage of the Earth and Mars: A subordinate role for carbonaceous chondrites. *Earth Planet. Sci. Lett., 311,* 93–100.

Watson D., Christensen L., Knudsen K. K., Richard J., Gallazzi A., and Michałowski M. J. (2015) A dusty, normal galaxy in the epoch of reionization. *Nature, 519,* 327–330.

Watson D. M., Leisenring J. M., Furlan E., Bohac C. J., Sargent B., Forrest W. J., Calvet N., Hartmann L., Nordhaus J. T., Green J. D., Kim K. H., Sloan G. C., Chen C. H., Keller L. D., d'Alessio P., Najita J., Uchida K. I., and Houck J. R. (2009) Crystalline silicates and dust processing in the protoplanetary disks of the Taurus young cluster. *Astrophys. J. Suppl., 180,* 84–101.

Watson W. D. (1974) Ion-molecule reactions, molecule formation, and hydrogen-isotope exchange in dense interstellar clouds. *Astrophys. J., 188,* 35–42.

Wdowiak T. J., Flickinger G. C., and Cronin J. R. (1988) Insoluble organic material of the Orgueil carbonaceous chondrite and unidentified infrared bands. *Astrophys. J. Lett., 328,* L75–L79.

Webb J. J., Price-Jones N., Bovy J., Portegies Zwart S., Hunt J. A. S., Mackereth J. T., and Leung H. W. (2020) Searching for solar siblings in APOGEE and Gaia DR2 with N-body simulations. *Mon. Not. R. Astron. Soc., 494,* 2268–2279.

Weidenschilling S. J. and Cuzzi J. N. (1993) Formation of planetesimals in the solar nebula. In *Protostars and Planets III* (E. H. Levy and J. I. Lunine, eds.), pp. 1031–1060. Univ. of Arizona, Tucson.

Weingartner J. C. and Draine B. T. (2001) Dust grain-size distributions and extinction in the Milky Way, large Magellanic cloud, and small Magellanic cloud. *Astrophys. J., 548,* 296–309.

Williams J. P. and Gaidos E. (2007) On the likelihood of supernova enrichment of protoplanetary disks. *Astrophys. J. Lett., 663,* L33–L36.

Wilson T. L. and Rood R. (1994) Abundances in the interstellar medium. *Annu. Rev. Astron. Astrophys., 32,* 191–226.

Winter A. J., Booth R. A., and Clarke C. J. (2018) Evidence of a past disc-disc encounter: HV and DO Tau. *Mon. Not. R. Astron. Soc., 479,* 5522–5531.

Wirström E. S. and Charnley S. B. (2018) Revised models of interstellar nitrogen isotopic fractionation. *Mon. Not. R. Astron. Soc., 474,* 3720–3726.

Wolfire M. G., McKee C. F., Hollenbach D., and Tielens A. G. G. M. (2003) Neutral atomic phases of the interstellar medium in the galaxy. *Astrophys. J., 587,* 278–311.

Woodward C. E., Wooden D. H., Harker D. E., Kelley M. S. P., Russell R. W., and Kim D. L. (2021) The coma dust of comet C/2013 US10 (Catalina): A window into carbon in the solar system. *Planet. Sci. J.,*

2, 25.
Yamamoto D., Kuroda M., Tachibana S., Sakamoto K., and Yurimoto H. (2018) Oxygen isotopic exchange between amorphous silicate and water vapor and its implications for oxygen isotopic evolution in the early solar system. *Astrophys. J.*, *865*, 98.
Yang L., Ciesla F. J., and Alexander C. M. O. D. (2013) The D/H ratio of water in the solar nebula during its formation and evolution. *Icarus*, *226*, 256–267.
Youdin A. N. and Goodman J. (2005) Streaming instabilities in protoplanetary disks. *Astrophys. J.*, *620*, 459–469.
Young E. D. (2016) Bayes' theorem and early solar short-lived radionuclides: The case for an unexceptional origin for the solar system. *Astrophys. J.*, *826*, 129.
Ysard N., Köhler M., Jones A., Dartois E., Godard M., and Gavilan L. (2016) Mantle formation, coagulation, and the origin of cloud/core shine. II. Comparison with observations. *Astron. Astrophys.*, *588*, A44.
Yurimoto H. and Kuramoto K. (2004) Molecular cloud origin for the oxygen isotope heterogeneity in the solar system. *Science*, *305*, 1763–1766.
Zhang M., Defouilloy C., Joswiak D. J., Brownlee D. E., Nakashima D., Siron G., Kitajima K., and Kita N. T. (2021) Oxygen isotope systematics of crystalline silicates in a giant cluster IDP: A genetic link to Wild 2 particles and primitive chondrite chondrules. *Earth Planet. Sci. Lett.*, *564*, 116928.
Zhukovska S., Dobbs C., Jenkins E. B., and Klessen R. S. (2016) Modeling dust evolution in galaxies with a multiphase, inhomogeneous ISM. *Astrophys. J.*, *831*, 147.
Zinner E. (2014) Presolar grains. In *Treatise on Geochemistry, second edition* (A. M. Davis, ed.), pp. 181–213. Elsevier, Oxford.
Zolensky M., Nakamura-Messenger K., Rietmeijer F., Leroux H., Mikouchi T., Ohsumi K., Simon S., Grossman L., Stephan T., Weisberg M., Velbel M., Zega T., Stroud R., Tomeoka K., Ohnishi I., Tomioka N., Nakamura T., Matrajt G., Joswiak D., Brownlee D., Langenhorst F., Krot A., Kearsley A., Ishii H., Graham G., Dai Z. R., Chi M., Bradley J., Hagiya K., Gounelle M., and Bridges J. (2008) Comparing Wild 2 particles to chondrites and IDPs. *Meteoritics & Planet. Sci.*, *43*, 261–272.
Zubko V., Dwek E., and Arendt R. G. (2004) Interstellar dust models consistent with extinction, emission, and abundance constraints. *Astrophys. J. Suppl.*, *152*, 211–249.

Aikawa Y., Okuzumi S., and Pontoppidan K. (2024) The physical and chemical processes in protoplanetary disks: Constraints on the composition of comets. In *Comets III* (K. J. Meech, M. R. Combi, D. Bockelée-Morvan, S. N. Raymond, and M. E. Zolensky, eds.), pp. 33–62. Univ. of Arizona, Tucson, DOI: 10.2458/azu_uapress_9780816553631-ch002.

The Physical and Chemical Processes in Protoplanetary Disks: Constraints on the Composition of Comets

Yuri Aikawa
Department of Astronomy, The University of Tokyo

Satoshi Okuzumi
Department of Earth and Planetary Sciences, Institute of Science Tokyo

Klaus Pontoppidan
Space Telescope Science Institute (now at the Jet Propulsion Laboratory)

We review the recent observations of protoplanetary disks together with relevant theoretical studies with an emphasis on the evolution of volatiles. In the last several years the Atacama Large Millimeter/submillimeter Array (ALMA) provided evidence of grain growth, gas-dust decoupling, and substructures such as rings and gaps in the dust continuum. Molecular line observations revealed radial and vertical distributions of molecular abundances and also provided significant constraints on the gas dynamics such as turbulence. While submillimeter and millimeter observations mainly probe the gas and dust outside the radius of several astronomical units, ice and inner warm gas are investigated at shorter wavelengths. Gas and dust dynamics are key to connect these observational findings. One of the emerging trends is inhomogeneous distributions of elemental abundances, most probably due to dust-gas decoupling.

1. INTRODUCTION

Planetary systems, including our solar system, are formed in protoplanetary disks, which are circumstellar disks around pre-main-sequence stars. Since *Comets II* was published in 2004, new astronomical instruments such as ALMA and the Herschel Space Observatory made dramatic progress in the observation of protoplanetary disks. In particular, ALMA made it possible to observe the thermal emission of dust grains with a spatial resolution of 5 au in the disks in the solar neighborhood (e.g., the Ophiuchus star-forming region). The discovery of the ring-gap structure stimulated theoretical studies on grain growth and planetesimal formation processes. Various gas molecules such as CO, on the other hand, are observed with a spatial resolution of ~15 au. Observations of these molecules provide an important clue to unveil the composition of ice, which, together with dust, is the ingredient of comets. In this chapter, we review the progress of these observational and theoretical studies of protoplanetary disks, which serve as references to investigate the comet formation in the solar nebula.

Before proceeding to the main sections, we briefly explain the process of star formation in order to set up the stage and to clarify the relationship with the chapter in this volume by Bergin et al. Low-mass stars like the Sun are formed by the gravitational collapse of a molecular cloud core. Since the core has non-zero angular momentum as a whole, the central star and its circumstellar disk are formed simultaneously. In the early stages, the star and disk are deeply embedded in dense gas, which we call an envelope. Since the envelope is cold, the spectral energy distribution (SED) of the object has a peak in the far-infrared. The objects in this evolutionary stage are called Class 0. In spite of its low temperature at the outermost radius of the envelope, the temperature of the forming disk would be high due to the high mass accretion rate inside the disk (i.e., release of gravitational energy) and shock heating caused by accretion from the envelope to the disk (e.g., *Offner and McKee*, 2011) (see also Fig. 1 and section 2.2). While the envelope gas accretes to the central star via disk, the high-velocity jet and outflow are launched from the vicinity of the central star. Eventually, the envelope gas decreases as it falls to the disk and star or is swept by the jet and outflow. When the central star (i.e., in near-infrared wavelengths) becomes apparent in the SED — which is, however, still dominated by mid- to far-infrared — the objects are called Class I. Eventually, when the envelope gas is dissipated, the central star becomes a T Tauri star and the disk becomes a protoplanetary disk, which is also called a Class II disk. More detailed explanations and quantitative definitions of Class 0, I, and II are found in *Evans et al.* (2009).

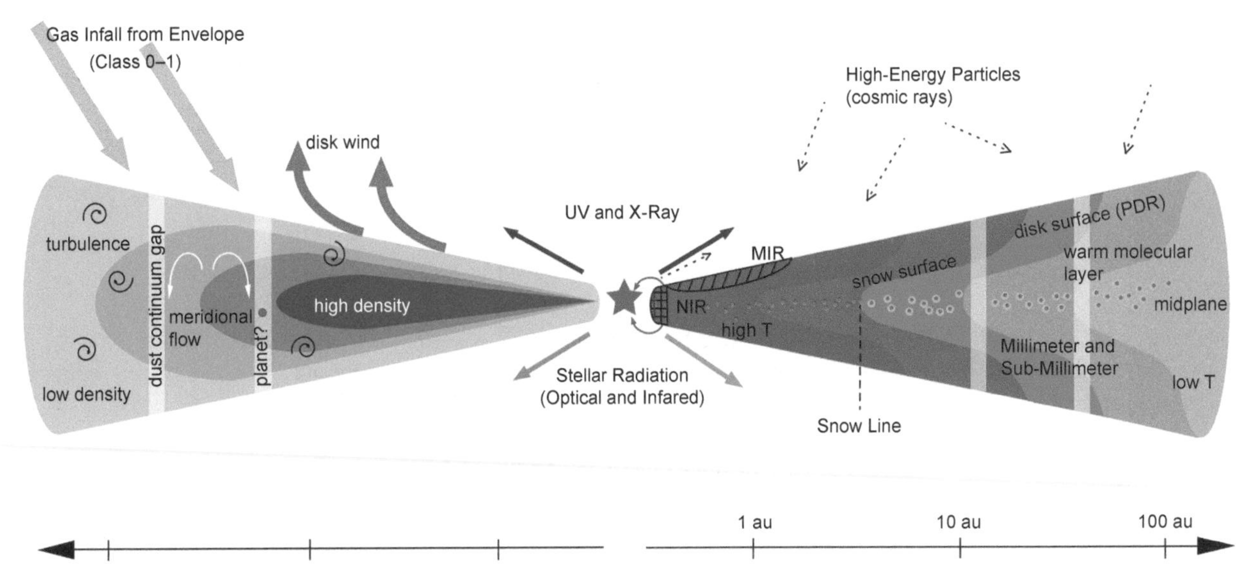

Fig. 1. Schematic view of the disk structure. The physical structures and processes are shown on the left: density distribution (blue contour), weak turbulence, disk wind, gaps seen in dust continuum emission, and meridional flow. In early disk-forming stages (Class 0 and I), we also have gas accretion from the envelope. The temperature distribution (red contour) is depicted on the right, together with the basic chemical structures: disk surface irradiated by UV radiation, warm molecular layer, and midplane. Snow surface is defined as the boundary outside which the dust temperature is lower than the sublimation temperature of a volatile molecule. Snow line is the intersection between the snow surface and the disk midplane. Infrared observations trace hot regions at the inner radius and disk surface, while millimeter and submillimeter observations probe the outer (≥5 au) regions.

While it is not easy to determine the age of pre-main-sequence stars, the typical age of the Class II objects is a few 10^6 yr. Statistical observations (i.e., number counts) indicate that the lifetime for Class 0 and Class I is ~0.1 –0.2 m.y. and ~0.4–0.5 m.y., respectively (e.g., *White et al.,* 2007; *Evans et al.,* 2009). In order to set solar system formation in an astronomical context, these timescales are compared with the range of absolute ages (i.e., formation interval) of calcium-aluminium-rich inclusions (CAIs) and chondrules in meteorites. CAIs have the highest condensation temperature among minerals in chondrites and are thus considered to be formed in the hottest stage and/or region. The formation interval of CAIs is estimated by isotope dating to be 0.16 m.y., which is comparable to the lifetime of the Class 0 stage. The age of the chondrules, on the other hand, varies over a few million years, which coincides with the lifetime of Class II disks (*Connelly et al.,* 2012). While Class II disks have been considered to be the birthplace of planetary systems, a ring-gap structure of dust thermal emission is recently found in some Class I disks as well. This suggests that planetary system formation may start early (see section 4). In summary, the solar nebula may correspond to the evolutionary stages as a whole, from Class 0, to I, and to II.

Having said that, in this chapter, we focus our attention on Class II disks. Chondrule fragments found in the sample return mission of Comet Wild 2 indicate the formation of comets during or after chondrule formation (*Bridges et al.,* 2012), i.e., in Class II disks. The physical and chemical processes and structures in Class II disks are better understood as depicted in Fig. 1 compared with the younger disks, which need more careful analysis to be distinguished from the accreting envelope gas and outflows (e.g., *van Dishoeck et al.,* 2021). In section 2, we describe the physical structure and processes in disks such as temperature distributions and gas dynamics, which are of fundamental importance for chemistry and comet formation in disks. Since comets contain significant amounts of volatiles such as water ice, they are considered to be formed in low-temperature regions outside the snowline. In section 3 we review the basic chemical structure of the disk and recent observations of molecular gas and ice at various wavelengths, as well as the studies on isotope fractionation in disks. The chemical properties of disk material provide a reference for investigations of the formation and evolution of comets based on their volatile composition and isotope ratios. In section 4 we overview two outstanding issues on disk chemistry: the chemical evolution of volatiles from earlier phases (Class 0 and I) to Class II disks, and how and what we can learn about volatiles in solids from line observations of disks and central stars.

2. PHYSICAL PROCESSES IN PROTOPLANETARY DISKS

In this section, we describe the basic physical structure and relevant observational results. The radial temperature distribution in disks determines snowlines, outside of which specific volatile species will be in ice to be incorporated into comets. Vertical temperature structure is also important to interpret the disk observations. The major heating source is stellar irradiation and gravitational energy released by the mass accretion toward the central star. Both the stellar luminosity and mass accretion rate are expected to vary with time, which results in temporal variation of the snowlines. Mass accretion, in turn, is determined by angular momentum transfer within the disk, which could be caused by turbulence and/or disk winds (Fig. 1). We note that angular momentum transfer and turbulence will also determine how the ingredients of comets — ices and grains — are mixed and distributed within the disk. We thus start this section with gas dynamics. Later in this section, we also provide the current overview of dust observations. Observations indicate that dust grains have grown to at least 100 μm and possibly to larger sizes. As grains grow, they decouple from the gas. Large grains, called pebbles, lose or gain angular momentum via gas drag to radially migrate, and are concentrated at a local gas pressure maximum. Such dust rings (and gaps) are found in many disks in the last several years.

2.1. Gas Dynamics

Pre-main-sequence stars accrete material from their surrounding disks. For solar-mass pre-main-sequence stars, the mass accretion rates estimated from observations are on average $\sim 10^{-8}$ $M_{\odot}$ yr^{-1} at a stellar age t of 1 m.y. and crudely scale inversely with t (*Hartmann et al.*, 2016). Some pre-main-sequence stars show months-long and decades-long luminosity eruptions, called Ex Ori and FU Ori outbursts, with estimated accretion rates of 10^{-7} $M_{\odot}$ yr^{-1} and 10^{-5}–10^{-4} $M_{\odot}$ yr^{-1}, respectively (*Hartmann et al.*, 2016; *Fischer et al.,* 2023).

The observed stellar accretion indicates that the inner region of disks loses angular momentum. Which mechanisms are responsible for the angular momentum transport is a longstanding question in the study of protoplanetary disk evolution. Classically, the accretion of protoplanetary disks as well as other astrophysical accretion disks was attributed to outward angular momentum transport within the disks by turbulence (*Lynden-Bell and Pringle*, 1974). However, as we detail below, neither theory nor observations support the picture that protoplanetary disks are strongly turbulent everywhere.

On the theoretical side, it was previously thought that the magnetorotational instability (MRI) (*Balbus and Hawley*, 1991) is a major cause of protoplanetary disk turbulence. The MRI is most likely to operate in the innermost disk regions where the gas is hot ($\gtrsim$1000 K) and thermally well-ionized (*Gammie*, 1996; *Desch and Turner*, 2015). However, farther out in the disks, the gas is only poorly ionized and the magnetohydrodynamics (MHD) is strongly subject to magnetic diffusion (*Sano et al.*, 2000; *Ilgner and Nelson*, 2006; *Wardle*, 2007; *Bai*, 2011). Recent theoretical studies (*Bai*, 2011; *Bai and Stone*, 2013; *Simon et al.*, 2013a,b; *Lesur et al.*, 2014; *Gressel et al.*, 2015; *Bai*, 2017) have shown that while Ohmic diffusion suppresses the MRI near the disk midplane, ambipolar diffusion (which is a type of magnetic diffusion occurring in low-density regions) also suppresses the MRI well above the midplane. In the inner few astronomical units of the disks, the combined effect of Ohmic and ambipolar diffusion can quench the MRI at all heights (*Bai*, 2013).

From the observational side, direct constraints on disk turbulence strength have been obtained from the measurements of nonthermal Doppler broadening of molecular emission lines (*Hughes et al.*, 2011; *Guilloteau et al.*, 2012). Recently, there have been several attempts to detect non-thermal gas motions of disks using ALMA, but most of them have resulted in non-detection with upper limits on the non-thermal velocity dispersion of several to ten percent of the local sound speed (*Flaherty et al.*, 2015, 2017, 2018; *Teague et al.*, 2016, 2018). To date, the disks around DM Tau (*Flaherty et al.,* 2020) and IM Lup (*Paneque-Carreño et al.,* 2024; *Flaherty et al.,* 2024) are the only cases for which turbulent motion has been detected with ALMA, albeit at a few gas scale heights above the midplane. There are indirect constraints on turbulence strength at the midplane from the morphology or azimuthal emission variation of dust rings and gaps seen in ALMA millimeter continuum images (*Pinte et al.*, 2016; *Dullemond et al.*, 2018; *Rosotti et al.*, 2020; *Doi and Kataoka*, 2021) (see section 2.5 for more detail about disk substructures found by millimeter observations). Some rings show an indication of strong vertical settling or radial dust concentration, suggesting that the turbulence at the midplane of the disks is too weak to diffuse the dust particles. However, constraining the level of turbulence quantitatively from these observations requires additional constraints on the dust particle size because large dust particles may settle and concentrate even in the presence of strong turbulence.

While the turbulence is found to be weak, we still need angular momentum transport to account for the mass accretion from the disk to the central star. Various mechanisms are proposed and investigated. The Keplerian shear of the gas disk amplifies large-scale horizontal magnetic fields even when Ohmic diffusion suppresses the MRI (*Turner and Sano*, 2008). The Keplerian shear, when coupled to the Hall drift of magnetic fields, also leads to exponential amplification of horizontal magnetic fields, the phenomenon called the Hall-shear instability (*Kunz*, 2008), which becomes prominent in inner disk regions (*Bai*, 2014, 2017; *Lesur et al.*, 2014). Importantly, MHD simulations show that these mechanisms produce coherent horizontal magnetic fields with no appreciable level of turbulence. In other words, they induce angular momentum transport leading to large-scale disk accretion but would not cause small-scale mixing of disk material.

Large-scale magnetic fields threading the disks induce another important dynamical phenomenon: disk winds. Protoplanetary disks are thought to inherit magnetic flux from their parent molecular clouds. Depending on the strength and inclination of the threading magnetic fields, the centrifugal force along the field lines or the magnetic field pressure in the vertical direction can accelerate the material on the disk surface to the escape velocity (*Bland-*

ford and Payne, 1982; *Shibata and Uchida*, 1986). The magnetically driven winds also induce gas accretion within the disks because the magnetically accelerated wind material takes away the disks' angular momentum. Recent MHD simulations have shown that the wind-driven accretion alone can account for the observed accretion rates for protoplanetary disks if the magnetic fields vertically threading the disks are sufficiently strong (*Bai and Stone*, 2013; *Bai*, 2013; *Simon et al.*, 2013a; *Lesur et al.*, 2014; *Gressel et al.*, 2015). However, lacking direct measurements of the strength of disks' large-scale magnetic fields [although there are some constraints on solar nebula magnetic field strengths from paleomagnetic measurements of meteorites; see *Fu et al.* (2014) and *Wang et al.* (2017)], it is unclear whether realistic protoplanetary disks can retain the required amount of the vertical magnetic flux. There are several recent theoretical studies in this direction (*Guilet and Ogilvie*, 2014; *Okuzumi et al.*, 2014; *Takeuchi and Okuzumi*, 2014; *Bai and Stone*, 2017; *Zhu and Stone*, 2018; *Leung and Ogilvie*, 2019).

There are purely hydrodynamical (i.e, non-MHD) instabilities that can produce weak turbulence, including the vertical shear instability (*Urpin and Brandenburg*, 1998; *Nelson et al.*, 2013), convective overstability (*Klahr and Hubbard*, 2014; *Lyra*, 2014), and zombie vortex instability (*Marcus et al.*, 2015). These hydrodynamical instabilities operate in different ranges of the gas cooling timescale and hence in different disk regions, with the vertical shear instability being most relevant to cold outer disk regions where comets form (*Malygin et al.*, 2017; *Lyra and Umurhan*, 2019). According to hydrodynamical simulations, turbulence driven by these instabilities does transfer the disk's angular momentum radially, but its transport efficiency appears to be low compared to strong MRI-driven turbulence and magnetically driven winds (*Lyra and Umurhan*, 2019). This does not mean that hydrodynamical instabilities are negligible, because they can play a significant role in the transport of disk material. For instance, the vertical shear instability produces vertically elongated turbulent eddies that strongly diffuse gas and dust in the vertical direction (*Flock et al.*, 2020). The hydrodynamical instabilities also cause the formation of gas pressure bumps and vortices that can efficiently trap dust particles (*Flock et al.*, 2020; *Raettig et al.*, 2021).

Massive planets also have interesting effects on disk gas dynamics. Early theoretical studies already predicted that a planet larger than Neptune can carve an annular density deficit called a gap in the background gas disk (*Lin and Papaloizou*, 1993). Recent three-dimensional simulations have shown that a massive planet also induces meridional (i.e., radial and vertical) flows of the disk gas both inside and outside the gap (*Morbidelli et al.*, 2014; *Fung and Chiang*, 2016) (Fig. 1). Signatures of the meridional flow are obtained from a detailed analysis of high-resolution molecular line data (*Teague et al.*, 2019). The gaps and meridional flows induced by massive planets have important implications for dust evolution (see section 2.3).

2.2. Temperature Distribution and Snowlines

As in the present-day solar system, the temperature in protoplanetary disks generally decreases with orbital radius. Therefore, each chemical species in a disk has a critical radius inside which its solid form is unstable. For volatile species, these critical radii are called the snowlines. Because different volatile species have different sublimation temperatures, each of them has its own snowline. (Considering the temperature gradient in the vertical direction, the boundary is actually a snow surface. The snowline is an intersection between the snowsurface and the disk midplane as depicted in Fig 1. We focus on the snowline here, because the midplane dominates in the mass distribution.)

The snowlines are generally expected to affect the radial distribution of volatiles in solids and also in the gas (*Hayashi*, 1981; *Öberg et al.*, 2011). If we fully understand where the snowlines are and how they move with time, we will be able to constrain where and when planets and comets of different compositions form in disks.

Themodynamically, the snowline of each volatile species can be defined as the location where its saturation vapor pressure equals its partial pressure in the gas phase. To illustrate how the locations of the snowlines are determined, we plot in Fig. 2 the saturation vapor pressure curves for pure CO, CO_2, NH_3, and H_2O ices as a function of radial distance r for the classical, optically thin minimum-mass solar nebula (MMSN) model of *Hayashi* (1981), which has radial temperature and midplane pressure profiles of $T_{MMSN} = 280(r/1\ au)^{-1/2}$ K and $P_{MMSN} = 1.4(r/1\ au)^{-13/4}$ Pa, respectively. Here, we adopt

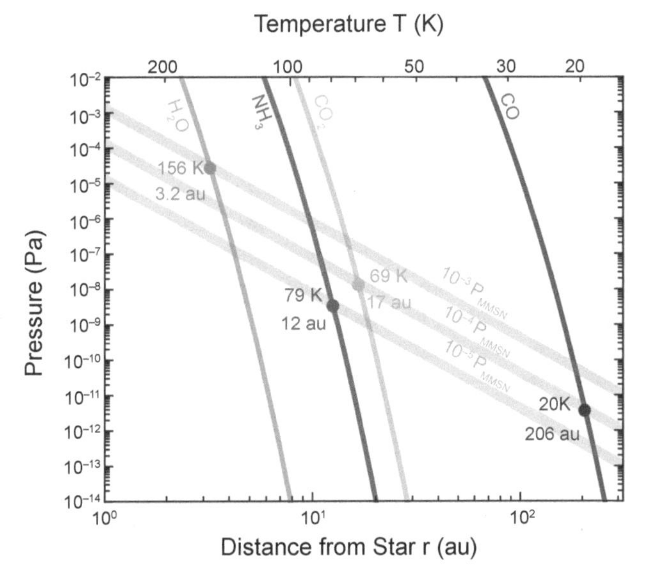

Fig. 2. Saturation vapor pressures for pure CO, CO_2, NH_3, and H_2O ices (thin lines) as a function of temperature T (indicated on the top edge). For reference, the thick lines show how the temperature and 0.1, 0.01, and 0.001% of the total midplane gas pressure vary with orbital radius (indicated on the bottom edge) in a radially extended version of the classical, optically thin minimum-mass solar nebula (*Hayashi*, 1981). The circles mark the locations of the snowlines.

analytic expressions for the saturation vapor pressures as a function of temperature provided by *Bauer et al.* (1997) for H_2O and by *Yamamoto et al.* (1983) for the other species and plot them as a function of r using the assumed temperature profile. Let us assume that the disk contains CO, CO_2, NH_3, and H_2O at spatially uniform molar abundances of $f = 10^{-4}$, 10^{-4}, 10^{-5}, and 10^{-3}, respectively, so that the relative abundances between the four volatiles are crudely consistent with the abundances in comets (*Mumma and Charnley,* 2011). If all the volatiles were in the gas phase, they would have partial pressures of f P_{MMSN}. By comparing the saturation vapor pressures and assumed partial pressures, we find that the snowlines of CO, CO_2, NH_3, and H_2O are located at $r \approx 206$, 17, 12, and 3.2 au, with sublimation temperatures of $\approx$20, 69, 79, and 156 K, respectively, in this particular disk model (see the circles in Fig. 2). [The abundance of H_2O relative to H_2 is set to 10^{-3} assuming that H_2O is the dominant reservoir of oxygen (*Lodders,* 2003). If water is mostly inherited from molecular clouds, H_2O abundance would be 10^{-4} (*Whittet,* 1993). Then the water snowline is slightly shifted outwards. We also note that the ices would be a mixture of various molecules, while we consider pure ices here for simplicity. Sublimation of mixed ice is more complicated than that of pure ice (e.g., see the review by *Hama and Watanabe*, 2013; *Minissale et al.*, 2022). For example, laboratory experiments show that the adsorption energy and thus the sublimation temperature of a molecule depends on the composition and surface structure (e.g., crystal or amorphous) of the substrate ice. Referring to the observations of comets and interstellar ice, H_2O is expected to be a dominant composition of ice in disks. Then a fraction of molecule with higher volatility, such as CO and CO_2, would be trapped in H_2O ice and desorb at the H_2O snowline.]

The strong dependence of the saturation vapor pressures on temperature illustrated in Fig. 2 clearly indicates that realistic modeling of the disk temperature structure is key to accurately inferring the snowline locations. Unfortunately, the simple optically thin disk model adopted above does not apply to protoplanetary disks, which are mostly optically thick to radiation from the central star; in the MMSN, for instance, an optical thickness to stellar radiation can be as high as $\sim 10^5$. Optically thick disks can receive stellar radiation only at their surfaces, and hence their interior temperatures tend to be lower than in optically thin disks unless any internal heat source is present. Specifically, for an optically thick disk passively heated by a central star of luminosity L_* and mass M_*, the temperature of dust grains at the optically thin surface and optically thick midplane can be estimated as (*Kusaka et al.*, 1970; *Chiang and Goldreich*, 1997)

$$T_{irr,surf} \approx 490 \left(\frac{\varepsilon}{0.1}\right)^{-1/4} \left(\frac{L_*}{L_\odot}\right)^{1/4} \left(\frac{r}{1\ au}\right)^{-1/2} K \qquad (1)$$

$$T_{irr,mid} \approx 120 \left(\frac{L_*}{L_\odot}\right)^{2/7} \left(\frac{M_*}{M_\odot}\right)^{1/7} \left(\frac{r}{1\ au}\right)^{-3/7} K \qquad (2)$$

respectively, where ε is the ratio of the infrared to visible absorption cross sections of the grains at the disk surface. The temperature profile T_{MMSN} of the MMSN disk model is equivalent to $T_{irr,surf}$ with $\varepsilon = 1$ and $L_* = 1\ L_\odot$, where the value $\varepsilon = 1$ applies to dust particles larger than micrometers. We here assume that only submicrometer-sized grains remain in the surface region and adopt $\varepsilon \sim 0.1$ (Isella and Natta, 2005). The estimate for $T_{irr,mid}$ depends weakly on the height of the surface where the starlight is finally absorbed (*Chiang and Goldreich*, 1997). The scale height, and thus $T_{irr,mid}$ depends on the mass of the central star. In equation (2), we have assumed that the starlight absorption surface lies at three scale heights above the midplane.

The large temperature difference between the midplane and surface regions of the disks has significant implications for the compositional distribution of solids in the disks. Figure 3a illustrates how the locations of the snowlines of some major volatiles at the surface (here referring to three scale heights above the midplane) and at the midplane evolve in an optically thick, passively irradiated disk (see also the snow surface in Fig. 1). Here, the central star is assumed to have a fixed mass of $M_* = 1\ M_\odot$, while the stellar luminosity is assumed to decrease from 10 $L_\odot$ at t = 0.1 m.y. to 0.5 $L_\odot$ at t = 10 m.y. in a power-law fashion, mimicking the evolution of solar-mass pre-main-sequence stars. For simplicity, the abundances of the volatiles are assumed to be spatially uniform and equal to the values used in Fig. 2, and the surface density profile of the MMSN is used to calculate the disk pressure distribution. We find from the lower panel of Fig. 3a that the H_2O snowline at the midplane would move inside 1 au — the radius of Earth's current orbit — within the first million years of star and planet formation. In contrast, the H_2O snowline in the optically thin surface region lies farther away, at more than 10 au from the central star (see the upper panel of Fig. 3a). This difference arises mainly from the vertical temperature (rather than pressure) gradient.

The midplane temperature given by equation (2) should be taken as a lower limit because we have neglected any internal heat sources. The classical accretion disk model that assumes vertically uniform turbulent viscosity (*Lynden-Bell and Pringle*, 1974) predicts that the gravitational energy liberated by mass accretion toward the star can significantly heat the disk interior, in particular in the early disk evolutionary phase where both the accretion rate and disk surface density are high. Figure 3b illustrates the evolution of the midplane temperature for a disk heated by internal viscous heating in addition to stellar radiation. We consider a solar-mass star and let its luminosity evolve as in Fig. 3a. The disk accretion rate is assumed to decay as $\dot{M} = 4 \times 10^{-8}(t/1\ m.y.)^{-1.07}\ M_\odot\ yr^{-1}$ (*Hartmann et al.*, 2016). The disk opacity is taken to be $2(T/100\ K)^2\ cm^2\ g^{-1}$ (*Bell and Lin*, 1994), and the disk's turbulent viscosity is taken to be 1% of the local sound velocity times the gas scale height. The figure shows that the viscous heating pushes the H_2O snowline out beyond 1 au during the entire Class II phase ($t \lesssim$ several million years). Note that the temperature evolution illustrated here depends on the disk opacity assumed, and hence on the evolution of dust grains

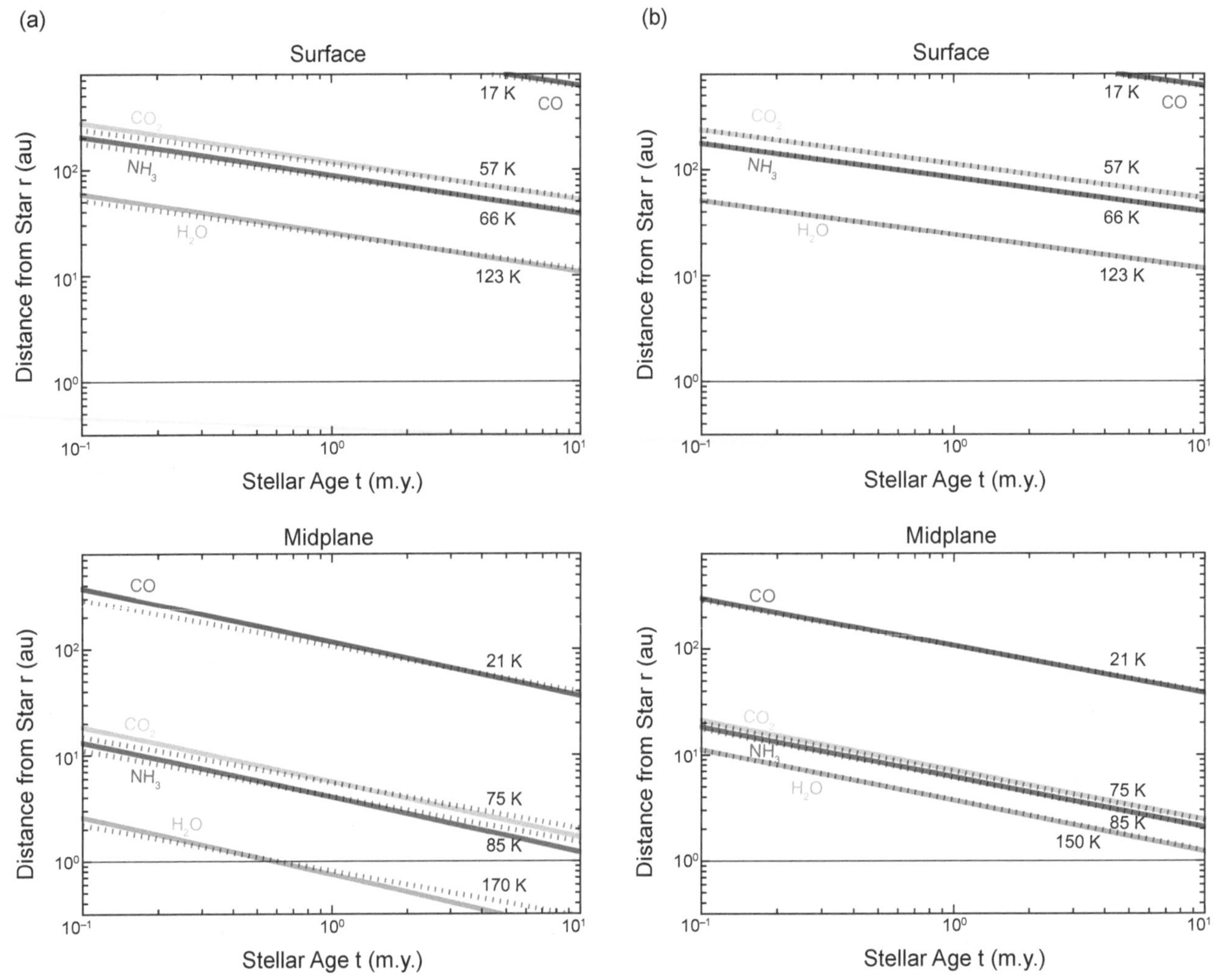

Fig. 3. Locations of the snowlines for pure CO, CO_2, NH_3, and H_2O ice at the optically thin surface (upper panels) and at the optically thick midplane (lower panels) as a function of stellar age t for disks around a solar-mass pre-main-sequence star. The molar abundances of CO, CO_2, NH_3, and H_2O are assumed to be 10^{-4}, 10^{-4}, 10^{-5}, and 10^{-3}, respectively, everywhere in the disks. **(a)** An optically thick MMSN heated by stellar radiation only, with the snowlines migrating as the stellar luminosity evolves. **(b)** An optically thick, viscous accretion disk with uniform accretion heating in addition to stellar irradiation. In this latter model, the evolution of both the stellar luminosity and accretion rate causes the migration of the snowlines. The dotted lines are contours of constant temperatures, and the thin horizontal line marks the location of the current Earth's orbit, r = 1 au.

that dominate the opacity. In the disk model adopted here, a factor of 10 decrease in the opacity leads to a factor of 2 decrease in the midplane temperature.

However, as already mentioned in section 2.1, theoretical studies call into question the applicability of the simple viscous accretion disk model. A more realistic picture of disk accretion may be that the accretion is mainly driven by magnetic fields. In this picture, disk heating still occurs through the Joule dissipation of electric currents associated with the magnetic fields. However, the Joule heating tends to predominantly occur near the disk surface because the low electric conductivity around the midplane prohibits the development of strong currents. Thermal radiation generated near the surface region can easily escape and is therefore inefficient at heating the midplane region. For this reason, the temperature structure of magnetically accreting disks tends to be close to that of passively irradiated disks as shown in Fig. 3a unless both the disk opacity and ionization fraction are sufficiently high (*Hirose and Turner*, 2011; *Béthune and Latter*, 2020; *Mori et al.*, 2019, 2021).

The disk temperature structure and the locations of the snowlines drastically change during episodic accretion outburst events introduced in section 2.1. For example, V883 Ori is a solar-mass FU Ori star with an elevated luminosity of ~200 $L_\odot$ (*Furlan et al.*, 2016) and a massive circumstellar disk. Equation (2) implies that the midplane of the V883 Ori disk in the outburst phase should be 4.5× hotter than those of disks around solar-luminosity stars. Assuming a sublimation temperature of 150 K for H_2O ice, we find that the strong irradiation heating should push the H_2O snowline out to r ~ 20 au. In reality, the bolometric luminosity of ~200 $L_\odot$ includes accretion heating, which could push the snowline further out by heating the disk from inside..

These theoretical estimates of disk temperature structure are compared with SED and spatially-resolved observations of dust continuum and molecular lines. *Cieza et al.* (2016) observed 1.3-mm dust continuum toward V883 Ori to find that the brightness and optical depth sharply rise inward at 40 au. A plausible interpretation is that there is a water snowline at 40 au and ice sublimation causes the abrupt change of the dust grain size and opacity [see also section 4.2 and *Schoonenberg et al.* (2017)]. More recently, *Law et al.* (2021b) derived the two-dimensional (radial and vertical) gas temperature distribution of five disks using high-resolution observations (~20 au) of CO and its optically thick isotopologs. The derived temperatures are roughly in agreement with the simple estimates presented in Fig. 3 (see also *Dutrey et al.*, 2017; *Flores et al.*, 2021).

2.3. Dust Growth: Theory

In studies of planet formation and protoplanetary disks, dust can refer to any solid particles smaller than planetesimals, not only (sub)micrometer-sized dust grains but also pebble- to boulder-sized solid particles. Although the solids only comprise a minor fraction of the mass in disks, they play significant roles in planet formation and disk evolution. The solids are the ultimate building blocks of all solid bodies in planetary systems including comets. Small dust grains govern the disks' thermal structure, and their size distribution can affect snowline locations (*Oka et al.*, 2011) and the disks' hydrodynamical instabilities (*Malygin et al.*, 2017; *Barranco et al.*, 2018; *Fukuhara et al.*, 2021). Thermal emission from micrometer- to millimeter-sized dust particles dominates the disks' continuum emission that we observe with infrared and radio telescopes. Small grains can even regulate the disk MHD by facilitating the recombination of ionized gas particles interacting with magnetic fields (*Sano et al.*, 2000; *Ilgner and Nelson*, 2006; *Wardle*, 2007; *Bai*, 2011). Therefore, to understand how solid bodies form in protoplanetary disks, we must fully understand how dust evolves and how it interacts with the disk gas.

From the theoretical point of view, there is no doubt that dust growth starts already in the earliest stage of disk evolution, at least in inner disk regions. Models of dust evolution in both laminar and turbulent disks (*Weidenschilling,* 1980; *Nakagawa et al.,* 1981; *Dullemond and Dominik*, 2005; *Tanaka et al.*, 2005; *Brauer et al.*, 2008; *Birnstiel et al.*, 2010) show that micrometer-sized grains grow into 0.1–1-mm-sized grains on a timescale of ≪1 m.y. In moderately turbulent disks, a useful simple estimate for the local growth timescale (size-doubling time) is available (*Takeuchi et al.*, 2005; *Brauer et al.*, 2008)

$$t_{\rm grow} \sim \frac{\Sigma_{\rm gas}}{\Sigma_{\rm dust}}\frac{1}{\Omega_{\rm K}} \sim 20\left(\frac{\Sigma_{\rm gas}/\Sigma_{\rm dust}}{100}\right)\left(\frac{r}{1\ {\rm au}}\right)^{3/2}\left(\frac{M_*}{M_\odot}\right)^{-1/2} {\rm yr} \qquad (3)$$

where $\Sigma_{\rm gas}$ and $\Sigma_{\rm dust}$ are the mass surface densities of gas and dust, respectively, and $\Omega_{\rm K}$ is the local Keplerian frequency. Equation (3) does not involve turbulence strength because the negative effect of turbulence on dust settling cancels its positive effect on particle collision speeds. Furthermore, equation (3) is independent of particle size, implying that the particle radius increases exponentially with time. For instance, micrometer-sized grains at 30 au would grow to millimeter-sized aggregates in ~2×10^4 yr. Strictly speaking, equation (3) overestimates the growth time of small grains that are vertically well mixed in the disk, so the timescale on which millimeter-sized aggregates form can be even shorter. Highly porous (or "fluffy") dust aggregates can also grow faster than estimated by equation (3) (*Okuzumi et al.*, 2012; *Kataoka et al.*, 2013; *Garcia and Gonzalez*, 2020). In any case, the growth timescale at 30 au is much shorter than the typical ages of Class II disks (~10^6 yr) and is even comparable to the ages of the youngest Class 0 sources (~10^4 yr).

Equation (3) assumes that grain collisions always result in sticking. This assumption is valid for micrometer-sized grains with low collisional velocities, unless they are strongly negatively charged in the partially ionized protoplanetary disks (*Okuzumi*, 2009). However, as the grain aggregates grow, they obtain a higher collision velocity induced by turbulence (*Ormel and Cuzzi*, 2007) and by radial drift on the background gas (*Whipple*, 1972; *Adachi et al.*, 1976; *Weidenschilling*, 1977). The increased collision velocity may prevent further growth by inducing collisional fragmentation (*Blum and Wurm*, 2000; *Güttler et al.*, 2010) and bouncing (*Güttler et al.*, 2010; *Zsom et al.*, 2010). Moreover, in a smooth gas disk with a negative radial pressure gradient, their radial drift is inward (see the chapter in this volume by *Simon et al.*). For millimeter- and centimeter-sized aggregates, the timescale of the radial inward drift can be shorter than the local growth timescale $t_{\rm grow}$, meaning that these pebbles would fall toward the central star before growing to meter-sized boulders. Therefore, a more realistic picture of dust growth is that grains rapidly grow initially and then reach the maximum size set by fragmentation, bouncing, or radial drift (*Birnstiel et al.*, 2012; *Drazkowska et al.*, 2022). These growth barriers complicate planetesimal formation (see also the chapter by Simon et al.).

In the context of this chapter, it is important to point out that the chemical composition of grains is key to understanding how they grow. Early models and experiments suggested that aggregates made of water ice are stickier than silicates (*Dominik and Tielens*, 1997) and may even overcome the fragmentation barrier (*Wada et al.*, 2009). Such sticky grains also tend to form highly porous aggregates, which is beneficial for overcoming the drift barrier because the porous aggregates grow rapidly (*Okuzumi et al.*, 2012; *Kataoka et al.*, 2013). However, this sticky water ice scenario is questioned by recent experiments showing that water ice is not so sticky at low temperatures (*Gundlach et al.*, 2018; *Musiolik and Wurm*, 2019). Instead, some recent studies suggest that silicates are stickier than previously thought (*Kimura et al.*, 2015; *Steinpilz et al.*, 2019). Some (if not all) types of organic matter are also sticky in a warm environment (*Kouchi et al.*, 2002; *Piani et al.*, 2017; *Bischoff et al.*, 2020). CO_2 ice appears to be less sticky than water ice (*Musiolik et al.*, 2016a,b; *Arakawa and Krijt*, 2021; *Fritscher and Teiser*, 2021). This implies that the fate of dust coagulation and planetesimal formation may depend on temperature and hence on distance from the central star (*Birnstiel et al.*, 2010; *Pinilla et al.*, 2017; *Homma et al.*, 2019; *Okuzumi and Tazaki*, 2019).

Ice sublimation, condensation, and sintering around snowlines can produce local pileups of solids (*Cuzzi and Zahnle*, 2004; *Saito and Sirono*, 2011; *Sirono*, 2011; *Okuzumi et al.*, 2016; *Schoonenberg and Ormel*, 2017; *Drążkowska and Alibert*, 2017; *Ida and Guillot*, 2016; *Ida et al.*, 2021; *Hyodo et al.*, 2019, 2021). These pileups can have important implications for planetesimal formation and the disks' observational appearance (see also section 2.5).

2.4. Dust Growth: Observations

Dust grains are the dominant source of disk opacities. Continuum emission and scattering thus provide information on the physical properties of grains (including size, porosity, and composition). Here, we briefly review important constraints on the degree of dust growth in disks obtained from radio observations. For more comprehensive reviews on disk dust observations, we refer to *Testi et al.* (2014) and *Miotello et al.* (2023).

The spectral index of dust continuum emission in the millimeter wavelength has been used to constrain the maximum grain size in disks; it is defined by

$$\alpha_\nu = \frac{d \ln I_\nu}{d \ln \nu} \tag{4}$$

where I_ν is the intensity of the emergent radiation at frequency ν. For a black body of uniform temperature T, I_ν is equal to the Planck function B_ν (T) for the temperature, and one has $\alpha_\nu = 2$ in the Rayleigh-Jeans limit ($B_\nu \propto \nu^2$). For an optically thin disk, one has $I_\nu \approx \kappa_\nu \Sigma_{dust} B_\nu$, where κ_ν is the dust absorption cross section per unit dust mass (*Miyake and Nakagawa*, 1993). In the latter case, α_ν reflects the frequency dependence of κ_ν, which in turn reflects the size distribution of opacity-dominating dust grains. For example, interstellar dust grains of maximum size 0.1 μm yield $\kappa_\nu \propto \nu^2$ and hence $\alpha_\nu \approx 4$ at radio wavelengths. As the grains grow and their maximum grain size exceeds the wavelengths, the ν dependence of κ_ν becomes weaker, leading to $\alpha_\nu < 4$ (e.g., *Miyake and Nakagawa*, 1993). Constraining the grain size from α_ν is also possible for disks that are optically thick but have a non-zero albedo ω_ν, for which $I_\nu \sim \sqrt{1-\omega_\nu} B_\nu$ (*Rybicki and Lightman*, 1979). The factor $\sqrt{1-\omega_\nu}$ (<1) represents the effect of multiple-scattering by dust particles in an optically thick disk suppressing the disk's thermal emission (*Rybicki and Lightman*, 1979; *Miyake and Nakagawa*, 1993; *Birnstiel et al.*, 2018; *Liu*, 2019; *Zhu et al.*, 2019; *Sierra and Lizano*, 2020).

Early millimeter and submillimeter surveys (*Weintraub et al.*, 1989; *Beckwith et al.*, 1990; *Beckwith and Sargent*, 1991; *Andrews and Williams*, 2005, 2007; *Ricci et al.*, 2010a,b) already showed that the spatially integrated continuum emission from T Tauri disks has spectral indices of 2–3 in the (sub)millimeter range. More recently, high-resolution millimeter interferometric observations have provided information on the spatial variation of α_ν in individual disks (*Pérez et al.*, 2012, 2015; *ALMA Partnership et al.*, 2015; *Guidi et al.*, 2016; *Tazzari et al.*, 2016; *Huang et al.*, 2018; *Cazzoletti et al.*, 2018a; *Dent et al.*, 2019; *Carrasco-González et al.*, 2019; *Soon et al.*, 2019; *Macías et al.*, 2021). Overall, local values of α_ν range between 1.5 and 4 and appear to be anticorrelated with millimeter intensity. This anticorrelation indicates either that particles larger than a millimeter are concentrated in the brighter regions or that the brighter regions simply have higher optical depths. In some disks, the spectral index approaches ~4 toward the disks' outer edges (*Pérez et al.*, 2012; *Guidi et al.*, 2016; *Tazzari et al.*, 2016; *Dent et al.*, 2019; *Carrasco-González et al.*, 2019), suggesting that the grains in these outermost regions are smaller than a millimeter in size (*Miyake and Nakagawa*, 1993). The values of $\alpha_\nu \lesssim 2$ can be explained by optically thick emission from dust grains that have higher albedos at shorter wavelengths. In particular, in the (sub)millimeter wavelength range, $\alpha_\nu \lesssim 2$ implies a population of dust grains with maximum grain sizes of 0.1–1 mm (*Liu*, 2019; *Zhu et al.*, 2019; *Sierra and Lizano*, 2020). An alternative explanation is that the disks with $\alpha_\nu < 2$ have temperatures that decrease with distance from the midplane (*Sierra and Lizano*, 2020).

If we assume that the entire disk is optically thin at the observed wavelengths, the spectral index of ~2–3 indicates grain growth to millimeters in radius (*Miyake and Nakagawa*, 1993; *Draine*, 2006). However, it is not evident that the disks, even their outer part, are always optically thin at submillimeter and millimeter wavelengths. Inferring the optical thicknesses of disks is intrinsically difficult because the disks' temperature distribution is unknown a priori. *Huang et al.* (2018) and *Dullemond et al.* (2018) used a simple model for the temperature of passively irradiated disks (similar to equation (2)) and found that some bright dust rings observed in ALMA millimeter images have similar millimeter absorption optical thicknesses of 0.2–1. Although one interpretation is that the rings are barely optically thin by coincidence, another possibility is that the rings are actually optically thick but appear to be darker than the blackbody because of scattering, $I_\nu \sim \sqrt{1-\omega_\nu B_\nu}$ (*Zhu et al.*, 2019).

Recently, radio polarimetric observations have been used to derive independent constraints on the grain size in the outer regions ($\gtrsim$10 au) of protoplanetary disks. The observations have shown that many inclined disks produce uniformly polarized submillimeter continuum emission whose polarization direction is parallel to the disks' minor axes (e.g., *Stephens et al.*, 2014, 2017; *Hull et al.*, 2018;*Harris et al.*, 2018; *Cox et al.*, 2018; *Sadavoy et al.*, 2018; *Dent et al.*, 2019). This polarization pattern can be explained if the polarized continuum emission is dust thermal radiation scattered by the dust particles themselves (*Kataoka et al.*,

2015; *Yang et al.*, 2016). In this interpretation, the observed degree of polarization gives a strong constraint on the size of the opacity-dominating dust particles because the scattering and polarization efficiencies depend strongly on grain size. Models assuming spherical, compact silicate particles (*Kataoka et al.*, 2015; *Yang et al.*, 2016) predict a maximum grain size of 100 μm for the disks with uniformly polarized submillimeter emission.

This new constraint has garnered considerable attention in recent years because the inferred grain size is an order of magnitude smaller than estimated from the spectral index. The cause of the discrepancy between the two grain size estimates is under debate. In fact, the estimates from disk polarized emission largely depend on the grain properties assumed, including grain shape (*Kirchschlager and Bertrang*, 2020), porosity (*Tazaki et al.*, 2019; *Brunngräber and Wolf*, 2021), and composition (*Yang and Li*, 2020). For disks that are optically thick at submillimeter wavelengths, the discrepancy may be resolved if grains larger than 100 μm have already settled to the optically thick midplane region and give no contribution to submillimeter emission, while still affecting the spectral index by contributing to millimeter emission that is optically thinner (*Brunngräber and Wolf*, 2020; *Ueda et al.*, 2021).

2.5. Dust Spatial Distribution

With ALMA, we can now observe dust emission from protoplanetary disks in nearby star-forming regions at a spatial resolution of 5 au. Arguably the most striking discovery from the ALMA disk observations is the prevalence of small-scale structures — rings, gaps, spirals, and crescents — in dust thermal emission (e.g., *ALMA Partnership et al.*, 2015; *Andrews et al.*, 2018; *Long et al.*, 2018; *Cieza et al.*, 2021). Substructures on similar or smaller scales are also found in near-infrared scattered light images that probe the disk surfaces (e.g., *Avenhaus et al.*, 2018; *Garufi et al.*, 2018). We refer to *Andrews* (2020) for a review of recent disk substructure observations at both radio and infrared wavelengths.

The most commonly observed substructures are axisymmetric rings and gaps. They were first discovered in the disk around HL Tau (*ALMA Partnership et al.*, 2015) and have since been found in many large, bright Class II disks (*Huang et al.*, 2018; *Long et al.*, 2018). Figure 4 shows ALMA dust continuum images of the well-studied protoplanetary disks around IM Lup, GM Aur, and AS 209. It illustrates the diversity of the ring/gap substructures: Whereas the disk around IM Lup only exhibits shallow gaps (plus spirals), the AS 209 disk features deeper, more extended gaps together with remarkably narrow rings. The disk around GM Aur has a ring connected to a fainter outer disk, a feature also visible in the dust images of some other disks [e.g., Elias 24 (*Huang et al.*, 2018)].

The origins of these dust substructures are a subject of active research (*Andrews*, 2020). A widely accepted explanation for narrow dust rings is dust trapping in ring-shape pressure maxima (bumps) (*Pinilla et al.*, 2012; *Dullemond et al.*, 2018; *Rosotti et al.*, 2020), where the radial inward drift of dust particles halts (*Whipple*, 1972). This scenario is tempting because dust trapping at pressure bumps has long been anticipated as a solution to the radial drift problem in planetesimal formation (*Johansen et al.*, 2014). There are a variety of physical mechanisms potentially yielding annular pressure bumps, including planet-disk interaction (*Paardekooper and Mellema*, 2006), MHD effects (*Johansen et al.*, 2009; *Suriano et al.*, 2018; *Flock et al.*, 2015), and dust-gas interaction (*González et al.*, 2017). Planets can also produce gaps and spirals.

In principle, snowlines could also produce dust rings and gaps as mentioned in section 2.3. *Zhang et al.* (2015) and *Okuzumi et al.* (2016) argued that the snowlines of major volatiles, including H_2O and CO, are responsible for the major dust rings and gaps in the HL Tau disk. However, the current understanding is that most of the rings and gaps that have been observed in a number of disks are not relevant to snowlines (*Huang et al.*, 2018; *Long et al.*, 2018; *van der Marel et al.*, 2019). The main reason is that the radial positions of the substructures do not appear to be correlated with the central star's luminosity, which should determine snowline locations. Nevertheless, there are a few disks (HD 163296 and

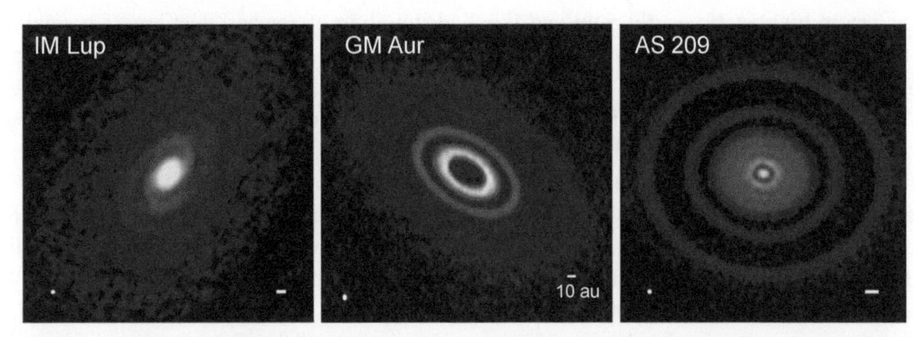

Fig. 4. ALMA millimeter continuum images of three selected protoplanetary disks [IM Lup and AS 209 adapted from *Andrews et al.* (2018); GM Aur from *Huang et al.* (2020)] showing ring/gap substructures in the disks' dust component. Beam sizes and 10-au scale bars are shown in the bottom left and right corners of each panel, respectively. The image of the IM Lup disk also exhibits spirals.

MWC 480 in addition to HL Tau) in which a dust ring is found close to the expected location of the CO snowline (*Zhang et al.*, 2021).

ALMA observations of dust emission have also provided important constraints on the dust surface density in the outer (r $\gtrsim$ 10 au) disk region where icy bodies like comets form. If we assume that this region is optically thin to its own millimeter emission, the dust surface density is derived from the millimeter intensity I_ν as $\Sigma_{dust} = I_\nu/(\kappa_\nu B_\nu)$ (see also section 2.4). For example, the dust surface densities of the IM Lup, GM Aur, and AS 209 disks are estimated to be comparable or an order of magnitude higher than that in the MMSN [Σ(ice + rock) = 30(r/1 au)$^{-3/2}$ g cm^{-2}] for a reasonable assumption about κ_ν for 0.1–1-mm-sized grains (see the bottom panels of Fig. 5 in section 3). We note, however, that these particular objects are among the largest, most massive disks; their dust surface densities may not represent those of more common, compact disks (*Miotello et al.*, 2022).

3. CHEMISTRY IN PROTOPLANETARY DISKS

Comets contain a significant amount of icy material, which has been investigated to reveal the thermal history and chemical environment in the cold outer regions of the solar nebula (e.g., see the chapter by *Biver et al.* in this volume). For example, the composition of major cometary ice is often similar to that of interstellar ice. The relative abundance of CO_2 and CH_3OH to water is 2–30% and 0.2–7% in comets, while these ratios are 15–44% and 5–12% in quiescent molecular clouds (*Mumma and Charnley*, 2011). Inheritance of interstellar ice in comets has thus long been discussed. Observations of protoplanetary disks, on the other hand, can reveal the spatial distribution and evolution of molecules at the sites of ongoing planetary system formation. These two kinds of information are complementary to elucidate the formation of comets and the solar system.

Theoretical and observational studies show that various chemical processes are going on within the disks (e.g., *Aikawa et al.*, 2002; *Henning and Semenov*, 2013; *Öberg et al.*, 2023). As we have seen in the previous section, there are large gradients of temperature and density in both the radial and vertical directions in a disk (Fig. 1). In the vertical directions, gases are in hydrostatic equilibrium and the density decreases toward the disk surface. The disk surface is directly irradiated by the stellar radiation and thus is warm. Molecules are dissociated to radicals and atoms by UV radiation from the central star and interstellar radiation field; the disk surface is a photon-dominated region (PDR). [While the surface temperature of T Tauri stars is ~4000 K, UV radiation is produced by the gas accretion from the disk to the star.] Beneath the surface layer, there is a warm molecular layer ($\gtrsim$20 K) that is moderately shielded from UV radiation and thus harbors various molecular gas such as CO. Ion-molecule reactions are triggered by X-rays from the central object and cosmic rays. (X-rays are emitted due to magnetic activity on the stellar surface.) In the midplane, the density is highest and dust grains could be further concentrated due to sedimentation. Since the temperature decreases with radial distance, snowlines are defined in the midplane as explained in section 2.2 (Fig. 2). Thermal energy could be high enough to drive chemical reactions in the inner hot regions [e.g., above several hundreds of K at <1 au (*Yang et al.*, 2013)]. Even in the outer cold midplane, reactions in the gas phase and within ices can be triggered by high-energy cosmic rays, X-rays, and decay of radioactive nuclei (e.g., ^{26}Al). Such layered structures, i.e., the PDR layer, warm molecular layer, and icy cold midplane, are confirmed by spatially and spectrally resolved observations (i.e., channel maps) of disks (*Dutrey et al.*, 2017; *Ruíz-Rodríguez et al.*, 2021; *Rosenfeld et al.*, 2013; *Law et al.*, 2021b). While the chemical composition of ice in the midplane would be most directly relevant to comets, observation of such ice is not easy (see section 3.3). Gaseous line observations in millimeter and submillimeter wavelengths are much more sensitive to low-abundance species. Chemistry in the gas and ice is linked via freeze-out, sublimation, and radial/vertical transport as we will discuss below.

In this section, we overview observations of cold molecular gas ($\lesssim$100 K) in millimeter and submillimeter wavelengths (section 3.1) and warm molecular gas ($\gtrsim$100 K) and ice in infrared and shorter wavelengths (section 3.2 and section 3.3; see also Fig. 1). Table 1 lists the molecular species so far detected in Class II disks in the order of increasing number of atoms and molecular mass. The observed wavelength is listed as well, since it tells where these molecules reside, i.e., cold or warm regions. In addition to the molecular abundances, we cover isotope fractionation and spin temperatures in section 3.4.

3.1. Molecules in Cold Gas

Figure 5 shows the radial distributions of the column densities of gaseous molecules and dust grains toward three disks around T Tauri stars (IM Lup, GM Aur, and AS 209) derived by ALMA observations (*Öberg et al.*, 2021b) with spatial resolutions of 18–48 au. In the following, we overview what these data tell us about chemistry in disks. Besides the overall chemical structure described above (i.e., the vertically layered structure and snowlines), coupling and decoupling of gas and dust are of special importance, since they could modify not only the molecular abundances but also the local elemental abundances.

3.1.1. CO and H_2O. In molecular clouds, CO and H_2O ice are major carbon and oxygen carriers with an abundance of ~10^{-4} relative to H_2, and thus are naively expected so in the disks. While H_2O snowlines are too close (less than several astronomical units) to the central star to be spatially resolved in the observations, the CO snowline is at r ~ 10–30 au around T Tauri stars (section 2.2). The observation of the CO snowline is, however, not straightforward, since CO gas is abundant in the warm molecular layer even outside the CO snowline. The N_2H^+ line is used to trace the CO snowline; its distribution anti-correlates with CO, since it is destroyed (N_2H^+ + CO $\rightarrow$ N_2 + HCO^+)

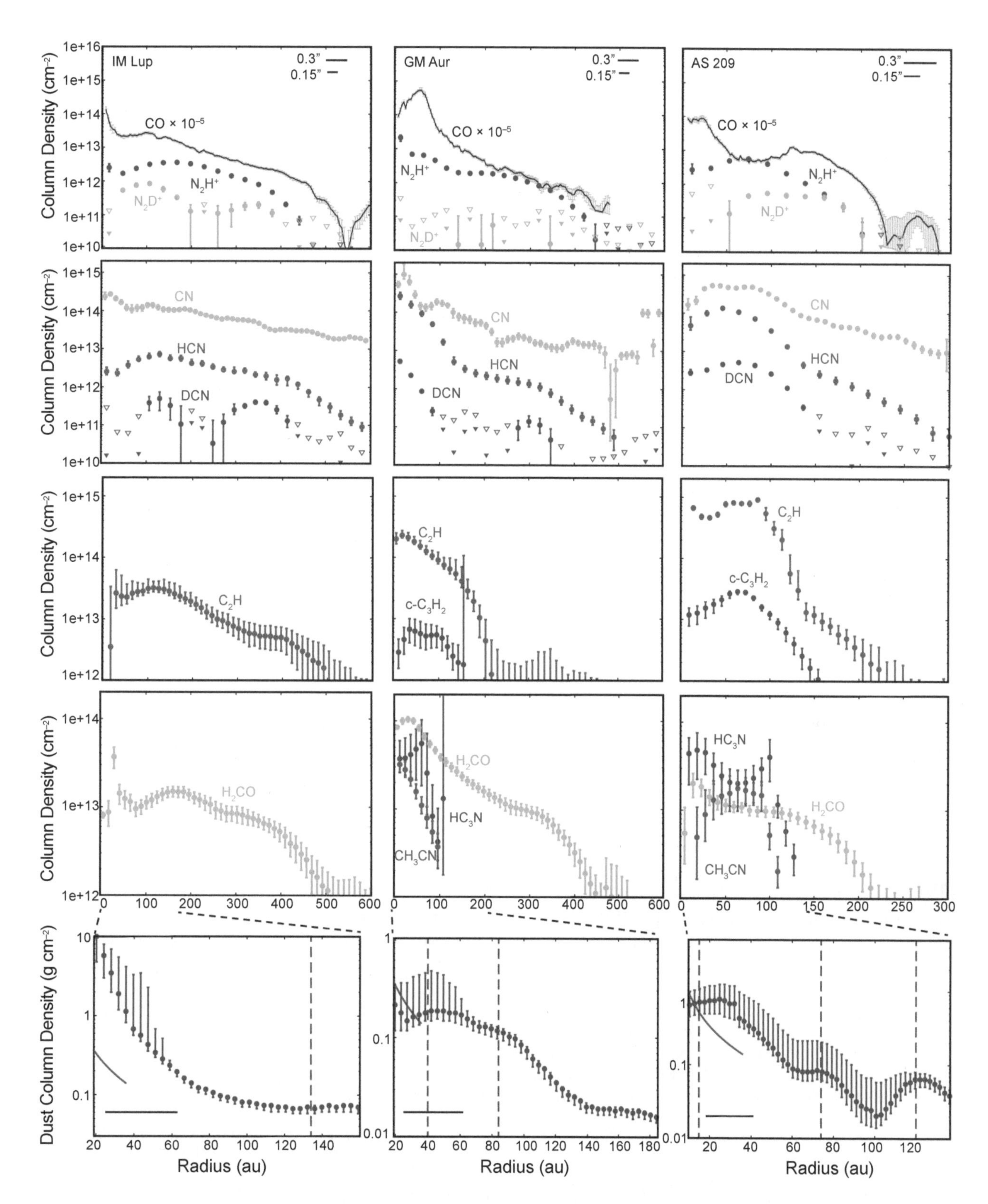

Fig. 5. Radial column density distributions of gaseous molecules and dust column densities in **(a)** IM Lup, **(b)** GM Aur, and **(c)** AS 209 from *Zhang et al.* (2021), *Bergner et al.* (2021), *Cataldi et al.* (2021), *Guzmán et al.* (2021), *Ilee et al.* (2021), and *Sierra et al.* (2021). The molecular column densities of C_2H and CO are derived from the data with a beam size of 0.15″, while the beam size is 0.3″ for other molecules. The error bars correspond to 1σ. For N_2H^+, N_2D^+, and DCN, the 1σ and 3σ upper limits are plotted at radii where the median value of the molecular column density is lower than the 1σ upper limits by a factor of >10. In the panels of dust column densities, the resolution of the dust continuum observations is shown with a horizontal bar, while the vertical dashed lines depict the positions of the rings. The red curves depict the dust column density of the MMSN (see section 2.5).

TABLE 1. Molecules detected in Class II disks.*

Species	Wavelength	Reference Examples	Species	Wavelength	Reference Examples
H_2	NIR,MIR	[1],[2]	HD	FIR	[3],[4]
CH^+	FIR	[5],[6]	OH	NIR,MIR,FIR	[6],[7]
CN	submm,mm	[8],[9]	^{13}CN	mm	[53]
$C^{15}N$	submm	[10]	CO	NIR,MIR,FIR,submm,mm	[8],[11]
^{13}CO	NIR,FIR,submm,mm	[8],[12]	$C^{17}O$	NIR,submm	[9],[13]
$C^{18}O$	NIR,submm,mm	[12],[14]	$^{13}C^{18}O$	submm	[14]
$^{13}C^{17}O$	submm	[15]	NO	submm	[54]
CS	mm	[8],[17]	^{13}CS	mm	[17],[18]
$C^{34}S$	mm	[8],[17]	SO	submm, mm	[19],[20]
H_2O	NIR,MIR,FIR,submm	[7],[21],[22]	HDO	mm	[55]
$H_2^{18}O$	mm	[55]	C_2H	mm	[8],[24]
C_2D	mm	[18]	HCN	NIR,MIR,submm,mm	[23],[24]
$H^{13}CN$	submm,mm	[25],[23]	$HC^{15}N$	submm	[23]
DCN	mm	[26],[27]	HNC	mm	[8],[28]
DNC	submm	[18]	HCO^+	submm,mm	[8],[29]
DCO^+	submm,mm	[30],[27]	$H^{13}CO^+$	submm,mm	[30],[25]
$HC^{18}O^+$	submm	[31]	N_2H^+	mm	[32],[33]
N_2D+	mm	[34],[35]	H_2S	mm	[36],[37]
CO_2	MIR	[16],[56]	$^{13}CO_2$	MIR	[56]
C_2S	mm	[53]	OCS	mm	[60]
$^{34}SO_2$	submm	[38]	$^{34}SO_2$	submm	[38]
NH_3	MIR,submm	[39],[58]	C_2H_2	NIR,MIR	[40],[56]
$^{13}C^{12}CH_2$	MIR	[57]	H_2CO	submm,mm	[8],[41]
H_2CS	submm,mm	[17],[18]	HC_3N	mm	[42],[43]
HCOOH	mm	[44]	c-C_3H_2	mm	[45],[43]
CH_4	NIR	[46]	CH_2CN	mm	[47]
CH_3CN	mm	[48],[43]	CH_3OH	submm,mm	[49],[50]
C_4H_2	MIR	[57]	CH_3CHO	submm	[52]
CH_3OCHO	submm	[51],[52]	$^{13}CH_3OCHO$	mm	[60]
$CH_3O^{13}CHO$	mm	[60]	CH_3OCH_3	submm	[51],[59]
CH_3COCH_3	submm	[52]	t–C_2H_3CHO	mm	[60]
s–C_2H_5CHO	mm	[60]	C_6H_6	MIR	[57]

*Adopted and revised from *McGuire* (2018). The table includes molecules detected in the disk of V883 Ori, which is a Class I object (section 4.2). The range of wavelengths are 2–5 μm for near-infrared (NIR), 5–40 μm for mid-infrared (MIR), 40–300 μm for far-infrared (FIR), 300–1000 μm for submillimeter (submm), and ≥1 mm for millimeter (mm). References are not exhaustive. In addition to the first detection papers, we list recent papers for readers to refer to references therein (see also text).

References: [1] *Weintraub et al.* (2000); [2] *Martin-Zaïdi et al.* (2007); [3] *Bergin et al.* (2013); [4] *McClure et al.* (2016); [5] *Thi et al.* (2011); [6] *Fedele et al.* (2013); [7] *Salyk et al.* (2008); [8] *Dutrey et al.* (1997); [9] *Guilloteau et al.* (2013); [10] *Hily-Blant et al.* (2017); [11] *Najita et al.* (2003); [12] *Ansdell et al.* (2016); [13] *Qi et al.* (2011); [14] *Zhang et al.* (2017); [15] *Booth et al.* (2019); [16] *Carr and Najita* (2008); [17] *Le Gal et al.* (2019); [18] *Loomis et al.* (2020); [19] *Fuente et al.* (2010); [20] *Booth et al.* (2018); [21] *Carr et al.* (2004); [22] *Hogerheijde et al.* (2011); [23] *Hily-Blant et al.* (2019); [24] *Guzmán et al.* (2021); [25] *Huang et al.* (2017); [26] *Qi et al.* (2008); [27] *Öberg et al.* (2021a); [28] *Long et al.* (2021); [29] *Aikawa et al.* (2021); [30] *van Dishoeck et al.* (2003); [31] *Furuya et al.* (2022b); [32] *Qi et al.* (2003); [33] *Qi et al.* (2019); [34] *Huang and Öberg* (2015); [35] *Cataldi et al.* (2021); [36] *Phuong et al.* (2018); [37] *Rivière-Marichalar et al.* (2021); [38] *Booth et al.* (2021a); [39] *Salinas et al.* (2016); [40] *Lahuis et al.* (2006); [41] *Pegues et al.* (2020); [42] *Chapillon et al.* (2012); [43] *Ilee et al.* (2021); [44] *Favre et al.* (2018); [45] *Qi et al.* (2013b); [46] *Gibb and Horne* (2013); [47] *Canta et al.* (2021); [48] *Öberg et al.* (2015); [49] *Walsh et al.* (2016); [50] *van der Marel et al.* (2021); [51] *Brunken et al.* (2022); [52] *Lee et al.* (2019); [53] *Phuong et al.* (2021); [54] *Leemker et al.* (2023); [55] *Tobin et al.* (2023); [56] *Grant et al.* (2023); [57] *Tabone et al.* (2023); [58] *Najita et al.* (2021); [59] *Yamato et al.* (2024a); [60] *Yamato et al.* (2024b).

when CO is abundant in the gas phase (*Qi et al.,* 2013a, 2019; *Aikawa et al.,* 2015). Alternative tracers are rare isotopes of CO (e.g., $^{13}C^{18}O$) whose emission is expected to be optically thin (*Zhang et al.,* 2017). In TW Hya, for example, the CO snowline is estimated to be 28–31 au and 20.5 au from N_2H^+ and $^{13}C^{18}O$, respectively (*Qi et al.,* 2013a; *Zhang et al.,* 2017).

The observations of rare isotopes of CO also suggest that the gaseous abundance of CO is actually lower than 10^{-4} even in the warm molecular layer. The CO column densities derived from the observations are compared with the disk models, which are constrained by dust continuum emission. As an example, Fig. 5 shows the radial distribution of CO and dust column densities derived from $C^{18}O$ (J = 2–1) and dust continuum (*Zhang et al.,* 2021; *Sierra et al.,* 2021). Assuming the gas/dust mass ratio is 100, the CO depletion factor, i.e., the ratio of the canonical CO abundance (10^{-4}) divided by the estimated CO abundance in the warm molecular layer, varies from 1 to 100 among disks in radii outside the CO snowline (*Miotello et al.,* 2017; *Zhang et al.,* 2019, 2021). While the CO abundance tends to increase inward, the CO depletion factor is 10–100 even inside the CO snowline in some disks. An alternative interpretation of these observations is that the gas/dust mass ratio is lower than 100. The submillimeter HD emission lines, which are observed by Herschel toward a limited number of disks, suggest CO depletion rather than gas depletion (*Bergin et al.,* 2013; *Favre et al.,* 2013; *McClure et al.,* 2016). It should however be noted that the evaluation of disk gas mass from HD emission also depends on the assumed thermal structure, i.e., disk models (*Miotello et al.,* 2023).

Herschel also observed a few H_2O lines in several disks (*Hogerheijde et al.,* 2011; *Salinas et al.,* 2016; *Du et al.,* 2017). While the beam size (e.g., 37″ at 557 GHz) is too large to provide any information on the spatial distribution, the emission lines of H_2O are significantly fainter than the predictions of disk models, in which H_2O is photodesorbed from ice outside its snowline. It suggests that H_2O ice-coated grains are mostly sedimented to the disk midplane where photodesorption is inefficient, and could even be radially drifted inward (section 2.3). The low CO abundance in the warm molecular layer can be explained similarly by the cold-finger effect; while CO is in the gas phase in the warm layer, it could be brought to the cold midplane by weak turbulence to freeze out onto grains. Conversion of CO to less volatile molecules (e.g., CO_2 ice and CH_3OH ice) would also help (e.g., *Furuya and Aikawa,* 2014; *Kama et al.,* 2016; *Krijt et al.,* 2020; *Furuya et al.,* 2022a).

It is noteworthy that H_2O ice and CO are expected to be the dominant reservoirs of oxygen and carbon in the initial condition (i.e., gas and solids accreted from the molecular clouds). While both can be depleted from the warm molecular layer as described above, lower volatility of H_2O would result in heavier depletion of oxygen than carbon, and an elemental abundance of C/O > 1 in the warm molecular layer. The high C/O ratio is supported by the observations of other molecules as explained below.

The similarity of the cometary ice with interstellar ice may indicate the sedimentation of ice-coated grains before ice composition is significantly affected by UV and X-rays in the upper layers of the solar nebula, while molecules formed in the upper layers can still be incorporated into the midplane ice via the cold finger effect.

3.1.2. Hydrocarbons: C_2H and c-C_3H_2. Unsaturated hydrocarbons, C_2H and c-C_3H_2, are expected to be abundant in the surface PDR layer, which is supported by the imaging observation of edge-on disks (*Ruíz-Rodríguez et al.,* 2021) and their relatively high excitation temperature (20 K–50 K) (*Guzmán et al.,* 2021; *Cleeves et al.,* 2021).

The abundance of C_2H is also sensitive to the C/O ratio in the gas phase; the comparison of its line brightness between observations and models indicates C/O > 1 (*Bergin et al.,* 2016; *Bergner et al.,* 2019; *Miotello et al.,* 2019). However, variation among disks should be noted; some disks are faint in C_2H emission (*Miotello et al.,* 2019). Spatially resolved observations show that the C_2H emissions often show a ring-like structure. Some rings coincide with the prominent gaps seen in the dust continuum, which is consistent with the expectation that C_2H abundance is enhanced by UV penetration. But not all dust gap correlates with the C_2H ring and vice versa (*Guzmán et al.,* 2021; *Law et al.,* 2021a). c-C_3H_2 is detected in several disks, although its emission lines are weaker than those of C_2H (*Qi et al.,* 2013b). The column density ratio of c-C_3H_2/C_2H is relatively constant with 5–10% in four disks where this ratio is derived (*Ilee et al.,* 2021) (Fig. 5). These observations and theoretical models suggest that c-C_3H_2 is formed in the gas phase together with C_2H.

3.1.3. HCN and CN. CN is also predicted to be abundant in the surface PDR layer, which is supported by the excitation analysis (*Teague and Loomis,* 2020; *Bergner et al.,* 2021). A ring-like distribution of CN emission is found in some disks, which is explained by CN formation via the reaction of N atom with UV-pumped vibrationally excited H_2 (*Cazzoletti et al.,* 2018b). HCN, on the other hand, is more susceptible to photodissociation than CN; it is dissociated by UV at longer wavelengths than for CN, and by Lyα emitted from T Tauri stars.

A recent high-resolution observation by *Bergner et al.* (2021) show that the column densities of CN and HCN are positively correlated (Fig. 5), which suggests that their formations are connected. For example, chemical models predict that both CN and HCN abundances are enhanced if the C/O ratio is >1 in the gas phase (*Cleeves et al.,* 2018; *Cazzoletti et al.,* 2018b). The column density ratio of CN/HCN also increases with radius (Fig. 5), which is consistent with the model prediction that CN is more abundant in the lower density UV-irradiated disk surface (*Bergner et al.,* 2021).

3.1.4. S-bearing molecules. As in molecular clouds, the abundances of gaseous S-bearing molecules in the disks are much lower than expected from model predictions assuming the solar elemental abundance, which suggests significant depletion of S from the gas to the solid phase (see section 4.2 and the chapter in this volume by Bergin et al.). Six S-bearing molecules are detected in disks: CS, SO, H_2S, H_2CS, C_2S, and SO_2. Among them, CS is the most readily detected, and its column density is estimated to be 10^{12}–10^{13} cm^{-2} in several disks (*Le Gal et al.*, 2019). Detection of other S-bearing molecules is rather limited. SO is first detected in the AB Aur disk by *Fuente et al.* (2010). *Guilloteau et al.* (2016) observed SO in ~20 disks using the Institut de Radioastronomie Millimétrique (IRAM) 30-m telescope and detected the emission only in four disks. Prevalence of CS over SO may suggest a high C/O ratio in disks (*Le Gal et al.*, 2021). H_2S, the dominant S-bearing molecule in comets, is first detected in disks by *Phuong et al.* (2018), and is recently detected in four additional disks (*Rivière-Marichalar et al.*, 2021). H_2CS is detected in the disk of the Herbig Ae star MWC 480 (*Le Gal et al.*, 2019).

Recently SO_2 was first detected in the disk of IRS 48, which is a transitional disk (i.e., a disk with a central hole in dust continuum emission) with an asymmetric dust emission peak (*Booth et al.*, 2021a). The emission peak would correspond to the local pressure maximum, at which dust grains are trapped (section 2). Both SO and SO_2 are bright only at this dust peak, while CS is not detected in this disk. The low CS/SO ratio indicates a low C/O ratio (<1) at the dust trap of IRS 48.

3.1.5. Complex organic molecules. In the astrochemistry community, organic molecules consist of six atoms and more are called complex organic molecules (COMs). Methanol, the most abundant and prototypical COM in molecular clouds, is so far detected only in TW Hya, HD 100546, IRS 48, and HD 169142 (*Walsh et al.*, 2016; *Booth et al.*, 2021b, 2023; *van der Marel et al.*, 2021) in spite of deep searches in some other disks (*Loomis et al.*, 2020; *Carney et al.*, 2019). It is also the key molecule to investigate the relation between ice and gas in disks; since its formation in the gas phase is known to be inefficient, gaseous CH_3OH should be mainly desorbed from ice, which in turn is formed by hydrogenation of CO on cold (≲20 K) ice surfaces. The CH_3OH emission in TW Hya is very weak and is in a ring region outside the CO snowline (~30 au) and inside the millimeter dust continuum edge. Since the dust temperature in the emitting region is below its sublimation temperature (~100 K), CH_3OH is considered to be desorbed by non-thermal processes. HD 100546, on the other hand, is a warm Herbig Ae disk with a central hole in the dust continuum. Most of the CH_3OH emission originates from the spatially unresolved central region (≲60 au) and thus could be tracing the thermally sublimated CH_3OH. The warm dust temperature (>20 K) in the disk of HD 100546 indicates that CH_3OH is not formed in the disk, but inherited from molecular clouds (*Booth et al.*, 2021b). In IRS 48, the CH_3OH emission spatially coincides with the dust trap and shows a high excitation temperature 100 K. Gaseous methanol abundance could be enhanced by a combination of ice-coated pebble concentration in the dust trap and irradiation heating at the edge of the central hole (*van der Marel et al.*, 2021). Recently, CH_3OCH_3 and CH_3OCHO are also detected at the same position as CH_3OH in IRS 48 (*Brunken et al.*, 2022).

Formic acid (HCOOH), which is often detected in the protostellar cores with bright CH_3OH emissions, is so far detected only in TW Hya (*Favre et al.*, 2018). H_2CO is another relevant molecule to CH_3OH. It is an intermediate product of CO hydrogenation to form CH_3OH on grain surfaces, while it can be formed via gas-phase reactions as well. H_2CO is also considered to be a precursor species of organics in meteorites (*Cody et al.*, 2011). *Pegues et al.* (2020) detected H_2CO in 13 disks out of 15, and derived an excitation temperature of 20–50 K and column density of ~5×10^{11}–5×10^{14} cm^{-2}. The fourth row of panels in Fig. 5 show the radial distribution of H_2CO in three T Tauri disks assuming the excitation temperature of 20 K (*Guzmán et al.*, 2021).

A nitrogen-bearing COM, CH_3CN, and another nitrile species HC_3N are detected in several disks (*Öberg et al.*, 2015; *Bergner et al.*, 2018). *Ilee et al.* (2021) spatially resolved the emission lines of these nitriles in disks of GM Aur, AS 209, HD 163296, and MWC 480; the emission is compact (≲100 au) comparable to the extent of the dust continuum. Unlike CH_3OH, these species can be formed in the gas phase. The weak correlation of their emission distributions with the dust continuum, however, may indicate the role of ice reservoir for their formation. The excitation temperatures (~30–50 K) are lower than their sublimation temperatures, indicating that they are non-thermally desorbed from ice or formed in the gas phase from the species with lower sublimation temperatures. Their column densities (10^{13}–10^{14} cm^{-2}) (Fig. 5) are higher than predicted in static disk models; either grain growth or turbulent mixing or a combination of the two would enhance their abundances (*Semenov and Wiebe*, 2011; *Furuya and Aikawa*, 2014; *Öberg et al.*, 2015).

3.2. Molecules in Warm and Hot Gas

Abundant warm molecular gas (T ≳ 200 K) is generally observed in typical protoplanetary disks surrounding stars of widely different masses (*Brittain et al.*, 2003; *Blake and Boogert*, 2004; *Brown et al.*, 2013). The regions of protoplanetary disks where such warm gas is present are often referred to as the "inner disk." The inner disk may similarly be defined as corresponding to disk radii inside the water snowline. While refractory carbon and silicate-dominated dust sublimate at 1500 K, i.e., at the radius of 0.01–0.1 au for typical protoplanetary disks, many chemically important volatile molecular species (water, CO_2, CH_3OH, CH_4, NH_3, etc.) persist in the gas phase down to dust temperatures of 100 K or below. Such temperatures are achieved at 1–2 au in the midplane for disks around young solar-mass stars, and at distances of up to 10 au in the superheated disk surface (*Sasselov and Lecar*, 2000; *Garaud and Lin*, 2007) (Fig. 3). Indeed, high-resolution spectroscopy and spectral imaging have demonstrated that the emission originates well inside their respective snowlines (*Goto et al.*, 2006; *Pontoppidan et al.*, 2008), and that the location of the snowline likely varies with the evolutionary stage of the star-disk system (*Hsieh et al.*, 2019) (see also section 2.2).

Warm molecular gas in disks is readily observed at mid-infrared wavelengths through their rovibrational transitions, through high-J pure rotational transitions in the case of water, and in the UV through electronic transitions (Fig. 6). It is also possible to observe warm and hot gas through rotational transitions between excited vibrational states in the (sub) millimeter regime, but beam dilution of the typically very small emitting inner disk regions makes such lines challenging to detect and image, even with facilities such as ALMA (*Notsu et al.*, 2019).

Given the theoretical prediction that the water snowline plays a key role in the formation of planets, multiple attempts have been made to observationally measure its location (and that of similar species), but with mixed results. While the CO snowline is located at disk radii large enough to allow for measurements using spatially resolved millimeter interferometric imaging (section 3.1.1), the water snowline typically only subtends an angle of 10–100 mas, below the imaging resolution limit of current facilities. Furthermore, while some hot water lines are detectable from the ground, water can only reliably and consistently be detected from space. Consequently, various techniques using purely spectral signatures have been used to constrain the location of the water snowline. *Zhang et al.* (2013) and *Blevins et al.* (2016) used the line spectral energy distribution, representing gas at a range of temperatures, to constrain the water sublimation radius in the surface layer in five protoplanetary disks, finding values of 3–10 au. *Notsu et al.* (2016) proposed a method using optically thin water lines from transitions with low spontaneous emission probability to probe the snowline deeper in the disk, but attempts at detecting the intrinsically weak lines have so far not been successful (*Notsu et al.*, 2019).

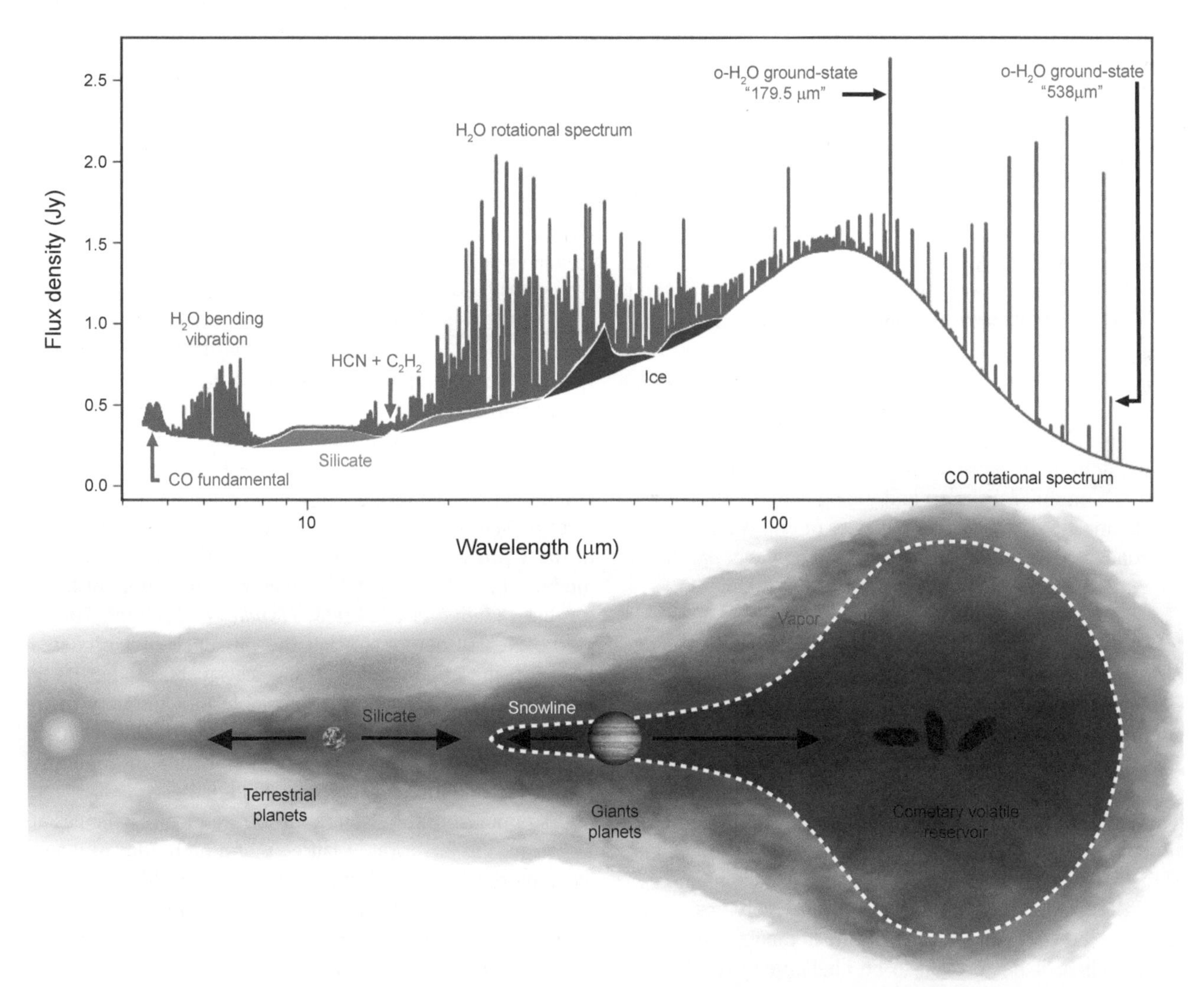

Fig. 6. Radiative transfer model infrared spectrum of a water-rich protoplanetary disk around a solar-mass star, based on observed Spitzer and Herschel spectra. The spectrum is rendered at high spectral resolving power (3 km s^{-1}). The spectrum includes lines from CO, water, and organics, and also includes the far-infrared solid-state bands due to water ice at 43 and 62 μm. The figure is reproduced from *Pontoppidan et al.* (2019a).

Because typical protoplanetary disks are primarily externally heated, except for the innermost regions (<1 au) or disks with very high accretion rates (>10^{-7} $M_\odot$ yr^{-1}) (section 2.2), warm molecular gas lines in the infrared overwhelmingly trace the surface layers of the disks (Kamp and Dullemond, 2004). This effect is exacerbated by the high vertical optical depths of most disks in the 1–10-au radius range. Consequently, the composition of the disk midplanes at these radii is generally not being observed, and can only be inferred under the assumption that some degree of vertical mixing is present (*Semenov and Wiebe*, 2011; *Anderson et al.*, 2021).

The relative abundances of warm gas-phase volatiles in inner disk surfaces appear to be different from those observed in ices in the cold interstellar medium and in solar system comets. This indicates that the observed inner disk chemistry is not directly inherited from cold ice chemistry, but is significantly altered by local processes. Concurrent measurements of H_2 and CO using UV spectroscopy suggest that inner disk CO abundances range from slightly depleted up to canonical ([CO/H_2] ~ 10^{-5}–10^{-4}) (*France et al.*, 2014; *Cauley et al.*, 2021). Other observable species, such as NH_3, CH_4, and CO_2, are further depleted relative to CO by orders of magnitude, clearly inconsistent with primordial ice chemistry (*Mandell et al.*, 2012; *Pontoppidan and Blevins*, 2014; *Bosman et al.*, 2017; *Pontoppidan et al.*, 2019b) but consistent with current predictions for warm gas-phase chemistry, which tends to destroy particularly those species, driving nitrogen into N_2 and carbon into CO (*Agúndez et al.*, 2008; *Walsh et al.*, 2015). Water, on the other hand, is commonly highly abundant in disks around low-mass and solar-mass young stars, as revealed by an extensive sample of disks observed by Spitzer (*Carr and Najita*, 2008, 2011; *Salyk et al.*, 2008; *Pontoppidan et al.*, 2010), with retrieved abundances consistent with [CO/H_2O] ~ 1 (*Salyk et al.*, 2011). JWST is currently confirming the ubiquitous presence of mid-infrared emission from warm gas in inner disks around young stars of all masses, and demonstrates a widening diversity of relative molecular abundances from water and organics (*Tabone et al.*, 2023; *Banzatti et al.*, 2023; *van Dishoeck et al.*, 2023).

3.3. Observations of Ices in Disks

During most of the evolution of a planetary system, the bulk of its volatile component is sequestered in the solid phase in the form of molecular ices. Indeed, most of the disk mass is typically located beyond the snowline (Fig. 6), with a few notable exceptions such as disks undergoing violent outbursts from instability-induced accretion events or disks that have been truncated by stellar companions.

While gas-phase volatiles can be relatively easily observed in the warm disk surface, or the hot inner disk inside the snowline, it has proven to be challenging to observe ices in disks, especially in their cold, outer midplanes where the formation of comets is presumably active. Ices are most readily identified through their strong mid-infrared (3–20 μm) resonance bands. However, these bands are nearly universally seen in absorption toward a background source of light, as the dust temperatures required to excite the bands in emission are too high (>150 K) to retain the ice on the emitting grains. Furthermore, the midplanes of protoplanetary disks during planetesimal formation are generally hidden beneath highly optically thick (at infrared wavelengths) layers of dust.

As a result, direct observations of ices in protoplanetary disks are sparse, and rely on particular geometric configurations that tend to be difficult to interpret. Even in cases where disk ices are detected, such ice is confined to the surface layers of the disk at τ_{IR} ~ 1, corresponding to a fraction of a percent of the total vertical column density.

Ices in disks have been observed when the disk is viewed close to edge-on, where ice in a flared outer disk absorbs thermal emission from hot dust in the inner disk, light scattered off the disk surface, or some combination thereof (*Pontoppidan et al.*, 2005; *Terada et al.*, 2007). Interpretation of such direct absorption spectroscopy is complicated by the complex geometry of the radiative transfer, making it difficult to establish unambiguously the location of the ices within the disk. Furthermore, the signal-to-noise ratios of edge-on disk spectra have often been low due to the inherent faintness of edge-on disks at most infrared wavelengths. In many cases, part of the absorption may also be due to cold material in a remnant protostellar envelope, rather than in the disk itself. In the most unambiguous cases, it appears that the most abundant ices have relative concentrations consistent with those observed in dense clouds and protostellar envelopes (*Aikawa et al.*, 2012), although more sensitive observations of rarer species, such as CO_2 and CH_3OH, will likely be needed to establish evolutionary patterns (*Ballering et al.*, 2021).

A complementary method for detecting the mid-infrared resonances of ice is through their detection in scattered light. This method is limited for use at the shortest wavelengths where scattering is the most efficient for the grain sizes present in disk surfaces (<5 μm). Thus far, detections of the 3-μm water ice stretching mode have been made in medium-band coronagraphic imaging of a number of face-on disks (*Inoue et al.*, 2008; *Honda et al.*, 2009, 2016). While clearly demonstrating the presence of ice-coated grains in the uppermost disk surface layers, model fits to ice band optical depths were not able to accurately measure the surface ice/rock ratio, and in particular whether photodesorption is playing a significant role in depleting ice in photodominated regions of the disks.

As opposed to the strong mid-infrared ice resonances at 3–15 μm, the far-infrared lattice vibration modes of water ice in particular near 43 and 62 μm (*Warren*, 1984; *Hudgins et al.*, 1993) appear in emission from a significant fraction of the disk dust mass in the comet-forming regions (*Malfait* et al., 1998). The excitation of these features in low-temperature dust means that they are potentially powerful tracers of the ice mass reservoirs. Furthermore, as they are related to vibrations of the solid lattice structure, rather than to vibrations within individual molecules, they are highly sensitive tracers of ice phase and crystallinity. The far-infrared water ice bands have been detected in at least four protoplanetary disks, and suggest high abundances of water ice, yielding estimates of ice/rock mass ratios of 0.36–1.6 (*McClure et al.*, 2015; *Min et al.*, 2016). Attempts have been made to detect far-infrared features in disks from ice species other than water, but these have thus far been unsuccessful, likely due to limitations of sensitivity (*McGuire et al.*, 2016; *Giuliano et al.*, 2016).

Observations of the molecular inventory in the inner disk may be used to constrain the efficiency of radial transport of ices. *Najita et al.* (2013) found a trend between the mid-infrared HCN/H_2O flux ratio and dust disk mass as measured by ALMA. The proposed interpretation of this apparent link between inner and outer disk properties is that more massive disks may have experienced more efficient buildup of large, water-rich planetesimals, leaving less water ice available in dust grains that migrate inward. In this case, the observed HCN/H_2O ratio is also a measure of the inner disk elemental C/O ratio. More recently, this scenario was supported by the observation of *Banzatti et al.* (2020), who found a similar relation of enhanced water emission for disks with small dust emission radii, suggesting that compact disks have experienced strong inward radial drift of water-ice rich dust, thereby enriching the inner disk with water vapor and decreasing the local C/O ratio (see also section 4.2).

3.4. Key Markers for Comets

3.4.1. Isotope fractionation. In the solar system material, isotope ratios of molecules are often different from the elemental abundance ratio. This is called isotope fractionation, and is used to investigate the origin of the planetary material. Isotope ratios in comets and other solar system materials are reviewed in the chapter in this volume by Biver et al. Here we summarize the observations and models of fractionation in disks. We refer the reader to *Nomura et al.* (2023) for a more detailed review. There are two mechanisms known to be responsible for the fractionation: (1) exothermic exchange reactions at low temperatures and (2) selective photodissociation.

The D/H ratio of Earth's ocean (1.56×10^{-4}) is higher than the elemental abundance of D/H in the interstellar matter (1.5×10^{-5}) or the protosolar value (2.1×10^{-5}) (see the chapter by Biver et al. and references therein). Such deuterium fractionation (enrichment) is caused by exchange reactions, e.g.,

$$H_3^+ + HD \rightarrow H_2D^+ + H_2 \tag{5}$$

$$CH_3^+ + HD \rightarrow CH_2D^+ + H_2 \tag{6}$$

Since these reactions are exothermic, the backward reactions are inefficient at low temperatures (less than or equal to several tens of Kelvins).

The high D/H ratio (e.g., H_2D^+/H_3^+) propagates to other molecules via chemical reactions; e.g., HDO is formed by grain surface reactions of O atom or OH with D atom, whose abundance is enhanced by the dissociative recombination of H_2D^+. The high D/H ratio of the ocean suggests that water is supplied to Earth from cold regions, at least partially, and has been motivating the measurement of D/H ratios in comets. The D/H ratio of cometary water, $(1.4–6.5) \times 10^{-4}$, overlaps with that of the ocean (*Mumma and Charnley*, 2011; *Ceccarelli et al.*, 2014; *Altwegg et al.*, 2015) (see also the chapter by Biver et al.). Recent observations of higher D_2O/HDO ratio than HDO/H_2O both in Comet 67P/Churyumov-Gerasimenko and protostellar cores indicate the inheritance of comet water from molecular clouds (*Furuya et al.*, 2016, 2017; *Altwegg et al.*, 2017), while the variation of HDO/H_2O ratio among comets indicates some reprocessing in disks.

In protoplanetary disks, five deuterated molecules, DCO^+, N_2D^+, DCN, C_2D, and HDO, are detected so far (e.g., *Huang et al.*, 2017; *Salinas et al.*, 2017; *Cataldi et al.*, 2021; *Öberg et al.*, 2021a; *Loomis et al.*, 2020; *Tobin et al.,* 2023) (Table 1). The disk-averaged D/H ratios of the former three species are summarized in Table 2. It should be noted that the recombination timescale of ions is much shorter than the typical ages of disks (i.e., a few million years). The high D/H ratio of molecular ions is thus clear evidence of active deuterium fractionation in disks. The D/H ratio of N_2H^+ is higher than that of HCO^+. It is consistent with the theoretical expectation that N_2H^+ traces the cold (<20 K) layer near the midplane where CO is frozen out, which enhances the deuterium fractionation (*Willacy*, 2007; *Cleeves et al.*, 2014; *Aikawa et al.*, 2015, 2018).

Recently, *Tobin et al.* (2023) detected HDO and $H_2{}^{18}O$ in the disk of FU Ori star V883 Ori, and derived the HDO/H_2O ratio to be $(2.23 \pm 0.63) \times 10^{-3}$. Since this ratio is comparable to those observed in the warm central protostellar cores, they conclude that water in the disk is inherited from molecular clouds.

Since temperature increases radially inward in disks, the molecular D/H ratio is expected to decrease inward, which is confirmed by the spatially resolved observations of DCN, DCO^+, and N_2D^+ (*Cataldi et al.,* 2021; *Öberg et al.,* 2021a). The observed peak D/H ratios and their radial positions are summarized in Table 2 (see also Fig. 5). The radial distribution of the molecular D/H ratio could also probe the deuteration pathways of each molecular species. The decline of D/H ratio toward the warm central region is expected to be less steep if the molecule is formed mainly via hydrocarbons rather than H_3^+, since the endothermicity of the reaction in equation (6) is higher than that of equation (5) (*Öberg et al.,* 2012). *Cataldi et al.* (2021) derived the DCN/HCN ratio as a function of the excitation temperature of HCN (i.e., proxy of gas temperature), which indicates the significant contribution of equation (6). They also found that the DCN/HCN ratio becomes comparable to the value measured in Hale-Bopp, $(2.3 \pm 0.4) \times 10^{-3}$ (*Meier et al.,* 1998; *Crovisier et al.,* 2004), in the regions of ~30–40 K.

The $^{14}N/^{15}N$ ratio of HCN, CN, NH_2 has been measured

TABLE 2. Molecular D/H ratios measured in Class II disks and comets.

Object	DCO^+/HCO^+	N_2D^+/N_2H^+	DCN/HCN
DM Tau	0.1 (50 au)–0.2 (450 au)*		
TW Hya	0.01 (30 au)–0.1 (70 au)†		0.17‡
AS 209	0.037–0.08§	0.3–0.5¶	0.028–0.059§
		0.67–0.94 (159 au)**	0.074–0.092 (118 au)**
IM Lup	0.023–0.037§		0.043–0.074§
		0.28–0.37 (76 au)**	0.17–0.24 (367 au)**
V4046 Sgr	0.014–0.029§		0.004–0.008§
LkCa15	0.019–0.038§		0.04–0.12§
GM Aur		<0.064 (216 au)**	0.079–0.13 (298 au)**
MWC 480	0.021–0.043§		0.006–0.018§
		0.40–0.66 (143 au)**	0.055–0.094 (133 au)**
HD 163296	0.039–0.075§		0.012–0.027§
	0.04–0.07††	0.19–0.66††	0.01–0.03††
		1.29–1.72 (152 au)**	0.048–0.062 (129 au)**
Comet Hale-Bopp			$(2.3 \pm 0.4) \times 10^{-3}$‡‡

**Teague et al.* (2015).
†*Qi et al.* (2008).
‡Disk average (*Öberg et al.,* 2012).
§Disk average (*Huang et al.,* 2017).
¶Disk average (*Huang and Öberg,* 2015).
**Peak value in the radial distribution (*Cataldi et al.,* 2021).
††Disk average (*Salinas et al.,* 2017).
‡‡*Meier et al.* (1998); *Crovisier et al.* (2004).

in comets. The ratio is lower than the elemental abundance in the local interstellar medium (~200–300) (e.g., *Ritchey et al.*, 2015) and the protosolar value (441) (*Marty et al.*, 2011). For example, the average $C^{14}N/C^{15}N$ ratio over 20 comets is 147.8 ± 5.7 (*Manfroid et al.*, 2009). The ^{15}N enrichment could be due to exchange reactions and/or selective photodissociation of N_2. The former would be less effective than previously thought, since some key exchange reactions are found to have activation barriers (*Roueff et al.*, 2015). In the latter mechanism, $^{14}N^{15}N$ is photodissociated in deeper layers of a disk or molecular cloud than the major isotope, producing excess ^{15}N atoms, which are incorporated to other molecules such as CN (*Liang et al.*, 2007; *Heays et al.*, 2014; *Furuya et al.*, 2017). *Visser et al.* (2018) calculated $^{14}N/^{15}N$ fractionation in disk models to find that the fractionation is fully dominated by selective photodissociation of N_2 (see also *Lee et al.*, 2021).

Guzmán et al. (2017) observed $H^{13}CN$ and $HC^{15}N$ to find the $HCN/HC^{15}N$ ratio of 80–160 in five disks, which are roughly consistent with the values in comets. In the disk of TW Hya (Table 3), *Hily-Blant et al.* (2019) found that the $HCN/HC^{15}N$ ratio increases with radius, which is consistent with the fractionation due to selective photodissociation (see also *Guzmán et al.*, 2017). The ratio of $CN/C^{15}N$, on the other hand, is 323 ± 30 (*Hily-Blant et al.*, 2019). The different $^{14}N/^{15}N$ ratio in HCN and CN could be a natural outcome of the vertical gradient of molecular abundance; i.e., CN has its abundance peak in the upper layer than that of HCN (*Cazzoletti et al.*, 2018b; *Teague and Loomis*, 2020) (section 3.1).

Carbon and oxygen isotope ratios in comets are mostly consistent with the solar abundance, with the notable exception of some molecules in 67P/Churyumov-Gerasimenko (see the chapter by Biver et al. for more details). On the theoretical side, selective photodissociation of CO, combined with isotope exchange reaction of $^{13}C^+ + CO \rightarrow {}^{13}CO + C^+$, could induce fractionation of $^{12}C/^{13}C$ and $^{16}O/^{17}O/^{18}O$ (*Yurimoto and Kuramoto*, 2004; *Lyons and Young*, 2005). *Miotello et al.* (2014) calculated the isotope fractionation of CO in disk models to show that the abundance and thus the line flux of $C^{18}O$ and $C^{17}O$ could be significantly reduced by the selective photodissociation, especially in disks with grain growth. Observational confirmation is challenging, since the lines of the major isotope of CO are optically thick, while the rare isotope lines are very weak. *Smith et al.* (2009) found that the $C^{18}O/CO$ and $C^{17}O/CO$ ratios are lower than the elemental abundances in the disk of VV CrA by observing absorption lines in the near-infrared with the binary star as a possible light source. As for carbon, *Hily-Blant et al.* (2019) found that the $HCN/H^{13}CN$ ratio is mostly flat with 77.9–88.7 at R = 20–55 au in the disk of TW Hya. *Yoshida et al.* (2022), on the other hand, derived $^{12}CO/^{13}CO = 21 \pm 5$ at 70–110 au,

TABLE 3. Molecular $^{14}N/^{15}N$ ratios measured in Class II disks and comets.

Object	$HC^{14}N/HC^{15}N$	$C^{14}N/C^{15}N$
TW Hya	223 ± 21*	323 ± 30*
	121 ± 11 (20 au)–339 ± 21 (45 au)†	
AS 209	156 ± 71‡	
V4046 Sgr	115 ± 35‡	
LkCa15	83 ± 32‡	
MWC 480	123 ± 45‡	
HD 163296	142 ± 59‡	
Comet Hale-Bopp	205 ± 70§	140 ± 35¶
Comet 17P/Holmes	139 ± 26§	165 ± 40§

* Disk average (*Hily-Blant et al.,* 2019).
† Radial distribution (*Hily-Blant et al.,* 2019).
‡ Disk average (*Guzmán et al.,* 2017).
§ *Bockelée-Morvan et al.* (2008).
¶ *Arpigny et al.* (2003).

which indicates the fractionation via the isotope fractionation (C^+ + CO) in the gas of high C/O (>1) elemental abundance. *Furuya et al.* (2022b) detected $HC^{18}O^+$ emission in TW Hya. Combining this detection with previous $H^{13}CO^+$ observation and disk chemical models, the $^{13}CO/C^{18}O$ ratio is estimated to be consistent with the elemental abundance ratio in the local ISM. More recently, *Yamato et al.* (2024b) found that the $^{12}C/^{13}C$ ratios of freshly sublimated complex organic molecules in V883 Ori is low (~20–30) (section 4.2).

In summary, we see at least a qualitative agreement in deuterium enrichment and nitrogen fractionation between disks and comets, in spite of the fact that we are observing gaseous components in disks rather than ice.

3.4.2. Spin temperature. Hydrogen atoms have nuclear spin angular momentum, and molecules that contain two or more H atoms in symmetrical locations have quantum states distinguished by the total spin angular momentum (I). In the case of H_2O, for example, I = 1 for ortho (o-H_2O) and I = 0 for para (p-H_2O). Spontaneous conversion of spin state (e.g., ortho to para) is strongly forbidden by the quantum mechanical selection rules. The rotational energy levels for each spin state form a discrete ladder, and the energy of the lowest-lying level is lower for p-H_2O than o-H_2O by 34.2 K. The relative population of the ortho-to-para ratio (OPR) of H_2O is described as

$$\mathrm{OPR} = \frac{3\Sigma(2J+1)\exp\left(\frac{-E_o(J_{K_a,K_c})}{k_B T_{spin}}\right)}{\Sigma(2J+1)\exp\left(\frac{-E_p(J_{K_a,K_c})}{k_B T_{spin}}\right)} \qquad (7)$$

where the statistical weight of ortho and para state are 3 and 1, respectively, T_{spin} is spin temperature, and K_a and K_c are the quantum numbers to describe the projected angular momentum. While the OPR is 3 in the high T_{spin} limit, the OPR is smaller than 3 due to the energy difference (34.2 K) of the lowest-lying state at <50 K (*Hama and Watanabe,* 2013).

The OPR of H_2O, NH_3, and some other hydrated molecules have been measured in comets, assuming that T_{spin} is equal to the temperature of the formation site of the molecule (e.g., *Mumma et al.,* 1987; *Mumma and Charnley,* 2011; *Shinnaka et al.,* 2020). The derived spin temperature range is 20–60 K with the typical value of ~30 K. For H_2O, the corresponding OPR is ~2.5 (*Mumma and Charnley,* 2011).

In protoplanetary disks, on the other hand, the OPR of warm water vapor (>300 K) detected in mid-infrared wavelength is consistent with OPR = 3 (*Pontoppidan et al.,* 2010; *van Dishoeck et al.,* 2013). Herschel detected emission lines of cold water vapor (<100 K) in TW Hya; the OPR is 0.77 ± 0.07, which corresponds to T_{spin} = 13.5 ± 0.5 K.

The interpretation of T_{spin} as the temperature of the molecular formation site, however, turned out to be too simplistic. A laboratory experiment by *Hama et al.* (2018) showed that the OPR of H_2O desorbed from ice is 3 regardless of the temperature of the ice formation (see also *Hama et al.,* 2016). While the energy difference of the lowest-lying state between ortho and para is 34.2 K in the gas phase, the rotation motion is constrained and the energy difference becomes significantly small in the ice phase. Then the magnetic dipole interaction of a proton with neighboring protons enables rapid spin conversion (*Limbach et al.,* 2006; *Hama et al.,* 2018).

The lower OPR than the statistical value observed in comets and outer protoplanetary disks would thus be due to the gas-phase reactions; the spin state changes via proton exchange reactions such as o-H_2O + H^+ → p-H_2O + H^+. The cold water vapor in protoplanetary disks is much less abundant compared with the value expected from the desorption of ice (section 3.1). The gas-phase formation of

H_2O could thus play a role, while it is much less efficient than the grain-surface reactions at <20 K. As for comets, the effect of sublimation processes and gas-phase reactions in the coma needs to be considered for the interpretation of the observed OPR ratios (*Hama and Watanabe,* 2013; *Hama et al.,* 2016).

4. OTHER OUTSTANDING ISSUES

4.1. From Molecular Clouds to Disks

It has long been discussed whether the molecular composition of volatiles in comets inherits ices in molecular clouds or is reset in the solar nebula (e.g., *Mumma and Charnley,* 2011). To pursue this question, we need to investigate not only Class II disks but also the disk formation stage (i.e., Class 0–I) (see section 1). Studies of forming disks have seen significant progress in the last decade.

On the theoretical side, disk formation has been studied by hydrodynamics simulations starting from the collapse of the parental cloud core, considering the effect of radiation transfer and magnetic fields to derive the physical structures (e.g., size, mass, temperatures) of the disks as a function of time (e.g., *Machida et al.*, 2011; *Tsukamoto et al.*, 2015, 2023; *Hennebelle et al.*, 2016). The molecular evolution of gas and ice is investigated by calculating the chemical reaction network along the gas flow from the cloud core to the disk, considering the temporal variation of density, temperature, and UV radiation flux along the trajectories (e.g., *Visser et al.*, 2009, 2011; *Yoneda et al.*, 2016; *Drozdovskaya et al.*, 2016; *Furuya et al.*, 2017; *Aikawa et al.*, 2020). Stable abundant ices such as H_2O and CO_2 are delivered to the disk without significant alteration, unless the trajectory is close to the outflow cavity where photodissociation is effective. [Outflows create a cavity (a pair of cone-shaped low-density regions) in the envelope. The wall of the outflow cavity could be irradiated by the UV radiation from the central object.] The abundance of COMs, on the other hand, could increase in the ice mantle as the thermal diffusion and reactions are activated in warm circumstellar conditions (e.g., *Garrod and Herbst*, 2006; *Garrod et al.*, 2022).

Observations of forming disks are more complicated than those of Class II, since the emission from the disk needs to be distinguished from the infalling envelope and outflow. Careful analysis of spatial distributions and velocity structures shows that emission lines of rare isotopes of CO (with optical depth lower than their normal isotope), H_2CO, and CS tend to probe the disk, while SO traces weak accretion shock onto the disk (*Sakai et al.*, 2014; *Aso et al.*, 2017; *Oya et al.*, 2018; *Harsono et al.*, 2021; *Tychoniec et al.*, 2021; *Garufi et al.*, 2021; *Yamato et al.,* 2023). The vertical stratification of molecules similar to that in Class II disks (section 3.1) is also found (*Podio et al.*, 2020). *van't Hoff et al.* (2020) show that the young disks tend to be warmer than Class II disks, with the CO snowline at several tens of astronomical units, possibly indicating a higher accretion rate (section 2.2). Disks with warped structures are also found (e.g., *Sakai et al.*, 2019; *Sai et al.*, 2020). Recent observations found large-scale (thousands of astronomical units) non-axisymmetric steamers onto young disks (*Yen et al.*, 2014; *Pineda et al.*, 2020; *Garufi et al.*, 2022). Such a flow provides chemically fresh interstellar material (e.g., carbon chains) and angular momentum that could perturb the disk structure to enhance mixing.

4.2. Estimation of Volatile Composition in Solids

At the time of writing this review, the direct observations of ices in Class II disks are relatively limited (section 3.3) compared with the line observations of gaseous species. Here we provide an overview and discuss how we can estimate the composition and evolution of volatiles in the solid phase based on gaseous observations.

Observations of fresh sublimates are the most straightforward to reveal the ice composition, and are possible during luminosity outbursts. The outbursts, which are caused by a sudden increase of mass accretion from the disk to the central star, are observed in Class I or the transition phase of Class I to Class II (section 2.1). While the water snowline is typically located at approximately a few astronomical units in Class II disks, it extends further away upon outburst. Around V883 Ori the dust continuum observation suggests that the water snowline is located at ~40 au (*Cieza et al.*, 2016). V883 Ori is a FU Ori-type star for which the duration of the outburst phase is ~100 years. Since it is shorter than typical timescales of the gas-phase chemistry, we can observe fresh sublimates. Various COMs (e.g., CH_3CHO and CH_3OCHO) are detected; their abundances relative to CH_3OH are comparable to those in comets (*Lee et al.,* 2019; *Yamato et al.,* 2024b). HDO is recently detected as well (*Tobin et al.,* 2023) (section 3.4.1).When the burst ceases, the COMs and water will again freeze on grain surfaces.

As described in section 3.1, observations of Class II disks outside the CO snowline indicate that carbon and oxygen are depleted in the warm molecular layer, most probably due to the sedimentation and radial drift of ice-coated dust grains. Observations of molecules such as C_2H suggest a high C/O ratio (>1) in the gas phase, which in turn suggests that the solid phase is oxygen-rich, e.g., with abundant water ice. The molecular composition of ice could then be determined by the competition of the timescales of dust growth, settling, and molecular evolution in the upper layers of the disk. If the grain growth and settling are fast enough, the ices inherited from molecular clouds might be stored and survive in the midplane, which is cold and shielded from UV and X-ray radiation. Interestingly, recent observations suggest that the growth and sedimentation of dust grains and CO depletion proceed over the short timescale from Class 0/I to II (*Bergner et al.*, 2020; *Zhang et al.*, 2020; *Harsono et al.*, 2021; *Ohashi et al.,* 2023).

It should be noted, however, that the freeze-out of gaseous molecules onto grain surfaces continues even after the dust sedimentation. While the turbulence in disks is weaker than previously expected (section 2.1), weak turbulence can bring molecules processed in the upper layers to the midplane, where they are frozen onto grains due to low temperature and high density (section 3.1). Chemical reactions inside the ice mantle also continue after the sedimentation. For example, impinging cosmic-ray particles cause radiolysis (*Shingledecker et al.*, 2018) and/or induce photochemistry (*Gredel et al.* 1989), although the flux of such energy source at the midplane is uncertain (*Cleeves et al.*, 2014; *Aikawa et al.*, 2021). As the grains migrate to inner radii, the gradual temperature rise enables the thermal diffusion and reactions of radicals inside cracks and pours of the ice mantle. *Krijt et al.* (2020) constructed numerical models that include pebble formation from small dust grains, radial drift of pebbles, diffusion of gas and dust, and simple gas-grain chemistry. They found that the heavy CO depletion observed in several disks is reproduced only in the model with the chemical conversion of CO to less volatile molecules and pebble sedimentation and radial drift. More recently, *Furuya et al.* (2022a) performed calculations with a larger chemical network to show that the depletion of nitrogen is less severe than that of carbon.

Inside the CO snowline, CO abundance varies among objects (*Zhang et al.*, 2019, 2021). In some disks, the CO abundance goes back to the canonical abundance, while it does not in other disks. Assuming that the gas/dust mass ratio is 100, there are two possibilities in the latter case: (1) CO is converted to less-volatile species or (2) CO ice is locked outside the CO snowline due to the planetesimal formation and/or dust trap at the local pressure maximum. For scenario (1), the product molecules would eventually sublimate when the ice-coated grains reach the inner warmer regions. For scenario (2), on the other hand, the elemental abundances of carbon and oxygen would be low even in the inner hot regions, ultimately at the inner edge of the disk and in the accreting material from the disk to the central star. *Bosman and Banzatti* (2019) and *McClure et al.* (2020) observed gas at the innermost radius of the TW Hya disk to find that both carbon and oxygen are depleted. This method works for other elements as well. *Kama et al.* (2019) analyzed the stellar spectra of young disk-hosting stars to find depletion of S, which suggests that 89 ± 8% of elemental S is in refractory form, e.g., FeS, and trapped in dust traps. The radial variation of elemental abundances should depend on the relative positions of snowlines of major ices and the dust trap or planetesimal formation. Ideally, observations of gases inside the possible comet-forming dust ring could tell us the amount of volatiles used to make comets.

5. BRIEF SUMMARY AND FUTURE DIRECTIONS

Studies on solar system material, including comets, and protoplanetary disks play complementary roles in exploring the formation of the solar system. While comets and asteroids are remnants or fragments of planetesimals, i.e., the building blocks of planetary systems, we can observe their ingredients, i.e., molecules, dust grains, and ices in disks. While the chemical composition of the solar system material contains rich but degenerated information about their formation environment, the statistical and spatially resolved observations of disks, including those around Class 0 and I objects, reveal the physical and chemical structures and their evolution. Variations of exoplanetary systems suggest that not all disks produce planetary systems like the solar system. Understanding basic physical and chemical processes in disks is essential to reveal the specific conditions for the solar system formation.

While ALMA has revealed the thermal emission of dust with subarcsecond resolution (~5 au), derivation of the mass distribution of dust grains depends on the opacity, which in turn depends on the size distribution of grains. The combination of the observations at various wavelengths, polarization, and theoretical studies on dust properties are crucial (section 2). In order to probe the dust in the inner radius, which tends to be optically thick even in millimeter wavelength, low-frequency observations by Band 1 ALMA and the Next Generation Very Large Array (ngVLA) are essential. These low-frequency instruments are also suited for observations of major volatile molecules such as NH_3 and CH_3OH. In recent years, various ring and linear molecules are newly detected in the line surveys of molecular cloud TMC-1 at the Green Bank Telescope and the Yabe 40-m telescope at ≤50 GHz (*Cernicharo et al.*, 2020; *McGuire et al.*, 2020). While the detection of new molecules in disks could be limited by sensitivity, since large molecules suitable for low-frequency observations tend to be frozen in the ice mantle, disks in outburst could be a good target.

Molecular line observations by ALMA are revealing the gas dynamics (section 2) and composition and isotope ratios of disk gas (section 3). Theoretical studies of disk chemistry and multi-wavelength observations are crucial to estimate the composition of solids, which is directly linked to comets (section 3 and section 4). It is clear that a significant fraction of dust grains are grown and decoupled from gas in Class II disks, which makes the elemental abundance heterogeneous. Combined analysis of various molecular lines is important to derive not only the chemical composition but also gas mass distributions in disks. One of the key probes of gas mass is the far-infrared line of HD (section 3.1), which would be observable with a new instrument on a suborbital platform, such as a balloon or a future far-infrared space mission (*Bergin et al.,* 2019).

In shorter wavelengths, the James Webb Space Telescope (JWST) is now providing a comprehensive view of inner disk chemistry in protoplanetary disks around stars of a wide range of masses. Active programs include the JWST Disk Infrared Spectroscopic Chemistry Survey (JDISCS) (*Pontoppidan et al.,* 2024) and the Mid-infrared Disk Survey (MINDS) (*Kamp et al.,* 2023). These surveys promise to yield inventories of a wide range of bulk volatile species and their isotopologs within the snowline, including water, organics, and nitrogen-bearing species. While the JWST observations will be able to survey disks around stars of all masses, its medium-resolution spectrometer will not be able to spectrally resolve individual lines. However, a new generation of high-resolution spectrometers on groundbased facilities will continue to offer complementary spectroscopy at high resolution. In particular, the new CRyogenic InfraRed Echelle Spectrograph (CRIRES) Upgrade project (CRIRES+) instrument (*Dorn* et al., 2014) on the European Southern Observatory Very Large Telescope (ESO-VLT), and in the future, the European Extremely Large Telescope's (ELT) Mid-Infrared ELT Imager and Spectrograph (METIS) instrument (*Brandl* et al., 2010) and the mid-infrared imager and spectrometer (MICHI) (*Packham et al.*, 2018) on the Thirty Meter Telescope will provide sensitive 3–12-μm spectroscopy of protoplanetary disk chemistry at high spectral and spatial resolution. Finally, the U.S. 2020 Decadal Survey recommended either a far-infrared or X-ray probe class misssion in the 2030s, and several concepts to trace the physics and chemistry of volatiles around the snowline using far-infrared spectroscopy are under consideration [e.g., the PRobe Far-Infrared Mission for Astrophysics (PRIMA) (*Glenn et al.,* 2023)].

Acknowledgments. Y.A. acknowledges support by NAOJ ALMA Scientific Research Grant Numbers 2019-13B, Grant-in-Aid for Scientific Research (S) 18H05222, (B) 24K00674, and Grant-in-Aid for Transformative Research Areas (A) 20H05844 and 20H05847. S.O. is supported by JSPS KAKENHI Grant Numbers JP18H05438 and JP20H00182. Part of this work was carried out at the Jet Propulsion Laboratory, California Institute of Technology, under a contract with the National Aeronautics and Space Administration (80NM0018D0004).

REFERENCES

Adachi I., Hayashi C., and Nakazawa K. (1976) The gas drag effect on the elliptical motion of a solid body in the primordial solar nebula. *Prog. Theor. Phys., 56,* 1756–1771.

Agúndez M., Cernicharo J., and Goicoechea J. R. (2008) Formation of simple organic molecules in inner T Tauri disks. *Astron. Astrophys., 483,* 831–837.

Aikawa Y., van Zadelhoff G. J., van Dishoeck E. F., et al. (2002) Warm molecular layers in protoplanetary disks. *Astron. Astrophys., 386,* 622–632.

Aikawa Y., Kamuro D., Sakon I., et al. (2012) AKARI observations of ice absorption bands towards edge-on young stellar objects. *Astron. Astrophys., 538,* A57.

Aikawa Y., Furuya K., Nomura H., et al. (2015) Analytical formulae of molecular ion abundances and the N_2H^+ ring in protoplanetary disks. *Astrophys. J., 807,* 120.

Aikawa Y., Furuya K., Hincelin U., et al. (2018) Multiple paths of deuterium fractionation in protoplanetary disks. *Astrophys. J., 855,* 119.

Aikawa Y., Furuya K., Yamamoto S., et al. (2020) Chemical variation among protostellar cores: Dependence on prestellar core conditions. *Astrophys. J., 897,* 110.

Aikawa Y., Cataldi G., Yamato Y., et al. (2021) Molecules with ALMA at Planet-forming Scales (MAPS). XIII. HCO^+ and disk ionization structure. *Astrophys. J. Suppl. Ser., 257,* 13.

ALMA Partnership, Brogan C. L., Pérez L. M., et al. (2015) The 2014 ALMA Long Baseline Campaign: First results from high angular resolution observations toward the HL Tau region. *Astrophys. J. Lett., 808,* L3.

Altwegg K., Balsiger H., Bar-Nun A., et al. (2015) 67P/Churyumov-Gerasimenko, a Jupiter family comet with a high D/H ratio. *Science, 347,* 1261952.

Altwegg K., Balsiger H., Berthelier J. J., et al. (2017) D_2O and HDS in the coma of 67P/Churyumov-Gerasimenko. *Philos. Trans. R. Soc. A, 375,* 20160253.

Anderson D. E., Blake G. A., Cleeves L. I., et al. (2021) Observing carbon and oxygen carriers in protoplanetary disks at mid-infrared wavelengths. *Astrophys. J., 909,* 55.

Andrews S. M. (2020) Observations of protoplanetary disk structures. *Annu. Rev. Astron. Astrophys., 58,* 483–528.

Andrews S. M. and Williams J. P. (2005) Circumstellar dust disks in Taurus-Auriga: The submillimeter perspective. *Astrophys. J., 631,* 1134–1160.

Andrews S. M. and Williams J. P. (2007) A submillimeter view of circumstellar dust disks in ρ Ophiuchi. *Astrophys. J., 671,* 1800–1812.

Andrews S. M., Huang J., Pérez L. M., et al. (2018) The Disk Substructures at High Angular Resolution Project (DSHARP). I. Motivation, sample, calibration, and overview. *Astrophys. J. Lett., 869,* L41.

Ansdell M., Williams J. P., van der Marel N., et al. (2016) ALMA survey of Lupus protoplanetary disks. I. Dust and gas masses. *Astrophys. J., 828,* 46.

Arakawa S. and Krijt S. (2021) On the stickiness of CO_2 and C_2O ice particles. *Astrophys. J., 910,* 130.

Arpigny C., Jehin E., Manfroid J., et al. (2003) Anomalous nitrogen isotope ratio in comets. *Science, 301,* 1522–1525.

Aso Y., Ohashi N., Aikawa Y., et al. (2017) ALMA observations of the protostar L1527 IRS: Probing details of the disk and the envelope structures. *Astrophys. J., 849,* 56.

Avenhaus H., Quanz S. P., Garufi A., et al. (2018) Disks around T Tauri Stars with SPHERE (DARTTS-S). I. SPHERE/IRDIS polarimetric imaging of eight prominent T Tauri disks. *Astrophys. J., 863,* 44.

Bai X.-N. (2011) Magnetorotational-instability-driven accretion in protoplanetary disks. *Astrophys. J., 739,* 50.

Bai X.-N. (2013) Wind-driven accretion in protoplanetary disks. II. Radial dependence and global picture. *Astrophys. J., 772,* 96.

Bai X.-N. (2014) Hall-effect-controlled gas dynamics in protoplanetary disks. I. Wind solutions at the inner disk. *Astrophys. J., 791,* 137.

Bai X.-N. (2017) Global simulations of the inner regions of protoplanetary disks with comprehensive disk microphysics. *Astrophys. J., 845,* 75.

Bai X.-N. and Stone J. M. (2013) Wind-driven accretion in protoplanetary disks. I. Suppression of the magnetorotational instability and launching of the magnetocentrifugal wind. *Astrophys. J., 769,* 76.

Bai X.-N. and Stone J. M. (2017) Hall effect-mediated magnetic flux transport in protoplanetary disks. *Astrophys. J., 836,* 46.

Balbus S. A. and Hawley J. F. (1991) A powerful local shear instability in weakly magnetized disks. I. Linear analysis. *Astrophys. J., 376,* 214–222.

Ballering N. P., Cleeves L. I., and Anderson D. E. (2021) Simulating observations of ices in protoplanetary disks. *Astrophys. J., 920,* 115.

Banzatti A., Pascucci I., Bosman A. D., et al. (2020) Hints for icy pebble migration feeding an oxygen-rich chemistry in the inner planet-forming region of disks. *Astrophys. J., 903,* 124.

Banzatti A., Pontoppidan K. M., Carr J. S., et al. (2023) JWST reveals excess cool water near the snow line in compact disks, consistent with pebble drift. *Astrophys. J. Lett., 957,* L22.

Barranco J. A., Pei S., and Marcus P. S. (2018) Zombie vortex instability. III. Persistence with nonuniform stratification and radiative damping. *Astrophys. J., 869,* 127.

Bauer I., Finocchi F., Duschl W. J., et al. (1997) Simulation of chemical reactions and dust destruction in protoplanetary accretion disks. *Astron. Astrophys., 317,* 273–289.

Beckwith S. V. W. and Sargent A. I. (1991) Particle emissivity in circumstellar disks. *Astrophys. J., 381,* 250–258.

Beckwith S. V. W., Sargent A. I., Chini R. S., et al. (1990) A survey for circumstellar disks around young stellar objects. *Astron. J., 99,* 924–945.

Bell K. R. and Lin D. N. C. (1994) Using FU Orionis outbursts to constrain self-regulated protostellar disk models. *Astrophys. J., 427,* 987–1004.

Bergin E. A., Cleeves L. I., Gorti U., et al. (2013) An old disk still capable of forming a planetary system. *Nature, 493,* 644–646.
Bergin E. A., Du F., Cleeves L. I., et al. (2016) Hydrocarbon emission rings in protoplanetary disks induced by dust evolution. *Astrophys. J., 831,* 101.
Bergin E., Pontoppidan K., Bradford C., et al. (2019) The disk gas mass and the far-IR revolution. In *Astro2020: Decadal Survey on Astronomy and Astrophysics, Science White Papers, No. 222, Bull. Am. Astron. Soc., 51,* id 222.
Bergner J. B., Guzmán V. G., Öberg K. I., et al. (2018) A survey of CH_3CN and HC_3N in protoplanetary disks. *Astrophys. J., 857,* 69.
Bergner J. B., Öberg K. I., Bergin E. A., et al. (2019) A survey of C_2H, HCN, and $C^{18}O$ in protoplanetary disks. *Astrophys. J., 876,* 25.
Bergner J. B., Öberg K. I., Bergin E. A., et al. (2020) An evolutionary study of volatile chemistry in protoplanetary disks. *Astrophys. J., 898,* 97.
Bergner J. B., Öberg K. I., Guzmán V. V., et al. (2021) Molecules with ALMA at Planet-forming Scales (MAPS). XI. CN and HCN as tracers of photochemistry in disks. *Astrophys. J. Suppl. Ser., 257,* 11.
Béthune W. and Latter H. (2020) Electric heating and angular momentum transport in laminar models of protoplanetary discs. *Mon. Not. R. Astron. Soc., 494,* 6103–6119.
Birnstiel T., Dullemond C. P., and Brauer F. (2010) Gas- and dust evolution in protoplanetary disks. *Astron. Astrophys., 513,* A79.
Birnstiel T., Klahr H., and Ercolano B. (2012) A simple model for the evolution of the dust population in protoplanetary disks. *Astron. Astrophys., 539,* A148.
Birnstiel T., Dullemond C. P., Zhu Z., et al. (2018) The Disk Substructures at High Angular Resolution Project (DSHARP). V. Interpreting ALMA maps of protoplanetary disks in terms of a dust model. *Astrophys. J. Lett., 869,* L45.
Bischoff D., Kreuzig C., Haack D., et al. (2020) Sticky or not sticky? Measurements of the tensile strength of microgranular organic materials. *Mon. Not. R. Astron. Soc., 497,* 2517–2528.
Blake G. A. and Boogert A. C. A. (2004) High-resolution 4.7 micron Keck/NIRSPEC spectroscopy of the CO emission from the disks surrounding Herbig Ae stars. *Astrophys. J. Lett., 606,* L73–L76.
Blandford R. D. and Payne D. G. (1982) Hydromagnetic flows from accretion disks and the production of radio jets. *Mon. Not. R. Astron. Soc., 199,* 883–903.
Blevins S. M., Pontoppidan K. M., Banzatti A., et al. (2016) Measurements of water surface snow lines in classical protoplanetary disks. *Astrophys. J., 818,* 22.
Blum J. and Wurm G. (2000) Experiments on sticking, restructuring, and fragmentation of preplanetary dust aggregates. *Icarus, 143,* 138–146.
Bockelée-Morvan D., Biver N., Jehin E., et al. (2008) Large excess of heavy nitrogen in both hydrogen cyanide and cyanogen from comet 17P/Holmes. *Astrophys. J. Lett., 679,* L49–L52.
Booth A. S., Walsh C., Kama M., et al. (2018) Sulphur monoxide exposes a potential molecular disk wind from the planet-hosting disk around HD 100546. *Astron. Astrophys., 611,* A16.
Booth A. S., Walsh C., Ilee J. D., et al. (2019) The first detection of $^{13}C^{17}O$ in a protoplanetary disk: A robust tracer of disk gas mass. *Astrophys. J. Lett., 882,* L31.
Booth A. S., van der Marel N., Leemker M., et al. (2021a) A major asymmetric ice trap in a planet-forming disk. II. Prominent SO and SO_2 pointing to C/O < 1. *Astron. Astrophys., 651,* L6.
Booth A. S., Walsh C., Terwisscha van Scheltinga J., et al. (2021b) An inherited complex organic molecule reservoir in a warm planet-hosting disk. *Nature Astron., 5,* 684–690.
Booth A. S., Law C. J., Temmink M., Leemker M., and Macías E. (2023) Tracing snowlines and C/O ratio in a planet-hosting disk. *Astron. Astrophys., 678,* A146.
Bosman A. D. and Banzatti A. (2019) The dry and carbon-poor inner disk of TW Hydrae: Evidence for a massive icy dust trap. *Astron. Astrophys., 632,* L10.
Bosman A. D., Bruderer S., and van Dishoeck E. F. (2017) CO_2 infrared emission as a diagnostic of planet-forming regions of disks. *Astron. Astrophys., 601,* A36.
Brandl B. R., Lenzen R., Pantin E., et al. (2010) Instrument concept and science case of the mid-IR E-ELT imager and spectrograph METIS. In *Ground-Based and Airborne Instrumentation for Astronomy III* (I. S. McLean et al., eds.), p. 77352G. SPIE Conf. Ser. 7735, Bellingham, Washington.
Brauer F., Dullemond C. P., and Henning T. (2008) Coagulation, fragmentation and radial motion of solid particles in protoplanetary disks. *Astron. Astrophys., 480,* 859–877.
Bridges J. C., Changela H. G., Nayakshin S., et al. (2012) Chondrule fragments from Comet Wild2: Evidence for high temperature processing in the outer solar system. *Earth Planet. Sci. Lett., 341,* 186–194.
Brittain S. D., Rettig T. W., Simon T., et al. (2003) CO emission from disks around AB Aurigae and HD 141569: Implications for disk structure and planet formation timescales. *Astrophys. J., 588,* 535–544.
Brown J. M., Pontoppidan K. M., van Dishoeck E. F., et al. (2013) VLT-CRIRES survey of rovibrational CO emission from protoplanetary disks. *Astrophys. J., 770,* 94.
Brunken N. G. C., Booth A. S., Leemker M., et al. (2022) A major asymmetric ice trap in a planet-forming disk. III. First detection of dimethyl ether. *Astron. Astrophys., 659,* A29.
Brunngräber R. and Wolf S. (2020) Self-scattering in protoplanetary disks with dust settling. *Astron. Astrophys., 640,* A122.
Brunngräber R. and Wolf S. (2021) Self-scattering on large, porous grains in protoplanetary disks with dust settling. *Astron. Astrophys., 648,* A87.
Canta A., Teague R., Le Gal R., et al. (2021) The first detection of CH_2CN in a protoplanetary disk. *Astrophys. J., 922,* 62.
Carney M. T., Hogerheijde M. R., Guzmán V. V., et al. (2019) Upper limits on CH_3OH in the HD 163296 protoplanetary disk. Evidence for a low gas-phase CH_3OH-to-H_2CO ratio. *Astron. Astrophys., 623,* A124.
Carr J. S. and Najita J. R. (2008) Organic molecules and water in the planet formation region of young circumstellar disks. *Science, 319,* 1504–1506.
Carr J. S. and Najita J. R. (2011) Organic molecules and water in the inner disks of T Tauri stars. *Astrophys. J., 733,* 102.
Carr J. S., Tokunaga A. T., and Najita J. (2004) Hot H_2O emission and evidence for turbulence in the disk of a young star. *Astrophys. J., 603,* 213–220.
Carrasco-González C., Sierra A., Flock M., et al. (2019) The radial distribution of dust particles in the HL Tau disk from ALMA and VLA observations. *Astrophys. J., 883,* 71.
Cataldi G., Yamato Y., Aikawa Y., et al. (2021) Molecules with ALMA at Planet-forming Scales (MAPS). X. Studying deuteration at high angular resolution toward protoplanetary disks. *Astrophys. J. Suppl. Ser., 257,* 10.
Cauley P. W., France K., Herzceg G. J., et al. (2021) A CO-to-H_2 ratio of $\approx 10^{-5}$ toward the Herbig Ae star HK Ori. *Astron. J., 161,* 217.
Cazzoletti P., van Dishoeck E. F., Pinilla P., et al. (2018a) Evidence for a massive dust-trapping vortex connected to spirals: Multi-wavelength analysis of the HD 135344B protoplanetary disk. *Astron. Astrophys., 619,* A161.
Cazzoletti P., van Dishoeck E. F., Visser R., et al. (2018b) CN rings in full protoplanetary disks around young stars as probes of disk structure. *Astron. Astrophys., 609,* A93.
Ceccarelli C., Caselli P., Bockelée-Morvan D., et al. (2014) Deuterium fractionation: The Ariadne's thread from the precollapse phase to meteorites and comets today. In *Protostars and Planets VI* (H. Beuther et al., eds.), pp. 859–882. Univ. of Arizona, Tucson.
Cernicharo J., Marcelino N., Agúndez M., et al. (2020) Discovery of HC_4NC in TMC-1: A study of the isomers of HC_3N, HC_5N, and HC_7N. *Astron. Astrophys., 642,* L8.
Chapillon E., Dutrey A., Guilloteau S., et al. (2012) Chemistry in disks. VII. First detection of HC_3N in protoplanetary disks. *Astrophys. J., 756,* 58.
Chiang E. I. and Goldreich P. (1997) Spectral energy distributions of T Tauri stars with passive circumstellar disks. *Astrophys. J., 490,* 368–376.
Cieza L. A., Casassus S., Tobin J., et al. (2016) Imaging the water snow-line during a protostellar outburst. *Nature, 535,* 258–261.
Cieza L. A., González-Ruilova C., Hales A. S., et al. (2021) The Ophiuchus DIsc Survey Employing ALMA (ODISEA) — III. The evolution of substructures in massive discs at 3-5 au resolution. *Mon. Not. R. Astron. Soc., 501,* 2934–2953.
Cleeves L. I., Bergin E. A., and Adams F. C. (2014) Exclusion of cosmic rays in protoplanetary disks. II. Chemical gradients and observational signatures. *Astrophys. J., 794,* 123.
Cleeves L. I., Öberg K. I., Wilner D. J., et al. (2018) Constraining gas-phase carbon, oxygen, and nitrogen in the IM Lup protoplanetary

disk. *Astrophys. J., 865*, 155.
Cleeves L. I., Loomis R. A., Teague R., et al. (2021) The TW Hya Rosetta Stone Project IV: A hydrocarbon-rich disk atmosphere. *Astrophys. J., 911*, 29.
Cody G. D., Heying E., Alexander C. M. O., et al. (2011) Cosmochemistry special feature: Establishing a molecular relationship between chondritic and cometary organic solids. *Proc. Natl. Acad. Sci. U. S. A., 108*, 19171–19176.
Connelly J. N., Bizzarro M., Krot A. N., et al. (2012) The absolute chronology and thermal processing of solids in the solar protoplanetary disk. *Science, 338*, 651–655.
Cox E. G., Harris R. J., Looney L. W., et al. (2018) ALMA's polarized view of 10 protostars in the Perseus molecular cloud. *Astrophys. J., 855*, 92.
Crovisier J., Bockelée-Morvan D., Colom P., et al. (2004) The composition of ices in comet C/1995 O1 (Hale-Bopp) from radio spectroscopy: Further results and upper limits on undetected species. *Astron. Astrophys., 418*, 1141–1157.
Cuzzi J. N. and Zahnle K. J. (2004) Material enhancement in protoplanetary nebulae by particle drift through evaporation fronts. *Astrophys. J., 614*, 490–496.
Dent W. R. F., Pinte C., Cortes P. C., et al. (2019) Submillimetre dust polarization and opacity in the HD163296 protoplanetary ring system. *Mon. Not. R. Astron. Soc., 482*, L29–L33.
Desch S. J. and Turner N. J. (2015) High-temperature ionization in protoplanetary disks. *Astrophys. J., 811*, 156.
Doi K. and Kataoka A. (2021) Estimate on dust scale height from the ALMA dust continuum image of the HD 163296 protoplanetary disk. *Astrophys. J., 912*, 164.
Dominik C. and Tielens A. G. G. M. (1997) The physics of dust coagulation and the structure of dust aggregates in space. *Astrophys. J., 480*, 647–673.
Dorn R. J., Anglada-Escude G., Baade D., et al. (2014) CRIRES+: Exploring the cold universe at high spectral resolution. *The Messenger, 156*, 7–11.
Draine B. T. (2006) On the submillimeter opacity of protoplanetary disks. *Astrophys. J., 636*, 1114–1120.
Drążkowska J. and Alibert Y. (2017) Planetesimal formation starts at the snow line. *Astron. Astrophys., 608*, A92.
Drążkowska J., Bitsch B., Lambrechts M., et al. (2023) Planet formation theory in the era of ALMA and Kepler: From pebbles to exoplanets. In *Protostars and Planets VII* (S.-i. Inutsuka et al., eds.), pp. 707–758. ASP Conf. Ser. 534, Astronomical Society of the Pacific, San Francisco.
Drozdovskaya M. N., Walsh C., van Dishoeck E. F., et al. (2016) Cometary ices in forming protoplanetary disc midplanes. *Mon. Not. R. Astron. Soc., 462*, 977–993.
Du F., Bergin E. A., Hogerheijde M., et al. (2017) Survey of cold water lines in protoplanetary disks: Indications of systematic volatile depletion. *Astrophys. J., 842*, 98.
Dullemond C. P. and Dominik C. (2005) Dust coagulation in protoplanetary disks: A rapid depletion of small grains. *Astron. Astrophys., 434*, 971–986.
Dullemond C. P., Birnstiel T., Huang J., et al. (2018) The Disk Substructures at High Angular Resolution Project (DSHARP). VI. Dust trapping in thin-ringed protoplanetary disks. *Astrophys. J. Lett., 869*, L46.
Dutrey A., Guilloteau S., and Guelin M. (1997) Chemistry of protosolar-like nebulae: The molecular content of the DM Tau and GG Tau disks. *Astron. Astrophys., 317*, L55–L58.
Dutrey A., Guilloteau S., Piétu V., et al. (2017) The flying saucer: Tomography of the thermal and density gas structure of an edge-on protoplanetary disk. *Astron. Astrophys., 607*, A130.
Evans Neal J. I., Dunham M. M., Jørgensen J. K., et al. (2009) The Spitzer c2d Legacy results: Star-formation rates and efficiencies: Evolution and lifetimes. *Astrophys. J. Suppl. Ser., 181*, 321–350.
Favre C., Cleeves L. I., Bergin E. A., et al. (2013) A significantly low CO abundance toward the TW Hya protoplanetary disk: A path to active carbon chemistry? *Astrophys. J. Lett., 776*, L38.
Favre C., Fedele D., Semenov D., et al. (2018) First detection of the simplest organic acid in a protoplanetary disk. *Astrophys. J. Lett., 862*, L2.
Fedele D., Bruderer S., van Dishoeck E. F., et al. (2013) DIGIT survey of far-infrared lines from protoplanetary disks. I. [O I], [C II], OH, H_2O, and CH^+. *Astron. Astrophys., 559*, A77.
Fischer W. J., Hillenbrand L. A., Herczeg G. J., Johnstone D., Kóspál Á., and Dunham M. M. (2023) Accretion variability as a guide to stellar mass assembly. In *Protostars and Planets VII* (S. Inutsuka et al., eds.), pp. 355–377. Astron. Soc. Pacific Conf. Ser. 534, San Francisco, California.
Flaherty K. M., Hughes A. M., Rosenfeld K. A., et al. (2015) Weak turbulence in the HD 163296 protoplanetary disk revealed by ALMA CO observations. *Astrophys. J., 813*, 99.
Flaherty K. M., Hughes A. M., Rose S. C., et al. (2017) A three-dimensional view of turbulence: Constraints on turbulent motions in the HD 163296 protoplanetary disk using DCO^+. *Astrophys. J., 843*, 150.
Flaherty K. M., Hughes A. M., Teague R., et al. (2018) Turbulence in the TW Hya disk. *Astrophys. J., 856*, 117.
Flaherty K., Hughes A. M., Simon J. B., et al. (2020) Measuring turbulent motion in planet-forming disks with ALMA: A detection around DM Tau and nondetections around MWC 480 and V4046 Sgr. *Astrophys. J., 895*, 109.
Flaherty K., Hughes A. M., Simon J. B., et al. (2024) Evidence for non-zero turbulence in the protoplanetary disc around IM Lup. *Mon. Not. R. Astron. Soc., 532*, 363.
Flock M., Ruge J. P., Dzyurkevich N., et al. (2015) Gaps, rings, and non-axisymmetric structures in protoplanetary disks: From simulations to ALMA observations. *Astron. Astrophys., 574*, A68.
Flock M., Turner N. J., Nelson R. P., et al. (2020) Gas and dust dynamics in starlight-heated protoplanetary disks. *Astrophys. J., 897*, 155.
Flores C., Duchêne G., Wolff S., et al. (2021) The anatomy of an unusual edge-on protoplanetary disk. II. Gas temperature and a warm outer region. *Astron. J., 161*, 239.
France K., Herczeg G. J., McJunkin M., et al. (2014) CO/H_2 abundance ratio $\approx 10^{-4}$ in a protoplanetary disk. *Astrophys. J., 794*, 160.
Fritscher M. and Teiser J. (2021) CO_2-ice collisions: A new experimental approach. *Astrophys. J., 923*, 134.
Fu R. R., Weiss B. P., Lima E. A., et al. (2014) Solar nebula magnetic fields recorded in the Semarkona meteorite. *Science, 346*, 1089–1092.
Fuente A., Cernicharo J., Agúndez M., et al. (2010) Molecular content of the circumstellar disk in AB Aurigae: First detection of SO in a circumstellar disk. *Astron. Astrophys., 524*, A19.
Fukuhara Y., Okuzumi S., and Ono T. (2021) Effects of dust evolution on the vertical shear instability in the outer regions of protoplanetary disks. *Astrophys. J., 914*, 132.
Fung J. and Chiang E. (2016) Gap opening in 3D: Single-planet gaps. *Astrophys. J., 832*, 105.
Furlan E., Fischer W. J., Ali B., et al. (2016) The Herschel Orion Protostar Survey: Spectral energy distributions and fits using a grid of protostellar models. *Astrophys. J. Suppl. Ser., 224*, 5.
Furuya K. and Aikawa Y. (2014) Reprocessing of ices in turbulent protoplanetary disks: Carbon and nitrogen chemistry. *Astrophys. J., 790*, 97.
Furuya K., van Dishoeck E. F., and Aikawa Y. (2016) Reconstructing the history of water ice formation from HDO/H_2O and D_2O/HDO ratios in protostellar cores. *Astron. Astrophys., 586*, A127.
Furuya K., Drozdovskaya M. N., Visser R., et al. (2017) Water delivery from cores to disks: Deuteration as a probe of the prestellar inheritance of H_2O. *Astron. Astrophys., 599*, A40.
Furuya K., Lee S., and Nomura H. (2022a) Different degrees of nitrogen and carbon depletion in the warm molecular layers of protoplanetary disks. *Astrophys. J., 938*, 29.
Furuya K., Tsukagoshi T., Qi C., et al. (2022b) Detection of $HC^{18}O^+$ in a protoplanetary disk: Exploring oxygen isotope fractionation of CO. *Astrophys. J., 926*, 148.
Gammie C. F. (1996) Layered accretion in T Tauri disks. *Astrophys. J., 457*, 355–362.
Garaud P. and Lin D. N. C. (2007) The effect of internal dissipation and surface irradiation on the structure of disks and the location of the snow line around Sun-like stars. *Astrophys. J., 654*, 606–624.
Garcia A. J. L. and Gonzalez J.-F. (2020) Evolution of porous dust grains in protoplanetary discs — I. Growing grains. *Mon. Not. R. Astron. Soc., 493*, 1788–1800.
Garrod R. T. and Herbst E. (2006) Formation of methyl formate and other organic species in the warm-up phase of hot molecular cores. *Astron. Astrophys., 457*, 927–936.
Garrod R. T., Jin M., Matis K. A., et al. (2022) Formation of complex

organic molecules in hot molecular cores through non-diffusive grain-surface and ice-mantle chemistry. *Astrophys. J. Suppl. Ser., 259*, 1.
Garufi A., Benisty M., Pinilla P., et al. (2018) Evolution of protoplanetary disks from their taxonomy in scattered light: Spirals, rings, cavities, and shadows. *Astron. Astrophys., 620*, A94.
Garufi A., Podio L., Codella C., et al. (2021) ALMA chemical survey of disk-outflow sources in Taurus (ALMA-DOT). V. Sample, overview, and demography of disk molecular emission. *Astron. Astrophys., 645*, A145.
Garufi A., Podio L., Codella C., et al. (2022) ALMA chemical survey of disk-outflow sources in Taurus (ALMA-DOT). VI. Accretion shocks in the disk of DG Tau and HL Tau. *Astron. Astrophys., 658*, A104.
Gibb E. L. and Horne D. (2013) Detection of CH_4 in the GV Tau N protoplanetary disk. *Astrophys. J. Lett., 776*, L28.
Giuliano B. M., Martín-Doménech R., Escribano R. M., et al. (2016) Interstellar ice analogs: H_2O ice mixtures with CH_3OH and NH_3 in the far-IR region. *Astron. Astrophys., 592*, A81.
Glenn J., Bradford C., Pope A., et al. (2023) PRIMA: The PRobe Infrared Mission for Astrophysics. American Astronomical Society Meeting #241, id. 160.08. *Bull. Am. Astron. Soc., 55*, e-id 2023n2i160p08.
Gonzalez J. F., Laibe G., and Maddison S. T. (2017) Self-induced dust traps: Overcoming planet formation barriers. *Mon. Not. R. Astron. Soc., 467*, 1984–1996.
Goto M., Usuda T., Dullemond C. P., et al. (2006) Inner rim of a molecular disk spatially resolved in iInfrared CO emission lines. *Astrophys. J., 652*, 758–762.
Grant S. L., van Dishoeck E. F., Tabone B., et al. (2023) MINDS. The detection of $^{13}CO_2$ with JWST-MIRI indicates abundant CO_2 in a protoplanetary disk. *Astrophys. J. Lett., 947*, L6.
Gredel R., Lepp S., Dalgarno A., et al. (1989) Cosmic-ray-induced photodissociation and photoionization rates of interstellar molecules. *Astrophys. J., 347*, 289–293.
Gressel O., Turner N. J., Nelson R. P., et al. (2015) Global simulations of protoplanetary disks with Ohmic resistivity and ambipolar diffusion. *Astrophys. J., 801*, 84.
Guidi G., Tazzari M., Testi L., et al. (2016) Dust properties across the CO snowline in the HD 163296 disk from ALMA and VLA observations. *Astron. Astrophys., 588*, A112.
Guilet J. and Ogilvie G. I. (2014) Global evolution of the magnetic field in a thin disc and its consequences for protoplanetary systems. *Mon. Not. R. Astron. Soc., 441*, 852–868.
Guilloteau S., Dutrey A., Wakelam V., et al. (2012) Chemistry in disks. VIII. The CS molecule as an analytic tracer of turbulence in disks. *Astron. Astrophys., 548*, A70.
Guilloteau S., Di Folco E., Dutrey A., et al. (2013) A sensitive survey for ^{13}CO, CN, H_2CO, and SO in the disks of T Tauri and Herbig Ae stars. *Astron. Astrophys., 549*, A92.
Guilloteau S., Reboussin L., Dutrey A., et al. (2016) Chemistry in disks. X. The molecular content of protoplanetary disks in Taurus. *Astron. Astrophys., 592*, A124.
Gundlach B., Schmidt K. P., Kreuzig C., et al. (2018) The tensile strength of ice and dust aggregates and its dependence on particle properties. *Mon. Not. R. Astron. Soc., 479*, 1273–1277.
Güttler C., Blum J., Zsom A., et al. (2010) The outcome of protoplanetary dust growth: Pebbles, boulders, or planetesimals? I. Mapping the zoo of laboratory collision experiments. *Astron. Astrophys., 513*, A56.
Guzmán V. V., Öberg K. I., Huang J., et al. (2017) Nitrogen fractionation in protoplanetary disks from the $H^{13}CN/HC^{15}N$ ratio. *Astrophys. J., 836*, 30.
Guzmán V. V., Bergner J. B., Law C. J., et al. (2021) Molecules with ALMA at Planet-forming Scales (MAPS). VI. Distribution of the small organics HCN, C_2H, and H_2CO. *Astrophys. J. Suppl. Ser., 257*, 6.
Hama T. and Watanabe N. (2013) Surface processes on interstellar amorphous solid water: Adsorption, diffusion, tunneling reactions, and nuclear-spin conversion. *Chem. Rev., 113*, 8783–8839.
Hama T., Kouchi A., and Watanabe N. (2016) Statistical ortho- to-para ratio of water desorbed from ice at 10 Kelvin. *Science, 351*, 65–67.
Hama T., Kouchi A., and Watanabe N. (2018) The ortho-to-para ratio of water molecules desorbed from ice made from para-water monomers at 11 K. *Astrophys. J. Lett., 857*, L13.
Harris R. J., Cox E. G., Looney L. W., et al. (2018) ALMA observations of polarized 872 μm dust emission from the protostellar systems VLA 1623 and L1527. *Astrophys. J., 861*, 91.
Harsono D., van der Wiel M. H. D., Bjerkeli P., et al. (2021) Resolved molecular line observations reveal an inherited molecular layer in the young disk around TMC1A. *Astron. Astrophys., 646*, A72.
Hartmann L., Herczeg G., and Calvet N. (2016) Accretion onto pre-main-sequence stars. *Annu. Rev. Astron. Astrophys., 54*, 135–180.
Hayashi C. (1981) Structure of the solar nebula, growth and decay of magnetic fields and effects of magnetic and turbulent viscosities on the nebula. *Prog. Theor. Phys. Suppl., 70*, 35–53.
Heays A. N., Visser R., Gredel R., et al. (2014) Isotope selective photodissociation of N_2 by the interstellar radiation field and cosmic rays. *Astron. Astrophys., 562*, A61.
Hennebelle P., Commerçon B., Chabrier G., et al. (2016) Magnetically self-regulated formation of early protoplanetary disks. *Astrophys. J. Lett., 830*, L8.
Henning T. and Semenov D. (2013) Chemistry in protoplanetary disks. *Chem. Rev., 113*, 9016–9042.
Hily-Blant P., Magalhaes V., Kastner J., et al. (2017) Direct evidence of multiple reservoirs of volatile nitrogen in a protosolar nebula analogue. *Astron. Astrophys., 603*, L6.
Hily-Blant P., Magalhaes de Souza V., Kastner J., et al. (2019) Multiple nitrogen reservoirs in a protoplanetary disk at the epoch of comet and giant planet formation. *Astron. Astrophys., 632*, L12.
Hirose S. and Turner N. J. (2011) Heating and cooling protostellar disks. *Astrophys. J. Lett., 732*, L30.
Hogerheijde M. R., Bergin E. A., Brinch C., et al. (2011) Detection of the water reservoir in a forming planetary system. *Science, 334*, 338–340.
Homma K. A., Okuzumi S., Nakamoto T., et al. (2019) Rocky planetesimal formation aided by organics. *Astrophys. J., 877*, 128.
Honda M., Inoue A. K., Fukagawa M., et al. (2009) Detection of water ice grains on the surface of the circumstellar disk around HD 142527. *Astrophys. J. Lett., 690*, L110–L113.
Honda M., Kudo T., Takatsuki S., et al. (2016) Water ice at the surface of the HD 100546 disk. *Astrophys. J., 821*, 2.
Hsieh T.-H., Murillo N. M., Belloche A., et al. (2019) Chronology of episodic accretion in protostars—An ALMA survey of the CO and H_2O snowlines. *Astrophys. J., 884*, 149.
Huang J. and Öberg K. I. (2015) Detection of N_2D^+ in a protoplanetary disk. *Astrophys. J. Lett., 809*, L26.
Huang J., Öberg K. I., Qi C., et al. (2017) An ALMA survey of DCN/$H^{13}CN$ and $DCO^+/H^{13}CO^+$ in protoplanetary disks. *Astrophys. J., 835*, 231.
Huang J., Andrews S. M., Dullemond C. P., et al. (2018) The Disk Substructures at High Angular Resolution Project (DSHARP). II. Characteristics of annular substructures. *Astrophys. J. Lett., 869*, L42.
Huang J., Andrews S. M., Dullemond C. P., et al. (2020) A multifrequency ALMA characterization of substructures in the GM Aur protoplanetary disk. *Astrophys. J., 891*, 48.
Hudgins D. M., Sandford S. A., Allamandola L. J., et al. (1993) Mid- and far-infrared spectroscopy of ices: Optical constants and integrated absorbances. *Astrophys. J. Suppl. Ser., 86*, 713.
Hughes A. M., Wilner D. J., Andrews S. M., et al. (2011) Empirical constraints on turbulence in protoplanetary accretion disks. *Astrophys. J., 727*, 85.
Hull C. L. H., Yang H., Li Z.-Y., et al. (2018) ALMA observations of polarization from dust scattering in the IM Lup protoplanetary disk. *Astrophys. J., 860*, 82.
Hyodo R., Ida S., and Charnoz S. (2019) Formation of rocky and icy planetesimals inside and outside the snow line: Effects of diffusion, sublimation, and back-reaction. *Astron. Astrophys., 629*, A90.
Hyodo R., Guillot T., Ida S., et al. (2021) Planetesimal formation around the snow line. II. Dust or pebbles? *Astron. Astrophys., 646*, A14.
Ida S. and Guillot T. (2016) Formation of dust-rich planetesimals from sublimated pebbles inside of the snow line. *Astron. Astrophys., 596*, L3.
Ida S., Guillot T., Hyodo R., et al. (2021) Planetesimal formation around the snow line. I. Monte Carlo simulations of silicate dust pile-up in a turbulent disk. *Astron. Astrophys., 646*, A13.
Ilee J. D., Walsh C., Booth A. S., et al. (2021) Molecules with ALMA at Planet-forming Scales (MAPS). IX. Distribution and properties of the large organic molecules HC_3N, CH_3CN, and $c\text{-}C_3H_2$. *Astrophys. J. Suppl. Ser., 257*, 9.

Ilgner M. and Nelson R. P. (2006) On the ionisation fraction in protoplanetary disks. I. Comparing different reaction networks. *Astron. Astrophys., 445,* 205–222.
Inoue A. K., Honda M., Nakamoto T., et al. (2008) Observational possibility of the "snow line" on the surface of circumstellar disks with the scattered light. *Publ. Astron. Soc. Japan, 60,* 557–563.
Isella A. and Natta A. (2005) The shape of the inner rim in protoplanetary disks. *Astron. Astrophys., 438,* 899–907.
Johansen A., Youdin A., and Klahr H. (2009) Zonal flows and long-lived axisymmetric pressure bumps in magnetorotational turbulence. *Astrophys. J., 697,* 1269–1289.
Johansen A., Blum J., Tanaka H., et al. (2014) The multifaceted planetesimal formation process. In *Protostars and Planets VI* (H. Beuther et al., eds.), pp. 547–570. Univ. of Arizona, Tucson.
Kama M., Bruderer S., van Dishoeck E. F., et al. (2016) Volatile-carbon locking and release in protoplanetary disks: A study of TW Hya and HD 100546. *Astron. Astrophys., 592,* A83.
Kama M., Shorttle O., Jermyn A. S., et al. (2019) Abundant refractory sulfur in protoplanetary disks. *Astrophys. J., 885,* 114.
Kamp I. and Dullemond C. P. (2004) The gas temperature in the surface layers of protoplanetary disks. *Astrophys. J., 615,* 991–999.
Kamp I., Henning T., Arabhavi A. M., et al. (2023) The chemical inventory of the inner regions of planet-forming disks — The JWST/MINDS program. *Faraday Discussions, 245,* 112–137.
Kataoka A., Tanaka H., Okuzumi S., et al. (2013) Fluffy dust forms icy planetesimals by static compression. *Astron. Astrophys., 557,* L4.
Kataoka A., Muto T., Momose M., et al. (2015) Millimeter-wave polarization of protoplanetary disks due to dust scattering. *Astrophys. J., 809,* 78.
Kimura H., Wada K., Senshu H., et al. (2015) Cohesion of amorphous silica spheres: Toward a better understanding of the coagulation growth of silicate dust aggregates. *Astrophys. J., 812,* 67.
Kirchschlager F. and Bertrang G. H. M. (2020) Self-scattering of non-spherical dust grains: The limitations of perfect compact spheres. *Astron. Astrophys., 638,* A116.
Klahr H. and Hubbard A. (2014) Convective overstability in radially stratified accretion disks under thermal relaxation. *Astrophys. J., 788,* 21.
Kouchi A., Kudo T., Nakano H., et al. (2002) Rapid growth of asteroids owing to very sticky interstellar organic grains. *Astrophys. J. Lett., 566,* L121–L124.
Krijt S., Bosman A. D., Zhang K., et al. (2020) CO depletion in protoplanetary disks: A unified picture combining physical sequestration and chemical processing. *Astrophys. J., 899,* 134.
Kunz M. W. (2008) On the linear stability of weakly ionized, magnetized planar shear flows. *Mon. Not. R. Astron. Soc., 385,* 1494–1510.
Kusaka T., Nakano T., and Hayashi C. (1970) Growth of solid particles in the primordial solar nebula. *Prog. Theor. Phys., 44,* 1580–1595.
Lahuis F., van Dishoeck E. F., Boogert A. C. A., et al. (2006) Hot organic molecules toward a young low-mass star: A look at inner disk chemistry. *Astrophys. J. Lett., 636,* L145–L148.
Law C. J., Loomis R. A., Teague R., et al. (2021a) Molecules with ALMA at Planet-forming Scales (MAPS). III. Characteristics of radial chemical substructures. *Astrophys. J. Suppl. Ser., 257,* 3.
Law C. J., Teague R., Loomis R. A., et al. (2021b) Molecules with ALMA at Planet-forming Scales (MAPS). IV. Emission surfaces and vertical distribution of molecules. *Astrophys. J. Suppl. Ser., 257,* 4.
Lee J.-E., Lee S., Baek G., et al. (2019) The ice composition in the disk around V883 Ori revealed by its stellar outburst. *Nature Astron., 3,* 314–319.
Lee S., Nomura H., Furuya K., et al. (2021) Modeling nitrogen fractionation in the protoplanetary disk around TW Hya: Model constraints on grain population and carbon-to-oxygen elemental abundance ratio. *Astrophys. J., 908,* 82.
Le Gal R., Öberg K. I., Loomis R. A., et al. (2019) Sulfur chemistry in protoplanetary disks: CS and H_2CS. *Astrophys. J., 876,* 72.
Le Gal R., Öberg K. I., Teague R., et al. (2021) Molecules with ALMA at Planet-forming Scales (MAPS). XII. Inferring the C/O and S/H ratios in protoplanetary disks with sulfur molecules. *Astrophys. J. Suppl. Ser., 257,* 12.
Leemker M., Booth A. S., van Dishoeck E. F., et al. (2023) A major asymmetric ice trap in a planet-forming disk. IV. Nitric oxide gas and a lack of CN tracing sublimating ices and a C/O ratio <1. *Astron. Astrophys., 673,* A7.
Lesur G., Kunz M. W., and Fromang S. (2014) Thanatology in protoplanetary discs: The combined influence of Ohmic, Hall, and ambipolar diffusion on dead zones. *Astron. Astrophys., 566,* A56.
Leung P. K. C. and Ogilvie G. I. (2019) Local semi-analytic models of magnetic flux transport in protoplanetary discs. *Mon. Not. R. Astron. Soc., 487,* 5155–5174.
Liang M.-C., Heays A. N., Lewis B. R., et al. (2007) Source of nitrogen isotope anomaly in HCN in the atmosphere of Titan. *Astrophys. J. Lett., 664,* L115–L118.
Limbach H.-H., Buntkowsky G., S. G., et al. (2006) Novel insights into the mechanism of the ortho/para spin conversion of hydrogen pairs: Implications for catalysis and interstellar water. *ChemPhysChem, 7,* 551–554.
Lin D. N. C. and Papaloizou J. C. B. (1993) On the tidal interaction between protostellar disks and companions. In *Protostars and Planets III* (E. H. Levy and J. I. Lunine, eds.), pp. 749–835. Univ. of Arizona, Tucson.
Liu H. B. (2019) The anomalously low (sub)millimeter spectral indices of some protoplanetary disks may be explained by dust self-scattering. *Astrophys. J. Lett., 877,* L22.
Lodders K. (2003) Solar system abundances and condensation temperatures of the elements. *Astrophys. J., 591,* 1220–1247.
Long F., Pinilla P., Herczeg G. J., et al. (2018) Gaps and rings in an ALMA survey of disks in the Taurus star-forming region. *Astrophys. J., 869,* 17.
Long F., Bosman A. D., Cazzoletti P., et al. (2021) Exploring HNC and HCN line emission as probes of the protoplanetary disk temperature. *Astron. Astrophys., 647,* A118.
Loomis R. A., Öberg K. I., Andrews S. M., et al. (2020) An unbiased ALMA spectral survey of the LkCa 15 and MWC 480 protoplanetary disks. *Astrophys. J., 893,* 101.
Lynden-Bell D. and Pringle J. E. (1974) The evolution of viscous discs and the origin of the nebular variables. *Mon. Not. R. Astron. Soc., 168,* 603–637.
Lyons J. R. and Young E. D. (2005) CO self-shielding as the origin of oxygen isotope anomalies in the early solar nebula. *Nature, 435,* 317–320.
Lyra W. (2014) Convective overstability in accretion disks: Three-dimensional linear analysis and nonlinear saturation. *Astrophys. J., 789,* 77.
Lyra W. and Umurhan O. M. (2019) The initial conditions for planet formation: Turbulence driven by hydrodynamical instabilities in disks around young stars. *Publ. Astron. Soc. Pac., 131,* 072001.
Machida M. N., Inutsuka S.-I., and Matsumoto T. (2011) Effect of magnetic braking on circumstellar disk formation in a strongly magnetized cloud. *Publ. Astron. Soc. Japan, 63,* 555–573.
Maćıas E., Guerra-Alvarado O., Carrasco-González C., et al. (2021) Characterizing the dust content of disk substructures in TW Hydrae. *Astron. Astrophys., 648,* A33.
Malfait K., Waelkens C., Waters L. B. F. M., et al. (1998) The spectrum of the young star HD 100546 observed with the Infrared Space Observatory. *Astron. Astrophys., 332,* L25–L28.
Malygin M. G., Klahr H., Semenov D., et al. (2017) Efficiency of thermal relaxation by radiative processes in protoplanetary discs: Constraints on hydrodynamic turbulence. *Astron. Astrophys., 605,* A30.
Mandell A. M., Bast J., van Dishoeck E. F., et al. (2012) First detection of near-infrared line emission from organics in young circumstellar disks. *Astrophys. J., 747,* 92.
Manfroid J., Jehin E., Hutsemékers D., et al. (2009) The CN isotopic ratios in comets. *Astron. Astrophys., 503,* 613–624.
Marcus P. S., Pei S., Jiang C.-H., et al. (2015) Zombie vortex instability. I. A purely hydrodynamic instability to resurrect the dead zones of protoplanetary disks. *Astrophys. J., 808,* 87.
Martin-Zäidi C., Lagage P. O., Pantin E., et al. (2007) Detection of warm molecular hydrogen in the circumstellar disk around the Herbig Ae star HD 97048. *Astrophys. J. Lett., 666,* L117–L120.
Marty B., Chaussidon M., Wiens R. C., et al. (2011) A ^{15}N-poor isotopic composition for the solar system as shown by Genesis solar wind samples. *Science, 332,* 1533–1536.
McClure M. K., Espaillat C., Calvet N., et al. (2015) Detections of trans-neptunian ice in protoplanetary disks. *Astrophys. J., 799,* 162.
McClure M. K., Bergin E. A., Cleeves L. I., et al. (2016) Mass measurements in protoplanetary disks from hydrogen deuteride. *Astrophys. J., 831,* 167.

McClure M. K., Dominik C., and Kama M. (2020) Measuring the atomic composition of planetary building blocks. *Astron. Astrophys., 642,* L15.

McGuire B. A. (2018) 2018 census of interstellar, circumstellar, extragalactic, protoplanetary disk, and exoplanetary molecules. *Astrophys. J. Suppl. Ser., 239,* 17.

McGuire B. A., Ioppolo S., Allodi M. A., et al. (2016) THz time-domain spectroscopy of mixed CO_2-CH_3OH interstellar ice analogs. *Phys. Chem. Chem. Phys., 18,* 20199–20207.

McGuire B. A., Burkhardt A. M., Loomis R. A., et al. (2020) Early science from GOTHAM: Project overview, methods, and the detection of interstellar propargyl cyanide ($HCCCH_2CN$) in TMC-1. *Astrophys. J. Lett., 900,* L10.

Meier R., Owen T. C., Jewitt D. C., et al. (1998) Deuterium in comet C/1995 O1 (Hale-Bopp): Detection of DCN. *Science, 279,* 1707–1710.

Meixner M., Cooray A., Leisawitz D., et al. (2019) Origins Space Telescope mission concept study report. *ArXiV e-prints,* arXiv:1912.06213.

Min M., Bouwman J., Dominik C., et al. (2016) The abundance and thermal history of water ice in the disk surrounding HD 142527 from the DIGIT Herschel key program. *Astron. Astrophys., 593,* A11.

Minissale M., Aikawa Y., Bergin E., et al. (2022) Thermal desorption of interstellar ices: A review on the controlling parameters and their implications from snowlines to chemical complexity. *ACS Earth Space Chem., 6,* 597–630.

Miotello A., Bruderer S., and van Dishoeck E. F. (2014) Protoplanetary disk masses from CO isotopologue line emission. *Astron. Astrophys., 572,* A96.

Miotello A., van Dishoeck E. F., Williams J. P., et al. (2017) Lupus disks with faint CO isotopologues: Low gas/dust or high carbon depletion? *Astron. Astrophys., 599,* A113.

Miotello A., Facchini S., van Dishoeck E. F., et al. (2019) Bright C_2H emission in protoplanetary discs in Lupus: High volatile $C/O > 1$ ratios. *Astron. Astrophys., 631,* A69.

Miotello A., Kamp I., Birnstiel T., et al. (2023) Setting the stage for planet formation: Measurements and implications of the fundamental disk properties. In *Protostars and Planets VII* (S.-i. Inutsuka et al., eds.), pp. 501–537. ASP Conf. Ser. 534, Astronomical Society of the Pacific, San Francisco.

Miyake K. and Nakagawa Y. (1993) Effects of particle size distribution on opacity curves of protoplanetary disks around T Tauri stars. *Icarus, 106,* 20–41.

Morbidelli A., Szulágyi J., Crida A., et al. (2014) Meridional circulation of gas into gaps opened by giant planets in three-dimensional low-viscosity disks. *Icarus, 232,* 266–270.

Mori S., Bai X.-N., and Okuzumi S. (2019) Temperature structure in the inner regions of protoplanetary disks: Inefficient accretion heating controlled by nonideal magnetohydrodynamics. *Astrophys. J., 872,* 98.

Mori S., Okuzumi S., Kunitomo M., et al. (2021) Evolution of the water snow line in magnetically accreting protoplanetary disks. *Astrophys. J., 916,* 72.

Mumma M. J. and Charnley S. B. (2011) The chemical composition of comets — Emerging taxonomies and natal heritage. *Annu. Rev. Astron. Astrophys., 49,* 471–524.

Mumma M. J., Weaver H. A., and Larson H. P. (1987) The ortho-para ratio of water vapor in comet P/ Halley. *Astron. Astrophys., 187,* 419–424.

Musiolik G. and Wurm G. (2019) Contacts of water ice in protoplanetary disks — Laboratory experiments. *Astrophys. J., 873,* 58.

Musiolik G., Teiser J., Jankowski T., et al. (2016a) Collisions of CO_2 ice grains in planet formation. *Astrophys. J., 818,* 16.

Musiolik G., Teiser J., Jankowski T., et al. (2016b) Ice grain collisions in comparison: CO_2, H_2O, and their mixtures. *Astrophys. J., 827,* 63.

Najita J., Carr J. S., and Mathieu R. D. (2003) Gas in the terrestrial planet region of disks: CO fundamental emission from T Tauri stars. *Astrophys. J., 589,* 931–952.

Najita J. R., Carr J. S., Pontoppidan K. M., et al. (2013) The HCN-water ratio in the planet formation region of disks. *Astrophys. J., 766,* 134.

Najita J. R., Carr J. S., Brittain S. D., et al. (2021) High-resolution mid-infrared spectroscopy of GV Tau N: Surface accretion and detection of NH_3 in a young protoplanetary disk. *Astrophys. J., 908,* 171.

Nakagawa Y., Nakazawa K., and Hayashi C. (1981) Growth and sedimentation of dust grains in the primordial solar nebula. *Icarus, 45,* 517–528.

Nelson R. P., Gressel O., and Umurhan O. M. (2013) Linear and non-linear evolution of the vertical shear instability in accretion discs. *Mon. Not. R. Astron. Soc., 435,* 2610–2632.

Nomura H., Furuya K., Cordiner M. A., et al. (2022) The isotopic links from planet forming regions to the solar system. In *Protostars and Planets VII* (S.-i. Inutsuka et al., eds.), pp. 1075–1099. ASP Conf. Ser. 534, Astronomical Society of the Pacific, San Francisco.

Notsu S., Nomura H., Ishimoto D., et al. (2016) Candidate water vapor lines to locate the H_2O snowline through high-dispersion spectroscopic observations. I. The case of a T Tauri star. *Astrophys. J., 827,* 113.

Notsu S., Akiyama E., Booth A., et al. (2019) Dust continuum emission and the upper limit fluxes of submillimeter water lines of the protoplanetary disk around HD 163296 observed by ALMA. *Astrophys. J., 875,* 96.

Öberg K. I., Murray-Clay R., and Bergin E. A. (2011) The effects of snowlines on C/O in planetary atmospheres. *Astrophys. J. Lett., 743,* L16.

Öberg K. I., Qi C., Wilner D. J., et al. (2012) Evidence for multiple pathways to deuterium enhancements in protoplanetary disks. *Astrophys. J., 749,* 162.

Öberg K. I., Guzmán V. V., Furuya K., et al. (2015) The comet-like composition of a protoplanetary disk as revealed by complex cyanides. *Nature, 520,* 198–201.

Öberg K. I., Cleeves L. I., Bergner J. B., et al. (2021a) The TW Hya Rosetta Stone Project. I. Radial and vertical distributions of DCN and DCO^+. *Astron. J., 161,* 38.

Öberg K. I., Guzmán V. V., Walsh C., et al. (2021b) Molecules with ALMA at Planet-forming Scales (MAPS). I. Program overview and highlights. *Astrophys. J. Suppl. Ser., 257,* 1.

Öberg K. I., Facchini S., and Anderson D. E. (2023) Protoplanetary disk chemistry. *Annu. Rev. Astron. Astrophys., 61,* 287–328.

Offner S. S. R. and McKee C. F. (2011) The protostellar luminosity function. *Astrophys. J., 736,* 53.

Ohashi N., Tobin J. J., Jørgensen J. K., et al. (2023) Early Planet Formation In Embedded Disks (eDisk). I. Overview of the program and first results. *Astrophys. J., 951,* 8.

Oka A., Nakamoto T., and Ida S. (2011) Evolution of snow line in optically thick protoplanetary disks: Effects of water ice opacity and dust grain size. *Astrophys. J., 738,* 141.

Okuzumi S. (2009) Electric charging of dust aggregates and its effect on dust coagulation in protoplanetary disks. *Astrophys. J., 698,* 1122–1135.

Okuzumi S. and Tazaki R. (2019) Nonsticky ice at the origin of the uniformly polarized submillimeter emission from the HL Tau disk. *Astrophys. J., 878,* 132.

Okuzumi S., Tanaka H., Kobayashi H., et al. (2012) Rapid coagulation of porous dust aggregates outside the snow line: A pathway to successful icy planetesimal formation. *Astrophys. J., 752,* 106.

Okuzumi S., Takeuchi T., and Muto T. (2014) Radial transport of large-scale magnetic fields in accretion disks. I. Steady solutions and an upper limit on the vertical field strength. *Astrophys. J., 785,* 127.

Okuzumi S., Momose M., Sirono S.-i., et al. (2016) Sintering-induced dust ring formation in protoplanetary disks: Application to the HL Tau disk. *Astrophys. J., 821,* 82.

Ormel C. W. and Cuzzi J. N. (2007) Closed-form expressions for particle relative velocities induced by turbulence. *Astron. Astrophys., 466,* 413–420.

Oya Y., Moriwaki K., Onishi S., et al. (2018) Chemical and physical picture of IRAS 16293-2422 source B at a sub-arcsecond scale studied with ALMA. *Astrophys. J., 854,* 96.

Paardekooper S. J. and Mellema G. (2006) Dust flow in gas disks in the presence of embedded planets. *Astron. Astrophys., 453,* 1129–1140.

Packham C., Honda M., Chun M., et al. (2018) The key science drivers for MICHI: A thermal-infrared instrument for the TMT. In *Ground-Based and Airborne Instrumentation for Astronomy VII* (C. J. Evans et al., eds.), p. 10702A0. SPIE Conf. Ser. 10702, Bellingham, Washington.

Paneque-Carreño T., Izquierdo A. F., Teague R., et al. (2024) High turbulence in the IM Lup protoplanetary disk. Direct observational constraints from CN and C_2H emission. *Astron. Astrophys., 684,* A174.

Pegues J., Öberg K. I., Bergner J. B., et al. (2020) An ALMA survey of

H_2CO in protoplanetary disks. *Astrophys. J., 890,* 142.
Pérez L. M., Carpenter J. M., Chandler C. J., et al. (2012) Constraints on the radial variation of grain growth in the AS 209 circumstellar disk. *Astrophys. J. Lett., 760,* L17.
Pérez L. M., Chandler C. J., Isella A., et al. (2015) Grain growth in the circumstellar disks of the young stars CY Tau and DoAr 25. *Astrophys. J., 813,* 41.
Phuong N. T., Chapillon E., Majumdar L., et al. (2018) First detection of H_2S in a protoplanetary disk: The dense GG Tauri A ring. *Astron. Astrophys., 616,* L5.
Phuong N. T., Dutrey A., Chapillon E., et al. (2021) An unbiased NOEMA 2.6 to 4 mm survey of the GG Tau ring: First detection of CCS in a protoplanetary disk. *Astron. Astrophys., 653,* L5.
Piani L., Tachibana S., Hama T., et al. (2017) Evolution of morphological and physical properties of laboratory interstellar organic residues with ultraviolet irradiation. *Astrophys. J., 837,* 35.
Pineda J. E., Segura-Cox D., Caselli P., et al. (2020) A protostellar system fed by a streamer of 10,500 au length. *Nature Astron., 4,* 1158–1163.
Pinilla P., Birnstiel T., Ricci L., et al. (2012) Trapping dust particles in the outer regions of protoplanetary disks. *Astron. Astrophys., 538,* A114.
Pinilla P., Pohl A., Stammler S. M., et al. (2017) Dust density distribution and imaging analysis of different ice lines in protoplanetary disks. *Astrophys. J., 845,* 68.
Pinte C., Dent W. R. F., Ménard F., et al. (2016) Dust and gas in the disk of HL Tauri: Surface density, dust settling, and dust-to-gas ratio. *Astrophys. J., 816,* 25.
Podio L., Garufi A., Codella C., et al. (2020) ALMA chemical survey of disk-outflow sources in Taurus (ALMA-DOT). II. Vertical stratification of CO, CS, CN, H_2CO, and CH_3OH in a Class I disk. *Astron. Astrophys., 642,* L7.
Pontoppidan K. M. and Blevins S. M. (2014) The chemistry of planet-forming regions is not interstellar. *Faraday Discussions, 168,* 49–60.
Pontoppidan K. M., Dullemond C. P., van Dishoeck E. F., et al. (2005) Ices in the edge-on disk CRBR 2422.8-3423: Spitzer spectroscopy and Monte Carlo radiative transfer modeling. *Astrophys. J., 622,* 463–481.
Pontoppidan K. M., Blake G. A., van Dishoeck E. F., et al. (2008) Spectroastrometric imaging of molecular gas within protoplanetary disk gaps. *Astrophys. J., 684,* 1323–1329.
Pontoppidan K. M., Salyk C., Blake G. A., et al. (2010) A Spitzer survey of mid-infrared molecular emission from protoplanetary disks. I. Detection rates. *Astrophys. J., 720,* 887–903.
Pontoppidan K., Banzatti A., Bergin E., et al. (2019a) The trail of water and the delivery of volatiles to habitable planets. *Bull. Am. Astron. Soc., 51,* 229.
Pontoppidan K. M., Salyk C., Banzatti A., et al. (2019b) The nitrogen carrier in inner protoplanetary disks. *Astrophys. J., 874,* 92.
Pontoppidan K. M., Salyk C., Banzatti A., et al. (2024) High-contrast JWST-MIRI spectroscopy of planet-forming disks for the JDISC survey. *Astrophys. J., 963,* 158.
Qi C., Kessler J. E., Koerner D. W., et al. (2003) Continuum and CO/HCO^+ emission from the disk around the T Tauri star LkCa 15. *Astrophys. J., 597,* 986–997.
Qi C., Wilner D. J., Aikawa Y., et al. (2008) Resolving the chemistry in the disk of TW Hydrae. I. Deuterated species. *Astrophys. J., 681,* 1396–1407.
Qi C., D'Alessio P., Öberg K. I., et al. (2011) Resolving the CO snow line in the disk around HD 163296. *Astrophys. J., 740,* 84.
Qi C., Öberg K. I., Wilner D. J., et al. (2013a) Imaging of the CO snow line in a solar nebula analog. *Science, 341,* 630–632.
Qi C., Öberg K. I., Wilner D. J., et al. (2013b) First detection of C_3H_2 in a circumstellar disk. *Astrophys. J. Lett., 765,* L14.
Qi C., Öberg K. I., Espaillat C. C., et al. (2019) Probing CO and N_2 snow surfaces in protoplanetary disks with N_2H^+ emission. *Astrophys. J., 882,* 160.
Raettig N., Lyra W., and Klahr H. (2021) Pebble trapping in vortices: Three-dimensional simulations. *Astrophys. J., 913,* 92.
Ricci L., Testi L., Natta A., et al. (2010a) Dust grain growth in ρ-Ophiuchi protoplanetary disks. *Astron. Astrophys., 521,* A66.
Ricci L., Testi L., Natta A., et al. (2010b) Dust properties of protoplanetary disks in the Taurus-Auriga star forming region from millimeter wavelengths. *Astron. Astrophys., 512,* A15.
Ritchey A. M., Federman S. R., and Lambert D. L. (2015) The $C^{14}N$/$C^{15}N$ ratio in diffuse molecular clouds. *Astrophys. J. Lett., 804,* L3.
Rivière-Marichalar P., Fuente A., Le Gal R., et al. (2021) H_2S observations in young stellar disks in Taurus. *Astron. Astrophys., 652,* A46.
Rosenfeld K. A., Andrews S. M., Hughes A. M., et al. (2013) A spatially resolved vertical temperature gradient in the HD 163296 disk. *Astrophys. J., 774,* 16.
Rosotti G. P., Teague R., Dullemond C., et al. (2020) The efficiency of dust trapping in ringed protoplanetary discs. *Mon. Not. R. Astron. Soc., 495,* 173–181.
Roueff E., Loison J. C., and Hickson K. M. (2015) Isotopic fractionation of carbon, deuterium, and nitrogen: A full chemical study. *Astron. Astrophys., 576,* A99.
Rúiz-Rodríguez D., Kastner J., Hily-Blant P., et al. (2021) Tracing molecular stratification within an edge-on protoplanetary disk. *Astron. Astrophys., 646,* A59.
Rybicki G. B. and Lightman A. P. (1979) *Radiative Processes in Astrophysics.* Wiley, New York. 382 pp.
Sadavoy S. I., Myers P. C., Stephens I. W., et al. (2018) Dust polarization toward embedded protostars in Ophiuchus with ALMA. I. VLA 1623. *Astrophys. J., 859,* 165.
Sai J., Ohashi N., Saigo K., et al. (2020) Disk structure around the Class I protostar L1489 IRS revealed by ALMA: A warped-disk system. *Astrophys. J., 893,* 51.
Saito E. and Sirono S.-i. (2011) Planetesimal formation by sublimation. *Astrophys. J., 728,* 20.
Sakai N., Oya Y., Sakai T., et al. (2014) A chemical view of protostellar-disk formation in L1527. *Astrophys. J. Lett., 791,* L38.
Sakai N., Hanawa T., Zhang Y., et al. (2019) A warped disk around an infant protostar. *Nature, 565,* 206–208.
Salinas V. N., Hogerheijde M. R., Bergin E. A., et al. (2016) First detection of gas-phase ammonia in a planet-forming disk. NH_3, N_2H^+, and H_2O in the disk around TW Hydrae. *Astron. Astrophys., 591,* A122.
Salinas V. N., Hogerheijde M. R., Mathews G. S., et al. (2017) DCO^+, DCN, and N_2D^+ reveal three different deuteration regimes in the disk around the Herbig Ae star HD 163296. *Astron. Astrophys., 606,* A125.
Salyk C., Pontoppidan K. M., Blake G. A., et al. (2008) H_2O and OH gas in the terrestrial planet-forming zones of protoplanetary disks. *Astrophys. J. Lett., 676,* L49–L52.
Salyk C., Pontoppidan K. M., Blake G. A., et al. (2011) A Spitzer survey of mid-infrared molecular emission from protoplanetary disks. II. Correlations and local thermal equilibrium models. *Astrophys. J., 731,* 130.
Sano T., Miyama S. M., Umebayashi T., et al. (2000) Magnetorotational instability in protoplanetary disks. II. Ionization state and unstable regions. *Astrophys. J., 543,* 486–501.
Sasselov D. D. and Lecar M. (2000) On the snow line in dusty protoplanetary disks. *Astrophys. J., 528,* 995–998.
Schoonenberg D. and Ormel C. W. (2017) Planetesimal formation near the snowline: In or out? *Astron. Astrophys., 602,* A21.
Schoonenberg D., Okuzumi S., and Ormel C. W. (2017) What pebbles are made of: Interpretation of the V883 Ori disk. *Astron. Astrophys., 605,* L2.
Semenov D. and Wiebe D. (2011) Chemical evolution of turbulent protoplanetary disks and the solar nebula. *Astrophys. J. Suppl. Ser., 196,* 25.
Shibata K. and Uchida Y. (1986) A magnetohydrodynamic mechanism for the formation of astrophysical jets. II. Dynamical processes in the accretion of magnetized mass in rotation. *Publ. Astron. Soc. Japan, 38,* 631–660.
Shingledecker C. N., Tennis J., Le Gal R., et al. (2018) On cosmic-ray-driven grain chemistry in cold core models. *Astrophys. J., 861,* 20.
Shinnaka Y., Kawakita H., and Tajitsu A. (2020) High-resolution optical spectroscopic observations of comet 21P/Giacobini-Zinner in its 2018 apparition. *Astron. J., 159,* 203.
Sierra A. and Lizano S. (2020) Effects of scattering, temperature gradients, and settling on the derived dust properties of observed protoplanetary disks. *Astrophys. J., 892,* 136.
Sierra A., Pérez L. M., Zhang K., et al. (2021) Molecules with ALMA at Planet-forming Scales (MAPS). XIV. Revealing disk substructures in multiwavelength continuum emission. *Astrophys. J. Suppl. Ser., 257,* 14.
Simon J. B., Bai X.-N., Armitage P. J., et al. (2013a) Turbulence in the

outer regions of protoplanetary disks. II. Strong accretion driven by a vertical magnetic field. *Astrophys. J.*, *775*, 73.
Simon J. B., Bai X.-N., Stone J. M., et al. (2013b) Turbulence in the outer regions of protoplanetary disks. I. Weak accretion with no vertical magnetic flux. *Astrophys. J.*, *764*, 66.
Sirono S.-i. (2011) Planetesimal formation induced by sintering. *Astrophys. J. Lett.*, *733*, L41.
Smith R. L., Pontoppidan K. M., Young E. D., et al. (2009) High-precision $C^{17}O$, $C^{18}O$, and $C^{16}O$ measurements in young stellar objects: Analogues for CO self-shielding in the early solar system. *Astrophys. J.*, *701*, 163–175.
Soon K.-L., Momose M., Muto T., et al. (2019) Investigating the gas-to-dust ratio in the protoplanetary disk of HD 142527. *Publ. Astron. Soc. Japan*, *71*, 124.
Steinpilz T., Teiser J., and Wurm G. (2019) Sticking properties of silicates in planetesimal formation revisited. *Astrophys. J.*, *874*, 60.
Stephens I. W., Looney L. W., Kwon W., et al. (2014) Spatially resolved magnetic field structure in the disk of a T Tauri star. *Nature*, *514*, 597–599.
Stephens I. W., Yang H., Li Z.-Y., et al. (2017) ALMA reveals transition of polarization pattern with wavelength in HL Tau's disk. *Astrophys. J.*, *851*, 55.
Suriano S. S., Li Z.-Y., Krasnopolsky R., et al. (2018) The formation of rings and gaps in magnetically coupled disc-wind systems: ambipolar diffusion and reconnection. *Mon. Not. R. Astron. Soc.*, *477*, 1239–1257.
Tabone B., Bettoni G., van Dishoeck E. F., et al. (2023) A rich hydrocarbon chemistry and high C to O ratio in the inner disk around a very low-mass star. *Nature Astron.*, *7*, 805–814.
Takeuchi T. and Okuzumi S. (2014) Radial transport of large-scale magnetic fields in accretion disks. II. Relaxation to steady states. *Astrophys. J.*, *797*, 132.
Takeuchi T., Clarke C. J., and Lin D. N. C. (2005) The differential lifetimes of protostellar gas and dust disks. *Astrophys. J.*, *627*, 286–292.
Tanaka H., Himeno Y., and Ida S. (2005) Dust growth and settling in protoplanetary disks and disk spectral energy distributions. I. Laminar disks. *Astrophys. J.*, *625*, 414–426.
Tazaki R., Tanaka H., Kataoka A., et al. (2019) Unveiling dust aggregate structure in protoplanetary disks by millimeter-wave scattering polarization. *Astrophys. J.*, *885*, 52.
Tazzari M., Testi L., Ercolano B., et al. (2016) Multiwavelength analysis for interferometric (sub-)mm observations of protoplanetary disks: Radial constraints on the dust properties and the disk structure. *Astron. Astrophys.*, *588*, A53.
Teague R. and Loomis R. (2020) The excitation conditions of CN in TW Hya. *Astrophys. J.*, *899*, 157.
Teague R., Semenov D., Guilloteau S., et al. (2015) Chemistry in disks. IX. Observations and modelling of HCO^+ and DCO^+ in DM Tauri. *Astron. Astrophys.*, *574*, A137.
Teague R., Guilloteau S., Semenov D., et al. (2016) Measuring turbulence in TW Hydrae with ALMA: Methods and limitations. *Astron. Astrophys.*, *592*, A49.
Teague R., Henning T., Guilloteau S., et al. (2018) Temperature, mass, and turbulence: A spatially resolved multiband non-LTE analysis of CS in TW Hya. *Astrophys. J.*, *864*, 133.
Teague R., Bae J., and Bergin E. A. (2019) Meridional flows in the disk around a young star. *Nature*, *574*, 378–381.
Terada H., Tokunaga A. T., Kobayashi N., et al. (2007) Detection of water ice in edge-on protoplanetary disks: HK Tauri B and HV Tauri C. *Astrophys. J.*, *667*, 303–307.
Testi L., Birnstiel T., Ricci L., et al. (2014) Dust evolution in protoplanetary disks. In *Protostars and Planets VI* (H. Beuther et al., eds.), pp. 339–362. Univ. of Arizona, Tucson.
Thi W. F., Ménard F., Meeus G., et al. (2011) Detection of CH^+ emission from the disc around HD 100546. *Astron. Astrophys.*, *530*, L2.
Tobin J. J., van't Hoff M. L. R., Leemker M., et al. (2023) Deuterium-enriched water ties planet-forming disks to comets and protostars. *Nature*, *615*, 227–230.
Tsukamoto Y., Takahashi S. Z., Machida M. N., et al. (2015) Effects of radiative transfer on the structure of self-gravitating discs, their fragmentation and the evolution of the fragments. *Mon. Not. R. Astron. Soc.*, *446*, 1175–1190.
Tsukamoto Y., Maury A., Commercon B., et al. (2023) The role of magnetic fields in the formation of protostars, disks, and outflows. In *Protostars and Planets VII* (S.-i. Inutsuka et al., eds.), pp. 317–354. ASP Conf. Ser. 534, Astronomical Society of the Pacific, San Francisco.
Turner N. J. and Sano T. (2008) Dead zone accretion flows in protostellar disks. *Astrophys. J. Lett.*, *679*, L131–L134.
Tychoniec Ł., van Dishoeck E. F., van't Hoff M. L. R., et al. (2021) Which molecule traces what: Chemical diagnostics of protostellar sources. *Astron. Astrophys.*, *655*, A65.
Ueda T., Kataoka A., Zhang S., et al. (2021) Impact of differential dust settling on the SED and polarization: Application to the inner region of the HL Tau disk. *Astrophys. J.*, *913*, 117.
Urpin V. and Brandenburg A. (1998) Magnetic and vertical shear instabilities in accretion discs. *Mon. Not. R. Astron. Soc.*, *294*, 399–406.
van der Marel N., Dong R., di Francesco J., et al. (2019) Protoplanetary disk rings and gaps across ages and luminosities. *Astrophys. J.*, *872*, 112.
van der Marel N., Booth A. S., Leemker M., et al. (2021) A major asymmetric ice trap in a planet-forming disk. I. Formaldehyde and methanol. *Astron. Astrophys.*, *651*, L5.
van Dishoeck E. F., Thi W. F., and van Zadelhoff G. J. (2003) Detection of DCO^+ in a circumstellar disk. *Astron. Astrophys.*, *400*, L1–L4.
van Dishoeck E. F., Herbst E., and Neufeld D. A. (2013) Interstellar water chemistry: From laboratory to observations. *Chem. Rev.*, *113*, 9043–9085.
van Dishoeck E. F., Kristensen L. E., Mottram J. C., et al. (2021) Water in star-forming regions: Physics and chemistry from clouds to disks as probed by Herschel spectroscopy. *Astron. Astrophys.*, *648*, A24.
van Dishoeck E. F., Grant S., Tabone B., et al. (2023) The diverse chemistry of protoplanetary disks as revealed by JWST. *Faraday Discussions*, *245*, 52–79.
van't Hoff M. L. R., Harsono D., Tobin J. J., et al. (2020) Temperature structures of embedded disks: Young disks in Taurus are warm. *Astrophys. J.*, *901*, 166.
Visser R., van Dishoeck E. F., Doty S. D., et al. (2009) The chemical history of molecules in circumstellar disks. I. Ices. *Astron. Astrophys.*, *495*, 881–897.
Visser R., Doty S. D., and van Dishoeck E. F. (2011) The chemical history of molecules in circumstellar disks. II. Gas-phase species. *Astron. Astrophys.*, *534*, A132.
Visser R., Bruderer S., Cazzoletti P., et al. (2018) Nitrogen isotope fractionation in protoplanetary disks. *Astron. Astrophys.*, *615*, A75.
Wada K., Tanaka H., Suyama T., et al. (2009) Collisional growth conditions for dust aggregates. *Astrophys. J.*, *702*, 1490–1501.
Walsh C., Nomura H., and van Dishoeck E. (2015) The molecular composition of the planet-forming regions of protoplanetary disks across the luminosity regime. *Astron. Astrophys.*, *582*, A88.
Walsh C., Loomis R. A., Öberg K. I., et al. (2016) First detection of gas-phase methanol in a protoplanetary disk. *Astrophys. J. Lett.*, *823*, L10.
Wang H., Weiss B. P., Bai X.-N., et al. (2017) Lifetime of the solar nebula constrained by meteorite paleomagnetism. *Science*, *355*, 623–627.
Wardle M. (2007) Magnetic fields in protoplanetary disks. *Astrophys. Space Sci.*, *311*, 35–45.
Warren S. G. (1984) Optical constants of ice from the ultraviolet to the microwave. *Appl. Opt.*, *23*, 1206–1225.
Weidenschilling S. J. (1977) Aerodynamics of solid bodies in the solar nebula. *Mon. Not. R. Astron. Soc.*, *180*, 57–70.
Weidenschilling S. J. (1980) Dust to planetesimals: Settling and coagulation in the solar nebula. *Icarus*, *44*, 172–189.
Weintraub D. A., Sandell G., and Duncan W. D. (1989) Submillimeter measurements of T Tauri and FU Orionis stars. *Astrophys. J. Lett.*, *340*, L69–L72.
Weintraub D. A., Kastner J. H., and Bary J. S. (2000) Detection of quiescent molecular hydrogen gas in the circumstellar disk of a classical T Tauri star. *Astrophys. J.*, *541*, 767.
Whipple F. L. (1972) On certain aerodynamic processes for asteroids and comets. In *From Plasma to Planet* (A. Elvius, ed.), pp. 211–232. Wiley, New York.
White R. J., Greene T. P., Doppmann G. W., et al. (2007) Stellar properties of embedded protostars. In *Protostars and Planets V* (B. Reipurth et al., eds.), pp. 117–132.
Whittet D. C. B. (1993) Observations of molecular ices. In *Dust and Chemistry in Astronomy* (T. J. Millar and D. A. Williams, eds.), p. 9.

Graduate Series in Astronomy, Institute of Physics, Philadelphia.
Willacy K. (2007) The chemistry of multiply deuterated molecules in protoplanetary disks. I. The outer disk. *Astrophys. J., 660*, 441–460.
Yamamoto T., Nakagawa N., and Fukui Y. (1983) The chemical composition and thermal history of the ice of a cometary nucleus. *Astron. Astrophys., 122*, 171–176.
Yamato Y., Aikawa Y., Ohashi N., et al. (2023) Early Planet Formation in Embedded Disks (eDisk). IV. The ringed and warped structure of the disk around the Class I protostar L1489 IRS. *Astrophys. J., 951,* 11.
Yamato Y., Aikawa Y., Guzmán V. V., et al. (2024a) Detection of dimethyl ether in the central region of the MWC 480 protoplanetary disk. *Astrophys. J., 974,* 83.
Yamato Y., Notsu S., Aikawa Y., et al. (2024b) Chemistry of complex organic molecules in the V883 Ori disk revealed by ALMA Band 3 observations. *Astron. J., 167,* 66.
Yang H. and Li Z.-Y. (2020) The effects of dust optical properties on the scattering-induced disk polarization by millimeter-sized grains. *Astrophys. J., 889*, 15.
Yang H., Li Z.-Y., Looney L., et al. (2016) Inclination-induced polarization of scattered millimetre radiation from protoplanetary discs: The case of HL Tau. *Mon. Not. R. Astron. Soc., 456*, 2794–2805.
Yang L., Ciesla F. J., and Alexander C. M. O. D. (2013) The D/H ratio of water in the solar nebula during its formation and evolution. *Icarus, 226*, 256–267.
Yen H.-W., Takakuwa S., Ohashi N., et al. (2014) ALMA observations of infalling flows toward the Keplerian disk around the class I protostar L1489 IRS. *Astrophys. J., 793*, 1.
Yoneda H., Tsukamoto Y., Furuya K., et al. (2016) Chemistry in a forming protoplanetary disk: Main accretion phase. *Astrophys. J., 833*, 105.
Yoshida T. C., Nomura H., Furuya K., et al. (2022) A new method for direct measurement of isotopologue ratios in protoplanetary disks: A case study of the $^{12}CO/^{13}CO$ ratio in the TW Hya disk. *Astrophys. J., 932,* 126.
Yurimoto H. and Kuramoto K. (2004) Molecular cloud origin for the oxygen isotope heterogeneity in the solar system. *Science, 305*, 1763–1766.
Zhang K., Pontoppidan K. M., Salyk C., et al. (2013) Evidence for a snow line beyond the transitional radius in the TW Hya protoplanetary disk. *Astrophys. J., 766*, 82.
Zhang K., Blake G. A., and Bergin E. A. (2015) Evidence of fast pebble growth near condensation fronts in the HL Tau protoplanetary disk. *Astrophys. J. Lett., 806*, L7.
Zhang K., Bergin E. A., Blake G. A., et al. (2017) Mass inventory of the giant-planet formation zone in a solar nebula analogue. *Nature Astron., 1*, 0130.
Zhang K., Bergin E. A., Schwarz K., et al. (2019) Systematic variations of CO gas abundance with radius in gas-rich protoplanetary disks. *Astrophys. J., 883*, 98.
Zhang K., Schwarz K. R., and Bergin E. A. (2020) Rapid evolution of volatile CO from the protostellar disk stage to the protoplanetary disk stage. *Astrophys. J. Lett., 891*, L17.
Zhang K., Booth A. S., Law C. J., et al. (2021) Molecules with ALMA at Planet-forming Scales (MAPS). V. CO gas distributions. *Astrophys. J. Suppl. Ser., 257*, 5.
Zhu Z. and Stone J. M. (2018) Global evolution of an accretion disk with a net vertical field: Coronal accretion, flux transport, and disk winds. *Astrophys. J., 857*, 34.
Zhu Z., Zhang S., Jiang Y.-F., et al. (2019) One solution to the mass budget problem for planet formation: Optically thick disks with dust scattering. *Astrophys. J. Lett., 877*, L18.
Zsom A., Ormel C. W., Güttler C., et al. (2010) The outcome of protoplanetary dust growth: Pebbles, boulders, or planetesimals? II. Introducing the bouncing barrier. *Astron. Astrophys., 513*, A57.

Simon J. B., Blum J., Birnstiel T., and Nesvorný D. (2024) Comets and planetesimal formation. In *Comets III* (K. J. Meech, M. R. Combi, D. Bockelée-Morvan, S. N. Raymond, and M. E. Zolensky, eds.), pp. 63–94. Univ. of Arizona, Tucson, DOI: 10.2458/azu_uapress_9780816553631-ch003.

Comets and Planetesimal Formation

Jacob B. Simon
Iowa State University

Jürgen Blum
Technische Universität Braunschweig

Til Birnstiel
Ludwig-Maximilians-Universität München

David Nesvorný
Southwest Research Institute

In this chapter, we review the processes involved in the formation of planetesimals and comets. We will start with a description of the physics of dust grain growth and how this is mediated by gas-dust interactions in planet-forming disks. We will then delve into the various models of planetesimal formation, describing how these planetesimals form as well as their resulting structure. In doing so, we focus on and compare two paradigms for planetesimal formation: the gravitational collapse of particle overdensities (which can be produced by a variety of mechanisms) and the growth of particles into planetesimals via collisional and gravitational coagulation. Finally, we compare the predictions from these models with data collected by the Rosetta and New Horizons missions and that obtained via observations of distant Kuiper belt objects.

1. INTRODUCTION

The formation of planetesimals from small submicrometer grains encompasses approximately 10–12 orders of magnitude increase in scale and numerous physical processes, ranging from particle-particle sticking to angular momentum exchange between disk gas and solid particles. While there are many unanswered questions in how exactly these physical processes work, recent years have seen substantial progress in addressing several key questions:

1. How do submicrometer-sized grains grow to the larger millimeter-to-centimeter solids (referred to as pebbles; see section 2.5 for a more complete definition) that have been observationally inferred to exist in disks around young stars?

2. How do these pebbles continue their growth to forming kilometer-sized planetesimals?

3. What are the properties of the resulting planetesimals and how do they compare with observations of solar system planetesimal populations, such as comets and Kuiper belt objects (KBOs)?

Along the theoretical front, a number of new insights have been gained. In particular, the powerful combination of analytical techniques and numerical simulations have identified a number of instabilities (e.g., *Youdin and Goodman,* 2005; *Youdin,* 2005) and mechanisms (e.g., *Cuzzi et al.,* 2008; *Hopkins,* 2016) that can concentrate small grains into regions of sufficient density to gravitationally collapse into planetesimals.

These techniques have directly led to predictions that can be tested with current and forthcoming observational campaigns, and along this front, observational data has also been abundant. With a number missions (e.g., OSIRIS-REx, Rosetta, New Horizons) to small solar system bodies (e.g., *Stern et al.,* 2021), as well as large surveys of small-body populations, such as the Outer Solar System Origins Survey (OSSOS), a plethora of new data has arrived on the physical characteristics of solar system planetesimals (e.g., *Kavelaars et al.,* 2021; *Fraser et al.,* 2021). Furthermore, the advent of next generation optical/infrared and radio telescopes, such as the Atacama Large Millimeter/submillimeter Array (ALMA) and the Very Large Telescope (VLT), have revolutionized our understanding of both gas and small-particle dynamics in planet-forming disks around nearby stars (*ALMA Partnership et al.,* 2015; *Isella et al.,* 2016; *Huang et al.,* 2018).

Finally, laboratory experiments have been very fruitful in addressing how exactly solid particles of different compositions interact when they collide and have thus served as a strong complement to both the observational and theoretical/numerical studies (*Blum and Wurm,* 2008). Moreover, the physical properties of planetesimals formed by different hypothesized mechanisms have also been studied in the laboratory (*Blum,* 2018), allowing a comparison to those of the most primitive bodies in the solar system.

Indeed, the combination of theory, observations, and laboratory work has proved to be very successful in improving our understanding of planetesimal and comet formation. In this chapter, we review such work and detail the latest understanding of how planetesimals are born. Our approach is primarily pedagogical and as such, much of our focus will be on the physics of planetesimal formation itself, while making connections to both observations and laboratory experiments when appropriate.

2. DUST GRAIN GROWTH

In this section, we describe in detail our current understanding of how small solids grow in protoplanetary disks (PPDs). Some discussion of dust growth is discussed in the chapter in this volume by Aikawa et al. However, here, we delve into the details and describe the processes that limit growth of small grains beyond millimeter-to-centimeter sizes, thus setting the stage for many of the mechanisms described in section 3.

The evolution of solids in PPDs is influenced by many effects, but they can be broadly categorized into the two areas of (1) the evolution of their composition, size, and morphology and (2) their transport and dynamics within PPDs. It is important to realize that one cannot be treated separately from the other because these two categories affect each other in many ways: The makeup, size, and morphology of a particle affects its aerodynamical properties, and the aerodynamic properties determine how a particle is transported throughout the disk and how its environment changes along the way. The environment, in turn, determines how collisions and composition evolve: It sets the collision speeds (see, e.g., *Birnstiel et al.,* 2016) and/or condensation/sublimation rates (e.g., *Stammler et al.,* 2019). In the following section, we will describe how dust is transported and how it evolves through collisions and will highlight how those aspects affect each other.

We begin by introducing a crucial quantity to describe the evolution from small grains to a planetesimal: the particle stopping time. It is defined as the timescale on which the particle velocity adapts to the gas velocity through drag forces and reads

$$t_{stop} = \frac{m\Delta v}{F_{drag}} = \frac{a\rho_s}{\rho_g v_{th}} \quad (1)$$

where m, a, ρ_s, Δv, and v_{th} are the particle mass, its radius, its material density, the absolute value of the velocity difference between gas and dust, and the mean thermal velocity of the gas (related to the gas sound speed through a factor on the order of unity), respectively. The right equality in equation (1) assumes the drag force to be in the Epstein regime (*Epstein,* 1924; *Weidenschilling,* 1977), where the particle size is smaller than the mean-free path of the gas molecules and furthermore that the dust particles are spherical and nonfractal. Often, the stopping time is expressed by the Stokes number

$$St = t_{stop}\,\Omega \quad (2)$$

with Ω being the Kepler frequency at the location in the disk. The Stokes number is thus a dimensionless number specifying the aerodynamic behavior of the particle. The Stokes number can therefore be seen as a "dimensionless particle size" or is sometimes also called the dimensionless stopping time as particles with $St \ll 1$ couple much faster to the gas flow than the orbital timescale, while particles with $St \gg 1$ require many orbits to couple. (A more accurate definition of the Stokes number is discussed in section 3.1.5, but for our current purposes, this definition will suffice.)

It is this coupling by drag forces that gives rise to systematic drift motion of particles (*Whipple,* 1972; *Nakagawa et al.,* 1986), determines how particles are affected by turbulence (*Voelk et al.,* 1980; *Ormel and Cuzzi,* 2007), and sets how strongly the radial gas flow drags particles along (*Takeuchi and Lin,* 2002). In the following, we will discuss some of these dynamical effects in more detail.

2.1. Vertical Settling

A particle on an inclined orbit (and in the absence of gas) would be oscillating vertically above and below a reference plane where the vertical coordinate has its origin, which we will refer to henceforth as the disk "mid-plane." This can be seen from the equation of motion of a particle near the mid-plane (*Nakagawa et al.,* 1986)

$$\ddot{z} + z\Omega^2 + \frac{\dot{z}}{t_{stop}} = 0 \quad (3)$$

where the first two terms describe a harmonic oscillator. The remaining term is the velocity-dependent drag term. The presence of the gas disk thus causes a deceleration of the particle if it moves vertically up and down through the gas, and the vertical motion of the particle can be regarded as a damped harmonic oscillator. Particles with $St \ll 1$ adapt quickly to the gas velocity, and in this limit, the particles are not oscillating; instead their velocity quickly tends to vertical gas velocity that we assume to be zero here. Due to the vertical component of gravity that accelerates the particle to the mid-plane, the particle velocity approaches a steady state between the height-dependent gravitational acceleration and the velocity dependent deceleration. Within a few stopping times, this equilibrium ($\ddot{z} = 0$) is reached at a terminal velocity of

$$v_{sett} = -z\,\Omega\,\text{St} \quad (\text{for St} \ll 1) \tag{4}$$

This is analogous to dropping a feather that quickly reaches a constant velocity.

If this vertical sedimentation were to proceed unhindered, then all particles would be accumulating at the midplane in a razor-thin disk, similar to Saturn's rings. This is, however, not the case, as current scattered-light images of PPDs spectacularly depict (e.g., *Avenhaus et al.,* 2018).

Some form of gas motion (likely vertically diffusive turbulence, although bulk flows are also possible) is believed to be at play that vertically mixes the particles away from the mid-plane. In this case, a balance is reached when settling leads to dust concentration gradients that in turn lead to a diffusive flux that becomes equal, but opposing to the settling flux

$$\rho_d v_{sett} - D\rho_g \frac{\partial \rho_d/\rho_g}{\partial z} = 0 \tag{5}$$

where the diffusion flux acts to smooth out gradients in the composition as the dust is mixed by and well coupled to the turbulent gas. The solution of the above equation is a steady-state dust distribution

$$\rho_d(z) = \rho_d(0)\frac{\rho_g(z)}{\rho_g(0)} \exp\left(\int_0^z \frac{v_{sett}}{D}\, dz'\right) \tag{6}$$

The Schmidt number Sc is the ratio of gas and dust diffusivities (see the discussion in *Youdin and Johansen,* 2007). All particles with St ≪ 1 are well enough coupled to the turbulent eddies of the gas, leading to a Schmidt number of effectively unity. *Youdin and Johansen* (2007) showed that it can be approximated by $\text{Sc} = 1 + \text{St}^2$. If the gas follows the isothermal hydrostatic solution of a Gaussian curve with scale height H, then around the mid-plane, the vertical density profile of particles of a given size also follows a Gaussian curve but with a scale height of

$$H_p = H\sqrt{\frac{\alpha}{\alpha + \text{St}}} \tag{7}$$

where α is a quantification of the turbulence. The history and various definitions of α are long and not within the scope of this chapter. Furthermore, α as used here should be distinguished from the so-called "turbulent viscosity" of *Shakura and Syunyaev* (1973). The latter quantity depends on correlations between radial and azimuthal velocity perturbations and includes turbulent magnetic stresses. The former is purely a dimensionless turbulent diffusion parameter, such as has been quantified by, e.g., *Johansen and Klahr* (2005) and *Fromang and Papaloizou* (2006). For our purposes, it suffices to assume that the turbulence is isotropic and define the turbulent diffusion as $\alpha = \delta v^2/c_s^2$, where δv and c_s are the turbulent velocity and sound speed, respectively. (More accurately, α should include the eddy turnover time in units of Ω^{-1}. However, at the typical driving length scales for disk turbulence, the eddy turnover time is $\sim\Omega^{-1}$, making the *dimensionless* turnover time equal to unity.)

The dust to gas ratio at the mid-plane is therefore enhanced for particles with a Stokes number St > α, since

$$\frac{\rho_d(0)}{\rho_g(0)} = \frac{\Sigma_d}{\Sigma_g}\sqrt{\frac{\text{St}}{\alpha} + 1} \tag{8}$$

where Σ_d/Σ_g is the ratio of the dust and gas column densities.

2.2. Radial Drift

A basic understanding of vertical settling now established, we can apply a similar approach for the radial and azimuthal velocity of dust and gas. The equation of motion of dust and gas are coupled by drag forces and this set of equations can be solved for a steady state, i.e., the terminal velocities (see *Nakagawa et al.,* 1986). Summarizing these results, the gas disk in a force balance between pressure, gravity, and centrifugal forces rotates slightly sub-Keplerian for a typical, radially decreasing pressure profile. This leads to the problem that the dust particle (that feels negligible pressure acceleration) needs to be on a Keplerian orbit to reach force balance. The resulting headwind on the particles decelerate them and remove some of their angular momentum, which in turn causes them to spiral inward radially. The gas, which has gained angular momentum from the particles, moves radially outward, and an equilibrium is reached, as derived by *Nakagawa et al.* (1986) (see also *Whipple,* 1972; *Weidenschilling,* 1977), where the dust radial velocity is

$$v_{r,dust} = -\frac{2}{(1+\varepsilon)^2\,\text{St}^{-1} + \text{St}}\eta v_K \tag{9}$$

$\varepsilon = \rho_d/\rho_g$ is the dust-to-gas mass ratio

$$\eta = -\frac{1}{2}\left(\frac{H}{r}\right)^2 \frac{\partial \ln P}{\partial \ln r} \tag{10}$$

r is the radial coordinate, and P is the gas pressure. The resulting timescale on which particles drift inward is therefore (for ε, St ≪ 1)

$$t_{drift} = \frac{1}{St\Omega}\left(\frac{H}{r}\right)^{-2}\left|\frac{\partial \ln P}{\partial \ln r}\right|^{-1} \simeq \frac{64 \text{ orbits}}{St}\left(\frac{H/r}{0.03}\right)^{-2} \quad (11)$$

which is long compared to the orbital timescale, but short in terms of the overall disk evolution timescale. This shows that growing particles can travel substantial distances during their evolution.

Both radial drift and vertical settling are a result of the terminal velocity approximation that more generally states (see *Youdin and Goodman,* 2005) that particles drift with respect to the gas toward higher pressure, i.e., that the dust-gas velocity difference is $-(\nabla P/\rho_g)t_{stop}$.

2.3. Collisional Growth and Destruction

When two dust grains collide, their fate is determined by the collision energy as well as the material and makeup of the particles. In general, three collisional outcomes are possible: sticking, bouncing, and fragmentation, although mixtures of and transitions between these outcomes are possible too (*Güttler et al.* 2010) (see also Fig. 1).

The initial grain sizes can be estimated by comparison with observations of meteorites, comets, or interstellar dust. *Vaccaro et al.* (2015) investigated the size distribution of matrix grains in primitive carbonaceous chondrites, using high-resolution electron microscopy, and found geometric mean diameters on the order of 200 nm. *Mannel et al.* (2019) determined very similar mean particle sizes for the dust particles collected from Comet 67P/Churyumov-Gerasimenko by the Rosetta/MIDAS instrument. Depending on the particular dust particle investigated, they found median values of the constituent grain diameters between ~100 nm and ~400 nm. In contrast to the collected samples of our own solar system, dust grains in the interstellar medium (ISM) seem not to possess a single characteristic size value. ISM grains rather follow a power-law size-frequency distribution, with the smallest grains in the 1-nm size range and the largest ones having diameters of ~500 nm (*Mathis et al.,* 1977). The slope of the power law is such that while most of the cross section resides in the smallest grains, most of the mass is in the largest grains. Thus, the largest effect on the mass growth will be from collisions between large grains; an appropriate range over which to investigate the collisional growth is 0.1–1 μm.

Laboratory experiments have shown that initially, dust and ice grains in PPDs generally stick upon mutual collisions (*Blum and Wurm,* 2008; *Gundlach and Blum,* 2015) because their sizes are small (typically $\lesssim$1 μm, see above) and their collision speeds are low (*Weidenschilling,* 1977; *Weidenschilling,* 1980). Two 1-μm solid silicate grains stick to each other when colliding with speeds $\lesssim$1 m s^{-1} (*Blum and Wurm,* 2008), whereas two 1-μm solid water-ice grains stick for collision velocities $\lesssim$10 m s^{-1} (*Gundlach and Blum,* 2015). The latter result is somewhat surprising, because the surface energy of water ice at low temperatures does not exceed that of silicates (*Gundlach et al.,* 2018; *Musiolik and Wurm,* 2019). Obviously, the surface energy is not the only material parameter with relevance to the collision and sticking behavior. *Krijt et al.* (2013) showed that the yield strength and the viscous relaxation time are two additional parameters that determine the collisional outcome. Recently, *Arakawa and Krijt* (2021) showed how the earlier and latest experimental data on the sticking threshold of H_2O-ice, CO_2-ice, and SiO_2 grains can be reconciled with the model by *Krijt et al.* (2013). From the various experimental and theoretical approaches it became clear that the smaller the grains, the higher their threshold velocity for sticking (*Blum and Wurm,* 2008). A recent study also showed that the static cohesion (a measure for the surface energy) between grains of carbon-bearing (mostly organic) substances is heavily material-dependent, with the highest values for graphite and paraffin and the smallest for humic acid and brown coal; the ratio in static cohesion between these two groups is as large as a factor of 1000 (*Bischoff et al.,* 2020).

Figure 1 shows examples of a state-of-the-art collision model for dust and ice grains in PPDs, based on laboratory collision experiments (*Kothe,* 2016). In the top row, the effect of monomer-grain size on the collision outcome is presented by a comparison between SiO_2 grains with 1.5 μm and 0.1 μm diameter, respectively. Green colors denote the regions in which growth, either by direct sticking or by mass transfer in collisions in which one of the aggregates fragments, dominates. In the yellow (bouncing), orange (erosion, abrasion, cratering), and red regions (fragmentation), growth is not possible. It is important to note that the collision outcome not only depends on the sizes of the two colliding aggregates, but also on their collision speed, which is in the example shown determined for a minimum-mass solar nebula model (dashed contours denote collision velocities in meters per second). From this comparison, it can be seen that for the more cohesive small grains, potential routes to planetesimal formation are present, namely by mass transfer in collisions between large bodies and smaller dust aggregates. However, these growth avenues are surrounded by orange and red regions so that an easy assessment of whether growth or destruction dominates is not possible. For large monomer-dust grains, the pathway to planetesimals is almost absent. Further out in the PPD, where water ice may be the dominant dust-grain material, is similar to the 0.1-μm SiO_2 grains at 1 au, even for ice particles of 1.5-μm diameter (bottom row in Fig. 1). This is due to the intrinsic higher collisional stickiness of water-ice particles in comparison to the SiO_2 grains (*Gundlach and Blum,* 2015).

It should be noted that the collision model by *Kothe* (2016) does not explicitly include the possible change in porosity during the growth of dust agglomerates. However, an earlier model by *Güttler et al.* (2010) showed that there is an effect of porosity on the collisional outcome. Whenever the colliding particles are in the sticking regime, this difference is small. However, for larger aggregate masses and/or collisions speeds, compact (low-porosity) particles

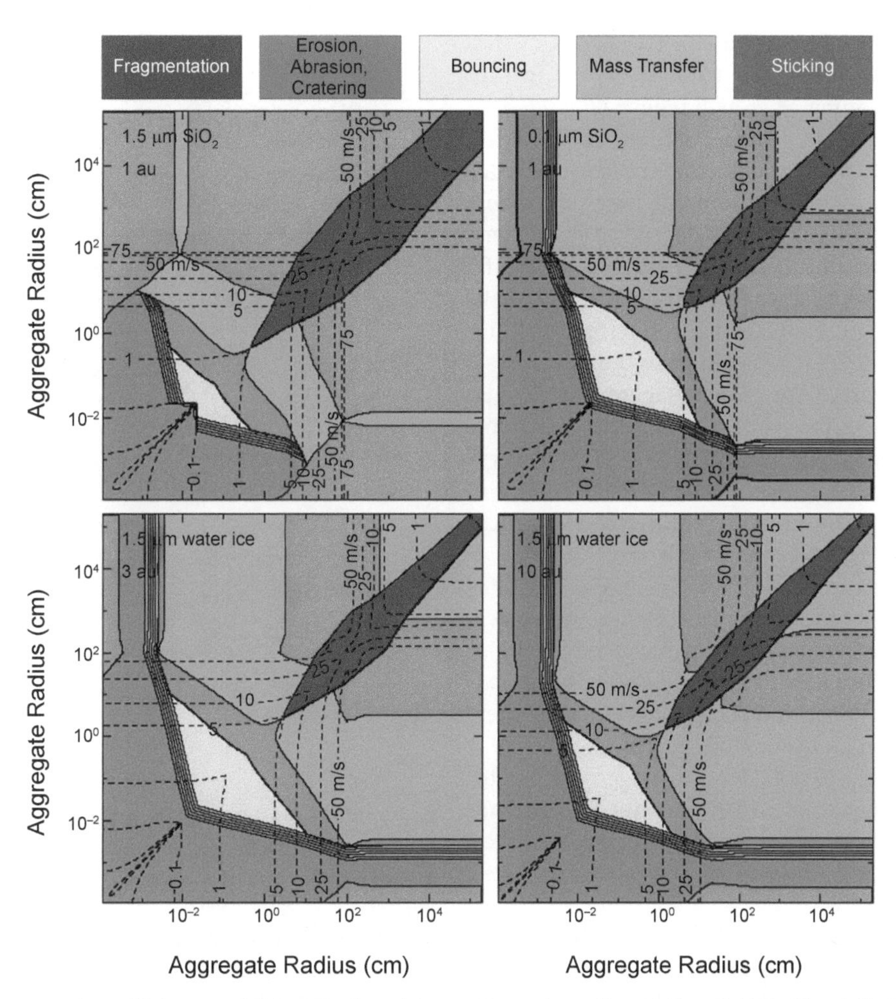

Fig. 1. A dust-aggregate collision model for dust and ice aggregates after *Kothe* (2016). Top left: for SiO_2 monomer grains of 1.5-μm diameter at 1 au; top right: for SiO_2 monomer grains of 0.1-μm diameter at 1 au; bottom left: for water-ice monomer grains of 1.5-μm diameter at 3 au; bottom right: for water-ice monomer grains of 1.5-μm diameter at 10 au. The underlying PPD model is a minimum-mass solar nebula model. Credit: Stefan Kothe.

enter the bouncing regions, whereas fluffy (high-porosity) particles can still grow. The transition between fluffy and compact particles is usually achieved when the collision energies exceed the rolling-friction threshold, as determined by *Dominik and Tielens* (1997). This threshold energy is required to overcome the cohesional friction against mutual rolling of the particles over their surfaces. It is apparent that a reliable modeling of dust growth has to take into account the change in porosity during the growth process. For example, *Zsom and Dullemond* (2008) show that initially the dust aggregates start in extremely fluffy (or even fractal, see below) configurations, but get compacted in mutual collisions when the collision energy becomes sufficiently high (*Blum and Wurm,* 2000; *Weidling et al.,* 2009).

Because of the initially slow collision speeds, sticking will be the dominant collisional outcome for the initial stages of grain growth in PPDs (see all four panels in Fig. 1). *Dominik and Tielens* (1997) predicted and *Blum and Wurm* (2000) experimentally confirmed that if the collision energy is much less than the energy for rolling or sliding (the latter is analogous to the rolling case and refers to the force required to slide two surfaces over each other against cohesional friction) of the grain contacts, the collisional adhesion will be of the hit-and-stick type, i.e., the grain contact will be frozen at the position of the first contact. In such a case, the growing aggregates can be described by a fractal description, $m \propto s^{D_f}$, between their mass m and their characteristic radius s, with the fractal dimension D_f. The role of the fractal dimension can best be understood if one calculates the average mass density and the collision cross section for fractal aggregates, respectively. The former can be estimated by $\rho_f \propto m/s^3 \propto m^{1-3/D_f}$, the latter by $\sigma_f \propto s^2 \propto m^{2/D_f}$. For $D_f = 3$, the aggregates behave like any solid material, i.e., ρ_f = const and $\sigma_f \propto m^{2/3}$. These aggregates are just porous, with the same porosity throughout the volume. For smaller fractal dimensions, however, e.g., $D_f = 2$ (see Fig. 2 for an example), the situation changes, because now $\rho_f \propto m^{-1/2}$ and $\sigma_f \propto m$. This means that the average density decreases with increasing aggregate mass and also from the center of mass radially outward if one considers a specific aggregate. On the other hand, the collision cross section increases in proportion with mass and thus much faster than for higher fractal dimensions, which can have considerable impact on the growth rate.

The exact value of D_f depends on the details of the growth process and can reach from $D_f \approx 1.5$ in the Brownian-motion-dominated growth regime (*Paszun and Dominik,* 2006) to $D_f = 3$ when a large particle drifts through a cloud of small grains (*Blum,* 2006). Fractal aggregates with $D_f < 2$ possess stopping times or Stokes numbers that are only weakly mass-dependent (see equations (1) and (2)) and thus behave dynamically very differently than those with larger fractal dimensions. Small fractal dimensions are equivalent to preferred sticking collisions among similar-sized aggregates, i.e., the mass-frequency distribution of the growing aggregates in this case is narrow at any given time. On the other hand, for the highest D_f values the size distribution may be very wide or even bidisperse (*Blum,* 2006) so that the largest aggregates preferentially grow by collisions with much smaller particles. Detailed studies on this will be discussed in the following subsection.

As the collision energy of protoplanetary dust aggregates increases with increasing mass of the particles, the criterion for hit-and-stick collisions will no longer be fulfilled and collisions will lead to the compaction of the aggregates (*Wada et al.,* 2009; *Güttler et al.,* 2010; *Zsom et al.,* 2010; *Okuzumi et al.,* 2012). Thus, dust aggregates with $D_f < 2$ might be converted into those with $D_f > 2$ and ultimately may reach $D_f = 3$. It should be mentioned that the compaction process depends on monomer-grain size, shape, and material.

For even higher collision speeds, the colliding aggregates will either bounce or even fragment upon collision (*Blum and Wurm,* 2008). While the former leads to an outside-in compaction of the aggregates (*Weidling et al.,* 2009; *Güttler et al.,* 2010), the latter changes the mass-frequency distribution dramatically (*Güttler et al.,* 2010; *Bukhari Syed et al.,* 2017). Laboratory studies have shown that the threshold speed for fragmentation of centimeter-sized aggregates of silica grains is very close to the sticking threshold of their constituent grains (*Blum and Wurm,* 2008; *Gundlach and Blum,* 2015; *Bukhari Syed et al.,* 2017), i.e., around 1 m s^{-1} for ~1-µm SiO_2 grains. However, recent experiments with centimeter-sized aggregates consisting of micrometer-sized water-ice particles at low temperatures showed that their fragmentation speed is also 1 m s^{-1} (*Landeck,* 2018) and thus much smaller than the sticking threshold of their constituent particles, which is around 1 m s^{-1} at low temperatures (*Gundlach and Blum,* 2015). In contrast to the collisional sticking process, fragmentation seems to be dominated by the surface energy. This is consistent with the findings by *Gundlach et al.* (2018) that the tensile strength of aggregates consisting of ~1-µm SiO_2 grains and ~1-µm H_2O grains is very similar. The conclusion of the limited empirical data available is that the survival of aggregates in collisions above a few meters per second is very unlikely.

In order to understand the evolution of the particle size in PPDs, given these collisional outcomes, one needs to track all possible grain-grain collisions and their resulting aggregates and fragments, a formidable task. Numerical models therefore have to simplify this evolution in one way or another. Most common approaches fall either into a mass-bin-based direct solution of the collisional evolution (e.g., *Weidenschilling and Ruzmaikina,* 1994; *Dullemond and Dominik,* 2005; *Brauer et al.,* 2008; *Birnstiel et al.,* 2010; *Okuzumi et al.,* 2012) or into Monte Carlo-based approaches (e.g., *Ormel and Spaans,* 2008; *Zsom and Dullemond,* 2008); see *Drążkowska et al.* (2014) for a discussion of these approaches. However, many of those detailed simulations can be understood in simpler, approximate terms following *Birnstiel et al.* (2012): Even in the most optimistic case, where every collision is assumed to result in perfect sticking, radial drift can usually remove particles faster than they form. A very simplified treatment of the growth timescale can be derived from monodisperse growth (e.g., *Kornet et al.,* 2004; *Brauer et al.,* 2008; *Birnstiel et al.,* 2012), which gives

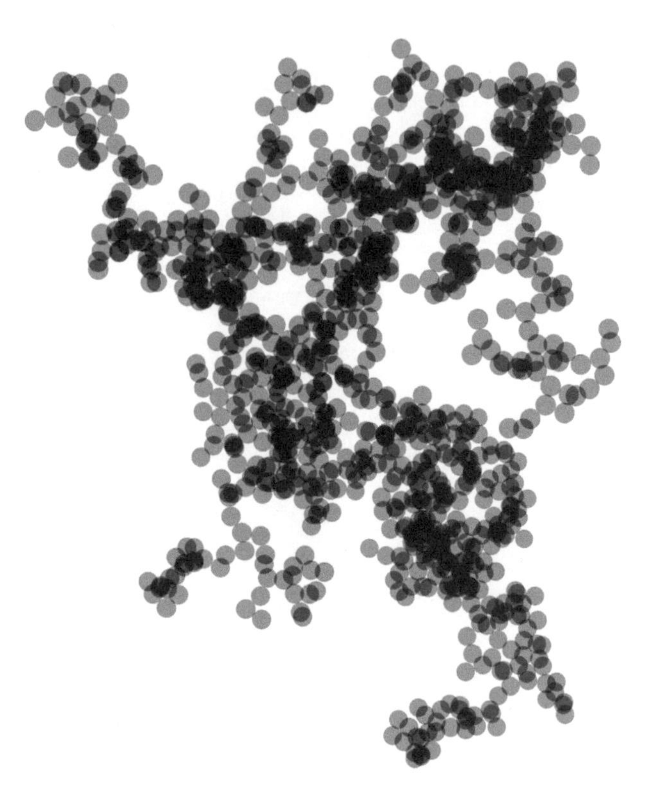

Fig. 2. Example of a simulated dust aggregate with a fractal dimension $D_f = 2.0$ consisting of 1024 monomer grains with a very narrow size distribution.

$$t_{grow} \sim \frac{1}{\varepsilon\Omega} \tag{12}$$

This is in part a factor of a few identical with the collisional timescale; see *Birnstiel et al.* (2016) for a derivation with a discussion of the approximations under which this is derived. This timescale (see the vertical arrows in Fig. 3) is roughly independent of particle size (but see *Powell et al.,* 2019), while the drift timescale (horizontal arrows in Fig. 3) does depend on particle size. Hence, small particles drift so slowly that they mainly grow *in situ,* while large particles can drift inward before they continue to grow substantially. Equating these particles gives us a limiting size, called the "drift limit," that particles approximately can reach before drift will move them to smaller distances from the star

$$a_{drift} \simeq 0.35 \frac{\Sigma_d}{\rho_s} \left(\frac{H}{r}\right)^{-2} \left|\frac{\partial \ln P}{\partial \ln r}\right|^{-1} \tag{13}$$

which is shown as an orange line in Fig. 3.

The resulting particle sizes are typically below millimeters in the outer disk but can be far beyond meters in the inner disk. This suggests that the outer disk may be limited by these dynamical effects, while in the inner disk, the above-mentioned collisional effects (bouncing, erosion, fragmentation) are more likely to stop particle growth. For those disk regions, a limiting particle size can be derived if the threshold velocity for bouncing or fragmentation is compared with the size-dependent collision speeds. For turbulent velocities (*Ormel and Cuzzi,* 2007), and, e.g., a fragmentation threshold velocity v_{frag}, the fragmentation limit becomes (e.g., *Birnstiel et al.,* 2012)

$$St_{frag} = \frac{1}{3\alpha} \left(\frac{v_{frag}}{c_s}\right)^2 \tag{14}$$

which shows a strong dependence on the fragmentation velocity. Similar limits can be derived for either different sources of relative velocities (e.g., drift velocity) or different threshold velocities (e.g., bouncing threshold velocity). Figure 3 only considers fragmentation and displays the fragmentation barrier as a purple line. Figure 3 shows how the simulated dust size distribution [blue contours, after *Birnstiel et al.* (2012)] is well contained below either the drift limit in the outer disk (orange line) or the fragmentation limit (purple line). The arrows display the approximate growth and drift rates. Particle growth is non-local in mass space. This means any particle size can interact with any other size and affect many other sizes, for example, in the case of a distribution of fragments. Still, the evolution of the bulk mass can often be understood by calculating trajectories in the mass/radius space (see black dashed lines in Fig. 3): Small grains grow but do not drift significantly, which results in a vertical motion in the figure. At 100 kiloyears (k.y.) the particles have reached the position marked with a cross, which shows that by that time, particles in the outer disk have barely grown while particles in the inner disk have since long grown to large sizes and have been lost toward the inner disk. This inward radial drift can be stopped if pressure maxima are present [we explain this in detail in section 3.1.5 below, but see also, e.g., *Pinilla et al.* (2012)]: At those positions, radial drift is stopped (see equations (9) and (10)) and dust can accumulate.

The black-dashed trajectories in Fig. 3 show that the

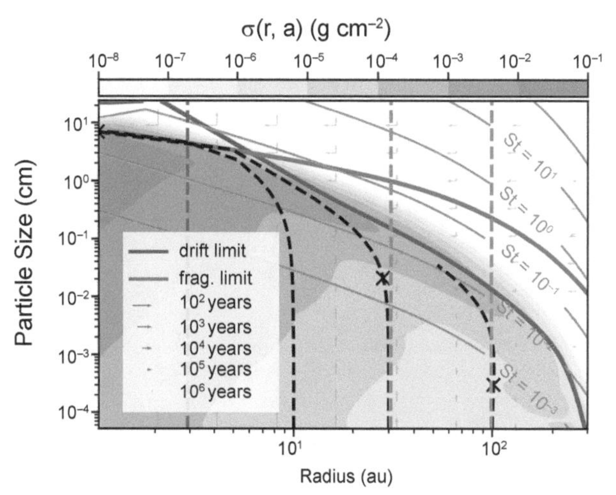

Fig. 3. Simulated dust size distribution after *Birnstiel et al.* (2010), shown as blue contours. Gray arrows show the drift and growth rates. The growth limits are displayed as brown/orange (drift limit) and purple (fragmentation limit) lines. Gray lines denote the corresponding Stokes numbers of the particles. Monodisperse growth/drift trajectories are shown as black dashed lines, ending at 700 k.y. and denoting the position at 100 k.y. with a cross. The vertical lines correspond to the size distributions shown in Fig. 4. Adapted from *Birnstiel et al.* (2016).

particles tend to grow toward and along the drift limit, and in the inner disk along the fragmentation limit. In the latter case, continuous fragmentation and coagulation balance each other, which maintains a constant population of both small and large particles. Figure 4 shows the particle size distributions at different positions in the disk. It can be seen that in all cases, most of the mass resides near the largest particles. This, coupled with the fact that the velocity also scales with the particle size, means that those largest sizes carry most of the radial mass flux (*Birnstiel et al.,* 2012). The resulting maximum particle size a_{max} determines the rate at which the dust surface density is transported inward and this explains why the two-population model of *Birnstiel et al.* (2012) reproduces more complex simulations well.

This inward-drifting, growing dust might feed dust pressure traps and/or accrete onto planets and it determines the local particle sizes that may eventually form planetesimals (*Lambrechts and Johansen,* 2014; *Lambrechts et al.,* 2019) (see section 3).

2.4. Condensational Growth

Collisional growth is not the only possible mode of grain growth in protoplanetary disks. Dust particles can also grow by deposition of vapor species on their surface. The key difference is, however, that condensation stops once the vapor is consumed, i.e., continuous condensation needs continuous resupply of condensible vapor. In contrast, collisional growth does not need to be sustained by adding mass. This means that significant growth by orders of magnitude in size can be easily achieved by collisions (e.g.,

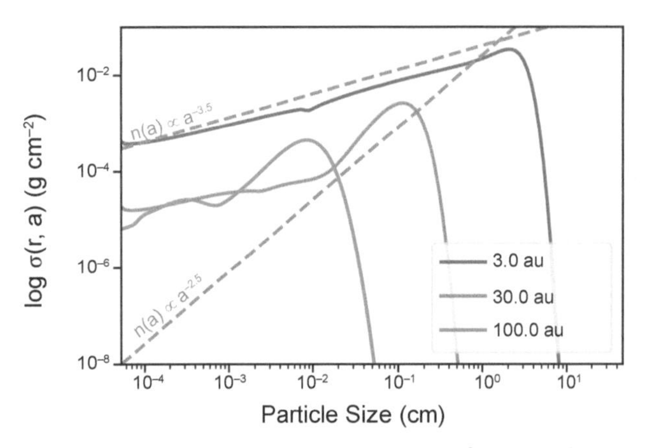

Fig. 4. Dust size density distributions (surface density per logarithmic mass) measured at 3, 30, and 100 au from the simulation in Fig. 3. The corresponding slope of the MRN size distribution as well as a top-heavier exponent ($\propto a^{-2.5}$) are shown as references. After *Birnstiel et al.* (2012).

Ormel et al., 2007; *Brauer et al.,* 2008; *Zsom and Dullemond,* 2008; *Okuzumi et al.,* 2009; *Birnstiel et al.,* 2010), but via condensation would require orders of magnitude of resupplied mass. A possible way around this limitation would be continuously repeating events of sublimation and recondensation. However, condensation/deposition is a surface effect and small particles usually dominate the dust surface area of a size distribution even if they constitute only a negligible fraction of the total dust mass. This means that condensational growth as described can only increase the sizes of small particles efficiently (*Hirashita,* 2012). This was also seen in *Stammler et al.* (2017), where both coagulation and sublimation/recondensation were simulated around the CO snow line. More recently, *Ros et al.* (2019) pointed out that heterogeneous ice nucleation on small silicate grains may be hindered. This, however, could at most offset the problem to larger sizes: Whenever there is a wide distribution of particle sizes, the smallest particles will grow more efficiently than the largest particles.

So while condensation cannot explain growth from ISM sizes to millimeter-centimeter sizes, it is still thought to be an important effect near the water snow line. As the most abundant volatile, water sublimation at the water snow line can deposit significant amounts of water vapor (e.g., *Cuzzi and Zahnle,* 2004; *Ciesla and Cuzzi,* 2006; *Gárate et al.,* 2020). Outward diffusion of this vapor can lead to deposition of water ice on the inward-drifting icy particles. This can locally increase both the particle size as well as the dust surface density and lead to conditions supporting the streaming instability (*Drążkowska et al.,* 2016; *Schoonenberg and Ormel,* 2017) (also, see below for a description of the streaming instability). The exact location where this takes place with respect to the snow line, however, depends also on the effects of backreaction of the dust onto the gas density distribution (*Hyodo et al.,* 2019; *Gárate et al.,* 2020).

2.5. Grain Size Distributions

Starting from an idealized monodisperse distribution of individual dust grains and assuming a high sticking probability in every occurring collision, the resulting mass-frequency distribution of the growing dust aggregates depend on (1) the velocity field among the particles and (2) the fractal dimension of the aggregates. If we approximate the mass dependency of the velocity field by a power law $v(m) = v_0\,(m/m_0)^\gamma$, with v_0 and m_0 being the velocity and mass of a monomer grain, and the collision cross section of the fractal aggregates by $\sigma = \sigma_0\,(m/m_0)^{2/D_f}$, with σ_0 being the collision cross section of the monomer grains, the approximate growth equation reads

$$\frac{d(m/m_0)(t)}{dt} = \frac{1}{\tau}\,(m/m_0)^{\gamma+(2/D_f)} \tag{15}$$

with $\tau = 1/(n_0\sigma_0 v_0)$ and n_0 being the collision timescale and the initial number density of monomer grains, respectively (*Blum,* 2006). For $\gamma + (2/D_f) < 1$, we are in the regime of orderly growth, for which the average mass grows with a power law of time and the mass-frequency distribution at any given time is quasi-monodisperse (i.e., narrow). For $\gamma + (2/D_f) > 1$, the growth is characterized by the so-called runaway process in which the mass-frequency distribution becomes bidisperse and the aggregates in the upper mass peak grow infinitely large in a finite time.

Thus, the growth behavior depends massively on the two parameters γ and D_f. The prior one can obtain values between $\gamma = -1/2$ for Brownian-motion-dominated growth and $\gamma = +1/3$ for drift-dominated velocities. The fractal dimension can obtain values between $D_f = 1.5$–2 for Brownian-motion-driven growth and $D_f = 3$ for either the pre-compaction stage (see above) or in the runaway-growth regime. Unfortunately, these two parameters are not independent and cannot easily be predicted ab initio. Thus, detailed numerical studies are required for the prediction of the growth speed, the morphology, and the mass-frequency distribution of the aggregates.

As pointed out in section 2.3, growth stops at the bouncing barrier or at the fragmentation barrier (*Güttler et al.,* 2010; *Zsom et al.,* 2010). As these barriers typically occur after the sticking-with-compaction stage has been reached, the aggregates have D = 3 and an internal porosity of ~60–70% (*Weidling et al.,* 2009; *Güttler et al.,* 2010; *Zsom et al.,* 2010). Their final size depends on the size and material of the monomer dust/ice grains as well as on the location in the PPD and can reach values of ~1 cm at 1 au (*Zsom et al.,* 2010) and ~1 mm at 30–50 au (*Lorek et al.,* 2018). These aggregates have been termed "pebbles."

While growth can proceed orderly (as a well defined peak), or in a runaway fashion, the outcome of many growth (and transport) simulations tend to reach a quasi-steady-state size distribution, the shape of which is determined by the collisional outcome.

First, if particles grow up to the fragmentation barrier, mass is effectively redistributed to all sizes smaller than the fragmenting particle. In that case dust mass is transported to larger sizes by growing collisions and to smaller sizes by fragmenting or eroding collisions. This quickly leads to a steady state where continuous growth and fragmentation balance each other. An example of this can be seen in Fig. 4 at 3 au, which shows slices of the dust distribution of Fig. 3. The resulting size distribution can depend on the way the particles fragment, particularly whether the most fragmented mass is in the smallest or the largest fragments. Analytic and semi-analytic models of these scenarios were derived in *Birnstiel et al.* (2011). Interestingly, for the most common regime of turbulent collision velocities and fragment distributions, the resulting size distribution is very close to $n(a) \propto a^{-3.5}$, which is the result for a fragmentation cascade, despite the different physics involved. Particles below roughly a micrometer in size, however, are even more strongly depleted, such that the particle surface area is not dominated by the smallest particles present, but instead by particles of about 1 μm (*Birnstiel et al.,* 2018).

Second, if particles do not fragment, but instead bounce off each other, most particle sizes tend to be around the bouncing barrier, leading to a very narrow size distribution with strongly reduced numbers of fine grains (*Zsom et al.,* 2010; *Windmark et al.,* 2012b).

Third, if growth is not halted by a collisional barrier, but instead by the dynamical removal of larger particles in the drift limit, then again the particle size distribution becomes "top-heavy" with the most mass in a narrow range around the maximum particle size, since small dust is not efficiently reproduced and instead swept up by the largest particles. The particle size distribution was measured to follow approximately $n(a) \propto a^{-2.5}$ in the drift-limited case [see *Birnstiel et al.* (2012) and Fig. 4]. The shallower slopes toward the smallest grains is caused either by radial mixing from the inner disk, where small grains are produced by fragmenting collisions [see the 30-au case in Fig. 4 and *Birnstiel et al.* (2015)], or are remnants of the initial condition that have not yet grown to larger sizes as is the case at 100 au.

2.6. Constraints from Observations

In recent years, much progress has been made toward testing the theoretical expectations above through observations. Traditionally, this was done by measuring the spectral index at millimeter wavelengths $F_\nu \propto \nu^{\alpha_{mm}}$, as α_{mm} (in the optically thin Rayleigh-Jeans limit) is linked to the grain opacity $\kappa_{abs} \propto \nu^\beta$ via $\alpha_{mm} = \beta + 2$. β in turn depends on the maximum particle size and to a lesser extent on the particle size distribution. While particles much smaller than the wavelength typically present values of $\beta \sim 1.7$, values of $\beta < 1$ tend to be only reached if the maximum particle size is larger than the observational wavelength (e.g., *Ricci et al.,* 2010; *Testi et al.,* 2014; and references therein).

Initial disk-integrated measurements already indicated low β values at millimeter wavelengths, indicative of millimeter-sized or larger particles, which are in agreement with the predicted maximum particle sizes of theoretical models. A large caveat, however, is that the lifetime of particles of such sizes should be much shorter than the disk age. Theoretical models produced the right sizes, but only for a short time of a few hundred thousand years (*Brauer et al.,* 2007, 2008; *Birnstiel et al.,* 2011). One possible solution, already envisioned by *Whipple* (1972), was the introduction of pressure perturbations that have a vanishing pressure gradient, thus locally halting radial drift. Strong enough pressure perturbations could locally trap the large particles long enough, as required by observations (*Pinilla et al.,* 2012).

The theoretically predicted size sorting of larger grains at smaller radii (*Birnstiel et al.,* 2012) was also confirmed by modeling the radially resolved spectral indices in *Pérez et al.* (2012) and *Tazzari et al.* (2016). However, it was not until ALMA's large baseline observations became available that substructures were imaged that seemed to agree with the idea of pressure traps (see *ALMA Partnership et al.,* 2015; *Andrews et al.,* 2016; *Huang et al.,* 2018). These observations also revived the idea of *Ricci et al.* (2012) that a significant fraction of the flux could be optically thick — then the observed low spectral index is not caused by large grains, but instead by optically thick emission (*Tripathi et al.,* 2017).

ALMA has also opened the way to using polarization to measure particle sizes via self-scattering (*Kataoka et al.,* 2015), which indicated particles not much larger than about 100 μm, a clear conflict with spectral-index based measurements that indicated millimeter- or centimeter-sized particles. While this conflict continues to linger [for a possible solution, see *Lin et al.* (2020)], the width of several of the imaged rings of DSHARP (*Andrews et al.,* 2018; *Huang et al.,* 2018) is so narrow that dust trapping, i.e., $\mathrm{St} \gtrsim \alpha$ is most likely required (*Dullemond et al.,* 2018), putting a constraint on the particle sizes. Furthermore, these rings might be providing the right conditions for planetesimal formation via the streaming instability and gravitational collapse (see section 3.1.3), as also argued for by *Stammler et al.* (2019), as this would naturally lead to the observed optical depths of around 0.4 in all the well-resolved substructures. The formation of planetesimals in pressure traps is briefly discussed further in section 3.1.5.

3. DUST TO PLANETESIMALS

Having described the formation and evolution as well as properties of small solids in PPDs, we now turn to a description of how pebbles transition to the next phase of planet formation: planetesimals. Currently, there are two schools of thought for how planetesimals form. The first relies on pebbles being concentrated somehow to the point where their mutual gravitational attraction overpower competing effects, such as stellar tides and turbulent diffusion. As we will see, this approach has provided the largest number of potential mechanisms, which we refer to as "gravitational collapse models."

The second school of thought takes the opposing view of bottom-up growth. These models, which we refer to as "coagulation models," rely on the imperfection of the various growth barriers described above; as described further below, there are routes toward sticking millimeter- to centimeter-sized pebbles together to further grow in size, such as the collision-induced mass transfer from smaller aggregates to larger bodies. Thus, coagulation models describe a route to planetesimal formation through purely collisional growth.

Most of the literature to date has focused on the former set of models, and consequently, we devote most of the remainder of this chapter to their description. However, in section 3.2, we describe in detail the coagulation models.

3.1. Gravitational Collapse Models

As mentioned above, the gravitational collapse models all provide a route for pebbles to be concentrated sufficiently well that their mass density ρ_d exceeds that which is required to counteract the destructive tidal forces from the central star, the so-called Roche density

$$\rho_R \equiv 3.5 \frac{M_*}{r^3} \tag{16}$$

where M_* is the mass of the central star, and r is the radial distance of the self-gravitating object from the star [for a derivation see *Armitage* (2007)].

In the remainder of this section, we describe a number of different mechanisms that all provide a route toward $\rho_d \gtrsim \rho_R$ locally, and thus for planetesimals to be born.

3.1.1. The Goldreich-Ward mechanism. The first such model put forth, and in some ways the simplest one, is that of a gravitationally unstable collisionless particle disk. This mechanism, often referred to as the Goldreich-Ward mechanism (*Goldreich and Ward,* 1973), is no longer thought to work for reasons we will address shortly. [Similar work can be found in *Safronov* (1972).] However, we include it here as a starting point for the discussion of other instabilities, especially as some of the ideas described here carry over to other mechanisms.

Following *Binney and Tremaine* (1987), our starting point with this discussion will be the consideration of a razor thin gaseous disk that extends infinitely, has a constant surface density Σ, has sound speed c_s, and is rotating uniformly at a rate Ω. By introducing small perturbations of the form $\exp[i(k_r r + k_\phi \phi - \omega t)]$ (where ω is the temporal frequency of the perturbed wave mode and k_r and k_ϕ are the wave vector components in the radial and azimuthal directions, respectively) to the governing equations and keeping only leading order (linear) terms [i.e., carrying out a "linear perturbation analysis"; we encourage the reader to read *Binney and Tremaine* (1987) for more details], one can show that the gravitational stability of this system is characterized by the so-called (*Toomre,* 1964) Q parameter,

$$Q_g \equiv \frac{c_s \Omega}{\pi G \Sigma} \tag{17}$$

where the subscript on Q denotes that it is relevant to the disk gas.

This key parameter quantifies the ratio of stabilizing to destabilizing parameters. More specifically, the gas disk is stabilized on small scales by thermal motions, hence the presence of c_s in the numerator, whereas on larger scales, shear acts to stabilize the fluid, as shown by Ω (also in the numerator). [This dependence of stability on scales can more directly be seen by examining the dispersion relation, as shown in *Binney and Tremaine* (1987).] For more massive disks (i.e., larger Σ in the denominator), the effect of the stabilizing parameters is diminished and the disk is more subject to gravitational instability.

More quantitatively, if $Q_g < Q_{g,crit}$, the system is gravitationally unstable, and for the setup considered here $Q_{g,crit} = 0.5$ (see *Binney and Tremaine,* 1987). However, gas is a collisional system, which is different than our collision*less* solid particle disk. Yet, even in this system, Q remains a very useful quantity. If we define σ as the one-dimensional velocity dispersion of particles (e.g., due to turbulent stirring by the gas) and replace Σ with Σ_p, the solid particle surface density, our "particle" Toomre parameter becomes

$$Q_p \equiv \frac{\sigma \Omega}{\pi G \ \Sigma_p} \tag{18}$$

The precise value of the critical Q for a differentially rotating Keplerian disk of particles (the system under consideration here), $Q_{p,crit}$, will be different than the gaseous disk considered above. However, as in the gas case, $Q_{p,crit} \approx 1$ and $Q_p \lesssim 1$ implies gravitational instability of the particle layer.

This is the crux of the Goldreich-Ward mechanism; for sufficiently large Σ_p and sufficiently low σ, $Q < 1$ and the particle layer fragments into bound objects.

But how viable is this mechanism given reasonable assumptions for disk parameters? Assuming the minimum mass solar nebula model [which is arguably not the most accurate model, but serves our purposes for now (*Hayashi,* 1981)], the gas surface density at 1 au is $\Sigma = 1700$ g cm^{-2}. With a standard dust-to-gas ratio of 1%, $\Sigma_p \approx 10$ g cm^{-2} at this location. Further assuming a solar-mass star at the center of the disk, $Q = 1$ corresponds to $\sigma \approx 10$ cm s^{-1}. Since $H_p = \sigma/\Omega$ and $c_s \sim 10^5$ cm s^{-1}, we find that for gravitational instability, the ratio of particle to gas scale heights is $H_p/H_g \sim 10^{-4}$.

For centimeter-sized particles at 1 au, St $\sim 10^{-3}$. Using equation (7), in order to maintain such a thin particle layer, $\alpha \sim 10^{-11}$ or less, whereas even very weak turbulent stirring often leads to a minimum of $\alpha \sim 10^{-6}$–10^{-5}! At this point, the problem with the Goldreich-Ward mechanism should be clear; turbulence produced by any number of magnetohydrodynamic or purely hydrodynamic instabilities (see *Balbus and Hawley,* 1998; *Lyra and Umurhan,* 2019) would *far* exceed this tiny number. Even absent these processes, the velocity shear induced by the particles pushing back on the gas near the mid-plane and the slower velocity away from the mid-plane (again, due to the radial pressure gradient) leads to the Kelvin-Helmholtz instability, which *easily* produces turbulence that prevents such a thin particle layer from forming (*Cuzzi et al.,* 1993).

3.1.2. Secular gravitational instability. While the Goldreich-Ward mechanism is very unlikely to produce planetesimals, there is a related phenomenon that acts similarly and *can* get around this thin layer problem. This process, known as the secular gravitational instability (SGI) (*Ward,* 1976) comes about by considering the gravitational instability problem we just described, but with *gas drag included.*

The physical mechanism of this process is as follows (see also *Goodman and Pindor,* 2000). Upon a steady-state particle layer with surface density Σ_p, we impose an axisymmetric particle overdensity. The width of this overdensity δr satisfies $H_p \ll \delta r \ll H_g$. Due to the effects of particle self-gravity, this overdensity will contract. However, as this contraction proceeds, the outer edge will move inward but will maintain its original angular momentum (i.e., there is no torque introduced by the self-gravity), thus particles increase their orbital velocity and resist the pull of the self-gravity. Up until now, what we have described is the stability of the system to pure gravitational instability, which can be overtaken if the mass of the overdensity is sufficiently strong. However, another route to overpower this stability is gas drag. Since $\delta r \ll H_g$, the gas's orbital speed is much less affected; thus, as the solid particles move inward, they feel an increased headwind, which removes their angular momentum and allows self-gravity to win and produce an even larger overdensity; from here, the whole process accelerates. Similar arguments can be applied to the inner part of the overdensity, and the result is runaway growth to eventually produce gravitationally bound objects (although there are caveats, as we will see shortly).

A number of works have examined different aspects of this mechanism, ranging from its linear growth (*Ward,* 2000; *Youdin,* 2005) to the behavior in the fully non-linear (i.e., no small perturbations assumed) regime (e.g., *Pierens,* 2021). *Takahashi and Inutsuka* (2014, 2016) carried out a linear analysis of the SGI but with particle feedback on the gas included, finding that this slightly modified SGI could be responsible for producing dust rings in protoplanetary systems, such as those observed in HL Tau (*ALMA Partnership et al.,* 2015).

Other works further explored this linear regime. *Youdin* (2011) and *Michikoshi et al.* (2012) found that for small ($\leq$1 mm) particles, radial drift timescales can limit the effectiveness of the SGI to regions with weak turbulent diffusion, namely $\alpha \lesssim 10^{-4}$. *Shadmehri et al.* (2019) found that non-axisymmetric modes of the SGI are more robust than axisymmetric modes, although are still limited by turbulent diffusion and are absent for $\alpha \geq 10^{-3}$.

More recently, a number of simulations of the non-linear SGI have been carried out. *Tominaga et al.* (2018) ran two-dimensional global simulations in the $r\phi$ plane and found that indeed the SGI can form dust rings, as was hypothesized by earlier work. More recently, *Pierens* (2021) carried out similar simulations and found that such dust rings can trigger the Rossby wave instability (RWI) (*Lovelace et al.,* 1999), which then gives rise to gravitationally bound clusters of particles via trapping in the RWI-induced vortices (see section 3.1.5 for more details on vortices trapping pebbles). *Tominaga et al.* (2020) followed up on their previous work by including radial drift and turbulent diffusion in their simulations, although in the more recent work, they assumed axisymmetry (i.e., their simulations were one-dimensional); they found that the SGI maintained its robustness in the nonlinear regime, even with these additional effects included.

Finally, *Abod et al.* (2019) carried out a number of local, shearing box simulations designed to study the streaming instability (see below), but one of their simulations produced gravitationally bound clumps in the presence of gas drag but with no radial drift (which requires that the clumps form via the SGI). The formed clumps had a initial mass function very similar to the streaming instability calculations, although with a slightly steeper power law toward low masses. Furthermore, the maximum clumps that formed had masses similar to that produced by the streaming instability (see below) and are thus roughly consistent with (although actually larger than) the largest solar system planetesimals. However, as pointed out in *Abod et al.* (2019), modeling the SGI in the setup that they used posed challenges, and as such, caution is warranted in interrupting these results.

Ultimately, studies of the SGI, while promising, are in their infancy and more work is needed to test the viability of this mechanism, particularly compared to others such as the streaming instability. Indeed, since streaming-instability clumping is now known to be triggered at relatively low dust abundances [see below and *Li and Youdin* (2021)], more work is needed to understand whether SGI could even be triggered before streaming instability clumping takes over. If the SGI can past these tests of robustness, then the next stage will be to quantify the properties of any planetesimals formed via this process.

3.1.3. Streaming instability. The streaming instability (hereafter SI) is the most well-studied of the mechanisms to produce gravitational collapse. As such, we will devote considerable attention to its description and the implications for planetesimal formation. We use the term "streaming instability" (SI) here to refer to a broad class of behaviors that have been observed in simulations that include at the minimum two key ingredients: a background pressure gradient (and thus radial drift; see below) and momentum feedback from solids to the gas. If these criterion are included, both linear analyses (e.g., *Youdin and Goodman,* 2005) and non-linear simulations (e.g., *Johansen and Youdin,* 2007) alike have demonstrated the unstable growth of perturbations that ultimately, as seen in the non-linear simulations, lead to clumping of solids.

However, as we describe in detail below, there has been considerable follow-up work to this, which has expanded our understanding of the behavior of this coupled solid-gas system both in the linear and nonlinear state with additional affects, such as vertical gravity, turbulence, and particle self-gravity. Some of these additional effects have led to the emergence of new routes to instability, such as the vertical shearing streaming instability of *Lin* (2021). We touch upon some of these extensions below, but for simplicity (and to be consistent with the literature), we collectively refer to these processes as the SI, deviating only when appropriate.

3.1.3.1. Simplest model: Linear, unstratified, and no self-gravity: Our starting point is the simplest model consisting of a uniform distribution of solid particles in a uniform gas, both subject to the radial component of the central star's gravity (but not vertical!) and with angular momentum exchange between solids and gas present. By carrying out a linear perturbation analysis on the governing equations, *Youdin and Goodman* (2005) and *Youdin and Johansen* (2007) showed that this system is unstable, with growth rates and unstable wavelengths that depend on ε, St, and η (as defined previously).

While a physically intuitive explanation for this instability remained elusive for many years, recent work by *Squire and Hopkins* (2018, 2020) has provided some clarity. This work led to the realization that in the limit of $\varepsilon < 1$, the SI is actually one of a broader class of "resonant drag instabilies." These instabilities arise when the drift velocity of solids equal the phase velocity of a specific wave mode in the gas. Such a configuration equates to a resonance in the frame of the solids. In the case of the SI, this resonance occurs between radial drift and epicyclic oscillations of the gas, which lead to slow growth of the dust concentration over time.

When $\varepsilon > 1$, on the other hand, the feedback from the dust overwhelms the gas, which accelerates the dust concentration drastically. Both regimes, however, operate on the same basic idea, as shown in Fig. 5: A dust overdensity drags gas inward (Fig. 5a). This inward motion is deflected into the azimuthal direction via the Coriolis force (Fig. 5b). Dust already present downstream of the overdensity then feels a boost from the deflected gas via drag, which causes this downstream dust to move outward radially, adding to the original dust overdensity (Fig. 5c). There are a number of subtleties associated with this picture, including the difference between the $\varepsilon < 1$ and $\varepsilon > 1$ regimes, which are described in detail in *Squire and Hopkins* (2020). However, for the purposes of this chapter, this picture captures the essence of the SI and demonstrates how at its core, it is a concentration-inducing instability. [While the original form of the SI, as derived by *Youdin and Goodman* (2005), is two-dimensional in the radial-vertical plane, this diagram sacrifices information about the vertical dimension and includes the azimuthal dimension in order to provide a more intuitive understanding of the mechanism. Furthermore, the mechanism as depicted here applies to the axisymmetric case as well.]

3.1.3.2. Nonlinear simulations: The linear phase of the SI elucidated, we now turn to a more in-depth discussion of the non-linear regime, probed with numerical simulations. While unstratified simulations have been run (*Johansen and Youdin,* 2007; *Bai and Stone,* 2010b), they have largely been used to test the employed code's ability to reproduce the linear growth phase. (For brevity, we will neglect a description of unstratified simulations beyond reproducing linear results.)

Most SI simulations have included the vertical component to the central star's gravity (the so-called "stratified" setup). Furthermore, with a few exceptions (see *Kowalik et al.,* 2013; *Mignone et al.,* 2019; *Schäfer et al.,* 2020; *Flock and Mignone,* 2021), these simulations have all employed the shearing box approximation. The shearing box treats a local, co-rotating patch of the disk of sufficiently small size to expand the relevant equations into Cartesian coordinates (see, e.g., *Hawley et al.,* 1995). This approach is generally

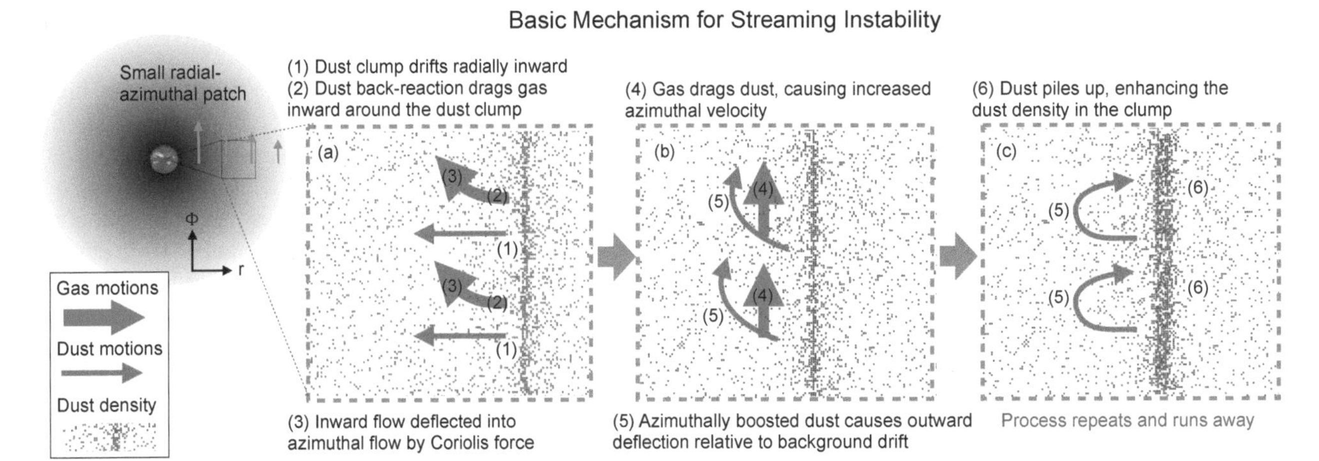

Fig. 5. Basic mechanism for the streaming instability. **(a)** A small dust overdensity brings gas inward via drag. **(b)** This inward gas motion is deflected azimuthally via the Coriolis force. Dust already present downstream of the overdensity then feels an azimuthal boost from the deflected gas via drag, which then **(c)** causes this downstream dust to move to larger radii (as it now has increased angular momentum), adding to the original dust overdensity. This entire process is in a frame corresponding to the average radial drift of the dust. Modified from *Squire and Hopkins* (2020).

done when small scales ($\ll H_g$) must be adequately resolved.

Within the context of the SI, such simulations have been done both in two dimensions (assuming axisymmetry) and fully three dimensions. The primary goal of these simulations has been to study the properties of and parameters responsible for "strong clumping," defined here as clumping sufficiently strong so as to locally reach the Roche density. Perhaps the most impactful of such studies have been a series of papers started by *Carrera et al.* (2015); the authors of that work ran a series of two-dimensional axisymmetric SI simulations finding an approximate "U-shaped" region of the St vs. Z (where $Z \equiv \Sigma_p/\Sigma$ is the dust concentration) parameter space where strong clumping occurs. To cover such a large parameter space, for each St, Z was slowly increased by removing gas from the domain.

Yang et al. (2017) expanded on this work to find that the SI can produce strong clumping at even smaller St than was determined by *Carrera et al.* (2015) if the numerical resolution is high enough and the simulation allowed to run for sufficiently long time. They also verified that for a subset of their parameters, the results are similar between two and three dimensions.

More recently, *Li and Youdin* (2021) reexamined the problem again, over a larger range of St values than explored by *Yang et al.* (2017) but also running two-dimensional and three-dimensional calculations for very long integration times. They found that the strong clumping criterion is even less restrictive than originally thought, allowing such clumping to occur for Z = 0.004, well below the standard value assumed to be inherited from the ISM. Figure 6 summarizes the results from these three works and indicates our current understanding of the parameters for which strong clumping is allowed.

Beyond St and ε, the latter of which can be related to Z via $Z = \varepsilon(H_g/H_p)$, the third parameter that controls the linear SI growth is the pressure gradient associated with stellar illumination. [The gas pressure decreases with radius to partially balance the disk against the radial component of gravity. The remainder of the balance comes from azimuthal flow (i.e., centrifugal balance). That pressure aids rotation in balancing gravity is the reason for the gas to be sub-Keplerian, albeit slightly.] This pressure gradient can be written in dimensionless form as

$$\Pi \equiv \frac{\Delta v}{c_s} = -\frac{1}{2}\left(\frac{c_s}{v_k}\right)\frac{\partial \ln P}{\partial \ln r} \tag{19}$$

where v_k is the Keplerian velocity, P is the gas pressure, and Δv is the headwind experienced by solid particles

$$\Delta v \equiv v_k - v_\phi = \eta v_k \tag{20}$$

where v_ϕ is the azimuthal gas velocity.

Carrera et al. (2015) found that lower Π decreased the threshold for strong clumping, consistent with earlier work by *Bai and Stone* (2010c). While this third parameter may make further parameter studies of the SI expensive, recent results (*Sekiya and Onishi* 2018) have demonstrated that Z/Π is actually the more relevant parameter for determining the clumping properties of the SI. That is, if one changes Z and Π separately but their ratio remain constant, the SI clumping appears to be very similar (*Sekiya and Onishi* 2018).

This important result allows us to expand the parameter surveys described above to define the critical concentration parameter Z_{crit} for any value of Π as [see equation (9) in *Li and Youdin* (2021)]

$$Z_{crit}(\Pi) = \frac{\Pi}{0.05} Z_{crit}(\Pi = 0.05) \tag{21}$$

This result has a clear physical interpretation when considering the result that larger Π is associated with more turbulence induced by the SI (*Carrera et al.*, 2015; *Abod et al.*, 2019). Higher turbulence equates to a larger H_p, which for a given Z will decrease ε (again because $Z = \varepsilon H_g/H_p$), which as has been shown from unstratified simulations is a control parameter for the fully non-linear SI.

This result led *Li and Youdin* (2021) to write an equation for the critical ε responsible for strong clumping by combining the *Sekiya and Onishi* (2018) result with their simulation suite. *Li and Youdin* (2021) also included the effect of externally driven turbulence (e.g., from the MRI) in their equation (although external turbulence was *not* included in their simulations through the traditional α parameter defined

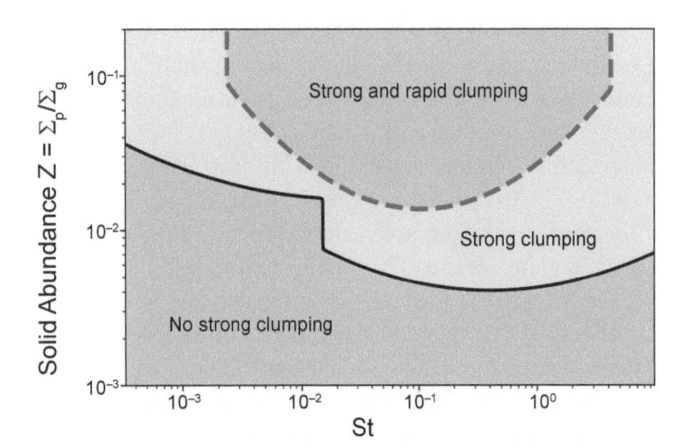

Fig. 6. Regions in St-Z space for which strong clumping does or does not occur, based on the work by *Carrera et al.* (2015), *Yang et al.* (2017), and *Li and Youdin* (2021) and assuming Π = 0.05. The pink region corresponds to no strong clumping, whereas both green regions correspond to strong clumping. The darker green region, outlined by the blue dashed line, indicates the work by *Carrera et al.* (2015) in which strong clumping was more rapid, which was made apparent by the continuous removal of gas, amounting to a 50% reduction of the remaining gas every 50 orbits of the local shearing box.

above), as this also sets H_p and thus ε. While we will return to the effect of externally driven turbulence shortly, it is worth writing this critical ε here as it is our *most current* understanding of the condition for strong clumping due to the SI. From *Li and Youdin* (2021)

$$\log_{10} \varepsilon_{crit} \simeq A'(\log_{10}St)^2 + B'(\log_{10}St) + C' \quad (22)$$

where

$$\begin{cases} A' = 0, B' = 0, C' = \log_{10}(2.5) & \text{if } St < 0.015 \\ A' = 0.48, B' = 0.87, C' = -0.11 & \text{if } St > 0.015 \end{cases}$$

This can then be included into a new equation for Z_{crit} that takes into account the effect of Π (see equation (21)) and externally driven turbulence parameterized via α

$$Z_{crit} \simeq \varepsilon_{crit}(St)\sqrt{\left(\frac{\Pi}{5}\right)^2 + \frac{\alpha}{(\alpha + St)}} \quad (23)$$

While the above set of equations represents our current best understanding of where in physical parameter space strong clumping occurs and planetesimal formation is likely, numerical effects must also be quantified when employing simulations. *Yang and Johansen* (2014) studied the effect of domain size, which is particularly important as the small scales over which the SI operate ($\ll H_g$) restrict simulations to small domain sizes, typically $\approx 0.2\ H_g$ on a side. The authors found an average separation of $\sim 0.2\ H_g$ between SI-induced clumps, suggesting that at least this domain size should be used in SI simulations. The authors also found that the maximum degree of clumping (measured as the maximum particle density throughout the simulation domain) increases with increasing resolution, consistent with prior work (*Johansen et al.,* 2007, 2012; *Bai and Stone,* 2010b), although *Yang and Johansen* (2014) did find tentative evidence for convergence in this quantity at 160–320 grid zones per H_g.

Li et al. (2018) followed up on this work, reexamining the effect of box size and resolution, while also exploring different vertical boundary conditions. Given the small domain size typically used in such simulations, previously used boundary conditions were either periodic or reflecting. *Li et al.* (2018) examined the effect of a modified open boundary [first developed for MHD simulations in *Simon et al.* (2011) and then modified by *Li et al.* (2018)], finding that such a boundary led to better convergence of results with box size, a smaller H_p (due to the absence of reflecting or periodic vertical gas flows), but no appreciable change in the maximum particle density.

More recently, *Li and Youdin* (2021) found that larger domains in the radial direction may be necessary for strong clumping when the relevant parameters lie close to the clumping boundary (the solid black line in Fig. 6). Furthermore, *Rucska and Wadsley* (2021) found evidence that scales larger than 0.2 H_g may be important in determining the properties of planetesimals born from the streaming instability.

Clearly, a picture of what parameters lead to strong clumping is emerging. However, there still remains more work to be done, exploring both physical parameters and numerical effects to fully flesh out the nature of strong clumping. For example, a major unanswered question is: *Why* does strong clumping occur for the parameter choices that it does? As pointed out by *Li and Youdin* (2021), large linear growth rates do not necessarily lead to strong clumping. Not only does this motivate a deeper understanding of how strong clumping works, but it also warrants caution in overinterpreting the results of linear analysis. Furthermore, improving upon other conditions used in previous work, such as assuming axisymmetry for most of the St, Z parameter exploration, as well as neglecting particle self-gravity and its influence on strong clumping, may prove to change (yet again) the parameters for which strong clumping occurs.

3.1.3.3. Extensions: Turbulence, multiple particle species: There have been a number of extensions applied to the SI problem, in the linear and nonlinear regimes alike. For instance, externally driven turbulence has been shown to significantly reduce the growth rate of the SI in the linear regime (*Umurhan et al.,* 2020; *Chen and Lin,* 2020) [see also preliminary comments on the matter in *Youdin and Goodman* (2005)]. However, as we just emphasized [and also mentioned in other works; see *Li and Youdin* (2021)], one should be cautious in interpreting the linear growth rates as having a direct influence on the degree of strong clumping.

Indeed, results from nonlinear simulations appear mixed. Simulations by *Gole et al.* (2020), for which turbulence was injected into a domain of sufficiently high resolution and small scale to resolve the SI, shows that strong clumping (and planetesimal formation; they included particle self-gravity) can be significantly reduced by turbulence.

A number of other works, on the other hand (*Johansen et al.,* 2007; *Yang et al.,* 2018; *Schäfer et al.,* 2020; *Xu and Bai,* 2022a) included the turbulence self-consistently (e.g., included magnetic fields to activate the MRI). These studies found that the sources of turbulence explored led to solid particle concentration on scales ~ 1–10 H_g, pushing the local concentration into the streaming-unstable regime.

While the obvious implication is that turbulence may actually *aid* in planetesimal formation, many of these studies were carried out at modest resolution. Given that the rate at which overdensities are diffused via turbulence scales with k^2, where $k = 2\pi/\lambda$ and λ is the typical size scale of an overdensity, resolving smaller scales may enhance the turbulent diffusion of small-scale clumping and thus reduce or quench the production of planetesimals. Ultimately, more work remains to answer the question of whether turbulence hinders or helps SI-induced planetesimal formation; such studies are already underway.

Another extension is the inclusion of multiple particle sizes. Of course, a non-monodisperse set of particles naturally result from collisional growth/fragmentation models, as outlined above (see section 2.5). However, for simplicity, many previous SI studies have assumed just a single size for all particles. Recently, a linear analysis carried out by *Krapp et al.* (2019) has shown that for $\varepsilon \lesssim 1$, the fastest growth rate of the SI decreases with increasing number of discrete particle sizes without any sign of convergence (i.e., the growth rate approaches zero as the continuum limit of particle sizes is reached). This surprising result was further explored by *Zhu and Yang* (2021) and *Paardekooper et al.* (2020), which found that the multispecies SI is largely controlled by ε and the maximum St of the distribution, St_{max}. Furthermore, for $\varepsilon \gtrsim 1$ and $St_{max} \gtrsim 1$, the SI growth rates are on the order of the dynamical time, regardless of the number of particle sizes.

More encouragingly, nonlinear, vertically stratified SI simulations with multiple particle sizes have also demonstrated the presence of the SI (*Bai and Stone,* 2010a; *Schaffer et al.,* 2018) and even the convergence of particle concentration via the SI with increasing number of particle sizes (*Schaffer et al.,* 2021). The latter work suggests that their results converge because the conditions near the disk mid-plane where the SI is active satisfy the convergent region of parameter space in the linear regime. While this result is encouraging, these considerations imply that careful tests of convergence should be carried out when running multiple-species SI simulations.

3.1.3.4. Formation of gravitationally bound clouds: Once strong clumping is achieved and the Roche density surpassed, gravitational forces overpower tidal and turbulent forces (see, e.g., *Gerbig et al.,* 2020) and gravitationally bound clouds form. (We use the term cloud here to distinguish these diffuse, spheroidal structures from the SI-induced clumps, which are not necessarily gravitationally bound.) This has been demonstrated numerically with a number of calculations that include particle-particle self-gravitational forces.

One such example is shown in Fig. 7; the SI produces elongated (nearly axisymmetric) filaments that continue to increase in concentration. Eventually, the Roche density is reached locally, and gravitationally bound clouds of particles spawn off of the SI-induced filaments. While the figure shows just one such simulation (from *Abod et al.,* 2019), a number of works have demonstrated that this is a robust process (e.g., *Johansen et al.,* 2007, 2015; *Simon et al.,* 2016, 2017; *Abod et al.,* 2019; *Li et al.,* 2019). Furthermore, these works have shown that the largest clouds have masses on roughly the same order as the largest solar system planetesimals and dwarf planets, such as Ceres (although as we will see shortly, this agreement is at best very approximate).

A more in-depth comparison of these clouds with observations of solar system planetesimals is addressed in section 4. However, at this point, a technical issue associated with many of the above calculations must be discussed. The algorithm to solve Poisson's equation for the particle gravitational potential, the particle-mesh (PM) method (which has been the only method employed in calculations of SI-induced planetesimal formation to date), limits the further collapse of such clouds to a factor of 10^3–10^4 larger than the scales of the largest observed planetesimals ($\approx$100 km). The specific details of why this happens are beyond the scope of this chapter. However, this is a purely numerical limitation, and in some cases, a workaround has already been achieved. For instance, recent work with the PENCIL code has included

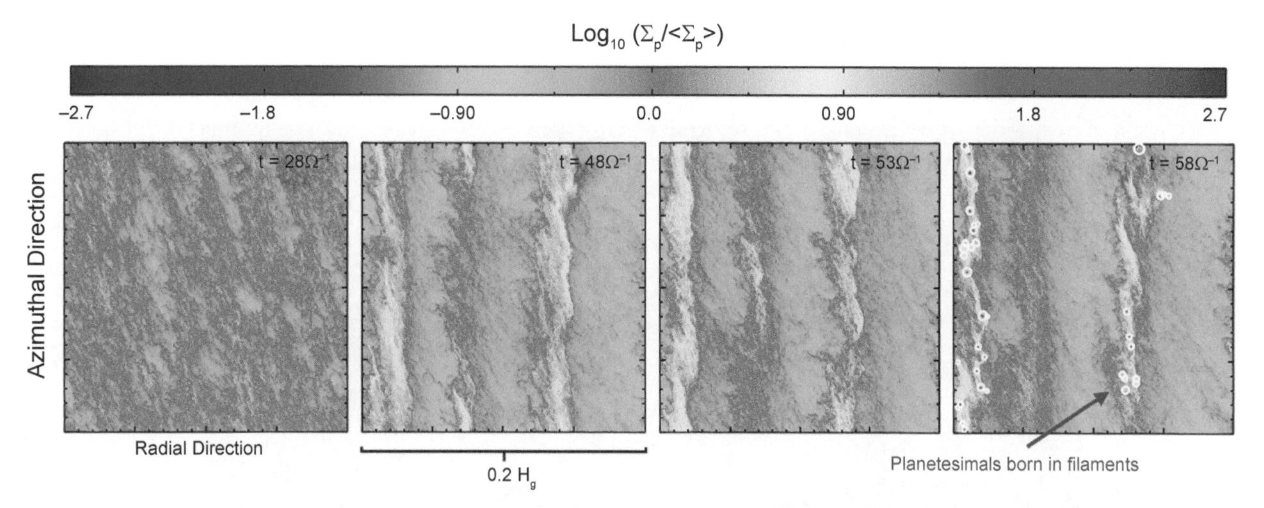

Fig. 7. Log of the pebble surface density (relative to the average value) in a series of SI simulation snapshots. The simulation domain is a small patch of a disk, and in this figure, the vertical axis corresponds to the azimuthal direction and the horizontal axis corresponds to the radial direction. As the SI proceeds, pebble overdensities begin to form, ultimately creating azimuthally elongated filaments. The mass density in these filaments eventually exceeds Roche (rightmost panel), leading to gravitationally bound clouds of pebbles. Modified from *Abod et al.* (2019). Ω as shown on the figure refers to the rotation rate at the center of the disk patch, and the white circles in the rightmost panel correspond to the Hill radius (i.e., the radius at which the gravity of the planetary body or planetesimal dominates over that from the Sun) of each bound cloud.

"sink particles" (*Johansen et al.,* 2015). Another workaround is an effective zoom-in on individual bound clouds with an improved treatment of gravity, as we now describe in more detail. (We should note that simulations including the entire range of scales from birth of planetesimals to final collapse are under development.)

3.1.4. Final stage of collapse: Planetesimals are born.

3.1.4.1. Planetesimal binaries: Recently, *Nesvorný et al.* (2021) carried out a series of N-body calculations with a modified version of PKDGRAV that solves particle gravity with a tree algorithm (see *Barnes and Hut,* 1986), includes collisions between particles, and properly accounts for energy dissipation via a new superparticle method (see *Nesvorný et al.,* 2020). Each of these calculations focused on individual particle clouds extracted from SI calculations. Furthermore, the particle collisions are expected to be inelastic: They extract the kinetic energy from the collapsing cloud, which allows it to collapse further toward denser and denser objects.

Nesvorný et al. (2021) found that the particle clouds rapidly collapse into short-lived disk structures from which planetesimals form (see Fig. 8). The planetesimal properties depend on the cloud's scaled angular momentum, $L' = L/(MR_H^2\Omega)$, where L and M are the angular momentum and mass, respectively, R_H is the Hill radius (i.e., the radius at which the gravity of the planetary body or planetesimal dominates over that from the Sun), and Ω is the orbital frequency. Low-L′ particle clouds produce tight (or contact) binaries and single planetesimals. Compact high-L′ clouds give birth to binary planetesimals with attributes that closely resemble the equally sized binaries found in the Kuiper belt (see section 4). Furthermore, 50–100% of the original mass of a particle cloud is converted to the final largest planetesimal or planetesimal binary. The remaining <50% of the original mass remains in particles and small planetesimals (*Robinson et al.,* 2020).

We will return shortly to binary production in collapsing particle clouds, particularly within the context of comparison with observations. However, at this point, it is worth describing studies of interior planetesimal structures, as this will also be directly relevant to the examination of actual solar system bodies.

3.1.4.2. Planetesimal structures — Predictions from theory: Wahlberg Jansson and Johansen (2014) and the subsequent works of these authors (see below) considered gravitational collapse of pebble clouds, although their setup was generic (i.e., not taken from an SI calculation) and had L ~ 0 (i.e., no global rotation). These authors developed statistical schemes to characterize the effects of collisions on the pebble size distribution, with different collisional regimes being considered, including sticking, bouncing, mass transfer, and fragmentation of pebbles (*Güttler et al.,* 2010). They found that the initial mass of a pebble cloud determines the interior structure of the resulting planetesimal.

The pebble collision speeds in low-mass clouds are below the threshold for fragmentation (~1–10 m s^{-1}; see section 2.3), forming pebble-pile planetesimals consisting of the primordial pebbles from the PPD. Planetesimals above 100 km in radius should consist of mixtures of dust (pebble fragments) and pebbles that have undergone substantial collisions with dust and other pebbles. If comet-sized planetesimals form by gravitational collapse of small pebble clouds, this model would predict that the recently visited solar system bodies 67P/Churyumov-Gerasimenko and Arrokoth (by Rosetta and New Horizons respectively) are pebble-pile planetesimals consisting of primordial pebbles from the solar nebula. If, instead, comets are collisional fragments of much larger planetesimals, the memory of original pebbles

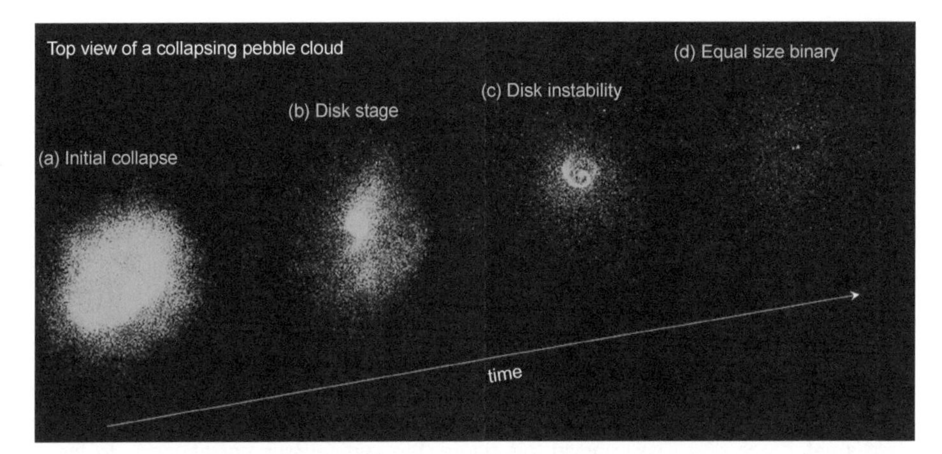

Fig. 8. Gravitational collapse of a particle cloud. The four snapshots show different stages of collapse at simulation time t = 0, 6, 13, and 50 yr (from left to right). The simulation was performed with the superparticle module of the PKDGRAV N-body code (*Nesvorný et al.,* 2020). The view is projected down the clump angular momentum axis. Each frame is 450,000 km across. The equal-sized binary in the righthand panel has primary diameter $D_1 \approx 110$ km, secondary diameter $D_2 \approx 78$ km, semimajor axis a_b = 10,400 km, and eccentricity e_b = 0.19. The binary represents 98% of the initial mass and 58% of the initial angular momentum of its parent cloud.

may have been lost.

These results were extended (*Wahlberg Jansson and Johansen,* 2017) by allowing the individual pebble subclouds to contract at different rates and including the effect of gas drag (on the contraction speed and energy dissipation). The results of this study yield comets that are porous pebble-piles with particle sizes varying with depth. The interior should consist of primordial pebbles with a narrow size distribution, yielding higher porosity. The surface layers should be a mixture of primordial pebbles and pebble fragments. The detailed predictions depend on the applied fragmentation model (*Wahlberg Jansson et al.,* 2017).

The physical properties of primordial pebble piles (ignoring pebble fragmentation during collapse) were calculated by *Skorov and Blum* (2012) and *Blum* (2018) and compared to known properties of comets (see section 4). *Skorov and Blum* (2012) calculated the tensile strength of the surface layers of pebble piles, which is many orders of magnitude lower than that of non-hierarchical, porous bodies consisting of (sub)-micrometer-sized dust/ice grains. The reason for the low strength is that the number of grain-grain contacts, which determines the cohesion of granular matter under low-gravity conditions, is greatly reduced. This in turn results from the contacts occurring only in the pebble-pebble contact area, which is much smaller than the pebble cross section. Other major differences to the non-hierarchical makeup of planetesimals are the vastly different heat conductivity and gas permeability, which are important properties to understand the thermal evolution of planetesimals and the activity of comets (see below).

Due to the hierarchical setup of these kinds of planetesimals (dust/ice grains–pebbles–planetesimals), a number of characteristics are present:

1. *Low mass density/high porosity.* Due to the effect of bouncing after the pebbles had been formed in the PPD, their internal volume filling factor was estimated to be in the range $\Phi_{int} \approx 0.33–0.58$ (*Güttler et al.,* 2009; *O'Rourke et al.,* 2020). The pebbles themselves form a pebble pile in the gravitational collapse that has a filling factor of $\Phi_{pp} \approx 0.55–0.64$ (*O'Rourke et al.,* 2020). Thus, small planetesimals ($\lesssim$100 km) as a whole should possess a volume filling factor of $\Phi_{pl} = \Phi_{int} \times \Phi_{pp} \approx 0.18–0.37$ and a mass density of $\rho_{pl} = \Phi_{pl} \times \rho_{solid}$, with ρ_{solid} being the average mass density of the solid monomer grains. For large planetesimals ($\gtrsim$100 km), their self-gravity is so large that the pebbles collapse and fill the macroscopic void spaces. Thus, these planetesimals have reduced porosities and increased densities and also possess other very different physical properties (for details see *Blum,* 2018).

2. *Low-tensile, compressive, and fragmentation strengths.* Due to the macroscopic size of the pebbles, the inner cohesion of planetesimals formed in a gentle gravitational collapse of a pebble cloud is extremely weak (*Skorov and Blum,* 2012). For millimeter- to centimeter-sized pebbles, tensile strength values for the planetesimals of $\lesssim$1 Pa are expected. Similar low values are expected for the compressive strength (*O'Rourke et al.,* 2020). In contrast, a homogeneous (i.e., non-hierarchical) packing of microscopic solid grains leads to tensile strengths $\gtrsim$1 kPa (*Blum,* 2018).

For the dynamical evolution of planetesimals, it is essential to consider their collisional behavior. A characteristic value often used in collisional evolution studies is the fragmentation strength, which is the collision energy per unit mass required to yield a largest fragment with half the mass of the original body. *Katsuragi and Blum* (2017) and *San Sebastián et al.* (2020) showed that the fragmentation strength scales with the static tensile strength. Thus, the energy required for the fragmentation of pebble-pile planetesimals can be many orders of magnitude smaller than in the non-hierarchical case (*Krivov et al.,* 2018; *San Sebastián et al.,* 2020).

3. *High gas permeability.* The macroscopic void spaces between the pebbles in a planetesimal that was formed by a gentle gravitational collapse of a pebble cloud play an important role in vapor transport [see *Lichtenberg et al.* (2016) for the influence of porosity on heating and gas transport]. Besides current solar system bodies exhibiting gas and/or dust activity, gas transport might be important in the first few million years after their formation if the abundance of short-lived radiogenic nuclei is high enough to vaporize volatile (e.g., water ice) or super-volatile (e.g., CO ice) materials (*Lichtenberg et al.,* 2021; *Golabek and Jutzi,* 2021; *Malamud et al.,* 2022). Due to a negative temperature gradient in the radial direction (in this context, radial means away from the center of the planetesimal), volatiles could be driven radially outward and could lead to their enhancement close to the surface of the planetesimal.

4. *Low solid-state and high radiative thermal conductivity.* Besides the gas permeability and the mechanical strengths, the heat conductivity is heavily dependent on the size of the building blocks of the planetesimals. For a homogeneous body consisting of microscopic dust or ice grains, the heat transport occurs through the network of grains. Thus, the heat conductivity decreases for increasing porosity and increasing grain size. In contrast, a pebble-pile makeup of the planetesimals may lead to an increased heat transport due to the influence of radiative heat transfer, in particular at higher temperatures and for larger pebble sizes (*Bischoff et al.,* 2021). The strong temperature dependency of the heat conductivity in the pebble case can lead to interesting diurnal effects, with high energy transport from the surface into the subsurface regions during periods of high-intensity insolation and low-energy transport at night times (*Bischoff et al.,* 2021).

As with the other subsections in section 3, we will return to planetesimal/comet structures shortly as we make a comparison with observations. However, for now, we conclude this section with a brief discussion of another route to producing planetesimals via gravitational collapse.

3.1.5. Concentration in gas features. The final gravitational collapse mechanisms we will discuss are the broad group of "concentration mechanisms." These mechanisms provide a route (alternative to the SI) toward concentrating pebbles to the point that their local density surpasses the Roche density. In what follows, we only briefly describe these mechanisms, as the literature is considerably less extensive compared with that of the SI.

The simplest concentration mechanism is an axisymmetric gas pressure enhancement. These can be induced by any number of things, including magnetohydrodynamical effects (e.g., *Johansen et al.,* 2009; *Simon and Armitage,* 2014; *Bai and Stone,* 2014; *Béthune et al.,* 2017), hydrodynamic structures from instabilities [e.g., the vertical shear instability (*Schäfer et al.,* 2020)], and planets themselves, which produce pressure maxima at the outer edge of planet-induced gaps (e.g., *Picogna and Kley,* 2015; *Fedele et al.,* 2017; *Hendler et al.,* 2018; *Baruteau et al.,* 2019; *Pérez et al.,* 2019; *Veronesi et al.,* 2020). Indeed, ALMA has observed the effect of dust concentration in such rings in a number of planet-forming disks (*Huang et al.,* 2018), although their origin has not yet been well constrained by such observations.

The process by which such concentration occurs is as follows. Pebbles on the outer part of a pressure enhancement drift toward increasing pressure due to the headwind they feel from the partially pressure supported sub-Keplerian gas (see section 2.2). However, on the other side of the pressure enhancement (i.e., at smaller radii), the opposite is true. In order for the pressure enhancement to be long lived, the Coriolis force and gas pressure gradient must balance (this is known as geostrophic balance). Thus, inward of the enhancement, gas is moving super-Keplerian, which provides a tailwind to pebbles, increasing their anguar momentum; the pebbles move toward pressure maxima and are trapped in the case where $dP/dr \geq 0$. (In the case in which a local maximum is not produced, the inward radial drift of pebbles continues, but their drift slows down as they enter the pressure enhancement, causing a "traffic jam.")

A related process is the concentration of pebbles in vortices. Gas vortices are another form of geostrophic balance, but without axisymmetry. In the non-inertial co-rotating shearing box frame, cyclonic vortices (i.e., with $\nabla \times \mathbf{v} > 0$) are sheared out by the anti-cyclonic shear flow. Only anti-cyclonic vortices survive, and in this configuration a high-pressure region exists within the vortex that balances the Coriolis force; geostrophy is maintained.

As the vortex orbits the central body (at Keplerian velocities), it sweeps up pebbles that are inwardly drifting from larger radii. These pebbles are captured in the outer region of the vortex and are swept along with the vortex streamlines. However, since they experience drag but no pressure gradient, they move within the vortex at a higher velocity relative to the gas, which causes them to slowly spiral toward the center of the vortex. An apt analogy is to think of the vortex as a miniature disk with the role of radial gravity being replaced by the Coriolis force. Thus the pebbles orbit the center of the vortex as they would a disk, at a higher velocity than the gas. The resulting headwind on the pebbles causes them to spiral inward to the center of the vortex.

Indeed, there is already evidence that dust can become entrapped by these vortices. For instance, early dust continuum observations with ALMA (e.g., *van der Marel et al.,* 2013; *Pérez et al.,* 2014) depicted the presence of large-scale asymmetric concentrations of dust grains, which have been interpreted as trapping of these grains in vortices.

A third mechanism, known as *turbulent clustering,* is the concentration of pebbles due to turbulent gas in planet-forming disks (*Klahr and Henning,* 1997; *Cuzzi et al.,* 2008; *Pan et al.,* 2011; *Hopkins,* 2016; *Hartlep et al.,* 2017; *Hartlep and Cuzzi,* 2020), namely, through the interaction between pebbles and turbulent eddies. Within this mechanism, pebbles that are coupled to the turbulent gas are centrifugally "flung" outward away from the centers of turbulent vorticies/eddies and are concentrated between the eddies to the point of gravitational collapse. However, for this to work effectively, several conditions must be satisfied:

- The pebble clump is of sufficiently high density to resist the destruction by fluid motion within the turbulent eddy as the clump and eddy interact.
- The pebble clump is of sufficiently high density that it can resist disruption from "ram pressure" (i.e., feeling a headwind as it moves through the gas).
- The pebble clump is of sufficiently *low* density that the feedback from pebbles does not destroy the turbulent eddies.

More quantitatively, if one considers pebbles with a specific stopping time t_{stop}, the Stokes number at length-scale ℓ is defined as

$$St_\ell \equiv \frac{t_{stop}}{t_{eddy,\ell}} \tag{24}$$

where $t_{eddy,\ell}$ is the eddy turnover time at scale ℓ. For scales smaller than H, $t_{eddy,\ell} < \Omega^{-1}$. (The ratio of stopping time to eddy turnover time is the *true* definition of the Stokes number and the quantity St defined in equation (2) is the Stokes number at scale H in which $t_{eddy,H} = \Omega^{-1}$. When planetary scientists and astronomers use the term Stokes number, they most often to refer to the stopping time in units of Ω^{-1}, which is equivalent to the Stokes number at the driving scale of the turbulence.)

In the case of a turbulent cascade of energy toward smaller scales [e.g., Kolmogorov turbulence (*Kolmogorov,* 1941)], t_{eddy} decreases with decreasing spatial scale ℓ. Thus, for a given pebble size, St_ℓ increases with decreasing ℓ.

The necessary conditions for collapse, as enumerated above, are best reached at scales where $St_\ell \sim 0.3$ (*Hartlep et al.,* 2017). At larger scales, the Stokes number is sufficiently small that the pebbles are strongly coupled to turbulent eddies and thus cannot be concentrated to meet the above collapse criteria. On smaller scales, the pebbles are only very weakly coupled to the gas and thus feel the turbulent motions as kicks that do not appreciably change the pebble density. The "sweet spot" for planetesimal formation happens at intermediate values, namely $St_\ell \sim 0.3$ (see *Hartlep et al.,* 2017; *Hartlep and Cuzzi,* 2020).

We will return to this mechanism later in this chapter within the context of predictions for planetesimal sizes. For now, however, it is worth mentioning that while this mechanism is promising, more work is needed to study its role in planetesimal formation and in particular in comparing predictions made by this model with solar system constraints.

Finally, it is essential to point out that these processes are not mutually exclusive with the SI. Indeed, it has recently been shown that pebble concentration in axisymmetric pressure bumps (*Taki et al.,* 2016; *Onishi and Sekiya,* 2017; *Carrera et al.,* 2021, 2022; *Xu and Bai,* 2022a,b) can provide a sufficient increase in localized dust-to-gas ratio such that the SI becomes active near the peak of the pressure bump. However, while this process may be robust for some parameters [e.g., the strength of the force generating the bumps (*Carrera et al.,* 2022)], the simulations by *Carrera and Simon* (2022) suggest that for sufficiently small pebbles, planetesimals are only formed when these pebbles are completely trapped in the pressure bump. In this case, which requires a bump of sufficiently high amplitude that it becomes unstable to the RWI [based on the conditions in *Ono et al.* (2016)], planetesimals are only formed by concentrating the pebbles beyond the Roche density; planetesimals were formed from direct gravitational collapse with no help from the SI.

Similarly, vortices can also provide sufficient concentration to kickstart the SI. In particular, *Raettig et al.* (2015, 2021) carried out local simulations of pebble concentration in vortices and found that the SI can be activated for dust-to-gas ratios much less than 1%.

Ultimately, however, these additional mechanisms coupled with the SI deserve further study. Encouragingly, such efforts are already underway.

3.2. Coagulation Models

Although the pebble-collapse model has had a great impact in the scientific community and can explain many observational facts (see below), there are alternative planetesimal-formation models that still merit study. In this section, we discuss the "coagulation models," of which there are two different types. The first involves the growth (via sticking) of particles beyond the various barriers discussed above, whereas the second involves the gravitational growth of small planetesimals into larger ones. We discuss both here for completeness, although given the limited literature in this area compared to collapse models, this section is necessarily short.

3.2.1. Growth past the barriers. The obstacles of the bouncing and fragmentation barriers discussed in section 2.3 are not absolute. The physical principles to overcome these barriers have been discussed before (see, e.g., *Blum,* 2018) and will only be briefly reviewed here.

The first possibility is to avoid reaching these barriers altogether. *Okuzumi et al.* (2012) and *Kataoka et al.* (2013) pointed out that extremely small (i.e., nanometer-sized) monomer grains possess a very high sticking-bouncing threshold (see Fig. 13 in *Blum and Wurm,* 2008) and a rather high tensile (cohesive) strength (see Fig. 4 in *Gundlach et al.,* 2018). This means that the growing aggregates may remain in the fractal growth regime (see section 2.5) long enough to grow to macroscopic sizes. *Okuzumi et al.* (2012) and *Kataoka et al.* (2013) show that aggregates can grow to centimeter sizes with a low fractal dimension $D_f \lesssim 2$ so that their gas-grain response time does not change by much and radial drift is slow. Although the fractal aggregates are very rigid, their mechanical stiffness is finite. Thus, collisions among centimeter-sized fractal aggregates lead to a compaction without changing the sticking efficiency. *Kataoka et al.* (2013) also took into account compaction through gas drag and self-gravity. Beyond the collisional-compaction size, the aggregates continue to grow with only slightly increasing densities (from $\sim 10^{-5}$–10^{-4} g cm^{-3} for ~centimeter-sized aggregates to $\sim 10^{-3}$ g cm^{-3} for ~100-m-sized bodies). Self-gravity is responsible for a rather steep increase in density for bodies above ~100 m in size. Ultimately, the resulting planetesimals have sizes of ~10 km and densities in the range ~0.1–1 g cm^{-3}. With this growth pattern, the radial-drift barrier (see section 2.2) can be avoided. Recently, *Kobayashi and Tanaka* (2021) showed that based on such a scenario even the cores of the giant planets can form in a relatively short time. Although the basic assumptions of high sticking threshold velocity and high tensile strength is certainly undoubted for very small monomer grains, the collision behavior of the extremely fluffy bodies above ~1 cm in size is speculative and cannot easily be tested in the lab. In particular, ignoring fragmentation as a possible collision outcome deserves a revisit in future lab experiments.

A second possibility is that of mass transfer during collisions. Although fragmentation in collisions among dust aggregates in principle leads to the mass loss of the colliding bodies, there are situations for which this is not the case for one of the collision partners. If the mass ratio of the two aggregates is large and the impact energy is moderate, only the smaller aggregate will fragment and may transfer some of its mass to the intact target aggregate (*Meisner et al.,* 2013). Based upon the mass-transfer process, *Windmark et al.* (2012a,b,c) and *Garaud et al.* (2013) showed that planetesimal formation is feasible, in spite of the dominance of fragmentation in collisions among similar-sized bodies. *Windmark et al.* (2012a,b,c) based their dust-evolution model on the collision model by *Güttler et al.* (2010) and took into account a variety of collisional outcomes. In order to overcome the bouncing barrier and enter the mass-transfer regime, *Windmark et al.* (2012a) artificially inserted 1-cm-sized seed particles, large enough to start the mass-transfer process. In contrast, *Windmark et al.* (2012b,c) used velocity-probability distribution functions to initiate the mass-transfer process. While *Windmark et al.* (2012a) needed 10^6 yr to grow planetesimals 100 m in size, *Windmark et al.* (2012b,c) achieved this result in $\sim 5 \times 10^4$ yr. *Garaud et al.* (2013) used a similar approach as *Windmark et al.* (2012b,c) and achieved aggregate sizes on the order of several hundred meters within 10^4 yr at 1 au, whereas further out, the aggregates remained much smaller. A question of critical importance for the credibility of the models is: Under which conditions does the mass-transfer growth process dominate over mass-loss effects such as fragmentation, cratering, or erosion? As reviewed by *Blum* (2018) and based upon the dust-aggregate collision model by *Güttler et al.* (2010) with later supplements from lab experiments, fragmentation is important as long as the size ratio between the colliding bodies is $\lesssim 10$ under solar-nebula conditions; cratering becomes the dominating process when the projectile sizes are $\gtrsim$1 cm, because in that case the impact energies are too large for mass transfer to occur; finally, erosion happens when the projectiles are $\lesssim$100 μm in size (*Schräpler et al.,* 2018). This leaves only a very small size window for the projectiles to reach the mass-transfer realm. Whether fragmentation, cratering, and erosion can completely prevent the formation of planetesimals by the mass-transfer process needs still to be clarified.

In conclusion, more empirical data on the collision behavior of fluffy aggregates are required to determine whether protoplanetary growth of solid bodies by coagulation inevitably stops at pebble sizes or continues toward planetesimal scales.

3.2.2. Formation of 100-km planetesimals. The second type of coagulation model has been examined in a number of works (e.g., *Stern and Colwell,* 1997; *Weidenschilling,* 1997; *Kenyon and Bromley,* 2004, 2012; *Schlichting and Sari,* 2011; *Schlichting et al.,* 2013) and is built on the assumption that smaller grains have already grown to some distribution of initial sizes, often ~ 1 km. These smaller planetesimals then collisionally grow to larger 10–100-km-sized bodies.

The basic physics of how this works is relatively straightforward. At the scales associated with the initial planetesimal populations (e.g., kilometer scales), St $\gg$ 1, and drag forces are negligible. Instead, small bodies grow larger when they collide with one another, "sticking" together primarily due to gravitational forces. Gravitational focusing also plays a role, although it is significantly more relevant for larger bodies as the square of the impact parameter is enhanced by a number that is linearly proportional to the mass [assuming equal mass bodies (see *Armitage,* 2007)].

Again, this mechanism for planetesimal formation has not been as extensively studied as the gravitational collapse models (although one could argue that because these models often *start* with planetesimals, albeit small ones, that it is more a mechanism to change the size distribution of planetesimals than it is a process to form them in the first place). However, later in this chapter, we will come back to it and the predictions that it makes compared with the SI model.

4. PLANETESIMALS IN THE OUTER SOLAR SYSTEM

Having discussed the various mechanisms thought to produce planetesimals, we now turn to a specific discussion of comets and Kuiper belt objects (KBOs). We focus in particular on comparing observed and/or measured characteristics of KBOs and comets to those predicted by models.

4.1. Mass and Size Distributions

The size distribution of planetesimals in the solar system has been one of the primary diagnostics employed to test planetesimal formation theory. In this section, we first describe the size distribution inferred from planetesimal formation calculations, followed by a discussion of the impact of collisional evolution, ending with a comparison with observational data (a brief discussion on this can also be found in the chapter in this volume by Fraser et al.).

4.1.1. Primordial mass distribution. As discussed previously, a number of works have explored the characteristics of SI-generated planetesimals, generally agreeing that the largest planetesimals have masses roughly consistent with large solar system planetesimals. In this section, we dive deeper into this topic: the initial mass and (equivalently) size distributions of planetesimals. As described above, many SI simulations prevent collapse below certain scales due to numerical limitations. Thus, in what follows, the size we reference for gravitationally bound pebble clouds is that which the total mass bound within a cloud would have if it were collapsed to material densities, generally assumed to be ~1 g cm^{-3}.

Johansen et al. (2015) carried out high-resolution simulations of SI-induced planetesimal formation in order to allow a large range of planetesimal sizes to form. Modeling the mass distribution of planetesimals as a simple power law

$$N(>M_p) \propto M_p^{-p+1} \tag{25}$$

they found that p $\approx$ 1.6. More recently, a number of works followed up on this and found that p $\approx$ 1.6 independent of relative strength of gravity (effectively the value of the Toomre Q — see section 3.1), numerical resolution (*Simon et al.,* 2016), St, Z (*Simon et al.,* 2017), and Π (*Abod et al.,* 2019); i.e., with this simple power law shape to the mass and size distributions, the so-called "slope" appeared to be nearly universal.

While *Abod et al.* (2019) did fit a simple power law to their planetesimal masses, they were able to increase the number of detected bound clouds with a tree-based algorithm constructed from halo finders in cosmological simulations [the PLAN code (*Li,* 2019)]. They found that an exponentially tapered distribution

$$N(>M_p) \propto M_p^{-p+1} \exp\left[-M_p/M_c\right] \tag{26}$$

where M_c is a characteristic "cut-off" mass, fits the planetesimal masses significantly better than the simple power law. In their fit, p $\approx$ 1.3, approximately independent of Π (other parameters were not explored in this work).

This work was roughly consistent with *Johansen et al.* (2015) and *Schäfer et al.* (2017), although both studies fit a slightly modified version of the exponentially tapered distribution, introducing a third parameter, β

$$N(>M_p) \propto M_p^{-p+1} \exp\left[\left(-M_p/M_c\right)^{\beta}\right] \tag{27}$$

Given their resolution limits, *Schäfer et al.* (2017) were unable to constrain p but found that β = 0.3–0.4, independent of their domain size. *Johansen et al.* (2015) carried out simulations in a smaller domain and at higher resolution (in terms of grid cells per H_g). They found best fit values p = 1.6 and $\beta = 4/3$; the slope of the power-law component was consistent with *Simon et al.* (2016, 2017) but the extent of exponential tapering was significantly larger than *Schäfer et al.* (2017). (We reserve discussion of the remaining parameter, M_c, to section 4.1.3 below.)

Most recently, *Li et al.* (2019) used PLAN and the highest-resolution SI simulations to date to further quantify the mass distribution. Analyzing two simulations with different St and Z values, they found that the best fit model differed for each simulation. Furthemore, when comparing *identical* models (e.g., the exponentially tapered power law), the best-fit parameters were different at a statistically significant level.

Taken together, these results call into question the originally hypothesized universality of the initial mass distribution. However, further work is required to determine what trends emerge with varying physical parameters. Furthermore, a more rigorous analysis is called for in which the same tools (e.g., advanced clump-finding with PLAN and improved statistics via Markov-Chain Monte Carlo techniques) are applied consistently across a large region of parameter space.

Despite this complexity, there are some consistent features that emerge from these calculations. The slope of the mass distribution toward smaller masses generally falls within the range 1–2 [although not in all models (see *Li et al.,* 2019)] and thus the mass function is dominated in number by small planetesimals but in mass by large planetesimals (see Fig. 4 of *Li et al.,* 2019). Furthermore, the mass distributions are in the broadest sense composed of a low-mass region (or regions) with shallow slopes and higher-mass regions that drop more rapidly with mass, either in the form of an expoential cut-off or a steeper power law.

Of course, the ultimate question is what do these results tell us, if anything, when compared to observations? This is a point to which we will return shortly. However, as these results only address the *initial* planetesimal mass distribution, we must briefly discuss the role of post-formation collisional effects and how these may affect our understanding of the mass and size distributions of planetesimal populations.

4.1.2. Collisional evolution. There are two main stages to the collisional evolution of comets and KBOs. During the first stage, bodies presumably resided in a massive (~20 $M_\oplus$) transneptunian disk [for further discussion of KBO dynamics, see *Morbidelli and Nesvorný* (2020) and the chapters in this volume by Fraser et al. and by Kaib and Volk] and experienced strong collisional grinding over the lifetime of the disk of solids ($t_{s,disk}$). They were subsequently implanted into the Kuiper belt, scattered disk, and Oort cloud when Neptune migrated to 30 au (e.g., *Dones et al.,* 2015). The collisional evolution during the subsequent 4+ G.y., the second stage of collisional evolution, is thought to be less significant; at these times, these populations were distant and/or did not contain much mass (*Singer et al.,* 2019; *Greenstreet et al.,* 2019; *Morbidelli et al.,* 2021).

The collisional evolution during the massive disk stage was studied in *Nesvorný et al.* (2018). They showed that the Patroclus-Menoetius (P-M) binary in the Jupiter Trojan population poses an important constraint on the disk lifetime. This is because the longer the P-M binary stays in the disk, the greater is the likelihood that its components will be stripped from each other (by impacts). This indicates $t_{s,disk}$ < 100 m.y. (*Clement et al.,* 2018; *de Sousa et al.,* 2020; *Morgan et al.,* 2021).

Different studies assumed different initial size distributions and impact scaling laws to model the collisional evolution of KBOs/comets (*Pan and Sari,* 2005; *Fraser,* 2009; *Campo Bagatin and Benavidez,* 2012). For example, *Nesvorný and Vokrouhlický* (2019) adopted the initial size distribution inspired by the SI simulations (e.g., *Simon et al.,* 2017) and impact scaling laws for weak ice to demonstrate that the collisional grinding would remove very large bodies and produce the KBO size-distribution break at D ~ 100 km (*Bernstein et al.,* 2004; *Fraser,* 2009). If that is the case, most comets with D < 10 km must be fragments of large planetesimals since in collisional grinding, the collisional lifetime increases with the size of the body (*Morbidelli and Rickman,* 2015). The small survivors would have suffered multiple reshaping collisions (*Jutzi et al.,* 2017). Bilobed comets should either be the product of subcatastrophic collisions (*Jutzi and Benz,* 2017) or fragments of catastrophic collisions (*Schwartz et al.,* 2018).

Jupiter Trojans have a well-defined cumulative power index −2.1 from below 10 to 100 km (*Emery et al.,* 2015), which is very close to the equilibrium slope expected for the gravity regime [D > 0.5 km (*O'Brien and Greenberg,* 2003)]. This represents a very important constraint on the collisional evolution. Specifically, if the Jupiter Trojan size distribution was established during the massive disk stage, the disk must have been dynamically cold [large impact velocities would produced waves in the size distribution that are not observed (e.g., *Fraser,* 2009; *Kenyon and Bromley,* 2020)]. It is also possible, however, that $t_{s,disk}$ = 0, which corresponds to the time immediately after gas dispersal, in which case the observed size distribution of Jupiter Trojans and KBOs above ~10 km would be primordial (and an important constraint on the formation processes). Even in this case, however, it is difficult to avoid the collisional disruption of comet-sized bodies [D < 10 km (e.g., *Jutzi et al.,* 2017; *Nesvorný and Vokrouhlický,* 2019)], unless comets are jocularly stronger that we think.

The first stage of evolution in the massive planetesimal disk, when most collisional grinding happened, only applies to bodies that formed in the massive disk at radial distances less than 30 au. This presumably includes most bodies in the dynamically hot populations of the Kuiper belt (hot classicals, resonant objects, etc.), and comets, making a direct comparison with SI planetesimal formation models difficult (as the SI simulations do not include collisional evolution). However, cold-classical KBOs are thought to have formed in a low-mass planetesimal disk at >40 au (*Parker and Kavelaars,* 2010; *Batygin et al.,* 2011; *Dawson and Murray-Clay,* 2012), not significantly collisionally evolved during stage two (stage one does not apply to them), and thus likely represent the most pristine planetesimals in the solar system. As such, we restrict our comparison to these objects in what follows.

4.1.3. Comparison between models and cold classical Kuiper belt objects. To best compare numerical models with observations of the cold-classicals (a discussion of the other KBO populations is covered in the chapter in this volume by Fraser et al.), we can rewrite the relevant parameters of our mass functions as equivalent size distribution parameters. Namely, the cumulative size distribution for bodies of diameter D in the three-parameter exponentially tapered power law (which we choose based on its common use in the literature thus far) is

$$N(>D) \propto D^{-q+1} \exp\left[\left(-D/D_c\right)^{\gamma}\right] \quad (28)$$

where $q = 3p-2$, $D = 2(M_c/4\pi\rho_s)^{1/3}$, and $\gamma = 3\beta$.

From the discussion in section 4.1.1 and using the above relation between q and p, q ranges from 1.9 in the tapered models of *Abod et al.* (2019) to 2.8 for the simple power law (*Simon et al.,* 2016, 2017) and the tapered model of *Johansen et al.* (2015).

On the observational side, *Fraser et al.* (2014) analyzed the H-band magnitude distribution of KBOs by combining the results of a number of independent KBO surveys. In fitting their data to two separate power laws (joined at a "break point") of the form

$$N(<H) \propto 10^{\alpha H} \quad (29)$$

they were able to determine $\alpha \approx 1.5$ for the largest planetesimals, translating to $q = 5\alpha + 1 = 8.5$, and $\alpha \approx 0.38$ for small planetesimals, equating to $q = 2.9$. The magnitude at the break corresponds to a KBO diameter of 140 km.

More recently, *Kavelaars et al.* (2021) reexamined this problem with new data collected from OSSOS as well as previous observations. They found that the cumulative magnitude distribution is best fit with an exponential cutoff corresponding to diameters ranging from 80 km to 130 km (assuming an albedo of 0.15) and a slope for small objects (the data on these objects were obtained from other sources) very similar to the *Fraser et al.* (2014) value.

While the $q \approx 2.8$ from the numerical models is temptingly close to the value inferred from these observations, we must reiterate that $q = 2.8$ is only the best fit for simpler power laws (which are clearly not an accurate representation of planetesimal size distributions). In fact, with the exception of the one simulation in *Johansen et al.* (2015), the models that are fit with an exponentially tapered power law [which at least agrees with the shape of the *Kavelaars et al.* (2021) fit in principle, athough see below] have much smaller q values. [While this simulation has a slope that agrees with observations at small sizes, there is significant mismatch between the simulations and observations at larger planetesimal sizes (not shown in Fig. 9).]

Beyond the slope, one can also compare characteristic

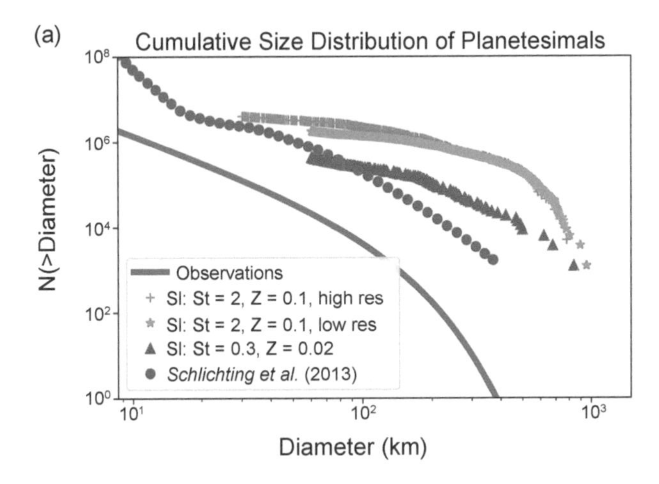

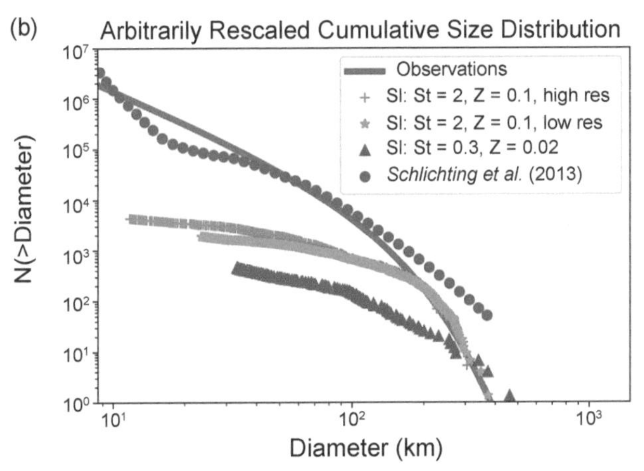

Fig. 9. A fit to the observed cumulative size distribution of KBOs from the Outer Solar System Origins Survey (solid blue) (*Kavelaars et al.,* 2021) compared with theoretical predictions made by SI simulations with different parameters (orange pluses, green stars, red triangles) and from the gravitational coagulation model of *Schlichting et al.* (2013) (purple circles). **(a)** Both the SI and coagulation calculations have been adjusted to account for the full extent of the cold-classical belt [e.g., we multiplied the data from *Schlichting et al.* (2013) by the projected area of the belt on the sky, which we assumed to be 180° × 5°). **(b)** The same data but with the models arbitrarily rescaled so that at least some of the predicted distribution lines up with the data. For the SI simulations, St, Z, and the resolution are varied. Neither the SI simulations, nor the coagulation model, match the observations, both in terms of numbers of planetesimals and the shapes of the distributions.

planetesimal sizes. A fit to a characteristic diameter was not analyzed explicitly in *Simon et al.* (2016, 2017), but the effective diameters of their largest planetesimals were ~200–800 km. Similarly, *Schäfer et al.* (2017) found large planetesimals with diameters ≈600–1200 km (when scaling their units to the Kuiper belt), and the maximum planetesimal diameters of *Li et al.* (2019) ranged from 320 km to 1200 km. On the other hand, *Abod et al.* (2019) found a narrower range of characteristic diameters: ~200–400 km, which is more consistent with observations, although still slightly larger than those inferred from observations.

Finally, one should consider the total number of planetesimals, as the debiased data in *Kavelaars et al.* (2021) put a constraint on the total number of cold-classical KBOs. When adjusting for the fact that local, shearing boxes represent only a very small fraction of the cold-classical belt, SI calculations generally produce too many planetesimals. For instance, the St = 0.3, Z = 0.02 model of *Simon et al.* (2017) produces a factor of ~10 more planetesimals (at least at the resolution probed by the simulations) compared to the OSSOS data.

To summarize the takeaway points of this comparison, Fig. 9 shows the cumulative size distribution from three SI simulations representing different regions of numerical and physical parameter space as well as a fit to the observed data as described in *Kavelaars et al.* (2021). Clearly, there remains a mismatch between the size distribution shape, the total number of planetesimals, and the maximum (or characteristic) planetesimal size, although, as shown in Fig. 9b, the slope for the largest planetesimals in the SI simulations with St = 2, Z = 0.1 do match the observations reasonably well when the theoretical curves have been manually rescaled to lie on top of the observational data.

These discrepancies between theory and observation are not yet resolved. One possibility regarding planetesimal sizes and the shape of their size distribution is that due to pebble clouds not collapsing below a numerically limited spatial scale in the ATHENA simulations (see above), such clouds may accrete pebbles at an unrealistically large rate or suffer from interactions with each other (e.g., tidal stripping, merging). However, this issue is not present with the sink particle runs of PENCIL, and there are still discrepancies between those simulations and KBO observations (e.g., *Schäfer et al.,* 2017).

Another possibility is that since the continued collapse of the pebble clouds leads to the formation of multiple smaller planetesimals, as discussed in section 3.1.4 and *Nesvorný et al.* (2021), the initial mass/size distributions will change. While the fact that many of these clouds put most of their mass into a binary seems to suggest that the distributions will not change appreciably, more work is required to resolve this issue. In particular, a larger exploration of parameter space (both for the SI and for the relevant parameters in the PKDGRAV calculations) is required, as well as a direct calculation of the size distribution from the final collapsed planetesimals.

It is also possible that a larger volume of the SI parameter space requires exploration. Even with the many publications on the SI mass function, there remain many parameter combinations that have not been explored. For the sake of computational expediency, most of the SI simulations have been carried out far from the border between strong clumping and no strong clumping (see Fig. 6). In reality though, as pebbles grow from smaller grains, regions of the disk may move from left to right in the Z-St parameter space, and thus the relevant St values may be smaller than what has been studied so far. While to the best of our knowledge small St values have only been explored within the context of the size distribution in *Simon et al.* (2017), a less-sophisticated clump-detection method in that work led to poorer sampling of planetesimals, ultimately resulting in a simple power law fit to the data (as discussed above). Similarly, the Z values may be closer to the border and not as high as Z = 0.1, unless there is a sudden swift increase in the dust-to-gas ratio in the disk such that the system is rapidly placed well within the strong clumping regime.

Indeed, simulations along this border have shown some indication of marginal behavior [e.g., transient excursions into the strong clumping regime (*Li and Youdin,* 2021)]. If true, then this would be a natural route toward reducing the number of planetesimals formed by the SI by, e.g., limiting the amount of time or spatial regions in which strong clumping occurs. How exactly this marginal behavior would affect the size distribution and maximum planetesimal sizes is less clear, but certainly remains a question worthy of study.

Finally, it is also possible that including a range of particle sizes will change the results, given the role that multiple particle sizes play in the linear regime of the SI (*Krapp et al.,* 2019; *Paardekooper et al.,* 2020; *Zhu and Yang,* 2021). However, as discussed above, simulations of the non-linear state of the SI that include the vertical component of stellar gravity suggest that the nature of strong clumping does not drastically change when multiple particle sizes are included, at least not to the extent suggested by linear analyses (*Bai and Stone,* 2010a; *Schaffer et al.,* 2018, 2021).

The coagulation models for planetesimal formation also make specific predictions for the size distribution of these bodies. *Schlichting and Sari* (2011) predict that $q \approx 4$ over a large range of planetesimal sizes, which is largely inconsistent with the most recent observational constraints. An updated model that includes collisional evolution (*Schlichting et al.,* 2013) (shown in Fig. 9) also does not agree with observations. As with the SI, there are too many planetesimals compared with that from *Kavelaars et al.* (2021). In particular, this model produced an increasingly large number of D < 2 km bodies (not shown on the graph). This excess of small bodies is inconsistent with recent measurements of cratering records on Pluto and Charon showing a turnover in the size distribution (*Singer et al.,* 2019) as well as even more recent work that, while suggesting a turnover does not exist, limits the slope to being shallower than predicted (*Morbidelli et al.,* 2021). Beyond this, the general shape of the *Schlichting et al.* (2013) size distribution is inconsistent with observations, as shown in the figure.

The numerical calculations of *Kenyon and Bromley* (2012) more broadly demonstrated that a wide range of size distribution slopes q (their results had different q values for different planetesimal size ranges) were possible depending on their initial conditions and chosen parameters. However, they generally found that $q \approx 3$ for D > 200 km; again, this is inconsistent with observations. However, it is worth noting that at smaller planetesimal diameters, $q \approx 3$–3.5, which *is* approximately consistent.

Finally, the turbulent clustering mechanism makes specific predictions for the initial size distribution of planetesimals. In particular, as outlined in *Hartlep and Cuzzi* (2020), the initial size distribution has a well-defined peak between 10 and 100 km in diameter, depending on the relevant model parameters, and does not have a power-law shape at smaller scales. [The total number of planetesimals formed via this mechanism is also largely dependent on these parameters (*Hartlep and Cuzzi,* 2020).] While the characteristic size of planetesimals formed via this mechanism is of approximately the same order of magnitude as observed objects, there is still disagreement in the shape of the distribution compared with observations.

In summary, the various gravitational collapse mechanisms that have quantified the size distribution as well as the coagulation model have yet to completely reproduce the typical sizes of planetesimals, their size distributions, and the total number of bodies formed, strongly motivating future work in this area. However, as we will now see, another diagnostic *has* proven to be a powerful discriminant between two of these models.

4.2. Constraints from Kuiper Belt Object Binaries

It is now known that many observed KBOs are actually in a binary configuration (e.g., *Noll et al.,* 2008). More recently, *Grundy et al.* (2019) analyzed 35 such binaries and found that ≈80% of them orbit each other in a prograde sense.

Nesvorný et al. (2019) showed that the SI mechanism produces *a nearly identical* distribution of binary orbital orientations: 80% of the simulated planetesimals orbit prograde. This astonishingly strong agreement is shown in the cumulative distribution (see Fig. 10) of orbital inclination for both SI models and observations. Even more encouragingly, *Nesvorný et al.* (2021) extended this and found that the angular momentum *magnitude* of SI-produced binaries also agrees with observed KBO binaries. [See also *Nesvorný et al.* (2010) and *Robinson et al.* (2020) for similar work on pebble cloud collapse.]

Future work to test the SI hypothesis for planetesimal formation includes a further exploration of parameter space as well as extending to a larger number of simulated particles in PKDGRAV. The latter is important to numerically resolving the tight and contact binaries, such as Arrokoth (*McKinnon et al.* 2020).

In contrast, the gravitational coagulation model described in section 4.1.3 makes predictions that are inconsistent with these observations. More specifically, if the KBO velocities during this coagulation stage are on the order of or greater than the so-called "Hill velocity" [see *Schlichting and Sari* (2008) for a definition], this model predicts roughly an equal number of prograde and retrograde orbits (this line is also included in Fig. 10). For KBO velocities less than 10% of the Hill velocity, retrograde orbits dominate by 97% (*Schlichting and Sari,* 2008). Clearly, these considerations taken together strongly support the SI-induced gravitational collapse paradigm.

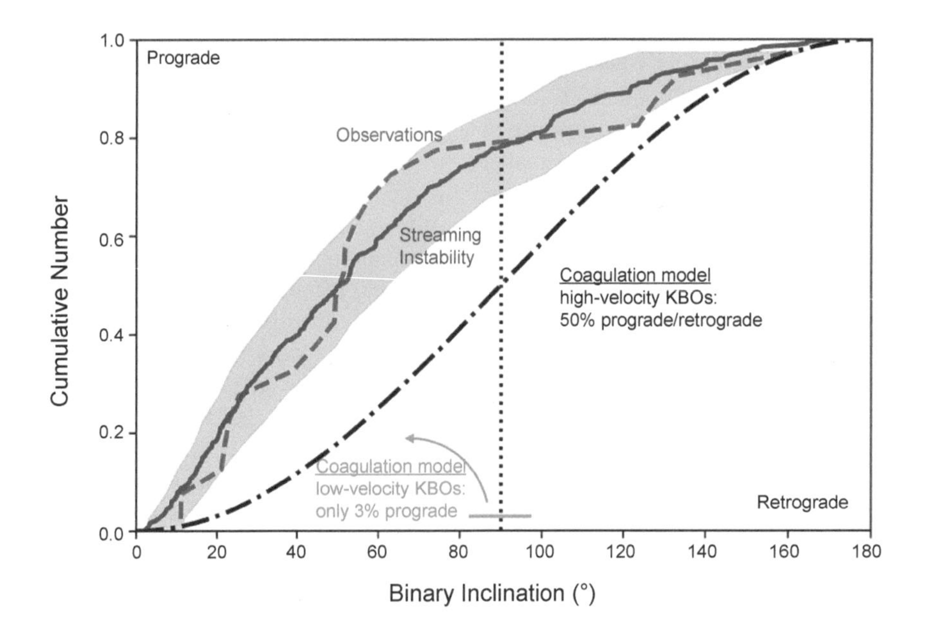

Fig. 10. The inclination distribution of binary orbits in a snapshot of the St = 2, Z = 0.1, higher-resolution run of *Nesvorný et al.* (2019) (blue solid line) and observations of binary KBOs (red dashed line). The shaded area is a 68% confidence envelope of the model, resulting from a sample of model orbits being randomly drawn from a sample size equal to that of real binaries with known inclinations. The binary orbits are predominantly prograde with approximately 80% having inclination <90°. The competing gravitational coagulation model predicts either a 50% split between prograde and retrograde (black dash-dot line, which was calculated assuming a uniform distribution of inclinations) or 3% prograde and 97% retrograde (see orange horizontal marker). The SI model is in excellent agreement with observations, in contrast to the coagulation model.

4.3. Comparison with Comet 67P/ Churyumov-Gerasimenko

Moving from constraints from KBOs to those from comets, we now consider recent observations of Comet 67P/ Churyumov-Gerasmineko (hereafter Comet 67P) and compare these with model predictions of planetesimal-formation and -evolution models. This comparison will include internal properties such as porosity, granularity, cohesive and tensile strength, and activity patterns.

Owing to the groundbreaking Rosetta mission of the European Space Agency, our knowledge about the inner makeup of cometary nuclei has increased enormously. We are now in a position to use these findings as diagnostic tools for the formation of these bodies (see *Blum et al.,* 2017; *Blum,* 2018; *Weissman et al.,* 2020). The most striking physical property of comets for which we now possess empirical evidence is their extremely low mechanical strength, claimed to be essential for sustaining the activity of comets (*Kuehrt and Keller,* 1994; *Skorov and Blum,* 2012; *Blum et al.,* 2014). Measurements of the tensile strength of the cometary material have shown that the tensile strength indeed is $\lesssim 1$ Pa (*Attree et al.,* 2018). Numerical studies suggest that such a requirement is mandatory for a sustained cometary activity (*Skorov and Blum,* 2012; *Gundlach et al.,* 2020). Extremely low compressive-strength values were also inferred from the bouncing of the Philae lander (*O'Rourke et al.,* 2020).

Rosetta also allowed precise mass and volume determinations of the nucleus of Comet 67P (*Pätzold et al.,* 2019). In combination with assumptions on the material composition, this allows the derivation of global porosity values. Different approaches result in values for the porosity of 73–76% (*Herique et al.,* 2019), 65–80% (*Pätzold et al.,* 2019), and 63–79% (*Fulle et al.,* 2016), which correspond to global volume filling factors of the nucleus of Comet 67P of $\Phi_{67P} = 0.24–0.27$, $\Phi_{67P} = 0.20–0.35$, and $\Phi_{67P} = 0.21–0.37$, respectively. As pointed out in section 3.1.4, the expected volume filling factor for small ($\lesssim 100$ km) planetesimals is $\Phi_{pl} \approx 0.18–0.37$, which is well matched by the values estimated for Comet 67P. *Güttler et al.* (2019) showed that such volume-filling factors indicate a makeup consisting of two hierarchical levels (e.g., solid dust grains and fluffy pebbles). Fewer levels lead to densities that are too high, whereas more levels render the planetesimals too porous.

A "smoking gun" for the primordial nature of Comet 67P could be the discovery of extremely fluffy dust particles by the Giada and Midas instruments of Rosetta (see the review by *Güttler et al.,* 2019, and original references therein). These particles possess porosities >95% and have pre-detection sizes on the order of 1 cm, but electrostatically fragment in the vicinity of the Rosetta spacecraft. Even under the most optimistic conditions, such particles could not have formed in the coma of Comet 67P. To see this, consider the maximum measured dust mass flux of Comet 67P: $F = 10^{-8}$ kg m^{-2} s^{-1} (*Merouane et al.,* 2016). With this result, along with the assumption that the monomer grains have sizes of 100 nm (see section 2.3) and masses of $m = 10^{-18}$ kg, a particle flux (per unit mass) of $F' = F/m = 10^{10}$ m^{-2} s^{-1} results. The dust monomers are so small that they couple to the gas outflow on short timescales and thus obtain typical outflow velocities of $v_{out} = 100$ m s^{-1}. Hence, a maximum number density of the monomer grains in the coma of $n = F'/v_{out} = 10^8$ m^{-3} results. To determine the collision timescale, we assume that all collisions occur at the sticking-bouncing threshold, ignoring that fractal growth requires much lower collision speeds. From the models described in section 2.3, the sticking-bouncing threshold velocity of 100-nm grains should be on the order of $v_r = 10$ m s^{-1}. With a collision cross section for 100-nm grains of $\sigma \approx 10^{-13}$ m^2, a collision timescale of $\tau_{coll} = 1/n\sigma v_r = 10^4$ s follows. At a distance of Rosetta from the surface of Comet 67P of d = 10–30 km (*Della Corte et al.,* 2015), the time between emission of the dust and detection by the Rosetta/Giada instrument was $\tau_{trans} = d/v_{out} = 100–300$ s, much too short to allow any collisional growth of the grains. Thus, the observed fluffy grains must have been present inside the nucleus of Comet 67P before they were ejected into the coma.

Fulle and Blum (2017) showed that these fluffy particles could be the remnants of the pre-pebble growth stage of protoplanetary dust, captured and preserved between the pebbles during the gravitational collapse. If that scenario is correct, the cometary nuclei in the solar system could not be rubble piles, reassembled after a catastrophic collision. If post-formation collisions were unavoidable, then they had to be subcatastrophic so that major parts of the colliding planetesimals would remain intact (see, e.g., *Schwartz et al.,* 2018).

Under the assumption that Comet 67P was formed by the gentle gravitational collapse of a pebble cloud, *Blum et al.* (2017) used a multi-Rosetta-instrument approach to estimate the pebble sizes. They found that pebble radii in the range of 3–6 mm were the most likely, with the upper limit constrained by the subsurface temperatures measured by the Miro instrument and the lower limit given by the size distribution of surface features identified by the Civa instrument on the Philae lander. These pebble sizes are in good agreement with estimates of the bouncing barrier at large heliocentric distances [see *Lorek et al.* (2018) and section 2.5].

We should note that there are alternative formation models for the nucleus of Comet 67P [see section 3.2 and *Weissman et al.* (2020) for a recent discussion on this issue] and that there are open questions with the formation of cometary nuclei by the gentle gravitational collapse of a pebble cloud. The most prominent questions to be answered in future work are: (1) What is the influence of radiogenic heating on the inner makeup of planetesimals (see section 3.1.4)? (2) Can the gravitational collapse form comet-sized planetesimals directly and what is their rate of binarity? (3) What is the role of collisions in the post-formation era and how can present-day cometary nuclei form as a result of these collisions without losing their volatile and fluffy-particle inventory? That being said, the evidence for a gravitational collapse origin of cometesimals is reasonably compelling, giving further support to the SI as the mechanism to form planetesimals (although, of course, other gravitational collapse mechanisms cannot be ruled out).

5. CONCLUDING REMARKS

Less than two decades ago, the formation of planetesimals was one of the biggest unanswered questions in planetary science and astrophysics. While this issue does remain open and of utmost importance to understanding the formation of the solar system (as well as other planetary systems), it is very encouraging that so much progress has been made in the time since then.

We now have a detailed understanding of how submicrometer-sized grains grow to pebble sizes, backed by both theoretical calculations and laboratory experiments, and we have *multiple* models to explain how pebbles then reach the next stage and form planetesimals.

Of course, there remain issues that must be solved in order to build toward a comprehensive understanding of and to determine which mechanisms are responsible for planetesimal formation. For instance, the largest issue of contention is the discrepancy in the size distribution between observations and theoretical models, those based on both the SI and gravitational coagulation. Will other gravitational collapse mechanisms (e.g., the SGI) show better agreement? Or is the discrepancy, at least in the SI simulations, the result of something simpler, such as a more extensive survey of relevant parameters or collapse below the numerically limited scales in the SI calculations?

Even with these outstanding issues, however, there are indications that the gravitational collapse paradigm, and specifically the SI-induced collapse model, is the correct one. The thermal and structural properties of planetesimals made available from the Rosetta mission do indeed support a gravitational collapse model for planetesimal formation. Furthermore, the excellent agreement in KBO binary properties between SI simulations and observations *strongly* support the SI as the mechanism for planetesimal formation, while strongly disfavoring the gravitational coagulation model.

Furthermore, as numerical models improve and computational resources continue to be readily accessible, the theoretical side of this question will likely see significant progress in the coming years. These enhancements will be further augmented by improved observational capabilities. For example, in addition to continued observations from OSSOS, the Vera C. Rubin observatory will drastically expand our database of detected KBO objects, while the James Webb Space Telescope will provide another way to peer into the earliest stages of planet formation in our solar system. Furthermore, new missions, such as Lucy and Psyche, are either being built or are already on their way to exploring solar system planetesimals.

In closing, with ever improving numerical models, new observational facilities coming online, and new missions set to explore solar system planetesimals, we live in a golden era of planetesimal and comet exploration. As such, there should be significant optimism that we as a community will solve the outstanding issues of planetesimal and comet formation in the coming years; the future of small body studies looks quite bright indeed.

Acknowledgments. J.B.S. acknowledges support from NASA under Emerging Worlds through grant 80NSSC21K0037. J.B. acknowledges the support of this work by the Deutsche Forschungsgemeinschaft (DFG) through grant BL 298/27-1 and thanks DFG and the Deutsches Zentrum für Luft-und Raumfahrt–German Space Agency for continuous support. T.B. acknowledges funding from the European Research Council (ERC) under the European Union's Horizon 2020 research and innovation programme under grant agreement No 714769 and funding by the Deutsche Forschungsgemeinschaft (DFG, German Research Foundation) under Ref no. FOR 2634/1 and under Germany's Excellence Strategy – EXC-2094 – 390783311. D.N. acknowledges support from the NASA Emerging Worlds program. We thank D. Carrera, W. Lyra, A. Youdin, and R. Li for useful discussions that greatly improved the quality of this chapter. We also thank the anonymous reviewers whose suggestions significantly strengthened the chapter. Furthermore, we thank J. Squire for providing the necessary materials and information to create Fig. 5. We also thank Ingo von Borstel for creating the fractal aggregate in Fig. 2. Finally, the writing of this document was supported in part by the Munich Institute for Astro- and Particle Physics (MIAPP), which is funded by the Deutsche Forschungsgemeinschaft (DFG, German Research Foundation) under Germany's Excellence Strategy – EXC-2094 – 390783311, and J.B.S. thanks MIAPP for their hospitality while he finished much of this chapter.

REFERENCES

Abod C. P., Simon J. B., Li R., Armitage P. J., Youdin A. N., and Kretke K. A. (2019) The mass and size distribution of planetesimals formed by the streaming instability. II. The effect of the radial gas pressure gradient. *Astrophys. J.*, *883*, 192.

ALMA Partnership, Brogan C. L., Pérez L. M., Hunter T. R., Dent W. R. F., Hales A. S., Hills R. E., Corder S., Fomalont E. B., Vlahakis C., Asaki Y., Barkats D., Hirota A., Hodge J. A., Impellizzeri C. M. V., Kneissl R., Liuzzo E., Lucas R., Marcelino N., Matsushita S., Nakanishi K., Phillips N., Richards A. M. S., Toledo I., Aladro R., Broguiere D., Cortes J. R., Cortes P. C., Espada D., Galarza F., Garcia-Appadoo D., Guzman-Ramirez L., Humphreys E. M., Jung T., Kameno S., Laing R. A., Leon S., Marconi G., Mignano A., Nikolic B., Nyman L. A., Radiszcz M., Remijan A., Rodón J. A., Sawada T., Takahashi S., Tilanus R. P. J., Vila Vilaro B., Watson L. C., Wiklind T., Akiyama E., Chapillon E., de Gregorio-Monsalvo I., Di Francesco J., Gueth F., Kawamura A., Lee C. F., Nguyen Luong Q., Mangum J., Pietu V., Sanhueza P., Saigo K., Takakuwa S., Ubach C., van Kempen T., Wootten A., Castro-Carrizo A., Francke H., Gallardo J., Garcia J., Gonzalez S., Hill T., Kaminski T., Kurono Y., Liu H. Y., Lopez C., Morales F., Plarre K., Schieven G., Testi L., Videla L., Villard E., Andreani P., Hibbard J. E., and Tatematsu K. (2015) The 2014 ALMA long baseline campaign: First results from high angular resolution observations toward the HL Tau region. *Astrophys. J. Lett.*, *808*, L3.

Andrews S. M., Wilner D. J., Zhu Z., Birnstiel T., Carpenter J. M., Pérez L. M., Bai X.-N., Öberg K. I., Hughes A. M., Isella A., and Ricci L. (2016) Ringed substructure and a gap at 1 au in the nearest protoplanetary disk. *Astrophys. J. Lett.*, *820*, L40.

Andrews S. M., Huang J., Pérez L. M., Isella A., Dullemond C. P., Kurtovic N. T., Guzmán V. V., Carpenter J. M., Wilner D. J., Zhang S., Zhu Z., Birnstiel T., Bai X.-N., Benisty M., Hughes A. M., Öberg K. I., and Ricci L. (2018) The Disk Substructures at High Angular Resolution Project (DSHARP). I. Motivation, sample, calibration,

and overview. *Astrophys. J. Lett.*, *869*, L41.
Arakawa S. and Krijt S. (2021) On the stickiness of CO_2 and H_2O ice particles. *Astrophys. J.*, *910*, 130.
Armitage P. J. (2007) Lecture notes on the formation and early evolution of planetary systems. *ArXiV e-prints*, arXiv:astroph/0701485.
Attree N., Groussin O., Jorda L., Nébouy D., Thomas N., Brouet Y., Kührt E., Preusker F., Scholten F., Knollenberg J., Hartogh P., Sierks H., Barbieri C., Lamy P., Rodrigo R., Koschny D., Rickman H., Keller H. U., A'Hearn M. F., Auger A. T., Barucci M. A., Bertaux J. L., Bertini I., Bodewits D., Boudreault S., Cremonese G., Da Deppo V., Davidsson B., Debei S., De Cecco M., Deller J., El-Maarry M. R., Fornasier S., Fulle M., Gutiérrez P. J., Güttler C., Hviid S., Ip W. H., Kovacs G., Kramm J. R., Küppers M., Lara L. M., Lazzarin M., Lopez Moreno J. J., Lowry S., Marchi S., Marzari F., Mottola S., Naletto G., Oklay N., Pajola M., Toth I., Tubiana C., Vincent J. B., and Shi X. (2018) Tensile strength of 67P/Churyumov-Gerasimenko nucleus material from overhangs. *Astron. Astrophys.*, *611*, A33.
Avenhaus H., Quanz S. P., Garufi A., Perez S., Casassus S., Pinte C., Bertrang G. H. M., Caceres C., Benisty M., and Dominik C. (2018) Disks around T Tauri stars with SPHERE (DARTTS-S). I. SPHERE/IRDIS polarimetric imaging of eight prominent T Tauri disks. *Astrophys. J.*, *863*, 44.
Bai X.-N. and Stone J. M. (2010a) Dynamics of solids in the mid-plane of protoplanetary disks: Implications for planetesimal formation. *Astrophys. J.*, *722*, 1437–1459.
Bai X.-N. and Stone J. M. (2010b) Particle-gas dynamics with Athena: Method and convergence. *Astrophys. J. Suppl. Ser.*, *190*, 297–310.
Bai X.-N. and Stone J. M. (2010c) The effect of the radial pressure gradient in protoplanetary disks on planetesimal formation. *Astrophys. J. Lett.*, *722*, L220–L223.
Bai X.-N. and Stone J. M. (2014) Magnetic flux concentration and zonal flows in magnetorotational instability turbulence. *Astrophys. J.*, *796*, 31.
Balbus S. A. and Hawley J. F. (1998) Instability, turbulence, and enhanced transport in accretion disks. *Rev. Mod. Phys.*, *70*, 1–53.
Barnes J. and Hut P. (1986) A hierarchical O(N log N) force-calculation algorithm. *Nature*, *324*, 446–449.
Baruteau C., Barraza M., Pérez S., Casassus S., Dong R., Lyra W., Marino S., Christiaens V., Zhu Z., Carmona A., Debras F., and Alarcon F. (2019) Dust traps in the protoplanetary disc MWC 758: Two vortices produced by two giant planets? *Mon. Not. R. Astron. Soc.*, *486*, 304–319.
Batygin K., Brown M. E., and Fraser W. C. (2011) Retention of a primordial cold classical Kuiper belt in an instability-driven model of solar system formation. *Astrophys. J.*, *738*, 13.
Bernstein G. M., Trilling D. E., Allen R. L., Brown M. E., Holman M., and Malhotra R. (2004) The size distribution of trans-neptunian bodies. *Astron. J.*, *128*, 1364–1390.
Béthune W., Lesur G., and Ferreira J. (2017) Global simulations of protoplanetary disks with net magnetic flux. I. Non-ideal MHD case. *Astron. Astrophys.*, *600*, A75.
Binney J. and Tremaine S. (1987) *Galactic Dynamics*. Princeton Univ., Princeton. 733 pp.
Birnstiel T., Dullemond C. P., and Brauer F. (2010) Gas- and dust evolution in protoplanetary disks. *Astron. Astrophys.*, *513*, A79.
Birnstiel T., Ormel C. W., and Dullemond C. P. (2011) Dust size distributions in coagulation/fragmentation equilibrium: Numerical solutions and analytical fits. *Astron. Astrophys.*, *525*, A11.
Birnstiel T., Klahr H., and Ercolano B. (2012) A simple model for the evolution of the dust population in protoplanetary disks. *Astron. Astrophys.*, *539*, A148.
Birnstiel T., Andrews S. M., Pinilla P., and Kama M. (2015) Dust evolution can produce scattered light gaps in protoplanetary disks. *Astrophys. J. Lett.*, *813*, L14.
Birnstiel T., Fang M., and Johansen A. (2016) Dust evolution and the formation of planetesimals. *Space Sci. Rev.*, *205*, 41–75.
Birnstiel T., Dullemond C. P., Zhu Z., Andrews S. M., Bai X.-N., Wilner D. J., Carpenter J. M., Huang J., Isella A., Benisty M., Pérez L. M., and Zhang S. (2018) The Disk Substructures at High Angular Resolution Project (DSHARP). V. Interpreting ALMA maps of protoplanetary disks in terms of a dust model. *Astrophys. J. Lett.*, *869*, L45.
Bischoff D., Kreuzig C., Haack D., Gundlach B., and Blum J. (2020) Sticky or not sticky? Measurements of the tensile strength of microgranular organic materials. *Mon. Not. R. Astron. Soc.*, *497*, 2517–2528.
Bischoff D., Gundlach B., and Blum J. (2021) A method to distinguish between micro- and macro-granular surfaces of small solar system bodies. *Mon. Not. R. Astron. Soc.*, *508*, 4705–4721.
Blum J. (2006) Dust agglomeration. *Adv. Phys.*, *55*, 881–947.
Blum J. (2018) Dust evolution in protoplanetary discs and the formation of planetesimals: What have we learned from laboratory experiments? *Space Sci. Rev.*, *214*, 52.
Blum J. and Wurm G. (2000) Experiments on sticking, restructuring, and fragmentation of preplanetary dust aggregates. *Icarus*, *143*, 138–146.
Blum J. and Wurm G. (2008) The growth mechanisms of macroscopic bodies in protoplanetary disks. *Annu. Rev. Astron. Astrophys.*, *46*, 21–56.
Blum J., Gundlach B., Mühle S., and Trigo-Rodriguez J. M. (2014) Comets formed in solar-nebula instabilities! — An experimental and modeling attempt to relate the activity of comets to their formation process. *Icarus*, *235*, 156–169.
Blum J., Gundlach B., Krause M., Fulle M., Johansen A., Agarwal J., von Borstel I., Shi X., Hu X., Bentley M. S., Capaccioni F., Colangeli L., Della Corte V., Fougere N., Green S. F., Ivanovski S., Mannel T., Merouane S., Migliorini A., Rotundi A., Schmied R., and Snodgrass C. (2017) Evidence for the formation of comet 67P/Churyumov-Gerasimenko through gravitational collapse of a bound clump of pebbles. *Mon. Not. R. Astron. Soc.*, *469*, S755–S773.
Brauer F., Dullemond C. P., Johansen A., Henning T., Klahr H., and Natta A. (2007) Survival of the mm-cm size grain population observed in protoplanetary disks. *Astron. Astrophys.*, *469*, 1169–1182.
Brauer F., Dullemond C. P., and Henning T. (2008) Coagulation, fragmentation and radial motion of solid particles in protoplanetary disks. *Astron. Astrophys.*, *480*, 859–877.
Bukhari Syed M., Blum J., Wahlberg Jansson K., and Johansen A. (2017) The role of pebble fragmentation in planetesimal formation. I. Experimental study. *Astrophys. J.*, *834*, 145.
Campo Bagatin A. and Benavidez P. G. (2012) Collisional evolution of trans-neptunian object populations in a Nice model environment. *Mon. Not. R. Astron. Soc.*, *423*, 1254–1266.
Carrera D. and Simon J. B. (2022) The streaming instability cannot form planetesimals from millimeter-size grains in pressure bumps. *Astrophys. J. Lett.*, *933*, L10.
Carrera D., Johansen A., and Davies M. B. (2015) How to form planetesimals from mm-sized chondrules and chondrule aggregates. *Astron. Astrophys.*, *579*, A43.
Carrera D., Simon J. B., Li R., Kretke K. A., and Klahr H. (2021) Protoplanetary disk rings as sites for planetesimal formation. *Astron. J.*, *161*, 96.
Carrera D., Thomas A. J., Simon J. B., Small M. A., Kretke K. A., and Klahr H. (2022) Resilience of planetesimal formation in weakly reinforced pressure bumps. *Astrophys. J.*, *927*, 52.
Chen K. and Lin M.-K. (2020) How efficient is the streaming instability in viscous protoplanetary disks? *Astrophys. J.*, *891*, 132.
Ciesla F. J. and Cuzzi J. N. (2006) The evolution of the water distribution in a viscous protoplanetary disk. *Icarus*, *181*, 178–204.
Clement M. S., Kaib N. A., Raymond S. N., and Walsh K. J. (2018) Mars' growth stunted by an early giant planet instability. *Icarus*, *311*, 340–356.
Cuzzi J. N. and Zahnle K. J. (2004) Material enhancement in protoplanetary nebulae by particle drift through evaporation fronts. *Astrophys. J.*, *614*, 490–496.
Cuzzi J. N., Dobrovolskis A. R., and Champney J. M. (1993) Particle-gas dynamics in the midplane of a protoplanetary nebula. *Icarus*, *106*, 102–134.
Cuzzi J. N., Hogan R. C., and Shariff K. (2008) Toward planetesimals: Dense chondrule clumps in the protoplanetary nebula. *Astrophys. J.*, *687*, 1432–1447.
Dawson R. I. and Murray-Clay R. (2012) Neptune's wild days: Constraints from the eccentricity distribution of the classical Kuiper belt. *Astrophys. J.*, *750*, 43.
Della Corte V., Rotundi A., Fulle M., Gruen E., Weissman P., Sordini R., Ferrari M., Ivanovski S., Lucarelli F., Accolla M., Zakharov V., Mazzotta Epifani E., Lopez-Moreno J. J., Rodriguez J., Colangeli L., Palumbo P., Bussoletti E., Crifo J. F., Esposito F., Green S. F., Lamy P. L., McDonnell J. A. M., Mennella V., Molina A., Morales R., Moreno F., Ortiz J. L., Palomba E., Perrin J. M., Rietmeijer F. J. M., Rodrigo R., Zarnecki J. C., Cosi M., Giovane F., Gustafson B., Herranz M. L., Jeronimo J. M., Leese M. R., Lopez-Jimenez A. C.,

and Altobelli N. (2015) GIADA: Shining a light on the monitoring of the comet dust production from the nucleus of 67P/Churyumov-Gerasimenko. *Astron. Astrophys.*, *583*, A13.
de Sousa R. R., Morbidelli A., Raymond S. N., Izidoro A., Gomes R., and Vieira Neto E. (2020) Dynamical evidence for an early giant planet instability. *Icarus*, *339*, 113605.
Dominik C. and Tielens A. G. G. M. (1997) The physics of dust coagulation and the structure of dust aggregates in space. *Astrophys. J.*, *480*, 647–673.
Dones L., Brasser R., Kaib N., and Rickman H. (2015) Origin and evolution of the cometary reservoirs. *Space Sci. Rev.*, *197*, 191–269.
Drążkowska J., Windmark F., and Dullemond C. P. (2014) Modeling dust growth in protoplanetary disks: The breakthrough case. *Astron. Astrophys.*, *567*, A38.
Drążkowska J., Alibert Y., and Moore B. (2016) Close-in planetesimal formation by pile-up of drifting pebbles. *Astron. Astrophys.*, *594*, A105.
Dullemond C. P. and Dominik C. (2005) Dust coagulation in protoplanetary disks: A rapid depletion of small grains. *Astron. Astrophys.*, *434*, 971–986.
Dullemond C. P., Birnstiel T., Huang J., Kurtovic N. T., Andrews S. M., Guzmán V. V., Pérez L. M., Isella A., Zhu Z., Benisty M., Wilner D. J., Bai X.-N., Carpenter J. M., Zhang S., and Ricci L. (2018) The Disk Substructures at High Angular Resolution Project (DSHARP). VI. Dust trapping in thin-ringed protoplanetary disks. *Astrophys. J. Lett.*, *869*, L46.
Emery J. P., Marzari F., Morbidelli A., French L. M., and Grav T. (2015) The complex history of Trojan asteroids. In *Asteroids IV* (P. Michel et al., eds.), pp. 203–220. Univ. of Arizona, Tucson.
Epstein P. S. (1924) On the resistance experienced by spheres in their motion through gases. *Phys. Rev.*, *23*, 710–733.
Fedele D., Carney M., Hogerheijde M. R., Walsh C., Miotello A., Klaassen P., Bruderer S., Henning T., and van Dishoeck E. F. (2017) ALMA unveils rings and gaps in the protoplanetary system HD 169142: Signatures of two giant protoplanets. *Astron. Astrophys.*, *600*, A72.
Flock M. and Mignone A. (2021) Streaming instability in a global patch simulation of protoplanetary disks. *Astron. Astrophys.*, *650*, A119.
Fraser W. C. (2009) The collisional divot in the Kuiper belt size distribution. *Astrophys. J.*, *706*, 119–129.
Fraser W. C., Brown M. E., Morbidelli A., Parker A., and Batygin K. (2014) The absolute magnitude distribution of Kuiper belt objects. *Astrophys. J.*, *782*, 100.
Fraser W. C., Benecchi S. D., Kavelaars J. J., Marsset M., Pike R. E., Bannister M. T., Schwamb M. E., Volk K., Nesvorny D., Alexandersen M., Chen Y.-T., Gwyn S., Lehner M. J., and Wang S.-Y. (2021) Col-OSSOS: The distinct color distribution of single and binary cold classical KBOs. *Planet. Sci. J.*, *2*, 90.
Fromang S. and Papaloizou J. (2006) Dust settling in local simulations of turbulent protoplanetary disks. *Astron. Astrophys.*, *452*, 751–762.
Fulle M. and Blum J. (2017) Fractal dust constrains the collisional history of comets. *Mon. Not. R. Astron. Soc.*, *469*, S39–S44.
Fulle M., Della Corte V., Rotundi A., Rietmeijer F. J. M., Green S. F., Weissman P., Accolla M., Colangeli L., Ferrari M., Ivanovski S., Lopez-Moreno J. J., Epifani E. M., Morales R., Ortiz J. L., Palomba E., Palumbo P., Rodriguez J., Sordini R., and Zakharov V. (2016) Comet 67P/Churyumov-Gerasimenko preserved the pebbles that formed planetesimals. *Mon. Not. R. Astron. Soc.*, *462*, S132–S137.
Gárate M., Birnstiel T., Drążkowska J., and Stammler S. M. (2020) Gas accretion damped by dust back-reaction at the snow line. *Astron. Astrophys.*, *635*, A149.
Garaud P., Meru F., Galvagni M., and Olczak C. (2013) From dust to planetesimals: An improved model for collisional growth in protoplanetary disks. *Astrophys. J.*, *764*, 146.
Gerbig K., Murray-Clay R. A., Klahr H., and Baehr H. (2020) Requirements for gravitational collapse in planetesimal formation — The impact of scales set by Kelvin-Helmholtz and nonlinear streaming instability. *Astrophys. J.*, *895*, 91.
Golabek G. J. and Jutzi M. (2021) Modification of icy planetesimals by early thermal evolution and collisions: Constraints for formation time and initial size of comets and small KBOs. *Icarus*, *363*, 114437.
Goldreich P. and Ward W. R. (1973) The formation of planetesimals. *Astrophys. J.*, *183*, 1051–1062.
Gole D. A., Simon J. B., Li R., Youdin A. N., and Armitage P. J. (2020) Turbulence regulates the rate of planetesimal formation via gravitational collapse. *Astrophys. J.*, *904*, 132.
Goodman J. and Pindor B. (2000) Secular instability and planetesimal formation in the dust layer. *Icarus*, *148*, 537–549.
Greenstreet S., Gladman B., McKinnon W. B., Kavelaars J. J., and Singer K. N. (2019) Crater density predictions for New Horizons flyby target 2014 MU_{69}. *Astrophys. J. Lett.*, *872*, L5.
Grundy W. M., Noll K. S., Roe H. G., Buie M. W., Porter S. B., Parker A. H., Nesvorný D., Levison H. F., Benecchi S. D., Stephens D. C., and Trujillo C. A. (2019) Mutual orbit orientations of transneptunian binaries. *Icarus*, *334*, 62–78.
Gundlach B. and Blum J. (2015) The stickiness of micrometer-sized water-ice particles. *Astrophys. J.*, *798*, 34.
Gundlach B., Schmidt K. P., Kreuzig C., Bischoff D., Rezaei F., Kothe S., Blum J., Grzesik B., and Stoll E. (2018) The tensile strength of ice and dust aggregates and its dependence on particle properties. *Mon. Not. R. Astron. Soc.*, *479*, 1273–1277.
Gundlach B., Fulle M., and Blum J. (2020) On the activity of comets: understanding the gas and dust emission from comet 67/Churyumov-Gerasimenko's south-pole region during perihelion. *Mon. Not. R. Astron. Soc.*, *493*, 3690–3715.
Güttler C., Krause M., Geretshauser R. J., Speith R., and Blum J. (2009) The physics of protoplanetesimal dust agglomerates. IV. Toward a dynamical collision model. *Astrophys. J.*, *701*, 130–141.
Güttler C., Blum J., Zsom A., Ormel C. W., and Dullemond C. P. (2010) The outcome of protoplanetary dust growth: Pebbles, boulders, or planetesimals? I. Mapping the zoo of laboratory collision experiments. *Astron. Astrophys.*, *513*, A56.
Güttler C., Mannel T., Rotundi A., Merouane S., Fulle M., Bockelée-Morvan D., Lasue J., Levasseur-Regourd A. C., Blum J., Naletto G., Sierks H., Hilchenbach M., Tubiana C., Capaccioni F., Paquette J. A., Flandes A., Moreno F., Agarwal J., Bodewits D., Bertini I., Tozzi G. P., Hornung K., Langevin Y., Krüger H., Longobardo A., Della Corte V., Tóth I., Filacchione G., Ivanovski S. L., Mottola S., and Rinaldi G. (2019) Synthesis of the morphological description of cometary dust at comet 67P/Churyumov-Gerasimenko. *Astron. Astrophys.*, *630*, A24.
Hartlep T. and Cuzzi J. N. (2020) Cascade model for planetesimal formation by turbulent clustering. *Astrophys. J.*, *892*, 120.
Hartlep T., Cuzzi J. N., and Weston B. (2017) Scale dependence of multiplier distributions for particle concentration, enstrophy, and dissipation in the inertial range of homogeneous turbulence. *Phys. Rev. E*, *95*, 033115.
Hawley J. F., Gammie C. F., and Balbus S. A. (1995) Local three-dimensional magnetohydrodynamic simulations of accretion disks. *Astrophys. J.*, *440*, 742–763.
Hayashi C. (1981) Structure of the solar nebula, growth and decay of magnetic fields and effects of magnetic and turbulent viscosities on the nebula. *Prog. Theor. Phys. Suppl.*, *70*, 35–53.
Hendler N. P., Pinilla P., Pascucci I., Pohl A., Mulders G., Henning T., Dong R., Clarke C., Owen J., and Hollenbach D. (2018) A likely planet-induced gap in the disc around T Cha. *Mon. Not. R. Astron. Soc.*, *475*, L62–L66.
Herique A., Kofman W., Zine S., Blum J., Vincent J. B., and Ciarletti V. (2019) Homogeneity of 67P/Churyumov-Gerasimenko as seen by CONSERT: Implication on composition and formation. *Astron. Astrophys.*, *630*, A6.
Hirashita H. (2012) Dust growth in the interstellar medium: How do accretion and coagulation interplay? *Mon. Not. R. Astron. Soc.*, *422*, 1263–1271.
Hopkins P. F. (2016) A simple phenomenological model for grain clustering in turbulence. *Mon. Not. R. Astron. Soc.*, *455*, 89–111.
Huang J., Andrews S. M., Dullemond C. P., Isella A., Pérez L. M., Guzmán V. V., Öberg K. I., Zhu Z., Zhang S., Bai X.-N., Benisty M., Birnstiel T., Carpenter J. M., Hughes A. M., Ricci L., Weaver E., and Wilner D. J. (2018) The Disk Substructures at High Angular Resolution Project (DSHARP). II. Characteristics of annular substructures. *Astrophys. J. Lett.*, *869*, L42.
Hyodo R., Ida S., and Charnoz S. (2019) Formation of rocky and icy planetesimals inside and outside the snow line: Effects of diffusion, sublimation, and back-reaction. *Astron. Astrophys.*, *629*, A90.
Isella A., Guidi G., Testi L., Liu S., Li H., Li S., Weaver E., Boehler Y., Carperter J. M., De Gregorio-Monsalvo I., Manara C. F., Natta A., Pérez L. M., Ricci L., Sargent A., Tazzari M., and Turner N. (2016) Ringed structures of the HD 163296 protoplanetary disk revealed by ALMA. *Phys. Rev. Lett.*, *117*, 251101.
Johansen A. and Klahr H. (2005) Dust diffusion in protoplanetary disks by magnetorotational turbulence. *Astrophys. J.*, *634*, 1353–1371.

Johansen A. and Youdin A. (2007) Protoplanetary disk turbulence driven by the streaming instability: Nonlinear saturation and particle concentration. *Astrophys. J.*, *662*, 627–641.

Johansen A., Oishi J. S., Mac Low M.-M., Klahr H., Henning T., and Youdin A. (2007) Rapid planetesimal formation in turbulent circumstellar disks. *Nature*, *448*, 1022–1025.

Johansen A., Youdin A., and Klahr H. (2009) Zonal flows and long-lived axisymmetric pressure bumps in magnetorotational turbulence. *Astrophys. J.*, *697*, 1269–1289.

Johansen A., Youdin A. N., and Lithwick Y. (2012) Adding particle collisions to the formation of asteroids and Kuiper belt objects via streaming instabilities. *Astron. Astrophys.*, *537*, A125.

Johansen A., Mac Low M.-M., Lacerda P., and Bizzarro M. (2015) Growth of asteroids, planetary embryos, and Kuiper belt objects by chondrule accretion. *Sci. Adv.*, *1*, e1500109.

Jutzi M. and Benz W. (2017) Formation of bi-lobed shapes by subcatastrophic collisions: A late origin of comet 67P's structure. *Astron. Astrophys.*, *597*, A62.

Jutzi M., Benz W., Toliou A., Morbidelli A., and Brasser R. (2017) How primordial is the structure of comet 67P?: Combined collisional and dynamical models suggest a late formation. *Astron. Astrophys.*, *597*, A61.

Kataoka A., Tanaka H., Okuzumi S., and Wada K. (2013) Fluffy dust forms icy planetesimals by static compression. *Astron. Astrophys.*, *557*, L4.

Kataoka A., Muto T., Momose M., Tsukagoshi T., Fukagawa M., Shibai H., Hanawa T., Murakawa K., and Dullemond C. P. (2015) Millimeter-wave polarization of protoplanetary disks due to dust scattering. *Astrophys. J.*, *809*, 78.

Katsuragi H. and Blum J. (2017) The physics of protoplanetesimal dust agglomerates. IX. Mechanical properties of dust aggregates probed by a solid-projectile impact. *Astrophys. J.*, *851*, 23.

Kavelaars J. J., Petit J.-M., Gladman B., Bannister M. T., Alexandersen M., Chen Y.-T., Gwyn S. D. J., and Volk K. (2021) OSSOS finds an exponential cutoff in the size distribution of the cold classical Kuiper belt. *Astrophys. J. Lett.*, *920*, L28.

Kenyon S. J. and Bromley B. C. (2004) The size distribution of Kuiper belt objects. *Astron. J.*, *128*, 1916–1926.

Kenyon S. J. and Bromley B. C. (2012) Coagulation calculations of icy planet formation at 15–150 AU: A correlation between the maximum radius and the slope of the size distribution for trans-neptunian objects. *Astron. J.*, *143*, 63.

Kenyon S. J. and Bromley B. C. (2020) Craters on Charon: Impactors from a collisional cascade among trans-neptunian objects. *Planet. Sci. J.*, *1*, 40.

Klahr H. H. and Henning T. (1997) Particle-trapping eddies in protoplanetary accretion disks. *Icarus*, *128*, 213–229.

Kobayashi H. and Tanaka H. (2021) Rapid formation of gas-giant planets via collisional coagulation from dust grains to planetary cores. *Astrophys. J.*, *922*, 16.

Kolmogorov A. (1941) The local structure of turbulence in incompressible viscous fluid for very large Reynolds' numbers. *Akad. Nauk SSSR Doklady*, *30*, 301–305.

Kornet K., Różyczka M., and Stepinski T. F. (2004) An alternative look at the snowline in protoplanetary disks. *Astron. Astrophys.*, *417*, 151–158.

Kothe S. (2016) Mikrogravitationsexperimente zur entwicklung eines empirischen stoßmodells für protoplanetare staubagglomerate. Ph.D. thesis, Technische Universität Braunschweig, Braunschweig. 141 pp.

Kowalik K., Hanasz M., Wóltański D., and Gawryszczak A. (2013) Streaming instability in the quasi-global protoplanetary discs. *Mon. Not. R. Astron. Soc.*, *434*, 1460–1468.

Krapp L., Benítez-Llambay P., Gressel O., and Pessah M. E. (2019) Streaming instability for particle-size distributions. *Astrophys. J. Lett.*, *878*, L30.

Krijt S., Güttler C., Heißelmann D., Dominik C., and Tielens A. G. G. M. (2013) Energy dissipation in head-on collisions of spheres. *J. Phys. D: Appl. Phys.*, *46*, 435303.

Krivov A. V., Ide A., Löhne T., Johansen A., and Blum J. (2018) Debris disc constraints on planetesimal formation. *Mon. Not. R. Astron. Soc.*, *474*, 2564–2575.

Kuehrt E. and Keller H. U. (1994) The formation of cometary surface crusts. *Icarus*, *109*, 121–132.

Lambrechts M. and Johansen A. (2014) Forming the cores of giant planets from the radial pebble flux in protoplanetary discs. *Astron. Astrophys.*, *572*, A107.

Lambrechts M., Morbidelli A., Jacobson S. A., Johansen A., Bitsch B., Izidoro A., and Raymond S. N. (2019) Formation of planetary systems by pebble accretion and migration: How the radial pebble flux determines a terrestrial-planet or super-Earth growth mode. *Astron. Astrophys.*, *627*, A83.

Landeck A. (2018) Aufbau und durchführung eines experiments für stöße zwischen eisagglomeraten. Master's thesis, Technische Universität Braunschweig, Braunschweig.

Li R. (2019) *PLAN: A Clump-Finder for Planetesimal Formation Simulations*. Astrophysics Source Code Library, ascl:1911.001.

Li R. and Youdin A. N. (2021) Thresholds for particle clumping by the streaming instability. *Astrophys. J.*, *919*, 107.

Li R., Youdin A. N., and Simon J. B. (2018) On the numerical robustness of the streaming instability: Particle concentration and gas dynamics in protoplanetary disks. *Astrophys. J.*, *862*, 14.

Li R., Youdin A. N., and Simon J. B. (2019) Demographics of planetesimals formed by the streaming instability. *Astrophys. J.*, *885*, 69.

Lichtenberg T., Golabek G. J., Gerya T. V., and Meyer M. R. (2016) The effects of short-lived radionuclides and porosity on the early thermo-mechanical evolution of planetesimals. *Icarus*, *274*, 350–365.

Lichtenberg T., Drazkowska J., Schönbächler M., Golabek G. J., and Hands T. O. (2021) Bifurcation of planetary building blocks during solar system formation. *Science*, *371*, 365–370.

Lin M.-K. (2021) Stratified and vertically shearing streaming instabilities in protoplanetary disks. *Astrophys. J.*, *907*, 64.

Lin Z.-Y. D., Li Z.-Y., Yang H., Looney L., Stephens I., and Hull C. L. H. (2020) Validating scattering-induced (sub)millimetre disc polarization through the spectral index, wavelength-dependent polarization pattern, and polarization spectrum: The case of HD 163296. *Mon. Not. R. Astron. Soc.*, *496*, 169–181.

Lorek S., Lacerda P., and Blum J. (2018) Local growth of dust- and ice-mixed aggregates as cometary building blocks in the solar nebula. *Astron. Astrophys.*, *611*, A18.

Lovelace R. V. E., Li H., Colgate S. A., and Nelson A. F. (1999) Rossby wave instability of Keplerian accretion disks. *Astrophys. J.*, *513*, 805–810.

Lyra W. and Umurhan O. M. (2019) The initial conditions for planet formation: Turbulence driven by hydrodynamical instabilities in disks around young stars. *Publ. Astron. Soc. Pac.*, *131*, 072001.

Malamud U., Landeck W. A., Bischoff D., Kreuzig C., Perets H. B., Gundlach B., and Blum J. (2022) Are there any pristine comets? Constraints from pebble structure. *Mon. Not. R. Astron. Soc.*, *514*, 3366–3394.

Mannel T., Bentley M. S., Boakes P. D., Jeszenszky H., Ehrenfreund P., Engrand C., Koeberl C., Levasseur-Regourd A. C., Romstedt J., Schmied R., Torkar K., and Weber I. (2019) Dust of comet 67P/Churyumov-Gerasimenko collected by Rosetta/MIDAS: Classification and extension to the nanometer scale. *Astron. Astrophys.*, *630*, A26.

Mathis J. S., Rumpl W., and Nordsieck K. H. (1977) The size distribution of interstellar grains. *Astrophys. J.*, *217*, 425–433.

McKinnon W. B., Richardson D. C., Marohnic J. C., Keane J. T., Grundy W. M., Hamilton D. P., Nesvorný D., Umurhan O. M., Lauer T. R., Singer K. N., Stern S. A., Weaver H. A., Spencer J. R., Buie M. W., Moore J. M., Kavelaars J. J., Lisse C. M., Mao X., Parker A. H., Porter S. B., Showalter M. R., Olkin C. B., Cruikshank D. P., Elliott H. A., Gladstone G. R., Parker J. W., Verbiscer A. J., Young L. A., and the New Horizons Science Team (2020) The solar nebula origin of (486958) Arrokoth, a primordial contact binary in the Kuiper belt. *Science*, *367*, aay6620.

Meisner T., Wurm G., Teiser J., and Schywek M. (2013) Preplanetary scavengers: Growing tall in dust collisions. *Astron. Astrophys.*, *559*, A123.

Merouane S., Zaprudin B., Stenzel O., Langevin Y., Altobelli N., Della Corte V., Fischer H., Fulle M., Hornung K., Silén J., Ligier N., Rotundi A., Ryno J., Schulz R., Hilchenbach M., Kissel J., and the COSIMA Team (2016) Dust particle flux and size distribution in the coma of 67P/Churyumov-Gerasimenko measured in situ by the COSIMA instrument on board Rosetta. *Astron. Astrophys.*, *596*, A87.

Michikoshi S., Kokubo E., and Inutsuka S.-i. (2012) Secular gravitational instability of a dust layer in shear turbulence. *Astrophys. J.*, *746*, 35.

Mignone A., Flock M., and Vaidya B. (2019) A particle module for the PLUTO code. III. Dust. *Astrophys. J. Suppl. Ser.*, *244*, 38.

Morbidelli A. and Rickman H. (2015) Comets as collisional fragments

of a primordial planetesimal disk. *Astron. Astrophys., 583,* A43.
Morbidelli A. and Nesvorný D. (2020) Kuiper belt: Formation and evolution. In *The Trans-Neptunian Solar System* (D. Prialnik et al., eds.), pp. 25–59. Elsevier, Amsterdam.
Morbidelli A., Nesvorny D., Bottke W. F., and Marchi S. (2021) A re-assessment of the Kuiper belt size distribution for subkilometer objects, revealing collisional equilibrium at small sizes. *Icarus, 356,* 114256.
Morgan M., Seligman D., and Batygin K. (2021) Collisional growth within the solar system's primordial planetesimal disk and the timing of the giant planet instability. *Astrophys. J. Lett., 917,* L8.
Musiolik G. and Wurm G. (2019) Contacts of water ice in protoplanetary disks — Laboratory experiments. *Astrophys. J., 873,* 58.
Nakagawa Y., Sekiya M., and Hayashi C. (1986) Settling and growth of dust particles in a laminar phase of a low-mass solar nebula. *Icarus, 67,* 375–390.
Nesvorný D. and Vokrouhlický D. (2019) Binary survival in the outer solar system. *Icarus, 331,* 49–61.
Nesvorný D., Youdin A. N., and Richardson D. C. (2010) Formation of Kuiper belt binaries by gravitational collapse. *Astron. J., 140,* 785–793.
Nesvorný D., Vokrouhlický D., Bottke W. F., and Levison H. F. (2018) Evidence for very early migration of the solar system planets from the Patroclus-Menoetius binary Jupiter Trojan. *Nature Astron., 2,* 878–882.
Nesvorný D., Li R., Youdin A. N., Simon J. B., and Grundy W. M. (2019) Trans-neptunian binaries as evidence for planetesimal formation by the streaming instability. *Nature Astron., 3,* 808–812.
Nesvorný D., Youdin A. N., Marschall R., and Richardson D. C. (2020) Superparticle method for simulating collisions. *Astrophys. J., 895,* 63.
Nesvorný D., Li R., Simon J. B., Youdin A. N., Richardson D. C., Marschall R., and Grundy W. M. (2021) Binary planetesimal formation from gravitationally collapsing pebble clouds. *Planet. Sci. J., 2,* 27.
Noll K. S., Grundy W. M., Chiang E. I., Margot J.-L., and Kern S. D. (2008) Binaries in the Kuiper belt. In *The Solar System Beyond Neptune* (M. A. Barucci et al., eds.), pp. 345–363. Univ. of Arizona, Tucson.
O'Brien D. P. and Greenberg R. (2003) Steady-state size distributions for collisional populations: Analytical solution with size-dependent strength. *Icarus, 164,* 334–345.
Okuzumi S., Tanaka H., and Sakagami M.-a. (2009) Numerical modeling of the coagulation and porosity evolution of dust aggregates. *Astrophys. J., 707,* 1247–1263.
Okuzumi S., Tanaka H., Kobayashi H., and Wada K. (2012) Rapid coagulation of porous dust aggregates outside the snow line: A pathway to successful icy planetesimal formation. *Astrophys. J., 752,* 106.
Onishi I. K. and Sekiya M. (2017) Planetesimal formation by an axisymmetric radial bump of the column density of the gas in a protoplanetary disk. *Earth Planets Space, 69,* 50.
Ono T., Muto T., Takeuchi T., and Nomura H. (2016) Parametric study of the Rossby wave instability in a two-dimensional barotropic disk. *Astrophys. J., 823,* 84.
Ormel C. W. and Cuzzi J. N. (2007) Closed-form expressions for particle relative velocities induced by turbulence. *Astron. Astrophys., 466,* 413–420.
Ormel C. W. and Spaans M. (2008) Monte Carlo simulation of particle interactions at high dynamic range: Advancing beyond the googol. *Astrophys. J., 684,* 1291–1309.
Ormel C. W., Spaans M., and Tielens A. G. G. M. (2007) Dust coagulation in protoplanetary disks: Porosity matters. *Astron. Astrophys., 461,* 215–232.
O'Rourke L., Heinisch P., Blum J., Fornasier S., Filacchione G., Van Hoang H., Ciarniello M., Raponi A., Gundlach B., Blasco R. A., Grieger B., Glassmeier K.-H., Küppers M., Rotundi A., Groussin O., Bockelée-Morvan D., Auster H.-U., Oklay N., Paar G., Perucha M. d. P. C., Kovacs G., Jorda L., Vincent J.-B., Capaccioni F., Biver N., Parker J. W., Tubiana C., and Sierks H. (2020) The Philae lander reveals low-strength primitive ice inside cometary boulders. *Nature, 586,* 697–701.
Paardekooper S.-J., McNally C. P., and Lovascio F. (2020) Polydisperse streaming instability — I. Tightly coupled particles and the terminal velocity approximation. *Mon. Not. R. Astron. Soc., 499,* 4223–4238.
Pan L., Padoan P., Scalo J., Kritsuk A. G., and Norman M. L. (2011) Turbulent clustering of protoplanetary dust and planetesimal formation. *Astrophys. J., 740,* 6.
Pan M. and Sari R. (2005) Shaping the Kuiper belt size distribution by shattering large but strengthless bodies. *Icarus, 173,* 342–348.
Parker A. H. and Kavelaars J. J. (2010) Destruction of binary minor planets during Neptune scattering. *Astrophys. J. Lett., 722,* L204–L208.
Paszun D. and Dominik C. (2006) The influence of grain rotation on the structure of dust aggregates. *Icarus, 182,* 274–280.
Pätzold M., Andert T. P., Hahn M., Barriot J.-P., Asmar S. W., Häusler B., Bird M. K., Tellmann S., Oschlisniok J., and Peter K. (2019) The nucleus of comet 67P/Churyumov-Gerasimenko — Part I: The global view — nucleus mass, mass-loss, porosity, and implications. *Mon. Not. R. Astron. Soc., 483,* 2337–2346.
Pérez L. M., Carpenter J. M., Chandler C. J., Isella A., Andrews S. M., Ricci L., Calvet N., Corder S. A., Deller A. T., Dullemond C. P., Greaves J. S., Harris R. J., Henning T., Kwon W., Lazio J., Linz H., Mundy L. G., Sargent A. I., Storm S., Testi L., and Wilner D. J. (2012) Constraints on the radial variation of grain growth in the AS 209 circumstellar disk. *Astrophys. J. Lett., 760,* L17.
Pérez L. M., Isella A., Carpenter J. M., and Chandler C. J. (2014) Large-scale asymmetries in the transitional disks of SAO *206462 and SR 21. Astrophys. J. Lett., 783,* L13.
Pérez S., Casassus S., Baruteau C., Dong R., Hales A., and Cieza L. (2019) Dust unveils the formation of a mini-Neptune planet in a protoplanetary ring. *Astron. J., 158,* 15.
Picogna G. and Kley W. (2015) How do giant planetary cores shape the dust disk?: HL Tauri system. *Astron. Astrophys., 584,* A110.
Pierens A. (2021) On the non-axisymmetric fragmentation of rings generated by the secular gravitational instability. *Mon. Not. R. Astron. Soc., 504,* 4522–4532.
Pinilla P., Birnstiel T., Ricci L., Dullemond C. P., Uribe A. L., Testi L., and Natta A. (2012) Trapping dust particles in the outer regions of protoplanetary disks. *Astron. Astrophys., 538,* A114.
Powell D., Murray-Clay R., Pérez L. M., Schlichting H. E., and Rosenthal M. (2019) New constraints from dust lines on the surface densities of protoplanetary disks. *Astrophys. J., 878,* 116.
Raettig N., Klahr H., and Lyra W. (2015) Particle trapping and streaming instability in vortices in protoplanetary disks. *Astrophys. J., 804,* 35.
Raettig N., Lyra W., and Klahr H. (2021) Pebble trapping in vortices: Three-dimensional simulations. *Astrophys. J., 913,* 92.
Ricci L., Testi L., Natta A., Neri R., Cabrit S., and Herczeg G. J. (2010) Dust properties of protoplanetary disks in the Taurus-Auriga star forming region from millimeter wavelengths. *Astron. Astrophys., 512,* A15.
Ricci L., Trotta F., Testi L., Natta A., Isella A., and Wilner D. J. (2012) The effect of local optically thick regions in the long-wave emission of young circumstellar disks. *Astron. Astrophys., 540,* A6.
Robinson J. E., Fraser W. C., Fitzsimmons A., and Lacerda P. (2020) Investigating gravitational collapse of a pebble cloud to form transneptunian binaries. *Astron. Astrophys., 643,* A55.
Ros K., Johansen A., Riipinen I., and Schlesinger D. (2019) Effect of nucleation on icy pebble growth in protoplanetary discs. *Astron. Astrophys., 629,* A65.
Rucska J. J. and Wadsley J. W. (2021) Streaming instability on different scales — I. Planetesimal mass distribution variability. *Mon. Not. R. Astron. Soc., 500,* 520–530.
Safronov V. S. (1972) *Evolution of the Protoplanetary Cloud and Formation of the Earth and Planets.* Keter, Jerusalem. 212 pp.
San Sebastián I. L., Dolff A., Blum J., Parisi M. G., and Kothe S. (2020) The tensile strength of compressed dust samples and the catastrophic disruption threshold of pre-planetary matter. *Mon. Not. R. Astron. Soc., 497,* 2418–2424.
Schäfer U., Yang C.-C., and Johansen A. (2017) Initial mass function of planetesimals formed by the streaming instability. *Astron. Astrophys., 597,* A69.
Schäfer U., Johansen A., and Banerjee R. (2020) The coexistence of the streaming instability and the vertical shear instability in protoplanetary disks. *Astron. Astrophys., 635,* A190.
Schaffer N., Yang C.-C., and Johansen A. (2018) Streaming instability of multiple particle species in protoplanetary disks. *Astron. Astrophys., 618,* A75.
Schaffer N., Johansen A., and Lambrechts M. (2021) Streaming instability of multiple particle species. II. Numerical convergence with increasing particle number. *Astron. Astrophys., 653,* A14.

Schlichting H. E. and Sari R. (2008) The ratio of retrograde to prograde orbits: A test for Kuiper belt binary formation theories. *Astrophys. J.*, *686*, 741–747.
Schlichting H. E. and Sari R. (2011) Runaway growth during planet formation: Explaining the size distribution of large Kuiper belt objects. *Astrophys. J.*, *728*, 68.
Schlichting H. E., Fuentes C. I., and Trilling D. E. (2013) Initial planetesimal sizes and the size distribution of small Kuiper belt objects. *Astron. J.*, *146*, 36.
Schoonenberg D. and Ormel C. W. (2017) Planetesimal formation near the snowline: In or out? *Astron. Astrophys.*, *602*, A21.
Schräpler R., Blum J., Krijt S., and Raabe J.-H. (2018) The physics of protoplanetary dust agglomerates. X. High-velocity collisions between small and large dust agglomerates as a growth barrier. *Astrophys. J.*, *853*, 74.
Schwartz S. R., Michel P., Jutzi M., Marchi S., Zhang Y., and Richardson D. C. (2018) Catastrophic disruptions as the origin of bilobate comets. *Nature Astron.*, *2*, 379–382.
Sekiya M. and Onishi I. K. (2018) Two key parameters controlling particle clumping caused by streaming instability in the dead-zone dust layer of a protoplanetary disk. *Astrophys. J.*, *860*, 140.
Shadmehri M., Oudi R., and Rastegarzadeh G. (2019) Linear analysis of the non-axisymmetric secular gravitational instability. *Mon. Not. R. Astron. Soc.*, *487*, 5405–5415.
Shakura N. I. and Syunyaev R. A. (1973) Black holes in binary systems: Observational appearance. *Astron. Astrophys.*, *24*, 337–355.
Simon J. B. and Armitage P. J. (2014) Efficiency of particle trapping in the outer regions of protoplanetary disks. *Astrophys. J.*, *784*, 15.
Simon J. B., Hawley J. F., and Beckwith K. (2011) Resistivity-driven state changes in vertically stratified accretion disks. *Astrophys. J.*, *730*, 94.
Simon J. B., Armitage P. J., Li R., and Youdin A. N. (2016) The mass and size distribution of planetesimals formed by the streaming instability. I. The role of self-gravity. *Astrophys. J.*, *822*, 55.
Simon J. B., Armitage P. J., Youdin A. N., and Li R. (2017) Evidence for universality in the initial planetesimal mass function. *Astrophys. J. Lett.*, *847*, L12.
Singer K. N., McKinnon W. B., Gladman B., Greenstreet S., Bierhaus E. B., Stern S. A., Parker A. H., Robbins S. J., Schenk P. M., Grundy W. M., Bray V. J., Beyer R. A., Binzel R. P., Weaver H. A., Young L. A., Spencer J. R., Kavelaars J. J., Moore J. M., Zangari A. M., Olkin C. B., Lauer T. R., Lisse C. M., Ennico K., the New Horizons Geology, Geophysics and Imaging Science Theme Team, the New Horizons Surface Composition Science Theme Team, and the New Horizons Ralph and LORRI Teams (2019) Impact craters on Pluto and Charon indicate a deficit of small Kuiper belt objects. *Science*, *363*, 955–959.
Skorov Y. and Blum J. (2012) Dust release and tensile strength of the non-volatile layer of cometary nuclei. *Icarus*, *221*, 1–11.
Squire J. and Hopkins P. F. (2018) Resonant drag instabilities in protoplanetary discs: The streaming instability and new, faster growing instabilities. *Mon. Not. R. Astron. Soc.*, *477*, 5011–5040.
Squire J. and Hopkins P. F. (2020) Physical models of streaming instabilities in protoplanetary discs. *Mon. Not. R. Astron. Soc.*, *498*, 1239–1251.
Stammler S. M., Birnstiel T., Panić O., Dullemond C. P., and Dominik C. (2017) Redistribution of CO at the location of the CO ice line in evolving gas and dust disks. *Astron. Astrophys.*, *600*, A140.
Stammler S. M., Drążkowska J., Birnstiel T., Klahr H., Dullemond C. P., and Andrews S. M. (2019) The DSHARP rings: Evidence of ongoing planetesimal formation? *Astrophys. J. Lett.*, *884*, L5.
Stern S. A. and Colwell J. E. (1997) Accretion in the Edgeworth-Kuiper belt: Forming 100–1000 km radius bodies at 30 AU and beyond. *Astron. J.*, *114*, 841–884.
Stern S. A., Keeney B., Singer K. N., White O., Hofgartner J. D., Grundy W., and the New Horizons Team (2021) Some new results and perspectives regarding the Kuiper belt object Arrokoth's remarkable, bright neck. *Planet. Sci. J.*, *2*, 87.
Takahashi S. Z. and Inutsuka S.-i. (2014) Two-component secular gravitational instability in a protoplanetary disk: A possible mechanism for creating ring-like structures. *Astrophys. J.*, *794*, 55.
Takahashi S. Z. and Inutsuka S.-i. (2016) An origin of multiple ring structure and hidden planets in HL Tau: A unified picture by secular gravitational instability. *Astron. J.*, *152*, 184.
Takeuchi T. and Lin D. N. C. (2002) Radial flow of dust particles in accretion disks. *Astrophys. J.*, *581*, 1344–1355.
Taki T., Fujimoto M., and Ida S. (2016) Dust and gas density evolution at a radial pressure bump in protoplanetary disks. *Astron. Astrophys.*, *591*, A86.
Tazzari M., Testi L., Ercolano B., Natta A., Isella A., Chandler C. J., Pérez L. M., Andrews S., Wilner D. J., Ricci L., Henning T., Linz H., Kwon W., Corder S. A., Dullemond C. P., Carpenter J. M., Sargent A. I., Mundy L., Storm S., Calvet N., Greaves J. A., Lazio J., and Deller A. T. (2016) Multiwave-length analysis for interferometric (sub-)mm observations of protoplanetary disks: Radial constraints on the dust properties and the disk structure. *Astron. Astrophys.*, *588*, A53.
Testi L., Birnstiel T., Ricci L., Andrews S., Blum J., Carpenter J., Dominik C., Isella A., Natta A., Williams J. P., and Wilner D. J. (2014) Dust evolution in protoplanetary disks. In *Protostars and Planets VI* (H. Beuther et al., eds.), pp. 339–362. Univ. of Arizona, Tucson.
Tominaga R. T., Inutsuka S.-i., and Takahashi S. Z. (2018) Non-linear development of secular gravitational instability in protoplanetary disks. *Publ. Astron. Soc. Japan*, *70*, 3.
Tominaga R. T., Takahashi S. Z., and Inutsuka S.-i. (2020) Secular gravitational instability of drifting dust in protoplanetary disks: Formation of dusty rings without significant gas substructures. *Astrophys. J.*, *900*, 182.
Toomre A. (1964) On the gravitational stability of a disk of stars. *Astrophys. J.*, *139*, 1217–1238.
Tripathi A., Andrews S. M., Birnstiel T., and Wilner D. J. (2017) A millimeter continuum size-luminosity relationship for protoplanetary disks. *Astrophys. J.*, *845*, 44.
Umurhan O. M., Estrada P. R., and Cuzzi J. N. (2020) Streaming instability in turbulent protoplanetary disks. *Astrophys. J.*, *895*, 4.
Vaccaro E., Wozniakiewicz P. J., Starkey N. A., Franchi I. A., and Russell S. S. (2015) Grain size distribution in the matrix of primitive meteorites. In *78th Annual Meeting of the Meteoritical Society*, Abstract #5258. LPI Contribution No. 1856, Lunar and Planetary Institute, Houston.
van der Marel N., van Dishoeck E. F., Bruderer S., Birnstiel T., Pinilla P., Dullemond C. P., van Kempen T. A., Schmalzl M., Brown J. M., Herczeg G. J., Mathews G. S., and Geers V. (2013) A major asymmetric dust trap in a transition disk. *Science*, *340*, 1199–1202.
Veronesi B., Ragusa E., Lodato G., Aly H., Pinte C., Price D. J., Long F., Herczeg G. J., and Christiaens V. (2020) Is the gap in the DS Tau disc hiding a planet? *Mon. Not. R. Astron. Soc.*, *495*, 1913–1926.
Voelk H. J., Jones F. C., Morfill G. E., and Roeser S. (1980) Collisions between grains in a turbulent gas. *Astron. Astrophys.*, *85*, 316–325.
Wada K., Tanaka H., Suyama T., Kimura H., and Yamamoto T. (2009) Collisional growth conditions for dust aggregates. *Astrophys. J.*, *702*, 1490–1501.
Wahlberg Jansson K. and Johansen A. (2014) Formation of pebble-pile planetesimals. *Astron. Astrophys.*, *570*, A47.
Wahlberg Jansson K. and Johansen A. (2017) Radially resolved simulations of collapsing pebble clouds in protoplanetary discs. *Mon. Not. R. Astron. Soc.*, *469*, S149–S157.
Wahlberg Jansson K., Johansen A., Bukhari Syed M., and Blum J. (2017) The role of pebble fragmentation in planetesimal formation. II. Numerical simulations. *Astrophys. J.*, *835*, 109.
Ward W. R. (1976) The formation of the solar system. In *Frontiers of Astrophysics* (E. H. Avrett, ed.), pp. 1–40. Harvard Univ., Cambridge.
Ward W. R. (2000) On planetesimal formation: The role of collective particle behavior. In *Origin of the Earth and Moon* (R. M. Canup et al., eds.), pp. 75–84. Univ. of Arizona, Tucson.
Weidenschilling S. J. (1977) Aerodynamics of solid bodies in the solar nebula. *Mon. Not. R. Astron. Soc.*, *180*, 57–70.
Weidenschilling S. J. (1980) Dust to planetesimals: Settling and coagulation in the solar nebula. *Icarus*, *44*, 172–189.
Weidenschilling S. J. (1997) The origin of comets in the solar nebula: A unified model. *Icarus*, *127*, 290–306.
Weidenschilling S. J. and Ruzmaikina T. V. (1994) Coagulation of grains in static and collapsing protostellar clouds. *Astrophys. J.*, *430*, 713–726.
Weidling R., Güttler C., Blum J., and Brauer F. (2009) The physics of protoplanetesimal dust agglomerates. III. Compaction in multiple collisions. *Astrophys. J.*, *696*, 2036–2043.
Weissman P., Morbidelli A., Davidsson B., and Blum J. (2020) Origin

and evolution of cometary nuclei. *Space Sci. Rev., 216,* 6.
Whipple F. L. (1972) On certain aerodynamic processes for asteroids and comets. In *From Plasma to Planet* (A. Elvius, ed.), pp. 211–232. Wiley, New York.
Windmark F., Birnstiel T., Güttler C., Blum J., Dullemond C. P., and Henning T. (2012a) Planetesimal formation by sweep-up: How the bouncing barrier can be beneficial to growth. *Astron. Astrophys., 540,* A73.
Windmark F., Birnstiel T., Ormel C. W., and Dullemond C. P. (2012b) Breaking through: The effects of a velocity distribution on barriers to dust growth. *Astron. Astrophys., 544,* L16.
Windmark F., Birnstiel T., Ormel C. W., and Dullemond C. P. (2012c) Breaking through: The effects of a velocity distribution on barriers to dust growth (corrigendum). *Astron. Astrophys., 548,* C1.
Xu Z. and Bai X.-N. (2022a) Dust settling and clumping in MRI-turbulent outer protoplanetary disks. *Astrophys. J., 924,* 3.
Xu Z. and Bai X.-N. (2022b) Turbulent dust-trapping rings as efficient sites for planetesimal formation. Turbulent dust-trapping rings as efficient sites for planetesimal formation. *Astrophys. J. Lett., 937,* L4.
Yang C.-C. and Johansen A. (2014) On the feeding zone of planetesimal formation by the streaming instability. *Astrophys. J., 792,* 86.
Yang C.-C., Johansen A., and Carrera D. (2017) Concentrating small particles in protoplanetary disks through the streaming instability. *Astron. Astrophys., 606,* A80.
Yang C.-C., Mac Low M.-M., and Johansen A. (2018) Diffusion and concentration of solids in the dead zone of a protoplanetary disk. *Astrophys. J., 868,* 27.
Youdin A. N. (2005) Planetesimal formation without thresholds. I: Dissipative gravitational instabilities and particle stirring by turbulence. *ArXiV e-prints,* arXiv:astro-ph/0508659.
Youdin A. N. (2011) On the formation of planetesimals via secular gravitational instabilities with turbulent stirring. *Astrophys. J., 731,* 99.
Youdin A. N. and Goodman J. (2005) Streaming instabilities in protoplanetary disks. *Astrophys. J., 620,* 459–469.
Youdin A. and Johansen A. (2007) Protoplanetary disk turbulence driven by the streaming instability: Linear evolution and numerical methods. *Astrophys. J., 662,* 613–626.
Zhu Z. and Yang C.-C. (2021) Streaming instability with multiple dust species — I. Favourable conditions for the linear growth. *Mon. Not. R. Astron. Soc., 501,* 467–482.
Zsom A. and Dullemond C. P. (2008) A representative particle approach to coagulation and fragmentation of dust aggregates and fluid droplets. *Astron. Astrophys., 489,* 931–941.
Zsom A., Ormel C. W., Güttler C., Blum J., and Dullemond C. P. (2010) The outcome of protoplanetary dust growth: Pebbles, boulders, or planetesimals? II. Introducing the bouncing barrier. *Astron. Astrophys., 513,* A57.

Part 2:

Orbital Dynamics of Comets

Kaib N. A. and Volk K. (2024) Dynamical population of comet reservoirs. In *Comets III* (K. J. Meech, M. R. Combi, D. Bockelée-Morvan, S. N. Raymond, and M. E. Zolensky, eds.), pp. 97–120. Univ. of Arizona, Tucson, DOI: 10.2458/azu_uapress_9780816553631-ch004.

Dynamical Population of Comet Reservoirs

N. A. Kaib
University of Oklahoma (now at Planetary Science Institute)

Kathryn Volk
Planetary Science Institute/University of Arizona

The Oort cloud and the scattered disk are the two primary reservoirs for long-period and short-period comets, respectively. In this review, we assess the known observational constraints on these reservoirs' properties and their formation. In addition, we discuss how the early orbital evolution of the giant planets generated the modern scattered disk from the early, massive planetesimal disk and how ~5% of this material was captured into the Oort cloud. We review how the Sun's birth environment and dynamical history within the Milky Way alters the formation and modern structure of the Oort cloud. Finally, we assess how the coming decade's anticipated observing campaigns may provide new insights into the formation and properties of the Oort cloud and scattered disk.

1. DISCERNING THE COMET RESERVOIRS

Comets make up one of the major populations of small bodies in the solar system, and their passages near Earth and the Sun have been documented by astronomers for millennia (*Kronk,* 1999). However, these bodies are known to have relatively short lifespans both dynamically, due to orbital instability, and physically, due to the volatile loss and surface dust deposition that occurs during each near-Sun perihelion passage (*Weissman,* 1980; *Rickman et al.,* 1990; *Levison and Duncan,* 1994). As a result, it has long been assumed that the short-lived modern population of comets are supplied from more populous distant reservoirs whose members have much longer stability timescales.

Until 1992, Pluto was the most distant known object in the solar system, and the nature of comet reservoirs had to be inferred solely from the properties of the more proximate but transient population of comets passing near Earth. Historically, the catalog of known comets has been broadly divided into two groups based on orbital period: short-period comets (SPCs) whose orbital periods are less than 200 yr, and long-period comets (LPCs) whose orbital periods are greater than 200 yr. It has long been recognized that the orbital inclinations of comets with large periods are nearly isotropic, while SPCs have an inclination distribution strongly concentrated toward prograde ecliptic orbits (e.g., *Everhart,* 1972; *Marsden,* 2009). This fundamental difference in the orbital distribution of SPCs and LPCs has fueled decades of dynamical studies attempting to understand how the supply of comets in the inner solar solar system is replenished and what the source reservoirs of this replenishment must look like (see, e.g., historical reviews by *Davies et al.,* 2008; *Fernández,* 2020).

1.1 An Oort Comet Cloud

Among the LPCs, the distribution of the inverse semimajor axes (1/a) of the comets is a representation of their orbital energy distribution, and *Oort* (1950) noted an overabundance of comets with $0 < 1 = a < 0:0001$ au^{-1}, or $a > 10^4$ au. *Oort* (1950) posited that the comets belonging to this overabundance (now known as the Oort spike) are bodies that are making their first perihelion passage near the Sun, while LPCs with smaller semimajor axes are on subsequent passages, their orbital energies having been drawn down via perturbations from Jupiter during previous perihelion passages. Jovian perturbations can also increase the orbital energies of new LPCs during their initial perihelion passage, but these comets are so weakly bound that this process permanently ejects them from the solar system on hyperbolic orbits.

To explain a continuous supply of new LPCs, *Oort* (1950) proposed that the solar system is surrounded by a vast cloud of ~10^{11} comet-sized bodies extending to

~10^5 au. In this scenario, LPCs passing near Earth on extreme eccentricities comprise a tiny fraction of the bodies within this cloud, whose orbits have semimajor axes of at least thousands of astronomical units but whose eccentricities range all the way from 0 to 1. If this cloud were dynamically inert, its high-eccentricity (low-perihelion) portion should have been depleted long ago due to hyperbolic ejection events during perihelion passages. However, at such large distances from the Sun, perturbations external to the solar system become important. *Öpik* (1932) noted that impulses from passing stars will alter the perihelia (and therefore eccentricities) of distant heliocentric orbits. This effect allows new Oort cloud bodies to be brought into the inner solar system as previous LPCs are lost through ejection by the giant planets, namely Jupiter. Moreover, these same perturbations will randomize the directions of orbital angular momenta, leading to an isotropic distribution of LPCs and a spherical structure to the cloud. Decades later it was realized that the tide of the Milky Way's disk (generated by the vertical distribution of matter within the disk) will also provide perihelion and inclination shifts in Oort cloud bodies that are, on average, larger than those due to passing stars (*Morris and Muller,* 1986; *Heisler and Tremaine,* 1986). This perturbative force will also steadily replenish the supply of LPCs from lower-eccentricity Oort cloud bodies.

1.2. The Edgeworth-Kuiper Belt and Scattered Disk

While the seminal work of *Oort* (1950) outlined most of the broad features of LPC dynamics and their distant reservoir, the origin of SPCs still remained a puzzle due to their lower-inclination orbits. *Everhart* (1972) argued that the observed SPC population could be produced from multiple rounds of jovian perturbations that lower the semimajor axes of low-inclination LPCs. However, such a process appeared to be orders of magnitude too inefficient (*Joss,* 1973; *Kresak and Pittich,* 1978) and would require an extremely massive Oort cloud (*Fernandez,* 1980b) to explain the abundance of SPCs.

Meanwhile, it had long been speculated that the falling surface density of solid material with increasing heliocentric distance in the early outer solar system would produce a belt of icy planetesimals that were never incorporated into fully formed planets (*Edgeworth,* 1943, 1949; *Kuiper,* 1951). While *Kuiper* (1951) assumed that perturbations from Pluto (whose mass was then thought to be comparable to Earth's) had dispersed this belt into the Oort cloud, *Edgeworth* (1943) proposed that members of this belt occasionally visit the inner solar system as comets. In another study, *Everhart* (1977) found that some Oort cloud bodies with perihelia near Neptune will eventually be scattered into Jupiter-crossing orbits and evolve into SPCs. *Fernandez* (1980b) realized that a belt of small bodies residing beyond Neptune would naturally supply near-Neptune perihelia and could generate a population of SPCs on its own. However, it was assumed that a reservoir completely comprised of Neptune-encountering orbits would be quickly depleted over the age of the solar system due to its inherent orbital instability.

Duncan et al. (1988) used numerical simulations to show that the inclinations of Oort cloud bodies are approximately conserved as their semimajor axes are decreased via encounters with the giant planets, and they concluded that the Oort cloud cannot be the main source of SPCs. An additional set of simulations showed that a low-inclination reservoir of Neptune-crossing bodies generates a low-inclination population of SPCs similar to the observed population, and these results were refined and confirmed in *Quinn et al.* (1990). Thus, it seemed that the population of SPCs required the existence of a second reservoir of bodies in a belt beyond Neptune.

The mechanism that injected members of this belt into the comet population was still unclear, however. Planetesimal-planetesimal scattering suggested by *Fernandez* (1980b) required such massive belt members that they would not have evaded detection. *Torbett* (1989) showed instead that chaos resulting from the gravitational effects of the four giant planets could raise the eccentricities of initially more circular orbits to slowly supply Neptune-encountering bodies, and therefore comets, from this region. Direct integrations showed that a similarly slow destabilization process takes place near the boundaries of mean-motion resonances with Neptune (*Duncan et al.,* 1995). However, additional work in *Levison and Duncan* (1997) found that most regions of orbital space of the conventional Kuiper belt (low-eccentricity orbits from ~40–50 au) either destabilize quickly ($\lesssim 10^8$ yr) or are stable over timescales longer than the age of the solar system. However, *Duncan and Levison* (1997) noted that a small subset (~5%) of Kuiper belt objects that have a close encounter with Neptune will attain ~billion-year dynamical lifetimes. They achieve these long dynamical lifetimes by occupying eccentric orbits whose perihelia have been temporarily lifted away from Neptune, typically within mean-motion resonances. With a dynamical lifetime comparable to the solar system's age, these "scattered disk" bodies are a more efficient producer of SPCs compared to the classical Kuiper belt. Thus, the subpopulation of transneptunian objects (TNOs) known as the scattered disk is widely thought to be the main source of SPCs.

Before TNOs make it all the way into the inner solar system, they must pass through the giant planet region as Centaurs, the intermediary small body population connecting SPCs and the scattered disk (e.g., *Duncan and Levison,* 1997; reviewed recently by *Peixinho et al.,* 2020). The exact orbital definition of Centaurs differs in the literature (e.g., *Gladman et al.,* 2008; *Jewitt,* 2009), but the general criteria is that they are small bodies on giant-planet-crossing orbits that are dynamically unstable on timescales much shorter than the age of the solar system.

The first discovered object widely recognized as belonging to the new class of objects later dubbed the Centaurs was (2060) Chiron (*Kowal and Gehrels,* 1977). It was quickly realized that Chiron's orbit is dynamically unstable (e.g., *Oikawa and Everhart,* 1979; *Scholl,* 1979). More general studies of orbital stability in the giant planet region reveal that Centaurs have typical dynamical lifetimes on the order of 1–10 m.y. (e.g., *Gladman and Duncan,* 1990; *Holman and Wisdom,* 1993; *Tiscareno and Malhotra,* 2003; *Di Sisto and Rossignoli,* 2020). Centaurs thus are a transition population for comets enroute from the transneptunian region (e.g., *Stern and Campins,* 1996) rather than a reservoir. However, not all Centaurs become comets. Centaurs spend most of their time in the outer, ice giant region, with only about half making it to the Jupiter-Saturn region and about a third evolving onto SPC orbits (e.g., *Tiscareno and Malhotra,* 2003; *Sarid et al.,* 2019). The chapter by Fraser et al. in this volume discusses the evolution of Centaurs into comets in more detail.

1.3. Bodies with Ambiguous Sources

There are a few groups of comets/comet predecessors that do not fall neatly along the long- or short-period divide and are not easily produced by models of Oort cloud or scattered disk delivery. In addition to orbital period, comets are often divided into subgroups based on their Tisserand parameter with respect to Jupiter, T_J

$$T_J = \frac{a_J}{a} + 2\sqrt{\frac{a}{a_J}(1-e^2)}\cos i \qquad (1)$$

where a_J and a are the semimajor axes of Jupiter and the comet, respectively, e is the eccentricity of the comet's orbit, and i is its orbital inclination. This parameter is related to a comet's Jacobi constant, which is a strictly conserved quantity in the restricted three-body problem (Sun + circular Jupiter + comet) and is approximately conserved on 1000-yr timescales for SPCs in the real solar system [*Levison and Duncan* (1994); see also, e.g., *Murray and Dermott* (1999) for a full discussion of the Jacobi and Tisserand parameters]. It has long been noted that the vast majority of SPCs have T_J slightly below 3, indicating a tight dynamical coupling with the gas giant (*Kresák,* 1972), and *Levison and Duncan* (1997) demonstrated how dynamical transfer from the scattered disk generates this tight distribution of T_J. Based on these results, *Levison* (1996) argued that the Tisserand parameter should be the main criterion upon which to base comet classification. In particular, *Levison and Duncan* (1997) argue that comets with $2 < T_J < 3$ should be considered Jupiter-family comets and are consistent with a Kuiper belt origin.

1.3.1. Halley-type comets. Figure 1 shows the current known population of SPCs, color coded by T_J; it is clear that not all comets are confined in the tight cluster of $T_J = 2–3$. In particular, there exists a subgroup of SPCs with $T_J < 2$,which *Carusi et al.* (1987) designated Halley-type comets (or HTCs). The driver for their lower $T_J < 2$ values are their generally significantly larger orbital inclinations (up to and including retrograde orbits) as well as moderately

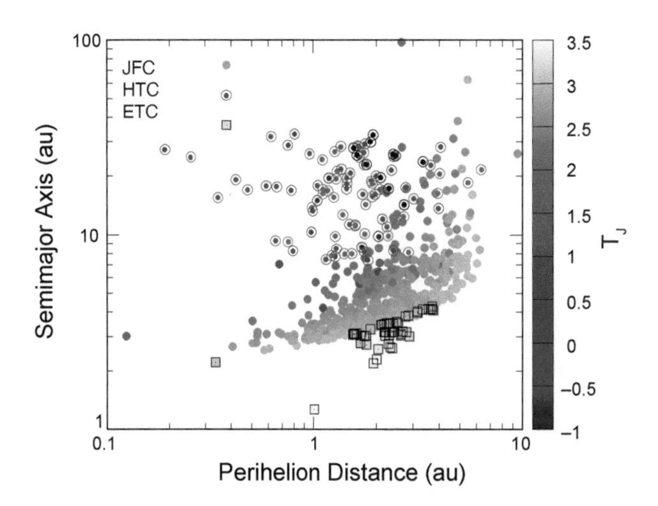

Fig. 1. Semimajor axis vs perihelion distance for all Jupiter-family (JFC; circles), Halley-type (HTC; triangles), and Encke-type (ETC; squares) comets listed by JPL (queried from *https://ssd.jpl.nasa.gov/sbdb query.cgi* on August 22, 2021). The comets are color-coded by Tisserand parameter with respect to Jupiter (T_J) to highlight the separations between the different dynamical classes of short-period comets.

larger semimajor axes. The ultimate source of these comets has been debated, with *Levison et al.* (2001) suggesting that HTCs are consistent with an origin from a flattened interior of the Oort cloud, *Levison et al.* (2006a) arguing for a scattered disk origin whose most distant members are pushed inward via the galactic tide (which also isoptropizes ecliptic inclinations), and *Nesvorný et al.* (2017) arguing that HTCs are consistent with returning LPCs from the Oort cloud whose semimajor axes have been drawn down through repeated planetary encounters. *Fernández et al.* (2018) find that HTC orbits can be produced from integrations of inactive Centaurs with perihelia in the Jupiter-Saturn region, suggesting that the lack of observed cometary activity could be due to a long residence time at these heliocentric distances that depleted surface volatiles. While *Fernández et al.* (2018) assume a scattered disk origin for all the Centaurs, perhaps this is also consistent with the *Nesvorný et al.* (2017) returning LPC scenario. Additional modeling and observations are likely necessary to test these different proposed HTC origins.

1.3.2. Encke-type comets. On the opposite side of the T_J divide from the JFCs are the Encke-type comets (ETCs); these comets typically have T_J just above 3 with low semimajor axes that slightly decouple the comets from Jupiter compared to JFCs (see Fig. 1). *Levison et al.* (2006b) investigated the origins of comet Encke and found that such orbits can be part of the standard model for JFC evolution, but that it is a rare outcome that typically required 1.5×10^5 perihelion passages in the inner solar system to achieve. This long timespan is problematic for producing active comets like the observed ETCs because comets are expected to fade (either via disruption or loss of volatiles) on timescales on the order of a few tens to a few hundred perihelion passages (e.g., *Levison and Duncan,* 1997; *Brasser and Schwamb,* 2015). *Levison et al.* (2006b) suggest that Encke could perhaps have built up an inactive surface layer during this time that was then shed after a dramatic drop in q, but it remains unclear whether this kind of evolution can plausibly explain the entire ETC population. It is possible that the production of ETCs depends on more than just purely gravitational dynamics. Most dynamical models of how SPCs are supplied from their source regions neglect nongravitational forces such as outgassing, but these might be important for the ETCs. *Fernández et al.* (2002) found a higher rate of ETC production when such forces were included in the model, although they found they needed to be sustained over a relatively long timeframe. Future models of ETC production with more realistic nongravitational forces would likely help shed light on this question.

1.3.3. Highly-inclined Centaurs. While the majority of Centaurs have low-to-moderate inclinations consistent with production from the scattered disk enroute to the JFC population, there exist a number of very-high-inclination and retrograde Centaurs and TNOs with perihelia in the giant planet region (e.g., *Gladman et al.,* 2009; *Chen et al.,* 2016). Much like the high-inclination HTCs, the origin of these highly inclined objects is unclear. It has been proposed that such bodies could originate from the Oort cloud (*Brasser et al.,* 2012b; *Kaib et al.,* 2019). In this scenario, galactic perturbations drive the perihelia of Oort cloud bodies (whose inclination distribution is nearly isotropic) into the giant planet region, at which point energy kicks from planetary perturbations can draw these bodies' semimajor axes to smaller values (*Emel'yanenko et al.,* 2005; *Kaib et al.,* 2009). An alternative explanation for high-inclination Centaurs is that they are a signature of a distant undetected planet that perturbs the inclinations of scattered bodies (*Gomes et al.,* 2015; *Batygin and Brown,* 2016b). However, such a mechanism may generate more high-inclination Centaurs than are actually observed (*Kaib et al.,* 2019) and has other, observable consequences for the distribution of TNOs (*Shankman et al.,* 2017b). Another recently proposed origin for this population is outward scattering of objects from the asteroid populations in the inner solar system, which can produce very high-inclination and retrograde orbits (*Greenstreet et al.,* 2020). This last hypothesis would have interesting (and testable) compositional implications for the highest-inclination outer solar system objects.

2. OBSERVATIONAL CONSTRAINTS ON RESERVOIRS

2.1. Short-Period Comet Population

Figure 2 shows the current inventory of solar system objects with orbital periods less than 200 yr classified as comets by JPL. There are obviously observational biases in this sample, but it highlights a few important features of the SPC population. First, note the large spread of inclinations

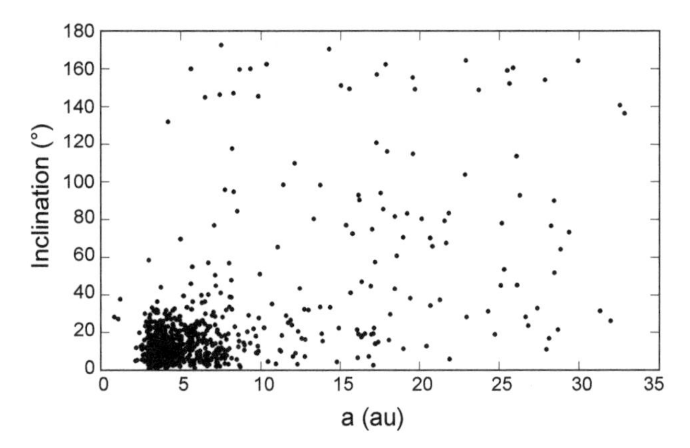

Fig. 2. Inclination vs. semimajor axis distribution of observed cometary objects with orbital periods shorter than 200 yr (the standard definition of SPCs; queried from JPL). The concentration of comets with low-a and low-i orbits highlight the JFC population.

for objects with a $\gtrsim$ 5 au; many of the high-inclination objects in this group are the HTCs discussed in section 1.3.1, whose dominant source population remains an active area of research. Second, note that the vast majority of comets with semimajor axes near or inside Jupiter's orbit (the JFCs) have inclinations smaller than ~30°. This concentration of inclinations relatively close to the ecliptic plane was the first notable evidence linking JFCs to the outer solar system's TNO populations (as discussed in section 1.2), which have a similar inclination distribution.

If we knew the exact number of SPCs in the inner solar system at the moment, we could use our dynamical models of the delivery of these comets to constrain the total number of comet-sized TNOs in the scattering population (the dominant supplier of these comets). However, this extrapolation is complicated by the fact that our observed sample of SPCs is biased toward active comets on easy-to-observe orbits. The bias toward active comets is particularly important because we have strong evidence that SPCs must "fade" (i.e., either become inactive and faint or be physically destroyed) on timescales much shorter than their dynamical lifetimes. The inclination distribution of the JFCs is the dominant evidence for this. Over their entire ~0.5-m.y. dynamical lifetimes in the inner solar system (*Levison and Duncan,* 1994), interactions with the planets tend to raise the orbital inclinations of JFCs to values larger than those seen in the observed population. To get the modeled JFC population to match the observed one in inclinations, *Brasser and Wang* (2015) estimate that comets could fade after as few as ~50 perihelion passages; *Nesvorný et al.* (2017) model fading as a size-dependent process and estimate that small comets remain visible for a few hundred perihelion passages (both models consider perihelion passages with q < 2.5 au). Our models of comet fading are thus still quite uncertain, making it difficult to extrapolate the current census of active JFCs into a total population of active and inactive JFCs. Additionally, the very fact that comets are active makes measuring their sizes more difficult because the gas and dust coma obscures the comet nucleus. Thus measurements of the size distribution of comets, necessary for connecting population estimates to their source region, can vary (e.g., *Lowry et al.,* 2008; *Snodgrass et al.,* 2011; *Fernández et al.,* 2013); this, combined with different assumptions about the ratio of active to dormant comets, means estimates of the current number of JFCs can vary by factors of several (see, e.g., discussion in *Nesvorný et al.,* 2017). These variable comet population estimates combined with variations in orbital models for their scattering population reservoir means that attempts to extrapolate from the number of observed JFCs to the number of comet-sized objects in the TNO region can differ enormously. Published estimates of the current population of the scattered disk based on supplying the current number of JFCs include 6×10^8 ["comet-sized" (*Duncan and Levison,* 1997)], $\sim 10^{10}$ [D > 1.4 km (*Emel'yanenko et al.,* 2004)], 0.5–1.1 $\times 10^9$ [D > 1 km (*Volk and Malhotra,* 2008)], $\sim 6 \times 10^9$ [D > 2.3 km (*Brasser and Wang,* 2015)], 4.4×10^8 [D > 2 km (*Nesvorný et al.,* 2017)]. While these estimates are not all for identical size ranges, it is clear that they vary by at least an order of magnitude.

2.2. Direct Observation of the Scattered Disk

Since the detection of the first scattering TNO (*Luu et al.,* 1997), hundreds of additional members of this population and the closely related high-a resonant and detached populations have been detected. Our picture of the high-a TNO population has evolved as observations built up. It became apparent fairly early that not all of the high-a TNOs had perihelion distances that brought them into direct gravitational range of Neptune; observations of objects with high perihelion distances ($q \gtrsim 40$ au) led to discussion of an "extended" scattered disk (e.g., *Gladman et al.,* 2002) or a dynamically "detached" population [the current more commonly used terminology (e.g., *Gladman et al.,* 2008)]. The important role of Neptune's resonances in the scattering population was noted in early dynamical models (*Levison and Duncan,* 1997), and their potential role in creating the detached population was also noted (e.g., *Gomes et al.,* 2005; *Gallardo,* 2006). As the orbits of distant TNOs have been measured more precisely, it is also apparent that there are large, stable populations of objects in many of Neptune's distant resonances (e.g., *Gladman et al.,* 2012; *Pike et al.,* 2015; *Volk et al.,* 2018) in addition to a large number of metastable objects temporarily stuck in resonances on a variety of timescales (e.g., *Lykawka and Mukai,* 2007; *Bannister et al.,* 2016; *Holman et al.,* 2018). Figure 3 shows the observed orbital distribution of TNOs in the scattering, a > 50 au, resonant, and detached populations.

It was apparent in the early TNO datasets that the higher-eccentricity TNOs (including the scattering objects) have a much wider range of orbital inclinations than the low-eccentricity "classical" KBOs belt objects, giving the total TNO population a bimodal inclination distribution (*Brown,* 2001). This gave rise to a division of the TNOs into dynamically "cold" and "hot" components, with the cold population being found only in the classical belt. The scattering, resonant, and detached populations all belong to the hot population. This division has been strengthened by subsequent surveys of the physical properties of TNOs. The photometric colors of hot and cold TNOs are distinct (e.g., *Tegler et al.,* 2003; *Pike et al.,* 2017; *Schwamb et al.,* 2019), implying that they have different surface compositions, perhaps reflecting different formation regions. The cold population also contains significantly more binary objects than the hot population; the few observed binary objects in the hot population tend to have a massive primary with small satellites rather than the prevalent equal-sized binaries in the cold population (see a recent review by *Noll et al.,* 2020). There was also early evidence that the hot population's size distribution differed from the cold population. *Bernstein et al.* (2004) found evidence for this difference, finding significantly fewer hot population objects at faint magnitudes than expected. Subsequent observational surveys have found additional evidence for differences between the hot and cold populations and for a "turnover" in the magnitude/size distribution from a steep slope for the bright/large objects to a shallower one at fainter/smaller objects (e.g., *Fraser and Kavelaars,* 2009; *Fuentes et al.,* 2009; *Fraser et al.,* 2010; *Shankman*

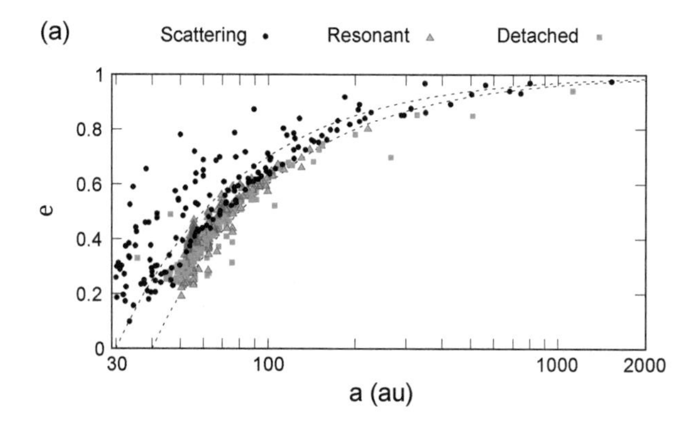

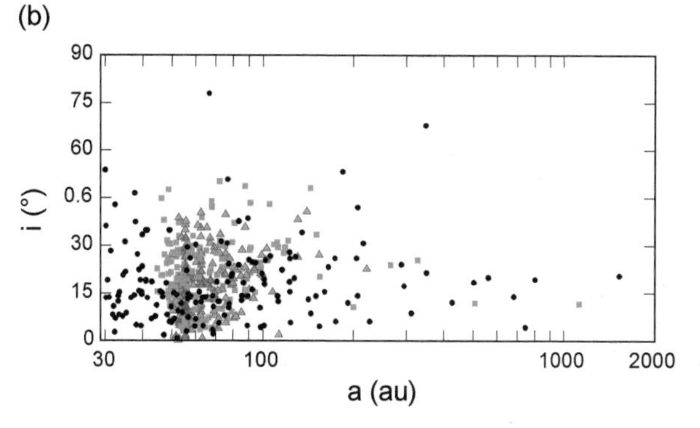

Fig. 3. Observed distribution of scattering, detached, and large-a (a > 50 au) resonant TNOs in **(a)** eccentricity and **(b)** inclination vs. semimajor axis. The dashed lines in **(a)** show constant perihelion distances of 30 and 40 au for reference. We include only objects with known dynamical classifications [we use the set of classified objects listed in *Smullen and Volk* (2020)] to highlight the overlaps between these three dynamical classes [see *Gladman et al.* (2008) for detailed definitions of these classes].

et al., 2013; *Fraser et al.,* 2014; *Kavelaars et al.,* 2021); however, the exact nature and location of this change has been difficult to constrain due to the strong observational biases and faintness of the TNOs. Recent results from the Outer Solar System Origins Survey (OSSOS) have shown that the scattering population of TNOs can be well fit with either a break or a "divot" (an actual decrease in the differential H magnitude distribution rather than just a change in slope) near $H_r \approx 8$–8.5 (e.g., *Shankman et al.,* 2013, 2017b; *Lawler et al.,* 2018b).

All these observations have led to the current view that the hot populations, which are the progenitors of the SPCs, formed closer to the Sun than the current cold classical region. They formed in the current giant planet region and were dispersed onto their current orbits during the epoch of giant planet migration (discussed in section 3). At present, observational estimates of the number of comet-sized objects in the scattering population are still fairly uncertain. In reflected sunlight, kilometer-sized bodies in the scattering population are far too faint for direct detection even at perihelion. Comet-sized population estimates thus must be extrapolated based on the size distribution of brighter bodies. But even the population estimates for these brighter bodies have large uncertainties because the observational biases in the scattering population are so strong [see, e.g., discussion of observational biases in *Gladman and Volk* (2021)]. Briefly, TNOs with large semimajor axes and large eccentricities spend most of their time at large heliocentric distances; they are typically only bright enough to be detected near their perihelia and thus over only a small portion of their orbits. Estimating the intrinsic number of scattering objects from the observed sample of objects requires either weighting each observed object by the inverse of its probability of detection based on its size and orbit (a traditional "debiasing" approach) or taking a forward-modeling approach and applying observational biases to a model of the scattering population and iterating that model until it gives you a match to the observed sample. The accuracy of both approaches rely to a large extent on how well the biases of a given observational survey are understood and measured as well as how well the observed objects sample the orbital distribution of the scattering population (this is discussed further in section 4.1). *Adams et al.* (2014) provide estimates of the scattered disk population based on debiasing the objects detected by the Deep Ecliptic Survey (*Elliot et al.,* 2005); they estimate that there are $\sim 10^4$ scattered objects with absolute magnitudes $H > 7.5$. Taking a forward-modeling approach, *Lawler et al.* (2018b) provide population estimates for the scattering population from OSSOS. They use the *Kaib et al.* (2011) model of the intrinsic orbital distribution of the scattering population [with a slightly modified inclination distribution to better match the observed distribution (see *Shankman et al.,* 2016)] with a variety of functional forms for the absolute magnitude distribution to find acceptable matches to the observed sample. From this, *Lawler et al.* (2018b) estimate that there are $(0.7–1) \times 10^5$ scattering objects with $H_r > 8.66$ (approximately $D > 100$ km; range covers all their H magnitude models); extrapolating down to $H_r < 12$ ($D \gtrsim 20$ km) yields population estimates of $\sim(2–3) \times 10^6$ objects. These are our current best estimates of the scattering population based on direct observations; while these estimates will likely improve with additional observations of the scattering population (see section 4.1), the inherent brightness limitations of TNO surveys will make it difficult to push the direct constraints on the *in situ* scattering population down to comet-sized bodies.

Two avenues of current research show promise for connecting the observational constraints on the size distribution of larger scattering objects to the population of comet-sized ones: occultation surveys (discussed in section 4.2) and analysis of cratering records in the outer solar system. In the latter case, data from the New Horizons mission provided a wealth of new data about the distribution of impact craters on TNOs Pluto and Arrokoth (see, e.g., *Singer et al.,* 2019; *Spencer et al.,* 2020). Impact crater distributions do not easily translate to *direct* constraints on the TNO size distribution. In addition to some uncertainties in crater scaling laws, converting crater sizes to impactor sizes requires knowing the impact speed distribution onto the surface in question, and this distribution depends on the assumed model for the orbital distributions and relative populations of the various dynamical classes of TNOs (see discussion in *Morbidelli et al.,* 2021); despite these uncertainties, the crater distributions offer our strongest current constraints on the distribution of the smallest TNOs. The scattering population does not dominate the impact flux at either Pluto or Arrokoth (*Greenstreet et al.,* 2015, 2019), so the crater size distribution does not directly constrain its total population at small sizes. However, the hot classical population of TNOs (which is expected to have similar origins as the scattering population; see section 3.2) does contribute to the cratering on both bodies, so the cratering record can offer insights into the size distribution of the scattering population even if not the absolute numbers; hopefully future detailed modeling can connect the implied combined TNO impactor distributions to population estimates for the comet-sized portion of the scattering population.

2.3. Direct Observation of Oort Cloud

With semimajor axes of thousands of astronomical units and perihelia decoupled from the solar system's planetary region, direct detection of Oort cloud objects is exceptionally difficult. To date, the TNO 2015 TG_{387} (*Sheppard et al.,* 2019) is the most likely candidate for a directly detected Oort cloud object (outside of the LPCs and other lower-perihelia bodies whose orbits have been substantially modified by recent planetary interactions). With a perihelion of 65 au, it is clearly very decoupled from scattering events with the giant planets. Its semimajor axis is 1170 au, meaning that it is sensitive to the stellar and galactic tidal perturbations that influence the rest of the Oort cloud (*Sheppard et al.,* 2019). With a sample of just one detected object, it is extremely difficult to make confident inferences about the nature of the Oort cloud. Given the steep dependence of discovery probability on semimajor axis, we do not have strong constraints on the semimajor axis distribution of the Oort cloud. In addition, while 2015 TG_{387} is estimated to have a diameter of 300 km, one cannot construct an accurate size distribution with a single object. Additional direct detections will be required to develop confident constraints on the Oort cloud's properties.

2.4. Long-Period Comet Flux

Because of the extreme distance to the Oort cloud, the vast majority of observational constraints come from observations of LPCs. In particular, the flux of dynamically new ($a > 10^{44}$ au) comets near Earth is one of the main observational parameters to constrain the cloud. The reason that comets with semimajor axes above 10^4 au are considered dynamically new is that a single passage through the gas giant region will typically deliver an energy kick strong enough to either eject such a comet onto a hyperbolic orbit or lower it to a much smaller semimajor axis.

For the last several decades, efforts have been made to measure the annual flux of dynamically new comets with the idea that such measurements will constrain the mass and population size of the Oort cloud. (A more massive or more populous cloud will generate larger numbers of LPCs.) While quantifying the flux, most works have sought to measure the number of new comets whose absolute magnitude, H_T, is brighter than some value. Here, H_T is given by

$$H_T = m - 2.5n \log r - 5 \log \Delta \tag{2}$$

where m is the total apparent magnitude, r is the distance to the Sun, Δ is the distance to Earth, and n is the comet's photometric index, or the power-law index defining the relationship between the comet's brightness and heliocentric distance. H_T is the absolute magnitude of a comet (nucleus and coma) when it is 1 au from Earth and the Sun. For a non-active body $n = 2$, but for a comet whose outgassing increases during a solar approach n is often higher. (If one assumes $n = 4$, as is often done, the absolute magnitude is noted as H_{10}.)

An oft-quoted metric is the flux of dynamically new comets per year with perihelia below 4 au (interior to Jupiter's orbit and potentially observable from Earth) and whose H_T is brighter than 11. *Everhart* (1967) found that ~20 new LPCs pass within 4 au of the Sun per year, but this result relied on the extrapolation of an uncertain comet magnitude distribution derived from historical comet searches with selection biases that were extremely difficult to quantify. Meanwhile, utilizing a greater number of detections from modern surveys *Francis* (2005) found a flux of only 2.9 per year, and *Fouchard et al.* (2017) found a flux of 3.6 per year.

2.4.1. Population of Oort cloud. To estimate the total population of the Oort cloud, many past works have examined the new LPC production rate within simulations of the Oort cloud and scaled this value to the observed LPC flux. *Kaib and Quinn* (2009) found that, on average, 1 in every $\sim 10^{11}$ Oort cloud bodies passes through perihelion within 4 au of the Sun per year. In contrast, *Fouchard et al.* (2014) found a lower rate of 1 in 5×10^{11}, but this may have been due to the higher level of central concentration in their Oort cloud semimajor axis distribution, which placed more bodies onto small semimajor axes with lower comet production rates (*Kaib and Quinn,* 2009). Taking a mean of the two results and assuming a real flux of three new comets per year with $q < 4$ au and $H_T < 11$ yields a total Oort cloud population of 7–8 $\times 10^{11}$ cometary bodies with $H_T < 11$.

2.4.2. Mass of Oort cloud. H_T is of course not an actual direct measure of a comet's nucleus, because a comet's brightness is dominated by the coma when it is at 1 au. To discuss the flux of comets of an actual size or mass, H_T has often been converted to a nuclear magnitude, which is then converted to a size/mass, but such a conversion is fraught with uncertainty (e.g., *Bailey and Stagg,* 1988; *Weissman,* 1996). Using mass- H_T relationships derived from observations of Comet Halley in combination with the LPC H_T distribution found in *Everhart* (1967), *Weissman* (1996) determined the average LPC mass to be 0.8–4 $\times 10^{16}$ g. This would imply an Oort cloud mass of 1–5 $M_\oplus$.

Alternatively, *Boe et al.* (2019) utilized Pan-STARRS1 LPC detections and applied a universal comet sublimation model to convert each cometary magnitude to nucleus radius. With this approach, they found that the LPC size distribution is a broken power-law with a steep bright-end index ($r^{-4.6}$) transitioning to a shallow faint-end index ($r^{-1.5}$) near an inferred comet radius of 2 km. This diminishes the relative contribution of subkilometer bodies to the mass of the Oort cloud, and *Boe et al.* (2019) estimated a total Oort cloud mass of 0.5–2 $M_\oplus$ in bodies larger than 1 km in radius.

Sosa and Fernández (2011) took yet a different approach and searched for a relationship between the observed water sublimation and non-gravitational forces altering comets' orbits to find the masses of LPCs. Using this relationship, they then derive one for the total visual magnitude and mass of a comet

$$\log R(\text{km}) = 0.9 - 0.13H_T \tag{3}$$

While this method was a unique and independent attempt to measure LPC nuclei, it relies upon several uncertain parameters, including an assumed bulk density of comets and a conversion of visual magnitude into water sublimation rate. The above mass-visual magnitude relationship was derived over a relatively narrow range of absolute magnitudes of 5–9, corresponding to radii of 0.5–1.8 km according to equation (3). Applying this conversion relation to a broader number of LPC discoveries, *Fernández and Sosa* (2012) inferred a comet size distribution comprised of three different power laws. However, extending their bright-end power law to infinity implies an infinite mass to the Oort cloud, so it must become steeper at some larger comet nucleus size beyond the range they study (r ~ 20 km).

2.4.3. Population ratio of the Oort cloud to the scattered disk. Historically, JFCs and LPCs have been used to estimate the population of comet-sized bodies within the scattered disk and Oort cloud, respectively. Taken at face value, these population estimates imply that there are nearly 3 orders of magnitude more such bodies in the Oort cloud than the scattered disk (*Levison et al.,* 2010). However, these estimates still largely rely on uncertain conversions of H_T to comet nuclear magnitudes. *Brasser and Morbidelli* (2013) attempted to limit the uncertainty by only considering LPCs and JFCs with inferred radii above 2.3 km, which largely avoided the previously discussed breaks in the LPC size distribution. With this approach, they found a substantially smaller Oort cloud-to-scattered disk population ratio of 44^{+54}_{-34}. In line with these results, *Nesvorný et al.* (2017) and *Vokrouhlický et al.* (2019) suggest that the sizes of the nuclei in LPCs may be overestimated and are typically just 300–400 m in diameter, rather than the kilometer sizes inferred in most works (e.g., *Francis,* 2005; *Fernández and Sosa,* 2012; *Boe et al.,* 2019). This would also lower the population of the Oort cloud and its ratio to the scattered disk (*Nesvorný et al.,* 2017).

3. BUILDING THE COMET RESERVOIRS

3.1. A Scattered Oort Cloud

3.1.1. A rough framework. Although the population of LPCs is very consistent with the existence of a distant spherical Oort cloud, they do not tell us how this cloud was formed. *Oort* (1950) realized, however, that the same stellar perturbations that produce LPCs from the Oort cloud can also explain the cloud's formation. If solid planetesimals are leftover after the formation of the planets, it is inevitable that many will eventually undergo close encounters with the giant planets, exchanging energy with the planets and driving a diffusion in the planetesimals' semimajor axes. As the planetesimals' semimajor axes diffuse, their perihelia effectively remain locked in the giant planet region, allowing energy exchanges to occur during each perihelion passage. Left unchecked, this process would eventually result in all the planetesimals either being ejected onto hyperbolic orbits or colliding with the Sun or a planet.

However, if planetesimals reach semimajor axes of thousands of astronomical units, the perturbations from passing stars begin to become significant (e.g., *Rickman,* 1976). If stellar perturbations can torque the planetesimals' perihelia beyond the planetary region this would save these planetesimals from planetary ejections onto hyperbolic orbits. These same perturbations would act to isotropize the orbital planes of planetesimals and could give rise to a distant spherical cloud of planetesimals similar to that inferred from LPC observations.

Thus, *Oort* (1950) hypothesized a basic framework for the formation of the Oort cloud, but at the time, the efficiency of such a process was unknown. While *Oort* (1950) assumed that the planetesimals occupying the Oort cloud originated between Mars and Jupiter, *Fernandez* (1978) noted that the probability that planetesimal-planet encounters result in a near-parabolic orbit rather than a hyperbolic one increases with planet-Sun distance. As a result, material scattering off Uranus and Neptune should be more likely to reach the Oort cloud than planetesimals closer to the Sun. This process was numerically verified in *Fernandez* (1980a), where particles were initiated on near parabolic orbits with perihelia between 20 and 30 au. They were numerically integrated as they scattered off of the giant planets while also being torqued by passing stars for 10^9 yr. In this timespan, over 50% of bodies were ejected while most survivors resided in a nearly isotropized cloud between 20,000 and 30,000 au. Extrapolating from these results, *Fernandez* (1980a) concluded that Uranus and Neptune implanted ~10% of the material they scattered into the Oort cloud.

Another early attempt to model the construction of the Oort cloud through these mechanisms was *Shoemaker and Wolfe* (1984). Shoemaker and Wolfe assumed the planetesimals were leftover from the formation of Uranus and Neptune and employed a distribution of initially circular orbits between 15 and 35 au for their planetesimals. The evolution of these planetesimals' orbits due to planetary encounters was modeled with a statistical approach (*Opik,* 1976), and the influence of passing stars was implemented as in *Oort* (1950). Largely in agreement with *Fernandez* (1980a), *Shoemaker and Wolfe* (1984) found that ~10% of planetesimals will avoid ejection or collision with the Sun. Of those survivors, the vast majority (~90%) will be retained on orbits with perihelia beyond the planets' orbits and with semimajor axes between 500 and 20,000 au. A smaller fraction (~10%) were predicted to orbit with semimajor axes beyond 20,000 au, comprising the Oort cloud that *Oort* (1950) predicted to explain LPC production. Thus, the survivors in the *Shoemaker and Wolfe* (1984) work tended to orbit the Sun at closer distances than the *Fernandez* (1980a) work.

3.1.2. A numerical model. This process of Oort cloud formation was then revisited with still more direct numerical techniques in *Duncan et al.* (1987). In this work, planetary perturbations on planetesimals were still applied with a Monte Carlo approach once per planetesimal orbit. However, the Monte Carlo sample was built from a distribution of energy kicks compiled from direct integrations of planetesimals through the giant planet region of the solar system. (For an example of such a distribution, see Fig. 5). As planetesimals reached large (>2000 au) semimajor axes, their orbits were directly integrated. During these direct integrations, planetary energy kicks were still applied at each perihelion passage, stellar perturbations consistent with the solar neighborhood kinematics were applied via the impulse approximation, and a vertical (with respect to the galactic plane) tidal force due to the distribution of matter within the Milky Way's disk was included (*Heisler and Tremaine,* 1986).

The approach of *Duncan et al.* (1987) revealed the key timescales involved that govern the formation of the Oort cloud. The semimajor axes of planetesimals diffuse due to planetary energy kicks received near perihelion. The timescale of this diffusion depends on the orbit's perihelion (which determines the giant planet that is primarily delivering the kicks) and the current semimajor axis (which determines how often perihelion passages occur as well as the orbit's binding energy). As semimajor axis increases, the diffusing timescale for further changes in semimajor axis decreases with $a^{-1/2}$ for a fixed q. Meanwhile, a competing timescale is the typical time for a significant change in perihelion due to perturbations from passing stars and the galactic tide. A torque large enough ($\Delta q \gtrsim 10$ au) to potentially pull an orbit's perihelion outside the planetary region will prevent future energy kicks from planetary encounters and halt diffusion in semimajor axis. This second timescale also decreases with semimajor axis since external perturbations become more significant as the distance from the Sun's gravitational pull increases, but it falls more steeply than the diffusion timescale, or as a^{-2}.

Both the timescale for semimajor axis diffusion due to planetary encounters and the timescale for perihelion torques due to external perturbations are plotted as a function of semimajor axis in Fig. 4. To avoid a case where all planetesimals are either ejected or collide with the Sun or planets, the timescale for a perihelion torquing must fall below the semimajor axis diffusion timescale at some value of semimajor axis. The semimajor axis for which the torquing timescale is shorter than the diffusion timescale depends on the planetesimal's initial perihelion distance because Jupiter and Saturn provide larger energy kicks than Uranus and Neptune. For planetesimals being scattered by Uranus and Neptune, the two timescales become comparable at a $\simeq$ 5000 au, and for planetesimals being scattered by Jupiter and Saturn, they become comparable at a $\simeq$ 20,000 au. These semimajor axis values determine the inner edge (i.e., minimum a) of the Oort cloud, interior to which objects' perihelia cannot be changed by external perturbations before they experience significant changes in semimajor axis; objects interior to this a boundary are part of the transneptunian scattering population rather than the Oort cloud. Exterior to this value, objects' perihelia can be lifted out of the planetary region, effectively locking their semimajor axes. Thus, bodies gravitationally scattering off the giant planets can enter the Oort cloud at or beyond the semimajor axis at which the two timescales become comparable. There is also a natural maximum semimajor axis to the Oort cloud that exists due to the tide of the Milky Way's disk, which overwhelms the Sun's gravitational pull around 0.5–1 pc or 1–2 × 10^5 au (*Heisler and Tremaine,* 1986).

With initial semimajor axes of 2000 au and bins of initial perihelia between 5 au and 35 au, *Duncan et al.* (1987) directly integrated ~7000 particles under the influence of the Sun, planets, passing stars, and galactic tide for 4.5 G.y.

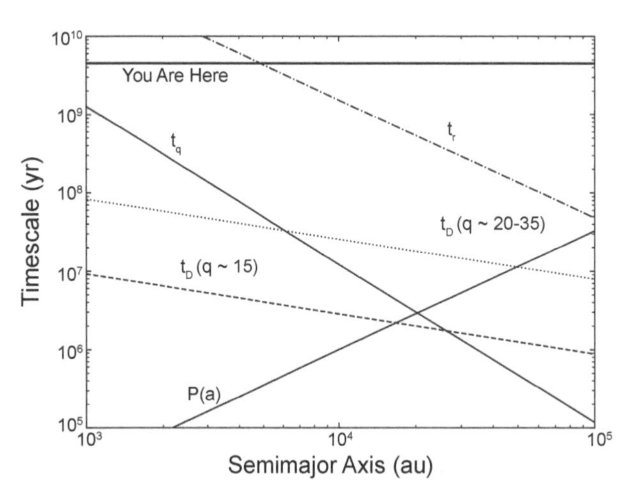

Fig. 4. Plot of key dynamical timescales as a function of particle semimajor axis. t_q is the timescale for a perihelion shift of 10 au (assuming initial q of 25 au) due to galactic tidal torquing, t_D is the diffusion timescale for the semimajor axis to change by order of its present value due to planetary energy kicks, t_r is the timescale for the galactic tide to cycle an orbit from q = 25 au to its maximum perihelion and back again, P(a) is the orbital period, and the horizontal line marks the age of the solar system. Adapted from *Duncan et al.* (1987).

At the end of this simulation, they found that a substantial fraction of particles were trapped in the Oort cloud with perihelia beyond the planets and semimajor axes between a few thousand astronomical units and ~10^5 au. However, the trapping fraction varied greatly depending on particles' initial perihelia. For those with perihelia of 5 au (near Jupiter), the trapping fraction was only 2%, while the fraction approached 40% for those with perihelia near Neptune (25–35 au). The reason for this is that the magnitude of energy kicks particles receive during perihelion passage sensitively depends on perihelion value (*Fernandez,* 1978).

These energy kicks were first numerically measured by *Fernandez* (1981), and the typical energy kick per perihelion passage for different orbits is shown in Fig. 5. Objects passing near Jupiter and Saturn receive energy kicks more than an order of magnitude larger than those passing near Uranus and Neptune. As already stated, for objects with perihelia near Jupiter and Saturn, their perihelia cannot be pulled from the planetary region until they attain semimajor axis $\gtrsim 10^4$ au. However, the entire orbital energy window of the Oort cloud ($10^4 \lesssim a \lesssim 10^5$ au) is smaller than the typical energy kick these objects receive on a single perihelion passage. Thus, most of them overshoot the Oort cloud and leave the solar system on hyperbolic orbits. Since the opposite is true for objects scattering off Uranus and Neptune (the weaker energy kicks of these planets are smaller than the Oort cloud's orbital energy window), Uranus and Neptune are much more efficient at placing bodies into the Oort cloud. Within the populations of particles trapped in the Oort cloud, *Duncan et al.* (1987) found that the same perturbations (passing stars and the galactic tide) that were

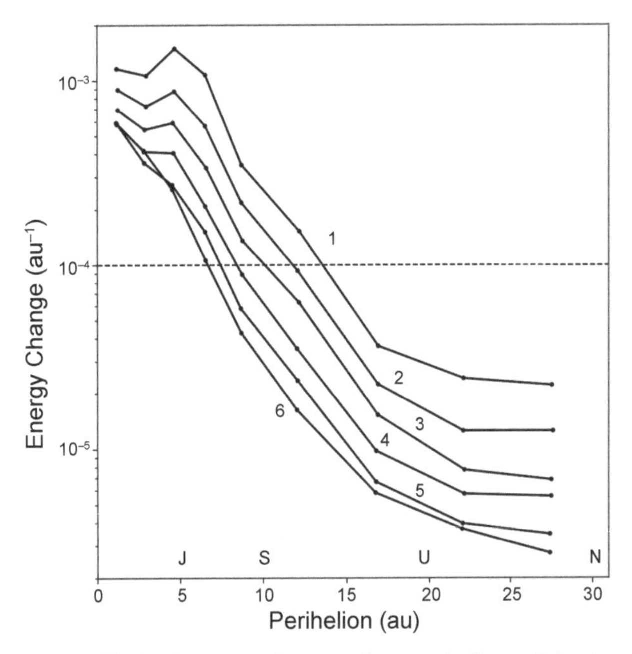

Fig. 5. Typical energy change of a parabolic particle during a single perihelion passage as a function of perihelion. Solid lines 1–6 mark 30° inclination bins from 0°–30° to 150°–180°. The dashed line coincides with the approximate width of the Oort cloud energy window for particles scattering off Jupiter and Saturn. Adapted from *Fernandez* (1981).

responsible for perihelion shifts were strong enough to complete isotropize orbital inclinations and eccentricities beyond a semimajor axis of ~5000 au.

Another prediction of the numerical simulations of *Duncan et al.* (1987) was that the Oort cloud is strongly centrally concentrated, with 80% of bodies orbiting on semimajor axes between 3000 and 20,000 au and the remaining 20% of bodies orbiting beyond a > 20,000 au. This finding was interpreted to imply that only the outer 20% of the Oort cloud was capable of supplying observed LPCs. The reason for this is that the perihelion shift timescale plotted in Fig. 4 is longer than the orbital period for semimajor axes a $\lesssim$ 20,000 au. Thus, objects on semimajor axes below 20,000 au must first make a perihelion passage near Jupiter and Saturn before they would be capable of making a perihelion passage near Earth. However, as shown in Fig. 5, the energy kick delivered during this first perihelion passage will typically be powerful enough to either eject the particle on a hyperbolic orbit or decrease its semimajor axis to a level where its perihelion evolution is extremely slow.

3.1.3. Lower central concentration and trapping efficiency. From the above logic it was thought that LPCs only provided constraints on the outer periphery of Oort cloud beyond 20,000 au, but the inner region could contain large numbers of bodies as well (*Hills,* 1981). Hills speculated that this inner portion could be ~100× more populous, yet would only supply comets near Earth during "comet showers" coinciding with rare, powerful stellar encounters (*Hills,* 1981; *Hut et al.,* 1987). The simulations of *Duncan et al.* (1987) suggested that these inner population estimates were far too high.

Based on the high Oort cloud trapping efficiencies of objects gravitationally scattering off Uranus and Neptune in *Duncan et al.* (1987), it was presumed that tens of percents of leftover planetesimals may have been captured into the Oort cloud after the giant planets formed. However, *Fernández* (1997) noted that many objects beginning on orbits near Uranus and Neptune evolve to Jupiter- and Saturn-crossing orbits before being scattered to large semimajor axes (*Fernandez and Ip,* 1984) and that this process may lower the fraction of material that is trapped in the Oort cloud (*Fernandez and Ip,* 1981). Numerical work by *Dones et al.* (2004) confirmed this. *Dones et al.* (2004) found that the actual fraction of initial bodies that are still trapped in the modern Oort cloud was much more likely no more than ~5%, and subsequent works have pointed to even lower fractions of 2–3% (*Kaib and Quinn,* 2008; *Kaib et al.,* 2011).

The reason for this difference from *Duncan et al.* (1987) was that *Dones et al.* (2004) initialized their particles on nearly circular orbits between the giant planets, whereas *Duncan et al.* (1987) had to begin their particles (due to computing limitations of the era) on highly eccentric orbits with semimajor axes of 2000 au. The choice of *Duncan et al.* (1987) effectively locked the perihelia of their particles unless the galactic tide or passing stars altered them. On the other hand, the approach of *Dones et al.* (2004) allowed the perihelia of bodies to significantly evolve before they were scattered to eccentric orbits.

Through earlier numerical experiments of particles scattering off the giant planets done by *Fernandez and Ip* (1984), it was found that particles beginning on circular orbits near the ice giants often first evolve to the Jupiter/Saturn region before being scattered to large semimajor axes. An example of this evolution is shown in Fig. 6. This is the dynamical behavior that lowered the Oort cloud trapping efficiency so much in *Dones et al.* (2004). Although objects scattering off Uranus or Neptune have a large probability of becoming trapped in the Oort cloud, most of the bodies that begin on circular orbits near these planets see their perihelia evolve toward Jupiter and Saturn, and, as discussed above, these planets are much more likely to eject objects than allow them to become trapped in the Oort cloud. The diminished role of Uranus and Neptune in placing objects into the Oort cloud also changes the cloud's structure. Since objects scattered by Jupiter and Saturn have little chance of becoming trapped in the Oort cloud until they attain semimajor axes above 10,000–20,000 au, the fraction of bodies retained with a < 20,000 au is nearly equal to that with a > 20,000 au (*Dones et al.,* 2004). Thus, the inner region of the cloud, thought to be unconstrained by observed LPCs, was predicted to be comparably populous to the outer region beyond 20,000 au.

3.2. Generating a Scattered Disk

The Oort cloud as described above is thought to be populated from an initially dynamically cold planetesimal disk in the current giant planet region. However, the ex-

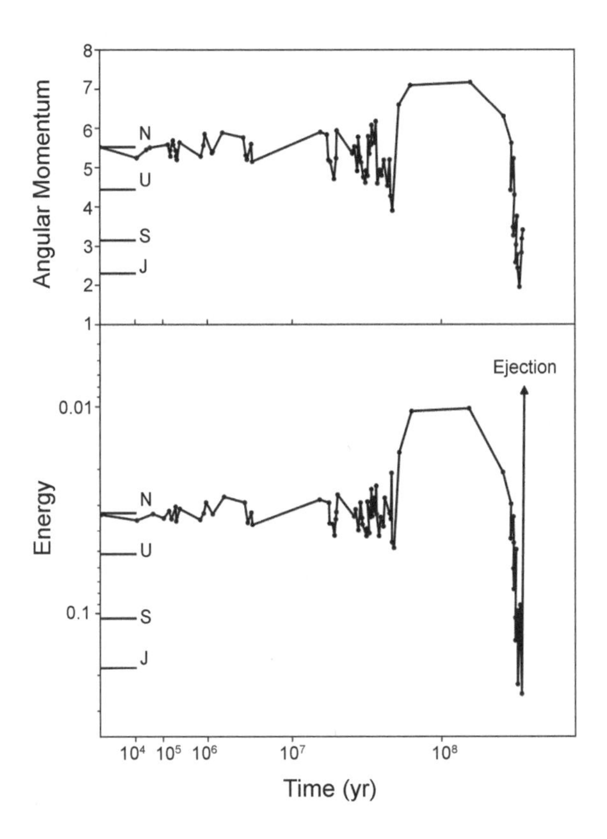

Fig. 6. Evolution of a particle's angular momentum and orbital energy over time as it interacts with the giant planets. The energies and angular momenta of the giant planets are marked with letters along the y-axis. Energy and angular momentum units are such that $GM_{\odot}$ = 1 and distance is in astronomical units. Adapted from *Fernandez and Ip* (1984).

istence of the SPCs revealed the need for an extension of this disk past Neptune to provide a low-inclination source of these non-isotropic comets (see section 1.2). It was then realized that the dynamical erosion of this disk that could feed the SPCs naturally leads to the formation of a scattered disk of objects with semimajor axes inside the inner Oort cloud boundary and perihelia near Neptune [see *Torbett* (1989) and *Duncan and Levison* (1997), although as discussed below much of the scattering population is now thought to originate from the initial planetesimal disk located interior to Neptune's current orbit]. Such objects were observational confirmed among the early TNO discoveries (*Luu et al.,* 1997).

However, as more TNOs were discovered, the complex dynamical structures revealed in the classical, resonant, and scattering populations (recently reviewed by *Gladman and Volk,* 2021) showed that the giant planets did not form on their current orbits. Early work on the interactions between the giant planets and the initially massive planetesimal disk showed the energy and angular momentum exchange can cause the planets' orbits to migrate (*Fernandez and Ip,* 1984). Outward migration of Neptune was then shown to lead to capture of objects into its external mean-motion resonances. The large eccentricity of Pluto's 3:2 resonant orbit was potentially explained by Neptune's outward migration by ~7 au (*Malhotra,* 1993, 1995). The large populations of objects in many of Neptune's resonances that have been found in subsequent surveys has provided strong evidence of migration-related resonant capture.

Our picture of the dynamical history of the giant planets has evolved significantly since the early proposals of so-called smooth planetary migration. To explain both the complex orbital distribution of the small bodies in the outer solar system and the dynamical state of the giant planets' orbits, more violent scenarios have been proposed. The Nice model of the giant planets' evolution was one such early model; *Tsiganis et al.* (2005) suggested that Jupiter and Saturn crossed a strong mutual mean-motion resonance while migrating, triggering an epoch of large eccentricities in all four giant planets' orbits that led to planet-planet encounters and even a swap in the order of the ice giants. This specific proposed resonance crossing has since been shown to not satisfactorily reproduce the orbits of Jupiter and Saturn (e.g., *Brasser et al.,* 2009), but the concept of excited planetary eccentricities, potential planet-planet gravitational encounters, or even ejection of an ice giant [see *Nesvorný* (2018) for a recent review of the giant planets' dynamical history] hold promise for explaining some of the TNO orbital distribution features.

The current picture of the giant planets' dynamical history is that a mix of violent and smooth migration likely took place (see, e.g., *Nesvorný,* 2018). Their dynamical evolution dispersed the massive disk of planetesimals that formed in the current giant planet region into the trans-neptunian and Oort cloud populations. A small portion of the observed TNOs (the cold classical population on low-eccentricity, low-inclination orbits in the 42–48-au region) represent objects that formed *in situ* beyond Neptune in a low-mass density disk [see, e.g., results from the New Horizons flyby of the classical TNO Arrokoth (*McKinnon et al.,* 2020)]. The majority of the TNO populations (the dynamically hot populations, including the scattering population) were scattered outward from closer-in, more massive disk with an outer edge near 30 au. A small fraction of these scattered bodies survived in metastable orbits within the modern TNO population. This same scattering process also populated the Oort cloud.

It is clear from modeling that smooth migration into an extended massive disk (one that extends out to ~50 au) significantly overpopulates Neptune's resonances compared to the observed populations (e.g., *Hahn and Malhotra,* 2005). The other motivation for an end to the massive initial planetesimal disk at ~30 au is also, of course, that Neptune stopped migrating where it did [i.e., ran out of fuel for migration (*Gomes et al.,* 2004)]. But even assuming a transition to a less-massive disk at ~30 au, something besides smooth migration is needed to produce the detailed distribution of objects in the classical Kuiper belt as well as the distributions of objects in the scattering, detached, and distant resonant/near-resonant populations.

The dynamical excitation from very violent scattering events (such as in the original Nice model) can still be consistent with retaining the cold classical population in certain circumstances (e.g., *Batygin and Brown,* 2010; *Dawson and Murray-Clay,* 2012; *Gomes et al.,* 2018), although more recent models investigated less dramatic instabilities. Smaller-scale "jumps" in Neptune's orbit caused, for example, by the ejection of an additional ice giant (see, e.g., *Nesvorný,* 2011; *Nesvorný and Morbidelli,* 2012) might help explain the "kernel" [an overdensity of objects near 45 au noted by *Petit et al.* (2011)] in the cold classical region (*Nesvorný,* 2015b); Fig. 7 shows an example of such an evolution for the giant planets.

In addition to planet-planet interactions, the number and mass of dwarf-planets in the planetesimal disk affects how "grainy" the planets' migration is; scattering events between Neptune and these largest objects in the disk cause small but significant jumps in Neptune's semimajor axis that can alter the capture rate into Neptune's resonances (e.g, *Murray-Clay and Chiang,* 2005; *Nesvorný and Vokrouhlický,* 2016). The speed of migration, likely determined by the mass distribution of the planetesimal disk, also affects the expected TNO distribution. There are models that predict that some period of slow migration, particularly during the last stage of migration after any planet-planet scattering has occurred, can help explain the high-perihelion objects in the scattering/detached TNO population by allowing for secular-resonant interactions to occur, which might also help excite the inclinations of the hot populations (e.g., *Nesvorný,* 2015a; *Lawler et al.,* 2019; *Fraser et al.,* 2017). Adding graininess to this slow

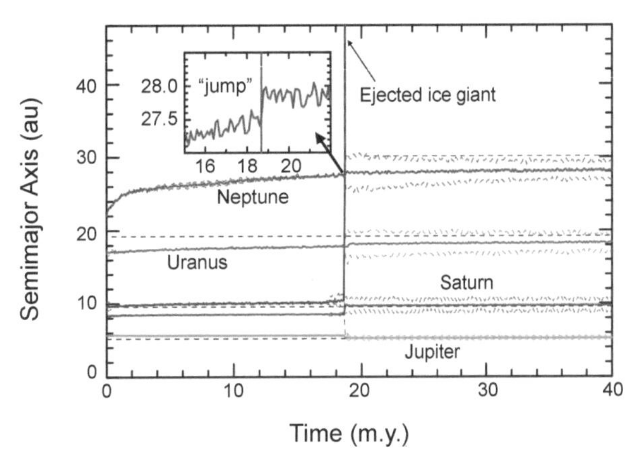

Fig. 7. One plausible version of the giant planets' orbital histories from *Nesvorný and Vokrouhlický* (2016) (modified from their Fig. 3). Semimajor axis vs. time for each planet is shown, with thinner dashed lines indicating the perihelion and aphelion distances of each planet. This migration scenario involves an additional ice giant that is ejected after ~18 m.y., causing moderate eccentricity excitation and small jumps in the semimajor axes of several planets, including Neptune (shown in the inset). Neptune slowly migrates to its final location in the final ~100 m.y. of the simulation.

migration can help strand objects out of the scattering/resonant population by dropping them out of resonances at a high-q portion of this secular evolution, building up a population of near-resonant detached objects (e.g., *Kaib and Sheppard,* 2016; *Lawler et al.,* 2019). Figure 8 shows an example of the scattering/detached/resonant populations from a $\simeq$ 50–100 au that are expected for a planetary migration scenario with a small instability followed by a long period of grainy migration (from *Nesvorný and Vokrouhlický,* 2016). Observations of the high-a, high-q TNOs remain relatively sparse (as indicated by the relatively small number of observed objects included in Fig. 8) due to the strong observational biases against detecting them, but future surveys will help provide more robust observational constraints on models of the scattering and detached populations (see section 4).

3.3. Concurrent Oort Cloud and Kuiper Belt Formation

Over the past three decades, evidence has mounted that planetesimal scattering drove the giant planets through substantial orbital evolution early in the solar system's history. However, Oort cloud formation, a product of this scattering, was not simulated in concert with giant planet orbital evolution until the work of *Brasser and Morbidelli* (2013). In this work, the authors replicated the post-instability behavior of the surviving giant planets within an early iteration of the Nice model (*Levison et al.,* 2008). Neptune and Uranus began with eccentricities above 0.2 and damped to their modern values while their semimajor axes migrated outward 2–3 au over the course of tens of millions of years. At the initiation of this sequence, test particles were started with semimajor axes between 29 and 34 au, eccentricities of 0.15, and coplanar inclinations. It was this outer belt from which *Brasser and Morbidelli* (2013) assumed the Oort cloud (and scattered disk) is formed.

After 4 G.y. of evolution *Brasser and Morbidelli* (2013) found an Oort cloud trapping efficiency of ~7%, modestly higher than *Dones et al.* (2004) and more than double that of *Kaib and Quinn* (2008). While giant planet orbital evolution may have contributed to this, the confinement of test particles to beyond a $\geq$ 29 au may also have been responsible. No particles were begun in the immediate vicinity of Jupiter and Saturn, which are very inefficient Oort cloud populators, while the contributions of Uranus and Neptune were maximized.

In addition to forming the Oort cloud, the simulations of *Brasser and Morbidelli* (2013) also generated a scattered disk. The population of this scattered disk was ~12× smaller than that of the Oort cloud. This result was in line with the ~10:1 population ratios found in *Dones et al.* (2004) and *Kaib and Quinn* (2008). Meanwhile, comparisons of LPC flux with the steady-state populations of JFCs have historically been interpreted as implying a ~100:1, or perhaps even higher ratio [see section 2.4.3 as

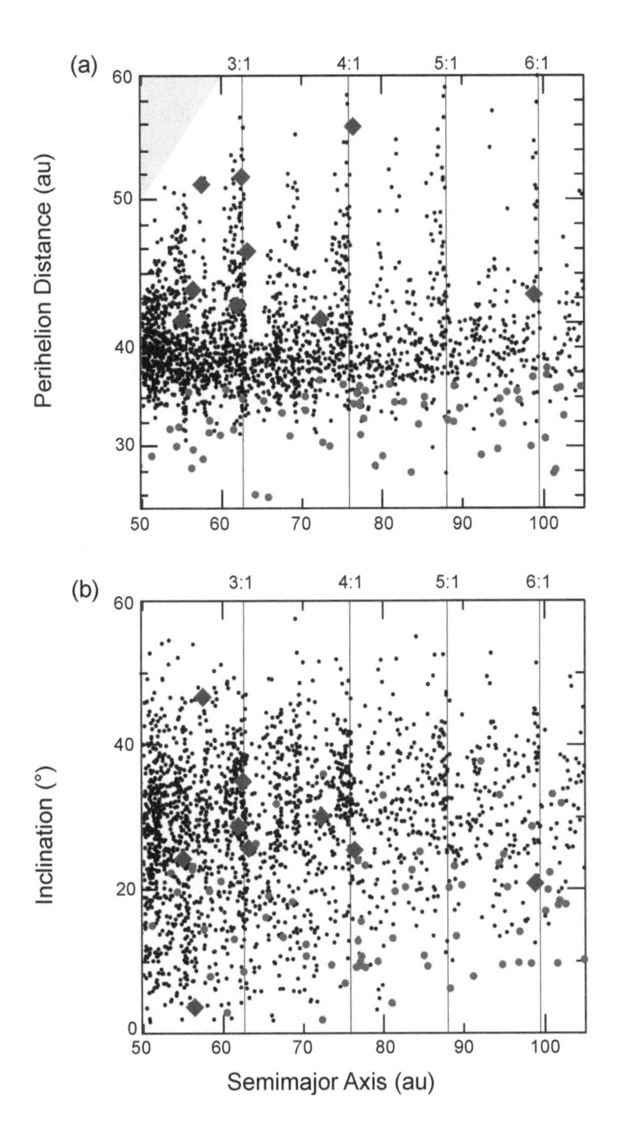

Fig. 8. An example simulated distribution of the TNO populations from a = 50–100 au from *Nesvorný et al.* (2016) (modified from their Fig. 1) produced from a giant planet history scenario similar to that shown in Fig. 7 that also included grainy migration of the planets due to interactions with Pluto-mass objects. **(a)** Perihelion distance and **(b)** inclination vs. semimajor axis are shown for the simulated particles that ended up in the scattering population (red dots) and the high-a detached and resonant populations (black dots); a subset of the then-observed TNOs with q > 40 au are shown (blue diamonds) for comparison.

well as *Levison et al.* (2010)].

The planetary orbital evolution of *Brasser and Morbidelli* (2013) utilized an early, violent version of the Nice model that was demonstrated to conflict with numerous aspects of the modern solar system's architecture (*Brasser et al.*, 2009; *Batygin et al.*, 2011; *Dawson and Murray-Clay*, 2012). As discussed in section 3.2, recent detailed comparisons of TNO surveys with simulations of Kuiper belt formation appear to favor a less-violent giant planet evolution in which Neptune's migration was relatively smooth from 24 to 30 au except for a brief interruption during the ejection of an additional ice giant (*Nesvorný*, 2011, 2015a,b). Incorporating this later iteration of early giant planet evolution, *Vokrouhlický et al.* (2019) again modeled the formation of the Oort cloud. In this instance, particles were initiated on circular orbits between 24 and 30 au while Uranus and Neptune migrated several astronomical units to their modern semimajor axes over ~100-m.y. timescales. Although this work focused on modeling the comet production from such a cloud, the authors found that ~6% of their bodies were trapped in the Oort cloud after 4.5 G.y. of evolution, in line with the results of *Brasser and Morbidelli* (2013), likely due to the backweighted distribution of initial test particle orbits. In addition, *Vokrouhlický et al.* (2019) found that the inner Oort cloud has nearly the same population size as the outer Oort cloud, although they divided the inner and outer clouds at a = 15,000 au.

For illustrative purposes, Fig. 9 shows the distribution of particle semimajor axes, eccentricities, and inclinations after 4 G.y. of evolution in an Oort cloud formation simulation. This simulation, taken from *Kaib et al.* (2019), forms an Oort cloud and scattered disk concurrently with very similar parameters to the work of *Vokrouhlický et al.* (2019). As one can see, the scattered disk generally has perihelia between ~30 and ~45 au and inclinations below ~45° until reaching semimajor axes above ~10^3 au. At this point, perturbations from the local galaxy begin

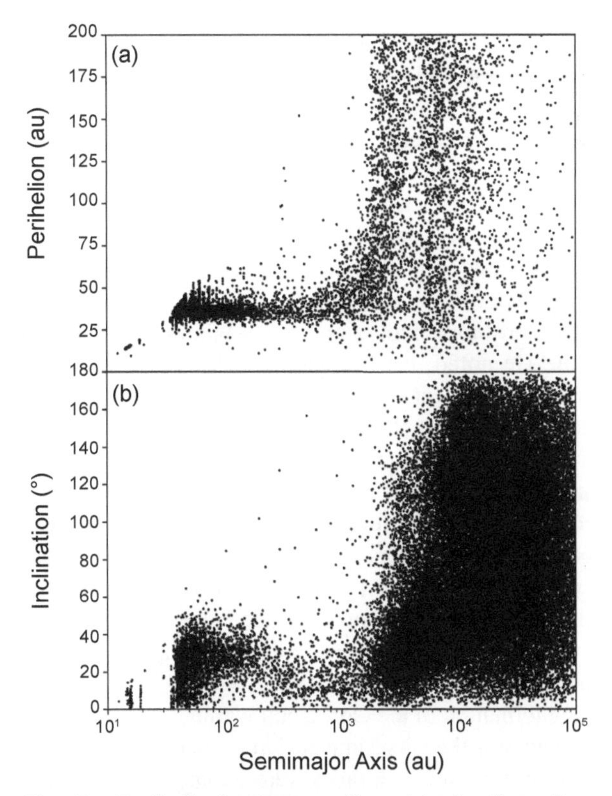

Fig. 9. Perihelia (upper panel) and inclinations (lower panel) vs. semimajor axes of particles at t = 4 G.y. in a simulation forming the Oort cloud and scattered disk concurrently. Plot is generated from the "OC" simulation of *Kaib et al.* (2019).

to isotropize inclinations and detach perihelia from the planetary region.

3.4. Influence of the Sun's Local Environment

3.4.1. Solar birth cluster. While the structure of the scattered disk should be relatively insensitive to the Sun's local galactic environment, perturbations from this environment are critical to the formation and evolution of the Oort cloud, and variations in this environment could impact the Oort cloud's formation. The numerical work of *Duncan et al.* (1987) and *Dones et al.* (2004) assumed that the Sun's local galactic environment did not significantly change with time. The strength of the galactic tide was based on the local density of matter in the Milky Way disk inferred from stellar motion (*Bahcall,* 1984), and the velocities and rates of passing stars were meant to mimic the population and kinematics of the solar neighborhood (e.g., *García-Sánchez et al.,* 2001; *Bailer-Jones et al.,* 2018).

This static galactic environment was, of course, an approximation, and there are reasons to expect it to have changed significantly over the Sun's lifetime. Perhaps most significantly, it is well-known that most stars form within clusters and spend some portion of their early lives within them (*Lada and Lada,* 2003). If this occurred for the Sun, the local stellar densities during this cluster phase would have been orders of magnitude higher than the modern solar neighborhood, and the tide of the cluster would have dwarfed the current galactic tide, perturbing the modern solar system. *Gaidos* (1995) realized that these enhanced external perturbations could have dramatic impacts on the formation of the Oort cloud, allowing objects to become trapped within the Oort cloud at a much lower semimajor axis than predicted for a galactic field environment. However, *Gaidos* (1995) argued that such a comet cloud would also be rapidly destroyed by the same cluster perturbations and that little material should remain at these distances.

Fernández (1997) further considered the scenario in which Oort cloud formation begins within a stellar birth cluster. He noted that the heightened perturbations of a cluster environment would dramatically widen the energy window of the Oort cloud and potentially allow Jupiter and Saturn to play more prominent roles in populating the cloud. Moreover, much of this material could reside at hundreds of astronomical units to a few thousand astronomical units, well interior to the semimajor axis range of the "classical" Oort cloud. Moreover, *Fernández* (1997) argued that the relatively short lifetimes of open and embedded clusters would have allowed many of these bodies to remain trapped, contrary to the conclusions of *Gaidos* (1995).

The formation of the Oort cloud within a star cluster environment was then first simulated in *Fernández and Brunini* (2000). In this work, particles were initiated on semimajor axes between 100 and 250 au with perihelia distributed among the giant planets. The entire solar system was then immersed within a star cluster environment that contained stars passing much more slowly (1 km s^{-1}) and closely than a galactic field environment. The stellar densities of clusters were varied between 10 and 100 stars pc^{-3} and linearly declined to zero over 10^8 yr. In addition, *Fernández and Brunini* (2000) included tidal torquing from a uniform-density spherical distribution of gas that dispersed at 10^7 yr meant to replicate the gaseous component of an embedded cluster. They found that an inner comet cloud readily formed under such conditions and the radial extent (semimajor axes from a couple hundred astronomical units to a few thousand astronomical units) depended on the stellar density of the star cluster, with denser clusters yielding more compact clouds. Moreover, as in *Duncan et al.* (1987), the fraction of material captured into this cloud depended sensitively on particles' initial perihelion. Those with perihelia in the Jupiter/Saturn region only had trapping efficiencies of a few percent, whereas tens of percents of material could be captured into the cloud for initial perihelia near the ice giants. Finally, the relatively short cluster lifetimes allowed these inner clouds to largely stay intact past cluster dissolution.

Oort cloud formation within a cluster was revisited by *Brasser et al.* (2006). This work specifically focused on the embedded cluster environment, in which most stars appear to form (*Lada and Lada,* 2003). The embedded cluster environments employed within *Brasser et al.* (2006) had higher central densities (in gas and stars) and shorter lifetimes than those used in *Fernández and Brunini* (2000). Due to the large energy window of the comet cloud and the short cluster lifetimes, roughly 10% of material scattered by Jupiter and Saturn could be caught in an inner comet cloud, whose range of semimajor axes again depended on the assumed density of the cluster environment (see Fig. 10). Nearly all this inner comet cloud material would be retained until the present epoch and very little would be lost or diffuse to the classical Oort cloud beyond a > 20,000 au (*Brasser et al.,* 2008).

Additional investigations by *Brasser et al.* (2007) studied the effect of the gaseous component of the solar nebula dur-

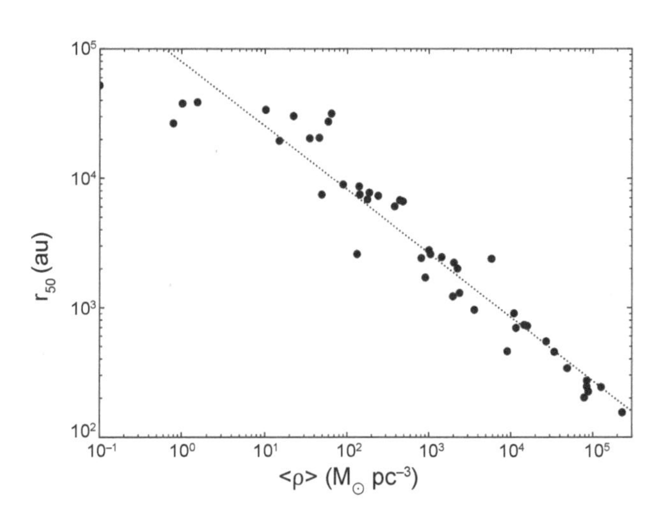

Fig. 10. Median position of bodies within the Oort cloud as a function of the mean density of a cluster environment for a given Oort cloud formation simulation. From *Brasser et al.* (2006).

ing the formation of the Oort cloud within a star cluster. For small kilometer-sized bodies, aerodynamic drag prevented objects from being scattered to large semimajor axes. Instead, they concentrated into nearly circular orbits interior to Jupiter. Meanwhile, objects with radii of tens of kilometers were decoupled from the gas and freely reached the Oort cloud. If the gaseous component of the Sun's protoplanetary disk outlived the Sun's birth cluster, this suggested that the centrally concentrated, cluster-generated comet cloud should only contain large ($r \gtrsim 10$ km) bodies.

Brasser et al. (2012a) repeated the embedded cluster scenario with improved simulations that used updated, observationally motivated cluster profiles and stellar velocity distributions. Furthermore, in this new work cluster stars were able to gravitationally interact with one another. While centrally concentrated Oort clouds still formed down to semimajor axes of several hundred astronomical units, the formation efficiency dropped to 1–2% due to the Oort cloud harassment th at occurred with a heightened number of stellar interactions in these new environments.

Complementing studies of Oort cloud formation within embedded clusters, *Kaib and Quinn* (2008) simulated the formation of the Oort cloud within longer-lived ($t \simeq 10^8$ yr) open cluster environments. Unlike the work of *Fernández and Brunini* (2000), particles were initiated on circular orbits. As expected, this allowed particles near Uranus and Neptune to evolve to Jupiter- and Saturn-crossing orbits before being scattered to large semimajor axes. This led to lower (<10%) trapping efficiencies than predicted by *Fernández and Brunini* (2000). Owing to the clusters' longer lifetimes and the more prominent role of powerful stellar encounters compared to the gas-rich environments of *Brasser et al.* (2006), the Oort clouds formed within open clusters displayed lower overall trapping efficiencies ($\lesssim$5%) and a greater level of stochasticity in the degree of central cloud concentration (which was largely dependent on the few most powerful stellar encounters). Nonetheless, this work showed that an Oort cloud formed within an open cluster should possess bodies on semimajor axes from several hundred to several thousand astronomical units whose numbers are comparable to those in the classical Oort cloud.

3.4.1.1. Discovered objects as possible birth cluster relics: Beginning with the pioneering works of *Gaidos* (1995) and *Fernández* (1997), it has been predicted that if the Sun formed within a star cluster, we should expect the inner edge of the Oort cloud to potentially reside at a semimajor axis of hundreds of astronomical units rather than the thousands of astronomical units suggested by models that only consider the modern solar neighborhood. These predictions seem extremely prescient, as in the early 2000s TNO surveys began discovering objects with semimajor axes of hundreds of astronomical units but perihelia that dynamically decoupled them from planetary energy kicks (*Gladman et al.,* 2002). Although some of these bodies may attain their high perihelia via Kozai-Lidov cycles within neptunian mean-motion resonances, this mechanism cannot generate all observed orbits (*Brasser and Schwamb,* 2015; *Gomes et al.,* 2005; *Kozai,* 1962; *Lidov,* 1962). In particular, the objects Sedna and 2012 VP_{113} both have perihelia beyond 70 au and semimajor axes beyond 250 au and are therefore not susceptible to Kozai-Lidov cycling (*Brown et al.,* 2004; *Trujillo and Sheppard,* 2014).

The discoveries of such objects appear consistent with the existence of a cluster-generated Oort cloud (*Morbidelli and Levison,* 2004). However, other explanations have been put forth, such as scattering off of an ~Earth-mass planetesimal or exchange with another young star system (*Gladman and Chan,* 2006; *Kenyon and Bromley,* 2004; *Silsbee and Tremaine,* 2018). Of particular note, it has been posited that these detached TNOs with large semimajor axes are the dynamical signature of a distant, yet-undetected planet (*Batygin and Brown,* 2016a). Such a planet has been shown to be capable of aligning the poles and perihelion directions of an ensemble of such orbits, and the detected orbits appear to display an anomalous anisotropy (*Batygin and Brown,* 2016a). However, the total sample size of objects remains small, and the observing biases associated with combining the results of many TNO surveys are complex (*Shankman et al.,* 2017a; *Kaib et al.,* 2019). Detections from well-characterized surveys do not find evidence of significant orbital clustering (*Shankman et al.,* 2017a; *Napier et al.,* 2021; *Bernardinelli et al.,* 2021).

3.4.1.2. Oort cloud capture within a birth cluster: Another aspect of Oort cloud formation within a cluster environment concerns the possibility of the Sun capturing exoplanetesimals, or bodies formed around another star cluster member and subsequently ejected into intracluster space. This possibility was first considered in *Eggers et al.* (1997) and then again in *Zheng et al.* (1990). Both works found that forming our entire Oort cloud through the capture of exoplanetesimals within a cluster is unlikely. Such a scenario would require 10^{16} planetesimals per star, which would be difficult to reconcile with the (at the time) non-detection of interstellar comets passing through the solar system (*Zheng et al.,* 1990).

However, the capture probabilities used in those early works may have been underestimates. *Levison et al.* (2010) directly simulated the dynamical evolution and dissolution of embedded clusters in which each star was surrounded by a disk of distant planetesimals. Over time, many planetesimals became unbound from their stars-of-origin via cluster tides and star-star encounters. A snapshot of one simulation is shown in Fig. 11. In these simulations, it was found that stars often captured other stars' planetesimals. As stars and intracluster planetesimals escaped the potential of the dissolving cluster, some star and planetesimal velocity vectors were inevitably similar. The escape from the potential decreased relative velocities even further, enabling capture. In this scenario, stars typically captured a couple percent of their original planetesimal number through this mechanism. Thus, the efficiency of such capture should generally not exceed the efficiency of the solar system capturing its own scattered material through the "classic" Oort cloud formation process. However, occasionally direct exchange

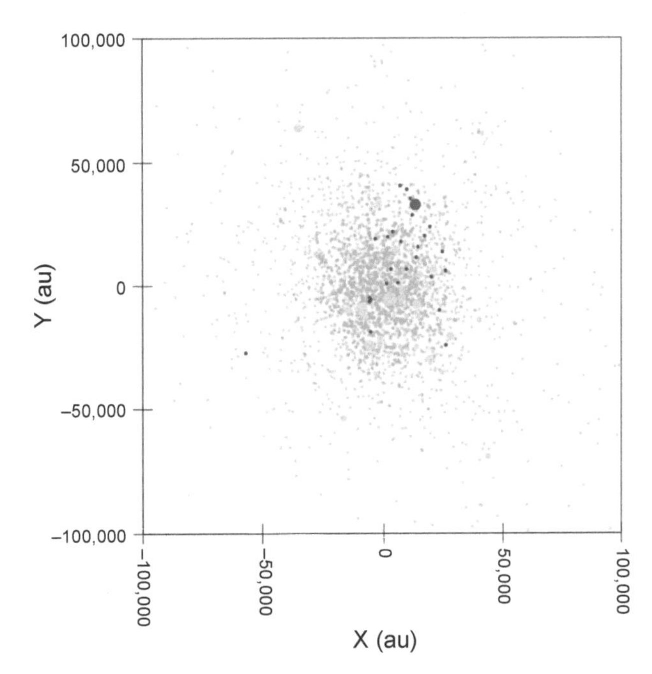

Fig. 11. Snapshot of a simulation of Oort cloud capture within an embedded cluster. Datapoints mark the Cartesian X and Y positions of simulation particles. Yellow and red points are cluster stars, while green and blue points mark planetesimals. Blue planetesimals become bound to the red cluster star after cluster dissolution. From *Levison et al.* (2010).

of planetesimals through a close star-star encounter led to capture fractions over 10%. From these results, *Levison et al.* (2010) argued that a significant fraction, and perhaps the majority, of bodies in the Oort cloud could have originated around other stars. However, many of these bodies were captured onto distant orbits with semimajor axes of tens of thousands of astronomical units, and their subsequent evolution and survival within the galaxy until the present epoch was not modeled.

3.4.2. Galactic migration. Even after the Sun's birth cluster disperses, there is ample reason to expect significant changes in the Sun's local galactic environment. Dynamical simulations of spiral galaxies indicate that the galactocentric orbits of most stars are not static with time. Instead, they often exchange angular momentum with the spiral arms, typically within corotation resonances, whose locations fluctuate as spiral wave pattern speed varies (*Sellwood and Binney,* 2002; *Roškar et al.,* 2008). On average, this transports disk stars outward over time (although inward transport is also possible). Importantly, this process does not necessarily increase the peculiar velocities of stars, and they can remain on nearly circular orbits about the galactic center with modest oscillations above and below the galactic midplane.

This stellar migration process obviously has large ramifications for our understanding of spiral galaxies and their properties, but it has also raised the prospect that the Sun may have undergone significant migration (likely outward) during its lifetime in the Milky Way. If this were the case, this may have significant implications for the formation of our Oort cloud. As the Sun's galactocentric distance changes, the mean density of its local environment will also change, which will modulate the rate of stellar encounters and the strength of the galactic tide (*Matese et al.,* 1995). *Brasser et al.* (2010) conducted simulations of Oort cloud formation at various galactocentric distances of 2, 4, 6, 8, 14, and 20 kpc using an analytic model for the Milky Way potential. Unsurprisingly, the different environmental densities varied the Oort cloud inner edge (which was defined as the 5th percentile in semimajor axis) from 30,000 au at r = 20 kpc to 2000 au at 4 = 2 kpc. In addition, the Oort cloud trapping efficiency varied between 2% and 4% and did not display a strong dependence on galactocentric distance.

However, the simulations of *Brasser et al.* (2010) still assumed a static (local and global) galactic environment and tide for the each entire simulation. In fact, stellar migration within galaxies implies that the local environment and tide should be changing with time. The work of *Kaib et al.* (2011) addressed this by merging Oort cloud formation simulations with data output from the evolution of a Milky Way analog spiral galaxy (*Roškar et al.,* 2008). Within the galactic simulation, an ensemble of 31 star particles were selected that possessed Sun-like galactic kinematics and ages. The histories of these particles were then backtraced throughout the simulation. These particles displayed a wide variety of behaviors. On average, their galactocentric distance varied by ~5 kpc throughout their lifetimes, and the closest approach to and furthest excursion from the galactic center was 2 kpc and 13 kpc, respectively. At each galactic simulation time output, the local density and galactic tidal field about each star particle was measured, and with this data, a time-varying set of external perturbations (galactic tide and passing field stars) was built for each star particle, or for each example of a galactic dynamical history of a Sun-like star. These perturbation sets were then used to model the formation of the Oort cloud for 31 hypothetical dynamical histories within the Milky Way.

Kaib et al. (2011) found that the galactic dynamical histories of their star particles had a significant impact on Oort cloud formation, with the median semimajor axis of the Oort cloud varying by nearly an order of magnitude across the sample of star particles. Generally, the minimum galactocentric distance attained, which typically determined the maximum local density of matter inhabited, had the largest impact on the Oort cloud structure. Excursions toward the galactic center resulted in more centrally concentrated clouds. Overall, Oort cloud formation models that do not account for the Sun's dynamical history within the galaxy likely overestimate the cloud's inner edge and median semimajor axis.

In addition, the time at which a star attains its minimum galactocentric distance affected the Oort cloud trapping efficiency. If the minimum galactocentric distance was attained early in the system's life, the trapping efficiency was enhanced (to 3–4%), because the Oort cloud energy window was widened early in the system's history, and it was subjected to somewhat weaker galactic perturbations later in time. Conversely, late excursions to small galactocentric distances decreased Oort cloud trapping efficiency to 1–2%, as the energy window of the Oort cloud remained small early in the system's history, and the late inward excursion drove the loss of weakly bound bodies.

3.4.3. Critical roles of stellar passages. As illustrated in Fig. 12, *Kaib et al.* (2011) also revealed that Oort cloud structure contains significant variation even after the Sun's potential migration history is considered, especially with respect to the Oort cloud's inner edge. The main source of this scatter is due to the effects of stellar passages. Since *Heisler and Tremaine* (1986) showed the galactic tide to be a more powerful mean perturbing force than stellar passages, works studying the formation of the Oort cloud have often focused more on galactic tidal effects and less on stars. However, the conclusions of *Heisler and Tremaine* (1986) are only true *on average*. Passages of field stars are inherently stochastic events that occur on very short ($<10^5$ yr) timescales, so there can be great variability in their significance from one epoch to another. Therefore, over the course of 4.5 G.y., there must have been short periods where the perturbation from stellar passages on the Oort cloud greatly exceeded the perturbation from the galactic tide. Because the time to be scattered into the Oort cloud is much shorter than the age of the solar system, these brief, stellar-dominated periods of perturbation are ultimately what sets the very minimum semimajor axis at which the perihelia of bodies are torqued into the Oort cloud.

When accounting for the role of stellar passages setting the Oort cloud's inner edge, there is a large amount of stochasticity since it depends on the few most powerful stellar passages over the Sun's history. To estimate the Oort cloud's very inner edge, *Kaib et al.* (2011) examined orbits that had only been weakly torqued from the planets ($60 < q < 100$ au) and measured the tenth percentile of semimajor axis. With

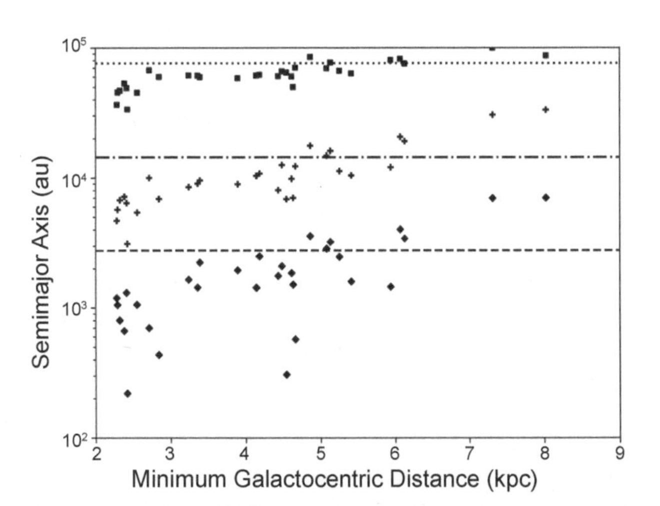

Fig. 12. Plot of maximum (squares, 95th percentile), median (crosses, 50th percentile), and minimum (diamonds, 5th percentile) semimajor axis of Oort cloud as a function of the minimum galactocentric distance attained by the solar analog within a galactic N-body simulation. Adapted from *Kaib et al.* (2011).

this parameterization, they found that the Oort cloud's inner edge can vary by a factor of several (even with a fixed local galactic environment) due to the stochastic nature of stellar passages. In fact, this inner edge was pushed interior to Sedna's orbit in ~20% of their simulations, raising the prospect of whether a solar birth cluster (or other alternative mechanism) is necessary to explain this orbit. These effects of the most powerful stellar encounters can also be seen in the smattering of low-semimajor-axis particles with detached perihelia in Fig. 9. An additional analysis of the stellar encounters experienced by the Sun for different galactic migration scenarios largely agreed with these results (*Martínez-Barbosa et al.*, 2017).

Stellar encounters also play a critical role in maintaining the Oort cloud's orbital isotropy. The vertical term of the galactic tide causes the orbital elements of Oort cloud bodies to fluctuate in a very regular manner (*Heisler and Tremaine,* 1986). Consequently, left unperturbed, the Oort cloud would exhaust the set of orbits that can attain very low ($q < 15$ au) perihelion under the action of the galactic tide, since these bodies are generally lost when they interact with Jupiter and Saturn. Before these regions of orbital space are completely depleted by the galactic tide, however, the randomizing perturbations of stellar encounters replenish them (*Rickman et al.,* 2008). Thus, although the galactic tide delivers most Oort cloud bodies into the planetary region as LPCs, stellar encounters are necessary to ensure that an abundant supply of Oort cloud bodies remain within the phase space that yields LPCs.

3.4.4. Recent stellar encounters with the Sun. As our census of the Sun's nearby stellar neighbors has improved, numerous works have searched for particularly close stellar passages in the recent past as well as the short-term future ($\lesssim\pm10$ m.y.). Using Hipparcos data, *García-Sánchez et al.* (2001) found several potential encounters within 1 pc of the Sun within ±10 m.y. In particular, they noted that the K star Gliese 710 should approach the Sun within 0.4 pc roughly 1.8 m.y. from now. None of the anticipated encounters were projected to deliver significant perturbations to the Oort cloud, but *Rickman et al.* (2012) noted that the Hipparcos catalog was perhaps only ~10–20% complete over this timeline and uncertainties within astrometry and radial velocity prevented confident periastron calculations for close stellar encounters (*García-Sánchez et al.,* 2001).

The ongoing Gaia mission is assembling the three-dimensional space velocities of over 1 billion stars, 4 orders of magnitude greater than Hipparcos, and its astrometric precision will be 2 orders of magnitude finer than Hipparcos. *Rickman et al.* (2012) concluded that the final data release of Gaia (June 2022) should reveal ~90% of the stellar encounters that have contributed to the flux of currently observable LPCs. With the second release of Gaia data, *Bailer-Jones et al.* (2018) found that ~20 stars m.y.$^{-1}$ approach the Sun within at least 1 pc. In addition, with the improved astrometry of Gaia, they found that Gliese 710 has a 95% probability of coming within 17,000 au of the Sun, and later refinements of this star's kinematics showed that it will come within 10^4 au 1.3 m.y. from now (*de la Fuente Marcos and de la Fuente Marcos,* 2020). The stellar encounter flux of *Bailer-Jones et al.* (2018) implies that such a close encounter should only be expected every ~10 m.y., and this is the closest known stellar encounter with the Sun. The improved stellar radial velocities coming with the third Gaia data release may still reveal other close passing stars in the near future or recent past.

3.4.5. Comet showers. Another well-studied effect that stellar encounters have on the Oort cloud dynamics are comet showers. Typically, the only bodies in the Oort cloud that can undergo a perihelion shift from beyond Saturn's orbit to near Earth's orbit within a single revolution about the Sun have semimajor axes ≳20,000 au (*Heisler and Tremaine,* 1986). However, as described in the last section, the perturbative strength of field star encounters varies wildly over time. During the most powerful stellar passages that have occurred in the solar system's history, the perturbations delivered on the Oort cloud are temporarily strong enough to enable this perihelion shift for virtually any semimajor axis in the Oort cloud. This led *Hills* (1981) to argue that most of the time, Earth is only exposed to LPCs from the outer periphery of the Oort cloud, but during powerful stellar passages, Earth is temporarily exposed to the entire Oort cloud, meaning that the flux of LPCs and potential Earth impactors could increase greatly during these events, which are called comet showers.

The potency of comet showers clearly depends on how well-populated the inner (a ≲ 20,000 au) portion of the Oort cloud is relative to the outer cloud. *Hills* (1981) assumed the inner Oort cloud could hold up to 100× as many bodies as the outer cloud, and under this assumption, the most powerful stellar encounter of the past 500 m.y. would temporarily increase the flux of comets near Earth by ~3 orders of magnitude. *Hut et al.* (1987) proposed that the impacts associated with such events could be responsible for one or more of Earth's mass extinctions. However, subsequent Oort cloud formation models predicted a much more modest inner Oort cloud population that is just 1–2× larger than the outer Oort cloud (*Dones et al.,* 2004).

Since it was presumed that Jupiter and Saturn prevented the inner Oort cloud from contributing to the known catalog of LPCs [the vast majority of which have perihelia of a few astronomical units or less (*Marsden and Williams,* 2008)], estimates of the inner Oort cloud population appeared to be completely model dependent. However, *Kaib and Quinn* (2009) showed that objects from the inner Oort cloud can also evolve to become observable LPCs. Illustrated in Fig. 13, as the perihelia of inner Oort cloud bodies slowly march sunward through the outer solar system, some can incur a set of planetary energy kicks that are too weak to eject them but sufficiently strong to inflate their semimajor axes above ~20,000 au. At this point, their perihelia can evolve rapidly toward Earth in a single orbital period, allowing them to become discovered with semimajor axis that obscure their inner Oort cloud source region.

This dynamical pathway was found to be roughly half as efficient at producing observable LPCs as the classic outer Oort cloud production (*Kaib and Quinn,* 2009). Thus, the current known population of LPCs implies that the inner Oort cloud population cannot be much greater than ~2× the number of comets that had previously been estimated to occupy the outer Oort cloud. Otherwise the comet flux would be higher than what is observed. This rather modest inner Oort cloud population calls into question the significance of comet showers' relationship to mass extinctions.

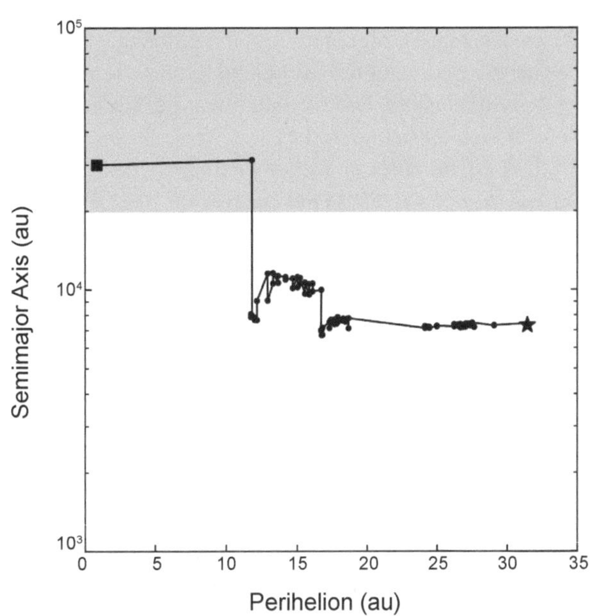

Fig. 13. Plot of the orbital evolution of an Oort cloud body as it enters the solar system's planetary region. Data points mark the semimajor axis vs. the perihelion each time the body enters the inner 50 au of the solar system. The evolution begins at the star data point and ends at the square data point. Letters mark the giant planet orbits along the x-axis. The shaded region marks Oort cloud semimajor axes historically assumed to supply observed LPCs, while unshaded semimajor axes had previously been assumed to be incapable of producing known LPCs.

4. FUTURE DEVELOPMENTS

4.1. Vera C. Rubin Observatory

With a projected first light in the early to mid-2020s, the Vera C. Rubin Observatory promises to revolutionize our understanding of the small-body populations of the solar system as it conducts the Legacy Survey of Space and Time (LSST) over 10 yr (*Ivezić et al.,* 2019). This observatory and survey is projected to discover tens of thousands of TNOs as well as provide the most extensive and uniform set of comet observations and discoveries in history. This will undoubtedly enhance our understanding of comets and their reservoirs.

With an expected limiting magnitude of ~24.5 covering all of the southern sky and the northern sky near the ecliptic plane, LSST's observations of the TNO populations will dwarf the present TNO catalog. To date, surveys reaching those limiting magnitudes have only covered small portions of the sky [e.g., the ~120 deg^2 OSSOS (*Bannister et al.,* 2018)], with all-sky surveys more typically limited to magnitudes brighter than ~22 [e.g., PanStarrs (*Chambers et al.,* 2016)]. LSST's combined wide-sky coverage and deep limiting magnitude will dramatically increase the number of observed TNOs in the scattering and detached populations. This is critical to improving our models of both the intrinsic orbital and size distributions of these populations. Current constraints on these populations have been limited by the fact that that these TNOs spend essentially all of their time at distances where they are unobservable. This limits the ability to directly debias observations (by weighting the observed objects by observability) to build a picture of the intrinsic population because the small current sample sizes make it likely we are missing examples from key regions of their orbital distributions. Likewise, forward modeling via the use of survey simulators to test models of the intrinsic scattering/detached TNO population against observations (see, e.g., *Lawler et al.,* 2018a) must be done carefully because portions of the model phase space will be invisible and thus untestable. The much larger observational sample expected from LSST, and the fact that it will come from a coherent survey with (hopefully) understood biases, will lead to more detailed models of the scattering and detached populations. The very wide latitude range of the observations (especially in the southern sky) will also provide strong new constraints on the inclination distribution of the TNOs. This will provide better tests of solar system formation and giant planet migration scenarios, which will also inform ideas of where different kinds of comets formed. The improved size distribution measurements along with improved orbital distributions will also yield higher-fidelity models of the rate at which comets are supplied to the inner solar system and perhaps help inform models of cometary physical evolution.

Rubin comet observations will dramatically improve our observational constraints on the Oort cloud. At present, our understanding of how cometary absolute magnitudes (H_T) relate to nuclear magnitudes is extremely uncertain. For each comet discovery, LSST will possess perhaps 50 separate temporally spaced observations made with the same instrument. This will allow us to precisely measure the photometric index of each comet as well as observe the comet at epochs when its brightness is less coma-dominated. In this manner, we can hope to measure nuclear magnitudes much more precisely and therefore estimate the size distribution of comets much more precisely. This in turn will enable much better mass estimates of the Oort cloud.

In addition, the Rubin observatory will build up a catalog of LPCs with larger perihelia than most of those that have been previously discovered. This offers a new dynamical window to the Oort cloud because in order for a LPC to attain a perihelion within the terrestrial planet region ($q \lesssim 5$ au), it must either already have possessed a large ($a \gtrsim 20{,}000$ au) semimajor axis, or it must have attained one through previous encounter(s) with the giant planets at larger perihelia (see Fig. 12). Thus, the true semimajor axis distribution of Oort cloud bodies is obscured within our current sample of comets. As the samples of observed cometary perihelia approach and even exceed 10 au, the semimajor axis distribution of LPCs should change, and these changes will depend on the orbital distribution of bodies within the Oort cloud (*Silsbee and Tremaine,* 2018).

Rubin's anticipated TNO discoveries will also be immensely helpful in furthering our understanding of the Oort cloud. As discussed previously, highly inclined Centaurs are a class of objects that potentially originate from the Oort cloud. Their larger sizes and more distant perihelia offer a different set of constraints on the Oort cloud. Currently, ~20 such bodies are known with perihelia beyond Saturn's orbit and inclinations over 45°, and these have been detected with an amalgam of different surveys, making it difficult to account for observational biases within the current catalog. We should expect hundreds of such objects detected by the single LSST survey, giving us a less-biased catalog of highly inclined Centaurs, making it more straightforward to infer their ultimate dynamical source.

LSST will also be capable of detecting Sedna-like bodies out to distances of over 100 au (*Ivezić et al.,* 2019). Thus, our sample of TNOs with large semimajor axes and perihelia very decoupled will increase by manyfold. This will help us discern whether the Sun's cluster environment, a distant planet, or some other mechanism placed these objects on their current orbits.

4.2. Occultation Searches

Observing ~kilometer-sized objects *in situ* within the scattered disk or Oort cloud is not viable due to their extreme faintness. Moreover, the uncertainty of converting cometary brightness into nucleus size leaves our estimates of the population of ~kilometer-sized, or comet-sized, bodies within these reservoirs poorly constrained. While the crater size distribution observed by the New Horizons mission may offer some clues for the scattered disk (*Singer et al.,* 2019; *Morbidelli et al.,* 2021) (see also section 2.2), stellar occultations may be the only tool to detect sizeable numbers of kilometer-sized bodies as they orbit within the scattered disk and Oort cloud (*Bailey,* 1976; *Nihei et al.,* 2007; *Ortiz et al.,* 2020). When such objects pass in front of a star, the star's brightness will dim for a fraction of a second (typically, for a kilometer-sized body at least tens of astronomical units from Earth).

At present only a handful of serendipitous stellar outer solar system occultations have been observed. Using high-cadence stellar observations taken with HST's fine guidance sensors, *Schlichting et al.* (2009) announced the discovery of a stellar occultation of a $r \simeq 500$ m object at ~45 au. With this single detection, they concluded that there was a deficiency of subkilometer-sized bodies within the Kuiper belt (and potentially the scattered disk) compared to a simple extrapolation of the power-law size distribution observed for TNO sizes above 100-km diameters. Analyzing additional HST guidance sensor data, *Schlichting et al.* (2012) discovered a second occultation by a similarly sized TNO at 40 au, which confirmed their previous conclusions about a deficiency of subkilometer-sized TNOs relative to an extrapolation of the ~100-km TNO size distribution. The Taiwanese-American Occultation Survey (TAOS) also searched for such events by monitoring ~500 stars at 5-Hz cadence for 7 yr (*Zhang et al.,* 2013). Although no occultation events were discovered, this non-detection placed an upper limit on the slope of the TNO size distribution below 90 km. A second-generation survey is set to begin operations in 2022 (TAOS-II) that will monitor over 10^4 stars at 20-Hz cadence (*Lehner et al.,* 2012). This occultation survey has a potential rate detection that is 100× greater than its predecessor and will of course provide much tighter constraints on the TNO size distribution between $0.5 \lesssim r \lesssim 30$ km. Although these constraints will largely be supplied by the classical Kuiper belt, they can likely be extrapolated to the scattered disk.

Owing to its much larger average distance, occultation events by kilometer-sized Oort cloud objects will generate much smaller decreases in stellar brightness (*Nihei et al.,* 2007). This may place the detection of kilometer-sized Oort cloud occultations beyond the realm of groundbased surveys (*Nihei et al.,* 2007).

5. SUMMARY

New SPCs and LPCs are steadily produced from the large, distant reservoirs of the scattered disk and the Oort cloud, respectively. The scattered disk was generated early in the solar system's history via the dispersal of a massive primordial belt of planetesimals exterior to the formation region of the giant planets. This dispersal occurred when Neptune migrated outward by several astronomical units and the giant planets likely passed through an orbital instability (e.g., *Malhotra,* 1993; *Nesvorný,* 2018). Less than 1% of these bodies were captured into the modern TNO populations, and an even smaller fraction now reside in the scattered disk. While some of these distant TNOs are actively scattering off the giant planets today (the scattering TNOs, a fraction of which are destined to become JFCs), many are stored in mean-motion resonances with Neptune and/or on orbits whose perihelia are dynamically detached from the planets. These detached orbits become unstable on timescales comparable to the solar system's age and seed the production of new scattering objects and JFCs. Our best current observational constraints imply a population of ~10^5 scattering TNOs larger than ~100 km in diameter (*Lawler et al.,* 2018b). Extrapolations down to comet-sized objects are still very uncertain; models of the delivery of SPCs imply a few times 10^8 scattering TNOs larger than ~2 km in diameter (*Nesvorný et al.,* 2017).

Meanwhile, up to ~5% percent of the dispersed primordial belt bodies are captured into the Oort cloud at semimajor axes of ~10^{3-5} au (*Dones et al.,* 2004). Most of this capture occurs during the first billion years of the solar system's history. As the giant planets scatter these primordial belt bodies to larger semimajor axes, perturbations from passing field stars and the galactic tide torque the perihelia of some of these bodies out of the planetary region, ending the scattering process and (at least temporarily) saving them from ejection. Although most dispersed bodies ultimately scatter off Jupiter, this planet is more prone to ejecting bodies than emplacing them into the Oort cloud because the total orbital energy window of the Oort cloud is smaller than the typical energy kick delivered by Jupiter during a single encounter (*Fernandez and Ip,* 1984; *Duncan et al.,* 1987). Neptune and Uranus are much more efficient populators of the Oort cloud but are less likely to retain dynamical control of bodies throughout the scattering process.

Simulations of the formation of the Oort cloud and scattered disk suggest that ~10× as many bodies should reside in the Oort cloud than in the scattering population (*Dones et al.,* 2004). However, LPC observations suggest that the Oort cloud contains 7–8 × 10^{11} comet-sized bodies (yielding an uncertain mass of 0.5–5 $M_{\oplus}$), which is a factor of ~50–1000× larger than scattered-disk population estimates (*Brasser and Morbidelli,* 2013; *Levison et al.,* 2010). This potentially represents a major discrepancy between formation models and comet observations, but this population ratio estimate relies on a highly uncertain conversion of cometary magnitude to nucleus size.

Simulations have shown that the structure of the Oort cloud is highly sensitive to the Sun's dynamical history within the galaxy, in particular its stellar birth cluster and the degree of radial migration it has undergone within the Milky Way (*Fernández,* 1997; *Kaib et al.,* 2011). Both of these effects can push the Oort cloud's range of semimajor axes closer to the Sun, potentially as close as several hundred astronomical units. The exact inner edge of the Oort cloud is also very dependent on the handful of closest stellar encounters that have occurred over the Sun's history, and the process of setting the inner edge is therefore quite stochastic (*Kaib et al.,* 2011). While it has been previously speculated that the number of comet-sized bodies in the interior 20,000 au of the Oort cloud could be many times the number that are exterior to 20,000 au, we now know that LPCs can be generated from Oort cloud semimajor axes as small as ~5000 au (*Kaib and Quinn,* 2009). This finding places an upper limit of ~10^{12} comet-sized bodies orbiting beyond ~5000 au in the Oort cloud.

To date, much of our understanding of the population and orbital structure of the Oort cloud relies upon observations of comets passing near Earth. In the coming decade, the Vera C. Rubin Observatory will increase the inventory of small solar system bodies by manyfold (*Ivezić et al.,* 2019). Its discoveries will include a sample of comets discovered and observed at large ($\gtrsim$10 au) perihelion and heliocentric distance that will have experienced less thermal processing (by the Sun) and dynamical processing (by Jupiter) compared to the historical comet catalog. In addition, we expect LSST to increase the inventory of observed TNOs by an order of magnitude, including discovering many more objects at high (≳45°) inclinations and with large (~40–100 au) perihelia. Finally, we anticipate the scope and sensitivity of stellar occultation searches to advance in the coming decade, and this should provide new, comet-independent constraints on the number of kilometer-sized bodies in the scattered disk (*Lehner et al.,* 2012). However, a similar effort for the Oort cloud may require a spacebased campaign (*Nihei et al.,* 2007). These anticipated discoveries will provide new windows into the population and structure of the scattered disk and the Oort cloud, whose properties offer clues to the giant planet's orbital evolution as well as the Sun's dynamical history.

Acknowledgments. We thank reviewer J. Fernández and another anonymous reviewer for comments and suggestions that greatly improved this review. N.A.K. thanks the Department of Astronomy at Case Western Reserve University for hosting him as a visitor during this work's preparation. N.A.K. acknowledges support from NSF CAREER award 1846388 and NASA Emerging Worlds grant 80NSSC18K0600. K.V. acknowledges support from NSF (grant AST-1824869) and NASA (grants 80NSSC19K0785 and 80NSSC21K0376).

REFERENCES

Adams E. R., Gulbis A. A. S., Elliot J. L., et al. (2014) De-biased populations of Kuiper belt objects from the Deep Ecliptic Survey. *Astron. J., 148,* 55.
Bahcall J. N. (1984) K giants and the total amount of matter near the Sun. *Astrophys. J., 287,* 926–944.
Bailer-Jones C. A. L., Rybizki J., Andrae R., et al. (2018) New stellar encounters discovered in the second Gaia data release. *Astron. Astrophys., 616,* A37.
Bailey M. E. (1976) Can 'invisible' bodies be observed in the solar system? *Nature, 259,* 290–291.
Bailey M. E. and Stagg C. R. (1988) Cratering constraints on the inner Oort cloud: Steady-state models. *Mon. Not. R. Astron. Soc., 235,* 1–32.
Bannister M. T., Alexandersen M., Benecchi S. D., et al. (2016) OSSOS. IV. Discovery of a dwarf planet candidate in the 9:2 resonance with Neptune. *Astron. J., 152,* 212.
Bannister M. T., Gladman B. J., Kavelaars J. J., et al. (2018) OSSOS. VII. 800+ trans-neptunian objects — The complete data release. *Astrophys. J. Suppl. Ser., 236,* 18.
Batygin K. and Brown M. E. (2010) Early dynamical evolution of the solar system: Pinning down the initial conditions of the Nice model. *Astrophys. J., 716,* 1323–1331.
Batygin K. and Brown M. E. (2016a) Evidence for a distant giant planet in the solar system. *Astron. J., 151,* 22.
Batygin K. and Brown M. E. (2016b) Generation of highly inclined trans-neptunian objects by Planet Nine. *Astrophys. J. Lett., 833,* L3.
Batygin K., Brown M. E., and Fraser W. C. (2011) Retention of a primordial cold classical Kuiper belt in an instability-driven model of solar system formation. *Astrophys. J., 738,* 13.
Bernardinelli P. H., Bernstein G. M., Sako M., et al. (2021) A search of the full six years of the Dark Energy Survey for outer solar system objects. *ArXiV e-prints,* arXiv:2109.03758.
Bernstein G. M., Trilling D. E., Allen R. L., et al. (2004) The size distribution of trans-neptunian bodies. *Astron. J., 128,* 1364–1390.
Boe B., Jedicke R., Meech K. J., et al. (2019) The orbit and size-frequency distribution of long period comets observed by Pan-STARRS1. *Icarus, 333,* 252–272.
Brasser R. and Morbidelli A. (2013) Oort cloud and scattered disc formation during a late dynamical instability in the solar system. *Icarus, 225,* 40–49.
Brasser R. and Schwamb M. E. (2015) Re-assessing the formation of the inner Oort cloud in an embedded star cluster — II. Probing the inner edge. *Mon. Not. R. Astron. Soc., 446,* 3788–3796.
Brasser R. and Wang J. H. (2015) An updated estimate of the number of Jupiter-family comets using a simple fading law. *Astron. Astrophys., 573,* A102.
Brasser R., Duncan M. J., and Levison H. F. (2006) Embedded star clusters and the formation of the Oort cloud. *Icarus, 184,* 59–82.
Brasser R., Duncan M. J., and Levison H. F. (2007) Embedded star clusters and the formation of the Oort cloud. II. The effect of the primordial solar nebula. *Icarus, 191,* 413–433.
Brasser R., Duncan M. J., and Levison H. F. (2008) Embedded star clusters and the formation of the Oort cloud. III. Evolution of the inner cloud during the galactic phase. *Icarus, 196,* 274–284.
Brasser R., Morbidelli A., Gomes R., et al. (2009) Constructing the secular architecture of the solar system II: The terrestrial planets. *Astron. Astrophys., 507,* 1053–1065.
Brasser R., Higuchi A., and Kaib N. (2010) Oort cloud formation at various galactic distances. *Astron. Astrophys., 516,* A72.
Brasser R., Duncan M. J., Levison H. F., et al. (2012a) Reassessing the formation of the inner Oort cloud in an embedded star cluster. *Icarus, 217,* 1–19.
Brasser R., Schwamb M. E., Lykawka P. S., et al. (2012b) An Oort cloud origin for the high-inclination, high-perihelion Centaurs. *Mon. Not. R. Astron. Soc., 420,* 3396–3402.
Brown M. E. (2001) The inclination distribution of the Kuiper belt. *Astron. J., 121,* 2804–2814.
Brown M. E., Trujillo C., and Rabinowitz D. (2004) Discovery of a candidate inner Oort cloud planetoid. *Astrophys. J., 617,* 645–649.
Carusi A., Kresak L., Perozzi E., et al. (1987) High-order librations of Halley-type comets. *Astron. Astrophys., 187,* 899–905.
Chambers K. C., Magnier E. A., Metcalfe N., et al. (2016) The Pan-STARRS1 surveys. *ArXiV e-prints,* arXiv:1612.05560.
Chen Y.-T., Lin H. W., Holman M. J., et al. (2016) Discovery of a new retrograde trans-neptunian object: Hint of a common orbital plane for low semimajor axis, high-inclination TNOs and Centaurs. *Astrophys. J. Lett., 827,* L24.
Davies J. K., McFarland J., Bailey M. E., et al. (2008) The early development of ideas concerning the transneptunian region. In *The Solar System Beyond Neptune* (M. A. Barucci et al., eds.), pp. 11–23. Univ. of Arizona, Tucson.
Dawson R. I. and Murray-Clay R. (2012) Neptune's wild days: Constraints from the eccentricity distribution of the classical Kuiper belt. *Astrophys. J., 750,* 43.
de la Fuente Marcos R. and de la Fuente Marcos C. (2020) An update on the future flyby of Gliese 710 to the solar system using Gaia EDR3: Slightly closer and a tad later than previous estimates. *Res. Notes Am. Astron. Soc., 4,* 222.
Di Sisto R. P. and Rossignoli N. L. (2020) Centaur and giant planet crossing populations: Origin and distribution. *Cel. Mech. Dyn. Astron., 132,* 36.
Dones L., Weissman P. R., Levison H. F., et al. (2004) Oort cloud formation and dynamics. In *Star Formation in the Interstellar Medium: In Honor of David Hollenbach, Chris McKee, and Frank Shu* (D. Johnstone et al., eds.), pp. 371–379. ASP Conf. Ser. 323, Astronomical Society of the Pacific, San Francisco.
Duncan M. J. and Levison H. F. (1997) A scattered comet disk and the origin of Jupiter family comets. *Science, 276,* 1670–1672.
Duncan M., Quinn T., and Tremaine S. (1987) The formation and extent of the solar system comet cloud. *Astron. J., 94,* 1330.
Duncan M., Quinn T., and Tremaine S. (1988) The origin of short-period comets. *Astrophys. J. Lett., 328,* L69.
Duncan M. J., Levison H. F., and Budd S. M. (1995) The dynamical structure of the Kuiper belt. *Astron. J., 110,* 3073.
Edgeworth K. E. (1943) The evolution of our planetary system. *J. Br. Astron. Assoc., 53,* 181–188.
Edgeworth K. E. (1949) The origin and evolution of the solar system. *Mon. Not. R. Astron. Soc., 109,* 600–609.
Eggers S., Keller H. U., Kroupa P., et al. (1997) Origin and dynamics of comets and star formation. *Planet. Space Sci., 45,* 1099–1104.
Elliot J. L., Kern S. D., Clancy K. B., et al. (2005) The Deep Ecliptic Survey: A search for Kuiper belt objects and Centaurs. II. Dynamical classification, the Kuiper belt plane, and the core population. *Astron. J., 129,* 1117–1162.
Emel'yanenko V. V., Asher D. J., and Bailey M. E. (2004) High-eccentricity trans-neptunian objects as a source of Jupiter-family comets. *Mon. Not. R. Astron. Soc., 350,* 161–166.
Emel'yanenko V. V., Asher D. J., and Bailey M. E. (2005) Centaurs from the Oort cloud and the origin of Jupiter-family comets. *Mon. Not. R. Astron. Soc., 361,* 1345–1351.

Everhart E. (1967) Comet discoveries and observational selection. *Astron. J., 72*, 716.
Everhart E. (1972) The origin of short-period comets. *Astrophys. J. Lett., 10*, 131–135.
Everhart E. (1977) The evolution of comet orbits as perturbed by Uranus and Neptune. In *Comets, Asteroids, Meteorites: Interrelations, Evolution and Origins* (A. H. Delsemme, ed.), pp. 99–104. IAU Coll. 39, Univ. of Toledo, Toledo.
Fernández J. A. (1978) Mass removed by the outer planets in the early solar system. *Icarus, 34*, 173–181.
Fernández J. A. (1980a) Evolution of comet orbits under the perturbing influence of the giant planets and nearby stars. *Icarus, 42*, 406–421.
Fernández J. A. (1980b) On the existence of a comet belt beyond Neptune. *Mon. Not. R. Astron. Soc., 192*, 481–491.
Fernández J. A. (1981) New and evolved comets in the solar system. *Astron. Astrophys., 96*, 26–35.
Fernández J. A. (1997) The formation of the Oort cloud and the primitive galactic environment. *Icarus, 129*, 106–119.
Fernández J. (2020) Introduction: The trans-neptunian belt — Past, present and future. In *The Trans-Neptunian Solar System* (D. Prialnik et al., eds.), pp. 1–22. Elsevier, Oxford.
Fernández J. A. and Brunini A. (2000) The buildup of a tightly bound comet cloud around an early Sun immersed in a dense galactic environment: Numerical experiments. *Icarus, 145*, 580–590.
Fernández J. A. and Ip W. H. (1981) Dynamical evolution of a cometary swarm in the outer planetary region. *Icarus, 47*, 470–479.
Fernández J. A. and Ip W. H. (1984) Some dynamical aspects of the accretion of Uranus and Neptune: The exchange of orbital angular momentum with planetesimals. *Icarus, 58*, 109–120.
Fernández J. A. and Sosa A. (2012) Magnitude and size distribution of long-period comets in Earth-crossing or approaching orbits. *Mon. Not. R. Astron. Soc., 423*, 1674–1690.
Fernández J. A., Gallardo T., and Brunini A. (2002) Are there many inactive Jupiter-family comets among the near-Earth asteroid population? *Icarus, 159*, 358–368.
Fernández Y. R., Kelley M. S., Lamy P. L., et al. (2013) Thermal properties, sizes, and size distribution of Jupiter-family cometary nuclei. *Icarus, 226*, 1138–1170.
Fernández J. A., Helal M., and Gallardo T. (2018) Dynamical evolution and end states of active and inactive Centaurs. *Planet. Space Sci., 158*, 6–15.
Fouchard M., Rickman H., Froeschlé C., et al. (2014) Planetary perturbations for Oort cloud comets: II. Implications for the origin of observable comets. *Icarus, 231*, 110–121.
Fouchard M., Rickman H., Froeschlé C., et al. (2017) Distribution of long-period comets: Comparison between simulations and observations. *Astron. Astrophys., 604*, A24.
Francis P. J. (2005) The demographics of long-period comets. *Astrophys. J., 635*, 1348–1361.
Fraser W. C. and Kavelaars J. J. (2009) The size distribution of Kuiper belt objects for D > 10 km. *Astron. J., 137*, 72–82.
Fraser W. C., Brown M. E., and Schwamb M. E. (2010) The luminosity function of the hot and cold Kuiper belt populations. *Icarus, 210*, 944–955.
Fraser W. C., Brown M. E., Morbidelli A., et al. (2014) The absolute magnitude distribution of Kuiper belt objects. *Astrophys. J., 782*, 100.
Fraser W. C., Bannister M. T., Pike R. E., et al. (2017) All planetesimals born near the Kuiper belt formed as binaries. *Nature Astron., 1*, 0088.
Fuentes C. I., George M. R., and Holman M. J. (2009) A Subaru pencil-beam search for m_R ~27 trans-neptunian bodies. *Astrophys. J., 696*, 91–95.
Gaidos E. J. (1995) Paleodynamics: Solar system formation and the early environment of the Sun. *Icarus, 114*, 258–268.
Gallardo T. (2006) The occurrence of high-order resonances and Kozai mechanism in the scattered disk. *Icarus, 181*, 205–217.
García-Sánchez J., Weissman P. R., Preston R. A., et al. (2001) Stellar encounters with the solar system. *Astron. Astrophys., 379*, 634–659.
Gladman B. and Chan C. (2006) Production of the extended scattered disk by rogue planets. *Astrophys. J. Lett., 643*, L135–L138.
Gladman B. and Duncan M. (1990) On the fates of minor bodies in the outer solar system. *Astron. J., 100*, 1680.
Gladman B. and Volk K. (2021) Transneptunian space. *Annu. Rev. Astron. Astrophys., 59*, 203–246.
Gladman B., Holman M., Grav T., et al. (2002) Evidence for an extended scattered disk. *Icarus, 157*, 269–279.
Gladman B., Marsden B. G., and Vanlaerhoven C. (2008) Nomenclature in the outer solar system. In *The Solar System Beyond Neptune* (M. A. Barucci et al., eds.), pp. 43–57. Univ. of Arizona, Tucson.
Gladman B., Kavelaars J., Petit J. M., et al. (2009) Discovery of the first retrograde transneptunian object. *Astrophys. J. Lett., 697*, L91–L94.
Gladman B., Lawler S. M., Petit J. M., et al. (2012) The resonant trans-neptunian populations. *Astron. J., 144*, 23.
Gomes R. S., Morbidelli A., and Levison H. F. (2004) Planetary migration in a planetesimal disk: Why did Neptune stop at 30 AU? *Icarus, 170*, 492–507.
Gomes R. S., Gallardo T., Fernández J. A., et al. (2005) On the origin of the high-perihelion scattered disk: The role of the Kozai mechanism and mean motion resonances. *Cel. Mech. Dyn. Astron., 91*, 109–129.
Gomes R. S., Soares J. S., and Brasser R. (2015) The observation of large semi-major axis Centaurs: Testing for the signature of a planetary-mass solar companion. *Icarus, 258*, 37–49.
Gomes R., Nesvorný D., Morbidelli A., et al. (2018) Checking the compatibility of the cold Kuiper belt with a planetary instability migration model. *Icarus, 306*, 319–327.
Greenstreet S., Gladman B., and McKinnon W. B. (2015) Impact and cratering rates onto Pluto. *Icarus, 258*, 267–288.
Greenstreet S., Gladman B., McKinnon W. B., et al. (2019) Crater density predictions for New Horizons flyby target 2014 MU_{69}. *Astrophys. J. Lett., 872*, L5.
Greenstreet S., Gladman B., and Ngo H. (2020) Transient Jupiter co-orbitals from solar system sources. *Astron. J., 160*, 144.
Hahn J. M. and Malhotra R. (2005) Neptune's migration into a stirred-up Kuiper belt: A detailed comparison of simulations to observations. *Astron. J., 130*, 2392–2414.
Heisler J. and Tremaine S. (1986) The influence of the galactic tidal field on the Oort comet cloud. *Icarus, 65*, 13–26.
Hills J. G. (1981) Comet showers and the steady-state infall of comets from the Oort cloud. *Astron. J., 86*, 1730–1740.
Holman M. J. and Wisdom J. (1993) Dynamical stability in the outer solar system and the delivery of short period comets. *Astron. J., 105*, 1987.
Holman M. J., Payne M. J., Fraser W., et al. (2018) A dwarf planet class object in the 21:5 resonance with Neptune. *Astrophys. J. Lett., 855*, L6.
Hut P., Alvarez W., Elder W. P., et al. (1987) Comet showers as a cause of mass extinctions. *Nature, 329*, 118–126.
Ivezić Ž., et al. (2019) LSST: From science drivers to reference design and anticipated data products. *Astrophys. J., 873*, 111.
Jewitt D. (2009) The active Centaurs. *Astron. J., 137*, 4296–4312.
Joss P. C. (1973) On the origin of short-period comets. *Astron. Astrophys., 25*, 271.
Kaib N. A. and Quinn T. (2008) The formation of the Oort cloud in open cluster environments. *Icarus, 197*, 221–238.
Kaib N. A. and Quinn T. (2009) Reassessing the source of long-period comets. *Science, 325*, 1234.
Kaib N. A. and Sheppard S. S. (2016) Tracking Neptune's migration history through high-perihelion resonant trans-neptunian objects. *Astron. J., 152*, 133.
Kaib N. A., Becker A. C., Jones R. L., et al. (2009) 2006 SQ_{372}: A likely long-period comet from the inner Oort cloud. *Astrophys. J., 695*, 268–275.
Kaib N. A., Roškar R., and Quinn T. (2011) Sedna and the Oort cloud around a migrating Sun. *Icarus, 215*, 491–507.
Kaib N. A., Pike R., Lawler S., et al. (2019) OSSOS. XV. Probing the distant solar system with observed scattering TNOs. *Astron. J., 158*, 43.
Kavelaars J., Petit J.-M., Gladman B., et al. (2021) OSSOS finds an exponential cutoff in the size distribution of the cold classical Kuiper belt. *ArXiV e-prints*, arXiv:2107.06120.
Kenyon S. J. and Bromley B. C. (2004) Stellar encounters as the origin of distant solar system objects in highly eccentric orbits. *Nature, 432*, 598–602.
Kowal C. T. and Gehrels T. (1977) Slow-moving object Kowal. *IAU Circular 3129.*
Kozai Y. (1962) Secular perturbations of asteroids with high inclination and eccentricity. *Astron. J., 67*, 591–598.
Kresák L. (1972) Jacobian integral as a classificational and evolutionary parameter of interplanetary bodies. *Bull. Astron. Inst. Czech., 23*, 1.
Kresák L. and Pittich E. M. (1978) The intrinsic number density of active long-period comets in the inner solar system. *Bull. Astron.*

Inst. Czech., 29, 299.
Kronk G. W. (1999) *Cometography: A Catalog of Comets, Volume 1: Ancient–1799.* Cambridge Univ., Cambridge. 579 pp.
Kuiper G. P. (1951) On the origin of the solar system. *Proc. Natl. Acad. Sci. U.S.A., 37,* 1–14.
Lada C. J. and Lada E. A. (2003) Embedded clusters in molecular clouds. *Annu. Rev. Astron. Astrophys., 41,* 57–115.
Lawler S. M., Kavelaars J. J., Alexandersen M., et al. (2018a) OSSOS: X. How to use a survey simulator: Statistical testing of dynamical models against the real Kuiper belt. *Front. Astron. Space Sci., 5,* 14.
Lawler S. M., Shankman C., Kavelaars J. J., et al. (2018b) OSSOS. VIII. The transition between two size distribution slopes in the scattering disk. *Astron. J., 155,* 197.
Lawler S. M., Pike R. E., Kaib N., et al. (2019) OSSOS. XIII. Fossilized resonant dropouts tentatively confirm Neptune's migration was grainy and slow. *Astron. J., 157,* 253.
Lehner M. J., Wang S.-Y., Alcock C. A., et al. (2012) The Transneptunian Automated Occultation Survey (TAOS II). In *Ground-Based and Airborne Telescopes IV* (L. M. Stepp et al., eds.), p. 84440D. SPIE Conf. Ser. 8444, Bellingham, Washington.
Levison H. F. (1996) Comet taxonomy. In *Completing the Inventory of the Solar System* (T. Rettig and J. M. Hahn, eds.), pp. 173–191. ASP Conf. Ser. 107, Astronomical Society of the Pacific, San Francisco.
Levison H. F. and Duncan M. J. (1994) The long-term dynamical behavior of short-period comets. *Icarus, 108,* 18–36.
Levison H. F. and Duncan M. J. (1997) From the Kuiper belt to Jupiter-family comets: The spatial distribution of ecliptic comets. *Icarus, 127,* 13–32.
Levison H. F., Dones L., and Duncan M. J. (2001) The origin of Halley-type comets: Probing the inner Oort cloud. *Astron. J., 121,* 2253–2267.
Levison H. F., Duncan M. J., Dones L., et al. (2006a) The scattered disk as a source of Halley-type comets. *Icarus, 184,* 619–633.
Levison H. F., Terrell D., Wiegert P. A., et al. (2006b) On the origin of the unusual orbit of Comet 2P/Encke. *Icarus, 182,* 161–168.
Levison H. F., Morbidelli A., Van Laerhoven C., et al. (2008) Origin of the structure of the Kuiper belt during a dynamical instability in the orbits of Uranus and Neptune. *Icarus, 196,* 258–273.
Levison H. F., Duncan M. J., Brasser R., et al. (2010) Capture of the Sun's Oort cloud from stars in its birth cluster. *Science, 329,* 187–190.
Lidov M. L. (1962) The evolution of orbits of artificial satellites of planets under the action of gravitational perturbations of external bodies. *Planet. Space Sci., 9,* 719–759.
Lowry S., Fitzsimmons A., Lamy P., et al. (2008) Kuiper belt objects in the planetary region: The Jupiter-family comets. In *The Solar System Beyond Neptune* (M. A. Barucci et al., eds.), pp. 397–410. Univ. of Arizona, Tucson.
Luu J., Marsden B. G., Jewitt D., et al. (1997) A new dynamical class of object in the outer solar system. *Nature, 387,* 573–575.
Lykawka P. S. and Mukai T. (2007) Resonance sticking in the scattered disk. *Icarus, 192,* 238–247.
Malhotra R. (1993) The origin of Pluto's peculiar orbit. *Nature, 365,* 819–821.
Malhotra R. (1995) The origin of Pluto's orbit: Implications for the solar system beyond Neptune. *Astron. J., 110,* 420.
Marsden B. G. (2009) Orbital properties of Jupiter-family comets. *Planet. Space Sci., 57,* 1098–1105.
Marsden B. G. and Williams G. V. (2008) *Catalogue of Cometary Orbits, 17th edition.* Smithsonian Astrophysical Observatory, Cambridge. 195 pp.
Martínez-Barbosa C. A., Jílková L., Portegies Zwart S., et al. (2017) The rate of stellar encounters along a migrating orbit of the Sun. *Mon. Not. R. Astron. Soc., 464,* 2290–2300.
Matese J. J., Whitman P. G., Innanen K. A., et al. (1995) Periodic modulation of the Oort cloud comet flux by the adiabatically changing galactic tide. *Icarus, 116,* 255–268.
McKinnon W. B., Richardson D. C., Marohnic J. C., et al. (2020) The solar nebula origin of (486958) Arrokoth, a primordial contact binary in the Kuiper belt. *Science, 367,* aay6620.
Morbidelli A. and Levison H. F. (2004) Scenarios for the origin of the orbits of the trans-neptunian objects 2000 CR_{105} and 2003 VB_{12} (Sedna). *Astron. J., 128,* 2564–2576.
Morbidelli A., Nesvorný D., Bottke W. F., et al. (2021) A re-assessment of the Kuiper belt size distribution for sub-kilometer objects, revealing collisional equilibrium at small sizes. *Icarus, 356,* 114256.
Morris D. E. and Muller R. A. (1986) Tidal gravitational forces: The infall of "new" comets and comet showers. *Icarus, 65,* 1–12.
Murray C. D. and Dermott S. F. (1999) *Solar System Dynamics.* Cambridge Univ., Cambridge. 592 pp.
Murray-Clay R. A. and Chiang E. I. (2005) A signature of planetary migration: The origin of asymmetric capture in the 2:1 resonance. *Astrophys. J., 619,* 623–638.
Napier K. J., Gerdes D. W., Lin H. W., et al. (2021) No evidence for orbital clustering in the extreme trans-neptunian objects. *Planet. Sci. J., 2,* 59.
Nesvorný D. (2011) Young solar system's fifth giant planet? *Astrophys. J. Lett., 742,* L22.
Nesvorný D. (2015a) Evidence for slow migration of Neptune from the inclination distribution of Kuiper belt objects. *Astron. J., 150,* 73.
Nesvorný D. (2015b) Jumping Neptune can explain the Kuiper belt kernel. *Astron. J., 150,* 68.
Nesvorný D. (2018) Dynamical evolution of the early solar system. *Annu. Rev. Astron. Astrophys., 56,* 137–174.
Nesvorný D. and Morbidelli A. (2012) Statistical study of the early solar system's instability with four, five, and six giant planets. *Astron. J., 144,* 117.
Nesvorný D. and Vokrouhlický D. (2016) Neptune's orbital migration was grainy, not smooth. *Astrophys. J., 825,* 94.
Nesvorný D., Vokrouhlický D., and Roig F. (2016) The orbital distribution of trans-neptunian objects beyond 50 au. *Astrophys. J. Lett., 827,* L35.
Nesvorný D., Vokrouhlicky D., Dones H. C. L., et al. (2017) Origin and evolution of short-period comets. *AAS/Division for Planetary Sciences Meeting Abstracts, 49,* #401.01.
Nesvorný D., Vokrouhlický D., Dones L., et al. (2017) Origin and evolution of short-period comets. *Astrophys. J., 845,* 27.
Nihei T. C., Lehner M. J., Bianco F. B., et al. (2007) Detectability of occultations of stars by objects in the Kuiper belt and Oort cloud. *Astron. J., 134,* 1596–1612.
Noll K., Grundy W. M., Nesvorný D., et al. (2020) Trans-neptunian binaries (2018). In *The Trans-Neptunian Solar System* (D. Prialnik et al., eds.), pp. 201–224. Elsevier, Oxford.
Oikawa S. and Everhart E. (1979) Past and future orbit of 1977 UB, object Chiron. *Astron. J., 84,* 134–139.
Oort J. H. (1950) The structure of the cloud of comets surrounding the solar system and a hypothesis concerning its origin. *Bull. Astron. Inst. Netherlands, 11,* 91–110.
Öpik E. (1932) Note on stellar perturbations of nearly parabolic orbits. *Proc. Am. Acad. Arts Sci., 67,* 169–183.
Öpik E. J. (1976) *Interplanetary Encounters: Close-Range Gravitational Interactions.* Elsevier, Amsterdam. 155 pp.
Ortiz J. L., Sicardy B., Camargo J. I. B., et al. (2020) Stellar occultation by TNOs: From predictions to observations. In *The Trans-Neptunian Solar System* (D. Prialnik et al., eds.), pp. 413–437. Elsevier, Oxford.
Peixinho N., Thirouin A., Tegler S. C., et al. (2020) From Centaurs to comets: 40 years. In *The Trans-Neptunian Solar System* (D. Prialnik et al., eds.), pp. 307–329. Elsevier, Oxford.
Petit J. M., Kavelaars J. J., Gladman B. J., et al. (2011) The Canada-France Ecliptic Plane Survey — Full data release: The orbital structure of the Kuiper belt. *Astron. J., 142,* 131.
Pike R. E., Kavelaars J. J., Petit J. M., et al. (2015) The 5:1 Neptune resonance as probed by CFEPS: Dynamics and population. *Astron. J., 149,* 202.
Pike R. E., Fraser W. C., Schwamb M. E., et al. (2017) Col-OSSOS: z-band photometry reveals three distinct TNO surface types. *Astron. J., 154,* 101.
Quinn T., Tremaine S., and Duncan M. (1990) Planetary perturbations and the origin of short-period comets. *Astrophys. J., 355,* 667.
Rickman H. (1976) Stellar perturbations of orbits of long-period comets and their significance for cometary capture. *Bull. Astron. Inst. Czech., 27,* 92–105.
Rickman H., Fernández J. A., and Gustafson B. A. S. (1990) Formation of stable dust mantles on short-period comet nuclei. *Astron. Astrophys., 237,* 524–535.
Rickman H., Fouchard M., Froeschlé C., et al. (2008) Injection of Oort cloud comets: The fundamental role of stellar perturbations. *Cel. Mech. Dyn. Astron., 102,* 111–132.
Rickman H., Fouchard M., Froeschlé C., et al. (2012) Gaia and the new comets from the Oort cloud. *Planet. Space Sci., 73,* 124–129.
Roškar R., Debattista V. P., Quinn T. R., et al. (2008) Riding the spiral waves: Implications of stellar migration for the properties of galactic

disks. *Astrophys. J. Lett.*, *684*, L79.

Sarid G., Volk K., Steckloff J. K., et al. (2019) 29P/Schwassmann-Wachmann 1, a Centaur in the gateway to the Jupiter-family comets. *Astrophys. J. Lett.*, *883*, L25.

Schlichting H. E., Ofek E. O., Wenz M., et al. (2009) A single sub-kilometre Kuiper belt object from a stellar occultation in archival data. *Nature*, *462*, 895–897.

Schlichting H. E., Ofek E. O., Sari R., et al. (2012) Measuring the abundance of sub-kilometer-sized Kuiper belt objects using stellar occultations. *Astrophys. J.*, *761*, 150.

Scholl H. (1979) History and evolution of Chiron's orbit. *Icarus, 40*, 345–349.

Schwamb M. E., Fraser W. C., Bannister M. T., et al. (2019) Col-OSSOS: The colors of the Outer Solar System Origins Survey. *Astrophys. J. Suppl. Ser.*, *243*, 12.

Sellwood J. A. and Binney J. J. (2002) Radial mixing in galactic discs. *Mon. Not. R. Astron. Soc.*, *336*, 785–796.

Shankman C., Gladman B. J., Kaib N., et al. (2013) A possible divot in the size distribution of the Kuiper belt's scattering objects. *Astrophys. J. Lett.*, *764*, L2.

Shankman C., Kavelaars J., Gladman B. J., et al. (2016) OSSOS. II. A sharp transition in the absolute magnitude distribution of the Kuiper belt's scattering population. *Astron. J.*, *151*, 31.

Shankman C., Kavelaars J. J., Bannister M. T., et al. (2017a) OSSOS. VI. Striking biases in the detection of large semimajor axis trans-neptunian objects. *Astron. J.*, *154*, 50.

Shankman C., Kavelaars J. J., Lawler S. M., et al. (2017b) Consequences of a distant massive planet on the large semimajor axis trans-neptunian objects. *Astron. J.*, *153*, 63.

Sheppard S. S., Trujillo C. A., Tholen D. J., et al. (2019) A new high perihelion trans-plutonian inner Oort cloud object: 2015 TG_{387}. *Astron. J.*, *157*, 139.

Shoemaker E. M. and Wolfe R. F. (1984) Evolution of the Uranus-Neptune planetesimal swarm. In *Lunar Planet. Sci. XV*, Abstract #1395, pp. 780–781. LPI Contribution No. 1581, Lunar and Planetary Institute, Houston.

Silsbee K. and Tremaine S. (2018) Producing distant planets by mutual scattering of planetary embryos. *Astron. J.*, *155*, 75.

Singer K. N., McKinnon W. B., Gladman B., et al. (2019) Impact craters on Pluto and Charon indicate a deficit of small Kuiper belt objects. *Science*, *363*, 955–959.

Smullen R. A. and Volk K. (2020) Machine learning classification of Kuiper belt populations. *Mon. Not. R. Astron. Soc.*, *497*, 1391–1403.

Snodgrass C., Fitzsimmons A., Lowry S. C., et al. (2011) The size distribution of Jupiter family comet nuclei. *Mon. Not. R. Astron. Soc.*, *414*, 458–469.

Sosa A. and Fernández J. A. (2011) Masses of long-period comets derived from non-gravitational effects — analysis of the computed results and the consistency and reliability of the non-gravitational parameters. *Mon. Not. R. Astron. Soc.*, *416*, 767–782.

Spencer J., Grundy W. M., Nimmo F., et al. (2020) The Pluto system after New Horizons. In *The Trans-Neptunian Solar System* (D. Prialnik et al., eds.), pp. 271–288. Elsevier, Oxford.

Stern A. and Campins H. (1996) Chiron and the Centaurs: Escapees from the Kuiper belt. *Nature*, *382*, 507–510.

Tegler S. C., Romanishin W., and Consolmagno G. J. (2003) Color patterns in the Kuiper belt: A possible primordial origin. *Astrophys. J. Lett.*, *599*, L49–L52.

Tiscareno M. S. and Malhotra R. (2003) The dynamics of known Centaurs. *Astron. J.*, *126*, 3122–3131.

Torbett M. V. (1989) Chaotic motion in a comet disk beyond Neptune: The delivery of short-period comets. *Astron. J.*, *98*, 1477–1481.

Trujillo C. A. and Sheppard S. S. (2014) A Sedna-like body with a perihelion of 80 astronomical units. *Nature*, *507*, 471–474.

Tsiganis K., Gomes R., Morbidelli A., et al. (2005) Origin of the orbital architecture of the giant planets of the solar system. *Nature*, *435*, 459–461.

Vokrouhlický D., Nesvorný D., and Dones L. (2019) Origin and evolution of long-period comets. *Astron. J.*, *157*, 181.

Volk K. and Malhotra R. (2008) The scattered disk as the source of the Jupiter family comets. *Astrophys. J.*, *687*, 714–725.

Volk K., Murray-Clay R. A., Gladman B. J., et al. (2018) OSSOS. IX. Two objects in Neptune's 9:1 resonance — Implications for resonance sticking in the scattering population. *Astron. J.*, *155*, 260.

Weissman P. R. (1980) Physical loss of long-period comets. *Astron. Astrophys.*, *85*, 191–196.

Weissman P. R. (1996) The Oort cloud. In *Completing the Inventory of the Solar System* (T. Rettig and J. M. Hahn, eds.), pp. 265–288. ASP Conf. Ser. 107, Astronomical Society of the Pacific, San Francisco.

Zhang Z. W., Lehner M. J., Wang J. H., et al. (2013) The TAOS Project: Results from seven years of survey data. *Astron. J.*, *146*, 14.

Zheng J.-Q., Valtonen M. J., and Valtaoja L. (1990) Capture of comets during the evolution of a star cluster and the origin of the Oort cloud. *Cel. Mech. Dyn. Astron.*, *49*, 265–272.

Fraser W. C., Dones L., Volk K., Womack M., and Nesvorný D. (2024) The transition from the Kuiper belt to the Jupiter-family comets. In *Comets III* (K. J. Meech, M. R. Combi, D. Bockelée-Morvan, S. N. Raymond, and M. E. Zolensky, eds.), pp. 121–151. Univ. of Arizona, Tucson, DOI: 10.2458/azu_uapress_9780816553631-ch005.

The Transition from the Kuiper Belt to the Jupiter-Family Comets

Wesley C. Fraser
Herzberg Astronomy and Astrophysics Research Centre, National Research Council of Canada

Luke Dones
Southwest Research Institute

Kathryn Volk
Planetary Science Institute/University of Arizona

Maria Womack
National Science Foundation and University of Central Florida

David Nesvorný
Southwest Research Institute

Kuiper belt objects (KBOs), or more generally transneptunian objects (TNOs), are planetesimals found beyond the orbit of Neptune. Some TNOs evolve onto Neptune-crossing orbits and become Centaurs. Many Centaurs in turn reach Jupiter-crossing orbits and become Jupiter-family comets (JFCs). TNOs are the main source of the JFCs. TNOs offer a different window than the JFCs, of more primordial bodies and over a different size and temperature range. It is in that context that this chapter is written. Here we discuss the dynamical pathways taken from the transneptunian region to the JFCs, and the most important properties of TNOs that relate to the JFC population, including considerations of their origins, compositions, morphologies, and size distributions. We relate these properties to the JFCs whenever possible. We reflect on a few key outstanding issues regarding our incomplete knowledge of TNOs as they pertain to the Centaurs and JFC populations. We finish with a short discussion of notable new and upcoming facilities and the impacts they will have regarding these outstanding questions.

1. INTRODUCTION

Oort (1950) proposed the first, and for some time the only, reservoir for comets. He noted that, of 19 long-period comets with well-determined orbits and semimajor axes $a \gtrsim 1000$ au, 10 had $a > 20,000$ au before they entered the planetary region. These "near-parabolic" comets had a wide range of inclinations to the ecliptic, both prograde and retrograde. Oort proposed that the solar system was surrounded by a roughly spherical cloud of $\approx 10^{11}$ comets that extended beyond $\approx 10^5$ au from the Sun, about half the distance to the nearest star.

In the wonderful words of *Kazimirchak-Polonskaya* (1972), the major planets can indeed serve as "powerful transformers of cometary orbits," with some evolving from near-parabolic orbits onto orbits like those of Jupiter-family comets (JFCs), which we discuss in section 1.2 (*Everhart,* 1972). However, the "capture" of Oort cloud comets into short-period orbits is far too inefficient to produce the observed number of JFCs (*Joss,* 1973). This mismatch suggested that there must be another source of short-period comets. Several astronomers, notably *Edgeworth* (1949) and *Kuiper* (1951), suggested the existence of a belt of low-inclination comet-like bodies beyond Neptune's orbit, ≈35–50 au from the Sun [see *Davies et al.* (2008) and *Fernández* (2020) for historical accounts of such ideas, which percolated even before the discovery of Pluto in 1930]. *Fernandez* (1980) calculated that such a belt could be the source of short-period comets if it contained bodies up to the mass of Ceres ($\approx 10^{24}$ g, or 7% the mass of Pluto). *Duncan et al.* (1988)

performed extensive numerical integrations and confirmed that only a low-inclination source could produce enough JFCs and proposed that this source be called the Kuiper belt. The first Kuiper belt object (KBO), 15760 Albion (1992 QB_1), was discovered four years later (*Jewitt and Luu,* 1993). (The community has referred to the objects beyond Neptune as KBOs and TNOs, largely interchangeably. We adopt the latter here as a more generalized and less historically charged term.) *Duncan et al.* (1988) also remarked that Chiron, the only Centaur known at the time, "may well be a bright member of the parent population of [short-period] comets." Hundreds of Centaurs are now known, including P/2019 LD_2 (ATLAS), an active body that passed within ≈0.1 au of Jupiter in 2017 and will do again in 2028 and 2063 (*Steckloff et al.,* 2020). The 2063 encounter, the closest of the three, is expected to place the comet on an orbit with a semimajor axis $a \approx 4$ au and a perihelion distance $q < 3$ au, making it a JFC.

Because of the inverse-fourth-power dependence on reflected flux density with distance, the known TNOs are typically much larger than the Centaurs, which, in turn, are generally larger than the nuclei of JFCs. Most known TNOs have diameters on the order of 100 km, ranging up to Pluto and Eris (2300–2400 km); Centaurs are generally in the range of 10 to 100 km, with Chariklo the largest at 300 km; the nuclei of JFCs range from ≈0.3 km up to more than 10 km. These size differences should be kept in mind when comparing different classes of bodies.

In the remainder of this section, we summarize basic orbital properties of TNOs, Centaurs, and comets, particularly the JFCs. Section 2 discusses how the Kuiper belt attained its complex structure and the orbital evolution of TNOs to Centaurs and JFCs. In section 3, we discuss the compositions of TNOs and Centaurs revealed by broadband colors and reflectance spectroscopy. Section 4 examines the active Centaurs, whose activity must be driven by substances more volatile than water ice; the issue of whether the frequent appearance of bilobed structures in JFCs reflects morphological evolution after leaving the Kuiper belt or dates to the formation of these bodies billions of years ago; and the size distributions of TNOs, Centaurs, and JFCs. Section 5 concludes with a discussion of outstanding questions and future observations that may clarify many of the issues we discuss. This chapter addresses a variety of different topics regarding TNOs. Rather than providing a detailed review of a specific topic, this chapter aims to provide an overview of the most relevant TNO science relating to the field of cometary astronomy.

Many chapters in this book discuss subjects that we do not have room to describe. We particularly note the chapters on the dynamical populations of comet reservoirs by Kaib and Volk; planetesimal and comet formation by Simon et al.; cometary nuclei by Guilbert-Lepoutre et al., Pajola et al., Filacchione et al., and Knight et al.; past and future comet missions by Snodgrass et al.; and transition objects chapter by Jewitt and Hsieh, especially their discussion of asteroids on cometary orbits.

1.1 Meet the Kuiper Belt

Three decades after the discovery of the first transneptunian object (TNO) (besides Pluto/Charon), the dynamical structure of the Kuiper belt has proven to be more complex than that of the asteroid belt (see *Gladman and Volk,* 2021, for a recent review). TNOs are typically classified based on 10-m.y. orbital integrations, as this timescale reveals the population's relevant key dynamical behaviors [a scheme that began with early observational surveys (e.g., *Elliot et al.,* 2005). *Gladman et al.* (2008) established a widely-used classification scheme (Fig. 1) that divides outer solar system objects into the following categories:

- Jupiter-coupled: Objects with semimajor axes exterior to Jupiter, but with Tisserand parameters with respect to Jupiter less than 3.05, consistent with cometary orbits (note that the Tisserand parameter T of a comet with respect to a planet with semimajor axis a_P is $T = a_P/a + 2 \cos i \sqrt{a(1-e^2)/a_P}$, where a, e, and i are the semimajor axis, eccentricity, and inclination of the comet with respect to the planet's orbital plane; the Tisserand parameter is an approximation to the Jacobi integral, which is a constant of motion for the circular restricted three-body problem, and for Jupiter-coupled comets, the Tisserand parameter varies more gradually than other orbital parameters, such as the comet's period of revolution around the Sun (*Kresák,* 1972; *Carusi and Valsecchi,* 1987; *Levison,* 1996)
- Centaurs: Non-Trojan objects with semimajor axes between those of Jupiter (5.2 au) and Neptune (30.1 au)
- Resonant: Objects in mean-motion resonances (MMRs) with Neptune
- Scattering: Objects whose semimajor axes vary by at least 1.5 au over 10 m.y., which typically coincides with perihelion distances ≲38–40 au
- Classical: Non-resonant, non-scattering objects with eccentricities <0.24, an arbitrary cutoff that restricts

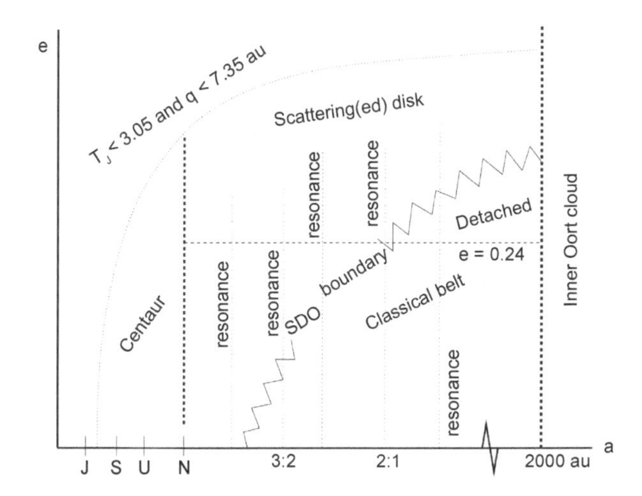

Fig. 1. Nomenclature scheme for objects with semimajor axes beyond Jupiter's orbit (from *Gladman et al.* 2008). This taxonomy primarily focuses on TNOs, and, by design, excludes most JFCs. We use a similar classification.

this population largely to semimajor axes between 30 and ≈50 au

- Detached: Objects on stable orbits that fall into none of the previous categories

The population of classical objects residing between Neptune's 3:2 and 2:1 resonances (39.4 < a < 47.8 au) is often referred to as the "main belt," with classical objects interior and exterior to those resonances referred to as inner and outer classical belt objects. Figure 2 shows a set of classified observed outer solar system objects that illustrates a few key dynamical features of this population: the prevalence of resonant objects, a bimodal inclination distribution extending to quite high i, and a mix of scattering and detached objects with a wide range of perihelion distances in the high-a population.

Populations of dynamically stable objects in Neptune's MMRs were apparent from early surveys (e.g., *Chiang and Jordan,* 2002; *Elliot et al.,* 2005) and predicted by early theoretical studies (e.g., *Duncan et al.,* 1995; *Malhotra,* 1995). Although the 3:2 MMR, which hosts Pluto and numerous "Plutinos," contains more *known* TNOs than other resonances, the raw numbers are misleading. This is because bodies in the 3:2 have shorter orbital periods and smaller perihelion distances than most TNOs, and so are easier to discover. Analyses accounting for observational biases imply that a wide range of Neptune's resonances in the classical belt and in more distant regions have large populations,

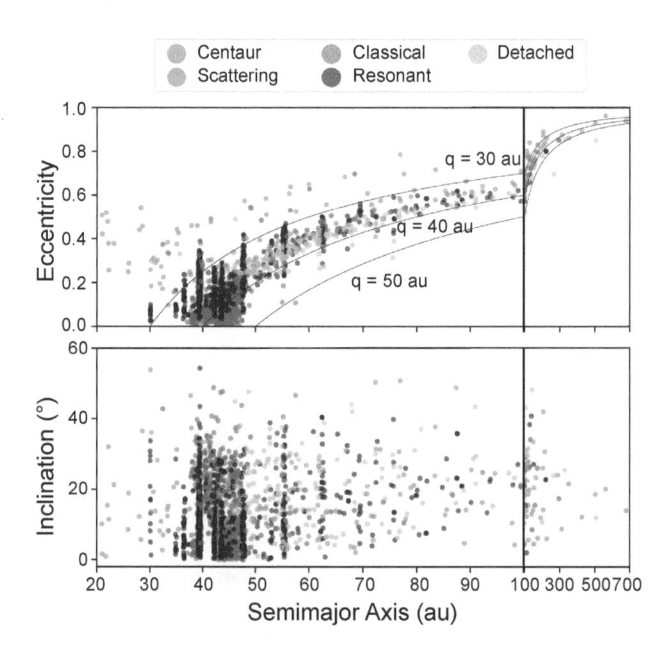

Fig. 2. Eccentricity and inclination vs. semimajor axis of ~1800 observed TNOs color-coded by dynamical class; the set of TNOs is the full OSSOS sample (*Bannister et al.,* 2018) combined with a subset of other TNOs in the Minor Planet Center database with classifications taken from *Smullen and Volk* (2020). The three solid lines in the top panel labeled with values between 30, 40, and 50 represent lines of constant perihelion distance. Modified from *Gladman and Volk* (2021). See Fig. 8 of *Bernardinelli et al.* (2022) for a similar plot of discoveries by the Dark Energy Survey.

comparable to that of the 3:2 MMR (e.g., *Gladman et al.,* 2012; *Adams et al.,* 2014; *Volk et al.,* 2016; *Chen et al.,* 2019; *Crompvoets et al.,* 2022). This has important implications for the dynamical history of the outer solar system (see section 2.1). By the late 1990s, it was clear that the orbits of observed TNOs extended to higher inclinations than did orbits of main-belt asteroids. However, there is a concentration of low-inclination orbits in the classical belt from a = 42–45 au. The bimodal nature of the inclination distribution in the classical Kuiper belt region was first demonstrated by *Brown* (2001). The low-inclination classical belt objects also tend to have low eccentricities; this sparked the division of TNO populations into categories of dynamically "hot" and "cold" subpopulations [meant to invoke thermodynamics because the "hot" population have larger radial and vertical velocities than "cold" objects (*Levison and Stern,* 2001)]. The cold population is essentially entirely constrained to a = 42–45 au [although with possible extensions just beyond 50 au (see, e.g., *Bannister et al.,* 2018)], while the scattering, detached, and majority of the resonant populations all fall into the dynamically hot category (the implications of which are discussed in section 2.1).

At large semimajor axes, the observed population is a mix of resonant, scattering, and detached objects spanning an unexpectedly wide range of perihelion distances from 30 to 80 au. The existence of a scattered disk (*Luu et al.,* 1997; *Duncan and Levison,* 1997; *Volk and Malhotra,* 2008) or scattering population (*Gladman et al.,* 2008; *Gladman and Volk,* 2021) with perihelion distances less than ≈37–40 au is a natural consequence of objects being fed onto Neptune-crossing orbits, either today or in the early solar system (e.g., *Levison and Duncan,* 1997; *Lykawka and Mukai,* 2007; also see the discussion in the chapter in this volume by Kaib and Volk). Objects with these smaller perihelion distances can experience perturbations from Neptune at perihelion that are strong enough to cause significant changes in orbital energy, resulting in a random walk in semimajor axis. This creates the fan-like structure in a–e space seen in Fig. 2, where objects are dispersed in semimajor axis along lines of constant perihelion distance (the dashed curves). Interactions with Neptune's resonances can cause some objects' perihelia to be raised and lowered, causing them to cycle between scattering, resonant, and detached states; other objects appear to be permanently stranded in high-perihelion detached orbits, some likely via resonant interactions during planet migration (reviewed in, e.g., *Gomes et al.,* 2008). There is a subset of detached objects with such large perihelia [90377 Sedna with q = 76 au, 2012 VP_{113} with q = 80 au (*Brown et al.,* 2004; *Trujillo and Sheppard,* 2014; *Sheppard et al.,* 2019)] that they require another explanation beyond the perturbations of the giant planets (see, e.g., *Gladman and Volk,* 2021). Additional large-a, large-q objects that might bridge gaps between the scattering, detached, and perhaps the inner Oort cloud populations are also beginning to be discovered (e.g., 541132 Leleakuhonua (2015 TG_{387}) with q = 65 au and a ≈ 1000 au). As these "extreme" objects do not contribute to the supply of short-period comets, we will not discuss them at length here (their implications for the solar system's history are touched on in section 2.1, and their possible relationship to the inner Oort cloud is discussed in the chapter by Kaib and Volk chapter).

1.2. Meet the Comets

Most comets likely formed beyond the water "snow line" ≈3 au from the Sun [however, the snow line's location changes as the protoplanetary disk evolves (*Öberg et al.,* 2011; *Marboeuf et al.,* 2014; *Harsono et al.,* 2015; *Cieza et al.,* 2016; *Drozdovskaya et al.,* 2016)] and within ≈30 au (*Nesvorný,* 2018) in the solar system's initial planetesimal disk.

Comets are commonly divided into groups based on their orbital periods. In the traditional taxonomy, long-period comets (LPCs) are defined as those with orbital periods P > 200 yr, corresponding to a > 34.2 au, and short-period, or periodic, comets have P < 200 yr. The 200-yr boundary has no fundamental significance dynamically or compositionally (see, e.g., *A'Hearn et al.,* 1995; *Mumma and Charnley,* 2011).

Long-period comets are thought to have been gravitationally scattered by interactions with the giant planets from their closer-in formation regions (≈3–30 au) out to distances between 2000 and ~100,000 au to form a collective known as the Oort cloud (see, e.g., *Duncan et al.,* 1987; *Vokrouhlický et al.,* 2019). At those distances, perturbations external to the solar system, including passing stars and galactic tides, raise and lower comets' perihelion distances in and out of the giant planet region, and cause their orbits to take on nearly random orientations (see, e.g., *Weissman,* 1996; *Dones et al.,* 2004) (see also the chapter by Kaib and Volk in this volume for further discussion of long-period comets and Oort cloud formation).

The short-period comets can be further divided into subpopulations (*Horner et al.,* 2003; *Seligman et al.,* 2021). Most are classified as JFCs, many of which have aphelion distances near Jupiter's distance from the Sun (Fig. 3). Most JFCs and Centaurs originate from the dynamically hot transneptunian populations (section 1.1) and orbit the Sun on prograde orbits with inclinations smaller than ~30°, which is why they are also sometimes referred to as ecliptic comets (e.g., *Levison and Duncan,* 1997). While most JFCs originate from the TNO populations, other populations likely contribute as well; section 2.1 discusses this in further detail. The smaller semimajor axes and inclinations of JFC orbits, which are not consistent with an Oort cloud origin, pointed the way to the existence of the Kuiper belt/TNO populations prior to their discovery (e.g., *Edgeworth,* 1949; *Fernandez,* 1980; *Duncan et al.,* 1988). This connection was later confirmed by models of the newly observed reservoir (e.g., *Levison and Duncan,* 1997; *Duncan and Levison,* 1997).

Historically, JFCs were often defined as comets with P < 20 yr (a < 7.4 au), although a variety of definitions appear in the literature. JFCs are now often taken to be comets with Tisserand parameters with respect to Jupiter, T_J, between 2 and 3 (*Carusi and Valsecchi,* 1987). This is the definition suggested by *Levison and Duncan* (1997) as being the most indicative of an origin in the Kuiper belt. Slight variations on the upper T_J limit for JFCs exist. We adopt the upper limit for JFCs of $T_J < 3.05$ from *Gladman et al.* (2008). By contrast, *Jewitt et al.* (2015) set the dividing line between JFCs and active asteroids, some of which are main-belt comets, at $T_J = 3.08$.

Levison (1996) defined "Encke-type comets" (ETCs) as comets with $a < a_J$ and $T_J > 3$ to distinguish "decoupled" comets like 2P/Encke, whose aphelion distance is ≈1 au interior to Jupiter's orbit, from typical JFCs that undergo close encounters with Jupiter. We do not find this category useful, as objects defined this way are a mix of asteroids and comets, as we discuss below. We consider Encke, which has $T_J = 3.025$, to be a JFC on an unusual orbit. Encke's perihelion distance q = 0.34 au (smaller than Mercury's semimajor axis) is unique among comets with $T_J > 3$.

Encke is generally assumed to originate in the Kuiper belt, but numerical simulations that reproduce the comet's decoupled orbit require it to remain active for an implausibly long time (*Valsecchi et al.,* 1995; *Fernández et al.,* 2002; *Levison et al.,* 2006b).

About half of the other "Encke-type comets" listed in the JPL Small Body Database are active asteroids (see the chapter in this volume by Jewitt and Hsieh). The other half are JFCs that do not cross Jupiter's orbit, but still come within 2 Hill radii of the planet and interact strongly with it. Many of these JFCs are quasi-Hilda comets (*Tancredi et al.,* 1990; *Di Sisto et al.,* 2005; *Toth,* 2006; *Ohtsuka et al.,* 2008; *Gil-Hutton and García-Migani,* 2016), which undergo slow, and thus strong encounters with Jupiter, sometimes leading to temporary captures by the planet. Indeed, D/1993 F2 (Shoemaker-Levy 9) may have been a quasi-Hilda before its capture by Jupiter and impactful demise (*Chodas and Yeomans,* 1996).

Halley-type comets (HTCs) were once defined as periodic comets with orbital periods between 20 and 200 yr (*Carusi et al.,* 1987), but now are typically classified as periodic comets with $T_J < 2$ (e.g., *Levison and Duncan,* 1994; *Levison,* 1996). HTCs generally have larger semimajor axes (and higher inclinations) than the JFCs, but the new classification includes "Sun-skirting" comets like 96P/Machholz with a ≈ 3 au and perihelion distances well within the orbit of Mercury (*Jones et al.,* 2018).

Halley-type comets have often been taken to be the shortest-period tail of the distribution of returning Oort cloud comets. However, the population of known HTCs is small, about 10% that of JFCs, in large part because HTCs' (generally) infrequent appearances provide few opportunities for discovery. The orbital distribution of Halley-type comets is therefore not very well constrained (e.g., *Wiegert and Tremaine,* 1999). HTCs can be produced both from the scattering TNO population (e.g., *Brasser et al.,* 2012; *Levison et al.,* 2006a) and from the Oort cloud (e.g., *Wang and Brasser,* 2014; *Nesvorný et al.,* 2017). More discussion of Encke- and Halley-type comets can be found in the chapter

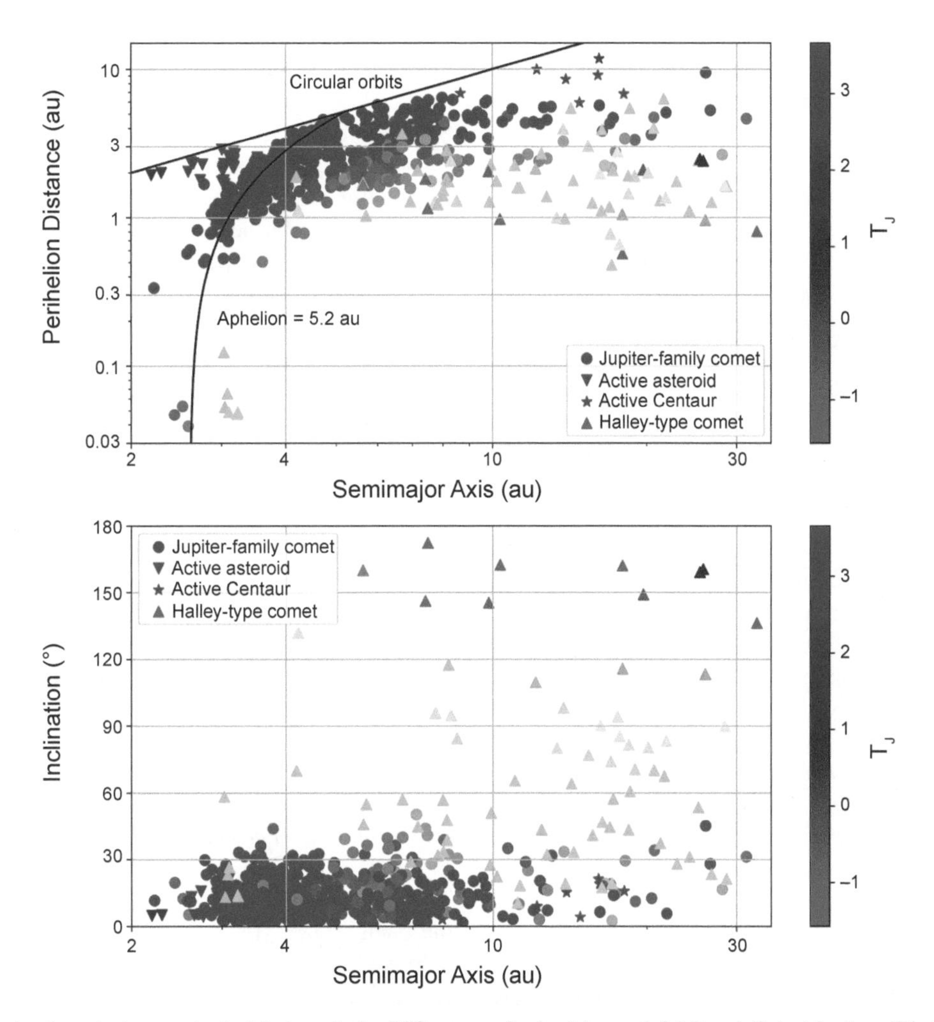

Fig. 3. Plot of short-period comets (orbital period <200 yr, equivalent to a $\lesssim$ 34.2 au) listed in the JPL Small-Body Database browser. We use the definition of *Gladman et al.* (2008) for Jupiter-family comets (JFCs), Tisserand parameter relative to Jupiter $2 < T_J < 3.05$. JFCs are denoted by circles; active asteroids ($T_J \geq 3.05$, $a < a_J$) by downward-pointing arrows); active Centaurs ($T_J \geq 3.05$, $a \geq a_J$) by stars; and Halley-type comets ($T_J \leq 2$) by upward-pointing arrows. The top panel shows perihelion distance vs. semimajor axis, while the bottom panel shows inclination vs. semimajor axis. The color scale shows T_J for each comet (right colorbars). The lines in the top panel denote circular orbits and aphelion distance equal to Jupiter's semimajor axis of 5.2 au. Objects between the lines have orbits entirely within Jupiter's orbit.

in this volume by Kaib and Volk.

The population of Centaurs, small bodies with orbits in the giant planet region, provides the link between the outer solar system TNO populations and the inner solar system short-period comets. Most Centaurs are former TNOs that evolved onto Neptune-encountering orbits and were scattered inward to the giant planet region. There they have relatively short dynamical lifetimes of ~1–10 m.y. (*Tiscareno and Malhotra,* 2003; *Horner et al.,* 2004; *Di Sisto and Brunini,* 2007; *Bailey and Malhotra,* 2009; *Fernández et al.,* 2018) and either evolve inward onto short-period comet orbits or outward onto highly eccentric orbits in the outer solar system; the details of this evolution are discussed in section 2.2. Cometary activity has been detected on some Centaurs with perihelion distances as large as ≈12 au (e.g., *Jewitt,* 2009; *Lilly et al.,* 2021) (see discussion in section 4.1), further demonstrating that they are the transition population between the TNOs and JFCs. However, the presence of activity does not by itself show that the populations are related. For instance, although some active asteroids have sublimation-driven activity and are thus *bona fide* comets, most are thought to have resided in the asteroid belt for billions of years.

2. ORIGINS OF TRANSNEPTUNIAN OBJECTS AND ECLIPTIC COMETS

2.1. Dynamical Origins of Transneptunian Objects

How was the Kuiper belt populated? The belt is thought to be a relic of an ancient massive planetesimal disk outside the orbits of the giant planets (*Nesvorný,* 2018). *Fernandez and Ip* (1984) showed that gravitational interactions of the giant planets with small bodies cause Jupiter to lose angular momentum (because it ultimately ejects most such bodies from the solar system) and Saturn, Uranus, and Neptune to gain angular momentum. If the planetesimal disk is massive enough (comparable to Uranus or Neptune), these interactions can change the planets' orbits substantially. *Malhotra* (1993, 1995) showed that the outward planetesimal-driven migration of Neptune would result in capture of Pluto and "Plutinos" into the 3:2 MMR. This model predicted that TNOs would also be captured into other MMRs, notably the 2:1 resonance near 48 au. TNOs in the 2:1 and many other MMRs were soon found by surveys such as the Dark Energy Survey (DES) (*Elliot et al.,* 2005), Canada-France Ecliptic Plane Survey (CFEPS) (*Petit et al.,* 2011), and Outer Solar System Origins Survey (OSSOS) (*Bannister et al.,* 2018) (see section 1.1).

In these migration scenarios, the original orbits of the giant planets would have been more closely spaced than they are now. In the Nice model (*Tsiganis et al.,* 2005), Jupiter and Saturn are initially closer together then cross a mutual 2:1 resonance as a result of interactions with the planetesimal disk, exciting their eccentricities and triggering an instability in the orbits of all the giant planets (*Tsiganis et al.,* 2005; *Levison et al.,* 2008; *Brasser et al.,* 2009; *Morbidelli et al.,* 2010; *Levison et al.,* 2011; *Nesvorný and Morbidelli,* 2012; *Izidoro et al.,* 2016; *Gomes et al.,* 2018; *Quarles and Kaib,* 2019; *Nesvorný et al.,* 2021). More recent models of planetary instabilities have addressed problems with the Nice model, such as the excitation of excessively large eccentricities and inclinations in the terrestrial planets (*Brasser et al.,* 2009; *Agnor and Lin,* 2012). Efforts to match the intricate orbital distributions of the transneptunian populations, asteroid belt, and other small-body reservoirs have also led to substantial revision of the Nice model (see reviews by *Dones et al.,* 2015; *Nesvorný,* 2018; *Malhotra,* 2019; *Morbidelli and Nesvorný,* 2020; *Raymond and Nesvorný,* 2022). For instance, most current models invoke a fifth giant planet that was later ejected from the solar system by Jupiter (*Nesvorný,* 2011; *Nesvorný and Morbidelli,* 2012; *Batygin et al.,* 2012; *Cloutier et al.,* 2015) and find that the instability probably took place early, in the first 10–100 m.y. of the solar system (*Deienno et al.,* 2017; *Mojzsis et al.,* 2019; *De Sousa et al.,* 2020), rather than after hundreds of millions of years (*Gomes et al.,* 2005a). Our evolving understanding of the timing of the instability has implications for the pre-migration evolution of the disk, including the likely collisional histories of the planetesimals that eventually become the TNOs that feed the short-period comet populations (e.g., *Morbidelli and Rickman,* 2015).

During the planetary instability, most bodies in the proto-transneptunian disk are ejected from the solar system, but some reach quasi-stable niches, where they survive to the present. Models predict that the Oort cloud and TNO populations contain the overwhelming majority of the survivors. Based on the known populations of small bodies, particularly Jupiter Trojans, and the capture efficiency found by modeling the instability, the disk is estimated to have originally contained about 6×10^9 bodies with diameters d $\gtrsim$ 10 km (*Morbidelli et al.,* 2009b; *Nesvorný,* 2018). The instability model of *Nesvorný et al.* (2019b), evolved to the present, predicts the number of Centaurs of this size that the Outer Solar System Origins Survey (OSSOS) discovered to within a factor of 2 (also see *Bottke et al.,* 2022). However, we know little about how many such objects reside in the TNO populations and especially the Oort cloud. The fraction of those objects that reach the Oort cloud and TNO populations and survive today are estimated to be $\eta = 0.05 \pm 0.01$ (*Vokrouhlickýet al.,* 2019) and 0.004 (*Nesvorný et al.,* 2017), respectively, corresponding to $6 \times 10^9 \times 0.05 = 3 \times 10^8$ d > 10 km bodies in the Oort cloud and 2×10^7 in the TNOs. Bodies from the disk are also implanted into the asteroid main belt [$\eta \approx 5–8 \times 10^{-6}$ (*Levison et al.,* 2009; *Vokrouhlickýet al.,* 2016)], as Hildas and Trojans of Jupiter [each with $\eta \approx (6 \pm 1) \times 10^{-7}$ (*Nesvorný et al.,* 2013; *Vokrouhlický et al.,* 2016)], and as irregular satellites of Jupiter, Uranus, and Neptune [each with $\eta \approx 2 - –3 \times 10^{-8}$) and Saturn [$\eta \approx 5 \times 10^{-8}$ (*Nesvorný et al.,* 2014)], yielding some 40,000 main-belt asteroids, 4000 Hildas, 4000 Trojans, and 100–300 irregular satellites of each giant planet.

Finding scenarios for the early evolution of the outer planets that are consistent with the complex orbital distribution of TNOs (section 1.1) is difficult and a topic of active research. In general, some amount of planetesimal-driven migration (e.g., *Fernandez and Ip,* 1984; *Malhotra,* 1993, 1995) is invoked to explain the large number of resonant TNOs, while some sort of dynamical instability among the giant planets (e.g., *Tsiganis et al.,* 2005) is invoked to help explain both the planets' orbits and help dynamically excite the TNOs. Three main migration models have been proposed, which depend on the value of Neptune's eccentricity e_N at that time: Model A: very low-eccentricity ($e_N \lesssim 0.01$) migration of Neptune from <25 au to 30 au (e.g., *Malhotra,* 1993, 1995; *Hahn and Malhotra,* 2005); Model B: instability-driven scattering of Neptune from <20 au to an eccentric orbit ($e_N \sim 0.3$) at ~30 au, and subsequent circularization of Neptune's orbit by dynamical friction from the planetesimal disk (e.g., *Tsiganis et al.,* 2005; *Levison et al.,* 2008); and Model C: an intermediate case in which Neptune's migration is interrupted by the instability, with its eccentricity reaching a peak of $e_N = 0.03–0.1$, and then dropping to $\simeq 0.01$ as Neptune slowly migrates toward 30 au (e.g., *Nesvorný and Morbidelli,* 2012; *Nesvorný and Vokrouhlický,* 2016). As discussed below, these different kinds of migration models result in different distributions of the TNO populations that feed into the current comet populations.

The *original* mass and radial span of the Kuiper belt are unknown, but models that initially put several Earth masses beyond 30 au run into problems with Neptune's migration, because Neptune continues to migrate past that distance (*Gomes et al.,* 2004). Migration beyond 30 au can also result in excess mass in the residual belt; today's TNOs with r < 50 au represent <0.1 $M_\oplus$ (*Di Ruscio et al.,* 2020). A sharp truncation of the original massive disk, an exponential cutoff of the disk surface density near 30 au, or some combination of the two is required to match Neptune's orbit (*Nesvorný et al.,* 2020). Specifically, the disk's mass density at 30 au must have been $\lesssim 1\ M_\oplus\ au^{-1}$ for Neptune to stop at 30 au (*Nesvorný,* 2018). The original disk most likely continued, with a low surface density, to 45 au, where the cold classical TNOs formed and survived (*Batygin et al.,* 2011). The current mass of cold TNOs is estimated to be only $\sim 3 \times 10^{-4}\ M_\oplus$ (*Fraser et al.,* 2014) to $(3 \pm 2) \times 10^{-3}\ M_\oplus$ (*Nesvorný et al.,* 2020). These values assume that TNOs have densities of 1 g cm^{-3}. The vast majority of the dynamically hot TNO populations that feed the Centaur and short-period-comet populations originate in the more massive, closer-in portion of this original planetesimal disk.

The three models outlined above have different implications for the hot TNO populations. A pure instability model (Model A) does not explain the wide orbital inclination distribution of hot TNOs (*Petit et al.,* 2011; *Nesvorný,* 2015) because the instability happens so fast that there is not enough time to sufficiently excite the orbital inclinations. The existence of cold classical TNOs also limits how large Neptune's eccentricity can be when it reaches its current semimajor axis [$e_N < 0.15$ (*Dawson and Murray-Clay,* 2012)]. In all three models, planetesimals originating from the <30-au portion of the disk are scattered by Neptune to higher-a and high-e orbits; to end up in today's metastable hot TNO populations, these orbits must subsequently be stabilized by some dynamical mechanism that causes their orbital eccentricities to drop. The Kozai resonance (*Kozai,* 1962) near and inside MMRs with Neptune is presumably the dominant implantation mechanism during periods of slow migration (*Nesvorný,* 2020). The Kozai resonance produces anti-correlated oscillations of e and i. As the TNO's orbital eccentricity decreases, its orbit can decouple from Neptune scattering (i.e., the TNO's perihelion distance can evolve to beyond Neptune's aphelion distance) or drop out of resonance while the TNO's orbital inclination increases. The ν_{18} secular resonance (which occurs when a TNO's orbital nodal precession rate matches one of the dominant nodal precession rates of the giant planets, and is the mode most associated with Neptune's nodal rate) can also influence the inclination distribution (*Volk and Malhotra,* 2019), but it does not wipe out the Kozai signature. When the Kozai implantation dominates for smooth migration (Model A), it is difficult to obtain implanted hot TNO orbits with $i < 10°$ (Fig. 4). The finer details of these stabilization mechanisms and how they affect the efficiency of implantation from the original planetesimal disk into the hot TNO populations is still not fully explored, even in the simplest planetesimal driven migration scenario (Model A) (see discussion in *Volk and Malhotra,* 2019).

In general, the results of migration models better match the observed hot TNO populations if Neptune's orbit is modestly excited by an instability at some point during migration (Model C). This is particularly true for matching the inclination distribution of the hot TNOs (Fig. 4). In this case, bodies are implanted in the hot TNO populations as the eccentricities of high-a and high-e orbits drop due to the ν_8 secular resonance (which occurs when a TNO's perihelion precession rate matches one of the dominant perihelion precession rates of the giant planets, the mode most associated with Neptune's perihelion precession rate), which is stronger in this case, because $e_N \neq 0$ (*Nesvorný,* 2021); here the contribution of Kozai cycles is relatively minor. As the ν_8 resonance does not affect inclinations, the inclination distribution is roughly preserved during implantation. The inclination distribution is then primarily controlled

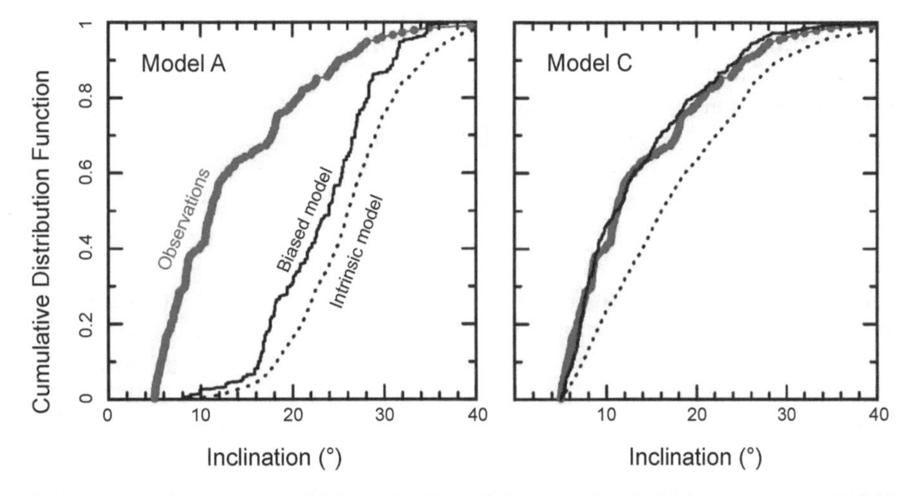

Fig. 4. The cumulative distribution function (CDF) of TNO orbital inclinations obtained in dynamical Models A (left panel) and C (right panel). The intrinsic-model, biased-model and observed distributions are shown by dotted, solid, and red lines, respectively. The observed distribution shows all detections from the Outer Solar System Origins Survey (OSSOS) with $40 < a < 47$ au, $q > 36$ au, and $I > 5°$ (*Bannister et al.,* 2018). The biased model distribution was obtained by applying the OSSOS survey simulator (*Lawler et al.,* 2018a) to the intrinsic model. In Model A, Neptune migrated from 24 au to 30 au with $e_N \simeq 0.01$ on a 10-Myr timescale. In Model C, Neptune's eccentricity was excited to $e_N \simeq 0.1$ when Neptune reached 28 au and slowly damped after that such that $e_N \simeq 0.01$ in the end. All planetesimals shown here started within 30 au of the Sun.

by Neptune's scattering and the ν_{18} secular resonance; in some simulations, the slower Neptune's migration is, the broader the inclination distribution becomes (*Nesvorný,* 2015), although this is not always a generic outcome of slow migration simulations (*Volk and Malhotra,* 2019). This model, assuming that Neptune's migration was long-range and slow (e-folding time $\gtrsim$10 m.y.) better matches the Kuiper belt's orbital structure, such as its inclination distribution (Fig. 4) and the fraction of TNOs in resonant orbits (*Nesvorný,* 2018).

The scattered disk is thought to be the main source of ecliptic comets (or JFCs) (*Duncan and Levison,* 1997; *Levison and Duncan,* 1997; *Volk and Malhotra,* 2008; *Brasser and Morbidelli,* 2013; *Nesvorný et al.,* 2017) [see section 2.2 and *Dones et al.* (2015) for a review of this topic]. The dynamical structure of the inner detached disk (50–100 au), with dropout bodies — objects that fell out of resonance during migration on the sunward side of MMRs (*Bernardinelli et al.,* 2022) — is an important constraint on Neptune's migration (*Kaib and Sheppard,* 2016; *Nesvorný et al.,* 2016; *Lawler et al.,* 2019). The efficiency of implantation into these populations and how the inner detached disk may feed the actively scattering population on long timescales depends on the details of migration. The scattered disk decayed by a factor of ~100–300 since its formation 4.5 G.y. ago [see, e.g., the scattered disk/Oort cloud formation and evolution models of *Brasser and Morbidelli* (2013) and *Nesvorný et al.* (2017)]. This is reflected by a rapidly decreasing number of ecliptic comets and planetary impactors soon after the instability, and a gradual decrease during the past ≈4 G.y.

2.2. Dynamical Routes from the Centaurs to the Jupiter-Family Comets

The short-period comets are fed into the inner solar system from the Neptune-crossing population of TNOs. This is predominantly the population of scattering objects (see the discussion in the chapter in this volume by Kaib and Volk), although other subpopulations of TNOs contribute as well (see section 2.3). Some objects on Neptune-crossing orbits will be scattered inward into the Centaur population, from which point their evolution is dominated by gravitational scattering by the giant planets. Objects can spend upwards of ~100 m.y. in the scattering TNO population with perihelion very near or interior to Neptune. Once an object has been transferred onto an orbit in the giant planet region, it takes ~1–10 m.y. for that object to either be ejected back into the scattering population or transferred into the JFC population (e.g., *Tiscareno and Malhotra,* 2003; *Di Sisto and Brunini,* 2007; *Bailey and Malhotra,* 2009; *Sarid et al.,* 2019; *Di Sisto and Rossignoli,* 2020).

Figure 5 illustrates the typical timespan objects spend traversing each part of the Centaur region [which in Fig. 5 is defined as an orbit entirely enclosed between Jupiter and Neptune, q > 5.2 au and Q < 30.1 au, which is slightly more restrictive than the *Gladman et al.* (2008) definition we typically use]. Most of the time an object spends in the

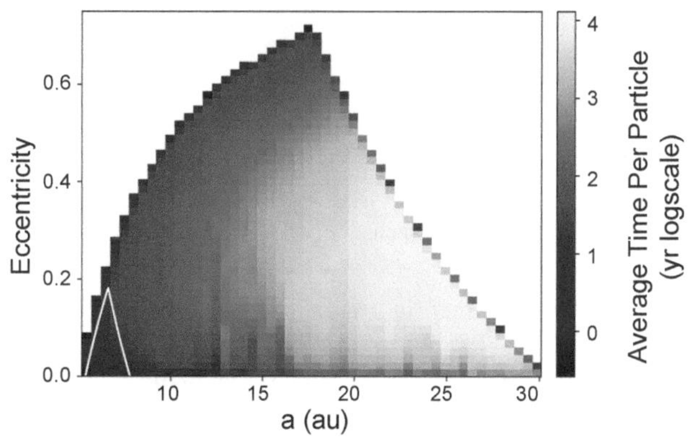

Fig. 5. The colormap shows the time-weighted orbital distribution of Centaurs (using a definition whereby Centaurs' orbits are entirely enclosed in the Jupiter-Neptune region) in a simulation of their evolution from the outer solar system into the JFC population. From *Sarid et al.* (2019).

Centaur population is spent in the Uranus-Neptune region because encounters with these planets are gentler than those with Jupiter and Saturn. Even so, 20% of Centaurs will be ejected back into the outer solar system before making it inside Uranus' orbit; only roughly half the Centaurs will be transferred onto orbits inside Saturn's, with roughly one-third making it past Jupiter to become JFCs (*Sarid et al.,* 2019; see also, e.g., *Tiscareno and Malhotra,* 2003; *Di Sisto and Rossignoli,* 2020). On the order of 1% of Centaurs will end their journeys via impact with one of the giant planets (e.g., *Levison et al.,* 2000; *Tiscareno and Malhotra,* 2003; *Raymond et al.,* 2018; *Wong et al.,* 2021). Dynamical evolution in the Centaur region is relatively insensitive to the exact source population in the outer solar system, although Centaurs with high orbital inclinations do tend to have longer lifetimes in the giant planet region due to the decreased probability of close encounters with the planets (e.g., *Di Sisto and Rossignoli,* 2020). Very high inclination and retrograde Centaurs are not well-explained by the observed populations of TNOs described in section 1.1 because the journey through the giant planet region does not radically alter the inclination distribution of Centaurs relative to their source population (e.g., *Brasser et al.,* 2012; *Volk and Malhotra,* 2013). The observed high-inclination Centaurs and scattering objects (e.g., *Gladman et al.,* 2009; *Chen et al.,* 2016) likely originate from non-TNO source regions (see discussion in the chapter by Kaib and Volk).

The approximately one-third of Centaurs that become JFCs (*Levison and Duncan,* 1997; *Tiscareno and Malhotra,* 2003; *Di Sisto and Brunini,* 2007; *Bailey and Malhotra,* 2009; *Fernández et al.,* 2018) typically transfer into the inner solar system via a low-eccentricity orbit just exterior to Jupiter like that of Comet 29P/Schwassmann-Wachmann 1; this orbital region, dubbed the JFC Gateway (*Sarid et al.,* 2019), represents an important phase in the transition from Centaur to JFC because it coincides with thermal conditions that allow for significant cometary activity from sublimation of water ice (see, e.g., *Steckloff et al.,* 2020). The Centaur to JFC transition is, of course, reversible, with objects passing back into the Centaur population (also typically through a Gateway orbit). How many JFCs survive, and in what state, to reenter the Centaur population remains an open question.

Dynamical models of JFC orbital evolution indicate that, after becoming JFCs, it can take up to ~0.5 m.y. for a comet to be dynamically removed (e.g., *Levison and Duncan,* 1994; *Di Sisto et al.,* 2009). However, during this extended orbital evolution, the JFCs' orbital inclinations are pumped up to larger and larger values, driving the overall simulated JFC inclination distribution to disagree with the observed one (e.g., *Levison and Duncan,* 1994; *Di Sisto et al.,* 2009; *Nesvorný et al.,* 2017). To solve this mismatch between simulated and observed JFC populations, it is usually assumed that a comet becomes inactive after some number of orbits with a perihelion distance below some threshold (often $q \lesssim 2.5$ au). Fits to the inclination distribution of JFCs with diameters between 1 and 10 km suggest physical lifetimes of ≈4000–40,000 yr (≈500–5000 orbits), which is a small fraction of the dynamical lifetimes (e.g., *Levison and Duncan,* 1997; *Di Sisto et al.,* 2009; *Brasser and Wang,* 2015; *Nesvorný et al.,* 2017). Subkilometer JFC nuclei must have much shorter physical lifetimes. The exact nature of this fading [i.e., whether it is due to buildup of a surface lag layer ("mantle") (e.g., *Rickman et al.* (1990) or physical destruction of small comet nuclei through spin-up or other mechanisms] is not well-understood; observations of dormant comet nuclei in various solar system regions (including the Centaur population) could help constrain this problem. An additional complication is that the activity of JFCs can turn on and off repeatedly. For instance, Comet Blanpain (now known as 289P) was discovered in 1819, observed for two months, and not seen again for almost two centuries until it was identified with the asteroid 2003 WY_{25} and found to have a weak coma (*Jewitt,* 2006).

2.3. Kuiper Belt Sources of Centaurs and Comets

Dynamical models of the evolution of transneptunian populations confirm that the dynamically hot populations, particularly the scattering population, dominate the influx of new Centaurs and then JFCs in the inner solar system. Figure 6, for example, highlights the original orbits of particles that evolve onto cometary orbits from a model of the TNOs (*Nesvorný et al.,* 2017). Most of these comet-supplying TNO orbits have perihelion distances below ~37–38 au, placing them in the scattering population, the one originally argued to be the dominant supplier of JFCs (e.g., *Duncan and Levison,* 1997). However, it is clear that a few other TNO populations also contribute to the influx. Figure 6 shows a number of JFCs sourced from Neptune's resonant populations (vertical features most evident in the top panel). Most notable are the close-in 3:2 and 2:1 resonances, which are known to have large populations (e.g., *Volk et al.,* 2016); chaotic diffusion within these resonances (and others) can slowly feed objects onto Neptune-crossing orbits that supply the JFCs (e.g., *Morbidelli,* 1997; *Tiscareno and Malhotra,* 2009). Figure 6 also shows a handful of high-perihelion

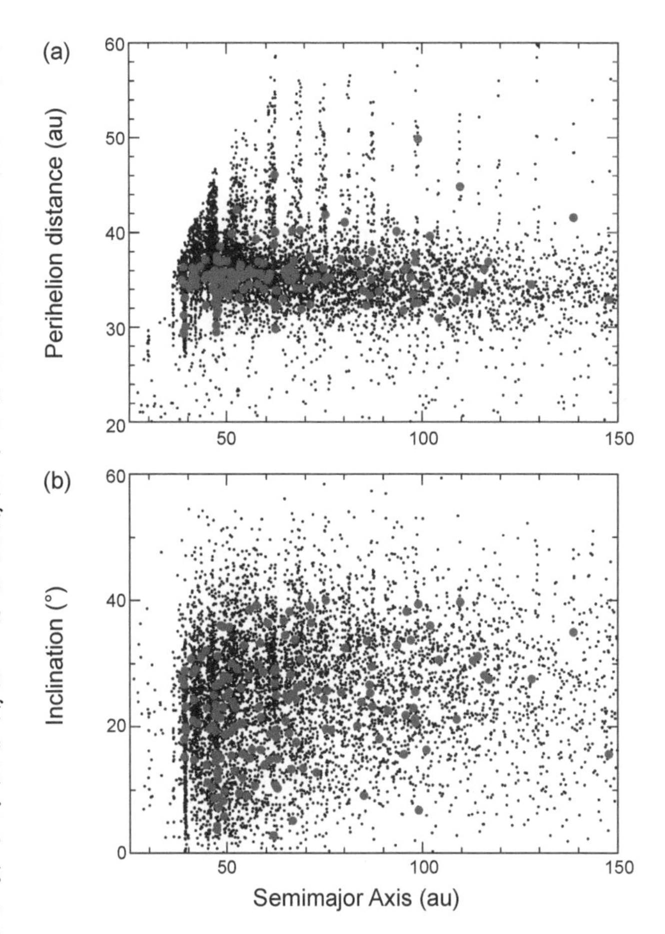

Fig. 6. A model of the TNO population (modeled from formation to the present day) with the source-region orbits of eventual ecliptic comets highlighted. The black dots show a snapshot of the TNO population from ~3 G.y. ago; objects from that snapshot that later become ecliptic comets (defined as $2 < T_J < 3$, $P < 20$ yr, $q < 2.5$ au) are shown as larger red dots. About 75% of TNOs that become short-period comets in this simulation originate in the scattered disk, with semimajor axes between 50 and 200 au; ~20% originate from TNOs with $a < 50$ au (some of which enter the 50–200-au scattering population en route); about 3% come from the Oort cloud. From *Nesvorný* (2018), which was adapted from *Nesvorný et al.* (2017).

TNO orbits that nonetheless evolve into JFCs. Most of these simulated objects appear to be in Neptune's more distant MMRs, where secular effects cause their perihelia to cycle in and out of Neptune's reach (e.g., *Gomes et al.,* 2005b). In the real TNO population, it is still not entirely clear how many of the so-called "detached" TNOs on high-perihelion orbits are still in Neptune's resonances (and thus could cycle back into the scattering population) and how many are truly stranded at high perihelia [for example, by being dropped out of resonance during the late stages of Neptune's migration (e.g., *Kaib and Sheppard,* 2016; *Lawler et al.,* 2019)]; this is because the orbits of very large semimajor axis TNOs are difficult to measure to high enough precision for this determination (discussed in, e.g., *Gladman and Volk,* 2021). The orbital distribution of the detached TNOs and how much that population can cycle in and out of the actively scattering population is one of the larger uncertainties in determining the exact delivery rate of JFCs from the transneptunian region [also relevant to this are the effects of any unknown perturbers in the distant solar system (see discussion in *Nesvorný et al.,* 2017)]. However, following the evolution of our best models of the TNO populations into the inner solar system (*Levison and Duncan,* 1997; *Di Sisto et al.,* 2009; *Brasser and Morbidelli,* 2013; *Nesvorný et al.,* 2017) results in an excellent match to the observed orbital distribution of comets (Fig. 7), confirming that the majority of JFCs likely originate as TNOs.

For completeness, we note that a few other solar system populations may contribute to the observed JFCs. These include the low-inclination Themis family in the outer main asteroid belt (*Hsieh et al.,* 2020) (also see the chapter by Jewitt and Hsieh in this volume), the Hildas (*Di Sisto et al.,* 2005) in the 3:2 resonance with Jupiter near 4 au, and the Jupiter and Neptune Trojans (*Horner and Lykawka,* 2010; *Di Sisto et al.,* 2019). These sources are likely minor suppliers of the JFC population compared to the TNOs. Bodies from the primordial Kuiper belt were implanted into all these populations (see section 2.1), so JFCs from different present-day sources may not show clear signs of their birthplaces. Indeed, two large asteroids near the middle of the asteroid belt, 203 Pompeja and 269 Justitia, have recently been proposed to have originated as TNOs, based on their colors, which are even redder than D-type Trojans and Hildas (*Hasegawa et al.,* 2021).

3. COMPOSITIONS AND PHYSICAL PROPERTIES

In this section, we discuss the compositions of TNOs, Centaurs, and JFCs as observed from Earth- and spacecraft-based telescopic observations. Measured properties of these objects are valuable for constraining solar system formation models (*Barucci et al.,* 2011; *van der Wiel et al.,* 2014). As Centaurs are thought to be an intermediate stage in the orbital evolution from TNOs to JFCs, intercomparison of their compositions can provide insights into how the journey from the scattering TNO population inward influences outgassing behavior and observed coma composition. Where available, we discuss the few compositional links between these dynamically connected populations.

3.1. Surface Compositions of Transneptunian Objects and Centaurs

Beyond atmospheric escape through processes such as Jeans loss (*Lisse et al.,* 2021), TNOs are generally inactive bodies. As such, specific material identification on surfaces

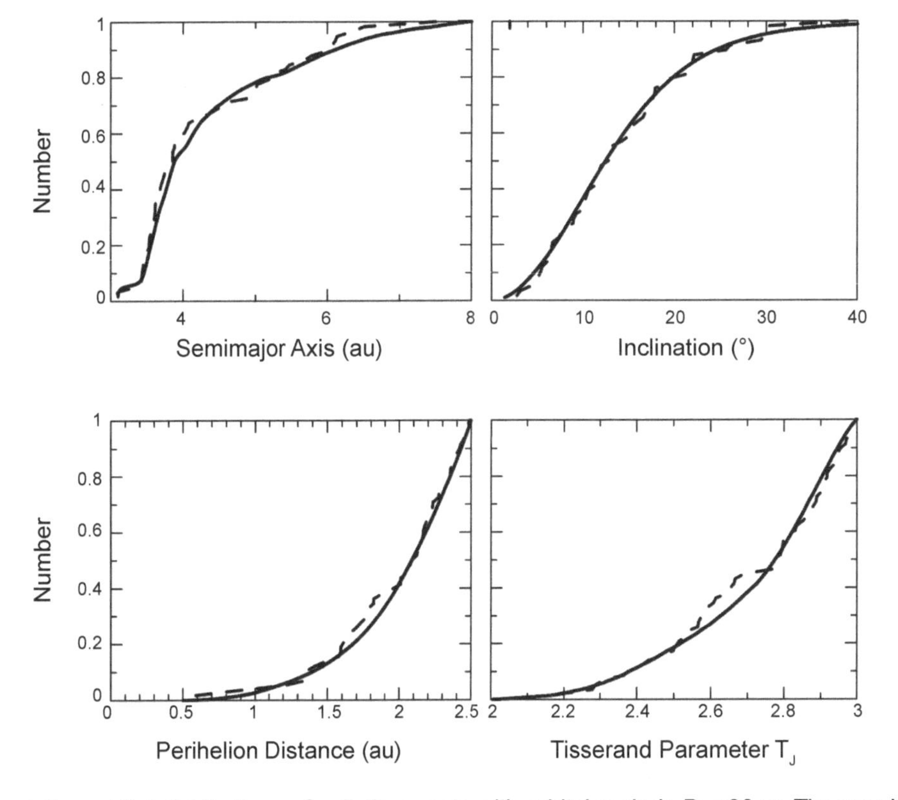

Fig. 7. The cumulative orbital distributions of ecliptic comets with orbital periods $P < 20$ yr, Tisserand parameter with respect to Jupiter $2 < T_J < 3$, and perihelion distance $q < 2.5$ au. The model results (solid lines) (*Nesvorný et al.,* 2017) are compared with the distribution of known JFCs (dashed lines). The model assumes that ecliptic comets remain active and visible for $N_p(2.5) = 500$ perihelion passages with $q < 2.5$ au. From *Nesvorný* (2018), which was adapted from *Nesvorný et al.* (2017). The panels are normalized so that the numbers of ecliptic comets with $q < 2.5$ and with $2 < T_J < 3 = 1$.

is made entirely through reflectance spectroscopic remote sensing. TNOs tend to have spectra that are largely devoid of identifying features (Fig. 8). The detection of specific materials on the surfaces of TNOs and Centaurs is so scarce that we can easily tabulate them in Table 1, which shows the limited list of surface ices detected on these objects.

The most frequently detected material on TNOs is water ice, in both amorphous and crystalline forms (*Zheng et al.,* 2009; *Barucci et al.,* 2011). Water ice is thought to be abundant in all TNOs and Centaurs (*Brown et al.,* 2012), and it is clearly one of the major building blocks of those bodies. Water ice absorption is most clearly seen in Haumea and its collisional family (*Barkume et al.,* 2006; *Trujillo et al.,* 2011). Methanol was first detected on the surface of the Centaur Pholus (*Cruikshank et al.,* 1998), and has been detected with varying levels of confidence on only a handful of TNOs through groundbased spectroscopy (*Barucci et al.,* 2011), and in the spectrum of Arrokoth by the New Horizon spacecraft (*Grundy et al.,* 2020). Its rarity of detection implies either that its spectral signatures are usually too weak to detect, either because it is not common on TNO surfaces, or because those signatures are masked by other materials. The former condition might imply that methanol is a by-product of some post-formation process, rather than a material that was abundant during the formation of TNOs.

The remaining materials are only found on the largest TNOs, i.e., on "dwarf planets," such as Eris, Pluto, Makemake, and Quaoar (for a recent summary, see *Barucci and Merlin,* 2020). These are the known volatile ices at TNO temperatures (≈30–50 K) that are unstable to sublimation on timescales shorter than the age of the solar system: N_2, CH_4, CO, and CO_2. It is only by virtue of the masses of the largest TNOs that atmospheric retention of these ices is sufficient to preserve a detectable abundance of these materials since formation (*Schaller and Brown,* 2007; *Brown et al.,* 2011; *Lisse et al.,* 2021). At the most basic level, the interplay between temperature and gravity governs the retention of each volatile, resulting in dramatic variations in relative abundances from object to object. This explains why some objects have spectra that are N_2 dominated (e.g., Eris), some are methane dominated (e.g., Quaoar), and some exhibit features of all the above volatile ices (e.g., Pluto). (We caution that dwarf planets should not be conflated with volatile-rich TNOs. For an object to be considered a dwarf planet requires only that it be massive enough for self-gravity to govern its shape. It is a coincidence of their formation and current surface temperatures that the dwarf-planet TNOs are the only bodies able to retain detectable levels of the volatile ices, which is the condition we are interested in here.) Figure 8 presents a small sample of spectra of the dwarf TNOs, exhibiting the sheer diversity of their spectra. Much like water on Earth, the presence of these materials in a semistable state drives interesting atmospheric and surface recirculation cycles that dominate the surface rheologies of these bodies (*Bertrand et al.,* 2018; *Hofgartner et al.,* 2019). These processes and the loss rates, which are highly sensitive to the surface and interior temperature histories of these bodies, greatly complicate estimates of the primordial ice abundances. This interesting topic is well beyond the scope of this chapter. For our purposes it is sufficient to emphasize that these volatile ices are primordial in nature, and are not fully depleted from the interiors of smaller TNOs, the interior presence of which is betrayed by the presence of these ices in the comae of JFCs (see section 3.2).

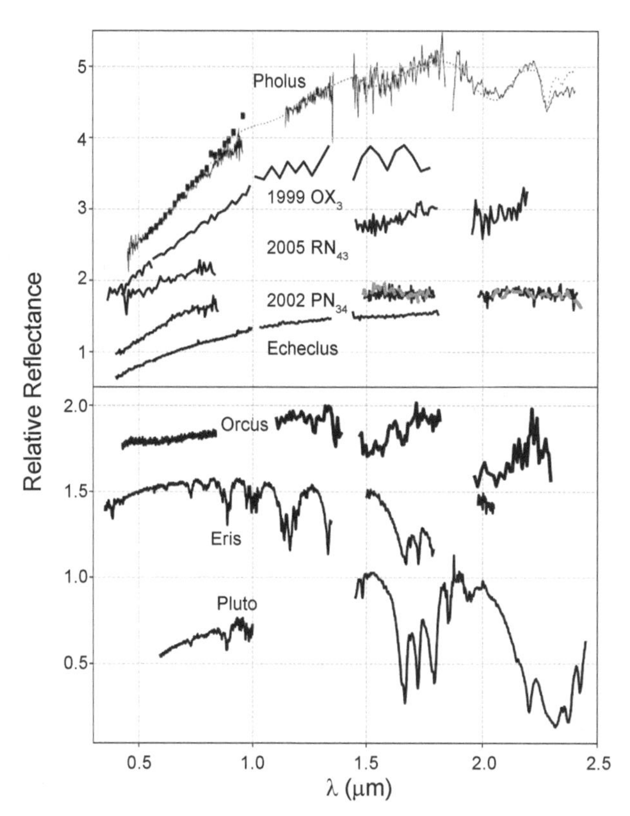

Fig. 8. Select TNO and Centaur reflectance spectra. Top: Spectra of (from top to bottom) the inactive Centaur 5145 Pholus (*Cruikshank et al.,* 1998), TNOs (44594) 1999 OX_3 (*Seccull et al.,* 2018), (145452) 2005 RN_{43} (*Alvarez-Candal et al.,* 2008; *Guilbert et al.,* 2009), and (73480) 2002 PN_{34} [black, *DeMeo et al.* (2010); gray, *Brown et al.* (2012)], and the active Centaur 60558 Echeclus, also known as Comet 174P (*Seccull et al.,* 2018). Spectra were normalized to unity at 0.65 μm and offset vertically in steps of 0.45. The NIR spectrum of 2005 RN_{43} has been renormalized to the optical spectrum using its observed (V-R) color. The dashed curve is a spectral model of Pholus, which includes contributions from olivine, tholin, water ice, and methanol. Bottom: Spectra of the dwarf planets Orcus [top, *DeMeo et al.* (2010)], Eris [middle, *Alvarez-Candal et al.* (2011)], and Pluto [bottom, *Merlin et al.* (2010)]. The spectra were normalized to unity at 1.5 μm, and those of Orcus and Eris have been offset by 0.8 and 1.0 vertically. In both panels, regions affected by telluric lines or instrumental artifacts have been removed, and data have been rebinned for clarity. This figure omits notable objects for which the published spectra were not available to the authors, including that of the methane-rich Quaoar (*Jewitt,* 2004) and the water-ice rich Haumea (*Trujillo et al.,* 2007).

Of course, the compositions of TNOs are not limited only to the few ices that have been spectrally identified.

TABLE 1. Representative listing of molecules detected/inferred on surface ices and in comae or atmospheres (volatiles) of TNOs, Centaurs, and comets.

TNO surface ices	H_2O	CH_3OH	CH_4	N_2	CO	CO_2	NH_3	C_2H_6
TNO volatiles	N_2	CO	CH_4					
Centaur surface ices	H_2O	CH_3OH						
Centaur volatiles	CO	H_2O	CO_2	HCN	CN	CO^+	N_2^+	
Comet ices	H_2O	CO_2	COOH-group					
Comet volatiles	H_2O	CO_2	CO	CH_4	HCN	CH_3OH	H_2CO	NH_3
	HNC	C_2H_2	C_2H_6	HCOOH	$HCOOCH_3$	HNCO	H_2S	OCS
	HC_3N	SO	SO_2	CS	CH_3CO	S_2	CN	C_2
	NH_2	C_3	CO^+	N_2^+	H_2O^+	CO_2^+	N_2	

TNOs typically have red colors at optical wavelengths. The most striking example of this is the New Horizons flyby imagery of the small (36-km-long) TNO Arrokoth (*Grundy et al.,* 2020). The dynamic range of TNO/Centaur optical colors is enormous, spanning from nearly neutral reflectors [e.g., the Haumea family members (*Brown et al.,* 2007)] to some of the reddest objects in the solar system [e.g., Gonggong — 2007 OR_{10} (*Fraser and Brown,* 2012)]. This red color is most commonly attributed to chemical processing of organic materials by cosmic rays (*Thompson et al.,* 1987; *Cruikshank et al.,* 1998; *Barucci et al.,* 2011; *Fraser and Brown,* 2012). Such materials exhibit the so-called optical gap absorption feature, driven by the C-H π-bond. Centered in the near-infrared (NIR), the shape and width of this feature depend on the amount of dehydrogenation of the organic, how disordered the molecular structure is, and the level of non-organic contaminant in the molecular chains, in particular (*Seccull et al.,* 2021). This feature is common to many organic materials that are considered suitable astrophysical analogs; compare the chemically simple polycyclic aromatic hydrocarbons (e.g., *Izawa et al.,* 2014) to the highly disordered laboratory materials called tholins (e.g., *Roush and Dalton,* 2004), both of which show this deep optical gap. If attribution to organic materials is correct, the red colors of TNOs imply that simple organics such as methane were abundant for enough time in the early solar system that red dehydrogenated crusts could develop (*Brunetto et al.,* 2006) before those volatile ices were depleted.

The spectrum of Arrokoth reveals a mostly featureless spectrum typical of most TNOs, except for two weak features at 1.8 and 2.2 μm. The absorption at 2.2 μm has been attributed to methanol ice (*Grundy et al.,* 2020). It has been suggested that the 1.8-μm feature can be driven by the presence of sulfur in the organic residue (*Mahjoub et al.,* 2021). Irradiation experiments on laboratory ice mixtures containing methane, ammonia, hydrogen sulfide, and water result in a red material bearing an absorption at 1.8 μm that is absent in mixtures with no sulfur (*Mahjoub et al.,* 2021). Such a feature has not been detected on any other TNO, possibly as a result of the low available signal-to-noise ratio of even the best spectra of small TNOs. It may be that the feature is unique to the so-called cold-classical TNOs (see section 2.1), which are the only TNO population presumed to have formed *in situ.* The search for sulfur in the comae of JFCs has taken on new importance, as the presence or absence of sulfur-bearing materials in JFCs would provide significant insights regarding the compositional variations in the protoplanetesimal disk. The recent strong detection of atomic sulfur at 1425 Å in the JFC 46P/Wirtanen rivals that of atomic hydrogen (*Noonan et al.,* 2021) and is consistent with emission directly from the nucleus or from grains very near the surface, similar to atomic sulfur and other sulfur-bearing species in 67P/Churyumov-Gerasimenko (*Calmonte et al.,* 2016). Additional searches are underway for other sulfur-bearing species in JFCs to test models of cometary ices and their subsequent processing history (*Presler-Marshall et al.,* 2020; *Saki et al.,* 2020; *Altwegg et al.,* 2022).

No discussion about TNO compositions would be complete without the mention of silicate materials. Two TNOs, the ≈700- and 300-km-diameter Plutinos (208996) 2003 AZ_{84} (*Fornasier et al.,* 2009) and (120216) 2004 EW_{95} (*Seccull et al.,* 2018), exhibit spectra that appear similar to C-type asteroids, with absorption features consistent with hydrated silicates. These two bodies are not icy, lacking even the absorptions due to water ice at 1.5 and 2.0 μm seen in the Haumea family (*Brown et al.,* 1999). Frustratingly, silicate materials have avoided spectroscopic detection for icy TNOs.

Their presence was first inferred by the need for silicate materials to account for the densities of the largest TNOs. Of the eleven TNOs with diameters d > 600 km and measured densities ρ, all but one, 55637 (2002 UX_{25}), have ρ > 1 g cm^{-3} (*Brown,* 2013; *Brown and Butler,* 2017) [see *Bierson and Nimmo* (2019) and *Grundy et al.* (2019a) for densities of other TNOs]. The largest density is 2.43 ± 0.05 g cm^{-3} for Eris/Dysnomia, the most massive TNO system known (*Holler et al.,* 2021). The presence of silicates has also been inferred from spectroscopic modeling, either of individual spectra (see *Barucci et al.,* 2011, for example) or of the continuum of TNO optical and NIR colors (e.g., *Fraser and Brown,* 2012). Recent modeling of spectrophotometry spanning ~0.5–4.5 μm implies the presence of silicate-rich surfaces among some of the brighter TNOs (*Fernández-Valenzuela et al.,* 2021). These results demonstrate that silicates *do* exist in icy TNOs (of course they do!), but simply are masked by the presence of other materials. It is likely that characterization of silicates in TNOs will have to await observations at longer wavelengths [$\lambda \gtrsim 3$ μm (*Parker et al.,* 2016)]. Spectral observations from the James Webb Space Telescope (JWST) are likely to be quite important in this regard (see section 5).

Beyond spectral studies, many insights regarding the compositions of TNOs have come from broadband photometric techniques. We review those here.

Significant effort has been devoted to developing a taxonomic system for TNOs, as even the simple act of determining the number of classes is likely to influence our interpretation of the early solar system. It is generally thought that the varied classes of TNOs reflect the compositional structure of the regions of the protoplanetary disk from which TNOs originated. Proposed mechanisms for compositional differences of planetesimals include ice lines (*Dalle Ore et al.,* 2015) and post-formation volatile loss (*Wong and Brown,* 2016; *Brown et al.,* 2011), which would lead to variable composition with distance from the Sun, prior to disk dispersal by the migrating gas giants. It is important to highlight that to date, TNO compositional measurements are all measured from reflected light, and are typically assumed to be indicative of primitive nucleus values.

Efforts to determine the number of TNO classes have been frustrated by their nearly featureless spectra. TNO reflectance spectra (e.g., *Fornasier et al.,* 2009; *Guilbert et al.,* 2009; *Barkume et al.,* 2008; *Barucci et al.,* 2011), like those of most icy bodies, including the Jupiter Trojans, Centaurs, and JFCs (*Emery et al.,* 2011; *Cruikshank et al.,* 1998; *Barucci et al.,* 2002), can be broadly described as linear in the optical and NIR, with a different spectral slope in each region, and a smooth rollover or transition between the two slopes, very roughly centered at ~0.9 μm, and sometimes an absorption band of water ice at 1.5 μm and other ices in the infrared (see section 3.1). A few examples are presented in Fig. 8. By comparison, asteroids generally exhibit distinct features between classes that aid taxonomic interpretation (*DeMeo et al.,* 2009); such features are generally unavailable for the classification of TNOs. Those few materials that have been confidently detected from the spectra of TNOs, such as water ice, do not seem to belong uniquely to certain TNO taxa. Rather, taxonomic systems have been largely generated from the variations in optical and NIR colors seen from object to object.

Centaurs and most dynamically excited TNOs exhibit a bifurcated optical color distribution (*Tegler and Romanishin,* 1998; *Peixinho et al.,* 2012; *Tegler et al.,* 2016; *Fraser and Brown,* 2012; *Peixinho et al.,* 2015; *Lacerda et al.,* 2014; *Marsset et al.,* 2019). The bimodal color distribution has provided a functional taxonomy of two populations for small TNOs, now colloquially referred to as red and very red. In this basic system, Pluto is a red member, and Arrokoth belongs to the very red population of TNOs (*Grundy et al.,* 2020). Intrinsically, red class TNOs outnumber the very red objects by at least 4:1 (*Wong and Brown,* 2017; *Schwamb et al.,* 2019). This ratio has been used to infer that the purported compositional line that divides the proto-red and proto-very red populations fell between 30 and 40 au (*Nesvorný et al.,* 2020; *Buchanan et al.,* 2022).

The red/very red taxonomy only considers optical spectral slope or color, and does not make use of any other wavelengths. Longward of ~0.9 μm, TNOs tend to exhibit NIR colors that are correlated with optical spectral slope (e.g., *Fraser and Brown,* 2012). Attempts have been made to use the optical-NIR color space to create more complex taxonomies. Early efforts applied principal component analysis to a sample of BVRIJ colors and found four classes (e.g., *Barucci et al.,* 2005). This four-taxon system is presented in Fig. 9, and essentially divides the red and very red classes into two subpopulations based on their IR behavior. For example, the IR and RR classes are very red, with the RR exhibiting red spectral slopes across the BVRIJ range, and the IR exhibiting bluer NIR spectral slopes than found in the optical. Generally, the number of taxa found would increase with the number of filters or the complexity of the analysis (*Dalle Ore et al.,* 2013). A recent effort to include visible albedo in the analysis expands the number of taxa to as many as six, with an additional four taxa each containing a single dwarf planet (Pluto, Eris, Makemake, and Quaoar). Alternative analyses have concluded that the optical-NIR color distribution should not be further subdivided. Rather, excited TNOs exhibit only two separate taxa, with each exhibiting a continuum of colors through the optical and NIR color spaces (*Fraser and Brown,* 2012; *Schwamb et al.,* 2019).

For completeness, we point out the so-called cold classical TNOs, which stand out in many ways, including their tight orbital distribution (see section 1.1) and high binary fraction (*Noll et al.,* 2008, 2020). They are found in a tight annulus between 42 < a < 48 with inclinations $i \lesssim 5°$. Considering their red optical colors (*Gulbis et al.,* 2006), cold

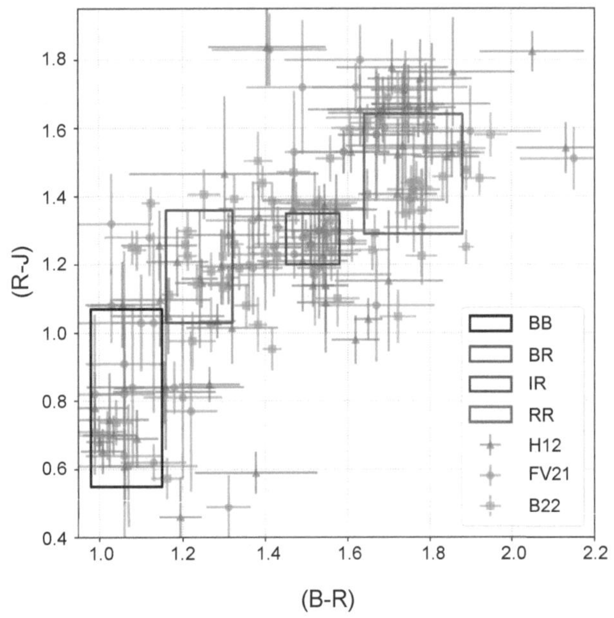

Fig. 9. Colors of TNOs and Centaurs from *Hainaut et al.* (2012) (H12), *Fernández-Valenzuela et al.* (2021) (F21), and *Buchanan et al.* (2022) (B22). The colors of the four-taxon system of *Barucci et al.* (2005) are shown by the rectangles (from *Hasegawa et al.,* 2021). Estimates of (B-R) of the *Buchanan et al.* (2022) measurements were done by determining the range of linear spectral slopes that match the reported (g-r) colors, and projecting those slopes onto (B-R). (R-J) was estimated from (r-J) in a similar fashion.

classicals mainly appear to belong to the very red taxon. Unlike most dynamically excited very red objects, however, cold classicals appear to exhibit NIR colors that tend to values closer to solar (*Pike et al.,* 2017) and have higher than usual visual albedos (*Brucker et al.,* 2009), suggesting cold classical objects may occupy a third taxon. Due to their stable orbits, however, they do not act as a significant source of Centaurs or JFCs.

Unsurprisingly, the Centaurs tend to exhibit similar optical-NIR color distributions as the TNOs that feed the Centaur population. A detailed comparison is limited by biases, both observational, such as inconsistent target selection in color surveys (*Schwamb et al.,* 2019), and physical, such as that colors seem to vary with size (*Benecchi et al.,* 2019), and surface alteration processes (*Seccull et al.,* 2019), all of which will likely affect the TNO and Centaur color distributions differently.

JFCs do not exhibit any of the color structure shown by TNOs and Centaurs. Instead, they consist entirely of objects with surfaces only slightly redder than solar (e.g., *Jewitt,* 2015). As the JFCs are predominantly directly fed from the excited TNO populations, the gray surfaces of the JFCs must be the result of alterations to their surfaces as objects migrate closer to the Sun. Cometary activity seems a likely culprit. Indeed, those Centaurs that have been found to be active all have surface colors that are closer to solar than the average TNO, with none of the Centaurs with very red colors exhibiting detectable levels of activity. This implicates activity as a main driver for surface color alteration, possibly through deposition of gray dust (e.g., *Seccull et al.,* 2018). Other alteration mechanisms might be at work, such as the thermal destruction of reddening agents.

3.2. Coma Compositions of Centaurs

At least 10–15% of known Centaurs exhibit dust comae and are deemed active (*Jewitt,* 2009; *Bauer et al.,* 2013). Such comae are composed primarily of dust grains with expansion velocities of 0.3 km s^{-1} or less and production rates ranging from ~1–1000 kg s^{-1} (Table 2) (*Whitney,* 1955; *Hughes,* 1991; *Trigo-Rodríguez et al.,* 2010; *Kulyk et al.,* 2016). Identifying the volatile components and ascertaining which are produced in sufficiently high amounts is an important first step in constraining models of Centaur activity and solar system formation and evolution.

A summary of volatile production rates in Centaurs is provided in Table 2. There are so few detections and significant limits that their production rates (in molecules per second and kilograms per second) can be listed succinctly in this table. Representative dust mass loss rates are also included in the table, with 29P leading the pack with the highest values, although low rates are sometimes seen in this Centaur.

Despite decades of observations, very few volatiles have been detected on Centaurs. This is largely due to their faintness and because most are inactive. Their distant orbits combined with only rare episodes of comae or outbursts make it difficult to discover when they are active and then carry out observations. Challenges also arise from the fact that some of the candidate species for driving activity, such as CO_2, CH_4, N_2, and O_2, are rotationally symmetric and thus have no pure rotational transitions at millimeter wavelengths, which is how most cometary volatiles are observed. Furthermore, telluric contamination is a substantial obstacle for detecting CO_2 and CH_4 emission via infrared rovibrational spectra, and thus observing these two volatiles directly has been restricted to a few instruments above Earth's atmosphere, such as the Infrared Space Observatory, AKARI Space Observatory, and *in situ* spacecraft mission measurements.

Highly volatile species, such as N_2 and O_2, may be abundant enough in nuclei to play a role in Centaur activity, but thus far have not been detected. N_2^+, a probable ionization product of N_2, has been detected in optical spectra of 29P's coma. N_2^+ is also a frequent night-sky emission feature; however, a telluric source can be ruled out with longslit spectra when N_2^+ emission is visible only in the tailward direction (e.g., *Ivanova et al.* (2019). Analysis of 29P's N_2^+ and CO^+ emission in such tailward spectra of 29P is consistent with an N_2 production rate of ~17 kg s^{-1}, which is much smaller than CO (*Wierzchos and Womack,* 2020). However, care should be used when interpreting either N_2^+ and CO^+ spectra, because their line strengths in 29P's coma appear strongly correlated with solar wind particle velocities, possibly indicating a charging mechanism of solar wind proton impact onto CO and N_2 in the coma (*Cochran et*

TABLE 2. Measured and derived properties of dust and gas in Centaurs.

Comet	Volatile	r (au)	Production rate 10^{27} mol s^{-1}	Mass loss rate* (kg s^{-1})	Reference
29P	Dust	5.8–6.3	–	30–4700	[1,2,3,4,5]
	CO	5.7–6.3	10–70	460–3200	[6,7,8,9,10,11]
	H_2O	6.2–6.3	4.1–6.8	120–188	[5,8]
	CO_2	6.2	0.5 ± 0.3	36 ± 22	[12]
	N_2 (N_2^+)†	6.3	0.4 ± 0.2	17 ± 8	[10,11,13]
	O_2‡	6.3	~0.24	~13	[11]
	HCN	6.3	0.048 ± 0.011	1.8 ± 0.4	[5]
	CN	5.8	0.008 ± 0.003	0.3 ± 0.1	[14]
	NH_3	6.3	<4.5	<128	[9]
	CH_3OH	6.3	<0.55	<29	[9]
	CH_4	6.3	<1.3	<34	[9]
	H_2CO	6.3	<0.1	<5	[9]
Chiron	Dust	9–18	—	1–45	[15,16,17]
	CO	8.5	13 ± 3	605 ± 151	[18]
	CN	11.3	0.037 ± 0.011	1.4 ± 0.3	[19]
Echeclus	Dust	13.1	—	10–700	[20,21]
	CO	6.1	0.8 ± 0.3	300 ± 130	[22]
	CN	12.9	<0.04	<2	[23]
P/2019 LD_2	Dust	4.5	—	40–60	[24]
	CO	4.58	<4.4	<200	[25]
	CN	4.5	<0.0014	<0.1	[23]

*Gaseous mass loss rates were calculated using production rates and appropriate atomic mass units.
†N_2 production rate is derived from CO+, N_2^+ and CO measurements; see *Womack et al.* (2017).
‡O_2 production rate is inferred by scaling the mixing ratio of O_2 and H_2O in 67P (a JFC) to the water production rate measured in 29P by AKARI; see *Wierzchos and Womack* (2020) for details.

References: [1] *Fulle* (1992); [2] *Ivanova et al.* (2011); [3] *Hosek et al.* (2013); [4] *Schambeau et al.* (2021); [5] *Bockelée-Morvan et al.* (2022); [6] *Senay and Jewitt* (1994); [7] *Festou et al.* (2001); [8] *Ootsubo et al.* (2012); [9] *Paganini et al.* (2013); [10] *Womack et al.* (2017); [11] *Wierzchos and Womack* (2020); [12] *Harrington Pinto et al.* (2022); [13] *Ivanova et al.* (2016); [14] *Cochran and Cochran* (1991); [15] *Luu and Jewitt* (1990); [16] *Romon-Martin et al.* (2003); [17] *Fornasieret al.* (2013); [18] *Womack and Stern* (1999); [19] *Bus et al.* (1991); [20] *Rousselot et al.* (2016); [21] *Bauer et al.* *(2008)*; [22] *Wierzchos et al.* (2017); [23] *Rousselot* (2008); [24] *Licandro et al.* (2021); [25] *Kareta et al.* (2021).

al., 1991; *Ivanova et al.,* 2019). An O_2 production rate was estimated by using measurements of its abundance in the Rosina mass spectroscopy of the JFC 67P and then scaling it to the measured water production rate (*Wierzchos and Womack,* 2020). These values suggest that N_2 and O_2 are not produced in sufficiently high amounts to drive much of the activity of 29P.

HCN emission was detected at millimeter wavelengths in only one Centaur — 29P in 2010 — but with relatively low production rates and with a spatial profile that indicates it may have been released from icy grains and not directly sublimating from the nucleus (*Bockelée-Morvan et al.,* 2022). Although seen in TNO and Centaur reflectance spectra, methanol (CH_3OH) emission has not yet been detected in Centaur comae, with the strongest upper limit of 29 kg s^{-1} set for 29P (*Biver,* 1997).

Evidence for water-ice grains has been observed in cometary comae (*Yang et al.,* 2009; *A'Hearn,* 2011; *Kelley et al.,* 2013; *Protopapa et al.,* 2018); by extension, it is reasonable to search for them in Centaur comae. Centaurs orbit at heliocentric distances too large for water ice on the nucleus surface to sublimate efficiently, but water vapor may still be detected in the coma due to water ice-rich grains that are carried off the nucleus by other outgassing mechanisms and then sublimate once bathed in the solar radiation field, as was reported for 29P for water (*Ootsubo et al.,* 2012) and in the aforementioned HCN. Recently, the signature of water ice grains was again reported during a significant outburst of 29P in September 2021 (*Kelley et al.,* 2021), and a water ice grain signature at the 1–10% level was seen in the near-infrared spectrum of the Centaur C/2019 LD_2 (Pan-STARRS)'s coma at 4.6 au (*Kareta et al.,* 2021).

There are also occasional reports of CN, C_2, and C_3 detections or upper limits. These are radicals and not capable of long-term storage in the nucleus, and so are probably daughter products from another cometary volatile and not directly responsible for driving activity (*Bus et al.,* 1991; *Cochran et al.,* 1991; *Womack et al.,* 2017; *Ivanova et al.,* 2019; *Kareta et al.,* 2021; *Licandro et al.,* 2021; *Bolin et al.,* 2021).

Although considered as a likely candidate for distant activity in Centaurs because of its relatively high abundance in many comets and the good match of CO_2's sublimation efficiency at Centaur distances, CO_2 emission has been searched for in Centaur comae, with no detections thus far. A significant limit was obtained only for 29P with the AKARI space telescope (*Ootsubo et al.,* 2012), implying a CO/CO_2 ratio >90, much higher than what is seen in other comets (*Harrington Pinto et al.,* 2022). The spacebased telescopes Spitzer and NEOWISE have been used to observe many Centaurs and were set up to detect the combined emission of CO and CO_2 in the same ~4.5-µm filter bandpass. Unfortunately, because the emission is combined, individual CO_2 or CO production rates cannot be derived from the data without independent and simultaneous measurements of one of the molecules. Recently, CO_2 production rates were inferred for 29P using Spitzer and NEOWISE imaging data in 2010 along with contemporaneously obtained CO millimeter-wavelength spectra, which confirms that CO_2 is produced in relatively small amounts in the 29P coma, as low as approximately 1% of CO and 14% of H_2O (*Harrington Pinto et al.,* 2022).

Consequently, CO is the most tractable volatile for direct measurement in Centaur comae that may play a significant role in activity, and these observations are possible using groundbased as well as spacebased instruments at millimeter and infrared wavelengths. Detection of CO emission is reported for three Centaurs: 29P, 95P/Chiron, and 174P/Echeclus (*Senay and Jewitt,* 1994; *Womack and Stern,* 1999; *Wierzchos et al.,* 2017; *Bockelée-Morvan et al.,* 2022). On a related note, CO emission is also detected in some long-period comets at "Centaur distances" from the Sun, such as C/1995 O1 (Hale-Bopp) and C/2017 K2 (PanSTARRS) (*Gunnarsson et al.,* 2003; *Yang et al.,* 2021), and CO appears to play a large role in distant comet activity (*Biver et al.,* 2002; *Womack et al.,* 2017).

The apparent absence, or at least very low abundance in the coma, of CO_2 emission in 29P's coma is instructive. If we assume that Centaurs and comets formed in similar environments and should have similar chemical compositions of their nuclei, then the available data on comets may be useful for predicting what we should see in other Centaurs. In a study of 25 comets for which CO and CO_2 were simultaneously measured, the mixing ratio of CO and CO_2 production rates showed a possible preference for CO_2 in the comae of most JFCs, but the opposite is true for some Oort cloud comets and the lone Centaur (29P) (see *A'Hearn et al.,* 2012; *Harrington Pinto et al.,* 2022). However, if JFCs, active Centaurs, and long-period comets are analyzed as an aggregate, the coma mixing ratio may be better explained as following a heliocentric trend out to at least 6 au. CO_2 is preferentially detected in most comets (not just JFCs) within 3 au, and CO is dominant beyond 3 au, although selection effects cannot be ruled out since JFCs are typically fainter than long-period comets and less likely to be seen at large distances. However, if this trend continues for other active Centaurs, then CO emission is likely to exceed CO_2 emission in the comae of active Centaurs, as is the case for 29P. This effect may be at least partly due to differences in the degree of thermal processing of the nucleus, rather than compositional differences (*A'Hearn et al.,* 2012; *Harrington Pinto and Womack,* 2019). Indeed, one of the problems remaining in cometary science is accurately determining comet nucleus composition from coma abundances.

4. ACTIVITY AND PHYSICAL EVOLUTION: COMET BEGINNINGS

4.1. Activity in Centaurs

The drivers of activity in Centaurs have implications for the nature and evolution of ice in small bodies and the cause of all activity in icy bodies at large heliocentric distances. Unfortunately, very little is known about the dominant mechanisms. Active Centaurs frequently display two types of activity: sustained comae and discrete outbursts. The first Centaur discovered, 95P/Chiron, undergoes periods of sustained activity as well as short-term brightening periods. Arguably, the best-known active Centaur is 29P, an object on a nearly circular orbit just beyond Jupiter's distance that exhibits continuous activity with occasional outbursts of 10–250× brightness superimposed. Another well-known active Centaur is 174P/Echeclus, which has undergone several outbursts, including a large one in which the activity was centered on an apparent fragment of the main nucleus, and a few longer-term periods of activity with lower-level dust comae.

Vigorous water-ice sublimation in comets releases many minor species in cometary comae, but Centaurs are too far from the Sun for water ice to sublimate efficiently. Thermal equilibrium temperatures of Centaur nuclei are most similar to the sublimation temperatures of CO_2 and NH_3 and hence they are favored by some models of Centaur activity, e.g., for the Centaur C/2014 OG_{392} at ~10 au (*Chandler et al.,* 2020). However, no emission from these volatiles has yet been detected in a Centaur, and studies of 29P place stringent limits on their mixing ratio relative to CO.

As discussed in the previous section, CO is the only molecule detected in the gaseous state in active Centaurs that is capable of driving activity. CO is also abundant in a few Oort cloud comets active beyond 5 au, but no CO detections or significant limits exist for JFCs this far out (this may not imply compositional differences between JFCs and Centaurs). There is evidence for water ice grains in the comae for two Centaurs: 29P and P/2019 LD_2 (ATLAS). These grains were most likely released from a surface component, similar to what has been reported for some Centaur and TNO surfaces in section 3.2.

Another model for Centaur activity invokes cosmogonically important volatiles with lower sublimation temperatures, such as CO, which can survive in the nucleus as ices just below the surface, and also be partially incorporated in the gaseous state within amorphous ice and then released when this ice undergoes the phase change to the crystalline state (*Prialnik et al.,* 1995, 2008; *De Sanctis et al.,* 2000; *Capria et al.,* 2009; *Guilbert-Lepoutre,* 2011). This phase change is optimized around ~120–140 K (equivalent to Centaurs and comets at 5–10 au), but it can proceed at farther distances (and colder temperatures) less efficiently.

The most popular model that explains the observations of Centaur (and distant comet) comae is that of a nucleus with pockets of amorphous ice that undergoes the crystallization process, which releases trapped volatiles. Another model invokes isolated and insulated pockets of frozen CO or CO_2 that sublimates once the surface is disrupted enough to expose a relatively small and previously protected frozen patch (e.g., *Prialnik et al.,* 2008). However, given the very few measurements we have on Centaurs, more observations are needed to settle between these two models.

Because of the lack of detections or low production rates and significant limits for other volatiles in Centaurs, our main clues to outgassing must come from CO in 29P. The spectral line profile of CO via its J=2–1 rotational transition at 230 GHz is very narrow and slightly blueshifted, consistent with sunward emission of CO from a very cold region and a secondary source more distributed over the nucleus surface (Fig. 10). This is also true for CO in Chiron (*Womack and Stern,* 1999) and Echeclus (*Wierzchos et al.,* 2017), as well as other distantly active comets like Hale-Bopp (*Biver,* 1997; *Womack et al.,* 1997) and C/2017 K2 (Pan-STARRS) (*Yang et al.,* 2021). The strikingly similar spectral shapes suggest there is a common outgassing mechanism for CO in active Centaurs and distant Oort cloud comets (*Womack and Stern,* 1999; *Biver et al.,* 2002;*Gunnarsson et al.,* 2003, 2008; *Wierzchos et al.,* 2017; *Yang et al.,* 2021) originating from a very cold (~4 K) gas (e.g., *Paganini et al.,* 2013). Thus, it is reasonable to plan for narrow emission from a very cold gas in the inner coma when searching for CO in other active Centaurs.

Interestingly, Hale-Bopp, 29P, and Echeclus appear to have approximately the same nucleus diameter and were all observed at 6 au, making an intercomparison possible that minimizes size and heliocentric distance contributions (*Wierzchos et al.,* 2017). Thus, 29P is an abundant producer of CO, even sometimes outproducing Oort cloud Comet Hale-Bopp at the same distance. In contrast, Echeclus and Chiron emitted CO very weakly, and it was only marginally detected. Searches for CO in many other Centaurs did not yield detections (*Drahus et al.,* 2017), but some, such as (10199) Chariklo, 342842 (2008 YB_3), (8405) Asbolus, and 95626 (2002 GZ_{32}) provide tight upper limits that eliminate the possibility of significant outgassing activity from CO at the time of observation (see Fig. 11). We do not know whether Centaurs like Chiron, Echeclus, and others that produce little to no CO formed in a different environment, or whether they are devolatilized, or have not yet started to become more active, partly due to spending more of their orbits well beyond 6 au, when compared to 29P. The difference in CO output in 29P and these other active Centaurs is particularly striking and may be useful to constrain Centaur models. Further accurate measurements of nucleus diameters and CO and CO_2 production rates are needed to better understand how these species contribute to activity and ultimately to determine the composition of their nuclei.

Centaur activity also appears to be related to the residence time in its orbital region, and a decrease in perihelion distance and/or semimajor axis may occur before the observed onset of activity (*Fernández et al.,* 2018; *Sarid et al.,* 2019; *Lilly et al.,* 2021). A Centaur's perihelion distance also apparently plays a large role in whether a

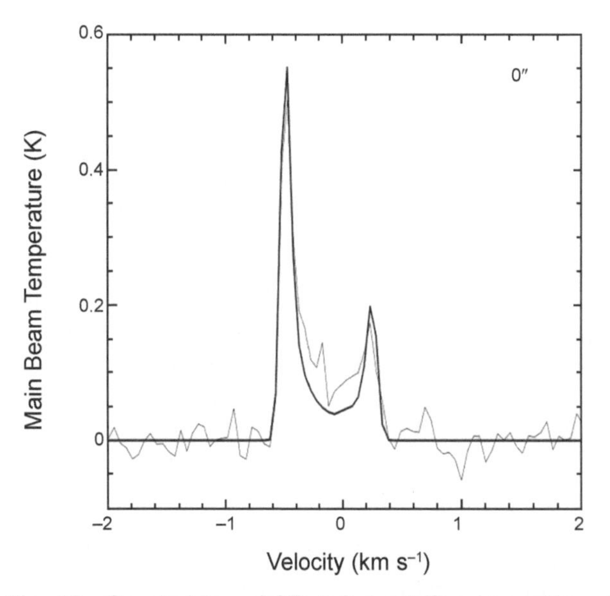

Fig. 10. Spectral line of CO J=2–1 rotational transition at 230 GHz for the Centaur 29P/Schwassmann-Wachmann 1 obtained with the IRAM 30-m telescope in January 2004. Superimposed is a model fit to the data, which includes a strong blue-shifted peak of sunward emission from a very cold nucleus source and a smaller contribution from the nightside. From *Gunnarsson et al.* (2008).

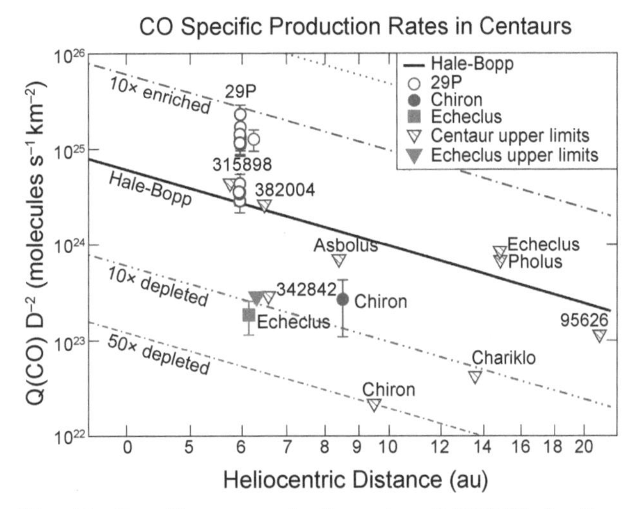

Fig. 11. Specific gas production rates, Q(CO)/D^2, for Centaurs. The solid line is a fit to data for Hale-Bopp provided for comparison. As the figure shows, after normalized by surface area, 29P produces far more CO than other Centaurs like Echeclus and Chiron, where CO was detected in lower amounts, and other Centaurs for which strong upper limits were set. From *Wierzchos et al.* (2017).

Centaur will become active, with lower values more likely to correlate with activity (*Jewitt,* 2009). *Rickman et al.* (1991) noted a similar trend for short-period comets whose perihelion distances had recently decreased. Such comets had larger values for their nongravitational parameters, suggesting that their mantles had been removed, allowing more of their surfaces to be active. The ongoing CO outgassing in Centaur 29P is recently proposed to be due to the nucleus responding via the crystallization phase change of water due to the relatively sudden (within ≈2000 yr) median residence time of the Gateway orbit (*Sarid et al.,* 2019), with a change in its external thermal environment produced by its dynamical migration from the Kuiper belt to the Gateway region where it maintains a nearly constant thermal environment at a heliocentric distance of ≈6 au (*Lisse et al.,* 2022). Thus, orbital history may play a large role in triggering the activity of Centaurs.

In addition to steady-state production, which may produce a coma lasting weeks or months, active Centaurs sometimes have discrete outbursts lasting a few days to weeks. 29P is the best-known example whose outbursts have been documented for more than a century, primarily with secular lightcurves of the visible magnitudes and images that capture morphology in the coma (e.g., *Trigo-Rodríguez et al.,* 2010; *Schambeau et al.,* 2017, 2019). One of the largest outbursts in decades began in September 2021, when 29P increased its brightness by a factor of ~250 during a series of four smaller outbursts occurring over a few days. Such outbursts give an opportunity to test models of chemical composition and physical mechanisms of ejection. Unlike 29P, Chiron and Echeclus typically do not maintain long-term quiescent comae, but both have exhibited outbursts or brightening episodes. In 2005 Echeclus underwent a ~7-mag outburst (from 21 to 14, corresponding to an increase in brightness by a factor of ~630) and was accompanied by a large, detached coma ~2 arcmin across, projected to be 1,000,000 km at the Centaur's distance and was visible for a few weeks (*Choi and Weissman,* 2006; *Tegler et al.,* 2006; *Rousselot,* 2008). More details about its observational timeline, orbital dynamics, and measured characteristics during outbursts are found in *Wierzchos et al.* (2017), *Seccull et al.* (2019), and *Kareta et al.* (2019).

29P's long-term behavior of continuously having a dust and gas coma, regularly punctuated by outbursts, is well-suited to dedicated observation over the years, and its light curve displays two types of outburst shapes: Most have an asymmetric sawtooth pattern, where the reflected light grows tremendously in a few hours and then decays in days to weeks. These outbursts must have an explosive trigger that releases a great deal of dust. Less frequently, the visible light curves of smaller outbursts appear to grow and decay symmetrically in time (cf. *Trigo-Rodríguez et al.,* 2010; *Wierzchos and Womack,* 2020; *Clements and Fernandez,* 2021). In contrast to the larger sawtooth-shaped outbursts, the smaller symmetric outbursts may originate from a source region that spans a larger surface area on the nucleus that releases material over a longer period of time, perhaps a few days. Still another mechanism contributing to the changing brightness of Chiron is the presence of rings or ring arcs [also detected around Chariklo (*Braga-Ribas et al.,* 2014)], which changes the Centaur's apparent brightness as the aspect of the rings varies from open to edge-on (*Sickafoose et al.,* 2020; *Fernández-Valenzuela,* 2022).

There are very little data about possible triggers for the outbursts in 29P. However, outbursts of CO and dust were both recorded during simultaneous observations of CO millimeter-wavelength spectral flux measurements and visible magnitudes, providing an opportunity to test the hypothesis that a strong CO outburst was needed to trigger the dust outbursts. Interestingly, the CO production rate doubled, but did not trigger a noticeable rise in dust production (*Wierzchos and Womack,* 2020). Similarly, two dust outbursts occurred without an accompanying increase in CO production. Two other dust outbursts may show CO gas involvement. These odd results may be explained if the CO is not always substantially incorporated with the dust component in the nucleus, or if CO is primarily released through a porous material.

4.2. Morphological Evolution and Transneptunian Object Binaries in Context

The prevalence of contact binaries found in the comet and TNO populations is suggestive of a formation mechanism that preferentially forms contact binary objects. Such a connection is not so obvious, however, without morphological and evolutionary considerations. This short discussion highlights only a few of the important processes one needs to consider when comparing the morphologies of JFCs to their precursor TNO populations, and the challenges faced when attempting to make inferences about their formation modes.

Most cometary nuclei imaged by spacecraft or radar have bilobed shapes; of the seven, only 9P/Tempel 1 and 81P Wild are nearly spheroidal (see the chapters in this volume by Pajola et al. and Knight et al. for further details). This suggests that comets are born bilobed or morphologically evolve to become bilobed during their lifetimes. Here we first consider the formation stage.

Formation of small planetesimals in the outer solar system favors binarity. For example, the streaming instability model (*Youdin and Goodman,* 2005; see also the chapter in this volume by Simon et al.), where rotating clouds of pebbles collapse under their own gravity, gives birth to binary systems with near-equal-sized components, with properties that match observations (*Grundy et al.,* 2019a; *Nesvorný et al.,* 2019a). The components of a newly formed close binary can be brought into contact by gas drag (*Lyra and Umurhan,* 2019), producing a contact binary. The early-stage low-speed collisions between similar size bodies can lead to mergers and bilobate shapes as well (*Jutzi and Asphaug,* 2015).

Contact binaries are ubiquitous in the Kuiper belt. For example, the New Horizons spacecraft revealed that Arrokoth is a contact binary, which likely resulted from a low-speed merger of two flattened, spheroidal components (*Stern et al.,* 2019). Additionally, lightcurve observations of TNOs indicate that the contact binary fraction can be 30% or higher (*Sheppard and Jewitt,* 2004; *Thirouin and Sheppard,* 2018, 2019, 2022; *Noll et al.,* 2020; *Showalter et al.,* 2021).

It is suggestive to draw connections between these observations and bilobed comets. Note, however, that (1) comets are much smaller than most known TNOs, and (2) Arrokoth, with its nearly-circular, nearly-ecliptic orbit 44 au from the Sun, formed well beyond the original formation region of most comets, in a sparsely populated region of the Kuiper belt. As discussed in section 2.1, present-day comets presumably formed in a massive disk ~20–30 au from the Sun, became part of the scattering TNO population (at distances of ~50–1000 au) for ~4.5 G.y., then evolved onto inner solar system orbits. The size and formation distance of TNOs affect their survival. Arrokoth is thought to be a pristine planetesimal that formed and survived essentially unchanged in the low-mass classical belt (*McKinnon et al.,* 2020). Small comets are instead less likely to survive intact, their first obstacle being the disruptive and shape-changing collisions during the massive disk stage (*Benavidez et al.,* 2022).

The overall significance of collisions during the disk stage mainly depends on the disk lifetime, t_{disk}. If $t_{disk} \gtrsim$ 10 m.y., the great majority of comet-sized bodies would be disrupted (*Morbidelli and Rickman,* 2015) and the size distribution of small bodies would approach the Dohnanyi slope (*Dohnanyi,* 1969; *O'Brien and Greenberg,* 2003). This could explain the size distribution break observed near D = 100 km (*Bernstein et al.,* 2004, 2006; *Fraser et al.,* 2014) (see also section 4.3). Any traces of the original surface morphology would be wiped out: most comets would be fragments of larger bodies. If $t_{disk} \ll$ 10 m.y. instead, 67P-sized comets would avoid being catastrophically disrupted. Smaller impacts, however, could still cause important shape changes (*Jutzi and Benz,* 2017).

Catastrophic disruptions (*Schwartz et al.,* 2018) and subcatastrophic impacts on elongated and rotating bodies have been modeled to demonstrate the formation of bilobate comets (*Jutzi and Benz,* 2017). As the massive disk was dispersed by Neptune, the collisional probabilities dropped, and the collision speeds increased from hundreds of meters per second to several kilometers per second (*Nesvorný and Vokrouhlický,* 2019). It is likely in this situation that each catastrophic disruption was followed by sub-catastrophic impacts, and each subcatastrophic impact — potentially capable of generating a bilobate shape — was followed by shape-changing impacts. The observed comet shapes could be a complex end product of this sequence.

As comets evolve into the inner solar system, they become affected by H_2O sublimation torques. Simulations suggest that 67P, and bilobed comets in general, should often spin up past the breakup limit, fission, and reconfigure (*Hirabayashi et al.,* 2016). Many new bilobate configurations can be produced by this process. CO or CO_2 sublimation-driven spin-up might be capable of disrupting a typical JFC even before it reaches the inner solar system (*Safrit et al.,* 2021). These results highlight the important relationship between spin and morphology. Adding to that, a spinning bilobed object also better resists reconfiguration into a more spheroidal body by small impacts (*Jutzi et al.,* 2017).

There are at least two basic possibilities: (1) small TNOs formed as contact binaries and these shapes survived, even if with some modifications, such that TNOs remained bilobed as they dynamically evolve into Centaurs/JFCs; or (2) the bilobed shapes of JFCs have nothing to do with the formation of TNOs, but instead were produced by other processes, such as impact disruption, activity (*Safrit et al.,* 2021), rotational fission, etc., and could reflect the reaccumulation of fragments. It may be hard to distinguish between these two possibilities. One option would be to measure the shapes of a large sample of small TNOs via occultations or lightcurve surveys, and, in the long run, via spacecraft imaging. This would give us a rough sense of the fraction of contact binaries in different TNO populations, including the cold classicals. For example, option (1) could be ruled out if small cold classicals show a very low contact binary fraction. This would indicate that small planetesimals in the outer solar system did not form bilobed, therefore pointing toward evolutionary processes as the primary cause. It is worth pointing out, however, that the only small TNO we have flown past happens to be bilobed. We could also expect some information from future surveys about the physical properties of TNO contact binaries — for example, about the relative size of individual components in each detected small contact binary — and compare that with bilobed JFCs and Centaurs. A match would be expected if small TNOs formed bilobed and the evolutionary processes have only a small effect on the overall shape.

4.3. Size Distributions of Transneptunian Objects, Centaurs, and Jupiter-Family Comets

Here we consider the size-frequency distribution (SFD) of TNOs, Centaurs, and JFCs. The SFD holds information about the formative, morphological, and destructive processes that have altered the sizes of these bodies.

The SFD of TNOs is almost always estimated from reflectance photometry of TNOs in well-characterized surveys. The cumulative distribution of apparent magnitudes m, which we will call the luminosity function (LF), is typically expressed as $\Sigma(< m) \propto 10^{\alpha m}$. Here $\Sigma(< m)$ represents the surface density in TNOs per square degree on the sky (usually on the ecliptic) of bodies with magnitudes less than (brighter than) m, and α is the slope of the distribution on a plot of $\log_{10} \Sigma$ vs. m. If the TNOs follow a power-law differential size distribution of the form $dN/dD \propto D^{-q}$, where D is the diameter of the TNO, and we assume constant albedos, then the differential SFD slope, q, and α are related by $q = 5\alpha + 1$. The cumulative size distribution is given by $N (> D) \propto D^{-\gamma}$, where $\gamma = 5\alpha = q{-}1$ (e.g., *Gladman et al.,* 2001). Correctly converting from a magnitude distribution to an SFD relies on a number of assumptions, including (1) the albedo and size distributions are constant and do not depend on heliocentric distance, (2) shape/lightcurve effects are not important, and (3) observational biases such as the limiting magnitude of the survey are accounted for. With these limitations in mind, surveys revealed a steep LF in the outer solar system for objects brighter than $m_R \sim 25$, with slope $\alpha \sim 0.7$ or $q \sim 4.5$ (*Jewitt et al.,* 1998; *Gladman et al.,* 1998; *Trujillo et al.,* 2001; *Fraser et al.,* 2008), significantly steeper than the value of q between 3 and 4 expected for a population in steady-state collisional cascade (*Dohnanyi,* 1969; *Matthews et al.,* 2014; *O'Brien and Greenberg,* 2003). Deep "pencil-beam" surveys revealed a shallower LF with slope $\alpha \sim 0.2$ at fainter magnitudes, with a transition between the bright and faint slopes at a brightness $m_R \sim 26$ (*Bernstein et al.,* 2004, 2006; *Fuentes et al.,* 2009; *Fraser and Kavelaars,* 2009). This transition is referred to as the "knee" or break magnitude in the LF.

As surveys and observational techniques improved, more of the discovered objects were tracked over time, allowing accurate distance measurements and even dynamical classification of the survey discoveries. We highlight the Canada-France-Ecliptic Plane Survey, which arguably was the first survey to provide 100% tracking for all survey discoveries (*Petit et al.,* 2011). This improvement enabled the first direct measurements of the differential absolute luminosity function (ALF), $\Sigma(H)$, where 1520 $H = m{-}2.5 \log(\Delta r){-}f(\alpha_p)$. Here m is the apparent magnitude of a TNO (often in R band), r and Δ are the TNO's heliocentric and geocentric distances, α_p is the phase angle, or observer-Sun-TNO angle, in degrees, and f is a function that describes the decrease in reflectance of a TNO with increasing α_p. For the low phase angles for TNOs accessible from Earth, f is usually linearly approximated as $f = 0.15\ \alpha_p$, where α_p is measured in degrees (see *Alvarez-Candal et al.,* 2016, for further details). The ability to measure the ALF provided higher fidelity toward inferring the true underlying SFD.

The shape of the ALF has been characterized by power-law slopes of the form $\Sigma(H) = 10^{\alpha_H (H-H_o)}$, and is known to exhibit three distinct regions. This functional form has been chosen as a matter of convenience, as it not only trivially translates to the power-law SFD discussed above, but also provides a statistically sufficient description of the observations (*Fraser et al.,* 2014). Formative and collisional processes do not necessarily favor the production of power-law SFDs (*Li et al.,* 2019; see also the chapter in this volume by Simon et al.).

The slopes of the ALF of the dynamically excited populations of TNOs are well described by

$$\alpha_H \sim \begin{cases} 0.2, \text{ for } H_r \lesssim 4 \\ 0.87, \text{ for } 4 \lesssim H_r \lesssim 8 \\ 0.2, \text{ for } H_r > 8 \end{cases} \tag{1}$$

We show a depiction of this shape in Fig. 12. This ALF translates to a size distribution that is shallow for the largest TNOs ($500 \lesssim D \lesssim 2100$ km), with SFD slope $q \sim 2$ for $D \gtrsim 800$ km for a 6% albedo (*Brown,* 2008; *Fraser et al.,* 2014; *Nesvorný et al.,* 2017; *Abedin and Kavelaars,* 2022).

This part of the size distribution is colloquially referred to as the *foot* of the SFD (*Fraser et al.,* 2012). The SFD has a steep slope $q \sim 5.25$ for objects with $100 \lesssim D \lesssim$ 600 km and then becomes shallow again, with $q \sim 2$, for

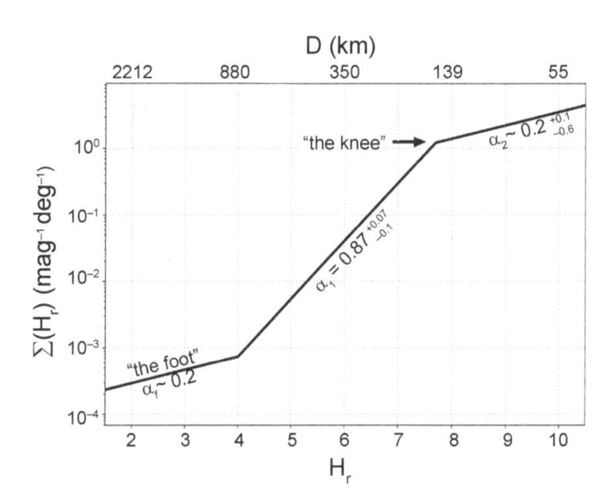

Fig. 12. A depiction of the measured differential absolute luminosity function of the dynamically excited TNOs. Each section of the ALF is approximated by the functional form $\Sigma(H) \propto 10^{\alpha(H-H_o)}$. Slopes and transition magnitudes are taken from *Fraser et al.* (2014). Object diameters are shown across the top, and assume a geometric albedo of 6%.

sizes smaller than the knee, D $\lesssim$ 100 km (*Fraser et al.,* 2014). Direct survey constraints on the SFD much below the knee (e.g., to sizes D $\ll$ 100 km) are difficult to gather due to the faintness of such small objects in reflected light.

Two separate scenarios for the creation of the SFD of the hot population have been postulated. Historically, the break at $H_r \approx 8$ has been associated with the size above which catastrophic collisions rarely occur. That is, since the epoch of formation, collisions have been largely disruptive due to the high relative velocities between TNOs (e.g., *Dell'Oro et al.,* 2013). Also, for most of the solar system's planetesimal populations, including TNOs, smaller bodies are more numerous, and also easier to disrupt down to subkilometer sizes (*Benz and Asphaug,* 1999). The net result is that smaller bodies are more likely to be collisionally disrupted than are relatively larger bodies (*Bottke et al.,* 2005). It follows that most of the largest bodies may have avoided disruption. The steep slope for large objects then is a result of their formation process, originally thought to occur by hierarchical accretion, and the break diameter reflects the size below which, on average, all objects have been collisionally disrupted. Numerical simulations bear out this idea (e.g., *Kenyon et al.,* 2008; *Benavidez and Campo Bagatin,* 2009), although recently, this interpretation has fallen out of favor. It is now thought that the observed break diameter is primordial, not collisional, in origin. If the break were collisional, the significant population of binary systems seen in the Kuiper belt could not have survived (*Parker and Kavelaars,* 2010; *Nesvorný and Vokrouhlický,* 2019).

The alternative idea to the collisional disruption scenario is the so-called born-big scenario, first put forth to explain the size distribution of the main-belt asteroids (*Morbidelli et al.,* 2009a), which has also been invoked to explain the break in the SFD of the Neptune Trojans (*Sheppard and Trujillo,* 2010). In this scenario, planetesimals form with a preferred size, D ~ 200 km, in a formation process that is much more rapid than can be achieved through hierarchical accretion. One possible mechanism rests with the streaming instability discussed in section 4.2. In this scenario, the break diameter reflects the preferred formation size, and the steep SFD at larger sizes reflects subsequent growth through classical hierarchical accretion. Notably, this scenario is broadly compatible with the observed properties of TNO binaries, including the frequency (*Nesvorný et al.,* 2011; *Robinson et al.,* 2020), orbital distribution (*Grundy et al.,* 2019b; *Nesvorný,* 2021), and diameters of TNOs themselves (*Li et al.,* 2019).

In either scenario, the *foot* of the SFD remains unexplained. This structure seems similar to expectations from models of runaway growth (*Lithwick,* 2014), although that has not been confirmed. It may also be that the foot is merely a signature of the largest objects that can form through the streaming instability.

Extension of the SFD to smaller diameters comes from interpretation of the Pluto, Charon, and Arrokoth cratering records. At diameters D $\gtrsim$ 1–2 km, the data are consistent with a slope $q \approx 3$ from the knee down to these sizes. For D $\lesssim$ 1–2 km, the SFD appears to break again at the "elbow" to an even shallower slope, $q = 1.7 \pm 0.3$, betraying a relative dearth of small TNOs with $0.1 < D < 1$ km, compared with an extrapolation from larger bodies (*Greenstreet et al.,* 2015, 2016; *Singer et al.,* 2019; *Parker,* 2021; *Robbins and Singer,* 2021; *Singer et al.,* 2021). A robust explanation of the change in the slope of the SFD near 1 km remains unavailable.

There is some hint of a so-called *divot* in the size distribution, just smaller than the knee diameter. The divot is a purported sudden downward deviation in the SFD, with fewer objects at sizes just below the break than above it (*Shankman et al.,* 2013; *Lawler et al.,* 2018b). Such a feature can result from a population of planetesimals that already has an SFD reminiscent of that observed (a steep slope followed by a break) that undergoes a sudden increase in velocity dispersion, such as that experienced during the onset of planet migration (*Fraser,* 2009). When the velocity dispersion rises, objects suddenly have enough kinetic energy to disrupt larger bodies than those they could shatter previously. Due to the transition from a steep to shallow distribution at the break, objects larger than the break see a relative increase in disruption rate than do objects smaller than the break, with the largest relative increase occurring for bodies of size just small enough to be disrupted by objects equal to the break radius. The result is preferential destruction of bodies ~10× larger than the initial break radius. While the presence of a divot is not statistically required to match the observed ALF, its presence is compatible with observations and, notably, is compatible with expectations of the formative and dynamical history of TNOs (*Fraser,* 2009).

For completeness, we point out that the cold population appears to exhibit a different SFD than do the dynamically excited populations. *Fraser et al.* (2014) found that the largest cold classical objects have D ~ 400 km and fall on the steep part of the SFD, with a slope q ~ 8. [In terms of absolute magnitude, the brightest cold classical TNO is 79360 Sila-Nunam (1997 CS_{29}), which has H = 5.29. Sila-Nunam is a roughly equal-mass binary in which each component has D $\approx$ 250 km (*Grundy et al.,* 2012).] Contrast this to the hot population SFD: The cold-classical SFD is missing the foot, and is much steeper brightward of the knee. This implies a significant difference in formation histories between those TNOs that appear to have formed *in situ*, and those that were scattered outward during Neptune's migration. The highest fidelity measure of the cold-classical SFD (*Kavelaars et al.,* 2021) demonstrates that this population is not well described by a broken power-law as in equation (1), but rather is better fit by a modified exponential function, $N(< H) = 10^{\alpha_S(H-H_o)} \times \exp^{10^{\beta_S(H-H_o)}}$, similar to the SFD produced in some simulations of formation via the streaming instability (*Li et al.,* 2019). For the hot populations, it is likely that future observations will clarify whether power laws are reasonable descriptions of the SFD of the hot population.

We also point out a tension in the populations of small, excited TNOs inferred from the ALF and the Pluto/Charon cratering record with that inferred from three serendipitous stellar occultations. Taken at face value, the occultations imply the existence of a population of small, D ~ 1 km excited bodies that is more than an order of magnitude larger than implied from the former techniques (*Schlichting et al.,* 2012; *Arimatsu et al.,* 2019; *Parker,* 2021). A solution to this tension may be the presence of a very large population of small TNOs beyond ≈50 au (*Shannon et al.,* 2023). It may also be the case that there is a population of small objects with extremely low albedos that have caused them to avoid optical detection. It may also be that some occultation events are not true detections, but rather are the result of unknown instrumental artifacts.

Centaurs and JFCs provide another opportunity to probe the small-end size distribution of their parent population in the Kuiper belt. However, there are challenges to interpreting observations of both Centaurs and JFCs. The first arises from the lack of dedicated, well-characterized surveys, particularly for Centaurs. This is primarily due to the Centaurs' small numbers, a result of their short dynamical lifetimes on planet-crossing orbits, and their wide range of ecliptic latitudes owing to their dynamically hot inclination distribution. Centaurs are thus much less dense on the sky than TNOs. Most observational surveys are focused on detecting either the close near-Earth-asteroid population or the more distant TNO populations, so the biases in the observed Centaur population are typically not well understood (see, e.g., discussion in *Peixinho et al.,* 2020). Another complication is that JFCs and some Centaurs are active. For such bodies, their observed magnitudes contain contributions from (and often are dominated by) the coma, rather than just the nucleus. Astronomers have tried to minimize coma signal by taking observations when the comets are at aphelion and thus less active or inactive; however, there is always the possibility of unresolved coma that nonetheless affects the photometry (*Hui and Li,* 2018).

With these limitations in mind, does the TNO size distribution match the JFC and Centaur size distributions? *Lamy et al.* (2004) review data on 65 ecliptic comets with effective diameters between 0.4 and 30 km, and infer a cumulative size distribution for $D \gtrsim 3$ km with $N(>D) \propto D^{-\gamma}$, where $\gamma = 1.9 \pm 0.3$, corresponding to $q \approx 2.6$–3.2 for the differential size distribution and $\alpha \approx 0.3$–0.4 for the luminosity function. *Lamy et al.* (2004) attribute the shallower size distribution seen for comets with $D \lesssim 3$ km to observational incompleteness and mass loss due to activity. *Meech et al.* (2004), on the other hand, find that the flatter slope for small comets is not explained by observational bias.

Snodgrass et al. (2011) performed Monte Carlo simulations of the JFC size distribution, accounting for uncertainties in photometry and the albedo, phase function, and shape of the nucleus. They inferred $\gamma = 1.92 \pm 0.20$ for nuclei with $D > 2.5$ km, consistent with *Lamy et al.* (2004).

The largest survey to date of JFCs was carried out by *Fernández et al.* (2013), who observed 89 JFCs with Spitzer and included nine other JFCs from the literature. They infer $\gamma \approx 1.9 \pm 0.2$ for $D > 3$ km, but note that, surprisingly, JFCs with $D > 6$ km and $q < 2$ au are still being found. *Fernández et al.* (2013) agree with *Meech et al.* (2004) that the rarity of small comets is real. *Jewitt* (2021) explains the scarcity of small short-period comets as the result of sublimation torques, which he estimates can spin up comets with perihelion distances between 1 and 2 au to rotational disruption in $25(D/1\ \text{km})^2$ yr, i.e., only a few orbits. The inferred SFD slopes for JFCs agree with those commonly adopted for the excited TNO parent populations, but the SFD of the Jupiter Trojans is more commonly used as a proxy for the TNO SFD for comet-sized bodies, as more Trojans are known than JFCs, and Trojans are inactive (*Jewitt et al.,* 2000; *Yoshida et al.,* 2019, 2020). [Since the year 2000, the number of known near-Earth asteroids (defined as inactive bodies with $q <$ 1.3 au) has increased by a factor of ≈30, while the number of near-Earth comets has less than doubled. These numbers are tabulated by the Center for Near Earth Object Studies at *https://cneos.jpl.nasa.gov/stats/totals.html*. Although biases in the discovery of asteroids and comets differ, the different discovery rates support the idea that small JFCs are intrinsically rare.]

Using data from the Deep Ecliptic Survey, *Adams et al.* (2014) found $\alpha = 0.42 \pm 0.02$ (i.e., $\gamma = 2.1 \pm 0.1$) for seven Centaurs with absolute magnitudes between 7.5 and 11, corresponding to 30 km $\lesssim D \lesssim$ 170 km for an assumed albedo of 0.06. OSSOS discovered 15 Centaurs with $a < 30$ au, $q >$ 7.5 au, and $H_r < 13.7$ (i.e., $D > 10$ km for an albedo of 0.06). *Nesvorný et al.* (2019b) found that his planetary instability models predicted 11 ± 4 such Centaurs if he assumed $\gamma =$ 2.1, similar to the slope inferred for small Trojans (*Wong and Brown,* 2015). In general, the debiased SFDs of the JFCs and Centaurs are consistent with those measured for the TNO populations that feed the JFCs, but a detailed comparison cannot yet be made. We discuss this in section 5.

5. OUTSTANDING QUESTIONS AND FUTURE PROSPECTS

In this section, we highlight three outstanding questions regarding TNOs and their link to cometary populations. We point out the anticipated contributions of certain new and upcoming telescope facilities that should provide significant leaps forward in our understanding of these questions. This list should in no way be considered impartial or complete, but rather reflects problems that the authors find particularly pertinent.

The first outstanding problem we wish to highlight regards the compositions of TNOs, and by extension the Centaurs and JFCs that are fed from the Kuiper belt. As expressed in section 3.1, despite more than 25 yr of observations by the astronomical community, very little is known about the surfaces or internal compositions of TNOs. We are still ignorant of the nature of the reddening agent responsible for the colors of these icy bodies. Moreover, beyond a couple of exceptions, no detection has been made of signatures of the silicate materials that must be present on and inside TNOs.

The main reason for this sad state of affairs is the lack of identifying absorption features in the optical and NIR atmospheric transmission regions. It seems that whatever materials are present on the surfaces of TNOs, they mutually mask any features that fall in this wavelength range. See, for example, the spectral model of the Centaur Pholus in Fig. 8, where tholin and olivine mask the strong absorption features each would exhibit on their own.

Longer wavelengths hold the potential to reveal some strong absorption features. For example, many of the ice species known or suspected to exist on TNOs exhibit deep and broad absorption features in the $2.5 \lesssim \lambda \lesssim 4.5$ µm range (see Fig. 13). This spectral region is very difficult to observe for these faint bodies with current technology, especially groundbased because of contamination from Earth's atmosphere. It is for this reason that JWST holds the massive potential to revolutionize our knowledge of the compositions of TNOs.

Most insight into the compositions of TNOs will come from spectral observations with JWST's Near-Infrared Spectrograph (NIRSpec), which will provide unprecedented sensitivity across a critical wavelength range out to 5 µm. In Fig. 14 we reproduce a figure from *Parker et al.* (2016) that presents three different spectral models, all of which are broadly compatible with the spectrum of the distant dwarf planet (90377) Sedna. Each model is wildly different in its compositional makeup, but nearly indistinguishable in reflectance spectra at wavelengths $\lambda \lesssim 2.5$ µm. At longer wavelengths, spectra from NIRSpec will be particularly useful in diagnosing a surface as organic-rich or organic-poor, as the C-H and C-N vibrational fundamental and overtone bands fall in this range (see *Roush and Dalton,* 2004; *Izawa et al.,* 2014, for recent discussions). Icy species, including H_2O, CH_3OH, CH_4, N_2, CO, and CO_2, also exhibit absorption features in this range that should be readily apparent in high-quality NIRSpec observations [see Fig. 13 of this chapter as well as Fig. 4 of *Fernández-Valenzuela et al.* (2021)].

Longer-wavelength spectra in the mid-infrared range $5 < \lambda < 28$ µm will come from JWST's Mid-Infrared Instrument (MIRI), where fine-grained silicate materials exhibit emission features (*Martin et al.,* 2022). While objects in the Kuiper belt are too distant, and therefore too faint, for observations with MIRI, many Centaurs should be bright enough. Spitzer observed 20 Centaurs in the mid-infrared (*Lisse et al.,* 2020), but obtained spectral data over the range 7.5–38 µm for only one, (8405) Asbolus (*Barucci et al.,* 2008). That spectrum, which suggests the presence of fine-grained silicates on the Centaur's surface, is broadly similar to the spectra of three Jupiter Trojans observed by Spitzer (*Dotto et al.,* 2008).

Unfortunately, the low signal-to-noise of Spitzer's spectrum of Asbolus makes it impossible to infer the nature of its putative silicates. The MIRI spectra should be of sufficient quality to not only definitively identify silicate emissions on many Centaurs, but also to identify specific properties of the silicates, such as whether those silicates are amorphous or crystalline in nature.

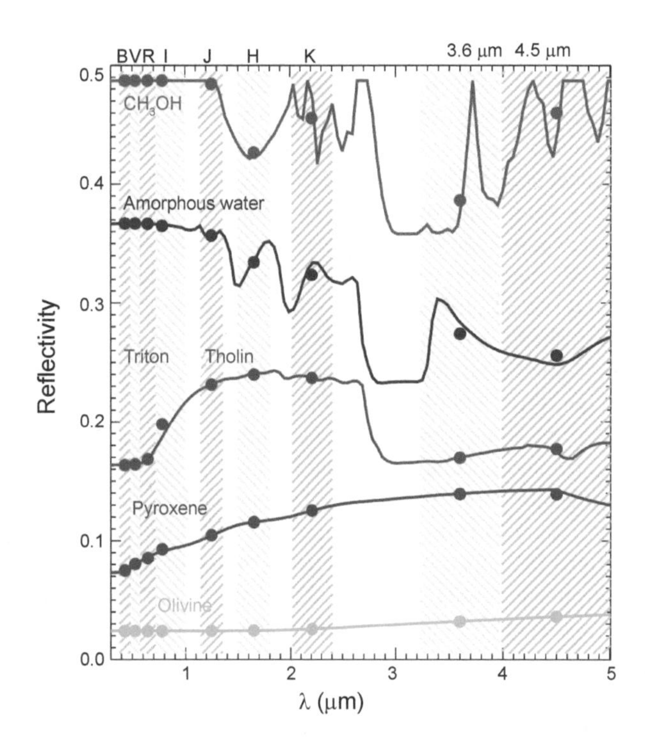

Fig. 13. Spectra of various laboratory materials through the optical and NIR spectral range, reproduced from *Fernández-Valenzuela et al.* (2021). Photometric bands of the Johnson-Cousins system, which are accessible from groundbased facilities, as well as the Spitzer-IRAC 3.6- and 4.5-µm bands, are shown.

With regard to the volatile component, the JWST NIRSpec-integral field unit (IFU) spectrograph is likely to either detect CO_2 emission in active Centaurs, or set strong upper limits, possibly up to 2 orders of magnitude less than the

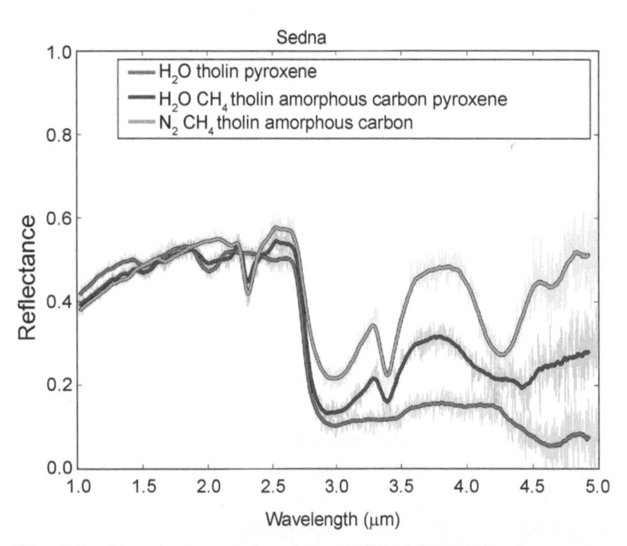

Fig. 14. Spectral models of the TNO (90377) Sedna, using the silicate pyroxene, the organic material Titan tholin, CH_4, and amorphous carbon, and N_2 and H_2O ices. Reproduced from *Parker et al.* (2016).

limit of 3.5 × 10^{26} for 29P achieved by AKARI (*Ootsubo et al.,* 2012; A. McKay, personal communication). In addition, CO and H_2O emission will also be reachable with the instrument (see Fig. 15). As discussed in section 4.1, CO_2 is reduced with respect to CO in the comae of objects beyond 3 au, a trend that increases with heliocentric distance (*Harrington Pinto et al.,* 2022). Thus, if other Centaurs are like 29P, then their comae may be also remarkably depleted in CO_2. JWST will be capable of detecting the 1.5, 2.0, and 3.0 μm bands of water, with the 3-μm band being the one most likely to be detected in the comae of active Centaurs, and the 2-μm band being the next likely.

An important, unanswered question about Centaur gaseous comae is whether comparing the CO, CO_2, and H_2O mixing ratios in cometary and Centaur comae is indeed a good match to compositional models of nuclei, or instead is more influenced by outgassing behavior at different heliocentric distances, due to differences in sublimation of the two species. This is a significant opportunity for the modeling community.

The second outstanding problem we wish to highlight is the discrepancy between the SFD of the JFCs and that predicted from our knowledge of the source populations feeding the JFCs (see sections 2.1 and 2.2). While the slopes of the SFDs between the two populations seem to align (section 4.3), there still appears to be disagreement between the availability of the source populations and the observed density of JFCs. This problem was highlighted by *Volk and Malhotra* (2008), who found that TNOs following the *Bernsteinet et al.* (2004, 2006) SFD for faint TNOs could produce only ≈1% of the observed JFCs. While modern models of the supply of JFCs from the TNOs match the orbital distributions of the comets well (e.g., *Di Sisto et al.,* 2009; *Brasser and Morbidelli,* 2013; *Nesvorný et al.,* 2017), the problem of directly, quantitatively linking the TNO and JFC population remains frustrated by several factors: the imprecise measurement of the SFDs of the scattered disk objects and other TNO populations, especially down to

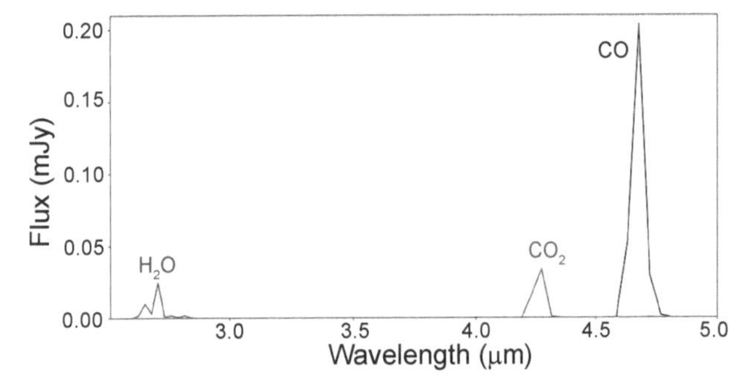

Fig. 15. Predicted JWST NIRSpec model gas emissions for an active Centaur with a similar coma compositional profile as 29P (A. McKay, personal communication). The model was calculated using the NASA Planetary Spectrum Generator [*https://psg.gsfc.nasa.gov* (*Villanueva et al.,* 2018)] assuming AKARI production rates for CO and H_2O from *Ootsubo et al.* (2012) and inferred $Q(CO_2)$ from *Harrington Pinto et al.* (2022).

comet-sized objects; the unknown ratio of active to inactive JFCs, i.e., the fading problem discussed in section 4.3; and the incomplete knowledge of the orbital structure of the most distant TNO populations and the interplay between the scattering and detached populations (discussed in section 2.3).

The Vera C. Rubin Telescope's Legacy Survey of Space and Time (LSST) will provide many key insights toward resolving this problem. One of the major advances from LSST will be a much improved census of the solar system's small bodies, down to a brightness threshold of r ~ 24.5, with well-understood observational biases. LSST will increase the inventory of known TNOs with well-determined orbits by at least an order of magnitude, detect large numbers of comet-sized (d ≲ 10 km) Centaurs, and increase the observed JFC population (*LSST Science Collaboration et al.,* 2009). These observations will yield important new constraints on the detailed orbital distributions of all three populations. LSST will also provide 10 years of monitoring for each discovered JFC and Centaur, on a 4–8-day cadence. This monitoring will be fundamental in detecting the onset and turnoff of activity as comets move between aphelion and perihelion as well as measuring the vigor of the activity. It will also yield improved insights into the beginning stages of activity in the Centaur region. Importantly, the monitoring will enable a robust derivation of the SFD of JFCs when they are inactive, which is what is required to make a robust comparison with the Centaur and TNO populations. A final issue is the possible presence of bodies more massive than Pluto and Eris (0.002–0.003 $M_\oplus$) in the distant Kuiper belt. Even with its sublunar mass, Pluto can destabilize bodies in the 3:2 MMR (*Nesvorný et al.,* 2000). *Gladman and Chan* (2006) investigated how an Earth-mass body, 10 Mars-mass bodies (each ≈0.1 $M_\oplus$), or both would raise the inclinations and perihelion distances of some objects in the primordial Kuiper belt and produce detached bodies like Sedna. *Lykawka and Mukai* (2008) proposed that a planet with mass ≈0.5 $M_\oplus$ excited the early Kuiper belt and evolved into a stable orbit with a > 100 au, q > 80 au, and an inclination between 20° and 40°. *Trujillo and Sheppard* (2014) discovered a TNO, 2012 VP_{113}, with a ≈ 266 au and a perihelion distance even larger than Sedna's, q = 80 au. They noted that the 12 TNOs with a > 150 au and q > 30 au appeared to cluster in argument of perihelion, and proposed that a 2–15-$M_\oplus$ body could produce the clustering. This idea is now known as the "Planet Nine" hypothesis (*Batygin et al.,* 2019; *Brown and Batygin,* 2021) and invokes a 5–10-$M_\oplus$ body with a perihelion distance ≈300 au to shape the orbits of TNOs with a > 250 au. Such a planet would even affect the orbital distributions of JFCs and HTCs (*Nesvorný et al.,* 2017). Well-characterized TNO surveys (*Shankman et al.,* 2017; *Kavelaars et al.,* 2020; *Napier et al.,* 2021) find that the orbital distribution of "extreme" TNOs does not require the existence of Planet Nine, but neither can it (or less massive perturbers) be ruled out. Assuming an albedo between 0.20 and 0.75, *Brown and Batygin* (2021) predict that Planet Nine's most likely R magnitude is ≈20, but it could be as faint as magnitude 25. Searches in the far-IR with archival

Infrared Astronomical Satellite (IRAS) and AKARI data have also been carried out recently (*Rowan-Robinson,* 2022; *Sedgwick and Serjeant,* 2022). Observations in the next decade with the Vera Rubin Observatory and other large telescopes such as Subaru, the Very Large Telescope (VLT), and Gemini should clarify the orbital distribution of distant TNOs and may finally detect the long-speculated transneptunian planet.

Acknowledgments. K.V. acknowledges support from NSF (grant AST-1824869) and NASA (grants 80NSSC19K0785, 80NSSC21K0376, and 80NSSC22K0512). L.D. thanks the Cassini Data Analysis Program for support. D.N. would like to acknowledge support from the Emerging Worlds program. This material is based in part on work done by M.W. while serving at the National Science Foundation. We would especially like to thank R. Kokotanekova and A. Fitzsimmons for their verbal contributions to this chapter. We thank A. McKay and G. Williams for useful discussions. We also thank S. Hasewaga for providing data tables.

REFERENCES

Abedin A. Y. and Kavelaars J. J. (2022) On the lack of catastrophic collisions in the present Kuiper belt. *Lunar and Planetary Science LIII,* Abstract #1998. Lunar and Planetary Institute, Houston.

Adams E. R., Gulbis A. A. S., Elliot J. L., et al. (2014) De-biased populations of Kuiper belt objects from the Deep Ecliptic Survey. *Astron. J., 148,* 55.

Agnor C. B. and Lin D. N. C. (2012) On the migration of Jupiter and Saturn: Constraints from linear models of secular resonant coupling with the terrestrial planets. *Astrophys. J., 745,* 143.

A'Hearn M. F. (2011) Comets as building blocks. *Annu. Rev. Astron. Astrophys., 49,* 281–299.

A'Hearn M. F., Millis R. C., Schleicher D. O., et al. (1995) The ensemble properties of comets: Results from narrowband photometry of 85 comets, 1976–1992. *Icarus, 118,* 223–270.

A'Hearn M. F., Feaga L. M., Keller H. U., et al. (2012) Cometary volatiles and the origin of comets. *Astrophys. J., 758,* 29.

Altwegg K., Combi M., Fuselier S. A., et al. (2022) Abundant ammonium hydrosulphide embedded in cometary dust grains. *Mon. Not. R. Astron. Soc., 516,* 3900–3910.

Alvarez-Candal A., Fornasier S., Barucci M. A., et al. (2008) Visible spectroscopy of the new ESO large program on trans-Neptunian objects and Centaurs: Part 1. *Astron. Astrophys., 487,* 741–748.

Alvarez-Candal A., Pinilla-Alonso N., Licandro J., et al. (2011) The spectrum of (136199) Eris between 350 and 2350 nm: Results with X-Shooter. *Astron. Astrophys., 532,* A130.

Alvarez-Candal A., Pinilla-Alonso N., Ortiz J. L., et al. (2016) Absolute magnitudes and phase coefficients of trans-Neptunian objects. *Astron. Astrophys., 586,* A155.

Arimatsu K., Tsumura K., Usui F., et al. (2019) A kilometre-sized Kuiper belt object discovered by stellar occultation using amateur telescopes. *Nature Astron., 3,* 301–306.

Bailey B. L. and Malhotra R. (2009) Two dynamical classes of Centaurs. *Icarus, 203,* 155–163.

Bannister M. T., Gladman B. J., Kavelaars J. J., et al. (2018) OSSOS. VII. 800+ trans-neptunian objects — The complete data release. *Astrophys. J. Suppl. Ser., 236,* 18.

Barkume K. M., Brown M. E., and Schaller E. L. (2006) Water ice on the satellite of Kuiper belt object 2003 EL61. *Astrophys. J. Lett., 640,* L87–L89.

Barkume K. M., Brown M. E., and Schaller E. L. (2008) Near-infrared spectra of Centaurs and Kuiper belt objects. *Astron. J., 135,* 55–67.

Barucci M. A. and Merlin F. (2020) Surface composition of trans-neptunian objects. In *The Trans-Neptunian Solar System* (D. Prialnik et al., eds.), pp. 109–126. Elsevier, Amsterdam.

Barucci M. A., Cruikshank D. P., Mottola S., et al. (2002) Physical properties of Trojan and Centaur asteroids. In *Asteroids III* (W. F. Bottke et al., eds.), pp. 273–287. Univ. of Arizona, Tucson.

Barucci M. A., Belskaya I. N., Fulchignoni M., et al. (2005) Taxonomy of Centaurs and trans-neptunian objects. *Astron. J., 130,* 1291–1298.

Barucci M. A., Brown M. E., Emery J. P., et al. (2008) Composition and surface properties of transneptunian objects and Centaurs. In *The Solar System Beyond Neptune* (M. A. Barucci et al., eds.), pp. 143–160. Univ. of Arizona, Tucson.

Barucci M. A., Alvarez-Candal A., Merlin F., et al. (2011) New insights on ices in Centaur and transneptunian populations. *Icarus, 214,* 297–307.

Batygin K., Brown M. E., and Fraser W. C. (2011) Retention of a primordial cold classical Kuiper belt in an instability-driven model of solar system formation. *Astrophys. J., 738,* 13.

Batygin K., Brown M. E., and Betts H. (2012) Instability-driven dynamical evolution model of a primordially five-planet outer solar system. *Astrophys. J. Lett., 744,* L3.

Batygin K., Adams F. C., Brown M. E., et al. (2019) The Planet Nine hypothesis. *Phys. Rept., 805,* 1–53.

Bauer J. M., Choi Y.-J., Weissman P. R., et al. (2008) The large-grained dust coma of 174P/Echeclus. *Publ. Astron. Soc. Pac., 120,* 393–404.

Bauer J. M., Grav T., Blauvelt E., et al. (2013) Centaurs and scattered disk objects in the thermal infrared: Analysis of WISE/NEOWISE observations. *Astrophys. J., 773,* 22.

Benavidez P. G. and Campo Bagatin A. (2009) Collisional evolution of trans-neptunian populations: Effects of fragmentation physics and estimates of the abundances of gravitational aggregates. *Planet. Space Sci., 57,* 201–215.

Benavidez P. G., Campo Bagatin A., Curry J., et al. (2022) Collisional evolution of the trans-neptunian region in an early dynamical instability scenario. *Mon. Not. R. Astron. Soc., 514,* 4876–4893.

Benecchi S. D., Borncamp D., Parker A. H., et al. (2019) The color and binarity of (486958) 2014 MU_{69} and other long-range New Horizons Kuiper belt targets. *Icarus, 334,* 22–29.

Benz W. and Asphaug E. (1999) Catastrophic disruptions revisited. *Icarus, 142,* 5–20.

Bernardinelli P. H., Bernstein G. M., Sako M., et al. (2022) A search of the full six years of the Dark Energy Survey for outer solar system objects. *Astrophys. J. Suppl. Ser., 258,* 41.

Bernstein G. M., Trilling D. E., Allen R. L., et al. (2004) The size distribution of trans-neptunian bodies. *Astron. J., 128,* 1364–1390.

Bernstein G. M., Trilling D. E., Allen R. L., et al. (2006) Erratum: The size distribution of trans-neptunian bodies (AJ, 128, 1364 [2004]). *Astron. J., 131,* 2364–2364.

Bertrand T., Forget F., Umurhan O. M., et al. (2018) The nitrogen cycles on Pluto over seasonal and astronomical timescales. *Icarus, 309,* 277–296.

Bierson C. J. and Nimmo F. (2019) Using the density of Kuiper belt objects to constrain their composition and formation history. *Icarus, 326,* 10–17.

Biver N. (1997) Molècules mères comètaires: Observations et modèlisations. Ph.D. thesis, Universitè Paris Diderot, Paris. 265 pp.

Biver N., Bockelée-Morvan D., Colom P., et al. (2002) The 1995–2002 long-term monitoring of Comet C/1995 O1 (Hale Bopp) at radio wavelength. *Earth Moon Planets, 90,* 5–14.

Bockelée-Morvan D., Biver N., Schambeau C. A., et al. (2022) Water, hydrogen cyanide, carbon monoxide, and dust production from distant Comet 29P/Schwassmann-Wachmann 1. *Astron. Astrophys., 664,* A95.

Bolin B. T., Fernandez Y. R., Lisse C. M., et al. (2021) Initial characterization of active transitioning Centaur, P/2019 LD_2 (ATLAS), using Hubble, Spitzer, ZTF, Keck, Apache Point Observatory, and GROWTH visible and infrared imaging and spectroscopy. *Astron. J., 161,* 116.

Bottke W. F., Durda D. D., Nesvorný D., et al. (2005) Linking the collisional history of the main asteroid belt to its dynamical excitation and depletion. *Icarus, 179,* 63–94.

Bottke W. F., Marschall R., Vokrouhlický D., et al. (2022) Collisional evolution of the primordial Kuiper belt, scattered disk, and Trojan populations. *Lunar and Planetary Science LIII,* Abstract #2638. Lunar and Planetary Institute, Houston.

Braga-Ribas F., Sicardy B., Ortiz J. L., et al. (2014) A ring system detected around the Centaur (10199) Chariklo. *Nature, 508,* 72–75.

Brasser R. and Morbidelli A. (2013) Oort cloud and scattered disc formation during a late dynamical instability in the solar system. *Icarus, 225,* 40–49.

Brasser R. and Wang J. H. (2015) An updated estimate of the number of Jupiter-family comets using a simple fading law. *Astron. Astrophys., 573,* A102.

Brasser R., Morbidelli A., Gomes R., et al. (2009) Constructing the secular architecture of the solar system II: The terrestrial planets. *Astron. Astrophys., 507*, 1053–1065.
Brasser R., Schwamb M. E., Lykawka P. S., et al. (2012) An Oort cloud origin for the high-inclination, high-perihelion Centaurs. *Mon. Not. R. Astron. Soc., 420*, 3396–3402.
Brown M. E. (2001) The inclination distribution of the Kuiper belt. *Astron. J., 121*, 2804–2814.
Brown M. E. (2008) The largest Kuiper belt objects. In *The Solar System Beyond Neptune* (M. A. Barucci et al., eds.), pp. 335–344. Univ. of Arizona, Tucson.
Brown M. E. (2013) The density of mid-sized Kuiper belt object 2002 UX25 and the formation of the dwarf planets. *Astrophys. J. Lett., 778*, L34.
Brown M. E. and Batygin K. (2021) The orbit of Planet Nine. *Astron. J., 162*, 219.
Brown M. E. and Butler B. J. (2017) The density of mid-sized Kuiper belt objects from ALMA thermal observations. *Astron. J., 154*, 19.
Brown M. E., Trujillo C., and Rabinowitz D. (2004) Discovery of a candidate inner Oort cloud planetoid. *Astrophys. J., 617*, 645–649.
Brown M. E., Barkume K. M., Ragozzine D., et al. (2007) A collisional family of icy objects in the Kuiper belt. *Nature, 446*, 294–296.
Brown M. E., Schaller E. L., and Fraser W. C. (2011) A hypothesis for the color diversity of the Kuiper belt. *Astrophys. J. Lett., 739*, L60.
Brown M. E., Schaller E. L., and Fraser W. C. (2012) Water ice in the Kuiper belt. *Astron. J., 143*, 146.
Brown R. H., Cruikshank D. P., and Pendleton Y. (1999) Water ice on Kuiper belt object 1996 TO_{66}. *Astrophys. J. Lett., 519*, L101–L104.
Brucker M. J., Grundy W. M., Stansberry J. A., et al. (2009) High albedos of low inclination classical Kuiper belt objects. *Icarus, 201*, 284–294.
Brunetto R., Barucci M. A., Dotto E., et al. (2006) Ion irradiation of frozen methanol, methane, and benzene: Linking to the colors of Centaurs and trans-neptunian objects. *Astrophys. J., 644*, 646–650.
Buchanan L. E., Schwamb M. E., Fraser W. C., et al. (2022) Col-OSSOS: Probing ice line/color transitions within the Kuiper belt's progenitor populations. *Planet. Sci. J., 3*, 9.
Bus S. J., A'Hearn M. F., Schleicher D. G., et al. (1991) Detection of CN emission from (2060) Chiron. *Science, 251*, 774–777.
Calmonte U., Altwegg K., Balsiger H., et al. (2016) Sulphur-bearing species in the coma of comet 67P/Churyumov-Gerasimenko. *Mon. Not. R. Astron. Soc., 462*, S253–S273.
Capria M. T., Coradini A., de Sanctis M. C., et al. (2009) Thermal modeling of the active Centaur P/2004 A1 (LONEOS). *Astron. Astrophys., 504*, 249–258.
Carusi A. and Valsecchi G. B. (1987) Dynamical evolution of short-period comets. *Publ. Astron. Inst. Czech. Acad. Sci., 2*, 21–28.
Carusi A., Kresak L., Perozzi E., et al. (1987) High-order librations of Halley-type comets. *Astron. Astrophys., 187*, 899–905.
Chandler C. O., Kueny J. K., Trujillo C. A., et al. (2020) Cometary activity discovered on a distant Centaur: A nonaqueous sublimation mechanism. *Astrophys. J. Lett., 892*, L38.
Chen Y.-T., Lin H. W., Holman M. J., et al. (2016) Discovery of a new retrograde trans-neptunian object: Hint of a common orbital plane for low semimajor axis, high-inclination TNOs and Centaurs. *Astrophys. J. Lett., 827*, L24.
Chen Y.-T., Gladman B., Volk K., et al. (2019) OSSOS. XVIII. Constraining migration models with the 2:1 resonance using the Outer Solar System Origins Survey. *Astron. J., 158*, 214.
Chiang E. I. and Jordan A. B. (2002) On the Plutinos and Twotinos of the Kuiper belt. *Astron. J., 124*, 3430–3444.
Chodas P. W. and Yeomans D. K. (2006) The orbital motion and impact circumstances of comet D/Shoemaker-Levy 9. In *The Collision of Comet Shoemaker-Levy 9 and Jupiter* (K. S. Noll et al., eds.), pp. 1–30. IAU Colloq. 156, Cambridge Univ., Cambridge.
Choi Y.-J. and Weissman P. (2006) Discovery of cometary activity for Centaur 174P/Echeclus (60558). *AAS/Division for Planetary Sciences Meeting Abstracts, 38*, #37.05.
Cieza L. A., Casassus S., Tobin J., et al. (2016) Imaging the water snow-line during a protostellar outburst. *Nature, 535*, 258–261.
Clements T. D. and Fernandez Y. (2021) Dust production from mini outbursts of Comet 29P/Schwassmann-Wachmann 1. *Astron. J., 161*, 73.
Cloutier R., Tamayo D., and Valencia D. (2015) Could Jupiter or Saturn have ejected a fifth giant planet? *Astrophys. J., 813*, 8.
Cochran A. L. and Cochran W. D. (1991) The first detection of CN and the distribution of CO^+ gas in the coma of Comet P/Schwassmann-Wachmann 1. *Icarus, 90*, 172–175.
Cochran A. L., Cochran W. D., Barker E. S., et al. (1991) The development of the CO^+ coma of comet P/Schwassmann-Wachmann 1. *Icarus, 92*, 179–183.
Crompvoets B. L., Lawler S. M., Volk K., et al. (2022) OSSOS XXV: Large populations and scattering-sticking in the distant trans-neptunian resonances. *Planet. Sci. J., 3*, 113.
Cruikshank D. P., Roush T. L., Bartholomew M. J., et al. (1998) The composition of Centaur 5145 Pholus. *Icarus, 135*, 389–407.
Dalle Ore C. M., Dalle Ore L. V., Roush T. L., et al. (2013) A compositional interpretation of trans-neptunian objects taxonomies. *Icarus, 222*, 307–322.
Dalle Ore C. M., Barucci M. A., Emery J. P., et al. (2015) The composition of "ultra-red" TNOs and Centaurs. *Icarus, 252*, 311–326.
Davies J. K., McFarland J., Bailey M. E., et al. (2008) The early development of ideas concerning the transneptunian region. In *The Solar System Beyond Neptune* (M. A. Barucci et al., eds.), pp. 11–23. Univ. of Arizona, Tucson.
Dawson R. I. and Murray-Clay R. (2012) Neptune's wild days: Constraints from the eccentricity distribution of the classical Kuiper belt. *Astrophys. J., 750*, 43.
De Sanctis M. C., Capria M. T., Coradini A., et al. (2000) Thermal evolution of the Centaur object 5145 Pholus. *Astron. J., 120*, 1571–1578.
De Sousa R. R., Morbidelli A., Raymond S. N., et al. (2020) Dynamical evidence for an early giant planet instability. *Icarus, 339*, 113605.
Deienno R., Morbidelli A., Gomes R. S., et al. (2017) Constraining the giant planets' initial configuration from their evolution: Implications for the timing of the planetary instability. *Astron. J., 153*, 153.
Dell'Oro A., Campo Bagatin A., Benavidez P. G., et al. (2013) Statistics of encounters in the trans-neptunian region. *Astron. Astrophys., 558*, A95.
DeMeo F. E., Binzel R. P., Slivan S. M., et al. (2009) An extension of the Bus asteroid taxonomy into the near-infrared. *Icarus, 202*, 160–180.
DeMeo F. E., Barucci M. A., Merlin F., et al. (2010) A spectroscopic analysis of Jupiter-coupled object (52872) Okyrhoe, and TNOs (90482) Orcus and (73480) 2002 PN_{34}. *Astron. Astrophys., 521*, A35.
Di Ruscio A., Fienga A., Durante D., et al. (2020) Analysis of Cassini radio tracking data for the construction of INPOP19a: A new estimate of the Kuiper belt mass. *Astron. Astrophys., 640*, A7.
Di Sisto R. P. and Brunini A. (2007) The origin and distribution of the Centaur population. *Icarus, 190*, 224–235.
Di Sisto R. P. and Rossignoli N. L. (2020) Centaur and giant planet crossing populations: Origin and distribution. *Cel. Mech. Dyn. Astron., 132*, 36.
Di Sisto R. P., Brunini A., Dirani L. D., et al. (2005) Hilda asteroids among Jupiter family comets. *Icarus, 174*, 81–89.
Di Sisto R. P., Fernández J. A., and Brunini A. (2009) On the population, physical decay and orbital distribution of Jupiter family comets: Numerical simulations. *Icarus, 203*, 140–154.
Di Sisto R. P., Ramos X. S., and Gallardo T. (2019) The dynamical evolution of escaped Jupiter Trojan asteroids, link to other minor body populations. *Icarus, 319*, 828–839.
Dohnanyi J. S. (1969) Collisional model of asteroids and their debris. *J. Geophys. Res., 74*, 2531–2554.
Dones L., Weissman P. R., Levison H. F., et al. (2004) Oort cloud formation and dynamics. In *Comets II* (M. C. Festou et al., eds.), pp. 153–174. Univ. of Arizona, Tucson.
Dones L., Brasser R., Kaib N., et al. (2015) Origin and evolution of the cometary reservoirs. *Space Sci. Rev., 197*, 191–269.
Dotto E., Emery J. P., Barucci M. A., et al. (2008) De Troianis: The Trojans in the planetary system. In *The Solar System Beyond Neptune* (M. A. Barucci et al., eds.), pp. 383–395. Univ. of Arizona, Tucson.
Drahus M., Yang B., Lis D. C., et al. (2017) New limits to CO outgassing in Centaurs. *Mon. Not. R. Astron. Soc., 468*, 2897–2909.
Drozdovskaya M. N., Walsh C., van Dishoeck E. F., et al. (2016) Cometary ices in forming protoplanetary disc midplanes. *Mon. Not. R. Astron. Soc., 462*, 977–993.
Duncan M. J. and Levison H. F. (1997) A scattered comet disk and the origin of Jupiter family comets. *Science, 276*, 1670–1672.
Duncan M., Quinn T., and Tremaine S. (1987) The formation and extent of the solar system comet cloud. *Astron. J., 94*, 1330–1338.

Duncan M., Quinn T., and Tremaine S. (1988) The origin of short-period comets. *Astrophys. J. Lett., 328*, L69–L73.

Duncan M. J., Levison H. F., and Budd S. M. (1995) The dynamical structure of the Kuiper belt. *Astron. J., 110*, 3073–3081.

Edgeworth K. E. (1949) The origin and evolution of the solar system. *Mon. Not. R. Astron. Soc., 109*, 600–609.

Elliot J. L., Kern S. D., Clancy K. B., et al. (2005) The Deep Ecliptic Survey: A search for Kuiper belt objects and Centaurs. II. Dynamical classification, the Kuiper belt plane, and the core population. *Astron. J., 129*, 1117–1162.

Emery J. P., Burr D. M., and Cruikshank D. P. (2011) Near-infrared spectroscopy of Trojan asteroids: Evidence for two compositional groups. *Astron. J., 141*, 25.

Everhart E. (1972) The origin of short-period comets. *Astrophys. Lett., 10*, 131–135.

Fernández J. A. (1980) On the existence of a comet belt beyond Neptune. *Mon. Not. R. Astron. Soc., 192*, 481–491.

Fernández J. A. (2020) Introduction: The trans-neptunian belt — Past, present, and future. In *The Trans-Neptunian Solar System* (D. Prialnik et al., eds.), pp. 1–22. Elsevier, Amsterdam.

Fernández J. A. and Ip W. H. (1984) Some dynamical aspects of the accretion of Uranus and Neptune: The exchange of orbital angular momentum with planetesimals. *Icarus, 58*, 109–120.

Fernández J. A., Gallardo T., and Brunini A. (2002) Are there many inactive Jupiter-family comets among the near-Earth asteroid population? *Icarus, 159*, 358–368.

Fernández J. A., Helal M., and Gallardo T. (2018) Dynamical evolution and end states of active and inactive Centaurs. *Planet. Space Sci., 158*, 6–15.

Fernández Y. R., Kelley M. S., Lamy P. L., et al. (2013) Thermal properties, sizes, and size distribution of Jupiter-family cometary nuclei. *Icarus, 226*, 1138–1170.

Fernández-Valenzuela E. (2022) Modeling long-term photometric data of trans-neptunian objects and Centaurs. *Front. Astron. Space Sci., 9*, 796004.

Fernández-Valenzuela E., Pinilla-Alonso N., Stansberry J., et al. (2021) Compositional study of trans-neptunian objects at $\lambda > 2.2$ *µm. Planet. Sci. J., 2*, 10.

Festou M. C., Gunnarsson M., Rickman H., et al. (2001) The activity of comet 29P/Schwassmann-Wachmann 1 monitored through its CO J=2→1 radio line. *Icarus, 150*, 140–150.

Fornasier S., Barucci M. A., de Bergh C., et al. (2009) Visible spectroscopy of the new ESO large programme on trans-neptunian objects and Centaurs: Final results. *Astron. Astrophys., 508*, 457–465.

Fornasier S., Lellouch E., Müller T., et al. (2013) TNOs are cool: A survey of the trans-neptunian region. VIII. Combined Herschel PACS and SPIRE observations of nine bright targets at 70–500 µm. *Astron. Astrophys., 555*, A15.

Fraser W. C. (2009) The collisional divot in the Kuiper belt size distribution. *Astrophys. J., 706*, 119–129.

Fraser W. C. and Brown M. E. (2012) The Hubble Wide Field Camera 3 Test of Surfaces in the Outer Solar System: The compositional classes of the Kuiper belt. *Astrophys. J., 749*, 33.

Fraser W. C. and Kavelaars J. J. (2009) The size distribution of Kuiper belt objects for D $\gtrsim$ 10 km. *Astron. J., 137*, 72–82.

Fraser W. C., Kavelaars J. J., Holman M. J., et al. (2008) The Kuiper belt luminosity function from $m_R = 21$ *to 26. Icarus, 195*, 827–843.

Fraser W., Brown M. E., Morbidelli A., et al. (2012) The absolute magnitude distributions of the cold and hot Kuiper belt populations. *AAS/Division for Planetary Sciences Meeting Abstracts, 44*, #502.03.

Fraser W. C., Brown M. E., Morbidelli A., et al. (2014) The absolute magnitude distribution of Kuiper belt objects. *Astrophys. J., 782*, 100.

Fuentes C. I., George M. R., and Holman M. J. (2009) A Subaru pencil-beam search for $m_R \sim 27$ trans-neptunian bodies. *Astrophys. J., 696*, 91–95.

Fulle M. (1992) Dust from short-period comet P/Schwassmann-Wachmann 1 and replenishment of the interplanetary dust cloud. *Nature, 359*, 42–44.

Gil-Hutton R. and García-Migani E. (2016) Comet candidates among quasi-Hilda objects. *Astron. Astrophys., 590*, A111.

Gladman B. and Chan C. (2006) Production of the extended scattered disk by rogue planets. *Astrophys. J. Lett., 643*, L135–L138.

Gladman B. and Volk K. (2021) Transneptunian space. *Annu. Rev. Astron. Astrophys., 59*, 203–246.

Gladman B., Kavelaars J. J., Nicholson P. D., et al. (1998) Pencil-beam surveys for faint trans-neptunian objects. *Astron. J., 116*, 2042–2054.

Gladman B., Kavelaars J. J., Petit J.-M., et al. (2001) The structure of the Kuiper belt: Size distribution and radial extent. *Astron. J., 122*, 1051–1066.

Gladman B., Marsden B. G., and Van Laerhoven C. (2008) Nomenclature in the outer solar system. In *The Solar System Beyond Neptune* (M. A. Barucci et al., eds.), pp. 43–57. Univ. of Arizona, Tucson.

Gladman B., Kavelaars J., Petit J. M., et al. (2009) Discovery of the first retrograde transneptunian object. *Astrophys. J. Lett., 697*, L91–L94.

Gladman B., Lawler S. M., Petit J. M., et al. (2012) The resonant trans-neptunian populations. *Astron. J., 144*, 23.

Gomes R. S., Morbidelli A., and Levison H. F. (2004) Planetary migration in a planetesimal disk: Why did Neptune stop at 30 AU? *Icarus, 170*, 492–507.

Gomes R., Levison H. F., Tsiganis K., et al. (2005a) Origin of the cataclysmic Late Heavy Bombardment period of the terrestrial planets. *Nature, 435*, 466–469.

Gomes R. S., Gallardo T., Fernández J. A., et al. (2005b) On the origin of the high-perihelion scattered disk: The role of the Kozai mechanism and mean motion resonances. *Cel. Mech. Dyn. Astron., 91*, 109–129.

Gomes R. S., Fernández J. A., Gallardo T., et al. (2008) The scattered disk: Origins, dynamics, and end states. In *The Solar System Beyond Neptune* (M. A. Barucci et al., eds.), pp. 259–273. Univ. of Arizona, Tucson.

Gomes R., Nesvorný D., Morbidelli A., et al. (2018) Checking the compatibility of the cold Kuiper belt with a planetary instability migration model. *Icarus, 306*, 319–327.

Greenstreet S., Gladman B., and McKinnon W. B. (2015) Impact and cratering rates onto Pluto. *Icarus, 258*, 267–288.

Greenstreet S., Gladman B., and McKinnon W. B. (2016) Corrigendum to "Impact and Cratering Rates onto Pluto" [Icarus 258 (2015) 267–288]. *Icarus, 274*, 366–367.

Grundy W. M., Benecchi S. D., Rabinowitz D. L., et al. (2012) Mutual events in the cold classical transneptunian binary system Sila and Nunam. *Icarus, 220*, 74–83.

Grundy W. M., Noll K. S., Buie M. W., et al. (2019a) The mutual orbit, mass, and density of transneptunian binary G!kúnǁ'hòmdímà (229762 2007 UK_{126}). *Icarus, 334*, 30–38.

Grundy W. M., Noll K. S., Roe H. G., et al. (2019b) Mutual orbit orientations of transneptunian binaries. *Icarus, 334*, 62–78.

Grundy W. M., Bird M. K., Britt D. T., et al. (2020) Color, composition, and thermal environment of Kuiper belt object (486958) Arrokoth. *Science, 367*, aay3705.

Guilbert A., Alvarez-Candal A., Merlin F., et al. (2009) ESO-Large Program on TNOs: Near-infrared spectroscopy with SINFONI. *Icarus, 201*, 272–283.

Guilbert-Lepoutre A. (2011) A thermal evolution model of Centaur 10199 Chariklo. *Astron. J., 141*, 103.

Gulbis A. A. S., Elliot J. L., and Kane J. F. (2006) The color of the Kuiper belt core. *Icarus, 183*, 168–178.

Gunnarsson M., Bockelée-Morvan D., Winnberg A., et al. (2003) Production and kinematics of CO in Comet C/1995 O1 (Hale-Bopp) at large post-perihelion distances. *Astron. Astrophys., 402*, 383–393.

Gunnarsson M., Bockelée-Morvan D., Biver N., et al. (2008) Mapping the carbon monoxide coma of Comet 29P/Schwassmann-Wachmann 1. *Astron. Astrophys., 484*, 537–546.

Hahn J. M. and Malhotra R. (2005) Neptune's migration into a stirred-up Kuiper belt: A detailed comparison of simulations to observations. *Astron. J., 130*, 2392–2414.

Hainaut O. R., Boehnhardt H., and Protopapa S. (2012) Colours of minor bodies in the outer solar system. II. A statistical analysis revisited. *Astron. Astrophys., 546*, A115.

Harrington Pinto O., Womack M., Fernandez Y., et al. (2022) A survey of CO, CO_2, and H_2O in comets and Centaurs. *Planet. Sci. J., 3*, 247.

Harsono D., Bruderer S., and van Dishoeck E. F. (2015) Volatile snowlines in embedded disks around low-mass protostars. *Astron. Astrophys., 582*, A41.

Hasegawa S., Marsset M., DeMeo F. E., et al. (2021) Discovery of two TNO-like bodies in the asteroid belt. *Astrophys. J. Lett., 916*, L6.

Hirabayashi M., Scheeres D. J., Chesley S. R., et al. (2016) Fission and reconfiguration of bilobate comets as revealed by 67P/Churyumov-Gerasimenko. *Nature, 534*, 352–355.

Hofgartner J. D., Buratti B. J., Hayne P. O., et al. (2019) Ongoing

resurfacing of KBO Eris by volatile transport in local, collisional, sublimation atmosphere regime. *Icarus, 334*, 52–61.
Holler B. J., Grundy W. M., Buie M. W., et al. (2021) The Eris/Dysnomia system I: The orbit of Dysnomia. *Icarus, 355*, 114130.
Horner J. and Lykawka P. S. (2010) Planetary Trojans — the main source of short period comets? *Intl. J. Astrobiol., 9*, 227–234.
Horner J., Evans N. W., Bailey M. E., et al. (2003) The populations of comet-like bodies in the solar system. *Mon. Not. R. Astron. Soc., 343*, 1057–1066.
Horner J., Evans N. W., and Bailey M. E. (2004) Simulations of the population of Centaurs — I. The bulk statistics. *Mon. Not. R. Astron. Soc., 354*, 798–810.
Hosek Jr. M. W., Blaauw R. C., Cooke W. J., et al. (2013) Outburst dust production of Comet 29P/Schwassmann-Wachmann 1. *Astron. J., 145*, 122.
Hsieh H. H., Novaković B., Walsh K. J., et al. (2020) Potential Themis-family asteroid contribution to the Jupiter-family comet population. *Astron. J., 159*, 179.
Hughes D. W. (1991) Possible mechanisms for cometary outbursts. In *Comets in the Post-Halley Era* (R. L. Newburn et al., eds.), pp. 825–851. IAU Colloq. 116, Cambridge Univ., Cambridge.
Hui M.-T. and Li J.-Y. (2018) Is the cometary nucleus-extraction technique reliable? *Publ. Astron. Soc. Pac., 130*, 104501.
Ivanova O. V., Skorov Y. V., Korsun P. P., et al. (2011) Observations of the long-lasting activity of the distant Comets 29P Schwassmann-Wachmann 1, C/2003 WT42 (LINEAR) and C/2002 VQ94 (LINEAR). *Icarus, 211*, 559–567.
Ivanova O. V., Luk'yanyk I. V., Kiselev N. N., et al. (2016) Photometric and spectroscopic analysis of Comet 29P/Schwassmann-Wachmann 1 activity. *Planet. Space Sci., 121*, 10–17.
Ivanova O., Agapitov O., Odstrcil D., et al. (2019) Dynamics of the CO^+ coma of Comet 29P/Schwassmann-Wachmann 1. *Mon. Not. R. Astron. Soc., 486*, 5614–5620.
Izawa M. R. M., Applin D. M., Norman L., et al. (2014) Reflectance spectroscopy (350–2500 nm) of solid-state polycyclic aromatic hydrocarbons (PAHs). *Icarus, 237*, 159–181.
Izidoro A., Raymond S. N., Pierens A., et al. (2016) The asteroid belt as a relic from a chaotic early solar system. *Astrophys. J., 833*, 40.
Jewitt D. C. (2004) From cradle to grave: The rise and demise of the comets. In *Comets II* (M. C. Festou et al., eds.), pp. 659–676. Univ. of Arizona, Tucson.
Jewitt D. (2006) Comet D/1819 W1 (Blanpain): Not dead yet. *Astron. J., 131*, 2327–2331.
Jewitt D. (2009) The active Centaurs. *Astron. J., 137*, 4296–4312.
Jewitt D. (2015) Color systematics of comets and related bodies. *Astron. J., 150*, 201.
Jewitt D. (2021) Systematics and consequences of comet nucleus outgassing torques. *Astron. J., 161*, 261.
Jewitt D. and Luu J. (1993) Discovery of the candidate Kuiper belt object 1992 QB_1. *Nature, 362*, 730–732.
Jewitt D., Luu J., and Trujillo C. (1998) Large Kuiper belt objects: The Mauna Kea 8K CCD survey. *Astron. J., 115*, 2125–2135.
Jewitt D. C., Trujillo C. A., and Luu J. X. (2000) Population and size distribution of small jovian Trojan asteroids. *Astron. J., 120*, 1140–1147.
Jewitt D., Hsieh H., and Agarwal J. (2015) The complex history of Trojan asteroids. In *Asteroids IV* (P. Michel et al., eds.), pp. 203–220. Univ. of Arizona, Tucson.
Jones G. H., Knight M. M., Battams K., et al. (2018) The science of sungrazers, sunskirters, and other near-Sun comets. *Space Sci. Rev., 214*, 20.
Joss P. C. (1973) On the origin of short-period comets. *Astron. Astrophys., 25*, 271–273.
Jutzi M. and Asphaug E. (2015) The shape and structure of cometary nuclei as a result of low-velocity accretion. *Science, 348*, 1355–1358.
Jutzi M. and Benz W. (2017) Formation of bi-lobed shapes by subcatastrophic collisions: A late origin of comet 67P's structure. *Astron. Astrophys., 597*, A62.
Kaib N. A. and Sheppard S. S. (2016) Tracking Neptune's migration history through high-perihelion resonant trans-neptunian objects. *Astron. J., 152*, 133.
Kareta T., Sharkey B., Noonan J., et al. (2019) Physical characterization of the 2017 December outburst of the Centaur 174P/Echeclus. *Astron. J., 158*, 255.
Kareta T., Woodney L. M., Schambeau C., et al. (2021) Contemporaneous multiwavelength and precovery observations of the active Centaur P/2019 LD2 (ATLAS). *Planet. Sci. J., 2*, 48.
Kavelaars J. J., Lawler S. M., Bannister M. T., et al. (2020) Perspectives on the distribution of orbits of distant trans-neptunian objects. In *The Trans-Neptunian Solar System* (D. Prialnik et al., eds.), pp. 61–77. Elsevier, Amsterdam.
Kavelaars J. J., Petit J.-M., Gladman B., et al. (2021) OSSOS finds an exponential cutoff in the size distribution of the cold classical Kuiper belt. *Astrophys. J. Lett., 920*, L28.
Kazimirchak-Polonskaya E. I. (1972) The major planets as powerful transformers of cometary orbits. In *The Motion, Evolution of Orbits, and Origin of Comets* (G. A. Chebotarev et al., eds.), pp. 373–397. Springer, Dordrecht.
Kelley M. S., Fernández Y. R., Licandro J., et al. (2013) The persistent activity of Jupiter-family comets at 3–7 AU. *Icarus, 225*, 475–494.
Kelley M. S. P., Sharma K., Kumar H., et al. (2021) Two outbursts of Comet 29P/Schwassmann-Wachmann 1. *Astron. Telegram*, 14543.
Kenyon S. J., Bromley B. C., O'Brien D. P., et al. (2008) Formation and collisional evolution of Kuiper belt objects. In *The Solar System Beyond Neptune* (M. A. Barucci et al., eds.), pp. 293–313. Univ. of Arizona, Tucson.
Kozai Y. (1962) Secular perturbations of asteroids with high inclination and eccentricity. *Astron. J., 67*, 591–598.
Kresák L. (1972) Jacobian integral as a classificational and evolutionary parameter of interplanetary bodies. *Bull. Astron. Inst. Czech., 23*, 1–34.
Kuiper G. P. (1951) On the origin of the solar system. In *Astrophysics: A Topical Symposium Commemorating the 50th Anniversary of the Yerkes Observatory and a Half Century of Progress in Astrophysics* (J. A. Hynek, ed.), pp. 357–424. McGraw-Hill, New York.
Kulyk I., Korsun P., Rousselot P., et al. (2016) P/2008 CL94 (Lemmon) and P/2011 S1 (Gibbs): Comet-like activity at large heliocentric distances. *Icarus, 271*, 314–325.
Lacerda P., Fornasier S., Lellouch E., et al. (2014) The albedo-color diversity of transneptunian objects. *Astrophys. J. Lett., 793*, L2.
Lamy P. L., Toth I., Fernandez Y. R., et al. (2004) The sizes, shapes, albedos, and colors of cometary nuclei. In *Comets II* (M. C. Festou et al., eds.), pp. 223–264. Univ. of Arizona, Tucson.
Lawler S. M., Kavelaars J. J., Alexandersen M., et al. (2018a) OSSOS: X. How to use a survey simulator: Statistical testing of dynamical models against the real Kuiper belt. *Front. Astron. Space Sci., 5*, 14.
Lawler S. M., Shankman C., Kavelaars J. J., et al. (2018b) OSSOS. VIII. The transition between two size distribution slopes in the scattering disk. *Astron. J., 155*, 197.
Lawler S. M., Pike R. E., Kaib N., et al. (2019) OSSOS. XIII. Fossilized resonant dropouts tentatively confirm Neptune's migration was grainy and slow. *Astron. J., 157*, 253.
Levison H. F. (1996) Comet taxonomy. In *Completing the Inventory of the Solar System* (T. Rettig and J. M. Hahn, eds.), pp. 173–191. ASP Conf. Ser. 107, Astronomical Society of the Pacific, San Francisco.
Levison H. F. and Duncan M. J. (1994) The long-term dynamical behavior of short-period comets. *Icarus, 108*, 18–36.
Levison H. F. and Duncan M. J. (1997) From the Kuiper belt to Jupiter-family comets: The spatial distribution of ecliptic comets. *Icarus, 127*, 13–32.
Levison H. F. and Stern S. A. (2001) On the size dependence of the inclination distribution of the main Kuiper belt. *Astron. J., 121*, 1730–1735.
Levison H. F., Duncan M. J., Zahnle K., et al. (2000) Note: Planetary impact rates from ecliptic comets. *Icarus, 143*, 415–420.
Levison H. F., Duncan M. J., Dones L., et al. (2006a) The scattered disk as a source of Halley-type comets. *Icarus, 184*, 619–633.
Levison H. F., Terrell D., Wiegert P. A., et al. (2006b) On the origin of the unusual orbit of Comet 2P/Encke. *Icarus, 182*, 161–168.
Levison H. F., Morbidelli A., Van Laerhoven C., et al. (2008) Origin of the structure of the Kuiper belt during a dynamical instability in the orbits of Uranus and Neptune. *Icarus, 196*, 258–273.
Levison H. F., Bottke W. F., Gounelle M., et al. (2009) Contamination of the asteroid belt by primordial trans-neptunian objects. *Nature, 460*, 364–366.
Levison H. F., Morbidelli A., Tsiganis K., et al. (2011) Late orbital instabilities in the outer planets induced by interaction with a self-gravitating planetesimal disk. *Astron. J., 142*, 152.
Li R., Youdin A. N., and Simon J. B. (2019) Demographics of planetesimals formed by the streaming instability. *Astrophys. J., 885*, 69.
Licandro J., de León J., Moreno F., et al. (2021) Activity of the Jupiter

co-orbital Comet P/2019 LD_2 (ATLAS) observed with OSIRIS at the 10.4 m GTC. *Astron. Astrophys., 650,* A79.
Lilly E., Hsieh H., Bauer J., et al. (2021) No activity among 13 Centaurs discovered in the Pan-STARRS1 detection database. *Planet. Sci. J., 2,* 155.
Lisse C., Bauer J., Cruikshank D., et al. (2020) Spitzer's solar system studies of comets, Centaurs and Kuiper belt objects. *Nature Astron., 4,* 930–939.
Lisse C. M., Young L. A., Cruikshank D. P., et al. (2021) On the origin and thermal stability of Arrokoth's and Pluto's ices. *Icarus, 356,* 114072.
Lisse C. M., Steckloff J. K., Prialnik D., et al. (2022) 29P/Schwassmann-Wachmann 1: A Rosetta Stone for amorphous water ice and CO ↔ CO_2 conversion in Centaurs and comets? *Planet. Sci. J., 3,* 251.
Lithwick Y. (2014) After runaway: The trans-Hill stage of planetesimal growth. *Astrophys. J., 780,* 22.
LSST Science Collaboration, Abell P. A., Allison J., et al. (2009) LSST Science Book, Version 2.0. *ArXiV e-prints,* arXiv:0912.0201.
Luu J. X. and Jewitt D. C. (1990) Cometary activity in 2060 Chiron. *Astron. J., 100,* 913–932.
Luu J., Marsden B. G., Jewitt D., et al. (1997) A new dynamical class of object in the outer solar system. *Nature, 387,* 573–575.
Lykawka P. S. and Mukai T. (2007) Dynamical classification of trans-neptunian objects: Probing their origin, evolution, and interrelation. *Icarus, 189,* 213–232.
Lykawka P. S. and Mukai T. (2008) An outer planet beyond Pluto and the origin of the trans-neptunian belt architecture. *Astron. J., 135,* 1161–1200.
Lyra W. and Umurhan O. M. (2019) The initial conditions for planet formation: Turbulence driven by hydrodynamical instabilities in disks around young stars. *Publ. Astron. Soc. Pac., 131,* 072001.
Mahjoub A., Brown M. E., Poston M. J., et al. (2021) Effect of H_2S on the near-infrared spectrum of irradiation residue and applications to the Kuiper belt object (486958) Arrokoth. *Astrophys. J. Lett., 914,* L31.
Malhotra R. (1993) The origin of Pluto's peculiar orbit. *Nature, 365,* 819–821.
Malhotra R. (1995) The origin of Pluto's orbit: Implications for the solar system beyond Neptune. *Astron. J., 110,* 420–429.
Malhotra R. (2019) Resonant Kuiper belt objects: A review. *Geosci. Lett., 6,* 12.
Marboeuf U., Thiabaud A., Alibert Y., et al. (2014) From stellar nebula to planetesimals. *Astron. Astrophys., 570,* A35.
Marsset M., Fraser W. C., Pike R. E., et al. (2019) Col-OSSOS: Color and inclination are correlated throughout the Kuiper belt. *Astron. J., 157,* 94.
Martin A. C., Emery J. P., and Loeffler M. J. (2022) Spectral effects of regolith porosity in the mid-IR — Forsteritic olivine. *Icarus, 378,* 114921.
Matthews B. C., Krivov A. V., Wyatt M. C., et al. (2014) Observations, modeling, and theory of debris disks. In *Protostars and Planets VI* (H. Beuther et al., eds.), pp. 521–544. Univ. of Arizona, Tucson.
McKinnon W. B., Richardson D. C., Marohnic J. C., et al. (2020) The solar nebula origin of (486958) Arrokoth, a primordial contact binary in the Kuiper belt. *Science, 367,* aay6620.
Meech K. J., Hainaut O. R., and Marsden B. G. (2004) Comet nucleus size distributions from HST and Keck telescopes. *Icarus, 170,* 463–491.
Merlin F., Barucci M. A., de Bergh C., et al. (2010) Chemical and physical properties of the variegated Pluto and Charon surfaces. *Icarus, 210,* 930–943.
Mojzsis S. J., Brasser R., Kelly N. M., et al. (2019) Onset of giant planet migration before 4480 million years ago. *Astrophys. J., 881,* 44.
Morbidelli A. (1997) Chaotic diffusion and the origin of comets from the 2/3 resonance in the Kuiper belt. *Icarus, 127,* 1–12.
Morbidelli A. and Nesvorný D. (2020) Kuiper belt: Formation and evolution. In *The Trans-Neptunian Solar System* (D. Prialnik et al., eds.), pp. 25–59. Elsevier, Amsterdam.
Morbidelli A. and Rickman H. (2015) Comets as collisional fragments of a primordial planetesimal disk. *Astron. Astrophys., 583,* A43.
Morbidelli A., Bottke W. F., Nesvorný D., et al. (2009a) Asteroids were born big. *Icarus, 204,* 558–573.
Morbidelli A., Levison H. F., Bottke W. F., et al. (2009b) Considerations on the magnitude distributions of the Kuiper belt and of the Jupiter Trojans. *Icarus, 202,* 310–315.
Morbidelli A., Brasser R., Gomes R., et al. (2010) Evidence from the asteroid belt for a violent past evolution of Jupiter's orbit. *Astron. J., 140,* 1391–1401.
Mumma M. J. and Charnley S. B. (2011) The chemical composition of comets — Emerging taxonomies and natal heritage. *Annu. Rev. Astron. Astrophys., 49,* 471–524.
Napier K. J., Gerdes D. W., Lin H. W., et al. (2021) No evidence for orbital clustering in the extreme trans-neptunian objects. *Planet. Sci. J., 2,* 59.
Nesvorný D. (2011) Young solar system's fifth giant planet? *Astrophys. J. Lett., 742,* L22.
Nesvorný D. (2015) Evidence for slow migration of Neptune from the inclination distribution of Kuiper belt objects. *Astron. J., 150,* 73.
Nesvorný D. (2018) Dynamical evolution of the early solar system. *Annu. Rev. Astron. Astrophys., 56,* 137–174.
Nesvorný D. (2020) Influence of Neptune's migration parameters on the inclination distribution of Kuiper belt objects (KBOs). *Res. Notes Am. Astron. Soc., 4,* 212.
Nesvorný D. (2021) Eccentric early migration of Neptune. *Astrophys. J. Lett., 908,* L47.
Nesvorný D. and Morbidelli A. (2012) Statistical study of the early solar system's instability with four, five, and six giant planets. *Astron. J., 144,* 117.
Nesvorný D. and Vokrouhlický D. (2016) Neptune's orbital migration was grainy, not smooth. *Astrophys. J., 825,* 94.
Nesvorný D. and Vokrouhlický D. (2019) Binary survival in the outer solar system. *Icarus, 331,* 49–61.
Nesvorný D., Roig F., and Ferraz-Mello S. (2000) Close approaches of trans-neptunian objects to Pluto have left observable signatures on their orbital distribution. *Astron. J., 119,* 953–969.
Nesvorný D., Vokrouhlický D., Bottke W. F., et al. (2011) Observed binary fraction sets limits on the extent of collisional grinding in the Kuiper belt. *Astron. J., 141,* 159.
Nesvorný D., Vokrouhlický D., and Morbidelli A. (2013) Capture of Trojans by jumping Jupiter. *Astrophys. J., 768,* 45.
Nesvorný D., Vokrouhlický D., and Deienno R. (2014) Capture of irregular satellites at Jupiter. *Astrophys. J., 784,* 22.
Nesvorný D., Vokrouhlický D., and Roig F. (2016) The orbital distribution of trans-neptunian objects beyond 50 au. *Astrophys. J. Lett., 827,* L35.
Nesvorný D., Vokrouhlický D., Dones L., et al. (2017) Origin and evolution of short-period comets. *Astrophys. J., 845,* 27.
Nesvorný D., Li R., Youdin A. N., et al. (2019a) Trans-neptunian binaries as evidence for planetesimal formation by the streaming instability. *Nature Astron., 3,* 808–812.
Nesvorný D., Vokrouhlický D., Stern A. S., et al. (2019b) OSSOS. XIX. Testing early solar system dynamical models using OSSOS Centaur detections. *Astron. J., 158,* 132.
Nesvorný D., Vokrouhlický D., Alexandersen M., et al. (2020) OSSOS XX: The meaning of Kuiper belt colors. *Astron. J., 160,* 46.
Nesvorný D., Roig F. V., and Deienno R. (2021) The role of early giant-planet instability in terrestrial planet formation. *Astron. J., 161,* 50.
Noll K. S., Grundy W. M., Stephens D. C., et al. (2008) Evidence for two populations of classical transneptunian objects: The strong inclination dependence of classical binaries. *Icarus, 194,* 758–768.
Noll K., Grundy W. M., Nesvorný D., et al. (2020) Trans-neptunian binaries. In *The Trans-Neptunian Solar System* (D. Prialnik et al., eds.), pp. 205–224. Elsevier, Amsterdam.
Noonan J. W., Harris W. M., Bromley S., et al. (2021) FUV observations of the inner coma of 46P/Wirtanen. *Planet. Sci. J., 2,* 8.
Öberg K. I., Murray-Clay R., and Bergin E. A. (2011) The effects of snowlines on C/O in planetary atmospheres. *Astrophys. J. Lett., 743,* L16.
O'Brien D. P. and Greenberg R. (2003) Steady-state size distributions for collisional populations: Analytical solution with size-dependent strength. *Icarus, 164,* 334–345.
Ohtsuka K., Ito T., Yoshikawa M., et al. (2008) Quasi-Hilda Comet 147P/Kushida-Muramatsu: Another long temporary satellite capture by Jupiter. *Astron. Astrophys., 489,* 1355–1362.
Oort J. H. (1950) The structure of the cloud of comets surrounding the solar system and a hypothesis concerning its origin. *Bull. Astron. Inst. Netherlands, 11,* 91–110.
Ootsubo T., Kawakita H., Hamada S., et al. (2012) AKARI near-infrared spectroscopic survey for CO_2 in 18 comets. *Astrophys. J., 752,* 15.
Paganini L., Mumma M. J., Boehnhardt H., et al. (2013) Ground-based

infrared detections of CO in the Centaur-comet 29P/Schwassmann-Wachmann 1 at 6.26 AU from the Sun. *Astrophys. J., 766*, 100.
Parker A. H. (2021) Transneptunian space and the post-Pluto paradigm. In *The Pluto System After New Horizons* (S. A. Stern et al., eds.), pp. 545–568. Univ. of Arizona, Tucson.
Parker A. H. and Kavelaars J. J. (2010) Destruction of binary minor planets during Neptune scattering. *Astrophys. J. Lett., 722*, L204–L208.
Parker A., Pinilla-Alonso N., Santos-Sanz P., et al. (2016) Physical characterization of TNOs with the James Webb Space Telescope. *Publ. Astron. Soc. Pac., 128*, 018010.
Peixinho N., Delsanti A., Guilbert-Lepoutre A., et al. (2012) The bimodal colors of Centaurs and small Kuiper belt objects. *Astron. Astrophys., 546*, A86.
Peixinho N., Delsanti A., and Doressoundiram A. (2015) Reanalyzing the visible colors of Centaurs and KBOs: What is there and what we might be missing. *Astron. Astrophys., 577*, A35.
Peixinho N., Thirouin A., Tegler S. C., et al. (2020) From Centaurs to comets: 40 years. In *The Trans-Neptunian Solar System* (D. Prialnik et al., eds.), pp. 307–329. Elsevier, Amsterdam.
Petit J. M., Kavelaars J. J., Gladman B. J., et al. (2011) The Canada-France Ecliptic Plane Survey — Full data release: The orbital structure of the Kuiper belt. *Astron. J., 142*, 131.
Pike R. E., Fraser W. C., Schwamb M. E., et al. (2017) Col-OSSOS: z-band photometry reveals three distinct TNO surface types. *Astron. J., 154*, 101.
Presler-Marshall B., Schleicher D. G., Knight M. M., et al. (2020) Diatomic sulfur in Jupiter family comets. *AAS/Division for Planetary Sciences Meeting Abstracts, 52*, #212.06.
Prialnik D., Brosch N., and Ianovici D. (1995) Modelling the activity of 2060 Chiron. *Mon. Not. R. Astron. Soc., 276*, 1148–1154.
Prialnik D., Sarid G., Rosenberg E. D., et al. (2008) Thermal and chemical evolution of comet nuclei and Kuiper belt objects. *Space Sci. Rev., 138*, 147–164.
Protopapa S., Kelley M. S. P., Yang B., et al. (2018) Icy grains from the nucleus of comet C/2013 US_{10} (Catalina). *Astrophys. J. Lett., 862*, L16.
Quarles B. and Kaib N. (2019) Instabilities in the early solar system due to a self-gravitating disk. *Astron. J., 157*, 67.
Raymond S. N. and Nesvorný D. (2022) Origin and dynamical evolution of the asteroid belt. In *Vesta and Ceres: Insights from the Dawn Mission for the Origin of the Solar System* (S. Marchi et al., eds.), pp. 227–249. Cambridge Univ., Cambridge.
Raymond S. N., Armitage P. J., Veras D., et al. (2018) Implications of the interstellar object 1I/'Oumuamua for planetary dynamics and planetesimal formation. *Mon. Not. R. Astron. Soc., 476*, 3031–3038.
Rickman H., Fernandez J. A., and Gustafson B. A. S. (1990) Formation of stable dust mantles on short-period comet nuclei. *Astron. Astrophys., 237*, 524–535.
Rickman H., Kamél L., Froeschlé C., et al. (1991) Nongravitational effects and the aging of periodic comets. *Astron. J., 102*, 1446–1463.
Robbins S. J. and Singer K. N. (2021) Pluto and Charon impact crater populations: Reconciling different results. *Planet. Sci. J., 2*, 192.
Robinson J. E., Fraser W. C., Fitzsimmons A., et al. (2020) Investigating gravitational collapse of a pebble cloud to form transneptunian binaries. *Astron. Astrophys., 643*, A55.
Romon-Martin J., Delahodde C., Barucci M. A., et al. (2003) Photometric and spectroscopic observations of (2060) Chiron at the ESO Very Large Telescope. *Astron. Astrophys., 400*, 369–373.
Roush T. L. and Dalton J. B. (2004) Reflectance spectra of hydrated Titan tholins at cryogenic temperatures and implications for compositional interpretation of red objects in the outer solar system. *Icarus, 168*, 158–162.
Rousselot P. (2008) 174P/Echeclus: A strange case of outburst. *Astron. Astrophys., 480*, 543–550.
Rousselot P., Korsun P. P., Kulyk I., et al. (2016) A long-term follow up of 174P/Echeclus. *Mon. Not. R. Astron. Soc., 462*, S432–S442.
Rowan-Robinson M. (2022) A search for Planet 9 in the IRAS data. *Mon. Not. R. Astron. Soc., 510*, 3716–3726.
Safrit T. K., Steckloff J. K., Bosh A. S., et al. (2021) The formation of bilobate comet shapes through sublimative torques. *Planet. Sci. J., 2*, 14.
Saki M., Gibb E. L., Bonev B. P., et al. (2020) Carbonyl sulfide (OCS): Detections in comets C/2002 T7 (LINEAR), C/2015 ER61 (PanSTARRS), and 21P/Giacobini-Zinner and stringent upper limits in 46P/Wirtanen. *Astron. J., 160*, 184.
Sarid G., Volk K., Steckloff J. K., et al. (2019) 29P/Schwassmann-Wachmann 1, a Centaur in the gateway to the Jupiter-family comets. *Astrophys. J. Lett., 883*, L25.
Schaller E. L. and Brown M. E. (2007) Volatile loss and retention on Kuiper belt objects. *Astrophys. J. Lett., 659*, L61–L64.
Schambeau C. A., Fernández Y. R., Samarasinha N. H., et al. (2017) Analysis of R-band observations of an outburst of Comet 29P/Schwassmann-Wachmann 1 to place constraints on the nucleus' rotation state. *Icarus, 284*, 359–371.
Schambeau C. A., Fernández Y. R., Samarasinha N. H., et al. (2019) Analysis of HST WFPC2 observations of Centaur 29P/Schwassmann-Wachmann 1 while in outburst to place constraints on the nucleus' rotation state. *Astron. J., 158*, 259.
Schambeau C. A., Fernández Y. R., Samarasinha N. H., et al. (2021) Characterization of thermal-infrared dust emission and refinements to the nucleus properties of Centaur 29P/Schwassmann-Wachmann 1. *Planet. Sci. J., 2*, 126.
Schlichting H. E., Ofek E. O., Sari R., et al. (2012) Measuring the abundance of sub-kilometer-sized Kuiper belt objects using stellar occultations. *Astrophys. J., 761*, 150.
Schwamb M. E., Fraser W. C., Bannister M. T., et al. (2019) Col-OSSOS: The colors of the outer solar system origins survey. *Astrophys. J. Suppl. Ser., 243*, 12.
Schwartz S. R., Michel P., Jutzi M., et al. (2018) Catastrophic disruptions as the origin of bilobate comets. *Nature Astron., 2*, 379–382.
Seccull T., Fraser W. C., Puzia T. H., et al. (2018) 2004 EW_{95}: A phyllosilicate-bearing carbonaceous asteroid in the Kuiper belt. *Astrophys. J. Lett., 855*, L26.
Seccull T., Fraser W. C., Puzia T. H., et al. (2019) 174P/Echeclus and its blue coma observed post-outburst. *Astron. J., 157*, 88.
Seccull T., Fraser W. C., and Puzia T. H. (2021) Near-UV reddening observed in the reflectance spectrum of high-inclination Centaur 2012 DR_{30}. *Planet. Sci. J., 2*, 239.
Sedgwick C. and Serjeant S. (2022) Searching for giant planets in the outer solar system with far-infrared all-sky surveys. *Mon. Not. R. Astron. Soc., 515*, 4828–4837.
Seligman D. Z., Kratter K. M., Levine W. G., et al. (2021) A sublime opportunity: The dynamics of transitioning cometary bodies and the feasibility of *in situ* observations of the evolution of their activity. *Planet. Sci. J., 2*, 234.
Senay M. C. and Jewitt D. (1994) Coma formation driven by carbon monoxide release from Comet Schwassmann-Wachmann 1. *Nature, 371*, 229–231.
Shankman C., Gladman B. J., Kaib N., et al. (2013) A possible divot in the size distribution of the Kuiper belt's scattering objects. *Astrophys. J. Lett., 764*, L2.
Shankman C., Kavelaars J. J., Bannister M. T., et al. (2017) OSSOS. VI. Striking biases in the detection of large semimajor axis trans-neptunian objects. *Astron. J., 154*, 50.
Shannon A., Doressoundiram A., Roques F., et al. (2023) Understanding the trans-neptunian solar system: Reconciling the results of serendipitous stellar occultations and the inferences from the cratering record. *Astron. Astrophys., 673*, A138.
Sheppard S. S. and Jewitt D. (2004) Extreme Kuiper belt object 2001 QG_{298} and the fraction of contact binaries. *Astron. J., 127*, 3023–3033.
Sheppard S. S. and Trujillo C. A. (2010) The size distribution of the Neptune Trojans and the missing intermediate-sized planetesimals. *Astrophys. J. Lett., 723*, L233–L237.
Sheppard S. S., Trujillo C. A., Tholen D. J., et al. (2019) A new high perihelion trans-plutonian inner Oort cloud object: 2015 TG387. *Astron. J., 157*, 139.
Showalter M. R., Benecchi S. D., Buie M. W., et al. (2021) A statistical review of light curves and the prevalence of contact binaries in the Kuiper belt. *Icarus, 356*, 114098.
Sickafoose A. A., Bosh A. S., Emery J. P., et al. (2020) Characterization of material around the Centaur (2060) Chiron from a visible and near-infrared stellar occultation in 2011. *Mon. Not. R. Astron. Soc., 491*, 3643–3654.
Singer K. N., McKinnon W. B., Gladman B., et al. (2019) Impact craters on Pluto and Charon indicate a deficit of small Kuiper belt objects. *Science, 363*, 955–959.
Singer K. N., Stern S. A., Elliott J., et al. (2021) A new spacecraft mission concept combining the first exploration of the Centaurs and an astrophysical space telescope for the outer solar system. *Planet.*

Space Sci., *205*, 105290.
Smullen R. A. and Volk K. (2020) Machine learning classification of Kuiper belt population. *Mon. Not. R. Astron. Soc.*, *497*, 1391–1403.
Snodgrass C., Fitzsimmons A., Lowry S. C., et al. (2011) The size distribution of Jupiter family comet nuclei. *Mon. Not. R. Astron. Soc.*, *414*, 458–469.
Steckloff J. K., Sarid G., Volk K., et al. (2020) P/2019 LD2 (ATLAS): An active Centaur in imminent transition to the Jupiter family. *Astrophys. J. Lett.*, *904*, L20.
Stern S. A., Weaver H. A., Spencer J. R., et al. (2019) Initial results from the New Horizons exploration of 2014 MU_{69}, a small Kuiper belt object. *Science*, *364*, aaw9771.
Tancredi G., Lindgren M., and Rickman H. (1990) Temporary satellite capture and orbital evolution of Comet P/Helin-Roman-Crockett. *Astron. Astrophys.*, *239*, 375–380.
Tegler S. C. and Romanishin W. (1998) Two distinct populations of Kuiper-belt objects. *Nature*, *392*, 49–51.
Tegler S., Consolmagno G., and Romanishin W. (2006) Comet 174P/Echeclus. *IAU Circular 8701*.
Tegler S. C., Romanishin W., Consolmagno G. J., et al. (2016) Two color populations of Kuiper belt and Centaur objects and the smaller orbital inclinations of red Centaur objects. *Astron. J.*, *152*, 210.
Thirouin A. and Sheppard S. S. (2018) The Plutino population: An abundance of contact binaries. *Astron. J.*, *155*, 248.
Thirouin A. and Sheppard S. S. (2019) Light curves and rotational properties of the pristine cold classical Kuiper belt objects. *Astron. J.*, *157*, 228.
Thirouin A. and Sheppard S. S. (2022) Lightcurves and rotations of trans-neptunian objects in the 2:1 mean motion resonance with Neptune. *Planet. Sci. J.*, *3*, 178.
Thompson W. R., Murray B. G. J. P. T., Khare B. N., et al. (1987) Coloration and darkening of methane clathrate and other ices by charged particle irradiation: Applications to the outer solar system. *J. Geophys. Res.–Space Phys.*, *92*, 14933–14947.
Tiscareno M. S. and Malhotra R. (2003) The dynamics of known Centaurs. *Astron. J.*, *126*, 3122–3131.
Tiscareno M. S. and Malhotra R. (2009) Chaotic diffusion of resonant Kuiper belt objects. *Astron. J.*, *138*, 827–837.
Toth I. (2006) The quasi-Hilda subgroup of ecliptic comets — An update. *Astron. Astrophys.*, *448*, 1191–1196.
Trigo-Rodríguez J. M., García-Hernández D. A., Sánchez A., et al. (2010) Outburst activity in comets — II. A multiband photometric monitoring of comet 29P/Schwassmann-Wachmann 1. *Mon. Not. R. Astron. Soc.*, *409*, 1682–1690.
Trujillo C. A. and Sheppard S. S. (2014) A Sedna-like body with a perihelion of 80 astronomical units. *Nature*, *507*, 471–474.
Trujillo C. A., Jewitt D. C., and Luu J. X. (2001) Properties of the trans-neptunian belt: Statistics from the Canada-France-Hawaii Telescope Survey. *Astron. J.*, *122*, 457–473.
Trujillo C. A., Brown M. E., Barkume K. M., et al. (2007) The surface of 2003 EL_{61} in the near-infrared. *Astrophys. J.*, *655*, 1172–1178.
Trujillo C. A., Sheppard S. S., and Schaller E. L. (2011) A photometric system for detection of water and methane ices on Kuiper belt objects. *Astrophys. J.*, *730*, 105.
Tsiganis K., Gomes R., Morbidelli A., et al. (2005) Origin of the orbital architecture of the giant planets of the solar system. *Nature*, *435*, 459–461.
Valsecchi G. B., Morbidelli A., Gonczi R., et al. (1995) The dynamics of objects in orbits resembling that of P/Encke. *Icarus*, *118*, 169–180.
van der Wiel M. H. D., Naylor D. A., Kamp I., et al. (2014) Signatures of warm carbon monoxide in protoplanetary discs observed with Herschel SPIRE. *Mon. Not. R. Astron. Soc.*, *444*, 3911–3925.
Villanueva G. L., Smith M. D., Protopapa S., et al. (2018) Planetary Spectrum Generator: An accurate online radiative transfer suite for atmospheres, comets, small bodies and exoplanets. *J. Quant. Spectrosc. Radiat. Transfer*, *217*, 86–104.
Vokrouhlický D., Bottke W. F., and Nesvorný D. (2016) Capture of trans-neptunian planetesimals in the main asteroid belt. *Astron. J.*, *152*, 39.
Vokrouhlický D., Nesvorný D., and Dones L. (2019) Origin and evolution of long-period comets. *Astron. J.*, *157*, 181.
Volk K. and Malhotra R. (2008) The scattered disk as the source of the Jupiter family comets. *Astrophys. J.*, *687*, 714–725.
Volk K. and Malhotra R. (2013) Do Centaurs preserve their source inclinations? *Icarus*, *224*, 66–73.
Volk K. and Malhotra R. (2019) Not a simple relationship between Neptune's migration speed and Kuiper belt inclination excitation. *Astron. J.*, *158*, 64.
Volk K., Murray-Clay R., Gladman B., et al. (2016) OSSOS III — Resonant trans-neptunian populations: Constraints from the first quarter of the Outer Solar System Origins Survey. *Astron. J.*, *152*, 23.
Wang J. H. and Brasser R. (2014) An Oort cloud origin of the Halley-type comets. *Astron. Astrophys.*, *563*, A122.
Weissman P. R. (1996) The Oort cloud. In *Completing the Inventory of the Solar System* (T. Rettig and J. M. Hahn, eds.), pp. 265–288. ASP Conf. Ser. 107, Astronomical Society of the Pacific, San Francisco.
Whitney C. (1955) Comet outbursts. *Astrophys. J.*, *122*, 190–195.
Wiegert P. and Tremaine S. (1999) The evolution of long-period comets. *Icarus*, *137*, 84–121.
Wierzchos K. and Womack M. (2020) CO gas and dust outbursts from Centaur 29P/Schwassmann-Wachmann. *Astron. J.*, *159*, 136.
Wierzchos K., Womack M., and Sarid G. (2017) Carbon monoxide in the distantly active Centaur (60558) 174P/Echeclus at 6 au. *Astron. J.*, *153*, 230.
Womack M. and Stern S. A. (1999) The detection of carbon monoxide gas emission in (2060) Chiron. *Solar System Res.*, *33*, 187–191.
Womack M., Festou M. C., and Stern S. A. (1997) The heliocentric evolution of key species in the distantly-active comet C/1995 O1 (Hale-Bopp). *Astron. J.*, *114*, 2789–2865.
Womack M., Sarid G., and Wierzchos K. (2017) CO in distantly active comets. *Publ. Astron. Soc. Pac.*, *129*, 031001.
Wong E. W., Brasser R., and Werner S. C. (2021) Early impact chronology of the icy regular satellites of the outer solar system. *Icarus*, *358*, 114184.
Wong I. and Brown M. E. (2015) The color-magnitude distribution of small Jupiter Trojans. *Astron. J.*, *150*, 174.
Wong I. and Brown M. E. (2016) A hypothesis for the color bimodality of Jupiter Trojans. *Astron. J.*, *152*, 90.
Wong I. and Brown M. E. (2017) The bimodal color distribution of small Kuiper belt objects. *Astron. J.*, *153*, 145.
Yang B., Jewitt D., and Bus S. J. (2009) Comet 17P/Holmes in outburst: The near infrared spectrum. *Astron. J.*, *137*, 4538–4546.
Yang B., Jewitt D., Zhao Y., et al. (2021) Discovery of carbon monoxide in distant comet C/2017 K2 (PANSTARRS). *Astrophys. J. Lett.*, *914*, L17.
Yoshida F., Terai T., Ito T., et al. (2019) A comparative study of size frequency distributions of Jupiter Trojans, Hildas and main belt asteroids: A clue to planet migration history. *Planet. Space Sci.*, *169*, 78–85.
Yoshida F., Terai T., Ito T., et al. (2020) A comparative study of size frequency distributions of Jupiter Trojans, Hildas and main belt asteroids: A clue to planet migration history (corrigendum). *Planet. Space Sci.*, *190*, 104977.
Youdin A. N. and Goodman J. (2005) Streaming instabilities in protoplanetary disks. *Astrophys. J.*, *620*, 459–469.
Zheng W., Jewitt D., and Kaiser R. I. (2009) On the state of water ice on Saturn's moon Titan and implications to icy bodies in the outer solar system. *J. Phys. Chem. A*, *113*, 11174–11181.

Part 3:

Missions, Campaigns, and Laboratory Experiments

Snodgrass C., Feaga L., Jones G. H., M. Küppers, and Tubiana C. (2024) Past and future comet missions. In *Comets III* (K. J. Meech, M. R. Combi, D. Bockelée-Morvan, S. N. Raymond, and M. E. Zolensky, eds.), pp. 155–191. Univ. of Arizona, Tucson, DOI: 10.2458/azu_uapress_9780816553631-ch006.

Past and Future Comet Missions

C. Snodgrass
University of Edinburgh

L. Feaga
University of Maryland

G. H. Jones
University College London (now at European Space Agency)

M. Küppers
European Space Agency

C. Tubiana
Istituto di Astrofisica e Planetologia Spaziali

We review the history of spacecraft encounters with comets, concentrating on those that took place in the recent past, since the publication of the *Comets II* book. This includes the flyby missions Stardust and Deep Impact, and their respective extended missions, the Rosetta rendezvous mission, and serendipitous encounters. While results from all these missions can be found throughout this book, this chapter focuses on the questions that motivated each mission, the technologies that were required to answer these questions, and where each mission opened new areas to investigate. There remain a large number of questions that will require future technologies and space missions to answer; we also describe planned next steps and routes forward that may be pursued by missions that have yet to be selected, and eventually lead to cryogenic sample return of nucleus ices for laboratory study.

1. INTRODUCTION

In the period since the publication of *Comets II* (*Festou et al.,* 2004) we have enjoyed a golden age of space exploration of comets. In that book, *Keller et al.* (2004) reviewed results from the relatively low-resolution images returned by the missions to Comet 1P/Halley (hereafter 1P) in the 1980s and Deep Space 1's flyby of Comet 19P/Borrelly (hereafter 19P). At the time, the next generation of missions had not yet returned any results, even though they were already on their journeys. Stardust, Deep Impact, and Rosetta would completely change our view of comets, with more detailed imaging (Fig. 1), spectroscopy, and *in situ* measurements, and making technological leaps forward (impacting or landing on the nucleus itself, and returning coma samples to Earth). This chapter reviews the motivations for each mission and the instruments and technologies that each required to answer their science questions. Major results are highlighted from each mission, although the detailed scientific results are described elsewhere in this book; nearly every chapter touches on results from space missions at some point, due to their significant impact on the field. For more details on the results of the Deep Impact/EPOXI and Stardust missions, see comprehensive reviews by *A'Hearn and Johnson* (2015) and *Sandford et al.* (2021), respectively. Rosetta results are both too extensive and too recent to have been captured in single reviews, but, alongside the many relevant chapters in this book, the reader could begin with the summary of key early results by *Taylor et al.* (2017).

All comet missions share some common goals: cross-cutting themes that motivated past missions and continue to be key questions today, as each subsequent mission has shed some light on the topic but also opened new avenues to explore. In very broad terms, these motivations can be described as wanting to understand:

- the formation and evolution of comets;
- the diversity of material in the solar system and how it has been distributed and transported;
- how the physical processes that drive cometary activity work and influence the first two topics.

Individual missions also explore specific and focused questions, but at their heart all are built around these topics.

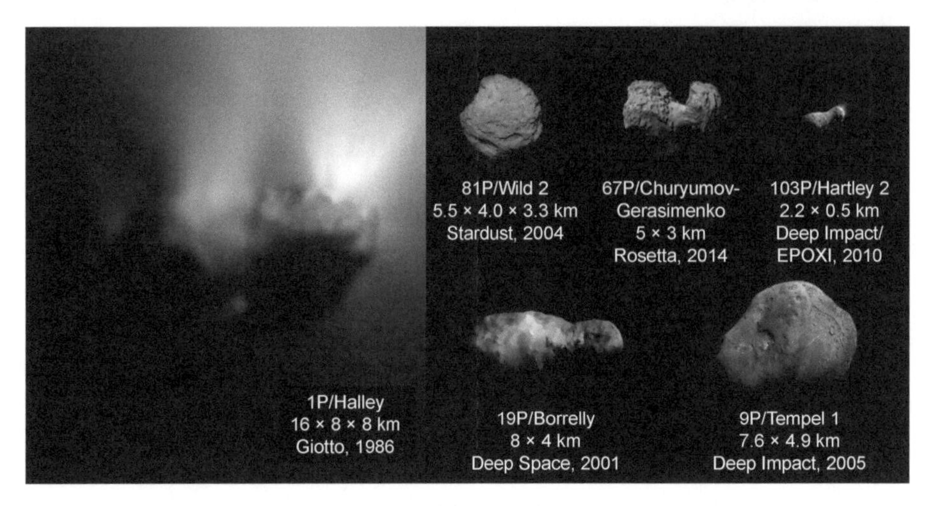

Fig. 1. All comet nuclei imaged by spacecraft, at approximately the same spatial scale. Adapted from a graphic by E. Lakdawalla, used with permission. Credit for individual images: 1P — ESA/MPS; 9P — NASA/JPL/UMD; 19P — NASA/JPL/Ted Stryk; 67P — ESA/Rosetta/NavCam/Emily Lakdawalla; 81P — NASA/JPL; 103P — NASA/JPL/UMD.

Reviews of our current understanding of the three themes can be found in the chapters by Simon et al., Guilbert-Lepoutre et al., Pajola et al., Filacchione et al., Knight et al., Marschall et al., Biver et al., Engrand et al., Agarwal et al., and Marty et al. in this volume. Missions have made significant contributions to the first topic by returning resolved nucleus images of increasing resolution, revealing morphological features, and by providing clues to interior structure (density, porosity) through indirect [Deep Impact (*A'Hearn et al.,* 2005b)] or direct [Rosetta (*Kofman et al.,* 2020)] measurements. Regarding the second topic, sampling of cometary dust and gas by *in situ* mass spectroscopy by Giotto and especially Rosetta has revealed a "zoo" of chemicals that cannot be detected by remote observation (*Rubin et al.,* 2020), while the return of a coma dust sample by Stardust has enabled incredibly detailed measurements of composition using laboratory equipment and revealed large-scale transport of material in the protoplanetary disk (*Brownlee et al.,* 2006). Finally, the question of how cometary activity works remains enigmatic, despite successive missions identifying various sources with increasing precision, from discrete areas [Giotto (*Keller et al.,* 1986)] to smooth flows and retreating scarps [Deep Impact (*Thomas et al.,* 2007)]. Deep Impact and EPOXI were able to associate dust and ice outflows with gas species (*A'Hearn et al.,* 2011), and finally Rosetta followed the changing patterns of activity with diurnal and seasonal cycles (*De Sanctis et al.,* 2015; *Fornasier et al.,* 2016). The approved future missions will further advance our understanding of all these topics by visiting comets with potentially more diverse regions of origin, and at different evolutionary states. Key technological steps required to make major further progress in each topic include, respectively, using ground-penetrating radar to understand the interior structure of nuclei, (cryogenic) sample return of nucleus material, and *in situ* investigation of the (micro-)physics of the near surface layers.

In this chapter we begin in section 2 by highlighting the synergy between missions and supporting observations. Our review of past missions includes descriptions of all those with a recognized comet as a named target, covering flyby missions in section 3 and Rosetta in more detail in section 4. We list all these missions in Table 1, and summarize their key goals, technologies, and results in Table 2. It is worth noting that the most important results of these exploration missions are often the unexpected ones, and not necessarily the ones described in the initial mission goals [*A'Hearn and Johnson* (2015) call this the "Harwit principle"].

We include a first review of distant *in situ* encounters by spacecraft with comets and their tails, whether targeted or not, in section 3.7, and briefly summarize the serendipitous observations of the close approach of Comet C/2013 A1 Siding Spring to Mars by the various probes in orbit (or on the surface) of that planet in section 3.8, but do not include remote observations of comets by space telescopes (these are described in the chapter in this volume by Bauer et al.). We also look ahead, in section 5, to mission concepts under development: the Japanese DESTINY$^+$ mission to Phaethon; the Chinese mission ZhengHe, which is expected to visit an active asteroid; and the European Space Agency (ESA) mission Comet Interceptor, whose objective is to visit an Oort cloud comet. We also suggest future possibilities of where and how we will continue to explore comets, in missions yet to be proposed or selected.

2. GROUNDBASED CAMPAIGNS

Most missions are accompanied by significant international campaigns of supporting observations from groundbased (or remote spacebased) telescopes. These serve multiple purposes: mission planning, providing the most complete and up-to-date reconnaissance on the comet before the spacecraft arrives; context, allowing comparison of spacecraft results to the comet on different spatial and timescales; and relation of the target comet to the wider population. The latter is particularly important as the number of comets visited by spacecraft will always be small

TABLE 1. Summary of all past comet missions (see Fig. 8 and sections 3.7 and 3.8 for unplanned encounters and tail crossings).

Mission	Comet	Dates		Agency*	Type†	Mass‡ (kg)	Speed (km s^{-1})	CA§ (km)	r¶ (au)
		Launch	Encounter						
ICE	21P	1978 Aug 12	1985 Sep 11	NASA/ESA	F	479	20.7	7862	+1.03
Vega 1	1P	1984 Dec 15	1986 Mar 06	SAS	F	4920**	79.2	8889	+0.79
Vega 2	1P	1984 Dec 21	1986 Mar 09	SAS	F	4920**	76.8	8030	+0.83
Sakigake	1P	1985 Jan 07	1986 Mar 11	ISAS	F	138	75.3	7×10^6	+0.86
Giotto	1P	1985 Jul 02	1986 Mar 14	ESA	F	583	68.4	605	+0.90
GEM	26P		1992 Jul 10	ESA	F		14.0	<200	–1.01
Suisei	1P	1985 Aug 19	1986 Mar 08	ISAS	F	139	73.0	152400	+0.82
Deep Space 1	19P	1998 Oct 24	2001 Sep 22	NASA	F	373	16.6	2171	+1.36
Stardust	81P	1999 Feb 07	2004 Jan 02	NASA	CSR	300	6.1	237	+1.86
Stardust-NExT	9P		2011 Feb 15	NASA	F		10.9	181	+1.55
CONTOUR	—	2002 Jul 03	—	NASA	MF	328	—	—	—
Rosetta	67P	2004 Mar 02	2014 Aug 06	ESA/NASA	R	1230	0	0	††
Philae	67P	2004 Mar 02	2014 Nov 12	ESA	L	100	0	0	–2.99
Deep Impact	9P	2005 Jan 12	2005 Jul 04	NASA	F + I	879	10.2	575	–1.51
EPOXI	103P		2010 Nov 04	NASA	F		12.3	694	+1.06

*NASA = National Aeronautics and Space Administration (USA); ESA = European Space Agency; SAS = Soviet Academy of Sciences (USSR); ISAS = Institute of Space and Astronautical Science (Japan).
†F = Flyby; CSR = Coma sample return; MF = Multiple flybys; R = Rendezvous; L = Lander; I = Impactor.
‡Dry mass in kilograms at launch.
§Closest approach distance to the nucleus.
¶Heliocentric distance at encounter (negative indicates pre-perihelion, positive post-perihelion).
**Includes Venus lander and balloon.
††Rosetta's extended operations are detailed in Table 3.

compared to the number observed with telescopes, and it is necessary to identify what features of a particular comet are unique, vs. characteristic of a class or comets in general.

The scale of various groundbased campaigns varied: The International Halley Watch that accompanied the 1980s missions defined, purchased, and distributed narrowband filters to isolate gas species in the coma that would be widely used for comet photometry for a decade (*Schleicher and Farnham,* 2004). The Deep Impact observing campaign (*Meech et al.,* 2005a) was an integral part of the mission, relied on to characterize the evolution of the ejecta plume after the spacecraft flyby, and as such is described in more detail in section 3.4. The Rosetta observing campaign continued throughout the 2.5 years that the spacecraft operated at the comet (*Snodgrass et al.,* 2017b).

A common goal of mission-supporting observation campaigns was monitoring the total brightness of the comet as a proxy for its activity levels — typically of dust brightness [or Afρ (*A'Hearn et al.,* 1984)] via broadband imaging, but also including spectroscopy or narrowband photometry to constrain gas production rates when the comets were bright enough. One key result from the Rosetta campaign was confirmation that the water production rates measured over months at the comet correlated well with the groundbased measurements of total dust brightness (*Hansen et al.,* 2016). This resulted in increased confidence in models that derive gas production from broadband photometry (e.g., *Meech and Svoren,* 2004), which were used to estimate long-term evolution of gas production rates around previous flybys [e.g., EPOXI (*Meech et al.,* 2011a)].

Intensive monitoring of 9P/Tempel 1 (hereafter 9P) ahead of the Deep Impact encounter revealed regular "mini-outbursts," although comparison between groundbased data and images taken by the spacecraft on approach showed many of these events were too small to affect the large-scale brightness of the comet (*Lara et al.,* 2006; *A'Hearn et al.,* 2005b). Similarly, the small-scale outbursts that punctuate 67P/Churyumov-Gerasimenko's (hereafter 67P) activity were seen by Rosetta but not the accompanying groundbased observations (*Vincent et al.,* 2016; *Snodgrass et al.,* 2017b). The long-term monitoring before and after spacecraft flybys possible from the ground was essential to understanding the rotation pole and period. The timing of the Stardust-NExT encounter with respect to rotational phasing and illumination of 9P was critical to meet mission objectives (see section 3.6) (*Veverka et al.,* 2013; *Belton et*

TABLE 2. Summary of key science goals, enabling technologies, and results for the missions described in detail in sections 3 and 4.

Mission	Goals	Technologies	Results	Relevant Chapter(s)*
Stardust	Return ~1 mg of coma dust sample to Earth; measure dust flux; image nucleus and inner coma	Aerogel dust capture; sample return (high-speed capsule, sample recovery, handling, curation); laboratory investigations	Highly pitted nucleus; fragmentation of coma particles; solid and fragile particles; composition of dust (mineralogy, petrology, atomic and isotopic composition of components), including high-temperature minerals	Bergin et al., Guilbert-Lepoutre et al., Pajola et al., Engrand et al.
Deep Impact	Excavate a ~100-m-diameter × ~25-m-depth crater; study interior composition, layering, heterogeneities; determine porosity, density, and strength of the nucleus; identify vertical/lateral structure; characterize nucleus and coma environment prior to impact	Releasable sub-spacecraft; autonomous navigation and targeting	Frequent mini-outbursts; morphological diversity, flows and layers, but overall homogeneous nucleus color and albedo; low strength, density, and thermal inertia; limited surface ice, possible frosts; compositional homogeneity with depth, but variation across surface (differing source regions of H_2O and CO_2); localized activity; presence of small subsurface H_2O ice grains, a hot (~1000 K) ejecta plume immediately following impact	Guilbert-Lepoutre et al., Pajola et al., Filacchione et al., Marschall et al.
EPOXI	Explore diversity of comets; determine the origin and significance of hyperactivity	Repurposed spacecraft; test of the deep space communications network	Bilobed morphology; rough terrain on lobes, a smooth "waist," and tall spires; decimeter-scale chunks of icy material in coma on bound orbits; spatially and temporally resolved distribution of gases; micrometer-sized ice associated with CO_2 driven activity providing explanation for "hyperactivity"	Pajola et al., Filacchione et al., Marschall et al., Agarwal et al.
Stardust-NExT	Image Deep Impact impact site; search for surface changes over one orbit; increase observed surface area of 9P	Repurposed spacecraft; precise prediction of rotational phase; targeting with very little fuel; recalibration of flying camera	Nucleus layering; localized erosion; scarp retreat by tens of meters; identification of crater and confirmation of low density and strength nucleus	Pajola et al.
Rosetta	Study changes in nucleus, gas, dust, and plasma around orbit; understand activity processes; provide detailed composition and structure measurements at unprecedented resolution	Rendezvous (inc. spacecraft hibernation, long-term operations at an active comet); lander targeting and release; extensive and novel payload (see Table 4 and section 4.2)	Complex and varied nucleus morphology, with clues on interior structure; mass, volume, density, and porosity of the nucleus; surface variation with time; frosts, ice revealed by cliff collapse; significant fallback of dust; detailed inventory of minor gas species, including complex organics and isotopic measurements; variation of activity and coma composition with time and location; strong seasonal effects; physical structure of dust particles; importance of electron impact dissociation; better understanding of solar wind interactions at low/intermediate activity	Bergin et al., Guilbert-Lepoutre et al., Pajola et al., Filacchione et al., Bodewits et al., Marschall et al., Biver et al., Beth et al., Goetz et al., Engrand et al., Agarwal et al.
Philae	Investigate nucleus surface composition, mechanical, electrical and physical properties; interior structure via sounding	Landing equipment for unknown surface; highly miniaturized instrumentation	Surface images and dust composition measurements; (debated) local surface strength assessment	Guilbert-Lepoutre et al., Filacchione et al., Goetz et al.

*See relevant chapters elsewhere in this volume for more details on the scientific results of each mission.

al., 2011). By monitoring 103P/Hartley 2 (hereafter 103P), significant changes in rotation rate through the encounter perihelion passage were revealed (*Knight et al.,* 2015). For 67P, groundbased measurements of the rotation rate together with Rosetta approach data revealed longterm spin-up, with the rotation period reducing by ~20 min per orbit, over a timescale much longer than the Rosetta mission (*Mottola et al.,* 2014). Combining these results, estimates of the non-gravitational acceleration of the comet from groundbased astrometry and orbit fitting, and the detailed shape model from the spacecraft data has enabled quite detailed modeling of the forces on 67P from outgassing (*Attree et al.,* 2019; *Kramer and Läuter,* 2019). Furthermore, findings from Rosetta confirmed that groundbased predictions of the pole position of 67P, via three independent methods, were correct (*Davidsson and Gutiérrez,* 2005; *Lowry et al.,* 2012; *Vincent et al.,* 2013). The resulting seasonal effects were used to explain the temporal differences in CN gas production rates measured from groundbased spectroscopy (*Opitom et al.,* 2017). However, Rosetta results also demonstrated the limitations of techniques for estimating nuclei shapes from groundbased observation (see section 4.3 and Fig. 11 for details).

In addition to complementing the temporal context around missions, groundbased observations are key to understanding the wider spatial scale of comets. Apart from the more distant encounters described in section 3.7, missions target the near-nucleus region, while the coma and tails extend over a region orders of magnitude larger. A prime example of this was on display at 103P where, prior to the EPOXI mission, the comet was classified as hyperactive (*Groussin et al.,* 2004), implying ~100% of its nucleus surface was active compared to a more typical few percent level. Earth-based observations during the 2010 apparition indicated extended and asymmetric distributions of secondary species (e.g., OH and H) enhanced in the anti-sunward direction (*Meech et al.,* 2011a). Spacecraft visible and spectral images revealed that the true source of the hyperactivity and extended water coma was an icy grain source that was being actively expelled from the nucleus during the encounter (*A'Hearn et al.,* 2011; *Protopapa et al.,* 2014; *Sunshine and Feaga,* 2021). The color, polarization, and radial brightness profile of the dust coma of 67P indicated evolution of particle properties as dust moved outward, interpreted as fragmentation of these particles at distances beyond where Rosetta was measuring their properties *in situ* (*Boehnhardt et al.,* 2016; *Rosenbush et al.,* 2017). A combination of these large-scale observations and inner-coma spacecraft measurements is required to build a full model of how the cometary coma develops from the nucleus (*Marschall et al.,* 2020). On the largest scales, cometary tails reveal both comet dust and gas properties and how their induced magnetospheres interact with the solar wind (see the chapter by Goetz et al. in this volume). The interpretation of remote ion tail observations combined with spacecraft measurements of magnetic fields would be particularly illuminating. Unfortunately, the majority of spacecraft targets to date [low-inclination Jupiter family comets (JFCs)] do not typically present strong ion tails with a favorable geometry; for example, any ion tail from 67P would have been projected behind the comet's dust coma as seen from Earth, and was not definitively detected. With favorable geometry, dust and ion tails can be well characterized by amateur astronomer observations, typically made with smaller telescopes with wide fields of view. There have been significant amateur contributions to mission-supporting observation campaigns, which, in addition to their direct scientific value, have been hugely successful in engaging a wider community and the general public in cometary exploration (*McFadden et al.,* 2005; *Usher et al.,* 2020).

Finally, groundbased observations are key to understanding how spacecraft targets relate to the wider comet population. Establishing a firm taxonomy of comets, and linking observed differences in composition to either formation or evolutionary processes, is an ongoing endeavour (see the chapter by Biver et al. in this volume), but observations have been able to broadly group comets into "typical" or "carbon-chain-depleted" classes (*A'Hearn et al.,* 1995). Groundbased observations show that three of the eight comets visited by missions to date (Table 1) fall into the typical range (1P, 9P, and 103P), with the others counting as depleted, as would be expected when we consider that most have a source region in the Kuiper belt, and the majority of comets from there were seen to be carbon-chain-depleted in the original taxonomy (*A'Hearn et al.,* 1995). More recently, a compilation of infrared spectroscopic data shows that both dynamical classes of comets are represented throughout the spread of compositional abundances, however, there is an overall emerging trend that JFCs are depleted in volatile species with respect to water as compared to Oort cloud comets (*Dello Russo et al.,* 2016). Smaller nuclei are often seen to be "hyperactive," another often-used classification of comets, including 103P and the original Rosetta target, 46P/Wirtanen (hereafter 46P). Of the spacecraft target comets, 21P/Giacobini-Zinner (hereafter 21P) and 103P have shown some evidence of hyperactivity (active fraction >50%) in remote observations with the Solar Wind Anisotropies (SWAN) instrument on the Solar and Heliospheric Observatory (SOHO) (*Combi et al.,* 2019). Recent work suggests a continuum of activity levels (*Sunshine and Feaga,* 2021) and that the degree of hyperactivity may correlate with the D/H isotopic ratio (*Lis et al.,* 2019).

3. FLYBY MISSIONS

3.1. Early Missions (pre-*Comets II*)

The first targeted cometary spacecraft encounter was an ion tail crossing, performed by the NASA/ESA ISEE-3 spacecraft, which had been monitoring the solar wind upstream of Earth since 1978. The spacecraft was renamed as the International Cometary Explorer (ICE), and repurposed to encounter 21P, which it achieved by crossing the comet's ion tail on September 11, 1985, as planned, 7800 km downstream. ICE's payload comprised instruments designed to study the solar wind only and had no cameras. The data returned were however extremely valuable: They confirmed the long-expected pattern of draped heliospheric magnetic field lines (*Alfvén,* 1957), forming an induced magnetotail ~10,000 km wide. Outside this region lay the ionosheath (or cometosheath) of weaker, loosely-draped magnetic fields, its outer boundary marked by a bow wave/shock (*Smith et al.,* 1986). ICE returned to Earth's vicinity in 2014, but attempts to adjust its trajectory were largely unsuccessful and any further targeted comet encounters are not possible (*Dunham et al.,* 2015).

The first missions dedicated to cometary science were the spacecraft encountering 1P, the most famous comet, during its perihelion passage in 1986. It was an opportunity to study the nucleus and the innermost coma, point sources in Earth-based observations, in spatial detail. A total of five spacecraft visited 1P within ~2 weeks (see Table 1). Due to 1P's retrograde orbit, all the spacecraft flew by the comet at very high velocities of 68–79 km s^{-1}. While ICE and the two Japanese missions Suisei and Sakigake measured the plasma environment of 1P, the two Vega missions and Giotto also provided the first detailed investigation of a comet nucleus, with most detail of the nucleus surface provided by Giotto. A composite of 68 images taken during the approach to 1P is shown in Fig. 1.

The missions to 1P and their results are described in detail in *Comets II* and will not be repeated here. However, after the rendezvous of Rosetta with 67P, several authors revisited the 1P data. Notably, the detection of molecular oxygen in 67P prompted a reanalysis of mass spectrometer data on Giotto. If 3.7 ± 1.7% of O_2 relative to water is included, similar to the abundance detected in 67P (*Rubin et al.,* 2015), the fit to the data is improved. This suggests that 67P is not special in terms of O_2 abundance and that molecular oxygen is a parent molecule with a typical abundance of a few percent relative to water in comets. Furthermore, plasma measurements from the Giotto and Vega missions were revisited to compare with Rosetta results on ion composition (*Heritier et al.,* 2017) and plasma processes and boundaries (see the chapter by Goetz et al. in this volume and references therein). The reuse of Giotto data 30+ years after the mission illustrates the value of long-term archiving of scientific data.

To change trajectories in order to intercept Comet 26P/Grigg-Skjellerup (hereafter 26P) on July 10, 1992, Giotto performed the first Earth flyby by a planetary spacecraft in July 1990. The spin-stabilized probe's high-gain antenna was oriented specifically for the 1P encounter; therefore, to maintain real-time communication with Earth, the spacecraft had to encounter 26P without full protection by its dust impact shield, but did so successfully (*Grensemann and Schwehm,* 1993). Despite not carrying an operational camera, much was learned about the comet's coma structure (*Levasseur-Regourd et al.,* 1993), the possible existence of nucleus fragments (*McBride et al.,* 1997), and the solar wind interaction of this relatively low-activity object (e.g., *Coates and Jones,* 2009). Giotto operations ceased on July 23, 1992, with insufficient fuel remaining for a further encounter (*Grensemann and Schwehm,* 1993).

Deep Space 1 was a NASA technology demonstration mission. Its extended mission included flybys of asteroid (9969) Braille and 19P (*Rayman and Varghese,* 2001; *Rayman,* 2002), which provided the first opportunity to identify surface morphological structures in sufficient detail to investigate a comet from a geological point of view (*Britt et al.,* 2004). Deep Space 1 paved the way for future comet and asteroid missions by pioneering new technologies (*Rayman et al.,* 2000). In particular, solar electric propulsion, autonomous navigation (including autonomous spacecraft maneuver execution), and miniaturized scientific instruments are now used by several spacecraft.

3.2. CONTOUR

The COmet Nucleus TOUR (CONTOUR) was a cometary flyby mission selected under the NASA Discovery program led by Principal Investigator (PI) J. Veverka. It successfully launched on July 3, 2002, but suffered a catastrophic failure on August 15, 2002, during the solid rocket burn that would have propelled the spacecraft out of Earth's orbit and on its trajectory to its first target (*NASA,* 2003). The mission's primary objective was to complete close flybys of two active cometary nuclei around 1 au from the Sun: 2P/Encke, a highly thermally evolved comet, and 73P/Schwassmann-Wachmann 3 (hereafter 73P), a recently fragmented comet (*Cochran et al.,* 2002).

CONTOUR's design was risk averse with a suite of heritage instruments (a visible and infrared imaging spectrograph, a visible imager, a mass spectrometer, and a dust analyzer) and a simple and compact spacecraft design with few movable parts (*Cochran et al.,* 2002). CONTOUR would have been one of the first missions to use only a tone to alert Earth that the spacecraft was still functioning during long periods of minimal power hibernation (A. Cochran, personal communication). The mission would have doubled the number of cometary nuclei imaged by spacecraft at the time, advanced our scientific knowledge of the characteristics of comet nuclei, and most importantly explored the diversity and evolution of comets (*Cochran et al.,* 2002). The most remarkable aspect of the CONTOUR mission was the flexibility in its trajectory design, implemented with many Earth flybys, commonly used now, to allow retargeting maneuvers as desired (*Cochran et al.,* 2002). This enabled options for multiple comet flybys, or even more scientifically compelling, rerouting the prime mission to a newly discovered long-period comet (LPC) from the Oort cloud, a class of comet then yet to be imaged by spacecraft.

3.3. Stardust

Stardust, a NASA Discovery mission led by PI D. Brownlee, is the only mission to date to return cometary dust of known origin to Earth for detailed analysis. The mission objectives included quantifying the nature and amount of dust released by the JFC 81P/Wild 2 (hereafter 81P), and relating cometary dust to meteoritic samples collected on Earth and interstellar dust particles also collected by Stardust (*Brownlee et al.,* 2003). A major component in the success of this mission was the comprehensive state of the art analysis that was conducted in the lab once the ~1-mg total sample was returned to Earth. These detailed measurements would not have been feasible *in situ*. See *Sandford et al.* (2021) for a detailed recent review of the mission results.

The biggest challenge of this mission was to capture a sample of dust particles without destroying them in the process nor harming the spacecraft. This was achieved thanks to the design of a spacecraft trajectory that produced a low encounter relative velocity of 6.1 km s^{-1} and the use of an aerogel capture medium (*Brownlee et al.,* 2003). Aerogel is a porous silica foam with density comparable to that of air: Its low density gradually slowed the impacting particles, avoiding melting and/or vaporization. Stardust was equipped with an ~40-cm-diameter tray of aerogel cells that was held outside the spacecraft and exposed to the dust flux, while a Whipple bumper protected the main body of the spacecraft (*Brownlee et al.,* 2003). Following the encounter, the whole tray was placed within a return capsule.

On January 2, 2004, the Stardust spacecraft flew by 81P at a distance of 234 km from the surface (*Brownlee et al.,* 2004) and collected more than 10,000 dust particles in the size range between 1 and 300 μm, which are expected to be a representative sample of the non-volatile component of the interior of the comet (*Brownlee et al.,* 2006). Particles impacting on the aerogel produced millimeter-sized tracks (Fig. 2); most of the material of the incoming particle was left as fragments on the wall of the track, but usually an intact terminal particle was found at the end of the track (*Burnett,* 2006). The shape of the deep tracks in the silica aerogel varied depending on the nature of the impacting particle. In addition, the aluminum frame used to hold the aerogel tiles showed bowl-shaped impact craters containing residues of the incoming particles (*Graham et al.,* 2006).

Another obstacle for sample return missions is to survive Earth atmospheric reentry and the environment of the Utah Test and Training Range (UTTR), or other such landscape, until recovery. Stardust's sample return capsule withstood peak temperatures from aerodynamic friction near 3000 K during the fastest Earth entry of a manmade object in history, while a mission requirement was levied to keep the temperature of the sample canister inside the capsule below 343 K until retrieval from the UTTR (*Brownlee et al.,* 2003). Stardust employed a heat shield constructed from a lightweight ceramic ablating material to meet these necessary conditions and to stay within mission mass and cost constraints. The 0.05-m-thick Phenolic-Impregnated Carbon Ablator (PICA) was an enabling technology with its first flight application on Stardust (*Tran et al.,* 1997).

The benefit of bringing samples back to Earth is to characterize and quantify small amounts of even smaller particles with instrumentation too large and expensive to send into space (*Zolensky et al.,* 2000). The Stardust samples underwent a wide range of analysis including, but not limited to, laser mass spectrometry, liquid chromatography, time-of-flight mass spectrometry, scanning transmission X-ray microscopy, infrared and Raman spectroscopy, ion chromatography, secondary ion mass spectrometry, and scanning and transmission electron beam microscopy (*Sandford et al.,* 2006). A key implication afforded by the detailed analysis of the returned 81P samples is the importance of large-scale mixing in the early solar system, based on the direct evidence for high-temperature phase materials like chondrules accreted within a cold icy comet (*Brownlee et al.,* 2006). Other significant results include the detection of glycine and other amino acids, the presence of amorphous and crystalline silicates, the absence of hydrous material, a heterogeneous organic distribution similar to interplanetary dust particles, and an isotopic makeup suggestive of interstellar origin (*Glavin et al.,* 2008; *Elsila et al.,* 2009; *Keller et al.,* 2006; *Zolensky et al.,* 2006; *Sandford et al.,* 2006). More details on the Stardust dust composition results are given in the chapter by Engrand et al. in this volume.

Additionally, while particles were collected during the flyby, the suite of onboard instruments also acquired data. Throughout the closest approach to 81P, the spacecraft maintained a constant attitude relative to the ram direction, to enable dust collection and to protect the spacecraft behind dust shields, while the navigation camera NAVCAM tracked the comet via a rotating mirror, which was itself protected by a "periscope" of two 45° mirrors to look forward around the shield (*Newburn et al.,* 2003). Stardust acquired 72 close up images of the nucleus (Fig. 3) that allowed for the determination of the size and shape of the comet as well as characterization of its surface and near-surface activity. Stardust observed an oblate nucleus, with a landscape of numerous flat-floored depressions and rounded central pits (*Duxbury et al.,* 2004; *Brownlee et al.,* 2004; *Sunshine et al.,* 2016; see also the chapter by Pajola et al. in this volume) and active jets of material emanating from it. The range of viewing geometries enabled by the rotating mirror approach, and the significant improvement in spatial resolution of the nucleus imaging compared to previous missions, allowed the source regions of activity (the pits) to be unambiguously identified and resolved for the first time. Although planned, multicolor imaging was not possible as the camera filter wheel stuck in the white-light position before arrival at the comet. 81P remains distinct from other nuclei imaged by spacecraft in appearing to be much more heavily pitted, and is relatively unusual in being a single comparatively round object, rather than bilobed.

The Stardust spacecraft also had two *in situ* dust experiments onboard, the Dust Flux Monitoring Instrument (DFMI), which measured the dust particle flux and size distribution (*Tuzzolino et al.,* 2003), and the Comet and Interstellar Dust Analyzer (CIDA), which measured dust grain composition (*Kissel et al.,* 2003) during various phases

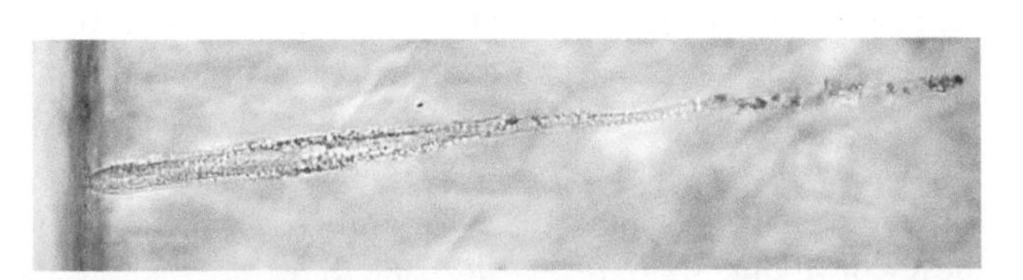

Fig. 2. Track created in the aerogel by a dust particle. From *Burnett* (2006).

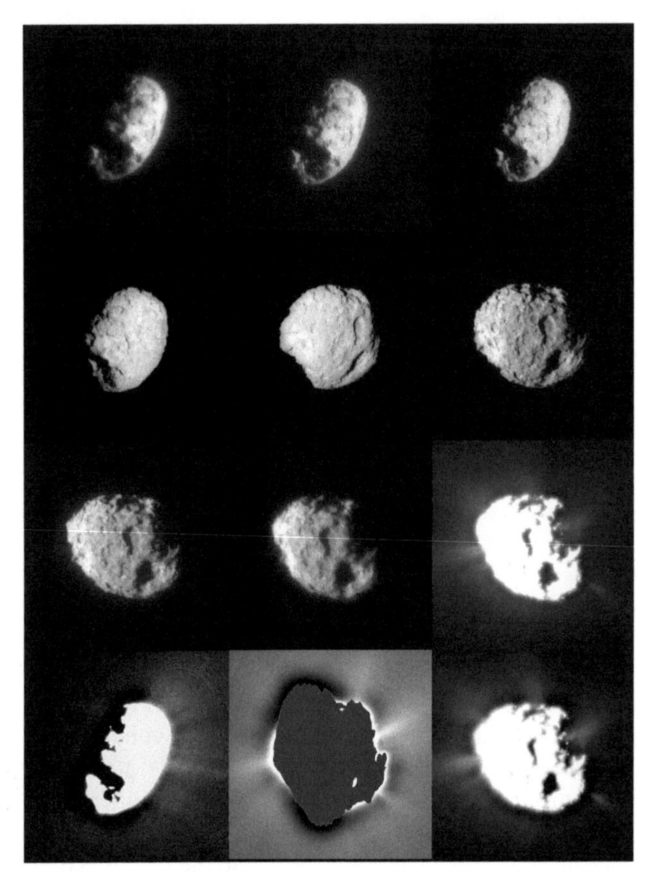

Fig. 3. Images of the nucleus of Comet 81P/Wild 2, acquired by the Stardust navigation camera during the comet flyby. Credit: NASA/JPL/Stardust Team.

of the mission. Measurements from CIDA unambiguously identified organic compounds within the dust (*Kissel et al.,* 2004). The DFMI measurements of non-uniform bursts of dust hits, with few impacts between, indicate that the nucleus of 81P releases large aggregate dust clumps that subsequently disintegrate once in the coma (*Tsou et al.,* 2004). DFMI had two different sensor subsystems: thin films of polyvinylidene fluoride sensitive to impacts by small particles, and acoustic detectors mounted on the dust shield to measure larger impacts. By carefully calibrating these, *Green et al.* (2004) showed that Stardust was much more sensitive than the dust detectors on Giotto, and that such dust fragmentation could be an important effect in all comets that had previously been missed; it would subsequently be seen again by Stardust-NExT and Rosetta (see sections 3.6 and 4).

Although the aerogel capture was highly successful, it modified all collected particles to some degree. Particles larger than 1 μm were generally well preserved due to their higher thermal inertia. Instead, submicrometer-sized dust particles were strongly modified and survived only when shielded by a larger particle (*Brownlee et al.,* 2006). A good understanding of this alteration process is important for understanding the properties of the cometary dust. Many laboratory calibration efforts were carried out to determine impactor properties from track characteristics (e.g., *Kearsley et al.,* 2012). This nonetheless motivates the need for future sample collection with little to no particle alteration, i.e., at essentially zero relative velocity to the comet, which can only realistically be achieved with a rendezvous mission.

3.4. Deep Impact

Deep Impact elevated the complexity of cometary science missions by including an experiment onboard that would interact with the nucleus of 9P. Envisioned first by *Belton and A'Hearn* (1999) and successfully proposed by PI M. A'Hearn in NASA's eighth round of Discovery missions in 1998, Deep Impact was the most risky cometary mission up to that time. The project comprised two fully functional spacecraft mated together, a flyby and an impactor. The goal was to explore beyond the highly evolved upper meters of the nucleus, exposed to cosmic radiation and solar insolation, by excavating less-altered and relatively pristine material at depth and comparing the properties and composition of the subsurface and coma components. This would be achieved via a hypervelocity impact with known kinetic parameters (*A'Hearn et al.,* 2005a). In order to be properly executed, the two spacecraft had to separate ~24 hr before impact and the flyby spacecraft had to divert and slow down to safely miss the nucleus by 500 km (Fig. 4) while providing an 800-s viewing window of the cratering event (*Blume,* 2005).

Target considerations for cometary missions are often determined by launch window and orbit feasibility and not chosen simply for the science case. In the case of Deep Impact, there were further constraints to consider. For purposes of both reliable targeting of the impactor during its autonomous navigation journey after release by the flyby spacecraft, and nucleus size and strength to sustain a crater of ~100 m in size, a comet nucleus with radius >2 km was required (*A'Hearn et al.,* 2005a). The approach phase angle was also restricted such that the nucleus would be illuminated as seen from the impactor. For successful autonomous navigation, the crater would be illuminated as seen from the flyby spacecraft before and after its creation, and the comet could be studied prior to the impact experiment for broader context (*Mastrodemos et al.,* 2005). The mission was to deliver a >350-kg impactor at a velocity >10 km s^{-1} in order to create a large-scale cratering event observable from the flyby spacecraft and Earth (*Blume,* 2005). An offset from the center of brightness, determined from inflight scene analysis, was programmed into the autonavigation system for both spacecraft for the greatest probability of impacting the comet in an observable and illuminated location and to avoid pointing confusion from a bright limb of the nucleus or the ejecta plume (*Mastrodemos et al.,* 2005). The event timing was also crucial as it needed to occur in darkness from at least one major astronomical observatory because of the important role of remote sensing (*Meech et al.,* 2005b). 9P was chosen because it met or exceeded all these constraints (*A'Hearn et al.,* 2005a).

The highly capable instrument suite on the flyby spacecraft included a high-resolution camera (HRI-VIS), a wide-

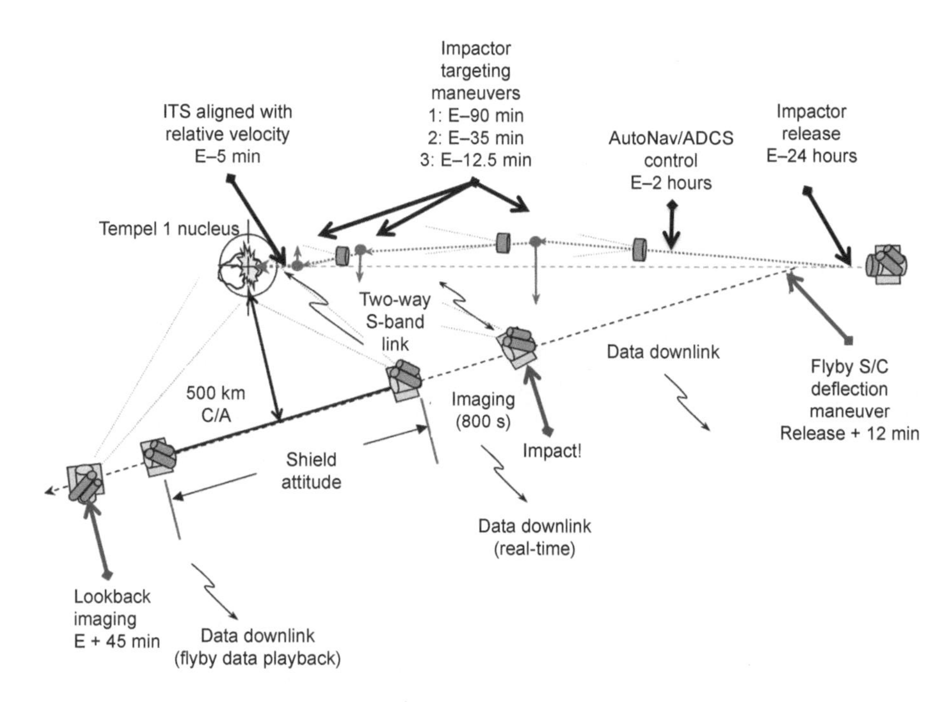

Fig. 4. Graphic adapted from *Mastrodemos et al.* (2005) depicting the timeline of the impactor separation from the Deep Impact flyby spacecraft and targeting maneuvers, deflection maneuver and shield mode of the flyby spacecraft, and various communication links.

field medium-resolution camera (MRI), and an infrared imaging spectrometer (HRI-IR) that were body mounted and co-aligned (*Hampton et al.,* 2005). The HRI-VIS and HRI-IR shared a single f/35 telescope with 10.5-m focal length, which was discovered to be out of focus after launch due to flawed pre-launch calibrations, resulting in spatial resolution degradation and blurred images that were improved with a deconvolution algorithm (*Lindler et al.,* 2007). The f/17.5 telescope feeding the MRI had a focal length of 2.1m. Both visible cameras utilized filter wheels with nine optimized filters for cometary science studies. Aside from two broadband clear filters for each camera, the HRI-VIS geology filters had a bandpass of ~100 nm for pre- and post-impact context imaging, observing the progression of the ejecta curtain and crater growth (Fig. 5), monitoring photometric variation with rotation, analyzing the surface color and reflectivity, and examining surface layers. The narrowband MRI filters were ~5–10 nm in bandwidth and were designed for detailed coma gas (e.g., OH, CN, and C_2) and dust studies, including morphology of faint structures and overall coma production. The HRI-VIS achieved 10 m pixel^{-1} or better spatial resolution over 25% of the nucleus during the flyby and the MRI provided stereoscopic information for the nucleus shape reconstruction (*A'Hearn et al.,* 2005b). The HRI-IR double-prism spectrometer covered a wavelength region of 1.05–4.85 μm with variable spectral resolution, peaking at the short- and long-wavelength ends of the spectrometer and having a minimum R ~200 at 2.5 μm. This wavelength region afforded the simultaneous detection and mapping of the three primary volatiles known to constitute the frozen interior of comets, namely H_2O, CO_2, and CO, as well as other important compositional components like carbon-based organics (*A'Hearn et al.,* 2005b). The spectral range also encompassed short- and long-wavelength continuum used to parameterize the reflective and thermal properties of the nucleus and ejected material.

In addition to the flyby spacecraft instrumentation, the impactor, made predominantly of copper to reduce contamination in the measurements of the cometary composition, had a complete attitude control system, propulsion system, and an S-band radio link to the flyby spacecraft (*Blume,* 2005). It carried a wide-field camera (ITS) identical to the MRI but without a filter wheel. The ITS captured images of the intended impact site on 9P up until a few seconds before impact (Fig. 5) and achieved spatial resolutions of ~1 m pixel^{-1} (*A'Hearn et al.,* 2005b), the highest resolution ever obtained on a cometary nucleus until Rosetta.

As previous missions encountered, in order to successfully study the diverse aspects of a comet whose nucleus is comparatively bright with respect to the dim diffuse coma, the instruments were designed to handle a large dynamic range. Before launch, a spectral attenuator that covered one third of the slit length was added at the entrance slit of the HRI-IR to reduce emission from the nucleus to account for the potential 10,000:1 brightness ratio between the nucleus and coma (*Hampton et al.,* 2005). In addition, observing sequences, exposure times, and data compression lookup tables for all the instruments were created to span the nucleus and coma signal, so as not to compromise low-signal details in the coma jets and ejecta. In the end, the ejecta was quite bright and optically thick and overwhelmed some of the images. The ejecta material was so bright that it biased the autonavigation and instrument pointing of the flyby spacecraft, which misinterpreted the ejecta as part of the

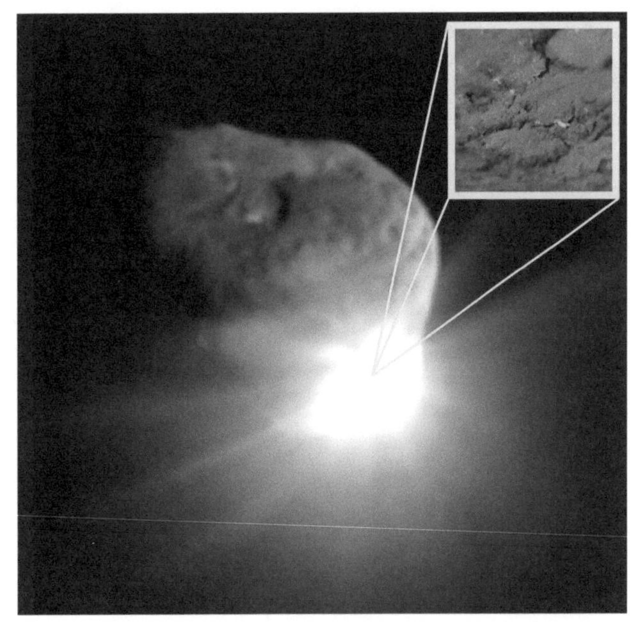

Fig. 5. Clear filter image of Comet 9P acquired with the Deep Impact HRI-VIS ~1 min after the impact experiment while the flyby spacecraft was ~8000 km away from the nucleus [PDS-SBN image ID 9000937 (*McLaughlin et al.,* 2014)]. The growing ejecta cloud is obvious in the image. The spatial resolution is ~15 m $pixel^{-1}$. The inset composite image is adapted from *Wellnitz et al.* (2013) and was derived after geometric and pointing corrections were performed on nested ITS images from its final approach before impacting the nucleus. The processed images were then carefully registered to preceding images to a small fraction of a pixel resulting in an image with scale of 1 m $pixel^{-1}$.

nucleus and thus pointed further from the impact site than the planned center of brightness offset (*Klaasen et al.,* 2008).

As with most cometary missions, 9P's nucleus properties were not well constrained at the time of mission design. The mission team had the challenge of designing to various physical parameters (e.g., composition, ice-to-silicate ratio, gas-to-dust ratio, mass, density, porosity, material strength, thermal properties), which often spanned an order of magnitude or more. Additionally, although modeled by *Richardson et al.* (2005) and *Schultz et al.* (2005), the exact impact scenario could not be assumed prior to the event and best-guess boundary conditions of low strength and low density were used to limit the expected situation in order to design the full mission. Regardless of the resulting cratering scenario, e.g., burrowing into very porous material or fragmenting a piece off the nucleus, the data from the impact experiment would lead to fundamental conclusions about cometary structure.

The Deep Impact flyby spacecraft implemented a 45° rotation with respect to the velocity vector 800 s after impact to protect the instruments and other critical subsystems from dust particle hits during closest approach and the comet's orbital plane crossing (Fig. 4). As an additional safety measure, the spacecraft design included dust shielding on the leading edge of the solar panels and instruments in this flyby orientation. Particle hits were detected and measured by the impactor's navigation system once released by the flyby spacecraft (neither the impactor nor flyby had dedicated dust impact monitoring instruments), and while some of the larger hits did cause a slew in pointing, none were fatal to the mission and the attitude control system always recovered the impactor pointing. This result has since been used to inform the assessment of dust hazards for subsequent mission proposals.

Major science results from the Deep Impact mission were first presented in *A'Hearn et al.* (2005b) and led to advances in our understanding of cometary nuclei (see also chapters in this volume by Filacchione et al. and Pajola et al.). For example, the results from pre-impact monitoring and mapping of 9P are extensive. The mini-outbursts observed on approach were nearly instantaneous, with many being cyclic and originating from regions on the limb coming into daybreak (*Farnham et al.,* 2007; *Belton et al.,* 2008). The diverse examples of surface terrain are suggestive of numerous geologic processes at work on 9P: The smooth surface flows are hypothesized to be cryogenic in nature; the rough regions have an assortment of circular features, where some may be attributed to cratering events but others are pits likely formed by sublimation and material collapse; the active scarps are suggestive of backwasting; and the thick layered ridge may be a relic of formation or a result of erosion (*Thomas et al.,* 2007). The single-scattering albedo was 0.039 at 550 nm wavelength and the photometric variations on 9P, showing little variegation, were relatively small compared to other comets (*Li et al.,* 2007). A low thermal inertia was sufficient to explain the temperature map of 9P's nucleus, which had a maximum of 336 K near the subsolar point and fell off toward the terminator, following the topography, and was indicative of cold temperatures just below the surface that may harbor ices (*Groussin et al.,* 2007). The surface ice was comprised of loose aggregates tens of micrometers in size and, like the mini-outbursts, appeared to be confined to small patches along the morning terminator (*Sunshine et al.,* 2006). Heterogeneous gas emission was observed for the first time: The H_2O outgassing was solar driven with enhancements in the sunlit hemisphere, while the CO_2 was released in nighttime jets and may be correlated with the smooth flows and eroding scarps (*Feaga et al.,* 2007).

On July 4, 2005, at 05:44:36 UTC, Deep Impact delivered 19 GJ of kinetic energy to the nucleus of 9P as designed, fulfilling a major mission objective and furthering the scientific return of the mission. The volatile and silicate composition of the ejecta was similar between the interior and exterior of the comet, suggesting that the primordial inventory may be reflected in ambient activity. The exception was that the first meter of excavated material was devoid of H_2O ice, while micrometer-sized H_2O-ice grains were present at depths as shallow as 1 m and extending tens of meters into the nucleus (*Sunshine et al.,* 2007). The impact heated and vaporized silicates, volatiles, and organics in the nucleus and created a self-luminous impact flash. The properties of the flash were indicative of a highly-porous substrate (*Ernst and Schultz,* 2007). Material properties of the nucleus were derived from analysis and modeling of the ejecta rays and curtain and their behavior. The results conclude that the surface of 9P is layered and has a high porosity (*Schultz et al.,* 2007). Additionally, the nucleus has low strength on the order of 1–10 kPa, bulk density of 400 kg m^{-3}, and the cratering event was controlled by gravity rather than material strength (*Richardson et al.,* 2007).

3.5. EPOXI

After a successful seven-month mission and impact experiment, the Deep Impact flyby spacecraft remained healthy, with all three scientific instruments functioning nominally, and had adequate consumables onboard to continue on for an extended mission (*A'Hearn et al.,* 2011). A second cometary mission, one to explore the diversity of JFCs, was proposed for the extension and selected as a Mission of Opportunity in the 2006 round of Discovery proposals with the caveat that the Deep Impact eXtended Investigation (DIXI) mission (PI M. A'Hearn) would be merged with the Extrasolar Planetary Observation and Characterization (EPOCh) mission (PI D. Deming), resulting in the EPOXI mission. The out-of-focus HRI telescope was advantageous to fulfilling the EPOCh objective of using the Deep Impact flyby spacecraft as a remote observatory for transiting extrasolar planet science during its long cruise to another comet (*Ballard et al.,* 2011). Photometric observations acquired by the HRI-VIS did not saturate the CCD, but rather the HRI focus issue spread the light from observations over many pixels, generating better data for the purpose of the study. The extended mission also incorporated a technology demonstration for NASA, testing the deep space communication network (*Vilnrotter et al.,* 2008).

The scientific value of repurposing the Deep Impact spacecraft was enormous. The extended mission utilized three Earth gravity assists to put it on a trajectory bringing the spacecraft within 700 km from 103P on November 4, 2010. During the Earth-Moon system flybys, hydration on the lunar surface was investigated (*Sunshine et al.,* 2009) and the Earth-Moon system was observed from afar as an analog extrasolar planet (*Robinson et al.,* 2011; *Crow et al.,* 2011). In addition, Deep Impact had a short six-month commissioning, calibration, and cruise phase from launch to approach of 9P during the primary mission, leaving much to improve upon for the extended mission, including extensive calibration campaigns throughout cruise and designing more effective observing sequences to be executed for the 103P encounter (*Klaasen et al.,* 2013a).

The EPOXI team was faced with a challenge leading up to the 103P encounter as a result of the spacecraft design for the primary mission, where the instruments were to point at the comet while the antenna was directed at Earth. During the 103P encounter, the flyby geometry was such that the instruments and antenna could not simultaneously point to their respective targets. The mitigation strategy was to manage data acquisition, data storage, data downlink, and thermal control of the spacecraft using a do-si-do maneuver where data were taken and stored, then the spacecraft slewed to point the antenna to Earth for data downlink, during which the spacecraft heated up, and then slewed back to the comet and cooled before data acquisition resumed. This programmed "dance" transpired daily on approach and caused spikes in the instrument temperatures but resulted in successful observing sequences.

103P was the fifth comet imaged up close by a spacecraft, but was not the original target of the extended mission. Comet 85D/Boethin was the proposed target for EPOXI, but was abandoned when it was not recovered in an extensive Earth-based campaign in 2005–2007 (*Meech et al.,* 2013), prior to a planned trajectory correction maneuver to put the spacecraft on the path to intercept the comet, again emphasizing the importance of mission support observational campaigns (section 2). The primary target became 103P, which resulted in paradigm-changing science due in part to its intrinsic hyperactivity (*Groussin et al.,* 2004). The cometary objectives of the EPOXI mission were to systematically determine the degree of diversity among comets by comparing comets of similar age and orbit, using the same suite of instruments, and to disentangle evolutionary vs. primordial aspects of cometary behavior and morphology. The resolved images of 103P showed an elongated bilobed nucleus with a smooth "waist," rough terrain on both lobes, and several tall spires (>40 m) (*Thomas et al.,* 2013b). The nucleus, which was in a state of complex rotation (*Belton et al.,* 2013b), was engulfed in icy particles (Fig. 6) (*Hermalyn et al.,* 2013). Reminiscent of 9P's heterogeneous coma, noncorrelated discrete concentrations of H_2O and CO_2 above the ubiquitous ambient coma were also detected at 103P (*A'Hearn et al.,* 2011). It was clear that the hyperactivity of 103P was due to the sublimation of H_2O-ice grains, including a population of small grains that were expelled from the small lobe of the nucleus in a jet of CO_2 gas (*A'Hearn et al.,* 2011; *Protopapa et al.,* 2014). These particles did not cause damage to the spacecraft.

The Deep Impact flyby spacecraft was subsequently used for the remote study of distant comets such as C/2009 P1 (Garradd) in early 2012, determining that the comet had a very high and increasing CO to H_2O abundance as it receded from the Sun (*Feaga et al.,* 2014), and C/2012 S1 (ISON) in early 2013, detecting several small spontaneous outbursts when it was approaching the Sun but still beyond the H_2O snowline (*Farnham et al.,* 2017). Communication with the Deep Impact spacecraft was lost in August 2013 after it was determined that the spacecraft computer entered an infinite fault protection loop due to a 64-bit time stamp being converted to and not fitting into a 32-bit value (*Farnham et al.,* 2017), thus bringing a close to the mission.

The combination of the Deep Impact and the EPOXI missions unveiled certain details about several extreme cometary behaviors and their manifestations, but generated just as many questions. In particular, it is still uncertain how deep below the surface the ice reservoirs are; what mechanism causes outbursts (*Belton and Melosh,* 2009);

Fig. 6. Deconvolved HRI-VIS image of 103P/Hartley 2 acquired November 4, 2010, when Deep Impact was ~1950 km away from the nucleus. There are large icy chunks surrounding the nucleus as well as small icy grains being expelled from the nucleus in the image. Every speck in the image is a particle. The image was downloaded from the PDS-SBN [image ID 5004008 (*Lindler et al.,* 2012)].

whether the surface flows are cryogenic as suggested (*Belton et al.,* 2008); what the true nature of the small vs. large ice particle populations is (*Sunshine et al.,* 2007; *Protopapa et al.,* 2014; *Kelley et al.,* 2013); and, most importantly, which morphological and compositional cometary properties reflect evolutionary vs. formational conditions (*Sunshine et al.,* 2016). Rosetta would shed light on many of these topics, but not completely close these knowledge gaps.

3.6. Stardust-NExT

Also selected as a Mission of Opportunity in the 2006 round of NASA Discovery proposals and led by PI J. Veverka, Stardust-NExT (New Exploration of Tempel) repurposed the healthy Stardust spacecraft to fly by 9P one perihelion passage after the Deep Impact experiment (*Veverka et al.,* 2013). The NAVCAM, which was built from spare parts from previous flight projects (*Newburn et al.,* 2003), and not well calibrated during the primary mission because of its supporting role during the Stardust sample collection, became the priority instrument for the extended Stardust-NExT mission. As such, substantial calibration improvements were made to the geometric correction, spatial resolution, and radiometric calibration during the extended mission (*Klaasen et al.,* 2013b). In addition, for optimal imaging quality, recurring camera contamination from dust was minimized by periodically heating the instrument via internal heaters or direct sunlight. The dust instruments, DFMI and CIDA, both collected data during the 9P encounter.

For the first time, Stardust-NExT imaged a comet previously visited by a spacecraft and documented surface changes after a full orbital period had passed, and after its surface was altered by an artificial impact. Stardust-NExT extended the nucleus surface coverage (Fig. 7) — and thus knowledge — of 9P from ~30% to ~70% and measured the Deep Impact experimental crater site (*Veverka et al.,* 2013).

Precise targeting and timing of the flyby were essential in order to realize these goals because the crater specifically, and other regions studied by Deep Impact more generally, needed to be illuminated and in the field of view of the spacecraft at the time of closest approach to make the intended measurements and comparisons. For this, 9P's spin state and spin rate had to be accurately predicted before the encounter, and thus the importance of synergistic observations was recognized by the community yet again. A worldwide observing campaign of 9P (*Meech et al.,* 2005a, 2011b), supporting both the Deep Impact and Stardust-NExT missions, was used to acquire enough data from which the evolution and current state of the rotational period and pole orientation could be derived from the light curve to feed to the navigation team a full year before closest approach (*Belton et al.,* 2011). A trajectory correction maneuver in 2010 was made to delay arrival by ~8.5 hr. Upon arrival, it was clear that the team met the challenge of accurately determining the rotation rate and spin axis orientation as the encounter occurred within ~20° of the targeted longitude (*Veverka et al.,* 2013; *Belton et al.,* 2011; *Chesley et al.,* 2013). This was the first time that a change in rotation rate of a comet was unambiguously measured from one perihelion to another.

At a speed of 10.9 km s^{-1} and closest approach distance of 178 km, Stardust-NExT arrived at 9P on February 15, 2011, and took 72 images with the NAVCAM (*Veverka et al.,* 2013). The highest-quality image of 9P and the artificial crater had a resolution of 11 m $pixel^{-1}$. The imaging frequency was constrained to one image every 6 s due to the maximum rate supported by the data system. During the flyby, the change in geometry and lighting was amenable to taking and creating stereo pairs of images. The data collected during closest approach allowed for the Deep Impact crater to be detected and measured: the images reveal a 50-m-diameter bowl-shaped crater within a larger (~180-m-diameter) depression, hinting at either a layered nucleus structure and/or a complex cratering process (*Schultz et al.,* 2013; *Vincent et al.,* 2015). Even more smooth terrain was discovered on 9P than had been seen with Deep Impact, amounting to ~30% of the surface, most likely new material that had erupted from the subsurface well after formation of the comet (*Veverka et al.,* 2013; *Thomas et al.,* 2013a). There was also additional evidence of nightside activity, jets emanating from steep slopes, extensive layering, and pitted terrain on 9P (*Veverka et al.,* 2013; *Farnham et al.,* 2013). Contrary to some expectations, the active regions of the surface identified by temporal variability comprised only ~10% of the surface, especially

Fig. 7. Clear NAVCAM image of Comet 9P acquired by Stardust-NExT on February 15, 2011, from ~200 km away from the nucleus. The illuminated face of the comet displayed here is territory not imaged by Deep Impact and shows a diversity in morphology including examples of layering, ridges, and circular depressions on the surface at a resolution of ~12 m pixel^{-1}. Downloaded from the PDS-SBN [image ID 30039 (*Veverka et al.,* 2011)].

evident in scarp retreat, with 1–10 m of material lost in one apparition (*Veverka et al.,* 2013; *Thomas et al.,* 2013a). Additional images were acquired on approach and departure, but the faint comet was not detected in the maximum commandable exposure time of 20 s until 27 days prior to encounter and was not resolved in the latter. The jet activity in 2011 was weaker than in 2005, which may be due to the pre- vs. post-perihelion behavior of the comet and the timing of the encounters. Deep Impact was 1 day pre-perihelion while Stardust-NExT was 34 days post-perihelion. There was also no evidence in the photometry of the periodic mini-outbursts seen by Deep Impact (*Belton et al.,* 2008, 2013a).

The dust instruments were only operational for a very short time (~hours) surrounding closest approach (*Veverka et al.,* 2013). Similar to the dust environment at 81P, the DFMI data were not uniform and were dominated by detections of clusters of particles fragmenting from aggregates rather than a steady stream of small particle hits originating from active areas of the nucleus. Up to 1000 particles were measured over kilometer scales followed by voids with no particle impacts (*Economou et al.,* 2013). Dust mass spectra collected with CIDA showed prominent peaks at mass 1 and 26, H and CN respectively, and long tails at high-mass numbers, suggesting complex molecules (*Veverka et al.,* 2013).

Although extended missions are challenging, as enumerated here and in section 3.5, recycled spacecraft have led to scientifically enlightening outcomes at a cost an order of magnitude less than that of the primary mission.

3.7. Other Encounters

In addition to planned ion tail crossings, spacecraft can also serendipitously traverse tails if they are at the right place and time to be in the flow of ions downstream of a comet. As many comets' ion tails have a significant width, this alignment does not need to be perfect for cometary ions to be detectable at the spacecraft. The first such tail crossing, found unexpectedly, was the Ulysses spacecraft's encounter with the ion tail of C/1996 B2 Hyakutake in 1996: *Jones et al.* (2000) recognized the magnetic field structure to be consistent with the comet's ion tail, and *Gloeckler et al.* (2000) reported the observation of cometary pickup ions in the craft's Solar Wind Ion Composition Instrument (SWICS) instrument data.

Several such unplanned comet tail crossings are now known (Fig. 8). There were two more by Ulysses: C/1999 T1 McNaught-Hartley on October 19–20, 2000 (*Gloeckler et al.,* 2004), and C/2006 P1 McNaught over several days centered on February 7, 2007 (*Neugebauer et al.,* 2007). Pickup ions from 73P were detected by the near-Earth Advanced Composition Explorer (ACE) and Wind spacecraft in May and June 2006 (*Gilbert et al.,* 2015), and ions from C/2010 X1 Elenin by the Solar Terrestrial Relations Observatory (STEREO)-B mission (*Galvin and Simunac,* 2015).

These unplanned comet encounters complement data from targeted comet missions, and provide *in situ* information on the solar wind's response to a comet's presence, including flow speed and direction; the magnetic field structure of a comet's ion tail; and the wind's composition and that of pickup ions that are carried by it. Despite not being designed for cometary encounters, general advancements in the sensitivity of *in situ* instruments have improved the quality of the serendipitous observations obtained.

As demonstrated by the case of C/2013 A1 (see section 3.8), the advance warning of a tail crossing can allow mission teams to select favorable instrument modes to maximize the science return of these serendipitous events. Such warnings were provided for ESA's Solar Orbiter crossing the tails of C/2019 Y4 ATLAS (*Jones et al.,* 2020) in June 2020 and C/2020 A1 Leonard in December 2021. *In situ* instruments detected these tails' draped magnetic field, pickup ions, and the presence of ion-scale waves excited by associated instabilities (e.g., *Matteini et al.,* 2021).

NASA's Parker Solar Probe (PSP) passed as close as 0.025 au, or 3.7 million km, to 322P/SOHO 1 on September 2, 2019. This encounter is particularly intriguing, given this object's possible asteroidal nature (*Knight et al.,* 2016). However, there was little unambiguous evidence of the comet in data gathered then by PSP (*He et al.,* 2021).

A remarkable aspect of several of the serendipitous tail crossings is the immense tail lengths sometimes involved. Pickup protons from 153P/Ikeya-Zhang were detected by the NASA-led Cassini spacecraft at ~6.5 au downwind of

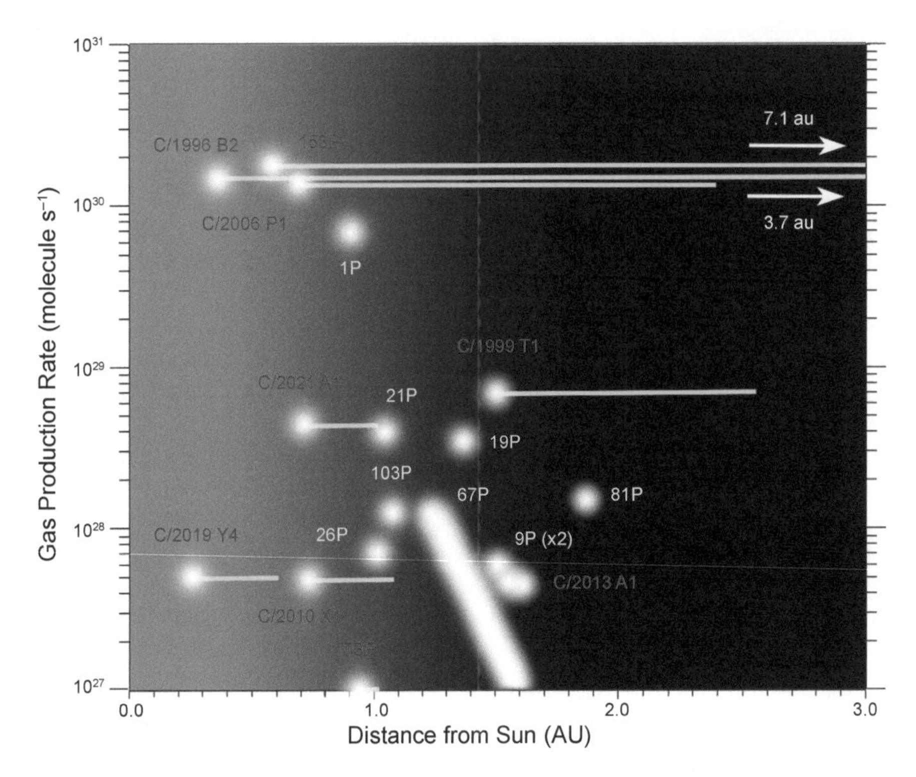

Fig. 8. All targeted comet encounters (yellow) and serendipitous ion tail crossings (red). Positions are plotted to show the heliocentric distance of each comet's nucleus against its gas production rate at the time. Tail crossings occurred at the end of each blue line; those outside the plot's range are as shown by the arrows. Sources of production rates, for H_2O unless indicated, if not cited in the text: 1P: (total) *Krankowsky et al.* (1986); 9P: *Schleicher et al.* (2006); 19P: *Young et al.* (2004); 21P: *Neugebauer et al.* (2007); 26P: *Johnstone et al.* (1993); 81P: *Fink et al.* (1999); 103P: *Dello Russo et al.* (2011); 153P, C/1999 T1: *Combi et al.* (2008); C/2019 Y4: *Combi et al.* (2021). Continuum of 67P values: *Hansen et al.* (2016); C/2021 A1: M. Combi, personal communication.

the nucleus (*Jones et al.,* 2022). Detections at such large distances imply that ion tails, although ultimately originating at bodies of a few kilometers across, can potentially remain as detectable structures all the way to the heliosphere's edge.

We also note the phenomenon of interplanetary field enhancements (IFEs). These are uncommon solar wind events, characterized by a heliospheric magnetic field magnitude increase, often with a thorn-shaped profile, that are usually accompanied by a discontinuity in the magnetic field direction. The first reported event of this kind (*Russell et al.,* 1983) has been followed by several surveys and case studies (e.g., *Russell et al.,* 1984; *Jones et al.,* 2003a). Although the exact cause of these signatures is still to be widely accepted, it appears clear from several associations between IFEs and cometary orbital plane crossings (e.g., *Lai et al.,* 2015; *Jones et al.,* 2003b) that they are somehow associated with interactions between the solar wind and dust trails that lie along the orbits of some comets and asteroids. Once the formation process for IFEs is better understood, routine solar wind observations by spacecraft may add to our understanding of dust trails.

3.8. C/2013 A1 at Mars

By far the closest known unplanned approach to a comet is the case of C/2013 A1 Siding Spring. This dynamically-new LPC's orbit took it to within 140,751 ± 175 km of the center of Mars on October 19, 2014 (*Farnocchia et al.,* 2016). All spacecraft then operating in the planet's orbit — NASA's Mars Odyssey, Mars Reconnaissance Orbiter (MRO), and MAVEN; ESA's Mars Express; and India's Mangalyaan — effectively performed a comet flyby. Pre-encounter studies of the comet suggested a not-insignificant dust risk for spacecraft near Mars, so these satellites' orbits were phased such that they were largely protected from dust impacts by Mars itself during the the comet's closest approach to the planet. Instruments on several of the spacecraft were also purposefully turned off for safety reasons.

The comet's ~0.5-km-wide nucleus was successfully resolved by the MRO HiRISE camera at a resolution of 138 m pixel^{-1} (*Farnham et al.,* 2015; *Lisse,* 2015). Images of the inner coma were also obtained using NASA's Curiosity and Opportunity rovers. Lyman-α observations by MAVEN indicated a water production rate of 0.5×10^{28} s^{-1} (*Mayyasi et al.,* 2020). Interpretations of the solar wind-comet-planet interactions were complicated by the arrival of the interplanetary counterpart of a coronal mass ejection shortly before the encounter. Nevertheless, the influence of the comet's own induced magnetosphere on the near-Mars environment was clear: Mars's much smaller induced magnetosphere underwent a rotation in magnetic field orientation as the comet approached, with particularly strong distortion of the planet's ionospheric magnetic field around closest approach, and increased planetary magnetosheath turbulence post-encounter (*Espley et al.,* 2015). A significant increase in energetic particle fluxes was observed during the comet's passage; these displaying similarities to features observed at other comets (*Sánchez-Cano et al.,* 2018).

Of particular interest was the interaction between Mars's atmosphere and the comet coma's dust and gas. The energetic particles mentioned above would have been deposited in the planet's atmosphere, as well as pickup ions such as O^+ that were detected for ~10 hr near the planet (*Sánchez-Cano et al.,* 2020). A transient ionized layer at altitudes of 80–100 km was present in the hours after closest approach, resulting from the shower of cometary dust (*Gurnett et al.,* 2015). An increase in the abundance of metallic ion species, consistent with cometary dust deposition, was also detected at altitudes of ~185 km (*Benna et al.,* 2015).

4. THE ROSETTA RENDEZVOUS MISSION

4.1. Introduction to the Rosetta Mission

The original concept that would become Rosetta began life in the 1980s as a hugely ambitious comet nucleus sample return mission. Despite the fact that the study began before anyone had even seen a nucleus (pre-1P encounters), it would conclude that a mission should drill a meters-long core and return the sample for analysis in laboratories on Earth, maintaining the ices at cryogenic temperatures throughout [see *Keller and Kührt* (2020) and *Thomas et al.* (2019) for short histories on the early evolution of the mission concept]. As described in section 5.6, the technology required for such a concept appears to be some way off, even today. However, the motivation for a sample return mission was, and is, compelling and obvious: By collecting ice from the interior of a comet, we would sample more-or-less completely unaltered material laid down during the formation of our solar system, and be able to subject it to the full range of investigative techniques possible with modern laboratory equipment. This would be revolutionary not only in the study of comets, but also in understanding the wider topic of planet formation processes. Motivated by a desire to understand the physics of how cometary activity really works, Rosetta would also study the changes on the surface of the comet over an extended period as it approached the Sun. This is important within the field in its own right, and would also be essential to determine just how "pristine" (or not) the returned sample really is, by measuring the effects of activity on the immediate subsurface layers.

Buoyed by the success of Giotto, ESA selected Rosetta as the only planetary probe among the four cornerstone missions for its Horizon 2000 program. An advanced set of *in situ* instruments was proposed as a more realistic alternative to sample return, including separate landers (originally two), but the key advance from previous missions was retained: Rosetta would rendezvous with a comet and follow its evolving behavior as it approached the Sun, instead of seeing it at only a single moment in time with a flyby. This two-year period of operation was the key to the Rosetta mission, enabling much more data to be collected on the comet and allowing for repeated measurements to be made, to see how the nucleus surface, coma, and interactions with the solar wind changed over time.

Rosetta's original cometary target was 46P, but a year-long delay following the failure of a previous Ariane 5 rocket meant that a new target was required. The mission launched toward 67P on March 2, 2004. To match orbits with the comet, Rosetta took a looping 10-year trajectory through the inner solar system (Table 3). This trajectory included two asteroid flybys en route as Rosetta passed through the main belt, the 5-km-diameter (2867) Šteins and the 100-km (21) Lutetia (*Barucci et al.,* 2015). Additionally, a number of "bonus" cruise phase science investigations were completed utilizing the Optical, Spectroscopic, and Infrared Remote Imaging System (OSIRIS) cameras for remote observations of comets C/2002 T7 LINEAR, C/2004 Q2 Machholz, and 9P (*Keller* 2006; *Rengel et al.,* 2007; *Küppers et al.,* 2005; *Keller et al.,* 2005, 2007b) and active asteroid 354P/LINEAR (*Snodgrass et al.,* 2010). Finally, while Rosetta was too far from the Sun to generate sufficient power from its solar panels, it entered a 31-month-long "hibernation" through its final aphelion before arrival.

Rosetta woke from hibernation in January 2014 and began recommissioning of its payload, with the first images of 67P being acquired by OSIRIS on March 23, 2014, from a distance of ~5 × 10^6 km (*Tubiana et al.,* 2015). The next five months saw a careful approach to the comet, followed by three months of mapping and preparation for the Philae landing, which occurred on November 12, 2014 (see section 4.3.3 for details). Rosetta then entered the "escort phase" of the mission, continuing through 67P's perihelion in August 2015, and an extended phase beyond the nominal operations period into 2016. During the escort phase the distance between Rosetta and the nucleus was typically between 10 and 200 km (increasing to more than 300 km with higher activity levels near perihelion; Fig. 9). "Excursions" to larger distances (1500 km toward the Sun and 1000 km in the ion tail direction) were made, primarily to better understand the plasma environment around the comet.

The Rosetta mission ended on September 30, 2016, with the controlled impact of the orbiter onto the smaller lobe of the nucleus. Figure 10 shows the approximate landing site. The spacecraft most likely bounced, but as contact was lost at impact time, we may never know its final location.

4.2. Instrumentation

The instrument suite on Rosetta was chosen, via a separate call for proposals after the mission was selected, to achieve as much of the original sample-return science as possible with *in situ* measurements. This led to a large collection of 11 instruments on the orbiter and 10 on the

TABLE 3. Timeline of key events in the Rosetta mission.

Dates	r*	Event
2004 Mar 02	1.0	Launch
2005 Mar 04	1.0	First Earth flyby
2007 Feb 25	1.5	Mars flyby
2007 Nov 13	1.0	Second Earth flyby
2008 Sep 05	2.1	Šteins flyby (800 km)
2009 Nov 13	1.0	Third Earth flyby
2010 Jul 10	2.7	Lutetia flyby (3200 km)
2011 Jun 08	4.5	Hibernation entry
2012 Oct 03	5.3	Aphelion
2014 Jan 20	4.5	Hibernation exit
2014 Aug 06	3.6	Arrival at comet
2014 Nov 12	3.0	Philae landing
2014 Nov 15	3.0	End of Philae science operations
2015 Feb 14	2.3	First close flyby (6 km)
2015 Mar 28	2.0	Second close flyby (14 km)
2015 May 10	1.7	Equinox
2015 Jul 09	1.3	Last signal from Philae
2015 Aug 13	1.2	Perihelion
2015 Sep 30	1.4	Sunward excursion to 1500 km
2016 Mar 21	2.6	Equinox
2016 Mar 30	2.7	Tailward excursion to 1000 km
2016 Sep 02	3.7	Philae found
2016 Sep 30	3.8	End of mission

*Heliocentric distance at event (au).

lander (Table 4), with the latter in particular focused on local composition measurements and including a drill to sample the subsurface. Some of the most important results from the mission would come from composition measurements, particularly via the Rosetta Orbiter Spectrometer for Ion and Neutral Analysis (ROSINA), which sampled neutral and ionized gas in the coma with high sensitivity and mass resolution (*Balsiger et al.,* 2007), while the Cometary Secondary Ion Mass Analyser (COSIMA) revealed both composition and structure of dust particles (*Kissel et al.,* 2007). Rosetta's instruments included a number of novel elements, including the first atomic force microscope in space [Micro-Imaging Dust Analysis System (MIDAS) (*Riedler et al.,* 2007)] to measure dust grain structure at nanometer scales, and an experiment to use the communication transmission between the orbiter and lander to explore the nucleus interior structure [Comet Nucleus Sounding Experiment by Radiowave Transmission (CONSERT) (*Kofman et al.,* 2007)]. Rosetta also carried remote sensing instruments covering ultraviolet through to microwave wavelength ranges, establishing a legacy that would see some of these instrument designs fly on other missions throughout the solar system. The UV spectrograph Alice (*Stern et al.,* 2007) would form the basis of the New Horizons instrument of the same name and the Lunar Reconnaissance Orbiter's Lyman Alpha Mapping Project (LAMP) instrument. The Visible and Infrared Thermal Imaging Spectrometer (VIRTIS) (*Coradini et al.,* 2007) would also be flown on Dawn and Venus Express. The two cameras of OSIRIS, the visible scientific imaging system (*Keller et al.,* 2007a), were equipped with 16-bit 2k × 2k CCD detectors and mechanical shutters, unique features that enabled the return of more than 80,000 detailed images. Two of the instruments were actually made up of suites of independent sensors provided by different teams but operated via a common interface to the spacecraft: Rosetta Plasma Consortium (RPC) and Surface Electric Sounding and Acoustic Monitoring Experiment (SESAME). The RPC plasma package (*Carr et al.,* 2007) was located on the orbiter and contained five sensors to study the comet's interaction with the solar wind (Ion Composition Analyser, Ion and Electron Sensor, Langmuir Probe, Fluxgate Magnetometer, and Mutual Impedance Probe), while the SESAME collection of sensors (*Seidensticker et al.,* 2007) was on the lander (Cometary Acoustic Surface Sounding Experiment, Permittivity Probe, and Dust Impact Monitor).

The selection of the Rosetta payload after the mission and its primary science goals had already been confirmed had some key advantages, primarily in allowing an "optimized" collection that achieved the best fit to the science goals based on what ESA member states and the international partner agency, NASA, would support. However, this progression led to some problems, as requirements for different instruments were often mutually exclusive with each other or with spacecraft limitations. For example, RPC measurements of the comet's interaction with the solar wind ideally needed to sample a wide range of distances from the comet, while *in situ* instruments had to be as close to the nucleus as possible for sensitivity considerations. Also, there were frequent conflicts between remote sensing and *in situ* payloads in pointing requirements. The ESA approach

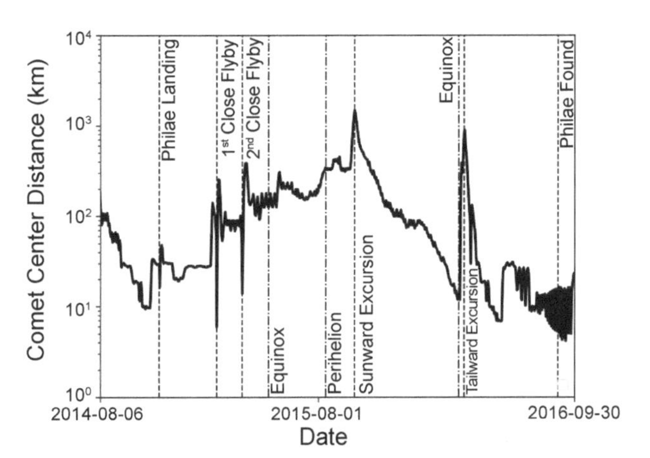

Fig. 9. Distance between Rosetta and the comet center from arrival at the comet until end of mission. The vertical dashed lines indicate spacecraft activities and the vertical dash-dotted lines indicate comet orbital milestones.

Fig. 10. Image from the OSIRIS Wide Angle Camera on Rosetta, showing the landing area, with the last image acquired by OSIRIS before the crash landing superimposed. The image of the Rosetta spacecraft is to scale. Credit: ESA/Rosetta/MPS for OSIRIS Team MPS/UPD/LAM/IAA/SSO/INTA/UPM/DASP/IDA.

of independent instrument teams encouraged an (occasionally fraught) atmosphere of competition over spacecraft resources rather than cooperation. It was only relatively late in the mission that coordinated interinstrument measurement campaigns were organized, and successful joint analyses of data from multiple instruments were carried out, e.g., in the analysis of serendipitous simultaneous observations of an outburst (*Grün et al.,* 2016), or in joint remote sensing and *in situ* study of dust (*Güttler et al.,* 2019).

4.3. Challenges for a Comet Rendezvous and Solutions

As the first mission to rendezvous with, and make a soft landing on, a comet, Rosetta faced several challenges. In what follows we describe how Rosetta coped with some of them. The treatment partially follows *Küppers* (2017).

4.3.1. Matching a comet orbit. Due to the high eccentricity of the orbit of 67P, matching the orbit of Rosetta with that of the comet required 10 years of interplanetary cruise, including 4 planetary flybys as gravity assists (three at Earth and one at Mars; see Table 3). The propellant required by Rosetta, most of it for the rendezvous maneuvers at comet arrival, constituted more than half of the spacecraft mass at launch. The orbit matching also required a heliocentric distance range of 0.89–5.3 au between Rosetta and the Sun. At low heliocentric distance, stringent attitude constraints, determined through inflight tests, had to be applied to avoid overheating of thrusters and radiators. At large heliocentric distance, Rosetta, with its 64 m^2 of solar panels, could only collect sufficient power to operate out to approximately 4.5 au, the largest heliocentric distance for a solar powered spacecraft at the time. For the 2.5 years spent outside 4.5 au the spacecraft had to be commanded into hibernation, meaning that only essential elements, in particular heaters and thermostats, were kept on to keep all spacecraft and payload components within their designed temperature range. All other elements were switched off and needed to survive 2.5 years without activation or communication. Power-driven constraints on simultaneous operation of payloads prevailed from hibernation exit until summer 2014 and again in the last weeks before Rosetta landed on 67P at a heliocentric distance of 3.83 au.

4.3.2. Rendezvous with a little known target. When 67P was chosen as Rosetta's new target in 2003, its orbit was certain but little else was known about it. An observing program was initiated to prepare for the Rosetta mission implementation by collecting as much information as possible about 67P. Here we summarize the results obtained and compare them to the ground truth from Rosetta.

4.3.2.1. Orbit and spin. 67P was discovered by Klim Ivanovich Churyumov and Svetlana Ivanova Gerasimenko in 1969. The comet was observed over 45 years between first detection and the arrival of Rosetta, meaning that the orbit was known well enough that it could be used to define the spacecraft trajectory until the onboard cameras detected the comet and images taken by Rosetta could be used for further trajectory refinements and navigation.

In a synthesis of data acquired between 2003 and 2007, *Lowry et al.* (2012) determined a rotation rate of 12.76137 ± 0.00006 hr. Somewhat surprisingly at the time, when Rosetta approached 67P in 2014, several light curves taken of the unresolved nucleus resulted in a spin period of 12.4043 ± 0.0007 hr (*Mottola et al.,* 2014). The difference was explained by a change in the rotation period during the perihelion passage in 2009. It was caused by the reaction torque from the non-radial component of cometary outgassing. This explanation was confirmed by monitoring the spin period of the comet during the perihelion passage in 2015, when the period was again observed to decrease by about 20 min (*Keller et al.,* 2015).

The spin axis direction of the comet as determined by *Lowry et al.* (2012) agreed with the solution obtained *in situ* by Rosetta to within approximately 10°. No deviation from principal axis rotation was found in groundbased data. Rosetta detected an excited rotational state, but with a very small amplitude [$\approx$0.15° (*Jorda et al.,* 2016)].

The spin period of 67P was known well enough to plan for onboard navigation. The only adaptation was a change of the frequency of navigation slots to once every 5 hr, from 4 or 6 hr, to avoid the resonance with the spin period of 67P of $\approx$12 hr after perihelion.

4.3.2.2. Size, shape, volume, mass, and density. Groundbased and Hubble Space Telescope observations of 67P arrived at an effective radius of 2 km and an albedo of 0.04–0.06 (*Lamy et al.,* 2006; *Lowry et al.,* 2012). Those values were later confirmed by the images acquired by Rosetta.

The determination of small-body shapes is difficult if only lightcurve data are available, in particular in the case of an object like 67P, as the technique is not sensitive to concavities. Cometary activity complicates the determination of a nucleus light curve close to the Sun, while the faintness of the nucleus and corresponding limitations in signal-to-noise, together with contamination of the data by any faint coma, complicate the analysis far from the Sun. Furthermore, 67P orbits the Sun close to the ecliptic, implying that light curves

TABLE 4. Rosetta instruments on the orbiter and lander.

Instrument	Description	Reference
Orbiter Instruments		
Alice (Ultraviolet Imaging Spectrometer)	R ~ 100, 70–210 nm FUV long-slit spectrograph	*Stern et al.* (2007)
CONSERT (Comet Nucleus Sounding Experiment by Radio wave Transmission)	Radio signal timing experiment between orbiter and lander	*Kofman et al.* (2007)
COSIMA (Cometary Secondary Ion Mass Analyser)	SIMS dust composition measurements over 1–3500 amu mass range, and microscope camera	*Kissel et al.* (2007)
GIADA (Grain Impact Analyser and Dust Accumulator)	Microbalance plate and laser curtain to measure mass/momentum of dust particles larger than ~15 μm	*Colangeli et al.* (2007)
MIDAS (Micro-Imaging Dust Analysis System)	Atomic force microscope to image dust in three dimensions, on nanometer to micrometer scales	*Riedler et al.* (2007)
MIRO (Microwave Instrument for the Rosetta Orbiter)	Microwave temperature measurements at 190 and 562 GHz, with 44-kHz-resolution spectroscopy around the latter band	*Gulkis et al.* (2007)
OSIRIS (Optical, Spectroscopic, and Infrared Remote Imaging System)	Narrow-angle (2°) and wide-angle (11°) CCD cameras, with a total of 26 filters covering 0.25–1 μm	*Keller et al.* (2007a)
ROSINA (Rosetta Orbiter Spectrometer for Ion and Neutral Analysis)	Separate high-resolution ($m/\Delta m > 3000$) ion and neutral mass spectrometers and pressure gauge	*Balsiger et al.* (2007)
RPC (Rosetta Plasma Consortium)	Sensors for the plasma environment: ion mass spectrometer, ion and electron sensors, Langmuir probe, magnetometer, and mutual impedance probe	*Carr et al.* (2007)
RSI (Radio Science Investigation)	Ultrastable oscillator to allow accurate ranging of Rosetta via radio signals	*Pätzold et al.* (2007)
VIRTIS (Visible and Infrared Thermal Imaging Spectrometer)	Medium- (R ~ 200) and high- (R ~ 2000) resolution spectrometers, covering 0.22–5 μm and 1.9–5 μm, respectively	*Coradini et al.* (2007)
Lander Instruments		
APXS (Alpha Proton X-ray Spectrometer)	Surface elemental composition measurements	*Klingelhöfer et al.* (2007)
CIVA (Comet Infrared and Visible Analyser)	Panoramic cameras (seven monochromatic visible cameras) and visible and infrared spectroscopic microscopes	*Bibring et al.* (2007a)
CONSERT (Comet Nucleus Sounding Experiment by Radio wave Transmission)	Radio signal timing experiment between orbiter and lander	*Kofman et al.* (2007)
COSAC (COmetary SAmpling and Composition)	Gas chromatograph and time-of-flight mass spectrometer focused on elemental and molecular composition	*Goesmann et al.* (2007)
Ptolemy (Mass spectrometer)	Gas chromatograph and ion trap mass spectrometer focused on isotopic composition	*Wright et al.* (2007)
MUPUS (MUlti-PUrpose Sensors for Surface and Subsurface Science)	Thermal and mechanical surface and subsurface properties probe	*Spohn et al.* (2007)
ROLIS (Rosetta Lander Imaging System)	Descent camera system, monochromatic visible camera	*Mottola et al.* (2007)
ROMAP (Rosetta Lander Magnetometer and Plasma Monitor)	Fluxgate magnetometer, ion, electron and pressure sensors	*Auster et al.* (2007)
SD2 (Sampling Drilling and Distribution)	Drill and system to pass samples to other instruments	*Finzi et al.* (2007)
SESAME (Surface Electric Sounding and Acoustic Monitoring Experiment)	Combination of three probes on landing legs; acoustic sounder, permittivity probe, and dust impact monitor	*Seidensticker et al.* (2007)

cannot cover the whole surface and the shape reconstruction is non-unique. Therefore it was not surprising that the shape models from Earth-based light curves did not resemble the very peculiar shape of 67P (see Fig. 11), although the presence of large flat facets in these models can be taken as an indication of significant concavities (*Devogèle et al.*, 2015). The highly concave shape of the comet caused some initial problems for shape modeling from resolved Rosetta imaging. However, the routines were adapted to reach accurate representations of the nucleus (e.g., *Jorda et al.*, 2016; *Preusker et al.*, 2017).

The mass of 67P was estimated from the modification of the cometary orbit by non-gravitational forces (the reaction force to outgassing) by *Davidsson and Gutiérrez* (2005). Their models provided nucleus masses in the range of 0.35–1.22 × 10^{13} kg. The measured mass of 67P gleaned from spacecraft data, 0.998 × 10^{13} kg (*Pätzold et al.*, 2016), is within this range. However, even with the size of 67P being approximately known from ground-based observations, the volume of the rather regular shape models is larger than that of the highly concave, bilobed nucleus. Therefore *Davidsson and Gutiérrez* (2005) overestimated the volume of the nucleus, underestimating its density. Nevertheless, their correct determination of a mass range for 67P provides some confidence in applying this method to other comets.

4.3.2.3. Activity. 67P's activity profile turned out to be relatively repeatable from one perihelion passage to the next; the global dust and gas production could be well predicted (*Snodgrass et al.*, 2013). However, the possibility of predicting conditions in the innermost coma, where Rosetta would spend most of its mission, was extremely limited due to the comparably low spatial resolution of Earth-based data. One valuable correct prediction from Earth-based data was that 67P would already be active at the time of Rosetta's arrival and the Philae landing.

4.3.3. Choosing a landing site and delivering a lander. The relative velocity between Rosetta and its target was successively reduced from 775 m s^{-1} to less than 1 m s^{-1} in a series of maneuvers taking place between May and August 2014. On August 6, 2014, Rosetta was inserted into the first of a series of hyperbolic arcs around 67P, representing the first comet rendezvous. As the Philae lander had to be delivered at a heliocentric distance not much below 3 au, due to thermal reasons, this left only three months for nucleus and gravity field characterization, moving to closed orbits, global mapping, landing site selection, and planning of the Philae delivery.

The characterization of the nucleus was achieved from hyperbolic orbit arcs, first at distances of 80–100 km, then moving in to 50–70 km. During this phase, five landing site candidates for Philae were selected by August 24, 2014. The main information sources for landing site selection were images by OSIRIS and the navigation cameras (for shape, surface roughness, etc.) and temperature information provided by MIRO and VIRTIS. Selection criteria included the scientific interest of the site and the following engineering criteria (*Ulamec et al.*, 2015):

- Illumination of the landing site at a Sun angle <60° from the time of landing until at least 40 min later
- Impact velocity between 0.3 m s^{-1} and 1.2 m s^{-1}
- Alignment of vertical axis of the lander, lander velocity vector, and surface normal within 30°, implying a constraint on average surface roughness
- Nominal separation velocity from the orbiter being commandable between 5 and 50 cm s^{-1}; a backup (emergency)

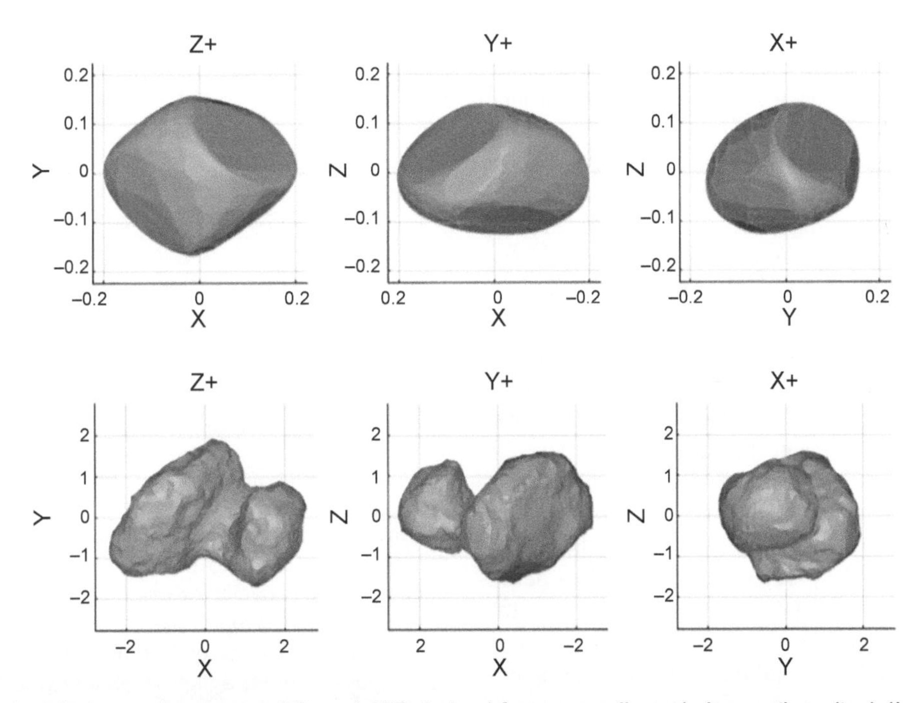

Fig. 11. Comparison between the shape of Comet 67P derived from groundbased observations (top) (*Lowry et al.*, 2012)) with a shape model determined by Rosetta (bottom) (*Preusker et al.*, 2017).

release mechanism would eject the lander with a separation velocity of ~18 cm s^{-1}, and preferred landing sites could be reached with a nominal separation velocity of ~18 cm s^{-1}, so that the nominal descent trajectory would be approximately followed using the backup mechanism and therefore a longer than necessary descent time was acceptable

On September 9, Rosetta entered a 30-km circular orbit around 67P, becoming the first comet orbiter. A few days later, the five landing sites were ranked and the prime and backup landing sites were determined. In subsequent weeks, Rosetta moved to orbits of 20 km and 9 km distance from the comet, and, as no showstopper (e.g., very high surface roughness at decimeter to meter scale) was found in the close observations, the landing site was confirmed on October 14, 2014. For the orbit changes during the mapping and close observation phases, the feasibility of getting closer was decided only a few days before the maneuvers were executed, requiring the maintenance of three different payload operations plans. More details of the operational aspects of landing preparation and landing are given in section 4.4 and by *Accomazzo et al.* (2016) and *Ashman et al.* (2016).

4.3.4. Operating a spacecraft around an active comet. Due to the low gravity and irregular shape of the nucleus, the gas drag from cometary outgassing, dust particles in the field of view of cameras resembling stars, and the changes of cometary activity over time, the spacecraft was exposed to an environment of limited predictability and strong variability. The resulting challenges for spacecraft operations were unprecedented.

Before lander delivery, the Rosetta mission focused largely on the preparations for landing. After Philae's first science sequence, the focus shifted to the needs of the orbiter instruments for scientific observations. This involved coordination of requests from 11 different instruments and adaptation to the operational constraints, such as the limitation on trajectories while navigating the spacecraft around the comet. The gas drag from the expanding cometary coma accelerated the spacecraft, and the amount of the drag (comet activity) was not fully predictable. In addition, inaccuracies of spacecraft maneuvers (maneuver error) contributed to an uncertainty in the motion of Rosetta relative to 67P. The uncertainty of the position of Rosetta translated into an error for spacecraft pointing (see Fig. 12). As navigation around the comet required regular images to be taken to confirm the relative position, a trajectory was defined to be possible if the pointing error did not exceed 2.5° during navigation periods, dictated by the field-of-view of the navigation cameras.

The feasibility of a trajectory therefore depended on the short-term predictability (a few days ahead) of the gas drag on the spacecraft, and therefore on the predictability of cometary activity. The problem was attacked by defining two cases: a preferred case, a trajectory defined by the science team based on a best guess on the activity (and its predictability), and a high activity case, based on very conservative assumptions. Rosetta flew trajectories according to the preferred case trajectory, and if that trajectory at some point was not feasible in terms of navigation, the spacecraft was commanded to change to the high activity case trajectory. Notably, as most of the cross section of the spacecraft was its 64 m^2 solar panels, gas drag on Rosetta was minimized by staying close to the terminator, where the gas flow was directed toward the edge of the solar panel. This trajectory design was leveraged for a large part of the mission, although it limited viewing geometry and sampling locations relative to the nucleus and its illuminated regions.

The remaining problem for the Rosetta Science Working Team (SWT) was to define a preferred case activity level that allowed trajectories close to the nucleus without violating navigational constraints. Observations from previous perihelion passages, measurements of the current conditions at 67P by Rosetta instruments, and theoretical models were used to propose an activity level that was then agreed on by the SWT and used for trajectory design. A comprehensive summary of the science planning activities in this phase is given by *Vallat et al.* (2017).

After the planning scheme had been executed for a few months, the concern was raised that the resulting trajectories were quite conservative in that Rosetta was significantly more distant from 67P than necessary, as suggested by the gas pressure measured by the COmet Pressure Sensor (COPS), part of the ROSINA instrument. Steps were prepared to improve the planning, but before implementation, events at Rosetta enforced a more dramatic change.

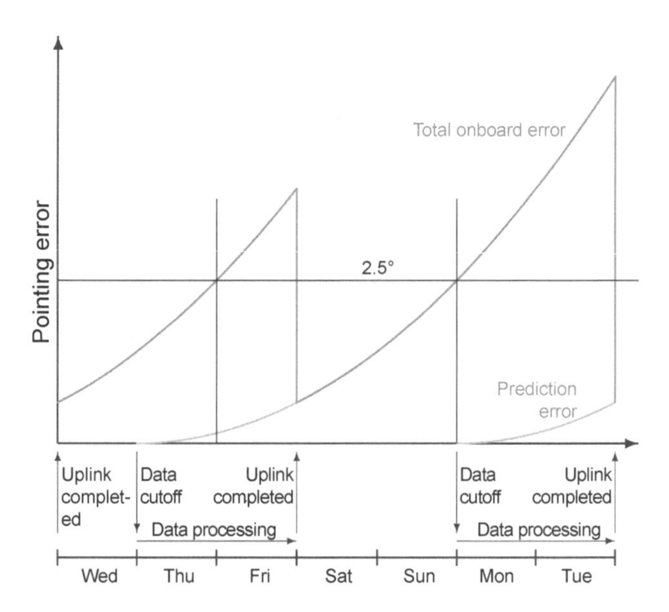

Fig. 12. The navigation cycle during the Rosetta mission. Navigation images were continuously taken every 4–6 hr. All data up to a cutoff (usually two days before uplink of the new commands to the spacecraft) were used for the next navigation cycle. At the time of uplink, the pointing error was reduced to the prediction error from the last data cutoff. Then, the error propagated to the next data uplink. The requirement for a flight trajectory was that the pointing error at data cutoff was not larger than 2.5°. The plot shows the limiting case. Credit: Björn Grieger.

The possibility of star trackers being confused between dust particles and stars, and thus not fully functional, was discussed in preparation for the Rosetta mission, but not considered to be a major constraint. However, the number of individually visible dust particles in the coma of 67P was higher than expected. During a close flyby on February 14, 2015, the Rosetta star trackers temporarily lost track due to such confusion. Operational measures were taken to mitigate the problem during the execution of a second close flyby on March 28, 2015. However, in spite of those measures, the star trackers again lost tracking over an extended period of time during that flyby, and while the spacecraft narrowly avoided a loss of attitude control, it ended up in safe mode. It became clear that neither the planned (preferred case) trajectory nor the high-activity trajectory, designed to be robust against navigation error, were safe in this situation and thus could not be flown.

Prior to this safing event, Rosetta had been flown with the typical planning scheme of an ESA planetary mission: In iterations between the Mission Operations Centre (MOC), the Science Ground Segment (SGS), and the PI teams, the trajectory was fixed in a long-term planning process 4–8 months in advance, the spacecraft pointing was fixed in a medium-term process 1–2 months in advance, and the details of resource (data volume, power) distribution and commanding were fixed 1–2 weeks in advance. As a consequence of the star tracker problems, the planning of the Rosetta mission had to be redesigned. The new scheme was used for most of the remaining mission operations. Some of the main changes in the planning process were:

- A large reduction in all turn-around times, so that planning could better consider the comet conditions
- Requests from the SWT and SGS for trajectories from MOC that met certain criteria, rather than specific trajectory implementations
- Using the long-term trajectory as indicative only, and waiting to define the actual trajectory until one day before execution

For the final ~7 weeks of the mission, a still-more-demanding and less-risk-averse scheme was designed. To get Rosetta even closer to the comet, elliptical orbits with a constant duration of three days were designed (corresponding to a semimajor axis of 10.5 km), implying an even shorter turnaround. In order to support this effort, a pre-defined attitude profile was flown, thereby somewhat reducing the flexibility in pointing. Starting with 13 × 8-km ellipses, the pericenter distances were successively reduced, reaching less than 2 km, or one nucleus radius, from the surface.

Overall, Rosetta was operated successfully over its more than two-year rendezvous with 67P. However, some of the mission goals were not fully achieved. With the benefit of hindsight, what could be improved for the future? In reviews of the Rosetta mission (*Thomas et al.,* 2019; *Keller and Kührt,* 2020), the main science objective that was identified as not being fully achieved is understanding the mechanism of cometary activity. High-resolution observations of the nucleus during the most active phase of 67P, around perihelion, would have been required to obtain the relevant data. The main obstacle to being close to the active nucleus was the attitude uncertainty, first due to maneuver error and uncertainties in the prediction of gas drag, then because of the star tracker failing when in the inner coma during high comet activity. While details of the operations scheme, chosen trajectories, and risk philosophy during Rosetta operations may be debated, the most straightforward way to overcome those constraints is autonomous attitude navigation of the spacecraft. It would eliminate the buildup of "pointing error" as shown in Fig. 12 and therefore the main constraint preventing the spacecraft from getting closer to the nucleus. Even the star tracker failure, while unlikely to reoccur in a future mission, may have been at least partly overcome with such a scheme. The position of the comet plus that of the Sun would provide at least a rough attitude solution. While a closed loop between the attitude control system and the navigation cameras was available to keep the Rosetta target asteroids in the field of view during the flybys, the algorithm was unfortunately not adequate for operations close to 67P, as it required the target object to underfill the field of view of the NAVCAMs.

Another shortcoming pointed out in the reviews is that there was no systematic approach to revisit, within a short timeframe, locations where interesting events were detected. The example most frequently mentioned is the fracture in the Aswan cliff that resulted in a (predictable) collapse (*Pajola et al.,* 2017). While most of those locations were observed repeatedly over the mission, in most cases those revisits were random during either comet center pointing attitude or pointings designed for other objectives, as opposed to the outcome of a specific targeted observing strategy. This may have allowed several objectives to be met in a single pass, but it was generally not optimal for following the development of changing features. The long leadtimes in the early comet phase of Rosetta were not ideal for such dedicated follow-up observations. In the later phases, the main issue was again the pointing uncertainty, which complicated targeted observations from close distances. Again, autonomous attitude guidance would have helped to increase the flexibility.

4.4. The Philae Lander

During the Rosetta mission study, the original two lander concepts for Rosetta were merged into the Philae lander. As some of the properties of the cometary surface, such as Young's modulus and strength, were unknown prior to Rosetta, Philae was designed against a large range of those properties, covering 3–4 orders of magnitude. This included several devices to keep it on the surface at touchdown (*Bibring et al.,* 2007b). During descent, the Active Descent System (ADS) was foreseen to fire thrusters in an upward direction to avoid rebound. Furthermore, anchoring harpoons were to be fired into the ground at landing. Finally, the feet of the lander contained screws to secure it on the surface after touchdown. The overall tensile and compressive strength of the surface of 67P turned out to be lower than the minimum assumed when Philae was built, although analysis of the Philae landing showed that the compressive strength at the original landing site Agilkia was within the Philae design range (*Groussin et al.,* 2019; *Biele et al.,* 2015; *Möhlmann et al.,* 2018). Furthermore, when the Rosetta target was changed from 46P to 67P, an update to the Philae landing gear was required to cope with the increased gravity of the new mission target, which is about 3× larger and therefore ≈30× more massive (*Ulamec et al.,* 2006).

On November 12, 2014, Philae separated from Rosetta at an altitude of 20.5 km and with a relative velocity of 19 cm s^{-1} (*Biele et al.,* 2015). After 7 hr of ballistic descent, Philae touched down on the Agilkia landing site at 15:34:03 UTC, only 120 m from the planned position, after having been imaged while flying over the surface of the comet (Fig. 13). At touchdown, the anchor harpoons failed to fire and the holddown thrust of the cold gas system did not work, leading to Philae bouncing and rising again from the surface before the screws could activate. Philae traveled across the nucleus for 2 hr, touching the surface twice before reaching its final resting position in Abydos, a shaded rocky area in the southern hemisphere (Fig. 14) at 17:31:17 UTC. It took almost two years to successfully find and undoubtedly image the lander in its final location.

Philae continued communicating with Rosetta during its entire flight. This provided an unexpected opportunity for some of the payload instruments to acquire data in two substantially different locations (Agilkia and Abydos), and the Rosetta Lander Magnetometer and Plasma Monitor (ROMAP) to collect data during most of Philae's voyage over the surface. The communication link was interrupted, for geometrical reasons, about 27 min after the final landing. It was reestablished four more times until the primary batteries ran out of power on November 15, 2014 (*Biele et al.,* 2015). In addition to OSIRIS and NAVCAM images, CONSERT played a key role in narrowing down the location of the final landing site to an area of 22 × 106 m^2 (*Herique et al.,* 2015).

The originally planned science activities of the Philae instruments had to be rapidly adapted to the illumination conditions of the final landing site and to the unknown attitude of the lander, but eventually they were all activated and collected invaluable *in situ* data. Unfortunately, the lander attitude ("hanging" sideways at a cliff) did not allow sample collection and drilling, and the Multi-Purpose Sensors for Surface and Subsurface Science (MUPUS) hammer failed to penetrate into the surface. It is still debated whether this was because of a harder than expected surface layer, or because the hammer did not deploy as expected due to the final lander attitude (*Spohn et al.,* 2015; *Heinisch et al.,* 2019). The MUPUS surface strength measurement should therefore be treated with caution, and may not be a useful constraint in future mission design; it is orders of magnitude higher than all other observational, experimental, and theoretical results (see *Groussin et al.,* 2019, in particular their Fig. 10, and the chapter in this volume by Guilbert-Lepoutre et al.). Communication between Rosetta and Philae was briefly reestablished on June 15, 2015, when Rosetta flew over the Abydos region at a distance of 200 km. Communications were irregular and unstable and ended on July 9, 2015 (*O'Rourke et al.,* 2019). The intermittent communications did not allow for a restart of science observations. Therefore, the Philae long-term science, expected to last weeks based on power from rechargeable batteries, could not be recovered.

From March 2016 the distance between Rosetta and the comet was regularly less than 20 km, after several

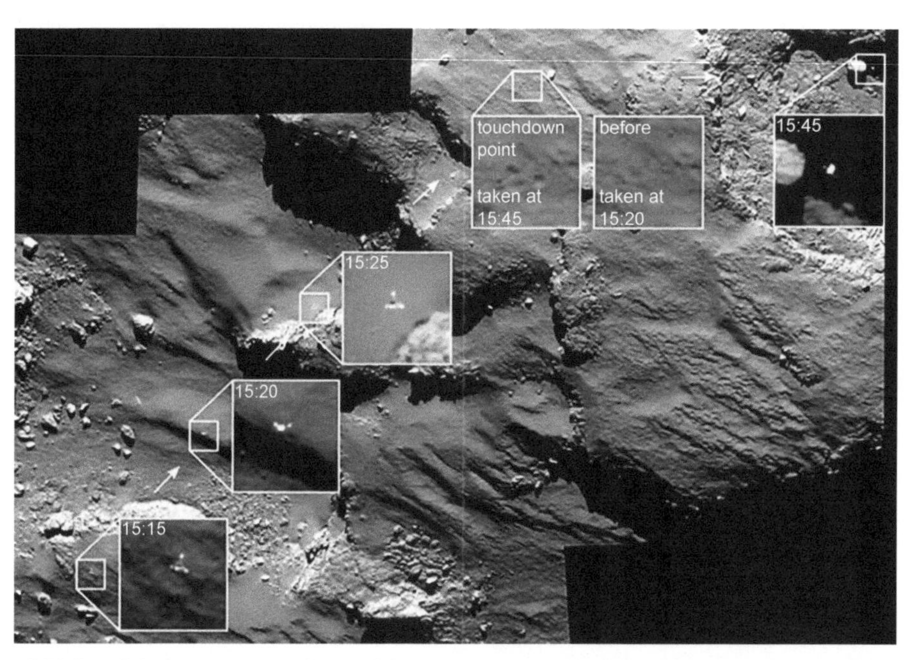

Fig. 13. Images of Philae flying over the surface of the comet on its way to the touchdown location on November 12, 2014. Credit: ESA/Rosetta/MPS for OSIRIS Team MPS/UPD/LAM/IAA/SSO/INTA/UPM/DASP/IDA.

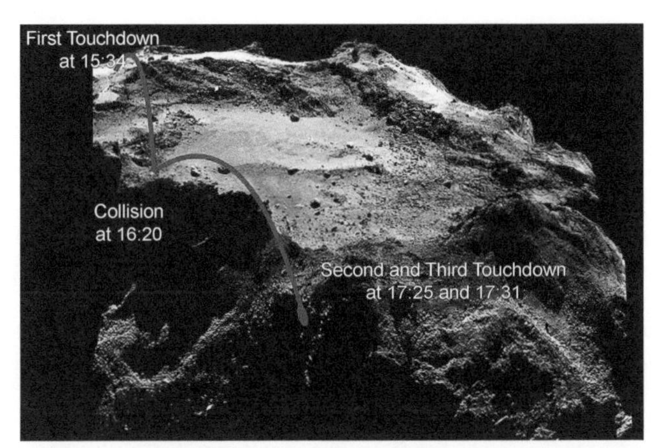

Fig. 14. Philae's flight to its final touch down. Credit: ESA/Rosetta/MPS for OSIRIS Team MPS/UPD/LAM/IAA/SSO/INTA/UPM/DASP/IDA.

months spent at large distances due to the high perihelion activity levels (Fig. 9), and the lander search restarted. Twenty-eight dedicated observing campaigns were carried out. A major difficulty was to be able to unambiguously identify the "bright dot" in the OSIRIS images as Philae (Fig. 15). Rosetta needed to be at a distance <5 km from the comet for Philae to be at least 10 pixels in diameter in OSIRIS images.

Despite the many challenges (e.g., limited observation slots at small enough distance, no pointing flexibility, non-optimal viewing geometries and illumination conditions, large pointing error, small projected fields of view), Philae was successfully and unmistakably imaged on September 2, 2016, at 19:59 UTC from a distance of 2.7 km from the surface. Imaging Philae in its final resting position (Fig. 16) was important for several reasons: It provided context for the measurements performed by the Philae instruments; it confirmed the location identified by CONSERT and allowed its measurements to be properly interpreted in terms of the nucleus interior (*Kofman et al.,* 2020); and it brought to a close the story, begun almost two years before, of the first spacecraft to land on a comet. The landing of Philae marked a milestone in space history.

4.5. Results and Open Questions for the Future

The most important results from Rosetta were arguably those on the composition of the comet, including the discovery of abundant O_2 and its correlation with H_2O (*Bieler et al.,* 2015; *Keeney et al.,* 2019; *Luspay-Kuti et al.,* 2022), complex organics [including the amino acid glycine (*Altwegg et al.,* 2016)], salts (*Altwegg et al.,* 2020), noble gases (*Rubin et al.,* 2018), and the role of electron impact dissociation in coma chemistry (*Feldman et al.,* 2015). The ROSINA and COSIMA instruments were also able to probe the comet's composition at an isotopic level, with implications for comet formation (including incorporation of pre-solar material), activity models, and delivery of various species to the early Earth (e.g., *Altwegg et al.,* 2015; *Marty et al.,* 2017; *Hoppe et al.,* 2018). The remote sensing instruments revealed, in exquisite detail, the morphology and composition of the nucleus and changes in these with time. Meanwhile, microscopic imaging of coma dust grains showed their structure and fragility on nanometer to micrometer scales (Fig. 17) (*Güttler et al.,* 2019).

Rosetta's contribution to cometary science is undoubted; in addition to the highlights summarized here (and listed in Table 2), results from the mission are quoted in nearly every chapter in this book. However, Rosetta is also not the final word in the field; some important questions remain, and some new ones that arose following Rosetta's findings. *A'Hearn* (2017) and *Thomas et al.* (2019) both consider what Rosetta achieved and what future missions could do to answer remaining questions. The highly critical review by *Keller and Kührt* (2020) also raises some valid points, the

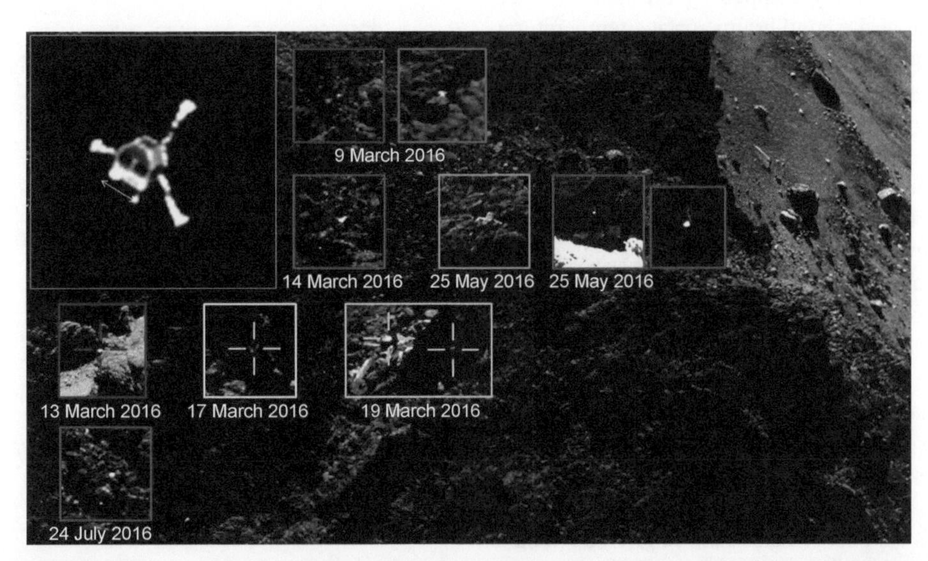

Fig. 15. Examples of OSIRIS images acquired during the search campaign for Philae, showing how difficult it was to unambiguously identify Philae in the images. Credit: ESA/Rosetta/MPS for OSIRIS Team MPS/UPD/LAM/IAA/SSO/INTA/UPM/DASP/IDA.

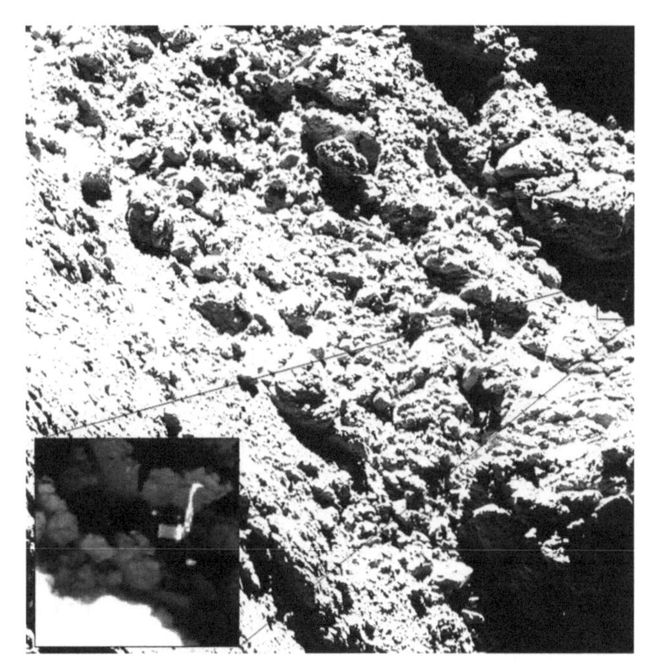

Fig. 16. Philae in its final resting position in Abydos in an OSIRIS image acquired on September 2, 2016, at 19:59 UTC. The full field of view is approximately 100 m across. Credit: ESA/Rosetta/MPS for OSIRIS Team MPS/UPD/LAM/IAA/SSO/INTA/UPM/DASP/IDA.

most important being that our understanding of the processes driving cometary activity is still far from complete. Since Rosetta followed a comet from approximately the onset of activity through perihelion and then outbound again, this does feel like a missed opportunity.

A key piece of missing information appears to be the temperature of the (near-)surface, which was constrained by near-IR and microwave observations by VIRTIS and MIRO, respectively, and locally from Philae measurements (*Groussin et al.,* 2019, and references therein), but would have been most effectively mapped by thermal IR (e.g., 5–20 μm) imaging at the equilibrium surface temperatures expected at 1–3 au from the Sun. Such an instrument was proposed for Rosetta, but not selected for budgetary reasons (*Thomas et al.,* 2019). The limited measurements from Philae, which should have provided details on the microscopic structure of the surface and range of temperatures in the immediate subsurface layers, also mean that there are still significant assumptions in the interpretation of the existing remote sensing data and uncertainties in where subsurface ice layers may exist (see the chapter in this volume by Guilbert-Lepoutre et al.).

In addition, the navigation issues described in section 4.3 meant that Rosetta was not at its closest to the nucleus during the more active phases, and therefore missed opportunities to directly sample the acceleration region *in situ,* and resolve centimeter-scale surface changes or lifted particles, when these effects were most significant. Improved onboard autonomy, which would allow safe spacecraft operation for extended periods at closer distances, as well as the ability to react and sample certain phenomena with a real-time "joystick" control option, seems a necessary technology for future comet rendezvous missions.

The fine-scale structure and composition of the nucleus surface remains a key missing piece of information, relevant not only for understanding activity, but also the formation processes of comets, and for interpreting unresolved data (e.g., photometric or spectroscopic behavior seen from groundbased telescopes). Remote sensing can only give broad assessments of nucleus composition [e.g., that it contains organics (*Capaccioni et al.,* 2015)], and Philae's composition measurements were unable to give any further detail to compare with measurements of dust particle composition in the inner coma [from COSIMA measurements (*Hilchenbach et al.,* 2016)]. One key question that remains the topic of energetic debate is what the volatile-to-refractory ratio is, both within particles lifted into the coma and originally within the nucleus (see *Choukroun et al.,* 2020, for a review). While models of comet formation that use coma measurements of this parameter have been proposed (*Blum et al.,* 2017), the lack of surface (or ideally subsurface) measurements limits our current knowledge. As well as how much ice is present, we would still like to know what and where it is, both locally and globally, e.g., changes with depth from the surface due to heating, or on the scale of the whole nucleus to test ideas of primordial (in)homogeneity. CONSERT could only operate for a short time and the detailed interior structure of 67P remains largely unknown (*Kofman et al.,* 2020). Ground-penetrating radar, operating from an orbiter without the need for a lander, could enable significant advances in future comet (or asteroid) missions (*Asphaug,* 2015; *Herique et al.,* 2018).

Finally, one of the surprising results from Rosetta was the revelation that much more dust falls back to the nucleus than was previously thought (*Thomas et al.,* 2015a; *Keller et al.,* 2017; *Hu et al.,* 2017). Together with the array of

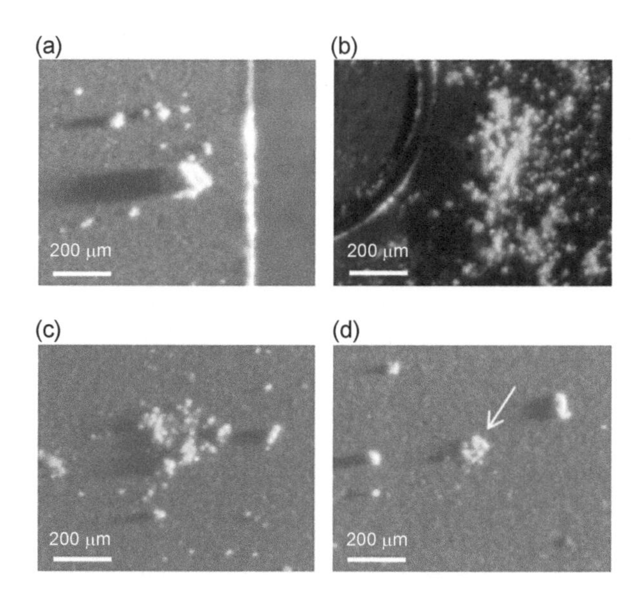

Fig. 17. Selection of dust particles imaged by Rosetta's COSIMA instrument. From *Merouane et al.* (2016).

surface morphology features seen on 67P [e.g., pits, terraces, fractures (*Thomas et al.,* 2015b)], it has become clear that the nucleus surface has been significantly eroded or covered by cometary activity. To access more pristine material, and really test planet formation models, will require either reaching subsurface layers (possibly at considerable depth), or visiting a comet that has not had previous close perihelion passes, and whose surface remains nearly unaltered.

5. FUTURE COMET MISSIONS

5.1. DESTINY$^+$

We include the Japanese Aerospace Exploration Agency's (JAXA) Demonstration and Experiment of Space Technology for INterplanetary voYage with Phaethon fLyby and dUst Science (DESTINY$^+$) (*Ozaki et al.,* 2022) as a future comet mission as its primary target, near-Earth asteroid (3200) Phaethon, has been demonstrated to have repeated activity at perihelion and is responsible for the Geminid meteor stream (*Jewitt and Li,* 2010; *Jewitt et al.,* 2013; *Li and Jewitt,* 2013). Phaethon has been dubbed a "rock comet" as its orbit is such that water ice is not expected to survive, even in the interior, but at perihelion (0.14 au) it is close enough to the Sun for thermal fracturing and electrostatic lifting of rock to form a dust coma without ice sublimation (*Jewitt and Li,* 2010). DESTINY$^+$ is primarily a technology demonstration mission that will show that a medium-sized (~500 kg wet mass) satellite launched into low-Earth orbit by a modest rocket can perform interplanetary missions, through patient use of ion engines to slowly raise the orbit to reach a lunar flyby, and then escape from the Earth-Moon system through a lunar gravity assist maneuver. It promises low-cost deep-space exploration via the enabling technologies of ion engines; large and very lightweight solar arrays; and advanced, compact, thermal and attitude control systems. It has a lot of flexibility in its trajectory and launch window (*Sarli et al.,* 2018). DESTINY$^+$ is currently expected to launch in 2028 and perform a fast (30–40 km s^{-1}) flyby of Phaethon in 2030 (*Ozaki et al.,* 2022). Its primary science goal is to measure dust composition: both interplanetary/interstellar dust during the long cruise and dust released from Phaethon. It will have an instrument based on the Cassini Cosmic Dust Analyzer, which relies on high-speed destructive particle impact with the back of the instrument, and mass spectroscopy of the resulting plasma (*Srama et al.,* 2004), similar to CIDA on Stardust. The mission will also carry narrow-angle panchromatic and wide-angle multi-color cameras to image the nucleus from an expected minimum distance of 500 km (*Ishibashi et al.,* 2020).

5.2. ZhengHe

The coming decade should also see the first dedicated small bodies mission from China, expected to be called ZhengHe, although the name and final approval for the mission remain to be confirmed at the time of writing (*Zhang et al.,* 2019, 2021). ZhengHe is ambitious and will demonstrate a wide range of technologies by first performing an autonomous touch-and-go surface sampling of a small near-Earth asteroid (2016 HO_3) in the late 2020s, returning the sample to Earth, and then redirecting the spacecraft to rendezvous with 311P/Pan-STARRS (hereafter 311P), an active asteroid. This is an interesting object that showed multiple tails at its discovery epoch in 2013, but has not yet shown the repeated activity from orbit-to-orbit that is associated with sublimation of ices and therefore marks main-belt objects as true comets (see the chapter in this volume by Jewitt and Hsieh). In any case, ZhengHe is expected to arrive at 311P in 2034 and operate there for a year (*Zhang et al.,* 2021), covering the approach to aphelion phase of the orbit, and not the true anomaly range in which activity was previously reported. The mission may therefore not encounter an active object, but in any case should provide detailed characterization of the nucleus, and help solve the puzzle of its strange activity pattern, which had been attributed to rotational disruption until lightcurve observations revealed a relatively slow rotation and possible binarity (*Jewitt et al.,* 2018). The ZhengHe payload will include cameras and spectroscopic capability covering the ultraviolet to thermal infrared range, radar (possibly with the capability to detect subsurface ice), a magnetometer, and *in situ* analysis of any gas (*Zhang et al.,* 2021). A proposed dust analysis instrument for ZhengHe presents some novel aspects; if flown and the mission encounters an active body, it will combine momentum and size measurements via piezo-electric microbalances and a laser curtain, as used by Rosetta's Grain Impact Analyser and Dust Accumulator (GIADA), with microscopic imaging and a first-order assessment of composition (volatile vs. refractory fraction) by heating (evaporating) collected particles and measuring the resulting change in mass (*Zhao et al.,* 2022). A subsurface penetrator has also been proposed for this mission (*Tang et al.,* 2021) (see section 5.6.3).

Note added in proof: This mission is now called Tianwen 2 and is confirmed for launch in 2025.

5.3. Comet Interceptor

The primary goal of the ESA-led Comet Interceptor mission is to visit a "pristine" comet (*Snodgrass and Jones,* 2019; *Jones et al.,* 2024). Due to launch in 2029, this will be the first of ESA's F-class missions, a relatively short-development-time (for ESA) program. Despite a budget an order of magnitude smaller and a wet launch mass less than ~1000 kg, Comet Interceptor will go beyond what ESA achieved with Giotto and Rosetta in some ways, such as the first multi-point *in situ* investigation of a cometary coma and the inclusion of thermal IR and more extensive polarimetric imaging.

Comet Interceptor will be a flyby mission, targeting a population that has yet to be visited: LPCs with a source region in the Oort cloud. If possible, it will target a dynamically new LPC (i.e., one that hasn't previously had a perihelion in the inner solar system), but in all cases the target comet will have experienced very little insolation-driven evolution on past perihelion passages, and is expected to have a relatively pristine surface with little or no devolatized layer. The main challenge associated with visiting a LPC is that the mission cannot be designed around a well-characterized target on a known orbit. Suitable LPCs have orbital periods of many thousands of years and are therefore only discovered months or years before their perihelion. The Comet Interceptor concept involves launch before the target is known. ISEE-3/ICE had arguably demonstrated this concept already, albeit with a spacecraft not specifically designed for comet characterization. ISEE-3 had operated at the Sun-Earth L1 for several years before being directed to 21P and then upstream of 1P. *Hewagama et al.* (2018) have proposed a similar "waypoint" approach for a CubeSat-sized comet mission. Comet Interceptor will wait in a halo orbit around the Sun-Earth L2 point, a favored location for modern space telescopes, and will depart from there to meet its chosen comet once a suitable one is found and a transfer orbit calculated. The number of possible targets will depend on the capability of the spacecraft (specifically, the maximum possible change of velocity, Δv, achievable by its propulsion system and available fuel), the length of time that the mission is prepared to wait, and the amount of warning time available for optimized transfer trajectories. Simulations suggest a high probability of a suitable target being reached within the nominal six-year lifetime of the mission (*Sánchez et al.,* 2021). Should no LPC target be found within an acceptable mission lifetime, Comet Interceptor will instead be sent to encounter one of eight possible backup targets [known JFCs (*Schwamb et al.,* 2020)], which will not achieve the primary mission goal of exploring a pristine surface, but are interesting in their own right (including comets thought to be at the end of their lifetimes, and either almost dormant, as in the case of 289P/Blanpain, or disintegrating via multiple fragmentation events, such as 73P). Calculations have also been made that suggest Comet Interceptor will have a low (but non-zero; ~5%) chance of encountering an interstellar comet (*Hoover et al.,* 2022) (see section 5.5).

Aside from its new target type and launch-before-discovery approach, Comet Interceptor will be novel as the first multi-point exploration of a coma, with three separate spacecraft: a "mothership" and two small-sat probes, one of which will be provided by JAXA. The probes (currently designated B1 and B2) will be released from the main spacecraft (designated A) ~24 hr before closest approach, and will pass closer to the nucleus than spacecraft A, which will perform a relatively distant flyby at around 1000 km closest approach distance. This approach allows for sampling the coma on three different trajectories, and also combines a low-risk flyby for A with "high-risk/high-reward" measurements from the B probes, which are designed to be short-lived and expendable, permitting a less risk-averse approach than is normally found in spacecraft operations. All three spacecraft will carry magnetometers to investigate the three-dimensional structure of the comet's interaction with the solar wind, and cameras to obtain different viewing geometries of the nucleus and inner coma, while there will be dust detectors on both ESA-provided spacecraft (A and B2) and ion and/or neutral mass spectrometers for *in situ* composition measurement on A and B1. These multi-point measurements will be complemented by a high-resolution visible color camera, infrared imaging/spectroscopy in the 1–25-μm range, and fields and energetic particle measurements on spacecraft A; wide- and narrow-angle visible and far UV (Ly-α) cameras on B1; and all-sky polarimetry and a forward looking visible camera on B2. It is expected that the highest-resolution images (~10 m pixel^{-1}) will come from spacecraft A, despite it being at the largest distance at closest approach, as the mothership will carry larger instruments, including the main science camera Comet Camera (CoCA), which is based on the Colour and Stereo Surface Imaging System (CaSSIS) camera onboard the ESA Mars orbiter Trace Gas Orbiter (*Thomas et al.,* 2017). The probes instead rely on advances in instrument miniaturization from the small-sat and CubeSat developments of recent years, particularly on the JAXA B1 probe, which reuses the UV camera developed for the PRoximate Object Close flYby with Optical Navigation (PROCYON) small-sat technology demonstration mission (*Shinnaka et al.,* 2017).

Comet Interceptor will encounter its comet near 1 au from the Sun and as it crosses the ecliptic plane; for both fuel and thermal design reasons, the spacecraft is not designed to travel very far from Earth's orbit. Since LPCs can have high inclinations (including retrograde orbits) and will be near their perihelion at the encounter, the relative velocity of the flyby could be high (up to 70 km s^{-1}), providing a number of technical challenges that the spacecraft design must meet, including adequate shielding and maintaining the nucleus in the fields of view of the various cameras. Spacecraft A is expected to employ a rotating mirror to track the nucleus while keeping shields toward the dust flow, as was done by Stardust. Probe B2 will be spin-stabilized, akin to a mini-Giotto, which it will take advantage of to scan the full sky as it rotates, with the Entire visible Sky (EnVisS) polarimeter looking out of one side. There is expected to be limited communication with Earth during the flyby itself, depending on the geometry of the encounter. Data from all three spacecraft will be stored onboard spacecraft A and transmitted in the weeks or months after encounter.

5.4. Future Concepts

Looking further into the future, there are two main routes to advancing the spacecraft exploration of comets: visiting different types of comets and performing different types of experiments, enabled by advances in spacecraft technology. Such missions will vary in cost and size depending on how challenging the targets are (mostly governed by the Δv required to reach them) and the complexity of the spacecraft and instruments, progressing from "simple" single-spacecraft flybys to the significant challenges of cryogenic sample return of nucleus ices. The latter remains a distant prospect but is widely regarded as the holy grail of comet exploration. Both the Ambition concept white paper presented to ESA's Voyage 2050 call (*Bockelée-Morvan et al.,* 2021) and the Cryogenic Comet Sample Return white paper presented to NASA's 2023–2032 Planetary Science and Astrobiology Decadal Survey call (*Westphal et al.,* 2021) explored the current state of the field and the technological steps needed to reach cryogenic sample return (see section 5.6). The Ambition paper also considered a wide range of potential mission and target types, identifying opportunities for different scales of mission that may be proposed in upcoming calls. In addition, over the past decade, NASA has funded two cometary Phase A concept studies, one in the Discovery class [Comet Hopper (*Clark et al.,* 2008)] and one in the New Frontiers class [Comet Astrobiology Exploration SAmple Return (CAESAR) (*Squyres et al.,* 2018)], emphasizing the key role that comets play in planetary exploration and providing insight into the origin of life. Figure 18 presents a grid of comet type and mission class, listing the past or planned missions described earlier in this chapter, and showing where the most promising future opportunities are. Some specific concepts in these areas are described in the following subsections. We note that the list of future missions we present is incomplete — there will always be concepts studied and proposed that we are not aware of, and we do not try to list all missions, but rather a selection covering the range of possibilities. We mention proposed missions by name where there are published descriptions in the literature, but note that there are very different norms regarding advertising competition-sensitive details of preselected mission concepts in different communities, due to differences in the selection processes of the various space agencies (e.g., NASA vs. ESA), and related commercial sensitivities. Inclusion or exclusion of any particular mission concept should not be read as endorsement or criticism.

5.5. Target Types

5.5.1. Jupiter-family comets and Centaurs. Of the two major comet classes — JFCs with orbits near the ecliptic and a source region in the Kuiper belt and scattered disk, and nearly-isotropic LPCs or Halley-types (HTCs) from the Oort cloud — the former present much easier spacecraft targets. [The distinction between LPCs and HTCs is on orbital period (longer or shorter than 200 years respectively), but from a mission point of view they present similar challenges (potentially high inclinations), so we group them together for the purposes of this chapter. We note that HTCs may not have an Oort cloud source after all (*Levison et al.,* 2006; *Fernández et al.,* 2016); see also the chapters by Kaib and Volk and Fraser et al. in this volume.] This is due to their typically low inclinations and short periods, with aphelia typically within the orbit of Jupiter. As demonstrated by Rosetta, it is feasible with current launchers to match these orbits and, with modern solar panels such as those used by Juno (*Matousek,* 2007) or Lucy (*Olkin et al.,* 2021), perform scientific operations all the way to aphelion. JFCs have well-characterized orbits and all classes of mission can be planned well in advance. This is reflected in the dominance of these comets in the list of previous missions. However, there remain interesting cases for future missions to JFCs: different types of mission (see section 5.6), or exploring the long-term evolution of these bodies by visiting split or fragmented comets or related populations, e.g., Centaurs or extinct comets.

Centaurs are thought to be precursors to JFCs as objects transitioning from outer solar system orbits. They have a wide range of orbits and sizes (see the chapter in this volume by Fraser et al.), and consequently vary considerably in how feasible a mission to one would be. Feasible medium-sized missions have been proposed (e.g., *Harris et al.,* 2019) for flyby or rendezvous with 29P/Schwassmann-Wachmann 1, which has a relatively circular orbit just beyond that of Jupi-

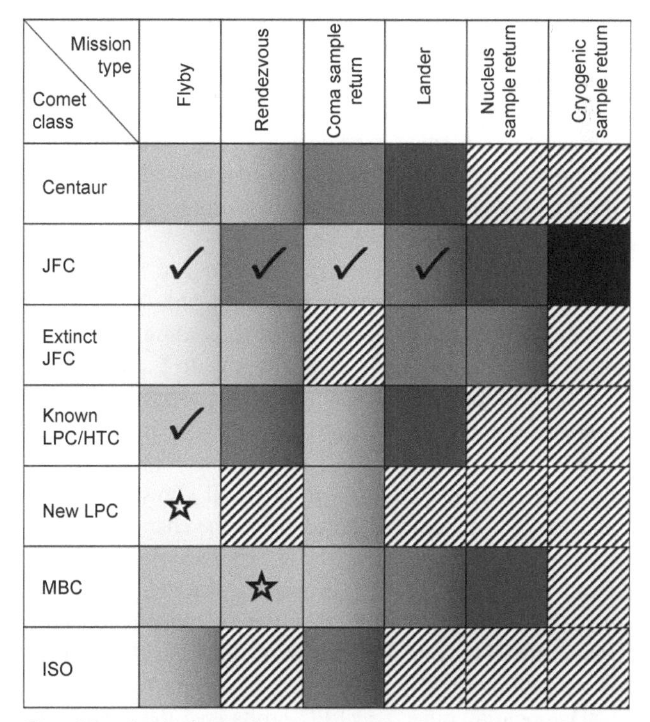

Fig. 18. Grid of mission type and comet class, indicating the approximate level of mission (cost/difficulty) needed to achieve the combination, from lighter shades (cheaper/easier) to darker (most challenging), including gradients between shades where a variety of approaches or targets give a wide range of possibilities. Hatched squares indicate that such a combination [e.g., nucleus sample return (SR) from an interstellar object] seems unfeasible with current or near future technology, or within the cost caps of the largest current interplanetary missions. Checks mark those combinations already tried (various JFC and Halley flybys described in section 3, JFC coma sample return by Stardust, and JFC rendezvous and landing by Rosetta and Philae) and stars indicate those selected for future missions (flyby of a new LPC and rendezvous with a MBC, by Comet Interceptor and ZhengHe, respectively). After *Bockelée-Morvan et al.* (2021).

ter, and is particularly interesting as it is likely in the process of transitioning from a Centaur to a JFC (*Sarid et al.,* 2019). More distant Centaurs, e.g., the large (~200-km-diameter) ringed worlds Chariklo or Chiron, with orbits between 8 and 18 au from the Sun, have been proposed as mission targets as more accessible Kuiper belt objects, rather than due to their relationship with comets as such, but would be more costly: e.g., NASA Frontiers class for a Deep Impact-style flyby and impactor combination (*Howell et al.,* 2018) or multi-Centaur tour (*Singer et al.,* 2021).

Comets at the other end of their active lifetimes can present more accessible targets. A large number of asteroids in cometary orbits (ACOs) have been identified, including many in near-Earth space, some of which may have low Δv transfers for flyby, rendezvous, or even sample return (see the chapter in this volume by Jewitt and Hsieh). As these objects are thought to be dormant or extinct comets, with their near-surface ices either exhausted or buried at sufficient depth to extinguish activity, they offer an opportunity to conduct a *relatively* easy comet mission, without the concerns for spacecraft safety caused by activity, but a scientifically compelling one, to understand what it is that causes comet activity to cease. However, the most significant uncertainty with such a proposal is in being sure of the identification of any individual ACO as definitely a previously active comet, rather than a body with an asteroidal origin but somehow scattered onto a comet-like orbit.

5.5.2. Long-period comets. As noted in section 5.3, LPCs or HTCs are much more challenging targets to reach, as they can have high inclinations (including retrograde orbits) and relatively high speeds relative to Earth and any spacecraft as they pass through the inner solar system. For this reason, only flyby missions to LPCs seem reasonably feasible in the short term. The 1P missions took advantage of a long-predicted return of a rare well known comet in this class, while Comet Interceptor has the chance to visit a new comet by implementing the use of a waypoint. Further sampling the diversity of LPCs may be rewarding, especially given the recent discoveries of activity at large heliocentric distances well beyond the water snow line [e.g., C/2017 K2 (*Meech et al.,* 2017)] and low-activity or inactive "Manx" or "Damocloid" objects in LPC orbits (see the chapter by Jewitt and Hsieh in this volume). Discussions of a return to 1P at its 2061 perihelion, to rendezvous or even return a sample, highlight the current need for significant technology development to achieve this step (*Ozaki et al.,* 2021). While another mission at the next return of this most famous comet has strong cultural and historic appeal, and its predictability decades in advance inspires some of the long-term planning needed to develop the necessary technology, there are perhaps easier LPCs/HTCs to consider. Although it comes close to Earth, 1P's highly retrograde orbit ($i = 162°$) will always mean very high energy requirements to match its velocity. A concept was investigated to address this issue for 1P's 1986 apparition through the use of a solar sail (*Friedman,* 1976); this technology deserves to be considered further.

5.5.3. Main-belt comets. In the decades since the last comet mission was launched, it has become apparent that there may be a third class of comet orbit, and source reservoir, within the solar system: the main-belt comets (MBCs), which suggest an unexpected significant population of ice-rich bodies in the outer asteroid belt (see the chapter by Jewitt and Hsieh in this volume). With very low activity levels and relatively circular orbits, MBCs are reasonably accessible for rendezvous missions, particularly for missions propelled by ion engines or other low thrust technology, as employed by the Dawn mission to Vesta and Ceres (*Thomas et al.,* 2011). As described in section 5.2, an upcoming Chinese mission may make the first visit to a MBC as a secondary target, but may not encounter an active body. A number of other missions have been proposed to other agencies, but not yet selected. As a potential guide to future proposals in this area, we identify some of the approaches and technologies proposed in those with publicly available information. These include, for example, the NASA Discovery class Proteus rendezvous mission concept (*Meech and Castillo-Rogez,* 2015); the Castalia rendezvous concept proposed as an ESA medium-sized (M-class) mission and described in detail by *Snodgrass et al.* (2018); and a Stardust-like coma sample return mission, Caroline, which was also previously proposed as an ESA M-class mission (*Jones et al.,* 2018). The main science drivers for most MBC missions rely on *in situ* sampling of their volatiles, to confirm that MBC activity is indeed driven by sublimating water ice, which has not yet been possible via remote observation (*Snodgrass et al.,* 2017a), and to reveal their composition and thus test solar system formation and evolution models, and whether or not the asteroid belt could have been a significant source of Earth's water. The very low activity levels of MBCs make this challenging — to build up the necessary signal-to-noise to make a sensitive detection via mass spectroscopy, Castalia proposed extended periods of automated close-proximity station keeping. This would have been a significant technological advance from Rosetta, where navigation with ground control in the loop meant reaction times of a few days, but automated onboard navigation is now a mature spacecraft technology, to the point of enabling precision landings on asteroids or Mars (*Ge et al.,* 2019). MBCs also differ from other comets in that their ices may be buried deeper under a refractory crust, or exposed in only a small and recently uncovered active area, and locating volatiles must constitute an important part of the mission. Surface (or subsurface) exploration with landers, drills, or penetrators would also be very interesting, if they can be targeted to volatile-rich areas, although estimates of the depth of buried ice vary from centimeters to tens of meters, depending on location and the individual MBC's age, orbit, and pole orientation (*Snodgrass et al.,* 2017a).

5.5.4. Interstellar objects. Finally, recent years have seen the discovery of the first interstellar comets, which would naturally be very-high-priority spacecraft targets, as they represent an opportunity to sample extrasolar material without the massive technological challenges or centuries-timescales of an interstellar probe. Our knowledge of interstellar objects (ISOs) and the motivations and concepts for a mission to one are discussed in the chapter by Fitzsimmons et al. in this volume, and were explored by a recent NASA Planetary Science Summer Seminar cohort (*Moore et al.,* 2021), but we briefly discuss the technology requirements for such a probe. In terms of mission design, ISOs are similar to new LPCs, with high velocities and potentially high inclinations, and consequently present similar challenges. Fast flyby missions, perhaps combined with a Deep Impact-style impactor (*Moore et al.,* 2021), appear to be the limit of what is feasible. A waiting spacecraft, similar to Comet Interceptor in concept, seems the most likely way to reach an ISO, but the current poor constraints on the total population size mean that it is hard to judge the necessary waiting times or Δv requirements. Given the limited heliocentric distance range in which Comet Interceptor will be able to operate, it seems that it will have to be exceedingly lucky to be able to encounter an ISO within its planned six-year lifetime and $\Delta v < 1$ km s^{-1}. A more capable version of a similar concept could be imagined for a larger-budget mission class (e.g., ESA M-class or NASA Discovery), as the true size of the ISO population becomes clearer following the expected decade of discovery with the Vera C. Rubin Observatory's Legacy Survey of Space and Time (LSST) (*Seligman and Laughlin,* 2018). Alternatively, "launch on discovery" concepts, rather than waiting in space, have been proposed to encounter ISOs (*Seligman and Laughlin,* 2018; *Moore et al.,* 2021), although these also require quite rapid reaction (see section 5.6.5). Clearly cometary ISOs like 2I/Borisov, if discovered inbound, present the best chances of success for interception, while the available warning times for another small 1I/'Oumuamua-like object, with no visible coma, remain impractically short even with LSST (one or two months at best). There are suggestions to launch missions to chase after departing ISOs, catching up to them at very large distance from the Sun (100 au or more) decades after their perihelion, but these are limited by both available spacecraft technology (propulsion, power, communications) and the navigation and operational practicalities of even locating such a small object so far from the Sun while approaching at tens of kilometers per second (*Hein et al.,* 2019, 2022). Such a mission does not appear feasible in the near term.

5.6. Technology for Future Comet Missions

There still remains much to be learned about the more accessible JFCs, and new missions to these, or even returns to previously visited comets, are well justified. In this section we review current and near-future technology that could enable the next comet missions to JFCs and beyond (see also *A'Hearn,* 2017; *Thomas et al.,* 2019).

5.6.1. Landing on the nucleus. As (cryogenic) sample return remains the ultimate goal of comet exploration, a key technology requirement is the ability to land on the nucleus safely: either in a brief "touch and go" maneuver, as used by the asteroid missions Hayabusa, Hayabusa 2 and OSIRIS-REx, and proposed at a comet by the Triple-F mission (*Küppers et al.,* 2009), or for a longer period of measurement/drilling. Landers also have a lot of potential to advance our knowledge of cometary activity, independent of any sample return attempt, through local measurements of the surface (micro)physics, as was planned for Philae. Sampling multiple locations on the nucleus would be particularly valuable for this, whether using one mobile lander or a network of static ones (the latter also having the advantage of redundancy and being able to perform seismology or CONSERT-like measurements between probes (*Thomas et al.,* 2019). Future mission architectures to consider include very low speed landers, rovers, and multi-point landers or hoppers, like the proposed implementation of NASA's Phase A Comet Hopper mission concept (*Clark et al.,* 2008), and a great deal of autonomous operations. Taking into account the microgravity environment at the surface of a comet and balancing it with the desire to minimize surface alteration upon arrival, various spacecraft designs coupled with landing and holding strategies have been evaluated in concept studies (*Clark et al.,* 2008; *Ulamec et al.,* 2011) or executed without success to date (section 4.4). With our knowledge of the surface strength and regolith properties still ranging an order of magnitude or more after Philae's inconclusive results (*Groussin et al.,* 2019), there remains a need for replicating and modeling porous regolith in microgravity conditions on Earth for landing tests, and to assess the particle movement near the output of spacecraft thrusters and the thermal and electrostatic response of the regolith to an approaching spacecraft. Low-speed landers will reduce the potential for bouncing on the surface and numerous small thrusters oriented at an angle from the subspacecraft point will undoubtedly prove to be beneficial.

Once in the vicinity of the comet nucleus or safely landed, interacting with the comet's surface is also technologically challenging, whether to rove or hop around to explore, to gather a surface sample, or to penetrate and sample the subsurface. There are still many unknowns in each individual comet's properties and orders of magnitude in possible strength regimes and surface topographies to design for. A significant difference between classes of comet (e.g., possibly asteroid-like MBCs vs "pristine" LPCs) could also be expected. Mission designs employing smaller articulating solar panels or nuclear power sources will enable agile traversal of the diverse terrains observed on the comets visited thus far. Additionally, missions powered by nuclear sources will enable operation at larger heliocentric distances with fewer limitations on shared spacecraft resources as well as in non-illuminated regions of a comet, the most likely location to find surface ice (*Sunshine et al.,* 2006; *De Sanctis et al.,* 2015; *Filacchione et al.,* 2016).

5.6.2. Sample return. The Stardust mission showed the immense value of laboratory analysis of comet samples. The next step in this field should be to collect a sample from the nucleus, although a more advanced coma sample return mission could also be imagined. Stardust and Rosetta showed that coma dust includes an easily fragmented and fragile component, which would have to be collected very gently (e.g., at near-zero relative velocity by a rendezvous mission) to obtain a sample that retains structural information. Measuring the refractory-to-ice abundance in the coma would also be an important, but challenging, goal for such a sample.

For nucleus sample return, there is a need for community guidelines and agreement on the depth to which a sample should be collected to have the greatest science return, from the exposed regolith (*Nakamura-Messenger et al.,* 2021) to reaching the least-altered high-volatility subsurface ice reservoirs. As the surface composition of comets remains uncertain even after Rosetta, there is value in the former approach, which formed the basis of the CAESAR proposal that was the runner-up in the 2019 NASA New Frontiers call (*Squyres et al.,* 2018). A surface regolith sample, which could be returned to Earth in non-cryogenic conditions, would allow for detailed investigation of thermally processed material and surface chemistry at a level beyond what Philae could have hoped to achieve, even if it had operated perfectly, and could be compared to similar asteroid samples from Hayabusa 2 and OSIRIS-REx. However, a surface sample return mission that scrapes, sucks, or brushes up regolith makes detailed chemistry measurements at the cost of knowledge of the physical structure of the surface material. A more ambitious sample return approach would be to collect a core that preserves both composition and its stratigraphy: information on the size and shape of grains, the bulk strength, porosity, and relative locations of different components (see *Thomas et al.,* 2019, for a more detailed discussion). The depth to which a core must reach to sample various ices is not well determined, but is expected to be on the order of meters or possibly considerably more to reach "pristine" layers (see discussion in the chapter by Guilbert-Lepoutre et al. in this volume). Comet missions can look to technological advances from ground-penetrating radar designed to characterize the subsurface structure 3–10 m deep at Mars (*Ciarletti et al.,* 2017) to first locate the true ice depths. This would be a useful measurement for future rendezvous missions whether or not they also attempt to land on or sample the nucleus [e.g., the CORE mission concept, an orbiter whose primary payload is a radar mapping experiment to study nucleus interior structure (*Asphaug,* 2015)]. Drilling into the ice, and extracting an unaltered core, would be technically challenging. The SD2 drill on Philae was designed to reach a depth of 23 cm and retrieve material (but not an intact core) for composition analysis (*Finzi et al.,* 2007), while Triple-F proposed to extract 50-cm cores in touch-and-go landings (*Küppers et al.,* 2009). Neither of these has been successfully demonstrated at a comet; obtaining meters-long ice cores will require significant advances of these technologies, but may benefit from current developments for ocean world applications (*Dachwald et al.,* 2020). The largest drill being developed for a near-future flight opportunity is the 1-m-long TRIDENT drill (*Zacny et al.,* 2021), expected to launch to the Moon no earlier than 2025 on a large lander or rover (450–2000 kg; a very different class of mission than the 100-kg Philae).

Cryogenic sample return missions face the challenge not only of acquiring a sample with context and retaining stratigraphy, but also preserving it during the return flight, Earth atmospheric entry, descent, landing, recovery, and curation. The latter is a significant challenge in its own right; we do not currently have suitable facilities, but progress is being made (*McCubbin et al.,* 2021). The sample would need to be protected to minimize and ideally prevent alteration that would affect interpretation (this includes shock, vibration, and heat). Spacebased retrieval (i.e., docking with a crewed space station, and subsequent "gentle" return of the sample with the astronauts) has even been considered to minimize some of these effects. There have been recent technological advances in cryogenic flight systems and instrumentation, as well as in cryogenic sample curation, handling, and analysis (*Westphal et al.,* 2021). *Küppers et al.* (2009) proposed a passive system that would keep the sample below 133 K during cruise and 163 K during reentry, which would return a valuable sample at relatively low cost, even if some ices were lost. *Bockelée-Morvan et al.* (2021) suggest that temperatures must be kept below 90 K, and pressure at 1 bar, to preserve most volatile ices (H_2O, CO_2, HCN). Even lower temperatures or higher pressures would be necessary to keep CO stable. *Westphal et al.* (2021) argue that sample collection and transport at 60 K is necessary to prevent an amorphous-amorphous phase transition in water ice that occurs at 80 K, and that this is considered achievable with current flight system technologies. While challenging, it appears that cryogenic sample preservation is technically possible in the near future. In the end, a small advantage that comet missions have once a sample is collected and contained is that operating in low-gravity environments will allow for easier departure.

5.6.3. Subsurface penetrators. Penetrator probes, which reach subsurface layers of a planetary body by burying themselves in a high-speed impact, rather than a gentle landing and then drilling, have been proposed for some time (*Lorenz,* 2011). Although typically suggested as a solution for larger bodies such as the Moon or icy moons of the outer planets (e.g., *Gowen et al.,* 2011), where gravity makes soft-landing far more challenging, they were studied in detail for comets as part of the NASA-led Comet Rendezvous Asteroid Flyby mission, which was canceled in the early 1990s (*Swenson et al.,* 1987; *Boynton and Reinert,* 1995). Penetrator probes rely on sufficiently robust instrumentation that can survive the sudden deceleration as the probe hits the ground, typically at $\sim$100 m s^{-1}, but can reveal physical structure of the surface by measuring this deceleration. They can measure subsurface composition, obtain (microscopic) images of the surrounding material, and measure physical properties such as temperatures, conductivity, etc., *in situ.* Although a penetrator has yet to be successfully flown on a planetary mission, they have been studied in detail, including high-speed-impact tests of instrumentation on Earth, and similar technology has been widely developed for military purposes;

penetrators are therefore a realistic and mature technology that would be of value on a suitable rendezvous mission. One may be included on the ZhengHe mission to 311P (*Tang et al.,* 2021), and, if it is, probably represents the best chance for this mission to detect any subsurface ice.

5.6.4. Small satellites and multi-point measurements. A relatively new mission architecture is proving to be beneficial in planetary sciences and has application for comets: the development of small, standardized spacecraft (e.g., CubeSats). The advantage of small, fast, and relatively cheap missions should not be overlooked. Small-sats can play a role in reconnaissance and rapid response, and could be employed as a fleet. They can also be used to address focused questions and may be the only way in the near future to explore an ISO or LPC up close, either as released probes as part of a larger mission (e.g., Comet Interceptor) or even as independent spacecraft (*Clark et al.,* 2018). A CONTOUR-like tour (*Birch et al.,* 2020) and Comet Interceptor-like "waiting in space" LPC mission (*Hewagama et al.,* 2018) have been proposed within NASA's Small Innovative Missions for Planetary Exploration (SIMPLEx) program.

The idea of having a large number of independent spacecraft at a comet has many advantages, particularly for coma sampling and measurements of the plasma environment. Comet Interceptor will perform the first multi-point sampling, but the only instrument common to all three spacecraft is a magnetometer, and it is a fast flyby mission. A plasma-science-dedicated rendezvous mission with at least four identical spacecraft measuring the interaction between a comet and the solar wind has been proposed as the next big step forward for this field (*Goetz et al.,* 2021). While this concept assumes four small- to medium-sized independent spacecraft, it is also interesting to consider the possibilities presented by advances in technology for very small platforms. The Comet Interceptor-released probes, while small, will still be bespoke spacecraft with rigorous design and build standards. The use of real CubeSats (i.e., standardized units built with relatively cheap off-the-shelf parts) in deep space has been demonstrated to be feasible by the MarCO mission to Mars (*Baker et al.,* 2019). Even smaller "ChipSats" have been demonstrated to be capable of making simple single measurements at many separate locations in Earth orbit and have been proposed for small-body missions (*Ledbetter et al.,* 2018). Many ($N \gg 1$) point measurements with a "coma swarm" would provide fascinating insights into the temporal and spatial variations within a coma, whether simply by tracing dynamics and variations in coma structure (dust or gas density, temperature, or magnetic fields), or even making multiple *in situ* composition measurements [e.g., with CubeSat-sized mass spectrometers based on Philae Ptolemy technology (*Wright et al.,* 2007)]. Having a large number of small probes also enables a paradigm shift in build and operation costs that is useful in a cometary environment — with sufficiently large N, failure *is* an option, and results are assured even if some of the probes fail or are lost soon after deployment.

5.6.5. Rapid reaction. One mission type that is at a true disadvantage in the current program architecture for most, if not all, space agencies is a rapid response mission (*Moore et al.,* 2021). These missions would undoubtedly result in high-reward science and could be in response to any number of phenomena or target type, e.g., a major splitting event of an JFC on its inbound perihelion leg, a newly discovered inbound LPC, or an ISO quickly passing through our solar system. Implementations of rapid response missions may have inherent schedule risk to mission success, or risk failing to achieve their full science potential due to launch vehicle, trajectory, and orbital constraints. Rapid response missions could be executed via a very short and rushed design, build, test, and launch phase, with the major disadvantages of high likelihood of schedule slip, ballooning costs, and potential lack of dedicated launch vehicle or timely ride-share launch. In contrast, rapid response missions could take advantage of ground- or spacebased storage and thus allow for a more typical and lengthy design and build phase. Having a pre-built spacecraft in storage or in a parking orbit in space waiting for a target has drawbacks, manifesting as having been designed with a target agnostic approach, requiring a minimum intact skeleton team for a long period of low to no mission activity, utilizing technology that is not state-of-the-art by the time of target discovery, degrading parts in orbit, and having the potential need for a dedicated launch vehicle. The key to justifying such a mission will be some reasonably reliable statistical knowledge of how long the wait may have to be, and that the (presumably rare) opportunity requiring rapid reaction is worth the associated cost.

6. SUMMARY

There has been fantastic progress in spacecraft exploration of comets since the beginning of this field around the time of 1P's last return. As this comet passed aphelion in 2023, it is instructive to look at what we have achieved and what we may look forward to in the next half of its orbit before its 2061 return. Reviewing the three major themes we described in the introduction (formation, composition, and activity), we see significant progress with each past mission and clear priorities for future missions and technology. Our understanding of nucleus formation and structure has primarily advanced by the increasing resolution of surface imaging with each mission to date, revealing an apparent preference for bilobed shapes and a wide variety of surface morphologies. The next step needs to be probing microphysical details of the nucleus and its interior structure. Composition measurements *in situ* have examined gas, dust, and ions, including isotopic abundances, and Stardust enabled incredibly detailed laboratory analysis of solid grains — but increasing the diversity of comets visited and sample return from the nucleus are promising next steps to take, especially to investigate how the volatile, organic, and refractory components are associated. Finally, processes linked to activity have been localized but fundamentally remain a puzzle that will likely require lander missions, capable of temporally studying the physical, thermal, and chemical properties of the nucleus (sub)surface at microscopic scales, to solve. There are still many knowledge gaps that require *in situ* measurement to explore: There are strong cases for missions to comets to better understand nucleus surfaces at small scales, their activity, and the diversity between comets from different source regions or at different evolutionary stages. The return of frozen ices from a comet nucleus remains the highest-priority mission objective, yet experiences from flyby, impactor, coma sample return, rendezvous, and landed missions in the past decades have shown that this will be highly challenging. Returning to 1P, and/or cryogenic sample return, in the distant future necessitates long-term planning and investment in technology now. In the meantime, there are still many questions that can, and should, be addressed with a range of smaller missions, taking advantage of current advances in small, fast, and cheap spacecraft. No matter what the mission design and target, the science objectives should be compelling and concise, and drive the spacecraft design and technological advances needed for future cometary exploration.

Acknowledgments. The authors thank the referees, J. Sunshine and H. U. Keller, for their helpful and constructive reviews of this chapter. We thank A. Donaldson for providing Fig. 11. C.S. and G.H.J. acknowledge support from the UK Science and Technology Research Council and UK Space Agency.

REFERENCES

Accomazzo A., Lodiot S., and Companys V. (2016) Rosetta mission operations for landing. *Acta Astronaut., 125,* 30–40.

A'Hearn M. F. (2017) Comets: Looking ahead. *Philos. Trans. R. Soc. A, 375,* 20160261.

A'Hearn M. F. and Johnson L. N. (2015) Deep Impact and related missions. In *Handbook of Cosmic Hazards and Planetary Defense* (J. Pelton and F. Allahdadi, eds.), pp. 513–534. Springer, Cham.

A'Hearn M. F., Schleicher D. G., Millis R. L., et al. (1984) Comet Bowell 1980b. *Astron. J., 89,* 579–591.

A'Hearn M. F., Millis R. C., Schleicher D. O., et al. (1995) The ensemble properties of comets: Results from narrowband photometry of 85 comets, 1976–1992. *Icarus, 118,* 223–270.

A'Hearn M. F., Belton M. J. S., Delamere A., et al. (2005a) Deep Impact: A large-scale active experiment on a cometary nucleus. *Space Sci. Rev., 117,* 1–21.

A'Hearn M. F., Belton M. J. S., Delamere W. A., et al. (2005b) Deep Impact: Excavating Comet Tempel 1. *Science, 310,* 258–264.

A'Hearn M. F., Belton M. J. S., Delamere W. A., et al. (2011) EPOXI at Comet Hartley 2. *Science, 332,* 1396–1340.

Alfvén H. (1957) On the theory of comet tails. *Tellus, 9,* 92–96.

Altwegg K., Balsiger H., Bar-Nun A., et al. (2015) 67P/Churyumov-Gerasimenko, a Jupiter family comet with a high D/H ratio. *Science, 347,* 1261952.

Altwegg K., Balsiger H., Bar-Nun A., et al. (2016) Prebiotic chemicals — amino acid and phosphorus — in the coma of Comet 67P/Churyumov-Gerasimenko. *Sci. Adv., 2,* e1600285.

Altwegg K., Balsiger H., Hänni N., et al. (2020) Evidence of ammonium salts in Comet 67P as explanation for the nitrogen depletion in cometary comae. *Nature Astron., 4,* 533–540.

Ashman M., Barthélémy M., O'Rourke L., et al. (2016) Rosetta science operations in support of the Philae mission. *Acta Astronaut., 125,* 41–64.

Asphaug E. and the CORE Science Team (2015) Comet Radar Explorer. In *Conference on Spacecraft Reconnaissance of Asteroid and Comet Interiors,* Abstract #6044. LPI Contribution No. 1829, Lunar and Planetary Institute, Houston.

Attree N., Jorda L., Groussin O., et al. (2019) Constraining models of activity on Comet 67P/Churyumov-Gerasimenko with Rosetta trajectory, rotation, and water production measurements. *Astron. Astrophys., 630,* A18.

Auster H. U., Apathy I., Berghofer G., et al. (2007) ROMAP: Rosetta Magnetometer and Plasma Monitor. *Space Sci. Rev., 128,* 221–240.

Baker J., Colley C. N., Essmiller J. C., et al. (2019) MarCo: The first interplanetary CubeSats. *EPSC-DPS Joint Meeting 2019,* 2009.

Ballard S., Christiansen J. L., Charbonneau D., et al. (2011) A search for additional planets in five of the exoplanetary systems studied by the NASA EPOXI mission. *Astrophys. J., 732,* 41.

Balsiger H., Altwegg K., Bochsler P., et al. (2007) Rosina — Rosetta Orbiter Spectrometer for Ion and Neutral Analysis. *Space Sci. Rev., 128,* 745–801.

Barucci M. A., Fulchignoni M., Ji J., et al. (2015) The flybys of asteroids 2867 Steins, 21 Lutetia, and 4179 Toutatis. In *Asteroids IV* (P. Michel et al., eds.), pp. 433–450. Univ. of Arizona, Tucson.

Belton M. J. S. and A'Hearn M. F. (1999) Deep sub-surface exploration of cometary nuclei. *Adv. Space Res., 24,* 1167–1173.

Belton M. J. S. and Melosh J. (2009) Fluidization and multiphase transport of particulate cometary material as an explanation of the smooth terrains and repetitive outbursts on 9P/Tempel 1. *Icarus, 200,* 280–291.

Belton M. J. S., Feldman P. D., A'Hearn M. F., et al. (2008) Cometary cryo-volcanism: Source regions and a model for the UT 2005 June 14 and other mini-outbursts on Comet 9P/Tempel 1. *Icarus, 198,* 189–207.

Belton M. J. S., Meech K. J., Chesley S., et al. (2011) Stardust-NExT, Deep Impact, and the accelerating spin of 9P/Tempel 1. *Icarus, 213,* 345–368.

Belton M. J. S., Thomas P., Carcich B., et al. (2013a) The origin of pits on 9P/Tempel 1 and the geologic signature of outbursts in Stardust-NExT images. *Icarus, 222,* 477–486.

Belton M. J. S., Thomas P., Li J.-Y., et al. (2013b) The complex spin state of 103P/Hartley 2: Kinematics and orientation in space. *Icarus, 222,* 595–609.

Benna M., Mahaffy P. R., Grebowsky J. M., et al. (2015) Metallic ions in the upper atmosphere of Mars from the passage of Comet C/2013 A1 (Siding Spring). *Geophys. Res. Lett., 42,* 4670–4675.

Bibring J. P., Lamy P., Langevin Y., et al. (2007a) CIVA. *Space Sci. Rev., 128,* 397–412.

Bibring J. P., Rosenbauer H., Boehnhardt H., et al. (2007b) The Rosetta lander ("Philae") investigations. *Space Sci. Rev., 128,* 205–220.

Biele J., Ulamec S., Maibaum M., et al. (2015) The landing(s) of Philae and inferences about comet surface mechanical properties. *Science, 349,* aaa9816.

Bieler A., Altwegg K., Balsiger H., et al. (2015) Abundant molecular oxygen in the coma of Comet 67P/Churyumov-Gerasimenko. *Nature, 526,* 678–681.

Birch S., Hayes A., Milam S., et al. (2020) NEAT: A multi-comet flyby mission that performs discovery-level science on a SIMPLEx budget. *AAS/Division for Planetary Sciences Meeting Abstracts, 52,* #001.04.

Blum J., Gundlach B., Krause M., et al. (2017) Evidence for the formation of Comet 67P/Churyumov-Gerasimenko through gravitational collapse of a bound clump of pebbles. *Mon. Not. R. Astron. Soc., 469,* S755–S773.

Blume W. H. (2005) Deep Impact mission design. *Space Sci. Rev., 117,* 23–42.

Bockelée-Morvan D., Filacchione G., Altwegg K., et al. (2021) AMBITION — Comet nucleus cryogenic sample return. *Exp. Astron.,* DOI: 10.1007/s10686-021-09770-4.

Boehnhardt H., Riffeser A., Kluge M., et al. (2016) Mt. Wendelstein imaging of the post-perihelion dust coma of 67P/Churyumov-Gerasimenko in 2015/2016. *Mon. Not. R. Astron. Soc., 462,* S376–S393.

Boynton W. V. and Reinert R. P. (1995) The cryo-penetrator: An approach to exploration of icy bodies in the solar system. *Acta Astronaut., 35,* 59–68.

Britt D. T., Boice D. C., Buratti B. J., et al. (2004) The morphology and surface processes of Comet 19/P Borrelly. *Icarus, 167,* 45–53.

Brownlee D. E., Tsou P., Anderson J. D., et al. (2003) Stardust: Comet and interstellar dust sample return mission. *J. Geophys. Res.–Planets, 108,* 8111.

Brownlee D. E., Horz F., Newburn R. L., et al. (2004) Surface of young Jupiter family Comet 81P/Wild 2: View from the Stardust spacecraft. *Science, 304,* 1764–1769.

Brownlee D., Tsou P., Aléon J., et al. (2006) Comet 81P/Wild 2 under a microscope. *Science, 314,* 1711–1716.

Burnett D. S. (2006) NASA returns rocks from a comet. *Science, 314,* 1709–1710.

Capaccioni F., Coradini A., Filacchione G., et al. (2015) The organic-rich surface of Comet 67P/Churyumov-Gerasimenko as seen by VIRTIS/Rosetta. *Science, 347,* aaa0628.

Carr C., Cupido E., Lee C. G. Y., et al. (2007) RPC: The Rosetta Plasma Consortium. *Space Sci. Rev., 128,* 629–647.

Chesley S. R., Belton M. J. S., Carcich B., et al. (2013) An updated rotation model for Comet 9P/Tempel 1. *Icarus, 222,* 516–525.

Choukroun M., Altwegg K., Kührt E., et al. (2020) Dust-to-gas and refractory-to-ice mass ratios of Comet 67P/Churyumov-Gerasimenko from Rosetta observations. *Space Sci. Rev., 216,* 44.

Ciarletti V., Clifford S., Plettemeier D., et al. (2017) The WISDOM radar: Unveiling the subsurface beneath the ExoMars rover and identifying the best locations for drilling. *Astrobiology, 17,* 565–584.

Clark B. C., Sunshine J. M., A'Hearn M. F., et al. (2008) Comet Hopper: A mission concept for exploring the heterogeneity of comets. In *Asteroids, Comets, Meteors 2008,* Abstract #8131. LPI Contribution No. 1405, Lunar and Planetary Institute, Houston.

Clark P., Hewagama T., Aslam S., et al. (2018) Overview of Primitive Object Volatile Explorer (PrOVE) CubeSat or SmallSat concept. In *CubeSats and NanoSats for Remote Sensing II* (T. S. Pagano and C. D. Norton, eds.), p. 107690J. SPIE Conf. Ser. 10769, Bellingham, Washington.

Coates A. J. and Jones G. H. (2009) Plasma environment of Jupiter family comets. *Planet. Space Sci., 57,* 1175–1191.

Cochran A., Veverka J., Bell J., et al. (2002) The Comet Nucleus Tour (Contour): A NASA Discovery mission. *Earth Moon Planets, 89,* 289–300.

Colangeli L., Lopez-Moreno J. J., Palumbo P., et al. (2007) The Grain Impact Analyser and Dust Accumulator (GIADA) experiment for the Rosetta mission: Design, performances and first results. *Space Sci. Rev., 128,* 803–821.

Combi M. R., Mäkinen J. T. T., Henry N. J., et al. (2008) Solar and Heliospheric Observatory/Solar Wind Anisotropies observations of five moderately bright comets: 1999–2002. *Astron. J., 135,* 1533–1550.

Combi M. R., Mäkinen T. T., Bertaux J. L., et al. (2019) A survey of water production in 61 comets from SOHO/SWAN observations of hydrogen Lyman-alpha: Twenty-one years 1996–2016. *Icarus, 317,* 610–620.

Combi M. R., Shou Y., Mäkinen T., et al. (2021) Water production rates from SOHO/SWAN observations of six comets: 2017–2020. *Icarus, 365,* 114509.

Coradini A., Capaccioni F., Drossart P., et al. (2007) VIRTIS: An imaging spectrometer for the Rosetta mission. *Space Sci. Rev., 128,* 529–559.

Crow C. A., McFadden L. A., Robinson T., et al. (2011) Views from EPOXI: Colors in our solar system as an analog for extrasolar planets. *Astrophys. J., 729,* 130.

Dachwald B., Ulamec S., Postberg F., et al. (2020) Key technologies and instrumentation for subsurface exploration of ocean worlds. *Space Sci. Rev., 216,* 83.

Davidsson B. J. R. and Gutiérrez P. J. (2005) Nucleus properties of Comet 67P/Churyumov Gerasimenko estimated from nongravitational force modeling. *Icarus, 176,* 453–477.

Dello Russo N., Vervack R. J. J., Lisse C. M., et al. (2011) The volatile composition and activity of Comet 103P/Hartley 2 during the EPOXI closest approach. *Astrophys. J. Lett., 734,* L8.

Dello Russo N., Kawakita H., Vervack R. J., et al. (2016) Emerging trends and a comet taxonomy based on the volatile chemistry measured in thirty comets with high-resolution infrared spectroscopy between 1997 and 2013. *Icarus, 278,* 301–332.

De Sanctis M. C., Capaccioni F., Ciarniello M., et al. (2015) The diurnal cycle of water ice on Comet 67P/Churyumov-Gerasimenko. *Nature, 525,* 500–503.

Devogèle M., Rivet J. P., Tanga P., et al. (2015) A method to search for large-scale concavities in asteroid shape models. *Mon. Not. R. Astron. Soc., 453,* 2232–2240.

Dunham D. W., Farquhar R. W., Loucks M., et al. (2015) The 2014 Earth return of the ISEE-3/ICE spacecraft. *Acta Astronaut., 110,* 29–42.

Duxbury T. C., Newburn R. L., and Brownlee D. E. (2004) Comet 81P/Wild 2 size, shape, and orientation. *J. Geophys. Res.–Planets, 109,* E12S02.

Economou T. E., Green S. F., Brownlee D. E., et al. (2013) Dust Flux Monitor Instrument measurements during Stardust-NExT flyby of Comet 9P/Tempel 1. *Icarus, 222,* 526–539.

Elsila J. E., Glavin D. P., and Dworkin J. P. (2009) Cometary glycine detected in samples returned by Stardust. *Meteorit. Planet. Sci., 44,* 1323–1330.

Ernst C. M. and Schultz P. H. (2007) Evolution of the Deep Impact flash: Implications for the nucleus surface based on laboratory experiments. *Icarus, 191,* 123–133.

Espley J. R., DiBraccio G. A., Connerney J. E. P., et al. (2015) A comet engulfs Mars: MAVEN observations of Comet Siding Spring's influence on the martian magnetosphere. *Geophys. Res. Lett., 42,* 8810–8818.

Farnham T. L., Wellnitz D. D., Hampton D. L., et al. (2007) Dust coma morphology in the Deep Impact images of Comet 9P/Tempel 1. *Icarus, 191,* 146–160.

Farnham T. L., Bodewits D., Li J. Y., et al. (2013) Connections between the jet activity and surface features on Comet 9P/Tempel 1. *Icarus, 222,* 540–549.

Farnham T. L., Delamere W. A., Kelley M. S. P., et al. (2015) The resolved nucleus and coma of Comet Siding Spring from MRO observations. *AAS/Division for Planetary Sciences Meeting Abstracts 47,* #415.04.

Farnham T. L., Kelley M. S. P., A'Hearn M. F., et al. (2017) Comet C/2012 S1 (ISON): Final observations from the Deep Impact spacecraft. *Icarus, 284,* 106–113.

Farnocchia D., Chesley S. R., Micheli M., et al. (2016) High precision comet trajectory estimates: The Mars flyby of C/2013 A1 (Siding Spring). *Icarus, 266,* 279–287.

Feaga L. M., A'Hearn M. F., Sunshine J. M., et al. (2007) Asymmetries in the distribution of H_2O and CO_2 in the inner coma of Comet 9P/Tempel 1 as observed by Deep Impact. *Icarus, 191,* 134–145.

Feaga L. M., A'Hearn M. F., Farnham T. L., et al. (2014) Uncorrelated volatile behavior during the 2011 apparition of Comet C/2009 P1 Garradd. *Astron. J., 147,* 24.

Feldman P. D., A'Hearn M. F., Bertaux J.-L., et al. (2015) Measurements of the near-nucleus coma of Comet 67P/Churyumov-Gerasimenko with the Alice far-ultraviolet spectrograph on Rosetta. *Astron. Astrophys., 583,* A8.

Fernández J. A., Gallardo T., and Young J. D. (2016) The end states of long-period comets and the origin of Halley-type comets. *Mon. Not. R. Astron. Soc., 461,* 3075–3088.

Festou M. C., Keller H. U., and Weaver H. A., eds. (2004) *Comets II.* Univ. of Arizona, Tucson. 780 pp.

Filacchione G., Raponi A., Capaccioni F., et al. (2016) Seasonal exposure of carbon dioxide ice on the nucleus of Comet 67P/Churyumov-Gerasimenko. *Science, 354,* 1563–1566.

Fink U., Hicks M. P., and Fevig R. A. (1999) Production rates for the Stardust mission target: 81P/Wild 2. *Icarus, 141,* 331–340.

Finzi A. E., Zazzera F. B., Dainese C., et al. (2007) SD2 — How to sample a comet. *Space Sci. Rev., 128,* 281–299.

Fornasier S., Mottola S., Keller H. U., et al. (2016) Rosetta's Comet 67P/Churyumov-Gerasimenko sheds its dusty mantle to reveal its icy nature. *Science, 354,* 1566–1570.

Friedman L. D. (1976) The Halley rendezvous via solar sailing mission description. In *Proceedings of the Shuttle-Based Cometary Science Workshop* (G. A. Gary and K. S. Clifton, eds.), pp. 251–256. NASA TM-82317, Huntsville.

Galvin A. B. and Simunac K. (2015) An extended spacecraft encounter with the tail of Comet Elenin: Plasma and suprathermal ion observations by STEREO-B. In *IAU General Assembly Meeting Abstracts, 29,* #2257907.

Ge D., Cui P., and Zhu S. (2019) Recent development of autonomous GNC technologies for small celestial body descent and landing. *Prog. Aerospace Sci., 110,* 100551.

Gilbert J. A., Lepri S. T., Rubin M., et al. (2015) In situ plasma measurements of fragmented Comet 73P Schwassmann-Wachmann 3. *Astrophys. J., 815,* 12.

Glavin D. P., Dworkin J. P., and Sandford S. A. (2008) Detection of cometary amines in samples returned by Stardust. *Meteorit. Planet. Sci., 43,* 399–413.

Gloeckler G., Geiss J., Schwadron N. A., et al. (2000) Interception of comet Hyakutake's ion tail at a distance of 500 million kilometres. *Nature, 404,* 576–578.

Gloeckler G., Allegrini F., Elliott H. A., et al. (2004) Cometary ions

trapped in a coronal mass ejection. *Astrophys. J. Lett., 604*, L121–L124.
Goesmann F., Rosenbauer H., Roll R., et al. (2007) COSAC, the Cometary Sampling and Composition experiment on Philae. *Space Sci. Rev., 128*, 257–280.
Goetz C., Gunell H., Volwerk M., et al. (2021) Cometary plasma science. *Exp. Astron.*, DOI: 10.1007/s10686-021-09783-z.
Gowen R. A., Smith A., Fortes A. D., et al. (2011) Penetrators for in situ subsurface investigations of Europa. *Adv. Space Res., 48*, 725–742.
Graham G. A., Teslich N., Dai Z. R., et al. (2006) Focused ion beam recovery of hypervelocity impact residue in experimental craters on metallic foils. *Meteorit. Planet. Sci., 41*, 159–165.
Green S. F., McDonnell J. A. M., McBride N., et al. (2004) The dust mass distribution of Comet 81P/Wild 2. *J. Geophys. Res.–Planets, 109*, E12S04.
Grensemann M. G. and Schwehm G. (1993) Giotto's second encounter: The mission to Comet P/Grigg-Skjellerup. *J. Geophys. Res.–Space Phys., 98*, 20907–20910.
Groussin O., Lamy P., Jorda L., et al. (2004) The nuclei of Comets 126P/IRAS and 103P/Hartley 2. *Astron. Astrophys., 419*, 375–383.
Groussin O., A'Hearn M. F., Li J. Y., et al. (2007) Surface temperature of the nucleus of Comet 9P/Tempel 1. *Icarus, 191*, 63–72.
Groussin O., Attree N., Brouet Y., et al. (2019) The thermal, mechanical, structural, and dielectric properties of cometary nuclei after Rosetta. *Space Sci. Rev., 215*, 29.
Grün E., Agarwal J., Altobelli N., et al. (2016) The 2016 Feb 19 outburst of Comet 67P/CG: An ESA Rosetta multi-instrument study. *Mon. Not. R. Astron. Soc., 462*, S220–S234.
Gulkis S., Frerking M., Crovisier J., et al. (2007) MIRO: Microwave Instrument for Rosetta Orbiter. *Space Sci. Rev., 128*, 561–597.
Gurnett D. A., Morgan D. D., Persoon A. M., et al. (2015) An ionized layer in the upper atmosphere of Mars caused by dust impacts from Comet Siding Spring. *Geophys. Res. Lett., 42*, 4745–4751.
Güttler C., Mannel T., Rotundi A., et al. (2019) Synthesis of the morphological description of cometary dust at Comet 67P/Churyumov-Gerasimenko. *Astron. Astrophys., 630*, A24.
Hampton D. L., Baer J. W., Huisjen M. A., et al. (2005) An overview of the instrument suite for the Deep Impact mission. *Space Sci. Rev., 117*, 43–93.
Hansen K. C., Altwegg K., Berthelier J. J., et al. (2016) Evolution of water production of 67P/Churyumov-Gerasimenko: An empirical model and a multi-instrument study. *Mon. Not. R. Astron. Soc., 462*, S491–S506.
Harris W., Woodney L., and Villanueva G. (2019) Chimera: A mission of discovery to the first Centaur. *EPSC-DPS Joint Meeting 2019*, 1094.
He J., Cui B., Yang L., et al. (2021) The encounter of the Parker Solar Probe and a comet-like object near the Sun: Model predictions and measurements. *Astrophys. J., 910*, 7.
Hein A. M., Perakis N., Eubanks T. M., et al. (2019) Project Lyra: Sending a spacecraft to 1I/'Oumuamua (former A/2017 U1), the interstellar asteroid. *Acta Astronaut., 161*, 552–561.
Hein A. M., Eubanks T. M., Lingam M., et al. (2022) Interstellar now! Missions to explore nearby interstellar objects. *Adv. Space Res., 69*, 402–414.
Heinisch P., Auster H. U., Gundlach B., et al. (2019) Compressive strength of Comet 67P/Churyumov-Gerasimenko derived from Philae surface contacts. *Astron. Astrophys., 630*, A2.
Herique A., Rogez Y., Pasquero O. P., et al. (2015) Philae localization from CONSERT/Rosetta measurement. *Planet. Space Sci., 117*, 475–484.
Herique A., Agnus B., Asphaug E., et al. (2018) Direct observations of asteroid interior and regolith structure: Science measurement requirements. *Adv. Space Res., 62*, 2141–2162.
Heritier K. L., Altwegg K., Balsiger H., et al. (2017) Ion composition at Comet 67P near perihelion: Rosetta observations and model-based interpretation. *Mon. Not. R. Astron. Soc., 469*, S427–S442.
Hermalyn B., Farnham T. L., Collins S. M., et al. (2013) The detection, localization, and dynamics of large icy particles surrounding Comet 103P/Hartley 2. *Icarus, 222*, 625–633.
Hewagama T., Aslam S., Clark P., et al. (2018) Primitive Object Volatile Explorer (PrOVE) — Waypoints and opportunistic deep space missions to comets. In *49th Lunar and Planetary Science Conference*, Abstract #2800. LPI Contribution No. 2083, Lunar and Planetary Institute, Houston.
Hilchenbach M., Kissel J., Langevin Y., et al. (2016) Comet 67P/Churyumov-Gerasimenko: Close-up on dust particle fragments. *Astrophys. J. Lett., 816*, L32.
Hoover D. J., Seligman D. Z., and Payne M. J. (2022) The population of interstellar objects detectable with the LSST and accessible for in situ rendezvous with various mission designs. *Planet. Sci. J., 3*, 71.
Hoppe P., Rubin M., and Altwegg K. (2018) Presolar isotopic signatures in meteorites and comets: New insights from the Rosetta mission to Comet 67P/Churyumov-Gerasimenko. *Space Sci. Rev., 214*, 106.
Howell S. M., Chou L., Thompson M., et al. (2018) Camilla: A Centaur reconnaissance and impact mission concept. *Planet. Space Sci., 164*, 184–193.
Hu X., Shi X., Sierks H., et al. (2017) Seasonal erosion and restoration of the dust cover on Comet 67P/Churyumov-Gerasimenko as observed by OSIRIS onboard Rosetta. *Astron. Astrophys., 604*, A114.
Ishibashi K., Hong P., Okamoto T., et al. (2020) Flyby observation of asteroid (3200) Phaethon to be conducted by cameras onboard DESTINY+ spacecraft. In *51st Lunar and Planetary Science Conference*, Abstract #1698. LPI Contribution No. 2326, Lunar and Planetary Institute, Houston.
Jewitt D. and Li J. (2010) Activity in Geminid parent (3200) Phaethon. *Astron. J., 140*, 1519–1527.
Jewitt D., Li J., and Agarwal J. (2013) The dust tail of asteroid (3200) Phaethon. *Astrophys. J. Lett., 771*, L36.
Jewitt D., Weaver H., Mutchler M., et al. (2018) The nucleus of active asteroid 311P/(2013 P5) PANSTARRS. *Astron. J., 155*, 231.
Johnstone A. D., Coates A. J., Huddleston D. E., et al. (1993) Observations of the solar wind and cometary ions during the encounter between Giotto and Comet Grigg-Skjellerup. *Astron. Astrophys., 273*, L1–L4.
Jones G. H., Balogh A., and Horbury T. S. (2000) Identification of Comet Hyakutake's extremely long ion tail from magnetic field signatures. *Nature, 404*, 574–576.
Jones G. H., Balogh A., McComas D. J., et al. (2003a) Strong interplanetary field enhancements at Ulysses — Evidence of dust trails' interaction with the solar wind? *Icarus, 166*, 297–310.
Jones G. H., Balogh A., Russell C. T., et al. (2003b) Possible distortion of the interplanetary magnetic field by the dust trail of Comet 122P/de Vico. *Astrophys. J. Lett., 597*, L61–L64.
Jones G. H., Agarwal J., Bowles N., et al. (2018) The proposed Caroline ESA M3 mission to a main belt comet. *Adv. Space Res., 62*, 1921–1946.
Jones G. H., Afghan Q., and Price O. (2020) Prospects for the in situ detection of Comet C/2019 Y4 ATLAS by Solar Orbiter. *Res. Notes Am. Astron. Soc., 4*, 62.
Jones G. H., Elliott H., McComas D., et al. (2022) Cometary ions detected by the Cassini spacecraft 6.5 au downstream of Comet 153P/Ikeya-Zhang. *Icarus, 388*, 115199.
Jones G. H., Snodgrass C., Tubiana C., et al. (2024) The Comet Interceptor mission. *Space Sci. Rev. 220, 9,* DOI: 10.1007/s11214-023-01035-0.
Jorda L., Gaskell R., Capanna C., et al. (2016) The global shape, density and rotation of Comet 67P/Churyumov-Gerasimenko from preperihelion Rosetta/OSIRIS observations. *Icarus, 277*, 257–278.
Kearsley A. T., Burchell M. J., Price M. C., et al. (2012) Experimental impact features in Stardust aerogel: How track morphology reflects particle structure, composition, and density. *Meteorit. Planet. Sci., 47*, 737–762.
Keeney B. A., Stern S. A., Feldman P. D., et al. (2019) Stellar occultation by Comet 67P/Churyumov-Gerasimenko observed with Rosetta's Alice far-ultraviolet spectrograph. *Astron. J., 157*, 173.
Keller H. U. (2006) Comet observations by OSIRIS onboard the Rosetta spacecraft. *IAU Joint Discussion, 26,* JD10-29.
Keller H. U., Arpigny C., Barbieri C., et al. (1986) First Halley Multicolour Camera imaging results from Giotto. *Nature, 321*, 320–326.
Keller H. U., Britt D., Buratti B. J., et al. (2004) In situ observations of cometary nuclei. In *Comets II* (M. C. Festou et al., eds.), pp. 211–222. Univ. of Arizona, Tucson.
Keller H. U., Jorda L., Küppers M., et al. (2005) Deep Impact observations by OSIRIS onboard the Rosetta spacecraft. *Science, 310*, 281–283.
Keller H. U., Barbieri C., Lamy P., et al. (2007a) OSIRIS — The scientific camera system onboard Rosetta. *Space Sci. Rev., 128*, 433–506.
Keller H. U., Küppers M., Fornasier S., et al. (2007b) Observations of Comet 9P/Tempel 1 around the Deep Impact event by the OSIRIS

cameras onboard Rosetta. *Icarus, 187,* 87–103.
Keller H. U., Mottola S., Skorov Y., et al. (2015) The changing rotation period of Comet 67P/Churyumov-Gerasimenko controlled by its activity. *Astron. Astrophys., 579,* L5.
Keller H. U., Mottola S., Hviid S. F., et al. (2017) Seasonal mass transfer on the nucleus of Comet 67P/Chuyumov-Gerasimenko. *Mon. Not. R. Astron. Soc., 469,* S357–S371.
Keller H. U. and Kührt E. (2020) Cometary nuclei — From Giotto to Rosetta. *Space Sci. Rev., 216,* 14.
Keller L. P., Bajt S., Baratta G. A., et al. (2006) Infrared spectroscopy of Comet 81P/Wild 2 samples returned by Stardust. *Science, 314,* 1728–1731.
Kelley M. S., Lindler D. J., Bodewits D., et al. (2013) A distribution of large particles in the coma of Comet 103P/Hartley 2. *Icarus, 222,* 634–652.
Kissel J., Glasmachers A., Grün E., et al. (2003) Cometary and interstellar dust analyzer for Comet Wild 2. *J. Geophys. Res.–Planets, 108,* 8114.
Kissel J., Krueger F. R., Silén J., et al. (2004) The cometary and interstellar dust analyzer at Comet 81P/Wild 2. *Science, 304,* 1774–1776.
Kissel J., Altwegg K., Clark B. C., et al. (2007) COSIMA — High resolution time-of-flight secondary ion mass spectrometer for the analysis of cometary dust particles onboard Rosetta. *Space Sci. Rev., 128,* 823–867.
Klaasen K. P., A'Hearn M. F., Baca M., et al. (2008) Invited article: Deep Impact instrument calibration. *Rev. Sci. Instrum., 79,* 091301.
Klaasen K. P., A'Hearn M., Besse S., et al. (2013a) EPOXI instrument calibration. *Icarus, 225,* 643–680.
Klaasen K. P., Brown D., Carcich B., et al. (2013b) Stardust-NExT NAVCAM calibration and performance. *Icarus, 222,* 436–452.
Klingelhöfer G., Brückner J., D'Uston C., et al. (2007) The Rosetta Alpha Particle X-Ray Spectrometer (APXS). *Space Sci. Rev., 128,* 383–396.
Knight M. M., Mueller B. E. A., Samarasinha N. H., et al. (2015) A further investigation of apparent periodicities and the rotational state of Comet 103P/Hartley 2 from combined coma morphology and light curve data sets. *Astron. J., 150,* 22.
Knight M. M., Fitzsimmons A., Kelley M. S. P., et al. (2016) Comet 322P/SOHO 1: An asteroid with the smallest perihelion distance? *Astrophys. J. Lett., 823,* L6.
Kofman W., Herique A., Goutail J. P., et al. (2007) The Comet Nucleus Sounding Experiment by Radiowave Transmission (CONSERT): A short description of the instrument and of the commissioning stages. *Space Sci. Rev., 128,* 413–432.
Kofman W., Zine S., Herique A., et al. (2020) The interior of Comet 67P/C-G: Revisiting CONSERT results with the exact position of the Philae lander. *Mon. Not. R. Astron. Soc., 497,* 2616–2622.
Kramer T. and Läuter M. (2019) Outgassing-induced acceleration of Comet 67P/Churyumov-Gerasimenko. *Astron. Astrophys., 630,* A4.
Krankowsky D., Lammerzahl P., Herrwerth I., et al. (1986) In situ gas and ion measurements at Comet Halley. *Nature, 321,* 326–329.
Küppers M. (2017) Rosetta: Experience of operating a spacecraft next to a comet for 2 years. In *From Giotto to Rosetta: 30 Years of Cometary Science from Space and Ground* (C. Barbieri and C. G. Someda, eds.), pp. 167–181. Accademia Galileiana di Scienze, Lettere ed Arti, Padova.
Küppers M., Bertini I., Fornasier S., et al. (2005) A large dust/ice ratio in the nucleus of Comet 9P/Tempel 1. *Nature, 437,* 987–990.
Küppers M., Keller H. U., Kührt E., et al. (2009) Triple F — A comet nucleus sample return mission. *Exp. Astron., 23,* 809–847.
Lai H. R., Russell C. T., Jia Y. D., et al. (2015) Momentum transfer from solar wind to interplanetary field enhancements inferred from magnetic field draping signatures. *Geophys. Res. Lett., 42,* 1640–1645.
Lamy P. L., Toth I., Weaver H. A., et al. (2006) Hubble Space Telescope observations of the nucleus and inner coma of Comet 67P/Churyumov-Gerasimenko. *Astron. Astrophys., 458,* 669–678.
Lara L. M., Boehnhardt H., Gredel R., et al. (2006) Pre-impact monitoring of Comet 9P/Tempel 1, the Deep Impact target. *Astron. Astrophys., 445,* 1151–1157.
Ledbetter W. G., Sood R., and Keane J. T. (2018) The interior structure of asteroids and comets revealed by ChipSat swarm gravimetry. In *49th Lunar and Planetary Science Conference,* Abstract #2136. LPI Contribution No. 2083, Lunar and Planetary Institute, Houston.
Levasseur-Regourd A. C., Goidet B., Le Duin T., et al. (1993) Short communication: Optical probing of dust in Comet Grigg-Skjellerup from the Giotto spacecraft. *Planet. Space Sci., 41,* 167–169.
Levison H. F., Duncan M. J., Dones L., et al. (2006) The scattered disk as a source of Halley-type comets. *Icarus, 184,* 619–633.
Li J. and Jewitt D. (2013) Recurrent perihelion activity in (3200) Phaethon. *Astron. J., 145,* 154.
Li J.-Y., A'Hearn M. F., Belton M. J. S., et al. (2007) Deep Impact photometry of Comet 9P/Tempel 1. *Icarus, 191,* 161–175.
Lindler D., Busko I., A'Hearn M. F., et al. (2007) Restoration of images of Comet 9P/Tempel 1 taken with the Deep Impact High Resolution Instrument. *Publ. Astron. Soc. Pac., 119,* 427–436.
Lindler D. J., A'Hearn M. F., and McLaughlin S. A. (2012) *EPOXI 103P/Hartley 2 Encounter — HRIV Deconvolved Images V1.0.* NASA Planetary Data System, DIF-C-HRIV-5-EPOXI-HARTLEY2-DECONV-V1.0.
Lis D. C., Bockelée-Morvan D., Güsten R., et al. (2019) Terrestrial deuterium-to-hydrogen ratio in water in hyperactive comets. *Astron. Astrophys., 625,* L5.
Lisse C. (2015) Comet Siding Spring, up close and personal. *Science, 350,* 277–278.
Lorenz R. D. (2011) Planetary penetrators: Their origins, history and future. *Adv. Space Res., 48,* 403–431.
Lowry S., Duddy S. R., Rozitis B., et al. (2012) The nucleus of Comet 67P/Churyumov-Gerasimenko: A new shape model and thermophysical analysis. *Astron. Astrophys., 548,* A12.
Luspay-Kuti A., Mousis O., Pauzat F., et al. (2022) Dual storage and release of molecular oxygen in Comet 67P/Churyumov-Gerasimenko. *Nature Astronomy, 6,* 724–730.
Marschall R., Skorov Y., Zakharov V., et al. (2020) Cometary comae-surface links. *Space Sci. Rev., 216,* 130.
Marty B., Altwegg K., Balsiger H., et al. (2017) Xenon isotopes in 67P/Churyumov-Gerasimenko show that comets contributed to Earth's atmosphere. *Science, 356,* 1069–1072.
Mastrodemos N., Kubitschek D. G., and Synnott S. P. (2005) Autonomous navigation for the Deep Impact mission encounter with Comet Tempel 1. *Space Sci. Rev., 117,* 95–121.
Matousek S. (2007) The Juno New Frontiers mission. *Acta Astronaut., 61,* 932–939.
Matteini L., Laker R., Horbury T., et al. (2021) Solar Orbiter's encounter with the tail of Comet C/2019 Y4 (ATLAS): Magnetic field draping and cometary pick-up ion waves. *Astron. Astrophys., 656,* A39.
Mayyasi M., Clarke J., Combi M., et al. (2020) Lyα observations of Comet C/2013 A1 (Siding Spring) using MAVEN IUVS echelle. *Astron. J., 160,* 10.
McBride N., Green S. F., Chantal Levasseur-Regourd A., et al. (1997) The inner dust coma of Comet 26P/Grigg-Skjellerup: Multiple jets and nucleus fragments? *Mon. Not. R. Astron. Soc., 289,* 535–553.
McCubbin F., Allton J. H., Barnes J. J., et al. (2021) Advanced curation of astromaterials for planetary science over the next decade. *Bull. Am. Astron. Soc., 53,* DOI: 10.3847/25c2cfeb.1c20e2ca.
McFadden L. A., Rountree-Brown M. K., Warner E. M., et al. (2005) Education and public outreach for NASA's Deep Impact mission. *Space Sci. Rev., 117,* 373–396.
McLaughlin S. A., Carcich B., Sackett S. E., et al. (2014) *Deep Impact 9P/Tempel Encounter — Reduced HRIV Images V3.0.* NASA Planetary Data System, DIF-C-HRIV-3/4-9P-ENCOUNTER-V3.0.
Meech K. J. and Castillo-Rogez J. C. (2015) Proteus — A mission to investigate the origins of Earth's water. In *IAU General Assembly Meeting Abstracts, 29,* #2257859.
Meech K. J. and Svoreň J. (2004) Using cometary activity to trace the physical and chemical evolution of cometary nuclei. In *Comets II* (M. C. Festou et al., eds.), pp. 317–335. Univ. of Arizona, Tucson.
Meech K. J., Ageorges N., A'Hearn M. F., et al. (2005a) Deep Impact: Observations from a worldwide Earth-based campaign. *Science, 310,* 265–269.
Meech K. J., A'Hearn M. F., Fernández Y. R., et al. (2005b) The Deep Impact Earth-based campaign. *Space Sci. Rev., 117,* 297–334.
Meech K. J., A'Hearn M. F., Adams J. A., et al. (2011a) EPOXI: Comet 103P/Hartley 2 observations from a worldwide campaign. *Astrophys. J. Lett., 734,* L1.
Meech K. J., Pittichová J., Yang B., et al. (2011b) Deep Impact, Stardust-NExT and the behavior of Comet 9P/Tempel 1 from 1997 to 2010. *Icarus, 213,* 323–344.
Meech K. J., Kleyna J., Hainaut O. R., et al. (2013) The demise of Comet 85P/Boethin, the first EPOXI mission target. *Icarus, 222,*

662–678.
Meech K. J., Kleyna J. T., Hainaut O., et al. (2017) CO-driven activity in Comet C/2017 K2 (PANSTARRS). *Astrophys. J. Lett., 849*, L8.
Merouane S., Zaprudin B., Stenzel O., et al. (2016) Dust particle flux and size distribution in the coma of 67P/Churyumov-Gerasimenko measured in situ by the COSIMA instrument on board Rosetta. *Astron. Astrophys., 596*, A87.
Möhlmann D., Seidensticker K. J., Fischer H.-H., et al. (2018) Compressive strength and elastic modulus at Agilkia on Comet 67P/Churyumov-Gerasimenko derived from the SESAME/CASSE touchdown signals. *Icarus, 303*, 251–264.
Moore K., Castillo-Rogez J., Meech K. J., et al. (2021) Rapid response missions to explore fast, high-value targets such as interstellar objects and long period comets. *Bull. Am. Astron. Soc., 53*, DOI: 10.3847/25c2cfeb.1d58e5af.
Moore K., Courville S., Ferguson S., et al. (2021) Bridge to the stars: A mission concept to an interstellar object. *Planet. Space Sci., 197*, 105137.
Mottola S., Arnold G., Grothues H.-G., et al. (2007) The ROLIS experiment on the Rosetta lander. *Space Sci. Rev., 128*, 241–255.
Mottola S., Lowry S., Snodgrass C., et al. (2014) The rotation state of 67P/Churyumov-Gerasimenko from approach observations with the OSIRIS cameras on Rosetta. *Astron. Astrophys., 569*, L2.
Nakamura-Messenger K., Hayes A. G., Sandford S., et al. (2021) The case for non-cryogenic comet nucleus sample return. *Bull. Am. Astron. Soc., 53*, DOI: 10.3847/25c2cfeb.441b2770.
NASA (2003) *Comet Nucleus Tour (CONTOUR): Mishap Investigation Board Report.* NASA, Washington, DC. 64 pp.
Neugebauer M., Gloeckler G., Gosling J. T., et al. (2007) Encounter of the Ulysses spacecraft with the ion tail of Comet McNaught. *Astrophys. J., 667*, 1262–1266.
Newburn R. L., Bhaskaran S., Duxbury T. C., et al. (2003) Stardust imaging camera. *J. Geophys. Res.–Planets, 108*, 8116.
Olkin C. B., Levison H. F., Vincent M., et al. (2021) Lucy mission to the Trojan asteroids: Instrumentation and encounter concept of operations. *Planet. Sci. J., 2*, 172.
Opitom C., Snodgrass C., Fitzsimmons A., et al. (2017) Ground-based monitoring of Comet 67P/Churyumov-Gerasimenko gas activity throughout the Rosetta mission. *Mon. Not. R. Astron. Soc., 469*, S222–S229.
O'Rourke L., Tubiana C., Güttler C., et al. (2019) The search campaign to identify and image the Philae lander on the surface of Comet 67P/Churyumov-Gerasimenko. *Acta Astronaut., 157*, 199–214.
Ozaki N., Yoshioka K., Kameda S., et al. (2021) Concept study of comet Halley revisiting missions. *43rd COSPAR Scientific Assembly,* 254.
Ozaki N., Yamamoto T., Gonzalez-Franquesa F., et al. (2022) Mission design of DESTINY+: Toward active asteroid (3200) Phaethon and multiple small bodies. *Acta Astronaut., 196*, 42–56.
Pajola M., Höfner S., Vincent J.-B., et al. (2017) The pristine interior of Comet 67P revealed by the combined Aswan outburst and cliff collapse. *Nature Astron., 1*, 0092.
Pätzold M., Häusler B., Aksnes K., et al. (2007) Rosetta Radio Science Investigations (RSI). *Space Sci. Rev., 128*, 599–627.
Pätzold M., Andert T., Hahn M., et al. (2016) A homogeneous nucleus for Comet 67P/Churyumov-Gerasimenko from its gravity field. *Nature, 530*, 63–65.
Preusker F., Scholten F., Matz K. D., et al. (2017) The global meter-level shape model of Comet 67P/Churyumov-Gerasimenko. *Astron. Astrophys., 607*, L1.
Protopapa S., Sunshine J. M., Feaga L. M., et al. (2014) Water ice and dust in the innermost coma of Comet 103P/Hartley 2. *Icarus, 238*, 191–204.
Rayman M. D. (2002) The Deep Space 1 extended mission: Challenges in preparing for an encounter with Comet Borrelly. *Acta Astronaut., 51*, 507–516.
Rayman M. D. and Varghese P. (2001) The Deep Space 1 extended mission. *Acta Astronaut., 48*, 693–705.
Rayman M. D., Varghese P., Lehman D. H., et al. (2000) Results from the Deep Space 1 technology validation mission. *Acta Astronaut., 47*, 475–487.
Rengel M., Jones G. H., Küppers M., et al. (2007) The ion tail of Comet Machholz observed by OSIRIS as a tracer of the solar wind velocity. *Geophys. Res. Abstr., 9,* Abstract #02744.
Richardson J. E., Melosh H. J., Artemeiva N. A., et al. (2005) Impact cratering theory and modeling for the Deep Impact mission: From mission planning to data analysis. *Space Sci. Rev., 117*, 241–267.
Richardson J. E., Melosh H. J., Lisse C. M., et al. (2007) A ballistics analysis of the Deep Impact ejecta plume: Determining Comet Tempel 1's gravity, mass, and density. *Icarus, 191*, 176–209.
Riedler W., Torkar K., Jeszenszky H., et al. (2007) MIDAS — The Micro-Imaging Dust Analysis System for the Rosetta mission. *Space Sci. Rev., 128*, 869–904.
Robinson T. D., Meadows V. S., Crisp D., et al. (2011) Earth as an extrasolar planet: Earth model validation using EPOXI Earth observations. *Astrobiology, 11*, 393–408.
Rosenbush V. K., Ivanova O. V., Kiselev N. N., et al. (2017) Spatial variations of brightness, colour and polarization of dust in Comet 67P/Churyumov-Gerasimenko. *Mon. Not. R. Astron. Soc., 469*, S475–S491.
Rubin M., Altwegg K., van Dishoeck E. F., et al. (2015) Molecular oxygen in Oort cloud Comet 1P/Halley. *Astrophys. J. Lett., 815*, L11.
Rubin M., Altwegg K., Balsiger H., et al. (2018) Krypton isotopes and noble gas abundances in the coma of Comet 67P/Churyumov-Gerasimenko. *Sci. Adv., 4*, eaar6297.
Rubin M., Engrand C., Snodgrass C., et al. (2020) On the origin and evolution of the material in 67P/Churyumov-Gerasimenko. *Space Sci. Rev., 216*, 102.
Russell C. T., Luhmann J. G., Barnes A., et al. (1983) An unusual interplanetary event — Encounter with a comet? *Nature, 305*, 612–615.
Russell C. T., Arghavani M. R., and Luhmann J. G. (1984) Interplanetary field enhancements in the solar wind: Statistical properties at 0.72 AU. *Icarus, 60*, 332–350.
Sánchez J. P., Morante D., Hermosin P., et al. (2021) ESA F-Class Comet Interceptor: Trajectory design to intercept a yet-to-be-discovered comet. *Acta Astronaut., 188*, 265–277.
Sánchez-Cano B., Witasse O., Lester M., et al. (2018) Energetic particle showers over Mars from Comet C/2013 A1 Siding Spring. *J. Geophys. Res.–Space Phys., 123*, 8778–8796.
Sánchez-Cano B., Lester M., Witasse O., et al. (2020) Mars' ionospheric interaction with Comet C/2013 A1 Siding Spring's coma at their closest approach as seen by Mars Express. *J. Geophys. Res.–Space Phys., 125*, e27344.
Sandford S. A., Aléon J., Alexander C. M. O. D., et al. (2006) Organics captured from Comet 81P/Wild 2 by the Stardust spacecraft. *Science, 314*, 1720–1724.
Sandford S. A., Brownlee D. E., and Zolensky M. E. (2021) The Stardust sample return mission. In *Sample Return Missions: The Last Frontier of Solar System Exploration* (A. Longobardo, ed.), pp. 79–104. Elsevier, Amsterdam.
Sarid G., Volk K., Steckloff J. K., et al. (2019) 29P/Schwassmann-Wachmann 1, a Centaur in the gateway to the Jupiter-family comets. *Astrophys. J. Lett., 883*, L25.
Sarli B. V., Horikawa M., Yam C. H., et al. (2018) DESTINY+ trajectory design to (3200) Phaethon. *J. Astronaut. Sci., 65*, 82–110.
Schleicher D. G. and Farnham T. L. (2004) Photometry and imaging of the coma with narrowband filters. In *Comets II* (M. C. Festou et al., eds.), pp. 449–469. Univ. of Arizona, Tucson.
Schleicher D. G., Barnes K. L., and Baugh N. F. (2006) Photometry and imaging results for Comet 9P/Tempel 1 and Deep Impact: Gas production rates, postimpact light curves, and ejecta plume morphology. *Astron. J., 131*, 1130–1137.
Schultz P. H., Ernst C. M., and Anderson J. L. B. (2005) Expectations for crater size and photometric evolution from the Deep Impact collision. *Space Sci. Rev., 117*, 207–239.
Schultz P. H., Eberhardy C. A., Ernst C. M., et al. (2007) The Deep Impact oblique impact cratering experiment. *Icarus, 191*, 84–122.
Schultz P. H., Hermalyn B., and Veverka J. (2013) The Deep Impact crater on 9P/Tempel-1 from Stardust-NExT. *Icarus, 222*, 502–515.
Schwamb M. E., Knight M. M., Jones G. H., et al. (2020) Potential backup targets for Comet Interceptor. *Res. Notes Am. Astron. Soc., 4*, 21.
Seidensticker K. J., Möhlmann D., Apathy I., et al. (2007) SESAME — An experiment of the Rosetta lander Philae: Objectives and general design. *Space Sci. Rev., 128*, 301–337.
Seligman D. and Laughlin G. (2018) The feasibility and benefits of in situ exploration of 'Oumuamua-like objects. *Astron. J., 155*, 217.
Shinnaka Y., Fougere N., Kawakita H., et al. (2017) Imaging observations of the hydrogen coma of Comet 67P/Churyumov-Gerasimenko in 2015 September by the PROCYON/LAICA. *Astron. J., 153*, 76.
Singer K. N., Stern S. A., Elliott J., et al. (2021) A new spacecraft

mission concept combining the first exploration of the Centaurs and an astrophysical space telescope for the outer solar system. *Planet. Space Sci., 205*, 105290.
Smith E. J., Slavin J. A., Bame S. J., et al. (1986) Analysis of the Giacobini-Zinner bow wave. In *Proceedings of the 20th ESLAB Symposium on the Exploration of Halley's Comet* (B. Battrick et al., eds.), p. 461. ESA SP-250, Noordwijk, The Netherlands.
Snodgrass C. and Jones G. H. (2019) The European Space Agency's Comet Interceptor lies in wait. *Nature Commun., 10*, 5418.
Snodgrass C., Tubiana C., Vincent J.-B., et al. (2010) A collision in 2009 as the origin of the debris trail of asteroid P/2010A2. *Nature, 467*, 814–816.
Snodgrass C., Tubiana C., Bramich D. M., et al. (2013) Beginning of activity in 67P/Churyumov-Gerasimenko and predictions for 2014–2015. *Astron. Astrophys., 557*, A33.
Snodgrass C., Agarwal J., Combi M., et al. (2017a) The main belt comets and ice in the solar system. *Astron. Astrophys. Rev., 25*, 5.
Snodgrass C., A'Hearn M. F., Aceituno F., et al. (2017b) The 67P/Churyumov-Gerasimenko observation campaign in support of the Rosetta mission. *Philos. Trans. R. Soc. A, 375*, 20160249.
Snodgrass C., Jones G. H., Boehnhardt H., et al. (2018) The Castalia mission to main belt Comet 133P/Elst-Pizarro. *Adv. Space Res., 62*, 1947–1976.
Spohn T., Seiferlin K., Hagermann A., et al. (2007) MUPUS — A thermal and mechanical properties probe for the Rosetta lander Philae. *Space Sci. Rev., 128*, 339–362.
Spohn T., Knollenberg J., Ball A. J., et al. (2015) Thermal and mechanical properties of the near-surface layers of Comet 67P/Churyumov-Gerasimenko. *Science, 349*, DOI: 10.1126/science.aab0464.
Squyres S. W., Nakamura-Messenger K., Mitchell D. F., et al. (2018) The CAESAR New Frontiers mission: 1. Overview. In *49th Lunar and Planetary Science Conference*, Abstract #1332. LPI Contribution No. 2083, Lunar and Planetary Institute, Houston.
Srama R., Ahrens T. J., Altobelli N., et al. (2004) The Cassini Cosmic Dust Analyzer. *Space Sci. Rev., 114*, 465–518.
Stern S. A., Slater D. C., Scherrer J., et al. (2007) ALICE: The Rosetta ultraviolet imaging spectrograph. *Space Sci. Rev., 128*, 507–527.
Sunshine J. M. and Feaga L. M. (2021) All comets are somewhat hyperactive and the implications thereof. *Planet. Sci. J., 2*, 92.
Sunshine J. M., A'Hearn M. F., Groussin O., et al. (2006) Exposed water ice deposits on the surface of Comet 9P/Tempel 1. *Science, 311*, 1453–1455.
Sunshine J. M., Groussin O., Schultz P. H., et al. (2007) The distribution of water ice in the interior of Comet Tempel 1. *Icarus, 191*, 73–83.
Sunshine J. M., Farnham T. L., Feaga L. M., et al. (2009) Temporal and spatial variability of lunar hydration as observed by the Deep Impact spacecraft. *Science, 326*, 565–568.
Sunshine J. M., Thomas N., El-Maarry M. R., et al. (2016) Evidence for geologic processes on comets. *J. Geophys. Res.–Planets, 121*, 2194–2210.
Swenson B. L., Squyres S. W., and Knight T. C. D. (1987) A proposed Comet Nucleus Penetrator for the Comet Rendezvous Asteroid Flyby mission. *Acta Astronaut., 15*, 471–479.
Tang J., Xiao J., Jiang S., et al. (2021) A robotic penetrator for the main belt Comet 133P/Elst-Pizarro exploration. *43rd COSPAR Scientific Assembly*, 153.
Taylor M. G. G. T., Altobelli N., Buratti B. J., et al. (2017) The Rosetta mission orbiter science overview: The comet phase. *Philos. Trans. R. Soc. A, 375*, 20160262.
Thomas N., Davidsson B., El-Maarry M. R., et al. (2015a) Redistribution of particles across the nucleus of Comet 67P/Churyumov-Gerasimenko. *Astron. Astrophys., 583*, A17.
Thomas N., Sierks H., Barbieri C., et al. (2015b) The morphological diversity of Comet 67P/Churyumov-Gerasimenko. *Science, 347*, aaa0440.
Thomas N., Cremonese G., Ziethe R., et al. (2017) The Colour and Stereo Surface Imaging System (CaSSIS) for the ExoMars Trace Gas Orbiter. *Space Sci. Rev., 212*, 1897–1944.
Thomas N., Ulamec S., Kührt E., et al. (2019) Towards new comet missions. *Space Sci. Rev., 215*, 47.
Thomas P. C., Veverka J., Belton M. J. S., et al. (2007) The shape, topography, and geology of Tempel 1 from Deep Impact observations. *Icarus, 191*, 51–62.
Thomas P., A'Hearn M., Belton M. J. S., et al. (2013a) The nucleus of Comet 9P/Tempel 1: Shape and geology from two flybys. *Icarus, 222*, 453–466.
Thomas P. C., A'Hearn M. F., Veverka J., et al. (2013b) Shape, density, and geology of the nucleus of Comet 103P/Hartley 2. *Icarus, 222*, 550–558.
Thomas V. C., Makowski J. M., Brown G. M., et al. (2011) The Dawn spacecraft. *Space Sci. Rev., 163*, 175–249.
Tran H. K., Johnson C. E., Rasky D. J., et al. (1997) *Phenolic Impregnated Carbon Ablators (PICA) as Thermal Protection Systems for Discovery Missions*. NASA TM-110440, Washington, DC.
Tsou P., Brownlee D. E., Anderson J. D., et al. (2004) Stardust encounters Comet 81P/Wild 2. *J. Geophys. Res.–Planets, 109*, E12S01.
Tubiana C., Snodgrass C., Bertini I., et al. (2015) 67P/Churyumov-Gerasimenko: Activity between March and June 2014 as observed from Rosetta/OSIRIS. *Astron. Astrophys., 573*, A62.
Tuzzolino A. J., Economou T. E., McKibben R. B., et al. (2003) Dust Flux Monitor Instrument for the Stardust mission to Comet Wild 2. *J. Geophys. Res.–Planets, 108*, 8115.
Ulamec S., Espinasse S., Feuerbacher B., et al. (2006) Rosetta lander — Philae: Implications of an alternative mission. *Acta Astronaut., 58*, 435–441.
Ulamec S., Kucherenko V., Biele J., et al. (2011) Hopper concepts for small body landers. *Adv. Space Res., 47*, 428–439.
Ulamec S., Biele J., Blazquez A., et al. (2015) Rosetta lander — Philae: Landing preparations. *Acta Astronaut., 107*, 79–86.
Usher H., Snodgrass C., Green S. F., et al. (2020) Seeing the bigger picture: Rosetta mission amateur observing campaign and lessons for the future. *Planet. Sci. J., 1*, 84.
Vallat C., Altobelli N., Geiger B., et al. (2017) The science planning process on the Rosetta mission. *Acta Astronaut., 133*, 244–257.
Veverka J. F., Klaasen K. P., Kissel J., et al. (2011) *Stardust RDR NAVCAM Images of 9P/Tempel 1 V1.0*. NASA Planetary Data System, SDU-C/CAL-NAVCAM-3-NEXT-TEMPEL1-V1.0.
Veverka J., Klaasen K., A'Hearn M., et al. (2013) Return to Comet Tempel 1: Overview of Stardust-NExT results. *Icarus, 222*, 424–435.
Vilnrotter V., Tsao P. C., Lee D. K., et al. (2008) EPOXI uplink array experiment of June 27, 2008. *IPN Prog. Rep., 42-174*, 1–25.
Vincent J.-B., Lara L. M., Tozzi G. P., et al. (2013) Spin and activity of Comet 67P/Churyumov-Gerasimenko. *Astron. Astrophys., 549*, A121.
Vincent J.-B., Oklay N., Marchi S., et al. (2015) Craters on comets. *Planet. Space Sci., 107*, 53–63.
Vincent J.-B., A'Hearn M. F., Lin Z. Y., et al. (2016) Summer fireworks on Comet 67P. *Mon. Not. R. Astron. Soc., 462*, S184–S194.
Wellnitz D. D., Collins S. M., A'Hearn M. F., et al. (2013) The location of the impact point of the Deep Impact impactor on Comet 9P/Tempel 1. *Icarus, 222*, 487–491.
Westphal A., Nittler L. R., Stroud R., et al. (2021) Cryogenic comet sample return. *Bull. Am. Astron. Soc., 53*, DOI: 10.3847/25c2cfeb.6197be94.
Wright I. P., Barber S. J., Morgan G. H., et al. (2007) Ptolemy — An instrument to measure stable isotopic ratios of key volatiles on a cometary nucleus. *Space Sci. Rev., 128*, 363–381.
Young D. T., Crary F. J., Nordholt J. E., et al. (2004) Solar wind interactions with Comet 19P/Borrelly. *Icarus, 167*, 80–88.
Zacny K., Chu P., Vendiola V., et al. (2021) TRIDENT drill for VIPER and PRIME1 missions to the Moon. In *52nd Lunar and Planetary Science Conference*, Abstract #2400. LPI Contribution No. 2548. Lunar and Planetary Institute, Houston.
Zhang T., Xu K., and Ding X. (2021) China's ambitions and challenges for asteroid-comet exploration. *Nature Astron., 5*, 730–731.
Zhang X., Huang J., Wang T., et al. (2019) ZhengHe — A mission to a near-Earth asteroid and a main belt comet. In *50th Lunar and Planetary Science Conference*, Abstract #1045. LPI Contribution No. 2132, Lunar and Planetary Institute, Houston.
Zhao C., Wang Y., Li D., et al. (2022) A design of dust analyzer for future main belt comet exploration mission. *Adv. Space Res., 69*, 3880–3890.
Zolensky M. E., Pieters C., Clark B., et al. (2000) Invited review — Small is beautiful: The analysis of nanogram-sized astromaterials. *Meteorit. Planet Sci., 35*, 9–29.
Zolensky M. E., Zega T. J., Yano H., et al. (2006) Mineralogy and petrology of Comet 81P/Wild 2 nucleus samples. *Science, 314*, 1735.

Bauer J. M., Fernández Y. R., Protopapa S., and Woodney L. M. (2024) Comet science with groundbased and spacebased surveys in the new millennium. In *Comets III* (K. J. Meech, M. R. Combi, D. Bockelée-Morvan, S. N. Raymond, and M. E. Zolensky, eds.), pp. 193–211. Univ. of Arizona, Tucson, DOI: 10.2458/azu_uapress_9780816553631-ch007.

Comet Science with Groundbased and Spacebased Surveys in the New Millennium

J. M. Bauer
University of Maryland

Y. R. Fernández
University of Central Florida

S. Protopapa
Southwest Research Institute

L. M. Woodney
California State University

We summarize the comet science provided by surveys. This includes surveys where the detections of comets are an advantageous benefit but were not part of the survey's original intent, as well as some pointed surveys where comet science was the goal. Many of the surveys are made using astrophysical and heliophysics assets. The surveys in our scope include those using groundbased as well as spacebased telescope facilities. Emphasis is placed on current or recent surveys, and science that has resulted since the publication of *Comets II*, although key advancements made by earlier surveys [e.g., Infrared Astronomical Satellite (IRAS), Cosmic Background Explorer (COBE), Near-Earth Asteroid Tracking (NEAT), etc.] will be mentioned. The proportionally greater number of discoveries of comets by surveys have in turn yielded larger samples of comet populations and subpopulations for study, resulting in better-defined evolutionary trends. While providing an array of remarkable discoveries, most of the survey data has been only cursorily investigated. It is clear that continuing to fund ground- and spacebased surveys of large numbers of comets is vital if we are to address science goals that can give us a population-wide picture of comet properties.

1. INTRODUCTION

Over the span of the last decade and a half, a more automated approach to the analysis of data has become common. This is in part owing to the arrival of datasets that are so large that each of the observations is not practically analyzed by human interaction, but rather is conducted by automated pipeline. Analysis routines are prototyped, tested, and applied to larger datasets, while outliers and diagnostic triggers indicate where special circumstances apply and further manipulation, or rejection, of the data are required. Much of this has been driven by the advent of the vast quantities of data provided by automated sky surveys. Additionally specialized datasets that are the product of targeted observations now have a certain expectation of providing statistically significant samples large enough for outliers to be identified and trends to be discerned.

Previous generation surveys generally had relatively small sample sizes. Many of these surveys, demonstrably the spacebased surveys, made robust discoveries with these smaller samples. *Lisse et al.* (1998) found temperature excess in dust comae from observations of five comets at perihelion distances ~1 au obtained by the Cosmic Microwave Background Explorer (COBE). Röntgen Satellite (ROSAT) observations of 6 (*Dennerl et al.,* 1997), and later 11 (cf. *Lisse et al.,* 2004), comets revealed charge exchange between highly charged heavy ions in the solar wind and cometary neutrals dominated cometary X-ray emissions. A subsequent survey by *Bodewits et al.* (2007) of eight comets with the Chandra observatory found that the characteristics of observed X-ray spectra mainly reflect the state of the local solar wind. The Infrared Astronomy Satellite (IRAS) mission data provided the first thermal dust trail measurements from eight identified comets (*Sykes and Walker,* 1992). Such surveys had large impacts on the cometary field but did not employ the more automated large-sample approaches, such as with astroinformatic techniques (cf. *Borne et al.,* 2009), now utilized for larger survey samples.

For the definition of survey within these pages, we include sample sizes of 20 or greater, owing to the conditions that even for simple statistical correlation tests, sample sizes ~20 or greater are required to achieve 95% confidence values even for strong correlations (cf. *Bonnet and Wright,* 2000). Here we concentrate on classical comet populations [long-period comets (LPCs) and short-period comets (SPCs)] and their dynamically defined subclassifications (cf. *Levison,* 1996). SPCs are defined to have orbital periods <200 years, and LPCs having orbit periods $\gtrsim$200 years. Jupiter-family comets (JFCs), for example, are a subclass of SPCs with orbital periods $\lesssim$20 years and prograde low orbital inclinations $\lesssim$40°, while dynamically new comets are a subclass of LPCs with original orbital semimajor axis values $\gtrsim 10^4$ au. Generally speaking, the source of LPCs is the Oort cloud while the Kuiper belt feeds the population of SPCs. Notably, Halley-type comets (HTCs) have historically often been lumped with SPCs, but most of them are likely to be highly-evolved (in the dynamic sense) objects from the Oort cloud. Thus they are more closely related to the LPCs. There are also different opinions on the meaning of the term "Oort-cloud comet" (OCC); e.g., it may only include dynamically new comets, or it may include all LPCs and HTCs that were in the Oort cloud any time in the past.

Measurements of large populations from single platforms and the same, or similar, instrumentation provide a basis for comparative samples, in contrast with compilations (cf. *A'Hearn et al.,* 1995, 2012; *Lisse et al.,* 2020). Such samples of cometary physical properties may be targeted, such as narrow-band filter surveys (cf. *Schleicher and Farnham,* 2004) or spectroscopic surveys (cf. *Dello Russo et al.,* 2016), or serendipitous observations, such as the data obtained with many groundbased or spacebased sky surveys. [Here "sky survey" refers to a survey that covers regions of the inertial frame, or background, sky with target coordinates fixed in the equatorial, ecliptic, or galactic coordinate frames, as opposed to moving targets (or solar system objects).] These two categories are significantly different in the selection of the objects observed, and how representative the samples are of the background populations.

In the first case of targeted samples, known solar-system object targets are selected based on their optical brightness, and were discovered often by sky surveys. They are often observed at preferred geometries (opposition, for example, for groundbased telescopic surveys) and detected while they are most active. As such, there are potential selection biases in the sample that may skew projection of behavior or physical properties of the base population. For targeted observations the observing time can be selected to sample the points through a comet's orbit where the expected levels of activity are best matched to the physical property of interest. For example, optical surveys of comets at aphelion may provide more accurate absolute magnitude values of the nucleus, leading to better-derived reflectances if the size of the body is known. Alternatively, a more comprehensive inventory of gas species may be derived at perihelion where so-called hypervolatiles and water-related species are released, and following a comet through its orbit may reveal when particular species dominate the activity.

1.1. Survey Discoveries of Comets

Prior to the 1990s, comets were generally discovered either in large photographic plate exposures or by individuals that visually scanned the sky, often employing specialized telescopes or binoculars with fast optics. In the late 1980s and early 1990s, digital cameras began to be employed in regular searches of the sky for solar system objects (cf. *Scotti et al.,* 1991) with a handful of early comet discoveries. In addition, astrophysical sky surveys were conceived to identify transient behavior, like supernova events, in extrasolar-system objects. With the advent of the earliest digital sky surveys, the automated surveys began to make significant contributions in the number of comet discoveries in the mid-1990s. These foundational surveys employed charge-coupled device (CCD) cameras with fields of view that by today's standards would be quite modest, on the order of 1° on each side (cf. *Pravdo et al.,* 1999), and rapidly began to outpace other means of discovery.

The efforts to detect solar system objects have been largely driven by the intent to discover and characterize near-Earth objects (NEOs). Observing cadences, pointing strategy, and approaches to analysis were therefore optimized or prioritized toward these NEO-related goals. These efforts have been remarkably effective, and discovery of more than 83% of the known NEOs has been the result of these efforts (*Landis and Johnson,* 2019). As a means of discovering comets, the NEO search programs have also been effective, not only with the comets that are a component of the NEO population, but also with more distant comets.

As of September 30, 2021, approximately 3586 comets had been discovered, as registered by the Jet Propulsion Laboratory (JPL) small bodies database (*https://ssd.jpl.nasa.gov/tools/sbdb query.html*). A summary of leading discovery platforms is provided in Table 1. According to the database, 41% of the comet discoveries listed were discovered by the Solar and Heliospheric Observatory (SOHO) [with the Solar Wind ANisotropy (SWAN) experiment and Large Angle and Spectrometric COronagraph (LASCO) instrument] or the Solar Terrestrial Relations Observatory (STEREO) spacecraft. In total, including these surveys, over 71% of comet discoveries up to October 2021 have been made by sky surveys. It is worth noting that the Minor Planet Center count (~4430 comets as of September 30, 2021), and future counts in the near term, are likely to be higher, as a significant remainder of the data from the SOHO spacecraft have yet to be processed and the JPL number includes only those comets that have been observed by other nonsolar-observing platforms in addition. Figure 1 shows the annual number of discoveries and observations of objects by sky

TABLE 1. Comet discoveries by selected sky surveys.

Survey	Location*	SPCs†	LPCs†	Total
Catalina	G96, 703, I52	200	193	393
Pan-STARRS	F51, F52	112	142	254
LINEAR	704	91	128	219
NEAT	566, 644	39	16	55
ATLAS	T05, T07, T08	16	35	51
NEOWISE	C51	16	23	39
Spacewatch	291, 691	16	12	28
LONEOS	699	17	5	22
ZTF/PTF	I41	2	16	18
SOHO/ SWAN‡	249	12	1466	1478
STEREO‡	C49	1	8	9

* The Minor Planet Center Observatory Code contributors.
† Short-period comets (SPCs) with orbital periods <200 years and long-period comets (LPCs) with orbital periods ≥200 years.
‡ Sun-looking survey total.

Note that the count of SOHO-discovered comets includes only those contributions with additional non-SOHO observations (see text).

surveys reported to the MPC and listed in the JPL database. The drop-off in the discoveries near 2010 coincides with the curtailment of Sun-pointing spacecraft survey data by the MPC, which has been recently resumed (*Battams and Boonplod,* 2020), although it has not yet encompassed the multi-year backlog (*Battams and Knight,* 2017).

1.2. Survey Observations

Along with a marked increase in the number of comet discoveries brought through groundbased surveys, the number of observations of comets has increased as well. Figure 1 shows that the number of observations closely tracks the number of objects observed by the surveys. (Note that the drop in 2021 in Fig. 1 is owing to the tally for that year being derived from the mid-year numbers.) On average, an object is observed on the order of 10× per year, during its range of detectability, e.g., while the comet passes through its perihelion. Table 2 lists the number of observations from each of the leading five surveys at five-year intervals back to 2000. The table shows that the number of comet observations is relatively small compared with the total observations of small bodies. It also reveals the slowly changing ranks (in order of total observations) in the lead surveys. The output of some very active programs are temporarily diminished [cf. Near-Earth Asteroid Tracking (NEAT) in the year 2000]; each program either upgrades and incorporates more sites or becomes outpaced by competing surveys, in which case existing survey programs or sites often shift to a priority from discovery to highly productive follow-up.

Both groundbased and spacebased sky surveys have been used to characterize cometary populations. However, the full and systematic utilization of the majority of data obtained by the surveys is in its early stages, with only a handful of instances of the data being used to quantitatively characterize the comet populations. Much of the initial exploration of these sky survey datasets are centered around characterization of particular comets of interest. *Dobson et al.* (2021) use Asteroid Terrestrial-impact Last Alert System (ATLAS) data to identify the longevity of 95P/Chiron's 2018 onset of activity. Zwicky Transient Factory (ZTF) survey (*Kelley et al.,* 2021), Transiting Exoplanet Survey Satellite (TESS) spacecraft (*Farnham et al.,* 2019), and Near-Earth Object Wide-field Infrared Survey Explorer (NEOWISE) survey (*Bauer et al.,* 2021) observations were used to monitor and characterize the behavior of 46P/Wirtanen during its 2019 perihelion approach. Investigations of statistically significant samples of comet populations (cf. *Farnham et al.,* 2021) are likewise facilitated by sky surveys, and are beginning to be analyzed using astroinformatic approaches. Larger surveys that compile cometary populations to constrain populations statistics are rarer still. *Hicks et al.* (2007) reported magnitudes and Afρ values (cf. *A'Hearn et al.,* 1984) for 52 comets observed by NEAT between 2001 and 2003, and produced estimates of nucleus size for 25 of the lowest-activity comets in the sample. Searches for cometary activity among asteroids are more common (see also the chapter by Jewitt and Hsieh in this volume). *Waszczak et al.* (2013) searched for undiscovered main-belt comets, but identified 115 comets in the Palomar Transient Factory (PTF) data taken from 2009 through 2012, listing the maximum and minimum magnitudes observed in the images. *Sonnett et al.* (2011) observed 924 asteroids, and *Hsieh et al.* (2015) conducted a large search of main-belt objects observed in Panoramic Survey Telescope and Rapid Response System (Pan-STARRS) data to find cometary activity, while *Martino et al.* (2019) and *Mommert et al.* (2020) searched for activity among asteroids in comet-like orbits.

1.3. Survey Biases

It is abundantly clear from the previous discussion that sky and sun-pointing surveys have had a remarkable impact on our statistical understanding of the comet populations. However, these data possess limitations, according to their sensitivities, coverage strategies, and cadences. Non-survey observations remain, therefore, highly valuable to the community, as demonstrated in the discovery of the second interstellar object (cf. *Borisov and Shustov,* 2021). Figures 2 and 3 are illustrative of the sample biases that can remain even when short-term biases, like those imposed by weather, are removed or averaged over, and how they can be convolved

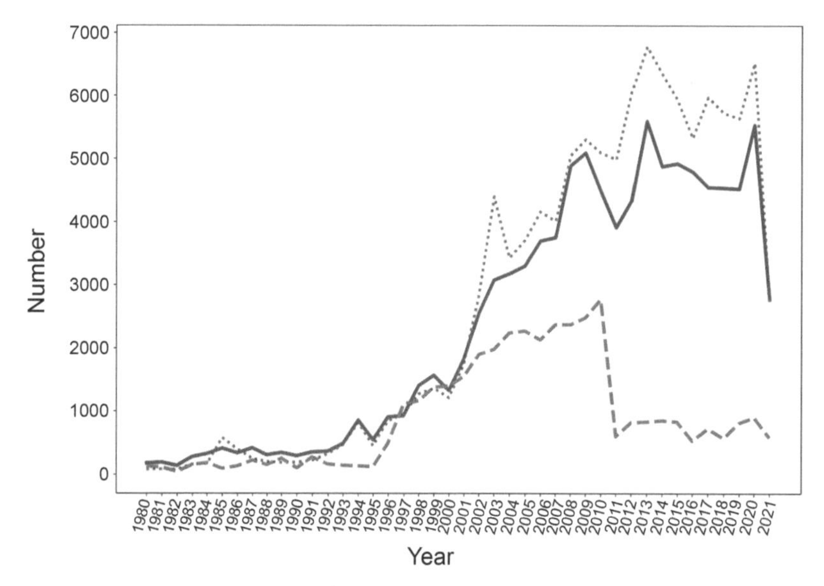

Fig. 1. Discoveries and reported observations of comets per year, 1980–2021. Green dashed line indicates discoveries per year (multiplied by 10) as listed in the JPL small bodies database (but see text for explanation of the 2010 drop-off). Blue solid line indicates the number of comets with observations reported to the MPC. Red dotted line indicates the number of observations reported to the MPC (divided by 10).

with real population features. The SPC inclination features are mostly real, and are dominated by the JFC population's clustering near low-inclination orbits. The outliers in inclination, around near-retrograde orbits, have contributions from the active Centaur and Halley-type comet populations. Eccentricity is nearly level, but falls off at near-zero values, corresponding to near-circular orbits; the high-eccentricity outliers on short-period orbits are in part strengthened by the near Sun-grazing comets seen by Sun-looking surveys (SLSs), or higher elongation, or terminator-pointing surveys (TPSs), like NEOWISE. [NEOWISE uses the repurposed Wide-field Infrared Survey Explorer (WISE) spacecraft to search for NEOs and other solar system bodies.] Comets tend to be most active as they approach toward and retreat from their perihelion distance, so the discoveries with the furthest perihelia are made first by the opposition-looking surveys, while the TPSs make the near-earth perihelion discoveries, and the remaining low-perihelion comets are found by SLSs. The dips near 80° and 270° in the SPC argument of perihelion (ω) distributions roughly correspond to where the tug of Jupiter disrupts SPC orbits.

For the LPCs, in contrast to the SPCs, the discoveries are dominated by SLSs, and thus a few high-inclination Sun-grazing comets, particularly the Kreutz-family comets. The NEOWISE weak cluster in inclination, near 105°, pointed out in *Bauer et al.* (2017), has become mildly more pronounced with additional survey data, and the peaks in ascending node (Ω) and ω are clustering from the noted Kreutz-family comets. In each survey approach, the success with particular populations show statistical outliers that can bias the derived distributions if naively extrapolated from the observed distributions or not carefully removed.

The large representative samples of comets discovered and observed by sky surveys facilitates analyses of their total populations that lead to constraints on their total numbers. Such derivations are common among other representative populations, for example, with NEOs (*Mainzer et al.,* 2012) and Centaurs (*Jedicke and Herron,* 1997). Accurate accounting of factors that affect the survey's detection efficiency, such as observing cadence, pointing pattern and viewing geometry, sensitivity, and weather (for groundbased surveys) are critical to the assessment of the underlying population numbers from the observations. Because these factors are intrinsic to each survey (or instrument/telescope combination), they must be considered for each separate contribution. Comets, however, are different from other populations in that they require an extra layer of accounting to derive the final total population numbers from the observed sample; the brightness variations from activity have to be accounted for as well. Comets tend to vary greatly in their brightness throughout their orbit, usually achieving peak brightness around, although often not precisely at, their perihelion. Even surveys with more predictable observing circumstances, e.g., spacebased surveys, have an additional significant level of uncertainty on any derived constraints of total populations.

The earliest estimates of background populations based on a modern sky survey was conducted by *Francis* (2005). The author used the Lincoln Near-Earth Asteroid Research (LINEAR) survey to assess the long-period comet population. *Francis* (2005) found a total population of $\sim 5 \times 10^{11}$ comets with a nucleus size of roughly 1 km in diameter, roughly a factor of ~2.5× that predicted by *Oort* (1950). An important detail is that because the small end of the comet size distribution is difficult to measure, many of the

TABLE 2. Yearly comet survey observations.*

Survey†	Total‡Detections§	Comets Observed	Comet Detections§
		2020	
Pan-STARRS	12181991	344	3606
ATLAS	10396137	284	11787
Catalina	10134103	335	4754
NEOWISE	152141	30	324
Spacewatch	70385	13	60
Yearly Total	32934757	1006	20531
Survey Fraction		*0.18*	*0.31*
		2015	
Pan-STARRS	7256500	176	2533
Catalina	3950145	206	1566
Spacewatch	512214	51	276
NEOWISE	158595	53	840
ATLAS	104495	189	5780
Yearly Total	11981950	514	5484
Survey Fraction		*0.10*	*0.09*
		2010	
Catalina	3296494	128	1119
NEOWISE¶	2410314	111	1477
LINEAR	2193193	78	1122
Spacewatch	852890	67	462
Pan-STARRS	597563	7	28
Yearly Total	9350456	391	4208
Survey Fraction		*0.09*	*0.08*
		2005	
Catalina	2325309	132	1342
LINEAR	2056210	98	1697
Spacewatch	1063276	46	287
NEAT	549313	37	207
LONEOS	803620	66	1513
Yearly Total	6797728	379	4046
Survey Fraction		*0.11*	*0.11*
		2000	
Spacewatch	2149917	14	134
LINEAR	2094140	84	1682
LONEOS	456892	50	343
Catalina	48878	7	36
NEAT	29	2	6
Yearly Total	2814856	157	2201
Survey Fraction		*0.09*	*0.11*

* Annual totals shown at five-year intervals.
† Non-solar-pointing surveys and follow-up programs.
‡ The total includes asteroids and comets.
§ Observations reported to the MPC; more complete summary available at *https://sbnmpc.astro.umd.edu*.
¶ *Bauer et al.* (2017) notes 164 comets observed by WISE/NEOWISE within the year, many retrieved by stacking. This number represents those detected by the automated detection pipeline.

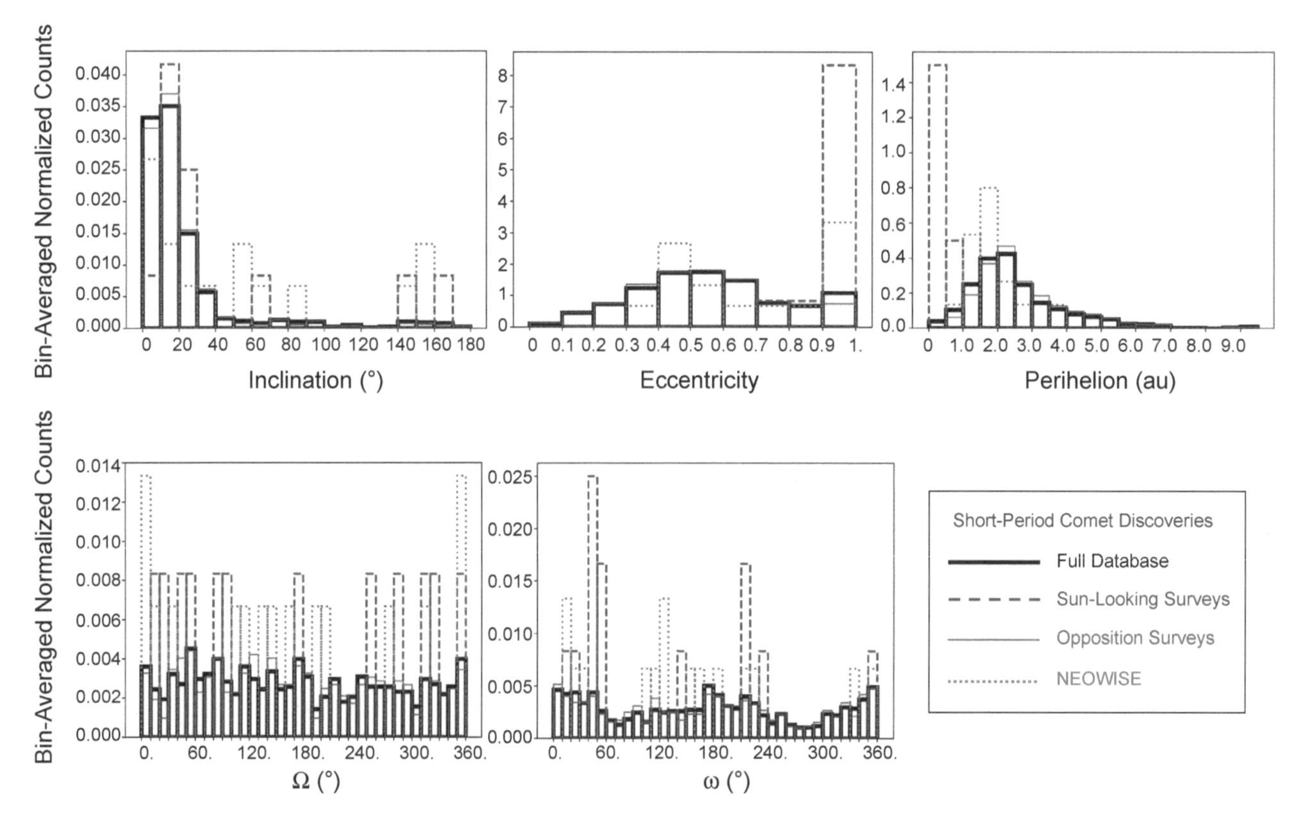

Fig. 2. The relative distributions of the orbital elements of short-period comet (SPC) discoveries from the JPL small bodies database are shown. The database's SPC populations are shown in total (black histogram) and compared with representative samples of contributions from Sun-looking surveys (blue dashed line; SOHO, SWAN, and STEREO comets), opposition-pointing surveys (red thin solid line), and terminator-pointing surveys (green dotted line; NEOWISE). Scaled distribution histograms are shown for orbital inclination (upper left), eccentricity (upper middle), perihelion (upper right), ascending node (lower left), and argument of perihelion (middle left). Note that each histogram is plotted with bin-averaged, normalized counts, such that the sum of the counts times the bin-width across all the bins equal 1.

population estimates are for lower-bounded size ranges. The difficulty in assessing the comet populations with effective diameters <1 km often results in values of $\gtrsim$1 km for the lower bound in size for population comparisons. The Survey of Ensemble Physical Properties of Cometary Nuclei (SEPPCoN) (*Fernández et al.,* 2013) provided constraints on the JFC population between 2000 and 10,000 objects with diameters of approximately 1 km or larger. *Bauer et al.* (2017), using the NEOWISE survey data, arrived at a number that fell within the lower end of that range, ~2100 JFCs. Applying a similar technique to the observed LPCs, *Bauer et al.* (2017) found a total population of 1.3×10^{12} OCCs, about twice that of the LINEAR-derived value by *Francis* (2005), and also found that the majority of LPCs, ~60%, were already detected by contemporary surveys. It is worth noting that since 2015, the rate of the discovery of non-Sun-grazing LPCs has held an average of 6.2 comets per year with perihelia within 1.5 au per year. Most recently, the PanSTARRS survey has been assessed and debiased to obtain JFC and LPC population constraints (*Boe et al.,* 2019). The comparative size constraints will be discussed in section 4, but population totals for JFC and LPC comets find similar numbers, with ~10^{12} Oort cloud objects as the speculative total.

2. SURVEYS OF COMETARY DUST

2.1. Broadband Visible Imaging

While spectroscopy in the visible and infrared wavelength ranges is the most diagnostic tool to investigate the composition of cometary dust, including both refractories and ice compounds, photometry, especially in the visible-wavelength range, has been the technique most used to characterize a large number of objects. Imaging with broadband filters in the visible-wavelength range, yielding color indices, enables measurement of the solar light scattered by cometary dust particles, from which it is possible to infer first-order compositional information such as particle size and ice to refractory ratios. Analysis of possible correlations between optical colors and orbital parameters can highlight composition diversity among subpopulations possibly attributable to variations at formation in the protoplanetary disk and/or evolutionary processes.

Solontoi et al. (2012) (see Table 3 for a broader context of surveys since *Comets II*) compiled u-, g-, r-, i-, and z-band photometry of 26 active comets (6 LPCs and 20 JFCs) observed by the Sloan Digital Sky Survey (SDSS) (*York et al.,*

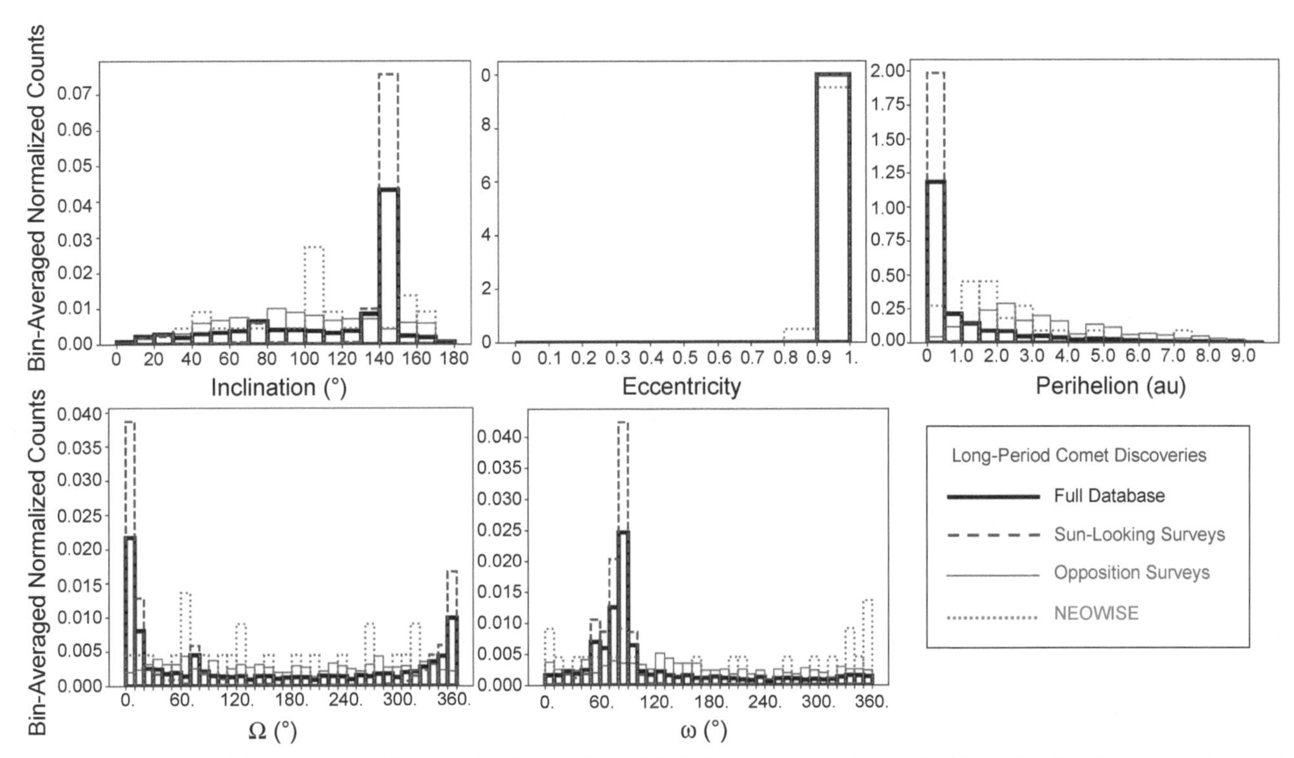

Fig. 3. Similar to Fig. 2, the relative distribution of the orbital elements are shown, but here for the long-period comet (LPC) discoveries from the JPL small bodies database. The database's LPC populations are shown in total (black histogram) and compared with representative samples of contributions from Sun-looking surveys (blue dashed line; SOHO, SWAN, and STEREO comets), opposition-pointing surveys (red thin solid line), and terminator-pointing surveys (green dotted line; NEOWISE). Scaled distribution histograms are shown for orbital inclination (upper left), eccentricity (upper middle), perihelion (upper right), ascending node (lower left), and argument of perihelion (middle left).

2000), spanning a range of heliocentric distances between ~1 and ~6 au (observations of unresolved comets are not considered). *Jewitt* (2015) extended the work by *Solontoi et al.* (2012) by presenting new B V R photometric measurements for 23 active LPCs obtained with the 10-m-diameter Keck I telescope at Mauna Kea and the Low Resolution Imaging Spectrometer (LRIS) camera (*Oke et al.,* 1995). This dataset not only quadrupled the number of LPCs for which colors are available, but also broadened the heliocentric distance range with measurements up to 18 au from the Sun. After transforming the measurements by *Solontoi et al.* (2012) in the Sloan filter system into B V R I photometry (*Ivezić et al.,* 2007), *Jewitt* (2015) investigated possible correlations between optical colors and orbital parameters for the combined SDSS + Keck dataset (see Fig. 4). No significant difference was found between the mean colors of active SPCs and LPCs (Fig. 4a), suggesting the lack of compositional variation between these two groups. The author pointed out the agreement between this finding and gas-phase studies reported by *A'Hearn et al.* (2012) and *Cochran et al.* (2015) and attributed it to the idea, already put forth by *A'Hearn et al.* (2012), that JFCs and LPCs formed in largely overlapping regions of the protoplanetary disk.

No trend was found between the B–V and V–R colors and heliocentric distance (R_h, Fig. 4b,c). *Jewitt* (2015) attributed this evidence to (1) the ice-to-dust ratio being on the order of only a few percent and (2) small particles not being abundant enough to dominate the scattering cross section. The first conclusion relies on the idea that solid-state water has been detected in cometary comae (*Davies et al.,* 1997; *Kawakita et al.,* 2004; *Yang et al.,* 2014; *Protopapa et al.,* 2018); it is stable at large heliocentric distances with sublimation rates varying inversely with heliocentric distance and it is bluer than refractory materials. Therefore, an ice-to-dust ratio larger than a few percent would lead to bluer colors with increasing heliocentric distance, contrary to what was observed. The second conclusion leans upon the expectation that, at large heliocentric distances, given the lower gas flow, the mean size of the ejected particles should fall into the Raleigh regime ($X = \pi D/\lambda \ll 1$, where D is the particle diameter and λ is the wavelength of observation), yielding to blue colors (*Bohren and Huffman,* 1983). However, *Gundlach et al.* (2015) found through numerical modeling and laboratory results that the size range of the dust aggregates able to escape from a comet nucleus into space widens when the comet approaches the Sun and narrows with increasing heliocentric distance. This is because the tensile strength of the dust aggregates decreases with increasing aggregate size. Therefore, at large heliocentric distances, given the lower gas flow, only large aggregates would be lifted off the nucleus. These arguments, which rely on the assumption that comets have formed by gravitational instability (*Skorov and Blum,* 2012; *Blum et al.,* 2014), weaken the conclusion that small particles are not abundant enough to dominate the scattering cross section.

TABLE 3. Summary of selected surveys since *Comets II*.

N		N_{obs}	λ	Location	Designated Asset	Technique	Telescope or Survey	Instru-ment	Reference
SPCs	LPCs								
0	150	150	735, 870 nm	G	A	I	Pan-STARRS 1	—	*Boe et al.* (2019)
95	56	~3000	3.5, 4.6, 11, 22 μm	S	A	I	NEOWISE	—	*Bauer et al.* (2015, 2017), *Kramer et al.* (2015)
100	0	~200	16, 22, 24 μm	S	A	I	Spitzer	IRS, MIPS	*Fernández et al.* (2013) *Kelley et al.* (2013)
34	0	34	24 μm	S	A	I	Spitzer	MIPS	*Reach et al.* (2007)
18	4	38	3, 6, 4, 5 μm	S	A	I	Spitzer	IRAC	*Reach et al.* (2013)
23	1	30	B,V, R, I	S	A	I	HST	WFPC2	*Lamy and Toth* (2009)
17	44	3700	Ly-α	S	H	I	SOHO	SWAN	*Combi et al.* (2019)
44	0	~200	R	G	A	I	various†	various†	*Snodgrass et al.* (2011)†
0	23	29	B,V, R, I	G	A	I	Keck I	LRIS	*Jewitt* (2015)
24	6	35	u, g, r, i, z	G	A	I, S	SDSS	CCD	*Solontoi et al.* (2012)
6	14	25	13–14 μm	G	P	S	IRTF	BASS	*Sitko et al.* (2004)
42	0	53	Johnson R	G	P, A	I	Kiso 1.05m	CCD	*Ishiguro et al.* (2009)
100	0	215	various	G, S	P	I	various	various	*Mazzotta Epifani et al.* (2009)
28	22	218	near-UV, visible	G	P	S	UA 1.54m	CCD	*Fink* (2009)
77	53	558	near-UV, visible	G	P	S	McDonald	CCD	*Cochran et al.* (2012)
11	19	152	2.5–5 μm	G	P, A	S	various*	various*	*Dello Russo et al.* (2016)
8	12	54	2.5–5 μm	G	A	S	Keck II	NIR-SPEC	*Lippi et al.* (2021)
—	50	152	Johnson R	G	P, A	I	0.6/0.9/1.8m Schmidt	CCD	*Sárneczky et al.* (2016)
~100	~100	~1000	near-UV, visible	G	P	I	Lowell		*Bair and Schleicher* (2021)

* NASA-IRTF with CSHELL, Keck II with NIRSPEC, Subaru with IRCS, and VLT with CRIRES.
† A variety of telescopes were used by the same group for the survey. The paper mentioned here compiles all the group's previous results from earlier papers.

We limit this table to surveys observing 20 or more comets. N refers to the number of comets (of SPCs and LPCs) in the survey; N_{obs} refers to the total number of observations made of those N comets; λ indicates the primary wavelength(s) of survey's observation. In the "Location" column, G = groundbased, S = spacebased. The "Designated Asset" column gives the original designation, which presumably drove the platform's design requirements; A = Astrophysics, P = Planetary, and H = Heliophysics. In the "Technique" column, I = imaging and S = spectroscopy.

Not all broadband comet surveys provide color information, but they still measure activity and dust production. *Sárneczky et al.* (2016) measured Afρ values and the coma slope parameters for 50 LPCs with known activity beyond 5 au. The Afρ quantity is the product of albedo (A), filling factor of the grains within the field of view (f), and the linear radius of the field of view at the comet (ρ), while the slope parameter is defined as d log Afρ/d log ρ (*A'Hearn et al.,* 1984). These parameters are diagnostic of the comet activity and are a proxy for the dust production rate and the morphological appearance of the coma. *Sárneczky et al.* (2016) divided the LPC sample into dynamically new ($a > 10^4$ au) and returning comets, and found that on average the Afρ of dynamically new comets significantly exceed those of the recurrent LPCs, similar to the earlier work of *Meech et al.* (2009) with a smaller sample. Furthermore, they found that new comets usually exhibit negative (shallow) slope (d log Afρ/d log ρ) parameters and symmetric comae. They found that the comets that were strongly active beyond 10 au tend to have a smaller increase in Afρ with decreasing heliocentric distance.

2.2. Thermal Dust

Analogously to the Afρ parameter introduced by *A'Hearn et al.* (1984) (see section 2.1) as a proxy for the dust production rate measured at visible wavelengths for scattered light observations, *Kelley et al.* (2013) used the εfρ parameter for

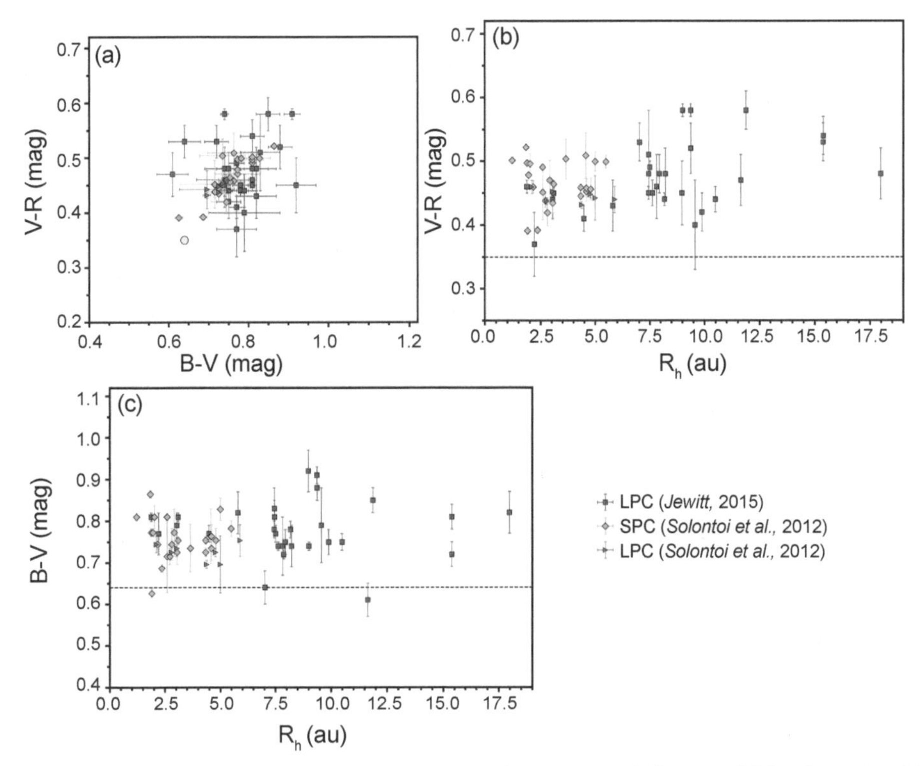

Fig. 4. (a) Color–color diagram comparing LPCs from *Jewitt* (2015) with LPCs and SPCs from the SLOAN survey reported by *Solontoi et al.* (2012). The color of the Sun is marked by a yellow circle. **(b)** V–R and **(c)** B–V colors of the same comets are shown as a function of heliocentric distance in astronomical units. The dashed horizontal line represents the color of the Sun.

observations of thermal emission from comet comae introduced by *Lisse et al.* (2002). The effective emissivity of the grains is parameterized by ε, while f is the areal filling factor within an observed aperture of radius ρ. *Bauer et al.* (2015) carried out a comparison between the log εfρ as derived from the WISE W3 (12 μm) and W4 (22 μm) channel dust emission and the log Afρ values derived from the W1 (3.4 μm) channel assuming dust signal was dominated by reflectance. For the comets observed at heliocentric distances exceeding 3 au, the difference between the values (~0.86 ± 0.1) were found to be consistent with dark dust with emissivity near 0.9. Comets observed at heliocentric distances inside 3 au deviated from this trend, possibly owing to the different size range of aggregates lifted by activity at smaller rather than larger heliocentric distances, given the higher gas flow, or to the more pronounced thermal emission component within the 3.4-μm band signal for the dust at small distances. Outside of 3 au, the sample was dominated by LPCs; only four SPCs were measured outside of 3 au.

The physical temperature of cometary dust grains is a function of composition, grain size, and morphology. An approximation for the true physical temperature of the grains is given by the color temperature T_c determined through analysis of the thermal spectral energy distribution over a defined wavelength range (e.g., *Wooden,* 2002; *Kolokolova et al.,* 2004). The temperature excess of a cometary coma over that of a blackbody is usually parameterized by the superheat parameter S_{heat} introduced by *Gehrz and Ney* (1992) and defined as the ratio of the color temperature and the temperature of an isothermal blackbody sphere in LTE at the same heliocentric distance (TBB = 278 $R_h^{-1/2}$). *Bauer et al.* (2015) reported color dust temperatures based on the WISE W3 and W4 band thermal fluxes for 24 SPCs and 14 LPCs while *Kelley et al.* (2013) reported color temperatures for 15 SPCs through analysis of the 16- and 22-μm Spitzer Space Telescope (SST) Multiband Imaging Photometer for Spitzer (MIPS) observations (Fig. 5). No differences were found between SPCs and LPCs (*Bauer et al.,* 2015). Overall, cometary comae color temperatures measured with WISE displayed a slight excess of 1.6 ± 0.1% [the error-weighted mean of all the superheat measurements is $\langle T_c/T_{BB} \rangle$ = 1.016 ± 0.001] and were found to be consistent with isothermal bodies with emissivity ~0.9 and albedo 0.1 at 3.4 μm (*Bauer et al.,* 2015). A more significant temperature excess was found by *Kelley et al.* (2013), who reported an error-weighted mean of all their color temperature measurements of $\langle T_c/T_{BB} \rangle$ = 1.074 ± 0.006. This translates in the color temperature of the dust to be 7.4 ± 0.6% warmer on average than an isothermal blackbody sphere in LTE. An important caveat to consider when assessing the validity of color temperatures obtained from broadband infrared photometry is the possible presence of emission features, specifically silicate features, above the continuum that could affect the thermal flux measurements and consequently the color temperature estimates. Therefore, to properly characterize the thermal properties of cometary grains, spectroscopic data of a large number of comets are

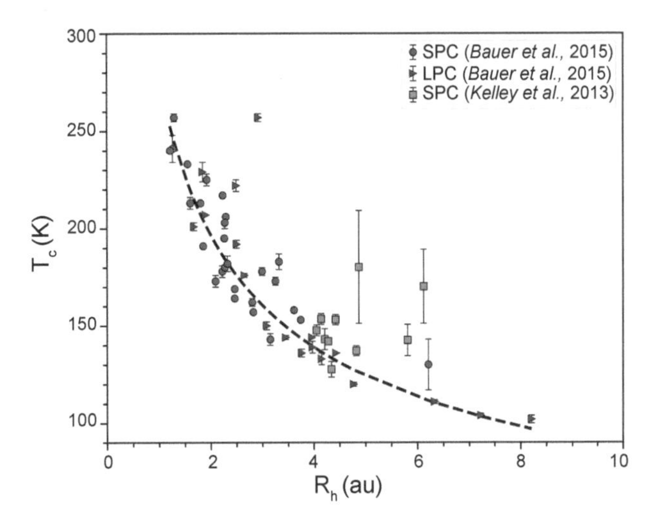

Fig. 5. Coma color temperatures as a function of heliocentric distance comparing LPCs from *Bauer et al.*(2015) with SPCs reported by *Bauer et al.* (2015) and *Kelley et al.* (2013) (only color temperatures of the dust centered on cometary nuclei are shown). The dashed line represents the temperature of an isothermal blackbody (278 $R_h^{-1/2}$).

required. *Sitko et al.* (2004) analyzed spectroscopic data over the wavelength range 3–14 μm of 20 comets belonging to different dynamical classes and found cometary grains radiating at temperatures in excess of that of a blackbody at the equilibrium temperature expected for their heliocentric distances. This effect is expected for a grain population that includes a significant fraction of the grains with sizes smaller than the wavelength of light being radiated, in this case from 3 to 14 μm. Additionally, *Sitko et al.* (2004) found a strong correlation between excess temperature and silicate band strength for dynamically new and long-period comets, confirming the results by *Gehrz and Ney* (1992) and *Williams et al.* (1997). The majority of Jupiter-family objects were found to deviate from this relation. To explain the different trend between JFCs and dynamically new comets, the authors put forth the idea of a radial gradient in the size distribution of silicate grains within the protoplanetary disk. Further observations are required to confirm this finding.

Mid-infrared broadband images of comets were found to be well suited not only for characterizing the properties of the dust grains but also to investigate the activity level of comets. *Kelley et al.* (2013), using SST images acquired as part of the SEPPCoN survey, investigated the activity of 89 JFCs at 3–7 au from the Sun and found that activity, detected in at least ≈24% of the comet sample, is significantly biased to post-perihelion epochs. Additionally, *Kelley et al.* (2013) suggested a bias in the discovered JFC population given that low-activity comets with large perihelion distances were found missing from the survey sample. The link between activity level and present and historical orbital parameters was also investigated by *Mazzotta Epifani et al.* (2009) through analysis of a sample of 90 SPCs as seen from the ground and space at heliocentric distances greater than 3 au. This analysis led to several findings including, but not limited to, the higher likelihood of SPCs being active post-perihelion rather than pre-perihelion, similarly to what was found by *Kelley et al.* (2013); the lack of a sharp cut-off in heliocentric distance marking the activity-fade; and a weak trend of comets with increasing perihelion distance to be more likely active at large heliocentric distance.

Not all analyses of thermal dust are based on flux measures. *Reach et al.* (2007) conducted a survey of 34 JFCs at 24 μm using SST's MIPS, and found that the majority (27) of the comets exhibited detectable trails. *Ishiguro et al.* (2009) confirmed the prevalence of JFC-associated dust trails, and found that 6 out of 42 JFCs exhibited trails detectable at visual-band wavelengths. By overplotting zero-velocity syndynes (cf. *Finson and Probstein,* 1968), *Reach et al.* (2007) found that the dust particles were dominated by millimeter- and centimeter-sized dust, and that the size distribution of the dust particles is not accurately modeled by a single power-law. *Kramer et al.* (2017) and *Kramer* (2014) demonstrated a novel technique of fitting the cometary dust in the thermal infrared observations using WISE/NEOWISE data. Subsequent application of the technique (*Kramer et al.,* 2015) to a sample of 89 comets shows that such techniques can elucidate the behavior of dust output of comets when they are most active. Using radiation pressure to sift the different ranges of particle sizes in the tail in combination with the dust thermal signal, they found that much of the dust is emitted preceding perihelion, and that the dust mass is primarily residing in millimeter- to centimeter-sized grains.

3. SURVEYS OF COMETARY GAS

By nature cometary gas surveys are slow, requiring multi-year commitments to observing the same sorts of objects as they become available one at a time. Whether waiting for a comet bright enough, close enough, or at the right phase angle, there may only be one or two comets available per year for which a detailed study of multiple species is feasible. Spacecraft visits may show us a snapshot, but the long-term surveys of composition have been critical to telling us what comets have in common, when a snapshot observation is surprising or different, and are beginning to help us understand how comets evolve. Surveys of cometary gas have been critical to the development of our understanding of what comets are, what they are made of, and how that composition relates to the origin of our solar system. The history of how surveys of cometary gas have shaped our understanding of comets is already well summarized by *Cochran et al.* (2015), *Bockelée-Morvan et al.* (2015), as well as in the chapter by Biver et al. in this volume, so here we shall only note a few key points and recent developments.

3.1. Groundbased Surveys

Early optical spectroscopic surveys are what told us that while all comets contained the same molecular species, that those species came in differing abundances. This led to what has become the grail of modern cometary gas studies: whether there are compositional classes of comets that can be tied to dynamical origin, and thus constrain our knowledge of the composition and temperatures in the protoplanetary disk. The promising early results of *A'Hearn et al.* (1995) demonstrated that there exists a class of comets, dominated by JFCs, that are depleted in carbon-chain molecules. They were followed by multiple groups confirming this result, although finding enough OCCs in the carbon-depleted class that the tie between depletion and dynamical origin is clearly not a simple one (*Cochran et al.*, 2012; *Fink*, 2009; *Langland-Shula and Smith* 2011; *Bair and Schleicher* 2021). Infrared and millimeter-wave surveys have made the picture more complicated with no consensus on a classification scheme [cf. the compilations of *Mumma and Charnley* (2011), *Bockelée-Morvan and Biver* (2017), and *Dello Russo et al.* (2016)], but new analyses of some of these observations such as that of *Lippi et al.* (2021) are beginning to find ways to disentangle the numerous abundances and find patterns.

Lippi et al. (2021) also demonstrate the necessity of continued surveys in the infrared and submillimeter-wavelength regimes. The authors admit significant limitations to the 20 comets within their survey [or the 33 in *Dello Russo et al.* (2016)]. *Harrington Pinto et al.* (2021) is working to compile matched observation of comets with CO and CO_2 production rate constraints. Yet *Lippi et al.* (2021) affirm that presently, the overlapping samples across optical through the submillimeter are too sparse to deconvolve compositional states from cometary origin and evolutionary effects. In order to successfully make sense of the compositional trends, the species detectable over the full range of wavelengths must be characterized for numbers of comets comparable to those where optical spectroscopic analyses are presently available.

While we generally think of gas surveys as a means to understand relative composition, the physical state of the gas species is a key detail in understanding comet comae as well. Outflow velocity is a critical characteristic in determining gas production rates, dissociation scales, and total mass loss. Line-of-sight velocity of the gas can be measured via spectroscopy done at sufficient resolution to detect doppler line broadening of emission lines. Since coma gas velocity is generally close to ~1 km s^{-1}, sufficient velocity resolution is currently achievable with groundbased observing via radio observations at submillimeter and longer wavelengths. Furthermore, when radio spectroscopy includes spatial mapping, and observations are made at a 90° phase angle, the survey can distinguish potential asymmetries in sunward and anti-sunward outgassing velocities. In their survey mapping the OH coma at 18 cm in 28 comets, *Lovell and Howell* (2015) found the gas outflow velocity beyond 1 au was 0.8 km s^{-1} regardless of size, production rate, or direction. While at this time the Atacama Large Millimeter/submillimeter Array (ALMA) is still too new for there to be enough observations of comets to qualify as a survey, its potential to spatially map both coma compositions and velocities will produce a unique survey to look forward to.

3.2. Spacecraft Surveys

The advantages of spacecraft are clear and are partially addressed in section 1, but for the purposes of studying gas production have additional advantages. The often higher resolution provided by space platforms can facilitate the measurement of product decay scales and associations with nucleus orientations and features (cf. *Fougere et al.*, 2016), However, spacebased telescopes and instruments are often designed with both solar-system and non-solar-system targets in mind. Hence bandpasses, resolution, and spatial scales may be only moderately suited to the measurements used to place constraints on the given species.

The Solar Wind ANisotropies (SWAN) instrument on the SOHO spacecraft, for example, was designed to measure H-alpha line emission associated with large-scale structures in the solar wind (*Bertaux et al.*, 1995). However, comets manifest h-alpha emission, 90% of which is produced by water dissociation mechanisms. Hence, SOHO's SWAN instrument has been effectively used to measure pre- and post-perihelion production of water in 61 comets (*Combi et al.*, 2019). These have provided power-law relationships for water production (Q_{H_2O}) vs. heliocentric distance (R_h) in 44 LP and 17 SP comets. *Combi et al.* (2011) demonstrated the methodology employed in the analysis. Power-law fits were provided for the pre- and post-perihelion approaches of the comets using the relation

$$Q_{H_2O} = Q_{1\,au} R_h{}^p \quad (1)$$

where $Q_{1\,au}$ is the water production rate extrapolated to when the comet is at R_h = 1 au, and p is the "slope" parameter (manifested as a slope in log units). Comparisons were made of the water production slopes according to dynamical subclasses of LPCs, dynamically new OCCs (with semimajor axis values greater than 20,000 au) and SPCs, and compositional subclasses of "carbon-depleted" and "typical" LPCs and SPCs (*A'Hearn et al.*, 1995). The dynamically new OCCs or younger comets tended to have less variation in slope values, clustering around values of ~–2 ± 1, and exhibited a possible steepening in slope as LPCs dynamically aged. For SPCs with measured nucleus effective radius values, a larger fractional active area correlated with comets with larger perihelion distances, consistent with less processing. The correlations with compositional classifications and LP pre- and post-perihelion observations were inconclusive.

Spacecraft observations can provide unique opportunities to assess gas species such as CO_2 that are not available to groundbased observations. CO_2 and CO are two drivers of cometary activity, outpacing water production in a limited range of circumstances, and their outgassing in these cases provides the dominant means of ejecting dust from the nucleus into the coma. CO can be detected from the ground at submillimeter and infrared wavelengths, but this is not the case for CO_2 emission, which is blocked by the absorption of CO_2 present in Earth's atmosphere. Alternative means of detecting CO_2 are under investigation (e.g., *Decock et al.,* 2013; *McKay et al.,* 2016, 2019) but direct detection, usually from the 4.26-μm infrared emission line, is the current means of assessing production rates. Both SST's Infrared Camera's (IRAC's) 4.5-μm imaging band and the WISE/NEOWISE 4.6-μm band contain both the infrared CO_2 and CO (4.67 μm) emission feature. The CO band relative to the CO_2 band is on the order of 11× weaker. However, without an accompanying CO observation from groundbased assets or spectroscopic data, there is no definitive way to determine which species causes excess in the ~4.5-μm channels of these two spacecraft. Furthermore, dust thermal emission signal dominates over CO + CO_2 excess at heliocentric distances within ~2 au. For comets at smaller heliocentric distances, the dust thermal signal must be well-characterized or the emission excess extremely pronounced (or both) for successful detections. That being said, *Reach et al.* (2013) produced measurements of 23 comets with CO or CO_2 emission using SST, and *Bauer et al.* (2015) measured 39 comets with CO or CO_2 excess using the WISE/NEOWISE (hereafter NEOWISE) survey data. *Reach et al.* (2013) attributed the majority of the excess to CO_2 production. Additionally, by comparing with literature measurements of Q_{H_2O} the SST survey concluded that water sublimation remained dominant out to ~2.8 au for most comets, and the highest-resolution IRAC images suggested more localized active regions for CO_2 production. The NEOWISE results suggested that approximately a quarter of comets observed had significant CO or CO_2 excess, and a Q_{CO_2} proportionality $\sim R_h^{-2}$ within 4 au. Outside 4 au LPCs tended to be the producers of CO or CO_2, and *Bauer et al.* (2021) attributed that to LPCs possibly being more CO-rich. Recent work by *Gicquel et al.* (2021) extends these analyses with an additional 52 comets observed by NEOWISE in 2014. It is also important to note that both the SST and NEOWISE surveys referenced the *Ootsubo et al.* (2012) Akari spacecraft results, although with a sample of 18 comets, smaller than the survey number threshold considered here, the attribution of CO_2 as the main species with the SST and NEOWISE surveys was at least in part based on these measurements. Furthermore, the Akari spacecraft spectra provided simultaneous water production comparisons and demonstrated that LPCs were generally producing CO_2 at greater distances and in some instances outpacing water production. About a fifth of the CO_2 producers had CO production that outpaced CO_2 production rates for the Akari-observed sample.

4. SURVEYS OF COMETARY NUCLEI

The basic parameters of nucleus observations have been described by, e.g., *Jewitt* (1991), but surveys of nuclei have historically been difficult to perform due to the problem of coma confusion. While we as a field have continued to make good progress on observing nuclei since the time of *Comets II*, we discuss here some of the effects that can fool us into misinterpreting nucleus photometry and thereby lead to systematic inaccuracies.

For a thorough review of the current understanding of nucleus ensemble properties, we refer the reader to the chapter in this volume by Knight et al., who discuss what we currently know about the sizes, shapes, spin states, scattering properties, and thermal properties of nuclei. One overarching result that is clear from such a compilation is that a survey that samples the full diversity of variation, and that samples enough nuclei to drive down the Poisson noise in each sampling bin, is crucial for being able to take the next step in interpreting the distribution in the context of origins and evolutionary processes.

4.1. Measurements

The most fundamental (and ongoing) problem is perhaps that of separating the coma's flux from the nucleus' flux. In many cases the comet has an extended coma within which a point-source is embedded, so it is obvious that the comet is not just showing us light from the nucleus. A significant step forward in handling these cases came with the development of empirical coma-fitting routines that could photometrically separate the contributions from nucleus and coma (*Lamy et al.,* 1996, 2004), and such techniques have been used in several nucleus surveys (e.g., *Fernández et al.,* 2013; *Bauer et al.,* 2017). The technique has proven to be successful as evinced by its success at finding the size of nuclei that were then observed directly with resolved imaging by visiting spacecraft (e.g., *Lamy et al.,* 1998; *Lisse et al.,* 2009). The limitation of this technique arises when the contrast between the nucleus and coma is too low, i.e., if a large fraction of the light in the central pixels is from the coma (*Hui and Li,* 2018). This can happen if the coma is particularly strong or if the comet is distant. While comets are still most often imaged at visible wavelengths, imaging in the infrared, if the coma is still sufficiently well-detected, is generally more likely to result in a robust extraction (*Bauer et al.,* 2020). For one thing, the most optically-active grains at some infrared wavelengths usually provide less total surface area (for a typical size distribution) than those in the visible. Also, at thermal wavelengths, the nucleus will generally be hotter than the surrounding dust grains (since an area on the nucleus only emits into 2π sr vs. the 4π sr dust grains emit into). Both of these effects would tend to increase the nucleus-coma contrast. The other scenario that can be problematic depends on grain outflow dynamics; if the coma's surface brightness profile at a given azimuth deviates significantly in the inner coma from what is measured in the outer coma, the extrapolation can yield incorrect results. An additional, but related, aspect to this is that the method assumes that the light from the extracted point-source is all from the nucleus, which may not be true. One example of a difficult case is that of C/1995 O1 (Hale-Bopp), where the extreme dust production complicated the extraction of the nucleus' signal (*Fernández,* 2002).

There is also the case of a comet that appears as a point-source — and hence one might assume that the comet is inactive — but yet the photometry indicates that there is excess light. The prototypical comet for this situation is Comet 2P/Encke (*Meech et al.,* 2001; *Fernández et al.,* 2005). Generally the more distant the comet, where the linear width at the comet of the (angular) point-spread function is larger, the easier it is to hide a dust coma within the seeing disk. However, it seems that this phenomenon alone cannot explain the specific situation with 2P/Encke; *Presler-Marshall* (2021) showed that for a particular dataset (and the observing conditions that went with it) where the comet looked entirely point-source like, only about ~20% of the flux could be from a steady-state 1/ρ dust coma. Any more than that would be revealed as wings in the comet's profile. The analysis did not assess other coma shapes, so it is possible that a large-grain coma, with particles moving below escape velocity around 2P/Encke's nucleus, could play a role.

Even when one is sure that the light from a comet is all or nearly all from the nucleus, the interpretation can still be muddled if the observation is just a snapshot. The rotational context is often necessary to be sure of what one is actually measuring. Fortunately, this problem is not as terrible as it may seem as first. For example, *Lamy et al.* (2004) show that most measurements even without rotational context will still often be within ~90% of the correct answer anyway. Furthermore, for survey data such rotational variations in profile often may be averaged over, as data may span days.

Ideally temporal coverage will extend all around the orbit so as to understand not only the spin period but also the spin axis direction and some shape information. That often presents a problem, since one may only be able to see one region of a nucleus from Earth when the comet is highly active. Of course we now also have much observational evidence that comets change their spin states on orbital (and shorter!) timescales as well, for example, so measurements obtained at multiple epochs may be challenging to fold into each other in the classical ways (e.g., *Stellingwerf,* 1978).

4.2. Sizes

Assessing the size distribution of cometary nuclei by definition requires an extensive survey, since it can only be measured by having a sufficient number of targets. One significant problem is that there is always a diameter above which the sample is sufficiently representative for a robust analysis. Although that size lower-limit is dependent on the survey's sensitivity, it is not often clear where that critical diameter is (see also section 1.3). A plot of the cumulative size distribution (CSD) of nuclei N(>D) (where N is the number of nuclei with diameter larger than D) always shows a flattening at small (~1–2 km) diameters; equivalently, a plot of the size-frequency distribution (SFD) n(D) dD (where n is the number of nuclei with diameter between D and D + dD) shows a dropoff toward zero at small diameters. This is at least partially due to the fact that (1) our discovery of such comets is less efficient to begin with, and (2) those are often the very comets for which it is hardest to determine accurate nucleus information because the coma obscures the nucleus signal and/or the nucleus is just too faint. In any case, clearly such a feature of the CSD or SFD adds difficulty when trying to fit, say, a power law.

A more robust solution is to assess what the observational biases are in the discovery of the comets. These observational biases will often lead to one or two effects that can accounted for in the CSD or SFD itself and so will yield a more realistic distribution. This is a challenging task, however, when comet discoveries are made by a wide range of facilities. For the smallest comets (~1 km diameter and below), it is especially difficult.

For the JFCs, ideally a survey would either sample a significant fraction of the known comets or make a thorough sweep of the sky to discover the population, including observing the known comets. There must be sufficiently robust software to identify an object as being active. At time of writing there are over 600 known JFCs (see *https://physics.ucf.edu/~yfernandez/cometlist.html*).

For the LPCs, the additional problem, in contrast to the JFCs, is that the comets are simply not visible for as long a period of time. A JFC will return again and again to provide (at least theoretically) multiple observing chances over several decades. An LPC is viewable throughout its perihelion only once over the lifetime of the survey. The LPCs are often more active than the JFCs as well, making it harder to extract nucleus properties.

Given the number of complications that come with studying nuclei, it may not be surprising that there is some divergence in results regarding the size distribution, and different methodologies complicate the picture. For example, the LPC nucleus distribution reported by *Bauer et al.* (2017) comes from NEOWISE observations in the infrared and makes use of the coma-extraction technique. Additionally, there is the LPC distribution reported by *Boe et al.* (2019), which instead comes from visible-wavelength imaging via the "nuclear" absolute magnitudes in JPL's database (see *https://ssd.jpl.nasa.gov/tools/sbdb_query.html*), and a model of activity to correct those absolute magnitudes — which generally are not representative of the nucleus alone — for coma contribution. Thus direct comparisons between the two studies could be difficult. Independent estimates of the particular LPCs in the two studies would be a useful check, but such estimates are sparse. It can be noted that a comparison of JFC (not LPC) nucleus diameters in the *Bauer et al.* (2017) work with those in the SEPPCoN survey (*Fernández et al.,* 2013) and with those from spacecraft encounters show a reasonable match to within 25%.

Another potentially useful approach to get around the problem of contaminating coma is to restrict a survey to objects that are known to be inactive or only very weakly active. For example, as part of the overall ExploreNEO survey, *Mommert et al.* (2015) report observations of several dormant or extinct comets. However, the connection between the size distribution of such highly-evolved comets and of the active JFCs is still to be determined. After all, extinct comets are, by definition, the survivors of an active lifetime that for many comets includes significant fragmentation (and thus potentially a large change in size) if not total disintegration.

5. FUTURE SURVEYS OF COMETS

Surveys will continue to be vital for us to probe the ensemble properties of comets and specifically to understand the full diversity of the population. While flybys and rendezvous of specific comets will of course provide detailed studies of such objects and phenomenological firsthand accounts, it is important that the comet community continue to take advantage of groundbased and spacebased telescopic assets that can shed light on a representative cometary sample. This includes making use of facilities whose original science drivers lie in the astrophysical or heliophysical realms.

Some facilities that we hope will become active in the 2020s have the potential to bring us a significant jump in the number of known, characterized comets. The Rubin Observatory (cf. *Jones et al.,* 2009) and NEO Surveyor (*Mainzer et al.,* 2021), which will both be scanning the skies for solar system objects in the near future, will provide us with the number statistics that would be incredibly helpful. Estimates of the surveys' efficiencies suggest that we could be finding thousands of new comets (*Solontoi et al.,* 2010; *Vera C. Rubin Observatory LSST Solar System Science Collaboration et al.,* 2020; *Sonnett et al.,* 2021). In particular, we can increase the number of LPCs that are discovered per year, and the number of such comets that are discovered beyond 5 and 10 au, expanding baselines of behavior before activity has ramped up. This will also make it easier for follow-up observations to assess the properties of their nuclei. Another important consideration that number statistics will help with is our understanding of the evolutionary paths of the JFCs. For example, both surveys are supposed to be sensitive enough to sample the subkilometer JFC population and we might expect such a population to exist as a result of mass loss and fragmentation over the comets' lifetimes. However, it is also possible such small comets quickly disintegrate all the way to dust. Rubin and NEO Surveyor will extend the number statistics into this size regime. Hence the surveys will be better able to determine how well small comets survive their active lifetimes.

Finally, the two surveys will be very complementary to each other since the combination of visible (reflected) and infrared (thermal) wavelength observations, and observations spanning several years at multiple epochs, will be tremendously helpful for gauging gas, dust, and nucleus properties. In a similar vein as WISE/NEOWISE, NEO Surveyor will provide measurements of nucleus sizes, dust characteristics, and CO and CO_2 production of manifold larger statistical samples.

The Spectro-Photometer for the History of the Universe, Epoch of Reionization, and Ices Explorer (SPHEREx) mission will provide us with near-infrared spectroscopic investigations of dozens of comets. Importantly, such data will extend to wavelengths where CO and CO_2 rovibrational bands emit, which means SPHEREx may build upon results of Akari with additional insight into these important species (*Doré et al.,* 2016).

Furthermore, low-cost mission concepts could, if brought to fruition, also address highly specific comet-related questions through a survey. In particular, cubesats can provide such survey work for relatively low cost. For example, a small but fast wide-field ultraviolet telescope in Earth orbit could let us make measurements of the OH electronic band near 309 nm in hundreds of cometary comae. Such a database, with observations covering all dynamical types and covering a range of heliocentric distances, used in concert with dust production measurements, could give a simple test of just how and when the water production is tied to dust. Cubesats could also be employed for an *in situ* survey of multiple comets. For example, equipping a fleet of cubesats with replicas of the MIRO instrument on Rosetta, and sending them out to a few dozen comets in the inner solar system, would give us unprecedented views of cometary near-surface interiors and let us assess how void space and consolidation evolve as a comet ages through its active lifetime.

A survey with the James Webb Space Telescope (JWST) has the potential to let us take the next step in our understanding of the comet population's nuclei, gas production, and dust production. It will provide more detailed gas production measurements at a larger range of heliocentric distances, and for a broader range of species (especially parent species), than ever before. We will be able to watch the changing release of various volatiles over time as a comet approaches and recedes from the Sun, and do it not just for exceptional comets such as Hale-Bopp, but for more typical comets and comets from all dynamical classes and ages. Similar synoptic coverage of the thermal and scattering properties of the dust as the comet moves in its orbit will likewise let us investigate how the grain properties change in response to the activity driver, giving clues about the nature of the ice-rock mixture in cometary subsurface layers. This will come from not only spectroscopic assessment of the dust spectral energy distribution (and its resulting decomposition into mineralogy) but also from resolved imaging of the dust coma. JWST's stable point-spread function and high spatial resolution will give us the best chance of overcoming the coma confusion problem, and let us do so at a range of wavelengths, thereby letting us have a better handle on overall thermal emission from the nucleus. Being able to do all this for 20 to 30 comets would be spectacular.

There are of course many additional, large-scale space telescopes in various levels of planning/concreteness that would theoretically arrive in the decades of the 2030s and 2040s. The Roman Space Telescope, for example, will provide multiband red and near-infrared imaging of cometary dust, as well as grism spectroscopy that covers the 1.5-μm water ice absorption band, thus potentially providing a survey of icy grains (*Holler et al.,* 2018). In the more distant future, observatories like the Large Ultraviolet Optical Infrared Surveyor (LUVOIR) or the Origins Space Telescope (cf. *National Academies of Sciences, Engineering, and Medicine* 2021) would further expand the samples. These observatories, combining significant improvements in sensitivity with high-resolution imaging and spectroscopy, may be used to get a statistical sense of the nature of low-level activity in comets at large heliocentric distances, as well as obtain large samples of surface constituents via spectroscopic studies and explore the presence of other possible drivers of activity, like methane, at larger distances (cf. *Meech and Svoren,* 2004; *Brown,* 2000). Such facilities, if capable of non-sidereal tracking, would certainly provide a new jump in our understanding of comets by taking us to the next level of detail on dozens to hundreds of these bodies.

Radio-wavelength surveys with existing high-spatial-resolution facilities like ALMA and with future facilities like the next-generation Very Large Array (ngVLA) and the Next Generation Arecibo Telescope (NGAT) (*Roshi et al.,* 2019) will provide great insight into energetics of the gas coma. The NGAT would also have a phased radar capability and thus provide a significantly higher power output that has been previously possible. This would allow us to obtain more detailed pictures of the large-grain (centimeter-scale) dust coma as well as the nucleus structure on centimeter scales. The number of radar-detected comets has slowly increased since the predecessor volume and is still fairly small, so a boost to the emitted power could drastically increase the number of available comets that could be sampled in this way. Radio continuum measurements (in passive mode) at a variety of submillimeter, millimeter, and centimeter wavelengths would let us sample different depths in a nucleus, down to approximately 1 m. This has already been demonstrated with ALMA with observations of Ganymede (*de Kleer et al.,* 2021). Again, an assessment of the surface and subsurface properties of a range of nuclei of varying ages and activity levels could be insightful.

There are many excellent surveys of cometary properties that are discussed throughout this volume and its predecessor, *Comets II.* But often the statistics could be improved to sharpen the conclusions by observing more comets or by observing the same comets in more detail (e.g., better temporal coverage, wavelength coverage, spatial resolution, or spectral resolution). With growing datasets, selecting the necessary qualities of the data to undertake analyses of physical properties, and tailor selections to particular subpopulations, will require improved metadata. Complex information models as the basis of metadata, like PDS4 (*Raugh and Hughes,* 2021) and the Minor Planet Center's new Astrometry Data Exchange Standard (ADES) (*Chesley et al.,* 2017), which are associated with archived data, will facilitate these applications and analyses. The metadata labels and automated tools will allow users to identify and extract the desired data for analysis. The question of when one has "enough" samples to draw a conclusion at a sufficient confidence level is not easy to answer ahead of time unless one has a good sense of the inherent diversity in the population.

6. SUMMARY

Surveys have had a large impact on our knowledge base of comets. They provide a systematical approach toward collecting large samples of data reflective of cometary physical characteristics and behavior and more revealing of the comprehensive cometary object populations. It is important to also acknowledge that comet science can advance as well with surveys done by telescopic assets that may have primarily astrophysical or heliophysical science drivers, and we advocate for the continued use of such facilities. We also would like to see that those facilities that are open to general observers for targeted observations accommodate at least some of the non-sidereal tracking capability that is so important for solar system science.

Some key features in the sky survey analysis approaches include:

- most comets are now discovered by all-sky surveys, SLSs and TPSs
- the majority of yearly detections are also made by such surveys
- critical margins remain outside the survey coverage, which creates opportunities for targeted single-object observations and non-survey discovery of unique cometary objects
- the data from sky surveys are well-explored for discovery and astrometric measurements, while only cursorily exploited as a resource for physical and behavior characterization

For targeted larger-sample surveys:

- larger sample sizes, the expanse of the sample, and broadening of wavelength regimes will lead to more comprehensive understanding of compositional variations across orbit classes and subclasses and separate original composition from evolutionary effects
- characterizing the physical states of comet components (e.g., gas phase, tail dust, nucleus spin, etc.) remains in the early stages of large-sample collection

Larger-scale surveys in general will lead to improved Earth-space situational awareness, understanding of the evolutionary processing of comets, and solar system volatile transport, as well as solar system formation. Finally, future larger-scale surveys will require the use of advanced astroinformatic techniques and incorporation of AI routines to realistically process the increased data volume.

Acknowledgments. J.M.B. and Y.R.F. acknowledge support from the NEO Surveyor mission, funded by NASA Contract 80MSFC20C0045. Y.R.F. also acknowledges support from the Center for Lunar and Asteroid Surface Science, funded by NASA Contract 80NSSC19M0214. We acknowledge helpful discussions with A. Lovell that improved this chapter.

REFERENCES

A'Hearn M. F., Schleicher D. G., Millis R. L., Feldman P. D., and Thompson D. T. (1984) Comet Bowell 1980b. *Astron. J., 89,* 579–591.

A'Hearn M. F., Millis R. C., Schleicher D. O., Osip D. J., and Birch P. V. (1995) The ensemble properties of comets: Results from narrowband photometry of 85 comets, 1976–1992. *Icarus, 118,* 223–270.

A'Hearn M. F., Feaga L. M., Keller H. U., Kawakita H., Hampton D. L., Kissel J., Klaasen K. P., McFadden L. A., Meech K. J., Schultz P. H., Sunshine J. M., Thomas P. C., Veverka J., Yeomans D. K., Besse S., Bodewits D., Farnham T. L., Groussin O., Kelley M. S., Lisse C. M., Merlin F., Protopapa S., and Wellnitz D. D. (2012) Cometary volatiles and the origin of comets. *Astrophys. J., 758,* 29.

Bair A. N. and Schleicher D. G. (2021) Water production rates in comets: Evidence for physical evolution of cometary surfaces. *AAS/Division for Planetary Sciences Meeting Abstracts, 53,* #210.03.

Battams K. and Boonplod W. (2020) C/2020 X3 (SOHO). *Minor Planet Electronic Circular 2020-Y19.*

Battams K. and Knight M. M. (2017) SOHO comets: 20 years and 3000 objects later. *Philos. Trans. R. Soc. A, 375,* 20160257.

Bauer J. M., Stevenson R., Kramer E., Mainzer A. K., Grav T., Masiero J. R., Fernández Y. R., Cutri R. M., Dailey J. W., Masci F. J., Meech K. J., Walker R., Lisse C. M., Weissman P. R., Nugent C. R., Sonnett S., Blair N., Lucas A., McMillan R. S., Wright E. L., and the WISE and NEOWISE Teams (2015) The NEOWISE-discovered comet population and the CO + CO_2 production rates. *Astrophys. J., 814,* 85.

Bauer J. M., Grav T., Fernández Y. R., Mainzer A. K., Kramer E. A., Masiero J. R., Spahr T., Nugent C. R., Stevenson R. A., Meech K. J., Cutri R. M., Lisse C. M., Walker R., Dailey J. W., Rosser J., Krings P., Ruecker K., Wright E. L., and the NEOWISE Team (2017) Debiasing the NEOWISE cryogenic mission comet populations. *Astron. J., 154,* 53.

Bauer J., Gicquel A., Kramer E., Mainzer A., Masiero J., Fernández Y., Schambeau C., Stevenson R., Lisse C., Meech K., and the NEOWISE Team (2020) Cometary coma extraction and nucleus size determination with NEOWISE. *AAS/Division for Planetary Sciences Meeting Abstracts, 52,* #316.04.

Bauer J. M., Gicquel A., Kramer E., and Meech K. J. (2021) NEOWISE observed CO and CO_2 production rates of 46P/Wirtanen during the 2018–2019 apparition. *Planet. Sci. J., 2,* 34.

Bertaux J. L., Kyrölä E., Quémerais E., Pellinen R., Lallement R., Schmidt W., Berthé M., Dimarellis E., Goutail J. P., Taulemesse C., Bernard C., Leppelmeier G., Summanen T., Hannula H., Huomo H., Kehlä V., Korpela S., Leppälä K., Strömmer E., Torsti J., Viherkanto K., Hochedez J. F., Chretiennot G., Peyroux R., and Holzer T. (1995) SWAN: A study of solar wind anisotropies on SOHO with Lyman alpha sky mapping. *Solar Phys., 162,* 403–439.

Blum J., Gundlach B., Mühle S., and Trigo-Rodriguez J. M. (2014) Comets formed in solar-nebula instabilities! An experimental and modeling attempt to relate the activity of comets to their formation process. *Icarus, 235,* 156–169.

Bockelée-Morvan D. and Biver N. (2017) The composition of cometary ices. *Philos. Trans. R. Soc. A, 375,* 20160252.

Bockelée-Morvan D., Calmonte U., Charnley S., Duprat J., Engrand C., Gicquel A., Hässig M., Jehin E., Kawakita H., Marty B., Milam S., Morse A., Rousselot P., Sheridan S., and Wirström E. (2015) Cometary isotopic measurements. *Space Sci. Rev., 197,* 47–83.

Bodewits D., Christian D. J., Torney M., Dryer M., Lisse C. M., Dennerl K., Zurbuchen T. H., Wolk S. J., Tielens A. G. G. M., and Hoekstra R. (2007) Spectral analysis of the Chandra comet survey. *Astron. Astrophys., 469,* 1183–1195.

Boe B., Jedicke R., Meech K. J., Wiegert P., Weryk R. J., Chambers K. C., Denneau L., Kaiser N., Kudritzki R. P., Magnier E. A., Wainscoat R. J., and Waters C. (2019) The orbit and size-frequency distribution of long period comets observed by Pan-STARRS1. *Icarus, 333,* 252–272.

Bohren C. F. and Huffman D. R. (1983) *Absorption and Scattering of Light by Small Particles.* Wiley, Weinheim. 530 pp.

Bonnet D. G. and Wright T. A. (2000) Sample size requirements for estimating Pearson, Kendall and Spearman correlations. *Psychometrika, 65,* 23–28.

Borisov G. V. and Shustov B. M. (2021) Discovery of the first interstellar comet and the spatial density of interstellar objects in the solar neighborhood. *Solar System Res., 55,* 124–131.

Borne K., Accomazzi A., Bloom J., Brunner R., Burke D., Butler N., Chernoff D. F., Connolly B., Connolly A., Connors A., Cutler C., Desai S., Djorgovski G., Feigelson E., Finn L. S., Freeman P., Graham M., Gray N., Graziani C., Guinan E. F., Hakkila J., Jacoby S., Jefferys W., Kashyap, Kelly B., Knuth K., Lamb D. Q., Lee H., Loredo T., Mahabal A., Mateo M., McCollum B., Muench A., Pesenson M., Petrosian V., Primini F., Protopapas P., Ptak A., Quashnock J., Raddick M. J., Rocha G., Ross N., Rottler L., Scargle J., Siemiginowska A., Song, Szalay A., Tyson J. A., Vestrand T., Wallin J., Wandelt B., Wasserman I. M., Way M., Weinberg M., Zezas A., Anderes E., Babu J., Becla J., Berger J., Bickel P. J., Clyde M., Davidson I., van Dyk D., Eastman T., Efron B., Genovese C., Gray A., Jang W., Kolaczyk E. D., Kubica J., Loh J. M., Meng X.-L., Moore A., Morris R., Park T., Pike R., Rice J., Richards J., Ruppert D., Saito N., Schafer C., Stark P. B., Stein M., Sun J., Wang D., Wang Z., Wasserman L., Wegman E. J., Willett R., Wolpert R., and Woodroofe M. (2009) Astroinformatics: A 21st century approach to astronomy. In *Astro2010: The Astronomy and Astrophysics Decadal Survey,* Position Paper #6. National Academies, Washington, DC.

Brown M. E. (2000) Near-infrared spectroscopy of Centaurs and irregular satellites. *Astron. J., 119,* 977–983.

Chesley S. R., Hockney G. M., and Holman M. J. (2017) Introducing ADES: A new IAU Astrometry Data Exchange Standard. *AAS/Division for Planetary Sciences Meeting Abstracts, 49,* #112.14.

Cochran A. L., Barker E. S., and Gray C. L. (2012) Thirty years of cometary spectroscopy from McDonald Observatory. *Icarus, 218,* 144–168.

Cochran A. L., Levasseur-Regourd A.-C., Cordiner M., Hadamcik E., Lasue J., Gicquel A., Schleicher D. G., Charnley S. B., Mumma M. J., Paganini L., Bockelée-Morvan D., Biver N., and Kuan Y.-J. (2015) The composition of comets. *Space Sci. Rev., 197,* 9–46.

Combi M. R., Lee Y., Patel T. S., Mäkinen J. T. T., Bertaux J. L., and Quémerais E. (2011) SOHO/SWAN observations of short-period spacecraft target comets. *Astron. J., 141,* 128.

Combi M. R., Mäkinen T. T., Bertaux J. L., Quémerais E., and Ferron S. (2019) A survey of water production in 61 comets from SOHO/SWAN observations of hydrogen Lyman-alpha: Twenty-one years 1996–2016. *Icarus, 317,* 610–620.

Davies J. K., Roush T. L., Cruikshank D. P., Bartholomew M. J., Geballe T. R., Owen T., and de Bergh C. (1997) The detection of water ice in Comet Hale-Bopp. *Icarus, 127,* 238–245.

Decock A., Jehin E., Hutsemékers D., and Manfroid J. (2013) Forbidden oxygen lines in comets at various heliocentric distances. *Astron. Astrophys., 555,* A34.

de Kleer K., Butler B., de Pater I., Gurwell M. A., Moullet A., Trumbo S., and Spencer J. (2021) Ganymede's surface properties from millimeter and infrared thermal emission. *Planet. Sci. J., 2,* 5.

Dello Russo N., Kawakita H., Vervack R. J., and Weaver H. A. (2016) Emerging trends and a comet taxonomy based on the volatile chemistry measured in thirty comets with high-resolution infrared spectroscopy between 1997 and 2013. *Icarus, 278,* 301–332.

Dennerl K., Englhauser J., and Trümper J. (1997) X-ray emissions from comets detected in the Röntgen X-ray satellite all-sky survey. *Science, 277,* 1625–1630.

Dobson M. M., Schwamb M. E., Fitzsimmons A., Kelley M. S. P., Lister T., Shingles L. J., Denneau L., Heinze A. N., Smith K. W., Tonry J. L., Weiland H., Young D. R., Benecchi S. D., and Verbiscer A. J. (2021) New or increased cometary activity in (2060) 95P/Chiron. *Res. Notes Am. Astron. Soc., 5,* 211.

Doré O., Werner M. W., Ashby M., Banerjee P., Battaglia N., Bauer J., Benjamin R. A., Bleem L. E., Bock J., Boogert A., Bull P., Capak P., Chang T.-C., Chiar J., Cohen S. H., Cooray A., Crill B., Cushing M., de Putter R., Driver S. P., Eifler T., Feng C., Ferraro S., Finkbeiner D., Gaudi B. S., Greene T., Hillenbrand L., Höflich P. A., Hsiao E., Huffenberger K., Jansen R. A., Jeong W.-S., Joshi B., Kim D., Kim M., Kirkpatrick D., Korngut P., Krause E., Kriek M., Leistedt B., Li A., Lisse C. M., Mauskopf P., Mechtley M., Melnick G., Mohr J., Murphy J., Neben A., Neufeld D., Nguyen H., Pierpaoli E., Pyo J., Rhodes J., Sandstrom K., Schaan E., Schlaufman C., Silverman J., Su K., Stassun K., Stevens D., Strauss M. A., Tielens X., Tsai C.-W., Tolls V., Unwin S., Viero M., Windhorst R. A., and Zemcov M. (2016) Science impacts of the SPHEREx all-sky optical to near-infrared spectral survey: Report of a community workshop

examining extragalactic, galactic, stellar and planetary science. *ArXiV e-prints*, arXiv:1606.07039.
Farnham T. L., Kelley M. S. P., Knight M. M., and Feaga M. (2019) First results from TESS observations of Comet 46P/Wirtanen. *Astrophys. J. Lett., 886*, L24.
Farnham T., Kelley M., and Bauer J. (2021) A survey of comets using TESS. *AAS/Division for Planetary Sciences Meeting Abstracts, 53*, #301.06.
Fernández Y. R. (2002) The nucleus of comet Hale-Bopp (C/1995 O1): Size and activity. *Earth Moon Planets, 89*, 3–25.
Fernández Y. R., Lowry S. C., Weissman P. R., Mueller B. E. A., Samarasinha N. H., Belton M. J. S., and Meech K. J. (2005) New near-aphelion light curves of Comet 2P/Encke. *Icarus, 175*, 194–214.
Fernández Y. R., Kelley M. S., Lamy P. L., Toth I., Groussin O., Lisse C. M., A'Hearn M. F., Bauer J. M., Campins H., Fitzsimmons A., Licandro J., Lowry S. C., Meech K. J., Pittichová J., Reach W. T., Snodgrass C., and Weaver H. A. (2013) Thermal properties, sizes, and size distribution of Jupiter-family cometary nuclei. *Icarus, 226*, 1138–1170.
Fink U. (2009) A taxonomic survey of comet composition 1985–2004 using CCD spectroscopy. *Icarus, 201*, 311–334.
Finson M. J. and Probstein R. F. (1968) A theory of dust comets. I. Model and equations. *Astrophys. J., 154*, 327–352.
Fougere N., Altwegg K., Berthelier J. J., Bieler A., Bockelée-Morvan D., Calmonte U., Capaccioni F., Combi M. R., De Keyser J., Debout V., Erard S., Fiethe B., Filacchione G., Fink U., Fuselier S. A., Gombosi T. I., Hansen K. C., Hässig M., Huang Z., Le Roy L., Leyrat C., Migliorini A., Piccioni G., Rinaldi G., Rubin M., Shou Y., Tenishev V., Toth G., and Tzou C. Y. (2016) Direct simulation Monte Carlo modelling of the major species in the coma of Comet 67P/Churyumov-Gerasimenko. *Mon. Not. R. Astron. Soc., 462*, S156–S169.
Francis P. J. (2005) The demographics of long-period comets. *Astrophys. J., 635*, 1348–1361.
Gehrz R. D. and Ney E. P. (1992) 0.7- to 23-μm photometric observations of P/Halley 1986 III and six recent bright comets. *Icarus, 100*, 162–186.
Gicquel A., Bauer J., Kramer E., Mainzer A., Masiero J., Fernández Y., Schambeau C., Lisse C., and Meech K. (2021) CO + CO_2 production rates of comets observed by NEOWISE in 2014. *AAS/Division for Planetary Sciences Meeting Abstracts, 53*, #210.17.
Gundlach B., Blum J., Keller H. U., and Skorov Y. V. (2015) What drives the dust activity of Comet 67P/Churyumov-Gerasimenko? *Astron. Astrophys., 583*, A12.
Harrington Pinto O., Womack M., Fernández Y., and Bauer J. (2021) CO and CO_2 production rates in comets and Centaurs. *AAS/Division for Planetary Sciences Meeting Abstracts, 53*, #210.05.
Hicks M. D., Bambery R. J., Lawrence K. J., and Kollipara P. (2007) Near-nucleus photometry of comets using archived NEAT data. *Icarus, 188*, 457–467.
Holler B. J., Milam S. N., Bauer J. M., Alcock C., Bannister M. T., Bjoraker G. L., Bodewits D., Bosh A. S., Buie M. W., Farnham T. L., Haghighipour N., Hardersen P. S., Harris A. W., Hirata C. M., Hsieh H. H., Kelley M. S. P., Knight M. M., Kramer E. A., Longobardo A., Nixon C. A., Palomba E., Protopapa S., Quick L. C., Ragozzine D., Reddy V., Rhodes J. D., Rivkin A. S., Sarid G., Sickafoose A. A., Simon A. A., Thomas C. A., Trilling D. E., and West R. A. (2018) Solar system science with the Wide-Field Infrared Survey Telescope. *J. Astron. Telesc. Instr. Syst., 4*, 034003.
Hsieh H. H., Denneau L., Wainscoat R. J., Schörghofer N., Bolin B., Fitzsimmons A., Jedicke R., Kleyna J., Micheli M., Vereš P., Kaiser N., Chambers K. C., Burgett W. S., Flewelling H., Hodapp K. W., Magnier E. A., Morgan J. S., Price P. A., Tonry J. L., and Waters C. (2015) The main-belt comets: The Pan-STARRS1 perspective. *Icarus, 248*, 289–312.
Hui M.-T. and Li J.-Y. (2018) Is the cometary nucleus-extraction technique reliable? *Publ. Astron. Soc. Pac., 130*, 104501.
Ishiguro M., Sarugaku Y., Nishihara S., Nakada Y., Nishiura S., Soyano T., Tarusawa K., Mukai T., Kwon S. M., Hasegawa S., Usui F., and Ueno M. (2009) Report on the Kiso cometary dust trail survey. *Adv. Space Res., 43*, 875–879.
Ivezić Ž., Smith J. A., Miknaitis G., Lin H., Tucker D., Lupton R. H., Gunn J. E., Knapp G. R., Strauss M. A., Sesar B., Doi M., Tanaka M., Fukugita M., Holtzman J., Kent S., Yanny B., Schlegel D., Finkbeiner D., Padmanabhan N., Rockosi C. M., Jurić M., Bond N., Lee B., Stoughton C., Jester S., Harris H., Harding P., Morrison H., Brinkmann J., Schneider D. P., and York D. (2007) Sloan Digital Sky Survey standard star catalog for stripe 82: The dawn of industrial 1% optical photometry. *Astron. J., 134*, 973–998.
Jedicke R. and Herron J. D. (1997) Observational constraints on the Centaur population. *Icarus, 127*, 494–507.
Jewitt D. (1991) Cometary photometry. In *Comets in the Post-Halley Era* (J. Newburn et al., eds.), pp. 19–66. Astrophysics and Space Science Library 167, Kluwer, Dordrecht.
Jewitt D. (2015) Color systematics of comets and related bodies. *Astron. J., 150*, 201.
Jones R. L., Chesley S. R., Connolly A. J., Harris A. W., Ivezic Z., Knezevic Z., Kubica J., Milani A., and Trilling D. E. (2009) Solar system science with LSST. *Earth Moon Planets, 105*, 101–105.
Kawakita H., Watanabe J., Ootsubo T., Nakamura R., Fuse T., Takato N., Sasaki S., and Sasaki T. (2004) Evidence of icy grains in comet C/2002 T7 (LINEAR) at 3.52 AU. *Astrophys. J. Lett., 601*, L191–L194.
Kelley M. S., Fernández Y. R., Licandro J., Lisse C. M., Reach W. T., A'Hearn M. F., Bauer J., Campins H., Fitzsimmons A., Groussin O., Lamy P. L., Lowry S. C., Meech K. J., Pittichová J., Snodgrass C., Toth I., and Weaver H. A. (2013) The persistent activity of Jupiter-family comets at 3–7 AU. *Icarus, 225*, 475–494.
Kelley M. S. P., Farnham T. L., Li J.-Y., Bodewits D., Snodgrass C., Allen J., Bellm E. C., Coughlin M. W., Drake A. J., Duev D. A., Graham M. J., Kupfer T., Masci F. J., Reiley D., Walters R., Dominik M., Jørgensen U. G., Andrews A. E., Bach-Møller N., Bozza V., Burgdorf M. J., Campbell-White J., Dib S., Fujii Y. I., Hinse T. C., Hundertmark M., Khalouei E., Longa-Peña P., Rabus M., Rahvar S., Sajadian S., Skottfelt J., Southworth J., Tregloan-Reed J., Unda-Sanzana E., and the Mindstep Collaboration (2021) Six outbursts of Comet 46P/Wirtanen. *Planet. Sci. J., 2*, 131.
Kolokolova L., Hanner M. S., Levasseur-Regourd A. C., and Gustafson B. Å. S. (2004) Physical properties of cometary dust from light scattering and thermal emission. In *Comets II* (M. C. Festou et al., eds.), pp. 577–604. Univ. of Arizona, Tucson.
Kramer E. (2014) Studying short-period comets and long-period comets detected by WISE/NEOWISE. Ph.D. thesis, University of Central Florida, Orlando. 218 pp.
Kramer E. A., Bauer J. M., Fernández Y. R., Mainzer A. K., Masiero J. R., Grav T., Nugent C. R., Sonnett S., Lisse C. M., Meech K. J., and the WISE Team (2015) Cometary dust tails in NEOWISE. In *46th Lunar and Planetary Science Conference*, Abstract #2820. LPI Contribution No. 1832, Lunar and Planetary Institute, Houston.
Kramer E. A., Bauer J. M., Fernández Y. R., Stevenson R., Mainzer A. K., Grav T., Masiero J., Nugent C., and Sonnett S. (2017) The perihelion emission of Comet C/2010 L5 (WISE). *Astrophys. J., 838*, 58.
Lamy P. and Toth I. (2009) The colors of cometary nuclei — Comparison with other primitive bodies of the solar system and implications for their origin. *Icarus, 201*, 674–713.
Lamy P. L., Toth I., Grün E., Keller H. U., Sekanina Z., and West R. M. (1996) Observations of Comet P/Faye 1991 XXI with the planetary camera of the Hubble Space Telescope. *Icarus, 119*, 370–384.
Lamy P. L., Toth I., and Weaver H. A. (1998) Hubble Space Telescope observations of the nucleus and inner coma of Comet 19P/1904 Y2 (Borrelly). *Astron. Astrophys., 337*, 945–954.
Lamy P. L., Toth I., Fernández Y. R., and Weaver H. A. (2004) The sizes, shapes, albedos, and colors of cometary nuclei. In *Comets II* (M. C. Festou et al., eds.), pp. 223–264. Univ. of Arizona, Tucson.
Landis R. and Johnson L. (2019) Advances in planetary defense in the United States. *Acta Astronaut., 156*, 394–408.
Langland-Shula L. E. and Smith G. H. (2011) Comet classification with new methods for gas and dust spectroscopy. *Icarus, 213*, 280–322.
Levison H. F. (1996) Comet taxonomy. In *Completing the Inventory of the Solar System* (T. Rettig and J. M. Hahn, eds.), pp. 173–191. ASP Conf. Ser. 107, Astronomical Society of the Pacific, San Francisco.
Lippi M., Villanueva G. L., Mumma M. J., and Faggi S. (2021) Investigation of the origins of comets as revealed through infrared high-resolution spectroscopy I. Molecular abundances. *Astron. J., 162*, 74.
Lisse C. M., A'Hearn M. F., Hauser M. G., Kelsall T., Lien D. J., Moseley S. H., Reach W. T., and Silverberg R. F. (1998) Infrared observations of comets by COBE. *Astrophys. J., 496*, 971–991.
Lisse C. M., Fernández Y. R., A'Hearn M. F., and Peschke S. B. (2002) A search for trends in cometary dust emission. In *COSPAR Colloquia Series, Vol. 15: Dust in the Solar System and Other*

Planetary Systems (S. F. Green et al., eds.), pp. 259–268. Elsevier, Amsterdam.

Lisse C. M., Cravens T. E., and Dennerl K. (2004) X-ray and extreme ultraviolet emission from comets. In *Comets II* (M. C. Festou et al., eds.), pp. 631–643. Univ. of Arizona, Tucson.

Lisse C. M., Fernández Y. R., Reach W. T., Bauer J. M., A'Hearn F., Farnham T. L., Groussin O., Belton M. J., Meech K. J., and Snodgrass C. D. (2009) Spitzer Space Telescope observations of the nucleus of Comet 103P/Hartley 2. *Publ. Astron. Soc. Pac., 121*, 968–975.

Lisse C., Bauer J., Cruikshank D., Emery J., Fernández Y., Fernandez-Valenzuela E., Kelley M., McKay A., Reach W., Pendleton Y., Pinilla-Alonso N., Stansberry J., Sykes M., Trilling D. E., Wooden D., Harker D., Gehrz R., and Woodward C. (2020) Spitzer's solar system studies of comets, Centaurs and Kuiper belt objects. *Nature Astron., 4*, 930–939.

Lovell A. J. and Howell E. S. (2015) Radio OH spectroscopic mapping survey of 26 comets: Trends in outflow velocity and collisional quenching. *AAS/Division for Planetary Sciences Meeting Abstracts, 47*, #415.14.

Mainzer A., Grav T., Masiero J., Bauer J., McMillan R. S., Giorgini J., Spahr T., Cutri R. M., Tholen D. J., Jedicke R., Walker R., Wright E., and Nugent C. R. (2012) Characterizing subpopulations within the near-Earth objects with NEOWISE: Preliminary results. *Astrophys. J., 752*, 110.

Mainzer A., Abell P., Bauer J., Bottke W., Grav T., Kelley M., Kramer E., Masci F., Masiero J., Reddy V., Reinhart L., Sonnett S., Wright E., and Wong A. (2021) Near-Earth Object Surveyor mission: Data products and survey plan. *AAS/Division for Planetary Sciences Meeting Abstracts, 53*, #306.16.

Martino S., Tancredi G., Monteiro F., Lazzaro D., and Rodrigues T. (2019) Monitoring of asteroids in cometary orbits and active asteroids. *Planet. Space Sci., 166*, 135–148.

Mazzotta Epifani E., Palumbo P., and Colangeli L. (2009) A survey on the distant activity of short period comets. *Astron. Astrophys., 508*, 1031–1044.

McKay A. J., Kelley M. S. P., Cochran A. L., Bodewits D., DiSanti A., Russo N. D., and Lisse C. M. (2016) The CO_2 abundance in Comets C/2012 K1 (PanSTARRS), C/2012 K5 (LINEAR) and 290P/Jäger as measured with Spitzer. *Icarus, 266*, 249–260.

McKay A. J., DiSanti M. A., Kelley M. S. P., Knight M. M., Womack M., Wierzchos K., Harrington Pinto O., Bonev B., Villanueva G. L., Dello Russo N., Cochran A. L., Biver N., Bauer J., Vervack J., Ronald J., Gibb E., Roth N., and Kawakita H. (2019) The peculiar volatile composition of CO-dominated Comet C/2016 R2 (PanSTARRS). *Astron. J., 158*, 128.

Meech K. J. and Svoren J. (2004) Using cometary activity to trace the physical and chemical evolution of cometary nuclei. In *Comets II* (M. C. Festou et al., eds.), pp. 317–335. Univ. of Arizona, Tucson.

Meech K. J., Fernández Y., and Pittichová J. (2001) Aphelion activity of 2P/Encke. *AAS/Division for Planetary Sciences Meeting Abstracts, 33*, #20.06.

Meech K. J., Pittichová J., Bar-Nun A., Notesco G., Laufer D., Hainaut O. R., Lowry S. C., Yeomans D. K., and Pitts M. (2009) Activity of comets at large heliocentric distances pre-perihelion. *Icarus, 201*, 719–739.

Mommert M., Harris A. W., Mueller M., Hora J. L., Trilling D. E., Bottke W. F., Thomas C. A., Delbo M., Emery J. P., Fazio G., and Smith H. A. (2015) ExploreNEOs. VIII. Dormant short-period comets in the near-Earth asteroid population. *Astron. J., 150*, 106.

Mommert M., Trilling D. E., Hora J. L., Lejoly C., Gustafsson A., Knight M., Moskovitz N., and Smith H. A. (2020) Systematic characterization of and search for activity in potentially active asteroids. *Planet. Sci. J., 1*, 10.

Mumma M. J. and Charnley S. B. (2011) The chemical composition of comets — Emerging taxonomies and natal heritage. *Annu. Rev. Astron. Astrophys., 49*, 471–524.

National Academies of Sciences, Engineering, and Medicine (2021) *Pathways to Discovery in Astronomy and Astrophysics for the 2020s*. National Academies, Washington, DC. 615 pp.

Oke J. B., Cohen J. G., Carr M., Cromer J., Dingizian A., Harris F. H., Labrecque S., Lucinio R., Schaal W., Epps H., and Miller J. (1995) The Keck low-resolution imaging spectrometer. *Publ. Astron. Soc. Pac., 107*, 375.

Oort J. H. (1950) The structure of the cloud of comets surrounding the solar system and a hypothesis concerning its origin. *Bull. Astron. Inst. Netherlands, 11*, 91–110.

Ootsubo T., Kawakita H., Hamada S., Kobayashi H., Yamaguchi M., Usui F., Nakagawa T., Ueno M., Ishiguro M., Sekiguchi T., Watanabe J.-i., Sakon I., Shimonishi T., and Onaka T. P. (2012) AKARI near-infrared spectroscopic survey for CO_2 in 18 comets. *Astrophys. J., 752*, 15.

Pravdo S. H., Rabinowitz D. L., Helin E. F., Lawrence K. J., Bambery R. J., Clark C. C., Groom S. L., Levin S., Lorre J., Shaklan S. B., Kervin P., Africano J. A., Sydney P., and Soohoo V. (1999) The Near-Earth Asteroid Tracking (NEAT) program: An automated system for telescope control, wide-field imaging, and object detection. *Astron. J., 117*, 1616–1633.

Presler-Marshall B. (2021) An aphelion analysis of Comet 2P/Encke. M.S. thesis, University of Central Florida, Orlando.

Protopapa S., Kelley M. S. P., Yang B., Bauer J. M., Kolokolova L., Woodward C. E., Keane J. V., and Sunshine J. M. (2018) Icy grains from the nucleus of Comet C/2013 US10 (Catalina). *Astrophys. J. Lett., 862*, L16.

Raugh A. and Hughes J. S. (2021) The road to an archival data format — Data structures. *Planet. Sci. J., 2*, 204.

Reach W. T., Kelley M. S., and Sykes M. V. (2007) A survey of debris trails from short-period comets. *Icarus, 191*, 298–322.

Reach W. T., Kelley M. S., and Vaubaillon J. (2013) Survey of cometary CO_2, CO, and particulate emissions using the Spitzer Space Telescope. *Icarus, 226*, 777–797.

Roshi A., Anderson L. D., Araya E., Balser D., Brisken W., Brum C., Campbell D., Chatterjee S., Churchwell E., Condon J., Cordes J., Cordova F., Fernández Y., Gago J., Ghosh T., Goldsmith P. F., Heiles C., Hickson D., Jeffs B., Jones K. M., Lautenbach J., Lewis B. M., Lynch R. S., Manoharan P. K., Marshall S., Minchin R., Palliyaguru N. T., Perera B. B. P., Perillat P., Pinilla-Alonso N., Pisano D. J., Quintero L., Raizada S., Ransom S. M., Fernandez-Rodriguez F. O., Salter C. J., Santos P., Sulzer M., Taylor P. A., Venditti F. C. F., Venkataraman A., Virkki A. K., Wolszczan A., Womack M., and Zambrano-Marin L. F. (2019) Arecibo Observatory in the next decade. *Bull. Am. Astron. Soc., 51*, 244.

Sárneczky K., Szabó G. M., Csák B., Kelemen J., Marschalkó G., Pál A., Szakáts R., Szalai T., Szegedi-Elek E., Székely P., Vida K., Vinkó J., and Kiss L. L. (2016) Activity of 50 long-period comets beyond 5.2 au. *Astron. J., 152*, 220.

Schleicher D. G. and Farnham T. L. (2004) Photometry and imaging of the coma with narrowband filters. In *Comets II* (M. C. Festou et al., eds.), pp. 449–469. Univ. of Arizona, Tucson.

Scotti J. V., Rabinowitz D. L., and Gehrels T. (1991) Automated detection of asteroids in real-time with the Spacewatch telescope. In *Asteroids, Comets, Meteors 1991* (A. W. Harris and E. Bowell, eds.), pp. 541–544. Lunar and Planetary Institute, Houston.

Sitko M. L., Lynch D. K., Russell R. W., and Hanner M. S. (2004) 3–14 micron spectroscopy of Comets C/2002 O4 (Hönig), C/2002 V1 (NEAT), C/2002 X5 (Kudo-Fujikawa), C/2002 Y1 (Juels-Holvorcem), and 69P/Taylor and the relationships among grain temperature, silicate band strength, and structure among comet families. *Astrophys. J., 612*, 576–587.

Skorov Y. and Blum J. (2012) Dust release and tensile strength of the non-volatile layer of cometary nuclei. *Icarus, 221*, 1–11.

Snodgrass C., Fitzsimmons A., Lowry S. C., and Weissman P. (2011) The size distribution of Jupiter family comet nuclei. *Mon. Not. R. Astron. Soc., 414*, 458–469.

Solontoi M., Ivezić Ž., West A. A., Claire M., Jurić M., Becker A., Jones L., Hall P. B., Kent S., Lupton R. H., Knapp G. R., Quinn T., Gunn J. E., Schneider D., and Loomis C. (2010) Detecting active comets in the SDSS. *Icarus, 205*, 605–618.

Solontoi M., Ivezić Ž., Jurić M., Becker A. C., Jones L., West A. A., Kent S., Lupton R. H., Claire M., Knapp G. R., Quinn T., Gunn J. E., and Schneider D. P. (2012) Ensemble properties of comets in the Sloan Digital Sky Survey. *Icarus, 218*, 571–584.

Sonnett S., Kleyna J., Jedicke R., and Masiero J. (2011) Limits on the size and orbit distribution of main belt comets. *Icarus, 215*, 534–546.

Sonnett S., Mainzer A., Grav T., Spahr T., Lilly E., and Masiero J. (2021) NEO Surveyor cadence and simulations. *7th IAA Planetary Defense Conference*, p. 56.

Stellingwerf R. F. (1978) Period determination using phase dispersion minimization. *Astrophys. J., 224*, 953–960.

Sykes M. V. and Walker R. G. (1992) Cometary dust trails I. Survey. *Icarus, 95*, 180–210.

Vera C. Rubin Observatory LSST Solar System Science Collaboration,

Jones R. L., Bannister M. T., Bolin B. T., Chandler C. O., Chesley S. R., Eggl S., Greenstreet S., Holt T. R., Hsieh H. H., Ivezić Z., Jurić M., Kelley M. S. P., Knight M. M., Malhotra R., Oldroyd W. J., Sarid G., Schwamb M. E., Snodgrass C., Solontoi M., and Trilling D. E. (2020) The scientific impact of the Vera C. Rubin Observatory's Legacy Survey of Space and Time (LSST) for solar system science. *ArXiV e-prints*, arXiv:2009.07653.

Waszczak A., Ofek E. O., Aharonson O., Kulkarni S. R., Polishook D., Bauer J. M., Levitan D., Sesar B., Laher R., Surace J., and the PTF Team (2013) Main-belt comets in the Palomar Transient Factory survey — I. The search for extendedness. *Mon. Not. R. Astron. Soc.*, *433*, 3115–3132.

Williams D. M., Mason C. G., Gehrz R. D., Jones T. J., Woodward C. E., Harker D. E., Hanner M. S., Wooden D. H., Witteborn F. C., and Butner H. M. (1997) Measurement of submicron grains in the coma of Comet Hale-Bopp C/1995 01 during 1997 February 15–20 UT. *Astrophys. J. Lett.*, *489*, L91–L94.

Wooden D. H. (2002) Comet grains: Their IR emission and their relation to ISM grains. *Earth Moon Planets*, *89*, 247–287.

Yang B., Keane J., Meech K., Owen T., and Wainscoat R. (2014) Multi-wavelength observations of Comet C/2011 L4 (Pan-STARRS). *Astrophys. J. Lett.*, *784*, L23.

York D. G., Adelman J., Anderson J., John E., Anderson S. F., Annis J., Bahcall N. A., Bakken J. A., Barkhouser R., Bastian S., Berman E., Boroski W. N., Bracker S., Briegel C., Briggs J. W., Brinkmann J., Brunner R., Burles S., Carey L., Carr M. A., Castander F. J., Chen B., Colestock P. L., Connolly A. J., Crocker J. H., Csabai I., Czarapata P. C., Davis J. E., Doi M., Dombeck T., Eisenstein D., Ellman N., Elms B. R., Evans M. L., Fan X., Federwitz G. R., Fiscelli L., Friedman S., Frieman J. A., Fukugita M., Gillespie B., Gunn J. E., Gurbani V. K., de Haas E., Haldeman M., Harris F. H., Hayes J., Heckman T. M., Hennessy G. S., Hindsley R. B., Holm S., Holmgren D. J., Huang C.-h., Hull C., Husby D., Ichikawa S.-I., Ichikawa T., Ivezić Ž., Kent S., Kim R. S. J., Kinney E., Klaene M., Kleinman A. N., Kleinman S., Knapp G. R., Korienek J., Kron R. G., Kunszt P. Z., Lamb D. Q., Lee B., Leger R. F., Limmongkol S., Lindenmeyer C., Long D. C., Loomis C., Loveday J., Lucinio R., Lupton R. H., MacKinnon B., Mannery E. J., Mantsch P. M., Margon B., McGehee P., McKay T. A., Meiksin A., Merelli A., Monet D. G., Munn J. A., Narayanan V. K., Nash T., Neilsen E., Neswold R., Newberg H. J., Nichol R. C., Nicinski T., Nonino M., Okada N., Okamura S., Ostriker J. P., Owen R., Pauls A. G., Peoples J., Peterson R. L., Petravick D., Pier J. R., Pope A., Pordes R., Prosapio A., Rechenmacher R., Quinn T. R., Richards G. T., Richmond M. W., Rivetta C. H., Rockosi C. M., Ruthmansdorfer K., Sandford D., Schlegel D. J., Schneider D. P., Sekiguchi M., Sergey G., Shimasaku K., Siegmund W. A., Smee S., Smith J. A., Snedden S., Stone R., Stoughton C., Strauss M. A., Stubbs C., SubbaRao M., Szalay A. S., Szapudi I., Szokoly G. P., Thakar A. R., Tremonti C., Tucker D. L., Uomoto A., Vanden Berk D., Vogeley M. S., Waddell P., Wang S.-i., Watanabe M., Weinberg D. H., Yanny B., Yasuda N., and the SDSS Collaboration (2000) The Sloan Digital Sky Survey: Technical summary. *Astron. J.*, *120*, 1579–1587.

Poch O., Pommerol A., Fray N., and Gundlach B. (2024) Laboratory experiments to understand comets. In *Comets III* (K. J. Meech, M. R. Combi, D. Bockelée-Morvan, S. N. Raymond, and M. E. Zolensky, eds.), pp. 213–245. Univ. of Arizona, Tucson, DOI: 10.2458/azu_uapress_9780816553631-ch008.

Laboratory Experiments to Understand Comets

Olivier Poch
Université Grenoble Alpes, Centre National de la Recherche Scientifique

Antoine Pommerol
Physikalisches Institut, University of Bern, Switzerland

Nicolas Fray
Université de Paris Est Créteil and Université Paris Cité, Centre National de la Recherche Scientifique

Bastian Gundlach
Institut für Planetologie, Universität Münster

In order to understand the origin and evolution of comets, one must decipher the processes that formed and processed cometary ice and dust. Cometary materials have diverse physical and chemical properties and are mixed in various ways. Laboratory experiments are capable of producing simple to complex analogs of comet-like materials, measuring their properties, and simulating the processes by which their compositions and structures may evolve. The results of laboratory experiments are essential for the interpretations of comet observations and complement theoretical models. They are also necessary for planning future missions to comets. This chapter presents an overview of past and ongoing laboratory experiments exploring how comets were formed and transformed, from the nucleus interior and surface, to the coma. Throughout these sections, the pending questions are highlighted, and the perspectives and prospects for future experiments are discussed.

1. INTRODUCTION

Comets contain clues about the physical and chemical processes that occurred during the early phases of solar system formation, and they evolve dramatically when approaching the Sun. They may have brought water, other volatiles, and organics to the terrestrial planets. Therefore, fundamental questions of cometary science concern the origin and evolution of comets: What are comets made of? What is their internal structure? How did comets form? How do they evolve, and what causes their activity?

To answer these questions, the cometary science community relies on observations of comets using telescopes (see the chapter by Bauer et al. in this volume), flybys or rendezvous space missions (see the chapter by Snodgrass et al.), as well as measurements of cometary samples collected in space or found within Earth's atmosphere and surface (see the chapter by Engrand et al.). But the proper preparation, analysis, and interpretation of these observations also requires theoretical models and experimental analogs, which are based on the current knowledge of comets obtained from previous studies. A strong synergy between observations, theories, and experiments is thus essential to understand comets, as well as other objects of the solar system.

The complexity of comets is due to their composition, structure, and evolution. Indeed, comets are made of multiple constituents of various origins and volatility (see the chapter by Bergin et al.): ices (H_2O, CO_2, CO, CH_3OH, NH_3, CH_4, etc.), semivolatile materials (e.g., heavier organic molecules, salts), and refractory materials (refractory organic matter, amorphous and crystalline silicates, carbonates, oxides, sulfides, and metals constituting the dust), the nature and proportion of which may vary depending on the object. Each of these individual constituents can have extremely different physical and chemical properties, and will thus play different roles in the evolution of comets. Of major importance is also the physical arrangement of this mixture of constituents, from the grain scale (≥10 nm to 100 μm) to the nucleus scale (≥0.1–70 km), to form a low-density agglomerate with a bulk porosity in excess of 50%. The initial composition and structure of cometary ice and dust depend on comet formation processes and on their place of birth. Both the composition and structure are modified with time, by slow evolutionary processes affecting comets during their long residence in the Oort cloud or the scattered disk, and during their short active phase when approaching the Sun (see the chapter by Guilbert-Lepoutre et al.).

When, where, and how did these cometary constituents form? How were they agglomerated together to form cometesimals? What are the fractions inherited from the interstellar medium, transformed, or produced in the protoplanetary disk? How do the composition and structure of comets generate cometary activity? How are they modified by various processes?

To answer these questions, we need laboratory experiments because of the limited ability of theoretical models to simulate the observed natural complexity (Fig. 1). First-principle physics or chemistry arguments cannot predict the evolution of such complex natural systems as cometary materials and structures. In the laboratory, we can produce and study analogs of these cometary constituents and simulate cometary processes.

The objectives of laboratory experiments are to (1) improve the general understanding of cometary formation, evolution, and activity, i.e., the physical and chemical processes that governed each of these stages; (2) interpret observations (light scattering, thermal, dielectric measurements, etc.) of cometary nuclei and comae performed by spacecraft instruments or by telescopes on Earth (or Earth's orbit), through direct comparison of observations with both theoretical and experimental simulations; and (3) prepare future cometary explorations via predictions of physical and chemical properties.

Two complementary approaches exist for laboratory experiments, depending on the initial complexity of the sample under study (Fig. 1). Experiments performed with relatively *simple* samples compared to actual cometary material (such as a nanometer-thick layer of a single constituent, or aggregates of pure silicate grains) are useful to determine physical constants and their dependencies on sample properties that can be implemented in theoretical models. On the other hand, experiments performed on more *complex* samples, often called cometary analogs, are used to simulate some aspects of cometary complexity (mixture of constituents, structure, combination of processes, etc.) and assess their influence on the processes (thermal, radiolytic, mechanical processes, etc.) modifying comets. The measurements done on these complex cometary analogs can be directly compared to observations, and/or can be used to test theoretical models. In all cases, it is crucial to characterize the composition and structure of the produced samples in quantitative detail before they are used for experiments.

This chapter presents an overview of laboratory experiments relevant to comets. Section 2 describes experiments used to understand cometary nuclei properties, from their formation to evolution during residence in the outer solar system and approach to the Sun. Section 3 covers experiments exploring the properties and evolution of comae particles, and section 4 provides some perspectives on experiments needed in the future.

2. COMETARY NUCLEI

2.1 Experiments on Comet Formation, from Interstellar and Circumstellar Grains to Cometesimals

2.1.1. Brief summary on comet origin and evolution. Comets formed in the young solar system when the protoplanetary disk was still present [see the review by *Weissman et al.* (2020) and the chapter by Aikawa et al. in this volume]. However, it is not yet known precisely where and when comets have formed. In this section, we refer to cometesimals as the building blocks that formed comets. After their formation, comets have been altered by cometary evolutionary processes (see Fig. 1 and section 2.2 and onward).

Due to the presence of water ice and more volatile species, called supervolatiles (i.e., having an equilibrium sublimation pressure higher than 10^{-9} bar at 100 K, like

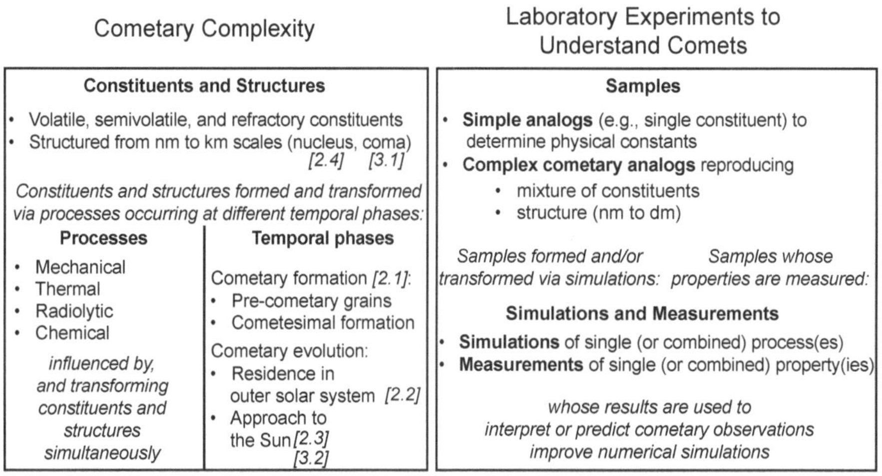

Cometary Complexity

Constituents and Structures
- Volatile, semivolatile, and refractory constituents
- Structured from nm to km scales (nucleus, coma) [2.4] [3.1]

Constituents and structures formed and transformed via processes occurring at different temporal phases:

Processes
- Mechanical
- Thermal
- Radiolytic
- Chemical

influenced by, and transforming constituents and structures simultaneously

Temporal phases

Cometary formation [2.1]:
- Pre-cometary grains
- Cometesimal formation

Cometary evolution:
- Residence in outer solar system [2.2]
- Approach to the Sun [2.3] [3.2]

Laboratory Experiments to Understand Comets

Samples
- **Simple analogs** (e.g., single constituent) to determine physical constants
- **Complex cometary analogs** reproducing
  - mixture of constituents
  - structure (nm to dm)

Samples formed and/or transformed via simulations: *Samples whose properties are measured:*

Simulations and Measurements
- **Simulations** of single (or combined) process(es)
- **Measurements** of single (or combined) property(ies)

whose results are used to interpret or predict cometary observations improve numerical simulations

Fig. 1. This diagram indicates how the natural complexity of comets can be studied and understood using laboratory experiments. Numbers in brackets refer to the sections of this chapter where the relevant experiments are described.

CO_2, NO, etc.) and hypervolatiles (i.e., higher than 10^{-9} bar at 50 K, like Kr, CO, etc.) (*Fray and Schmitt,* 2009), we know that cometesimal formation must have taken place beyond the snow line, which describes at which distance from the central star water ice (or other frozen volatile species) are stable in the protoplanetary disk (*Stevenson and Lunine,* 1988). The average low density of cometary nuclei (below 1 g cm^{-3} whenever estimated) favors two possible scenarios for their formation, namely the gravitational instability scenario that predicts a cometary material that consists of pebbles and fractal agglomerates, and the mass transfer scenario that predicts an homogeneous material that consists of (sub-)micrometer-sized grains (see the chapter by Simon et al. in this volume). Laboratory experiments are performed with the aim of studying these growth processes via coagulation of grains and collision of aggregates having different compositions. The formation scenario may also influence the composition of cometesimals, because the direct growth scenario requires more time than the gravitational instability one. Indeed, if cometesimals formed relatively fast after the first solid components of the solar system [the calcium-aluminum-rich inclusions (CAIs) found in chondrite meteorites], they must have experienced significant heating due to the decay of short-lived radionuclides (see, e.g., *Prialnik et al.,* 2008), leading to thermally-induced chemical reactions and depletion of the volatiles in the interior and probably in the whole object. In contrast, a delayed formation of cometesimals would imply a lower thermal evolution induced by radioactive decay than an early formation.

Observations and analyses of cometary materials indicate that grains of ice and dust of very different origins were mixed to form cometesimals (*Brownlee,* 2014). Some of these grains might have formed in molecular clouds or dense cores [prestellar environments, also called the interstellar medium (ISM)] and incorporated intact in cometesimals (see the chapter by Bergin et al. in this volume), and many others have been modified or (re-)formed in various regions of the protostellar and protoplanetary disk before being transported and incorporated in cometesimals (see the chapters by Aikawa et al. and Engrand et al. in this volume). Moreover, possible later accretion of refractory materials after the icy cometesimal formed cannot be excluded (see the chapter by Engrand et al.). The paragraphs below present the different pathways to synthesize and study analogs of cometary volatile molecules and carbonaceous and organic dust, as well as minerals via laboratory experiments.

Important open questions about the chemical composition of cometesimals are the degree of inheritance vs. reset of the volatile and refractory materials from prestellar and protostellar phases to cometesimals, and the elemental budgets throughout this evolution, i.e., the identification of the materials/molecules carrying each atom C, O, N, and S to account for their cosmic abundance (see reviews in *Caselli and Ceccarelli,* 2012; *Boogert et al.,* 2015; *Öberg and Bergin,* 2021; *Boogert,* 2019). Laboratory experiments, in conjunction with theoretical works, play important roles in answering these questions, by providing a means to identify and quantify the volatile and refractory species from observational data of precometary environments and by studying the processes affecting them before their incorporation in cometesimals.

2.1.2. Chemistry of pre-cometary grains. *2.1.2.1. Volatile molecules:* Volatile molecules found in comets are mainly H_2O, CO, CO_2, and other minor constituents such as CH_3OH, NH_3, CH_4, etc., including larger carbon-bearing molecules that are called organic molecules or complex organic molecules (COM), the latter denomination being used in ISM studies to refer to all molecules containing at least six atoms including carbon [see *Herbst and Van Dishoeck* (2009) and the chapter by Bergin et al. in this volume]. These molecules are formed by chemical reactions in the gas phase, at the surfaces of bare dust particles, or in the ice mantles that cover these particles in cold dense interstellar clouds and possibly in later stages of star formation (see reviews in *Herbst and Van Dishoeck,* 2009; *Herbst,* 2014). To predict or interpret observations and to complete numerical models, laboratory experiments are performed to obtain parameters (rate constants, branching ratios) of these chemical reactions in the gas (*Smith,* 2011), gas-solid, and solid phases (*Öberg et al.,* 2009). Other experiments also provide means to identify and quantify these molecules, via gas phase molecular excitation, and spectroscopy of gases and ices at infrared wavelengths (see references in *Herbst and Van Dishoeck,* 2009; *Herbst,* 2014; *Öberg and Bergin,* 2021). These experiments dedicated to the chemistry of volatile molecules in pre- and protostellar environments are crucial to understand their inheritance and/or processing between the ISM and comets (see reviews in *Caselli and Ceccarelli,* 2012; *Boogert et al.,* 2015; *Öberg and Bergin,* 2021; and the chapters by Bergin et al. and Aikawa et al. in this volume).

Of particular interest to understand the composition of cometesimals are the experiments investigating the evolution of ice mantles coating interstellar grains from quiescent clouds to young stellar objects envelopes and disks (*Caselli and Ceccarelli,* 2012). Chemical reactions in and on these ices take place through neutral-neutral atom and radical addition reactions following the accretion of species, or via the interaction with ultraviolet photons and impacting particles (protons, ions, electrons from galactic cosmic rays), and also via heating. Laboratory studies aiming at reproducing the astrophysical conditions experienced by ice mantles are described below (more detailed reviews can be found in *Linnartz et al.,* 2015; *Öberg,* 2016; *Gudipati et al.,* 2015).

Setups for ice mantle chemistry. Figure 2 shows setups consisting of an ultra-high vacuum chamber (with pressure down to 10^{-10} mbar) containing a metal or infrared-transparent salt surface (~2 cm diameter) mounted on the tip of a helium cryostat in order to be cooled down to about 10 K and controlled by resistance heaters. This cold substrate is exposed to a flow of gas of controlled composition, either pure or a mixture (H_2O, NH_3, etc.), to build a layer of ice few nanometers to micrometers thick. This ice surface is then exposed to heat by changing the temperature of the

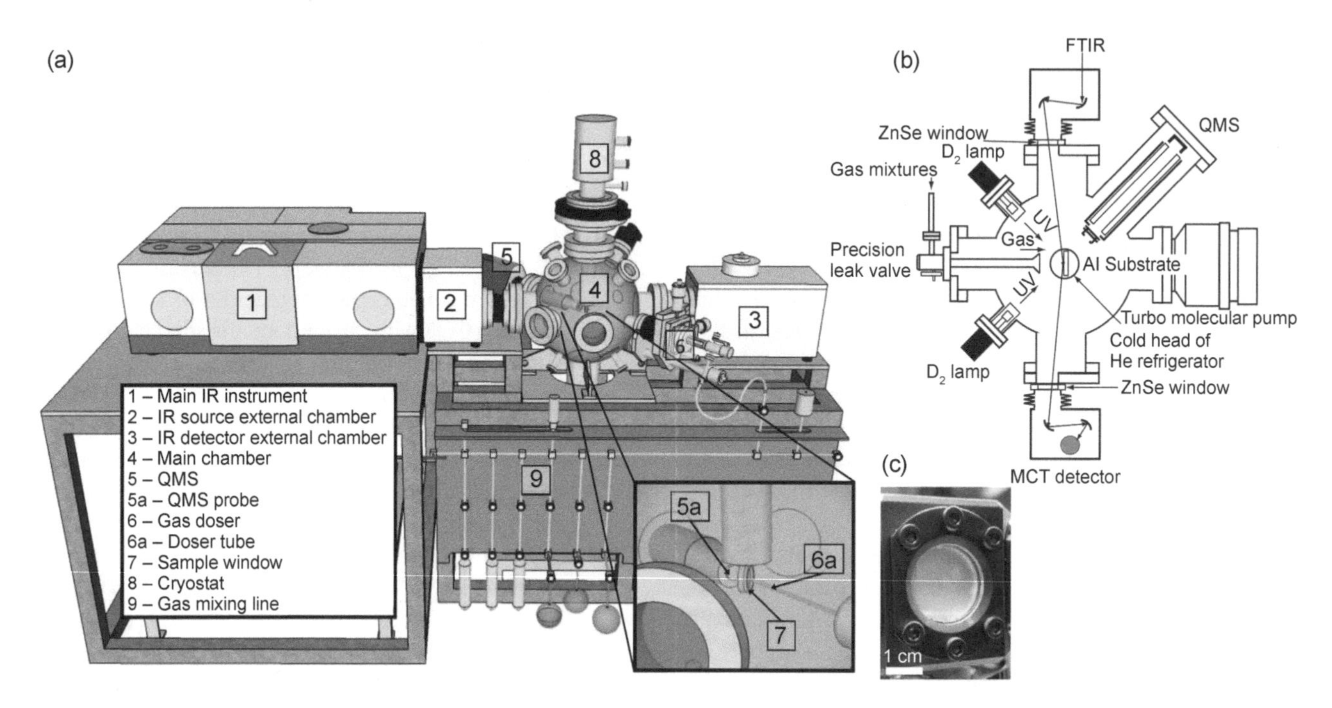

Fig. 2. Setups for astrochemistry of (pre-)cometary ices analogs, deposited as thin (nm-μm) layers. **(a)** Ultra-high vacuum and cryogenic setup, with sample monitoring by infrared spectroscopy and mass spectrometry (*Lauck et al.*, 2015). **(b)** Another setup, with UV irradiation of the sample (*Oba et al.*, 2017). Various irradiation and monitoring techniques can be used. **(c)** Refractory organic residue on a sample substrate (*Augé et al.*, 2016). Reproduced with permission of the AAS.

cryostat and/or to vacuum ultraviolet photons (from 115 to 200 nm, using a microwave-driven H_2 discharge lamp), X-rays (from a synchrotron), energetic particles (neutral or ionized atoms, electrons, e.g., from an accelerator) inducing desorption, structural changes, and chemical reactions. These changes are monitored *in situ*, most often via infrared spectroscopy [transmission or reflectance Fourier transform infrared (FTIR)] and mass spectrometry [quadrupole mass spectrometer (QMS) or time-of-flight mass spectrometer (TOFMS)] during the irradiation and/or during progressive heating of the substrate [called temperature programmed desorption (TPD)]. Infrared spectroscopy monitors the composition of the solid phase by identifying absorption bands at frequencies specific to molecular vibrations in the midinfrared (MIR) (2.5–25 μm). In some experiments, *in situ* ultraviolet-visible (UV-Vis) (190–700 nm), vacuum ultraviolet (VUV) (100–200 nm), Raman, far-infrared, and THz time-domain spectroscopies are also performed (*Allodi et al.*, 2013). Mass spectrometry techniques use an ionization source, a mass-selective analyzer, and an ion detector to identify individual molecules produced in the gas phase and/or in the ice. During progressive TPD heating (rate from 0.1 to 10 K min^{-1}), each molecule has its maximum sublimation rate at a specific temperature, facilitating its identification by mass spectrometry (*Öberg et al.*, 2009). The temperature at which the molecule sublimates also depends on its mixing, trapping, or segregation, providing information not only on the composition but also on the structure of the solid phase. Some experiments implement *in situ* analysis of the ice via non-thermal laser desorption/ablation and ionization mass spectrometry (LD-TOFMS) or Matrix Assisted Laser Desorption Ionization (MALDI-TOFMS) (*Henderson and Gudipati,* 2015).

Results of ice mantle chemistry experiments. Such experiments have shown that the exposure of ices to ionizing radiation (UV, energetic particles) breaks molecular bonds and creates radicals and ions, which can diffuse, meet, and react in or on the ice to form more complex molecules. Photolysis or radiolysis induces similar products, suggesting that the overall chemical evolution mainly depends on the amount — not the kind — of energy delivered to the ice mantle (*Gerakines et al.,* 2001, 2004; *Rothard et al.,* 2017). Heating in star-forming regions (from 10 to 100 K or more) facilitates the mobility of these species produced by photolysis or radiolysis, allowing their reaction. Moreover, mixtures of neutral molecules (NH_3, CO_2, etc.) can produce new molecules by purely thermal reactions (acid-base, nucleophilic addition and elimination, or condensation), without any other source of energy, providing that the initial abundance of the reactants, the temperature, and the time available are adequate (see review in *Theulé et al.,* 2013). The association of ionizing radiations followed by warming of a mixture of ices results in an impressive molecular diversity (Fig. 3). Water, which is the most abundant molecule in pre-cometary grains and comets, play an important role in this chemistry, as reactant, as catalyst, or as trapping matrix and diffusion surface allowing reactions to occur (*Fresneau et al.,* 2014, and references therein; *Ghesquière et al.,* 2018). Newly formed molecules of larger size are often less volatile (more refractory) than their precursors. After warm-up and complete sublimation of the ice at the end of the experiment, the molecules remaining in solid phase at

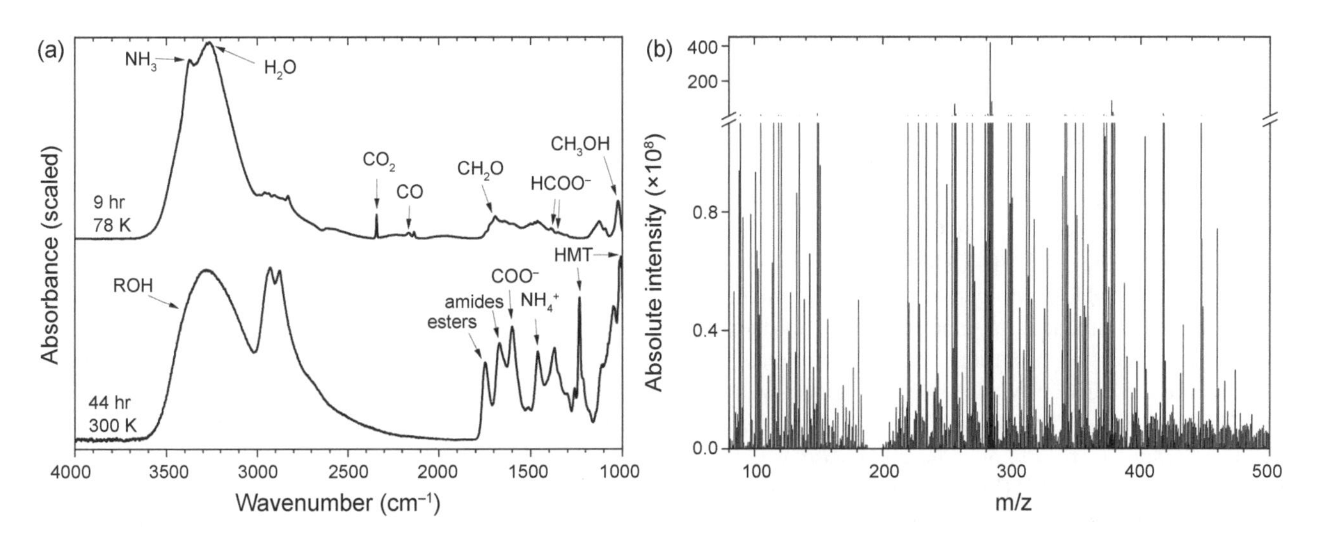

Fig. 3. Molecular diversity after irradiation of ices and warming. **(a)** FTIR spectra of (top) H_2O:NH_3:CH_3OH in a ratio 3:1:1 after about 9 h of UV irradiation at 78 K, and (bottom) the resulting organic residue after 44 h of UV at 78 K and warmed up to 300 K. **(b)** Mass spectrum of the methanol-soluble fraction of this organic residue, showing thousands of peaks. From *Danger et al.* (2013).

room temperature (Fig. 2c) are thus called the "refractory residue" [historically called "yellow stuff" in *Greenberg et al.* (1995) and references therein]. These are semi-volatile molecules, such as salts (ionic solids), including molecules of prebiotic interest, as well as polymers up to refractory carbon-bearing compounds. *Ex situ* analyses, outside the simulation chamber, are performed via various techniques to reveal the nature of this residue. Experiments on the chemistry of refractory residues are presented in section 2.1.2.2.

Diffusion and reaction parameters. Laboratory experiments with single-component or binary ices enable the determination of reaction pathways and physical constants, such as reaction rates, cross-sections, desorption rates, and branching ratios needed to complete existing chemical networks used to compute the evolution of the ice/gas composition [for a review of these models, see section 1.3.2 in *Öberg and Bergin* (2021)]. These parameters are significantly influenced by the structure of the ice (porosity, morphology, etc.), so experiments on various ice structures (produced by varying the deposition method on the cold substrate or the temperature) should be performed (*Isokoski et al.,* 2014; *Noble et al.,* 2020; *Kouchi et al.,* 2021). Moreover, the simultaneous presence of several molecules in the ice also changes the reaction kinetics compared to pure or simple ice mixtures. Other important parameters to obtain via experiments are diffusion rates of different radicals, ions, or neutrals in ices at different temperatures (*Öberg,* 2016; *Mispelaer et al.,* 2013; *Minissale et al.,* 2019) (see section 2.3.4.2).

Influence of dust on the ice chemistry. Pre-cometary ices have condensed on refractory particles, which are made of minerals (mainly silicates, etc.) and/or carbonaceous materials. The nature of the refractory particle (mineral/carbonaceous) and its structure (porosity, etc.) will affect the electrostatic or bonding interactions with the ice molecules, the energy dissipation of exothermic reactions, and the active area in contact with the molecules, influencing reaction rates, diffusion, and concentration of species. Only a limited number of studies have addressed the effects of dust substrates on the astrophysical ice chemistry, but some studies demonstrated the catalytic effect of dust surface on processes in ices [see, e.g., *Potapov et al.* (2019) and the review in *Jäger et al.* (2019)]. In the setup described above (Fig. 2), ices are deposited as a layer on a substrate. However, observations suggest that cometary dust particles possess a large range of porosity, from highly porous to compact aggregates (*Wooden,* 2008; *Levasseur-Regourd et al.,* 2018). Ices condensing on such aggregates probably fill the pores, having a much larger surface of contact with the dust than in typical experiments, and consequently enhanced interactions. Future laboratory experiments should thus develop methods to prepare such ice/dust mixtures, with various composition and structure, and analyze their influences on ice chemistry and spectroscopy.

2.1.2.2. From ices to refractory organic dust: The organic refractory residues, formed after energetic irradiation followed by warming and sublimation of ices (Fig. 2c), comprise an impressive diversity of molecules, such as salts (ionic solids), individual molecules, polymers and other macromolecules, including some of prebiotic interest, which are analyzed by different techniques available in the laboratory.

Analysis of the entire molecular diversity. After energetic processing, warming, and subsequent sublimation of the ice sample inside the experimental setup, the refractory residue remaining on the substrate can be analyzed *in situ*, most often via infrared spectroscopy, and/or *ex situ* via mass spectrometry following desorption or extraction techniques (Fig. 3). Other *ex situ* analyses such as X-ray absorption near-edge structure spectroscopy (XANES), energy-dispersive X-ray spectroscopy (EDX), and transmission and scanning electron microscopy (TEM and SEM, respectively) are also sometimes used for the elemental (N/C, O/C, etc.), chemical (bonds, chemical functions), and structural (macromolecular organization) characterization of residues. Infrared spectroscopy of residues has revealed the presence of several molecular species and chemical groups in the residues, in particular hexamethylenetetramine (HMT, $[(CH_2)_6N_4]$), ammonium salts of carboxylic acids $[R{-}COO^- \; NH_4^+]$, amides $[H_2NC(=O){-}R]$, esters $[R{-}C(=O){-}O{-}R]$, and species related to polyoxymethylene (POM) $[(-CH_2O-)_n]$ (*Muñoz Caro and Schutte,* 2003). HMT, a nitrogen-bearing cyclic molecule, is an abundant component of some refractory residues obtained after processing of various initial ice mixtures: For example, after irradiation (via UV or protons) and warming of H_2O:NH_3:CH_3OH ices (*Bernstein et al.,* 1995; *Cottin et al.,* 2001; *Muñoz Caro and Schutte,* 2003; *Muñoz Caro et al.,* 2004; *Oba et al.,* 2017) (Fig. 3a), or after only warming of H_2CO:NH_3:HCOOH or CH_2NH:HCOOH ices (*Vinogradoff et al.,* 2013). HMT can reach up to 50 wt% of the total organic products, especially if methanol (CH_3OH) is present in the initial ice mixture (*Danger et al.,* 2013). Depending on the relative concentration of NH_3 or CN^- and formaldehyde (H_2CO) in the ice mixture, H_2CO can polymerize to form POM or POM-like polymers that are then the most abundant components of the residues (*Butscher et al.,* 2019; *Duvernay et al.,* 2014). These experiments have suggested the presence of HMT or POM in comets. Although HMT is found at part-per-billion level concentration in carbonaceous chondrites (*Oba et al.,* 2020), it has not been detected in a comet up to now, and the presence of POM in Giotto (*Huebner,* 1987) and Rosetta (*Wright et al.,* 2015) observations is debated (*Altwegg et al.,* 2017). These compounds, although abundant in some laboratory refractory residues, are not observed in comets yet, either because they are mainly decomposed, or because their precursors are not formed or destroyed, under cometary conditions. Future experiments should investigate these questions and suggest when/where these compounds could form and be preserved, or under which conditions they decompose. Other abundant components of refractory residues primarily characterized by their infrared absorption spectra are ammonium salts ($NH_4^+X^-$, where X^- is the anion of a generic acid molecule XH). They can be produced by acid-base or nucleophilic addition reactions activated by heat (*Theulé et al.,* 2013). If NH_3 and an acid molecule (e.g., HCOOH, HCN, HNCO, etc.) can diffuse close enough to each other to exchange a proton, the pair of ions NH_4^+ (ammonium) and $HCOO^-$ (formate) is formed. Alternatively, additions of NH_3 on CO_2 can end up in the formation of ammonium carbamates (*Theulé et al.,* 2013). Laboratory experiments performed for many years had predicted the presence of ammonium salts as a likely component of comets (*Colangeli et al.,* 2004), which was confirmed in the dust of Comet 67P/Churyumov-Gerasimenko (hereafter 67P) by the Visible and Infrared Thermal Imaging Spectrometer (VIRTIS) and Rosetta Orbiter Spectrometer for Ion and Neutral Analysis (ROSINA) instruments of Rosetta (*Altwegg et al.,* 2020; *Poch et al.,* 2020). Moreover, NH_4^+ is suspected to be present in interstellar ices, as a counter-ion of isocyanide OCN^-, which has been identified in interstellar ices (*Öberg et al.,* 2011; *Boogert et al.,* 2015, 2022; *McClure et al.,* 2023) and in different protostar environments (*Schutte and Khanna,* 2003; *van Broekhuizen et al.,* 2005). Therefore, ammonium salts might be inherited from interstellar ices, and/or they may form at later cometary stages (see section 2.3.4.2).

For *ex situ* analyses, the organic refractory residue is collected and placed in the chamber of a microprobe laser-desorption laser-ionization mass spectrometer (*Dworkin et al.,* 2004) or a laser-desorption time-of-flight mass spectrometer (LD-TOFMS) (*Modica et al.,* 2012). LD-TOFMS analyses have revealed the presence of macromolecular components between 1000 and 3000 amu in residues (*Modica et al.,* 2012). Other mass spectrometers include (Very) High Resolution Mass Spectrometry (HRMS or VHRMS) techniques, such as Fourier Transform Ion Cyclotron Resonance (FT-ICR) or Fourier Transform Orbitrap (FT-Orbitrap). For these analyses, the refractory residues are dissolved (often in ultrapure methanol) before being injected into the mass spectrometer (Fig. 3). VHRMS provides sufficient mass resolution to retrieve the chemical formulae of the molecules, enabling the sorting of molecules in families according to their content in specific atoms, group of atoms, and even aromaticity (*Danger et al.,* 2013, 2016; *Fresneau et al.,* 2017; *Gautier et al.,* 2020). VHRMS is a promising technique to determine how the overall chemistry of analogs is affected by different production conditions, and how it compares to that of cosmomaterial organic matter, especially future comet samples that will be returned on Earth (see the chapter by Snodgrass et al. in this volume).

Identification of specific molecules. To search for specific molecules in refractory organic residues, chromatography techniques are used in order to separate the molecules according to their chemical properties, before they are weighed by mass spectrometry. After dissolution, residues are analyzed by high performance liquid chromatrography coupled to a mass spectrometer (HPLC-MS). Residues can also be analyzed via gas chromatography coupled to mass spectrometry (GCMS), providing that they are treated chemically to increase the volatility of some molecules (a step called derivatization or functionalization). These techniques allowed for the confirmation of the presence of HMT (*Cottin et al.,* 2001; *Muñoz Caro et al.,* 2004), amines, and various nitrogen- and oxygen-bearing linear or cyclic organic compounds including precursors of molecules of prebiotic interest such as urea, hydantoin, and carbamic acid (*Chen et al.,* 2007; *Marcellus et al.,* 2011; *Nuevo et al.,* 2010). Molecules of prebiotic interest identified in refractory residues are amino acids (*Bernstein et al.,* 2002a; *Nuevo et al.,* 2008), amphiphiles (*Dworkin et al.,* 2001), nucleobases (*Oba et al.,* 2019), and sugars (*Meinert et al.,* 2016; *Nuevo et al.,* 2018). Moreover, studies using enantioselective multidimensional gas chromatography coupled to time-of-flight mass spectrometry (GC×GC-TOFMS) have shown that the circular polarization of the VUV light used to irradiate the ices induces the production of a small enantiomeric excess in amino acids (*Marcellus et al.,* 2011; *Modica et al.,* 2014). This result suggests that an enantiomeric excess brought to the early Earth by cometary materials might have been amplified via prebiotic chemistry, potentially explaining the specific amino acids enantiomers chosen by terrestrial life forms whose homochirality is a universal property. Among the molecules of prebiotic interest, only glycine and precursors such as ammonium salts have been firmly identified on comets (*Altwegg et al.,* 2016; *Poch et al.,* 2020), but neither of these molecules are chiral. These techniques employed on analogs will allow searching for these and other molecules in future samples returned from comets, and some of them may be miniaturized to be run onboard future space missions. The main drawback of *ex situ* analyses is the exposure of the organic residue to air and ambient temperature, and the possible alteration of the most reactive species it contains with water vapor or O_2, for example. Future laboratory experiments should address this problem by developing new methods to transfer the samples for *ex situ* analyses or by developing novel techniques for *in situ* analyses, maybe inspired by other research fields (*Fulvio et al.,* 2021).

From refractory ice residues to even more refractory compounds. The carbon in 67P, and possibly in other comets, appears to be mainly present as high-molecular-weight organic matter, analogous to the insoluble organic matter (IOM) found in the carbonaceous chondrite meteorites (*Fray et al.,* 2016; *Bardyn et al.,* 2017). The refractory organic matter of 67P has a higher hydrogen/carbon ratio than IOM, suggesting that it could be more pristine (*Isnard et al.,* 2019). How this refractory organic matter was formed is a major question of cosmochemistry. Organic refractory residues obtained after sublimation of the ices may have been exposed to irradiation in pre- and protostellar environments, as well as heat in the forming protoplanetary disk. These subsequent processings of organic residues are studied by dedicated laboratory experiments, described in this paragraph. Moreover, irradiations and heat could also transform some of the organic compounds after their incorporation in cometesimals, as discussed in sections 2.2.3 and 2.3.4.2. Experiments have shown that partial thermal processing at 300°–400°C is required to convert organic residues made from ice photoprocessing and warm-up into the amorphous carbon with low heteroatom content found in IDPs (*Muñoz Caro et al.,* 2006). Organic residues subjected to further UV irradiation produced a thin altered dark crust insoluble on top of the initially soluble residue (*de Marcellus et al.,* 2017; *Piani et al.,* 2017). This insoluble material shows spectral similarities with natural samples of IOM extracted from carbonaceous chondrites (*de Marcellus et al.,* 2017). TEM and SEM morphological observations showed amorphous nanospherules similar to organic nanoglobules observed in the least-altered chondrites, chondritic porous interplanetary dust particles (CP-IDPs), and cometary samples (*Piani et al.,* 2017). However, to date no experiment has been able to reproduce all the structural, chemical and isotopic properties of IOM. *Faure et al.* (2021) performed ion irradiation experiments of various organic polymers, revealing a dramatic precursor effect on the final chemistry of the residues below a nuclear dose of ~10 eV atom^{-1}. Above this dose, any precursor transforms into amorphous carbon. Other experiments exposed refractory organics residues, previously prepared in the laboratory, to solar UV from low-Earth orbit (*Baratta et al.,* 2019; *Greenberg et al.,* 1995). Future exposure experiments in outer space (e.g., on the planned "Lunar Gateway" space station) may allow simultaneous exposure of many different samples to the solar radiations on long time scales (*Cottin et al.,* 2017). However, these experiments on thin layers only produce very low amounts of organic residues, preventing their use in complex cometary analogs for larger-scale experiments (see sections 2.3.2 and 2.4). For such experiments, an alternative is to produce several hundreds of milligrams of analogs of these carbonized organic residues by thermal degradation of *tholins* produced from gases, or HCN polymers produced from aqueous chemistry (*Bonnet et al.,* 2015).

2.1.2.3. Amorphous carbon grains: In the diffuse ISM, a significant portion of carbon is expected to be in the dust (*van Dishoeck,* 2014). Infrared spectra of the diffuse interstellar medium suggest that this carbon dust is much more similar to carbon-rich hydrocarbon materials than energetically processed ice residues containing heteroatoms (*Pendleton and Allamandola,* 2002). *Henning et al.* (2004) and *Ehrenfreund and Cami* (2010) provide reviews of laboratory studies on the formation, evolution, and means of characterization of different forms of carbon-rich dust expected in the ISM. This carbon dust could be in the form of various cross-linked three-dimensional networks of "amorphous carbon" that contain only C and H, and "hydrogenated amorphous carbon" containing higher level of H. Molecules such as polycyclic aromatic hydrocarbons (PAHs) and their clusters, widespread in the ISM, could contribute to the formation of this carbon dust and/or be produced from their decomposition. Finally, "macromolecular organic matter" containing additional atoms such as O, N, and S may be produced through processing of these carbon dust precursors, or from ice chemistry as discussed above (section 2.1.2.2). It is unclear if this interstellar carbonaceous dust was preserved, modified, or destroyed in the protoplanetary disk, and how much could have been incorporated in cometesimals.

Laboratory experiments have been carried out to study amorphous carbon with different degrees of hydrogenation. Different amorphous carbon grains are produced by condensation of carbon vapor obtained by laser pyrolysis of various hydrocarbons (*Jäger et al.,* 2006) or laser ablation of carbon surfaces, or arc discharge between two carbon rods (*Mennella et al.,* 2003, and references therein), or ion and photon irradiation of hydrocarbon-rich (e.g., pure CH_4) ice layers (*Dartois et al.,* 2005; *Strazzulla,* 1999). Carbon soots, produced by controlled combustion of acetylene or ethylene in a low-pressure flat flame burner, are of special interest because their precursors may be PAHs and fullerene-like molecules (*Carpentier et al.,* 2012). These carbon dusts are mainly characterized by infrared (*Gavilan et al.,* 2017), but also by Raman spectroscopy (*Brunetto et al.,* 2009), which is sensitive to the carbon backbone structure and to its degree of order.

Laboratory experiments investigating how this ISM-inherited carbon dust could have been processed in the disk are of interest for cometary science. Once produced and characterized, these carbon dusts can be submitted to energetic processing encountered in space. Galactic cosmic ray and solar energetic particle irradiations are simulated by ion irradiation experiments, causing their progressive amorphization (*Brunetto et al.,* 2009; *Mennella et al.,* 2003; *Pino et al.,* 2019). The impact of these irradiations on PAHs or amorphous/hydrogenated carbon dusts embedded in ices is also of interest to understand if these carbon networks could break, incorporate heteroatoms (O, N, S), and link via aliphatic chains to form a macromolecular network similar to IOM. Up to now, few laboratory experiments have started to address this question, showing oxygenation/hydroxylation of PAH molecules embedded in water ice after UV or proton irradiation (*Bernstein et al.,* 2002b; *Yang and Gudipati,* 2014).

2.1.2.4. Gas phase organic synthesis: Another way proposed to form pre-cometary organic dust is from gas phase or gas-solid phase chemistry in the hot (~1000 K) innermost region close to the protostar, followed by radial outward transport before incorporation into cometesimals. Gas-mineral Haber-Bosch (HB) catalytic reduction of N_2 by hydrogen to make NH_3, and Fischer-Tropsch type (FTT) catalytic reduction of CO by hydrogen in the presence of NH_3, were early proposed as plausible sources of organic dust. Laboratory experiments have demonstrated their feasibility under nebular conditions, on various mineral grains (amorphous silicates, pure silica smokes, etc.), and at low pressure (*Nuth et al.,* 2008, and references therein; *Nuth et al.,* 2016). These experiments take place in a glass apparatus where the mineral powder is subjected to a flow of gas at temperatures from 500 to 900 K (Fig. 4). A carbonaceous coating is formed on the grains, and this coating is itself an efficient catalytic surface for the reaction to continue.

More recently, other experiments have been conducted to simulate radical-rich plasma environments of the protoplanetary disk (inner and irradiated upper layers of the disk, around 1 to 10 au from the protostar), by ionizing gas-phase mixtures (*Bekaert et al.,* 2018; *Biron et al.,* 2015; *Kuga et al.,* 2015). These experimental setups, coined "Nebulotron," consist of a glass line in which a gas mixture (CO, N_2, H_2) at about 1 mbar is flowed through a microwave generator. The microwaves trigger a plasma discharge in which gases are dissociated and ionized at 800–1000 K. Ions and/or radicals condense progressively on the glass surface, forming a refractory organic solid, which is analyzed *ex situ.* This material reproduces some features of the chondritic IOM: hydrocarbon backbone structure (*Biron et al.,* 2015) and elementary and isotopic signatures of noble gases (*Kuga et al.,* 2015).

2.1.2.5. Hydrothermal organic synthesis: Refractory organics are also produced via condensation and polymerization reactions of small molecules in liquid water. Such reactions may occur in the interior of sufficiently large comets (≥50 km) (*Gounelle et al.,* 2008) or asteroids. Moreover, because cometesimals have accreted some materials such as crystalline silicates that could have been formed in the innermost region of the protoplanetary disk, they might also have accreted some organic matter produced from hydrothermal reactions in parent bodies that were destroyed and whose materials were transported to the cometesimal formation region (*Cody et al.,* 2011).

Several experiments have shown that insoluble refractory organics, as well as soluble ones, can be synthesized in alkaline liquid water at ~90°–250°C from formaldehyde (H_2CO) and ammonia (NH_3) (*Isono et al.,* 2019; *Kebukawa*

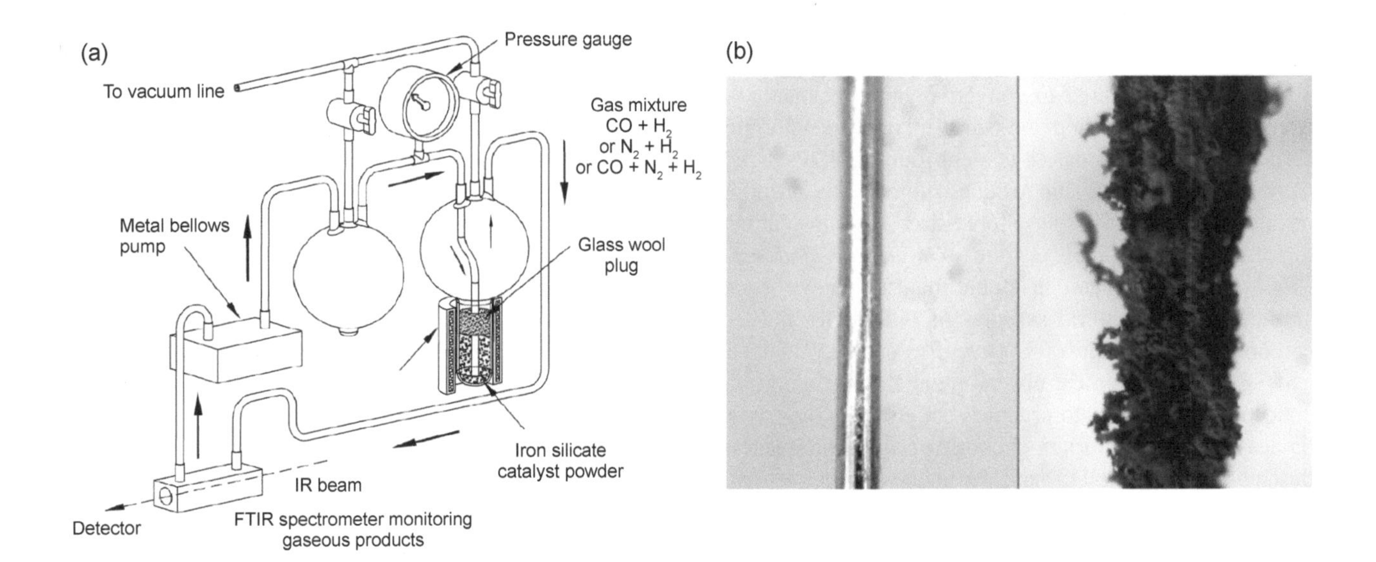

Fig. 4. Production of organic dust via catalytic reduction of N_2 and CO by H_2 on mineral grains or on an iron wire (~200 μm in diameter) (*Nuth et al.,* 2008, 2016). Such dust is used as a cometary analog in light-scattering experiments (see Fig. 12). Figures reproduced from *Nuth et al.* (2020) with permission.

et al., 2017, 2013), but also from HMT (*Vinogradoff et al.,* 2018) (see section 2.1.2.2). The reactions leading to these compounds likely include formose and Maillard-type reactions (*Vinogradoff et al.,* 2018, and references therein). Further works could address how various minerals can influence this hydrothermal chemistry (*Vinogradoff et al.,* 2020) and provide estimates of how much of these organic solids could have been produced and incorporated in cometesimals.

2.1.2.6. Minerals: Silicates observed in the diffuse interstellar medium are mostly amorphous (*Do-Duy et al.,* 2020), but comets contain a significant portion of crystalline silicates and refractory minerals (see the chapter by Engrand et al. in this volume). These crystalline or refractory minerals have been formed at high temperatures, most probably in inner regions of the protoplanetary disk, before being transported and incorporated in cometesimals (*Wooden,* 2008). Open questions are how they formed, where they came from, and when and how they were transported (see the chapter by Engrand et al. in this volume). Laboratory experiments have been conducted to study the formation of amorphous silicates and their thermal processing to form crystalline silicates, as reviewed by *Colangeli et al.* (2003) and *Jäger et al.* (2009).

Production. Analogs of cosmic/cometary silicates are produced by gas-phase condensation, in a furnace at 500–1500 K containing atmospheres of SiO-H_2-O_2 or Mg/Fe-SiO-H_2-O_2 (*Nuth et al.,* 2000, 2002) at ~100 or 10^{-7} mbar (*Nagahara et al.,* 2009), or via laser vaporization/ablation of a silicate pellet, also called pulsed laser deposition (*Brucato et al.,* 2002; *Sabri et al.,* 2013). In these experiments, after vaporization, the cooling of the gas phase induces the nucleation of clusters growing into macroscopic solid grains, which are deposited on a substrate for *in situ* (or collected, for *ex situ*) analyses. Analogs of cometary silicates are also produced from sol-gel inorganic or organic chemical reactions, where a colloidal solution evolves toward a solid gel. After drying of the solution, amorphous silicates are obtained (*Thompson et al.* (2019) and references therein). Once produced, their composition and structure are characterized by a variety of usual techniques: scanning or transmission electron microscopy (SEM, TEM), energy dispersive X-ray (EDX), X-ray diffraction (XRD), etc., as detailed in section 7 of *Colangeli et al.* (2003). The silicates produced are individual spherical grains 10–100 nm in diameter, arranged in fluffy chain-like agglomerates (Fig. 5), which are usually amorphous. Experiments producing silicates are constantly evolving with the aim to improve the production speed (*Thompson et al.,* 2019), the reproducibility, and the homogeneity of the dust, and to control experimental parameters such as oxygen fugacity (i.e., the amount of oxygen in the gas phase) representative of different pre- or protoplanetary disk environments (*Wooden et al.,* 2017).

Post-formation processing. After production, the amorphous silicates can be subjected to energetic processes simulating those encountered in the ISM and in protoplanetary disks, before their incorporation in cometesimals. Annealing is the process by which heating of amorphous silicates (e.g., after shocks in protoplanetary disks) results in the diffusion and rearrangement of structural units to form higher-order structures, i.e., crystallization. Laboratory experiments are performed to determine the temperature at which crystallization can be triggered and the kinetic of the annealing process (determination of the activation energy of crystallization) (*Colangeli et al.,* 2003; *Jäger et al.,* 2009). In the ISM, the silicate dust is irradiated by cosmic-ray ions, which lead to sputtering (erosion), amorphization, and possible implantation of protons (H^+). The annealing properties of such ion-irradiated silicate grains may be different from the original grains. Moreover, H^+ implantation may form hydroxyl OH groups in the silicates before their incorporation in cometesimals, as shown by laboratory experiments (*Djouadi et al.,* 2011; *Jin and Bose,* 2021; *Mennella et al.,* 2020). More laboratory experiments are needed to determine the ionization rate, implantation efficiency, and diffusion rate of hydrogen and deuterium in silicate minerals (*Jin and Bose,* 2021).

Fig. 5. SEM images for **(a)** MgSiO and **(b)** FeSiO smokes produced by gas-phase condensation (*Hadamcik et al.,* 2007).

Future experiments should also address formation and evolution of other refractory minerals present in comets, such as FeS, Fe-Ni alloys, iron oxides, and carbonates (see the chapter by Engrand et al. in this volume). For example, experiments have shown that Mg-carbonates are produced with amorphous silicates during the non-equilibrium condensation of a silicate gas in a H_2O-CO_2-rich vapor (*Toppani et al.,* 2005). Furthermore, there are tentative hints for the presence of small amounts of aqueously altered minerals in comets (*Rubin et al.,* 2020, and references therein; see also the chapter by Engrand et al. in this volume). Some of them might have been accreted from dust of aqueously altered bodies formed in the inner solar system and transported into cometesimals, or they have been produced in comets (*Suttle et al.,* 2020; *Gounelle et al.,* 2008). Laboratory experiments on these aqueous alteration processes are described in section 2.3.4.2.

2.1.3. Physical formation of a cometesimal, from grains to a cometary nucleus. Laboratory experiments with the aim of studying the growth of planets and other objects in the solar system have been extensively carried out in the last few decades. The experiments can be classified by the material used or by the process to be studied. While experiments with non-volatile materials are easier to perform, experiments with water ice and other super- or hypervolatiles are rare because of the complex handling of the icy materials. The growth of micrometer-sized particles to larger, millimeter-sized objects can be studied in so-called coagulation experiments. The further growth and evolution of the pebbles that were formed by the coagulation process can be studied in collision experiments to learn about the first growth phases in the young solar system.

Grains production and dispersion. One major problem that occurs when performing laboratory experiments with micrometer-sized particles is that the sample material has to be dispersed prior to the execution of the experiments. This is often done by a particle dispenser. Depending on the material type, different methods for the dispersion process can be used. The best solution to create non-volatile micrometer-sized monomers is to fill the desired material (often silica, SiO_2, is used) into a piston and the open end is pressed against a rotating cogwheel that disperses the material into monomers (*Blum et al.,* 2006). Then, the produced particles are carried to the desired location by a gas stream. Often, a siphon is introduced into the pipe system to ensure that aggregates cannot follow the gas stream, and only single dust particles are collected on a filter at the output of the dispenser.

To create micrometer-sized water ice particles another method is required. Typically, a droplet dispenser is used to create a fog of micrometer-sized water droplets that are then forced into liquid nitrogen by a dry, cold gas stream (*Gundlach et al.,* 2011a; *Jost et al.,* 2013). After evaporation of the liquid nitrogen, the sample material is accessible. For using water ice particles in dynamical collision experiments, the fog can be directly introduced into a cold pipe system. The droplets then freeze during flight through the pipes and are directed into the vacuum chamber (*Gundlach and Blum,* 2015). The creation of CO_2-ice particles can be realized by adiabatic expansion of compressed CO_2 gas. The expansion process cools the gas and CO_2 particles are then created with typical grain size of 2 μm. Another technique to create larger CO_2 particles/aggregates is to first create a solid layer of CO_2 ice onto a cold surface that is then mechanically scraped by a rotating gearwheel (see Fig. 6) (*Musiolik et al.,* 2016). This method produces larger particles with a mean radius of about 90 μm; intramixtures can be realized by using several gas species for the same sample creation on the cold target, e.g., CO_2 in combination with H_2O gas.

Many experiments have been performed to study the first stage of planetesimal/cometesimal formation and the main findings are summarized in the following paragraphs.

Specific surface energy. The specific surface energy is an important material parameter that describes the outcome of particle-particle collisions as well as the strength of the cohesion of materials. It is therefore an important physical property that has to be known to simulate coagulation scenarios and the collisional evolution of objects, as well as cometary activity. Different types of experiments were performed to measure this physical property.

The first method was only used on SiO_2 dust particles with radii ranging from 0.5 to 2.5 μm, but provides a precise measurement for the specific surface energy. Therefore, the adhesion force between micrometer-sized grains was determined by gluing single-spherical SiO_2 particles to the cantilever of an atomic force microscope (AFM) and on the respective substrate (*Heim et al.,* 1999). Then, the particles were brought into contact and were separated afterward. To derive the specific surface energy from the force curves, the Johnson, Kendall, and Roberts (JKR) theory derived by *Johnson et al.* (1971) was used, which yielded a specific surface energy $\gamma = 0.019$ J m^{-2} for SiO_2.

Another technique to measure the specific surface energy was introduced by *Blum and Wurm* (2000). The idea was to observe coagulating micrometer-sized particles when they formed a growing fractal agglomerate on a thin needle. At some point, parts of the grown agglomerate became too heavy and gravitational restructuring of these parts occurred. These restructuring events were observed with a camera to determine the mass of the restructuring part of the agglomerate. A determination of the specific surface energy was possible if the restructuring event could be described by rolling of the aggregate part around one particle. Under the assumption that the rolling friction force equals the gravitational force, these events were

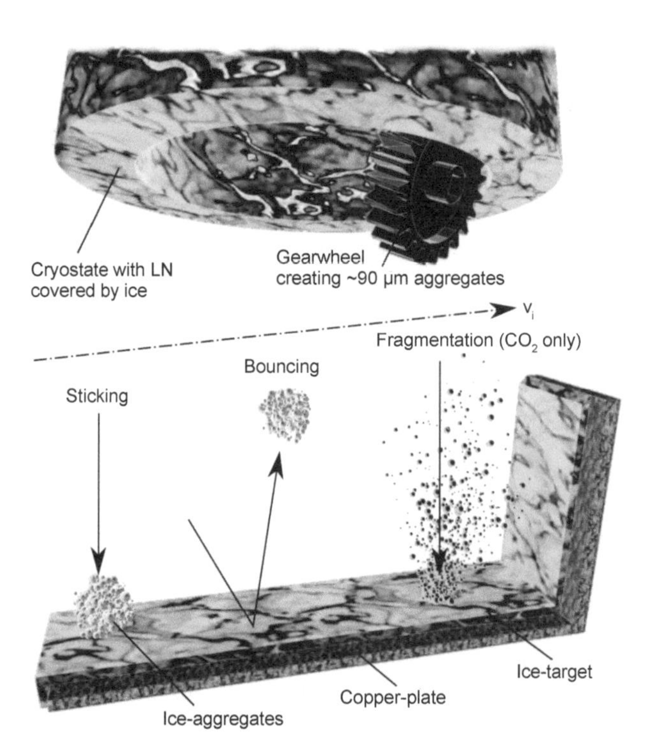

Fig. 6. Setup to study the collision of H_2O/CO_2 ice aggregates under 80 K and 0.1 to 1 mbar. Depending on the impact velocity (v_i) of the aggregates colliding with the ice layer, either sticking, bouncing, or fragmentation is observed (*Musiolik et al.,* 2016).

used to calculate specific surface energies for SiO_2 [γ = 0.02 J m^{-2}; (*Blum and Wurm,* 2000)] and for H_2O [γ = 0.19 J m^{-2} (*Gundlach et al.,* 2011a)].

The experiments discussed above indicate that the specific surface energy of water ice is an order of magnitude higher than the value for SiO_2 and this result is not in agreement with the idea that SiO_2 particles possess a surrounding water layer under terrestrial atmospheric conditions and also in vacuum experiments (*Kimura et al.,* 2015). Thus, it is expected that SiO_2 should have the same specific surface energy as water ice. A possible explanation for the deviation seen in the experiments discussed above is that the water ice was not kept cold enough during these experiments. Recent studies have investigated the specific surface energy indirectly by measuring the tensile strength of cylindrical samples composed of either SiO_2 or micrometer-sized water ice particles (*Gundlach et al.,* 2018). The advantage of these experimental studies was that the environmental temperature was controlled, so that water ice temperatures below 140 K have been secured. These experiments have shown that water ice at low temperatures possesses the same specific surface energy compared to SiO_2.

Sticking threshold velocity. Direct collision experiments have been utilized to determine the sticking threshold velocity of the particles (*Poppe et al.,* 2000; *Gundlach and Blum,* 2015). A particle jet was created and directed onto a quartz target under vacuum conditions. The interaction of the particle jet with the target was studied with a long-distance microscope and a laser, which illuminated the particle jet in opposition. To avoid the direct laser light reaching the long-distance microscope, a small absorption plate was mounted directly in front of the imaging system. With this setup, only the forward-scattered light of the particles was detected by the microscope. Because the laser light was turned on and off with a very high frequency (stroboscopic laser light) the flight direction and the speed of the particles were measured. *Poppe et al.* (2000) used this technique to measure the critical velocity at which the particles stop to stick to the surface, the so-called sticking threshold velocity. At higher speeds, the particles are not able to stick to the surface and bounce off. *Gundlach and Blum* (2015) refined this technique for use with water ice. The produced micrometer-sized ice particles were transferred into the vacuum chamber by a pipe system and the collision occurred with a cold ice target. Both works have shown that the sticking threshold velocity of water ice at low temperature (below ~210 K) is 10× higher compared to the value for SiO_2: v_{St,H_2O} = 9.6 m s^{-1} and v_{St,SiO_2} = 1.0 m s^{-1}. At higher ice temperatures, the sticking threshold increases up to ~50 m s^{-1}, which can be explained by the formation of a liquid-like layer on the surface of the grains (*Gärtner et al.,* 2017) that could be the cause for the enhanced specific surface energy value measured by *Gundlach et al.* (2011a).

Musiolik et al. (2016) performed an additional set of experiments to study the collisional properties of CO_2 ice as well as different ice mixtures. As shown in Fig. 6, the authors created different ice samples that were scratched off the surface by a rotating gearwheel. These particles were then attracted by Earth gravity and hit an ice target of the same composition underneath. The results of these experiments indicate lower values for the sticking threshold velocity, namely 0.04 m s^{-1} for pure CO_2 particles and 0.43 m s^{-1} for a 1:1 mixture of H_2O and CO_2.

Aggregation regimes. Blum and Wurm (2000) performed microgravity experiments to study the aggregation behavior of fractal agglomerates (fractal dimension of 1.9) under realistic conditions. For this purpose, a turbomolecular pump was modified to produce the fractal agglomerates during flight. The collisions of the formed agglomerates with a thin Si_3N_4 target occurred at velocities of 0.07–0.5 m s^{-1} and were observed with a camera. The authors identified four aggregation regimes dependent on the impact velocity. Hit-and-stick collisions occur at the lowest velocities (~0.2 m s^{-1}). Impact restructuring was observed for intermediate-low velocities (~0.65 m s^{-1}), and at intermediate-high collision speeds (~1.2 m s^{-1}) compact growth was observed. High-velocity (~1.9 m s^{-1}) impacts led to fragmentation events. These experimental findings have confirmed the model developed by *Dominik and Tielens* (1997).

The outcome of pebble collisions. The collisional growth of small particles to larger objects (agglomerates and pebbles) was extensively studied in many experiments in the past years and their results were used as input for a very detailed collision model [see Fig. 1 in *Blum* (2018) and references therein] that describes the different regimes for agglomerate-agglomerate collisions in the protoplanetary disk. In general two factors determine the collisional outcome: (1) the size ratio and (2) the speed of the collision partners. Sticking collisions (please note that different aggregation regimes exist; see above) occur at low speeds and if the size ratio is close to unity (see, e.g., *Wurm and Blum,* 1998; *Blum and Wurm,* 2000; *Kothe et al.,* 2013). Bouncing collisions take place when the energy dissipation during collision is insufficient to allow sticking (*Weidling et al.,* 2012; *Brisset et al.,* 2016, 2017) and causes compaction of the pebbles (*Weidling et al.,* 2009). Fragmentation always occurs at larger speeds if the mass ratio is about unity (*Beitz et al.,* 2011; *Schräpler et al.,* 2012; *Deckers and Teiser,* 2013; *Bukhari Syed et al.,* 2017).

In the case of different aggregate sizes, other collision results can occur. Mass transfer describes the process by which the smaller impacting aggregate loses mass by fragmentation during the collision (see, e.g., *Wurm et al.,* 2005b; *Teiser and Wurm,* 2009; *Güttler et al.,* 2010; *Beitz et al.,* 2011; *Deckers and Teiser,* 2014; *Bukhari Syed et al.,* 2017). Typically 1–50% of the mass is transferred to the larger object. Cratering may occur if the projectile aggregate is larger than in the previous case. Then the target loses material, because more material is excavated than transferred (*Wurm et al.,* 2005a; *Paraskov et al.,* 2007). Cratering is a transition process to the fragmentation regime. Erosion is another process that can occur when very small (<0.1 mm) aggregates collide with a larger object at higher speeds (>50 m s^{-1}) (*Bukhari Syed et al.,* 2017).

All these experiments can be used to improve simulations that study the coagulation and collisional evolution in the protoplanetary disk in the early stages of planet formation (see, e.g., *Windmark et al.,* 2012a,b; *Zsom et al.,* 2010). The collision behavior of larger objects (from millimeter to meter) have been studied in laboratory experiments reviewed in *Güttler et al.* (2013). These experiments measured the coefficient of restitution (i.e., the ratio of the final to initial relative speed between two colliding objects) for spherical particles made of various materials (including water ice) up to meter size (*Durda et al.,* 2011).

Once formed, cometesimals and comets can experience collisions. Hydrodynamical models are used to predict the outcome of the collisions in terms of the general structure of the object and material properties such as porosity as well as their thermal history (*Jutzi and Michel,* 2020; *Schwartz et al.,* 2018). These models use various experimental data as inputs, e.g., the so-called "crush-curves" (the relation between porosity and pressure) or the thermal conductivity and diffusivity of analog material (*Güttler et al.,* 2009, 2010). While these properties are well known for the various refractory components of putative cometary dust, ices and complex organic molecules (pure or mixed) are more difficult to study. Systematic measurements with such materials would be very beneficial for future modeling studies. In the meantime, it is mandatory to derive high-quality scaling laws to apply the experimental findings on small icy and organic-rich particles to larger objects representative of the later stages of the formation process.

2.2. Experiments on Comet Evolution During Residence in the Outer Solar System

2.2.1. Brief summary of the processes. After formation, the structure and composition of comets continue to evolve due to several processes, mostly irradiation by energetic particles and heating (see review in *Weissman et al.,* 2020, and the chapter by Guilbert-Lepoutre et al. in this volume).

Through the long residence time of comets in the Oort cloud or Kuiper belt/scattered disk, galactic cosmic rays (GCRs) and solar energetic particles (SEPs) can deposit enough energy to transform the cometary materials chemical and/or physical properties down to tens of meters (*Gronoff et al.,* 2020; *Maggiolo et al.,* 2020). The surface and subsurface temperature of comets may also increase by 10 K or more due to supernovae (down to 2 m) or stars passing through the Oort cloud (down to 50 m) (*Stern,* 2003). In the case of a sufficiently large comet with a high dust-to-ice mass ratio, the decomposition of short-lived radio nuclides (^{26}Al) could provide internal heating (*Prialnik et al.,* 2004, 2008). Collisions with other bodies can also heat the surface and interior of comets (*Schwartz et al.,* 2018), changing their composition and structure. Finally, other mechanical processes such as accretion or erosion caused by the impact of interstellar grains (reviewed in *Mumma et al.,* 1993) can also modify the surface or subsurface of comets.

Laboratory simulations of these processes could provide clues to distinguish the physical and chemical characteristics that are primordial from the results of later processes or secondary evolution. This is particularly important to interpret correctly the observations of a comet approaching the Sun for the first time, which is the goal of the Comet Interceptor mission (see the chapter by Snodgrass et al. in this volume).

2.2.2. Irradiations by energetic particles. Cosmic rays are particles (mostly protons, but also electrons and more massive ions) accelerated at energies up to several giga-electron volts (*Bringa and Johnson,* 2003). They can penetrate the surface of cometary nuclei down to tens of meters; the more energetic, the deeper. When they go through a material, such energetic ions can move atomic nuclei (elastic collisions) or produce nuclear reactions and/or ionizations and excitations (inelastic collisions), inducing rearrangements of the material's structure and chemical reactions. Moreover, secondary particles produced from a single incident particle can also lead to a cascade of nuclear and ionizing reactions in the material. Molecules can dissociate and new ones can form that are more or less volatile than the initial ones. These energetic projectiles can also cause sputtering, i.e., the expulsion of atoms or molecules from the surface, and changes of material structure (amorphisation, compaction). Finally, ions can be implanted down to a certain depth in the surface material and contribute to the formation of new molecules.

Setups and products. Laboratory experiments simulating these structural and chemical effects of GCRs usually use electrons, protons, and heavier ions in the kilo-electron-volt to mega-electron-volt energy range, so with lower energy but higher particle fluxes than GCRs, to bombard a sample that is most often very thin (nanometers to micrometers thick) compared to the penetration depth of the radiation. The projectiles are generated by sources, such as electron guns (up to kilo-electron volts), ion accelerators including cyclotrons (up to mega-electron volts to giga-electron volts) or synchrotrons (up to giga-electron volts). Varying the mass and the energy of the ions allows for the investigation of either elastic or inelastic interactions and for varying the stopping power (i.e., the energy loss during collision) of the projectile in the material. From these experiments with multiple projectiles, the relation between the sputtering or radiolysis cross sections and the stopping power can be calculated. Knowing the flux of GCRs as a function of their energy in the Oort cloud or Kuiper belt, the temporal evolution of cometary materials (via sputtering, radiolysis, amorphization, etc.) can be estimated (see *Rothard et al.,* 2017, for details). Sophisticated numerical models taking into account many parameters (generation of secondary species by projectiles, diffusion of products, etc.) can estimate how deep the nucleus material is affected by these processes (*Gronoff et al.,* 2020; *Maggiolo et al.,* 2020).

Ion irradiation on ices. In section 2.1.2, we described the laboratory experiments studying the chemical effects of photolysis and radiolysis in pre-cometary ices. The same chemistry continues to occur in cometary ices after comet formation. *Rothard et al.* (2017) and *Allodi et al.* (2013) have provided detailed reviews of the laboratory experiments simulating the modifications of ices caused by cosmic rays and solar wind. A large number of ion or electron irradiation experiments have been performed on single-component ices (H_2O, CH_4, N_2, etc.), H_2O-rich ices (H_2O being the dominant ice in comets), or N_2-rich ices [because N_2 is largely present on transneptunian objects (TNOs) and so possibly on distant comets], as reviewed in *Hudson et al.* (2008). More recently, experiments on methanol (CH_3OH)-rich ices have also been performed (*Urso et al.,* 2020). Experiments performed with doses in the range of about 1–20 eV/16 amu allow the monitoring of the progressive formation of new molecules, analyzed mostly via infrared spectroscopy and mass spectrometry as described previously in section 2.1.2.1 (Fig. 2). But at higher doses on the order of 100 eV/16 amu, carbon-containing ices (such as CH_4, CH_3OH, or C_6H_6) progressively lose H_2 and increase their carbon-to-hydrogen ratio: The produced molecules bind together, forming colorized refractory organics and finally an organic crust, masking the ice below it (*Brunetto et al.,* 2006; *Strazzulla et al.,* 2003, 1991). The colorization of ices by ion irradiation is especially studied in the case of TNOs (*Dalle Ore et al.,* 2011). Experiments have shown that the total deposited energy (elastic plus inelastic collisions) is the most important parameter in the reddening process (*Brunetto et al.,* 2006). Such laboratory experiments also tested the hypothesis of the formation of the N-rich organic matter observed in ultracarbonaceous Antarctic micrometeorites (UCAMMs) by GCR irradiation of N_2-CH_4-rich ices at the surface of a cometary body in the Oort cloud (*Augé et al.,* 2016). At high irradiation dose, these organic residues evolve toward a hydrogenated amorphous carbon (*Baratta et al.,* 2008). The carbonization process is efficient only for bodies accumulating a sufficient dose of GCRs such as comets in the scattered disk or in the Oort cloud, where this crust of highly processed organic material may extend down to several meters (*Strazzulla et al.,* 2003, 1991). GCRs may change not only the chemistry but also the structure of the ices in comets, inducing amorphization of crystalline ices and compaction of amorphous ice with time. These phenomena have been described by experiments performed in the context of interstellar water ice mantles (*Dartois et al.,* 2015), or icy satellite surfaces (*Strazzulla,* 2013), but they may apply in the context of comets. In comets, the amorphization induced by GCRs may compete with the crystallization induced by thermal waves (see sections 2.2.3 and 2.3.4.1).

Ion irradiation on refractories. Studies were also dedicated to the effects induced by ion irradiation on refractory materials mainly representative of the Moon and of asteroids, but some of these results are relevant to comets as well. Ion irradiation of minerals causes their amorphization, and the formation of nanophase reduced iron (np-Fe^0) if they are Fe-bearing minerals (for reviews, see *Bennett et al.,* 2013; *Brunetto et al.,* 2015). These nanoparticles cause the darkening and spectral reddening of irradiated silicates (*Hapke,* 2001). However, complex mineral mixtures, such as carbonaceous chondrites, are bluing or reddening upon ion irradiation (*Lantz et al.,* 2017). Ion irradiation was also performed on materials whose chemical structures can be considered analogous to some cometary refractory carbonaceous compounds, such as natural hydrocarbons (asphaltite, kerite) (*Baratta et al.,* 2008; *Moroz et al.,* 2004) and soots produced in a low-pressure flame (*Brunetto et al.,* 2009). These experiments have shown the progressive carbonization of the samples, evolving toward amorphous phases. Irradiated soots exhibit Raman spectra similar to those of some meteorites, IDPs, and Comet Wild 2 grains collected by the Stardust mission (*Brunetto et al.,* 2009). Visible and near-infrared reflectance spectra of the natural hydrocarbons evolve from red to neutral after ion irradiation (*Moroz et al.,* 2004). These changes were mainly caused by elastic collisions, in contrast with ices whose evolution is controlled by the total of elastic and inelastic collisions.

Toward experiments with thicker and mixed samples. Most of these experiments have been done on nanometer- to micrometer-thick samples of ice or dust. However, highly penetrating energetic particles interact with a much thicker layer of cometary materials (up to several meters) having some porosity and being a mixture of ice and dust components. These large-scale structural and compositional properties certainly induce effects that are not apparent in small-scale irradiation experiments performed up to now. These effects can be heating, charging, and diffusion of radiolysis products on the surface or in the volume of the ice, inducing new/enhanced chemical reactions and structural changes. For example, experiments performed on thick and porous water ice samples have revealed the charging and sublimation of the ice upon irradiation by ions (*Galli et al.,* 2018).

2.2.3. Thermal evolution. The heat experienced by cometary nuclei during their residence in the outer solar system, due to the decay of short-lived radio nuclides (^{26}Al) or the influence of nearby stars or supernovae or to impacts, can change the composition and structure of a significant fraction of their surface and interior (*Stern,* 2003; *Prialnik et al.,* 2004, 2008; *Schwartz et al.,* 2018; *Mumma et al.,* 1993). Moreover, when comets approach the Sun for the first time, their surface is heated by the solar radiation. Due to the very low thermal conductivity of granular materials at their surface, the heat cannot efficiently penetrate into the interior of the nucleus. This also implies that cometary nuclei should experience alteration of their orbital elements due to the Yarkowski effect and rotational spin up by the Yarkovsky-O'Keefe-Radzievskii-Paddack (YORP) effect (*Bottke et al.,* 2006). Both effects strongly depend on the thermal conductivity of the cometary materials.

Laboratory experiments studying these thermal properties and processes are described in section 2.3 for comets penetrating the inner solar system. Thermal processes are exacerbated during the approach to the Sun but some of them may also occur to some extent during/after the formation of comets and during their residence in the scattered disk or Oort cloud. Experimentalists should keep in mind that these thermal properties and processes are strongly influenced by the differences of composition and structure of the body considered, inherited from its history, and whether this body is a cometesimal, a pristine or slighly heated comet (depending on its mass and formation history), or a comet significantly more altered after multiple passages close to the Sun. Future experiments should consider each of these cases, and clearly differentiate between them, to improve our understanding of the thermal evolution of these bodies.

2.3. Experiments on Comet Evolution During Approach to the Sun

2.3.1. Brief summary of cometary activity processes. When a comet penetrates the inner solar system, the absorption of the incident solar light by the surface dust and/or ice components causes the cometary surface to heat. The heat is then transported to deeper layers of the nucleus (see the chapter by Guilbert-Lepoutre et al. in this volume). Cometary nuclei being extremely porous, we can distinguish three mechanisms of heat transport: (1) the conduction through the solid network of particles, (2) the radiative heat transport originating from thermal radiation through void spaces between particles, and (3) the gaseous heat transport due to the gas present in the void spaces. The effective thermal conductivity of porous materials is then given by the combination of these three effects. The heat wave can cause various physical and chemical transformations: phase changes and segregation of the solid ices or salts, sublimation, release of trapped gases, decomposition of clathrates, and activation of chemical reactions. In return, these transformations can release or absorb energy inside the nucleus. For instance, the phase transition of pure water ice from amorphous to crystalline releases a significant amount of energy but the same process can be endothermic if the ice contains impurities (*Kouchi and Sirono,* 2001). Moreover, the inward-flowing gas from the sublimation front to the nucleus interior leads to the recondensation of ices in deeper and cooler layers, acting as a transport of latent heat of condensation toward the interior (see Fig. 9b). Layers made of components having similar volatility can thus form in the nucleus, leading to what is called nucleus chemical differentiation. These layers could get reduced porosity. Chemical transformations (isotopic fractionation, gas-dust reactions, etc.) could also occur during diffusion. At the surface, the sublimation of the ices leads to the formation of a dessicated surface layer made of the most refractory components. Gases flowing from the sublimation fronts at depth to the surface can break fragments of this layer, or of subsurface layers, and eject dust/ice particles in the coma.

All these processes are responsible for not only shaping the nucleus interior and surface, but also the coma. Indeed, a good knowledge of the mechanisms producing the gases and ejecting the ice/dust particles is essential for a correct interpretation of the observations of outgassing comets. Furthermore, the planning of any future cometary nucleus material sampling for *in situ* analysis or sample-return will require a good knowledge of the (sub)surface nucleus properties. Numerical models taking into account most of these multiple processes have been developed to simulate the nucleus and coma evolution of periodic comets (*Marboeuf et al.,* 2012; *Prialnik et al.,* 2008). Experiments carried out to understand the different steps of this energy flow, and the numerous parameters and processes controlling these physical and chemical transformations of comets, are described in the paragraphs below.

2.3.2. Large-scale insolation experiments. With the aim to simulate experimentally the evolution of a cometary nucleus during its approach to the Sun, several experiments have studied the consequences of incident solar-like (or infrared) light illuminating the surface of complex cometary analog samples made of porous granular mixtures of ices and non-volatile constituents. These experiments are performed in simulation chambers, allowing the maintainance of such samples, generally several centimeters large and thick, under high vacuum (10^{-5}–10^{-6} mbar) and cometary-like temperatures (100–250 K). Typically, a solar simulator irradiates the sample surface, and the transformations of the sample surface and interior are monitored using various instruments (temperature sensors, cameras, hyperspectral imaging system, penetrometer, scale, etc.) (see Fig. 7a,b). A mass spectrometer can also be used to monitor the released gases. In the 1970s, at the Ioffe Institute in Leningrad and at the Institute of Astrophysics in Dushanbe, E. Kajmakov and co-workers prepared mixtures of ice and minerals and/or organics and observed the formation — and outbursts — of a porous ice-free matrix after sublimation of ices at 10^{-5}–10^{-6} mbar and 180–240 K (*Dobrovolsky and Kajmakov,* 1977). In the 1980s, at the Jet Propulsion Laboratory in Pasadena, *Saunders et al.* (1986) studied the sublimation of water ice samples prepared from a suspension of various silicates and organic matter (*Storrs et al.,* 1988). They noticed that some classes of phyllosilicates and organics were able to form fluffy filamentary sublimate residues, whereas most classes of silicates were not. Later, in the 1980s and 1990s, the Kometen Simulation (KOSI) project at the Deutsches Zentrum für Luft- und Raumfahrt (DLR) conducted a total of 11 experimental simulations over 7 years, considerably enhancing our understanding of cometary activity (*Sears et al.,* 1999; *Grün et al.,* 1991; *Kochan et al.,* 1999). Many different properties of the samples (made of H_2O, CO_2 ices, and various silicates and organics) as well as the evolved gases were studied in detail. *Bar-Nun and Laufer* (2003) constructed a simulation chamber in which large (200 cm^2 × 10 cm) samples made of fluffy agglomerates of 200-μm gas-laden amorphous ice grains were prepared at 80 K and 10^{-5} mbar. These samples were irradiated from above using infrared radiation in order to study the evolution of gases released and the thermal and mechanical properties. More recently, the Comets Physics Laboratory (CoPhyLab) team at the Technical University of Braunschweig constructed a new laboratory hosting several small-scale experiments as well as a large-scale comet-simulation chamber (L-Chamber). The latter, equipped with 14 different scientific instruments,

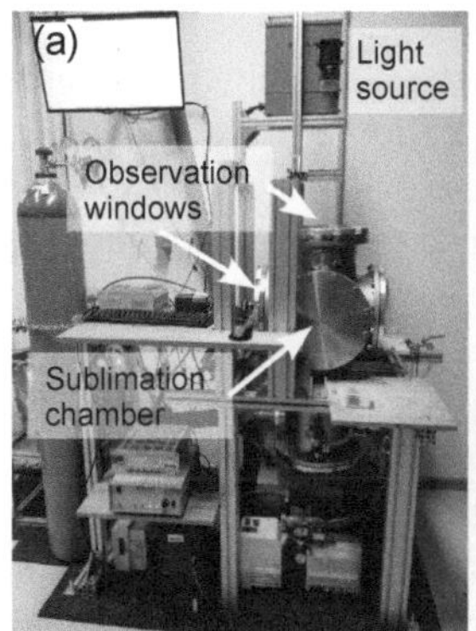

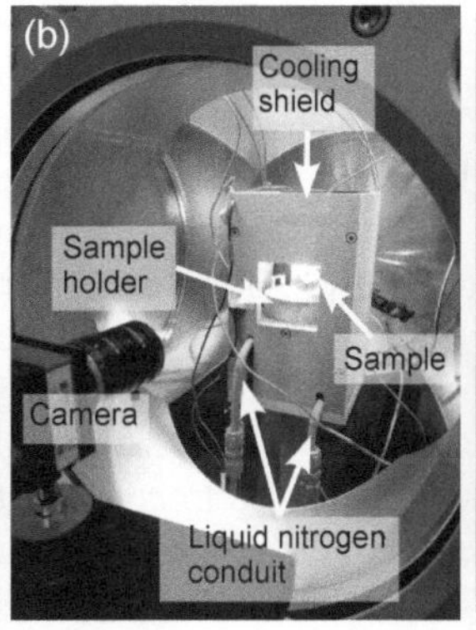

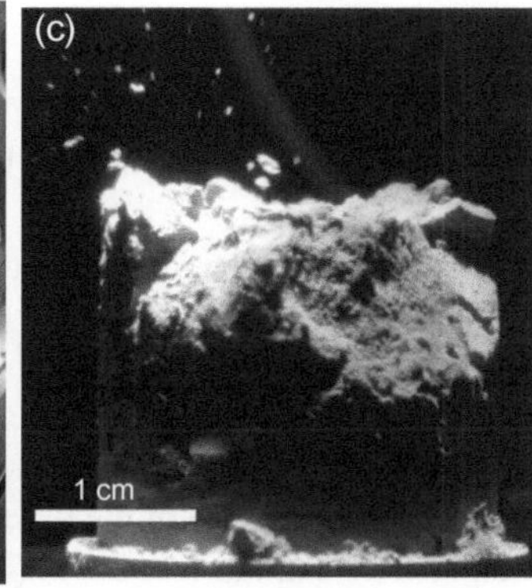

Fig. 7. Setup for sublimation experiments at the Technical University of Braunschweig. **(a),(b)** External and internal views of the vacuum chamber equipped to monitor the sublimation of several-cm-large cometary analogue samples. **(c)** Ejections of icy-dust fragments during the sublimation of a sample made of water ice (20 vol%) mixed with fly ash (mostly SiO_2 and Al_2O_3) particles (*Haack et al.,* 2021a).

allows uninterrupted experiments such as solar illumination cycles on 10–30-cm thick and 25-cm-diameter dust-ice samples at low temperatures and pressures for up to several weeks (*Kreuzig et al.,* 2021). The main results of these large-scale insolation experiments are described in the paragraphs below. They allowed the study of the numerous structural and compositional transformations occuring after insolation (for examples, see Figs. 7 and 9a). However, to understand how each experimental parameter (initial sample properties, temperature, etc.) influences these transformations, it is necessary to perform a series of experiments in which a single parameter is changed at a time. This was not always the case in past experiments, such as KOSI. Moreover, most of these experiments also used initial ice-dust samples having relatively low refractory-to-ice mass ratio compared to the range of values (from 0.3 to 7) estimated from the Rosetta mission instruments (*Choukroun et al.,* 2020). Future experiments of this kind should thus be more systematic and investigate samples having larger refractory-to-ice mass ratio. Other smaller-scale experiments, focused on peculiar processes or properties using simpler samples, are complementary to the large-scale ones, and equally essential to provide physical constants as input to numerical models.

2.3.3. Experiments on heat transport. In cometary nuclei, heat can be transported due to the conduction through the solid network, the radiation inside the pores, and the gas diffusion (*Gundlach and Blum,* 2012). Numerical models of comet nuclei thermal evolution need the input of numerous physical parameters including the thermal conductivity (in watts per meter Kelvin, or W m^{-1} K^{-1}), which determines the amount of heat transported from the insolated surface into the interior of cometary nuclei and thus the temperature gradient. In the laboratory, samples deposited on a cold plate can be irradiated by a laser or a light source. The temperature evolution of the illuminated surface and bottom surface of the sample can be measured using an infrared camera and temperature sensors, respectively. The thermal conductivity can then be derived from this temperature evolution (*Krause et al.,* 2011; *Gundlach and Blum,* 2012). Such experiments can be performed under vacuum or not and the sample size can be of several centimeters or larger. Numerous experimental measurements of thermal conductivity of porous media have been performed to test models that allow the calculation of the thermal conductivity from a set of parameters such as the particle size and composition, porosity, and temperature (*Gundlach and Blum,* 2012; *Sakatani et al.,* 2017; *Wood,* 2020). In a porous medium, the thermal conductivity can be reduced by several orders of magnitude compared to the conductivity of the bulk material. Indeed, due to the small contact areas between the particles, which determine the efficiency of the heat exchange between the particles, the thermal conductivity decreases with increasing porosity or with decreasing filling factor (*Arakawa et al.,* 2017; *Krause et al.,* 2011; *Sakatani et al.,* 2017). The radiative thermal conductivity depends on the particle size and on the mean free path of the photons inside the pores (*Gundlach and Blum,* 2012; *Sakatani et al.,* 2017). If the porous medium is not under vacuum, the thermal conductivity depends also on the static gas pressure inside the pores. *Seiferlin et al.* (1996) measured a matrix conductivity close to 0.02 W m^{-1} K^{-1}, but maximum values for the effective (= matrix + vapor) thermal conductivity exceeding 0.25 W m^{-1} K^{-1} at large pressures close to 1 bar. We can also note that most of the experiments designed to measure the thermal conductivity of cometary analog samples have used pure water ice, whereas the gas trapped in amorphous water ice can have an impact on the thermal conductivity of the water ice. Indeed, the thermal conductivity of gas-laden ice is much higher than for pure ice, and increases with the concentration of trapped species (*Bar-Nun and Laufer,* 2003). This effect of trapped species in water ice on its thermal conductivity is poorly known and new experiments, with relevant concentration of trapped species, are desirable.

Moreover, to model the temperature gradient in cometary nuclei, one should also consider the inward flowing of gas from the sublimation front to the nucleus interior, leading to the recondensation of ices in the deeper and cooler interior (Fig. 9b). This inward flow of gas and the ice recondensation can be seen as a transport of latent heat toward the interior (*Benkhoff and Huebner,* 1995). This latter mechanism can be significant. Indeed, an experiment performed on a pure, porous water ice sample allows us to estimate that the energy transport by the inward flowing of gaseous water represents 40% of the total heat flux available for heating (*Benkhoff et al.,* 1995).

The thermal conductivity of cometary nuclei is generally assumed to be homogeneous and constant in time. Nevertheless, some KOSI experiments have provided experimental evidence of evolution of the studied sample, such as the recondensation of ices and the formation of a dust mantle (Fig. 9a), which could lead to a variation of the thermal conductivity as a function of time and nucleus depth (*Huebner et al.,* 2006). For example, the sintering of ice grains leads to a growth of the grain to grain contact area, enhancing the thermal conductivity (*Kossacki et al.,* 1994). The thermal conductivity of the dust mantle seems to depend on the presence of organics. *Kömle et al.* (1996) studied the evolution of an originally homogeneous multi-component sample (containing water ice, organics, and minerals) to a residuum containing (finally) only minerals and organics. During this evolution the thermal properties changed dramatically. The heat conductivity of the cohesive residuum was found to be at least an order of magnitude larger than the typical value for a loose dust mantle containing no organic material.

While the effect of numerous parameters, such as the grain size and composition, porosity, and temperature, on the thermal conductivity of porous samples are well known (*Gundlach and Blum,* 2012; *Sakatani et al.,* 2017; *Wood,* 2020), most of the experiments aimed to measure the thermal conductivity of cometary analogs consider mixtures made of only minerals and water ice. The impact of refractory organics and ammonium salts, which could be important components of the cometary nuclei, on the thermal conductivity are almost unknown. In a similar way, if the thermal conductivity of pure water ice is well known, that of gas-laden amorphous water ice should be studied further.

2.3.4. Experiments on ice thermal transformations.

2.3.4.1. Thermally induced physical transformations: Different forms of ices. The ice found in comets is mainly water (70–90%), along with other minor species possibly mixed with water. Depending on its place of formation and evolution before it approaches the Sun, different types of water-rich ice mixtures may be found in a comet, apart from pure crystalline or amorphous water ice: mixtures of water ice with other compounds, or molecules forming strong hydrate complexes, clathrate hydrates, or a mixture of these different structures. Clathrate hydrates are crystalline solids where water molecules are organized in the form of individual cages, stabilized by hydrogen bonds, trapping a single molecule other than water. The thermodynamic properties of each of these structures are different, and need to be measured via experiments.

Crystallization. The crystallization of amorphous water ice can occur during the heating of cometary nuclei. The rate of this phase transition has been experimentally measured by *Schmitt et al.* (1989). If the amorphous water ice does not contain impurities, this process is exothermic and the heat release is about 9×10^4 J kg^{-1} (*Ghormley et al.,* 1968). Nevertheless, for sufficiently high impurity content, the process is endothermic (*Kouchi and Sirono,* 2001). The negative energy balance during crystallization may be caused by the energy lost in expelling the guest molecules from the water lattice.

Sublimation. The heat experienced by the ices contained in the cometary nuclei leads to the production of gases and cometary activity. The first step to model cometary activity is to consider that cometary nuclei contain only pure ices. The temperature dependence of the sublimation vapor pressure for pure ices has been reviewed by *Fray and Schmitt* (2009), who have compiled more than 1800 experimental measurements of vapor pressure for 53 molecules and propose some vapor pressure relations that can be included in models. Nevertheless, the reality is much more complex and most of the cometary molecules could be trapped in water ice. Some volatile species, such as CO, N_2, or Ar, can be partially trapped during the condensation of amorphous water ice even if the condensation temperature is too high to allow the direct condensation of these volatile species. This process has been intensively studied by Bar-Nun and coworkers (*Bar-Nun et al.,* 1988, 2007; *Bar-Nun and Owen,* 1998; *Greenberg et al.,* 2017).

The sublimation of binary mixtures has also been extensively studied. When water ice trapping a hypervolatile species such as CO, N_2, or Ar is slowly heated, the release of the more-volatile components occurred in several discrete temperature regions. Thus, a fraction of the hypervolatile species can remain trapped in water ice until its complete sublimation (see Fig. 8) (*Bar-Nun et al.,* 1985, 1988; *Bar-Nun and Owen,* 1998; *Hudson and Donn,* 1991; *Kouchi,* 1990; *Kouchi and Yamamoto,* 1995). An extensive experimental study of the vaporization of binary mixtures has been performed by *Collings et al.* (2004), allowing separation of the different species into three categories based on their desorption behavior. The sublimation of more complex ice mixtures, containing numerous species, has been studied by *Kouchi and Yamamoto* (1995), *Martín-Doménech et al.* (2014), and *Potapov et al.* (2019). In that case, the desorption of the different species as a function of the temperature is rather complex and some effects caused by the presence of CH_3OH in the mixture cannot be reproduced in experiments with binary mixtures. Such experiments show the complexity of the sublimation processes that take place in cometary nuclei.

When heated, the clathrate hydrates release the molecules that were trapped. The experimental data available on the equilibrium between clathrate hydrates, water ice, and gas have been reviewed by *Lunine and Stevenson* (1985) and *Choukroun et al.* (2013). For some species, like O_2 or N_2, experimental data are available only at pressures larger than 10 bar. To be applied to the cometary nuclei, these data must be extrapolated at lower pressures. As the empirical laws, such as $\ln P = A + \frac{B}{T}$, are generally applicable only on moderate pressures ranges, the validity of such extrapolation is questionable. For some other

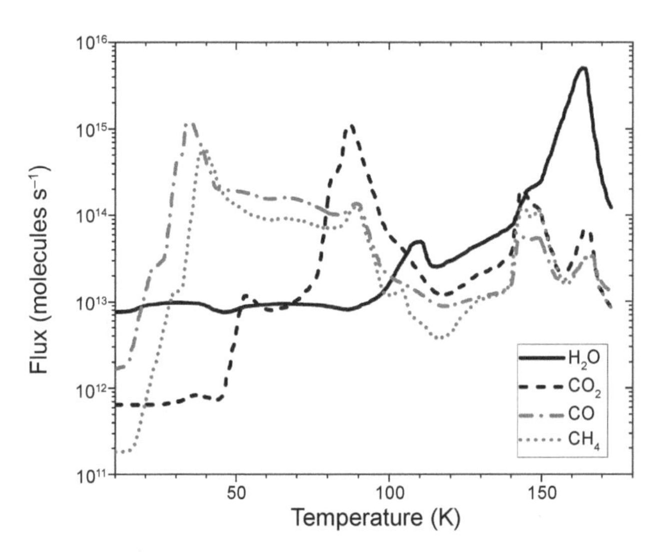

Fig. 8. Molecular fluxes of CO, CH_4, CO_2, and H_2O as a function of the temperature from an ice mixture condensed at 10 K (*Kouchi and Yamamoto,* 1995). The initial ratio at 10 K is H_2O:CO_2:CH_4:CO = 65:10:10:15. The sublimation behavior is similar to that of a H_2O-X binary mixture, but the very broad peak ranging from 20 K to 90 K is due to the release of CO and CH_4 trapped in an amorphous CO_2 ice. Such experiments show the necessity of studying the sublimation from complex ice mixtures, and not only from H_2O-X binary mixtures, to understand the processes taking place in cometary nuclei.

species, like CO, there are no experimental data for the equilibrium at low pressure and temperature. Moreover, the available experimental data are relevant only for pure clathrate hydrates; the stability of mixed clathrate hydrates at low pressure and temperature is very poorly known. For all these reasons, new experimental measurements on the stability of clathrate hydrates would be welcome.

2.3.4.2. Thermally induced chemical transformations: Thermally induced ice chemistry. The volatile molecules constituting ices are primarily affected by heating processes, as explained in section 2.3.4.1. The increase of ice temperature leads to the rearrangement of the molecules, from amorphous to crystalline phase, and to the separation of mixed molecules, i.e., segregation. Heating can also lead to the sublimation of molecules, passing from the solid to the gas phase. Simultaneously to these structural changes, volatile molecules initially trapped in the amorphous water ice can be released. This enhanced molecular mobility favors chemical reactions in the solid phase, and possibly the gas phase, inside cometary nuclei.

The dynamics of solid-state chemical reactions at the surface and inside the ice grains is controlled by both the diffusion coefficients of the reactants and the reaction rate constants. Laboratory experiments are performed to measure these parameters, serving as inputs to numerical models of chemical reactions networks (see review in *Theulé,* 2019). Although these experiments are primarily done in the context of pre-cometary interstellar grains, they are also relevant to the chemistry happening in cometary nuclei ice mixtures. Using setups similar to those described in section 2.1.2.1 and Fig. 2, these experiments are studying reactions between neutral molecules [neutral-neutral, such as condensation, acid-base, and nucleophilic addition (*Theulé et al.,* 2013)], or reactions with radicals formed from interactions with GCRs/SEPs [neutral-radical and radical-radical reactions (*Linnartz et al.,* 2015)]. Recent experiments have shown that diffusion of molecules on the surfaces of ice (external or internal surfaces such as cracks or pores) is much faster than inside the bulk ice (*Minissale et al.,* 2019; *Mispelaer et al.,* 2013). Structural changes induced by heating result in the formation of many cracks in the ice (*May et al.,* 2012), allowing the diffusion of reactants along them (*Ghesquière et al.,* 2018). Barrier-less or low-activation-energy reactions then form many new molecules, resulting in the so-called "organic refractory dust" (Fig. 3). Laboratory investigations of this organic dust were detailed in section 2.1.2.2. An open question is how much of the salts, polymers, or macromolecules constituting this organic dust are formed during the residence in cometary reservoirs and during the active phase of comets, compared to those that were incorporated in cometesimals. Another question to investigate is how these reactions, triggered by heating, modify the isotopic composition of cometary ices via exchanges of D/H atoms (*Faure et al.,* 2015; *Lamberts et al.,* 2015).

Liquid water-induced chemistry? If temperature and pressure conditions allow the melting of the water ice in the subsurface of cometary nuclei, even transiently, this liquid water may induce additional chemistry, affecting mineral and organic phases. Analogs of pre-cometary amorphous silicates produced by gas-phase condensation experiments (see section 2.1.2.6) have been reacted with liquid water at different temperatures (331–378 K) and reaction times (1–142 h) (*Nelson et al.,* 1987; *Rietmeijer et al.,* 2004). The porous amorphous silicate particles were transformed into phyllosilicates, and it was observed that the swelling of phyllosilicates decreased the porosity of the dust, limiting further propagation of the liquid water and leading to localized alteration (*Rietmeijer et al.,* 2004). Other experiments conducted on CP-IDPs suspended in liquid water showed that their amorphous silicates are altered into phyllosilicates 2 orders of magnitude faster than laboratory analogs (*Nakamura-Messenger et al.,* 2011). Based on these studies, in addition to numerical modeling and observations, *Suttle et al.* (2020) proposed that for some comets passing within <1.5 au of the Sun, the solar radiation would generate temperatures high enough in their immediate subsurface (<40 cm) for the water ice to melt, potentially forming the minor secondary minerals observed in some IDPs and comets. In addition to phyllosilicates, future laboratory experiments could also investigate the formation of other minerals, as well as the organic chemistry potentially produced by transient liquid water on comets.

2.3.5. Experiments on gas diffusion from the nucleus interior to the surface and coma. *2.3.5.1. Experiments on gas diffusion through the dust: Stratification.* Large-scale cometary insolation experiments, especially KOSI, have shown that the propagation of the heat wave inside an initially homogeneous mixture of ice and dust particles results in the vertical stratification of the sample (see Fig. 9a). The loss of water and other volatiles from the surface leads to its erosion and to the formation of a refractory dust mantle on top of the sample. This mantle, also called sublimation residue, is often quite porous and loosely consolidated with a microstructure of filaments made of aggregated dust grains (*Saunders et al.,* 1986; *Poch et al.,* 2016b; *Sears et al.,* 1999). Below this dust mantle, a system of ice layers formed by chemical differentiation have been identified. The sublimation of ices generates gases that diffuse inward and outward of the sublimation front. These gases can then recondense on sufficiently cold regions (recondensation fronts), especially at greater depth. After most of the KOSI experiments, a consolidated layer formed by recrystallization of H_2O and CO_2 was found just below the dust mantle, and 200-μm pure ice crystals were also observed (*Hsiung and Roessler,* 1989; *Roessler et al.,* 1992a; *Sears et al.,* 1999).

Permeability. The degree of permeability of the medium in which gases diffuse will influence these internal sublimation/recondensation processes, but also the ejection in the coma of gases and dust particles. Experiments on gas diffusion have been performed by *Gundlach et al.* (2011b) to study the permeability of dust layers. Two approaches were used. The first approach utilized a gas flow streaming through a sample placed inside a glass tube. By measuring the gas flow rate and the pressure difference between the two pressure reservoirs above and below the sample, the permeability was derived. The second approach used a sublimating ice surface that was covered by a dust layer. Both experiments yielded the same results, namely that the escaping rate of molecules is reduced by the dust layers. The thicker the layer, the stronger the reduction. In addition, *Schweighart et al.* (2021) studied the gas diffusion through granular materials in great detail. They varied grain size as well as the material type. The main result is that the density as well as the grain size distribution of the sample play a major role for the permeability of the sample, with larger grains allowing for larger permeability values. Additional measurements in low- and hypergravity conditions have recently been collected by *Capelo et al.* (2022). They show that for samples with irregularly shaped particles the deposition and compaction of the dust layer in conditions of variable gravity have a strong influence on the measured gas permeability.

2.3.5.2. Chemistry involving the gas and the dust: When the gaseous molecules produced by the sublimation of ices flow through the cometary nucleus before expanding in the coma, collisions between them and/or with the surface of ice or dust grains may lead to their adsorption followed by chemical reactions, changing the composition of the gas, and possibly forming new molecules in the ice or dust.

Isotopic fractionation. The influence of the sublimation process on the partition of isotopes between the solid and gas phases was studied in several laboratory experiments. During some of the KOSI cometary simulation experiments, isotopic analyses of the ice remaining at different depths af-

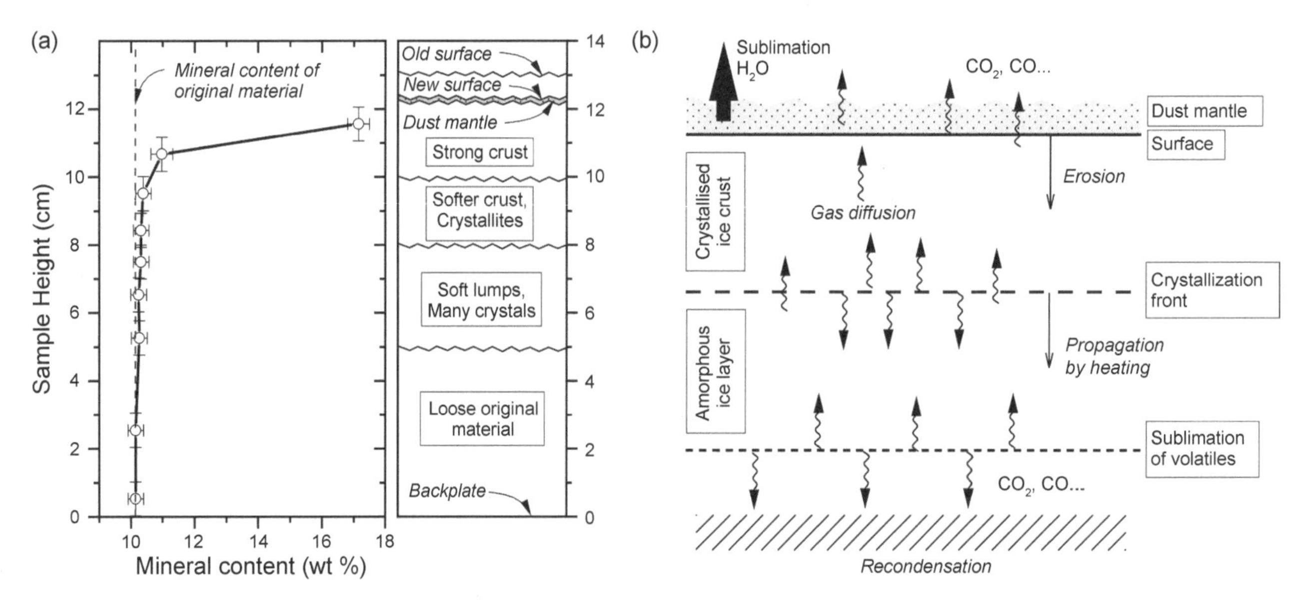

Fig. 9. Possible stratification in a comet nucleus approaching the Sun. **(a)** Cometary analog sample mineral content and macroscopic stratigraphy after a large-scale insolation experiment (KOSI 9) (*Grün et al.,* 1993) with a sample made of 10% dust and 90% crystalline water ice (see section 2.3.2). **(b)** Possible processes in a short-period comet nucleus (*Rickman,* 1991; *Sears et al.,* 1999). The insolation of the surface is followed by heat propagation processes inside the nucleus (see section 2.3.1). At the surface, the sublimation of the ice causes erosion and the formation of a dust mantle. Deeper, the water ice initially amorphous (contrary to the experiment on the left) becomes crystalline. During crystallization, volatile molecules initially trapped in water are released. Some gas molecules diffuse upward through warmer material, other diffuse inward and recondense into volatile-enriched layers.

ter insolation under vacuum of mixtures of water, CO_2 ices, and silicate dust were performed (*Roessler et al.,* 1992a,b). The upper layers of water ice were found to be enriched in heavy isotopes of hydrogen (deuterium, D) and oxygen (^{18}O), whereas the lower layers kept their initial isotopic composition. *Moores et al.* (2012) measured the isotopic composition of the water vapor produced continuously by the sublimation of water ice mixed with 1–25 wt% of various types of dust grains larger than 1 µm. The D/H of water vapor was observed to decrease with time, particularly for samples of highest dust contents, with isotopic fractionation factor ($(D/H)_{ice}/(D/H)_{gas}$) up to 2.5. The preferential adsorption of HDO on dust grains is the proposed explanation for the depletion of deuterium in the gas phase. By contrast, experiments studying the sublimation of pure water ice have reported much lower fractionation factors, varying from 0.969 to 1.123 (*Lécuyer et al.,* 2017; *Mortimer et al.,* 2018). The D/H of H_2O in comae has been shown to vary between 1 and 3× Earth's ocean value. Based on observational data, *Lis et al.* (2019) suggested that D/H of H_2O in comae may be correlated with the active fraction of the comet nuclei: The more active, the lower the D/H. *Lis et al.* (2019) proposed that hyperactive comets eject icy grains from deeper in the nucleus than less-active comets such as 67P, outgassing higher D/H water from shallower depths. Future experiments on isotopic fractionation of gases emitted by comets are needed to test this scenario and explain these observations, which are of fundamental importance to constrain the origin of Earth's water (*O'Brien et al.,* 2018). Experiments should be conducted with aggregates of submicrometer-sized dust grains and with relatively large dust-to-ice ratio simulating the upper layer of cometary nuclei. Moreover, they should investigate not only the D/H fractionation of water, but also $^{18}O/^{16}O$, $^{17}O/^{16}O$, $^{13}C/^{12}C$, and $^{14}N/^{15}N$ fractionations of H_2O, CO_2, CO, and NH_3 molecules, for example.

Alteration of mineral and organic dust. The gas constantly diffusing through the cometary dust may also react with the dust and change its composition. *Takigawa et al.* (2019) exposed amorphous silicates to D_2O and D_2O + 0.15% NH_3 ices and vapors for 10 to 120 days, at temperatures from 246 to 323 K. Hydration of the amorphous silicates was only observed in experiments above 298 K. Experiments performed at low temperatures did not show any modification of the silicates, or of organic molecules that were exposed to the same D_2O and NH_3 vapors, suggesting that alteration of dust by vapors should not be significant on comets.

2.3.5.3. Experiments on ejections of dust/ice particles and nucleus erosion: Large-scale insolation experiments have reported the ejection of dust/ice particles from cometary analog samples irradiated under vacuum (*Thiel,* 1989; *Poch et al.,* 2016b; *Bischoff et al.,* 2019; *Haack et al.,* 2021a). In these experiments, as the dust mantle left by the sublimation of the surface ices becomes thicker, it imposes more resistance to the gas flow. As the tortuosity of the path followed by the gas to exit the mantle layer increases, the force exerted by the gas on the mantle increases up to the point when it reaches the tensile strength of the mantle and breaks it in fragments that are ejected away by the gas flow (see Fig. 7c). The sudden release of pressure within the ice results in an increase of its sublimation rate and a decrease of its temperature because of latent heat absorption. After the ejection event, the mantle is progressively rebuilt on the surface of the ice.

In recent years, the group at the University of Warsaw has performed an extensive series of experiments to better understand how the sublimation of water ice is influenced by various parameters such as the presence, nature, concentration, and location of contaminants including silicate dust and organics of variable volatility (*Kossacki and Leliwa-Kopystynski,* 2014; *Kossacki et al.,* 2017, 2019, 2022; *Kossacki,* 2019, 2021). The experiments are conducted in a vacuum chamber equipped with a cooled plate where the temperature gradients within the sample (5.5 × 8 cm) and the rate of recession of the surface are measured continuously. The results of these experiments are used to constrain and validate a thermophysical model of the surface and subsurface evolution, which can then be used to predict the evolution of the nucleus topography (*Kossacki and Czechowski,* 2018; *Kossacki and Jasiak,* 2019; *Kossacki et al.,* 2020a). The case of the sliding of the silicate dust at the surface of the sublimating ice was also investigated. Experiments showed that the reduction in friction caused by the drag force from the gas results in sliding at angles much lower than the angle of repose in static conditions (*Kossacki et al.,* 2020b, 2022).

Other recent experiments on outgassing and material ejection have been performed by the CoPhyLab team at the Technical University of Braunschweig. Pure granular water ice samples were illuminated by an artificial Sun (Kreuzig et al., in preparation). The energy input into the sample material caused the ejection of water ice particles from the water ice sample. The first main result of this investigation is that water ice can eject itself, which means that the pressure increase inside the ice sample occurs before the sublimation front is able to erode the overlying particle layers. In addition, the particle trajectories have been analyzed and it was concluded that the ejected material possesses a non-zero starting velocity, which implies that the particles either start from within the porous sample, or have experienced a rapid initial acceleration by the ejection mechanism. The larger the temperature of the ice sample, the larger the initial velocity of the grains.

In a previous work, *Bischoff et al.* (2019) studied the ejection of dust pebbles by sublimation from an icy surface. A pure, solid ice surface was produced and dust aggregates (composed of silica) were placed on top of the ice surface. Then, the temperature was sequentially increased by a heater installed inside the ice. At temperatures of about 200 K the material was ejected from the ice surface by the pressure buildup at the ice-dust interface. The thicker the aggregate layer, the larger the ejected aggregate clusters are. A tentative explanation of this trend is that as the layer thickness increases, the force exerted by the gas on the dust layer increases, causing some rearrangements of dust particles that increase the layer internal cohesion, resulting in the ejection of larger chunks.

Spadaccia et al. (2022) have investigated experimentally whether the sublimation of the ice contained in millimeter-sized pebbles resulted in the ejection of silicate dust grains. They found that even under the harsh conditions of the laboratory (Earth gravity, accelerated sublimation), most of the initially icy pebbles retained their integrity upon sublimation of the ice while pebbles made of coarser silicate particles disrupted.

Experiments measuring the mechanical properties of various cometary surfaces analogs, described in more details in section 2.4.3, also provide knowledge on the ejection of particles from comets. The tensile strength (i.e., the pressure required to separate one particle from another) is orders of magnitudes lower for surfaces made of dust/ice pebbles (pascals) (*Blum et al.*, 2014) compared to homogeneous dust/ice powders (kilopascals) (*Gundlach et al.*, 2018; *Bischoff et al.*, 2020). If we assume that the gas pressure of volatile compounds are typically less than a pascal, then ejection of materials by sublimation seems to be only possible if the surface consists of pebbles.

Finally, the comet size may influence its activity and erosion rate. *Blum et al.* (2014) have studied the compaction of pebble assemblies under simulated load of the material. A gas stream was used to compress samples positioned inside a glass-vacuum tube. In a next step, the gas stream was inverted to destroy the pebble assemblage in order to measure the low tensile strength of the packing. The main result of these experiments is that the pebbles do "remember" ~3% of the applied compression after load reduction ("memory effect"). This effect has a major implication for understanding cometary activity. Excavated aggregate layers that have experienced gravitational compression require an enhanced gas pressure to be released. Hence, sufficiently large comets should not be able to sustain dust activity until the whole nucleus has been eroded. *Gundlach and Blum* (2016) argued that the activity stops after a threshold tensile strength is reached, because the gas pressure buildup will then not be able to overcome this critical value. The interiors of smaller comets have not experienced significant gravitational compression so that the memory effect can be neglected and, in this case, the whole object can in principle be eroded by the gas pressure buildup (if enough energy is supplied). Asteroids in cometary orbits (ACOs) are non-active bodies on cometary-like orbits and *Gundlach and Blum* (2016) state that these objects are extinguished cometary nuclei that were born larger, were eroded by sublimation until the threshold value was reached, and are now inactive.

Despite recent progress on these questions, the exact mechanisms of ejection and the initial velocities of the ejected particles deserve additional investigations with a wide range of experimental parameters. The role of super- and hypervolatiles should also be studied, starting with carbon dioxide mixed with water and dust. Such experiments have just been initiated within the CoPhyLab project.

2.4. Experiments to Constrain the Composition and Structure of Cometary Nuclei from Observations

Laboratory experimentation is crucial for the interpretation of observational data. Laboratory investigations serve two distinct and complementary purposes: (1) Laboratory measurements provide values of the fundamental physical quantities of relevant materials in their pure forms, and (2) laboratory setups provide the opportunity to prepare complex but well-controlled samples to mitigate the lack of "ground truth".

2.4.1. Surface light scattering properties: Spectroscopy, photometry. Over the reflected solar spectrum, cometary nuclei display an overall low reflectance, a red slope in the visible and a strong absorption around 3 μm (see the chapter by Filacchione et al. in this volume). A number of laboratory experiments on complex analogs have attempted to reproduce these properties to decipher the chemical nature and physical properties of the darkening and red component(s) as well as the ices with which they are mixed. This empirical approach is complementary to modeling and simulation, where the measured optical constants of the pure materials are used as inputs to calculate the spectrophotometric properties of parameterized surfaces.

The optical constants of the two types of ice observed on cometary surfaces, water and carbon dioxide, have been measured at the temperatures relevant for cometary nuclei on thin films of ice deposited on cryostat windows (*Warren,* 1986; *Hansen,* 1997; *Grundy and Schmitt,* 1998; *Schmitt et al.,* 1998). These values can be used to model the reflectance of macroscopic icy surfaces using a parametric representation of their properties (*Hapke,* 2012; *Shkuratov et al.,* 1999), which permits the quantitative inversion of remote-sensing observations (*Filacchione et al.,* 2019).

Efforts have been made to produce well-characterized and reproducible particles of water ice (*Gundlach et al.,* 2011b; *Jost et al.,* 2013) and mix them with contaminants in different ways (*Yoldi et al.,* 2015; *Poch et al.,* 2016a; *Pommerol et al.,* 2019; *Ciarniello et al.,* 2021) (Fig. 10). The same work should now be done for CO_2 ice and later other volatiles. While carbon and carbon-based compounds have often been used to lower the albedo of cometary analogs (*Oehler and Neukum,* 1991; *Stephens and Gustafson,* 1991; *Moroz et al.,* 1998), opaque minerals such as nanoparticle Fe-Ni alloys or iron sulfides are also relevant (*Quirico et al.,* 2016; *Rousseau et al.,* 2018; *Sultana et al.,* 2023). Experimenting with complex organics and nanoparticles of metal is very challenging as the materials are difficult to procure in large amounts, might be unstable in ambient laboratory conditions, and/or might be hazardous for the experimenters. These difficulties currently limit the experimental investigations on the exact nature and properties of cometary material.

Beside the exact nature of the components, the way they are mixed with bright ices and in what quantities are also of primary interest. Intimate mixtures of transparent ices and fine-grained dark material show that the absorbing material has an influence disproportionate to its abundance, and

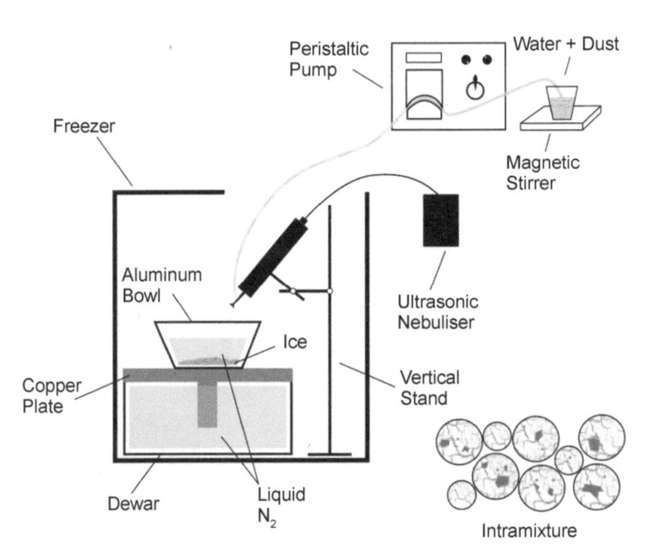

Fig. 10. The Setup for the Preparation of Icy Planetary Analogues (SPIPA-B) is used to produce comet analog mixtures of ice and dust (see Fig. 11) (*Pommerol et al.,* 2019).

minute amounts are sufficient to lower the albedo enough to hide the ice (*Yoldi et al.,* 2015; *Jost et al.,* 2017a). Unfortunately, this makes the inversion of albedo into dust-to-ice ratio highly challenging without any *a priori* knowledge of the physical properties of the compounds.

In addition to the overall albedo, the spectral red slope and the dependence of the reflectance to phase angle have been investigated, as they provide additional constraints on the nature and properties of the surface material. The first attempts at reproducing the observed properties of the 67P nucleus with mixtures of ice, organics, and minerals have shown that while it is possible to find mixtures of ice, minerals, and organics that mimic each individual property, satisfying combinations of constraints is more challenging and no analog could be produced that fits all observed properties (*Rousseau et al.,* 2018; *Jost et al.,* 2017a).

Sublimation experiments with icy analogs produce desiccated textures that probably resemble closely the surface of cometary nuclei. Depending on the initial composition of the dust and how it is mixed with the ice, sublimation can produce high-porosity cohesive mantles that display unique spectrophotometric properties (*Poch et al.,* 2016a,b). The strength of the absorption bands is reduced by the decrease in size and the deaggregation of the dust grains, to a point where spectral features are barely recognizable despite a high abundance of the material (*Sultana et al.,* 2021, 2023). The strong backscattering behaviour of such porous mantle is the best match to the phase dependence of reflectance observed at the nucleus of 67P (*Jost et al.,* 2017b). In addition to the microporosity, the submicrometer grain size of components having contrasted optical indexes (such as mixtures of silicates and opaque minerals or organics) could together explain the low albedo, visible and near-infrared spectral slope, mid-infrared spectral emissivity, and visible polarimetric phase curves of the most primitive small bodies, as shown in *Sultana et al.* (2023). While laboratory measurements are useful to understand the photometry of the nucleus at low phase angle, large-scale roughness and topographic effects likely dominate the photometry at high phase angle, which cannot easily be studied in the laboratory. Models and/or very large-scale experiments would be required to make progress on this question.

Beside the visible red slope, the strong 3-μm band of 67P is the most prominent spectral feature identified on a cometary nucleus. Its interpretation is challenging as different components contribute to it. Here as well, sublimation experiments have proven crucial to measure spectra influenced by both the chemical nature of the compounds and the physical texture of the surface. *Poch et al.* (2020) were able to prove in this way for the first time that ammonium salts are a key component of the surface of 67P (Fig. 11), while *Raponi et al.* (2020) have showed the additional contribution to the 3-μm band of aliphatic organics, and *Mennella et al.* (2020) the possible contribution of hydroxylated silicates.

To be useful to the large community, the laboratory data (optical constants, reflectance spectra, sample descriptions, etc.) need to be available readily and freely. Recent efforts have resulted in the development, update, and modernization of numerical databases to store and distribute these data and associated documentation (*Schmitt et al.,* 2018; *Milliken et al.,* 2016; *Kokaly et al.,* 2017).

During approach to the Sun, cometary surfaces are irradiated by solar wind and UV light, and interactions of coma gases with the solar wind generate accelerated molecular ions that impact and sputter the nucleus surface, as seen on 67P (*Nilsson et al.,* 2015; *Wurz et al.,* 2015). The influence of energetic particles on the optical properties of cometary surfaces are studied by the experiments described in section 2.2.2. The irradiation dose received during approach to the Sun may be lower than during their residence in the outer solar system, because of the higher resurfacing rate due to cometary activity, and lower energy of SEPs compared to GCRs. Finally, we note that the charging of the cometary surface may induce the levitation of dust particles as shown in the laboratory by *Wang et al.* (2016), possibly influencing surface optical properties.

2.4.2. Subsurface and interior electrical properties. The quantitative interpretation of radar data relies on the knowledge of the complex dielectric constants of the materials encountered at the surface and in the subsurface of cometary nuclei. While the dielectric constants of many materials of interest in their pure form have been determined for decades (see Table A1 in *Herique et al.,* 2016), recent experiments have focused on the measurements of dielectric properties of complex synthetic or natural analogs for cometary material. Indeed, as for other spectral domains, mixing laws exist and are readily used (*Sihvola,* 1999), but their applicability to complex high-porosity mixtures of ices, minerals, and organics must be verified.

Motivated by the interpretation of multi-instrumental Rosetta and Philae measurements at 67P, *Brouet et al.* (2016b) have measured systematically and over a wide spectral range the dielectric permittivity of porous mixtures of water ice

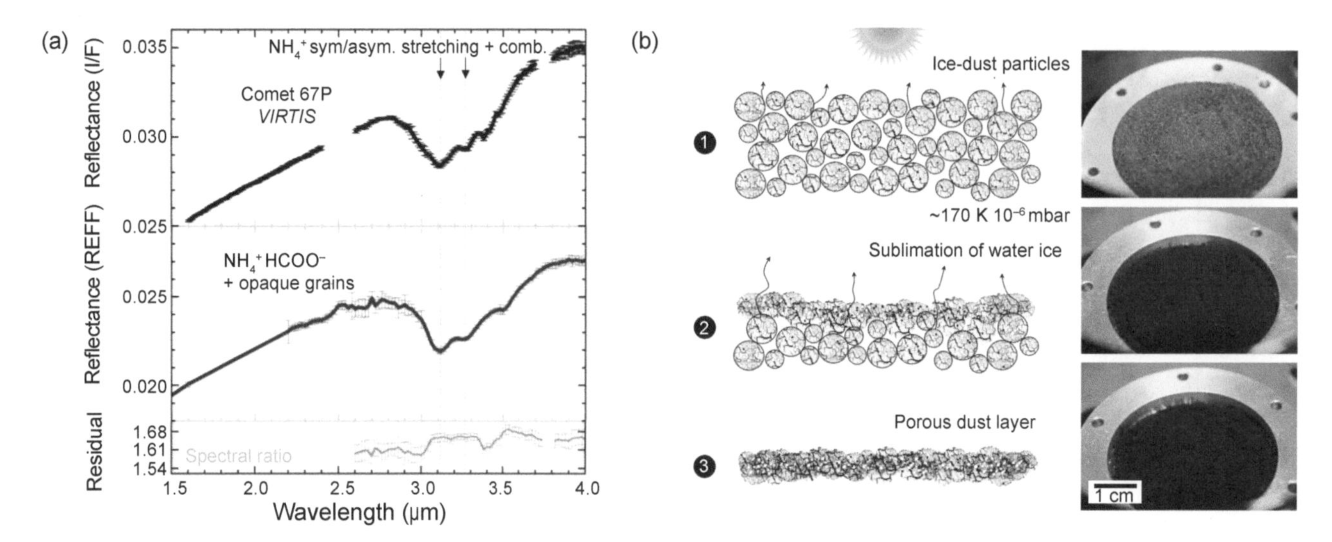

Fig. 11. Identification of NH_4^+ salts on the surface of Comet 67P. **(a)** Comparison of Rosetta/VIRTIS average 67P spectrum with the one of a sublimation residue produced by **(b)** mixing water, $NH_4^+HCOO^-$, and opaque submicrometer grains to make icy-dust particles (Fig. 10), followed by the sublimation of the water in a vacuum chamber (*Poch et al.,* 2020).

and dust. The application of these measurements to Rosetta data indicates a significant vertical gradient of porosity in the 67P nucleus (*Brouet et al.,* 2016a). To interpret the data from the COmet Nucleus Sounding Experiment by Radiowave Transmission (CONSERT) instrument, *Heggy et al.* (2012) and *Kofman et al.* (2015) reported on new measurements of the dielectric properties of some carbonaceous chondrites, thought to be reasonable analogs for the dust found in cometary nuclei. The values are notably lower than values for ordinary chondrites found in databases, but the authors also note the need to better study experimentally the potential effect of metal oxidation, which might bias the results (*Herique et al.,* 2016). Beside Titan, the dielectric measurements of tholins performed by *Lethuillier et al.* (2018) are also relevant for cometary material in which relatively heavy and complex organic molecules are abundant. The measurements revealed that both the real and imaginary parts of their dielectric constant are sensitive to the composition of the gas from which tholins are produced.

2.4.3. Experiments on mechanical properties of dust/ice surfaces. Laboratory studies have been performed to measure the tensile strength of granular materials. Two different material types have to be distinguished. The first type comprises all samples whose structure can be described as a homogeneous dust layer. These experiments can be used to understand the tensile strength of dust layers that were formed by the mass transfer scenario (see section 2.1.1 and the chapter by Simon et al. in this volume), and can be used to investigate the internal tensile strength of the dust/ice pebbles. *Gundlach et al.* (2018) have performed experiments (so-called "Brazilian disk tests") to study the tensile strength of cylindrical samples a few centimeters large, composed of granular silica or water ice. A loading platen is lowered on to a cylindrical sample by a stepper motor and exerts force and thus pressure onto the cylinder. The sample stays intact until the tensile strength of the material is reached. These measurements have shown that the tensile strength of homogeneous samples is in the kilopascal range and no difference between water ice and silica was detected. This implies that the internal tensile strength of pebbles as well as the tensile strength of cometary surface formed by the mass transfer process are in the kilopascal range. Also, different organic materials have been studied with the result that the exact tensile strength values deviate from sample to sample, but all investigations yield values of about 10^{-1} kPa to 10^1 kPa (*Bischoff et al.,* 2020). In contrast, *Blum et al.* (2014) studied the tensile strength of surfaces made of pebbles made of ice and dust and found that their tensile strength is orders of magnitude lower, namely in the pascal range. These experiments are crucial for our understanding of cometary activity discussed in section 2.3.5.3, because if we assume that the gas pressure of volatile compounds is typically less than a pascal, then ejection of materials by sublimation seems to be only possible if the surface consists of pebbles. In other words, comets that show activity (ejection of dust) must possess a pebble-like surface structure with a very low tensile strength.

Following many large-scale insolation experiments (described in section 2.3.2), a hardening of the remaining ice was observed underneath the dehydrated dust mantle at the surface (*Kochan et al.,* 1989; *Ratke and Kochan,* 1989; *Seiferlin et al.,* 1995; *Thiel,* 1989; *Poch et al.,* 2016b; *Kaufmann and Hagermann,* 2018) (Fig. 9a). This consolidation can be explained by the recondensation of H_2O/CO_2 between the individual ice/dust particles and/or by sintering: Both processes build bonds connecting the ice particles together and providing solidification of the sample. Measurements of the hardness of these ice-dust crusts, performed after the KOSI experiments, yielded values of 0.15 to 5 MPa (*Grün et al.,* 1993; *Kochan et al.,* 1989). Experiments also showed that the hardening depends on the amount of dust (*Kaufmann and Hagermann,* 2018) and on the way the ice and the dust are initially mixed together (*Poch et al.,* 2016b).

While a number of experiments with cometary analogs have provided observations on the formation and evolution of morphologic features upon sublimation of the ice, few laboratory simulations have focused entirely on this aspect. *Haack et al.* (2021a,b) have recently investigated the evolution of mixtures of water ice and dust (fly ash) under simulated cometary conditions and using scaling laws to select adequate particle sizes and energy input taking into account the effect of terrestrial gravity. Figure 7c shows an example of a sample affected by the ejection of large agglomerates. The morphologies observed are diverse, some of them reminiscent of the high-resolution images obtained at 67P such as fractures, overhangs, and collapses. Both the dust-to-ice mass ratios and the direction of the incident light strongly affect the morphology resulting from the sublimation. The introduction of salt (sodium acetate) and an amino acid (glycine) in the sample (*Haack et al.*, 2021b) changes the outcome of the experiments by inhibiting the collapse events, leading to thick, cohesive, and stable but fractured mantles at the surface of the sample.

The influence of composition on the sublimation coefficient, and hence the erosion rate of cometary analogs, has been studied experimentally by the Warsaw group [see *Kossacki and Leliwa-Kopystynski* (2014) and other references listed in section 2.3.5.3]. Based on these experimental results, they have developed a thermal model, which they then use to interpret the topography observed on cometary nuclei (*Kossacki and Czechowski,* 2018; *Kossacki and Jasiak,* 2019; *Kossacki et al.*, 2020a).

3. COMAE

3.1. Experiments to Constrain the Composition and Structure of Comae

Comae are made of gases dragging solid particles of dust and ice. They are observed remotely via different techniques. Ultraviolet-visible, infrared or radio/mm-wave spectroscopy allow to determine the composition of the gas phase. Broad-band visible photometry, polarimetry, and infrared spectroscopy are used to infer the composition and physical properties of the solid particles.

3.1.1. Coma gases. Gases produced by the sublimation of ices inside the nucleus may undergo several processes modifying their composition before they reach the coma, as discussed in section 2.3.5.2. Once in the coma, these molecules collide with each other, with electrons, and with solar wind ions. All these processes and their resulting spectral emission need to be studied in the laboratory and computationally to understand fully and correctly the light emitted by comae. This requires the determination of physical parameters such as frequencies and strengths of lines, collisional cross-sections, photodissociation rates, etc. for many molecules and their isotopologs, which are reported in databases. These laboratory experiments and databases on gas coma properties are reported in the chapters in this volume by Bodewits et al. and Biver et al. In the following paragraphs, we will focus on experiments on solid particles in the coma.

3.1.2. Coma particles' light scattering properties. Dust in the coma can be studied from a distance by analyzing thermally emitted light at long infrared wavelengths and solar scattered light in the near-UV to near-IR spectral range (see the chapter by Kolokolova et al. in this volume). In the first case, the spectral energy distribution (SED) of the emitted light is the primary source of information on the grains properties. At shorter wavelengths, the analysis of the angular distribution of scattered light can provide additional information on the physical characteristics of the dust beside the spectral dimension. The study of the light polarization ideally complements total light analyses for that purpose.

The interpretation of infrared SEDs relies on laboratory measurements of the thermal emissivity of fine powdered materials relevant for comets, such as minerals, carbonaceous compounds, and ices, essential to infer the composition, temperature, size distribution, and porosity of cometary dusts (*Lisse,* 2007; *Wooden,* 2008). The numerous challenges related to the selection and procurement of suitable analog materials discussed in the context of the nucleus are also relevant here.

Analyses of the scattering of light by dust particles in the coma can provide quantitative information on a number of properties of the dust, but the inversion of observational data into physical characterization of the particles is a complicated process that relies on scattering models and parameterizations of the particles that should be tested (see the chapter by Kolokolova et al. in this volume). As unpolarized sunlight gets partially polarized when scattered by particles in the coma, it carries information on the complex refractive index, i.e., the composition, as well as on the size, shape, and structure of the dust particles, complementary to SED observations (*Kolokolova et al.*, 2007). The variations of polarization with wavelength and phase angle as comets orbit the Sun provide information on these properties (*Kiselev et al.*, 2015; *Hines and Levasseur-Regourd,* 2016) and are therefore extensively studied in the laboratory to compare with the observations (*Levasseur-Regourd et al.*, 2015). Light-scattering properties of particles randomly oriented in a gas stream are measured at the Cosmic Dust Laboratory (CoDuLab) of the Instituto de Astrofisica de Andalucia (IAA) in Granada (*Hovenier and Muñoz,* 2009; *Muñoz et al.*, 2010, 2011), where the whole Mueller scattering matrixes (*Muñoz et al.*, 2012) have been obtained for a variety of analogs. Phase functions of millimeter-sized irregular dust grains have been compared to numerical models with the aim to improve the interpretation of comets and protoplanetary disk observations (*Muñoz et al.*, 2017; *Escobar-Cerezo et al.*, 2017). PROGRA2 instruments were developed in France in 1998 and have been in use since that time in order to compare linear polarization phase functions of small body surfaces and cometary comae with those measured on a wide set of samples (*Levasseur-Regourd et al.*, 1998; *Worms et al.*, 1999). Measurements of interest for comae may be obtained in the laboratory for particles smaller than 10 μm through an air-draught technique, while more systematic measurements are obtained under microgravity conditions (0 ± 0.05 g) during parabolic flight campaigns organized by the CNES or ESA space agencies (*Hadamcik et al.,* 2011) (Fig. 12). The PROGRA2-Vis instrument can provide measurements between 5° and 165° phase angles at 543.5 nm and 632.8 nm, while the PROGRA2-IR instrument operates in the IR at 1.5 μm (*Levasseur-Regourd et al.,* 2015). The polarimetric phase curves of cometary analogs made of porous aggregates of submicrometer-sized Mg-silicates, Fe-silicates, and carbon black grains mixed with compact Mg-silicate grains appear similar to observations of comae (*Hadamcik et al.,* 2007). Aggregates of submicrometer-sized monomers provide the best fits for the higher polarization observed in cometary jets, while a mixture of porous aggregates and compact grains is needed to fit whole comae observations (*Hadamcik et al.,* 2006). The influence of the size of individual dust grains (from nanometer- to micrometer-sized) and agglomerates (from micrometer- to centimeter-sized) on the maximum polarization of phase curves has been studied (*Hadamcik et al.,* 2009), allowing the interpretation of polarimetric images of the coma of 103P/Hartley 2 in term of progressive sublimation of ice, leading to particle fragmentations with increasing distance from the nucleus (*Hadamcik et al.,* 2013). Finally, a light scattering unit, derived from PROGRA2-Vis, had been developed as a tentative precursor experiment for aggregation of microspheres in microgravity conditions and flown on the ESA sounding rocket flight MASER-8 (*Levasseur-Regourd,* 2003).

3.2. Experiments on the Processes Transforming Coma Particles

The dust particles ejected in the coma are subjected to different processes (photolysis, heating, etc.), which can modify their composition and generate additional gaseous products. As a consequence, the distribution of some gaseous species in comae correspond not only to their direct sublimation from the nucleus, but also to their production from larger molecules or dust grains. Such additional sources of gas in the coma are called "extended sources" or "distributed sources" (*Cottin and Fray,* 2008). Distributed sources are observed for molecules and radicals such as H_2CO, CO, HNC, CN, OCS, HF, HCl, C_2, C_3, and even glycine (see the chapter by Biver et al. in this volume).

The absorption of solar radiation by molecules in the dust can cause their photodegradation by UV photons, and it also induces the warming of the dust. Volatile and semivolatile species still present in the particles can undergo photolysis or sublimation, and even refractory compounds can decompose into gases due to irradiation or heating. Indeed, the submicrometer-sized particles made of absorptive materials can reach temperatures higher than a blackbody, up to 500 K or more for 0.2-μm grains (*Kolokolova et al.,* 2004). In addition to solar radiation, dust particles could also be irradiated by energetic particles from the cometary plasma and solar wind (*Mendis and Horányi,* 2013).

Laboratory experiments are performed to identify the nature of the solid precursors responsible for these distributed sources, and retrieve qualitative and quantitative data

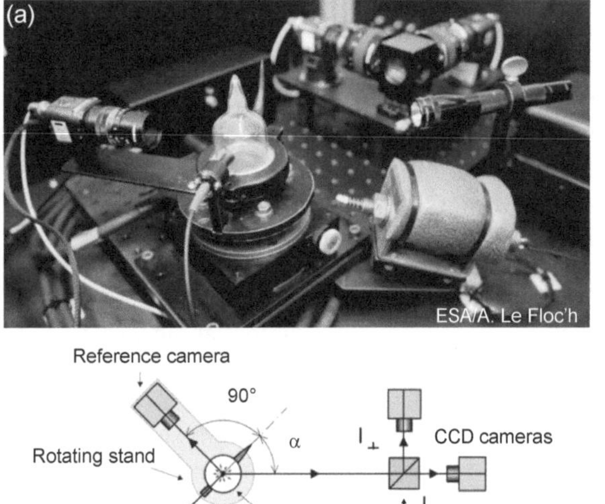

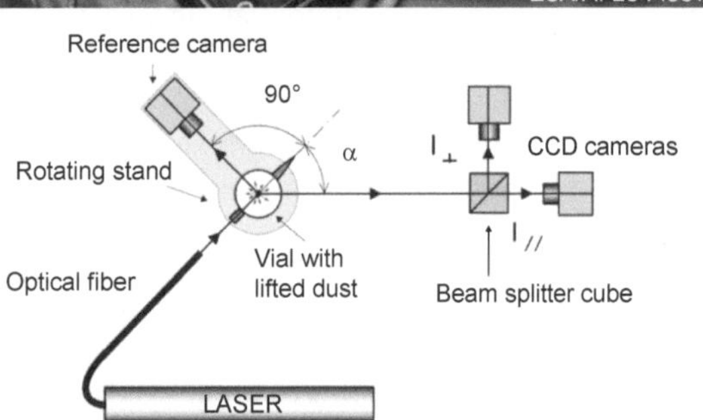

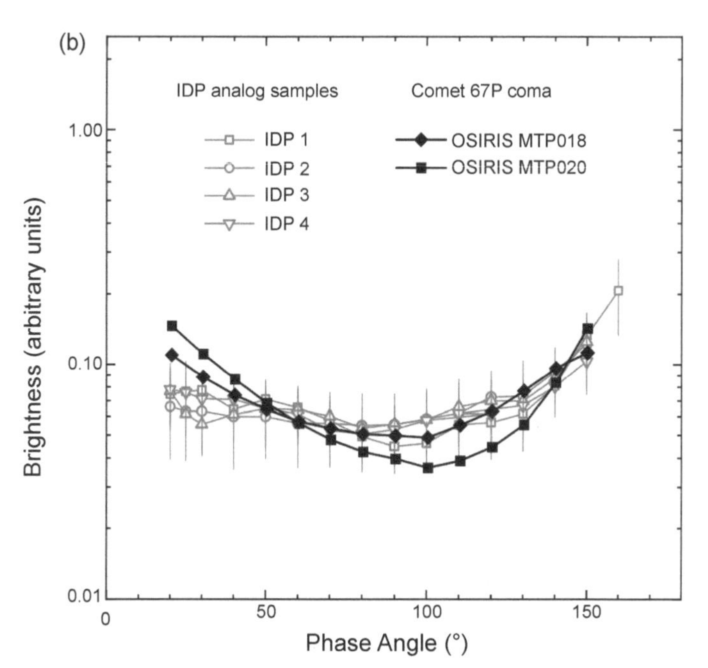

Fig. 12. Light-scattering properties of cometary dust analogs in suspension. **(a)** PROGRA2-Vis instrument; (top) picture of the setup under microgravity, (bottom) scheme of laboratory setup (*Hadamcik et al.,* 2009). **(b)** Comparison between phase functions of Comet 67P (Rosetta/OSIRIS data) and measurements of IDP analogs (aggregates of submicrometer grains of carbonaceous compounds and minerals, such as those shown in Figs. 4 and 5) (*Levasseur-Regourd et al.,* 2019).

on the processes by which they are transformed in gases. Experiments investigating photodegradation and thermal degradation of compounds relevant for comets enable the determination of absorption cross sections, quantum yields of photodegradation, Arrhenius constants and activation energies of thermal degradation, or vapor pressures of sublimation, which serve as inputs into numerical models of gas distributions observed in comae (*Cottin and Fray,* 2008; *Hadraoui et al.,* 2019). Photolysis of solid compounds stable at room temperature can be performed in a temperature-controlled (300 K) reactor illuminated with a UV lamp, maintained under vacuum (10^{-4} mbar) and connected to a system analyzing the gaseous products, such as an infrared spectrometer or a mass spectrometer possibly coupled with a gas chromatograph (*Cottin and Fray,* 2008; *Fray et al.,* 2004b). Thermal degradation of the same kind of compounds can be performed in a heated reactor or a pyrolyzer connected to similar analytic systems. To study the degradation of compounds formed after irradiation and heating of cometary ices, high vacuum chambers such as the ones described in section 2.1.2.1 and Fig. 2 are used to produce these compounds and degrade them *in situ* via UV and/or controlled heating from 20 to 800 K (*Briani et al.,* 2013). Several experiments have studied the photodegradation and/or thermal degradation of refractory organic compounds such as polyoxymethylene (POM) [($-CH_2O-$)n]) (*Fray et al.,* 2004a, 2006) (Fig. 13) and hexamethylenetetramine (HMT) [$(CH_2)_6N_4$]) (*Briani et al.,* 2013; *Fray et al.,* 2004b) produced from ice chemistry (see section 2.1.2.2), HCN-polymers (*Fray et al.,* 2004b), and C_3O_2-polymers, which have been proposed as a source of CO (see references in *Cottin and Fray,* 2008). However, to date the presence of these compounds in cometary dust

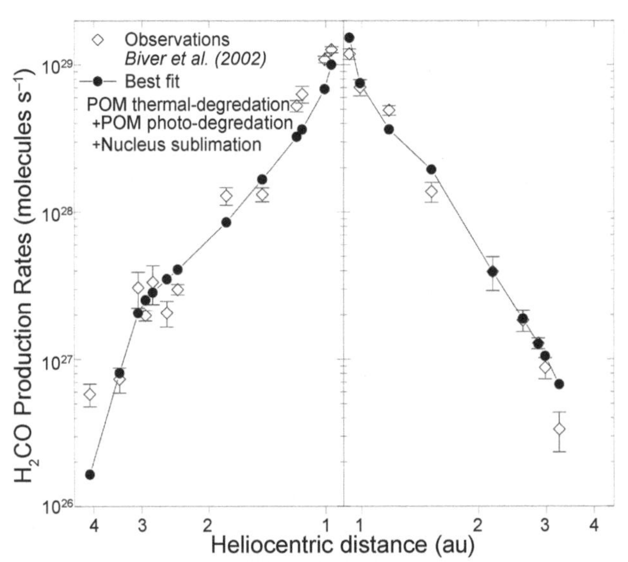

Fig. 13. H_2CO production rate as a function of heliocentric distance for Comet C/1995 O1 Hale-Bopp as observed (open squares), and as computed from experimental data on POM (black circles), assuming a mass fraction in the grains of 3.1% and H_2CO production at the surface of the nucleus equal to 3% of HCN production (*Fray et al.,* 2006).

is uncertain. The Rosetta mission has shown the presence of high-molecular-weight organic matter sharing similarities with meteoritic IOM (*Fray et al.,* 2016; *Raponi et al.,* 2020) and ammonium salts (*Altwegg et al.,* 2020; *Poch et al.,* 2020) in the dust of 67P. Therefore, future experiments should investigate the photo- and thermal degradation of these types of compounds. The thermal decomposition of ammonium salts ($NH_4^+Cl^-$ and $NH_4^+HCOO^-$) and the detectability of their products after ionization in the Rosetta mass spectrometer ROSINA was studied in the laboratory, enabling their identification (*Hänni et al.,* 2019). Moreover, NH_4^+ salts thermal degradation at relatively high temperature could explain the apparent depletion of comae in nitrogen, the increase of NH_3 in comets closer to the Sun (*Altwegg et al.,* 2020; *Dello Russo et al.,* 2016), and possibly several distributed sources (*Hänni et al.,* 2020). To confirm this hypothesis, future experiments of photo-/thermal degradation on pure and mixed ammonium salts should be performed.

The dust in comae may be subjected to other processes not yet investigated in experiments. In particular, the solar wind interaction with the coma, responsible for the cometary ionosphere, could induce radiolysis of dust particles by energetic particles or charging processes, causing erosion and disruption of dust particles (*Mendis and Horányi,* 2013).

3.3. Link with Circumstellar Disks

Our knowledge of circumstellar disks has evolved considerably in the last decades thanks to fast progress in observations at optical and radio wavelengths. These data provide new and important observational constraints to better understand planetary formation and complement knowledge gained in our solar system by studying primordial objects such as comets. Indeed, circumstellar disks are made of objects whose physicochemical properties could be similar to cometesimals or comets depending on the disk age, and studies of circumstellar disks and comets share the same scientific questions on the origin and evolution of matter around stars. For these reasons, experiments on comets are also useful to better interpret observations of circumstellar disks (see the chapters by Aikawa et al. and Fitzsimmons et al. in this volume). *Levasseur-Regourd et al.* (2020) show that the properties of cometary dust, in particular their high porosity, are compatible with modeling of light scattering in protoplanetary and debris disks. For example, *Hunziker et al.* (2021) extract the polarimetric properties of dust particles in the protoplanetary disk HD 142527 and conclude that large (> 1 μm) and porous aggregates are present. Future laboratory efforts to further characterize light-scattering properties of analog samples relevant for circumstellar disks will be necessary to better constrain the physical models on which these interpretations are based. While the initial steps can largely be based on similar experiments related to comets, some specificity of disks will also need to be taken into account.

4. PERSPECTIVES FOR FUTURE EXPERIMENTS

Laboratory experiments provide essential reference data to understand what comets are made of, how they formed, how they evolve, and what causes their activity. Experiments extend to all the cometary materials (ice, dust, gas) and all the processes they undergo from their formation/processing in pre-cometary environments to their incorporation and possible transformation in comet nuclei and comae. In this chapter, we have provided an overview of the diversity of laboratory experiments supporting the scientific investigations of comets.

Over the last two decades, cometary missions (Stardust sample-return, Deep Impact, Rosetta) and astronomical observations have provided new constraints on comet structure and composition. These data should drive the preparation of future experiments, indicating specific structures, compounds, and conditions of interest to study with "simple" or "complex" samples. In particular, experiments performed on "complex cometary analogs" need to be continued with samples more representative of the current knowledge on both the physical structure and the chemical composition of cometary materials. For example, experiments with mixtures of ices and dust are essential to study the chemistry (formation, transformation) and the spectroscopy of cometary materials. The dust comprises not only silicates and other minerals, but also a large fraction of high-molecular-weight organic matter, as well as semivolatile components such as lower molecular weight organic molecules and salts. These components appear to be arranged in the form of porous or compact aggregates of submicrometer-sized grains. Cometary analogs consisting of such realistic mixtures, in terms of structure and composition, will provide a better understanding of various physical processes such as coagulation, accretion, thermal evolution, outgassing, light scattering, etc. Furthermore, the realization of these experiments in the most realistic conditions (low temperatures, vacuum, microgravity) will also be beneficial. Closer collaborations between cometary physicists and chemists should be encouraged to perform such experiments.

Large-scale cometary laboratory experiments can also be used to explore and test new ideas of cometary processes still unidentified, such as the effects of cosmic-ray irradiation in thick sections of cometary analogs of various porosity, the processes related to the diffusion of gas through the nucleus material (transformations, isotopic fractionation), the role of super- and hypervolatiles in the ejection of dust/ice particles, or the photo/thermal decomposition and interaction with plasma of grains in comae.

Future experiments, either refining already known cometary processes or revealing new ones, will be complementary to theoretical models simulating these processes. Many open questions about comets would only receive reasonable answers via the interplay of models and experiments. For example, to understand comet formation processes it is essential to have good scaling laws to extrapolate results from experiments to large objects. Moreover, models of (pre-) cometary ice chemistry using networks of reactions (whose parameters are determined by theories and/or experiments) could allow answering the questions of how, when, and where some cometary materials were formed. Furthermore, models of comet thermal evolution and outgassing would benefit from measurements of the thermal conductivity of refractory organics, salts, and gas-laden amorphous ice, as well as measurements of the solid-gas equilibrium of ices and clathrate hydrates at low pressure and temperature. Finally, future experiments could also benefit from new technologies or methods of physical or chemical measurements, providing better characterization of the samples and of their processing.

The next cometary space missions (see the chapter by Snodgrass et al. in this volume) will also drive the realization of future experiments. First, experiments are needed to prepare and interpret these observations. In particular, one of the challenges of observing a dynamically-new comet, as planned by the Comet Interceptor mission, will be to disentangle the origins of the observed compounds or structures (already formed in cometesimals, or produced later). Second, and reciprocally, new observations will provide knowledge to take into account in future experiments. A long-term challenge of cometary science is to plan and execute future analyses of cometary nucleus material, either *in situ* or via sample return. A potential future sample return of cometary nucleus to Earth would require various experimental and theoretical simulations to define how to best sample, preserve, and analyze the cometary material. Future large-scale experiments could play a fundamental role in refining our knowledge of the depth profile of a cometary nucleus (composition and interrelationship of volatiles and refractories, physical structure, thermal and mechanical properties), crucial for all steps of the mission from sampling to interpreting the results (*Thomas et al.,* 2019). In return, such analyses of cometary material will provide excellent data to improve experimental and theoretical models of comets.

Astronomical observatories such as the James Webb Space Telescope, Extremely Large Telescope, Atacama Large Millimeter/submillimeter Array, etc., will also provide observations of a multitude of comae and nuclei and allow comparisons with pre- and protostellar environments, protoplanetary or debris disks, but also TNOs and asteroids of our own solar system. Cosmomaterials and cosmochemistry studies could provide interesting insights as well related to comets. Laboratory experiments will be instrumental in understanding the evolutionary links between these environments/objects and comets.

Acknowledgments. The authors dedicate this chapter to the memory of Anny-Chantal Levasseur-Regourd (1945–2022). She was a renowned specialist of comets, especially of the optical properties of cometary and interplanetary dusts she studied via observations, laboratory, and numerical simulations. The first version of this chapter benefited from her thorough reading and advice. Her inexhaustible enthusiasm will continue to inspire us. The work of O.P. related to comets was supported by the Center for Space and Habitability of the University of Bern, the National Centre of Competence in Research (NCCR) PlanetS supported by the Swiss National Science Foundation, the Centre National d'Etudes Spatiales (CNES), the French Agence Nationale de la Recherche (program Classy, ANR-17-CE31-0004), and the European Research Council (under the SOLARYS grant ERC-CoG2017-771691). O.P. acknowledges B. Schmitt for discussions about cometary physics and M. Choukroun for discussions on the chapter's structure. The authors are grateful to J. A. Nuth and P. A. Gerakines for their reviews of this chapter.

REFERENCES

Allodi M., Baragiola R., Baratta G., et al. (2013) Complementary and emerging techniques for astrophysical ices processed in the laboratory. *Space Sci. Rev., 180,* 101–175.

Altwegg K., Balsiger H., Bar-Nun A., et al. (2016) Prebiotic chemicals — amino acid and phosphorus — in the coma of Comet 67P/Churyumov-Gerasimenko. *Sci. Adv., 2,* 1600285.

Altwegg K., Balsiger H., Berthelier J., et al. (2017) Organics in Comet 67P — A first comparative analysis of mass spectra from ROSINA-DFMS, COSAC and Ptolemy. *Mon. Not. R. Astron. Soc., 469,* 130–141.

Altwegg K., Balsiger H., Hänni N., et al. (2020) Evidence of ammonium salts in Comet 67P as explanation for the nitrogen depletion in cometary comae. *Nature Astron., 4,* 533–540.

Arakawa S., Tanaka H., Kataoka A., et al. (2017) Thermal conductivity of porous aggregates. *Astron. Astrophys., 608,* L7.

Augé B., Dartois E., Engrand C., et al. (2016) Irradiation of nitrogen-rich ices by swift heavy ions: Clues for the formation of ultracarbonaceous micrometeorites. *Astron. Astrophys., 592,* A99.

Baratta G., Brunetto R., Leto G., et al. (2008) Raman spectroscopy of ion-irradiated astrophysically relevant materials. *J. Raman Spectrosc., 39,* 211–219.

Baratta G., Accolla M., Chaput D., et al. (2019) Photolysis of cometary organic dust analogs on the EXPOSE-R2 mission at the International Space Station. *Astrobiology, 19,* 1018–1036.

Bardyn A., Baklouti D., Cottin H., et al. (2017) Carbon-rich dust in Comet 67P/Churyumov-Gerasimenko measured by COSIMA/Rosetta. *Mon. Not. R. Astron. Soc., 469,* S712–S722.

Bar-Nun A. and Owen T. (1998) Trapping of gases in water ice and consequences to comets and the atmospheres of the inner planets. In *Solar System Ices* (B. Schmitt et al., eds.), pp. 353–366. Astrophys. Space Sci. Library, Vol. 227, Kluwer, Dordrecht.

Bar-Nun A. and Laufer D. (2003) First experimental studies of large samples of gas-laden amorphous "cometary" ices. *Icarus, 161,* 157–163.

Bar-Nun A., Herman G., Laufer D., et al. (1985) Trapping and release of gases by water ice and implications for icy bodies. *Icarus, 63,* 317–332.

Bar-Nun A., Kleinfeld I., and Kochavi E. (1988) Trapping of gas mixtures by amorphous water ice. *Phys. Rev. B, 38,* 7749–7754.

Bar-Nun A., Notesco G., and Owen T. (2007) Trapping of N_2, CO and Ar in amorphous ice — Application to comets. *Icarus, 190,* 655–659.

Beitz E., Güttler C., Blum J., et al. (2011) Low-velocity collisions of centimeter-sized dust aggregates. *Astrophys. J., 736,* 34.

Bekaert D., Derenne S., Tissandier L., et al. (2018) High-temperature ionization-induced synthesis of biologically relevant molecules in the protosolar nebula. *Astrophys. J., 859,* 142.

Benkhoff J. and Huebner W. (1995) Influence of the vapor flux on temperature, density, and abundance distributions in a multicomponent, porous, icy body. *Icarus, 114,* 348–354.

Benkhoff J., Seidensticker K., Seiferlin K., et al. (1995) Energy analysis of porous water ice under space-simulated conditions: Results from the KOSI-8 experiment. *Planet. Space Sci., 43,* 353–361.

Bennett C., Pirim C., and Orlando T. (2013) Space-weathering of solar system bodies: A laboratory perspective. *Chem. Rev., 113,* 9086–9150.

Bernstein M. P., Sandford S., Allamandola L., et al. (1995) Organic compounds produced by photolysis of realistic interstellar and cometary ice analogs containing methanol. *Astrophys. J., 454,* 327–344.

Bernstein M. P., Dworkin J., Sandford S., et al. (2002a) Racemic amino acids from the ultraviolet photolysis of interstellar ice analogues. *Nature, 416,* 401–403.

Bernstein M. P., Moore M., Elsila J., et al. (2002b) Side group addition to the polycyclic aromatic hydrocarbon coronene by proton irradiation in cosmic ice analogs. *Astrophys. J. Lett., 582,* L25–L29.

Biron K., Derenne S., Robert F., et al. (2015) Toward an experimental synthesis of the chondritic insoluble organic matter. *Meteorit. Planet. Sci., 50,* 1408–1422.

Bischoff D., Gundlach B., Neuhaus M., et al. (2019) Experiments on cometary activity: Ejection of dust aggregates from a sublimating water-ice surface. *Mon. Not. R. Astron. Soc., 483,* 1202–1210.

Bischoff D., Kreuzig C., Haack D., et al. (2020) Sticky or not sticky? Measurements of the tensile strength of microgranular organic materials. *Mon. Not. R. Astron. Soc., 497,* 2517–2528.

Blum J. (2018) Dust evolution in protoplanetary discs and the formation of planetesimals. What have we learned from laboratory experiments? *Space Sci. Rev., 214,* 52.

Blum J. and Wurm G. (2000) Experiments on sticking, restructuring, and fragmentation of preplanetary dust aggregates. *Icarus, 143,* 138–146.

Blum J., Schräpler R., Davidsson B. J. R., et al. (2006) The physics of protoplanetesimal dust agglomerates. I. Mechanical properties and relations to primitive bodies in the solar system. *Astrophys. J., 652,* 1768–1781.

Blum J., Gundlach B., Mühle S., et al. (2014) Comets formed in solar-nebula instabilities! — An experimental and modeling attempt to relate the activity of comets to their formation process. *Icarus, 235,* 156–169.

Bonnet J.-Y., Quirico E., Buch A., et al. (2015) Formation of analogs of cometary nitrogen-rich refractory organics from thermal degradation of tholin and HCN polymer. *Icarus, 250,* 53–63.

Boogert A. C. A. (2019) Questions about the evolution of ices, from diffuse molecular clouds to comets. *Proc. Intl. Astron. Union, 15,* 15–20.

Boogert A. C. A., Gerakines P., and Whittet D. (2015) Observations of the icy universe. *Annu. Rev. Astron. Astrophys., 53,* 541–581.

Boogert A. C. A., Brewer K., Brittain A., et al. (2022) Survey of ices toward massive young stellar objects. I. OCS, CO, OCN^-, and CH_3OH. *Astrophys. J., 941,* 32.

Bottke W. F., Vokrouhlický D., Rubincam D. P., et al. (2006) The Yarkovsky and YORP effects: Implications for asteroid dynamics. *Annu. Rev. Earth Planet. Sci., 34,* 157–191.

Briani G., Fray N., Cottin H., et al. (2013) HMT production and sublimation during thermal process of cometary organic analogs: Implications for its detection with the Rosetta instruments. *Icarus, 226,* 541–551.

Bringa E. and Johnson R. (2003) Ion interactions with solids: Astrophysical applications. In *NATO Science Series, Vol. 120: Solid State Astrochemistry* (V. Pirronello et al., eds.), pp. 357–393, Springer, Dordrecht.

Brisset J., Heißelmann D., Kothe S., et al. (2016) Submillimetre-sized dust aggregate collision and growth properties: Experimental study of a multi-particle system on a suborbital rocket. *Astron. Astrophys., 593,* A3.

Brisset J., Heißelmann D., Kothe S., et al. (2017) Low-velocity collision behaviour of clusters composed of sub-millimetre sized dust aggregates. *Astron. Astrophys., 603,* A66.

Brouet Y., Levasseur-Regourd A.-C., Sabouroux P., et al. (2016a) A porosity gradient in 67P/C-G nucleus suggested from CONSERT and SESAME-PP results: An interpretation based on new laboratory permittivity measurements of porous icy analogues. *Mon. Not. R. Astron. Soc., 462,* S89–S98.

Brouet Y., Neves L., Sabouroux P., et al. (2016b) Characterization of the permittivity of controlled porous water ice-dust mixtures to support the radar exploration of icy bodies. *J. Geophys. Res.–Planets, 121,* 2426–2443.

Brownlee D. (2014) The Stardust mission: Analyzing samples from the edge of the solar system. *Annu. Rev. Earth Planet. Sci., 42,* 179–205.

Brucato J., Mennella V., Colangeli L., et al. (2002) Production and processing of silicates in laboratory and in space. *Planet. Space Sci., 50,* 829–837.

Brunetto R., Barucci M., Dotto E., et al. (2006) Ion irradiation of frozen methanol, methane, and benzene: Linking to the colors of Centaurs and trans-neptunian objects. *Astrophys. J., 644,* 646–650.

Brunetto R., Pino T., Dartois E., et al. (2009) Comparison of the Raman spectra of ion irradiated soot and collected extraterrestrial carbon.

Icarus, 200, 323–337.
Brunetto R., Loeffler M. J., Nesvorný D., et al. (2015) Asteroid surface alteration by space weathering processes. In *Asteroids IV* (P. Michel et al., eds.), pp. 597–616. Univ. of Arizona, Tucson.
Bukhari Syed M., Blum J., Wahlberg Jansson K., et al. (2017) The role of pebble fragmentation in planetesimal formation. I. Experimental study. *Astrophys. J., 834*, 145.
Butscher T., Duvernay F., Danger G., et al. (2019) Radical-assisted polymerization in interstellar ice analogues: Formyl radical and polyoxymethylene. *Mon. Not. R. Astron. Soc., 486*, 1953–1963.
Capelo H. L., Kühn J., Pommerol A., et al. (2022) TEMPus VoLA: The timed Epstein multi-pressure vessel at low accelerations. *Rev. Sci. Instrum., 93*, 104502.
Carpentier Y., Féraud G., Dartois E., et al. (2012) Nanostructuration of carbonaceous dust as seen through the positions of the 6.2 and 7.7 μm AIBs. *Astron. Astrophys., 548*, A40.
Caselli P. and Ceccarelli C. (2012) Our astrochemical heritage. *Astron. Astrophys. Rev., 20*, 56.
Chen Y.-J., Nuevo M., Hsieh J.-M., et al. (2007) Carbamic acid produced by the UV/EUV irradiation of interstellar ice analogs. *Astron. Astrophys., 464*, 253–257.
Choukroun M., Kieffer S. W., Lu X., et al. (2013) Clathrate hydrates: Implications for exchange processes in the outer solar system. In *The Science of Solar System Ices* (M. S. Gudipati and J. Castillo-Rogez, eds.), pp. 409–454. Astrophys. Space Sci. Library, Vol. 356, Kluwer, Dordrecht.
Choukroun M., Altwegg K., Kührt E., et al. (2020) Dust-to-gas and refractory-to-ice mass ratios of Comet 67P/Churyumov-Gerasimenko from Rosetta observations. *Space Sci. Rev., 216*, 44.
Ciarniello M., Moroz L. V., Poch O., et al. (2021) VIS-IR spectroscopy of mixtures of water ice, organic matter, and opaque mineral in support of small body remote sensing observations. *Minerals, 11*, 1222.
Cody G., Heying E., Alexander C., et al. (2011) Establishing a molecular relationship between chondritic and cometary organic solids. *Proc. Natl. Acad. Sci. U.S.A., 108*, 19171–19176.
Colangeli L., Henning T., Brucato J., et al. (2003) The role of laboratory experiments in the characterisation of silicon-based cosmic material. *Astron. Astrophys. Rev., 11*, 97–152.
Colangeli L., Brucato J. R., Bar-Nun A., et al. (2004) Laboratory experiments on cometary materials. In *Comets II* (M. C. Festou et al., eds.), pp. 695–717. Univ. of Arizona, Tucson.
Collings M., Anderson M., Chen R., et al. (2004) A laboratory survey of the thermal desorption of astrophysically relevant molecules. *Mon. Not. R. Astron. Soc., 354*, 1133–1140.
Cottin H. and Fray N. (2008) Distributed sources in comets. *Space Sci. Rev., 138*, 179–197.
Cottin H., Szopa C., and Moore M. (2001) Production of hexamethylenetetramine in photolyzed and irradiated interstellar cometary ice analogs. *Astrophys. J., 561*, 139–142.
Cottin H., Kotler J., Billi D., et al. (2017) Space as a tool for astrobiology: Review and recommendations for experimentations in Earth orbit and beyond. *Space Sci. Rev., 209*, 83–181.
Dalle Ore C., Fulchignoni M., Cruikshank D., et al. (2011) Organic materials in planetary and protoplanetary systems: Nature or nurture? *Astron. Astrophys., 533*, A98.
Danger G., Orthous-Daunay F.-R., de Marcellus P., et al. (2013) Characterization of laboratory analogs of interstellar/cometary organic residues using very high resolution mass spectrometry. *Geochim. Cosmochim. Acta, 118*, 184–201.
Danger G., Fresneau A., Abou Mrad N., et al. (2016) Insight into the molecular composition of laboratory organic residues produced from interstellar/pre-cometary ice analogues using very high resolution mass spectrometry. *Geochim. Cosmochim. Acta, 189*, 184–196.
Dartois E., Muñoz Caro G., Deboffle D., et al. (2005) Ultraviolet photoproduction of ISM dust: Laboratory characterisation and astrophysical relevance. *Astron. Astrophys., 432*, 895–908.
Dartois E., Augé B., Boduch P., et al. (2015) Heavy ion irradiation of crystalline water ice: Cosmic ray amorphisation cross-section and sputtering yield. *Astron. Astrophys., 576*, A125.
Deckers J. and Teiser J. (2013) Colliding decimeter dust. *Astrophys. J., 769*, 151.
Deckers J. and Teiser J. (2014) Macroscopic dust in protoplanetary disks — From growth to destruction. *Astrophys. J., 796*, 99.
Dello Russo N., Kawakita H., Vervack R., et al. (2016) Emerging trends and a comet taxonomy based on the volatile chemistry measured in thirty comets with high-resolution infrared spectroscopy between 1997 and 2013. *Icarus, 278*, 301–332.
de Marcellus P., Fresneau A., Brunetto R., et al. (2017) Photo and thermochemical evolution of astrophysical ice analogues as a source for soluble and insoluble organic materials in solar system minor bodies. *Mon. Not. R. Astron. Soc., 464*, 114–120.
Djouadi Z., Robert F., d'Hendecourt L., et al. (2011) Hydroxyl radical production and storage in analogues of amorphous interstellar silicates: A possible "wet" accretion phase for inner telluric planets. *Astron. Astrophys., 531*, A96.
Dobrovolsky O. V. and Kajmakov E. (1977) Surface phenomena in simulated cometary nuclei. In *Comets, Asteroids, Meteoroids* (A. H. Delsemme, ed.), pp. 37–46. IAU Colloquium 39, Cambridge Univ., Cambridge.
Do-Duy T., Wright C., Fujiyoshi T., et al. (2020) Crystalline silicate absorption at 11.1 μm: Ubiquitous and abundant in embedded YSOs and the interstellar medium. *Mon. Not. R. Astron. Soc., 493*, 4463–4517.
Dominik C. and Tielens A. G. G. M. (1997) The physics of dust coagulation and the structure of dust aggregates in space. *Astrophys. J., 480*, 647–673.
Durda D. D., Movshovitz N., Richardson D. C., et al. (2011) Experimental determination of the coefficient of restitution for meter-scale granite spheres. *Icarus, 211*, 849–855.
Duvernay F., Danger G., Theulé P., et al. (2014) Formaldehyde chemistry in cometary ices: On the prospective detection of NH_2CH_2OH, $HOCH_2OH$, and POM by the on-board ROSINA instrument of the Rosetta mission. *Astrophys. J., 791*, 75.
Dworkin J., Deamer D., Sandford S., et al. (2001) Self-assembling amphiphilic molecules: Synthesis in simulated interstellar/precometary ices. *Proc. Natl. Acad. Sci. U.S.A., 98*, 815–819.
Dworkin J., Seb Gillette J., Bernstein M., et al. (2004) An evolutionary connection between interstellar ices and IDPs? Clues from mass spectroscopy measurements of laboratory simulations. *Adv. Space Res., 33*, 67–71.
Ehrenfreund P. and Cami J. (2010) Cosmic carbon chemistry: From the interstellar medium to the early Earth. *Cold Spring Harb. Perspect. Biol., 2*, a002097.
Escobar-Cerezo J., Palmer C., Muñoz O., et al. (2017) Scattering properties of large irregular cosmic dust particles at visible wavelengths. *Astrophys. J., 838*, 74.
Faure M., Quirico E., Faure A., et al. (2015) Kinetics of hydrogen/deuterium exchanges in cometary ices. *Icarus, 261*, 14–30.
Faure M., Quirico E., Faure A., et al. (2021) A radiolytic origin of organic matter in primitive chondrites and trans-neptunian objects? New clues from ion irradiation experiments. *Icarus, 364*, 114462.
Filacchione G., Groussin O., Herny C., et al. (2019) Comet 67P/CG nucleus composition and comparison to other comets. *Space Sci. Rev., 215*, 19.
Fray N. and Schmitt B. (2009) Sublimation of ices of astrophysical interest: A bibliographic review. *Planet. Space Sci., 57*, 2053–2080.
Fray N., Bénilan Y., Cottin H., et al. (2004a) New experimental results on the degradation of polyoxymethylene: Application to the origin of the formaldehyde extended source in comets. *J. Geophys. Res.–Planets, 109*, E07S12.
Fray N., Bénilan Y., Cottin H., et al. (2004b) Experimental study of the degradation of polymers: Application to the origin of extended sources in cometary atmospheres. *Meteorit. Planet. Sci., 39*, 581–587.
Fray N., Benilan Y., Biver N., et al. (2006) Heliocentric evolution of the degradation of polyoxymethylene: Application to the origin of the formaldehyde (H_2CO) extended source in Comet C/1995 O1 (Hale-Bopp). *Icarus, 184*, 239–254.
Fray N., Bardyn A., Cottin H., et al. (2016) High-molecular-weight organic matter in the particles of Comet 67P/Churyumov-Gerasimenko. *Nature, 538*, 72–74.
Fresneau A., Danger G., Rimola A., et al. (2014) Trapping in water — An important prerequisite for complex reactivity in astrophysical ices: The case of acetone $(CH_3)_2C = O$ and ammonia NH_3. *Mon. Not. R. Astron. Soc., 443*, 2991–3000.
Fresneau A., Mrad N., d'Hendecourt L., et al. (2017) Cometary materials originating from interstellar ices: Clues from laboratory experiments. *Astrophys. J., 837*, 168.
Fulvio D., Potapov A., He J., et al. (2021) Astrochemical pathways to complex organic and prebiotic molecules: Experimental perspectives for *in situ* solid-state studies. *Life, 11*, 568.

Galli A., Vorburger A., Wurz P., et al. (2018) 0.2 to 10 keV electrons interacting with water ice: Radiolysis, sputtering, and sublimation. *Planet. Space Sci., 155,* 91–98.
Gärtner S., Gundlach B., Headen T. F., et al. (2017) Micrometer-sized water ice particles for planetary science experiments: Influence of surface structure on collisional properties. *Astrophys. J., 848,* 96.
Gautier T., Danger G., Mousis O., et al. (2020) Laboratory experiments to unveil the molecular reactivity occurring during the processing of ices in the protosolar nebula. *Earth Planet. Sci. Lett., 531,* 116011.
Gavilan L., Le K., Pino T., et al. (2017) Polyaromatic disordered carbon grains as carriers of the UV bump: Far-UV to mid-IR spectroscopy of laboratory analogs. *Astron. Astrophys., 607,* A73.
Gerakines P. A., Moore M. H., and Hudson R. L. (2001) Energetic processing of laboratory ice analogs: UV photolysis versus ion bombardment. *J. Geophys. Res.–Planets, 106,* 33381–33385.
Gerakines P. A., Moore M. H., and Hudson R. L. (2004) Ultraviolet photolysis and proton irradiation of astrophysical ice analogs containing hydrogen cyanide. *Icarus, 170,* 202–213.
Ghesquière P., Ivlev A., Noble J., et al. (2018) Reactivity in interstellar ice analogs: Role of the structural evolution. *Astron. Astrophys., 614,* A107.
Ghormley J. A. (1968) Enthalpy changes and heat-capacity changes in the transformations from high-surface-area amorphous ice to stable hexagonal ice. *J. Chem. Phys., 48,* 503–508.
Gounelle M., Morbidelli A., Bland P. A., et al. (2008) Meteorites from the outer solar system? In *The Solar System Beyond Neptune* (M. A. Barucci et al., eds.), pp. 525–541. Univ. of Arizona, Tucson.
Greenberg A., Laufer D., and Bar-Nun A. (2017) The effect of CO_2 on gases trapping in cometary ices. *Mon. Not. R. Astron. Soc., 469,* S517–S521.
Greenberg J., Li A., Mendoza-Gómez C., et al. (1995) Approaching the interstellar grain organic refractory component. *Astrophys. J. Lett., 455,* L177.
Gronoff G., Maggiolo R., Cessateur G., et al. (2020) The effect of cosmic rays on cometary nuclei. I. Dose deposition. *Astrophys. J., 890,* 89.
Grün E., Kochan H., and Seidensticker K. J. (1991) Laboratory simulation, a tool for comet research. *Geophys. Res. Lett., 18,* 245–248.
Grün E., Gebhard J., Bar-Nun A., et al. (1993) Development of a dust mantle on the surface of an insolated ice-dust mixture: Results from the KOSI-9 experiment. *J. Geophys. Res.–Planets, 98,* 15091–15104.
Grundy W. M. and Schmitt B. (1998) The temperature-dependent near-infrared absorption spectrum of hexagonal H_2O ice. *J. Geophys. Res.–Planets, 103,* 25809–25822.
Gudipati M., Abou Mrad N., Blum J., et al. (2015) Laboratory studies towards understanding comets. *Space Sci. Rev., 197,* 101–150.
Gundlach B. and Blum J. (2012) Outgassing of icy bodies in the solar system — II: Heat transport in dry, porous surface dust layers. *Icarus, 219,* 618–629.
Gundlach B. and Blum J. (2015) The stickiness of micrometer-sized water-ice particles. *Astrophys. J., 798,* 34.
Gundlach B. and Blum J. (2016) Why are Jupiter-family comets active and asteroids in cometary-like orbits inactive? How hydrostatic compression leads to inactivity. *Astron. Astrophys., 589,* A111.
Gundlach B., Kilias S., Beitz E., et al. (2011a) Micrometer-sized ice particles for planetary-science experiments — I. Preparation, critical rolling friction force, and specific surface energy. *Icarus, 214,* 717–723.
Gundlach B., Skorov Y. V., and Blum J. (2011b) Outgassing of icy bodies in the solar system — I. The sublimation of hexagonal water ice through dust layers. *Icarus, 213,* 710–719.
Gundlach B., Schmidt K. P., Kreuzig C., et al. (2018) The tensile strength of ice and dust aggregates and its dependence on particle properties. *Mon. Not. R. Astron. Soc., 479,* 1273–1277.
Güttler C., Krause M., Geretshauser R. J., et al. (2009) The physics of protoplanetesimal dust agglomerates. IV. Toward a dynamical collision model. *Astrophys. J., 701,* 130–141.
Güttler C., Blum J., Zsom A., et al. (2010) The outcome of protoplanetary dust growth: Pebbles, boulders, or planetesimals?* — I. Mapping the zoo of laboratory collision experiments. *Astron. Astrophys., 513,* A56.
Güttler C., Heißelmann D., Blum J., et al. (2013) Normal collisions of spheres: A literature survey on available experiments. *ArXiV e-prints,* arXiv:1204.0001.
Haack D., Kreuzig C., Gundlach, B., et al. (2021a) Sublimation of organic-rich comet analog materials and their relevance in fracture formation. *Astron. Astrophys., 653,* A153.
Haack D., Lethuillier A., Kreuzig C., et al. (2021b) Sublimation of ice-dust mixtures in cooled vacuum environments to reproduce cometary morphologies. *Astron. Astrophys., 649,* A35.
Hadamcik E., Renard J.-B., Levasseur-Regourd A.-C., et al. (2006) Light scattering by fluffy particles with the PROGRA2 experiment: Mixtures of materials. *J. Quant. Spectrosc. Radiat. Transfer, 100,* 143–156.
Hadamcik E., Renard J.-B., Rietmeijer F., et al. (2007) Light scattering by fluffy Mg-Fe-SiO and C mixtures as cometary analogs (PROGRA2 experiment). *Icarus, 190,* 660–671.
Hadamcik E., Renard J.-B., Levasseur-Regourd A.-C., et al. (2009) Light scattering by agglomerates: Interconnecting size and absorption effects (PROGRA2 experiment). *J. Quant. Spectrosc. Radiat. Transfer, 110,* 1755–1770.
Hadamcik E., Renard J.-B., Levasseur-Regourd A.-C., et al. (2011) Laboratory measurements of light scattered by clouds and layers of solid particles using an imaging technique. In *Polarimetric Detection, Characterization and Remote Sensing* (M. I. Mishchenko et al., eds.), pp. 137–176. Springer, Dordrecht.
Hadamcik E., Sen A., Levasseur-Regourd A.-C., et al. (2013) Dust in Comet 103P/Hartley 2 coma during EPOXI mission. *Icarus, 222,* 774–785.
Hadraoui K., Cottin H., Ivanovski S., et al. (2019) Distributed glycine in Comet 67P/Churyumov-Gerasimenko. *Astron. Astrophys., 630,* A32.
Hänni N., Gasc S., Etter A., et al. (2019) Ammonium salts as a source of small molecules observed with high-resolution electron-impact ionization mass spectrometry. *J. Phys. Chem. A, 123,* 5805–5814.
Hänni N., Altwegg K., Pestoni B., et al. (2020) First in situ detection of the CN radical in comets and evidence for a distributed source. *Mon. Not. R. Astron. Soc., 498,* 2239–2248.
Hansen G. B. (1997) The infrared absorption spectrum of carbon dioxide ice from 1.8 to 333 μm. *J. Geophys. Res.–Planets, 102,* 21569–21588.
Hapke B. (2001) Space weathering from Mercury to the asteroid belt. *J. Geophys. Res.–Planets, 106,* 10039–10073.
Hapke B. (2012) *Theory of Reflectance and Emittance Spectroscopy, 2nd edition.* Cambridge Univ., Cambridge. 513 pp.
Heggy E., Palmer E. M., Kofman W., et al. (2012) Radar properties of comets: Parametric dielectric modeling of Comet 67P/Churyumov-Gerasimenko. *Icarus, 221,* 925–939.
Heim L.-O., Blum J., Preuss M., et al. (1999) Adhesion and friction forces between spherical micrometer-sized particles. *Phys. Rev. Lett., 83,* 3328–3331.
Henderson B. and Gudipati M. (2015) Direct detection of complex organic products in ultraviolet (Ly-alpha) and electron-irradiated astrophysical and cometary ice analogs using two-step laser ablation and ionization mass spectrometry. *Astrophys. J., 800,* 66.
Henning T., Jäger C., and Mutschke H. (2004) Laboratory studies of carbonaceous dust analogs. In *Astrophysics of Dust* (A. N. Witt et al., eds.), pp. 603–628. ASP Conf. Ser. 309, Astronomical Society of the Pacific, San Francisco.
Herbst E. (2014) Three milieux for interstellar chemistry: gas, dust, and ice. *Phys. Chem. Chem. Phys, 16,* 3344–3359.
Herbst E. and Van Dishoeck E. (2009) Complex organic interstellar molecules. *Annu. Rev. Astron. Astrophys., 47,* 427–480.
Herique A., Kofman W., Beck P., et al. (2016) Cosmochemical implications of CONSERT permittivity characterization of 67P/CG. *Mon. Not. R. Astron. Soc., 462,* S516–S532.
Hines D. and Levasseur-Regourd A.-C. (2016) Polarimetry observations of comets: Status, questions, future pathways. *Planet. Space Sci., 123,* 41–50.
Hovenier J. and Muñoz O. (2009) Light scattering in the solar system: An introductory review. *J. Quant. Spectrosc. Radiat. Transfer, 110,* 1280–1292.
Hsiung P. and Roessler K. (1989) CO_2 depth profiles on cometary model substances of KOSI. In *Proceedings of an International Workshop on Physics and Mechanics of Cometary Materials* (J. Hunt and T. D. Guyenne, eds.), pp. 191–196. ESA SP-302, Noordwijk, The Netherlands.
Hudson R. L. and Donn B. (1991) An experimental study of the sublimation of water ice and the release of trapped gases. *Icarus, 94,* 326–332.
Hudson R. L., Palumbo M. E., Strazzulla G., et al. (2008) Laboratory studies of the chemistry of transneptunian object surface materials.

In *The Solar System Beyond Neptune* (A. M. Barucci et al., eds.), pp. 507–523. Univ. of Arizona, Tucson.
Huebner W. (1987) First polymer in space identified in Comet Halley. *Science, 237*, 628–630.
Huebner W., Benkhoff J., Capria M.-T., et al., eds. (2006) *Heat and Gas Diffusion in Comet Nuclei.* International Space Science Institute, Bern. 258 pp.
Hunziker S., Schmid H. M., Ma J., et al. (2021) HD 142527: Quantitative disk polarimetry with SPHERE. *Astron. Astrophys., 648*, A110.
Isnard R., Bardyn A., Fray N., et al. (2019) H/C elemental ratio of the refractory organic matter in cometary particles of 67P/Churyumov-Gerasimenko. *Astron. Astrophys., 630*, A27.
Isokoski K., Bossa J.-B., Triemstra T., et al. (2014) Porosity and thermal collapse measurements of H_2O, CH_3OH, CO_2, and H_2O:CO_2 ices. *Phys. Chem. Chem. Phys., 16*, 3456–3465.
Isono Y., Tachibana S., Naraoka H., et al. (2019) Bulk chemical characteristics of soluble polar organic molecules formed through condensation of formaldehyde: Comparison with soluble organic molecules in Murchison meteorite. *Geochem. J., 53*, 41–51.
Jäger C., Krasnokutski S., Staicu A., et al. (2006) Identification and spectral properties of polycyclic aromatic hydrocarbons in carbonaceous soot produced by laser pyrolysis. *Astrophys. J. Suppl. Ser., 166*, 557–566.
Jäger C., Mutschke H., Henning T., et al. (2009) Analogs of cosmic dust. In *Cosmic Dust — Near and Far* (T. Henning et al., eds.), pp. 319–337. ASP Conf. Ser. 414, Astronomical Society of the Pacific, San Francisco.
Jäger C., Potapov A., Rouillé G., et al. (2019) Laboratory experiments on cosmic dust and ices. *Proc. Intl. Astron. Union, 15*, 27–34.
Jin Z. and Bose M. (2021) Hydration of nebular minerals through the implantation-diffusion process. *Astrophys. J., 913*, 116.
Johnson K. L., Kendall K., and Roberts A. D. (1971) Surface energy and the contact of elastic solids. *Proc. R. Soc. A, 324*, 301–313.
Jost B., Gundlach B., Pommerol A., et al. (2013) Micrometer-sized ice particles for planetary-science experiments — II. Bidirectional reflectance. *Icarus, 225*, 352–366.
Jost B., Pommerol A., Poch O., et al. (2017a) Bidirectional reflectance of laboratory cometary analogues to interpret the spectrophotometric properties of the nucleus of Comet 67P/Churyumov-Gerasimenko. *Planet. Space Sci., 148*, 1–11.
Jost B., Pommerol A., Poch O., et al. (2017b) Bidirectional reflectance and VIS-NIR spectroscopy of cometary analogues under simulated space conditions. *Planet. Space Sci., 145*, 14–27.
Jutzi M. and Michel P. (2020) Collisional heating and compaction of small bodies: Constraints for their origin and evolution. *Icarus, 350*, 113867.
Kaufmann E. and Hagermann A. (2018) Constraining the parameter space of comet simulation experiments. *Icarus, 311*, 105–112.
Kebukawa Y., David Kilcoyne A., and Cody G. (2013) Exploring the potential formation of organic solids in chondrites and comets through polymerization of interstellar formaldehyde. *Astrophys. J., 771*, 19.
Kebukawa Y., Chan Q., Tachibana S., et al. (2017) One-pot synthesis of amino acid precursors with insoluble organic matter in planetesimals with aqueous activity. *Sci. Adv., 3*, 1602093.
Kimura H., Wada K., Senshu H., et al. (2015) Cohesion of amorphous silica spheres: Toward a better understanding of the coagulation growth of silicate dust aggregates. *Astrophys. J., 812*, 67.
Kiselev N., Rosenbush V., Levasseur-Regourd A.-C., and Kolokolova L. (2015) Comets. In *Polarimetry of Stars and Planetary Systems* (L. Kolokolova et al., eds.), pp. 379–404. Cambridge Univ., Cambridge.
Kochan H., Feuerbacher B., Joo F., et al. (1989) Comet simulation experiments in the DFVLR space simulators. *Adv. Space Res., 9*, 113–122.
Kochan H. W., Huebner W. F., and Sears D. W. G. (1999) Simulation experiments with cometary analogous material. In *Laboratory Astrophysics and Space Research* (P. Ehrenfreund et al., eds.), pp. 623–665. Astrophys. Space Sci. Library, Vol. 236, Springer, Dordrecht.
Kofman W., Herique A., Barbin Y., et al. (2015) Properties of the 67P/Churyumov-Gerasimenko interior revealed by CONSERT radar. *Science, 349*, aab0639.
Kokaly R. F., Clark R. N., Swayze G. A., et al. (2017) *USGS Spectral Library Version 7: U. S. Geological Survey Data Series 1035.* U. S. Geological Survey, Reston. 61 pp.
Kolokolova L., Hanner M. S., Levasseur-Regourd A.-C., et al. (2004) Physical properties of cometary dust from light scattering and thermal emission. In *Comets II* (M. C. Festou et al., eds.), pp. 577–604. Univ. of Arizona, Tucson.
Kolokolova L., Kimura H., Kiselev N., et al. (2007) Two different evolutionary types of comets proved by polarimetric and infrared properties of their dust. *Astron. Astrophys., 463*, 1189–1196.
Kömle N., Kargl G., Thiel K., et al. (1996) Thermal properties of cometary ices and sublimation residua including organics. *Planet. Space Sci., 44*, 675–689.
Kossacki K. J. (2019) Sublimation of cometary ices in the presence of organic volatiles II. *Icarus, 319*, 470–475.
Kossacki K. J. (2021) Sublimation of porous granular ice in vacuum. *Icarus, 368*, 114613.
Kossacki K. J. and Leliwa-Kopystynski J. (2014) Temperature dependence of the sublimation rate of water ice: Influence of impurities. *Icarus, 233*, 101–105.
Kossacki K. J. and Czechowski L. (2018) Comet 67P/Churyumov-Gerasimenko, possible origin of the depression Hatmehit. *Icarus, 305*, 1–14.
Kossacki K. J. and Jasiak A. (2019) The evolution of gently sloping mantled deposits on Comet 67P/Churyumov-Gerasimenko. *Icarus, 319*, 381–391.
Kossacki K. J., Koemle N. I., Kargl G., et al. (1994) The influence of grain sintering on the thermoconductivity of porous ice. *Planet. Space Sci., 42*, 383–389.
Kossacki K. J., Leliwa-Kopystynski J., Witek P., et al. (2017) Sublimation of cometary ices in the presence of organic volatiles. *Icarus, 294*, 227–233.
Kossacki K. J., Misiura K., and Czechowski L. (2019) Sublimation of buried cometary ice. *Icarus, 329*, 72–78.
Kossacki K. J., Czechowski L., and Skóra G. (2020a) Influence of landslides on the erosion of slopes on Comet 9P/Tempel 1: Laboratory experiments and numerical simulations. *Icarus, 340*, 113529.
Kossacki K. J., Skóra G., and Czechowski L. (2020b) Comets, sliding of surface dust. *Icarus, 348*, 113781.
Kossacki K. J., Wesołowski M., Skóra G., et al. (2022) Comets, sliding of surface dust II. *Icarus, 379*, 114946.
Kothe S., Blum J., Weidling R., et al. (2013) Free collisions in a microgravity many-particle experiment. III. The collision behavior of sub-millimeter-sized dust aggregates. *Icarus, 225*, 75–85.
Kouchi A. (1990) Evaporation of H_2O-CO ice and its astrophysical implications. *J. Cryst. Growth, 99*, 1220–1226.
Kouchi A. and Yamamoto T. (1995) Cosmoglaciology — Evolution of ice in interstellar space and the early solar-system. *Prog. Cryst. Growth Charact. Mater., 30*, 83–108.
Kouchi A. and Sirono S. (2001) Crystallization heat of impure amorphous H_2O ice. *Geophys. Res. Lett., 28*, 827–830.
Kouchi A., Tsuge M., Hama T., et al. (2021) Transmission electron microscopy study of the morphology of ices composed of H_2O, CO_2, and CO on refractory grains. *Astrophys. J., 918*, 45.
Krause M., Blum J., Skorov Y., et al. (2011) Thermal conductivity measurements of porous dust aggregates: I. Technique, model and first results. *Icarus, 214*, 286–296.
Kreuzig C., Kargl G., Pommerol A., et al. (2021) The CoPhyLab comet-simulation chamber. *Rev. Sci. Instrum., 92*, 115102.
Kuga M., Marty B., Marrocchi Y., et al. (2015) Synthesis of refractory organic matter in the ionized gas phase of the solar nebula. *Proc. Natl. Acad. Sci. U.S.A., 112*, 7129–7134.
Lamberts T., Ioppolo S., Cuppen H., et al. (2015) Thermal H/D exchange in polar ice — Deuteron scrambling in space. *Mon. Not. R. Astron. Soc., 448*, 3820–3828.
Lantz C., Brunetto R., Barucci M., et al. (2017) Ion irradiation of carbonaceous chondrites: A new view of space weathering on primitive asteroids. *Icarus, 285*, 43–57.
Lauck T., Karssemeijer L., Shulenberger K., et al. (2015) CO diffusion into amorphous H_2O ices. *Astrophys. J., 801*, 118.
Lethuillier A., Le Gall A., Hamelin M., et al. (2018) Electrical properties of tholins and derived constraints on the Huygens landing site composition at the surface of Titan. *J. Geophys. Res.–Planets, 123*, 807–822.
Levasseur-Regourd A.-C. (2003) Cosmic dust physical properties and the ICAPS facility on board the ISS. *Adv. Space Res., 31*, 2599–2606.
Levasseur-Regourd A.-C., Cabane M., Haudebourg V., et al. (1998) Light scattering by dust under microgravity conditions. *Earth Moon*

Planets, 80, 343–368.
Levasseur-Regourd A.-C., Renard J.-B., Shkuratov Y., et al. (2015) Laboratory studies. In *Polarimetry of Stars and Planetary Systems* (L. Kolokolova et al., eds.), pp. 62–80. Cambridge Univ., Cambridge.
Levasseur-Regourd A.-C., Agarwal J., Cottin H., et al. (2018) Cometary dust. *Space Sci. Rev., 214,* 64.
Levasseur-Regourd A.-C., Renard J.-B., Hadamcik E., et al. (2019) Interpretation through experimental simulations of phase functions revealed by Rosetta in 67P/Churyumov-Gerasimenko dust coma. *Astron. Astrophys., 630,* A20.
Levasseur-Regourd A.-C., Baruteau C., Lasue J., et al. (2020) Linking studies of tiny meteoroids, zodiacal dust, cometary dust and circumstellar disks. *Planet. Space Sci., 186,* 104896.
Linnartz H., Ioppolo S., and Fedoseev G. (2015) Atom addition reactions in interstellar ice analogues. *Intl. Rev. Phys. Chem., 34,* 205–237.
Lis D., Bockelée-Morvan D., Güsten R., et al. (2019) Terrestrial deuterium-to-hydrogen ratio in water in hyperactive comets. *Astron. Astrophys., 625,* L5.
Lisse C. (2007) Comparison of the composition of the Tempel 1 ejecta to the dust in Comet C/Hale-Bopp 1995 O1 and YSO HD 100546. *Icarus, 187,* 69–86.
Lécuyer C., Royer A., Fourel F., et al. (2017) D/H fractionation during the sublimation of water ice. *Icarus, 285,* 1–7.
Lunine J. I. and Stevenson D. J. (1985) Thermodynamics of clathrate hydrate at low and high pressures with application to the outer solar system. *Astrophys. J. Suppl. Ser., 58,* 493–531.
Maggiolo R., Gronoff G., Cessateur G., et al. (2020) The effect of cosmic rays on cometary nuclei — II. Impact on ice composition and structure. *Astrophys. J., 901,* 136.
Marboeuf U., Schmitt B., Petit J.-M., et al. (2012) A cometary nucleus model taking into account all phase changes of water ice: Amorphous, crystalline, and clathrate. *Astron. Astrophys., 542,* A82.
Marcellus P., Meinert C., Nuevo M., et al. (2011) Non-racemic amino acid production by ultraviolet irradiation of achiral interstellar ice analogs with circularly polarized light. *Astrophys. J. Lett., 727,* L27.
Martín-Doménech R., Caro G., Bueno J., et al. (2014) Thermal desorption of circumstellar and cometary ice analogs. *Astron. Astrophys., 564,* A8.
May R., Smith R., and Kay B. (2012) The molecular volcano revisited: Determination of crack propagation and distribution during the crystallization of nanoscale amorphous solid water films. *J. Phys. Chem. Lett., 3,* 327–331.
McClure M. K., Rocha W. R. M., Pontoppidan K. M., et al. (2023) An Ice Age JWST inventory of dense molecular cloud ices. *Nature Astron., 7,* 431–443.
Meinert C., Myrgorodska I., Marcellus P., et al. (2016) Ribose and related sugars from ultraviolet irradiation of interstellar ice analogs. *Science, 352,* 208–212.
Mendis D. and Horányi M. (2013) Dusty plasma effects in comets: Expectations for Rosetta. *Rev. Geophys., 51,* 53–75.
Mennella V., Baratta G., Esposito A., et al. (2003) The effects of ion irradiation on the evolution of the carrier of the 3.4 micron interstellar absorption band. *Astrophys. J., 587,* 727–738.
Mennella V., Ciarniello M., Raponi A., et al. (2020) Hydroxylated Mg-rich amorphous silicates: A new component of the 3.2 μm absorption band of Comet 67P/Churyumov-Gerasimenko. *Astrophys. J. Lett., 897,* L37.
Milliken R. E., Hiroi T., and Patterson W. (2016) The NASA Reflectance Experiment LABoratory (RELAB) facility: Past, present, and future. In *47th Annual Lunar and Planetary Science Conference,* Abstract #2058. LPI Contribution No. 1903, Lunar and Planetary Institute, Houston.
Minissale M., Nguyen T., and Dulieu F. (2019) Experimental study of the penetration of oxygen and deuterium atoms into porous water ice. *Astron. Astrophys., 622,* A148.
Mispelaer F., Theulé P., Aouididi H., et al. (2013) Diffusion measurements of CO, HNCO, H_2CO, and NH_3 in amorphous water ice. *Astron. Astrophys., 555,* A13.
Modica P., de Marcellus P., Baklouti D., et al. (2012) Organic residues from ultraviolet irradiation of interstellar ice analogs. In *Setting a New Standard in the Analysis of Binary Stars* (C. Stehlé et al., eds.), pp. 343–347. EAS Publ. Series, Vol. 58, Cambridge Univ., Cambridge.
Modica P., Meinert C., Marcellus P., et al. (2014) Enantiomeric excesses induced in amino acids by ultraviolet circularly polarized light irradiation of extraterrestrial ice analogs: A possible source of asymmetry for prebiotic chemistry. *Astrophys. J., 788,* 79.
Moores J., Brown R., Lauretta D., et al. (2012) Experimental and theoretical simulation of sublimating dusty water ice with implications for D/H ratios of water ice on comets and Mars. *Planet. Sci., 1,* 2.
Moroz L. V., Arnold G., Korochantsev A. V., et al. (1998) Natural solid bitumens as possible analogs for cometary and asteroid organics: 1. Reflectance spectroscopy of pure bitumens. *Icarus, 134,* 253–268.
Moroz L., Baratta G., Strazzulla G., et al. (2004) Optical alteration of complex organics induced by ion irradiation: 1. Laboratory experiments suggest unusual space weathering trend. *Icarus, 170,* 214–228.
Mortimer J., Lécuyer C., Fourel F., et al. (2018) D/H fractionation during sublimation of water ice at low temperatures into a vacuum. *Planet. Space Sci., 158,* 25–33.
Mumma M. J., Weissman P. R., and Stern S. A. (1993) Comets and the origin of the solar system — Reading the Rosetta Stone. In *Protostars and Planets III* (E. Levy and J. I. Lunine, eds.), pp. 1177–1252. Univ. of Arizona, Tucson.
Muñoz Caro G. and Schutte W. (2003) UV-photoprocessing of interstellar ice analogs: New infrared spectroscopic results. *Astron. Astrophys., 412,* 121–132.
Muñoz Caro G., Meierhenrich U., Schutte W., et al. (2004) UV-photoprocessing of interstellar ice analogs: Detection of hexamethylenetetramine-based species. *Astron. Astrophys., 413,* 209–216.
Muñoz Caro G., Matrajt G., Dartois E., et al. (2006) Nature and evolution of the dominant carbonaceous matter in interplanetary dust particles: Effects of irradiation and identification with a type of amorphous carbon. *Astron. Astrophys., 459,* 147–159.
Muñoz O., Moreno F., Guirado D., et al. (2010) Experimental determination of scattering matrices of dust particles at visible wavelengths: The IAA light scattering apparatus. *J. Quant. Spectrosc. Radiat. Transfer, 111,* 187–196.
Muñoz O., Moreno F., Guirado D., et al. (2011) The IAA cosmic dust laboratory: Experimental scattering matrices of clay particles. *Icarus, 211,* 894–900.
Muñoz O., Moreno F., Guirado D., et al. (2012) The Amsterdam-Granada Light Scattering Database. *J. Quant. Spectrosc. Radiat. Transfer, 113,* 565–574.
Muñoz O., Moreno F., Vargas-Martín F., et al. (2017) Experimental phase functions of millimeter-sized cosmic dust grains. *Astrophys. J., 846,* 85.
Musiolik G., Teiser J., Jankowski T., et al. (2016) Ice grain collisions in comparison: CO_2, H_2O, and their mixtures. *Astrophys. J., 827,* 63.
Nagahara H., Ogawa R., Ozawa K., et al. (2009) Laboratory condensation and reaction of silicate dust. In *Cosmic Dust — Near and Far* (T. Henning et al., eds.), pp. 403–410. ASP Conf. Ser. 414, Astronomical Society of the Pacific, San Francisco.
Nakamura-Messenger K., Clemett S., Messenger S., et al. (2011) Experimental aqueous alteration of cometary dust. *Meteorit. Planet. Sci., 46,* 843–856.
Nelson R., Nuth J. A., and Donn B. (1987) A kinetic study of the hydrous alteration of amorphous MgSiO smokes: Implications for cometary particles and chondrite matrix. *J. Geophys. Res.–Solid Earth, 92,* 657–662.
Nilsson H., Wieser G., Behar E., et al. (2015) Evolution of the ion environment of Comet 67P/Churyumov-Gerasimenko — Observations between 3.6 and 2.0 AU. *Astron. Astrophys., 583,* A20.
Noble J. A., Michoulier E., Aupetit C., et al. (2020) Influence of ice structure on the soft UV photochemistry of PAHs embedded in solid water. *Astron. Astrophys., 644,* A22.
Nuevo M., Auger G., Blanot D., et al. (2008) A detailed study of the amino acids produced from the vacuum UV irradiation of interstellar ice analogs. *Origins Life Evol. Biospheres, 38,* 37–56.
Nuevo M., Bredehöft J., Meierhenrich U., et al. (2010) Urea, glycolic acid, and glycerol in an organic residue produced by ultraviolet irradiation of interstellar/pre-cometary ice analogs. *Astrobiology, 10,* 245–256.
Nuevo M., Cooper G., and Sandford S. (2018) Deoxyribose and deoxysugar derivatives from photoprocessed astrophysical ice analogues and comparison to meteorites. *Nature Commun., 9,* 5276.
Nuth J. A. III, Hallenbeck S., and Rietmeijer F. J. M. (2000) Laboratory studies of silicate smokes: Analog studies of circumstellar materials. *J. Geophys. Res.–Space Phys., 105,* 10387–10396.

Nuth J. A. III, Rietmeijer F., and Hill H. (2002) Condensation processes in astrophysical environments: The composition and structure of cometary grains. *Meteorit. Planet. Sci., 37,* 1579–1590.
Nuth J. A. III, Johnson N., and Manning S. (2008) A self-perpetuating catalyst for the production of complex organic molecules in protostellar nebulae. *Astrophys. J. Lett., 673,* L225–L228.
Nuth J. A. III, Johnson N. M., Ferguson F. T., et al. (2016) Gas/solid carbon branching ratios in surface-mediated reactions and the incorporation of carbonaceous material into planetesimals. *Meteorit. Planet. Sci., 51,* 1310–1322.
Nuth J. A. III, Ferguson F. T., Hill H. G. M., et al. (2020) Did a complex carbon cycle operate in the inner solar system? *Life, 10,* 206.
Oba Y., Takano Y., Naraoka H., et al. (2017) Deuterium fractionation upon the formation of hexamethylenetetramines through photochemical reactions of interstellar ice analogs containing deuterated methanol isotopologues. *Astrophys. J., 849,* 122.
Oba Y., Takano Y., Naraoka H., et al. (2019) Nucleobase synthesis in interstellar ices. *Nature Commun., 10,* 4413.
Oba Y., Takano Y., Naraoka H., et al. (2020) Extraterrestrial hexamethylenetetramine in meteorites — A precursor of prebiotic chemistry in the inner solar system. *Nature Commun., 11,* 6243.
Öberg K. I. (2016) Photochemistry and astrochemistry: Photochemical pathways to interstellar complex organic molecules. *Chem. Rev., 116,* 9631–9663.
Öberg K. I. and Bergin E. A. (2021) Astrochemistry and compositions of planetary systems. *Phys. Rept., 893,* 1–48.
Öberg K. I., Garrod R. T., Dishoeck E. F., et al. (2009) Formation rates of complex organics in UV irradiated CH_3OH-rich ices: I. Experiments. *Astron. Astrophys., 504,* 891–913.
Öberg K. I., Boogert A. C. A., Pontoppidan K. M., et al. (2011) The Spitzer Ice Legacy: Ice evolution from cores to protostars. *Astrophys. J., 740,* 109.
O'Brien D., Izidoro A., Jacobson S., et al. (2018) The delivery of water during terrestrial planet formation. *Space Sci. Rev., 214,* 47.
Oehler A. and Neukum G. (1991) Visible and near IR albedo measurements of ice/dust mixtures. *Geophys. Res. Lett., 18,* 253–256.
Paraskov G. B., Wurm G., and Krauss O. (2007) Impacts into weak dust targets under microgravity and the formation of planetesimals. *Icarus, 191,* 779–789.
Pendleton Y. and Allamandola L. (2002) The organic refractory material in the diffuse interstellar medium: Mid-infrared spectroscopic constraints. *Astrophys. J. Suppl. Ser., 138,* 75–98.
Piani L., Tachibana S., Hama T., et al. (2017) Evolution of morphological and physical properties of laboratory interstellar organic residues with ultraviolet irradiation. *Astrophys. J., 837,* 35.
Pino T., Chabot M., Béroff K., et al. (2019) Release of large polycyclic aromatic hydrocarbons and fullerenes by cosmic rays from interstellar dust — Swift heavy ion irradiations of interstellar carbonaceous dust analogue. *Astron. Astrophys., 623,* A134.
Poch O., Pommerol A., Jost B., et al. (2016a) Sublimation of ice-tholins mixtures: A morphological and spectro-photometric study. *Icarus, 266,* 288–305.
Poch O., Pommerol A., Jost B., et al. (2016b) Sublimation of water ice mixed with silicates and tholins: Evolution of surface texture and reflectance spectra, with implications for comets. *Icarus, 267,* 154–173.
Poch O., Istiqomah I., Quirico E., et al. (2020) Ammonium salts are a reservoir of nitrogen on a cometary nucleus and possibly on some asteroids. *Science, 367,* aaw7462.
Pommerol A., Jost B., Poch O., et al. (2019) Experimenting with mixtures of water ice and dust as analogues for icy planetary material: Recipes from the Ice Laboratory at the University of Bern. *Space Sci. Rev., 215,* 37.
Poppe T., Blum J., and Henning T. (2000) Analogous experiments on the stickiness of micron-sized preplanetary dust. *Astrophys. J., 533,* 454–471.
Potapov A., Theulé P., Jäger C., et al. (2019) Evidence of surface catalytic effect on cosmic dust grain analogs: The ammonia and carbon dioxide surface reaction. *Astrophys. J. Lett., 878,* L20.
Prialnik D., Benkhoff J., and Podolak M. (2004) Modeling the structure and activity of comet nuclei. In *Comets II* (M. C. Festou et al., eds.), pp. 359–387. Univ. of Arizona, Tucson.
Prialnik D., Sarid G., Rosenberg E. D., et al. (2008) Thermal and chemical evolution of comet nuclei and Kuiper belt objects. *Space Sci. Rev., 138,* 147–164.
Quirico E., Moroz L. V., Schmitt B., et al. (2016) Refractory and semi-volatile organics at the surface of Comet 67P/Churyumov-Gerasimenko: Insights from the VIRTIS/Rosetta imaging spectrometer. *Icarus, 272,* 32–47.
Raponi A., Ciarniello M., Capaccioni F., et al. (2020) Infrared detection of aliphatic organics on a cometary nucleus. *Nature Astron., 4,* 500–505.
Ratke L. and Kochan H. (1989) Fracture mechanical aspects of dust emission processes from a model comet surface. In *Proceedings of an International Workshop on Physics and Mechanics of Cometary Materials* (J. Hunt and T. D. Guyenne, eds.), pp. 121–128. ESA SP-302, Noordwijk, The Netherlands.
Rickman H. (1991) The thermal history and structure of cometary nuclei. In *Comets in the Post-Halley Era* (R. L. Newburn Jr. et al., eds.), pp. 733–760. IAU Colloq. 116, Cambridge Univ., Cambridge.
Rietmeijer F., Nuth J., and Nelson R. (2004) Laboratory hydration of condensed magnesiosilica smokes with implications for hydrated silicates in IDPs and comets. *Meteorit. Planet. Sci., 39,* 723–746.
Roessler K., Benit J., and Sauer M. (1992a) Chemical and physical effects in the bulk of cometary analogs. In *Asteroids, Comets, Meteors 1991* (A. W. Harris and E. Bowell, eds.), pp. 521–524. LPI Contribution No. 1087, Lunar and Planetary Institute, Houston.
Roessler K., Eich G., Klinger E., et al. (1992b) Changes of natural isotopic abundances in the KOSI comet simulation experiments. *Ann. Geophys., 10,* 232–234.
Rothard H., Domaracka A., Boduch P., et al. (2017) Modification of ices by cosmic rays and solar wind. *J. Phys. B: Atom. Mol. Phys., 50,* 062001.
Rousseau B., Érard S., Beck P., et al. (2018) Laboratory simulations of the Vis-NIR spectra of Comet 67P using sub-µm sized cosmochemical analogues. *Icarus, 306,* 306–318.
Rubin M., Engrand C., Snodgrass C., et al. (2020) On the origin and evolution of the material in 67P/Churyumov-Gerasimenko. *Space Sci. Rev., 216,* 102.
Sabri T., Gavilan L., Jäger C., et al. (2013) Interstellar silicate analogs for grain-surface reaction experiments: Gas-phase condensation and characterization of the silicate dust grains. *Astrophys. J., 780,* 180.
Sakatani N., Ogawa K., Iijima Y., et al. (2017) Thermal conductivity model for powdered materials under vacuum based on experimental studies. *AIP Adv., 7,* 015310.
Saunders S., Fanale F., Parker T., et al. (1986) Properties of filamentary sublimation residues from dispersions of clay in ice. *Icarus, 66,* 94–104.
Schmitt B., Espinasse S., Grim R., et al. (1989) Laboratory studies of cometary ice analogues. In *Proceedings of an International Workshop on Physics and Mechanics of Cometary Materials* (J. Hunt and T. D. Guyenne, eds.), pp. 65–69. ESA SP-302, Noordwijk, The Netherlands.
Schmitt B., Quirico E., Trotta F., et al. (1998) Optical properties of ices from UV to infrared. In *Solar System Ices* (B. Schmitt et al., eds.), pp. 199–240. Astrophys. Space Sci. Library, Vol. 227, Kluwer, Dordrecht.
Schmitt B., Bollard P., Garenne A., et al. (2018) SSHADE: The European solid spectroscopy database infrastructure. *European Planet. Sci. Congr. 2018,* 529.
Schräpler R., Blum J., Seizinger A., et al. (2012) The physics of protoplanetesimal dust agglomerates. VII. The low-velocity collision behavior of large dust agglomerates. *Astrophys. J., 758,* 35.
Schutte W. and Khanna R. (2003) Origin of the 6.85 µm band near young stellar objects: The ammonium ion (NH_4^+) revisited. *Astron. Astrophys., 398,* 1049–1062.
Schwartz S., Michel P., Jutzi M., et al. (2018) Catastrophic disruptions as the origin of bilobate comets. *Nature Astron., 2,* 379–382.
Schweighart M., Macher W., Kargl G., et al. (2021) Viscous and Knudsen gas flow through dry porous cometary analogue material. *Mon. Not. R. Astron. Soc., 504,* 5513–5527.
Sears D., Kochan H., and Huebner W. (1999) Invited review: Laboratory simulation of the physical processes occurring on and near the surfaces of comet nuclei. *Meteorit. Planet. Sci., 34,* 497–525.
Seiferlin K., Spohn T., and Benkhoff J. (1995) Cometary ice texture and the thermal evolution of comets. *Adv. Space Res., 15,* 35–38.
Seiferlin K., Kömle N., Kargl G., et al. (1996) Line heat-source measurements of the thermal conductivity of porous H_2O ice, CO_2 ice and mineral powders under space conditions. *Planet. Space Sci., 44,* 691–704.

Shkuratov Y., Starukhina L., Hoffmann H., et al. (1999) A model of spectral albedo of particulate surfaces: Implications for optical properties of the Moon. *Icarus, 137,* 235–246.
Sihvola A. (1999) *IEEE Electromagnetic Waves, Vol. 47: Electromagnetic Mixing Formulas and Applications.* Institution of Engineering and Technology, London. 296 pp.
Smith I. (2011) Laboratory astrochemistry: Gas-phase processes. *Annu. Rev. Astron. Astrophys., 49,* 29–66.
Spadaccia S., Capelo H. L., Pommerol A., et al. (2022) The fate of icy pebbles undergoing sublimation in protoplanetary discs. *Mon. Not. R. Astron. Soc., 509,* 2825–2835.
Stephens J. R. and Gustafson B. A. S. (1991) Laboratory reflectance measurements of analogues to "dirty" ice surfaces on atmosphereless solar system bodies. *Icarus, 94,* 209–217.
Stern A. (2003) The evolution of comets in the Oort cloud and Kuiper belt. *Nature, 424,* 639–642.
Stevenson D. J. and Lunine J. I. (1988) Rapid formation of Jupiter by diffusive redistribution of water vapor in the solar nebula. *Icarus, 75,* 146–155.
Storrs A., Fanale F., Saunders R., et al. (1988) The formation of filamentary sublimate residues (FSR) from mineral grains. *Icarus, 76,* 493–512.
Strazzulla G. (1999) Ion irradiation and the origin of cometary materials. *Space Sci. Rev., 90,* 269–274.
Strazzulla G. (2013) Crystalline and amorphous structure of astrophysical ices. *Low Temp. Phys., 39,* 430–433.
Strazzulla G., Baratta G., Johnson R., et al. (1991) Primordial comet mantle: Irradiation production of a stable organic crust. *Icarus, 91,* 101–104.
Strazzulla G., Cooper J., Christian E., et al. (2003) Ion irradiation of TNOs: From the fluxes measured in space to the laboratory experiments. *Compt. Rend. Phys., 4,* 791–801.
Sultana R., Poch O., Beck P., et al. (2021) Visible and near-infrared reflectance of hyperfine and hyperporous particulate surfaces. *Icarus, 357,* 114141.
Sultana R., Poch O., Beck P., et al. (2023) Reflection, emission, and polarization properties of surfaces made of hyperfine grains, and implications for the nature of primitive small bodies. *Icarus, 395,* 115492.
Suttle M., Folco L., Genge M., et al. (2020) Flying too close to the Sun — The viability of perihelion-induced aqueous alteration on periodic comets. *Icarus, 351,* 113956.
Takigawa A., Furukawa Y., Kimura Y., et al. (2019) Exposure experiments of amorphous silicates and organics to cometary ice and vapor analogs. *Astrophys. J., 881,* 27.
Teiser J. and Wurm G. (2009) High-velocity dust collisions: Forming planetesimals in a fragmentation cascade with final accretion. *Mon. Not. R. Astron. Soc., 393,* 1584–1594.
Theulé P. (2019) Chemical dynamics in interstellar ice. *Proc. Intl. Astron. Union, 15,* 139–143.
Theulé P., Duvernay F., Danger G., et al. (2013) Thermal reactions in interstellar ice: A step towards molecular complexity in the interstellar medium. *Adv. Space Res., 52,* 1567–1579.
Thiel K., et al. (1989) Dynamics of crust formation and dust emission of comet nucleus analogues under isolation. In *Proceedings of an International Workshop on Physics and Mechanics of Cometary Materials* (J. Hunt and T. D. Guyenne, eds.), pp. 221–225. ESA SP-302, Noordwijk, The Netherlands.
Thomas N., Ulamec S., Kührt E., et al. (2019) Towards new comet missions. *Space Sci. Rev., 215,* 47.
Thompson S., Herlihy A., Murray C., et al. (2019) Amorphous Mg-Fe silicates from microwave-dried sol-gels: Multi-scale structure, mid-IR spectroscopy and thermal crystallization. *Astron. Astrophys., 624,* A136.
Toppani A., Robert F., Libourel G., et al. (2005) A 'dry' condensation origin for circumstellar carbonates. *Nature, 437,* 1121–1124.
Urso R., Baklouti D., Djouadi Z., et al. (2020) Near-infrared methanol bands probe energetic processing of icy outer solar system objects. *Astrophys. J. Lett., 894,* L3.
van Broekhuizen F. A., Pontoppidan K. M., Fraser H. J., et al. (2005) A 3–5 μm VLT spectroscopic survey of embedded young low mass stars II: Solid OCN^-. *Astron. Astrophys., 441,* 249–260.
van Dishoeck E. F. (2014) Astrochemistry of dust, ice and gas: Introduction and overview. *Faraday Discussions, 168,* 9–47.
Vinogradoff V., Fray N., Duvernay F., et al. (2013) Importance of thermal reactivity for hexamethylenetetramine formation from simulated interstellar ices. *Astron. Astrophys., 551,* A128.
Vinogradoff V., Bernard S., Le Guillou C., et al. (2018) Evolution of interstellar organic compounds under asteroidal hydrothermal conditions. *Icarus, 305,* 358–370.
Vinogradoff V., Le Guillou C., Bernard S., et al. (2020) Influence of phyllosilicates on the hydrothermal alteration of organic matter in asteroids: Experimental perspectives. *Geochim. Cosmochim. Acta, 269,* 150–166.
Wang X., Schwan J., Hsu H.-W., et al. (2016) Dust charging and transport on airless planetary bodies. *Geophys. Res. Lett., 43,* 6103–6110.
Warren S. G. (1986) Optical constants of carbon dioxide ice. *Appl. Opt., 25,* 2650–2674.
Weidling R., Güttler C., Blum J., et al. (2009) The physics of protoplanetesimal dust agglomerates. III. Compaction in multiple collisions. *Astrophys. J., 696,* 2036–2043.
Weidling R., Güttler C., and Blum J. (2012) Free collisions in a microgravity many-particle experiment. I. Dust aggregate sticking at low velocities. *Icarus, 218,* 688–700.
Weissman P., Morbidelli A., Davidsson B., et al. (2020) Origin and evolution of cometary nuclei. *Space Sci. Rev., 216,* 6.
Windmark F., Birnstiel T., Güttler C., et al. (2012a) Planetesimal formation by sweep-up: How the bouncing barrier can be beneficial to growth. *Astron. Astrophys., 540,* A73.
Windmark F., Birnstiel T., Ormel C. W., et al. (2012b) Breaking through: The effects of a velocity distribution on barriers to dust growth. *Astron. Astrophys., 544,* L16.
Wood S. (2020) A mechanistic model for the thermal conductivity of planetary regolith: 1. The effects of particle shape, composition, cohesion, and compression at depth. *Icarus, 352,* 113964.
Wooden D. (2008) Cometary refractory grains: Interstellar and nebular sources. *Space Sci. Rev., 138,* 75–108.
Wooden D., Ishii H., and Zolensky M. (2017) Cometary dust: The diversity of primitive refractory grains. *Philos. Trans. R. Soc. A, 375,* 20160260.
Worms J., Renard J., Hadamcik E., et al. (1999) Results of the PROGRA^2 experiment: An experimental study in microgravity of scattered polarized light by dust particles with large size parameter. *Icarus, 142,* 281–297.
Wright I., Sheridan S., Barber S., et al. (2015) CHO-bearing organic compounds at the surface of 67P/Churyumov-Gerasimenko revealed by Ptolemy. *Science, 349,* aab0673.
Wurm G. and Blum J. (1998) Experiments on preplanetary dust aggregation. *Icarus, 132,* 125–136.
Wurm G., Paraskov G., and Krauss O. (2005a) Ejection of dust by elastic waves in collisions between millimeter- and centimeter-sized dust aggregates at 16.5 to 37.5 m/s impact velocities. *Phys. Rev. E, 71,* 021304.
Wurm G., Paraskov G., and Krauss O. (2005b) Growth of planetesimals by impacts at 25 m/s. *Icarus, 178,* 253–263.
Wurz P., Rubin M., Altwegg K., et al. (2015) Solar wind sputtering of dust on the surface of 67P/Churyumov-Gerasimenko. *Astron. Astrophys., 583,* A22.
Yang R. and Gudipati M. (2014) Novel two-step laser ablation and ionization mass spectrometry (2S-LAIMS) of actor-spectator ice layers: Probing chemical composition of D_2O ice beneath a H_2O ice layer. *J. Chem. Phys., 140,* 104202.
Yoldi Z., Pommerol A., Jost B., et al. (2015) VIS-NIR reflectance of water ice/regolith analogue mixtures and implications for the detectability of ice mixed within planetary regoliths. *Geophys. Res. Lett., 42,* 6205–6212.
Zsom A., Ormel C. W., Güttler C., et al. (2010) The outcome of protoplanetary dust growth: Pebbles, boulders, or planetesimals? II. Introducing the bouncing barrier. *Astron. Astrophys., 513,* A57.

Part 4:

Nucleus

Guilbert-Lepoutre A., Davidsson B. J. R., Scheeres D. J., and Ciarletti V. (2024) Comet nucleus interiors. In *Comets III* (K. J. Meech, M. R. Combi, D. Bockelée-Morvan, S. N. Raymond, and M. E. Zolensky, eds.), pp. 249–287. Univ. of Arizona, Tucson, DOI: 10.2458/azu_uapress_9780816553631-ch009.

Comet Nucleus Interiors

Aurélie Guilbert-Lepoutre
Centre National de la Recherche Scientifique (CNRS)–Laboratoire de Géologie de Lyon

Björn J. R. Davidsson
Jet Propulsion Laboratory, California Institute of Technology

Daniel J. Scheeres
University of Colorado

Valérie Ciarletti
Université de Versailles Saint-Quentin-en-Yvelines–Université Paris-Saclay–Laboratoire Atmosphères, Milieux, Observations Spatiales/Institut Pierre-Simon-Laplace

---

This chapter addresses the current knowledge of the internal properties of comet nuclei. We discuss observations and methods to access these characteristics, together with processes, thermal or mechanical in nature, known to be at the origin of their evolution through time. We thus discuss how comet nuclei may have been affected by collisions, tidal and centrifugal forces, radiogenic heating, or solar irradiation, assessing in particular how such processing may release, absorb, and transport energy throughout nuclei. We discuss how on long timescales, the complex interplay between these processes, acting at various stages during a comet's history, affecting its nucleus in varying degrees, can produce a complex internal structure. Finally, we try to connect activity patterns observed today, and the degree of variability between nuclei, to the internal structure of comet nuclei and their past history.

1. INTRODUCTION

Comets provide among the most valuable clues and insights into the formative stages of our planetary system. They may be the closest we can get to understanding the conditions prevailing in the early solar system. As detailed in the following sections, comets contain an abundance of volatile icy materials that are highly sensitive to temperature changes. The refractory solid phase includes organic material, which might have a presolar origin, and may have played a role in the emergence of life on Earth. Their inferred high internal porosities and low strengths suggest that comet nuclei could be the most primitive bodies in the solar system. However, comet nuclei are affected by collisions, tidal and centrifugal forces, radiogenic heating, and solar irradiation. Over long timescales, the complex interplay between these thermal and mechanical processes, acting at various stages during a nucleus' history, could be modifying their bulk properties to varying degrees. They ultimately shape a nucleus' activity pattern, internal structure, and surface morphology as they are observed today.

Only in rare occasions can comet nuclei be studied directly, as these objects are generally observed while active. The coma they produce approaching the Sun shields comet nuclei from any close Earth-based scrutiny, making them inherently difficult objects to detect and characterize. Space missions have brought some ground truth (see the chapter by Snodgrass et al. in this volume for a comprehensive discussion), required to interpret observations made from the Earth, as they observed through the dust and gas coma to bring constraints on some nuclei properties. In addition to 1P/Halley being the target of several space missions [including the European Space Agency's (ESA) Giotto mission] during its last perihelion passage in 1986, five nuclei of Jupiter-family comets (JFCs) have been targeted by space missions: 19P/Borrelly [NASA's Deep Space 1 (*Soderblom et al.,* 2002)], 81P/Wild 2 [NASA's Stardust

(*Brownlee et al.,* 2004)], 9P/Tempel 1 [NASA's Deep Impact (*A'Hearn et al.,* 2005a) and Stardust NexT (*Veverka et al.,* 2013)], 103P/Hartley 2 [NASA's EPOXI (*A'Hearn et al.,* 2011)], and 67P/Churyumov-Gerasimenko (ESA's Rosetta). In addition, the Goldstone Solar System Radar in California's Mojave Desert and the Arecibo Observatory in Puerto Rico obtained detailed radar images of some comet nuclei passing very close to the Earth. These nuclei are shown in Fig. 1.

Comet nuclei observed directly appear diverse in size, shape, and surface morphology and produce various activity levels. This diversity needs to be interpreted as an expression of both their formation and evolution (Vincent et al., 2017; Steckloff and Samarasinha, 2018), i.e., the complex interplay between processes mentioned earlier. Indeed, the ongoing modification of comet nuclei cannot be ignored, after two years of close examination and the varied and complementary observations of 67P that Rosetta has provided. A revealing picture of how activity develops from the nucleus with respect to seasons, for instance, depending on heliocentric distance but also on latitude, spin axis orientation, and subsolar latitude, has been brought to our knowledge. Two chapters of this volume are dedicated to the properties of surface layers of comet nuclei and processes, both thermal and mechanical in nature, actively shaping them on diurnal and seasonal timescales (see the chapters in this volume by Pajola et al. and Filacchione et al.). This is where the reader will find detailed information on topics such the cyclic release of gas and dust, the spatial distribution and transport of material across the surface, and the morphology and topography of cometary surfaces. A third chapter by Knight et al. is dedicated to the current knowledge on comet nuclei as a population, gained from remote observations of large samples, where comprehensive discussions on the size distribution of comet nuclei or their rotational properties are described. This chapter, dedicated to the nuclei of comets, provides a review of the internal properties of comet nuclei and processes responsible for their evolution over the age of the solar system. Most of our current knowledge on these topics is built on rare direct measurements and mostly derived or inferred from surface properties.

1.1. Origin of Cometary Material

We briefly describe the origin of cometary material and comet nuclei (for more comprehensive discussions, see the chapters in this volume by Bergin et al., Aikawa et al., and Simon et al.), because these early stages dictate initial internal properties such as strength, porosity, or thermal characteristics, which in turn influence the evolution of comet nuclei to the present day. Comet nuclei are made of a mixture of refractory material, organics, and frozen volatile species: H_2O, CO_2, and CO are the most abundant ices, but multiple minor species have be detected (*Altwegg et al.,* 2019; see also the chapter by Biver et al. in this volume). The isotopic, elemental, and molecular abundances in these phases may be used to gain insights into the origin and evolution of cometary material.

1.1.1. Refractory material. The composition of cometary dust (see the chapter by Engrand et al. in this volume) is constrained for only a handful of comets: 1P (*Jessberger et al.,* 1988; *Kissel et al.,* 1986; *Bregman et al.,* 1988; *Campins and Ryan,* 1989), 81P (*Brownlee et al.,* 2006; *Brownlee,* 2014, and references therein), 9P (*Lisse et al.,* 2006), C/1995 O1 (*Crovisier et al.,* 1997), and 67P (*Wurz et al.,* 2015; *Bardyn et al.,* 2017; *Fray et al.,* 2016, 2017; *Isnard et al.,* 2019). Although minerals show some variations in nature, they mostly consist of amorphous or crystalline silicates (such as olivines and pyroxenes), metal, and sulfides. Amorphous silicates observed in comets (*Kelley and Wooden,* 2009) appear similar or identical to the glass with embedded metal and sulfides (GEMS) found in anhydrous interplanetary dust particles (IDPs), considered to have formed in the interstellar medium (*Wooden et al.,* 2017). Crystalline minerals were manifestly formed at high temperature, as grains recovered in the Stardust sample resemble calcium-aluminum-rich inclusions (CAIs), chondrules, and large ($\lesssim$50 μm) crystalline silicate grains (*Zolensky et al.,* 2006; *Nakamura et al.,* 2008; *Westphal et al.,* 2009). The coexistence of minerals forming at high and low temperature suggests a substantial mixing of different types of material, occurring in the early phases of the solar system formation (*Bockelée-Morvan et al.,* 2002; *Ciesla,* 2007). In terms of physical characteristics, the Stardust and Rosetta missions

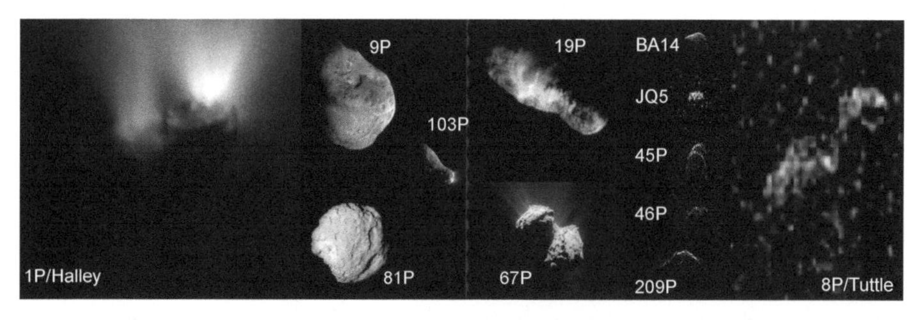

Fig. 1. Comet nuclei observed by spacecraft missions: 1P — ESA/MPS, 9P — NASA/JPL/UMD, 103P — NASA/JPL/UMD, 81P — NASA/JPL, 19P — NASA/JPL, 67P — ESA/Rosetta/NavCam. Comet nuclei observed by radar facilities: P/2016 BA14 — Goldstone/NASA/JPL/GSSR, P/2005 JQ5, 45P, 46P, 209P, 8P — Arecibo/NASA/NFS. We note that radar images are different in essence, as they plot the Doppler shift vs. delay, rather than the spatial dimensions.

demonstrated that nanometer- to micrometer-sized mineral grains can assemble in larger, sometimes fractal aggregates (*Brownlee et al.,* 2006; *Bentley et al.,* 2016; *Mannel et al.,* 2016; *Fulle et al.,* 2016; *Hilchenbach et al.,* 2017). In particular, the Rosetta dust collectors captured a wide variety of grain and aggregate types of potential diverse origins, as presolar and protoplanetary disk grains are expected to have different mineralogies and sizes and may differ in terms of porosity and structure (*Güttler et al.,* 2019).

1.1.2. Organic molecules. The gas coma of comets contains an abundance of carbon-based molecules, as evidenced by emission lines of carbonated small molecules visible in their optical spectra (e.g., *Cochran and Cochran,* 2002). These could be photodissociation products of organic parent molecules escaping the nucleus (see the chapter by Biver et al. in this volume). Owing to submillimeter and radio observations, a growing number of organic molecules — defined as having at least one carbon atom and one hydrogen atom — has been detected (e.g., *Bockelée-Morvan et al.,* 2000). Space missions have become key to identifying organic material on comet nuclei: The first detection of organic matter in cometary dust was made during the encounter of space probes with 1P/Halley in 1986. Giotto and Vega data allowed the detection of carbon, hydrogen, oxygen, and nitrogen (CHON)-rich particles (*Kissel et al.,* 1986; *Langevin et al.,* 1987) and phenantrene (*Moreels et al.,* 1994), a typical polycyclic aromatic hydrogenated (PAH) molecule containing three benzene rings. Dust particles returned to Earth from Comet 81P/Wild 2 by the Stardust mission show unambiguous evidence for glycine (*Elsila et al.,* 2009), amines (*Glavin et al.,* 2008), and complex macromolecular material rich in amorphous carbon and containing aromatic, aliphatic, and carboxylic functional groups (*Sandford et al.,* 2006; *Muñoz Caro et al.,* 2008). In the coma of 67P, a number of organic molecules have been identified by the Rosetta Orbiter Spectrometer for Ion and Neutral Analysis (ROSINA), such as methylamine (CH_3NH_2), acetonitrile (CH_3CN), acetaldehyde (CH_3CHO), or ethylamine ($C_2H_5NH_2$) (*Altwegg et al.,* 2016, 2017). The near-infrared spectrum of the comet surface shows a feature around in the 2.9–3.6-μm wavelength range compatible with nonvolatile organic macromolecular material (*Capaccioni et al.,* 2015). The origin of such organic molecules could be traced back to the interstellar medium (e.g., *Bertaux and Lallement,* 2017), assuming they survived the radiation of the early Sun in the outer regions of the solar system where comet nuclei probably formed. Indeed, the degree of processing of interstellar material sustained after the solar nebula phase is the subject of debate, ranging from zero processing (at least in the outer parts of the protoplanetary disk where comets were formed) to full destruction of interstellar grains (see the chapter by Bergin et al. in this volume for a comprehensive discussion).

1.1.3. Volatile species. The study of the gas phase chemistry of cometary volatiles began in earnest in 1910 with the detection of a CN optical line in the tail of 1P/Halley (see *Altwegg et al.,* 2019, for a review). By the time *Comets II* was published in 2004, almost 30 parent molecules had been detected, and taxonomies based on the chemical diversity of comets had started (*A'Hearn et al.,* 1995; *Mumma and Charnley,* 2011; *Dello Russo et al.,* 2016). More recently, observations in the infrared and submillimeter wavelengths targeting the rotational and vibrational modes of these parent molecules have been made possible, owing to new space- and groundbased instruments such as the Atacama Large Millimeter/submillimeter Array (ALMA) (*Bockelée-Morvan and Biver,* 2017). These advances allowed Rosetta to make the first-ever measurements of the abundance of noble gases (Ar, Kr, Xe) on a comet, detect unexpected molecules, and complete detailed studies of isotopic ratios for many elements. For example, the presence of S_2 molecules (*Calmonte et al.,* 2016), the xenon isotope ratios (*Marty et al.,* 2017), the HDO/H_2O and D_2O/HDO ratios, and the high O_2 abundance (*Rubin et al.,* 2020) suggest that even the icy component of 67P's nucleus is at least partially of interstellar origin.

1.1.4. Mixing of the constituents. As of today, it is unclear in which manner refractories, organics, and ices are mixed. On a microscopic scale, condensation sequences and radiative processing in molecular clouds and other parts of the interstellar medium suggest that ice would coat a grain with a silicate core and organic mantle (*Greenberg,* 1982). The icy and refractory phases may also exist as separate components, so that mixing can also occur at a macroscopic level, to create larger units with distinct mineralogy or volatile abundance. Potentially, such mixing may explain the existence of meter-sized ice-rich blocks found on the surface of 67P (*Pommerol et al.,* 2015; *Oklay et al.,* 2016; *Deshapriya et al.,* 2018). Such blocks might be swept-up debris from collisionally disrupted differentiated objects with icy mantles (*Oklay et al.,* 2016). Alternatively, they could originate from, and be representative of, the nucleus interior [fresh cliff collapse sites expose bright and ice-rich material with a 40% visual geometric albedo (*Pajola et al.,* 2017)] and have not had time to develop the dust mantle yet, similar to that covering most of the nucleus' surface (*Fornasier et al.,* 2015). Ultimately, the relative proportion of refractories and ices within comet nuclei is one key to tracing their formation location and studying the evolution processes that subsequently shaped them (*Weissman and Lowry,* 2008; *Rubin et al.,* 2020). Such comet nuclei characteristics as the degree and type of compositional heterogeneity — from grain level to global size scales, or the refractory-to-ice mass ratio — are also important to constrain for understanding how comets work.

1.2. Formation of Comet Nuclei

1.2.1. Growing comet nuclei. The theory of planetesimal growth, including comet nuclei, has undergone a paradigm shift since the publication of *Comets II.* The dominating view that had emerged during the preceding 30 years was that comet nuclei grew through gentle hit-and-stick collisions between "cometesimals," which are themselves the products of similar previous mergers, a process known as "hierarchical growth" or "coagulation" (*Weidenschilling, 2004*). Since then, laboratory experiments have demonstrated that micrometer-sized grains of silica and/or crystalline water ice only form millimeter- to centimeter-sized assemblages through hit-and-stick, before collisions become destructive (*Blum and Wurm, 2008*). In addition, the development of a theoretical framework for streaming instabilities (*Youdin and Goodman, 2005*) has enabled three-dimensional magneto-hydrodynamic simulations, which have demonstrated that such instabilities could create pebble swarms capable of direct gravitational collapse into large planetesimals (*Johansen et al., 2007*). Finally, ultra-wide binaries were discovered in the Kuiper belt (*Parker et al., 2011*), whose existence is difficult to explain unless pebble-swarm collapse took place (*Nesvorný et al., 2010*). This "pebble-collapse" model may be considered the currently dominating theory of comet formation (see the chapter by Simon et al. in this volume), although hierarchical growth scenarios still might be possible under certain conditions (see their section 3.2).

We note that a first generation of pebble-collapse nuclei may have continued to grow through gravity-assisted mergers: Gravitational focusing becomes important for 1-km bodies at the low relative velocities initially prevailing in the solar nebula (e.g., *Hartmann and Davis, 1975*; *Weidenschilling, 1997*). Such runaway growth (e.g., *Kokubo and Ida, 1996*) may be necessary to explain the Kuiper belt size distribution, crowned by Eris- and Pluto-sized bodies (e.g., *Schlichting and Sari, 2011*; *Kenyon and Bromley, 2012*), and may have contributed to the formation of bilobate objects (see sections 4.1.1 and 4.2.1). Furthermore, it has been suggested that comet nuclei may have grown hierarchically from the material remaining after the larger Kuiper belt objects formed through pebble collapse (*Davidsson et al., 2016*). Both pebble-collapse formation and hierarchical growth scenarios therefore remain relevant for now.

1.2.2. Predicting comet nuclei characteristics. To advance our knowledge of comet formation, any proposed formation scenario must make predictions about the physical and structural properties of comet nuclei that can be tested *in situ* by spacecraft missions or through the modeling of observed activity. Indeed, different formation scenarios may lead to distinct thermal or mechanical properties, which then influence the evolution of comet nuclei between their formation and the present day. Ideally, it would also be necessary to identify quantitative differences between primordial nuclei and objects that experienced high-velocity collisional processing (*Jutzi et al., 2017*; *Jutzi and Benz, 2017*; *Schwartz et al., 2018*; *Steckloff et al., 2023*). Unfortunately, such predictions are currently few. Furthermore, because evolutionary processes that alter comet nuclei after formation are still poorly understood (e.g., their collisional evolution, their sublimation evolution during an orbital transition through the giant planet region), these initial structures and properties may be significantly altered, and therefore make the task of using comet observations to constrain these formation scenarios much more difficult.

Hierarchical growth modeling has primarily focused on calculating the size-frequency distribution of planetesimal populations and how it evolved with time and heliocentric distance. The physical and structural properties of the comet nuclei resulting from such growth have rarely been considered in detail. Nonetheless, the large-scale layering observed on 9P/Tempel 1 has led to the "talps layered pile" model (*Belton et al., 2007*), in which cometesimals are smeared to sheets during accretion. It is still unclear, however, how a hierarchically grown nucleus would look like in terms of shape, density, magnitude, and size-scale of internal density variations, including the size and spatial density of voids. Nuclei formed by pebble collapse are less likely to display significant large-scale heterogeneities because of the small size of the constituent pebbles, although clumping during gravitational collapse (e.g., *Michel and Richardson, 2013*) may introduce substructures within such nuclei as well.

The recent modeling efforts toward predicting comet nuclei properties from formation scenarios are encouraging. However, observational constraints on the internal structure of 67P placed by the Rosetta mission in terms of nucleus formation and evolution mechanisms clearly show the pressing need for further work. Indeed, 67P's nucleus is bilobed, highly porous, internally homogeneous on $\gtrsim$10-m scales but with a potential porosity increase with depth, having exposed structures (pits, accumulation basins, layering, positive relief features) on $\gtrsim$100-m scales (see the chapter by Pajola et al. in this volume). These characteristics have been taken as support for primordial hierarchical growth (*Davidsson et al., 2016*), primordial pebble collapse (*Blum et al., 2017*), and severe high-velocity collisional processing (*Rickman et al., 2015*; *Jutzi et al., 2017*). Clearly, all three scenarios cannot be simultaneously correct.

1.3. Overview

In this chapter, we start by describing the bulk properties of comet nucleus interiors as they are observed today: their strength, density, composition, and dielectric and thermal properties. We then address multiple aspects of the thermal processing of comet nuclei (section 3): We discuss effects such as the formation of a stratified structure, the development of non-uniform layers, and activity in the outer solar system. We emphasize again that surface processes at the origin of activity and evolution of the surface itself are discussed in detail by two other chapters (see the chapters in this volume by Filacchione et al. and Pajola et al.), so we do not discuss these aspects in great detail here. We then move to the mechanical processing of comet nuclei (section 4), sustained at various stages of their lives: This is where the outcomes of collisional or tidal evolution are described, and how these processes are at the origin of further evolution of internal properties. Finally, we present in section 5 current challenges to our understanding of comet nuclei evolution. We propose relevant strategies for improving this understanding of the quantitative physical and structural differences among nuclei, built in various formation and evolutionary scenarios. We also propose observations that would certainly make a different in the future for making significant advances in this field.

2. BULK PROPERTIES OF COMET NUCLEI

2.1 Strength

Mechanical properties, such as the strength of a nucleus, are important parameters for understanding the evolution and activity of comets. In particular, observable physical processes such as cometary activity, impacts, or the transport of material across the surface depend on such properties. For instance, strength dictates the response of a comet nucleus to internal or external stresses, such as those involved in accretion, encounters with giant planets, or collisions. In this section, we discuss three types of strength relevant for understanding comet nuclei: the tensile strength σ_T (Pa), which corresponds to the maximal mechanical tension that can build up inside a material before it breaks; the shear strength σ_S (Pa) that corresponds to the maximum shear load that can be applied to a material without failure; and the compressive strength σ_C (Pa), indicative of the maximum load that can be applied to a material without changing the porosity of a sample. These are notoriously difficult to measure outside of a laboratory, i.e., for the larger spatial scales. It is important to remember that strength is a scale-dependent parameter, with a general trend of decreasing strength with increasing spatial scale [see Fig. 2 and the discussion by *Biele et al.* (2022)], a notion to keep in mind when comparing measurements for diverse sample sizes of cometary material, from local to global comet nucleus scales.

2.1.1. Small spatial scales: Grains and aggregates. Recently, measurements have been performed in the laboratory for silica grains, icy grains, and mixtures of dust and ice grains. These experiments on cometary material analogs have provided valuable estimates of the strength for water ice and dust grains, with σ_T in the range of 2–4 kPa for pure 200-µm water ice grains (*Bar-Nun et al.,* 2007), and σ_T in the range of 1–10 kPa for micrometer-sized silicate grains (*Blum et al.,* 2006). These grains typically form aggregates for which various configurations can be inferred based on a combination of formation and evolution processes acting on comets. At least two structures of cometary material are currently under debate: homogeneous dust layers (e.g., *Davidsson et al.,* 2016) and aggregate dust layers (e.g., *Blum et al.,* 2017). The tensile strength of homogeneous dust layers has been measured in various studies, and is roughly on the order of 1 kPa (*Blum and Schräpler,* 2004; *Güttler et al.,* 2009; *Gundlach et al.,* 2011, 2018; *Meisner et al.,* 2012). These results are in good agreement with measurements of the breakup of cometary dust particles observed with Rosetta's Cometary Secondary Ion Mass Analyser (COSIMA) [~1 kPa (*Hornung et al.,* 2016)] and with the derived tensile strength from the breakup of meteor streams in Earth's atmosphere: 0.4–150 kPa (*Blum et al.,*

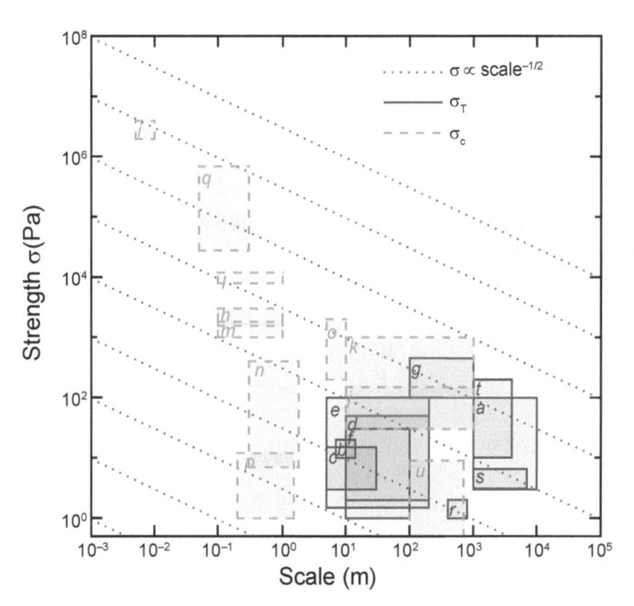

Fig. 2. Summary of tensile (solid blue line) and compressive (dashed yellow line) strength for cometary materials, adapted from *Biele et al.* (2022). Each box represents a domain of scale/strength from a specific study. The dotted gray lines illustrate the qualitative relationship between strength and scale. References: (a) *Biele et al.* (2009), (b) *Thomas et al.* (2015a), (c) *Groussin et al.* (2015), (d) *Vincent et al.* (2015, 2017), (e) *Basilevsky et al.* (2016), (f) *Attree et al.* (2018), (g) *Matonti et al.* (2019), (h) *Biele et al.* (2015), (i) *Spohn et al.* (2015), (j) *Groussin et al.* (2015), (k) *Basilevsky et al.* (2016), (l) *Möhlmann et al.* (2018), (m) *Roll and Witte* (2016), (n) *Heinisch et al.* (2019), (o) *Vincent et al.* (2019), (p) *O'Rourke et al.* (2020), (q) *Biele et al.* (2022), (r) *Sekanina and Yeomans* (1985), (s) *Asphaug and Benz* (1996), (t) *Hirabayashi et al.* (2016), (u) *Steckloff et al.* (2015).

2014) and 0.04–1 kPa (*Trigo-Rodríguez and Llorca,* 2006). Thus, the tensile strength of homogeneous dust layers can be used to describe the internal tensile strength of dust aggregates. Measurements of the tensile strength of packings of aggregates have revealed much lower values, on the order of 1 Pa (*Blum et al.,* 2014). It is therefore important to note that the tensile strength of a granular material can vary by at least 3 orders of magnitude, just by different arrangements of the material (homogeneous dust layers vs. aggregate packings). It is therefore critical to understand the structure of dust aggregates, as it not only has a strong influence on the thermal properties of cometary material, but also on the strength of the surface material. Both are subject to further modifications, for example, when sintering occurs to modify the contact area between grains.

2.1.2. Meter to 100-meter spatial scales. While laboratory experiments and theoretical studies provide reliable values for the strength of cometary material at the smallest spatial scales, most values reported for larger spatial scales are derived from measurements outside the controlled laboratory environment and are typically lower and upper boundaries (see Fig. 2). On scales up to ~100 m, cometary topography and impact events can be used to constrain the local strength of material. The Deep Impact mission inferred σ_C < 65 Pa (*A'Hearn et al.,* 2005a); however, *Holsapple and Housen* (2007) argue that this controlled impact experiment is compatible with σ_C as large as 12 kPa. Origins, Spectral Interpretation, Resource Identification, and Security–Regolith Explorer (OSIRIS-REx) images of 67P have shown a varied and complex morphology including surface features suggestive of nonzero material strength, such as pits, cliffs, overhangs, and fractured consolidated material (see the chapter by Pajola et al. in this volume). *Groussin et al.* (2015) estimated the strength of collapsed surface features: they found σ_T = 3–15 Pa and σ_C = 30–150 Pa at the 5–30-m scales. They also estimated σ_S = 4–30 Pa for the surface material to hold meter-sized boulders on slope, and σ_S > 30 Pa to resist the lateral pressure at the bottom of the 900-m-high Hathor cliffs. Further observational constraints are discussed in detail in *Groussin et al.* (2019) and *Biele et al.* (2022), including constraints from the bouncing of the lander Philae on 67P's surface in the Agilkia region. *Basilevsky et al.* (2016) and *Vincent et al.* (2017) analyzed the properties of overhangs, landslides, and cliff collapses respectively to find comparable strength estimates. Despite these low strengths, fractures on 67P indicate that some areas of the surface consist of material with sufficient strength to be brittle. Indeed, fractures are present at all scales in OSIRIS-REx images, from hundreds of meters down to tens of centimeters (*Thomas et al.,* 2015b; *El-Maarry et al.,* 2015; *Pajola et al.,* 2015; *Matonti et al.,* 2019), as well as smaller scales (centimeter and below) as evidenced by images from Philae (*Poulet et al.,* 2016). These may be caused by stresses associated with rotational and shape effects, tidal forces, collisions, and thermal cycling (see the chapter by Pajola et al. in this volume for a comprehensive discussion on that topic).

2.1.3. Strength of comet nuclei. At the largest scales, it is possible to estimate the strength of a whole nucleus (not just the surface material) if and when it experiences a mechanical stress, for example, when cometary material has to resist centrifugal forces due to the nucleus' rotation, or when cometary material has to resist tidal forces during the encounter of a nucleus with a giant planet or the Sun. From the observed rotational period of several nuclei, *Davidsson* (2001) and *Toth and Lisse* (2006) estimated that $\sigma_T \sim$ 1–100 Pa. Investigations of the rotational evolution of 67P suggest a bulk cohesive strength of 10–200 Pa (*Hirabayashi et al.,* 2016). The close encounter of Comet Shoemaker-Levy 9 (SL9) with Jupiter allowed *Asphaug and Benz* (1996) to estimate σ_T < 6.5 Pa. Finally, *Klinger et al.* (1989) estimated $\sigma_T \sim$ 100 Pa for a 1-km-radius sungrazing comet.

Overall, cometary material seems to follow the general expected trend of an increase in strength with decreasing spatial scale [i.e., $\sigma \propto L^{-1/2}$ with L the length scale; see Fig. 2 (*Steckloff and Jacobson,* 2016; *Biele et al.,* 2022)]: Individual grains are stronger than aggregates, which are stronger than boulders, which are themselves stronger than large topographic features such as cliffs or the nucleus itself. Unconsolidated materials and fine deposits that cover a large fraction of the surface have a lower strength than consolidated materials from which boulders and cliffs are made. It should be noted that at any scale, strength values can span several orders of magnitude, which could be explained by intrinsic differences between the material considered in various studies (e.g., fallback or consolidated material, ice-rich or ice-poor material), but can also be the result of the evolution of comet nuclei. Indeed, water-ice sublimation/recondensation cycles, as well as sintering close to the surface, can produce layers that are significantly harder. Theoretical studies and laboratory experiments indicate that strengths as high as a few millipascals can be reached by sintered icy material, although values 1 order of magnitude lower are deemed more probable (*Grün et al.,* 1991; *Kossacki et al.,* 1994, 2015; *Pommerol et al.,* 2015; *Gundlach et al.,* 2018). Philae data indicate that the top layer of the Abydos site on 67P consists of a hard layer not exceeding a few centimeters in thickness, strong enough to deflect the Multi Purpose Sensors for Surface and Subsurface Science (MUPUS) penetrator. We note that the existence of this layer is inferred from the nonpenetration of MUPUS only (*Knapmeyer et al.,* 2018). Below that, the authors report a rigid layer of at least 10 cm, possibly as thick as 50 cm [with a shear modulus ranging from 3.6 to 346 MPa, a Young's modulus from 7.2 to 980 MPa, and a bulk porosity below 0.74 (*Knapmeyer et al.,* 2018)]. The picture that emerges from these measurements and constraints is one of weak comet nuclei, with a bulk tensile strength on the order of less than 100 Pa, with strong variations across the surface and with depth.

2.2. Density

To estimate the bulk density ρ_{bulk} (kg m^{-3}) of a comet nucleus, the mass and volume of the nucleus must be determined, when both of these characteristics remain very difficult to measure. The volume itself can be constrained from direct observations of a nucleus by spacecraft missions or radar observations. Indirect assessments can be made, from rotational lightcurves constraining a global shape (see the chapter by Knight et al. in this volume), or from the brightness of inactive nuclei, which can be translated into an approximate radius. Assessing the mass of a comet nucleus is more arduous. We detail below the main methods to reach this constraint.

2.2.1. From nongravitational forces. First, it is possible to assess the mass of a comet nucleus by modeling the expected nongravitational force (NGF) sustained by a nucleus with thermophysical models. Comparisons with the observed nongravitational acceleration (NGA) thus yield an estimate of the nucleus mass. This approach was suggested by *Wallis and MacPherson* (1981) and successfully applied by *Rickman* (1986, 1989) for Comet 1P/Halley. Combined with an estimate of the nucleus volume from Giotto images, the nucleus bulk density was estimated as ρ_{bulk} = 280–650 kg m^{-3} (*Rickman,* 1989). NGF analysis had been performed for dozens of comets, systematically suggesting that comets are underdense compared to their main compacted components (i.e., ice with $\rho \sim 1000$ kg m^{-3}, organics with $\rho \sim 2000$ kg m^{-3}, or silicates with ~3500 kg m^{-3}). A bulk density $\rho_{bulk} \lesssim 600$ kg m^{-3} seems typical (*Weissman et al.,* 2004), suggesting that the volume occupied by voids — or porosity — is $\psi \gtrsim 70\%$ (*Davidsson,* 2006; *Weissman and Lowry,* 2008).

Spacecraft flyby of specific comets (and intensified groundbased observations of such targets) have constrained several parameters that are usually poorly known in NGF analysis, and contribute to the typically large uncertainties of that method: the nucleus volume, shape, spin axis orientation, and surface activity pattern, as well as the water production rate curve before and after perihelion. With flyby missions, only a fraction of the comet nucleus is directly imaged, so that the uncertainty on the volume can remain relatively large. In contrast, Rosetta has provided a precise measurement of the volume of 67P's nucleus. Based on these volume estimates, the bulk density of 19P, 81P, 9P, and 67P has been constrained (see Fig. 3). Comet 67P constitutes an important test-case for NGF analysis because its mass and bulk density are indeed well-known. Pre-Rosetta mass estimates include 0.1–1.3 × 10^{13} kg (*Rickman et al.,* 1987); 0.35–1.3 × 10^{13} kg in total, but 1.1–0.2 × 10^{13} kg for the particular spin axis orientation that later turned out to be correct (*Davidsson and Gutiérrez,* 2005); 0.72–1.1 × 10^{13} kg (*Kossacki and Szutowicz,* 2006); and 1.5 ± 0.6 × 10^{13} kg (*Sosa and Fernández,* 2009). These estimates are consistent with the actual mass, M = 0.9982 ± 0.0003 × 10^{13} kg (*Pätzold et al.,* 2016). It thus seems that NGF-based comet masses can be rather reliable (albeit delicate to work with), and that the largest uncertainty remains in the measurement of a comet nucleus' volume. Finally, we note that *Sosa and Fernández* (2010, 2011) extended the NGF method of mass estimates to 15 long-period comets. Instead of attempting to estimate nucleus volumes and bulk densities, they assumed $\rho_{bulk} = 400$ kg m^{-3} and calculated the corresponding nucleus radii. By doing so, they found that long-period comets have higher active area fractions and/or lower bulk density than short-period comets.

2.2.2. From centrifugal forces. Another method to estimate the bulk density of a comet nucleus is to calculate the minimum value needed in order for self-gravity to prevent its disruption due to the centrifugal force, assuming a strengthless object (e.g., *Davidsson,* 2001; *Snodgrass et al.,* 2006; *Weissman and Lowry,* 2008). This requires some knowledge of the spin period (see the chapter by Knight et al. in this volume for further discussion on spin periods) and shape of the object. Most known JFCs do not spin faster than 5.5 hr (*Weissman and Lowry,* 2008), so they would not require a bulk density $\rho_{bulk} \gtrsim 600$ kg m^{-3} in order to remain stable. There are, however, a few examples of faster-rotating objects — e.g., 96P/Machholz 1 (*Eisner et al.,* 2019), 322P/SOHO (*Knight et al.,* 2016), and 323P/SOHO (*Hui et al.,* 2022) — that seem to challenge the traditional assumptions. It is possible that they have a bulk density larger than 600 kg m^{-3}, at least 1000 kg m^{-3} in the

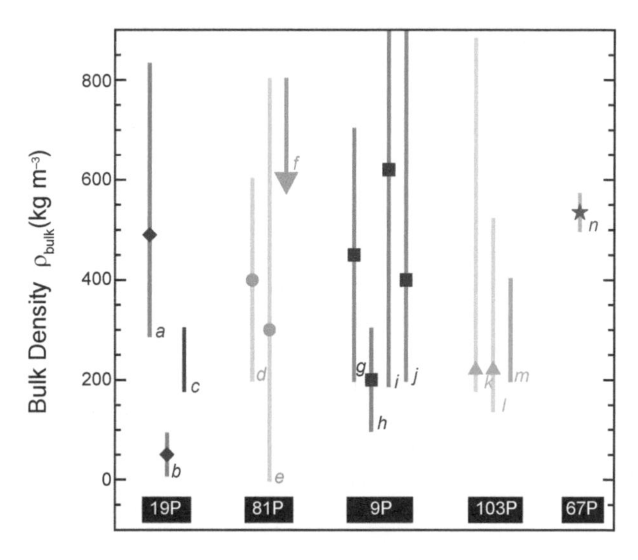

Fig. 3. Summary of bulk densities derived for comet nuclei with a volume measured by spacecraft missions. References: for 19P/Borrelly — (a) 490^{+340}_{-200} kg m^{-3} (*Farnham and Cochran,* 2002), (b) 50 ± 40 kg m^{-3} (*Sosa and Fernández,* 2009), (c) 180–300 kg m^{-3} (*Davidsson and Gutiérrez,* 2004), for 81P/Wild 2 — (d) 400 ± 200 kg m^{-3} (*Szutowicz et al.,* 2008), (e) 300^{+500}_{-300} kg m^{-3} (*Sosa and Fernández,* 2009), (f) ≲600–800 kg m^{-3} (*Davidsson and Gutiérrez,* 2006), for 9P/Tempel 1 — (g) 450 ± 250 kg m^{-3} (*Davidsson et al.,* 2007), (h) 200 ± 100 kg m^{-3} (*Sosa and Fernández,* 2009), (i) 620^{+470}_{-330} kg m^{-3} (*A'Hearn et al.,* 2005b), (j) 400 kg m^{-3} (200–1000 kg m^{-3}) (*Richardson et al.,* 2007), for 103P/Hartley 2 — (k) 220 kg m^{-3} (180–880 kg m^{-3}) (*A'Hearn et al.,* 2011), (l) 220 kg m^{-3} (140–520 kg m^{-3}) (*Richardson and Bowling,* 2014), (m) 200–400 kg m^{-3} (*Thomas et al.,* 2013), for 67P/Churyumov-Gerasimenko — (n) 535 ± 35 kg m^{-3} (*Preusker et al.,* 2015).

case of 322P, which would require a lower internal porosity and/or a dust-rich composition. Alternatively, these nuclei could have a nonzero internal strength, as evidence for such internal strength, especially for the upper layers of a comet nucleus, has been laid out. We should note that both 96P and 323P were observed to fragment, which could potentially invalidate the measurements of their rotation period.

Fractures in stronger cometary material can allow for a more rapid splitting of nuclei, as rotation periods can change over time due to sublimation-induced torques (*Steckloff and Jacobson,* 2016; *Kramer et al.,* 2019; *Kramer and Läuter,* 2019). For example, on 67P, *Sierks et al.* (2015) and *Matonti et al.* (2019) noted evidence of fractures near the "neck" of the bilobate nucleus. *Hirabayashi et al.* (2016) computed the stresses required to propagate a fracture on the order of 100 m between the two lobes, resulting in the split of the nucleus. They found that the rotation period would need to reach values as low as 7 hr: It would take only about 15 orbits, almost 100 years, for 67P to reach this point, provided its rotation period maintains its current trend of linear decrease with every orbit. If rotational spin-up is one leading cause of disruptive events among comet nuclei, other mechanisms may be involved, as discussed in section 4. For example, it seems unlikely that rotational instability is responsible for the fragmenting episodes of 73P/Schwassmann-Wachmann 3 (*Graykowski and Jewitt,* 2019). Instead, internal thermal stresses due to strong temperature gradients in the subsurface may play a role, as Rosetta did indeed provide some evidence of thermal stressing and cracking at the surface of 67P (*Auger et al.,* 2018). Ultimately, a more accurate understanding of the interplay between density, strength, and other internal characteristics are key to assess the effect of evolution mechanisms (thermal or mechanical). This is evidenced by the fact that dynamically new comets, with a very different thermal evolution in particular, are found to split significantly more often than typical JFCs (*Weissman,* 1980; *Boehnhardt,* 2004; *Jewitt,* 2022).

2.2.3. From the equipotential surface method. Finally, we briefly mention a method available for comet nuclei observed by spacecraft missions: the "equipotential surface" method of density measurement. If the surface of the nucleus contains a topographic feature (see the chapter by Pajola et al. in this volume for a comprehensive discussion), which can be assumed to lie along an equipotential surface; the analysis of the shape of that nucleus can be used to find the density that best fits this equipotential. This technique was used, for instance, to estimate the bulk density of 103P/Hartley 2 to 220 kg m^{-3} (*A'Hearn et al.,* 2011). Later, *Richardson and Bowling* (2014) refined this technique and found a density for 103P of 220 kg m^{-3}, and in the 140–520-kg m^{-3} range. By allowing the rotation period of the nucleus to vary, *Thomas et al.* (2013) finally found a best fit density of 200–400 kg m^{-3} (see Fig. 3) for this comet.

2.3. Composition

2.3.1. Uniform vs. non-uniform. Splitting events offer a rare opportunity to probe the internal structure and composition of comet nuclei. Observing multiple components of a fragmented comet indeed sets constraints on whether its nucleus is homogeneous in composition, a key constraint to understand its formation mechanism and the effect of subsequent evolution. Famous splitting comet nuclei include D/1993 F2 (SL9), which did not resolve the matter as no gas was detected (*Crovisier,* 1996), and 73P/Schwassmann-Wachmann 3, which has been releasing multiple components. All 73P's fragments have exhibited the same chemical composition (*Villanueva et al.,* 2006; *Dello Russo et al.,* 2007; *Bertini et al.,* 2009), which is consistent with the composition of the progenitor nucleus measured prior to fragmentation (*Fink,* 2009; *Schleicher and Bair,* 2011). Most comet nuclei are not observed to fragment, however; their internal composition must be inferred from the composition of the coma they produce.

Connecting the release of dust and gas in the coma to the nucleus is a strenuous task (see the chapter by Marschall et al. in this volume). Space missions thus offer a rare opportunity to directly link coma compositions to nucleus properties. For 67P, both showed morphological heterogeneity. 67P consists (as do most comet nuclei observed directly; see section 4) of two lobes (*Sierks et al.,* 2015), proposed to have formed as two individual bodies (*Massironi et al.,* 2015; *Penasa et al.,* 2017; *Tognon et al.,* 2019) that later merged. Alternatively, the two lobes may have formed from the same original body: *Schwartz et al.* (2018) argue that catastrophic disruptions due to impacts could form bilobate comet nuclei shapes from the same original body, while *Safrit et al.* (2021) argues that sublimative torques in the giant planet region could spin comet nuclei into these bilobate shapes. This could explain the compositional similarities between the two lobes. Indeed, while some properties are identical for the two lobes, such as the D/H ratio (*Schroeder et al.,* 2019), others are different with a statistical significance. Among those, the small lobe has fewer and smaller bright spots (*Fornasier et al.,* 2021), fewer and less-frequent spectrally blue regions, larger "goosebumps" [i.e., meter-sized spheroids of disputed origin: primordial cometesimals according to *Davidsson et al.* (2016), or thermal contraction crack polygons according to *El-Maarry et al.* (2015) and *Auger et al.* (2018)], and substantially fewer and smaller morphological changes, compared to similarly illuminated regions on the large lobe. This suggests a potential large-scale heterogeneity in terms of water ice abundance, mechanical, and thermal properties. Ultimately, the origin and implications of bilobate comet nuclei is not settled, but rather is an active area of research.

Non-uniform outgassing was also observed by EPOXI at 103P (*A'Hearn et al.,* 2011), where CO_2 was mainly outgassed from one lobe, while H_2O was coming from the neck, i.e., the junction between the two lobes of the nucleus. Nucleus heterogeneity has thus been proposed in order to explain the factor of 2 differences in the CO_2/H_2O outgassing ratio between the two lobes of this comet (*A'Hearn et al.,* 2011, 2012), as well as variability in regional H_2O, CO_2, and CO activity levels that are not readily explained by differences in illumination (*Fougere et al.,* 2016a). This heterogeneity could be limited to the superficial layers though, as *Steckloff et al.* (2016) argued that the observed CO_2 activity could be the result of a large mass-wasting event that exposed CO_2-rich material, while the H_2O from the waist could correspond to fallback material. In this scenario, 103P would be displaying activity differences that result from different evolutionary processes acting on the different parts of its nucleus. On 9P/Tempel 1, asymmetries in the spatial distribution of H_2O and CO_2 in the coma have been taken as an indication of chemical heterogeneities in the nucleus (*Feaga et al.,* 2007). A comprehensive discussion on these aspects can be found in the chapter by Filacchione et al. in this volume.

Current groundbased observation facilities and techniques allow for coma observations that can monitor compositional variability as a function of time. For instance, variations in the production of CH_3OH from Comet 46P/Wirtanen are correlated with the 9-hr rotation period of its nucleus (*Roth et al.,* 2021b). The outgassing is therefore likely related to the illumination of active areas on the nucleus. Comet 46P also showed significant heterogeneity in dust production

(*Lejoly et al.,* 2022). Overall, we may conclude that the layer of the comet nucleus that contributes to the activity is heterogeneous in composition, possibly as a result of evolution.

2.3.2. ***Relative abundances.*** Beside knowing whether comet nuclei are homogeneous or heterogeneous in composition, their bulk refractory-to-ice (RTI) mass ratio is used to trace back to their formation conditions and for assessing the material's thermal properties. The water abundance in chondritic meteorites is 15–28 wt% (*Garenne et al.,* 2014), and the Pluto-Charon system has 35 wt% (*McKinnon et al.,* 2017). Comets ought to have similar water abundances: For most comets studied this far, the RTI ratio has been assumed to be equivalent to the dust-to-ice mass ratio (DTI) measured in their comae. Two major issues can thus be raised: (1) those measurements are usually made at one given moment in the orbit, and (2) it is assumed that the coma is directly representative of the bulk nucleus. *Choukroun et al.* (2020) have shown that individual measurements cannot constrain the RTI ratio of Comet 67P, based on a thorough review of the Rosetta data. They showed that the DTI ratio is an evolving number, both along the orbit and possibly across multiple perihelion passages. For the breadth of information gathered on 67P, they estimate that an RTI of 1 is a good approximation, a value consistent with that derived by *Davidsson et al.* (2022b) by reproducing the long-term H_2O production rates.

The same caveat applies to the average composition in minor volatile species, as measured with respect to water ice. Historically, two taxonomic classes have been identified: one typical, and one carbon-depleted class with low C_2 and C_3 production rates relative to CN (*A'Hearn et al.,* 1995). When including major volatile species CO, CO_2, and H_2O, *A'Hearn et al.* (2012) found no evidence for major differences between comets originating from the Kuiper belt or the Oort cloud, which would suggest a similar formation location for these two populations and negligible modification due to subsequent thermal processing. However, recent facilities such as the Transiting Planets and Planetesimals Small Telescope (TRAPPIST) have enabled cometary activity to be tracked over a wide range of heliocentric distances. As a result, *Opitom et al.* (2016) showed that production rates vary significantly across the orbit. This effect is best demonstrated by the Rosetta observations: When the spacecraft arrived at Comet 67P (in 2014, around 3 au), H_2O was originating mostly from the illuminated northern hemisphere, while CO_2 was mainly outgassed from the southern one (*Hässig et al.,* 2015). CO was more evenly distributed over the nucleus (*Läuter et al.,* 2019). During this period, the ratio of most observed minor species was larger when emitted from the southern hemisphere (*Le Roy et al.,* 2015). After the first equinox in May 2015, both molecules were released from the same regions in the south. The period from May to June 2015 has been found to be the most representative of the nucleus composition (*Calmonte et al.,* 2016). The reason is that all volatile species including water were coming predominantly from the southern hemisphere, which was experiencing summer at that time. Besides, all species were correlated independently of the Rosetta spacecraft position. After that, 67P showed many short-lived outbursts of activity, peaking at perihelion, which changed the relative abundances in the coma quite significantly (e.g., *Läuter et al.,* 2019). If most of these measurements can be ascribed to seasonal variations of the composition, or heterogeneities in the layers producing the coma, they do show that taxonomies based on a limited number of observations ought to be used with caution. They remain a useful tool to quickly assess the nature of a comet nucleus, especially when the inferred composition is far from the average composition. This is, for instance, the case of N_2-rich Comet C/2016 R2 (*Biver et al.,* 2018; *McKay et al.,* 2019; *Opitom et al.,* 2019): Its unique composition may relate to a precise formation location in the disk; alternatively, C/2016 R2 could be a debris from a disrupted, differentiated, 1000-km object.

2.4. Dielectric Properties

The dielectric properties of a comet nucleus provide information on the nature of cometary material, such as its roughness, porosity, or density. These internal characteristics are mostly inaccessible to direct observations, but can be constrained by radar and permittivity probes. The interaction of an electromagnetic wave with a nonhomogeneous body is a complex process that includes macroscopic phenomena such as reflections on structures significantly larger than the wavelength, refraction at the interfaces between two different units, volume and surface scattering on features of the size of the wavelength, or propagation with attenuation in the medium. For nonmagnetic bodies such as comet nuclei (*Auster et al.,* 2015), the phenomena mentioned above are essentially driven by a parameter called effective permittivity that combines the electrical permittivity (i.e., the capacity of migration or reorientation of charged particles when submitted to an electric field) and the electric conductivity (i.e., the capacity of motion of free charges within the material) of the medium. Within their spatial resolution (which depends on the frequency bandwidth for a radar and the distance between electrodes for a permittivity probe), instruments see an averaged, uniform medium: The retrieved bulk permittivity needs to be compared to laboratory measurements performed in controlled conditions to derive constraints on the composition (mainly the RTI ratio) or porosity of the sounded medium (*Heggy et al.,* 2012; *Brouet et al.,* 2015, 2016). These techniques have made key contributions to the exploration of comet nuclei in the recent years, most notably as part of the Rosetta mission and with observations by Earth-based radar.

2.4.1. ***Earth-based observations.*** From Earth, radar observations are based on sending an unmodulated (monochromatic) microwave and measuring the power of the Doppler-broadened echo. They are operated at frequencies of the order of a few gigahertz. The detectability of comet nuclei with such techniques is limited by the Δ^{-4} dependence of the received signal (where Δ is the geocentric distance): Earth-based radars can only study close-approaching comets, typically with $\Delta \lesssim 0.3$ au. These events are rare: Half a dozen comet nuclei have been imaged by radar antennas in the past 20 years (see Fig. 1). Radar albedo measurements can be interpreted in terms of shape, surface roughness, and permittivity value of the shallow subsurface of the target (see *Harmon et al.,* 2004), which are consistent with this subsurface layer being highly porous. It is indeed important to remember that radar measurements are sensitive to the top layer of the nucleus, with a thickness on the order of ~10× λ (where λ is the wavelength), which is commensurate with the penetration depth of the radar wave δ, which also depends on the properties of the probed material.

In addition to the integrated received power, the frequency shift observed in the Doppler spectra may be used to estimate the relative velocity of the target with respect to the radar. Hence, radar images can be produced and velocity values retrieved from the Doppler shift can allow observers to disentangle the response of the nucleus from that of the coma and identify bilobate nuclei (*Harmon et al.,* 2010). Even a nondetection can provide constraints on the size of a nucleus, as was demonstrated in the case of 67P (*Kamoun et al.,* 1998): The lack of radar detection provided an upper limit of the nucleus radius of around 3 km. Based on information provided by other instruments, additional quantitative results were extracted from the Arecibo radar data, such as a maximum value of 0.05 for the radar albedo, consistent with a low permittivity value (2 ± 0.1), which was interpreted as a bulk porosity between 55% and 65% in the first 2.5 m beneath the surface (*Kamoun et al.,* 2014).

2.4.2. Observations from Rosetta. Rosetta provided the unique opportunity to obtain *in situ* data for comparison with permittivity values inferred from Earth-based distant radar measurements. The Philae lander operated at the surface of the 67P nucleus at the Abydos landing site, with the Surface Electric Sounding and Acoustic Monitoring Experiment/Permittivity Probe (SESAME-PP) to measure the permittivity of the shallow subsurface (down to 1 m). The amount of data collected by SESAME-PP was very limited, due to a combination of unfortunate events (*Brouet et al.,* 2016; *Lethuillier et al.,* 2016). Nevertheless, the SESAME-PP *in situ* measurements were consistent with the Arecibo radar measurements, although local variations at the surface could not be excluded.

To complement these measurements on the shallow subsurface, a unique dataset on the deep subsurface has been collected by the Comet Nucleus Sounding Experiment by Radio wave Transmission (CONSERT) tomographic radar (*Kofman et al.,* 2007), specifically designed for Rosetta to probe the interior of 67P using lower frequencies that can penetrate deeper inside the nucleus. CONSERT used a frequency around 90 MHz (λ = 3.33 m in vacuum) and, to further reduce the attenuation of the radar signal inside the nucleus, was operating in transmission between the Rosetta orbiter and the Philae lander at the nucleus surface rather than in reflection. The final non-optimal attitude and position of Philae at the surface of 67P limited the number of surveys carried out by the radar and caused a significant antenna mismatch that affected the quality of the received signals. However, two datasets of excellent quality were collected for the smaller lobe of the nucleus over depths reaching a maximum of 150 m beneath the surface. The results of these experiments can be summarized as follows. First, the retrieved permittivity is consistent with a bulk porosity ranging between 75% and 85% (*Kofman et al.,* 2015). There is evidence that the permittivity value decreases with depth, which is consistent with an increase in porosity and/or a decrease in the refractory-to-ice content with depth [permittivity values 1.7 to 1.95 for the <25-m subsurface, and between 1.2 and 1.32 at depths up to 150 m (*Herique et al.,* 2016; *Ciarletti et al.,* 2015; *Kofman et al.,* 2020)].

2.4.3. Implications for the bulk porosity. This variability mentioned above is likely an indication that near the surface, the nucleus is less porous. It is possible that the surface material has been compacted due to sintering during sublimation (*Kossacki et al.,* 2015), or compacted due to the CO_2 gas pressure gradient (*Davidsson et al.,* 2022c), or has a higher RTI ratio. It is interesting to mention that more generally, radar surface density estimates range between 500 and 1500 kg m^{-3} (*Harmon et al.,* 2004), larger than bulk density estimates derived from NGF. This reinforces the idea that harder and/or denser materials are present in the top layer (*Davidsson et al.,* 2009). Second, the analysis of the radar pulse width, which gets larger when volume scattering is significant, has allowed the confirmation that the nucleus is relatively homogeneous at a few meters scale (*Ciarletti et al.,* 2017). In other words, CONSERT observations did not find any substantial macroscopic voids, a finding confirmed by the Radio Science Investigation (RSI) (*Pätzold et al.,* 2016). From the constraints on internal properties described above, we can generally infer that comet nuclei should be highly porous, with voids occupying 70–80% of the total volume of a nucleus. With simple estimates of the bulk density and the composition, it is difficult to disentangle the relative contributions of micro- and macroporosity, i.e., the porosity inside dust aggregates vs. the porosity on a global scale. Therefore, radar observations aimed at determining the permittivity of the nucleus interior have thus provided important constraints on how porosity is distributed, although we stress again that porosity has never been measured directly. The fact that Rosetta did not observe macroscopic voids or cavities within the small lobe (at or above a scale of 10 m) is an important indication that porosity might primarily be confined to smaller scales.

2.5. Thermal Properties

Thermal properties dictate the temperature distribution throughout the nucleus, and are thus key to describe physical and chemical processes occurring in response to solar illumination or radiogenic heating. For instance, the thermal inertia of a comet nucleus drives its ability to adapt its temperature to a change in local insolation. A material with a large thermal inertia takes longer to adapt its temperature to changing illumination conditions compared to a material with low thermal inertia.

2.5.1. Thermal inertia of the surface layer. Before the era of spacecraft flybys, only basic considerations on the porosity and component mixing ratios were made to assess the thermal inertia of a comet nucleus. Estimates have then been derived from remote observations, either in the infrared (e.g., Infrared Space Observatory, Spitzer) or the millimeter (e.g., Institut de radioastronomie millimétrique) wavelength range. Targeted spatially unresolved observations of comet nuclei allow for the derivation of the minimum temperature, the spectral energy distribution from that of a blackbody, and sometimes a temporal shift in its thermal infrared light curve (e.g., *Fernández et al.,* 2013). The current dataset obtained through these methods indicates a thermal inertia lower than 50 J K^{-1} m^{-2} $s^{-1/2}$.

With space missions, the thermal inertia has been derived from radiance measurements from which temperature can be inferred, on the surface with spatially resolved maps of Comets 9P, 103P, and 67P. They also point toward a low thermal inertia, between 50 and 200 J K^{-1} m^{-2} $s^{-1/2}$ for 9P (*Davidsson et al.,* 2013), and less than 250 J K^{-1} m^{-2} $s^{-1/2}$ for 103P (*Groussin et al.,* 2013). The suite of instruments onboard Rosetta indicates that 67P's nucleus reaches its maximum temperature close to the subsolar point, and that the maximum temperature is close to the blackbody temperature. These two effects are indicative of a thermal inertia low enough not to cool the surface temperature significantly, or to dramatically shift it in time relative to insolation. Thermal inertia is thus estimated to be between 5 and 350 J K^{-1} m^{-2} $s^{-1/2}$ (*Schloerb et al.,* 2015; *Leyrat et al.,* 2015; *Marshall et al.,* 2018; *Leyrat,* 2019; *Davidsson et al.,* 2022c).

Thermal inertia varies across the surface (*Attree et al.,* 2018), which could be due to variations in material properties such as density or porosity between consolidated and unconsolidated terrains. There is indeed a correlation between the distribution of local gravitational slopes (*Groussin et al.,* 2015) and the thermal inertia derived for 67P (*Leyrat et al.,* 2015; *Marshall et al.,* 2018), where dusty terrains with very flat surfaces present a low thermal inertia, while consolidated terrains with very steep slopes have a high thermal inertia. The MUPUS measurement at the landing site of Philae suggests a local thermal inertia of 85 ± 35 J K^{-1} m^{-2} $s^{-1/2}$ (*Spohn et al.,* 2015), although parameters used in the model to derive this value have recently been revised, yielding a thermal inertia on the order of 120 J K^{-1} m^{-2} $s^{-1/2}$ (*Groussin et al.,* 2019). It varies also with depth, as estimates in the near-subsurface, down to 1 and 4 cm below the surface, were made possible by the Microwave Instrument for the Rosetta Orbiter (MIRO) (*Schloerb et al.,* 2015). They suggest a thermal inertia lower than that of the surface, on the order of 10 to 60 J K^{-1} m^{-2} $s^{-1/2}$, and lower than 80 J K^{-1} m^{-2} $s^{-1/2}$ (*Gulkis et al.,* 2015; *Choukroun et al.,* 2015; *Schloerb et al.,* 2015; *Marshall et al.,* 2018). It is possible that this tendency only reflects the temperature dependence of thermal properties.

2.5.2. Observational limitations. It is worth emphasizing that thermal inertia estimates discussed previously strongly depend on models used to derive them. We must take them for what they are: an indication that the material on the (near-)surface has a low thermal inertia (*Groussin et al.,* 2019; *Thomas et al.,* 2019). First, ground- and spacebased observations provide assessments of the thermal inertia for the top surface layer of a comet nucleus: It is generally assumed that these surface measurements can be relevant for the whole nucleus, although we have no evidence that this should be the case, and variations are not only plausible but indeed observed. Several other limitations arise from the observational technique itself. With remote sensing instruments, temperatures are not directly measured, but rather the IR flux, or equivalently the brightness temperature, is detected. The kinetic temperature, a basic physical quantity that gives information on thermal properties of a surface, has to be retrieved through models, taking the emissivity and roughness of the surface into account. Measurements in the near-infrared can be contaminated by reflected solar radiation (*Keihm et al.,* 2012). In addition, due to the poor resolution compared to optical images [several tens of meters, for instance, for the Visible and Infrared Thermal Imaging Spectrometer (VIRTIS)-M onboard Rosetta], the instrument ultimately detects a nonlinear average of potentially very different temperatures in the field of view, with large- and small-scale topographic features and perhaps compositional heterogeneities. Measurements from MIRO were affected by both the thermal and the optical properties of the material, which makes the interpretation challenging. Yet, a larger restriction comes from the lack of a thermal infrared instrument on the Rosetta payload. The lower limit for the temperature derived from VIRTIS-M effectively restricts the measurements to the dayside of the nucleus (*Tosi et al.,* 2019), while *Choukroun et al.* (2015) present MIRO measurements of the polar night side of the comet, getting temperatures in the 25–50-K range. Nonetheless, kinetic temperatures over complete diurnal cycles and for the same layer could not be retrieved, and thermal inertia maps were derived with large error bars (*Groussin et al.,* 2019). Finally, thermal properties are sensitive to the composition of cometary material, its structure (e.g., porosity), and often on the temperature itself. Variations in thermal inertia $\Gamma = \sqrt{\kappa \rho c}$ (J K^{-1} m^{-2} $s^{-1/2}$) may be ascribed to different variations in each of its components: the thermal conductivity κ (W K^{-1} m^{-1}), the density ρ (kg m^{-3}), or the heat capacity c (J kg^{-1} K^{-1}). Direct measurements of the thermal inertia, or the conductivity and heat capacity, cannot be made without a sample of cometary material; sample return missions would therefore fill a critical gap in our understanding of comet nuclei in this context. Until such a mission is selected, we have to rely on the goodness of our cometary analogs studied through laboratory experiments and modeling to provide estimates of the thermal properties of comet nuclei.

2.5.3. Modeling of thermal properties. The chapter in this volume by Poch et al. is dedicated to laboratory experiments providing critical measurements of the thermal properties relevant to comet nuclei. Here we discuss some methods or models developed over the years to assess thermal properties, which are of paramount importance when assessing the thermal processing of these objects. For example, the effective specific heat capacity c (J kg^{-1} K^{-1}) of the bulk material is typically given by a mass-weighted average of all constituents' specific heat capacities in the mixture [often measured in laboratory experiments on cometary analogs or natural material such as meteorites (e.g., *Piqueux et al.,* 2021)]. Like for the bulk density, the detailed knowledge of the composition is thus required to assess this parameter, which can additionally vary throughout the nucleus and evolve through time. The effective thermal conductivity κ (W K^{-1} m^{-1}) of cometary material is perhaps the most crucial yet the most complex parameter to assess; it needs to account for the composition, but also the porosity of the material. Heat transfer can occur through the solid phase: The way grains are in contact with each other thus matters critically. Besides, heat can also be transported through radiation in the empty pores, as well as by gas advection through these pores. Heat transfers by solid conduction, radiation, and gas diffusion are usually considered to follow parallel paths through the material, so that thermal conductivities add up and yield an effective thermal conductivity

$$\kappa = \kappa_{sol} + \kappa_{rad} + \kappa_{gas} \quad (1)$$

with κ_{sol} the effective conductivity of the solid phase, κ_{rad} the conductivity associated with the radiative heat transport, and κ_{gas} the conductivity associated with heat transport due to gas diffusion (all in units of W m^{-1} K^{-1}). Many studies have been performed in an effort to constrain the value of the thermal conductivity, to account for the highly porous structure of the material and its temperature-dependence, as well as to understand processes responsible for its evolution, such as sintering (e.g., *Molaro et al.,* 2019).

In many models, κ_{gas} is usually neglected, which is justified when the pressure remains low. The thermal conductivity associated with radiation in spherical pores is given by

$$\kappa_{rad} = 4\varepsilon\sigma r_p T^3 \quad (2)$$

with ε the emissivity of the solid material, σ the Stefan-Boltzmann constant, r_p (m) the effective size of pores, and T (K) the temperature. We note that *Ferrari and Lucas* (2016) and *Gundlach and Blum* (2012) provide more general formulae, which give the same value for the conductivity within a factor of a few. From equation (2), it is clear that the size of pores plays a significant role; this is where the formation mechanism of comet nuclei may affect their thermal properties. Recent estimates for the expected pore size cover a range from submicrometer- (*Keller et al.,* 2015a; *Skorov et al.,* 2017), which result in limited transport of heat through radiation, to centimeter-sized pores (*Gundlach et al.,* 2015), which dominate the heat transport at high temperatures (see Fig. 4 for millimeter-sized pores). Assessing the effective thermal conductivity of the solid phase κ_{sol} is not straightforward, as it requires a knowledge of how individual materials are mixed. For example, if grains have a layered structure (a silicate core, an organic mantle, and an ice coating), their effective conductivity would be dominated by the conductivity of the ice coating (e.g., *Haruyama et al.,* 1993). Alternatively, icy grains could co-exist with dust grains, so that their effective conductivity κ_{mix} should account for all individual contributions.

Finally, we need to account for the aggregation of these grains into the final structure of cometary material, with a method consistent with the estimation of κ_{rad}. This often leads to correction factors (described in section 2.5.4) that reduce the effective conductivity by the ratio between the contact area between grains and the mean cross section of grains. The correction factor commonly known as the Hertz factor f_H can be substantial (*Squyres et al.,* 1985; *Kossacki et al.,* 1994), and additionally varies due to sintering (*Molaro et al.,* 2019). It should be included as a correction factor to κ_{mix} before the resulting conductivity is further modified by a factor f_ψ to take into account the porous structure. The advantage of having two separate correction factors allows for the possibility to independently study processes (such as sintering) and to assess the effect of each correction factor. Some studies have, however, lumped the two correction factors into one correction factor ϕ. Thus, the effective thermal conductivity of the solid phase may be written as

$$\kappa_{sol} = f_\psi \, f_H \, \kappa_{mix} = \phi\kappa_{mix} \quad (3)$$

2.5.4. Critical considerations. Various methods exist for computing correction factors to be applied when deriving the thermal properties of cometary material: We briefly discuss three main methods, and how they compare to measurements in laboratory experiments or observations of the thermal properties of cometary material, as illustrated in Fig. 4. First, *Shoshany et al.* (2002) derived a formula for ϕ for a fractal structure of grains. They compute a single correction factor with no further correction required as

$$\phi = \left(1 - \frac{p_0(\psi)}{0.7}\right)^{(4.1 p_0(\psi) + 0.22)^2} \quad (4)$$

where $p_0(\psi) = 1 - \sqrt{1-\psi}$, and ψ the porosity. The benefit of this formula is that it accounts for a distribution of pore sizes, on purely geometrical considerations. However, this formula is both extremely sensitive to porosity and is applicable to only a limited a range of porosities (see area shown with blue lines in Fig. 4).

Second, *Gundlach and Blum* (2012, 2013) consider not only the possibility for individual grains to form the overall porous structure, but also the possibility for these monomers to be assembled in fluffy aggregates that then form the overall structure. They provide formulae for computing

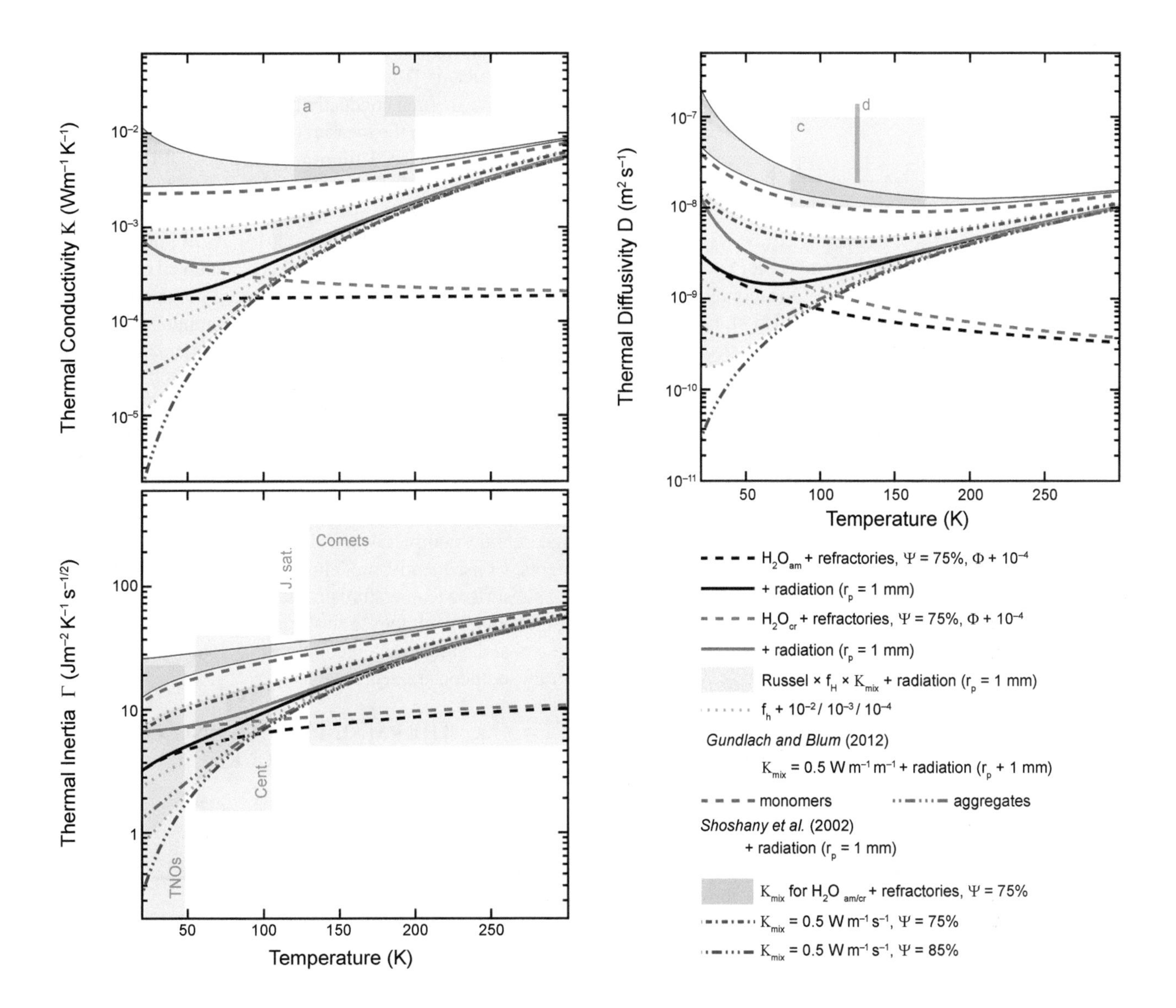

Fig. 4. *Top left:* Thermal conductivity of cometary material as a function of temperature, modeled under various assumptions (see legend and text for references). Green boxes show values measured by laboratory experiments in the relevant range of temperature: (a) *Steiner and Koemle* (1991), *Spohn et al.* (1989), and *Seiferlin et al.* (1996); (b) *Steiner and Koemle* (1991). *Bottom left:* Corresponding thermal inertia. Green boxes provided measurements of the surface thermal inertia for various populations of icy objects throughout the solar system, compiled by *Lellouch et al.* (2013), *Fornasier et al.* (2013), and *Ferrari* (2018). *Top right:* Corresponding thermal diffusivity. Green boxes show values measured in the laboratory: (c) *Spohn et al.* (1989) and *Spohn and Benkhoff* (1990); (d) *Benkhoff and Spohn* (1991). Blue lines correspond to the formula given by *Shoshany et al.* (2002) with ψ = 75%. The blue area shows variations due to the use of amorphous or crystalline water ice in κ_{mix}. The dotted line considers a constant temperature independent value for κ_{mix}, with ψ = 75%. The dashed blue line gives the effective thermal conductivity for ψ = 85%. Red lines correspond to the correction factors from *Gundlach and Blum* (2012), both for monomers (dashed line) and for aggregates (three-dotted-dashed line). Yellow lines correspond to Russel's formula with ψ = 75%, and a Hertz factor f_H with three different values: 10^{-2}, 10^{-3}, and 10^{-4}. For all these curves, κ_{rad} is added to the resulting effective conductivity of solid material, computed from equation (2) with ε = 0.9 and r_p = 1 mm. When needed, κ_{mix} is computed with 80 wt% refractories and 20 wt% water ice. Finally, black and gray dashed lines show the simplest way to write the effective thermal conductivity: $\kappa_{sol} = \phi\kappa_{mix}$, with $\phi = 10^{-4}$ and either amorphous water ice (gray; no crystallization is accounted for in this formula) or crystalline water ice (black). For the corresponding solid lines, κ_{rad} is added. Corresponding thermal inertias $\Gamma = \sqrt{\kappa\rho c}$ use ρ = 500 kg m^{-3} and c computed with the same mass fractions previously mentioned. For individual heat capacities, we used c = 7.49 T + 90 J kg^{-1} K^{-1} for water ice (*Giauque and Stout,* 1936; *Klinger,* 1980), and adopted an exponential expression for refractories similar to *Malamud et al.* (2022) based on *Bertoldi et al.* (2005): c = 1260 × (1–exp (–3.5 × 10^{-3}/T)) J kg^{-1} K^{-1}. We note that other expressions exist for meteoritic material (e.g., *Piqueux et al.,* 2021) that are relevant for cometary material.

correction factors for both components (monomers and aggregates) of this complex structure. For a porous structure of monomers (red dashed line in Fig. 4), the correction factor is given by

$$f_m = \left(\frac{9\pi}{4} \frac{1-\mu_m^2}{E_m} \frac{\gamma_m(T)}{r} \right)^{1/3} \quad (5)$$

where μ_m, E_m, and $\gamma_m(T)$ are Poisson's ratio, Young's modulus, and the specific surface energy of the monomer material respectively (see *Gundlach and Blum,* 2013). For the more complex aggregate structure, some further correction applies (red three-dotted-dashed line in Fig. 4), given by

$$f_{agg} = \left(\frac{9\pi}{4} \frac{1-\mu_{agg}^2}{E_{agg}} \frac{\gamma_{agg}(T)}{r} \right)^{1/3} \quad (6)$$

where μ_{agg}, E_{agg}, and $\gamma_{agg}(T)$ are Poisson's ratio, Young's modulus, and the specific surface energy of the aggregate material. For both types of structure, a correction factor f_ψ can be computed as a function of a filling factor, a parameter that depends on the packing structure of the material, either among monomers or aggregates (see *Gundlach and Blum,* 2012, for a complete discussion). However, we note that the aggregate model is only valid for shallow depths close to the surface of the nucleus, since the calculation of the contact between aggregates is only valid when gravitational compression is negligible with respect to the Van der Waals adhesive force (see *Gundlach and Blum,* 2012).

Finally, some models consider Russel's formula for f_ψ (*Prialnik et al.,* 2004) and apply a constant value for f_H, which is more or less equivalent to considering a constant value of ϕ (see area in yellow lines in Fig. 4). The correction factor is computed as

$$f_\psi = \frac{\psi^{2/3} f + (1-\psi^{2/3})}{\psi - \psi^{2/3} + 1 - \psi^{2/3}(\psi^{1/3} - 1) f} \quad (7)$$

where $f = \kappa_{rad}/\kappa_{mix}$, and ψ the porosity.

The comparison with actual measurements or observations is not meant to highlight the most appropriate correction factors, since the underlying assumptions on compositional and structural effects are not properly taken into account. Besides, the observed trend of thermal inertia dropping with increasing heliocentric distance (see *Ferrari and Lucas,* 2016; *Ferrari,* 2018) could be due to a strong dependence of thermophysical properties with the temperature, and thermal inertia measurements are made for the uppermost layer of surface material. Such measurements thus produce a biased view of the cometary material thermal properties. On the contrary, Fig. 4 is meant to illustrate how measurements of the surface thermal inertia (including those derived from Rosetta data) do not allow a critical assessment of how much heat is transported through the upper layers of a comet nucleus. This would require a direct constraint on the effective thermal conductivity itself, which was supposed to be done through the measurement of a temperature profile below 67P's surface. Unfortunately, the MUPUS penetrator onboard Philae could not provide such measurements, because the thermal sensors were not hammered into the ground as planned (*Spohn et al.,* 2015). As of now, only limited laboratory experiments have been attempted on adequate cometary analogs; values of the thermal conductivity vary depending on the assumptions used to compute it and span at least 2 to 3 orders of magnitude at low temperatures. In this figure, thermal properties converge beyond 200 K because heat transport through radiation inside the millimeter-sized pores starts to dominate in this temperature range (i.e., $\kappa_{rad} > \kappa_{sol}$). Despite these large differences, the conversion from thermal conductivity to thermal inertia (assuming the same density and effective heat capacity) provides values that remain completely consistent with current measurements. Consequently, modeling work aimed at constraining the subsurface temperature profile and the behavior of active comet nuclei continues to rely on the assumptions described above, and simulation outcomes remain strongly dependent on these unconstrained parameters.

3. THERMAL PROCESSING OF COMET NUCLEI

3.1. Internal Properties Derived from Activity Patterns

The ubiquitous characteristic shared by all comet nuclei is their mass-losing behavior ("activity") with its wealth of manifestations, sometimes erratic and unexpected. The general framework for understanding the thermal processing of these objects is thus tightly linked to the understanding of cometary activity itself. For instance, the depth of a volatile species can be assessed by correlating its production in the coma with heliocentric distance, as such correlation clearly depends on volatility. Typically, the production of water in a coma is the most strongly and directly correlated to solar heating, since water ice sublimates mostly from the surface or a thin subsurface layer: It thus quickly responds to changes in the surface temperature. It is thought that as volatility increases, so to does the depth where the corresponding ice sublimates. If volatile ices are located within the orbital skin depth, they can still correlate with insolation, although a shift in their production peak with respect to perihelion would be expected, due to the lag of the heat wave propagating below the surface. A similar effect is expected on diurnal cycles.

In this framework, the most volatile species could originate from deep layers as they require less heat to sublimate, or could be released from amorphous water ice crystallization or from CO_2 segregation. Therefore, their production rates should be independent of diurnal variations in insolation. Evidence for such behavior can be found in VIRTIS observations of 67P. Indeed, *Bockelée-Morvan et al.* (2015) and *Migliorini et al.* (2016) observed that H_2O vapor does follow the local insolation: It is produced by detectable well-illuminated active areas and is only weakly emitted from areas with low solar illumination. This confirms that water ice is indeed present very close to the surface. However, CO_2 vapor appears to be produced by both illuminated and non-illuminated areas of the nucleus, and shows a more uniform distribution in the coma. This is consistent with CO_2 sublimating from layers deeper than at least the diurnal skin depth [with sublimation fronts ~3.8 m below the surface in the northern hemisphere of 67P, and 1.9 m below the surface in the south according to model reproduction of the CO_2 production according to *Davidsson et al.* (2022b)]. From a structural point of view, Rosetta provided additional constraints on subsurface properties, from MIRO and Philae, which are consistent with 67P's top meter surface layer being more compacted than the rest of the interior (*Ciarletti et al.,* 2015; *Lethuillier et al.,* 2016; *Brouet et al.,* 2016), which could be the result of sintering, the process that compacts porous material after heating (*Kossacki et al.,* 2015).

Whether the internal structure and composition of a comet nucleus is uniform may be assessed from the variability of its activity, or possibly outbursts. Internal heterogeneities, whether primordial or formed by post-formation processing, indeed have an impact on a comet's general activity pattern (*Rosenberg and Prialnik,* 2007, 2009, 2010; *Guilbert-Lepoutre et al.,* 2016). For instance, *Rosenberg and Prialnik* (2010) studied the effect of internal heterogeneities in the form of talps (*Belton et al.,* 2007), which in the model consist of extended, 10-m-thick layers, with characteristics differing considerably from the "reference" material in dust to ice ratio, density, and thermal conductivity. Internal layers of different structure and composition affect the overall evolution of the internal structure itself, and the formation of a surface dust mantle, in different ways. They also affect the progression of phase transitions, including surface erosion due to water ice sublimation: The addition of non-uniform layers in the subsurface results in varying production rates, even when surface temperatures remain the same as for a reference, internally uniform nucleus. Overall, heterogeneous internal properties lead to a somewhat erratic activity, with peaks of water and dust production at all heliocentric distances, even close to aphelion. Cometary outbursts of activity, diverse in both intensity and duration, have been observed over a wide range of heliocentric distances for a long time: They may simply arise from internal heterogeneities. Of course, Rosetta has further complicated this picture, as short-lived, local outbursts have been observed to originate from the surface processes throughout the mission (see the chapter by Filacchione et al. in this volume).

3.2 Early Thermal Alteration

3.2.1. Effect of radiogenic heating. Assessing the early thermal evolution of comet nuclei remains very speculative, since we only have limited constraints on key aspects of this phase of their life, e.g., their formation mechanism, formation time, formation timescale, or formation location, which cannot be directly assessed from their properties as observed at present. At present there is a general consensus that if early processing of cometary nuclei took place, it would be dominated by the effect of radiogenic heating, induced by the decay of short-lived radiogenic isotopes such as ^{26}Al, the most efficient radiogenic source, or ^{60}Fe and ^{53}Mn. The corresponding heating rate $\mathcal{Q}_{rad}$ is given by

$$\mathcal{Q}_{rad} = \sum_{rad} \varrho_d \, X_{rad}(0) \, \Delta H_{rad} \frac{1}{\tau_{rad}} \exp\left(\frac{-t}{\tau_{rad}}\right) \tag{8}$$

with ϱ_d (kg m^{-3}) the mass per unit volume of refractory dust, $X_{rad,0}$ the initial mass fraction, ΔH_{rad} (J kg^{-1}) the heat released per unit mass, and τ_{rad} (s) the mean lifetime of each radiogenic isotope rad. The total heating rate is found by adding individual contribution from all short-lived elements. From equation (8) it is immediately apparent that depending on the formation time, timescale, or initial composition of cometary nuclei, the resulting amount of energy delivered to their internal parts may vary greatly (see Fig. 5 for examples of central temperature evolutions).

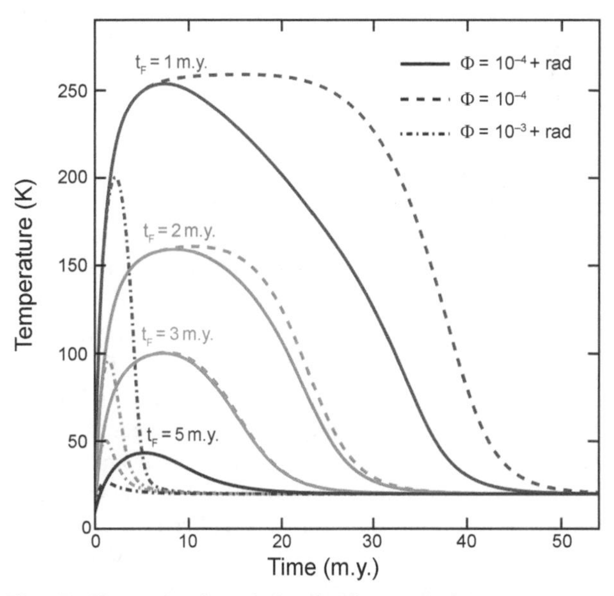

Fig. 5. Example of evolution for the central temperature of a comet nucleus due to radiogenic heating, with a radius R = 5 km and a refractory-to-ice ratio of 1. Different colors correspond to different formation times from 1 to 5 m.y. after CAI formation. Line styles correspond to different thermal conductivities, expressed in the form of a correction factor applied to the thermal conductivity of the solid material, and the addition of heat transferred by radiation in 1-mm pores.

Different studies have used varied initial abundances of radionuclides, based on the choice of chondritic sample supposed to best represent cometary material within our meteoritic collections. The literature thus offers a range of initial mass fractions (e.g., *De Sanctis et al.,* 2001; *Choi et al.,* 2002; *Guilbert-Lepoutre et al.,* 2011; *Davidsson,* 2021), and effects of radiogenic heating may thus result in a range of outcomes. If comet nuclei are formed early (and quickly) enough, short-lived radioactive nuclide ^{26}Al should be effective in heating their interior, although only for a very limited period after the formation of CAIs (believed to be the first solids to condense in the solar system). Whether this nuclide was actually common in the comet-forming region as in the region forming carbonaceous chondrites, and thus accreted in comet nuclei, is still a matter of debate (*Prialnik and Podolak,* 1995). Indeed, the Stardust dust samples show that although minerals do contain CAIs, the inclusions have no radiogenic excess of ^{26}Mg, the decay product of ^{26}Al (at least the inclusions for which an excess of ^{26}Mg has been searched for). CAIs with an initial inferred low ratio of ^{26}Al/^{27}Al [$\sim 5 \times 10^{-6}$, compared to the canonical value of 5×10^{-5} (*MacPherson et al.,* 1995)] have been reported for some meteorites (*Kunihiro et al.,* 2004; *Makide et al.,* 2009) and used to invoke a non-uniform distribution of ^{26}Al within the solar system (*Makide et al.,* 2011; *Krot et al.,* 2012; *Ogliore et al.,* 2012). However, whether this is the result of a bias in our samples remains to be understood, as other studies argue for a uniform distribution of ^{26}Al in the early solar system (*Huss et al.,* 2001). A comet sample return mission could address this point more specifically. Additional sources of uncertainties arise from the nature of cometary material and the thermophysical properties resulting from the formation mechanism (e.g., *Malamud et al.,* 2022). We illustrate this effect in Fig. 6, where different thermal conductivities drive different internal heating patterns.

As a consequence, studying the impact of early radiogenic heating on the internal properties of comet nuclei remains a difficult task. Assuming an initial value of 5×10^{-5} for the ^{26}Al/^{27}Al ratio, and depending on the choice of initial characteristics such as thermophysical parameters, sizes, or formation time, the early evolution of comet nuclei could lead to distinct configurations (*Prialnik et al.,* 1987; *Haruyama et al.,* 1993; *Prialnik and Podolak,* 1995; *De Sanctis et al.,* 2001; *Choi et al.,* 2002; *Mousis et al.,* 2017). Pristine structures could be thoroughly preserved, especially if the formation time of comet nuclei is delayed by several million years with respect to the formation of CAIs. Alternatively, the interior could be completely crystallized except for an outer layer that might remain primitive, assuming water ice is initially amorphous. Finally, some nuclei might develop a differentiated, stratified structure with a crystallized core and layer of condensed volatiles (i.e., during crystallization, trapped volatile species are released, diffuse, and condense in colder regions away from the center of the nucleus) and more pristine outer layers made of unaltered material. The occurrence of high internal temperatures, leading to the production of liquid water (which additionally requires high pressure, above the triple point), has also been reported. Such high internal temperatures are more likely when accounting for the effect of accretional heating, affecting the early evolution of comets concurrently with radiogenic heating during the short lifetime of ^{26}Al. *Merk and Prialnik* (2006) showed that the occurrence of liquid water in 2- to 32-km-radius bodies may be a very common phenomenon, extending from 10% to 90% of the overall interior. Regardless, early thermal evolution is thought to significantly alter the volatile species initially present in comet nuclei (e.g., *Lisse*

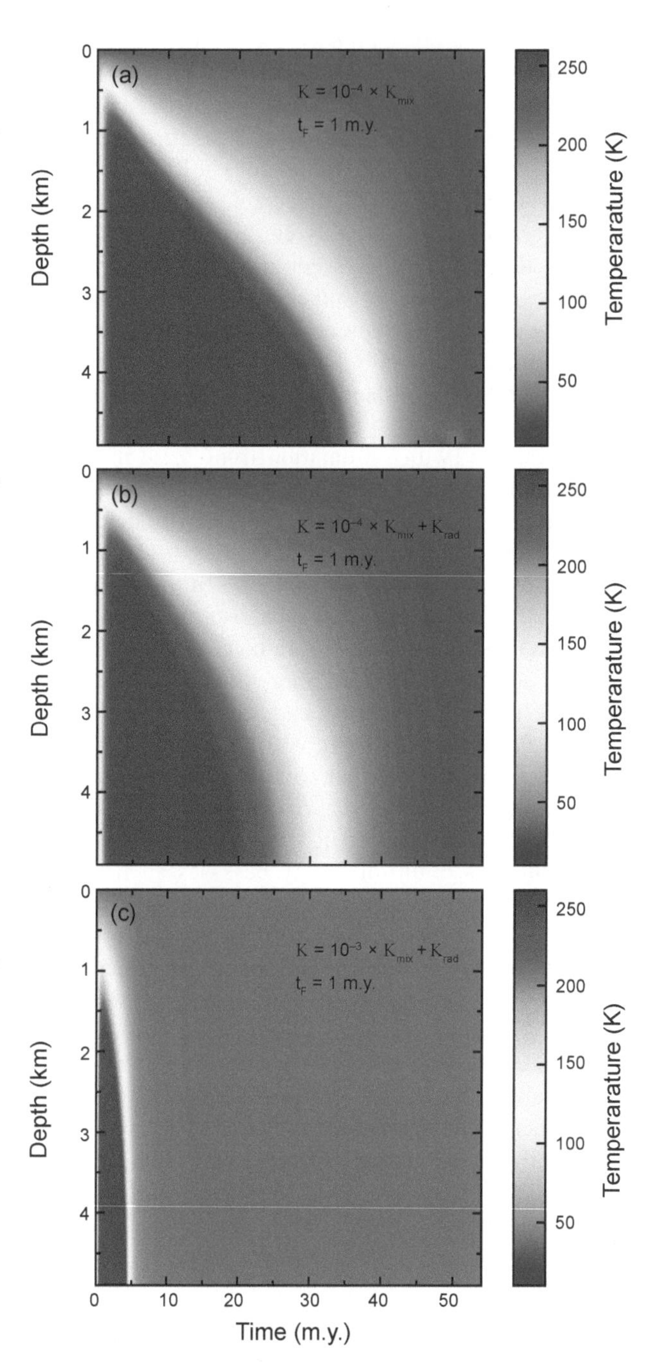

Fig. 6. Examples of evolution of the internal temperature distribution for a 5-km-radius comet nucleus formed 1 m.y. after CAI formation. Panels correspond to different thermal conductivities: **(a)** $\kappa = 10^{-4}\ \kappa_{mix} + \kappa_{rad}$, **(b)** $\kappa = 10^{-4}\ \kappa_{sol}$, **(c)** $\kappa = 10^{-3}\ \kappa_{mix} + \kappa_{rad}$. κ_{rad} is computed for 1-mm pores.

et al., 2022; *Steckloff et al.*, 2023). With a pebble structure for cometary material, *Malamud et al.* (2022) further argue that pristine comet nuclei could be rare. Nonetheless, the fact that hypervolatile species are observed today in most comets suggests that these nuclei have not been significantly heated before their arrival in the inner solar system, or formed after they could have incorporated much ^{26}Al.

3.2.2. Effect of surface irradiation. During the early stages of comet nuclei evolution, superficial heating sources can additionally contribute to the alteration of surface layers, left relatively untouched by radiogenic heating. Indeed, the luminosity of the proto-Sun was more intense than today's, which would lead to more intense heating of comet nuclei. However, the presence of the Sun's protoplanetary disk could dampen this effect in the comet-forming region. Such effects are only starting to be considered in models aimed at assessing the impact of early thermal processing, so there is still a large degree of discrepancy between studies. For instance, *Davidsson* (2021) suggests that nuclei of any size exposed to the intense heating of the early proto-Sun at their putative birth location would lose not only hypervolatiles condensed as pure ices, but also CO_2 down to 30 m. In addition, partial crystallization could occur in the upper 200 m. *Lisse et al.* (2022) argue that all comet nuclei would have lost all untrapped volatiles. As a consequence, the near-surface layers of a comet nucleus may be significantly processed: Even dynamically new comets are likely not pristine, and most would have lost a significant amount of hypervolatiles prior to their scattering to the Oort cloud. Through a combination of protosolar and long-lived radionuclide heating, comet nuclei in the protoplanetary disk with diameters ranging from 4 to 200 km might lose all their condensed CO ice on timescales smaller than the minimum time necessary to eject them in the reservoirs (*Davidsson,* 2021; *Lisse et al.,* 2022). Complete loss of condensed hypervolatiles is confirmed by *Prialnik* (2021) in concurrent heat and gas diffusion simulations similar to those of *Davidsson* (2021), and is also supported by simplified models (*De Sanctis et al.,* 2001; *Choi et al.,* 2002; *Steckloff et al.,* 2021; *Lisse et al.,* 2021). The magnitude of CO_2 loss depends on assumptions about the clearing time of the solar nebula and the protosolar luminosity at that point in time (*Davidsson,* 2021; *Steckloff et al.,* 2021; *Lisse et al.,* 2021).

Consequently, the objects stored in the outer solar system reservoirs (the transneptunian region or the Oort cloud) probably do not preserve a truly pristine inventory of hypervolatiles or mechanical structure. Once in the Kuiper belt (or the transneptunian region in general), the equilibrium temperature can range from 30 to 50 K, so that all hypervolatiles initially present as pure ice condensates might sublimate during the time of residence in this reservoir. This is typically what is expected for objects smaller than 4 km (*De Sanctis et al.,* 2001; *Choi et al.,* 2002; *Jewitt,* 2004; *Davidsson,* 2021; *Lisse et al.,* 2021, 2022; *Steckloff et al.,* 2021). In the Oort cloud, comet nuclei are subject to irradiation by galactic cosmic rays or passing stars, which would produce a volatile-depleted and hardened superficial crust of several meters (see also section 2.1). The extent of alteration depends on the energy of the impacting particles and the duration of these events. As a result, up to 20% of comet nuclei stored in the Oort cloud could have been heated to at least 30 K, down to several dozen meters below the surface, due to the passage of luminous stars during the history of the solar system (*Stern and Shull,* 1988). Most of them might have been heated to 45 K in the uppermost 1-m layer due to stochastic supernovae events. *Stern* (2003) further reports that passing stars and supernovae heating events could modify the primordial composition of comet nuclei down to 5 to 50 m (for heating due to passing stars), and 0.1 to 2 m (for heating due to supernovae events).

3.3. Processing Beyond the Water Ice Line

3.3.1. Mechanisms at play in the giant planet region. When comet nuclei leave the outer solar system reservoirs to transit toward the inner solar system, insolation, which is a function of d_H^{-2}, with d_H the heliocentric distance, is the dominant energy source. Phase transitions can then be induced locally that also act as secondary heat sources and sinks. As long as comet nuclei remain located beyond 4–6 au from the Sun, water ice is stable and does not sublimate. Other phase transitions (e.g., CO_2 sublimation or amorphous water ice crystallization) can nonetheless be triggered, and contribute to modifications of the internal structure and composition. In addition, the release of material drives cometary activity in these outer regions of the solar system; therefore, by studying the activity patterns and the diversity of behaviors of Centaurs in the giant planet region, we may infer constraints on the degree of thermal processing sustained by cometary nuclei before.

3.3.1.1. Sublimation/condensation: Because the internal structure of comet nuclei is very porous, sublimation and condensation of volatiles can occur in any volume that sustains the appropriate conditions to trigger these phase transitions. Overall, this process efficiently transports energy in the form of latent heat, with a heat source term written as [derived from *Langmuir* (1913) for application to comet nuclei by *Weissman and Kieffer* (1981)]

$$\mathcal{Q}_{\mathrm{subl}} = \mathcal{S}\left(\mathcal{P}_{\mathrm{sat,i}} - \mathcal{P}_{\mathrm{i}}\right)\sqrt{\frac{m_i}{2\pi k_B T}}\,\Delta H_i \qquad (9)$$

with $\mathcal{S}$ (m^{-1}) the surface to volume ratio of pores [see *Prialnik et al.* (2004) for various formulations], m_i (kg) the mass of a molecule of species i, k_B the Boltzmann constant, and ΔH_i (J kg^{-1}) the latent heat of sublimation. The gas pressure $\mathcal{P}_i$ (Pa) is estimated with the ideal gas law, while the saturated vapor pressure $\mathcal{P}_{sat,i}$ (Pa) is computed from the Clausius-Clapeyron equation. *Fray and Schmitt* (2009) propose vapor pressure relations for more than 50 volatile species, either empirical or theoretical, valid over a large range of temperatures, including low temperatures representative of cometary interiors. Finally, we note that if $\mathcal{P}_i > \mathcal{P}_{sat,I}$, then equation (9) describes the heat released due to recondensation.

We note that local sublimation and condensation events drive a form of ice metamorphism, also known as sintering (mentioned in previous sections), which influences the microstructural evolution of the grain aggregates forming cometary material. Sintering is driven by the thermodynamic requirement to decrease surface energy, with various diffusion mechanisms dominating the process under different temperatures, pressures, depths, or material (*Molaro et al.,* 2019). As a result, the recondensation of gas molecules on cometary grains tend to form sintering necks, which — by increasing the contact area between grains — result in the increase of the material's thermal conductivity, strength, and the overall evolution of local thermophysical properties (*Kossacki et al.,* 1994, 2015; *Molaro et al.,* 2019).

3.3.1.2. Crystallization of water ice: The transition from amorphous water ice to its crystalline form has traditionally been recognized as a key transformation that could explain some comet nuclei behaviors. However, any direct detection of amorphous water ice remains to be performed. This volume includes a chapter dedicated to this topic (see the chapter by Prialnik and Jewitt) and its potential role in driving cometary activity and behavior throughout the solar system, so we only provide here some simple analytical considerations to understand the process. Crystallization of amorphous water ice as a heat source can be described by

$$\mathcal{Q}_{cryst} = \lambda(T)\, \varrho_a\, \Delta H_{ac} \tag{10}$$

with ϱ_a (kg m^{-3}) the amorphous water ice mass per unit volume. The phase transition releases a latent heat ΔH_{ac} = 9 × 10^4 J kg^{-1} (*Klinger,* 1981) at a rate (in s^{-1}) determined by *Schmitt et al.* (1989)

$$\lambda(T) = 1.05 \times 10^{13}\, e^{-5370/T} \tag{11}$$

There is a debate, however, as to how much energy may actually be released upon crystallization, in particular when amorphous water ice is hosting other species or impurities (*González et al.,* 2008; *Kouchi and Sirono,* 2001). Beside its local heating effect, this phase transition has a number of consequences on cometary material; e.g., it changes its overall thermal conductivity. More importantly, it releases occluded gases, which can escape or condense in colder volumes inside the nucleus, thereby modifying the local inventory and distribution of volatiles.

3.3.1.3. Segregation of hypervolatiles from carbon dioxide ice: Amorphous water ice is not necessarily the only storage medium of hypervolatile species and possibly not the most important one. Laboratory experiments show that CO_2 ice efficiently stores CH_4 (*Luna et al.,* 2008), N_2 (*Satorre et al.,* 2009), and CO (*Cooke et al.,* 2018; *Simon et al.,* 2019) at temperatures substantially above the sublimation temperatures of these guest species. Intimate mixtures of CO_2 and CO ices are common among interstellar ices (e.g., *Pontoppidan et al.,* 2008), and could potentially be present within comet nuclei to this day. Correlations between the local CO_2 production rate and those of CO, C_2H_2 (*Luspay-Kuti et al.,* 2015), CH_4, HCN, and H_2S (*Gasc et al.,* 2017) are present in the Rosetta/ROSINA dataset for 67P. *Gasc et al.* (2017) therefore suggested that these species are stored in CO_2 ice. From our current understanding of the process, substantial segregation (i.e., release of hypervolatile vapor trapped in the CO_2 ice) can take place near T = 75 K, after the CO_2 ice goes through a compaction of its structure that substantially reduces its microporosity, which was acquired during its low-temperature formation (*Luna et al.,* 2008; *Satorre et al.,* 2009). At such temperatures, amorphous water ice is stable on ≥10^5-yr timescales (*Schmitt et al.,* 1989). Carbon dioxide segregation therefore provides a mechanism of hypervolatile release that is intermediate between CO sublimation (20–40 K) and amorphous water ice crystallization (starting effectively around 100–110 K). *Davidsson* (2021) proposed a segregation rate μ (s^{-1})

$$\mu(T) = \frac{3}{2}(1-\psi)\Lambda e^{-E_s/T} \tag{12}$$

where ψ is the porosity, Λ is the pre-exponential factor of the segregation activation law, E_s (K) is the segregation activation energy, and T (K) is temperature. The energy consumed during segregation is taken as the latent energy of sublimation for the released species. The consequences of CO_2 entrapment for the long-term storage of hypervolatiles in comets, as well as for distant activity in such bodies, deserves further study both in the form of laboratory experiments and thermal evolution modeling.

3.3.2. Evidence for ongoing processing. *3.3.2.1. Onset of activity for Oort cloud comets:* By the time *Comets II* was written, cometary activity had been observed for objects at great heliocentric distances, in the 10–15 au range (e.g., *Meech and Jewitt,* 1987; *Meech and Belton,* 1990; *Roemer,* 1962; *Sekanina,* 1975). Activity at larger heliocentric distance (>20 au) had thus been postulated for a long time. Recently, a combination of more sensitive instruments and dedicated survey facilities have provided clear detections of activity in very distant comets (e.g., *Jewitt,* 2009; *Meech et al.,* 2009, 2017), which suggests that further progress in sensitivity will inform us about whether this type of activity is exceptional or common. For instance, Comet C/2010 U3 (Boattini) was reported to be active at d_H =25.8 au (*Hui et al.,* 2019), and Comet C/2014 UN271 (Bernardinelli-Bernstein) was reported to be active beyond 20 au [at d_H ~ 29 au in 2014 and d_H ~ 20 au (*Bernardinelli et al.,* 2021)]. The current record is held by long-period Comet C/2017 K2 (PANSTARRS), which was discovered at an inbound heliocentric distance d_H = 15.9 au, and was found to be active in prediscovery data at d_H = 23.7 au (*Jewitt et al.,* 2017a; *Meech et al.,* 2017; *Hui et al.,* 2018). By studying the development of activity since its discovery, *Jewitt et al.* (2021) suggest that dust production could have started at $d_H \sim 35$ au. With a surface temperature of roughly ~70 K at 16 au, and ~45 K at 35 au, water ice is not only thermodynamically stable, but the crystallization of amorphous water ice would require a timescale larger than the age of the solar system to occur. Such extreme cometary activity can be sustained for example by hypervolatile sublimation: CO was indeed detected in the coma of C/2017 K2, albeit at 6.72 au (*Yang et al.,* 2021).

3.3.2.2. Active Centaurs: Given the known dynamical cascade between Kuiper belt objects, Centaurs, and JFCs, studying the origin of cometary activity among Centaurs is very relevant in constraining the overall evolution and stratification pattern of the nuclei of short-period comets. As of today, the source of the activity of Centaurs has not been definitively identified, and different processes may be involved for different individual objects (*Prialnik et al.,* 1995; *Capria et al.,* 2000; *De Sanctis et al.,* 2000). Our current impediments for further understanding of the origin of active Centaurs come, on the one hand, from the lack of volatile detection in their comae (maybe due to limited observational sensitivity). For instance, no strong detection of gaseous CO has been made to date, except for Centaur 29P/ Schwassmann-Wachmann 1 (*Senay and Jewitt,* 1994; *Crovisier et al.,* 1995; *Gunnarsson et al.,* 2008). CO was marginally detected in the coma of 2060 Chiron at d_H =8.5 au (*Womack and Stern,* 1999), as well as on the coma of 174P/ Echeclus during an outburst at 6.1 au (*Wierzchos et al.,* 2017). Otherwise, the most sensitive observations of CO to date are consistent with the activity of Centaurs not being driven by the sublimation of CO (*Drahus et al.,* 2017). We may be further limited by the current sensitivity of observational facilities, as some objects may be active below our current detection limits.

Jewitt (2009) noticed that the physical and orbital properties of active Centaurs are such that their activity should be driven by a thermal process: To understand the activity of Centaurs, we can thus rely on thermal evolution models. Currently, no active Centaur with a perihelion distance beyond ~12 au (Fig. 7) has been detected. Given these constraints, processes able to drive the activity of Centaurs could be related to the transition from amorphous to crystalline water ice [(efficient from 5 to 10 au, possibly up to 12 au (*Guilbert-Lepoutre,* 2012)] or phase transitions of CO_2 [sublimation or segregation, efficient in the 8–12-au region where crystallization rates strongly decline if CO_2 is present (*Davidsson,* 2021)]. Due to the long-term, inward propagation of phase transition fronts below the surface, cometary activity can only be sustained for a limited time, typically hundreds to thousands of years (*Guilbert-Lepoutre,* 2012). This implies that active Centaurs should have suffered from a recent orbital change: After a change in orbital parameters, the nucleus would need to adjust to new thermal conditions, at which point phase transitions could be triggered and produce some observable cometary activity. Empirical data confirm that active Centaurs are prone to drastic drops in

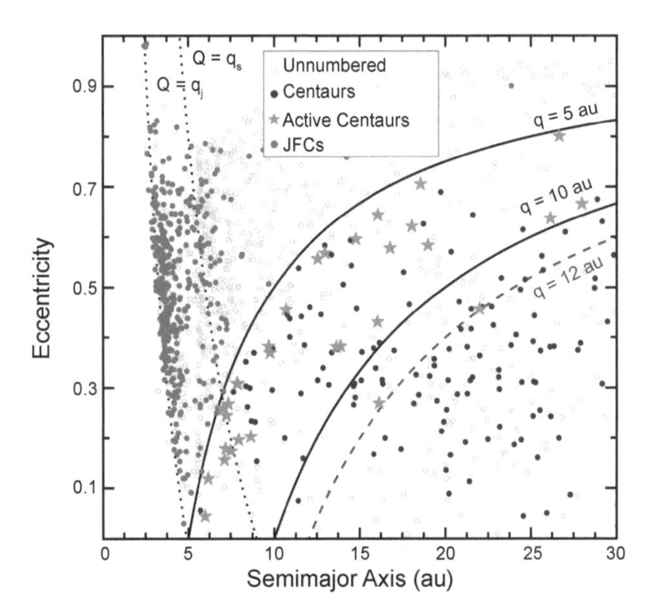

Fig. 7. Orbital distribution of Centaurs and JFCs in the semimajor axis vs. eccentricity plane. The locus of perihelion distances at 5, 10, and 12 au is shown with solid black lines and dashed red lines respectively. The locus of aphelion distance at Jupiter and Saturn is shown with dotted black lines. Active Centaurs are marked with orange stars. Orbital elements come from the JPL Solar System Dynamics database (*https://ssd.jpl.nasa.gov*).

their perihelion distances, with a timescale of 10^2 to 10^3 yrs (*Fernández et al.,* 2018). *Cabral et al.* (2019) argued that crystallization as the source of cometary activity among Centaurs would favor activity on objects dynamically new to the Jupiter-Saturn region, as others having previously stayed in this region would have exhausted their amorphous water ice in the near-surface layers. We note that searches for activity among recently discovered Centaurs have failed (*Cabral et al.,* 2019; *Li et al.,* 2020; *Lilly et al.,* 2021), which is to be expected as they target Centaurs on relatively stable orbits beyond Saturn.

3.3.3. Transitioning to the inner solar system. In general, Oort cloud comets are injected into the inner solar system on a "direct" path, whereas JFCs spend a significant amount of time [typically 10 m.y. (*Levison and Duncan,* 1997; *Tiscareno and Malhotra,* 2003; *Volk and Malhotra,* 2008; *Bailey and Malhotra,* 2009)] as Centaurs. We can anticipate that such an indirect pathway toward the Sun would result in modifications of the internal composition and structure of observed JFCs, possibly quite extensive, in marked contrast to the evolution resulting from the direct injection of Oort cloud comets. Pathways from the outer to the inner solar system can be very complex (see the chapter by Fraser and Dones in this volume): Most JFCs come from the scattered disk in the transneptunian region (*Nesvorný et al.,* 2017). The classical Kuiper belt, including various resonant populations below 50 au, is also an important source of short-period comets (*Emel'yanenko et al.,* 2013). Both *Emel'yanenko et al.* (2013) and *Nesvorný et al.* (2017) noticed that a few JFCs could originate in the Oort cloud: The orbital evolution of these comet nuclei is similar to Halley-type comets, i.e., dynamically evolved, returning LPCs.

The extent of the thermal modifications sustained during the dynamical evolution of comets is only starting to be properly quantified. When comet nuclei are observed at the same heliocentric distance, models tell us that we should expect some similarities: (1) surface temperatures, driven by the heat balance at the surface; and (2) water and CO_2 gas productions, which are controlled by erosion that keeps both water and CO_2 ices close to the surface (see *Huebner et al.,* 2006; *Steckloff and Samarasinha,* 2018). Yet distinct orbital evolutions should result in significant differences on the internal composition and structure of comet nuclei, resulting from the inward propagation of the heat wave during the very different timescales comet nuclei reside in the giant planet region, or more generally closer to the Sun, and reflected in the depth of key phase transition fronts. *Guilbert-Lepoutre et al.* (2016) suggested that typical JFC nuclei could have their subsurface altered down to a few hundred meters before entering the inner solar system. *Gkotsinas et al.* (2022) found that due to the stochastic nature of JFC trajectories toward the inner solar system, they experience multiple heating events, resulting in substantial chemical alteration of their upper layers, down to several hundred meters.

The dynamical cascade between transneptunian objects (TNOs), Centaurs, and JFCs could be characterized by two modes of migration through the giant planet region: (1) the well-known dynamical chaos and (2) mean-motion resonance hopping (*Bailey and Malhotra,* 2009; *Seligman et al.,* 2021). As a result, the individual orbital track that any comet nuclei follows to reach the orbit where it is currently observed is impossible to constrain beyond the last close passage to Jupiter (see Fig. 8 for one example). Recently, there has been a large effort to identify Centaurs on the edge of transitioning to a JFC-like orbit, as those would allow an investigation of how dynamical and thermal evolution alters comet nuclei before they become JFCs (*Sarid et al.,* 2019; *Steckloff et al.,* 2020; *Kareta et al.,* 2021; *Seligman et al.,* 2021). These studies rely on the transition from the Centaur to the JFC region through a restricted area in the orbital elements space, called the Gateway. However, by coupling a thermal evolution model to the orbital evolution of JFC dynamical clones, *Guilbert-Lepoutre et al.* (2023) showed that most comet nuclei found in this region are altered due to previously sustained thermal processing. This is because when the orbits of comet nuclei are controlled by interactions with Jupiter, multiple passages from the giant planet region to the inner solar system are possible (*Sarid et al.,* 2019; *Seligman et al.,* 2021; *Guilbert-Lepoutre et al.,* 2023) (see also Fig. 9 for an example). It is, however, impossible to know whether any given Centaur is on its first pass to the inner solar system. As a result, comet nuclei in the gateway are statistically more prone to being thermally processed than other Centaurs crossing the orbit of Jupiter for the first time, because a higher fraction of gateway objects have already been JFCs.

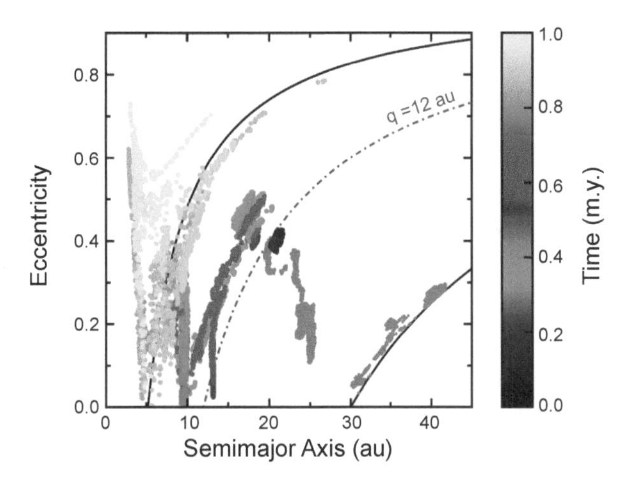

Fig. 8. Example of one possible orbital evolution from the transneptunian region, in the eccentricity vs. semimajor axis plane. One data point is given every 100 yr of dynamical evolution: the color code provides the time evolution for the last million years, while the prior orbital evolution is shown in gray. The black solid lines show orbits with a perihelion distance of 5.2 and 30 au. The blue dotted-dashed line shows orbits with a perihelion distance of 12 au. Data from *Nesvorný et al.* (2017).

Depending on thermophysical parameters, compositions, and other poorly constrained internal properties, it is possible that the effects of thermal processing sustained during the Centaur phase could be limited to a modest near-surface layer. Our current understanding of such characteristics indicates that the layer actively contributing to the production of cometary activity in the inner solar system (discussed in the next section) will reflect the composition and structure inherited from their previous stages of evolution described here. This layer is likely very different from its initial state; consequently, for a significant number of comet nuclei, outgassing observed today might still occur from a layer thermally altered during the Centaur stage.

3.4. Activity in the Inner Solar System

The processing of comet nuclei in the inner solar system is so complex that it is the focus of two separate chapters in this volume (see the chapters in this volume by Filacchione et al. and Pajola et al. for the composition and morphology, respectively, of comet nuclei and their evolution during this phase). Here we briefly discuss the main features of the thermally-induced alteration of comet nuclei, focusing on the effect on their internal characteristics. Once comet nuclei get closer than 4–6 au to the Sun, variations in the temperature across their surface start to become significant. Since phase transition rates depend exponentially on the temperature and vary widely between volatile species, the local inducement of such transitions is expected to give rise to distinct sublimation fronts below the surface as well as complex lateral patterns (e.g., *Guilbert-Lepoutre and Jewitt,* 2011).

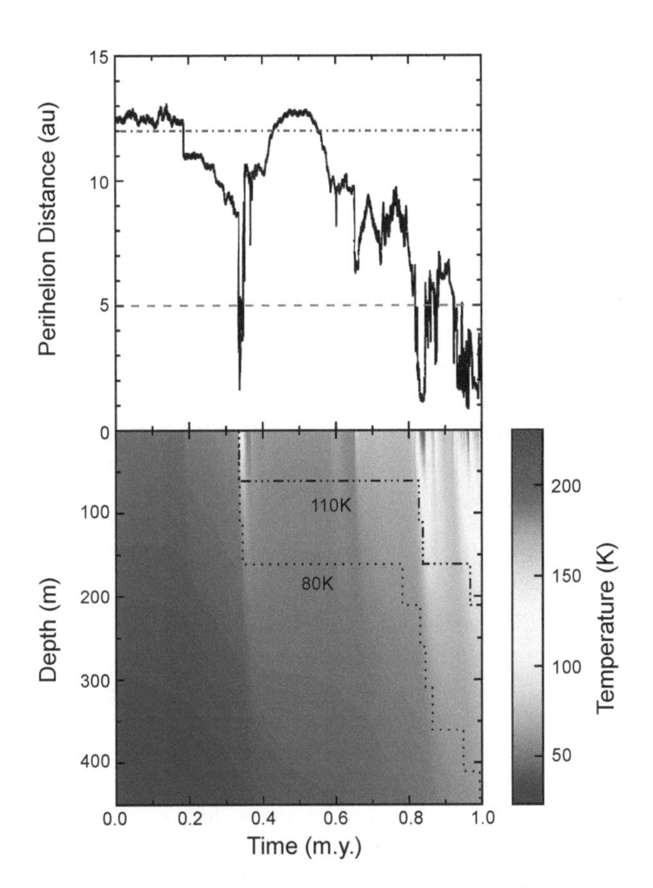

Fig. 9. *Top panel:* Evolution of the perihelion distance as a function of time, resulting from the orbital evolution shown in Fig. 8 (only the last million years are shown). *Bottom panel:* Distribution of the internal temperature as a function of depth and time, resulting from this orbital evolution. The maximum depths of the 80 and 110 K isotherms are shown with black dotted and dash-dotted lines respectively. Adapted from *Guilbert-Lepoutre et al.* (2023).

Therefore, once cometary activity properly sets in (i.e., inside the orbit of Jupiter), characteristics such as the rotational period of a nucleus, the orientation of its spin axis (the properties of which are detailed in the chapter in this volume by Knight et al.), or variations of the surface composition and local topography (see the chapters by Filacchione et al. and Pajola et al. respectively), become instrumental in shaping the evolution of comet nuclei. Indeed, these influence how much energy is provided to the surface and how it is provided. As *Benseguane et al.* (2022) showed that both the peak energy received close to perihelion and the integrated energy received over one orbit can drive the evolution of 67P's surface features, we illustrate this effect in Fig. 10, which shows how the latitude of the subsolar point — the point at the surface with the largest temperature — evolves across one orbital revolution. *Kossacki* (2016) further showed that all locations on 67P's surface are not processed to the same degree, with some areas where a pristine unconsolidated ice-dust mixture may have been preserved, and other areas subject to considerable processing of both composition and structure. In addition, *Keller et al.* (2015a) showed that the energy balance at the surface of 67P can be extremely varied, especially when accounting for self-heating and shadowing: Its concave neck area receives 50% more energy during the northern summer due to self-heating, for example.

These complex heating patterns have an influence on how energy is transferred to the nucleus surface because of subsurface energy storage. For example, when Rosetta arrived at 67P, the nucleus was already active (e.g., *Snodgrass et al.,* 2016), which can be explained by the subsurface retention of heat after a perihelion passage. This same energy storage was noticed in the EPOXI mission: The heliocentric light curve showed that there was still activity from the nucleus after the nucleus passed perihelion and was on its way in (*Meech et al.,* 2011). Furthermore, the characteristics of an outburst of 67P observed in 2016 were incompatible with the free sublimation of water ice under solar illumination, so that additional energy stored near the surface could have been involved (*Agarwal et al.,* 2017a). After several orbital revolutions, we may expect that non-uniform compositions and structures resulting from seasonal and diurnal cycles (described below) could be fairly common in the subsurface (e.g., *Guilbert-Lepoutre et al.,* 2016).

3.4.1. Seasonal cycle. Generally, seasonal variations of the activity are expected to occur due to variations of the insolation along the highly elliptical orbits of comet nuclei. In the case of 67P, the irregular shape combined with the relatively high rotational obliquity to further complicate this picture. The concept of cometary obliquity playing a role in the evolution of both the coma and the surface of comet nuclei was extensively discussed after the Giotto encounter with Comet 1P (e.g., *Schleicher et al.,* 1990, 1998). However, it is Rosetta that has been instrumental in demonstrating the importance of seasons as a main contributor for shaping comet nuclei. Due to the high inclination of 67P's rotation axis (~55°), the distribution of solar illumination across the surface is extremely asymmetric. The northern hemisphere is mildly heated during a long summer close

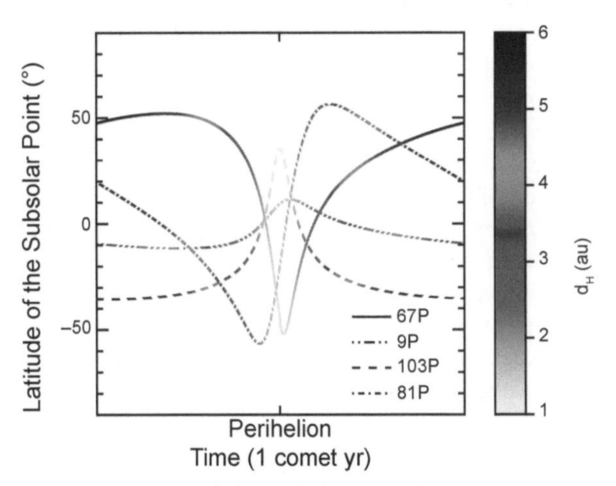

Fig. 10. Latitude of the subsolar point as a function of time (normalized to one comet year, i.e., one orbital revolution around the Sun) for Comets 67P, 9P, 103P, and 81P. The corresponding heliocentric distance is color-coded.

to aphelion, while the southern hemisphere suffers from an intense short-lived summer at perihelion (e.g., *Choukroun et al.,* 2015; *Keller et al.,* 2015a). This leads to complex patterns of surface processes detailed in the chapters by Filacchione et al. and Pajola et al. The analysis of Rosetta data throughout the mission has shown how gas species were released from source locations that varied with both a diurnal and a seasonal cycle (*Läuter et al.,* 2019; *Biver et al.,* 2019), and the changes in the relative abundance of major species when the Sun crossed 67P's equator (*Gasc et al.,* 2017) also point to complex relationships between gas production and insolation. For instance, strong differences in the coma composition above the northern and southern hemispheres were reported, with CO varying from 2.7% in the north to 20% in the south, and CO_2 from 2.5% in the north to 80% in the south [with respect to H_2O (*Le Roy et al.,* 2015)]. Once the southern hemisphere became more active during its short and intense summer months, CO and CO_2 ratios with respect to water measured at 1.5 au (*Rubin et al.,* 2019) were similar to the ones measured above the northern hemisphere at 3.1 au. This is consistent with the heterogeneity initially measured at 3.1 au being driven by temperature effects on a seasonal level (*Hässig et al.,* 2015). We note that a difference was also reported for the global CN abundance from groundbased measurements (*Opitom et al.,* 2017).

3.4.2. Diurnal cycle. Rotational variations of the activity can be observed from groundbased facilities. For example, *Bockelée-Morvan et al.* (2000) reported strong variations in the spectral line profile of CO in Comet C/1995 O1 (Hale-Bopp), consistent with the presence of a rotating jet, active both on the day- and nightside of the nucleus. More recently, *Roth et al.* (2021b) reported variations in the CH_3OH outgassing from Comet 46P/Wirtanen, probably due to diurnal variations in the insolation of the surface. For comets visited by spacecraft, a careful examination of groundbased observations obtained during the missions have been instrumental at constraining this diurnal cycle of activity. For example, EPOXI data coupled with groundbased observations of Comet 103P found that variations in insolation of the nucleus' small lobe was the cause for variations in the outgassing (*A'Hearn et al.,* 2011; *Drahus et al.,* 2012; *Boissier et al.,* 2014). Rosetta observations have revealed the full extent of the diurnal cycle active on 67P. Variations in the release of CO_2, CH_4, CO, and C_2H_6 (*Luspay-Kuti et al.,* 2015; *Fink et al.,* 2016) have been connected to day-night variations of the surface insolation. The implication is that the related ices cannot be too far from the surface, in order to react to insolation without too much delay.

One key issue is thus that direct observations of cometary nuclei have revealed very limited amounts of surface ice, not only on 67P (*Capaccioni et al.,* 2015; *Sierks et al.,* 2015; *Gulkis et al.,* 2015), but also more generally on other comets such as 19P (*Soderblom et al.,* 2002), 9P (*Sunshine et al.,* 2006; *Feaga et al.,* 2007), or 103P (*A'Hearn et al.,* 2011). The limited amounts of ice observed at the surface cannot sustain the observed outgassing, in particular the water-driven outgassing. Consequently, the subsurface layer can be involved in producing the observed coma (e.g., *De Sanctis et al.,* 2015; *Hu et al.,* 2017; *Hoang et al.,* 2020; *Herny et al.,* 2021; *Davidsson et al.,* 2022b, for 67P), where heat gets transferred on a cyclic basis. Such a mechanism was suggested to take place on 9P, where it was used to explain the outbursts of activity occurring near sunrise, observed by Deep Impact (*Prialnik et al.,* 2008): It is thus very likely that this is a general process acting on all cometary nuclei. An illustration of the diurnal cycle is given in Fig. 11. *Filacchione et al.* (2016) further suggested that exposed water ice on the surface of 67P could have a grain size consistent with grain growth by vapor diffusion in ice-rich layers, or by sintering, processes typically affecting the uppermost layers of a nucleus while the comet orbits close to perihelion.

3.5. Summary

We provide in Fig. 12 a visual summary of the extent of thermal processing below the surface of a nucleus, derived on a statistical manner from the discussion in this section. The actual consequences of both early radiogenic heating and early surface irradiation on the initial budget of volatiles, and hypervolatiles like CO, N_2, or O_2 in particular, or the mechanical properties of comet nuclei, remain mostly unconstrained. The fact that comet nuclei have retained hypervolatiles argues in favor of a delayed formation with respect to CAI formation, so as to avoid substantial heating from short-lived radiogenic nuclides. The entrapment of the most volatile species either in amorphous water ice, CO_2 ice, or as clathrate hydrates in crystalline water ice could secure their survival on long timescales, in particular for JFCs, which experience temperatures high enough for pure hypervolatile condensates to sublimate within a significant fraction of their nuclei.

As comet nuclei leave the outskirts of the solar system, insolation becomes the dominant source of energy. Cometary activity starts in the outer solar system, eventually leading to a complicated, stratified pattern progressively emerging below the surface (even if the composition was initially uniform), with possibly some lateral differentiation. Such thermal processing may induce some mechanical changes too, in particular when sintering of the grains occurs (*Kossacki et al.,* 2015) or if CO_2 pressure gradients are strong enough to displace solids (*Davidsson et al.,* 2022c). The extent of the thermally-processed layer broadly depends on the timescale, so that differences between Oort cloud comets and JFCs are expected. While the former preferentially follow direct, rapid paths to the Sun, most short-period comets leave their storage location on chaotic orbits through the giant planet region over several million years. For these comet nuclei, the stratified structure may thus extend down to a few hundred meters below the surface, but is most prevalent in the upper 10 to 100 m, which is easily processed during their Centaur phase. As a result, we could expect these nuclei to show activity variations both along their orbit and from one orbit to another. However,

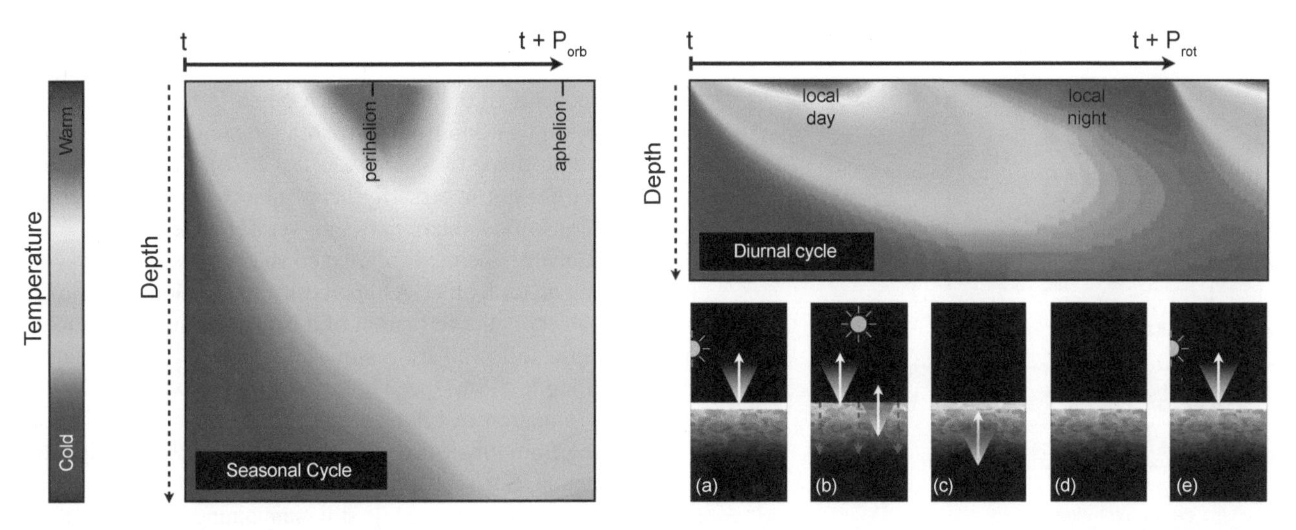

Fig. 11. Qualitative illustration of the seasonal and diurnal cycles showing the evolution of subsurface temperature distributions as a function of depth and time. For the diurnal cycle, the schematics illustrate the sublimation/recondensation pattern: **(a)** the surface heats due to insolation in the morning, and water ice sublimates from the surface (white arrow); **(b)** heat gets transferred to subsurface layers (red arrows) where sublimation can take place (white arrow); **(c)** the surface goes into shadow so a temperature inversion occurs between the cold surface and the hot subsurface layers: ice sublimates in the warmer subsurface (white arrow) and vapor flows toward the surface and recondenses; **(d)** cold nighttime; **(e)** the diurnal cycle starts again the next morning. This description of diurnal activity is simplified, as more complex processes such as related to mass waiting events can additionally deliver volatiles to the surface. The depth ranges from a few centimeters to ~40 cm for the diurnal cycle, and from several tens of centimers to a few meters for the seasonal cycle.

when looking at orbits prior the ones where comet nuclei are observed, we remain blind to the past dynamical evolution of these objects beyond the last close passage to Jupiter. Inevitably, we maintain a poor understanding as to how significant the thermally-induced physical and chemical alteration was before a given nucleus is observed.

Finally, the impact of the diurnal and seasonal processing of comet nuclei is not trivial. Because the energy balance across the surface can be strongly non-uniform, local alteration depends both on the duration of the day/night cycle and orbital period. The diurnal cycle depends not only on the complex rotation of nuclei, but also on complex shadowing effects in the case of extreme shapes such as 67P's. For the seasonal processing, the orientation of the spin axis and the orbital parameters are crucial to assess the breadth of consequences on the surface and subsurface. Since erosion is typically on the order of a few meters per revolution, i.e., on the same order as the extent of the seasonal processing (*Keller et al.,* 2015a, for 67P), seasonal and diurnal cycles could lead to local differential erosion and variations in the thermally-induced processing of the surface itself.

4. MECHANICAL PROCESSING OF COMET NUCLEI

4.1. Morphology of Comet Nuclei

With multiple comet flybys, a rendezvous mission, and multiple radar observations over the last few decades, we have come to an increased awareness of the unique morphologies that comet nuclei present and what role their shapes play in the overall evolution of these bodies. Key questions explored in the literature are how primordial nuclei shapes and morphologies are formed and how they become altered once they enter their active phase. In particular, issues that have been explored are whether comet nuclei take on their characteristic shapes and morphologies due to active processes once they leave the outer solar system reservoirs (i.e., the Oort cloud, the Kuiper belt, and scattered disk), or whether these active processes preserve their fundamental initial morphologies, even as the nucleus goes through drastic transformations due to outgassing and other interactions in the solar system. As discussed earlier in the chapter, the general consensus is that comet nuclei formed through an accretion process in the early solar system. Beyond that, there is debate as to whether they are then significantly influenced by collisional processes, or whether their subsequent evolution and emergence of key morphological features are related to their active phase once they depart their reservoir in the Oort cloud and the Kuiper belt. A key morphology indicator that has emerged from decades of space missions to cometary nuclei, culminating with the Rosetta rendezvous with 67P, are that nuclei are commonly found to have a bilobate structure (*Keller et al.,* 2015a; *Safrit et al.,* 2021). This bilobate structure can be described as a bifurcated shape, with two apparent lobes that are connected by a more narrow waist. While not all comet nuclei have such a morphology, a sizable fraction of them do. Out of seven reported cometary nuclei whose shapes have been detected, five out of them are considered to have bilobate shapes with the secondary lobe of size greater than 10% of the primary lobe (*Hirabayashi et al.,* 2016). Given the inherently random selection of comet mission targets,

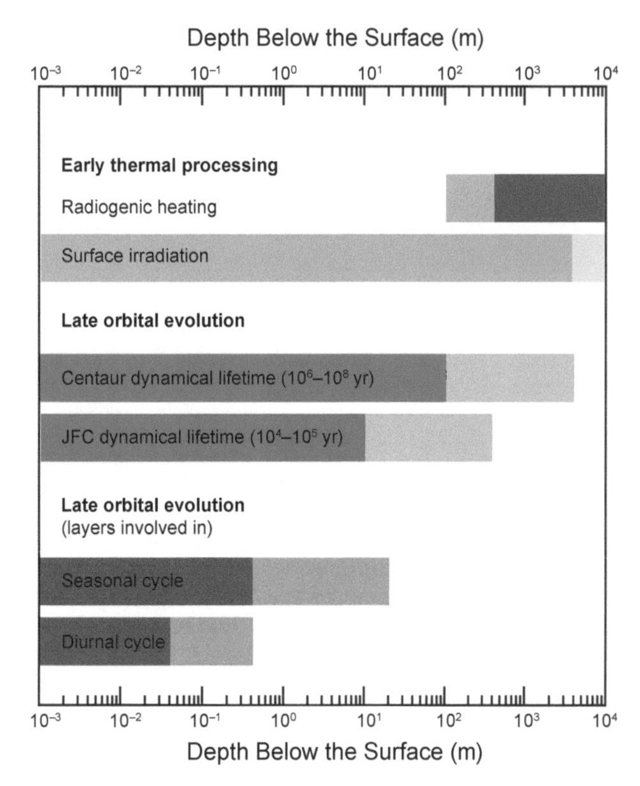

Fig. 12. Layering expected due to thermal processing sustained through the lifetime of a comet nucleus. Solid color areas provide the minimum depth reached by each process; shaded areas give the uncertainty related to thermophysical parameters and timescales.

this ratio is significant and indicates that this should be a common feature of comets. An important aspect of a body with a bifurcated shape is that the globally weakest point of such shapes will be at the neck/waist region that joins the lobes (*Hirabayashi and Scheeres,* 2019). The implication is that when a nucleus is subject to stresses due to impacts, tides, or rapid rotation, it would preferentially fail in this region if strength were distributed homogeneously. Conversely, the nucleus would require significant strength for the waist region to remain connected when subjected to extreme stresses — i.e., stronger than the rest of the nucleus. Thus, any changes in a nucleus morphology during its active phase should apparently preserve or create such bilobate shapes. On the other hand, if such lobes are prone to split when subjected to stresses, a valid question is why they do not mutually escape and become separate comets about the Sun, similar to asteroid pairs (*Pravec et al.,* 2010).

4.1.1. Properties inherited from primordial formation. There exist a few different proposed mechanisms for the formation of this morphology, with the components either forming due to their initial accretion process, forming after accretion via some process in the reservoir itself, or following the migration of the nuclei out of the reservoir. Starting at the earliest epochs, the formation of multilobed structures in comet nuclei has been posited as a hallmark of the formation process during accretion, as the gravitational collapse of planetesimals can frequently lead to the formation of binary objects (*Nesvorný et al.,* 2010). However, these are generally not formed as contact binaries, but as separate components in orbit about each other. In this context, it is significant to point out Arrokoth's distinct bifurcated and primitive shape (*McKinnon et al.,* 2020). This Kuiper belt object has not necessarily been subject to an active phase (in terms of cometary activity), yet has a clearly (and even more sharply defined) bifurcated shape of two components resting on each other. A hypothesized formation scenario for this object is the formation of a binary during gravitational collapse followed by a slow inward spiraling ultimately resulting in a low-speed collision (*McKinnon et al.,* 2020). The formation of bilobate nuclei has also been posited as arising from low-speed impacts between nuclei shortly after formation. Such relatively low-speed impacts can retain intact lobate structures while still maintaining interior volatiles (*Davidsson et al.,* 2016; *Jutzi et al.,* 2017; *de Niem et al.,* 2018; *McKinnon et al.,* 2020). Other work, described later, has proposed that these characteristic morphologies arise later during a comet's active phase.

4.1.2. Further observational constraints. Additional observations on the structure of cometary nuclei have emerged from earlier space missions, groundbased observations of active and splitting comets, and the detailed observations of 67P by Rosetta. In addition to their bilobate structure, the individual components of a nucleus when observed *in situ* have shown a degree of subsurface layering. The observation of interior stratification was first identified from images of 9P/Tempel 1 from the Deep Impact mission, and motivated some early models of the accretion process for these bodies (*Belton et al.,* 2007, 2018). Imaging and analysis of Rosetta images of 67P have clearly identified such structures (*Massironi et al.,* 2015; *Davidsson et al.,* 2016; *Penasa et al.,* 2017), and have led to claims that these strata even follow local gravitational equipotential surfaces (*Massironi et al.,* 2015). It should be noted that the local gravitational geopotential is highly sensitive to mass and density distribution and nucleus rotation rate, which is known to vary significantly over the lifetime of a comet nucleus (*Jewitt,* 2021). Intriguingly, deeper seated layers are brighter than more shallow ones (*Ferrari et al.,* 2018; *Tognon et al.,* 2019; *Davidsson et al.,* 2022a), for reasons that currently are not understood.

Gravity field and mass determination from radiometric tracking of the Rosetta spacecraft has provided the first insight into a comet's interior, and has supported the hypothesis that the interior density distribution of the body is both highly porous and homogenous (*Pätzold et al.,* 2016). However, the gravity constraints are not strong, given the low degree and order of the gravity field estimation, only up to second degree and order. These few coefficients that are consistent with a homogeneous distribution are not conclusive, as an accreted body could follow a homeoid density distribution and thus mask its density heterogeneity even if it exists (similar to an onion-like structure of a spherical body). A homeoid is the ellipsoidal equivalent to a set of nesting spheres of different radii. Thus, a homeoid density distribution for a triaxial ellipsoid is one that could be layered in such a way that layers fit above or below each other, allowing them to nest.

Finally, recent space- and groundbased observations of comets have enabled constraints on the overall strength of these bodies to be estimated. As discussed above, surface strength is defined by nearly strengthless covering with a harder, stronger undersurface (*Heinisch et al.,* 2019). Interior strength has been evaluated with FEM codes, using the crack in the Hathor region on the surface of 67P as a definitive feature, providing an upper bound of 200 Pa for the contact region between the 67P lobes (*Hirabayashi et al.,* 2016). Other observations of splitting rubble piles has yielded similar strengths (*Jewitt and Luu,* 2019; *Jewitt et al.,* 2015, 2017b).

There are still significant data analysis tasks that could be performed using the Rosetta data to further explore the mass distribution of 67P and its interior properties. As one example, it should be possible to identify and track dust ejecta during the mission to carry out a similar analysis as was done for Bennu, which yielded a ninth degree and order gravity field estimate (*Chesley et al.,* 2020). There are significant challenges with such an approach at a comet, however, driven by the outgassing environment and the multitude of particles that would need to be tracked. Another investigation could carefully track the estimated outgassing torques acting on the nucleus and its resulting spin rate changes (*Kramer et al.,* 2019), as such comet nucleus rotational excitation should also cause a nucleus to tumble and spin about a nonprincipal axis, as has been seen for some cometary nuclei (*Samarasinha et al.,* 2004). However, the nucleus of 67P was seen to not be tumbling prior to perihelion passage, indicating that excess spin energy was dissipated from its previous perihelion passage. Given the detailed models of the 67P nucleus and torques due to outgassing it should be possible to study this effect in more detail, which would provide strong constraints on the degree of energy dissipation for this particular comet.

4.2. Processing During the Active Phase

The main question during this active phase is how the primordial nuclei shapes and morphologies become altered. Issues explored in the literature relate to whether they take on their characteristic shapes and morphologies due to these active processes, or whether these active processes preserve their fundamental initial morphologies, even as the nucleus goes through drastic transformations due to outgassing and sublimation. *Keller and Kührt* (2020) argued that comet nuclei observed from space missions document different stages of activity-driven evolution. In this discussion, a potentially relevant observation is of the recent discovery of a comet binary body, 288P, associated with an impact family (*Agarwal et al.,* 2017b), although its formation mechanism is not fully constrained, and could be due to rotational fission as well (*Agarwal et al.,* 2020). The existence of this body provides some evidence of continued collisional or rotational processing of cometary nuclei. While the lifetime of this object as a binary is likely to be short, due to expected nongravitational effects on the orbit, such effects could just as easily evolve the components inward to form a contact binary with a bilobate shape.

4.2.1. Effects of collisions. Comets may have formed through primordial accretion or collisionally-induced disintegration of larger parent bodies. On the one hand, *Jutzi and Asphaug* (2015) showed that low densities and high porosities can be preserved in subcatastrophic collisional mergers. However, *Jutzi et al.* (2017) and *Jutzi and Benz* (2017) point out that Kuiper belt objects larger than 5 km can suffer from multiple shape changing or catastrophic disruptions throughout the lifetime of the solar system. Once comets enter into the inner solar system, or even after their orbits have become randomized following their early formation, higher impact speeds that could catastrophically shatter a nucleus can occur. Following our current understanding of how asteroid families are formed, such impacts would generate rubble pile bodies, which could reaccrete into the currently seen bilobate structures as explored in *Jutzi et al.* (2017) and *Schwartz et al.* (2018). These impact evolutions are to be contrasted with the low-speed collisional processing hypothesized to occur shortly after formation. A key question to discriminate between these two models is whether the disruptive impacts would be consistent with the observed interior layering structure that has been documented and discussed above, and the extent to which interior volatiles can be preserved.

Because comet nuclei contain volatile compounds, sensitive to any increase of the internal or superficial temperature, constraining the heating effect of collisions is a legitimate issue. Laboratory experiments show, for example, that collisions under present day dynamical conditions would result in the melting of water ice at the point of impact (*Stewart and Ahrens,* 2004). However, the theoretical study of collisions among asteroids suggests that impacts would have been unable to drive a global-scale, significant thermal processing, with *Keil et al.* (1997) reporting that the global temperature increase would be approximately 10 K. *Davison et al.* (2010) showed that collisions on porous objects would result in higher temperatures than on nonporous objects, on the order of 10 K. Such estimates may depend significantly on model parameters used in collision codes, particularly the assumed pressure p_s needed to remove porosity. Increasing p_s from typically considered values to those measured for water ice may enable both segregation and crystallization, with associated hypervolatile loss (*Davidsson,* 2023; *Steckloff et al.,* 2023). *Schwartz et al.* (2018) investigated the amount of impact heating caused by catastrophic collisions. They suggest that cometary material that is subsequently reaccreted should be only marginally heated, while the material that is more significantly processed (albeit heated up by several tens of Kelvin) should be mainly found in the ejecta. Therefore, it is entirely possible that hypervolatile species can be retained in cometary material during a catastrophic collision. As a corollary, this result suggests that the presence of hypervolatiles, as well as the low densities, do not alone guarantee that a nucleus was not subject to any collisional processing in the past. Hence the presence of highly volatile molecules and 67P's low density (*Pätzold et al.,* 2016) alone is not proof for the nucleus itself to be primordial. Nevertheless, *Jutzi et al.* (2017) predict that a comet like 67P undergoes many shape changing collisions, critically depending on the differential size distribution of the objects in the disk. Depending on the index of the size distribution [shallow in *Davidsson et al.* (2016), steeper in *Jutzi and Benz* (2017)], comet nuclei could have an increased probability of surviving in their primordial form, or they could be the result of many collisions.

Collisions may be particularly relevant to explain the observed activity among main-belt comets, where impact-induced occasional outgassing could be at the origin of some dormant nuclei's activity. Here, a comet nucleus that shows no other activity is occasionally subjected to impacts that expose underlying volatile material, which then sustain observable activity. The evolution of a cometary nucleus to such an end state has been predicted previously (*Jewitt,* 2004), with detailed information provided about the mechanism given by Rosetta observations. In this situation, however, the impacts are not as important for the overall evolution of the nuclei, although they may provide an indication of which bodies were previously active cometary nuclei. Furthermore, not all impacts can generate a sustained activity, as revealed by the Deep Impact experiment, which failed to produce such sustained activity.

4.2.2. Tidal effects. The effect of tidal forces when a cometary nucleus has a close passage to a planet can be quite dramatic, as evidenced by the spectacular break-up of SL9 in 1992 (*Asphaug and Benz,* 1994; *Boehnhardt,* 2004). It has been estimated that cometary disruptions at Jupiter occur every 200–400 years (*Schenk et al.,* 1996), and it has been noted that such infrequent flybys can split nuclei and create additional comets from a population of fewer bodies (*Jewitt,* 2004). Such close flybys could mimic the effect of catastrophic impacts, creating a disaggregated body that could accrete and form the observed morphologies mentioned above. However, a disaggregated cometary body at the distance of Jupiter may be subject to loss of volatiles, if it were fully separated such as SL9. If, instead, it were a composite body and were separated into individual lobes due to a close passage, this presumes the existence of a multilobe structure susceptible to separation.

The observation of the dramatic effect that tidal disruption had on the SL9 comet motivated significant research on the overall effect of close planetary flybys on different solar system bodies. While it is clear that such close passages can significantly evolve a comet nucleus' shape, these effects are still quite occasional and would be difficult to create a globally common morphology across all cometary nuclei. For these extreme effects, the cometary nucleus would need to come close to or within the Roche lobe of Jupiter without impacting. Simple computations can show that the probability of a comet with a density of 500 kg m^{-3} coming within Jupiter's Roche lobe without impacting is just twice that of the comet impacting with Jupiter. Thus, the probability of any single comet nuclei having exactly the flyby distance necessary to be pulled apart into a bifurcated shape is likely very low. In fact, such an analysis does not even apply to SL9, since that comet was first captured, increasing its odds of the close passage that disrupted it. Such an occasional process argues against this being a ubiquitous process that should shape the overall morphologies of the comet nuclei we see, as it is frequent (e.g., *Safrit et al.,* 2021), but not frequent enough to necessarily be relevant for the reshaping and evolution of most individual cometary shapes. Tidal disruption undoubtedly serves as a sink for comets, but is not as ubiquitous as the sublimation and spin evolution of these bodies.

4.2.3. Effects of outgassing on shape. An additional process that has been proposed for the formation of the bilobate structures of cometary nuclei is the global process of mass loss through sublimation (*El-Maarry et al.,* 2017; *Vavilov et al.,* 2019). Here, the proposed interactions between insolation and outgassing, along with possible heterogeneities within the nucleus, can lead to a non-uniform loss of surface material from a nucleus. Over repeated apparitions, this can lead to the emergence of more complex, bilobate structures (*Vavilov et al.,* 2019). The effect of outgassing and loss of mass on a cometary nucleus is their defining characteristic, and is clearly the fundamental driver of a comet nucleus' evolution. The total mass of 67P decreased by approximately 0.1% during the perihelion passage observed by Rosetta (*Pätzold et al.,* 2019). When averaged over the entire body, this corresponds to an average loss on the order of 0.5 m of its surface. In specific regions, however, 67P was seen to have lost an order of magnitude more in depth (*El-Maarry et al.,* 2017). Nonetheless, despite this significant loss, the overall global shape and morphology of the comet was not seen to be strongly affected (*El-Maarry et al.,* 2017), so that if the current shape were due solely to outgassing and sublimation effects, the comet would have had to orbit significantly closer to the Sun, in particular at perihelion, earlier in its life (*El-Maarry et al.,* 2017).

While the loss of surface material clearly has an effect on the shape of a nucleus, it is important to note that only a fraction of the levitated dust is seen to escape from it. From Rosetta observations, it was estimated that 1.8 to 4.3 times the lost mass was levitated, but fell back to the surface, while the mass that actually escaped was mostly gas molecules (*Pätzold et al.,* 2019). These infalling dust components were seen to be of centimeter to decimeter size, and can contain volatiles that can be released at later apparitions, although presumably at a decreased intensity. This poses a challenge for current modeling. A few critical notes can be made here. First, the infall of dust is expected to be relatively uniform; however, when mapped to the body of the nucleus, it is expected that there will be preferential regions where material may accumulate (e.g., on the leading edges of a rotating nucleus), or in geopotential lows on the surface of the nucleus (*Geissler et al.,* 1996). Second, the preferential loss of volatiles and infall of refractory dust will clearly transform the surface of the nucleus over multiple apparitions. Simply applying mechanical models to this loss of surface material predicts a limited lifetime for such nuclei (*Jewitt,* 2004). If the dust insulates the nucleus from the effect of solar heating, it also can clearly lead to reduced activity on a nucleus surface. Such insulation can lead to diminished and delayed outgassing activity, potentially culminating in a defunct or dead comet. Finally, the accumulation of dust in the gravitational lows would also work against the preferential creation of bifurcated regions. For a waist-shaped body, the gravitational low will in general lie around the waist structure (at the relatively slow spin rates that cometary nuclei have). This would work against these regions being preferentially active, and would create a competition between the sublimation reshaping effect and the overall insulation of these regions due to infall.

4.2.4. Effects of outgassing on rotation. Outgassing has a particularly strong effect on the rotation of nuclei. This fact has been long understood and studied in detail over the previous decades (*Samarasinha et al.,* 2004; *Neishtadt et al.,* 2002; *Chesley et al.,* 2013). Recent analyses have also clearly demonstrated the role of rotational fission in the short lifetimes of smaller cometary bodies (*Jewitt,* 2021). The Rosetta observations have provided the first detailed view of this process over an entire perihelion passage, and has verified that a comet spin rate is definitely affected by its outgassing during such perihelion passage (*Kramer et al.,* 2019; *Mottola et al.,* 2020). Detailed modeling of 67P provides clear insight into how this process works. A key observation is that a Yarkovsky-O'Keefe-Radzievskii-Paddack (YORP)-like effect due to outgassing drives the spin-up and spin-down characteristics (*Keller et al.,* 2015b). By this, we mean that the torques acting on the nucleus are coupled with the surface area and orientation of the comet nucleus surface to the incident sunlight. The use of the classical YORP model is further motivated by the observed low thermal inertia on the cometary nuclei, which is a key assumption of the so-called "normal YORP" effect (*Vokrouhlický et al.,* 2015). Despite this, the cometary YORP effect is expected to be different and more effective than what is experienced by asteroidal bodies, mainly driven by the much higher efficiency of mass loss in placing larger toques on the nucleus, giving it an ability to significantly change a body's rotation rate over short time spans (*Chesley et al.,* 2013; *Steckloff and Jacobson,* 2016; *Bodewits et al.,* 2018).

Given the strong variation of outgassing strength with solar distance, the bulk of the spin rate change will occur around the perihelion passage, with the total change in nucleus angular momentum being a function of the surface morphology and solar illumination geometry at perihelion. This is fundamentally controlled by the subsolar latitude of the Sun in a comet's nucleus-fixed frame. If we assume that the comet rotates, even at a modest rate, it will average this effect out over all of its longitudes. The identification of the subsolar latitude's importance was shown in *Hirabayashi et al.* (2016). In the same study, it was also noted that a comet's subsolar latitude will vary chaotically over time due to distant flybys of Jupiter. This migration can cause a variable change in spin rate for each passage, leading to a random walk in a comet's spin rate over multiple apparitions. This effect can delay the nucleus reaching disruptive spin rates, yet may also eventually lead a nucleus to a spin rate at which fission occurs. For 67P, it is seen that disruption of the lobes would not occur until a spin period of ~7 hr, compared to its current spin period of ~12 hr or less, would be reached (see Fig. 13).

If rapid rotation does lead to the separation of lobes, or fission within a body, it is an interesting question whether the separated components would escape from each other or would remain bound and eventually reimpact. In *Jacobson and Scheeres* (2011) and *Scheeres* (2009), it is shown that a bilobate body that is spun to fission cannot undergo escape in the absence of exogenous effects if the two lobes are of similar size. For constant density bodies, if the size of the smaller lobe is ~0.1 or larger relative to the primary, the system does not have enough energy for the two components to escape, and thus may recollapse and preserve its bilobate structure. In *Hirabayashi et al.* (2016), it was noted that all observed bilobate nuclei have a size ratio larger than this ratio. This indicates that these bodies may have been subject to such rotational fission and recollapse in the past.

If fission occurs and the body has a bilobate structure with the secondary having a size less than 0.1 of the primary, the system will be energetic enough to lose this secondary. There may be biases associated with clearly identifying smaller bilobate components, although for 67P there was a clear distinction between lobe sizes and other surface components. If a cometary nucleus were composed of multiple smaller components, one can imagine a rotational cascade starting from larger bodies and accelerating for smaller components, as has been hypothesized for weakly cohesive rubble-pile asteroids (*Scheeres,* 2018). Under this scenario, a nucleus subjected to rapid rotation may lose components and mass and thus be increasingly sensitive to future torques. Conversely, if a body initially has a strongly bilobate structure with components of similar size, it may survive rotational fission as a single system, and be preserved as a larger cometary system, as proposed by *Hirabayashi et al.* (2016).

Finally, we note that a nucleus spun up by outgassing torques could experience mass flows that increase its rotational inertia, slowing down its rotation state. *Steckloff et al.* (2016) found that the bifurcated nucleus of Comet 103P/Hartley 2 would actually experience massive landslides, which would redistribute material toward the ends of the nucleus. Indeed, spun-up comet nuclei would tend to experience mass-wasting events that redistribute material toward an equipotential surface, generating a negative feedback that counters changes to their rotation state (*Richardson and Bowling,* 2014). Overall, sublimation torques would also need to overcome this process in order to spin up comet nuclei to disruption. *Jewitt* (2021) studied the effect of anisotropic outgassing applying a torque significant enough to change the angular momentum of the nucleus,

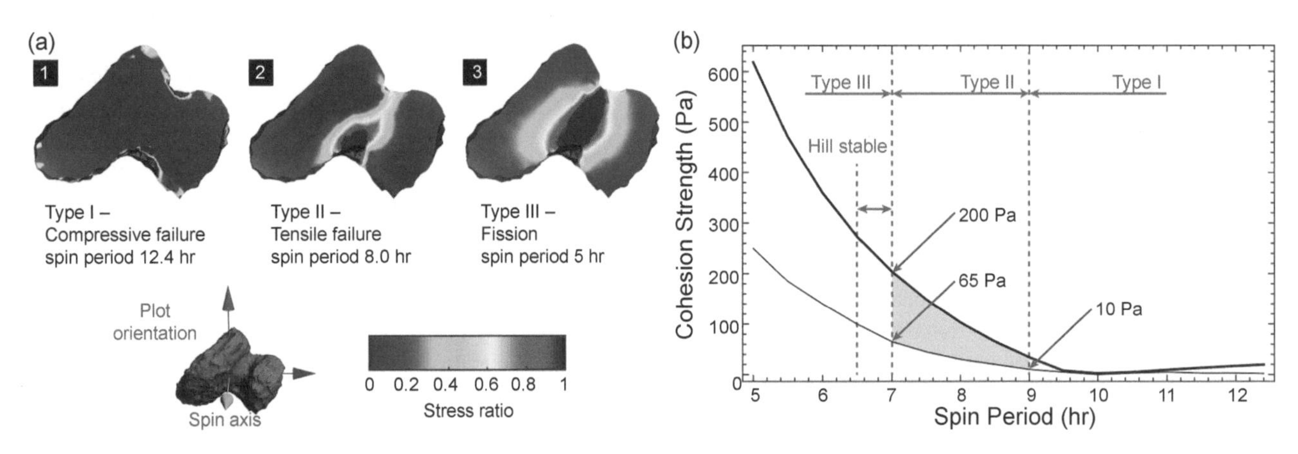

Fig. 13. Failure analysis of 67P at different spin periods. **(a)** Stress analysis of 67P at different spin rates, showing the compressive regime at (1) the nucleus' current period, (2) the crack-forming regime at moderate spin, and (3) the fission regime at fast spin. The stress ratio is normalized to failure for each case. **(b)** Diagram showing the relationship between the cohesive strength of 67P, failure regime, and spin period. The thick and thin black lines show results obtained from elastic and plastic finite-element-model analyses respectively. If the cohesive strength is below the thick line, cracks should appear at the stress peaks. If cohesive strength is below the narrow line, the nucleus splits into two. Given that the cracks on 67P have already formed, the lines delineate upper and lower bounds to the bulk cohesive strength at a given spin period. The boundary between types I and II failure is determined by when stress peaks first appear in the neck. The boundary between types II and III is determined by when the gravitational and centrifugal forces are balanced across the neck. The type I compressive failure condition is less than 10 Pa and much lower than the reported compressive strength on the order of 1 kPa (*Biele et al.,* 2015; *Groussin et al.,* 2015). In a type III failure, the existing cracks would experience tension in excess of its strength and fail catastrophically. Therefore, the shaded region indicates the bulk cohesive strength of the nucleus. Results shown here are relatively insensitive to variations in density and mass distribution. The Hill stability condition is determined by finding when the spin rate at fission when the two components have sufficient energy to escape each other. Modified from *Hirabayashi et al.* (2016).

potentially leading to some rotational instability. Since the characteristic spin-up timescale strongly depends on the nucleus radius, they report that comet nuclei smaller than 1 km have spin-up timescales comparable to their orbital periods, so that outgassing torques can rapidly lead them to rotational disruption. They further argue that torque-induced rotational instability could explain why subkilometer nuclei are scarcely observed.

5. OBSERVATIONS CHALLENGING MODELS FOR FUTURE ADVANCES

5.1. Fallback Material

Fallback material poses a challenge to classical thermophysical evolution models, in addition to further challenges to the classical model of dust particle ejection (*Jewitt et al.,* 2019). Evidence for fallback material was first put forward in the case of Comet 103P/Hartley 2 during the EPOXI flyby. Comet 103P has a bilobate nucleus, with a smooth but variegated region in between the two lobes, called the "waist" (*Thomas et al.,* 2013). This waist is covered by deposited icy grains: At the time of the flyby, water outgassing was coming mostly from this area. This material did not seem to contain any CO_2 ice, probably because this species can sublimate easily enough even from the innermost part of these chunks [see *Davidsson et al.* (2021) for timescales for CO_2 loss from fallback material]. The importance of fallback has definitively been revealed by Rosetta. The detailed investigation of 67P's nucleus revealed large smooth plains in the northern hemisphere, covered by a coating of dust originating elsewhere on the comet, which provides evidence for dust redistribution on large scales over the nucleus (*Thomas et al.,* 2015a). The depth of this coating is unknown and almost certainly non-uniform: Estimates of the fallback material thickness have been provided by *Hu et al.* (2017) and *Davidsson et al.* (2021), ranging from a fraction of a meter to several meters. Fallback is now known to be a dominant mechanism for transferring mass from the southern to the northern hemispheres of 67P, following ejection processes occurring around perihelion (*Keller et al.,* 2015a, 2017; *Lai et al.,* 2016). Finally, *Cambianica et al.* (2020, 2021) measured the thickness of the layer that was removed on the inbound part of the orbit, and the thickness of the layer deposited on the outbound part of the orbit, by looking at the lengths of boulder shadows present at the surface of 67P. They find that both thicknesses are on the order of 1 to 2 m.

In addition, fallback is significant as a process in connection with activity itself. Indeed, tracking back the origin of the water distribution in the coma of 67P during the early, pre-perihelion phase of the Rosetta mission, *Fougere et al.* (2016b) found that H_2O originated predominantly from the northern regions. It may thus be directly linked to the fallback material that ejected during the previous perihelion passage of the comet and settled down slowly during the post-perihelion portion of the orbit (*Bertini et al.,* 2019). This fallback material is likely dominated in mass by centimeter- to decimeter-sized particles, which have been observed in the inner coma of 67P (*Rotundi et al.,* 2015). This large size would allow them to preserve ices within their interior and enable their reactivation later on (*Fulle et al.,* 2019). Indeed, pre-perihelion CO/H_2O and CO_2/H_2O ratios are not so different when originating from the north or the south (*Le Roy et al.,* 2015; *Rubin et al.,* 2019). This suggests that either ices more volatile than water have been retained in the fallback material, or the fallback layer in the north is not thick enough to dampen the release of CO and other volatile species from below. The high CO/H_2O and CO_2/H_2O in the south early in the mission can be explained by low water emission, directly linked to low insolation (*Biver et al.,* 2015; *Choukroun et al.,* 2015). However, because the layer processed on a seasonal timescale in the southern hemisphere of 67P is of the same scale as the layer eroded, fallback material originating from the south has likely been processed to some extent prior to their ejection in the coma (*Fulle et al.,* 2019).

5.2. Extended Sources of Activity and Hyperactivity

Icy grains sublimating once released in a comet's coma can contribute to the enhanced production of various species, befuddling the link between gas production and volatile abundance in the nucleus. As such, hyperactive comets have more water vapor released in their coma than can be produced by water ice sublimation over the entire surface of their nucleus, a concept that was first introduced by *A'Hearn et al.* (1995). For instance, the presence of icy grains in the coma may help reconcile some dust observations made by Rosetta (*Agarwal et al.,* 2016; *Gicquel et al.,* 2016), and have been suggested on other comets as well. This is the case of 103P/Hartley 2, for which extended sources of activity are abundant (*Meech et al.,* 2011; *A'Hearn et al.,* 2011). For 103P, most of the activity is driven by intense CO_2 sublimation, which snatches icy chunks from the surface and near-subsurface. The coma observed by EPOXI is therefore spatially heterogeneous, with icy grains and CO_2 mostly present above the illuminated small lobe and water vapor emitted from the waist. From the ground, *Drahus et al.* (2012) observed variations in the production of HCN and CH_3OH: Their outgassing patterns suggest that HCN was released from both icy grains and a jet, while CH_3OH was released only from icy grains. *Boissier et al.* (2014) further confirmed this interpretation, based on offset production curves between CO_2 being released by the nucleus only, while HCN, CH_3OH, and H_2O are also released by icy grains. A similar behavior is observed for 46P/Wirtanen, another hyperactive comet, with a high production of water relative to the small size of its nucleus (*Lis et al.,* 2019). A recent and thorough investigation of its coma composition has found significant enhancements in water emission, which supports the large (~40%) contribution of icy grains as an extended source of activity (*Bonev et al.,* 2021; *Khan et al.,* 2021; *Roth et al.,* 2021a). These icy grains are also invoked to explain the production of CH_3OH in the coma (*Roth et al.,* 2021b). However, icy grains are not directly detected in 46P's coma (*Protopapa et al.,* 2021).

Recently, the isotopic composition of hyperactive comets has challenged our understanding of the formation and evolution of comet nuclei (see the chapter by Marty et al. in this volume for an extensive discussion). Due to the relative high mass difference between deuterium (D) and hydrogen (H), the D/H ratio is prone to fractionation effects, including from sublimation (*Brown et al.,* 2012). The D/H ratio has been used as a key measurement to infer the origin of water on Earth, for instance, since the degree of fractionation depends on external conditions under which ices were formed. For example, D/H is expected to increase with heliocentric distance (*Aikawa and Herbst,* 2001; *Ceccarelli et al.,* 2014). This ratio is notoriously difficult to measure in comets, as it requires either a spacebased facility for faint comets or a very bright comet. Most early measurements were thus available for Oort cloud comets only: They showed a larger D/H ratio compared to Earth [the Vienna Standard Mean Ocean Water (VSMOW), with $D/H = (1.558 \pm 0.001) \times 10^{-4}$]. However, and most notably, the Herschel space observatory measured a terrestrial D/H ratio in Comet 103P/Hartley 2 (*Hartogh et al.,* 2011). As of today, D/H has been measured in a dozen comets coming from both the Oort cloud and the Kuiper belt (*Paganini et al.,* 2017; *Lis et al.,* 2019). If the measured D/H ratio was initially thought to reflect the formation location of comet nuclei, a more recent study has changed this picture. Indeed, D/H in 46P/Wirtanen is measured to be compatible with VSMOW. As both comets with terrestrial D/H ratios are also hyperactive comets, and since a correlation can be found between the so-called "active surface area" and the D/H ratio, *Lis et al.* (2019) suggested that the observed differences in the isotopic composition of comets may be due to the current evolutionary state of their nucleus rather than its formation location.

5.3. Semivolatile Phase and Volatile Mixtures

Current thermal models are challenged by the coupling between the gas phases and a rigorous treatment of ice mixtures. We have seen that thermal evolution models use gas production rates in order to reproduce the observed activity of comet nuclei. Assumptions must be made as to whether the observed volatile species come from pure ice condensates, or are released upon amorphous water ice crystallization, CO_2 segregation, or clathrate destabilization [the existence of which in cometary conditions still needs to be proven (e.g., *Choukroun et al.,* 2019)]. Since layers at different depths may contribute to cometary activity observed at any given time due to different volatilities of the associated material, we need to decipher whether an activity pattern is better reproduced by ice mixtures and/or a stratified structure. However, we are still lacking critical laboratory experiments to properly understand, and model, ice mixtures. Further ice trapping experiments of hypervolatiles and noble gases in comet-like ices, containing not only H_2O but also CO_2, have started to be carried out (see the chapter in this volume by Poch et al. for an extensive discussion).

Accounting for a mix of ices is especially important since Rosetta observations suggest that the behavior of volatiles cannot be represented by one single phase of water ice. For example, noble gases were better correlated with CO_2 than with H_2O. Minor volatile and hypervolatile species show better correlation with either H_2O or CO_2, with *Luspay-Kuti et al.* (2015) reporting maxima in the intensity profiles of the apolar species (CO, C_2H_6, and CO_2) occurring independently of H_2O (whose behavior is more closely followed by HCN, CH_3OH). Methane shows a different diurnal pattern from all other species analyzed in this work, which argues for a complex internal mixture, including the trapping of guest molecules inside an icy matrix of H_2O or CO_2. Besides, even if crystallization of amorphous water ice is thought to play a key role in the thermal evolution of comet nuclei, it becomes necessary to carry out experiments to quantify the amount of trapped gas, the outgassing process during the warming up of the sample (in particular with regards to the exothermic vs. endothermic nature of the process), or the possibility of clathrate hydrates forming when the amorphous ice reorganizes and when a mixture of gases recondenses in the nucleus (in a co-deposit manner).

Further complexity can arise: HCN and NH_3 were found to better correlate with the production of C_2H_6 than H_2O in 30 comets (*Dello Russo et al.,* 2016), and both showed enrichments in comets within 1 au of the Sun. These contribute to an emerging view that at least some volatiles observed in comae may come from the dissociation of semivolatile ammonium salts (e.g., $NH_4^+CN^-$) (*Hänni et al.,* 2020; *Mumma et al.,* 2019; *Altwegg et al.,* 2020; *Poch et al.,* 2020). Recent studies even suggest that the "missing" nitrogen in comets may be hidden in such salts (*Altwegg et al.,* 2020; *Poch et al.,* 2020). These semivolatile salts could form in the interstellar medium at low temperatures and be accreted in cometary ices. They could then sublimate in the inner solar system, releasing volatiles such as NH_3, HCN, organic acids, and other byproducts into the comae (*Hänni et al.,* 2020). Such a phase is not yet accounted for in any thermal evolution model.

5.4. Continuous Monitoring of Activity

We have known for quite some time that gas production rates observed at a given time do not reflect the corresponding abundance of a volatile species in the nucleus (*Benkhoff and Huebner,* 1995; *Prialnik,* 2006). As we have discussed, volatile species cover a wide range of sublimation temperatures, so that relative abundances in the coma necessarily depend on the heliocentric distance. This effect has been well documented by Rosetta for 67P, where relative abundances of the major species H_2O, CO, CO_2, and O_2 changed over the course of the comet's orbit (*Biver et al.,* 2019; *Bockelée-Morvan et al.,* 2016; *Fougere et al.,* 2016a; *Keeney et al.,* 2017). In addition, a nucleus can be heterogeneous in composition, either by formation or as a result of the non-uniform structure that inevitably develops with thermal evolution, so that gas production rates at any given time of the diurnal or seasonal cycle should not be taken to reflect the composition of the nucleus (*Huebner and Benkhoff,* 1999). Finally, some comets can exhibit an erratic behavior from one perihelion passage to another. These variations can be illustrated by the recent observations of Comet 2P/Encke, for which extreme differences in the mixing ratios of several primary volatiles were found during the 2017 passage compared to data obtained in 2003 (*Roth et al.,* 2018). These observations were nonetheless not obtained during the same orbital phase and could possibly be more reflective of seasonal changes. However, this reinforces the difficulty of constraining a cometary nucleus' composition from limited observations of gas production rates obtained at a given time. *Prialnik* (2006) further tested this effect by comparing instantaneous production rates as well as integrated production rates to the initial volatile abundance ratios fed into a nucleus thermal evolution model. They showed that if we integrate production rates over time to obtain the total amount of material ejected during one orbit for each species, differences between the composition of the coma and the nucleus become smaller. After repeated orbital revolutions, volatile ratios in the coma converge to values closely representative of the nucleus abundance ratios. This modeling result is of great significance for building a strategy for comet observations. Indeed, in order to access volatile relative abundances representative of the nucleus — or more appropriately, representative of the layer of the nucleus contributing to cometary activity — a comet should be monitored for a sufficiently large portion of its orbit where it is active, and gas production rates should be integrated over that time.

5.5. Survey of Centaurs

Centaurs are connected to JFCs through a dynamical cascade that brings cometary nuclei from the Kuiper belt to the inner solar system. Understanding the mechanisms at the origin of their activity is paramount to fully comprehend the extent of the post-formation thermal processing of the short-period comets. By analyzing the observations available in 2018, as well as thermal modeling outputs and dynamical simulations, *Cabral et al.* (2019) argued that we currently do not have an appropriate dataset to constrain the source of Centaurs' activity. Compared to TNOs, Centaurs have different detectability when survey detections are motion-rate dependent. With an upper limit for the motion rate set at 15″/hr, the Outer Solar System Origins Survey (OSSOS) (see the chapter by Fraser et al. in this volume) detected mostly objects located beyond 10 au. As such, OSSOS, like most TNO surveys, has an observation cadence biased toward dynamically stable orbits found beyond Saturn (*Tiscareno and Malhotra,* 2003; *di Sisto et al.,* 2010). Surveys targeting objects closer to the Sun, i.e., asteroid surveys, typically have their sensitivity drop beyond 5 au. More generally, this sensitivity bias is likely to be inherent in the Centaur discoveries by all existing surveys. As of today, there have been no surveys that adequately target the motion rate for objects in the 5–10-au region, where Centaurs with more unstable orbits can be found, and most active Centaurs are observed today, in a well-characterized manner. This is a weakness of the entire known Centaur dataset, and an impediment toward understanding cometary activity in the giant planet region. We thus argue that a dedicated survey is required to actually understand the origin of Centaurs' activity.

5.6. Sample of Cometary Material

The usefulness of a cometary sample return mission is discussed in several chapters in this volume. Certainly, many advances in our understanding of cometary material, comet nuclei, their internal properties, and their past history (both thermal and mechanical) can only come from sampling a comet. For example, we have mentioned that early radiogenic heating of comet nuclei is directly dependent upon the amount of nuclides: Collecting CAIs to measure their ^{26}Mg excess could address this issue more specifically. Cometary samples could lift the degeneracy between porosity and bulk composition: Both additionally vary with depth and location across the nucleus, as well as in time due to evolution. Direct measurements of the thermal inertia, the conductivity and heat capacity, cannot be made without a sample of cometary material. Besides, the mixing of refractory, organic, and ice components, as well as their stratification, could be studied if the petrography was maintained during the sample collection and atmospheric transfer.

We stress again that the effects of the diurnal and seasonal processing of comet nuclei are not trivial. Their extent on the surface and subsurface depends both on the duration of the local day/night cycle (modulated by the orientation of the spin axis or the shape, for example), orbital parameters, and local thermophysical characteristics of cometary material. A superficial sample would most likely only collect refractories as well as some water ice depending on local thermophysical properties. It would not necessarily require a cryogenic sample return. For JFCs, which would be the likely targets for such missions, erosion is on the same order as seasonal processing [for 67P, see *Keller et al.* (2015a)]. Accessing a CO_2 layer or other species with similar volatility, or an amorphous water ice layer (with a cryogenic sample return at 80 K), would thus require at least a few meters of digging, which increases the challenge from an engineering point of view [however, *Davidsson et al.* (2022c) argue that CO_2 may have been as shallow as ~0.5 m at a pit-forming region in Hapi on 67P, based on analysis of MIRO observations]. Finally, the subsurface layer contributing to the observed activity of a JFC is likely to have been processed during the nucleus transition from the Kuiper belt to the inner solar system as a Centaur. It is therefore unlikely that pure hypervolatile condensates, if any remain within JFCs observed today, can be accessed. We suggest that sample return missions should be prepared for assessing whether such species are present during the collection of samples, but overall do not require a cryogenic collection at temperature below 30 K, which is prohibitive from many aspects (sample collection, atmospheric transfer, curation, budget). Collecting several samples at different locations, although complex, would address the heterogeneity of comet nuclei, both at and below the surface.

5.7. Deployment of Radar Tomography

The internal characteristics of comet nuclei provide crucial clues about their formation mechanisms and their collisional, rotational, or thermal evolution. They are, however, extremely difficult to constrain: Long-wavelength radar is one of the few remote-sensing techniques that can directly probe the deep interior of a small bodies, to retrieve constraints on the structure and distribution of a nucleus' density, porosity, or composition. Indeed, the radar exploration of 67P by Rosetta/CONSERT, although limited by the unfavorable landing position of Philae, has proved that significant results on the deep composition and structure of the nucleus can be retrieved, unveiling accretion and evolution mechanisms. Radar tomography uses long-wavelength radio waves to image the three-dimensional internal structure of an object. With this technique, radar echoes are acquired at many viewing angles, processed using inverse scattering techniques, and used to reconstruct variations of nucleus dielectric properties. The deployment of such technique is thus costly on a space mission dedicated to a specific target. However, tomographic observations have the potential of revealing internal characteristics otherwise inaccessible by any other method, and should thus be a priority for future exploration of comet nuclei.

Acknowledgments. We warmly thank J. Steckloff, K. Meech, D. Jewitt, and A. Gkotsinas for comments and discussions that greatly improved this chapter. A.G.-L. acknowledges support from the European Research Council (ERC) under the European Union's Horizon 2020 research and innovation programme (Grant agreement No. 802699), and from the PSMN (Pôle Scientifique de Modélisation Numérique) of the ENS de Lyon for computing resources. Parts of the research were carried out at the Jet Propulsion Laboratory, California Institute of Technology, under a contract with the National Aeronautics and Space Administration. B.J.R.D. acknowledges "raise the bar" funding from the Jet Propulsion Laboratory.

REFERENCES

Agarwal J., A'Hearn M. F., Vincent J. B., et al. (2016) Acceleration of individual, decimetre-sized aggregates in the lower coma of Comet 67P/Churyumov-Gerasimenko. *Mon. Not. R. Astron. Soc., 462,* S78–S88.
Agarwal J., Della Corte V., Feldman P. D., et al. (2017a) Evidence of sub-surface energy storage in Comet 67P from the outburst of 2016 July 03. *Mon. Not. R. Astron. Soc., 469,* S606–S625.
Agarwal J., Jewitt D., Mutchler M., et al. (2017b) A binary main-belt comet. *Nature, 549,* 357–359.
Agarwal J., Kim Y., Jewitt D., et al. (2020) Component properties and mutual orbit of binary main-belt Comet 288P/(300163)

2006 VW139. *Astron. Astrophys., 643,* A152.
A'Hearn M. F., Millis R. C., Schleicher D. O., et al. (1995) The ensemble properties of comets: Results from narrowband photometry of 85 comets, 1976–1992. *Icarus, 118,* 223–270.
A'Hearn M. F., Belton M. J. S., Delamere A., et al. (2005a) Deep Impact: A large-scale active experiment on a cometary nucleus. *Space Sci. Rev., 117,* 1–21.
A'Hearn M. F., Belton M. J. S., Delamere W. A., et al. (2005b) Deep Impact: Excavating Comet Tempel 1. *Science, 310,* 258–264.
A'Hearn M. F., Belton M. J. S., Delamere W. A., et al. (2011) EPOXI at Comet Hartley 2. *Science, 332,* 1396–1400.
A'Hearn M. F., Feaga L. M., Keller H. U., et al. (2012) Cometary volatiles and the origin of comets. *Astrophys. J., 758,* 29.
Aikawa Y. and Herbst E. (2001) Two-dimensional distributions and column densities of gaseous molecules in protoplanetary disks. II. Deuterated species and UV shielding by ambient clouds. *Astron. Astrophys., 371,* 1107–1117.
Altwegg K., Balsiger H., Bar-Nun A., et al. (2016) Prebiotic chemicals — amino acid and phosphorus — in the coma of Comet 67P/Churyumov-Gerasimenko. *Sci. Adv., 2,* e1600285.
Altwegg K., Balsiger H., Berthelier J.-J., et al. (2017) Organics in Comet 67P – A first comparative analysis of mass spectra from ROSINA-DFMS, COSAC and Ptolemy. *Mon. Not. R. Astron. Soc., 469,* S130–S141.
Altwegg K., Balsiger H., and Fuselier S. A. (2019) Cometary chemistry and the origin of icy solar system bodies: The view after Rosetta. *Annu. Rev. Astron. Astrophys., 57,* 113–155.
Altwegg K., Balsiger H., Hänni N., et al. (2020) Evidence of ammonium salts in Comet 67P as explanation for the nitrogen depletion in cometary comae. *Nature Astron., 4,* 533–540.
Asphaug E. and Benz W. (1994) Density of Comet Shoemaker-Levy 9 deduced by modelling breakup of the parent 'rubble pile.' *Nature, 370,* 120–124.
Asphaug E. and Benz W. (1996) Size, density, and structure of Comet Shoemaker-Levy 9 inferred from the physics of tidal breakup. *Icarus, 121,* 225–248.
Attree N., Groussin O., Jorda L., et al. (2018) Tensile strength of 67P/Churyumov-Gerasimenko nucleus material from overhangs. *Astron. Astrophys., 611,* A33.
Auger A.-T., Groussin O., Jorda L., et al. (2018) Meter-scale thermal contraction crack polygons on the nucleus of Comet 67P/Churyumov-Gerasimenko. *Icarus, 301,* 173–188.
Auster H.-U., Apathy I., Berghofer G., et al. (2015) The nonmagnetic nucleus of Comet 67P/Churyumov-Gerasimenko. *Science, 349,* aaa5102.
Bailey B. L. and Malhotra R. (2009) Two dynamical classes of Centaurs. *Icarus, 203,* 155–163.
Bardyn A., Baklouti D., Cottin H., et al. (2017) Carbon-rich dust in Comet 67P/Churyumov-Gerasimenko measured by COSIMA/Rosetta. *Mon. Not. R. Astron. Soc., 469,* S712–S722.
Bar-Nun A., Pat-El I., and Laufer D. (2007) Comparison between the findings of Deep Impact and our experimental results on large samples of gas-laden amorphous ice. *Icarus, 187,* 321–325.
Basilevsky A. T., Krasil'nikov S. S., Shiryaev A. A., et al. (2016) Estimating the strength of the nucleus material of Comet 67P Churyumov-Gerasimenko. *Solar System Res., 50,* 225–234.
Belton M. J. S., Thomas P., Veverka J., et al. (2007) The internal structure of Jupiter family cometary nuclei from Deep Impact observations: The "talps" or "layered pile" model. *Icarus, 187,* 332–344.
Belton M. J., Zou X.-D., Li J.-Y., et al. (2018) On the origin of internal layers in comet nuclei. *Icarus, 314,* 364–375.
Benkhoff J. and Huebner W. F. (1995) Influence of the vapor flux on temperature, density, and abundance distributions in a multicomponent, porous, icy body. *Icarus, 114,* 348–354.
Benkhoff J. and Spohn T. (1991) Thermal histories of the KOSI samples. *Geophys. Res. Lett., 18,* 261–264.
Benseguane S., Guilbert-Lepoutre A., Lasue J., et al. (2022) Evolution of pits at the surface of 67P/Churyumov-Gerasimenko. *Astron. Astrophys., 668,* A132.
Bentley M. S., Schmied R., Mannel T., et al. (2016) Aggregate dust particles at Comet 67P/Churyumov-Gerasimenko. *Nature, 537,* 73–75.
Bertaux J.-L. and Lallement R. (2017) Diffuse interstellar bands carriers and cometary organic material. *Mon. Not. R. Astron. Soc., 469,* S646–S660.
Bertini I., Lara L. M., Vincent J. B., et al. (2009) Activity evolution, outbursts, and splitting events of Comet 73P/Schwassmann-Wachmann 3. *Astron. Astrophys., 496,* 235–247.
Bertini I., La Forgia F., Fulle M., et al. (2019) The backscattering ratio of Comet 67P/Churyumov-Gerasimenko dust coma as seen by OSIRIS onboard Rosetta. *Mon. Not. R. Astron. Soc., 482,* 2924–2933.
Bertoldi C., Dachs E., Cemic L., et al. (2005) The heat capacity of the serpentine subgroup mineral berthierine $(Fe_{2.5}Al_{0.5})[Si_{1.5}Al_{0.5}O_5](OH)_4$. *Clays Clay Miner., 53,* 380–388.
Biele J., Ulamec S., Richter L., et al. (2009) The putative mechanical strength of comet surface material applied to landing on a comet. *Acta Astronaut., 65,* 1168–1178.
Biele J., Ulamec S., Maibaum M., et al. (2015) The landing(s) of Philae and inferences about comet surface mechanical properties. *Science, 349,* aaa9816.
Biele J., Vincent J.-B., and Knollenberg J. (2022) Mechanical properties of cometary surfaces. *Universe, 8,* 487.
Biver N., Hofstadter M., Gulkis S., et al. (2015) Distribution of water around the nucleus of Comet 67P/Churyumov-Gerasimenko at 3.4 AU from the Sun as seen by the MIRO instrument on Rosetta. *Astron. Astrophys., 583,* A3.
Biver N., Bockelée-Morvan D., Paubert G., et al. (2018) The extraordinary composition of the blue comet C/2016 R2 (PanSTARRS). *Astron. Astrophys., 619,* A127.
Biver N., Bockelée-Morvan D., Hofstadter M., et al. (2019) Long-term monitoring of the outgassing and composition of Comet 67P/Churyumov-Gerasimenko with the Rosetta/MIRO instrument. *Astron. Astrophys., 630,* A19.
Blum J. and Schräpler R. (2004) Structure and mechanical properties of high-porosity macroscopic agglomerates formed by random ballistic deposition. *Phys. Rev. Lett., 93,* 115503.
Blum J. and Wurm G. (2008) The growth mechanisms of macroscopic bodies in protoplanetary disks. *Annu. Rev. Astron. Astrophys., 46,* 21–56.
Blum J., Schräpler R., Davidsson B. J. R., et al. (2006) The physics of protoplanetesimal dust agglomerates. I. Mechanical properties and relations to primitive bodies in the solar system. *Astrophys. J., 652,* 1768–1781.
Blum J., Gundlach B., Mühle S., et al. (2014) Comets formed in solar-nebula instabilities! — An experimental and modeling attempt to relate the activity of comets to their formation process. *Icarus, 235,* 156–169.
Blum J., Gundlach B., Krause M., et al. (2017) Evidence for the formation of Comet 67P/Churyumov-Gerasimenko through gravitational collapse of a bound clump of pebbles. *Mon. Not. R. Astron. Soc., 469,* S755–S773.
Bockelée-Morvan D. and Biver N. (2017) The composition of cometary ices. *Philos. Trans. R. Soc. A, 375,* 20160252.
Bockelée-Morvan D., Lis D. C., Wink J. E., et al. (2000) New molecules found in Comet C/1995 O1 (Hale-Bopp): Investigating the link between cometary and interstellar material. *Astron. Astrophys., 353,* 1101–1114.
Bockelée-Morvan D., Gautier D., Hersant F., et al. (2002) Turbulent radial mixing in the solar nebula as the source of crystalline silicates in comets. *Astron. Astrophys., 384,* 1107–1118.
Bockelée-Morvan D., Debout V., Erard S., et al. (2015) First observations of H_2O and CO_2 vapor in Comet 67P/Churyumov-Gerasimenko made by VIRTIS onboard Rosetta. *Astron. Astrophys., 583,* A6.
Bockelée-Morvan D., Crovisier J., Erard S., et al. (2016) Evolution of CO_2, CH_4, and OCS abundances relative to H_2O in the coma of Comet 67P around perihelion from Rosetta/VIRTIS-H observations. *Mon. Not. R. Astron. Soc., 462,* S170–S183.
Bodewits D., Farnham T. L., Kelley M. S., et al. (2018) A rapid decrease in the rotation rate of Comet 41P/Tuttle-Giacobini-Kresák. *Nature, 553,* 186–188.
Boehnhardt H. (2004) Split comets. In *Comets II* (M. C. Festou et al., eds.), pp. 301–316. Univ. of Arizona, Tucson.
Boissier J., Bockelée-Morvan D., Biver N., et al. (2014) Gas and dust productions of Comet 103P/Hartley 2 from millimetre observations: Interpreting rotation-induced time variations. *Icarus, 228,* 197–216.
Bonev B. P., Dello Russo N., DiSanti M. A., et al. (2021) First comet observations with NIRSPEC-2 at Keck: Outgassing sources of parent volatiles and abundances based on alternative taxonomic compositional baselines in 46P/Wirtanen. *Planet. Sci. J., 2,* 45.

Bregman J. D., Tielens A. G. G. M., Witteborn F. C., et al. (1988) 3 micron spectrophotometry of Comet Halley: Evidence for water ice. *Astrophys. J., 334,* 1044–1048.
Brouet Y., Levasseur-Regourd A. C., Sabouroux P., et al. (2015) Permittivity measurements of porous matter in support of investigations of the surface and interior of 67P/Churyumov-Gerasimenko. *Astron. Astrophys., 583,* A39.
Brouet Y., Levasseur-Regourd A. C., Sabouroux P., et al. (2016) A porosity gradient in 67P/C-G nucleus suggested from CONSERT and SESAME-PP results: An interpretation based on new laboratory permittivity measurements of porous icy analogues. *Mon. Not. R. Astron. Soc., 462,* S89–S98.
Brown R. H., Lauretta D. S., Schmidt B., et al. (2012) Experimental and theoretical simulations of ice sublimation with implications for the chemical, isotopic, and physical evolution of icy objects. *Planet. Space Sci., 60,* 166–180.
Brownlee D. (2014) The Stardust mission: Analyzing samples from the edge of the solar system. *Annu. Rev. Earth Planet. Sci., 42,* 179–205.
Brownlee D. E., Horz F., Newburn R. L., et al. (2004) Surface of young Jupiter family Comet 81P/Wild 2: View from the Stardust spacecraft. *Science, 304,* 1764–1769.
Brownlee D., Tsou P., Aléon J., et al. (2006) Comet 81P/Wild 2 under a microscope. *Science, 314,* 1711–1714.
Cabral N., Guilbert-Lepoutre A., Fraser W. C., et al. (2019) OSSOS. XI. No active Centaurs in the Outer Solar System Origins Survey. *Astron. Astrophys., 621,* A102.
Calmonte U., Altwegg K., Balsiger H., et al. (2016) Sulphur-bearing species in the coma of Comet 67P/Churyumov-Gerasimenko. *Mon. Not. R. Astron. Soc., 462,* S253–S273.
Cambianica P., Fulle M., Cremonese G., et al. (2020) Time evolution of dust deposits in the Hapi region of Comet 67P/Churyumov-Gerasimenko. *Astron. Astrophys., 636,* A91.
Cambianica P., Cremonese G., Fulle M., et al. (2021) Long-term measurements of the erosion and accretion of dust deposits on Comet 67P/Churyumov-Gerasimenko with the OSIRIS instrument. *Mon. Not. R. Astron. Soc., 504,* 2895–2910.
Campins H. and Ryan E. V. (1989) The identification of crystalline olivine in cometary silicates. *Astrophys. J., 341,* 1059–1066.
Capaccioni F., Coradini A., Filacchione G., et al. (2015) The organic-rich surface of Comet 67P/Churyumov-Gerasimenko as seen by VIRTIS/Rosetta. *Science, 347,* aaa0628.
Capria M.-T., Coradini A., De Sanctis M. C., et al. (2000) Chiron activity and thermal evolution. *Astron. J., 119,* 3112–3118.
Ceccarelli C., Caselli P., Bockelée-Morvan D., et al. (2014) Deuterium fractionation: The Ariadne's thread from the precollapse phase to meteorites and comets today. In *Protostars and Planets VI* (H. Beuther et al., eds.), pp. 859–882. Univ. of Arizona, Tucson.
Chesley S., Belton M., Carcich B., et al. (2013) An updated rotation model for Comet 9P/Tempel 1. *Icarus, 222,* 516–525.
Chesley S., French A., Davis A., et al. (2020) Trajectory estimation for particles observed in the vicinity of (101955) Bennu. *J. Geophys. Res.–Planets, 125,* e2019JE006363.
Choi Y.-J., Cohen M., Merk R., et al. (2002) Long-term evolution of objects in the Kuiper belt zone — Effects of insolation and radiogenic heating. *Icarus, 160,* 300–312.
Choukroun M., Keihm S., Schloerb F. P., et al. (2015) Dark side of Comet 67P/Churyumov-Gerasimenko in Aug.–Oct. 2014: MIRO/Rosetta continuum observations of polar night in the southern regions. *Astron. Astrophys., 583,* A28.
Choukroun M., Vu T. H., and Fayolle E. C. (2019) No compelling evidence for clathrate hydrate formation under interstellar medium conditions over laboratory time scales. *Proc. Natl. Acad. Sci. U. S. A., 116,* 14407–14408.
Choukroun M., Altwegg K., Kührt E., et al. (2020) Dust-to-gas and refractory-to-ice mass ratios of Comet 67P/Churyumov-Gerasimenko from Rosetta observations. *Space Sci. Rev., 216,* 44.
Ciarletti V., Levasseur-Regourd A. C., Lasue J., et al. (2015) CONSERT suggests a change in local properties of 67P/Churyumov-Gerasimenko's nucleus at depth. *Astron. Astrophys., 583,* A40.
Ciarletti V., Herique A., Lasue J., et al. (2017) CONSERT constrains the internal structure of 67P at a few metres size scale. *Mon. Not. R. Astron. Soc., 469,* S805–S817.
Ciesla F. J. (2007) Outward transport of high-temperature materials around the midplane of the solar nebula. *Science, 318,* 613–615.
Cochran A. L. and Cochran W. D. (2002) A high spectral resolution atlas of Comet 122P/de Vico. *Icarus, 157,* 297–308.
Cooke I. R., Öberg K. I., Fayolle E. C., et al. (2018) CO diffusion and desorption kinetics in CO_2 ices. *Astrophys. J., 852,* 75.
Crovisier J. (1996) Observational constraints on the composition and nature of Comet D/Shoemaker-Levy 9. In *The Collision of Comet Shoemaker-Levy 9 and Jupiter* (K. S. Noll et al., eds.), pp. 31–54. IAU Colloq. 156, Cambridge Univ., Cambridge.
Crovisier J., Biver N., Bockelée-Morvan D., et al. (1995) Carbon monoxide outgassing from Comet P/Schwassmann-Wachmann 1. *Icarus, 115,* 213–216.
Crovisier J., Leech K., Bockelée-Morvan D., et al. (1997) The spectrum of Comet Hale-Bopp (C/1995 O1) observed with the infrared space observatory at 2.9 astronomical units from the Sun. *Science, 275,* 1904–1907.
Davidsson B. J. R. (2001) Tidal splitting and rotational breakup of solid biaxial ellipsoids. *Icarus, 149,* 375–383.
Davidsson B. J. R. (2006) The bulk density of cometary nuclei. In *Advances in Geosciences, Vol. 3: Planetary Science* (W.-H. Ip and A. Bhardwaj, eds.), pp. 155–169. World Scientific, Singapore.
Davidsson B. J. R. (2021) Thermophysical evolution of planetesimals in the primordial disc. *Mon. Not. R. Astron. Soc., 505,* 5654–5685.
Davidsson B. J. R. (2023) Collisional heating of icy planetesimals — I. Catastrophic collisions. *Mon. Not. R. Astron. Soc., 521,* 2484–2503.
Davidsson B. J. R. and Gutiérrez P. J. (2004) Estimating the nucleus density of Comet 19P/Borrelly. *Icarus, 168,* 392–408.
Davidsson B. J. R. and Gutiérrez P. J. (2005) Nucleus properties of Comet 67P/Churyumov-Gerasimenko estimated from non-gravitational force modeling. *Icarus, 176,* 453–477.
Davidsson B. J. R. and Gutiérrez P. J. (2006) Non-gravitational force modeling of Comet 81P/Wild 2. I. A nucleus bulk density estimate. *Icarus, 180,* 224–242.
Davidsson B. J. R., Gutiérrez P. J., and Rickman H. (2007) Nucleus properties of Comet 9P/Tempel 1 estimated from non-gravitational force modeling. *Icarus, 187,* 306–320.
Davidsson B. J. R., Gutiérrez P. J., and Rickman H. (2009) Physical properties of morphological units on Comet 9P/Tempel 1 derived from near-IR Deep Impact spectra. *Icarus, 201,* 335–357.
Davidsson B. J. R., Gutiérrez P. J., Groussin O., et al. (2013) Thermal inertia and surface roughness of Comet 9P/Tempel 1. *Icarus, 224,* 154–171.
Davidsson B. J. R., Sierks H., Güttler C., et al. (2016) The primordial nucleus of Comet 67P/Churyumov-Gerasimenko. *Astron. Astrophys., 592,* A63.
Davidsson B. J. R., Birch S., Blake G. A., et al. (2021) Airfall on Comet 67P/Churyumov-Gerasimenko. *Icarus, 354,* 114004.
Davidsson B. J. R., Buratti B. J., and Hicks M. D. (2022a) Albedo variegation on Comet 67P/Churyumov-Gerasimenko. *Mon. Not. R. Astron. Soc., 516,* 5125–5142.
Davidsson B. J. R., Samarasinha N., Farnocchia D., et al. (2022b) Modeling the water and carbon dioxide production rates of Comet 67P/Churyumov-Gerasimenko. *Mon. Not. R. Astron. Soc., 509,* 3065–3085.
Davidsson B. J. R., Schloerb F. P., Fornasier S., et al. (2022c) CO_2-driven surface changes in the Hapi region on Comet 67P/Churyumov-Gerasimenko. *Mon. Not. R. Astron. Soc., 516,* 6009–6040.
Davison T. M., Collins G. S., and Ciesla F. J. (2010) Numerical modelling of heating in porous planetesimal collisions. *Icarus, 208,* 468–481.
Dello Russo N., Vervack R. J., Weaver H. A., et al. (2007) Compositional homogeneity in the fragmented Comet 73P/Schwassmann-Wachmann 3. *Nature, 448,* 172–175.
Dello Russo N., Kawakita H., Vervack R. J., et al. (2016) Emerging trends and a comet taxonomy based on the volatile chemistry measured in thirty comets with high-resolution infrared spectroscopy between 1997 and 2013. *Icarus, 278,* 301–332.
de Niem D., Kührt E., Hviid S., et al. (2018) Low velocity collisions of porous planetesimals in the early solar system. *Icarus, 301,* 196–218.
De Sanctis M. C., Capria M.-T., Coradini A., et al. (2000) Thermal evolution of the Centaur object 5145 Pholus. *Astron. J., 120,* 1571–1578.
De Sanctis M. C., Capria M.-T., and Coradini A. (2001) Thermal evolution and differentiation of Edgeworth-Kuiper belt objects. *Astron. J., 121,* 2792–2799.
De Sanctis M. C., Capaccioni F., Ciarniello M., et al. (2015) The diurnal cycle of water ice on Comet 67P/Churyumov-Gerasimenko. *Nature,*

525, 500–503.
Deshapriya J. D. P., Barucci M. A., Fornasier S., et al. (2018) Exposed bright features on the Comet 67P/Churyumov-Gerasimenko: Distribution and evolution. *Astron. Astrophys.*, *613*, A36.
di Sisto R. P., Brunini A., and de Elía G. C. (2010) Dynamical evolution of escaped plutinos, another source of Centaurs. *Astron. Astrophys.*, *519*, A112.
Drahus M., Jewitt D., Guilbert-Lepoutre A., et al. (2012) The sources of HCN and CH_3OH and the rotational temperature in Comet 103P/Hartley 2 from time-resolved millimeter spectroscopy. *Astrophys. J.*, *756*, 80.
Drahus M., Yang B., Lis D. C., et al. (2017) New limits to CO outgassing in Centaurs. *Mon. Not. R. Astron. Soc.*, *468*, 2897–2909.
Eisner N. L., Knight M. M., Snodgrass C., et al. (2019) Properties of the bare nucleus of Comet 96P/Machholz 1. *Astron. J.*, *157*, 186.
El-Maarry M. R., Thomas N., Garcia Berná A., et al. (2015) Fractures on Comet 67P/Churyumov-Gerasimenko observed by the Rosetta/OSIRIS camera. *Geophys. Res. Lett.*, *42*, 5170–5178.
El-Maarry M. R., Groussin O., Thomas N., et al. (2017) Surface changes on Comet 67P/Churyumov-Gerasimenko suggest a more active past. *Science*, *355*, 1392–1395.
Elsila J. E., Glavin D. P., and Dworkin J. P. (2009) Cometary glycine detected in samples returned by Stardust. *Meteorit. Planet. Sci.*, *44*, 1323–1330.
Emel'yanenko V. V., Asher D. J., and Bailey M. E. (2013) A model for the common origin of Jupiter family and Halley type comets. *Earth Moon Planets*, *110*, 105–130.
Farnham T. L. and Cochran A. L. (2002) A McDonald Observatory study of Comet 19P/Borrelly: Placing the Deep Space 1 observations into a broader context. *Icarus*, *160*, 398–418.
Feaga L. M., A'Hearn M. F., Sunshine J. M., et al. (2007) Asymmetries in the distribution of H_2O and CO_2 in the inner coma of Comet 9P/Tempel 1 as observed by Deep Impact. *Icarus*, *190*, 345–356.
Fernández J. A., Helal M., and Gallardo T. (2018) Dynamical evolution and end states of active and inactive Centaurs. *Planet. Space Sci.*, *158*, 6–15.
Fernández Y. R., Kelley M. S., Lamy P. L., et al. (2013) Thermal properties, sizes, and size distribution of Jupiter-family cometary nuclei. *Icarus*, *226*, 1138–1170.
Ferrari C. (2018) Thermal properties of icy surfaces in the outer solar system. *Space Sci. Rev.*, *214*, 111.
Ferrari C. and Lucas A. (2016) Low thermal inertias of icy planetary surfaces: Evidence for amorphous ice? *Astron. Astrophys.*, *588*, A133.
Ferrari S., Penasa L., La Forgia F., et al. (2018) The big lobe of 67P/Churyumov-Gerasimenko comet: Morphological and spectrophotometric evidences of layering as from OSIRIS data. *Mon. Not. R. Astron. Soc.*, *479*, 1555–1568.
Filacchione G., de Sanctis M. C., Capaccioni F., et al. (2016) Exposed water ice on the nucleus of Comet 67P/Churyumov-Gerasimenko. *Nature*, *529*, 368–372.
Fink U. (2009) A taxonomic survey of comet composition 1985–2004 using CCD spectroscopy. *Icarus*, *201*, 311–334.
Fink U., Doose L., Rinaldi G., et al. (2016) Investigation into the disparate origin of CO_2 and H_2O outgassing for Comet 67/P. *Icarus*, *277*, 78–97.
Fornasier S., Lellouch E., Müller T., et al. (2013) TNOs are cool: A survey of the trans-neptunian region. VIII. Combined Herschel PACS and SPIRE observations of nine bright targets at 70–500 μm. *Astron. Astrophys.*, *555*, A15.
Fornasier S., Hasselmann P. H., Barucci M., et al. (2015) Spectrophotometric properties of the 67P/Churyumov-Gerasimenko's nucleus from the OSIRIS instrument onboard the Rosetta spacecraft. *Astron. Astrophys.*, *583*, A30.
Fornasier S., Bourdelle de Micas J., Hasselmann P. H., et al. (2021) Small lobe of Comet 67P: Characterization of the Wosret region with Rosetta-OSIRIS. *Astron. Astrophys.*, *653*, A132.
Fougere N., Altwegg K., Berthelier J.-J., et al. (2016a) Direct simulation Monte Carlo modelling of the major species in the coma of Comet 67P/Churyumov-Gerasimenko. *Mon. Not. R. Astron. Soc.*, *462*, S156–S169.
Fougere N., Altwegg K., Berthelier J.-J., et al. (2016b) Three-dimensional direct simulation Monte-Carlo modeling of the coma of Comet 67P/Churyumov-Gerasimenko observed by the VIRTIS and ROSINA instruments on board Rosetta. *Astron. Astrophys.*, *588*, A134.
Fray N. and Schmitt B. (2009) Sublimation of ices of astrophysical interest: A bibliographic review. *Planet. Space Sci.*, *57*, 2053–2080.
Fray N., Bardyn A., Cottin H., et al. (2016) High-molecular-weight organic matter in the particles of Comet 67P/Churyumov-Gerasimenko. *Nature*, *538*, 72–74.
Fray N., Bardyn A., Cottin H., et al. (2017) Nitrogen-to-carbon atomic ratio measured by COSIMA in the particles of Comet 67P/Churyumov-Gerasimenko. *Mon. Not. R. Astron. Soc.*, *469*, S506–S516.
Fulle M., Della Corte V., Rotundi A., et al. (2016) Comet 67P/Churyumov-Gerasimenko preserved the pebbles that formed planetesimals. *Mon. Not. R. Astron. Soc.*, *462*, S132–S137.
Fulle M., Blum J., and Rotundi A. (2019) How comets work. *Astrophys. J. Lett.*, *879*, L8.
Garenne A., Beck P., Montes-Hernandez G., et al. (2014) The abundance and stability of "water" in type 1 and 2 carbonaceous chondrites (CI, CM and CR). *Geochim. Cosmochim. Acta*, *137*, 93–112.
Gasc S., Altwegg K., Balsiger H., et al. (2017) Change of outgassing pattern of 67P/Churyumov-Gerasimenko during the March 2016 equinox as seen by ROSINA. *Mon. Not. R. Astron. Soc.*, *469*, S108–S117.
Geissler P., Petit J.-M., Durda D. D., et al. (1996) Erosion and ejecta reaccretion on 243 Ida and its moon. *Icarus*, *120*, 140–157.
Giauque W. F. and Stout J. W. (1936) The entropy of water and third law of thermodynamics: The heat capacity of ice from 15 to 273 K. *J. Am. Chem. Soc.*, *58*, 1144–1144.
Gicquel A., Vincent J. B., Agarwal J., et al. (2016) Sublimation of icy aggregates in the coma of Comet 67P/Churyumov-Gerasimenko detected with the OSIRIS cameras on board Rosetta. *Mon. Not. R. Astron. Soc.*, *462*, S57–S66.
Gkotsinas A., Guilbert-Lepoutre A., Raymond S. N., et al. (2022) Thermal processing of Jupiter-family comets during their chaotic orbital evolution. *Astrophys. J.*, *928*, 43.
Glavin D. P., Dworkin J. P., and Sandford S. A. (2008) Detection of cometary amines in samples returned by Stardust. *Meteorit. Planet. Sci.*, *43*, 399–413.
González M., Gutiérrez P. J., Lara L. M., et al. (2008) Evolution of the crystallization front in cometary models: Effect of the net energy released during crystallization. *Astron. Astrophys.*, *486*, 331–340.
Graykowski A. and Jewitt D. (2019) Fragmented Comet 73P/Schwassmann-Wachmann 3. *Astron. J.*, *158*, 112.
Greenberg J. M. (1982) What are comets made of? A model based on interstellar dust. In *Comets* (L. L. Wilkening, ed.), pp. 131–163. Univ. of Arizona, Tucson.
Groussin O., Sunshine J. M., Feaga L. M., et al. (2013) The temperature, thermal inertia, roughness and color of the nuclei of Comets 103P/Hartley 2 and 9P/Tempel 1. *Icarus*, *222*, 580–594.
Groussin O., Jorda L., Auger A. T., et al. (2015) Gravitational slopes, geomorphology, and material strengths of the nucleus of Comet 67P/Churyumov-Gerasimenko from OSIRIS observations. *Astron. Astrophys.*, *583*, A32.
Groussin O., Attree N., Bouet Y., et al. (2019) The thermal, mechanical, structural, and dielectric properties of cometary nuclei after Rosetta. *Space Sci. Rev.*, *215*, 29.
Grün E., Benkhoff J., Heidrich R., et al. (1991) Energy balance of the KOSI 4 experiment. *Geophys. Res. Lett.*, *18*, 257–260.
Guilbert-Lepoutre A. (2012) Survival of amorphous water ice on Centaurs. *Astron. J.*, *144*, 97.
Guilbert-Lepoutre A. and Jewitt D. (2011) Thermal shadows and compositional structure in comet nuclei. *Astrophys. J.*, *743*, 31.
Guilbert-Lepoutre A., Lasue J., Federico C., et al. (2011) New 3D thermal evolution model for icy bodies application to trans-neptunian objects. *Astron. Astrophys.*, *529*, A71.
Guilbert-Lepoutre A., Rosenberg E. D., Prialnik D., et al. (2016) Modelling the evolution of a comet subsurface: Implications for 67P/Churyumov-Gerasimenko. *Mon. Not. R. Astron. Soc.*, *462*, S146–S155.
Guilbert-Lepoutre A., Gkotsinas A., Raymond S. N., et al. (2023) The gateway from Centaurs to Jupiter-family comets: Thermal and dynamical evolution. *Astrophys. J.*, *942*, 92.
Gulkis S., Allen M., von Allmen P., et al. (2015) Subsurface properties and early activity of Comet 67P/Churyumov-Gerasimenko. *Science*, *347*, aaa0709.
Gundlach B. and Blum J. (2012) Outgassing of icy bodies in the solar system — II: Heat transport in dry, porous surface dust layers. *Icarus*, *219*, 618–629.

Gundlach B. and Blum J. (2013) A new method to determine the grain size of planetary regolith. *Icarus, 223*, 479–492.
Gundlach B., Skorov Y. V., and Blum J. (2011) Outgassing of icy bodies in the solar system — I. The sublimation of hexagonal water ice through dust layers. *Icarus, 213*, 710–719.
Gundlach B., Blum J., Keller H. U., et al. (2015) What drives the dust activity of Comet 67P/Churyumov-Gerasimenko? *Astron. Astrophys., 583*, A12.
Gundlach B., Schmidt K. P., Kreuzig C., et al. (2018) The tensile strength of ice and dust aggregates and its dependence on particle properties. *Mon. Not. R. Astron. Soc., 479*, 1273–1277.
Gunnarsson M., Bockelée-Morvan D., Biver N., et al. (2008) Mapping the carbon monoxide coma of Comet 29P/Schwassmann-Wachmann 1. *Astron. Astrophys., 484*, 537–546.
Güttler C., Krause M., Geretshauser R. J., et al. (2009) The physics of protoplanetesimal dust agglomerates. IV. Toward a dynamical collision model. *Astron. J., 701*, 130–141.
Güttler C., Mannel T., Rotundi A., et al. (2019) Synthesis of the morphological description of cometary dust at Comet 67P/Churyumov-Gerasimenko. *Astron. Astrophys., 630*, A24.
Hänni N., Altwegg K., Pestoni B., et al. (2020) First *in situ* detection of the CN radical in comets and evidence for a distributed source. *Mon. Not. R. Astron. Soc., 498*, 2239–2248.
Harmon J. K., Nolan M. C., Ostro S. J., et al. (2004) Radar studies of comet nuclei and grain comae. In *Comets II* (M. C. Festou et al., eds.), pp. 265–279. Univ. of Arizona, Tucson.
Harmon J. K., Nolan M. C., Giorgini J. D., et al. (2010) Radar observations of 8P/Tuttle: A contact-binary comet. *Icarus, 207*, 499–502.
Hartmann W. K. and Davis D. R. (1975) Satellite-sized planetesimals and lunar origin. *Icarus, 24*, 504–515.
Hartogh P., Lis D. C., Bockelée-Morvan D., et al. (2011) Ocean-like water in the Jupiter-family Comet 103P/Hartley 2. *Nature, 478*, 218–220.
Haruyama J., Yamamoto T., Mizutani H., et al. (1993) Thermal history of comets during residence in the Oort cloud: Effect of radiogenic heating in combination with the very low thermal conductivity of amorphous ice. *J. Geophys. Res.–Planets, 98*, 15079–15090.
Hässig M., Altwegg K., Balsiger H., et al. (2015) Time variability and heterogeneity in the coma of 67P/Churyumov-Gerasimenko. *Science, 347*, aaa0276.
Heggy E., Palmer E. M., Kofman W., et al. (2012) Radar properties of comets: Parametric dielectric modeling of Comet 67P/Churyumov-Gerasimenko. *Icarus, 221*, 925–939.
Heinisch P., Auster H.-U., Gundlach B., et al. (2019) Compressive strength of Comet 67P/Churyumov-Gerasimenko derived from Philae surface contacts. *Astron. Astrophys., 630*, A2.
Herique A., Kofman W., Beck P., et al. (2016) Cosmochemical implications of CONSERT permittivity characterization of 67P/CG. *Mon. Not. R. Astron. Soc., 462*, S516–S532.
Herny C., Mousis O., Marschall R., et al. (2021) New constraints on the chemical composition and outgassing of 67P/Churyumov-Gerasimenko. *Planet. Space Sci., 200*, 105194.
Hilchenbach M., Fischer H., Langevin Y., et al. (2017) Mechanical and electrostatic experiments with dust particles collected in the inner coma of Comet 67P by COSIMA onboard Rosetta. *Philos. Trans. R. Soc. A, 375*, 20160255.
Hirabayashi M. and Scheeres D. J. (2019) Rotationally induced failure of irregularly shaped asteroids. *Icarus, 317*, 354–364.
Hirabayashi M., Scheeres D. J., Chesley S. R., et al. (2016) Fission and reconfiguration of bilobate comets as revealed by 67P/Churyumov-Gerasimenko. *Nature, 534*, 352–355.
Hoang M., Garnier P., Lasue J., et al. (2020) Investigating the Rosetta/RTOF observations of Comet 67P/Churyumov-Gerasimenko using a comet nucleus model: Influence of dust mantle and trapped CO. *Astron. Astrophys., 638*, A106.
Holsapple K. A. and Housen K. R. (2007) A crater and its ejecta: An interpretation of Deep Impact. *Icarus, 187*, 345–356.
Hornung K., Merouane S., Hilchenbach M., et al. (2016) A first assessment of the strength of cometary particles collected *in-situ* by the COSIMA instrument onboard Rosetta. *Planet. Space Sci., 133*, 63–75.
Hu X., Shi X., Sierks H., et al. (2017) Seasonal erosion and restoration of the dust cover on Comet 67P/Churyumov-Gerasimenko as observed by OSIRIS onboard Rosetta. *Astron. Astrophys., 604*, A114.
Huebner W. F. and Benkhoff J. (1999) From coma abundances to nucleus composition. *Space Sci. Rev., 90*, 117–130.
Huebner W. F., Benkhoff J., Capria M.-T., et al. (2006) *Heat and Gas Diffusion in Comet Nuclei.* International Space Science Institute, Bern. 284 pp.
Hui M.-T., Jewitt D., and Clark D. (2018) Pre-discovery observations and orbit of Comet C/2017 K2 (PANSTARRS). *Astron. J., 155*, 25.
Hui M.-T., Farnocchia D., and Micheli M. (2019) C/2010 U3 (Boattini): A bizarre comet active at record heliocentric distance. *Astron. J., 157*, 162.
Hui M.-T., Tholen D. J., Kracht R., et al. (2022) The lingering death of periodic near-Sun comet 323P/SOHO. *Astron. J., 164*, 1.
Huss G. R., MacPherson G. J., Wasserburg G. J., et al. (2001) Aluminum-26 in calcium-aluminum-rich inclusions and chondrules from unequilibrated ordinary chondrites. *Meteorit. Planet. Sci., 36*, 975–997.
Isnard R., Bardyn A., Fray N., et al. (2019) H/C elemental ratio of the refractory organic matter in cometary particles of 67P/Churyumov-Gerasimenko. *Astron. Astrophys., 630*, A27.
Jacobson S. A. and Scheeres D. J. (2011) Dynamics of rotationally fissioned asteroids: Source of observed small asteroid systems. *Icarus, 214*, 161–178.
Jessberger E. K., Christoforidis A., and Kissel J. (1988) Aspects of the major element composition of Halley's dust. *Nature, 332*, 691–695.
Jewitt D. C. (2004) From cradle to grave: The rise and demise of the comets. In *Comets II* (M. C. Festou et al., eds.), pp. 659–676. Univ. of Arizona, Tucson.
Jewitt D. (2009) The active Centaurs. *Astron. J., 137*, 4296–4312.
Jewitt D. (2021) Systematics and consequences of comet nucleus outgassing torques. *Astron. J., 161*, 261.
Jewitt D. (2022) Destruction of long-period comets. *Astron. J., 164*, 158.
Jewitt D. and Luu J. (2019) Disintegrating inbound long-period Comet C/2019 J2. *Astrophys. J. Lett., 883*, L28.
Jewitt D., Hsieh H., Agarwal J., et al. (2015) The active asteroids. In *Asteroids IV* (P. Michel et al., eds.), pp. 221–241. Univ. of Arizona, Tucson.
Jewitt D., Agarwal J., Li J., et al. (2017a) Anatomy of an asteroid breakup: The case of P/2013 R3. *Astron. J., 153*, 223.
Jewitt D., Hui M.-T., Mutchler M., et al. (2017b) A comet active beyond the crystallization zone. *Astrophys. J. Lett., 847*, L19.
Jewitt D., Agarwal J., Hui M.-T., et al. (2019) Distant comet C/2017 K2 and the cohesion bottleneck. *Astron. J., 157*, 65.
Jewitt D., Kim Y., Mutchler M., et al. (2021) Cometary activity begins at Kuiper belt distances: Evidence from C/2017 K2. *Astron. J., 161*, 188.
Johansen A., Oishi J. S., Low M.-M. M., et al. (2007) Rapid planetesimal formation in turbulent circumstellar disks. *Nature, 448*, 1022–1025.
Jutzi M. and Asphaug E. (2015) The shape and structure of cometary nuclei as a result of low-velocity accretion. *Science, 348*, 1355–1358.
Jutzi M. and Benz W. (2017) Formation of bi-lobed shapes by sub-catastrophic collisions: A late origin of Comet 67P's structure. *Astron. Astrophys., 597*, A62.
Jutzi M., Benz W., Toliou A., et al. (2017) How primordial is the structure of Comet 67P? Combined collisional and dynamical models suggest a late formation. *Astron. Astrophys., 597*, A61.
Kamoun P., Campbell D., Pettengill G., et al. (1998) Radar observations of three comets and detection of echoes from one: P/Grigg-Skjellerup. *Planet. Space Sci., 47*, 23–28.
Kamoun P., Lamy P. L., Toth I., et al. (2014) Constraints on the subsurface structure and density of the nucleus of Comet 67P/Churyumov-Gerasimenko from Arecibo radar observations. *Astron. Astrophys., 568*, A21.
Kareta T., Woodney L. M., Schambeau C., et al. (2021) Contemporaneous multiwavelength and precovery observations of the active Centaur P/2019 LD2 (ATLAS). *Planet. Sci. J., 2*, 48.
Keeney B. A., Stern S. A., A'Hearn M. F., et al. (2017) H_2O and O_2 absorption in the coma of Comet 67P/Churyumov-Gerasimenko measured by the Alice far-ultraviolet spectrograph on Rosetta. *Mon. Not. R. Astron. Soc., 469*, S158–S177.
Keihm S., Tosi F., Kamp L., et al. (2012) Interpretation of combined infrared, submillimeter, and millimeter thermal flux data obtained during the Rosetta fly-by of asteroid (21) Lutetia. *Icarus, 221*, 395–404.
Keil K., Stoeffler D., Love S. G., et al. (1997) Constraints on the role of impact heating and melting in asteroids. *Meteorit. Planet. Sci., 32*, 349–363.

Keller H. U. and Kührt E. (2020) Cometary nuclei: From Giotto to Rosetta. *Space Sci. Rev.*, *216*, 1–26.
Keller H. U., Mottola S., Davidsson B., et al. (2015a) Insolation, erosion, and morphology of Comet 67P/Churyumov-Gerasimenko. *Astron. Astrophys.*, *583*, A34.
Keller H. U., Mottola S., Skorov Y., et al. (2015b) The changing rotation period of Comet 67P/Churyumov-Gerasimenko controlled by its activity. *Astron. Astrophys.*, *579*, L5.
Keller H. U., Mottola S., Hviid S. F., et al. (2017) Seasonal mass transfer on the nucleus of Comet 67P/Churyumov-Gerasimenko. *Mon. Not. R. Astron. Soc.*, *469*, S357–S371.
Kelley M. S. and Wooden D. H. (2009) The composition of dust in Jupiter-family comets inferred from infrared spectroscopy. *Planet. Space Sci.*, *57*, 1133–1145.
Kenyon S. J. and Bromley B. (2012) Coagulation calculations of icy planet formation at 15–150 AU: A correlation between the maximum radius and the slope of the size distribution for trans-neptunian objects. *Astron. J.*, *143*, 63.
Khan Y., Gibb E. L., Bonev B. P., et al. (2021) Testing short-term variability and sampling of primary volatiles in Comet 46P/Wirtanen. *Planet. Sci. J.*, *2*, 20.
Kissel J., Sagdeev R. Z., Bertaux J. L., et al. (1986) Composition of Comet Halley dust particles from Vega observations. *Nature*, *321*, 280–282.
Klinger J. (1980) Influence of a phase transition of ice on the heat and mass balance of comets. *Science*, *209*, 271–272.
Klinger J. (1981) Some consequences of a phase transition of water ice on the heat balance of comet nuclei. *Icarus*, *47*, 320–324.
Klinger J., Espinasse S., and Schmidt B. (1989) Some considerations on cohesive forces in Sun-grazing comets. In *Proceedings of an International Workshop on Physics and Mechanics of Cometary Materials* (J. Hunt and T. D. Guyenne, eds.), pp. 197–200. ESA SP-302, Noordwijk, The Netherlands.
Knapmeyer M., Fischer H. H., Knollenberg J., et al. (2018) Structure and elastic parameters of the near surface of Abydos site on Comet 67P/Churyumov-Gerasimenko, as obtained by SESAME/CASSE listening to the MUPUS insertion phase. *Icarus*, *310*, 165–193.
Knight M. M., Fitzsimmons A., Kelley M. S. P., et al. (2016) Comet 322P/SOHO 1: An asteroid with the smallest perihelion distance? *Astrophys. J. Lett.*, *823*, L6.
Kofman W., Herique A., Goutail J. P., et al. (2007) The Comet Nucleus Sounding Experiment by Radiowave Transmission (CONSERT): A short description of the instrument and of the commissioning stages. *Space Sci. Rev.*, *128*, 413–432.
Kofman W., Herique A., Barbin Y., et al. (2015) Properties of the 67P/Churyumov-Gerasimenko interior revealed by CONSERT radar. *Science*, *349*, aab0639.
Kofman W., Zine S., Herique A., et al. (2020) The interior of Comet 67P/C-G: Revisiting CONSERT results with the exact position of the Philae lander. *Mon. Not. R. Astron. Soc.*, *497*, 2616–2622.
Kokubo E. and Ida S. (1996) On runaway growth of planetesimals. *Icarus*, *123*, 180–191.
Kossacki K. J. (2016) Comet 67P/Churyumov-Gerasimenko, location of pristine material? *Planet. Space Sci.*, *125*, 96–104.
Kouchi A. and Sirono S. (2001) Crystallization heat of impure amorphous H_2O ice. *Geophys. Res. Lett.*, *28*, 827–830.
Kossacki K. J. and Szutowicz S. (2006) Comet 67P/Churyumov-Gerasimenko: Modeling of orientation and structure. *Planet. Space Sci.*, *54*, 15–27.
Kossacki K. J., Koemle N. I., Kargl G., et al. (1994) The influence of grain sintering on the thermoconductivity of porous ice. *Planet. Space Sci.*, *42*, 383–389.
Kossacki K. J., Spohn T., Hagermann A., et al. (2015) Comet 67P/Churyumov-Gerasimenko: Hardening of the sub-surface layer. *Icarus*, *260*, 464–474.
Kramer T. and Läuter M. (2019) Outgassing-induced acceleration of Comet 67P/Churyumov-Gerasimenko. *Astron. Astrophys.*, *630*, A4.
Kramer T., Läuter M., Hviid S., et al. (2019) Comet 67P/Churyumov-Gerasimenko rotation changes derived from sublimation-induced torques. *Astron. Astrophys.*, *630*, A3.
Krot A. N., Makide K., Nagashima K., et al. (2012) Heterogeneous distribution of ^{26}Al at the birth of the solar system: Evidence from refractory grains and inclusions. *Meteorit. Space Sci.*, *47*, 1948–1979.
Kunihiro T., Rubin A. E., McKeegan K. D., et al. (2004) Initial ^{26}Al/^{27}Al in carbonaceous-chondrite chondrules: Too little ^{26}Al to melt asteroids. *Geochim. Cosmochim. Acta*, *68*, 2947–2957.
Lai I.-L., Ip W.-H., Su C.-C., et al. (2016) Gas outflow and dust transport of Comet 67P/Churyumov-Gerasimenko. *Mon. Not. R. Astron. Soc.*, *462*, S533–S546.
Langevin Y., Kissel J., Bertaux J. L., et al. (1987) First statistical analysis of 5000 mass spectra of cometary grains obtained by PUMA 1 (Vega 1) and PIA (Giotto) impact ionization mass spectrometers in the compressed modes. *Astron. Astrophys.*, *187*, 761–766.
Langmuir I. (1913) The effect of space charge and residual gases on thermionic currents in high vacuum. *Phys. Rev.*, *2*, 450–486.
Läuter M., Kramer T., Rubin M., et al. (2019) Surface localization of gas sources on Comet 67P/Churyumov-Gerasimenko based on DFMS/COPS data. *Mon. Not. R. Astron. Soc.*, *483*, 852–861.
Lejoly C., Harris W., Samarasinha N., et al. (2022) Radial distribution of the dust comae of comets 45P/Honda-Mrkos-Pajdušáková and 46P/Wirtanen. *Planet. Sci. J.*, *3*, 17.
Lellouch E., Santos-Sanz P., Lacerda P., et al. (2013) "TNOs are Cool": A survey of the trans-neptunian region. IX. Thermal properties of Kuiper belt objects and Centaurs from combined Herschel and Spitzer observations. *Astron. Astrophys.*, *557*, A60.
Le Roy L., Altwegg K., Balsiger H., et al. (2015) Inventory of the volatiles on Comet 67P/Churyumov-Gerasimenko from Rosetta/ROSINA. *Astron. Astrophys.*, *583*, A1.
Lethuillier A., Le Gall A., Hamelin M., et al. (2016) Electrical properties and porosity of the first meter of the nucleus of 67P/Churyumov-Gerasimenko as constrained by the Permittivity Probe SESAME-PP/Philae/Rosetta. *Astron. Astrophys.*, *591*, A32.
Levison H. F. and Duncan M. J. (1997) From the Kuiper belt to Jupiter-family comets: The spatial distribution of ecliptic comets. *Icarus*, *127*, 13–32.
Leyrat C. (2019) Mapping of thermal properties over CG-67/P surface. *EPSC-DPS Joint Meeting 2019*, 2052.
Leyrat C., Tosi F., Capaccioni F., et al. (2015) Investigations of 67/P-CG surfaces thermal properties at southern latitudes and variations with heliocentric distances with VIRTIS/Rosetta. *European Planet. Sci. Congr., 2015*, 435.
Li J., Jewitt D., Mutchler M., et al. (2020) Hubble Space Telescope search for activity in high-perihelion objects. *Astron. J.*, *159*, 209.
Lilly E., Hsieh H., Bauer J., et al. (2021) No activity among 13 Centaurs discovered in the Pan-STARRS1 detection database. *Planet. Sci. J.*, *2*, 155.
Lis D. C., Bockelée-Morvan D., Güsten R., et al. (2019) Terrestrial deuterium-to-hydrogen ratio in water in hyperactive comets. *Astron. Astrophys.*, *625*, L5.
Lisse C. M., VanCleve J., Adams A. C., et al. (2006) Spitzer spectral observations of the Deep Impact ejecta. *Science*, *313*, 635–640.
Lisse C. M., Young L. A., Cruikshank D. P., et al. (2021) On the origin and thermal stability of Arrokoth's and Pluto's ices. *Icarus*, *356*, 114072.
Lisse C. M., Gladstone G. R., Young L. A., et al. (2022) A predicted dearth of majority hypervolatile ices in Oort cloud comets. *Planet. Sci. J.*, *3*, 112.
Luna R., Millán C., Domingo M., et al. (2008) Thermal desorption of CH_4 retained in CO_2 ice. *Astrophys. Space Sci.*, *314*, 113–119.
Luspay-Kuti A., Hässig M., Fuselier S. A., et al. (2015) Composition-dependent outgassing of Comet 67P/Churyumov-Gerasimenko from ROSINA/DFMS: Implications for nucleus heterogeneity? *Astron. Astrophys.*, *583*, A4.
MacPherson G. J., Davis A. M., and Zinner E. K. (1995) The distribution of aluminum-26 in the early solar system — A reappraisal. *Meteoritics*, *30*, 365–386.
Makide K., Nagashima K., Krot A. N., et al. (2009) Oxygen- and magnesium-isotope compositions of calcium-aluminum-rich inclusions from CR2 carbonaceous chondrites. *Geochim. Cosmochim. Acta*, *73*, 5018–5050.
Makide K., Nagashima K., Krot A. N., et al. (2011) Heterogeneous distribution of ^{26}Al at the birth of the solar system. *Astrophys. J. Lett.*, *733*, L31.
Malamud U., Landeck W. A., Bischoff D., et al. (2022) Are there any pristine comets? Constraints from pebble structure. *Mon. Not. R. Astron. Soc.*, *514*, 3366–3394.
Mannel T., Bentley M. S., Schmied R., et al. (2016) Fractal cometary dust — A window into the early solar system. *Mon. Not. R. Astron.*

Soc., 462, S304–S311.
Marshall D., Groussin O., Vincent J. B., et al. (2018) Thermal inertia and roughness of the nucleus of Comet 67P/Churyumov-Gerasimenko from MIRO and VIRTIS observations. *Astron. Astrophys., 616*, A122.
Marty B., Altwegg K., Balsiger H., et al. (2017) Xenon isotopes in 67P/Churyumov-Gerasimenko show that comets contributed to Earth's atmosphere. *Science, 356*, 1069–1072.
Massironi M., Simioni E., Marzari F., et al. (2015) The two independent and primitive envelopes of the bilobate nucleus of Comet 67P/Churyumov-Gerasimenko. *Nature, 526*, 402–405.
Matonti C., Attree N., Groussin O., et al. (2019) Bilobate comet morphology and internal structure controlled by shear deformation. *Nature Geosci., 12*, 157–162.
McKay A. J., DiSanti M. A., Kelley M. S. P., et al. (2019) The peculiar volatile composition of CO-dominated comet C/2016 R2 (PanSTARRS). *Astron. J., 158*, 128.
McKinnon W. B., Stern S. A., Weaver H. A., et al. (2017) Origin of the Pluto-Charon system: Constraints from the New Horizons flyby. *Icarus, 287*, 2–11.
McKinnon W., Richardson D., Marohnic J., et al. (2020) The solar nebula origin of (486958) Arrokoth, a primordial contact binary in the Kuiper belt. *Science, 367*, eaay6620.
Meech K. J. and Jewitt D. (1987) Comet Bowell at record heliocentric distance. *Nature, 328*, 506–509.
Meech K. J. and Belton M. J. S. (1990) The atmosphere of 2060 Chiron. *Astron. J., 100*, 1323–1338.
Meech K. J., Pittichová J., Bar-Nun A., et al. (2009) Activity of comets at large heliocentric distances pre-perihelion. *Icarus, 201*, 719–739.
Meech K. J., A'Hearn M. F., Adams J. A., et al. (2011) EPOXI: Comet 103P/Hartley 2 observations from a worldwide campaign. *Astrophys. J. Lett., 734*, L1.
Meech K. J., Kleyna J. T., Hainaut O., et al. (2017) CO-driven activity in Comet C/2017 K2 (PANSTARRS). *Astrophys. J. Lett., 849*, L8.
Meisner T., Wurm G., and Teiser J. (2012) Experiments on centimeter-sized dust aggregates and their implications for planetesimal formation. *Astron. Astrophys., 544*, A138.
Merk R. and Prialnik D. (2006) Combined modeling of thermal evolution and accretion of trans-neptunian objects — Occurrence of high temperatures and liquid water. *Icarus, 183*, 283–295.
Michel P. and Richardson D. C. (2013) Collisions and gravitational reaccumulation: Possible formation mechanism of the asteroid Itokawa. *Astron. Astrophys., 554*, L1.
Migliorini A., Piccioni G., Capaccioni F., et al. (2016) Water and carbon dioxide distribution in the 67P/Churyumov-Gerasimenko coma from VIRTIS-M infrared observations. *Astron. Astrophys., 589*, A45.
Möhlmann D., Seidensticker K. J., Fischer H.-H., et al. (2018) Compressive strength and elastic modulus at Agilkia on Comet 67P/Churyumov-Gerasimenko derived from the SESAME/CASSE touchdown signals. *Icarus, 303*, 251–264.
Molaro J. L., Choukroun M., Phillips C. B., et al. (2019) The microstructural evolution of water ice in the solar system through sintering. *J. Geophys. Res.–Planets, 124*, 243–277.
Moreels G., Clairemidi J., Hermine P., et al. (1994) Detection of a polycyclic aromatic molecule in Comet P/Halley. *Astron. Astrophys., 282*, 643–656.
Mottola S., Attree N., Jorda L., et al. (2020) Nongravitational effects of cometary activity. *Space Sci. Rev., 216*, 1–20.
Mousis O., Drouard A., Vernazza P., et al. (2017) Impact of radiogenic heating on the formation conditions of Comet 67P/Churyumov-Gerasimenko. *Astrophys. J. Lett., 839*, L4.
Mumma M. J. and Charnley S. B. (2011) The chemical composition of comets — Emerging taxonomies and natal heritage. *Annu. Rev. Astron. Astrophys., 49*, 471–524.
Mumma M., Charnley S., Cordiner M., et al. (2019) The relationship of HCN, NH_3, C_2H_6, H_2O, and ammoniated salts in comets: A key clue to origins. *EPSC-DPS Joint Meeting 2019*, 1916.
Muñoz Caro G. M., Dartois E., and Nakamura-Messenger K. (2008) Characterization of the carbon component in cometary Stardust samples by means of infrared and Raman spectroscopy. *Astron. Astrophys., 485*, 743–751.
Nakamura T., Noguchi T., Tsuchiyama A., et al. (2008) Chondrule-like objects in short-period Comet 81P/Wild 2. *Science, 321*, 664–667.
Neishtadt A., Scheeres D., Sidorenko V., et al. (2002) Evolution of comet nucleus rotation. *Icarus, 157*, 205–218.
Nesvorný D., Youdin A. N., and Richardson D. C. (2010) Formation of Kuiper belt binaries by gravitational collapse. *Astron. J., 140*, 785–793.
Nesvorný D., Vokrouhlický D., Dones L., et al. (2017) Origin and evolution of short-period comets. *Astrophys. J., 845*, 27.
Ogliore R. C., Huss G. R., Nagashima K., et al. (2012) Incorporation of a late-forming chondrule into Comet Wild 2. *Astrophys. J. Lett., 745*, L19.
Oklay N., Sunshine J.-M., Pajola M., et al. (2016) Comparative study of water ice patches on cometary nuclei using multispectral imaging data. *Mon. Not. R. Astron. Soc., 462*, S394–S414.
Opitom C., Guilbert-Lepoutre A., Jehin E., et al. (2016) Long-term activity and outburst of Comet C/2013 A1 (Siding Spring) from narrow-band photometry and long-slit spectroscopy. *Astron. Astrophys., 589*, A8.
Opitom C., Snodgrass C., Fitzsimmons A., et al. (2017) Ground-based monitoring of Comet 67P/Churyumov-Gerasimenko gas activity throughout the Rosetta mission. *Mon. Not. R. Astron. Soc., 469*, S222–S229.
Opitom C., Hutsemékers D., Jehin E., et al. (2019) High resolution optical spectroscopy of the N_2-rich Comet C/2016 R2 (PanSTARRS). *Astron. Astrophys., 624*, A64.
O'Rourke L., Heinisch P., Blum J., et al. (2020) The Philae lander reveals low-strength primitive ice inside cometary boulders. *Nature, 586*, 697–701.
Paganini L., Mumma M. J., Gibb E. L., et al. (2017) Ground-based detection of deuterated water in Comet C/2014 Q2 (Lovejoy) at IR wavelengths. *Astrophys. J. Lett., 836*, L25.
Pajola M., Vincent J.-B., Güttler C., et al. (2015) Size-frequency distribution of boulders ≥7 m on Comet 67P/Churyumov-Gerasimenko. *Astron. Astrophys., 583*, A37.
Pajola M., Höfner S., Vincent J. B., et al. (2017) The pristine interior of Comet 67P revealed by the combined Aswan outburst and cliff collapse. *Nature Astron., 1*, 0092.
Parker A. H., Kavelaars J. J., Petit J.-M., et al. (2011) Characterization of seven ultra-wide trans-neptunian binaries. *Astrophys. J., 743*, 1.
Pätzold M., Andert T., Hahn M., et al. (2016) A homogeneous nucleus for Comet 67P/Churyumov-Gerasimenko from its gravity field. *Nature, 530*, 63–65.
Pätzold M., Andert T. P., Hahn M., et al. (2019) The nucleus of Comet 67P/Churyumov-Gerasimenko —Part I: The global view — Nucleus mass, mass-loss, porosity, and implications. *Mon. Not. R. Astron. Soc., 483*, 2337–2346.
Penasa L., Massironi M., Naletto G., et al. (2017) A three-dimensional modelling of the layered structure of Comet 67P/Churyumov-Gerasimenko. *Mon. Not. R. Astron. Soc., 469*, S741–S754.
Piqueux S., Vu T. H., Bapst J., et al. (2021) Specific heat capacity measurements of selected meteorites for planetary surface temperature modeling. *J. Geophys. Res.–Planets, 126*, e2021JE007003.
Poch O., Istiqomah I., Quirico E., et al. (2020) Ammonium salts are a reservoir of nitrogen on a cometary nucleus and possibly on some asteroids. *Science, 367*, eaaw7462.
Pommerol A., Thomas N., El-Maarry M. R., et al. (2015) OSIRIS observations of meter-sized exposures of H_2O ice at the surface of 67P/Churyumov-Gerasimenko and interpretation using laboratory experiments. *Astron. Astrophys., 583*, A25.
Pontoppidan K. M., Boogert A. C. A., Fraser H. J., et al. (2008) The c2d Spitzer spectroscopic survey of ices around low-mass young stellar objects. II. CO_2. *Astrophys. J., 678*, 1005–1031.
Poulet F., Lucchetti A., Bibring J. P., et al. (2016) Origin of the local structures at the Philae landing site and possible implications on the formation and evolution of 67P/Churyumov-Gerasimenko. *Mon. Not. R. Astron. Soc., 462*, S23–S32.
Pravec P., Vokrouhlický D., Polishook D., et al. (2010) Formation of asteroid pairs by rotational fission. *Nature, 466*, 1085–1088.
Preusker F., Scholten F., Matz K.-D., et al. (2015) Shape model, reference system definition, and cartographic mapping standards for Comet 67P/Churyumov-Gerasimenko — Stereophotogrammetric analysis of Rosetta/OSIRIS image data. *Astron. Astrophys., 583*, A33.
Prialnik D. (2006) What makes comets active? In *Asteroids, Comets, Meteors* (D. Lazzaro et al., eds.), pp. 153–170. IAU Symp. 229, Cambridge Univ., Cambridge.
Prialnik D. (2021) Modeling sublimation of ices during the early evolution of Kuiper belt objects. *AAS/Division for Planetary Sciences Meeting Abstracts, 53*, #307.10.

Prialnik D. and Podolak M. (1995) Radioactive heating of porous comet nuclei. *Icarus, 117,* 420–430.
Prialnik D., Bar-Nun A., and Podolak M. (1987) Radiogenic heating of comets by ^{26}Al and implications for their time of formation. *Astrophys. J., 319,* 993–1002.
Prialnik D., Brosch N., and Ianovici D. (1995) Modelling the activity of 2060 Chiron. *Mon. Not. R. Astron. Soc., 276,* 1148–1154.
Prialnik D., Benkhoff J., and Podolak M. (2004) Modeling the structure and activity of comet nuclei. In *Comets II* (M. C. Festou et al., eds.), pp. 359–387. Univ. of Arizona, Tucson.
Prialnik D., A'Hearn M. F., and Meech K. J. (2008) A mechanism for short-lived cometary outbursts at sunrise as observed by Deep Impact on 9P/Tempel 1. *Mon. Not. R. Astron. Soc., 388,* L20–L23.
Protopapa S., Kelley M. S. P., Woodward C. E., et al. (2021) Nondetection of water-ice grains in the coma of Comet 46P/Wirtanen and implications for hyperactivity. *Planet. Sci. J., 2,* 176.
Richardson J. E. and Bowling T. J. (2014) Investigating the combined effects of shape, density, and rotation on small body surface slopes and erosion rates. *Icarus, 234,* 53–65.
Richardson J. E., Melosh H. J., Lisse C. M., et al. (2007) A ballistics analysis of the Deep Impact ejecta plume: Determining Comet Tempel 1's gravity, mass, and density. *Icarus, 191,* 176–209.
Rickman H. (1989) The nucleus of Comet Halley: Surface structure, mean density, gas and dust production. *Adv. Space Res., 9,* 59–71.
Rickman H. and Fernandez J. A. (1986) Masses and densities of comets Halley and Kopff. In *The Comet Nucleus Sample Return Mission* (O. Melita, ed.), pp. 195–205. ESA SP-249, Noordwijk, The Netherlands.
Rickman H., Kamél L., Festou M. C., et al. (1987) Estimates of masses and densities of short-period comet nuclei. In *Symposium on the Diversity and Similarity of Comets* (M. Nicolet et al., eds.), pp. 471–481. ESA SP-278, Noordwijk, The Netherlands.
Rickman H., Marchi S., A'Hearn M. F., et al. (2015) Comet 67P/Churyumov-Gerasimenko: Constraints on its origin from OSIRIS observations. *Astron. Astrophys., 583,* A44.
Roemer E. (1962) Activity in comets at large heliocentric distance. *Publ. Astron. Soc. Pac., 74,* 351–365.
Roll R. and Witte L. (2016) Rosetta lander Philae: Touch-down reconstruction. *Planet. Space Sci., 125,* 12–19.
Rosenberg E. D. and Prialnik D. (2007) A fully 3-dimensional thermal model of a comet nucleus. *New Astron., 12,* 523–532.
Rosenberg E. D. and Prialnik D. (2009) Fully 3-dimensional calculations of dust mantle formation for a model of Comet 67P/Churyumov-Gerasimenko. *Icarus, 201,* 740–749.
Rosenberg E. D. and Prialnik D. (2010) The effect of internal inhomogeneity on the activity of comet nuclei — Application to Comet 67P/Churyumov-Gerasimenko. *Icarus, 209,* 753–765.
Roth N. X., Gibb E. L., Bonev B. P., et al. (2018) A tale of "two" comets: The primary volatile composition of Comet 2P/Encke across apparitions and implications for cometary science. *Astron. J., 156,* 251.
Roth N. X., Bonev B. P., DiSanti M. A., et al. (2021a) The volatile composition of the inner coma of Comet 46P/Wirtanen: Coordinated observations using iSHELL at the NASA-IRTF and Keck/NIRSPEC-2. *Planet. Sci. J., 2,* 54.
Roth N. X., Milam S. N., Cordiner M. A., et al. (2021b) Rapidly varying anisotropic methanol (CH_3OH) production in the inner coma of Comet 46P/Wirtanen as revealed by the ALMA Atacama Compact Array. *Planet. Sci. J., 2,* 55.
Rotundi A., Sierks H., Della Corte V., et al. (2015) Dust measurements in the coma of Comet 67P/Churyumov-Gerasimenko inbound to the Sun. *Science, 347,* aaa3905.
Rubin M., Altwegg K., Balsiger H., et al. (2019) Elemental and molecular abundances in Comet 67P/Churyumov-Gerasimenko. *Mon. Not. R. Astron. Soc., 489,* 594–607.
Rubin M., Engrand C., Snodgrass C., et al. (2020) On the origin and evolution of the material in 67P/Churyumov-Gerasimenko. *Space Sci. Rev., 216,* 102.
Safrit T. K., Steckloff J. K., Bosh A. S., et al. (2021) The formation of bilobate comet shapes through sublimative torques. *Planet. Sci. J., 2,* 14.
Samarasinha N. H., Mueller B. E., Belton M. J., et al. (2004) Rotation of cometary nuclei. In *Comets II* (M. C. Festou et al., eds.), pp. 281–299. Univ. of Arizona, Tucson.
Sandford S. A., Aléon J., Alexander C. M. O., et al. (2006) Organics captured from Comet 81P/Wild 2 by the Stardust spacecraft. *Science, 314,* 1720–1724.
Sarid G., Volk K., Steckloff J. K., et al. (2019) 29P/Schwassmann-Wachmann 1, a Centaur in the gateway to the Jupiter-family comets. *Astrophys. J. Lett., 883,* L25.
Satorre M. Á., Luna R., Millán C., et al. (2009) Volatiles retained in icy surfaces dominated by CO_2 after energy inputs. *Planet. Space Sci., 57,* 250–258.
Scheeres D. J. (2009) Stability of the planar full 2-body problem. *Cel. Mech. Dyn. Astron., 104,* 103–128.
Scheeres D. J. (2018) Disaggregation of small, cohesive rubble pile asteroids due to YORP. *Icarus, 304,* 183–191.
Schenk P. M., Asphaug E., McKinnon W. B., et al. (1996) Cometary nuclei and tidal disruption: The geologic record of crater chains on Callisto and Ganymede. *Icarus, 121,* 249–274.
Schleicher D. G. and Bair A. N. (2011) The composition of the interior of Comet 73P/Schwassmann-Wachmann 3: Results from narrowband photometry of multiple components. *Astron. J., 141,* 177.
Schleicher D. G., Millis R. L., Thompson D. T., et al. (1990) Periodic variations in the activity of Comet P/Halley during the 1985/1986 apparition. *Astron. J., 100,* 896–912.
Schleicher D. G., Millis R. L., and Birch P. V. (1998) Narrowband photometry of Comet P/Halley: Variation with heliocentric distance, season, and solar phase angle. *Icarus, 132,* 397–417.
Schlichting H. E. and Sari R. (2011) Runaway growth during planet formation: Explaining the size distribution of large Kuiper belt objects. *Astrophys. J., 728,* 68.
Schloerb F. P., Keihm S., von Allmen P., et al. (2015) MIRO observations of subsurface temperatures of the nucleus of 67P/Churyumov-Gerasimenko. *Astron. Astrophys., 583,* A29.
Schmitt B., Espinasse S., Grim R. J. A., et al. (1989) Laboratory studies of cometary ice analogues. In *Proceedings of an International Workshop on Physics and Mechanics of Cometary Materials* (J. Hunt and T. D. Guyenne, eds.), pp. 65–69. ESA SP-302, Noordwijk, The Netherlands.
Schroeder I. R. H. G., Altwegg K., Balsiger H., et al. (2019) A comparison between the two lobes of Comet 67P/Churyumov-Gerasimenko based on D/H ratios in H_2O measured with the Rosetta/ROSINA DFMS. *Mon. Not. R. Astron. Soc., 489,* 4734–4740.
Schwartz S. R., Michel P., Jutzi M., et al. (2018) Catastrophic disruptions as the origin of bilobate comets. *Nature Astron., 2,* 379–382.
Seiferlin K., Kömle N. I., Kargl G., et al. (1996) Line heat-source measurements of the thermal conductivity of porous H_2O ice, CO_2 ice and mineral powders under space conditions. *Planet. Space Sci., 44,* 691–704.
Sekanina Z. (1975) A study of the icy tails of the distant comets. *Icarus, 25,* 218–238.
Sekanina Z. and Yeomans D. K. (1985) Orbital motion, nucleus precession, and splitting of periodic comet Brooks 2. *Astron. J., 90,* 2335–2352.
Seligman D. Z., Kratter K. M., Levine W. G., et al. (2021) A sublime opportunity: The dynamics of transitioning cometary bodies and the feasibility of *in situ* observations of the evolution of their activity. *Planet. Sci. J., 2,* 234.
Senay M. C. and Jewitt D. (1994) Coma formation driven by carbon monoxide release from Comet Schwassmann-Wachmann 1. *Nature, 371,* 229–231.
Shoshany Y., Prialnik D., and Podolak M. (2002) Monte Carlo modeling of the thermal conductivity of porous cometary ice. *Icarus, 157,* 219–227.
Sierks H., Barbieri C., Lamy P. L., et al. (2015) On the nucleus structure and activity of Comet 67P/Churyumov-Gerasimenko. *Science, 347,* aaa1044.
Simon A., Öberg K. I., Rajappan M., et al. (2019) Entrapment of CO in CO_2 ice. *Astrophys. J., 883,* 21.
Skorov Y. V., Rezac L., Hartogh P., et al. (2017) Is near-surface ice the driver of dust activity on 67P/Churyumov-Gerasimenko. *Astron. Astrophys., 600,* A142.
Snodgrass C., Lowry S. C., and Fitzsimmons A. (2006) Photometry of cometary nuclei: Rotation states, colours and a comparison with Kuiper belt objects. *Mon. Not. R. Astron. Soc., 373,* 1590–1602.
Snodgrass C., Jehin E., Manfroid J., et al. (2016) Distant activity of 67P/Churyumov-Gerasimenko in 2014: Ground-based results during the Rosetta pre-landing phase. *Astron. Astrophys., 588,* A80.
Soderblom L. A., Becker T. L., Bennett G., et al. (2002) Observations

of Comet 19P/Borrelly by the Miniature Integrated Camera and Spectrometer aboard Deep Space 1. *Science, 296,* 1087–1091.
Sosa A. and Fernández J. A. (2009) Cometary masses derived from non-gravitational forces. *Mon. Not. R. Astron. Soc., 393,* 192–214.
Sosa A. and Fernández J. A. (2010) Non-gravitational forces and masses of some long-period comets: The cases of Hale-Bopp and Hyakutake. In *Icy Bodies in the Solar System* (J. A. Fernández et al., eds.), pp. 85–88. Proc. Int. Astron. Union Symp. 263, Cambridge Univ., Cambridge.
Sosa A. and Fernández J. A. (2011) Masses of long-period comets derived from non-gravitational effects — Analysis of the computed results and the consistency and reliability of the non-gravitational parameters. *Mon. Not. R. Astron. Soc., 416,* 767–782.
Spohn T. and Benkhoff J. (1990) Thermal history models for KOSI sublimation experiments. *Icarus, 87,* 358–371.
Spohn T., Benkhoff J., Klinger J., et al. (1989) Thermal modelling of two KOSI comet nucleus simulation experiments. *Adv. Space Res., 9,* 127–131.
Spohn T., Knollenberg J., Ball A. J., et al. (2015) Thermal and mechanical properties of the near-surface layers of Comet 67P/Churyumov-Gerasimenko. *Science, 349,* aab0464.
Squyres S. W., McKay C. P., and Reynolds R. T. (1985) Temperatures within comet nuclei. *J. Geophys. Res.–Solid Earth, 90,* 12381–12392.
Steckloff J. K. and Jacobson S. A. (2016) The formation of striae within cometary dust tails by a sublimation-driven YORP-like effect. *Icarus, 264,* 160–171.
Steckloff J. K. and Samarasinha N. H. (2018) The sublimative torques of Jupiter family comets and mass wasting events on their nuclei. *Icarus, 312,* 172–180.
Steckloff J. K., Johnson B. C., Bowling T., et al. (2015) Dynamic sublimation pressure and the catastrophic breakup of Comet ISON. *Icarus, 258,* 430–437.
Steckloff J. K., Graves K., Hirabayashi M., et al. (2016) Rotationally induced surface slope-instabilities and the activation of CO_2 activity on Comet 103P/Hartley 2. *Icarus, 272,* 60–69.
Steckloff J. K., Sarid G., Volk K., et al. (2020) P/2019 LD2 (ATLAS): An active Centaur in imminent transition to the Jupiter family. *Astrophys. J. Lett., 904,* L20.
Steckloff J. K., Lisse C. M., Safrit T. K., et al. (2021) The sublimative evolution of (486958) Arrokoth. *Icarus, 356,* 113998.
Steckloff J. K., Sarid G., and Johnson B. C. (2023) The effects of early collisional evolution on amorphous water ice bodies. *Planet. Sci. J., 4,* 4.
Steiner G. and Koemle N. I. (1991) A model of the thermal conductivity of porous water ice at low gas pressures. *Planet. Space Sci., 39,* 507–513.
Stern S. A. (2003) The evolution of comets in the Oort cloud and Kuiper belt. *Nature, 424,* 639–642.
Stern S. A. and Shull J. M. (1988) The influence of supernovae and passing stars on comets in the Oort cloud. *Nature, 332,* 407–411.
Stewart S. T. and Ahrens T. J. (2004) A new H_2O Hugoniot: Implications for planetary impact events. In *Shock Compression of Condensed Matter — 2003* (M. D. Furnish et al., eds.), pp. 1478–1483. AIP Conf. Ser. 706, American Institute of Physics, Melville, New York.
Sunshine J. M., A'Hearn M. F., Groussin O., et al. (2006) Exposed water ice deposits on the surface of Comet 9P/Tempel 1. *Science, 311,* 1453–1455.
Szutowicz S., Królikowska M., and Rickman H. (2008) Non-gravitational motion of the Jupiter-family Comet 81P/Wild 2. II. The active regions on the surface. *Astron. Astrophys., 490,* 393–402.
Thomas N., Davidsson B., El-Maarry M. R., et al. (2015a) Redistribution of particles across the nucleus of Comet 67P/Churyumov-Gerasimenko. *Astron. Astrophys., 583,* A17.
Thomas N., Sierks H., Barbieri C., et al. (2015b) The morphological diversity of Comet 67P/Churyumov-Gerasimenko. *Science, 347,* aaa0440.
Thomas N., Ulamec S., Kührt E., et al. (2019) Towards new comet missions. *Space Sci. Rev., 215,* 47.
Thomas P. C., A'Hearn M. F., Veverka J., et al. (2013) Shape, density, and geology of the nucleus of Comet 103P/Hartley 2. *Icarus, 222,* 550–558.
Tiscareno M. S. and Malhotra R. (2003) The dynamics of known Centaurs. *Astron. J., 126,* 3122–3131.
Tognon G., Ferrari S., Penasa L., et al. (2019) Spectrophotometric variegation of the layering in Comet 67P/Churyumov-Gerasimenko as seen by OSIRIS. *Astron. Astrophys., 630,* A16.
Tosi F., Capaccioni F., Capria M.-T., et al. (2019) The changing temperature of the nucleus of Comet 67P induced by morphological and seasonal effects. *Nature Astron., 3,* 649–658.
Toth I. and Lisse C. M. (2006) On the rotational breakup of cometary nuclei and Centaurs. *Icarus, 181,* 162–177.
Trigo-Rodríguez J. M. and Llorca J. (2006) The strength of cometary meteoroids: Clues to the structure and evolution of comets. *Mon. Not. R. Astron. Soc., 372,* 655–660.
Vavilov D., Eggl S., Medvedev Y. D., et al. (2019) Shape evolution of cometary nuclei via anisotropic mass loss. *Astron. Astrophys., 622,* L5.
Veverka J., Klaasen K., A'Hearn M., et al. (2013) Return to Comet Tempel 1: Overview of Stardust-NExT results. *Icarus, 222,* 424–435.
Villanueva G. L., Bonev B. P., Mumma M. J., et al. (2006) The volatile composition of the split ecliptic Comet 73P/Schwassmann-Wachmann 3: A comparison of fragments C and B. *Astrophys. J. Lett., 650,* L87–L90.
Vincent J.-B., Bodewits D., Besse S., et al. (2015) Large heterogeneities in Comet 67P as revealed by active pits from sinkhole collapse. *Nature, 523,* 63–66.
Vincent J.-B., Hviid S. F., Mottola S., et al. (2017) Constraints on cometary surface evolution derived from a statistical analysis of 67P's topography. *Mon. Not. R. Astron. Soc., 469,* S329–S338.
Vincent J.-B., Birch S., Hayes A., et al. (2019) Bouncing boulders on Comet 67P. *EPSC-DPS Joint Meeting 2019,* 502.
Vokrouhlický D., Bottke W., Chesley S., et al. (2015) The Yarkovsky and YORP effects. In *Asteroids IV* (P. Michel et al., eds.), pp. 509–532. Univ. of Arizona, Tucson.
Volk K. and Malhotra R. (2008) The scattered disk as the source of the Jupiter family comets. *Astrophys. J., 687,* 714–725.
Wallis M. K. and MacPherson A. K. (1981) On the outgassing and jet thrust of snowball comets. *Astron. Astrophys., 98,* 45–49.
Weidenschilling S. J. (1997) The origin of comets in the solar nebula: A unified model. *Icarus, 127,* 290–306.
Weidenschilling S. J. (2004) From icy grains to comets. In *Comets II* (M. C. Festou et al., eds.), pp. 97–104. Univ. of Arizona, Tucson.
Weissman P. R. (1980) Physical loss of long-period comets. *Astron. Astrophys., 85,* 191–196.
Weissman P. R. and Kieffer H. H. (1981) Thermal modeling of cometary nuclei. *Icarus, 47,* 302–311.
Weissman P. R. and Lowry S. C. (2008) Structure and density of cometary nuclei. *Meteorit. Planet. Sci., 43,* 1033–1047.
Weissman P. R., Asphaug E., and Lowry S. C. (2004) Structure and density of cometary nuclei. In *Comets II* (M. C. Festou et al., eds.), pp. 337–357. Univ. of Arizona, Tucson.
Westphal A. J., Fakra S. C., Gainsforth Z., et al. (2009) Mixing fraction of inner solar system material in Comet 81P/Wild 2. *Astrophys. J., 694,* 18–28.
Wierzchos K., Womack M., and Sarid G. (2017) Carbon monoxide in the distantly active Centaur (60558) 174P/Echeclus at 6 au. *Astron. J., 153,* 230.
Womack M. and Stern S. A. (1999) The detection of carbon monoxide gas emission in (2060) Chiron. *Solar System Res., 33,* 187.
Wooden D. H., Ishii H. A., and Zolensky M. E. (2017) Cometary dust: The diversity of primitive refractory grains. *Philos. Trans. R. Soc. A, 375,* 20160260.
Wurz P., Rubin M., Altwegg K., et al. (2015) Solar wind sputtering of dust on the surface of 67P/Churyumov-Gerasimenko. *Astron. Astrophys., 583,* A22.
Yang B., Jewitt D., Zhao Y., et al. (2021) Discovery of carbon monoxide in distant comet C/2017 K2 (PANSTARRS). *Astrophys. J. Lett., 914,* L17.
Youdin A. N. and Goodman J. (2005) Streaming instabilities in protoplanetary disks. *Astrophys. J., 620,* 459–469.
Zolensky M. E., Zega T. J., Yano H., et al. (2006) Mineralogy and petrology of Comet 81P/Wild 2 nucleus samples. *Science, 314,* 1735–1739.

Pajola M., Vincent J.-B., El-Maarry M. R., and Lucchetti A. (2024) Cometary nuclei morphology and evolution. In *Comets III* (K. J. Meech, M. R. Combi, D. Bockelée-Morvan, S. N. Raymond, and M. E. Zolensky, eds.), pp. 289–313. Univ. of Arizona, Tucson, DOI: 10.2458/azu_uapress_9780816553631-ch010.

Cometary Nuclei Morphology and Evolution

Maurizio Pajola
National Institute for Astrophysics–Astronomical Observatory of Padova

Jean-Baptiste Vincent
Institute of Planetary Research, Deutches Zentrum für Luft- und Raumfahrt

Mohamed Ramy El-Maarry
Space and Planetary Science Center and Department of Earth Sciences, Khalifa University

Alice Lucchetti
National Institute for Astrophysics–Astronomical Observatory of Padova

---

Among all bodies of our solar system, cometary nuclei show some of the most diverse morphological features on their surfaces. Indeed, comets are the result of the sublimation activity that repeatedly rises and falls along the heliocentric orbit, coupled with both diurnal and seasonal thermal stresses that widely affect their surface and subsurface layers. During the last 35 years of space exploration, our knowledge of the cometary nuclei morphology has widely improved, thanks to many dedicated flyby or rendezvous missions. Moreover, long-term *in situ* observations have provided us with the possibility of witnessing how such peculiar morphologies form and evolve with time, providing evidence that comets are some of the most geologically active bodies in the solar system. This chapter fully focuses on the cometary nuclei surface properties and structure, together with the activity-driven surface evolution, as well as the seasonal and orbital morphological changes.

1. INTRODUCTION

The year 1986 marks the beginning of spacecraft close exploration of cometary nuclei, with the Giotto flyby mission of Comet 1P/Halley, i.e., the most prominent representative of comets that are believed to have been members of the Oort cloud (*Oort,* 1950; *Wang and Brasser,* 2014). It is indeed thanks to more than 2000 images of 1P taken by the Halley Multicolour Camera (HMC) (*Keller et al.,* 1986, 1987) that our vision and interpretation of comets changed from purely unresolved astronomical objects observed from afar, to resolved morphological surfaces characterized by, e.g., depressions, hills, smooth terrains, and boulder fields. Moreover, such observations confirmed the *Whipple* (1950) interpretation that a single, small solid body was at the center of the cometary activity, while the flying sandbank model [before *Whipple* (1950) there was a general consensus that comets consisted of loose aggregations called "swarm" of widely separated dust particles (*Lyttleton,* 1948)] was discarded (*Keller et al.,* 2004).

320- to 60-m-resolution Halley Multicolor Camera (HMC) images revealed that Halley's nucleus shape and morphology are characterized by an irregular-shaped bilobed surface with a large central depression that laces the nucleus almost like a waist. In addition, a chain of four ~500-m-sized hills regularly spaced is present, together with a roundish feature interpreted to be a "crater" that is ~2000 m wide and 150 m deep and a "mountain" that peaks roughly 0.5–1 km above the surface (*Keller et al.,* 1988) (Fig. 1).

The next spacecraft flyby of a comet took place in 2001, when the Deep Space 1 probe observed the Jupiter-family comet (JFC) (*Duncan and Levison,* 1997) 19P/Borrelly (*Soderblom et al.,* 2002). Like 1P/Halley, the surface morphology of 19P proved to be particularly complex, being characterized by dark spots, mottled terrains, bright terrains and mesas, and possibly ridges and fractures (*Britt et al.,* 2004) (Fig. 2).

From 2004 to 2011 a surge of cometary flybys occurred (Fig. 2), extensively broadening our knowledge of JFC nuclei morphology and evolution, with the closest approach of:

- Comet 81P/Wild 2, observed by the Stardust spacecraft in 2004. Its surface was covered by circular features that resemble impact craters characterized by nearly vertical walls, but that are thought to be a combination

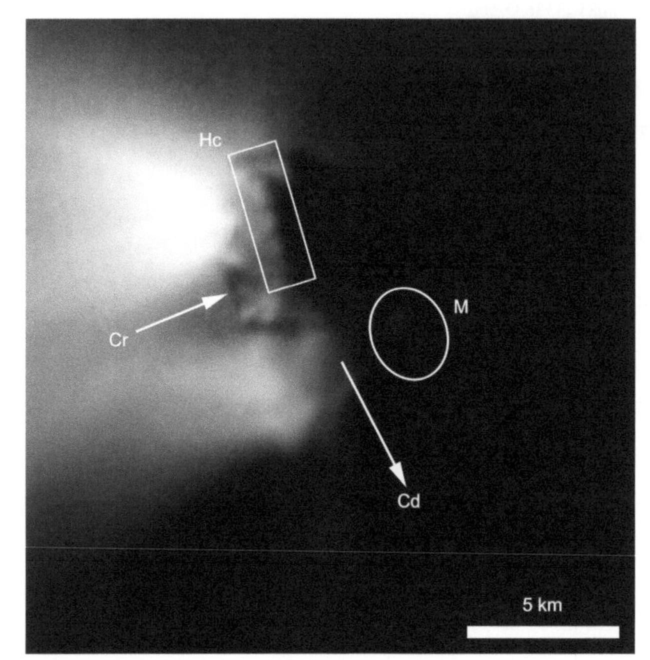

Fig. 1. One of the highest-spatial-scale HMC images obtained of Comet 1P/Halley. The central depression (Cd), the "mountain" (M), the hills chain (Hc), and the "crater" (Cr) are indicated.

of impact, sublimation, mass wasting, and ablation (*Brownlee et al.,* 2004);

- Comet 9P/Tempel 1, impacted in 2005 by the Deep Impact mission probe, while a companion spacecraft witnessed the event from close proximity (*A'Hearn et al.,* 2005). During the flyby several high-resolution images taken of the 9P nucleus revealed a surface morphology with several pitted and smooth terrains, mesas, a terraced area, a long concave eroded scarp, and a ~3-km-long flow with an estimated thickness of 10 to 15 m (*Thomas et al.,* 2007). In 2011 the Stardust-NExT spacecraft performed a second flyby of 9P, extending the previous image coverage of the comet to almost two-thirds of the surface. This closest approach led to the discovery of a large-scale layering of the nucleus (*Belton et al.,* 2007; *Veverka et al.,* 2013) as well as the first morphological change ever witnessed on a comet, i.e., a 50-m cliff retreat of the prominent smooth flow observed 5.5 years before (*Thomas et al.,* 2013a; *Veverka et al.,* 2013);
- Comet 103P/Hartley 2, observed by the EPOXI mission in 2011 (*A'Hearn et al.,* 2011). [The name "EPOXI" is derived from a combination of two science investigations: Extrasolar Planet Observation and Characterization (EPOCH) and Deep Impact eXtended Investigation (DIXI).] This bilobed hyperactive comet showed a surface morphology characterized by irregularly-shaped smooth areas, a smooth but variegated region forming a "waist" between the lobes, and several small mounds/boulders with sizes <40 m across (*Thomas et al.,* 2013b).

All the above observations showed the strong morphological and shape variability existing among all nuclei and eventually confirmed that comets are all but simple "dirty snowballs." Nevertheless, it has been thanks to the Rosetta mission to the JFC 67P/Churyumov-Gerasimenko (*Glassmeier et al.,* 2007) that we finally had the chance not only to observe the surface morphology of a nucleus with a centimeter-to-millimeter scale resolution, but also to observe how it seasonally evolves during its orbit, particularly at and near perihelion.

After a 10-year journey, the Rosetta spacecraft arrived at Comet 67P in August 2014 (*Taylor et al.,* 2017). Following an intense period of mapping and nucleus characterization (*Sierks et al.,* 2015; *Ulamec et al.,* 2015), it successfully deployed the lander Philae (*Glassmeier et al.,* 2007), which landed on the comet's surface on 12 November 2014 (*Boehnhardt et al.,* 2017; *Mottola et al.,* 2015, and references herein). The huge multi-instrument, multi-spacecraft observation of the 67P nucleus revealed a surface morphology characterized by widespread diversity (*Thomas et al.,* 2015a) where dust-covered terrains, brittle material, large-scale depression structures, smooth terrains, exposed consolidated surface, boulder fields and taluses, mass movements, high-reflective particle clusters, as well as fracturing are present (Fig. 2). In addition, thanks to more than 76,000 high-resolution images from the Optical, Spectroscopic, and Infrared Remote Imaging System (OSIRIS) (*Keller et al.,* 2007) it has been possible to prepare the most detailed morphological and geological maps (*El-Maarry et al.,* 2015a, 2016; *Giacomini et al.,* 2016; *Lee et al.,* 2016) ever obtained for a cometary nucleus so far, even leading to higher-scale morphological studies of 67P (e.g., *Auger et al.,* 2015; *La Forgia et al.,* 2015; *Lucchetti et al.,* 2016; *Pajola et al.,* 2019).

It is noteworthy that several morphological features that have been previously observed only on one or few visited nuclei appear to all be present with different frequency and dimensions on 67P (*Thomas et al.,* 2015a; *Sunshine et al.,* 2016); nevertheless, they are observed with much higher spatial and temporal resolution. For this reason, everything we have learned and are still understanding from Comet 67P can be used to reinterpret unclear morphologies and their evolution previously identified on other comets, together with groundbased observed events (e.g., *Altenhoff et al.,* 2009) that are traceable back to the surface.

This chapter first reviews in detail the surface properties and structure of the cometary nuclei observed so far (section 2). It then focuses on the activity-driven evolution of comets (section 3) as well as their seasonal evolution (section 4), highlighting not only the cometary changes over many orbits, but also discussing the geological evolution that is believed to occur for bodies in their journey from the Kuiper belt (KB) or Oort cloud to an evolved JFC that orbits the inner solar system.

2. COMET NUCLEI SURFACE PROPERTIES AND STRUCTURE

Cometary surface properties and structures are key to understanding both the nuclei origin and evolution. In this

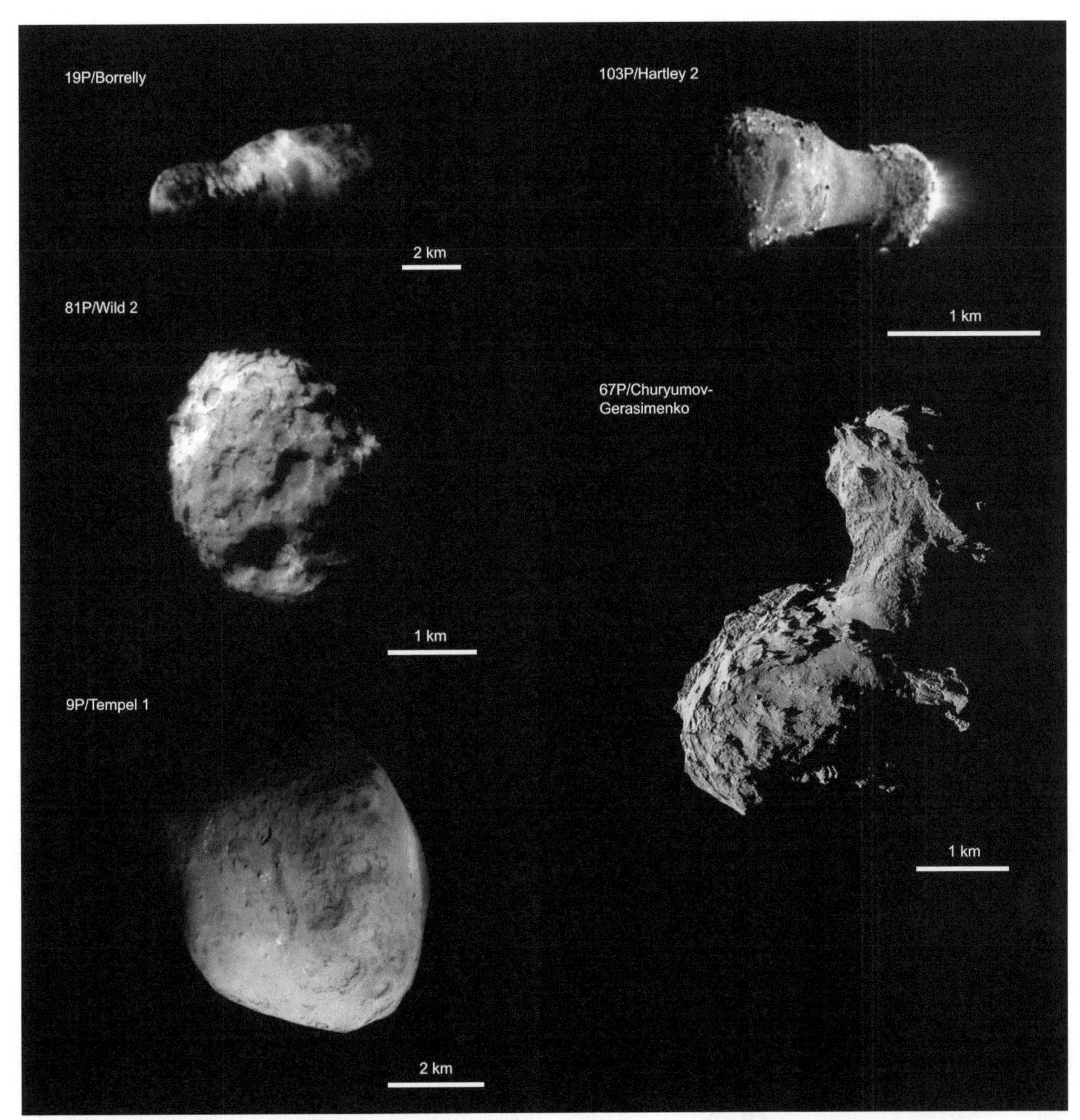

Fig. 2. The nuclei of the Jupiter-family comets visited so far.

section, global shapes and morphologies will be detailed for the six comets that have been closely observed by space missions in the last 35 years.

2.1. Global Shapes

Cometary shapes have been commonly presented in terms of the best-fit triaxial ellipsoid (*Thomas,* 2020) in order to provide hints on the main-body dimensions (Table 1). Nevertheless, since Giotto's observation of Comet 1P/Halley (*Keller et al.,* 1986), it has become clear that such bodies may be characterized by irregular shapes, such as prolate (1P/Halley) or elongated (19P/Borrelly) ellipsoids, a spheroidal shape (81P/Wild 2), a rough pyramidal shape (9P/Tempel 1), and an axially symmetric (103P/Hartley 2) or an axially asymmetric (67P/Churyumov-Gerasimenko) bilobate shape.

After overcoming the hardly suitable ellipsoid definition, two different techniques provide the possibility to derive the true shape of a resolved irregular nucleus, called a shape model. On one side, the stereo photogrammetry technique (SPG) combines images of the same area that have been taken under different viewing conditions, hence deriving the elevation with respect to an image plane (e.g., *Preusker et al.,* 2015, 2017). After repeatedly iterating this approach on different images of the target, the complete tridimensional shape model can be reconstructed. SPG maximizes its return on areas where the topographic relief is large; however, if flat

TABLE 1. The sizes of the comets that have been currently visited by spacecraft (the three axes of the best-fitting ellipsoids are given, when available; alternatively, the mean radius is indicated).

Comet	Size	Reference
1P/Halley	15.3 × 7.8 × 7.4 km ± 0.5 km	*Keller et al.* (1987)
19P/Borrelly	4.0 ± 0.1 × 1.60 ± 0.02 × 1.60 ± 0.02 km	*Buratti et al.* (2004)
81P/Wild 2	1.65 × 2.00 × 2.75 km ± 0.05 km	*Duxbury et al.* (2004)
9P/Tempel 1	Mean radius: 2.83 ± 0.1 km	*Thomas et al.* (2013a)
103P/Hartley 2	Mean radius: 0.58 ± 0.018 km	*Thomas et al.* (2013b)
67P/Churyumov-Gerasimenko	4.34 ± 0.02 × 2.60 ± 0.02 × 2.12 ± 0.06 km	*Jorda et al.* (2016)

areas are solely considered, a second technique called stereo photoclinometry (SPC) works better. Indeed, by assuming that the reflectance properties of the observed surface elements are identical, SPC makes use of brightness changes to evaluate changes in photometric angles, hence providing the local slope and elevation (e.g., *Jorda et al.,* 2016). If both techniques are complementarily used, then the final dataset should be the most reliable (*Thomas,* 2020).

Such techniques have been used on most of the cometary nuclei observed at close range, though; depending on the resolution and quality of the images, the viewing and illumination conditions, and the number of images obtained of the target, the resulting shape models can present finer- or coarser-scale details, as presented in Fig. 3. It is worth pointing out that with the exception of Comets 81P and 9P, all the remaining nuclei show elongated shapes. In addition, for 19P, 103P, and 67P, i.e., 50% of the total number of nuclei observed so far, a clear bilobate shape is visible with a narrowing region connecting the two lobes. This is true if we restrict it only to nuclei observed at close distance with cameras; nevertheless, a few more comets have been observed by Arecibo with a resolution good enough to get the shape and some large features. Two such objects — 45P/Honda-Mrkos-Pajdušáková and 8P/Tuttle (*Harmon et al.,* 2010; *Springmann et al.,* 2022) — are bilobate as well. As presented by *Davidsson et al.* (2016), the frequent bilobate appearance of the cometary nuclei may well be the result of the hierarchical formation theory of comets, which says that comets slowly grow out of the material that remained after the formation of a transneptunian object (TNO) during the primordial phase of the solar system (*Nesvorný et al.,* 2010; *Wahlberg Jansson and Johansen,* 2014). Such ice-rich remnants could have then merged with slow relative velocity, forming the shapes we see today. Another possibility is that the slow accretion forming the bilobate or elongated nuclei may occur at later epochs as part of the reaccumulation process resulting from the collisional disruption of larger bodies, but still keeping their volatiles and low density throughout the process (*Schwartz et al.,* 2018).

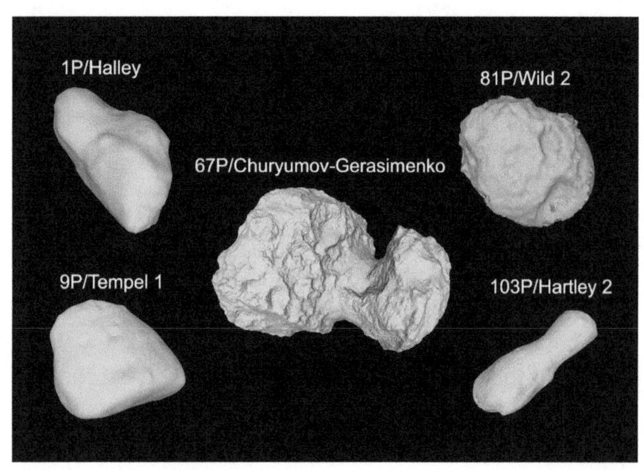

Fig. 3. Shape models of the cometary nuclei visited by spacecraft [1P/Halley (*Stooke,* 2002); 81P (*Farnham et al.,* 2005); 9P (*Farnham et al.,* 2013a); 103P (*Farnham et al.,* 2013b); 67P (*Preusker et al.,* 2017)]. For Comet 19P/Borrelly no shape model has been published yet.

2.2. Surface Features and Deposits

The morphology and global surface texture of the nuclei observed at close range might be different at first sight; nevertheless, it is worth pointing out that different surface features such as cliffs, circular depressions, and fractures, as well as dusty deposits and boulders, are not unique and reoccur in some if not all visited comets.

2.2.1. Cliffs, terraces, and layering. Scarps underlying terraces and mesas are common features on cometary nuclei, even if in some cases the limited resolution of the images may prevent their observation, as in the case of Comet 1P/Halley (*Keller et al.,* 1988). Flat-topped mesas were recognized on the surface of 19P/Borrelly with an elevation of typically 100 m above the nearby terrains (Fig. 4a) (*Britt et al.,* 2004). The tops of the mesas display an albedo similar to the close mottled terrain, while the steep slopes surrounding the mesa tops are brighter and may be near the angle of repose (*Britt et al.,* 2004). Such features are mainly located in the central part of the comet and may be associated with jet activity, suggesting that gas emission occurs from ablating cliff faces that retreat with time with estimated rate of several meters per perihelion passage (*Britt et al.,* 2004). Mesas have also been observed on 81P/Wild 2, such as the one identified close to the south rim of crater Rahe (*Brownlee et al.,* 2004) (Fig. 4b). They have a typical elevation of 100 m above the nearby terrain as in the 19P/Borrelly case, but are smaller in extension possibly due to

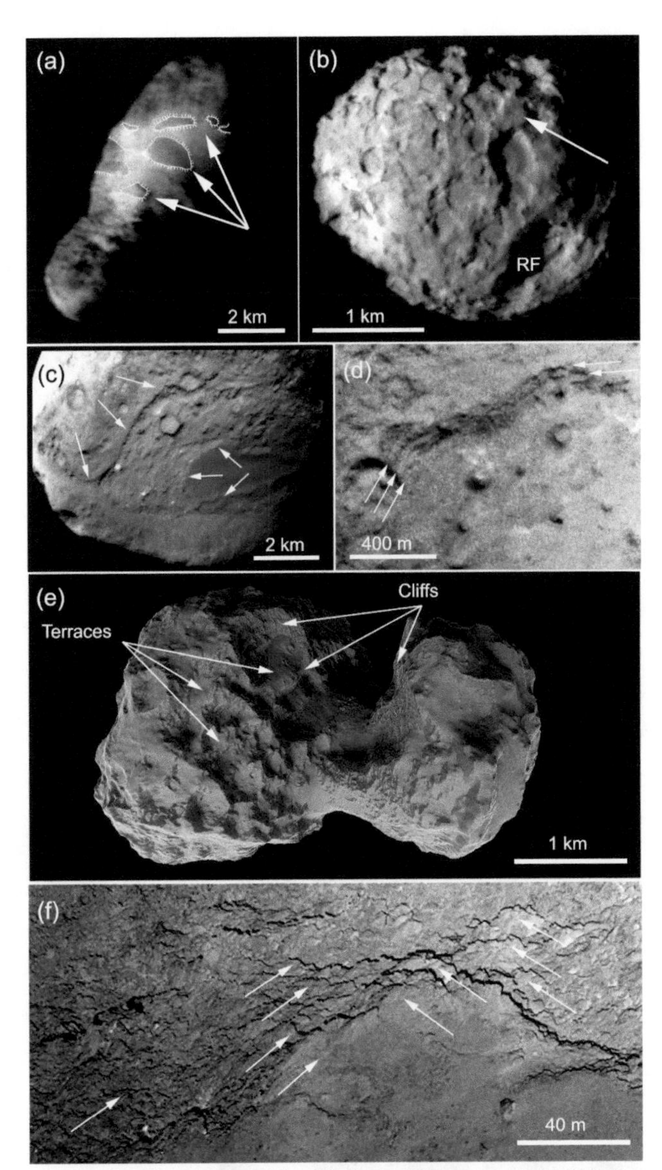

Fig. 4. (a) The mesas identified on 19P surrounded by the 100-m steep slopes, indicated by white arrows. **(b)** The mesa on 81P located close to the south rim of crater Rahe. RF stands for Right-Foot crater, i.e., where terracing and layers have been identified. **(c)** A >2-km-long concave eroding scarp and a sharply bounded flow edge identified on the surface of Comet 9P. **(d)** Kilometer-long terraced terrain on Tempel 1, with a height of 50 m. **(e)** Cliffs (colored in red) and terraces identified in the northern heminucleus of 67P. **(f)** Decimeter- to meter-size layering observed on 67P at the boundary between Imhotep and Ash regions (modified from *Feller et al.,* 2016).

the fact that 19P/Borrelly is at an earlier stage of ablation evolution (*Brownlee et al.,* 2004). Moreover, there are two lines of evidence to suggest that 81P/Wild 2 is stratified at different scales: the 150-m outcrops located on the southwest rim of Right Foot crater, indicating modest terracing with layers tens of meters thick (Fig. 4b), and pits, suggesting possible discontinuities in the comet's physical properties at a depth of 100 m.

Comet P/Tempel 1 displays a large variety of scarp morphology whose edges are sharp, concave, terraced, and scalloped, highlighting that slope erosion and retreat are the dominant processes modeling the cometary surface (Fig. 4c) (*Thomas et al.,* 2007; *Sunshine et al.,* 2016). Such scarp variety, in terms of slopes and morphology, implies a strong heterogeneity in material and/or compositional properties (*Veverka et al.,* 2013). In this case, the evidence of layering (Fig. 4d) on global scales (*Thomas et al.,* 2007) consists of two different type of layers: (1) thin layers (10–15 m thick) associated with the smooth flows [in *Belton and Melosh* (2009), the authors discuss the fluidization and multiphase transport of cometary material made of particles as the possible explanation of flows located on the smooth terrains of 9P] and (2) thick layers (50–100 m thick) that are being exposed in places by erosion (*Veverka et al.,* 2013). The pervasive layering affecting the surface of 9P/Tempel 1 has been explained as a relict of the primordial process of the nucleus formation or as the result of thermal processing on a homogeneous and larger body that led to a "onion-like" stratification (*Thomas et al.,* 2007).

Among all observed nuclei, that of 67P is the one that more clearly shows the widespread presence of almost-vertical, laterally contiguous cliffs stacked on top of each other (*Thomas et al.,* 2015b; *El-Maarry et al.,* 2015a, 2016), leading to a series of planar terraces in a repetitive staircase pattern (*Massironi et al.,* 2015) (Fig. 4e). Moreover, the vertical scarps show a parallelism of linear traces, identified as layers (*Penasa et al.,* 2017), that are consistently aligned with the nearby terraces and also continue inside the deep circular pits (*Vincent et al.,* 2015a).

Through the OSIRIS high-resolution images of the full comet, coupled with the m-scale shape models prepared, it has been highlighted that stratification is a widespread structural aspect of Comet 67P (Fig. 4f) both in the main body and in the head lobe. Indeed, by reconstructing the full comet nucleus through the strata distribution and orientation, *Massironi et al.* (2015) concluded that the two lobes once independently formed from primordial pebbles (*Davidsson et al.,* 2016) in an onion-like stratification that is several hundred meters thick. At the time of merging, the two cometesimals joined together involving reciprocal motion with dextral strike-slip kinematics that bent the layers in the contact area without obliterating them (*Franceschi et al.,* 2020). The terrace-cliff-terrace series that is observable on 67P derives from differential sublimation of the stratified sequences, hence implying changes in the relative volatile abundances inside the cometary strata (*Massironi et al.,* 2015). The most prominent scarp of 67P is Hathor (*Thomas et al.,* 2015a), which is located on the comet's small lobe. This cliff is characterized by a height of ~900 m and it perpendicularly rises from the comet's equatorial plane. This region shows evidence of different small terraces (*El-Maarry,* 2015a), suggesting the inner presence of layering not only in the main lobe, but also in the small one. A similar feature, but with lower relief with respect to the neck underneath, is the Neith cliff (*El-Maarry et al.,* 2016), which shows a clearly ridged morphology and multiple aligned terraces.

The presence of scarps and cliffs located on all nuclei observed so far supports the interpretation that comets are characterized by some degree of layering of their surface and subsurface. Moreover, this suggests that scarp erosion and retreat, most likely triggered by volatile sublimation (*Britt et al.,* 2004; *Veverka et al.,* 2013; *Massironi et al.,* 2015; *Keller et al.,* 2015a; *Pajola et al.,* 2016b), is a common process that shapes these objects, hence modifying the above terraces and mesas and the global shapes of the nuclei in general.

*2.2.2. **Circular features.*** Spacecraft closeup observations of comet nuclei revealed the presence of circular depressions on their surface, sometimes resembling impact structures. On planetary solid surfaces, the presence of circular depressions is usually related to the impact cratering process, providing hints of the collisional history of the body. However, cometary nuclei surfaces do not necessarily keep records of impact events due to the strong erosion that affects them. Indeed, cometary surface landscapes are usually modified by volatile sublimation, suggesting that circular depressions may be also linked to cometary activity. Despite the extensive characterization of the depressions located on several comet nuclei, the determination of their formation process is still a challenge (*Vincent et al.,* 2015a).

Comet 1P/Halley's nucleus was observed by the Halley Multicolour Camera (HMC) onboard the Giotto spacecraft (*Keller et al.,* 1988), revealing the absence of clear impact features on its surface (Fig. 1). However, two depressions were discovered and linked to coma dust jets, even if their origin remains unclear. The nucleus of 19P/Borrelly was imaged by the Miniature Integrated Camera and Spectrometer (MICAS) (*Soderblom et al.,* 2004) onboard the Deep Space 1 mission, highlighting a variety of terrains and morphological features (*Soderblom et al.,* 2002). Many rounded depressions were identified on the mottled terrain of 19P/Borrelly (Fig. 5a), but no fresh impact craters down to a scale of 200 m, hence indicating a young and active surface (*Soderblom et al.,* 2002, 2004). In addition, all the circular pits observed have a similar diameter, suggesting that they may be the product of sublimation of subsurface volatiles and collapse rather than of impact cratering processes. Images taken by the Stardust mission during its flyby of 81P/Wild 2 exhibited surface depressions of different sizes (from a few hundred meters to 2 km in diameter) and morphologies (Fig. 5b). Such depressions were classified as (1) pit halos, which are circular pits surrounded by an irregular halo of excavated material, and (2) flat-floored features rounded by steep, nearly vertical walls and without a rim (*Brownlee et al.,* 2004). Even if both morphologies do not show typical signs of impact craters, such as raised rims and ejecta, they are interpreted as impact features based on hypervelocity impact laboratory experiments on a porous target (*Brownlee et al.,* 2004). The morphological difference between pit halos and flat-floored features has been attributed to the different physical properties of the target again based on laboratory experiments: Pit halo craters originate in a target characterized by some weak surface layer overlying a harder material, while the flat-floored ones originate in a target with the opposite setup (*Brownlee et al.,* 2004). The surface of 81P/Wild 2 also shows the presence of depressions that are not circular: Some of them have steep walls like the flat-floored craters, while other have rounded edges

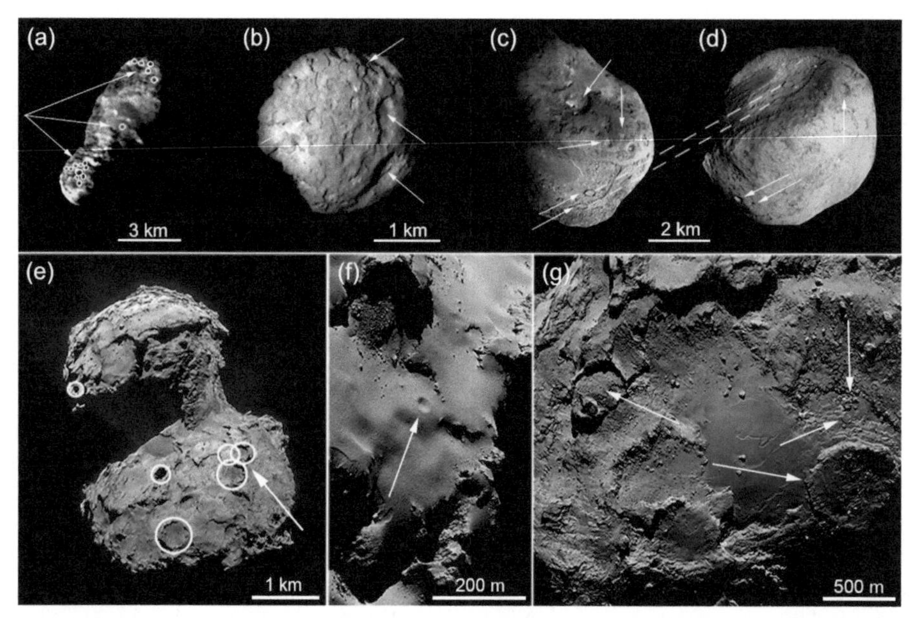

Fig. 5. (a) The location of pits on Comet 19P (modified from *Britt et al.,* 2004) indicated with white circles. **(b)** Circular and irregularly shaped features on Comet 81P. **(c),(d)** Two different views of the circular features identified on Comet 9P observed by the **(c)** Deep Impact and **(d)** Stardust-NExT missions. Two rounded features present in both images are indicated with dashed lines. **(e)** Some of the circular features identified on the northern heminucleus of 67P. The white arrow points at the ~200–350-m circular features located in the Seth region. **(f)** The ~35-m bowl-shaped crater observed in the Seth region of 67P. **(g)** Circular features indicated with white arrows located in the Imhotep region of 67P.

that may have been modified by loose material sliding into the depressions. Such features seem to not be directly linked to impacts, but may be formed by other processes, such as sublimation, mass wasting, ablation, or a combination of these processes (*Brownlee et al.,* 2004).

Comet 9P/Tempel 1 has been imaged twice by the Deep Impact experiment (Fig. 5c) and Stardust/NExT mission (Fig. 5d), revealing a pitted terrain covering half of the nucleus surface. Some of the depressions observed are tens of meters to 100 m across, while larger depressions extend up to 400 m with flat floors, no raised rims, and a variety of roughly concentric albedo markings (*Thomas et al.,* 2013a). The depressions observed on 9P/Tempel 1 are less prominent with respect to the ones located on 81P/Wild 2, relative to nucleus size. In particular, 380 pits with sizes ranging from 25 m to 940 m were identified on the comet's surface (*Belton et al.,* 2013), far more than what is expected from the calculation of impact rates on a typical JFC during its lifetime (*Belton et al.,* 2013; *Vincent et al.,* 2015b). The large amounts of pits per surface area and their spatial clustering in areas prone to outbursts makes it unlikely that the impact-cratering process could be the responsible for their formation. They may rather originate from volatile outburst and sublimation erosion. *Belton et al.* (2013) suggest that most pits on 9P/Tempel 1 are due to activity, originating during "mini-outbursts." The only feature clearly originating from an impact is the artificial one made by the projectile sent by the Deep Impact experiment on the 9P/Tempel 1 surface that was observed 5.5 years after its formation by the Stardust/NExT mission. The Deep Impact site shows a 50-m-diameter pit surrounded by a collapse region of 200 m (*Richardson and Melosh,* 2013). Numerical simulations (*Vincent et al.* 2015b, and references therein) are able to reproduce this peculiar morphology with the currently known range of cometary material mechanical properties.

A few pits have also been observed on Comet 103P/Hartley 2 that are similar to 9P/Tempel 1's features and are located in active jet regions, suggesting that their formation process is directly related to the activity of the nucleus and not to impact craters (*Bruck Syal et al.,* 2013). It should be noted that 103P/Hartley 2 is a hyperactive comet where sublimation and dust fall back constantly erase the topography and would prevent the long-term survival of crater-like morphologies.

67P/Churyumov-Gerasimenko displays several circular and quasicircular features on its surface (Fig. 5e), although very few are unambiguously interpreted as impact craters. The largest among them is the ~35-m bowl-shaped depression located in the Ash region (Fig. 5f). Other examples are depressions caused by the fall of boulders from cliffs or the footprints left by the Philae lander upon its first touchdown. Both examples are low-velocity impacts. On the other hand, several pits have been observed on the 67P surface (*Ip et al.,* 2015), mainly in the northern hemisphere (*Vincent et al.,* 2015a), that are suggested to be a signature of endogenic activity. In particular, *Ip et al.* (2015) obtained the size-frequency distribution of such circular depressions, deriving a power-law index of -2.3 ± 0.2 for sizes >200 m, which is comparable to the ones obtained for 9P and 81P, hence implying a similar formation process. Since orbital integration calculations show that the surface erosion histories of 81P and 9P could be shorter than those of 67P, the circular depressions could be dated back to the pre-JFC phase or the TNO phase of these comets.

A few pits were observed by Rosetta as being newly formed, e.g., Hapi, Imhotep, and Anubis (*Groussin et al.,* 2015; *El-Maarry et al.,* 2017), within a few months around perihelion in 2015. The high-resolution images provided by Rosetta allow the identification of deep circular pits with different sizes that tend to cluster in small groups. The pits were classified as active and currently inactive with a depth-to-diameter ratio (d/D) of 0.78 and 0.26, respectively (*Vincent et al.,* 2015a). When compared to the depressions observed on other comet nuclei (*Kirk et al.,* 2005), the d/D of active pits is much higher. 67P active pits probably originated by a sinkhole collapse mechanism, possibly associated with outbursts, which may be followed by sublimation-driven retreat of the walls (*Vincent et al.,* 2015a; *Pajola et al.,* 2016b). In addition, *Vincent et al.* (2015a) suggested that an age-to-depth relationship could exist, with deep pits being younger, such as those located in the Ma'at and Seth regions.

Another type of roundish feature has been observed on the surface of 67P, but only in the gravitational lowest area of the Imhotep region (Fig. 5g). They are characterized by a circular, elliptical, or irregular roundish shape, with sizes between 2 m and 59 m, resembling the circular depressions observed on 9P/Tempel 1 (*Auger et al.,* 2015). Roundish features are located close to terraces, hinting at a possible link between them (Fig. 5g). It has been proposed that such features can be interpreted as ancient degassing conduits that have been revealed by the differential erosion of preexisting layers (*Auger et al.,* 2015; *Brownlee et al.,* 2004). Finally, other circular depressions were identified on Comet 67P, such as those located in the Seth region [with sizes ranging from ~200 to 350 m (white arrow in Fig. 5e)], which are characterized by a huge number of boulders on their floor and by steep walls. These features could be the result of landslide events or retreat of cliffs through progressive erosion on 67P (*Lucchetti et al.,* 2017).

2.2.3. Fractures. Prior to the Rosetta mission to 67P/Churyumov-Gerasimenko, it was difficult to undoubtedly identify fractures at a high enough confidence level given the high spatial resolutions needed to resolve these predominately small features [see, e.g., the uncertain fractures identification on 19P/Borrelly (*Soderblom et al.,* 2002)]. Nevertheless, given the ice-rich composition of cometary materials, diurnal and seasonal temperature fluctuations, and variable dust-to-ice ratios, there was always the possibility that tensional fractures could develop on cometary surfaces as thermal stresses build up in the near-subsurface of the cometary nucleus.

OSIRIS images have shown that fractures are ubiquitous on the surface of Comet 67P, with variable origins inferred (Fig. 6) (*El-Maarry et al.* 2015b). Follow-up global mapping by *Auger et al.* (2018) has also verified the ubiquity of meter-scale polygonal fracture patterns. Based on these findings, a seasonal contraction mechanism akin to similar processes on Earth and Mars was suggested for the formation of the polygonal patterns. In addition, fractures were observed on cliffs and boulders, hinting at further evolution and fragmentation through insolation weathering (*El-Maarry et al.*, 2015b; *Cambianica et al.*, 2019). Finally, a notable large-scale linear fracture system was observed in the neck region of 67P.

This fracture might have formed due to activity-induced tectonic forces that develop as the comet undergoes seasonal variations in its rotational speed (*Thomas et al.*, 2015a; *Keller et al.*, 2015b): This might occur in response to variations in the global torques caused by seasonal activity variations (*Matonti et al.*, 2019). Indeed, modeling efforts by *Hirabayashi et al.* (2016) focused on this phenomenon and predicted that bilobed comets may undergo repeated splitting and reconfiguration because of this effect. Furthermore, long-term monitoring of surface changes on 67P (*El-Maarry et al.*, 2017) (see also section 3) suggested that the neck fractures extended in length during the mission at a time that coincided with changes in rotation speed (*El-Maarry et al.*, 2017). Finally, it is worth noting that development of fractures generated by the shear deformation of the bodies colliding at slow relative velocity (*Davidsson et al.*, 2016; *Schwartz et al.*, 2018; *Franceschi et al.*, 2020) has been suggested, which could have propagated several hundred meters below the nucleus, hence facilitating heat penetration and consequent sublimation and resulting in surface erosion (*Matonti et al.*, 2019).

Fig. 6. An OSIRIS NAC image taken on 30 September 2016 showing consolidated terrains partly covered by smooth deposits on the small lobe of 67P. Pervasive fracturing is visible in the consolidated materials forming in locations polygonal patterns, which have been suggested to be indicative of insolation weathering.

*2.2.4. **Deposits.*** Comets may display deposits on their surfaces that are a result of weathering, mass wasting, and transport of materials under the influence of gravity and surface processes (particularly sublimation-related events, such as jets and outbursts). Deposits may appear smooth when compared to rougher consolidated terrains and landforms, but overall their texture would ultimately depend on surface composition, material strength, and the formation mechanism(s) of the deposits (*Sunshine et al.*, 2016; *El-Maarry et al.*, 2019). For instance, deposits may form *in situ* through cliff/scarp collapse with the resulting debris-forming deposits of various sizes at the foot of the cliffs (*Pajola et al.*, 2015; *Lucchetti et al.*, 2019). Examples of such deposits have been observed on 9P/Tempel 1 and at a higher resolution (both spatially and temporally) on 67P/Churyumov-Gerasimenko (*Giacomini et al.*, 2016). At 67P, diamicton (poorly-sorted) deposits are observed on the bases of scarps and cliffs, most notably in the northern hemisphere's Ash region [for all 67P regions mentioned in the manuscript we refer to *El-Maarry et al.*, 2015a, 2016)].

In other cases, deposits may form through *in situ* weathering, for instance, as a byproduct of sublimation events, but later transported by being entrained in the escaping gas to other parts of the comet falling under the influence of gravity (*Thomas et al.*, 2015b; *Cambianica et al.*, 2021). Comet 103P/Hartley 2 is another bilobate comet displaying large smooth areas. These terrains are predominantly located around the comet "waist," at the interface between the two lobes (Fig. 7a). Spectroscopy and morphological analyses (*A'Hearn et al.*, 2011) reveal that smooth areas result from a global cycle of sublimation/transport/deposition on the comet. The ends of both lobes are rich in CO_2 ice, which starts sublimating at a lower temperature than H_2O ice. Upon release, the gas ejects grains of dust and water ice from the surface that travel around the comet and fall back in the regions of higher gravity, thus covering these areas with a smooth dust blanket enriched in H_2O. These grains would eventually sublimate, leading to distinct water-rich plumes from the comet waist, in contrast with the CO_2-rich plumes emitted from the lobes (*Kelley et al.*, 2013).

Activity-related deposits were clearly observed as well on Comet 67P and partly account for the hemispherical dichotomy observed at 67P (*El-Maarry et al.*, 2016; *Lee et al.*, 2016). Indeed, modeling efforts have shown that intense activity in Comet 67P's southern hemisphere results in the production of friable deposits that are ejected during the activity and transported by gravity to the northern hemisphere where they settle in local gravitational lows (*Keller et al.*, 2017).

On 67P, smooth deposits tend to appear with very fine texture (*Thomas et al.*, 2015b) not fully resolved in OSIRIS images taken at a distance of around 30 km from the surface [corresponding to a spatial resolution of ~0.5 m pixel^{-1} (Fig. 7b)]. However, higher-resolution images from orbit, the Philae lander during its landing sequence, and Rosetta itself during its final touchdown in the Sais area (Fig. 7c) of the Ma'at region all show that these "smooth" deposits

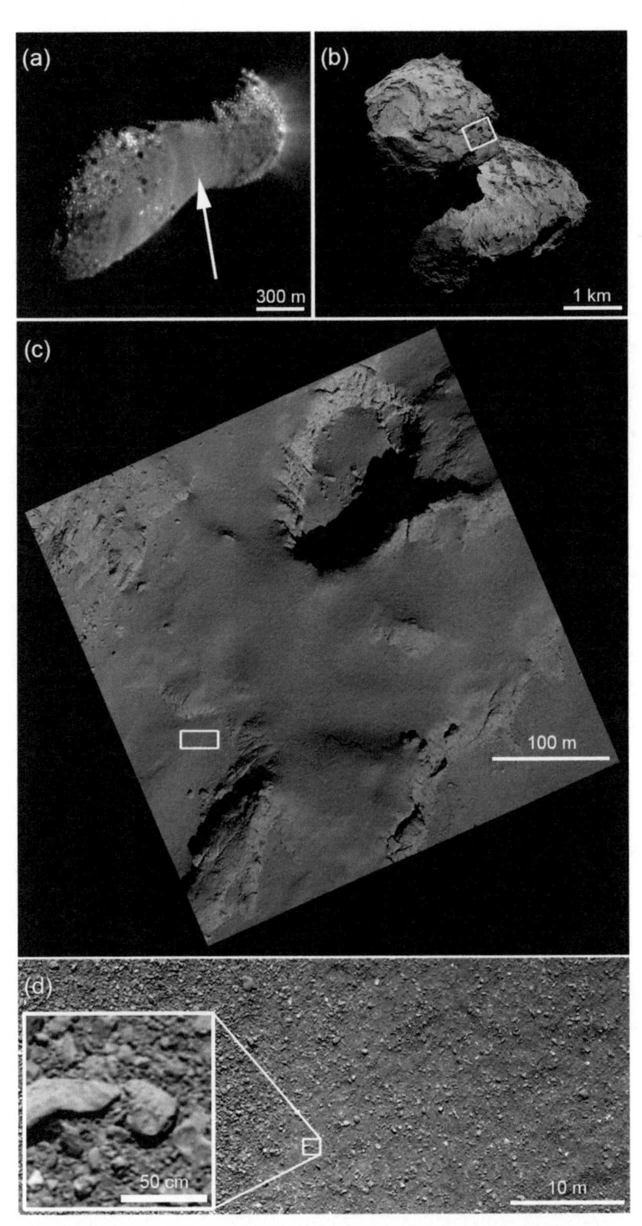

Fig. 7. (a) Comet 103P as observed with a resolution of 3.9 m. The white arrow points at the waist between the two lobes, i.e., where a large smooth area is present. **(b)** Comet 67P as observed with a resolution of 2.4 m. The white box shows the location of **(c)**. **(c)** The dusty deposits located in the Ma'at region of 67P. The white box shows the location of **(d)**. **(d)** One of the final images taken by Rosetta during its touchdown on 30 September 2016. The previously identified "smooth" deposits are characterized by boulder-sized (tens of centimeters to meter-sized boulders) as observed in the high-resolution closeup.

comprise pebble- to boulder-sized (tens of centimeters to meter-sized) clasts (e.g., *Mottola et al.,* 2015; *Pajola et al.,* 2016a) (Fig. 7d). An interesting byproduct of the comet's weak gravity is that meter-sized boulders on the surface can act dynamically like sand-sized (typically tens of micrometers up to a couple of millimeters) particles on Earth under comparable processes. Notably, "winds" may develop on the surface as gases are accelerated between the colder and warmer (illuminated) regions (e.g., *Thomas et al.,* 2015b), leading to aeolian-like transport mechanisms and landforms as shown by the ripple-like formations observed in the northern neck region of 67P (*Jia et al.,* 2017).

The evidence that at 67P such landforms underwent substantial changes in the neck region [called Hapi (*Pajola et al.,* 2019)] and redeposition of new materials took place in the northern hemisphere (*El-Maarry et al.,* 2017) clearly shows that weathering and transportation can play a major and continuous role in the evolution of cometary surfaces.

2.2.5. Boulders. The first cometary nucleus where positive reliefs with sizes <40 m across have been detected is Comet 103P/Hartley 2 (*A'Hearn et al.,* 2011; *Thomas et al.,* 2013b). Such features, later called "boulders" since they resemble the shapes of the terrestrial boulders, are characterized by the constant presence of an elongated shadow identifiable on multiple images and appear to protrude from the surface where they stand (*Pajola et al.,* 2016c). The identification of such features and their size-frequency distribution (SFD) fitting analysis is of pivotal importance, since it can provide important information on how they may have formed and/or degraded (see, e.g., *Pajola et al.,* 2021; *Schröder et al.,* 2021, and references therein), hence potentially indicating the processes that affected the surfaces of the studied targets.

The images obtained by the Deep Impact impactor camera a few seconds ahead of impacting Comet 9P/Tempel 1 also show the possible presence of boulders close to the impact area (*A'Hearn et al.,* 2005); nevertheless, their identification is still debated despite the high-resolution acquisition, and their absolute number is too low to extract any possible SFD. Instead, thanks to the Deep Impact/EPOXI High Resolution Visible CCD (HRIV) images (*Klaasen et al.,* 2008) of 103P with a spatial scale ranging from 5 to 2 m pixel^{-1}, it has been possible to derive on the entire illuminated side of the nucleus (50%) the SFD of boulders larger than 10 m, getting a best-fit power-law index of -2.7 ± 0.2 and a spatial density of 140 boulders $\geq$10 m km^{-2} (Figs. 8a,b). *Pajola et al.* (2016c) justified such a trend as the result of the concurring disintegration of the smallest boulders that make them disappear due to the strong nucleus sublimation, as well as by their lifting due to the drag force produced by the outflowing gas triggered by the cometary activity. Both processes lead to the decrease of smaller boulders with respect to the larger ones, therefore lowering the cumulative SFD power-law index. Similar analysis has been performed on the two 103P lobes separately, getting power-law indices that overlap [-2.7 ± 0.2 for Lobe 1 (L1) and -2.6 $+0.2/-0.5$ for Lobe 2 (L2)] with the global one, hence pointing toward similar fracturing/disintegration and lifting phenomena that may occur both in L1 and L2. Nevertheless, the 2× difference in the number of boulders per square kilometer between L1 and L2 suggests that the more diffuse H_2O sublimation on L1 produce twice the boulders per square kilometer with respect to those produced on L2, whose activity is CO_2-driven.

At 67P, *Pajola et al.* (2015, 2016d) identified all boulders in the size range 7–38 m. By taking into account

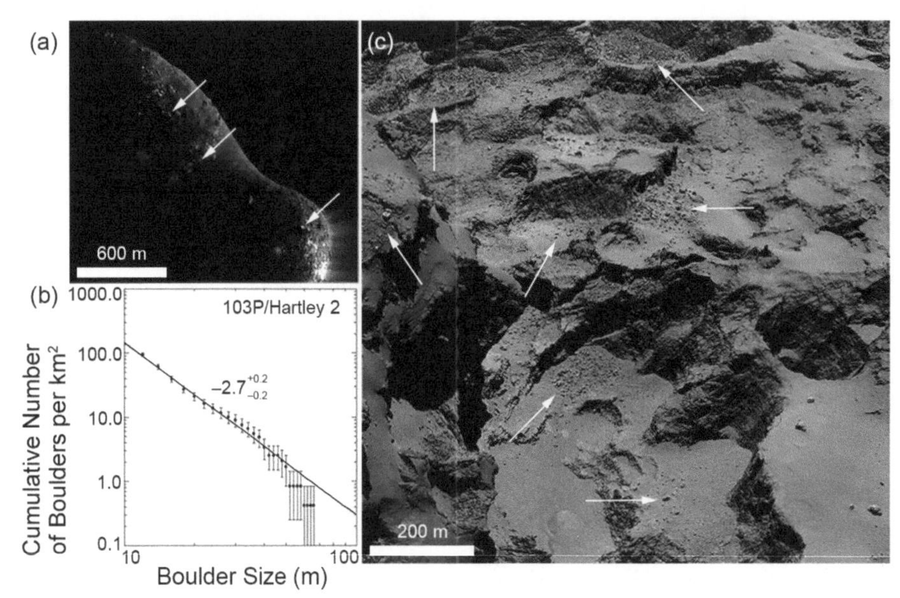

Fig. 8. (a) Comet 103P/Hartley 2 with some boulders highlighted with white arrows. **(b)** 103P boulders SFD with the power-law best-fitting curve presenting an index of −2.7 ± 0.2. **(c)** The northern heminucleus of Comet 67P, as observed through the OSIRIS Narrow Angle Camera. Some of the boulder fields located below the cliffs are indicated with white arrows.

those counts belonging to similar morphological settings it has been possible to identify that (1) pit formations and creation of depressions with subsequent escape of high-pressure volatiles and consequent high fracturing are characterized by power-law indices of about –5 to –6.5 (*Pajola et al.,* 2015); (2) gravitational events triggered by sublimation and/or thermal fracturing causing regressive erosion present power-law indices of about –3.5 to –4.5 (*Pajola et al.,* 2016b); and (3) the evolution of the original material formed during both (1) or (2) were not particularly renewed or present in areas where repeated, high-sublimation rates occurred showing power-law indices of about –1 to –2 (*Pajola et al.,* 2019). We highlight that a useful schematic figure showing the different power-laws obtained on 67P and how they evolve from steeper values (associated with younger gravitational deposits induced by sublimation) to shallower distributions (associated with mature deposits that underwent a prolonged sublimation activity) is presented in *Schmedemann et al.* (2018, Fig. 13.15).

Since Comet 67P is globally dominated by the presence of boulders located below receding cliffs (*El-Maarry et al.,* 2015a, 2016; *Giacomini et al.,* 2016), and the global SFD power-law index derived on >7000 counts is –3.6 ± 0.2 (both in the northern and the southern heminucleus), it was suggested that sublimation and thermal fracturing causing regressive erosion and gravitational collapse is the dominant process for the boulder genesis at 67P (Fig. 8c). In contrast to 103P, Comet 67P does not show strong volatile differences between the two lobes (*Hässig et al.,* 2015; *Gulkis et al.,* 2015), whose main activity is H_2O-driven; nevertheless, the much larger thermal stresses (*Keller et al.,* 2015a,b) and higher level of activity of the southern heminucleus, due to the much stronger insolation occurring during southern perihelion (*Keller et al.,* 2017), result in a 3× increase of the southern boulder density per square kilometer with respect to the northern one.

Beside Comets 103P, 67P, and 9P, it is likely that all other observed nuclei may also be characterized by the presence of boulders since sublimation, lifting, and thermal processes (with different intensities and strengths) occur on all visited targets. However, since the sizes of the identified boulders range from meters to few decameters, the highest resolution of the images acquired for 1P [~50 m (*Keller et al.,* 1988)], 19P [47 to 58 m (*Soderblom et al.,* 2002)] and 81P [maximum scale of 14 m pixel^{-1} (*Brownlee et al.,* 2004)] hampered their identification.

3. ACTIVITY-DRIVEN SURFACE EVOLUTION

Cometary nuclei show ever-changing surfaces that are heavily affected by diurnal and seasonal thermal stresses, increasing and decreasing sublimation activity along the orbit, as well as sudden outbursts. Such factors all favor mass-wasting processes as cliff collapses and landslides, dusty deposit disappearance and formation, and boulder/pebble lifting and ejection.

3.1. Cliff Collapses, Mass Wasting, and Landslides

Mass-wasting events such as landslides, cliff collapses, or scarp retreats have been postulated or observed on the nucleus of almost every comet visited so far (*A'Hearn et al.,* 2005; *Belton et al.,* 2013), suggesting that they can be important processes in reshaping the cometary surface. Such processes are often also correlated with jet activity and strong outbursts (*Vincent et al.,* 2016; *Agarwal et al.,* 2017). *Britt et al.* (2004) suggested that mass wasting heavily contributed to the formation of mesas on Comet 19P/Borrelly, proposing that this process removed non-volatile material from steep slopes, allowing scarps to retreat through sublimation. This was also supported by the observation of jet activity mainly concentrating in the areas of such mesas. Such a volatile-driven process has been also observed on the surface of 9P/Tempel 1, where it was possible to quantitively estimate one scarp retreat by more than 50 m (*Thomas et al.,* 2013a; *Sunshine et al.,* 2016). In addition, *Farnham et al.* (2013c) interpreted the long-terraced scarp on the side of 9P observed with Stardust NExT to be the source of the seven jets observed. On Comet 103P/Hartley, *Steckloff and Samarasinha* (2018) proposed that avalanches may have been induced by rotational instabilities activating the small lobe of the nucleus and producing the observed CO_2 outgassing. In this context, mass-wasting events have been considered important in maintaining the sublimation activity of comets, but also in reactivating dormant comets (*Steckloff and Samarasinha,* 2018).

More recently, the Rosetta spacecraft observed many cliff-like structures whose faces or tops are often characterized by the presence of fractures, displaying talus deposits at their base (e.g., *Pajola et al.,* 2015; *Bouquety et al.,* 2021), consistent with the occurrence of mass-wasting processes (*Thomas et al.,* 2015a). The evidence for 67P activity associated with cliffs has been presented by *Vincent et al.* (2016), while the role of fractures in driving cliff collapse and mass wasting has been suggested by *El-Maarry et al.* (2015b) and *Ali-Lagoa et al.* (2015). In particular, thanks to multiple OSIRIS images taken pre-perihelion, *Pajola et al.* (2016b) identified a peculiar fracture near the edge of the Aswan escarpment in the northern heminucleus of 67P (Figs. 9a,b). The fracture, 70 m long and 1 m wide, propagated almost perpendicularly from the scarp edge at its two ends and inward in a semicircular fashion ~12 m away from the edge at its farthest point. On 10 July 2015, after constantly monitoring this location throughout the mission, the Rosetta Navigation Camera captured a large plume of dust that could be traced back to the Aswan escarpment (Fig. 9c), hence revealing that a cliff collapse had occurred (*Pajola et al.,* 2017a). Out of the total collapsed volume (2.20×10^4 m^3, with a 1σ uncertainty of 0.34×10^4 m^3;

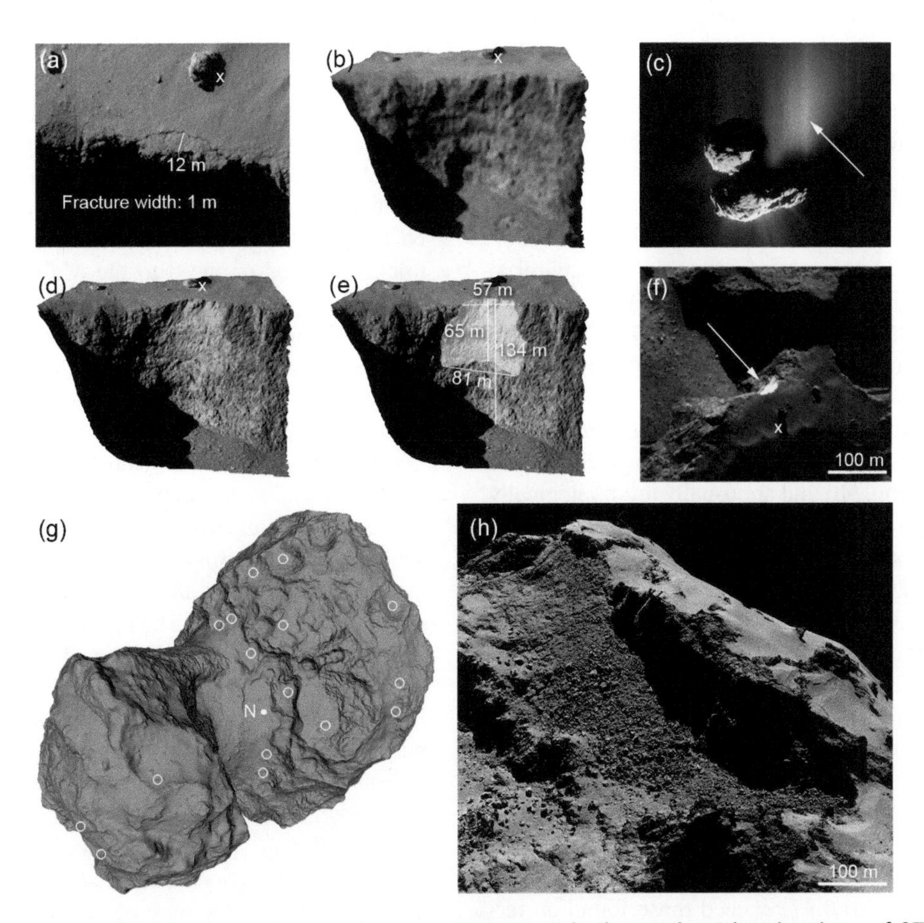

Fig. 9. (a) The 70-m-long fracture located in the Aswan escarpment, in the northern heminucleus of 67P. The x feature marks a big boulder used for reference in **(b)**, **(d)**, and **(f)**. **(b)** Three-dimensional shape model prepared using the SPG technique of the pre-collapse Aswan cliff. **(c)** The large plume of dust coming from the Aswan cliff (marked with a white arrow) as observed through the Rosetta NAVCAM. **(d)** Three-dimensional shape model of the post-collapse Aswan escarpment. **(e)** Same as **(d)**, but with the dimensions of the volume that detached from the cliff. **(f)** The bright, sharp scarp edge showing the pristine interior of 67P, as observed on 26 December 2015. **(g)** The location of the landslide deposits (white, voided circles) on 67P northern heminucleus. The north pole is indicated. **(h)** High-resolution OSIRIS image showing a wide landslide deposit located in the Ash region.

Figs. 9d,e), 1% of it was lost to space during the event, while the remaining part formed the boulder talus at the foot of the scarp. A few days later, an OSIRIS image revealed a fresh, sharp scarp edge, where the cliff collapse exposed the pristine and fresh interior of Comet 67P, characterized by water ice with an albedo >0.4 (Fig. 9f). Modeling temperature behavior (*Höfner et al.,* 2017) indicated that during the July 2015 timeframe (close to perihelion), the temperature of the surface at the cliff face increased from 130 K to 320 K in less than 20 minutes, with a maximum of 30 K per minute shortly after local sunrise, while the nearby areas were still remaining cold. The thermal cracking of the already highly ruptured and fractured cliff led to the propagation of fractures at the surface into the interior of the cliff, inducing sublimation and erosion in deeper layers of the cometary nucleus, hence favoring the collapse itself (*Pajola et al.,* 2017a).

The exposure of fresher, brighter material has been also identified at the base of another 67P cliff between Khepry and Imhotep, where the material has been spectrophotometrically interpreted as water-ice-rich material thanks to OSIRIS and the Visible and InfraRed Thermal Imaging Spectromete (VIRTIS) measurements (*Pommerol et al.,* 2015; *Barucci et al.,* 2016; *Filacchione et al.,* 2016a). Once more, such observations strengthen the idea that cliff collapse is a pivotal process actively affecting the cometary surface.

From a global perspective, multiple landslide deposits have been observed on the surface of 67P, not all correlated with fractured cliffs or bright boulders, but associated with well-defined scarps from which the material has been detached (*Lucchetti et al.,* 2019) (Fig. 9g). A morphological detailed characterization of such features was only possible on 67P due to higher-resolution images, even if talus deposits were also observed at the base of many cliffs on Comet 81P/Wild 2 (*Brownlee et al.,* 2004). The 67P landslides deposits are made of fragments in the form of boulders (Fig. 9h) and are usually morphologically considered as the result of a single collapse, even if there are a few cases that resembled multiple rock-fall events (*Lucchetti et al.,* 2019). Moreover, landslide deposits are identified on consolidated terrains and not in the southern heminucleus of the comet, where a smoother topography is present. The sizes of 67P landslide deposits range from tens to hundreds of meters, being small with respect to the kilometer-scale landslides observed on other solar system bodies. To better investigate the occurrence of such events on a comet surface, the height to run-out length ratio (H/L) has been calculated for 67P, which is an approximation for the friction coefficient of landslide material. The high H/L found (from 0.5 to 0.97) exceed those found on terrestrial dry landslides, suggesting a mechanically rocky-type behavior of cometary material that, once collapsed, assumes a rock avalanche mobilization (*Lucchetti et al.,* 2019). Finally, the different H/L of 67P landslides may be an indicator of the different volatile content of the material, fostering it to slide closer or farther from the scarp face (*Lucchetti et al.,* 2019).

The appearance of multiple cliffs and mass-wasting deposits on several nuclei, coupled with the direct witness of scarp retreats both on 9P and 67P, strongly supports the interpretation that comets are some of the most geologically active bodies in the whole solar system, with landslides, collapses, and the associated dust-gas plumes that are major processes causing their surface evolution.

3.2. Dusty Deposit Disappearance and Formation

As already discussed so far, non-consolidated materials (NCMs) undergo sublimation-related activity as observed in all comet nuclei that have been imaged during or close to perihelion by flyby or orbiting missions. A natural consequence of this activity is the ultimate desiccation of the top layers of the NCMs. However, *De Sanctis et al.* (2015) described the possibility of a diurnal volatile cycle that would replenish volatile inventory of the upper layers. Nevertheless, the activity itself and the mobilization of volatiles (through sublimation and possibly diffusion from below) may lead to textural changes in the cometary deposits. Long-term monitoring of the surface of 67P, particularly around the time of perihelion, has demonstrated that NCMs undergo indeed substantial, if localized, morphological changes in two main ways:

- Development of circular patterns and retreating scarps: Circular patterns from tens to hundreds of meters in diameter were observed to develop in certain regions where NCMs are thought to have accumulated to considerable thicknesses (as opposed to the Ma'at and Ash regions where the NCMs appear to drape the underlying terrain with a thin mantle), particularly the Imhotep, Anubis, and Hapi regions (*El-Maarry et al.,* 2015a). Some of the first examples of such features were observed in the Imhotep region (*Groussin et al.,* 2015). Scarps were seen to develop and retreat in the smooth deposits in the central part of the region, accompanied by bright albedo changes in the scarps, suggesting that the scarps were undergoing sublimation (*Groussin et al.,* 2015). Similar features were observed in the Hapi and Anubis regions (*El-Maarry et al.,* 2017) (Figs. 10a,b). The timing of activity in the three regions is consistent with seasonal illumination patterns and how they are related to different local regions and their latitudes (*El-Maarry et al.,* 2017). However, the calculated rate of expansion of the developing patterns and the scarp retreats is considerably faster than what would be expected from water-ice sublimation (*El-Maarry et al.,* 2017).
- Development of "honeycomb"-like surface textures: The Ma'at and Ash regions displayed numerous patches that developed around the time of perihelion ranging from a few tens of meters to 100 m (*Hu et al.,* 2017; *El-Maarry et al.* 2017) (Figs. 10c,d). Morphometric information derived from local elevation models suggests that these features were 0.5 m deep on average (*Hu et al.,* 2017). The patches were observed to fade away (Fig. 10e) as they were most probably covered by new fall deposits from the activity in the southern hemisphere (*El-Maarry et al.,* 2017).

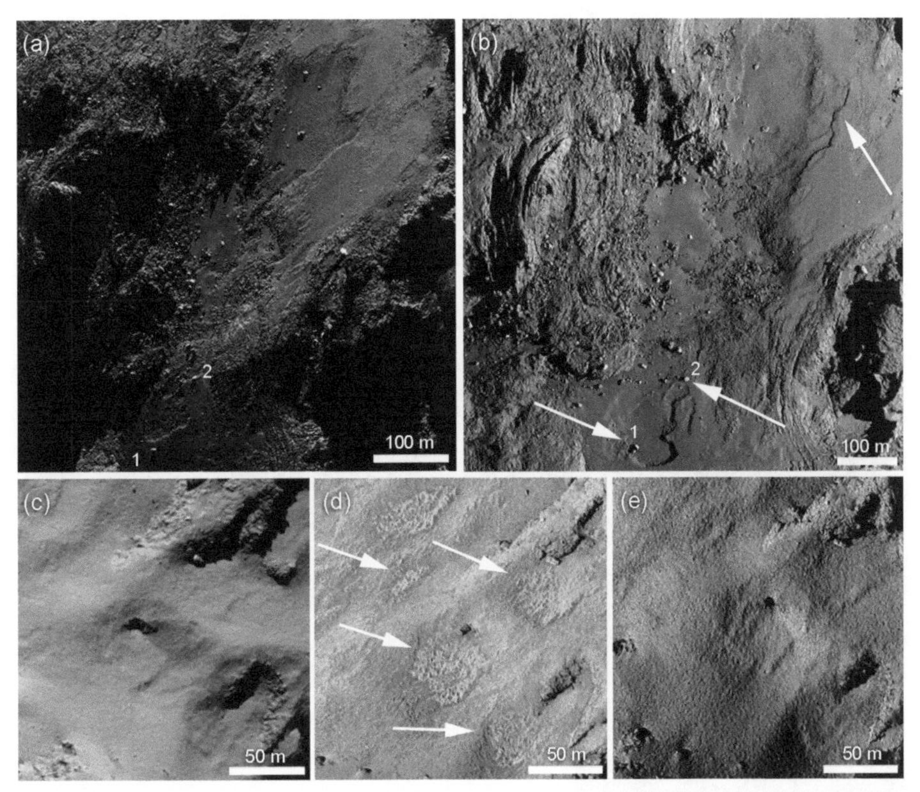

Fig. 10. (a),(b) A view of the Anubis region before (December 2014) and after (June 2016) perihelion passage (in August 2015) showing scarp retreats or modifications (white arrows). Note the boulders that are marked 1 and 2 for orientation. **(c)–(e)** An area in the Ma'at region tracking the appearance of the so-called honeycomb features a few months before perihelion passage [**(c)**, September 2014; **(d)**, March 2015] and consequent disappearance after perihelion as new NCM deposits cover the features (March 2016).

Despite the sublimation of volatiles being a clear agent in the development of changes, the actual physics and mechanisms of change remain poorly understood. In the Ma'at and Ash regions, it is possible that the textural changes that occur, which resemble honeycomb structures, are caused by the loss of ice from the upper layers of the deposits. However, it is not well understood why the patches do not grow in size or increase in number with time. Similarly, the circular patterns were observed solely in certain locations on the comet, suggesting that compositional heterogeneities coupled with varying illumination patterns may account for the isolated features. For the specific case of the Imhotep region, for example, a vertical loss of materials of ~4 m in thickness was recorded in one notable location (*El-Maarry et al.,* 2017), which is a factor of 4 more than the inferred global loss (*Keller et al.,* 2017). As for the honeycomb textures, it is not clear why this occurred only in this specific area, although it is possible that the thickness of the deposits plays an important role by favoring circular patterns in thicker deposits.

3.3. Boulder/Dust Lifting and Ejection

One of the most important consequences that the increasing sublimation activity has on comets while approaching perihelion is not only the lifting of dust (commonly considered in the 0.1–100-μm size range) but also of millimeter- to meter-sized particles in general (*Sunshine et al.,* 2016). This process has been clearly identified for the hyperactive nucleus of 103P, where high-resolution images showed a coma of large particles and boulders reaching dimensions between 0.1 and 2.21 m, with some of the largest cases reaching effective radii ~4 m (*Kelley et al.,* 2013) (Fig. 11a). The icy chunks detected from the Deep Impact/EPOXI were dragged from the surface by the strong activity of the nucleus, with their consequent sublimation providing the largest part of the total gaseous output of the comet (*Lisse et al.,* 2009; *A'Hearn et al.,* 2011).

By taking into account the nucleus local surficial gravity field, the centrifugal force, and the drag force generated by the outflowing gas triggered by sublimation processes, it is possible to derive the particle largest dimension that can be lifted from the surface. As seen in *Fulle* (1997), the gas drag force, called $F_G D$, can be computed from Stokes' formula as

$$F_G D = M_b \times a = \frac{C_D}{2} s^2 v_g \left(\frac{dm}{dt}\right)_g \frac{1}{r^2} \quad (1)$$

where M_b is the boulder mass, a is the acceleration, C_D is the drag coefficient, s is the boulder radius, v_g is the gas velocity, $(dm/dt)_g$ is the gas-loss rate, and r is the cometocentric distance. If the drag force is similar to the local gravity, the

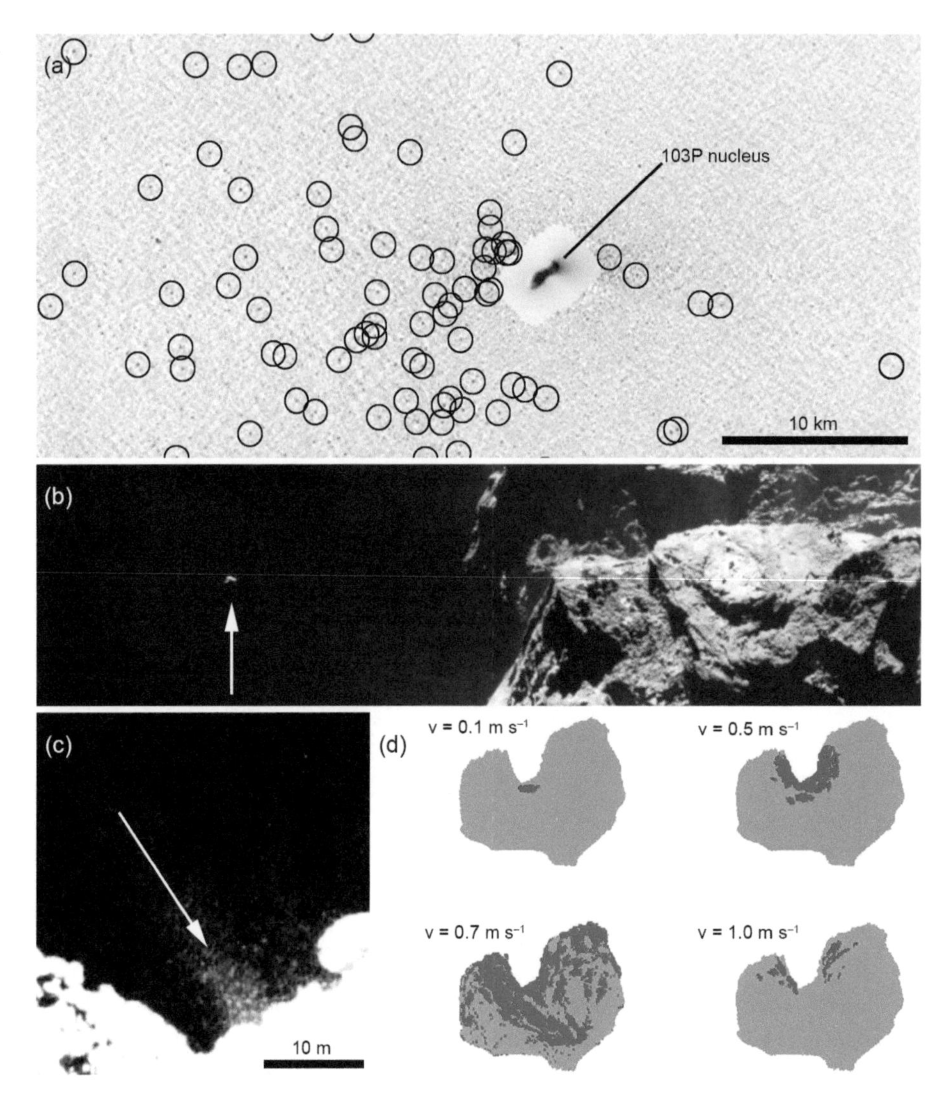

Fig. 11. (a) The large particles and boulders (circled with black circles) identified inside the coma of Comet 103P/Hartley 2. **(b)** The irregular ~1-m-size boulder that had been ejected from the surface of 67P, as observed by the OSIRIS NAC camera on July 30, 2015. **(c)** Individual, resolved particles ejected from the surface of 67P with a velocity of >2 m s^{-1}. **(d)** The surface location (red areas) of reimpacting particles ejected from Hapi at different velocities. The particles with ejection speeds >1.0 m s^{-1} leave 67P, with the exception of those that collide with the surfaces of Hapi and Seth before escaping. Particles with ejection velocities <0.5 m s^{-1} remain in the neck (*Thomas et al.,* 2015a).

boulder can take off, and depending on the ejection velocity it can consequently escape the cometary nucleus.

Pajola et al. (2015) used the above equation to estimate the largest boulder size that could be lifted by the outgassing of 67P, taking into consideration the highest and smallest surface gravity of the comet as well as its gas-loss rate range (*Bertaux et al.,* 2014). This was done because several isolated boulders appear on the surface of 67P, not related to any cliff collapse nor localized fracturing, and hence could have landed there after being lifted from other areas of the nucleus. The resulting largest liftable size obtained ranges between 2 and ~6 m and agrees with the satellite analysis of *Bertini et al.* (2015) that evidenced no detections of boulders larger than a few meters orbiting or in close proximity to 67P. Moreover, on 30 July 2015 (i.e., close to 67P perihelion) OSIRIS/Narrow Angle Camera (NAC) observations of 67P showed the presence of a resolved ~1-m irregular boulder that had been ejected from the surface (*Fulle et al.,* 2016) (Fig. 11b), hence confirming that the lifting process of decimeter- to meter-sized particles can commonly occur on cometary nuclei (Fig. 11c).

An important aspect to keep in mind in order to understand how a cometary surface evolves along its orbit is a direct consequence of the dust/boulder-lifting process. Any particle not accelerated to escape velocity will fall back to the surface, as initially suggested by *Möhlmann* (1994), hence causing the production of a loosely packed regolith with centimeter-sized particles as the main contributors to the surface coverage (*Kührt et al.,* 1997). For 67P, this was modeled by *Thomas et al.* (2015a), which explained the occurrence of the smooth thin deposits located in the northern heminucleus of the comet as the result of the redistribution of particles ejected at different velocities and sizes (Fig. 11d) from the Hapi region at the neck (*Pajola et al.,* 2019). This "fallback" process occurs not only while Hapi is active, but also when the 4× more active southern heminucleus ejects millimeter- to meter-sized material, hence transferring mass to the northern inactive regions (*Keller et al.,* 2015a; *Lai et al.,* 2016) and producing smooth-texture surfaces (*Pajola et al.,* 2017b; *Hu et al.,* 2017).

3.4. Outburst Correlation

Establishing a clear correlation between outburst events and specific morphology is a challenging task. Outbursts are defined as a transient release of dust and gas, generally much more material than what is typically ejected by perennial activity (*Vincent et al.,* 2020). Such events can often be detected in groundbased observations and have been also characterized by spacebased observations of several comets, e.g., 9P/Tempel 1 in *Farnham et al.* (2007) and 67P/Churyumov-Gerasimenko in *Vincent et al.* (2016). They lead to the ejection of tens to hundreds of tons of material and should leave a recognizable signature on the surface (Fig. 12).

However, the transient nature of such events makes it so that most, if not all, observations are unique and cannot easily be traced back to the nucleus; typical activity source inversion techniques require multiple observations of the same event from different angles (e.g., *Knollenberg et al.,* 2016). It should also be noted that all missions before Rosetta were flybys, and therefore do not have before/after images of the surface around the time of an outburst. Such observations exist for Rosetta but are very limited. Because of safety concerns, the spacecraft retreated from the comet to a safe distance of several hundred kilometers from the nucleus during perihelion (*Taylor et al.,* 2017), when most outbursts occurred on Comet 67P (*Vincent et al.,* 2016). The resulting low resolution means that it is often impossible to unambiguously find the source location of an outburst, and high-resolution images of the source could not be acquired until many months after the event. Therefore, any association of morphological change with outbursts should be considered cautiously. Examples of the determination of outburst sources and their connection to morphology can be found in *Belton et al.* (2013), *Vincent et al.* (2016), *Grün et al.* (2016), *Agarwal et al.,* (2017), and *Pajola et al.* (2017a).

Keeping these limitations in mind, we can nonetheless argue that some morphology is very likely related to an outburst event. Circular features of non-impact origin found on all comets suggest that the surface can cave in locally due to explosive sublimation (*Belton et al.,* 2013) or sink-hole formation (*Vincent et al.,* 2015a). Avalanches are known to take place on comets (*Steckloff et al.,* 2016; *Lucchetti et al.,* 2019), as well as catastrophic cliff collapses (*Pajola et al.,* 2017a). All these events lead to the mobilization and acceleration of a massive amount of cometary material, which can easily be ejected in space due to the low gravity and gas expansion.

On both Comets 9P/Tempel 1 and 67P/Churyumov-Gerasimenko, outbursts sources that could be reliably inverted from the dust plume observations lead to locations on the nuclei that are rich in small pits (*Belton et al.,* 2013), or that display cliffs with signs of regressive erosion [e.g., cracks on the cliff face, boulders fields at the bottom, avalanche runs (see *Vincent et al.,* 2016; *Fornasier et al.,* 2019), indicating that local resurfacing events have taken place]. Two of the most striking and least ambiguous, events observed on Comet 67P are the Aswan cliff collapse (*Pajola et al.,* 2017a) and a series of micro-outbursts originating from newly formed pits in the Imhotep region (*Agarwal et al.,* 2017; *Oklay,* 2018) (Fig. 12a). In both cases, we can constrain the area from which the outburst arose, and we have unambiguous evidence for significant morphological changes (Figs. 12b,c) occurring in the same location and at the same epoch. In both cases, the scale of surface modifications is compatible with the amount of material ejected during the event.

Hence, although the processes behind cometary outbursts remain challenging to establish, we do have evidence for significant morphological transformations of cometary nuclei in areas prone to outbursting. What remains to be understood is the precise chronology of the events and whether different types of outbursts can be associated with different surface transformations.

4. SEASONAL AND ORBITAL SURFACE EVOLUTION

Comets are seasonally active geological bodies whose energy source is primarily solar insolation; hence, nuclei morphology can locally change while approaching perihelion. Moreover, such changes accumulate over time, evolving over many orbits. This section focuses on the nuclei seasonal surface evolution, as well as the possible geological changes that can occur from the Kuiper belt object (KBO) phase to an evolved JFC.

4.1. Seasonal Surface Evolution

Cometary nuclei are characterized by unpredictable behavior since sudden outbursts or fragmentation can occur at any time (e.g., *Boehnhardt,* 2004; *Tubiana et al.,* 2015; *El-Maarry et al.,* 2019). Such events can lead to a change in nucleus period, spin orientation, and activity (e.g., *Bertini et al.,* 2009; *Bodewits et al.,* 2018). The source for comet activity is mainly driven by solar insolation, hence making them seasonally active geological bodies. Some of this energy is re-radiated back into space as thermal emission (*Groussin et al.,* 2019), but part of it initiates the sublimation of ices from the surface of the nucleus (*Skorov et al.,* 2016). Finally, (1) non-uniform gravitational forces affecting material movements and (2) spatially heterogeneous outgassing affecting orbital dynamics and leading to tidal stresses are both consequences of irregular shapes of comet nuclei (*Lamy et al.,* 2004; *Thomas,* 2020). To observe seasonal and orbital nucleus changes, it is therefore pivotal to have spacecraft remote sensing observations that follow the entire

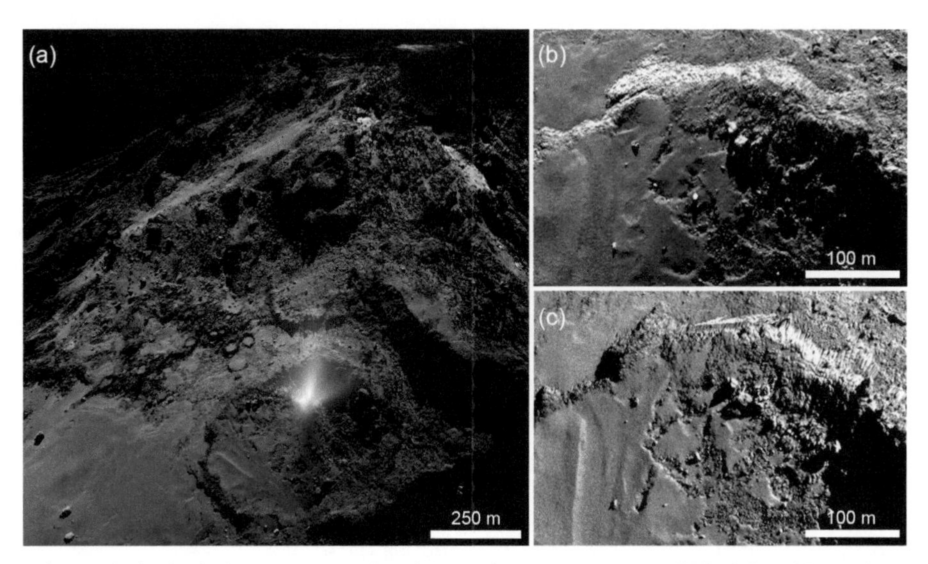

Fig. 12. Example of morphological changes associated to outbursts on comet 67P. **(a)** A big outburst observed in the Imhotep region of 67P on July 3, 2016 (*Agarwal et al.,* 2017). **(b)** The area where the outburst of **(a)** originated. This region is known to have experienced many small outbursts in the 2014 to 2016 period and even beyond (*Oklay,* 2018). **(c)** Same as **(b)**, but with the newly formed several small pits. Figure adapted from *Vincent et al.* (2020).

orbit of a comet (or at least a substantial period before and after perihelion passage), or observe a comet at different revolutions around the Sun.

9P/Tempel 1 is the first case in which a study of surface changes was carried out because it was visited by two space missions during two different orbits in 2005 and 2011, respectively. The Deep Impact and Stardust-NExT mission flybys were separated by one 9P orbit, i.e., 5.5 years. Specifically, the Deep Impact spacecraft observed the comet one day before perihelion passage, while the Stardust-NExT mission imaged it 34 days after the next perihelion. The region observed by the Stardust-NExT mission covered about 20 km^2 of the area previously imaged by Deep Impact, specifically the area of the impact site and the northern part of the southernmost smooth flow discovered by Deep Impact (*Veverka et al.,* 2013). The observations made by the two encounters have different viewing and illumination conditions, making it difficult to assess and evaluate the apparent surface changes. No large-scale changes in albedo occurred, but several small brighter albedo spots with a size less than 30 m in the Deep Impact region disappeared, while others appeared (*Thomas et al.,* 2013a). However, due to the small size of such features, it remains unclear if these changes can be considered real or due to different illumination conditions of the images analyzed. The absence of changes in the pit features observed on the surface of Tempel 1 suggest that their formation timescales are longer than one orbit, probably at least tens of perihelion passages.

The only significant morphological change of 9P is visible along parts of the scarp bounding a smooth flow that has been observed at a resolution of 9 m and 11 m by Deep Impact and Stardust-NExT, respectively (Figs. 13a,b). As shown in Figs. 13c,d, the scarp retreated by more than 50 m in six years at least in two places, corresponding to the disappearance of an area of 1.6×10^4 m^2. Moreover, it was possible to estimate the thickness at the flow front between 8 and 15 m, corresponding to a total volume loss of $2–4 \times 10^5$ m^3 and a total loss of material of $8–16 \times 10^7$ kg, considering the mean average density of the comet (*Thomas et al.,* 2013a). The observed scarp retreat seems to operated more at its brighter part, possibly being the result of removing darker lag material by scarp collapse or small-scale sublimation pressure removing crust (*Thomas et al.,* 2013a). The area of eroding scarp is at 40°S, thus receiving moderate heating and never being in total darkness throughout a rotation. In addition, no jets were detected in that region, but the source of most jets occurred at lower altitude, along a terraced scarp imaged only by Stardust-NExT, hence making it impossible to discover any possible change. Even if surface change and the source of some jets are correlated with scarps, the cause-and-effect relation was not clear, as the influence of solar heating in jet activity (see *Farnham et al.,* 2013c).

On the contrary, Comet 67P orbits around the Sun in 6.44 years, on an elliptical orbit with an aphelion distance of 5.68 au, a perihelion of 1.24 au, and an inclination of 7.04°, typical of JFCs. Its shape is characterized by two lobes, with a strong constriction between them [the neck of 67P, called Hapi (*El-Maarry et al.,* 2015a)]. The axis of rotation of the comet is close to parallel to the long axes of the lobes and almost perpendicular to the neck (*Sierks et al.,* 2015; *Preusker et al.,* 2017). Such a complex shape, coupled with the 52° obliquity of the comet's rotational axis, lead to strong diurnal irradiation changes between the lobes and the connecting region, as well as strong seasonal thermal changes between the northern heminucleus, which is mostly illuminated farther out from the Sun, and the southern heminucleus, which completely receives the full perihelion irradiation in only few months (*Keller et al.,* 2015a).

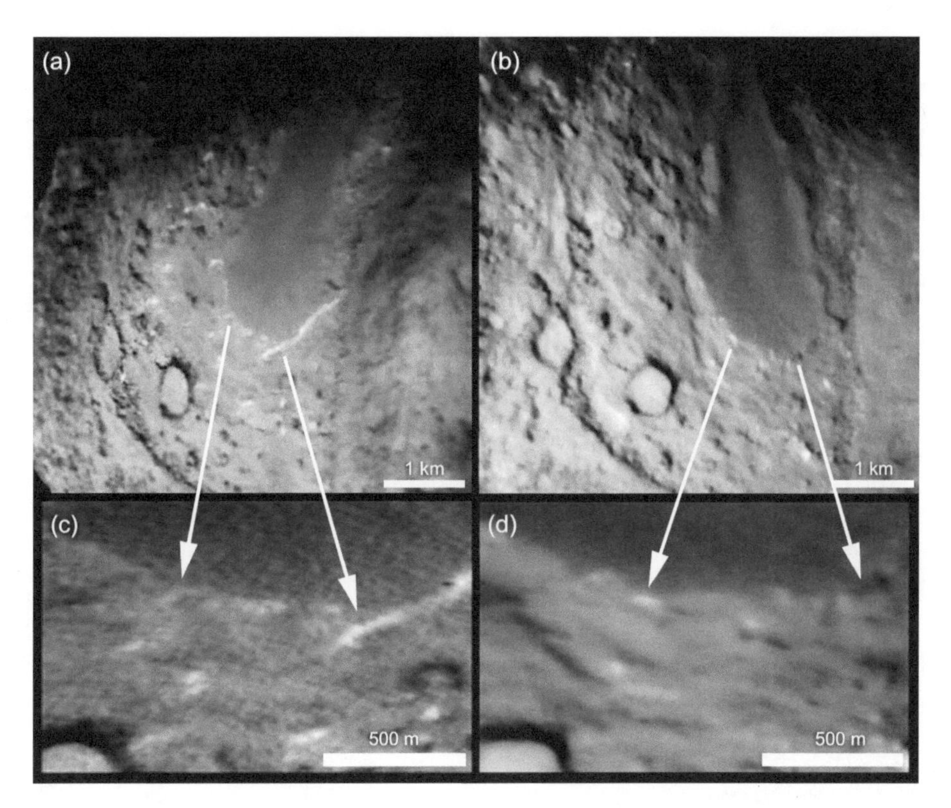

Fig. 13. (a) The scarp bounding the smooth flow as observed by Deep Impact on Comet 9P/Tempel 1. **(b)** Same as **(a)**, but as observed by Stardust-NExT ~6 years later. **(c)** Closeup view of the scarp in **(a)**. The white arrows point to two roughly triangular segments (extent of ~50 m) that disappeared in **(d)** the Stardust-NExT observation. Figure modified from *Thomas et al.* (2013a).

So far, Comet 67P is the only nucleus that has been continuously observed at different resolutions along most of its orbit, inward from ~4.4 au, reaching perihelion, and then outward up to 3.8 au. Therefore, the Rosetta instrumentation has permitted us to observe how a cometary surface seasonally evolves while approaching and leaving the Sun.

By using a two-layer thermophysical model where a thin dust layer is located on top of the cometary dust/ice matrix, *Keller et al.* (2015a) first estimated the insolation and ensuing erosion of the surface of 67P, deriving that the typical erosion caused by the water ice sublimation is 4× stronger on the southern heminucleus rather than in the northern regions (Fig. 14), with maximum, localized erosions between 15 and 20 m. Moreover, the models showed that the erosion over one cometary day could vary between a fraction of a millimeter on the northern hemisphere, when the comet is ~3 au from the Sun before perihelion, to several centimeters on the southern regions around perihelion. Based on such computations, *Keller et al.* (2015a) predicted that the heminuclei morphology would have showed important differences that accumulated over the last eight orbits since 67P's orbit changed after a close Jupiter encounter in 1959 (*Krolikowska,* 2003).

The first direct hint of activity on 67P was observed at ~3.7 au, in the form of jets that were unambiguously coming from the neck region (*Lara et al.,* 2015; *Sierks et al.,* 2015; *Lin et al.,* 2015) of the comet (Fig. 15a). As supported by the thermophysical analysis (*Sierks et al.,* 2015), in the August–December 2014 timeframe Hapi was characterized by rapid temperature and illumination changes, causing strong thermal gradients and stresses (*Ali-Lagoa et al.,* 2015) with an additional self-heating effect (*Pajola et al.,* 2019). As pointed out by *Keller et al.* (2017), the ice-rich aggregates coming from the southern heminucleus and deposited on Hapi's surface during the previous perihelion passage were "shock frozen" while entering the seasonal shadow, keeping their water ice content pristine. Shortly after aphelion, when Hapi started to be illuminated, such deposits began to sublimate, hence causing the exceptional activity level that was observed early in the mission. No morphological changes were observed during the first five months in orbit, but on 30 December 2014, at 2.66 au from the Sun, two spots appeared in the neck, later growing into a 100 × 70-m depression (Figs. 15b,c), with a depth of 0.5 m (*Davidsson et al.,* 2017; *El-Maarry et al.,* 2017).

In the following months while approaching the Sun at 1.52 au, the level of sublimation activity largely increased and it was possible to observe another morphological change. This occurred in June 2015 in the Imhotep region (*Groussin et al.,* 2015; *Auger et al.* 2015), which is a gravitational low that is almost always illuminated along 67P orbit and that receives 70–80% of the maximum energy received by the nucleus (*Keller et al.,* 2015a). In the smooth terrains of Imhotep, a collapse of 5 m thickness of the upper surface occurred in a circular fashion (with a slightly visible shallow bulging at the center of the collapse), propagating and

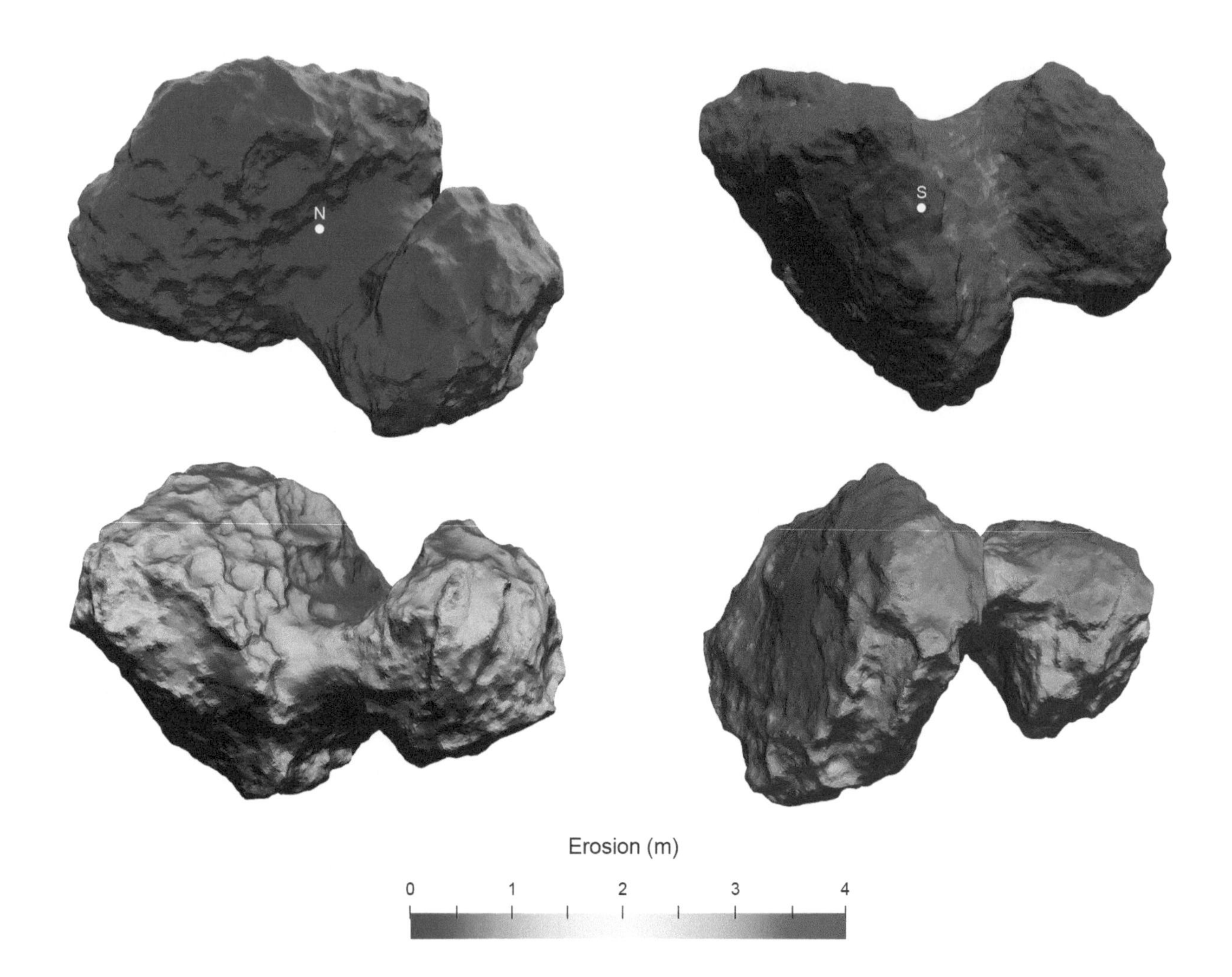

Fig. 14. The water ice sublimation based on the two-layer thermodynamic model B of *Keller et al.* (2015a).

expanding across the surface, hence exposing fresh ices on the surface in a timeframe of days. Despite the different process duration, such changes may well be interpreted as those observed on Comet 9P, i.e., the retreat of the scarps due to progressive sublimation of the ice (*Veverka et al.,* 2013).

An almost-continuous set of OSIRIS images taken around perihelion revealed also an important scarp retreat with a rate ~5.4 m per day occurring in the Anubis region, consistent with a similar range of values observed in Imhotep (*El-Maarry et al.,* 2017), hence supporting the interpretation that such a process is a common one seasonally shaping the cometary nuclei.

In addition to such changes, several other modifications of 67P surface and localized texture have been observed, all around perihelion, such as the erosional transport of dust deposits located in Imhotep, where diverse surface characterized by smooth terrains, isolated boulders, and circular features were exhumed from a depth of ~4 m (*El-Maarry et al.,* 2017). As mentioned in section 3.3, such dusty deposits have been lifted up by gas drag due to sublimation, which locally exceeded the nucleus local surficial gravity field. In one particular case, a 30-m boulder located in the Khonsu region was observed 140 m further away from its initial location (Fig. 15d) (*El-Maarry et al.,* 2017). As noted before, the extension of a preexisting >500-m-long fracture in the neck region has been observed as well, along with the possible development of new fractures parallel to it (Fig. 15e). 67P's spin rate steadily increased from May 2015 until perihelion (*Keller et al.,* 2015b), hence it is possible that such fracture extension is the result of activity-induced torques and the development of tensile stresses in the neck region. Since Churyumov-Gerasimenko is not the only bilobate comet observed so far, it is possible that the mutual readjustment due to sublimation activity of the two lobes may have seasonally developed fractures on other observed comets; nevertheless, new observations would be necessary in order to confirm such an interpretation. Cliff collapses in Aswan (*Pajola et al.,* 2017a) (see also section 3.1), as well as in the boundary between the Ma'at and Wosret regions (*El-Maarry and Driver,* 2019) and the Ash region (*Bouquety et al.,* 2021), have also been observed around perihelion. Eventually, the appearance of numerous patches resembling honeycomb structures (Fig. 15f), previously non-visible on the surface of Ma'at dust-covered terrains, showed an increase of surface roughness, with activity departing from such spots (*Hu et al.,* 2017). Such decameter-sized features

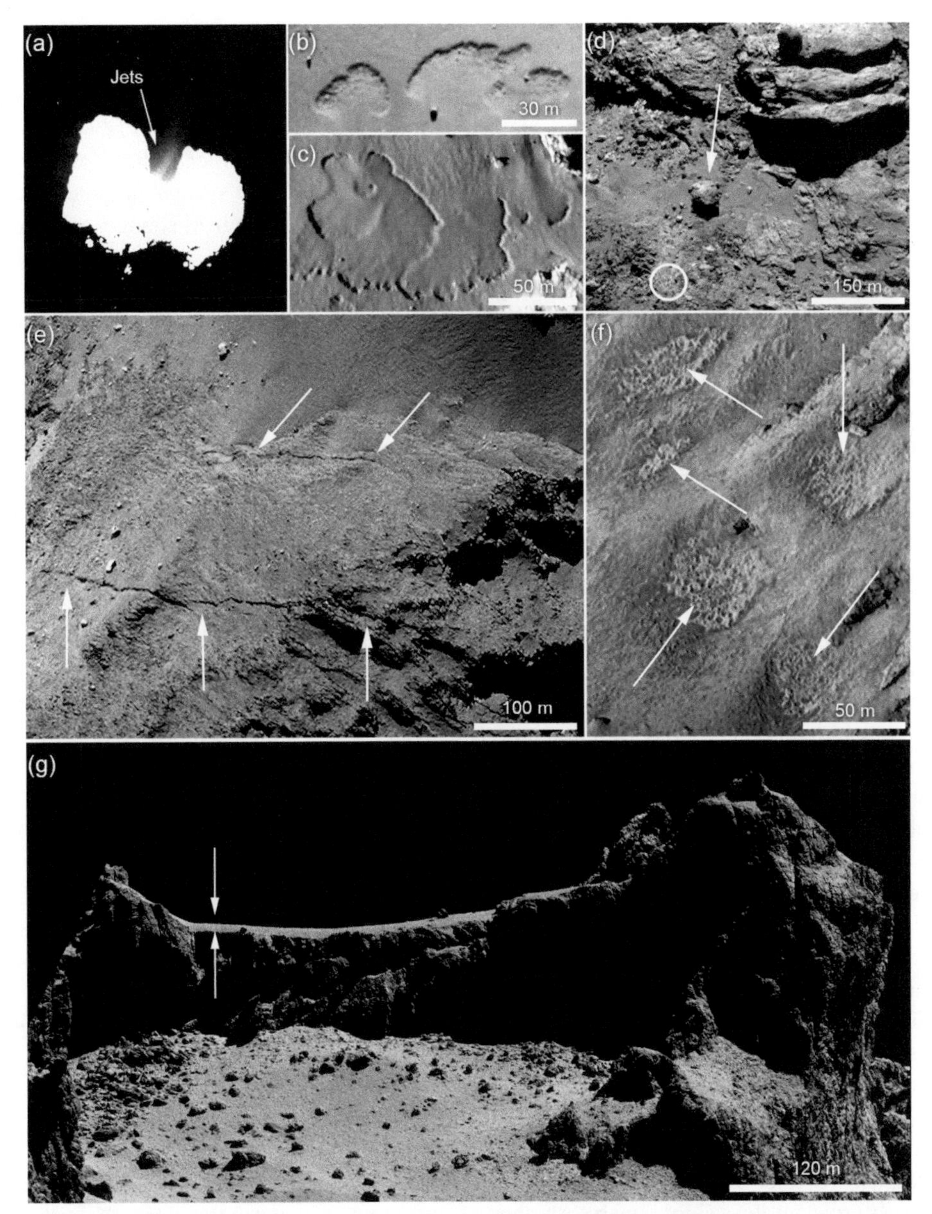

Fig. 15. (a) The jets coming from the Hapi region of 67P. **(b)** The two depressions appearing on the surface of Hapi in December 2014. **(c)** Same as **(b)**, but on January 2015. **(d)** The 30-m-size boulder located in the Khonsu region that was moved 140 m from its original location, indicated with the white circle. **(e)** The extension of the neck fracture (below), and the appearance of a new fracture parallel to it (above). **(f)** The honeycomb structures that appeared in the Ma'at region as observed in March 2015. **(g)** The meter-size thickness of a dusty deposit located in the northern heminucleus of 67P.

have never been observed on other nuclei before, most probably because they are too small to see in the datasets of the flyby missions. Nevertheless, their occurrence below dusty deposits close to perihelion suggests that their visibility may only be limited in time. This is above all true because a few months after passing perihelion, at a heliocentric distance >2 au, a general resurfacing event occurred where most of the dust and pebbles lifted from the southern hemisphere that did not reach escape velocity redeposited in the unilluminated northern heminucleus of 67P, making the pits in Hapi and Imhotep and the honeycomb in Ma'at and other regions disappear again under a blanket locally a few meters thick. This process suggests that backfall and redistribution of backfall (Fig. 15g) between the heminucleus that is illuminated during perihelion and the one that is inactive in that timeframe might be a common process occurring on all cometary nuclei and highly variable with the seasons (*Lai et al.,* 2016; *Hu et al.,* 2017; *Pajola et al.,* 2017b; *Keller et al.,* 2017).

4.2. Geological Evolution from the Surface of a Kuiper Belt Object to an Evolved Jupiter-Family Comet

In the current paradigm, most small icy bodies formed at large heliocentric distances, beyond the orbit of Neptune. After formation, gravitational perturbations induced by the giant planets of our solar system or external galactic events in the relative vicinity perturbed this population of small bodies (*Morbidelli and Nesvorný,* 2020, and references therein), scattering objects in different dynamical populations [i.e., TNOs, scattered disk objects (SDOs), Oort cloud, etc.]. The majority of these objects remain far from the Sun until further perturbations transform into elliptical orbits with a perihelion in the inner solar system (i.e., inside the orbit of Jupiter). Close encounters with Jupiter may modify the small body orbit and bring down the aphelion significantly, effectively trapping the object in a short-period orbit, within ~5 au from the Sun. Such objects are referred to as JFCs. While they represent only a minority population within the realm of comets and other SDOs, JFCs are the best-studied comets as their short orbital period and heliocentric distance range make them relatively easy to reach with a spacecraft. Indeed, all cometary missions so far have visited a short-period comet, most with an orbital period of less than 10 years with the exception of Comet 1P/Halley (~75-year period), imaged by the European Space Agency's (ESA) Giotto mission and the Soviet Vega missions in 1986, which can nonetheless be considered a short-period comet for the purpose of this discussion. This means that all observations of cometary surfaces reported in this chapter describe morphologies that have been affected by significant exposure to solar insolation. As detailed above, the heating drives erosion by sublimating volatile material present in the body. Such erosion can be gentle and locally remove a few tens of centimeters in each orbit, or it can lead to more dramatic changes with sudden outbursts carving holes in scales of tens to hundreds of meters or even the entire surface [e.g., Comet 17P/Holmes (*Montalto et al.,* 2007)]. By contrast, objects that remain far from the Sun do not experience such erosion and their surface may evolve on completely different timescales than JFCs. This was confirmed by NASA's flyby of (486958) Arrokoth with the New Horizons mission in 2019 (*Stern et al.,* 2019). Indeed, at comparable spatial resolutions, imaging data revealed a surface much smoother than what had been observed on JFCs (Figs. 16a,b). Another important observation at Arrokoth was the apparent paucity of features of unambiguous impact origin with only a few circular depressions that could be attributed to impacts or possibly sublimation events (*Stern et al.,* 2019; *Spencer et al.,* 2020). These observations confirmed the inferred low impact probability in the Kuiper belt due to the lower population densities (e.g., *Singer et al.,* 2019; *Morbidelli et al.,* 2021), and further indicate that KBOs, or bodies sourced from the outer solar system in general, can only undergo substantial morphological changes when pushed into orbits closer to the Sun.

Therefore, we currently have data from the end members of cometary surface evolution: a frozen, mostly pristine KBO, and highly eroded JFCs. While this provides reference points for modeling evolutionary processes, the timeline of surface transformations remains challenging to establish. One particular area of interest is the transition period from KBO to JFC, during which the comet perihelion and aphelion are brought within the inner solar system. This transition is colloquially known as the Centaur phase, Centaurs being dynamically unstable objects with orbits in between the outer planets (*Di Sisto and Brunini,* 2007). Evolutionary studies looking at the transformations taking place on short-period comets with different dynamical age (number of orbits in the inner solar system) show that erosion mostly driven by the sublimation of water ice tends to smoothen the topography by collapsing large features into boulders, pebbles, and dust (*Vincent et al.,* 2020; *Birch et al.,* 2017). While this explains the current evolution, such erosion cannot reproduce the larger topographic changes such as deep pits and tall cliffs observed on JFCs (*El-Maarry et al.,* 2017). Because such features are not detected on KBOs either, it is assumed that significant surface transformations must have taken place in the transition period. The uneven spatial distribution of pits and cliffs, often found in clusters, argues in favor of transformations driven by localized pockets of supervolatile material [e.g., CO or CO_2 ice (*Filacchione et al.,* 2016b)] rather than impact events. It is postulated that such material would rapidly sublimate in the Centaur phase, as the comet orbit semimajor axis is lowered, leading to massive outbursts that can transform the surface on a large scale. The fact that many centaurs are known to outburst regularly [e.g., 29P/Schwassmann-Wachmann 1 (*Trigo-Rodríguez et al.,* 2008)] supports this hypothesis. However, it should be

Fig. 16. (a) (486958) Arrokoth imaged by the New Horizons spacecraft. **(b)** Comet 67P/Churyumov-Gerasimenko imaged by Rosetta. Objects are not to scale; Arrokoth is about 6× larger than 67P. Although Arrokoth data has lower spatial scale (30 m px^{-1} vs. 5 m px^{-1} for the 67P image shown here), the rugged comet morphology displays erosional morphologies at scales that would be easily detected on Arrokoth, should such features exist on a KBO (cliffs or pits hundreds of meters wide).

noted that cometary evolution is an active area of research with no clear conclusion at the time of the writing of this chapter. Future missions like ESA's Comet Interceptor (*Snodgrass and Jones,* 2019), to be launched in 2029 with the intent of investigating a long-period comet, preferably a dynamically new one entering the inner solar system for the first time, are indeed necessary to constrain the evolutionary processes taking place in that critical phase of a comet's life.

5. CONCLUSIONS AND OPEN QUESTIONS — OUTLOOK FOR FUTURE MISSIONS

The six visited cometary nuclei were revealed to be among the most geologically active bodies in the solar system. *In situ* observations have revealed the occurrence of multiple physical processes that act on the cometary surfaces, hence modifying their landscape, such as pit formation, crack development, scarp retreat, cliff collapses, or redistribution of backfall material previously ejected through outbursts. In addition, the variability of morphological landforms detected on both individual bodies and among different nuclei have widened our perspective of comets at large. Nevertheless, the wealth of existing observations has led us to answering some of the questions we previously had on the morphology of such bodies, how cometary surfaces might evolve on their approach to perihelion, or if dust exchange occurs between the illuminated or unilluminated parts of the nucleus, but has definitely raised a series of new questions, well establishing how enigmatic these intriguing worlds are.

Most of the observed nuclei exhibit bilobate shapes, hence providing observables to test different cometary formation hypotheses, such as the slow primordial growth and merging of such bodies, vs. the reaccumulation process resulting from the collisional disruption of larger nuclei. Moreover, layering has been considered as a pervasive characteristic of comets and, in particular, a widespread structural aspect of Comet 67P. Coupled with the bilobate shape, this has helped to suggest the formation and evolution of such bodies; nevertheless, definitive proof about the origin of such layers is still lacking.

It is clear that the occurrence of various morphological landforms on several scales, such as flows on 9P or flat floor depressions on 81P, differ from one comet to another. However, the reason why certain geological processes act on some comets and not on others remains unclear. It is indeed well known that insolation, which depends on a comet's rotation, pole position, and topography, is expected to be the major contributor in modifying the cometary geology landscape on diurnal, seasonal, and orbital timescales. Specifically, the sublimation driven by insolation has been considered as the source of several cliff collapses and mass-wasting events documented on comets, such as 9P and 67P. However, the link between such a process and the roughness of surfaces, the local heating, and its association with outgassing activity is not yet fully understood.

Another open question about the nuclei surface is the occurrence of ubiquitous fracturing at all scales, as clearly visible on the surface of 67P, as a result of the thermal stresses affecting such bodies. More experiments are definitely needed to fully understand the control on fracturing mechanisms in cometary analog materials in space conditions.

Finally, surface changes on orbital and seasonal timescales, such as those observed on 9P and 67P respectively, have allowed us to quantify and describe the surface evolution and should provide information about the various process and timescales acting on such peculiar bodies. However, the physical process that govern and control these changes merit further exploration.

Unveiling the origin and evolution of cometary surfaces will require further synthesis of existing data acquired by previous missions, coupled with new cometary missions that will widen and deepen our knowledge of cometary surfaces. Questions about the evolutionary or primordial origin of layers; the occurrence of specific morphological structures on some comets and not on others (even similar); the quantification and modeling of active geological processes; the connection between geological processes and activity; the processes controlling the seasonal and orbital changes; and the link between insolation, erosion, and sublimation require additional investigations on both visited and yet-to-be-explored bodies. Specifically, high-resolution images acquired during the entire orbit of a comet will be pivotal to assess the morphology and changes happening on a comet, as largely demonstrated by Rosetta on 67P, coupled with observations at different orbits that will help us in understanding comet evolution at large. Furthermore, it is evident that with comets having gone through nearly all stages of space exploration encompassing remote observation, close flybys, orbiting, and landing [we recall that the work done by the Philae lander was necessarily not effective as planned (*Boehnhardt et al.,* 2017), hence the possibility of a new landing and/or roving with the main aim of onsite analyses still remains an important task to be accomplished], the natural progressive step in our exploration would need to be sample return. Samples [preferably with the means to cryogenically store and deliver them to Earth in a pristine condition (e.g., *Hayes et al.,* 2018; *Bockelée-Morvan et al.,* 2021)] can yield accurate information about the mechanical and chemical characteristics of cometary materials. This phase has already been achieved for the Moon (e.g., *Che et al.,* 2021) and near-Earth asteroids [e.g., Itokawa (*Chan et al.,* 2021), Ryugu (*Morota et al.,* 2020), and Bennu (*Lauretta et al.,* 2024)], and there are plans to progress in that direction on Mars. It is clear, and almost inevitable, that we should reach this stage with cometary missions too.

Beyond sample return, another area of exploration that remains to be pursued is the *in situ* investigation of cometary objects beyond JFCs. Indeed, in order to constrain the timeline of physical processes that transform cometary nuclei, is it necessary to visit such targets that are entering the inner solar system for the first time (i.e., ESA's Comet Interceptor mission) or that are currently in their Centaur phase: transitioning from a remote orbit in the Kuiper belt or beyond to a perihelion distance between Jupiter and Neptune. This would give us the opportunity to study activity and evolution driven mostly by the sublimation of supervolatile elements rather than water. Several known Centaurs (2060 Chiron, 60558 Echeclus, and 29P/Schwassmann-Wachmann 1) have detectable emission of gas and dust (*Bus et al.,* 1991; *Wierzchos et al.,* 2017), with 29P known for outbursting regularly while orbiting around 6 au (*Trigo-Rodríguez et al.,* 2010), while others like 10199 Chariklo have a system of icy rings (*Braga-Ribas et al.,* 2014) that may result from large-scale resurfacing events. On the other side of evolution, it would also be valuable to consider missions to main-belt comets: objects in asteroid-like orbits that display a dust tail but no gas coma [within detection limits (*Snodgrass et al.,* 2017)]. Such targets could be depleted comets, or a new class of objects in the small-bodies continuum. As of 2024, the Japan Aerospace Exploration Agency's (JAXA) Destiny+ mission is scheduled to visit the active asteroid 3200 Phaeton in 2028 (*Arai et al.,* 2018), and the China National Space Administration's (CNSA) ZhengHe mission will rendezvous with the main-belt comet 133P/Elst-Pizarro (*Zhang et al.,* 2019) in the coming decade.

Acknowledgments. The authors are grateful to W.-H. Ip and M. Massironi for the constructive comments and suggestions that highly improved the quality of the chapter. We also thank L. Penasa for preparing Fig. 5e. This work is dedicated to all doctors who gave and are still giving their energy, time, and lives during the ongoing COVID-19 pandemic.

REFERENCES

Agarwal J., Della Corte V., Feldman P. D., et al. (2017) Evidence of sub-surface energy storage in Comet 67P from the outburst of 2016 July 03. *Mon. Not. R. Astron. Soc., 469,* S606–S625.

A'Hearn M. F., Belton M. J. S., Delamere W. A., et al. (2005) Deep Impact: Excavating Comet Tempel 1. *Science, 310,* 258–264.

A'Hearn M. F., Belton M. J. S., Delamere W. A., et al. (2011) EPOXI at Comet Hartley 2. *Science, 332,* 1396–1400.

Ali-Lagoa V., Delbo M., and Libourel G. (2015) Rapid temperature changes and the early activity on Comet 67P/Churyumov-Gerasimenko. *Astrophys. J. Lett., 810,* L22.

Altenhoff W. J., Kreysa E., Menten K. M., et al. (2009) Why did Comet 17P/Holmes burst out? Nucleus splitting or delayed submilation? *Astron. Astrophys., 495,* 975–978.

Arai T., Kobayashi M., Ishibashi K., et al. (2018) Destiny+ mission: Flyby of geminids parent asteroid (3200) Phaethon and in-situ analyses of dust accreting on the Earth. In *49th Lunar and Planetary Science Conference,* Abstract #2570. LPI Contribution No. 2083, Lunar and Planetary Institute, Houston.

Auger A. T., Groussin O., Jorda L., et al. (2015) Geomorphology of the Imhotep region on Comet 67P/Churyumov-Gerasimenko from OSIRIS observations. *Astron. Astrophys., 583,* A35.

Auger A. T., Groussin O., Jorda L., et al. (2018) Meter-scale thermal contraction crack polygons on the nucleus of Comet 67P/Churyumov-Gerasimenko. *Icarus, 301,* 173–188.

Barucci M. A., Filacchione G., Fornasier S., et al. (2016) Detection of exposed H_2O ice on the nucleus of Comet 67P/Churyumov-Gerasimenko, as observed by Rosetta OSIRIS and VIRTIS instruments. *Astron. Astrophys., 595,* A102.

Belton M. J. and Melosh J. (2009) Fluidization and multiphase transport of particulate cometary material as an explanation of the smooth terrains and repetitive outbursts on 9P/Tempel 1. *Icarus, 200,* 280–291.

Belton M. J., Thomas P., Veverka J., et al. (2007) The internal structure of Jupiter family cometary nuclei from Deep Impact observations: The "talps" or "layered pile" model. *Icarus, 187,* 332–344.

Belton M. J., Thomas P., Carcich B., et al. (2013) The origin of pits on 9P/Tempel 1 and the geologic signature of outbursts in Stardust-NExT images. *Icarus, 222,* 477–486.

Bertaux J.-L., Combi M. R., Quémerais E., et al. (2014) The water production rate of Rosetta target Comet 67P/Churyumov-Gerasimenko near perihelion in 1996, 2002 and 2009 from Lyman α observations with SWAN/SOHO. *Planet. Space Sci., 91,* 14–19.

Bertini I., Lara L. M., Vincent J. B., et al. (2009) Activity evolution, outbursts, and splitting events of Comet 73P/Schwassmann-Wachmann 3. *Astron. Astrophys., 496,* 235–247.

Bertini I., Gutiérrez P. J., Lara L. M., et al. (2015) Search for satellites near Comet 67P/Churyumov-Gerasimenko using Rosetta/OSIRIS images. *Astron. Astrophys., 583,* A19.

Birch S., Tang Y., Hayes A. G., et al. (2017) Geomorphology of Comet 67P/Churyumov-Gerasimenko, *Mon. Not. R. Astron. Soc., 469,* S50–S67.

Bockelée-Morvan D., Filacchione G., Altwegg K., et al. (2021) AMBITION — Comet nucleus cryogenic sample return. *Exp. Astron.,* DOI: 10.1007/s10686-021-09770-4.

Bodewits D., Farnham T. L., Kelley M. S. P., et al. (2018) A rapid decrease in the rotation rate of Comet 41P/Tuttle-Giacobini-Kresák. *Nature, 553,* 186–189.

Boehnhardt H. (2004) Split comets. In *Comets II* (M. C. Festou et al., eds.), pp. 301–316. Univ. of Arizona, Tucson.

Boehnhardt H., Bibring J.-B., Apathy I., et al. (2017) The Philae lander mission and science overview. *Phil. Trans. R. Soc. A, 375,* 20160248.

Bouquety A., Jorda L., Groussin O., et al. (2021) Ancient and present surface evolution processes in the Ash region of Comet 67P/Churyumov-Gerasimenko. *Astron. Astrophys., 649,* A82.

Braga-Ribas F., Sicardy B., Ortiz J. L., et al. (2014) A ring system detected around the Centaur (10199) Chariklo. *Nature, 508,* 72–75.

Britt D. T., Boice D. C., Buratti B. J., et al. (2004) The morphology and surface processes of Comet 19/P Borrelly. *Icarus, 167,* 45–53.

Brownlee D. E., Horz F., Newburn R. L., et al. (2004) Surface of young Jupiter family Comet 81P/Wild 2: View from the Stardust spacecraft. *Science, 304,* 1764–1768.

Bruck Syal M., Schultz P. H., Sunshine J. M., et al. (2013) Geologic control of jet formation on Comet 103P/Hartley 2. *Icarus, 222,* 610–624.

Buratti B. J., Hicks M. D., Soderblom L. A., et al. (2004) Deep Space 1 photometry of the nucleus of Comet 19P/Borrelly. *Icarus, 167,* 16–29.

Bus S. J., A'Hearn M. F., Schleicher D. G., et al. (1991) Detection of CN emission from (2060) Chiron. *Science, 251,* 774–777.

Cambianica P., Cremonese G., Naletto G., et al. (2019) Quantitative analysis of isolated boulder fields on Comet 67P/Churyumov-Gerasimenko. *Astron. Astrophys., 630,* A15.

Cambianica P., Cremonese G., Fulle M., et al. (2021) Long-term measurements of the erosion and accretion of dust deposits on Comet 67P/Churyumov-Gerasimenko with the OSIRIS instrument. *Mon. Not. R. Astron. Soc., 504,* 2895–2910.

Chan Q. H. S., Stephant A., Franchi I. A., et al. (2021) Organic matter and water from asteroid Itokawa. *Sci. Rept., 11,* 5125.

Che X., Nemchin A., Liu D., et al. (2021) Age and composition of young basalts on the Moon, measured from samples returned by Chang'e-5. *Science, 374,* 887–890.

Davidsson B. J. R., Sierks H., Güttler C., et al. (2016) The primordial nucleus of Comet 67P/Churyumov-Gerasimenko. *Astron. Astrophys., 592,* A63.

Davidsson B. J. R., Seungwon L., von Allmen P., et al. (2017) Large scale morphological changes in the Hapi region on Comet 67P/Churyumov-Gerasimenko. *AAS/Division for Planetary Sciences Meeting Abstracts, 49,* #415.02.

De Sanctis M. C., Capaccioni F., Ciarniello M., et al. (2015) The diurnal cycle of water ice on Comet 67P/Churyumov-Gerasimenko. *Nature, 525,* 500–503.

Di Sisto R. P. and Brunini A. (2007) The origin and distribution of the Centaur population. *Icarus, 190,* 224–235.

Duncan M. J. and Levison H. F. (1997) A scattered comet disk and the origin of Jupiter family comets. *Science, 276,* 1670–1672.

Duxbury T. C., Newburn R. L., and Brownlee D. E. (2004) Comet 81P/Wild 2 size, shape, and orientation. *J. Geophys. Res.–Planets, 109,* E12S02.

El-Maarry M. and Driver G. (2019) Cliff collapses on Comet 67P/Churyumov-Gerasimenko following outbursts as observed by the

Rosetta mission. *EPSC-DPS Joint Meeting 2019,* 1727.
El-Maarry M. R., Thomas N., Giacomini L., et al. (2015a) Regional surface morphology of Comet 67P/Churyumov-Gerasimenko from Rosetta/OSIRIS images. *Astron. Astrophys., 583,* A26.
El-Maarry M. R., Thomas N., Gracia-Berná A., et al. (2015b) Fractures on Comet 67P/Churyumov-Gerasimenko observed by Rosetta/OSIRIS. *Geophys. Res. Lett., 42,* 5170–5178.
El-Maarry M. R., Thomas N., Giacomini L., et al. (2016) Regional surface morphology of Comet 67P/Churyumov-Gerasimenko from Rosetta/OSIRIS images: The southern hemisphere. *Astron. Astrophys.,593,* A110.
El-Maarry M. R., Groussin O., Thomas N., et al. (2017) Surface changes on Comet 67P/Churyumov-Gerasimenko suggest a more active past. *Science, 355,* 1392–1395.
El-Maarry M. R., Groussin O., Keller H. U., et al. (2019) Surface morphology of comets and associated evolutionary processes: A review of Rosetta's observations of 67P/Churyumov-Gerasimenko. *Space Sci. Rev., 215,* 36.
Farnham T. L. and Thomas P. C. (2013a) *Plate Shape Model of Comet 9P/Tempel 1 V2.0.* NASA Planetary Data System, DIF-C-HRIV/ITS/MRI-5-TEMPEL1-SHAPE-MODEL-V2.0.
Farnham T. L. and Thomas P. C. (2013b) *Plate Shape Model of Comet 103P/Hartley 2 V1.0.* NASA Planetary Data System, DIF-C-HRIV/MRI-5-HARTLEY2-SHAPE-V1.0.
Farnham T. L., Duxbury T., and Li J.-Y. (2005) *Shape Models of Comet Wild 2.* NASA Planetary Data System, SDU-C-NAVCAM-5-WILD2-SHAPE-MODEL-V2.1.
Farnham T. L., Wellnitz D. D., Hampton D. L, et al. (2007) Dust coma morphology in the Deep Impact images of Comet 9P/Tempel 1. *Icarus, 187,* 26–40.
Farnham T. L., Bodewits D., Li J. Y., et al. (2013c) Connections between the jet activity and surface features on Comet 9P/Tempel 1. *Icarus, 222,* 540–549.
Feller C., Fornasier S., Hasselmann P. H., et al. (2016) Decimetre-scaled spectrophotometric properties of the nucleus of Comet 67P/Churyumov-Gerasimenko from OSIRIS observations. *Mon. Not. R. Astron. Soc., 462,* S287–S303.
Filacchione G., De Sanctis M. C., Capaccioni F., et al. (2016a) Exposed water ice on the nucleus of Comet 67P/Churyumov-Gerasimenko. *Nature, 529,* 368–372.
Filacchione G., Raponi A., Capaccioni F., et al. (2016b) Seasonal exposure of carbon dioxide ice on the nucleus of Comet 67P/Churyumov-Gerasimenko. *Science, 354,* 1563–1566.
Fornasier S., Feller C., Hasselmann P. H., et al. (2019) Surface evolution of the Anhur region on Comet 67P/Churyumov-Gerasimenko from high-resolution OSIRIS images. *Astron. Astrophys., 630,* A13.
Franceschi M., Penasa L., Massironi M., et al. (2020) Global-scale brittle plastic rheology at the cometesimals merging of Comet 67P/Churyumov-Gerasimenko. *Proc. Natl. Acad. Sci. U.S.A., 117,* 10181–10187.
Fulle M. (1997) Injection of large grains into orbits around comet nuclei. *Astron. Astrophys., 325,* 1237–1248.
Fulle M., Marzari M., Della Corte V., et al. (2016) Evolution of the dust size distribution of Comet 67P/Churyumov-Gerasimenko from 2.2 au to perihelion. *Astrophys. J., 821,* 19.
Giacomini L., Massironi M., El-Maarry M. R., et al. (2016) Geologic mapping of the Comet 67P/Churyumov-Gerasimenko's northern hemisphere. *Mon. Not. R. Astron. Soc., 462,* S352–S367.
Glassmeier K.-H., Boehnhardt H., Koschny D., et al. (2007) The Rosetta mission: Flying towards the origin of the solar system. *Space Sci. Rev., 128,* 1–21.
Groussin O., Sierks H., Barbieri C., et al. (2015) Temporal morphological changes in the Imhotep region of Comet 67P/Churyumov-Gerasimenko. *Astron. Astrophys., 583,* A36.
Groussin O., Attree N., Brouet Y., et al. (2019) The thermal, mechanical, structural, and dielectric properties of cometary nuclei after Rosetta. *Space Sci. Rev., 215,* 29.
Grün E., Agarwal J., Altobelli N, et al. (2016) The 2016 Feb 19 outburst of Comet 67P/CG: An ESA Rosetta multi-instrument study. *Mon. Not. R. Astron. Soc., 462,* S220–S234.
Gulkis S., Allen M., von Allmen, P., et al. (2015) Subsurface properties and early activity of Comet 67P/Churyumov-Gerasimenko. *Science, 347,* aaa0709.
Harmon J. K., Nolan M. C., Giorgini J. D., et al. (2010) Radar observations of 8P/Tuttle: A contact-binary comet. *Icarus, 207,* 499–502.
Hässig M., Altwegg K., Balsiger H., et al. (2015) Time variability and heterogeneity in the coma of 67P/Churyumov-Gerasimenko. *Science, 347,* aaa0276.
Hayes A., Glavin D., Lauretta D., et al. (2018) The CAESAR New Frontiers mission. *42nd COSPAR Scientific Assembly,* B1.1-9-18.
Hirabayashi M., Scheeres D. J., Chelsey S. R., et al. (2016) Fission and reconfiguration of bilobate comets as revealed by 67P/Churyumov-Gerasimenko. *Nature, 534,* 352–355.
Höfner S., Vincent J. B., Blum J., et al. (2017) Thermophysics of fractures on Comet 67P/Churyumov-Gerasimenko. *Astron. Astrophys., 608,* A121.
Hu X., Shi X., Sierks H., et al. (2017) Seasonal erosion and restoration of the dust cover on Comet 67P/Churyumov-Gerasimenko as observed by OSIRIS onboard Rosetta. *Astron. Astrophys., 604,* A114.
Ip W.-H., Lai I.-L., Lee J.-C., et al. (2015) Physical properties and dynamical relation of the circular depressions on Comet 67P/Churyumov-Gerasimenko. *Astron. Astrophys., 591,* A132.
Jia P., Andreotti B., and Claudin P. (2017) Giant ripples on Comet 67P/Churyumov-Gerasimenko sculpted by sunset thermal wind. *Proc. Natl. Acad. Sci. U.S.A., 114,* 2509–2514.
Jorda L., Gaskell R., Capanna C., et al. (2016) The global shape, density and rotation of Comet 67P/Churyumov-Gerasimenko from preperihelion Rosetta/OSIRIS observations. *Icarus, 277,* 257–278.
Keller H. U., Arpigny C., Barbieri C., et al. (1986) First Halley Multicolour Camera imaging results from Giotto. *Nature, 321,* 320–326.
Keller H. U., Delamere W. A., Huebner W. F., et al. (1987) Comet P/Halley's nucleus and its activity. *Astron. Astrophys., 187,* 807–823.
Keller H. U., Kramm R., and Thomas N. (1988) Surface features on the nucleus of Comet Halley. *Nature, 331,* 227–231.
Keller H. U., Britt D., Buratti B. J., et al. (2004) In situ observations of cometary nuclei. In *Comets II* (M. C. Festou et al., eds.), pp. 211–222. Univ. of Arizona, Tucson.
Keller H. U., Barbieri C., Lamy P., et al. (2007) OSIRIS — The scientific camera system onboard Rosetta. *Space Sci. Rev., 128,* 433–506.
Keller H. U., Mottola S., Davidsson B., et al. (2015a) Insolation, erosion, and morphology of Comet 67P/Churyumov-Gerasimenko. *Astron. Astrophys., 583,* A34.
Keller H. U., Mottola S., Skorov Y., et al. (2015b) The changing rotation period of Comet 67P/Churyumov-Gerasimenko controlled by its activity. *Astron. Astrophys., 579,* L5.
Keller H. U., Mottola S., Hviid S. F., et al. (2017) Seasonal mass transfer on the nucleus of Comet 67P/Churyumov-Gerasimenko controlled by its activity. *Mon. Not. R. Astron. Soc., 469,* S357–S371.
Kelley M. S., Lindler D. J., Bodewits D., et al. (2013) A distribution of large particles in the coma of Comet 103P/Hartley 2. *Icarus, 222,* 634–652.
Kirk R. L., Duxbury T. C., Horz F., et al. (2005) Topography of the 81P/Wild 2 nucleus derived from Stardust stereoimages. In *36th Lunar and Planetary Science Conference,* Abstract #2244. LPI Contribution No. 1234, Lunar and Planetary Institute, Houston.
Klaasen K., A'Hearn M. F., Baca M., et al. (2008) Deep Impact instrument calibration. *Rev. Sci. Instrum., 79,* 77.
Knollenberg J., Lin Z. Y., Hviid S. F., et al. (2016) A mini outburst from the nightside of Comet 67P/Churyumov-Gerasimenko observed by the OSIRIS camera on Rosetta. *Astron. Astrophys., 569,* A89.
Krolikowska M. (2003) 67P/Churyumov-Gerasimenko — Potential target for the Rosetta mission. *Acta Astron., 53,* 195–209.
Kührt E., Knollenberg J., and Keller H. U. (1997) Physical risks of landing on a cometary nucleus. *Planet. Space Sci., 45,* 665–680.
La Forgia F., Giacomini L., Lazzarin M., et al. (2015) Geomorphology and spectrophotometry of Philae's landing site on Comet 67P/Churyumov-Gerasimenko. *Astron. Astrophys., 583,* A41.
Lai I-L., Ip W.-H., Su C.-C., et al. (2016) Gas outflow and dust transport of Comet 67P/Churyumov-Gerasimenko. *Mon. Not. R. Astron. Soc., 462,* S533–S546.
Lamy P. L., Toth I., Fernández Y. R., et al. (2004) The sizes, shapes, albedos, and colors of cometary nuclei. In *Comets II* (M. C. Festou et al., eds.), pp. 223–264. Univ. of Arizona, Tucson.
Lara L. M., Lowry S., Vincent J. B., et al. (2015) Large-scale dust jets in the coma of 67P/Churyumov-Gerasimenko as seen by the OSIRIS instrument onboard Rosetta. *Astron. Astrophys., 583,* A9.
Lauretta D. S., Connolly H. C. Jr., Aebersold J. E., et al. (2024) Asteroid (101955) Bennu in the laboratory: Properties of the sample collected

by OSIRIS-REx. *Meteorit. Planet. Sci., 59,* 2453–2486. DOI: 10.1111/maps.14227.

Lee J.-C., Massironi M., Ip W.-H., et al. (2016) Geologic mapping of the Comet 67P/Churyumov-Gerasimenko's southern hemisphere. *Mon. Not. R. Astron. Soc., 462,* S573–S592.

Lin Z.-Y., Ip W.-H., Lai I.-L., et al. (2015) Morphology and dynamics of the jets of Comet 67P/Churyumov-Gerasimenko: Early-phase development. *Astron. Astrophys., 583,* A11.

Lisse C. M., Fernández Y. R., Reach W. T., et al. (2009) Spitzer Space Telescope observations of the nucleus of Comet 103P/Hartley 2. *Publ. Astron. Soc. Pac., 121,* 968–975.

Lucchetti A., Cremonese G., Jorda L., et al. (2016) Characterization of the Abydos region through OSIRIS high-resolution images in support of CIVA measurements. *Astron. Astrophys., 585,* L1.

Lucchetti A., Pajola M., Fornasier S., et al. (2017) Geomorphological and spectrophotometric analysis of Seth's circular niches on Comet 67P/Churyumov-Gerasimenko using OSIRIS images. *Mon. Not. R. Astron. Soc., 469,* S238–S251.

Lucchetti A., Penasa L., Pajola M., et al. (2019) The rocky-like behavior of cometary landslides on 67P/Churyumov-Gerasimenko. *Geophys. Res. Lett., 46,* 14336–14346.

Lyttleton R. A. (1948) On the origin of comets. *Mon. Not. R. Astron. Soc., 108,* 465–475.

Massironi M., Simioni E., Marzari F., et al. (2015) Two independent and primitive envelopes of the bilobate nucleus of Comet 67P. *Nature, 526,* 402–405.

Matonti C., Attree N., Groussin O., et al. (2019) Bilobate comet morphology and internal structure controlled by shear deformation. *Nature Geosci., 12,* 157–162.

Möhlmann D. (1994) Surface regolith and environment of comets. *Planet. Space Sci., 42,* 933–937.

Montalto M., Riffeser A., Hopp U., et al. (2007) The Comet 17P/Holmes 2007 outburst: The early motion of the outburst material. *Astron. Astrophys., 479,* L45–L49.

Morbidelli A. and Nesvorný D. (2020) Kuiper belt: Formation and evolution. In *The Trans-Neptunian Solar System* (D. Prialnik et al., eds.), pp. 25–59. Elsevier, Amsterdam.

Morbidelli A., Nesvorný D., Bottke W. F., et al. (2021) A re-assessment of the Kuiper belt size distribution for sub-kilometer objects, revealing collisional equilibrium at small sizes. *Icarus, 356,* 114256.

Morota T., Sugita S., Cho Y., et al. (2020) Sample collection from asteroid (162173) Ryugu by Hayabusa2: Implications for surface evolution. *Science, 368,* 654–659.

Mottola S., Arnold G., Grothues H.-G., et al. (2015) The structure of the regolith on 67P/Churyumov-Gerasimenko from ROLIS descent imaging. *Science, 349,* aab0232.

Nesvorný D., Youdin A. N., and Richardson D. C. (2010) Formation of Kuiper belt binaries by gravitational collapse. *Astron. J., 140,* 785–793.

Oklay N. and the OSIRIS Team (2018) Large sub-surface volatile reservoirs of Comet 67P. In *49th Lunar and Planetary Science Conference*, Abstract #1282. LPI Contribution No. 2083, Lunar and Planetary Institute, Houston.

Oort J. H. (1950) The structure of the cloud of comets surrounding the solar system and a hypothesis concerning its origin. *Bull. Astron. Inst. Netherlands, 11,* 91–110.

Pajola M., Vincent J.-B., Güttler C., et al. (2015) Size-frequency distribution of boulders ≥7 m on Comet 67P/Churyumov-Gerasimenko. *Astron. Astrophys., 583,* A37.

Pajola M., Lucchetti A., Bertini I., et al. (2016c) Size-frequency distribution of boulders ≥10 m on Comet 103P/Hartley 2. *Astron. Astrophys., 585,* A85.

Pajola M., Lucchetti A., Bertini I., et al. (2016d) The southern hemisphere of 67P/Churyumov-Gerasimenko: Analysis of the preperihelion size-frequency distribution of boulders ≥7 m. *Astron. Astrophys., 592,* L2.

Pajola M., Lucchetti A., Fulle M., et al. (2016a) The Agilkia boulders/pebbles size-frequency distributions: OSIRIS and ROLIS joint observations of 67P surface. *Mon. Not. R. Astron. Soc., 462,* S242–S252.

Pajola M., Oklay N., La Forgia F., et al. (2016b) Aswan site on Comet 67P/Churyumov-Gerasimenko: Morphology, boulder evolution, and spectrophotometry. *Astron. Astrophys., 592,* A69.

Pajola M., Höfner S., Vincent J. B., et al. (2017a) The pristine interior of Comet 67P revealed by the combined Aswan outburst and cliff collapse. *Nature Astron., 1,* 0092.

Pajola M., Lucchetti A., Fulle M., et al. (2017b) The pebbles/boulders size distributions on Sais: Rosetta's final landing site on Comet 67P/Churyumov-Gerasimenko. *Mon. Not. R. Astron. Soc., 469,* S636–S645.

Pajola M., Lee J.-C., Oklay N., et al. (2019) Multidisciplinary analysis of the Hapi region located on Comet 67P/Churyumov-Gerasimenko. *Mon. Not. R. Astron. Soc., 485,* 2139–2154.

Pajola M., Lucchetti A., Senter L., et al. (2021) Blocks size frequency distribution in the Enceladus tiger stripes area: Implications on their formative processes. *Universe, 7,* 82.

Penasa L., Massironi M., Naletto G., et al. (2017) A three-dimensional modelling of the layered structure of Comet 67P/Churyumov-Gerasimenko. *Mon. Not. R. Astron. Soc., 469,* S741–S754.

Pommerol A., Thomas N., El-Maarry M. R., et al. (2015) OSIRIS observations of meter-sized exposures of H_2O ice at the surface of 67P/Churyumov-Gerasimenko and interpretation using laboratory experiments. *Astron. Astrophys., 583,* A25.

Preusker F., Scholten F., Matz K.-D., et al. (2015) Shape model, reference system definition, and cartographic mapping standards for Comet 67P/Churyumov-Gerasimenko — Stereo-photogrammetric analysis of Rosetta/OSIRIS image data. *Astron. Astrophys., 583,* A33.

Preusker F., Scholten F., Matz K.-D., et al. (2017) The global meter-level shape model of Comet 67P/Churyumov-Gerasimenko. *Astron. Astrophys., 607,* L1.

Richardson J. E. and Melosh H. J. (2013) An examination of the Deep Impact collision site on Comet Tempel 1 via Stardust-NExT: Placing further constraints on cometary surface properties. *Icarus, 222,* 492–501.

Schmedemann N., Massironi M., Wagner R., et al. (2018) Small bodies and dwarf planets. In *Planetary Geology* (A. P. Rossi and S. van Gasselt, eds.), pp. 311–343. Springer Praxis, Cham.

Schröder S. E., Carsenty U., Hauber E., et al. (2021) The brittle boulders of dwarf planet Ceres. *Planet. Sci. J., 2,* 111.

Schwartz S. R., Michel P., Jutzi M., et al. (2018) Catastrophic disruptions as the origin of bilobate comets. *Nature Astron., 2,* 379–382.

Sierks H., Barbieri C., Lamy P. L., et al. (2015) On the nucleus structure and activity of Comet 67P/Churyumov-Gerasimenko. *Science, 347,* aaa1044.

Singer K. N., McKinnon W. B., Gladman B., et al. (2019) Impact craters on Pluto and Charon indicate a deficit of small Kuiper belt objects. *Science, 363,* 955–959.

Skorov Y. V., Rezac L., Hartogh P., et al. (2016) A model of short-lived outbursts on the 67P from fractured terrains. *Astron. Astrophys., 593,* A76.

Snodgrass C. and Jones G. (2019) The European Space Agency's Comet Interceptor lies in wait. *Nature Commun., 10,* 5418.

Snodgrass C., Agarwal J., Combi M., et al. (2017) The main belt comets and ice in the solar system. *Astron. Astrophys. Rev., 25,* 5.

Soderblom L. A., Becker T. L., Bennett G., et al. (2002) Observations of Comet 19P/Borrelly by the Miniature Integrated Camera and Spectrometer aboard Deep Space 1. *Science, 296,* 1087–1091.

Soderblom L. A., Boice D. C., Britt D. T., et al. (2004) Imaging Borrelly. *Icarus, 167,* 4–15.

Spencer J. R., Stern S. A., Moore J. M., et al. (2020) The geology and geophysics of Kuiper belt object (486958) Arrokoth. *Science, 367,* aay3999.

Springmann A., Harris W. M., Ryana E. L., et al. (2022) Repeating gas ejection events from Comet 45P/Honda-Mrkos-Pajdušáková. *Planet. Sci. J., 3,* 15.

Steckloff J. K. and Samarasinha N. H. (2018) The sublimative torques of Jupiter family comets and mass wasting events on their nuclei. *Icarus, 312,* 172–180.

Steckloff J., Graves K., Hirabayashi M., et al. (2016) Rotationally induced surface slope-instabilities and the activation of CO_2 activity on Comet 103P/Hartley 2. *Icarus, 272,* 60–69.

Stern S. A., Weaver H. A., Spencer J. R., et al. (2019) Initial results from the New Horizons exploration of 2014 MU_{69}, a small Kuiper belt object. *Science, 364,* aaw9771.

Stooke P. (2002) *Small Body Shape Models.* NASA Planetary Data System, EAR-A-5-DDR-STOOKE-SHAPE-MODELS-V1.0.

Sunshine J. M., Thomas N., El-Maarry M. R., et al. (2016) Evidence for geologic processes on comets. *J. Geophys. Res.–Planets, 121,* 2194–2210.

Taylor M. G. G. T., Altobelli N., Buratti B. J., et al. (2017) The Rosetta

mission orbiter science overview: The comet phase. *Phil. Trans. R. Soc. A, 375,* 20160262.
Thomas N. (2020) *An Introduction to Comets. Post-Rosetta Perspectives.* Springer, Cham. 503 pp.
Thomas N., Sierks H., Barbieri C., et al. (2015a) The morphological diversity of Comet 67P/Churyumov-Gerasimenko. *Science, 347,* aaa0440.
Thomas N., Davidsson B., El-Maarry M. R., et al. (2015b) Redistribution of particles across the nucleus of Comet 67P/ Churyumov-Gerasimenko. *Astron. Astrophys., 583,* A17.
Thomas P. C., Veverka J., Belton M. J., et al. (2007) The shape, topography, and geology of Tempel 1 from Deep Impact observations. *Icarus, 191,* 51–62.
Thomas P. C., A'Hearn M., Belton M. J. S., et al. (2013a) The nucleus of Comet 9P/Tempel 1: Shape and geology from two flybys. *Icarus, 222,* 453–466.
Thomas P. C., A'Hearn M. F., Veverka J., et al. (2013b) Shape, density, and geology of the nucleus of Comet 103P/Hartley 2. *Icarus, 222,* 550–558.
Trigo-Rodríguez J. M., García-Melendo E., Davidsson B. J. R., et al. (2008) Outburst activity in comets: I. Continuous monitoring of Comet 29P/Schwassmann-Wachmann 1. *Astron. Astrophys., 485,* 599–606.
Trigo-Rodríguez J. M., García-Hernandez D. A., Sanchez A., et al. (2010) Outburst activity in comets: II. A multiband photometric monitoring of Comet 29P/Schwassmann-Wachmann 1. *Mon. Not. R. Astron. Soc., 409,* 1682–1690.
Tubiana C., Snodgrass C., Bertini I., et al. (2015) 67P/Churyumov-Gerasimenko: Activity between March and June 2014 as observed from Rosetta/OSIRIS. *Astron. Astrophys., 573,* A62.
Ulamec S., Biele J., Blazquez A., et al. (2015) Rosetta lander — Philae: Landing preparations. *Acta Astronaut., 107,* 79–86.
Veverka J., Klaasen K., A'Hearn M., et al. (2013) Return to Comet Tempel 1: Overview of Stardust-NExT results. *Icarus, 222,* 424–435.
Vincent J.-B., Bodewits D., Besse S., et al. (2015a) Large heterogeneities in Comet 67P as revealed by active pits from sinkhole collapse. *Nature, 523,* 63–66.
Vincent J.-B., Oklay N., Marchi S., Höfner S., et al. (2015b) Craters on comets. *Planet. Space Sci., 107,* 53–63.
Vincent J.-B., A'Hearn M. F., Lin Z. Y., et al. (2016) Summer fireworks on Comet 67P. *Mon. Not. R. Astron. Soc., 462,* S184–S194.
Vincent J.-B., Farnham T., Kührt E., et al. (2020) Local manifestations of cometary activity. *Space Sci. Rev., 215,* 30.
Wahlberg Jansson K. and Johansen A. (2014) Formation of pebble-pile planetesimals. *Astron. Astrophys., 570,* A47.
Wang J.-H. and Brasser R. (2014) An Oort cloud origin of the Halley-type comets. *Astron. Astrophys., 563,* A122.
Whipple F. L. (1950) A comet model I. The acceleration of Comet Encke. *Astrophys. J., 111,* 375–394.
Wierzchos K., Womack M., Sarid G. (2017) Carbon monoxide in the distantly active Centaur (60558) 174P/Echeclus at 6 au. *Astron. J., 153,* 230.
Zhang X., Huang J., Wang T., et al. (2019) ZhengHe — A mission to a near-Earth asteroid and a main belt comet. In *50th Lunar and Planetary Science Conference,* Abstract #1045. LPI Contribution No. 2132, Lunar and Planetary Institute, Houston.

Filacchione G., Ciarniello M., Fornasier S., and Raponi A. (2024) Comet nuclei composition and evolution. In *Comets III* (K. J. Meech, M. R. Combi, D. Bockelée-Morvan, S. N. Raymond, and M. E. Zolensky, eds.), pp. 315–360. Univ. of Arizona, Tucson, DOI: 10.2458/azu_uapress_9780816553631-ch011.

Comet Nuclei Composition and Evolution

G. Filacchione and M. Ciarniello
INAF-IAPS, Institute for Space Astrophysics and Planetology

S. Fornasier
LESIA, Université Paris Cité, Observatoire de Paris/Université PSL, CNRS/Institut Universitaire de France

A. Raponi
INAF-IAPS, Institute for Space Astrophysics and Planetology

Thanks to data from the Rosetta orbiter and the Philae lander our knowledge of the composition of cometary nuclei has experienced tremendous advancement. The properties of the nucleus of 67P/Churyumov-Gerasimenko are discussed and compared with other comets explored in the past by space missions. Cometary nuclei are made by a collection of ices, minerals, organic matter, and salts resulting in very dark and red-colored surfaces. When relatively far from the Sun (>3 au), exposed water and carbon dioxide ices are found only in few locations of 67P/CG where the exposure of pristine subsurface layers or the recondensation of volatile species driven by the solar heating and local terrain morphology can sustain their temporary presence on the surface. The nucleus surface appears covered by a dust layer of variable thickness caused by the backfall flux of coma grains. Depending on local morphology, some regions are less influenced by the backfall and show consolidated terrains. Dust grains appear mostly dehydrated and are made by an assemblage of minerals, organic matter, and salts mixed together. Spectral analysis shows that the mineral phase is dominated by silicates and fine-grained opaques (possibly including Fe-sulfides such as troilite or pyrrhotite) and ammoniated salts. Aliphatic and aromatic groups, with the presence of the strong hydroxyl group, are identified within the organic matter. The surface composition and physical properties of cometary nuclei evolve with heliocentric distance and seasonal cycling: Approaching perihelion, the increase of the solar flux boosts the activity through the sublimation of volatile species, which in turn causes the erosion of surface layers, the exposure of ices, the activity in cliffs and pits, the collapse of overhangs and walls, and the mobilization and redistribution of dust. The evolution of color, composition, and texture changes occurring across different morphological and geographical regions of the nucleus are correlated with these processes. In this chapter we discuss 67P/CG nucleus composition and evolutionary processes as observed by the Rosetta mission in the context of other comets previously explored by space missions or observed from Earth.

1. INTRODUCTION

Cometary nuclei are among the most pristine bodies currently populating our solar system. Comets and planetesimals both formed through the accretion of dust grains, volatiles, and organic matter in the outer regions of the primordial solar nebula (see the chapter by Simon et al. in this volume). Whatever the origin of cometary matter, whether it is prior to the formation of the solar system or whether it contains materials formed in the solar nebula disk is still a subject of debate. The analysis of the dust samples collected by the Stardust mission show a mixing of presolar grains and minerals formed at high temperature, like forsterite (Mg_2SiO_4), and enstatite (Mg_2SiO_3) as well as calcium-aluminium inclusions (CAIs) formed in the hottest regions (>1400 K) of the primordial solar nebula (*Brownlee et al.,* 2006), thus suggesting the presence of radial mixing due to turbulence (*Bockelée-Morvan et al.,* 2002) and ballistic transport (*Shu et al.,* 2001). Conversely, other measurements suggest a different formation process: (1) the elemental abundances of many species (C, O, etc.) are close to the solar values; (2) the organic matter abundance is similar to molecular clouds and star-forming regions; (3) the nonsolar values of the isotopic ratios of Si, Xe, and S; and (4) the high $[D_2O/HDO]/[HDO/H_2O]$ value measured on 67P/CG are compatible with a scenario in which comets were formed very far from the Sun from presolar matter mixed with never-sublimated (amorphous) ice, at temperature as low as 20 K and in a poorly mixed protoplanetary disk (*Altwegg et al.,* 2019). In contrast to larger bodies orbiting closer to the Sun that have sustained thermal processing, impacts, and chemical alterations (*de Niem et al.,* 2018), comets preserve their original composition by being relegated to the outer solar system for the greater part of their lifetimes. Comets are currently stored in two large reservoirs — the Oort cloud (heliocentric distance between 20,000 and 100,000 au) and the Kuiper belt (30–50 au) — where they have been orbiting since the time of their formation. Gravitational perturbations and resonances with external planets are capable of deflecting them on orbits closer to the Sun. Jupiter-family comets (JFCs), the class of comets to which 67P/Churyumov-Gerasimenko (hereafter 67P/CG) belongs, have orbital periods shorter than 20 years, low inclinations ($\leq 30°$), aphelia located approximately beyond Jupiter's orbit, and perihelia between 1 and 2 au from the Sun.

During their orbital evolution, comets cross different snow lines, corresponding to the equilibrium temperature at which a given ice sublimates in vacuum. In the current solar system the frost line of water ice (H_2O) is located at heliocentric distance of about 3 au, for carbon dioxide (CO_2) is at about 14 au, for carbon monoxide (CO) is at 28 au, and for nitrogen (N_2) at 32 au. Each time a comet orbits inbound one of these snow lines the corresponding ice starts to sublimate. Comets orbiting in the Kuiper belt are kept at cryogenic temperatures continuously during their lifetimes and for this reason their composition remains unaltered until today (*Mumma and Charnley,* 2011). This equilibrium is broken when comets are perturbed and injected on orbits closer to the Sun (see the chapter by Fraser et al. in this volume). During their inbound orbits toward the perihelion, all comets show gaseous activity due to the triggering of ices sublimation when they cross the frost lines (*Podolak and Zucker,* 2004), resulting in the formation of a coma made of accelerated gas and dust grains mainly released from the illuminated areas of the nucleus. Due to the small dimension (typically a few kilometers) and irregular shape of the nucleus, the gaseous activity is not homogeneous but is driven by complex relationships among local morphology, solar heating, and availability of volatile species in the subsurface layers (see the chapter by Marschall et al. in this volume). This activity steadily increases when the comet moves toward the Sun, reaches maximum efficiency immediately after the perihelion passage, and then decreases during the outbound orbit. A similar behavior is common among comets having been observed on both 9P/Tempel 1 (*Meech et al.,* 2011) and 67P/CG (*Hansen et al.,* 2016). After recrossing the frost line along the outbound orbit, the water-ice sublimation subsides and the coma becomes more tenuous, sustained only by the sublimation of more-volatile, less-abundant species like CO_2 and CO.

In recent years the classical distinction between comets and asteroids has been reconsidered thanks to the discovery of ice-rich and active asteroids orbiting in the outer part of the main belt (see the discussion in *Hsieh and Jewitt,* 2006; *Snodgrass et al.,* 2017; and the chapter by Jewitt and Hsieh in this volume). In addition, in the last decade a certain number of inactive objects moving on orbits similar to those of comets has been observed (*Tancredi,* 2014). The presence of this population is very relevant because these objects share surface properties comparable with comet nuclei, making them likely dormant objects (*Licandro et al.,* 2016, 2018). Moreover, cometary dust samples returned from Comet 81P/Wild 2 by the Stardust mission (*Brownlee et al.,* 2006) or 67P/CG coma grains characterized by Rosetta's Cometary Secondary Ion Mass Analyser (COSIMA) (*Fray et al.,* 2016) indicate a direct composition link between comets and asteroids as a consequence of the mixing of inner solar system material that has migrated toward the outer regions. Apart from keeping a record of their primordial composition, comets have played a primary role in planetary evolution, being the major carriers of water and C-bearing matter on internal planets: Ocean water and organic matter, the two main ingredients necessary to activate biological processes on Earth, have been provided by cometary and asteroidal impacts taking place during the early phases of the solar system evolution (*Delsemme,* 1984; *Mayo Greenberg and Mendoza-Gómez,* 1992). For these reasons, the exploration of comets and primitive asteroids is an unavoidable step in understanding the processes of solar system formation and evolution.

Modern exploration of comets began in 1986 thanks to the European Space Agency's (ESA's) Giotto mission, which performed a close flyby (minimum distance $\approx$ 600 km) of Comet 1P/Halley. Thanks to the data collected by the instruments onboard, it was for the first time possible to resolve the very dark surface of a cometary nucleus (*Keller et al.,* 1986), to observe seven active jets releasing matter at a rate of about 3 ton s^{-1} (*McDonnell et al.,* 1986), and to characterize organic-rich composition of dust particles in the coma (*Kissel et al.,* 1986). Our knowledge of comets has advanced greatly through NASA missions Deep Space 1, Stardust, and Deep Impact/EPOXI, which targeted comets 19P/Borrelly, 81P/Wild 2, 9P/Tempel 1, and 103P/Hartley, respectively. With these missions, we have discovered the variability of nuclei and activity processes, including the presence of red-colored dehydrated nuclei surfaces (*Soderblom et al.,* 2002), their very low bulk density ($\approx$600 kg m^{-3}) (*A'Hearn et al.,* 2005), the presence of chunks of ice ejected from the nucleus by sublimation of CO_2 ice (*A'Hearn et al.,* 2011), and the presence of both presolar and solar system origin materials embedded in dust particles (*Brownlee et al.,* 2006). More recently, ESA's Rosetta mission (*Taylor et al.,* 2017) has accomplished a thorough exploration of JFC 67P/CG by orbiting at a close distance from the nucleus during an entire perihelion passage, e.g., from August 2014 (heliocentric distance 3.54 au), to perihelion in August 2015 (1.23 au) to the end of the mission in September 2016 (3.82 au). By performing continuous observations for more than 2 years, the 10 scientific instruments onboard Rosetta were able to study 67P/CG's nucleus, coma, and solar wind interactions from a vantage viewpoint, allowing the comet's evolution to be followed during perihelion passage with both remote sensing and *in situ* techniques. In the following, we'll discuss the main results from the Rosetta mission on the

67P/CG nucleus composition and evolutionary processes in the context of other comets explored by space missions or observed from Earth.

2. COMET NUCLEI PROPERTIES

2.1. Photometry

The photometric investigation of cometary nuclei is relatively difficult because these bodies are small, irregularly shaped and extremely dark, and thus very faint, spatially unresolved by groundbased telescopes and masked by their coma when approaching the Sun. Cometary nuclei were originally believed to be bright because of their high volatile content deduced from coma measurements. However, the first *in situ* images of Comet 1P/Halley acquired by the Giotto mission revealed an extremely dark nucleus reflecting only 4% of the incoming solar light (*Keller,* 1987). All subsequent measurements of the geometric albedo (A_{geo}) from groundbased telescopes and space missions confirmed that cometary nuclei are among the darker objects in the solar system, with 3% < A_{geo} < 7% [Table 1; see also the compendia of comet nuclei albedo by *Lamy et al.* (2004)], although they are rich in volatiles. Comets 28P/Neujmin 1 and 19P/Borrelly show the darkest surfaces with average albedo values of 2.5–3% (*Buratti et al.,* 2004; *Soderblom et al.,* 2002; *Delahodde et al.,* 2001). However, Comet 19P/Borrelly displays huge variation in the local photometric parameters and *Li et al.* (2007b) found a geometric albedo of 7 ± 2%.

The unprecedented observations of the 67P/CG nucleus during more than two years at different scales of spatial resolution have allowed the Rosetta scientists to perform the most detailed study of a comet ever attempted. *In situ* measurements by the Optical, Spectroscopic, and Infrared Remote Imaging System (OSIRIS) (see *Keller et al.,* 2007) and Visible, InfraRed, and Thermal Imaging Spectrometer (VIRTIS) (see *Coradini et al.,* 2007) instruments have provided a A_{geo} value of 5.9 ± 0.3% at 535 nm (*Fornasier et al.,* 2015) and 6.2 ± 0.2% at 550 nm (*Ciarniello et al.,* 2015) respectively from global photometry, indicating that Comet 67P/CG nucleus is among the brightest explored so far by space missions, together with Comets 9P/Tempel 1 and 81P/Wild 2.

The β coefficient giving the slope of the linear fit of the relation between magnitude and phase angle α is reported in Table 1 for several comets, and it ranges between 0.01 and 0.076 mag/°. It is usually computed for phase angles >5°–7° in order to avoid the opposition surge contribution. For Comet 67P/CG, *Fornasier et al.* (2015) and *Ciarniello et al.* (2015) reported a β value of 0.047 ± 0.002 mag/° and 0.041 ± 0.001 mag/° for α > 7° and α > 15°, respectively. The linear slope strongly increases when extending the fit to small phase angles reaching values of 0.074 mag/° (*Fornasier et al.* 2015) and 0.082 mag/° (*Ciarniello et al.,* 2015), thus indicating for Comet 67P/CG the presence of a strong opposition effect, which is a steep increase of the reflectance at small phase angles.

The linear slope value of Comet 67P/CG is similar to the average value of JFCs [β = 0.053 ± 0.016 mag/° (*Snodgrass et al.,* 2011)], close to the values found for Comets 1P/Halley, 9P/Tempel 1, 19P/Borelly, and 103P/Hartley 2 (Table 1), and similar to the value reported in the literature for low-albedo asteroids (*Belskaya and Shevchenko,* 2000).

Hapke's semi-empirical models based on the radiative transfer theory (*Hapke,* 1993, 2002, 2012) are commonly applied to the global and spatially-resolved photometry of minor bodies to constrain their photometric properties. These models use the parameters listed in the following:

- SSA is the particle's single-scattering albedo;
- b is the asymmetric factor, which describes the direction in which the medium spreads most of the radiation: backward if b < 0 and forward if b > 0;
- $\bar{\theta}$ is the angle indicating the average macroscopic roughness slope of the surface medium;
- B_{sh} and h_{sh} are respectively the amplitude and angular width of the peak of the opposition effect (OE) due to the shadow hiding mechanism;x
- B_{cb} and h_{cb} are respectively the amplitude and angular width of the peak of the opposition effect due to the coherent backscattering mechanism.

An additional parameter K, which takes into account the porosity of the surface, is also included in the latest versions of the Hapke model (*Hapke,* 2008, 2012). The coherent-backscattering mechanism is expected to have a negligible contribution on dark surfaces like those of cometary nuclei, where multiple scattering should have a small contribution relative to single scattering.

We report in Table 1 the SSA, b, and $\bar{\theta}$ parameters for the comets observed by space missions. Cometary nuclei have very low single-scattering-albedo values (2–6%), strongly backscattering surfaces (–0.52 < b < –0.42), and moderately rough surfaces (15 < $\bar{\theta}$ < 27°). However, spatially resolved data from space missions revealed not only the complex shape and geomorphology of cometary nuclei but also some colors and photometric properties heterogeneity. For instance, for Comet 19P/Borrelly, Li et al. (2007b) reported variations of the SSA by a factor of 2.5, of the asymmetric factor from almost isotropic (–0.1) to a strongly backscattering (–0.7) behavior, and a roughness parameter ranging from 13° to 55°. Also at ultraviolet wavelengths the single-scattering albedo of cometary nuclei is very low: Feaga et al. (2015) have measured a value of 0.031 ± 0.003 at 1425–1525 Å for 67P/CG. For Comet 67P/CG a number of features with distinct albedo differences from the average terrain, including bright spots, blue veins, and dark boulders, have also been observed and investigated (Hasselmann et al., 2017; Feller et al., 2016).

Among comets explored by space missions 67P/CG is the only one for which the opposition surge has been characterized, thanks to the observations at low phase angles acquired with the OSIRIS imaging system onboard Rosetta. *Fornasier et al.* (2015) applied the *Hapke* (1993) model on

TABLE 1. Spectrophotometric properties and Hapke parameters of 67P/CG compared to other cometary nuclei (from *Filacchione et al.*, 2019).

Comet	B–V	V–R	R–I	SSA	b	θ[°]	A_{geo}	β[mag/°]	α(°)
67P/CG[*]	0.73 ± 0.07	0.57 ± 0.03	0.59 ± 0.04	0.052 ± 0.013	−0.42	$19°^{+4}_{-9}$	0.062 ± 0.002	0.041 ± 0.001[b]	1.2°–111.5°
67P/CG[†]	—	—	—	0.037 ± 0.002	−0.42 ± 0.03	15°	0.059 ± 0.003	0.047 ± 0.002	1.3°–53.9°
67P/CG[‡]	—	—	—	0.042	−0.37	15°	0.065 ± 0.002	0.047 ± 0.002	1.3°–53.9°
67P/CG[§] bright spots	—	—	—	0.067 ± 0.004	−0.26 ± 0.07	15°	0.077	—	1°–30°
67P/CG[¶] bright spots	—	—	—	0.047 ± 0.001	−0.34 ± 0.02	15°	0.073 ± 0.05	—	0.1°–68°
67P/CG[§] dark boulders	—	—	—	0.029 ± 0.004	−0.41 ± 0.06	15°	0.064	0.045 ± 0.002	1°–30°
67P/CG[**]	0.83 ± 0.08	0.54 ± 0.05	0.46 ± 0.04	—	—	—	—	0.061 – 0.076	0.5°–10.6°
1P/Halley[††]	0.72 ± 0.04	0.41 ± 0.03	0.39 ± 0.06	—	—	—	—	0.04[‡‡]	107°
1P/Halley[§§]	—	—	—	—	—	—	—	$0.05^{+0.03}_{-0.01}$	—
1P/Halley[¶¶]	—	—	—	—	—	—	—	$0.04^{+0.01}_{-0.02}$	—
2P/Encke[***]	0.78 ± 0.02	0.48 ± 0.02	—	—	—	—	0.046 ± 0.023	0.06	—
9P/Tempel1[†††]	0.84 ± 0.01	0.50 ± 0.01	0.49 ± 0.02	0.039 ± 0.005	−0.49 ± 0.02	16° ± 8°	0.056 ± 0.009	0.046 ± 0.007	63°–117°
19P/Borrelly[‡‡‡]	—	—	—	0.057 ± 0.009	−0.43 ± 0.07	22° ± 5°	0.072 ± 0.020	0.043	51°–75°
19P/Borrelly[§§§]	—	0.25 ± 0.78[¶¶¶]	—	0.020 ± 0.004	−0.45 ± 0.05	20° ± 5°	0.029 ± 0.002	—	52°–89°
28P/Neujmin 1[¶¶¶]	—	0.45 ± 0.05	—	—	—	—	0.025	0.006 ± 0.006	1°–19°
45P/Honda-Mrkos-Pajdušáková[****]	1.12 ± 0.03	0.44 ± 0.03	0.20 ± 0.03	—	—	—		0.06	—
48P/Johnson[††††]	—	—	—	—	—	—		0.059	6°–16°
55P/Tempel-Tuttle[****]	—	—	—	—	—	—	0.060 ± 0.025	0.041	3°–55°
81P/Wild 2[‡‡‡‡]	—	—	—	0.038 ± 0.04	−0.52 ± 0.04	27° ± 5°	0.059 ± 0.004	0.0513 ± 0.0002	11°–100°
103P/Hartley 2[§§§§]	0.75 ± 0.05	0.43 ± 0.04	—	0.036 ± 0.006	−0.46 ± 0.06	15° ± 10°	0.045 ± 0.009	0.046 ± 0.002	79°–93°
143P/Kowal-Mrkos[¶¶¶¶]	0.84 ± 0.02	0.58 ± 0.02	0.55 ± 0.02	—	—	—	—	0.043 ± 0.014	4.8°–12.7°

[*] From *Ciarniello et al.* (2015), β computed for phase angle $\alpha > 15°$.

[†] Values for w, b, θ, and β from *Fornasier et al.* (2015), Hapke global photometric model at $\lambda = 0.535\ \mu m$, β computed for phase angle $\alpha > 7°$.

[‡] From *Fornasier et al.* (2015); *Hapke* (2002) photometric model at $\lambda = 0.649\ \mu m$.

[§] From *Feller et al.* (2016); *Hapke* (2012) photometric model at $\lambda = 0.649\ \mu m$.

[¶] From *Hasselmann et al.* (2017); *Hapke* (2012) photometric model at $\lambda = 0.612\ \mu m$.

[**] From *Tubiana et al.* (2011). B–V has been computed from B–R and V–R index reported in *Tubiana et al.* (2011).

[††] From *Thomas and Keller* (1989).

[‡‡] From *Keller* (1987).

[§§] From *Hughes* (1985).

[¶¶] From *Sagdeev et al.* (1986).

TABLE 1. (continued)

*** From *Fernández et al.* (2000); (B–V) and (V–R) colors derived from *Luu and Jewitt* (1990) spectrophotometry.
††† Photometric parameters derived in V band from *Li et al.* (2007a); groundbased and HST observations with solar phase angle down to 4° have been included to compute the slope parameter.
‡‡‡ Photometric parameters derived in R band from *Li et al.* (2007b); groundbased observations with solar phase angle down to 13° have been included to compute the slope parameter.
§§§ From *Buratti et al.* (2004).
¶¶¶ From *Campins et al.* (1987) and *Delahodde et al.* (2001).
**** Colors from *Lamy et al.* (1999); β from *Lamy et al.* (2004).
†††† From *Jewitt and Sheppard* (2004).
‡‡‡‡ Photometric parameters derived in R band by *Li et al.* (2009).
§§§§ From *Li et al.* (2013); Gemini and HST observations at low phase angle have been included to constrain Hapke modeling and derive the slope parameter.
¶¶¶¶ From *Jewitt et al.* (2003).

disk-average photometry of OSIRIS observations acquired with eight filters covering the 325–990-nm range during the comet approach phase in 2014, and having a spatial resolution up to 2.1 m/pixel. They found an amplitude of the opposition effect larger than 1.8 and a width larger than 0.021 and no clear wavelength dependence of these parameters, as expected when the shadow-hiding effect is the main cause of the opposition surge. They also model 67P/CG resolved data acquired with the orange filter centered at 649 nm with the latest versions of the Hapke model (*Hapke,* 2002, 2012), finding a strong opposition peak dominated by the shadow-hiding effect (B_{sh} = 2.5, h_{sh} = 0.079) with a small contribution coming from the coherent backscattering (B_{sh} = 0.19, h_{sh} = 0.017). They also estimate the top surface layer porosity at 87%, suggesting the presence of a very weak and porous mantle with irregular dust particles having structures similar to fractal aggregates.

These results were further confirmed at greater spatial resolution during two Rosetta flybys designed to observe the nucleus at low phase angles. The pre-perihelion flyby took place in February 2015 and permitted observation of an area near the borders of the Ash, Apis, and Imhotep regions at decimeter scale [see *El-Maarry et al.* (2015) and *El-Maarry et al.* (2016) for a definition of the regions on Comet 67P/CG]. The second flyby took place in April 2016 (after perihelion), and it covered the Imhotep-Khepry transition region at a spatial resolution reaching up to 0.5 m/pixel. These higher spatial-scale-resolution observations confirmed that the opposition surge is dominated by the shadow-hiding effect and that there is no evidence of a sharp opposition peak at small phase angles associated with the coherent backscattering mechanism. The area observed during the February 2015 (*Feller et al.,* 2016) flyby has Hapke parameters indistinguishable from those of the overall nucleus presented in *Fornasier et al.* (2015); it shares the same high value of superficial porosity (85–87%) and has a phase function similar to that observed in laboratory for carbon soot samples, or for intramixtures composed of tholins and carbon black (*Jost et al.,* 2017), suggesting the presence of a fluffy and dark regolith coverage in this area (*Masoumzadeh et al.,* 2017).

Feller et al. (2016) also investigated the photometric properties of some features showing higher (bright spots) or darker (dark boulders) than average albedo values (Table 1). In fact, several meter-sized boulders were identified, and all have both colors redder than those of the average nucleus, as well as a smaller opposition effect. Some bright spots were also observed, but it is worth noting that they are only ~50% brighter than the average terrain and do not have a spectral behavior consistent with exposures of water-ice-rich material as found elsewhere on the 67P/CG nucleus and discussed in the next sections.

These features have very different SSA and b parameter values: The bright spots have high SSA (6.7%) and low b in absolute value (–0.29), while dark boulders have a very low SSA (2.9%) and they are more backscattering.

The April 2016 flyby was investigated by *Hasselmann et al.* (2017) and *Feller et al.* (2019) using the WAC and the NAC cameras of the OSIRIS imaging system respectively. While the NAC data permitted a high-resolution investigation of the flyby area, highlighting the presence of boulders 20% darker than the average terrain as well as of several spots at least 2× brighter than the average terrain, and having colors compatible with a local enrichment in water-ice (*Feller et al.,* 2019), the WAC observations covered the whole Imhotep and surrounding regions, catching the opposition surge swiping across the Imhotep region (*Hasselmann et al.* 2017). The photometric properties of the whole imaged surface during this flyby as well as of some particular features including bright spots and spectrally blue veins were investigated by *Hasselmann et al.* (2017), who confirmed the presence of a strong opposition peak produced by the shadow-hiding mechanism, and deduced that the cometary surface is dominated by opaque and dehydrated grains with irregular shape and fairy castle structures. Bright spots and blue veins display a sharper opposition surge consistent with a local enrichment of a few percent water ice mixed with the dark compounds of the comet nucleus. Comparing the area observed both during the February 2015 (pre-perihelion) and April 2016 (post-perihelion) flybys, *Hasselmann et al.* (2017) find that this area became 7% darker post-perihelion compared to the pre-perihelion observations and explain this albedo evolution as due to a local air falling deposit of dust.

The presence on 67P/CG of a surface layer of opaque and dehydrated grains organized in underdense fairy castle structures is in line with the findings reported by *Emery et al.* (2006) on Trojan asteroids. Trojans appear covered by a layer of small silicate (likely dehydrated) grains embedded in a relatively transparent matrix, or in very underdense surface structures. Moreover, the emissivity spectra of Trojans more closely resemble the emission spectra of cometary comae. These results further support the existence of a common link between comets and asteroids.

Overall, the photometric properties of the 67P/CG nucleus are very similar to those of dark asteroids. Both its linear slope and opposition surge are consistent with the values published in the literature for the majority of dark asteroids (*Belskaya and Shevchenko,* 2000), including C and P classes, with the notable exception of D-type asteroids, which are the closest to cometary nuclei in terms of surface colors. In fact, *Shevchenko et al.* (2012) investigate the phase function of three Trojan D-type asteroids, finding no opposition effect, proposing that this could be due to their extremely dark surfaces, where only single light-scattering is important and the coherent-backscattering enhancement is completely negligible. They also assumed that asteroids showing no opposition surge should be the darkest, an assumption strengthened by the results of *Belskaya and Shevchenko* (2000), who found a strong correlation between the amplitude of the opposition effect and the bodies' albedos, with the lowest albedo asteroids showing the smallest opposition effect.

However, this assumption does not work for the very dark surfaces (albedo of ~0.044) of the near-Earth asteroids Ryugu and Bennu, recently visited by the Hayabusa 2 and Origins, Spectral Interpretation, Resource Identification, and Security-Regolith Explorer (OSIRIS-REX) missions, respectively. In fact, even if dark, they show a moderate opposition surge. Their opposition surge value, defined as the reflectance at 0.3° over that at 5° phase angle, ranges from 1.2 for Bennu (*Golish et al.,* 2021) to 1.26 for Ryugu (*Domingue et al.,* 2021). For Comet 67P/CG, the opposition ratio is comprised of reflectance phase curves between 1.25 and 1.34 as extrapolated from *Hasselmann et al.* (2017) and *Feller et al.* (2016), thus consistent with that found for P- and C-type asteroids [1.25–1.39 from *Belskaya and Shevchenko* (2000)]. The three large Trojans investigated by *Shevchenko et al.* (2012) thus look peculiar among dark bodies, and the opposition surge is not solely related to the surface albedo but very likely to the surface physical properties such as porosity, grain size, and composition.

Figure 1 displays some color variations and the normal albedo map at 649 nm of Comet 67P/CG's northern hemisphere obtained from pre-perihelion images acquired with the OSIRIS instrument (*Fornasier et al.,* 2015). Hapi, the region in between the two lobes of the 67P/CG nucleus, is the brightest one (*Fornasier et al.,* 2015), followed by Imhotep. The same results are found also by VIRTIS measurements (*Ciarniello et al.,* 2015). Both these regions are characterized by a smooth surface, suggesting that their brightness may be at least partially due to the texture of the surface, even if Hapi clearly also shows evidence of local enrichment of water ice (*De Sanctis et al.,* 2015; *Filacchione et al.,* 2016a). The Seth region results to be the darkest of the northern hemisphere. Very similar results are visible on SSA maps from VIRTIS observations (*Ciarniello et al.,* 2015; *Filacchione et al.,* 2016a). *Ciarniello et al.* (2015) derived for the 67P/CG nucleus a SSA spectral slope of 0.20/kÅ in the visible range and a lower value, 0.033/kÅ, in the near-infrared range. The SSA maps at 0.55 and 1.8 μm derived from VIRTIS data show a SSA distribution consistent with the geometric albedo at 0.649 μm observed with OSIRIS in the visible range (Fig. 1) during a similar time period. The VIRTIS near-infrared 1.8-μm SSA map highlights the Imhotep region as the brightest one on the surface of 67P/CG. This result is in contrast with the 0.55-μm albedo distribution, which shows a maximum across Hapi (see maps in the bottom row of Fig. 1). *Longobardo et al.* (2017) derived the photometrically-corrected reflectance at phase angle α = 30° at several wavelengths for the 67P/CG nucleus, finding values ranging from 0.013 at 0.55 μm to 0.041 at 2.8 μm and 4 μm. Examining the correlation between reflectance and SSA as a function of the wavelength, they deduced that the single scattering, even if dominating at all wavelengths, is larger at shorter wavelengths than at longer ones. *Longobardo et al.* (2017) also derived the photometrically corrected reflectance at α = 30° for the other comets explored by space missions, finding values consistent with those of Comet 67P/CG and confirming the very low reflectance of cometary nuclei.

2.2. Colors of Cometary Nuclei

As shown in Fig. 2, cometary nuclei are characterized by red colors (Table 1) and their spectrophotometric properties are similar to those of primitive D-type asteroids like the Jupiter Trojans (*Fornasier et al.,* 2004, 2007) and the moderately red transneptunians objects (TNO) (see *Sierks et al.,* 2015; *Capaccioni et al.,* 2015). While visible color data on resolved cometary nuclei and asteroid are still limited to the few objects explored by space missions, much better statistics are available from telescopic observations, which allow tracing of the color gradients among different families of primordial bodies. As discussed by *Jewitt* (2015), a color gradient exists across minor-body groups that increases from C-complex asteroids (C, B, F, G classes), to D-types and Trojans, to active comets and Centauri up to the reddest populations of Kuiper belt objects (KBOs) (see Fig. 2), including some inactive Centaurs (*Lacerda et al.,* 2014). 67P/CG colors are similar to the average values of active JFCs and are the reddest among the comets (a redder color has been observed only on 143P). During the pre-Rosetta era visible and infrared telescopic observations have provided a large dataset of spectrophotometric observations of disk-integrated cometary nuclei, including Comets 162P/Siding

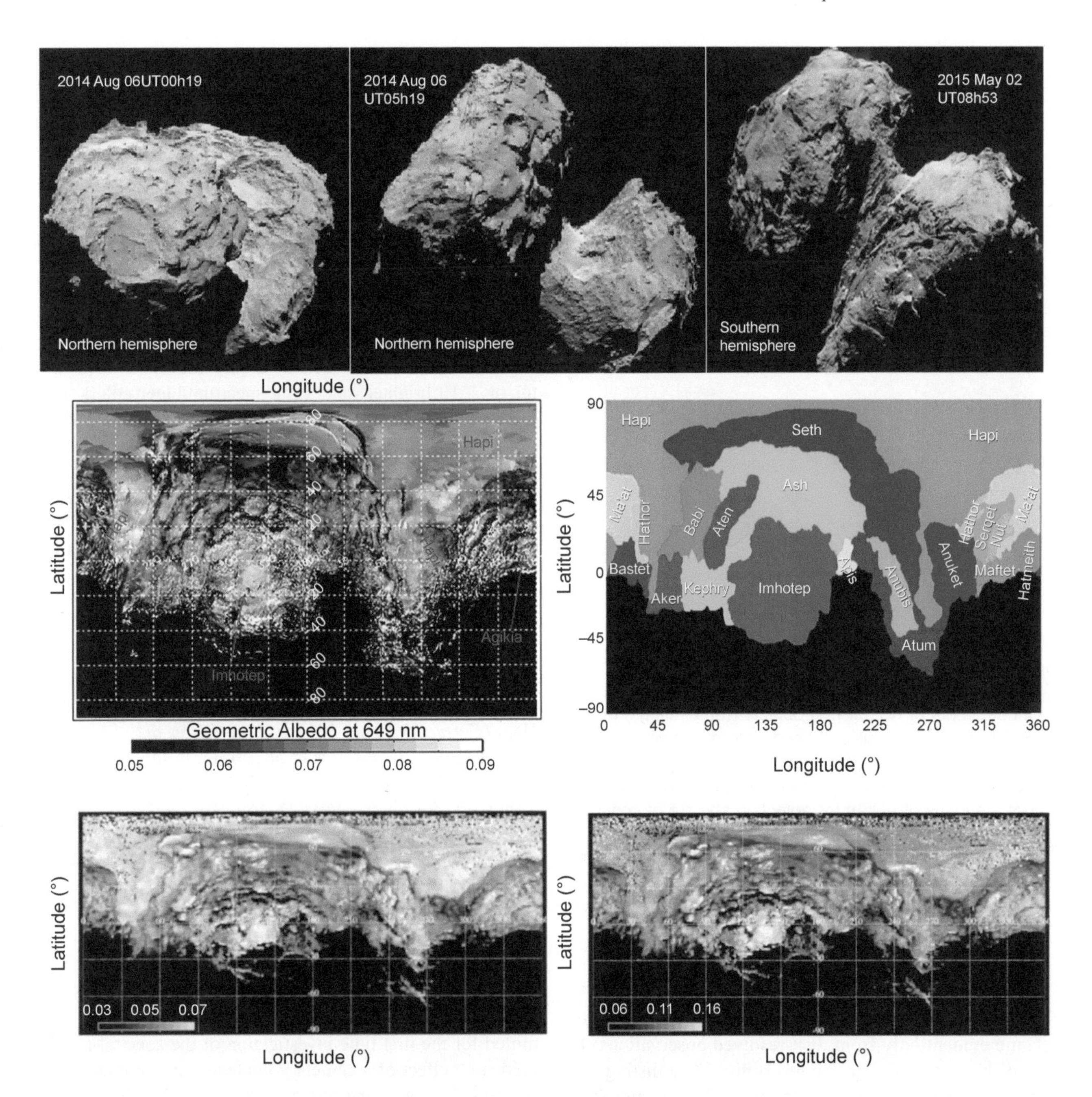

Fig. 1. Colors and albedo map of 67P/CG nucleus. Top panel: OSIRIS-REx color images (filters B: 480 nm, G: 649 nm, R: 882 nm) for the north [left and center image (from *Fornasier et al.,* 2015)] and south hemispheres [right image (from *Fornasier et al.,* 2016)]. Spatial resolution is about 2.2 m/px. Center left panel: Cylindrical map of the normal albedo at 649 nm by OSIRIS-REx (adapted from *Fornasier et al.,* 2015) derived from July–August 2014 observations. Center right panel: Boundaries of 67P/CG's 19 geomorphological regions of the northern hemisphere in cylindrical projection (from *El-Maarry et al.,* 2015). Note that the regions' boundaries have evolved in more recent works by *Thomas et al.* (2018) and *Leon-Dasi et al.* (2021). Missing data represents the unilluminated southern hemisphere at the given epochs. Bottom panel: Cylindrical maps of the single scattering albedo at 0.55 μm (left) and 1.8 μm derived from VIRTIS data (*Ciarniello et al.,* 2015). Note that all cylindrical maps of 67P/CG show degeneration in certain regions due to the irregular shape of the nucleus.

Spring (*Fernández et al.,* 2006; *Campins et al.,* 2006), 124P/Mrkos (*Licandro et al.,* 2003), and 28P/Neujmin 1 (*Campins et al.,* 2007). Moreover, to understand the end-states of comets and the sizes of the comet population, infrared data from the Wide-field Infrared Survey Explorer (WISE) mission were used by *Licandro et al.* (2016) to derive the absolute magnitude H of different classes of asteroids in cometary orbits and assess if they are dormant or extinct comet nuclei. All these telescopic observations were not able to distinguish water ice nor other volatile species on the surfaces of cometary nuclei that share similar low albedos and red-colored surfaces.

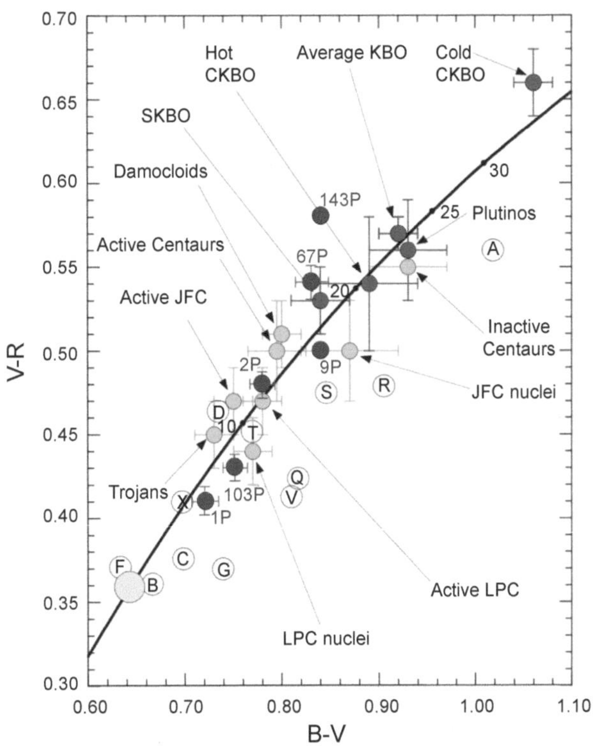

Fig. 2. B-V vs. V-R color index diagram showing the colors of various small-body populations including dynamically distinct subsets of Kuiper belt objects (red circles), comet-related bodies (cyan circles), and asteroid spectral types (black circled letters), as defined in the *Tholen* (1984) classification system (data from *Dandy et al.*, 2003). The solid line shows the locus of points for reflection spectra of constant gradient, S'; numbers give the slope in units of % $kÅ^{-1}$. The large yellow circle shows the color of the Sun. Some overlapping points have been displaced (by 0.005 mag) for clarity. Error bars show the uncertainties on the respective means. Data for comets (including 67P/CG) from Table 1 are shown in magenta. Figure adapted from *Jewitt* (2015).

The correlations between colors and surface properties become evident only from disk-resolved observations: In fact, disk-integrated data are not sufficient to distinguish surface regional variability due to the local rough morphology typical of cometary nuclei, nor to take into account their diurnal and seasonal evolution. On 67P/CG different kind of terrains, from the spectrally bluer and water-ice-enriched (like Hapi), to the redder ones associated mostly with dusty regions, have been identified by OSIRIS and VIRTIS since the first resolved images acquired in July–August 2014 (*Fornasier et al.*, 2015; *Filacchione et al.*, 2016a), covering mostly the northern hemisphere of the nucleus. No color variability between the small and big lobes nor evidence of vertical diversity in the composition for the first tens of meters was seen on the OSIRIS images at a spatial resolution of about 2 m/pixel. The southern hemisphere, visible from Rosetta since March 2015, shows a lack of spectrally red regions compared to the northern hemisphere. This was interpreted as due to the absence of widespread smooth or dust-covered terrains (*Fornasier et al.*, 2016; *El-Maarry et al.*, 2016) in the highly active southern hemisphere regions, which are exposed to high insolation during the perihelion passage.

The heterogeneity in colors and albedo was confirmed at a higher (submeter) resolution scale (*Feller et al.*, 2016; *Hasselmann et al.*, 2017; *Pommerol et al.*, 2015; *Fornasier et al.*, 2019a) by OSIRIS, and brightness variations at centimeter and millimeter scale were reported by the Comet Infrared and Visible Analyser (CIVA) instrument onboard the Philae lander (*Bibring et al.*, 2015).

These color changes have been correlated with composition heterogeneities and active areas across the nucleus. In fact, areas characterized by a very high albedo (a factor of 2–10 brighter) and a flat spectrum/bluer colors than the average dark terrain have been proven to be water ice enriched by joint observations carried out with the OSIRIS and VIRTIS instruments (*Barucci et al.*, 2016; *Filacchione et al.*, 2016b). Therefore, bright and relatively blue regions observed in high-resolution images acquired with OSIRIS permitted the identification of composition inhomogeneities across the nucleus surface. Notable examples are water-ice-enriched areas observed in the northern hemisphere (*Pommerol et al.*, 2015; *Barucci et al.*, 2016), in the Anhur-Bes regions (*Fornasier et al.*, 2016, 2019a), with the detection also of CO_2 ice by VIRTIS (*Filacchione et al.*, 2016c); in the Khonsu (*Deshapriya et al.*, 2016; *Hasselmann et al.*, 2019), Hapi (*Fornasier et al.*, 2015; *De Sanctis et al.*, 2015), and Wosret regions (*Fornasier et al.*, 2021); in the Aswan cliff after its collapse (*Pajola et al.*, 2017a); in several areas in Imhotep (*Filacchione et al.*, 2016b; *Agarwal et al.*, 2017; *Oklay et al.*, 2017); in the Agilkia and Abydos landing sites (*La Forgia et al.*, 2015; *Lucchetti et al.*, 2016; *O'Rourke et al.*, 2020), and in the circular niches of the Seth region (*Lucchetti et al.*, 2017).

2.3. Phase Reddening

The detailed *in situ* investigation of Comet 67P/CG permitted for the first time investigation of the spectral-phase-reddening effect of a cometary nucleus, i.e., the increase in spectral slope with the phase angle. This effect is a consequence of the small-scale surface roughness and multiple scattering in the surface medium at extreme geometries (high phase angles) and it is commonly observed on several solar system bodies. A strong spectral-phase-reddening effect has been noticed for the 67P/CG nucleus since the very first observations with both OSIRIS and VIRTIS instruments covering the northern hemisphere (*Fornasier et al.*, 2015; *Ciarniello et al.*, 2015; *Longobardo et al.*, 2017), and later confirmed also in the southern hemisphere (*Fornasier et al.*, 2016) for specific areas observed during the 2015 and 2016 flybys (*Feller et al.*, 2016, 2019), for the Wosret and Anhur regions (*Fornasier et al.*, 2017, 2019a, 2021), and for the Abydos landing site (*Hoang et al.*, 2020). The phase-reddening coefficients derived from these works for Comet 67P/CG are summarized in Table 2, as well as those for other dark asteroids for comparison. This effect is interpreted as due to the presence of fine particles, namely, of approximately micrometer size, or to the irregular surface structure of larger grains, having micrometer-scale surface roughness (*Schröder et al.,* 2014; *Fornasier et al.,* 2015; *Ciarniello et al.,* 2015).

The 67P/CG phase reddening coefficient is a monotonic and wavelength-dependent function (Table 2), with values of 0.04–0.1 × 10^{-4} nm^{-1} deg^{-1} in the 535–882-nm range and lower values, 0.015–0.018 × 10^{-4} nm^{-1} deg^{-1} in the 1–2-µm range. Moreover, the phase-reddening effect also varies over time, showing a seasonal cycle linked to the cometary activity. The highest phase-reddening effect is observed pre- and post-perihelion at relatively large heliocentric distances when the cometary activity is low or moderate. Conversely, this effect greatly diminished during the peak of the cometary activity close to the perihelion passage (*Fornasier et al.,* 2016), when ungoing surface erosion exposes the underlying layers enriched in volatiles. This seasonal variation of the phase-reddening effect is observed globally and in some individual regions like Anhur, Wosret, and the Abydos landing site (*Fornasier et al.,* 2019a, 2021; *Hoang et al.,* 2020), which is located in Wosret but close to the boundaries with the Bastet and Hatmehit regions.

The phase-reddening effect is lower in the southern hemisphere regions like Wosret and Anhur, which lack widespread dust coating (*El-Maarry et al.,* 2016), and stronger in the northern regions, which are overall richer in dust. In fact, *Keller et al.* (2017) found that the dust particles ejected from the southern hemisphere of Comet 67P/CG during the peak of activity partially fall back and are then deposited in the northern hemisphere.

Compared to other dark minor bodies, the phase reddening effect on the 67P/CG nucleus is comparable to that of D-type asteroids and the dwarf planet Ceres, and it is a factor of 2 or 4 higher compared to the dusty depleted surfaces of Ryugu and Bennu Apollo asteroids, recently visited by the Hayabusa 2 and OSIRIS-REx missions, respectively (Table 2). This different reddening behavior is probably a result of the fact that the Apollo family asteroids orbit much closer to the Sun than comets, resulting in a surface regolith made by smaller and more thermal processed particles showing a fractal structure affecting their micro- and submicro roughness (*Fornasier et al.,* 2020).

2.4. Ultraviolet-Visible-Infrared Spectral Properties from Optical Remote Sensing

During the approximately two-year-long (August 2014–September 2016) escort phase of the Rosetta mission, the nucleus of Comet 67P/CG has been extensively imaged by the Alice, OSIRIS, and VIRTIS experiments, characterizing the ultraviolet-visible-infrared spectral properties of the surface.

TABLE 2. Phase reddening coefficients evaluated for Comet 67P/CG and for other dark small bodies (adapted from *Fornasier et al.,* 2021).

Body and Conditions	Wavelength Range (nm)	α (°)	γ (10^{-4} nm^{-1} deg^{-1})	Y_0 (10^{-4} nm^{-1})	References
Northern hemisphere pre-perihelion (2014)	535–882	0–55	0.1040 ± 0.0030	11.3 ± 0.2	*Fornasier et al.* (2015)
Northern hemisphere pre-perihelion (2014–Feb 2015)	1000–2000	25–110	0.018	2.3	*Ciarniello et al.* (2015)
Northern hemisphere pre-perihelion (2014)	1100–2000	12–100	0.015 ± 0.001	2.0	*Longobardo et al.* (2017)
Northern and southern hemisphere at perihelion (2015)	535–882	60–90	0.0410 ± 0.0120	12.8 ± 1.0	*Fornasier et al.* (2016)
Small region in Imhotep, Ash, Apis (Feb. 2015 flyby)	535–743	0–34	0.065 ± 0.001	17.9 ± 0.1	*Feller et al.* (2016)
Small region in Imhotep-Khepry (April 2016 flyby)	535–743	0–62	0.064 ± 0.001	17.99 ± 0.01	*Feller et al.* (2019)
Abydos and surroundings (post-perihelion, 2016)	535–882	20–106	0.0486 ± 0.0075		*Hoang et al.* (2020)
Wosret around perihelion (2015)	535–882	60–90	0.0396 ± 0.0067	12.7 ± 0.5	*Fornasier et al.* (2021)
Wosret post-perihelion (2016)	535–882	20–106	0.0546 ± 0.0042	13.6 ± 0.4	*Fornasier et al.* (2021)
D-type asteroids	450–2400	10–73	0.05 ± 0.03		*Lantz et al.* (2017)
(1) Ceres	550–800	7–132	0.046	–0.2	*Ciarniello et al.* (2017)
(101955) Bennu	550–860	7–90	0.014 ± 0.001	–1.3 ± 0.1	*Fornasier et al.* (2020)
(162173) Ryugu	550–860	0–40	0.020 ± 0.007		*Tatsumi et al.* (2020)

The quantity γ is the phase reddening coefficient; Y_0 is the estimated spectral slope at zero phase angle from the linear fit of the data and α the phase angle range used for computing the phase reddening coefficients.

The average ultraviolet spectrum of the nucleus' surface by Alice (700–2050 Å) shows a very dark regolith characterized by a blue spectral slope without significant evidences of exposed water-ice absorption (*Stern et al.,* 2015). The average ultraviolet spectral reflectance best-fit is compatible with a model made of 99.5% tholins, 0.5% water ice plus a neutral darkening agent in a fluffy, light-trapping medium (Fig. 3). The observed blue spectral slope could be the consequence of a surface rich in tholins or be the result of Rayleigh scattering on very fine particles in the regolith. While the fit is non-unique, it is compatible with very dehydrated tholin-rich material.

Hyperspectral images (0.25–5 μm) of the nucleus acquired by VIRTIS during the pre-perihelion phase, when the comet was orbiting at a heliocentric distance of 3.6–3.3 au (*Capaccioni et al.,* 2015; *Filacchione et al.,* 2016a), are characterized by a red slope on I/F visible and infrared spectra and display a broad absorption feature between 2.9 μm and 3.6 μm (referred to as 3.2-μm absorption) superimposed over the thermal emission, which extends down to approximately 3 μm (Fig. 4). The 3.2-μm absorption has been interpreted as carried by a combination of ammoniated salts (*Poch et al.,* 2020), organic materials (*Capaccioni et al.,* 2015; *Quirico et al.,* 2016; *Raponi et al.,* 2020), and hydroxylated silicates (*Mennella et al.,* 2020), as discussed in greater detail in sections 3.2 and 3.3. The I/F spectra of most of the surface do not display the diagnostic absorption features of water ice at 1.5 and 2.0 μm, a result that is compatible with the lack of extended ice-rich patches with sizes much larger than a few tens of meters (see section 3.1).

Representative spectra at standard observation geometry

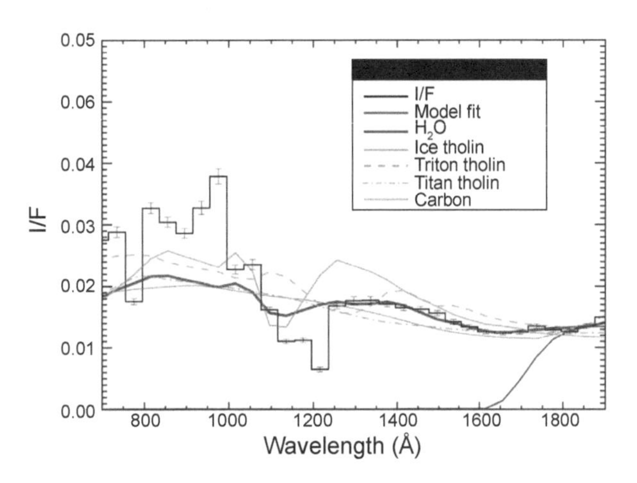

Fig. 3. UV average I/F of 67P/CG nucleus acquired by ALICE on August 14–15, 2014 (black curve) and best-fit model (red). The fit corresponds to a mixture of 99.5% tholins (orange curves), 0.5% water ice (blu), plus a neutral darkening agent like carbon (gray). The tholins are modeled as an intimate mixture made of Triton tholin (abundance 79.9%, grain size 3.5 μm), Titan tholin (0.9%, grain size 5.0 μm), and ice tholin (19.2%, grain size 1.7 μm). Calibration residuals are the cause of the abrupt changes visible at 800, 975, and 1225 Å. From *Stern et al.* (2015).

(incidence = 30°, emission = 0°, phase angle = 30°) of four nucleus macroregions (referred to as "head," "neck," "body," and "bottom" in the following text), as computed by using the corresponding Hapke parameters derived by spectrophotometric modeling of VIRTIS observations acquired from August 2014 (~3.6 au) to February 2015 (~2.4 au) (*Ciarniello et al.,* 2015), reveal a limited spectral variability across the surface at global scale (Fig. 5). This result is a consequence of the dust backfall, which homogenizes the spectral response on great percentage of the nucleus surface. For this reason, despite the richness of morphological features visible on the surface, from a spectral perspective the nucleus appear quite homogeneous before perihelion passage. Nonetheless, the "neck," connecting the two main lobes of the comet and corresponding to the Hapi geomorphological region defined by *El-Maarry et al.* (2015), stands out, displaying reduced visible (*Capaccioni et al.,* 2015; *Fornasier et al.,* 2015; *Filacchione et al.,* 2016b) and infrared spectral slopes (*Capaccioni et al.,* 2015; *Filacchione et al.,* 2016b), and a deeper 3.2-μm absorption feature.

The different spectral response of Hapi with respect to the rest of the 67P/CG nucleus surface clearly emerges from the mapping of comet spectral indicators (CSI) (*Filacchione et al.,* 2016a; *Ciarniello et al.,* 2016), aimed at characterizing the spectral variability across the nucleus. In Fig. 6, maps of the visible (0.5–0.8 μm) and infrared (1.0–2.5 μm) spectral slopes and of the 3.2-μm absorption band depth and position are displayed, with the Hapi region again showing visible and infrared slopes lower than the rest of the surface, and a deeper 3.2-μm band depth characterized by a shift of the feature toward smaller wavelengths. These concomitant variations of the CSI are compatible with an average water-ice enrichment in Hapi of approximately 1 vol% (*Ciarniello et al.,* 2016).

Along with 67P/CG, visible-infrared spectral properties of cometary nuclei have been investigated in detail for Comets 9P/Tempel 1 (explored by the Deep Impact mission), 103P/Hartley 2 (EPOXI mission), and 19P/Borrelly (Deep Space 1). For Tempel 1, disk-integrated observations of the nucleus at visible wavelengths by means of the High-Resolution Instrument (HRI) and the Medium-Resolution Instrument (MRI) cameras (*Hampton et al.,* 2005) reveal a linear red-sloped spectrum (spectral slope of 12.5 ± 1%/kÅ) lacking any feature at the spectral resolution of the multispectral observations of the nucleus (*Li et al.,* 2007a). In addition, disk-resolved observations show limited color variations across the surface (~3% with respect to the nucleus average), as displayed in the color ratio maps (Fig. 7) obtained by rationing images acquired at 750 nm and 550 nm (*Li et al.,* 2007a), with the exception of water-ice-rich regions [color ratio of about 80% of the nucleus average (from *Sunshine et al.,* 2006)].

Similar results are provided by observations at infrared wavelengths from the High Resolution Instrument-InfraRed spectrometer (HRI-IR) (*Hampton et al.,* 2005), which again indicate minor spatial variabilities in the spectral slope in the 1.5–2.2-μm interval (normalized

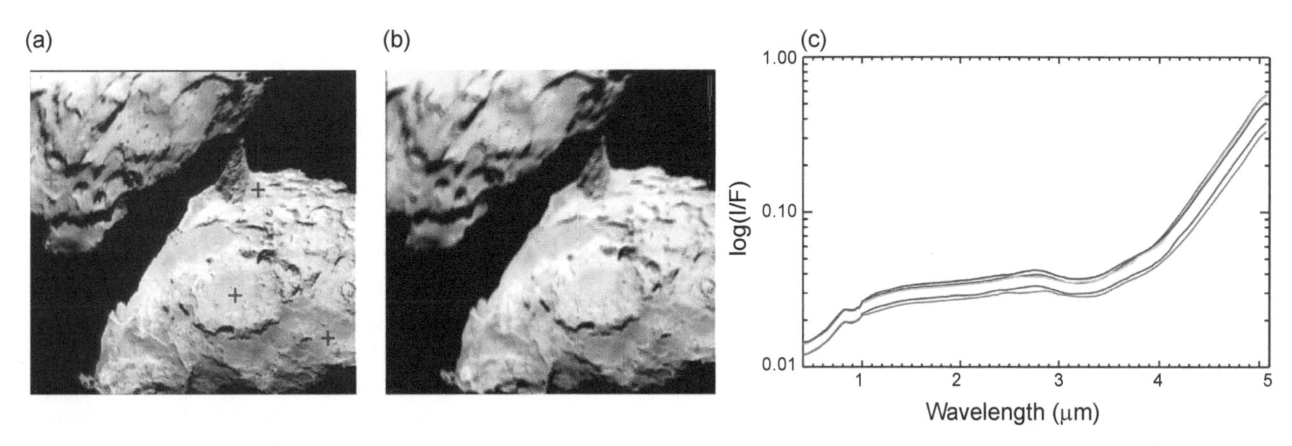

Fig. 4. Hyperspectral images (at ~13 m/pix resolution) of the nucleus of Comet 67P/CG from VIRTIS. **(a)** I/F color image (blue = 0.44 μm, green = 0.55 μm, red = 0.7 μm). **(b)** I/F color image (blue = 1.5 μm, green = 3 μm, red = 4.5 μm). **(c)** Average I/F spectra of 5 × 5 pixel regions indicated by the colored crosses in **(a)** and **(b)**. The feature common to all the spectra in the 0.8–1-μm interval is a calibration artifact corresponding to the junction of the instrument visible and infrared channels. Adapted from *Filacchione et al.* (2016a).

at 1.8 μm), with typical values of 3.0%/kÅ to 3.5%/kÅ, and smaller spectral slopes (2.0%/kÅ) corresponding to the water-ice-rich areas (*Groussin et al.,* 2013) (Fig. 8, right panel). In addition, *Davidsson et al.* (2013) found a correlation between 9P/Tempel 1 surface morphology and the 1.5–2.2-μm spectral slope, which appears redder on thick layered terrains (median value 3.4%/kÅ) and more neutral on smooth and water-ice-enriched terrains (median value 3.1%/kÅ).

For the nucleus of Comet Hartley 2, HRI and MRI disk-integrated observations at visible wavelengths revealed a linear featureless spectrum in the 400–850-nm interval, with an overall spectral slope of 7.6 ± 3.6%/kÅ, while color ratio maps indicate relatively larger color variations with respect to Tempel 1, with a full width of half maximum (FWHM) of 12% (*Li et al.,* 2013). In particular, images acquired during the inbound leg of the Deep Impact spacecraft flyby of Hartley 2 showed areas near the terminator characterized by a smaller spectral slope, indicating a higher fraction of water ice on the surface (*Li et al.,* 2013; *Sunshine et al.,* 2011).

A similar spatial distribution is revealed for the 1.5–2.2-μm spectral slope as inferred from HRI-IR inbound images (*Groussin et al.,* 2013), with red areas (spectral slope of 3.0%/kÅ to 3.5%/kÅ) close to the nucleus limb, and infrared-bluer regions (average spectra slope values of 0%/kÅ to 1.5%/kÅ) near the terminator (Fig. 8).

Comet 19P/Borrelly's nucleus surface has been investigated by means of the Miniature Integrated Camera and Spectrometer (MICAS) (*Rodgers et al.,* 2007), which provided both visible broadband images and infrared (1.3–2.6-μm) spectral-spatial images. In particular, these latter evidenced a linear red-sloped infrared spectrum, possibly exhibiting some variability between the large and small ends of the nucleus. The spectrum lacks significant features (including water-ice bands), with the exception of a recurrent 0.1-μm-wide absorption at 2.39 μm that remains unassigned (*Soderblom et al.,* 2004).

In Fig. 9 we compare representative infrared spectra of cometary nuclei explored by spacecraft: 67P/CG, 19P/Borrelly, 103P/Hartley 2, and 9P/Tempel 1. For 67P/CG we report spectra at two different phase angles (40° and 90°), providing a reference for possible phase reddening (see section 2.3) affecting the overall spectral slope of the other comets, which were observed at different phase angles (from 41° to 84°). Whereas the above-mentioned distinct absorptions at 3.2 μm on 67P/CG and at 2.39 μm on 19P/Borrelly are visible, for Comets 103P/Hartley 2 and 9P/Tempel 1, no relevant features have been reported so far, with the exception of spectra acquired over water-ice-rich areas (*Sunshine et al.,* 2006, 2011). Comets 67P/CG, 103P/Hartley 2, and 9P/Tempel 1 display a similar infrared spectral slope, comparable to D-type objects, which are among the reddest asteroids, whereas 19P/Borrelly displays an even larger infrared spectral slope, standing out as the reddest nucleus. Assuming that phase reddening, if present, produces effects with magnitude comparable to those observed for 67P/CG, 19P/Borrelly spectrum (phase angle = 41°) appears to be intrinsically redder than the others. In this respect, we also note that the resulting infrared spectral slope can depend on heliocentric distance, as shown for Comet 67P/CG (*Fornasier et al.,* 2016; *Ciarniello et al.,* 2016, 2022; *Filacchione et al.,* 2016c, 2020), for which a reduction of the visible and infrared slope has been observed while approaching the Sun, as the effect of an increase of water-ice surface content. The observations of 19P/Borrelly discussed here were acquired at an heliocentric distance of 1.36 au (*Soderblom et al.,* 2004), significantly smaller than the ones of 67P/CG (3.6–2.4 au), and intermediate between the cases of Tempel 1 (1.51 au) and Hartley 2 (1.06 au) (*Groussin et al.,* 2013). From this, one might speculate that 19P/Borrelly would appear even redder, if observed at heliocentric distances comparable to the ones of the presented 67P/CG spectrum; however, any firm conclusion in this respect would require an orbital monitoring of the surface spectral properties.

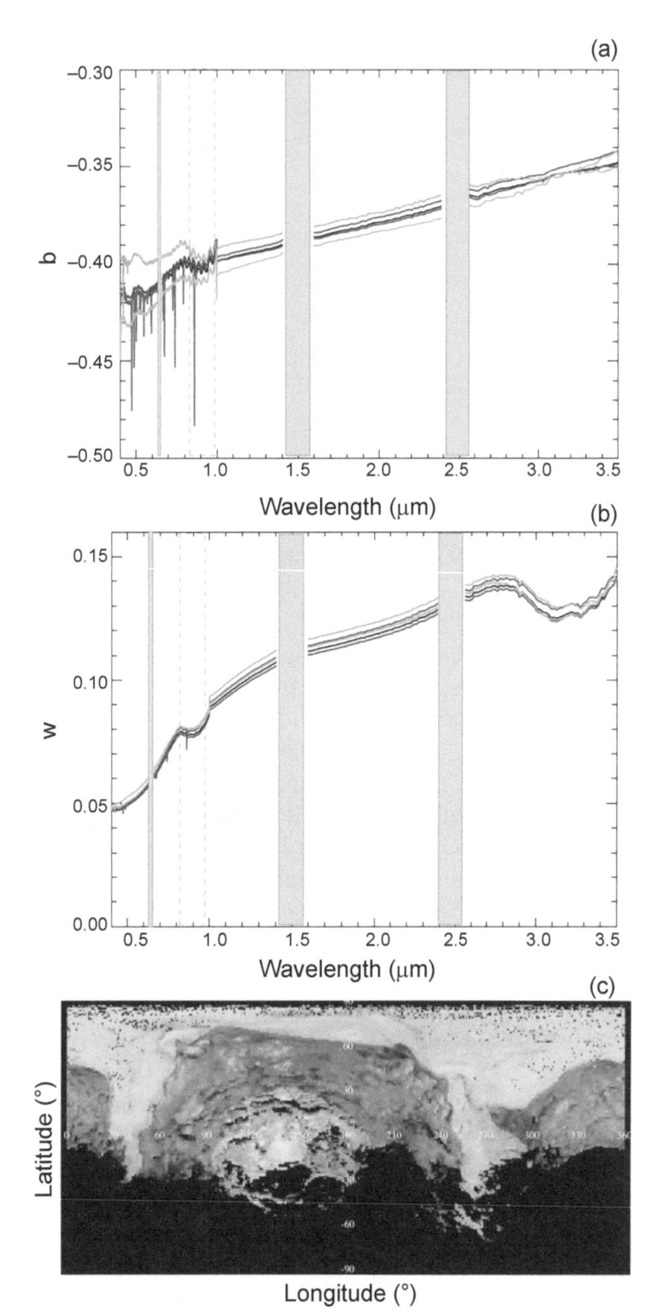

Fig. 5. 67P/CG average photometric parameters: **(a)** asymmetry factor b and **(b)** single scattering albedo w derived for four macro-regions: "head" (orange), "neck" (blue), "body" (green), and "bottom" (red) as defined on the map in **(c)**. Black curves correspond to the average b and w values. Gray bands on the spectral plots indicate the positions of the junctions of instrument order-sorting filters; the dashed box corresponds to the junction of the visible and infrared spectral channels and instrument straylight. Adapted from *Ciarniello et al.* (2015).

2.5. Composition from Surface Measurements

Prior to the Rosetta-Philae space mission, the main challenge in studying the cometary dust was in the collection method adopted on the spacecraft. As an example, the Stardust mission during the flyby with Comet 81P/Wild 2 was moving with a high relative velocity (~6 km s^{-1}) with respect to the nucleus. The spacecraft successfully collected dust grains trapped in the aerogel target exposed on the sample return capsule (*Brownlee et al.,* 2003). The main bias of this method is the high relative motion because at such velocities a fraction of the dust material, especially its organic components, is thermally destroyed or evaporated when impacting the collection target (*Brownlee et al.,* 2006).

Conversely, the Rosetta orbiter reduced the relative velocity to ~10 m s^{-1}; consequently, the chemical composition and physical structure of a substantial number of particles collected by the Grain Impact Analyser and Dust Accumulator (GIADA) dust experiment remained mainly unchanged (*Rotundi et al.,* 2015). Another advantage of the Rosetta mission is the presence of the Philae lander aiming for the first time to sample and analyze the cometary material directly on the nucleus surface. The scientific payload of Philae included two mass spectrometers devoted to the analysis of volatile and refractory material. The task of Philae was a direct analysis of the nucleus composition, other than the physical properties of the cometary surface and interior, bypassing the complex relationship between coma and nucleus compositions (*de Almeida et al.,* 1996; *Huebner and Benkhoff,* 1999). The contributions of *in situ* instruments onboard the Philae lander and remote sensing payloads on the Rosetta orbiter are valuable to obtain a unique picture of the comet composition. On November 12, 2014, at 15:34:04 UTC, Philae touched down at Agilkia (335.69° longitude, 12.04° latitude), just 112 m away from the selected landing site (*Biele et al.,* 2015). However, the shooting of the harpoons, necessary to secure the lander to the nucleus' surface, did not happen when the touchdown signal trigger arrived. The lander feet sunk into the soft regolith on the surface, ejecting a cloud of dust grains over the touchdown area imaged by OSIRIS (*Keller et al.,* 2007) and the navigation cameras onboard the Rosetta orbiter. After a timespan of about 20 s, Philae bounced back from the surface, starting an almost two-hour-long hopping tour across the surface of the comet. All available orbiter and lander observations and housekeeping data from the Philae touchdown at Agilkia were analyzed and permitted the reconstruction of the sequence of events the lander experienced on the ground, and even allowed the determination of important physical parameters of the cometary surface at the landing site and at the second touchdown. In the latter, Philae spent almost two minutes of its cross-comet journey, exposing primitive water ice that was analyzed by instruments onboard the orbiter, allowing important findings on the content of pristine water ice in the nucleus material (*O'Rourke et al.,* 2020), which are discussed in section 3.1. Lastly, the lander Philae arrived under an overhanging cliff in the Abydos region, where its journey ended. Here the lack of solar illumination prevented the batteries to be further recharged by the solar cells and caused the end of the Philae mission.

Thus, information on the chemical composition coming from the Philae lander was mainly obtained by the Cometary Sampling and Composition (COSAC) (*Goesmann et*

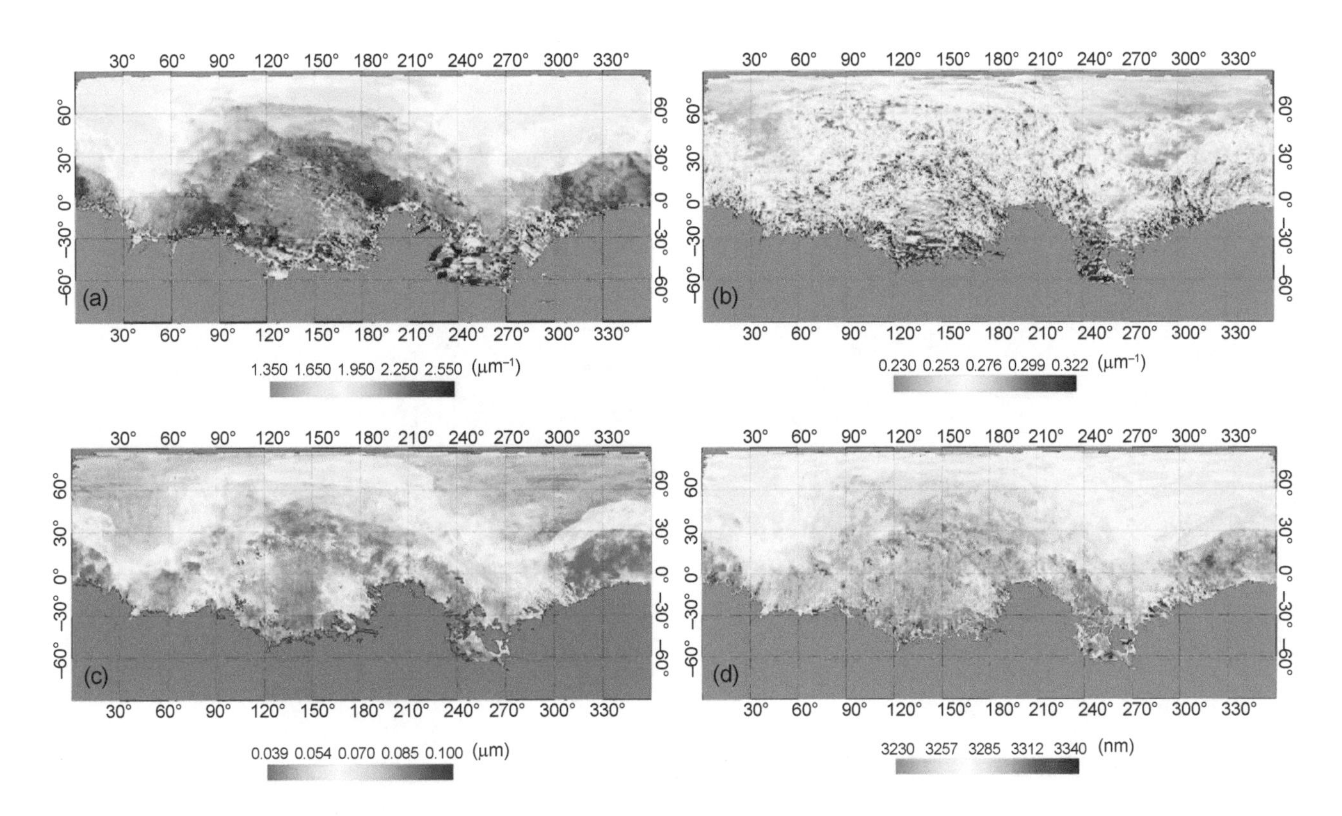

Fig. 6. Comet Spectral Indicator maps from VIRTIS-M observations. Shown are the **(a)** VIS and **(b)** IR spectral slopes, and the **(c)** band depth and **(d)** band center position of the 3.2-μm absorption feature, for observations performed in the timeframe August 8–September 2, 2014, when 67P/CG was at a heliocentric distance of 3.62–3.44 au on the inbound orbit. Adapted from *Ciarniello et al.* (2016).

al., 2007) and Ptolemy (*Wright et al.,* 2007) instruments, operating in "sniffing mode" during the early hopping phase of the lander and after landing at the Abydos site. Besides species probably released directly from the nucleus (H_2O, CO_2, CO), most of the peaks in the measured low-resolution mass spectra in both instruments are due to fragment byproducts partly created with or after the release of the species, but mostly by the ionization process inside the instruments. These fragments are associated with organic species. In this respect, the COSAC data were interpreted (*Goesmann et al.,* 2015) as being dominated by gas released from cometary dust particles unintentionally collected via the instrument gas exhaust pipe during the formation of the dust cloud following the lander touchdown at Agilkia. Twenty-five minutes after Philae's initial comet touchdown, the COSAC mass spectrometer took a spectrum that displayed a suite of 16 compounds, including many nitrogen-bearing species, and four compounds — methyl isocyanate, acetone, propionaldehyde, and acetamide — that had not previously been reported in comets. Sulfur compounds, if any, are below the detection limits for both the COSAC and Ptolemy instruments (*Wright et al.,* 2015; *Goesmann et al.,* 2015).

Fig. 7. Color ratio maps of the nucleus of Comet Tempel 1 from HRI and MRI images. The darkest areas correspond to the water-ice-rich (bluest) regions. From *Li et al.* (2007a).

The interpretation of the mass spectra peaks measured with Ptolemy (*Wright et al.,* 2015) suggests a fractionation pattern that might be related to chain-forming species. Short-chained polyoxymethylenes, first proposed to explain ion mass spectra during the Giotto flyby at Comet 1P/Halley (*Huebner et al.,* 1987), were considered possible candidates. Ptolemy measurements also indicated an apparent absence of aromatic compounds. Such Ptolemy preliminary results were partially revised in a successive analysis by comparing Ptolemy, COSAC, and Rosetta Orbiter Spectrometer for Ion and Neutral Analysis (ROSINA) data, being the latter onboard the orbiter; the peaks on masses 91 and 92 Da in the Ptolemy spectrum are most probably due to the aromatic molecule toluene (CH_3-C_6H_5) and not to polyoxymethylene (*Altwegg et al.,* 2017). Ptolemy measurements also indicate very low concentrations of nitrogenous material. The low m/z 14 signal

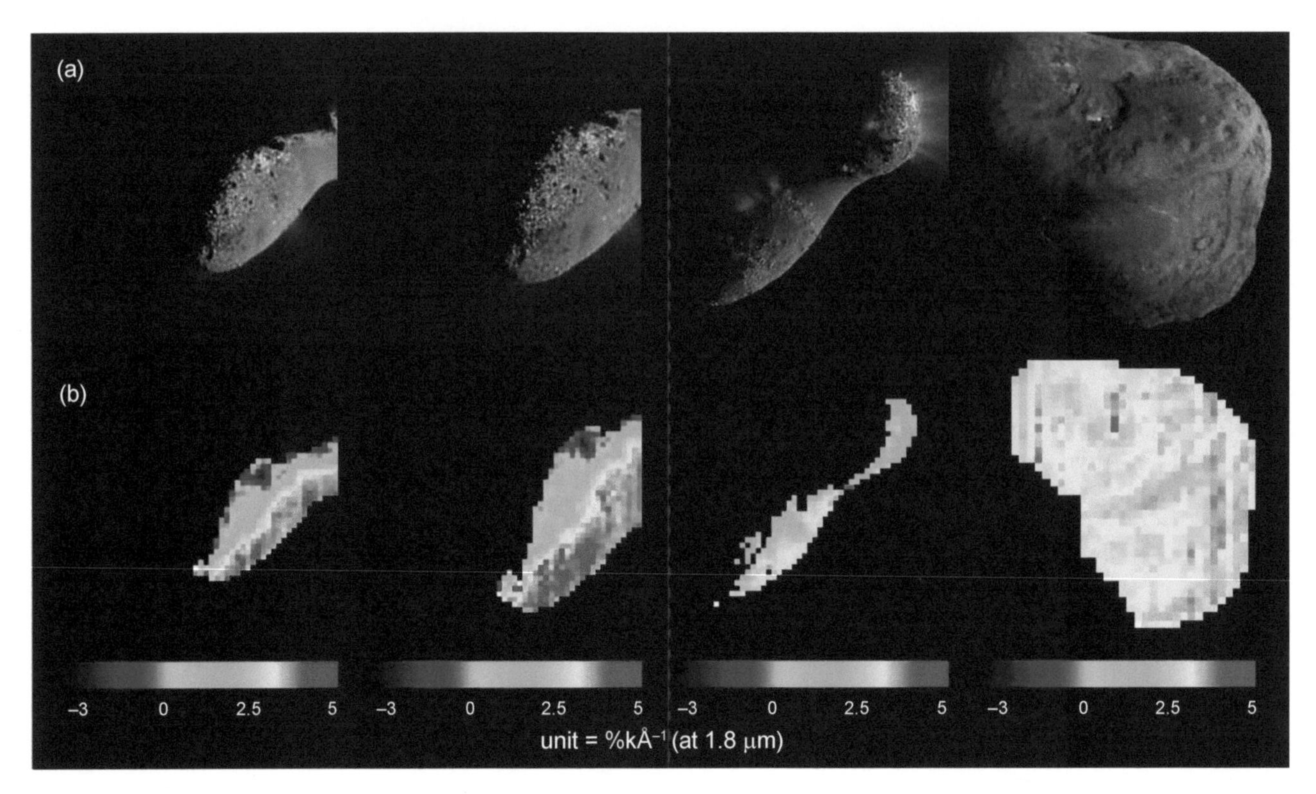

Fig. 8. (a) Context images of Comet Hartley 2 (three left panels) and Tempel 1 (right panel) at visible wavelengths from the HRI camera. **(b)** Maps of the 1.5–2.2-μm spectral slope normalized at 1.8 μm. Adapted from *Groussin et al.* (2013).

indicates that N_2 is not in high abundance, which agrees with the average N_2/CO ratio of 5.7×10^{-3} obtained by ROSINA (*Rubin et al.,* 2015). However, COSAC detected far more nitrogen-bearing compounds than Ptolemy. The COSAC findings differ from those of Ptolemy because COSAC sampled particles excavated by Philae's impact on the surface (*Goesmann et al.,* 2015) that entered the warm exhaust tubes located on the bottom of the lander, where they pointed toward the surface, whereas Ptolemy sampled ambient coma gases entering exhaust tubes located on top of the lander, where they pointed toward the sky. This was confirmed by the peak at m/z 44 detected by both instruments, which decays much slower in the COSAC measurements than in the Ptolemy data: That measured by COSAC was likely dominated by organic species, whereas the peak measured by Ptolemy was interpreted to be mostly due to CO_2 (*Krüger et al.,* 2017). Even if unambiguous detection of NH_3 was not possible because of the presence of H_2O and other potential compounds, such as CH_4, a nitrogen source such as NH_3 must originally have been abundant to form the many N-bearing species. All organic molecules detected by COSAC can be formed by ultraviolet irradiation and/or radiolysis of ices due to the incidence of galactic and solar cosmic rays: alcohols and carbonyls derived from CO and H_2O ices (*Goesmann et al.,* 2015), and amines and nitriles from CH_4 and NH_3 ices (*Kim and Kaiser,* 2011). Some of the compounds detected by these onsite measurements, especially those containing carbon-nitrogen bonds, play a key role in the synthesis of amino acids, sugars, and nucleins (*Dorofeeva,* 2020).

Ammonia, combining with many acids such as HCN, HNCO, and HCOOH encountered in the interstellar medium as well as in cometary ice, would also be the source of ammonium salts ($NH_4^+ X^-$) that two independent analyses from ROSINA (*Altwegg et al.,* 2020) and VIRTIS (*Poch et al.,* 2020) data suggested being abundant in the cometary dust of 67P/CG (see discussion in section 3.3). In this respect, cometary dust of 67P/CG analyzed by COSIMA shows a ratio N/C = 0.035 ± 0.011, which is 3× higher than the value observed in carbonaceous chondrites (*Fray et al.,* 2017), as also found for earlier observations of Comet 1P/Halley. Conversely, the content of oxygen in the dust fraction of both Comets 67P/CG and 1P/Halley is found to be lower than that in primitive chondrites. The explanation of these results is that the cometary material contains only dehydrated minerals, while all silicates in primitive carbonaceous CI-type chondrites are hydrated. The lack of hydrated minerals on the 67P/CG surface is also supported by analysis performed with VIRTIS instrument data (*Capaccioni et al.,* 2015; *Quirico et al.,* 2016). This is a probable indication of the absence of post-accretional internal heating of the 67P/CG nucleus (*Dorofeeva,* 2020).

2.6. Composition from Coma Measurements

By orbiting the Sun, a cometary nucleus undergoes cyclic sublimation and gas release resulting in the formation of the coma, a tenuous atmosphere made of accelerated gas and dust/ice particles encircling the nucleus (for further discussion about coma properties, see the chapters by Bodewits

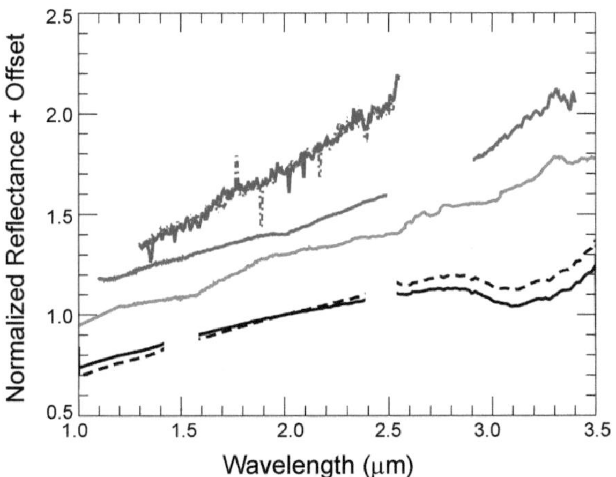

Fig. 9. Comparison among IR spectra of Comets 19P/Borrelly (blue curves, phase angle = 41°), 103P/Hartley 2 (red, phase angle = 84°), 9P/Tempel 1 (cyan, phase angle = 63°), and 67P/CG (solid and dashed black curves corresponding to phase angles of 40° and 90°, respectively). Spectra are normalized at 2 μm and offset for clarity. For 19P/Borrelly two spectra of the small end of the nucleus are superimposed (dashed and solid lines), which are affected by negligible thermal contribution (*Soderblom et al.,* 2004). For 67P/CG the two spectra are computed by applying the Hapke model from *Ciarniello et al.* (2015) at 40° and 90° phase angle, respectively, after the application of a corrective spectral factor resulting from the absolute calibration with star observations described in *Raponi et al.* (2020). In this case, a minor contribution of thermal emission is expected longward of ~3 μm. Missing parts of the spectrum for 67P/CG correspond to the junctions of instrumental order-sorting filters. The spectra of 9P/Tempel 1 and 103P/Hartley 2 have been derived by averaging spectra avoiding water-ice-rich spots from HRII hyperspectral images 9000036 and 5005001, respectively, after removal of the thermal contribution (*Raponi,* 2015). The missing part of the spectrum of 103P/Hartley 2 at ~2.7 μm corresponds to the wavelength interval of the H_2O-vapor emission in the foreground coma, contaminating the nucleus signal (*A'Hearn et al.,* 2011).

et al. and Biver et al. in this volume). As a consequence, the composition of the gas and dust particles dispersed in the coma allows an indirect characterization of the nucleus composition. The large diversity of volatile ices composition and relative abundances in comet populations (*Bockelée-Morvan and Biver,* 2017) can be traced back to formation environments in the outskirts of the protosolar disk; to the chemical, thermal, and radiation evolution during their orbital permanence in Oort and Kuiper clouds far from the Sun; and to recent thermal processing during their orbital evolution closer to the Sun.

Gaseous species are the result of the sublimation of solid ice fractions in the nucleus, mainly H_2O, CO_2, CO, while the dust grains are made of refractory and organic matter lifted by the gas flux from the surface's regolith. The genetic link between surface and coma composition is complicated for two reasons:

1. The volatile species are available on the nucleus as condensed ice or as trapped gases (*de Almeida et al.,* 1996; *Huebner and Benkhoff,* 1999). Each ice species is characterized by its own thermodynamical properties, resulting in different sublimation temperatures, pressures, and depths within the nucleus (*Huebner et al.,* 2006; *Fray and Schmitt,* 2009; *Marboeuf and Schmitt,* 2014). Instead, the availability of trapped gases depends on the molecular structure of the ice matrix (amorphous, crystalline, or clathrate hydrate), which is driven by thermophysical conditions occurring during the gravitational collapse from which the nucleus has been formed (*Espinasse et al.,* 1991; *De Sanctis et al.,* 2005; *Huebner et al.,* 2006; *Marboeuf and Schmitt,* 2014; *Blum et al.,* 2017). This means that the sublimation of each volatile species occurs at different heliocentric distances, with the more volatile species (CO_2, CO) still active at far distances from the Sun while water ice becomes active while the comet is orbiting inside the frost line (about 3.2 au from the Sun). The formation of the dust coma is in general associated with the diurnal sublimation of water ice whose flux is capable of lifting small grains from active areas (*Vincent et al.,* 2019), whereas larger icy chunks are associated with CO_2 activity (*A'Hearn et al.,* 2011).

2. The dust mantle blanketing the nucleus surface has a primary role in driving the intensity of the outgassing through its thickness and porosity (*Marboeuf and Schmitt,* 2014). Dust grains, once lifted from the surface by the gas flux, are illuminated by the Sun and experience rapid heating as a result of their low albedo and low heat capacity (*Bockelée-Morvan et al.,* 2019). Moreover, dust grains and gas molecules experience intense solar ultraviolet photon irradiation and electronic bombardment, causing the formation of ionized particles (*Cravens,* 1987; *Heritier et al.,* 2018): Erosion, aggregation, and chemical reactions of particles can alter the original properties of dust grains. All these mechanisms develop during the time of flight of each dust particle and are related to the heliocentric distance, the orbital obliquity, the irregular shape of the nucleus, and the period of rotation (*Huebner and Benkhoff,* 1999; *Fulle et al.,* 2016a).

The combined effects of these processes alter the coma properties, which could not represent directly the initial abundances of volatiles on the nucleus (*Espinasse et al.,* 1991; *Benkhoff,* 1995; *de Almeida et al.,* 1996; *Huebner,* 2008; *Marboeuf and Schmitt,* 2014). Several studies tried to link the spatial distribution of the coma composition with either a homogeneous or heterogeneous nucleus composition, taking advantage of the few observations available. Their results show very different cases: The analysis of two fragments of Comet 73P/Schwassmann-Wachmann 3 revealed very similar properties, suggesting a homogeneous composition of the nucleus (*Dello Russo et al.,* 2007). On the contrary, the Deep Impact mission found that the fresh material excavated from Comet 9P/Tempel 1 was remarkably different from the rest of the nucleus' surface (*Feaga et al.,* 2007). As shown in Fig. 10, Comet 103P/Hartley 2 (characterized by a bilobate nucleus shape similar to 67P/CG) also

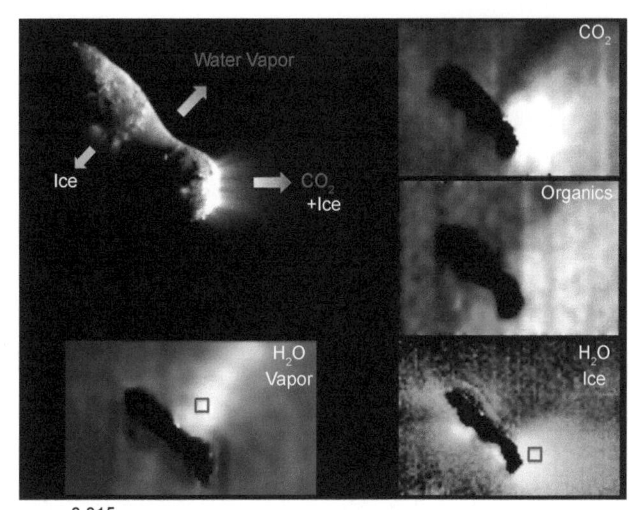

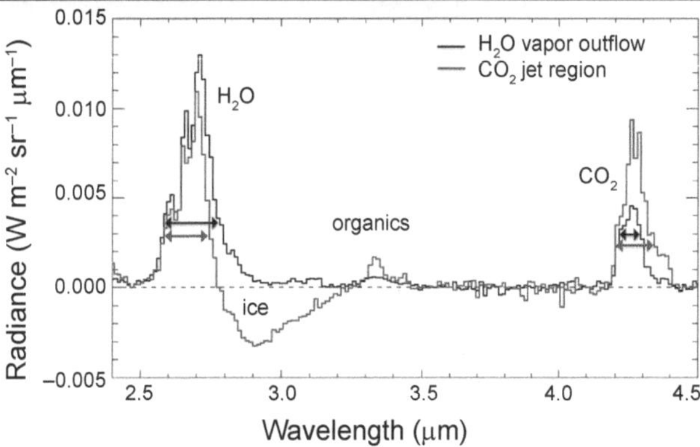

Fig. 10. Top panels: Distribution of Hartley 2 coma emissions due to water vapor, CO_2, and organic matter, in addition to a map of water ice particle absorption. These species are identified on spectral radiances (bottom panel) measured on H_2O vapor outflow and CO_2 jet regions. These two regions are localized within the two red squares shown in the images in the top panels. Water ice particles, identified through the 2.9-μm absorption band visible on the red spectrum, are associated with the CO_2 gas emission, which drags them out of the nucleus surface. From *A'Hearn et al.* (2011).

shows a coma with a heterogeneous spatial distribution as a consequence of intrinsic variations of composition within the nucleus (*A'Hearn et al.,* 2011; *Protopapa et al.,* 2014).

Before the Rosetta mission, many of the coma observations were too sparse and biased by the limited time series and surface coverage to allow a reliable estimate of the nucleus composition from coma properties. This explains the difficulties in interpreting cometary outgassing data and the resulting vivacious ongoing debate on the measurement of the dust-to-gas and refractory-to-ice mass ratios in comets and in 67P/CG in particular (*Choukroun et al.,* 2020): These ratios are the key parameters necessary to understand if the observed heterogeneity is primordial or a consequence of the evolutionary processes.

For 67P/CG, the ROSINA instrument (*Balsiger et al.,* 2007) has provided the time series of the H_2O, CO, and CO_2 fluxes, allowing exploration of the coma properties from different distances, nucleus orientations, and illumination conditions (Fig. 11). The time series collected along the inbound orbit show large fluctuations of the gases flux (*Hässig et al.,* 2015; *Luspay-Kuti et al.,* 2015). By linking gas flux, Rosetta's relative position, and nucleus orientation, *Läuter et al.* (2019) has inverted H_2O and CO_2 flux measured by ROSINA in the coma with localized gas emission sources on the nucleus surface. A similar study by *Hoang et al.* (2019) demonstrated that during the inbound orbit the outgassing of CO_2 and CO was more pronounced in the southern than in the northern hemisphere, and that the water vapor maximum sublimation was taking place on the Hapi active region. These results are in agreement with (1) the cyclic condensation of water frost in Hapi occurring at each nucleus' rotation observed at infrared wavelengths (*De Sanctis et al.,* 2015) and (2) the detection of a transient CO_2 ice deposit in the southern Anhur region (*Filacchione et al.,* 2016c). Both results will be discussed in the following sections. Near perihelion, several water-ice-rich locations placed on steep scarps and cliffs were identified as the sources of short-lived outbursts repeating across several nucleus rotations (*Vincent et al.,* 2016a): These were the major sources of water vapor activity measured on the coma by ROSINA at that time. Finally, carbon dioxide distribution in the coma appears compatible with the outgassing taking place above the transient CO_2 ice layer observed in the Anhur region while transiting from seasonal night to dayside (*Filacchione et al.,* 2016c; *Fornasier et al.,* 2016, 2017). The complete inventory of gases and parent species measured by ROSINA on 67P/CG is discussed in *Altwegg et al.* (2019). The gas production rates of 14 gas species (including H_2O, CO_2, CO, H_2S, O_2, C_2H_6, CH_3OH, H_2CO, CH_4, NH_3, HCN, C_2H_5OH, OCS, and CS_2) inferred from a two-year-long (2014–2016) time series are reported by *Läuter et al.* (2020). In general, from ROSINA data it appears evident that 67P/CG is enriched not only in CO and CO_2 with respect to other JFCs, but also in organic and highly volatile molecules (*Le Roy et al.,* 2015). Finally, mass spectroscopy has been successful in identifying other prebiotic species, including phosphorous, simple amino acid glycine (*Altwegg et al.,* 2016), ammoniated salts (*Altwegg et al.,* 2020), and cyano radicals (CN) (*Hänni et al.,* 2021).

Dust particles can be characterized with dedicated *in situ* instruments designed to collect and analyze them while the spacecraft is orbiting near the nucleus. On Rosetta, a suite of three payloads is employed, each of them dedicated to a specific range of particle sizes: (1) the Micro-Imaging Dust Analysis System (MIDAS) is an atomic-force microscope designed to measure grains from nanometers to micrometer sizes (*Riedler et al.,* 2007); (2) COSIMA is a time-of-flight secondary-ion mass spectrometer (TOF-SIMS) (*Kissel et al.,* 2007) sensitive to particles ranging from tens of micrometers to millimeters in size and able to image them after impacting on exposed targets (*Langevin et al.,* 2016); (3) GIADA measures the dust flux, including the velocity and mass of single grains, for particles $\geq$100 μm (*Colangeli et al.,* 2007).

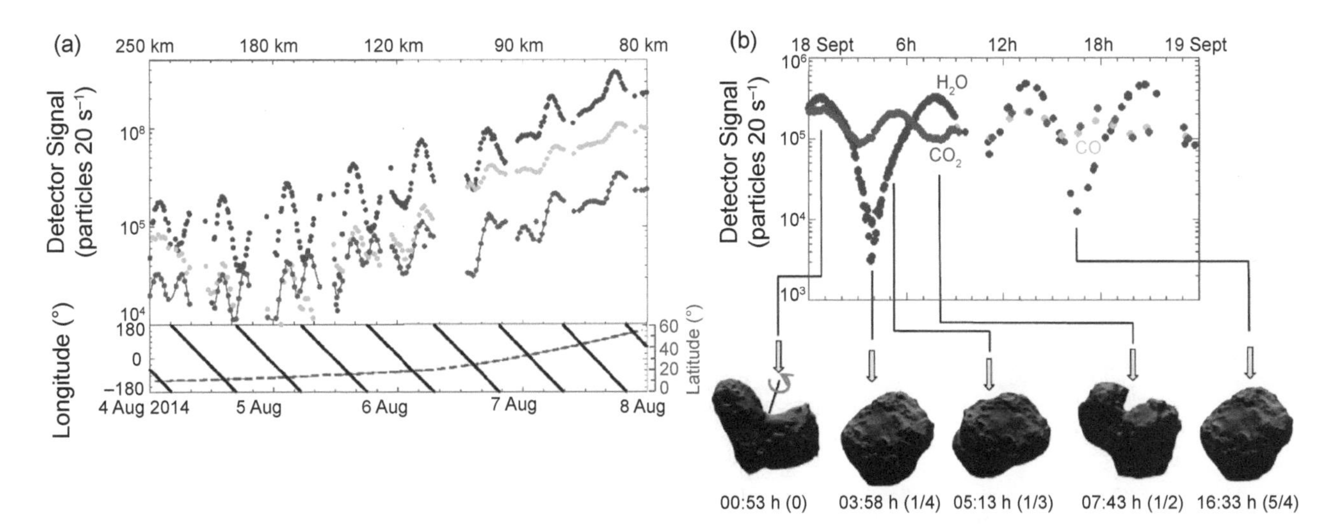

Fig. 11. ROSINA time series of the flux of H_2O, CO, and CO_2 in the 67P/CG coma. **(a)** Fluxes measured between August 4–8, 2014 (top plot). The corresponding position of the Rosetta spacecraft in longitude and latitude is shown in the bottom plot. (b) Fluxes (top plot) correlated with the relative orientation of the nucleus (bottom) for September 18, 2014. From *Hässig et al.* (2015).

The smallest dust grains observed by MIDAS are made by hierarchical agglomerates of compact and fractal particles (*Mannel et al.,* 2016; *Bentley et al.,* 2016). The elemental composition of collected grains is determined through secondary ion mass spectroscopy by COSIMA, which has inferred anhydrous mineral phases for 55% of the mass and 45% of the organic matter (*Bardyn et al.,* 2017) (see Table 3). This retrieval is based on certain assumptions on the H/C and O/C ratios, which cannot be determined by the instrument. Grains collected by Rosetta's dust instruments appear in general depleted in ices because they undergo rapid heating after ejection from the nucleus' surface.

A remarkable result concerns the purity of the individual grains: Within COSIMA's dust size range sensitivity (from 50 to 1000 μm), all collected particles contain both minerals and organics. This means that the non-volatile matter is intimately mixed up to tens of micrometers in scale, suggesting the presence of macromolecular carbonaceous material as an assemblage made of the elements listed in Table 3.

The dust flux of grains larger than 100 μm is measured by GIADA. *Rotundi et al.* (2015) and *Fulle et al.* (2016c) have determined that the dust-to-water-ice mass ratio is equal to 6 during the inbound orbit (from 3.6 au heliocentric distance to 1.23 au perihelion) and that the mass distribution of the dust particles is dominated by compact grains with a density of about 2000 kg m^{-3} and with sizes ≥1 mm. Moreover, GIADA data indicate that when measuring dust grains mass two different families appear: compact particles and fluffy porous aggregates having masses from 10^{-10} to 10^{-7} kg (*Della Corte et al.,* 2015).

By merging Rosetta's *in situ* and remote sensing data, *Güttler et al.* (2019) have obtained a comprehensive classification of 67P/CG's dust particles (Fig. 12), including (1) solid particles (e.g., irregular grains, roundish monomers, dense aggregate of grains); (2) fluffy particles (fractal dendritic agglomerates); and (3) porous particles (agglomerates and cluster of agglomerates). Apart from these classes, combinations of them are possible, e.g., porous aggregates with fluffy chains attached.

Apart from Rosetta, we have few insights into cometary dust composition. The Stardust mission collected and returned to Earth several dust samples from Comet 81P/Wild 2. As said, the high relative velocity during the sample collection (≈6.1 km s^{-1}) caused the loss of volatiles and organic matter fraction (*Brownlee et al.,* 2006). Elemental analysis of collected grains shows that 81P's particles are made by a matrix containing sulfide pyrrhotite, enstatite, and fine-grained porous chondritic aggregate. The existence of crystalline and high-temperature minerals in the comet's dust (*Bockelée-Morvan et al.,* 2002) indicates some transport mechanism of these materials from the inner regions of the solar nebula to the outer fringes (*Cuzzi et al.,* 2003). A large portion of the dust grains larger than 1 μm returned by the Stardust mission are made of olivine and pyroxene: Similar silicates have been observed on infrared spectra of Hale-Bopp (*Wooden et al.,* 1999) and Tempel 1 (*Lisse et al.,* 2006). Moreover, the presence of chondritic material suggests a further evolutionary link between comets and asteroids (*Gounelle,* 2011).

TABLE 3. Averaged composition of 67P/CG's dust particles as measured by COSIMA (from *Bardyn et al.,* 2017).

Element	Atomic Fraction	Weight Fraction
Oxygen	30%	41%
Carbon	30%	30%
Hydrogen	30%	2.5%
Silicon	5.5%	13%
Iron	1.6%	7.5%
Magnesium	0.6%	1.3%

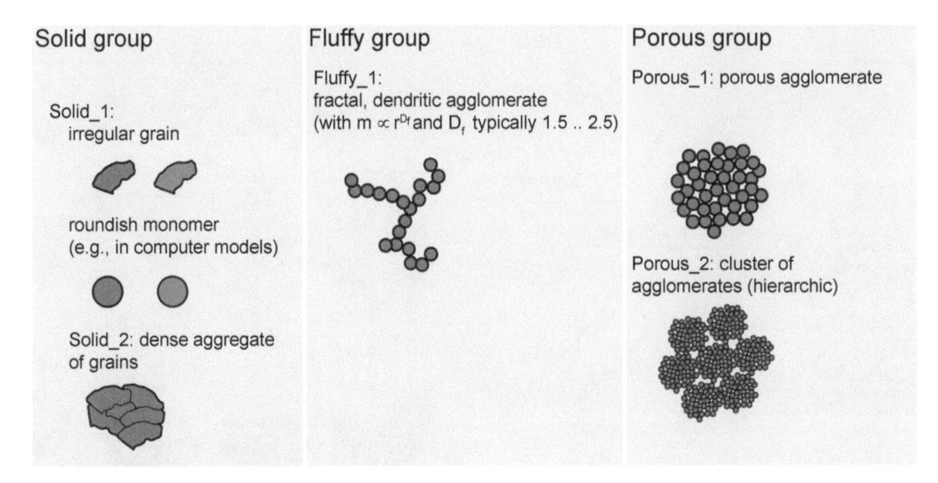

Fig. 12. 67P/CG's dust grain classifications. The gray/blue color indicates different compositions, as refractories and ices. From *Güttler et al.* (2019).

3. COMPOSITION END MEMBERS

Materials assembled in cometary nuclei preserve the composition of the presolar disk at the time of their formation. There is a general consensus among scientists that comets nuclei are dark and dirty snowballs are as described by the *Whipple* (1950) model. The cometary matter is made by a macromolecular assemblage containing a mixture of ices, organic matter rich in carbon, hydrogen, oxygen, nitrogen, and sulfur (referred to as CHONS), and minerals. In the following, we review recent results by space missions that have allowed us to better constrain the composition and relative abundances of end members.

3.1. Water and Carbon Dioxide Ices

Optical remote sensing observations at visible and near-infrared wavelengths allow sampling the surface composition of nuclei within a shallow layer (corresponding to about 10× the wavelength, i.e., 10 μm at $\lambda = 1$ μm) assuming a uniform slab. In porous media, like the surface of comets, photons can be scattered multiple times among nearby grains allowing in principle the sampling of larger depths. However, due to the extremely low albedo of the grains, multiple scattering is very limited in the cometary regolith. Moreover, the presence of dust particles, e.g., dehydrated fine and dark material, falling back on the nucleus' surface, is blanketing and coating underneath pristine layers, effectively hiding ices' spectral signatures. These effects must be taken into serious consideration when trying to extrapolate the nucleus' internal composition from surface composition measurements. While the spectral identification of cometary volatile species in extended comas is possible from Earth, surface ices are much more difficult to resolve due to the small dimensions of the nuclei and small amount of exposed ices and for these reasons have been recognized only through remote sensing observations performed by space missions able to navigate near the nuclei. So far, the only two space missions able to accomplish these measurements were Deep Impact at 9P/Tempel 1 and Rosetta at 67P/CG. Both comets have shown the presence of crystalline water ice while only on 67P/CG was it possible to identify carbon dioxide ice. In general, the presence of ices on cometary surfaces is regulated by solar illumination and by local topography.

3.1.1. Water ice. Despite the fact that water ice is the dominant volatile species in cometary nuclei, its presence has been assessed by visible-infrared spectroscopy only in very limited locations on 9P/Tempel 1 and 67P/CG where specific processes are active at the time of the observations. The following paragraphs discuss the main modalities in which water ice has been detected.

3.1.1.1. Recently exposed landslides and debris fields: Due to the rough surface topography, is not uncommon that elevated and overhang structures can collapse due to erosion and thermal stresses. This is the case of at least two specific areas located on 67P/CG's Imhotep region (see Fig. 13) where recent collapses have exposed bright material within the debris fields located at their bases (*Filacchione et al.,* 2016b). The study of these areas, called bright area patches (BAPs), allows therefore the derivation of the characteristics of more fresh ice as present in the outer layers of the nucleus. The best spectral fit to VIRTIS data evidence that the pixel's area located on the debris fields contain up to 1.2% of water-ice grains with a diameter ≈2 mm while the residual 98.8% of the pixel area consists of an intimate mixture made of 3.4% water-ice grains of 56 μm and average dark terrain for the remaining 95.4% (Fig. 13g). On average, the total amount of exposed water ice on the debris fields is ≈4.6% given a 2.5 m/pixel spatial resolution. The surface temperature of the water-ice-rich areas is T < 160–180 K, about 20–25 K colder than in the surrounding dehydrated areas. Despite these low temperatures, the debris fields are still warm enough to justify the presence of water ice in the crystalline state. The presence of a millimeter-sized grain population can be explained by the recent sintering of smaller grains or by the growth of ice crystals from vapor diffusion in cold layers of the subsurface.

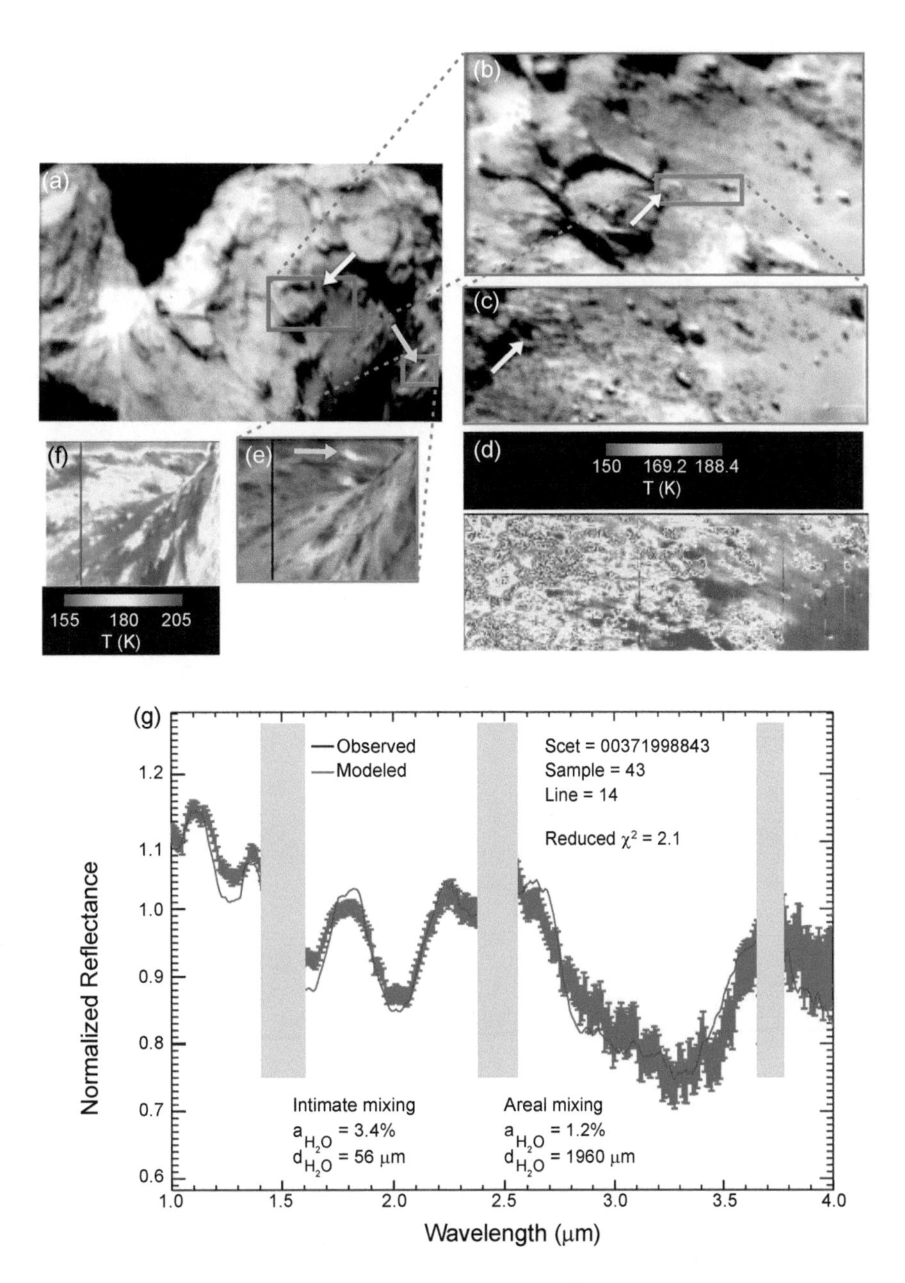

Fig. 13. Exposed water ice on recently disrupted terrains, like landslides and debris fields seen by Rosetta-VIRTIS on 67P/CG. **(a)** Color image showing BAPs 1–2 (white and yellow arrows, respectively) in the Imhotep region. **(b)** Water-ice-rich BAP 1 (blue area) is on the right side of a circular elevated structure. **(c)** High-resolution image of debris field 1 better shows the distribution of the water-ice-rich unit (bluish color). **(d)** Temperature image map shows that the water-ice-rich terrain is at $T < 160$ K, colder than the neighboring dark terrain. **(e)** Color image showing BAP 2. **(f)** Temperature map of BAP 2 showing the colder water-ice-rich terrain. All previous color images are rendered from 1.3 μm (blue), 2.0 μm (green), and 2.9 μm (red) wavelengths. **(g)** Water-ice-rich VIRTIS spectrum (black curve with error bars in red) from BAP 1 compared with the best-fit spectral model (blue curve) showing the percentage of crystalline water ice (a_{H_2O}) and the water-ice grain diameter (d_{H_2O}). The spectral ranges colored in gray correspond to positions of instrumental order-sorting filters where responsivity is uncertain. From *Filacchione et al.* (2016b).

3.1.1.2. Localized icy patches: These features are preferentially located on equatorial regions and in dust-free areas. High-albedo icy patches have been observed on high-resolution visible images by OSIRIS (*Pommerol et al.,* 2015) and for eight of them VIRTIS has confirmed the presence of water (*Barucci et al.,* 2016) (see Fig. 14). The icy patch size spans from a few to about 60 m and appears distributed in clusters of small bright spots or in much larger individual ones. Some of them are probable sources of activity and outbursts (*Vincent et al.,* 2016a). The surface water ice localized in the patches is almost stable for months on some of them, while in others it appears more variable and correlated with the illumination conditions (*Raponi et al.,* 2016). Co-located spectral analysis by VIRTIS shows extremely variable water-ice abundances and grain sizes: Mixing with the ubiquitous dark terrain is in both areal

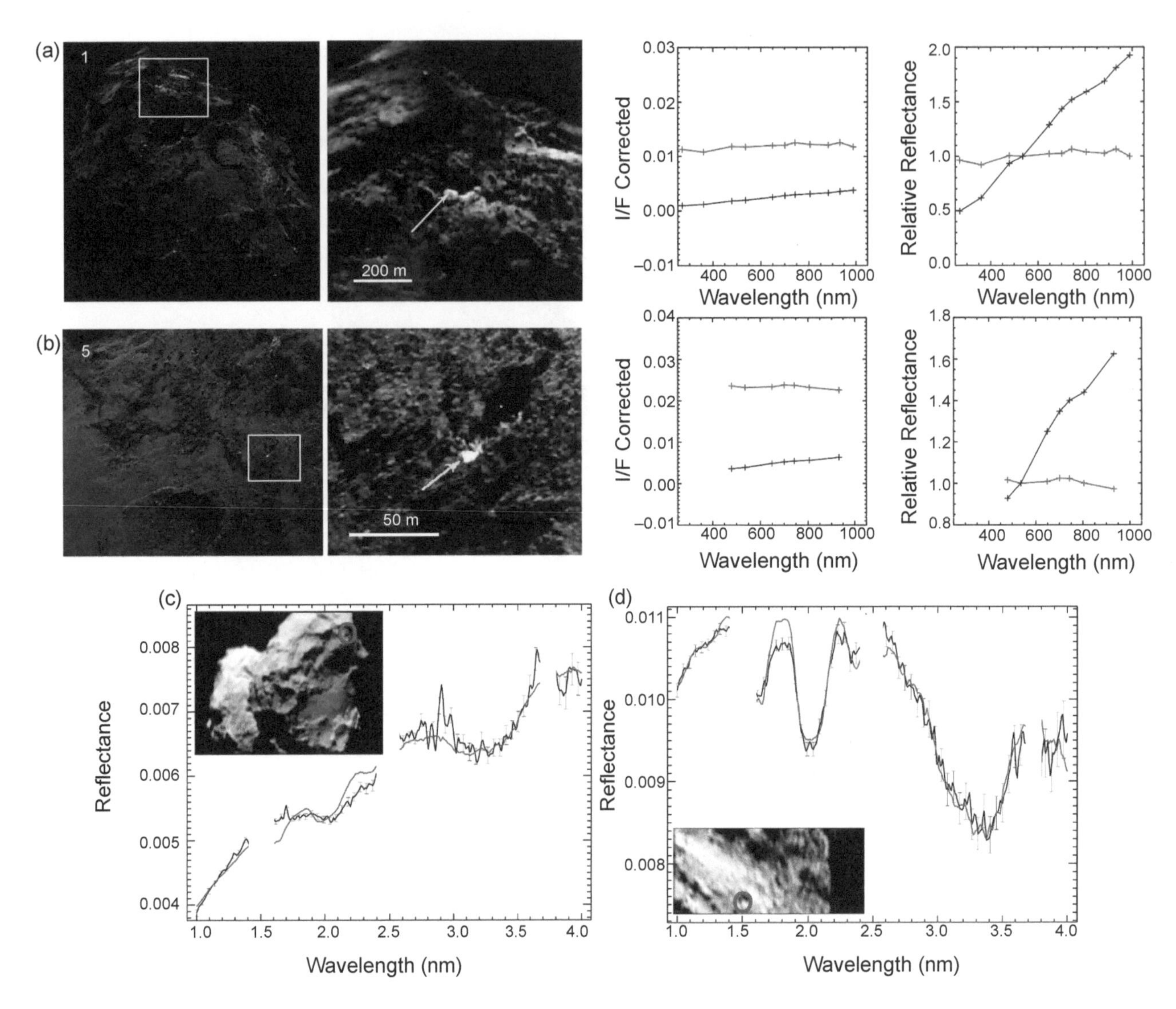

Fig. 14. Examples of localized icy patches observed by Rosetta, OSIRIS, and VIRTIS on 67P/CG. **(a)** OSIRIS images and VIS I/F spectra for patch #1. Spectrum in red is the one on the icy patch, the one in black is taken on the nearby non-icy terrain. The plot in column 3 shows I/F corrected spectra with higher values on the icy patch while in column 4 the same data are normalized at 0.55 μm to enhance the flat slope of the ice at visible wavelengths. **(b)** Same as **(a)** but for patch #5. **(c)** VIRTIS context image, reflectance spectrum, and best-fit solution for spot #1; **(d)** same as **(c)** but for spot #5. Patches #1 and #5 are located in Imhotep at lat. –5.8°, lon. 189.4° and at lat. –22°, lon. 182.8° respectively. From *Barucci et al.* (2016).

and intimate modalities. The derived water-ice abundance is in the 0.3–4% range with the presence of both tens of micrometer-sized grains in intimate mixing with the dark terrain or very large millimeter-sized grains in areal mixing.

Although water is the dominant volatile observed in the coma, exposed water ice on the cometary surface has been detected in relatively small amounts (a few percent) in several regions of the 67P/CG comet (*De Sanctis et al.,* 2015; *Filacchione et al.,* 2016b; *Pommerol et al.,* 2015; *Barucci et al.,* 2016; *Oklay et al.,* 2016; *Fornasier et al.,* 2017) and in higher amounts (>20%) in localized areas observed by the OSIRIS camera in the Anhur, Bes, Khonsu, Wosret, and Imhotep regions (*Fornasier et al.,* 2016, 2019a, 2021; *Deshapriya et al.,* 2016; *Oklay et al.,* 2017; *Hasselmann et al.,* 2019; *Leon-Dasi et al.,* 2021), in the Aswan site (*Pajola et al.,* 2017a), and in the Abydos landing site and surrounding area (*Hoang et al.,* 2020; *O'Rourke et al.,* 2020). Unfortunately, for many of those last detections infrared spectroscopic data are not available and ice identification and abundance are based on modeling albedo and visible colors.

The wider surface exposures of water ice were detected in the boundary between the Anhur and Bes regions in the southern hemisphere (*Fornasier et al.,* 2016). These two patches were 4–7× brighter than the surrounding regions, and about 1500 m^2 wide each. As shown in Fig. 15 they were observed for the first time on April 27, 2015, and survived on the surface for at least 10 days. They were located on a flat terrace bounded by two 150-m-high scarps and were formed on a smooth terrain made up of a thin seam of fine deposits covering the consolidated material.

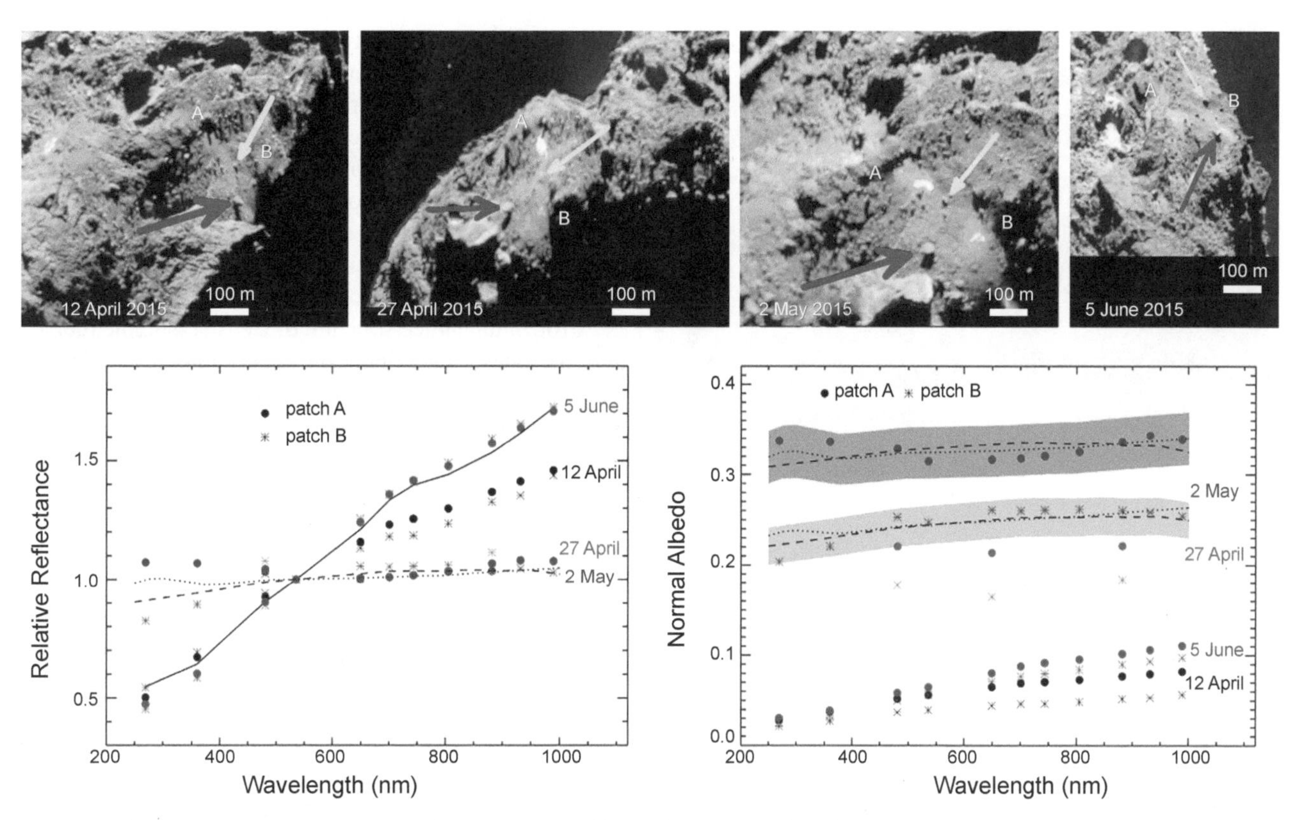

Fig. 15. Composite images (882-nm, 649-nm, and 480-nm filters) showing the appearance and evolution of the bright patches in the Anhur/Bes regions of 67P/CG (the arrows indicate two common boulders for reference). The reflectance relative to 535 nm and the normal albedo are represented in the bottom panels. The black line represents the mean spectrum of the comet from a region close to the patches. Dashed and dotted lines show the best-fit spectral models to the patches (gray shows the associated uncertainty), produced by the linear mixture of the comet dark terrain enriched with 21 ± 3% of water ice (dashed line) or 23 ± 3% of water frost (dotted line) for patch B, and with 29 ± 3% of water ice (dashed line) or 32 ± 3% of water frost (dotted line) for patch A. Simulations are performed assuming a water ice grain of 30 μm. From *Fornasier et al.* (2016).

Figure 15 shows an example of the spectrophotometric evolution of these bright patches: On April 12, 2015, the terrain looks spectrally bluer, a hint for higher water-ice abundance, but at this time the bright patches were not exposed yet. Instead they become clearly visible on April 27 and May 2 images when they show a relatively flat spectrum, consistent with the direct exposure of water ice. On June 5, 2015, that region was observed again and the two bright patches had fully sublimated (*Fornasier et al.,* 2016, 2017), leaving a layer spectrally indistinguishable from the average dark terrain. A water-ice abundance of 20–30% was estimated in these bright areas from linear mixing models. This corresponds to an ice loss rate between 1.4 and 2.5 kg day^{-1} m^{-2} for an intimate mixture and the estimated solid ice-equivalent thickness layer was 1.5–27 mm (*Fornasier et al.,* 2016). Notably, the presence of a small amount of CO_2 ice was detected by the VIRTIS spectrometer in an 80 m × 60 m area corresponding to the location of patch A (*Filacchione et al.,* 2016c), about one month before the detection of the bright patches with OSIRIS (see discussion in section 3.1.2).

3.1.1.3. Diurnal condensation cycle: This process is driven by the alternate illumination conditions occurring during each comet's rotation that are able to induce cyclic sublimation during the day and condensation during the night or when a given area moves into shadows (*De Sanctis et al.,* 2015). The net effect of this process is the condensation of water vapor, released from the subsurface when reached by the propagation of the diurnal thermal wave, on the cold surface during the night. In these conditions, in fact, the temperature inversion occurs between the colder surface regolith and the warmer internal layer. On 67P/CG, this cycle has been observed on the active Hapi region during the inbound orbit. Since the 67P/CG rotation period is about 12.4 hr (*Jorda et al.,* 2016), and due to the bilobate shape of the nucleus, the Hapi region was experiencing two Sun exposures, one shadow transit (caused by the small lobe) and one night period for each comet rotation. Infrared spectral analysis performed on the areas close to the transition line between nightside and dayside (see Fig. 16) shows that during the cold night very fine (<3 μm) water-ice grains were condensing and accumulating on the surface, reaching a maximum abundance of 14% and resulting in intimate mixing with the ubiquitous dark terrain (*De Sanctis et al.,* 2015). The volatile condensation-sublimation process has been suggested by *Prialnik et al.* (2008), who correlated the diurnal cycle with the presence of outbursts occurring during the morning hours on 9P/Tempel 1.

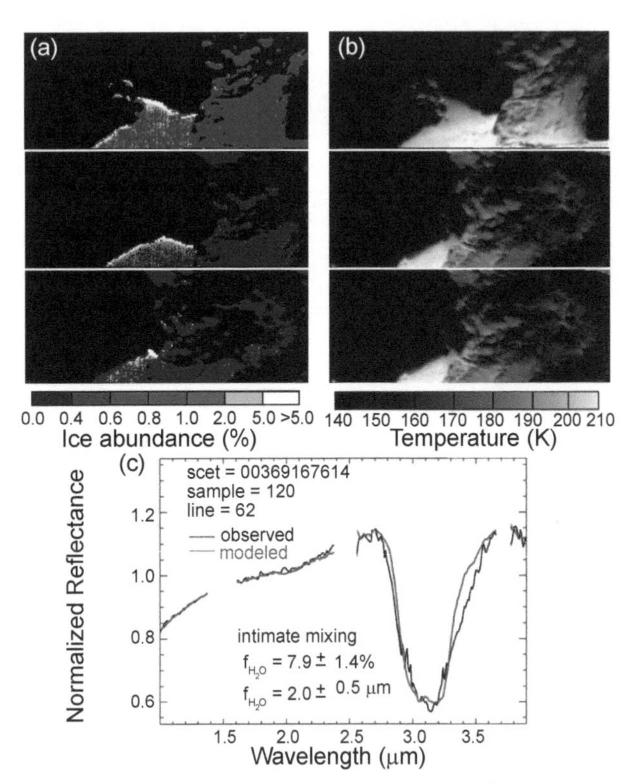

Fig. 16. The diurnal water-ice cycle on 67P/CG. **(a)** Images showing the ice abundance (by volume) across the Hapi active area while transiting across dawn hours. **(b)** Corresponding temperature images derived from thermal emission show that close to the shadow line the minimum temperature is about 160–170 K. **(c)** VIRTIS spectrum on a specific location compared with the best-spectral-fit matches to a mixture of average dark terrain and water ice grains with an abundance of 8% and grain size of 2 μm. From *De Sanctis et al.* (2015).

3.1.1.4. Dust-covered icy boulders: During the Philae landing process, the lander collided two times with the nucleus surface before stopping under an overhanging cliff located in the Abydos region.

The touchdown (TD) sites are shown in Figs. 17a,b,c. While Philae was bouncing on the TD2 site (area marked with the green dashed line in Fig. 17d), the lander impacted in a crevice located between two boulders, causing the removal of their dust-covered surface and the exposure of an area of ≈3.5 m^2 showing their pristine interiors (*O'Rourke et al.,* 2020). A multi-instrument campaign based on both Philae and Rosetta data shows that the boulders' interior material is very bright (up to 6× more than the average dark terrain nearby), enriched in water ice (46.4% with grain sizes on the order of 30–100 μm), and has a very low compressive strength (<12 Pa). The measurements show that the local dust-to-ice mass ratio is 2.3:1 (*O'Rourke et al.,* 2020). So far, this very peculiar measurement is the one that has allowed us to measure the highest abundance of water ice on the surface of 67P/CG and it suggests that pristine ice-rich localizations are available in the subsurface.

3.1.1.5. Water ice on 67P/CG: A summary: Thanks to infrared observations performed by Rosetta it is possible to understand the distribution of the superficial water-ice grains. A first result is that the water ice appears only in crystalline form, a result derived from the observation of the diagnostic properties of the 1.5–1.65- and 2.05-μm bands. Remote sensing observations cannot resolve the question about the presence of primordial amorphous ice in the deep nucleus since it cannot be reached at optical wavelengths. Moreover, at the typical surface temperature measured on water-ice-rich locations [from 160 to 180 K (*Filacchione et al.,* 2016b)], the lattice structure of ice is crystalline (cubic) since the phase transition from amorphous to crystalline form occurs at lower temperature [137 K (*Lynch,* 2005)]. Crystalline surface ice signals its presence in three distinct grain size populations whose existence is related to the processes behind the ice formation: (1) Micrometer-sized grains are characteristic of the more active areas where diurnal variations occurring on timescales of few hours preclude the condensation of larger grains (*De Sanctis et al.,* 2015). This presence of micrometer-sized water-ice grains is evidenced by the very absorbing H_2O stretch band between 2.7 and 3.0 μm and by the contemporary absence of the other diagnostic bands at 1.05, 1.25, 1.5, 1.65, and 2.05 μm that instead are associated with larger grains. Such micrometer-sized grains are also probably contributing to the 1–2% abundance of water ice (upper limit) associated with the dark and red dehydrated material ubiquitously seen across the surface of 67P/CG (*Capaccioni et al.,* 2015). (2) Tens of micrometers, or intermediate grains, are instead associated with intimate mixtures of water ice and dark material localized in bright icy patches. (3) Coarse millimeter-sizes grains are probably the consequence of condensation of ice crystals from the vapor diffusion occurring in ice-rich colder layers or by ice grain coalescing and sintering caused by thermal evolution. Due to their larger dimensions, this family of grains appears preferentially distributed in areal mixtures together with dark terrain grains.

3.1.1.6. Ices on other comets: Before 67P/CG exploration by Rosetta, other space missions have searched for hints of ices on comet nuclei. The Deep Space 1 mission to Comet 19P/Borrelly (*Nelson et al.,* 2004) was not able to detect any water-ice signatures in the 1.3–2.6-μm spectral range with its near-infrared spectrometer (*Soderblom et al.,* 2002, 2004). On the contrary, three large blue-colored high-visible-albedo areas were identified on Comet 9P/Tempel 1 by the HRI and MRI onboard the Deep Impact mission (*Sunshine et al.,* 2006). Concurrent observations of the same areas by the onboard infrared spectrometer confirmed the presence of water ice thanks to the detection of the diagnostic absorption bands at 1.5, 2.0, and 3.0 μm (see Fig. 18). The ice-rich areas correspond to three depressions reaching about 80 m below the neighborhood. The overall area occupied by these depressions is estimated in about 0.5 km^2 or 0.5% of the total nucleus surface.

Spectral modeling points out that in the depressions the water ice is mixed with the average dark terrain, reaching an average abundance of 6 ± 3% and grain sizes of 30 ± 20 μm (Fig. 19). Moreover, the depressions are acting like cold

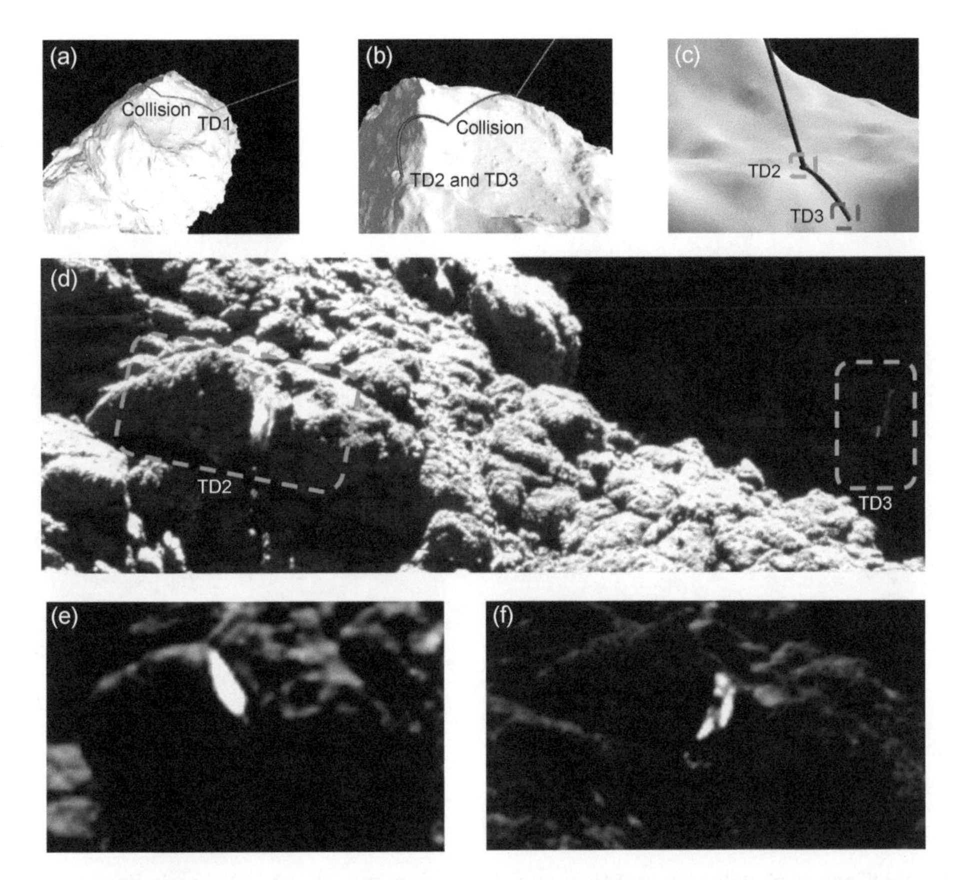

Fig. 17. Discovery of dust-covered icy boulders on the surface of 67P/CG. **(a)–(c)** Philae's landing trajectory and touchdown locations. **(d)** High-resolution image (4.9 cm/pixel) showing TD2 (green dashed line) and Philae final position on TD3 (blue dashed line). Note the crevice in the TD2 place. **(e)–(f)** Images of the high-albedo icy interiors of the boulders in the crevice. From *O'Rourke et al.* (2020).

traps, having diurnal temperatures of 280–290 K, much colder than nearby terrains (≥300 K).

Finally, as already discussed in Fig. 10, water-ice grains emitted from the surface of Hartley 2 associated with the emission of CO_2 have been observed in the coma (*A'Hearn et al.*, 2011).

3.1.2. Carbon dioxide ice. After water ice, carbon dioxide is the second most abundant gaseous species observed on cometary comae (*Bockelée-Morvan and Biver*, 2017). Despite its high abundance as gas, the observation of the solid form on nuclei surfaces is extremely difficult due to the high volatility: In vacuum, carbon dioxide rapidly sublimates starting at 70 K (*Huebner et al.*, 2006). So far, the only spectroscopic detection of carbon dioxide ice on a cometary nucleus has been reported by VIRTIS on 67P/CG (*Filacchione et al.*, 2016c) on a 60 × 80-m-wide area located in the Anhur region. This specific area was transiting outside the four-year-long winter-night season and returning on the dayside, an event occurring only in the proximity of the perihelion passage. During a great part of the orbit far from the Sun, including the aphelion passage, the Anhur area is constantly in the nightside. As a consequence of the lack of solar illumination it reaches very low surface temperatures, probably on the order of <50 K. Thanks to Rosetta's close orbit to the nucleus and trajectory's excursion, on March 21, 2015, it was possible to observe the areas close to the terminator where VIRTIS has detected the diagnostic 2.0-μm triplet feature and the 2.6–2.75-μm bands characteristic of CO_2 ice (Fig. 20).

The best spectral fits to VIRTIS data give an abundance of CO_2 ice of <1% in both areal or intimate mixing with the dark terrain (Fig. 20a). Subsequent observations confirm that the CO_2 ice deposit was rapidly sublimating over the following weeks, until completely disappearing when VIRTIS observed the same area in Anhur on April 12–13, 2015 (Fig. 20b). The thermal history experienced by this area indicates that the surface underwent a loss of 56 kg of CO_2 ice corresponding to an erosion of a layer of 9 cm. To explain this process one could presume that the ice patch was formed during the previous orbit when the carbon dioxide outgassing from the nucleus' deeper layers reached by the perihelion's thermal heatwave was condensing on the cold winter surface. The Anhur region in fact is placed on the nucleus' southern part, which after a continuous and intense insolation at perihelion, enters a four-year-long winter night. This occurs with a fast drop in illumination after the outbound equinox, as a consequence of which the outgassing of CO_2 coming from the inner warm layers finds a rapidly cooling surface acting as a point of condensation and accumulation for the ice. When the comet orbits toward the

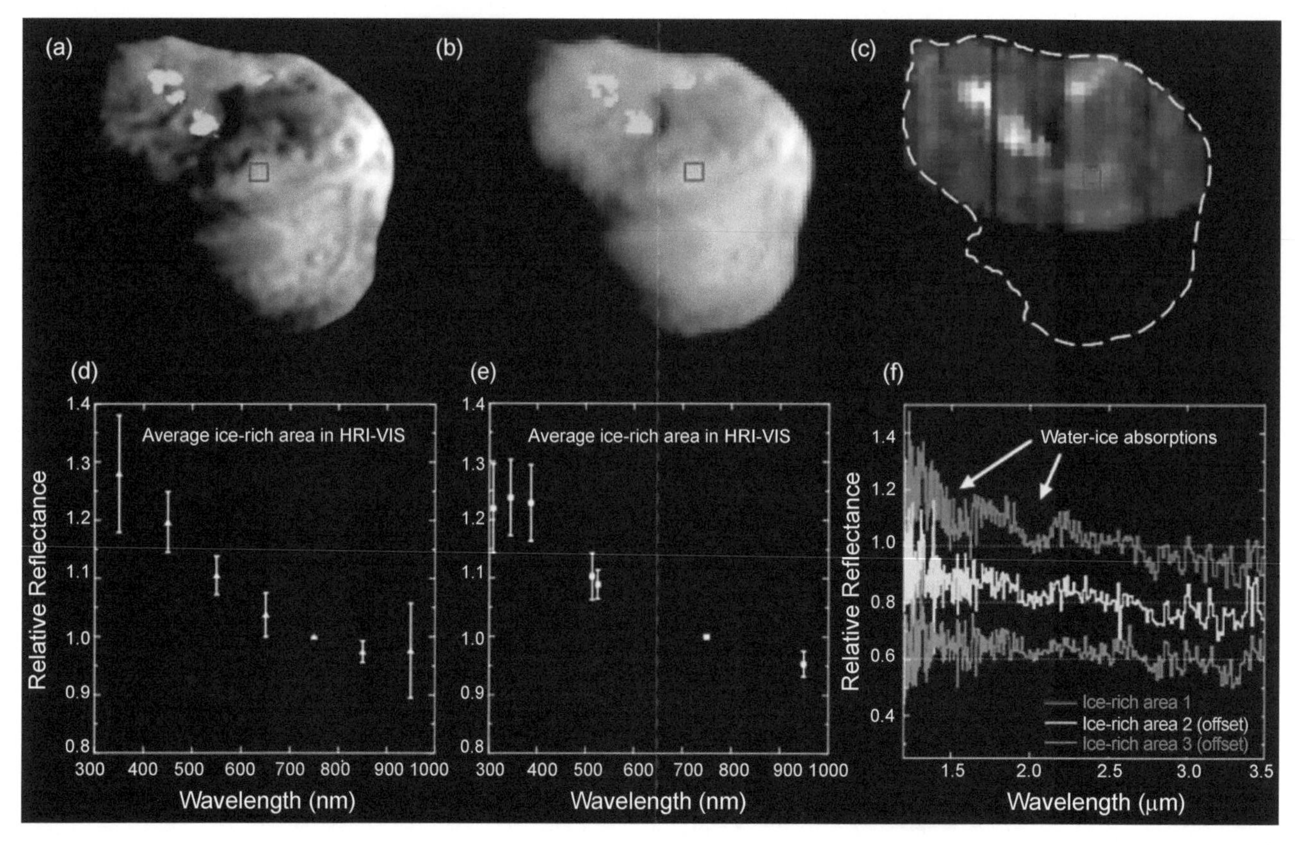

Fig. 18. Spectral identification of water ice on 9P/Tempel 1 by Deep Impact. **(a)** HRI visible image at 16 m/pixel spatial resolution showing the three blue-colored high-albedo depressions (marked in light blue color). **(b)** MRI visible image. **(c)** IR image showing the 2.0-μm water-ice band depth at 120 m/pixel spatial resolution. **(d)** Relative spectral reflectance of the inset shown in **(a)**. **(e)** Spectrum of the area shown in the inset in **(b)** at 82 m/pixel confirms HRI data. **(f)** Relative reflectances of the three water-ice-rich areas. Note the presence of the water-ice diagnostic bands on the three spectra shown. From *Sunshine et al.* (2006).

perihelion, the Anhur region moves out from winter night to local summer day: The abrupt increase of the solar illumination causes the rapid disappearance of carbon dioxide ice from the surface.

3.2. Organic Matter

Prior to the Rosetta mission, direct information on the refractory component of the organic material on a comet were provided by 81P/Wild (Wild 2) grains gathered by the Stardust sample return mission and by meteorites of probable cometary origin, such as interplanetary dust particles (IDPs) and Antarctic micrometeorites (AMMs). Analysis performed on Wild 2 grains demonstrates the presence of complex aromatic organic matter. N-rich material appears to be present in the form of aromatic nitriles. However, aromatic organics consist of a small fraction of the total organic matter present (*Clemett et al.,* 2010). Infrared spectra of Wild 2 grains reveal indigenous aliphatic hydrocarbons with longer chain lengths than those observed in the diffuse interstellar medium (*Keller et al.,* 2006). Stratospherically collected IDPs and AMMs show a close match with organic species in the Stardust samples, relatively to the above-mentioned properties (*Keller et al.,* 2006; *Clemett et al.,* 2010). In turn, refractory organic matter in stratospheric IDPs and AMMs of cometary origin shares similarities with chondritic insoluble organic matter (IOM). Raman spectroscopy analyses have revealed that these particles also contain a disordered polyaromatic structure (*Quirico et al.,* 2005, 2016). However, H/C values reported in stratospheric IDPs display a larger range of values than reported in chondritic IOM (*Aléon et al.,* 2001).

The scientific payload onboard the Rosetta spacecraft and the lander Philae provided an unprecedented amount of data on the organics of 67P/CG. The refractory dust component analyzed *in situ* by COSIMA revealed a mixture of carbonaceous matter and anhydrous mineral phases, consistent with the composition of chondritic porous IDPs and anhydrous AMMs (*Bardyn et al.,* 2017). The elemental composition of 67P/CG dust shares similarities with the macromolecular IOM extracted from primitive chondrites (*Fray et al.,* 2016), with the exception of carbon, which greatly exceeds the abundance measured in carbonaceous chondrites (*Bardyn et al.,* 2017). The mass spectrometers of ROSINA, COSAC, and Ptolemy investigated the semivolatile compounds, finding a relatively high abundance of the aromatic compound toluene (*Altwegg et al.,* 2017).

The organic-rich nature of 67P/CG has also been highlighted by the reflectance spectra measured by the VIRTIS

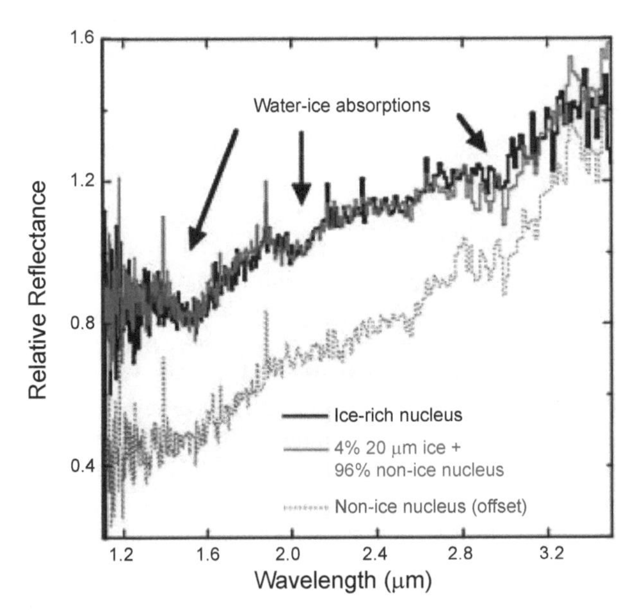

Fig. 19. Comparison of the water-ice-rich spectra on 9P/Tempel 1 (black curve) with the best spectral fit corresponding to a mixture of 4% water ice grains of 20-μm size and 96% of nucleus dark terrain (blue curve). The average dehydrated dark terrain spectrum is shown in red. All curves are relative reflectances. From *Sunshine et al.* (2006).

instrument in the 1–4-μm range (*Capaccioni et al.,* 2015), which showed a dark and red surface (*Ciarniello et al.,* 2015), similar to other spectra of cometary nuclei although with slight differences (see section 2.4). Moreover, VIRTIS detected a broad absorption at 2.8–3.6 μm, centered at 3.2 μm, never detected on other cometary nuclei nor on meteorite spectra. Analysis on laboratory analogs show that the reddish slope of 67P/CG should be attributed to a significant fraction of a dark refractory polyaromatic carbonaceous component mixed with opaque minerals (*Quirico et al.,* 2016; *Rousseau et al.,* 2018). However, most of the spectral information on the nature of 67P/CG organics comes from the broad 3.2-μm absorption. It is compatible with a complex mixture of various types of carbon–hydrogen (C–H) and/or oxygen–hydrogen (O–H) chemical groups, with contribution from nitrogen–hydrogen (N–H) groups such as ionized ammonium (NH_4^+). A firm identification of the exact compounds was made possible thanks to a calibration refinement and performing an average spectrum from a few million surface spectra, producing a spectrum of unprecedented quality in terms of signal-to-noise ratio (*Raponi et al.,* 2020). It revealed that the complex spectral structure of the broad 3.2-μm band is made up of weaker and ubiquitous spectral features (Fig. 21). The strongest ones are centered at 3.10 μm, 3.30 μm, 3.38 μm, 3.42 μm, and 3.47 μm. The whole broad 2.8–3.6-μm absorption feature can be affected by the presence of O–H stretching in compounds such as carboxylic acids, alcohols, phenols (*Quirico et al.,* 2016), and O–H bearing minerals. The 3.3-μm as well as the 3.1-μm subfeature have been attributed to a significant abundance of ammonium salts by *Poch et al.* (2020) (see section 3.3). Micrometer-sized water-ice grains also contribute to the short-wavelength part of the absorption band, consistently with the findings about temporal variability of the broad absorption (see section 3.1).

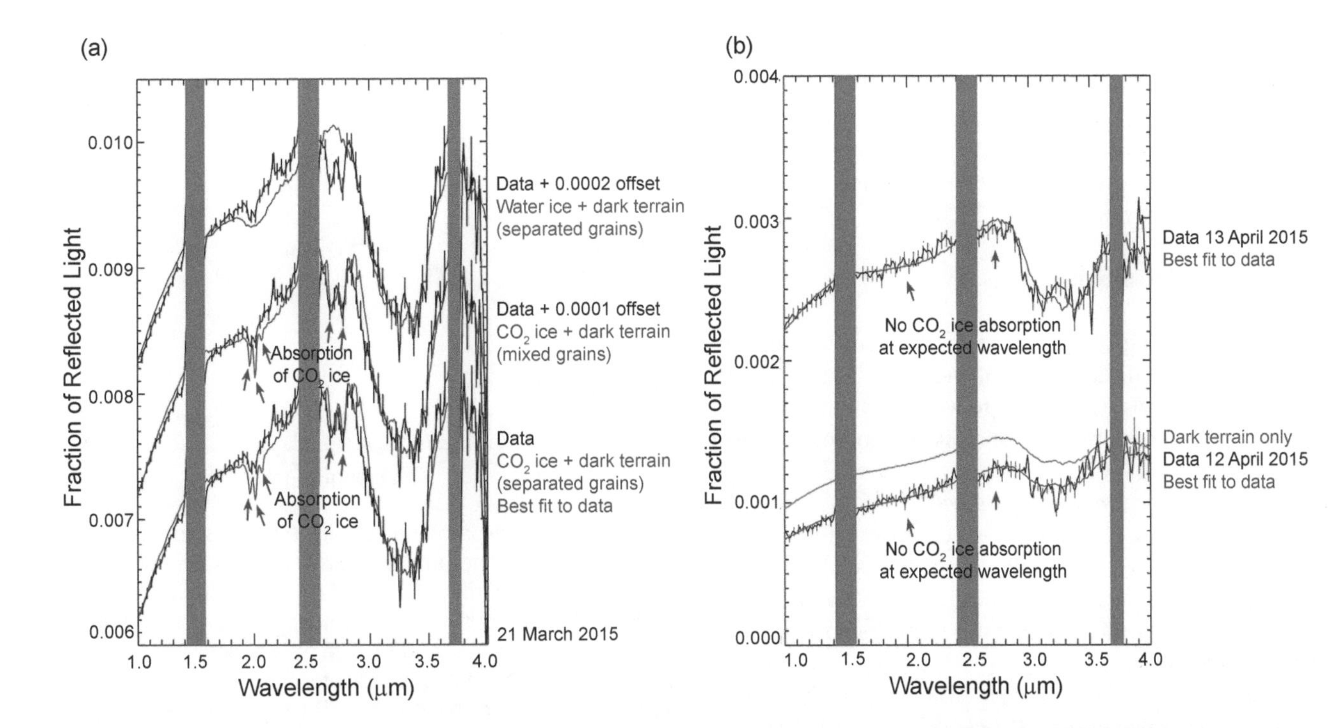

Fig. 20. Spectral identification and modeling of transient CO_2 ice in the Anhur region by Rosetta/VIRTIS. **(a)** In observations taken on March 21, 2015, IR spectra are showing CO_2 ice signatures. **(b)** In observations of the same area taken on April 12–13, 2015, the CO_2 ice features disappeared because the ice had been sublimating in the meanwhile. The spectral ranges shown in gray correspond to instrumental order-sorting filters. From *Filacchione et al.* (2016c).

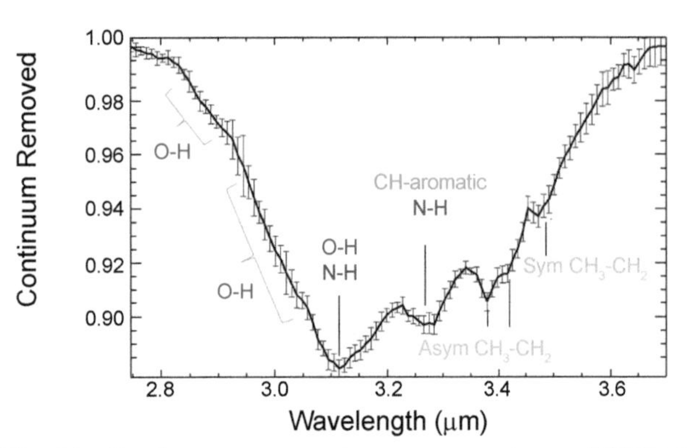

Fig. 21. Continuum-removed average spectrum of 67P/CG across the broad absorption at 2.8–3.6 μm. From *Raponi et al.* (2020).

The absorption bands at 3.38 μm and 3.42 μm identify the asymmetric C–H stretching modes of the methyl (CH_3) and methylene (CH_2) aliphatic groups, respectively. The absorption at 3.47 μm can be assigned to the symmetric modes of CH_3 and CH_2 groups (*Moroz et al.,* 1998), which in 67P/CG spectrum are blended for the possible presence of perturbing groups such as aromatic molecules (*Sandford et al.,* 1991; *Keller et al.,* 2004).

The strength of the features detected in the spectrum of the surface of 67P/CG is peculiar: The asymmetric stretching mode of CH_3 (3.38 μm) seems more intense than the asymmetric stretching of CH_2 (3.42 μm). Such a configuration is unusual for most of the other investigated extraterrestrial materials, except for the interstellar medium (ISM) (*Pendleton et al.,* 1994; *Dartois et al.,* 2004) and for IOM extracted from primitive carbonaceous chondrites (*Kaplan et al.,* 2018; *Kebukawa et al.,* 2011; *Orthous-Daunay et al.,* 2013). This suggests both a possible evolutionary link between hydrocarbons in the diffuse ISM and comets and a link in carbon composition among chondrites and comets (*Mennella,* 2010; *Raponi et al.,* 2020).

3.3. Ammoniated Salts

Remote sensing spectroscopic observations of cometary gaseous comae indicate that the abundance of nitrogen-bearing volatile species with respect to water is about 1%, being mainly represented by NH_3 and HCN, ranging between 0.3–0.7% and 0.08–0.25%, respectively (*Bockelée-Morvan and Biver,* 2017; *Dello Russo et al.,* 2016; *Lippi et al.,* 2021). A deficiency of nitrogen in comets was first pointed out by *Geiss* (1987) from the composition analysis of the gas and dust of the coma of Comet 1P/Halley by *in situ* and groundbased data. In particular, for the refractory phase, a nitrogen-to-carbon atomic ratio (N/C) of 0.052 ± 0.028 was reported by *Jessberger et al.* (1988), significantly smaller than the solar value of 0.3 ± 0.1 (*Lodders,* 2010). A similar result was obtained for Comet 67P/CG by the analysis of the cometary dust particles collected by COSIMA, providing N/C = 0.035 ± 0.011 (*Fray et al.,* 2017).

Although different formation scenarios have been proposed to explain the puzzling nitrogen deficiency in comets (*Willacy et al.,* 2015, and references therein), a possible answer comes only recently with the discovery of significant amounts of ammoniated salts on 67P/CG, resulting from the separate analysis of the data collected by the VIRTIS experiment (*Poch et al.,* 2020) and the ROSINA instrument (*Altwegg et al.,* 2020) onboard the Rosetta mission.

Analysis of the surface visible-infrared spectrum of Comet 67P/CG pointed out that the low-reflectance level and red slope were compatible with refractory polyaromatics organics mixed with opaque mineral, while a semivolatile component with low molecular weight, along with carboxylic (-COOH)-bearing molecules or NH_4^+ ions, were considered as the best candidates to explain the broad 3.2-μm feature (*Capaccioni et al.,* 2015; *Quirico et al.,* 2016). Dedicated laboratory spectral reflectance measurements carried out by *Poch et al.* (2020) provided further insight on the nature of the carrier of the 3.2-μm feature by showing that the main characteristic of the 67P/CG average spectrum are reproduced by the spectrum of sublimate residues of mixtures originally containing water ice, fine-grained iron sulfides, and ammonium salts (Fig. 22).

In particular, the spectrum of the sublimate residue (a porous mixture of submicrometer grains of pyrrhotite and

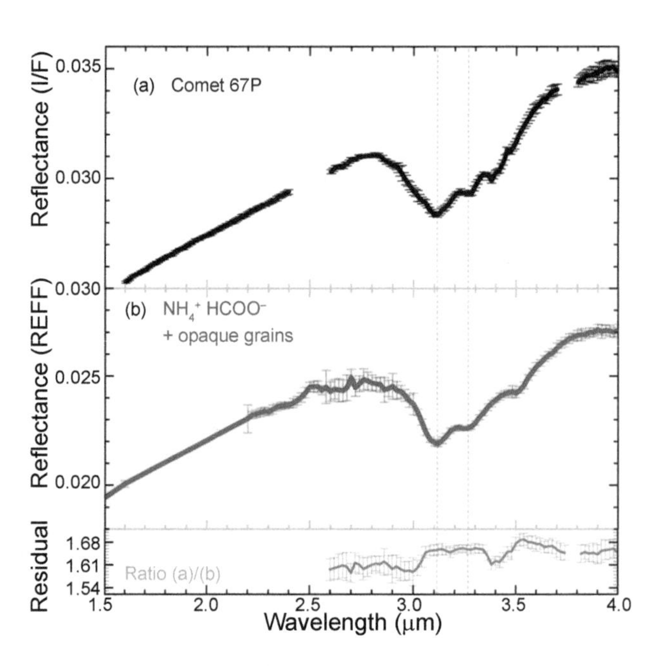

Fig. 22. Comparison of the average 67P/CG spectrum from VIRTIS-M (black line) and a sublimation residue containing ≤17 wt% ammonium formate and pyrrhotite submicrometric grains (red curve). The 67P/CG spectrum is obtained from the refined calibration of VIRTIS-M data described in *Raponi et al.* (2020). Dashed lines indicate the 3.1-μm and 3.3-μm spectral minima in the broad 67P/CG 3.2-μm absorption, corresponding to the spectral minima observed in the sublimation residue. The bottom panel reports the spectral ratio between the comet spectrum and sublimation residue spectrum. Adapted from *Poch et al.* (2020).

ammonium salt) reproduces the position and asymmetry of the cometary 3.2-*μm* absorption, along with the 3.1-μm and 3.3-μm spectral minima resulting from the N-H vibration modes in NH_4^+ (Fig. 22). The comet spectrum exhibits additional features in the 3.35–3.60-μm spectral interval that are not observed in the ammonium-salt-bearing mixture spectra, and are assigned by *Raponi et al.* (2020) to the C–H modes in aliphatic organics (see section 3.2), while further differences can be ascribed to the presence of minor amounts of water ice in the comet spectrum, along with the occurrence of other compounds, including hydroxylated Mg-silicates (see section 3.4) and possible different properties of the ammonium salts on the comet surface (e.g., counter-ions, abundance, mixing modality). Depending on the nature of the counter-ions, the position of the NH^+-related absorptions may shift and vary in relative intensity. *Poch et al.* (2020) investigated the spectral effects of different ammonium salts, and found that ammonium formate, ammonium sulfate, and ammonium citrate are characterized by spectral features similar to those observed in 67P/CG, making it possible that a mixture of different NH^+-bearing salts is present on the comet surface.

Further evidence for ammonium salts was provided by *Altwegg et al.* (2020), taking advantage of measurements carried out by the ROSINA experiment during a short-lived outburst event that occurred on September 5, 2016, at 18:00 UTC. At this time the Rosetta spacecraft and ROSINA sensors were directly hit by dust particles that were fragmented by the impact. A comparison of the mass spectra before (around September 5, 2016, 17:00 UTC) and after (around September 5, 2016, 20:00 UTC) the impact showed a marked increment of species that are possible products of ammonium salts (NH_4Cl, NH_4CN, NH_4OC, NH_4HCOO, and NH_4CH_3COO). This is shown in Fig. 23, where the relative abundances normalized to H_2O of the different species, after and before the dust impact event, are reported. The post-event increase of ammonium-salt-related products has been interpreted as the result of ongoing sublimation of ammonium salts in the grains that entered the double focusing mass spectrometer (DFMS) of the ROSINA instrument, at a temperature of 273 K.

This is compatible with the fact that ammonium salts are more volatile than the typical cometary refractory materials while being generally less volatile than water ice (see Supplementary Table 1 of *Altwegg et al.,* 2020, and references therein). It also suggests that the N/C ratio inferred by COSIMA is likely a lower limit of the real value of 67P/CG, as most of the ammonium salts in the collected particles would have sublimated during the pre-analysis storage phase in the instrument at 283 K (*Fray et al.,* 2017).

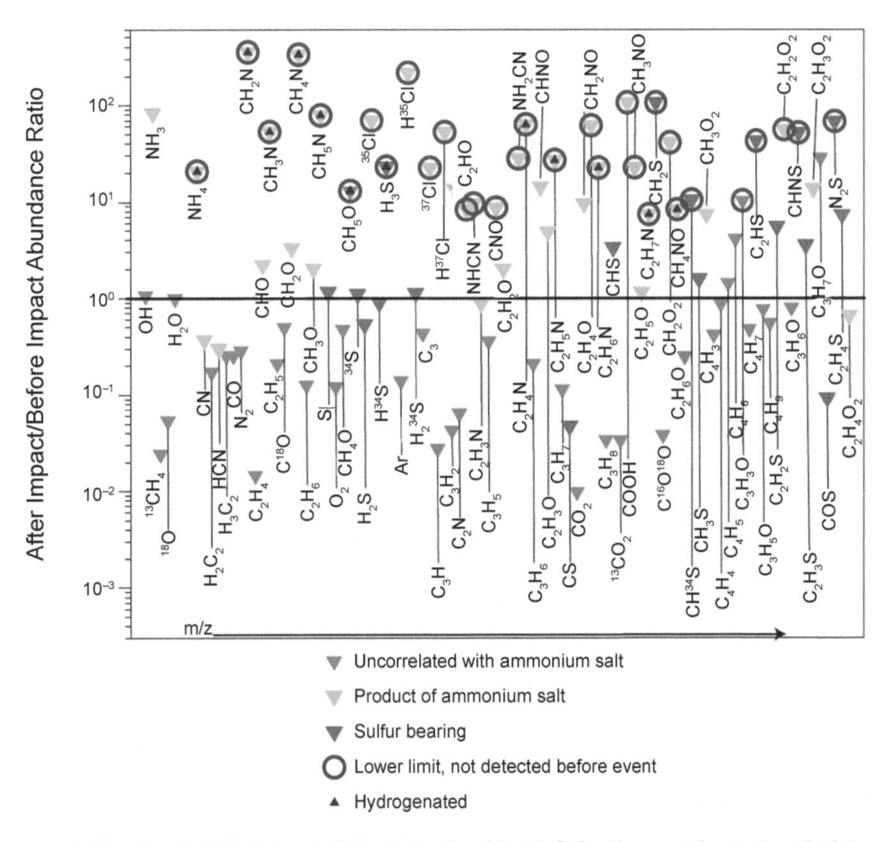

Fig. 23. ROSINA measurements of abundance ratios normalized to H_2O for the post-impact period (around September 5, 2016, 20:00 UTC) compared with the pre-impact period (around September 5, 2016, 17:00 UTC). Although ammonium salt sublimation can produce H_2O, CO, and CO_2, the corresponding contribution to their abundance is considered small, as these are dominant species in the undisturbed coma. Thus, H_2O, CO, and CO_2 are considered to be uncorrelated with ammonium salts. From *Altwegg et al.* (2020), where further details can be found.

The detection of abundant ammonium salts on the nucleus and in the dust of Comet 67P/CG provides an explanation for the apparent nitrogen depletion in cometary comae (*Poch et al.,* 2020; *Altwegg et al.,* 2020). Ammonium salts sequester nitrogen in a semivolatile form, not accessible to spectroscopic investigations of the coma unless the cometary dust is exposed to temperatures high enough to trigger ammonium salts sublimation. This could be the case for comets reaching small heliocentric distances, which in fact tend to display a larger abundance of NH_3 (*Dello Russo et al.,* 2016), suggesting that a significant fraction of the ammonium salts would sublimate.

By assuming that the refractory component of 67P/CG dark material is composed of ~45 wt% organics and ~55 wt% minerals, as indicated by COSIMA measurements (*Bardyn et al.,* 2017), *Poch et al.* (2020) estimate an upper limit of ≲40 wt% ammonium salts in the cometary dust. Furthermore, depending on the nature of NH^+ counter-ions, they find that amounts of ammonium salts ranging between 10 wt% and 30 wt% would provide a cometary N/C ratio matching the solar values. In addition, *Altwegg et al.* (2020) show that the abundance of nitrogen in comets 67P/CG and 1P/Halley gets close to solar abundances if one assumes that their actual ammonia content is the one observed in comets reaching the smallest heliocentric distance [~4% relative to H_2O (*Dello Russo et al.,* 2016)], which means accounting also for the ammonia hidden in ammonium salts.

Ammonium salts may represent an important reservoir of nitrogen in comets and, as suggested by similarities in the infrared spectrum with 67P/CG, possibly on other solar system small bodies (*Poch et al.,* 2020). This is shown in Fig. 24, where the 67P/CG spectrum is compared with the ones from a few main-belt asteroids (Themis, Cybele, Europa, and Bononia) and Jupiter trojans. All these bodies evidence a relatively broad absorption feature at 3.1–3.2 μm, which could be compatible with ammonium salts if we take into account that the ammonium-salt-related features show some variability in intensity and position depending on the counter-ions, the matrix materials, and surface temperature (*Poch et al.,* 2020).

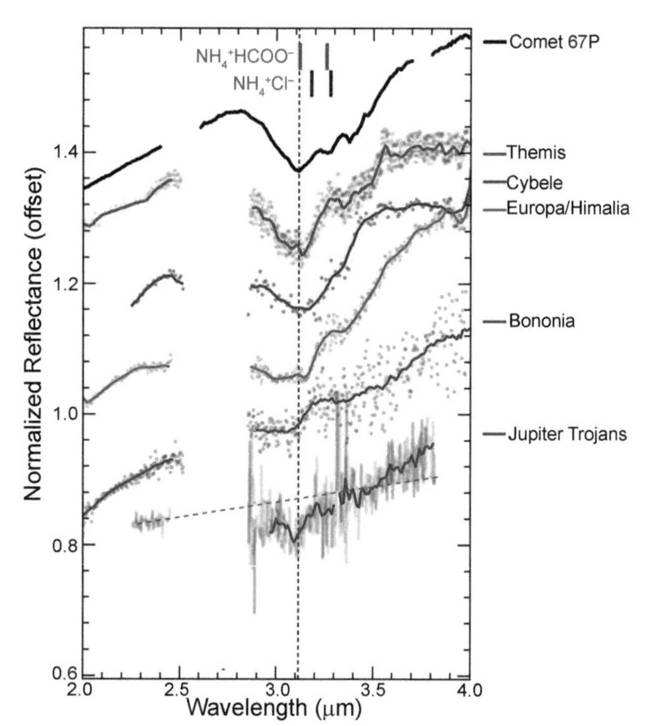

Fig. 24. IR spectrum of Comet 67P/CG compared to the main-belt asteroids 24 Themis (*Rivkin and Emery,* 2010), 65 Cybele (*Licandro et al.,* 2011), 52 Europa (*Takir and Emery,* 2012), and 361 Bononia (*Takir and Emery,* 2012), and the average spectrum of six Jupiter trojans [the "less red" group from *Brown* (2016)]. The spectrum of the Jupiter trojans is divided by 3. Himalia is also indicated, as its spectrum is indistinguishable from Europa (*Brown and Rhoden,* 2014). Running averages (solid lines) are superimposed over the observation data (dots). The blue dashed line indicates the average extrapolation of the K-band spectra of six Jupiter trojan (*Emery et al.,* 2011). The vertical dashed line indicates the position of the 3.11-μm feature in the 67P/CG spectrum. The positions of the two main absorptions for pyrrhotite-ammonium formate (red vertical marks) and pyrrhotite-ammonium chloride (green vertical marks) sublimate residue, respectively, are indicated. The plot is adapted from *Poch et al.* (2020).

3.4. Inorganic Refractories

Together with volatiles and organic matter, inorganic refractories are the third major composition end member observed in comets. Inorganic refractories include different mineral species, among which silicates are the more abundant. Silicates were first observed in Comet 1P/Halley by means of infrared spectrophotometric measurements by *Bregman et al.* (1987) and *Campins and Ryan* (1989) and later confirmed in C/1995 O1 Hale-Bopp by *Crovisier et al.* (1997). Also, the samples of dust collected by Stardust on 81P/Wild 2 evidenced a large fraction of Mg-silicates (*Brownlee et al.,* 2006). There is a general consensus about the origin of silicates that were synthesized around evolved stars and then went through evolution in the ISM where the irradiation with ions resulted in amorphization, implantation, and sputtering (*Demyk et al.,* 2001; *Carrez et al.,* 2002; *Jäger et al.,* 2003; *Brucato et al.,* 2004; *Djouadi et al.,* 2005). The effects caused by ion-induced processing on silicates have also been observed on IDPs by *Bradley et al.* (1999, 2005). The composition of silicates in the ISM is dominated by the amorphous form (98%) with only a minor presence of the crystalline form (2%). As reported by *Sugita et al.* (2005), the amount of amorphous silicates is between 70% and 98% in the JFCs where the Mg-rich species are the dominant ones (*Brownlee et al.,* 2006; *Kelley et al.,* 2017). During their evolution in the ISM, Mg-rich silicates undergo irradiation with H atoms resulting in the formation of OH bonds. As reported by *Zeller et al.* (1966), *Djouadi et al.* (2011), and *Schaible and Baragiola* (2014), the implantation of protons within the kiloelectron volt to megaelectron volt energy range in silicates with different compositions and structures causes the formation of an

OH bond recognizable by means of its spectral signature around 2.8 μm.

A similar mechanism has been explored through laboratory experiments to synthesize hydroxylated Mg-rich amorphous silicates by *Mennella et al.* (2020), interpreting the wide absorption signature visible on 67P/CG spectra at around 3.2 μm. In Fig. 25a the appearance of the broad 3.2-μm absorption band on an Mg-rich silicate is shown after a cycle of annealing at 300°C and irradiation with a flux of 10^{18} H atoms 1 cm^{-2}. The processed sample (SIL3) is derived from natural olivine with Mg/Si = 1.7 and Fe/Si = 0.2 ratios. The 3.2-μm band is reproduced also on other low-Mg (Mg/Si = 1.2) types of silicates after H atom irradiation.

The measured optical depth allows computing the best-fit I/F to the spectrum of observed comets, as shown in Fig. 25b. The two figures show that the hydroxylated magnesium silicate component is able to reproduce the general shape of the band, suggesting a contribution to observed absorption in addition to aromatic and aliphatic organic matter (section 3.2) and ammonium salts (section 3.3).

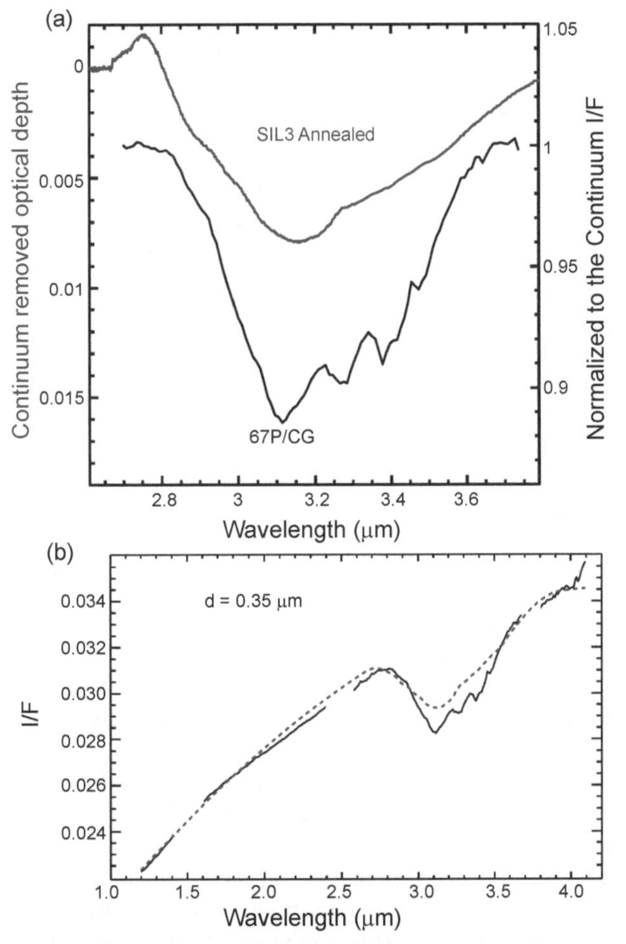

Fig. 25. (a) The 3.2-μm absorption band observed in SIL3 sample (blue line), which underwent annealing at 300°C and was then exposed to a flux of 10^{18} H atoms 1 cm^{-2}, is compared to the observed main absorption band visible in the 67P/CG I/F spectrum (black line). **(b)** The best-fit solution from Hapke's model corresponds to a mixture of 15% hydroxylated amorphous silicate and 85% comet's average dark-red end member [blue dashed line (from *Raponi et al.*, 2020)] with 5-μm grain size distribution. For comparison, the black line corresponds to the observed average 67P/CG reflectance spectrum. From *Mennella et al.* (2020).

Supplementary clues about the mineral composition of 67P/CG can be derived from the analysis of the visible–near-infrared spectral range. *Rousseau et al.* (2018) measured several ternary mixtures of cometary analogs with the aim of reproducing the low reflectance, red color lope, and featureless continuum observed in the 0.4–2.5-μm range. The investigated mixtures are made by three components: carbon (lignite coal PSCO1532), iron sulfide (pyrrhotite), and silicate (dunite, 95% olivine). At the initial sample, made by a 67:33 wt% proportion of carbon and pyrrhotite, an increasing amount of dunite from 5 wt% to 30 wt% is added (Fig. 26) to explore variations in color and albedo. The mineral phase composition and grain size are selected to match the values measured by dust instruments (GIADA, COSIMA) on Rosetta. The experiment, repeated in both intimate mixing and submicrometer grains coating larger clumps, shows that these solutions are capable of reproducing the low albedo level and featureless continuum but fail to match the red slope observed in VIRTIS data, which is higher than the simulated one. This experiment shows that additional reddening can be introduced by adding complex organic matter to the mineral phase mixtures.

Moroz et al. (2017) reports the spectral properties of mixtures of organic matter and minerals, including pyrrhotites and troilites. The reflectance of these sulfides shows a decrease when grains become small (e.g., submicrometer-sized), a behavior described by *Hunt et al.* (1971). The low albedo of 67P/CG could be therefore explained by a mixture of fine-grained minerals mixed with organic matter. The additional presence of silicates appears less relevant to model spectral slopes and albedo (*Moroz et al.*, 2017; *Rousseau et al.*, 2018). Spectral measurements on several kerite-troilite and kerite-pyrrhotite mixtures show that iron-sulfides mixed with organic matter can modify the fine structure of the aromatic-aliphatic group absorption bands within the wide 3.2-μm band (*Moroz et al.*, 2017).

4. ACTIVITY-DRIVEN COMPOSITION EVOLUTION

The composition of the nuclei of comets is driven by the level of gaseous activity, which in turn is a complex result of heliocentric distance, surface morphology, rotational state, and volatile abundance. In the text that follows we describe the nucleus evolution caused by diurnal (section 4.1) and seasonal cycles (section 4.2) as well as the changes occurring at local scales where activity events are observed (section 4.3). Finally, in section 4.4, a model of cometary nucleus internal structure and composition derived from the surface seasonal evolution is discussed.

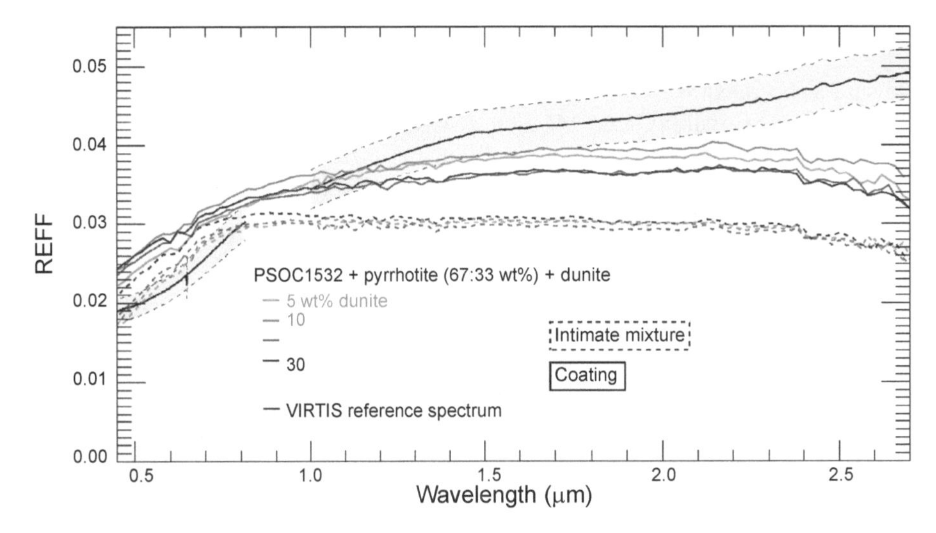

Fig. 26. Comparison of 67P/CG VIRTIS reflectance with laboratory-measured spectra of intimate mixtures of carbon (PSOC 1532) and pyrrhotite in 67:33 wt% proportion plus dunite (from 5 wt% to 30 wt%) in intimate (dot line) or coating mixture (solid line). From *Rousseau et al.* (2018).

4.1. Diurnal Composition Variation

The first evidence of the diurnal cycle of water on Comet 67P/CG was reported by *De Sanctis et al.* (2015) who detected an area close to shadows in the Hapi region showing variations in the absorption band near 3 μm, being deeper and shifted at shorter wavelengths with respect to the average comet spectrum as measured by VIRTIS. They explain these features as a local enrichment of water frost formed when water vapor released from the warm subsurface recondenses on the cold surface after sunset and then rapidly sublimates when exposed to the Sun. Molecules in the inner coma may also be backscattered to the nucleus surface and recondense on cold areas, contributing to the frost formation (*Davidsson and Skorov,* 2004). After this first evidence of the diurnal cycle of water relatively far from the Sun (heliocentric distance ~3 au at the time of the observations), no other evidence of surface frost was reported until the comet approached perihelion. Starting from June 2015, i.e., a couple of months before 67P/CG's perihelion passage, the nucleus showed repeated diurnal color variations on extended areas and the occurrence of water frost close to morning shadows on both lobes of the comet, as reported in *Fornasier et al.* (2016). An example of these phenomena is illustrated in the middle panel of Fig. 27, showing the color changes on large areas in the Imhotep region from images acquired on June 27, 2015, with a cadence of only 40 minutes. Areas just emerging from the shadows are spectrally bluer than their surroundings (Fig. 27, left side of the middle panel) and their spectral slope increased, i.e., the regions got relatively redder, in the images acquired 40 minutes later (Fig. 27, right side of the middle panel). *Fornasier et al.* (2016) interpreted the relatively blue surface at dawn as the presence of additional water frost that condensed during the previous night and that suddenly sublimates when the region is illuminated by the Sun. These rapid color changes are accompanied by the presence of bright material close to the shadows, proven to be morning water frost that sublimates within few minutes once illuminated (see the bottom panel of Fig. 27). This material is 6× brighter than the mean comet reflectance and shows a spectrum that is flat in the visible and near-infrared range but with a flux enhancement in the ultraviolet region, consistent with the spectra of laboratory water frost (*Wagner et al.,* 1987). *Fornasier et al.* (2016) reproduced the spectral behavior by linear mixing of the cometary dark terrain with 17 ± 4% of water frost, and estimated the frost layer is extremely thin, with a depth of only 10–15 μm.

4.2. Seasonal Composition Evolution

By orbiting in formation with the nucleus of 67P/CG, Rosetta had the unique opportunity to follow the evolution of the surface and coma properties as a function of the heliocentric distance and seasonal changes. Due to the 52.4° obliquity of the rotation axis (*Jorda et al.,* 2016), the summer season on the north and south hemispheres of the nucleus is reached in different orbital periods: The northern regions are continuously illuminated by the Sun during the 5.5-year-long northern summer season. This includes the aphelion passage, which occurs at a heliocentric distance of 5.68 au (beyond Jupiter's orbit). The inbound equinox was reached in May 2015 when the comet was orbiting at 1.7 au from the Sun: This date marks the beginning of the 10-month-long and intense summer on the southern hemisphere while the north hemisphere was moving in night and was no more illuminated by the Sun. The perihelion passage occurred on August 13, 2015, when the comet reached the minimum heliocentric distance of 1.24 au. During this phase the southern hemisphere was experiencing its brief summer season, when the Sun becomes circumpolar (never going below the line of the horizon during a comet's day), and the surface was receiving the maximum solar flux resulting in the maximum activity and erosion of a surface's

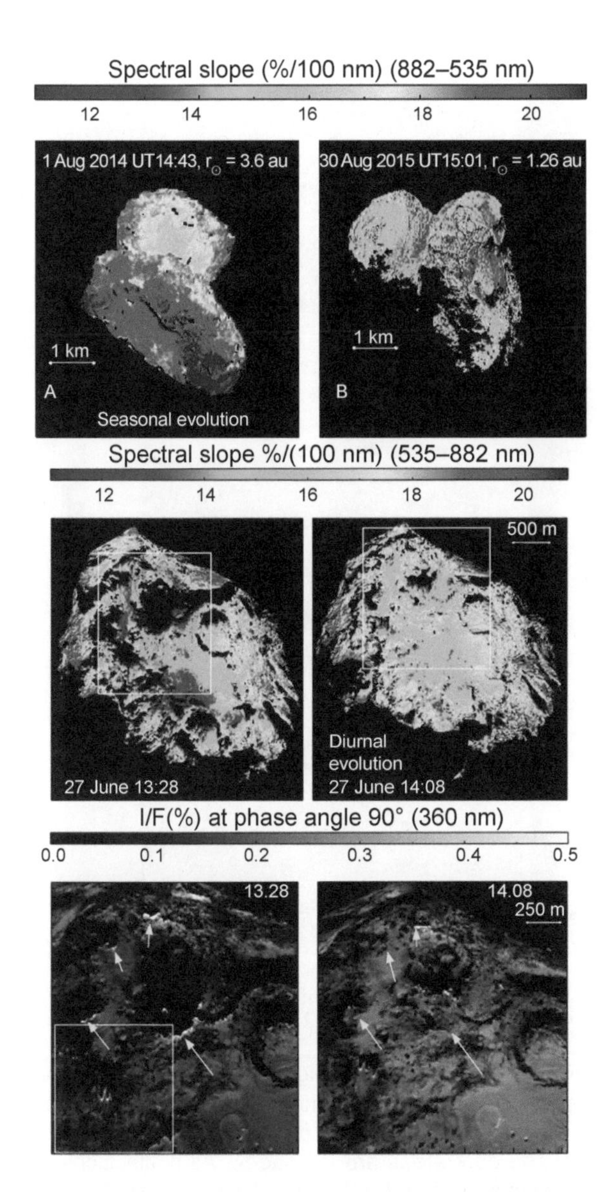

Fig. 27. Seasonal (top) and diurnal (center and bottom panels) color evolution of Comet 67P/CG from OSIRIS observations (*Fornasier et al.,* 2016). Top panel: Seasonal evolution of the spectral slope from August 2014 (when the comet was at 3.6 au inbound) to August 2015 (just after the perihelion passage). Middle panel: Diurnal evolution of the spectral slope in the Imhotep region. The two images shown were acquired on June 27, 2015, and only 40 min apart. The corresponding heliocentric distance was 1.37 au, the spatial resolution 3.2 m/pixel, and the phase angle 90°. The Sun is toward the top. Bottom panel: Radiance factor (I/F) of the regions indicated by the white rectangle on the middle panels showing morning frost (yellow arrows) disappearing and moving with shadows.

layer of thickness ≈4 m (*Fulle et al.,* 2019). Depending on the model, somewhere between 20% (*Keller et al.,* 2017) and 50% (*Hu et al.,* 2017) of the dust flux lifted from the southern hemisphere at perihelion falls back on the northern hemisphere. The grain size distribution of the fallback flux is dominated by tens of centimeter-sized dehydrated aggregates containing only a very small fraction of water ice (*Keller et al.,* 2017). Along the outbound orbit, the equinox is repeating again in March 2016 when the southern hemisphere returned in nightside and the northern in dayside. In this phase, with the progressive settling of the activity on the southern regions, the layering of the dust on the surface blankets again exposed pristine icy material and the nucleus surface returns darker and redder. This condition is the same observed at the beginning of the Rosetta mission.

The surface colors of the nucleus have been routinely monitored by VIRTIS, allowing catching the changing color and brightness occurring on different regions as more pristine, water-ice-rich material was progressively exposed on the surface (*Filacchione et al.,* 2016a; *Ciarniello et al.,* 2016). On average, the comet's surface appears dark and red where dehydrated dust made out of a mixture of minerals and organic matter prevails. Localized active regions (as discussed in the chapter by Pajola et al. in this volume) and the occasional ice-rich exposures are characterized by higher albedo and more neutral colors. VIRTIS and OSIRIS observations show that global changes are noticeable, with an overall trend of the comet becoming bluer and more water-ice-rich while approaching perihelion (*Fornasier et al.,* 2016; *Filacchione et al.,* 2020; *Ciarniello et al.,* 2022).

The heliocentric distance is also driving the long-term stability of the water-ice-rich BAPs discussed in section 3.1, which starts to sublimate when the solar exposure becomes substantial (*Raponi et al.,* 2016). These patches are buried below a dust layer when the comet is at relatively large heliocentric distances ($\gtrsim$3.4 au) and for this reason, they are scarcely recognizable at 1.5–2 μm where the diagnostic absorption bands of water ice are located. Approaching the Sun, the activity rapidly starts to remove dust particles from the surface, gradually exposing the ice beneath it. The patch keeps sublimating thanks to the steady increase of the solar flux until the ice deposit is completely lost. This trend is shown in Fig. 28 where we report the six-month temporal sequence of the water-ice-band area for heliocentric distances between 3.44 and 2.23 au along the inbound orbit. During this period the Imhotep region is moving from winter to spring. Long-term ice stability on the surface of 67P/CG has been reported for many other ice deposits observed by OSIRIS (*Oklay et al.,* 2017).

Spectral modeling of synthetic spectra of water ice and dark terrain mixtures, in both areal and intimate mixing, has been exploited by *Raponi et al.* (2016) to correlate visible slope and the 2-μm-band area (Fig. 29), which are the two widely-used spectral indicators sensitive to water-ice abundance on 67P/CG spectra.

Theoretical trends from models are compared with experimental data collected at 95° phase angle. The choice of the phase angle reflects the fact that Rosetta was spending a great part of its time orbiting along terminator orbits. A similar relationship is useful to infer water-ice abundance by visible colors in the absence of infrared data. It is noteworthy that since both slope and band area are phase-dependent, this correspondence is valid only at the specific 95° phase

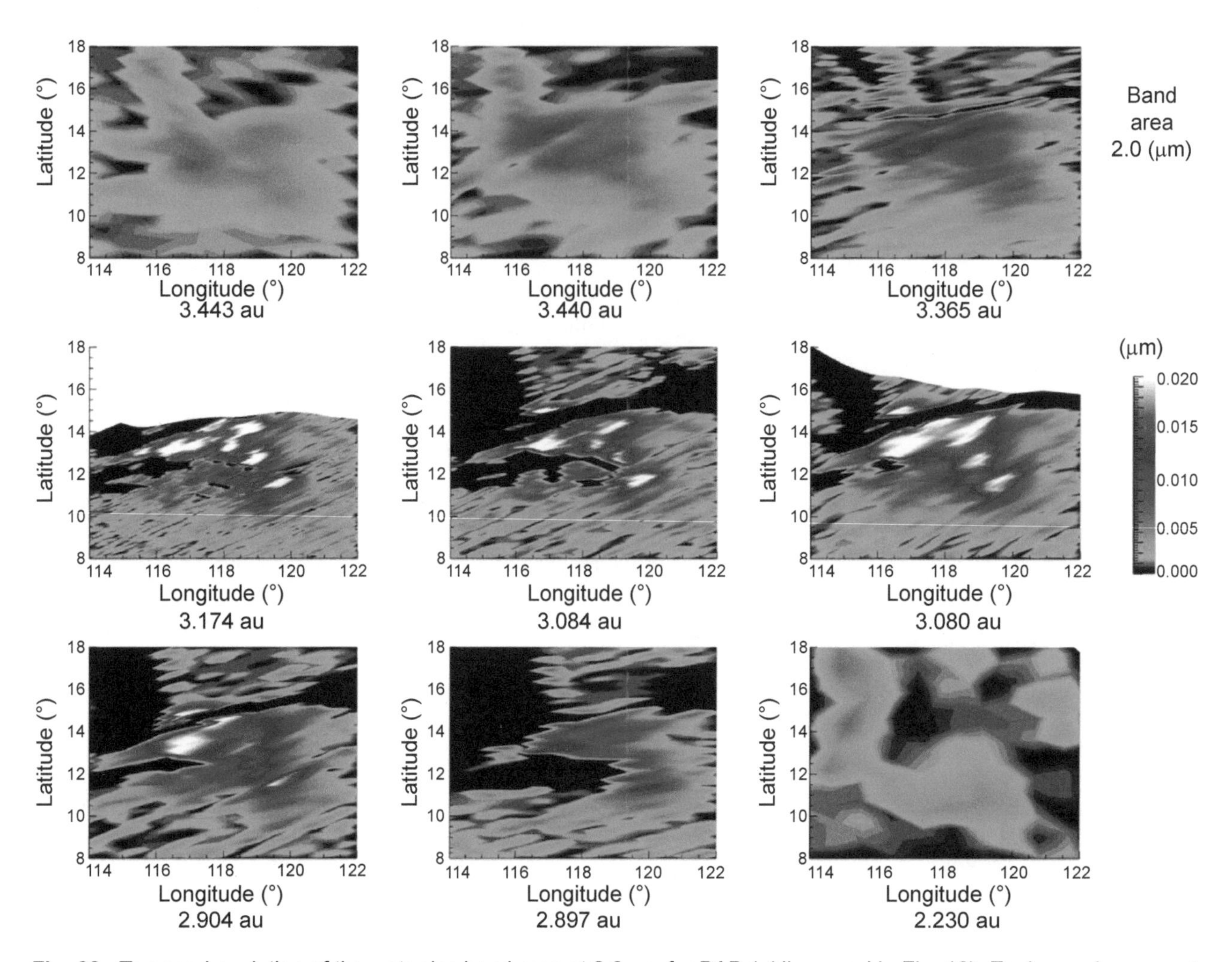

Fig. 28. Temporal evolution of the water-ice band area at 2.0 μm for BAP 1 (discussed in Fig. 13). Each panel represents the extension and intensity of the water-ice band for heliocentric distances between 3.443 and 2.230 au along the inbound orbit corresponding to a period of 6 months. The feature steadily increases on the first six panels and then decreases in the last three. In the last panel, the feature is very weak or absent. The dark areas are in shadow. All acquisitions were taken with similar illumination conditions (average incidence angle between 53° and 67°, corresponding to early morning local time). The white areas outside the edge of the map are out of the field of view of the acquisitions. From *Raponi et al.* (2016).

angle. As computed by *Ciarniello et al.* (2015), the slope changes with phase by +0.007 μm^{-1} deg^{-1}.

Few studies have focused their analyses on the variability of the comet's environment at different heliocentric distances with the aim of correlating the spectral changes occurring on the nucleus and in the coma. The evolution of the nucleus and coma spectral properties demonstrates the presence of an orbital cycle of water ice, which can be observed by measuring multiple spectral indicators such as integrated radiance, spectral slopes, and wavelength of the radiance's peak measured across the visible spectral range (*Filacchione et al.,* 2020). The gaseous activity of comets is strongly dependent on solar heating, which provides the energy input to sublimate volatile species. The sublimation of volatile species is responsible for the removal of the dust grains from the surface, a process that depends on multiple factors such as heliocentric distance, local orientation of the surface with respect to the solar direction, composition, and physical properties of the subsurface (*Gundlach et al.,* 2015; *Tubiana et al.,* 2019). As a consequence of the activity, both the nucleus (*Filacchione et al.,* 2016a; *Ciarniello et al.,* 2016, 2022; *Fornasier et al.,* 2016) and the coma (*Hansen et al.,* 2016; *Bockelée-Morvan et al.,* 2019) change their appearance along the orbit. The light scattered by dust grains in the coma, taking into account for grain size distribution, shapes (*Güttler et al.,* 2019), and viewing geometry, allows the inference of their composition: Carbonaceous-organic grains are recognizable by their red color (*Jewitt and Meech,* 1986), while magnesium silicate (*Zubko et al.,* 2011; *Hadamcik et al.,* 2014) and water ice (*Fernández et al.,* 2007) grains are recognizable by their distinctive blue color.

By modeling the light-scattering properties of the dust particles in the coma of 67P/CG, *Filacchione et al.* (2020) have found different populations of grains along the orbit: During the inbound orbit water-ice grains are the dominant group, at perihelion submicrometer-sized organic material

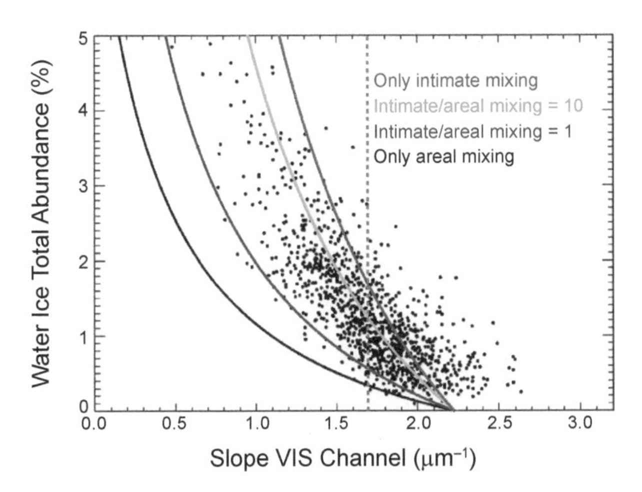

Fig. 29. The theoretical abundance of water ice as a function of the slope at a phase angle of 95° for different mixing modalities (color lines) compared with VIRTIS data (points in black). From *Raponi et al.* (2016).

and carbon-rich particles are more abundant, and along the outbound orbit water ice (or possibly magnesium silicate, which has similar color) is dominant. Such dust composition end members and grain sizes are reconcilable with other Rosetta studies: (1) During the emission of outbursts, *Agarwal et al.* (2017) have observed the release of submicrometer-sized water-ice grains mixed with hundreds of micrometers sized refractory grains; (2) at the same time, *Bockelée-Morvan and Biver* (2017) have reported a broad emission feature centered at 3.5 μm visible on the outbursts that is compatible with the sublimation of organic materials caused by the grains' heating by the solar flux; (3) similar dust grain composition has been inferred by *Frattin et al.* (2017), who modeled resolved dust grain color images by OSIRIS with three composition classes: organic material grains corresponding to high spectral slope, mixtures of silicate and organic material grains with intermediate slope, and water-ice grains with flat slope; (4) the presence of elemental carbon, silicon, and magnesium in dust grains has been inferred by COSIMA (see Table 3 of *Bardyn et al.*, 2017); and (5) predictions on the dust properties (composition and size) are found in the theoretical "cometary model" of *Fulle* (2021).

When comparing spectral properties derived from two years of continuous VIRTIS observations of the dust dispersed in the coma and nucleus surface of 67P/CG (Fig. 30), two opposite seasonal color cycles become evident through its perihelion passage (*Filacchione et al.*, 2020): When far from the Sun, the nucleus appears redder than the dust particles in the bluish coma, which are dominated by water-ice grains of about 100 μm in size. Moving on the inbound orbit toward perihelion passage, the nucleus shows a progressive exposure of water-ice-rich and pristine materials; as a consequence of this process, the nucleus reaches the maximum blueing at perihelion (Fig. 31) and concurrently, the large quantities of dust lifted in the coma become progressively redder orbiting closer to the Sun due to an enrichment of organic matter and amorphous carbon-rich submicrometer grains (Fig. 30). The high temperatures reached by the grains cause the rapid sublimation of any residual ice within them, a mechanism through which the grains are rapidly dehydrated once lifted off from the nucleus surface (*Bockelée-Morvan et al.*, 2019). This process develops in reverse order after the perihelion passage (plus about one month of delay) along the outbound orbit when the activity settles and the redeposition of dehydrated dust particles on the nucleus make it red-colored again while water-ice particles start again to populate the coma. The overall cycle of colors developing on the nucleus and coma of 67P/CG along its orbit is summarized in Fig. 32.

The trends observed by VIRTIS were also confirmed by OSIRIS observations: *Fornasier et al.* (2016) monitored the color evolution of the nucleus from 3.6 au to perihelion (1.24 au) and beyond by measuring changes in the spectral slope in the 535–882-nm range (Fig. 27). They found that the whole nucleus became relatively bluer near perihelion, when the increasing level of activity provides a progressive exposure of water ice. This color evolution is accompanied by a change in the phase reddening effect that decreases at perihelion, as reported in section 2.1.

4.3. Local Activity-Driven Composition Evolution

The Rosetta spacecraft has been orbiting Comet 67P/CG for more than two years, providing the unique opportunity to study the surface evolution over time. It has been therefore possible for the very first time not only to identify a number of morphological changes in several regions of the nucleus, but also to follow the composition variations produced by the cometary activity and by the diurnal and seasonal cycles of water (see section 4.2). A number of jets and outbursts have been identified and investigated in the literature: The very first outburst was caught by OSIRIS on April 2014, when the comet was far from the Sun (~4 au) and unresolved (*Tubiana et al.*, 2015), followed shortly thereafter by the detection of water vapor with the Microwave Instrument for the Rosetta Orbiter (MIRO) instrument (*Gulkis et al.*, 2015). Hapi was the most active region during the first resolved observations (*Pajola et al.*, 2019) and characterized by brighter-than-average and relatively blue colors (*Fornasier et al.*, 2015), indicating a local enrichment of water ice that was in fact detected by VIRTIS (*Capaccioni et al.*, 2015), but it was during the cometary summer close to the perihelion passage that a number of outbursts and jets were individually identified and their source position on the nucleus retrieved through geometrical tracing (*Vincent et al.*, 2016a; *Knollenberg et al.*, 2016) or direct imaging (*Vincent et al.*, 2016b; *Shi et al.*, 2016; *Pajola et al.*, 2017a; *Fornasier et al.*, 2019b). Finally, during the Rosetta extended mission from January to September 2016, some plumes/faint jets were observed in the equatorial and southern regions in the high-resolution images (*Fornasier et al.*, 2017, 2019a; *Agarwal et al.*, 2017; *Deshapriya et al.*, 2018; *Hasselmann et al.*, 2019).

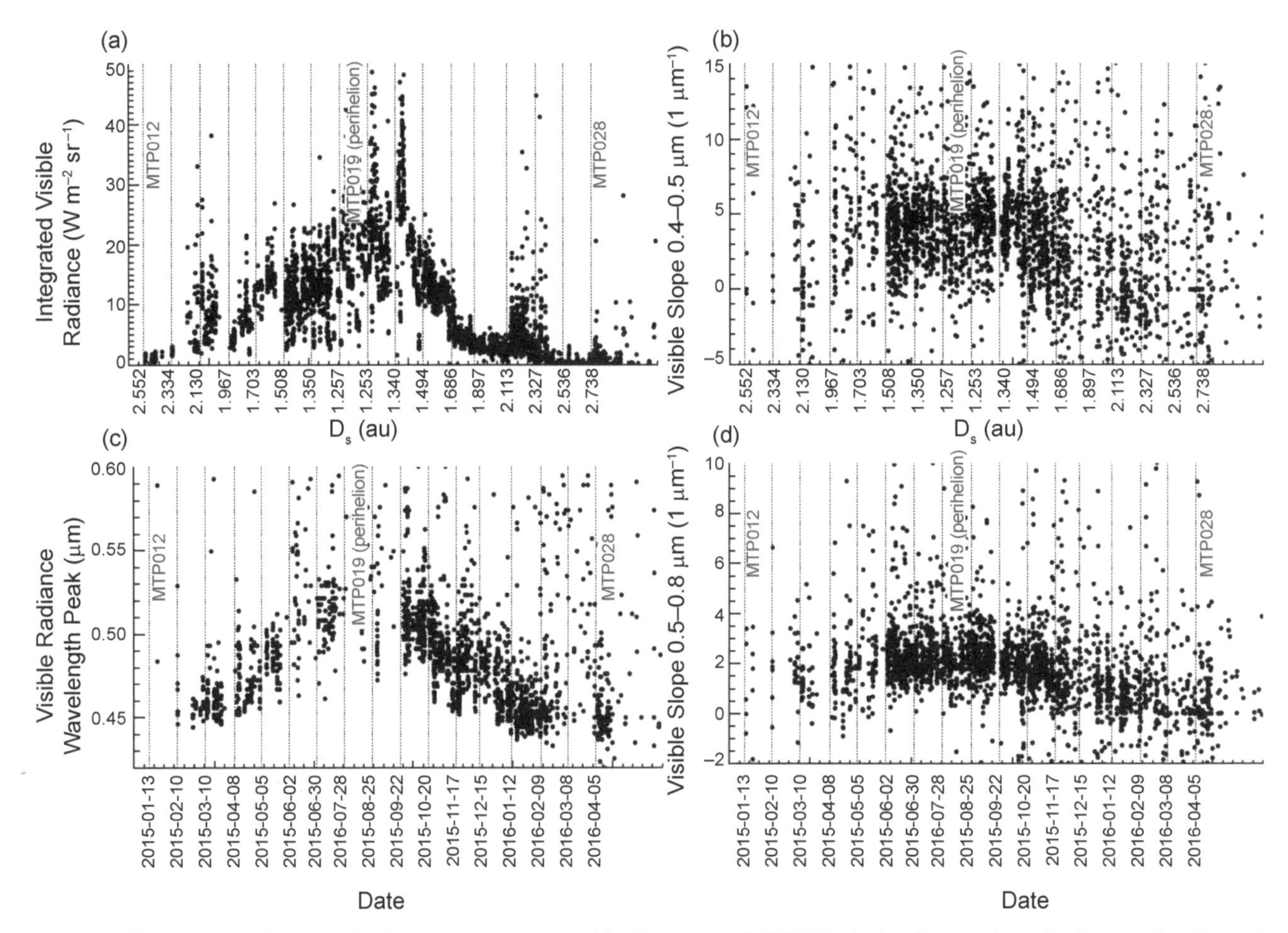

Fig. 30. Time-series of spectral indicators as measured in the coma of 67P/CG during inbound, perihelion, and outbound orbital phases. **(a)** Visible integrated radiance; **(b)** 0.4–0.5-µm spectral slope; **(c)** visible radiance wavelength peak; **(d)** 0.5–0.8-µm spectral slope. Each point corresponds to the average value of the spectral indicators as computed on an annulus defined by a tangent altitude between 1 and 2.5 km from the nucleus' surface on a single observation. For comparison, the temporal changes occurring on the nucleus are shown in Fig. 31. From *Filacchione et al.* (2020).

Thanks to the direct identification of activity sources on the nucleus, a number of regions were carefully investigated to look for morphological and composition changes produced by the activity. The main processes originating composition variations on the nucleus can be categorized as the following.

4.3.1. Dust removal and redeposition by volatile sublimation. The surface dust is mobilized by means of the sublimation of volatile species [mainly CO_2 (*Gundlach et al.,* 2020) and H_2O]. The cometary activity progressively removes dust and exposes pristine material, including water ice able to survive on the surface until sublimation occurs. This is causing the composition changes observed on the overall nucleus, or on extended areas of it, where the seasonal and diurnal cycle of water (as detailed in section 4.2) are manifested, as well as the spectral phase reddening effect variations (see section 2.1).

The transfer and redeposition of the dust are strongly related to the sublimation of volatiles and thus to the activity level, lift pressure, and local topography (*Thomas et al.,* 2015; *Keller et al.,* 2017; *Hu et al.,* 2017; *Pajola et al.,* 2017b). Because the southern hemisphere experiences strong thermal effects, being intensely illuminated during and close to the perihelion passage, it is highly eroded. By modeling only water-ice sublimation, *Keller et al.* (2017) has found that dust ejection occurs from the south and a significant fraction of it is redeposited in the northern regions. *Lai et al.* (2016) estimated a dust loss of about 0.4 m in depth per orbit in Hapi and up to 1.8 m in Wosret, where the highest water production rate was measured by MIRO during the perihelion passage (*Marshall et al.,* 2017). Concurrent OSIRIS measurements during the inbound orbit show an erosion of about 1.7 ± 0.2 m in the Hapi region that is nearly balanced by a 1.4 ± 0.8 m fallout layer from the southern hemisphere during perihelion (*Cambianica et al.,* 2020).

The first consequence of dust mobilization is the seasonal variation of colors, with the nucleus becoming bluer approaching perihelion because of the exposure of underlying layers enriched in water ice (Fig. 27).

Secondly, the morphological landscape changed locally: A local removal of ~4 m of dust coating was observed in part of Imhotep, with the exhumation of boulders and roundish features (*El-Maarry et al.,* 2017) and up to ~14 m in a cavity of the Khonsu region (*Hasselmann et al.,* 2019)

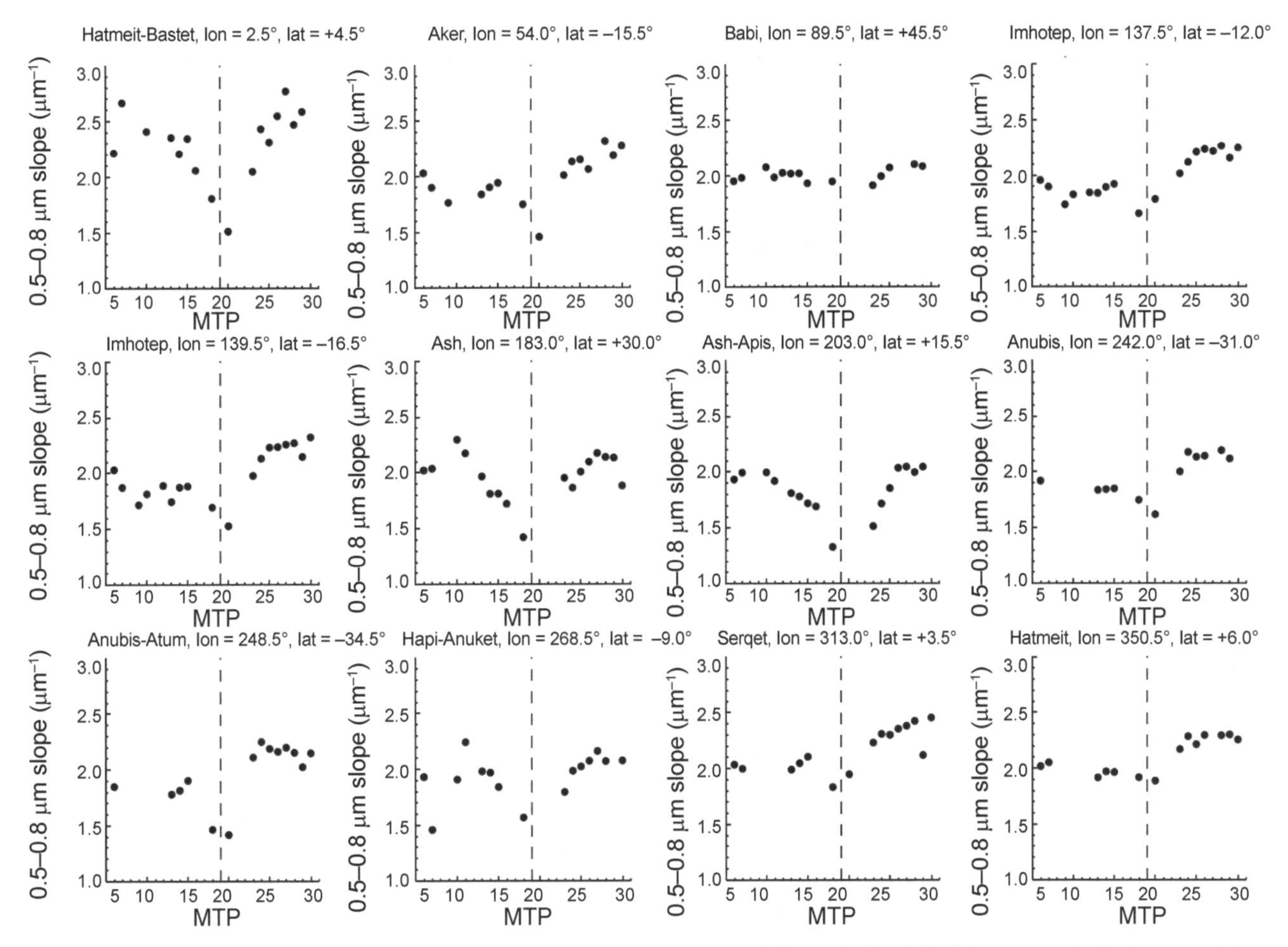

Fig. 31. Time-series of 67P/CG nucleus colors evolution as measured through the 0.5–0.8-μm spectral slope on 12 areas of the surface. MTPs are Rosetta's Medium Term Planning phases starting from MTP5 (July 2014) and ending with MTP28 (May 2016). Each MTP is one month long. Orbital passage at perihelion is during MTP19 (marked by the vertical dashed line on each plot). The region name and position (long., lat.) of the 12 test areas are indicated in each plot. For comparison, the temporal changes occurring on the coma are shown in Fig. 30. From *Filacchione et al.* (2020).

and within the canyon-like structure in Anhur (*Fornasier et al.,* 2019a). Both these southern hemisphere regions originated several jets; in particular, the canyon in Anhur was the source of the so-called perihelion outburst, one of the strongest-activity events reported for Comet 67P/CG. Moreover, several ice exposures were observed in the Anhur and Khonsu regions (*Deshapriya et al.,* 2018; *Fornasier et al.,* 2019a; *Hasselmann et al.,* 2019), and water frost was repeatedly observed within the Anhur canyon (see Fig. 33b).

Finally, the dust that fell back onto the nucleus is preserving some water ice, as suggested by *Keller et al.* (2017) and observed by *Fornasier et al.* (2019a). In fact, these authors noticed an asymmetry in the presence of frost/water ice on 67P/CG by comparing pre- and post-perihelion images acquired at similar spatial resolutions and at similar heliocentric distances. While during the inbound orbits water frost/ice was observed only in Hapi, in the outbound orbits it was observed in several regions, including Seth, Anhur, Khonsu, Imhotep, and, in a minor amount, in Khonsu (*Lucchetti et al.,* 2017; *Fornasier et al.,* 2019a, 2021; *Hasselmann et al.,* 2019; *Deshapriya et al.,* 2016, 2018; *Oklay et al.,* 2017). All these pieces of evidence proved that the falling-back dust still preserves some ice.

4.3.2. Local surface deflation. This mechanism is associated with the sublimation of subsurface reservoirs of volatiles, resulting in a local surface erosion. An example of manifestation of this mechanism has occurred in the large smooth central area of the Imhotep region with the appearance of two roundish depressions with sizes increasing in time up to several hundred meters and depth of 5 ± 2 m when the comet was approaching perihelion (*Groussin et al.,* 2015b). These authors invoked exothermic processes such as the crystallization of water ice and/or the clathrate destabilization to explain the expansion rate of these features, which was higher than 18 cm hr^{-1}. Tiny bright spots, as well as relatively bluer colors/spectral slopes near the rims/edges of these roundish features, were observed, indicating a local enrichment in volatiles.

4.3.3. Cliff/overhang collapse. Cliff and overhang collapses may be triggered by episodic and explosive activity events, producing the mechanical disruption of old terrains (*Groussin et al.,* 2015a; *Pommerol et al.,* 2015). A

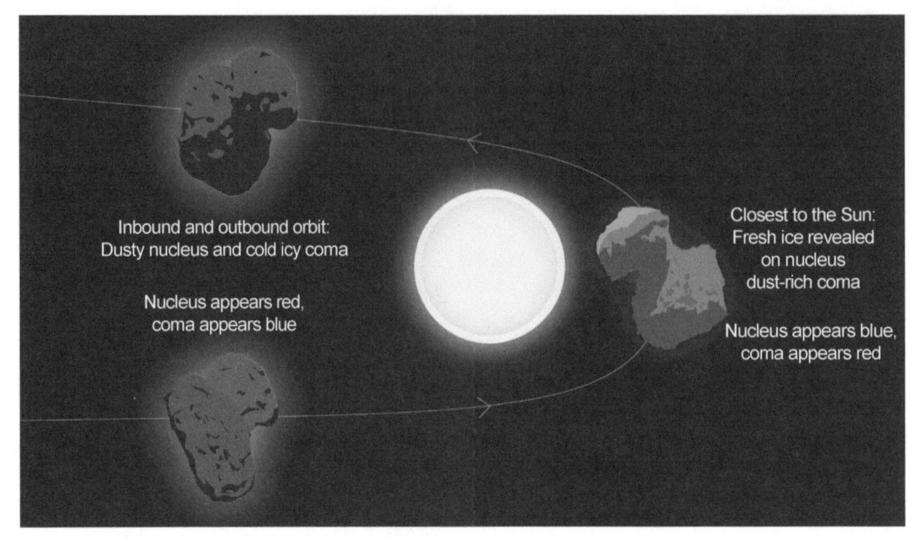

Fig. 32. A synthesis of the color cycling occurring on the nucleus surface and inner coma of 67P/CG. From ESA's Science and Technology Blog (available online at *https://sci.esa.int/s/8kaR0bw*).

notable example is the July 2015 outburst associated with the cliff's collapse in the Aswan site in the Seth region (*Pajola et al.,* 2017a). Following the cliff's collapse, an inner layer enriched in fresh and bright icy material (with albedo beyond 40%) has been exposed. This layer has survived for several months: few examples of this kind of event are shown in Figs. 33a,d. Several jets were identified by *Fornasier et al.* (2019b) directly on the nucleus and they were found arising below or close to cliffs or scarps. Often, the walls or the debris fields of cliffs and overhangs are spectrally bluer and brighter, indicating the local enrichment of water ice, from the freshly exposed underneath layers (*Pommerol et al.,* 2015; *Filacchione et al.,* 2016b; *Vincent et al.,* 2016b; *Oklay et al.,* 2017; *Fornasier et al.,* 2017, 2019a).

4.3.4. Local activity. Outburst and jets arise from cliffs, pits, bright spots, and fractures, or from cavities that cast shadows and favor the recondensation of volatiles showing periodic activity. These activity sources are often observed to be brighter and relatively bluer in colors/spectral slope, showing a local enrichment in volatiles (*Vincent et al.,* 2016b; *Agarwal et al.,* 2017; *Fornasier et al.,* 2017, 2019a,b; *Hasselmann et al.,* 2019; *Deshapriya et al.,* 2018; *Hoang et al.,* 2020). Examples are reported in Figs. 33b,c. The investigation of activity events from *Vincent et al.* (2016a) and *Fornasier et al.* (2019b) shows that cometary activity is triggered by the local illumination conditions and it is stable between several nucleus rotations. *Fornasier et al.* (2019b) identified the source areas of more than 200 jets and found that the activity events are not correlated with the nucleus morphology, originating both from consolidated and smooth terrains. Interestingly, *Fornasier et al.* (2019b) found that some faint jets may have an extremely short duration time, less than a couple of minutes, so it may be difficult to catch this kind of event, which is probably common in comets.

4.3.5. Composition heterogeneities. Local color and composition heterogeneities have been identified in different regions of the 67P/CG nucleus and several of them are associated with the local exposure of volatiles. A notable example is Anhur, which, as discussed in section 3.1, showed a first exposure of CO_2 ice (*Filacchione et al.,* 2016c), followed by the exposure of two large water-ice-rich patches (*Fornasier et al.,* 2016) on the same area. These authors deduced compositional heterogeneities in the subsurface composition on a scale of tenths of meters in the Anhur region. The sublimation of the surface and subsurface volatile reservoir here observed is at the origin of the fragile terrain where a new scarp formed as shown in Fig. 33a, just near the location of the two bright patches.

Some compositional inhomogeneities in Comet 9P/Tempel 1 linked to the detection of dirty water-ice rich material have also been reported by *Sunshine et al.* (2006). Moreover, the comparison of Tempel 1 images from Deep Impact and the Stardust flybys show morphological changes (*Veverka et al.,* 2013), interpreted as the progressive sublimation and depletion of volatiles.

Comparing the physical and composition properties of two southern hemisphere regions, Anhur and Wosret, subject to the same strong thermal heating during the cometary summer but located in the large and small lobes, respectively, *Fornasier et al.* (2021) have noticed a number of differences in the physical and mechanical properties among the two lobes, although no appreciable differences were reported in the literature on the global surface composition (*Capaccioni et al.,* 2015) and on the D/H ratio (*Schroeder et al.,* 2019). Wosret, in the small lobe, shows larger goosebump features, very few morphological changes compared to Anhur and Khonsu regions despite the fact that they all experienced the same high insolation, and less-frequent and smaller frost and water-ice-enriched areas. Considering that Wosret is highly eroded and thus exposes the inner layers of the small lobe, *Fornasier et al.* (2021) deduced that the small lobe has a

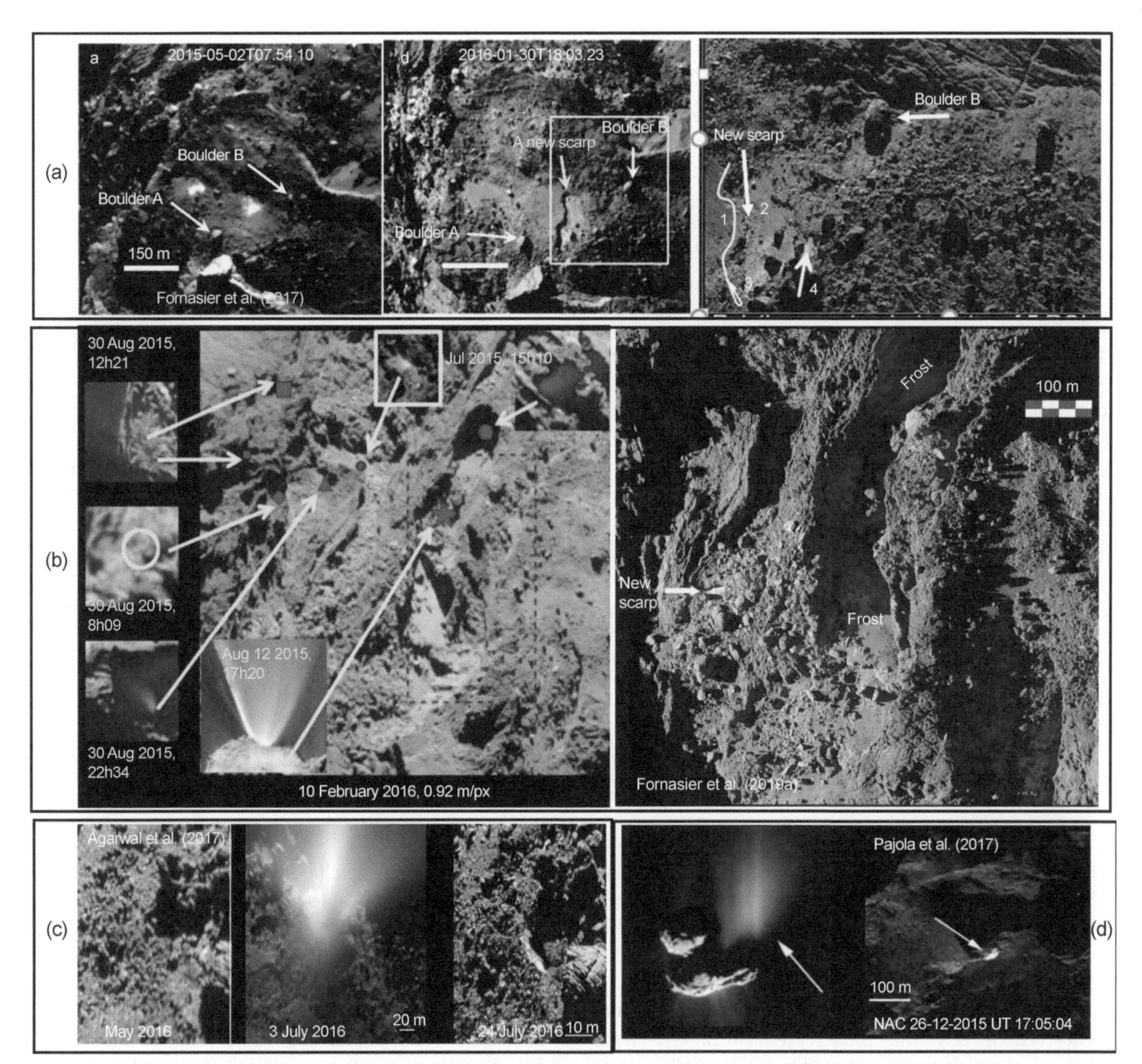

Fig. 33. Examples of composition changes associated with activity events. **(a)** The Anhur region showing the two larger water-ice-rich patches (left), the new scarp formed between August and December 2015 near the patches (center), and the corresponding RGB map from images acquired on May 7, 2016, UT 04:15, showing the new scarp and the bluer and water-ice-rich material at its feet (from *Fornasier et al.,* 2017). **(b)** Example of jet locations (left) and some associated activity events in the Anhur region, including the perihelion outburst, which took place on August 12, 2015 (from *Fornasier et al.,* 2019b). The right side of **(b)** shows a RGB composite image (from the images acquired with filters centered at 882, 649, and 480 nm) of the Anhur region from observations acquired on June 25, 2016, at UT 01h37 at a resolution of 0.35 m/pixel. Several exposures of water ice are also detected, notably at the base of a new scarp, and frost observed within the canyon structure (adapted from *Fornasier et al.,* 2019a). **(c)** The plume and associated new cavity formation with exposure of water ice identified in the Imhotep region by *Agarwal et al.* (2017). **(d)** The outburst and cliff collapse in Aswan site with the exposure of the fresh water-ice-rich layer at the wall of the cliff (from *Pajola et al.,* 2017a).

lower volatile content, at least in its top layers, than the big one, and different mechanical properties, as already noticed by *El-Maarry et al.* (2016).

A comparison of the observed surface properties on the active regions of the small and big lobes of 67P/CG (Table 4) allows appreciation of how the latter is more eroded and processed by solar heating during the perihelion passages.

4.4. From Surface Evolution to the Internal Structure of Cometary Nuclei

Even though the blueing evolution hints a global water-ice enrichment toward perihelion passage (see section 4.2), the water ice is observed in a patchy distribution across the nucleus. As we have discussed in section 3.1.1, high spatial resolution observations of 67P/CG have revealed a rich

TABLE 4. Summary of the observed behavior and physical properties of active areas located on the small and big lobe of Comet 67P/CG from *Fornasier et al.* (2021); properties are derived from the investigation of southern hemisphere regions submitted to similar high heating at perihelion.

	Small Lobe (Wosret)	Big Lobe	References
Incoming solar flux	~550 W m^{-2}	~550 W m^{-2} (Anhur and Khonsu)	*Marshall et al.* (2017)
Morphology	Consolidated material that appears highly fractured with occasional pits	Consolidated material with significant intermediate-scale roughness (Anhur)	*Thomas et al.* (2018)
Exposed water ice	in a few and small bright spots (~m^2)	in many bright spots and some in big patches (1500 m^2)	*Fornasier et al.* (2021); *Fornasier et al.* (2016, 2019a)
Relatively blue area enriched in frost/ice	in very spatially limited area, periodic, not frequently observed	in extended area, periodic often observed	*Fornasier et al.* (2016, 2017); *Fornasier et al.* (2019a, 2021); *Hasselmann et al.* (2019)
Average goose-bump diameter	the largest in 67P/CG: 4.7 ± 1.5 m	2.2–3.2 m (Seth, Imhothep, Anubis, Atum)	*Fornasier et al.* (2021); *Sierks et al.* (2015); *Davidsson et al.* (2016)
Level of activity	very high	very high	*Fornasier et al.* (2021, 2019b); *Hasselmann et al.* (2019)
Surface morphology changes	few and not so important: a relatively small cavity, local dust removal (~1 m depth) revealing a cluster of outcrops	many and important: local dust removal up to 14 m in depth, new relatively big scarps and cavities, big vanishing structures, boulders displacements and fragmentation	*Fornasier et al.* (2021); *Fornasier et al.* (2017, 2019a); *Hasselmann et al.* (2019)
Surface mass loss	~1.2 × 10^6 kg	>50 × 10^6 kg in Anhur; ~2 × 10^8 kg in Khonsu	*Fornasier et al.* (2016, 2019a); *Hasselmann et al.* (2019)

phenomenology of localized water-ice-rich spots, ranging in size from <1 m up to tens of meters. These spots appear distributed across the nucleus surface and are recognizable thanks to their intrinsic high reflectance and blue-colored visible spectra, which distinguish them from the average dark and red-colored terrain [we refer to such spots as blue patches (BPs)]. A similar property is also in agreement with the dust ejection rates reported by *Fulle et al.* (2016b) derived by GIADA at perihelion (*Fulle et al.,* 2020).

Starting from this observational evidence, *Ciarniello et al.* (2022) have explained the color changes occurring on the surface of 67P in the framework of the water-driven activity model of *Fulle et al.* (2020) for a nucleus made of centimeter-sized pebbles (*Blum et al.,* 2017) by defining a new model of the comet's internal structure. According to the proposed scenario (Fig. 34), the perihelion blueing of 67P is interpreted as an increase of the areal fraction covered in BPs. In this respect, the BPs represent the exposed counterparts of a population of water-ice-enriched blocks (WEBs), formed of pebbles with a high content of water ice, immerse in a matrix of drier pebbles (see details in Fig. 34 and caption). The WEBs are exposed as a consequence of the surface erosion through chunk ejection induced by CO_2 ice sublimation (*Gundlach et al.,* 2020), and are eroded by H_2O ice sublimation ejecting subcentimeter dust (*Fulle et al.,* 2020) when directly reached by sunlight. The observed color evolution of the nucleus is the result of these two competing mechanisms. By matching the temporal variation of the surface color while taking into account the constraints on the surface erosion by CO_2-driven chunk emission (*Fulle et al.,* 2019, 2020; *Cambianica et al.,* 2020) and the modeled water-driven erosion rates (*Fulle et al.,* 2020), *Ciarniello et al.* (2022) infer a dominant size for the WEBs of 0.5–1 m, with a volumetric abundance in the nucleus from 9.5 to 5.5%. The presence of submeter WEBs in the nucleus of 67P/CG is in agreement with Comet Nucleus Sounding Experiment by Radiowave Transmission (CONSERT) radar observations, which found a homogeneous internal structure down to a few meters scale (*Ciarletti et al.,* 2017).

A uniform distribution of WEBs in the interior of the 67P/CG nucleus explains both the observed seasonal spectral changes and the measured dust ejection rates, making the macroscopic compositional dishomogeneities observed in comets, e.g., BPs, compatible with the homogeneous structure at small scale (the centimeter-sized pebbles) and with the sublimation processes occurring at microscopic (subpebble) scales. If this is the case, WEBs must be of primordial origin: In fact, during the orbital evolution water ice can recondensate only within a very shallow layer (*De Sanctis et al.,* 2015; *Fornasier et al.,* 2016) but not in the deep internal parts. With similar characteristics, WEBs are imposing strict constraints on the formation scenarios for comets. The detection of crystalline water ice in protoplanetary disks (PPDs) by *Min et al.* (2016) would be compatible

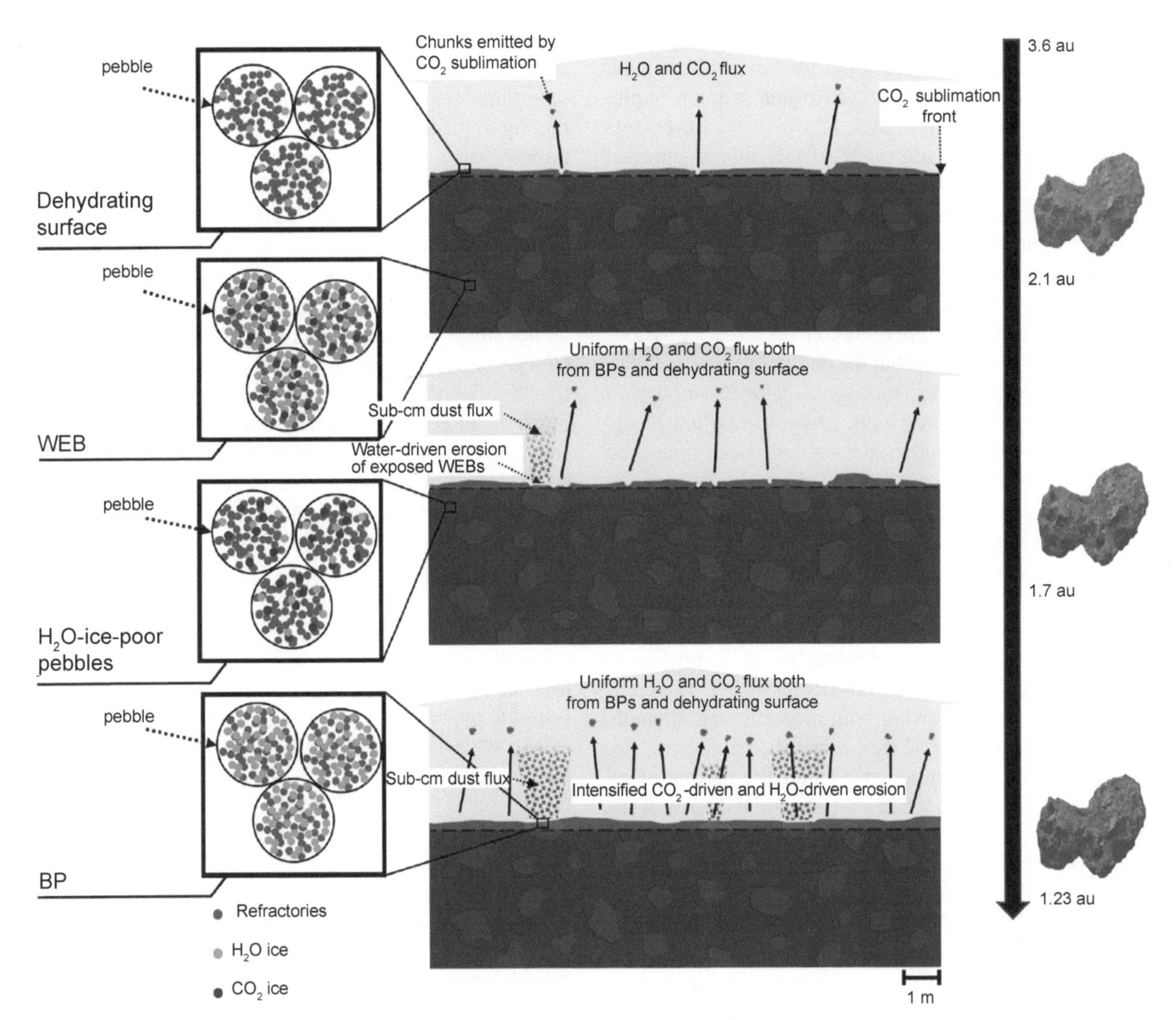

Fig. 34. The progressive blueing of 67P/CG surface orbiting toward perihelion is caused by the progressive exposure to sunlight of water-ice-enriched blocks (WEBs). In this model the comet nucleus is made up of two populations of pebbles composed of refractories (organic matter and minerals) and CO_2 ice with, respectively, (1) a high abundance of H_2O ice (composing the WEBs) and (2) a low abundance of H_2O ice (H_2O-ice poor pebbles). The composition of the H_2O-ice-poor pebbles corresponds to the material observed during the occasional exposure of CO_2 ice (*Filacchione et al.,* 2016c), making most of the nucleus. The CO_2 ice is stable at depths >0.1 m where its condensation front is localized (*Gundlach et al.,* 2020). When approaching perihelion, the sublimation rate of CO_2 steadily increases, causing the erosion of the surface through the ejection of chunks and exposing the subsurface WEBs. Once exposed, the WEBs rapidly lose their CO_2 content and appear as blue patches (BPs). After this, the solar flux activates the sublimation of the H_2O ice, which erodes the BPs while removing subcentimeter-sized dust from their surface and preventing the formation of a dry crust (*Fulle et al.,* 2020). The BPs can survive on the surface until their water-ice fraction is gone. In the meanwhile, their presence is the origin of the observed surface blueing. While WEBs evolve along this path, H_2O-ice-poor pebbles experience dehydration rather than erosion due to their intrinsic lower abundance of water ice (*Fulle et al.,* 2020), whereas the formation of a crust is hampered by the concurrent CO_2 driven erosion. The same water flux is released by the BPs and the dehydrating surface units as both contain some water ice and are assumed to share the same temperature (*Fulle et al.,* 2020). The resulting blueing trend when 67P/CG approaches perihelion is shown along the right axis. Adapted from *Ciarniello et al.* (2022), where further details can be found.

with the formation of water-ice-rich pebbles from local ice recondensation, likely stickier than the water-poor ones, and possibly favoring the accretion into WEBs during the gravitational collapse originated by the streaming instabilities in PPDs (*Blum et al.,* 2017).

5. CONCLUSIONS

The nuclei of comets are repositories of primordial materials synthesized in the ISM and then processed in the protosolar disk. As such, they are cold time capsules able to preserve primordial matter until the present day and, through exploring them, offer us the possibility of understanding the conditions that occurred during the formation of our solar system. To achieve this goal, the Rosetta mission implemented an innovative approach in which measurements performed through a wide range of techniques aimed at inferring the composition of 67P/CG from the orbiter and from the Philae lander at different heliocentric distances. In this respect, the Rosetta mission has been a game changer for cometary science.

So far, the analysis of Rosetta data discussed in this chapter in the context of previous cometary space missions provides important clues about the nature of cometary matter that is compatible with a *macromolecular assemblage in which ices, organic matter, salts, and minerals* are bound together. While the volatile species have been measured and characterized as ices on the nucleus and as gases in the coma by multiple instruments, the determination of the non-volatile fraction is much more complicated to achieve. As of the present time, it is still unclear how to reconcile the elemental composition of gaseous species and dust grains achieved by *in situ* instruments (ROSINA, COSIMA, Ptolemy, COSAC) with the optical properties of the nucleus at visible-infrared wavelengths (VIRTIS, OSIRIS). The complex organic matter, showing both aromatic and aliphatic bonds, appears well-mixed with minerals, as silicates, Fe-sulfides, carbon, and ammoniated salts. The mixing of all these materials results in the low albedo and red color of the dust covering the whole nucleus. When far from the Sun, the majority of the surface appears very dehydrated, with a water-ice residual of no more than 1%.

The composition of the non-icy materials allows back-tracing the origins of comets: The nature of the refractories, including organic aromatic and aliphatic matter and silicates, shares similarities with presolar matter. The stronger intensity of the CH_3 asymmetric stretch (3.38 μm) with respect to the CH_2 (3.42 μm) resembles the properties observed in the ISM and IOM in primitive carbonaceous condrites. The 2.8 μm OH band in hydroxylated-magnesium silicates also indicates a possible genetic link between cometary refractories and ISM, where amorphous silicates are irradiated by keV-MeV energy protons. Moreover, Rosetta's results can explain the apparent nitrogen depletion observed in cometary comae being sequestered in semivolatile ammoniated salts. In this respect, comets show analogies with other asteroids and outer solar system objects observed at visible-infrared wavelengths, suggesting an evolutionary link among all these objects. This evidence shows that to understand the composition of comets, it will be necessary to improve statistics (so far only six comets have been explored by space missions) and comparisons with asteroids, outer solar system objects, and ISM data.

The Rosetta mission faced multiple (and unexpected) challenges in exploring 67P/CG: The very low albedo, the complex shape and morphology of the nucleus, the evolution of the solar heating, and the changing activity responsible for the dust and gas environment in the coma were factors playing against a straightforward determination of the nucleus' composition. Moreover, the less than ideal landing of Philae further complicated the mission planning and the scientific interpretation of the results. Similar difficulties will remain in place for any future cometary mission relying on remote sensing and *in situ* payloads. In order to overcome them we need a paradigm shift in cometary exploration by implementing a more ambitious exploration approach, such as by adopting a cryogenic sample return, in which a sample containing original cometary material, keeping the ices and refractory parts unaltered, is returned to Earth to conduct composition analyses with multiple analytical techniques (*Bockelée-Morvan et al.,* 2021). A similar sample will be the "holy grail" for the cosmochemistry community, allowing light to be shed on many unresolved questions: Apart from determining the composition of cometary material and whether they have a solar or presolar origin, it would permit an understanding of cometary formation mechanisms, internal structure, dust-to-ice ratio, and the presence of prebiotic molecules.

Acknowledgments. The authors gratefully thank S. Besse (ESA-ESAC, Madrid, Spain) and J. Licandro (Instituto de Astrofísica de Canarias, Spain) for their thorough reviews, which helped to improve the quality of the manuscript. G.F., M.C., and A.R. acknowledge support from INAF-IAPS and ASI, Italian Space Agency. S.F. acknowledges support from LESIA, the France Agence Nationale de la Recherche (programme Classy, ANR-17-CE31-0004), and from the Institut Universitaire de France. This research has made use of NASA's Astrophysics Data System (NASA-ADS).

REFERENCES

Agarwal J., Della Corte V., Feldman P. D., et al. (2017) Evidence of sub-surface energy storage in Comet 67P from the outburst of 2016 July 03. *Mon. Not. R. Astron. Soc., 469,* S606–S625.

A'Hearn M. F., Belton M. J. S., Delamere W. A., et al. (2005) Deep Impact: Excavating Comet Tempel 1. *Science, 310,* 258–264.

A'Hearn M. F., Belton M. J. S., Delamere W. A., et al. (2011) EPOXI at Comet Hartley 2. *Science, 332,* 1396–1400.

Aléon J., Engrand C., Robert F., et al. (2001) Clues to the origin of interplanetary dust particles from the isotopic study of their hydrogen-bearing phases. *Geochim. Cosmochim. Acta, 65,* 4399–4412.

Altwegg K., Balsiger H., Bar-Nun A., et al. (2016) Prebiotic chemicals — amino acid and phosphorus — in the coma of Comet 67P/Churyumov-Gerasimenko. *Sci. Adv., 2,* e1600285.

Altwegg K., Balsiger H., Berthelier J. J., et al. (2017) Organics in Comet 67P — a first comparative analysis of mass spectra from ROSINA — DFMS, COSAC and Ptolemy. *Mon. Not. R. Astron. Soc., 469,* S130–S141.

Altwegg K., Balsiger H., and Fuselier S. A. (2019) Cometary chemistry and the origin of icy solar system bodies: The view after Rosetta. *Annu. Rev. Astron. Astrophys., 57,* 113–155.

Altwegg K., Balsiger H., Hänni N., et al. (2020) Evidence of ammonium salts in Comet 67P as explanation for the nitrogen depletion in cometary comae. *Nature Astron., 4,* 533–540.

Balsiger H., Altwegg K., Bochsler P., et al. (2007) Rosina — Rosetta Orbiter Spectrometer for Ion and Neutral Analysis. *Space Sci. Rev., 128,* 745–801.

Bardyn A., Baklouti D., Cottin H., et al. (2017) Carbon-rich dust in Comet 67P/Churyumov-Gerasimenko measured by COSIMA/Rosetta. *Mon. Not. R. Astron. Soc., 469,* S712–S722.

Barucci M. A., Filacchione G., Fornasier S., et al. (2016) Detection of exposed H_2O ice on the nucleus of Comet 67P/Churyumov-

Gerasimenko as observed by Rosetta OSIRIS and VIRTIS instruments. *Astron. Astrophys., 595,* A102.

Belskaya I. N. and Shevchenko V. G. (2000) Opposition effect of asteroids. *Icarus, 147,* 94–105.

Benkhoff J. (1995) Numerical simulations of the gas flux at the surface of 2060 Chiron. *Bull. Am. Astron. Soc., 27,* 1338.

Bentley M. S., Schmied R., Mannel T., et al. (2016) Aggregate dust particles at Comet 67P/Churyumov-Gerasimenko. *Nature, 537,* 73–75.

Bibring J.-P., Langevin Y., Carter J., et al. (2015) 67P/Churyumov-Gerasimenko surface properties as derived from CIVA panoramic images. *Science, 349,* aab0671.

Biele J., Ulamec S., Maibaum M., et al. (2015) The landing(s) of Philae and inferences about comet surface mechanical properties. *Science, 349,* aaa9816.

Blum J., Gundlach B., Krause M., et al. (2017) Evidence for the formation of Comet 67P/Churyumov-Gerasimenko through gravitational collapse of a bound clump of pebbles. *Mon. Not. R. Astron. Soc., 469,* S755–S773.

Bockelée-Morvan D. and Biver N. (2017) The composition of cometary ices. *Philos. Trans. R. Soc.* A, *375,* 20160252.

Bockelée-Morvan D., Gautier D., Hersant F., et al. (2002) Turbulent radial mixing in the solar nebula as the source of crystalline silicates in comets. *Astron. Astrophys., 384,* 1107–1118.

Bockelée-Morvan D., Leyrat C., Erard S., et al. (2019) VIRTIS-H observations of the dust coma of Comet 67P/Churyumov-Gerasimenko: Spectral properties and color temperature variability with phase and elevation. *Astron. Astrophys., 630,* A22.

Bockelée-Morvan D., Filacchione G., Altwegg K., et al. (2021) AMBITION — Comet nucleus cryogenic sample return. *Exp. Astron.,* DOI: 10.1007/s10686-021-09770-4.

Bradley J. P., Keller L. P., Snow T. P., et al. (1999) An infrared spectral match between GEMS and interstellar grains. *Science, 285,* 1716–1718.

Bradley J., Dai Z. R., Erni R., et al. (2005) An astronomical 2175 Å feature in interplanetary dust particles. *Science, 307,* 244–247.

Bregman J. D., Witteborn F. C., Allamandola L. J., et al. (1987) Airborne and groundbased spectrophotometry of Comet P/Halley from 5–13 micrometers. *Astron. Astrophys., 187,* 616–620.

Brown M. E. (2016) The 3–4 μm spectra of Jupiter Trojan asteroids. *Astron. J., 152,* 159.

Brown M. E. and Rhoden A. R. (2014) The 3 μm spectrum of Jupiter's irregular satellite Himalia. *Astrophys. J. Lett., 793,* L44.

Brownlee D. E., Tsou P., Anderson J. D., et al. (2003) Stardust: Comet and interstellar dust sample return mission. *J. Geophys. Res.–Planets, 108,* 8111.

Brownlee D., Tsou P., Aléon J., et al. (2006) Comet 81P/Wild 2 under a microscope. *Science, 314,* 1711–1716.

Brucato J. R., Strazzulla G., Baratta G., et al. (2004) Forsterite amorphisation by ion irradiation: Monitoring by infrared spectroscopy. *Astron. Astrophys., 413,* 395–401.

Buratti B. J., Hicks M. D., Soderblom L. A., et al. (2004) Deep Space 1 photometry of the nucleus of Comet 19P/Borrelly. *Icarus, 167,* 16–29.

Cambianica P., Fulle M., Cremonese G., et al. (2020) Time evolution of dust deposits in the Hapi region of Comet 67P/Churyumov-Gerasimenko. *Astron. Astrophys., 636,* A91.

Campins H. and Ryan E. V. (1989) The identification of crystalline olivine in cometary silicates. *Astrophys. J., 341,* 1059.

Campins H., A'Hearn M. F., and McFadden L.-A. (1987) The bare nucleus of Comet Neujmin 1. *Astrophys. J., 316,* 847.

Campins H., Ziffer J., Licandro J., et al. (2006) Nuclear spectra of Comet 162P/Siding Spring (2004 TU12). *Astron. J., 132,* 1346–1353.

Campins H., Licandro J., Pinilla-Alonso N., et al. (2007) Nuclear spectra of Comet 28P Neujmin 1. *Astron. J., 134,* 1626–1633.

Capaccioni F., Coradini A., Filacchione G., et al. (2015) The organic-rich surface of Comet 67P/Churyumov-Gerasimenko as seen by VIRTIS/Rosetta. *Science, 347,* aaa0628.

Carrez P., Demyk K., Cordier P., et al. (2002) Low-energy helium ion irradiation-induced amorphization and chemical changes in olivine: Insights for silicate dust evolution in the interstellar medium. *Meteorit. Planet. Sci., 37,* 1599–1614.

Choukroun M., Altwegg K., Kührt E., et al. (2020) Dust-to-gas and refractory-to-ice mass ratios of Comet 67P/Churyumov-Gerasimenko from Rosetta observations. *Space Sci. Rev., 216,* 44.

Ciarletti V., Herique A., Lasue J., et al. (2017) CONSERT constrains the internal structure of 67P at a few metres size scale. *Mon. Not. R. Astron. Soc., 469,* S805–S817.

Ciarniello M., Capaccioni F., Filacchione G., et al. (2015) Photometric properties of Comet 67P/Churyumov-Gerasimenko from VIRTIS-M onboard Rosetta. *Astron. Astrophys., 583,* A31.

Ciarniello M., Raponi A., Capaccioni F., et al. (2016) The global surface composition of 67P/Churyumov-Gerasimenko nucleus by Rosetta/VIRTIS. II) Diurnal and seasonal variability. *Mon. Not. R. Astron. Soc., 462,* S443–S458.

Ciarniello M., De Sanctis M. C., Ammannito E., et al. (2017) Spectrophotometric properties of dwarf planet Ceres from the VIR spectrometer on board the Dawn mission. *Astron. Astrophys., 598,* A130.

Ciarniello M., Fulle M., Raponi A., et al. (2022) Macro and micro structures of pebble-made cometary nuclei reconciled by seasonal evolution. *Nature Astron., 6,* 546–553.

Clemett S. J., Sandford S. A., Nakamura-Messenger K., et al. (2010) Complex aromatic hydrocarbons in Stardust samples collected from Comet 81P/Wild 2. *Meteorit. Planet. Sci., 45,* 701–722.

Colangeli L., Lopez-Moreno J. J., Palumbo P., et al. (2007) The Grain Impact Analyser and Dust Accumulator (GIADA) experiment for the Rosetta mission: Design, performances and first results. *Space Sci. Rev., 128,* 803–821.

Coradini A., Capaccioni F., Drossart P., et al. (2007) VIRTIS: An imaging spectrometer for the Rosetta mission. *Space Sci. Rev., 128,* 529–559.

Cravens T. E. (1987) Theory and observations of cometary ionospheres. *Adv. Space Res., 7,* 147–158.

Crovisier J., Leech K., Bockelée-Morvan D., et al. (1997) The spectrum of Comet Hale-Bopp (C/1995 O1) observed with the infrared space observatory at 2.9 astronomical units from the Sun. *Science, 275,* 1904–1907.

Cuzzi J. N., Davis S. S., and Dobrovolskis A. R. (2003) Blowing in the wind. II. Creation and redistribution of refractory inclusions in a turbulent protoplanetary nebula. *Icarus, 166,* 385–402.

Dandy C. L., Fitzsimmons A., and Collander-Brown S. J. (2003) Optical colors of 56 near-Earth objects: Trends with size and orbit. *Icarus, 163,* 363–373.

Dartois E., Marco O., Muñoz-Caro G. M., et al. (2004) Organic matter in Seyfert 2 nuclei: Comparison with our galactic center lines of sight. *Astron. Astrophys., 423,* 549–558.

Davidsson B. J. R. and Skorov Y. V. (2004) A practical tool for simulating the presence of gas comae in thermophysical modeling of cometary nuclei. *Icarus, 168,* 163–185.

Davidsson B. J. R., Gutiérrez P. J., Groussin O., et al. (2013) Thermal inertia and surface roughness of Comet 9P/Tempel 1. *Icarus, 224,* 154–171.

Davidsson B. J. R., Sierks H., Güttler C., et al. (2016) The primordial nucleus of Comet 67P/Churyumov-Gerasimenko. *Astron. Astrophys., 592,* A63.

de Almeida A. A., Huebner W. F., Benkhoff J., et al. (1996) The three sources of gas in the comae of comets. *Rev. Mex. Astron. Astrofis. Ser. Conf., 4,* 110.

Delahodde C. E., Meech K. J., Hainaut O. R., et al. (2001) Detailed phase function of Comet 28P/Neujmin 1. *Astron. Astrophys., 376,* 672–685.

Della Corte V., Rotundi A., Fulle M., et al. (2015) GIADA: Shining a light on the monitoring of the comet dust production from the nucleus of 67P/Churyumov-Gerasimenko. *Astron, Astrophys., 583,* A13.

Dello Russo N., Kawakita H., Vervack R. J., et al. (2016) Emerging trends and a comet taxonomy based on the volatile chemistry measured in thirty comets with high-resolution infrared spectroscopy between 1997 and 2013. *Icarus, 278,* 301–332.

Dello Russo N., Vervack R. J., Weaver H. A., et al. (2007) Compositional homogeneity in the fragmented Comet 73P/Schwassmann-Wachmann 3. *Nature, 448,* 172–175.

Delsemme A. H. (1984) The cometary connection with prebiotic chemistry. *Origins Life, 14,* 51–60.

Demyk K., Carrez P., Leroux H., et al. (2001) Structural and chemical alteration of crystalline olivine under low energy He^+ irradiation. *Astron. Astrophys., 368,* L38–L41.

de Niem D., Kührt E., Hviid S., et al. (2018) Low velocity collisions of porous planetesimals in the early solar system. *Icarus, 301,* 196–218.

De Sanctis M. C., Capria M. T., and Coradini A. (2005) Thermal evolution model of 67P/Churyumov-Gerasimenko, the new Rosetta

target. *Astron. Astrophys., 444,* 605–614.
De Sanctis M. C., Capaccioni F., Ciarniello M., et al. (2015) The diurnal cycle of water ice on Comet 67P/Churyumov-Gerasimenko. *Nature, 525,* 500–503.
Deshapriya J. D. P., Barucci M. A., Fornasier S., et al. (2016) Spectrophotometry of the Khonsu region on the Comet 67P/Churyumov-Gerasimenko using OSIRIS instrument images. *Mon. Not. R. Astron. Soc., 462,* S274–S286.
Deshapriya J. D. P., Barucci M. A., Fornasier S., et al. (2018) Exposed bright features on the Comet 67P/Churyumov-Gerasimenko: Distribution and evolution. *Astron. Astrophys., 613,* A36.
Djouadi Z., D'Hendecourt L., Leroux H., et al. (2005) First determination of the (re)crystallization activation energy of an irradiated olivine-type silicate. *Astron. Astrophys., 440,* 179–184.
Djouadi Z., Robert F., Le Sergeant D'Hendecourt L., et al. (2011) Hydroxyl radical production and storage in analogues of amorphous interstellar silicates: a possible "wet" accretion phase for inner telluric planets. *Astron. Astrophys., 531,* A96.
Domingue D., Kitazato K., Matsuoka M., et al. (2021) Spectrophotometric properties of 162173 Ryugu's surface from the NIRS3 opposition observations. *Planet. Sci. J., 2,* 178.
Dorofeeva V. A. (2020) Chemical and isotope composition of Comet 67P/Churyumov-Gerasimenko: The Rosetta-Philae mission results reviewed in the context of cosmogony and cosmochemistry. *Solar System Res., 54,* 96–120.
El-Maarry M. R., Thomas N., Giacomini L., et al. (2015) Regional surface morphology of Comet 67P/Churyumov-Gerasimenko from Rosetta/OSIRIS images. *Astron. Astrophys., 583,* A26.
El-Maarry M. R., Thomas N., Gracia-Berná A., et al. (2016) Regional surface morphology of Comet 67P/Churyumov-Gerasimenko from Rosetta/OSIRIS images: The southern hemisphere. *Astron. Astrophys., 593,* A110.
El-Maarry M. R., Groussin O., Thomas N., et al. (2017) Surface changes on Comet 67P/Churyumov-Gerasimenko suggest a more active past. *Science, 355,* 1392–1395.
Emery J. P., Burr D. M., and Cruikshank D. P. (2011) Near-infrared spectroscopy of Trojan asteroids: Evidence for two compositional groups. *Astron. J., 141,* 25.
Emery J. P., Cruikshank D. P., and Van Cleve J. (2006) Thermal emission spectroscopy (5.2–38 μm) of three Trojan asteroids with the Spitzer Space Telescope: Detection of fine-grained silicates. *Icarus, 182,* 496–512.
Espinasse S., Klinger J., Ritz C., et al. (1991) Modeling of the thermal behavior and of the chemical differentiation of cometary nuclei. *Icarus, 92,* 350–365.
Feaga L. M., A'Hearn M. F., Sunshine J. M., et al. (2007) Asymmetries in the distribution of H_2O and CO_2 in the inner coma of Comet 9P/Tempel 1 as observed by Deep Impact. *Icarus, 191,* 134–145.
Feaga L. M., Protopapa S., Schindhelm R., et al. (2015) Far-UV phase dependence and surface characteristics of Comet 67P/Churyumov-Gerasimenko as observed with Rosetta Alice. *Astron. Astrophys., 583,* A27.
Feller C., Fornasier S., Hasselmann P. H., et al. (2016) Decimetre-scaled spectrophotometric properties of the nucleus of Comet 67P/Churyumov-Gerasimenko from OSIRIS observations. *Mon. Not. R. Astron. Soc., 462,* S287–S303.
Feller C., Fornasier S., Ferrari S., et al. (2019) Rosetta/OSIRIS observations of the 67P nucleus during the April 2016 flyby: High-resolution spectrophotometry. *Astron. Astrophys., 630,* A9.
Fernández Y. R., Lisse C. M., Ulrich Käufl H., et al. (2000) Physical properties of the nucleus of Comet 2P/Encke. *Icarus, 147,* 145–160.
Fernández Y. R., Campins H., Kassis M., et al. (2006) Comet 162P/Siding Spring: A surprisingly large nucleus. *Astron. J., 132,* 1354–1360.
Fernández Y. R., Lisse C. M., Kelley M. S., et al. (2007) Near-infrared light curve of Comet 9P/Tempel 1 during Deep Impact. *Icarus, 187,* 220–227.
Filacchione G., Capaccioni F., Ciarniello M., et al. (2016a) The global surface composition of 67P/CG nucleus by Rosetta/VIRTIS. (I) Prelanding mission phase. *Icarus, 274,* 334–349.
Filacchione G., de Sanctis M. C., Capaccioni F., et al. (2016b) Exposed water ice on the nucleus of Comet 67P/Churyumov-Gerasimenko. *Nature, 529,* 368–372.
Filacchione G., Raponi A., Capaccioni F., et al. (2016c) Seasonal exposure of carbon dioxide ice on the nucleus of Comet 67P/Churyumov-Gerasimenko. *Science, 354,* 1563–1566.
Filacchione G., Groussin O., Herny C., et al. (2019) Comet 67P/CG nucleus composition and comparison to other comets. *Space Sci. Rev., 215,* 19.
Filacchione G., Capaccioni F., Ciarniello M., et al. (2020) An orbital water-ice cycle on Comet 67P from colour changes. *Nature, 578,* 49–52.
Fornasier S., Dotto E., Marzari F., et al. (2004) Visible spectroscopic and photometric survey of L5 Trojans: Investigation of dynamical families. *Icarus, 172,* 221–232.
Fornasier S., Dotto E., Hainaut O., et al. (2007) Visible spectroscopic and photometric survey of Jupiter Trojans: Final results on dynamical families. *Icarus, 190,* 622–642.
Fornasier S., Hasselmann P. H., Barucci M. A., et al. (2015) Spectrophotometric properties of the nucleus of Comet 67P/Churyumov-Gerasimenko from the OSIRIS instrument onboard the Rosetta spacecraft. *Astron. Astrophys., 583,* A30.
Fornasier S., Mottola S., Keller H. U., et al. (2016) Rosetta's Comet 67P/Churyumov-Gerasimenko sheds its dusty mantle to reveal its icy nature. *Science, 354,* 1566–1570.
Fornasier S., Feller C., Lee J.-C., et al. (2017) The highly active Anhur-Bes regions in the 67P/Churyumov-Gerasimenko comet: Results from OSIRIS/Rosetta observations. *Mon. Not. R. Astron. Soc., 469,* S93–S107.
Fornasier S., Feller C., Hasselmann P. H., et al. (2019a) Surface evolution of the Anhur region on Comet 67P/Churyumov-Gerasimenko from high-resolution OSIRIS images. *Astron. Astrophys., 630,* A13.
Fornasier S., Hoang V. H., Hasselmann P. H., et al. (2019b) Linking surface morphology, composition, and activity on the nucleus of 67P/Churyumov-Gerasimenko. *Astron. Astrophys., 630,* A7.
Fornasier S., Hasselmann P. H., Deshapriya J. D. P., et al. (2020) Phase reddening on asteroid Bennu from visible and near-infrared spectroscopy. *Astron. Astrophys., 644,* A142.
Fornasier S., Bourdelle de Micas J., Hasselmann P. H., et al. (2021) Small lobe of Comet 67P: Characterization of the Wosret region with Rosetta-OSIRIS. *Astron. Astrophys., 653,* A132.
Frattin E., Cremonese G., Simioni E., et al. (2017) Post-perihelion photometry of dust grains in the coma of 67P Churyumov-Gerasimenko. *Mon. Not. R. Astron. Soc., 469,* S195–S203.
Fray N. and Schmitt B. (2009) Sublimation of ices of astrophysical interest: A bibliographic review. *Planet. Space Sci., 57,* 2053–2080.
Fray N., Bardyn A., Cottin H., et al. (2016) High-molecular-weight organic matter in the particles of Comet 67P/Churyumov-Gerasimenko. *Nature, 538,* 72–74.
Fray N., Bardyn A., Cottin H., et al. (2017) Nitrogen-to-carbon atomic ratio measured by COSIMA in the particles of Comet 67P/Churyumov-Gerasimenko. *Mon. Not. R. Astron. Soc., 469,* S506–S516.
Fulle M. (2021) Water and deuterium-to-hydrogen ratio in comets. *Mon. Not. R. Astron. Soc., 505,* 3107–3112.
Fulle M., Altobelli N., Buratti B., et al. (2016a) Unexpected and significant findings in Comet 67P/Churyumov-Gerasimenko: An interdisciplinary view. *Mon. Not. R. Astron. Soc., 462,* S2–S8.
Fulle M., Della Corte V., Rotundi A., et al. (2016b) Comet 67P/Churyumov-Gerasimenko preserved the pebbles that formed planetesimals. *Mon. Not. R. Astron. Soc., 462,* S132–S137.
Fulle M., Marzari F., Della Corte V., et al. (2016c) Evolution of the dust size distribution of Comet 67P/Churyumov-Gerasimenko from 2.2 au to perihelion. *Astrophys. J., 821,* 19.
Fulle M., Blum J., Green S. F., et al. (2019) The refractory-to-ice mass ratio in comets. *Mon. Not. R. Astron. Soc., 482,* 3326–3340.
Fulle M., Blum J., Rotundi A., et al. (2020) How comets work: Nucleus erosion versus dehydration. *Mon. Not. R. Astron. Soc., 493,* 4039–4044.
Geiss J. (1987) Composition measurements and the history of cometary matter. *Astron. Astrophys., 187,* 859–866.
Goesmann F., Rosenbauer H., Roll R., et al. (2007) COSAC, The Cometary Sampling and Composition Experiment on Philae. *Space Sci. Rev., 128,* 257–280.
Goesmann F., Rosenbauer H., Bredehöft J. H., et al. (2015) Organic compounds on Comet 67P/Churyumov-Gerasimenko revealed by COSAC mass spectrometry. *Science, 349,* aab0689.
Golish D. R., DellaGiustina D. N., Li J. Y., et al. (2021) Disk-resolved photometric modeling and properties of asteroid (101955) Bennu. *Icarus, 357,* 113724.
Gounelle M. (2011) The asteroid-comet continuum: In search of lost

primitivity. *Elements, 7,* 29–34.
Groussin O., Sunshine J., Feaga L., et al. (2013) The temperature, thermal inertia, roughness and color of the nuclei of Comets 103P/Hartley 2 and 9P/Tempel 1. *Icarus, 222,* 580–594.
Groussin O., Jorda L., Auger A. T., et al. (2015a) Gravitational slopes, geomorphology, and material strengths of the nucleus of Comet 67P/Churyumov-Gerasimenko from OSIRIS observations. *Astron. Astrophys., 583,* A32.
Groussin O., Sierks H., Barbieri C., et al. (2015b) Temporal morphological changes in the Imhotep region of Comet 67P/Churyumov-Gerasimenko. *Astron. Astrophys., 583,* A36.
Gulkis S., Allen M., von Allmen P., et al. (2015) Subsurface properties and early activity of Comet 67P/Churyumov-Gerasimenko. *Science, 347,* aaa0709.
Gundlach B., Blum J., Keller H. U., et al. (2015) What drives the dust activity of Comet 67P/Churyumov-Gerasimenko? *Astron. Astrophys., 583,* A12.
Gundlach B., Fulle M., and Blum J. (2020) On the activity of comets: Understanding the gas and dust emission from Comet 67/Churyumov-Gerasimenko's south-pole region during perihelion. *Mon. Not. R. Astron. Soc., 493,* 3690–3715.
Güttler C., Mannel T., Rotundi A., et al. (2019) Synthesis of the morphological description of cometary dust at Comet 67P/Churyumov-Gerasimenko. *Astron. Astrophys., 630,* A24.
Hadamcik E., Renard J., Buch A., et al. (2014) Linear polarization of light scattered by cometary analogs: New samples. In *Asteroids, Comets, Meteors 2014* (K. Muinonen et al., eds.), p. 215. Univ. of Helsinki, Helsinki.
Hampton D. L., Baer J. W., Huisjen M. A., et al. (2005) An overview of the instrument suite for the Deep Impact mission. *Space Sci. Rev., 117,* 43–93.
Hänni N., Altwegg K., Balsiger H., et al. (2021) Cyanogen, cyanoacetylene, and acetonitrile in Comet 67P and their relation to the cyano radical. *Astron. Astrophys., 647,* A22.
Hansen K. C., Altwegg K., Berthelier J. J., et al. (2016) Evolution of water production of 67P/Churyumov-Gerasimenko: An empirical model and a multi-instrument study. *Mon. Not. R. Astron. Soc., 462,* S491–S506.
Hapke B. (1993) *Theory of Reflectance and Emittance Spectroscopy.* Cambridge Univ., Cambridge. 455 pp.
Hapke B. (2002) Bidirectional reflectance spectroscopy. 5. The coherent backscatter opposition effect and anisotropic scattering. *Icarus, 157,* 523–534.
Hapke B. (2008) Bidirectional reflectance spectroscopy. 6. Effects of porosity. *Icarus, 195,* 918–926.
Hapke B. (2012) Bidirectional reflectance spectroscopy 7. The single particle phase function hockey stick relation. *Icarus, 221,* 1079–1083.
Hasselmann P. H., Barucci M. A., Fornasier S., et al. (2017) The opposition effect of 67P/Churyumov-Gerasimenko on post-perihelion Rosetta images. *Mon. Not. R. Astron. Soc., 469,* S550–S567.
Hasselmann P. H., Barucci M. A., Fornasier S., et al. (2019) Pronounced morphological changes in a southern active zone on Comet 67P/Churyumov-Gerasimenko. *Astron. Astrophys., 630,* A8.
Hässig M., Altwegg K., Balsiger H., et al. (2015) Time variability and heterogeneity in the coma of 67P/Churyumov-Gerasimenko. *Science, 347,* aaa0276.
Heritier K. L., Galand M., Henri P., et al. (2018) Plasma source and loss at Comet 67P during the Rosetta mission. *Astron. Astrophys., 618,* A77.
Hoang M., Garnier P., Gourlaouen H., et al. (2019) Two years with Comet 67P/Churyumov-Gerasimenko: H_2O, CO_2, and CO as seen by the ROSINA/RTOF instrument of Rosetta. *Astron. Astrophys., 630, A33.*
Hoang H. V., Fornasier S., Quirico E., et al. (2020) Spectrophotometric characterization of the Philae landing site and surroundings with the Rosetta/OSIRIS cameras. *Mon. Not. R. Astron. Soc., 498,* 1221–1238.
Hsieh H. H. and Jewitt D. (2006) A population of comets in the main asteroid belt. *Science, 312,* 561–563.
Hu X., Shi X., Sierks H., et al. (2017) Seasonal erosion and restoration of the dust cover on Comet 67P/Churyumov-Gerasimenko as observed by OSIRIS onboard Rosetta. *Astron. Astrophys., 604,* A114.
Huebner W. F. (2008) Origins of cometary materials. *Space Sci. Rev., 138,* 5–25.
Huebner W. F. and Benkhoff J. (1999) From coma abundances to nucleus composition. *Space Sci. Rev., 90,* 117–130.
Huebner W. F., Boice D. C., Sharp C. M., et al. (1987) Evidence for first polymer in Comet Halley: Polyoxymethylene. In *Proceedings of the International Symposium on the Diversity and Similarity of Comets* (E. J. Rolfe et al., eds.), pp. 163–167. ESA-SP-278, Noordwijk, The Netherlands.
Huebner W. F., Benkhoff J., Capria M.-T., et al., eds. (2006) *Heat and Gas Diffusion in Comet Nuclei.* International Space Science Institute, Bern. 258 pp.
Hughes D. W. (1985) The size, mass, mass loss and age of Halley's comet. *Mon. Not. R. Astron. Soc., 213,* 103–109.
Hunt G., Salisbury J., and Lenhoff C. (1971) Visible and near-infrared spectra of minerals and rocks: IV. Sulphides and sulphates. *Mod. Geol., 3,* 1–14.
Jäger C., Dorschner J., Mutschke H., et al. (2003) Steps toward interstellar silicate mineralogy. VII. Spectral properties and crystallization behaviour of magnesium silicates produced by the sol-gel method. *Astron. Astrophys., 408,* 193–204.
Jessberger E. K., Christoforidis A., and Kissel J. (1988) Aspects of the major element composition of Halley's dust. *Nature, 332,* 691–695.
Jewitt D. (2015) Color systematics of comets and related bodies. *Astron. J., 150,* 201.
Jewitt D. and Meech K. J. (1986) Cometary grain scattering versus wavelength, or, "What color is comet dust?" *Astrophys. J., 310,* 937.
Jewitt D. and Sheppard S. (2004) The nucleus of Comet 48P/Johnson. *Astron. J., 127,* 1784–1790.
Jewitt D., Sheppard S., and Fernández Y. (2003) 143P/Kowal-Mrkos and the shapes of cometary nuclei. *Astron. J., 125,* 3366–3377.
Jorda L., Gaskell R., Capanna C., et al. (2016) The global shape, density and rotation of Comet 67P/Churyumov-Gerasimenko from preperihelion Rosetta/OSIRIS observations. *Icarus, 277,* 257–278.
Jost B., Pommerol A., Poch O., et al. (2017) Bidirectional reflectance of laboratory cometary analogues to interpret the spectrophotometric properties of the nucleus of Comet 67P/Churyumov-Gerasimenko. *Planet. Space Sci., 148,* 1–11.
Kaplan H. H., Milliken R. E., and Alexander C. M. O. (2018) New constraints on the abundance and composition of organic matter on Ceres. *Geophys. Res. Lett., 45,* 5274–5282.
Kebukawa Y., Alexander C. M. O. D., and Cody G. D. (2011) Compositional diversity in insoluble organic matter in type 1, 2 and 3 chondrites as detected by infrared spectroscopy. *Geochim. Cosmochim. Acta, 75,* 3530–3541.
Keller H. U. (1987) The nucleus of Comet Halley. In *Proceedings of the International Symposium on the Diversity and Similarity of Comets* (E. J. Rolfe et al., eds.), pp. 447–454. ESA-SP-278, Noorrdwijk, The Netherlands.
Keller H. U., Arpigny C., Barbieri C., et al. (1986) First Halley multicolour camera imaging results from Giotto. *Nature, 321,* 320–326.
Keller H. U., Barbieri C., Lamy P., et al. (2007) OSIRIS — The scientific camera system onboard Rosetta. *Space Sci. Rev., 128,* 433–506.
Keller H. U., Mottola S., Hviid S. F., et al. (2017) Seasonal mass transfer on the nucleus of Comet 67P/Chuyumov-Gerasimenko. *Mon. Not. R. Astron. Soc., 469,* S357–S371.
Keller L. P., Messenger S., Flynn G. J., et al. (2004) The nature of molecular cloud material in interplanetary dust. *Geochim. Cosmochim. Acta, 68,* 2577–2589.
Keller L. P., Bajt S., Baratta G. A., et al. (2006) Infrared spectroscopy of Comet 81P/Wild 2 samples returned by Stardust. *Science, 314,* 1728–1731.
Kelley M. S. P., Woodward C. E., Gehrz R. D., et al. (2017) Mid-infrared spectra of comet nuclei. *Icarus, 284,* 344–358.
Kim Y. S. and Kaiser R. I. (2011) On the formation of amines (RNH_2) and the cyanide anion (CN^-) in electron-irradiated ammonia-hydrocarbon interstellar model ices. *Astrophys. J., 729,* 68.
Kissel J., Brownlee D. E., Buchler K., et al. (1986) Composition of Comet Halley dust particles from Giotto observations. *Nature, 321,* 336–337.
Kissel J., Altwegg K., Clark B. C., et al. (2007) COSIMA — High resolution time-of-flight secondary ion mass spectrometer for the analysis of cometary dust particles onboard Rosetta. *Space Sci. Rev., 128,* 823–867.
Knollenberg J., Lin Z. Y., Hviid S. F., et al. (2016) A mini outburst from the nightside of Comet 67P/Churyumov-Gerasimenko observed by the OSIRIS camera on Rosetta. *Astron. Astrophys., 596,* A89.

Krüger H., Goesmann F., Giri C., et al. (2017) Decay of COSAC and Ptolemy mass spectra at Comet 67P/Churyumov-Gerasimenko. *Astron. Astrophys., 600,* A56.
Lacerda P., Fornasier S., Lellouch E., et al. (2014) The albedo-color diversity of transneptunian objects. *Astrophys. J. Lett., 793,* L2.
La Forgia F., Giacomini L., Lazzarin M., et al. (2015) Geomorphology and spectrophotometry of Philae's landing site on Comet 67P/Churyumov-Gerasimenko. *Astron. Astrophys., 583,* A41.
Lai I.-L., Ip W.-H., Su C.-C., et al. (2016) Gas outflow and dust transport of Comet 67P/Churyumov-Gerasimenko. *Mon. Not. R. Astron. Soc., 462,* S533–S546.
Lamy P. L., Toth I., A'Hearn M. F., et al. (1999) Hubble Space Telescope observations of the nucleus of Comet 45P/Honda-Mrkos-Pajdusakova and its inner coma. *Icarus, 140,* 424–438.
Lamy P. L., Toth I., Fernández Y. R., et al. (2004) The sizes, shapes, albedos, and colors of cometary nuclei. In *Comets II* (M. C. Festou et al., eds.), pp. 223–264. Univ. of Arizona, Tucson.
Langevin Y., Hilchenbach M., Ligier N., et al. (2016) Typology of dust particles collected by the COSIMA mass spectrometer in the inner coma of 67P/Churyumov Gerasimenko. *Icarus, 271,* 76–97.
Lantz C., Brunetto R., Barucci M. A., et al. (2017) Ion irradiation of carbonaceous chondrites: A new view of space weathering on primitive asteroids. *Icarus, 285,* 43–57.
Läuter M., Kramer T., Rubin M., et al. (2019) Surface localization of gas sources on Comet 67P/Churyumov-Gerasimenko based on DFMS/COPS data. *Mon. Not. R. Astron. Soc., 483,* 852–861.
Läuter M., Kramer T., Rubin M., et al. (2020) The gas production of 14 species from Comet 67P/Churyumov-Gerasimenko based on DFMS/COPS data from 2014 to 2016. *Mon. Not. R. Astron. Soc., 498,* 3995–4004.
Leon-Dasi M., Besse S., Grieger B., et al. (2021) Mapping a duck: Geological features and region definitions on Comet 67P/Churyumov-Gerasimenko. *Astron. Astrophys., 652,* A52.
Le Roy L., Altwegg K., Balsiger H., et al. (2015) Inventory of the volatiles on Comet 67P/Churyumov-Gerasimenko from Rosetta/ROSINA. *Astron. Astrophys., 583,* A1.
Li J.-Y., A'Hearn M. F., Belton M. J. S., et al. (2007a) Deep Impact photometry of Comet 9P/Tempel 1. *Icarus, 187,* 41–55.
Li J.-Y., A'Hearn M. F., McFadden L. A., et al. (2007b) Photometric analysis and disk-resolved thermal modeling of Comet 19P/Borrelly from Deep Space 1 data. *Icarus, 188,* 195–211.
Li J.-Y., A'Hearn M. F., Farnham T. L., et al. (2009) Photometric analysis of the nucleus of Comet 81P/Wild 2 from Stardust images. *Icarus, 204,* 209–226.
Li J.-Y., Besse S., A'Hearn M. F., et al. (2013) Photometric properties of the nucleus of Comet 103P/Hartley 2. *Icarus, 222,* 559–570.
Licandro J., Campins H., Hergenrother C., et al. (2003) Near-infrared spectroscopy of the nucleus of Comet 124P/Mrkos. *Astron. Astrophys., 398,* L45–L48.
Licandro J., Campins H., Kelley M., et al. (2011) (65) Cybele: Detection of small silicate grains, water-ice, and organics. *Astron. Astrophys., 525,* A34.
Licandro J., Alí-Lagoa V., Tancredi G., et al. (2016) Size and albedo distributions of asteroids in cometary orbits using WISE data. *Astron. Astrophys., 585,* A9.
Licandro J., Popescu M., de León J., et al. (2018) The visible and near-infrared spectra of asteroids in cometary orbits. *Astron. Astrophys., 618,* A170.
Lippi M., Villanueva G. L., Mumma M. J., et al. (2021) Investigation of the origins of comets as revealed through infrared high-resolution spectroscopy I. Molecular abundances. *Astron. J., 162,* 74.
Lisse C. M., VanCleve J., Adams A. C., et al. (2006) Spitzer spectral observations of the Deep Impact ejecta. *Science, 313,* 635–640.
Lodders K. (2010) Solar system abundances of the elements. In *Principles and Perspectives in Cosmochemistry* (A. Goswami and E. B. Reddy, eds.), pp. 379–417. Astrophysics and Space Science Proceedings Vol. 16, Springer-Verlag, Berlin.
Longobardo A., Palomba E., Capaccioni F., et al. (2017) Photometric behaviour of 67P/Churyumov-Gerasimenko and analysis of its pre-perihelion diurnal variations. *Mon. Not. R. Astron. Soc., 469,* S346–S356.
Lucchetti A., Cremonese G., Jorda L., et al. (2016) Characterization of the Abydos region through OSIRIS high-resolution images in support of CIVA measurements. *Astron. Astrophys., 585,* L1.
Lucchetti A., Pajola M., Fornasier S., et al. (2017) Geomorphological and spectrophotometric analysis of Seth's circular niches on Comet 67P/Churyumov-Gerasimenko using OSIRIS images. *Mon. Not. R. Astron. Soc., 469,* S238–S251.
Luspay-Kuti A., Hässig M., Fuselier S. A., et al. (2015) Composition-dependent outgassing of Comet 67P/Churyumov-Gerasimenko from ROSINA/DFMS. Implications for nucleus heterogeneity? *Astron. Astrophys., 583,* A4.
Luu J. and Jewitt D. (1990) The nucleus of Comet P/Encke. *Icarus, 86,* 69–81.
Lynch D. K. (2005) *The Infrared Spectral Signature of Water Ice in the Vacuum Cryogenic AI&T Environment.* Report TR-2006(8570)-1, Aerospace Corporation, El Segundo.
Mannel T., Bentley M. S., Schmied R., et al. (2016) Fractal cometary dust — A window into the early solar system. *Mon. Not. R. Astron. Soc., 462,* S304–S311.
Marboeuf U. and Schmitt B. (2014) How to link the relative abundances of gas species in coma of comets to their initial chemical composition? *Icarus, 242,* 225–248.
Marshall D. W., Hartogh P., Rezac L., et al. (2017) Spatially resolved evolution of the local H_2O production rates of Comet 67P/Churyumov-Gerasimenko from the MIRO instrument on Rosetta. *Astron. Astrophys., 603,* A87.
Masoumzadeh N., Oklay N., Kolokolova L., et al. (2017) Opposition effect on Comet 67P/Churyumov-Gerasimenko using Rosetta-OSIRIS images. *Astron. Astrophys., 599,* A11.
Mayo Greenberg J. and Mendoza-Gómez C. X. (1992) The seeding of life by comets. *Adv. Space Res., 12,* 169–180.
McDonnell J. A. M., Alexander W. M., Burton W. M., et al. (1986) Dust density and mass distribution near Comet Halley from Giotto observations. *Nature, 321,* 338–341.
Meech K. J., Pittichová J., Yang B., et al. (2011) Deep Impact, Stardust-NExT and the behavior of Comet 9P/Tempel 1 from 1997 to 2010. *Icarus, 213,* 323–344.
Mennella V. (2010) H atom irradiation of carbon grains under simulated dense interstellar medium conditions: The evolution of organics from diffuse interstellar clouds to the solar system. *Astrophys. J., 718,* 867–875.
Mennella V., Ciarniello M., Raponi A., et al. (2020) Hydroxylated Mg-rich amorphous silicates: A new component of the 3.2 μm absorption band of Comet 67P/Churyumov-Gerasimenko. *Astrophys. J. Lett., 897,* L37.
Min M., Bouwman J., Dominik C., et al. (2016) The abundance and thermal history of water ice in the disk surrounding HD 142527 from the DIGIT Herschel Key Program. *Astron. Astrophys., 593,* A11.
Moroz L. V., Arnold G., Korochantsev A. V., et al. (1998) Natural solid bitumens as possible analogs for cometary and asteroid organics: 1. Reflectance spectroscopy of pure bitumens. *Icarus, 134,* 253–268.
Moroz L. V., Markus K., Arnold G., et al. (2017) Laboratory spectral reflectance studies aimed at providing clues to composition of refractory phases of Comet 67P/CG's nucleus. *European Planet. Sci. Congr., 2017,* EPSC2017-266.
Mumma M. J. and Charnley S. B. (2011) The chemical composition of comets — Emerging taxonomies and natal heritage. *Annu. Rev. Astro. Astrophys., 49,* 471–524.
Nelson R. M., Rayman M. D., and Weaver H. A. (2004) The Deep Space 1 encounter with Comet 19P/Borrelly. *Icarus, 167,* 1–3.
Oklay N., Vincent J.-B., Fornasier S., et al. (2016) Variegation of Comet 67P/Churyumov-Gerasimenko in regions showing activity. *Astron. Astrophys., 586,* A80.
Oklay N., Mottola S., Vincent J. B., et al. (2017) Long-term survival of surface water ice on Comet 67P. *Mon. Not. R. Astron. Soc., 469,* S582–S597.
O'Rourke L., Heinisch P., Blum J., et al. (2020) The Philae lander reveals low-strength primitive ice inside cometary boulders. *Nature, 586,* 697–701.
Orthous-Daunay F. R., Quirico E., Beck P., et al. (2013) Mid-infrared study of the molecular structure variability of insoluble organic matter from primitive chondrites. *Icarus, 223,* 534–543.
Pajola M., Höfner S., Vincent J. B., et al. (2017a) The pristine interior of Comet 67P revealed by the combined Aswan outburst and cliff collapse. *Nature Astron., 1,* 0092.
Pajola M., Lucchetti A., Fulle M., et al. (2017b) The pebbles/boulders size distributions on Sais: Rosetta's final landing site on Comet 67P/Churyumov-Gerasimenko. *Mon. Not. R. Astron. Soc., 469,* S636–S645.
Pajola M., Lee J. C., Oklay N., et al. (2019) Multidisciplinary analysis

of the Hapi region located on Comet 67P/Churyumov-Gerasimenko. *Mon. Not. R. Astron. Soc., 485,* 2139–2154.
Pendleton Y. J., Sandford S. A., Allamandola L. J., et al. (1994) Near-infrared absorption spectroscopy of interstellar hydrocarbon grains. *Astrophys. J., 437,* 683.
Poch O., Istiqomah I., Quirico E., et al. (2020) Ammonium salts are a reservoir of nitrogen on a cometary nucleus and possibly on some asteroids. *Science, 367,* aaw7462.
Podolak M. and Zucker S. (2004) A note on the snow line in protostellar accretion disks. *Meteorit. Planet. Sci., 39,* 1859–1868.
Pommerol A., Thomas N., El-Maarry M. R., et al. (2015) OSIRIS observations of meter-sized exposures of H_2O ice at the surface of 67P/Churyumov-Gerasimenko and interpretation using laboratory experiments. *Astron. Astrophys., 583,* A25.
Prialnik D., A'Hearn M. F., and Meech K. J. (2008) A mechanism for short-lived cometary outbursts at sunrise as observed by Deep Impact on 9P/Tempel 1. *Mon. Not. R. Astron. Soc., 388,* L20–L23.
Protopapa S., Sunshine J. M., Feaga L. M., et al. (2014) Water ice and dust in the innermost coma of Comet 103P/Hartley 2. *Icarus, 238,* 191–204.
Quirico E., Borg J., Raynal P.-I., et al. (2005) A micro-Raman survey of 10 IDPs and 6 carbonaceous chondrites. *Planet. Space Sci., 53,* 1443–1448.
Quirico E., Moroz L. V., Schmitt B., et al. (2016) Refractory and semi-volatile organics at the surface of Comet 67P/Churyumov-Gerasimenko: Insights from the VIRTIS/Rosetta imaging spectrometer. *Icarus, 272,* 32–47.
Raponi A. (2015) Spectrophotometric analysis of cometary nuclei from in situ observations. Ph.D. thesis, Univ. of Rome II, Rome. 131 pp. *ArXiV e-prints,* arXiv:1503.08172.
Raponi A., Ciarniello M., Capaccioni F., et al. (2016) The temporal evolution of exposed water ice-rich areas on the surface of 67P/Churyumov-Gerasimenko: Spectral analysis. *Mon. Not. R. Astron. Soc., 462,* S476–S490.
Raponi A., Ciarniello M., Capaccioni F., et al. (2020) Infrared detection of aliphatic organics on a cometary nucleus. *Nature Astron., 4,* 500–505.
Riedler W., Torkar K., Jeszenszky H., et al. (2007) MIDAS — The Micro-Imaging Dust Analysis System for the Rosetta mission. *Space Sci. Rev., 128,* 869–904.
Rivkin A. S. and Emery J. P. (2010) Detection of ice and organics on an asteroidal surface. *Nature, 464,* 1322–1323.
Rodgers D. H., Beauchamp P. M., Soderblom L. A., et al. (2007) Advanced technologies demonstrated by the Miniature Integrated Camera and Spectrometer (MICAS) aboard Deep Space 1. *Space Sci. Rev., 129,* 309–326.
Rotundi A., Sierks H., Della Corte V., et al. (2015) Dust measurements in the coma of Comet 67P/Churyumov-Gerasimenko inbound to the Sun. *Science, 347,* aaa3905.
Rousseau B., Érard S., Beck P., et al. (2018) Laboratory simulations of the Vis-NIR spectra of Comet 67P using sub-µm sized cosmochemical analogues. *Icarus, 306,* 306–318.
Rubin M., Altwegg K., Balsiger H., et al. (2015) Molecular nitrogen in Comet 67P/Churyumov-Gerasimenko indicates a low formation temperature. *Science, 348,* 232–235.
Sagdeev R. Z., Avanesov G. A., Ziman Y. L., et al. (1986) TV experiment of the Vega mission: Photometry of the nucleus and the inner coma. In *Proceedings of the 20th ESLAB Symposium on the Exploration of Halley's Comet Vol. 2: Dust and Nucleus* (B. Battrick et al., eds.), pp. 317–326. ESA-SP-250, Noordwijk, The Netherlands.
Sandford S. A., Allamandola L. J., Tielens A. G. G. M., et al. (1991) The interstellar C-H stretching band near 3.4 microns: Constraints on the composition of organic material in the diffuse interstellar medium. *Astrophys. J., 371,* 607.
Schaible M. J. and Baragiola R. A. (2014) Hydrogen implantation in silicates: The role of solar wind in SiOH bond formation on the surfaces of airless bodies in space. *J. Geophys. Res.–Planets, 119,* 2017–2028.
Schröder S. E., Grynko Y., Pommerol A., et al. (2014) Laboratory observations and simulations of phase reddening. *Icarus, 239,* 201–216.
Schroeder I. R. H. G., Altwegg K., Balsiger H., et al. (2019) A comparison between the two lobes of Comet 67P/Churyumov-Gerasimenko based on D/H ratios in H_2O measured with the Rosetta/ROSINA DFMS. *Mon. Not. R. Astron. Soc., 489,* 4734–4740.
Shevchenko V. G., Belskaya I. N., Slyusarev I. G., et al. (2012) Opposition effect of Trojan asteroids. *Icarus, 217,* 202–208.
Shi X., Hu X., Sierks H., et al. (2016) Sunset jets observed on Comet 67P/Churyumov-Gerasimenko sustained by subsurface thermal lag. *Astron. Astrophys., 586,* A7.
Shu F. H., Shang H., Gounelle M., et al. (2001) The origin of chondrules and refractory inclusions in chondritic meteorites. *Astrophys. J., 548,* 1029–1050.
Sierks H., Barbieri C., Lamy P. L., et al. (2015) On the nucleus structure and activity of Comet 67P/Churyumov-Gerasimenko. *Science, 347,* aaa1044.
Snodgrass C., Fitzsimmons A., Lowry S. C., et al. (2011) The size distribution of Jupiter family comet nuclei. *Mon. Not. R. Astron. Soc., 414,* 458–469.
Snodgrass C., Agarwal J., Combi M., et al. (2017) The main belt comets and ice in the solar system. *Astron. Astrophys. Rev., 25,* 5.
Soderblom L. A., Becker T. L., Bennett G., et al. (2002) Observations of Comet 19P/Borrelly by the Miniature Integrated Camera and Spectrometer aboard Deep Space 1. *Science, 296,* 1087–1091.
Soderblom L., Britt D., Brown R., et al. (2004) Short-wavelength infrared (1.3–2.6 µm) observations of the nucleus of Comet 19P/Borrelly. *Icarus, 167,* 100–112.
Stern S. A., Feaga L. M., Schindhelm R., et al. (2015) First extreme and far ultraviolet spectrum of a comet nucleus: Results from 67P/Churyumov-Gerasimenko. *Icarus, 256,* 117–119.
Sugita S., Ootsubo T., Kadono T., et al. (2005) Subaru Telescope observations of Deep Impact. *Science, 310,* 274–278.
Sunshine J. M., A'Hearn M. F., Groussin O., et al. (2006) Exposed water ice deposits on the surface of Comet 9P/Tempel 1. *Science, 311,* 1453–1455.
Sunshine J. M., Feaga L. M., Groussin O., et al. (2011) Water ice on Comet 103P/Hartley 2. *EPSC-DPS Joint Meeting 2011,* 1345.
Takir D. and Emery J. P. (2012) Outer main belt asteroids: Identification and distribution of four 3-µm spectral groups. *Icarus, 219,* 641–654.
Tancredi G. (2014) A criterion to classify asteroids and comets based on the orbital parameters. *Icarus, 234,* 66–80.
Tatsumi E., Domingue D., Schröder S., et al. (2020) Global photometric properties of (162173) Ryugu. *Astron. Astrophys., 639,* A83.
Taylor M. G. G. T., Altobelli N., Buratti B. J., et al. (2017) The Rosetta mission orbiter science overview: The comet phase. *Philos. Trans. R. Soc. A, 375,* 20160262.
Tholen D. J. (1984) Asteroid taxonomy from cluster analysis of photometry. Ph.D. thesis, University of Arizona, Tucson. 150 pp.
Thomas N. and Keller H. U. (1989) The colour of Comet P/Halley's nucleus and dust. *Astron. Astrophys., 213,* 487–494.
Thomas N., Sierks H., Barbieri C., et al. (2015) The morphological diversity of Comet 67P/Churyumov-Gerasimenko. *Science, 347,* aaa0440.
Thomas N., El Maarry M. R., Theologou P., et al. (2018) Regional unit definition for the nucleus of Comet 67P/Churyumov-Gerasimenko on the SHAP7 model. *Planet. Space Sci., 164,* 19–36.
Tubiana C., Böhnhardt H., Agarwal J., et al. (2011) 67P/Churyumov-Gerasimenko at large heliocentric distance. *Astron. Astrophys., 527,* A113.
Tubiana C., Snodgrass C., Bertini I., et al. (2015) 67P/Churyumov-Gerasimenko: Activity between March and June 2014 as observed from Rosetta/OSIRIS. *Astron. Astrophys., 573,* A62.
Tubiana C., Rinaldi G., Güttler C., et al. (2019) Diurnal variation of dust and gas production in Comet 67P/Churyumov-Gerasimenko at the inbound equinox as seen by OSIRIS and VIRTIS-M on board Rosetta. *Astron. Astrophys., 630,* A23.
Veverka J., Klaasen K., A'Hearn M., et al. (2013) Return to Comet Tempel 1: Overview of Stardust-NExT results. *Icarus, 222,* 424–435.
Vincent J.-B., A'Hearn M. F., Lin Z.-Y., et al. (2016a) Summer fireworks on Comet 67P. *Mon. Not. R. Astron. Soc., 462,* S184–S194.
Vincent J.-B., Oklay N., Pajola M., et al. (2016b) Are fractured cliffs the source of cometary dust jets? Insights from OSIRIS/Rosetta at 67P/Churyumov-Gerasimenko. *Astron. Astrophys., 587,* A14.
Vincent J.-B., Farnham T., Kührt E., et al. (2019) Local manifestations of cometary activity. *Space Sci. Rev., 215,* 30.
Wagner J. K., Hapke B. W., and Wells E. N. (1987) Atlas of reflectance spectra of terrestrial, lunar, and meteoritic powders and frosts from 92 to 1800 nm. *Icarus, 69,* 14–28.
Whipple F. L. (1950) A comet model. I. The acceleration of Comet *Encke. Astrophys. J., 111,* 375–394.

Willacy K., Alexander C., Ali-Dib M., et al. (2015) The composition of the protosolar disk and the formation conditions for comets. *Space Sci. Rev., 197,* 151–190.

Wooden D. H., Harker D. E., Woodward C. E., et al. (1999) Silicate mineralogy of the dust in the inner coma of Comet C/1995 01 (Hale-Bopp) pre- and postperihelion. *Astrophys. J., 517,* 1034–1058.

Wright I. P., Barber S. J., Morgan G. H., et al. (2007) Ptolemy an instrument to measure stable isotopic ratios of key volatiles on a cometary nucleus. *Space Sci. Rev., 128,* 363–381.

Wright I. P., Sheridan S., Barber S. J., et al. (2015) CHO-bearing organic compounds at the surface of 67P/Churyumov-Gerasimenko revealed by Ptolemy. *Science, 349,* aab0673.

Zeller E. J., Ronca L. B., and Levy P. W. (1966) Proton-induced hydroxyl formation on the lunar surface. *J. Geophys. Res., 71,* 4855–4860.

Zubko E., Furusho R., Kawabata K., et al. (2011) Interpretation of photo-polarimetric observations of Comet 17P/Holmes. *J. Quant. Spec. Radiat. Transf., 112,* 1848–1863.

Knight M. M., Kokotanekova R., and Samarasinha N. H. (2024) Physical and surface properties of comet nuclei from remote observations. In *Comets III* (K. J. Meech, M. R. Combi, D. Bockelée-Morvan, S. N. Raymond, and M. E. Zolensky, eds.), pp. 361–403. Univ. of Arizona, Tucson, DOI: 10.2458/azu_uapress_9780816553631-ch012.

Physical and Surface Properties of Comet Nuclei from Remote Observations

Matthew M. Knight
United States Naval Academy

Rosita Kokotanekova
Institute of Astronomy and National Astronomical Observatory, Bulgarian Academy of Sciences/European Southern Observatory

Nalin H. Samarasinha
Planetary Science Institute

We summarize the collective knowledge of physical and surface properties of comet nuclei, focusing on those that are obtained from remote observations. We now have measurements or constraints on effective radius for over 200 comets, rotation periods for over 60, axial ratios and color indices for over 50, geometric albedos for over 25, and nucleus phase coefficients for over 20. The sample has approximately tripled since the publication of *Comets II*, with infrared surveys using Spitzer and the Near-Earth Object Wide-field Infrared Survey Explorer responsible for the bulk of the increase in effective radii measurements. Advances in coma morphology studies and long-term studies of a few prominent comets have resulted in meaningful constraints on rotation period changes in nearly a dozen comets, allowing this to be added to the range of nucleus properties studied. The first delay-Doppler radar and visible light polarimetric measurements of comet nuclei have been made since *Comets II* and are considered alongside the traditional methods of studying nuclei remotely. We use the results from recent *in situ* missions, notably Rosetta, to put the collective properties obtained by remote observations into context, emphasizing the insights gained into surface properties and the prevalence of highly elongated and/or bilobate shapes. We also explore how nucleus properties evolve, focusing on fragmentation and the likely related phenomena of outbursts and disintegration. Knowledge of these behaviors has been shaped in recent years by diverse sources: high-resolution images of nucleus fragmentation and disruption events, the detection of thousands of small comets near the Sun, regular photometric monitoring of large numbers of comets throughout the solar system, and detailed imaging of the surfaces of mission targets. Finally, we explore what advances in the knowledge of the bulk nucleus properties may be enabled in coming years.

1. INTRODUCTION

Comets are among the best known and most accessible of astronomical phenomena. They have undoubtedly been noticed — and quite often feared — as long as there have been humans to do so. Records of comet observations are extant from well over 2000 years ago (cf. *Xi,* 1984; *Kronk,* 1999). And yet, an accurate understanding of the physical properties of cometary nuclei only came about in the last few decades of the twentieth century, far later than one might have assumed given their prominent places in the sky.

The modern concept of a comet as a small, consolidated nucleus was established by *Whipple* (1950), although holdouts in support of the earlier "sandbank" model (e.g., *Lyttleton,* 1953) forced continued debate in the literature for over a decade (e.g., *Whipple,* 1963). Whipple's "dirty snowball" model was primarily motivated by the desire to explain the non-gravitational forces evident on comet orbits as well as observed gas production rates, but a natural consequence was that comet nuclei were quite small, on the order of a few kilometers in radius rather than the tens of kilometers that had been inferred observationally from the apparent sizes of central condensation of comets.

The first conclusive measurements of nucleus properties came in the 1980s due to the proliferation of modern instrumentation over the previous two decades. Key advances included the ability to observe at wavelengths beyond the near-ultraviolet (UV) and visible, bigger telescopes at higher-altitude sites around the world, the introduction of charge-coupled-device (CCD) cameras, and the ability to observe from space. Notably, *Millis et al.* (1985, 1988) made the first convincing measurement of nucleus size and albedo (of 49P/Arend-Rigaux), while *Cruikshank et al.* (1985) concluded that 1P/Halley's albedo was <10%. When the Vega 1, Vega 2, and Giotto spacecraft reached Halley and unequivocally measured its size and albedo (e.g., *Keller et al.,* 1986; *Sagdeev et al.,* 1986), they confirmed that observations from Earth could successfully determine nucleus properties.

Despite technological advances, knowledge of nucleus properties has remained elusive due to the often quixotic nature of their study. When active, comets' gas and dust comae typically obscure the nucleus, but when inactive, their small and dark nuclei are often too faint to detect. What is more, comets' elliptical orbits bring them into the inner solar system only infrequently, on the order of 10 years for the short-period, low-inclination Jupiter-family comets (JFCs), a few decades for the higher-inclination Halley-type comets (HTCs), and hundreds to thousands of years or more for long-period comets (LPCs) from the Oort cloud. Even during these infrequent passages, an individual comet can only be studied well if it happens to pass close to Earth. The result is that our knowledge of comets often comes in bursts, via either predictable and long-anticipated favorable apparitions of individual JFCs and HTCs or unexpected arrivals of LPCs.

Comets are renowned as "fossils" left over from the formation of the solar system, with their study motivated by the insight they provide into the conditions in the Sun's protoplanetary disk. However, they are not static, and their properties, individually or as a population, must be properly contextualized. An individual comet will have formed between approximately 15–30 au (e.g., *Gomes et al.,* 2004) and will have been been scattered to a more distant orbit during the solar system's early evolution where it remained in relative stasis, although not completely unmodified, before being eventually perturbed into the inner solar system (see the chapters by Kaib and Volk and by Fraser et al. in this volume). Some comets will settle into relatively stable orbits in the inner solar system while others will experience large orbital changes due to gravitational perturbations. Comets are subject to a host of fates, ranging from complete disappearance from sublimation of volatile ices, quiescence due to loss or burial of accessible volatiles near the surface, spontaneous disruption for a variety of reasons, impact with the Sun or planets, or ejection from the solar system. Furthermore, each comet is at a different place in its evolution, and its specific history is unknown. The current comet population is thought to be in a quasi-steady state, with roughly equal numbers of comets newly arriving from the outer solar system reservoirs as being lost via the various mechanisms just discussed. While the studies conducted today merely give a snapshot in time of an individual comet, they sample numerous members of various populations at different times in their evolution. By combining observations and dynamical modeling, researchers seek to assemble all these snapshots into coherent stories.

1.1. Nomenclature

What is a "comet" has become difficult to precisely define as more objects with unusual or ambiguous properties are discovered. For this chapter we will concentrate on objects that are traditionally cometary; i.e., they contain volatile ices that sublimate to produce a gas and/or dust coma. We further consider primarily those objects whose dynamics bring them into the inner solar system where sublimation can readily occur. Except as noted, we are generally excluding active asteroids (e.g., *Jewitt et al.,* 2015), main-belt comets (e.g., *Hsieh and Jewitt,* 2006), asteroidal objects on comet-like orbits (ACOs) (e.g., *Weissman et al.,* 2002), and outer solar system objects like Centaurs and Kuiper belt objects (KBOs).

A common means of assigning taxonomic classifications to orbits is the Tisserand parameter with respect to Jupiter

$$T_J = \frac{a_J}{a} + 2\sqrt{\frac{a}{a_J}(1-e^2)}\ \cos(i) \tag{1}$$

where a, e, and i are the orbit's semimajor axis, eccentricity, and inclination, respectively, and a_J is the semimajor axis of Jupiter's orbit (*Kresák,* 1972; *Carusi et al.,* 1987). From a dynamical standpoint, comets are objects that have $T_J \lesssim 3.0$. We follow traditional definitions (e.g., *Levison,* 1996) to distinguish between JFCs having $2 < T_J < 3$, and HTCs and LPCs having $T_J < 2$. A stricter method of orbit classification to distinguish between comets and asteroids was presented by *Tancredi* (2014) but is not needed in this chapter.

At times, we will distinguish between returning Oort cloud comets and dynamically new comets (DNC) that are statistically likely to be on their first passage through the inner solar system. DNCs are identified dynamically by their reciprocal original semimajor axes ($1/a_0$). We use $1/a_0 < 5 \times 10^{-5}$ au^{-1} to identify comets that are unlikely to have previously passed close enough to the Sun for substantial outgassing to occur (e.g., *A'Hearn et al.,* 1995; see also *Oort,* 1950). Note, however, that more stringent requirements are needed to ensure a high likelihood that an object is genuinely new (cf. *Dybczyński and Królikowska,* 2015). Unless explicitly stated that we are discussing DNCs, we include DNCs in the term LPCs.

Other populations to which we will refer on occasion are Centaurs, damocloids, and Manx comets. Centaurs are small bodies on orbits that are intermediate between JFCs and objects residing entirely in the transneptunian region. Note that a strict classification of the Centaur population on orbital grounds causes some overlap with other populations (e.g., *Gladman et al.,* 2008). Damocloids, a term coined by *Jewitt* (2005), are objects on HTC or LPC orbits that do not exhibit cometary activity and are presumed to be inert (or nearly so) nuclei. Introduced by *Meech et al.* (2016), Manx comets are objects on Oort cloud and DNC orbits with no activity — suggesting (especially for DNCs) that they may never have had much (or any) ice. It is thought that Centaurs and damocloids, along with other minor bodies in the outer solar system such as KBOs [a.k.a. transneptunian objects (TNOs)], Jupiter Trojans (asteroids that share Jupiter's orbit, but librate around its L4 or L5 Lagrange points), Hildas (asteroids on a 3:2 resonance with Jupiter, and residing within its orbit), and irregular satellites formed in the same region as traditional comets (*Dones et al.,* 2015, and references therein). Manx comets may have formed with little or no water (e.g., asteroidal material); some may represent early solar system building blocks that formed near the water-ice line in our solar system. See the chapters by Fraser et al. and Jewitt and Hsieh in this volume for additional discussion of these objects.

1.2. Overview and Related Chapters

Our understanding of the nucleus properties of comets has grown substantially since the publication of *Comets II* (*Festou et al.,* 2004), driven in large part by an unprecedented string of successful space missions (see the chapter by Snodgrass et al. in this volume). As a result, several chapters in this volume deal with insights into comet nuclei gained from these missions including interior and global structure and density (chapter by Guilbert-Lepoutre et al. in this volume), surface properties (chapter by Pajola et al. in this volume), and nucleus activity and surface evolution (chapter by Filacchione et al. in this volume). Although we will touch on many of the following topics, readers should consult other chapters for more detailed discussion of planetesimal/comet formation (see the chapter by Simon et al. in this volume), dynamical population of comet reservoirs (chapter by Kaib and Volk in this volume), the journey from the Kuiper belt and TNOs to comets (chapter by Fraser et al. in this volume), asteroid/comet transition objects (chapter by Jewitt and Hsieh in this volume), and comet science with astrophysical assets (chapter by Bauer et al. in this volume).

The current chapter will largely focus on the knowledge gained from remote observations — made using ground- and spacebased telescopes in the vicinity of Earth, as opposed to *in situ* studies by dedicated spacecraft — of a larger number of comets. As such, it builds on a number of comprehensive papers published in the last two decades, notably *Boehnhardt* (2004), *Lamy et al.* (2004), *Samarasinha et al.* (2004), *Weissman et al.* (2004), *Snodgrass et al.* (2006), *Lowry et al.* (2008), *Fernández* (2009), *Fernández et al.* (2013), *Bauer et al.* (2017), and *Kokotanekova et al.* (2017). As will be discussed later, remote studies must contend with a variety of limitations in order to ascertain the properties of the nuclei under study, and a broad understanding of an individual comet typically requires the synthesis of many different investigations. Due to page limitations, this chapter cannot provide a comprehensive list of citations for all nucleus properties. We will cite individual papers whenever possible and will frequently refer to earlier review papers, but readers are encouraged to seek out the original sources when citing in future work.

This chapter is laid out as follows. Section 2 gives a brief summary of the observational techniques by which nucleus properties are measured or constrained and section 3 discusses the ensemble properties of comet nuclei. For each property, we first describe how observations are translated to the relevant measurements, then review the known properties and discuss what insights were learned from them. We conclude with a discussion of future advances that are likely to shape our understanding in section 4 and a brief summary in section 5.

2. OBSERVATIONAL TECHNIQUES

2.1. Optical Studies

By far the most common method for determining comet nucleus properties is optical studies, which we define to broadly include the near-UV to mid-infrared (IR) wavelengths that are available to groundbased observers. We introduce these techniques first before moving on to radar studies and space missions in the following subsections.

2.1.1. The difficulty with studying comet nuclei. Comet nuclei are small — as will be discussed below, typically a few kilometers — so the vast majority of remote observations are not capable of resolving the nucleus. At a distance of 1 au, 1″ corresponds to ~725 km. Thus, a 5-km-diameter nucleus would need to pass within 0.07 au of Earth to extend 0.1″ and appear two pixels wide with 0.05″/pix resolution typical of Hubble Space Telescope (HST) or adaptive optics images. More realistically, it would need to be several times closer in order for it to be resolved sufficiently for a meaningful constraint on the size to be made. Approaches to Earth of even 0.07 au are extremely rare. Since HST's launch in 1990, just four comets have been observed passing this close; of these only 252P/LINEAR was observed with high-resolution imaging (*Li et al.,* 2017). Thus, comet nucleus sizes and other physical properties are generally not directly measurable, but must be deduced via other means.

2.1.2. (Mostly) direct detections of the nucleus. Cometary activity tends to obscure the nucleus, and much of this section deals with how this obscuration can be mitigated. However, in certain cases, it is possible to detect the nucleus directly. Direct detection occurs when the nucleus is inactive or so weakly active that its signal dominates that from the coma and tail. When the nucleus signal can be assumed to dominate, its physical properties can be investigated.

The techniques were first successfully applied to the largest comets, since they could be detected when much further from the Sun and thus less active. Prominent early examples include 28P/Neujmin 1 (*Campins et al.,* 1987), 10P/Tempel 2 (*Jewitt and Meech,* 1988; *A'Hearn et al.,* 1989), and the aforementioned 49P, all of which were initially characterized by the late 1980s. Comets with very low dust-to-gas ratios in their coma (cf. *A'Hearn et al.,* 1995) can also have their nuclei directly detected even when active, e.g, 2P/Encke (see *Fernández et al.,* 2000, and references therein) and

162P/Siding Spring (*Fernández et al.,* 2006). Technological advancements have allowed smaller comets to be reliably observed at larger heliocentric distances, enabling subsequent studies to include far more objects. Notable surveys have been published by *Lowry et al.* (1999, 2003), *Licandro et al.* (2000), *Lowry and Fitzsimmons* (2001, 2005), *Lowry and Weissman* (2003), *Meech et al.* (2004), *Snodgrass et al.* (2006), and *Kokotanekova et al.* (2017).

2.1.3. Nucleus detection via coma fitting. Since most comets have coma contribution too prominent for a direct detection, techniques have been developed that can successfully remove the coma under certain circumstances. When there is sufficiently high spatial resolution that the coma can reliably be extrapolated all the way to the nucleus, the coma signal at the center can be removed and the excess signal attributed to the nucleus. The technique has been used most effectively for spacebased observations, but can be utilized in groundbased observations of comets coming extremely close to Earth or for facilities having exceptional angular resolution. Lamy and collaborators have utilized HST extensively in such studies (e.g., *Lamy and Toth,* 1995; *Lamy et al.,* 2009, 2011), and the technique is described in more detail in *Lamy et al.* (2004).

A similar procedure can be applied to mid-IR observations of comets despite lower spatial resolution when there is high enough contrast between the nucleus and the dust coma, typically at heliocentric distances beyond ~3 au. A tweak to the approach removes a range of scaled point spread functions (PSFs), allowing the nucleus signal to be extracted when the coma cannot be fit with a single power law. *Lamy et al.* (2004) and *Fernández et al.* (2013) describe the IR nucleus extraction processes in detail; the latter processed their sample independently using both techniques and found good agreement. The approach is limited to the few facilities capable of making thermal IR (roughly 5–25 μm) observations such as the groundbased Infrared Telescope Facility (IRTF) (*Lisse et al.,* 1999) and spacebased facilities including Infrared Space Observatory (ISO) (*Jorda et al.,* 2000; *Lamy et al.,* 2002; *Groussin et al.,* 2004), Spitzer Space Telescope (e.g., *Lisse et al.,* 2005; *Groussin et al.,* 2009), and Wide-field Infrared Survey Explorer (WISE), which operated from 2009–2010 and was henceforth known as the Near-Earth Object Wide-field Infrared Survey Explorer (NEOWISE) (e.g., *Bauer et al.,* 2011, 2012). Two major surveys have been conducted in the thermal IR: Survey of Ensemble Physical Properties of Cometary Nuclei (SEPPCoN), a targeted survey of JFCs with Spitzer (*Fernández et al.,* 2013), and a compilation of all comets observed during the cryogenic phase of WISE/NEOWISE (*Bauer et al.,* 2017).

2.1.4. Coma morphology. A method that has been utilized more frequently since *Comets II* is studying varying structures in the coma to infer nucleus properties. If repeating features can be identified, their temporal spacing and/or rate of motion can allow a rotation period to be determined or constrained. The observed morphology will vary as the viewing geometry changes, so three-dimensional modeling of activity can yield the pole orientation and, frequently, estimates of the location and extent of the active regions on the surface producing the observed coma features. Variations in the observed morphology can indicate seasonal changes and/or be diagnostic of the extent of any non-principal-axis (NPA) rotation (see section 3.3 for additional details).

This method is best applied to bright comets making reasonably close approaches to Earth ($\lesssim$1.0 au) since the spatial resolution and signal-to-noise are highest. The ability to resolve features in the coma is dependent on the nature of those structures, including their contrast relative to the ambient coma, the number and projected velocity of the features, and the rotation period and spin state of the nucleus [see the reviews by *Schleicher and Farnham* (2004) and *Farnham* (2009) for detailed discussion]. Asymmetries in the coma can often be identified by eye in raw images, but a variety of image-enhancement techniques are used to accentuate the small differences, often just a few percent in brightness, between the structures of interest and the ambient coma [see the reviews by *Schleicher and Farnham* (2004) and *Samarasinha and Larson* (2014)].

Coma morphology studies require much brighter comets than the direct detections described above, but they have several advantages. Most critically, coma morphology can be used to infer the rotation period and spin state of active comets when the nucleus is heavily obscured. Many comet nuclei are only accessible in this manner, and the technique is particularly helpful for expanding our knowledge of LPC nucleus properties since these nuclei often remain active until the nucleus is too faint to detect. Another advantage is that morphology studies are much less sensitive to observing conditions than photometric studies, and often can be conducted without the need for absolute calibrations. Meaningful constraints can often be set with sparse sampling of just one or two visits per night over several nights.

Coma structures can be due to gas or dust. Although easier to detect, dust is generally harder to interpret because it is slower moving than gas (requiring better spatial resolution to resolve), it has large velocity dispersion (thus smearing out features), and its trajectory is altered by solar radiation pressure. Gas is harder to detect than dust, but travels at a much higher velocity (~1 km s^{-1} vs. ~0.1 km s^{-1}) and is much less affected by solar radiation pressure. The gases seen at optical wavelengths are so called "fragment species," daughter or granddaughter gases descended from the parent ices that left the nucleus (e.g., *Feldman et al.,* 2004). Although these fragment species gain excess velocities in random directions during their production, the excess velocities are essentially randomized and the bulk outward motion of the parents leaving the nucleus is approximately preserved. As a result, gas species have proven to be far more useful for rotational studies.

In order to study the gas coma, the gas must be isolated from the dust. This is most commonly accomplished using specialized narrowband filters (*Farnham et al.,* 2000; *Schleicher and Farnham,* 2004), with the cyanogen (CN) filter being by far the most heavily utilized due to the CN's bright emission band and high contrast relative to reflected solar continuum at the same wavelengths. In principle, the hydroxyl (OH) filter is also effective for such studies, but it suffers from severe atmospheric extinction, and many telescopes have very low throughput at the relevant wavelengths. With recent improvements in integral field unit (IFU) spectroscopy, it is becoming possible to study coma morphology in gas "images" constructed from spectra (*Vaughan et al.,* 2017; *Opitom et al.,* 2019), with the added possibility of studying lines that are too spread out for conventional narrowband filters, like NH_2. Coma morphology techniques are not limited to near-UV and visible wavelengths. However, few other methods result in images with sufficient signal-to-noise, spatial resolution, and temporal coverage to conduct such studies. CO, CO_2, and/or dust features are detectable in the comae of some comets imaged by Spitzer's 4.5-μm channel (e.g., *Reach et al.,* 2013), so similar studies should be possible for some comets observed by NEOWISE, the James Webb Space Telescope (JWST), and future spacebased IR telescopes if image duration and cadence permit.

Attempts to interpret coma morphology in order to infer properties of the nucleus date back many decades (e.g., *Whipple,* 1978; *Sekanina,* 1979). These images were generally dominated by dust, and more recent work has shown that the assumptions in these early modeling efforts yielded results that are incompatible with modern solutions (cf. *Sekanina,* 1991a; *Knight et al.,* 2012).

The first results using gas filters were achieved by studying CN in 1P (*A'Hearn et al.,* 1986; *Samarasinha et al.,* 1986; *Hoban et al.,* 1988). With the improvement in CCDs and the production of the ESA and HB filters in the 1990s (*Farnham et al.,* 2000; *Schleicher and Farnham,* 2004), coma morphology studies have now been applied to many comets. Work has largely been concentrated among a few groups with access to narrowband filters and sufficient telescope time to constrain rotation periods: Schleicher and collaborators at Lowell Observatory (e.g., *Schleicher et al.,* 1998; *Farnham and Schleicher,* 2005; *Knight and Schleicher,* 2011; *Bair et al.,* 2018), Jehin and collaborators using the TRAnsiting Planets and PlanetesImals Small Telescope (TRAPPIST) telescopes (e.g., *Jehin et al.,* 2010; *Opitom et al.,* 2015; *Moulane et al.,* 2018), Samarasinha and colleagues using the Kitt Peak National Observatory (*Mueller et al.,* 1997; *Farnham et al.,* 2007a; *Samarasinha et al.,* 2011), and Waniak and collaborators at Rozhen National Observatory in Bulgaria (*Waniak et al.,* 2009, 2012). The technique has matured to the point that off-the-shelf filters with a quasi-CN bandpass have been successfully employed (*Ryske,* 2019).

2.1.5. Other methods of constraining nucleus properties. Coma lightcurves have been used for decades to constrain rotation periods (e.g., *Millis and Schleicher,* 1986; *Feldman et al.,* 1992), with the technique becoming much more prevalent in the modern era (e.g., *Anderson,* 2010; *Santos-Sanz et al.,* 2015; *Manzini et al.,* 2016). Aperture photometry is typically employed, with a signal that is assumed to be dominated by dust and/or gas in the coma as opposed to the nucleus. Except in rare cases, such lightcurves tend to have very small amplitudes (often just a few 0.01 mag from peak to trough) that can be easily affected by seeing variations, background contamination, calibration systematics, etc. When phased to a "best" period, interpretation is dependent on an assumption about the source of the variations (frequently assumed to be a single source), but unless this can be constrained, e.g., by coma morphology, it can lead to aliasing problems. Even in comets for which much information is known *a priori*, conclusive interpretation of a coma lightcurve can be challenging (cf. *Schleicher et al.,* 2015), so we generally consider periods from coma lightcurves to be less reliable than those obtained by the means discussed above.

Various efforts have been made to extract nucleus information from observations not necessarily designed for that purpose. Compilations of multi-wavelength constraints on individual comets to determine their nucleus radius include C/1995 O1 (Hale-Bopp) (*Weaver and Lamy,* 1997; *Fernández,* 2002) and C/1983 H1 (IRAS-Araki-Alcock) (*Groussin et al.,* 2010). Non-gravitational accelerations (caused by momentum transfer to the nucleus from outgassing materials) from orbit calculations and overall brightness have been used to infer mass (e.g., *Rickman,* 1986, 1989; *Sosa and Fernández,* 2009, 2011). When sizes are also known, this can also yield estimates of the bulk density of the nucleus (e.g., *Farnham and Cochran,* 2002; *Davidsson and Gutiérrez,* 2006). For a detailed discussion of nucleus densities and how they are estimated, see the chapter by Guilbert-Lepoutre et al. in this volume.

Secular lightcurves (brightness/activity as a function of heliocentric distance) imply seasonal variations in activity of some comets; these can provide constraints on pole orientations and spin state (e.g., *Schleicher et al.,* 2003; *Farnham and Schleicher,* 2005). *Boe et al.* (2019) combined a cometary activity model with a survey simulator to statistically characterize the size distribution of LPCs. *Tancredi et al.* (2000, 2006), *Ferrín* (e.g., 2010), and *Weiler et al.* (2011) have assembled data from a variety of published observations to attempt to extract nucleus magnitudes from observations at large distances. The comets studied are primarily JFCs near aphelion, but are occasionally LPCs when they are assumed to be inactive. As acknowledged by these authors, such observations can be problematic since many JFCs still show distant activity near aphelion (e.g., *Mazzotta Epifani et al.,* 2007, 2008; *Kelley et al.,* 2013). *Hui and Li* (2018) showed that the nucleus must account for more than 10% of the total signal for a reliable extraction from typical groundbased observations, effectively ruling out nucleus extraction for more active JFCs and LPCs.

2.2. Radar Studies

Radar contributes unique insight into comet nuclei, being capable of imaging nuclei at spatial scales only achieved otherwise by space missions and of measuring the rotation rate of the nucleus directly. Since radar observations are conducted by sending a burst of microwaves toward a target and measuring the power of the returned echo, the received signal varies as Δ^{-4}, where Δ is the geocentric distance. This effectively limits the detectable population to only those approaching within 0.1 au of Earth unless their nuclei are especially large (*Harmon et al.,* 1999). A detailed review of radar observations was provided by *Harmon et al.* (2004); we discuss the technique briefly and provide updates of key observations since then in section 3.1.3.

The highest-quality radar observations use delay-Doppler imaging, which measures both the echo Doppler spectrum as well as the time delay of the echo, resulting in spatial and rotational rate information about different positions on the nucleus. Delay-Doppler data can be inverted to form a three-dimensional model of the nucleus. While common for asteroids, this has only been achieved for a handful of comets. Doppler-only detections yield a radar cross section, which has a degeneracy between nucleus size and radar albedo unless additional information is considered. The Doppler signal can be interpreted to give constraints on the rotation period, polarization by the surface, surface roughness, and density of the surface layer. Interestingly, radar albedos have been found to be similar to optical albedos despite probing meters into the surface (*Harmon et al.,* 2004). Although beyond the scope of this chapter, radar observations can also provide information about the properties of large grains in the inner coma and even their position relative to the nucleus.

All radar detections of comet nuclei to date have been achieved using Arecibo and/or Goldstone. Bistatic observations, where the emitter and receiver are located at different telescopes, with Greenbank Observatory as the receiver have been made occasionally. With the loss of Arecibo in 2020 and few known upcoming close-approaching comets, there are no foreseeable opportunities for radar comet observations until the mid-2030s. New comet radar detections in the next decade will require as yet undiscovered comets passing close to Earth (E. Howell, personal communication, 2021).

2.3. Space Missions

Dedicated space missions to comets provide definitive measurements of comet nucleus properties since they fully resolve the nucleus and are so close that the dust and gas along the line of sight are negligible. Six comet nuclei have been imaged *in situ*: 1P by Vega 1 and 2 and Giotto, 19P/Borrelly by Deep Space 1, 81P/Wild 2 by Stardust, 9P/Tempel 1 by Deep Impact and Stardust, 103P/Hartley 2 by Deep Impact/EPOXI, and 67P/Churyumov-Gerasimenko by Rosetta. These are discussed extensively in the chapter by Snodgrass et al. in this volume, but we note that some properties of these nuclei were constrained by remote observations before the mission encounters and were in good agreement with mission findings. Successful examples include the surprisingly small size of 103P for its activity level (*Lisse et al.,* 2009) and the rotation period of 67P (*Lowry et al.,* 2012). The consistency between the remote observations and spacecraft results is a validation that the remote techniques are reliable. The spacecraft results provide a ground truth for the selected objects, allowing us to trust that, under the proper conditions, the same techniques give good results for other objects as well.

3. THE RANGES OF PHYSICAL AND SURFACE PROPERTIES

The catalog of measured nucleus properties is now approximately three times as large as when they were reviewed in *Comets II.* We also now have detailed information from several new space missions that influence our interpretation of these measurements. We tabulate the physical (Table 1) and surface (Table 2) properties for all comets having reliable determinations of any of the following: rotation period, axial ratio, albedo, nucleus phase function, or nucleus color. The distributions of these properties are plotted in Fig. 1. For any comet with one of the other properties measured, we tabulate what we conclude to be the most reliable effective radius in Table 1. This is not a complete review of all measured sizes; it aims to enable searching for the relation between nucleus size and the other parameters. Due to the limited space, it is not possible to list all measured effective radii and provide the necessary context to evaluate the accuracy of the measurements. Instead, we plot in Fig. 1 all effective radii determined from the two major thermal IR surveys (*Fernández et al.,* 2013; *Bauer et al.,* 2017). Although this is not a complete set of effective nucleus measurements, it encompasses 213 comets and the methodology was the same throughout, allowing more meaningful comparisons. Owing to the still sparse collection of properties for HTCs and LPCs, these are grouped together in the figure for comparison with the JFCs, but are discussed individually below as warranted.

In the following subsections we discuss these cumulative properties and provide updated mean, standard deviation, and median values when it is reasonable to do so. Section 3.1 discusses nucleus sizes, the nucleus size-frequency distribution, and nucleus shapes. Section 3.2 discusses the cumulative surface properties, with separate subsections for albedo (section 3.2.1), phase function and phase reddening (section 3.2.2), and nucleus colors/spectroscopy (section 3.2.3). Rotation periods and changes in rotation period are discussed in section 3.3.

This extensive compilation of nucleus properties provides an opportunity to explore correlations between properties. We conducted Spearman rank correlations for all comets and for only the JFC population (the HTC plus LPC data are too sparse for their own test) for all combinations of the data compiled in Tables 1 and 2. The pairings resulting in the highest Spearman rank correlations (ρ) were albedo and phase coefficient [ρ = 0.52 for all comets and ρ = 0.43 for JFCs; previously identified by *Kokotanekova et al.* (2017)], period and axial ratio (ρ = 0.33), and phase coefficient with radius (ρ = −0.31). The first can be considered to have

TABLE 1. Physical properties of comet nuclei.

Comet	r_n†(km)	Ref*	P (h)	Ref*	a/b‡	Ref*
Jupiter-family comets						
2P/Encke	2.4 ± 0.3	F00	11.0830 ± 0.0030[a,b?]	L07, B05	≥1.44 (0.06)	L07
4P/Faye	1.77 ± 0.04	L09			≥1.25	L04
6P/d'Arrest	$2.23^{+0.13}_{-0.15}$	F13	6.67 ± 0.03[b?]	G03	≥1.08	G03
7P/Pons-Winnecke	2.64 ± 0.17	F13	$7.9^{+1.6}_{-1.1}$	S05	≥1.3 (0.1)	S05
9P/Tempel 1	2.83 ± 0.10	T13a	41.335 ± 0.005[a]	B11	1.28[c]	T13a
10P/Tempel 2	5.98 ± 0.04	L09	8.948 ± 0.001[a]	S13	≥1.9	J89
14P/Wolf	2.95 ± 0.19	F13	9.07 ± 0.01	K18	≥1.41 (0.06)	K17
17P/Holmes	2.41 ± 0.53	BG17	7.2/8.6/10.3/12.8	S06	≥1.3 (0.1)	S06
19P/Borrelly	2.5 ± 0.1	B04	26.0 ± 1.0[a]	M02	2.5 (0.07)[c]	B04
21P/Giacobini-Zinner	1	T00	7.39 ± 0.01; 10.66 ± 0.01[a]	G23	≥1.5	M92
22P/Kopff	2.15 ± 0.17	F13	12.30 ± 0.8	L03a	≥1.66 (0.11)	L03a
26P/Grigg-Skjellerup	1.3	L04			≥1.1	B99
28P/Neujmin 1	~10.7	L04	12.75 ± 0.03	D01	≥1.51 (0.07)	D01
31P/Schwassmann-Wachmann 2	$1.65^{+0.11}_{-0.12}$	F13	5.58 ± 0.03	L92	≥1.6 (0.15)	L92
36P/Whipple	2.31 ± 0.29	BG17	~40	S08	≥1.9 (0.1)	S08
37P/Forbes	$1.23^{+0.08}_{-0.09}$	F13				
41P/Tuttle-Giacobini-Kresák	0.7	T00	19.75–20.05[a,b?]	BF18, H18	≥2.1	BR20
44P/Reinmuth	2.55 ± 0.15	BG17				
45P/Honda-Mrkos-Pajdušáková	0.6–0.65	LH22	7.6 ± 0.5	S22	≥1.3	LT99
46P/Wirtanen	0.56 ± 0.04	B02	$8.94^{+0.02}_{-0.01}$[a]	F21	≥1.4 (0.1)	B02
47P/Ashbrook-Jackson	$3.11^{+0.20}_{-0.21}$	F13	15.6 ± 0.1	K17	≥1.36 (0.07)	K17
48P/Johnson	$2.97^{+0.19}_{-0.20}$	F13	29.00 ± 0.04	J04	≥1.34 (0.06)	J04
49P/Arend-Rigaux	4.57 ± 0.06	KW17	13.450 ± 0.005	E17	≥1.63 (0.07)	M88
50P/Arend	1.49 ± 0.13	F13				
53P/Van Biesbroeck	3.33–3.37	M04				
59P/Kearns-Kwee	0.79 ± 0.03	L09				
61P/Shajn-Schaldach	2.28 ± 0.64	BG17	4.9 ± 0.2	L11	≥1.3	L11
63P/Wild 1	1.46 ± 0.03	L09	14 ± 2	BdS20	≥2.19 (0.02)	BdS20
67P/Churyumov-Gerasimenko	1.649 ± 0.007	J16	12.055 ± 0.001[a]	Rosetta	1.67 (0.01)[c]	J16
70P/Kojima	1.84 ± 0.09	L11			≥1.1	L11
71P/Clark	0.79 ± 0.03	L09				
73P/Schwassmann-Wachmann 3-C[f]	0.4 ± 0.1	GJ19	20.76 ± 0.08	GJ19	≥1.8 (0.3)	T06
74P/Smirnova-Chernykh	$3.31^{+0.60}_{-0.69}$	F13			≥1.14	L11
76P/West-Kohoutek-Ikemura	0.31 ± 0.01	L11	6.6 ± 1.0	L11	≥1.45	L11
81P/Wild 2	1.98 ± 0.05	SB04	13.5 ± 0.1	M10	1.38 (0.04)	D04
82P/Gehrels 3	0.59 ± 0.04	L11	≥24 ± 5	L11	≥1.59	L11
84P/Giclas	0.90 ± 0.05	L09				
86P/Wild 3	0.42 ± 0.02	L11			≥1.35	L11
87P/Bus	0.26 ± 0.01	L11	32 ± 9	L11	≥2.2	L11
92P/Sanguin	2.08 ± 0.01	S05	6.22 ± 0.05	S05	≥1.7 (0.1)	S05
93P/Lovas 1	2.59 ± 0.26	F13	$18.2^{+1.5}_{-15}$	K17	≥1.21(0.06)	K17
94P/Russell 4	$2.27^{+0.13}_{-0.15}$	F13	20.70 ± 0.07	K17	≥2.8 (0.2)	K17

TABLE 1. (continued)

Comet	r_n†(km)	Ref*	P (h)	Ref*	a/b‡	Ref*
Jupiter-family comets (continued)						
96P/Machholz 1[f]	3.40 ± 0.20	E19	4.096 ± 0.002	E19	≥1.6 (0.1)	E19
103P/Hartley 2	0.580 ± 0.018	T13b	16.4 ± 0.1[a,b]	M11	3.11[c]	T13b
106P/Schuster	0.94 ± 0.03	L09				
110P/Hartley 3	2.20 ± 0.10	L11	10.153 ± 0.001	K17	≥1.20 (0.03)	K17
112P/Urata-Niijima	0.90 ± 0.05	L09				
113P/Spitaler	1.70 ± 0.10	F13				
114P/Wiseman-Skiff	0.78 ± 0.05	L09				
121P/Shoemaker-Holt 2	$3.87^{+0.26}_{-0.27}$	F13	10^{+8}_{-2}	S08	≥1.15 (0.03)	S08
123P/West-Hartley	2.18 ± 0.23	F13			≥1.6 (0.1)	K17
131P/Mueller 2	$1.11^{+0.09}_{-0.07}$	F13				
137P/Shoemaker-Levy 2	$4.04^{+0.31}_{-0.32}$	F13			≥1.18 (0.05)	K17
143P/Kowal-Mrkos	$4.79^{+0.32}_{-0.33}$	F13	17.20 ± 0.02	K18	≥1.49 (0.05)	L04
147P/Kushida-Muramatsu	0.21 ± 0.02	L11	10.5 ± 1.0/ 4.8 ± 0.2	L11	≥1.53	L11
149P/Mueller 4	$1.42^{+0.09}_{-0.10}$	F13				
162P/Siding Spring	$7.03^{+0.47}_{-0.48}$	F13	32.864 ± 0.001	D23	1.56 ($^{+0.44}_{-0.16}$)[c]	D23
169P/NEAT	$2.48^{+0.13}_{-0.14}$	F13	8.4096 ± 0.0012	K10	≥1.74 (0.03)[d]	W06
209P/LINEAR	~1.53	H14	10.93 ± 0.020	H14, S16	≥1.55	H14
249P/LINEAR	1.0–1.3	FL17				
252P/LINEAR	0.3 ± 0.03	L17	5.41 ± 0.07/ 7.24 ± 0.07[b?]	L17		
260P/McNaught	$1.54^{+0.09}_{-0.08}$	F13	8.16 ± 0.24	M14	≥1.07[d]	M14
280P/Larsen	$1.23^{+0.15}_{-0.17}$	F13				
322P/SOHO[x]	0.075–0.16	K16	2.8 ± 0.3	K16	≥1.3	K16
323P/SOHO[f,x]	0.086 ± 0.003	H22	0.522024 ± 0.000002	H22	~1.25[c]	H22
333P/LINEAR	3.04	S21				
P/2016 BA14 (PANSTARRS)	0.55–0.8	K22	~40	N16		
Halley-type comets						
1P/Halley	5.5	L04	~68[b]	S04	2.0 (0.1)[c]	K87
8P/Tuttle	2.9	G19	11.4	H10	1.41 (0.07)[cb]	G19
55P/Tempel-Tuttle	1.8	L04			≥1.5	H98
109P/Swift-Tuttle	15 ± 3	F95	69.4	L04		
C/2001 OG108 (LONEOS)	7.6 ± 1.0	A05	57.2 ± 0.5	A05	≥1.3	A05
C/2002 CE10 (LINEAR)	8.95 ± 0.45	S18	8.19 ± 0.05	S18	≥1.2 (0.1)	S18
P/1991 L3 (Levy)	5.8 ± 0.1	F94	8.34	F94	≥1.3	F94
P/2006 HR30 (Siding Spring)	11.95–13.55	BI17				
Long-period comets						
C/1983 H1 (IRAS-Araki-Alcock)	3.4 ± 0.5	G10	51.3 ± 0.3	S88		
C/1990 K1 (Levy)			17.0 ± 0.1[a]	F92		
C/1995 O1 (Hale-Bopp)	37	SK12	11.35 ± 0.04	J02	≥1.72(0.07)	SK12
C/1996 B2 (Hyakutake)	2.4 ± 0.5	LF99	6.273 ± 0.007	S02		
C/2002 VQ94 (LINEAR)	40.7	J05				
C/2004 Q2 (Machholz)			17.60 ± 0.05	F07		
C/2007 N3 (Lulin)	6.10 ± 0.25	BG17	41.45 ± 0.05	BS18		

TABLE 1. (continued)

Comet	r_n†(km)	Ref*	P (h)	Ref*	a/b‡	Ref*
Long-period comets (continued)						
C/2009 P1 (Garradd)	13.50 ± 2.50	BG17	11.1 ± 0.8	I17		
C/2012 F6 (Lemmon)			9.52 ± 0.05	O15		
C/2012 K1 (PANSTARRS)	8.67 ± 0.08	BdS20	9.4 ± 0.4	BdS20	≥1.44 (0.03)	BdS20
C/2013 A1 (Siding Spring)	~1	F17	8.00 ± 0.08	L16		
C/2014 Q2 (Lovejoy)			17.89 ± 0.17	SR15		
C/2014 S2 (PANSTARRS)	1.3	BdS18	68 ± 2	BdS18	≥1.45 (0.13)[d]	BdS18
C/2014 UN271 (Bernardinelli-Bernstein)	69 ± 9/60 ± 7	LM22, HJ22				
C/2020 F3 (NEOWISE)	2.5	J. Bauer (unpubl. data)	7.8 ± 0.2	M21		

Comets are grouped by dynamical classification, and not all properties are known for a given comet. Column 1 gives the comet name. Columns 2, 4, and 6 give the effective radius (r_n), rotation period (P), and axial ratio (a/b) respectively. Columns 3, 5, and 7 give the corresponding reference code(s).

*Additional references are available for some comets. The table lists the most recent and/or most precise measurements. All other literature values have been omitted due to space limitations. Published precision has been preserved, resulting in differing numbers of significant digits.
[x]Excluded from bulk calculations and Figs. 1–4 for the following reasons: 322P: suspected asteroid only active near extremely small perihelion of 0.071 au (*Knight et al.,* 2016); 323P: disintegrating, only active near extremely small perihelion of 0.039 au (*Hui et al.,* 2022b).
†Nucleus radius measurements are given for any comet with P, a/b, β, p_V, (V–R) measurements (see text for additional details).
‡The uncertainty of the axial ratio value is listed in brackets whenever available.
[a]Period changes observed; this is the minimum known value measured with sufficient precision.
[b]NPA rotation (suspected NPA rotators indicated with [b?]).
[c]A shape model was quoted in the cited paper; the provided axial ratio is obtained by dividing the largest shape model axis by the second largest axis.
[cb]A shape model consisting of two spheres with radii $R_1 > R_2$ was presented in the cited paper; the axial ratio was approximated by $(R_1 + R_2)/R_1$.
[d]Calculated from the brightness variation in the cited paper.
[f]Ongoing fragmentation observed.

References: A05: *Abell et al.* (2005); B02: *Boehnhardt et al.* (2002); B04: *Buratti et al.* (2004); B05: *Belton et al.* (2005); B11: *Belton et al.* (2011); B99: *Boehnhardt et al.* (1999); BdS18: *Betzler et al.* (2018); BdS20: *Betzler et al.* (2020); BF18: *Bodewits et al.* (2018); BG17: *Bauer et al.* (2017); BI17: *Bach et al.* (2017); BR20: *Boehnhardt et al.* (2020). BS18: *Bair et al.* (2018); D01: *Delahodde et al.* (2001); D04: *Duxbury et al.* (2004); D23: *Donaldson et al.* (2023); E17: *Eisner et al.* (2017); E19: *Eisner et al.* (2019); F00: *Fernández et al.* (2000); F07: *Farnham et al.* (2007a); F13: *Fernández et al.* (2013); F17: *Farnham et al.* (2017); F21: *Farnham et al.* (2021); F92: *Feldman et al.* (1992); F94: *Fitzsimmons and Williams* (1994); F95: *Fomenkova et al.* (1995); FL17: *Fernández et al.* (2017); G03: *Gutierrez et al.* (2003a); G10: *Groussin et al.* (2010); G19: *Groussin et al.* (2019); G23: *Goldberg et al.* (2023); GJ19: *Graykowski and Jewett* (2019); H10: *Harmon et al.* (2010); H14: *Howell et al.* (2014); H18: *Howell et al.* (2018); H22: *Hui et al.* (2022b); H98: *Hainaut et al.* (1998); HJ22: *Hui et al.* (2022a); I17: *Ivanova et al.* (2017); J02: *Jorda and Gutiérrez* (2002); J04: *Jewitt and Sheppard* (2004); J05: *Jewitt* (2005); J16: *Jorda et al.* (2016); J89: *Jewitt and Luu* (1989); K10: *Kasuga et al.* (2010); K16: *Knight et al.* (2016); K17: *Kokotanekova et al.* (2017); K18: *Kokotanekova et al.* (2018); K22: *Kareta et al.* (2022); K87: *Keller et al.* (1987); KW17: *Kelley et al.* (2017); L03a: *Lowry and Weissman* (2003); L04: *Lamy et al.* (2004); L07: *Lowry and Weissman* (2007); L09: *Lamy et al.* (2009); L11: *Lamy et al.* (2011); L16: *Li et al.* (2016); L17: *Li et al.* (2017); L92: *Luu and Jewett* (1992); LF99: *Lisse et al.* (1999); LH22: *Lejoly et al.* (2022); LM22: *Lellouch et al.* (2022); LT99: *Lamy et al.* (1999); M02: *Mueller and Samarasinha* (2002); M04: *Meech et al.* (2004); M10: *Mueller et al.* (2010); M11: *Meech et al.* (2011); M14: *Manzini et al.* (2014); M21: *Manzini et al.* (2021); M88: *Millis et al.* (1988); M92: *Mueller* (1992); N16: *Naidu et al.* (2016); O15: *Opitom et al.* (2015); S02: *Schleicher and Osip* (2002); S04: *Samarasinha et al.* (2004); S05: *Snodgrass et al.* (2005); S06: *Snodgrass et al.* (2006); S08: *Snodgrass et al.* (2008); S13: *Schleicher et al.* (2013); S16: *Schleicher and Knight* (2016); S18: *Sekiguchi et al.* (2018); S21: Simion et al., personal communication, 2021; S22: *Springmann et al.* (2022); S88: *Sekanina* (1988); SB04: *Sekanina et al.* (2004); SK12: *Szabó et al.* (2012); SR15: *Serra-Ricart and Licandro* (2015); T00: *Tancredi et al.* (2000); T06: *Toth et al.* (2006); T13a: *Thomas et al.* (2013a); T13b: *Thomas et al.* (2013b); W06: *Warner et al.* (2006).

moderate correlation, while the remainder have low correlation. We caution that for some properties there are still few enough measurements that Spearman ranks can change substantially with the addition of a single new measurement. The lack of correlation between some quantities that might reasonably be assumed to correlate such as color and albedo ($\rho = -0.09$) or color and phase function ($\rho = -0.07$) may be informative. The scatter in some properties (albedo, phase coefficient, color) is larger for smaller radii. While this may be an indication of a greater diversity of surface properties in small nuclei, more data are needed to rule out observational bias since most have large uncertainties. We discuss these and other interesting comparisons between properties in the relevant subsections. We also performed Spearman rank correlations to check for any possible correlations between the nucleus properties in Tables 1 and 2 and orbital parameters. The intriguing possibility for a correlation between (V–R) and perihelion distance is discussed in section 3.2.3.

3.1. Nucleus Sizes and Shapes

3.1.1. Effective radius. Except in rare instances of radar observations or space missions, the actual width of the nucleus cannot be measured directly. Instead, the effective radius (r_n; the radius of a sphere having the same cross-section as the comet nucleus) is used to quantify its size.

TABLE 2. Surface properties of comet nuclei.

	p_V	Ref*	β (mag/°)	Ref*	(V–R)	Ref*
Jupiter-family comets						
2P/Encke	0.04 ± 0.03^{R}	F00	0.053 ± 0.003^{WM}	F00, B08	0.44 ± 0.06	LT09
4P/Faye					0.45 ± 0.04	L09
6P/d'Arrest					0.51±0.10	LT09
7P/Pons-Winnecke					0.49 ± 0.03	S05
9P/Tempel 1	0.056 ± 0.007	L07a	0.046 ± 0.007	L07a	0.50 ± 0.01	L07a
10P/Tempel 2	$0.022^{+0.004}_{-0.006}$	AH89	0.037 ± 0.004	S91	0.54 ± 0.03	LT09
14P/Wolf	0.046 ± 0.007^{rPS}	K17	0.060 ± 0.005	K17	0.57 ± 0.07	S05
17P/Holmes					0.53 ± 0.07	S06
19P/Borrelly	0.06 ± 0.02^{R}	L07b	0.043 ± 0.009	L07b	0.25 ± 0.78	L03b
21P/Giacobini-Zinner					0.50 ± 0.02	L93
22P/Kopff	0.042 ± 0.006	L02			0.50 ± 0.05	LT09
26P/Grigg-Skjellerup					0.3 ± 0.1	B99
28P/Neujmin 1	0.03 ± 0.01^{R}	J88	0.025 ± 0.006	D01	0.47 ± 0.05	LT09
36P/Whipple			0.060 ± 0.019	S08	0.47 ± 0.02	S08
37P/Forbes					0.29 ± 0.03	L09
44P/Reinmuth					0.62 ± 0.08	L09
45P/Honda-Mrkos-Pajdušáková			0.06	L04	0.44 ± 0.05	LT09
46P/Wirtanen					0.45 ± 0.06	LT09
47P/Ashbrook-Jackson	0.054 ± 0.008^{rPS}	K17			0.43 ± 0.09	LT09
48P/Johnson			0.059 ± 0.002	J04	0.5 ± 0.3	L00
49P/Arend-Rigaux	0.04 ± 0.01	C95			0.47 ± 0.05	LT09
50P/Arend					0.81 ± 0.10	L09
53P/Van Biesbroeck					0.34 ± 0.08	M04
59P/Kearns-Kwee					0.62 ± 0.07	L09
63P/Wild 1					0.50 ± 0.05	L09
67P/Churyumov-Gerasimenko	0.059 ± 0.002	S15	0.047 ± 0.002	F15	0.57 ± 0.03	C15
70P/Kojima					0.60 ± 0.09	L11
71P/Clark					0.64 ± 0.07	L09
73P/Schwassmann-Wachmann 3-C					0.45 ± 0.20	LT09
81P/Wild 2	0.059 ± 0.004	LAH09	0.0513 ± 0.0002	LAH09		
84P/Giclas					0.32 ± 0.03	L09
86P/Wild 3					0.86 ± 0.10	L11
92P/Sanguin					0.54 ± 0.04	S05
94P/Russell 4	0.043 ± 0.007^{rPS}	K17	0.039 ± 0.002	K17	0.62 ± 0.05	S08
96P/Machholz 1					0.41 ± 0.04	E19
103P/Hartley 2	0.045 ± 0.009	L13	0.046 ± 0.002	L13	0.43 ± 0.04	L13
106P/Schuster					0.52 ± 0.06	L09
110P/Hartley 3			0.069 ± 0.002	K17	0.67 ± 0.09	L11
112P/Urata-Niijima					0.53 ± 0.04	L09
113P/Spitaler					0.58 ± 0.08^{C}	SI12
114P/Wiseman-Skiff					0.46 ± 0.02	L09
121P/Shoemaker-Holt 2					0.53 ± 0.03	S08
131P/Mueller 2					0.45 ± 0.12	S08
137P/Shoemaker-Levy 2	0.030 ± 0.005^{rPS}	K17	0.035 ± 0.004	K17	0.71 ± 0.18	S06

TABLE 2. (continued)

	p_V	Ref*	β (mag/°)	Ref*	(V–R)	Ref*
Jupiter-family comets (continued)						
143P/Kowal-Mrkos	0.044 ± 0.008 [rPS]	K18	0.043 ± 0.014	J03	0.58 ± 0.02	LT09
149P/Mueller 4	0.030 ± 0.005 [rPS]	K17	0.03 ± 0.02	K17		
162P/Siding Spring	0.021 ± 0.002	D23	0.051 ± 0.002	D23	0.45 ± 0.01	C06
169P/NEAT	0.03 ± 0.01	DM08			0.43 ± 0.02	K10
249P/LINEAR					0.37 ± 0.01[S']	K21
280P/Larsen					0.49 ± 0.03	S08
322P/SOHO[x]			0.031 ± 0.004	K16	0.41 ± 0.04	K16
323P/SOHO[x]			0.0326 ± 0.0004	H22	0.05 ± 0.06; 0.13 ± 0.09[D]	H22
333P/LINEAR					0.44 ± 0.01[S']	S21
P/2016 BA14 (PANSTARRS)	0.01–0.03	K22				
Halley-type comets						
1P/Halley	$0.04^{+0.02}_{-0.01}$	S86			0.41 ± 0.03	T89
8P/Tuttle	0.04 ± 0.01[R]	G19	0.033–0.04	L12	0.53 ± 0.04	LT09
55P/Tempel-Tuttle	0.05 ± 0.02[R]	C02	0.041	L04	0.51 ± 0.05	LT09
109P/Swift-Tuttle	0.02–0.04[R]	L04				
C/2001 OG108 (LONEOS)	0.04 ± 0.01	A05	0.034[TW]	A05	0.46 ± 0.02	A05
C/2002 CE10 (LINEAR)	0.03 ± 0.01	S18			0.568 ± 0.039	S18
P/2006 HR30 (Siding Spring)	0.035–0.045	BI17			0.45 ± 0.01	H07
Long-period comets						
C/1983 H1 (IRAS-Araki-Alcock)	0.04 ± 0.01	G10	0.04	G10		
C/1995 O1 (Hale-Bopp)	0.04 ± 0.03[var]	C02				
C/2002 VQ94 (LINEAR)					0.50 ± 0.02	J05
C/2014 UN271 (Bernardinelli-Bernstein)	0.034±0.008	LM22			0.46 ± 0.04[C]	B21

Comets are grouped by dynamical classification, and not all properties are known for a given comet. Column 1 gives the comet name. Columns 2, 4, and 6 give the V-band albedo (p_V), phase coefficient (β), and (V–R) color index, respectively. Columns 3, 5, and 7 give the corresponding reference code(s).

*Additional references are available for some comets. The table lists the most recent and/or most precise measurements. All other literature values have been omitted due to space limitations. Published precision has been preserved, resulting in differing numbers of significant digits.
[x]Excluded from bulk calculations and Figs. 1–4 for the following reasons: 322P: suspected asteroid only active near extremely small perihelion of 0.071 au (*Knight et al.*, 2016); 323P: disintegrating, only active near extremely small perihelion of 0.039 au (*Hui et al.*, 2022b).
[C]Converted to (V–R) using *Jester et al.* (2005).
[D]Different colors before and after perihelion; converted to (V–R) using *Jester et al.* (2005).
[R]Original geometric albedo in R-band converted to V-band using the (V–R) color of the nucleus in this table. Due to the large uncertainty in the color index of 19P, the average (V–R) = 0.50 ± 0.03 for JFC nuclei (*Lamy and Toth*, 2009; *Jewitt*, 2015) was used instead.
[S']Using the expression in equation (2) in *Luu and Jewitt* (1990b) can be approximated to a color index (V–R), using a (V–R) = 0.354 ± 0.010 mag for the Sun (*Holmberg et al.*, 2006).
[rPS]Original geometric albedo in r_{PS}-band converted to V-band using $p_V = p_{rPS} \times 0.919$ for the mean color index (B–V) = 0.87 ± 0.05 mag (*Lamy and Toth*, 2009).
[TW]Linear fit derived from data in *Abell et al.* (2005).
[var]Albedo variations observed (*Szabó et al.*, 2012).
[WM]Weighted mean.

References: A05: *Abell et al.* (2005); AH89: *A'Hearn et al.* (1989); B08: *Boehnhardt et al.* (2008); B21: *Bernardinelli et al.* (2021); B99: *Boehnhardt et al.* (1999); BI17: *Bach et al.* (2017); C02: *Campins and Fernández* (2002); C06: *Campins et al.* (2006); C15: *Ciarniello et al.* (2015); C95: *Campins et al.* (1995); D01: *Delahodde et al.* (2001); D23: *Donaldson et al.* (2023); DM08: *DeMeo and Binzel* (2008); E19: *Eisner et al.* (2019); F00: *Fernández et al.* (2000); F15: *Fornasier et al.* (2015); G10: *Groussin et al.* (2010); G19: *Groussin et al.* (2019); H07: *Hicks and Bauer* (2007); H22: *Hui et al.* (2022b); J03: *Jewitt et al.* (2003); J04: *Jewitt and Sheppard* (2004); J05: *Jewitt* (2005); J88: *Jewitt and Meech* (1988); K10: *Kasuga et al.* (2010); K16: *Knight et al.* (2016); K17: *Kokotanekova et al.* (2017); K18: *Kokotanekova et al.* (2018); K21: *Kareta et al.* (2021); K22: *Kareta et al.* (2022); L00: *Licandro et al.* (2000); L02: *Lamy et al.* (2002); L03b: *Lowry et al.* (2003); L04: *Lamy et al.* (2004); *(continued on next page)* L07a: *Li et al.* (2007a); L07b: *Li et al.* (2007b); L09: *Lamy et al.* (2009); L11: *Lamy et al.* (2011); L12: *Lamy et al.* (2012); L13: *Li et al.* (2013); L93: *Luu* (1993); LAH09: *Li et al.* (2009); LM22: *Lellouch et al.* (2022); LT09: *Lamy and Toth* (2009); M04: *Meech et al.* (2004); S05: *Snodgrass et al.* (2005); S06: *Snodgrass et al.* (2006); S08: *Snodgrass et al.* (2008); S15: *Sierks et al.* (2015); S18: *Sekiguchi et al.* (2018); S21: Simion et al., personal communication, 2021; S86: *Sagdeev et al.* (1986); S91: *Sekanina* (1991a); SI12: *Solontoi et al.* (2012); T89: *Thomas and Keller* (1989).

More nuanced measurements of the radius are sometimes used [$r_{n,a}$ and $r_{n,v}$ (see *Lamy et al.,* 2004)], but since many authors do not specify which they have measured, we generically use the term r_n in this chapter. As shown by *Lamy et al.* (2004), $r_{n,a}$ and $r_{n,v}$ agree to within 10% for axial ratios up to 3, so any discrepancies between nomenclature are minimal.

The apparent magnitude, m, can be translated into r_n (in meters) from reflected-light observations by the following equation, which was originally derived by *Russell* (1916)

$$p\Phi(\alpha)r_n^2 = 2.238 \times 10^{22} r_H^2 \Delta^2 10^{0.4(m_\odot - m)} \quad (2)$$

Here, p is the geometric albedo, $\Phi(\alpha)$ is the phase function of the nucleus at solar phase angle α (solar phase angle, henceforth simply "phase angle," is the Sun-comet-observer angle), r_H is the heliocentric distance in astronomical units, Δ is the distance from the observer to the nucleus in astronomical units, and $m_\odot$ is the apparent magnitude of the Sun. The quantities p, $\Phi(\alpha)$, $m_\odot$, and m need to be taken in the same spectral band. The phase function is typically assumed to be linear with phase angle (see section 3.2.2), and is given by

$$-2.5 \log[\Phi(\alpha)] = \alpha\beta \quad (3)$$

where β is the linear phase coefficient in mag/°. Traditionally, this phase coefficient is taken to be independent of wavelength; however, this should be investigated with additional photometric data taken at multi-bandpasses in the future.

Although not discussed further in this chapter, a common method of comparing comet nucleus sizes and in searching for unresolved activity contaminating the photometric aperture is via the absolute magnitude, H, given by

$$H = m - 5 \log(r_H\Delta) - \alpha\beta \quad (4)$$

This is the magnitude the nucleus would have at $r_H = \Delta = 1$ au and $\alpha = 0°$, a physically impossible scenario, but one that is convenient for normalizations.

Thermal IR observations are usually dominated by thermal re-radiation from the nucleus and, when paired with thermophysical modeling, can be used to determine nucleus sizes. The process utilizes the size, shape, rotation period, spin axis orientation, and surface properties (thermal inertia, surface roughness) to predict the observed thermal flux. This is more mathematically complex than the reflected light procedure just discussed. Detailed explanations are given in *Lamy et al.* (2004) and *Fernández et al.* (2005); the key result of relevance here is that it yields an effective nucleus radius. This is best constrained with concurrent visible wavelength observations, but single-wavelength thermal IR observations yield small enough uncertainties (*Fernández et al.,* 2013; *Bauer et al.,* 2017) that they are generally considered reliable.

Together, these methods have revealed a large diversity of comet sizes, ranging from effective radii of a few hundred meters up to tens of kilometers (see Table 1 and Fig. 1). The largest known nucleus is C/2014 UN271 (Bernardinelli-Bernstein) [r_n ~60 km (*Hui et al.,* 2022a; *Lellouch et al.,* 2022)], with C/2002 VQ94 and Hale-Bopp also having $r_n \gtrsim$ 30 km. As noted earlier, we exclude active Centaurs like Chiron from consideration. Based on the existence of objects with radius of approximately hundreds of kilometers in the Centaur (*Stansberry et al.,* 2008) and scattered-disk-object (SDO) populations (*Müller et al.,* 2020), it is conceivable that comparably large comet nuclei are present in the Oort cloud but have not yet been discovered due to their lower abundance. Large ($r_n \gtrsim$ 10 km) ACOs also exist, including (3552) Don Quixote, which has shown recurrent activity (*Mommert et al.,* 2020). It is noteworthy that the debiased average size of LPCs has been found to be 1.6 times larger than the debiased average JFC nucleus size (*Bauer et al.,* 2017) (see the next subsection).

3.1.2. Size-frequency distribution. Collecting the r_n of a large sample of objects and studying their cumulative size-frequency distribution (SFD) provides an invaluable tool to probe the formation and subsequent evolution of comets. The SFDs of minor-planet populations in the solar system are expressed as a power law of the form

$$N(> r_n) \propto r_n^{-a} \quad (5)$$

where N is the number of objects with radius larger than r_n. According to analytical models, collisionally relaxed populations of self-similar bodies with identical physical parameters have a power-law SFD with a = 2.5 (*Dohnanyi,* 1969). In contrast, a shallower slope, a = 2.04 is predicted for collisionally relaxed populations of strengthless bodies (*O'Brien and Greenberg,* 2003). Observations of asteroids reveal that their SFD shows characteristic "breaks," or changes in the slope of the power-law distribution, which can be used to probe the material strength and the population evolutionary processes (e.g., *O'Brien and Greenberg,* 2005; *Bottke et al.,* 2005).

Traditionally comets are presumed to have had a very different evolutionary history than the asteroid belt (e.g., they are not expected to have reached a collisionally steady state). Moreover, they undergo sublimation-driven mass loss, which is expected to result in significant differences between the SFDs of cometary populations and their source populations. However, recent work reviewed in *Weissman et al.* (2020) and *Morbidelli et al.* (2021) suggests that the source population of comets and TNOs in the primordial transneptunian disk may have evolved to reach collisional equilibrium prior to being dispersed. While the debate about this possibility remains open, some of the deciding evidence may come from a better understanding of the small-end of the SFD of today's remnants from the primordial transplanetary disk and specifi-

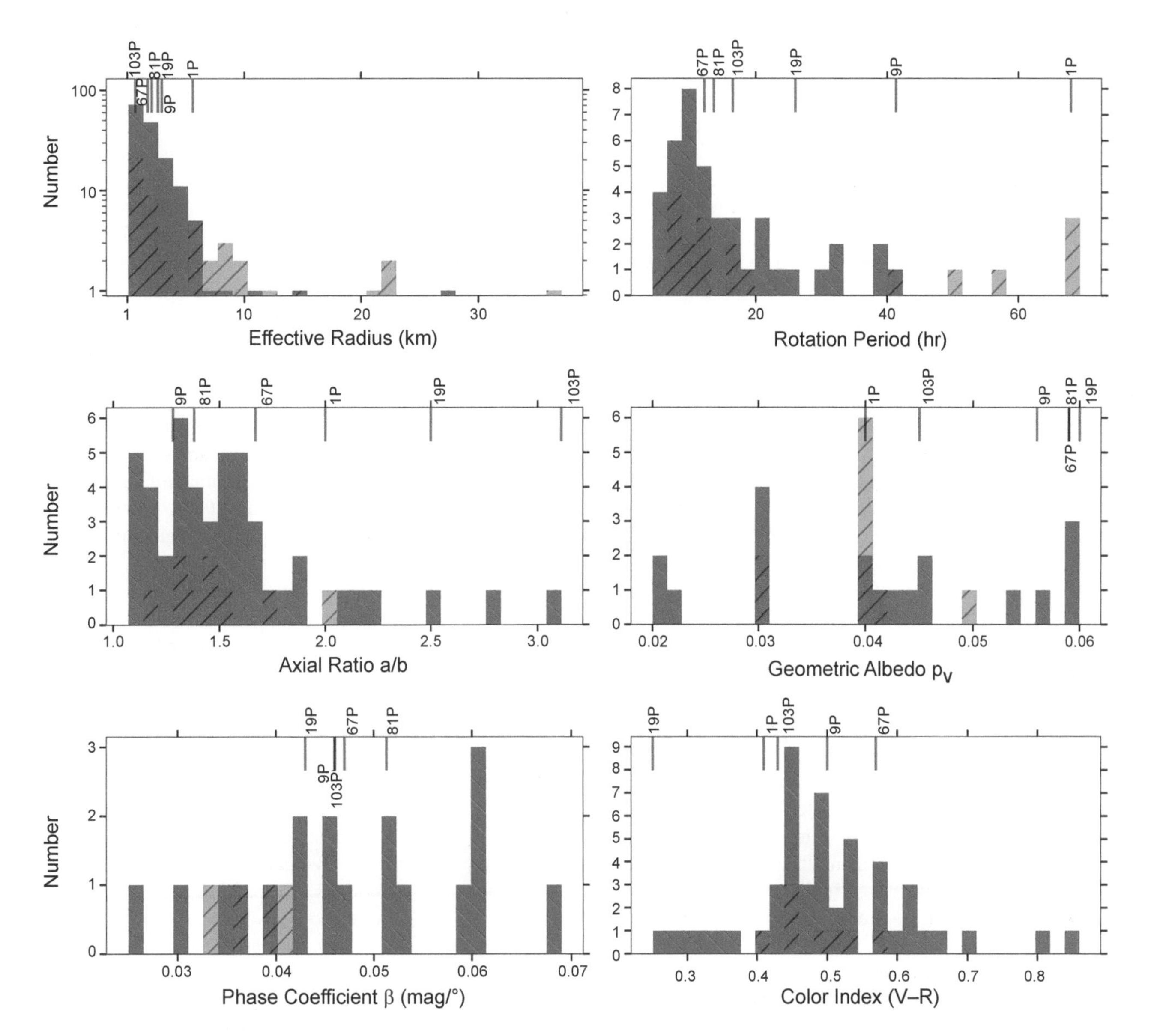

Fig. 1. Histogram of the measured nucleus properties for JFCs (blue backward hashmarks) and HTCs plus LPCs (green forward hashmarks). The values for comets with *in situ* spacecraft measurements are indicated by vertical lines at the top axis. This figure aims to provide an overview of each parameter's range and the sample size for which it has been measured. The property being plotted is given in the abscissa label for each plot. There are different numbers of comets in each sample because only a few comets have had all properties measured. As discussed in the text, the effective radius values are taken only from the thermal-IR studies of *Fernández et al.* (2013) and *Bauer et al.* (2017), while all other quantities represent our best effort to include all published values. In the cases when a comet had multiple measurements of a certain property, we have displayed the most recent sufficiently precise measurement, therefore only plotting a given comet's property once. Note that most axial ratios plotted are lower limits. See the text for additional details.

cally from comets. It is therefore informative to examine the observational evidence on the SFDs of short-period comets (SPCs) and LPCs in an attempt to distinguish the signatures of recent activity-driven evolution from those of planetesimal formation and/or early collisional history.

Earlier attempts to derive the SFD of JFCs from optical observations resulted in slightly different slopes, a (e.g., *Lowry et al.*, 2003; *Lamy et al.*, 2004; *Meech et al.*, 2004; *Tancredi et al.*, 2006; *Weiler et al.*, 2011). This highlighted the need to assess the uncertainty of the power-law slope determination by assessing the contribution of the various assumptions of the size estimates (e.g., on the albedo, phase function, and shape of the nucleus, as well as photometric uncertainty). This was addressed by *Snodgrass et al.* (2011) and yielded a SFD with a = 1.92 ± 0.20 for JFCs with $r_n \geq 1.25$ km. This result is comparable to the SFD slopes determined for JFCs from thermal-IR observations [1.92 ± 0.23 (*Fernández et al.*, 2013)]; both are consistent with the expected slope for a collisionally relaxed population of strengthless bodies (*O'Brien and Greenberg*, 2003), although as just discussed, the implications of this are not yet settled.

These studies, however, did not take into account the observational biases influencing the SFD, the most prominent being the bias against detecting small comet nuclei. *Bauer et al.* (2017) therefore performed careful debiasing of the NEOWISE comet size estimates and determined a slope a = 1.0 ± 0.1 for LPCs and a steeper slope, a = 2.3 ± 0.2, for JFCs. This analysis, however, has left a few prominent questions unresolved. Theory suggests that there is an underabundance of subkilometer comets (e.g., *Samarasinha,* 2007; *Jewitt,* 2021); it is important for future observations to determine whether there is a paucity of small JFCs and what it reveals about comet disruption and the primordial SFD of outer solar system planetesimals (*Fernández et al.,* 2013; *Bauer et al.,* 2017). Other features, such as the small bump in the SFD of JFCs between 3 and 6 km (*Fernández et al.,* 2013), also remain to be confirmed and explained in the context of planetesimal formation or sublimation evolution (*Kokotanekova et al.,* 2018). For more details about these debates, we refer the reader to the chapter by Bauer et al. in this volume, where the details of the telescope surveys used to derive the comet SFD are covered, and to the chapter by Fraser et al., where the comet size distribution is discussed in the context of other outer solar system populations.

3.1.3. Nucleus shapes. Three main observational techniques can be employed to study the shapes of comet nuclei: rotational lightcurves, radar observations, and spacecraft data. The least complex and most easily available are rotational lightcurves in which the nucleus signal dominates the flux in the photometric aperture. Lightcurves in which coma flux dominates yield fundamentally different information and are discussed in section 3.3.

Observations that are frequent enough, ideally on the same night or over several consecutive nights, can be combined to create a lightcurve in which the magnitude is plotted as a function of time or rotational phase (if known). Lower limits to the projected nucleus axial ratio can be determined from the the peak-to-trough amplitude ($\Delta m = m_{min} - m_{max}$) as

$$\frac{a}{b} \geq 10^{0.4(m_{min} - m_{max})} \tag{6}$$

where a and b are the semi-long and and semi-intermediate axes in a triaxial ellipsoid in simple rotation, and m_{min} and m_{max} are the magnitudes at lightcurve minimum and maximum, respectively.

Broadly available for a large number of comet nuclei, Δm allows the study of a large sample of comets (e.g., *Lamy et al.,* 2004) under the assumption that they are approximately triaxial ellipsoids. *Kokotanekova et al.* (2017) updated the sample of well-constrained JFC lightcurves and estimated a median axial ratio of a/b = 1.5, in agreement with the previous estimate from *Lamy et al.* (2004). We show in Fig. 1 an updated version that includes all known comet axial ratios; these are also tabulated in Table 1. The known range extends from 1.07 to 3.11 for the elongated nucleus of Comet 103P. The axial ratios tabulated in Table 1 have a mean of 1.55 ± 0.40 and a median of 1.45. Similar results are obtained when considering only JFCs vs. HTCs plus LPCs.

It is important to keep in mind that the projected axial ratio derived from rotational lightcurves is just a lower limit unless the spin axis is normal to the observer's line of sight. If the lightcurve is observed at an unfavorable geometry or when the nucleus is surrounded by an undetected coma, the actual nucleus elongation can be significantly underestimated. This limitation becomes evident when the axial ratios of comets observed *in situ* by spacecraft are compared to the total population average. Most spacecraft targets have axial ratios (≥1.5), with Comets 103P and 19P reaching some of the largest a/b of 3.1 and 2.5 (Fig. 1).

Rotational lightcurves taken at a wide variety of different observing geometries can also be analyzed using the convex lightcurve inversion (CLI) technique (*Kaasalainen and Torppa,* 2001; *Kaasalainen et al.,* 2001). This technique has been successfully applied to produce shape models for thousands of asteroids (*Durech et al.,* 2010); a few comets have thus far been modeled, including 67P (*Lowry et al.,* 2012), 162P (*Donaldson et al.,* 2023), and 323P/SOHO (*Hui et al.,* 2022b) as well as Don Quixote (*Mommert et al.,* 2020). This method is limited to producing convex shape models and cannot recreate the concavities now known to be characteristic for comet nuclei (see below). Despite this limitation, CLI can accurately determine the object's pole orientation and axial ratio with great precision. Another challenge posed by this method is that it requires a lot of observing time on comparatively large (>2-m) telescopes, given the faintness of bare comet nuclei. Additionally, only a small number of comets are inactive at large portions of their orbits, limiting the possibility to probe different observing geometries. However, an increasing number of bare nuclei have well-observed rotational lightcurves collected mainly to study changes in their rotation rates (see section 3.3 below). The addition of data from future all-sky surveys (see section 4) may enable the shapes of additional comets to be constrained. Moreover, large flat surfaces on the asteroid convex shape models can be used to infer the existence of concavities (*Devogèle et al.,* 2015). Combined with large elongations, this could reveal more contact binaries among the known JFC population and has the potential to improve our understanding of the binary fraction among comet nuclei (see below).

On the rare occasions when a comet passes sufficiently close to Earth, radar delay-Doppler imaging can be performed to constrain the shapes of comet nuclei (see section 2.2). At the time of *Harmon et al.* (2004)'s writing, nine comets had Doppler-only detection with radar and none had delay-Doppler imaging. Thanks to technological improvements and a confluence of close-approaching comets, at least eight now have delay-Doppler imaging of their nucleus: 300P/Catalina (*Harmon et al.,* 2006), 73P/Schwassmann-Wachmann 3 fragments B and C (*Nolan et al.,* 2006), 8P/Tuttle (*Harmon et al.,* 2010), 103P (*Harmon et al.,* 2011), 209P/LINEAR (*Howell et al.,* 2014), 460P/PANSTARRS (2016 BA14) (*Naidu et al.,* 2016), and 45P/Honda-Mrkos-Pajdušáková (*Lejoly and Howell,* 2017). Doppler-only detections since *Harmon et al.* (2004) include 252P, 289P/Blanpain, 41P/Tuttle-Giacobini-Kresák (*Howell et al.,* 2017), and 46P/Wirtanen (*Lejoly et al.,* 2019). As evidenced by these studies, high-resolution radar imaging is occasionally possible. In such cases, precise shape modeling can be achieved by combining radar observations with optical lightcurve modeling (see *Ostro et al.,* 2002).

Finally, owing to the space missions equipped with onboard cameras (see the chapter by Snodgrass et al. in this volume), the shapes of six comets have been studied in great detail (1P, 9P, 19P, 67P, 81P, and 103P). Four were found to be highly elongated or potentially bilobed: 1P (*Keller et al.,* 1986), 19P (*Britt et al.,* 2004; *Oberst et al.,* 2004), 103P (*Thomas et al.,* 2013b), and 67P (*Sierks et al.,* 2015). Additionally, radar observations of Comet 8P were consistent with a contact binary shape (*Harmon et al.,* 2010).

The progress in characterizing comet nucleus shapes in the last decade revealed a striking overabundance of highly-elongated/bilobate objects in comparison to other minor planets. The comparison with other populations is somewhat complicated by the different definitions used by the different communities. Works focusing on near-Earth asteroids (NEAs) often use a strict contact binary definition that sets a limit on the components' mass ratio and implies that the objects might have been separate in the past (see *Benner et al.,* 2015). TNO studies, on the other hand, are now making the first steps in understanding the shapes of individual objects and use a less-restrictive definition (e.g., *Thirouin and Sheppard,* 2019). It is nevertheless informative to outline the contrasting findings for the different populations. Out of the six comets visited by spacecraft, four are highly-elongated/bilobed. If 8P is also considered, this sample of well-constrained comet shapes contains more than two-thirds highly-elongated/bilobed shapes. In comparison, only 14% of the almost 200 radar-imaged NEAs are bilobate (*Taylor and Margot,* 2011; *Benner et al.,* 2015). The large abundance of bilobate shapes cannot be traced to the Centaur region where no contact binary has been identified (*Peixinho et al.,* 2020), while the contact-binary fraction among TNOs is estimated as 10–25% for cold classicals or up to 50% for Plutinos (*Thirouin and Sheppard,* 2018, 2019). However, recently *Showalter et al.* (2021) presented evidence that the contact-binary fraction in the Kuiper belt can even be higher if the shapes and the directional distribution of the objects' rotation poles are accounted for.

The unusually large abundance of highly-elongated/bilobate objects among comets prompted a number of works to investigate which combination of the comet's physical properties and/or evolutionary processes have led to the formation of a large number of contact binaries. This motivation was further enhanced by the finding that New Horizons' target in the cold classical Kuiper belt, Arrokoth, is also a contact binary (*Stern et al.,* 2019). Arrokoth's shape is consistent with formation by merger of a collapsed binary system (*McKinnon et al.,* 2020). However, unlike cold classical KBOs, the progenitors of JFCs have undergone giant-planet encounters and possibly a significant collisional evolution (*Morbidelli and Nesvorný,* 2020), which have most likely destroyed any distant binary systems early on. Instead, the formation of bilobate comet nuclei is better explained by the reaccretion of material ejected from catastrophic collisions in the early solar system (*Jutzi et al.,* 2017; *Schwartz et al.,* 2018) or even from multiple fission and reconfiguration cycles (*Hirabayashi et al.,* 2016). Alternatively, modeling work by *Safrit et al.* (2021) shows that bilobate shapes can also form after comet nuclei experience rotational disruption caused by sublimation-driven torques soon after the onset of activity (e.g., during the Centaur phase). As shown by *Zhao et al.* (2021), Arrokoth and by extension other icy bodies as well could have evolved their shapes through sublimation to produce objects with enhanced elongated shapes if the conditions were right. It is remarkable that these scenarios not only reproduce the highly-elongated/bilobate shapes of comet nuclei but can also preserve the cometary physical properties and volatile content (*Schwartz et al.,* 2018).

3.2. Cumulative Surface Properties of the Comet Population

The merit in exploring the cumulative surface properties lies in its potential to reveal dependencies between the comets' surface properties and the physical properties or orbital characteristics, which could, in turn, shed light on cometary evolution. On the other hand, comparing the bulk properties of comet nuclei with other minor planet populations could be used to establish the dynamical and evolutionary links among the diverse small-body populations in the solar system.

3.2.1. Albedo. The energy balance on the surface of a comet can be described using several photometric properties. The most frequently constrained is the geometric albedo p_λ for a given wavelength λ, defined as the ratio between the disk-integrated reflectance at opposition and that of a perfectly reflective flat disk with the same size. If the phase function of the object at wavelength λ is denoted by $\Phi_\lambda(\alpha)$, where α is the phase angle, and observations cover a large phase-angle range [typically covering a phase angle range of 70° and above (*Verbiscer and Veverka,* 1988)], its phase integral q_λ can be derived from:

$$q_\lambda = 2\int \Phi_\lambda(\alpha) \sin(\alpha)\, d\alpha \quad (7)$$

In such cases, the spherical albedo at wavelength λ, A_λ (sometimes referred to as the Bond albedo), can be calculated as

$$A_\lambda = p_\lambda q_\lambda \quad (8)$$

Physically, this is the fraction of power of the incident radiation that is scattered back into space over all angles and wavelengths. See *Hanner et al.* (1981) for further discussion of terminology.

The phase integral has been constrained by *in situ* spacecraft observations of five JFCs: 19P (*Li et al.,* 2007b), 9P (*Li et al.,* 2007a), 81P (*Li et al.,* 2009), 103P (*Li et al.,* 2013), and 67P (*Ciarniello et al.,* 2015). Besides 81P, which has an exceptionally low phase integral of 0.16, the other JFCs have small phase integrals in the range 0.2–0.3, similar to small low-albedo asteroids (*Verbiscer et al.,* 2019).

As indicated by equation (2), determination of reliable nucleus sizes requires knowledge of the geometric albedo. Albedo can be determined by coupling simultaneous observations of sunlight reflected by the nucleus with observations that depend on the nucleus size but *not* its ability to reflect sunlight (most commonly via thermal re-radiation in the mid-IR). The geometric albedo can be determined using ground- and spacebased telescopes (see *Lamy et al.,* 2004, for details) and therefore the sample of comet nuclei with well-constrained albedo in the visible range is comparatively large. The geometric albedos of JFCs were most recently reviewed by *Snodgrass et al.* (2011) and *Kokotanekova et al.* (2017), while the HTCs and LPCs were last summarized in *Lamy et al.* (2004).

As seen in Table 2 and Fig. 1, this sample of 29 comet nuclei (19 JFCs and 10 HTCs plus LPCs) contains V-band albedos (p_V) between 0.02 and 0.06. Even though the absolute range is small, albedos vary by a factor of ~3 from darkest to brightest. This distribution clearly identifies comets as some of the darkest objects in the solar system. The unweighted average albedo of the whole sample of comets is 0.040 ± 0.011 (median 0.040). Subdividing into JFCs or HTC/LPC yields virtually identical values. Thus, the common practice of assuming p = 0.04 when albedo is unknown (cf. *Lamy et al.,* 2004) continues to be reasonable. Since $p_\lambda \sim 0.04$ and $q_\lambda \sim 0.2$–0.3, this means that $A_\lambda \sim 0.01$, and thus ~99% of the incident solar energy is absorbed by the comet.

Interestingly, as pointed out by *Kokotanekova et al.* (2018), the objects with the largest geometric albedos are all comets with albedo estimates from spacecraft observations (9P, 19P, 67P, 81P). On the other hand, the darkest surfaces belong mostly to the largest (*Fernandez et al.,* 2016; *Kokotanekova et al.,* 2017) and least-active comets whose nuclei are easiest to characterize with telescope observations (*Kokotanekova et al.,* 2018). The latter result may be an observational selection effect since small and dark objects will be more difficult to discover, but the former almost certainly is not — spacecraft targets were primarily selected for orbital accessibility.

As compared to other populations, comet nuclei span a narrow range of geometric albedos. The average albedo of comet nuclei is comparable to that of C-, D-, and P-type asteroids (*Mainzer et al.,* 2011), but unlike the asteroids of the respective classes, the comet population lacks objects with larger albedos. The consistently low albedo of comet nuclei has been utilized as a criterion to distinguish extinct comets from asteroids in the ACO populations identified by using dynamical criteria (*Fernández et al.,* 2001; *Licandro et al.,* 2016). Centaurs and scattered disk objects, thought to be the progenitors of current SPCs, contain objects with albedos as low as those of comet nuclei. However, the mean geometric albedo of these populations is somewhat higher (0.056 for Centaurs and 0.057 for SDOs) and they contain objects with albedos of up to 0.25. The known high-albedo surfaces of Centaurs and TNOs are also associated with redder colors (larger spectral slopes in the visible) and form the "bright red" surface type with $18 < S' < 58\%/100$ nm (S′ is the normalized reflectivity gradient; see section 3.2.3), and albedo >0.06 that is thought to disappear with the onset of Centaur activity (see *Jewitt,* 2015). On the other hand, the "dark gray" Centaurs have albedos similar to these of Jupiter Trojans and Hildas (*Romanishin and Tegler,* 2018), which has been interpreted as evidence of the common origins of these populations and consequently between comet nuclei, Jupiter Trojans, and Hildas.

An interesting question that has been difficult to answer broadly from telescope observations is whether comet surfaces have large-scale albedo variations. Synchronous visible/IR lightcurves [e.g., 10P by *A'Hearn et al.* (1989)] suggest they do not, but these observations can only be accomplished for a small number of low-activity nuclei. It was, however, possible to search for surface areas with different albedos on the comets visited by spacecraft. While the brightness variations on the surface of 19P have been found to be up to a factor of 2 (*Buratti et al.,* 2004; *Li et al.,* 2007b), they are thought to be caused by surface roughness variations rather than differences in the albedo. All other comets that have been measured have smaller albedo variations, $\lesssim 16\%$; 81P (*Li et al.,* 2009), 9P (*Li et al.,* 2007a), 103P (*Li et al.,* 2013), and 67P (*Fornasier et al.,* 2015).

3.2.2. Phase function and phase reddening. The observed spectrophotometric properties of a reflecting surface are known to change depending on the phase angle, α, of the observations. The spectral slope (as a function of wavelength) increase with α is referred to as phase reddening while the phase darkening, or phase function, describes the decrease of an object's brightness with increasing α. Moreover, at small phase angles $\alpha \leq 10°$, the phase function can undergo a sharp nonlinear increase, known as the opposition effect (e.g., *Gehrels,* 1956).

These patterns are characteristic for airless regolith surfaces of solar system bodies. They can be analyzed using physically motivated photometric models, which relate the reflectance changes with the varying geometry and the properties of the object's surface layer. The currently most widely used models follow the Hapke formalism (e.g., *Hapke,* 1981) and provide an opportunity to characterize the surface properties on small scales below the resolution of the observing instruments available on telescopes and most spacecraft. For example, careful modeling of the opposition effect or of the phase function beyond 90° can be used to derive properties of the surface regolith at small scales. The moderate angle phase function, on the other hand, can be used to study the surface roughness and topography (see *Verbiscer et al.,* 2013). In many cases, however, remote telescope observations of minor planets provide insufficient coverage to constrain the analytic photometric models. Instead, the phase function is approximated by a simpler parametric formalism, such as the International Astronomical Union (IAU) H-G phase function (*Bowell et al.,* 1989), which generally works well for asteroids.

The phase functions of comet nuclei present an observational challenge because they require relatively large telescopes that are able to detect the bare nuclei and a substantial amount of observing time in order to characterize the rotation rate and correct for rotational variability. Despite the diversity of comet orbits, in practice, all nucleus phase functions observed from the ground span a narrow phase angle range of less than 20° (see *Kokotanekova et al.,* 2018) with the exception of 2P [2.5°–177° (*Fernández et al.,* 2000)]. None of the nucleus phase functions (reviewed in *Lamy et al.,* 2004; *Snodgrass et al.,* 2011; *Kokotanekova et al.,* 2017) provide evidence for the presence of an opposition effect and are in excellent agreement with linear phase function fits, hence its use in equation (3). Note that a different, non-linear phase function is used to characterize coma dust [e.g., Schleicher-Marcus model (*Schleicher and Bair,* 2011)].

In situ observations during the flyby missions covered a broad phase-angle range but were insufficient to model the phase function close to opposition (*Li et al.,* 2007a,b, 2009, 2013). Rosetta's rendezvous with 67P was the first opportunity to observe a comet nucleus close to opposition. Even though the phase function between 0.5° and 12.5° of 67P from groundbased observations was well described by a linear function with a coefficient $\beta = 0.059 \pm 0.006$ (*Lowry et al.,* 2012), the disk-integrated phase function of 67P observed by OSIRIS between 1.3° and 54° exhibited a strong opposition effect and was better characterized by an IAU H-G model with $G = -0.13$ (*Fornasier et al.,* 2015). The shape of the phase curve during the opposition effect derived from disk-resolved observations of different nucleus areas also enabled studies of the structure and properties of 67P's surface (e.g., *Masoumzadeh et al.,* 2017; *Hasselmann et al.,* 2017).

Table 2 lists the phase coefficients of 24 comets (20 JFCs and 4 HTCs and LPCs) ranging from 0.025 to 0.07 mag/°. From this sample, the average phase coefficient for JFCs is 0.047 ± 0.012 mag/° (with median 0.046 mag/°). As shown in Fig. 1, the phase function slopes measured for LPCs and HTCs are smaller than the JFC average, although with the caveat that there are only five combined LPC plus HTC measurements. All phase coefficients determined from *in situ* measurements, presumed to provide the most reliable estimates, are larger than the commonly used phase coefficient of 0.04 mag/° (e.g., *Lamy et al.,* 2004). We therefore recommend an updated value of 0.047 mag/° to be assumed as an approximation when the phase coefficient of a nucleus is unknown.

Analyzing disk-resolved observations with the Visible and Infrared Thermal Imaging Spectrometer (VIRTIS) onboard Rosetta, *Longobardo et al.* (2017) found that terrains with larger roughness have steeper phase functions. At the same time, a comparison with the phase functions from previous missions indicated that nuclei known to have rougher surfaces (81P and 9P) have steeper phase functions than 103P and 19P, characterized by smoother surfaces. Interestingly, dynamical studies of these five comets by *Ip et al.* (2016) suggested that the comets with smoother surfaces and shallower phase functions have spent more time in the inner solar system and have experienced more sublimation-driven erosion.

The notion that the phase functions of comet nuclei can reveal the extent of their surface evolution was strengthened when ground observations of JFC nuclei were included in the sample. *Kokotanekova et al.* (2017) identified a possible correlation of increasing phase coefficient with increasing albedo for the 14 JFCs with both of these parameters constrained (updated version shown in Fig. 2). This was interpreted as a possible evolutionary path of JFCs where the least-evolved comet nuclei have bright surfaces with steep phase possibly associated with volatile-rich surface layers and topography dominated by deep pits and cliffs (see *Vincent et al.,* 2017; *Kokotanekova et al.,* 2018; *Vincent,* 2019). This hypothesis suggests that sublimation-driven evolution is responsible for gradually eroding their surfaces, which in turn decreases their albedos and phase coefficients. If confirmed by future observations, this idea provides a compelling possibility to characterize the evolutionary state of comets and to distinguish between extinct comets and objects of asteroidal origin (*Kokotanekova et al.,* 2018). Alternatively, it is also conceivable that better phase-angle coverage and therefore better-characterized nucleus phase functions in the future will challenge this hypothesis. For example, the phase coefficient of 162P was updated to a significantly larger value of 0.051 ± 0.002 mag/° (*Donaldson et al.,* 2023), which leads to a weaker Spearman rank correlation ($\rho = 0.43$) as compared to the sample in *Kokotanekova et al.* (2018). Additionally, the smallest known JFC phase coefficient, that of 28P, is also likely to be much larger, ~0.05 mag/° (*Schleicher et al.,* 2022).

Nevertheless, it is interesting to note that asteroids follow a reverse correlation of decreasing phase coefficient for increasing geometric albedo (*Belskaya and Shevchenko,* 2000). However, their dataset does not allow conclusive results for objects with albedo similar to the darkest comet nuclei (smaller than 0.05). Another comparison between comet nuclei and asteroids worth considering is the presence of an opposition effect. The opposition effect amplitude (defined as the magnitude difference between the phase function at 0.3° and the extrapolation of the linear part of the phase curve) is found to decrease for small albedos (*Belskaya and Shevchenko,* 2000). Moreover, some low-albedo Hilda asteroids and Jupiter Trojans do not show evidence for the presence of an opposition effect down to very small phase angles [0.1° (*Shevchenko et al.,* 2008)]. It is conjectured that the strong opposition effect found for 67P can be related to its comparatively large albedo. It is therefore essential to probe the phase functions of more comet nuclei of different albedos in order to verify how good of an approximation the linear phase function is.

Although attempts were made to determine the phase reddening of earlier flyby mission targets, they were not successful. The long duration of the Rosetta rendezvous allowed the first detections of phase-reddening of a nucleus in the visible and near-IR range (*Ciarniello et al.,* 2015; *Fornasier et al.,* 2015). The spectral slopes at different phase angles were analyzed for both disk-integrated and disk-resolved data and show a significant phase reddening in the 1.3°–54° phase angle range [from spectral slope S′ of 11%/100 nm to 16%/100 nm (*Fornasier et al.,* 2015)]. In addition to this, the phase reddening was observed to vary with the changing heliocentric distance, first decreasing toward perihelion and then increasing again in the outbound orbit (*Fornasier et al.,* 2016), indicating a significant seasonal variability.

Although the opposition effect and phase reddening have not yet been detected in groundbased data, both may be feasible to detect in the coming era (section 4), and future investigators should be mindful of the possibility when analyzing their data. As shown by recent findings, the spectrophotometric properties of comet nuclei can be used to probe not only the physical characteristics of the surface layers of comets but also their sublimation-driven evolution. This highlights the need for future photometric and spectroscopic observations of bare nuclei both at different geometries and heliocentric distances.

3.2.3. Color and spectra. Provided that sufficiently sensitive instruments are used, spectroscopy is the most desirable means of characterizing the nucleus surface as, in principle, it can reveal both precise color determinations and the presence of any absorption features due to surface ice or minerals. This compelling possibility motivated the early comet nucleus spectroscopy study by *Luu* (1993), which, however, revealed mostly featureless spectra. The absence of absorption features in the visible and near-IR has been confirmed by all subsequent spectroscopic observations (previously reviewed by, e.g., *Lamy et al.,* 2004; *Kelley et al.,* 2017; *Licandro et al.,* 2018). As a result, the normalized reflectivity gradient S′ (% per 100 nm) defined by *A'Hearn et al.* (1984) is the most commonly used parameter to characterize nucleus spectra. For visible wavelengths, reflectance is often normalized to 1 at some reference wavelength, frequently chosen to be 550 nm.

Although less informative than spectroscopy, the color index is determined from photometry by measuring the bare nucleus brightness in two filters and subtracting the magnitude at the longer wavelength from the shorter, e.g., V–R. Since photometric techniques are generally more sensitive than spectroscopy, this allows the colors of fainter nuclei to be measured with the caveat that they risk contamination by emission lines if unrecognized coma is present. This technique has been used to obtain the color indices of a wide variety of comet nuclei, building statistically significant samples covering the different classes of comets. A direct expression provided in *Luu and Jewitt* (1990b) allows an easy conversion between S′ and the color index in the corresponding wavelength range (e.g., V–R, B–R, R–I, g–r, etc.) and allows for comparison of the surface colors of comets observed with either spectroscopy or photometry to other solar system populations. An important confirmation of the utility of groundbased spectroscopy was the finding that the spectrum of 67P near aphelion (*Tubiana et al.,* 2011) was consistent with the visible and near-IR spectrum from the Optical, Spectrocopic and Infrared Remote Imaging System (OSIRIS) onboard Rosetta (*Fornasier et al.,* 2015).

In the visible to near-IR, comet surface spectra have spectral gradients up to 20% per 1000 Å (see *Campins et al.,* 2007; *Jewitt,* 2002; *Kelley et al.,* 2017). While some spectra have been classified as closer to T- and X-type asteroids (*DeMeo and Binzel,* 2008), most comet spectra have been identified to be closest to those of the primitive D-type asteroids, which have similar spectral slopes [9.1 ± 1.1% per 1000 Å (*Fitzsimmons et al.,* 1994; *Fornasier et al.,* 2007)] and to some moderately red Centaurs and TNOs (*Fornasier et al.,* 2009). In the near-IR, comet nuclei exhibit spectral diversity similar to that known for Jupiter Trojan asteroids (*Emery and Brown,* 2003; *Campins et al.,* 2007, and references therein). The similarity between comets and Jupiter Trojans extends to further spectral features such as the 10-μm emission plateau attributed to a surface layer of fine dust (*Kelley et al.,* 2017; *Licandro et al.,* 2018) and has been used to investigate the possible dynamical links between SPCs and Jupiter Trojans proposed by dynamical studies (e.g., *Morbidelli et al.,* 2005).

Since the existing comet surface spectra probe mainly weakly active or extinct comets (e.g., 28P, 49P, 162P, 249P/LINEAR, 364P/PANSTARRS, 196256 (2003 EH1), P/2006 HR30 Siding Spring), the spectra of these objects have been used to study the transition to dormancy. It is a notoriously difficult problem to distinguish inactive objects with a cometary origin in the outer solar system (ACOs) from asteroids using purely dynamical criteria (see *Fernández et al.,* 2005; *Tancredi,* 2014). Probing the spectra of ACOs is therefore considered to provide important clues on understanding the differences between these populations (*Licandro et al.,* 2018; *Simion et al.,* 2021). Additionally, increasing numbers of so-called "active asteroids" [objects on typically asteroidal orbits that exhibit cometary activity (cf. *Hsieh,* 2017)] and SPCs whose dynamics suggest an origin in the asteroid belt (*Fernández and Sosa,* 2015; *Hsieh et al.,* 2020) are being discovered. Considering this complexity, attributing any observed spectral signatures to the objects' origin and evolution can be enhanced if combined with dynamical studies. This approach was assumed in the recent work by *Kareta et al.* (2021), who compared the reflectance spectra of three different dormant comets and inactive solar system small body 2003 EH1 in an attempt to study the diversity among the spectra of dormant comets

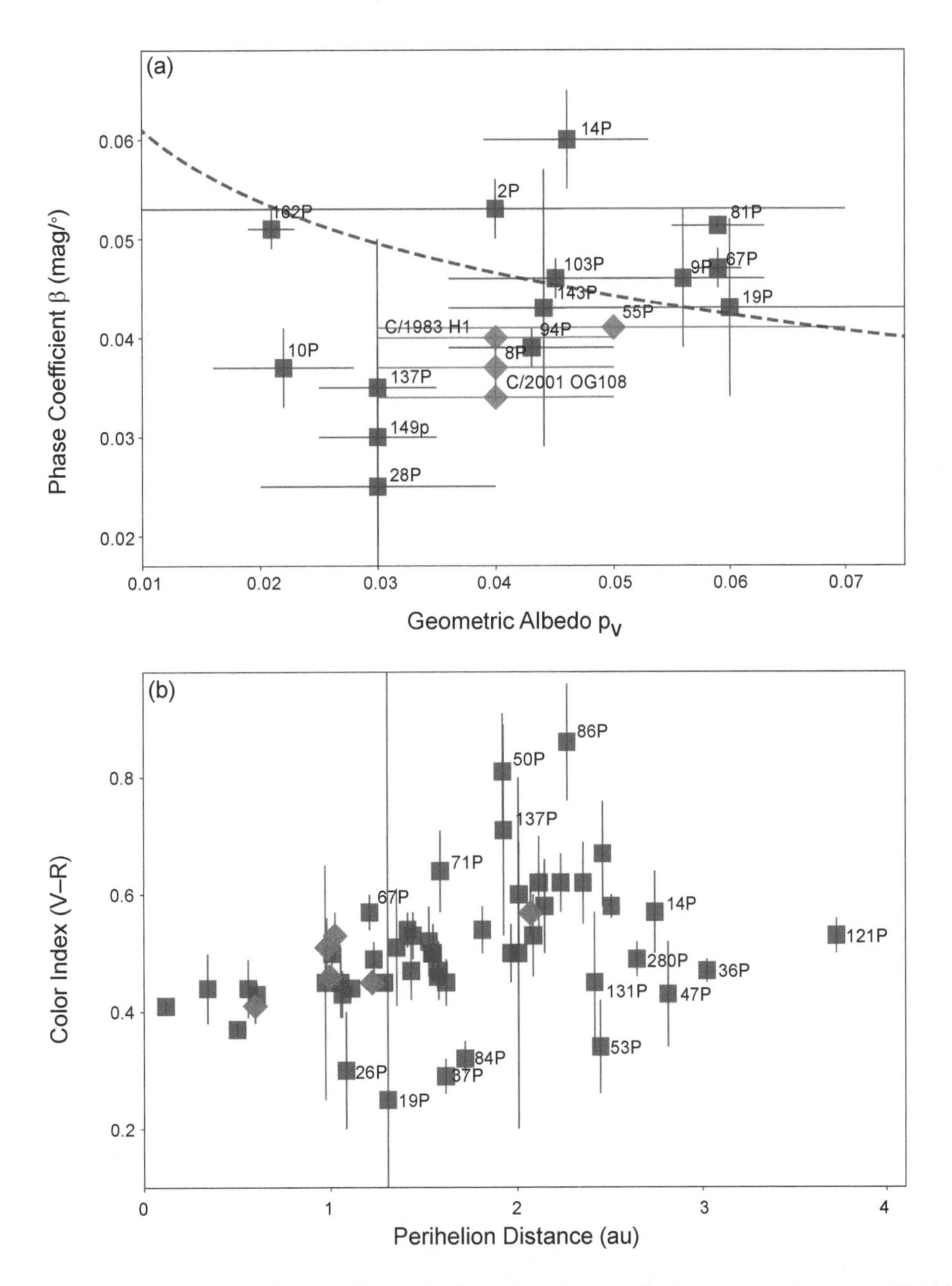

Fig. 2. Plots of the two highest correlations we found in the datasets compiled here: **(a)** phase coefficient as a function of geometric albedo and **(b)** (V–R) color vs. perihelion distance. The blue squares are JFCs and the green diamonds are HTCs and LPCs. The purple dashed line in the top panel is the correlation between phase coefficient and albedo for asteroids (*Belskaya and Shevchenko,* 2000). A recent update of 162P's phase coefficient to a higher value has made the trend with albedo first identified in *Kokotanekova et al.* (2017) less pronounced. As identified by *Lamy and Toth* (2009) and discussed further in section 3.2.3, there is a group of 10 comets that run parallel to but below the main cluster in the color vs. perihelion distance plot.

and objects identified as meteor shower sources.

Previously, the surface colors of comet nuclei were reviewed by *Lamy et al.* (2004), *Lamy and Toth* (2009), and *Solontoi et al.* (2012); we have updated these reviews with subsequently published colors in Table 2 and Fig. 1. *Jewitt* (2015) calculated the average color indices B–V, V–R, R–I, and B–R, distinguishing between nucleus and coma observations of JFCs and LPCs, and comparing them to the different TNO populations, Centaurs, Jupiter Trojans, and damocloids. This study reinforced previous findings that the average colors of the coma of JFCs and LPCs and Centaurs are indistinguishable from the nucleus colors within the uncertainties. Additionally, the comparison between the nucleus colors of JFCs, LPCs, damocloids, and Trojans are indistinguishable within the uncertainties for each class (*Jewitt,* 2015).

One of the main motivations behind comparing the surface colors of comet nuclei with Centaurs and TNOs is understanding the absence of the so called "ultrared" material on comet nuclei (*Jewitt,* 2002). While a significant fraction of Centaurs and TNOs contain the very red surfaces of spectral slopes S′ > 25% (*Lacerda et al.,* 2014, and references therein) and B–R > 1.5 (see *Tegler et al.,* 2008, 2016; *Peixinho et al.,* 2012; *Fraser and Brown,* 2012; *Wong and Brown,* 2017) alongside more neutral surfaces, none of the observed comet nuclei exhibit such extreme surface colors [with an average B–R of 1.22 ± 0.03 and 1.37 ± 0.08 for LPC and JFC nuclei, respectively (*Jewitt,* 2015)]. According to *Jewitt* (2015), the disappearance of the ultrared matter coincides with a perihelion distance of ~10 au where Centaur activity is observed to begin and is hypothesized to be due to blanketing of the primordially red material by fallback material ejected during the onset of activity.

Despite the effort to link surface color to different orbital and nucleus properties (perihelion distance, inclination, nucleus size, and active fraction) of comets, no statistically significant correlations have been found to date (cf. *Lamy and Toth,* 2009; *Jewitt,* 2015). We investigated correlations between our expanded collection of nucleus properties and various dynamical parameters (perihelion distance, aphelion distance, semimajor axis, eccentricity, inclination, and T_J). The only relationship evincing some correlation is color with perihelion distance (plotted in Fig. 2), with bluer nuclei having smaller perihelion distances (and a coupled trend of higher-eccentricity JFCs having bluer nuclei), mimicking the behavior seen in the asteroid belt (e.g., *Gradie and Tedesco,* 1982) and among near-Earth asteroids (e.g., *Marchi et al.,* 2006). Previously noticed by *Lamy and Toth* (2009), this apparent trend is not statistically significant (Spearman rank 0.47 for JFCs or 0.44 for all comets in our sample). However, similarly to *Lamy and Toth* (2009), we identify two distinct groups in the (V–R) vs. q distribution, with the main group consisting of 39 JFCs. Considering only comets from this group results in a Spearman rank of 0.8, suggestive of a moderate to high correlation. Notably, all HTCs from the table also agree with the observed trend, while both LPCs have perihelia much larger than 3 au and do not follow the observed trend. The second group consists of 10 JFCs, which are bluer than the other group and follow a similar pattern of increasing redness with increasing heliocentric distance. We hypothesize that the existence of the second group can be connected with recent orbital changes, but interpreting these results is beyond the scope of this chapter.

As discussed briefly at the beginning of section 3, we found no statistically significant correlations between surface color and the geometric albedo or phase coefficient for the JFCs. Owing to the paucity of polarization data (see section 3.2.4), its correlation with other surface properties remains unexplored. The above links should be explored further with the precise color determinations expected from all-sky surveys such as the Vera Rubin Observatory's Legacy Survey of Space and Time (LSST).

3.2.4. Polarization. Polarimetry is the study of the degree of polarization of light. Light becomes polarized — its electromagnetic wave having a preferred plane of oscillation — by scattering off of particles. The degree of linear polarization can be easily calculated (see *Kolokolova et al.,* 2004, for details), and its variation reveals information about the particles from which it scattered. The degree of linear polarization of airless solar system bodies changes with phase angle and wavelength, producing a phase-polarization curve with a negative branch between 0° and 20° and a positive branch for higher phase angles that peaks in the range 90°–100° (cf. *Kolokolova et al.,* 2004). Characterizing the phase-polarization curve of comet nuclei is limited by the faintness and the narrow phase angle range when comets are sufficiently far from the Sun to be inactive. Although polarimetric observations of comets date to the early 1800s (reviewed by *Kiselev et al.,* 2015), it has only been applied to comet nuclei since 2004.

The first polarization studies of a comet nucleus were by *Jewitt* (2004a), who studied 2P at phase angles of 22° as well as 93°–100°. Later, *Boehnhardt et al.* (2008) observed the coma-free nucleus of 2P between 2.7 and 2.1 au, constraining its linear phase-polarization curve between 4° and 28° in the R- and V-band. In this range, the polarization in both bands increases linearly with increasing phase angle, suggesting an inversion angle ≤4°. Similar low values of the inversion angle are characteristic for F-type asteroids (*Belskaya et al.,* 2005) and have been found for active asteroid 107P/Wilson-Harrington (see *Belskaya et al.,* 2019) and main-belt comet 133P/Elst-Pizarro (*Bagnulo et al.,* 2010).

Kuroda et al. (2015) observed the linear polarization of the highly anemic 209P at large phase angles from 92.2° to 99.5°. This was possible because 209P was very weakly active even at heliocentric distance ~1 au and the nucleus polarization could be determined separately from that of the coma. They found no significant difference in 209P's R- and J-band polarization and estimated that maximum polarization P_{max} = 30.8% is reached around the phase angles of the observations. Interestingly, the phase angle at which maximum polarization occurs coincides with that of asteroids, but the value of the maximum polarization of 209P is significantly higher. *Boehnhardt et al.* (2008) and *Kuroda et al.* (2015) explored whether existing empirical phase-polarization relationships for asteroids can be applied to comet nuclei. This subject remains to be explored further when the polarization of more comet nuclei is characterized in detail.

3.3. Spin State

3.3.1. Overview and explanation of terminology. The spin state, or the rotational state, of a comet fully describes the rotational properties of the nucleus. In this chapter, we use the terms "spin state" and "rotational state" interchangeably. A comet could be in (1) the least-rotational kinetic energy "simple rotation" around its maximum moment of inertia (the latter is physically identified with the short axis, or more precisely the principal short axis defined by the moment of inertia ellipsoid), or (2) in an excited rotational state. An "excited" rotational state, by definition, has more rotational kinetic energy for a given rotational angular momentum than the least-energy state described earlier.

A simple (i.e., a principal-axis) rotation requires three independent parameters and an initial condition to uniquely define it (the rotation period, two parameters defining the direction of the rotational angular momentum vector such as RA and Dec, and the orientation of a reference longitude of the nucleus at a specific time). In theory, a simple rotation could represent a rotation around any one of the three principal axes. However, a simple rotation around the intermediate principal axis or that around the long principal axis is extremely rare and both can be changed due to mechanical damping of energy. Thus, in this chapter, the term "simple rotation" means the least-kinetic-energy rotation around the short principal axis. The observational techniques for identifying and analyzing properties of simple rotations are discussed in the review by *Samarasinha et al.* (2004) in *Comets II* and thus not repeated here.

Except for the principal-axis (PA) rotational states around the intermediate principal axis or the long principal axis, a rotationally excited nucleus undergoes three concurrent component rotations, of which two are dynamically coupled, and such a spin state is known as an NPA rotation. In general, these component rotations occur at rates that vary with time but are periodic and therefore they can be expressed by time-averaged periods (e.g., *Landau and Lifshitz,* 1976).

To uniquely define an NPA spin state, one requires six independent parameters and two initial conditions. These specific parameters and other details of the NPA rotation are described in *Comets II*. The NPA rotational states are the most general spin states and PA spin states can be considered limiting cases. The vast majority of cometary nuclei are observationally determined to be PA rotators in the least-energy state; however, it is likely that many of them might be in slightly excited NPA states and groundbased observations are not sensitive enough to detect them [such as 67P (*Gutiérrez et al.,* 2016)].

3.3.2. Detection of non-principal-axis rotation. As remarked earlier, many cometary nuclei appear to be simple rotators. However, delving deeper into the characteristics of NPA rotation helps us comprehend the observations of cometary spin and spin evolution of comets coherently. As described in the previous subsection, an NPA spin state is extremely complicated. Now, consider the possible NPA states of a nucleus that is near but not exactly prolate. This nuclear shape is considered since different space mission images show many cometary nuclei are like that; lightcurves of nuclei, too, provide additional supporting evidence for this prevalent shape. Such a nucleus, if only slightly excited, is likely to be in a short axis mode (SAM), whereas a highly excited nucleus will be in a long axis mode (LAM) NPA state. The SAM and LAM NPA states are described in detail in *Samarasinha et al.* (2004) of *Comets II* (see their Fig. 1 for visualizations of the NPA rotations described here).

The near-prolate nucleus undergoing a SAM NPA rotation has its long axis (1) rotating around the rotational angular momentum vector at a time-averaged angle of 90°, (2) a roll back-and-forth motion around the long axis itself with a back-and-forth half-rolling angle between 0° and 90°, and (3) a low-amplitude (i.e., a few degree) nodding motion. For a distant observer, the resultant lightcurve would be similar to that of a simple rotator within the observational uncertainties. Furthermore, if coma morphology is due to source region(s) near the end of the long axis, the coma morphology too would mimic a simple rotation. Only if the source region is located on the waist of the nucleus and the half-rolling angle is $\gg 0°$ might the coma morphology hint at an NPA rotation.

Similarly, if the nucleus is in a LAM NPA state, the long axis has (1) a precessional motion around the rotational angular momentum vector and making an angle $<90°$, (2) a full rolling motion around itself, and (3) a low-amplitude (i.e., a few degrees) nodding motion. Again, the lightcurve may not be distinguishable from a simple rotator nor would the coma morphology caused by a source region(s) near the end of the long axis indicate an NPA rotation. However, a source region on the waist of the nucleus could point to an NPA state.

These examples point to the difficulty that observers face in conclusively identifying NPA rotators, especially if the signal-to-noise of the data is not high or there are not extensive data from multiple observing geometries. It would not be surprising if future observations indicate that some comets that we now consider to be simple rotators are in NPA spin states, or if discrepancies are found between simple/NPA spin states for different observational techniques.

3.3.3. Observational evidence for non-principal-axis spin and excitation into and damping of non-principal-axis spin. So far, only two comets — 1P (e.g., *Schleicher et al.,* 1990; *Belton et al.,* 1991; *Samarasinha and A'Hearn,* 1991) and 103P (e.g., *A'Hearn et al.,* 2011; *Samarasinha et al.,* 2011; *Knight and Schleicher,* 2011; *Drahus et al.,* 2011) — have been definitively shown to be in NPA rotation from remote observations. In both cases, extensive observations were required, a necessity that makes NPA rotation difficult to conclusively establish. Rosetta observations of 67P demonstrated that it is also excited, albeit at a level too low to detect from Earth (*Gutiérrez et al.,* 2016; *Kramer et al.,* 2019). Several other suspected NPA rotators are indicated by "b?" in Table 1. Possible NPA spin states were suggested to explain conflicting observations or observations that were deemed incompatible with simple rotations. Clearly additional observations and analyses are needed to confirm NPA states for these comets.

An NPA spin state has been proposed for active Centaur 29P/Schwassmann-Wachmann 1 with component periods on the order of a day (*Meech et al.,* 1993). However, recent modeling based on multiple disparate datasets suggests a much longer rotation period for that comet (e.g., *Miles,* 2016a; *Schambeau et al.,* 2017, 2019).

The fraction of periodic comets observed to be in NPA states is much smaller than what the theory suggests based on the timescale for rotation period change of a simple rotator, τ_P, due to outgassing torques (*Samarasinha et al.,* 1986;

Jewitt, 1997). This timescale is indeed a lower limit to the timescale for rotational excitation; it will also depend on the relative component moments of inertia of the nucleus and the components of the torque. The rotational excitation of a nucleus is a complicated process and *Gutiérrez et al.* (2003b) cite the following as a necessary (but not sufficient) condition for a nucleus to get rotationally excited and to remain that way

$$\frac{N_l}{N_s} > \frac{(I_s - I_i)\,|\Omega_s|}{(I_i - I_l)\,|\Omega_l|} \qquad (9)$$

where N, I, and Ω represent torque, moment of inertia, and angular velocity of the nucleus, respectively, whereas l, i, and s represent respective components along long, intermediate, and short principal axes. From this condition, it can be seen that as near-prolate nuclei have nearly equal I_s and I_i, such nuclei are comparatively easy to excite. Also, slow rotators (i.e., small Ω_s) are easier to excite than fast rotators (i.e., large Ω_s) and the former have comparatively small rotational kinetic energies. An explanation is presented in section 3.3.6 as to why small nuclei are comparatively easier to excite. The reader is referred to *Gutiérrez et al.* (2003b) for an indepth discussion on the process of rotational excitation.

Even among the periodic comet nuclei, which experience recurrent torques due to outgassing, only a very small fraction of comets are observed to be in NPA states. In addition to the observational biases against discovering NPA states discussed earlier, a nucleus may be prevented from being excited if the damping timescale is shorter than the excitation timescale. A flexible nucleus (e.g., rubble-pile-like) will have shorter damping timescales and most comets are likely to be structurally flexible objects with low tensile strengths. The loss of mechanical energy of a non-rigid body in an excited spin state causes it to ultimately rotate around the maximum moment of inertia. The corresponding damping timescale, τ_{damp}, was derived by *Burns and Safronov* (1973) to be $\tau_{damp} \propto r_n^{-2}\,\Omega^{-3}$, where r_n is the effective nuclear radius and Ω is the effective angular velocity. An immediate conclusion is large comets that are fast rotators have comparatively short damping timescales and are thus less likely to be in NPA spin states. *Samarasinha* (2008) discusses difficulties associated with deriving accurate damping timescales primarily due to our poor understanding of the values for structural parameters of comets. For example, for a 1-km-radius nucleus with an effective rotation period of 1 day, τ_{damp} can easily vary from $\sim 10^4$ to $\sim 10^{10}$ yr depending on the structural flexibility of the nucleus. Therefore, τ_{damp} could be on the same order as the orbital period for a few kilometer-sized fast rotating nuclei that is extremely flexible. On the other hand, τ_{damp} could be as large as the age of the solar system if the nucleus is comparatively more rigid. As most nuclei appear to be simple rotators (or close to that), it is likely that this is another pointer to the low rigidness of cometary nuclei.

3.3.4. Range of measured rotation periods. Among the rotational parameters that define the spin state of a comet, the rotation period of the nucleus is the parameter that has measurements for the largest number of comets, albeit a relatively small number when compared with the asteroids with known rotation periods. This paucity of data is primarily due to impediments caused by cometary comae when deriving nuclear lightcuves of comets close to us (see section 2).

Rotation periods can be determined from rotational lightcurves, coma morphology, or radar. As the latter two techniques were described in section 2, we only discuss lightcurves here. Ideally, the flux in the photometric aperture would be dominated by the nucleus, but techniques that measure the coma flux have been successful for some comets. Common techniques for determining the correct rotation period from a lightcurve include phase dispersion minimization [PDM (*Stellingwerf,* 1978)], Fourier transformations (e.g., *Belton,* 1990), and simply aligning the lightcurves by eye. When the correct rotation period has been identified, data phased to this period will have their features (peaks and troughs) from different rotation cycles aligned with each other. Although rotational lightcurves are ideally acquired close together in time so that the viewing geometry is nearly constant, recent work by *Kokotanekova et al.* (2017) has shown that bare nucleus data acquired at widely separated epochs can reliably be combined in some circumstances.

The range of measured rotation periods of unambiguous comets varies from approximately four hours to multiple days (see Table 1 and Fig. 1). Clearly this distribution is affected by observational selection effects. For example, to confidently derive a particular rotation period, typically an observational window longer than the rotation period is required, and therefore there is a bias against deriving long rotational periods. Similarly, groundbased lightcurve observations have a bias against detecting rotation periods that are multiples of diurnal observing windows (e.g., periods near 12 and 24 hr) since nearly the same portion of the lightcurve is seen from night to night at the same location. Rotation periods based on the repetition of coma morphology have a distinct disadvantage against detecting short rotation periods because of the higher likelihood of coma structures getting smeared. The smearing occurs when the spatial separation of repeating features is less than the image resolution. Short rotation periods and low outflow speeds (e.g., of dust when compared with that of gas) cause closely spaced repetitive features, whereas large geocentric distances cause low-spatial-resolution images.

In addition, there are a number of observational biases against measuring accurate rotation periods. They include lightcurve observations with small amplitude variability, either because of the intrinsic nuclear shapes or due to the specific observing geometries (for a nucleus lightcurves) or due to the diurnal activity pattern (for a coma lightcurve), and difficulties associated with detecting periods of NPA rotations discussed in the previous subsection.

Finally, different observational techniques could yield

different manifestations of nuclear rotation (e.g., a nuclear lightcurve indicating a double-peaked lightcurve vs. coma morphology showing a periodic coma structure originating from a particular location on the nucleus vs. radar observations of the nucleus) and care must be made to accurately interpret them and provide a consistent picture.

3.3.5. A comet spin barrier? Figure 3 plots rotation period vs. axial ratio. This can be used to derive a lower limit to the bulk density of the nucleus under the strengthless body assumption. *Lowry and Weissman* (2003) derived a lower limit to the density of ~600 kg m^{-3} from such a plot — corresponds to a minimum rotation period near 6 hr (also see *Weissman et al.,* 2004; *Lamy et al.,* 2004; *Kokotanekova et al.,* 2017) — a value consistent with direct density measurements for Comet 67P by the Rosetta mission (*Jorda et al.,* 2016). A review by *Groussin et al.* (2019) found a mean bulk density of 480 ± 220 kg m^{-3} from 20 published estimates using a variety of techniques (see also the chapter by Guilbert-Lepoutre et al. in this volume). A lower limit to the density estimated from a plot such as that in Fig. 3 is derived for an ensemble and is extremely sensitive to a small number of comets spinning at periods near their rotational breakup periods. Thus this estimate should be considered only as a gross ensemble property but not as a density estimate for a particular comet.

In contrast to this "minimum" rotation period of ~6 hr mentioned above for comets, the corresponding spin barrier for comparatively-sized asteroids occurs at ~2.2 hr (*Warner et al.,* 2009; also see *Hu et al.,* 2021). For comparatively-sized Jupiter Trojans the barrier occurs at ~4 hr (*Chang et al.,* 2021) due to the fact that their densities are higher than those of comets.

As noted earlier, there is a slight correlation between period and axial ratio. We suspect that this is an observational bias, due to the difficulty in detecting both low-amplitude lightcurves and long rotation periods. However, fast rotators could also be missed, regardless of amplitude, if observations are not frequent enough. Future analyses with more robust observational data, e.g., from LSST, should investigate this further, but observational biases should be accounted for prior to interpreting the results.

3.3.6. Evidence and implications of detected rotation period changes. The idea that the rotation of comets could be altered by sublimation of gases dates back to the seminal work of *Whipple* (1950) in which he introduced the "icy conglomerate" model of the nucleus. In *Whipple and Sekanina* (1979), this concept was further expanded to examine how the torque caused by the jet force due to sublimation can result in the precession of the rotation axis of a non-rigid body nucleus rotating around its maximum moment of inertia. However, the actual situation could be much more complex. The net torque due to outgassing could cause rotational excitation to an NPA spin state as well as changes to its period(s) of rotation, in addition to the forced precession of the direction of the rotational angular momentum vector (e.g., *Samarasinha et al.,* 2004, and references therein). As the rotation period has the largest spin parameter database, including the most number of determinations on their temporal changes, we now concentrate on the current state of known rotation period changes.

Rotation period changes require precise measurements and typically (but not always) observations spanning multiple apparitions, making them challenging to determine. Just over a dozen have been measured or meaningfully constrained (see Table 3), most since *Comets II.*

Samarasinha and Mueller (2013), based on four periodic comets, pointed out that the rotation period changes seem to be independent of the active fraction of a nucleus where active fraction is the fractional surface area of a solid water ice nucleus needed to reproduce the observed water production of the comet. They introduced a parameter X given by

$$X = \frac{|\delta P| r_n^2}{P^2 \zeta} \tag{10}$$

where P is the rotation period, $|\delta P|$ is the absolute value of the change of rotation period per orbit, r_n is the effective nuclear radius, and ζ is the total water production per unit surface area of the nucleus per orbit. It was found that X was nearly constant for all comets despite a large scatter in the range of active fractions. They suggest highly active comets undergo more cancellation of torques. *Mueller and Samarasinha* (2018) added two more comets and found X is still nearly constant. This near-constancy enables one to estimate $|\delta P|$ for various comets. *Steckloff and Samarasinha* (2018) proved that the sublimative torque model of *Steckloff and Jacobson* (2016) is consistent with the X parameter model.

In Table 3, X is confined to approximately an order of magnitude and the geometrical mean of $\frac{X}{X_E}$ of the 11 JFCs is ~2 where X_E is the value of X for Comet 2P/Encke. The observations cover only portions of the active orbit instead of the entire active orbit for C/1990 K1 (Levy) (*Schleicher et al.,* 1991; *Feldman et al.,* 1992) and C/2001 K5 (*Drahus and Waniak,* 2006); for the latter, only the rate of spin change at the middle of the observing window is quoted (instead of the net change), so there is insufficient information for inclusion in Table 3. Due to these reasons, we cannot properly estimate the X-parameters for these two comets. Upper limits exist for rotation period changes of three more JFCs — 14P/Wolf, 143P/Kowal-Mrkos, and 162P (*Kokotanekova et al.,* 2018). However, as they do not meaningfully constrain the X-parameter, they are not listed in Table 3.

Jewitt (2021) demonstrated that current observational data show that the active fraction, f_A, and the effective dimensionless moment arm corresponding to sublimative torques, k_T, approximately follow $f_A \propto r_n^{-2}$ and $k_T \propto r_n^2$ and therefore $f_A k_T$ is nearly constant. They also showed $f_A k_T$ is proportional to the X parameter of *Samarasinha and Mueller* (2013). They argued these relationships for f_A and k_T are affected by significant observational biases and a debiased sample would make the r_n dependencies shallower for both f_A and k_T. Therefore the net result would still be $f_A k_T$ (and therefore the X parameter) is nearly constant.

The most direct conclusion from this spin-evolution analysis is comets that have large active fractions show similar rotational evolution to comets with small active fractions. Comet nuclei have a range of active fractions spanning nearly three orders of magnitude (e.g., *A'Hearn et al.,* 1995); the near-constant X spans a factor of ~500 in active fraction (*Jewitt,* 2021). The exceptions appear to be comets with very low active fractions (≪0.1%), where the sublimative torques are apparently too weak to cause any spin evolution.

The rotation period change during an orbit is not necessarily a monotonic function of time [e.g., 9P (*Belton et al.,* 2011; *Chesley et al.,* 2013), 67P (*Jorda et al.,* 2016; *Kramer et al.,* 2019), and 46P (*Farnham et al.,* 2021b)]. So far, 46P is the most extreme case where rotation-period changes during one segment of the orbit get canceled during the subsequent segment of the orbit (see *Farnham et al.,* 2021b) and this is likely due to the comet's high obliquity of 70° (*Knight et al.,* 2021) as different torques dominate over different parts of the orbit. In fact, among the four comets whose intraorbit spin changes are well characterized, the percentage cancellation of δP per orbit in increasing order are for Comets 103P, 67P, 9P, and 46P. The obliquity (Table 3) increases in order: 9P, 103P, 67P, and 46P; however, as 103P is an NPA rotator, its obliquity-induced seasonal effects will be moderated and presumably the cancellation of torques too. This hints at percentage cancellation of spin change during an orbit correlating with obliquity. Could nuclear shape also play a part in this? Additional observations during the coming decades may help answer this.

The timescale for rotation period change for a simple rotator, τ_P, can be expressed by (see *Samarasinha et al.,* 1986; *Jewitt,* 2021)

$$\tau_P = k \frac{\rho_n r_n^2}{k_T f_A P \zeta} \tag{11}$$

where k is approximately a constant. Since $f_A k_T$ is nearly constant, $\tau_P \propto r_n^2$. Therefore, smaller nuclei will change their rotation periods more rapidly. As can be seen in *Kokotanekova et al.* (2017) and *Mueller and Samarasinha* (2018), shorter rotation periods are preferentially seen among the smaller nuclei. This is consistent with sublimation-driven spin evolution being the primary cause of spin changes among comets. For SPCs, τ_P is smaller than their dynamical lifetime and small nuclei in particular should have undergone extreme alterations to their spin sometimes causing potential splitting events. As a result, particularly for periodic comets, the spin we observe today is not primordial. However, for extremely large comets (e.g., $r_n \gg 10$ km) or for DNCs, the spin alterations during their lifetimes are much smaller and they may retain some residual evidence of the primordial distribution. This interpretation is also consistent with $\tau_P \sim 10^2\, r_n^2$ suggested by *Jewitt* (2021) for comets with perihelia around 1–2 au, where r_n is in kilometers and τ_P is in years.

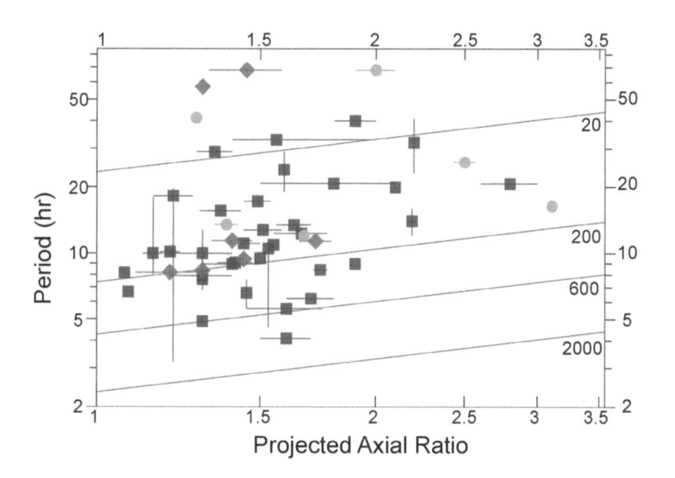

Fig. 3. Rotation period against projected axial ratio for comet nuclei, updated from *Kokotanekova et al.* (2017). The blue squares are JFCs and the green diamonds are HTCs and LPCs. For these points, the axial ratio is a lower limit and the uncertainties are plotted when they were stated by the authors. When authors provided multiple equally likely periods, the shortest rotation period is plotted since this is the most constraining. Orange circles are comets having shape models derived from spacecraft data. The diagonal lines indicate the minimum density (denoted in kg m^{-3} to the right), which a strengthless body of the given axial ratio and spin period requires to remain intact. Apart from the unusual case of 96P, which is undergoing continued fragmentation (*Eisner et al.,* 2019) potentially related to rotational spinup (see section 3.4.3), no comet requires a density greater than ~600 kg m^{-3} to remain stable against rotational breakup.

The changes to spin states due to outgassing torques can result in either a net spinup or spindown during an orbit. The nuclei that undergo spindown may ultimately undergo spinup, i.e., they will spin down to extremely long periods and then reverse course and start spinning up. During this epoch near extremely long rotation periods (with minimal rotational kinetic energies), if the conditions for rotational excitation are satisfied [see section 3.3.3 and *Gutiérrez et al.* (2003b)], the torques can put the nucleus into an excited spin state and it starts to spin up. As $\tau_P \propto r_n^{-2}$, the spindown process too (similar to spinup) is more efficient for smaller nuclei. Consequently, the rotational excitation is also more likely for smaller nuclei. The spinup of nuclei is likely to be a cause of splitting events and is discussed in section 3.4.3.

As sublimation-induced forces are the cause for the non-gravitational forces that change the orbits of comets, *Rafikov* (2018) investigated the likely changes in the rotation period with corresponding non-gravitational forces and suggested they are correlated. As sublimation is the primary driver for both the rotational- and non-gravitational-force-induced orbital changes of comets, simultaneous modeling of both these motions may be used to derive better constrained bulk densities for comets than just using orbital changes. Rafikov further determined that the dimensionless lever arm responsible for torquing was small, with a typical value around 1%, implying that most outgassing induced torques cancel themselves. *Jewitt* (2021) arrived at a similar conclusion from order of magnitude estimates of spinup timescales.

3.3.7. Orientation of spin axes and their changes. Typically, the term spin axis refers to the axis defined by the angular velocity. For a simple rotator, the spin axis of the body aligns with the direction of the rotational angular momentum vector. For an NPA rotator, the instantaneous spin axis is not fixed either in the body frame or in the inertial frame. In this chapter, as most comet nuclei are simple rotators, the term "spin axis" is used for simplicity; however, in the case of an NPA rotator, the discussion implicitly refers to the direction of the rotational angular momentum vector despite the usage of the term "spin axis."

Other than by space missions to comets, the primary means of determining spin axes of comets is using (1) coma morphology observations taken preferentially at different observing geometries and (2) multi-epoch observations of the bare-nucleus lightcurves corresponding to different observing geometries. Unlike for asteroids, such multi-epoch bare-nucleus lightcurves are difficult to carry out and expensive in terms of telescope time. Therefore, most spin-axis orientations of comets determined to date are via morphological observations; however, the lightcurve method may become practical for many more comets in the LSST era (see section 4).

Initial morphology-based spin-axis determinations were based on broadband continuum features in the coma representative of scattering of sunlight by dust grains (e.g., *Sekanina,* 1991b; *Samarasinha and Mueller,* 2002; *Farnham and Cochran,* 2002; *Schleicher et al.,* 2003; *Schleicher and Woodney,* 2003). With the relatively widespread use of narrowband filters within the last 2–3 decades, using primarily the morphology of CN features in the coma, many investigators were able to derive spin-axis orientations of multiple comets or place useful constraints (e.g., *Farnham et al.,* 2007a; *Woodney et al.,* 2007; *Samarasinha et al.,* 2011; *Bair et al.,* 2018; *Knight et al.,* 2021).

As mentioned earlier (section 3.3.6), *Whipple and Sekanina* (1979) argued that the spin axis of a non-rigid oblate nucleus would "precess" due to non-gravitational torques. Numerical simulations by *Samarasinha* (1997) point to evolution of the spin axis due to sublimative torques, irrespective of whether the spin state is excited or not. The spin axis gradually evolves either toward the orbital direction of peak outgassing (typically near perihelion) or the one diametrically opposite to that. Such a configuration is comparatively stable since that will provide the least net torque over time in the inertial frame. Subsequent work by *Neishtadt et al.* (2002) using a highly idealized model obtained the same result as one of the main evolutionary paths for the spin axis. Therefore, statistically, more spin axis directions may tend to align closer to the major axis of the orbit. As the current database for spin axis directions of comets is sparse, only future observations (e.g., from the LSST era) could test this prediction.

3.3.8. Properties and processes associated with spin. Spin studies provide insight into how spin is related to activity and other physical and structural parameters of the nucleus. The discussions above make it clear how activity drives the spin state changes; it is appropriate to note that the spin state, too, is responsible for determining the activity of a comet, the seasonal effects of outgassing caused by a specific spin state being a well-known example. As remarked earlier, since the timescale for rotation period change is larger for large nuclei (e.g., *Jewitt,* 2021), their spin changes are difficult to measure and there is an observational selection effect against detecting spin changes of large nuclei.

Spin studies also provide insights into the structural properties of the nucleus (see Fig. 4). The existence of a potential spin barrier and the disruption of Comet D/1993 F2 (Shoemaker-Levy 9) due to jovian tidal effects are consistent with low strengths for cometary nuclei [on the order of 10 Pa, about a million times weaker than water ice (*Weissman et al.,* 2004)]. In addition, the paucity of NPA spin states is a pointer to highly flexible nuclei with short damping timescales. However, both these structural traits should be considered as ensemble properties rather than specific to a particular comet. Here, the "flexibility of the nucleus" is meant to suggest that it is not a "perfect" rigid body but consists of a cohesive amalgamation of quasi subunits held together weakly. This flexibility of the nucleus is also consistent with the frequent splitting events observed in comets (see next section). Although the activity and other activity-caused effects (e.g., spinup) could be the primary trigger responsible for most splitting events, the low tensile strengths of comets could facilitate effortless splitting.

It is possible that the cometary surface layer could be somewhat more cohesive and stronger than the interior but it is still weak when compared to water ice or even some forms of snow (see the chapter by Guilbert-Lepoutre et al. in this volume). Naturally, it is likely that the low strengths for comets mean that the surface constituents (e.g., boulders) as well as the interior could respond to the stresses and resultant strains caused by rotation, including that due to NPA spin, by even altering cometary shapes. This possibility of mass redistribution and also the initiation of localized activity on the nucleus facilitated by a combination of low strengths of the nucleus and the rotation is an investigation worth pursuing in the future.

Spin-state changes can cause cometary outbursts in at least two ways: (1) the spinup-caused splitting can trigger outbursts due to exposure of new surface areas, and (2) the landslides triggered by rapid spin and/or spin changes can result in outbursts (*Steckloff and Samarasinha,* 2018, and references therein). However, the majority of these outbursts are unlikely to be major contributors to changes in the spin state as the net torques due to outbursts are comparatively small (since most outbursts are of short duration).

TABLE 3. Rotation period change per orbit (δP), effective nuclear radius (r_n), rotation period (P), total water production per unit surface area per orbit (ζ), and X-parameter of comets (with respect to those of 2P/Encke, denoted by subscript E) followed by the spin axis direction rounded to the nearest degree in RA, Dec, orbital longitude, and obliquity.

						Spin Axis Direction			
Comet	$\frac{\lvert\delta P\rvert}{\lvert\delta P_E\rvert}$	$\frac{r_n}{r_{nE}}$	$\frac{P}{P_E}$	$\frac{\zeta}{\zeta_E}$	$\frac{X}{X_E}$	RA (°)	Dec (°)	Orb Long (°)	Obliq (°)
2P/Encke*,†	1	1	1	1	1	218	8	48	58
	(4)	(2.4)	(11)	(1.66 × 10^4)	(5.3 × 10^{-5})				
9P/Tempel 1*,‡	3.5	1.18	3.73	0.35	1.0	255	64	285	16
10P/Tempel 2*,§	0.07	2.50	0.82	0.37	1.7	162	58	193	49
19P/Borrelly*,¶	>10	1.04	2.55	0.35	>4.8	214	–5	146	103
21P/Giacobini-Zinner**	3.6, 6.3	0.42	0.86	0.43	1.9, 3.4	169	73	199	10
41P/Tuttle-Giacobini-Kresák††	390	0.29	3.2	0.46	7.0				
46P/Wirtanen‡‡	0.6, 3.0	0.25	0.82	0.45	0.12, 0.62	319	–5	240	70
49P/Arend-Rigaux§§	<0.06	1.92	1.2	0.34	<0.45				
67P/Churyumov-Gerasimenko*,¶¶	5.3	0.69	1.09	0.38	5.5	70	64	21	52
96P/Machholz 1***	<0.54	1.42	0.37	1.67	<4.8				
103P/Hartley 2*,†††	37.5	0.24	1.64	0.43	1.9	8	54	340	48
C/1990 K1 (Levy)‡‡‡	57		3.3	0.38					

In the row corresponding to 2P, the absolute values for 2P are given within parentheses in units of minutes, km, hours, kg m^{-2}, and m^4 kg^{-1} s^{-1}, respectively. The orbital longitude is the angle measured prograde from the Sun-perihelion direction in the orbital plane. The sense of rotation is not known for some spin axis directions and thus could be the diametrically opposite directions.

**Mueller and Samarasinha* (2018).
†*Woodney et al.* (2007); also *Sekanina* (1988), *Festou and Barale* (2000).
‡*Thomas et al.* (2013a).
§*Knight et al.* (2012).
¶*Farnham and Cochran* (2002), *Schleicher et al.* (2003).
**There are two solutions for the spin period change; *Goldberg et al.* (2023).
††As the change in rotation period is large, the mean value is considered for the calculations. The possibility of NPA spin is suggested for this comet (see text) and if that is the case, this X-parameter estimate is highly uncertain; *Bodewits et al.* (2018), *Howell et al.* (2018), *Schleicher et al.* (2019), *Jewitt et al.* (2021).
‡‡The first entry for $\frac{\lvert\delta P\rvert}{\lvert\delta P_E\rvert}$ and $\frac{X}{X_E}$ represent the net change per orbit and the second entry for those correspond to the difference between the maximum and minimum rotation periods during the orbit (see text); *Farnham et al.* (2021), *Knight et al.* (2021).
§§*Eisner et al.* (2017).
¶¶*Preusker et al.* (2015), *Jorda et al.* (2016).
****Eisner et al.* (2019).
†††This comet is in an evolving NPA spin state and a representative average based on the observations during the 2010 apparition is chosen as the rotation period; *Belton et al.* (2013), *Knight et al.* (2015).
‡‡‡Observations cover only a segment of the active orbit; *Schleicher et al.* (1991), *Feldman et al.* (1992).

3.4. Evolution of Physical Properties

Formation mechanisms and early evolution of comet nuclei are the focus of the chapter by Simon et al. in this volume and were also the subject of excellent recent reviews by *Dones et al.* (2015) and *Weissman et al.* (2020). Briefly, the two leading formation theories are hierarchical agglomeration, whereby smaller cometesimals gradually merge into larger bodies through low-velocity collisions with minimal influence of gravity (cf. *Weidenschilling,* 1997; *Davidsson et al.,* 2016), and pebble accretion, in which millimeter- to centimeter-sized aggregates ("pebbles") are formed out of dust and ice grains and then concentrated by the streaming instability until coalescing in a gravitational collapse (cf. *Weidenschilling,* 1977; *Blum et al.,* 2017). Formation presumably occurred between 15 and 30 au (cf. *Gomes et al.,* 2004), and the comets likely remained in this region for long enough that the population was significantly modified by collisional evolution (cf. *Morbidelli and Rickman,* 2015; *Jutzi et al.,* 2017) before being scattered by giant planet migration to the Oort cloud or perturbed outward into the Kuiper belt [e.g., the "Nice model" (*Gomes et al.,* 2005; *Tsiganis et al.,* 2005)].

Once in these reservoirs, comets were likely further modified. As reviewed by *Weissman et al.* (2020), contributing processes are thought to include internal heating from short-lived radionuclides, heating by supernovae and passing stars, bombardment by high-energy particles, and accretion of and/or erosion by interstellar gas and dust grains. The first two processes would have heated nuclei at depth, affecting the volatile inventory and distribution. This has implications for the internal structure and composition (see the chapters in this volume by Guilbert-Lepoutre et al. and Filacchione et al.). The last two processes are estimated to alter the top layers to perhaps 2 m [high-energy particles; see review by *Strazzulla* (1999)] or ≲0.1 m [accretion/erosion (e.g., *O'Dell,* 1973; *Stern,* 1986)]. The above processes likely cannot appreciably change the size or shape of comet nuclei, although recent work by *Zhao et al.* (2021) suggests that sublimation could have significantly altered the shape of Arrokoth with additional assistance due to its specific spin orientation. The properties of the uppermost layers are undoubtedly altered by bombardment and accretion/erosion. This may result in changes to the color, albedo, phase function, and polarization, among other things. However, these are hard to discern observationally because the outermost layer is lost as soon as activity begins and typically before the comet is observable. These macroscopic surface properties could theoretically be measured in incoming dynamically new comets before significant activity has begun, but such observations are extraordinarily challenging (see section 4).

Subsequent perturbation from the Kuiper belt or Oort cloud into smaller perihelion distance orbits will result in the initiation of activity (see the chapter by Fraser et al. in this volume). Changes in activity are likely gradual as comets diffuse inward from the Kuiper belt and scattered disk via the Centaur region into JFCs, or rapidly for Oort cloud comets, which can have their perihelion distance change by many astronomical units in a single orbit (*Dybczyński and Królikowska,* 2011). The more rapid perihelion changes of Oort cloud comets and corresponding dramatic increase in insolation may trigger some of the fragmentation scenarios discussed below. Activity removes material to a depth on the order of 1 m per orbit for a typical comet, although this is not necessarily uniform across the surface due to seasons, topography, and fallback. This is small compared to the size of typical comet nuclei, so it will not change the size or shape appreciably in a single orbit. However, the cumulative effects of activity-driven mass loss may modify the sizes and shapes of comet nuclei over their active lifetimes, which is likely on the order of 10^3–10^5 yr (e.g., *Levison and Duncan,* 1994, 1997) and dependent on both size and number of close perihelion passages (*Nesvorný et al.,* 2017).

At the other end of their lifetimes, comets will be lost or destroyed by several mechanisms. *Levison and Duncan* (1997) estimate that the median dynamical lifetime of "ecliptic comets" (comets having $T_J > 2$) is 4.5×10^7 yr from the time of their first encounter with Neptune until they collide with the Sun or a planet or are ejected from the solar system or into the Oort cloud. Thus, most comets will cease to be active long before they are ejected or destroyed by dynamical means. Inactivity may come about by loss of accessible volatiles or the formation of an inert crust; specific mechanisms and estimates of their timescales were reviewed by *Jewitt* (2004b) in *Comets II.* Such inactive comets could represent ~10% of the near-Earth asteroid population (*Fernández et al.,* 2001). (For additional discussion, see the chapter by Jewitt and Hsieh in this volume.) Subkilometer nuclei can be destroyed by mass loss due to sublimation alone (cf. *Samarasinha,* 2007). Fragmentation (or splitting) and disintegration are also common end states. The frequency of these events is not well constrained, but is estimated to be a few percent per orbit (e.g., *Weissman,* 1980; *Chen and Jewitt,* 1994). More than any other mechanisms just discussed, fragmentation has the potential to modify the nucleus properties discussed in the preceding subsections, so we explore it in more detail below.

3.4.1. Fragmentation overview. The modern history of comet fragmentation begins with 3D/Biela, which was observed as a pair of fragments in 1846 following regular observations as a single nucleus since its discovery in 1772. It was recovered again in 1852 as a single comet, but has never been recovered since, with heightened meteor activity in 1872, 1885, 1892, and 1899 calculated to have originated in Biela's orbit (see *Kronk,* 1999, 2003, and references therein). High-profile split comets in recent decades include Shoemaker-Levy 9, which was discovered as a string of nuclei in 1993 and subsequently observed to impact Jupiter in 1994 (cf. *Hammel et al.,* 1995; *Weaver et al.,* 1995), and 73P, which has had several apparently spontaneous fragmentation events since 1995 (cf. *Sekanina,* 2005) and whose fragments were observed in exquisite detail during a remarkable close approach to Earth in 2005 (e.g., *Weaver et al.,* 2006; *Reach et al.,* 2009). See *Weissman* (1980), *Sekanina* (1982), and *Boehnhardt* (2004) for lists of other split comets and more detailed discussion of their splitting scenarios.

Despite the relatively high frequency of comet splitting, only a few groups of dynamically related comets are known. Most of the persistent groups are near-Sun comets, almost exclusively seen only by solar observatories. Best known is the Kreutz sungrazing group (perihelion distance ~0.005 au), which includes a number of historically bright, ground-observed comets including C/1882 R1 (Great September Comet) and C/1965 S1 (Ikeya-Seki) (cf. *Marsden,* 1967, 1989; *Sekanina and Chodas,* 2004), and thousands of minor fragments seen arriving in SOHO images every few days (cf. *Biesecker et al.,* 2002; *Knight et al.,* 2010). The minor fragments are seen only near the Sun for hours to days before they are destroyed, but are thought to have been produced by cascading fragmentation from one of a few major fragments since the previous perihelion passage (e.g., *Sekanina,* 2002). The Marsden and Kracht groups have ~10 times larger perihelion distances (q ~ 0.05 au) and are thought to be descended from 96P/Machholz 1 as part of the "Machholz complex" (*Ohtsuka et al.,* 2003; *Sekanina and Chodas,* 2005). Little is known about the Meyer group, which also has much larger perihelion distances than the Kreutz group (q ~ 0.04 au); it has not been linked to any other known objects but its members are thought to be on long-period orbits. See *Marsden* (2005), *Battams and Knight* (2017), and *Jones et al.* (2018) for reviews of these near-Sun groups.

Few other groups of persistent families are known. *Boehnhardt* (2004) lists several likely split pairs/families, but uncertainties in orbital integrations, worsened by nongravitational accelerations, make firm linkages difficult unless they are recently split; *Tancredi* (1995) estimated

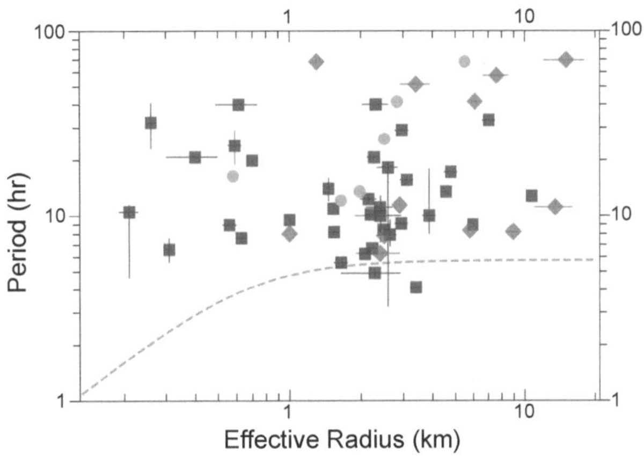

Fig. 4. Rotation period against effective radius of comet nuclei, updated from *Kokotanekova et al.* (2017). The blue squares are JFCs, green diamonds are HTCs and LPCs, and orange circles are comets visited by spacecraft. The dashed line is the model for density 532 kg m^{-3}, axial ratio 2, and tensile strength 15 Pa, which corresponds to the parameters measured for 67P from Rosetta (*Groussin et al.,* 2015; *Jorda et al.,* 2016). The two comets below the line are 61P, which also has two plausible longer-period solutions (*Lamy et al.,* 2011), and 96P, which is undergoing continued fragmentation (*Eisner et al.,* 2019) potentially related to rotational spinup (see section 3.4.3).

that even well-known comet orbits cannot be integrated reliably for more than ~1000 yr. The only strong linkage of comets on typical JFC orbits seems to be 42P/Neujmin 3 and 53P/Van Biesbroeck (*Carusi et al.,* 1985), while several likely split pairs of Oort cloud comets have been identified.

Most fragmentation events result in an initial brightening of the inner coma. As the new material expands, the inner coma brightness gradually returns to pre-outburst levels and one or more fragments may be discernible. There is generally a primary fragment having brightness approximately unchanged from before the event and fainter fragments that diminish in brightness or simply disappear on timescales of days to weeks. Frequently, but not always, the fainter fragments do not persist long enough to gain individual designations, and it is rare that they survive to be reobserved on subsequent apparitions. It is speculated (e.g., *Jewitt,* 2021) that these short lifetimes are due to rapid spinup (see below), although other mechanisms such as rapid depletion of volatiles could also be responsible.

3.4.2. Outburst, fragmentation, and disintegration. Fragmentation may be part of a continuum of phenomena, including outbursts and disruption/disintegration (see Fig. 5), that can modify the size, shape, and spin of a comet nucleus by progressively larger amounts. Most benign are outbursts, where a comet suddenly brightens due to a rapid, and usually short-lived, increase in activity. Historically, outbursts needed to be large to be detected (e.g., *Ishiguro et al.,* 2016; *Miles,* 2016b) unless a comet was already under regular observation. However, with the improved cadences and depth of modern surveys, smaller outbursts are being discovered with much higher frequency than previously known. For instance, M. Kelley (U. of Maryland) and collaborators have reported outbursts of more than a dozen comets in Zwicky Transient Facility images since 2019 (e.g., *Kelley et al.,* 2019; *Lister et al.,* 2022). Deep Impact and Rosetta showed that outbursts too small to be detected from the ground happen frequently (*Farnham et al.,* 2007b; *Vincent et al.,* 2016), so it is reasonable to assume that the behavior is widespread. *Belton* (2010) proposed a taxonomy to explain a range of activity/outbursts of different strengths observed in spacecraft target comets.

The largest outburst recorded is that of Comet 17P/Holmes in 2007 where it brightened by nearly a million times just within two days (e.g., *Montalto et al.,* 2008; *Lin et al.,* 2009; *Reach et al.,* 2010). However, except in the largest outbursts, the mass of material involved is likely very small compared to the whole nucleus. Estimates of the masses of the large outbursts reported in *Ishiguro et al.* (2016) are 10^8–10^{11} kg, considerably less than the ~10^{16}-kg mass of 67P (*Pätzold et al.,* 2019). For an assumed density of 600 kg m^{-3}, a 10^8-kg outburst corresponds to a sphere of equivalent radius ~34 m. Thus, the volume of material involved in most outbursts is likely comparable to the sizes of the pits and boulders seen on 67P (e.g., *Vincent et al.,* 2015; *Pajola et al.,* 2016) and is insufficient, individually, to make appreciable changes in a comet's size, shape, or rotation period.

Outbursts are occasionally accompanied by discrete brightness condensations in the coma that are identified as fragments. Fragments need not be associated with outbursts, and many are only discovered a considerable time after they were presumably produced. Identification of fragments and measurements of their sizes are limited by viewing geometry as well as the activity level of the coma. *Graykowski and Jewitt* (2021) identified fragments as small as 1 m in radius in HST observations of the debris field around the B and G fragments of 73P at 0.23 au from Earth in 2006, while the smallest discrete fragments identified by *Weaver et al.* (2001) of C/1994 S4 (LINEAR) at ~0.7 au from Earth were ~25 m in radius. We surmise that many outbursts that appear devoid of discrete fragments likely contain fragments either too small or too heavily enshrouded in coma to be detected.

Fragmentations can result in significant alteration of the original nucleus, particularly when the resulting pieces are large enough to persist. The primary surviving fragment of 73P is fragment C, which was estimated to have an equivalent radius of ~0.5 km in 2006 (*Toth et al.,* 2005; *Graykowski and Jewitt,* 2019) as compared to the pre-breakup radius of ~1.1 km for the full comet (*Boehnhardt et al.,* 1999). Sizes have not been published for the other fragments of 73P, but presuming they make up the bulk of the lost material, such mass loss could carry enough angular momentum and changes to the moments of inertia to significantly affect the rotation period and rotation state, in addition to the obvious changes to the size and shape.

The most extreme phenomenon in this continuum is

disintegration, when the nucleus completely disappears. This may be accompanied by discrete fragments (e.g., C/1999 S4) or simply a diffuse cloud of material devoid of brightness condensations like C/2010 X1 (Elenin), in which no fragments were detected down to ~40 m in radius (*Li and Jewitt,* 2015). Spontaneous disintegration seems to be far more common among dynamically new comets, an idea that we will revisit in the next subsection.

3.4.3. Causes of fragmentation. A variety of mechanisms have been proposed to explain why comets split. In a few cases, the cause can be determined with some confidence, but most are conjecture. The most commonly accepted explanations for comet fragmentation are discussed below and illustrated in Fig. 6.

3.4.3.1. Tidal forces: The differential gravitational force from a large mass felt on opposite sides of a less-massive body can cause disruption of the latter if the differential gravitational force exceeds the forces acting to hold the body together. The splitting of sungrazing comets C/1882 R1 and Ikeya-Seki when near perihelion inspired *Whipple* (1963) and *Opik* (1966) to think of comet nuclei as (nearly) strengthless bodies, prone to split due to tidal forces based on their radius and density. The dramatic discovery of the string of fragments in Shoemaker-Levy 9 motivated a host of studies, which led to the robust conclusion that it had been disrupted during a close approach to Jupiter in 1992 (e.g., *Sekanina et al.,* 1994; *Asphaug and Benz,* 1996). Subsequent work has shown that the distance at which tidal splitting occurs can be modified to some extent by other properties of the nucleus such as the axial ratio, sense of rotation, and rotation period (see *Knight and Walsh,* 2013, and references therein). However, such modifications are relatively small, and to first order, tidal splitting can be assumed to occur at the Roche limit.

Using the classical definition of the Roche limit as $\approx 2.45\ R_p\ (\frac{\rho_p}{\rho_c})^{\frac{1}{3}}$, where R_p is the radius of the primary body, comet, a comet with a density of 600 kg m^{-3} is in danger of tidal splitting if it has perihelion within ~3.2 solar radii [the common upper limit for orbits deemed to be "sungrazing" (see review by *Jones et al.,* 2018)], perijove within ~3.2 jovian radii, or perigee within ~5.1 Earth radii. Encounters sufficiently close to induce tidal splitting are infrequent, with the only clear examples being several bright Kreutz sungrazing comets, Shoemaker-Levy 9, and 16P/Brooks 2 (*Sekanina and Yeomans,* 1985).

3.4.3.2. Rotational spinup: A comet nucleus can split if the centrifugal forces due to rotation exceed the forces holding it together. As previously discussed, torques caused by the asymmetric outgassing of the nucleus can lead to changes in the rotation period and spin state. Although NPA rotation states are known to exist, for simplicity, we consider the case of simple rotation here. Torques can act to either decrease ("spinup") or increase ("spindown") the rotation period. Spindown decreases the centrifugal force and thus does not lead to splitting, while spinup can lead to destruction, as has been discussed extensively in the literature (e.g., *Sekanina,* 1982; *Jewitt,* 1997; *Jorda and Gutiérrez,* 2002).

As discussed in section 3.3.6, smaller nuclei can spin up more rapidly than larger nuclei. Order-of-magnitude calculations (e.g., *Jewitt,* 2021) suggest that this can lead to catastrophic breakup of very small comet nuclei on surprisingly short timescales consistent with observations: 20–200 days at 1.6 au for 20-m-radius fragments of 332P/Ikeya-Murakami (*Jewitt et al.,* 2016) and ~1.5 days at ~0.06 au for 50-m-radius sungrazing comets (*Knight et al.,* 2010). *Jewitt* (2021) further argues that, for typical SPCs of up to 10 km in radius, the timescale for destruction by spinup is on the order of 10^4 yr or less, and is shorter than the devolatilization timescale. Due, presumably, to the difficulty of making rotation-period measurements for LPCs as well as no possibility of measuring a change in rotation period on subsequent orbits, only two Oort cloud comets have had a change in period detected, so a similar analysis cannot be conducted on the LPC population.

Even if these timescales are underestimated, observational evidence suggests spinup is a frequent mechanism in the apparently spontaneous demise of comets. For example, size distributions of JFCs reveal an underabundance of subkilometer nuclei (e.g., *Fernández et al.,* 2013; *Bauer et al.,* 2017) that can be explained by spinup, while *Jewitt* (2022) used a sample of 27 LPCs to argue that rotational breakup can explain the destruction of near-Sun LPCs. However, note that spinup is not necessarily the controlling factor in the destruction of small nuclei. As demonstrated by *Jewitt* (2021) and consistent with earlier work by *Samarasinha* (2007), either mass loss or spinup can be the main driver of fragmentation.

A possibly related phenomenon is the disappearance of faint Oort cloud comets near the Sun. *Bortle* (1991) found that about 70% of Oort cloud comets for whom the absolute magnitude H_{10} in the *Vsekhsviatskii* (1964) system was greater than 7.0 + 6.0 q, where q is the perihelion distance in astronomical units, did not survive their perihelion passage. This equation is often referred to as the "Bortle limit." *Sekanina* (2019) found even less-frequent survival using a high-quality set of modern orbits. On the assumption that fainter comets have smaller nuclei, this behavior can naturally be explained by sublimation-driven spinup, with increasingly larger nuclei being destroyed as perihelion distance decreases. In the most extreme case of a sungrazing comet, simulations by *Samarasinha and Mueller* (2013) predicted significant changes to the rotational state for C/2012 S1 (ISON) due to sublimation.

3.4.3.3. Thermal effects: We use "thermal effects" to broadly encompass several processes related to the heating of the nucleus. The most commonly proposed mechanism is the buildup of subsurface gas pressure, which is eventually released catastrophically as an outburst or fragmentation. A variety of processes have been proposed including the transition of amorphous to crystalline H_2O ice (e.g., *Patashnick,* 1974; *Prialnik and Podolak,* 1999) and exothermic dissolution of gases like CO and CO_2 (*Miles,* 2016b). Such processes are likely to be most important at large

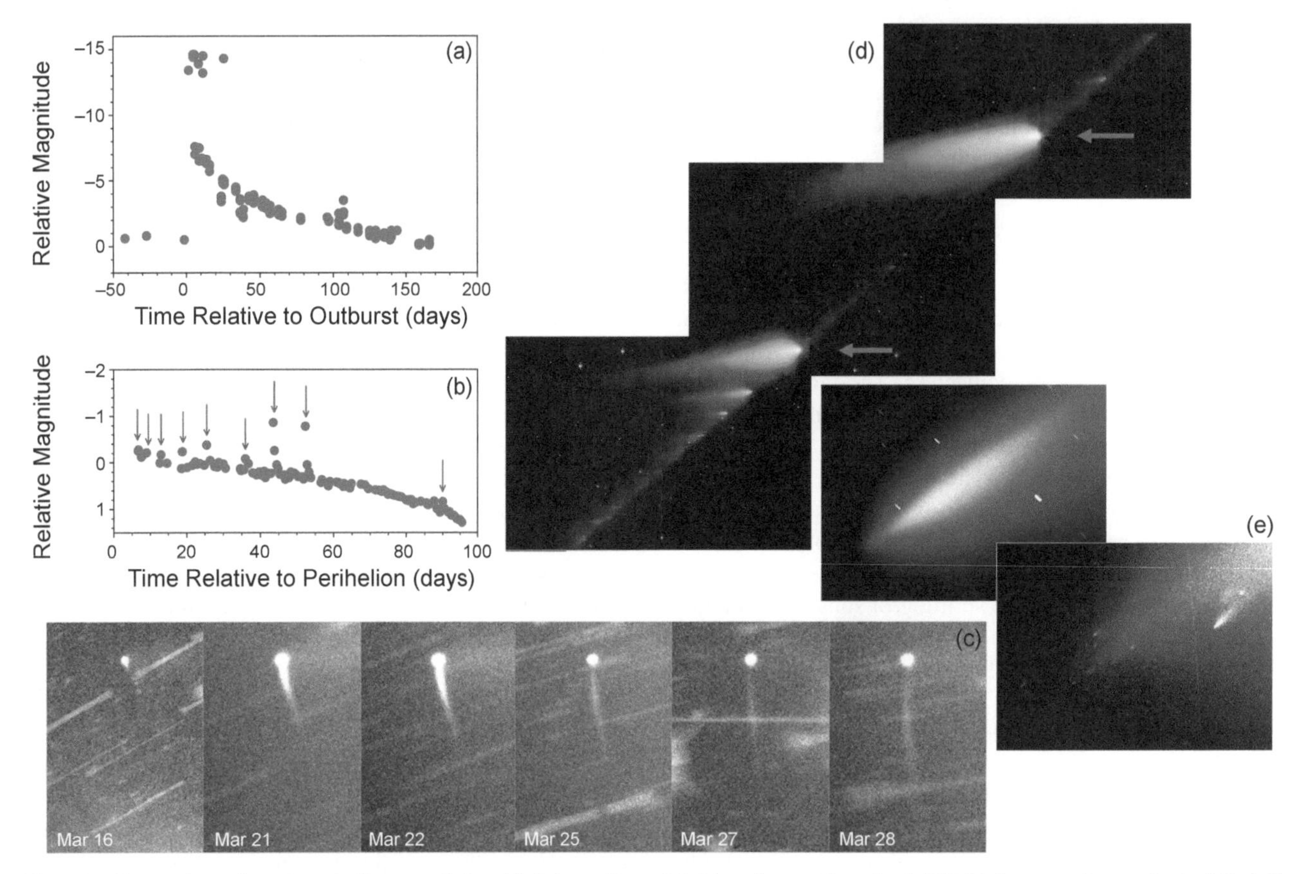

Fig. 5. Examples of outbursts/fragmentation/disintegration. **(a)** Massive outburst of 17P/Holmes using selected "total" coma magnitudes archived with the Minor Planet Center. Time is measured from 2007 October 23.0. **(b)** Recurring minor outbursts of 7P/Pons-Winnecke adapted from *Lister et al.* (2022). Time is measured from perihelion (2021 May 27.1) and outbursts are indicated with vertical arrows. **(c)** Evolution of minor outburst morphology of 49P/Arend-Rigaux in 2012 from *Eisner et al.* (2017). **(d)** Extensive fragmentation of 73P/Schwassman-Wachmann 3 imaged 2006 May 4–6 by Spitzer. Persistent fragments are indicated with horizontal arrows. Credit: NASA/JPL-Caltech/W. Reach (SSC/Caltech). **(e)** Disintegration of C/1999 S4 (LINEAR) as seen on 2000 August 5. The lower-right image by HST shows numerous discrete fragments that were not visible in the near-simultaneous upper left image acquired from the ground. Credit: H. Weaver (The Johns Hopkins University) and the HST Comet LINEAR Investigation Team, University of Hawaii, and NASA/ESA.

heliocentric distances; see reviews by *Hughes* (1991) and *Gronkowski and Wesołowski* (2016) for additional details. Thermal stress has often been cited as a likely cause of breakups (e.g., *Boehnhardt,* 2004). *Fernández* (2009) pointed out that the results from Deep Impact suggest that the thermal pulse from diurnal heating can only penetrate a few centimeters, so is unlikely to create large amounts of thermal stress.

Another thermal effect is the complete loss of the volatiles holding a comet together. As envisioned by *Sekanina and Kracht* (2018), a comet experiencing this might remain intact as a fluffy aggregate, but would be subject to extreme non-gravitational accelerations and would be expected to dissipate rapidly as the grains crumble. This loss of the volatile "glue" would be more likely to occur at small heliocentric distances and might, therefore, help explain the "Bortle limit" discussed above. In this scenario, there would be a survivor bias in which comets not prone to such behavior would survive, thus decreasing the frequency of disintegration by returning comets.

3.4.3.4. Shearing: The Rosetta team identified fractures that were centered in the neck region of 67P and propagated during the mission. *Matonti et al.* (2019) concluded that these expanding fractures were due to shearing deformation caused by stresses from 67P's two lobes and speculated that this could cause outbursts at large distances or ultimately lead to 67P breaking up. Given the apparent prevalence of highly elongated/bilobate nuclei, this could be an important evolutionary mechanism across the comet population.

3.4.3.5. Impacts: Although collisional evolution of the comet population undoubtedly occurred in the early stages of the solar system, spatial densities throughout the solar system are too low for impacts onto comets to trigger fragmentation with any appreciable regularity today. However, small impacts undoubtedly occur on occasion, and may be capable of triggering increased activity by exposing fresh, subsurface ice.

3.4.4. Potential insight gained. Splitting provides the unique opportunity to study the freshly exposed interior of a nucleus. If each fragment is sufficiently separated spatially to

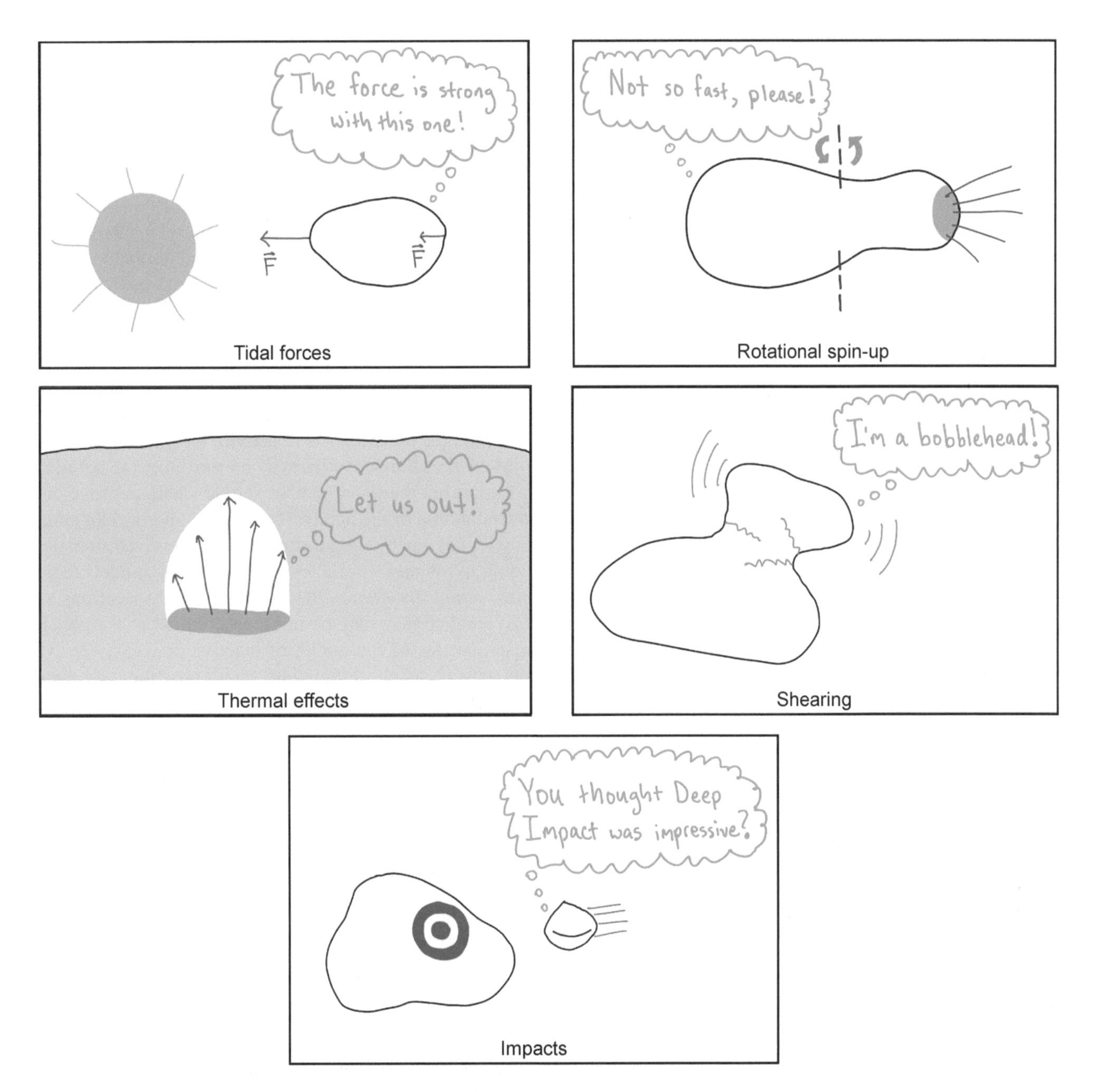

Fig. 6. Causes of fragmentation (see text for additional details).

allow its chemical composition to be measured, the compositions of fragments can be compared to investigate whether or not the parent nucleus was heterogeneous. To date, this has only been accomplished for 73P, whose fragments had similar compositions (e.g., *Dello Russo et al.,* 2007; *Schleicher and Bair,* 2011). Measuring compositions of more split comets is highly desirable, as the results should help discriminate between nucleus formation scenarios, notably how bilobate nuclei come to be.

Rotationally resolved observations of recently split nuclei while they are weakly active or inactive would permit searches for color, albedo, and/or polarimetric variations. Since no comet has been shown to have appreciable macroscopic variations in any of these properties, the detection of variations in a recently split comet would suggest that fresh materials have different properties from older surface material.

If comets split along paths through their interiors with less structural integrity, the resulting more consolidated fragments may be the cometesimals out of which the original comet formed. It is interesting to note that the sizes of the fragments sampled in occasional high-resolution images of recent fragmentations — e.g., 73P, C/1999 S4, 332P (all discussed above) — are of similar sizes to the inferred sizes of the SOHO-observed Kreutz comets (*Knight et al.,* 2010). Furthermore, all of these are consistent with *Weidenschilling* (1997)'s preferential sizes of cometesimals and roughly comparable in size to the boulders and pits on 67P (cf. *Vincent et al.,* 2015; *Pajola et al.,* 2016). If the fragments produced in splitting events are the fundamental units of comet formation, then measuring their size distribution will be valuable for constraining formation models. At present, the measured sizes of fragmented comets are thought to be limited by detector sensitivity. Future observations with greater survey depth should be capable of testing the small end of this size distribution.

4. FUTURE ADVANCES

We conclude this chapter by considering what lies ahead for the study of comet nuclei. We limit this discussion to the discoveries that will advance the topics discussed in the preceding sections via remote observations. Future advances are obviously difficult to predict, but we shall make educated guesses based on the technological advances already underway or planned as well as well-established long term goals of the community. We supplement this with some wishful thinking if cost and/or luck were no impediment.

4.1. Near-Term Telescope Advances

The easiest future insights to foresee are those that result from existing technology and soon-to-be completed facilities. The two major facilities expected to revolutionize planetary science in the near future are JWST and LSST.

4.1.1. James Webb Space Telescope. The JWST, a 6.5-m spacebased telescope located at the Earth-Sun L2 Lagrange point, began operations in mid-2022 and is capable of acquiring images and spectra from 0.6 to 28.3 μm. Its unprecedented thermal IR capabilities provide fascinating possibilities for nucleus studies. Most known JFC nuclei should be detectable at aphelion, meaning that a dedicated survey could theoretically measure the nucleus sizes of essentially the entire JFC population. In reality, many are likely to be active at such distances (*Kelley et al.,* 2013), making nucleus detections more challenging.

A more compelling advance would come from JWST detections of the nuclei of LPCs at large heliocentric distances post-perihelion, when activity has died down or is sufficiently low to allow the coma-nucleus separation techniques of *Fernández et al.* (2013). The efficiency is likely to be dependent on the dust-to-nucleus ratio, which would not be known in advance, but given the paucity of direct detections of the nuclei of comets coming from the Oort cloud, would be well worth exploring. A study comparable in scope to SEPPCoN (*Fernández et al.,* 2013) would have enormous value for shaping our understanding of the properties of LPC nuclei. Furthermore, JWST will enable surface spectroscopy of many more nuclei than have ever been studied in this manner. This may lead to the first remote detection of exposed ices and will make possible studies of spectral changes with rotation.

4.1.2. Legacy Survey of Space and Time. The LSST is expected to begin full operations in 2025. Surveying the entire sky visible from its southern hemisphere site in Chile every few nights to a limiting magnitude of $r \approx 24.7$, LSST is predicted to increase the number of known comets by about an order of magnitude (*Silsbee and Tremaine,* 2016; *Ivezić et al.,* 2019). Specific operational details are still being worked out and will have some effect on the number and nature of discoveries — a proposed twilight survey would find many new comets at small solar elongations (*Seaman et al.,* 2018), while the proposed northern ecliptic survey would detect more distantly active bodies in the main asteroid belt, Kuiper belt, and (possibly) inner Oort cloud (*Schwamb et al.,* 2018) — but it is clear that LSST will discover fainter and more distant objects than current surveys. The distant discoveries are especially enticing for the possibility of detecting bare nuclei of DNCs, allowing us to monitor the initiation of activity for the first time.

Most LSST discoveries will be too faint for detailed followup, but the sheer number should vastly improve our understanding of the population of small and/or faint comets with large perihelion distances. The improved statistics should allow meaningful tests of population models (e.g., *Silsbee and Tremaine,* 2016). Frequent observations of individual comets may permit measurement of the physical properties of the nuclei of inactive or weakly active comets such as phase function, rotation period, and axial ratios from lightcurves, color, and size for an assumed albedo. Regular monitoring will also allow the detection of unexpected activity or activity changes, both at individual epochs and by stacking multiple epochs. In addition to discovering many new weakly active comets, this will likely reveal many more fragmentation events than are currently known. The rapid identification of activity changes (e.g., *Kelley et al.,* 2019) will enable the triggering of followup observations on specialized instruments that should yield even better data, such as size distributions of newly formed fragments.

4.2. Longer-Term Telescope Advances

The Near-Earth Object Surveyor (NEO Surveyor), the planned followup to WISE/NEOWISE, is slated to launch in 2027 and operate for at least five years (*Sonnett et al.,* 2021). As proposed, it will have roughly 50% more collective power than WISE/NEOWISE and contain two passively cooled channels at mid-IR wavelengths, 4–5 *μm* and 6–10 *μm*. The longer-wavelength channel should be virtually free of reflected sunlight, making it more effective than the post-cryogen phase NEOWISE at obtaining thermal measurements of comet nuclei. However, it will likely be less effective than the cryogen phase of the WISE mission, since the 4–5-*μm* region will suffer CO and CO_2 gas contamination, meaning the results will depend only on the longer-wavelength channel (WISE had two gas-free bandpasses during its cryogen phase). On the whole, we expect NEO Surveyor to be capable of detecting the nuclei of smaller and more distant comets than WISE/NEOWISE but to produce results that are more model dependent. NEO

Surveyor's much longer planned duration than the cryogen phase of WISE means it will detect many nuclei at multiple epochs; the repeat sampling should increase confidence in marginal results and may yield insight into the axial ratio of higher signal-to-noise ratio targets.

A rebuilt Arecibo would reinstate our ability to directly measure nucleus properties of close-approaching comets. If it could be made more powerful, the volume of space in which such observations can be made greatly increases.

One or more 30-m-class telescopes are expected to become operational by the late 2020s: the Extremely Large Telescope (39 m diameter in Chile), the Thirty Meter Telescope (30 m; Hawaii), and Giant Magellan Telescope (24.5 m; Chile). After JWST, the next major spacebased observatory is expected to be the 2.4-m Nancy Grace Roman telescope with a launch date in the mid-2020s. Primarily planned to study dark energy and hunt for exoplanets, its primary value for comet studies is likely via serendipitous observations at large distances (*Holler et al.,* 2021). The true successor to HST as a flexible, multiwavelength spacebased observatory is likely to be the Large Ultraviolet Optical Infrared Surveyor (LUVOIR). The LUVOIR concept is still being developed, but it is likely to have a 10-m-class mirror and support imaging and spectroscopic capabilities from UV to IR wavelengths. Thirty-meter-class telescopes and next-generation space telescopes will possess collecting power roughly an order of magnitude greater than the most powerful existing telescopes, thereby permitting detection and characterization of smaller and/or more distant nuclei. Those that overlap with the LSST era could enable detailed studies that are impossible with current technology. The high angular resolution of 30-m-class telescopes could resolve nuclei of moderately sized comets without requiring extreme close approaches, might allow jets to be traced down to the active areas on the nucleus, or might detect submeter-scale fragments.

4.3. Serendipitous Advances

Advances that come from clever use of new technologies or simply luck are harder to predict. Entirely beyond our control are comets on orbits that bring them close enough to Earth for their nucleus properties to be measured. In general, encounters within 0.1 au yield outstanding data via numerous techniques. With diverse and often flexible missions in operation throughout the solar system, there are now more opportunities for close approaches of some kind than ever before. Indeed, the closest known approach of a comet to any planet was not a flyby of Earth, but C/2013 A1 (Siding Spring)'s 0.0009-au pass by Mars in October 2014, which yielded the first spatially resolved images of a dynamically new comet (*Farnham et al.,* 2017).

Occultations are another opportunity that depend to a great extent on luck. To date, no comet nucleus has conclusively been detected by an occultation, although *Fernández et al.* (1999) constrained Hale-Bopp to ≤30 or ≤48 km depending on model parameters. The technique depends on measuring a shadow on Earth that is only the size of the occulting body, and whose position on Earth is uncertain in proportion to the positional uncertainty of the orbit and its distance from Earth. Comet occultations are more challenging than asteroids and even KBOs due to their combination of small sizes, vastly larger positional uncertainty owing to their non-gravitational forces from outgassing, and optical depth effects. However, there is no technical reason preventing the observation of comet nucleus occultations. The highly successful campaign for New Horizons extended mission target Arrokoth (*Buie et al.,* 2020), which correctly identified it as a contact binary and provided accurate estimates of the size of each lobe as well as the albedo, engenders optimism that a similar campaign could be mounted for the right comet. In principle, a relatively low-cost program could be developed to attempt campaigns for promising targets. The logistics are daunting — tens to hundreds of observers spaced evenly across tens of kilometers — but with persistence, chords across the nucleus of multiple LPCs might be obtained, unequivocally constraining their sizes at a fraction of the cost of a dedicated space mission.

Unexpected results from new technology provide another possible advance. A prime example of relevance to this chapter is the discovery in SOlar Heliospheric Observatory (SOHO) images that there exists a nearly continuous stream of small comets approaching the Sun. Small sungrazing comets belonging to the Kreutz family were known prior to SOHO (*Marsden,* 1989), but the discovery of thousands of such comets (cf. *Battams and Knight,* 2017) was unexpected, as was the existence of several other groups of dynamically related comets (*Marsden,* 2005). There is no facility currently planned that will rival SOHO's surveying power, but the Parker Solar Probe (*Fox et al.,* 2016) and Solar Orbiter (*Müller et al.,* 2013), which are on trajectories with perihelion distance decreasing to 0.05 au and 0.28 au, respectively, might discover previously unknown groups of small comets near the Sun by fortuitously passing close to their orbits.

Moving further from luck to clever design is the opportunity to use instrumentation in untraditional ways. JWST will allow parallel observations for many of its instruments. Parallel observations will have no control over telescope orientation, but careful planning for serendipitous pointing that include known comets or simply data mining to search for unknown objects might prove a low-cost way to explore comet nucleus properties. Data mining for serendipitous observations of known solar system objects has become relatively common [e.g., asteroids in Gaia data by *Carry et al.* (2021)]; similar work with JWST data could be particularly compelling if it can detect the nuclei of distant comets. *Schlichting et al.* (2012) constrained the size distribution of subkilometer KBOs by searching for occultations of stars in HST images; perhaps future work might be capable of probing the size distribution of the Oort cloud (e.g., *Nihei et al.,* 2007; *Ofek and Nakar,* 2010).

Finally, computationally intensive data mining of ongoing and future surveys will likely prove fruitful. Analyses that can combine many epochs of data at a multitude of possible rates of motion and then efficiently search the results may be capable of discovering much smaller and/or more distant comets than currently known. These will not only yield new discoveries, but could be used to determine many of the properties discussed in this chapter. The proliferation of stable, spacebased datasets — Kepler/K2, the Transiting Exoplanet Survey Satellite (TESS), Gaia — as well as deep groundbased surveys from larger telescopes like the Dark Energy Camera (DECam) on the Blanco 4-m telescope at the Cerro Tololo Inter-American Observatory (CTIO) and Hypersuprime Cam on the 8-m Subaru telescope, along with, of course, LSST, should provide ample input for such analyses over the next decade.

4.4. Open Questions

4.4.1. What is the range of comet nucleus properties? As discussed previously, the size, axial ratio, albedo, color, phase slope, and rotation period have been measured for enough comet nuclei that we have a sense of their likely ranges. However, there are still few enough data points that individual measurements continue to have value in fleshing out the group statistics. A very reachable objective of the next decade or so is to measure each of these properties for enough comets to draw statistically significant conclusions, with the goal of making assessments akin to the revelation of the "spin barrier" among near-Earth asteroids (*Pravec and Harris,* 2000). Hints of this are already evident in the inference that comets had low densities prior to a direct measurement being made by Rosetta (*Lowry and Weissman,* 2003). However, might a clear limit be established for all comets, by dynamical class, or for a subset that share some common feature?

With sufficient data, exploration of the bounds of different properties might be especially insightful. Some questions that might be answerable in the relatively near future include: Is there a minimum size for comet nuclei? Is there a minimum and/or maximum albedo? What is the distribution of rotation periods? How common are highly-elongated/bilobate nuclei and what is their long-term stability if they outgas continuously? Are there binary nuclei similar to main-belt Comet 288P (300163) (*Agarwal et al.,* 2017)? Can phase function be used to discriminate between traditional asteroids and inactive comet nuclei on near-Earth object orbits?

4.4.2. What are the bulk properties of the nuclei of long-period comets? Tied closely to understanding the range of nucleus properties is the desire to understand the bulk properties of the comets originating from the Oort cloud. As discussed previously, few nucleus properties of LPCs are known, but a dedicated program to detect the nuclei of LPCs outbound at large heliocentric distances in the thermal IR with JWST may be capable of detecting sufficient numbers of LPC nuclei to determine bulk properties of this class of comets.

Another possibility is distant detections of inbound LPCs prior to the onset of activity. The recent discovery of comet C/2014 UN271 illustrates how challenging this will be. C/2014 UN271 was discovered in archival DECam images at ~29 au (*Bernardinelli et al.,* 2022). It is apparently exceptionally large and/or bright, so detection of a meaningful number of comets inbound requires being sensitive to smaller sizes. Since C/2014 UN271 was discovered at $V \approx 22.5$, surveys would need to go to $V \approx 27$ to detect 10-km objects at comparable distances. C/2014 UN271 exhibited clear activity in TESS data by a heliocentric distance of 23.8 au (*Farnham et al.,* 2021a) and likely activity at ~25 au in DECam images (*Bernardinelli et al.,* 2021), so discoveries must be made at even larger distances if the nucleus is to be detected outright. Alternatively, discovering that only a small fraction of LPCs activate at large heliocentric distances would support the recently proposed lack of hypervolatile-rich comets in the Oort cloud (*Lisse et al.,* 2022).

Such observations might be feasible with one of the proposed 30-m-class telescopes, although they would likely require a large field of view and dedicated time to survey a sufficiently large area of sky to get meaningful statistics about the population. This could be approached in a manner similar to the New Horizons team's search for an extended mission target. Many epochs of deep imaging of the same fields eventually yielded the discovery of Arrokoth and compelling insights into physical properties of the faintest KBOs (*Benecchi et al.,* 2015). Perhaps a 30-m-class telescope "deep field" akin to the famous Hubble Deep field could achieve such a feat.

5. SUMMARY

Since the publication of *Comets II*, our knowledge of comet nucleus properties has grown considerably and a few hundred comet nuclei, dominated by JFCs but spanning across all dynamical types, have one or more physical or surface properties measured. The vast majority of this growth has come via remote observations. Several dedicated missions have provided ground-truth measurements and confirmed that remote observations can yield accurate measurements of bulk properties. Thus, it is becoming possible to use the cumulative nucleus properties to meaningfully constrain models of solar system formation and evolution. Highlights include:

- The effective radius has been measured or constrained for a few hundred comets. Most of these (>200) were included in two large-scale IR surveys whose data permit a debiased assessment of the size distribution.
- Ever-improving observational capabilities and, in some instances, multi-orbit observations have yielded more than 50 axial ratio measurements and more than 60 rotation period measurements. *In situ* imaging from missions and radar observations have

demonstrated that highly elongated and/or bilobate nuclei are common, motivating new investigations into how this might come to be.

- The number of comets with changes in rotation period measured or meaningfully constrained has increased from just one to about a dozen, leading to the surprising finding that rotation-period changes seem to be nearly independent of the active fraction of the nucleus.
- The first nucleus polarization and delay-Doppler radar measurements have been made.
- Albedo and nucleus phase coefficients have each been measured for more than 20 comets, while nucleus color indices have been measured for more then 50 comets. A sufficient number of comets have had multiple nucleus surface properties measured to allow investigations into correlations between properties as well as comparisons with other small bodies in the solar system. When remote observations are combined with detailed mission data, studies into the "age" of comet surfaces are becoming viable.
- Detailed observations of several comets with highly favorable apparitions as well as greatly expanded surveying capabilities have led to advances in our understanding of nucleus spin states, outbursts, and disintegration/fragmentation.

We are on the cusp of the JWST and LSST eras, which are expected to provide incredible new sources of data, potentially making every currently known JFC accessible and yielding direct detections of fainter and more distant comets than previously possible. These telescopes, along with several 30-m-class groundbased facilities and eventual successors to HST and NEOWISE, should enable the first widespread studies of Oort cloud comet nuclei and undoubtedly result in paradigm-shifting breakthroughs in our understanding of small bodies in the solar system.

Acknowledgments. M.M.K. acknowledges support from NASA Solar System Observations program grant NNX17AK15G/80HQTR20T-0060. R.K. acknowledges support through the ESO Fellowship Programme. Funding via past NASA Planetary Science Division grants were instrumental in N.H.S.'s contributions and this support is gratefully acknowledged. We thank E. Howell, Y. Fernandez, M. S. P. Kelley, A. Donaldson, and J.-Y. Li for helpful discussions.

REFERENCES

Abell P. A., Fernández Y. R., Pravec P., et al. (2005) Physical characteristics of comet nucleus C/2001 OG 108 (LONEOS). *Icarus, 179,* 174–194.

Agarwal J., Jewitt D., Mutchler M., et al. (2017) A binary main-belt comet. *Nature, 549,* 357–359.

A'Hearn M. F., Schleicher D. G., Millis R. L., et al. (1984) Comet Bowell 1980b. *Astron. J., 89,* 579–591.

A'Hearn M. F., Hoban S., Birch P. V., et al. (1986) Cyanogen jets in Comet Halley. *Nature, 324,* 649–651.

A'Hearn M. F., Campins H., Schleicher D. G., et al. (1989) The nucleus of Comet P/Tempel 2. *Astrophys. J., 347,* 1155–1166.

A'Hearn M. F., Millis R. C., Schleicher D. O., et al. (1995) The ensemble properties of comets: Results from narrowband photometry of 85 comets, 1976–1992. *Icarus, 118,* 223–270.

A'Hearn M. F., Belton M. J. S., Delamere W. A., et al. (2011) EPOXI at Comet Hartley 2. *Science, 332,* 1396–1400.

Anderson W. M. Jr. (2010) Infrared observations of oxidized carbon in Comet C/2002 T7 (LINEAR). Ph.D. thesis, Catholic University, Washington, DC. 193 pp.

Asphaug E. and Benz W. (1996) Size, density, and structure of Comet Shoemaker-Levy 9 inferred from the physics of tidal breakup. *Icarus, 121,* 225–248.

Bach Y. P., Ishiguro M., and Usui F. (2017) Thermal modeling of comet-like objects from AKARI observation. *Astron. J., 154,* 202.

Bagnulo S., Tozzi G. P., Boehnhardt H., et al. (2010) Polarimetry and photometry of the peculiar main-belt object 7968 = 133P/Elst-Pizarro. *Astron. Astrophys., 514,* A99.

Bair A. N., Schleicher D. G., and Knight M. M. (2018) Coma morphology, numerical modeling, and production rates for Comet C/Lulin (2007 N3). *Astron. J., 156,* 159.

Battams K. and Knight M. M. (2017) SOHO comets: 20 years and 3000 objects later. *Philos. Trans. R. Soc. A, 375,* 20160257.

Bauer J. M., Walker R. G., Mainzer A. K., et al. (2011) WISE/NEOWISE observations of Comet 103P/Hartley 2. *Astrophys. J., 738,* 171.

Bauer J. M., Kramer E., Mainzer A. K., et al. (2012) WISE/NEOWISE preliminary analysis and highlights of the 67P/Churyumov-Gerasimenko near nucleus environs. *Astrophys. J., 758,* 18.

Bauer J. M., Grav T., Fernández Y. R., et al. (2017) Debiasing the NEOWISE cryogenic mission comet populations. *Astron. J., 154,* 53.

Belskaya I. N. and Shevchenko V. G. (2000) Opposition effect of asteroids. *Icarus, 147,* 94–105.

Belskaya I. N., Shkuratov Y. G., Efimov Y. S., et al. (2005) The F-type asteroids with small inversion angles of polarization. *Icarus, 178,* 213–221.

Belskaya I., Cellino A., Levasseur-Regourd A.-C., et al. (2019) Optical polarimetry of small solar system bodies: From asteroids to debris disks. In *Astronomical Polarisation from the Infrared to Gamma Rays* (R. Mignani et al., eds.), pp. 223–246. Astrophys. Space Sci. Library, Vol. 460, Springer, Cham.

Belton M. J. S. (1990) Rationalization of Comet Halley's periods. *Icarus, 86,* 30–51.

Belton M. J. S. (2010) Cometary activity, active areas, and a mechanism for collimated outflows on 1P, 9P, 19P, and 81P. *Icarus, 210,* 881–897.

Belton M. J. S., Julian W. H., Anderson A. J., et al. (1991) The spin state and homogeneity of Comet Halley's nucleus. *Icarus, 93,* 183–193.

Belton M. J. S., Samarasinha N. H., Fernández Y. R., et al. (2005) The excited spin state of Comet 2P/Encke. *Icarus, 175,* 181–193.

Belton M. J. S., Meech K. J., Chesley S., et al. (2011) Stardust-NExT, Deep Impact, and the accelerating spin of 9P/Tempel 1. *Icarus, 213,* 345–368.

Belton M. J. S., Thomas P., Li J.-Y., et al. (2013) The complex spin state of 103P/Hartley 2: Kinematics and orientation in space. *Icarus, 222,* 595–609.

Benecchi S. D., Noll K. S., Weaver H. A., et al. (2015) New Horizons: Long-range Kuiper belt targets observed by the Hubble Space Telescope. *Icarus, 246,* 369–374.

Benner L. A. M., Busch M. W., Giorgini J. D., et al. (2015) Radar observations of near-Earth and main-belt asteroids. In *Asteroids IV* (P. Michel et al., eds.), pp. 165–182. Univ. of Arizona, Tucson.

Bernardinelli P. H., Bernstein G. M., Montet B. T., et al. (2021) C/2014 UN271 (Bernardinelli-Bernstein): The nearly spherical cow of comets. *Astrophys. J. Lett., 921,* L37.

Bernardinelli P. H., Bernstein G. M., Sako M., et al. (2022) A search of the full six years of the Dark Energy Survey for outer solar system objects. *Astrophys. J. Suppl. Ser., 258,* 41.

Betzler A. S., de Sousa O. F., and Betzler L. B. S. (2018) Photometric study of Comet C/2014 S2 (PANSTARRS) after the perihelion. *Earth Moon Planets, 122,* 53–71.

Betzler A. S., de Sousa O. F., Diepvens A., et al. (2020) BVR photometry of Comets 63P/Wild 1 and C/2012 K1 (PANSTARRS). *Astrophys. Space Sci., 365,* 102.

Biesecker D. A., Lamy P., St. Cyr O. C., et al. (2002) Sungrazing comets discovered with the SOHO/LASCO coronagraphs 1996–1998. *Icarus, 157,* 323–348.

Blum J., Gundlach B., Krause M., et al. (2017) Evidence for the

formation of Comet 67P/Churyumov-Gerasimenko through gravitational collapse of a bound clump of pebbles. *Mon. Not. R. Astron. Soc., 469*, S755–S773.
Bodewits D., Farnham T. L., Kelley M. S. P., et al. (2018) A rapid decrease in the rotation rate of Comet 41P/Tuttle-Giacobini-Kresák. *Nature, 553*, 186–188.
Boe B., Jedicke R., Meech K. J., et al. (2019) The orbit and size-frequency distribution of long period comets observed by Pan-STARRS1. *Icarus, 333*, 252–272.
Boehnhardt H. (2004) Split comets. In *Comets II* (M. C. Festou et al., eds.), pp. 301–316. Univ. of Arizona, Tucson.
Boehnhardt H., Rainer N., Birkle K., et al. (1999) The nuclei of Comets 26P/Grigg-Skjellerup and 73P/Schwassmann-Wachmann 3. *Astron. Astrophys., 341*, 912–917.
Boehnhardt H., Delahodde C., Sekiguchi T., et al. (2002) VLT observations of Comet 46P/Wirtanen. *Astron. Astrophys., 387*, 1107–1113.
Boehnhardt H., Tozzi G. P., Bagnulo S., et al. (2008) Photometry and polarimetry of the nucleus of Comet 2P/Encke. *Astron. Astrophys., 489*, 1337–1343.
Boehnhardt H., Riffeser A., Ries C., et al. (2020) Mt. Wendelstein imaging of Comet 41P/Tuttle-Giacobini-Kresak during the 2017 perihelion arc. *Astron. Astrophys., 638*, A8.
Bortle J. E. (1991) Post-perihelion survival of comets with small q. *Intl. Comet Q., 13*, 89–91.
Bottke W. F., Durda D. D., Nesvorný D., et al. (2005) Linking the collisional history of the main asteroid belt to its dynamical excitation and depletion. *Icarus, 179*, 63–94.
Bowell E., Hapke B., Domingue D., et al. (1989) Application of photometric models to asteroids. In *Asteroids II* (R. P. Binzel et al., eds.), pp. 524–556. Univ. of Arizona, Tucson.
Britt D. T., Boice D. C., Buratti B. J., et al. (2004) The morphology and surface processes of Comet 19/P Borrelly. *Icarus, 167*, 45–53.
Buie M. W., Porter S. B., Tamblyn P., et al. (2020) Size and shape constraints of (486958) Arrokoth from stellar occultations. *Astron. J., 159*, 130.
Buratti B. J., Hicks M. D., Soderblom L. A., et al. (2004) Deep Space 1 photometry of the nucleus of Comet 19P/Borrelly. *Icarus, 167*, 16–29.
Burns J. A. and Safronov V. S. (1973) Asteroid nutation angles. *Mon. Not. R. Astron. Soc., 165*, 403–411.
Campins H. and Fernández Y. (2002) Observational constraints on surface characteristics of comet nuclei. *Earth Moon Planets, 89*, 117–134.
Campins H., A'Hearn M. F., and McFadden L.-A. (1987) The bare nucleus of Comet Neujmin 1. *Astrophys. J., 316*, 847–857.
Campins H., Osip D. J., Rieke G. H., et al. (1995) Estimates of the radius and albedo of comet-asteroid transition object 4015 Wilson-Harrington based on infrared observations. *Planet. Space Sci., 43*, 733–736.
Campins H., Ziffer J., Licandro J., et al. (2006) Nuclear spectra of Comet 162P/Siding Spring (2004 TU12). *Astron. J., 132*, 1346–1353.
Campins H., Licandro J., Pinilla-Alonso N., et al. (2007) Nuclear spectra of Comet 28P Neujmin 1. *Astron. J., 134*, 1626–1633.
Carry B., Thuillot W., Spoto F., et al. (2021) Potential asteroid discoveries by the ESA Gaia mission: Results from follow-up observations. *Astron. Astrophys., 648*, A96.
Carusi A., Kresák L., Perozzi E., et al. (1985) First results of the integration of motion of short-period comets over 800 years. In *Dynamics of Comets: Their Origin and Evolution* (A. Carusi and G. B. Valsecchi, eds.), pp. 319–340. IAU Colloq. 83, Cambridge Univ., Cambridge.
Carusi A., Kresak L., Perozzi E., et al. (1987) High-order librations of Halley-type comets. In *Explorations of Halley's Comet* (M. Grewing et al., eds)., pp. 899–905. Springer, Berlin.
Chang C.-K., Chen Y.-T., Fraser W. C., et al. (2021) FOSSIL. I. The spin rate limit of Jupiter Trojans. *Planet. Sci. J., 2*, 191.
Chen J. and Jewitt D. (1994) On the rate at which comets split. *Icarus, 108*, 265–271.
Chesley S. R., Belton M. J. S., Carcich B., et al. (2013) An updated rotation model for Comet 9P/Tempel 1. *Icarus, 222*, 516–525.
Ciarniello M., Capaccioni F., Filacchione G., et al. (2015) Photometric properties of Comet 67P/Churyumov-Gerasimenko from VIRTIS-M onboard Rosetta. *Astron. Astrophys., 583*, A31.
Cruikshank D. P., Hartmann W. K., and Tholen D. J. (1985) Colour, albedo and nucleus size of Halley's comet. *Nature, 315*, 122–124.
Davidsson B. J. R. and Gutiérrez P. J. (2006) Non-gravitational force modeling of Comet 81P/Wild 2: I. A nucleus bulk density estimate. *Icarus, 180*, 224–242.
Davidsson B. J. R., Sierks H., Güttler C., et al. (2016) The primordial nucleus of Comet 67P/Churyumov-Gerasimenko. *Astron. Astrophys., 592*, A63.
Delahodde C. E., Meech K. J., Hainaut O. R., et al. (2001) Detailed phase function of Comet 28P/Neujmin 1. *Astron. Astrophys., 376*, 672–685.
Dello Russo N., Vervack R. J., Weaver H. A., et al. (2007) Compositional homogeneity in the fragmented Comet 73P/Schwassmann-Wachmann 3. *Nature, 448*, 172–175.
DeMeo F. and Binzel R. P. (2008) Comets in the near-Earth object population. *Icarus, 194*, 436–449.
Devogèle M., Rivet J. P., Tanga P., et al. (2015) A method to search for large-scale concavities in asteroid shape models. *Mon. Not. R. Astron. Soc., 453*, 2232–2240.
Dohnanyi J. S. (1969) Collisional model of asteroids and their debris. *J. Geophys. Res., 74*, 2531–2554.
Donaldson A., Kokotanekova R., and Sondgrass C. (2023) The nucleus of 162P/Siding Spring from ground-based photometry. *Mon. Not. R. Astron. Soc., 521*, 1518–1531.
Dones L., Brasser R., Kaib N., et al. (2015) Origin and evolution of the cometary reservoirs. *Space Sci. Rev., 197*, 191–269.
Drahus M. and Waniak W. (2006) Non-constant rotation period of Comet C/2001 K5 (LINEAR). *Icarus, 185*, 544–557.
Drahus M., Jewitt D., Guilbert-Lepoutre A., et al. (2011) Rotation state of Comet 103P/Hartley 2 from radio spectroscopy at 1 mm. *Astrophys. J. Lett., 734*, L4.
Durech J., Sidorin V., and Kaasalainen M. (2010) DAMIT: A database of asteroid models. *Astron. Astrophys., 513*, A46.
Duxbury T. C., Newburn R. L., and Brownlee D. E. (2004) Comet 81P/Wild 2 size, shape, and orientation. *J. Geophys. Res.–Planets, 109*, E12S02.
Dybczyński P. A. and Królikowska M. (2015) Near-parabolic comets observed in 2006-2010 — II. Their past and future motion under the influence of the galaxy field and known nearby stars. *Mon. Not. R. Astron. Soc., 448*, 588–600.
Dybczyński P. A. and Królikowska M. (2011) Where do long-period comets come from? Moving through the Jupiter-Saturn barrier. *Mon. Not. R. Astron. Soc., 416*, 51–69.
Eisner N., Knight M. M., and Schleicher D. G. (2017) The rotation and other properties of Comet 49P/Arend-Rigaux, 1984–2012. *Astron. J., 154*, 196.
Eisner N. L., Knight M. M., Snodgrass C., et al. (2019) Properties of the bare nucleus of Comet 96P/Machholz 1. *Astron. J., 157*, 186.
Emery J. P. and Brown R. H. (2003) Constraints on the surface composition of Trojan asteroids from near-infrared (0.8–4.0 μm) spectroscopy. *Icarus, 164*, 104–121.
Farnham T. L. (2009) Coma morphology of Jupiter-family comets. *Planet. Space Sci., 57*, 1192–1217.
Farnham T. L. and Cochran A. L. (2002) A McDonald Observatory study of Comet 19P/Borrelly: Placing the Deep Space 1 observations into a broader context. *Icarus, 160*, 398–418.
Farnham T. L. and Schleicher D. G. (2005) Physical and compositional studies of Comet 81P/Wild 2 at multiple apparitions. *Icarus, 173*, 533–558.
Farnham T. L., Schleicher D. G., and A'Hearn M. F. (2000) The HB narrowband comet filters: Standard stars and calibrations. *Icarus, 147*, 180–204.
Farnham T. L., Samarasinha N. H., Mueller B. E. A., et al. (2007a) Cyanogen jets and the rotation state of Comet Machholz (C/2004 Q2). *Astron. J., 133*, 2001–2007.
Farnham T. L., Wellnitz D. D., Hampton D. L., et al. (2007b) Dust coma morphology in the Deep Impact images of Comet 9P/Tempel 1. *Icarus, 187*, 26–40.
Farnham T., Kelley M. S., Bodewits D., et al. (2017) The resolved nucleus of Comet Siding Spring (C/2013 A1) in MRO HiRISE images. *AAS/Division for Planetary Sciences Meeting Abstracts, 49*, #403.01.
Farnham T. L., Kelley M. S. P., and Bauer J. M. (2021a) Early activity in Comet C/2014 UN271 Bernardinelli-Bernstein as observed by TESS. *Planet. Sci. J., 2*, 236.
Farnham T. L., Knight M. M., Schleicher D. G., et al. (2021b) Narrowband observations of Comet 46P/Wirtanen during its exceptional apparition of 2018/19. I. Apparent rotation period and

outbursts. *Planet. Sci. J., 2,* 7.
Feldman P. D., Budzien S. A., Festou M. C., et al. (1992) Ultraviolet and visible variability of the coma of Comet Levy (1990c). *Icarus, 95,* 65–72.
Feldman P. D., Cochran A. L., and Combi M. R. (2004) Spectroscopic investigations of fragment species of the coma. In *Comets II* (M. C. Festou et al., eds.), pp. 425–447. Univ. of Arizona, Tucson.
Fernández J. A. and Sosa A. (2015) Jupiter family comets in near-Earth orbits: Are some of them interlopers from the asteroid belt? *Planet. Space Sci., 118,* 14–24.
Fernández J. A., Licandro J., Moreno F., et al. (2017) Physical and dynamical properties of the anomalous comet 249P/LINEAR. *Icarus, 295,* 34–45.
Fernández Y. R. (2009) That's the way the comet crumbles: Splitting Jupiter-family comets. *Planet. Space Sci., 57,* 1218–1227.
Fernández Y. R. (2002) The nucleus of Comet Hale-Bopp (C/1995 O1): Size and activity. *Earth Moon Planets, 89,* 3–25.
Fernández Y. R., Wellnitz D. D., Buie M. W., et al. (1999) The inner coma and nucleus of Comet Hale-Bopp: Results from a stellar occultation. *Icarus, 140,* 205–220.
Fernández Y. R., Lisse C. M., Ulrich Käufl H., et al. (2000) Physical properties of the nucleus of Comet 2P/Encke. *Icarus, 147,* 145–160.
Fernández Y. R., Jewitt D. C., and Sheppard S. S. (2001) Low albedos among extinct comet candidates. *Astrophys. J. Lett., 553,* L197–L200.
Fernández Y. R., Jewitt D. C., and Sheppard S. S. (2005) Albedos of asteroids in comet-like orbits. *Astron. J., 130,* 308–318.
Fernández Y. R., Campins H., Kassis M., et al. (2006) Comet 162P/Siding Spring: A surprisingly large nucleus. *Astron. J., 132,* 1354–1360.
Fernández Y. R., Kelley M. S., Lamy P. L., et al. (2013) Thermal properties, sizes, and size distribution of Jupiter-family cometary nuclei. *Icarus, 226,* 1138–1170.
Fernández Y. R., Weaver H. A., Lisse C. M., et al. (2016) The distribution of geometric albedos of Jupiter-family comets from SEPPCoN and visible-wavelength photometry. *AAS Meeting Abstracts, 227,* #141.22.
Ferrín I. (2010) Atlas of secular light curves of comets. *Planet. Space Sci., 58,* 365–391.
Festou M. C. and Barale O. (2000) The asymmetric coma of comets. I. Asymmetric outgassing from the nucleus of Comet 2P/Encke. *Astron. J., 119,* 3119–3132.
Festou M. C., Keller H. U., and Weaver H. A., eds. (2004) *Comets II.* Univ. of Arizona, Tucson. 745 pp.
Fitzsimmons A. and Williams I. P. (1994) The nucleus of Comet P/Levy 1991 XI. *Astron. Astrophys., 289,* 304–310.
Fitzsimmons A., Dahlgren M., Lagerkvist C. I., et al. (1994) A spectroscopic survey of D-type asteroids. *Astron. Astrophys., 282,* 634–642.
Fomenkova M. N., Jones B., Pina R., et al. (1995) Mid-infrared observations of the nucleus and dust of Comet P/Swift-Tuttle. *Astron. J., 110,* 1866–1874.
Fornasier S., Dotto E., Hainaut O., et al. (2007) Visible spectroscopic and photometric survey of Jupiter Trojans: Final results on dynamical families. *Icarus, 190,* 622–642.
Fornasier S., Barucci M. A., de Bergh C., et al. (2009) Visible spectroscopy of the new ESO large programme on trans-neptunian objects and Centaurs: Final results. *Astron. Astrophys., 508,* 457–465.
Fornasier S., Hasselmann P. H., Barucci M. A., et al. (2015) Spectrophotometric properties of the nucleus of Comet 67P/Churyumov-Gerasimenko from the OSIRIS instrument onboard the Rosetta spacecraft. *Astron. Astrophys., 583,* A30.
Fornasier S., Mottola S., Keller H. U., et al. (2016) Rosetta's Comet 67P/Churyumov-Gerasimenko sheds its dusty mantle to reveal its icy nature. *Science, 354,* 1566–1570.
Fox N. J., Velli M. C., Bale S. D., et al. (2016) The Solar Probe Plus mission: Humanity's first visit to our star. *Space Sci. Rev., 204,* 7–48.
Fraser W. C. and Brown M. E. (2012) The Hubble Wide Field Camera 3 test of surfaces in the outer solar system: The compositional classes of the Kuiper belt. *Astrophys. J., 749,* 33.
Gehrels T. (1956) Photometric studies of asteroids. V. The light-curve and phase function of 20 Massalia. *Astrophys. J., 123,* 331–338.
Gladman B., Marsden B. G., and Vanlaerhoven C. (2008) Nomenclature in the outer solar system. In *The Solar System Beyond Neptune* (M. A. Barucci et al., eds.), pp. 43–57. Univ. of Arizona, Tucson.
Goldberg C., Lejoly C., and Samarasinha N. (2023) Analysis of CN coma morphology features of Comet 21P/Giacobini-Zinner. *Planet. Sci. J., 4,* 28.
Gomes R. S., Morbidelli A., and Levison H. F. (2004) Planetary migration in a planetesimal disk: Why did Neptune stop at 30 AU? *Icarus, 170,* 492–507.
Gomes R., Levison H. F., Tsiganis K., et al. (2005) Origin of the cataclysmic Late Heavy Bombardment period of the terrestrial planets. *Nature, 435,* 466–469.
Gradie J. and Tedesco E. (1982) Compositional structure of the asteroid belt. *Science, 216,* 1405–1407.
Graykowski A. and Jewitt D. (2019) Fragmented Comet 73P/Schwassmann-Wachmann 3. *Astron. J., 158,* 112.
Graykowski A. and Jewitt D. (2021) Influence of small bodies on the evolution of the solar system: Fragmenting comets and irregular satellites. *AAS/Division for Planetary Sciences Meeting Abstracts, 53,* #201.01D.
Gronkowski P. and Wesołowski M. (2016) A review of cometary outbursts at large heliocentric distances. *Earth Moon Planets, 119,* 23–33.
Groussin O., Lamy P., Jorda L., et al. (2004) The nuclei of Comets 126P/IRAS and 103P/Hartley 2. *Astron. Astrophys., 419,* 375–383.
Groussin O., Lamy P., Toth I., et al. (2009) The size and thermal properties of the nucleus of Comet 22P/Kopff. *Icarus, 199,* 568–570.
Groussin O., Lamy P. L., and Jorda L. (2010) The nucleus of Comet C/1983 H1 IRAS-Araki-Alcock. *Planet. Space Sci., 58,* 904–912.
Groussin O., Jorda L., Auger A. T., et al. (2015) Gravitational slopes, geomorphology, and material strengths of the nucleus of Comet 67P/Churyumov-Gerasimenko from OSIRIS observations. *Astron. Astrophys., 583,* A32.
Groussin O., Attree N., Brouet Y., et al. (2019) The thermal, mechanical, structural, and dielectric properties of cometary nuclei after Rosetta. *Space Sci. Rev., 215,* 29.
Gutiérrez P. J., de León J., Jorda L., et al. (2003a) New spin period determination for Comet 6P/d'Arrest. *Astron. Astrophys., 407,* L37–L40.
Gutiérrez P. J., Jorda L., Ortiz J. L., et al. (2003b) Long-term simulations of the rotational state of small irregular cometary nuclei. *Astron. Astrophys., 406,* 1123–1133.
Gutiérrez P. J., Jorda L., Gaskell R. W., et al. (2016) Possible interpretation of the precession of Comet 67P/Churyumov-Gerasimenko. *Astron. Astrophys., 590,* A46.
Hainaut O. R., Meech K. J., Boehnhardt H., et al. (1998) Early recovery of Comet 55P/Tempel-Tuttle. *Astron. Astrophys., 333,* 746–752.
Hammel H. B., Beebe R. F., Ingersoll A. P., et al. (1995) HST imaging of atmospheric phenomena created by the impact of Comet Shoemaker-Levy 9. *Science, 267,* 1288–1296.
Hanner M. S., Giese R. H., Weiss K., et al. (1981) On the definition of albedo and application to irregular particles. *Astron. Astrophys., 104,* 42–46.
Hapke B. (1981) Bidirectional reflectance spectroscopy. I — Theory. *J. Geophys. Res.–Solid Earth, 86,* 3039–3054.
Harmon J. K., Campbell D. B., Ostro S. J., et al. (1999) Radar observations of comets. *Planet. Space Sci., 47,* 1409–1422.
Harmon J. K., Nolan M. C., Ostro S. J., et al. (2004) Radar studies of comet nuclei and grain comae. In *Comets II* (M. C. Festou et al., eds.), pp. 265–279. Univ. of Arizona, Tucson.
Harmon J. K., Nolan M. C., Margot J. L., et al. (2006) Radar observations of Comet P/2005 JQ5 (Catalina). *Icarus, 184,* 285–288.
Harmon J. K., Nolan M. C., Giorgini J. D., et al. (2010) Radar observations of 8P/Tuttle: A contact-binary comet. *Icarus, 207,* 499–502.
Harmon J. K., Nolan M. C., Howell E. S., et al. (2011) Radar observations of Comet 103P/Hartley 2. *Astrophys. J. Lett., 734,* L2.
Hasselmann P. H., Barucci M. A., Fornasier S., et al. (2017) The opposition effect of 67P/Churyumov-Gerasimenko on post-perihelion Rosetta images. *Mon. Not. R. Astron. Soc., 469,* S550–S567.
Hicks M. D. and Bauer J. M. (2007) P/2006 HR30 (Siding Spring): A low-activity comet in near-Earth space. *Astrophys. J. Lett., 662,* L47–L50.
Hirabayashi M., Scheeres D. J., Chesley S. R., et al. (2016) Fission and reconfiguration of bilobate comets as revealed by 67P/Churyumov-Gerasimenko. *Nature, 534,* 352–355.
Hoban S., Samarasinha N. H., A'Hearn M. F. A., et al. (1988) An investigation into periodicities in the morphology of CN jets in

Comet P/Halley. *Astron. Astrophys., 195*, 331–337.
Holler B., Milam S. N., Bauer J. M., et al. (2021) Minor body science with the Nancy Grace Roman Space Telescope. *Bull. Am. Astron. Soc., 53*, 30.
Holmberg J., Flynn C., and Portinari L. (2006) The colours of the Sun. *Mon. Not. R. Astron. Soc., 367*, 449–453.
Howell E. S., Nolan M. C., Taylor P. A., et al. (2014) Radar images of Comet 209P/LINEAR: Constraints on shape and rotation. *AAS/Division for Planetary Sciences Meeting Abstracts, 46*, #209.24.
Howell E. S., Lejoly C., Taylor P. A., et al. (2017) Arecibo radar observations of 41P/Tuttle-Giacobini-Kresák constrain the nucleus size and rotation. *AAS/Division for Planetary Sciences Meeting Abstracts, 49*, #414.24.
Howell E., Belton M. J., Samarasinha N. H., et al. (2018) Arguments for NPA rotation of Comet 41P/Tuttle-Giacobini-Kresák. *AAS/Division for Planetary Sciences Meeting Abstracts, 50*, #106.06.
Hsieh H. H. (2017) Asteroid-comet continuum objects in the solar system. *Philos. Trans. R. Soc. A, 375*, 20160259.
Hsieh H. H. and Jewitt D. (2006) A population of comets in the main asteroid belt. *Science, 312*, 561–563.
Hsieh H. H., Novaković B., Walsh K. J., et al. (2020) Potential Themis-family asteroid contribution to the Jupiter-family comet population. *Astron. J., 159*, 179.
Hu S., Richardson D. C., Zhang Y., et al. (2021) Critical spin periods of sub-km-sized cohesive rubble-pile asteroids: Dependences on material parameters. *Mon. Not. R. Astron. Soc., 502*, 5277–5291.
Hughes D. W. (1991) Possible mechanisms for cometary outbursts. In *Comets in the Post-Halley Era* (J. Newburn et al., eds.), pp. 825–851. IAU Colloq. 167, Cambridge Univ., Cambridge.
Hui M.-T. and Li J.-Y. (2018) Is the cometary nucleus-extraction technique reliable? *Publ. Astron. Soc. Pac., 130*, 104501.
Hui M.-T., Jewitt D., Yu L.-L., et al. (2022a) Hubble Space Telescope detection of the nucleus of Comet C/2014 UN271 (Bernardinelli-Bernstein). *Astrophys. J. Lett., 929*, L12.
Hui M.-T., Tholen D. J., Kracht R., et al. (2022b) The lingering death of periodic near-Sun Comet 323P/SOHO. *Astron. J., 164*, 1.
Ip W. H., Lai I. L., Lee J. C., et al. (2016) Physical properties and dynamical relation of the circular depressions on Comet 67P/Churyumov-Gerasimenko. *Astron. Astrophys., 591*, A132.
Ishiguro M., Kuroda D., Hanayama H., et al. (2016) 2014–2015 multiple outbursts of 15P/Finlay. *Astron. J., 152*, 169.
Ivanova O., Rosenbush V., Afanasiev V., et al. (2017) Polarimetry, photometry, and spectroscopy of Comet C/2009 P1 (Garradd). *Icarus, 284*, 167–182.
Ivezić Ž., Kahn S. M., Tyson J. A., et al. (2019) LSST: From science drivers to reference design and anticipated data products. *Astrophys. J., 873*, 111.
Jehin E., Manfroid J., Hutsemekers D., et al. (2010) Comet 103P/Hartley. *Central Bureau Electronic Telegram 2589*.
Jester S., Schneider D. P., Richards G. T., et al. (2005) The Sloan Digital Sky Survey view of the Palomar-Green Bright Quasar Survey. *Astron. J., 130*, 873–895.
Jewitt D. (1997) Cometary rotation: An overview. *Earth Moon Planets, 79*, 35–53.
Jewitt D. C. (2002) From Kuiper belt object to cometary nucleus: The missing ultrared matter. *Astron. J., 123*, 1039–1049.
Jewitt D. (2004a) Looking through the HIPPO: Nucleus and dust in Comet 2P/Encke. *Astron. J., 128*, 3061–3069.
Jewitt D. C. (2004b) From cradle to grave: The rise and demise of the comets. In *Comets II* (M. C. Festou et al., eds.), pp. 659–676. Univ. of Arizona, Tucson.
Jewitt D. (2005) A first look at the Damocloids. *Astron. J., 129*, 530–538.
Jewitt D. (2015) Color systematics of comets and related bodies. *Astron. J., 150*, 201.
Jewitt D. (2021) Systematics and consequences of comet nucleus outgassing torques. *Astron. J., 161*, 261.
Jewitt D. (2022) Destruction of long-period comets. *Astron. J., 164*, 158.
Jewitt D. and Luu J. (1989) A CCD portrait of Comet P/Tempel 2. *Astron. J., 97*, 1766–1790.
Jewitt D. C. and Meech K. J. (1988) Optical properties of cometary nuclei and a preliminary comparison with asteroids. *Astrophys. J., 328*, 974–986.
Jewitt D. and Sheppard S. (2004) The nucleus of Comet 48P/Johnson. *Astron. J., 127*, 1784–1790.
Jewitt D., Sheppard S., and Fernández Y. (2003) 143P/Kowal-Mrkos and the shapes of cometary nuclei. *Astron. J., 125*, 3366–3377.
Jewitt D., Hsieh H., and Agarwal J. (2015) The active asteroids. In *Asteroids IV* (P. Michel et al., eds.), pp. 221–241. Univ. of Arizona, Tucson.
Jewitt D., Mutchler M., Weaver H., et al. (2016) Fragmentation kinematics in Comet 332P/Ikeya-Murakami. *Astrophys. J. Lett., 829*, L8.
Jones G. H., Knight M. M., Battams K., et al. (2018) The science of sungrazers, sunskirters, and other near-Sun comets. *Space Sci. Rev., 214*, 20.
Jorda L. and Gutiérrez P. (2002) Rotational properties of cometary nuclei. *Earth Moon Planets, 89*, 135–160.
Jorda L., Lamy P., Groussin O., et al. (2000) ISOCAM observations of cometary nuclei. In *ISO Beyond Point Sources: Studies of Extended Infrared Emission* (R. J. Laureijs et al., eds.), pp. 61–66. ESA SP-455, Noordwijk, The Netherlands.
Jorda L., Gaskell R., Capanna C., et al. (2016) The global shape, density and rotation of Comet 67P/Churyumov-Gerasimenko from preperihelion Rosetta/OSIRIS observations. *Icarus, 277*, 257–278.
Jutzi M., Benz W., Toliou A., et al. (2017) How primordial is the structure of Comet 67P? Combined collisional and dynamical models suggest a late formation. *Astron. Astrophys., 597*, A61.
Kaasalainen M. and Torppa J. (2001) Optimization methods for asteroid lightcurve inversion: I. Shape determination. *Icarus, 153*, 24–36.
Kaasalainen M., Torppa J., and Muinonen K. (2001) Optimization methods for asteroid lightcurve inversion: II. The complete inverse problem. *Icarus, 153*, 37–51.
Kareta T., Hergenrother C., Reddy V., et al. (2021) Surfaces of (nearly) dormant comets and the recent history of the Quadrantid meteor shower. *Planet. Sci. J., 2*, 31.
Kareta T., Reddy V., Sanchez J. A., et al. (2022) Near-infrared spectroscopy of the nucleus of low-activity Comet P/2016 BA14 during its 2016 close approach. *Planet. Sci. J., 3*, 105.
Kasuga T., Balam D. D., and Wiegert P. A. (2010) Comet 169P/NEAT (=2002 EX_{12}): The parent body of the α-Capricornid meteoroid stream. *Astron. J., 140*, 1806–1813.
Keller H. U., Arpigny C., Barbieri C., et al. (1986) First Halley multicolour camera imaging results from Giotto. *Nature, 321*, 320–326.
Keller H. U., Delamere W. A., Reitsema H. J., et al. (1987) Comet P/Halley's nucleus and its activity. *Astron. Astrophys., 187*, 807–823.
Kelley M. S., Fernández Y. R., Licandro J., et al. (2013) The persistent activity of Jupiter-family comets at 3–7 AU. *Icarus, 225*, 475–494.
Kelley M. S. P., Woodward C. E., Gehrz R. D., et al. (2017) Mid-infrared spectra of comet nuclei. *Icarus, 284*, 344–358.
Kelley M. S. P., Bodewits D., Ye Q., et al. (2019) ZChecker: Finding cometary outbursts with the Zwicky Transient Facility. In *Astronomical Data Analysis Software and Systems XXVII* (P. J. Teuben et al., eds.), pp. 471–474. ASP Conf. Ser. 523, Astronomical Society of the Pacific, San Francisco.
Kiselev N., Rosenbush V., Levasseur-Regourd A.-C., et al. (2015) Comets. In *Polarimetry of Stars and Planetary Systems* (L. Kolokolova et al., eds.), pp. 379–404. Cambridge Univ., Cambridge.
Knight M. M. and Schleicher D. G. (2011) CN morphology studies of Comet 103P/Hartley 2. *Astron. J., 141*, 183.
Knight M. M. and Walsh K. J. (2013) Will Comet ISON (C/2012 S1) survive perihelion? *Astrophys. J. Lett., 776*, L5.
Knight M. M., A'Hearn M. F., Biesecker D. A., et al. (2010) Photometric study of the Kreutz comets observed by SOHO from 1996 to 2005. *Astron. J., 139*, 926–949.
Knight M. M., Schleicher D. G., Farnham T. L., et al. (2012) A quarter-century of observations of Comet 10P/Tempel 2 at Lowell Observatory: Continued spin-down, coma morphology, production rates, and numerical modeling. *Astron. J., 144*, 153.
Knight M. M., Mueller B. E. A., Samarasinha N. H., et al. (2015) A further investigation of apparent periodicities and the rotational state of Comet 103P/Hartley 2 from combined coma morphology and light curve data sets. *Astron. J., 150*, 22.
Knight M. M., Fitzsimmons A., Kelley M. S. P., et al. (2016) Comet 322P/SOHO 1: An asteroid with the smallest perihelion distance? *Astrophys. J. Lett., 823*, L6.
Knight M. M., Schleicher D. G., and Farnham T. L. (2021) Narrowband observations of Comet 46P/Wirtanen during its exceptional apparition of 2018/19. II. Photometry, jet morphology, and modeling results. *Planet. Sci. J., 2*, 104.
Kokotanekova R., Snodgrass C., Lacerda P., et al. (2017) Rotation of

cometary nuclei: New light curves and an update of the ensemble properties of Jupiter-family comets. *Mon. Not. R. Astron. Soc., 471*, 2974–3007.

Kokotanekova R., Snodgrass C., Lacerda P., et al. (2018) Implications of the small spin changes measured for large Jupiter-family comet nuclei. *Mon. Not. R. Astron. Soc., 479*, 4665–4680.

Kolokolova L., Hanner M. S., Levasseur-Regourd A.-C., et al. (2004) Physical properties of cometary dust from light scattering and thermal emission. In *Comets II* (M. C. Festou et al., eds.), pp. 577–604. Univ. of Arizona, Tucson.

Kramer T., Läuter M., Hviid S., et al. (2019) Comet 67P/Churyumov-Gerasimenko rotation changes derived from sublimation-induced torques. *Astron. Astrophys., 630*, A3.

Kresák L. (1972) Jacobian integral as a classificational and evolutionary parameter of interplanetary bodies. *Bull. Astron. Inst. Czech., 23*, 1–34.

Kronk G. W. (1999) *Cometography: A Catalog of Comets, Volume 1, Ancient–1799.* Cambridge Univ., New York. 580 pp.

Kronk G. W. (2003) *Cometography: A Catalog of Comets, Volume 2, 1800–1899.* Cambridge Univ., New York. 852 pp.

Kuroda D., Ishiguro M., Watanabe M., et al. (2015) Optical and near-infrared polarimetry for a highly dormant Comet 209P/LINEAR. *Astrophys. J., 814*, 156.

Lacerda P., Fornasier S., Lellouch E., et al. (2014) The albedo-color diversity of transneptunian objects. *Astrophys. J. Lett., 793*, L2.

Lamy P. L. and Toth I. (1995) Direct detection of a cometary nucleus with the Hubble Space Telescope. *Astron. Astrophys., 293*, L43–L45.

Lamy P. and Toth I. (2009) The colors of cometary nuclei — Comparison with other primitive bodies of the solar system and implications for their origin. *Icarus, 201*, 674–713.

Lamy P. L., Toth I., A'Hearn M. F., et al. (1999) Hubble Space Telescope observations of the nucleus of Comet 45P/Honda-Mrkos-Pajdusakova and its inner coma. *Icarus, 140*, 424–438.

Lamy P. L., Toth I., Jorda L., et al. (2002) The nucleus of Comet 22P/Kopff and its inner coma. *Icarus, 156*, 442–455.

Lamy P. L., Toth I., Fernandez Y. R., et al. (2004) The sizes, shapes, albedos, and colors of cometary nuclei. In *Comets II* (M. C. Festou et al., eds.), pp. 223–264. Univ. of Arizona, Tucson.

Lamy P. L., Toth I., Weaver H. A., et al. (2009) Properties of the nuclei and comae of 13 ecliptic comets from Hubble Space Telescope snapshot observations. *Astron. Astrophys., 508*, 1045–1056.

Lamy P. L., Toth I., Weaver H. A., et al. (2011) Properties of the nuclei and comae of 10 ecliptic comets from Hubble Space Telescope multi-orbit observations. *Mon. Not. R. Astron. Soc., 412*, 1573–1590.

Lamy P., Toth I., Jorda L., et al. (2012) Comet 8P/Tuttle: A portrait of a contact-binary nucleus from Hubble and Spitzer Space Telescopes observations. *EGU General Assembly Abstracts, 14*, 10506.

Landau L. D. and Lifshitz E. M. (1976) *Mechanics, 3rd edition.* Butterworth-Heinemann, Oxford. 224 pp.

Lejoly C. and Howell E. (2017) Comet 45P/Honda-Mrkos-Pajdusakova. *Central Bureau Electronic Telegram 4357.*

Lejoly C., Harris W., Samarasinha N., et al. (2019) Radial dust profiles of three close approach Jupiter family comets. *EPSC-DPS Joint Meeting 2019*, 718.

Lejoly C., Harris W., Samarasinha N., et al. (2022) Radial distribution of the dust comae of comets 45P/Honda-Mrkos-Pajdušáková and 46P/Wirtanen. *Planet. Sci. J., 3*, 17.

Lellouch E., Moreno R., Bockelée-Morvan D., et al. (2022) Size and albedo of the largest detected Oort-cloud object: Comet C/2014 UN271 (Bernardinelli-Bernstein). *Astron. Astrophys., 659*, L1.

Levison H. F. (1996) Comet taxonomy. In *Completing the Inventory of the Solar System* (T. Rettig and J. M. Hahn, eds.), pp. 173–191. ASP Conf. Ser. 107, Astronomical Society of the Pacific, San Francisco.

Levison H. F. and Duncan M. J. (1994) The long-term dynamical behavior of short-period comets. *Icarus, 108*, 18–36.

Levison H. F. and Duncan M. J. (1997) From the Kuiper belt to Jupiter-family comets: The spatial distribution of ecliptic comets. *Icarus, 127*, 13–32.

Li J. and Jewitt D. (2015) Disappearance of Comet C/2010 X1 (Elenin): Gone with a whimper, not a bang. *Astron. J., 149*, 133.

Li J.-Y., A'Hearn M. F., Belton M. J. S., et al. (2007a) Deep Impact photometry of Comet 9P/Tempel 1. *Icarus, 187*, 41–55.

Li J.-Y., A'Hearn M. F., McFadden L. A., et al. (2007b) Photometric analysis and disk-resolved thermal modeling of Comet 19P/Borrelly from Deep Space 1 data. *Icarus, 188*, 195–211.

Li J.-Y., A'Hearn M. F., Farnham T. L., et al. (2009) Photometric analysis of the nucleus of Comet 81P/Wild 2 from Stardust images. *Icarus, 204*, 209–226.

Li J.-Y., Besse S., A'Hearn M. F., et al. (2013) Photometric properties of the nucleus of Comet 103P/Hartley 2. *Icarus, 222*, 559–570.

Li J.-Y., Samarasinha N. H., Kelley M. S. P., et al. (2016) Seasonal evolution on the nucleus of Comet C/2013 A1 (Siding Spring). *Astrophys. J. Lett., 817*, L23.

Li J.-Y., Kelley M. S. P., Samarasinha N. H., et al. (2017) The unusual apparition of Comet 252P/2000 G1 (LINEAR) and comparison with Comet P/2016 BA14 (PanSTARRS). *Astron. J., 154*, 136.

Licandro J., Tancredi G., Lindgren M., et al. (2000) CCD photometry of cometary nuclei, I: Observations from 1990–1995. *Icarus, 147*, 161–179.

Licandro J., Alí-Lagoa V., Tancredi G., et al. (2016) Size and albedo distributions of asteroids in cometary orbits using WISE data. *Astron. Astrophys., 585*, A9.

Licandro J., Popescu M., de León J., et al. (2018) The visible and near-infrared spectra of asteroids in cometary orbits. *Astron. Astrophys., 618*, A170.

Lin Z.-Y., Lin C.-S., Ip W.-H., et al. (2009) The outburst of Comet 17P/Holmes. *Astron. J., 138*, 625–632.

Lisse C. M., Fernández Y. R., Kundu A., et al. (1999) The nucleus of Comet Hyakutake (C/1996 B2). *Icarus, 140*, 189–204.

Lisse C. M., A'Hearn M. F., Groussin O., et al. (2005) Rotationally resolved 8–35 micron Spitzer Space Telescope observations of the nucleus of Comet 9P/Tempel 1. *Astrophys. J. Lett., 625*, L139–L142.

Lisse C. M., Fernandez Y. R., Reach W. T., et al. (2009) Spitzer Space Telescope observations of the nucleus of Comet 103P/Hartley 2. *Publ. Astron. Soc. Pac., 121*, 968–975.

Lisse C. M., Gladstone G. R., Young L. A., et al. (2022) A predicted dearth of majority hypervolatile ices in Oort cloud comets. *Planet. Sci. J., 3*, 112.

Lister T., Kelley M. S. P., Holt C. E., et al. (2022) The LCO Outbursting Objects Key Project: Overview and year 1 status. *Planet. Sci. J., 3*, 173.

Longobardo A., Palomba E., Capaccioni F., et al. (2017) Photometric behaviour of 67P/Churyumov-Gerasimenko and analysis of its pre-perihelion diurnal variations. *Mon. Not. R. Astron. Soc., 469*, S346–S356.

Lowry S. C. and Fitzsimmons A. (2001) CCD photometry of distant comets II. *Astron. Astrophys., 365*, 204–213.

Lowry S. C. and Fitzsimmons A. (2005) William Herschel Telescope observations of distant comets. *Mon. Not. R. Astron. Soc., 358*, 641–650.

Lowry S. C. and Weissman P. R. (2003) CCD observations of distant comets from Palomar and Steward Observatories. *Icarus, 164*, 492–503.

Lowry S. C. and Weissman P. R. (2007) Rotation and color properties of the nucleus of Comet 2P/Encke. *Icarus, 188*, 212–223.

Lowry S. C., Fitzsimmons A., Cartwright I. M., et al. (1999) CCD photometry of distant comets. *Astron. Astrophys., 349*, 649–659.

Lowry S. C., Fitzsimmons A., and Collander-Brown S. (2003) CCD photometry of distant comets. III. Ensemble properties of Jupiter-family comets. *Astron. Astrophys., 397*, 329–343.

Lowry S., Fitzsimmons A., Lamy P., et al. (2008) Kuiper belt objects in the planetary region: The Jupiter-family comets. In *The Solar System Beyond Neptune* (M. A. Barucci et al., eds.), pp. 397–410. Univ. of Arizona, Tucson.

Lowry S., Duddy S. R., Rozitis B., et al. (2012) The nucleus of Comet 67P/Churyumov-Gerasimenko: A new shape model and thermophysical analysis. *Astron. Astrophys., 548*, A12.

Luu J. X. (1993) Spectral diversity among the nuclei of comets. *Icarus, 104*, 138–148.

Luu J. X. and Jewitt D. C. (1990a) Charge-coupled device spectra of asteroids. I. Near-Earth and 3:1 resonance asteroids. *Astron. J., 99*, 1985–2011.

Luu J. X. and Jewitt D. C. (1990b) Cometary activity in 2060 Chiron. *Astron. J., 100*, 913–932.

Luu J. X. and Jewitt D. C. (1992) Near-aphelion CCD photometry of Comet P/Schwassmann-Wachmann 2. *Astron. J., 104*, 2243–2249.

Lyttleton R. A. (1953) *The Comets and Their Origin.* Cambridge Univ., Cambridge. 173 pp.

Mainzer A., Grav T., Bauer J., et al. (2011) NEOWISE observations of near-Earth objects: Preliminary results. *Astrophys. J., 743*, 156.

Manzini F., Oldani V., Crippa R., et al. (2014) Comet McNaught

(260P/2012 K2): Spin axis orientation and rotation period. *Astrophys. Space Sci.*, *351*, 435–450.
Manzini F., Oldani V., Behrend R., et al. (2016) Comet C/2013 X1 (PanSTARRS): Spin axis and rotation period. *Planet. Space Sci.*, *129*, 108–117.
Manzini F., Oldani V., Ochner P., et al. (2021) Coma morphology and dust emission pattern of Comet C/2020 F3 (NEOWISE). *Mon. Not. R. Astron. Soc.*, *506*, 6195–6202.
Marchi S., Magrin S., Nesvorný D., et al. (2006) A spectral slope versus perihelion distance correlation for planet-crossing asteroids. *Mon. Not. R. Astron. Soc.*, *368*, L39–L42.
Marsden B. G. (1967) The sungrazing comet group. *Astron. J.*, *72*, 1170–1183.
Marsden B. G. (1989) The sungrazing comet group. II. *Astron. J.*, *98*, 2306–2321.
Marsden B. G. (2005) Sungrazing comets. *Annu. Rev. Astron. Astrophys.*, *43*, 75–102.
Masoumzadeh N., Oklay N., Kolokolova L., et al. (2017) Opposition effect on Comet 67P/Churyumov-Gerasimenko using Rosetta-OSIRIS images. *Astron. Astrophys.*, *599*, A11.
Matonti C., Attree N., Groussin O., et al. (2019) Bilobate comet morphology and internal structure controlled by shear deformation. *Nature Geosci.*, *12*, 157–162.
Mazzotta Epifani E., Palumbo P., Capria M. T., et al. (2007) The distant activity of short-period comets — I. *Mon. Not. R. Astron. Soc.*, *381*, 713–722.
Mazzotta Epifani E., Palumbo P., Capria M. T., et al. (2008) The distant activity of short period comets — II. *Mon. Not. R. Astron. Soc.*, *390*, 265–280.
McKinnon W. B., Richardson D. C., Marohnic J. C., et al. (2020) The solar nebula origin of (486958) Arrokoth, a primordial contact binary in the Kuiper belt. *Science*, *367*, eaay6620.
Meech K. J., Belton M. J. S., Mueller B. E. A., et al. (1993) Nucleus properties of P/Schwassmann-Wachmann 1. *Astron. J.*, *106*, 1222–1236.
Meech K. J., Hainaut O. R., and Marsden B. G. (2004) Comet nucleus size distributions from HST and Keck telescopes. *Icarus*, *170*, 463–491.
Meech K. J., A'Hearn M. F., Adams J. A., et al. (2011) EPOXI: Comet 103P/Hartley 2 observations from a worldwide campaign. *Astrophys. J. Lett.*, *734*, L1.
Meech K. J., Yang B., Kleyna J., et al. (2016) Inner solar system material discovered in the Oort cloud. *Sci. Adv.*, *2*, e1600038.
Miles R. (2016a) Discrete sources of cryovolcanism on the nucleus of Comet 29P/Schwassmann-Wachmann and their origin. *Icarus*, *272*, 387–413.
Miles R. (2016b) Heat of solution: A new source of thermal energy in the subsurface of cometary nuclei and the gas-exsolution mechanism driving outbursts of Comet 29P/Schwassmann–Wachmann and other comets. *Icarus*, *272*, 356–386.
Millis R. L. and Schleicher D. G. (1986) Rotational period of Comet Halley. *Nature*, *324*, 646–649.
Millis R. L., A'Hearn M. F., and Campins H. (1985) The nucleus and coma of Comet P/Arend-Rigaux. *Bull. Am. Astron. Soc.*, *17*, 688.
Millis R. L., A'Hearn M. F., and Campins H. (1988) An investigation of the nucleus and coma of Comet P/Arend-Rigaux. *Astrophys. J.*, *324*, 1194–1209.
Mommert M., Hora J. L., Trilling D. E., et al. (2020) Recurrent cometary activity in near-Earth object (3552) Don Quixote. *Planet. Sci. J.*, *1*, 12.
Montalto M., Riffeser A., Hopp U., et al. (2008) The Comet 17P/Holmes 2007 outburst: The early motion of the outburst material. *Astron. Astrophys.*, *479*, L45–L49.
Morbidelli A. and Nesvorný D. (2020) Kuiper belt: Formation and evolution. In *The Trans-Neptunian Solar System* (D. Prailnik et al., eds.), pp. 25–59. Elsevier, Amsterdam.
Morbidelli A. and Rickman H. (2015) Comets as collisional fragments of a primordial planetesimal disk. *Astron. Astrophys.*, *583*, A43.
Morbidelli A., Levison H. F., Tsiganis K., et al. (2005) Chaotic capture of Jupiter's Trojan asteroids in the early solar system. *Nature*, *435*, 462–465.
Morbidelli A., Nesvorny D., Bottke W. F., et al. (2021) A re-assessment of the Kuiper belt size distribution for subkilometer objects, revealing collisional equilibrium at small sizes. *Icarus*, *356*, 114256.
Moulane Y., Jehin E., Opitom C., et al. (2018) Monitoring of the activity and composition of comets 41P/Tuttle-Giacobini-Kresak and 45P/Honda-Mrkos-Pajdusakova. *Astron. Astrophys.*, *619*, A156.
Mueller B. E. A. (1992) CCD-photometry of comets at large heliocentric distances. In *Asteroids, Comets, Meteors 1991* (A. W. Harris and E. Bowell, eds.), pp. 425–428. Lunar and Planetary Institute, Houston.
Mueller B. E. A. and Samarasinha N. H. (2018) Further investigation of changes in cometary rotation. *Astron. J.*, *156*, 107.
Mueller B. E. A. and Samarasinha N. H. (2002) Visible lightcurve observations of Comet 19P/Borrelly. *Earth Moon Planets*, *90*, 463–471.
Mueller B. E. A., Samarasinha N. H., and Belton M. J. S. (1997) Imaging of the structure and evolution of the coma morphology of Comet Hale-Bopp (C/1995 O1). *Earth Moon Planets*, *77*, 181–188.
Mueller B. E. A., Farnham T. L., Samarasinha N. H., et al. (2010) Determination of the rotation period of Comet 81P/Wild 2. *AAS/Division for Planetary Sciences Meeting Abstracts*, *42*, #28.31.
Müller D., Marsden R. G., St. Cyr O. C., et al. (2013) Solar Orbiter: Exploring the Sun-heliosphere connection. *Solar Phys.*, *285*, 25–70.
Müller T., Lellouch E., and Fornasier S. (2020) Trans-neptunian objects and Centaurs at thermal wavelengths. In *The Trans-Neptunian Solar System* (D. Prailnik et al., eds.), pp. 153–181. Elsevier, Amsterdam.
Naidu S. P., Benner L. A. M., Brozovic M., et al. (2016) High-resolution Goldstone radar imaging of Comet P/2016 BA14 (Pan-STARRS). *AAS/Division for Planetary Sciences Meeting Abstracts*, *48*, #219.05.
Neishtadt A. I., Scheeres D. J., Sidorenko V. V., et al. (2002) Evolution of comet nucleus rotation. *Icarus*, *157*, 205–218.
Nesvorný D., Vokrouhlický D., Dones L., et al. (2017) Origin and evolution of short-period comets. *Astrophys. J.*, *845*, 27.
Nihei T. C., Lehner M. J., Bianco F. B., et al. (2007) Detectability of occultations of stars by objects in the Kuiper belt and Oort cloud. *Astron. J.*, *134*, 1596–1612.
Nolan M. C., Harmon J. K., Howell E. S., et al. (2006) Radar observations of Comet 73P/Schwassmann-Wachmann 3. *AAS/Division for Planetary Sciences Meeting Abstracts*, *38*, #12.06.
Oberst J., Giese B., Howington-Kraus E., et al. (2004) The nucleus of Comet Borrelly: A study of morphology and surface brightness. *Icarus*, *167*, 70–79.
O'Brien D. P. and Greenberg R. (2003) Steady-state size distributions for collisional populations: Analytical solution with size-dependent strength. *Icarus*, *164*, 334–345.
O'Brien D. P. and Greenberg R. (2005) The collisional and dynamical evolution of the main-belt and NEA size distributions. *Icarus*, *178*, 179–212.
O'Dell C. R. (1973) A new model for cometary nuclei. *Icarus*, *19*, 137–146.
Ofek E. O. and Nakar E. (2010) Detectability of Oort cloud objects using Kepler. *Astrophys. J. Lett.*, *711*, L7–L11.
Ohtsuka K., Nakano S., and Yoshikawa M. (2003) On the association among periodic Comet 96P/Machholz, Arietids, the Marsden comet group, and the Kracht comet group. *Publ. Astron. Soc. Japan*, *55*, 321–324.
Oort J. H. (1950) The structure of the cloud of comets surrounding the solar system and a hypothesis concerning its origin. *Bull. Astron. Inst. Netherlands*, *11*, 91–110.
Opik E. J. (1966) Sun-grazing comets and tidal disruption. *Irish Astron. J.*, *7*, 141–161.
Opitom C., Jehin E., Manfroid J., et al. (2015) TRAPPIST monitoring of Comet C/2012 F6 (Lemmon). *Astron. Astrophys.*, *574*, A38.
Opitom C., Yang B., Selman F., et al. (2019) First observations of an outbursting comet with the MUSE integral-field spectrograph. *Astron. Astrophys.*, *628*, A128.
Ostro S. J., Hudson R. S., Benner L. A. M., et al. (2002) Asteroid radar astronomy. In *Asteroids III* (W. F. Bottke et al., eds.), pp. 151–168. Univ. of Arizona, Tucson.
Pajola M., Lucchetti A., Vincent J.-B., et al. (2016) The southern hemisphere of 67P/Churyumov-Gerasimenko: Analysis of the preperihelion size-frequency distribution of boulders ≥7 m. *Astron. Astrophys.*, *592*, L2.
Patashnick H. (1974) Energy source for comet outbursts. *Nature*, *250*, 313–314.
Pätzold M., Andert T. P., Hahn M., et al. (2019) The nucleus of Comet 67P/Churyumov-Gerasimenko — Part I: The global view — nucleus mass, mass-loss, porosity, and implications. *Mon. Not. R. Astron. Soc.*, *483*, 2337–2346.
Peixinho N., Delsanti A., Guilbert-Lepoutre A., et al. (2012) The bimodal colors of Centaurs and small Kuiper belt objects. *Astron. Astrophys.*, *546*, A86.

Peixinho N., Thirouin A., Tegler S. C., et al. (2020) From Centaurs to comets — 40 years. In *The Trans-Neptunian Solar System* (D. Prailnik et al., eds.), pp. 307–329. Elsevier, Amsterdam.

Pravec P. and Harris A. W. (2000) Fast and slow rotation of asteroids. *Icarus, 148*, 12–20.

Preusker F., Scholten F., Matz K. D., et al. (2015) Shape model, reference system definition, and cartographic mapping standards for Comet 67P/Churyumov-Gerasimenko — Stereophotogrammetric analysis of Rosetta/OSIRIS image data. *Astron. Astrophys., 583*, A33.

Prialnik D. and Podolak M. (1999) Changes in the structure of comet nuclei due to radioactive heating. *Space Sci. Rev., 90*, 169–178.

Rafikov R. R. (2018) Non-gravitational forces and spin evolution of comets. *ArXiV e-prints*, arXiv:1809.05133.

Reach W. T., Vaubaillon J., Kelley M. S., et al. (2009) Distribution and properties of fragments and debris from the split Comet 73P/Schwassmann-Wachmann 3 as revealed by Spitzer Space Telescope. *Icarus, 203*, 571–588.

Reach W. T., Vaubaillon J., Lisse C. M., et al. (2010) Explosion of Comet 17P/Holmes as revealed by the Spitzer Space Telescope. *Icarus, 208*, 276–292.

Reach W. T., Kelley M. S., and Vaubaillon J. (2013) Survey of cometary CO_2, CO, and particulate emissions using the Spitzer Space Telescope. *Icarus, 226*, 777–797.

Rickman H. (1986) Masses and densities of comets Halley and Kopff. In *Proceedings of the Workshop on the Comet Nucleus Sample Return Mission* (O. Melita, ed.), pp. 195–205. ESA SP-249, Noordwijk, The Netherlands.

Rickman H. (1989) The nucleus of Comet Halley: Surface structure, mean density, gas and dust production. *Adv. Space Res., 9*, 59–71.

Romanishin W. and Tegler S. C. (2018) Albedos of Centaurs, jovian Trojans, and Hildas. *Astron. J., 156*, 19.

Russell H. N. (1916) On the albedo of the planets and their satellites. *Astrophys. J., 43*, 173–196.

Ryske J. (2019) Cometary CN cyanogen jet observations using small telescopes with narrowband UV filter. *EPSC-DPS Joint Meeting 2019*, 1980.

Safrit T. K., Steckloff J. K., Bosh A. S., et al. (2021) The formation of bilobate comet shapes through sublimative torques. *Planet. Sci. J., 2*, 14.

Sagdeev R. Z., Szabo F., Avanesov G. A., et al. (1986) Television observations of Comet Halley from Vega spacecraft. *Nature, 321*, 262–266.

Samarasinha N. H. (1997) Preferred orientations for the rotational angular momentum vectors of periodic comets. *Bull. Am. Astron. Soc.*, 29, 743.

Samarasinha N. H. (2007) Rotation and activity of comets. *Adv. Space Res., 39*, 421–427.

Samarasinha N. H. (2008) Rotational excitation and damping as probes of interior structures of asteroids and comets. *Meteorit. Planet. Sci., 43*, 1063–1073.

Samarasinha N. H. and A'Hearn M. F. (1991) Observational and dynamical constraints on the rotation of Comet P/Halley. *Icarus, 93*, 194–225.

Samarasinha N. H. and Larson S. M. (2014) Image enhancement techniques for quantitative investigations of morphological features in cometary comae: A comparative study. *Icarus, 239*, 168–185.

Samarasinha N. H. and Mueller B. E. A. (2013) Relating changes in cometary rotation to activity: Current status and applications to Comet C/2012 S1 (ISON). *Astrophys. J. Lett., 775*, L10.

Samarasinha N. H. and Mueller B. E. A. (2002) Spin axis direction of Comet 19P/Borrelly based on observations from 2000 and 2001. *Earth Moon Planets, 90*, 473–482.

Samarasinha N. H., A'Hearn M. F., Hoban S., et al. (1986) CN jets of Comet P/Halley: Rotational properties. In *ESLAB Symposium on the Exploration of Halley's Comet* (B. Battrick et al., eds.), pp. 487–491. ESA SP-250, Noordwijk, The Netherlands.

Samarasinha N. H., Mueller B. E. A., Belton M. J. S., et al. (2004) Rotation of cometary nuclei. In *Comets II* (M. C. Festou et al., eds.), pp. 281–299, Univ. of Arizona, Tucson.

Samarasinha N. H., Mueller B. E. A., A'Hearn M. F., et al. (2011) Rotation of Comet 103P/Hartley 2 from structures in the coma. *Astrophys. J. Lett., 734*, L3.

Santos-Sanz P., Ortiz J. L., Morales N., et al. (2015) Short-term variability of Comet C/2012 S1 (ISON) at 4.8 AU from the Sun. *Astron. Astrophys., 575*, A52.

Schambeau C. A., Fernández Y. R., Samarasinha N. H., et al. (2017) Analysis of R-band observations of an outburst of Comet 29P/Schwassmann-Wachmann 1 to place constraints on the nucleus' rotation state. *Icarus, 284*, 359–371.

Schambeau C. A., Fernández Y. R., Samarasinha N. H., et al. (2019) Analysis of HST WFPC2 observations of Centaur 29P/Schwassmann-Wachmann 1 while in outburst to place constraints on the nucleus' rotation state. *Astron. J., 158*, 259.

Schleicher D. G. and Bair A. N. (2011) The composition of the interior of Comet 73P/Schwassmann-Wachmann 3: Results from narrowband photometry of multiple components. *Astron. J., 141*, 177.

Schleicher D. G. and Farnham T. L. (2004) Photometry and imaging of the coma with narrowband filters. In *Comets II* (M. C. Festou et al., eds.), pp. 449–469. Univ. of Arizona, Tucson.

Schleicher D. G. and Knight M. M. (2016) The extremely low activity Comet 209P/LINEAR during its extraordinary close approach in 2014. *Astron. J., 152*, 89.

Schleicher D. G. and Osip D. J. (2002) Long- and short-term photometric behavior of Comet Hyakutake (1996 B2). *Icarus, 159*, 210–233.

Schleicher D. G. and Woodney L. M. (2003) Analyses of dust coma morphology of Comet Hyakutake (1996 B2) near perigee: Outburst behavior, jet motion, source region locations, and nucleus pole orientation. *Icarus, 162*, 190–213.

Schleicher D. G., Millis R. L., Thompson D. T., et al. (1990) Periodic variations in the activity of Comet P/Halley during the 1985/1986 apparition. *Astron. J., 100*, 896–912.

Schleicher D. G., Millis R. L., Osip D. J., et al. (1991) Comet Levy (1990c): Groundbased photometric results. *Icarus, 94*, 511–523.

Schleicher D. G., Millis R. L., Osip D. J., et al. (1998) Activity and the rotation period of Comet Hyakutake (1996 B2). *Icarus, 131*, 233–244.

Schleicher D. G., Woodney L. M., and Millis R. L. (2003) Comet 19P/Borrelly at multiple apparitions: Seasonal variations in gas production and dust morphology. *Icarus, 162*, 415–442.

Schleicher D. G., Knight M. M., and Levine S. E. (2013) The nucleus of Comet 10P/Tempel 2 in 2013 and consequences regarding its rotational state: Early science from the Discovery Channel Telescope. *Astron. J., 146*, 137.

Schleicher D. G., Bair A. N., Sackey S., et al. (2015) The evolving photometric lightcurve of Comet 1P/Halley's coma during the 1985/86 apparition. *Astron. J., 150*, 79.

Schleicher D. G., Knight M. M., Eisner N. L., et al. (2019) Gas jet morphology and the very rapidly increasing rotation period of Comet 41P/Tuttle-Giacobini-Kresák. *Astron. J., 157*, 108.

Schleicher D., Knight M., Skiff B., et al. (2022) Physical properties of Comet 28P/Neujmin 1: Nucleus rotation, shape, and phase function. *AAS/Division for Planetary Sciences Meeting Abstracts, 54*, #309.03.

Schlichting H. E., Ofek E. O., Sari R., et al. (2012) Measuring the abundance of sub-kilometer-sized Kuiper belt objects using stellar occultations. *Astrophys. J., 761*, 150.

Schwamb M. E., Volk K., Wen H., et al. (2018) A Northern Ecliptic Survey for solar system science. *ArXiV e-prints*, arXiv:1812.01149.

Schwartz S. R., Michel P., Jutzi M., et al. (2018) Catastrophic disruptions as the origin of bilobate comets. *Nature Astron., 2*, 379–382.

Seaman R., Abell P., Christensen E., et al. (2018) A near-Sun Solar System Twilight Survey with LSST. *ArXiV e-prints*, arXiv:1812.00466.

Sekanina Z. (1979) Fan-shaped coma, orientation of rotation axis, and surface structure of a cometary nucleus I. Test of a model on four comets. *Icarus, 37*, 420–442.

Sekanina Z. (1982) The problem of split comets in review. In *Comets* (L. L. Wilkening, ed.), pp. 251–287. Univ. of Arizona, Tucson.

Sekanina Z. (1988) Outgassing asymmetry of periodic Comet Encke. II. Apparitions 1868–1918 and a study of the nucleus evolution. *Astron. J., 96*, 1455–1475.

Sekanina Z. (1991a) Comprehensive model for the nucleus of periodic Comet Tempel 2 and its activity. *Astron. J., 102*, 350–388.

Sekanina Z. (1991b) Randomization of dust-ejecta motions and the observed morphology of cometary heads. *Astron. J., 102*, 1870–1878.

Sekanina Z. (2002) Statistical investigation and modeling of sun-grazing comets discovered with the Solar and Heliospheric Observatory. *Astrophys. J., 566*, 577–598.

Sekanina Z. (2005) Comet 73P/Schwassmann-Wachmann: Nucleus

fragmentation, its light-curve signature, and close approach to Earth in 2006. *Intl. Comet Q.*, *27*, 225–240.
Sekanina Z. (2019) 1I/'Oumuamua and the problem of survival of Oort cloud comets near the Sun. *ArXiV e-prints*, arXiv:1903.06300.
Sekanina Z. and Chodas P. W. (2004) Fragmentation hierarchy of bright sungrazing comets and the birth and orbital evolution of the Kreutz system. I. Two-superfragment model. *Astrophys. J.*, *607*, 620–639.
Sekanina Z. and Chodas P. W. (2005) Origin of the Marsden and Kracht groups of sunskirting comets. I. Association with Comet 96P/Machholz and its interplanetary complex. *Astrophys. J. Suppl. Ser.*, *161*, 551–586.
Sekanina Z. and Kracht R. (2018) Preperihelion outbursts and disintegration of Comet C/2017 S3 (Pan-STARRS). *ArXiV e-prints*, arXiv:1812.07054.
Sekanina Z. and Yeomans D. K. (1985) Orbital motion, nucleus precession, and splitting of periodic Comet Brooks 2. *Astron. J.*, *90*, 2335–2352.
Sekanina Z., Chodas P. W., and Yeomans D. K. (1994) Tidal disruption and the appearance of periodic Comet Shoemaker-Levy 9. *Astron. Astrophys.*, *289*, 607–636.
Sekanina Z., Brownlee D. E., Economou T. E., et al. (2004) Modeling the nucleus and jets of Comet 81P/Wild 2 based on the Stardust encounter data. *Science*, *304*, 1769–1774.
Sekiguchi T., Miyasaka S., Dermawan B., et al. (2018) Thermal infrared and optical photometry of asteroidal Comet C/2002CE_{10}. *Icarus*, *304*, 95–100.
Serra-Ricart M. and Licandro J. (2015) The rotation period of C/2014 Q2 (Lovejoy). *Astrophys. J.*, *814*, 49.
Shevchenko V. G., Chiorny V. G., Gaftonyuk N. M., et al. (2008) Asteroid observations at low phase angles. III. Brightness behavior of dark asteroids. *Icarus*, *196*, 601–611.
Showalter M. R., Benecchi S. D., Buie M. W., et al. (2021) A statistical review of light curves and the prevalence of contact binaries in the Kuiper belt. *Icarus*, *356*, 114098.
Sierks H., Barbieri C., Lamy P. L., et al. (2015) On the nucleus structure and activity of Comet 67P/Churyumov-Gerasimenko. *Science*, *347*, aaa1044.
Silsbee K. and Tremaine S. (2016) Modeling the nearly isotropic comet population in anticipation of LSST observations. *Astron. J.*, *152*, 103.
Simion N. G., Popescu M., Licandro J., et al. (2021) Spectral properties of near-Earth objects with low-jovian Tisserand invariant. *Mon. Not. R. Astron. Soc.*, *508*, 1128–1147.
Snodgrass C., Fitzsimmons A., and Lowry S. C. (2005) The nuclei of Comets 7P/Pons-Winnecke, 14P/Wolf and 92P/Sanguin. *Astron. Astrophys.*, *444*, 287–295.
Snodgrass C., Lowry S. C., and Fitzsimmons A. (2006) Photometry of cometary nuclei: Rotation rates, colours and a comparison with Kuiper belt objects. *Mon. Not. R. Astron. Soc.*, *373*, 1590–1602.
Snodgrass C., Lowry S. C., and Fitzsimmons A. (2008) Optical observations of 23 distant Jupiter family comets, including 36P/Whipple at multiple phase angles. *Mon. Not. R. Astron. Soc.*, *385*, 737–756.
Snodgrass C., Fitzsimmons A., Lowry S. C., et al. (2011) The size distribution of Jupiter family comet nuclei. *Mon. Not. R. Astron. Soc.*, *414*, 458–469.
Solontoi M., Ivezić Ž., Jurić M., et al. (2012) Ensemble properties of comets in the Sloan Digital Sky Survey. *Icarus*, *218*, 571–584.
Sonnett S., Mainzer A., Grav T., et al. (2021) NEO Surveyor cadence and simulations. In *52nd Lunar and Planetary Science Conference*, Abstract #2705. LPI Contribution No. 2548, Lunar and Planetary Institute, Houston.
Sosa A. and Fernández J. A. (2009) Cometary masses derived from non-gravitational forces. *Mon. Not. R. Astron. Soc.*, *393*, 192–214.
Sosa A. and Fernández J. A. (2011) Masses of long-period comets derived from non-gravitational effects — Analysis of the computed results and the consistency and reliability of the non-gravitational parameters. *Mon. Not. R. Astron. Soc.*, *416*, 767–782.
Springmann A., Harris W. M., Ryan E. L., et al. (2022) Repeating gas ejection events from Comet 45P/Honda-Mrkos-Pajdušáková. *Planet. Sci. J.*, *3*, 15.
Stansberry J., Grundy W., Brown M., et al. (2008) Physical properties of the Kuiper belt and Centaur objects: Constraints from the Spitzer Space Telescope. In *The Solar System Beyond Neptune* (M. A. Barucci et al., eds.), pp. 161–179. Univ. of Arizona, Tucson.
Steckloff J. K. and Jacobson S. A. (2016) The formation of striae within cometary dust tails by a sublimation-driven YORP-like effect. *Icarus*, *264*, 160–171.
Steckloff J. K. and Samarasinha N. H. (2018) The sublimative torques of Jupiter family comets and mass wasting events on their nuclei. *Icarus*, *312*, 172–180.
Stellingwerf R. F. (1978) Period determination using phase dispersion minimization. *Astrophys. J.*, *224*, 953–960.
Stern S. A. (1986) The effects of mechanical interaction between the interstellar medium and comets. *Icarus*, *68*, 276–283.
Stern S. A., Weaver H. A., Spencer J. R., et al. (2019) Initial results from the New Horizons exploration of 2014 MU69, a small Kuiper belt object. *Science*, *364*, aaw9771.
Strazzulla G. (1999) Ion irradiation and the origin of cometary materials. *Space Sci. Rev.*, *90*, 269–274.
Szabó G. M., Kiss L. L., Pál A., et al. (2012) Evidence for fresh frost layer on the bare nucleus of Comet Hale-Bopp at 32 AU distance. *Astrophys. J.*, *761*, 8.
Tancredi G. (1995) The dynamical memory of Jupiter family comets. *Astron. Astrophys.*, *299*, 288–292.
Tancredi G. (2014) A criterion to classify asteroids and comets based on the orbital parameters. *Icarus*, *234*, 66–80.
Tancredi G., Fernández J. A., Rickman H., et al. (2000) A catalog of observed nuclear magnitudes of Jupiter family comets. *Astron. Astrophys. Suppl. Ser.*, *146*, 73–90.
Tancredi G., Fernández J. A., Rickman H., et al. (2006) Nuclear magnitudes and the size distribution of Jupiter family comets. *Icarus*, *182*, 527–549.
Taylor P. A. and Margot J.-L. (2011) Binary asteroid systems: Tidal end states and estimates of material properties. *Icarus*, *212*, 661–676.
Tegler S. C., Bauer J. M., Romanishin W., et al. (2008) Colors of Centaurs. In *The Solar System Beyond Neptune* (M. A. Barucci et al., eds.), pp. 105–114. Univ. of Arizona, Tucson.
Tegler S. C., Romanishin W., Consolmagno G. J., et al. (2016) Two color populations of Kuiper belt and Centaur objects and the smaller orbital inclinations of red Centaur objects. *Astron. J.*, *152*, 210.
Thirouin A. and Sheppard S. S. (2018) The Plutino population: An abundance of contact binaries. *Astron. J.*, *155*, 248.
Thirouin A. and Sheppard S. S. (2019) Light curves and rotational properties of the pristine cold classical Kuiper belt objects. *Astron. J.*, *157*, 228.
Thomas N. and Keller H. U. (1989) The colour of Comet P/Halley's nucleus and dust. *Astron. Astrophys.*, *213*, 487–494.
Thomas P., A'Hearn M., Belton M. J. S., et al. (2013a) The nucleus of Comet 9P/Tempel 1: Shape and geology from two flybys. *Icarus*, *222*, 453–466.
Thomas P. C., A'Hearn M. F., Veverka J., et al. (2013b) Shape, density, and geology of the nucleus of Comet 103P/Hartley 2. *Icarus*, *222*, 550–558.
Toth I., Lamy P., and Weaver H. A. (2005) Hubble Space Telescope observations of the nucleus fragment 73P/Schwassmann-Wachmann 3-C. *Icarus*, *178*, 235–247.
Toth I., Lamy P., Weaver H., et al. (2006) HST observations of the nucleus fragment 73P/Schwassmann-Wachmann 3-C during its close approach to Earth in 2006. *AAS/Division for Planetary Sciences Meeting Abstracts*, *38*, #06.01.
Tsiganis K., Gomes R., Morbidelli A., et al. (2005) Origin of the orbital architecture of the giant planets of the solar system. *Nature*, *435*, 459–461.
Tubiana C., Böhnhardt H., Agarwal J., et al. (2011) 67P/Churyumov-Gerasimenko at large heliocentric distance. *Astron. Astrophys.*, *527*, A113.
Vaughan C. M., Pierce D. M., and Cochran A. L. (2017) Jet morphology and coma analysis of Comet 103P/Hartley 2. *Astron. J.*, *154*, 219.
Verbiscer A. J. and Veverka J. (1988) Estimating phase integrals: A generalization of Russell's law. *Icarus*, *73*, 324–329.
Verbiscer A. J., Helfenstein P., and Buratti B. J. (2013) Photometric properties of solar system ices. In *The Science of Solar System Ices* (M. S. Gudipati and J. Castillo-Rogez, eds.), pp. 47–72. Astrophys. Space Sci. Library, Vol. 356, Springer, New York.
Verbiscer A. J., Porter S., Benecchi S. D., et al. (2019) Phase curves from the Kuiper belt: Photometric properties of distant Kuiper belt objects observed by New Horizons. *Astron. J.*, *158*, 123.
Vincent J.-B. (2019) Cometary topography and phase darkening. *Astron. Astrophys.*, *624*, A5.
Vincent J.-B., Bodewits D., Besse S., et al. (2015) Large heterogeneities in Comet 67P as revealed by active pits from sinkhole collapse.

Nature, 523, 63–66.
Vincent J.-B., A'Hearn M. F., Lin Z. Y., et al. (2016) Summer fireworks on Comet 67P. *Mon. Not. R. Astron. Soc., 462*, S184–S194.
Vincent J.-B., Hviid S. F., Mottola S., et al. (2017) Constraints on cometary surface evolution derived from a statistical analysis of 67P's topography. *Mon. Not. R. Astron. Soc., 469*, S329–S338.
Vsekhsviatskii S. K. (1964) *Physical Characteristics of Comets (Fizicheskie Kharakteristiki Komet)*. Israel Program for Scientific Translations, Jerusalem. 596 pp. (Originally published in Russian, 1958, Gos. izd-vo fiziko-matematicheskoĭ lit-ry, Moscow).
Waniak W., Borisov G., Drahus M., et al. (2009) Rotation of the nucleus, gas kinematics and emission pattern of Comet 8P/Tuttle: Preliminary results from optical imaging of the CN coma. *Earth Moon Planets, 105*, 327–342.
Waniak W., Borisov G., Drahus M., et al. (2012) Rotation-stimulated structures in the CN and C_3 comae of Comet 103P/Hartley 2 close to the EPOXI encounter. *Astron. Astrophys., 543*, A32.
Warner B. D. (2006) Asteroid lightcurve analysis at the Palmer Divide Observatory: July–September 2005. *Minor Planet Bull., 33*, 35–39.
Warner B. D., Harris A. W., and Pravec P. (2009) The asteroid lightcurve database. *Icarus, 202*, 134–146.
Weaver H. A. and Lamy P. L. (1997) Estimating the size of Hale-Bopp's nucleus. *Earth Moon Planets, 79*, 17–33.
Weaver H. A., A'Hearn M. F., Arpigny C., et al. (1995) The Hubble Space Telescope (HST) observing campaign on Comet Shoemaker-Levy 9. *Science, 267*, 1282–1288.
Weaver H. A., Sekanina Z., Toth I., et al. (2001) HST and VLT investigations of the fragments of Comet C/1999 S4 (LINEAR). *Science, 292*, 1329–1334.
Weaver H. A., Lisse C. M., Mutchler M. J., et al. (2006) Hubble Space Telescope investigation of the disintegration of 73P/Schwassmann-Wachmann 3. *AAS/Division for Planetary Sciences Meeting Abstracts, 38*, #06.02.
Weidenschilling S. J. (1977) Aerodynamics of solid bodies in the solar nebula. *Mon. Not. R. Astron. Soc., 180*, 57–70.
Weidenschilling S. J. (1997) The origin of comets in the solar nebula: A unified model. *Icarus, 127*, 290–306.
Weiler M., Rauer H., and Sterken C. (2011) Cometary nuclear magnitudes from sky survey observations. *Icarus, 212*, 351–366.
Weissman P. R. (1980) Physical loss of long-period comets. *Astron. Astrophys., 85*, 191–196.
Weissman P. R., Bottke W. F. J., and Levison H. F. (2002) Evolution of comets into asteroids. In *Asteroids II* (R. P. Binzel et al., eds.), pp. 669–686. Univ. of Arizona, Tucson.
Weissman P. R., Asphaug E., and Lowry S. C. (2004) Structure and density of cometary nuclei. In *Comets II* (M. C. Festou et al., eds.), pp. 337–357. Univ. of Arizona, Tucson.
Weissman P., Morbidelli A., Davidsson B., et al. (2020) Origin and evolution of cometary nuclei. *Space Sci. Rev., 216*, 6.
Whipple F. L. (1950) A comet model. I. The acceleration of Comet Encke. *Astrophys. J., 111*, 375–394.
Whipple F. L. (1963) On the structure of the cometary nucleus. In *The Solar System, Vol. 4: The Moon, Meteorites and Comets* (B. M. Middlehurst and G. P. Kuiper, eds.), pp. 639–664. Univ. of Chicago, Chicago.
Whipple F. L. (1978) Rotation period of Comet Donati. *Nature, 273*, 134–135.
Whipple F. L. and Sekanina Z. (1979) Comet Encke — Precession of the spin axis, nongravitational motion, and sublimation. *Astron. J., 84*, 1894–1909.
Wong I. and Brown M. E. (2017) The bimodal color distribution of small Kuiper belt objects. *Astron. J., 153*, 145.
Woodney L., Schleicher D. G., Reetz K. M., et al. (2007) Rotational properties of Comet 2P/Encke based on nucleus lightcurves and coma morphology. *AAS/Division for Planetary Sciences Meeting Abstracts, 39*, #36.03.
Xi Z.-z. (1984) The cometary atlas in the silk book of the Han tomb at Mawangdui. *Chinese Astron. Astrophys., 8*, 1–7.
Zhao Y., Rezac L., Skorov Y., et al. (2021) Sublimation as an effective mechanism for flattened lobes of (486958) Arrokoth. *Nature Astron., 5*, 139–144.

Part 5:

Gas Coma

Bodewits D., Bonev B. P., Cordiner M. A., and Villanueva G. L. (2024) Radiative processes as diagnostics of cometary atmospheres. In *Comets III* (K. J. Meech, M. R. Combi, D. Bockelée-Morvan, S. N. Raymond, and M. E. Zolensky, eds.), pp. 407–431. Univ. of Arizona, Tucson, DOI: 10.2458/azu_uapress_9780816553631-ch013.

Radiative Processes as Diagnostics of Cometary Atmospheres

D. Bodewits
Auburn University

B. P. Bonev
American University

M. A. Cordiner
NASA Goddard Space Flight Center and Catholic University of America

G. L. Villanueva
NASA Goddard Space Flight Center

In this chapter, we provide a review of radiative processes in cometary atmospheres spanning a broad range of wavelengths, from radio to X-rays. We focus on spectral modeling, observational opportunities, and anticipated challenges in the interpretation of new observations, based on our current understanding of the atomic and molecular processes occurring in the atmospheres of small, icy bodies. Close to the surface, comets possess a thermalized atmosphere that traces the irregular shape of the nucleus. Gravity is too low to retain the gas, which flows out to form a large, collisionless exosphere (coma) that interacts with the heliospheric radiation environment. As such, cometary comae represent conditions that are familiar in the context of planetary atmosphere studies. However, the outer comae are tenuous, with densities lower than those found in vacuum chambers on Earth. Comets therefore provide us with unique natural laboratories that can be understood using state-of-the-art theoretical treatments of the relevant microphysical processes. Radiative processes offer direct diagnostics of the local physical conditions, as well as the macroscopic coma properties. These can be used to improve our understanding of comets and other astrophysical environments such as icy moons and the interstellar medium.

1. INTRODUCTION

The spectroscopic analysis of cometary comae dates back to *Huggins* (1868), who matched the optical spectrum of Comet 2P/Encke to the carbon emission signatures observed in flames from an organic gas (for historical reviews, see *Swings,* 1965; *Feldman et al.,* 2004). As shown in Fig. 1, the optical spectra of comets are typically dominated by the emission of several radicals (such as C_2, CN, and OH), superimposed on the emission of a blackbody spectrum (sunlight reflected by dust in the coma). Since these first observations, the spectroscopic investigation of comets has greatly expanded.

An important driver for these studies is that studying cometary compositions allows us to probe the physics and chemistry of our early solar system (see the chapter by Biver et al. in this volume). Observations of coma emission provide measurements that are diagnostic for various properties of both the nucleus and the coma. These measurements include column densities, production rates, relative abundances (also referred to as "mixing ratios"), rotational and kinetic temperatures, photodissociation scale lengths and lifetimes, gas outflow velocities, plasma structures, and the rotational state of the nucleus, among others. This large suite of "metrics" is then interpreted in terms of chemical composition and isotopic abundances, as well as the long-term storage, physicochemical evolution, and release of volatiles in small protobodies, both directly from the nucleus and from secondary sources. Comets are natural laboratories for studying a wide variety of chemical and physical phenomena owing to the range of processes at work (such as ice sublimation, volatile release and acceleration, thermal processing of grains, photochemistry, ionospheric processes) and their variability

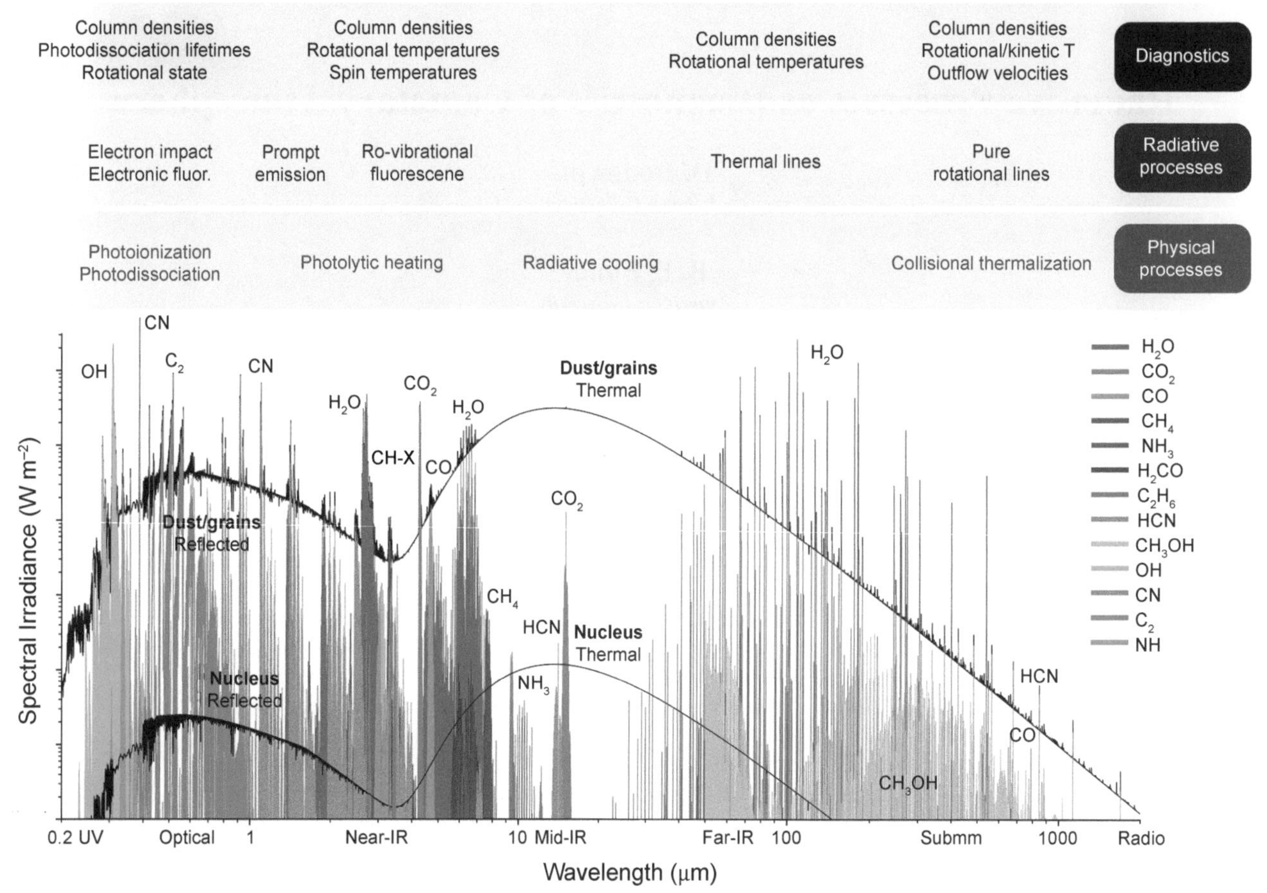

Fig. 1. Example (synthetic) spectrum of cometary emissions and their relationship to retrievable diagnostics, radiative processes, and the associated physical processes. This synthetic spectrum includes only a subset of cometary species, and was synthesized using the Planetary Spectrum Generator [PSG (*Villanueva et al.,* 2018); *https://psg.gsfc.nasa.gov*], capturing the dominant radiative processes at a resolving power of 5000. The broad cometary continuum at the assumed conditions (Q = 10^{29} s^{-1}, diameter = 5 km, heliocentric and geocentric distance = 1 au, aperture = 100″) is usually dominated by dust grains, yet for distant and low activity comets, the nucleus can become the main thermal continuum source.

over the comet's orbit about the Sun. Understanding these processes provides insights into the complex interactions between matter and radiation, knowledge of which informs other areas of astrophysics, astrochemistry, and planetary science. For instance, studies of cometary ice abundances help to understand the impact of radiation on the evolution of primordial ices (*Garrod,* 2019) and help constrain theories for the gas and ice-phase chemistry occurring in interstellar clouds and protoplanetary disks (*Eistrup et al.,* 2019; *Price et al.,* 2021; see also the chapter by Bergin et al. in this volume). The low densities and sublimative outflows of cometary comae offer an accessible analog to planetary exospheres. With volatile mass loss rates of tens to thousands of kilograms per second, comets can also provide a baseline for the study of outgassing from (and the chemistry of) icy moons, asteroids, and transneptunian objects. X-ray observations of comets paved the way to remotely study plasma interactions in a multitude of astrophysical environments (*Dennerl,* 2010). Last but not least, comets provide a road map for the detection of volatiles (including molecules that are possibly relevant to the origin of life) in low-density environments within the solar system and beyond (*Strøm et al.,* 2020; *Cordiner et al.,* 2020; *Bodewits et al.,* 2020; *Villanueva et al.,* 2021).

Cometary comae can span several hundreds of thousands kilometers in size, and their ion tails can stretch to over an astronomical unit in some cases (*Neugebauer et al.,* 2007). Remote sensing is a prime method for studying such large structures, yet interpreting the light emitted by atoms and molecules requires a detailed understanding of the physical processes and how they operate under the specific conditions of cometary atmospheres. These processes operate in the coma across all energies and wavelengths, but certain wavelengths provide more sensitivity to specific phenomena. For instance, fluorescence typically dominates the infrared (IR)/optical emission, but such processes also impact the ground rotational levels sampled at submillimeter/radio wavelengths. Specific wavelength ranges are better suited to probe certain processes in the coma, and observations at specific wavelengths can thus be tailored to obtain key diagnostics that distinguish and connect physical and chemical processing and primordial properties (see Fig. 1). This direct link between the diagnostic radiative emissions and the associated physical processes also simplifies the interpretation of spectroscopic data, in that only a specific set of processes needs to be included in the modeling and interpretation of the data.

A key challenge in interpreting cometary emission is the disequilibrium between the excitation and ambient temperature. This is due to the intrinsically tenuous nature of the atmosphere (because of a lack of equilibrating and thermalizing collisions), combined with a deeply penetrating solar radiation field that sublimates icy grains, breaks molecular bonds (photodissocation), ionizes atoms/molecules (photoionization), and also leads to non-equilibrated emissions via solar pumping (fluorescence).

Low-density regimes (such as the atmospheres of main-belt comets, Centaurs, distant comets) are challenging to observe, but with the launch of the James Webb Space Telescope (JWST) and many other large telescopes and space observatories planned (or in construction), the observational exploration of comets will soon enter a new era (see section 4). These new observational facilities will further demonstrate the limits of our current understanding of the processes that occur in these tenuous regimes. Their interpretation will require more detailed theoretical treatments and advanced models that incorporate accurate laboratory data, molecular constants, and parameterizations obtained under conditions that represent those found in the coma.

A review of excitation of cometary molecules and the resulting spectra was provided by *Bockelée-Morvan et al.* (2004) for parent volatiles and by *Feldman et al.* (2004) for fragment species. Building on this, in this chapter we discuss the processes that are responsible for the most notable emissions in comets, and which yield diagnostics that allow the remote characterization of physical conditions and chemical parameters of the nucleus and the coma. We organize the discussion by process rather than wavelength, since cometary processes span a broad range of energies and are therefore generally not restricted to a particular wavelength regime. We also explore the limitations of how we currently model and interpret these diagnostics and discuss the challenges and possible steps to be taken to advance the field of coma remote sensing.

2. RADIATIVE TRANSFER IN THE COMA

2.1. Physical Structure of the Coma

As ices in the nucleus sublimate, the resulting gases drag icy particles and refractory materials along and form an expanding atmosphere of gas and dust, referred to as the coma (Fig. 2). The thermal velocity of the gas molecules exceeds the low escape velocity of comet nuclei (typically $\lesssim$0.1 m s^{-1}), so instead of forming a hydrostatic atmosphere as in planets, an expanding atmosphere/exosphere is established, the dynamics of which are dominated by outgassing and acceleration processes. The gas and dust decouple within roughly 10 km of the surface (*Gerig et al.,* 2018). As the gas continues to expand, the density drops rapidly (approximately as ρ^{-2}, where ρ is the distance from the nucleus) and the neutral coma begins to depart from thermal and chemical equilibrium. This ultimately results in the formation of a well-defined ionosphere that interacts with the solar wind (see the chapter by Beth et al. in this volume).

The outflow velocity of the expanding coma (v_{out}) is governed by the balance of heating and cooling processes (*Rodgers et al.,* 2004), which vary with coma position, potentially leading to measurable variations in v_{out} with radius. However, for most observational applications (integrating over large spatial apertures), a constant value of v_{out} is assumed. Spatially averaged gas outflow velocities are commonly derived at radio wavelengths using heterodyne spectroscopy to measure spectrally resolved (Doppler) line profiles (*Biver et al.,* 1999; *Roth et al.,* 2021b). In the optical and IR, information on the outflow velocity can also be obtained from temporally resolved imaging of outward-moving coma density structures (*Schleicher and Farnham,* 2004; *Knight et al.,* 2021).

There are multiple different approaches to modeling the density, velocity, and structure of cometary atmospheres (*Combi et al.,* 2004; *Rodgers et al.,* 2004). *Haser* (1957) provided an analytical basis for understanding the large-scale chemical structure of cometary comae, under the assumption of an isotropic outflow of gases at a constant radial velocity away from the nucleus. Photodissociation by the solar radiation field leads to exponential decay of the gas radial density on top of the ρ^{-2} spherical expansion (see *Haser et al.,* 2020, for an English translation). For parent species, volume densities following Haser are described as

$$N(\rho) = \frac{Q}{4\pi\rho^2 v_{out}} e^{-\frac{\rho\beta}{v_{out}}}$$

where Q is the production rate, ρ is the distance to the nucleus, v_{out} is the expansion velocity, and β is the photodissociation rate.

Due to its simplicity and broad applicability, the Haser model continues to be the primary method used for linking column densities of atoms and molecules to the production rates. However, the Haser model has several limitations. It is time-independent, with the assumption of uniform, isotropic outgassing, and it cannot reproduce observed column density profiles with experimentally determined lifetimes of fragment species in the solar radiation field (*Combi and Delsemme,* 1980). To work around this, studies of daughter fragments adopt empirical scale lengths (*Cochran and Schleicher,* 1993; *A'Hearn et al.,* 1995) that are no longer based on physical outflow velocities and photodissocation rates. The column density profiles of some species, most notably C_2, require multi-generation Haser models (*O'Dell et al.,* 1988). Departures from the molecular distributions predicted by the spherically-symmetric Haser model become more evident with increasing telescope resolution and sensitivity. Anisotropic outflow is commonly seen in spectral line profiles observed at millimeter/submillimeter wavelengths (*Cordiner et al.,* 2014; *Gulkis et al.,* 2015; *Roth et al.,* 2021b). This results from diurnal modulation of the sublimation rates, jets,

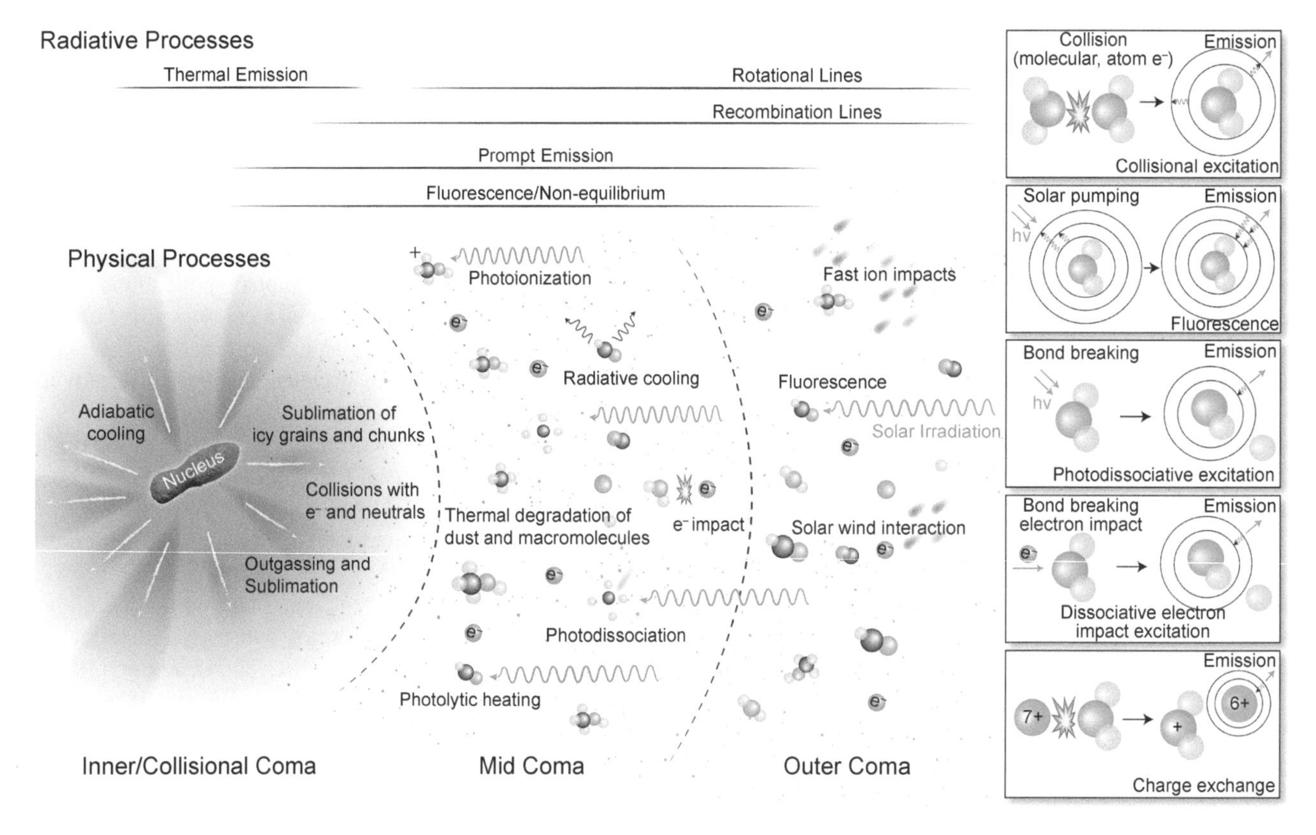

Fig. 2. Overview of physical regimes and structure of the coma (not to scale) and the dominant physical/chemical processes that factor into the molecular excitation and emission processes considered in this chapter. Key emission processes and their associated energetic driving mechanisms are shown on the right. Horizontal bars at the top indicate the approximate spatial ranges over which these emission processes occur.

and other heterogeneous outgassing effects related to surface/subsurface inhomogeneities and non-uniform illumination of the nucleus. Other well-known phenomena not described by the Haser model include asymmetric coma heating and photolysis rates, radial outflow acceleration (*Biver et al.,* 2011), and solar wind pickup of molecular cations.

To provide a more physically realistic description of the outer coma, *Festou* (1981) and *Combi and Delsemme* (1980) introduced models in which fragments are emitted with an isotropic velocity kick upon their dissociation. Because these models are rooted in photochemistry, it requires that the formation pathway and reaction dynamics of the coma species are known, which is not the case for most fragment species except for OH (*Feldman et al.,* 2004; *Combi et al.,* 2004). Both approaches are time-dependent but are only valid for the collisionless part of the coma. More numerically intensive, direct simulation Monte Carlo models are required for the most accurate determinations of molecular densities (and kinetic temperatures) across the complete range of coma scales (*Combi et al.,* 2004). Such models can capture the complex interplay between nucleus shape, insolation, and outgassing patterns in three dimensions (*Combi et al.,* 2020). Less complex, three-dimensional fluid dynamic models are also applicable for a wide range of conditions (*Shou et al.,* 2016). Nevertheless, because of their simplicity and ease of parameterization, ad hoc multi-component-type models, or semianalytic treatments, remain a favored approach for interpretation of high-resolution groundbased and *in situ* observations (e.g., *Biver et al.,* 2019; *Zhao et al.,* 2020).

In the outermost coma, solar wind interactions begin to dominate the gas dynamics, resulting in the formation of an ion tail containing cometary particles accelerated away in the anti-solar direction. Interpretation of coma ion observations can therefore require the use of models incorporating magnetohydrodynamics (*Rauer et al.,* 1997), although simpler, Monte Carlo treatments of the ion distribution can be appropriate in some cases (*Lovell et al.,* 1998). We refer to the chapter by Beth et al. in this volume for a more complete discussion of coma ion distributions.

2.2. Solar Radiation Field

The solar radiation field plays an important role in determining the excitation state of the atoms and molecules in the coma. Therefore, the quality of the input spectrum of the solar radiation field is a critical component when modeling cometary emissions. The tenuous nature of the coma environment permits solar radiation to penetrate deeply into the coma, where it plays an important role (e.g., sublimation of grains, solar pumping of molecules). The widespread presence of strong Fraunhofer lines across the solar spectrum, together with the typically narrow nature of cometary lines, makes the effects of the Sun on the coma dependent on the comet's heliocentric velocity and the outflow velocity of the gas (known as the Swings and Greenstein effects, respectively). Therefore, solar pumping, fluorescence, and maser emissions in comets are highly susceptible to the characterization of the detailed structure of the solar spectrum and the specific velocities at the time of the observations.

Our knowledge of the solar spectrum has greatly improved in the past few decades due to spacecraft measurements [Atmospheric Trace Molecule Spectroscopy Experiment (ATMOS) (*Abrams et al.,* 1996); Advanced Composition Explorer (ACE) (*Hase et al.,* 2010)], and also via solar surveys performed with groundbased observatories (*Wallace and Livingston,* 2003). One of the biggest limitations of some of these databases is that they are not flux-calibrated, and the spectra can only be used to extract transmittance information. Although theoretical models (*Tobiska et al.,* 2000; *Kurucz,* 2000) have been extremely successful in calculating a flux-calibrated solar continuum, their predictions of solar spectral features are still not optimum at high resolutions. To determine accurate solar optical/IR templates, a general solution has been to combine theoretical and empirical solar databases (*Bromley et al.,* 2021; *Villanueva et al.,* 2011b; *Fiorenza and Formisano,* 2005). At short wavelengths, the far-ultraviolet (FUV) solar model by *Fontenla et al.* (2014) can be used to model the typical hard energetic radiation (λ < 170 nm), which can be complemented by Solar Stellar Irradiance Comparison Experiment (SOLSTICE) measurements of the Sun during solar minimum in the 170–200-nm range (*Rottman et al.,* 1993) (*https://lasp.colorado.edu/home/sorce/data/ssi-data*).

The solar spectrum varies significantly at shorter wavelengths [e.g., FUV, X-rays (*Huebner and Mukherjee,* 2015)]. In particular, for ultraviolet (UV) observations (λ < 200 nm) of atomic features, it is important to scale the reference spectrum of the Sun to the exact conditions during the observations. An approach to capture the temporal evolution of the UV radiation field is to scale the solar spectrum by the daily averaged flux as measured with the Thermosphere Ionosphere Mesosphere Energetics and Dynamics (TIMED) Solar EUV Experiment (SEE) instrument (*Woods et al.,* 1998, 2000) for a given day of observations at 68 nm by the Solar Ultraviolet Measurements of Emitted Radiation (SUMER)-averaged relative spectrum to estimate a high-resolution UV solar spectrum (*https://www.swpc.noaa.gov/*). In addition, the daily 10.7-cm solar fluxes can be useful to estimate molecular lifetimes (*Crovisier,* 1989).

2.3. Energy Level Populations of Coma Molecules

The wavelengths and mechanisms for absorption and emission of radiation in cometary atmospheres are determined by the energy level populations of the coma gases. According to quantum mechanics, the internal energy of a molecule can be separated into rotational, vibrational, and electronic modes of excitation (*Herzberg,* 1971). Each of these modes is divided into a set of discrete energy levels, numbered "i" in order of ascending energy.

For a given molecule in local thermodynamic equilibrium (LTE) at temperature T, the energy level populations (P_i — the fraction of the total number of molecules in each level) follow a Boltzmann distribution

$$P_i = \frac{g_i e^{-\frac{E_i}{kT}}}{\Sigma_i g_i e^{-\frac{E_i}{kT}}}$$

where E_i are the energies of the levels (with respect to the ground state), g_i are their statistical weights, and k is Boltzmann's constant. In LTE, the excitation temperature T is equivalent to the kinetic temperature T_{kin}, which relates to the collisional and velocity/thermal spectral profile of the molecule.

Even in the general non-LTE case often encountered in the mid- or outer coma, analyses of molecular spectra (which, in reality, are likely to be based on a subset of the complete set of molecular energy levels) can be facilitated by considering the Boltzmann-equivalent temperature for that subset of levels (i'). Within a given vibronic level, the set of rotational level populations $P_{i'}$ can often be described using a single "rotational temperature," such that $T = T_{rot}$. Similarly, a "vibrational temperature" T_{vib} can be used to characterize a (quasi-Boltzmann) distribution of vibrational level populations.

2.4. The Equation of Radiative Transfer

The density in a typical cometary atmosphere spans an extremely broad range [from less than 1 cm^{-3} to more than 10^{22} cm^{-3} (*Tenishev et al.,* 2008)], and the coma is exposed to strong and variable fluxes of solar radiation and particles. Therefore, the distribution of energy levels depends on a detailed balance of collisional and radiative excitation (and de-excitation) processes (*Crovisier and Encrenaz,* 1983; *Weaver and Mumma,* 1984; *Crovisier,* 1987). Departure of the level populations from LTE can occur in rotational, vibrational, or electronic excitation modes, and will depend on many factors, including (1) the level of cometary activity (i.e., the density of collisional partners such as water and electrons), (2) the distance from the nucleus (since the density decreases exponentially with radius, and particle/radiation fluxes also vary with position), (3) the particular quantum characteristics of the species in question (including Einstein A and B coefficients; collisional cross sections), and (4) the wavelength of the observations, which often depends on the excitation energy of the system and therefore the source of excitation (e.g., thermal/non-thermal particle collisions, fluorescent pumping, energy balance of chemical reactions, and solar wind particles).

The total radiation field in a cometary coma is established by a comprehensive balance of emission and absorption processes (Fig. 2), which is summarized in the equation of radiative transfer [equation (1); *Liou,* 2002]. Specifically, the intensity of radiation (I_ν) propagating through the cometary coma is calculated by integrating the emission and absorption of radiation as a function of frequency (ν):

$$\frac{dI_\nu}{ds} = \gamma_\nu - \alpha_\nu I_\nu \tag{1}$$

where γ_ν and α_ν are the emission and absorption coefficients of the gas, respectively, and s is the distance through the coma. These coefficients are derived from the Einstein coefficients of the gas in question (A_{ij}, B_{ij}, and B_{ji}, for a transition between the upper energy level i and lower level j)

$$\gamma_\nu = \frac{h\nu}{4\pi} N P_i A_{ij} \phi_\nu \tag{2}$$

$$\alpha_\nu = \frac{Nh\nu}{4\pi}(P_j B_{ji} - P_i B_{ij})\phi_\nu \tag{3}$$

where N is the number of gas particles per unit volume, P_i, P_j are the relative populations in levels i and j, respectively, and ϕ_ν is a (normalized) function representing the frequency dispersion of the spectral line of interest (typically a Gaussian for individual, thermally-broadened lines). The B_{ij} and B_{ji} coefficients can be derived from the A_{ij} coefficients of spontaneous emission and/or from the line intensities (*Šimečková et al.,* 2006). In order to relate an observed spectral line intensity (from energy level i) to a column density, the relative population of level i, P_i, needs to be known. This can be obtained as a function of time (t) in the outflowing coma gas by solving the differential equation

$$\frac{dP_i}{dt} = -P_i\left[\sum_{j<i} A_{ij} + \sum_{j\neq i}(B_{ij}J_\nu + C_{ij}n_{gas})\right] + \sum_{k>i} P_k A_{ki} + \sum_{k\neq i} P_k(B_{ki}J_\nu + C_{ki}n_{gas}) \tag{4}$$

In this equation, C_{ij} are the temperature-dependent rate coefficients for transitions between levels i and j due to collisions between the gas particles (see section 3.1.1), and n_{gas} is the number density of the colliding gas (typically H_2O). The $C_{ij}n_{gas}$ terms can be replaced by a sum over all significant colliding gases, if necessary. J_ν is the local total radiation field, which includes contributions from gas emission, solar radiation, and thermal emission from coma dust particles and the nucleus (the latter two contributions are usually small enough to be neglected). Atoms and molecules can be produced (arising spontaneously in the coma, for example, as a result of dissociative excitation by photons or electrons) in an energized (non-thermally excited) state. Such processes can be accounted for, if needed, by adding a further source term to equation (4).

A given astronomical observation of a comet, restricted to a particular wavelength regime, is usually sensitive to emission from a small subset of energy levels. In practice, it is therefore typical to solve for a restricted subset of energy level populations $P_{i'}$ of the gas in question, thus simplifying the calculation. For instance, in radio/millimeter-wave spectroscopy in a near-thermal regime, only the rotational levels of interest for a molecule can be considered. For the case of some atomic emission lines, such as hydrogen and oxygen fluorescence in UV wavelengths (*Feldman et al.,* 2004; *Noonan et al.,* 2021), only a limited number of electronic states needs to be considered, although in more complex systems (such as nickel and iron), metastable states require a more complete treatment of the energy level structure (*Bromley et al.,* 2021).

In broadband photometry and low-resolution spectroscopy of rovibronic emission at optical and IR wavelengths, the rotational structure is unresolved. The rotational populations in the ground state do impact the total band pumping rates, since the solar spectrum at each rovibronic state will be different — this is particularly notable for CN and OH (*Crovisier and Encrenaz,* 1983). Simplified hybrid schemes are commonly adopted, for example, in the case of a non-thermal distribution of rotational levels modified by pumping through a few of the most populated vibrational levels. For IR spectroscopy, a full treatment of the relevant set of vibrational levels is required (pumped by solar radiation), while the ground rotational manifold remains "frozen" as a Boltzmann distribution (*Crovisier and Encrenaz,* 1983). As we will discuss further in section 3.3, such simplifications facilitate efficient modeling and interpretation of cometary spectra. Care is needed, however, when deciding which levels can be safely excluded (or approximated) in the calculation of $P_{i'}$. For example, given the low coma kinetic temperatures (typically less than ~100 K), one might erroneously consider only a few rotational levels of the ground vibronic state.Great complexity arises, however, as a result of the 5777 K solar radiation pump, which can excite highly energetic vibrational modes of a molecule. Other important sources of molecular excitation include photodissociative excitation (section 3.4), impacts with hot electrons (section 3.5), and solar wind charge exchange (section 3.7), resulting in the population of highly excited energy levels and giving rise to non-thermal emission at a variety of wavelengths.

2.5. Solving for the Coma Radiation Field

Non-thermal (disequilibrium) excitation occurs even close to the surface of the comet, where densities are high and thermalizing collisions are frequent. Evidence for this is observed at IR wavelengths, where fluorescent emission from solar-pumped rovibrational levels is so strong and prevalent across the whole coma that it becomes a prime physical and chemical diagnostic. As a result of large Einstein A values in the IR, vibrational fluorescent transitions occur rapidly, so despite traversing a manifold of vibrational states (ν), a molecule spends most of its time in the ground state ($\nu = 0$), where the rotational levels are thermalized by collisions.

In addition to the various excitation mechanisms already mentioned, the energy level population at a given point in the coma depends on the local radiation field, J_ν, which is calculated by summing the incident radiant energy received at that point from all solid angles. Since J_ν depends on the integrated emission over the entire spatial domain, equation (4) should formally be solved iteratively until convergence of the level populations is achieved, which can be computationally demanding. When optical thickness is important, the physical conditions of the coma need to be computed radially along different trajectories (radial paths through the coma, starting at the nucleus) for multiple gas parcels. In this way, coma asymmetries can also be properly taken into account.

In the less-dense parts of the coma where collisions are less frequent, non-LTE effects become increasingly important across all (rovibronic) levels, but the optical depth for photons leaving the coma is usually low (i.e., $\tau_\nu \lesssim 1$). In that case, the stimulated emission and absorption terms ($B_{ij}J_\nu$ and $B_{ji}J_\nu$) tend to be small for most gases in their vibronic ground states, and can often be neglected. On the other hand, for some transitions of abundant atoms and molecules, τ_ν can still be large in the non-LTE zone, so that photon-trapping effects can have a significant impact on the energy level populations. In this regime, to avoid the computational burden of iteratively solving for J_ν, the "escape probability" approximation is commonly used (Sobolev's method); see *Bockelée-Morvan* (1987) for the interpretation of radio/submillimeter pure rotational observations.

The solution to equation (4) requires a time-dependent integration of the coupled set of differential equations describing how the population of each energy level changes as a parcel of gas moves outward through the coma. Once the energy level populations are known as a function of radial distance to the nucleus, they can be mapped into three dimensions and ray-traced as a function of frequency using equation (1) to produce a model coma (spectral) image for comparison with observations. In practice, this is solved differently for each process/wavelength, taking into account the prevailing mechanism of excitation (Figs. 1 and 2). However, solving for solar fluorescence equilibrium for some molecules would require accounting for disequilibrium in millions of transitions, which would be an unsolvable system of millions of differential equations. For those cases, one can only solve for a single pump process and determine the cascade products (with the assumption that the ground rovibrational state is thermalized), which is the prevailing method to compute IR molecular fluorescence efficiencies for many species (*Crovisier and Encrenaz,* 1983; *Villanueva et al.,* 2012).

Similarly, when solving the rotational level populations, one can mostly concentrate on the ground state and only a few of the most relevant vibrational levels, permitting realistic treatments of the three-dimensional opacities, pumps, and other excitation processes. A further simplification to the (time-dependent) solution of equation (4) is to set $dP_x/dt = 0$, invoking the steady-state approximation and solving for the energy-level populations at each individual coma position (in up to three dimensions). The resulting equations of statistical equilibrium can be efficiently solved through matrix-inversion methods, and Monte Carlo photon propagation can be employed for a physically accurate, self-consistent calculation of the coma radiation field J_ν (*van der Tak et al.,* 2007; *Zakharov et al.,* 2007). The steady-state approximation is only strictly applicable when the radiative and collisional excitation timescales are much shorter than the dynamical timescale of the outflowing gas (*Cordiner et al.,* 2022). However, it has the benefit of allowing arbitrary geometries in three dimensions, including the possibility of rapidly varying or even discontinuous physical parameters in the radial dimension.

3. PROCESSES AND THEIR DIAGNOSTICS

The coma spans a broad range of excitation regimes and is subject to strong spatial variations in the radiation and collisional conditions. In the inner regions of the coma, collisions with neutrals and electrons provide a thermalizing influence. As we approach the more tenuous outer regions, or regions with strong chemical and/or solar pumping, additional, non-thermal excitation terms are required. For the example case of H_2O, some of the main processes leading to observable emission from cometary comae are summarized in Table 1. These different processes can lead to distinctly different spectra for the same molecule, as is illustrated for the OH $A^2\Sigma^+$–$X^2\Pi$ rovibronic emission in the near-UV (NUV) (Fig. 3). Fluorescence of OH radicals leads to relatively narrow emission features from the (0-0), (1-0), and (1-1) vibrational bands (*Schleicher and A'Hearn,* 1982), whereas the direct production of excited OH by the dissociation of water molecules leads to the population of higher vibrational states.

In the following sections, we describe the dominant excitation and radiative processes in detail for the general case of coma molecules, atoms, and ions, with an emphasis on recent developments in theory, analysis methods, and observational data.

3.1. Thermal (Collisional) Excitation

The distribution of energy level populations for coma gases is governed by an intricate balance of radiative and collisional processes, the understanding of which starts with two basic assumptions. First, molecules sublimate with vibrational and rotational temperatures equal to the sublimation temperature of the gas source (nucleus, icy grains, or possibly larger icy chunks ejected into the coma), i.e., ~150–200 K at a heliocentric distance $R_h \sim 1$ au. However, collision rates are generally too small to maintain the initial vibrational population, so this population decays radiatively to the ground vibrational state. Any further vibrational and/or electronic excitation usually occurs via non-thermal processes. Second, in the inner (often referred to as "collisional")

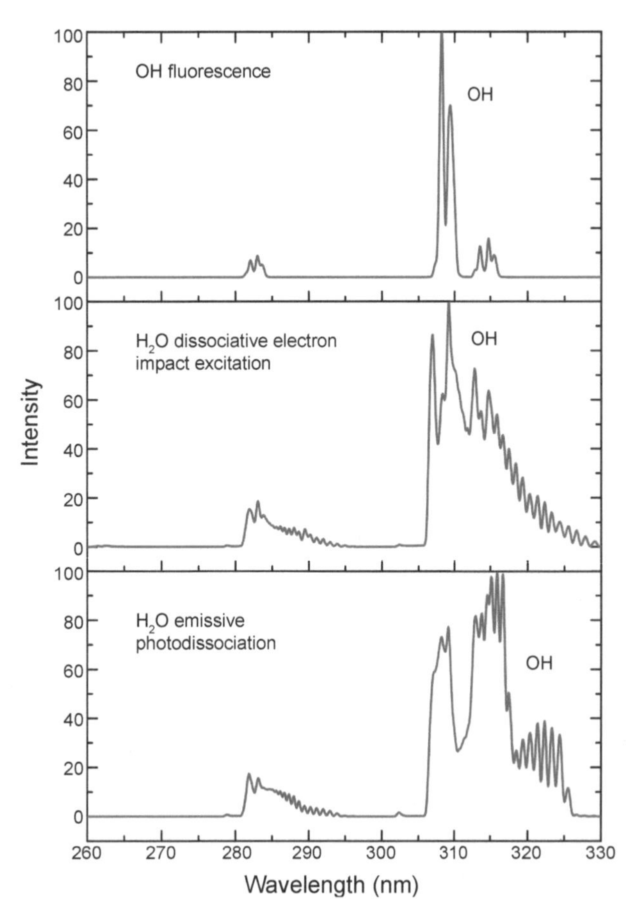

Fig. 3. Emission from transitions between the first excited electronic band of hydroxyl (OH) and its ground state is a prominent feature of cometary spectra in the near-UV. Different excitation processes lead to distinct spectra. The top panel shows the fluorescence spectrum of OH radicals already present in the coma, whereas the middle and bottom panels show the emission resulting from the production of excited OH by the dissociation of water molecules by electrons and photons respectively. Adapted from *Bodewits et al.* (2019).

coma, densities decrease with distance from the nucleus due to near-adiabatic expansion, but remain high enough for collisions to efficiently excite and thermalize the rotational energy level populations in the ground vibrational state. As a result, a Boltzmann distribution can be maintained with a rotational temperature similar to the kinetic temperature of the gas. Pure rotational transitions from the thermally excited rotational levels give rise to emission lines at millimeter/submillimeter wavelengths. In some cases, these rotational transitions may be optically thick, which reduces the efficiency of radiative de-excitation, thereby helping maintain the Boltzmann populations (*Bockelée-Morvan et al.,* 2004).

3.1.1. Collisional excitation and de-excitation rates. Interpreting rotational spectra and understanding the intricacies of collisional excitation among various coma species requires knowledge of the state-to-state collisional excitation/de-excitation rate coefficients (C_{ij}), which are generally less well known (and more difficult to derive) than the radiative decay coefficients. Thanks to studies of rotational excitation in interstellar clouds, collision rates between polyatomic molecules and atomic/molecular hydrogen/helium are now known for many species (*Roueff and Lique,* 2013; *van der Tak et al.,* 2020). In contrast, comparatively little data exists for collisions of known coma molecules with the most abundant cometary gases (H_2O, CO, and CO_2). Experimental collision rates are lacking due to the difficulty of quantum-state-resolved laboratory cross section measurements. Dedicated theoretical efforts, however, are beginning to address this knowledge gap, incorporating various simplifying assumptions to allow the molecular dynamics to be solved with sufficient accuracy for application to cometary spectra.

For example, *Buffa et al.* (2000) used a semiclassical method to calculate C_{ij} rates for the H_2O–H_2O collisional system, which was recently revisited in more detail by *Boursier et al.* (2020). Collision rates for para-H_2O–HCN were calculated by *Dubernet and Quintas-Sánchez* (2019) using a partially-converged, coupled-states calculation, whereas for the H_2O–CO system, *Faure et al.* (2020) used the statistical adiabatic channel method (SACM) (*Loreau et al.,* 2018). Collision rates for CO–CO (up to J = 5) were recently published by *Cordiner et al.* (2022), which are of particular use for comets at large heliocentric distances where the coma CO/H_2O ratio can exceed unity. Theoretical calculations for other collision systems relevant to cometary comae are expected in the coming years, using mixed quantum/classical and SACM approaches. Presently, however, it is common to assume that the required collision rates with H_2O are similar to the already-known rates with H_2 (or He) (e.g., *Roth et al.,* 2021a), or with the additional assumption that the collision rates scale in proportion to the "reduced mass" of the colliding system (*Hogerheijde et al.,* 2009). If

TABLE 1. Summary of the main processes leading to emission from water molecules and its fragments.

Collisional excitation	$H_2O + H_2O$	→	$2H_2O^*$
Radiative pumping and fluorescence	$\gamma + H_2O$	→	H_2O^*
Photodissociative excitation	$\gamma + H_2O$	→	$OH^* + H$; $O^* + H_2$
Dissociative electron impact excitation	$e^- + H_2O$	→	$OH^* + H$; H_2O^{+*}; $2H + O^*$
Charge exchange	$O^{7+} + H_2O$	→	$O^{6+,*} + H_2O^+$

quantum-state-specific C_{ij} values are unavailable, an alternative approach is to assume each molecular collision has a thermalizing effect on the distribution of rotational levels, employing a nominal estimate for the gases collisional cross sections (*Bockelée-Morvan et al.,* 1994; *Biver et al.,* 1999; *Boissier et al.,* 2014).

For atomic species with metastable states that have long lifetimes (e.g., O I, C I, N I; section 3.4), it is also critical to assess the relative importance of collisional quenching and radiative decay.

3.1.2. Rotational temperatures: measurements and interpretation. Gas rotational temperatures are commonly measured using radio and near-IR spectroscopy. To derive T_{rot} from pure rotational lines, the individual, spectrally integrated rotational line intensities are plotted with respect to energy of the upper state of the corresponding transition, E_u (e.g., *Bockelée-Morvan et al.,* 1994). The slope of the linear relation between these two is then directly related to the rotational temperature (see Fig. 4). At the high spectral resolution typically available in radio, modeling the profiles of individual velocity-resolved lines in a rotational band can provide further insight into opacity and line blending effects, allowing more robust determinations of rotational temperatures (*Cordiner et al.,* 2017).

Bonev et al. (2014) summarize methods and sources of uncertainty in the rotational temperature (usually in the ground vibrational state) from rovibrational data at IR wavelengths. Methods include (1) global fits finding the best agreement between measured and synthetic spectra (Levenberg–Marquardt χ^2 minimization and correlation analysis) and (2) plotting the ratio of observed-to-modeled line fluxes on a diagram as a function of rotational excitation energy. The model T_{rot} value is then varied until the points on the diagram lie along a flat line, also minimizing the line-by-line variance (zero slope excitation analysis and the ratio between flux and the g-factor $g(T_{rot})$ variance minimization). These iterative approaches for determining rotational temperatures are applicable to cases where lines from several bands are analyzed together.

In the case of high angular-resolution observations [e.g., with optical/IR spectroscopy or submillimeter interferometry at long baselines with the Atacama Large Millimeter/submillimeter Array (ALMA)], the small ($\lesssim 1''$) beam sizes (corresponding to distances $\lesssim 360$ km from the nucleus at a geocentric distance of $\Delta = 1$ au) are mostly sensitive to the collisionally dominated, inner coma (Fig. 4). Because rotational levels in the ground vibrational state can often be characterized by a single rotational temperature, the measured T_{rot} reflects the thermal state of the gas and, as such, provides a powerful diagnostic into the coma environment and processes responsible for coma heating [e.g., photochemical heating, in which fast dissociation products transfer kinetic energy to the ambient coma gas (*Combi and Smyth,* 1988)] and near-adiabatic expansion cooling.

We note that when several molecules are measured simultaneously across the same region of the inner coma, they all tend to share a similar rotational temperature (see, e.g., *Lippi et al.,* 2021). This further demonstrates the link between T_{kin}

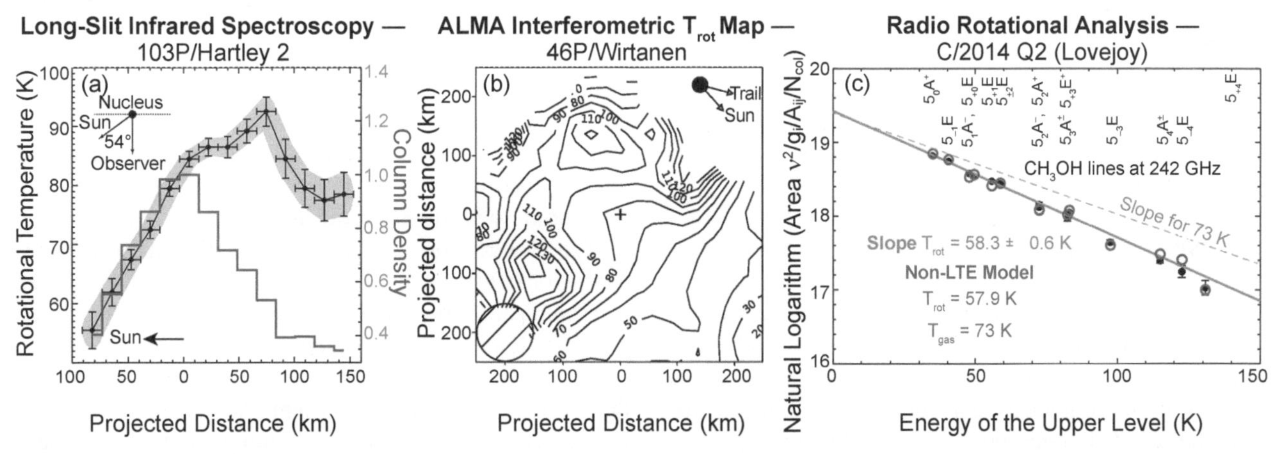

Fig. 4. **(a)** H_2O gas column density (N_{col}, blue histogram) and rotational temperature (T_{rot}, black circles) distributions in 103P/Hartley 2 (*Bonev et al.,* 2013). Such spatially-resolved measurements provide a testbed for coma thermodynamics models because the distributions of the measured parameters are diagnostic for the competition between near-adiabatic expansion cooling and various heating processes in the coma. **(b)** Contour map of CH_3OH coma rotational temperature (T_{rot}) in the inner coma of Comet 46P/Wirtanen based on observations with the Atacama Large Millimimeter/submillimeter Array (*Cordiner et al.,* 2019a); the coma center is denoted with "+." The angular resolution (hatched circle) is shown lower left. Contour labels are in units of Kelvin. **(c)** Rotational diagram for CH_3OH lines observed around 242 GHz in the coma of Comet C/2014 Q2 (Lovejoy) by *Biver et al.* (2015) (black filled circles), with line of best fit (orange) denoting an inferred slope consistent with a rotational temperature of 58.3 K. Open blue circles show the calculated fluxes for individual lines based on a non-LTE radiative transfer model at a kinetic temperature of T_{kin} = 73 K, including radiative processes and collisions with H_2O and electrons — radiative cooling results in $T_{rot} < T_{kin}$. The model nicely predicts a rotational temperature matching the one derived from the slope.

and T_{rot} in the collisional coma, which, depending mainly on gas production rates, can extend from tens to thousands of kilometers from the nucleus.

Measuring T_{rot} is critical for deriving column densities from an incomplete set of molecular line observations, as is typically the case given the limited wavelength coverage and sensitivity of astronomical observations. In that case, T_{rot} provides a useful approximation for the populations of unobserved levels, so that the total molecular number density can still be obtained.

3.1.3. Departure from local thermodynamic equilibrium in the outer coma. As the coma density falls with increasing distance from the nucleus, the reduced collision rates become insufficient to maintain LTE, so the rotational level populations in the ground vibrational state start to deviate strongly from a thermal (Boltzmann) distribution. The interpretation of measured T_{rot} becomes more complex, and its close relationship with T_{kin} breaks down. This transition from LTE to non-LTE regimes is commonly encountered in millimeter/submillimeter non-interferometric (i.e., "single-dish") rotational spectroscopy, with telescope beam sizes larger than a few arcseconds (corresponding to radial distances $\gtrsim$1000 km from the nucleus, for a geocentric distance of 1 au). As a result of relatively low telluric opacities and the high (milli-Kelvin) sensitivity of detectors, such observations represent the primary groundbased technique for the detection of new, complex coma molecules (*Biver et al.,* 2015; see also the chapter by Biver et al. in this volume). In contrast to near-IR spectroscopy and submillimeter interferometry at long baselines, these single dish measurements represent an average signal from the collisional and the extended coma, hence, interpretation of pure rotational lines must account for the departure from LTE of the rotational populations with increasing nucleocentric distance. In addition to a knowledge of the collisional excitation rates (section 3.1.1), understanding this departure from LTE involves the processes of radiative excitation (pumping) and fluorescence, discussed in the next two sections.

3.2. Radiative Excitation (Pumping)

Coma gases interact with various radiation fields. The most important one is the solar radiation field. Additional, minor sources of radiation include scattered and thermal emission from the nucleus and from the coma dust, emission from excited coma molecules, and the 2.7-K cosmic background radiation. This non-thermal excitation process is often referred to as radiative pumping. Absorption of solar photons is by far the major process leading to vibrational pumping (caused by IR photons) and electronic excitation (caused by UV and optical photons). Vibrational excitation from thermal dust emission should also be considered at wavelengths longer than ~10 μm (*Crovisier and Encrenaz,* 1983). Unlike vibrational and electronic excitation, pure rotational pumping by solar photons is negligible because the solar flux is too weak at the long (millimimeter/submillimeter) wavelengths of the rotational transitions. On the other hand, rotational levels (in the ground vibrational state) can be excited by the 2.7-K cosmic background (*Bockelée-Morvan et al.,* 2004).

3.3. Fluorescence

Radiative excitation is the first step in the fluorescence process. An atom or molecule is "pumped" to an excited level due to absorption of a solar photon (section 3.2), followed by radiative decay via spontaneous emission. Fluorescence impacts the interpretation of all types of spectra from coma gases because (1) this process explains the redistribution of rotational level populations in the extended (non-LTE) coma, as needed to interpret rotational spectra, especially from single-dish observations; and (2) solar-induced fluorescence is the main process leading to emissions from virtually all comet gases in a broad wavelength range spanning from the UV to the IR.

3.3.1. Transition from thermal to fluorescence equilibrium. As molecules sequentially absorb solar photons and radiatively decay via spontaneous emission they undergo transitions, in which the total angular momentum (J) can change by an amount equal to $\Delta J = 0$ or ± 1. This leads to a change in the degree of rotational excitation. For an initial distribution of rotational level populations $P_{i,v'}$ in vibrational state v′, following a vibrational (or electronic) transition, the ΔJ selection rules result in characteristic P, Q, and R branches in the spectrum, corresponding to $\Delta J = -1$, 0, and +1, respectively. The rovibrational transition probabilities vary as a function of J, ΔJ, the dipole moment, and the Franck-Condon factor of the vibrational transition (*Herzberg,* 1971). Therefore, after undergoing repeated absorption and emission of photons as a gas parcel travels outward through the coma, the distribution of rotational levels $P_{i,v'}$ changes from the initial, Boltzmann distribution prevalent in the inner coma. At large distances from the nucleus a balance between absorption and spontaneous emission is ultimately realized and the distribution of rotational levels attains fluorescence equilibrium with the solar radiation field.

The change from LTE to non-LTE distributions is evident in observed rotational temperatures. Spatially resolved measurements of the rotational temperature of H_2O acquired with the Spitzer Space Telescope showed a decrease with distance from the nucleus, best described as a transition from thermal to fluorescence equilibrium (Woodward et al., 2007). Importantly, this transition occurs more rapidly for strongly polar molecules, such as HCN, with higher rotational transition rates (larger Einstein A values), whereas molecules with smaller dipole moments, such as CO, maintain LTE out to much greater distances from the nucleus. Figure 5 shows the theoretical evolution of CO and HCN rotational temperatures with distance from the nucleus for comets with increasingly large water production rates (Q_{H_2O} = 10^{27}–10^{29} molecules s^{-1}). The smaller Einstein A values for rotational transitions of CO (for a given J level) lead to a larger population of the higher J states as a result of solar pumping. Consequently, at fluorescence equilibrium in the

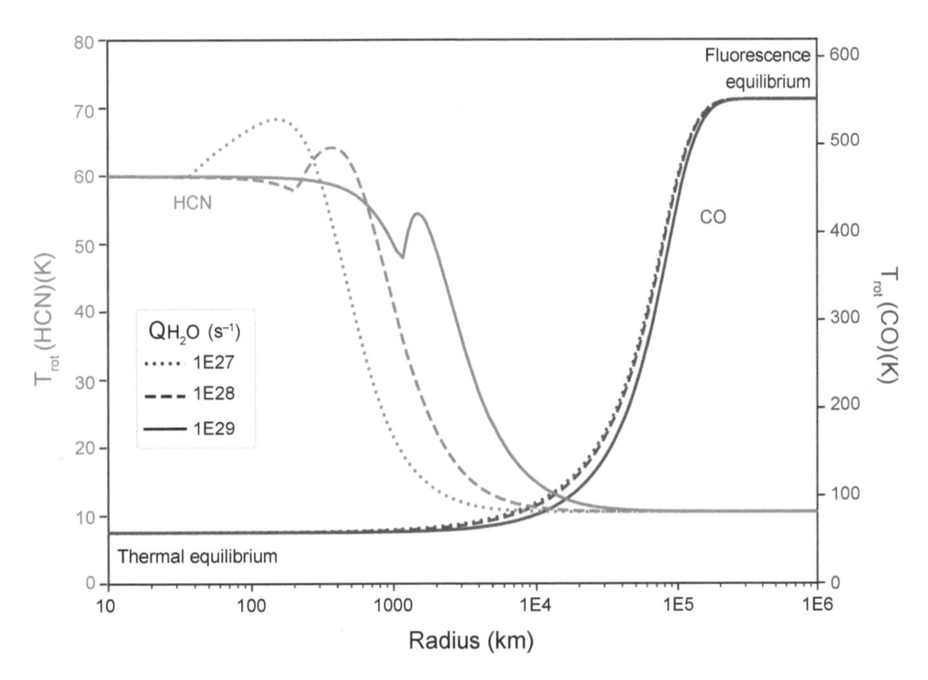

Fig. 5. Simulated rotational temperatures (T_{rot}) of HCN (orange, left ordinate) and CO (blue, right ordinate) as a function of distance from the nucleus for three different water production rates (Q_{H_2O}; 10^{27} molecules s^{-1}, dotted; 10^{28} molecules s^{-1}, dashed; 10^{29} molecules s^{-1}, solid), based on the populations of the lowest seven energy levels for each molecule. This assumes a spherically-symmetric outflow, with constant gas kinetic temperature T_{kin} = 60 K and a heliocentric distance of 1 au. Rotational energy level populations were calculated using equation (4), with CO–H_2O collision rates from *Faure et al.* (2020), and HCN–H_2O collision rates from *Dubernet and Quintas-Sánchez* (2019). Rotational temperatures evolve from thermal equilibrium at high density (on the left) to fluorescence equilibrium in the low-density outer coma (on the right). For additional model details, see *Cordiner et al.* (2022).

outer coma of a comet at 1 au from the Sun, T_{rot}(CO) reaches 550 K, whereas T_{rot}(HCN) reaches only 11 K. The relatively large rate of collisions between HCN and electrons is also responsible for significant departures of T_{rot}(HCN) from the kinetic temperature of the neutral gas, seen as humps in the T_{rot}(r) curve in Fig. 5, between 100 and 2000 km from the nucleus, where the densities of (hot) photoelectrons in the coma begin to become significant. The dramatically different behavior of these two commonly observed coma molecules demonstrates the importance of considering the detailed microphysics when using limited observational data for the derivation of production rates.

3.3.2. Advancements in fluorescence models. Fluorescence gives rise to emission lines of many cometary volatiles of significance to both cosmogony and astrobiology, including water (H_2O, see Fig. 6, and HDO), symmetric hydrocarbons (CH_4, C_2H_2, C_2H_4, C_2H_6), oxidized carbon compounds (CO, H_2CO, CH_3OH, OCS), nitrogen species (NH_3, HCN), and daughter species (e.g., C_2, CN, OH, NH). In particular, symmetric hydrocarbons can only be detected through their fluorescence IR emission, due to the lack of a permanent dipole moment. High-resolution ($\lambda/\Delta\lambda > 20{,}000$) optical and near-IR spectrographs at ground-based facilities can detect many volatiles simultaneously via multiple lines of each species (*Cochran et al.,* 2012; *Villanueva et al.,* 2011a).

The interpretation of observed spectra requires the development of fluorescence models. The first fluorescence calculations focused on the electronic excitation of radicals such as OH, CN, and CO^+ at optical and NUV wavelengths (*Schleicher and A'Hearn,* 1982; *Schleicher,* 1983; *Magnani and A'Hearn,* 1986). The characterization of the IR region in comets has grown tremendously in the last decade, thanks to the recent advent of extensive ab initio high-energy spectroscopic linelists [ExoMol (*Tennyson et al.,* 2020)], and more complete experimental linelists of trace species and hydrocarbons [HITRAN (*Gordon et al.,* 2017)]. Because fluorescence involves pumping and cascades from all rovibrational levels, the calculations require handling of a large number of high-energy transitions (due to solar pumping) as well as low-energy (thermal) transitions. This is particularly relevant for non-resonance fluorescence, which dominates the IR emissions of H_2O. Modern fluorescence models ingest information from many spectroscopic sources and handle the billions of transitions needed for accurate branching ratio calculations (*DiSanti et al.,* 2006; *Villanueva et al.,* 2012, 2013; *Gibb et al.,* 2013; *Kawakita and Mumma,* 2011). For instance, a recent high-energy water database contained 5×10^8 transitions for H_2O and 7×10^8 for HDO (*Villanueva et al.,* 2012). Figure 6 presents the main elements in a fluorescence calculation (pumps, cascades, and branching ratios), all underpinned by a complete molecular Hamiltonian and an equilibrium non-LTE model. These new models have become more accessible to the community by the develop-

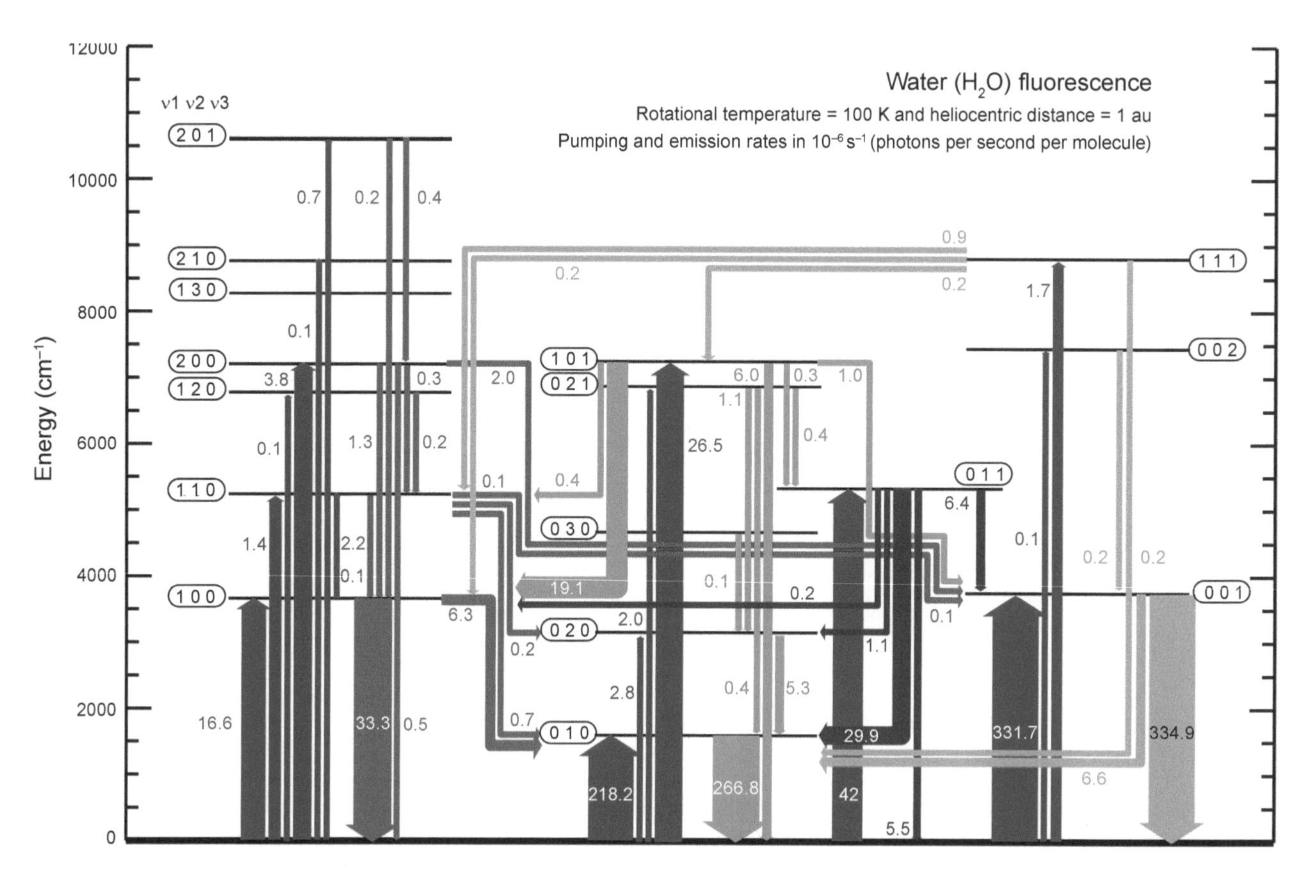

Fig. 6. Diagram showing the full non-resonance fluorescence tree for H_2O in a comet at 1 au from the Sun with a rotational population at 100 K. The pumping rates (in units of photons per second per molecule, shown in blue) were calculated considering a realistic solar model, and the emission rates (shown in red/green/purple/yellow colors) were calculated by subsequent cascade down to the ground-vibrational level and considering line-by-line and level-by-level branching ratios, which take into account all 500 million transitions. Adapted from *Villanueva et al.* (2012).

ment of open and public repositories and the inclusion of non-LTE modeling and fluorescence capabilities via public online tools (*Villanueva et al.,* 2018).

The interpretation of fluorescence spectra requires a hybrid approach, starting with the generally good approximation that in the inner coma rotational populations in the ground vibrational state are collisionally equilibrated to a Boltzmann distribution, and then incorporating the complexity of non-LTE physics. Solar pumping raises a molecule to a highly excited vibrational state (section 3.2) through rovibrational transitions, based on the impinging solar flux (J_v) and the Einstein B_{ji} coefficients of those transitions. Then, the molecule quickly decays radiatively to intermediate vibrational levels (non-resonant fluorescence or "hot-bands") or directly back to the ground vibrational state (fundamental bands). Each cascade gives rise to observable optical/IR lines, which can be characterized by a corresponding fluorescence efficiency. The cascade process is followed sequentially for all rovibrational levels, until the molecule returns to the ground vibrational state. At each level, the direction and proportionally of the cascade is calculated based on the branching ratios, which are determined from the spontaneous emission coefficients (A_{ij}) for transitions from that level. Fundamental bands with direct counterparts in absorption in Earth's atmosphere (such as CH_4, CO, HDO) require geocentric Doppler shifts for their detection in groundbased studies of comets. H_2O vibrational hot-bands are the easiest accessible groundbased diagnostic for water, because these transitions originate from levels not significantly populated in Earth's atmosphere (*Mumma et al.,* 1986). The final product of IR fluorescence models is the emission efficiency (g-factor, in units of photons s^{-1} molecule^{-1}, or W molecule^{-1}) for each rovibrational line across all possible vibrational modes.

A challenge in fluorescence calculations is the inclusion of opacity effects and non-Boltzmann rotational level populations, which require solving the radiative transfer equilibrium states iteratively. In most cases, the pumping transitions are optically thin due to the transient and sporadic nature of high-energy rovibrational transitions, yet for highly active comets, the coma can be quite opaque for the pumps of some abundant species. A hybrid solution to this issue can be achieved by scaling the line fluxes based on the probability of the corresponding pump being opaque, which can be approximated by exploring the Einstein B_{ji} coefficient and average densities in the atmosphere. Integrating this approach into the Q-curve formalism [i.e., extracting production rates based on flux collected in successive apertures along the slit (*Dello Russo et al.,* 1998)] has helped in addressing optically thick pumps in the derivation of densities and rotational temperatures in Comet C/2006 W3 (Christensen) (*Bonev et al.,* 2017).

Bockelée-Morvan et al. (2010) demonstrated the importance of considering optical depth effects in interpreting spatial distributions from IR observations in exceptionally bright comets, such as C/1995 O1 (Hale-Bopp), which exhibited extraordinary high CO production rates (~10^{29}–10^{30} molecules s^{-1}). This work presented radiative transfer calculations of CO fluorescent line intensities from the 1-0 band, taking into account (1) opacity in the solar pumps; (2) radiation trapping caused by self-absorption of IR photons emitted by CO molecules in coma, which acts as an additional source of vibrational excitation, counterbalancing the reduced direct pumping by solar photons; and (3) attenuation of the observed IR emission, also due to self-absorption. The authors showed that the spatial profiles of CO in Hale-Bopp can be explained by these optical depths effects rather than indicating the presence of extended coma sources of CO, as inferred by earlier analyses reviewed by *Bockelée-Morvan et al.* (2004).

Treating opacity effects in the coma is also particularly relevant when interpreting cometary spacecraft data, since the measured bands tend to be the strong fundamentals, the densities sampled of the inner coma are particularly large, and the effects regarding the directionality of the excitation source become notable. As such, dedicated models that combine modern numerical and radiative transfer methods have been developed to model such spectra (*Gersch and A'Hearn,* 2014; *Gersch et al.,* 2018; *Debout et al.,* 2016; *Cheng et al.,* 2022).

3.3.3. Coma diagnostics from fluorescence spectra. In the IR, comparison of measured spectra with fluorescence models enables the determination of gas rotational temperatures (section 3.1.2) and column densities (N_{col}), which in turn can be a testbed for coma thermodynamic models. *Fougere et al.* (2012) compared synthetic T_{rot} and N_{col} spatial profiles predicted by a kinetic Direct Simulation Monte Carlo model (*Tenishev et al.,* 2008) with the long-slit spatial distributions of these parameters obtained from IR fluorescence spectra in 73P/Schwassmann-Wachmann 3B. Their work suggested that water sublimated from icy grains in a rotationally hot state imparts energy through collisions with the ambient coma, thereby offsetting the cooling due to near-adiabatic expansion. This scenario also explains the elevated T_{rot} in the projected anti-sunward direction in 103P/Hartley 2 (Fig. 4a), where the EPOXI mission observed that icy grains were a significant source of water in the coma (*A'Hearn et al.,* 2011; *Protopapa et al.,* 2014).

Comparing the spatial distributions of column density among different volatiles addresses whether these species are associated with common or distinct outgassing sources (see the chapter by Biver et al. in this volume). Major differences in spatial distributions point to heterogeneous outgassing reflecting entirely distinct sources or one or more additional source(s) for some volatiles (*Kawakita et al.,* 2013; *Dello Russo et al.,* 2022). *DiSanti et al.* (2018) showed that short-term (hours) temporal variations in coma relative abundances can be directly linked to heterogenous outgassing, as revealed by spatial distributions of fluorescent emission measured in Comet C/2013 V5 (Oukaimeden). Another notable example is C/2007 W1 (Boattini), for which *Villanueva et al.* (2011a) interpreted the emission profiles in terms of release from two distinct moieties of ice: (1) clumps of mixed ice and dust released from the nucleus into the sunward hemisphere, and (2) small grains of nearly pure polar ice (water and methanol, without dark material or apolar volatiles).

Improvements in fluorescence models have also resulted in more accurate measurements of spin ratios in cometary comae. An intrinsic property of molecules is the existence of distinct symmetries of nuclear spin species, which do not interact radiatively under electric-dipole selection rules. Interconversion among these spin species is strongly forbidden in the gas, leading to long persistence times in the coma. An often overlooked aspect is that the fluorescent g-factors have separate dependencies on rotational temperature and on spin ratio. For example, even at the same T_{rot}, spectra of water will look substantially different for ortho-H_2O to para-H_2O abundance ratios of 1.5 and 3.0. Without factoring the effects of rotational temperature and spin ratio into the modeled g-factors, the observed line strengths cannot be accurately reproduced, and this can influence calculated column densities and production rates.

Infrared fluorescence spectra are particularly suitable for obtaining spin ratios because of the simultaneous sampling of numerous lines from different spin species and molecules provided by modern high-resolution spectrometers. Because the lowest energy levels in the various spin ladders differ slightly, the ratio between the total populations of different spin states would be temperature-dependent if we assume that the given spin ratio was realized in thermal equilibrium. Thus, a nuclear spin temperature, T_{spin}, can formally be derived from measured spin ratios. Although a spin temperature of 30 K was suggested by observational studies of H_2O and NH_3 in some comets (*Dello Russo et al.,* 2005; *Kawakita et al.,* 2006; *Mumma and Charnley,* 2011), subsequent groundbased retrievals of T_{spin} employing the latest models typically indicate spin temperatures above 50 K, consistent with statistical equilibrium (*Villanueva et al.,* 2012; *Bonev et al.,* 2013). *Cheng et al.* (2022) investigated the impact of optical depth on the derived spin temperatures from water fluorescence emissions observed by the Visible and Infrared Thermal Imaging Spectrometer (VIRTIS-H) onboard the Rosetta spacecraft. Their analysis also concluded that the spin temperature in statistical equilibrium (T_{spin}) provides the best explanation for their measurements.

Currently, there is a strong debate over whether T_{spin} preserves a record of the formation temperatures of cometary volatiles. Recent laboratory measurements do not support this scenario and explore the possibility that phase transition phenomena (*Hama et al.,* 2016), energetic particle irradiation (*Sliter et al.,* 2011), and/or formation of clusters (*Manca Tanner et al.,* 2013) may reset the ortho/para ratio of cometary water. The situation for ammonia is not well characterized experimentally, while the exchange of H-atoms in CH_4 and other hydrocarbons (C_2H_6, CH_3OH) should be extremely slow, even in ices.

3.4. Photodissociative Excitation

In this non-LTE process, parent species (H_2O, H_2CO, CO_2, CO) are photodissociated by solar UV photons, producing fragments (OH, CO, H_2, O I) in excited states. For example, photolysis in the second absorption band of H_2O (primarily by Ly-α radiation) can lead to OH in electronically excited states ($A^2\Sigma^+$) or in vibrationally and rotationally excited levels within the ground electronic state ($X^2\Pi$), while photolysis in the first absorption band ($\lambda >$ 136 nm) leads predominantly to rovibrationally excited OH ($X^2\Pi$) (*Crovisier,* 1989). These states of OH cannot be efficiently populated by fluorescence or collisions and generally have short lifetimes (~10 ms). As a result, the dissociatively excited fragment decays radiatively, giving rise to prompt emission.

This emission provides two main diagnostics. (1) Because of the short radiative lifetimes of dissociation products, optically-thin prompt emission is a good tracer for the spatial distribution of the precursor molecule (see Fig. 7), and can also be used to approximate the production rate of the precursor (*Bertaux,* 1986). However, analysis of prompt emission requires evaluating optical depth effects in the solar UV for exceptionally active comets (*Bonev et al.,* 2006). (2) The relative intensities of prompt emission lines may be quite different from those expected from fluorescence (Fig. 3) and can help understand the dissociation dynamics in cometary comae (*Bonev and Mumma,* 2006; *Bodewits et al.,* 2016). The quantum state distributions of dissociation products, inferred from comet observations, can be compared with those from laboratory studies (*Feldman et al.,* 2009; *A'Hearn et al.,* 2015). These distributions are governed by the exit channels from the dissociative parent state, whose excitation in turn depends on the initial state population of the parent (see section 3.1) and especially on the energy of the UV photon (e.g., first or second absorption band of H_2O).

Mumma (1982) and *Bertaux* (1986) predicted that prompt emission would be an important component in the IR and UV spectra of comets respectively. *Crovisier* (1989) modeled the integrated prompt emission rates for several OH vibrational bands. *Bockelée-Morvan and Crovisier* (1989) showed that OH prompt emission will contribute to low resolution ($\lambda/\Delta\lambda$ ~ 160) IR cometary spectra near 2.8 μm. Early detections of prompt emission were reported by *Budzien and Feldman* (1991), *Brooke et al.* (1996), *Mumma et al.* (2001), and references therein. Here we complement the recent summary by *Thomas* (2020), highlighting several post-*Comets II* developments.

A'Hearn et al. (2015) reported the first spectrally resolved detection of OH prompt emission in the NUV 0-0 band of the $A^2\Sigma^+$–$X^2\Pi$ electronic transitions. They analyzed spectroscopic observations of Comet C/1996 B2 (Hyakutake), whose small geocentric distance (0.11 au) allowed probing the near-nucleus region and to detection of prompt emission from the ground. The "rotationally-hot" quantum-state distribution of OH measured in Hyakutake was in excellent agreement with laboratory studies of the second absorption band of H_2O.

LaForgia et al. (2017) observed the NUV OH prompt emission in narrowband images taken during the Deep Impact spacecraft flyby of Comet 103P/Hartley 2. Demonstrating that the prompt emission closely tracked the maps of water vapor in the inner coma, the authors proposed a dedicated OH prompt emission filter, which could be used to directly image the distribution of H_2O in the inner coma

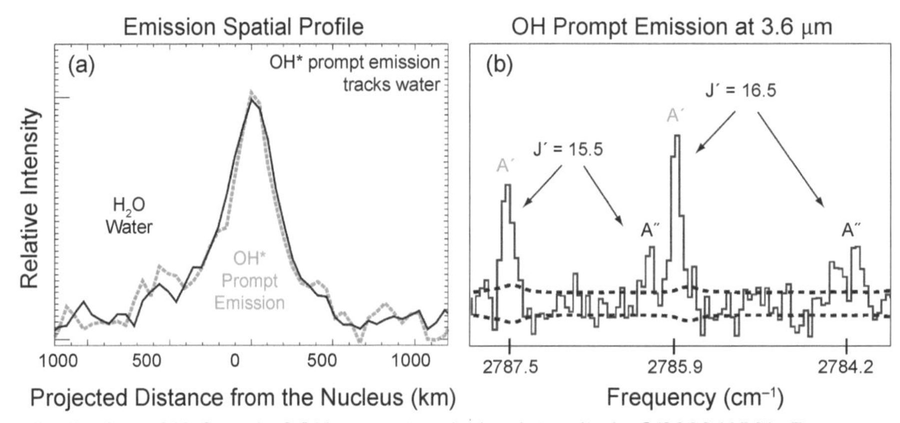

Fig. 7. (a) Spatial distribution of H_2O and of OH prompt emission intensity in C/2000 WM1. Prompt emission commonly peaks at the nucleus and tracks the spatial distribution of the dissociative precursor. **(b)** Example of OH prompt emission lines often detected in comets. Near-IR measurements typically use frequency in wavenumber (cm^{-1}) units (2785.9 cm^{-1} ≈ 3.6 *μm*). J′ is the rotational quantum number of the upper state for a transition. The ratio between the emission efficiencies of the Π(A′) and Π(A″) Λ-doublet components (marked as A′ and A″ on the plot) is an important diagnostic for the H_2O dissociation dynamics leading to rovibrationally excited OH states (after *Bonev et al.,* 2006; *Bonev and Mumma,* 2006).

when the first 200 km are resolved. Centered at ~318.8 nm, such a filter would be suitable for wavelengths commonly accessible by CCDs on both groundbased telescopes and platforms in space.

Many P-branch lines from vibrational prompt emission bands in OH ($X^2\Pi$) are detectable in groundbased near-IR observations of even moderately bright comets. Using simultaneous measurements of H_2O and OH, *Bonev et al.* (2006) empirically calibrated emission efficiencies (OH photons s^{-1} per H_2O molecule) for a suite of OH lines (2.8–3.6 μm). This has helped isolate the contribution of OH in spectrally crowded regions throughout the L-band in near-IR studies of cometary volatiles (see the chapter by Biver et al. in this volume).

The rotational distribution of OH ($X^2\Pi$) cannot be approximated by a single Boltzmann distribution. The dissociation of H_2O can lead to Λ-doublet states of OH with the same rotational quantum number J′, but with a different symmetry of the electronic wave function: $\Pi(A')$ and $\Pi(A'')$, as defined by *Alexander et al.* (1988). These states generally have vastly different populations, which affects the relative intensities of prompt emission lines (Fig. 7b). *Bonev and Mumma* (2006) showed that the ratio of OH line intensities from transitions originating from $\Pi(A')$ and $\Pi(A'')$ levels has a strong dependence on J′. This ratio may be diagnostic for the particular dissociation channel, reflecting preferential population of $\Pi(A'')$ states at low-J′ (dissociation in the first absorption band), and for $\Pi(A')$ states for $J' > 9.5$ (second absorption band).

The $a^3\Pi$ states of CO can be produced directly by the dissociation of CO_2 and the ensuing prompt emission has long been considered one of the main sources of the CO Cameron bands in UV spectra of comets (*Weaver et al.,* 1994). However, *Raghuram and Bhardwaj* (2012) evaluate the importance of both photodissociatve excitation and electron impact excitation (see section 3.5) in populating CO ($a^3\Pi$) and they concluded the latter is the main production mechanism in 1P/Halley.

Kalogerakis et al. (2012) showed that CO ($a^3\Pi$) can be efficiently populated via cascades from higher-energy levels of CO. These authors identified visible and near-IR prompt emission in laboratory spectra of CO resulting from CO_2 photodissociation. CO was produced in $a'^3\Sigma^+$, $d^3\Delta$, and $e^3\Sigma^-$ states. Cascades from these states can populate $a^3\Pi$ and lead to prompt emission in the visible and near-IR (wavelengths above 500 nm). *Kalogerakis et al.* (2012) suggested that if found in comets, these emissions would be an indirect "marker" for the presence on CO_2 in the coma, potentially enabling the estimation of upper limits for CO_2 abundances.

Dissociative excitation can lead to the production of radicals in states with longer radiative lifetimes (metastable states), which therefore do not favor prompt emission. These products can then be further excited by fluorescence. In the FUV, *Liu et al.* (2007) identified many lines pumped by solar Ly-α from rovibrational levels of H_2 produced by photodissociation of H_2O. *Feldman et al.* (2009) and *Feldman* (2015) discussed CO and H_2 fluorescence respectively, pumped from dissociatively excited states of H_2CO.

Several observational and modeling studies have focused on better understanding dissociative excitation leading to metastable atomic states. *McKay et al.* (2015), *Decock et al.* (2015), *Raghuram et al.* (2020), and references therein discuss both the usefulness and the uncertainties in an indirect method to obtain CO_2 production rates based on the intensity ratio of the green (^1S-^1D; $\lambda = 557.7$ nm) and red doublet (^1D-^3P, $\lambda = 630.0$ and 636.4 nm) lines of O I. This is a potentially important application, because CO_2 is a major volatile in comet nuclei that cannot be detected from the ground, unlike the easily detectable lines of O I. Two key challenges for this methodology are that atomic oxygen can be produced by the dissociation of various parents (such as H_2O, CO_2, O_2, and CO), and that the relevant dissociative excitation rates are not well known. These rates are difficult to constrain in the laboratory owing to the long radiative lifetime of metastable atomic states. Nevertheless, based on laboratory studies of H_2O photodissociation leading to O^1S and vibrationally-excitated H_2, *Kawakita* (2022) derived H_2O photodissociation rates leading to both O^1S and O^1D.

Raghuram et al. (2020) utilized the H_2O-depleted coma of C/2016 R2 (PanSTARSS) as a natural laboratory to study dissociative excitation of CO_2, CO, and N_2, leading to O I, C I, and N I, respectively. Collisional quenching is being evaluated in interpreting the line ratios and line widths of forbidden transitions, measured for different projected distances from the nucleus. These ratios help constrain dissociation yields and also identify photolysis of O_2 as an additional source of the O I red and green lines in comets. The line ratio for the N I doublet ($\lambda = 519.8$ and 520.0 nm, respectively) measured in C/2016 R2 is useful for obtaining the intrinsic transition probability ratio of the two sub-levels of N^2D. Further investigations are needed to more fully investigate these complex processes.

3.5. Electron Impact Excitation

As solar radiation impinges on the coma, photoelectrons arise in the gas with a distribution of energies that peaks between 1 and 30 eV (*Gan and Cravens,* 1990; *Engelhardt et al.,* 2018). These electrons redistribute their energy via collisions with atoms and molecules (*Cravens and Korosmezey,* 1986), and in doing so, contribute toward various coma excitation and emission processes, the nature of which depends strongly on electron energy.

Collisions with thermalized electrons in the inner coma help maintain LTE, but outside the contact surface, where the degree of ionization increases and electron temperatures and densities rise steeply with radius (*Körösmezey et al.,* 1987; *Eberhardt and Krankowsky,* 1995), impacts with hot (~10^4 K, or a few electron-volt) electrons can strongly affect the excitation of neutral coma gases (*Xie and Mumma,* 1992; *Lovell et al.,* 2004). Similar to the case of neutral-neutral molecular collisions (section 3.1.1), the state-to-state collision rates involving electrons and neutral molecules can be obtained from detailed quantum mechanical calculations

(*Faure et al.,* 2007). However, such calculations are yet to be performed for most molecules of interest over the range of electron energies found in cometary comae. Consequently, the simplified formula of *Itikawa* (1972) (using the Born approximation) is frequently used to determine state-to-state electron impact rates (*Biver,* 1997; *Zakharov et al.,* 2007; *Cordiner et al.,* 2019b).

3.6. Dissociative Electron Impact Excitation

Electrons with energies above 10 eV that collide with neutral molecules in the coma can produce excited fragments. Rosetta found that outside 2 au pre-perihelion, when outgassing rates were low, atomic and molecular emission features of 67P/Churyumov-Gerasimenko at UV/optical wavelengths were predominantly caused by dissociative electron impact excitation (*Feldman et al.,* 2015, 2018; *Bodewits et al.,* 2016; *Galand et al.,* 2020). Because gas production rates increased much faster than the ionizing solar radiation, a collisionopause was formed (*Nilsson et al.,* 2017). Within this boundary, collisions between water molecules and lower-energy electrons (below the threshold for dissociative impact excitation reactions) become dominant. For example, the formation of excited OH ($A^2\Sigma^+$) by electron impact on H_2O has an electron energy threshold of 9.1 eV, H Ly-α has a threshold of 15.4 eV, and the production of O I 130.4 nm emission requires electron temperatures above 23.5 eV (*Beenakker et al.,* 1974; *Bodewits et al.,* 2019). At electron temperatures below 9 eV, collisions result in rovibrational excitation rather than dissociation of H_2O (*Itikawa and Mason,* 2005) and energy is radiated out of the inner coma in the form of IR emission.

The Alice instrument onboard Rosetta recorded the emission of mostly atomic fragments in the FUV, including atomic hydrogen, carbon, and oxygen. It was noted that the relative intensities of lines (i.e., line ratios) were different from FUV spectra of previous comet observations (*Feldman et al.,* 2015), which could not resolve the inner coma. The largest surprise was the O I] line at 135.6 nm, which is a forbidden intercombination multiplet that is rarely seen in coma spectra. These line ratios indicated that the dominant excitation process was dissociative electron impact excitation, rather than photofluorescence. In addition, the relative strength of the electron impact induced emission lines of H I, O I, and C I in the FUV are highly sensitive to the parent molecule they are produced from, such as H_2O, O_2, and CO_2 (see Fig. 8). Electron impact emission thus provides a sensitive diagnostic probe of the local plasma environment in the

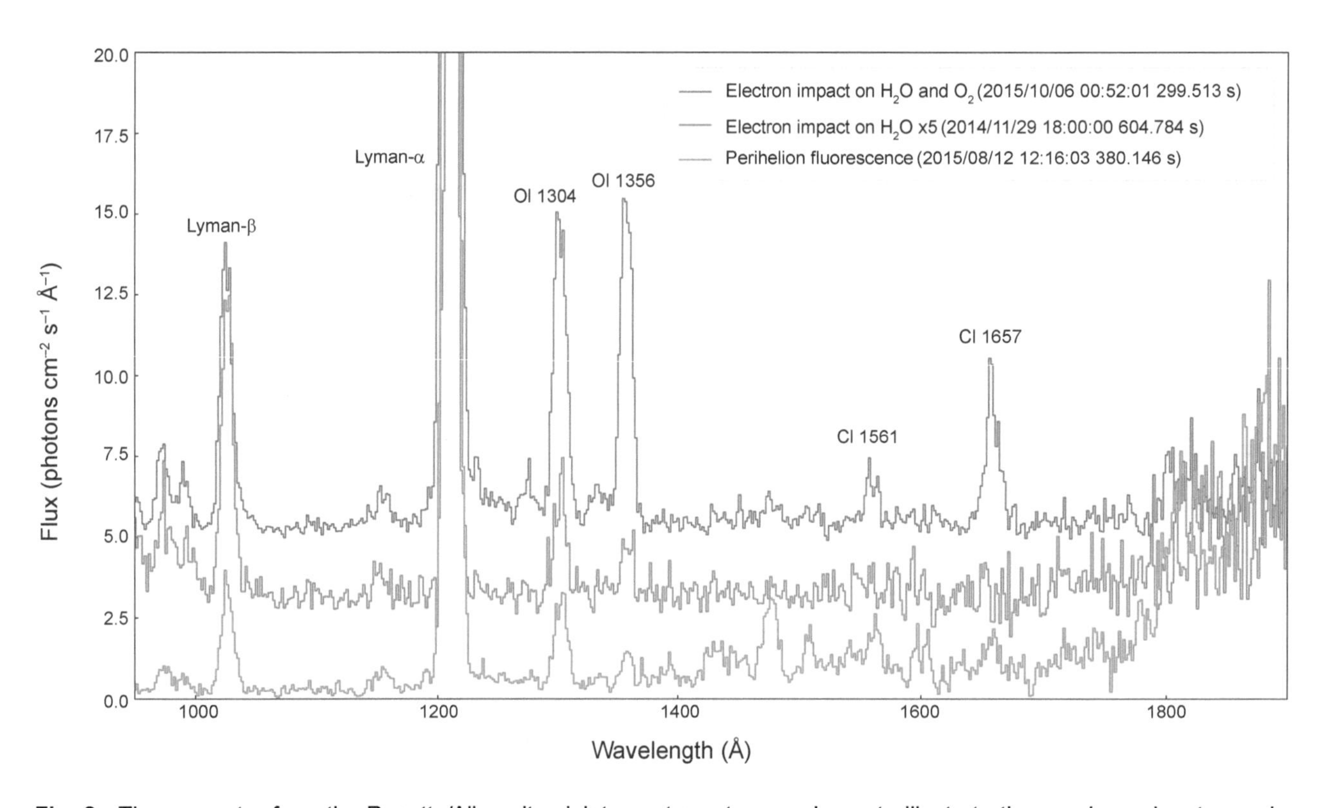

Fig. 8. Three spectra from the Rosetta/Alice ultraviolet spectrometer are shown to illustrate three unique signatures observed in the near-nucleus coma around 67P/Churyumov-Gerasimenko. All spectra are created using the narrow middle section of the Alice slit. Electron impact on molecular oxygen and water (blue) produces the clearest signature of electron impact at the O I] 1356 Å emission feature. Electron impact on water (orange) is clearest in the early days of the mission when 67P was beyond 2.7 au. It has a much weaker signal in that spectral region. For comparison, a spectrum taken at perihelion, in green, shows a richer spectrum of CO Fourth Positive group features and resonance fluorescence of the Lyman series, but little in the way of O I] 1356 Å (adapted from *Feldman et al.,* 2018).

coma and of the chemical composition of the atmospheres of small bodies. Remote studies using the Hubble Space Telescope have used this to characterize the composition of the atmospheres of Europa (*Hall et al.,* 1995; *Roth et al.,* 2014), Callisto (*Cunningham et al.,* 2015), and Ganymede (*Feldman et al.,* 2000; *Roth et al.,* 2021). However, Hubble Space Telescope remote studies of 46P/Wirtanen found no evidence of dissociative electron impact in the inner coma (*Noonan et al.,* 2021), likely because the local densities were high enough to suppress them, as was observed by Rosetta as 67P approached the Sun (*Bodewits et al.,* 2016; *Galand et al.,* 2020). Under which coma conditions dissociative electron impact can be the dominant process of emission remains an open question.

3.7. Solar Wind Charge Exchange

Comets emit up to 1 GW in extreme UV and X-ray radiation (*Lisse et al.,* 2004; *Krasnopolsky et al.,* 1997), despite the relatively low temperature of the gas in the coma (tens of Kelvins; Fig. 9). This emission is produced mostly by charge exchange between heavy ions in the solar wind (e.g., He^{2+}, $O^{7,8+}$, $C^{5,6+}$) and neutral molecules in the cometary atmosphere (*Cravens,* 1997; *Krasnopolsky,* 1998). The incoming ions capture one or more electrons directly into a highly excited state, which results in X-ray emission when the ion decays to its ground state.

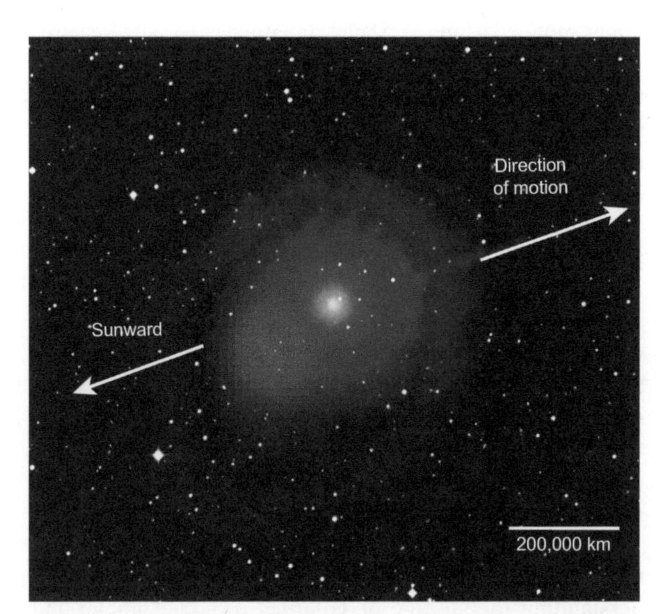

Fig. 9. Simultaneous observations of Comet C/2007 N3 (Lulin) by the Neil Gehrels-Swift observatory in soft X-rays (photon energies below 1 keV; red) and in the ultraviolet (200–350 nm; blue). In the UV, it sampled the fluorescent emission of OH radicals formed by the photodissociation of H_2O molecules. X-rays are emitted by solar wind ions coming from the direction of the Sun (bottom left), which capture electrons as they penetrate the neutral coma. Because the coma is collisionally thick to these incoming ions, the X-ray emission is asymmetric and offset toward the Sun.

In the energy regime accessible by current X-ray satellites (typically above 0.25 keV), cometary charge exchange spectra consist mostly of emission lines of hydrogen- and helium-like ions. In particular, they are characterized by strong dipole forbidden lines, such as the O VII features around 565 eV, as metastable states are populated through cascades originating from the highly excited states electrons are captured into.

Different from most optical telescopes, detectors onboard X-ray facilities such as Chandra, XMM-Newton, and the Neil Gehrels-Swift Observatory simultaneously provide spectral, spatial, and temporal information of incoming photons that can be used to probe properties of both the comet and the solar wind. For example, X-ray spectra can be used to measure the solar wind elemental composition and the charge state distribution of its ions, and thus of the conditions of its source region on the Sun (*Schwadron and Cravens,* 2000; *Bodewits et al.,* 2007). Because charge exchange emission is a quasi-resonant process, the angular momentum distribution of the captured electrons depends strongly on the velocity of the incoming ion and on electron donor species (*Beiersdorfer et al.,* 2003).

Since the discovery of charge exchange emission in comets, it has been detected in many different astrophysical environments where a hot gas collides with a cold, neutral gas, including Mars, galaxies, and supernova remnants (*Dennerl,* 2010). Charge exchange reactions with the dominant coma gases (such as H_2O, CO_2, and CO) typically have cross sections of approximately 10^{-15} cm^2. This implies that for moderately active comets (gas production rates of less than a few times 10^{28} molecules s^{-1}), solar wind ions will fly through most of the coma unhindered, and the charge exchange emission will map the neutral coma and jets, independent of the species of neutral gas. In this collisionally thin scenario, variations in the X-ray brightness depend on the heavy ion flux and neutral gas production rate of the comet. For comets with higher production rates, more of the coma will become collisionally thick to charge exchange and when observed under a sufficiently large phase angle, the X-ray morphology will take on a characteristic crescent shape (Fig. 9), first seen around Comet C/1996 B2 (Hyakutake) (*Lisse et al.,* 1996). Because the solar wind ions that are capable of emitting X-rays are depleted before they get close to the nucleus, the X-ray luminosity becomes decoupled from the gas production rate (*Dennerl et al.,* 1997) but not from variations in the solar wind (*Bonamente et al.,* 2021). XMM-Newton imaging of the particularly active Comet C/2001 WM_1 (LINEAR) allowed for the first remote characterization of a comet's plasma bowshock, which is formed when the solar wind is decelerated by picking up heavy cometary ions (*Wegmann and Dennerl,* 2005).

X-ray astronomers are highly anticipating the launch of calorimeter instruments, which will provide a spectral resolution between 2 and 5 eV and good sensitivity to low energies. These upcoming observations will challenge our current understanding of charge exchange processes. For example, modeling suggests clear differences between spectra resulting from water group molecules and CO and CO_2 (*Mullen et al.,* 2017). The spectral resolution of these instruments will allow testing of whether other plasma processes, such as electron bremsstrahlung, contribute to cometary X-ray emission (*Krasnopolsky,* 1998). Finally, these observations will allow using comets as natural laboratories and address the question of whether the state selective population by charge exchange results favors triplet or singlet states of helium-like ions (*Mullen et al.,* 2017). This question is very hard to address in the laboratory owing to the long lifetimes of the relevant metastable states, yet the resulting emission lines are among the strongest in astrophysical charge exchange spectra.

4. FUTURE DIRECTIONS

4.1 Spectrospatial Coma Observations

The main observational breakthroughs in cometary exploration expected in the coming decades relate to the increased sensitivity and instantaneous bandwidth (across all wavelengths) afforded by improved instrumentation and larger telescope collecting areas. These developments will result in improved signal-to-noise and fidelity for atomic and molecular imaging and spectroscopy that will provide new insights into coma physics and chemistry through a better understanding of the gas macroscopic and microscopic motions and compositions.

Most modern research on cometary objects has been restricted to comets in the inner solar system, and typically to a limited number of species. This originates from the dramatic drop in sensitivity of comets with heliocentric distance r_h and challenges in probing weak spectral signatures among many other bright primary emissions. For instance, fluorescence efficiencies scale by r_h^{-2}, global integrated flux densities scale by Δ^{-2} with distance, and if we consider insolation-driven activity, production rates would scale by r_h^{-2}. This would mean a dramatic dependence of r_h^{-6} in the expected IR fluxes of distant objects, or r_h^{-4} for radio/thermal emissions. Furthermore, the integration time needed for the same signal-to-noise ratio scales quadratically with the drop of flux in background dominated regimes (typically for $\lambda > 3$ μm for groundbased observatories), and linearly in the source's noise-dominated regimes (typically at shorter wavelengths). Some trace species (such as Na, Ni, Fe, C_2H_6) have very strong signatures due to large fluorescence efficiencies, even though they have 6 orders of magnitude lower abundances than the primary volatiles. Large collecting area observatories (e.g., E-ELT with a collecting area 16× greater than Keck and much improved in spatial sampling); spaceborne, cold observatories (e.g., JWST, with orders of magnitude improved sensitivity at thermal IR wavelengths); and *in situ* probes are therefore essential to explore new frontiers in cometary research.

We anticipate the spectroscopic detection and characterization of new atomic and molecular species, as well as expanded data and statistics on species for which detailed spatial/spectral/large survey-sample information has thus far been unavailable. In the era of JWST, the frontier of cometary studies will be expanded to fainter and more distant targets ($r_h > 5$ au), particularly at IR wavelengths. This will allow the elucidation of novel physical and chemical regimes in the less well-explored cometary populations, such as main-belt comets, active asteroids, Centaurs, transneptunian objects, interstellar objects, and more distant/pristine Oort cloud comets for which volatile sublimation may not yet be fully activated (*Kelley et al.,* 2016).

Such advances will permit major discoveries and new insights regarding the fundamental chemical building blocks of comets. As we continue to explore deeper at every wavelength, new important markers are being identified, such as the recent discovery of ammonium salt absorption (*Poch et al.,* 2020) and atomic metals fluorescing in the NUV/visible (*Manfroid et al.,* 2021). Future studies of coma isotope ratios have the potential to revolutionize our understanding of the origins and history of our solar system's most primitive materials (see the chapter by Biver et al. in this volume). More detailed mapping of coma density structures will help elucidate cometary activity and outgassing mechanisms, and expanded, more spatially complete, higher-resolution surveys of coma rotational temperatures will provide new theoretical challenges toward a complete understanding of coma heating and cooling processes.

Increased spectral resolution will diminish spectral confusion, which is particularly severe in the optical/IR, where many molecular and isotopic signatures overlap. Spectroscopic surveys at high spectral resolution ($\lambda/\delta\lambda > 50,000$) will permit the proper separation of isomeric and isotopic signatures and open new avenues for better understanding the excitation of fragment species. For example, *Nelson and Cochran* (2019) found that spectra of the C_2 Swan bands ($\lambda/\delta\lambda \sim 60,000$) in two comets are best described by a bimodal rotational temperature distribution. The underlying mechanism can be tested by high-resolution optical studies of comets spanning a range in heliocentric and cometocentric distances. The authors point out that such studies can first test if two-temperature distributions are a common feature in fluorescence spectra of cometary C_2, and second, test hypotheses of the underlying mechanisms, e.g., competing formation pathways from the dissociation of parent molecules.

The chapter by Biver et al. in this volume emphasizes the importance of spatially-resolved measurements in understanding the storage of volatiles in comet nuclei and their release in the coma. *Bonev et al.* (2021) discuss how clues about associations (or differences) in volatile release deduced from groundbased spatial studies can be linked to the detailed findings from the Rosetta and EPOXI mission to comets. As discussed in this chapter, spatial distributions are also informative of various excitation coma processes, including the effects of icy grain vaporization on coma temperatures. The basic parameters that limit spatial studies are the spatial scale (arcsec/pix) and the dimension(s) of the field-of-view (arcseconds, respectively kilometers in projected distance on the sky plane, the latter dependent on geocentric distance Δ). In the IR, the optimal combination of detector sensitivity and long-slit capability (~0.13 arcsec/pix) is currently offered by the Near Infrared Spectrometer (NIRSPEC) at Keck II (*Martin et al.,* 2018). The total slit length (24″) of NIRSPEC allows for significant extension in projected distance (>~2000–3000 km at Δ = 1 au), as needed for detecting molecular emission farther from the comet nucleus, including from extended coma sources. A key challenge in optimizing the design and cost of future IR high-resolution spectrographs may be incorporating increased spectral coverage, while maintaining slit length comparable to or exceeding that of NIRSPEC.

Looking beyond present-day instrumentation, a future space observatory offering high-resolution IR spectroscopy (at $\lambda/\delta\lambda \gtrsim 50{,}000$; see Fig. 10) would enable sampling of narrow cometary emission lines with unprecedented sensitivity and precision, therefore allowing measurement of many isotopic and spin species of H_2O and CO_2, CO, and trace organic species. Future and planned telescopes (e.g., European Extremely Large Telescope, Thirty Meter Telescope, Giant Magellan Telescope) with large collecting areas will permit exploration of ever fainter objects at increasingly large heliocentric distances and tracking cometary activity as they enter the inner solar system. Integral field unit (IFU) instrumentation with two-dimensional NUV/optical/IR spatial-spectral mapping capabilities would allow identification of heterogeneous regions of activity in the nucleus and secondary sources in the coma, as well as mapping of inner coma temperatures in a large number of comets with a wide range of production rates.

At longer wavelengths, multi-beam radio/millimeter/submillimeter receiver technology has opened up the field

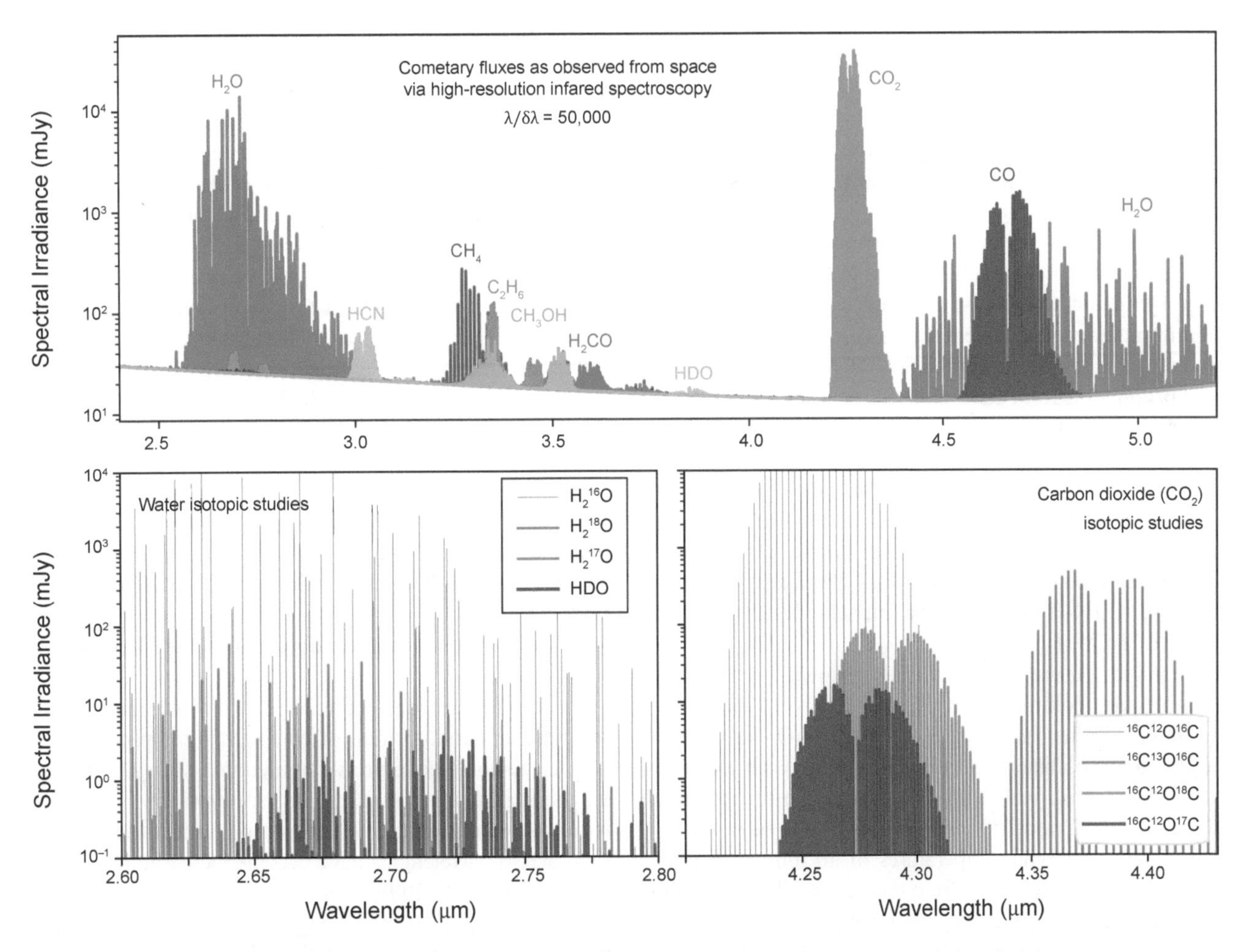

Fig. 10. High-resolution spectroscopy of comets from space will enable probing of the organic, isotopic and isomeric composition of these primordial bodies with unique accuracy, in particular considering that many of these bright emissions are not available to groundbased observations due to telluric absorption. High-resolution spectroscopy also maximizes the contrast between the cometary gases and continuum levels, reduces background noise, and removes the spectral confusion impacting low-resolution studies (e.g., JWST). The simulations presented here are for a typical comet (5–10 per year) as observed with a 1-m space observatory, and were calculated employing the Planetary Spectrum Generator (*Villanueva et al.,* 2018).

of high-spectral-resolution coma mapping, enabling snapshot molecular imaging in a small fraction of the time it takes to perform a conventional raster-style map using a single-pixel receiver (*Coulson et al.,* 2017, 2020). Inclusion of larger focal plane receiver arrays in future generations of millimeter/submillimeter telescopes will facilitate more detailed spatial studies of coma physical and chemical properties. High-sensitivity, spatially complete coma maps are difficult to obtain with current telescopes, even with ALMA, due to the interferometer's inherent lack of sensitivity to extended structures. Therefore, science related to the mapping of less-abundant molecules (such as isotopologs or complex organics) and to the determination of kinematics and production/release mechanisms will benefit strongly from next-generation IFU-style heterodyne focal plane arrays. Improved sensitivity for molecular mapping will be ensured thanks to ongoing (and planned) upgrades to currently available radio interferometers.

4.2. Fundamental Spectroscopic Data Requirements

The availability of sufficiently accurate theoretical and laboratory data limits the accuracy with which astronomical observations can be interpreted. New observational discoveries and improved spectroscopic sensitivity therefore drives the need for improved gas-phase experimental and spectroscopic data.

The capabilities of state-of-the-art radio interferometers, such as ALMA, have made it possible to probe a broader range of spatial scales than previously possible via radio/submillimeter molecular emissions (particularly in the inner coma). Accurate interpretation of rotational emission lines originating from such a broad range of nucleocentric radii places strict demands on the reliability of the underlying physical model. Major uncertainties still remain in our knowledge of state-to-state collisional excitation/de-excitation rate coefficients (C_{ij}), applicable to the calculation of rotational excitation of coma gases (section 3.1.1). More accurate collision rates between less abundant molecules and water $C_{ij}(X\text{-}H_2O)$, where the impactor X can be HDO, CH_3OH, H_2CO, HNC, HC_3N, CH_3CN, NH_2CHO, and CS, are required. Future studies will also benefit from C_{ij} data for molecules colliding with CO and CO_2 (for application to comets farther from the Sun where H_2O may not be the dominant collision partner), as well as with electrons (*Ahmed and Acharyya,* 2021). To realize the full potential of such detailed, non-LTE calculations, it will be necessary to track the rotational temperatures of all molecular collision partners as they depart from thermal equilibrium, since in general, the C_{ij} values depend on the rotational state of the impactor (characterized by T_{rot}) as well as the impact velocity (from T_{kin}).

The high spatial resolutions expected at shorter wavelengths owing to future extreme adaptive optics and ELTs will pose strong challenges to the interpretation of the resulting UV/optical/IR emission. Modeling the coma distributions of emitting molecules requires knowledge of the photodissociation lifetimes in the solar radiation field, and these are still not well known for many molecules. Understanding the spatial distributions of emitters will also require the incorporation of heterogeneous outgassing, accounting for multiple native and extended sources of major species (H_2O, CO_2, CO) and trace species (C_2H_2, HCN, etc.). It is not yet clear what the excitation state is (or how it is being affected) for molecules released from different extended sources such as dust, from icy grains, or in some photochemical reactions.

Similarly, many electron impact emission features observed by Rosetta's Alice and Optical, Spectrocopic and Infrared Remote Imaging System (OSIRIS) instruments at 67P/Churyumov-Gerasimenko have only been measured at a fixed impact energy, but not between their threshold energy and approximately 100 eV, the energy range that includes the majority of electrons observed at the comet. In the case of solar wind charge exchange, laboratory data are needed for reactions with the main relevant molecules. Such data include multi-electron capture processes and velocity-dependent charge-exchange cross sections.

As discussed in the previous sections, the completeness and accuracy of spectroscopic line lists play a critical role in the calculation of model fluxes for the processes discussed in this chapter. In the last decade, there has been a dramatic increase in the availability of high-temperature/energy ab initio line lists, yet there is still a lack of databases addressing species that are prevalent in cometary atmospheres but less common in planetary atmospheres [e.g., C_2H_6, CH_3OH, various cations such as H_2O^+, CO^+ (*Fortenberry et al.,* 2021)]. Such line lists with detailed transition level identifications (needed for fluorescence calculations) and high-energy range will potentially allow identification of the tens of thousands of currently unassigned lines (*Cochran and Cochran,* 2002; *Dello-Russo et al.,* 2016), enable sensitive probes of isotopic and isomeric ratios, and allow better characterization of prompt emission efficiencies (including their variations driven by variable solar activity) with proper assessment of dissociation yields in photochemistry (e.g., excited oxygen atoms from the photodissociation of CO_2 vs. H_2O).

4.3. Comprehensive Probing of Cometary Processes

In the last decade, our capabilities to characterize the structure and composition of comets and the processes active in the coma have grown tremendously. *In situ* probing via the Rosetta mission (e.g., mass spectrometry, inner coma remote sensing) and coordinated near-simultaneous observations across wavelengths have become increasingly common (e.g., *Jehin et al.,* 2009; *Meech et al.,* 2011). Observations that bridge multiple wavelength regimes and techniques not only benefit compositional studies but are also critical to exploration of novel excitation processes. For instance, coordinated IR and radio observations of parent species (e.g., HCN, CH_3OH, H_2CO) enable a deeper investigation of the origins of these species, which may be both precursor molecules or dissociation fragments themselves. Furthermore, coordinated optical/IR/radio observations better elucidate parent-fragment relationships, such as C_2H_2–C_2, HCN–CN, NH_3–NH_2–NH. Similarly, the derived kinematics, composition, and isotopic ratios from *in situ* mass spectrometry provide new challenges to the interpretation of (and consistency with) remote sensing observations. For instance, multi-wavelength observations can resolve the disagreements in the retrieved columns of species at different wavelengths, and provide a better understanding of the overall excitation processes and the impact of new modeling/analysis methods (*Lippi et al.,* 2021). A primary example of this is the long-standing disagreement in the production rates of hydrogen cyanide and other species when retrieved through IR and radio techniques (*Villanueva et al.,* 2013). The production rates of HCN, determined from rovibrational transitions (IR fluorescence), often exceed those obtained from observations of pure rotational transitions by about a factor of 2. Evaluating the role of cascades in IR fluorescence (section 3.3) did not bring results closer, so the challenge remains to critically examine the assumptions regarding HCN excitation made by each technique and the physical and chemical processes involved in the storage and release of cometary volatiles, together with their subsequent processing in the coma.

Decades-long observations of the CN red- and violet-band systems have been important in compositional classifications of comets (*A'Hearn et al.,* 1995; *Fink,* 2009). It is of note that the source of CN is not entirely clear (*Fray et al.,* 2005; *Altwegg et al.,* 2020). There are conflicting reports on whether the photodissociative excitation of HCN can be a significant source of electronically excited CN (*Bockelée-Morvan and Crovisier,* 1985; *Huebner and Mukherjee,* 2015), and this might be different for other sources of the molecule. *Paganini and Mumma* (2016) recently proposed a new excitation model for CN fluorescence and tested it against high-resolution IR spectra of Comet C/2014 Q2 (Lovejoy). This model includes both electronic transitions in the $A^2\Pi$–$X^2\Sigma^+$ and $B^2\Sigma^+$–$X^2\Sigma^+$ systems and rovibrational transitions within the CN ground electronic state. The IR can therefore sample both CN and its likely precursor HCN. Furthermore, contemporaneous IR and optical observations of CN at high spectral resolution are now feasible with facilities that allow flexible scheduling, like the McDonald Observatory and NASA's Infrared Telescope Facility (IRTF). Such observations would facilitate intercomparison of the assumptions and parameters of existing models for CN fluorescence, thereby allowing the excitation of this frequently observed radical to be traced from the inner collisional coma (IR) to the extended coma (optical).

Integrating analyses across wavelength regimes, with localized *in situ* mass-spectrometry measurements and magneto-hydrodynamic probing, would further advance ongoing studies of non-LTE and dissociative excitation in cometary comae. For example, there are systematic uncertainties in the release rates from a particular precursor (CO_2 or H_2O) leading to metastable atomic states (O ^{1}D and O ^{1}S) whose long lifetimes present a challenge to laboratory measurements (section 3.4). *McKay et al.* (2015) demonstrated that comparisons of CO_2 production rates from O I (groundbased optical observations) and from direct CO_2 detections (IR observations from space) can constrain dissociation release rates empirically.

5. OUTLOOK

Current spectroscopic studies have, out of necessity, focused on the brightest comets. The increased detector sensitivity of spectrographs creates opportunities for remote molecular detections in comets that could not be previously observed. This will open in-depth exploration of new radiation and density regimes. Interpreting volatile emissions from the extremely low-density environments of comets with marginal activity, or with activity driven by volatiles other than H_2O, will require reevaluating the balance between collisional and radiative excitation processes. Such regimes are expected in distant comets. In addition, *Nuth et al.* (2020) suggested that active comets, volatile-rich active asteroids, and "dormant" small bodies in the solar system represent a single evolutionary sequence, motivating the need to bridge radiative diagnostics of these objects, including intrinsically weak comets observed closer to the Sun.

Acknowledgments. D.B. acknowledges support from NASA as part of the Rosetta Data Analysis Program (No. 80NSSC19K1304). M.A.C. acknowledges support from the National Science Foundation (Grant No. AST-2009253) and the NASA Planetary Science Division Internal Scientist Funding Program, through the Fundamental Laboratory Research work package (FLaRe). G.L.V. acknowledges support from NASA as part of the U.S. participation in the Comet Interceptor mission. B.P.B. acknowledges support from the National Science Foundation (Grant No. AST-2009398) and NASA Solar System Workings Program (Grant No. 80NSSC20K0651).

REFERENCES

Abrams M. C., Goldman A., Gunson M. R., et al. (1996) Observations of the infrared solar spectrum from space by the ATMOS experiment. *Appl. Opt., 35,* 2747–2751.

A'Hearn M. F., Millis R. C., Schleicher D. O., et al. (1995) The ensemble properties of comets: Results from narrowband photometry of 85 comets, 1976–1992. *Icarus, 118,* 223–270.

A'Hearn M. F., Belton M. J. S., Delamere W. A., et al. (2011) EPOXI at Comet Hartley 2. *Science, 332,* 1396–1400.

A'Hearn M. F., Krishna Swamy K. S., Wellnitz D. D., et al. (2015) Prompt emission by OH in Comet Hyakutake. *Astron. J., 150,* 5.

Ahmed S. and Acharyya K. (2021) Gas-phase modeling of the cometary coma of interstellar Comet 2I/Borisov. *Astrophys. J., 923,* 91.

Alexander M. H., Andresen P., Bacis R., et al. (1988) A nomenclature for Λ-doublet levels in rotating linear molecules. *J. Chem. Phys., 89,* 1749–1753.

Altwegg K., Balsiger H., Hänni N., et al. (2020) Evidence of ammonium salts in Comet 67P as explanation for the nitrogen depletion in cometary comae. *Nature Astron., 4,* 533–540.

Beenakker C. I. M., Heer F. J. D., Krop H. B., et al. (1974) Dissociative excitation of water by electron impact. *Chem. Phys., 6,* 445–454.

Beiersdorfer P., Boyce K. R., Brown G. V., et al. (2003) Laboratory simulation of charge exchange-produced X-ray emission from comets. *Science, 300,* 1558–1559.

Bertaux J.-L. (1986) The UV bright spot of water vapor in comets. *Astron. Astrophys., 160,* L7–L10.

Biver N. (1997) Molécules mères cométaires: Observations et modélisations. Ph.D. thesis, Université Paris Diderot, Paris. 265 pp.

Biver N., Bockelée-Morvan D., Crovisier J., et al. (1999) Spectroscopic monitoring of Comet C/1996 B2 (Hyakutake) with the JCMT and IRAM radio telescopes. *Astron. J., 118,* 1850–1872.

Biver N., Bockelée-Morvan D., Colom P., et al. (2011) Molecular investigations of Comets C/2002 X5 (Kudo-Fujikawa), C/2002 V1 (NEAT), and C/2006 P1 (McNaught) at small heliocentric distances. *Astron. Astrophys., 528,* A142.

Biver N., Bockelée-Morvan D., Moreno R., et al. (2015) Ethyl alcohol and sugar in Comet C/2014 Q2 (Lovejoy). *Sci. Adv., 1,* 1500863.

Biver N., Bockelée-Morvan D., Hofstadter M., et al. (2019) Long-term monitoring of the outgassing and composition of Comet 67P/Churyumov-Gerasimenko with the Rosetta/MIRO instrument. *Astron. Astrophys., 630,* A19.

Bockelée-Morvan D. (1987) A model for the excitation of water in comets. *Astron. Astrophys., 181,* 169–181.

Bockelée-Morvan D. and Crovisier J. (1985) Possible parents for the cometary CN radical — Photochemistry and excitation conditions. *Astron. Astrophys., 151,* 90–100.

Bockelee-Morvan D. and Crovisier J. (1989) The nature of the 2.8-micron emission feature in cometary spectra. *Astron. Astrophys., 216,* 278–283.

Bockelée-Morvan D., Crovisier J., Colom P., et al. (1994) The rotational lines of methanol in Comets Austin 1990 V and Levy 1990 XX. *Astron. Astrophys., 287,* 647–665.

Bockelée-Morvan D., Crovisier J., Mumma M. J., et al. (2004) The composition of cometary volatiles. In *Comets II* (M. C. Festou et al., eds.), pp. 391–423. Univ. of Arizona, Tucson.

Bockelée-Morvan D., Boissier J., Biver N., et al. (2010) No compelling evidence of distributed production of CO in Comet C/1995 O1 (Hale-Bopp) from millimeter interferometric data and a re-analysis of near-IR lines. *Icarus, 210,* 898–915.

Bodewits D., Christian D. J., Torney M., et al. (2007) Spectral analysis of the Chandra comet survey. *Astron. Astrophys., 469,* 1183–1195.

Bodewits D., Lara L. M., A'Hearn M. F., et al. (2016) Changes in the physical environment of the inner coma of 67P/Churyumov–Gerasimenko with decreasing heliocentric distance. *Astron. J., 152,* 130.

Bodewits D., Orszagh J., Noonan J., et al. (2019) Diagnostics of collisions between electrons and water molecules in near-ultraviolet and visible wavelengths. *Astrophys. J., 885,* 167.

Bodewits D., Noonan J. W., Feldman P. D., et al. (2020) The carbon monoxide-rich interstellar Comet 2I/Borisov. *Nature Astron., 4,* 867–871.

Boissier J., Bockelée-Morvan D., Biver N., et al. (2014) Gas and dust productions of Comet 103P/Hartley 2 from millimetre observations: Interpreting rotation-induced time variations. *Icarus, 228,* 197–216.

Bonamente E., Christian D. J., Xing Z., et al. (2021) Variable X-ray emission of Comet 46P/Wirtanen. *Planet. Sci. J., 2,* 224.

Bonev B. P. and Mumma M. J. (2006) A comprehensive study of infrared OH prompt emission in two comets. II. Implications for unimolecular dissociation of H_2O. *Astrophys. J., 653,* 788–791.

Bonev B. P., Mumma M. J., DiSanti M. A., et al. (2006) A comprehensive study of infrared OH prompt emission in two comets. I. Observations and effective g-factors. *Astrophys. J., 653,* 774–787.

Bonev B. P., Villanueva G. L., Paganini L., et al. (2013) Evidence for two modes of water release in Comet 103P/Hartley 2: Distributions of column density, rotational temperature, and ortho-para ratio. *Icarus, 222,* 740–751.

Bonev B. P., DiSanti M. A., Villanueva G. L., et al. (2014) The inner coma of Comet C/2012 S1 (ISON) at 0.53 AU and 0.35 AU from the Sun. *Astrophys. J. Lett., 796,* L6.

Bonev B. P., Villanueva G. L., DiSanti M. A., et al. (2017) Beyond 3 au from the Sun: The hypervolatiles CH_4, C_2H_6, and CO in the distant Comet C/2006 W3 (Christensen). *Astron. J., 153,* 241.

Bonev B. P., Dello Russo N., DiSanti M. A., et al. (2021) First comet observations with NIRSPEC-2 at Keck: Outgassing sources of parent volatiles and abundances based on alternative taxonomic compositional baselines in 46P/Wirtanen. *Planet. Sci. J., 2,* 45.

Boursier C., Mandal B., Babikov D., et al. (2020) New H_2O-H_2O collisional rate coefficients for cometary applications. *Mon. Not. R. Astron. Soc., 498,* 5489–5497.

Bromley S. J., Neff B., Loch S. D., et al. (2021) Atomic iron and nickel in the coma of C/1996 B2 (Hyakutake): Production rates, emission mechanisms, and possible parents. *Planet. Sci. J., 2,* 228.

Brooke T. Y., Tokunaga A. T., Weaver H. A., et al. (1996) Detection of acetylene in the infrared spectrum of Comet Hyakutake. *Nature, 383,* 606–608.

Budzien S. A. and Feldman P. D. (1991) OH prompt emission in Comet IRAS-Araki-Alcock (1983 VII). *Icarus, 90,* 308–318.

Buffa G., Tarrini O., Scappini F., et al. (2000) H_2O-H_2O collision rate coefficients. *Astrophys. J. Suppl. Ser., 128,* 597–601.

Cheng Y. C., Bockelée-Morvan D., Roos-Serote M., et al. (2022) Water ortho-to-para ratio in the coma of Comet 67P/Churyumov-Gerasimenko. *Astron. Astrophys., 663,* A43.

Cochran A. L. and Cochran W. D. (2002) A high spectral resolution atlas of Comet 122P/de Vico. *Icarus, 157,* 297–308.

Cochran A. L. and Schleicher D. G. (1993) Observational constraints on the lifetime of cometary H_2O. *Icarus, 105,* 235–253.

Cochran A. L., Barker E. S., and Gray C. L. (2012) Thirty years of cometary spectroscopy from McDonald Observatory. *Icarus, 218,* 144–168.

Combi M. R. and Delsemme A. H. (1980) Neutral cometary atmospheres. I. An average random walk model for photodissociation in comets. *Astrophys. J., 237,* 633–640.

Combi M. R. and Smyth W. H. (1988) Monte Carlo particle trajectory models for neutral cometary gases. I. Models and equations. *Astrophys. J., 327,* 1026–1043.

Combi M. R., Harris W. M., and Smyth W. H. (2004) Gas dynamics and kinetics in the cometary coma: Theory and observations. In *Comets II* (M. C. Festou et al., eds.), pp. 523–552. Univ. of Arizona, Tucson.

Combi M., Shou Y., Fougere N., et al. (2020) The surface distributions of the production of the major volatile species, H_2O, CO_2, CO and O_2, from the nucleus of Comet 67P/Churyumov-Gerasimenko throughout the Rosetta mission as measured by the ROSINA double focusing mass spectrometer. *Icarus, 335,* 113421.

Cordiner M. A., Remijan A. J., Boissier J., et al. (2014) Mapping the release of volatiles in the inner comae of Comets C/2012 F6 (Lemmon) and C/2012 S1 (ISON) using the Atacama Large Millimeter/Submillimeter Array. *Astrophys. J. Lett., 792,* L2.

Cordiner M. A., Biver N., Crovisier J., et al. (2017) Thermal physics of the inner coma: ALMA studies of the methanol distribution and excitation in Comet C/2012 K1 (PanSTARRS). *Astrophys. J., 837,* 177.

Cordiner M. A., Biver N., Milam S., et al. (2019a) ALMA spectral imaging of volatiles and dust from Jupiter family comets: 21P/Giacobini-Zinner and 46P/Wirtanen. *EPSC-DPS Joint Meeting 2019,* EPSC-DPS2019-1131-2.

Cordiner M. A., Palmer M. Y., de Val-Borro M., et al. (2019b) ALMA autocorrelation spectroscopy of comets: The HCN/$H^{13}CN$ ratio in C/2012 S1 (ISON). *Astrophys. J. Lett., 870,* L26.

Cordiner M. A., Milam S. N., Biver N., et al. (2020) Unusually high CO abundance of the first active interstellar comet. *Nature Astron., 4,* 861–866.

Cordiner M. A., Coulson I. M., Garcia-Berrios E., et al. (2022) A SUBLIME 3D model for cometary coma emission: The hypervolatile-rich Comet C/2016 R2 (PanSTARRS). *Astrophys. J., 929,* 38.

Coulson I. M., Cordiner M. A., Kuan Y.-J., et al. (2017) JCMT spectral and continuum imaging of Comet 252P/LINEAR. *Astron. J., 153,* 169.

Coulson I. M., Liu F.-C., Cordiner M. A., et al. (2020) James Clerk Maxwell Telescope spectral and continuum imaging of hyperactive Comet 46P/Wirtanen. *Astron. J., 160,* 182.

Cravens T. E. (1997) Comet Hyakutake X-ray source: Charge transfer of solar wind heavy ions. *Geophys. Res. Lett., 24,* 105–108.

Cravens T. and Korosmezey A. (1986) *Vibrational and rotational cooling of electrons by water vapor. Planet. Space Sci., 34,* 961–970.

Crovisier J. (1987) Rotational and vibrational synthetic spectra of linear parent molecules in comets. *Astron. Astrophys., 68,* 223–258.

Crovisier J. (1989) The photodissociation of water in cometary atmospheres. *Astron. Astrophys., 213,* 459–464.

Crovisier J. and Encrenaz T. (1983) Infrared fluorescence of molecules in comets: The general synthetic spectrum. *Astron. Astrophys., 126,* 170–182.

Cunningham N. J., Spencer J. R., Feldman P. D., et al. (2015) Detection of Callisto's oxygen atmosphere with the Hubble Space Telescope. *Icarus, 254,* 178–189.

Debout V., Bockelée-Morvan D., and Zakharov V. (2016) A radiative transfer model to treat infrared molecular excitation in cometary atmospheres. *Icarus, 265,* 110–124.

Decock A., Jehin E., Rousselot P., et al. (2015) Forbidden oxygen lines at various nucleocentric distances in comets. *Astron. Astrophys., 573,*

A1.
Dello Russo N., DiSanti M. A., Mumma M. J., et al. (1998) Carbonyl sulfide in Comets C/1996 B2 (Hyakutake) and C/1995 O1 (Hale-Bopp): Evidence for an extended source in Hale-Bopp. *Icarus, 135*, 377–388.
Dello Russo N., Bonev B. P., DiSanti M. A., et al. (2005) Water production rates, rotational temperatures, and spin temperatures in Comets C/1999 H1 (Lee), C/1999 S4, and C/2001 A2. *Astrophys. J., 621*, 537–544.
Dello-Russo N., Kawakita H., Vervack R. J., et al. (2016) Emerging trends and a comet taxonomy based on the volatile chemistry measured in thirty comets with high-resolution infrared spectroscopy between 1997 and 2013. *Icarus, 278*, 301–332.
Dello Russo N., Vervack R. J., Kawakita H., et al. (2022) Volatile abundances, extended coma sources, and nucleus ice associations in Comet C/2014 Q2 (Lovejoy). *Planet. Sci. J., 3*, 6.
Dennerl K. (2010) Charge transfer reactions. *Space Sci. Rev., 157*, 57–91.
Dennerl K., Englhauser J., and Trumper J. (1997) X-ray emissions from comets detected in the Röntgen X-ray Satellite all-sky survey. *Science, 277*, 1625–1630.
DiSanti M. A., Bonev B. P., Magee-Sauer K., et al. (2006) Detection of formaldehyde emission in Comet C/2002 T7 (LINEAR) at infrared wavelengths: Line-by-line validation of modeled fluorescent intensities. *Astrophys. J., 650*, 470–483.
DiSanti M. A., Bonev B. P., Gibb E. L., et al. (2018) Comet C/2013 V5 (Oukaimeden): Evidence for depleted organic volatiles and compositional heterogeneity as revealed through infrared spectroscopy. *Astron. J., 156*, 258.
Dubernet M. L. and Quintas-Sánchez E. (2019) First quantum study of the rotational excitation of HCN by para-H_2O: Convergence of quantum results, influence of the potential energy surface, and approximate rate coefficients of interest for cometary atmospheres. *Mol. Astrophys., 16*, 100046.
Eberhardt P. and Krankowsky D. (1995) The electron temperature in the inner coma of Comet P/Halley. *Astron. Astrophys., 295*, 795–806.
Eistrup C., Walsh C., and van Dishoeck E. F. (2019) Cometary compositions compared with protoplanetary disk midplane chemical evolution: An emerging chemical evolution taxonomy for comets. *Astronomy Astrophys. Suppl., 629*, A84.
Engelhardt I. A. D., Eriksson A. I., Vigren E., et al. (2018) Cold electrons at Comet 67P/Churyumov-Gerasimenko. *Astron. Astrophys., 616*, A51.
Faure A., Varambhia H. N., Stoecklin T., et al. (2007) Electron-impact rotational and hyperfine excitation of HCN, HNC, DCN and DNC. *Mon. Not. R. Astron. Soc., 382*, 840–848.
Faure A., Lique F., and Loreau J. (2020) The effect of CO-H_2O collisions in the rotational excitation of cometary CO. *Mon. Not. R. Astron. Soc., 493*, 776–782.
Feldman P. D. (2015) The photodissociation of formaldehyde in comets. *Astrophys. J., 812*, 115.
Feldman P. D., McGrath M. A., Strobel D. F., et al. (2000) HST/STIS ultraviolet imaging of polar aurora on Ganymede. *Astrophys. J., 535*, 1085–1090.
Feldman P. D., Cochran A. L., and Combi M. R. (2004) Spectroscopic investigations of fragment species in the coma. In *Comets II* (M. C. Festou et al., eds.), pp. 425–447. Univ. of Arizona, Tucson.
Feldman P. D., Lupu R. E., McCandliss S. R., et al. (2009) The far-ultraviolet spectral signatures of formaldehyde and carbon dioxide in comets. *Astrophys. J., 699*, 1104–1112.
Feldman P. D., A'Hearn M. F., Bertaux J.-L., et al. (2015) Measurements of the near-nucleus coma of Comet 67P/Churyumov-Gerasimenko with the Alice far-ultraviolet spectrograph on Rosetta. *Astron. Astrophys., 583*, A8.
Feldman P. D., A'Hearn M. F., Bertaux J.-L., et al. (2018) FUV spectral signatures of molecules and the evolution of the gaseous coma of Comet 67P/Churyumov-Gerasimenko. *Astron. J., 155*, 9.
Festou M. C. (1981) The density distribution of neutral compounds in cometary atmospheres. I. Models and equations. *Astron. Astrophys., 95*, 69–79.
Fink U. (2009) A taxonomic survey of comet composition 1985–2004 using CCD spectroscopy. *Icarus, 201*, 311–334.
Fiorenza C. and Formisano V. (2005) A solar spectrum for PFS data analysis. *Planet. Space Sci., 53*, 1009–1016.
Fontenla J. M., Landi E., Snow M., et al. (2014) Far- and extreme-UV solar spectral irradiance and radiance from simplified atmospheric physical models. *Solar Phys., 289*, 515–544.
Fortenberry R. C., Bodewits D., and Pierce D. M. (2021) Knowledge gaps in the cometary spectra of oxygen-bearing molecular cations. *Astrophys. J. Suppl. Ser., 256*, 6.
Fougere N., Combi M. R., Tenishev V., et al. (2012) Understanding measured water rotational temperatures and column densities in the very innermost coma of Comet 73P/Schwassmann-Wachmann 3 B. *Icarus, 221*, 174–185.
Fray N., Bénilan Y., Cottin H., et al. (2005) The origin of the CN radical in comets: A review from observations and models. *Planet. Space Sci., 53*, 1243–1262.
Galand M., Feldman P. D., Bockelée-Morvan D., et al. (2020) Far-ultraviolet aurora identified at Comet 67P/Churyumov-Gerasimenko. *Nature Astron., 4*, 1084–1091.
Gan L. and Cravens T. E. (1990) Electron energetics in the inner coma of Comet Halley. *J. Geophys. Res., 95*, 6285–6303.
Garrod R. T. (2019) Simulations of ice chemistry in cometary nuclei. *Astrophys. J., 884*, 69.
Gerig S. B., Marschall R., Thomas N., et al. (2018) On deviations from free-radial outflow in the inner coma of Comet 67P/Churyumov-Gerasimenko. *Icarus, 311*, 1–22.
Gersch A. M. and A'Hearn M. F. (2014) Coupled escape probability for an asymmetric spherical case: Modeling optically thick comets. *Astrophys. J., 787*, 36.
Gersch A. M., A'Hearn M. F., and Feaga L. M. (2018) Modeling the Deep Impact near-nucleus observations of H_2O and CO_2 in Comet 9P/Tempel 1 using asymmetric spherical coupled escape probability. *Astrophys. J., 856*, 104.
Gibb E. L., Bonev B. P., Villanueva G., et al. (2013) Solar fluorescence model of CH_3D as applied to comet emission. *J. Mol. Spectrosc., 291*, 118–124.
Gordon I. E., Rothman L. S., Hill C., et al. (2017) The HITRAN2016 molecular spectroscopic database. *J. Quant. Spectrosc. Radiat. Transfer, 203*, 3–69.
Gulkis S., Allen M., von Allmen P., et al. (2015) Subsurface properties and early activity of Comet 67P/Churyumov-Gerasimenko. *Science, 347*, aaa0709.
Hall D. T., Strobel D. F., Feldman P. D., et al. (1995) Detection of an oxygen atmosphere on Jupiter's moon Europa. *Nature, 373*, 677–679.
Hama T., Kouchi A., and Watanabe N. (2016) Statistical ortho-to-para ratio of water desorbed from ice at 10 Kelvin. *Science, 351*, 65–67.
Hase F., Wallace L., McLeod S. D., et al. (2010) The ACE-FTS atlas of the infrared solar spectrum. *J. Quant. Spectrosc. Radiat. Transfer, 111*, 521–528.
Haser L. (1957) Distribution d'intensité dans la tête d'une comète. *Bull. Soc. R. Sci. Liege, 43*, 740–750.
Haser L., Oset S., and Bodewits D. (2020) Intensity distribution in the heads of comets. *Planet. Sci. J., 1*, 83.
Herzberg G. (1971) *The Spectra and Structures of Simple Free Radicals: An Introduction to Molecular Spectroscopy.* Cornell Univ., Ithaca.
Hogerheijde M. R., Qi C., de Pater I., et al. (2009) Simultaneous observations of Comet C/2002 T7 (LINEAR) with the Berkeley-Illinois-Maryland Association and Owens Valley Radio Observatory interferometers: HCN and CH_3OH. *Astron. J., 137*, 4837–4845.
Huebner W. F. and Mukherjee J. (2015) Photoionization and photodissociation rates in solar and blackbody radiation fields. *Planet. Space Sci., 106*, 11–45.
Huggins W. (1868) Spectrum analysis of comet II. *Astron. Reg., 6*, 169–170.
Itikawa Y. (1972) Rotational transition in an asymmetric-top molecule by electron collision: Applications to H_2O and H_2CO. *J. Phys. Soc. Japan, 32*, 217–226.
Itikawa Y. and Mason N. (2005) Cross sections for electron collisions with water molecules. *J. Phys. Chem. Ref. Data, 34*, 1–22.
Jehin E., Bockelée-Morvan D., Dello Russo N., et al. (2009) A multi-wavelength simultaneous study of the composition of the Halley family Comet 8P/Tuttle. *Earth Moon Planets, 105*, 343–349.
Kalogerakis K. S., Romanescu C., Ahmed M., et al. (2012) CO prompt emission as a CO_2 marker in comets and planetary atmospheres. *Icarus, 220*, 205–210.
Kawakita H. (2022) Photodissociation rate, excess energy, and kinetic total energy release for the photolysis of H_2O producing $O(^1S)$ by solar UV radiation field. *Astrophys. J., 931*, 24.
Kawakita H. and Mumma M. J. (2011) Fluorescence excitation models of ammonia and amidogen radical (NH_2) in comets: Application to

Comet C/2004 Q2 (Machholz). *Astrophys. J., 727,* 91.
Kawakita H., Dello Russo N., Furusho R., et al. (2006) Ortho-to-para ratios of water and ammonia in Comet C/2001 Q4 (NEAT): Comparison of nuclear spin temperatures of water, ammonia, and methane. *Astrophys. J., 643,* 1337–1344.
Kawakita H., Kobayashi H., Dello Russo N., et al. (2013) Parent volatiles in Comet 103P/Hartley 2 observed by Keck II with NIRSPEC during the 2010 apparition. *Icarus, 222,* 723–733.
Kelley M. S. P., Woodward C. E., Bodewits D., et al. (2016) Cometary science with the James Webb Space Telescope. *Publ. Astron. Soc. Pac., 128,* 018009.
Knight M. M., Schleicher D. G., and Farnham T. L. (2021) Narrowband observations of Comet 46P/Wirtanen during its exceptional apparition of 2018/19. II. Photometry, jet morphology, and modeling results. *Planet. Sci. J., 2,* 104.
Körösmezey A., Cravens T. E., Gombosi T. I., et al. (1987) A new model of cometary ionospheres. *J. Geophys. Res.–Space Phys., 92,* 7331–7340.
Krasnopolsky V. A. (1998) Excitation of X rays in Comet Hyakutake (C/1996 B2). *J. Geophys. Res., 103,* 2069–2075.
Krasnopolsky V. A., Mumma M. J., Abbott M. J., et al. (1997) Detection of soft X-rays and a sensitive search for noble gases in Comet Hale-Bopp (C/1995 O1). *Science, 277,* 1488–1491.
Kurucz R. (2000) Solar spectroscopy: Visible emission. In *Encyclopedia of Astronomy and Astrophysics* (P. Murdin, ed.), p. 2232. Institute of Physics, Bristol.
LaForgia F., Bodewits D., A'Hearn M. F., et al. (2017) Near-UV OH prompt emission in the innermost coma of 103P/Hartley 2. *Astron. J., 154,* 185.
Liou K.-N. (2002) *An Introduction to Atmospheric Radiation, 2nd edition.* Academic, San Diego. 608 pp.
Lippi M., Villanueva G. L., Mumma M. J., et al. (2021) Investigation of the origins of comets as revealed through infrared high-resolution spectroscopy I. Molecular abundances. *Astron. J., 162,* 74.
Lisse C. M., Dennerl K., Englhauser J., et al. (1996) Discovery of X-ray and extreme ultraviolet emission from Comet C/1996 B2 Hyakutake. *Science, 274,* 205–209.
Lisse C. M., Cravens T. E., and Dennerl K. (2004) X-ray and extreme ultraviolet emission from comets. In *Comets II* (M. C. Festou et al., eds.), pp. 631–643. Univ. of Arizona, Tucson.
Liu X., Shemansky D. E., Hallett J. T., et al. (2007) Extreme non-LTE H_2 in Comets C/2000 WM1 (LINEAR) and C/2001 A2 (LINEAR). *Astrophys. J. Suppl. Ser., 169,* 458–471.
Loreau J., Faure A., and Lique F. (2018) Scattering of CO with H_2O: Statistical and classical alternatives to close-coupling calculations. *J. Chem. Phys., 148,* 244308.
Lovell A. J., Schloerb F. P., Dickens J. E., et al. (1998) HCO^+ imaging of Comet Hale-Bopp (C/1995 O1). *Astrophys. J. Lett., 497,* L117–L121.
Lovell A. J., Kallivayalil N., Schloerb F. P., et al. (2004) On the effect of electron collisions in the excitation of cometary HCN. *Astrophys. J., 613,* 615–621.
Magnani L. and A'Hearn M. F. (1986) CO^+ fluorescence in comets. *Astrophys. J., 302,* 477–487.
Manca Tanner C., Quack M., and Schmidiger D. (2013) Nuclear spin symmetry conservation and relaxation in water ${}^1H_2{}^{16}O$ studied by cavity ring-down (CRD) spectroscopy of supersonic jets. *J. Phys. Chem. A, 117,* 10105–10118.
Manfroid J., Hutsemékers D., and Jehin E. (2021) Iron and nickel atoms in cometary atmospheres even far from the Sun. *Nature, 593,* 372–374.
Martin E. C., Fitzgerald M. P., McLean I. S., et al. (2018) An overview of the NIRSPEC upgrade for the Keck II telescope. In *Ground-Based and Airborne Instrumentation for Astronomy VII* (C. J. Evans et al., eds.), Proc. SPIE 10702, DOI: 10.1117/12.2312266.
McKay A. J., Cochran A. L., DiSanti M. A., et al. (2015) Evolution of H_2O, CO, and CO_2 production in Comet C/2009 P1 Garradd during the 2011–2012 apparition. *Icarus, 250,* 504–515.
Meech K. J., A'Hearn M. F., Adams J. A., et al. (2011) EPOXI: Comet 103P/Hartley 2 observations from a worldwide campaign. *Astrophys. J. Lett., 734,* L1.
Mullen P. D., Cumbee R. S., Lyons D., et al. (2017) Line ratios for solar wind charge exchange with comets. *Astrophys. J., 844,* 7.
Mumma M. J. (1982) Speculations on the infrared molecular spectra of comets. In *Vibrational-Rotational Spectroscopy for Planetary Atmospheres, Volume II* (M. J. Mumma et al., eds.), pp. 717–742. NASA CP-2223-VOL-2, Washington, DC.
Mumma M. J. and Charnley S. B. (2011) The chemical composition of comets — Emerging taxonomies and natal heritage. *Annu. Rev. Astron. Astrophys., 49,* 471–524.
Mumma M. J., Weaver H. A., Larson H. P., et al. (1986) Detection of water vapor in Halley's comet. *Science, 232,* 1523–1528.
Mumma M. J., McLean I. S., DiSanti M. A., et al. (2001) A survey of organic volatile species in Comet C/1999 H1 (Lee) using NIRSPEC at the Keck Observatory. *Astrophys. J., 546,* 1183–1193.
Nelson T. and Cochran A. L. (2019) Investigating the temperature distribution of diatomic carbon in comets using the Swan bands. *Astron. J., 158,* 221.
Neugebauer M., Gloeckler G., Gosling J. T., et al. (2007) Encounter of the Ulysses spacecraft with the ion tail of Comet McNaught. *Astrophys. J., 667,* 1262–1266.
Nilsson H., Wieser G. S., Behar E., et al. (2017) Evolution of the ion environment of Comet 67P during the Rosetta mission as seen by RPC-ICA. *Mon. Not. R. Astron. Soc., 469,* S252–S261.
Noonan J. W., Harris W. M., Bromley S., et al. (2021) FUV observations of the inner coma of 46P/Wirtanen. *Planet. Sci. J., 2,* 8.
Nuth J. A. I., Abreu N., Ferguson F. T., et al. (2020) Volatile-rich asteroids in the inner solar system. *Planet. Sci. J., 1,* 82.
O'Dell C. R., Robinson R. R., Krishna Swamy K. S., et al. (1988) C_2 in Comet Halley: Evidence for its being third generation and resolution of the vibrational population discrepancy. *Astrophys. J., 334,* 476–488.
Paganini L. and Mumma M. J. (2016) A solar-pumped fluorescence model for line-by-line emission intensities in the B–X, A–X, and X–X band systems of ${}^{12}C^{14}N$. *Astrophys. J. Suppl. Ser., 226,* 3.
Poch O., Istiqomah I., Quirico E., et al. (2020) Ammonium salts are a reservoir of nitrogen on a cometary nucleus and possibly on some asteroids. *Science, 367,* aaw7462.
Price E. M., Cleeves L. I., Bodewits D., et al. (2021) Ice-coated pebble drift as a possible explanation for peculiar cometary CO/H_2O ratios. *Astrophys. J., 913,* 9.
Protopapa S., Sunshine J. M., Feaga L. M., et al. (2014) Water ice and dust in the innermost coma of Comet 103P/Hartley 2. *Icarus, 238,* 191–204.
Raghuram S. and Bhardwaj A. (2012) Model for the production of CO Cameron band emission in Comet 1P/Halley. *Planet. Space Sci., 63,* 139–149.
Raghuram S., Hutsemékers D., Opitom C., et al. (2020) Forbidden atomic carbon, nitrogen, and oxygen emission lines in the water-poor Comet C/2016 R2 (Pan-STARRS). *Astron. Astrophys., 635,* A108.
Rauer H., Roesler F., Scherb F., et al. (1997) Ion emission line profiles in cometary plasma tails. *Astron. Astrophys., 325,* 839–846.
Rodgers S. D., Charnley S. B., Huebner W. F., et al. (2004) Physical processes and chemical reactions in cometary comae. In *Comets II* (M. C. Festou et al., eds.), pp. 505–522. Univ. of Arizona, Tucson.
Roth L., Saur J., Retherford K. D., et al. (2014) Transient water vapor at Europa's south pole. *Science, 343,* 171–174.
Roth L., Ivchenko N., Gladstone G. R., et al. (2021) A sublimated water atmosphere on Ganymede detected from Hubble Space Telescope observations. *Nature Astron., 5,* 1043–1051.
Roth N. X., Milam S. N., Cordiner M. A., et al. (2021a) Leveraging the ALMA Atacama Compact Array for cometary science: An interferometric survey of Comet C/2015 ER61 (PanSTARRS) and evidence for a distributed source of carbon monosulfide. *Astrophys. J., 921,* 14.
Roth N. X., Milam S. N., Cordiner M. A., et al. (2021b) Rapidly varying anisotropic methanol (CH_3OH) production in the inner coma of Comet 46P/Wirtanen as revealed by the ALMA Atacama Compact Array. *Planet. Sci. J., 2,* 55.
Rottman G. J., Woods T. N., and Sparn T. P. (1993) Solar-Stellar Irradiance Comparison Experiment 1: 1. Instrument design and operation. *J. Geophys. Res.–Atmos., 98,* 10667–10677.
Roueff E. and Lique F. (2013) Molecular excitation in the interstellar medium: Recent advances in collisional, radiative, and chemical processes. *Chem. Rev., 113,* 8906–8938.
Schleicher D. G. (1983) The fluorescence of cometary OH and CN. Ph.D. thesis, University of Maryland, College Park.
Schleicher D. G. and A'Hearn M. F. (1982) OH fluorescence in comets: Fluorescence efficiency of the ultraviolet bands. *Astrophys. J., 258,* 864–877.
Schleicher D. G. and Farnham T. L. (2004) Photometry and imaging of the coma with narrowband filters. In *Comets II* (M. C. Festou et al.,

eds.), pp. 449–469. Univ. of Arizona, Tucson.
Schwadron N. A. and Cravens T. E. (2000) Implications of solar wind composition for cometary X-rays. *Astrophys. J., 544,* 558–566.
Shou Y., Combi M., Toth G., et al. (2016) A new 3D multi-fluid model: A study of kinetic effects and variations of physical conditions in the cometary coma. *Astrophys. J., 833,* 160.
Šimečková M., Jacquemart D., Rothman L. S., et al. (2006) Einstein *A*-coefficients and statistical weights for molecular absorption transitions in the HITRAN database. *J. Quant. Spectrosc. Radiat. Transfer, 98,* 130–155.
Sliter R., Gish M., and Vilesov A. F. (2011) Fast nuclear spin conversion in water clusters and ices: A matrix isolation study. *J. Phys. Chem. A, 115,* 9682–9688.
Strøm P. A., Bodewits D., Knight M. M., et al. (2020) Exocomets from a solar system perspective. *Publ. Astron. Soc. Pac., 132,* 101001.
Swings P. (1965) Cometary spectra. *Q. J. R. Astron. Soc., 6,* 28–69.
Tenishev V., Combi M., and Davidsson B. (2008) A global kinetic model for cometary comae: The evolution of the coma of the Rosetta target Comet Churyumov-Gerasimenko throughout the mission. *Astrophys. J., 685,* 659–677.
Tennyson J., Yurchenko S. N., Al-Refaie A. F., et al. (2020) The 2020 release of the ExoMol database: Molecular line lists for exoplanet and other hot atmospheres. *J. Quant. Spectrosc. Radiat. Transfer, 255,* 107228.
Thomas N. (2020) *An Introduction to Comets: Post-Rosetta Perspectives.* Springer, Cham. 503 pp.
Tobiska W. K., Woods T., Eparvier F., et al. (2000) The SOLAR2000 empirical solar irradiance model and forecast tool. *J. Atmos. Terr. Phys., 62,* 1233–1250.
van der Tak F. F. S., Black J. H., Schöier F. L., et al. (2007) A computer program for fast non-LTE analysis of interstellar line spectra: With diagnostic plots to interpret observed line intensity ratios. *Astron. Astrophys., 468,* 627–635.
van der Tak F. F. S., Lique F., Faure A., et al. (2020) The Leiden Atomic and Molecular Database (LAMDA): Current status, recent updates, and future plans. *Atoms, 8,* 15.
Villanueva G. L., Mumma M. J., DiSanti M. A., et al. (2011a) The molecular composition of Comet C/2007 W1 (Boattini): Evidence of a peculiar outgassing and a rich chemistry. *Icarus, 216,* 227–240.
Villanueva G. L., Mumma M. J., and Magee-Sauer K. (2011b) Ethane in planetary and cometary atmospheres: Transmittance and fluorescence models of the ν_7 band at 3.3 μm. *J. Geophys. Res.–Planets, 116,* E08012.
Villanueva G. L., Mumma M. J., Bonev B. P., et al. (2012) Water in planetary and cometary atmospheres: H_2O/HDO transmittance and fluorescence models. *J. Quant. Spectrosc. Radiat. Transfer, 113,* 202–220.
Villanueva G. L., Magee-Sauer K., and Mumma M. J. (2013) Modeling of nitrogen compounds in cometary atmospheres: Fluorescence models of ammonia (NH_3), hydrogen cyanide (HCN), hydrogen isocyanide (HNC) and cyanoacetylene (HC_3N). *J. Quant. Spectrosc. Radiat. Transfer, 129,* 158–168.
Villanueva G. L., Smith M. D., Protopapa S., et al. (2018) Planetary Spectrum Generator: An accurate online radiative transfer suite for atmospheres, comets, small bodies and exoplanets. *J. Quant. Spectrosc. Radiat. Transfer, 217,* 86–104.
Villanueva G. L., Liuzzi G., Crismani M. M. J., et al. (2021) Water heavily fractionated as it ascends on Mars as revealed by ExoMars/NOMAD. *Sci. Adv., 7,* eabc8843.
Wallace L. and Livingston W. (2003) *An Atlas of the Solar Spectrum in the Infrared from 1850 to 9000 cm^{-1} (1.1 to 5.4 micrometer), revised.* National Solar Observatory, National Optical Astronomy Observatory, NSO Technical Report, Tucson.
Weaver H. A. and Mumma M. J. (1984) Infrared molecular emissions from comets. *Astrophys. J., 276,* 782–797.
Weaver H. A., Feldman P. D., McPhate J. B., et al. (1994) Detection of CO Cameron band emission in Comet P/Hartley 2 (1991 XV) with the Hubble Space Telescope. *Astrophys. J., 422,* 374–380.
Wegmann R. and Dennerl K. (2005) X-ray tomography of a cometary bow shock. *Astron. Astrophys. Suppl., 430,* L33–L36.
Woods T. N., Bailey S. M., Eparvier F. G., et al. (1998) TIMED solar EUV experiment. In *Missions to the Sun II* (C. M. Korendyke, ed.), pp. 180–191. SPIE Conf. Ser. 3442, Bellingham, Washington.
Woods T. N., Bailey S., Eparvier F., et al. (2000) TIMED solar EUV experiment. *Phys. Chem. Earth C, 25,* 393–396.
Woodward C. E., Kelley M. S., Bockelée-Morvan D., et al. (2007) Water in Comet C/2003 K4 (LINEAR) with Spitzer. *Astrophys. J., 671,* 1065–1074.
Xie X. and Mumma M. J. (1992) The effect of electron collisions on rotational populations of cometary water. *Astrophys. J., 386,* 720–728.
Zakharov V., Bockelée-Morvan D., Biver N., et al. (2007) Radiative transfer simulation of water rotational excitation in comets: Comparison of the Monte Carlo and escape probability methods. *Astron. Astrophys., 473,* 303–310.
Zhao Y., Rezac L., Hartogh P., et al. (2020) Constraining spatial pattern of early activity of Comet 67P/C-G with 3D modelling of the MIRO observations. *Mon. Not. R. Astron. Soc., 494,* 2374–2384.

Marschall R., Davidsson B. J. R., Rubin M., and Tenishev V. (2024) Neutral gas coma dynamics: Modeling of flows and attempts to link inner coma structures to properties of the nucleus. In *Comets III* (K. J. Meech, M. R. Combi, D. Bockelée-Morvan, S. N. Raymond, and M. E. Zolensky, eds.), pp. 433–458. Univ. of Arizona, Tucson, DOI: 10.2458/azu_uapress_9780816553631-ch014.

Neutral Gas Coma Dynamics: Modeling of Flows and Attempts to Link Inner Coma Structures to Properties of the Nucleus

Raphael Marschall
Southwest Research Institute/Observatoire de la Côte d'Azur

Björn J. R. Davidsson
Jet Propulsion Laboratory/California Institute of Technology

Martin Rubin
University of Bern

Valeriy Tenishev
University of Michigan

Deriving properties of cometary nuclei from coma data is of significant importance for our understanding of cometary activity and has implications beyond. Groundbased data represent the bulk of measurements available for comets. Yet, to date these observations only access a comet's gas and dust coma at rather large distances from the surface and do not directly observe its surface or even the outgassing layer. In contrast, spacecraft flyby and rendezvous missions are one of the only tools that gain direct access to surface measurements. However, these missions are limited to roughly one per decade. We can overcome these challenges by recognizing that the coma contains information about the nucleus's properties. In particular, the near-surface gas environment is most representative of the nucleus. It can inform us about the composition, regionality of activity, and sources of coma features and how they link to the topography, morphology, or other surface properties. The inner coma data is a particularly good proxy because it has not yet, or has only marginally, been contaminated by coma chemistry or secondary gas sources (e.g., from icy grains released into the coma) and can retain fine structure that needs time to dissipate. Additionally, when possible, the simultaneous observation of the innermost coma with the surface provides the potential to make a direct link between coma measurements and the nucleus. If we hope to link outer coma measurements obtained by Earth-based telescopes to the surface, we must first understand how the inner coma measurements are linked to the surface. Numerical models that describe the flow from the surface into the immediate surroundings are needed to make this connection. This chapter focuses on the advances made to understand the flow of the neutral gas coma from the surface to distances up to a few tens of nuclei radii. The current state of research on linking the inner gas coma properties and structures to the nucleus is explored, describing both simple/heuristic models and state-of-the-art physically consistent models. The model limitations and what they each are best suited for is discussed. In the end, the different approaches are compared to spacecraft data, and the remaining knowledge gaps and how best to address them in the future are presented.

1. INTRODUCTION

Comets are thought to be icy leftovers from planet formation, either planetesimal themselves or direct descendants of the former. For that reason, they are widely considered to have retained information about the early solar system and can inform our understanding of planet formation. While their interiors have likely retained their primordial properties, the same cannot be said for their surfaces (e.g., *Jutzi and Michel,* 2020). Cometary surfaces are heavily evolved by numerous processes such as, e.g., irradiation, impacts, thermal processing, and sublimation-driven activity. For more on the structure and properties of the surface, see the chapter by Pajola et al. in this volume.

Because the pristine interior of comets is not easily accessible directly, we turn our gaze to the gas and dust comae, which can be studied with spacecraft and groundbased telescopes. The Deep Impact mission (*A'Hearn et al.,* 2005) stands out for probing the subsurface of 9P/Tempel 1 with an impactor and visiting the first hyperactive comet

(103P/Hartley 2). But this bridge, from the interior/surface to the comae, requires us to devise methods to link the coma properties to the surface/interior. We need to understand the dynamics of the gas and dust from the surface to a spacecraft or the distances observed with groundbased telescopes. This chapter will describe the current state of the art in modeling the gas dynamics within the first few nucleus radii above the surface (corresponding to a few tens of kilometers in the case of a nucleus with a typical radius of a few kilometers) and critical insights from the past decade of research and spacecraft missions.

To date, only six comets (1P/Halley, 19P/Borrelly, 9P/Tempel 1, 67P/Churyumov-Gerasimenko, 81P/Wild 2, and 103P/Hartley 2) have been visited by spacecraft, which resolved their nuclei. Spacecraft observations provide a detailed, high spatial and temporal resolution of the surface and surrounding coma but are limited to a few target comets (~1–2 per decade). The European Space Agency's (ESA's) Rosetta mission (*Glassmeier et al.,* 2007) has given us the most recent and detailed picture of cometary evolution by following Comet 67P/Churyumov-Gerasimenko (hereafter 67P) through its perihelion for over two years. Apart from Rosetta, all previous comet missions have been flybys and thus did not cover the baseline to study temporal changes. Only Rosetta provided a long-time dataset to study the temporal variability and evolution of the coma in detail. Previous missions were able to observe comets for several rotation periods of the nucleus and were able to find periodicity in their activity. For example, the combination of Stardust NExT's exploration of 9P/Tempel 1 one full apparition after the Deep Impact experiment showed significant changes on the surface (*Veverka et al.,* 2013).

Comets are more easily and frequently observed using ground- and spacebased telescopes. In contrast to spacecraft, the spatial resolution is much lower, but we can observe many more comets and thus sample their diversity. Although the nuclei are not resolved in telescopic observations, the comae and tails are.

Both spacecraft and ground-/spacebased telescopes thus provide complementary datasets that require consolidation. This chapter focuses on the near-nucleus coma (within the first few nucleus radii of the surface). The three main reasons that motivate the study of this region are:

- We may link spacecraft measurements from the coma to the surface.
- Understanding the innermost coma is a prerequisite to understanding groundbased observations and linking those measurements to the nucleus, i.e., we first need to understand the near nucleus coma to interpret measurements at larger distances
- The gained knowledge of this region allows us to make predictions for future comet missions and assess hazards for spacecraft operating in that region.

Although we will touch on the issue of dust in the gas flow, dust dynamics is not the main focus of this chapter. Instead, we refer the reader to the chapter by Agarwal et al. in this volume and *Marschall et al.* (2020c) for detailed reviews of the state of the art in dust coma modeling. We will, however, discuss how dust can alter the properties of the gas flow but will leave the rest to the two references above.

Spacecraft- and Earth-based telescopes measure gas column densities along the line of sight. This is done indirectly through the measurement of emission lines of different gas species (e.g., *Feaga et al.,* 2007; *Biver et al.,* 2019) or absorption of starlight during occultations when in orbit with the comet (e.g., *Keeney et al.,* 2019). When embedded in a coma, a spacecraft can also measure the local gas densities (e.g., *Hässig et al.,* 2015). Both quantities, local gas densities, and line-of-sight column densities can be used to derive the parameters of the gas flux at the surface. This includes the global gas production rate and the distribution of sources at the surface. Furthermore, the relative abundances in the coma bear information on the composition of the ices in the nucleus (e.g., *Marboeuf and Schmitt,* 2014; *Prialnik,* 1992; *Herny et al.,* 2021).

In this chapter we will focus on two crucial questions of linking inner coma measurements to the surface:

- How can we confidently derive the gas production rate of different species and thus the volatile mass loss from coma measurements?
- Can we determine if coma structures (inhomogeneities in density, often referred to as "jets") are reflective of a heterogeneous nucleus, or are mere emergent phenomena in the gas flow due to, e.g., the complex shape of the nucleus?

We will only focus on the inner coma/near environment for this review. There is no strict definition of the inner coma, but here we consider it the region within which the major gas species (H_2O, CO_2, and CO) accelerate and do not yet experience any substantial loss through chemical reactions (ionization, ion-neutral reactions, etc.). These chemical processes act on tens of thousands of kilometers and will make a notable dent in the neutral gas profile (e.g., *Shou et al.,* 2016). The typical extent of the acceleration region is on the order of 10 nucleus radii [a few tens of kilometers for a typical comet (*Tenishev et al.,* 2008; *Shou et al.,* 2016; *Zakharov et al.,* 2018b)]. This region is typically only accessible with spacecraft missions and not by groundbased observations.

We ultimately want to understand how measurements at larger distances to the nucleus obtained with ground- or spacebased telescopes can be linked to the nucleus. But before we can understand the link between those measurements and the nucleus, we first need to understand how the near environment can be linked to the surface. Therefore we dedicate this chapter to the advances of the latter.

We also focus here on the inner coma because of the recent wealth of spacecraft data from Rosetta and Deep Impact. The close distances to the gas' source region also provide the biggest chance to link the coma to the surface unambiguously.

The second question posed above is controversial as it has been known for some time that it is theoretically possible to produce structures in the coma from a homogeneous but non-spherical nucleus. Moreover, inhomogeneous spherical and homogeneous aspherical nuclei may lead to similar structures in the gas coma (e.g., *Zakharov et al.,* 2008). The chapter by *Crifo et al.* (2004) in *Comets II* left us at that crossroads. At the time, the only modeling including an actual comet shape and data comparison had been done for 1P/Halley. Since then, we have added five more comets (19P, 81P, 9P, 103P, and 67P) to help us understand the gas flow from cometary nuclei. At the time of the previous book, the modeling of the inner coma was still primarily theoretical. There was a large amount of work done that explored active spots on or inhomogeneous outgassing from spherical nuclei (e.g., *Komle and Ip,* 1987; *Kitamura,* 1990; Kn*ollenberg,* 2017; *Crifo et al.,* 1995; *Crifo and Rodionov,* 1997) as well as homogeneous nuclei with complex shapes such as ellipsoids and beans (e.g., *Crifo and Rodionov,* 1997, 2000; *Crifo et al.,* 1999, 2002b). *Crifo et al.* (2004) had to leave the question of what drives inner coma structures open, and thus we intend to revisit this question in this chapter and provide some answers.

We will show in section 3 that the gas production rates can be reasonably safely estimated using heuristic models, at least for some comets. In section 4 we will present the state of the art of physical gas coma models and argue in section 5 that the evidence points to the fact that observed coma structures do not require a heterogeneous nucleus. Instead, the redistribution of material across the nucleus surface is sufficient to explain most regional heterogeneity of the observed activity. Overlain on these regional levels of activity is topography and the irregular shape of the nucleus that affect the flows through focusing and defocusing. We will conclude this chapter by giving an outlook and discussing open questions (section 6).

2. COMA STRUCTURE DEFINITIONS

The main property to differentiate structures in the gas flow is the spatial scale. There are large/global-scale (larger than the nucleus) structures and fine structures (much smaller than the scale of the nucleus).

A good example of the former large-scale gas flow structure is the CO_2 column density distribution in the coma of Comet 103P/Hartley 2 as observed during the Deep Impact eXtended Investigation (DIXI) (*Protopapa et al.,* 2014) shown in shown in Fig. 3 and Fig. 1c for Comet 67P. This feature does not appear to have a confined source region but rather covers a significant fraction of the smaller lobe of Comet 103P. These larger-scale structures reflect the global parameters, such as the total gas production rate, asymmetry and composition in gas production, and large-scale geometry of the nucleus (among others, its shape and rotation state).

In contrast, the fine structures reflect very local features of the nucleus including its topography. A good example, although observed indirectly through the reflectance of dust particles in the gas flow, are highly collimated features seen at Comet 67P/Churyumov-Gerasimenko during the Rosetta mission (*Vincent et al.,* 2016a). These events can be associated with outbursts. In such cases, we might refer to these features as a "jets." The use of the word "jet" is controversial, though, mainly because it has a strict physical interpretation. It is often used very liberally to describe any collimated feature in the coma (see, e.g., *Vincent et al.,* 2019). Therefore, in some parts of the literature, any inhomogeneity in the coma that appears to be collimated will be identified as a "jet." We argue that a "jet" should have a narrower definition, which is closer to a physical understanding of the word. At least the following two properties should be satisfied. A "jet" represents a gas stream with (1) a clear boundary to ambient flow and (2) outflowing from a source much smaller than the size of the nucleus (Fig. 1a/c). If the source region covers, e.g., an entire hemisphere, it would not be "jet" even though the resulting feature might appear bounded. In this example, we suggest the less-implicating term "stream" (e.g., the CO_2 feature in Fig. 1b nicely fits that). A counterexample to a bounded "jet" is the expansion into a solid angle of 2π, which would rather be regarded as a "plume." We will discuss in section 5 why this nomenclature can be extremely misleading and that, except for outbursts, most features in the coma do not warrant the label "jet." We should also note that for historical reasons the word "jet" is often used to describe features observed in the outer coma of ground-based data and is used descriptively. Here, we specifically encourage a more specific use for inner coma structures because there we have at least the possibility of more accurately distinguishing between these terms.

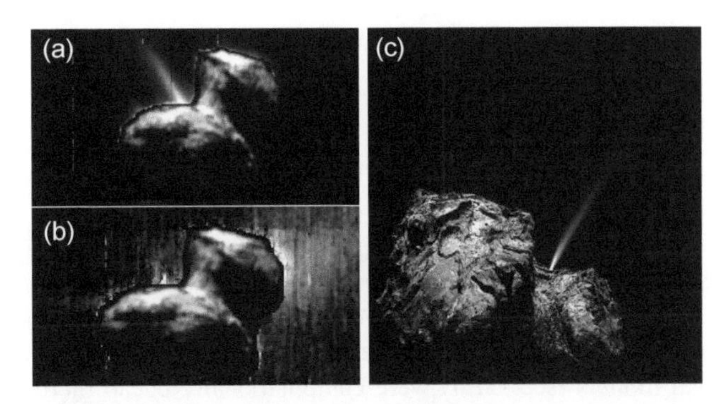

Fig. 1. (a),(b) Composite images of the Comet 67P/Churyumov-Gerasimenko's nucleus at 1.095 μm, superimposed with the **(a)** water column density and **(b)** CO_2 column density observed by Rosetta/VIRTIS-M (from *Migliorini et al.,* 2016). **(c)** Comet nucleus observed by Rosetta/OSIRIS (from *Vincent et al.,* 2016a). It shows the nucleus and the scattered light of a collimated feature, a "jet" of dust grains.

With increasing distance from the surface, the flow expands and the local density decreases. The mean free path (MFP) of the molecules therefore becomes large, and thus the flow gradients become smoothed. At large distances to the nucleus the increased rarefaction causes the fine structures of the flow to vanish. This is why coma structures on large distances will only reflect the global characteristics of the neutral gas flow. Nevertheless, even these global characteristics are of interest since they allow capturing the general properties of gas emission (e.g., total gas production rate) and therefore those of the nucleus.

3. HEURISTIC MODELS

A heuristic model by its nature sacrifices physical rigor for simplicity, and therefore gain computation speed. It attempts to simplify a problem by neglecting complexities deemed unnecessary to derive specific properties. If physical models (section 4) were computationally cheap there would not be any justification at all to turn to heuristic models. They help obtain rough estimates and thus provide a "sanity check" on physical models. Some inconsistencies or uncertainties of heuristic models are easy to spot; other limitations are not immediately obvious. It is therefore essential to understand the limitations of heuristic models and not apply them to inappropriate situations or the resulting interpretation of the respective data is likely to be wrong. As we will see, some heuristic models can be useful, but they also immediately show the need for physically accurate models, which we will discuss in section 4.

The most famous heuristic model in cometary comae research is the so-called "Haser model" (*Haser,* 1957). It assumes free molecular (i.e., collisionless) radial flow and is based on the conservation of the number of particles. It also takes into account chemical processes like photodissociation. The model becomes much more straightforward in its simplified form, where chemical processes are neglected. In this case it links the total gas production rate, Q, to the local gas density n via

$$Q = 4\pi r^2 nv \tag{1}$$

where r is the distance to the nucleus, and v is the gas speed. The equation above also assumes isotropic expansion. Nonisotropic flows will also reach radial expansion at a constant speed, and at that point the above expression is valid in a directional sense, conserving the directional mass flow. In its full form, the model includes the photochemical destruction of parent molecules and can be rewritten to track daughter species (*Combi et al.,* 2004). For a comet at 1 au these chemical processes act on tens of thousands of kilometers and will make a notable dent in the neutral gas profiles only on those scales (e.g., *Shou et al.,* 2016). These processes do not dominate on the short timescales of the near-nucleus environment. The flow can also be confined to, e.g., a half sphere (for instance, the sunward side) by modifying the solid angle from 4π to 2π steradian. In this sense, this is the simplest gas model one might think of. Although it is still the most commonly used approximation, owing to its apparent simplicity, it is physically adequate only at distances when the flow is expanding radially. To be precise, the flow generally expands radially and reaches 90% of terminal velocity at around 10 nucleus radii (*Zakharov et al.,* 2018b; *Gerig et al.,* 2018). This model cannot capture the dynamics close to any nucleus while the gas accelerates and simultaneously cools due to the associated expansion. At these short distances to the surface ($<10\ R_N$), effects from the nucleus shape also still play an important role. It should therefore not be used in the immediate vicinity of the nucleus. But equation (1) can be rather safely used between distances of 10 nucleus radii and 10,000 km. Beyond the latter, chemical reactions need to be accounted for.

For early data from the Rosetta mission, this model seemed to provide reasonable estimates of the gas production rate using the relationship in equation (1) and variations in the solar zenith angle (*Bieler et al.,* 2015). At that point in the mission, 67P was still beyond 3 au from the Sun, and Rosetta at ≥ 10 km cometocentric distance thus outside the gas acceleration region (*Tenishev et al.,* 2008; *Zakharov et al.,* 2018b, 2023). We will come back to the question of whether the "Haser model" or other heuristic models are useful to at least estimate the gas production rate.

Recently, two more models have appeared combining physics-based and heuristic approaches. In the first model, *Fougere et al.* (2016b) used the local gas densities from the Rosetta Orbiter Spectrometer for Ion and Neutral Analysis (ROSINA) (*Balsiger et al.,* 2007) instrument to perform a spherical harmonics fit of the data and constrain the surface gas emission distribution. Regions that overlap due to the concavities of the shape of 67P are ignored. These surface distributions in conjunction with local illumination were the initial conditions for a three-dimensional kinetic modeling of the dusty gas coma. The modeling was done using the Adaptive Mesh Particle Simulator (AMPS), which is a general-purpose Direct Simulation Monte Carlo (DSMC) model (*Tenishev et al.,* 2021). We refer interested readers to *Bird* (1994, 2013) for more detail on the DSMC method.

This validation step with a physical model is important because it gives some confidence in the result. We cannot be confident that the solution is unique because the surface-emission distribution is prescribed a "spherical" form. This approach will show large-scale (e.g., north-south) distribution of gas sources but does not seem adequate to link the surface activity to morphological differences on the surface.

This "spherical harmonics model" has not only been used to estimate the surface-emission distribution but also the global gas production rate of the major species along the orbit of 67P (*Fougere et al.,* 2016a; *Combi et al.,* 2020). While the surface-emission distribution contains considerable ambiguity, the global gas production rate overlaps with the results from other approaches (e.g., *Läuter et al.,* 2020; *Marschall et al.,* 2020b).

The second model (*Kramer et al.,* 2017; *Läuter et al.,* 2019, 2020) assumes that each surface facet of a shape model (here 67P) is an independent gas source. The gas outflow from each facet is described with an opening angle and then follows essentially collisionless outflow according to *Narasimha* (1962). The different sources from neighboring facets do not interact and simply contribute linearly to the gas densities at the spacecraft. In this sense, it is a "Haser"-type model that also takes into account the shape and we shall refer to it as a "Haser + shape model" [to some extent similar to *Bieler et al.* (2015)]. As with any heuristic model, it results in some nonphysical results, e.g., extremely high gas speeds (*Kramer et al.,* 2017). The model included an assumption on the coma temperature (200 K, 100 K, and 50 K). In contrast, *Tenishev et al.* (2008) obtained much lower temperatures in the 30 to 10 K range at distances between 10 km and 100 km, which may in part be responsible for this discrepancy. Later models (*Läuter et al.,* 2019) then used the modeled velocities by *Hansen et al.* (2016) as input.

Gas flows from different sources cannot be assumed to be independent. Although this can be correct in very rarefied cases, it is rarely true, even for a comet with comparably weak activity such as 67P. It has long been known that gas sources close to each other, e.g., two jets, interact with each other and the local gas density in the coma is not simply a linear combination of the gas densities of the two isolated jets (e.g., *Dankert and Koppenwallner,* 1984). Furthermore, the "Haser + shape model" cannot reproduce surface-emission maps from physical coma models (see appendix A of *Marschall et al.,* 2020a). Although this makes it unclear if the activity maps from this model are reliable, the global gas production rates from this model (*Läuter et al.,* 2019, 2020) should be fairly good estimates. Importantly, *Marschall et al.* (2020a) point out that there is a physical resolution limit to detecting heterogeneous emission distributions. This limit stems from the flow viscosity (which is connected with the MFP of the molecules) close to the surface. Viscous dissipation blurs fine structures of the flow and therefore the underlying information of the boundary conditions that caused these structures. For a comet like 67P, this resolution limit lies at several hundred meters. This resolution limit also prevents physical models from determining the activity map to arbitrary accuracy. Finally, by not taking illumination into account, the "Haser + shape model" is also much less suited to reproduce Microwave Instrument for Rosetta Orbiter (MIRO) (*Gulkis et al.,* 2007) and Visual IR Thermal Imaging Spectrometer (VIRTIS) (*Coradini et al.,* 2007) line-of-sight observations compared to ROSINA data, which were obtained predominantly in a terminator orbit.

Marschall et al. (2020b) used a physical model similar to *Fougere et al.* (2016a). Instead of applying spherical harmonics to parameterize the active area fraction (AAF) of the nucleus surface, they assumed a homogeneous nucleus composition (i.e., constant AAF). The gas production rate was modulated by the illumination conditions only, i.e., there was no regional heterogeneity as in the works by *Fougere et al.* (2016a), *Combi et al.* (2020), and *Läuter et al.* (2020), but remained calibrated with Rosetta/ROSINA data as in the other studies. They show that matching daily averaged measurements is sufficient to estimate the global mass loss. Reproducing the precise diurnal variation, with the associated regional heterogeneity, is not necessary to estimate the total mass loss. All the above-mentioned approaches predict the same global mass loss within error bars.

This indicates that the shape and regional heterogeneity do not significantly contribute to variations in the production rate of 67P. This is in line with findings from *Marshall et al.* (2019), who have shown that on average 67P behaves almost like a spherical nucleus. They also show that this is not true in general and different shapes and spin states can have a significant influence on the gas production rate. Importantly, so long as the respective model preserves mass conservation at large distances, it will correctly characterize the total gas production rate.

The rough global surface distributions found by *Fougere et al.* (2016a) and *Combi et al.* (2020) with the "spherical harmonics model," *Läuter et al.* (2019) with the "Haser + shape model," and *Zakharov et al.* (2018a) and *Marschall et al.* (2019) (the latter two both for the northern hemisphere) with physical models are in agreement in the following sense. There is, e.g., enhanced water emission from the northern hemisphere and CO_2 from the southern hemisphere of 67P during northern summer. The water production rate follows to first order the subsolar latitude. Although these are important first insights, they are inadequate to link activity with surface morphology and evolutionary history including erosion, which should be the ultimate goal. We should point out, though, that a diverse surface morphology might be the result of activity and not the driver of it.

Put another way, using a heuristic model, whichever it may be, instead of a physical model is sufficient to estimate the global production rate for 67P and other comets. *Combi et al.* (2019) used a semi-analytical model called the time-resolved model (TRM) (*Mäkinen and Combi,* 2005) to calculate global water production rates for 61 comets. This illustrates the strength of such approaches to determine the global properties of comets. A comet with a spin state and shape such that it behaves as a sphere [in the sense described in *Marshall et al.* (2019)] should be similarly suitable for these heuristic models as 67P appears to be.

We hope to have convinced the reader of the usefulness of heuristic models in some cases but have also shown that they are somewhat inadequate to link structures in the coma to emission distributions at the surface and through that to surface morphology. This link requires physically consistent models, which we will discuss in the next section. But, as mentioned above, even physically consistent models have spatial-resolution limits, and the appropriate error propagation needs to be accounted for (*Marschall et al.,* 2020a).

4. STATE-OF-THE-ART PHYSICAL MODELS

The comet nucleus is the primary source of vapor and refractory particles in the coma (coma solids that emit gas and dust may constitute a secondary extended or distributed source). The nucleus surface also acts as a boundary that may scatter or adsorb coma molecules. From a coma modeling perspective, the type of input information needed regarding the nucleus source depends on the coma model. Kinetic models based on the Boltzmann equation require the emission flux, temperature, and velocity distribution function for each species specified at the nucleus/coma interface. They also require the nucleus surface temperature when dealing with scattering or adsorption of coma molecules (*Bird,* 1994, 2013). Hydrodynamic models based on Euler (EE) or Navier-Stokes (NS) equations require the the flux, temperature, and drift speed on top of the Knudsen layer (the boundary between the non-equilibrium near-surface layer and the equilibrium fluid flow). We refer an interested reader to *Hirsch* (2007) and *Rodionov et al.* (2002) for more details on the theory of the fluid methods. Thermophysical nucleus models, discussed in section 4.1.1, provide outgassing rates, surface temperatures, and near-surface temperature gradients. This constitutes necessary, but not sufficient, information needed to calculate transmission velocity distribution functions, discussed in section 4.1.2, and Knudsen layer properties, mentioned in section 4.1.3. In general, the thermophysical model of the surface and gas environment model are coupled in both directions (i.e., they are interdependent). In practice, though, these are treated independently.

4.1. Boundary Conditions

4.1.1. Thermophysical modeling of the nucleus. A comet nucleus has a porous interior consisting of refractories, crystalline and/or amorphous water ice, and secondary highly volatile species such as CO_2 and CO. The vast majority of the surface is covered by an ice-free dust mantle that absorbs solar radiation and emits thermal infrared radiation. For 9P, 103P, and 67P very small exposed icy patches (H_2O and/or CO_2) have been observed on the surface (*Sunshine et al.,* 2006; *Pommerol et al.,* 2015; *Raponi et al.,* 2016; *Fornasier et al.,* 2016). State-of-the-art thermophysical models (e.g., the chapter by Guilbert-Lepoutre et al. in this volume; *Groussin et al.,* 2007, 2013; *Rosenberg and Prialnik,* 2010; *Davidsson et al.,* 2013, 2021; *Herny et al.,* 2021; *Marboeuf and Schmitt,* 2014, and references therein) solve the coupled differential equations for energy and mass conservation of the nucleus, which attempt to describe how such a system evolves as solar energy is transported by solid-state and radiative conduction, ice sublimates while consuming energy, vapor diffuses according to local temperature and pressure gradients while transporting energy by advection, and gas eventually escapes to space through the dust mantle or recondenses at depth while releasing latent energy. The solutions to these equations provide temperature, partial gas pressures, porosity, and abundances of solids as functions of latitude, time, and depth for the rotating and orbiting nucleus, as well as outgassing rates for each considered volatile.

Numerical coma models are generally too computationally demanding to allow for the usage of a coupled state-of-the-art thermophysical nucleus model. Therefore, simplified thermophysical models are employed that typically balance solar energy absorption, thermal reradiation, and energy consumption by surface water ice sublimation (e.g., *Crifo et al.,* 2005; *Zakharov et al.,* 2008; *Marschall et al.,* 2019). This simplification has at least three important consequences that may affect the accuracy of coma models.

First, simplified nucleus models with surface ice become substantially cooler than realistic nucleus models with dust mantles [during strong subsolar sublimation near perihelion, the former typically have surface temperatures of 200 K, while the latter have 350 K (*Groussin et al.,* 2007)]. This means that transmission velocity distributions are biased toward low molecular initial speeds, while initial translational temperatures and drift speeds are underestimated. This first issue could partially be mitigated by calculating the radiative equilibrium temperature, i.e., omitting sublimation cooling altogether. This would approximately account for the gas heating taking place as it diffuses through the hot dust mantle on its way to the surface. Another time-efficient option is to apply lookup tables generated by more advanced thermophysical models. A rudimentary version of that approach was employed by, e.g., *Tenishev et al.* (2008), *Fougere et al.* (2016b), and *Combi et al.* (2020).

Second, simplified models typically produce 1–2 orders of magnitude more gas compared to realistic models, for which a finite diffusivity of the dust mantle quenches the flow. This problem is typically handled by introducing an AAF that reduces the production rate (and thus the near-surface gas number density) to the observed level by assuming that only parts of the surface is covered by ice. We will come back to this issue in section 4.1.4.

Third, simplified nucleus models lack thermal inertia effects, caused by non-zero heat conductivity and heat capacity. Consequently, those models predict peak outgassing at local noon, while realistically modeled activity is strongest in the afternoon. This makes the modeled coma too axis-symmetric about the Sun-comet line, at least for a spherical comet. Complex shapes, such as the one of 67P, introduce additional complexity in the pattern of activity. Furthermore, nighttime activity artificially goes to zero, which typically is dealt with by setting a low but arbitrary background outgassing. While a small nighttime activity is mostly a good approximation for H_2O, it is not for more volatile species. Nighttime activity of CO_2 was observed both for 9P (*Feaga et al.,* 2007) and 103P (*Feaga et al.,* 2014). Even for 67P, which is dominated by H_2O, nighttime activity (likely of CO_2) needed to be invoked to understand the dust coma dynamics (*Gerig et al.,* 2020). Also, at 67P the coma above the southern hemisphere showed strongly enhanced CO_2/H_2O ratios during the poorly illuminated winter months early on in the Rosetta mission (*Hässig et al.,* 2015). More on this follows in section 5.3.4. These observed instances of nighttime activity need to be reflected in thermophysical modeling that goes into the boundary condition of coma models (e.g., *Pinzón-Rodríguez et al.,* 2021). This third issue is interesting because it has so far not seemed to hinder the modeling of H_2O. This might indicate that water ice is very close to the surface, thus making thermal inertia effects small enough that they cannot be picked up by coma models. Or the effects are so nuanced that they have simply not been discovered yet. Compared to H_2O, thermal inertia effects need to be properly addressed with a thermal model that includes the thermal lag (*Pinzón-Rodríguez et al.,* 2021). The measurements of CO (*Hässig et al.,* 2015) above the southern winter hemisphere of 67P show even less diurnal variation and a more uniform outgassing pattern than even CO_2. This suggests that CO comes from even deeper layers. Given that the CO is much more volatile than CO_2, this observation is not surprising.

4.1.2. Transmission velocity distribution functions. Most kinetic coma models postulate the semi-Maxwellian velocity distribution function (SMVDF) (e.g., *Huebner and Markiewicz,* 2000) as a boundary condition. To ensure that the gas has a semi-Maxwellian velocity distribution the initial expansion of the gas into the first cell needs to be taken into account. It turns out that one cannot directly draw the velocity vectors for the molecules at the surface from an SMVDF because by the time they have expanded into the first cell above the surface their velocity distribution function (VDF) has been altered. Therefore, there is a need to define a transmission semi-Maxwellian velocity distribution (TSMVD) to draw from (*Huebner and Markiewicz,* 2000). The SMVD and the TSMVD differ by a factor $\cos(\theta)$, where θ is the angle between the emission direction and the surface normal. Using the TSMVD when drawing initial velocities at the surface will — as molecules with higher v_z component (v_z being the component of the velocity in the direction of the surface normal) start to overtake molecules with a lower v_z component — establish an SMVD distribution inside the volume.

Whether a SMVDF is the proper VDF can of course be debated. *Skorov and Rickman* (1995) calculated the velocity distribution of molecules emerging from a cylindrical channel with a sublimating floor. They found that the emerging distribution function was Maxwellian if the channel was isothermal, but noted strong deviations when a temperature gradient was present. *Davidsson and Skorov* (2004) used a Monte Carlo approach to study molecular migration within a granular medium, as well as the distribution function of molecules exiting the medium and entering an empty half-space. They too found that the distribution function was semi-Maxwellian for isothermal media. However, when the temperature falls with depth (typical midday conditions) the outflow is less collimated, and when the temperature increases with depth (typical of late afternoon and night) the outflow is more collimated, compared to the semi-Maxwellian. *Liao et al.* (2016) investigated the consequences of changing the degree of collimation for the global coma properties. They found that increased initial collimation tended to lower the number density and translational temperature and increase the drift speed. Reducing the collimation would presumably have the opposite effect.

Therefore a fully self-consistent nucleus/coma model requires a thermophysical model that provides near-surface temperature gradients and outflow rates, as well as a detailed kinetic model of the molecular velocity distribution function of emerging gas, that can be fed to the kinetic coma model. We should add, though, that we currently lack critical information on the details of the pores, such as their physical dimensions.

Coma molecules that impact the comet surface either scatter or are adsorbed. Scattering occurs either through specular reflection or diffusively (in the latter case, thermalization on the nucleus surface may modify the molecular speeds). Adsorption gives rise to a (sub)monolayer of volatiles on top of the dust mantle that may have a short residence time. Because of the low surface density, their desorption rates should ideally be calculated with the first-order Polanyi-Wigner equation (e.g., *Suhasaria et al.,* 2017), using an activation energy that is suitable for vapor molecules attached to a silicate or organics surface. This is different from the zeroth-order sublimation that typically is considered for multilayer ice deposits. Scattered and thermally desorbed molecules form a separate population that should be added to obtain the complete transmission velocity distribution function. We have ample evidence that this population exists and can be adsorbed from the ambient coma (*Liao et al.,* 2018) or accumulate when the interior is warmer than the surface (*De Sanctis et al.,* 2015).

4.1.3. The Knudsen layer. The quasi-semi-Maxwellian velocity distribution emerging from a sublimating medium relaxes to a drifting Maxwell-Boltzmann velocity distribution over a finite distance due to molecular collisions. Within this "Knudsen layer" (e.g., *Cercignani,* 2000) the gas properties evolve according to the Boltzmann equation with a non-zero collision integral. At the upper boundary of the Knudsen layer (if it has a finite thickness), the collision integral goes to zero, so that the first three moments of the Boltzmann equation approach the NS and, furthermore, the EE, i.e., a hydrodynamic formulation becomes valid in the downstream flow.

The gas number density, translational temperature, and drift speed at the boundary (i.e., the top of the Knudsen layer) are connected to the nucleus surface temperature and near-nucleus number density through the "jump conditions." These were defined by *Anisimov* (1968), who assumed Mach number $M = 1$ at the upper boundary, while *Ytrehus* (1977) demonstrated that the problem does not form a closed set of equations so that the solution becomes a function of an assumed downstream Mach number. As described by *Crifo* (1987), the boundary conditions and the downstream hydrodynamic solutions therefore need to be brought in agreement through an iterative procedure. For further discussions about cometary Knudsen layers and application of jump conditions, see, e.g., *Davidsson* (2008) and *Davidsson et al.* (2021). If the Knudsen layer is thin, the jump conditions allow for the specification of inner coma boundary conditions in hydrodynamic coma models. Comparisons between the kinetic and hydrodynamic versions of coma models have been made by, e.g., *Crifo et al.* (2002a, 2003) and *Zakharov et al.* (2008).

4.1.4. Active area fraction. As described above, instead of a full thermophysical model, a simplified thermal balance is often employed to determine the boundary conditions for the dynamical models. This leads to the introduction of an

AAF to reduce the flux, usually 1–2 orders of magnitude too high, to realistic values. The AAF is simply a linear term multiplied with the gas production rate.

In recent years this AAF, although referred to differently by various groups, has been the main parameter used to fit the observed number/column densities (e.g., *Fougere et al.,* 2016b; *Marschall et al.,* 2016; *Zakharov et al.,* 2018a). Naturally, the question arises as to the physical interpretation of the found AAFs. Given that they are based on exposed pure ice surfaces, which we do not observe apart from very few isolated patches (e.g., *Sunshine et al.,* 2006; *Pommerol et al.,* 2015), the absolute values of the AAF should not be taken literally (i.e., having AAF = 0.05 does not mean that 95% of the surface is dust and 5% is literally exposed water ice). If that were true then an instrument such as Rosetta's VIRTIS would have seen 3-μm water ice absorption over vast swaths of the surface. This was not the case on 67P where VIRTIS saw the absorption only at a few specific locations (e.g., *Barucci et al.,* 2016; *Raponi et al.,* 2016). The relative values of AAF reveal actual differences in the response of the surface to solar illumination. The cause of these differences is not captured in the AAF. However, apart from isolated patches of water ice, the surface is covered by a dry layer. This layer quenches the flow from the sublimation front beneath it and thus provides a better stand-in for understanding the AAF.

Although the initial surface number density, temperature, and velocity are the common input parameters for gas kinetic models, we have seen that the simplified model leads to surfaces that are significantly cooler than realistic nucleus models with dust mantles and thus lead to slower initial molecular speed. The scaling through AAF ensures that the total flux is of the correct order of magnitude, but the temperatures and speed of the gas flows may not be. Observations of the gas number/column density are rather insensitive to this but measurements by Rosetta/MIRO indicate that deviations from the model temperatures and velocities are observed (*Marschall et al.,* 2019). In essence, the model production rates are to a large extent solid but the gas speeds and temperatures are not.

The AAF is nevertheless useful because it highlights real differences in how "active" different regions are. Presenting only local gas production rates would make it almost impossible to determine if two respective regions show distinct levels of activity because of different illumination conditions or because of, e.g., different ice contents or variations in the dust cover. Ideally, the AAF is independent of any diurnal and seasonal variations caused by the local illumination conditions.

The AAF is an important quantity to parameterize the asymmetry of surface activity but can be problematic if used to fit data far from a comet or set of comets that have a significant secondary contribution of gas from icy grains in the coma (extended source; see section 4.4.3 for more details). In the case of an extended gas source, the AAF will combine the effects of surface and coma sublimation terms and is no longer representative of the surface property.

4.2. Kinetic and Fluid Models

In most cases of practical interest, studying cometary comae involves the consideration of rarefied gas flows under strong non-equilibrium conditions.

Kinetic modeling of cometary comae is based on solving the Boltzmann equation (Fig. 2)

$$\frac{\partial f}{\partial t} + \mathbf{v}\frac{\partial f}{\partial x} + \dot{\mathbf{v}}\frac{\partial f}{\partial v} = I(\mathbf{v}) \qquad (2)$$

where $f = f(\vec{x}, \vec{v}, t)$ is a distribution function of atoms or molecules of the gas in phase space where $\vec{x}$, $\vec{v}$, t are the coordinate, velocity and time respectively. $I(\mathbf{v})$ is a collision integral describing the interaction of these particles.

Solving the Boltzmann equation is a challenging problem that is further complicated by the need to account for the complexity of the nucleus shape. Typically, kinetic models

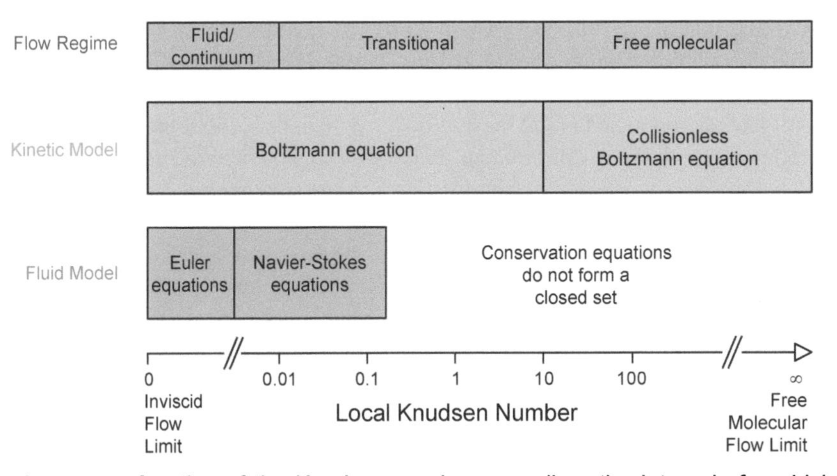

Fig. 2. Different flow regimes as a function of the Knudsen number as well as the intervals for which different models (fluid and kinetic) are valid. This figure has been adapted from its original in *Bird* (1994).

developed within the DSMC method are used for kinetic modeling of the dusty gas dynamics in cometary comae. This modeling approach allows the inclusion of important processes such as inelastic intermolecular collisions, photochemical reactions, and interaction with dust (e.g., *Combi,* 1996; *Tenishev et al.,* 2008, 2021; *Crifo et al.,* 2003, 2005; *Zakharov et al.,* 2008). Compared to fluid models, the main advantage of a kinetic model is that it is valid at any degree of non-equilibrium and/or rarefaction (i.e., conditions typical in the coma; Fig. 2). Another approach to model gas dynamics in cometary comae is based on solving the NS equation — a fluid model (Fig. 2) — as detailed by, e.g., *Crifo et al.* (2002a).

The Rosetta mission delivered a large volume of new data. This data includes, among others, almost continuous monitoring over two years of (1) the gas density and composition at the location of the spacecraft (ROSINA) and (2) spectral imaging of the gas coma (VIRTIS), coma line-of-sight (LOS) spectra (VIRTIS, MIRO), and imaging of the nucleus and dust coma by the Optical, Spectroscopic, and Infrared Remote Imaging System (OSIRIS) (*Keller et al.,* 2007). Kinetic models based on the DSMC method demonstrated robustness when they were applied for interpretation of these new data (e.g., *Fougere et al.,* 2016b; *Marschall et al.,* 2016; *Liao et al.,* 2016).

In the most general case, the flow in the coma covers regions with widely differing conditions — from fully collisionless to fluid. The degree of rarefaction is characterized by the Knudsen number, Kn

$$Kn = \lambda/L \tag{3}$$

where λ is the MFP of the molecules and L is a characteristic length. The choice of L is not immediately clear and depends on the problem under consideration. When describing the entire flow within the near nucleus environment by a single global Knudsen number, the equivalent radius of the nucleus is traditionally used for L as the characteristic scale of the flow. When characterizing the rarefaction of the flow locally, the scale length of the macroscopic gradient can be used as L (e.g., using the gas density, n_g, such that $L = n_g/|\nabla n_g|$). Depending on the local Kn three flow regimes can be roughly distinguished (illustrated in Fig. 2):

1. continuum/fluid, $Kn < 0.01$
2. transitional, $0.01 \leq Kn \leq 100$
3. free molecular, $Kn > 100$

The expansion of the flow leads to a decrease in collisions with radial distance, and therefore even if an equilibrium flow was established at the top of the Knudsen layer, the flow becomes non-equilibrium again at larger distances due to an insufficient collision rate to maintain the equilibrium distribution.

For a more detailed description of the numerical methods, we refer the reader to *Hirsch* (2007) and *Bird* (1994), as well as *Bird* (2013) for the DSMC method specifically. For a recent deeper description of the different models as well as a historical review of their development we refer to *Marschall et al.* (2020c).

4.3. Flow Regime Estimations

To select the appropriate approach (fluid or kinetic) we need to know which flow regime our cometary coma covers. To get a more quantitative understanding of the scales involved and which flow regimes can be expected, a simple order-of-magnitude estimation can be made using equation (1) and the fact that the MFP is

$$\lambda = \frac{1}{\sqrt{2}n\sigma} = \frac{4\pi r^2 v}{\sqrt{2}Q\sigma} \tag{4}$$

where σ is the collisional cross-section of the molecules. The MFP thus scales with the square of the distance to the center of the nucleus but only linearly with the gas speed and inversely proportional to the gas production rate.

Figure 3a shows the MFP as a function of the distance to the center of the nucleus. As a reminder, for equation (4) to be valid we assume an isotropic coma that expands at a constant speed. Here we have set the expansion speed, v, to 1000 m s^{-1}. In reality, v is not independent of the gas production rate, but 1000 m s^{-1} represents a reasonably realistic speed for the order-of-magnitude estimation we perform here. In general, the collisional cross-section, σ, is a function of temperature (or the relative velocity of colliding particles). It is constant only in the hard sphere model. For simplicity we assume $\sigma_{H_2O} = 10^{-19}$ m^2.

For very low water production rates (10^{26} molecules s^{-1} = 3 kg s^{-1}) the MFP is large even close to the nucleus. It quickly expands from 1 km at 1 km from the nucleus center to 10,000 km at 100 km from the nucleus center (Fig. 3a). Even for the highest water activity case (10^{29} molecules s^{-1} = 3000 kg s^{-1}) the MFP is on the order of 10 km at a distance of 100 km from the nucleus center.

As a reference, the nucleus of 67P had a water production rate of 10^{26} molecules s^{-1} = 3 kg s^{-1} at a heliocentric distance of 3 au inbound a few months after Rosetta arrived (*Marschall et al.,* 2020b; *Combi et al.,* 2020). 67P reached a peak production rate on the order of 1000 kg s^{-1} shortly after perihelion at a heliocentric distance of 1.25 au [estimates range between 500 and 1500 kg s^{-1}, depending on the instrument used and given that H_2O accounts for 80% of the total volatile mass loss (*Hansen et al.,* 2016; *Fougere et al.,* 2016a; *Marshall et al.,* 2017; *Kramer et al.,* 2017; *Combi et al.,* 2020; *Marschall et al.,* 2020b)]. During the 1P/Halley encounters *in situ* measurements by *Krankowsky et al.* (1986) showed a total gas production rate of 6.9 × 10^{29} molecules s^{-1} ≈ 20,000 kg s^{-1}.

Another useful concept is the distance beyond which we expect the final collision. We shall call this distance R_{fin} and can find it by

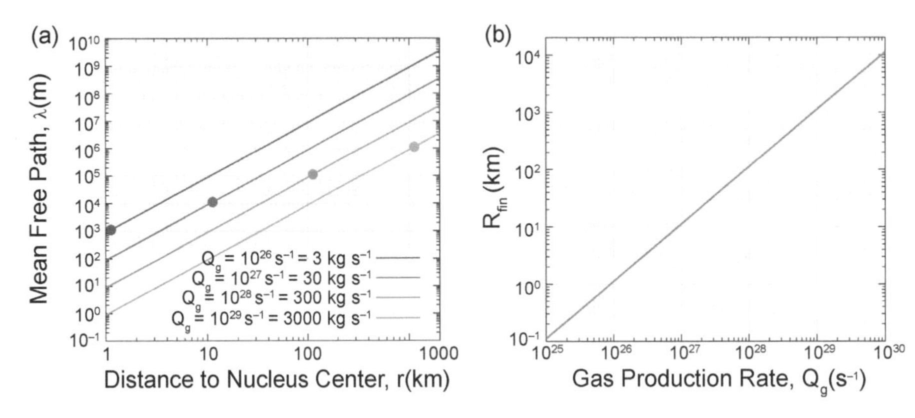

Fig. 3. (a) Mean free path, λ, of H_2O as a function of the distance to the center of the nucleus for four different gas production rates spanning four orders of magnitude according to equation (4). The coma is assumed to be isotropic and expanding at a constant speed. Here we assume v = 1000 m s^{-1} and $\sigma_{H_2O} = 10^{-19}$ m^2. The dots indicate the distances beyond which we expect the last collision. **(b)** Distance, R_{fin}, as a function of the gas production rate. R_{fin} is the distance beyond which we expect the last collision of a molecule with another. The respective expression is given in full in equation (6). All other assumptions are the same as in **(a)**.

$$1 \doteq \int_{R_{fin}}^{\infty} \frac{dr}{\lambda} = \frac{\sqrt{2}Q\sigma}{4\pi v R_{fin}} \tag{5}$$

and hence

$$R_{fin} = \frac{\sqrt{2}Q\sigma}{4\pi v} \tag{6}$$

Thus, for the isotropic coma expanding at constant speed the distance beyond which we expect only one last collision is proportional to the production rate and inversely proportional to the expansion speed of the gas. Of course, this is a big simplification. The assumption that — after its initial expansion — the gas speed is constant is a fairly good one for typical comets up to a distance of at least 10^4 km when considering H_2O (see, e.g., Fig. 2 in *Shou et al.,* 2016). The distance at which the gas seems to accelerate again is on the order of 10^4 km and is the result of a selection effect where dissociation and ionization preferentially removes the slower molecules from the phase space (*Tseng et al.,* 2007). Therefore, this is a fairly good order of magnitude estimate. Figure 3b shows R_{fin} as a function of the gas production rate. For very low water production rates (10^{26} molecules s^{-1} = 3 kg s^{-1}) R_{fin} is only 1 km. This means that for a nucleus larger than 1 km and with such a low water production rate the molecules will only experience one collision. The flow is essentially in the free molecular regime from the surface on. For the higher water-activity cases (e.g., 10^{29} molecules s^{-1} = 3000 kg s^{-1}) the last collision can only be expected beyond 1000 km.

Furthermore, we would like to know at which distance we transition from one flow regime to another (for example, from the fluid to the transitional). If we search for the distance, R_{Kn}, where the Knudsen number has a certain value and the characteristic length is the nucleus radius, $L = R_N$, we can combine equations (3) and (4) and find that

$$R_{Kn} = \sqrt{\frac{\sqrt{2}Q\sigma R_N Kn}{4\pi v}} \tag{7}$$

Two cases of this equation are particularly interesting. First, at which point is the MFP as large as the radius of the nucleus, R_N? It turns out that this is equivalent to finding R_{Kn} for Kn = 1. This means that when we model a coma environment that has the size $R_{Kn=1}$ then the global Knudsen number is unity and we are squarely in the transitional flow regime and therefore likely need to apply DSMC to capture the physics correctly. Figure 4a shows $R_{Kn=1}$ for different nucleus radii between 1 and 10 km and four gas production rates spanning 4 orders of magnitude. Any domain larger than the given $R_{Kn=1}$ will have a global Kn ≥ 1.

Second, we would like to find the approximate location of flow transitions. For example, we'd like to know when our system is between the fluid and transitional regime, i.e., the regime that can be safely studied using the more efficient fluid solvers (EE/NS) compared to the regime that will require a kinetic approach (DSMC). To be conservative, this transition occurs around Kn = 0.01. Thus, Fig. 4b shows $R_{Kn=0.01}$. Note that the shaded red area marks the interior of the nucleus up to the surface. For all production rates below Q = 10^{28} (~300 kg s^{-1}) the respective lines fall below the size of the nucleus. This means that the flow will be in the transitional regime above the surface for any nucleus larger than 1 km. Simulation domains that are larger than $R_{Kn=0.01}$ will have a Kn ≥ 0.01, i.e., the global regime shifts further toward free molecular flow. Only for much higher production rates, and smaller nuclei, will we find several meters to kilometers of fluid regions above the surface. When one

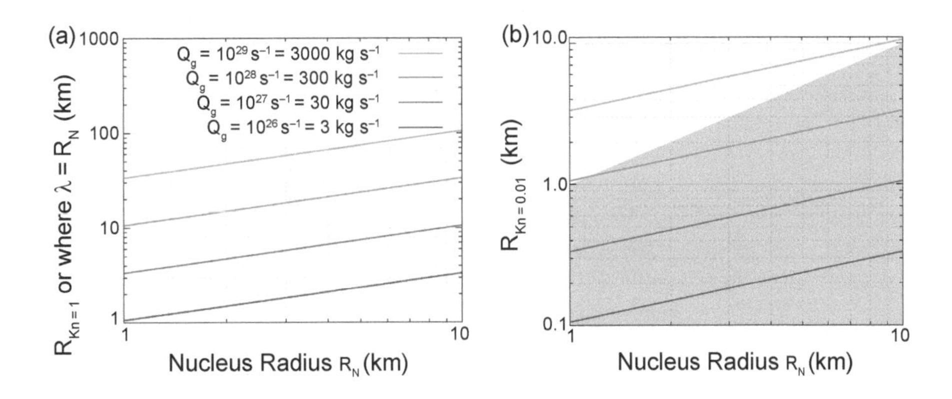

Fig. 4. (a) Distance at which the mean free path, λ, is equal to the radius of the nucleus, R_N, for four different production rates spanning four orders of magnitude and as a function of the R_N. The other assumptions are as in Fig. 3. This distance is at the location where Kn = 1. Therefore, the lines shown are given by equation (7) when Kn = 1. **(b)** Same as **(a)** but for Kn = 0.01 in equation (7). The value of Kn = 0.01 corresponds to the transition between the fluid and transitional regimes. The red shaded area denotes distances to the center of the nucleus, which lies below the current nucleus surface.

considers a larger Kn as the transition between fluid and transitional flow, e.g., 0.1 instead of 0.01, then the respective R_{Kn} distances increase by a factor of $\sqrt{10} \sim 3$. Therefore, for all production rates below Q = 10^{27} s^{-1} (~30 kg s^{-1}) the transition from fluid to transitional flow happens "below" the nucleus surface.

Note also that in equation (7) all terms (R_N, Q, Kn, and v) enter with the same weight. For the same Kn, any increase of the nucleus radius is compensated by the respective decrease in gas production rate (see Fig. 4b).

4.4. Dusty-Gas Flow

While we have focused here on the gas flow and, to a large extent, neglected the presence of dust particles in the flow, we would be remiss not to discuss some fundamental aspects of dusty-gas flows. For the most part, dust particles are treated as passive objects that act as test particles within the gas flow. Of course, this is not true in general. For a detailed discussion of the dust dynamics, we refer the reader to the chapter by Agarwal et al. in this volume.

In the most ideal and straightforward case, we are dealing with a coma containing dry dust and a dust-to-gas mass flux ratio much smaller than 1 (i.e., low dust content). In this case — where the backcoupling from the dust to the gas can be neglected — the gas flow can be treated independently from the dust flow. Separate models for the gas and dust flows can be run sequentially. This not only allows for more flexibility to explore the parameter space but is also much less computationally expensive because of the different timescales of gas molecule and dust particle motion. It is thus a convenient scheme to employ.

Here we will highlight how the presence of the dust can alter the gas flow, thus deviating from this ideal case and therefore requiring a coupled dust-gas coma model. This is on top of shadowing the nucleus from solar insolation in the case of a very dusty coma.

4.4.1. Momentum transfer. If the dusty-gas flow has a high dust-to-gas mass ratio the two flows cannot be treated independently. When the dust particles are accelerated by the gas flow, momentum is transferred from the gas to the dust. As the dust-to-gas ratio increases so do the kinetic energy and total momentum of the dust flow. When momentum transferred from gas to dust becomes a significant fraction of the gas flow momentum, the presence of dust becomes noticeable for the gas flow as well and the gas flow will be slowed down. There is an important caveat though; if the dust is hotter than the gas, it will be able to accelerate the molecules with which it interacts (see section 4.4.2).

Whether or not the dust flow impedes the expansion of the gas coma depends not only on the dust-to-gas ratio but also on the dust size distribution. A coma with large slow-moving particles will — for a given total dust mass — impact the gas flow significantly less than the case where the same mass is distributed in small particles that accelerate to a significant fraction of the gas speed. As far as we are aware, no study has quantified the combined effect of the dust size distribution and dust-to-gas ratio, but the reader shall be aware of this potential pitfall.

4.4.2. Energy transfer. A further effect of the presence of dust particles in the gas flow occurs from the fact that dust particles can be significantly hotter than the surrounding gas. The gas cools as it expands into space while dust particles heat up when exposed to the Sun (*Lien,* 1990). This makes the temperature gap between gas and dust larger with increasing cometocentric distances. Additionally, sufficient molecule-dust collisions need to occur, which typically happen close to the surface. Therefore, collisions of the gas molecules with the dust particles will increase the gas temperature. The presence of small (micrometer and submicrometer) particles can increase the gas temperature by as much as a factor of 3 (*Kitamura,* 1987; *Markelov et al.,* 2006). A mass-loaded gas flow will thus initially be slowed down and then subsequently heated in the inner part of the coma (*Crifo et al.,* 2002a).

Radiative heating of the gas by the hotter dust can become a dominant effect.

As in the case of momentum transfer (section 4.4.1) the effect of energy transfer becomes of particular concern when the coma is dominated by very small particles, which are more easily heated to very high temperatures and also constitute a much larger cross section to the gas for the same dust mass. Superheated dust particles have been observed at various comets [e.g., at 1P, Hale-Bopp, and 67P (*Gehrz and Ney,* 1992; *Williams et al.,* 1997; *Bockelée-Morvan et al.,* 2019)].

Although both momentum transfer and heating from dust can significantly alter the gas flow, we currently do not know the extent to which such particles have altered an observed gas flow. To date, no observational constraints are available to evaluate the effect of momentum and heat transfer within the gas flow due to mass loading.

4.4.3. Mass transfer. Another backcoupling from the dust to the gas comes from "wet" dust. When dust particles consist of ice or contain a large fraction of ice, they will begin to sublimate and contribute to the gas coma in the form of an extended/distributed gas source. The sublimation of icy or ice-rich dust particles directly alters the dynamics of the dust particles (*Kelley et al.,* 2013; *Agarwal et al.,* 2016; *Güttler et al.,* 2017).

The contribution from such icy particles can be significant and it might explain hyperactive comets (*Sunshine and Feaga,* 2021). But a recent non-detection of icy grains in the coma of hyperactive Comet 46P/Wirtanen by *Protopapa et al.* (2021) challenges this interpretation of hyperactivity.

Hyperactive Comet 103P/Hartley 2 — visited by the extended Deep Impact mission EPOXI (*A'Hearn et al.,* 2011) — shows an abundance of icy particles [bottom panel in Fig. 5 (*Protopapa et al.,* 2014)]. The icy grains, driven by CO_2 activity (bottom two panels in Fig. 5), sublimate in the coma while larger chunks can redeposit in the neck region where they could cause the observed water "jet" (second panel from the top in Fig. 5).

Sublimating icy grains lead to a slower expansion but warmer/hotter gas compared to a comet nucleus source only. This has been observed and modeled for Comet 73P/Schwassmann-Wachmann 3 (*Fougere et al.,* 2012) and Hartley 2 (*Fougere et al.,* 2013).

4.4.4. Modeling dusty gas flows. The first multidimensional models (axially symmetric and three-dimensional) of a dusty gas coma with the physically consistent description of a flow as a fluid were presented in *Kitamura* (1986, 1987, 1990), and *Korosmezey and Gombosi* (1990). These models were based on the numerical solution of the coupled hydrodynamic equations (representing mass, momentum, and energy conservation). The dust was treated as one of the components of the fluid consisting of single-sized spherical grains (<1 μm).

However, aspherical grains affect the maximum liftable sizes and velocity distribution. Unlike spherical particles, which experience only drag force, aspherical particles have also transversal (i.e., lift) force and torques (see *Ivanovski et al.,* 2017). As a consequence they may start to rotate, thus altering the trajectories with respect to their spherical counterparts. Aspherical particles may also have rotational energy and therefore act as an additional sink of energy from the gas. The chapter by Agarwal et al. in this volume goes into more detail about dust dynamics.

Modern numerical models of the dusty gas flow (e.g., *Tenishev et al.,* 2011) were constructed in the spirit of DSMC for the gas. The dust phase in the coma is represented by a large but finite number of model particles that represent real dust grains. The motion of a spherical, isothermal particle in the inertial frame is assumed to be governed by the gas drag and nucleus gravity force

$$\frac{4\pi}{3}\rho_d a_d^{\ 3}\frac{d\mathbf{v}_d}{dt} = \frac{1}{2}a_d^{\ 2}C_D\rho\left|\mathbf{u}-\mathbf{v}_d\right|\left(\mathbf{u}-\mathbf{v}_d\right)-\mathbf{F}_g \quad (8)$$

Here, v_d is the velocity of a spherical dust particle with radius a_d and bulk density ρ_d; $C_D = C_D(\mathbf{u}, \mathbf{v}_d, T_g, T_d)$ is the drag coefficient, which depends on the gas and dust flow parameters; F_g is the gravity force; and u and ρ are the gas velocity and mass density (gas density times molecular mass), respectively.

The drag coefficient is a function of relative velocity of the gas and dust and of their temperatures. Typically, the MFP in the coma is much larger than the size of the dust grains (meters vs. millimeters to micrometers), and therefore C_D can be taken for a particle in a free molecular flow (*Bird,* 1994). Because the drag coefficient asymptotically approaches 2 for most dust particle sizes, $C_D = 2$ is often assumed rather than the more complicated and complete description. The choice of $C_D = 2$ implies, though, that the acceleration of particles will in general be underestimated (for small particles it is strongly underestimated). See the chapter by Agarwal et al. in this volume for more details on this topic.

The dynamics of the dust grains are significantly affected by the geometry and mass distribution of the nucleus. The complexity of the nucleus surface geometry, in turn, complicates the calculation of the gravity force acting upon a dust particle. Realistic gravity fields must therefore be incorporated into the dynamical models (*Tenishev et al.,* 2016; *Marschall et al.,* 2016).

In the vicinity of the nucleus (distances of a few tens of nucleus radii), additional forces such as solar radiation pressure and solar gravity can be neglected and are therefore absent in equation (8). Additionally, equation (8) assumes spherical particles and thus does not include terms related to the rotation of particles (*Ivanovski et al.,* 2017).

The distinctive feature of a state-of-the-art model, like the one presented in *Tenishev et al.* (2011), is the self-consistent kinetic treatment of both the gas and the dust phases of the coma. These numerical experiments with that model suggested that the effect of dust on the gas flow in the coma is minimal. Thus the effect of the dust phase on the gas can in most cases be neglected. However, we must add that this

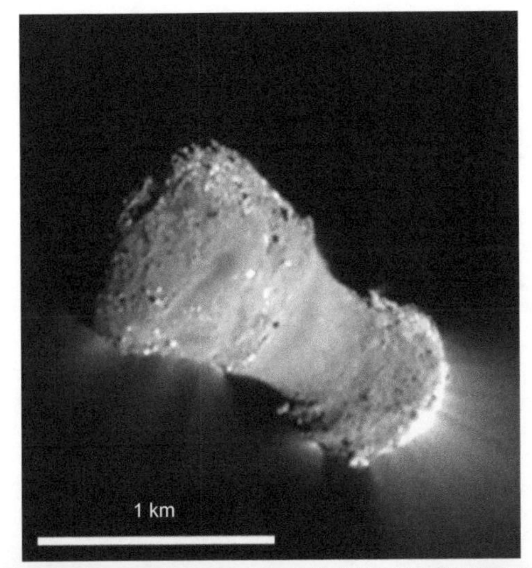

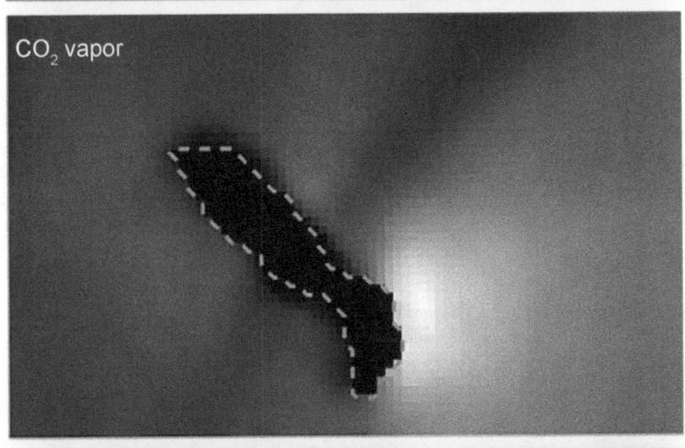

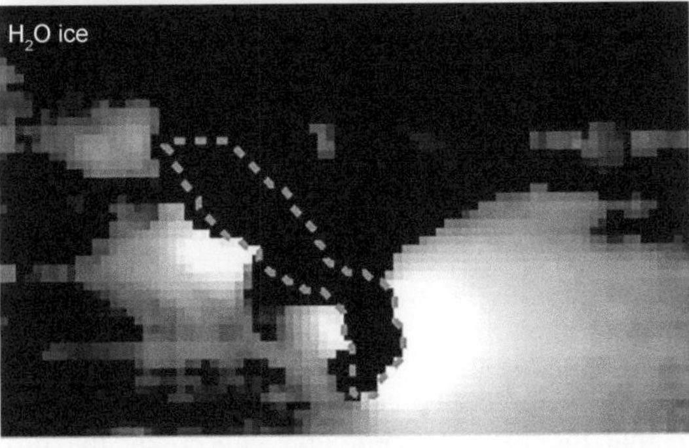

Fig. 5. The top panel shows the nucleus of 103P/Hartley 2 taken with the MRI instrument for context. The two panels labeled "H_2O/CO_2 vapor" are maps of the total flux in the relevant emission bands. The bottom panel labeled "H_2O ice" shows a map of the depth of the water ice absorption feature at 3 μm. The three lower panels were taken 7 minutes after the closest approach. The figure was adapted from *A'Hearn et al.* (2011) for the top panel and from *Protopapa et al.* (2014) for the lower three panels.

statement should be checked for a wider range of parameters in the future.

Importantly, within a region of roughly 10–100 R_N the assumption that the gas flow is in steady state can generally be used. In contrast, this is not true for the dust — at least not for all dust sizes. For a more precises estimate of the steady-state distance we need to take into account the rotation period of the nucleus, P_{rot}, which modulates the surface emission distribution (i.e., the boundary condition of the flow). There is no absolute threshold but let us suppose the nucleus rotation can be neglected, e.g., when it rotates by less than 5°. In this case the flow is steady on the length scale of $L_S = vP_{rot}\frac{5°}{360°}$. For a nucleus with a rotation period of 12 hours and a water expansion speed of v_{gas} = 1000 m s^{-1} we get L_S = 600 km. This is a much shorter scale than, e.g., the scale for photodissociation. For slow dust particles moving at, e.g., $v_{dust} \sim 10$ m s^{-1} we get L_S = 6 km. For typical nuclei sizes this is within the typical dust acceleration region of 10 R_N.

5. EMERGENT COMA STRUCTURES

Often gas structures are more difficult to observe than the dust. Because the dust is coupled to the gas (see equation (8)) dust particles — to some extent — trace the gas flow. There are meaningful differences between the two flows that we will highlight. Nevertheless, dust observations are often used to also inform our understanding of the gas flows.

Since the first up-close observations of 1P/Halley in 1986, collimated gas and/or dust features have been detected in the coma of all comets visited by spacecraft (Fig. 6). As described in section 2, we will refer to the observed structures in the inner coma as "filaments" or "collimated features" rather than "jets" unless the latter is clearly warranted.

The simplest explanation for the observed inner coma structures is the following. One might imagine that a surface is composed of active and inactive areas. Therefore the active areas produce high-density areas in the coma above, which are contrasted by low-density areas over inactive surface patches. An observed difference in coma density should thus be interpreted as the result of heterogeneity of the nucleus. Although a compelling story, it has been shown to be wrong — at least in general — for quite some time (e.g., *Crifo et al.*, 2004).

There are at least two mechanisms that can produce collimated features in the absence of a heterogeneous nucleus. One important thing to note is that the gas structures are generally broader than the intricate dust structures (Fig. 6 and 7) (see also *Combi et al.*, 2012). Dust particles [unless they are very small (submicrometer) or in a very dense gas flow] are only weakly coupled to the gas flow. Their primary flow direction is governed by the gas environment very close to the surface (see equation (8)) where the gas flow is dominantly perpendicular to the surface normal. With increasing distance from the nucleus and the associated decrease in gas density,

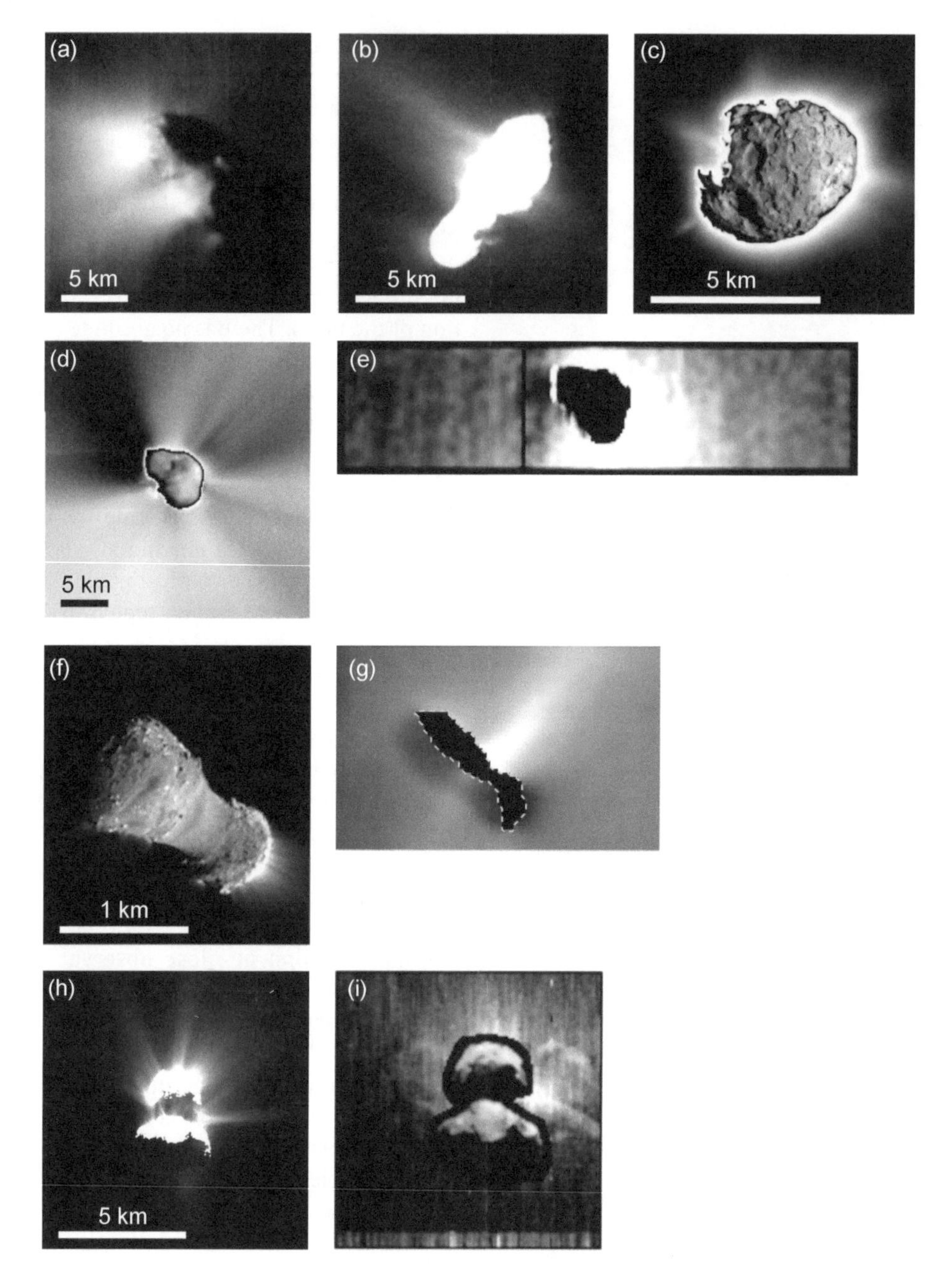

Fig. 6. In chronological order of comets visited: The inner dust structures of comets **(a)** 1P/Halley (*Keller et al.,* 1987), **(b)** 19P/Borrelly (*Boice et al.,* 2002), **(c)** 81P/Wild 2 (*Tsou et al.,* 2004), **(d)** 9P/Tempel 1 (*Feaga et al.,* 2007), **(f)** 103P/Hartley 2 (*A'Hearn et al.,* 2011), and **(h)** 67P/Churyumov-Gerasimenko (*Tubiana et al.,* 2019). Maps of the water column density are shown for the three most recently visited comets by the Deep Impact/EPOXI and Rosetta mission: **(e)** 9P (*Feaga et al.,* 2007), **(g)** 103P (*Protopapa et al.,* 2014), and **(i)** 67P (*Fink et al.,* 2016), respectively.

the dust particles decouple from the gas flow rather quickly. This enables the dust to better preserve the near-surface conditions. For non-uniform outgassing the dust may also return and fall back to the nucleus surface (*Thomas et al.,* 2015a; *Davidsson et al.,* 2021). See the chapter by Agarwal et al. in this volume for more detail on the dust dynamics.

In the following we will discuss two main causes leading to "collimated features" in the inner coma that have been identified: (1) a heterogeneous nucleus and (2) nonsphericity of the homogeneous nucleus and local topography. In addition to these two causes the reader should also be aware that the viewing geometry can create the illusion of structures (e.g., *Shi et al.,* 2018; *Tenishev et al.,* 2016; Agarwal et al. in this volume).

The two end members of the models we will discuss are (1) spherical nuclei with active areas and (2) homogeneous nuclei with complex shapes. The former case with a spherical nucleus is agnostic as to what causes the differences between active and inactive areas. These differences can be caused by a heterogeneous nucleus or evolutionary processes that alter the (near-)surface properties but leavethe bulk nucleus properties unaltered.

5.1. Heterogeneous Nucleus Surfaces

A significant amount of work was performed to explore active spots on or inhomogeneous outgassing from spherical nuclei (e.g., *Komle and Ip,* 1987; *Kitamura,* 1990; Kn*ollenberg,* 2017; *Crifo et al.,* 1995; *Crifo and Rodionov,* 1997). In these models, a spherical nucleus would be divided into active and inactive surface elements. Although early models (e.g., *Kitamura,* 1986, 1987, 1990; *Korosmezey and Gombosi,* 1990) did not use an underlying thermophysical model to determine the gas production rate at the surface, their conclusions still hold. These studies showed the formation of the shock structures due to interactions of several "jets." Even a single pure gas jet expanding into a co-current flow produces a shock structure.

Thus, as supported by our intuition, an active area will produce a gas structure in the inner coma and the simple story we discussed above is one way of explaining the inhomogeneous features. Areas of enhanced activity have indeed been identified. Regional variations of the surface activity have been invoked to explain the gas coma features at comets 9P (e.g., *Finklenburg et al.,* 2014), 103P (e.g., *Fougere et al.,* 2013), and 67P (e.g., *Fougere et al.,* 2016b; *Marschall et al.,* 2016). For example, *Finklenburg et al.* (2014) were unable to reproduce the observed data from Comet 9P/Tempel 1 with a homogeneous nucleus model.

Heterogeneous surface activity does not imply an inhomogeneous nucleus. On the contrary, there is ample evidence for regional heterogeneity resulting from, e.g., dust redeposition. Both 103P (*A'Hearn et al.,* 2011) and 67P (*Thomas et al.,* 2015a) showed clear evidence of dust airfall, i.e., dust that was ejected into the coma but did not reach escape speed and thus fell back to the surface. In the case of 103P, the dust redeposited in the neck region while 67P saw large deposits on the entire northern hemisphere in addition to the neck region called Hapi (*Thomas et al.,* 2015b). In both cases, an increased water production from the neck region, where dust was deposited, points to the fact that the deposited chunks contained a significant amount of water ice but were otherwise largely depleted in the hypervolatiles (e.g., *Davidsson et al.,* 2021). The enhanced activity of Hapi and reduced activity of other dusty deposits is also reflected in the AAF of coma models (e.g., *Marschall et al.,* 2016; *Fougere et al.,* 2016b; *Marschall et al.,* 2017). That the other northern dusty deposits are not active on the inbound leg comes from their different thermal history compared to Hapi (*Keller et al.,* 2017). Hapi only reenters the Sun around aphelion and therefore retains its water ice content until the next inbound leg. The other northern dust deposits are already illuminated on the outbound leg where they lose their water ice and are subsequently largely inactive on the following inbound leg (*Keller et al.,* 2017). It cannot be ruled out though that the distinct heterogeneity between the two lobes of 103P/Hartley 2 is caused by a heterogeneous nucleus; at least there are no telltale signs for that from the gas phase.

The most probable scenario is thus the emission of water-rich dust chunks from one part of the comet (the small lobe in the case of 103P and the southern hemisphere in the case of 67P) and subsequent redeposition on other areas of the surface. If we begin with a homogeneous nucleus, then this redistribution of material across the surface results in a natural alteration of the (near-)surface properties. These in turn result in regional differences in the outgassing.

5.2. Irregular Shape and Topography

As soon as multi-dimensional models were available, the effect of non-spherical shapes was examined. In contrast to the spherical models — where surface heterogeneities were studied — a homogeneous nucleus was assumed. For example, a "bean"-shaped nucleus was used to model 1P/Halley (*Crifo and Rodionov,* 1997) and 46P/Wirtanen (*Fulle et al.,* 1999; *Crifo and Rodionov,* 1999). A homogeneously outgassing nucleus with a "Haser"-type gas model of 67P was used in *Kramer and Noack* (2016) to fit dust structures.

These early models demonstrated very clearly that the focusing of the gas and dust flow from an irregular shape of the nucleus results in a strongly inhomogeneous inner coma and "collimated structures." The effect of a realistic shape for 1P/Halley was studied in *Crifo et al.* (2002b). In this case, the study included homogeneous and inhomogeneous (set of spots) emissions. This study showed that a more detailed description of the shape leads to a considerably more complicated flow structure near the surface. Most notably, the geometrical effects of the surface can be stronger than the effect of a surface inhomogeneity.

We can further illustrate this effect using a toy model of Comet 67P (Fig. 7). In this model, the emission of gas and dust on the entire surface is uniform (i.e., the production rate per unit area is constant across the surface). Gas "jets" are visible flowing upward and downward from the neck (Fig. 7b) due to the focus of the flow between the two lobes. The fine structures in the inner dust coma (Fig. 7c) are even more striking. None of the many dust filaments or gas jets has a "source" in any meaningful sense of the word. The observed structures are on the scale of the nucleus size and therefore only accessible to spacecraft.

This implies that absent any modeling of the gas and dust flow, we cannot and should not draw any conclusions as to the "origin" or "source" of any structure observed in the inner coma, as tempting as it may seem. Unfortunately, this also suggests that inverting or tracing back of features from the coma to the surface is bound to be futile because it makes the implicit assumption that sources exist in the first place. Only a forward-modeling approach can properly evaluate the different scenarios — heterogeneous surface vs. topographic focusing — and thus, e.g., exclude the possibility that features are produced by the irregular shape of the nucleus.

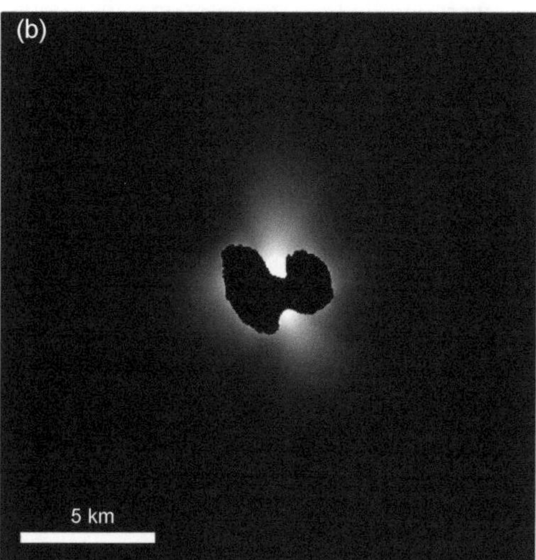

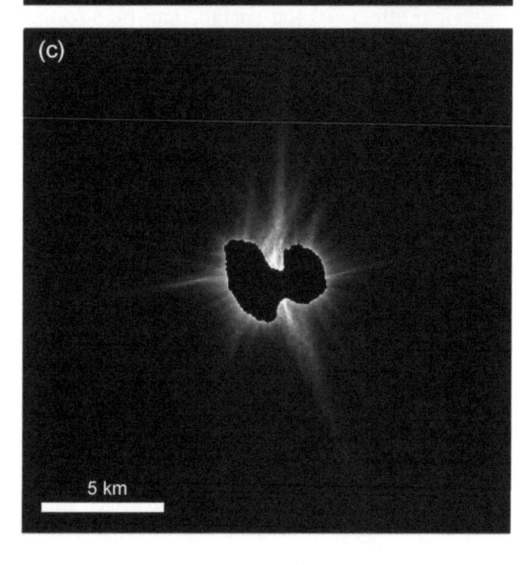

Fig. 7. (a) Shape model (*Preusker et al.*, 2017; *Jorda et al.*, 2016) and orientation of Comet 67P/Chryumov-Gerasimenko depicted in **(b)** and **(c)**. **(b)** Water column density of 67P if it were uniformly outgassing, i.e., equal gas production rate per unit surface over the entire surface. **(c)** Dust brightness of particles with a diameter of 4 μm also injected uniformly in the flow of **(b)**. From *Marschall* (2017).

5.3. Interpretation of Spacecraft Data

The previous two sections have left us in a sort of limbo. We have identified two end-member scenarios, both of which are capable of explaining collimated inner coma structures. But which scenario dominates in comets? At the time of the *Comets II* book (*Festou et al.,* 2004), insufficient data was available to clearly answer this question and thus *Crifo et al.* (2004) left it open.

In the following paragraphs, we will review what has been found for the four comets visited by spacecraft for which such an analysis has been done. We will discuss the results in chronological order, starting with ESA's Giotto, Japan's Sakigake, and the Russian Vega flybys at 1P/Halley in 1986, followed by NASA's Deep Impact/EPOXI encounters of 9P/Tempel 1 and 103P/Hartley 2 in 2010 and 2014 respectively, and finally the escort of 67P/Churyumov-Gerasimenko by ESA's Rosetta mission from 2014 to 2016.

For details about the composition of comets we refer the reader to the chapter by Biver et al. in this volume.

5.3.1. 1P/Halley. *Knollenberg et al.* (1996) studied the overall Halley dust coma appearance during the 1986 Giotto flyby (Fig. 6a). They found a gas and dust distribution with two circular active areas that result in the two observed main jet-like features roughly directed toward the Sun. They argue that data is well explained by three "jets" superimposed on a weak background. As mentioned above, this is not strictly self-consistent because the "jets" would interact with each other and could form interaction zones or even shock fronts depending on the level of activity. Furthermore, this model relies on axis-symmetric solutions, thus prohibiting the exploration of shape effects.

Crifo et al. (2002b) improved on previous work by using the nucleus shape model derived from the Vega 1 probe (*Szegö et al.,* 1995). They pointed out that the observed distribution can be explained primarily due to the effects of the shape. The orientation of the nucleus was not known well enough to fully constrain the model, thus introducing some ambiguity. Although the direction of some features was not completely matched, it did show, for the first time, that the nucleus shape can be sufficient to explain the coma morphology without the need for any ad hoc active spots on the surface.

This difference in interpretation highlights the difficulty in determining which of the two end members is the driving mechanism. We will see, though, that the data from 9P, 103P, and 67P suggest that comets do not fall in one of the two extreme cases. Rather, they appear to be driven by both regional differences in the surface activity and significant topographic modifications of the flow fields.

Another notable observation at 1P/Halley is the rapid transition to free radial outflow, within roughly 100 km (~20 nucleus radii), suggesting no notable extended source of the dominant water molecule or another altering process in the immediate vicinity of the comet (*Thomas et al.,* 1988). There is a substantial discussion on extended/distributed sources at Comet Halley, especially for CO (*Eberhardt et al.,* 1986); however, the measured distribution can also be explained by a change of the production rate (*Rubin et al.,* 2009) given that the flyby covered a large range of cometocentric distances and phase angles. A good review can be found by *Cottin and Fray* (2008). Distributed sources in minor species may hence not be at odds with H_2O coming mostly from the nucleus.

5.3.2. 9P/Tempel 1. To date, Comet 9P/Tempel 1 is by far the most spherical comet observed up close. It thus provides a more favorable opportunity to disentangle shape from surface composition effects. *Finklenburg et al.* (2014) found that a homogeneous surface composition was not sufficient to explain the distribution of H_2O vapor in the inner coma. It overpredicted the amount of vapor in the coma. Instead, they required active areas that were broadly in line with *Farnham et al.* (2013) and *Kossacki and Szutowicz* (2008) but need additional nightside activity at the northern pole. Although no direct link with morphology is made, the active areas identified by *Farnham et al.* (2013) correspond to steep slopes and the edges of smooth areas. *Kossacki and Szutowicz* (2008) argue for a varying dust mantle thickness. These two interpretations are not mutually exclusive. Steeper slopes might simply have thinner dust covers because dust cannot accumulate on them. In this sense, the conclusions of the three above-mentioned studies are in agreement.

Gersch et al. (2018) used an asymmetric model to improved the production rates of H_2O and CO_2. They were unable to fit the spectra assuming the coma was optically thin. This indicates that the coma needs to be treated as optically thick. Using the optically thick assumption, they found improved production rates that were almost 50% larger than those derived under the assumption of optically thin conditions.

5.3.3. 103P/Hartley 2. Comet 103P/Hartley 2 is a particularly interesting case because of its hyperactivity. The measurements obtained by the EPOXI mission (*A'Hearn et al.,* 2011) indicate that large chunks, rich in water ice, are ejected and then redeposited in the topographically low region of the neck (*Kelley et al.,* 2013). Such large chunks can retain much of their water ice (e.g., *Davidsson et al.,* 2021) and thus serve as the source of the water vapor above the neck. This interpretation is in line with coma models, which have found good agreement with the data assuming a regionally changing surface composition. Additionally, *Protopapa et al.* (2014) observed micrometer-sized icy grains ejected from the small lobe. There is a strong correlation between the water ice particles, dust particles, and the CO_2 spatial distribution. This suggests that CO_2 drives both the activity of dust particles and icy grains, which subsequently sublimate in the coma (*Protopapa et al.,* 2014).

Fougere et al. (2013) found that emission driven solely by the variation in the solar incidence angle did not fit the data. Rather, local enhancements of roughly 1 order of magnitude over the background gas emission were needed for both H_2O and CO_2. To fit the data, the emission of H_2O from the neck needed to be increased. For CO_2 the same was true for the small lobe. Furthermore, from the areas where CO_2 is emitted, the sublimation temperature for CO_2 was to be assumed as the surface temperature. Most of the icy grains get pushed to the nightside by radiation pressure and lateral gas expansion, which further leads to a drop of the gas density above active spots faster than $1/r^2$. Furthermore, nucleus gravity supports this process by pulling particles back from the subsolar direction of emission. Most of the water contribution to the extended source occurs toward the nightside. This is consistent with observations by *Knight and Schleicher* (2013) who reported an enhancement in OH in the anti-sunward direction. Similar observations were made by *Combi et al.* (2011) using Lyα emission of hydrogen by the Solar Wind ANisotropy (SWAN) instrument on the Solar and Heliospheric Observatory (SOHO), by *Bonev et al.* (2013) using long-slit spectra of H_2O emission acquired with the Near-Infrared Spectrometer (NIR-SPEC) on Keck II, and by *Meech et al.* (2011).

5.3.4. 67P/Churyumov-Gerasimenko. Because of the extended coverage provided by the Rosetta mission we have a very detailed picture of Comet 67P/Churyumov-Gerasimenko. The comet revealed a strong seasonal cycle caused by the obliquity of 56° (*Mottola et al.,* 2014). For most of the orbit, the Sun illuminates the northern hemisphere, resulting in cold northern summer. But, most of the activity occurs during southern summer when the comet is close to perihelion (southern solstice occurs only 14 days after perihelion).

Overall, the water production rate follows closely the movement of the subsolar latitude (*Fougere et al.,* 2016a; *Combi et al.,* 2020). Furthermore, the diurnal water activity appears to also follow the subsolar point with little thermal lag. Dust features, driven by that activity, are sustained for tens of minutes to 1 hour after local sunset (*Shi et al.,* 2016). This also suggests that water ice, although not directly observed at the surface [except for a few exposed patches (*Pommerol et al.,* 2015)], resides close to the surface, likely within a few millimeters of it (*Marboeuf and Schmitt,* 2014; *Herny et al.,* 2021; *Davidsson et al.,* 2022a).

CO_2 and CO, in contrast to water, have much shallower diurnal cycles and more uniform emission distributions (*Hässig et al.,* 2015; *Combi et al.,* 2020). This behavior is consistent with an expected longer thermal lag corresponding to deeper sublimation fronts (*Herny et al.,* 2021; *Davidsson,* 2021).

The water surface distribution has the strongest regional variation. During northern summer there is a prominent dust feature originating from the northern neck region (Hapi). Studies by *Fougere et al.* (2016b) and *Marschall et al.* (2016) both find a regional enhancement of water from that area (Figs. 8a,b). *Zakharov et al.* (2018a), on the other hand, finds the enhancement at the top of the neck valley rather than deep in it (Fig. 8c). But it is unclear if such a source — located outside of Hapi — would be consistent with the dust features or the VIRTIS gas maps (*Migliorini et al.,* 2016). This illustrates that the derivation of emission maps solely from *in situ* gas density data will retain a degeneracy of the solution above the previously mentioned theoretical resolution limit. Any reconstruction method suffers from limitations, be it the iterative approach by *Zakharov et al.* (2018a), the spherical harmonics one by *Fougere et al.* (2016b) and *Combi et al.* (2020), or the use of morphological regions as "basis" vectors by *Marschall et al.* (2016). Therefore, even with state-of-the-art forward models a multi-instrument approach with complementary data [e.g., *in situ* gas density from an instrument like Rosetta's ROSINA in combination with column-density maps from an instrument like Rosetta's VIRTIS or Deep Impact's High Resolution Instrument (HRI)] has to be considered to break this ambiguity (e.g., *Fougere et al.,* 2016b; *Marschall et al.,* 2019). Figure 8 also illustrates the correlation between activity and morphology. In particular, *Fougere et al.* (2016b) and *Marschall et al.* (2019) place the maximum of activity in the Hapi region, while the other regions that are dominated by dust deposits (such as Ash, Seth, and Ma'at) are at least 1 order of magnitude less active.

Vincent et al. (2016b) hypothesized that the water activity and by extension the dust features originate primarily from cliffs. This seems to be confirmed by coma models (*Marschall et al.,* 2017). Interestingly, the dusty deposits outside the Hapi region do not seem to contribute to water activity (*Marschall et al.,* 2017). This implies that the airfall material in those regions mostly loses its volatile content after deposition on the outbound leg of the comet's orbit after perihelion. The deposits in the Hapi region, on the other hand, have preserved their water ice because they remain continuously shadowed. These differences in activity are thus not due to inherent differences in the composition of the nucleus but stem from material transport across the surface combined with a different thermal history that alters the near-surface composition of the deposits.

Similar to the pattern observed at 103P, activity shapes the surface composition and thus in turn the observed activity. There are regions on comets that do not simply erode continuously and thus are not purely driven by the nucleus composition but rather by the feedback of redeposited material.

Most of the dust features overlaying this regional variation of activity appear purely driven by local topography (e.g., *Marschall et al.,* 2016, 2017) or the viewing geometry (e.g., *Shi et al.,* 2018).

5.3.5. What causes the observed coma structures? We thus return to the question of what causes the observed coma structures. There is clear evidence from all visited comets that there are regional differences in the strength of activity. However, these regional differences are consistent with evolutionary processes and material transport across the surface.

Large dust particles or chunks that do not reach escape speed will redeposit as airfall on the surface filling regions of gravitational lows. The angle of repose seems to be at roughly 30° (*Vincent et al.,* 2016b; *Marschall et al.,* 2017), corresponding to granular material. This leaves cliffs free of such insulating airfall and are thus active (e.g., *Farnham et al.,* 2013; *Vincent et al.,* 2016b; *Marschall et al.,* 2017).

The local topography and overall shape of the comet significantly influence the flow (e.g., *Crifo et al.,* 2002b; *Shi et al.,* 2018; *Marschall et al.,* 2016). The effect of topography is even more pronounced in the dust, which does not seem to require any additional small-scale (i.e., below the resolution limit of the gas) localized sources. A series of works (e.g., *Crifo et al.,* 2003; *Zakharov et al.,* 2008, 2009) even showed that flow structures generated due to surface topography and inhomogeneity can be identical, making it impossible to derive the reason for the structure formation just from how they appear. Therefore, it is crucial that additional information, like surface morphology and composition, is taken into account when assessing the cause of the structures in the coma. Features arising from a "flat" topography would indicate inhomogeneous sources. In contrast, a source region that is rather uniform in morphology (or, e.g., spectral properties) but has significant topography points to the topography as the source of those features.

Currently, most results point to the fact that no heterogeneous nucleus is needed to explain the heterogenous coma. Furthermore, apart from outbursts, there is no evidence for jets in the traditional sense of having a confined source area. Rather, there appears to be smooth activity originating from large regions. The absence of any observable shock features also points in this direction.

We should remind the reader of the fact that determining the surface-emission distribution from *in situ* gas measurements that are tens to hundreds of nuclei radii away from the surface has a physical resolution limit of several hun-

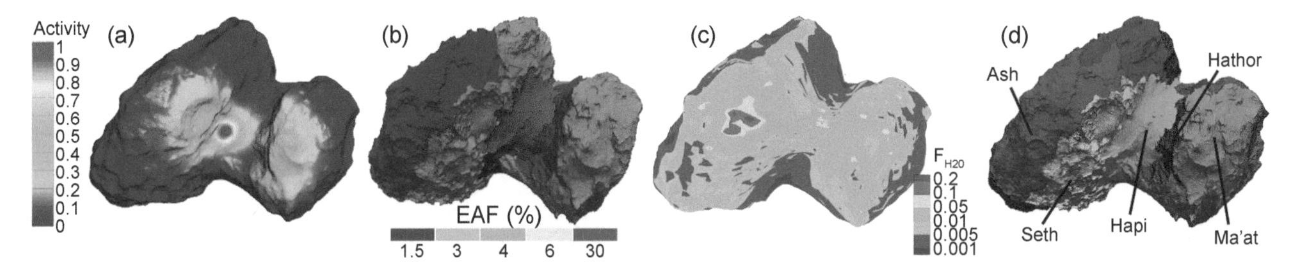

Fig. 8. (a)–(c) Variation of the active area fraction (AAF) (see section 4.1.4) from three different research groups modeling the same epoch of Rosetta data (Sept.–Dec. 2014). Although the scales cannot be directly compared the relative differences are. To derive production rates from these maps one needs to convolve them with the local illumination condition and respective thermal model of the surface. Results in **(a)** are by *Fougere et al.* (2016b), results in **(b)** are by *Marschall et al.* (2016, 2019), and results in **(c)** are by *Zakharov et al.* (2018a). **(d)** Morphological regions defined by *Thomas et al.* (2015b).

dreds of meters (*Marschall et al.,* 2020a). This can hide smaller-scale heterogeneities in the activity maps derived from such measurements.

From an Occam's razor argument, we can be satisfied with the above conclusions but future measurements should nevertheless probe the issue of sources below the physical resolution limit mentioned above. We will discuss which approaches can address this issue in section 6.

It appears — at least at this point — that there is no need for more refined activity maps to explain the coma data. Regional differences of the strength of outgassing (i.e., AAF) in combination with local changes in illumination and topography are sufficient to predict the coma structures.

5.3.6. Gas temperature and speed observations. As discussed above, most models have used either local number densities or integrated column densities to constrain surface parameters. The other flow properties — temperature and flow speed — have somewhat been neglected in inner coma models. The main reason is that there are limited spacecraft observational constraints with one notable exception.

MIRO has provided information in addition to the number or column density of the gas flow. Due to its ability to detect the $H_2^{16}O$ and $H_2^{18}O$ absorption lines information about the water speed and temperature can be retrieved (*Biver et al.,* 2019; *Rezac et al.,* 2021). *Biver et al.* (2019) observed general agreement between the terminal velocity and theoretical expectations from *Hansen et al.* (2016) over a large range of heliocentric distances (from 3.8 au to perihelion at 1.24 au). The terminal gas speeds increased from ~600 m s^{-1} at 3.6 au inbound to ~900 m s^{-1} at perihelion and then subsequently decreased to ~400 m s^{-1} at 3.8 au outbound. Notably, the lower terminal gas speed at similar heliocentric distances might be enhanced by the higher fraction of CO_2 in the coma outbound vs. inbound (*Biver et al.,* 2019; *Combi et al.,* 2020).

Using VIRTIS data, *Cheng et al.* (2022) found rotational temperatures for water of ~150 K close to the surface. At distances of 10 km from the center of the comet the temperatures had dropped to 60–80 K. These observations were taken at heliocentric distances between 1.4 and 1.8 au.

Self-consistent retrievals of the LOS profiles of water number density, temperature, and velocity from the spectral lines are also possible. For data from around the inbound equinox (May 2015), *Marschall et al.* (2019) found a notable deviation from the expected model profiles compared to the profiles retrieved from the data. Although several observations were found where the gas model and the LOS retrievals were in good agreement there were others where models failed (e.g., see Fig. 9). In some of these cases, the gas velocity along the LOS was up to 100 m s^{-1} faster in the data than in the model (bottom middle panel of Fig. 9). Conversely, the retrieved gas temperature from the spectrum was warmer than in the model. This is intriguing because the model assumed that the surface gas temperature is at the free sublimation temperature (~200 K) and therefore *Marschall et al.* (2019) speculated that in reality, the gas must have been significantly warmer upon emission than assumed. This could therefore imply that the gas sublimates from the subsurface and thus first travels through a much hotter desiccated surface layer. That gas can be efficiently heated when flowing through a porous layer has been shown for different porous surfaces (e.g., *Skorov et al.,* 2011; *Christou et al.,* 2018, 2020).

What this demonstrates is the potential information contained in the gas speed and temperature in the near-nucleus environment. These kinds of measurements contained, e.g., in the Rosetta/MIRO dataset remain largely unexploited. In a recent work using a thermophysical model, *Davidsson et al.* (2022b) demonstrates that CO_2 ice is fairly close to the surface for its effects to be detected by MIRO (*Davidsson et al.,* 2022a).

The determination of the gas temperature can also be done from ground. For Comet 73P-B/Schwassmann-Wachmann 3 *Bonev et al.* (2008) detected multiple H_2O emission lines in non-resonant fluorescence near 2.9 μm using the Subaru telescope. They retrieved a decrease in the rotational temperature from ~110 to ~90 K as the projected distance from the nucleus increased from ~5 to ~30 km. These measurements were taken when the comet was at a heliocentric distance of 1.027 au. *Fougere et al.* (2012) used this data to compare it to their kinetic coma model. Importantly, *Fougere et al.* (2012) find that the comparison with a model using pure water emission from the surface cannot account for the observed rotational temperatures. Such a model would predict a much steeper drop of the water column density and temperature with distance to the nucleus than was observed for Comet 73P. By introducing icy particles into the model — acting as an extended gas source — both the temperature and column density of the gas could be increased at large cometocentric distances to match the observations. With the extended source model *Fougere et al.* (2012) was able to conclude that the water coma of 73P is dominated by sublimation from icy grains in the coma rather than surface sublimation.

6. OUTLOOK AND OPEN QUESTIONS

As described above, *in situ* measurements taken beyond several nucleus radii from the surface are limited by the physical resolution limit stemming from the gas dynamics itself (*Marschall et al.,* 2020a). This is made even worse by data primarily from terminator orbits, as was often the case during the Rosetta mission, because of the lateral day-to-nightside flow of the gas in this region. This naturally leads to the question: Which observations can better constrain the gas source distribution at the surface?

First, *in situ* coma measurements at small phase angles, where the gas flow is mostly radial after only a few kilometers even for an irregularly shaped nucleus, would provide stronger constraints on the surface source distribution. Second, *in situ* coma measurements at much lower altitudes (a few hundred meters) above the surface would break the degeneracy stemming from the resolution limit. Such measurements have the added benefit of being within the acceleration region of the gas, thus providing valuable constraints on the velocity distribution function at the surface (if the gas temperature and speed can be measured in addition to the gas density) and any additional processes within that region, such as the sublimation of small icy particles or significant mass loading. Third, high-spatial-resolution (<10 m) spectral imaging could probe the near-surface structure of the gas flow without the need to fly a spacecraft close to the surface. In this case, a terminator orbit would be favored to show the emission into the subsolar direction. Fourth, measurements at a high phase angle would allow the determination of the amount of nightside activity, which is currently rather poorly constrained. Such data could also shed more light on the fading of activity after local sunset and thus hold valuable information about the depth of the sublimation front(s) and thermal properties (e.g., thermal inertia) of the subsurface.

The next open question pertains to the surface boundary conditions of coma models, i.e., the nucleus surface. What is the gas temperature and velocity distribution function at the nucleus surface? While the gas number density at tens of nucleus radii is rather insensitive to the surface gas temperature and velocity (e.g., *Liao et al.,* 2016) the temperatures and velocities themselves are not. There is some hope that existing data (e.g., from Rosetta/MIRO) can still provide

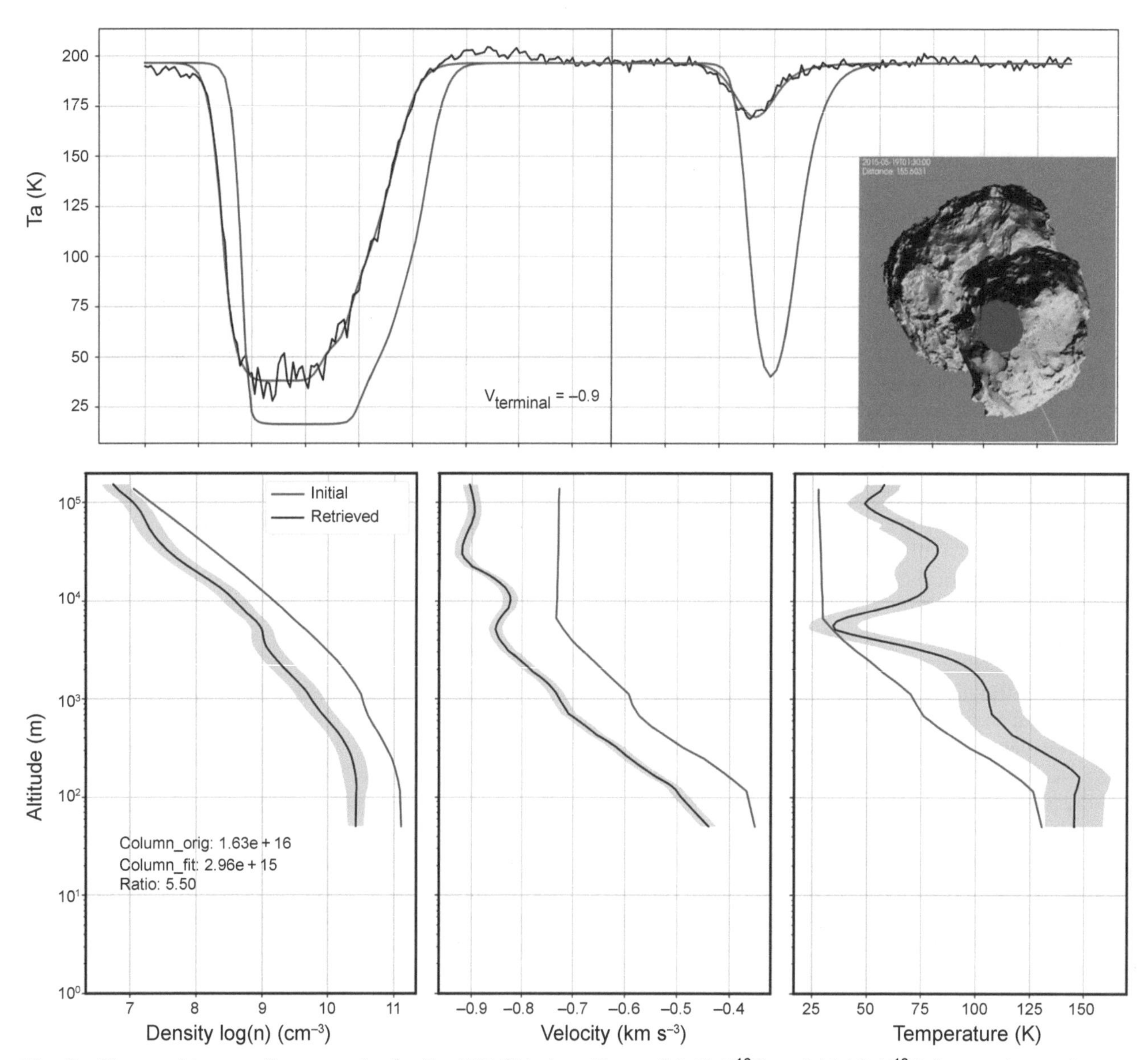

Fig. 9. Shown at top are three spectra for the 557-GHz transitions of (left) $H_2^{16}O$ and (right) $H_2^{18}O$ for an observation on 2015-05-19T01:30: black is the MIRO measurement, red is the retrieved synthetic spectrum, and blue is the synthetic spectrum of the gas model. The inset image shows the shape model of 67P and the pointing geometry with the MIRO footprint (blue circle) on the nucleus surface. Shown at bottom are (left) the vertical profiles of the number density, (middle) the expansion velocity along the MIRO line-of-sight, and (right) the kinetic temperature for the gas model in blue. Black lines are the retrieved profiles. The shaded region represents a 2σ component of uncertainty due to measurement random error propagation. From *Marschall et al.* (2019).

important information (e.g., *Marschall et al.,* 2019; *Pinzón-Rodríguez et al.,* 2021) about the near nucleus structure of the flow and by extension of the surface. But such advances will require the implementation of an actual non-idealized thermal model of the subsurface to inform the gas coma model boundary conditions. Such a thermal model will be crucial to properly remove diurnal/seasonal variability of the production rates and retrieve (sub)surface properties.

While the above-mentioned measurements illustrate a way forward from the standpoint of *in situ* and remote sensing data of the coma, they also highlight the greater problem that is only addressed indirectly by these measurements. It is the big overarching question: What drives and sustains cometary activity? It has become clear that even the Rosetta mission, with its extensive dataset, was not able to answer this question (*Thomas et al.,* 2019; *Keller and Kührt,* 2020). And it is not clear whether further *in situ* and remote sensing data from the coma, even with better instruments and/or different observation geometries, will be able to adequately answer this question. The lesson for this "failure" has to be that we need to understand the physicochemical structure of the subsurface. This has already been pointed out by *Thomas et al.* (2019) but is worth reiterating here. The straightforward way of understanding how activity works is by observing it *in situ* at multiple locations on the surface and determining how volatiles and refractories are mixed on the microscopic level. We thus deem it inevitable that a comet lander or hopper is needed to comprehensively address this question. We should also mention that it is unlikely that any sample return mission would be able to answer this question. Any sample returned to Earth would not be able to retain the physical structure of the sample during the high accelerations during reentry. Furthermore, retaining the ices of highly volatile species is very challenging. Sublimation of such molecules may further alter the physical structure. And it is this physical structure and how volatiles are embedded within the refractory components that seem to hold the key to understanding cometary activity.

A final open question we want to highlight touches on the outer coma and tail. Although these were not the focus of this chapter it is worth pondering them. Now that we know there are regional differences in surface activity, modeled without the need for any small-scale localized sources (in the sense of a classical "jet"), it is worth asking what the consequences are for the structures we see in the outer coma. First, it is more obvious than ever that the concept of a "jet" in the outer coma is a complete misnomer (*Crifo et al.,* 2004). If there are only large regional differences in the level of activity but no source in the traditional sense, then such features in the outer coma structures cannot be "jets." Second, the apparent mismatch between the intricate and rather small-scale structures and the comparably large spatial extents of outer coma "jets" needs to be resolved. How do these small-scale structures of the inner coma connect to the outer coma? For this we have yet to understand how coma structures connect from the surface to the inner coma (accessible to spacecraft) before they can ultimately be linked to structures in the outer coma (accessible with groundbased telescopes).

7. CONCLUSIONS

In this chapter we have focused on two crucial questions of linking inner coma measurements to the surface of cometary nuclei:

- How can we derive the gas production rate of different species and thus the volatile mass loss from coma measurements?
- Can we determine if coma structures (inhomogeneities in density, often referred to as "jets") are reflective of a heterogeneous nucleus, or are mere emergent phenomena in the gas flow due to, e.g., the complex shape of the nucleus?

Answering these questions allows us to (1) link spacecraft measurements from the inner coma to the surface, which is also a prerequisite to understanding groundbased observations and linking those measurements to the nucleus, and (2) make predictions for future comet missions and assess hazards for spacecraft operating in that region.

Deriving properties of a cometary nucleus from coma data is of significant importance for our understanding of cometary activity and has implications beyond. For example, whether a cometary nucleus is homogeneous or heterogeneous in composition will influence how we understand planetesimal formation. A homogeneous nucleus would indicate that the material was formed in close proximity and within a rather short time. A mechanism such as the streaming instability (e.g., *Goldreich and Ward,* 1973; *Youdin and Goodman,* 2005; *Johansen et al.,* 2007; *Simon et al.,* 2016; *Blum et al.,* 2017) followed by gravitational collapse would likely result in rather homogeneous nuclei. In contrast, a heterogeneous nucleus would support longer formation times, which allow for more mixing and therefore might be indicative of hierarchical formation scenarios (e.g., *Weidenschilling,* 1997; *Kenyon and Luu,* 1998; *Windmark et al.,* 2012a,b; *Davidsson et al.,* 2016).

To understand the coma-surface link, different types of models have been developed over the past decades. These have been used to derive properties of the nucleus surface from measurements of *in situ* or remote sensing instruments of spacecraft. We find that common heuristic models such as the "Haser model" (and more complex variations thereof) are useful to derive properties such as the global gas production rate. We caution the reader though that they are not a substitute for real physically self-consistent models. Heuristic models are — by their nature of neglecting physical processes — limited and therefore not well suited to, e.g., derive the gas emission distribution from the surface. Such models cannot give us answers as to the emergence of structures in the coma and properties of the (sub)surface.

To retrieve more detailed information about the dynamical structure of the gas coma current state-of-the-art physical models are needed. We described the need for accurate thermophysical models of the (sub)surface that serve as input for state-of-the-art kinetic coma models. Typically, kinetic models developed within the DSMC method are used to model the dusty-gas dynamics in cometary comae. Although they are computationally more expensive, they are our best tool to derive the near-surface properties. This includes, e.g., the determination of the gas temperature and speed, which hold information about the physiochemical subsurface structure that the gas flows through before leaving the nucleus. Therefore, we encourage the reader to opt for forward models whenever possible, because they are less likely to be misinterpreted and their limitations forgotten.

Complex models, which account for all relevant processes, are particularly useful when there are non-linear effects (such as molecular collisions) that cannot easily be parameterized. If a flow contains only linear features, or if observations only capture global properties of the flow, then complex models do not offer a benefit over simplified models. This is the case, for instance, when determining the global gas production rate. At the same time, we should always keep in mind that all models have some limitations. A complex model, such as DSMC, is more widely applicable but usually comes at a computational cost. Thus, understanding these limitations prevents us from misapplying the model to inappropriate situations (see, e.g., discussion in sections 4.3 and 4.2 with respect to the flow regime).

One key lesson to be learned from physical models is that there is a physical resolution limit to determining the distribution of source regions on the nucleus surface, especially from *in situ* data taken at tens to hundreds of kilometers away from the surface. Sources at the surface do not independently/linearly contribute to the gas density in the coma. This limits the accuracy to which we can resolve the heterogeneity at the surface, both with physical and heuristic models, even if the latter are mathematically valid to smaller spatial scales.

One of the main open questions from the previous volume (*Crifo et al.,* 2004) was what drove "jet"-like structures in the coma. Are they the result of (1) a heterogeneous nucleus or (2) features emerging from the focusing of flows from a complex nucleus shape and local topography? The past decades of spacecraft missions to comets 1P/Halley, 19P/Borrelly, 9P/Tempel 1, 67P/Churyumov-Gerasimenko, 81P/Wild 2, and 103P/Hartley 2 have given important insights to this question. The nuclei show a vast diversity in shape with both concave and convex terrain. It has become clear that all the nuclei visited show regional differences in the strength of outgassing beyond simple variations in local illumination. However, these regional differences do not appear to be driven by a heterogeneous nucleus itself but rather by surface processes that alter the near-surface composition. One of the main drivers of this surface evolution is airfall — the redeposition of large dust chunks that are depleted in ices more volatile than water. In addition to the regional variations of activity, local topography and the global complex shape of the nuclei is sufficient to explain the observed inhomogeneous structures of the inner coma. In most examples, there does not appear to be any need for small-scale sources that would link to "jets." This excludes, of course, the phenomena of outbursts that have a clear confined source location at the surface.

Thus, while there is no strong evidence for heterogeneous nuclei, heterogeneous nucleus surfaces are a common feature of comets. Real comets thus do not neatly fall into one of the two extremes outlined above. Rather, they are an intermediate case, i.e., regionally heterogeneous surfaces with corresponding topography shaping an inhomogeneous inner coma.

This also illustrates further what has been known for some time, that the term "jet," which has a strict physical meaning, is a misnomer (*Crifo et al.,* 2004; *Vincent et al.,* 2019). Because the "jet" is not an adequate description of the emergent structures in the inner coma, it also loses its meaning in the outer coma. Thus there is a need for a consistent, widely accepted nomenclature that more accurately describes the structures and features in cometary comae, near and far.

To improve the resolution of surface emission maps we propose that orbits much closer to the surface (several hundred meters), including low ($<30°$) and high ($>150°$) phase angles, are needed to derive more accurate surface properties from *in situ* coma measurements. Such a dataset would also allow us to probe the acceleration region of the gas and determine the level and nature of the nightside activity. Alternatively, high-spatial-resolution spectral imagers that can probe the coma structure very close to the surface from a terminator orbit would also significantly improve our understanding of the link between coma structures and the surface while remaining further away from the surface. Measuring the gas temperature and speed close to the surface is also crucial to deriving (sub)surface properties. This can be done with an instrument like Rosetta/MIRO, or *in situ* on low-altitude orbits.

We are now at a point where we start to understand the near-nucleus environment well enough that a crucial missing element can be tackled. How do coma structures on a spatial but also temporal scale trace from the surface out to the scales observable with groundbased telescopes? Answering this question will allow us to understand the nature of the large-scale structures in the outer coma, their origin, and what they tell us about the nucleus.

Finally, we suggest that observations of activity at the surface, in addition to subsurface measurements, are likely the key avenue to address how comets work by probing the physiochemical structure of the subsurface. In addition, thermal measurements, radar sounding, and subsurface sampling and imaging (to determine the physicochemical properties), for example, will all play an important role in understanding comets (*Thomas et al.,* 2019). These goals can be achieved by dedicated orbiter and lander infrastructures. While gas coma measurements close to the surface can shed important insights into the subsurface structure, degeneracies likely will remain. A lander or hopper that can study the near-surface physiochemical properties is thus the most promising next step to address this issue.

Acknowledgments. We thank the two reviewers, L. Feaga and V. Zakharov, for taking the time to proving detailed constructive suggestions that have greatly improved this chapter. We thank D. Bockelée-Morvan for taking time, as editor, to read the chapter and give us important feedback and suggestions. Furthermore, we thank S. Protopapa and A. Migliorini for their help with the water and CO_2 column density maps of 103P and 67P respectively and for providing high-resolution versions of the respective figures. R.M. acknowledges the support from NASA's Emerging Worlds program, grant NNX17AE83G, and funding from the European Research Council (ERC) under the European Union's Horizon 2020 research and innovation programme (Grant agreement No. 101019380). A portion of this research was carried out at the Jet Propulsion Laboratory, California Institute of Technology (internal "raise the bar" funding is acknowledged), under a contract with the National Aeronautics and Space Administration.

REFERENCES

Agarwal J., A'Hearn M. F., Vincent J.-B., et al. (2016) Acceleration of individual, decimetre-sized aggregates in the lower coma of Comet 67P/Churyumov-Gerasimenko. *Mon. Not. R. Astron. Soc., 462*, S78–S88.

A'Hearn M. F., Belton M. J. S., Delamere A., et al. (2005) Deep Impact: A large-scale active experiment on a cometary nucleus. *Space Sci. Rev., 117*, 1–21.

A'Hearn M. F., Belton M. J. S., Delamere W. A., et al. (2011) EPOXI at Comet Hartley 2. *Science, 332*, 1396–1400.

Anisimov S. I. (1968) Vaporization of metal absorbing laser radiation. *Sov. J. Exp. Theor. Phys., 27*, 182–183.

Balsiger H., Altwegg K., Bochsler P., et al. (2007) ROSINA — Rosetta Orbiter Spectrometer for Ion and Neutral Analysis. *Space Sci. Rev., 128*, 745–801.

Barucci M. A., Filacchione G., Fornasier S., et al. (2016) Detection of exposed H_2O ice on the nucleus of Comet 67P/Churyumov-Gerasimenko as observed by Rosetta OSIRIS and VIRTIS instruments. *Astron. Astrophys., 595*, A102.

Bieler A., Altwegg K., Balsiger H., et al. (2015) The role of numerical models in data analysis for the Rosetta mission. *European Planet. Sci. Congr. 2015*, 301.

Bird G. A. (1994) *Molecular Gas Dynamics And The Direct Simulation Of Gas Flows.* Clarendon, Oxford. 458 pp.

Bird G. A. (2013) *The DSMC Method.* CreateSpace Independent Publishing Platform, Scotts Valley, California. 300 pp.

Biver N., Bockelée-Morvan D., Hofstadter M., et al. (2019) Long-term monitoring of the outgassing and composition of Comet 67P/Churyumov-Gerasimenko with the Rosetta/MIRO instrument. *Astron. Astrophys., 630*, A19.

Blum J., Gundlach B., Krause M., et al. (2017) Evidence for the formation of Comet 67P/Churyumov-Gerasimenko through gravitational collapse of a bound clump of pebbles. *Mon. Not. R. Astron. Soc., 469*, S755–S773.

Bockelée-Morvan D., Leyrat C., Erard S., et al. (2019) VIRTIS-H observations of the dust coma of Comet 67P/Churyumov-Gerasimenko: Spectral properties and color temperature variability with phase and elevation. *Astron. Astrophys., 630*, A22.

Boice D. C., Soderblom L. A., Britt D. T., et al. (2002) The Deep Space 1 encounter with Comet 19P/Borrelly. *Earth Moon Planets, 89*, 301–324.

Bonev B. P., Mumma M. J., Kawakita H., et al. (2008) IRCS/Subaru observations of water in the inner coma of Comet 73P-B/Schwassmann-Wachmann 3: Spatially resolved rotational temperatures and ortho-para ratios. *Icarus, 196*, 241–248.

Bonev B. P., Villanueva G. L., Paganini L., et al. (2013) Evidence for two modes of water release in Comet 103P/Hartley 2: Distributions of column density, rotational temperature, and ortho-para ratio. *Icarus, 222*, 740–751.

Cercignani C. (2000) *Rarefied Gas Dynamics: From Basic Concepts to Actual Calculations.* Cambridge Univ., Cambridge. 340 pp.

Cheng Y. C., Bockelée-Morvan D., Roos-Serote M., et al. (2022) Water ortho-to-para ratio in the coma of Comet 67P/Churyumov-Gerasimenko. *Astron. Astrophys., 663*, A43.

Christou C., Dadzie S. K., Thomas N., et al. (2018) Gas flow in near surface comet like porous structures: Application to 67P/Churyumov-Gerasimenko. *Planet. Space Sci., 161*, 57–67.

Christou C., Dadzie S. K., Marschall R., et al. (2020) Porosity gradients as a means of driving lateral flows at cometary surfaces. *Planet. Space Sci., 180*, 104752.

Combi M. R. (1996) Time-dependent gas kinetics in tenuous planetary atmospheres: The cometary coma. *Icarus, 123*, 207–226.

Combi M. R., Harris W. M., and Smyth W. H. (2004) Gas dynamics and kinetics in the cometary coma: Theory and observations. In *Comets II* (M. C. Festou et al., eds.), pp. 523–552. Univ. of Arizona, Tucson.

Combi M. R., Bertaux J. L., Quémerais E., et al. (2011) Water production by Comet 103P/Hartley 2 observed with the SWAN instrument on the SOHO spacecraft. *Astrophys. J. Lett., 734*, L6.

Combi M. R., Tenishev V. M., Rubin M., et al. (2012) Narrow dust jets in a diffuse gas coma: A natural product of small active regions on comets. *Astrophys. J., 749*, 29.

Combi M. R., Mäkinen T. T., Bertaux J. L., et al. (2019) A survey of water production in 61 comets from SOHO/SWAN observations of hydrogen Lyman-alpha: Twenty-one years 1996–2016. *Icarus, 317*, 610–620.

Combi M., Shou Y., Fougere N., et al. (2020) The surface distributions of the production of the major volatile species, H_2O, CO_2, CO and O_2, from the nucleus of Comet 67P/Churyumov-Gerasimenko throughout the Rosetta mission as measured by the ROSINA double focusing mass spectrometer. *Icarus, 335*, 113421.

Coradini A., Capaccioni F., Drossart P., et al. (2007) VIRTIS: An imaging spectrometer for the Rosetta mission. *Space Sci. Rev., 128*, 529–559.

Cottin H. and Fray N. (2008) Distributed sources in comets. *Space Sci. Rev., 138*, 179–197.

Crifo J. F. (1987) Improved gas-kinetic treatment of cometary water sublimation and recondensation — Application to Comet P/Halley. *Astron. Astrophys., 187*, 438–450.

Crifo J. F. and Rodionov A. V. (1997) The dependence of the circumnuclear coma structure on the properties of the nucleus. *Icarus, 127*, 319–353.

Crifo J. F. and Rodionov A. V. (1999) Modelling the circumnuclear coma of comets: Objectives, methods and recent results. *Planet. Space Sci., 47*, 797–826.

Crifo J. F. and Rodionov A. V. (2000) The dependence of the circumnuclear coma structure on the properties of the nucleus. IV. Structure of the night-side gas coma of a strongly sublimating nucleus. *Icarus, 148*, 464–478.

Crifo J. F., Itkin A. L., and Rodionov A. V. (1995) The near-nucleus coma formed by interacting dusty gas jets effusing from a cometary nucleus: I. *Icarus, 116*, 77–112.

Crifo J. F., Rodionov A. V., and Bockelée-Morvan D. (1999) The dependence of the circumnuclear coma structure on the properties of the nucleus: III. First modeling of a CO-dominated coma, with application to Comets 46 P/Wirtanen and 29 P/Schwassmann-Wachmann I. *Icarus, 138*, 85–106.

Crifo J. F., Lukianov G. A., Rodionov A. V., et al. (2002a) Comparison between Navier-Stokes and direct Monte-Carlo simulations of the circumnuclear coma. I. Homogeneous, spherical source. *Icarus, 156*, 249–268.

Crifo J. F., Rodionov A. V., Szegö K., et al. (2002b) Challenging a paradigm: Do we need active and inactive areas to account for near-nuclear jet activity? *Earth Moon Planets, 90*, 227–238.

Crifo J. F., Loukianov G. A., Rodionov A. V., et al. (2003) Navier-Stokes and direct Monte Carlo simulations of the circumnuclear coma II. Homogeneous, aspherical sources. *Icarus, 163*, 479–503.

Crifo J. F., Fulle M., Kömle N. I., et al. (2004) Nucleus-coma structural relationships: Lessons from physical models. In *Comets II* (M. C. Festou et al., eds.), pp. 471–503. Univ. of Arizona, Tucson.

Crifo J. F., Loukianov G. A., Rodionov A. V., et al. (2005) Direct Monte Carlo and multifluid modeling of the circumnuclear dust coma: Spherical grain dynamics revisited. *Icarus, 176*, 192–219.

Dankert C. and Koppenwallner G. (1984) Experimental study of the interaction between two rarefied free jets. In *14th Symposium on Rarefied Gas Dynamics, Volume 1* (H. Oguchi, ed.), pp. 477–484. Univ. of Tokyo, Tokyo.

Davidsson B. J. R. (2008) Comet Knudsen layers. *Space Sci. Rev., 138,* 207–223.

Davidsson B. J. R. (2021) Thermophysical evolution of planetesimals in the primordial disc. *Mon. Not. R. Astron. Soc., 505,* 5654–5685.

Davidsson B. J. R. and Skorov Y. V. (2004) A practical tool for simulating the presence of gas comae in thermophysical modeling of cometary nuclei. *Icarus, 168,* 163–185.

Davidsson B. J. R., Gutiérrez P. J., Groussin O., et al. (2013) Thermal inertia and surface roughness of Comet 9P/Tempel 1. *Icarus, 224,* 154–171.

Davidsson B. J. R., Sierks H., Güttler C., et al. (2016) The primordial nucleus of Comet 67P/Churyumov-Gerasimenko. *Astron. Astrophys., 592,* A63.

Davidsson B. J. R., Birch S., Blake G. A., et al. (2021) Airfall on Comet 67P/Churyumov-Gerasimenko. *Icarus, 354,* 114004.

Davidsson B. J. R., Samarasinha N. H., Farnocchia D., et al. (2022a) Modelling the water and carbon dioxide production rates of Comet 67P/Churyumov-Gerasimenko. *Mon. Not. R. Astron. Soc., 509,* 3065–3085.

Davidsson B. J. R., Schloerb F. P., Fornasier S., et al. (2022b) CO_2-driven surface changes in the Hapi region on Comet 67P/Churyumov-Gerasimenko. *Mon. Not. R. Astron. Soc., 516,* 6009–6040.

De Sanctis M. C., Capaccioni F., Ciarniello M., et al. (2015) The diurnal cycle of water ice on Comet 67P/Churyumov-Gerasimenko. *Nature, 525,* 500–503.

Eberhardt P., Krankowsky D., Schulte W., et al. (1986) On the CO and N_8 abundance in Comet Halley. In *20th ESLAB Symposium on the Exploration of Halley's Comet* (B. Battrick et al., eds.), pp. 383–386. ESA SP-250, Noordwijk, The Netherlands.

Farnham T. L., Bodewits D., Li J. Y., et al. (2013) Connections between the jet activity and surface features on Comet 9P/Tempel 1. *Icarus, 222,* 540–549.

Feaga L. M., A'Hearn M. F., Sunshine J. M., et al. (2007) Asymmetries in the distribution of H_2O and CO_2 in the inner coma of Comet 9P/Tempel 1 as observed by Deep Impact. *Icarus, 191,* 134–145.

Feaga L., Sunshine J., Protopapa S., et al. (2014) Comet 103P/Hartley's volatiles within 100 kilometers: Sources of water and volatile dependence on illumination. In *Asteroids, Comets, Meteors 2014, Book of Abstracts* (K. Muinonen et al., eds.), p. 178. Univ. of Helsinki, Finland.

Festou M. C., Keller H. U., and Weaver H. A., eds. (2004) *Comets II.* Univ. of Arizona, Tucson. 745 pp.

Fink U., Doose L., Rinaldi G., et al. (2016) Investigation into the disparate origin of CO_2 and H_2O outgassing for Comet 67/P. *Icarus, 277,* 78–97.

Finklenburg S., Thomas N., Su C. C., et al. (2014) The spatial distribution of water in the inner coma of Comet 9P/Tempel 1: Comparison between models and observations. *Icarus, 236,* 9–23.

Fornasier S., Mottola S., Keller H. U., et al. (2016) Rosetta's Comet 67P/Churyumov-Gerasimenko sheds its dusty mantle to reveal its icy nature. *Science, 354,* 1566–1570.

Fougere N., Combi M. R., Tenishev V., et al. (2012) Understanding measured water rotational temperatures and column densities in the very innermost coma of Comet 73P/Schwassmann-Wachmann 3 B. *Icarus, 221,* 174–185.

Fougere N., Combi M. R., Rubin M., et al. (2013) Modeling the heterogeneous ice and gas coma of Comet 103P/Hartley 2. *Icarus, 225,* 688–702.

Fougere N., Altwegg K., Berthelier J. J., et al. (2016a) Direct simulation Monte Carlo modelling of the major species in the coma of Comet 67P/Churyumov-Gerasimenko. *Mon. Not. R. Astron. Soc., 462,* S156–S169.

Fougere N., Altwegg K., Berthelier J.-J., et al. (2016b) Three-dimensional direct simulation Monte-Carlo modeling of the coma of Comet 67P/Churyumov-Gerasimenko observed by the VIRTIS and ROSINA instruments on board Rosetta. *Astron. Astrophys., 588,* A134.

Fulle M., Crifo J. F., and Rodionov A. V. (1999) Numerical simulation of the dust flux on a spacecraft in orbit around an aspherical cometary nucleus — I. *Astron. Astrophys., 347,* 1009–1028.

Gehrz R. D. and Ney E. P. (1992) 0.7- to 23-μm photometric observations of P/Halley 1986 III and six recent bright comets. *Icarus, 100,* 162–186.

Gerig S.-B., Marschall R., Thomas N., et al. (2018) On deviations from free-radial outflow in the inner coma of Comet 67P/Churyumov-Gerasimenko. *Icarus, 311,* 1–22.

Gerig S.-B., Pinzón-Rodríguez O., Marschall R., et al. (2020) Dayside-to-nightside dust coma brightness asymmetry and its implications for nightside activity at Comet 67P/Churyumov-Gerasimenko. *Icarus, 351,* 113968.

Gersch A. M., A'Hearn M. F., and Feaga L. M. (2018) Modeling the Deep Impact near-nucleus observations of H_2O and CO_2 in Comet 9P/Tempel 1 using asymmetric spherical coupled escape probability. *Astrophys. J., 856,* 104.

Glassmeier K.-H., Boehnhardt H., Koschny D., et al. (2007) The Rosetta mission: Flying towards the origin of the solar system. *Space Sci. Rev., 128,* 1–21.

Goldreich P. and Ward W. R. (1973) The formation of planetesimals. *Astrophys. J., 183,* 1051–1062.

Groussin O., A'Hearn M. F., Li J. Y., et al. (2007) Surface temperature of the nucleus of Comet 9P/Tempel 1. *Icarus, 187,* 16–25.

Groussin O., Sunshine J. M., Feaga L. M., et al. (2013) The temperature, thermal inertia, roughness and color of the nuclei of Comets 103P/Hartley 2 and 9P/Tempel 1. *Icarus, 222,* 580–594.

Gulkis S., Frerking M., Crovisier J., et al. (2007) MIRO: Microwave Instrument for Rosetta Orbiter. *Space Sci. Rev., 128,* 561–597.

Güttler C., Hasselmann P. H., Li Y., et al. (2017) Characterization of dust aggregates in the vicinity of the Rosetta spacecraft. *Mon. Not. R. Astron. Soc., 469,* S312–S320.

Hansen K. C., Altwegg K., Berthelier J. J., et al. (2016) Evolution of water production of 67P/Churyumov-Gerasimenko: An empirical model and a multi-instrument study. *Mon. Not. R. Astron. Soc., 462,* S491–S506.

Haser L. (1957) Distribution d'intensité dans la tête d'une comète. *Bull. Soc. R. Sci. Liege, 43,* 740–750.

Hässig M., Altwegg K., Balsiger H., et al. (2015) Time variability and heterogeneity in the coma of 67P/Churyumov-Gerasimenko. *Science, 347,* aaa0276.

Herny C., Mousis O., Marschall R., et al. (2021) New constraints on the chemical composition and outgassing of 67P/Churyumov-Gerasimenko. *Planet. Space Sci., 200,* 105194.

Hirsch C. (2007) *Numerical Computation of Internal and External Flows: The Fundamentals of Computational Fluid Dynamics, 2nd edition.* Butterworth-Heinemann, Oxford. 656 pp.

Huebner W. F. and Markiewicz W. J. (2000) Note: The temperature and bulk flow speed of a gas effusing or evaporating from a surface into a void after reestablishment of collisional equilibrium. *Icarus, 148,* 594–596.

Ivanovski S., Zakharov V., Della Corte V., et al. (2017) Dynamics of aspherical dust grains in a cometary atmosphere: I. Axially symmetric grains in a spherically symmetric atmosphere. *Icarus, 282,* 333–350.

Johansen A., Oishi J. S., Mac Low M.-M., et al. (2007) Rapid planetesimal formation in turbulent circumstellar disks. *Nature, 448,* 1022–1025.

Jorda L., Gaskell R., Capanna C., et al. (2016) The global shape, density and rotation of Comet 67P/Churyumov-Gerasimenko from preperihelion Rosetta/OSIRIS observations. *Icarus, 277,* 257–278.

Jutzi M. and Michel P. (2020) Collisional heating and compaction of small bodies: Constraints for their origin and evolution. *Icarus, 350,* 113867.

Keeney B. A., Stern S. A., Feldman P. D., et al. (2019) Stellar occultation by Comet 67P/Churyumov-Gerasimenko observed with Rosetta's Alice far-ultraviolet spectrograph. *Astron. J., 157,* 173.

Keller H. U. and Kührt E. (2020) Cometary nuclei — From Giotto to Rosetta. *Space Sci. Rev., 216,* 14.

Keller H. U., Delamere W. A., Reitsema H. J., et al. (1987) Comet P/Halley's nucleus and its activity. *Astron. Astrophys., 187,* 807–823.

Keller H. U., Barbieri C., Lamy P., et al. (2007) OSIRIS — The scientific camera system onboard Rosetta. *Space Sci. Rev., 128,* 433–506.

Keller H. U., Mottola S., Hviid S. F., et al. (2017) Seasonal mass transfer on the nucleus of Comet 67P/Churyumov-Gerasimenko. *Mon. Not. R. Astron. Soc., 469,* S357–S371.

Kelley M. S., Lindler D. J., Bodewits D., et al. (2013) A distribution of large particles in the coma of Comet 103P/Hartley 2. *Icarus, 222,* 634–652.

Kenyon S. J. and Luu J. X. (1998) Accretion in the early Kuiper belt. I. Coagulation and velocity evolution. *Astron. J., 115,* 2136–2160.

Kitamura Y. (1990) A numerical study of the interaction between two cometary jets: A possibility of shock formation in cometary

atmospheres. *Icarus, 86,* 455–475.
Kitamura Y. (1986) Axisymmetric dusty gas jet in the inner coma of a comet. *Icarus, 66,* 241–257.
Kitamura Y. (1987) Axisymmetric dusty gas jet in the inner coma of a comet: II. The case of isolated jets. *Icarus, 72,* 555–567.
Knight M. M. and Schleicher D. G. (2013) The highly unusual outgassing of Comet 103P/Hartley 2 from narrowband photometry and imaging of the coma. *Icarus, 222,* 691–706.
Knollenberg J. (1993) Modellrechnung zur staubverteilung in der inneren koma von kometen unter spezieller berücksichtigung der HMC-daten de Giotto mission. Ph.D. thesis, Georg-August Universität, Göttingen. 139 pp.
Knollenberg J., Kührt E., and Keller H. U. (1996) Interpretation of HMC images by a combined thermal and gasdynamic model. *Earth Moon Planets, 72,* 103–112.
Komle N. I. and Ip W.-H. (1987) Anisotropic non-stationary gas flow dynamics in the coma of Comet P/Halley. *Astron. Astrophys., 187,* 405–410.
Korosmezey A. and Gombosi T. I. (1990) A time-dependent dusty gas dynamic model of axisymmetric cometary jets. *Icarus, 84,* 118–153.
Kossacki K. J. and Szutowicz S. (2008) Comet 9P/Tempel 1: Sublimation beneath the dust cover. *Icarus, 195,* 705–724.
Kramer T. and Noack M. (2016) On the origin of inner coma structures observed by Rosetta during a diurnal rotation of Comet 67P/Churyumov-Gerasimenko. *Astrophys. J. Lett., 823,* L11.
Kramer T., Läuter M., Rubin M., et al. (2017) Seasonal changes of the volatile density in the coma and on the surface of Comet 67P/Churyumov-Gerasimenko. *Mon. Not. R. Astron. Soc., 469,* S20–S28.
Krankowsky D., Lammerzahl P., Herrwerth I., et al. (1986) *In situ* gas and ion measurements at Comet Halley. *Nature, 321,* 326–329.
Läuter M., Kramer T., Rubin M., et al. (2019) Surface localization of gas sources on Comet 67P/Churyumov-Gerasimenko based on DFMS/COPS data. *Mon. Not. R. Astron. Soc., 483,* 852–861.
Läuter M., Kramer T., Rubin M., et al. (2020) The gas production of 14 species from Comet 67P/Churyumov-Gerasimenko based on DFMS/COPS data from 2014 to 2016. *Mon. Not. R. Astron. Soc., 498,* 3995–4004.
Liao Y., Su C. C., Marschall R., et al. (2016) 3D direct simulation Monte Carlo modelling of the inner gas coma of Comet 67P/Churyumov-Gerasimenko: A parameter study. *Earth Moon Planets, 117,* 41–64.
Liao Y., Marschall R., Su C. C., et al. (2018) Water vapor deposition from the inner gas coma onto the nucleus of Comet 67P/Churyumov-Gerasimenko. *Planet. Space Sci., 157,* 1–9.
Lien D. J. (1990) Dust in comets. I. Thermal properties of homogeneous and heterogeneous grains. *Astrophys. J., 355,* 680–692.
Mäkinen J. T. T. and Combi M. R. (2005) Temporal deconvolution of the hydrogen coma I. A hybrid model. *Icarus, 177,* 217–227.
Marboeuf U. and Schmitt B. (2014) How to link the relative abundances of gas species in coma of comets to their initial chemical composition? *Icarus, 242,* 225–248.
Markelov G., Skorov Y., and Keller H. (2006) DSMC modeling of dusty innermost cometary atmosphere around non-spherical nucleus. In *9th AIAA/ASME Joint Thermophysics and Heat Transfer Conference,* AIAA 2006-3392. American Institute of Aeronautics and Astronautics, Reston.
Marschall R. (2017) Inner gas and dust comae of comet: Building a 3D simulation pipeline to understand multi-instrument results from the Rosetta mission to Comet 67P/Churyumov-Gerasimenko. Ph.D. thesis, Universität Bern, Bern. 256 pp.
Marschall R., Su C. C., Liao Y., et al. (2016) Modelling observations of the inner gas and dust coma of Comet 67P/Churyumov-Gerasimenko using ROSINA/COPS and OSIRIS data: First results. *Astron. Astrophys., 589,* A90.
Marschall R., Mottola S., Su C. C., et al. (2017) Cliffs versus plains: Can ROSINA/COPS and OSIRIS data of Comet 67P/Churyumov-Gerasimenko in autumn 2014 constrain inhomogeneous outgassing? *Astron. Astrophys., 605,* A112.
Marschall R., Rezac L., Kappel D., et al. (2019) A comparison of multiple Rosetta data sets and 3D model calculations of 67P/Churyumov-Gerasimenko coma around equinox (May 2015). *Icarus, 328,* 104–126.
Marschall R., Liao Y., Thomas N., et al. (2020a) Limitations in the determination of surface emission distributions on comets through modelling of observational data — A case study based on Rosetta observations. *Icarus, 346,* 113742.
Marschall R., Markkanen J., Gerig S.-B., et al. (2020b) The dust-to-gas ratio, size distribution, and dust fall-back fraction of Comet 67P/Churyumov-Gerasimenko: Inferences from linking the optical and dynamical properties of the inner comae. *Front. Phys., 8,* 227.
Marschall R., Skorov Y., Zakharov V., et al. (2020c) Cometary comae-surface links. *Space Sci. Rev., 216,* 130.
Marshall D. W., Hartogh P., Rezac L., et al. (2017) Spatially resolved evolution of the local H_2O production rates of Comet 67P/Churyumov-Gerasimenko from the MIRO instrument on Rosetta. *Astron. Astrophys., 603,* A87.
Marshall D., Rezac L., Hartogh P., et al. (2019) Interpretation of heliocentric water production rates of comets. *Astron. Astrophys., 623,* A120.
Meech K. J., A'Hearn M. F., Adams J. A., et al. (2011) EPOXI: Comet 103P/Hartley 2 observations from a worldwide campaign. *Astrophys. J. Lett., 734,* L1.
Migliorini A., Piccioni G., Capaccioni F., et al. (2016) Water and carbon dioxide distribution in the 67P/Churyumov-Gerasimenko coma from VIRTIS-M infrared observations. *Astron. Astrophys., 589,* A45.
Mottola S., Lowry S., Snodgrass C., et al. (2014) The rotation state of 67P/Churyumov-Gerasimenko from approach observations with the OSIRIS cameras on Rosetta. *Astron. Astrophys., 569,* L2.
Narasimha R. (1962) Collisionless expansion of gases into vacuum. *J. Fluid Mech., 12,* 294–308.
Pinzón-Rodríguez O., Marschall R., Gerig S.-B., et al. (2021) The effect of thermal conductivity on the outgassing and local gas dynamics from cometary nuclei. *Astron. Astrophys., 655,* A20.
Pommerol A., Thomas N., El-Maarry M. R., et al. (2015) OSIRIS observations of meter-sized exposures of H_2O ice at the surface of 67P/Churyumov-Gerasimenko and interpretation using laboratory experiments. *Astron. Astrophys., 583,* A25.
Preusker F., Scholten F., Matz K. D., et al. (2017) The global meter-level shape model of Comet 67P/Churyumov-Gerasimenko. *Astron. Astrophys., 607,* L1.
Prialnik D. (1992) Crystallization, sublimation, and gas release in the interior of a porous comet nucleus. *Astrophys. J., 388,* 196–202.
Protopapa S., Sunshine J. M., Feaga L. M., et al. (2014) Water ice and dust in the innermost coma of Comet 103P/Hartley 2. *Icarus, 238,* 191–204.
Protopapa S., Kelley M. S. P., Woodward C. E., et al. (2021) Nondetection of water-ice grains in the coma of Comet 46P/Wirtanen and implications for hyperactivity. *Planet. Sci. J., 2,* 176.
Raponi A., Ciarniello M., Capaccioni F., et al. (2016) The temporal evolution of exposed water ice-rich areas on the surface of 67P/Churyumov-Gerasimenko: Spectral analysis. *Mon. Not. R. Astron. Soc., 462,* S476–S490.
Rezac L., Zorzi A., Hartogh P., et al. (2021) Gas terminal velocity from MIRO/Rosetta data using neural network approach. *Astron. Astrophys., 648,* A21.
Rodionov A. V., Crifo J. F., Szegö K., et al. (2002) An advanced physical model of cometary activity. *Planet. Space Sci., 50,* 983–1024.
Rosenberg E. D. and Prialnik D. (2010) The effect of internal inhomogeneity on the activity of comet nuclei — Application to Comet 67P/Churyumov-Gerasimenko. *Icarus, 209,* 753–765.
Rubin M., Hansen K. C., Gombosi T. I., et al. (2009) Ion composition and chemistry in the coma of Comet 1P/Halley — A comparison between Giotto's ion mass spectrometer and our ion-chemical network. *Icarus, 199,* 505–519.
Shi X., Hu X., Sierks H., et al. (2016) Sunset jets observed on Comet 67P/Churyumov-Gerasimenko sustained by subsurface thermal lag. *Astron. Astrophys., 586,* A7.
Shi X., Hu X., Mottola S., et al. (2018) Coma morphology of Comet 67P controlled by insolation over irregular nucleus. *Nature Astron., 2,* 562–567.
Shou Y., Combi M., Toth G., et al. (2016) A new 3D multi-fluid model: A study of kinetic effects and variations of physical conditions in the cometary coma. *Astrophys. J., 833,* 160.
Simon J. B., Armitage P. J., Li R., et al. (2016) The mass and size distribution of planetesimals formed by the streaming instability. I. The role of self-gravity. *Astrophys. J., 822,* 55.
Skorov Y. V. and Rickman H. (1995) A kinetic model of gas flow in a porous cometary mantle. *Planet. Space Sci., 43,* 1587–1594.
Skorov Y. V., van Lieshout R., Blum J., et al. (2011) Activity of comets: Gas transport in the near-surface porous layers of a cometary nucleus. *Icarus, 212,* 867–876.

Suhasaria T., Thrower J. D., and Zacharias H. (2017) Thermal desorption of astrophysically relevant molecules from forsterite(010). *Mon. Not. R. Astron. Soc., 472,* 389–399.
Sunshine J. M. and Feaga L. M. (2021) All comets are somewhat hyperactive and the implications thereof. *Planet. Sci. J., 2,* 92.
Sunshine J. M., A'Hearn M. F., Groussin O., et al. (2006) Exposed water ice deposits on the surface of Comet 9P/Tempel 1. *Science, 311,* 1453–1455.
Szegö K., Sagdeev R. Z., Whipple F. L., et al. (1995) *Images of the Nucleus of Comet Halley, Volume 2: Obtained by the Television System (TVS) On Board the Vega Spacecraft.* European Space Agency, Paris. 255 pp.
Tenishev V., Combi M., and Davidsson B. (2008) A global kinetic model for cometary comae: The evolution of the coma of the Rosetta target comet Churyumov-Gerasimenko throughout the mission. *Astrophys. J., 685,* 659–677.
Tenishev V., Combi M. R., and Rubin M. (2011) Numerical simulation of dust in a cometary coma: Application to Comet 67P/Churyumov-Gerasimenko. *Astrophys. J., 732,* 104.
Tenishev V., Fougere N., Borovikov D., et al. (2016) Analysis of the dust jet imaged by Rosetta VIRTIS-M in the coma of Comet 67P/Churyumov-Gerasimenko on April 12, 2015. *Mon. Not. R. Astron. Soc., 462,* S370–S375.
Tenishev V., Shou Y., Borovikov D., et al. (2021) Application of the Monte Carlo method in modeling dusty gas, dust in plasma, and energetic ions in planetary, magnetospheric, and heliospheric environments. *J. Geophys. Res.–Space Phys., 126,* e2020JA028242.
Thomas N., Boice D. C., Huebner W. F., et al. (1988) Intensity profiles of dust near extended sources on Comet Halley. *Nature, 332,* 51–52.
Thomas N., Davidsson B., El-Maarry M. R., et al. (2015a) Redistribution of particles across the nucleus of Comet 67P/Churyumov-Gerasimenko. *Astron. Astrophys., 583,* A17.
Thomas N., Sierks H., Barbieri C., et al. (2015b) The morphological diversity of Comet 67P/Churyumov-Gerasimenko. *Science, 347,* aaa0440.
Thomas N., Ulamec S., Kührt E., et al. (2019) Towards new comet missions. *Space Sci. Rev., 215,* 47.
Tseng W. L., Bockelée-Morvan D., Crovisier J., et al. (2007) Cometary water expansion velocity from OH line shapes. *Astron. Astrophys., 467,* 729–735.
Tsou P., Brownlee D. E., Anderson J. D., et al. (2004) Stardust encounters Comet 81P/Wild 2. *J. Geophys. Res.–Planets, 109,* E12S01.
Tubiana C., Rinaldi G., Güttler C., et al. (2019) Diurnal variation of dust and gas production in Comet 67P/Churyumov-Gerasimenko at the inbound equinox as seen by OSIRIS and VIRTIS-M on board Rosetta. *Astron. Astrophys., 630,* A23.
Veverka J., Klaasen K., A'Hearn M., et al. (2013) Return to Comet Tempel 1: Overview of Stardust-NExT results. *Icarus, 222,* 424–435.
Vincent J. B., A'Hearn M. F., Lin Z. Y., et al. (2016a) Summer fireworks on Comet 67P. *Mon. Not. R. Astron. Soc., 462,* S184–S194.
Vincent J. B., Oklay N., Pajola M., et al. (2016b) Are fractured cliffs the source of cometary dust jets? Insights from OSIRIS/Rosetta at 67P/Churyumov-Gerasimenko. *Astron. Astrophys., 587,* A14.
Vincent J.-B., Farnham T., Kührt E., et al. (2019) Local manifestations of cometary activity. *Space Sci. Rev., 215,* 30.
Weidenschilling S. J. (1997) The origin of comets in the solar nebula: A unified model. *Icarus, 127,* 290–306.
Williams D. M., Mason C. G., Gehrz R. D., et al. (1997) Measurement of submicron grains in the coma of Comet Hale-Bopp C/1995 01 during 1997 February 15–20 UT. *Astrophys. J. Lett., 489,* L91–L94.
Windmark F., Birnstiel T., Güttler C., et al. (2012a) Planetesimal formation by sweep-up: How the bouncing barrier can be beneficial to growth. *Astron. Astrophys., 540,* A73.
Windmark F., Birnstiel T., Ormel C. W., et al. (2012b) Breaking through: The effects of a velocity distribution on barriers to dust growth. *Astron. Astrophys., 544,* L16.
Youdin A. N. and Goodman J. (2005) Streaming instabilities in protoplanetary disks. *Astrophys. J., 620,* 459–469.
Ytrehus T. (1977) Theory and experiments on gas kinetics in evaporation. In *Rarefied Gas Dynamics* (J. L. Potter, ed.), pp. 1197–1212. *Prog. Astronaut. Aeronaut., Vol. 51,* American Institute of Aeronautics and Astronautics, Washington, DC.
Zakharov V. V., Rodionov A. V., Lukyanov G. A., et al. (2008) Navier Stokes and direct Monte-Carlo simulations of the circumnuclear gas coma: III. Spherical, inhomogeneous sources. *Icarus, 194,* 327–346.
Zakharov V. V., Rodionov A. V., Lukianov G. A., et al. (2009) Monte-Carlo and multifluid modelling of the circumnuclear dust coma: II. Aspherical-homogeneous, and spherical-inhomogeneous nuclei. *Icarus, 201,* 358–380.
Zakharov V. V., Crifo J. F., Rodionov A. V., et al. (2018a) The near-nucleus gas coma of Comet 67P/Churyumov-Gerasimenko prior to the descent of the surface lander Philae. *Astron. Astrophys., 618,* A71.
Zakharov V. V., Ivanovski S. L., Crifo J. F., et al. (2018b) Asymptotics for spherical particle motion in a spherically expanding flow. *Icarus, 312,* 121–127.
Zakharov V. V., Rotundi A., Bockelée-Morvan D., et al. (2023) Stationary expansion of gas mixture from a spherical source into vacuum. *Icarus, 395,* 115453.

Image Credits:

Figs. 1a,b. Courtesy of Alessandra Migliorini through private communication. The figure was adapted from Migliorini et al., *A&A, 589,* A45, 2016, reproduced with permission © ESO.
Fig. 1c. NAC image taken on 2015-07-29 13:25:28 UTC; Credit: ESA/Rosetta/MPS for OSIRIS Team MPS/UPD/LAM/IAA/SSO/INTA/UPM/DASP/IDA published under the Creative Commons license CC BY-SA 4.0.
Fig. 5a. Courtesy NASA/JPL-Caltech (https://photojournal.jpl.nasa.gov/jpeg/PIA13570.jpg).
Figs. 5b,c,d. Courtesy of Silvia Protopapa through private communications. The panels represent composition maps of the coma of Comet 103P/Hartley 2 published by *Protopapa et al.* (2014) and derived from data publicly available through the Small Bodies Node (SBN) of NASA's Planetary Data System (PDS) (*McLaughlin et al.,* 2013).
Fig. 6a. ESA/MPS under Creative Commons Attribution-Share Alike 3.0 igo (https://creativecommons.org/licenses/by-sa/3.0/igo/deed.en).
Fig. 6b. Courtesy NASA/JPL-Caltech (https://photojournal.jpl.nasa.gov/catalog/PIA03501).
Fig. 6c. Courtesy NASA/JPL-Caltech (https://photojournal.jpl.nasa.gov/catalog/PIA05578).
Fig. 6d. Adapted from *A'Hearn et al.* (2008) under a Creative Commons Attribution 4.0 International License (https://creativecommons.org/licenses/by/4.0/).
Fig. 6e. Courtesy of Lori Feaga through private communication. The panel represents a map derived from the current state of calibration of the data first published by *Feaga et al.* (2007) for 9P/Tempel 1. The data from which the map is derived is publicly available through the Small Bodies Node (SBN) of NASA's Planetary Data System (PDS) (*McLaughlin et al.,* 2014).
Fig. 6f. Courtesy NASA/JPL-Caltech (https://photojournal.jpl.nasa.gov/jpeg/PIA13570.jpg).
Fig. 6g. Courtesy of Silvia Protopapa through private communications. The panels represent composition maps of the coma of Comet 103P/Hartley 2 published by *Protopapa et al.* (2014) and derived from data publicly available through the Small Bodies Node (SBN) of NASA's Planetary Data System (PDS) (*McLaughlin et al.,* 2013).
Fig. 6h. Rosetta/OSIRIS/WAC image taken on 2015-04-27 18.17.57.683 UTC; Credit: ESA/Rosetta/MPS for OSIRIS Team MPS/UPD/LAM/IAA/SSO/INTA/UPM/DASP/IDA published under the Creative Commons license CC BY-SA 4.0.
Fig. 6i. Courtesy of David Kappel through private communication. The panel represents a map derived from the current state of calibration of the data (cube I1 00388776027), which is publicly available through the European Space Agency's Planetary Science Archive (PSA, https://www.cosmos.esa.int/web/psa/rosetta). This data was first published by *Fink et al.* (2016).
Fig. 7. Courtesy of Raphael Marschall through private communication. The figure was adapted from data presented in his thesis (*Marschall,* 2017).
Fig. 8a. Fougere et al., *A&A, 588,* A134, 2016, reproduced with permission © ESO.
Fig. 8b. Reprinted from *Icarus, 328,* 104–126, Marschall R., Rezac L., Kappel D. et al., Copyright (2019), with permission from Elsevier.
Fig. 8c. Zakharov et al., *A&A, 618,* A71, 2018, © ESO, published by EDP Sciences, under the terms of the Creative Commons Attribution License (http://creativecommons.org/licenses/by/4.0).
Fig. 9. Reprinted from *Icarus, 328,* 104–126, Marschall R., Rezac L., Kappel D. et al., Copyright (2019), with permission from Elsevier.

Biver N., Dello Russo N., Opitom C., and Rubin M. (2024) Chemistry of comet atmospheres. In *Comets III* (K. J. Meech, M. R. Combi, D. Bockelée-Morvan, S. N. Raymond, and M. E. Zolensky, eds.), pp. 459–498. Univ. of Arizona, Tucson, DOI: 10.2458/azu_uapress_9780816553631-ch015.

Chemistry of Comet Atmospheres

N. Biver
Observatoire de Paris/Paris Sciences and Lettres Research University/Centre National de la Recherche Scientifique (CNRS)/ Sorbonne Université/Université Paris Cité

N. Dello Russo
Johns Hopkins University Applied Physics Laboratory

C. Opitom
University of Edinburgh Royal Observatory

M. Rubin
Space Research and Planetary Sciences, Physics Insitute, University of Bern

---

The composition of cometary ices provides key information on the thermal and chemical properties of the outer parts of the protoplanetary disk where they formed 4.6 G.y. ago. This chapter reviews our knowledge of composition of cometary comae based on remote spectroscopy and *in situ* investigation techniques. Cometary comae can be dominated by water vapor, CO, or CO_2. The abundances of several dozen molecules, with a growing number of complex organics, have been measured in comets. Many species that are not directly sublimating from the nucleus ices have also been observed and traced out into the coma in order to determine their production mechanisms. Chemical diversity in the comet population and compositional heterogeneity of the coma are discussed. With the completion of the Rosetta mission, isotopic ratios, which hold additional clues on the origin of cometary material, have been measured in several species. Finally, important pending questions (e.g., the N deficiency in comets) and the need for further work in certain critical areas are discussed in order to answer questions and resolve discrepancies between techniques.

1. INTRODUCTION

We discuss the current knowledge of the composition of comets as derived from observations of molecular species in their comae. These species either sublimate directly into the coma from the nucleus ices (parent molecules) or they are secondary products. These secondary products can arrive in the coma from either the dissociation of a parent molecule or as a distributed source contained in released ice and dust. We review the investigative techniques that have made considerable progress in determining the composition of cometary comae over the last two decades, including remote spectroscopic observation [from ultraviolet (UV) to radio wavelengths] and *in situ* mass spectrometry (Giotto and Rosetta missions). The measured column density (remote observation) or local density (mass spectrometry) of the given molecule characterizes the local state of the cometary coma at a given time. Observational techniques with a range of spatial coverage and resolution provide information on how molecules are distributed in the coma. The basic assumption of steady-state continuous production of gas or dust from the nucleus is generally used to retrieve the molecular production rates Q_{molec}. We discuss the molecules that dominate the gas flow of the nucleus and drive cometary activity. We then present our current knowledge of all molecules that have been detected in the coma and their abundances relative to other volatiles, most often water (Q_{molec}/Q_{H_2O}), and the source of possible discrepancies between various measurements. Composition can vary significantly from one comet to another, and we review this variability and link it to possible comet origins. Spatial, short, and long-term monitoring of comets, especially as undertaken by the Rosetta mission to Comet 67P/ Churyumov-Gerasimenko (hereafter 67P), have also shown that the composition of the coma can vary both locally and with time or heliocentric distance. Isotopic abundances also provide important insights into the origin of cometary matter. Several key isotopic measurements, such as the D-to-H and $^{14}N/^{15}N$ ratios, which are fundamental to understanding the delivery of cometary material to planets, are presented here. We conclude with needs and perspectives that should improve our current understanding of comet composition.

1.1. Radio Spectroscopy of Rotational Lines

Radio spectroscopy is a powerful technique to probe the composition of cometary atmospheres. All asymmetric molecules have a dipole moment and can be observed in the millimeter to the submillimeter domain via their rotational lines. At the temperature of coma gas (typically 5–200 K), most molecules have their peak rotational emission in the 80–800-GHz frequency domain, probed by ground- and spacebased radio telescopes. In addition, the excitation process is mostly spontaneous emission due to radiative decay of the low-energy rotational levels and can be efficient up to large heliocentric distances, such as in the case of CO detected in at least three comets [C/1995 O1 (Hale-Bopp), 29P, and C/2016 R2 (PanSTARRS)] beyond 5 au from the Sun. Of the ~30 molecular species detected by remote spectroscopy in comets, 80% have been detected in the radio (*Bockelée-Morvan et al.,* 2004; *Biver et al.,* 2015). Radio spectroscopy of comets has the additional unique capability of resolving the line Doppler velocity profile due to its very high spectral resolution ($\nu/\delta\nu$ = 10^6–10^8). Cometary lines are typically 1–4 km s^{-1} wide due to Doppler broadening (see example in Fig. 1) and resolving the line enables measurement of the gas expansion velocity, asymmetry in outgassing, and characterization of opacity effects, such as in Comet 67P with the Microwave Instrument for the Rosetta Orbiter (MIRO) (*Biver et al.,* 2019).

1.1.1. Wide-band high-resolution spectroscopy. Since the advent of millimeter and submillimeter astronomy in the late 1980s, there has been steady progress in receiver and spectrometer performance. For the last 10 years, new wide-band receivers (typically 4–16 GHz instantaneous bandwidth) have equipped many facilities such as the Institut de Radioastronomie Millimetrique (IRAM) [i.e., Eight Mixer Receiver (EMIR); *Carter et al.* (2012)], NOrthern Extended Millimeter Array (NOEMA), or Atacama Large Millimeter Array (ALMA). High spectral resolution ($\nu/\delta\nu > 10^6$) is also necessary to resolve cometary lines in spectra and benefit from the full sensitivity of the receiver. This has become possible with the most recent correlators or Fourier transform spectrometers (FTS) (*Klein et al.,* 2012), which can offer a spectral resolution better than 200 kHz over several gigahertz of bandwidth. An example of the combined (instantaneous) frequency coverage (2 × 8 GHz) and high-resolution (200 kHz) spectrum of Comet 46P is shown in Fig. 2 (*Biver et al.,* 2021a).

1.1.2. Using a large number of lines to detect molecules. Another advantage of the wide frequency coverage of radio receivers and spectrometers is the ability to detect

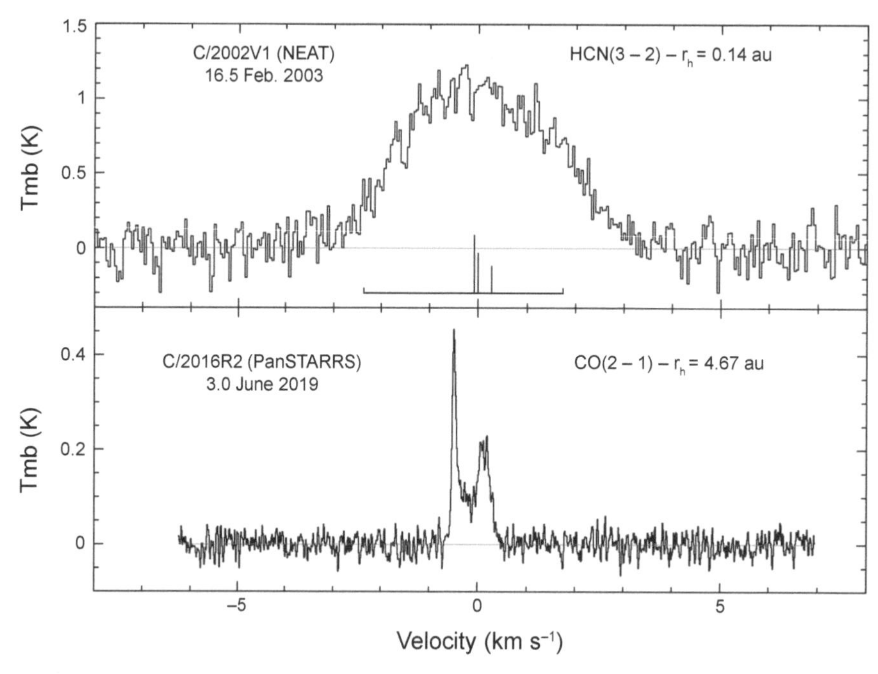

Fig. 1. Cometary lines observed at 1 mm wavelength at high spectral resolution (10–40 kHz). The horizontal axis is the Doppler velocity relative to the comet reference frame. The vertical axis is the line intensity in unit of Kelvins in the main-beam-antenna temperature scale. The top spectrum is the HCN(3–2) line at 265886.434 MHz showing a broad line width due to the large expansion velocity (2.0 km s^{-1}) at a very small heliocentric distance (r_h) (*Biver et al.,* 2011). The positions and intensities of the hyperfine components are indicated below the line. The bottom spectrum is the CO(2–1) line at 230538.000 MHz showing a narrow line in Comet C/2016 R2 far from the Sun. It also exhibits a very narrow blue-shifted component (full width at half maximum = 0.08 km s^{-1}) indicative of a very cold (~3 K) sunward jet.

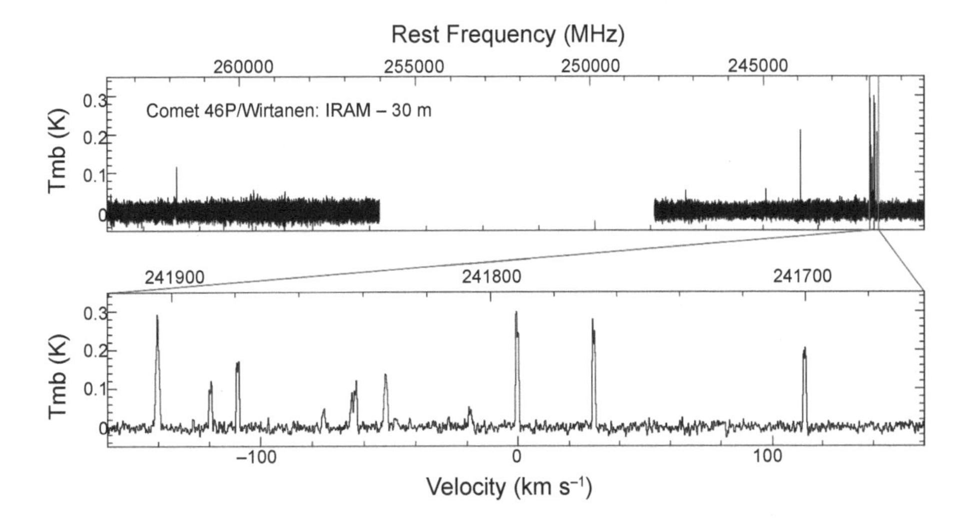

Fig. 2. Spectrum of Comet 46P/Wirtanen observed in December 2018 with the IRAM EMIR230 receiver coupled to the FTS spectrometer. With such a setup 2 × 8 GHz, on each vertical and horizontal polarization (averaged here), is instantaneously covered, with a 200-kHz resolution (zoom on one series of methanol lines at 242 GHz that are resolved).

more complex molecules, which have line emission spectra that encompass a large number of lines with similar intensities. The number of lines increases rapidly with molecular complexity; for example, in the 210–270-GHz domain, CO has one line, H_2CO three lines, and CH_3CHO 64 lines, with intensities comparable within a factor of 3. Averaging many lines of similar expected intensity (weighted according to their signal to noise, or S/N) for a given species increases its detectability; for example, the ability to average 100 lines of similar strength (and S/N) improves detection sensitivity by a factor of 10 over a single line. This method has been applied recently in the detection of complex organics in comets even when individual lines are too weak by themselves to be seen in spectra (e.g., *Biver et al.,* 2015). This technique is also applied to infrared echelle spectra that sample many rovibrational transitions of a given molecule (*Paganini et al.,* 2017). The high spectral resolution of radio spectroscopy and the relatively narrow width of cometary lines also limits confusion between lines of different molecules.

1.2. Infrared Spectroscopy of Vibrational Bands

Many parent molecules have strong vibrational bands in the near-infrared (IR) in the 2–5 μm spectral region. Because there are several windows of high atmospheric transmittance in the near-IR, groundbased observations have been the primary driver of molecular identifications in this spectral region in comets. Since many species have IR vibrational bands in overlapping spectral regions, high spectral resolving power ($\lambda/\delta\lambda > 10^4$) is needed to detect rovibrational lines within vibrational bands (Fig. 3). Even relatively simple molecules can have complex rovibrational spectra creating a forest of lines with multiple contributing species, making high-resolution necessary in order to disentangle these highly diagnostic emissions in IR spectra (Fig. 3).

The IR spectral region provides the only means to detect certain species. For example, symmetric hydrocarbons (e.g., CH_4, C_2H_6, C_2H_4, and C_2H_2) can only be investigated in comets with remote sensing techniques at IR wavelengths. This is also the case for CO_2, one of the main cometary volatiles that under many circumstances drives activity but can only be directly detected from spacebased platforms. Additionally, IR wavelengths provide the most efficient and straightforward method of studying H_2O, the most abundant volatile in comets, through multiple non-resonance fluorescence emissions from groundbased observatories or through its strong fundamental vibrational bands from space. Other molecules typically sampled at IR wavelengths (HCN, CH_3OH, H_2CO, NH_3, CO, and OCS) can also be detected at radio wavelengths (and in the case of CO both radio and UV wavelengths), providing additional independent methods for studying these volatiles in comets:

Derived volatile production rates are determined by temperature-dependent fluorescence efficiencies (g-factors) for individual rovibrational emissions for the given species. Over the years IR fluorescence models have been developed and improved for all molecules that have been detected in comets as well as additional molecules with potentially detectable IR bands (e.g., *Villanueva et al.,* 2018). The temperature-dependent strengths and positions of rovibrational emissions for H_2O, OH, CO, OCS, CH_4, C_2H_6, C_2H_2, CH_3OH, H_2CO, HCN, NH_3, and NH_2 are well enough understood to enable spectral blends to be interpreted and corrected for and production rates to be accurately determined with uniform methodology.

Many additional more complex molecules have strong IR bands that are theoretically detectable even at their lower abundances with the sensitivities of modern groundbased high-resolution IR spectrometers but are in practice largely inaccessible. The reasons for this are related to spectral complexity, which increases fast for larger molecules. First, many parameters needed for developing accurate fluorescence models are unavailable for these complex molecules. Second, spectral confusion from many lines of simpler but

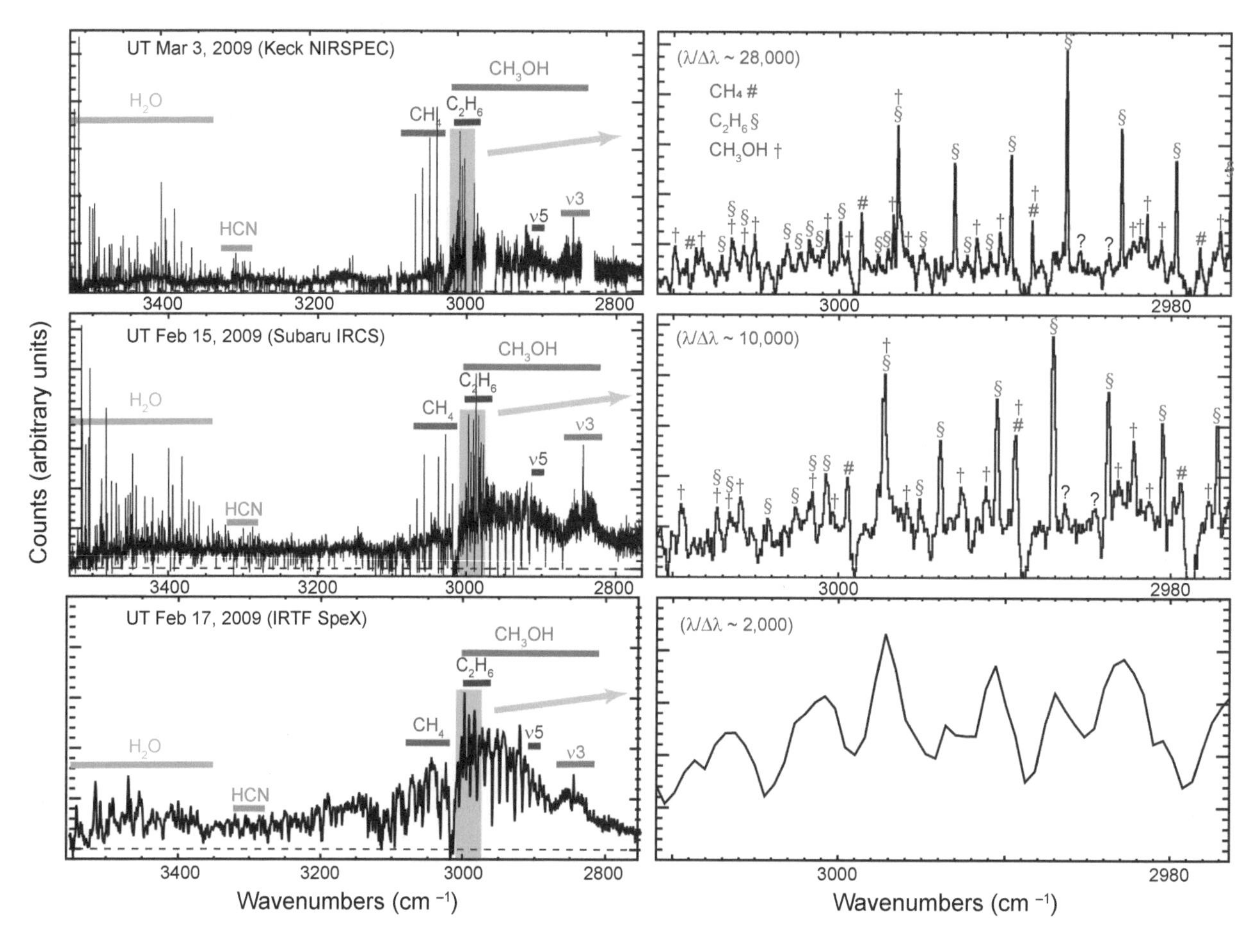

Fig. 3. Near-IR spectra of Comet C/2007 N3 Lulin obtained at different spectral resolving powers. This illustrates the density of rovibrational emissions from multiple species with overlapping vibrational bands in some spectral regions. High spectral resolving power is essential for disentangling and interpreting molecular contributions at IR wavelengths.

more abundant molecules makes definitive identification of more complex but less abundant molecules very difficult. For example, spectral complexity in the 3.35-μm region of comets even at high spectral resolution is dominated by multiple dense emissions of primarily C_2H_6 and CH_3OH (Fig. 3), which interferes with the detection of more complex hydrocarbons (e.g., C_3H_8, C_4H_{10}). The spectral complexity near 3.0 μm is dominated by the presence of many simple molecules (e.g., H_2O, OH, NH_3, NH_2, HCN, C_2H_2), which makes detection of more complex long-chain molecules such as HC_3N and C_4H_2 problematic.

Detection of species from groundbased IR observations can also be hindered by extinction from the terrestrial atmosphere. An example of this is CO_2, an abundant species in comets that has a very strong band near 4.26 μm but is completely obscured in groundbased observations by terrestrial CO_2. Detection of CO and CH_4 in comets depends on a sufficiently large relative geocentric velocity to shift cometary lines from their terrestrial counterparts. Strong lines from other species (e.g., C_2H_2, C_2H_4, and NH_3) can coincidentally be in spectral regions of poor transmittance, making their detection difficult unless atmospheric water vapor is low.

Groundbased observations of comets at mid- to far-IR wavelengths (~5–100 μm) are similarly hindered by terrestrial atmospheric extinction in addition to increasing and rapidly changing thermal emission from telescope and atmosphere. Thus, studies in this wavelength region are mostly done from airborne observatories [e.g., the Stratospheric Observatory for Infrared Astronomy (SOFIA)] or spacebased telescopes (e.g., Spitzer). While most volatile coma molecules are best detected at near-IR and radio wavelengths, the mid- to far-IR can provide information on the composition of the less-volatile dusty coma components (e.g., *Lisse et al.,* 2020).

1.3. Ultraviolet-Visible Spectroscopy of Electronic Transitions

Spectroscopic observations of cometary atmospheres at UV and visible wavelengths sample mostly electronic transitions from secondary products such as OH, NH, CH, CN, C_2, C_3, or NH_2, as illustrated in Fig. 4. Some parent species such as CO or S_2 can also be probed at UV wavelengths. In addition to purely electronic transitions, some prompt emission lines (see, e.g., *Bodewits et al.,* 2022) from O, C, or OH are also detected at visible wavelengths, which can give information on their parent molecules like H_2O and

CO_2. X-ray emissions have also been seen in comets, but this reveals the composition of the solar wind rather than the cometary coma (*Bodewits et al.,* 2022).

1.3.1. Low-resolution spectroscopy and narrow-band imaging. Comets have been observed with low-resolution spectroscopy at visible wavelengths for over a century, focusing on secondary products as illustrated in Fig.4. Even though spectra often have a high density of lines due to the number of molecular emission bands in the visible spectral region, narrow-band imaging and low-resolution spectroscopy were among the first tools available to constrain the composition of cometary atmospheres. Narrow-band filters isolate the light emitted by several gas species as well as the dust-reflected sunlight. When combined with a camera, narrow-band filters enable compositional and morphological studies of the coma. Narrow-band imaging and low-resolution spectroscopy of comets at visible wavelengths are among the most sensitive and routine observations available to study the composition of comets. Visible databases contain several hundred comets observed over decades, providing a means to study comet composition and its diversity statistically (*A'Hearn et al.,* 1995; *Schleicher,* 2008; *Fink,* 2009; *Langland-Shula and Smith,* 2011; *Cochran et al.,* 2012). Those two techniques are covered in detail by *Schleicher and Farnham* (2004) and *Feldman et al.* (2004) and we refer the reader to these publications for an extended discussion.

Because of terrestrial atmospheric extinction, observations blueward of 300 nm can only be performed from space. However, they are instrumental in constraining coma abundances of the main constituents of cometary ices. The main features observable in the UV spectra of comets from parent and fragment species are discussed by *Bockelée-Morvan et al.* (2004) and *Feldman et al.* (2004). An important example is H, one of the main photodissociation products of water, which is detectable through its Lyman-α emission at 121.6 nm.

The last decade has seen a significant increase in the use of integral field unit spectrographs (IFUs), particularly on large telescopes. This technique is especially advantageous for the observation of extended objects because it provides simultaneous spatial and spectral information. Observing comets with an IFU combines the advantages of spectroscopy and narrow-band imaging by enabling mapping of the spatial distributions of several gas species and the dust continuum at different wavelengths simultaneously. Given the rotational variability of comets on timescales as brief as a few hours, IFU observations have the advantage of simultaneous measurements of multiple species in contrast to narrow-band observations where different filters need to be cycled to sample each species. The first observations of comets with IFUs at visible wavelengths are relatively recent (*Dorman et al.,* 2013; *Vaughan et al.,* 2017; *Opitom et al.,* 2019b) but they have demonstrated the potential of this type of observation to study the spatial distributions of secondary products and their production mechanisms in the coma.

1.3.2. High-resolution spectroscopy. The last 30 years have seen the advent of high-resolution spectroscopic observations of comets. Fabry-Perot interferometers, providing high spectral resolutions but covering very limited wavelength ranges, were among the first instruments used to perform high-spectral-resolution observations of comets (*Roesler et al.,* 1985). Crossed-dispersed spectrographs with spectral resolutions from a few tens of thousands to as high as 200,000 and covering larger wavelength ranges are now available on large telescopes [e.g., Ultraviolet and Visual Echelle Spectrograph (UVES) at the Very Large Telescope, Hobby-Eberly Telescope (HET) at McDonald Observatory, High Resolution Echelle Spectrometer (HIRES) at Keck, or High Dispersion Spectrograph (HDS) at the Subaru Telescope]. This new generation of optical instrumentation allows emission lines composing molecular bands to be

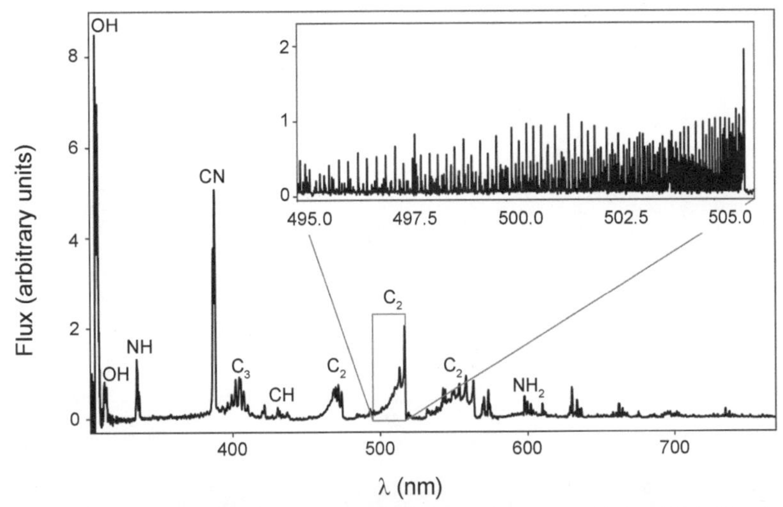

Fig. 4. Optical spectrum of Comet C/2015 ER61 (PANSTARRS) obtained with the ISIS spectrograph at the William Herschel Telescope (A. Fitzsimmons and M. Hyland, personal communication). The inset shows the region of the C_2 ($\Delta v = 0$) band observed with a much higher spectral resolution using the UVES spectrograph at the VLT (*Yang et al.,* 2018).

resolved and has opened a new era of visible observations of comets, as illustrated in Fig. 4.

A resolving power of ~40,000 is sufficient to separate emission lines from isotopologs of the same species. UV and visible high-resolution spectroscopic observations provide the opportunity to measure the isotopic abundance ratios of several key elements (H, C, N, and O). High-resolution spectroscopic observations of comets at optical wavelengths can also resolve ortho and para spin isomers, enabling estimates of the ortho-to-para abundance ratios (OPR) of water and NH_2 (*Kawakita et al.,* 2001; *Shinnaka et al.,* 2012) (see also the chapter in this volume by Aikawa et al. and section 7.6).

High-resolution spectroscopic observations are also necessary to resolve telluric and cometary emission lines. For example, forbidden O lines [O I] around 557, 630, and 636 nm and the sodium D doublet at 589.0 and 589.6 nm are blended with telluric lines at lower spectral resolution. High-resolution spectroscopy is currently only feasible for relatively active comets (with high gas production rates) that are close to the Sun (typically within 3 au); however, high-resolution spectrographs mounted on the next generation of extremely large telescopes, such as HIRES on the Extremely Large Telescope, will more routinely allow these studies in a larger subset of comets.

1.4. In Situ Mass Spectrometry

Another way to investigate the neutral gas environment around comets is *in situ* mass spectrometry. This method provides neutral gas abundances along the trajectory of the spacecraft and is therefore limited to dedicated missions. Neutral gas mass spectrometry has been carried out at two comets by Giotto and Rosetta; the profiles of these missions are discussed below.

1.4.1. Comet mission profiles. In 1986 the Giotto spacecraft flew by Comet 1P/Halley (*Reinhard,* 1986). During the inbound portion of the fast, 68.4-km-s^{-1} flyby the Neutral Mass Spectrometer (NMS) (*Krankowsky et al.,* 1986) carried out continuous measurements of the neutral gas abundances (*Eberhardt,* 1999) until, near closest approach, impacting dust grains led to the failure of the sensor.

The second comet visited by neutral gas mass spectrometers was 67P (*Glassmeier et al.,* 2007). Rosetta carried the Rosetta Orbiter Spectrometer for Ion and Neutral Analysis (ROSINA) (*Balsiger et al.,* 2007) Double Focusing Mass Spectrometer (DFMS) and Reflectron-type Time-Of-Flight (RTOF) on the orbiter. Furthermore, the two lander mass spectrometers, Ptolemy (*Wright et al.,* 2007) and the COmetary SAmpling and Composition experiment (COSAC) (*Goesmann et al.,* 2007) were carried by Philae. Rosetta carried out an extended two-year investigation following 67P from beyond 3 au, through perihelion at 1.24 au, and out again past 3 au. It was at times gravitationally bound to the comet and hence typical relative velocities between Rosetta and 67P were on the order of 1 m s^{-1}, much smaller compared to typical neutral gas velocities on the order of 1 km s^{-1}. Figure 5 shows an example mass spectrum containing parent and fragment species as well as minor isotopologs.

1.4.2. Instrument capabilities and mass resolution. Neutral gas mass spectrometry has some advantages and disadvantages. For instance, the observed atoms and molecules do not require a strong electric dipole moment and hence volatile species such as O_2 and N_2 can be investigated. Furthermore, being *in situ*, Earth's atmosphere will not interfere and there are no optical depth issues requiring reversion to a minor isotopolog with only limited information on the specific isotope ratio (cf. section 7.3). On the downside, the ability to distinguish isomers is limited and is only possible

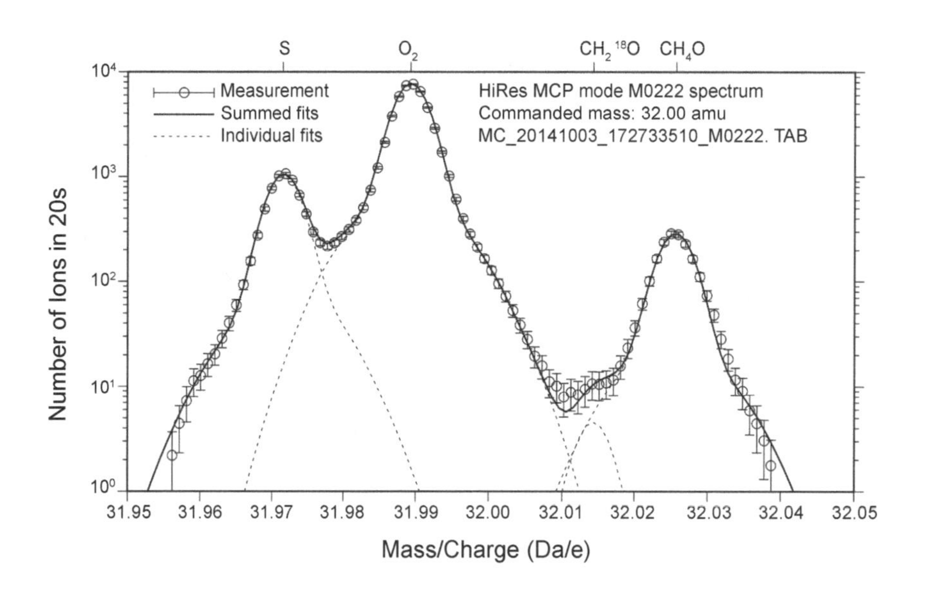

Fig. 5. ROSINA DFMS mass spectra (*Balsiger et al.,* 2007) of mass/charge 32 u/e with sulfur S (in parts a fragment of S-bearing species), molecular oxygen O_2, and methanol CH_3OH. $H_2C^{18}O$ due to formaldehyde and fragmentation of methanol, both with the heavy oxygen isotope, can also be found (cf. *Altwegg et al.,* 2020a).

when the differences associated to the fragmentation occurring during ionization (e.g., $H_2O \rightarrow OH^+$, $H_2O \rightarrow O^+$, etc.) are larger than the measurement uncertainties.

Table 1 lists the mass spectrometers that have carried out neutral gas measurements at comets. The instruments either employed electric and magnetic fields or measured time-of-flight to separate ions by their mass-to-charge. In order to resolve ionized atoms and molecules on the same integer mass line, e.g., O^+ and CH_4^+ at 15.9944 and 16.0308 u/e, respectively, a mass resolution on the order of $m/\Delta m > 1000$ (FWHM) is required. If the resolution does not allow separation of the different contributors to a given mass line, e.g., O^+, NH_2^+, and CH_4^+ on mass/charge 16 u/e, it may still be possible to disentangle their relative proportions: Based on the calibrated fragmentation patterns, the contributions to a given mass line, e.g., $H_2O \rightarrow O^+$ and $NH_3 \rightarrow NH_2^+$ can be subtracted from the total signal on mass/charge 16 u/e leaving only CH_4^+. The signal of CH_4^+ can then be related to the abundance of the cometary parent species methane, CH_4, in the local coma. This approach quickly becomes very complex in gas mixtures due to the large number of involved fragments and isomers.

1.4.3. Ion mass spectrometry. The composition of the neutral gas coma can also be constrained from *in situ* plasma measurements (see the chapter in this volume by Beth et al.). However, this approach requires additional modeling of the involved ionization, dissociation, and chemical reactions between neutrals, ions, and electrons. A suite of processes, such as photoionization, photodissociation, ion-neutral reactions, and charge exchange, as well as ion-electron recombination, have to be included.

For instance, *Allen et al.* (1987) found evidence for the presence of ammonia and methane in Comet 1P/Halley based on Giotto's Ion Mass Spectrometer (IMS) (*Balsiger et al.,* 1987) measurements. Furthermore, the D/H ratio in the water of the same comet was derived from IMS (*Balsiger et al.,* 1995) and ion mode measurements of NMS (*Eberhardt et al.,* 1987). At 67P, the total outgassing activity was estimated from the increasing He^+/He^{2+} ratio of solar wind helium crossing the gas coma and charge-exchanging with neutrals (*Wedlund et al.,* 2016; *Hansen et al.,* 2016). The same process, when high charge state solar wind minor ions undergo charge exchange excitation with neutrals, led to the X-ray emission observed at Comet C/1996 B2 (Hyakutake) (*Häberli et al.,* 1997) (cf. section 1.3).

2. FROM OBSERVATIONS TO ABUNDANCES

For estimating relative abundances in ices that sublimate near the surface of the nucleus, either for the parent molecules directly observed or the assumed parent of a secondary product species, we convert the measured quantities, a line intensity converted into a column density or a local density (*in situ* measurements) into production rate Q. The first step for spectroscopic observations requires modeling the excitation of considered energy levels of the molecule and radiative transfer. In all cases a knowledge of the spatial distribution of the molecule is needed. Most studies use the Haser model (*Haser,* 1957) to describe the density of a sublimating parent molecule. Remote spectroscopic observations nowadays are often two-dimensional or three-dimensional data cubes that combine spectral information at high resolution and one-dimensional (long-slit spectroscopy) to two-dimensional (data cubes, interferometric maps) spatial information. This is used to characterize the local density of the molecules and reveal evidence for departure from the basic Haser model: outgassing asymmetry and radial distribution for non-nucleus sources.

2.1. Deriving Production Rates

The processes that lead to excitation of molecular energy levels are multiple: collisions with neutral gas that tend to establish a local thermal equilibrium (LTE) at the local gas temperature, collisions with electrons created in the coma and from the solar wind, and radiative processes. The radiative environment for the inner coma is dominated by the solar radiation in the UV to IR range, the cosmic black-body background at 2.73 K in the radio range, and dust and nucleus thermal emission in the IR. Self-absorption by optically thick lines in the IR to submillimeter range can also play an important role. Generally, UV to IR emission will result from fluorescence excitation (absorption followed by rapid emission of photons due to the short lifetime of the electronic and vibrational states excited in this process) while rotational transitions in the radio domain result from

TABLE 1. Neutral gas mass spectrometers operated at comets.

Instrument	Craft	Mass Separation Technique	Mass Range (u/e)	Resolution m/Δm	Reference
NMS	Giotto	EM-fields	1–56*	unit†	*Krankowsky et al.* (1986)
ROSINA DFMS	Rosetta	EM-fields	12 to 180	3000 at 1%	*Balsiger et al.* (2007)
ROSINA RTOF	Rosetta	Time of Flight	1 to >300	>500 at 50%	*Balsiger et al.* (2007)
COSAC	Philae	Time of Flight	1 to 1500	300 at 50%	*Goesmann et al.* (2007)
Ptolemy	Philae	Time of Flight	12 to 150	unit	*Wright et al.* (2007)

* 1–38 u/e in nominal double focusing mode.
† Unit mass resolution in the range 12–56 u/e.

spontaneous emission due to slow decay of the rotational levels (lifetimes of 10^2–10^6 s) populated by collisions or radiative processes, including pumping via vibrational bands. Dissociation products of molecules can also radiate via prompt emission when they are created in an excited electronic state, such as CO Cameron bands in the UV or OH in the IR. All processes have been reviewed in detail in *Bockelée-Morvan et al.* (2004) or *Bodewits et al.* (2022). State-of-the-art models adapted to comets need to take into account non-steady-state calculation specific for cometary comae in which the gas is flowing away from the nucleus and the density seen by a given molecule decreases with time (*Marschall et al.,* 2022).

Several parameters, such as the gas temperature and the expansion velocity, can be inferred directly from observations: rotational temperatures measured in the IR or the radio from series of rovibrational or rotational lines and analysis of the line profiles resolved in the radio where width is directly related to the gas expansion velocity.

2.2. Spatial Distribution of Molecules in the Coma

While imaging cometary dust is often the easiest way to study coma morphology, it provides an incomplete picture as gas and dust spatial distributions are generally distinct. Determining and comparing the spatial distributions of volatile gases in the coma of comets can provide information both about their sources (direct sublimation from ice and/or extended coma sources) and about how ices are associated or separated in the nucleus. Both spacecraft and groundbased studies have provided abundant evidence that some molecules are principally released directly from their sublimating ices while others are released from the dissociation of other molecules in the coma; however, the contribution from these sources for a given molecule often differs from comet to comet and the progenitor species of secondary products are often difficult to identify. When no spatial information is available, it is difficult to assess whether some molecules are primarily parents or products. This can affect the accuracy of derived production rates if the parentage is wrongly assumed. Spatial studies of comets span a large range in scale and resolution, from the highest-resolution studies of the inner coma by *in situ* spacecraft to groundbased studies with various aperture sizes and different comet geocentric distances. The addition of spacecraft data and increasing groundbased capabilities has allowed more detailed studies of molecular spatial distributions in the coma.

2.2.1. Narrowband imaging and interferometric maps. Studies of cometary coma features have been conducted for many years at optical wavelengths and extending from the near-UV to the near-IR. Modern telescopes and imagers allow extensive spatial coverage in the coma while also providing subarcsecond spatial resolution over large-format CCD arrays. Broadband continuum filters have provided detailed information on the dust structure in the comae of comets. Using targeted narrowband filters and continuum removal techniques or IFUs has allowed the isolation of coma gas structure, which is needed to provide complete information on coma morphology because coma gas and dust distributions are generally distinct. Gas phase species sampled at these wavelengths are product species or ions (e.g., OH, NH, CN, C_2, C_3, CO^+, H_2O^+) (e.g., *Schleicher and Farnham* 2004). For certain bright comets the consistency of assumed parent scale lengths for these species can be tested, which is important as many of these values are poorly constrained. With increasing capabilities at radio and IR wavelengths, coordinated observations of the spatial distributions of parents and secondary products can be performed, providing a more direct comparison of parent-product associations in individual comets.

Interferometric maps at radio wavelengths provide information on the spatial and velocity distributions of gas molecules and dust. Modern heterodyne receivers allow the determination of high-resolution line-of-sight velocities that enable three-dimensional spatial structures of measured molecules to be obtained (e.g., *Qi et al.,* 2015). The importance of radio interferometry in obtaining spatial maps for molecules released from different sources has been demonstrated (e.g., *Boissier et al.,* 2007; *Cordiner et al.,* 2014, 2017; *Roth et al.,* 2021). Maps for molecular species show dramatic differences in their spatial distributions, suggesting heterogeneity in the way ices are stored and associated within the nucleus. Under favorable circumstances radio interferometery with facilities such as NOEMA, ALMA, or the Submillimeter Array (SMA) has determined the spatial distribution of parent and product species such as H_2CO, HCN, HNC (e.g., Fig 6), CH_3OH, CO, or CS. Future capabilities with facilities such as the Next-Generation Very Large Array (ngVLA) will allow higher sensitivities and cover longer wavelengths, allowing NH_3 and OH to be mapped in the comae of comets at high spatial resolution (e.g., *Cordiner and Qi,* 2018).

2.2.2. Infrared spatial profiles. The long-slit capabilities of IR groundbased observing enables the determination and comparison of spatial distributions of molecules in the inner coma. These comparisons address whether species observed simultaneously relate to a common or different outgassing sources and suggest potential nucleus associations. IR spatial analysis of comets has routinely determined the spatial distributions of H_2O, C_2H_6, CH_3OH, and HCN in many comets observed to date. Spatial distributions for CO and CH_4 are also readily obtained when the comet's geocentric velocity is sufficient to Doppler-shift cometary emissions from their atmospheric counterparts. For some of the brightest comets, spatial distributions have also been obtained for H_2CO, C_2H_2, NH_3, and OCS. Previous groundbased IR studies have suggested separate polar and apolar ice phases in the nucleus of some comets (e.g., *Mumma and Charnley,* 2011); however, this association by ice polarity is not seen in some cases (e.g., *Dello Russo et al.,* 2022). IR spatial analysis has revealed surprising evidence for extended sources for some species (Fig. 7a), as well as showing possible volatile associations of ices in the nucleus (Fig. 7b) (e.g., *Dello Russo et al.,* 2022).

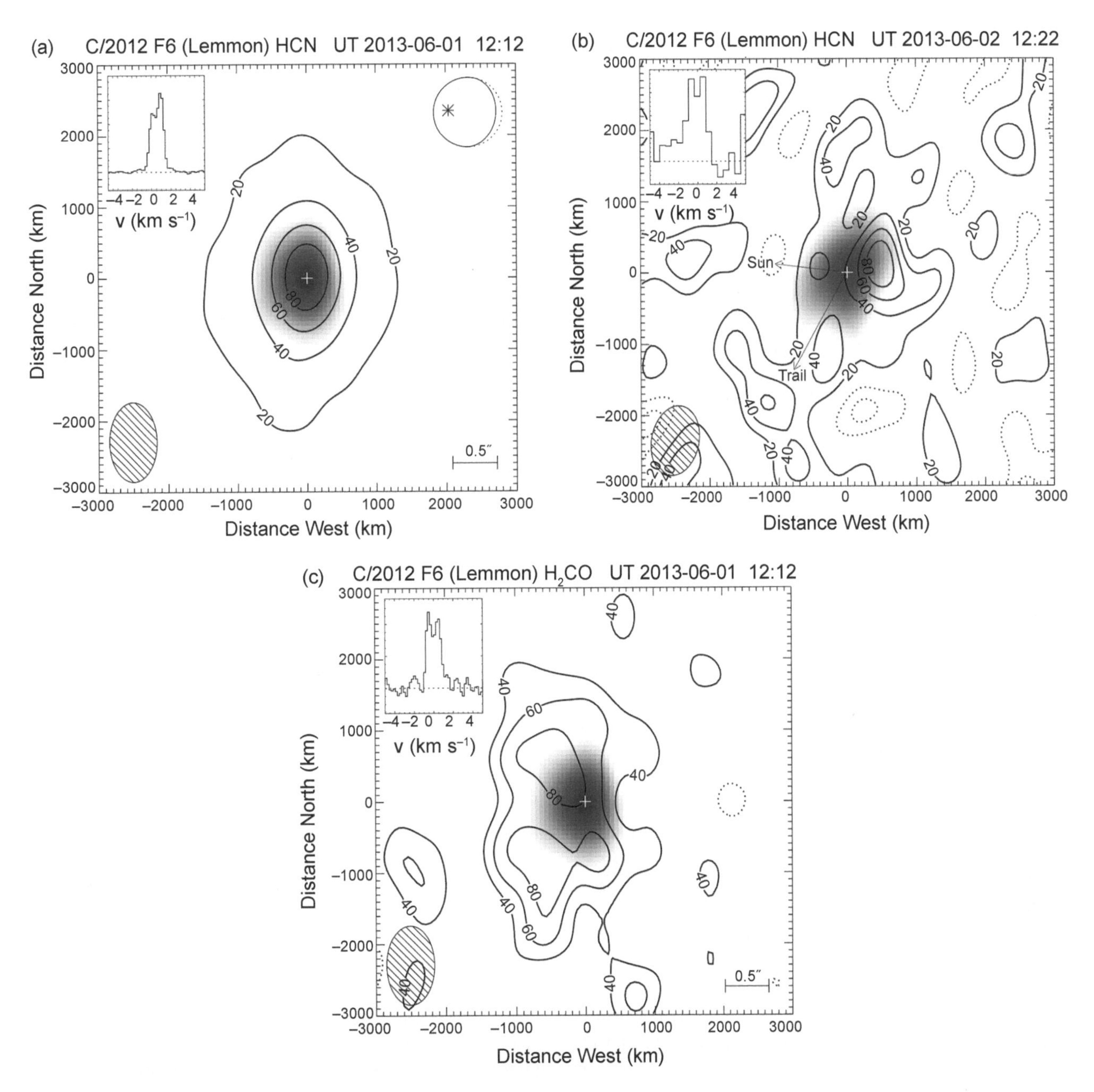

Fig. 6. ALMA interferometric contour maps of **(a)** HCN, **(b)** HNC, and **(c)** H_2CO in C/2012 F6 (Lemmon). The signal distributions of HNC and H_2CO are much less peaked due to their production in the coma in contrast to HCN coming mostly from the nucleus (*Cordiner et al.,* 2014). In the lower left corners the hatched ellipse shows the shape and size of the synthetic beam. The spectrum at the central point is shown in the upper left and the continuum flux is displayed in blue.

Technological advances in groundbased IR instrumentation have allowed spatial analysis to be done in more detail and on a rapidly increasing number of comets. This work is beginning to address which spatial properties are specific to individual comets and which are more general properties of all comets.

*2.2.3. **In situ radial profiles.*** Spacecraft missions have performed the highest-resolution studies of how volatiles are released into the inner comae of comets. In contrast to remote sensing observations, which derive spatial properties over the global coma, *in situ* studies can reveal volatile and dust release from specific areas on the nucleus. The Extrasolar Planet Observation and Deep Impact Extended Investigation (EPOXI) flyby revealed a two-lobed nucleus for 103P/Hartley 2 with the bulk of the activity emanating from the smaller lobe (*A'Hearn et al.,* 2011). Release from the small lobe was dominated by the spectral signature of CO_2 gas with entrained H_2O ice chunks, whereas H_2O vapor primarily sublimated from a different region along the waist between the lobes (*A'Hearn et al.,* 2011). EPOXI spatial studies of volatiles were based on low-

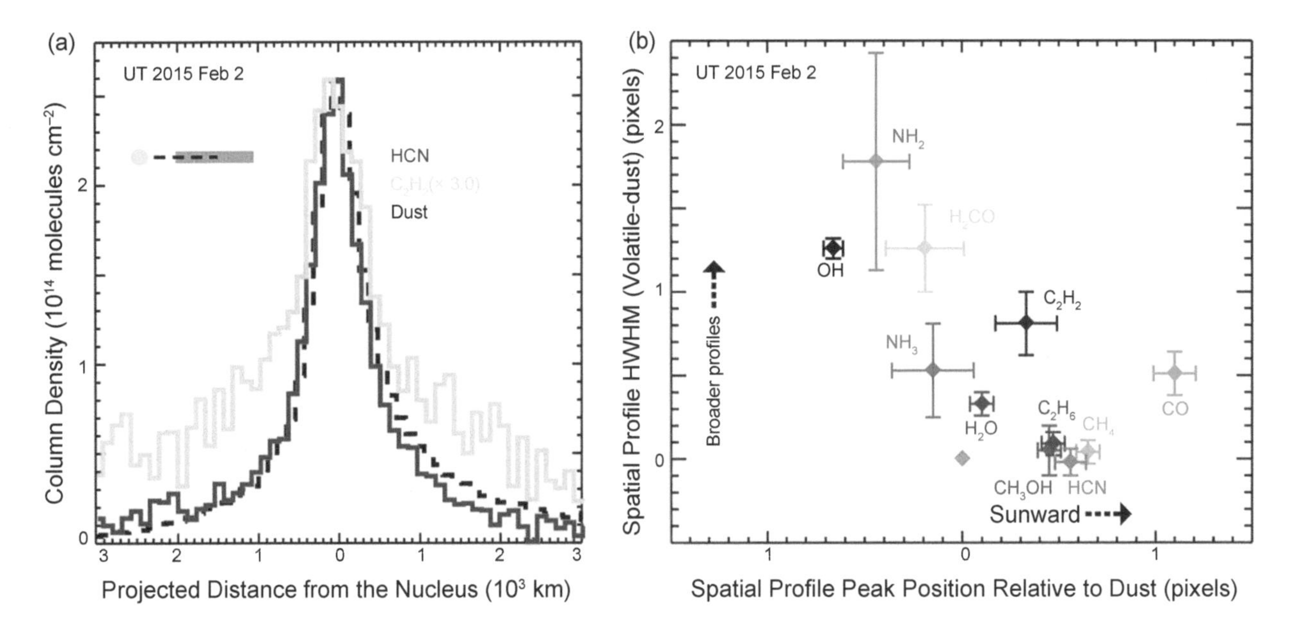

Fig. 7. **(a)** Spatial profiles of C_2H_2, HCN, and dust in C/2014 Q2 (Lovejoy) showing evidence for a C_2H_2 extended source due to excess flux on the profile wings compared to HCN. **(b)** Spatial profile properties for volatiles and dust in C/2014 Q2 (Lovejoy) showing the distinct characteristics of spatial profiles based on profile peak position and width (*Dello Russo et al.,* 2022).

resolution IR spectroscopic measurements during a flyby and therefore were a brief snapshot of the volatile coma structure of 103P and mainly sensitive to H_2O and CO_2, the strongest IR molecular bands. Rosetta studies of 67P enabled tracking of long-term molecular associations in the coma of many species with multiple instruments. In contrast to 103P, Rosetta observed dominant H_2O outgassing directly from ices in the nucleus of 67P and not from coma icy grains (e.g., *Luspay-Kuti et al.,* 2015), suggesting a more direct association between coma gas and nucleus ice distributions.

Processes like sublimating icy grains and photochemistry are imprinted in the gas densities along radial profiles from the nucleus. Marked deviations from $1/r^2$ (assuming constant expansion velocity) indicate contribution from a distributed source, either through sublimation from grains in the coma or dissociation of larger molecules. At 67P, *Altwegg et al.* (2016) reported that the total gas density dropped with cometocentric distance as expected but glycine was most likely released from water ice on dust particles (*Hadraoui et al.,* 2019). Based on Giotto NMS results at 1P/Halley, substantial distributed sources were reported for H_2CO, CO (*Meier et al.,* 1993; *Eberhardt,* 1999), and possibly H_2S, while CH_3OH behaved differently (*Eberhardt et al.,* 1994). During the flyby, the Giotto spacecraft covered a wide range in phase angle and cometocentric distances (*Reinhard,* 1986), and hence NMS measured gas released during different times from distinct regions on the nucleus with varying illumination. Therefore, temporally and spatially inhomogeneous outgassing could result in similar spatial profiles (cf. Fig. 3 of *Rubin et al.,* 2009).

3. DRIVERS OF COMET ACTIVITY

Water is the most abundant species in cometary ices and the main driver of activity for comets within 2–3 au from the Sun. At larger heliocentric distances, the water sublimation rate decreases, and water alone is not sufficient to sustain cometary activity. More volatile ices, like CO and CO_2, are believed to drive the distant activity of comets.

3.1. CO and Distant Activity

CO was first detected in comets close to the Sun in the UV via resonant fluorescence in the Fourth Positive Group ($A^1\Pi-X^1\Sigma^+$) with sounding rockets and satellites (*Bockelée-Morvan et al.,* 2004, and references therein). Since its first direct cometary detection in 29P/Schwassmann-Wachmann 1 in 1993 by *Senay and Jewitt* (1994) and *Crovisier et al.* (1995), CO has been detected in the radio and in the IR in several comets at large heliocentric distances, and up to 14 au in Comet C/1995 O1 (Hale-Bopp) (*Biver et al.,* 2002a). CO is one of the most volatile molecules, along with N_2, detected in comets. Due to its volatility, it can sublimate as far out as the Kuiper belt [T_{sub} ~25 K (*Fray and Schmitt,* 2009)]. Hence its signature has been searched for in many icy objects, including transneptunian objects (TNOs) and Centaurs (*Bockelée-Morvan et al.,* 2001; *Jewitt et al.,* 2008; *Womack et al.,* 2017). CO sublimation might have been responsible for transient or outburst activity in Centaurs like 95P/Chiron (*Womack et al.,* 2017) or 174P/Echeclus (*Wierzchos et al.,* 2017), but this has not been confirmed due to the difficulty to detect CO in these distant objects. Nevertheless, in several comets observed beyond 2–3 au (Table 2), a substantial outgassing of CO has been detected and often dominates the activity of the comet ($Q_{CO}/Q_{H_2O} > 1$). In those comets, CO sublimation can lift dust into the coma even at distances more than 20 au from the Sun. Radio lines also often show a strong blueshift (e.g., the bottom spectrum of Fig. 1), suggesting that CO sublimation is enhanced at the warmer subsolar point. This does not necessarily mean that the ices of these comets are dominated by CO: Comets like C/1995 O1 (Hale-Bopp) or C/2009 P1 (Garradd), when they were closer to the Sun (1.5 au or less), had comae dominated by water vapor with CO abundances around 5–25%. In fact, the abundance of CO relative to water in comets varies by at least 2 orders of magnitude (0.3 to 35%; Table 3 and Figs. 9 and 10) in comets observed within 1.5 au from the Sun. JFCs [e.g., 67P (*Läuter et al.,* 2020)] tend to be depleted in CO with abundances generally not exceeding 3% relative to water. Comets with low CO abundances may have most of their CO trapped in water ice and CO_2 ices (*Rubin et al.,* 2023), so even beyond the sublimation of water, CO generally doesn't control activity in JFCs. Finally, some comets like C/2016 R2 (*Biver et al.,* 2018; *McKay et al.,* 2019) (see also section 6.1.2) seem to be dominated by CO outgassing even at a distance where H_2O outgassing becomes significant in most comets.

3.2. Carbon Dioxide as Driver of Activity

The CO_2 abundance is very difficult to estimate from groundbased observations and most measurements of the CO_2 abundance in comets come from infrared space observatories [Infrared Space Observatory (ISO), Spitzer, AKARI] or spacecraft (Deep Impact, EPOXI, Rosetta). Indirect measurements can be made from observation of the CO Cameron band in the UV and forbidden O lines in the optical. The AKARI survey (*Ootsubo et al.,* 2012) has revealed that in most cases CO_2 is a major constituent of cometary atmospheres, with abundances typically ranging from 5% to 30% with no clear differences based on dynamical origin. Beyond 2.5 au from the Sun the CO_2 abundance relative to water increases and CO_2 can become the major constituent of the coma, driving the activity of the comet. Even closer to the Sun, the outgassing of CO_2 can dominate the activity of the nucleus, as in the case of (3552) Don Quixote [Spitzer (*Mommert et al.,* 2020)] or 103P/Hartley [EPOXI (*A'Hearn et al.,* 2011)].

The intricate pattern of activity in 67P is a result of the peculiar shape of its nucleus and associated illumination together with the obliquity of its rotation axis of 52° (*Sierks et al.,* 2015) and the different volatility of the involved gases (*Fray and Schmitt,* 2009). The gas activity induced a decrease in the rotation period (from 12.4 to 12.0 hr) as the comet passed perihelion (*Kramer et al.,* 2019; *Keller et al.,* 2015b). Furthermore, the transport and redeposition of (icy) grains around the nucleus (*Keller et al.,* 2017) leads to very different surface morphologies (*El-Maarry et al.,* 2015) leaving their imprint on the structures of the gas coma.

Figure 8 shows a time series of H_2O and CO_2 measurements late in the Rosetta mission at a heliocentric distance of ~3 au, just after the outbound equinox. While the spacecraft orbited the comet at close distances in the terminator plane, it passed over regions in the south where CO_2 was the main driver of activity while H_2O dominated above the northern hemisphere. At that time the subsolar latitude climbed from southern to northern latitudes; however, the southern hemisphere did remain more active for the

TABLE 2. Comets dominated by CO outgassing.

Comet	Date	r_h (au)	Q_{CO} (molec.s^{-1})	Q_{CO}/Q_{H_2O}	Reference
C/1995 O1 (Hale-Bopp)*	May 1996	4.6	6×10^{28}	4.5	*Biver et al.* (2002)
	Sep. 1996	3.3	15×10^{28}	~1	*Biver et al.* (2002)
	Aug. 1997	2.3	24×10^{28}	~1	*Biver et al.* (2002)
	Dec. 1997	3.9	10×10^{28}	3.8	*Biver et al.* (2002)
C/1997 J2 (Meunier-Dupouy)	Mar. 1998	3.1	0.39×10^{28}	†	*Biver et al.* (2018)
29P/Schwassmann-Wachmann 1	May 2010	6.2	4×10^{28}	~10	*Bockelée-Morvan et al.* (2010)
C/2006 W3 (Christensen)	Nov. 2009	3.3	3×10^{28}	>2.2	*Bockelée-Morvan et al.* (2010)
	Aug. 2010	4.9	1.2×10^{28}	~5	*Bonev et al.* (2017), *de Val-Borro et al.* (2014)
C/2009 P1 (Garradd)	Apr. 2012	2.1	3×10^{28}	0.6	*Feaga et al.* (2014)
C/2016 R2 (PanSTARRS)	Feb. 2018	2.8	8×10^{28}	~300	*McKay et al.* (2019)
C/2017 K2 (PanSTARRS)	Feb. 2021	6.7	0.16×10^{28}	†	*Yang et al.* (2021)
174P/Echeclus	Jun. 2016	6.1	0.08×10^{28}	†	*Wierzchos et al.* (2017)

*For Comet Hale-Bopp, for which activity was monitored from 7 au pre-perihelion to 14 au post-perihelion, we provide the CO/H_2O ratio for the most distant detections of H_2O and the time when it crossed the $Q_{CO} = Q_{H_2O}$. Water outgassing dominated between September 1996 and August 1997.

†Q_{H_2O} not measured, likely low.

TABLE 3. Molecular abundances in cometary atmospheres.

Molecule	Name	Abundance Relative to Water (%)		
		From Radio	From IR	*In Situ* in 67P*
CO	carbon monoxide	<1.23–35	0.3–26	0.3–3†
CO_2	carbon dioxide	—	4–30	7.0†
CH_4	methane	—	0.15–2.7	0.4†
C_2H_6	ethane	—	0.1–2.7	0.8†
C_2H_2	acetylene	—	0.03–0.37	—
C_2H_4	ethylene	—	0.2	
C_3H_8	propane	—	—	0.018 ± 0.004
C_6H_6	benzene	—	—	0.00069 ± 0.00014
C_7H_8	toluene	—	—	0.0062 ± 0.0012
CH_3OH	methanol	0.7–6.1	<0.13–4.3	0.5–1.5†
H_2CO	formaldehyde	0.13–1.4§	<0.02–1.1	0.5
HCOOH	formic acid	0.04–0.58	—	0.013
CH_3CHO	acetaldehyde	0.05–0.08	—	0.047
c-C_2H_4O	ethylene oxide	<0.006	—	
$(CH_2OH)_2$	ethylene glycol	0.07–0.35	—	0.011
CH_3OCH_2OH	methoxymethanol	<9	—	
$HCOOCH_3$	methyl formate	0.06–0.08	—	0.0034
CH_2OHCHO	glycolaldehyde	0.016–0.039	—	
CH_3COOH	acetic acid	<0.026	—	
C_2H_5OH	ethanol	0.11–0.19	—	0.10†
CH_3OCH_3	dimethyl ether	<0.025	—	
CH_3COCH_3	acetone	≤0.011	—	0.0047
C_2H_5CHO	propanal	—	—	
CH_2CO	ketene	≤0.0078	—	
N_2	molecular nitrogen	(<0.002–1000 from N_2^+)‡		0.089 ± 0.024
NH_3	ammonia	0.18–0.60	0.1–3.6	0.4†
HCN	hydrogen cyanide	0.05–0.25	0.03–0.5	0.20†
HNC	hydrogen isocyanide	0.0015–0.035	—	
CH_3CN	methyl cyanide	0.008–0.054	—	0.0059
HC_3N	cyanoacetylene	0.002–0.068	—	0.0004
HNCO	isocyanic acid	0.01–0.62	—	0.027
NH_2CHO	formamide	0.015–0.022	—	0.004
C_2H_3CN	vinyl cyanide	<0.0027		
C_2H_5CN	ethyl cyanide	<0.0036		
C_2N_2	cyanogen	—	—	0.0004±0.0002
H_2S	hydrogen sulfide	0.09–1.5	—	2.0†
SO	sulfur monoxide	0.04–0.30	—	0.071
SO_2	sulfur dioxide	0.03–0.23	—	0.127
CS	carbon monosulfide	0.03–0.20	—	—
CS_2	carbon disulfide	—	—	0.02†
CH_3SH	methyl mercaptan	<0.023	—	0.038
HF	hydrogen fluoride	0.018	—	0.003–0.048
HCl	hydrogen chloride	<0.011	—	0.002–0.059

TABLE 3. (continued)

Molecule	Name	Abundance Relative to Water (%)		
		From Radio	From IR	*In Situ* in 67P*
HBr	hydrogen bromide	—	—	0.00012–0.00083
PH_3	phosphine	<0.07	—	<0.003
PN	phosphorus nitride	<0.003	—	<0.001
PO	phosphorus oxide	<0.013	—	0.011
CH_3NH_2	methylamine	<0.055		
NH_2CH_2COOH	glycine I	<0.18	—	0.000017
Ar	argon			0.00058 ± 0.00022
Kr	krypton			0.000049 ± 0.000022
Xe	xenon			0.000024 ± 0.000011
O_2	molecular oxygen	—	—	2.0†

Table updated from information in *Bockelée-Morvan and Biver* (2017), *Biver et al.* (2021a), *Rubin et al.* (2019), *Dello Russo et al.* (2016a), *Lippi et al.* (2021), and *Biver et al.* (2024).

*Based on *Biver et al.* (2019), *Rubin et al.* (2019), *Läuter et al.* (2020), and *Hänni et al.* (2021).
†Value based on the integrated gas loss over the whole Rosetta mission (2 years).
‡From observations in the visible-UV.
§Assuming a daughter distribution.

remaining part of the mission out to almost 4 au despite the subsolar latitude increasing again above 20° north. CO_2 remained the dominant molecule and a similar picture was obtained on a global scale from studies of the H_2O and CO_2 gas production rates of 67P at this distance (*Combi et al.,* 2020; *Läuter et al.,* 2020). The same behavior was also observed very early on in the mission, beyond 3 au on the inbound path of the comet. *Hässig et al.* (2015) reported an elevated CO_2/H_2O ratio above the southern hemisphere from which pristine ices are sublimating due to the substantial erosion rates (*Keller et al.,* 2015a) observed during the intense but short summer months (*Capria et al.,*

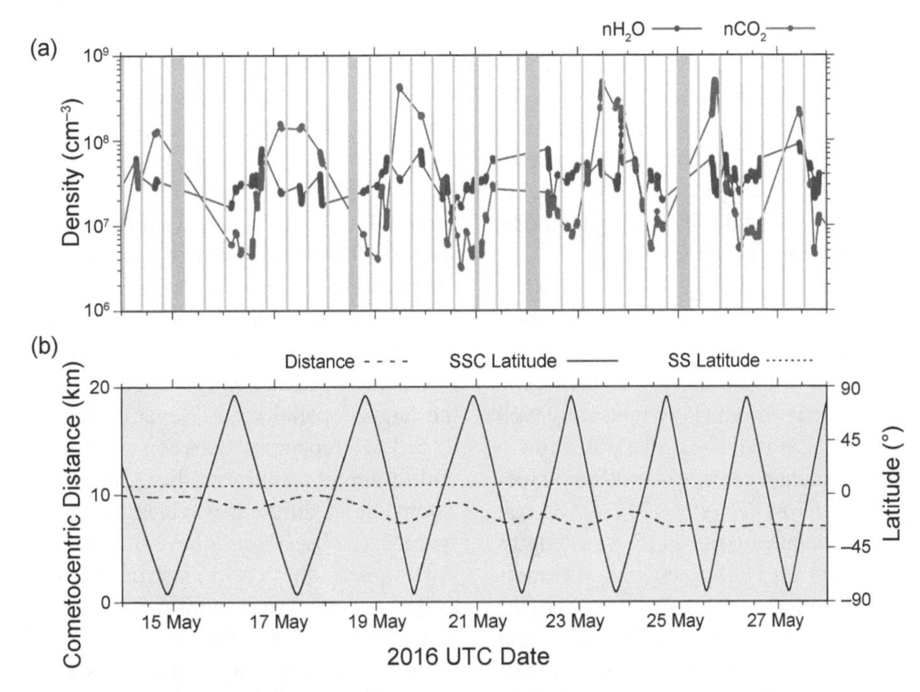

Fig. 8. (a) H_2O and CO_2 densities at 67P measured in May 2016 (post outbound equinox) at ~3 au by ROSINA DFMS (cf. *Gasc et al.,* 2017) during close terminator orbits. Gray areas mark times of thruster operations when the instrument was off, other gaps due to different measurement mode. **(b)** Cometocentric distance (<10 km), subspacecraft latitude (red: north, summer; blue: south, winter) and subsolar latitude.

2017). Local enhancements in CO_2/H_2O ratio have also been identified by remote sensing observations of the [O I] emission (*Bodewits et al.,* 2016).

3.3. Monitoring of H_2O Production

H_2O is the dominant volatile ice in comets and as such determining its production rate provides a measure of the overall comet volatile productivity when the comet is within about 2–3 au from the Sun. H_2O is also the molecule to which the abundances of all other volatile species are generally compared. Thus, determining H_2O production rates and coma spatial distribution is an important goal in overall investigations of comet composition. H_2O production in comets can be directly measured or inferred through its secondary product species by several techniques over a range of wavelengths from both groundbased and spacebased observatories.

3.3.1. Direct observations of H_2O. H_2O has strong IR vibrational bands near 2.7 and 6.3 μm, associated with the O-H stretching and bending modes respectively that can be easily detected from remote or *in situ* spacebased observations (e.g., *Weaver et al.,* 1987; *Crovisier,* 1997; *A'Hearn et al.,* 2011; *Bockelée-Morvan et al.,* 2015). H_2O can also be detected from space through rotational line transitions at radio wavelengths (e.g., *Neufeld et al.,* 2000; *Lecacheux et al.,* 2003; *Hartogh et al.,* 2011). Because of extinction from atmospheric H_2O, IR fundamental bands and rotational lines are generally inaccessible in comets from groundbased observing platforms. However, groundbased IR observations have routinely detected H_2O in comets through the years via non-resonance fluorescence emissions that are not opaque in Earth's atmosphere (e.g., *Mumma et al.,* 1996; *Dello Russo et al.,* 2000; *Villanueva et al.,* 2012; *Faggi et al.,* 2016). The spectral coverage of modern groundbased IR instruments allows the simultaneous sampling of H_2O with many other volatile species, providing a direct measure of relative volatile abundances and comparison of coma spatial distributions that eliminate uncertainties due to temporal variability in volatile release and coma dynamics.

3.3.2. Tracers of H_2O: OH, H, O. Because H_2O is by far the dominant source of OH, H, and O in the coma of comets close to the Sun, H_2O production rates can be inferred from measurements of these species, as branching ratios of H_2O to these secondary products are generally well known. Observations of OH at radio wavelengths allow a determination of H_2O production rates, gas outflow velocity, and asymmetries (e.g., *Crovisier et al.,* 2013). OH can also be detected through prompt emission at IR wavelengths and has been used as a tracer for H_2O production in comets (e.g., *Bonev et al.,* 2006). Optical and near-UV wavelengths enable the detection of H, O, and OH from both ground- and spacebased observatories (e.g., *Combi et al.,* 2019; *McKay et al.,* 2013; *Opitom et al.,* 2015). Because of the sensitivity of radio, optical, and near-UV techniques for observing H_2O product species, these have been the main methods over the years for long-term monitoring of volatile production in comets within an apparition. IR observatories and instrumentation now also have the sensitivity for long-term monitoring of H_2O production in comets, but are generally limited by telescope availability for such studies.

3.3.3. Water from icy grains. H_2O is characterized by its low volatility compared to other measured coma species, which has implications for its outgassing behavior. First, H_2O sublimation is generally not fully activated in a comet until it is between about 2 and 3 au from the Sun. Second, while H_2O sublimates into the coma directly from nucleus ices, H_2O icy grains can also be dragged into the coma by more volatile gases and survive as ice in the coma longer than other measured volatiles. The prevalence of H_2O icy grains in the coma may depend on the abundances of more volatile species such as CO_2 or CO that can also drive activity. For example, the correlation of CO_2 gas and H_2O ice in the inner coma of 103P/Hartley 2 suggests that areas of gas sublimation rich in CO_2 may have dragged H_2O ice into the coma from below the nucleus surface (*A'Hearn et al.,* 2011). From groundbased observatories, the presence of icy grains in the coma can often be inferred by the spatial distribution of H_2O molecules in the global coma compared to other volatiles. Because H_2O icy grains can survive longer in the coma than other volatiles, care must be taken when (1) determining abundances of volatiles relative to H_2O especially when the aperture size is small when projected on the comet, and (2) comparing H_2O production rates determined by techniques with significantly different aperture sizes.

3.3.4. Discrepancies in derived water production rates. Measurements of water abundances from different tracers are not always consistent, even for measurements close in time. Water production rates measured at 2 au pre-perihelion in Comet C/2009 P1 (Garradd) ranged from about 8×10^{28} molecules s^{-1} from high-resolution IR observation of H_2O with the CRyogenic high-resolution InfraRed Echelle Spectrograph (CRIRES) on the Very Large Telescope (VLT) (*Paganini et al.,* 2012) to $2–2.9 \times 10^{29}$ molecules s^{-1} from H_2O observations with the Herschel Heterodyne Instrument for the Far Infrared (HIFI) instrument, OH observations with Swift's UltraViolet and Optical Telescope (UVOT) in the near-UV (*Bockelée-Morvan et al.,* 2012; *Bodewits et al.,* 2014), and from H I observation with the Solar Wind Anisotropies (SOHO) instrument onboard the Solar and Heliospheric Observatory (SOHO) (*Combi et al.,* 2013) on the largest spatial scale. Several factors could play a role in the discrepancies between production rates measured with different techniques. First, instruments used for measurements at different wavelengths and from ground- and spacebased observatories have different fields of view (FoV). For example, the SWAN instrument onboard SOHO, used to derive water production rates from H I emission, has a FoV of 1° × 1°. On the other end, near-IR spectrographs [such as VLT's CRIRES or Keck's Near Infrared Echelle Spectrograph (NIRSPEC)] tend to have small slits on the order of a few arcseconds. Second, the use of different coma density models [e.g., the Haser model (*Haser,* 1957)] vs. the vectorial model (*Festou and Feldman,* 1981a) can also lead to differences in the derived production rates (*Cochran and Schleicher,* 1993). Finally, model parameter assumptions, such as scale lengths or gas velocities, can also impact derived production rates (*Cochran and Schleicher,* 1993; *Fink,* 2009).

4. VOLATILE COMPOSITION OF THE COMA

Around 30 molecules have been identified remotely in cometary atmospheres with an abundance relative to water often varying by 1 order of magnitude. Figure 9 shows the range of abundances measured and the number of comets in which they have been detected. The Rosetta mission has nearly doubled this number of molecules. The present status of abundances measured in comets and in 67P is summarized in Table 3 and detailed hereafter.

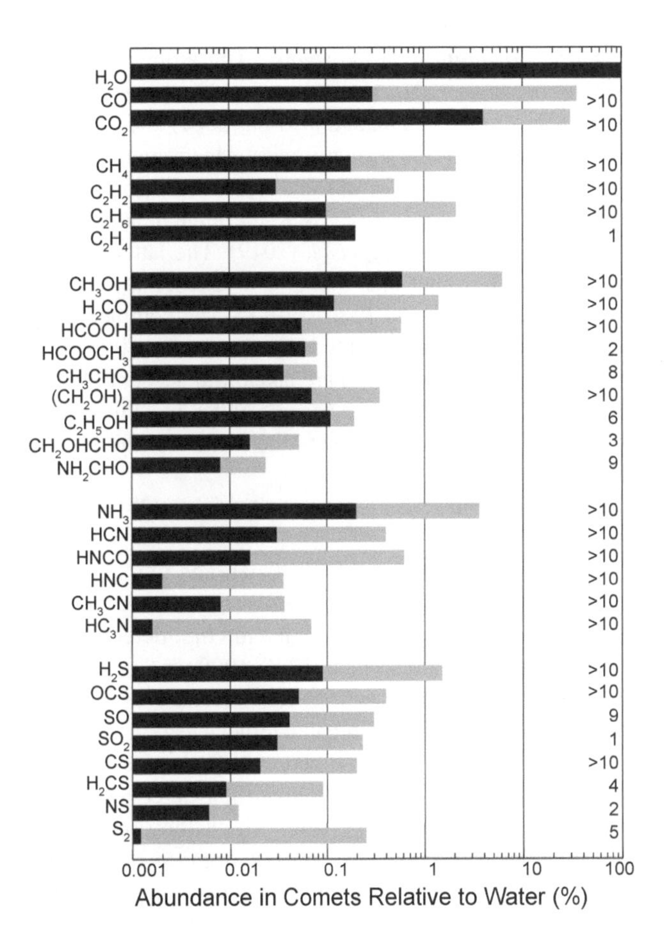

Fig. 9. The range of abundances relative to water for the molecules observed remotely in cometary comae. The number of comets in which the molecule has been detected is indicated to the right (in light gray). From *Bockelée-Morvan and Biver* (2017).

4.1. Hydrocarbons

Symmetric hydrocarbons such as CH_4, C_2H_6, C_2H_2, and C_2H_4 are investigated remotely in the IR. Based on the growing database of comet observations there is evidence that CH_4, C_2H_6, and C_2H_2 are depleted in Jupiter-family comets (JFCs) compared to long-period comets from the Oort cloud (OCCs). These and more complex hydrocarbons were detected in the coma of 67P using the ROSINA mass spectrometer during the Rosetta mission.

4.1.1. Methane (CH_4). CH_4 is the simplest and most volatile saturated hydrocarbon (C_nH_{2n+2}), so its relative abundance in comets may provide clues about the region of formation and processing history of cometary ices (*Gibb et al.,* 2003). CH_4 has a strong IR fundamental band (ν_3) near 3.3 μm; however, its detection in comets from Earth-based observatories requires a sufficient cometary geocentric velocity ($|d\delta/dt| > \sim 15$ km s^{-1}) to shift comet lines away from corresponding terrestrial counterparts. CH_4 is typically one of the most abundant hydrocarbons in comets with mixing ratios CH_4/H_2O ranging from ~0.15% to 3%, with typical abundances ~0.8% (*Dello Russo et al.,* 2016a, and references therein).

4.1.2. Ethane (C_2H_6). The C_2H_6 ν_7 band near 3.35 μm is favorable for detection in comets owing to the large intrinsic strength and a pile-up of rovibrational lines within Q-branches. Lines from the ν_5 band of C_2H_6 near 3.45 μm are also generally detected although the strongest lines are several times weaker than the ν_7 Q-branches. The terrestrial C_2H_6 component is very weak so no specific geocentric Doppler-shift is required for its detection in comets from the ground. Mixing ratios C_2H_6/H_2O range from ~0.1 to 2%, with typical abundances ~0.6% (*Dello Russo et al.,* 2016a, and references therein), so mixing ratios of C_2H_6/CH_4 ~1 in comets are typical.

4.1.3. Acetylene (C_2H_2). Rovibrational lines from the ν_3 and $\nu_2 + \nu_4 + \nu_5$ vibrational bands of C_2H_2 near 3 μm have been detected in many comets. However, detection of C_2H_2 is not routine owing to its low abundances, presence of lines in regions of low atmospheric transmittance, and blending with stronger lines of other species. Mixing ratios C_2H_2/H_2O range from <0.5%, with typical abundances ~0.1% (*Dello Russo et al.,* 2016a, and references therein). The typically low abundances of C_2H_2 relative to C_2H_6 in comets may indicate the importance of H-atom addition reactions on pre-cometary ices in the nebula.

4.1.4. Other hydrocarbons. Although C_2H_4 emissions are numerous in the ~3.2–3.35-μm region, they are relatively weak, located in many areas of generally low atmospheric transmittance, and are subject to blends from stronger and more abundant species. For these reasons C_2H_4 has been detected in only two comets: 67P from Rosetta ROSINA measurements (*Altwegg et al.,* 2017b) and C/2014 Q2 (Lovejoy) from groundbased IR measurements with C_2H_4/H_2O ~0.2% (*Dello Russo et al.,* 2022). Other more complex hydrocarbons also have strong IR bands but they have not been detected from the ground because of various factors including low abundances, spectral confusion with more abundant species (mostly CH_3OH), and lack of adequate fluorescence models. Many long carbon-chain molecules were detected for the first time in 67P with ROSINA (e.g., *Altwegg et al.,* 2017b), showing the complexity of hydrocarbons stored in comets.

4.1.5. Aromatic hydrocarbons. The aromatic hydrocarbons benzene, C_6H_6, and toluene, C_7H_8, were detected by ROSINA in 67P (*Schuhmann et al.,* 2019) and the latter also tentatively by the Ptolemy mass spectrometer on the Philae lander (*Altwegg et al.,* 2017b). Furthermore, tentative detections of xylene (C_8H_{10}) and naphtalene ($C_{10}H_8$) were reported by *Altwegg et al.* (2019). The latter is most likely associated with dust grains and released at elevated temperatures (*Lamy and Perrin,* 1988). The complexity of detected hydrocarbon species increases when sublimation from dust grains in the coma are taken into account (*Hänni et al.,* 2022). Such grains can become hotter by up to several hundreds of Kelvins compared to the surface of the comet.

4.2. CHO-Species and Complex Organics

Organic molecules containing C, H, and O (CHO) are significant constituents of cometary comae comprising about 3–5% relative to water. Since the first detections of methanol and formaldehyde in 1986–1990 via *in situ* mass spectrometry and radio and IR remote observations, the number and complexity of organic molecules discovered in cometary comae has steadily increased (*Bockelée-Morvan et al.,* 2004). They generally follow their detection in the interstellar medium (ISM), from which they may have inherited their diversity. Complex organic molecules (COMs) are often defined by astrophysicists as CHO-bearing molecules with six atoms or more, but there is no strict definition.

4.2.1. Methanol (CH_3OH). Methanol is the most abundant COM in cometary comae. It was first suggested and subsequently shown to be mainly responsible for the 3.52-μm feature in the low-resolution IR spectrum of Comet 1P/Halley in 1986 and in spectra of subsequently observed comets such as C/1989 X1 (Austin) and C/1990 V1 (Levy), in which its detection in the millimeter range via rotational lines was unambiguous (*Bockelée-Morvan et al.,* 2004, and references therein). Since then, methanol has been routinely detected in comets, both in the radio via its rotational lines and with high-resolution IR spectrometers via its rovibrational P, Q, R branches of the ν_2, ν_3, ν_9 bands in the 3.3–3.5-μm range. At radio wavelengths, multiple lines (up to 70 in bright comets) are often observed and used to probe the gas temperature (*Biver et al.,* 2015, 2021a). As of 2024, methanol has been detected in about 76 comets (70 in the radio and 40 in the IR, including two with *in situ* mass spectrometry) with an abundance relative to water ranging from 0.5% to 6%, within 2 au from the Sun. IR and radio observation often yield similar production rates, with methanol-rich comets exhibiting 3–4% methanol relative to water, and methanol-poor comets around 1%, but the distribution is not clearly bimodal (Fig. 10).

4.2.2. Formaldehyde (H_2CO). Formaldehyde was first unambiguously identified in a comet on 1P/Halley with the NMS instrument on the Giotto spacecraft. Subsequently H_2CO was detected in the radio in C/1989 X1 (Austin) and C/1990 K1 (Levy) (*Colom et al.,* 1992) and in the IR via its ν_1 band at 3.59 μm in 153P/Ikeya-Zhang (*DiSanti et al.,* 2002). Measured abundances range from 0.12% to 1.4% relative to water for the total production of H_2CO including distributed sources.

4.2.3. Complex organics. Including HCOOH, complex CHO-species have been observed in several comets in the radio, for example, in C/1996 B2 (Hyakutake), C/1995 O1 (Hale-Bopp), and C/2014 Q2 (Lovejoy). *In situ* mass spectrometry in Comet 67P has identified COMs, some not yet detected in the ISM.

- *Formic acid (HCOOH)* has been detected in 10 comets with an estimated abundance between less than 0.04% and 0.58% relative to water. But abundances can be uncertain by a factor of 2 or more owing to poor constraints on its photodissociation rate *(Biver et al.,* 2021a, 2024).
- *Acetaldehyde (CH_3CHO)* exhibits multiple lines of similar intensities that helps its detection in wide-band spectroscopic surveys. It has been detected in seven comets with very similar abundances of 0.05–0.08% relative to water.
- *Ethylene-glycol* in its lowest energy conformer *aGg'-(CH_2OH_2)* has been identified in the radio spectra of seven comets with an abundance of 0.07–0.35% relative to water, which is a substantial percentage for such a large molecule. It is notably higher than measurements by Rosetta (*Rubin et al.,* 2019) (but it was not measured at the time of peak activity) and measured values in star-forming regions (*Biver and Bockelée-Morvan,* 2019).
- *Methyl formate ($HCOOCH_3$)* has only been detected in two comets in the radio (~0.07% relative to water). It is the most abundant isomer of mass 60, with *glycoladelhyde (CH_2OHCHO)* 4× less abundant, and *acetic acid (CH_3COOH)* at least 2× less abundant in comet C/2014 Q2 (Lovejoy) (*Biver et al.,* 2015). These molecules appear depleted in 67P compared to previous groundbased measurements (Table 3).
- *Alcohols: Ethanol (C_2H_5OH)* has been identified in four comets with an abundance around 0.1%, 1 order of magnitude lower than methanol. More complex alcohols have been identified in 67P by Rosetta's ROSINA instrument (*Altwegg et al.,* 2019).

The more complex CHO-molecules, especially those with three carbons, have not been detected with remote observation due to their complex spectra and decreasing abundance with molecular complexity. Only ROSINA had the sensitivity to detect species such as C_3H_6O, C_3H_8O, C_4H_8O, $C_4H_{10}O$, and $C_5H_{12}O$ (*Altwegg et al.,* 2019), but other techniques are required to distinguish isomers. Again, the complexity increased when dusty periods were taken into account (*Hänni et al.,* 2023).

4.3. Nitrogen-Bearing Species

Nitrogen-bearing volatiles are generally detectable both at IR and radio wavelengths. Evidence from both remote

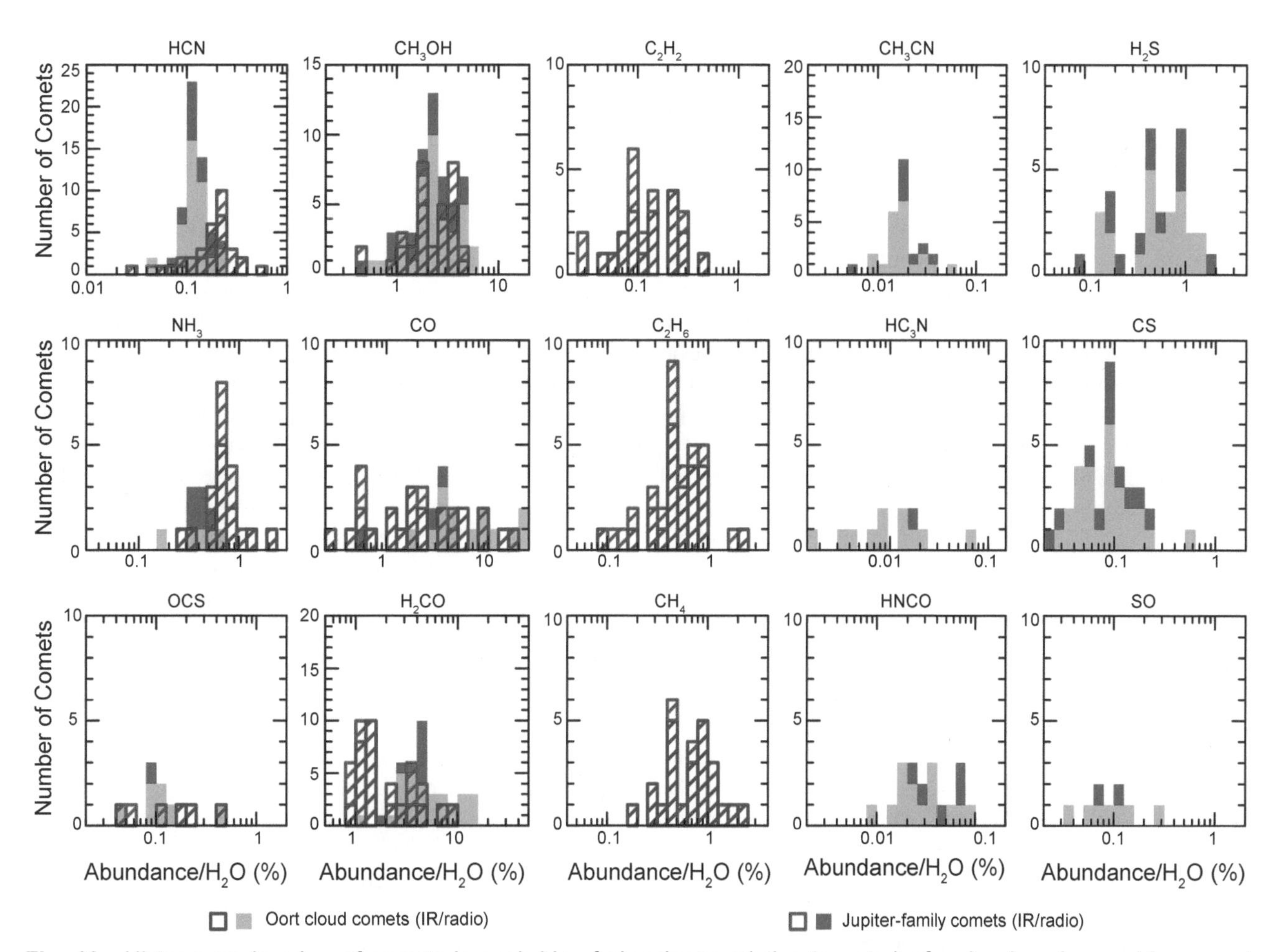

Fig. 10. Histograms (number of comets in each bin of abundance relative to water) of molecules observed in several comets both in the radio and the infrared. Radio data correspond to filled bars and infrared are hatched. Blue-green colors are for comets originating from the Oort cloud (OCC), on periodic to parabolic orbits of any inclination. Red-pink colors are for JFCs in low-inclination orbit with a short period. HCN shows more systematic scatter from infrared data while in the other cases similar behavior is observed from the two techniques, infrared being able to sample comets with lower CO abundance. Excepted for CO, which is less abundant on average in JFC [by a factor of 4 (*Dello Russo et al.*, 2016a)], there is no significant difference between JFCs and OCCs, with a scatter in abundance that can reach 1 order of magnitude.

sensing and *in situ* missions have established that comets contain many of the necessary ingredients for life from simple HCN and NH_3 to the amino acid glycine. Some N-bearing species show spatial distributions consistent with extended sources in the coma, providing further evidence for complex N molecules in comets.

4.3.1. Molecular nitrogen (N_2). The presence of N_2 in comets has until recently been a matter of debate as its direct detection from radio or IR techniques is not possible. However, N_2^+ can be observed at near-UV wavelengths to trace N_2 abundances in comets. Detections of the $B^2\Sigma_u^+$–$X^2\Sigma_g^+$ First Negative Band (0,0) band of N_2^+ at 391.4 nm have been claimed in the tails of comets from low-resolution spectroscopic observations, first in photographic spectra (*Swings and Haser,* 1956; *Cochran et al.,* 2000) and subsequently in the coma of 1P/Halley (*Wyckoff and Theobald,* 1989) and C/1987 P1 (Bradfield). However, these early detections have been questioned because at low spectral resolution N_2^+ emissions from the comet and the terrestrial atmosphere can be difficult to differentiate. More reliable detections of N_2^+ have only been reported in the last decade: in the coma of C/2002 VQ94 (LINEAR) and C/2016 R2 (PANSTARRS) using high spectral resolution (*Korsun et al.,* 2014; *Cochran and McKay,* 2018; *Opitom et al.,* 2019a) and in the coma of the centaur 29P/Schwassmann-Wachmann 1 from low-resolution spectroscopic observations (*Ivanova et al.,* 2016). In addition to the (0,0) band, the (1,1) and (0,1) bands were also detected in Comet C/2016 R2 (*Opitom et al.,* 2019a). The presence of N_2 in comets was confirmed by its detection in 67P by ROSINA onboard Rosetta (*Rubin et al.,* 2015a), showing this highly-volatile molecule is retained in the nucleus of an evolved JFC, most of it embedded in H_2O and CO_2 ices (*Rubin et al.,* 2023).

While N_2 is surely the dominant N-bearing volatile in some comets, its extremely high volatility makes it susceptible to depletion by evolutionary effects, so its abundance is probably highly variable in comets, e.g., in 67P NH_3 contributed more to the N reservoir than N_2 (*Rubin et al.,* 2019).

The N_2/CO ratio in cometary ices can be estimated from groundbased detection of CO^+ and the N_2^+ emission band at optical wavelengths, as was done for three comets (*Korsun et al.,* 2014; *Cochran and McKay,* 2018; *Ivanova et al.,* 2016). Inferring the relative abundance of neutrals from the ion emission intensity ratio is not always straightforward but *Raghuram et al.* (2021) showed that this can be done when the ion production is controlled by photon and photoelectron impact ionization of the neutrals (rather than ion-neutral chemistry). Rosetta provided a direct measure of the relative abundance [$N_2/CO = 0.029 \pm 0.012$ (*Rubin et al.,* 2019) (see Table 3)], about 5× higher than the value measured earlier in the mission at 3.1 au (*Rubin et al.,* 2015a). The evolutionary difference between the northern and southern hemispheres (*Keller et al.,* 2015a) coupled with the different illumination conditions during the two observation phases led to differences in the measured N_2/CO ratio, similar to the observations of the ratios of other species (*Läuter et al.,* 2020; *Combi et al.,* 2020). The presence of both species puts constraints on the maximum temperature of the material inside 67P since its formation. When comparing the measured N_2/CO ratio to the estimated primordial solar system ratio (*Lodders,* 2010) a formation temperature well below 30 K for amorphous ices (cf. *Rubin et al.,* 2015b, 2020) and <50 K for crystalline ices was derived (*Mousis et al.,* 2016a). These results may be further modified when taking into account how these species may not be present in their pure ice forms but trapped in either CO_2 or H_2O ices (*Kouchi and Yamamoto,* 1995; *Rubin et al.,* 2023).

4.3.2. Ammonia (NH_3). Ammonia is an important biogenic molecule as it was likely an intermediary for synthesis of amino acids on the early Earth (*Bernstein et al.,* 2002; *Muñoz Caro et al.,* 2002). NH_3 can be targeted at both radio and IR wavelengths in comets but it has been measured in fewer comets than HCN owing to the relative difficulty of its detection. Mixing ratios NH_3/H_2O range from <0.1% to 5%, with typical abundances ~0.5–1% (*Dello Russo et al.,* 2016a, and references therein).

4.3.3. Hydrogen cyanide (HCN). HCN is the simplest nitrile parent volatile as well as an important intermediary for synthesis of biochemical compounds and can be routinely detected in comets at both millimeter and IR wavelengths. HCN/H_2O mixing ratios in comets are typically ~0.1–0.2%, with no obvious differences between JFCs and long-period comets. Notably, production rates of HCN in comets derived at millimeter wavelengths are typically about a factor of 2 smaller than those derived at IR wavelengths (*Magee-Sauer et al.,* 2002; *Biver et al.,* 2002). The reason for this difference has not been discovered. Modern capabilities at IR and radio wavelengths can provide sensitive spatial maps of HCN in the coma of comets. Results generally suggest that HCN in the coma comes predominately from direct sublimation from nucleus ices.

4.3.4. Hydrogen isocyanide (HNC). Hydrogen isocyanide has been detected in several comets at millimeter wavelengths. HNC also has strong vibrational bands near 5 μm, but these transitions have not been detected in comets to date. The HNC/HCN mixing ratio has been shown to increase with decreasing heliocentric distance, with a mean value around 7% at 1 au (*Lis et al.,* 2008).

4.3.5. Other N-bearing molecules. Additional N-bearing molecules (e.g., NH_2CHO, HNCO, CH_3CN, and HC_3N) have been detected in a few comets at radio wavelengths with mixing ratios on the order of 10–30% with respect to HCN (Table 3) (*Bockelée-Morvan and Biver,* 2017). More complex N-bearing molecules have also been searched for in the radio and some detected by ROSINA, but in all cases they are minor contributors to the N budget in cometary atmospheres.

- *CH_3CN* and *HC_3N* were first identified in Comets Hyakutake and Hale-Bopp and have been detected in over 10 comets during the last 20 years. CH_3CN is routinely detected in cometary comae, typically representing 10–30% of the HCN abundance. On the other hand, the abundance of HC_3N can be more variable: Generally between only 2% and 13% relative to HCN, its abundance in Comet C/2002 X5 at 0.2 au from the Sun was found to be up to 40% relative to HCN (*Biver et al.,* 2011). A variation with heliocentric distance of the HC_3N abundance is not excluded.
- *Isocyanic acid (HNCO)* has also been detected in over 10 comets since 1996. Its abundance (Table 3) can vary between about 0.2 and 4 times that of HCN. This acid could also be associated with ammonium salts and its spatial distribution, and the heliocentric variation of its abundance needs to be investigated.
- *Formamide (NH_2CHO)* has only been detected in a handful of comets (Fig. 9), with some uncertainties on its abundance due to an unknown lifetime (*Biver et al.,* 2021a). Nevertheless, it does not exceed 2×10^{-4} relative to water in comets and formamide could also be produced in the coma (*Cordiner and Charnley,* 2021).

More complex N-bearing molecules have also been identified in comets (section 4.5.3).

4.3.6. The apparent nitrogen depletion in comets. *Geiss* (1988) reported an apparent lack of N in the coma of Comet 1P/Halley when compared to the Sun, meteorites, and terrestrial samples. These results were based on the abundances of the major N-bearing molecules NH_3 and HCN (including HNC) obtained during the Giotto flyby. Similarly, abundances of NH_3, HCN, and other minor N-bearing species in most comets measured with remote sensing techniques are consistent with this observed N deficiency (Table 3). Gases in the coma of 67P were analyzed (*Rubin et al.,* 2019), this time also including N_2 (*Rubin et al.,* 2015a), which revealed a similar underabundance of N.

The heliocentric distance dependence of NH_3/H_2O in comets shows evidence for increasing ratios for smaller heliocentric distances, up to >3% for Comet D/2012 S1 (ISON) (*Altwegg et al.,* 2020b; *Dello Russo et al.,* 2016a). This is consistent with the additional release of NH_3 from grains when a thermal threshold is reached. It has been suggested that ammonium salts may be present on the surface of 67P, similar to the surface of some observed asteroids, based on a broad spectral reflectance feature (*Quirico et al.,* 2016; *Poch et al.,* 2020). Ammonium salts are made up of the NH_3 base and an acid providing an H^+ to the base. Mass spectrometric measurements in the coma, most likely due to dust grains entering the ion source of ROSINA, revealed all possible sublimation products of the six different ammonium salts $NH_4^+Cl^-$, $NH_4^+CN^-$, $NH_4^+OCN^-$, $NH_4^+HCOO^-$, $NH_4^+CH_3COO$, and NH^+SH^- (*Altwegg et al.,* 2020b, 2022). Direct identification of ammonium salts is difficult as they mostly dissociate into the base and corresponding acid upon sublimation of the salt. Ammonium salts may hence be released from distributed grain sources that subsequently release simpler volatiles into the coma. Assuming that the NH_4/H_2O ratio of comets close to the Sun is more representative of the actual content of N, the apparent lack of cometary N could be an observational bias related to the difficulty in directly measuring N contained in less-volatile sources.

4.4. Sulfur-Bearing Species

Eight sulfureted molecular species (or radicals) have been identified in multiple comets from remote observations primarily at radio wavelengths: H_2S, CS, SO, SO_2, OCS, H_2CS, NS, and S_2 (Table 3, Fig. 9). The OCS ν_3 vibrational band has also been detected in the infrared at 4.85 μm (e.g., *Dello Russo et al.,* 1998; *Saki et al.,* 2020). S_2 has only been detected in the UV via electronic transitions around 295 nm (e.g. *Reylé and Boice,* 2003). Atomic S has also been observed in the UV (e.g., *Meier and A'Hearn,* 1997) but is interpreted as mostly a secondary photodissociation product of the other sulfureted species. The total S/O ratio observed in the volatile phase ranges from 0.5% to 2%, in agreement with *in situ* measurements in the coma of Comet 67P by ROSINA (*Rubin et al.,* 2019; *Calmonte et al.,* 2016).

4.4.1. Hydrogen sulfide (H_2S). H_2S, the dominant sulfur-bearing molecule in comets, has been detected in 34 comets since 1990 (*Bockelée-Morvan et al.,* 1991), almost exclusively by its rotational lines at 168.8 and 216.7 GHz or by *in situ* neutral gas mass spectrometry in Comet 67P and ion mass spectrometry at Comet 1P/Halley (*Eberhardt et al.,* 1994). Its abundance relative to water can vary by a factor of 16 from 0.09% in the carbon-chain-depleted JFC comet 21P to 1.5% in Comet Hale-Bopp. However, no systematic differences in H_2S abundances are seen between JFCs and OCCs. Rosetta measured an average abundance of 1.1% relative to water (*Calmonte et al.,* 2016) typical of a relatively H_2S-rich comet. In H_2S-depleted comets, the total abundance of other sulfur-bearing species can be larger than H_2S itself.

4.4.2. Sulfur monoxide and dioxide (SO and SO_2). The abundance of SO_2 has been measured at radio wavelengths in two comets and abundances relative to water are in the range <0.009 to 0.23%. SO_2 was also detected by Rosetta in Comet 67P (0.13%). The linear molecule SO has stronger millimeter lines and has been detected in eight comets. Although it is often assumed that SO is produced by the photolysis of SO_2 (β_{0,SO_2} = 2.5 × 10^{-4} s^{-1} at 1 au), there is evidence of a secondary source of SO, as suggested from interferometric maps (*Boissier et al.,* 2007). Rosetta's ROSINA found 0.07% of SO relative to water in 67P, with evidence of direct release from the sublimation of SO ice and not as a secondary product of SO (*Calmonte et al.,* 2016).

4.4.3. Carbon monosulfide (CS) and carbon disulfide (CS_2). CS has been detected in comets remotely via UV and radio spectroscopy for decades (*Meier and A'Hearn,* 1997; *Biver et al.,* 1999) with abundances between 0.03% and 0.2%. CS is a radical for which there is little information on its photolysis properties. A photodissociation rate around 2.5 × 10^{-5} s^{-1} at 1 au from the Sun was estimated from radio observations (*Boissier et al.,* 2007; *Biver et al.,* 2011). It is usually assumed that CS is a product species coming from the photodissociation of CS_2 [photodissociation rate β_{0,CS_2} = 1.7 × 10^{-3} s^{-1} at 1 au (*Jackson et al.,* 1986)]. However, some observations suggest a longer lifetime for the parent of CS (section 5.3.1). Rosetta' ROSINA detected CS_2 at very low abundances in 67P (*Calmonte et al.,* 2016; *Rubin et al.,* 2019), but CS could not be identified as a parent species.

4.4.4. Carbonyl sulfide (OCS) and thioformaldehyde (H_2CS). Including IR and radio techniques, OCS has been detected in 10 comets with an abundance relative to water ranging from 0.04% to 0.4%. An average value of 0.07% (*Läuter et al.,* 2020) was measured in Comet 67P by Rosetta. Thioformaldehyde (H_2CS), because of its likely short lifetime and low abundance (0.009–0.09% relative to water), has been detected remotely in only two comets. ROSINA measurements in the coma of 67P yield an even lower abundance of 0.003% (*Rubin et al.,* 2019).

4.4.5. Disulfur (S_2), trisulfur (S_3), and tetrasulfur (S_4). Sulfur has also been detected in the form of disulfur (S_2) in cometary comae via UV spectroscopy (*Reylé and Boice,* 2003, and references therein), and even in the form of S_3 and S_4 (*Calmonte et al.,* 2016) in the coma of 67P by Rosetta. UV observations yield abundances of S_2 ranging from 0.001% to 0.02%, comparable to ROSINA findings (0.002%). But some S_2 and most of the S_3 and S_4 detected by ROSINA are estimated to be secondary products not coming from nucleus ices but from a likely warmer dust component due to their low volatility (*Calmonte et al.,* 2016).

4.4.6. Nitrogen sulfide (NS), methanethiol (CH_3SH), and other sulfur-bearing molecules. Several other sulfur-bearing species have been identified in the coma of 67P such as methyl mercaptan or methanetiol (CH_3SH) and C_2H_6S (*Calmonte et al.,* 2016) and more complex species such as CH_4S_2, CH_4OS, C_2H_6OS, and C_3H_8OS according to *Altwegg et al.* (2019). CH_3SH has not yet been detected in a comet in the radio domain, but the abundance found in 67P (0.04% relative to water) is comparable to upper limits found in other comets (0.02–0.06%). NS was not detected in the coma of 67P but its abundance was estimated to be 0.006–0.012% in Comets C/2014 Q2 and C/1995 O1. The origin of this radical is still unknown (*Irvine et al.,* 2000).

4.5. Other Minor Species

4.5.1. Halogens. *Dhooghe et al.* (2017) reported the detection of three main halogen-bearing molecules HF, HCl, and HBr in the coma of 67P; the measured fragmentation pattern (cf. section 1.4.1) indicates at least one additional Cl-bearing compound aside from the already identified HCl, CH_3Cl, and NH_4Cl [cf. section 5.3.2 and *Fayolle et al.* (2017), *Altwegg et al.* (2020b)], and an increasing relative proportion of chlorine with distance indicates the likely presence of a distributed source (*De Keyser et al.,* 2017).

4.5.2. Phosphorus-bearing molecules. Phosphorous has been identified in the gas phase around 67P (*Altwegg et al.,* 2016) with PO being the main carrier while minor contributions of PN and PH_3 cannot be ruled out (*Rivilla et al.,* 2020). Phosphorous is an important element for life and its presence supports the theory that comets may have contributed key molecules related to the emergence of life on the early Earth (*Oró,* 1961; *Schwartz,* 2006).

4.5.3. Glycine and amines. The presence of the most simple amino acid glycine has been established in the dust samples returned by the Stardust mission to Comet 81P/Wild 2 (*Elsila et al.,* 2009). Glycine ($C_2H_5NO_2$) and two of its precursor molecules, methylamine (CH_3NH_2) and ethylamine ($C_2H_5NH_2$), have also been detected in the coma of 67P (*Altwegg et al.,* 2016). Together with the suite of C-, H-, O-, N-, and S-bearing organic compounds (cf. Table 3 and *Altwegg et al.,* 2017b) as well as phosophorous (cf. section 4.5.2), comets contain elements key in prebiotic chemistry and molecules identified as possible biomarkers in the search for life elsewhere (*Seager et al.,* 2016). More complex amino acids may form through aqueous alteration (*Burton et al.,* 2012) but were not identified in either 81P or 67P, which limits how their material is reprocessed in larger parent bodies.

4.5.4. Noble gases argon, krypton, and xenon. The noble gases Ar, Kr, and Xe have been identified in the coma of 67P (*Balsiger et al.,* 2015; *Marty et al.,* 2017; *Rubin et al.,* 2018) with Ar abundances much lower than reported in C/1995 O1 (Hale-Bopp) by *Stern et al.* (2000). Neon, on the other hand, was below the detection limit (*Rubin et al.,* 2018).

The measured relative abundances of noble gases (Table 3) may indicate trapping at elevated temperatures if the original abundances were solar or subsequent loss of the more volatile Ar and Ne during the comet's transition from the scattered disk, through the Centaur stage, and to its current orbit (*Maquet,* 2015; *Guilbert-Lepoutre et al.,* 2016; *Ligterink et al.,* 2024). Another possibility, suggested by *Marty et al.* (2017), is the addition of an exotic component of Xe required to explain its isotopic composition.

4.6. Molecular Oxygen (O_2)

The measured O_2 abundance in the coma of Comet 67P (3.1 ± 1.1)% relative to H_2O measured at ~1.5 au pre-perihelion (*Bieler et al.,* 2015; *Rubin et al.,* 2019) made it the dominant coma species after H_2O, CO_2, and CO. Furthermore, *Bieler et al.* (2015) reported a strong correlation with water despite the very different sublimation temperatures of the corresponding pure ices (*Fray and Schmitt,* 2009). Based on these results the presence of O_2 has also been inferred in the coma of 1P/Halley from measurements by Giotto's NMS (*Rubin et al.,* 2015b). At 67P, during suitable periods when CO_2 was at times even the dominant molecule in the coma (*Läuter et al.,* 2020), O_2 was found to be correlated to CO_2 (*Luspay-Kuti et al.,* 2022).

There are several possible mechanisms for the formation of O_2 in comets (cf. review by *Luspay-Kuti et al.,* 2018). *Dulieu et al.* (2017) proposed the dismutation of H_2O_2 during the sublimation process, where H_2O_2 would be co-produced with the water ice explaining the correlation of H_2O and O_2. On the other hand, a high conversion of $2\ H_2O_2 \rightarrow 2\ H_2O + O_2$ would be required. *Mousis et al.* (2016b) is consistent with radiolysis of water ice as the origin of O_2 with subsequent trapping in the ice, which is also consistent with the correlation between the two species. However, it is not clear these different scenarios can reproduce the observed amounts of O_2. Yet another mechanism was proposed by *Yao and Giapis* (2017) based on Eley-Rideal reactions forming O_2 involving energetic water ions. However, this mechanism is inconsistent with the low fluxes of energetic H_2O^+ ions and their poor correlation with neutral water in the coma (*Heritier et al.,* 2018). Additionally, all these production scenarios are at odds with the different O isotope ratios between the two species (cf. section 7.3).

To date, the most promising scenario remains a primordial origin of O_2 as proposed by *Taquet et al.* (2016) where the O_2 is formed through ice grain surface chemistry and/or in the gas phase (*Rawlings et al.,* 2019). A primordial origin is also favored by *Luspay-Kuti et al.* (2022).

5. FRAGMENT SPECIES

The first molecule identified in the coma of comets, C_2, is a secondary product. Secondary products are not present in cometary ices but are produced in the coma by the photodissociation of more complex molecules. For example, the photodissociation of water can lead to the production of OH, H, H_2, O, OH^+, O^+, or H^+ in the coma. Most secondary products can be observed at optical wavelengths and are among the easiest species to detect in the coma of comets. They are commonly used to infer the composition of cometary ices when observations of parent species are lacking. However, inferring the composition of cometary ices from the abundance of secondary products in the coma is not straightforward, as the parents of some radicals have not been identified. Some secondary products may also have multiple parents or be released from organic-rich grains. Species parentage in the coma of comets is a complex problem requiring further investigation and coordination of observations with modern techniques in multiple spectral regions.

5.1. Radicals

The main radicals observed in the coma of comets are OH, CN, C_2, C_3, CH, CS, NH, and NH_2. The NS radical was also detected in the coma of a handful of comets. CO is a special case as it can be both a parent or secondary product. *Feldman et al.* (2004) explores in detail the main emissions of these radicals. They are typically observed through electronic transitions at optical wavelengths, but some like OH also have transitions in the UV, radio, and IR. Most of these radicals are easy to observe at optical wavelengths and their abundances have been measured in the coma of hundreds of comets. The first taxonomic classifications of comets based on their composition relied on the systematic observations of these radicals.

The most abundant of these radicals is OH, which is produced by the photodissociation of water. Because the OH (0-0) band around 308 nm is one of the brightest emissions in the optical spectrum of comets, OH is often used as a water tracer and abundances of other radicals observed at optical wavelengths are often measured relative to OH. Just like for parent species, the relative abundances of radicals with respect to water (or OH) varies among comets, by more than 1 order of magnitude (*A'Hearn et al.,* 1995; *Schleicher,* 2008; *Cochran et al.,* 2012; *Langland-Shula and Smith,* 2011; *Fink,* 2009). CN abundances relative to OH vary between less than 0.1% to almost 1% while C_2 exhibits a wider range of abundances: In carbon-chain-depleted comets (see section 6.1.1), C_2 abundance can be lower than 0.01%, but it can reach maximum values similar to CN for typical comets. C_3 is generally less abundant than C_2 and CN (C_3/OH = 0.005–0.2%). NH and NH_2 have relatively similar abundances as they are both thought to originate from the photodissociation of NH_3 ranging between less than 0.1% to a few percent Abundances of CH are the least well constrained of the radicals observed at optical wavelengths, partly because it isn't isolated by any of the narrowband filters commonly used and it is usually faint in most comets. Estimates of its abundance range from about 0.3% to typically a few percent.

High-resolution spectroscopy has allowed the detailed characterization of the molecular band structure of several of these radicals. It has also paved the way for new modeling and laboratory measurements. For example, new models or laboratory measurements of the OH and NH_2 emission bands at optical wavelengths have been used to derive isotopic ratios of H, C, N, and O (*Rousselot et al.,* 2014; *Hutsemékers et al.,* 2008). Significant progress has also been made in the observations and modeling of CN, and especially the (A-X) red system, whose (0-0) band has a band head around 1100 nm (*Paganini and Mumma,* 2016; *Shinnaka et al.,* 2017).

5.2. Atoms

In addition to radicals, emissions from atoms are also detected at UV and visible wavelengths. H I, O I, C I, N I, S I, and the ions C II and O II have all been observed in the coma of comets, in the 80–200-nm range via solar fluorescence (*Feldman et al.,* 2004) and electron impact dissociative excitation of their parent in the inner coma of 67P (*Bodewits et al.,* 2022; *Feldman et al.,* 2015). H I Lyman-α emission at 121.6 nm, which totals almost 98% of far-UV emission in comets, has been used to estimate the water abundance in the coma of several dozen comets, in particular using the SWAN instrument onboard SOHO (*Combi et al.,* 2019). At optical wavelengths, the most prominent atomic features are the three forbidden O I emission lines at 557.73 nm, 630.03 nm, and 636.38 nm. These lines have been used for a number of years to estimate the water production rate and more recently as an indirect way to estimate the ratio between H_2O and CO_2 abundances in the coma of comets (*Festou and Feldman,* 1981b; *Schultz et al.,* 1992; *Morgenthaler et al.,* 2001; *McKay et al.,* 2013; *Decock et al.,* 2013). Forbidden [C I] lines at 872.7, 982.4, and 985.0 nm, coming from the photodissociation of neutral C-bearing species, have also been detected in the coma of C/1995 O1 (Hale-Bopp) and C/2016 R2 (PANSTARRS) (*Oliversen et al.,* 2002; *Opitom et al.,* 2019a). Recently, lines at 519.79 and 520.03 nm in the coma of C/2016 R2 (PANSTARRS) were identified as forbidden N lines (*Opitom et al.,* 2019a). To this day, [N I] lines have only been detected in the coma of a single comet.

The sodium D-line doublet was identified in comets over a century ago (*Newall,* 1910). This emission is difficult to observe from groundbased observatories due to telluric sodium lines. It has been mostly observed in the coma of comets well within 1 au from the Sun, manifesting as a neutral tail. Sodium detection in several comets is discussed by *Feldman et al.* (2004). There is no present consensus on the origin of neutral sodium in comets as both nucleus ice and dust grain sources have been proposed (*Cremonese et al.,* 2002). However, the detection of sodium in dust grains and in gas phase in the coma of 67P was reported from *Rosetta* data (*Schulz et al.,* 2015; *Wurz et al.,* 2015; *Rubin et al.,* 2022).

In exceptional cases when a comet passes very close to the Sun, other heavier elements are detected in the coma. Emission lines from Ca, K, V, Cr, Mn, Fe, Co, Ni, and Cu were detected in the coma of sungrazing comet C/1965 S1 Ikeya-Seki (*Preston,* 1967; *Slaughter,* 1969). These elements are typically only detected in the coma of sungrazing comets because of the extremely high temperatures needed to release them through the sublimation of dust grains. Si was detected in the coma of Comet C/2011 W3 (Lovejoy) with the Ultraviolet Coronagraph Spectrometer (UVCS) onboard SOHO (*Raymond et al.,* 2018). *Fulle et al.* (2007) also reported the discovery of an atomic iron tail for Comet C/2006 P1 (McNaught) at a heliocentric distance of 0.3 au from observations with the Solar Terrestrial Relations Observatory (STEREO) spacecraft. More recently *Manfroid et al.* (2021) showed that Fe I and Ni I emission lines are present in the coma of comets at a variety of heliocentric distances, from 0.7 to >3 au. This discovery was supported by the detection of Na, Si, K, Ca, and Fe detected by Rosetta (*Wurz et al.*, 2015; *Rubin et al.*, 2022). Their detection at large distances from the Sun rules out a production from the sublimation of dust grains as is the case for sungrazing comets. *Manfroid et al.* (2021) suggest that Fe and Ni could be contained in cometary ices within organometallic complexes {e.g., $[Fe(PAH)]^+$} or carbonyls such as $Fe(CO)_5$ and $Ni(CO)_4$.

5.3. Species Parentage and Release Mechanisms

While some species like OH have well-identified production mechanisms, it is not the case of all radicals. Some radicals have several potential parents, while for others no parent has been identified to date. Understanding the parentage of secondary products in the coma of comets provides a window into the composition of the nucleus.

5.3.1. Identifying parent molecules. *5.3.1.1. NH and NH_2:* It is generally believed that both NH and NH_2 result from the photodissociation of ammonia. NH would be a third-generation product, with NH_3 being first dissociated into NH_2, and subsequently into NH. Comparison of the abundances of NH, NH_2, and NH_3 to confirm this parentage is limited by a lack of simultaneous measurements of these species, the difficulty in detecting NH_3, and the uncertainty of NH scale lengths (*Fink et al.*, 1991; *Feldman et al.*, 2004).

5.3.1.2. CN: It has long been postulated that photodissociation of HCN is the dominant source of CN in the comae of comets. If this is true, then CN and HCN production rates measured simultaneously should be consistent with each other. However, for some comets, derived HCN abundances are significantly lower than those reported for CN (*Fray et al.*, 2005; *Dello Russo et al.*, 2009, 2016b). Inconsistencies between HCN and CN production rates were also observed in coma of 67P, from measurements with the ROSINA mass spectrometer (*Hänni et al.*, 2020). *Bockelée-Morvan and Crovisier* (1985) argued that the spatial profile of CN in the coma did not match a Haser model assuming release by the photodissociation of HCN. This is strong evidence that another source other than HCN plays a role in the production of CN in the coma of at least some comets. Other parent molecules have been investigated by *Fray et al.* (2005) who find that C_2N_2, HC_3N, and CH_3CN have lifetimes consistent with the parent of CN, even though their abundances in comets are likely insufficient to explain CN abundances. Sublimation of C-rich dust grains, CHON particles, or other macromolecules in the coma have also been suggested as potential extended sources of CN.

5.3.1.3. C_2 and C_3: The formation pathways of C_2 and C_3 are even less clear. Molecules like C_2H_2 and C_2H_6 were suspected to be potential parents of C_2 (e.g., *Jackson*, 1976) even before they were first detected at IR wavelengths in the 1990s (*Brooke et al.*, 1996; *Mumma et al.*, 1996). The photodissociation of C_3 itself is also thought to contribute to the formation of C_2. C_3 parents are less obvious but C_4H_2 and C_3H_4 were proposed in the 1960s–1970s by *Swings* (1965) and *Stief* (1972). Other potential parents for the two radicals include HC_3N and C_2H_4. Because the spatial distribution of C_2 in the inner coma of comets is often flatter than what would be expected from the simple photodissociation of a parent molecule, *Combi and Fink* (1997) suggested that C_2 might be released from a distributed source like CHON grains or by a two-step process. To this day, attempts to link C_2 and C_3 to potential parents through photochemical modeling have had mixed results: *Helbert et al.* (2005) concluded that C_3 could have C_3H_4 as main parent in C/1995 O1 (Hale-Bopp) while *Weiler* (2012) found that the breakdown of C_2H_6 does not produce C_2 efficiently enough to be a major parent in most comets. As pointed out by *Hölscher* (2015), some of the reactions in the chemical models used to estimate the origin of C_2 and C_3 have poorly known rate coefficients. Additionally, only some of the parents considered in the chemical networks mentioned above have been detected from groundbased observations of comets: HC_3N, C_2H_6, C_2H_2 (see, e.g., *Bockelée-Morvan and Biver*, 2017). Rosetta detected many saturated molecules and even more unsaturated ones containing at least two C atoms that could be potential C_2 and C_3 parents (*Schuhmann et al.*, 2019). Among all potential parents, C_3H_4 is not the most abundant, so it is likely not the only (or even primary) C_3 source in the coma of 67P. *Altwegg et al.* (2019) conclude that there are probably many parents leading to the production of these two radicals, with their exact contributions likely variable from comet to comet.

5.3.1.4. CS: Another radical with uncertain parent is CS. It has been assumed to generally come from the photodissociation of CS_2. Most observations show evidence of production in the coma at least compatible with its production from CS_2, but a few constraints on the parent scale length yield 2–4× longer lifetimes [$\tau_0 = 1/\beta_0 = 1000–2500$s (*Biver et al.*, 1999; *Roth et al.*, 2021; respectively) for the parent of CS than expected if the parent is CS_2. The low CS_2 abundances measured at 67P by Rosetta's ROSINA compared to CS abundances measured in other comets (Table 3) suggests that it is not the main parent of CS.

5.3.2. Distributed sources and production vs. r_h. Few comets are observed spectroscopically at small heliocentric distances (r_h) and information on long-term variability of composition from small heliocentric distances to greater than 1 au is lacking. However, there is emerging evidence from global molecular associations within the comet population that production rates of some molecules increase at smaller heliocentric distances relative to others (e.g., *Dello Russo et al.*, 2016a). These species, including C_2H_2, H_2CO, and NH_3, have been traditionally considered as parents released from nucleus ices. From infrared observations of D/2012 S1 (ISON), *DiSanti et al.* (2016) report an increase of the NH_3, NH_2, HCN, H_2CO, and C_2H_2 to H_2O ratio within 0.5 au while from optical observations, *McKay et al.* (2014) report increase of CN, C_2, CH, and C_3 abundances relative to H_2O. From Rosetta measurements, *De Keyser et al.* (2017) found changes of the relative abundance of the H halides HF and HCl with decreasing heliocentric distance in the coma of 67P. Observations also show evidence of a strong dependency of the CS abundance with heliocentric distance in the coma of comets (*Biver et al.,* 2006, 2011, and references therein), roughly as $Q_{CS}/Q_{H2O} \propto r_h^{-1}$, down to heliocentric distances as low as 0.1 au. A possible explanation for this behavior is significant release of these species from a distributed source of less-volatile grains once a thermal threshold is reached. Because of the relative difficulty of detecting species like C_2H_2 and NH_3 in comets, confirming extended release through spatial distributions is challenging; however, there is evidence of extended release of these from spatial studies in bright comets (e.g., *Dello Russo et al.,* 2016b, 2022; *DiSanti et al.,* 2016). The presence of ammonium salts ($NH^+{}_4X^-$) in cometary ices was inferred based on observations with ROSINA (*Altwegg et al.,* 2020b) and laboratory experiments (*Poch et al.,* 2020). Sublimation of ammonium salts, which happens at higher temperature than for species like HCN or even water, could account for the additional source of HCN and NH_3 at small heliocentric distances.

Other evidence also seems to indicate that some volatiles usually considered as parent volatiles could be released from extended sources even at typical heliocentric distances beyond 1 au. OCS showed a broad spatial distribution that is evidence of a possible extended source in C/1995 O1 (Hale-Bopp) (*Dello Russo et al.,* 1998), but high signal-to-noise measurements of OCS are rare so it is unclear if this behavior is typical. Chemical models also suggest that much of the HC_3N and NH_2CHO seen in comets at radio wavelengths could be produced from reactions in the coma (e.g., *Cordiner and Charnley,* 2021). Some comets have an inferred upper limit for the SO_2 abundance that is lower than the measured abundance for SO [e.g., C/2014 Q2 (*Biver et al.,* 2015)], which also suggests that there is another source of SO or that SO is a parent molecule. Further interferometric observations are needed to test the sources of SO in a larger subset of comets.

Radio observations have also shown that the abundance of H_2CO relative to water (e.g., *Biver et al.,* 2002) and the HNC/HCN ratio (*Lis et al.,* 2008) increase with decreasing heliocentric distance. While infrared observations indicate significant nucleus sources of formaldehyde (H_2CO) (*Dello Russo et al.,* 2016a), measurements from the Giotto flyby (*Meier et al.,* 1993) and radio and interferometric data point to an extended emission of formaldehyde for several comets (*Colom et al.,* 1992; *Biver et al.,* 1999; *Roth et al.,* 2021). Millimeter studies and ALMA observations suggest that the unknown formaldehyde parent would have a scale length between 1000 and 6800 km (*Bockelée-Morvan et al.,* 2000; *Cordiner et al.,* 2014, 2017), which is shorter than the dissociation scale length of CH_3OH, which mostly dissociates into H_2CO according to *Huebner and Mukherjee* (2015). Thermal degradation of polyoxymethylene (POM) (*Fray et al.,* 2006) has been proposed as parent source of formaldehyde. Similarly, radio interferometric data indicate that HNC is mostly produced in the coma within a few thousands of kilometers (*Cordiner et al.,* 2014; *Roth et al.,* 2021) and could be released from the degradation of N-rich grains (*Cordiner et al.,* 2017) or come from the photodissociation of a more complex nitrile parent (*Rodgers and Charnley,* 2001). More studies are necessary to better quantify the importance of extended sources for the production of species like OCS, H_2CO, HNC, or NH_3 and their variation among comets.

6. VARIATIONS OF COMA COMPOSITION

Cometary composition is often deduced by measurements obtained at a few (or one) points in time. These snapshots are frequently the only information that is available on the chemistry of a particular comet. Recently, longer timescale measurements of comets have been increasingly obtained. It is clear in many cases from groundbased observations obtained over long time periods and Rosetta measurements of 67P over much of its orbit that measured relative abundances of volatile species often vary with time. This variability within a comet must be accounted for when comparing measured compositions between comets. A key question is whether these temporal effects seen in individual comets due to heterogeneous active areas are signatures preserved from comet formation (natal) or have evolved through various processes over time.

6.1. Differences in Composition Between Comets

6.1.1. Taxonomic classes. Following the work done by *A'Hearn et al.* (1995) based on narrowband visible photometry of product species in 85 comets and a subsequent update on over 153 comets (*Schleicher,* 2008), IR and radio spectroscopy based datasets also enable classification of comets according to the composition of parent volatiles. *Dello Russo et al.* (2016a) and *Lippi et al.* (2021) have established a taxonomy based on the composition of 20 to 30 comets observed at IR wavelengths, using the abundances of HCN, CH_4, C_2H_6, C_2H_2, CH_3OH, H_2CO, CO, and NH_3. Earlier attempts based on radio spectroscopy have not established significant subgrouping based on, e.g., principal component analysis of the abundances of four to eight molecules. Difficulties are often linked to the consistency of the dataset, dealing with the largest number of comets vs. dealing with a more limited number of objects with more precise abundance determinations for a larger number of molecules. Histograms showing the distribution of abundances relative to water (or OH) based on visible, IR, and radio data show that for most molecules the distribution of abundances is typically Gaussian with a total width smaller than a factor of ~10 (Figs. 10 and 11). Evidence from the optical taxonomy from *A'Hearn et al.* (1995) shows that a group of comets is depleted in carbon-chain (C_2 and C_3) molecules, by typically 1 order of magnitude (Fig. 11). *Dello Russo et al.* (2016a) has established that arbitrary groupings of comets can be made based on their hydrocarbon, CH_3OH, HCN, NH_3, CO, and H_2CO content and, as established by *A'Hearn et al.* (1995) for product species, also showed that

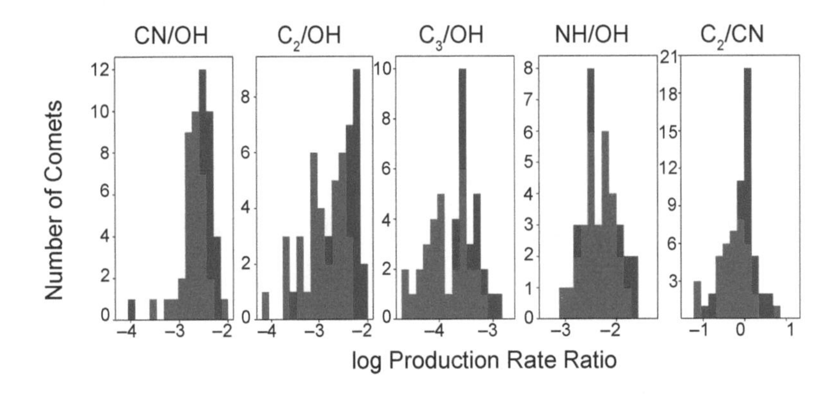

Fig. 11. Histograms of abundances relative to OH or CN based on observations of radicals in the visible range from *Osip et al.* (2010), based on the dataset of *A'Hearn et al.* (1995). Blue color is comets originating from the Oort cloud (OCC), on periodic to parabolic orbits of any inclination. Red color is for JFCs in low-inclination orbit with a short period.

the heliocentric dependence of parent abundances are also variable. Comets depleted in C_2H_2 in the IR are not necessarily correlated with the carbon-chain-depleted group of comets from optical measurements. There is, however, some evidence that carbon-chain-depleted comets have a rather low abundance of methanol in radio investigations (*Biver et al.*, 2021b). However, for some molecules the distribution shown in Figs. 10 and 11 might be better fitted by two Gaussians, which suggests that additional data are needed to characterize molecular trends in comets.

6.1.2. Outliers. Even considering the dispersion in molecular abundances seen within the comets observed over the last decades, a few comets show more extreme differences of one or two orders of magnitude in some molecular abundances. Comet C/2016 R2 (PanSTARRS) seems to be a member of a class of "blue" comets also comprising C/2002 VQ_{94} (LINEAR), C/1961 R1 (Humason), C/1908 R1 (Morehouse), and possibly 29P/Schwassmann-Wachmann 1. A few other cases based on visible spectroscopy are given in *Cochran et al.* (2000). As observed at 2.7 au from the Sun, the coma of Comet C/2016 R2 shows a very unusual composition, strongly enriched in CO, N_2, and CO_2, compared to other comets even at the same heliocentric distance (Table 2) (see also *Biver et al.*, 2018; *McKay et al.*, 2019; *Korsun et al.*, 2014). The CO:N_2:CH_3OH:HCN ratios are typically orders of magnitude different from other comets. Explanations for such differences could be that either these comets are fragments of a differentiated TNO that were formed mostly from the volatile part of its surface, or that they formed further away than most of the comets, beyond the CO/N_2 ice lines (*Mousis et al.*, 2021). The interstellar comet 2I/Borisov is also relatively enriched in CO compared to HCN (*Cordiner et al.*, 2020), but still much closer to the tail of the distribution observed in solar system comets.

Optical observations have also revealed potential compositional outliers such as comets 96P/Machholz and C/1988 Y1 (Yanaka): *Schleicher* (2008) found them very depleted in CN and C_2 with abundances relative to OH typically 2 orders of magnitude lower than the average value measured in comets, while *Fink* (2009) found them typical regarding their NH_2 content.

6.2. Heterogeneity and Evolution of the Coma

6.2.1. Temporal variability, outbursting, and split comets. Secular evolution in composition with time can also be investigated either on a short or longer timescales for periodic comets. For example, 2P/Encke (*Radeva et al.*, 2013; *Roth et al.*, 2018) showed large compositional changes between two apparitions (2003 and 2017), showing a substantial decrease of C_2H_6 and CH_3OH by factors of 8 and 4 respectively, whereas H_2CO abundances relative to water increased by a factor of 3. However, these observations were not obtained during the same orbital phase and might be related to seasonal changes on the nucleus of 2P. On the other hand, Comet 21P/Giacobini-Zinner was observed at three apparitions close to its perihelion time, in 1998, 2005, and 2018, and showed only minor short-term daily variations in abundances, with about the same average coma composition during each apparition (*Weaver et al.*, 1999; *Mumma et al.*, 2000; *DiSanti et al.*, 2013; *Roth et al.*, 2020; *Biver et al.*, 2021b; *Moulane et al.*, 2020). Generally, the carbon-chain-depleted comets within the visible taxonomy (see the previous section) show this characteristic behavior at each perihelion, which is indicative that a specific coma composition is generally characterizing each comet.

Comet compositional heterogeneity can often be rigorously tested when a comet breaks up or fragments. First, fragmenting comets release volatiles from the previously protected interior of the nucleus that is likely more indicative of primitive material. Second, when a comet completely disintegrates or loses a significant amount of its mass, subsequent measurements of coma material should be representative of the bulk nucleus. Third, when comets fragment into pieces that are large enough to separately investigate, the chemistry of different parts of the original nucleus can be sampled and compared. 73P/Schwassmann-Wachmann 3 split in several pieces in 1995, and subsequently came within

0.08 au of Earth in 2006, allowing observations with high sensitivity. Fragments C, B, and G showed a very similar composition from radio (*Biver et al.,* 2008), IR (*Dello Russo et al.,* 2007), and visible investigations (*Schleicher and Bair,* 2011). Measurements so far suggest that the degree of heterogeneity within comet nuclei is variable, and the likely cause of time variation (selective sublimation due to temperature threshold, seasonal variation due heat wave propagation, or modification of near surface composition after multi-perihelion passages) might not exclude a more homogeneous composition seen in many comets.

6.2.2. Local differences in the coma. Seasonal and diurnal variations in the (relative) abundances of volatile parent species have been observed in H_2O and CO_2 (cf. section 3.2) but also in many other species (*Feaga et al.,* 2007; *Hässig et al.,* 2015). Non-homogeneity has also been reported for the radical species OH, [O I], CN, NH, and NH_2, albeit more diffuse given their origin from dissociation of cometary parent molecules (*Bodewits et al.,* 2016). *Le Roy et al.* (2015) reported relative abundances for a suite of species measured at Comet 67P at 3.1 au inbound. In this comparative study strong differences in the coma composition above the northern summer vs. the southern winter hemisphere were obtained. CO varied from 2.7% to 20% and CO_2 from 2.5% to 80% with respect to H_2O north to south. *Bockelée-Morvan et al.* (2015) reported a range of 2–30% for CO_2 using the high-resolution (-H) optical subsystem of Rosetta's Visible and Infrared Thermal Imaging Spectrometer (VIRTIS), but this may be explained by the difference in the measurement technique, taking into account that local measurements can lead to much higher variations compared to remote sensing observations representing global integrated densities along a column (cf. section 1). When the southern winter hemisphere was poorly illuminated, increased ratios of CO, CO_2, CH_4, C_2H_2, C_2H_6, HCOOH, $HCOOCH_3$, HCN, OCS, and CS_2 with respect to H_2O were measured, whereas for some other organics the differences were less striking (*Le Roy et al.,* 2015). A difference was also noted from groundbased measurements of the global CN abundance (*Opitom et al.,* 2017). This was expected due to the low volatility of water relative to these species. However, CO, CO_2, CH_4, C_2H_2, C_2H_6, HCN, OCS, and CS_2 were also more abundant in absolute numbers above the southern winter hemisphere at 3.1 au inbound. This hints at a heterogeneous nucleus, most likely caused by evolutionary processes as the southern hemisphere is subject to significantly higher rates of erosion per orbit (*Keller et al.,* 2015a). A plausible explanation is that erosion exposes fresh material from the comet's interior to sublimation. However, once the southern hemisphere became more active during the short and intense summer months, the measured ratios with respect to water above the south at ~1.5 au (*Rubin et al.,* 2019) were similar to within a factor of a few to the ones measured above the northern hemisphere at ~3.1 au (*Le Roy et al.,* 2015). This indicates that the measured heterogeneity is also illumination dependent and hence temperature driven and occurs on seasonal as well as diurnal timescales (*Hässig et al.,* 2015), further modified by recondensation of H_2O (*De Sanctis et al.,* 2015). It may also provide clues as to how different species are embedded inside the ices of the nucleus.

6.2.3. Groups of molecules sharing similar behaviors. Cometary volatile species exhibit different outgassing patterns when it comes to seasonal (see, e.g., *Biver et al.,* 2002) and diurnal patterns (cf. *Hässig et al.,* 2015). It is no surprise that the volatility of the different ices can play a crucial role in the outgassing behavior of the nucleus. Hence, the relative abundances of the major volatiles CO and CO_2 with respect to H_2O tend to be lower near the comet's perihelion compared to larger heliocentric distances (cf. *Ootsubo et al.,* 2012).

Rosetta allowed an investigation of the relative abundances of species on diurnal and seasonal timescales. Very early in the mission, substantial amounts of O_2 were detected [cf. section 4.6 and *Bieler et al.* (2015)]. Interestingly, there was also a high correlation between the production of O_2 and H_2O despite the vastly different sublimation temperatures of the corresponding pure ices (*Fray and Schmitt,* 2009). On the other hand, O_2 showed a much lower correlation to CO and N_2. A similar picture was also obtained for other volatiles, e.g., NH_3 and H_2O (*Läuter et al.,* 2020; *Biver and Bockelée-Morvan,* 2019) and among the noble gases Ar, Kr, and Xe together with N_2 (*Balsiger et al.,* 2015; *Rubin et al.,* 2018).

Gasc et al. (2017) investigated the change in outgassing of H_2O, CO_2, CO, H_2S, CH_4, HCN, O_2, and NH_3 as a function of heliocentric distance during the outbound path past the second equinox. As mentioned above, H_2O, O_2, and NH_3 dropped rapidly as the comet moved away from the Sun. CO_2, CO, H_2S, CH_4, and HCN, on the other hand, dropped much more gradually and furthermore exhibited strong heterogeneity where the region of peak outgassing did not follow the subsolar latitude. Little correlation to the pure-ice sublimation temperatures were observed, i.e., O_2 (~30 K) behaved very similar to H_2O (~144 K) but very different from CH_4 (~36 K). The strongest decrease in outgassing was observed for water while the CO_2 activity dropped by the smallest amount of the species listed above.

Volatile production rates measured over large timescales suggested two distinct ice phases in 67P, associated with either H_2O or CO_2 release (*Hässig et al.,* 2015; *Fink et al.,* 2016; *Gasc et al.,* 2017; *Rubin et al.,* 2023). Two types of ices, a polar phase dominated by H_2O and an apolar phase dominated by CO and CO_2, have also been observed in the ISM and young stellar objects (*Boogert et al.,* 2015; *Mumma and Charnley,* 2011). At 67P, CO_2 appeared correlated with C_2H_6, CO, H_2S, and CH_4 (*Luspay-Kuti et al.,* 2015; *Hässig et al.,* 2015; *Gasc et al.,* 2017), whereas H_2O appeared correlated with CH_3OH, NH_3, and O_2 (*Luspay-Kuti et al.,* 2015; *Gasc et al.,* 2017). Some relationships varied with time; for example, HCN was sometimes correlated with CO_2 (*Gasc et al.,* 2017) and at other times with H_2O (*Luspay-Kuti et al.,* 2015). The correlation of O_2 with H_2O furthermore suggests that polarity was not the sole reason for this behavior. On the other hand, the different formation and processing mechanisms of some of these species, such as grain surface, gas phase chemistry, radiolysis, and thermal processing are likely key. Because the drop in outgassing of all these cometary parent molecules is bracketed by the two major species (*Rubin et al.,* 2019; *Läuter et al.,* 2020; *Combi et al.,* 2020), H_2O and CO_2, *Gasc et al.* (2017) suggested that all minor species are trapped in different proportions within H_2O and CO_2 and then released during phase transitions or co-desorbed during sublimation (*Kouchi and Yamamoto,* 1995; *Mousis et al.,* 2016b). It is possible that at one time minor species were present in their pure ices but that they are not anymore (or only to a very limited degree), with cometary activity governed by H_2O and CO_2 (*Rubin et al.,* 2023). In this scenario, outgassing may be further modified by recondensation and resublimation processes (*De Sanctis et al.,* 2015).

For both comets and the ISM, laboratory measurements are key to understanding this behavior (*Kouchi and Yamamoto,* 1995; *Collings et al.,* 2004; *Laufer et al.,* 2017). However, mixed ice experiments in the laboratory are quite challenging for species present in trace amounts, i.e., $\ll$1% with respect to H_2O and/or CO_2. But this corresponds to exactly the situation encountered at 67P. Furthermore, as the volatile coma structure in comets is explored by spacecraft it becomes important to connect these high-resolution spatial results with the global coma results obtained for a large number of comets from remote sensing observations.

6.2.4. Present-day composition: Natal or evolved? Ices within comet nuclei have been protected since formation by an overburden of material; thus, cometary volatiles observed today likely retain signatures from their birth. Yet comets are not perfectly preserved early solar system relics, as each nucleus has experienced a unique processing history over its long life, especially (but not only) during close passages to the Sun. Determining the degree to which comets retain their natal character is a fundamental question, but one that is difficult to answer. First, JFCs and OCCs have quite different dynamical histories; however, comet-forming regions were vast, which leads to significant expected and observed compositional diversity within each dynamical class. Second, while some systematic compositional differences have been seen between comets from different dynamical classes on average, this could plausibly reflect either different evolutionary histories or distinct formative regions.

As volatile abundances have been determined in an increasing number of comets, there are several avenues of investigation that allow natal vs. evolutionary effects in comets to be tested: (1) Determining abundances of the most volatile parents (e.g., N_2, CO, CH_4) shows the proficiency of comets in preserving these low-sublimation-temperature (hypervolatile) species. (2) Observations of JFCs over multiple apparitions with modern instrumentation that allow detailed compositional information to be obtained is now feasible. (3) Spatial distributions of parent molecules in the coma provide evidence for how ices are associated or sequestered in the nucleus. Determining spatial distributions in many comets can test which spatial properties and possible ice associations in the nucleus are comet-specific and which are global among the population of comets.

7. ISOTOPIC AND ORTHO-TO-PARA RATIOS

Isotopic ratios provide key information on the origin of cometary material. The abundances of the heavier elements in a molecule were determined by the reservoir of molecules containing heavy isotopes such as HD, atomic D, or H_2D^+, present during formation of these ices and the chemical reactions that happened during this time (e.g., *Taquet et al.,* 2013). Reactions in the gas phase or in grain-gas interactions lead to various fractionation mechanisms in the ISM. Then when the planets formed in a warmer environment in the protoplanetary disk, further fractionation occurred due to sublimation, photodissociation screening, mixing, and recondensation at various distances and temperatures. The results of these temperature and processing conditions, including isotopolog abundances, may have been frozen in cometary ices. So, isotopic ratios measured in the sublimating gas phase today may reveal their original value 4.6 G.y. ago. However, the impact of isotopic exchange in the ice phase and during sublimation is a subject of debate.

The ortho-to-para ratio (OPR) or spin temperature may be an estimate of the temperature of the molecules when they condensed into the icy grains that formed cometary nuclei. The change of spin state of a molecule is in principle strictly forbidden during collisions or radiative transitions and can only occur through chemical reactions, so present-day spin temperatures may be unchanged since cometary ices formed.

7.1. Deuterium/Hydrogen Ratios

D/H is the isotopic ratio of greatest interest as H is the most abundant element in the universe and the difference in the atomic mass (a factor of 2) is the largest, causing significant differences in D/H ratios in various molecules. The main molecular reservoir of D in the cold universe and solar system is HD, but in comets it is HDO. Strong enrichment in D in water (D/H = 10^{-4}–10^{-3}) and other molecules is observed in star-forming regions (e.g., *Drozdovskaya et al.,* 2021; *Jensen et al.,* 2021) and in the solar system. The reference D/H on Earth, the Vienna Standard Mean Ocean Water (VSMOW) value, is 1.558×10^{-4}, while in the local ISM the D/H value in molecular hydrogen is 2×10^{-5}. The HDO abundance in cometary ices is of key interest, especially when considering that comets may have contributed a significant amount of water to the young Earth. The D/H in water is half the HDO/H_2O, since there are two H atoms that can be substituted by D: D/H = 0.5 × (HDO/H_2O or DHO/H_2O).

D/H in cometary water has been measured in a dozen comets (Fig. 12) with sensitive upper limits obtained in a few others. Cometary D/H varies between 1 and 4× the VSMOW value as measured by different techniques. Radio spectroscopy samples HDO in the coma generally over thousands of kilometers around the nucleus via the J_{KaKc} = 1_{01}–0_{00} line at 464.9 GHz, the 1_{10}–1_{01} line at 509.3 GHz,

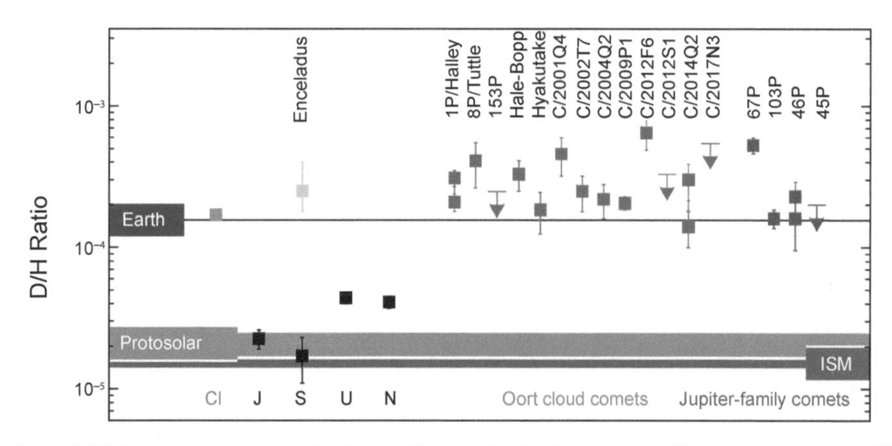

Fig. 12. Values of the D/H in solar system objects, in H_2 for giant planets, and the protosolar and ISM reservoirs or in water for Earth, CI meteorites, Enceladus, and comets. Adapted from *Lis et al.* (2019), with data from *Waite et al.* (2009), *Hartogh et al.* (2011), *Altwegg et al.* (2019), *Biver et al.* (2016), *Paganini et al.* (2017), Biver et al. (in preparation), and *Lis et al.* (2019, and references therein).

or the 2_{11}–2_{12} line at 241.6 GHz. HDO production rates determined at radio wavelengths are subject to uncertainty in the excitation model. IR spectroscopy (as well as UV spectroscopy of OD) is in theory less sensitive to coma temperature since it integrates several rovibrational lines within a vibrational band (electronic band in the UV). In both cases precise determination of the HDO/H_2O requires the detection of several HDO lines as well as the simultaneous detection of the water. However, H_2O and HDO are typically sampled through observations of lines in different spectral regions that require additional instrument settings. Values of the D/H in water measured in comets have no apparent correlation to the dynamical origin (JFCs vs. long-period OCCs). It has been suggested recently by *Lis et al.* (2019) that the D/H could be anti-correlated with hyperactivity of the comet. For example, comets with an equivalent active surface larger than 100% would have terrestrial D/H while those with a small active surface like Comet 67P have higher D/H. More measurements of D/H in comets are needed to test if this effect is real and whether this could be dictated by differences in formation conditions between hyperactive and low-activity comets.

D/H has also been measured in a handful of other molecules in the volatile phase: DCN was remotely detected in C/1995 O1 (Hale-Bopp) (*Meier et al.,* 1998) with a D/H 8× the VSMOW value. Measurements with ROSINA onboard the Rosetta spacecraft have found even higher enrichment in D in other molecules such as D_2O (D_2O/HDO =1.8 ± 0.9%), HDS, and CH_3OD/CH_2DOH. These are often expected as fractionation processes in the ISM and star-forming regions yield similar enrichments (*Drozdovskaya et al.,* 2021). D/H measurements in cometary comae are presented in Table 4.

7.2. Abundance of Nitrogen-15 in Comets

The ratio of $^{14}N/^{15}N$ across the solar system shows significant diversity, ranging from the protosolar value of 441 (*Marty et al.,* 2011), to the terrestrial value of 272 (*Anders and Grevesse,* 1989), to material very enriched in ^{15}N ($^{14}N/^{15}N$ ~50) in insoluble organic matter in carbonaceous chondrites (*Bonal et al.,* 2010). The origin of the diversity in ^{15}N across the solar system, and in comets, is still poorly understood.

The $^{14}N/^{15}N$ isotopic ratio has been measured in several comets from groundbased observations using HCN, CN, and NH_2 (*Bockelée-Morvan et al.,* 2015). Initial measurements made for Comet C/1995 O1 (Hale-Bopp) using submillimeter detection of HCN (*Jewitt et al.,* 1997; *Ziurys et al.,* 1999) derived a value of 323 ± 46, consistent with the terrestrial value while the value measured for CN was over 2× lower: 140 ± 35 (*Arpigny et al.,* 2003). Subsequent measurements for a dozen comets made from high-resolution spectroscopy of CN (*Jehin et al.,* 2009), together with almost simultaneous measurements in Comet 17P/Holmes from both HCN and CN and a reanalysis of the Hale-Bopp data (*Bockelée-Morvan et al.,* 2008), found an isotopic ratio consistent with the original CN measurement. For a sample of 20 comets of different origins and observed at different distances from the Sun, *Manfroid et al.* (2009) measured an average $^{14}N/^{15}N$ = 147.8 ± 5.7. They also point out that the isotopic ratios are remarkably constant (within the uncertainties) for all comets, irrespective of their origin.

Ammonia is another major N reservoir in cometary ices but the lack of sensitivity of current instrumentation has prevented the measurement of $^{15}NH_3$ so far. The $^{14}N/^{15}N$ was only recently indirectly measured in NH_2, assumed to be produced by the photodissociation of NH_3. The measurement performed both on co-added spectra of several comets and on individual comets is consistent within the uncertainties with that measured for $C^{14}N/C^{15}N$ (*Rousselot et al.,* 2014; *Shinnaka et al.,* 2016), as can be seen in Fig. 13. The first detection of N_2 in the coma of a comet by Rosetta's ROSINA and the subsequent measurement of a $^{14}N/^{15}N$ revealed a value consistent with the ratio measured for CN and NH_3 (*Altwegg et al.,* 2019).

TABLE 4. D/H in cometary molecules.

Deuterated Molecule	D/H Value	Comet	Reference
HDO	$1.4–6.5 \times 10^{-4}$	several	
HDCO	$<70 \times 10^{-4}$	C/2014 Q2	[2]
HDS	$12 \pm 3 \times 10^{-4}$	67P*	[1]
	$<170 \times 10^{-4}$	C/2014 Q2	[2]
	$<80 \times 10^{-4}$	17P/Holmes	[8]
DCN	$23 \pm 4 \times 10^{-4}$	Hale-Bopp	[3,4]
NH_2D	$11 \pm 2 \times 10^{-4}$	67P	[1]
	$<400 \times 10^{-4}$	Hale-Bopp	[6]
CH_3D	$24 \pm 3 \times 10^{-4}$	67P	[1]
	$<64 \times 10^{-4}$	C/2004 Q2	[5]
	$<50 \times 10^{-4}$	C/2004 Q2	[6]
C_2H_5D	$24 \pm 3 \times 10^{-4}$	67P	[1]
CH_3OD	$140 \pm 30 \times 10^{-4}$	67P	[7]
CH_2DOH	(for same deuteration on each H)		
CH_3OD	$<300 \times 10^{-4}$	Hale-Bopp	[4]
	$<100 \times 10^{-4}$	C/2014 Q2	[8]
CH_2DOH	$<80 \times 10^{-4}$	Hale-Bopp	[4]
	$<35 \times 10^{-4}$	C/2014 Q2	[8]
Doubly deuterated species (see text for D_2O)			
CHD_2OH	$110 \pm 10 \times 10^{-4}$	67P	[7]
CH_2DOD	(for same deuteration on each H)		

References: [1] *Müller et al.* (2022); [2] *Biver et al.* (2016); [3] *Meier et al.* (1998); [4] *Crovisier et al.* (2004); [5] *Kawakita and Kobayashi* (2009); [6] *Bonev et al.* (2009); [7] *Drozdovskaya et al.* (2021); [8] Biver, personal communication.

*Note that the value in *Altwegg et al.* (2019) and in the abstract of *Altwegg et al.* (2017a) is wrong by a factor of 2.

The origin of the large variation of N fraction across the solar system and the enrichment of ^{15}N in comets is still not understood. The existence of two distinct N reservoirs, one atomic (likely leading to the formation of HCN and NH_3) and one molecular (N_2), has been suggested (*Hily-Blant et al.,* 2017). We refer the reader to *Marty et al.* (2022) for a more in-depth discussion of N fractionation in the early solar system.

7.3. Other Isotopic Ratios: Sulfur, Oxygen, and Carbon

Isotopic ratios of C, O, and S are available for a suite of comets (cf. review by *Bockelée-Morvan et al.,* 2015). Table 5 provides $^{12}C/^{13}C$ ratios measured in comets and corresponding references. Carbon isotope ratios are often consistent within uncertainties with the the terrestrial Vienna Pee Dee Belemnite (V-PDB) reference. As discussed in section 5.1, a subset of the ratios provided in Table 5 are measured in radicals (C_2, CN) and may hence reflect a mixture of different parent species. This further complicates the identification of any species-dependent differences based on the chemical origin. Compared to the case of N (cf. section 7.2) only limited isotopic anomalies have been identified so far in comets. One notable exception is the case of H_2CO, which shall be discussed later.

Table 6 shows measured $^{16}O/^{18}O$ isotopic ratios including $^{16}O/^{17}O$ in H_2O and O_2 and corresponding references. Most $^{16}O/^{18}O$ ratios have been measured in cometary water and are in most cases consistent within 1σ with the terrestrial VSMOW reference. Rosetta revealed O isotope ratios in a few additional species (*Altwegg et al.,* 2020a). These initial results reveal a remarkable heterogeneity among the different chemical groups. Both the $^{16}O/^{17}O$ and $^{16}O/^{18}O$ ratios in H_2O and O_2 differ; on the other hand, the ratios $^{18}O/^{17}O = 4.5 \pm 1.0$ (O_2) and 4.9 ± 0.6 (H_2O) are both consistent with each other and within 1σ of the VSMOW reference (5.3). CO_2 and CH_3OH have $^{16}O/^{18}O$ ratios comparable to H_2O, but the ratios obtained for H_2CO, SO, SO_2, and OCS are about half while that for O_2 remains somewhere in between. Furthermore, *Schroeder et al.* (2019) analyzed O isotopes over the course of the full Rosetta mission and revealed no variation in the $^{16}O/^{18}O$ ratio as a function of cometary activity.

Table 7 lists the measured $^{32}S/^{33}S$ and $^{32}S/^{34}S$ ratios and corresponding references. The number of observed comets is limited, and the observed species are radicals such as CS or the S^+ ion originating from fragmentation of various S-

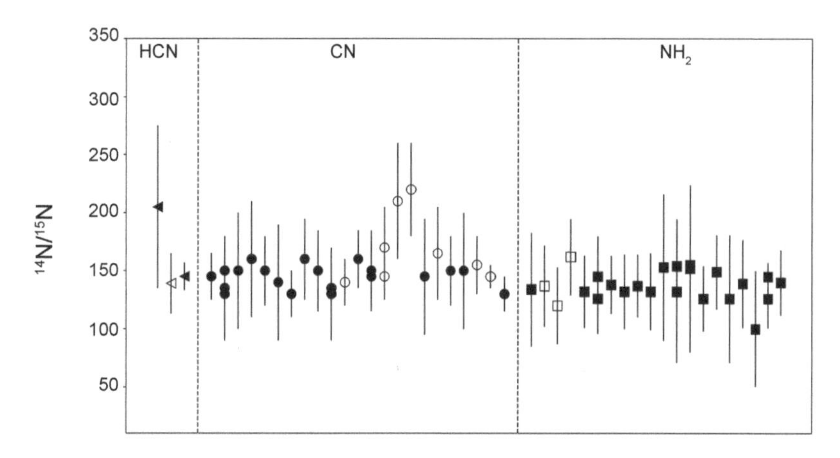

Fig. 13. $^{14}N/^{15}N$ isotopic ratios in comets measured using HCN, CN, and NH_2. Full symbols represent OCCs while open symbols represent JFCs from the Kuiper belt. Data from *Bockelée-Morvan et al.* (2008), *Manfroid et al.* (2009), *Jehin et al.* (2011), *Moulane et al.* (2020), *Yang et al.* (2018), *Shinnaka et al.* (2016), *Biver et al.* (2016), *Shinnaka et al.* (2014), *Rousselot et al.* (2015), and *Shinnaka and Kawakita* (2016).

bearing species inside the Giotto NMS (cf. section 1.4.1). Nevertheless, these results are consistent with the $^{32}S/^{34}S$ measurements of cometary parent species obtained at 67P: The $^{32}S/^{34}S$ ratios also align with the terrestrial reference samples within uncertainties and also the ratio $^{32}S/^{34}S$ = 21.6 ± 2.7 (*Paquette et al.,* 2017) measured in cometary dust by the COmetary Secondary Ion Mass Analyzer (COSIMA) onboard Rosetta (*Kissel et al.,* 2007). However, deviations from the terrestrial and solar system references have been observed in the $^{32}S/^{33}S$ ratios, namely the ^{33}S isotope was found to be depleted in H_2S, OCS, and CS_2. Similar to the O isotopes, the deviations are notable and in excess of what is typically found in the material in the inner solar system aside from the much less abundant presolar grains (*Hoppe et al.,* 2018).

Altwegg et al. (2020a) reported significant ^{18}O and ^{13}C enrichment in H_2CO compared to CH_3OH and terrestrial/solar system standards, while the $^{18}O/^{13}C$ ratios are in agreement. This is remarkable as both molecules were thought to originate from the same pathway, i.e., the hydrogenation of CO (*Watanabe and Kouchi,* 2002). There are observations in the ISM that reveal a similar picture, comparable C isotope ratios in CO and CH_3OH but significant deviations in H_2CO (*Wirström et al.,* 2011a,b). Large isotopic fractionation has also been modeled and found in other O-bearing species in the gas phase of cold clouds (10 K) in the ISM (*Loison et al.,* 2019). Hence, despite the resemblance to the findings at 67P, the question remains as to what extent these results can be related to each other.

Furthermore, the difference in both the $^{16}O/^{17}O$ and $^{16}O/^{18}O$ ratios in H_2O and O_2 is important and suggests that O_2 does not directly originate from H_2O (cf. section 4.6). This favors a primordial origin of O_2-based chemistry occurring on grain surfaces (*Taquet et al.,* 2016). The $^{16}O/^{18}O$ isotope ratio varies by up to a factor of 2 in the observed species. However, for the S-bearing subset, SO, SO_2, and OCS, the $^{32}S/^{34}S$ isotope ratio is consistent with terrestrial and solar reference material.

7.4. Noble Gas Isotopes

The relative abundances of a suite of noble gas isotopes reported in section 4.5.4 were obtained by Rosetta. Ar and Kr isotope ratios seem to be in agreement with the solar and terrestrial reference material listed in Table 8 within measurement uncertainties. For Xe, however, deviations from these standards were measured in 67P. In particular, ^{129}Xe was enriched while both ^{134}Xe and ^{136}Xe were considerably depleted. Taking cometary Xe, mixing it with solar wind Xe, and taking mass-dependent isotopic fractionation into account, *Marty et al.* (2017, 2022) estimated a contribution of 22.5 ± 5% of cometary Xe to the terrestrial atmosphere.

7.5. Halogen Isotopes

The measured isotopic ratios of $^{35}Cl/^{37}Cl$ and $^{79}Br/^{81}Br$ in 67P were consistent with solar system values when taking the uncertainties into account, which points to molecular cloud chemistry. The isotope ratio $^{35}Cl/^{37}Cl$ did not change throughout the mission (*Dhooghe et al.,* 2021).

7.6. Ortho-to-Para and Spin State Ratios

Molecules that have identical nuclei having non-zero nuclear spin, especially H atoms having a spin of 1/2, can exist in different energy levels due to their total spin value I. Molecules such as water, formaldehyde, or ammonia can be in two different spin states, (I = 1 ortho, or 3/2) or (I = 0 para, or 1/2), or A and E states and can be characterized by their OPR or A/E abundance ratio. CH_4 can be in three different spin states (I = 0, 1, or 2). H_2O, the most common species where OPRs are measured, achieves a statistical equilibrium

TABLE 5. Carbon isotope ratios in comets.

Species	Comet Reference	$^{12}C/^{13}C$ (89[1]/89[2])	Reference
C_2	C/1975 V1 (West)	60 ± 15	[3]
	four comets	93 ± 10	[4]
	C/2002 T7	85 ± 20	[5]
	C/2001 Q4	80 ± 20	[5]
	C/2012 S1	94 ± 33	[6]
CN	1P/Halley	65 ± 9	[7]
		89 ± 17	[8]
		95 ± 12	[9]
	C/1995 O1 (Hale-Bopp)	90 ± 15	[10]
		165 ± 40	[11]
	C/1990 K1 Levy	90 ± 10	[4]
	C/1989 X1 Austin	85 ± 20	[4]
	C/1989 Q1 O-L-R	93 ± 20	[4]
	122P/de Vico	90 ± 10	[12]
	88P/Howell	90 ± 10	[13]
	9P/Tempel 1	95 ± 15	[14]
	17P/Holmes	90 ± 20	[15]
	21 comets	91 ± 4	[16]
	103P/Hartley 2	95 ± 15	[17]
	C/2012 F6	95 ± 25	[18]
	C/2015 ER_{61}	100 ± 15	[19]
	21P/ Giacobini-Zinner	100 ± 10	[20]
HCN	C/1995 O1	111 ± 12	[21]
		94 ± 8	[15]
	17P/Holmes	114 ± 26	[15]
	C/2014 Q2	109 ± 14	[22]
	C/2012 F6	124 ± 64	[22]
	C/2012 S1	88 ± 18	[23]
CO	67P	86 ± 9	[24]
CO_2	67P	84 ± 4	[25]
CH_4	67P	88 ± 10	[26]
C_2H_6	67P	93 ± 10	[26]
H_2CO	67P	40 ± 14**	[27]
CH_3OH	67P	91 ± 10**	[27]

References: [1] V-PDB terrestrial (*Meija et al.,* 2016) and [2] solar (*Lodders* 2010) and in comets: [3] *Lambert and Danks* (1983); [4] *Wyckoff et al.* (2000); [5] *Rousselot et al.* (2012); [6] *Shinnaka et al.* (2014); [7] *Wyckoff et al.* (1989); [8] *Jaworski and Tatum* (1991); [9] *Kleine et al.* (1995); [10] *Lis et al.* (1997); [11] *Arpigny et al.* (2003); [12] *Jehin et al.* (2004); [13] *Hutsemékers et al.* (2005); [14] *Jehin et al.* (2006); [15] *Bockelée-Morvan et al.* (2008); [16] *Manfroid et al.* (2009); [17] *Jehin et al.* (2011); [18] *Bockelée-Morvan and Biver* (2017); [19] *Yang et al.* (2018); [20] *Moulane et al.* (2020); [21] *Jewitt et al.* (1997); [22] *Biver et al.* (2016); [23] *Cordiner et al.* (2019); [24] *Rubin et al.* (2017); [25] *Hässig et al.* (2017); [26] *Müller et al.* (2022); [27] *Altwegg et al.* (2020a).

value of 3/1 for temperatures above ~50 K, whereas the para species is increasingly favored as the temperature decreases below 50 K. OPRs have shown variability in measured values in comets (e.g., *Faure et al.,* 2019), with values for T_{spin} consistent with statistical equilibrium above 50 K in some cases, while others are sometimes clustered close to 30 K (*Bockelée-Morvan et al.,* 2004; *Kawakita and Kobayashi,* 2009, and references therein). To what extent differences in OPRs among comets relate to differences in the formation temperatures of ices in the solar nebula or sublimation and coma processes remains an open question (e.g., *Bonev et al.,* 2007, 2013; *Faure et al.,* 2019; *Bodewits et al.,* 2022; see also the chapter by Aikawa et al. in this volume). There is some evidence that once OPRs are locked into H_2O ice

TABLE 6. Oxygen isotope ratios in comets.

Species	Comet Reference	$^{16}O/^{17}O$ (2632[1]/2798[2])	$^{16}O/^{18}O$ (498.7[1]/530[2])	Reference
H_2O	1P/Halley		518 ± 45	[3]
			470 ± 40	[4]
	153P		530 ± 60	[5],[6]
	C/2001 Q4		~530 ± 60	[6]
	C/2002 T7		~550 ± 75	[6]
			425 ± 55	[7]
	C/2004 Q2		508 ± 33	[6]
	C/2012 F6		300 ± 150	[8]
	C/2009 P1		523 ± 32	[9]
	67P	2347 ± 191	445 ± 45	[10],[13]
O_2	67P	1544 ± 308	345 ± 40	[11]
CO_2	67P		494 ± 8	[12]
CH_3OH	67P		495 ± 40	[11]
H_2CO	67P		256 ± 100	[11]
SO	67P		239 ± 52	[11]
SO_2	67P		248 ± 88	[11]
OCS	67P		277 ± 70	[11]

References: [1] VSMOW terrestrial (*Clayton,* 2003) and [2] solar (*McKeegan et al.,* 2011) and in comets: [3] *Balsiger et al.* (1995); [4] *Eberhardt et al.* (1995); [5] *Lecacheux et al.* (2003); [6] *Biver et al.* (2007); [7] *Hutsemékers et al.* (2008); [8] *Bockelée-Morvan and Biver* (2017); [9] *Bockelée-Morvan et al.* (2012); [10] *Schroeder et al.* (2019); [11] *Altwegg et al.* (2020a); [12] *Hässig et al.* (2017); [13] *Müller et al.* (2022).

upon its formation, conversion is difficult (e.g., *Miani and Tennyson,* 2004), so it is possible that OPRs have remained stable since ices were incorporated into the nucleus and represent the chemical formation temperature of cometary ices (e.g., *Kawakita,* 2005). However, other laboratory results suggest OPRs are not ancient but instead are reset to their high-temperature equilibrium value after sublimation, independent of formation processes (*Hama et al.,* 2018).

8. PERSPECTIVES

Over the last 15 years since *Comets II* was published, increasing remote sensing capabilities have greatly increased the number of comets where the volatile composition and spatial distributions in the coma have been determined. Additionally, *in situ* analysis through comet missions have allowed the study of volatile release and coma spatial distributions with unprecedented detail. The EPOXI flyby provided the highest-spatial-resolution measurements of how volatiles were released from the nucleus of 103P/Hartley 2 (*A'Hearn et al.,* 2011) near perihelion and supporting groundbased studies showed that spatial distributions in the global coma were consistent with spacecraft observations (e.g., *Dello Russo et al.,* 2011; *Mumma et al.,* 2011; *Bonev et al.,* 2013; *Kawakita et al.,* 2013). The two-year Rosetta rendezvous with 67P allowed the chemical composition of a comet to be determined with unprecedented detail and many new and more complex volatiles were detected through observations with the ROSINA mass spectrometer (e.g., *Rubin et al.,* 2019; *Altwegg et al.,* 2017b; *Hänni et al.,* 2022, 2023). Rosetta and EPOXI observations also revealed distinct outgassing behavior for 67P and 103P/Hartley 2.

8.1. The State of Molecular Line Databases

Knowledge of the synthetic spectrum of a molecule or radical is essential to identify the species and compute its abundance in cometary comae. Knowing the fragmentation patterns in mass spectrometers is also necessary to evaluate the amount of the considered molecular species observed. There are still many unidentified lines in cometary spectra, especially in the high-spectral-resolution optical and IR spectra where the number of lines can be large. Laboratory measurements and ab initio calculations are regularly providing new catalogs of lines, available in databases such as those available from the Jet Propulsion Laboratory [*https://spec.jpl.nasa.gov* (*Pickett et al.,* 1998)] or the Cologne Database for Molecular Spectroscopy (CDMS) [*https://cdms.astro.uni-koeln.de* (*Müller et al.,* 2005)] for rotational transitions, and the High-Resolution Transmission Molecular Absorption Database (HITRAN) [*https://lweb.cfa.harvard.*

TABLE 7. Sulfur isotope ratios in comets.

Species	Comet Reference	$^{32}S/^{33}S$ (126.9[1]/126.7[2])	$^{32}S/^{34}S$ (22.6[1]/22.5[2])	Reference
S^+	1P/Halley		23 ± 6	[3]
CS	C/1995 O1		27 ± 3	[4]
	C/2014 Q2		24.7 ± 3.5	[5]
	C/2012 F6		20 ± 5	[5]
H_2S	C/1995 O1		16.5 ± 3.5	[6]
	46P/Wirtanen		20.6 ± 2.9	[7]
	67P			[8]
	(10/2014)	187 ± 9	23.6 ± 0.4	[8]
	(05/2016)	132 ± 3	22.4 ± 1.6	[8]
OCS	67P			[9]
	(10/2014)		25.1 ± 1.3	[8]
	(03/2016)		21.7 ± 4.0	[9]
	(05/2016)	165 ± 12	22.8 ± 0.3	[8]
CS_2	67P			
	(10/2014)	157 ± 7	24.3 ± 0.7	[8]
	(05/2016)	151 ± 8	25.3 ± 0.6	[8]
SO	67P		23.5 ± 2.5	[9]
SO_2	67P		21.3 ± 2.1	[9]

References: [1] V-CDT terrestrial (*Ding et al.,* 2001) and [2] solar (*Lodders,* 2010) and in comets: [3] *Altwegg* (1995); [4] *Jewitt et al.* (1997); [5] *Biver et al.* (2016); [6] *Crovisier et al.* (2004); [7] *Biver et al.* (2021a); [8] *Calmonte et al.* (2017); [9] *Altwegg et al.* (2020a). Note that the 1P/Halley value represents a mixture of species fragmenting into S (cf. section 1.4.1). Also note that *Calmonte et al.* (2017) analyzed two periods for H_2S, OCS, and CS_2, one early (10/2014) and the other late (05/2016) in the Rosetta mission.

edu/HITRAN (*Gordon et al.,* 2017)] or Gestion et Etude des Informations Spectroscopiques Atmosphériques (GEISA) [*https://geisa.aeris-data.fr* (*Jacquinet-Husson et al.,* 2016)] for rovibrational lines. Other databases are available but there is always a need for new data to be able to search for, detect, and quantify new molecular species or radicals in cometary atmospheres. For example, acetaldehyde was present in spectra of C/1995 O1 (Hale-Bopp) but could only be identified later when accurate frequencies where available (*Crovisier et al.,* 2004). Several molecular species identified by ROSINA in the coma of 67P, such as CH_3OSH, CH_3S_2H, and $C_3H_7NH_2$ (*Altwegg et al.,* 2019) (see Table 4), do not yet have published spectra. Similarly, data on line frequencies and strengths are also needed to extract new species from already existing spectra such as in the case of $^{15}NH_2$ (*Rousselot et al.,* 2014). There is also an ongoing need for basic chemical information such as provided by the National Institute of Standards and Technology (NIST) (see the NIST Chemistry WebBook at *https://webbook.nist.gov/chemistry*) for the fragmentation patterns of molecules in mass spectrometers.

In all cases the progress in the detection and determination of abundances of more and more complex species in cometary atmospheres, enabled by increased performance of observatories, cannot come without laboratory work and the release of new molecular data (see *Poch et al.,* 2022).

8.2. Comparison to the Interstellar Medium and Pathway for New Searches

Comets are considered as possible samples of interstellar material from which the solar system formed. Molecular complexity found in comets has allowed comparison to the composition of nearby star-forming regions, especially the low mass protostar IRAS 16293-2422(B) (*Biver et al.,* 2015; *Biver and Bockelée-Morvan,* 2019; *Drozdovskaya et al.,* 2019), in which abundances of COMs relative to CH_3OH measured in several comets (C/2014 Q2, 46P, 67P) are of the same order of magnitude. Inner regions of solar-type protostars (e.g., *Taquet et al.,* 2015) and those with strong polar jets like L1157-B1 (*Lefloch et al.,* 2017) reveal a complex chemistry that likely results from a process similar to the sublimation of comet ices. The heritage of cometary ices from the ISM is also reviewed in the chapters in this volume by Bergin et al. and Aikawa et al.. Such similarities in composition drives our search for new molecules in cometary comae, since many more complex species have been already detected in the ISM (*McGuire,* 2018). This requires

TABLE 8. Noble gas isotope ratios in Comet 67P and reference material.

Species	Isotopes						Reference
Argon	$^{36}Ar/^{38}Ar$						
67P	5.4 ± 1.4						*Balsiger et al.* (2015)
Solar wind	5.470 ± 0.003						*Heber et al.* (2012)
Terr. atm.	5.319 ± 0.008						*Lee et al.* (2006)
Krypton	$^{84}Kr/^{80}Kr$	$^{84}Kr/^{82}Kr$	$^{84}Kr/^{83}Kr$	$^{84}Kr/^{86}Kr$			
67P	23 ± 14	4.9 ± 0.4	5.3 ± 0.4	3.4 ± 0.2			*Rubin et al.* (2018)
Solar	24.4	4.88	4.93	3.31			*Lodders* (2010)
Terr. atm.	24.9	4.92	4.96	3.30			*Aregbe et al.* (2001)
Xenon	$^{132}Xe/^{128}Xe$	$^{132}Xe/^{129}Xe$	$^{132}Xe/^{130}Xe$	$^{132}Xe/^{131}Xe$	$^{132}Xe/^{134}Xe$	$^{132}Xe/^{136}Xe$	
67P	13 ± 4	0.7 ± 0.1	5.4 ± 0.9	1.2 ± 0.1	4.2 ± 0.9	8.6 ± 3.0	*Marty et al.* (2017)
Solar	11.8	0.96	6.02	1.21	2.73	3.35	*Lodders* (2010)
Terr. atm.	14.1	1.02	6.62	1.27	2.58	3.04	*Valkiers et al.* (1998)

more and more sensitivity as abundances generally decrease with molecular complexity (e.g., number of C atoms in the molecule) (*Crovisier et al.,* 2004). On the other hand, addressing the bulk abundances in cometary comae and ices is probably easier and less model dependent than in the ISM, due to the proximity of the source, optical thinness of the lines, and access to *in situ* observations via space missions.

8.3. Open Questions and Future Prospects

Despite the substantial insights gained in comet composition over the years, many fundamental questions remain unanswered (*Thomas et al.,* 2019). For instance, what are the processes governing outbursts and activity in general and their associated time and size scales? This includes how volatiles are distributed within the nucleus and the relationship between the composition of the nucleus and measurements in the coma. Another question is how comets evolve as they are processed in the inner solar system. The stratigraphy of the outgassing layer, affected by erosion and fallback material, and the interaction of the gas with the dust cover and how this sets the dynamics of the neutral gas in the coma are also open questions. What is the interrelation between the neutral gas, dust, and plasma (e.g., the lifting and acceleration of dust by neutral gas and the formation of a diamagnetic cavity, the innermost region around an active comet devoid of solar wind particles and fields)?

Some of these questions will be addressed by future missions (*Snodgrass et al.,* 2022) such as the Chinese ZengHe mission (*Zhang et al.,* 2021), which will visit MBC 311P/PanSTARRS, or ESA's Comet Interceptor (CI) mission (*Snodgrass and Jones,* 2019; *Jones et al.,* 2024), which proposes to fly by a long-period comet (LPC), or, if possible, a dynamically new comet (DNC) or an interstellar object (ISO). Further mission concepts have been proposed (*Thomas et al.,* 2019) and are discussed in the chapter in this volume by Snodgrass et al. A high priority in the coming decades is the collection and return of a comet nucleus sample and eventually a cryogenic sample that includes volatiles. This will allow analysis with the most advanced laboratory instrumentation on Earth, providing the next leap in our knowledge of the volatile content and structure of ices in a comet.

Despite these likely future advances, mission studies will remain rare and confined to a very few objects within a diverse population. Therefore, remote sensing techniques will continue to be important in studying the composition and diversity of comets as a population. Technological advances in ground- and spacebased facilities and instrumentation will continue to push toward improved sensitivity, efficiency, spectral coverage, and spatial resolution, which increases the number and variety of comets available for both compositional and coma spatial studies. In addition to improvements to existing facilities, new telescopes including the James Webb Space Telescope (JWST) (*Kelley et al.,* 2016), the Vera C. Rubin Observatory (*Jones et al.,* 2009; *Schwamb et al.,* 2018), the Thirty Meter Telescope (*Skidmore et al.,* 2015), and the European Extremely Large Telescope (ELT) (*Jehin et al.,* 2009; *Brandl et al.,* 2008) are expected to become available in the coming years and further advance our understanding of the composition and activity of comets.

Finally, we are now starting to be able to compare the composition of solar system comets to that of comets formed around other stars. Two interstellar objects have been detected crossing through our solar system. One of them was active and its composition could be compared to that of solar system comets. The existence of comets around other stars, inferred for the first time in the late 1980s, also provides us with a way to assess the composition of comets in other planetary systems. Comparing the composition of exocomets to solar system comets remains difficult, as they are observed under different circumstances, either when they pass very close to their stars (like sungrazing comets) or directly in the debris disk. Observations of interstellar comets and the study of exocomets with upcoming facilities will give us the opportunity to understand how different formation conditions influence the composition of comets and the nature of our solar system compared to its neighbors.

Acknowledgments. M.R. was funded by the State of Bern and the Swiss National Science Foundation (200020 182418, 200020 207312).

REFERENCES

A'Hearn M. F., Millis R. C., Schleicher D. O. et al. (1995) The ensemble properties of comets: Results from narrowband photometry of 85 comets, 1976–1992. *Icarus, 118,* 223–270.

A'Hearn M. F., Belton M. J., Delamere W. A. et al. (2011) EPOXI at Comet Hartley 2. *Science, 332,* 1396–1400.

Allen M., Delitsky M., Huntress W. et al. (1987) Evidence for methane and ammonia in the coma of Comet P/Halley. In *Exploration of Halley's Comet* (M. Grewing et al., eds.), pp. 502–512. Springer, Berlin.

Altwegg K. (1995) Sulfur in the coma of Comet Halley: Results from in situ measurements. Habilitation thesis, Univ. of Bern, Switzerland.

Altwegg K., Balsiger H., Bar-Nun A. et al. (2016) Prebiotic chemicals amino acid and phosphorus in the coma of Comet 67P/Churyumov-Gerasimenko. *Sci. Adv., 2,* e1600285.

Altwegg K., Balsiger H., Berthelier J.-J. et al. (2017a) D_2O and HDS in the coma of 67P/Churyumov-Gerasimenko. *Philos. Trans. R. Soc. A, 375,* 20160253.

Altwegg K., Balsiger H., Berthelier J.-J. et al. (2017b) Organics in Comet 67P — a first comparative analysis of mass spectra from ROSINA–DFMS, COSAC and Ptolemy. *Mon. Not. R. Astron. Soc., 469,* S130–S141.

Altwegg K., Balsiger H., and Fuselier S. A. (2019) Cometary chemistry and the origin of icy solar system bodies: The view after Rosetta. *Annu. Rev. Astron. Astrophys., 57,* 113–155.

Altwegg K., Balsiger H., Combi M. et al. (2020a) Molecule-dependent oxygen isotopic ratios in the coma of Comet 67P/Churyumov-Gerasimenko. *Mon. Not. R. Astron. Soc., 498,* 5855–5862.

Altwegg K., Balsiger H., Hänni N. et al. (2020b) Evidence of ammonium salts in Comet 67P as explanation for the nitrogen depletion in cometary comae. *Nature Astron., 4,* 533–540.

Altwegg K., Combi M., Fuselier S. A., Hänni N., et al. (2022) Abundant ammonium hydrosulphide embedded in cometary dust grains. *Mon. Not. R. Astron. Soc., 516,* 3900–3910.

Anders E. and Grevesse N. (1989) Abundances of the elements: Meteoritic and solar. *Geochim. Cosmochim. Acta, 53,* 197–214.

Aregbe Y., Valkiers S., Poths J. et al. (2001) A primary isotopic gas standard for krypton with values for isotopic composition and molar mass traceable to the Système International d'Unités. *Intl. J. Mass Spectr., 206,* 129–136.

Arpigny C., Jehin E., Manfroid J. et al. (2003) Anomalous nitrogen isotope ratio in comets. *Science, 301,* 1522–1525.

Balsiger H., Altwegg K., Benson J. et al. (1987) The ion mass spectrometer on Giotto. *J. Phys. E., 20,* 759.

Balsiger H., Altwegg K., and Geiss J. (1995) D/H and O-18/O-16 ratio in the hydronium ion and in neutral water from in situ ion measurements in Comet Halley. *J. Geophys. Res., 100,* 5827–5834.

Balsiger H., Altwegg K., Bochsler P. et al. (2007) ROSINA–Rosetta orbiter spectrometer for ion and neutral analysis. *Space Sci. Rev., 128,* 745–801.

Balsiger H., Altwegg K., Bar-Nun A. et al. (2015) Detection of argon in the coma of Comet 67P/Churyumov-Gerasimenko. *Sci. Adv., 1,* e1500377.

Bernstein M. P., Dworkin J. P., Sandford S. A. et al. (2002) Racemic amino acids from the ultraviolet photolysis of interstellar ice analogues. *Nature, 416,* 401–403.

Bieler A., Altwegg K., Balsiger H. et al. (2015) Abundant molecular oxygen in the coma of Comet 67P/Churyumov-Gerasimenko. *Nature, 526,* 678–681.

Biver N. and Bockelée-Morvan D. (2019) Complex organic molecules in comets from remote-sensing observations at millimeter wavelengths. *ACS Earth Space Chem., 3,* 1550–1555.

Biver N., Bockelée-Morvan D., Crovisier J. et al. (1999) Spectroscopic monitoring of Comet C/1996 B2 (Hyakutake) with the JCMT and IRAM radio telescopes. *Astron. J., 118,* 1850–1872.

Biver N., Bockelée-Morvan D., Colom P. et al. (2002a) The 1995–2002 long-term monitoring of Comet C/1995 O1 (Hale-Bopp) at radio wavelength (CP). In *Cometary Science after Hale-Bopp, Vol. 2* (H. Boehnhardt et al., eds.), pp. 5–14. Kluwer, Dordrecht.

Biver N., Bockelée-Morvan D., Crovisier J. et al. (2002b) Chemical composition diversity among 24 comets observed at radio wavelengths. *Earth Moon Planets, 90,* 323–333.

Biver N., Bockelée-Morvan D., Crovisier J. et al. (2006) Radio wavelength molecular observations of Comets C/1999 T1 (McNaught-Hartley), C/2001 A2 (LINEAR), C/2000 WM1 (LINEAR) and 153P/Ikeya-Zhang. *Astron. Astrophys., 449,* 1255–1270.

Biver N., Bockelée-Morvan D., Crovisier J. et al. (2007) Submillimetre observations of comets with Odin: 2001–2005. *Planet. Space Sci., 55,* 1058–1068.

Biver N., Bockelée-Morvan D., Crovisier J. et al. (2008) In-depth investigation of the fragmenting Comet 73P/Schwassmann-Wachmann 3 at radio wavelengths with the Nançay, IRAM, CSO, APEX and Odin radio telescopes. In *Asteroids, Comets, Meteors 2008*, Abstract #8149. Lunar and Planetary Institute, Houston.

Biver N., Bockelée-Morvan D., Colom P. et al. (2011) Molecular investigations of Comets C/2002 X5 (Kudo-Fujikawa), C/2002 V1 (NEAT), and C/2006 P1 (McNaught) at small heliocentric distances. *Astron. Astrophys., 528,* A142.

Biver N., Bockelée-Morvan D., Moreno R. et al. (2015) Ethyl alcohol and sugar in Comet C/2014 Q2 (Lovejoy). *Sci. Adv., 1,* 1500863.

Biver N., Moreno R., Bockelée-Morvan D. et al. (2016) Isotopic ratios of H, C, N, O, and S in comets C/2012 F6 (Lemmon) and C/2014 Q2 (Lovejoy). *Astron. Astrophys., 589,* A78.

Biver N., Bockelée-Morvan D., Paubert G. et al. (2018) The extraordinary composition of the blue Comet C/2016 R2 (PanSTARRS). *Astron. Astrophys., 619,* A127.

Biver N., Bockelée-Morvan D., Hofstadter M. et al. (2019) Long-term monitoring of the outgassing and composition of Comet 67P/Churyumov-Gerasimenko with the Rosetta/MIRO instrument. *Astron. Astrophys., 630,* A19.

Biver N., Bockelée-Morvan D., Boissier J. et al. (2021a) Molecular composition of Comet 46P/Wirtanen from millimetre-wave spectroscopy. *Astron. Astrophys., 648,* A49.

Biver N., Bockelée-Morvan D., Lis D. C. et al. (2021b) Molecular composition of short-period comets from millimetre-wave spectroscopy: 21P/Giacobini-Zinner, 38P/Stephan-Oterma, 41P/Tuttle-Giacobini-Kresák, and 64P/Swift-Gehrels. *Astron. Astrophys., 651,* A25.

Biver N., Bockelée-Morvan D., Handzlik B., et al. (2024) Chemical composition of comets C/2021 A1 (Leonard) and C/2022 E3 (ZTF) from radio spectroscopy and the abundance of HCOOH and HNCO in comets. *Astron. Astrophys., 690,* A271.

Bockelée-Morvan D. and Biver N. (2017) The composition of cometary ices. *Philos. Trans. R. Soc. London Ser. A, 375,* 20160252.

Bockelée-Morvan D. and Crovisier J. (1985) Possible parents for the cometary CN radical — Photochemistry and excitation conditions. *Astron. Astrophys., 151,* 90–100.

Bockelée-Morvan D., Colom P., Crovisier J. et al. (1991) Microwave detection of hydrogen sulphide and methanol in Comet Austin (1989c1). *Nature, 350,* 318–320.

Bockelée-Morvan D., Lis D. C., Wink J. E. et al. (2000) New molecules found in Comet C/1995 O1 (Hale-Bopp): Investigating the link between cometary and interstellar material. *Astron. Astrophys., 353,* 1101–1114.

Bockelée-Morvan D., Biver N., Moreno R. et al. (2001) Outgassing behavior and composition of Comet C/1999 S4 (LINEAR) during its disruption. *Science, 292,* 1339–1343.

Bockelée-Morvan D., Crovisier J., Mumma M. J. et al. (2004) The composition of cometary volatiles. In *Comets II* (M. C. Festou et al., eds.), pp. 391–423. Univ. of Arizona, Tucson.

Bockelée-Morvan D., Biver N., Jehin E. et al. (2008) Large excess of heavy nitrogen in both hydrogen cyanide and cyanogen from Comet 17P/Holmes. *Astrophys. J. Lett., 679,* L49.

Bockelée-Morvan D., Biver N., Crovisier J., et al. (2010) Comet 29P/

Schwassmann-Wachmann observed with the Herschel Space Observatory: Detection of water vapour and dust far-IR thermal emission. *Bull. Am. Astron. Soc., 42,* 946.
Bockelée-Morvan D., Hartogh P., Crovisier J. et al. (2010) A study of the distant activity of Comet C/2006 W3 (Christensen) with Herschel and ground-based radio telescopes. *Astron. Astrophys., 518,* L149.
Bockelée-Morvan D., Biver N., Swinyard B. et al. (2012) Herschel measurements of the D/H and $^{16}O/^{18}O$ ratios in water in the Oort-cloud Comet C/2009 P1 (Garradd). *Astron. Astrophys., 544,* L15.
Bockelée-Morvan D., Calmonte U., Charnley S. et al. (2015) Cometary isotopic measurements. *Space Sci. Rev., 197,* 47–83.
Bockelée-Morvan D., Debout V., Erard S. et al. (2015) First observations of H_2O and CO_2 vapor in Comet 67P/Churyumov-Gerasimenko made by VIRTIS onboard Rosetta. *Astron. Astrophys., 583,* A6.
Bodewits D., Farnham T. L., A'Hearn M. F. et al. (2014) The evolving activity of the dynamically young Comet C/2009 P1 (Garradd). *Astrophys. J., 786,* 48.
Bodewits D., Lara L. M., A'Hearn M. F. et al. (2016) Changes in the physical environment of the inner coma of 67P/Churyumov-Gerasimenko with decreasing heliocentric distance. *Astron. J., 152,* 130.
Boissier J., Bockelée-Morvan D., Biver N. et al. (2007) Interferometric imaging of the sulfur-bearing molecules H_2S, SO, and CS in Comet C/1995 O1 (Hale-Bopp). *Astron. Astrophys., 475,* 1131–1144.
Bonal L., Huss G. R., Krot A. N. et al. (2010) Highly ^{15}N-enriched chondritic clasts in the CB/CH-like meteorite Isheyevo. *Geochim. Cosmochim. Acta, 74,* 6590–6609.
Bonev B. P., Mumma M. J., DiSanti M. A. et al. (2006) A comprehensive study of infrared OH prompt emission in two comets. I. Observations and effective *g*-factors. *Astrophys. J., 653,* 774–787.
Bonev B. P., Mumma M. J., Villanueva G. L. et al. (2007) A search for variation in the H_2O ortho-para ratio and rotational temperature in the inner coma of Comet C/2004 Q2 (Machholz). *Astrophys. J. Lett., 661,* L97–L100.
Bonev B. P., Mumma M. J., Gibb E. L. et al. (2009) Comet C/2004 Q2 (Machholz): Parent volatiles, a search for deuterated methane, and constraint on the CH_4 spin temperature. *Astrophys. J., 699,* 1563–1572.
Bonev B. P., Villanueva G. L., Paganini L. et al. (2013) Evidence for two modes of water release in Comet 103P/Hartley 2: Distributions of column density, rotational temperature, and ortho-para ratio. *Icarus, 222,* 740–751.
Bonev B. P., Villanueva G. L., DiSanti M. A. et al. (2017) Beyond 3 au from the Sun: The hypervolatiles CH_4, C_2H_6, and CO in the distant Comet C/2006 W3 (Christensen). *Astron. J., 153,* 241.
Boogert A. A., Gerakines P. A., and Whittet D. C. (2015) Observations of the icy universe. *Annu. Rev. Astron. Astrophys., 53,* 541–581.
Brandl B. R., Lenzen R., Pantin E. et al. (2008) METIS: The mid-infrared E-ELT imager and spectrograph. In *Ground-based and Airborne Instrumentation for Astronomy II* (I. S. McLean and M. M. Casali, eds.), Abstract #70141N. SPIE Conf. Series 7014, Bellingham, Washington.
Brooke T. Y., Tokunaga A. T., Weaver H. A. et al. (1996) Detection of acetylene in the infrared spectrum of Comet Hyakutake. *Nature, 383,* 606–608.
Burton A. S., Stern J. C., Elsila J. E. et al. (2012) Understanding prebiotic chemistry through the analysis of extraterrestrial amino acids and nucleobases in meteorites. *Chemical Soc. Rev., 41,* 5459–5472.
Calmonte U., Altwegg K., Balsiger H. et al. (2016) Sulphur-bearing species in the coma of Comet 67P/Churyumov-Gerasimenko. *Mon. Not. R. Astron. Soc., 462,* S253–S273.
Calmonte U., Altwegg K., Balsiger H. et al. (2017) Sulphur isotope mass-independent fractionation observed in Comet 67P/Churyumov-Gerasimenko by Rosetta/ROSINA. *Mon. Not. R. Astron. Soc., 469,* S787–S803.
Capria M. T., Capaccioni F., Filacchione G. et al. (2017) How pristine is the interior of the comet 67P/Churyumov-Gerasimenko? *Mon. Not. R. Astron. Soc., 469,* S685–S694.
Carter M., Lazareff B., Maier D. et al. (2012) The EMIR multi-band mm-wave receiver for the IRAM 30-m telescope. *Astron. Astrophys., 538,* A89.
Clayton R. N. (2003) Oxygen isotopes in the solar system. In *Solar System History from Isotopic Signatures of Volatile Elements* (R. Kallenbach et al., eds.), pp. 19–32. Springer, New York.
Cochran A. L. and Schleicher D. G. (1993) Observational constraints on the lifetime of cometary H_2O. *Icarus, 105,* 235–253.
Cochran A. L., Cochran W. D., and Barker E. S. (2000) N^+ and CO^+ in Comets 122P/1995 S1 (deVico) and C/1995 O1 (Hale-Bopp). *Icarus, 146,* 583–593.
Cochran A. L., Barker E. S., and Gray C. L. (2012) Thirty years of cometary spectroscopy from McDonald Observatory. *Icarus, 218,* 144–168.
Cochran A. L. and McKay A. J. (2018) Strong CO^+ and N^+ emission in Comet C/2016 R2 (Pan-STARRS). *Astrophys. J. Lett., 854,* L10.
Collings M. P., Anderson M. A., Chen R. et al. (2004) A laboratory survey of the thermal desorption of astrophysically relevant molecules. *Mon. Not. R. Astron. Soc., 354,* 1133–1140.
Colom P., Crovisier J., Bockelée-Morvan D. et al. (1992) Formaldehyde in comets. I — Microwave observations of P/Brorsen-Metcalf (1989 X), Austin (1990 V) and Levy (1990 XX). *Astron. Astrophys., 264,* 270–281.
Combi M. R. and Fink U. (1997) A critical study of molecular photodissociation and CHON grain sources for cometary C_2. *Astrophys. J., 484,* 879–890.
Combi M. R., Mäkinen J. T. T., Bertaux J. L. et al. (2013) Water production rate of Comet C/2009 P1 (Garradd) throughout the 2011–2012 apparition: Evidence for an icy grain halo. *Icarus, 225,* 740–748.
Combi M. R., Mäkinen T. T., Bertaux J. L. et al. (2019) A survey of water production in 61 comets from SOHO/SWAN observations of hydrogen Lyman-alpha: Twenty-one years 1996–2016. *Icarus, 317,* 610–620.
Combi M., Shou Y., Fougere N. et al. (2020) The surface distributions of the production of the major volatile species, H_2O, CO_2, CO and O_2, from the nucleus of Comet 67P/Churyumov-Gerasimenko throughout the Rosetta mission as measured by the ROSINA double focusing mass spectrometer. *Icarus, 335,* 113421.
Cordiner M. A. and Charnley S. B. (2021) Neutral-neutral synthesis of organic molecules in cometary comae. *Mon. Not. R. Astron. Soc., 504,* 5401–5408.
Cordiner M. A. and Qi C. (2018) High-resolution imaging of comets. In *Science with a Next-Generation Very Large Array* (E. J. Murphy and the Next-Generation Very Large Array Science Advisory Council, eds.), p. 73. ASP Conf. Ser. 517, Astronomical Society of the Pacific, San Francisco.
Cordiner M. A., Remijan A. J., Boissier J. et al. (2014) Mapping the release of volatiles in the inner comae of Comets C/2012 F6 (Lemmon) and C/2012 S1 (ISON) using the Atacama Large Millimeter/Submillimeter Array. *Astrophys. J. Lett., 792,* L2.
Cordiner M. A., Boissier J., Charnley S. B. et al. (2017) ALMA mapping of rapid gas and dust variations in Comet C/2012 S1 (ISON): New insights into the origin of cometary HNC. *Astrophys. J., 838,* 147.
Cordiner M. A., Palmer M. Y., de Val-Borro M. et al. (2019) ALMA autocorrelation spectroscopy of comets: The HCN/$H^{13}CN$ ratio in C/2012 S1 (ISON). *Astrophys. J. Lett., 870,* L26.
Cordiner M. A., Milam S. N., Biver N. et al. (2020) Unusually high CO abundance of the first active interstellar comet. *Nature Astron., 4,* 861–866.
Cremonese G., Huebner W. F., Rauer H. et al. (2002) Neutral sodium tails in comets. *Adv. Space Res., 29,* 1187–1197.
Crovisier J. (1997) Infrared observations of volatile molecules in Comet Hale-Bopp. *Earth Moon Planets, 79,* 125–143.
Crovisier J., Biver N., Bockelée-Morvan D. et al. (1995) Carbon monoxide outgassing from Comet P/Schwassmann-Wachmann 1. *Icarus, 115,* 213–216.
Crovisier J., Bockelée-Morvan D., Colom P. et al. (2004) The composition of ices in Comet C/1995 O1 (Hale-Bopp) from radio spectroscopy: Further results and upper limits on undetected species. *Astron. Astrophys., 418,* 1141–1157.
Crovisier J., Colom P., Biver N. et al. (2013) Observations of the 18-cm OH lines of Comet 103P/Hartley 2 at Nançay in support to the EPOXI and Herschel missions. *Icarus, 222,* 679–683.
De Keyser J., Dhooghe F., Altwegg K. et al. (2017) Evidence for distributed gas sources of hydrogen halides in the coma of Comet 67P/Churyumov-Gerasimenko. *Mon. Not. R. Astron. Soc., 469,* S695–S711.
De Sanctis M. C., Capaccioni F., Ciarniello M. et al. (2015) The diurnal cycle of water ice on Comet 67P/Churyumov-Gerasimenko. *Nature, 525,* 500–503.
de Val-Borro M., Bockelée-Morvan D., Jehin E. et al. (2014) Herschel observations of gas and dust in Comet C/2006 W3 (Christensen) at

5 AU from the Sun. *Astron. Astrophys., 564,* A124.
Decock A., Jehin E., Hutsemékers D. et al. (2013) Forbidden oxygen lines in comets at various heliocentric distances. *Astron. Astrophys., 555,* A34.
Dello Russo N., DiSanti M. A., Mumma M. J. et al. (1998) Carbonyl sulfide in Comets C/1996 B2 (Hyakutake) and C/1995 O1 (Hale-Bopp): Evidence for an extended source in Hale-Bopp. *Icarus, 135,* 377–388.
Dello Russo N., Mumma M. J., DiSanti M. A. et al. (2000) Water production and release in Comet C/1995 O1 Hale-Bopp. *Icarus, 143,* 324–337.
Dello Russo N., Vervack R. J., Weaver H. A. et al. (2007) Compositional homogeneity in the fragmented Comet 73P/Schwassmann-Wachmann 3. *Nature, 448,* 172–175.
Dello Russo N., Vervack R. J. J., Weaver H. A. et al. (2009) The parent volatile composition of 6P/d'Arrest and a chemical comparison of Jupiter-family comets measured at infrared wavelengths. *Astrophys. J., 703,* 187–197.
Dello Russo N., Vervack R. J. J., Lisse C. M. et al. (2011) The volatile composition and activity of Comet 103P/Hartley 2 during the EPOXI closest approach. *Astrophys. J. Lett., 734,* L8.
Dello Russo N., Kawakita H., Vervack R. J. et al. (2016a) Emerging trends and a comet taxonomy based on the volatile chemistry measured in thirty comets with high-resolution infrared spectroscopy between 1997 and 2013. *Icarus, 278,* 301–332.
Dello Russo N., Vervack R. J., Kawakita H. et al. (2016b) The compositional evolution of C/2012 S1 (ISON) from ground-based high-resolution infrared spectroscopy as part of a world-wide observing campaign. *Icarus, 266,* 152–172.
Dello Russo N., Vervack R. J., Kawakita H. et al. (2022) Volatile abundances, extended coma sources, and nucleus ice associations in bright comet C/2014 Q2 (Lovejoy). *Planet. Sci. J., 3,* 46.
Dhooghe F., De Keyser J., Altwegg K. et al. (2017) Halogens as tracers of protosolar nebula material in Comet 67P/Churyumov-Gerasimenko. *Mon. Not. R. Astron. Soc., 472,* 1336–1345.
Dhooghe F., De Keyser J., Hänni N. et al. (2021) Chlorine-bearing species and the $^{37}Cl/^{35}Cl$ isotope ratio in the coma of Comet 67P/Churyumov-Gerasimenko. *Mon. Not. R. Astron. Soc., 508,* 1020–1032.
Ding T., Valkiers S., Kipphardt H. et al. (2001) Calibrated sulfur isotope abundance ratios of three IAEA sulfur isotope reference materials and V-CDT with a reassessment of the atomic weight of sulfur. *Geochim. Cosmochim. Acta, 65,* 2433–2437.
DiSanti M. A., Dello Russo N., Magee-Sauer K. et al. (2002) CO, H_2CO, and CH_3OH in Comet 2002 C1 Ikeya-Zhang. In *Asteroids, Comets, and Meteors: ACM 2002* (B. Warmbein, ed.), pp. 571–574. ESA SP-500, Noordwijk, The Netherlands.
DiSanti M. A., Bonev B. P., Villanueva G. L. et al. (2013) Highly depleted ethane and mildly depleted methanol in Comet 21P/Giacobini-Zinner: Application of a new empirical ν_2-band model for CH_3OH near 50 K. *Astrophys. J., 763,* 1.
DiSanti M. A., Bonev B. P., Gibb E. L. et al. (2016) En route to destruction: The evolution in composition of ices in Comet D/2012 S1 (ISON) between 1.2 and 0.34 AU from the Sun as revealed at infrared wavelengths. *Astrophys. J., 820,* 34.
Dorman G., Pierce D. M., and Cochran A. L. (2013) The spatial distribution of C_2, C_3, and NH in Comet 2P/Encke. *Astrophys. J., 778,* 140.
Drozdovskaya M. N., van Dishoeck E. F., Rubin M. et al. (2019) Ingredients for solar-like systems: Protostar IRAS 16293-2422 B versus Comet 67P/Churyumov-Gerasimenko. *Mon. Not. R. Astron. Soc., 490,* 50–79.
Drozdovskaya M. N., Schroeder I. R. H. G., Rubin M. et al. (2021) Prestellar grain-surface origins of deuterated methanol in Comet 67P/Churyumov-Gerasimenko. *Mon. Not. R. Astron. Soc., 500,* 4901–4920.
Dulieu F., Minissale M., and Bockelée-Morvan D. (2017) Production of O_2 through dismutation of H_2O_2 during water ice desorption: A key to understanding comet O_2 abundances. *Astron. Astrophys., 597,* A56.
Eberhardt P. (1999) Comet Halley's gas composition and extended sources: Results from the neutral mass spectrometer on Giotto. *Space Sci. Rev., 90,* 45–52.
Eberhardt P., Dolder U., Schulte W. et al. (1987) The D/H ratio in water from Comet P/Halley. In *Exploration of Halley's Comet* (M. Grewing et al., eds.), pp. 435–437. Springer, Berlin.
Eberhardt P., Meier R., Krankowsky D. et al. (1994) Methanol and hydrogen sulfide in Comet P/Halley. *Astron. Astrophys., 288,* 315–329.
Eberhardt P., Reber M., Krankowsky D. et al. (1995) The D/H and $^{18}O/^{16}O$ ratios in water from Comet P/Halley. *Astron. Astrophys., 302,* 301.
El-Maarry M. R., Thomas N., Giacomini L. et al. (2015) Regional surface morphology of Comet 67P/Churyumov-Gerasimenko from Rosetta/OSIRIS images. *Astron. Astrophys., 583,* A26.
Elsila J. E., Glavin D. P., and Dworkin J. P. (2009) Cometary glycine detected in samples returned by Stardust. *Meteorit. Planet. Sci., 44,* 1323–1330.
Faggi S., Villanueva G. L., Mumma M. J. et al. (2016) Detailed analysis of near-IR water (H_2O) emission in Comet C/2014 Q2 (Lovejoy) with the GIANO/TNG spectrograph. *Astrophys. J., 830,* 157.
Faure A., Hily-Blant P., Rist C. et al. (2019) The ortho-to-para ratio of water in interstellar clouds. *Mon. Not. R. Astron. Soc., 487,* 3392–3403.
Fayolle E. C., Öberg K. I., Jørgensen J. K. et al. (2017) Protostellar and cometary detections of organohalogens. *Nature Astron., 1,* 703–708.
Feaga L., A'Hearn M., Sunshine J. et al. (2007) Asymmetries in the distribution of H_2O and CO_2 in the inner coma of Comet 9P/Tempel 1 as observed by Deep Impact. *Icarus, 191,* 134–145.
Feaga L. M., A'Hearn M. F., Farnham T. L. et al. (2014) Uncorrelated volatile behavior during the 2011 apparition of Comet C/2009 P1 Garradd. *Astron. J., 147,* 24.
Feldman P. D., Cochran A. L., and Combi M. R. (2004) Spectroscopic investigations of fragment species in the coma. In *Comets II* (M. C. Festou et al., eds.), pp. 425–447. Univ. of Arizona, Tucson.
Feldman P. D., A'Hearn M. F., Bertaux J.-L. et al. (2015) Measurements of the near-nucleus coma of Comet 67P/Churyumov-Gerasimenko with the Alice far-ultraviolet spectrograph on Rosetta. *Astron. Astrophys., 583,* A8.
Festou M. and Feldman P. D. (1981a) The forbidden oxygen lines in comets. *Astron. Astrophys., 103,* 154–159.
Fink U. (2009) A taxonomic survey of comet composition 1985–2004 using CCD spectroscopy. *Icarus, 201,* 311–334.
Fink U., Combi M. R., and Disanti M. A. (1991) Comet P/Halley: Spatial distributions and scale lengths for C_2, CN, NH_2, and H_2O. *Astrophys. J., 383,* 356.
Fink U., Doose L., Rinaldi G. et al. (2016) Investigation into the disparate origin of CO_2 and H_2O outgassing for Comet 67/P. *Icarus, 277,* 78–97.
Fray N. and Schmitt B. (2009) Sublimation of ices of astrophysical interest: A bibliographic review. *Planet. Space Sci., 57,* 2053–2080.
Fray N., Bénilan Y., Cottin H. et al. (2005) The origin of the CN radical in comets: A review from observations and models. *Planet. Space Sci., 53,* 1243–1262.
Fray N., Bénilan Y., Biver N. et al. (2006) Heliocentric evolution of the degradation of polyoxymethylene: Application to the origin of the formaldehyde (H_2CO) extended source in Comet C/1995 O1 (Hale-Bopp). *Icarus, 184,* 239–254.
Fulle M., Leblanc F., Harrison R. A. et al. (2007) Discovery of the atomic iron tail of Comet McNaught using the heliospheric imager on STEREO. *Astrophys. J. Lett., 661,* L93–L96.
Gasc S., Altwegg K., Balsiger H. et al. (2017) Change of outgassing pattern of 67P/Churyumov-Gerasimenko during the March 2016 equinox as seen by ROSINA. *Mon. Not. R. Astron. Soc., 469,* S108–S117.
Geiss J. (1987) Composition measurements and the history of cometary matter. In *Exploration of Halley's Comet* (M. Grewing et al., eds.), pp. 859–866. Springer, Berlin.
Gibb E. L., Mumma M. J., Dello Russo N. et al. (2003) Methane in Oort cloud comets. *Icarus, 165,* 391–406.
Glassmeier K.-H., Boehnhardt H., Koschny D. et al. (2007) The Rosetta mission: Flying towards the origin of the solar system. *Space Sci. Rev., 128,* 1–21.
Goesmann F., Rosenbauer H., Roll R. et al. (2007) COSAC, the cometary sampling and composition experiment on Philae. *Space Sci. Rev., 128,* 257–280.
Gordon I. E., Rothman L. S., Hill C. et al. (2017) The HITRAN2016 molecular spectroscopic database. *J. Quant. Spectrosc. Radiat. Transfer, 203,* 3–69.
Guilbert-Lepoutre A., Rosenberg E. D., Prialnik D. et al. (2016) Modelling the evolution of a comet subsurface: Implications for 67P/Churyumov-Gerasimenko. *Mon. Not. R. Astron. Soc., 462,* S146–

S155.
Häberli R. M., Gombosi T. I., De Zeeuw D. L. et al. (1997) Modeling of cometary X-rays caused by solar wind minor ions. *Science, 276,* 939–942.
Hadraoui K., Cottin H., Ivanovski S. et al. (2019) Distributed glycine in Comet 67P/Churyumov-Gerasimenko. *Astron. Astrophys., 630,* A32.
Hama T., Kouchi A., and Watanabe N. (2018) The ortho-to-para ratio of water molecules desorbed from ice made from para-water monomers at 11 K. *Astrophys. J. Lett., 857,* L13.
Hänni N., Altwegg K., Pestoni B. et al. (2020) First in situ detection of the CN radical in comets and evidence for a distributed source. *Mon. Not. R. Astron. Soc., 498,* 2239–2248.
Hänni N., Altwegg K., Balsiger H. et al. (2021) Cyanogen, cyanoacetylene, and acetonitrile in Comet 67P and their relation to the cyano radical. *Astron. Astrophys., 647,* A22.
Hänni N., Altwegg K., Combi M., et al. (2022) Identification and characterization of a new ensemble of cometary organic molecules. *Nature Commun., 13,* 3639.
Hänni N., Altwegg K., Baklouti D., et al. (2023) Oxygen-bearing organic molecules in comet 67P's dusty coma: First evidence for abundant heterocycles. *Astron. Astrophys., 678,* A22.
Hansen K. C., Altwegg K., Berthelier J.-J. et al. (2016) Evolution of water production of 67P/Churyumov-Gerasimenko: An empirical model and a multi-instrument study. *Mon. Not. R. Astron. Soc., 462,* S491–S506.
Hartogh P., Lis D. C., Bockelée-Morvan D. et al. (2011) Ocean-like water in the Jupiter-family Comet 103P/Hartley 2. *Nature, 478,* 218–220.
Haser L. (1957) Distribution d'intensité dans la tête d'une comète. *Bull Soc. R. Sci. Liege, 43,* 740–750.
Hässig M., Altwegg K., Balsiger H. et al. (2015) Time variability and heterogeneity in the coma of 67P/Churyumov-Gerasimenko. *Science, 347,* DOI: 10.1126/science.aaa0276.
Hässig M., Altwegg K., Balsiger H. et al. (2017) Isotopic composition of CO_2 in the coma of 67P/Churyumov-Gerasimenko measured with ROSINA/DFMS. *Astron. Astrophys., 605,* A50.
Heber V. S., Baur H., Bochsler P. et al. (2012) Isotopic mass fractionation of solar wind: Evidence from fast and slow solar wind collected by the Genesis mission. *Astrophys. J., 759,* 121.
Helbert J., Rauer H., Boice D. C. et al. (2005) The chemistry of C_2 and C_3 in the coma of Comet C/1995 O1 (Hale-Bopp) at heliocentric distances $r_h \geq 2.9$ AU. *Astron. Astrophys., 442,* 1107–1120.
Heritier K., Altwegg K., Berthelier J.-J. et al. (2018) On the origin of molecular oxygen in cometary comae. *Nature Commun., 9,* 1–4.
Hily-Blant P., Magalhaes V., Kastner J. et al. (2017) Direct evidence of multiple reservoirs of volatile nitrogen in a protosolar nebula analogue. *Astron. Astrophys., 603,* L6.
Hölscher A. (2015) Formation of C_3 and C_2 in cometary comae. Ph.D. thesis, Technische Univ. Berlin. 263 pp.
Hoppe P., Rubin M., and Altwegg K. (2018) Presolar isotopic signatures in meteorites and comets: New insights from the Rosetta mission to Comet 67P/Churyumov-Gerasimenko. *Space Sci. Rev., 214,* 1–28.
Huebner W. F. and Mukherjee J. (2015) Photoionization and photodissociation rates in solar and blackbody radiation fields. *Planet. Space Sci., 106,* 11–45.
Hutsemékers D., Manfroid J., Jehin E. et al. (2005) Isotopic abundances of carbon and nitrogen in Jupiter-family and Oort cloud comets. *Astron. Astrophys., 440,* L21–L24.
Hutsemékers D., Manfroid J., Jehin E. et al. (2008) The $^{16}OH/^{18}OH$ and OD/OH isotope ratios in Comet C/2002 T7 (LINEAR). *Astron. Astrophys., 490,* L31–L34.
Irvine W. M., Senay M., Lovell A. J. et al. (2000) Note: Detection of nitrogen sulfide in Comet Hale-Bopp. *Icarus, 143,* 412–414.
Ivanova O. V., Luk'yanyk I. V., Kiselev N. N. et al. (2016) Photometric and spectroscopic analysis of Comet 29P/Schwassmann-Wachmann 1 activity. *Planet. Space Sci., 121,* 10–17.
Jackson W. M. (1976) Laboratory observations of the photochemistry of parent molecules: A review. In *The Study of Comets, Part 2* (B. Donn et al., eds.), pp. 679–704. NASA SP-393, Washington, DC.
Jackson W. M., Butterworth P. S., and Ballard D. (1986) The origin of CS in Comet IRAS-Araki-Alcock 1983d. *Astrophys. J., 304,* 515.
Jacquinet-Husson N., Armante R., Scott N. A. et al. (2016) The 2015 edition of the GEISA spectroscopic database. *J. Mol. Spectrosc., 327,* 31–72.
Jaworski W. A. and Tatum J. B. (1991) Analysis of the swings effect and Greenstein effect in Comet P/Halley. *Astrophys. J., 377,* 306.
Jehin E., Manfroid J., Cochran A. L. et al. (2004) The anomalous $^{14}N/^{15}N$ ratio in Comets 122P/1995 S1 (de Vico) and 153P/2002 C1 (Ikeya-Zhang). *Astrophys. J. Lett., 613,* L161–L164.
Jehin E., Manfroid J., Hutsemékers D. et al. (2006) Deep Impact: High-resolution optical spectroscopy with the ESO VLT and the Keck I telescope. *Astrophys. J. Lett., 641,* L145–L148.
Jehin E., Manfroid J., Hutsemékers D. et al. (2009) Isotopic ratios in comets: Status and perspectives. *Earth Moon Planets, 105,* 167–180.
Jehin E., Hutsemékers D., Manfroid J. et al. (2011) A multi-wavelength study with the ESO VLT of Comet 103P/Hartley 2 at the time of the EPOXI encounter. *EPSC Abstracts,* p. 1463.
Jensen S. S., Jørgensen J. K., Furuya K. et al. (2021) Modeling chemistry during star formation: Water deuteration in dynamic star-forming regions. *Astron. Astrophys., 649,* A66.
Jewitt D., Matthews H. E., Owen T. et al. (1997) The $^{12}C/^{13}C$, $^{14}N/^{15}N$ and $^{32}S/^{34}S$ isotope ratios in Comet Hale-Bopp (C/1995 O1). *Science, 278,* 90–93.
Jewitt D., Garland C. A., and Aussel H. (2008) Deep search for carbon monoxide in cometary precursors using millimeter wave spectroscopy. *Astron. J., 135,* 400–407.
Jones G. H., Snodgrass C., Tubiana C., et al. (2024) The Comet Interceptor mission. *Space Sci. Rev., 220,* 9.
Jones R., Chesley S., Connolly A. et al. (2009) Solar system science with LSST. *Earth Moon Planets, 105,* 101–105.
Kawakita H. (2005) Comets as fossils of our solar system. *Astron. Her.,* 98, 757–763.
Kawakita H. and Kobayashi H. (2009) Formation conditions of icy materials in Comet C/2004 Q2 (Machholz). II. Diagnostics using nuclear spin temperatures and deuterium-to-hydrogen ratios in cometary molecules. *Astrophys. J., 693,* 388–396.
Kawakita H., Watanabe J.-I., Ando H. et al. (2001) The spin temperature of NH_3 in Comet C/1999 S4 (LINEAR). *Science, 294,* 1089–1091.
Kawakita H., Kobayashi H., Dello Russo N. et al. (2013) Parent volatiles in Comet 103P/Hartley 2 observed by Keck II with NIRSPEC during the 2010 apparition. *Icarus, 222,* 723–733.
Keller H. U., Mottola S., Davidsson B. et al. (2015a) Insolation, erosion, and morphology of Comet 67P/Churyumov-Gerasimenko. *Astron. Astrophys., 583,* A34.
Keller H. U., Mottola S., Skorov Y. et al. (2015b) The changing rotation period of Comet 67P/Churyumov-Gerasimenko controlled by its activity. *Astron. Astrophys., 579,* L5.
Keller H. U., Mottola S., Hviid S. F. et al. (2017) Seasonal mass transfer on the nucleus of Comet 67P/Churyumov-Gerasimenko. *Mon. Not. R. Astron. Soc., 469,* S357–S371.
Kelley M. S., Woodward C. E., Bodewits D. et al. (2016) Cometary science with the James Webb Space Telescope. *Publ. Astron. Soc. Pac., 128,* 018009.
Kissel J., Altwegg K., Clark B. et al. (2007) COSIMA — high-resolution time-of-flight secondary ion mass spectrometer for the analysis of cometary dust particles onboard Rosetta. *Space Sci. Rev., 128,* 823–867.
Klein B., Hochgürtel S., Krämer I. et al. (2012) High-resolution wide-band fast Fourier transform spectrometers. *Astron. Astrophys., 542,* L3.
Kleine M., Wyckoff S., Wehinger P. A. et al. (1995) The carbon isotope abundance ratio in Comet Halley. *Astrophys. J., 439,* 1021.
Korsun P. P., Rousselot P., Kulyk I. V. et al. (2014) Distant activity of Comet C/2002 VQ94 (LINEAR): Optical spectrophotometric monitoring between 8.4 and 16.8 au from the Sun. *Icarus, 232,* 88–96.
Kouchi A. and Yamamoto T. (1995) Cosmoglaciology: Evolution of ice in interstellar space and the early solar system. *Prog. Cryst. Growth Charact. Mater., 30,* 83–107.
Kramer T., Läuter M., Hviid S. et al. (2019) Comet 67P/Churyumov-Gerasimenko rotation changes derived from sublimation-induced torques. *Astron. Astrophys., 630,* A3.
Krankowsky D., Lämmerzahl P., Herrwerth I. et al. (1986) In situ gas and ion measurements at Comet Halley. *Nature, 321,* 326–329.
Lambert D. L. and Danks A. C. (1983) High-resolution spectra of C_2 Swan bands from Comet West 1976 VI. *Astrophys. J., 268,* 428–446.
Lamy P. L. and Perrin J.-M. (1988) Optical properties of organic grains: Implications for interplanetary and cometary dust. *Icarus, 76,* 100–109.
Langland-Shula L. E. and Smith G. H. (2011) Comet classification with new methods for gas and dust spectroscopy. *Icarus, 213,* 280–322.
Laufer D., Bar-Nun A., and Ninio Greenberg A. (2017) Trapping

mechanism of O_2 in water ice as first measured by Rosetta spacecraft. *Mon. Not. R. Astron. Soc., 469,* S818–S823.

Läuter M., Kramer T., Rubin M. et al. (2020) The gas production of 14 species from Comet 67P/Churyumov-Gerasimenko based on DFMS/COPS data from 2014 to 2016. *Mon. Not. R. Astron. Soc., 498,* 3995–4004.

Le Roy L., Altwegg K., Balsiger H. et al. (2015) Inventory of the volatiles on Comet 67P/Churyumov-Gerasimenko from Rosetta/ROSINA. *Astron. Astrophys., 583,* A1.

Lecacheux A., Biver N., Crovisier J. et al. (2003) Observations of the 557 GHz water line in comets with the Odin satellite. *Astron. Astrophys., 402,* L55–L58.

Lee J.-Y., Marti K., Severinghaus J. P. et al. (2006) A redetermination of the isotopic abundances of atmospheric Ar. *Geochim. Cosmochim. Acta, 70,* 4507–4512.

Lefloch B., Ceccarelli C., Codella C. et al. (2017) L1157-B1, a factory of complex organic molecules in a solar-type star-forming region. *Mon. Not. R. Astron. Soc., 469,* L73–L77.

Ligterink N. F. W., Kipfer K. A., Rubin M., et al. (2024) Sublimation of volatiles from H_2O:CO_2 bulk ices in the context of comet 67P/Churyumov-Gerasimenko. II. Noble gases. *Astron. Astrophys., 687,* A78.

Lippi M., Villanueva G. L., Mumma M. J. et al. (2021) Investigation of the origins of comets as revealed through infrared high-resolution spectroscopy I. Molecular abundances. *Astron. J., 162,* 74.

Lis D. C., Keene J., Young K. et al. (1997) Spectroscopic observations of Comet C/1996 B2 (Hyakutake) with the Caltech Submillimeter Observatory. *Icarus, 130,* 355–372.

Lis D. C., Bockelée-Morvan D., Boissier J. et al. (2008) Hydrogen isocyanide in Comet 73P/Schwassmann-Wachmann (fragment B). *Astrophys. J., 675,* 931–936.

Lis D. C., Bockelée-Morvan D., Güsten R. et al. (2019) Terrestrial deuterium-to-hydrogen ratio in water in hyperactive comets. *Astron. Astrophys., 625,* L5.

Lisse C., Bauer J., Cruikshank D. et al. (2020) Spitzer's solar system studies of comets, Centaurs and Kuiper belt objects. *Nature Astron., 4,* 930–939.

Lodders K. (2010) Solar system abundances of the elements. In *Principles and Perspectives in Cosmochemistry* (A. Goswami and B. E. Reddy, eds.), pp. 379–417. Springer-Verlag, Berlin.

Loison J.-C., Wakelam V., Gratier P. et al. (2019) Oxygen fractionation in dense molecular clouds. *Mon. Not. R. Astron. Soc., 485,* 5777–5789.

Luspay-Kuti A., Hässig M., Fuselier S. A. et al. (2015) Composition-dependent outgassing of Comet 67P/Churyumov-Gerasimenko from ROSINA/DFMS: Implications for nucleus heterogeneity? *Astron. Astrophys., 583,* A4.

Luspay-Kuti A., Mousis O., Lunine J. I. et al. (2018) Origin of molecular oxygen in comets: Current knowledge and perspectives. *Space Sci. Rev., 214,* 1–24.

Luspay-Kuti A., Mousis O., Pauzat F. et al. (2022) Dual storage and release of molecular oxygen in Comet 67P/Churyumov-Gerasimenko. *Nature Astron., 6,* 724–730.

Magee-Sauer K., Mumma M. J., DiSanti M. A. et al. (2002) Hydrogen cyanide in Comet C/1996 B2 Hyakutake. *J. Geophys. Res.–Planets, 107,* 5096.

Manfroid J., Jehin E., Hutsemékers D. et al. (2009) The CN isotopic ratios in comets. *Astron. Astrophys., 503,* 613–624.

Manfroid J., Hutsemékers D., and Jehin E. (2021) Iron and nickel atoms in cometary atmospheres even far from the Sun. *Nature, 593,* 372–374.

Maquet L. (2015) The recent dynamical history of Comet 67P/Churyumov-Gerasimenko. *Astron. Astrophys., 579,* A78.

Marty B., Chaussidon M., Wiens R. C. et al. (2011) A ^{15}N-poor isotopic composition for the solar system as shown by Genesis solar wind samples. *Science, 332,* 1533.

Marty B., Altwegg K., Balsiger H. et al. (2017) Xenon isotopes in 67P/Churyumov-Gerasimenko show that comets contributed to Earth's atmosphere. *Science, 356,* 1069–1072.

McGuire B. A. (2018) 2018 census of interstellar, circumstellar, extragalactic, protoplanetary disk, and exoplanetary molecules. *Astrophys. J. Suppl., 239,* 17.

McKay A. J., Chanover N. J., Morgenthaler J. P. et al. (2013) Observations of the forbidden oxygen lines in DIXI target Comet 103P/Hartley. *Icarus, 222,* 684–690.

McKay A., Cochran A., Dello Russo N. et al. (2014) Evolution of fragment-species production in Comet C/2012 S1 (ISON) from 1.6 au to 0.04 au. In *Asteroids, Comets, Meteors 2014* (K. Muinonen et al., eds.), p. 365. Univ. of Helsinki, Helsinki.

McKay A. J., DiSanti M. A., Kelley M. S. P. et al. (2019) The peculiar volatile composition of CO-dominated Comet C/2016 R2 (PanSTARRS). *Astron. J., 158,* 128.

McKeegan K., Kallio A., Heber V. et al. (2011) The oxygen isotopic composition of the Sun inferred from captured solar wind. *Science, 332,* 1528–1532.

Meier R. and A'Hearn M. F. (1997) Atomic sulfur in cometary comae based on UV spectra of the S I triplet near 1814 Å. *Icarus, 125,* 164–194.

Meier R., Eberhardt P., Krankowsky D. et al. (1993) The extended formaldehyde source in Comet P/Halley. *Astron. Astrophys., 277,* 677.

Meier R., Owen T. C., Jewitt D. C. et al. (1998) Deuterium in Comet C/1995 O1 (Hale-Bopp): Detection of DCN. *Science, 279,* 1707.

Meija J., Coplen T. B., Berglund M. et al. (2016) Atomic weights of the elements 2013 (IUPAC Technical Report). *Pure Appl. Chem., 88,* 265–291.

Miani A. and Tennyson J. (2004) Can ortho-para transitions for water be observed? *J. Chem. Phys., 120,* 2732–2739.

Mommert M., Hora J. L., Trilling D. E. et al. (2020) Recurrent cometary activity in near-Earth object (3552) Don Quixote. *Planet. Sci. J., 1,* 12.

Morgenthaler J. P., Harris W. M., Scherb F. et al. (2001) Large-aperture [O I] 6300 Å photometry of Comet Hale-Bopp: Implications for the photochemistry of OH. *Astrophys. J., 563,* 451–461.

Moulane Y., Jehin E., Rousselot P. et al. (2020) Photometry and high-resolution spectroscopy of Comet 21P/Giacobini-Zinner during its 2018 apparition. *Astron. Astrophys., 640,* A54.

Mousis O., Lunine J., Luspay-Kuti A. et al. (2016a) A protosolar nebula origin for the ices agglomerated by Comet 67P/Churyumov-Gerasimenko. *Astrophys. J. Lett., 819,* L33.

Mousis O., Ronnet T., Brugger B. et al. (2016b) Origin of molecular oxygen in Comet 67P/Churyumov-Gerasimenko. *Astrophys. J. Lett., 823,* L41.

Mousis O., Aguichine A., Bouquet A. et al. (2021) Cold traps of hypervolatiles in the protosolar nebula at the origin of the peculiar composition of Comet C/2016 R2 (PanSTARRS). *Planet. Sci. J., 2,* 72.

Müller H. S. P., Schlöder F., Stutzki J. et al. (2005) The Cologne Database for Molecular Spectroscopy, CDMS: A useful tool for astronomers and spectroscopists. *J. Mol. Structure, 742,* 215–227.

Müller D. R., Altwegg K., Berthelier J. J. et al. (2022) High D/H ratios in water and alkanes in Comet 67P/Churyumov-Gerasimenko measured with Rosetta/ROSINA DFMS. *Astron. Astrophys., 662,* A69.

Mumma M. J. and Charnley S. B. (2011) The chemical composition of comets emerging taxonomies and natal heritage. *Annu. Rev. Astron. Astrophys., 49,* 471–524.

Mumma M. J., DiSanti M. A., Russo N. D. et al. (1996) Detection of abundant ethane and methane, along with carbon monoxide and water, in Comet C/1996 B2 Hyakutake: Evidence for interstellar origin, *Science, 272,* 1310–1314.

Mumma M. J., DiSanti M. A., Dello Russo N. et al. (2000) Detection of CO and ethane in Comet 21P/Giacobini-Zinner: Evidence for variable chemistry in the outer solar nebula. *Astrophys. J. Lett., 531,* L155–L159.

Mumma M. J., Bonev B. P., Villanueva G. L. et al. (2011) Temporal and spatial aspects of gas release during the 2010 apparition of Comet 103P/Hartley 2. *Astrophys. J. Lett., 734,* L7.

Muñoz Caro G. M., Meierhenrich U. J., Schutte W. A. et al. (2002) Amino acids from ultraviolet irradiation of interstellar ice analogues. *Nature, 416,* 403–406.

Neufeld D. A., Stauffer J. R., Bergin E. A. et al. (2000) Submillimeter wave astronomy satellite observations of water vapor toward Comet C/1999 H1 (Lee). *Astrophys. J. Lett., 539,* L151–L154.

Newall H. F. (1910) On the spectrum of the daylight Comet 1910a. *Mon. Not. R. Astron. Soc., 70,* 459.

Oliversen R. J., Doane N., Scherb F. et al. (2002) Measurements of [C I] emission from Comet Hale-Bopp. *Astrophys. J., 581,* 770–775.

Ootsubo T., Kawakita H., Hamada S. et al. (2012) AKARI near-infrared spectroscopic survey for CO_2 in 18 comets. *Astrophys. J., 752,* 15.

Opitom C., Jehin E., Manfroid J. et al. (2015) TRAPPIST photometry and imaging monitoring of Comet C/2013 R1 (Lovejoy): Implications for the origin of daughter species. *Astron. Astrophys., 584,* A121.

Opitom C., Snodgrass C., Fitzsimmons A. et al. (2017) Ground-based monitoring of Comet 67P/Churyumov-Gerasimenko gas activity throughout the Rosetta mission. *Mon. Not. R. Astron. Soc., 469,* S222–S229.

Opitom C., Hutsemékers D., Jehin E. et al. (2019a) High resolution optical spectroscopy of the N_2-rich Comet C/2016 R2 (PanSTARRS). *Astron. Astrophys., 624,* A64.

Opitom C., Yang B., Selman F. et al. (2019b) First observations of an outbursting comet with the MUSE integral-field spectrograph. *Astron. Astrophys., 628,* A128.

Oró J. (1961) Comets and the formation of biochemical compounds on the primitive Earth. *Nature, 190,* 389–390.

Osip D. J., A'Hearn M., and Raugh A. C. (2003) *Lowell Observatory Cometary Database — Production Rates.* NASA Planetary Data System, EAR-C-PHOT-5-RDR-LOWELL-COMET-DB-PR-V1.0.

Paganini L. and Mumma M. J. (2016) A solar-pumped fluorescence model for line-by-line emission intensities in the B-X, A-X, and X-X band systems of $^{12}C^{14}N$. *Astrophys. J. Suppl., 226,* 3.

Paganini L., Mumma M. J., Villanueva G. L. et al. (2012) The chemical composition of CO-rich Comet C/2009 P1 (Garradd) at R_h = 2.4 and 2.0 AU before perihelion. *Astrophys. J. Lett., 748,* L13.

Paganini L., Mumma M. J., Gibb E. L. et al. (2017) Ground-based detection of deuterated water in Comet C/2014 Q2 (Lovejoy) at IR wavelengths. *Astrophys. J. Lett., 836,* L25.

Paquette J., Hornung K., Stenzel O. J. et al. (2017) The $^{34}S/^{32}S$ isotopic ratio measured in the dust of Comet 67P/Churyumov-Gerasimenko by Rosetta/COSIMA. *Mon. Not. R. Astron. Soc., 469,* S230–S237.

Pickett H. M., Poynter R. L., Cohen E. A. et al. (1998) Submillimeter, millimeter, and microwave spectral line catalog. *J. Quant. Spectrosc. Radiat. Transfer, 60,* 883–890.

Poch O., Istiqomah I., Quirico E. et al. (2020) Ammonium salts are a reservoir of nitrogen on a cometary nucleus and possibly on some asteroids. *Science, 367,* DOI: 10.1126/science.aaw7462.

Preston G. W. (1967) The spectrum of Ikeya-Seki (1965f). *Astrophys. J., 147,* 718–742.

Qi C., Hogerheijde M. R., Jewitt D. et al. (2015) Peculiar near-nucleus outgassing of Comet 17P/Holmes during its 2007 outburst. *Astrophys. J., 799,* 110.

Quirico E., Moroz L., Schmitt B. et al. (2016) Refractory and semi-volatile organics at the surface of Comet 67P/Churyumov-Gerasimenko: Insights from the VIRTIS/Rosetta imaging spectrometer. *Icarus, 272,* 32–47.

Radeva Y. L., Mumma M. J., Villanueva G. L. et al. (2013) High-resolution infrared spectroscopic measurements of Comet 2P/Encke: Unusual organic composition and low rotational temperatures. *Icarus, 223,* 298–307.

Raghuram S., Bhardwaj A., Hutsemékers D. et al. (2021) A physico-chemical model to study the ion density distribution in the inner coma of Comet C/2016 R2 (Pan-STARRS). *Mon. Not. R. Astron. Soc., 501,* 4035–4052.

Rawlings J., Wilson T. G., and Williams D. A. (2019) A gas-phase primordial origin of O_2 in Comet 67P/Churyumov-Gerasimenko. *Mon. Not. R. Astron. Soc., 486,* 10–20.

Raymond J. C., Downs C., Knight M. M. et al. (2018) Comet C/2011 W3 (Lovejoy) between 2 and 10 solar radii: Physical parameters of the comet and the corona. *Astrophys. J., 858,* 19.

Reinhard R. (1986) The Giotto encounter with Comet Halley. *Nature, 321,* 313–318.

Reylé C. and Boice D. C. (2003) An S_2 fluorescence model for interpreting high-resolution cometary spectra. I. Model description and initial results. *Astrophys. J., 587,* 464–471.

Rivilla V., Drozdovskaya M. N., Altwegg K. et al. (2020) ALMA and ROSINA detections of phosphorus-bearing molecules: The interstellar thread between star-forming regions and comets. *Mon. Not. R. Astron. Soc., 492,* 1180–1198.

Rodgers S. D. and Charnley S. B. (2001) On the origin of HNC in Comet Lee. *Mon. Not. R. Astron. Soc., 323,* 84–92.

Roesler F. L., Scherb F., Magee K. et al. (1985) High spectral resolution line profiles and images of Comet Halley. *Adv. Space Res., 5,* 279–282.

Roth N. X., Gibb E. L., Bonev B. P. et al. (2018) A tale of "two" comets: The primary volatile composition of Comet 2P/Encke across apparitions and implications for cometary science. *Astron. J., 156,* 251.

Roth N. X., Gibb E. L., Bonev B. P. et al. (2020) Probing the evolutionary history of comets: An investigation of the hypervolatiles CO, CH_4, and C_2H_6 in the Jupiter-family comet 21P/Giacobini-Zinner. *Astron. J., 159,* 42.

Roth N. X., Milam S. N., Cordiner M. A. et al. (2021) Leveraging the ALMA Atacama Compact Array for cometary science: An interferometric survey of Comet C/2015 ER61 (PanSTARRS) and evidence for a distributed source of carbon monosulfide. *Astrophys. J., 921,* A14.

Rousselot P., Jehin E., Manfroid J. et al. (2012) The $^{12}C2/^{12}C^{13}C$ isotopic ratio in comets C/2001 Q4 (NEAT) and C/2002 T7 (LINEAR). *Astron. Astrophys., 545,* A24.

Rousselot P., Pirali O., Jehin E. et al. (2014) Toward a unique nitrogen isotopic ratio in cometary ices. *Astrophys. J. Lett., 780,* L17.

Rousselot P., Decock A., Korsun P. P. et al. (2015) High-resolution spectra of Comet C/2013 R1 (Lovejoy). *Astron. Astrophys., 580,* A3.

Rubin M., Hansen K. C., Gombosi T. I. et al. (2009) Ion composition and chemistry in the coma of Comet 1P/Halley: A comparison between Giotto's Ion Mass Spectrometer and our ion-chemical network. *Icarus, 199,* 505–519.

Rubin M., Altwegg K., Balsiger H. et al. (2015a) Molecular nitrogen in Comet 67P/Churyumov-Gerasimenko indicates a low formation temperature. *Science, 348,* 232–235.

Rubin M., Altwegg K., van Dishoeck E. F. et al. (2015b) Molecular oxygen in Oort cloud Comet 1P/Halley. *Astrophys. J. Lett., 815,* L11.

Rubin M., Altwegg K., Balsiger H. et al. (2017) Evidence for depletion of heavy silicon isotopes at Comet 67P/Churyumov-Gerasimenko. *Astron. Astrophys., 601,* A123.

Rubin M., Altwegg K., Balsiger H. et al. (2018) Krypton isotopes and noble gas abundances in the coma of Comet 67P/Churyumov-Gerasimenko. *Sci. Adv.,4,* eaar6297.

Rubin M., Altwegg K., Balsiger H. et al. (2019) Elemental and molecular abundances in Comet 67P/Churyumov-Gerasimenko. *Mon. Not. R. Astron. Soc., 489,* 594–607.

Rubin M., Engrand C., Snodgrass C. et al. (2020) On the origin and evolution of the material in 67P/Churyumov-Gerasimenko. *Space Sci. Rev., 216,* 1–43.

Rubin M., Altwegg K., Berthelier J.-J. et al. (2022) Refractory elements in the gas phase for Comet 67P/Churyumov-Gerasimenko. Possible release of atomic Na, Si, and Fe from nanograins. *Astron. Astrophys., 658,* A87.

Rubin M., Altwegg K., Berthelier J.-J., et al. (2023) Volatiles in the H_2O and CO_2 ices of comet 67P/Churyumov-Gerasimenko. *Mon. Not. R. Astron. Soc., 526,* 4209–4233.

Saki M., Gibb E. L., Bonev B. P. et al. (2020) Carbonyl sulfide (OCS): Detections in Comets C/2002 T7 (LINEAR), C/2015 ER61 (PanSTARRS), and 21P/Giacobini-Zinner and stringent upper limits in 46P/Wirtanen. *Astron. J., 160,* 184.

Schleicher D. G. (2008) The extremely anomalous molecular abundances of Comet 96P/Machholz 1 from narrowband photometry. *Astron. J., 136,* 2204–2213.

Schleicher D. G. and Bair A. N. (2011) The composition of the interior of Comet 73P/Schwassmann-Wachmann 3: Results from narrowband photometry of multiple components. *Astron. J., 141,* 177.

Schleicher D. G. and Farnham T. L. (2004) Photometry and imaging of the coma with narrowband filters. In *Comets II* (M. C. Festou et al., eds.), pp. 449–469. Univ. of Arizona, Tucson.

Schroeder I. R., Altwegg K., Balsiger H. et al. (2019) $^{16}O/^{18}O$ ratio in water in the coma of Comet 67P/Churyumov-Gerasimenko measured with the Rosetta/ROSINA double-focusing mass spectrometer. *Astron. Astrophys., 630,* A29.

Schuhmann M., Altwegg K., Balsiger H. et al. (2019) Aliphatic and aromatic hydrocarbons in Comet 67P/Churyumov-Gerasimenko seen by ROSINA. *Astron. Astrophys., 630,* A31.

Schultz D., Li G. S. H., Scherb F. et al. (1992) Comet Austin (1989c1) $O(^1D)$ and H_2O production rates. *Icarus, 96,* 190–197.

Schulz R., Hilchenbach M., Langevin Y. et al. (2015) Comet 67P/Churyumov-Gerasimenko sheds dust coat accumulated over the past four years. *Nature, 518,* 216–218.

Schwamb M. E., Jones R. L., Chesley S. R. et al. (2018) Large synoptic survey telescope solar system science roadmap. *ArXiV e-prints,* arXiv:1802.01783.

Schwartz A. W. (2006) Phosphorus in prebiotic chemistry, *Phil. Trans. R. Soc. London Ser. B, 361,* 1743–1749.

Seager S., Bains W., and Petkowski J. (2016) Toward a list of molecules as potential biosignature gases for the search for life on exoplanets and applications to terrestrial biochemistry. *Astrobiology, 16,* 465–485.

Senay M. C. and Jewitt D. (1994) Coma formation driven by carbon monoxide release from Comet Schwassmann-Wachmann 1. *Nature, 371,* 229–231.

Shinnaka Y. and Kawakita H. (2016) Nitrogen isotopic ratio of cometary ammonia from high-resolution optical spectroscopic observations of C/2014 Q2 (Lovejoy). *Astron. J., 152,* 145.

Shinnaka Y., Kawakita H., Kobayashi H. et al. (2012) Ortho-to-paraabundance ratio of water ion in Comet C/2001 Q4 (NEAT): Implication for ortho-to-para abundance ratio of water. *Astrophys. J., 749,* 101.

Shinnaka Y., Kawakita H., Nagashima M. et al. (2014) High-dispersion spectroscopic observations of Comet C/2012 S1 (ISON) with the Subaru Telescope. In *AAS/Division for Planetary Sciences Meeting Abstracts, 46,* #209.14.

Shinnaka Y., Kawakita H., Jehin E. et al. (2016) Nitrogen isotopic ratios of NH_2 in comets: Implication for ^{15}N-fractionation in cometary ammonia. *Mon. Not. R. Astron. Soc., 462,* S195–S209.

Shinnaka Y., Kawakita H., Kondo S. et al. (2017) Near-infrared spectroscopic observations of Comet C/2013 R1 (Lovejoy) by WINERED: CN red-system band emission. *Astron. J., 154,* 45.

Sierks H., Barbieri C., Lamy P. L. et al. (2015) On the nucleus structure and activity of Comet 67P/Churyumov-Gerasimenko. *Science, 347,* DOI: 10.1126/science.aaa1044.

Skidmore W., TMT International Science Development Teams, and TMT Science Advisory Committee (2015) Thirty Meter Telescope Detailed Science Case: 2015. *Res. Astron. Astrophys., 15,* 1945.

Slaughter C. D. (1969) The emission spectrum of Comet Ikeya-Seki 1965-f at perihelion passage. *Astron. J., 74,* 929.

Snodgrass C. and Jones G. H. (2019) The European Space Agency's comet interceptor lies in wait. *Nature Commun., 10,* 1–4.

Stern S., Slater D., Festou M. et al. (2000) The discovery of argon in Comet C/1995 O1 (Hale-Bopp). *Astrophys. J. Lett., 544,* L169.

Stief L. J. (1972) Origin of C_3 in comets. *Nature, 237,* 29.

Swings P. (1965) Cometary spectra. *Q. J. R. Astron. Soc., 6,* 28.

Swings P. and Haser L. (1956) *Atlas of Representative Cometary Spectra.* Impr. Ceuterick, Louvain. 37 pp.

Taquet V., Peters P. S., Kahane C. et al. (2013) Water ice deuteration: A tracer of the chemical history of protostars. *Astron. Astrophys., 550,* A127.

Taquet V., López-Sepulcre A., Ceccarelli C. et al. (2015) Constraining the abundances of complex organics in the inner regions of solar-type protostars. *Astrophys. J., 804,* 81.

Taquet V., Furuya K., Walsh C. et al. (2016) A primordial origin for molecular oxygen in comets: A chemical kinetics study of the formation and survival of O_2 ice from clouds to discs. *Mon. Not. R. Astron. Soc., 462,* S99–S115.

Thomas N., Ulamec S., Kührt E. et al. (2019) Towards new comet missions. *Space Sci. Rev., 215,* 1–35.

Valkiers S., Aregbe Y., Taylor P. et al. (1998) A primary xenon isotopic gas standard with SI traceable values for isotopic composition and molar mass. *Intl. J. Mass Spec. Ion Process., 173,* 55–63.

Vaughan C. M., Pierce D. M., and Cochran A. L. (2017) Jet morphology and coma analysis of Comet 103P/Hartley 2. *Astron. J., 154,* 219.

Villanueva G., Mumma M., DiSanti M. et al. (2012) A multi-instrument study of Comet C/2009 P1 (Garradd) at 2.1 AU (pre-perihelion) from the Sun. *Icarus, 220,* 291–295.

Villanueva G. L., Smith M. D., Protopapa S. et al. (2018) Planetary spectrum generator: An accurate online radiative transfer suite for atmospheres, comets, small bodies and exoplanets. *J. Quant. Spectrosc. Radiat. Transfer, 217,* 86–104.

Waite J. H. J., Lewis W. S., Magee B. A. et al. (2009) Liquid water on Enceladus from observations of ammonia and ^{40}Ar in the plume. *Nature, 460,* 487–490.

Watanabe N. and Kouchi A. (2002) Efficient formation of formaldehyde and methanol by the addition of hydrogen atoms to CO in H_2O-CO ice at 10 K. *Astrophys. J. Lett., 571,* L173.

Weaver H. A., Mumma M. J., and Larson H. P. (1987) Infrared investigation of water in Comet P/Halley. *Astron. Astrophys., 187,* 411–418.

Weaver H. A., Chin G., Bockelée-Morvan D. et al. (1999) An infrared investigation of volatiles in Comet 21P/Giacobini-Zinner. *Icarus, 142,* 482–497.

Wedlund C. S., Kallio E., Alho M. et al. (2016) The atmosphere of Comet 67P/Churyumov-Gerasimenko diagnosed by charge-exchanged solar wind alpha particles. *Astron. Astrophys., 587,* A154.

Weiler M. (2012) The chemistry of C_3 and C_2 in cometary comae. I. Current models revisited. *Astron. Astrophys., 538,* A149.

Wierzchos K., Womack M., and Sarid G. (2017) Carbon monoxide in the distantly active Centaur (60558) 174P/Echeclus at 6 au. *Astron. J., 153,* 230.

Wirström E., Geppert W. D., Hjalmarson Å. et al. (2011a) Observational tests of interstellar methanol formation. *Astron. Astrophys., 533,* A24.

Wirström E., Geppert W. D., Persson C. M. et al. (2011b) The $^{12}C/^{13}C$ ratio as a chemistry indicator. In *The Molecular Universe: Posters from the Proceedings of the 280th Symposium of the International Astronomical Union* (J. Cernicharo and R. Bachiller, eds.), Poster #384. IAU Symp. 280, Cambridge Univ., Cambridge.

Womack M., Sarid G., and Wierzchos K. (2017) CO in distantly active comets. *Publ. Astron. Soc. Pac., 129,* 031001.

Wright I., Barber S., Morgan G. et al. (2007) Ptolemy — an instrument to measure stable isotopic ratios of key volatiles on a cometary nucleus. *Space Sci. Rev., 128,* 363–381.

Wurz P., Rubin M., Altwegg K. et al. (2015) Solar wind sputtering of dust on the surface of 67P/Churyumov-Gerasimenko. *Astron. Astrophys., 583,* A22.

Wyckoff S. and Theobald J. (1989) Molecular ions in comets. *Adv. Space Res., 9,* 157–161.

Wyckoff S., Lindholm E., Wehinger P. A. et al. (1989) The $^{12}C/^{13}C$ abundance ratio in Comet Halley. *Astrophys. J., 339,* 488.

Wyckoff S., Kleine M., Peterson B. A. et al. (2000) Carbon isotope abundances in comets. *Astrophys. J., 535,* 991–999.

Yang B., Hutsemékers D., Shinnaka Y. et al. (2018) Isotopic ratios in outbursting Comet C/2015 ER61. *Astron. Astrophys., 609,* L4.

Yang B., Jewitt D., Zhao Y. et al. (2021) Discovery of carbon monoxide in distant Comet C/2017 K2 (PANSTARRS). *Astrophys. J. Lett., 914,* L17.

Yao Y. and Giapis K. P. (2017) Dynamic molecular oxygen production in cometary comae. *Nature Commun., 8,* 1–8.

Zhang T., Xu K., and Ding X. (2021) China's ambitions and challenges for asteroid-comet exploration. *Nature Astron., 5,* 730–731.

Ziurys L. M., Savage C., Brewster M. A. et al. (1999) Cyanide chemistry in Comet Hale-Bopp (C/1995 O1). *Astrophys. J. Lett., 527,* L67–L71.

Part 6:

Plasma

Beth A., Galand M., Simon Wedlund C., and Eriksson A. (2024) Cometary ionospheres: An updated tutorial. In *Comets III* (K. J. Meech, M. R. Combi, D. Bockelée-Morvan, S. N. Raymond, and M. E. Zolensky, eds.), pp. 501–541. Univ. of Arizona, Tucson, DOI: 10.2458/azu_uapress_9780816553631-ch016.

Cometary Ionospheres: An Updated Tutorial

Arnaud Beth
Umeå Universitet

Marina Galand
Imperial College London

Cyril Simon Wedlund
Österreichische Akademie der Wissenschaften, Institut für Weltraumforschung

Anders Eriksson
Institutet för Rymdfysik

This chapter aims at providing the tools and knowledge to understand and model the plasma environment surrounding comets in the innermost part near the nucleus. In particular, our goal is to give an updated post-Rosetta "view" of this ionized environment: what we knew, what we confirmed, what we overturned, and what we still do not understand.

1. INTRODUCTION

Comets have aroused the curiosity of humankind for millennia. Reported in different ways throughout history — for instance, through texts and artwork — a few of them have passed Earth close enough to be witnessed with the naked eye. But what makes comets visible and so bright? For those lucky enough to see them in their lifetime, comets in the sky display very singular shapes and colors: a very broad white tail and a thinner blue slightly transparent one, both originating from the same point where the nucleus is located. The "white tail" is partly directed anti-sunward (i.e., opposed to the Sun's direction), with a component along the comet's trajectory. It originates from the dust reflecting sunlight that is continuously expelled from the comet's surface and pushed away, swept by the solar radiation pressure. This radiation pressure is induced by solar photons, which, once they hit the dust, transfer a part of their momentum continuously to it, equivalent to a force. Depending on the size of dust grains, this force, in addition to inertial ones, is the main driver of the dust dynamics, which as a result may overcome the Sun's gravity [Finson-Probstein model; see *Finson and Probstein* (1968).]

The "blue" tail itself originates from ions present in the coma. This was first evidenced in the band spectra of Comet C/1907 L2 (Daniel) (*Deslandres and Bernard,* 1907; *Evershed,* 1907), although its true origin was unknown at the time (*Larsson et al.,* 2012). Subsequently, *Fowler* (1907, 1910) investigated these bands by means of laboratory experiments and electric discharges in different gases and concluded that carbon monoxide was the most likely cause. Fowler was *almost* right. The blue color displayed by this tail comes in fact from the fluorescence of CO^+ ions. Although the cometary gas is mainly composed of water (H_2O), other species are present such as carbon dioxide (CO_2), carbon monoxide (CO), and formaldehyde (methanal, H_2CO). As they leave the nucleus, these molecules may break into smaller fragments, neutral atoms/molecules (through photodissociation), be ionized (photoionization), or undergo both processes at the same time (dissociative photoionization). The newly formed CO^+ may in turn be excited by absorbing solar photons at specific wavelengths and is deexcited by reemitting a photon at the same wavelength (resonance fluorescence). Among cometary species, CO^+ is the only one that efficiently emits in the visible, especially in the blue part (so-called "comet tail" A-X bands emitting between 308 and 720 nm).

In almost a century, even though substantial progress has been made in astronomy and instrumentation, only a few other ions have been observed and confirmed at comets through remote sensing from Earth, for example, HO^+, CO_2^+, and H_2O^+ (see the chapter by Bodewits et al. in this volume). Early on, another mystery was also associated with the "ion" tail: its direction. Indeed, through the comet's course around the Sun, and unlike the dust tail, the blue ion tail is exactly anti-sunward. The discovery of the solar wind was motivated by the observations of cometary ion tails from Earth. Following early works by *Hoffmeister* (1943), *Biermann* (1951) was the first to suggest then that there is a flow of ionized particles from the Sun, "a stream of solar matter" (*Biermann,* 1952), although why this stream could arise was another unanswered question. In 1958, Eugene Parker followed on Biermann's lead and published his seminal paper on the generation and existence of the so-called "solar wind" (*Parker,* 1958), an idea that was initially ill received (*Obridko and Vaisberg,* 2017).

Interestingly, contemporarily to these works, the true nature of comets and of their nucleus remained unknown. Only in 1950 did Fred Whipple formulate the idea that comets were made of a conglomerate of ices combined with a conglomerate of meteoric materials (*Whipple,* 1950), known as the "dirty snowball" nucleus model. Our modern view based on recent observations of cometary nuclei shows that comets are rather *dirt balls with some snow*. Based on this model, *Haser* (1957) (see *Haser et al.,* 2020, for a modern, accessible, and translated version) later developed the mathematical framework to describe the cometary neutral environment, referred to as the "Haser model" in the current literature. Haser derived the mathematical formulation describing the spatial distribution and number density for neutral species, either those coming from the ice sublimation at the (near-)surface of the nucleus, the "parent" species, or those coming from the dissociation of the parent species, the "daughter" species. For "parent" molecules, such as H_2O and CO_2, primarily released from the (near) surface, the flux is conserved; the number density profile of the escaping gas of species p is thus given in m^{-3} by

$$n_p(r) = \frac{Q_p}{4\pi V_p r^2} \exp\left(-\frac{r}{\mathcal{T}_p V_p}\right) \quad (1)$$

where Q_p is the total number of "parent" (p) molecules expelled from the surface per unit time (molecules s^{-1} or simply s^{-1} for short), V_p is the radial speed of these molecules assumed constant (m s^{-1}) estimated around 0.4–0.9 km s^{-1} (see the chapter by Biver et al. in this volume and *Hansen et al.,* 2016; *Biver et al.,* 2019), r is the distance from the nucleus (m), and $\mathcal{T}_p$ is the lifetime of these molecules against dissociation (s). This equation remains valid for assymmetric outgassing (i.e., Q_p varies with latitude and longitude at a given r as long as the gas velocity remains radial). Usually, molecules are characterized, in a cometary context, by their so-called characteristic scale length $L_p = V_p\mathcal{T}_p$. This model is adequate for species and molecules that are mainly photodissociated by extreme ultraviolet (EUV) solar radiation, for instance, H_2O into HO + H. It also provides an easy way to retrieve the outgassing rate Q_p of the different cometary neutral parent species based on remote sensing observations. L_p is usually on the order of 10^4–10^5 km corresponding to a typical lifetime of 10^4–10^5 s (*Huebner and Mukherjee,* 2015); the exponential correction in equation (1) should be accounted for when comets are observed remotely from Earth (as the full extent of the coma is probed) or when numerical simulations of the whole cometary environment are developed. Furthermore, not all neutral species may come directly from the nucleus. Some species, such as H_2CO and CO, may also originate from dust and photodissociation of larger molecules already present in the coma, referred to as "extended sources," and thus do not follow equation (1) (*Eberhardt,* 1999). These extended sources must be accounted for as well in large-scale simulations beyond thousands of kilometers (see the chapter by Biver et al. in this volume). However, in the rest of this chapter, the exponential term may be dropped in equation (1) as we are focusing on major neutral species (originating primarily from the nucleus) and on *in situ* observations at cometocentric distances $r \ll L_p$. In these regions, *in situ* observations performed at 1P/Halley [hereafter 1P (*Krankowsky et al.,* 1986)] and at 67P/Churyumov-Gerasimenko [hereafter 67P (*Hässig et al.,* 2015)] showed that $n_{H_2O} \propto 1/r^2$ and equation (1) reduces to

$$n_p(r) \approx \frac{Q_p}{4\pi V_p r^2} \text{ for } r \ll L_p \quad (2)$$

Although there are neutral species produced in part within the coma and linked to the so-called *extended sources* (*Eberhardt,* 1999), such as CO and O, such a source can often be neglected in plasma models of the inner coma.

Similarly to equation (2), the total neutral number density n_n in the inner coma is given by

$$n_n(r) = \sum_p n_p \approx \frac{Q}{4\pi V_n r^2} \quad (3)$$

where

$$V_n = \frac{1}{n_n}\sum_p V_p n_p \quad (4)$$

Q is the total outgassing rate (total number of neutral species released by unit time from the (near) surface) and V_n, the mean bulk radial speed of the neutral species.

The neutral gas released by sublimation of ices at the surface of the nucleus is primarily made of H_2O, CO_2, and CO (see the chapter by Biver et al. in this volume). Many other molecules are also present in the coma, such as dioxygen (O_2), ammonia (NH_3), and glycine (NH_2CH_2COOH), unambiguously detected *in situ* at 67P (*Le Roy et al.,* 2015; *Altwegg et al.,* 2016; *Gasc et al.,* 2017) thanks to the Rosetta Orbiter Spectrometer for Ion and Neutral Analysis (ROSINA) DFMS Double Focusing Mass Spectrometer (DFMS) (*Balsiger et al.,* 2007). Along their path in the interplanetary medium, molecules may undergo different fates: They can be excited, dissociated into neutral fragments, or ionized by means of absorption of solar EUV radiation or impact of energetic particles. Although dissociation is more efficient than ionization and excitation, a fraction of the molecules still undergoes ionization; this forms a region made of plasma, a mixture of ions and electrons, surrounding the nucleus: the *ionosphere.*

Ionospheres are commonly present around any solar and extrasolar system body possessing a neutral gas layer, either an atmosphere, like at Earth, Mars, Jupiter and Titan, or an exosphere, like at Mercury or Ganymede. It plays a critical role in the interaction of the Sun (or the star) with the body. Indeed, none of these solar system bodies is isolated: They all bathe in an external plasma, either directly the solar wind for bodies without an intrinsic magnetic field or the magnetospheric plasma (and indirectly the solar wind) for magnetized bodies. Around comets, two broad types of plasma are present (see Fig. 1):

- The *ionosphere* proper, made of cometary electrons and ions born from outgassed neutral molecules, is dense (10^2–10^9 particles m^{-3}). Cometary ions are heavy [above 12 unified atomic mass units (u) or Daltons (Da)], slow (from 0.5 to a few kilometers per second), and cold.
- In contrast, the *solar wind* is rarefied (1–10 particles m^{-3} at 1 au), made of protons (H^+), alpha particles (He^{2+}), and electrons (e^-); it is fast ($|\mathbf{V}_{SW}|$ = 400–800 km s^{-1}) and hot [ion temperature $T_i \sim 0.8 \times 10^5$ K, electron temperature $T_e \sim 1.5 \times 10^5$ K at 1 au typically scaling to the heliocentric distance r_h as $r_h^{-2/3}$ and $r_h^{-1/3}$, respectively; see *Slavin and Holzer* (1981)].

The solar wind transports with itself electromagnetic fields. It carries the solar magnetic field $\mathbf{B}_{SW}$, which is "frozen-in"; i.e., the magnetic field is dragged into space by the expanding solar wind. Moreover, even though there is no electric field in the solar wind rest frame, as it moves, an observer in a moving frame with respect to the solar wind experiences an electric field from the advected magnetic field by means of a relativistic Lorentz transformation. This field is the so-called convective or motional electric field defined as $\mathbf{E}_{SW} = -\mathbf{V}_{SW} \times \mathbf{B}_{SW}$. A simplistic view of the relationship between a star and comets is shown in Fig. 1.

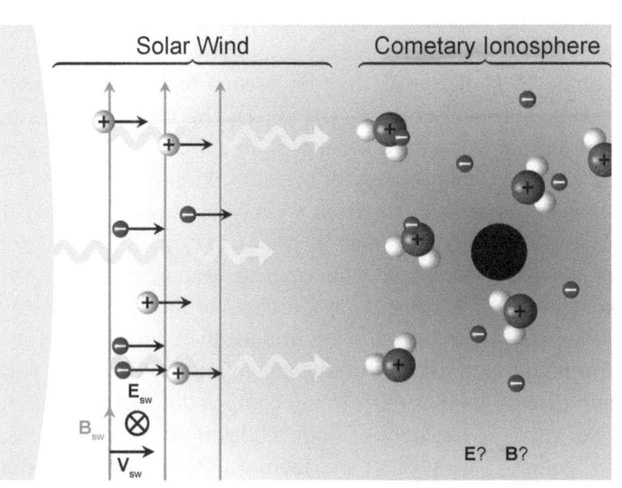

Fig. 1. Schematic of the actors of the interaction of the comet (black disk) with its space environment. They include (1) the cometary ionosphere, here represented by water H_2O^+ ions (red/white) and electrons (blue around the coma); (2) the solar extreme ultraviolet (EUV) radiation (yellow), responsible in part for ionizing the neutral coma; and (3) the solar wind plasma, made of H^+ (white) and electrons (blue) carrying the interplanetary magnetic field (gray vertical arrows), filling the interplanetary medium and traveling at speeds typically between 400 and 800 km s^{-1}. The transition between the solar wind and cometary ionosphere and the nature of this transition depend on many factors (e.g., outgassing activity, composition, heliocentric distance, etc.), discussed in detail in the chapter by Götz et al. in this volume.

The first spacecraft to visit a comet and its plasma environment was NASA's International Cometary Explorer (ICE), which flew by Comet 21P/Giacobini-Zinner (21P) on September 11, 1985, at a closest approach of 7800 km through the comet's tail (see *Farquhar,* 2001). This mission was mainly dedicated to plasma measurements and was the first to provide *in situ* information on the interaction of the cometary ionosphere with the solar wind (*von Rosenvinge et al.,* 1986; *Cowley et al.,* 1987). ICE focused on the tail and could not investigate the innermost coma. Although ions of cometary origin were observed up to 4×10^6 km from 21P's nucleus, they were already significantly accelerated, picked up by the solar wind convective electric field, at speeds around 150 km s^{-1} (*Ogilvie et al.,* 1986). At the time, mass spectrometers, which discriminate and separate ions according to their mass per charge (or simply their mass if ions are assumed singly ionized), had a very limited mass resolution. The resolution in *mass spectrometry* is defined by $R = m/\Delta m$ where m is the mass at which ions are measured and Δm is the width of the peak typically at half height [see the International Union of Pure and Applied Chemistry (IUPAC) recommendation and nomenclature (*IUPAC,* 1978)]. With ICE, the mass resolution was relatively poor, with measurements separated by 1–2 u q^{-1} (mass unit per charge, q being the elementary charge in Coulomb), making it extremely difficult to separate species with relatively close molecular masses, for example, HO^+, H_2O^+, and H_3O^+.

The following year, the Halley Armada, a fleet of five spacecraft from several space agencies [one from the European Space Agency (ESA) (Giotto), two from the Soviet Union and France (Vega 1 and Vega 2), and two from the Institute of Space and Astronomical Science, now JAXA, in Japan (Suisei and Sakigake)] flew by the most famous comet of all and one of the brightest. As such, Halley's comet was the first one to be identified as periodic (1P) with its 76-year period determined by Edmond Halley in 1705. In succession, Sakigake, Vega 1, Suisei, Vega 2, and finally Giotto flew by Comet 1P, each probe providing a more precise location of 1P and therefore helping the next probe to get closer to its nucleus. On March 13–14, 1986, the last spacecraft of the fleet, Giotto, achieved a closest approach of 600 km at a heliocentric distance r_h = 0.89 au from the Sun (*Reinhard,* 1986). The first detected "cometary" ion was O^+ (most likely candidate at the peak at 16 u q^{-1}) at 5.5×10^5 km from the nucleus (*Krankowsky et al.,* 1986). Most likely, it resulted from the dissociation of H_2O into O followed by ionization (*Balsiger et al.,* 1986). The mass resolution was high enough to separate HO^+ from H_2O^+, albeit not O^+ from NH_2^+ and CH_4^+ [$R \gtrsim 20$ (*Balsiger et al.,* 1987)].

As Giotto got closer to 1P's nucleus, the ion composition changed. Between 1.5×10^5 km and 7×10^4 km, the cometary plasma composition was dominated by ion mass

16 u q^{-1} (most likely O^+), followed by ion masses 17 u q^{-1} (most likely HO^+) and 18 u q^{-1} (most likely H_2O^+). From 6×10^4 km to 3×10^4 km, the order changed: mass 18 u q^{-1} dominated, followed by 17 u q^{-1} and 16 u q^{-1}. From 3×10^4 km to the closest approach, mass 19 u q^{-1} (H_3O^+) exceeded in counts mass 16 u q^{-1}, and even dominated the overall composition below 2×10^4 km. However, the reader should keep in mind that the main contributor at a given u q^{-1} may change during the flyby. For instance, close to the nucleus, the signal at 18 u q^{-1} may be associated with NH_4^+ while further away, it is more likely H_2O^+ due to the ion chemistry. For this reason, ion measurements from Giotto had to be combined with photochemical models to infer the neutral and ion composition of the coma (e.g., *Allen et al.,* 1987; *Geiss et al.,* 1991). Furthermore, not only the composition but also the spatial distribution of the plasma changed. The ion count associated with cometary ions exhibited a $1/r^2$-dependence for cometocentric distances r beyond 16,000 km, whereas in the innermost part of the coma, ion counts varied in 1/r.

Giotto was a great mission for exploring large spatial scales around a comet in a very limited time, taking a veritable snapshot of the cometary plasma down to 600 km, the best achievement until the ESA cometary mission Rosetta, which escorted Comet 67P over a two-year period (2014–2016), from $r_h \sim 3.6$ au to $r_h \sim 1.24$ au (perihelion) and back to 3.8 au at the end of the mission (*Glassmeier et al.,* 2007). Nevertheless, several important aspects should be kept in mind when we compare the Giotto mission to Rosetta. Comet 1P was very active ($Q \approx 6.9 \times 10^{29}$ s^{-1}), close to the Sun (0.89 au), and, in that respect, its flyby by Giotto and the entire armada is far from being representative of the whole cometary environment and development.

As shown in Fig. 2, Giotto's closest approach was quite far compared with Rosetta's ranges of cometocentric distances, even when scaling these distances to their vastly different outgassing rates. Rosetta not only explored a wide range of distances, from 500 km to the surface, but also more than 3 orders of magnitude in terms of outgassing rate, from $Q \approx 10^{25}$ to $\sim 3 \times 10^{28}$ s^{-1}, as 67P gradually became more active on its journey toward the inner solar system and then faded again on its outward journey. During the mission, the scientific community thus had the unique opportunity to monitor the cometary coma, the ionosphere and its relationship with the solar wind, and their combined evolution over time, providing a wealth of unprecedented scientific data. In the future, we may not have many opportunities to visit other comets with such a dedication; in this way, escorting 67P constitutes a major step forward in our understanding of solar wind-comet interactions.

This chapter proposes to look into the lessons learned from the combined past cometary flybys and the recent two years of escort by Rosetta, from the formation to the composition and evolution of a cometary ionosphere. It follows the example of the previous books on the matter, *Comets* (*Wilkening,* 1982) and *Comets II* (*Festou et al.,* 2004). *Comets*, published before *in situ* exploration of cometary

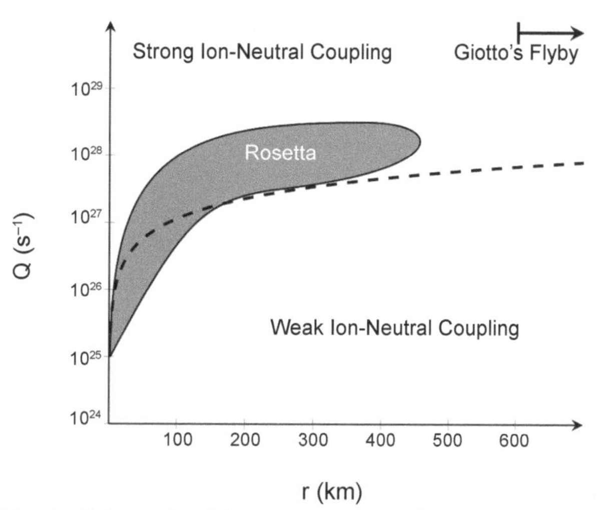

Fig. 2. Schematic of the cometocentric distance covered by Giotto and Rosetta as a function of the outgassing activity. Excursions have been excluded. The dashed line defines the theoretical limit where ions and neutrals become collisionally decoupled (*Gombosi,* 2015), referred to as $R_{i,n}$ in this chapter.

environments (prior to ICE and the Halley Armada), made use of remote sensing observations from Earth and possible theories (*Ip and Axford,* 1982; *Huebner et al.,* 1982), describing the solar wind assimilation into the cometary environment with the formation of plasma boundaries, and the emergence of photochemical models of the inner coma including ion-neutral gas chemistry. *Comets II* focused on the advances brought by the contemporary exploration of a very active comet, 1P/Halley, with the first attempt to compare models against *in situ* observations (*Ip,* 2004). This chapter is a tentative harmonization and consensus on our current knowledge of cometary ionospheres in order to provide some sort of a "continuous" picture from weakly active to very active comets. This chapter must be seen and read as a "tutorial" or "manual" for those who want to become familiar with cometary ionospheres; it sets the stage for future explorations of comets. In section 2, we introduce the theory behind the formation of a cometary ionosphere and its chemical loss and discuss its plasma balance, composition, and evolution over time, in the light of both modeling and observations. In section 3, we present and focus on the different electron populations and their role observed at comets, while in section 4 we present and focus on the different ion populations and their behavior. Section 5 is dedicated to the role of dust on cometary plasmas. Finally, in section 6, we summarize the contemporary progress and discoveries made in cometary physics followed by a non-exhaustive list of open questions.

2. BIRTH OF A COMETARY IONOSPHERE

2.1. Ion Continuity Equations

After being ejected from the nucleus' surface via sublimation, desorption, or other mechanisms (see the chapter by Filacchione et al. in this volume), the cometary neutral gas forming the coma is partially ionized by solar radiation ($\lambda \lesssim 100$ nm; see section 2.2.1), and by energetic electrons (section 2.2.2) and ions (section 2.2.3), leading to the formation of an ionosphere. The resulting ionosphere can be described by macroscopic, fundamental plasma quantities, such as the number density, mean velocity, and temperature, all *in situ* observables from instruments onboard a spacecraft:

- The plasma number density is the number of ions (or electrons assuming that ions only carry one positive charge +q) per unit volume, often expressed in m^{-3}
- The ion (or electron) mean (bulk) velocity is the statistical averaged velocity over all ions (or electrons) at a specific location, expressed in m s^{-1}
- The ion (or electron) temperature represents the variance of the velocity around the mean ion (or electron) velocity; it can be expressed in Kelvin or in electron volts (where 1 eV $\approx 1.6 \times 10^{-19}$ J $\approx k_B \times 11604$ K, k_B being the Boltzmann constant).

Note that by virtue of quasi-neutrality of the plasma, ion and electron number densities (noted n_i and n_e) must be "almost" equal ($n_i \approx n_e$). However, neither the mean velocities ($\mathbf{V}_i$ for ions and $\mathbf{V}_e$ for electrons) nor the temperatures (T_i and T_e) are necessarily so.

Let us look at the ion number density. The fate of newly-born ions is manifold (see section 2.2 for a detailed description). For instance, ions can be transported near or away from the cometary nucleus by means of electromagnetic forces. They can also collide along their path with neutral molecules. Through such a collision, the ion may lose its charge to the benefit of the neutral species, the latter itself becoming an ion. Finally, ions may recombine with free electrons and neutralize/dissociate, creating a net loss of ions and electrons. Such processes need to be taken into account when assessing the ion number densities. This can be achieved with the use of a mathematical model based on solving a set of differential equations.

The time evolution of the number density of cometary ion species s is captured by the continuity equation, which describes the evolution of the number of particles over time and space

$$\frac{\partial n_s(\mathbf{r},t)}{\partial t} = -\underbrace{\nabla \cdot \left(n_s(\mathbf{r},t)\mathbf{V}_s(\mathbf{r},t)\right)}_{\text{transport}} + \underbrace{S_s(\mathbf{r},t)}_{\text{source}} - \underbrace{L_s(\mathbf{r},t)}_{\text{chemical loss}} \qquad (5)$$

where $\mathbf{r}$ and t are the spatial and temporal coordinates respectively (the spatial reference frame is the comet's inertial frame centered on its nucleus), and where:

- $\partial n_s(\mathbf{r}, t)/\partial t$ is the instantaneous temporal variation of the ion number density in the infinitesimal volume $d\mathcal{V} = d^3\mathbf{r}$ at the position vector $\mathbf{r}$ and time t;
- $\nabla \cdot (n_s(\mathbf{r}, t)\mathbf{V}_s(\mathbf{r}, t))$ is the transport term and quantifies the rate at which plasma enters and leaves the infinitesimal volume $d\mathcal{V}$ at ($\mathbf{r}$, t). It is the net outward flux (here $n_s\mathbf{V}_s$) through the surface enveloping $d\mathcal{V}$. If there is no production and no chemical loss of ions s within $d\mathcal{V}$, the number density $n_s(\mathbf{r}, \mathbf{t})$ depends only on the amount of ions s passing through the surface dS at ($\mathbf{r}$, t), the outer boundary of $d\mathcal{V}$. Beware that unlike some fluids, plasma is compressible (i.e., $\nabla \cdot \mathbf{V} \neq 0$);
- $S_s(\mathbf{r}, t)$ is the local and instantaneous source (commensurable to a production rate) of the ion species s in the infinitesimal volume $d\mathcal{V}$ at ($\mathbf{r}$, t) through ionization or ion-neutral collisions (see sections 2.2 and 2.3);
- $L_s(\mathbf{r}, t)$ is the local and instantaneous chemical loss rate of the species s in the infinitesimal volume $d\mathcal{V}$ at ($\mathbf{r}$, t) through collisions (see section 2.3).

All four terms in equation (5) are expressed in m^{-3} s^{-1}. One should note that there are as many continuity equations as there are ion species. This constitutes a challenging system to solve and therefore assumptions should be made. Physicists tend to reduce the number of independent variables, here four: one for time and three for space.

The most practical and well-justified assumption is to model the system at/near steady state, i.e., in the limit of very slow variations/low frequencies and long timescales: The lefthand side of equation (5) becomes zero. Even if the system is disturbed, it will naturally converge and recover toward its steady state after a certain typical timescale, ruled here by transport and chemistry.

The second assumption is regarding the symmetry of the system. Although comets have various shapes, it has been shown that the neutral number density follows a $1/r^2$ law, at least for the major species, such as H_2O and CO_2 [e.g., at 67P (*Hässig et al.,* 2015)]: This is the same as for a point source or a sphere emitting gas with a constant radial speed, as in equation (2). Close to the nucleus and in the first few hundred kilometers from it, this assumption appears reasonable. Farther away from the nucleus, the loss of neutrals needs to be taken into account (see equation (1)).

Source and loss terms of cometary ions, S_s and L_s in equation (5), are described in detail in sections 2.2 (net production of charge in the coma) and 2.3 (chemical production and loss of cometary ions through ion collisions with neutrals and with electrons). Solution of equation (5), along with the relative importance of the different terms, is discussed in section 2.4.1 when applied to the full plasma and in section 2.4.2 when applied to individual ion species.

2.2. Source of the Cometary Plasma

We consider three net sources of charge for the cometary ionosphere: photoionization by solar EUV radiation (see section 2.2.1), ionization by energetic electrons (see section 2.2.2), and ionization and charge exchange by solar wind ions (see section 2.2.3). The relative importance of each of these sources is reviewed in section 2.2.4. Complementarily, the reader is directed to other textbooks on the matter (*Schunk and Nagy,* 2009).

2.2.1. Photoionization. Photoionization is the process of absorption of a sufficiently energetic photon by an atom/molecule A and the subsequent ejection of one or several of its bound electrons. This is a main plasma-production process leading to the formation of an ionosphere at any solar system body with a gas envelope

$$\begin{aligned} A + \text{photon (or } h\nu) &\rightarrow A^{m+(*)} + m\, e^- \\ &\rightarrow B^{m+(*)} + C^{(*)} + m\, e^- \end{aligned} \quad (6)$$

where m is the number of electrons released in the process (in most cases m = 1) and the star denotes a possible excited state of the ion A^{m+}, B^{m+}, and/or the neutral species C. Photoionization can produce new neutral atoms or molecules (here C) if A is a molecule: The atoms constituting B and C together constituted A. The produced ion and/or the neutral may be in an excited state and emissions are produced when the de-excitation is radiative (see the chapter by Bodewits et al. in this volume). The efficiency of an photoionization is linked to the *cross section* associated with the absorbing neutral species and produced ion species. The cross section depends on the energy (or wavelength λ) of the photon absorbed, i.e., $E = h\nu = hc/\lambda$ where h is the Planck constant, and the targetted neutral species.

For the most common molecules encountered around comets, photoionization is triggered by photons of energy $E \gtrsim 12$ eV ($\lambda \lesssim 100$ nm) corresponding to the EUV range. The minimum energy can be lower in the case of alkali metals. Photoionization depends on the distribution in energy of the incoming photons. First, the EUV spectrum significantly departs from the blackbody emission (see Fig. 3). Second, this spectrum is composed of emission "lines" in the EUV/soft X-ray range originating mainly from the chromosphere, the transition region, and the solar corona: They are emitted by excited (often highly ionized) species, such as H, He, C, O, Si, Fe, etc. (*Lean,* 1991) (see Fig. 3). The EUV photon energy distribution strongly depends on the solar activity and solar cycle.

In order to obtain an expression of the photoionization source term in equation (5), several steps are necessary. These are detailed below. The starting point is to compute the resulting efficiency of the solar EUV flux in ionizing neutrals. It is determined by the photoionization frequency of neutral species p leading to the production of ion species s

$$\nu_{p,s}^{h\nu,ioni}(\mathbf{r}) = \int_{\lambda_{min}}^{\lambda_{p,s}^{th}} \sigma_{p,s}^{h\nu,ioni}(\lambda) F_{h\nu}(\lambda,\mathbf{r})\, d\lambda \quad (7)$$

where

- $F_{h\nu}(\lambda, \mathbf{r})$ (photons m^{-2} s^{-1} nm^{-1}) is the spectral flux of the solar radiation at the position vector **r** (where the origin is taken at the center of the nucleus, $|\mathbf{r}| = r$ corresponds to the cometocentric distance). It is derived from equation (11);
- $\sigma_{p,s}^{h\nu,ioni}(\lambda)$ (m^2) is the photoionization cross section proportional to the probability to have the neutral species p absorbing a photon of wavelength λ and forming the ion species s (see Fig. 3);

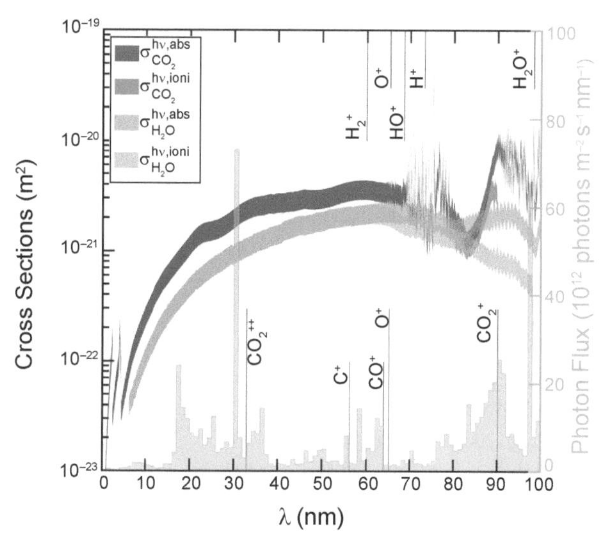

Fig. 3. Left axis: Total photoabsorption ($\sigma^{h\nu,abs}$) and photoionization ($\sigma^{h\nu,ioni}$) cross sections for H_2O and CO_2 as a function of wavelength. Associated wavelength appearances for ionized products are indicated (top for H_2O, bottom for CO_2). Right axis: EUV solar photon flux on 16 July 2016 between 0 and 100 nm from TIMED/SEE measured at Earth, normalized at 1 au, integrated over 1-nm-wide bins. Photoionization and photoabsorption cross sections are from *Heays et al.* (2017) and references therein. Wavelength appearances are from *Zavilopulo et al.* (2005).

- $\lambda_{p,s}^{th}$ is the maximum wavelength triggering the process; it corresponds to a threshold energy typically around 12 eV, referred to as the *ionization energy* $E_{p,s}^{th}$ (or archaically *ionization potential*). For example, water has an ionization energy of 12.5 eV (99.2 nm; see Fig. 3); if there is dissociation and/or excitation during the ionization, the threshold energy has a value higher than for a single, non-dissociative ionization generating all products in the ground state;
- λ_{min} is the minimum wavelength observable in the solar spectrum, approximately around 0.1 nm. Below this limit, the solar photon flux becomes negligible, even during flares.

As photons penetrate the dense coma, the amount of absorbed photons by cometary molecules becomes more and more significant. This results in the solar spectral flux being attenuated as the coma gets thicker and thicker, a process known as *photoabsorption*. Photoabsorption at the position vector **r** depends on the total column density N_p of neutral species p crossed by photons from a position $\mathbf{r}_\infty$ along the line of sight (where the distance r_∞ is large enough such that there is no significant absorption of solar photons sunward of $\mathbf{r}_\infty$ along the line of sight) to **r**. It is defined as

$$N_p(\mathbf{r}) = \int_{l=0}^{l=l_\infty} n_p(\mathbf{r}'(l))\, dl \quad (8)$$

where l represents the curvilinear abscissa along the Sun-comet line of sight between the point of interest **r** and $\mathbf{r}_\infty$ (see Fig. 4). The number density n_p decreases fast enough for N_p to be finite. For a comet, the neutral number density decreases as $1/r^2$ (see equation (2)) and therefore the neutral column density is given by

$$N_p(r,\chi) = \frac{Q_p}{4\pi V_p} \frac{\chi}{r \sin \chi} \tag{9}$$

where χ is the solar zenith angle between **r** and the direction of the Sun ($\chi = 0°$ in the subsolar direction; see Fig. 4). Equation (9) neither accounts for asymmetric outgassing nor adiabatic acceleration of the neutral gas. However, it gives an idea of how strong photoabsorption is.

To quantify the "thickness" of a coma regarding the penetration of photons of wavelength λ, the dimensionless "optical depth" τ is defined as:

$$\tau(\lambda,\mathbf{r}) = \sum_p \sigma_p^{h\nu,abs}(\lambda) N_p(\mathbf{r}) \tag{10}$$

where the subscript p refers to a specific neutral species and $\sigma_p^{h\nu,ioni}(\lambda)$ is the total photoabsorption cross section of the cometary neutral species p representing the probability of a photon of wavelength λ to be absorbed by the neutral species p, regardless of the process (ionization, dissociation, or excitation; see Fig. 3). The solar photon spectral flux $F_{h\nu}(\lambda,\mathbf{r})$ at the location **r** in the coma is then given by a simple Beer-Lambert law

$$F_{h\nu}(\lambda,\mathbf{r}) = F_{h\nu,1\,au}(\lambda)\exp[-\tau(\lambda,\mathbf{r})]/r_h^2 \tag{11}$$

where $F_{h\nu,1\,au}$ is the observed spectral flux measured near Earth, for example, by the Solar EUV Experiment (SEE) instrument on the NASA Thermosphere Ionosphere Mesosphere Energetics and Dynamics (TIMED) mission (*Woods et al.,* 2005) and normalized to 1 au, and r_h is the heliocentric distance given in astronomical units. Equation (11) does not take into account the temporal variation of the solar EUV flux nor the phase shift correction. The latter stems from the fact that the solar EUV radiation varies with ecliptic solar longitude and the comet is not aligned with Earth. It is then necessary to take into account the phase shift between Earth and the comet from the Sun in terms of days. For example, if the phase angle (i.e., the angle formed by Earth, the Sun, and the comet in the solar equatorial plane) is +90°, the reader interested in the solar EUV radiation at the comet on day "d" should look for the solar EUV radiation measured at Earth, ~6.5 days before (one-fourth of the solar synodic period). Applying equation (11), equation (7) becomes

$$\nu_{p,s}^{h\nu,ioni}(\mathbf{r}) = \int_{\lambda_{min}}^{\lambda_{p,s}^{th}} \sigma_{p,s}^{h\nu,ioni}(\lambda) \frac{F_{h\nu,1\,au}(\lambda)}{r_h^2} \exp[-\tau(\lambda,\mathbf{r})]\, d\lambda \tag{12}$$

For $\tau(\lambda,\mathbf{r}) \lesssim 0.1$, the coma is optically thin at **r** for photons of wavelength λ. Photoabsorption is thus negligible and the photoionization frequency is constant (independent of **r**), reducing to

$$\nu_{p,s}^{h\nu,ioni} = \int_{\lambda_{min}}^{\lambda_{p,s}^{th}} \sigma_{p,s}^{h\nu,ioni}(\lambda) \frac{F_{h\nu,1au}(\lambda)}{r_h^2} d\lambda \tag{13}$$

For $\tau(\lambda,\mathbf{r}) \gtrsim 0.1$–1, the coma is optically thick at **r** for photons of wavelength λ. In this case, photoabsorption needs to be taken into account as the solar flux is attenuated at **r**. The coma was optically thick in the vicinity of the nucleus but not necessarily at the spacecraft location during the flyby of 1P by Giotto and during the escort of 67P by Rosetta near perihelion, although in both cases the spacecraft was far enough from the nucleus such that the ionospheric density at the location of the spacecraft was not affected by the photoabsorption occurring closer to the nucleus (*Heritier et al.,* 2018; *Beth et al.,* 2019). We can now calculate the production rate, through photoionization, for the ion species s contributing to the source term in equation (5)

$$S_s^{h\nu}(\mathbf{r}) = \sum_p S_{p,s}^{h\nu}(\mathbf{r}) \tag{14}$$

where $S_{p,s}^{h\nu}$ is the production of an ion s from photoionization of the neutral species p given by

$$S_{p,s}^{h\nu}(\mathbf{r}) = \nu_{p,s}^{h\nu,ioni}(\mathbf{r}) n_p(\mathbf{r}) \tag{15}$$

The total photoionization rate or photoelectron production rate (as only one electron is usually produced per ionization process) is then simply given by

$$S_{e^-}^{h\nu}(\mathbf{r}) = \sum_{p,s} S_{p,s}(\mathbf{r}) \tag{16}$$

What are the typical behavior and spatial distribution of this photoelectron production rate in the coma? For

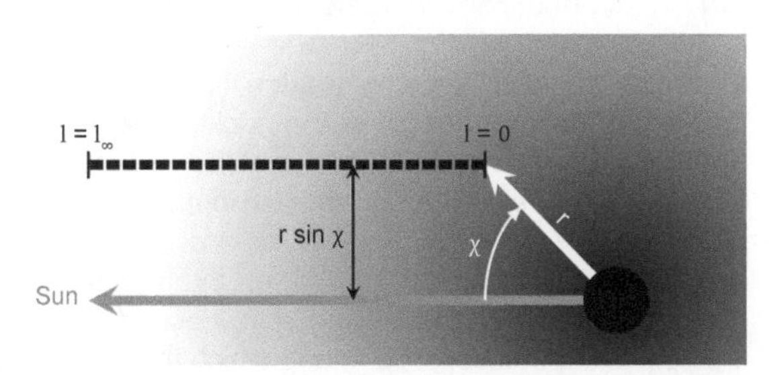

Fig. 4. Schematic of the geometry of solar photon beams near the cometary nucleus (black disk) illustrating different quantities introduced in equations (8) and (9); see text for further explanation.

$\tau(\lambda,\mathbf{r}) < 0.1$, $S_{e^-}^{h\nu}$ is proportional to n, i.e., to $1/r^2$. For $\tau \gtrsim 0.1$–1, $S_{e^-}^{h\nu}$ has additionally a spatial dependence associated with the optical depth. One might be interested in the location where most of the EUV radiation is absorbed within the coma and deposits its energy. This maximum absorption, which corresponds to a maximum in the photoelectron production rate, is taking place as a result of the neutral number density increasing with decreasing r, while the solar flux decreases. For a monochromatic radiation, single-species coma, and for a fixed χ, the photoelectron production rate induced by the absorption of photons at wavelength λ is maximum where (*Beth et al.,* 2019)

$$\tau(\lambda,\mathbf{r}) = 2 \tag{17}$$

The value of 2 at the maximum production differs from what is usually encountered at planetary objects (including moons with a dense atmosphere such as Titan). At these bodies, the atmosphere can be assumed to be in hydrostatic equilibrium and hence varies exponentially with decreasing altitude: The maximum of absorption and of production rate thus occurs at $\tau = 1$ (e.g., *Schunk and Nagy,* 2009) instead of 2.

From equations (9) and (10), equation (17) is reached at a cometocentric distance corresponding to

$$r_{p,s}^{max,abs}(\lambda,\chi) = \frac{\sigma_p^{h\nu,abs}(\lambda)Q_p}{8\pi V_p}\frac{\chi}{\sin\chi} \tag{18}$$

As shown in Fig. 5, the spatial variation in the Sun-comet plane of the photoelectron production rate for a monochromatic solar radiation can be very asymmetric as the comet becomes more active (increasing outgassing rates lead to an increased optical depth). For an optically thin coma ($\tau \ll 2$), the photoelectron production only depends on the cometocentric distance: Photoelectrons are produced uniformly, only as a function of the local neutral density. However, for $\tau \geq 0.1$–1, this changes. The neutral column density becomes high enough to significantly attenuate the photon flux, reducing the ionization efficiency. Photoelectrons are thus produced in most numbers along the Sun-comet axis. As the outgassing rate increases, this maximum is located at a gradually increasing cometocentric distance upstream of the nucleus.

The photoelectron production rate is not constant along the lines of constant τ as, for a given value of τ and λ, both the neutral number density and the column density of absorption depend on $\mathbf{r}$. This is of importance as, when considering all wavelengths, the total photoelectron production rate will result from the sum of different production profiles through the coma. Each profile is associated with a different energy distribution of the photoelectrons that varies with cometocentric distance. The energy of the newborn photoelectron is given by $E_{e^-_{h\nu}} = E_{h\nu} - E_{p,s}^{th}$ where $E_{h\nu}$ is the solar photon energy.

Above 20 eV, $\sigma_{H_2O}^{h\nu,abs}$ decreases with energy, meaning that the maximum of absorption occurs closer to the nucleus for photons at 100 eV than those at 20 eV. Whereas photons at 100 eV are efficiently absorbed and ionize the neutral species, photons at 20 eV cannot do so as they have been absorbed upstream due to their larger photoabsorption cross sections. Therefore the energy distribution of photoelectrons is depleted around 20 eV when the coma starts to be optically thick. Nevertheless, the 100-eV photons produce highly energetic electrons, which in turn impact neutral molecules to produce secondary electrons. This accounts for the main ionization source close to the nucleus for highly active comets, above $Q = 10^{29}$–10^{30} s^{-1} (*Bhardwaj,* 2003).

For more information regarding photoionization and photoabsorption, the reader is invited to read the historic work of *Chapman* (1931), initially developed for planetary atmospheres under hydrostatic equilibrium (exponentially

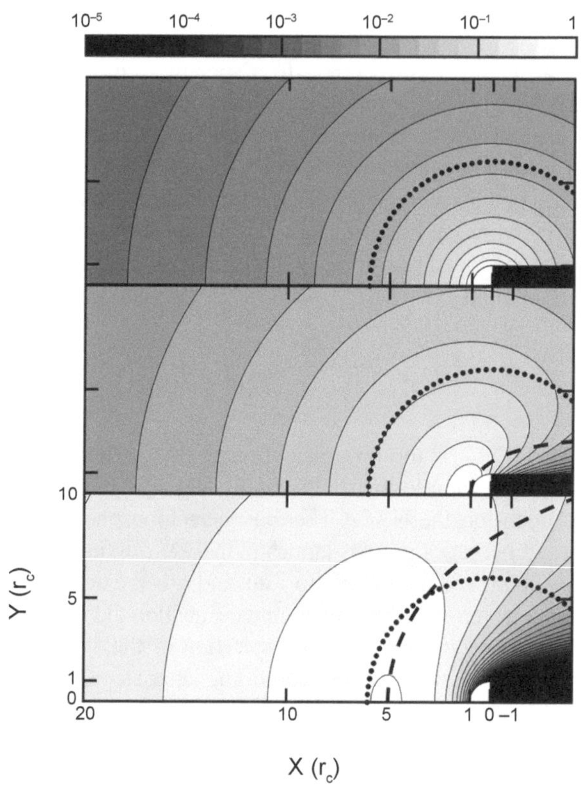

Fig. 5. Normalized photoelectron production rate for a monochromatic radiation in a pure water coma for different cometary activities using equation (16). The Sun-comet axis is along X, with origin at the nucleus' center, whereas the terminator plane is along Y. Instead of setting the outgassing rate, we set three different optical depths on the surface at the subsolar point ($X = r_c$ at $\chi = 0$): 0.01 (top panel, low activity), 2 (center panel, intermediate activity), 10 (bottom panel, high activity). The dashed line refers to the optical depth level $\tau = 2$, along which the neutral column density is constant. The neutral number density described by equation (2) is constant along the dotted lines (spherical symmetry). Cometocentric distances in the X–Y plane are expressed in terms of the nucleus' radius, r_c.

decreasing with height) and monochromatic radiation. The theoretical application to comets hosting an expanding coma is presented in full by *Beth et al.* (2019).

2.2.2. Electron-impact ionization. Electron-impact ionization (sometimes refers as electron ionization) is another critical process in the birth of a cometary ionosphere (e.g., *Heritier et al.,* 2018). Like photons, free electrons, with an energy above the ionization energy $E^{th}_{p,s}$ ($\approx$ 12 eV) efficiently tear off electrons attached to an atom/molecule A following

$$A + e^-_{prim} \rightarrow A^{m+(*)} + e^-_{prim} + (m-1)e^-_{sec}$$
$$\rightarrow B^{m+(*)} + C^{(*)} + e^-_{prim} + (m-1)e^-_{sec} \quad (19)$$

where subscripts "prim" (or "sec") refer to the primary impacting electron (or the secondary electron, freed during the ionization). The impacting electron can be a photoelectron, a solar wind electron, or a secondary electron (see section 3.1.3). It loses energy through the ionization: It loses the ionization energy to the neutral target, and eventually more whether one or more products are left in excited states and/or if A is dissociated, plus a fraction of it that is passed on to the secondary electron as kinetic energy. The corresponding electron-impact ionization frequency is given by

$$\nu^{e^-,ioni}_{p,s}(\mathbf{r}) = \int_{E^{th}_{p,s}}^{+\infty} \sigma^{e^-,ioni}_{p,s}(E) F_{e^-}(\mathbf{r},E)\, dE \quad (20)$$

where F_{e-} (E) is the electron flux (also referred sometimes as differential flux) given in (electrons) m^{-2} s^{-1} eV^{-1} and $\sigma^{e^-,ioni}_{p,s}$ is the electron-impact cross section of the neutral species p leading to the ion species s (see Fig. 6). This formula is very similar to equation (7) except that (1) solar spectral flux is replaced by electron flux and (2) the problem is defined in terms of energy instead of wavelength.

We can now calculate the electron-impact production rate for the ion species s contributing to the source term in equation (5)

$$S^{e^-}_s(\mathbf{r}) = \sum_p S^{e^-}_{p,s}(\mathbf{r}) \quad (21)$$

where $S^{e^-}_{p,s}$ is the electron-impact ionization rate of cometary neutral p leading to the production of a new cometary ion s and given by

$$S^{e^-}_{p,s}(\mathbf{r}) = \nu^{e^-,ioni}_{p,s}(\mathbf{r})\, n_p(r) \quad (22)$$

Although it is an endothermic reaction (the ionization energy is lost), the repartition of the remaining energy between the primary and secondary electrons varies. In extreme cases, when the incident electron is very energetic, above 200–300 eV, two electrons can be kicked out through the ionization process. This can also occur with energetic photons of the same energy. Such high-energy primaries may lead to a process known as the *Auger effect* (*Auger,* 1923), caused by the ejection of an electron from the innermost atomic or molecular shells (K-shell) and the subsequent filling of that gap with a higher-shell electron, itself leading to photoemission and/or emission of a so-called *Auger electron* (*Fox et al.,* 2008).

It is important to note that the produced ion A^+ in reaction (19) can be in an excited state A^{+*}, in a similar way to photoionization: Its radiative de-excitation toward a lower excited state (ultimately ground state) leads to the emission of a photon in the far ultraviolet (FUV)-visible range, a process at play most famously in cometary ion tails, but also in aurora-like structures throughout the solar system and recently discovered at comets (see *Galand et al.,* 2020, and the chapter in this volume by Bodewits et al. for more detail).

2.2.3. Charge-exchange and ionization by solar wind ions. In addition to solar photons and energetic electrons, energetic ions emitted by the Sun, such as protons (H^+) and alpha particles (He^{2+}), may ionize the neutral species surrounding the comet. This generally occurs either through (*Simon Wedlund et al.,* 2019b, 2020):

- direct impact ionization, as for energetic electrons; we refer to it as *solar wind ionization* (SWI);
- *charge-exchange processes* (with no *net* production of charge), referred to as SWCX in the following.

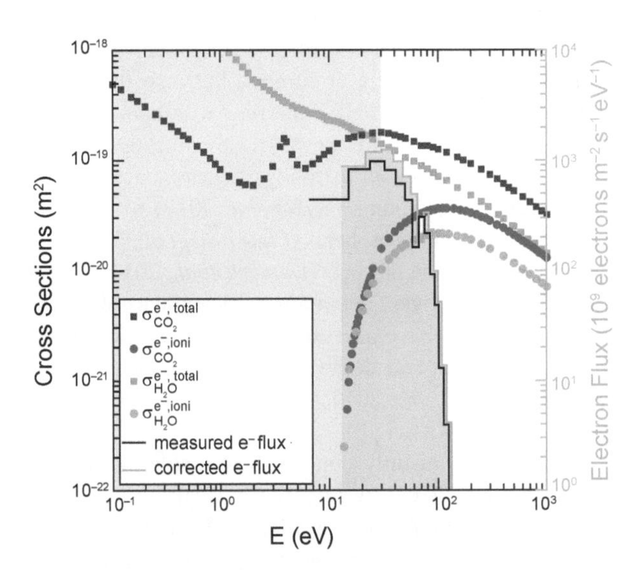

Fig. 6. Left axis: Total and ionization cross sections resulting from the impact of electrons on H_2O and CO_2. Right axis: Raw and corrected (from the spacecraft potential) electron flux by the electron spectrometer onboard Rosetta on 14 January 2015 00:06:59 (*Stephenson et al.,* 2021). The gray region represents values of $-qV_{SC}$ during Rosetta's mission (see section 3.1.1 for fuller explanation). It represents the minimal energy required by an electron to reach the spacecraft. Cross sections are from *Itikawa* (2002) for CO_2 and *Itikawa and Mason* (2005) for H_2O.

As solar wind energetic ions pass through the coma, they may strip electrons from the neutral atoms and molecules they encounter, thereby lowering their charge state. The first evidence of such a phenomenon was reported by *Lisse et al.* (1996), who observed at Comet Hyakutake C/1996 B2 a strong sunward emission of the coma in the X-ray and EUV ranges. Quickly after, *Cravens* (1997a) identified the mechanism behind these emissions. Solar wind plasma is not only formed of protons and alphas, but also of heavier ions amounting to a small proportion (<0.1%) of the total solar wind density. These ions, highly charged because of their origin in the deep solar corona, are, for example, O^{m+}, C^{m+}, and N^{m+} (noted X^{m+} in the following), where $m \geq 4$ is their charge number (*von Steiger et al.,* 2000). Such ion species have large and very different first ionization energies E_i, i.e., the minimum energy required to eject an electron from an isolated atom/molecule in the gaseous phase (*Muller,* 1994). As they stream through the coma, multiply-charged heavy ions undergo collisions with cometary molecules, such as water

$$X^{m+} + H_2O \rightarrow X^{(m-1)+*} + HO + H^+$$

a dissociative ionization process followed by radiative de-excitation

$$X^{(m-1)+*} \rightarrow X^{(m-1)+} + h\nu$$

The emitted photons have typical wavelengths below about 12 nm (~100 eV), placing them in the soft X-ray (roughly 0.1–10 nm) and EUV ranges (*Cravens,* 2002; *Krasnopolsky et al.,* 2004). SWCX is responsible for global X-ray and EUV emissions not only at comets, but also throughout the universe (*Dennerl,* 2010). Starting with comets, it has only recently become a rich subject of investigation and neutral gas diagnostic in astrophysics and space physics, from supernova remnants (*Lallement,* 2012) to planetary magnetospheres and exospheres (*Bhardwaj et al.,* 2007) with applications in fusion plasmas (*Dennerl et al.,* 2012; *Sembay et al.,* 2012). Moreover, laboratory measurements of charge-transfer processes have also experienced an unprecedented boom over the last two decades (see *Wargelin et al.,* 2008; *Simon Wedlund et al.,* 2019c; see also the chapter in this volume by Bodewits et al.).

SWCX does not only concern heavy solar wind ions: Protons and alphas constituting the bulk of the solar wind are also subject to it (*Simon Wedlund et al.,* 2019a), although they do not produce such energetic photons. In this case, the main effect of SWCX is to neutralize the solar wind [creating H and He energetic neutral atoms (ENAs) from incoming protons alphas, respectively] and implant heavier cometary ions into the solar wind flow so that, for example, for alpha particles

$$He^{2+}_{fast} \xrightarrow{H_2O} He^{+}_{fast}(+H_2O^+) \xrightarrow{H_2O} He_{fast}(+H_2O^+) \quad (23)$$

This in turn partakes of the so-called *mass/momentum-loading* of the solar wind as the newly-born H_2O^+ molecules are accelerated by the convective electric field (see section 4.2 and the chapter by Götz et al. in this volume).

For a solar wind ion species sw (e.g., H^+) charge-exchanging with a heavy neutral species p (e.g., H_2O), thereby creating heavy ion species s (e.g., H_2O^+), the charge-exchange frequency $\nu^{sw,CX}_{p,s}$ can be expressed as

$$\nu^{sw,CX}_{p,s}(\mathbf{r}) = \int_{E^{th,CX,sw}_{p,s}}^{+\infty} \sigma^{sw,CX}_{p,s}(E) F_{sw}(\mathbf{r},E)\, dE \quad (24)$$

where F_{sw} (m^{-2} s^{-1} eV^{-1}) is the attenuated flux of solar wind ions sw at the position **r**, $\sigma^{sw,CX}_{p,s}$ the corresponding charge-exchange cross section, and $E^{th,CX,sw}_{p,s}$ the threshold energy of the reaction involving solar wind ion species sw reacting with the neutral species p to form ion species s. SWCX is a cumulative process, hence the flux of solar wind ions is progressively attenuated the deeper they penetrate in the coma, whereas keeping most of its initial energy (at least to a first approximation). Assuming straight-line trajectories with no energy change of the solar wind along the Sun-comet axis even after charge-exchange, this translates to (*Simon Wedlund et al.,* 2016, 2019a)

$$F_{sw}(\mathbf{r},E) = F^{\infty}_{sw}(E) \exp\left(-\sum_{p,s} \sigma^{sw,CX}_{p,s}(E)\, N_p(r,\chi) \right)$$

where F^{∞}_{sw} is the upstream "undisturbed" solar wind flux, $N_p(r,\chi)$ the neutral column density at distance r, assuming cylindrical symmetry along the Sun-comet axis of the neutral coma, and solar zenith angle χ, defined in equation (9).

Similarly to photoionization and electron ionization processes, and summing through all solar wind ion species sw, the source term for charge-exchange reactions leading to the production of a new cometary ion s

$$S^{CX}_{p,s}(\mathbf{r}) = \sum_{sw} \nu^{sw,CX}_{p,s}(\mathbf{r})\, n_p(r) \quad (25)$$

Compared to photoionization and electron ionization, SWCX is expected to be a minor source of new plasma in the inner coma. That said, large-scale three-dimensional numerical simulations of the cometary plasma environment have shown that it is one of the most important processes responsible for the formation of boundaries upstream of the inner ionosphere (*Simon Wedlund et al.,* 2017; see also the chapter in this volume by Götz et al.).

Because charge-exchange processes have large interaction cross sections (and thus efficiencies), SWCX production rates usually dominate those of SWI. This is especially true outside transient solar wind phenomena such as coronal mass ejections (CME) or co-rotating interaction regions(CIR), which carry away particles at much larger bulk speeds (up to ~1–3 × 10^3 km s^{-1}) than the nominal "slow" solar wind (~400 km s^{-1} or ~833 eV u^{-1}). The explanation is straightforward: SWCX cross sections peak below 1 keV u^{-1}, whereas proton and He ionization cross sections in H_2O and CO_2 have their maximum above that limit (*Simon Wedlund et al.,* 2019c).

2.2.4. Which ionization sources matter at comets? The relative importance of each separate source of cometary ions, namely photoionization (see section 2.2.1), electron-impact ionization (see section 2.2.2), and solar wind charge-exchange (see section 2.2.3), depends on outgassing activity and cometocentric distance (*Heritier et al.,* 2018; *Simon Wedlund et al.,* 2019b, 2020). Whereas solar wind charge exchange plays an important role at large cometocentric distances, typically far upstream of the comet, the production of ions in the inner ionosphere is heavily driven by electron-impact ionization and photoionization. For example, at Comet 67P, Rosetta made it possible to monitor the ionization frequencies of each of these sources individually from low to high outgassing rates. Outside of the so-called *solar wind ion cavity* (see section 4.3, the chapter by Götz et al. in this volume, and Fig. 13), the electron-impact ionization frequency was on average 5–10 times larger than that of photoionization, the latter a factor of 10 (respectively, 100) or so larger than solar wind charge-exchange (ionization) frequencies, except for exceptional solar transient events, when charge exchange could briefly rival the other two main ionization sources. Inside the *solar wind ion cavity*, photoionization frequencies at the cometocentric distances probed by the spacecraft steadily increased with decreasing heliocentric distance, to progressively dominate over electron-impact ionization frequencies by the time the comet reached perihelion (*Heritier et al.,* 2018).

2.3. Center of Chemical Reactions

Beside the net production of charge in the coma presented in section 2.2, there are additional chemical processes that contribute to the source and loss terms, S_s and L_s, in equation (5) (see section 2.1). First, there are ion-neutral collisions, which heavily influence the cometary ion composition. Through such processes, there is no net change in charge in the coma, but there is a change in cometary ion species: It is a loss for the reacting ion and a production for the new ion species formed. Second, there are electron-ion dissociative recombination reactions, through which there is a net loss of charge. These two types of processes are presented hereafter.

At low outgassing activity (at large heliocentric distances), a coma can be described as a quasi-collisionless medium. In the case of Comet 67P, farther than 3 au, the outgassing rate was lower than 10^{26} s^{-1}, which corresponds to two 1.5-l bottles of water released to space each second. Hence, the probability of collisions between ions and neutrals was low. As the comet gets closer to the Sun and the outgassing activity grows, ion-neutral collisions occur more often and chemistry becomes increasingly effective within the inner coma, generating a rich zoo of ion species (*Beth et al.,* 2020).

What are the different key chemical processes contributing to the sources S_s and losses L_s in equation (5)? These are:

- Charge exchange: Through the collision of a cometary ion A^+ with an atom/molecule B, A^+ captures an electron from B, such that

$$A^+ + B \rightarrow A + B^+$$

For example

$$O^+ + H_2O \rightarrow O + H_2O^+$$

This is similar to an oxidation-reduction reaction but here in the context of gases: B is oxidized and A is reduced. Mathematically, this translates into the ion source/loss rates

$$S_{B^+} = L_{A^+} = k_{A^+,B}(T)\, n_{A^+} n_B \tag{26}$$

where k(T) (m^3 s^{-1}) is the reaction rate coefficient, which depends on the temperature of the gas T, n_B (m^{-3}) is the local number density of neutral species B, and $n_A{}^+$ (m^{-3}) is that of ion species A^+. This process is a chemical source for B (S_{B^+}), while it is a chemical loss for A^+ (L_{A^+}). For single electron capture, there is no net production of ions since we start with one ion and end up with another.

- Proton transfer: Through the collision of a cometary ion AH^+ with an atom/molecule B, a proton is transferred from AH^+ to B such that

$$AH^+ + B \rightarrow A + BH^+$$

For example

$$H_3O^+ + NH_3 \rightarrow H_2O + NH_4^+$$

Mathematically, this translates into the ion source/loss rates

$$S_{BH^+} = L_{AH^+} = k_{AH^+,B}(T)\, n_{AH^+} n_B \tag{27}$$

This process is a chemical source for BH^+ (S_{BH^+}), while it is a chemical loss for AH^+ (L_{AH^+}). Compared with charge transfer, this reaction produces "unique" ions. For example, whereas H_2O^+ can be produced both through ionization of H_2O and charge transfer, NH_4^+ can only be produced through protonation of NH_3. Thus, the existence of BH^+ (e.g., NH_4^+) does not imply that of BH (e.g., NH_4). This process is governed by the proton affinity (PA) of both neutral species A and B. This reaction occurs only if B has a higher PA than A. In the context of cometary ionospheres, this is of importance and pointed out initially by *Aikin* (1974), the first to apply ion-neutral gas chemistry at comets. Through ionization, the main ion produced is H_2O^+ (which may be loosely seen as protonated hydroxyl radical HO-H^+). Any neutral with a higher proton affinity than HO [PA(HO) = 6.15 eV] may steal the proton from H_2O^+ to be in turn protonated. Among cometary neutrals, candidates include H_2O (7.16 eV), H_2S (7.30 eV), H_2CO (7.39 eV), HCN/HNC (7.39/8.00 eV), and

NH_3 (8.84 eV) (*Hunter and Lias,* 1998). Ammonia, the latter, is thus at the top of the "proton food" chain. In the coma, once protonated ammonia NH_4^+ (ammonium ion) is formed, it is only destroyed through recombination with electrons or transport, and therefore it should persist for a long time (as such, it is referred to as terminal ion). The mass resolution of the ROSINA/DFMS mass spectrometer onboard Rosetta was high enough to unambiguously identify it for the first time at a comet, differentiating it from H_2O^+ (*Fuselier et al.,* 2016; *Beth et al.,* 2017) (see also Fig. 10).

Ion-neutral chemistry affects ion composition but, as there is no net production of charge, the total ion and electron number densities remain unchanged. Only one known process affects the electron density, reduces the net amount of charges, and efficiently removes ions and electrons from a plasma. It is the so-called

- Electron-ion dissociative recombination (DR): A molecular ion recombines with an electron, which yields fragmented neutral species. For example

$$H_3O^+ + e^- \rightarrow OH + H_2$$

"Dissociative" refers to the breaking of the molecular ion into neutral fragments in order to dissipate the exceeding energy. The temperature-dependent loss rate for the DR reaction is given by

$$L_{DR,s}(T_e) = \alpha_s(T_e)\, n_s\, n_e \tag{28}$$

where α_s stands for the DR reaction rate coefficient (a function of the electron temperature T_e), n_s for the number density of ion species s, and n_e for the electron number density ($n_e \approx \Sigma_s n_s$).

Some words of caution regarding the DR reaction. First, it depends on (1) the electron temperature (the colder the electrons are, the more efficient the DR is), and (2) the molecular ion species (*Heritier et al.,* 2017b). In fact, $\alpha_s(T_e)$ exhibits very different T_e-dependency. $\alpha_s(T_e)$ is often assumed to behave like $\alpha_s(T_e) \propto 1/\sqrt{T_e}$ as theoretically derived (*Wigner,* 1948), an approximation that works well with many ion species (*Florescu-Mitchell and Mitchell,* 2006). In reality, this dependency might only hold for a specific electron temperature range whether the molecular ions are diatomic or polyatomic (*McGowan and Mitchell,* 1984). For instance, $\alpha_{H_2O^+} = 4.3 \times (300/T_e)^{0.74} \times 10^{-13}$ ($m^3\ s^{-1}$) (*Rosén et al.,* 2000) for $T_e < 1000$ K and $\alpha_{CO_2^+} = 4.2 \times (300/T_e)^{0.75} \times 10^{-13}$ ($m^3\ s^{-1}$) (*Viggiano et al.,* 2005) without constraints on T_e, where T_e is expressed in Kelvin. Nevertheless, these recombination rates are of the same order of magnitude for similar $T_e \lesssim 10^3$ K so that assuming initially a common value is reasonable. When the range of T_e encountered at comets is considered, α_s may span several orders of magnitude (see section 2.4.1).

Second, DR introduces nonlinearity into the continuity equation. For ion-neutral chemistry in the inner coma, one may safely assume that the neutral density profile n_p is decreasing as $1/r^2$ (see equation (2)). DR introduces nonlinear terms, such as n_s^2 and $n_s n_e$. Last but not least, α_s is often greater than the coefficient rate k associated with ion-neutral chemistry. Despite ions being much less abundant than neutrals in the inner part of the coma by a factor of $\sim\nu^{ioni}\, r/V_n$ (see equations (2) and (42)), $\nu^{h\nu,ioni}$ (a main contributor of ionization for $r_h < 2$ au) increases as the comet gets closer to the Sun and the gap reduces. In addition, as the activity rises, collisions occur more frequently, cooling the electrons: The electron temperature decreases, the loss through DR increases, potentially overcoming the loss through transport (see section 2.4.1). This appears to be the case for outgassing rates $\gtrsim 10^{28}\ s^{-1}$ (*Heritier et al.,* 2018; *Beth et al.,* 2019).

2.4. The Ionosphere: Putting It All Together

2.4.1. Total ionospheric density. As shown in section 2.1, one has to solve a complex system made of s′ continuity equations, one for each ion species s, in order to model the cometary ion composition under steady-state conditions. This can be summarized by combining equations (26), (27), and (28) for the chemical reactions where

$$\begin{array}{llcllll}
 & \overbrace{\nabla\cdot(n_{A^+}\vec{V}_{A^+}\vec{V}^i)}^{\text{transport}} & = & \overbrace{\nu^{ioni}_{A^+} n_n \vec{V}^i}^{\text{ionization}} & \overbrace{-\cancel{k_{A^+,B}\, n_{A^+} n_B} \pm \ldots \vec{V}^i}^{\text{ion-neutral chemistry}} & \overbrace{-\alpha_{A^+} n_i n_{A^+} \vec{V}^i}^{\text{electron-ion recombination}} & \\
s'eqs. \left\{ + \right. & \nabla\cdot(n_{B^+}\vec{V}_{B^+}) & = & \nu^{ioni}_{B^+} n_n & +\cancel{k_{A^+,B}\, n_{A^+} n_B} \pm \ldots & -\alpha_{B^+} n_i n_{B^+} & (29) \\
 & & & \vdots & & & \\
\hline
\text{sum} & \nabla\cdot(n_i \vec{V}_i) & = & \nu^{ioni}_i n_n & +0 & -\left(\sum_{s=1}^{s'} \alpha_s n_s\right) n_i &
\end{array}$$

$$\mathbf{V}_i = \sum_s n_s \mathbf{V}_s / n_i \tag{30}$$

$$\nu_s^{ioni} = \sum_p n_p \left(\nu_{p,s}^{h\nu,ioni} + \nu_{p,s}^{e-,ioni} \right) / n_n \tag{31}$$

$$\nu_i^{ioni} = \sum_s \nu_s^{ioni} \tag{32}$$

n_n stands for the total neutral density, n_i for the total ion number density (equal to the electron density, assuming no doubly-charged ions), $\mathbf{V}_i$ for the mean ion velocity, ν_i^{ioni} for a mean/effective ionization rate (via photoionization, see section 2.2.1, and electron impact, see section 2.2.2), and α_s for the electron-ion dissociative recombination rate coefficient associated with ion species s. Indices p and s refer to a specific neutral and a specific ion species, respectively. Note that in equation (31), only photoionization and electron-impact ionization terms are included, as solar wind ionization and charge-exchange terms are usually negligible in the inner coma (see sections 2.2, 2.2.3, and 2.2.4).

Once all ion continuity equations are summed, the ion-neutral chemistry terms (second set of terms on the righthand side of equation (29)) cancel out between lost and produced ion species. From equation (29), the continuity equation ruling the total ion number density is weakly depending on the ion composition through the last term on the righthand side (electron-ion recombination) as the kinetic rates are different between ion species (see section 2.3).

Let us look more closely at the transport term (lefthand side of equation (29)). Its contribution in non-cometary environments, such as a planetary ionosphere in the plane-parallel approximation (*Schunk and Nagy,* 2009), is limited: Vertical transport between atmospheric layers remains small and is dominated by eddy or molecular diffusion. In contrast, at comets, the neutrals, and therefore the newly-born ions, are moving at speeds V_n ranging from 300 m s^{-1} close to the surface, up to 1 km s^{-1} farther away, assuming adiabatic expansion. The radial velocity thus plays a significant role at comets. In spherical symmetry and under steady-state conditions, physical quantities depend only on the cometocentric distance r and the bulk cometary ion velocity $\mathbf{V}_i(r)$ being mainly radial, although these assumptions are only valid for the inner cometary ionosphere and may significantly break at larger cometocentric distances (see the chapter by Götz et al. in this volume). Under such conditions, the transport term is reduced to

$$\nabla \cdot (n_i \mathbf{V}_i) = \frac{1}{r^2}\frac{d(n_i V_i r^2)}{dr} = V_i \frac{dn_i}{dr} + n_i \frac{dV_i}{dr} + \frac{2 n_i V_i}{r} \tag{33}$$

The spherical shell enclosed in [r, r + dr] has a volume $4\pi r^2$ dr such that ions move into shells of increasingly larger volumes the more outward they move. The last term $2n_iV_i/r$ in equation (33) accounts for this geometrical effect.

As a whole, equation (29) represents the balance between the source of the ionospheric plasma through ionization of the cometary neutrals and ionospheric loss terms, namely radial transport and electron-ion dissociative recombination (net chemical loss). Unfortunately, it is not possible to solve equation (29) analytically or without making assumptions/approximations as one needs the radial ion velocity and electron temperature profiles.

Before going further, the first question should be, which of these terms are the most relevant to consider and are there any that can be neglected? One approach is to look at a typical physical quantity characterizing each term. If equation (29), combined with equation (33), is now divided by n_i

$$\underbrace{-\frac{1}{n_i r^2}\frac{d(n_i V_i r^2)}{dr}}_{1/\mathcal{T}_{transport}} + \underbrace{\frac{\nu_i^{ioni} n_n}{n_i}}_{1/\mathcal{T}_{ionization}} - \underbrace{\sum_s \alpha_s n_s}_{1/\mathcal{T}_{recombination}} \approx 0 \tag{34}$$

an equation whose dimension is the inverse of time. Large (or small) terms in equation (34) are associated with fast (or slow) processes, i.e., short (or long) timescales.

Let us review each loss term first (terms with negative signs in equation (34), which represent a physical sink), i.e., transport and dissociative recombination, and assess their respective timescale. The transport timescale is given by

$$\frac{1}{\mathcal{T}_{transport}} = V_i \frac{d \log n_i}{dr} + \frac{dV_i}{dr} + \frac{2V_i}{r} \tag{35}$$

independent of the cometary activity. From observations, the plasma density in the inner cometary ionosphere of 1P and 67P alike varies in 1/r (*Balsiger et al.,* 1986; *Edberg et al.,* 2015; *Heritier et al.,* 2017a) such that d log n_i/dr ~ 1/r. Regarding the variation of the radial speed, there is no evidence of a strong velocity gradient (acceleration or deceleration) except near the contact surface at 1P (*Goldstein et al.,* 1989). However, within the diamagnetic cavity of 1P, the ion velocity was relatively constant (*Schwenn et al.,* 1987). At 67P, there is no evidence of strong velocity gradients either. This term can thus be neglected and we end up with

$$\frac{1}{\mathcal{T}_{transport}} \sim \frac{V_i}{r} \text{ or } \frac{V_n}{r} \tag{36}$$

In fact, equation (36) is an approximation that only holds beyond a few cometary radii, beyond the location where the plasma density peaks, where $n_i \propto 1/r$. Below and up to the location of this maximum, this is an underestimate and equation (35) must be used (see Fig. 7).

By comparison, the electron-ion recombination time scale for the whole plasma population (electron sink term) is given by

$$\frac{1}{\mathcal{T}_{\text{recombination}}} \approx \alpha_i(T_e)n_i(r) \sim \frac{\alpha_i(T_e)n_{i,\text{obs}}r_{\text{obs}}}{r} \quad (37)$$

where

$$\alpha_i = \frac{\sum_s \alpha_s n_s}{n_i}$$

and the subscript "obs" corresponds to a location of reference where the cometocentric distance and number plasma density are known, for instance, probed and observed at a spacecraft. The numerator in equation (37) has the dimension of a speed and should be compared with the ion speed. The recombination timescale may span over several orders of magnitude because it depends both on electron temperature (through the recombination coefficient α_i; see section 2.3) and plasma density.

Experimentally, α_i ranges between 10^{-12} and 10^{-15} m^3 s^{-1} for $10^2 < T_e < 10^4$ K (*Gombosi et al.,* 1996; *Rosén et al.,* 2000). The electron temperature may be very different throughout the coma or even between comets. The lowest range for T_e typically corresponds to the inner cometary ionosphere at 1P under active outgassing conditions. As further discussed in section 3.1.2, frequent electron-neutral collisions are expected to bring the electron temperature close to the nucleus down to that of the neutral gas, which is consistent with indirect observations at Comet 1P (see *Eberhardt and Krankowsky,* 1995). The highest range has been observed at 67P under low outgassing conditions when the bulk of the thermal population is not thermalized (*Wattieaux et al.,* 2019, 2020) (see section 3.1.1). During Giotto's flyby, ion number densities reached almost 10^{10} m^{-3} at $r_{\text{obs}} \sim 10^4$ km (*Altwegg et al.,* 1993) such that $n_{i,\text{obs}}r_{\text{obs}} \approx 10^{17}$ m^{-2}. Because $\alpha_i \sim 10^{-14}$ m^3 s^{-1} in those conditions, the dissociative recombination timescale was $\mathcal{T}_{\text{recombination}} \sim 10^4$ s [or ~2 h 45 min (*Gombosi et al.,* 1996)]. Compared with the transport timescale of about $r/V_i \sim 10^4$ s at the same location, transport and dissociative recombination were contributing equally to the ion sink at that location at 1P (*Beth et al.,* 2019). However, for $r \lesssim 10^4$ km, T_e fell drastically by 2–3 orders of magnitude: The recombination timescale was then shorter than the transport timescale and hence dissociative recombination was the main ion sink in the innermost of 1P's ionosphere.

Unlike the transport timescale, the recombination timescale depends on cometary activity: The higher the outgassing activity, the larger the ion number density, the more efficient the recombination. We can then try to find out for which activity we should consider/neglect the recombination. Let us consider a very simplistic case in the coma, with a uniform electron temperature and therefore a constant α_i. Under steady-state conditions and spherical symmetry, the ionospheric continuity equation (equation (29)), combined with equations (2) and (33), reduces to

$$\frac{1}{r^2}\frac{d(n_iV_ir^2)}{dr} = \frac{\nu_i^{\text{ioni}}Q}{4\pi V_n r^2} - \alpha_i n_i^2 \quad (38)$$

where Q is the total outgassing rate for the comet (all major neutral species included). Assuming constant and radial ion speed V_i, ionization frequency ν_i^{ioni}, and recombination rate coefficient α_i to be independent of r, the solution of equation (38) is given by (*Beth et al.,* 2019)

$$n_i(r) = \frac{V_i}{2\alpha_i r}(\delta-1)\frac{r^\delta - r_c^\delta}{r^\delta + \frac{\delta-1}{\delta+1}r_c^\delta} \quad (39)$$

where

$$\delta = \sqrt{1+\frac{Q}{Q_0}}, \quad Q_0 = \frac{\pi V_n V_i^2}{\nu_i^{\text{ioni}}\alpha_i}$$

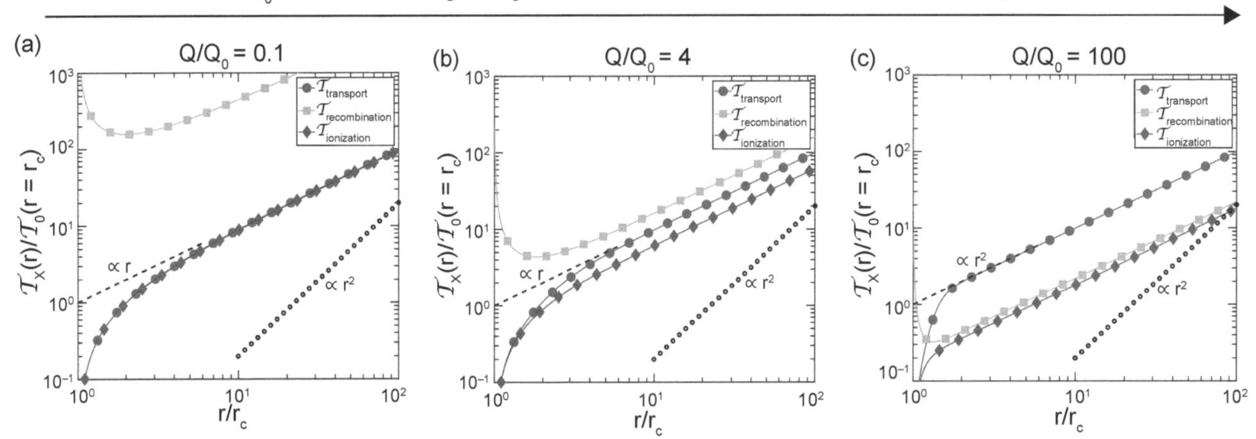

Fig. 7. Timescales for the ion number density profile from equation (39) for different activities normalized to Q_0: (a) Q/Q_0 = 0.1 (loss dominated by transport), (b) Q/Q_0 = 4 (losses through transport and recombination are similar), and (c) Q/Q_0 = 100 (loss dominated by recombination) as a function of cometocentric distances normalized to r_c and expressed in terms of the approximated transport timescale $\mathcal{T}_0$ at the nucleus' surface (equation (36)). Timescale dependencies in r and r^2 (losses in 1/r and $1/r^2$) are indicated for reference. Photoabsorption is ignored for simplicity.

and r_c is the nucleus' radius. Q_0 is homogeneous to an outgassing rate (s^{-1}). The different timescales (namely transport, recombination, and ionization) associated with this profile are given for illustration in Fig. 7. Even though ion-neutral chemical reactions do not affect the total ion density, their typical timescale (varying in $\propto r^2$) is indicated.

Regarding the assessment of Q_0, and unlike α_i, the order of magnitude for ν_i^{ioni}, V_n, and V_i^2 can be reasonably estimated (see section 2, the chapter by Biver et al. in this volume, and section 4 respectively). In the case of active comets, ν_i^{ioni} is not uniform (even when assuming spherical symmetry for the neutral density) and can be significantly reduced closer to the nucleus in the inner coma due to photoabsorption (see section 2.2.1). This has also the consequence of limiting the effect of recombination near the nucleus (*Beth et al.,* 2019): As the ion number density is near zero at the nucleus' surface, the recombination is negligible there.

Interestingly, equation (38) gives a constraint on the flux. By integrating equation (38) over radial distance r, one gets

$$n_i(r)V_i(r) \leq \frac{1}{r^2}\int_{r_c}^{r} \frac{\nu_i^{ioni}Q}{4\pi V_n}dr = \frac{\nu_i^{ioni}Q}{4\pi V_n}\frac{r-r_c}{r^2} < \frac{\nu_i^{ioni}Q}{4\pi V_n r} \quad (40)$$

Equation (40) may be used in two ways:

- If the ion number density is known based on *in situ* observations, it provides an upper limit for the ion radial velocity.
- When recombination is negligible compared to transport, equation (40) reduces to the equality (see equation (38)).

$$n_i(r)V_i(r) = \frac{\nu_i^{ioni}Q}{4\pi V_n}\frac{r-r_c}{r^2} \quad (41)$$

This corresponds to a solution to the ion continuity equation reduced to the simple balance between transport (loss) and ionization (source). In contrast with equation (39), equation (41) is valid for $V_i(r)$ varying with r but still purely radial.

For weakly active comets, such as 67P, the recombination timescale is much longer than the transport timescale (*Galand et al.,* 2016). This corresponds to the "equality" case given by equation (41). Applying equation (2), equation (41) can be rewritten as

$$n_i(r) = \frac{\nu_i^{ioni} n_n (r-r_c)}{V_i(r)} \quad (42)$$

This equation was successfully used at 67P at low activity to retrieve the electron number density observed by the Mutual Impedance Probe and Langmuir Probe (see section 3.1.1) onboard Rosetta (*Galand et al.,* 2016; *Heritier et al.,* 2018). A multi-instrument approach was applied to cometocentric distances from 10 km to 80 km, for heliocentric distances larger than 2 au. In those conditions, photoabsorption and dissociative recombination were negligible (*Heritier et al.,* 2018). The ions were assumed not to have undergone any significant acceleration at Rosetta's position and to have a bulk velocity similar to that of the neutrals ($V_i(r) \approx V_n$ at $r = r_{Rosetta}$). Electron-impact ionization frequency was derived from the electron spectrometer onboard Rosetta (see section 3.1.3), whereas photoionization frequency was extrapolated from solar flux measurements from Earth. Neutral density and neutral composition were derived from Rosetta pressure gauge and double-focusing mass spectrometer (*Balsiger et al.,* 2007; see the chapter by Biver et al. in this volume). Equation (42) captured very well the observed electron density from the balance between transport of nonaccelerated ions and the major sources of ionization, photoionization, and electron-impact ionization. The latter source was found to dominate at times, especially post-perihelion for large heliocentric distances [>2 au (see *Heritier et al.,* 2018)]. Equation (42) held even under solar-wind-disturbed conditions, such as CMEs and CIRs (*Galand et al.,* 2016; *Heritier et al.,* 2018).

This multi-instrument approach was also applied all the way down to the surface at the end of the Rosetta mission (*Heritier et al.,* 2017a). In that case, the model was refined by taking into account adiabatic expansion and acceleration of the cometary neutrals following sublimation near the surface (*Huebner and Markiewicz,* 2000). Excellent agreement was found between observed and modelled ionospheric densities, as shown in Fig. 8. The only exception was around 3 UTC near the neck of the comet, which had a concave shape and which is a region where the local outgassing has a complex structure and the gas may be collimated (see the chapter by Marschall et al. in this volume).

2.4.2. A rich diversity of ions: A real ion zoo. After the focus on the total ionospheric density (section 2.4.1), let us now have a look at the ion composition in the coma. We consider here individual ion species, whose number density is governed by the continuity equation given in equation (5). In terms of loss, three processes are in competition depending on the outgassing activity: transport, ion-neutral collisions, and electron-ion dissociative recombination (sections 2.3 and 2.4.1). Note that the recombination timescale for ion species s is simply given by $\mathcal{T}_{s,recombination} \approx [\alpha_s n_i(r)]^{-1}$ (equations (34) and (37)). Because it involves the electron density (or total ion density, n_i), it is almost the same for the different ion species. A given ion species also undergoes loss through ion-neutral collisions (see section 2.3 and equation (5)). The loss timescale for an ion species s through ion-neutral collisions is given by

$$\frac{1}{\mathcal{T}_{s,ion\text{-}neut\ loss}} = \sum_p k_{s,p}(T)\, n_p(r) = k_{s,n}\, n_n(r) \propto \frac{1}{r^2} \quad (43)$$

where

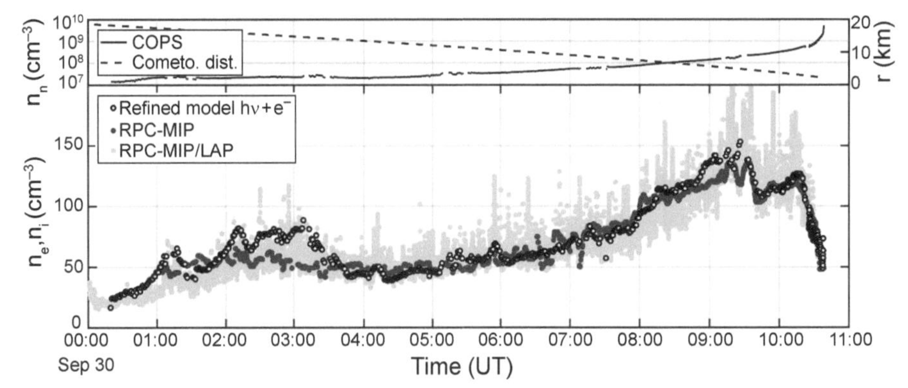

Fig. 8. Top panel: Time series of neutral number density from ROSINA pressure gauge (solid line) and of cometocentric distance (dashed line). Bottom panel: Time series of the modeled (black dots) and observed (magenta) ionospheric number densities at the end of Rosetta's mission, from 20 km down to the surface on 30 September 2016. The outgassing rate was around 4×10^{25} s^{-1}. The dark magenta dots correspond to RPC Mutual Impedance Probe measured electron density, while the light magenta dots represent the high-time-resolution RPC Langmuir Probe data cross-calibrated with RPC Mutual Impedance Probe. From Fig. 9 of *Heritier et al.* (2017a).

$$k_{s,n} = \frac{1}{n_n}\sum_p k_{s,p} n_p(r) \tag{44}$$

is the average reaction rate coefficient between ion species s and the set of neutral species n. The subscript p refers to the neutral species p and is incremented over all neutral species chemically reacting with ion species s. Excluding extended sources, the neutral density of a neutral species follows a spatial dependence in $1/r^2$ (equation (2)), thus the timescale $\mathcal{T}_{s,\text{ion-neut loss}}$ varies in r^2, whereas dissociative recombination and transport timescales vary in r (see Fig. 7). Therefore, the loss timescale through ion-neutral chemistry increases much faster with cometocentric distance r than that through transport or recombination. The ion-neutral chemistry and coupling are relevant in the region where $\mathcal{T}_{s,\text{ion-neut loss}} \leq \mathcal{T}_{\text{transport}}$ (*Gombosi,* 2015), i.e., within a region around the nucleus bounded by

$$R_{s,n} = \frac{k_{s,n} Q}{4\pi V_s V_n} \lesssim \frac{k_{s,n} Q}{4\pi V_n^2} \tag{45}$$

where the distance $R_{s,n}$ corresponds to $\mathcal{T}_{\text{transport}}$ (combining equations (2), (36), and (43)). Ions and neutrals are assumed to flow radially outward at the same speed (see sections 2.4.1 and 4.1). However, recombination may shrink this region of coupling (i.e., reduce $R_{s,n}$) if $\mathcal{T}_{\text{recombination}} \leq \mathcal{T}_{\text{transport}}$.

Ion species can be separated into three broad categories whose r-dependency of their number density is summarized in Table 1 and exemplified in Fig. 9:

- (A) ions produced from ionization but not (or barely) reacting with neutrals, i.e., $R_{s,n} < r_c$ even for high outgassing activity (see equation (47))
- (B) ions produced from ionization that are able to react efficiently with the main neutral cometary species for high enough outgassing activity ($r_c < R_{s,n}$)
- (C) ions only produced through ion-neutral chemistry, such as protonated molecules (except H_2O^+), and lost through mainly transport or chemistry (section 2.3); they are born from collisions between neutrals and ions from category B

When $R_{s,n} \lesssim r_c \leq r$, ions from category A and ions from category B for weak outgassing activity are mainly produced through ionization and are primarily lost through transport, which implies

$$\frac{1}{r^2}\frac{d(n_s V_s r^2)}{dr} \approx \nu_s^{\text{ioni}} n_n(r) \tag{46}$$

Applying equation (2) leads to

$$n_s(r) \approx \frac{\nu_s^{\text{ioni}} Q}{4\pi V_n V_s(r)} \frac{r - r_c}{r^2} \approx \frac{\nu_s^{\text{ioni}} (r - r_c)}{V_s(r)} n_n(r) \tag{47}$$

When $R_{s,n} > r_c$ (moderate to high activity), for $r_c \lesssim r \lesssim R_{s,n}$, ion species s from category B, produced through ionization, is lost primarily through ion-neutral chemistry (note that usually such an ion species is not terminal and hence not lost through electron-ion dissociative recombination); it is said to be in *photochemical equilibrium*, meaning that the ion number density is only governed by chemical reactions

$$0 \approx \nu_s^{\text{ioni}} n_n(r) - k_{s,n} n_n(r)\, n_s(r) \tag{48}$$

This implies that

$$n_s(r) \approx \frac{\nu_s^{\text{ioni}}}{k_{s,n}} \approx \text{constant} \tag{49}$$

This value is a plateau and an upper limit for the ion species s which cannot be surpassed. In addition, photoabsorption may reduce this value in an optically thick coma (see section 2.2.1).

TABLE 1. Different radial number density dependency for each category below and above the ion-neutral coupling limit $R_{s,n}$ (equation (45)) when transport dominates the ion loss ($Q/Q_0 \lesssim 1$).

Cat.	Ion density n_s when:		Ref. Eqs.
	$r_c < r < R_{s,n}$	$r_c < R_{s,n} < r$	
A	N/A ($R_{s,n} \ll r_c$)	$\propto \frac{r-r_c}{r^2}$*	equation (47)
B	$\approx$ constant $\leq \frac{v}{k}$	$\propto \frac{1}{r}$	equations (47), (49)
C†	$\propto \frac{r-r_c}{r^2}$	$\propto \frac{\log r}{r^2}$	equations (51), (52)
A + B + C	$\propto \frac{r-r_c}{r^2}$	$\propto \frac{r-r_c}{r^2}$	—

The r-dependencies given are for an optically thin coma.

* For category A, $R_{s,n} < r_c$, so third column refers to $r_c < r$.

† For category C, s in $R_{s,n}$ is not associated with the ion itself but to its parent ion, for example, s = H_2O^+ and n = H_2O for H_3O^+ and s = H_2O^+/H_3O^+ and n = NH_3 for NH_4^+. For category C, transport is here assumed to be the ion loss process.

Ions from category C, so-called *protonated ions,* are produced through proton-transfer reactions, either from H_2O^+ or through the production of other protonated ions, such as H_3O^+. Neglecting dissociative recombination in equation (29), the total ion density of protonated molecules AH^+ (i.e., the sum of the number densities of the protonated molecules such that $n_{AH^+} = n_{H_3O^+} + n_{NH_4^+}$ + etc.) obeys approximately the equation

$$\frac{1}{r^2}\frac{d\left(n_{AH^+}V_{AH^+}r^2\right)}{dr} \approx k_{H_2O^+,A}\, n_A\, n_{H_2O^+} \tag{50}$$

The loss through dissociative recombination did not play a key role to assess ion densities at the location of Rosetta during its escort phase (*Heritier et al.,* 2018), hence this process was disregarded in equation (50). However, in specific conditions, such as high enough outgassing rates at a certain range of cometocentric distances coupled with a high sensitivity of the recombination rate coefficient α_i with T_e, dissociative recombination may represent a key loss for protonated ions: This was, for example, the case during Giotto's flyby of 1P (*Rubin et al.,* 2009).

To solve equation (50), one needs to know the dependence of $n_{H_2O^+}$ with r as it plays the role of precursor for the formation of protonated molecules (see, e.g., *Aikin,* 1974). Indeed, the protonated molecules dominate the ion composition near the nucleus only when H_2O^+ is close to photochemical equilibrium and triggers the formation of H_3O^+ (see, e.g., Fig. 9). On the one hand, in the region $r_c \lesssim r \lesssim R_{H_2O^+,n}$ occurring only for high enough outgassing rates, H_2O^+ is in photochemical equilibrium and thus the lefthand side of equation (50) is $\propto 1/r^2$. Therefore, we have

$$n_{AH^+}(r) \propto \frac{r-r_c}{r^2} \text{ for } r \lesssim R_{H_2O^+,n} \tag{51}$$

On the other hand, for $r > R_{H_2O^+,n}$, H_2O^+ dominates again the ion composition and behaves $\propto 1/r$. The righthand side of equation (50) is $\propto 1/r^3$ and thus

$$n_{AH^+}(r) \propto \frac{\log r}{r^2} \text{ for } r \gtrsim R_{H_2O^+,n} \tag{52}$$

At low activity ($Q \leq 10^{25}$ s^{-1}), ion species from category B exhibit an ion density profile (blue, solid line) similar to that of ions from category A (black, dashed line) and given by equation (47), as shown in Figs. 9a,d. Ions from category B, although reacting with H_2O and CO_2 like H_2O^+, have their associated ion-neutral coupling limit $R_{B,n}$ smaller than r_c due to the low outgassing rate. So effectively, they behave as ions from category A. Ions from category C are hence barely produced as they require at least one ion-neutral collision and their density decreases very quickly with increasing r (see red profile in Figs. 9a,d). A-type and B-type ion species are dominating the ion composition.

As the outgassing activity increases ($Q \gtrsim 10^{25}$ s^{-1}), so does the number density of the neutral species (e.g., H_2O and CO_2), and $R_{B,n}$ passes above the surface ($R_{B,n} > r_c$), meaning there exists a region where ion-neutral chemistry dominates over transport. Within this region, ions from category B are efficiently reacting with neutrals and their density profile is near photochemical equilibrium (equation (49), see blue profile in Figs. 9b,c,e,f). In the same region, ions from category C are efficiently produced, such as H_3O^+ (and NH_4^+), and may dominate the composition region with a profile following the red solid line in Figs. 9b,c,e,f. Beyond $R_{B,n}$, transport or electron-ion recombination govern the loss. Ions from category B [or, C] are barely lost (or, produced) through ion-neutral collisions. A-type and C-type ions are dominating the ion composition at low r distances, whereas A-type and B-type ions are dominating at high r.

Regarding the detection of ion species in a cometary environment, Rosetta has revolutionized the field (*Beth et al.,* 2020). If the mass resolution of the ion mass spectrometer onboard a cometary probe is high enough, the mass, or mass-per-charge, may work as a unique identifier of ion species. For example, the mass of H_2O^+ is 18.015 u while that of NH_4^+ is 18.039 u. We thus need an instrument with a minimum resolving power of $\Delta m \approx 0.025$ u to separate both species. During the Giotto flyby of Comet 1P, a snapshot of the ion density profiles was obtained at different integer masses (*Altwegg et al.,* 1993); however, ion species of the same integer mass could not be distinguished due to the too low mass resolution of the ion mass spectrometer. In contrast, at Comet 67P, ion species were identified individually over a two-year period and a wide range of outgassing activities thanks to the ROSINA/DFMS onboard Rosetta (see Fig. 10a). This spectrometer was one of the most

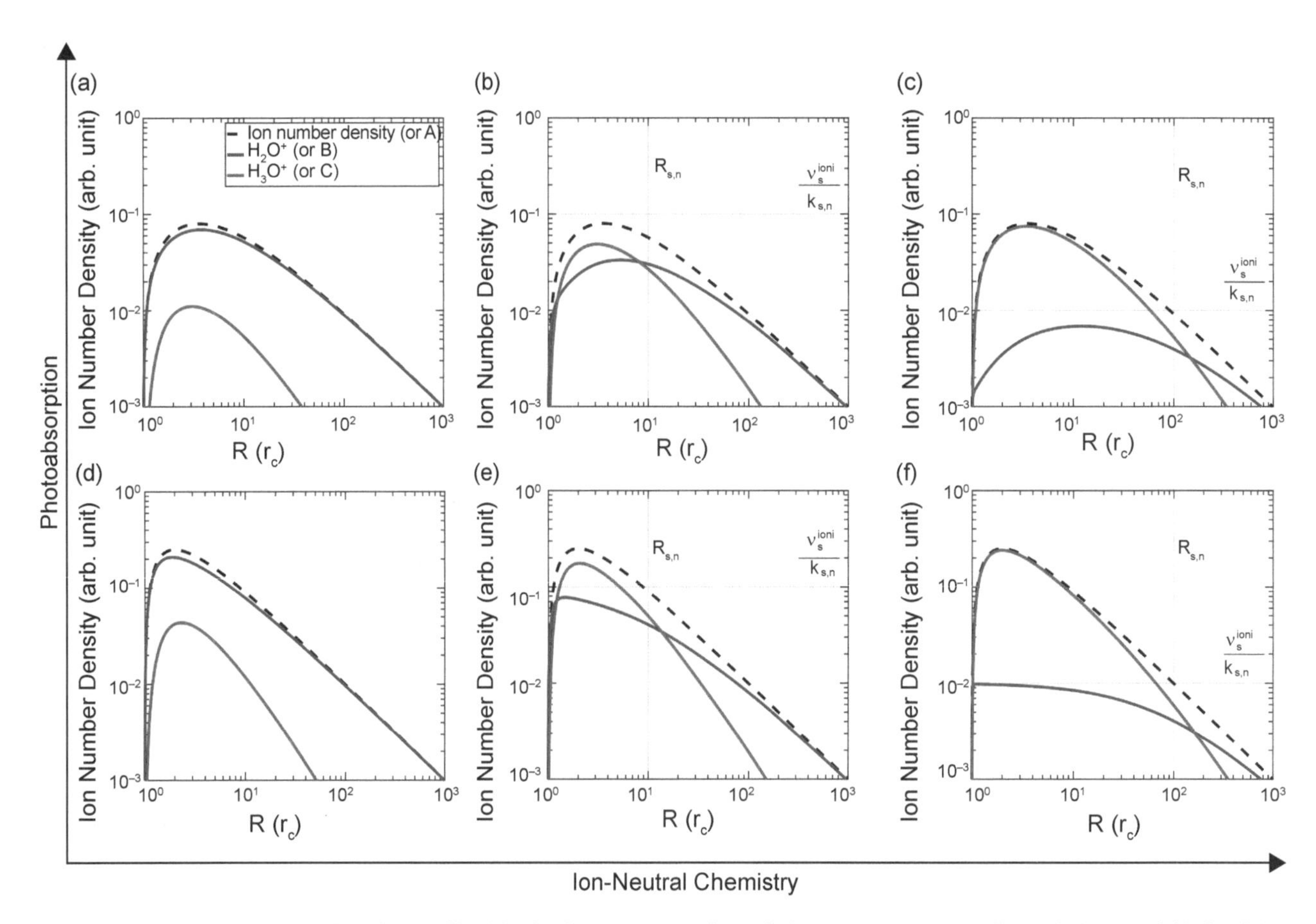

Fig. 9. Total ion number density profile (dashed, representative of the category as well) and those of H_2O^+ (blue, representative of category B) and H_3O^+ (red, representative of category C) for a coma made of H_2O only, in arbitrary unit using the analytical model from *Beth et al.* (2020). Electron-ion dissociative recombination is ignored. Although photoabsorption and ion-neutral chemistry are both linked to the outgassing rate, different conditions are achieved by increasing either photoabsorption (from bottom to top) or ion-neutral chemistry (from left to right) separately. The H_2O^+ number density at photochemical equilibrium (equation (49), horizontal dotted line) and the ion-neutral coupling limit (equation (45), vertical dotted line) are indicated.

powerful spectrometers ever flown on a planetary probe in terms of mass resolution, reaching up to $R \approx 3000$. Thanks to its high accuracy, more than 20 ion species were unambiguously identified in Comet 67P's ionosphere (*Beth et al.,* 2020) (see Figs. 10b,c), with the serendipitous detection of the CO_2^{++} dication.

3. ELECTRON POPULATION AND ITS IMPACT

3.1. Different Electron Populations

At any given time, the electrons in a cometary environment can be classified into three broad populations: *cold, warm,* and *hot.* Although the energy range of each population may slightly vary between published studies, this classification allows to distinguish different electron sources and processes affecting electrons. The cold and warm electron populations, which are primarily of cometary origin with some contribution from the solar wind to the warm electrons, as discussed below, represent a dense population, compared with the rarefied hot electron population. Note that, while cold and warm populations cohabit, one of them may be greatly dominating over the other, depending on the outgassing activity, neutral composition, and cometocentric distance. These three electron populations were first reported from *in situ* observations at around 12,000 km of Comet 21P during its flyby by ICE (*Zwickl et al.,* 1986). During the course of the Rosetta escort mission, these three populations were also identified. They were differentiated by their signatures as detected by different Rosetta Plasma Consortium (RPC) sensors onboard Rosetta (*Glassmeier et al.,* 2007). The characteristics, prime origin (photoelectrons, section 2.2.1, solar wind electrons or their secondaries, section 2.2.2), and detection of each of the three electron populations are discussed in more details hereafter:

- the newly-born, warm, cometary electrons (5–15 eV), which dominate the plasma density at large heliocentric distances, but are always present (section 3.1.1). They include photoelectrons (for coma optically thin to EUV radiation) and electrons from electron-impact ionization

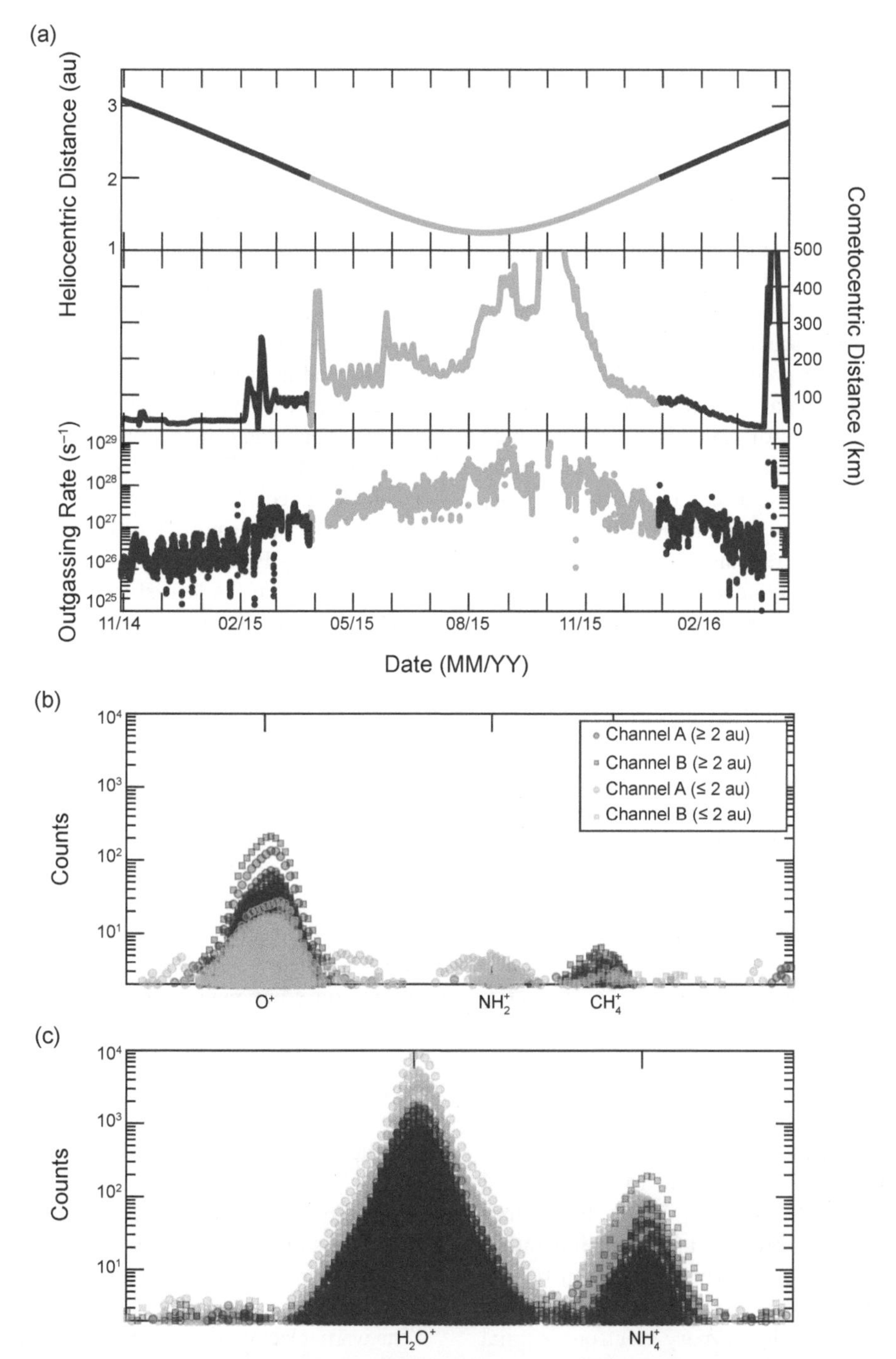

Fig. 10. **(a)** Overview of the heliocentric and cometocentric distances of Rosetta, and of the outgassing rate during the operational phase of ROSINA-DFMS mass spectrometer in ion mode. **(b),(c)** Spectra at 16 u and 18 u from ROSINA/DFMS mass spectrometer during Rosetta's mission. Both colors are associated with different heliocentric distances: above (blue) or below (yellow) 2 au. Adapted from *Beth et al.* (2020).

by hot electrons;

- the cold, cometary electrons (<1 eV), which are energy-degraded warm electrons and whose relative importance increases with outgassing activity (section 3.1.2);
- the more rarefied, hot electrons (>20 eV). When the coma is optically thin to EUV at large heliocentric distances, they are mainly accelerated, solar wind electrons, and are responsible for the bulk ionization of the neutral coma and for the auroral FUV emissions (section 3.1.3). When the coma is optically thick (e.g., during the flyby of 1P by Giotto), they are mainly photoelectrons in the inner coma (*Bhardwaj,* 2003; *Beth et al.,* 2019).

3.1.1. Newly born warm, cometary electrons. The warm electron population refers to electrons produced through (1) photoionization in an optically thin coma (see section 2.2.1) and (2) electron-impact ionization (see section 2.2.2). Photoionization of a neutral species p by solar radiation of incident energy $E_{h\nu}$ above ionization energy

$E^{th}_{p,s}$ yields the production of a photoelectron and an ion (see equation (6)). The rate at which neutrals p are ionized per unit time is given by the photoionization frequency, $\nu_p^{h\nu,ioni}$ (see equation (12) and Fig. 4). The total photoelectron production rate, $S^{h\nu}_{e^-}$, depends on the neutral composition and total neutral density (equation (16)). During photoionization of a neutral p yielding the production of a photoelectron and an ion s, the excess energy, $\Delta E = E_{h\nu} - E^{th}_{p,s}$, goes mainly to the released photoelectron, in the form of kinetic energy. However, as this process is largely isotropic, the electrons are uniformly scattered in every direction. The energy distribution of the produced photoelectrons is as structured as the incident solar radiation flux, some strong solar lines producing a peak of photoelectrons at a given energy, like at 27–28 eV [photoionization of H_2O or CO_2 by the He II 30.4 nm (40.8 eV) solar line (*Körösmezey et al.,* 1987; *Bhardwaj,* 2003; *Vigren and Galand,* 2013)]. Nevertheless, in an optically thin coma, the average photoelectron energy is around 10–15 eV (*Huebner et al.,* 1992). These electrons are referred to as *warm*, prior to efficient cooling in the coma through collisions (see section 3.1.2). In an optically thick coma, the mean photoelectron energy increases in regions of large neutral column densities (*Bhardwaj,* 2003). Indeed, as illustrated in Fig. 3, photoabsorption cross sections have the largest values above 20 nm. Hence, it is this part of the solar spectrum that is going to be absorbed first, at an optical depth $\tau = 2$ (equation (17)). Inward of the boundary corresponding to $\tau = 2$ at 20 nm, photoelectrons are produced by the remaining solar photons of wavelength less than 20 nm (energy larger than 60 eV). Subtracting the ionization energy (typically 10–15 eV), this means that all photoelectrons in this region are born with an energy larger than 45 eV. In such a case, photoelectrons are *hot*, no longer warm.

Electrons themselves can be energetic enough to ionize the neutral species. This is the case if their energy is above the ionization energy (see section 2.2.2). As energetic electrons at 67P were found to have typical energies of 30–40 eV (*Myllys et al.,* 2019), the energy of the secondary electron produced through ionization is less than 15 eV, because the primary electron retains at least half of the remaining energy (where the remaining energy corresponds to the primary electron original energy minus the ionization energy). Thus, electrons produced by electron-impact ionization also contribute to the warm population.

The warm electron population was detected throughout the Rosetta escort phase by several instruments. First, the Ion and Electron Sensor (IES) (*Burch et al.,* 2007), part of the RPC consortium, measured energy distributions of negatively-charged particles (primarily electrons) above the noise level over an energy range from 4.3 eV q^{-1} to a few 100 eV q^{-1} with an energy resolution $\Delta E/E \sim 8\%$. The electron velocity/energy distribution, fitted by a suprathermal double-kappa function, exhibited signatures from two populations: one warm and one hot (*Broiles et al.,* 2016b; *Myllys et al.,* 2019). Whereas the hot population was rarefied (see section 3.1.3), the warm population was dense (~10–200 cm^{-3}), with temperatures typically below 10 eV (*Myllys et al.,* 2019). At the beginning of Rosetta's mission, on November 14, 2014, *Broiles et al.* (2016a) identified two dense populations in the RPC-IES dataset: One with temperatures above 8.6 eV was observed when the local neutral number density at Rosetta was high ($n_n > 8 \times 10^6$ cm^{-3}) and one with temperature below 8.6 eV when n_n was generally low ($n_n < 8 \times 10^6$ cm^{-3}). *Myllys et al.* (2019) did not find that the electron temperature of the warm population T_w decreased with heliocentric distance, but the spread in electron temperature estimates was found to range from 1 eV to 15 eV. It should be pointed out that the warm population is not thermalized, its distribution is not following a Maxwellian, and the double-kappa fit is often not suitable, especially at large heliocentric distances (*Myllys et al.,* 2019). Furthermore, part of the warm population is unseen by RPC-IES (*Madanian et al.,* 2016), as a result of the combination of the lowest energy of detection (4.3 eV for electrons) and of the precarious effect of the spacecraft potential. Prior to launch, it was not anticipated that the spacecraft would "charge" so negatively. In space, there is no ground/Earth and the spacecraft may charge positively or negatively, depending on the material it is made of and the interaction with the surrounding plasma environment. At 1P, the spacecraft potential of Vega-1 and Vega-2 was only a few volts, negative ($V_{SC} < 0$) or positive ($V_{SC} > 0$) (*Pedersen et al.,* 1986). In contrast, at 67P, the warm and dense cometary electron population was responsible for the negative spacecraft potential, often between –15 V and –5 V, rarely below –20 V (*Odelstad et al.,* 2017). Biased elements on solar panels attracting cold electrons also contributed to the negative spacecraft potential (*Johansson et al.,* 2020). The presence of a negative spacecraft potential prevented low-energy electrons ($0 < E_{k,e^-} < -qV_{SC}$, where E_{k,e^-} is the electron kinetic energy and $q > 0$) to reach the spacecraft itself. In contrast, those with energies $E_{k,e^-} > -qV_{SC}$ could reach the spacecraft with a reduced kinetic energy, $E_{k,e^-} + qV_{SC}$ (see Fig. 6). This should be accounted for and corrected for when analyzing the RPC-IES electron spectrometer dataset (*Broiles et al.,* 2016a; *Galand et al.,* 2016).

The warm electron population was detected at 67P by two other instruments less sensitive to the spacecraft potential: the spherical LAngmuir Probe (LAP) (*Eriksson et al.,* 2007) and the Mutual Impedance Probe (MIP) (*Trotignon et al.,* 2007), both part of the RPC consortium. The RPC-LAP, based on a century-old technique, provides insight on both the ion and electron populations. The principle is relatively simple. A metallic sphere is set at different voltages, scanning from –30 V to +30 V in a short period of time, and the total current from charged particles in the plasma collected by the probe is measured. It results in a characteristic I–V curve (collected current I vs. applied voltage V) that, upon strong assumptions on the velocity distribution function of the ions and electrons and on the ion mass, may give access to macroscopic properties of the surrounding plasma, including number density, electron temperature, and ion speed (*Mott-Smith and Langmuir,* 1926). The second instrument, RPC-MIP, is able to probe the dielectric characteristics of the plasma to derive electron number density and temperature. In active mode, it consists in two emitting/transmitting electrodes excited by an intensity source of very high impedance. The excited electrodes inject an alternating current $I(\omega)$ (where ω is the angular frequency) into the medium, here the cometary plasma environment. This induces a voltage difference $V(\omega)$ between the two other receiving electrodes, itself a function of the frequency, ω. Expanding the concept of resistance to alternating currents, the ratio $Z(\omega) = V(\omega)/I(\omega)$ is therefore an impedance containing information on both amplitude and phase (complex number). Between emitting and receiving electrodes, the surrounding plasma plays the role of a "filter" as its dielectric constant ε depends on the frequency: $Z(\omega)$ is maximum near the electron plasma frequency $\omega_{p,e}$, i.e., max $Z(\omega) \approx Z(\omega_{p,e})$, where

$$\omega_{p,e} = \sqrt{\frac{q^2 n_e}{m_e \varepsilon_0}} \tag{53}$$

(with m_e the electron mass and ε_0 the vacuum permittivity) as the plasma is in resonance. Indeed, $\omega_{p,e}$ corresponds to the natural frequency at which electrons oscillate in response to a small charge separation. Beside this simple approach to derive the electron density in a nonmagnetized plasma characterized by only one electron temperature, more elaborated processing taking advantage of the full RPC-MIP mutual impedance spectra as a function of frequency can be applied to the dataset in order to do any or all of the following: (1) Derive electron density and temperature for two different electron populations, here warm and cold (when the densities are high enough and Debye length suitable for the operation mode). Indeed, the Debye length is defined by

$$\lambda_{Debye} = \sqrt{\frac{\varepsilon_0 k_B T_e}{q^2 n_e}} = \frac{v_{th,e}}{\omega_{p,e}} \tag{54}$$

where $v_{th,e} = \sqrt{k_B T_e/m_e}$ is the electron thermal speed. λ_{Debye} is a characteristic plasma length that corresponds to the typical distance at which electromagnetic waves are screened and damped by the surrounding plasma (here mainly electrons): RPC-MIP operates efficiently when the emitter-receiver distance is greater than $2\lambda_{Debye}$. (2) Take into account the magnetization of the environment. (3) Include the presence of an ion sheath surrounding the spacecraft and the probe as a result of the negative spacecraft charging (*Gilet et al.,* 2017, 2020; *Wattieaux et al.,* 2019, 2020). In passive mode, the emitting antennas are turned off and the plasma plays the role of a natural emitter of electromagnetic waves. The spectra measured corresponds to the Fourier Transform of the voltage in this case.

At 67P at low activity (> 2 au), the cometary plasma is dominated by warm electrons, with a mean energy typically around 5–10 eV but varying between 2 and 20 eV (*Wattieaux et al.,* 2020) (see Fig. 11). Observations from RPC-LAP (*Edberg et al.,* 2015), RPC-MIP (*Heritier et al.,* 2017a), and RPC-IES (*Myllys et al.,* 2019) showed that the (warm) electron number density $n_{e,warm}$ followed a 1/r dependence with cometocentric distance. *Galand et al.* (2016) and *Heritier et al.* (2017a, 2018) demonstrated that the warm electrons were produced by electron-impact ionization and, to a lesser extent, photoionization, and are lost through transport: The plasma density is well captured by equation (42) following a 1/r dependence (see section 2.4.1). The warm electron number density was found to be slightly decreasing with increasing heliocentric distance at Rosetta's position, a result of the increase in r with heliocentric distance throughout the mission. It is more pronounced when the electron number density is normalized to a given cometocentric distance r (*Myllys et al.,* 2019). Moreover, $n_{e,warm}$ increased during solar events due to an increase in the ionizing electron population (*Hajra et al.,* 2018). Its dependence with season and hemisphere is discussed in *Heritier et al.* (2018).

Near perihelion, the warm population around 67P was primarily driven by photoionization (*Heritier et al.,* 2018). The newborn, warm photoelectrons are quickly cooled through collisions with neutral species and the cometary electron population becomes *cold* (*Gilet et al.,* 2020). The relative importance of the warm versus cold electron populations observed in the inner coma of 67P during the Rosetta mission is reviewed in section 3.1.2.

3.1.2. From warm to cold cometary electrons. Since the newborn electrons have average energies (10 eV; see section 3.1.1) well above the cometary neutrals [0.01 eV near the surface; see *Gombosi et al.* (1986) and the chapters by Marschall et al. and Biver et al. in this volume], if the coma is dense enough, electrons may undergo efficient energy degradation through collisions with neutrals. The lowest electron temperature which can be reached is the temperature of the cometary neutrals.

Evidence of cold electrons was provided by electron spectrometer measurements from the ICE spacecraft in the tail of 21P (*Meyer-Vernet et al.,* 1986; *Zwickl et al.,* 1986). Indirect evidence during 1P's flyby by Giotto near 1 au from the analysis of observed ion composition by the use of an ionospheric model and observed neutral densities and composition (*Eberhardt and Krankowsky,* 1995): They derived electron temperature as low as 10^2 K (~0.1 eV) at a cometocentric distance of 2000 km.

At 67P, the RPC-LAP and RPC-MIP sensors detected cold electrons (*Eriksson et al.,* 2017; *Gilet et al.,* 2017). Both instruments have their own limitations in the cold electron detectability [i.e., the highly negative potential of Rosetta when the cold electron density is high can completely hide cold electrons from RPC-LAP, while assumptions used in the complex model used for RPC-MIP analysis of cold electrons may not always be valid, for example, when the ratio of warm and cold temperatures is too small (see

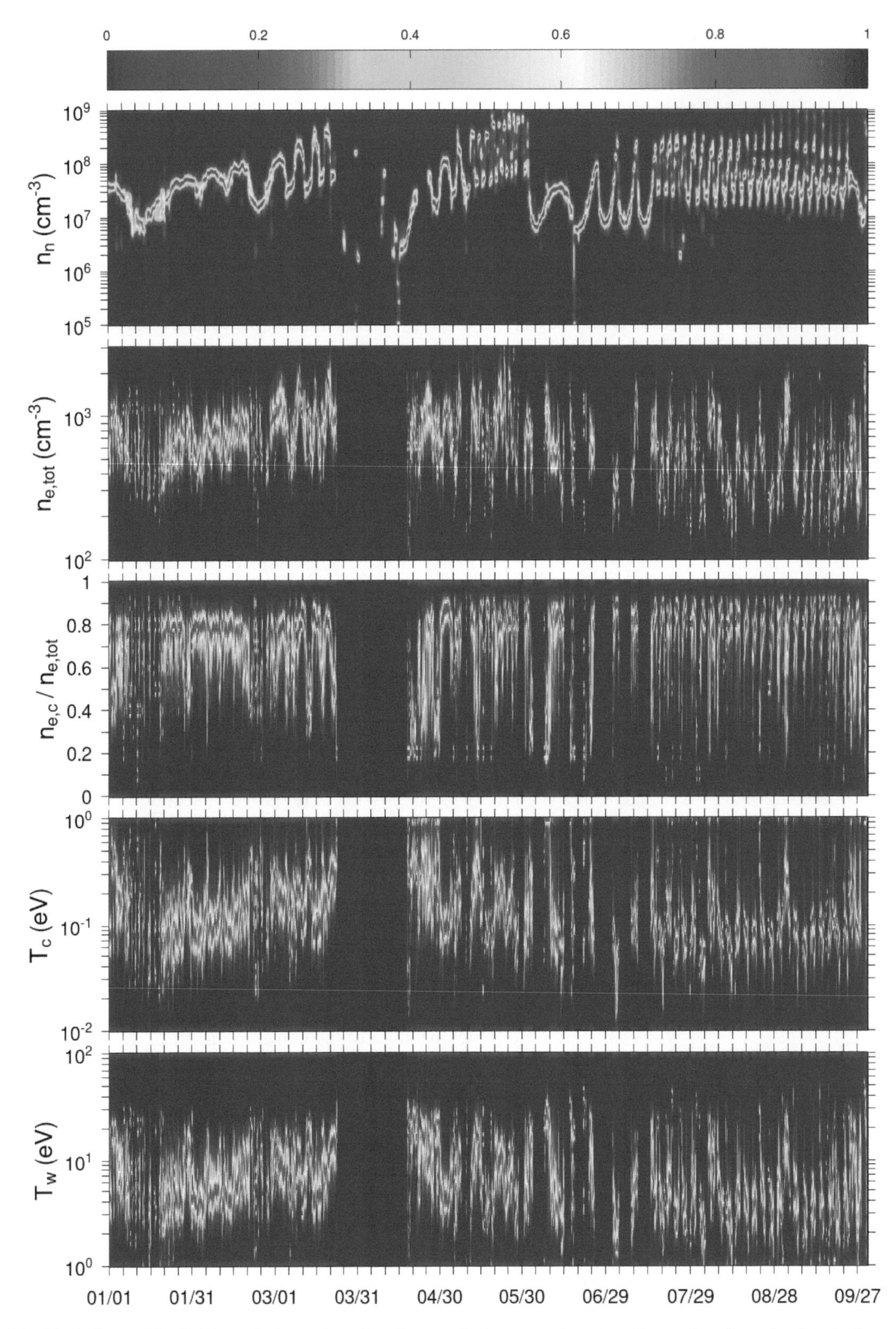

Fig. 11. Evolution of the total neutral number density n_n, electron number density n_e, fraction of cold electrons $n_{e,cold}/n_{e,total}$, and temperatures of both populations T_{cold} and T_{warm} from January to September 2016 (DD/MM) at Rosetta. The dispersion of the results was computed and normalized over time intervals of 6 h. Color bar corresponds to the occurrence. Adapted from *Wattieaux et al.* (2020).

Gilet et al., 2020; *Wattieaux et al.*, 2019)]. Despite these limitations, general trends were inferred (*Engelhardt et al.*, 2018; *Odelstad et al.*, 2018; *Gilet et al.*, 2020; *Wattieaux et al.*, 2020) (see Fig. 11). Although largely fluctuating, the contribution of the cold electrons to the total electron population was observed to mainly increase with local outgassing (hence latitude through season) and to decrease with cometocentric distance r and heliocentric distance r_h. Cold electrons were always detected when Rosetta was within the diamagnetic cavity where cold electrons made up close to 100% of the total electron density. Post-perihelion, the largest contribution of cold electrons was found to be over the southern, summer/autumn hemisphere, reaching values up to 70% to 90% of the total electron population, with temperatures as low as 0.05 eV and number densities as high as 1000 cm^{-3}. *Gilet et al.* (2020) ruled out the influence of the neutral composition as H_2O (abundant in the northern hemisphere) and CO_2 (more abundant in the southern hemisphere) have similar momentum-transfer cross sections. However, one may note that H_2O has higher total cross-sections than CO_2 (see Fig. 6): Electrons, on average, lose more momentum by colliding with CO_2 than with H_2O, although collisions with the former are scarcer.

At 1P, the presence of cold electrons was predicted by ionospheric models and showed that, for high outgassing activity, the bulk of the cometary plasma is cold, with electron temperatures reaching values as low as the neutral temperature within the diamagnetic cavity (see section 3.3), whereas warm electrons, constantly produced, but in minority, continue to be present (*Mendis et al.,* 1985; *Körösmezey et al.,* 1987; *Gan and Cravens,* 1990; *Eberhardt and Krankowsky,* 1995). At larger cometocentric distances, the electron temperature is expected to increase as (1) electron and neutral densities decrease, hence their collision probability decreases, and (2) there is efficient heating of the cold electrons through Coulomb collisions between the cold electrons and the newborn, more energetic, electrons.

In order to assess where electrons may undergo significant collisions, a heuristic approach may be applied. Similar to planetary atmospheres, one may estimate the position of the so-called *exobase* [also *critical level* (*Jeans,* 1923; *Spitzer,* 1949; *Chamberlain,* 1963)], i.e., the boundary where the gas (here electrons) transitions from a continuum flow (collisional) to a free-molecular regime (collisionless). It is also defined as the critical limit where the atom/molecule undergoes one collision on average along its path outward to infinity. For this, one may evaluate the Knudsen number

$$\mathrm{Kn} \sim \frac{\ell_{e^-,\mathrm{mfp}}}{L}$$

where $\ell_{e^-,mfp}$ stands for the mean free path [i.e., the mean distance covered by a particle between two successive collisions, here $(n_n\sigma_{e,n})^{-1}$ where $\sigma_{e,n}$ is the electron-neutral total cross section as electrons mainly collide with neutrals] and L stands for a characteristic length (e.g., the electron number density scale height $|\nabla \log n_i|^{-1}$, here ~ r). The electron exobase is expected to be located at a cometocentric distance

$$\mathrm{Kn} \sim 1 \rightarrow r_{e^-,\mathrm{exo}} = L \sim \frac{1}{n_n \sigma_{e,n}} \quad (55)$$

Cold electrons were expectedly detected near perihelion [around 1.24 au, with Q ~ 10^{28} s^{-1} (*Hansen et al.,* 2016; *Engelhardt et al.,* 2018; *Gilet et al.,* 2020)] when the electron exobase predicted by equation (55) can reach a few hundred kilometers.

Surprisingly, cold electrons were also detected away from perihelion beyond heliocentric distances of 3 au [Q < 10^{26} s^{-1} (*Gilet et al.,* 2020)], when the electron exobase predicted by equation (55) is located below the surface (*Engelhardt et al.,* 2018; *Gilet et al.,* 2020). Hence, no efficient cooling of the electrons by collisions with neutrals was anticipated. The simple approach based on the Knudsen number implies a radial flow for the electrons, which is not sufficient to generate a cold electron population as observed with Rosetta (*Gilet et al.,* 2020). Three-dimensional modeling from particle-in-cell (PiC) simulations, applied to low outgassing comets, have shown that electrons are trapped in the close environment of the nucleus (*Deca et al.,* 2017, 2019; *Sishtla et al.,* 2019). This region is set up by an ambipolar electric field produced by the cometary plasma. Based on a three-dimensional kinetic test particle model, *Stephenson et al.* (2022) demonstrated that the trapping increases significantly the path of electrons in the dense coma close to the nucleus: This allows electrons to undergo substantial collisions and become cold despite low neutral densities. In the light of the complex, electromagnetic environment at low outgassing, the electron exobase needs to be moved closer to the nucleus than originally predicted through the Knudsen number.

3.1.3. Accelerated electrons. The most energetic electron population is often referred to as *suprathermal*, either as their energy exceeds k_BT_e, where T_e is the mean temperature of the whole electron population, or as their energy distribution departs from a Maxwellian distribution at high energies (i.e., $f_e(E)\, dE \not\propto \sqrt{E} \exp(-E/k_BT_e)\, dE$). Note that the warm electron distribution in energy also departs from a Maxwellian distribution. The electron population under focus in this section corresponds to electrons with energies above the mean energy of the whole electron population dominated by cold and warm electrons (and above the upstream solar wind electrons, i.e., around 10 eV). Their energy distribution tends not to decrease exponentially [i.e., $f_e(E) \not\propto \sqrt{E} \exp(-E/k_BT_e)$] but rather follows a power law [i.e., $f_e(E) \propto E^{-\kappa}$, as in kappa distributions]. The reason for such a behavior in a plasma and its possible ubiquity in nature are still debated and beyond the scope of this chapter (*Pierrard et al.,* 2016).

Let us review the findings on the *hot* electron population detected by Rosetta by the RPC-IES electron spectrometer (*Broiles et al.,* 2016a,b; *Myllys et al.,* 2019). It was derived by fitting a suprathermal double-kappa function on the electron velocity distribution. From such a fitting, characteristics of the warm and hot populations were derived. While the bulk of the electron population is dominated by warm and/or cold electrons (see section 3.1.1 and section 3.1.2), the hot electrons with temperatures of a few tens of electron volts are rarefied with typical densities below a few cubic centimeters (*Myllys et al.,* 2019). Their density was found to decrease significantly from 3–10 cm^{-3} near perihelion down to less than

0.1 cm^{-3} at large heliocentric distances [3.0–3.8 au (*Broiles et al.*, 2016a; *Myllys et al.*, 2019)]. Their temperature does not exhibit any strong dependence with heliocentric distance or cometocentric distance. The dependence of the hot population with r is significantly less than for the warm population, with a dependence more in r^{-2} than in r^{-1}, although the uncertainty is large and the dependence breaks below 50 km. The energy distribution measured by RPC-IES needs to be corrected to take into account the spacecraft potential, which typically ranges between –20 and –5 eV (see Fig. 6 and section 3.1.1).

The source of the hot electron population has been under debate. It was first proposed by *Broiles et al.* (2016a) to be suprathermal halo solar wind electrons (*Pierrard et al.*, 2016). *Myllys et al.* (2019), however, showed that the hot electron population observed at Rosetta is denser (by almost an order of magnitude) and colder than the solar wind halo component. Furthermore, they found an increase in the electron temperature with increasing invariant kappa index, an opposite trend to the one derived for the halo solar wind electrons (*Pierrard et al.*, 2016). Hence, the bulk of the hot electrons has a different origin.

At large heliocentric distances, *Madanian et al.* (2016) suggested that electrons could be accelerated toward the comet by an ambipolar electric field and that this field was the result of the electron pressure gradient due to plasma inhomogeneity observed close to the comet. When an electron is born far from the comet or comes from the space environment, such as the solar wind electrons, it is accelerated toward the cometary nucleus when falling into the potential well set up by the ambipolar electric field. Using a self-consistent collisionless PiC model applicable to weakly outgassing comets, *Deca et al.* (2017) showed that the suprathermal electrons present close to the comet are originating from the solar wind and accelerated by the ambipolar electric field set up by cometary plasma. This explanation is consistent with the large-scale acceleration process for the source of the suprathermal electrons responsible for auroral emissions (*Galand et al.*, 2020; *Stephenson et al.*, 2022). These atomic emissions observed in the far ultraviolet (FUV) by the UV spectrograph Alice onboard Rosetta were shown to result from the dissociative excitation of molecules, such as water and CO_2, by energetic (>15–20 eV) electrons. Through a multi-instrument analysis, it was found that the variation in the FUV brightnesses correlated very closely with the variation in the electron-impact emission frequency, driven by the RPC-IES electron flux. This remained true even during solar events and when the FUV emissions were observed at the limb, far from the location of Rosetta where the RPC-IES was measuring the suprathermal electrons. These accelerated solar wind electrons play a critical role in the cometary environment, as they are responsible not only for the auroral emissions, but also for the bulk of the ionization, and hence of the cometary plasma, at large heliocentric distances (see *Galand et al.*, 2016; *Heritier et al.*, 2018) (see also section 2.2.2).

Near perihelion, the source of the suprathermal electrons is still under debate. Indeed, with the formation of a dense ionosphere under solar illumination, a diamagnetic cavity (see section 3.3) is formed around the cometary nucleus and solar wind electrons do not have (easy) access to the inner ionosphere anymore. *Madanian et al.* (2017) observed a modest but consistent drop in the RPC-IES electron fluxes over the 40 eV to 100 eV range between outside and inside the diamagnetic cavity. However, they showed that the lower fluxes measured inside the cavity are too high to be explained by photoelectrons alone. They proposed two mechanisms to explain the observations: a trapping mechanism of the photoelectrons inside the cavity and/or the penetration of part of the solar wind electrons into it.

3.2. Effect on the Ion Dynamics

Electrons play a critical role in the plasma and ion dynamics. Indeed, as their mass is much lower than that of positive charges, such as protons, they have a greater mobility and they react quicker to changes in the electromagnetic fields.

Ion dynamics is largely driven by macroscopic electric fields. In cometary and classical (i.e., not relativistic) plasmas, quasi-neutrality holds ($n_i \approx n_e$) and hence the charge density $[q(n_i–n_e)]$ is very small at large temporal and spatial scales. This means that the Maxwell-Gauss equation is reduced to $\nabla \cdot \mathbf{E} \approx 0$ and it cannot be used to derive the electric field (no unique solution in absence of boundary conditions). An alternative equation needs to be used: the generalized Ohm's law (GOL), which can be derived from the electron momentum equation as follows. For the electron momentum density $n_e m_e \mathbf{V}_e$, its temporal evolution in conservation form is given by

$$m_e \frac{\partial(n_e \mathbf{V}_e)}{\partial t} = - \overbrace{\nabla \cdot \mathsf{P}_e}^{\mathsf{P}_e \text{ gradient}} - \overbrace{\nabla \cdot (n_e m_e \mathbf{V}_e \otimes \mathbf{V}_e)}^{\text{Electron dynamic pressure gradient}} \underbrace{- q n_e (\mathbf{E} + \mathbf{V}_e \times \mathbf{B})}_{\text{Lorentz force}} + \underbrace{n_e m_e \mathbf{a}_{ext}}_{\text{External forces}} + \frac{\delta \mathcal{P}_e}{\delta t} \tag{56}$$

where $\mathbf{a}_{ext} = \mathbf{F}_{ext}/m_e$ with $\mathbf{F}_{ext}$ the set of forces acting upon electrons that are not electromagnetic and independent of the environment, e.g., gravity, P_e and $\mathbf{V}_e \otimes \mathbf{V}_e$ are tensors of rank 2, which are 3 × 3 matrixes in three dimensions, and $\delta \mathcal{P}_e/\delta t$ encompasses all microscopic source and loss of electron momentum, e.g., frictional interactions between electrons and neutrals/ions (*Cravens*, 1997b; *Szegö et al.*, 2000).

The electron number density n_e (a scalar), the electron mean (bulk) velocity $\mathbf{V}_e$ (a vector), and the electron thermal pressure P_e (a matrix) are mathematically defined respectively as the zeroth, first, and second central moments of the electron distribution function over velocity space and are given by

$$n_e(\mathbf{r}) = \int_{\mathbb{R}^3} f_e(\mathbf{r},\mathbf{v}) d^3\mathbf{v} \tag{57}$$

$$V_{e,j}(\mathbf{r}) = \frac{1}{n_e(\mathbf{r})}\int_{\mathbb{R}^3} v_j f_e(\mathbf{r},\mathbf{v})\, d^3\mathbf{v} \quad (58)$$

$$P_{e,jk}(\mathbf{r}) = m_e \int_{\mathbb{R}^3} [v_j - V_{e,j}(\mathbf{r})][v_k - V_{e,k}(\mathbf{r})] f_e(\mathbf{r},\mathbf{v})\, d^3\mathbf{v} \quad (59)$$

$$[\mathbf{V}_e \otimes \mathbf{V}_e]_{jk} = V_{e,j}(\mathbf{r}) V_{e,k}(\mathbf{r}) \quad (60)$$

where **v** is the particle's velocity vector, a microscopic quantity to distinguish from the particle's bulk velocity **V(r)**, a macroscropic (averaged) quantity at the location **r**. j and k are indexes for the different components (e.g., j, k = x, y, z and $d^3\mathbf{v} = dv_x dv_y dv_z$ in Cartesian coordinates, nine possible combinations in three dimensions, but there are different for other geometries). $f_e(\mathbf{r},\mathbf{v})$, in electrons m^{-6} s^3, is a fundamental function that contains all the properties of the electrons: It is the so-called velocity distribution function and corresponds to the density of particles (here electrons) in the volume element $d^3\mathbf{r}$ $d^3\mathbf{v}$ around the position (**r**,**v**) in the space-velocity domain (six dimensions). f_e gives a continuous statistical description of the behavior of the electrons with their number at a given velocity and at a given position. $f_e(\mathbf{r},\mathbf{v})/n_e(\mathbf{r})$ may be interpreted as a *probability distribution function* in the mathematical sense. As such, f_e can be reconstructed from moments at all orders. However, this is impossible in practice: Only a limited set of equations can be solved and moments can be known or approximated. In addition, the temporal evolution of the moment of order n always involves the divergence of the moment of order n + 1 (equation (56)). This is the reason why most mathematical models [e.g., magnetohydrodynamics (MHD) approach] apply closures that approximate the expressions of high-order moments in terms of low-order ones (e.g., see *Grad,* 1949; *Chapman and Cowling,* 1970, for a mathematical description).

Two simplifying assumptions may be done here: steady state and massless electrons. As electrons have a very low mass, they do not have inertia nor contribute to the dynamic pressure. In addition, at comets, forces such as gravity are irrelevant. Within a cometary plasma, the thermal energy is mainly carried by the electrons, while the dynamic pressure is carried by the ions. If we ignore microscopic source and loss terms of electron momenta and with the aforementioned assumptions, equation (56) becomes

$$0 \approx -\nabla \cdot \mathsf{P}_e - qn_e\mathbf{E} - qn_e\mathbf{V}_e \times \mathbf{B} \quad (61)$$

Instead of writing in terms of electron velocities, it is more common to use **V**, the mean plasma velocity, and **J**, the current density, given by

$$\mathbf{V} = \frac{n_i m_i \mathbf{V}_i + n_e m_e \mathbf{V}_e}{n_i m_i + n_e m_e}$$

$$\mathbf{J} = qn_i\mathbf{V}_i - qn_e\mathbf{V}_e$$

and conversely

$$\mathbf{V}_i \approx \mathbf{V} \text{ and } \mathbf{V}_e \approx \mathbf{V} - \frac{\mathbf{J}}{qn_e}$$

if $n_i \approx n_e$ and $m_e \ll m_i$. Equation (61) becomes

$$0 \approx -\nabla \cdot \mathsf{P}_e - qn_e\mathbf{E} - qn_e\mathbf{V} \times \mathbf{B} + \mathbf{J} \times \mathbf{B} \quad (62)$$

that is

$$\mathbf{E} \approx \overbrace{-\mathbf{V} \times \mathbf{B}}^{\text{Motional/ convective}} \overbrace{-\frac{1}{qn_e}\nabla \cdot \mathsf{P}_e}^{\text{Ambipolar/ polarization}} + \overbrace{\frac{1}{qn_e}\mathbf{J} \times \mathbf{B}}^{\text{Hall term}} \quad (63)$$

known as the GOL. It is not the most rigorous demonstration and, for more details, the reader is referred to, e.g., *Somov* (2007) or *Valentini et al.* (2007). This equation is fundamental as it provides a relatively good, unique, and physical estimate of the electric field without solving the Maxwell-Gauss equation at large temporal and spatial scales. The contributions of the different terms have been assessed at low outgassing activity by *Deca et al.* (2019) and are detailed in the chapter by Götz et al. in this volume.

In equation (63), one term is dominating the ion acceleration in the inner coma, within the diamagnetic cavity, and along the magnetic field: the ambipolar/polarization electric field, defined as $\mathbf{E}_{\mathsf{P}_e} \approx -\nabla \cdot \mathsf{P}_e/(qn_e)$. In the inner coma, electron and ion densities are dominated by the population of cometary origin, which only depends on the cometocentric distance (assuming spherical symmetry). Let us extend this assumption to the electron pressure and temperature. In such a case, $\mathbf{E}_{\mathsf{P}_e}$ is only radial and its amplitude is given by

$$E_{\mathsf{P}_e} \approx -\frac{1}{qn_e}\frac{\partial \mathsf{P}_{e,rr}}{\partial r} - \frac{2\mathsf{P}_{e,rr} - \mathsf{P}_{e,\theta\theta} - \mathsf{P}_{e,\phi\phi}}{r} \quad (64)$$

where $\mathsf{P}_{e,rr}$ stands for the electron pressure component along r and $\mathsf{P}_{e,\theta\theta}$ and $\mathsf{P}_{e,\phi\phi}$ stand for both perpendicular components. In spherical symmetry, $\mathsf{P}_{e,\theta\theta} = \mathsf{P}_{e,\phi\phi}$ with θ and ϕ the polar and azimuthal angles in spherical, polar coordinates. If the electron temperature is isotropic with a temperature $T_e(r)$ [i.e., $\mathsf{P}_e = \mathsf{P}_e(r)\, \mathsf{I}_3 = n_e(r)\, k_B\, T_e(r)\, \mathsf{I}_3$ where I_3 is the identity matrix and $\mathsf{P}_e(r) = n_e(r)\, k_B\, T_e(r)$ is a scalar], equation (64) is reduced to (*Cravens et al.,* 1984)

$$E_{\mathsf{P}_e} \approx -\frac{1}{qn_e}\frac{d\mathsf{P}_e}{dr} = -\frac{k_B T_e}{q}\frac{d \log n_e}{dr} - \frac{k_B}{q}\frac{dT_e}{dr} \quad (65)$$

Neglecting the electron temperature gradient and as $n_e(r) \propto 1/r$ from theory and observations (e.g., see *Balsiger et al.,* 1986; *Edberg et al.,* 2015; *Beth et al.,* 2019), one gets (*Vigren et al.,* 2015)

$$E_{P_e} \approx \frac{k_B T_e}{qr} \tag{66}$$

As the electrons often have more energy in the form of thermal agitation (although it may not be valid in the inner coma of 1P), electrons can leave the coma faster than the ions. However, quasi-neutrality should be maintained and $\mathbf{E}_{P_e}$ ensures to keep it that way, in particular along the magnetic field lines as other components act perpendicular to **B**. $\mathbf{E}_{P_e}$ appears mainly radial, oriented outward near the nucleus: It decelerates escaping cometary electrons inward and accelerates cometary ions outward. Conversely, ions coming from outside would be deflected and repelled from the nucleus, while electrons would be attracted and accelerated as they approach the nucleus (see Fig. 12).

Equation (66) should be considered with great caution as it is based upon some assumptions (e.g., constant and isotropic T_e through the coma) difficult to verify with a single spacecraft, above all a negatively charged one. It provides insight on how the electric field and associated potential might behave in the inner ionosphere but should not be applied over large scales (e.g., from the surface to the infinity in terms of cometocentric distances) or at low outgassing activity. Indeed, the associated potential would be in log(r) trapping all cometary electrons, preventing them to reach infinity while cometary ions could escape, leading to a charge imbalance. In addition, PiC simulations performed by *Deca et al.* (2019) at low outgassing activity show that $\mathbf{E}_{P_e}$ is not spherically symmetric near the nucleus (despite having assumed a spherically symmetric outgassing) and hence it cannot be described by equation (66).

From this approach, one would expect ions to be accelerated and exceed the local neutral speed at which they are born in the first tens or hundreds kilometers above the surface. However, this was not the case at 1P, for example. *Schwenn et al.* (1987) showed that the radial component of the ion velocity at 1P remained steady, close to the neutral speed between the closest approach and 20,000 km. This means either that ions and neutrals are collisionally coupled or that ions might not experience any radial acceleration by means of electromagnetic forces. As previously mentioned, $\mathbf{E}_{P_e}$ depends on the electron/plasma temperature. *Eberhardt and Krankowsky* (1995) showed that the plasma temperature in the same region was cold, increasing gradually from 100 K (0.01 eV) at 2000 km to 900 K (0.09 eV) at 8500 km. This must be compared with the kinetic energy of newborn water ions, which is around 0.1 eV. Therefore, at 1P, near the nucleus, newborn ions are slightly supersonic [their initial speed exceeds $\sqrt{k_B(T_i + T_e)/m_i}$]. This translates into the presence of an inner shock ahead of the contact surface (*Cravens,* 1989; *Goldstein et al.,* 1989). As ions decelerate when approaching the contact surface, supersonic ions become subsonic and a shock forms before being deflected. The existence of this shock was anticipated before Giotto (see *Houpis and Mendis,* 1980). However, it was difficult and hard to observe as Giotto was too fast.

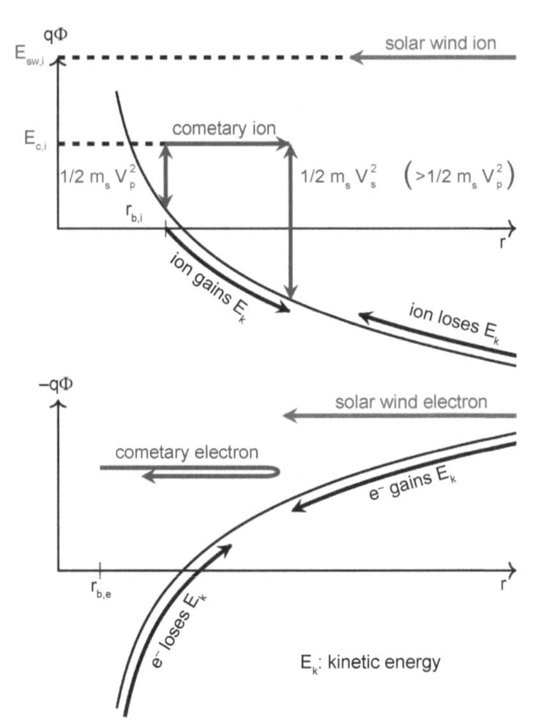

Fig. 12. Schematic of how ions and electrons may gain/lose energy by means of the electric potential ϕ (black line). This potential is driven by the electron pressure and is plotted as a function of the cometocentric distance r. Solar wind electrons gain energy as they fall and dive towards the nucleus, while cometary electrons born close to the nucleus at $r_{b,e}$ (<10–$100\ r_c$) are trapped by the potential well. In contrast, solar wind ions are barely decelerated and cometary ions are accelerated, although the gain in kinetic energy will depend on their birthplace $r_{b,i}$ and the potential profile. Magnetic field and collisions have been ignored here. The red and blue arrows show the direction of motion of the particles.

The situation is different at 67P. Although cold electrons were observed during Rosetta's escort phase, the mean electron temperature was mostly around 5–15 eV. Nevertheless, *Gilet et al.* (2020) showed that within the diamagnetic cavity the electron population was dominated by the cold component (<0.1 eV), consistent with previous observations by *Odelstad et al.* (2018). It is then not clear whether or not ions were subsonic/supersonic at birth, if that depends on the cometary activity and location (e.g., within and outside the diamagnetic cavity) or on the ability of solar wind ions and electrons to access the inner ionosphere. This aspect should be further investigated in the future.

3.3. Diamagnetic Cavity

Among the different regions met at comets and the boundaries separating the different environments covered in the chapter by Götz et al. in this volume, one in particular draws our attention regarding the inner cometary ionosphere: the *diamagnetic cavity*. The adjective "diamagnetic" comes from an analogy to materials that, if subject to an external magnetic field, generate in response an induced magnetic field in the opposite direction in order to reduce or nullify the total magnetic field strength. The diamagnetic cavity is characterized by an almost null magnetic strength (below instrumental detection) and a steep magnetic field gradient at its edge, the latter having different nicknames in the literature: *ionopause*, *contact surface*, or the *diamagnetic cavity boundary layer* (DCBL). This boundary is a tangential discontinuity whose existence was already anticipated by *Biermann et al.* (1967). The diamagnetic cavity was observed at two comets. The first time was during Giotto's flyby on March 13–14, 1986 (*Neubauer et al.,* 1986), when the spacecraft entered into the cavity for two minutes crossing the boundary at around 4500–5000 km (*Neubauer,* 1987). The second time was during the Rosetta mission on July 26, 2015 (*Goetz et al.,* 2016b), followed later by the identification of more than 600 crossings (*Goetz et al.,* 2016a). The detections were performed a few months on either side of the perihelion passage when 67P was most active; however, no detection at the location of Rosetta could be achieved at large heliocentric distances, missing the anticipated shrinking and/or disappearance of the diamagnetic cavity.

The origin of the boundary, the physics at play at the boundary, and how to maintain it were ardently debated in the 1990s following the detection of the diamagnetic cavity at 1P during the Giotto flyby. Debates converged in one idea: As the coma of 1P was extremely dense, the force induced by the magnetic field gradient (pressure and tension) may be balanced by the force exerted by the neutrals on the ions by means of collisions, the so-called *ion-neutral drag*, of which the strength depends on $|\mathbf{V}_i - \mathbf{V}_n|$ (*Cravens,* 1986; *Ip and Axford,* 1987). Indeed, at the microscopic level, as ions and neutrals may have very different velocities and neutrals are much more abundant, collisions tend to reduce the gap between both speeds and force the velocity of ions to be closer to that of neutrals. However, to balance the magnetic pressure in the right direction, ions are required to be slower than neutrals at the boundary and beyond, with, e.g., a null or even inward ion velocity. This strong velocity requirement was in general not fulfilled at 67P during Rosetta's many crossings of the boundary, questioning the validity of the simple ion-neutral drag model at low outgassing conditions. The question of whether or not this region is associated with collisions is open. Nevertheless, statistical studies have provided valuable insight. For instance, at 67P, *Henri et al.* (2017) showed that the DCBL crossings occurred close to the electron-neutral exobase (see section 3.1.2), meaning that collisions between electrons and neutrals might play a critical role in the formation of this cavity. Further evidence that collisions and electron cooling might have a significant role in its formation was the presence of a shock on the inner edge of the diamagnetic cavity at 1P (*Gombosi et al.,* 1996; *Rubin et al.,* 2009). The cause is quite well understood: Ions are born supersonic at 1P (i.e., $V_n \gtrsim \sqrt{k_B T_e/m_i}$) and as they are slowed down by the magnetic field gradient, they undergo deceleration from supersonic to subsonic speeds and a shock forms at the transition. Although V_n is not expected to vary between comets, T_e does and was extremely low at 1P compared with 67P. Such a low value is associated with high outgassing and cooling through collisions with the neutrals.

For a modeling perspective, it is critical to predict whether a diamagnetic cavity is formed and which shape and size it has, as the plasma dynamics in a magnetized and an unmagnetized medium are very different. On the one hand, in an unmagnetized medium, the plasma dynamics is driven by $\mathbf{E}_{P_e}$ (see section 3.2) and symmetries may exist. On the other hand, in a magnetized medium, the symmetry is broken (in a spherically symmetric problem, there is no magnetic field) and the dynamics of ions and electrons become more complex. For instance, ions may travel radially in the former case but not in the latter. In addition, the ability to accurately model the transition between collisionless and collisional media with the same approach is challenging, computationally and mathematically. In the collisional case, kinetic physics including elastic and inelastic collisions is required. Three-dimensional modeling of the DCBL remains a key challenge to tackle in the future.

4. ION POPULATION

As for electrons, several ion populations, discriminated in energy and origin, were identified during past cometary flybys and during the course of the Rosetta mission. They consist of pickup ions, solar wind ions, and cold cometary ions; each may predominate depending on cometocentric distance and/or cometary activity. Instruments of choice to characterize these populations are ion mass spectrometers. Onboard Rosetta, these were (1) ROSINA-DFMS, which provided detailed ion composition of the cometary ionosphere thanks to its high mass resolution mode (see section 2.4.2 for details); (2) RPC-Ion Composition Analyzer (ICA), able to determine the ion distribution function and distinguish in energy, mass, and direction between different populations (e.g., solar wind protons, solar wind alpha particles, pickup ions, cometary ions). In addition, there were the Langmuir probe, RPC-LAP, with the ability to measure the ion current, and the electron-ion analyzer, RPC-IES, which, although lacking the mass determination, measured the total ion fluxes throughout the mission. This section is dedicated to the different ion populations present within the cometary ionosphere in terms of energy and the processes that might affect the latter.

4.1. Collisions vs. Acceleration

In a cometary ionosphere, energy is stored under different forms, either in the electromagnetic fields (B^2/μ_0

and $\varepsilon_0 E^2$) or within the particles, kinetic or thermal. For electrons ($T_e \sim 10$ eV at most; see section 3.1.1) that are light species, the energy is mainly thermal, whereas for ions, it is mainly kinetic.

Cometary ions are produced from neutrals (see section 3.1.1). During an ionization, most of the energy goes to the newborn electron and hence the ion keeps the original momentum of the incident neutral. Thus, newborn ions have an initial speed of 0.5–1 km s^{-1}, namely a kinetic energy of 0.1–0.2 eV for ion masses spanning from 18 to 44 u. If most of the available energy is stored by the electrons in absence of electron and ion collisions, ions should undergo acceleration and "drain" energy from the electrons by means of electromagnetic fields. However, when ion collisions are frequent and ion-neutral chemistry is dominant (see section 2.4.1), ions struggle to accelerate: An ion may collide with a neutral (e.g., charge exchange or proton transfer); the ion, which may have accelerated, becomes a neutral, and the neutral a cold ion (or one of its fragments), reducing the energy of the ion present albeit not affecting the total plasma charge. Although the produced ion may have a different mass from the original one, the total plasma momentum is reduced overall. In a collisional coma (high outgassing rate), another reason why ions have difficulty being accelerated by electromagnetic fields is because the latter is weak: The region usually corresponds to the diamagnetic cavity (i.e., low or null magnetic field), dominated by cold electrons, yielding a weak electric field from the weak electron pressure gradient (see section 3.3).

At Comet 1P during the Giotto flyby at 0.89 au, ions were not accelerated in the inner ionosphere (*Schwenn et al.,* 1987), either because the coma was dense and/or cometary electrons were too cold for causing a large enough electron pressure-gradient-driven field. Acceleration by ions was observed beyond the diamagnetic cavity, especially from 20,000 km (*Altwegg et al.,* 1993; *Rubin et al.,* 2009). What is the situation at 67P? During low outgassing at large heliocentric distances (beyond 2 au), a multi-instrument analysis linking datasets from the Rosetta Plasma Consortium (RPC-IES, RPC-MIP, RPC-LAP) and ROSINA [Cometary Pressure Sensor (COPS), DFMS] by a physics-based model (see section 2.4.1) gave excellent agreement in terms of ionospheric density within the instrument uncertainties (*Galand et al.,* 2016; *Heritier et al.,* 2017a, 2018; *Vigren et al.,* 2019). The model assumes that the ions are moving radially at the speed of neutrals, i.e., they have not suffered from any acceleration between their birth place and the location of Rosetta.

Why the models work so well at large heliocentric distances/low outgassing activity (e.g., *Heritier et al.,* 2018) is not entirely clear, as there are theoretical as well as observational indications for the acceleration of newborn ions, flowing at speeds around an order of magnitude above the neutral gas speed. Based on theory and models, $\mathbf{E}_{P_e}$ must accelerate the ions away from the coma as they are born at subsonic speeds in the case of 67P. Therefore, are the ion-neutral collisions present at 67P, as discussed in section 2.3, sufficient to limit/prevent this acceleration? Simulations by *Vigren and Eriksson* (2019) indicate that this was not the case at the location of Rosetta even at perihelion, the most active outgassing phase. From RPC-ICA observations, *Bercic et al.* (2018) estimated the outward flow of low-energy cometary ion population during a one-month period when Rosetta was at 28 km from the nucleus at 2.5–2.7 au (low outgassing rate), around 6 km s^{-1}. Similar results were also reported by *Nilsson et al.* (2020). Ion velocities from combined analysis of RPC-LAP and RPC-MIP data (and by two different approaches) have also shown similar ion speeds (*Johansson et al.,* 2021).

While these observations appear to be consistent with the expected effect of the ambipolar electric field, one has to be aware of the technical problems with observing low energy ions (a few electron volts) by detectors on a spacecraft charging down to negative potentials, typically –5 to –20 V (see section 3.1.1), accelerating the ions to similar energies. The negative spacecraft potential is believed to be somehow beneficial for RPC-ICA measurements: All the ions must reach the spacecraft and be detectable as the minimum energy threshold of RPC-ICA was below $-qV_{SC}$. Simulations of the ion measurements based on the Spacecraft Plasma Interaction System (SPIS) did not clearly identify the reason(s) why low-energy ions should not reach the RPC-ICA detector (*Bergman et al.,* 2020, 2021). Some possibilities include the obstruction of its 360° × 90° field of view, the limited angular coverage in some operational modes, and/or large uncertainties regarding the geometric factor at low energies, i.e., the factor that allows conversion of counts from the instrument to physical quantities such as the ion differential flux in this case. However, these simulations have also shown that ions, particularly low-energy ones, had very distorted trajectories around Rosetta prior to their detection. Regardless, it is clear that the plasma density (the zeroth order moment) from the ion distribution measured by RPC-ICA is 1 to 2 orders of magnitude lower than those measured by RPC-MIP and RPC-LAP, considered as baselines, such that the direct measurements of the low-energy ions on and at Rosetta is still an open question.

The ion speed derived from combining both measurements by RPC-LAP and RPC-MIP was more indirect. The ion current measured by RPC-LAP is combined with the plasma density from RPC-MIP, assuming only one ion species at 18 or 19 u [an assumption far from being true, especially near perihelion, when many ions coexist and might have different speeds (*Beth et al.,* 2020)], to derive an effective speed, regardless of the direction (*Johansson et al.,* 2021). The only indication of flow direction by RPC-LAP was the observation of a dropout in the ion current observed when the spacecraft itself obstructed the ion flow, preventing the ions from reaching the probe. It occurred a few times when 67P, the spacecraft, and one of the RPC-LAP probes were aligned (*Odelstad et al.,* 2018), as expected for a radial outflow. While completely excluding effects of the spacecraft potential on RPC-LAP is difficult, errors sufficient to explain the large difference from the speed of the neutral gas are equally difficult to conceive. Some might be still inherent to the analysis of the measurements of the Langmuir probes. Indeed, deriving plasma properties from I–V curves has never been a straightforward and easy task and relies on assumptions. For instance, it is common to derive the effective ion speed (thermal and bulk speeds are indistinguishable) based on strong assumptions (see the seminal paper by *Mott-Smith and Langmuir* 1926): a *single* ion population and specific distribution functions (e.g., ring or Maxwellian distribution), drifted or not, in the orbital motion limited (OML) regime (the Debye length is infinite compared with the probe's size, which is not always the case), and for an unmagnetized plasma (while at low outgassing, the environment is magnetized). Depending on the ratio between the probe's radius and the ambient Debye length, the estimates of the ion effective speed may be very different (*Laframboise,* 1966) and can differ significantly from the bulk ion speed.

The multi-instrumental approach proposed at large heliocentric distances is not applicable near perihelion due to higher outgassing rates. Additional processes have to be taken into account. For instance, chemical loss through electron-ion dissociative recombination needs to be included (besides loss through transport) as well as attenuation of the EUV solar flux. Although the effect on the electron density of photoabsorption by the neutral gas is ruled out at the location of Rosetta (*Heritier et al.,* 2018), *Johansson et al.* (2017) observed attenuation of the EUV solar flux near perihelion. This was attributed to dust, lessening the ionization rate and causing in part this overestimation as models did not take into account dust absorption. Finally, large ion speeds were also reported (*Vigren et al.,* 2017; *Odelstad et al.,* 2018) implying that acceleration takes place. If confirmed, this would reduce the electron density locally (*Vigren et al.,* 2019) as the models are based on flux conservation.

The apparent discrepancy at low activity between the success of the multi-instrument analysis, assuming $V_i \approx V_n$ to reproduce the measured ion density, and the observations, indicating much higher ion speeds, needs to be solved. The physics of the interplay of ion-neutral collisions in an inhomogeneous gas with an inhomogeneous and dynamic plasma, bathed in electromagnetic fields varying in space and time, is a very rich and complex matter, with much still to be explored. On the one hand, hybrid simulations may include ion-neutral collisions, but as they lack the electron kinetics, much of the detailed physics causing acceleration is partially lost (e.g., *Koenders et al.,* 2015). On the other hand, PiC simulations may include collisions either self-consistently or by application of them as an extra layer somewhat along the lines of *Stephenson et al.* (2022). This approach is expected to be more and more pertinent in the coming years.

4.2. Pickup Ions

In a region where the solar wind flows, the newborn, cometary ions, initially at rest with respect to the ambient plasma, may feel the solar wind convective electric field $\mathbf{E}_{sw}$ and start being accelerated by it. This process, called pickup, results, in a kinetic sense, in these accelerated ions of cometary origin having cycloid trajectories in the $\mathbf{E} \times \mathbf{B}$-drift direction, where $\mathbf{B}$ is the local magnetic field. The implicit assumption is that solar wind flow and magnetic field directions are not aligned, so that the cross product is non-zero, otherwise the pickup process would be rendered negligible. In velocity space, the accelerated pickup ions form so-called "ring"-shaped velocity distribution functions (VDFs), with a maximum velocity of the picked-up particles on the order of $2V_{sw} \sin \xi$, where ξ is the angle between the local solar wind magnetic field $\mathbf{B}$ and solar wind velocity $\mathbf{V}_{sw}$. Pitch-angle scattering further distributes the energy in phase space, transforming the initial ideal pickup ring or partial ring VDFs into bispherical shell VDFs (*Coates,* 2004). Such VDFs, from ring-beam to bispherical shells, have been observed at comets including 1P and 67P (e.g., *Coates et al.,* 2015; *Nilsson et al.,* 2015; *Goldstein et al.,* 2015), denoting several stages of evolution in the pickup ion process. In turn, these distributions are unstable and generate waves heating the ambient plasma (*Wu and Davidson,* 1972). A more detailed outlook of pickup ions, their effect on large plasma structures, and wave-particle interactions are given in the chapter by Götz et al. in this volume.

The pickup process has a drastic effect on the solar wind flow (*Glassmeier,* 2017). As the energy and momentum are directly transferred from *light* solar wind to *heavy* cometary ions, two effects simultaneously take place: To conserve energy, the solar wind is slowed down upstream of the cometary nucleus, and, to conserve momentum, is deflected in the direction opposite to the pickup ions' motion. Moreover, due to the instability of the VDFs created in the pickup process, part of the transferred energy is diverted toward the excitation of low-frequency waves (ion cyclotron waves, Alfvén waves, etc.), which in turn heat the plasma and produce local turbulence. Momentum transfer between solar wind and accelerated cometary ions can be particularly encouraged when VDFs are non-gyrotropic, further feeding local instabilities. Because of the implantation of heavy ions into the solar wind flow, the pickup process thus contributes to the mass-loading of the solar wind (*Szegö et al.,* 2000), an idea first put forward by *Biermann et al.* (1967).

4.3. Solar Wind vs. Ionospheric Plasma

The interaction between the solar wind and the cometary neutral environment gives rise to several coexisting populations of ions, which may take precedence over one another in the inner coma depending on heliocentric distance (and hence, nucleus activity) and on cometocentric distance. What are the conditions for cometary ions (at high or low energy) to prevail over solar wind ions? Historical missions composing the Halley Armada or more recently Rosetta give us clues to answer this question.

The transition region where cometary ions gradually become dominant is usually called the *cometopause* (*Cravens and Gombosi,* 2004), and sometimes *protonopause* (*Sauer et al.,* 1995). It has historically been defined in two ways (see also the chapter by Götz et al. in this volume):

1. the *charge-exchange collisionopause,*
2. the inner region delimited by $n_{sw} \approx n_{ci}$ (or $F_{sw} \approx F_{ci}$), where n (respectively, F) are the total number densities (energy and/or momentum fluxes) of solar wind ions (sw) and of cometary ions (ci, referred to as solely i in section 2).

In definition 2, the cometopause may not be a sharp boundary but rather a smooth crossover transition between dominating species in the plasma; in the case when solar wind plasma is shocked and thermal speeds of cometary and solar wind ions are similar, density and flux ratio definitions are equivalent. In fact, the presence or not in the data of a cometopause at 1P and its physical relevance were highly debated at the time of its first discovery (*Gringauz et al.,* 1986; *Reme et al.,* 1994). *Gombosi* (1987) introduced a typical ion-neutral collision scale linked to the charge-exchange collisionopause (definition 1)

$$\ell_{CX} = \frac{\sigma_{p,s}^{sw,CX}(E_{sw})Q}{4\pi V_p} \tag{67}$$

resulting, from the point of view of the ion composition, in the conversion of fast (but already decelerated) light solar wind ions sw to slow heavy cometary ions. For a solar wind speed of 300 km s^{-1}, a charge-exchange cross section of protons in water of 2×10^{-19} m^2 (*Simon Wedlund et al.,* 2019c), Q ~ 10^{26}–10^{30} s^{-1} (representative of activity levels of 67P and 1P), and $V_p \sim 600$ m s^{-1}, $r_c \lesssim \ell_{CX} \lesssim 2.5 \times 10^4$ km. In turn, the subsolar position of the cometopause was defined by *Gombosi* (1987) as

$$R_{CX} \sim \sqrt{\ell_{CX} r_0} \tag{68}$$

where r_0 is the distance at which the flow speed of the shocked solar wind ions decreases down to the mean solar wind proton thermal speed. This is due to solar wind mass-loading behind the bow shock once the latter has formed (and which is situated thousands of kilometers upstream of the nucleus). In the case of 1P, $r_0 \sim 1.6 \times 10^5$ km, so that $R_{CX} \sim 6 \times 10^4$ km, possibly consistent with off-Sun-comet-axis Giotto and Vega 2 observations around 1.5×10^5 km along their respective path (*Gombosi,* 1987). At 67P around perihelion (Q ~ 5×10^{27} s^{-1}), numerical hybrid simulations predicted $R_{CX} \geq 800$ km using definition 2 to image the density crossover (*Simon Wedlund et al.,* 2017; *Alho et al.,* 2021) with $\ell_{CX} \sim 130$ km (from equation (67)), which implies $r_0 \sim 5000$ km (calculated by inverting equation (68)).

Two intimately intertwined aspects drive the evolution of the total plasma ion composition: cometocentric distance on the one hand, and outgassing rate on the other (for a description of the cometary ion composition, see section 2.4.2). The more the coma shrinks as the result of a smaller outgassing rate, the deeper into the coma must a given spacecraft venture to observe a relatively comparable ion composition. As a rule, H_2O^+ and CO_2^+ are the most prominent cometary ion species in the inner ionosphere whatever the outgassing rate, provided that we are at a sufficient distance from the nucleus, typically above $R_{s,n} \sim R_{CX}$ (see equation (45) and section 2.4.2). Let us illustrate this with two examples from the Rosetta mission: one at large heliocentric distance (low outgassing), one near perihelion (high outgassing). What would a mass spectrometer see? As a guide, typical ion spectrograms, captured during the Rosetta mission, of the solar wind ions and of the cometary ions are displayed in Fig. 13.

From the start of the activity to medium activities, the neutral outgassing rate of a typical Jupiter-family comet (*Lowry et al.,* 2008), such as 67P, is on the order of 10^{25}–10^{26} s^{-1} (see Fig. 2) (*Coates and Jones,* 2009; *Hansen et al.,* 2016). Under such conditions, $R_{CX} \sim r_c$ and solar wind ions are expected to have energy and momentum fluxes (although not necessarily densities) at least equal to those of cometary ions, as they pass almost undisturbed through the comet environment, with minimal mass-loading effects such as deflection and slowing down. However, the onset of magnetospheric-like behaviors, with the production of cometary pickup ions and sporadic solar wind charge-exchanged products, such as He^+, can already be seen from $Q \geq 10^{26}$ s^{-1}, as demonstrated by Rosetta's ion energy spectrograms (from RPC-ICA and RPC-IES) during both inbound and outbound conditions (*Nilsson et al.,* 2015, 2017; *Goldstein et al.,* 2017; *Simon Wedlund et al.,* 2019b). Of note, negatively charged H^- ions, thought to be arising from double charge exchange of solar wind protons, were also observed in the electron channel of RPC-IES (*Burch et al.,* 2015a). Thus, a lot of the total ion energy and momentum flux is still contained in the solar wind ions at relatively small cometocentric distances (*Williamson et al.,* 2020), such as those probed by Rosetta ($r \leq 150$ km). From Rosetta's point of view, the cometopause transition, seen as a gradual flux crossover between cometary and solar wind ions, occurred for $Q \approx 2$–5×10^{26} s^{-1}, i.e., in and around January 2015 (inbound leg of 67P's orbit around the Sun) and February 2016 (outbound leg) (see the chapter by Götz et al. in this volume).

At medium to high outgassing activity, typically for $Q \gtrsim 10^{27}$ s^{-1}, corresponding to closing in on perihelion conditions for 67P, $R_{CX} \gtrsim 500$ km, and the angular deflection of the solar wind ahead of the obstacle may become so large that the solar wind is effectively kept from entering the inner coma. This region, where no solar wind ions can be detected outside of occasional solar-wind-driven events, such as CMEs, was dubbed by the Rosetta team the *solar wind ion cavity* (SWIC). Rosetta stayed in this

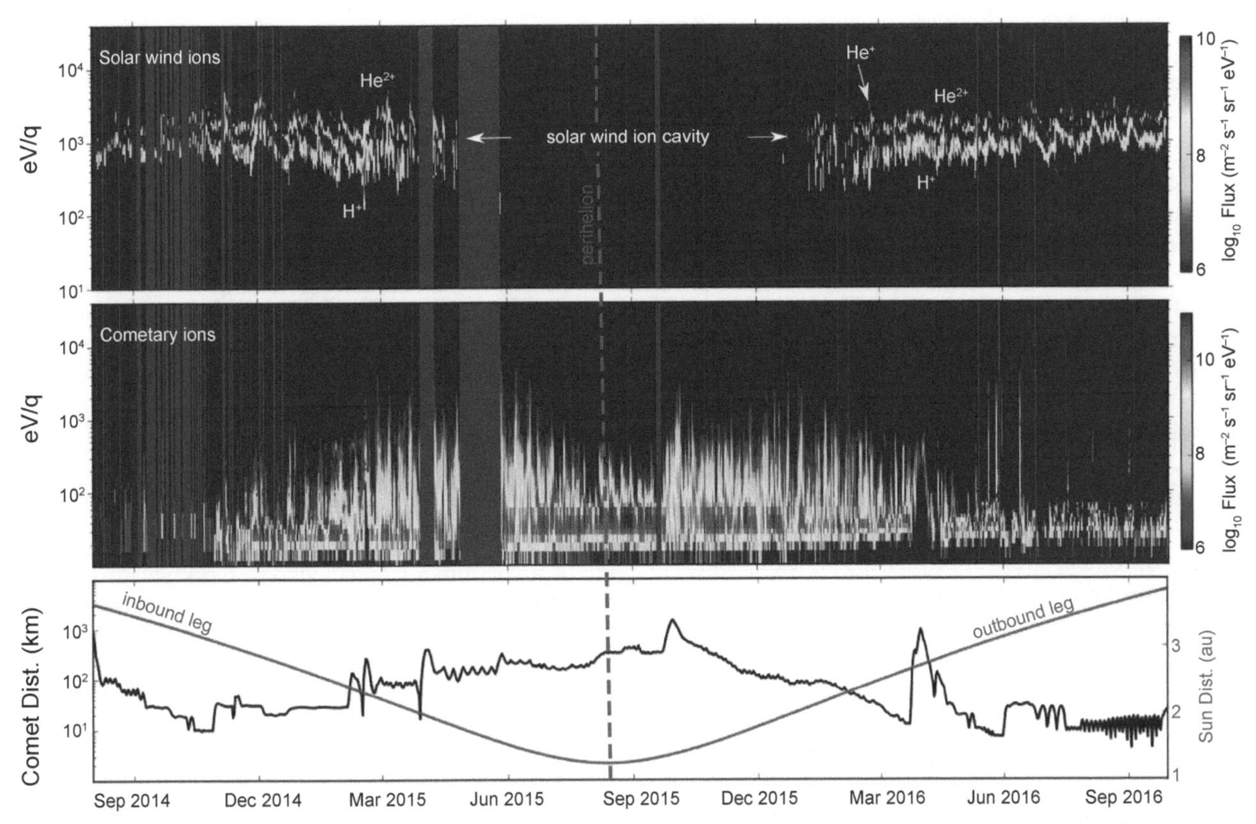

Fig. 13. Overview of the solar wind (top) and cometary ion (middle) energy spectrograms during the Rosetta mission, as measured by RPC-ICA and angle-integrated over 1 hr. Heliocentric and cometocentric distances are shown for reference on the bottom panel. At this low temporal resolution, solar wind protons (around $E_p/q \sim 1$ keV q^{-1}) and alphas (around $2E_p/q$ energy, fourfold heavier but doubly charged) can be clearly identified, whereas charge-exchange products He^+ (around $4E_p/q$) are more seldom seen. The solar wind ion cavity spans April–December 2015, with no detection of solar wind ions during that period. Data gaps are highlighted as gray zones. Note that the color bars on the first two panels do not have the same upper bounds. Adapted from Fig. 1 of *Nilsson et al.* (2017).

region from around April to December 2015 (*Behar et al.,* 2017) (see also Fig. 13). In it, the heavier, slower plasma of cometary origin, produced locally by photoionization and electron impact ionization, dominated the total ion composition and fluxes. Hence, for these relatively high outgassing rates, Rosetta was always inward of the cometopause, with the SWIC possibly being its inner edge (*Mandt et al.,* 2019). This is mainly because (1) Rosetta orbited close to the comet at any given time (<500 km outside temporary excursions) and (2) RPC-MIP and RPC-LAP, instruments measuring n_{ci}, could not probe plasma densities below ~50 cm^{-3}. As the solar wind density is typically $n_{sw} \approx 1$–10 cm^{-3}, it was not possible to locate where $n_{sw} \approx n_{ci}$. Thus, in a way similar to onions, the cometopause successively encompasses the ion-neutral and electron-neutral collisionopauses, where plasma-neutral collisions start to play a major role in the coma, but also the diamagnetic cavity once it forms. For more discussion on the formation of large-scale boundaries including the cometopause, the diamagnetic cavity, the collisionopauses, and the solar wind ion cavity, the reader is directed to the chapter by Götz et al. in this volume.

5. DUSTY (DIRTY) COMPLEX PLASMA

When we look at a comet in the night sky, our eyes mostly detect sunlight reflected by dust grains emitted from the comet nucleus, forming the coma and the dust tail (usually the more visible of the two tails; see the chapter by Agarwal et al. in this volume). This obvious presence of dust in the same volume of space as the cometary plasma immediately suggests that the coma and dust tail may be regions where there are lots of interactions between the dust and the plasma. The plasma may charge the dust, which in turn can influence the motion of the plasma, coupling them together in what is known as a *dusty* or *complex plasma.*

In this section, we will only provide a very limited introduction to the range of possible effects of dust-plasma effects at comets (section 5.1) and the mechanics of dust charging (section 5.2). The main reason for spending so little space on this is very simple: In contrast to the dust-free physics of a cometary ionosphere, presented in this chapter until now, and against many expectations before the mission arrival at 67P, the analysis of the Rosetta data has as yet not provided any entirely new picture of dust-plasma interactions at comets. As we will see in section 5.3, some dust-plasma interaction signatures have been identified in the Rosetta data. But the same section will show that the rich Rosetta datasets on plasma and cometary dust actually suggest such interactions probably have only a small impact on the plasma, at least for most of the mission and for the regions visited around this particular comet. For a more extensive treatment of dust charging and its possible effects on the comet plasma environment, we refer to previous literature on the subject (see, e.g., *Mendis and Horányi,* 2013).

5.1. Why Care About Dust Charging?

Consider free electrons moving around in the coma. If there are lots of dust grains, an electron might now and then hit one of the grains, perhaps knocking out one electron (in the process known as *secondary emission*) and thereby increasing the number of *free* electrons and charging the dust grain positively. The dust grain would then effectively become an enormously heavy positive ion. Alternatively, the impacting electron might stick to a dust grain, decreasing the number of free electrons and charging the dust negatively. The charge acquired on the dust grain suddenly couples it to the electromagnetic fields around it, and it will itself influence these fields and thereby other charges. Both plasma and dust interact as seen in Fig. 14.

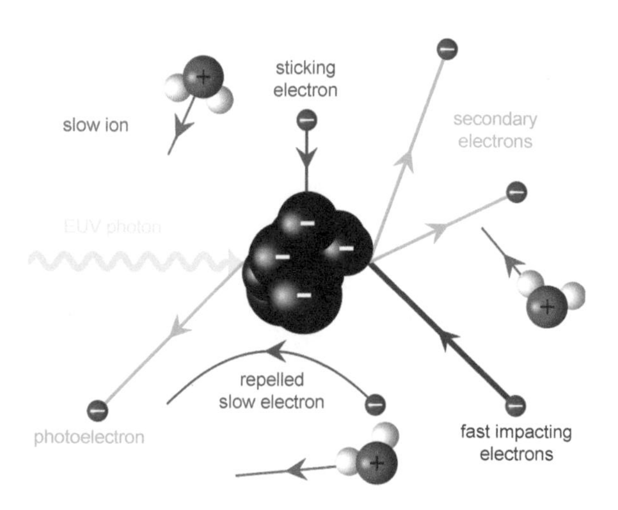

Fig. 14. Schematic of various effects influencing the charging of a dust grain (conglomerated black spheres). n_i and n_e are assumed equal within the figure (not every ion or electron is represented) but the higher thermal speed of the electrons leads to more hits on a dust grain thereby charging the grain negatively when impacting electrons stick to the grain. Sufficiently energetic free electrons (magenta line) and solar EUV photons (yellow) can kick out electrons from the grain (green lines), gradually shifting the total charge toward positive values and leading to a non-uniform charging of the grain.

Two main effects may ensue from dust charging and variations in the number density of free electrons. The first one affects the collective behavior of the plasma and its wave modes, such as the frequency at which it naturally pulsates, $\omega_{pe} \propto \sqrt{n_e}$ (see equation (53)). One can show that the decrease in the free-electron number density due to attachment to dust grains changes the dispersion relation of *ion acoustic waves* in such a way as to push them to higher frequency, also decreasing their Landau damping (*Barkan et al.,* 1996). It also allows a completely new wave mode, known as *dust acoustic waves*, which, because of the huge dust grain inertia, propagates much slower than usual plasma waves (*Merlino,* 2014). Moreover, electromagnetic modes can be influenced as the charged dust provides the plasma with a higher inertia and with a continuous distribution of dust masses: This may be interpreted as a continuum of cyclotron resonances. However, due to the large dust grain mass (see the chapter by Engrand et al. in this volume for a discussion of dust size distribution) and weak magnetic field (on the order of 10 nT) at a comet outside 1 au, the effects on electromagnetic waves are expected only at very low frequency such as to not be of practical relevance. Consequently, acoustic modes are the most relevant wave modes at comets. For a discussion of the physics related to cometary plasma waves (albeit not dust-related), see the chapter by Götz et al. in this volume.

But dust charging can also affect the dynamics of the dust. From the point of view of cometary evolution, the most important of these effects is the so-called *levitation,* in which dust grains can be expelled from the cometary nucleus by electrostatic repulsion if they and their immediate neighborhood are charged negatively when bombarded by electrons from the surrounding plasma. This effect could potentially be important as it may increase the nucleus dust production from the surface. We refer to the chapter by Agarwal et al. in this volume for further discussion on processes that lift off dust grains from the nucleus. In contrast, farther out from the nucleus' surface, a charged dust grain is subject not only to the forces of gravity and radiation pressure (i.e., a momentum transfer between the impacting solar photons and atoms, molecules, and dust grains in the coma) acting on any grain, but also to the electromagnetic fields in the plasma. Due to the large dust grain inertia, fields acting over large scales in space and time are most significant: We will see below that this effect actually was observed *in situ* by Rosetta. In addition, the attachment of charges (electrons) to a dust grain causes a repelling electrostatic force between various parts of the grain itself. This force may contribute to erosion of the grain, either by more or less continuous ablation with small fractions of the grain splitting off or fragmentation, when the grain breaks in two (or more) pieces of approximately the same size. Note, however, that other processes may also lead to the same result (see the chapter by Engrand et al. in this volume).

5.2. Dust Charging

What are the effects influencing the charge of a dust grain? As discussed earlier, a plasma consists of free charges, positive and negative, moving around at random velocities distributed in accordance with their temperatures (T_s for ion species s or T_e for free electrons) and masses (m_s and m_e). One of their typical speeds is the thermal speed $v_{th} = \sqrt{k_B T/m}$ corresponding to the standard deviation around their mean bulk velocity **V**. Ions and electrons have masses that differ by about 4 orders of magnitude ($m_s/m_e \sim 3.3 \times 10^4$ for water-group ions, typical for a cometary coma), while their temperatures are usually alike. In addition, in the inner cometary ionosphere, $|\mathbf{V}_s| \gg v_{th,s}$ for ions whereas $|\mathbf{V}_e| \ll v_{th,e}$. As a consequence, electrons typically move much faster (typically a few hundred times faster) than ions such that in a given time span, a dust grain (or any "body" in the ionized cometary gas) will be hit more frequently by electrons than by ions. The grain thus preferentially charges negatively, until reaching an equilibrium potential V_d sufficient to stop further charging by repulsion of electrons and catching of ions. This potential can be estimated as

$$V_d \sim -\frac{3k_B T_e}{q} \tag{69}$$

It can be noted that these considerations for dust grains apply similarly to a spacecraft, so that the spacecraft potential (discussed in section 3.1.1) is in principle set by a similar equation. However, analysis and modeling have shown that due to the design of the Rosetta spacecraft, its electrostatic potential in the inner coma was instead determined by the plasma density (*Johansson et al.,* 2021).

From the grain potential V_d, we can calculate its charge

$$q_d = C_d V_d \tag{70}$$

where C_d is the dust's capacitance. The latter can be estimated, or at least a minorant may be found, by considering a spherical grain of radius r_d in vacuum such that

$$C_d = 4\pi\varepsilon_0 r_d \tag{71}$$

Therefore, if all grains have the same size and are all spherical, the total charge density ρ_d (in A s m^{-3}) carried by the dust is obtained by combining equations (69)–(71)

$$\rho_d = q_d n_d \sim -12\pi\varepsilon_0 r_d n_d k_B T_e/q \tag{72}$$

where n_d is the number of dust grains per unit volume.

In reality, the attachment of free electrons is not the only factor influencing the grain potential (Fig. 14). The emission of electrons from sunlit dust grains due to the photoelectric effect will tend to charge the grains positively whereas, if the impacting flux of free electrons is sufficiently high, the negative charging by electrons attaching to grains will dominate. In practice, the limiting density is on the order of 10^2 cm^{-3} at 1 au (scaling with r_h^{-2}), depending on the dust grain properties. In the dense parts of a coma, we expect to find mainly negatively-charged dust grains. Another major effect is secondary emission, occurring when an electron of sufficiently high energy kicks out one or more electrons from a grain. It is typically important when the incoming electron has an energy of approximately hundreds of electron volts and does not matter much below 10 eV, although this depends on the detailed dust properties, often not very well known (size and mass distributions, shape, composition).

A further complicating factor is the time it takes to charge a dust grain to its equilibrium potential V_d (equation (69)). The total electric current collected by the grain I_e (A), due to the randomly moving plasma and mainly driven by the free electrons, is approximately given by $I_e \approx J_e S_d$ where $S_d = 4\pi r_d^2$ (m^2) is the grain surface and J_e (A m^{-2}) is the electron thermal current density, driven by free electrons moving toward the grain. J_e is given by (e.g., *Mott-Smith and Langmuir,* 1926)

$$J_e = -q n_e \sqrt{\frac{k_B T_e}{2\pi m_e}} \tag{73}$$

Determining q_d as a function of time requires solving a differential equation, exactly as for a resistor-capacitor circuit. The characteristic time, or time constant, for charging a dust grain from 0 to V_d is given by

$$\mathcal{T}_d \sim \frac{C_d V_d}{I_e} = 3\varepsilon_0 \frac{\sqrt{2\pi m_e k_B T_e}}{n_e q^2 r_d} = 3\sqrt{2\pi}\frac{\lambda_{Debye}}{r_d \omega_{p,e}} \tag{74}$$

where λ_{Debye} and $\omega_{p,e}$ are defined by equations (53)–(54). Each dust grain can be seen as a capacitor and a resistor in series with its characteristic capacitance C_d and resistance $R_d = V_d/I_e$ (Ω), depending not only on their size and shape but also on the local electron number density and temperature. The current is provided to this circuit by means of free electrons in the plasma falling on the dust grain. From equation (74), we note that large grains charge up faster than small ones (as $C \propto r_d$ and $I_e \propto r_d^2$; bigger grains collect more electrons). Only one of the most relevant dust charging processes at comets has been described here and a more exhaustive description is given in *Meyer-Vernet* (2013).

How long does a dust grain take to reach $\sim V_d$ for typical cometary plasma parameters? For a grain of radius $r_d = 10$ μm in a plasma with $n_e = 10^3$ cm^{-3} and $T_e = 10$ eV, we get $\mathcal{T}_d \sim 10$ s, while a 100-nm "nanograin" requires $\sim 10^3$ s (more than 15 minutes). During this time, the grain may travel through plasmas of quite varying characteristics, particularly in the inner coma [see the chapter by Agarwal et al. in this volume and *Marschall et al.* (2020) for discussions of dust grain velocities]. At some distances, the different $\mathcal{T}_d$ between grains associated with the outward dust flow through a plasma such that $n_e(r) \propto 1/r$ can lead to complex situations where small grains may conserved the negative charge acquired closer to the nucleus (long $\mathcal{T}_d$) while larger grains may quickly charge positively by means of photoemission (short $\mathcal{T}_d$). In addition, the secondary emission, which tends to be more efficient for small grains, can lead to similar results. As dust grains may travel at different speeds and take different trajectories through the plasma, they experience different charging along their path. All this leads to an expectation of a distribution of charge states even for similar dust grains — and, as is clear from the chapter by Engrand et al. in this volume, there is a large variation among dust grain properties.

In these circumstances, is it possible to estimate the fraction of the free electrons that will attach to dust grains? This is an important question, because this fraction must be significant in order for the dust charging to have any considerable effect on the plasma. Clearly, the answer must depend on n_e and n_d. However, the dust size distribution is also crucial in this respect. For example, let us assume that all dust grains have the time to reach their equilibrium potential as given by equation (69). The charge carried by a grain depends on its capacitance, which scales linearly with its size (as seen in the example of a sphere; equation (71)). Contrarily, the mass of a grain scales with its volume $\mathcal{V}_d \propto r_d^3$. As a consequence, for a fixed total dust mass, the plasma is more depleted in free electrons and dust charging is more efficient if the dust is mostly made of lots of small grains than of a few large ones (*Vigren et al.,* 2015). For a given dust size distribution, detailed calculations can be used to estimate the total amount of dust charging.

As the size distribution of dust in the innermost coma of Comet 67P is quite well constrained by the dust instruments (see the chapter by Engrand et al. in this volume), *Vigren et al.* (2021) carried out such calculations and constrained the possible electron depletion in this environment. Their model, combining observed plasma densities and electron energies with dust size distributions derived from the dust observations at Rosetta by means of equation (72), shows that because of the preponderance of large grains, the fraction of electrons attaching to dust grains must have been small or even negligible during most parts of the Rosetta mission. Quite extreme assumptions on the dust distribution had to be taken even to push the electron depletion to a few percent. A similar negative conclusion was reached by *Vigren et al.* (2022), using analogous arguments, for positive charging of dust grains.

There may be exceptions at specific locations and times, as in outbursts [see, e.g., the February 19, 2016, event studied by *Grün et al.* (2016)], which could lead to a locally high density of small grains, although no clear signatures of obvious electron depletion have yet been identified in the Rosetta dataset. An apparent absorption of solar EUV radiation has also been used to argue that the dust size distribution at large distances from Rosetta (several thousand kilometers or more) may include larger amounts of grains down to the size of tens of nanometers (*Johansson et al.,* 2017). As few such grains have been observed close to the nucleus, they have presumably resulted from one or several erosion and fragmentation processes (see discussion in section 5.1 and the chapter by Engrand et al. in this volume), which could possibly lead to stronger dust-plasma coupling farther from the nucleus than the regions investigated by Rosetta. One may note that very small dust grains, with masses down to 10^{-23} kg, were observed by the Vega spacecraft $\sim 10^5$ km from the nucleus of Comet Halley (*Sagdeev et al.,* 1989).

While there may still be much to learn from the Rosetta dataset, the combination of the observed average properties of dust and plasma quite clearly indicates that the typical electron depletion in the innermost coma of 67P is small, with correspondingly weak impact of the dust on the overall plasma dynamics. However, charged dust grains of very small size were indeed observed in some events, as discussed in section 5.3.

5.3. Charged Dust Dynamics at Comet 67P

How do charged dust grains move in the coma? Let us start by considering neutral dust grains. Close to the nucleus, the grain is dragged by the outflowing cometary gas. As the neutral gas density decreases with cometocentric distance (see equation (3)), the neutral drag force decays rapidly with distance and is hence stronger close to the nucleus. The very weak gravitational force of a comet nucleus can also mostly be neglected for most dust grains farther outside [however, see *Davidsson et al.* (2015) for examples of gravitationally bound dust motion around Comet 67P]. In addition, the solar radiation pressure impresses a constant force on every grain in the anti-sunward direction. Grains emitted from the nucleus in the direction of the Sun can therefore be turned back by the radiation pressure and return toward the nucleus and farther out downstream of the comet. The radiation pressure force on a grain depends on r_d^2, while the inertia scales with its volume, through its mass, which is $\propto r_d^3$. Therefore the radiation pressure affects more efficiently the dynamics of small/light grains than that of large/heavy grains. Other effects, such as Mie scattering and Poynting-Robertson drag, are not discussed here: Fuller details are given in *Burns et al.* (1979).

Charged grains are, in addition, subject to electromagnetic forces arising in the comet-plasma interaction, discussed in the chapter by Götz et al. in this volume. Due to the large mass of dust grains, compared with cometary ions and electrons (in the gas phase) [even small 10-nm grains are estimated to have a mass on the order of 10^5 u by *Gombosi et al.* (2015)], the gyroradius of dust grains generally becomes so large that the grain's gyromotion can be neglected within the inner coma explored by Rosetta at 67P. Nevertheless, the magnetic field indirectly leads to the appearance of a convection electric field, which may efficiently affect the dust dynamics. If this force is constant over the region traversed by the dust grain, the dust's trajectory will describe a parabola of which its axis of symmetry is along this force. However, electric and magnetic fields change in time, and, as mentioned in section 5.2, the charge of a grain may evolve as well, as it travels through plasma regions of different $J_e(\mathbf{r})$, while grains eventually erode and/or fragment, which yields changes in the dust's mass, size, charge, and number. Therefore the picture of motion in a conservative field must be applied with caution.

The RPC-IES spectrometer was designed for the measurement of (positive) ions and electrons. However, it was able to detect charged dust grains. As there was no possibility for direct mass separation, charged dust signatures could not be unambiguously separated from those of ions and electrons by IES. That said, some signatures stand out from ordinary charged particle signatures and better fit expectations for charged dust. *Burch et al.* (2015b) found a few events with negatively charged grains moving outward from the nucleus at kinetic energies of a few hundreds of electron volts (if assumed to be singly-charged; higher charge states would have correspondingly higher energy) as well as much more energetic grains (1–17 keV q^{-1}), also negatively charged, arriving from the approximate Sun's direction. All these detections were from the very early phase of the Rosetta escort of 67P, from August 23 to September 1, 2014, when Rosetta was at 50–60 km sunward of the nucleus and the outgassing activity was still very low. *Gombosi et al.* (2015) analyzed the observations of outward moving grains in terms of neutral gas drag acceleration. The anti-sunward moving grains were attributed to radiation pressure acting on small grains likely created from larger grains by erosion at large distances from the nucleus. For both types of grains, consistency between model and data resulted for grains' sizes within the 30–80-nm range. The convection electric field, changing in time in response to the interplanetary magnetic field, was discussed as a possible reason for the intermittency of the observations, deviating grains from the densest regions away from or toward Rosetta.

LLera et al. (2020) could identify another event, for a similar period (September 19, 2014), with positively-charged and negatively-charged grains over a broad energy range, as expected from the competing charging processes discussed in section 5.1. Both showed signatures very similar to the negatively charged anti-sunward moving grains previously reported. Interestingly, while grains of both signs had an anti-sunward velocity component, their trajectory was symmetric with respect to the anti-sunward component. This is what is to be expected for charged dust grains subject to a combination of the solar radiation pressure, same for both signs, and the solar wind convection electric field in the comet reference frame, which accelerates grains of opposite charge in opposite directions.

So far, these are the only events with signatures that have been attributed to charged dust grains in the tens of nanometers size range, all from the initial part of the Rosetta mission. Why are there so few? Regarding the outflowing grains discussed by *Burch et al.* (2015b), their signature is quite weak and may be swamped with electron fluxes in the same range, which were significant throughout most of the mission. The higher-energy signature of the (roughly) anti-sunward grains should not suffer as much, and here it seems likely they are just not present in significant numbers in the inner coma. According to the interpretation by *Johansson et al.* (2017) of anomalously low flux of solar radiation in the EUV observed at Rosetta around perihelion when 67P's activity peaked, nanometer-size dust grains were mainly present at cometocentric distances of at least a few thousand kilometers during this period. This may point to the fragmentation of dust grains occurring farther out at high activity than at low activity. Alternatively, the dust itself may already have different properties depending on Q and r_h.

Rosetta has as yet only showed some tantalizing glimpses of dust-plasma interactions, concentrated on the early phase of very low activity, while mission-wide modeling indicated that the dust impact on the plasma mostly is small or negligible. Nevertheless, the Rosetta dataset is far from exhausted and may still hold more clues to the presence and importance of dust charging.

6. SUMMARY AND OPEN QUESTIONS

Prior to the 1980s, comets and their plasma environment had relied only on speculation and simulation work, either in the laboratory or starting with the first numerical simulations. Thanks to the historical missions that followed, from the first flybys of comets in the 1980s to Rosetta's two-year mission escorting Comet 67P around the Sun 30 years later, many questions regarding the ionosphere of comets and its formation, structure, and evolution at different outgassing activities have now been answered. Table 2 presents highlights of our current understanding sorted by the themes structuring this chapter, from the physics presiding at the creation of a cometary ionosphere (section 2, presented as a tutorial-style toolbox), the description of the plasma population, electrons (section 3) and ions (section 4), and the puzzling and still largely unexplored role of charged dust in the overall dynamics and composition of the cometary ionosphere (section 5).

Many open questions remain, some of the more prominent ones are summarized in Table 3. With the advent of new missions, several of these questions will hopefully be answered. Ideally, multi-point, multi-instrument, *in situ* exploration of the cometary coma, either through a flyby or

TABLE 2. Highlights of this chapter on cometary ionospheres.

Topic	Highlights
Birth of cometary ionosphere	Identification of ionospheric sources and losses, and of their relative contribution, of different outgassing activities (section 2.2.4)
	CMEs/CIRs impact on the ionospheric sources and densities (sections 2.2.4 and 2.4.1)
	Plasma density profile observed all the way down to the surface (section 2.4.1)
	First detection and confirmation of ion species (up to 39 u q^{-1}); evolution of the ion composition over heliocentric distance and season (section 2.4.2)
Electron population	Three populations confirmed: cold, warm, and hot (section 3.1)
	Symbiosis between the three populations: coexist and are dependent on each other (section 3.1)
	Hot electrons: similarities with auroral physics at Earth (section 3.1.3)
	Detection of cold electrons even when the coma is quasi-collisionless (section 3.1.2)
	Evidence of the role played by electrons in the location of the diamagnetic cavity boundary (section 3.3)
Ion population	Three populations: slow cometary ions, early picked-up cometary ions, solar wind ions (sections 4.2 and 4.3)
	Evidence of chemistry (ion-neutral collisions) taking place in the coma (sections 2.4.2 and 4.1)
	Evidence for accelerated cometary ions (sections 3.2 and 4.1)
	Cometary ion density (but not flux) dominating over that of solar wind ions even at very low outgassing rate (section 2.2.4)
Dusty plasma	No evidence yet of a dusty plasma at Comet 67P with Rosetta (section 5.3)
	Observations of a few events of charged dust grains, positive and negative (section 5.3)

an extended escort phase of the comet nucleus, à la Rosetta, are necessary to address them with more than one spacecraft if possible as already proposed by the community for the ESA Voyage 2050 program call for white papers (*Goetz et al.,* 2021). This is among the goals of the highly anticipated ESA/JAXA Comet Interceptor mission (*Snodgrass and Jones,* 2019; *Jones et al.,* 2024), currently in the works as an international endeavor: Boasting simultaneous three-point measurements thanks to a mother spacecraft A complemented by two smaller probes, B1 and B2, the mission is planned to be launched in 2029 and to fly by a "dynamically new comet" yet to be discovered (i.e., visiting our inner solar system for the first time). Of particular interest for cometary ionospheric physics, probe A, farthest from the cometary nucleus, hosts a Dust, Field, and Plasma (DFP) set of sensors to probe the dust, electromagnetic fields, (dusty) plasma, and energetic particles (electrons, ions, and neutral atoms). This set is complemented by a neutral pressure gauge and mass spectrometer probing the major volatile species (MANIaC). On probe B2, which will get closest to the cometary nucleus, the DFP package is reduced to a dust impact sensor and counter and a magnetometer. Probe B1 hosts a plasma suite, which includes a time-of-flight mass spectrometer and a magnetometer. All these sensors will be probing the plasma environment around the target to deepen our understanding of cometary ionospheres, including formation, loss, and interaction with the space environment.

No new dedicated cometary mission, with sample return or not, is envisaged in the decade following Comet Interceptor, despite a strong call from the community, as proposed for the ESA Voyage 2050 program call (*Bockelée-Morvan et al.,* 2021; *Goetz et al.,* 2021). Because of their complexity, costs, and necessary multidisciplinary approach, future cometary missions will likely be international collaborations between space agencies (as for Comet Interceptor) and address a broad range of physics including plasma physics. By escorting Comet 67P around perihelion for two years, Rosetta is a successful proof of concept showing that a spacecraft orbiting a comet for a long period of time can be designed and flown and can harvest an amazingly rich dataset, despite the weak nucleus gravity. With our increasing capacity to detect more and more new targets of interest, interstellar objects or not, one can envision multiple flybys during an extended mission phase, with several small-scale missions, like Comet Interceptor or along the lines of the original Halley Armada. The next passage of Comet 1P/Halley in our vicinity in 2061, together with other comets from different reservoirs, may provide an interesting window of opportunity.

Acknowledgments. We would like to warmly thank P. Stephenson (Imperial College London, UK) and G. Wattieaux (Université of Toulouse, France) for their help and support, in particular in providing the un/processed dataset shown in Fig. 6 and in generating Fig. 11 respectively. We warmly thank Z. Lewis for her

TABLE 3. Non-exhaustive list of open questions regarding cometary ionospheres.

Open Questions
Birth of cometary ionosphere
What is the plasma balance near perihelion? What are the key plasma source and loss?
What is the three-dimensional distribution of the ionospheric densities under different outgassing activities?
What is the composition of heavy ions (>40 u q^{-1})?
What is the change, if any, of the ion composition in different plasma regions and boundaries?
What is the role played by the dust regarding EUV absorption under high activity?
Electron population
What is the energy distribution of electrons at high temporal and energy resolutions? How does it vary in space?
What is the exact origin of cold electrons in a quasi-collisional coma?
What is the role of cold electron in the presence of a diamagnetic cavity?
What is the source of hot electrons at high activity, when a diamagnetic cavity is formed?
Ion population
What is the energy distribution of cometary, slow ions?
Are ions accelerated near the surface at high activity and, if so, what is the process responsible for it?
What is driving the cometary ion dynamics?
What is the exact role of the electrons on the ion acceleration?
How can the observed high ion bulk speeds and the observed electron densities explained by a model excluding ion acceleration be reconciled?
How can the observed high ion bulk speeds and the observed ion composition attesting of ion-neutral chemistry in the coma be reconciled?
Dusty plasma
Does dust have significant impact on the cometary ionosphere for more active comets?
Are dusty plasma interactions the same at all comets and, if not, what are the conditions driving those effects?

careful reading of the chapter and her suggestions. A.B. and his work at Umeå University was supported by the Swedish National Space Agency (SNSA) grant 108/18. C.S.W. thanks the Austrian Science Fund (FWF) projects P32035-N36 and 10.55776/P35954. Work by M.G. at Imperial College London was supported in part by STFC of UK under grant ST/W001071/1.

REFERENCES

Aikin A. C. (1974) Cometary coma ions. *Astrophys. J., 193,* 263–264.

Alho M., Jarvinen R., Simon Wedlund C., et al. (2021) Remote sensing of cometary bow shocks: Modelled asymmetric outgassing and pickup ion observations. *Mon. Not. R. Astron. Soc., 506,* 4735–4749.

Allen M., Delitsky M., Huntress W., et al. (1987) Evidence for methane and ammonia in the coma of Comet P/Halley. *Astron. Astrophys., 187,* 502–512.

Altwegg K., Balsiger H., Geiss J., et al. (1993) The ion population between 1300 km and 230000 km in the coma of Comet P/Halley. *Astron. Astrophys., 279,* 260–266.

Altwegg K., Balsiger H., Bar-Nun A., et al. (2016) Prebiotic chemicals — amino acid and phosphorus — in the coma of Comet 67P/Churyumov-Gerasimenko. *Sci. Adv., 2,* e1600285.

Auger P. (1923) Sur les rayons β secondaires produits dans un gaz par des rayons x. *Compt. Rend. Hebd. Acad. Sci., 177,* 169–171.

Balsiger H., Altwegg K., Bühler F., et al. (1986) Ion composition and dynamics at Comet Halley. *Nature, 321,* 330–334.

Balsiger H., Altwegg K., Benson J., et al. (1987) The ion mass spectrometer on Giotto. *J. Phys. E: Sci. Instr., 20,* 759–767.

Balsiger H., Altwegg K., Bochsler P., et al. (2007) ROSINA — Rosetta Orbiter Spectrometer for Ion and Neutral Analysis. *Space Sci. Rev., 128,* 745–801.

Barkan A., D'Angelo N., and Merlino R. L. (1996) Experiments on ion-acoustic waves in dusty plasmas. *Planet. Space Sci., 44,* 239–242.

Behar E., Nilsson H., Alho M., et al. (2017) The birth and growth of a solar wind cavity around a comet — Rosetta observations. *Mon. Not. R. Astron. Soc., 469,* S396–S403.

Bercic L., Behar E., Nilsson H., et al. (2018) Cometary ion dynamics observed in the close vicinity of Comet 67P/Churyumov-Gerasimenko during the intermediate activity period. *Astron. Astrophys., 613,* A57.

Bergman S., Stenberg Wieser G., Wieser M., et al. (2020) The influence of varying spacecraft potentials and Debye lengths on *in situ* low-energy ion measurements. *J. Geophys. Res.–Space Phys., 125,* e2020JA027870.

Bergman S., Stenberg Wieser G., Wieser M., et al. (2021) Ion bulk speeds and temperatures in the diamagnetic cavity of Comet 67P from RPC-ICA measurements. *Mon. Not. R. Astron. Soc., 503,* 2733–2745.

Beth A., Altwegg K., Balsiger H., et al. (2017) First *in situ* detection of the cometary ammonium ion NH^+ (protonated ammonia NH_3) in the coma of 67P/C-G near perihelion. *Mon. Not. R. Astron. Soc., 462,* S562–S572.

Beth A., Galand M., and Heritier K. L. (2019) Comparative study of photo-produced ionosphere in the close environment of comets. *Astron. Astrophys., 630,* A47.

Beth A., Altwegg K., Balsiger H., et al. (2020) ROSINA ion zoo at Comet 67P. *Astron. Astrophys., 642,* A27.

Bhardwaj A. (2003) On the solar EUV deposition in the inner comae

of comets with large gas production rates. *Geophys. Res. Lett., 30,* 2244.
Bhardwaj A., Elsner R. F., Randall Gladstone G., et al. (2007) X-rays from solar system objects. *Planet. Space Sci., 55,* 1135–1189.
Biermann L. (1951) Kometenschweife und solare korpuskularstrahlung. *Z. Astrophys., 29,* 274–286.
Biermann L. (1952) Physical processes in comet tails and their relation to solar activity. In *La Physique des Cometes, Fourth Liège International Astrophysical Colloquia* (P. Swings, ed.), pp. 251–262. Institut d'Astrophysique et de Géophysique, Liège.
Biermann L., Brosowski B., and Schmidt H. U. (1967) The interactions of the solar wind with a comet. *Solar Phys., 1,* 254–284.
Biver N., Bockelée-Morvan D., Hofstadter M., et al. (2019) Long-term monitoring of the outgassing and composition of Comet 67P/Churyumov-Gerasimenko with the Rosetta/MIRO instrument. *Astron. Astrophys., 630,* A19.
Bockelée-Morvan D., Filacchione G., Altwegg K., et al. (2021) AMBITION — Comet nucleus cryogenic sample return. *Exp. Astron.,* DOI: 10.1007/s10686-021-09770-4.
Broiles T. W., Burch J. L., Chae K., et al. (2016a) Statistical analysis of suprathermal electron drivers at 67P/Churyumov-Gerasimenko. *Mon. Not. R. Astron. Soc., 462,* S312–S322.
Broiles T. W., Livadiotis G., Burch J. L., et al. (2016b) Characterizing cometary electrons with kappa distributions. *J. Geophys. Res.–Space Phys., 121,* 7407–7422.
Burch J. L., Goldstein R., Cravens T. E., et al. (2007) RPC-IES: The Ion and Electron Sensor of the Rosetta Plasma Consortium. *Space Sci. Rev., 128,* 697–712.
Burch J. L., Cravens T. E., Llera K., et al. (2015a) Charge exchange in cometary coma: Discovery of H^- ions in the solar wind close to Comet 67P/Churyumov-Gerasimenko. *Geophys. Res. Lett., 42,* 5125–5131.
Burch J. L., Gombosi T. I., Clark G., et al. (2015b) Observation of charged nanograins at comet 67P/Churyumov-Gerasimenko. *Geophys. Res. Lett., 42,* 6575–6581.
Burns J. A., Lamy P. L., and Soter S. (1979) Radiation forces on small particles in the solar system. *Icarus, 40,* 1–48.
Chamberlain J. W. (1963) Planetary coronae and atmospheric evaporation. *Planet. Space Sci., 11,* 901–960.
Chapman S. (1931) The absorption and dissociative or ionizing effect of monochromatic radiation in an atmosphere on a rotating earth. *Proc. Phys. Soc., 43,* 26–45.
Chapman S. and Cowling T. G. (1970) *The Mathematical Theory of Non-Uniform Gases: An Account of the Kinetic Theory of Viscosity, Thermal Conduction and Diffusion in Gases, 3rd edition.* Cambridge Univ., Cambridge. 448 pp.
Coates A. J. (2004) Ion pickup at comets. *Adv. Space Res., 33,* 1977–1988.
Coates A. J. and Jones G. H. (2009) Plasma environment of Jupiter family comets. *Planet. Space Sci., 57,* 1175–1191.
Coates A. J., Burch J. L., Goldstein R., et al. (2015) Ion pickup observed at Comet 67P with the Rosetta Plasma Consortium (RPC) particle sensors: Similarities with previous observations and AMPTE releases, and effects of increasing activity. *J. Phys. Conf. Ser., 642,* 012005.
Cowley S. W. H., McCrea W. H., Turner G., et al. (1987) ICE observations of Comet Giacobini-Zinner. *Philos. Trans. R. Soc. A, 323,* 405–420.
Cravens T. E. (1986) The physics of the cometary contact surface. In *ESLAB Symposium on the Exploration of Halley's Comet* (B. Battrick et al., eds.), pp. 241–246. ESA SP-250, Noordwijk, The Netherlands.
Cravens T. E. (1989) A magnetohydrodynamical model of the inner coma of Comet Halley. *J. Geophys. Res.–Space Phys., 94,* 15025–15040.
Cravens T. E. (1997a) Comet Hyakutake X-ray source: Charge transfer of solar wind heavy ions. *Geophys. Res. Lett., 24,* 105–108.
Cravens T. E. (1997b) *Physics of Solar System Plasmas.* Cambridge Univ., Cambridge. 477 pp.
Cravens T. E. (2002) X-ray emission from comets. *Science, 296,* 1042–1046.
Cravens T. E. and Gombosi T. I. (2004) Cometary magnetospheres: A tutorial. *Adv. Space Res., 33,* 1968–1976.
Cravens T. E., Gombosi T. I., Gribov B. E., et al. (1984) *The Role of Electric Fields in the Cometary Environment.* Tech. Rep. KFKI-1984-41. Hungarian Academy of Sciences, Budapest.
Davidsson B. J. R., Gutiérrez P. J., Sierks H., et al. (2015) Orbital elements of the material surrounding Comet 67P/Churyumov-Gerasimenko. *Astron. Astrophys., 583,* A16.
Deca J., Divin A., Henri P., et al. (2017) Electron and ion dynamics of the solar wind interaction with a weakly outgassing comet. *Phys. Rev. Lett., 118,* 205101.
Deca J., Henri P., Divin A., et al. (2019) Building a weakly outgassing comet from a generalized Ohm's Law. *Phys. Rev. Lett., 123,* 055101.
Dennerl K. (2010) Charge transfer reactions. *Space Sci. Rev., 157,* 57–91.
Dennerl K., Lisse C. M., Bhardwaj A., et al. (2012) Solar system X-rays from charge exchange processes. *Astron. Nachr., 333,* 324–334.
Deslandres H. and Bernard A. (1907) Étude spectrale de la comète Daniel d 1907: Particularités de la queue. *Compt. Rend. Hebd. Acad. Sci., 145,* 445–448.
Eberhardt P. (1999) Comet Halley's gas composition and extended sources: Results from the Neutral Mass Spectrometer on Giotto. *Space Sci. Rev., 90,* 45–52.
Eberhardt P. and Krankowsky D. (1995) The electron temperature in the inner coma of Comet P/Halley. *Astron. Astrophys., 295,* 795–806.
Edberg N. J. T., Eriksson A. I., Odelstad E., et al. (2015) Spatial distribution of low-energy plasma around Comet 67P/CG from Rosetta measurements. *Geophys. Res. Lett., 42,* 4263–4269.
Engelhardt I. A. D., Eriksson A. I., Vigren E., et al. (2018) Cold electrons at Comet 67P/Churyumov-Gerasimenko. *Astron. Astrophys., 616,* A51.
Eriksson A. I., Boström R., Gill R., et al. (2007) RPC-LAP: The Rosetta Langmuir probe instrument. *Space Sci. Rev., 128,* 729–744.
Eriksson A. I., Engelhardt I. A. D., André M., et al. (2017) Cold and warm electrons at Comet 67P/Churyumov-Gerasimenko. *Astron. Astrophys., 605,* A15.
Evershed J. (1907) Spectrum of Comet 1907 d (Daniel). *Mon. Not. R. Astron. Soc., 68,* 16–18.
Farquhar R. W. (2001) The flight of ISEE-3/ICE: Origins, mission history, and a legacy. *J. Astronaut. Sci., 49,* 23–73.
Festou M. C., Keller H. U., and Weaver H. A., eds. (2004) Comets II. Univ. of Arizona, Tucson. 745 pp.
Finson M. J. and Probstein R. F. (1968) A theory of dust comets. I. Model and equations. *Astrophys. J., 154,* 327–352.
Florescu-Mitchell A. and Mitchell J. (2006) Dissociative recombination. *Phys. Rept., 430,* 277–374.
Fowler A. (1907) The origin of certain bands in the spectra of sun-spots. *Mon. Not. R. Astron. Soc., 67,* 530–534.
Fowler A. (1910) Investigations relating to the spectra of comets. *Mon. Not. R. Astron. Soc., 70,* 484–496.
Fox J. L., Galand M. I., and Johnson R. E. (2008) Energy deposition in planetary atmospheres by charged particles and solar photons. *Space Sci. Rev., 139,* 3–62.
Fuselier S. A., Altwegg K., Balsiger H., et al. (2016) Ion chemistry in the coma of comet 67P near perihelion. *Mon. Not. R. Astron. Soc., 462,* S67–S77.
Galand M., Héritier K. L., Odelstad E., et al. (2016) Ionospheric plasma of Comet 67P probed by Rosetta at 3 au from the Sun. *Mon. Not. R. Astron. Soc., 462,* S331–S351.
Galand M., Feldman P. D., Bockelée-Morvan D., et al. (2020) Far-ultraviolet aurora identified at Comet 67P/Churyumov-Gerasimenko. *Nature Astron., 4,* 1084–1091.
Gan L. and Cravens T. E. (1990) Electron energetics in the inner coma of Comet Halley. *J. Geophys. Res.–Space Phys., 95,* 6285–6303.
Gasc S., Altwegg K., Balsiger H., et al. (2017) Change of outgassing pattern of 67P/Churyumov-Gerasimenko during the March 2016 equinox as seen by ROSINA. *Mon. Not. R. Astron. Soc., 469,* S108–S117.
Geiss J., Altwegg K., Anders E., et al. (1991) Interpretation of the ion mass spectra in the mass per charge range 25–35 amu/e obtained in the inner coma of Halley's comet by the HIS-sensor of the Giotto IMS experiment. *Astron. Astrophys., 247,* 226–234.
Gilet N., Henri P., Wattieaux G., et al. (2017) Electrostatic potential radiated by a pulsating charge in a two-electron temperature plasma. *Radio Sci., 52,* 1432–1448.
Gilet N., Henri P., Wattieaux G., et al. (2020) Observations of a mix of cold and warm electrons by RPC-MIP at 67P/Churyumov-Gerasimenko. *Astron. Astrophys., 640,* A110.
Glassmeier K.-H. (2017) Interaction of the solar wind with comets: A Rosetta perspective. *Philos. Trans. R. Soc. A, 375,* 20160256.
Glassmeier K.-H., Boehnhardt H., Koschny D., et al. (2007) The Rosetta

mission: Flying towards the origin of the solar system. *Space Sci. Rev., 128,* 1–21.

Goetz C., Koenders C., Hansen K. C., et al. (2016a) Structure and evolution of the diamagnetic cavity at Comet 67P/Churyumov-Gerasimenko. *Mon. Not. R. Astron. Soc., 462,* S459–S467.

Goetz C., Koenders C., Richter, I., et al. (2016b) First detection of a diamagnetic cavity at Comet 67P/Churyumov-Gerasimenko. *Astron. Astrophys., 588,* A24.

Goetz C., Gunell H., Volwerk M., et al. (2021) Cometary plasma science. *Exp. Astron.,* DOI: 10.1007/s10686-021-09783-z.

Goldstein B. E., Altwegg K. Balsiger H., Fuselier S. A., et al. (1989) Observations of a shock and a recombination layer at the contact surface of Comet Halley. *J. Geophys. Res.–Space Phys., 94,* 17251–17257.

Goldstein R., Burch J. L., Mokashi P., et al. (2015) The Rosetta Ion and Electron Sensor (IES) measurement of the development of pickup ions from Comet 67P/Churyumov-Gerasimenko. *Geophys. Res. Lett., 42,* 3093–3099.

Goldstein R., Burch J. L., Mokashi P., et al. (2017) Two years of solar wind and pickup ion measurements at Comet 67P/Churyumov-Gerasimenko. *Mon. Not. R. Astron. Soc., 469,* S262–S267.

Gombosi T. I. (1987) Charge exchange avalanche at the cometopause. *Geophys. Res. Lett., 14,* 1174–1177.

Gombosi T. I. (2015) Physics of cometary magnetospheres. In *Magnetotails in the Solar System* (A. Keiling et al., eds.), pp. 169–188. AGU Geophys. Monogr. 207, American Geophysical Union, Washington, DC.

Gombosi T. I., Nagy A. F., and Cravens T. E. (1986) Dust and neutral gas modeling of the inner atmospheres of comets. *Rev. Geophys., 24,* 667–700.

Gombosi T. I., De Zeeuw D. L., Häberli R. M., et al. (1996) Three-dimensional multiscale MHD model of cometary plasma environments. *J. Geophys. Res.–Space Phys., 101,* 15233–15253.

Gombosi T., Burch, J.L., and Horanyi, M. (2015) Negatively charged nano-grains at 67P/Churyumov-Gerasimenko. *Astron. Astrophys., 583,* A23.

Grad H. (1949) On the kinetic theory of rarefied gases. *Commun. Pure Appl. Math., 2,* 331–407.

Gringauz K. I., Gombosi T. I., Tátrallyay M., et al. (1986) Detection of a new "chemical" boundary at Comet Halley. *Geophys. Res. Lett., 13,* 613–616.

Grün E., Agarwal J., Altobelli N., et al. (2016) The 2016 Feb 19 outburst of Comet 67P/CG: An ESA Rosetta multi-instrument study. *Mon. Not. R. Astron. Soc., 462,* S220–S234.

Hajra R., Henri P., Myllys M., et al. (2018) Cometary plasma response to interplanetary corotating interaction regions during 2016 June–September: A quantitative study by the Rosetta Plasma Consortium. *Mon. Not. R. Astron. Soc., 480,* 4544–4556.

Hansen K. C., Altwegg K., Berthelier J. J., et al. (2016) Evolution of water production of 67P/Churyumov-Gerasimenko: An empirical model and a multi-instrument study. *Mon. Not. R. Astron. Soc., 462,* S491–S506.

Haser L. (1957) Distribution d'intensité dans la tête d'une comète. *Bull. Acad. R. Belgique, 43,* 740–750.

Haser L., Oset S., and Bodewits D. (2020) Intensity distribution in the heads of comets. *Planet. Sci. J., 1,* 83.

Hässig M., Altwegg K., Balsiger H., et al. (2015) Time variability and heterogeneity in the coma of 67P/Churyumov-Gerasimenko. *Science, 347,* aaa0276.

Heays A. N., Bosman A. D., and van Dishoeck E. F. (2017) Photodissociation and photoionisation of atoms and molecules of astrophysical interest. *Astron. Astrophys., 602,* A105.

Henri P., Vallières X., Hajra R., et al. (2017) Diamagnetic region(s): Structure of the unmagnetized plasma around Comet 67P/CG. *Mon. Not. R. Astron. Soc., 469,* S372–S379.

Heritier K. L., Henri P., Vallières X., et al. (2017a) Vertical structure of the near-surface expanding ionosphere of Comet 67P probed by Rosetta. *Mon. Not. R. Astron. Soc., 469,* S118–S129.

Heritier K. L., Altwegg K., Balsiger H., et al. (2017b) Ion composition at Comet 67P near perihelion: Rosetta observations and model-based interpretation. *Mon. Not. R. Astron. Soc., 469,* S427–S442.

Heritier K. L., Galand, M., Henri, P., et al. (2018) Plasma source and loss at Comet 67P during the Rosetta mission. *Astron Astrophys., 618,* A77.

Hoffmeister C. (1943) Physikalische untersuchungen an kometen. I. Die beziehungen des primären schweifstrahls zum radiusvektor. *Z. Astrophys., 22,* 265.

Houpis H. L. F. and Mendis D. A. (1980) Physicochemical and dynamical processes in cometary ionospheres. I. The basic flow profile. *Astrophys. J., 239,* 1107–1118.

Huebner W. and Markiewicz W. (2000) The temperature and bulk flow speed of a gas effusing or evaporating from a surface into a void after reestablishment of collisional equilibrium. *Icarus, 148,* 594–596.

Huebner W. and Mukherjee J. (2015) Photoionization and photodissociation rates in solar and blackbody radiation fields. *Planet. Space Sci., 106,* 11–45.

Huebner W. F., Giguere P. T., and Slattery W. L. (1982) Photochemical processes in the inner coma. In *Comets* (L. L. Wilkening, ed.), pp. 496–515. Univ. of Arizona, Tucson.

Huebner W. F., Keady J. J., and Lyon S. P. (1992) Solar photo rates for planetary atmospheres and atmospheric pollutants. *Astrophys. Space Sci., 195,* 1–294.

Hunter E. P. L. and Lias S. G. (1998) Evaluated gas phase basicities and proton affinities of molecules: An update. *J. Phys. Chem. Ref. Data, 27,* 413–656.

Ip W.-H. (2004) Global solar wind interaction and ionospheric dynamics. In *Comets II* (M. C. Festou et al., eds.), pp. 605–629. Univ. of Arizona, Tucson.

Ip W.-H. and Axford W. I. (1982) Theories of physical processes in the cometary comae and ion tails. In *Comets* (L. L. Wilkening, ed.), pp. 588–634. Univ. of Arizona, Tucson.

Ip W.-H. and Axford W. I. (1987) The formation of a magnetic-field-free cavity at Comet Halley. *Nature, 325,* 418–419.

Itikawa Y. (2002) Cross sections for electron collisions with carbon dioxide. *J. Phys. Chem. Ref. Data, 31,* 749–767.

Itikawa Y. and Mason N. (2005) Cross sections for electron collisions with water molecules. *J. Phys. Chem. Ref. Data, 34,* 1–22.

IUPAC (1978) Recommendations for symbolism and nomenclature for mass spectroscopy. *Pure Appl. Chem., 50,* 65–74.

Jeans J. (1925) *The Dynamical Theory of Gases.* Cambridge Univ., Cambridge. 444 pp.

Johansson F. L., Odelstad E., Paulsson J. J. P., et al. (2017) Rosetta photoelectron emission and solar ultraviolet flux at Comet 67P. *Mon. Not. R. Astron. Soc., 469,* S626–S635.

Johansson F. L., Eriksson A. I., Gilet N., et al. (2020) A charging model for the Rosetta spacecraft. *Astron. Astrophys., 642,* A43.

Johansson F. L., Eriksson A. I., Vigren E., et al. (2021) Plasma densities, flow and solar EUV flux at Comet 67P: A cross-calibration approach. *ArXiV e-prints,* arXiv:2106.15491.

Jones G. H., Snodgrass C., Tubiana C., et al. (2024) The Comet Interceptor mission. *Space Sci. Rev., 220,* 9.

Koenders C., Glassmeier K.-H., Richter I., et al. (2015) Dynamical features and spatial structures of the plasma interaction region of 67P/Churyumov-Gerasimenko and the solar wind. *Planet. Space Sci., 105,* 101–116.

Körösmezey A., Cravens T. E., Gombosi T. I., et al. (1987) A new model of cometary ionospheres. *J. Geophys. Res.–Space Phys., 92,* 7331–7340.

Krankowsky D., Lämmerzahl P., Herrwerth I., et al. (1986) *In situ* gas and ion measurements at Comet Halley. *Nature, 321,* 326–329.

Krasnopolsky V. A., Greenwood J. B., and Stancil P. C. (2004) X-ray and extreme ultraviolet emissions from comets. *Space Sci. Rev., 113,* 271–373.

Laframboise J. G. (1966) Theory of spherical and cylindrical Langmuir probes in a collisionless, Maxwellian plasma at rest. Ph.D. thesis, University of Toronto, Toronto. 210 pp.

Lallement R. (2012) Charge-exchange X-ray emission at hot/cool gas interfaces. *Astron. Nachr., 333,* 347–350.

Larsson M., Geppert W. D., and Nyman G. (2012) Ion chemistry in space. *Rept. Prog. Phys., 75,* 066901.

Lean J. (1991) Variations in the Sun's radiative output. *Rev. Geophys., 29,* 505–535.

Le Roy L., Altwegg K., Balsiger H., et al. (2015) Inventory of the volatiles on Comet 67P/Churyumov-Gerasimenko from Rosetta/ROSINA. *Astron. Astrophys., 583,* A1.

Lisse C. M., Dennerl K., Englhauser J., et al. (1996) Discovery of X-ray and extreme ultraviolet emission from Comet C/Hyakutake 1996 B2. *Science, 274,* 205–209.

Llera K., Burch J. L., Goldstein R., et al. (2020) Simultaneous observation of negatively and positively charged nanograins at Comet 67P/Churyumov-Gerasimenko. *Geophys. Res. Lett., 47,*

e2019GL086147.
Lowry S., Fitzsimmons A., Lamy P., et al. (2008) Kuiper belt objects in the planetary region: The Jupiter-family comets. In *The Solar System Beyond Neptune* (M. A. Barucci et al., eds.), pp. 397–410. Univ. of Arizona, Tucson.
Madanian H., Cravens T. E., Rahmati A., et al. (2016) Suprathermal electrons near the nucleus of Comet 67P/Churyumov-Gerasimenko at 3 AU: Model comparisons with Rosetta data. *J. Geophys. Res.–Space Phys., 121,* 5815–5836.
Madanian H., Cravens T. E., Burch J., et al. (2017) Plasma environment around Comet 67P/Churyumov-Gerasimenko at perihelion: Model comparison with Rosetta data. *Astron. J., 153,* 30.
Mandt K. E., Eriksson A., Beth A., et al. (2019) Influence of collisions on ion dynamics in the inner comae of four comets. *Astron. Astrophys., 630,* A48.
Marschall R., Markkanen J., Gerig S.-B., et al. (2020) The dust-to-gas ratio, size distribution, and dust fall-back fraction of Comet 67P/Churyumov-Gerasimenko: Inferences from linking the optical and dynamical properties of the inner comae. *Front. Phys., 8,* 227.
McGowan J. W. and Mitchell J. B. A. (1984) Electron-molecular positive-ion recombination. In *Electron-Molecule Interactions and Their Applications, Vol. 2* (L. Christophorou, ed.), pp. 65–88. Academic, Orlando.
Mendis D. A. and Horányi M. (2013) Dusty plasma effects in comets: Expectations for Rosetta. *Rev. Geophys., 51,* 53–75.
Mendis D. A., Houpis H. L. F., and Marconi M. L. (1985) *The Physics of Comets.* Gordon and Breach, New York. 380 pp.
Merlino R. L. (2014) 25 years of dust acoustic waves. *J. Plasma Phys., 80,* 773–786.
Meyer-Vernet N. (2013) On the charge of nanograins in cold environments and Enceladus dust. *Icarus, 226,* 583–590.
Meyer-Vernet N., Couturier P., Hoang S., et al. (1986) Plasma diagnosis from thermal noise and limits on dust flux or mass in Comet Giacobini-Zinner. *Science, 232,* 370–374.
Mott-Smith H. M. and Langmuir I. (1926) The theory of collectors in gaseous discharges. *Phys. Rev., 28,* 727–763.
Muller P. (1994) Glossary of terms used in physical organic chemistry (IUPAC recommendations 1994). *Pure Appl. Chem., 66,* 1077–1184.
Myllys M., Henri P., Galand M., et al. (2019) Plasma properties of suprathermal electrons near Comet 67P/Churyumov-Gerasimenko with Rosetta. *Astron. Astrophys., 630,* A42.
Neubauer F. M. (1987) Giotto magnetic-field results on the boundaries of the pile-up region and the magnetic cavity. *Astron. Astrophys., 187,* 73–79.
Neubauer F. M., Glassmeier K. H., Pohl M., et al. (1986) First results from the Giotto magnetometer experiment at Comet Halley. *Nature, 321,* 352–355.
Nilsson H., Stenberg Wieser G., Behar E., et al. (2015) Birth of a comet magnetosphere: A spring of water ions. *Science, 347,* aaa0571.
Nilsson H., Wieser G. S., Behar E., et al. (2017) Evolution of the ion environment of Comet 67P during the Rosetta mission as seen by RPC-ICA. *Mon. Not. R. Astron. Soc., 469,* S252–S261.
Nilsson H., Williamson H., Bergman S., et al. (2020) Average cometary ion flow pattern in the vicinity of Comet 67P from moment data. *Mon. Not. R. Astron. Soc., 498,* 5263–5272.
Obridko V. N. and Vaisberg O. L. (2017) On the history of the solar wind discovery. *Solar System Res., 51,* 165–169.
Odelstad E., Stenberg-Wieser G., Wieser M., et al. (2017) Measurements of the electrostatic potential of Rosetta at Comet 67P. *Mon. Not. R. Astron. Soc., 469,* S568–S581.
Odelstad E., Eriksson A. I., Johansson F. L., et al. (2018) Ion velocity and electron temperature inside and around the diamagnetic cavity of Comet 67P. *J. Geophys. Res.–Space Phys., 123,* 5870–5893.
Ogilvie K. W., Coplan M. A., Bochsler P., et al. (1986) Ion composition results during the international cometary explorer encounter with Giacobini-Zinner. *Science, 232,* 374–377.
Parker E. N. (1958) Dynamics of the interplanetary gas and magnetic fields. *Astrophys. J., 128,* 664–676.
Pedersen A., Grard R., Trotignon J. G., et al. (1986) Measurements of low energy electrons and spacecraft potentials near Comet P/Halley. In *ESLAB Symposium on the Exploration of Halley's Comet* (B. Battrick et al., eds.), pp. 425–431. ESA SP-250, Noordwijk, The Netherlands.
Pierrard V., Lazar M., Poedts S., et al. (2016) The electron temperature and anisotropy in the solar wind: Comparison of the core and halo populations. *Solar Phys., 291,* 2165–2179.
Reinhard R. (1986) The Giotto encounter with Comet Halley. *Nature, 321,* 313–318.
Reme H., Mazelle C., D'Uston C., et al. (1994) There is no "cometopause" at Comet Halley. *J. Geophys. Res.–Space Phys., 99,* 2301–2308.
Rosén S., Derkatch A., Semaniak J., et al. (2000) Recombination of simple molecular ions studied in storage ring: Dissociative recombination of H_2O^+. *Faraday Discussions, 115,* 295–302.
Rubin M., Hansen K. C., Gombosi T. I., et al. (2009) Ion composition and chemistry in the coma of Comet 1P/Halley — A comparison between Giotto's Ion Mass Spectrometer and our ion-chemical network. *Icarus, 199,* 505–519.
Sagdeev R. Z., Evlanov E. N., Fomenkova M. N., et al. (1989) Small-size dust particles near Halley's comet. *Adv. Space Res., 9,* 263–267.
Sauer K., Bogdanov A., and Baumgartel K. (1995) The protonopause — An ion composition boundary in the magnetosheath of comets, Venus and Mars. *Adv. Space Res., 16,* 153–158.
Schunk R. and Nagy A. (2009) *Ionospheres: Physics, Plasma Physics, and Chemistry, 2nd edition.* Cambridge Univ., Cambridge. 628 pp.
Schwenn R., Ip W. H., Rosenbauer H., et al. (1987) Ion temperature and flow profiles in Comet P/Halley's close environment. *Astron. Astrophys., 187,* 160–162.
Sembay S., Branduardi-Raymont G., Eastwood J. P., et al. (2012) AXIOM: Advanced X-ray imaging of the magnetosheath. *Astron. Nachr., 333,* 388–392.
Simon Wedlund C., Kallio E., Alho M., et al. (2016) The atmosphere of Comet 67P/Churyumov-Gerasimenko diagnosed by charge-exchanged solar wind alpha particles. *Astron. Astrophys., 587,* A154.
Simon Wedlund C., Alho M., Gronoff G., et al. (2017) Hybrid modelling of cometary plasma environments: I. Impact of photoionisation, charge exchange, and electron ionisation on bow shock and cometopause at 67P/Churyumov-Gerasimenko. *Astron. Astrophys., 604,* A73.
Simon Wedlund C., Behar E., Kallio E., et al. (2019a) Solar wind charge exchange in cometary atmospheres: II. Analytical model. *Astron. Astrophys., 630,* A36.
Simon Wedlund C., Behar E., Nilsson H., et al. (2019b) Solar wind charge exchange in cometary atmospheres: III. Results from the Rosetta mission to Comet 67P/Churyumov-Gerasimenko. *Astron. Astrophys., 630,* A37.
Simon Wedlund C., Bodewits D., Alho M., et al. (2019c) Solar wind charge exchange in cometary atmospheres: I. Charge-changing and ionization cross sections for He and H particles in H_2O. *Astron. Astrophys., 630,* A35.
Simon Wedlund C., Behar E., Nilsson H., et al. (2020) Solar wind charge exchange in cometary atmospheres: III. Results from the Rosetta mission to Comet 67P/Churyumov-Gerasimenko [Corrigendum to A&A 630, A37]. *Astron. Astrophys., 640,* C3.
Sishtla C. P., Divin A., Deca J., et al. (2019) Electron trapping in the coma of a weakly outgassing comet. *Phys. Plasmas, 26,* 102904.
Slavin J. A. and Holzer R. E. (1981) Solar wind flow about the terrestrial planets: 1. Modeling bow shock position and shape. *J. Geophys. Res.–Space Phys., 86,* 11401–11418.
Snodgrass C. and Jones G. H. (2019) The European Space Agency's Comet Interceptor lies in wait. *Nature Commun., 10,* 5418.
Somov B. V. (2007) The generalized Ohm's law in plasma. In *Astrophysics and Space Science Library, Vol. 341: Plasma Astrophysics* (G. P. Kuiper, ed.), pp. 193–204. Springer, New York.
Spitzer L. (1949) The terrestrial atmosphere above 300 km. In *The Atmospheres of the Earth and Planets* (G. P. Kuiper, ed.), pp. 211–247. Univ of Chicago, Chicago.
Stephenson P., Galand M., Feldman P. D., et al. (2021) Multi-instrument analysis of far-ultraviolet aurora in the southern hemisphere of Comet 67P/Churyumov-Gerasimenko. *Astron. Astrophys., 647,* A119.
Stephenson P., Galand M., Deca J., et al. (2022) A collisional test-particle model of electrons at a comet. *Mon. Not. R. Astron. Soc., 511,* 4090–4108.
Szegö K., Glassmeier K.-H., Bingham R., et al. (2000) Physics of mass loaded plasmas. *Space Sci. Rev., 94,* 429–671.
Trotignon J. G., Michau J. L., Lagoutte D., et al. (2007) RPC-MIP: The Mutual Impedance Probe of the Rosetta Plasma Consortium. *Space Sci. Rev., 128,* 713–728.
Valentini F., Trávníček P., Califano F., et al. (2007) A hybrid-Vlasov model based on the current advance method for the simulation of collisionless magnetized plasma. *J. Comp. Phys., 225,* 753–770.
Viggiano A. A., Ehlerding A., Hellberg F., et al. (2005) Rate constants

and branching ratios for the dissociative recombination of CO_2^+. *J. Chem. Phys., 122,* 226101.
Vigren E. and Eriksson A. I. (2019) On the ion-neutral coupling in cometary comae. *Mon. Not. R. Astron. Soc., 482,* 1937–1941.
Vigren E. and Galand M. (2013) Predictions of ion production rates and ion number densities within the diamagnetic cavity of Comet 67P/Churyumov-Gerasimenko at perihelion. *Astrophys. J., 772,* 33.
Vigren E., Galand M., Eriksson A. I., et al. (2015) On the electron-to-neutral number density ratio in the coma of Comet 67P/Churyumov-Gerasimenko: Guiding expression and sources for deviations. *Astrophys. J., 812,* 54.
Vigren E., André M., Edberg N. J. T., et al. (2017) Effective ion speeds at ~200–250 km from Comet 67P/Churyumov-Gerasimenko near perihelion. *Mon. Not. R. Astron. Soc., 469,* S142–S148.
Vigren E., Edberg N. J. T., Eriksson A. I., et al. (2019) The evolution of the electron number density in the coma of Comet 67P at the location of Rosetta from 2015 November through 2016 March. *Astrophys. J., 881,* 6.
Vigren E., Eriksson A. I., Johansson F. L., et al. (2021) A case for a small to negligible influence of dust charging on the ionization balance in the coma of Comet 67P. *Planet. Sci. J., 2,* 156.
Vigren E., Eriksson A. I., and Bergman S. (2022) On positively charged dust in the coma of Comet 67P. *Mon. Not. R. Astron. Soc., 513,* 536–540.
von Rosenvinge T. T., Brandt J. C., and Farquhar R. W. (1986) The international cometary explorer mission to Comet Giacobini-Zinner. *Science, 232,* 353–356.
von Steiger R., Schwadron N. A., Fisk L. A., et al. (2000) Composition of quasi-stationary solar wind flows from Ulysses/Solar Wind Ion Composition Spectrometer. *J. Geophys. Res., 105,* 27217–27238.
Wargelin B. J., Beiersdorfer P., and Brown G. V. (2008) EBIT charge-exchange measurements and astrophysical applications. *Canad. J. Phys., 86,* 151–169.
Wattieaux G., Gilet N., Henri P., et al. (2019) RPC-MIP observations at Comet 67P/Churyumov-Gerasimenko explained by a model including a sheath and two populations of electrons. *Astron. Astrophys., 630,* A41.
Wattieaux G., Henri P., Gilet N., et al. (2020) Plasma characterization at Comet 67P between 2 and 4 AU from the Sun with the RPC-MIP instrument. *Astron. Astrophys., 638,* A124.
Whipple F. L. (1950) A comet model. I. The acceleration of Comet Encke. *Astrophys. J., 111,* 375–394.
Wigner E. P. (1948) On the behavior of cross sections near thresholds. *Phys. Rev., 73,* 1002–1009.
Wilkening L. L., ed. (1982) *Comets.* Univ. of Arizona, Tucson. 766 pp.
Williamson H. N., Nilsson H., Stenberg Wieser G., et al. (2020) Momentum and pressure balance of a comet ionosphere. *Geophys. Res. Lett., 47,* e88666.
Woods T. N., Eparvier F. G., Bailey S. M., et al. (2005) Solar EUV Experiment (SEE): Mission overview and first results. *J. Geophys. Res.–Space Phys., 110,* A01312.
Wu C. S. and Davidson R. C. (1972) Electromagnetic instabilities produced by neutral-particle ionization in interplanetary space. *J. Geophys. Res.–Space Phys., 77,* 5399–5406.
Zavilopulo A. N., Chipev F. F., and Shpenik O. B. (2005) Ionization of nitrogen, oxygen, water, and carbon dioxide molecules by near-threshold electron impact. *Tech. Phys., 50,* 402–407.
Zwickl R. D., Baker D. N., Bame S. J., et al. (1986) Three component plasma electron distribution in the intermediate ionized coma of Comet Giacobini-Zinner. *Geophys. Res. Lett., 13,* 401–404.

Götz C., Deca J., Mandt K., and Volwerk M. (2024) Solar wind interaction with a comet: Evolution, variability, and implication. In *Comets III* (K. J. Meech, M. R. Combi, D. Bockelée-Morvan, S. N. Raymond, and M. E. Zolensky, eds.), pp. 543–574. Univ. of Arizona, Tucson, DOI: 10.2458/azu_uapress_9780816553631-ch017.

Solar Wind Interaction with a Comet: Evolution, Variability, and Implication

C. Götz
European Space Agency/European Space Research and Technology Centre (now at Northumbria University Newcastle)

J. Deca
Laboratory for Atmospheric and Space Physics, University of Colorado Boulder

K. Mandt
Johns Hopkins Applied Physics Laboratory

M. Volwerk
Space Research Institute, Austrian Academy of Sciences

Once a cometary plasma cloud has been created through ionization of the cometary neutrals, it presents an obstacle to the solar wind and the magnetic field within it. The acceleration and incorporation of the cometary plasma by the solar wind is a complex process that shapes the cometary plasma environment and is responsible for the creation of boundaries such as a bow shock and diamagnetic cavity boundary. It also gives rise to waves and electric fields, which in turn contribute to the acceleration of the plasma. This chapter aims to provide an overview of how the solar wind is modified by the presence of the cometary plasma and how the cometary plasma is incorporated into the solar wind. We will also discuss models and techniques widely used in the investigation of the plasma environment in the context of recent findings by Rosetta. In particular, this chapter highlights the richness of the processes and regions within this environment and how processes on small scales can shape boundaries on large scales. It has been 15 years since *Comets II* was published and since then we have made great advances in the field of cometary research. But many open questions remain, which are listed and discussed with particular emphasis on how to advance the field of cometary plasma science through future space missions.

1. INTRODUCTION

In the preceding chapter, the reader is introduced to the sources and losses of the plasma at a comet. In the following, we will describe what happens to the cometary plasma cloud as it interacts with the solar wind.

Several works have summarized and reviewed the extensive work done in the field of cometary plasma environments since the 1980s with the first artificial comet experiments and the 1P/Halley flyby. The reader is referred to, for example, summaries by *Gombosi* (2015) for an overview or *Szegö et al.* (2000) for the details of the physics in these plasmas. Most new discoveries since the *Comets II* book (*Combi et al.,* 2004) was written are due to the European Space Agency's Rosetta mission, and we therefore highlight some peculiarities of this mission here. The spacecraft arrived at Comet 67P/Churyumov-Gerasimenko (67P) in August 2014 and explored the plasma environment for over two years until the end of the mission at the end of September 2016. This is in stark contrast to all previous missions to comets that were equipped with plasma instruments, which were all single flybys with different closest approach distances (for more details, see the chapter in this volume by Snodgrass et al.). The particular advantage of the long-lasting rendezvous is that the observations cover a large range of activity levels. This allows us to study the cometary environment and its interaction with the solar wind at different stages and observe how the increase and decrease in gas production rate and consequently the cometary ion density affects the solar wind. Rosetta also covered much lower activity levels than other missions to comets and therefore expands our knowledge significantly. On the other hand, this means that comparisons with previous results always need to take into account the different situations and are usually non-trivial.

Rosetta's instruments were able to measure the magnetic field vector, the plasma density, the electron temperature, the ion energy distribution with mass resolution, and the electron energy distribution. For a few days in November 2014, data by the lander Philae were available for two-point measurements of the magnetic field. For more details, the reader is referred to the documentation for the Rosetta Plasma Consortium (RPC) (*https://www.cosmos.esa.int/web/psa/rosetta*). There, one can also find an extensive list of available data products and the limits and error sources associated with them.

As already described in the chapter in this volume by Beth et al., the primary source of the plasma at the comet is the neutral gas that is produced by the sublimation of the ices on the comet's surface. Processes like photoionization, charge exchange, and electron impact ionization produce ions and electrons from those neutrals. The resulting ion coma can extend millions of kilometers from the nucleus, interacting with the plasma and magnetic field of the solar wind in which it mostly resides. This forms the plasma environment of the comet.

As Comet 67P journeys through the solar system, the heliocentric distance decreases and increases and therefore the gas production rate increases and decreases as well. This is shown in the top and middle panels of Figs. 1a,b. At the same time, the solar wind parameters change as well. Figure 1c shows the estimated dynamic pressure of the solar wind at 67P from a model to illustrate this. The closer the comet is to the Sun the higher the solar wind density (and therefore the dynamic pressure) and magnetic field strength. On shorter timescales such as days and hours, solar wind transients like corotating interaction regions (CIRs) and interplanetary coronal mass ejections (ICMEs) can also cause the solar wind parameters to change significantly. Unfortunately, no solar wind monitor was included in the Rosetta mission and models need to suffice to infer the solar wind parameters at the comet. They usually utilize some sort of propagation model and either observations of the Sun or *in situ* measurements at other solar system locations to estimate the solar wind conditions with spatial and temporal uncertainties. This is one of the major drawbacks of a single-spacecraft mission and should be taken into consideration for future missions (*Götz et al.,* 2019).

Therefore, depending on the gas production rate and solar wind conditions, the comet influenced plasma environment may extend from as little as 100 km up to millions of kilometers from the nucleus. As a consequence, the collisionality of the plasma, the gyroradii of the particles within it, and the presence of different electron and ion populations also vary significantly and systematically over time.

In the following, the reader is given an overview of the solar wind-comet interaction, starting with a description of the induced cometosphere, the magnetic field structure near the comet and in the comet plasma tail, followed by a list of the boundaries that form. At the end, various wave phenomena are discussed.

2. INDUCED COMETOSPHERE

2.1. Introduction

Whereas the cometary environment might appear quite uneventful near aphelion, as a comet travels closer to the Sun, the increasing availability of neutral gas leads to the formation of a coma and dust and plasma tails. Observations of the latter phenomena revealed the existence of the solar wind and the interplanetary magnetic field (*Biermann et al.,* 1967). Important is the realization that the presence of neutral gas in the solar wind leads to mass-loaded plasmas of various natures (*Szegö et al.,* 2000; *Gombosi,* 2015). One such example is an induced cometosphere, a term analogous to a magnetosphere at planets, that forms when the magnetized solar wind plasma interacts with the cometary plasma environment (*Goetz et al.,* 2021a).

Mass-loading describes the process whereby slow, heavy ions like water, carbon dioxide, or oxygen are accelerated and incorporated into the fast solar wind flow. It is therefore an energy and momentum exchange process between different plasma populations (*Galeev et al.,* 1985; *Huddleston et al.,* 1993). What distinguishes the cometary neutral environment from that of, e.g., Venus or Mars is the lack of significant gravitational acceleration of the central body: While Venus and Mars have an atmosphere that is for the most part gravitationally bound to the planet, the neutral exosphere of a comet is radially expanding into space without much gravitational influence. As a consequence, neutral gas can be ionized anywhere in the environment, even millions of kilometers upstream of the source body. This then gives a size of the cometosphere that is dependent on the outgassing and ionization conditions. For example, the first signatures of a cometosphere at Comet 1P/Halley at about 1 au, which had a gas production rate of $Q = 10^{30}$ s^{-1} and a water photoionization rate of $\nu_{H_2O} = 10^{-6}$ s^{-1}, were found at 2×10^6 km upstream of the nucleus (*Neubauer et al.* 1986). On the other end of the scale, we find that the cometosphere at Comet 67P at 3.4 au only expands about 10^2 km at a gas production rate of 4×10^{25} s^{-1} and a water photoionization rate of 0.5×10^{-7} s^{-1} (*Richter et al.,* 2015). This impressive difference in scale of 4 orders of magnitude demonstrates the richness of the cometary plasma environment. Many of the processes and structures in the cometary plasma environment depend on the scale size and therefore the outgassing rate of the comet, which is why it is useful to talk about different activity stages of a comet. Following *Goetz et al.* (2021a), we will use the following definition: weakly active for $Q < 10^{26}$ s^{-1}, intermediately active for $10^{26} \leq Q < 5 \times 10^{27}$ s^{-1}, and strongly active for $Q \geq 5 \times 10^{27}$ s^{-1}. These values are useful guidelines, but should not be taken too strictly, as the transition between stages is smooth.

In addition to the outgassing rate, the state of the solar wind plasma is an important parameter shaping the cometosphere. Figure 1c shows the solar wind dynamic pressure at Comet 67P over two years at different helio-

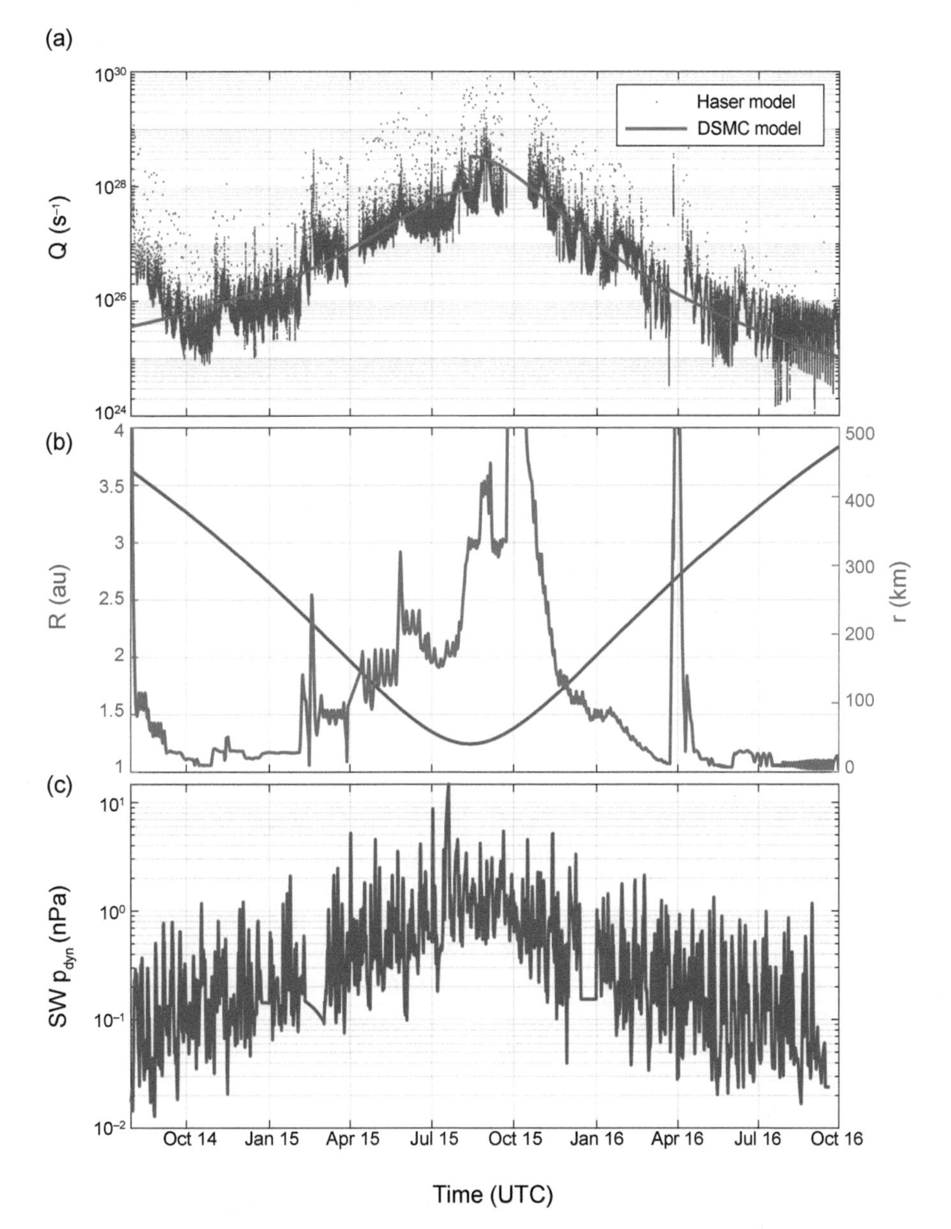

Fig. 1. (a) Gas production rate of Comet 67P derived from *in situ* measurments using the Haser model (red) and averaged gas production rate from a DSMC model (blue). **(b)** Heliocentric distance of Comet 67P (red) and cometocentric distance of Rosetta at 67P (blue). The dayside excursion in October 2015 and the nightside excursion in April 2015 went out to 1500 km and 1000 km respectively. **(c)** Solar wind dynamic pressure at 67P derived from Earth-based data and a Tao model (*Tao et al.*, 2005).

centric distances. The closer the comet is to the Sun, the higher the dynamic pressure ρv_{SW}^2, because the density ρ decreases approximately as r_h^{-2}, where r_h is the heliocentric distance. The interplanetary magnetic field magnitude is also higher closer to the Sun so that, as a comet follows its orbit, the solar wind parameters vary on timescales of weeks to months due to the heliocentric distance change. Without knowing the exact solar wind conditions that a comet encounters, it is difficult (but not impossible!) to distinguish between intrinsic changes of the plasma environment and changes due to external triggers. The large timescale changes are therefore usually absorbed in the definitions of the comet's activity stages, as they happen on the same timescale. The shorter timescales are then discussed on a case-by-case basis.

While most of this chapter will concern itself with the interaction of different plasma populations with each other, the solar wind can also interact with the nucleus directly. This is only possible if the activity of the nucleus is negligible, because if there is a significant atmosphere, the solar wind will not be able to reach the nucleus. Solar wind sputtering causes dust particles to be freed from the nucleus. This process is useful to investigate surface composition without having a surface probe, but so far only Comet 67P was suitable to observe sputtering (*Wurz et al.,* 2015). There, silicon was the most abundant species, with traces of sodium, calcium, and potassium all detected at 67P. Models showed that indeed the observed signal was due to sputtering, not gas or dust release due to thermal input into the surface. The ratio of sodium to potassium was on the same order as that of the lunar surface, the solar system, and meteorites.

2.2. Shaping the Induced Cometosphere

When intercepting Comet 67P at about 3.5 au ($Q \simeq 10^{26}$ s^{-1}), the Rosetta plasma instruments reported the first signatures of a plasma environment significantly affected by cometary matter at a few hundreds of kilometers from the cometary nucleus (*Clark et al.,* 2015; *Yang et al.,* 2016). At this time, the outgassing rate of 67P was estimated to be 4 orders of magnitude smaller than Comet 1P/Halley's outgassing rate (*Combi and Feldman,* 1993; *Hansen et al.,* 2016). Hence, even during its weakly outgassing phases, 67P was actively mass-loading the solar wind plasma and maintaining an induced cometosphere (*Nilsson et al.,* 2015; *Richter et al.,* 2015; *Deca et al.,* 2017).

2.2.1. Plasma sources and sinks. Fundamentally, there are two populations of particles that contribute to shaping the near-cometary environment, or better, the induced cometosphere: the incoming solar wind plasma and the radially-expanding neutral gas coma that is ionized through a variety of processes, such as photoionization, charge exchange, and electron-impact ionization (see the chapter in this volume by Beth et al.).

67P's neutral coma consists primarily of H_2O, CO_2, CO, and O_2 (see the chapter by Beth et al.). Historically, the Haser model, which assumes a spherically-symmetric homogeneous outgassing for each species at a constant neutral radial velocity, has been applied to estimate the production rates of cometary volatiles (*Haser,* 1957). While the model can include different species and loss through ionization, a reasonable approximation of the overall, momentary gas production rate Q can be obtained by

$$Q = 4\pi n_n u_n r^2 \tag{1}$$

where n_n is the local, measured neutral density, u_n is the neutral velocity of 500–1000 m s^{-1} and r is the cometocentric distance of the measurement point. Within a couple of kilometers of the comet a more accurate model should be used, but for the distances covered by Rosetta during most of its mission, this approximation is reasonable to disentangle the effects of the radial variation of the trajectory from that of the temporal variation.

Rosetta's advanced instruments were able to capture for the first time the details of the non-uniform time-varying volatile distribution in which the cometary nucleus is embedded (*Marshall et al.,* 2019). The complexity of the latter is caused by the rotation and orbital motion of the comet. In other words, Rosetta's high temporal and spatial resolution measurements allowed to develop an accurate shape model that allows solar illumination and shadowing effects to be taken into account to predict the evolution of the production rates as a function of heliocentric distance (*Bieler et al.,* 2016; *Fougere et al.,* 2016a,b; *Hansen et al., 2016).*

The cometary plasma density is determined by a continuous interplay of five main processes: (1) When a neutral molecule absorbs sunlight, a positive ion and photoelectron may be created. Photoelectrons are believed to be a significant contributor to the warm electron population (see below). (2) At larger heliocentric distances/lower outgassing rates, accelerated solar wind electrons (see section 2.2.3) lead to elevated electron-impact ionization rates, because the electron ionization thresholds for most cometary species are significantly higher than the typical local photoelectron energies (*Itikawa and Mason,* 2005; *Bodewits et al.,* 2016; *Deca et al.,* 2017; *Heritier et al.,* 2018; *Divin et al.,* 2020). (3) At very large cometocentric distances, fast and light solar wind H^+ and He^{2+} ions may capture electrons from slow heavy neutral species through charge exchange (*Fuselier et al.,* 1991; *Bodewits et al.,* 2004; *Simon Wedlund et al.,* 2019b). Although this process does not produce any net ionization, momentum is transferred from the fast solar wind ions to the slow neutral coma and mass-loads the cometary environment (*Simon Wedlund et al.,* 2017). (4) Fast solar wind ions that impact cometary neutrals can provide an additional source of charged particles to the cometary plasma density. (5) Finally, the primary process causing a loss of plasma in the induced cometosphere is electron-ion dissociative recombination. When the plasma number density is sufficiently high and contains cold electrons, ions and electrons can merge back together and with the excess energy create a neutral fragment (see also the chapter in this volume by Beth et al. and *Simon Wedlund et al.,* 2019a).

2.2.2. Ion dynamics. In previous books on cometary plasma processes, the discussion on the behavior of ions was often based on a (single-)fluid dynamics approach. This was for two reasons: (1) Simulations in a magnetohydrodynamic (MHD) regime were the predominant simulation method, as hybrid and fully kinetic simulations required computational resources out of reach of the previous generations of supercomputers; and (2) for the comets visited until then, a fluid dynamics approach was in reasonable agreement with the measurements (partially because fine-scale particle dynamics could not be resolved by the onboard instruments). However, even then it was clear that weakly active comets necessitate a different approach, because ratios between the relevant plasma scales are different from more active comets. A useful parameter that highlights these differences is the particle gyroradius

$$r_g = \frac{m v_\perp}{|q| B} \tag{2}$$

where m and q are the particle mass and charge, B is the magnetic field strength and $v_\perp$ is the particle velocity component that is perpendicular to the field. For B = 5 nT and a water group ion with a relative velocity to the magnetic field of 400 km s^{-1}, this results in a gyroradius of ~15,000 km. Compared to the scale size of Comet 1P/Halley's interaction region (2 × 10^6 km) this is small and thus gyroradius effects on the large-scale structure of the environment should be negligible. On the other hand, the interaction region at 67P only extends 100 km (at low activity) to ~7000 km (high activity) and thus the cometary ion gyroradius is larger and cannot be neglected.

These gyroradius effects are observable in the data collected by Rosetta's instruments. As was speculated in *Comets II*, the cometary ions are virtually at rest in the comet's frame of reference, which means they move at v_{sw} in the solar wind frame. Therefore, they are subject to the solar wind convective electric field

$$\vec{E}_{conv} = -\vec{v}_{sw} \times \vec{B} \quad (3)$$

and are accelerated along the field. This is illustrated in Fig. 2 by the purple arrows. *Rubin et al.* (2014) showed that these gyroradius effects can be modeled reasonably well with a hybrid as well as a multi-fluid MHD approach. The inclusion of the cometary ions as a separate fluid was shown

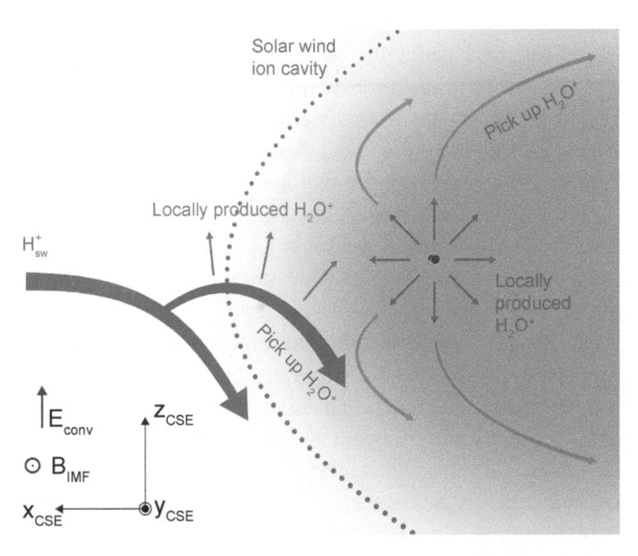

Fig. 2. Sketch of the bulk ion motion at an intermediately active comet. The Sun is to the left. The solar wind protons (red) are deflected to conserve momentum as the locally produced H_2O^+ (purple) is accelerated along the solar wind convective electric field E_{conv}. Pickup ions (dark blue) that were produced far upstream eventually take over the role of the solar wind protons in the flow toward the comet. Within the solar wind ion cavity, the convective electric field is shielded. Within this region, the locally produced water ions (light blue) expand radially and are eventually accelerated and picked up by various processes (electric fields, wave-particle interaction). Adapted from *Nilsson et al.* (2020).

to be a good method to simulate the large-scale gyroradius structures, expanding the applicability of the MHD models. If all cometary ions are moving in one direction, momentum conservation dictates that the solar wind ions have to move in the opposite direction; they are deflected away from the comet-Sun line. This has been observed at low activity at 67P as well as in the Active Magnetospheric Particle Tracer Explorers (AMPTE) artificial comet experiments (*Coates et al.,* 2015; *Behar et al.,* 2016b). This in turn modifies the plasma flow, as the addition of the heavy cometary ions leads to a reduction in overall flow velocity. Incoming solar wind ions are thus faster than the bulk flow and start to gyrate in the magnetic field, similar to the cometary ions. This leads to an additional component to the solar wind ion deflection that can add up to 180° in the denser part of the coma, where the magnetic field is higher. In effect this means that under certain outgassing conditions, solar wind ions can flow sunward (see also blue and red boxes in Fig. 10). At this point, the solar wind ions are so deflected that they cannot flow toward the inner coma anymore (red arrow in Fig. 2). This is the approximate position of the solar wind ion cavity. Within that region, no solar wind ions can be observed (see section 3). In the aftermath of the Halley flybys, the region outside the solar wind ion cavity was often referred to as the cometosheath, i.e., the region where cometary and solar wind ions are mixed and flow around the inner coma (*Mandt et al.,* 2016).

For any investigation of the ion flow pattern (as, e.g., shown in Fig. 2) at low gas production rates, or far from the nucleus, it can be advantageous to transform the data into a cometocentric solar electric (CSE) coordinate system. In this system the comet is at the origin and the x-axis points toward the Sun. The y-axis is aligned with the solar wind convective electric field and the z-axis completes the triad.

Ignoring the electron dynamics for now (see section 2.2.3), an ion plasma flow toward the nucleus remains, as also evidenced by the presence of an interplanetary magnetic field inside the solar wind ion cavity. This means that accelerated (pickup) cometary ions must have taken over the role of the solar wind ions in this flow. Indeed, observations show two distinct populations of cometary ions at this stage (*Berčič et al.,* 2018): one accelerated population flowing approximately anti-sunward (dark blue arrows in Fig. 2) and a radially expanding, slow-moving population (blue locally produced H_2O^+ in Fig. 2). This flow pattern is consistently observed in the innermost coma, where accelerated ions are often observed in pulses, while the slower, radially expanding population is ubiquitous. An open question remains: How do the kinetic aspects of the ion dynamics around a low-to-medium activity comet continuously evolve toward this more fluid-like behavior at high outgassing activity, close to the Sun? And, related, what is the speed of the bulk of those low-energy ions accelerated near the nucleus? What is the role of the ambipolar, polarization, and Hall electric field for these acceleration processes?

This transfer of energy and momentum from the solar wind to the cometary ions can also be surveyed in terms of momentum flux. The momentum flux in a cometary environment is the sum of the momentum flux of the cometary ions, solar wind ions, electron pressure, and magnetic pressure. The higher the gas production rate (and the solar wind dynamic pressure) the higher the total momentum flux measured by the ion instruments. From recent observations it becomes clear that the electron pressure gradient is the dominant source of momentum at all activity levels, except those that are very high (close to perihelion of 67P). While the solar wind ions carry some momentum into the coma at low activity, this momentum is taken up by the cometary ions within the solar wind ion cavity (*Williamson et al.,* 2020). This indicates an efficient momentum and energy transfer from the solar wind ions to the pickup cometary ions via the convective electric field.

Within the inner coma, the solar wind convective electric field is not present anymore; instead, other fields such as an ambipolar or a polarization electric field gain importance and lead to more complicated flow patterns (*Nilsson et al.,* 2018; *Gunell et al.,* 2019; *Deca et al.,* 2019). Close to the nucleus, where collisions are important, the dynamics of the ions are coupled to the radially expanding neutral molecules (*Nicolaou et al.,* 2017; *Berčič et al.,* 2018; *Nilsson et al.,* 2020). Closer to the Sun, more complex flow patterns were observed around the diamagnetic cavity as collisional effects become important in a greater area of the interaction region. In general, simulations and observations show that the accelerated ions have a flow pattern that guides them around the diamagnetic cavity most of the time, but in some instances the accelerated ions can go through the boundary and also flow tailward within the diamagnetic cavity (*Masunaga et al.,* 2019).

2.2.3. A four-fluid coupled system. Close to the nucleus of Comet 67P, Rosetta's Langmuir probe measured a 1/r decay of the plasma density with cometocentric distance (*Edberg et al.,* 2015; *Galand et al.,* 2016). Consequently, it indicates that the ratio between the cometary and solar wind plasma density, and by extension the full set of plasma parameters throughout 67P's plasma environment, continuously changes within a comet's plasma interaction region (*Nilsson et al.,* 2017; *Eriksson et al.,* 2017; *Berčič et al.,* 2018). In other words, the global configuration of the induced cometosphere is driven by changes in the locally dominating physical processes and the relative ratios of the local plasma scales, such as the electron/ion-neutral collisional mean-free-path and the inertial lengths of the plasma species. A variety of permanent and transient boundaries can be defined [section 3 and *Mandt et al.* (2016)]. In addition, (the existence of) these boundaries evolve(s) with heliocentric distance, or better, with the activity regime of the comet.

During weakly outgassing regimes, i.e., well before a bow shock or diamagnetic cavity forms upstream of the cometary nucleus, ionization of the outflowing neutral gas from the nucleus compresses the incoming solar wind. A magnetic pileup region forms and the magnetic field lines drape around the nucleus (see section 2.4) (see also *Koenders et al.,* 2013, 2015, 2016a; *Behar et al.,* 2016a,b; *Volwerk et al.,* 2016).

In addition to section 2.2.2, here we include the electron dynamics. The discussion is based on several fully kinetic particle-in-cell simulations that self-consistently model four plasma populations: solar wind protons, solar wind electrons, cometary water ions, and cometary electrons (*Deca et al.,* 2017, 2019; *Sishtla et al.,* 2019; *Divin et al.,* 2020). Assuming a collisionless plasma, the dynamical interaction that determines the general structure of the induced cometosphere during the weakly outgassing phases can be interpreted as a four-fluid coupled system, where the solar wind electrons move to neutralize the cometary ions, and the cometary electrons organize themselves to neutralize the solar wind ions (illustrated in Fig. 3) (*Deca et al.,* 2017). More precisely, ions of cometary origin accelerate along the local solar wind convective electric field in the cross-magnetic field direction, whereas electrons of cometary origin are initially accelerated in the opposite direction. This results in a net (Hall) current that locally decouples the solar wind protons and electrons as the solar wind plasma becomes more and more mass-loaded by cometary ions. It is important to note that the convective electric field carries an opposite sign in the solar wind and cometary ion reference frame and transfers momentum between the two species. As the solar wind protons are deflected and are no longer frozen in, the interplanetary magnetic field is carried close to the comet through the solar wind (and cometary) electrons. Note that a crucial component of the mechanism described above is provided by the electron dynamics. One of the revelations of the Rosetta mission and subsequent numerical modeling is that key interaction mechanisms based on fluid concepts, such as momentum and energy transfer between the solar wind and cometary plasma, can be explained by self-consistently including the electron dynamics. For example, the deflection of cometary electrons creates the electron current that induces the magnetic pileup region as the incoming solar wind is compressed (*Deca et al.,* 2019). Figure 3 presents an overview of the four-fluid behavior of the solar wind interaction with a weakly outgassing comet.

Although modeling both the electron and ion species is a necessity to unveil the interaction discussed above, less-complete approaches (from an electron point-of-view) often present different benefits. For example, more complex ion-interaction models can be implemented as less-computational resources are needed when keeping the electron populations as a massless charge-neutralizing fluid as it is done in a hybrid approach (*Simon Wedlund et al.,* 2019a; *Koenders et al.,* 2016a; *Alho et al.,* 2020). This has the advantage of reducing simulation time while retaining the ion kinetics. To resolve larger interaction regions, simulate longer timescales and for parameter studies that require a short turnover time, a multi-fluid approach can be useful (e.g., *Huang et al.,* 2018). However, this comes at the cost of losing the electron and ion kinetics and many small-scale structures. Compared to fully kinetic treatments, hybrid and multi-fluid (Hall) MHD models often assume a a suitable electron closure relation combined with a generalized Ohm's

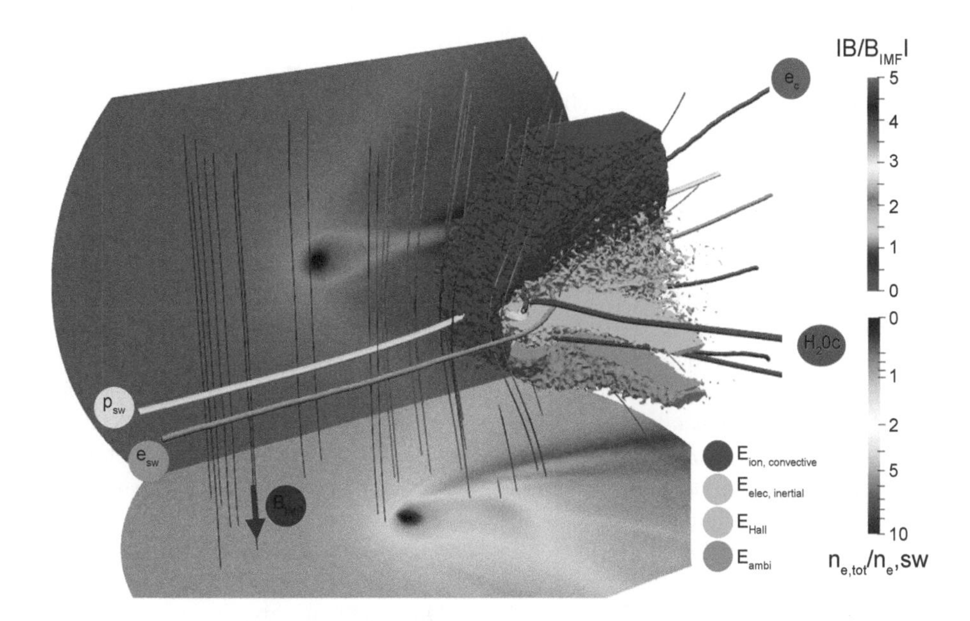

Fig. 3. Illustration of the four-fluid coupled system, including various isovolumes that indicate the significant components of a generalized Ohm's law. See text for details. Figure adapted from *Deca et al.* (2019).

law (GOL) to approximate the electric field in the system (*Huang et al.,* 2018; *Deca et al.,* 2019)

$$\mathbf{E} = -(\mathbf{u}_i \times \mathbf{B}) + \frac{1}{en}(\mathbf{j} \times \mathbf{B}) - \frac{1}{en}\nabla \times \mathbf{\Pi}_e + \frac{m_e}{e}\nabla \times (\mathbf{u}_e\mathbf{u}_e) \tag{4}$$

where e and m_e are the electron electric charge and mass, respectively, **B** the magnetic field, $\mathbf{u}_i$ the ion mean velocity, **j** the current density, n the plasma total number density defined as the sum of the solar wind and cometary densities, $n = n_{sw} + n_c$, and $\mathbf{\Pi}_e$ the electron pressure tensor derived from the electron momentum equation. The various terms decompose the total electric field in (from left to right) the convective electric field, the Hall electric field, the ambipolar electric field, and the contribution associated with the electron inertia. Using the results from their fully kinetic model, *Deca et al.* (2019) were able to compute the various contributions using the output particle data from the simulation, in this way identifying where an isotropic single-electron fluid Ohm's law approximation can be adopted and where it cannot. To illustrate, Fig. 3 includes various isovolumes that indicate the regions where specific terms of the GOL described above are significant. Consistent with Rosetta's measurements, they conclude that near the nucleus the electron pressure gradient is dominant, and that at spatial scales smaller than the ion plasma scales the total electric field is primarily a combination of the solar wind convective electric field and the ambipolar electric field. Interestingly, throughout most of the region the electron inertial term was found negligible. This also implies that multi-fluid and hybrid simulations with an electron pressure equation are valid most of the time, except when a distinction between electron populations is necessary. While cometary missions and simulations have helped us understand the complex interplay of these four fluids to a greater degree, the exact plasma distribution with respect to cometocentric distance (in three dimensions) and how it varies in time is still unclear. In addition, it remains to be investigated what role the (quiet) solar wind dynamics (e.g., turbulence) play in the inner coma dynamics, where the plasma is shielded from direct influence of solar wind variation but could be affected indirectly, via pickup ion dynamics.

2.2.4. Electron populations and the ambipolar electric field. The electron distributions observed in the inner coma of 67P can be traced back to two main categories: cometary electrons and solar wind electrons. The cometary electron population originates from ionizing neutral molecules of cometary origin through solar extreme ultraviolet radiation [photoelectrons (*Vigren and Galand,* 2013)] or secondary particle processes as described above (*Simon Wedlund et al.,* 2017; *Heritier et al.,* 2018). The undisturbed solar wind electron population typically consists of a thermal core ($E <$ 50 eV) and a suprathermal tail ($70 < E < 1000$ eV) (*Pierrard et al.,* 2016).

A variety of physical mechanisms in the cometary plasma environment can affect both these source populations, such as ambipolar electric fields (*Madanian et al.,* 2016; *Deca et al.,* 2019; *Divin et al.,* 2020), adiabatic compression and expansion (*Nemeth et al.,* 2016; *Broiles et al.,* 2016; *Deca et al.,* 2019; *Madanian et al.,* 2020), collisional cooling (*Engelhardt et al.,* 2018a), and wave-particle interactions (*Goldstein et al.,* 2019; *Lavorenti et al.,* 2021).

In reality, it is not feasible to tell which group an electron belongs to without additional information, e.g., from numerical modeling. Most generally, three distinct electron populations were observed: a warm (5–10 eV), cold (0.01–1 eV), and hot (or suprathermal) population (a few tens to 100 eV).

In order to understand the origins of these three distinct populations, we need to have a closer look at the electromagnetic environment surrounding the cometary nucleus. Whereas ionization of the cometary gas results in electrons that have thermal speeds on the order of 1000 km s^{-1} (*Vigren and Galand,* 2013), the much heavier ion counterparts tend to retain the speed of their parent neutral molecule (typically on the order of 1 km s^{-1}). A strong electron pressure gradient forms, which results in a potential well and ambipolar electric field that surrounds the cometary nucleus (*Madanian et al.,* 2016; *Deca et al.,* 2019; *Divin et al.,* 2020).

Note that efficient electron cooling during high outgassing phases may neutralize the ambipolar electric field component, such as was the case for the Giotto flyby of Comet 1P/Halley (*Gan and Cravens,* 1990), but the importance of the ambipolar electric field for low-activity comets has been one of the most intriguing discoveries of the Rosetta mission. For example, *Galand et al.* (2020) found that this acceleration of the electrons in an ambipolar field can produce extreme ultraviolet (EUV) emissions, similar to those associated with auroral emissions at other solar system bodies.

The ambipolar electric field and its associated potential well can temporarily trap cometary electrons and accelerate solar wind electrons near the cometary nucleus, leading to the observed warm and hot electron populations (*Deca et al.,* 2017). Studying the trajectories of trapped electrons in the ambipolar potential well surrounding the nucleus, a clear boundary in velocity space can be defined that separates temporarily trapped cometary electrons from passing solar wind electrons (*Sishtla et al.,* 2019). It shows that electrons may stay much longer in the region of dense neutral gas around the nucleus than previously believed, which may also lead to increased cooling of electrons in non-collisionless plasma regimes (*Engelhardt et al.,* 2018a; *Divin et al.,* 2020). Using fully kinetic particle-in-cell simulations it was found that cometary electrons exhibit an apparent isotropic and almost isothermal behavior, whereas the solar wind electrons, on the other hand, exhibit an anisotropic and apparent polytropic behavior (*Deca et al.,* 2019). In addition, at a few tens of kilometers from the nucleus, the electron sensor observed a highly variable bi-Maxwellian electron distribution with energies ranging from tens up to several hundred electron volts (*Clark et al.,* 2015), consistent with a dense warm and a rarefied hot electron population (*Madanian et al.,* 2016; *Deca et al.,* 2017). *Broiles et al.* (2016) constrained the kappa values of the warm population ranging from 10 to 1000, suggesting the warm electrons are near-thermal equilibrium [consistent with *Deca et al.* (2019)], whereas values between 1 and 10 were found for the hot electron component. The warm population showed a 1/r density dependence up to about 900 km from the nucleus and a $r^{-0.7}$ dependence farther out (*Edberg et al.,* 2015; *Myllys et al.,* 2019), where r is the cometocentric distance. In addition, *Myllys et al.* (2019) corroborated on a $1/r^2$ dependence with density for the hot electrons.

Taking into account the radial cometocentric dependence of the warm population, both the warm and hot components manifest an increase in density with decreasing heliocentric distance, consistent with the observed increase in neutral gas density and photoionization frequency. The density of the warm and hot population increased from about 30 to 100 cm^{-3} and 0.1–3 cm^{-3} between 3.5 and 1.3 au, respectively. The hot electron population showed a 33% increase in temperature with heliocentric distance, whereas the warm population seemed not affected by the latter parameter. *Myllys et al.* (2019) concluded as well that the ambipolar acceleration process could only explain part of the observed hot electron distribution, suggesting that part of the population may have its origin in the already accelerated suprathermal component of the solar wind electron distribution (*Broiles et al.* 2016).

Quite intriguingly, the accelerated hot electron population was found to be correlated with far ultraviolet (FUV) emissions observed near Comet 67P, showing for the first time that cometary aurora can be driven by the interaction of the solar wind with the local environment (*Galand et al.,* 2016).

Although Giotto could not directly measure cold electrons, it was assumed that the dense inner coma of 1P/Halley could efficiently cool electrons through collisions with the neutral gas (*Feldman et al.,* 1975; *Ip and Mendis,* 1975; *Eberhardt and Krankowsky,* 1995). Electron cooling is thought to be most efficient within the electron exobase (or collisionopause; see the chapter in this volume by Beth et al.). However, Rosetta's Langmuir probe and Mutual Impedance Probe (MIP) detected cold electron signatures significantly outside the electron exobase, suggesting that once cooled in the inner coma, there is no mechanism that is able to reheat cold electrons when they travel away from the nucleus (*Eriksson et al.,* 2017). At this point, it is not known what processes contribute to cooling newborn, warm electrons. The cold electron temperature observed ranged between 0.05 and 0.3 eV (*Wattieaux et al.,* 2020). In addition, cold electrons measurements were only significant in regions dominated by (warm) cometary plasma (Fig. 4). It was shown that there is a clear correlation between electron cooling and the observations of the diamagnetic cavity, suggesting that both have to be in close proximity (*Henri et al.,* 2017; *Odelstad et al.,* 2018). Finally, cold electrons have as of yet not been observed without the presence of a significant warm electron population (*Eriksson et al.,* 2017; *Myllys et al.,* 2019; *Gilet et al.,* 2020; *Wattieaux et al.,* 2020). So far it is unclear what the energy distribution of electrons below the spacecraft potential is.

2.2.5. Cometary outbursts. Up to this point, we have focused on a steady-state solar wind plasma interaction with the cometary environment. Throughout the Rosetta mission, however, at least one internal outburst was detected that significantly affected the near-cometary plasma environment (*Grün et al.,* 2016). During the outburst, the neutral density

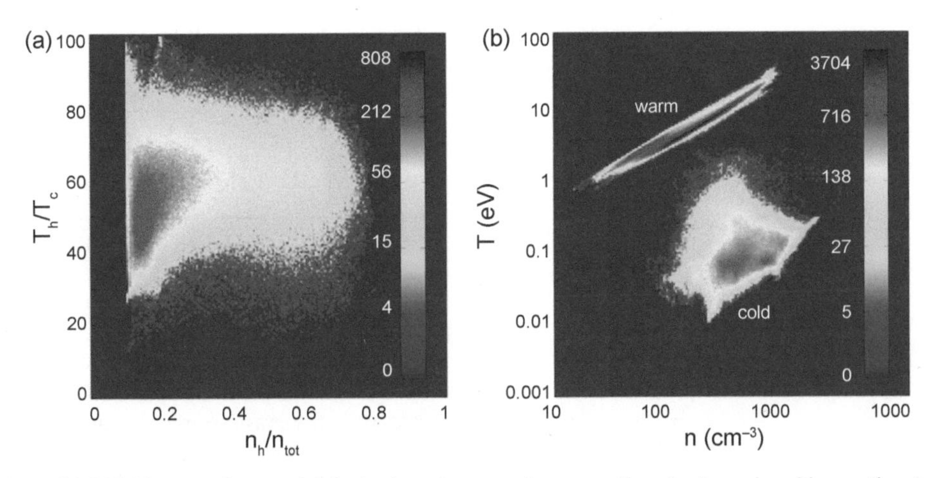

Fig. 4. Distribution of **(a)** T_h/T_c vs. n_h/n_{tot} and **(b)** electron temperatures vs. the electron densities estimated at the Rosetta orbiter locations in the cometary plasma of Comet 67P from January to September 2016 (570,000 samples), where the subscripts h, c, and tot indicate hot, cold, and total, respectively. Colors correspond to the number of responses (the color map is logarithmically scaled in both panels). Figure adapted from *Wattieaux et al.* (2020).

doubled, but interestingly, the plasma density quadrupled, which was attributed to an increased efficiency of the electron impact ionization process as well as more efficient ion-neutral coupling in the denser neutral gas, which reduces the ionization length scales (*Hajra et al.,* 2017). Rosetta also locally measured a decreased suprathermal electron flux and solar wind ion density, as well as a reconfigured magnetic field and a change in the singing comet wave activity [see section 4 and *Breuillard et al.* (2019)].

2.3. External Drivers to the Induced Cometosphere

Above we discussed the major internal processes that shape the near-cometary plasma environment. Solar wind (SW) transients, flares, CIRs, and ICMEs, on the other hand, are external drivers to the induced cometosphere. Transient phenomena change the solar wind parameters on relatively short timescales and, as a result, have the potential to significantly reconfigure/disturb the near-cometary plasma environment. The effects can be extremely variable and differ from event to event, depending on the strength and type of the transient and the heliocentric location and outgassing rate of the comet. Nevertheless, several characteristics can be identified that are consistently observed for specific classes of transient phenomena.

Solar flares are short-lived and temporarily increase the local EUV flux. However, during the weakly outgassing stages of a comet, photoionization is only a minor contributing plasma source. During the strongly outgassing stages, on the other hand, the induced cometosphere is inherently very variable even without external drivers. As a result, solar flares are not observed to significantly disturb the plasma interactions that shape the induced cometosphere (*Edberg et al.,* 2019).

CIRs, on the other hand, spiral outward in the solar system and are often accompanied by a forward shock at their leading edge [and a reverse shock at the trailing edge (*Balogh et al.,* 1999)]. This has been observed to lead to an increase in the local solar wind velocity, proton and cometary plasma density, temperature, magnetic field strength, and suprathermal (hot) electron flux (*Edberg et al.,* 2016b; *Hajra et al.,* 2018). The latter results in an increased electron-impact ionization rate, which, on its turn, may become the largest contributor to the total plasma density, rather than increased compression or acceleration due to the ion pickup process (*Galand et al.,* 2016; *Heritier et al.,* 2018; *Hajra et al.,* 2018). The magnetic compression resulting from the CIR impact on the induced cometosphere, on the other hand, is responsible for the local increase in proton density and magnetic field strength.

Specific for 67P, *Hajra et al.* (2018) observed greater increases on the southern (summer) hemisphere of the comet vs. its northern (winter) hemisphere. The effects of CIRs are typically observed for about 24 hr, with clear signatures of the forward and reverse shock wave visible even in the coma region. Langmuir waves can be excited near the forward and reverse shocks of a CIR in the solar wind, where electrons can be accelerated such that they counterstream in the solar wind flow (*Pierrard et al.,* 2016). Locally generated Langmuir waves have been observed during times of CIR (and ICME) interaction with the cometary environment, indicating that backstreaming electrons exist even in the coma of the comet. However, the exact mechanism that causes the electrons to backstream along the field lines from the shock interface is not yet understood (*Myllys et al.,* 2021).

ICMEs comprise the most variable class of external drivers to the induced cometosphere. However, they tend to accelerate most within 1 au, after which they gradually decelerate travelling outward into the solar system (*Zhao et al.,* 2017). Their interaction with the cometosphere can be characterized by an increase in magnetic field and plasma density, an increase in energetic electron flux (e.g., Fig. 5), and a rapid (Forbush) decrease in the observed galactic cosmic ray intensity (*Witasse et al.,* 2017). ICMEs may also merge with other transients and form more complex

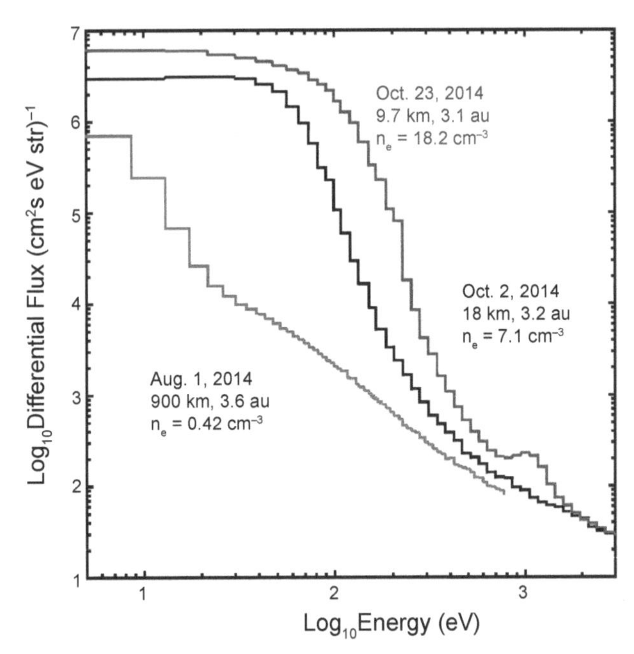

Fig. 5. The daily averaged electron differential flux as measured by Rosetta's Ion and Electron Spectrometer. The spectrum from August 1, 2014, shows a typical solar wind type electron flux. The spectrum from October 2, 2014, and October 23, 2014, show a typical quiet and ICME-impacted electron flux profile near the comet, respectively. Adapted from *Madanian et al.* (2016).

structures (*Witasse et al.,* 2017; *Wellbrock et al.,* 2018). Although the Rosetta spacecraft was not often in an ideal position to characterize the impacts of ICMEs throughout the induced cometosphere, an ICME impact detected close to perihelion led at 170 km from the cometary nucleus to the highest magnetic field strength (300 nT) measured through the mission (*Goetz et al.,* 2018).

2.4. Magnetic Field Pileup

The interaction of the interplanetary magnetic field (IMF) (sometimes also referred to as the heliospheric magnetic field) carried by the solar wind and an outgassing comet leads to the formation of an induced magnetosphere with magnetic field pileup and field line draping, an overview of which is shown in Fig. 6. This happens as soon as the comet gets active, usually when it enters Jupiter's orbit, i.e., 5 au radial distance from the Sun. The convective electric field of the solar wind will accelerate the newborn cometary ions, and these start to gyrate around the magnetic field lines. This is called ion pickup or mass-loading of the solar wind. As most of the earlier work on cometary magnetospheres is discussed in *Ip* (2004), this section will mainly concentrate on work after 2004. Because of conservation of momentum the mass-loaded solar wind will have to decelerate. *Biermann et al.* (1967) presented a simple one-dimensional MHD model of this solar wind-comet interaction by combining the continuity, momentum, and energy equations. The solar wind velocity $u_x(x)$ is in the x-direction and the IMF B_y is in the y-direction. This results in a solar wind velocity

$$u_x(x) = \left(2(f+1)(M(x)+\rho_\infty u_\infty)\right)\Big[(f+2)\rho_\infty u_\infty^2 \pm \sqrt{(f+2)\rho_\infty^2 u_\infty^4 - 4(f+1)(M(x)+\rho_\infty u_\infty)\rho_\infty u_\infty^3}\Big] \tag{5}$$

where ρ_∞ and u_∞ are the undisturbed solar wind density and velocity, f is the number of degrees of freedom of the ions, and M(x) is the mass source. Using, e.g., the *Haser* (1957) model for cometary outgassing, one can solve the continuity equation for the magnetic flux, which leads to $B_y(x) \to \infty$ at the location where the bow shock is formed. This can be solved by adding cooling of the plasma through charge exchange processes to obtain finite field strengths downstream of the bow shock (for details, see *Galeev et al.,* 1985; *Goetz et al.,* 2017). However, there are some limiting assumptions to this model: The solar wind magnetic field should be perpendicular to the plasma flow direction (but is in reality on average at the Parker spiral angle), single-fluid MHD is assumed (although the ion gyroradius can be larger than the interaction region), and the model is only valid on the Sun-comet line. Nevertheless, a comparison of the modeled magnetic field and the measured field at Comet 67P gives reasonable agreement (*Goetz et al.,* 2017).

2.5. Magnetic Field Draping

The one-dimensional model described above delivers the magnetic field strength in the subsolar cometary environ-

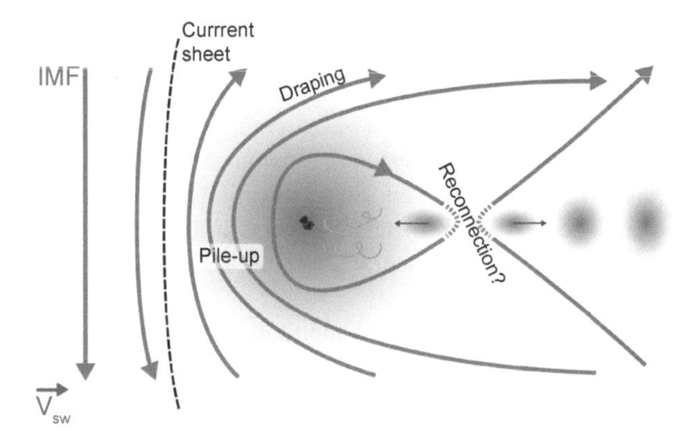

Fig. 6. An illustration of the large-scale magnetic field at a comet, including processes in the tail. The solar wind with the IMF approaches from the left. The magnetic field lines are shown in green, and a heliospheric current sheet is marked by the black dashed line. In the near-comet tail there are spiraling magnetic field waves in orange. A possible reconnection region in the tail produces plasma density enhancements (blue) that travel toward and away from the reconnection region. For further information and scale sizes see text.

ment but does not describe the full structure of the induced magnetosphere. Therefore, it needs to be expanded with a second radial dimension, perpendicular to the comet-Sun line. This results in the magnetic field line being draped around the active nucleus, as was first described by *Alfvén* (1957). Through the finite conductivity, the magnetic field will diffuse through the plasma cloud around the nucleus, thereby avoiding another magnetic catastrophe, i.e., an infinite magnetic field strength. The IMF, however, varies in direction and the radial component of the magnetic field changes sign regularly, therefore a differently-directed magnetic field can be draped around the active nucleus. When the diffusion speed through the plasma is smaller than the flow velocity of the mass-loaded solar wind, a layering of a differently-directed magnetic field will be created upstream of the comet, which is referred to as "nested draping." This phenomenon was first observed at Comet 1P/Halley during the flybys of Vega 1 (*Riedler et al.,* 1986) and Giotto (*Raeder et al.,* 1987). The boundaries between these different magnetic field directions will have to support current sheets. Indeed, during the early phase of Rosetta, during the unbound pyramidal orbits, many rotations of the field were found with a minimum magnetic field strength and a maximum in electron density in the center of the rotation (*Volwerk et al.,* 2017). Again, this would argue for nested draping of the IMF if this were a "snapshot" flyby like Giotto. However, Rosetta moves at a speed of only a few meters per second, which needs a different interpretation. On the other hand, if the diffusion speed is fast enough, no such pileup will occur, apart from a regular increase of the magnetic field strength created by the mass-loading and slowing down of the solar wind.

Another situation occurred at Comet 67P, during the dayside excursion by Rosetta. The spacecraft took an upstream excursion of ~20 days up to a distance of ~1500 km from the nucleus. In order to find nested draping, *Volwerk et al.* (2019) determined the cone angle of the magnetic field, defined as

$$\theta_{co} = \tan^{-1}\left\{\frac{\sqrt{B_y^2 + B_z^2}}{B_x}\right\} \tag{6}$$

where $\theta_{co} = 0°$ is a purely sunward-directed field and $\theta_{co} = 180°$ purely anti-sunward. (B_x, B_y, and B_z are defined in a CSEQ coordinate system, with the nucleus in the origin; $\vec{e}_x$ points to the Sun, $\vec{e}_z$ is parallel to the solar rotational axis, and $\vec{e}_y$ completes the righthanded system.) Indeed, regions of a differently-directed magnetic field were found, but when the spacecraft returned back after apoapsis, a region was entered that did not exist during the outbound leg (see Fig. 7). This indicates that the draping moves faster toward the nucleus than the spacecraft is moving, which is not difficult as Rosetta moved at ~1 m s^{-1}. The authors called this "dynamic draping." The question of how dynamic and nested field line draping are related is still open, as there is either a snapshot observation by a fast moving flyby or there is only a local long-term observation by an orbiting spacecraft. A multi-spacecraft mission could allow us to get a better grip on this problem: Different trajectories could be flown, like it is planned for the upcoming Comet Interceptor mission (*Snodgrass and Jones,* 2019).

Recently, *Delva et al.* (2014) reinvestigated the Vega 1 magnetometer data taken during the flyby of Comet 1P/Halley to study the magnetic pileup boundary (MPB) and the draping pattern of the magnetic field. The MPB is the boundary between the cometosheath (where there is no draping) and the pileup region (where there is strong draping). In order to study the draping, an electromagnetic reference frame was calculated from the aberrated cometocentric coordinate system, in which x_{IMF} is along $\mathbf{v}_{SW}$ and the new y_{IMF}-axis direction is given by $-\mathbf{v}_{SW} \times \mathbf{B}_{IMF}$. The draping is then studied by the relationship between $B_{x,IMF}$ and B_{rad}, the field component pointing radially away from the cometary nucleus. There should be a strong negative correlation between the two variables if the spacecraft is inside the MPB.

One effect has not been taken into account in this discussion, and that is conservation of momentum related to the pickup process through the acceleration of the heavy cometary ions. The newly-created ions are accelerated by the convective electric field of the solar wind. Through Newton's second law, this action will need an opposite reaction, and thus the solar wind ions will have to move in the direction opposite to the convective electric field (see also section 2). As the IMF is frozen into the solar wind plasma flow, this will result in the magnetic field draping in this direction. *Broiles et al.* (2015) first showed that indeed the solar wind and pickup ions moved in opposite directions, and the magnetic field draping effect was shown for March 28, 2015 (low to medium activity), when Rosetta had a closest approach of 15 km (see, e.g., Fig. 5 in *Koenders et al.,* 2016a). This effect starts to become very important when the pickup ion density starts to become similar to the solar wind density.

The comet-solar wind interaction has influence on the structure of the induced cometary magnetosphere. This influence can be used to obtain information about the solar wind for missions, like Rosetta, that do not have easy access to the undisturbed solar wind. Indeed, the interaction of an ICME and CIR with Comet 67P, on July 3, 2015, led to the unprecedented high magnetic field strength of almost 300 nT measured by Rosetta's magnetometer (*Goetz et al.,* 2018). But not only are such extreme field measurements necessary, *Timar et al.* (2019) derived a proxy by using the magnetic pressure just outside the boundary of the diamagnetic cavity. Assuming that the magnetic pressure in the induced magnetosphere needs to balance the dynamic pressure of the solar wind, they found a good correlation between the proxy and the propagated solar wind from Earth to the comet.

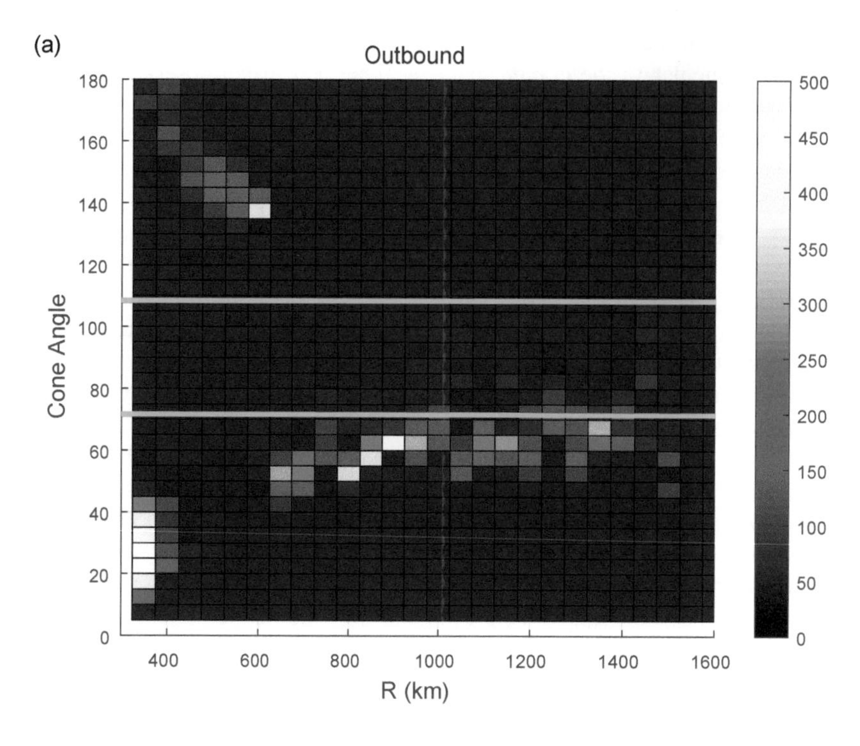

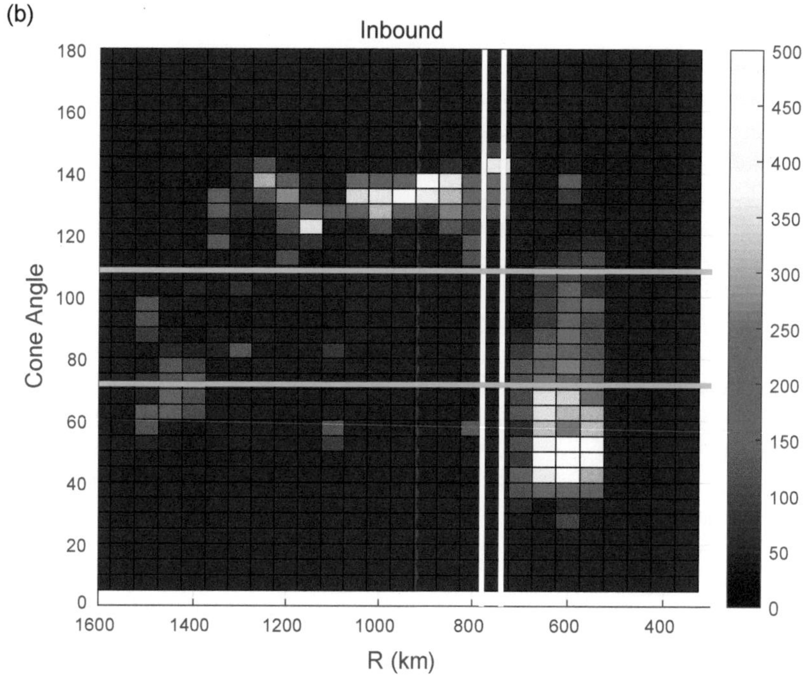

Fig. 7. Two-dimensional histograms of the cone angle of the magnetic field along the Rosetta orbit for the **(a)** outbound and **(b)** inbound leg. The green lines show the cone angles for the Parker spirals. The two white vertical lines show the interval during which the ICME interaction took place. From *Volwerk et al.* (2019).

2.6. Ion/Plasma/Magnetotail

The most notable structure of a comet is its tail, or rather tails: the dust tail, curved along the comet's orbit, and the ion tail, which points almost radially away from the Sun. Historically, the cometary tails were categorized by *Bredikhin* (1879) into three types: type I, a straight tail pointing away from the Sun (the ion/plasma tail), and types II and III, tails curved toward the orbit of the comet (the dust tails) with type II consisting of medium-heavy elements and type III consisting of heavy metals. With the available new observations, this classification is no longer valid. The following sections will focus on the plasma tail itself, but there is also some evidence that the interaction of a comet with the solar wind can affect the dust tail through Lorentz force acceleration of the charged dust particles (e.g., *Kramer et al.,* 2014; *Price et al.,* 2019).

2.6.1. Remote sensing of the plasma tail. Remote sensing, for example with groundbased observations of comets, gives a large-scale view of the ion tail, which, because of its extensive size, cannot be studied by spacecraft missions. These observations give insight in the structure and the dynamics of the tail. However, as we know that the cometary tail is created through the interaction of the solar wind and IMF, observations of the tail can also provide information about the ambient plasma (i.e., the solar wind) conditions.

Photometric observations of Comet 1P/Halley in 1986 showed many structures in the tail. *Brandt* (1982) already talked about rays, streams, kinks, knots, helixes, condensations, and disconnection events. *Saito and Saito* (1986) added arcades in the collection of structures. *Mozhenkov and Vaisberg* (2017) studied the large-scale structure of more than 1500 photographs of comets to categorize two different plasma tail types: double structures — two bright diverging rays from the coma — and outflow — a single large-scale thin tail. The double structure is assumed to be created by a line-of-sight effect of a cylindrical distribution of plasma in the tail created by mass loading of the solar wind. The outflow tails are posited to be created by outflow of plasma from the coma driven by the dynamic pressure of the solar wind, similar to martian ionospheric plasma outflow (*Lundin,* 2011).

2.6.2. Structure of the tail. *Biermann* (1951) and *Alfvén* (1957) laid the foundations for the generation of the ion tail through mass loading and field line draping described above. On September 11, 1986, the first evidence for this magnetic field draping in the comet's magnetotail was established through observations by the International Cometary Explorer (ICE) spacecraft. It passed Comet 21P/Giacobini-Zinner at the downstream side, crossing the induced magnetotail (*Smith et al.,* 1986; *Slavin et al.,* 1986b).

Assuming that the IMF is always perpendicular to the comet-Sun line is not realistic, although it makes the modeling easier. However, the IMF statistically follows the Parker spiral (*Parker,* 1958) and thus arrives at the comet at an angle, making the draping pattern asymmetric (*Volwerk et al.,* 2014), as also observed at Venus (*Delva et al.,* 2017).

Sometimes chance encounters with cometary tails by spacecraft happen, such as in the case of Ulysses traversing Comet Hyakutake's (C/1996 B2) ion tail (*Jones et al.,* 2000). It was detected at a distance of more than 3.8 au from its nucleus and had a diameter of at least 7,000,000 km.

It took 30 years for a planned investigation of a cometary tail. In the period from March 24 through April 10, 2016, Rosetta was sent on a tail excursion. In this case the spacecraft moved up to only ~1000 km from the nucleus so the very near-region was studied. *Volwerk et al.* (2018) looked at the draping of the field, the cone-angle θ_{co} was calculated, and the distribution peaked between 60° and 80°. This means that the "draped" magnetic field is more cross-tail than along the Sun-comet line as one would expect. This direction of the magnetic field is most likely caused by the deflected solar wind magnetic field (*Koenders et al.,* 2016a) as discussed above, which is transported to the downstream side of the comet. Eventually the field stretches out to the more regular tail structure.

There is a non-radial component to the orientation of the plasma tail, as it will be aberrated through the motion of the comet itself. The plasma tail will make an angle ε, with the radial direction to the Sun given by

$$\varepsilon \approx \tan^{-1}\left\{\frac{|V_c \sin(\gamma)|}{|V_{sw}|}\right\} \tag{7}$$

where V_c is the cometary orbital velocity, γ is the angle between V_c and the anti-solar radial direction, and V_{sw} is the radial component of the solar wind (*Mendis,* 2007), where typical values are at $\varepsilon < 6°$.

It was found that ion tails often show acceleration of condensations (i.e., bright blobs of plasma in the tail), which exceed the gravitational pull of the Sun by 1 to 2 orders of magnitude. *Öpik* (1964) used *Bobrovnikoff*'s (1930) observations of Comet 1P/Halley from 1910 to show that the usage of a correct coordinate system leads to better estimates of the acceleration, as sometimes values >1000 were published. The general acceptance was that the acceleration was caused by radiation pressure (see, e.g., *Wurm,* 1943).

2.6.3. Plasma acceleration. One of the open questions in (cometary) tail physics is how fast the plasma is accelerated. Knots are bright patches of plasma that are created by disconnection events in the tail. Studies by *Niedner* (1981), *Saito et al.* (1987), *Rauer and Jockers* (1993), *Kinoshita et al.* (1996), and *Buffington et al.* (2008) gave a broad range of velocities of these knots varying from 20 to 100 km s^{-1}. Acceleration of the knots were measured as 21 cm s^{-2} (*Niedner,* 1981) and 17 cm s^{-2} (*Saito et al.,* 1987).

Yagi et al. (2015) used the Subaru Telescope on Mauna Kea to observe the tail of Comet C/2013 R1 (Lovejoy) and study the short time behavior through short exposures, and thereby were able to determine the initial velocity of the knots to be between 20 and 25 km s^{-1}. This is smaller than the earlier studies, but there the initial velocity was determined by interpolation and may be inaccurate. There are reasons why the initial velocity can differ, e.g., the radial distance from the Sun or the heliospheric latitude. Significant changes in the width of the tail were also found over a timescale of 7 min, which makes high-time resolution imaging of cometary tails a necessity to understand the influence of solar wind variations on the tail.

2.6.4. Waves and oscillations. An interesting effect was found at distances ≥500 km from the nucleus of Comet 67P/Churyumov-Gerasimenko. *Volwerk et al.* (2018) calculated the clock angle, defined as

$$\phi_{cl} = \tan^{-1}\left\{\frac{B_z}{B_y}\right\} \quad (8)$$

The unwrapped clock angle showed a continuously increasing angle ϕ_{cl}, even when the spacecraft turns around at apoapsis and moves in the opposite direction. This was interpreted as a traveling rotational wave along the magnetic field. The angular frequency of this wave while Rosetta moved outbound(inbound) was $\omega_h \approx 5.65(5.83)° hr^{-1}$, and the spacecraft speed was $v_{ros} \approx 1.3(3)$ m s^{-1}. Assuming Rosetta observed a Doppler-shifted wave, a phase velocity for the wave was determined at $v_h \approx 136$ m s^{-1}. With an average magnetic field of $B_m \approx 10$ nT, this velocity does not correspond any known specific velocity, nor does the frequency correspond to any gyrofrequency.

Nisticò et al. (2018) studied oscillations of cometary tails to investigate the interaction of the solar wind with a cometary plasma tail, especially for Sun-grazing comets. Thereby, insight can be obtained in the solar wind conditions. Using images of Comets Encke and ISON by the Solar TErrestrial RElations Observatory (STEREO)-A/HI-2 telescope they found oscillations of the two cometary plasma tails, which they interpreted as so called "vortex shedding" (see, e.g., *Johnson et al.,* 2004). Modeling the tail as a driven and damped harmonic oscillator, the authors obtained information about the local solar wind velocity. Indeed, *Ramanjooloo* (2014) makes a case that ion tails can be used to investigate the inner structure of the whole heliosphere. This opens a new window in solar wind studies from historical and contemporary observations (see, e.g., *Ramanjooloo and Jones,* 2022).

Tail dynamics, such as waves, kinks, detachments, and also cometary ion rays, need to be studied through well-planned campaigns, not in the least because of the enormous size of a cometary tail. What actually causes detachments, what accelerates the condensations in the tail, what drives the kinks? To answer these questions, detailed *in situ* observations combined with ground observation campaigns are necessary.

2.6.5. Historical solar wind. Historical observations of cometary tails can also be used to investigate the historic solar wind conditions. One interesting historic period is the so-called Maunder Minimum (1645–1715). *Gulyaev* (2015) studied drawings of cometary plasma tails (type I tails) made during the Maunder Minimum. More than 20 comets were observed during that time, including Comet 1P/Halley. The presence of plasma tails is directly related to the presence of a solar wind. *Gulyaev* (2015) states that, in principle, it should be possible to obtain the solar wind velocity by looking at shifts of individual features in consecutive drawings; however, these are very hard to distinguish.

Zolotova et al. (2018) studied descriptions of comets from the eleventh to the eighteenth century but found from color descriptions that mainly dust tails were described. Using the determination of the aberration ε as determined by *Bredikhin* (1879), an average value of ~10° was found, also indicating that mainly dust tails were seen, because the plasma tail is much less intense and thus more difficult to observe with the unaided eye. *Zolotova et al.* (2018) concludes that probably the first identification of a plasma tail was done for the great Comet C/1769 P1. However, *Hayakawa et al.* (2021) discuss three cases, earlier than 1769, in which historical sources talk about two tails at Comets C/1577 V1 and 1P/Halley in both the years 760 and 837. Comparing these descriptions with simulations of the cometary tails showed that it was indeed plausible that the plasma tail was observed already centuries before 1769. Indeed, *Silverman and Limor* (2021) show that observations of the great comet of 1577 is clearly described as having two tails in various sources.

2.6.6. Sounding the plasma tail. *Iju et al.* (2015) used distant radio sources and scintillation of the signal to determine the electron density in the tail of Comet C/2021 S1 ISON. This interplanetary scintillation method has been regularly used to obtain the electron density in the solar wind and its structures such as ICMEs (*Hewish et al.,* 1964; *Gapper et al.,* 1982). This effect was first seen in a cometary tail (C/1973 E1 Kohoutek) by *Ananthakrishnan et al.* (1975), and *Lee* (1976) used that signal to estimate the electron density fluctuations in the tail with a value of $\Delta N_e \approx 80$ cm^{-3}.

Using five consecutive nights, November 13–17, 2013, *Iju et al.* (2015) obtained scintillation measurements of one radio source through the cometary tail from which they could derive the local electron density. There were no ICMEs passing through the field of view on November 13, 16, and 17, but the electron density was determined for all nights. Normally, the number density increases toward the center of the cometary tail (see, e.g., *Meyer-Vernet et al.,* 1986); however, what was measured was that the electron density increased as the radio source moved toward the edge of the cometary tail. This increase is thought to be created by an outburst of the comet (*Combi et al.,* 2014) or by turbulence on the boundary between the tail and the solar wind.

2.7. Induced Magnetospheric Activity

A notable feature of comets are the near-linear regions of enhanced brightness emerging from the coma around the nucleus, which bend backward toward the tail (e.g., *Wurm,* 1963). These are so-called cometary or ion rays (*Eddington,* 1910; *Rahe,* 1968). Attempts to understand the physics of ray formation did not start until *Alfvén*'s (1957) paper on tail formation; however, this explains the bend-back of the rays and not the brightness variations. Observations of Comet Morehouse, 1P/Halley, and others showed that the rays are most likely created by ionization of H_2O, CO, and CO_2 close to the nucleus, which happens in a discontinuous way (*Rahe and Donn,* 1969). Abrupt currents closing through the "head" of the comet, generated by tail reconnection (see section 2.8), can act as an extra source for ionization in the coma. *Wolff et al.* (1985) also considered cometary rays to be channeled outflow of cometary ions along magnetic flux tubes. However, in this case, the flux tubes are entering the cometary ionosphere through the Kelvin-Helmholtz instability on the ionopause. *Ip* (1994) expanded on this model by positing that modification of the electron heat conduction is cause for variations along the flux tubes.

Estimates for the tail magnetic field strength were performed by *Ip and Mendis* (1975), who derived a magnetic field of ≥100 γ (now nT) for Comet C/1973 E1 Kohoutek, which is similar to the field measured at Comet 21P/Giacobini-Zinner (*Slavin et al.,* 1986b) and around 67P (*Goetz et al.,* 2017). This value was determined through observations of a helical structure that moved down the cometary tail (*Hyder et al.,* 1974). Assuming that this wave moves at the Alfvén velocity, and that it was created by a kink-mode instability, *Hyder et al.* (1974) came up with a necessary tail current of $I \approx 2 \times 10^7$ A, which was later adjusted by *Ip and Mendis* (1975) to $I \approx 2 \times 10^6$ A.

Similar helical waves were also observed in the tails of Comets C/1973 E1 Kohoutek, C/1908 R1 Morehouse, and C/1956 R1 Arend-Roland by *Ershkovich and Heller* (1977). These authors modeled the tail as a cylinder that is separated from the solar wind by tangential discontinuities. They determined the dispersion relation for helical waves for an ideal compressible plasma, and for typical parameters the phase velocity of these waves was found to be close to the Alfvén velocity. The authors conclude that the origin of the waves lies in the Kelvin-Helmholtz instability.

Interest in tail currents was generated through the temporal variations in the observed brightness, e.g., in CO^+ close to the nucleus (*Wurm,* 1961) and the cometary rays. *Ip and Mendis* (1976) discussed how the folding of the cometary rays generates a strong tail field through accumulation of magnetic flux. Flux conservation leads to a field strength B_t of the tail

$$B_t \approx \frac{B_0 v_s t_f}{h_t} \tag{9}$$

where $B_0 \approx 5$ nT and $v_s \approx 300$ km s^{-1} are the solar wind magnetic field and velocity and $h_t \approx 5 \times 10^4$ km is the half-width of the tail. These assumptions lead to rather strong magnetic fields, $B_t \approx 1500$ nT, but it was realized that this folding leads to, similar as in Earth's magnetotail, two regions of oppositely directed magnetic field, as in the model by *Alfvén* (1957). Between these two lobes there needs to be a cross-tail current sheet, given by Ampère's law

$$\nabla \times \mathbf{B} = \mu_0 \mathbf{J} + \frac{1}{c^2}\frac{\partial \mathbf{E}}{\partial t} \tag{10}$$

Ip (1979) assumed that (part) of this current can suddenly be channeled through the cometary atmosphere by a process to avoid a magnetic catastrophe. This could happen in a similar way as in Earth's magnetotail during a substorm (*Boström,* 1974). This current will then flow along the field lines toward the coma, and can be used to increase the ionization rate near the nucleus.

Israelevich and Ershkovich (2014) studied the magnetotails of Titan, Venus, and Comet 1P/Halley. Using the Giotto flyby, they showed that the magnetic tension of the draped magnetic field lines, given by $T = (\mathbf{B} \cdot \nabla)\mathbf{B}$ is always pointing tailward because of the specific geometry of the field. A mean tension of $T \approx 1600$ nT^2/R$_{dc}$, with $R_{dc} = 4000$ km, the radius of the diamagnetic cavity. With $B \approx 80$ nT they estimate a curvature radius of the field lines of $R_c \approx 4R_{dc}$. This is important for tail detachment events discussed below.

The connection between the solar wind and the folding of cometary ion rays was clearly shown by *Degroote et al.* (2008), studying ion rays at Comet C/2004 Q2 Machholz. They combined telescopic U-band observations with the Solar and Heliospheric Observatory (SOHO) solar wind measurements and suggested that the ion rays were formed and folded back through a sudden change of the solar wind from slow to fast.

2.8. Magnetic Reconnection

The presence of oppositely directed magnetic field throughout the induced cometary magnetosphere, in the nested/dynamic draping upstream and the two lobes downstream of the nucleus, gives rise to the assumption that reconnection events should be ubiquitous.

As mentioned in section 2.5, changing directions of the IMF over time leads to a nested draping of the magnetic field around the outgassing nucleus. The rotations of the magnetic field create current sheets that can be observed when sufficiently high-resolution data are available (see, e.g., *Volwerk et al.,* 2017). These current sheets are usually the location where reconnection will occur.

There was much less nested draping observed by Vega 1 (see, e.g., Fig. 3 of *Volwerk et al.,* 2014), and inside the coma the plasma instrument did not measure any solar wind ions (see section 3.4). But near closest approach the plasma instrument showed bursts of 100–1000 eV ions over a period of 5 min. *Verigin et al.* (1987) showed the magnetic topology of the site at which the bursts were observed and proposed a reconnection region, which accelerates the ions.

Kirsch et al. (1989) showed that during the magnetic field rotations there is a peak in field-aligned high-energy ions and a drop in the low-energy ions. They calculated the energy of the ions assuming that they are accelerated over 50,000 km through reconnection and found a maximum energy of $E_i \approx 22$ keV, but this is well below the observed energies of 97–145 keV.

These are indirect indications that magnetic reconnection could be taking place in the nested draping region, and only a multi-probe mission could give a more precise answer. However, one should also realize the reconnection rate as determined by *Sweet* (1956) and *Parker* (1957) is probably very much reduced because of the high plasma densities in the coma. The rate is proportional to $1/\sqrt{\rho}$, where ρ is the plasma density, for slow reconnection, and $\propto 1/\ln S$ for fast reconnection (*Petschek,* 1964), where S is the Lundquist (or magnetic Reynolds) number.

The magnetic field measurements by ICE showed a clear draping pattern of the magnetic field around the comet, reminiscent of Earth's magnetotail. The spacecraft passed through two regions of oppositely directed field, which were separated by a current sheet of ~200 km thickness (*Slavin et al.,* 1986a,b). The current density was determined by *McComas et al.* (1987) and with that the **J** × **B**-force was calculated, which was negative, reaccelerating the plasma tailward. Such a magnetotail, comparable with Earth's, will experience instabilities driven by external (e.g., by the solar wind) forces or internally through, e.g., flux accumulation. These instabilities can give rise to magnetic reconnection and in the most drastic version a tail disconnection event can take place (e.g., *Niedner and Brandt,* 1978).

Solar wind sector boundaries that interact with a comet's induced magnetosphere, as well as other structures such as ICMEs (see, e.g., *Möstl et al.,* 2014), will influence the magnetospheric activity. *Vourlidas et al.* (2007) studied Comet 2P/Encke interacting with an ICME using STEREO-A's Sun Earth Connection Coronal and Heliospheric Investigation (SECCHI) HI-1 instrument. As an ICME passes over the comet, part of the tail is disconnected and transported away by the solar wind. After the tail disconnection, the magneto-tail slowly grows again to "full length."

To further study tail detachment events one has to turn to numerical simulations to model the interaction of a cometary tail and the solar wind and its structures. *Jia et al.* (2009) modeled the tail detachment event of Comet 2P/Encke. This interaction was modeled with an MHD code, which showed that a sudden 180° rotation of the IMF caused a reconnection event at the upstream side of the comet. As the model was based on only a single species of ions, the speed of the evolution was slower than observed. A change to multiple ion species, cometary and solar wind ions, might improve the agreement with the observations. Also, this model only looks at one type of ICME-tail interaction.

Comet 153/Ikeya-Zhang experienced multiple interactions with ICMEs (*Jones and Brandt,* 2004). These interactions showed evidence of possible shocks traveling along the length of the tail, which created a disruption and a disconnection of part of the tail.

Edberg et al. (2016b) found that with the CIR interaction (see section 2.3) there was a sudden reduction in the piled-up magnetic field, combined with an increase in energetic electrons. This could be evidence that a tail disconnection event is the comet's ionospheric response.

Similarly, there was an interaction with an ICME (see section 2.3) that compressed the magnetosphere (*Edberg et al.* 2016a). Large spikes on the magnetic field data were observed and these were assumed to be created through reconnection of the magnetic field lines of the ICME with the draped field lines around the nucleus. Due to the lack of plasma tail observations for 67P, it is not clear whether this interaction led to a tail disconnection event, and it need not have done.

Even though IMF reversals, through whatever means (sector boundaries, ICMEs, CIRs), or shocks can generate tail disconnection events, there is not a one-to-one correlation between the two. *Delva et al.* (1991) used the data from Vega 1/2 for the period of December 1, 1985, to May 1, 1986 (which includes the flybys of Comet 1P/Halley on March 6 and 9, 1986) and disconnection events observed by ground-based observatories. The authors correlated the crossing of sector boundaries of the solar wind with tail disconnection events. If changes in the solar wind (*Jockers,* 1985) generate dayside reconnection (*Niedner and Brandt,* 1978), then one should find a good correlation. In all, *Delva et al.* (1991) determined that 50% of the events showed a correlation between sector boundary crossing and tail disconnection. The authors also state that density enhancements in the solar wind are connected to events in the tail.

The first direct observation of an ICME interacting with a comet (but not the first one published in the literature) was done with the Solar Mass Ejection Imager (SMEI) looking at Comet C/2001 Q4 (NEAT) in 2004 (*Kuchar et al.,* 2008). The ICME first created a kink in the tail, which moved anti-sunward, which developed into knots. This seems to be a general behavior as seen in a handful of other events. The generation of the knots, evidence for the disruption of the tail, can result from various effects like shocks or polarity reversals of the magnetic field in the ICME.

Comparison of events at Comet C/2001 Q4 (NEAT) with those of Comet C/2002 T7 (LINEAR), and tracing them back along the Parker spiral, shows that these occur preferably when the comet is near the heliospheric current sheet (*Kuchar et al.,* 2008). *Brandt et al.* (1999) and *Brandt and Snow* (2000) also concluded this effect using observations of Comets 1P/Halley, 122P/de Vico, C/1996 B2 Hyakutake, and C/1995 O1 Hale-Bopp.

Li et al. (2018) performed laboratory experiments to study tail detachment events. They used a laser-driven plasma that collides with a cylindrical obstacle. Behind the obstacle detached tails are measured. At the same time numerical simulations of this interaction were performed. However, it should be noted that this was an unmagnetized interaction. The experiment showed that an electrostatic field is created in the plasma when the density is high, because of a temperature difference between the ions and the electrons. This field leads the plasma ions to converge in a tail and move away from the obstacle, giving the impression of a detached tail. The authors argue that this is a viable process to explain a detached tail, because the physical sizes related to cometary effects are shorter than the gyroradius of cometary ions in the solar wind, invalidating an MHD approach.

3. BOUNDARIES/REGIONS

3.1. Overview

Comet interactions with the solar wind begin at great distances from the comet because the high neutral outgassing rate leads to mass-loading of the solar wind (*Galeev and Sagdeev,* 1988). Closer to the nucleus, this interaction leads to the formation of large-scale structures in the form of boundaries that separate regions characterized by plasmas with differing parameters. Prior to spacecraft flybys of comets, computer simulations of the solar wind interaction with the comet predicted the existence of two permanent boundaries: a bow shock and a contact surface (*Schmidt and Wegmann,* 1982; *Ip and Axford,* 1982). These boundaries established three regions of comet-solar wind interaction: an upstream region outside the bow shock, a region between the bow shock and contact surface termed the cometosheath, and a diamagnetic cavity located between the contact surface and the nucleus. After spacecraft made *in situ* observations during several flybys and the escort of Comet 67P by Rosetta, several other boundaries were also observed: the cometopause, an ion pileup boundary, and ion-neutral and electron-neutral collisionopauses. Plasma interaction boundaries observed at a comet can be permanent features, solar wind and interplanetary magnetic field boundaries, or small-scale transient features created by waves or instabilities (*Cravens,* 1989). Permanent features include shocks and collisionopause boundaries. Shocks form when the relative velocity of the plasma and an obstacle, in this case the cometary plasma, exceed the characteristic speed of waves in the plasmas. As discussed in section 3.5, a collisionopause forms when collisions between neutrals and ions or electrons change characteristics of the plasma such as composition or velocity (*Mendis et al.,* 1986; *Cravens,* 1989, 1991). *Cravens* (1991) outlined various types of collisionopause for both ions and electrons depending on the collision processes and the reactions that may occur, including ion-neutral charge transfer and ion-neutral chemistry. Figure 8 illustrates the boundaries that are discussed in this section. We review here observations of five boundaries that were determined to be permanent features (*Gringauz and Verigin,* 1991).

3.2. Bow Shock

The bow shock is the location where the solar wind flow transitions from supersonic to subsonic as a result of mass loading. Simulations predicted, and spacecraft observations confirmed, that this boundary would be broad and weak compared to planetary bow shocks because the slowing of the solar wind begins well upstream of the shock (see also section 2.4). Spacecraft observations have noted sudden changes in the plasma density and velocity across the bow shock and an enhancement of the magnetic field (*Galeev and Sagdeev,* 1988). The detailed physics of the cometary bow shock formation and early modeling efforts are described in section 3 of *Ip* (2004) while spacecraft observations prior to Rosetta are outlined in section 4.1 of *Ip* (2004). Rosetta's exploration of the plasma environment around 67P has significantly advanced our understanding of cometary bow shocks for weakly outgassing comets. Hybrid modeling of the comet found that including electron impact ionization and charge-exchange processes in addition to photoionization increased the standoff distance of the bow shock along the Sun-comet line by more than a factor of 6, as illustrated in Fig. 9. Furthermore, simulations also showed that asymmetric outgassing and illumination-driven outgassing can extend the standoff distance even further (*Huang et al.,* 2016; *Alho et al.,* 2020). The Rosetta spacecraft did not travel far enough from the comet to cross the bow shock during outgassing rates relevant to these simulations.

The greatest advantage of Rosetta in advancing our understanding of cometary bow shocks was thanks to the extended amount of time spent escorting the comet around the Sun. This allowed detection of the bow shock as it began to form, or an infant bow shock (*Gunell et al.,* 2018). Statistical analyses of the full Rosetta dataset showed that the spacecraft encountered the infant bow

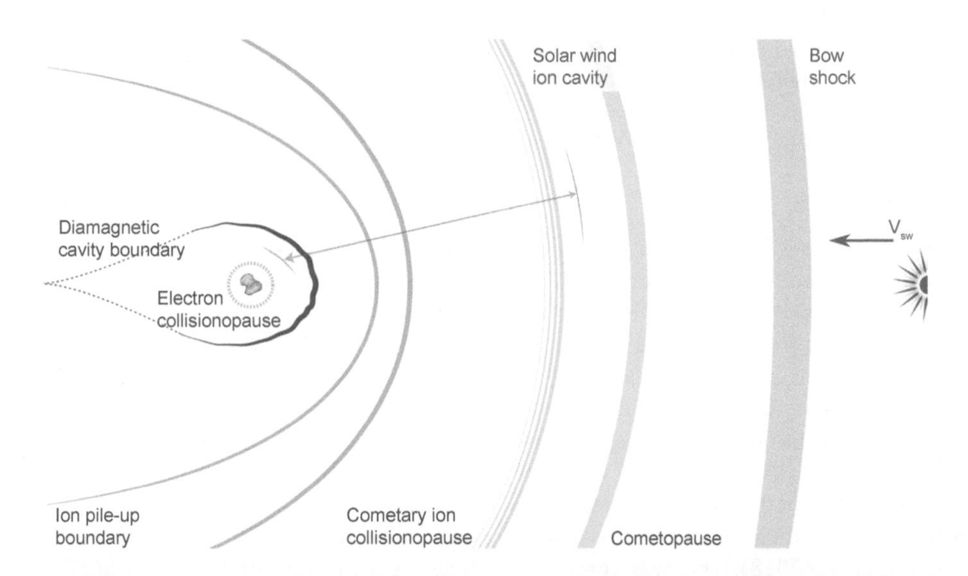

Fig. 8. Overview of large-scale boundaries observed by spacecraft at comets (not to scale). The gray arrow indicates the boundaries covered by Rosetta. Credit: E. Behar, adapted from *Goetz et al.* (2021a).

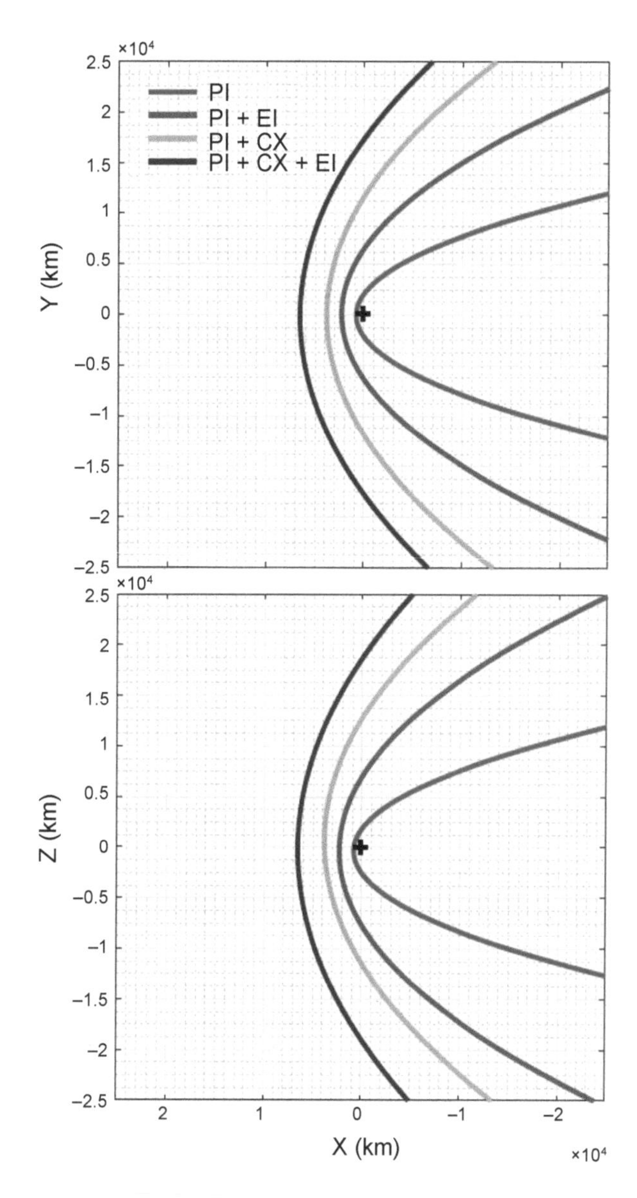

Fig. 9. Illustration of how the simulated distance of the bow shock from the nucleus increases compared to simulations with only photoionization (PI) when the influence of electron-impact ionization (EI) and charge-exchange processes (CX) are included. Adapted from Fig. 11 of *Simon Wedlund et al.* (2017).

shock multiple times between 3.0 and 1.7 au inbound to the Sun and then again outbound starting at around 1.8 au (*Goetz et al.,* 2021b). Additionally, a method for detecting the existence of a bow shock from inside the shock was discovered. The energy of ions at the spacecraft location that were accelerated by a constant electric field upstream of the spacecraft is proportional to the distance from the place where the ions were initially created. This was used to determine that a bow shock had formed at a distance of ~4000 km from the nucleus when the comet was 1.4 au from the Sun (*Nilsson et al.,* 2018). This was confirmed with hybrid simulations (*Alho et al.,* 2019) that were later used to determine the spectral features in ion energy that could indicate the presence of a bow shock upstream of the spacecraft (*Alho et al.,* 2020). Similarly, a bimodal ion distribution found in the environment of Comet 1P/Hall indicated that ions produced before and after the bow shock are observed as two distinct populations, indicating that the acceleration history of the ions is preserved (*Thomsen et al.,* 1987).

Although we have learned a great deal about cometary bow shocks, we are lacking observations that cover sufficient space and production rate to fully understand the transition between an infant bow shock and a full shock as well as the implications of asymmetric outgassing on the location and shape of the bow shock. Future observations should include simultaneous measurements at multiple points in space (*Snodgrass and Jones,* 2019) and multi-spacecraft missions over long time periods (*Götz et al.,* 2019), which should be combined with simulations that treat electrons and ions kinetically resolving all relevant scales (*Balogh and Treumann,* 2013).

3.3. Cometopause

The concept of the cometopause is introduced in section 4.3 of *Ip* (2004), which discusses ion properties inside the bow shock over a wide distance range. We have learned a great deal more from Rosetta about the interaction between the solar wind and the coma inside the cometopause. We outline here what new understanding we have gained about the cometopause and introduce new regions and boundaries that were discovered by Rosetta.

The cometopause is described as the boundary where the ion composition changes from predominantly solar wind ions to predominantly cometary ions (*Gringauz et al.,* 1986; *Mendis et al.,* 1989; *Coates,* 1997). Although some dispute the existence of this boundary (*Rème et al.,* 1994), it was determined by several researchers to be a permanent feature (*Gringauz and Verigin,* 1991; *Sauer et al.,* 1995). Several authors have proposed that the cometopause is the location where collisions causing charge exchange between solar wind protons and cometary neutrals become dominant (*Gringauz et al.,* 1986; *Gombosi,* 1987; *Cravens,* 1989; *Ip,* 1989), a form of collisionopause boundary (see Table 1 and section 3.5). However, this is not an appropriate explanation for the cometopause as explained in section 3.5. Instead, as theoretical simulations suggest, the cometopause is best explained by deflection of solar wind protons (*Sauer et al.,* 1995). Giotto observed plasma flux deflection that indicated plasma flow forced around a comet (*Perez-de Tejada,* 1989), suggesting that the cometopause could be the beginning of the solar wind ion cavity (see section 3.4). Previous work stated that Rosetta's instruments did not observe the cometopause (*Mandt et al.,* 2016), but if this is the boundary of the solar wind ion cavity, then it may have been observed several times as it was forming (*Behar et al.,* 2017). Further work is needed to better understand the physics of this region of the cometosphere.

TABLE 1. Summary of collisional processes that can take place in the cometary plasma.

Number	Type	Example
R1	Ion-neutral charge transfer	$H_2O^+_{fast} + H_2O \rightarrow H_2O_{fast} + H_2O^+$
R2	Ion-neutral chemistry	$H_2O^+ + H_2O \rightarrow H_3O^+ + OH$
R3	Ionization	$H^+_{fast} + H_2O \rightarrow H_{fast} + H_2O^+ + e^-$
R4	Electron removal	$O + H_2O \rightarrow O^+ + H_2O + e^-$
R5	Ion-electron recombination	$H_3O^+ + e^- \rightarrow OH + 2H$
R6	Ion-ion coulomb interaction	$H_2O^+ + H_2O^+ \rightarrow H_2O^+ + H_2O^+$

3.4. Solar Wind Ion Cavity

From May to December 2015 Rosetta was located in a region that was mostly free from solar wind ions, a region termed the solar wind ion cavity (*Nilsson et al.,* 2017; *Behar et al.,* 2017). This cavity was determined to have formed as a result of the deflection of the solar wind ions by magnetic pileup that results from mass loading of the solar wind. Prior to the disappearance of the solar wind ions, the ion instruments observed deflection of the solar wind that increased over time as shown in Fig. 10. When the ion deflections were greater than 90°, ion observations became less frequent and had lower densities until they disappeared. Deflections were observed to be very large near the boundary of the cavity, possibly in the region of the cometopause. A simple analytical model that estimates the global dynamics of solar wind protons for a given heliocentric distance provided strong agreement with observed deflection angles during Rosetta's nightside excursion, validating understanding of how the solar wind ions gyrate withing the coma and how they are repelled from the innermost region to form an ion cavity (*Behar et al.,* 2018). Although the solar wind ions are deflected away from the solar wind ion cavity, cometary ions picked up by the solar wind and the solar wind magnetic field are still present even though the pickup ions are deflected in a similar manner to the solar wind ions (*Nilsson et al.,* 2020). The boundary also does not constitute a discontinuity in the momentum budget, but just a change in the composition of the plasma flow (*Williamson et al.,* 2020). *Edberg et al.* (2016a) showed that the impact of an ICME can compress the solar wind ion cavity, indicating that the solar wind dynamic pressure regulates the size of this region.

3.5. Ion-Neutral Collisionopause

Collisions within an atmosphere or coma influence the dynamics and chemistry of neutrals and ions. The term collisionopause is used to define a boundary inside which collisions play a dominant role. We describe in this section ion-neutral collisionopause boundaries and cover the electron-neutral collisionopause in section 3.6.

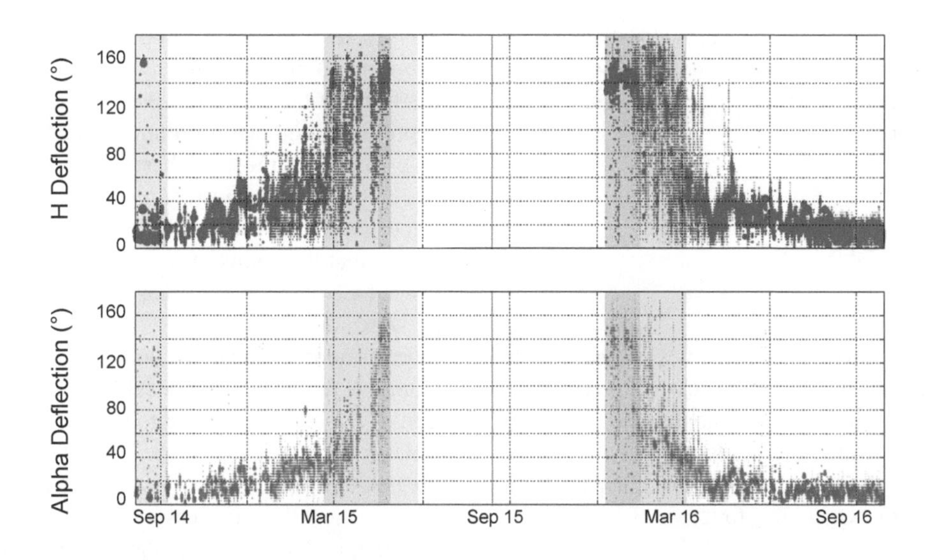

Fig. 10. Observations of the increasing deflection of solar wind protons and alpha particles as Comet 67P approached the Sun from September 2014 to May 2015, the formation of the solar wind ion cavity between May and December 2015, and the decreasing deflection of solar wind ions as the comet moved away from the Sun after December 2015. Adapted from Fig. 1 of *Behar et al.* (2017).

The location of an ion-neutral collisionopause is calculated in a manner similar to determine the location of the exobase in aeronomy, or the study of neutral atmospheres. In the case of a steady state atmosphere, the exobase is the location is where the Knudsen number, or the ratio of the mean free path, λ to the scale height, is one.

$$\lambda = \frac{1}{n_n \sigma} \tag{11}$$

where n_n is the neutral density and σ is the collision cross section. In aeronomy, when determining the location of the neutral exobase, the cross-section is the species-specific neutral collision cross-section. This means that each species has its own exobase. For an ion-neutral collisionopause, the cross-section would be the collision cross-section for the relevant ion species with water. In the coma, the neutral density, n_n, as a function of distance from the comet, can be approximated as

$$n_n(r) = n_{s/c}\left(\frac{r_{s/c}}{r}\right)^2 \tag{12}$$

where r is the distance from the nucleus, $n_{s/c}$ is the density at the spacecraft, and $r_{s/c}$ is the distance of the spacecraft from the nucleus.

The scale height is the distance over which the density of a neutral species or the plasma reduces by a factor of e. Rosetta observations suggested that the plasma scale height at the location of Rosetta when escorting 67P could be approximated by the distance of the spacecraft from the comet, r (*Edberg et al.,* 2015). Therefore, the location of both ion-neutral and electron-neutral collisionopause boundaries could be found by setting the mean free path equal to the distance from the comet

$$\lambda = r = n_{s/c} \sigma r_{s/c}^2 \tag{13}$$

In the case of Rosetta, the neutral density was measured at the spacecraft location, making this a reasonable equation to use for determining where a collisionopause would be located based on local measurements. A collisionopause distance can also be calculated by substituting for $n_{s/c}$ with the outgassing rate, Q, which is a function of $n_{s/c}$ and the neutral outflow velocity, v_n

$$Q = 4\pi r^2 n_{s/c} v_n \tag{14}$$

Cravens (1991) outlined various types of collisionopause for both ions and electrons depending on the collision processes involved, which are outlined in Table 1. Reactions of type R1 are charge transfer reactions where the charge is transferred from a fast (<300 km s^{-1}) ion in the mass-loaded solar wind flow (e.g., H^+ or H_2O^+) to a neutral traveling away from the comet at velocities of 1 km s^{-1} or less. As a result of this reaction, the ion becomes an energetic neutral, and the bulk velocity of the ions is reduced. Chemical reactions like R2 not only reduce the bulk velocity of the ions, but also alter the relative composition of the ion population. For example, in reaction R2, a proton is transferred from H_2O^+ to H_2O, creating an H_3O^+ ion, increasing the bulk mass 19/18 ratio of the ion population. If H_2O^+ is fast it will end as an energetic OH neutral while H_3O^+ will have the same energy as the neutrals. Another chemical reaction that should be noted is the proton transfer from H_3O^+ to NH_3 producing NH_4^+ in the dense coma (*Beth et al.,* 2016). This reaction depends on the volume mixing ratio of NH_3 in the coma and would reduce the mass 19/18 ratio, countering the effect of producing H_3O^+.

As outlined in *Mandt et al.* (2019), the collision cross-section is the greatest source of uncertainty for calculating ion-neutral collisionopause distances. In early studies, the ion-neutral cross-section was estimated to be 2×10^{-15} cm^2 for solar wind ions and 8×10^{-15} cm^2 for mass-loaded solar wind with a bulk composition of H_2O^+ (*Mendis et al.,* 1986). We illustrate in the shaded blue region of Fig. 11 the calculated location of the collisionopause throughout the Rosetta mission based on these cross-sections. Note that because the solar wind ion cross-section is smaller than the cross-section for water ions, the collisionopause location calculated with equation (13) for charge transfer of solar wind ions will be closer to the nucleus than the collisionopause for picked up cometary ions. Rosetta made several observations of a boundary where the bulk ion velocity transitioned between values greater than 10 km s^{-1} outside the boundary to velocities too low to measure directly because the ions were seen at the value of the spacecraft potential that were interpreted to represent a cometary ion collisionopause (*Mandt et al.,* 2016). However, most of these observations were within the solar wind ion cavity (*Mandt et al.,* 2016), meaning that the spacecraft was inside the cometopause. This demonstrates that the cometopause is not a collisionopause and is instead a boundary formed as a result of deflection of the solar wind ions.

Furthermore, although the boundary was observed within the range of predicted locations illustrated in the shaded blue region of Fig. 11, it appeared to vary in distance depending on solar wind dynamic pressure (*Mandt et al.,* 2016), similar to the cometopause (*Edberg et al.,* 2016a). Laboratory measurements show that the cross-sections for reactions (R1) and (R2) depend on the energy of the ions (*Lishawa et al.,* 1990; *Fleshman et al.,* 2012). A comparison of these cross-sections as a function of energy with the Rosetta observations of cometary ion collisionopause crossings (*Mandt et al.,* 2016), observations made by Giotto at 1P/Halley (*Schwenn et al.,*

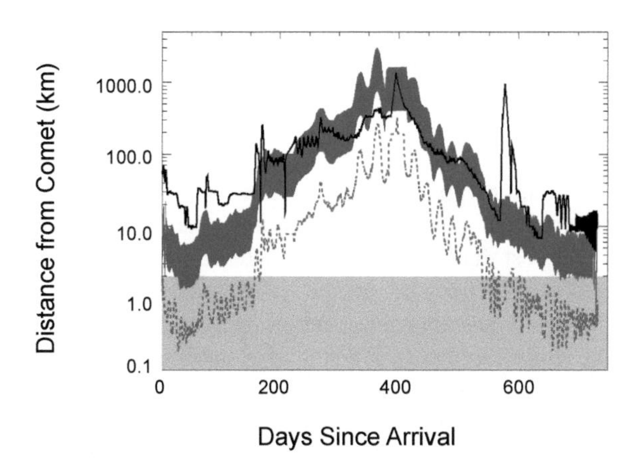

Fig. 11. Calculated location of the cometary ion collisionopause (blue shaded region) and the electron exobase (red dashed line) compared to the distance of Rosetta (black line) from the comet nucleus (gray shaded region).

1988; *Altwegg et al.,* 1993), and observations made by Deep Space 1 at 19P/Borelly (*Young et al.,* 2004) found that the collisionopause had been observed at all three comets and appeared to form as a result of a combination of reactions (R1) and (R2) (*Mandt et al.,* 2019).

Many questions remain about this region of the cometosphere and the role of collisions in influencing the dynamics of cometary ions that have been picked up by the solar wind. A greater understanding of cross sections is needed and more observations of ion and neutral composition would be of high value. Additionally, measurements of the electron temperatures are needed to constrain the electron recombination rates in this region and inside the collisionopause.

3.6. Electron-Neutral Collisionopause

Collisions between electrons and neutrals in the coma will cool the electrons to temperatures similar to the neutral gas temperature (*Eberhardt and Krankowsky,* 1995). The electron-neutral collisionopause, also called the electron exobase, represents the boundary outside of which collisions do not efficiently influence the electron temperatures. The location of the electron exobase calculated using equation (13) and a cross-section of 5×10^{-16} cm^2 is illustrated in Fig. 11. As shown here, Rosetta was not expected to cross inside the electron exobase. However, additional processes can enhance the electron collision rate, and thus electron cooling can extend the location where electrons thermalize to the same temperature as the neutrals beyond the calculated exobase location (*Henri et al.,* 2017). These processes include rotational and vibrational cooling for electrons originating close to the nucleus where electrons pass through the densest gas (*Engelhardt et al.,* 2018b), and possibly through an ambipolar electric field, either trapping electrons in the dense neutral region (*Madanian et al.,* 2016; *Vigren and Eriksson,* 2019) or increasing the electron collision cross-section by slowing their outward movement (*Engelhardt et al.,* 2018b).

At several points during the mission Rosetta's instruments observed a population of cold electrons that at times made up as much as 90% of the electron population (*Engelhardt et al.,* 2018b; *Gilet et al.,* 2020). *Henri et al.* (2017) found a relationship between the location of the electron exobase and the diamagnetic cavity boundary (DCB) (section 3.7). These observations confirmed that electron cooling is more efficient than suggested by the simple approximation shown in equation (13) and that a more comprehensive cooling model is required to estimate the location of the electron exobase (see also section 2). Furthermore, because Rosetta did not explore the coma at a close enough distance to the nucleus, *in situ* measurements of the electron neutral collisionopause are lacking. Future exploration of this region, measuring the ion and neutral composition and the electron densities and temperatures, would be of great value.

3.7. Diamagnetic Cavity

The mass-loading of the solar wind by cometary ions was predicted to decelerate the solar wind until the velocity reached zero upstream of the comet under high outgassing rates (*Biermann et al.,* 1967). Because the solar wind magnetic field is tied to the solar wind plasma, the magnetic field was predicted to come to a stop, creating a magnetic-field-free region around the nucleus (*Schmidt and Wegmann,* 1982; *Ip and Axford,* 1982). The first confirmation that a diamagnetic cavity could form in the conditions provided by a coma was during the AMPTE. The magnetic field briefly dropped to zero after release of a barium cloud, confirming the formation of a field-free region (*Luehr et al.,* 1988). The boundary of this region was found to be where the electron thermal pressure equaled the magnetic field pressure.

The Giotto spacecraft flew close enough to Comet 1P/Halley to pass within the predicted region where the diamagnetic cavity was expected to exist. The magnetic field measurements showed that the field magnitude dropped to and remained at close to 0 nT around 5000 km from the comet nucleus. The observations suggested that the boundary of the diamagnetic cavity was not symmetric about the nucleus and may have ripples across its surface (*Neubauer,* 1988). The position of the boundary was expected to be determined by a balance of outward ion-neutral drag and inward magnetic gradient forces (*Gringauz and Verigin,* 1991; *Coates,* 1997). Besides Giotto at 1P/Halley, no other spacecraft directly observed the diamagnetic cavity previous to the Rosetta mission. However, a decrease in HCO^+ emissions in the innermost coma of Comet Hale-Bopp was speculated to be associated to the diamagnetic cavity. As Hale-Bopp is much more active than 1P/Halley, the diamagnetic cavity was predicted to be around 500,000 km in size, which coincided with a drop in emissions (*Womack et al.,* 1997). It should be noted that in early works the solar wind ion cavity boundary and the diamagnetic cavity boundary were identified to be in the same place. However, Rosetta observations have clearly shown that the solar wind ion cavity is most of the time much larger than the diamagnetic cavity.

This is because the solar wind ions are deflected upstream and the pickup ions take over the role of the solar wind flow and carry the magnetic field into the inner coma where it is eventually stopped at the diamagnetic cavity boundary.

Prior to the Rosetta mission, several modeling studies for 67P were conducted to project the location of the diamagnetic cavity. Single-fluid MHD simulations predicted that the cavity would form around 50 km from the nucleus within two months of perihelion, while multi-fluid simulations showed that the boundary could be extended on the sunward side by an asymmetric outgassing profile (*Huang et al.,* 2016). Hybrid simulations agreed with these results. However, clear signatures of a magnetic-field-free region were observed at distances as great as 400 km from the nucleus much earlier in the mission when the comet was within 2 au of the Sun inbound and continued until the comet had reached 2.4 au from the Sun outbound (*Goetz et al.,* 2016a, 2018). In fact, the diamagnetic cavity was crossed into by the spacecraft more than 700×, indicating a highly variable boundary. The extension of the diamagnetic cavity was shown to be dependent on the gas production rate, with higher gas production rates leading to larger sizes of the diamagnetic cavity.

The ions inside the diamagnetic cavity are usually found to be quite constant in density and velocity, except for sporadic enhancements (see below). The ion velocity of the constant low-velocity population is around 5–10 km s^{-1} (*Bergmanet al.,* 2021). This is significantly higher than the neutral gas velocity [$\leq$1 km s^{-1} (*Hansen et al.,* 2016)], which indicates that there must be an acceleration mechanism for those ions. It is speculated that an ambipolar field caused by the electron pressure gradient of the expanding plasma accelerates these ions (*Vigren and Eriksson,* 2019). At the same time, the ions cannot collide frequently with the neutral gas, as that would cause ion cooling and a reduction in velocity.

This is quite interesting since for Comet 1P/Halley, the collisions between neutral gas and ions were identified as the process that prevents the magnetic field from diffusing or convecting into the diamagnetic cavity (*Cravens,* 1987). Therefore, another process must be at play at Comet 67P, but which one it is is still an open question.

A series of asymmetric, steepened waves were visible in the magnetic field and plasma density outside the boundary (*Goetz et al.,* 2016a; *Stenberg Wieser et al.,* 2017; *Hajra et al.,* 2018; *Ostaszewski et al.,* 2020b) and in the shape of the boundary itself (*Goetz et al.,* 2016b). The unmagnetized plasma density inside the cavity scales well with the neutral density (*Henri et al.,* 2017). Observations of dense plasma events when the spacecraft was inside the boundary (*Hajra et al.,* 2018; *Masunaga et al.,* 2019) provided indications that plasma from outside the diamagnetic cavity boundary could possibly penetrate into the cavity.

The electron environment changes at the diamagnetic cavity boundary: Electrons in an energy range of 100–200 eV are ubiquitous outside the diamagnetic cavity, but depleted inside the diamagnetic cavity (*Nemeth et al.,* 2016). These electrons probably originate from the solar wind (strahl electrons) and are tied to the magnetic field. As the magnetic field decreases into the diamagnetic cavity, the electrons are adiabatically transported and become field aligned. Therefore, without a perpendicular component, they cannot cross field lines and enter the diamagnetic cavity (*Madanian et al.,* 2020).

The relationship between observations of the electron exobase and the diamagnetic cavity boundary (*Henri et al.,* 2017) suggest that the mechanism determining the distance of the boundary from the nucleus is related to electron neutral collisions and that the location of the boundary could change quite rapidly as a result of instabilities. *Huang et al.* (2018) showed that the introduction of a Hall term in an MHD multi-fluid model could also lead to an extension of the diamagnetic cavity as well as the formation of filaments extending away from the main diamagnetic cavity boundary, similar to an instability as described above. In general, the Rosetta observations were unexpected, and many questions remain about the mechanisms involved in forming and determining the location of the dimagnetic cavity boundary. In particular, the pressure balance at the boundary is not well understood, nor the role of the changing plasma environment in the movement of the boundary. Why is there evidence for ion acceleration at 67P within the cavity but not at 1P/Halley? The processes that cool and/or accelerate electrons and ions in and near the cavity are poorly known, as well as the process for transmitting plasma enhancements through the boundary and inside the cavity. Finally, we do not know the origin of the asymmetry in the inbound and outbound crossing of the boundary. Answering these questions would likely require multi-point measurements of the diamagnetic cavity, its boundary, and the upstream plasma conditions (*Goetz et al.,* 2018).

4. WAVES

In general, waves in plasmas are oscillations in the properties of a coupled system. Since any oscillations have to adhere to the plasma equations, wave modes are discrete. Different approximations of the plasma will result in different wave dispersion relations, thus it is important to always check the underlying assumption of any approximation and make sure it is applicable to the situation.

Usually, a plasma that is initially in equilibrium becomes unstable because a source of free energy is added; at a comet, this source of free energy is the presence of newly ionized cometary ions in the solar wind flow. *Goetz et al.* (2017) found that, in general, the wave activity, or overall power spectral density of the magnetic field at 67P, is modulated by the neutral gas production rate, demonstrating that the addition of more ions leads to more free energy that needs to be distributed in the plasma in order to reach equilibrium again. This addition of free energy will induce wave-like disturbances at a multitude of frequencies. Most of them are damped quickly and only if the disturbances are at a frequency near a wave mode (e.g., ion cyclotron mode) can a wave actually develop and propagate. Due to various processes, the wave is dispersed and dissipated until the energy contained in the original instability is evenly distributed. For a wave to be detectable by instruments, its amplitude needs to be larger than the underlying thermal fluctuations. In addition, plasma wave excitation and propagation depends on the direction of the magnetic field.

Plasma waves in the cometary environment contribute to the heating and cooling of the plasma and couple fields and particles as well as different particle populations. Therefore, the study of these waves is important in understanding the energy and momentum transfer as well as the behavior of particles in the environment. Hereunder, the reader may find a list of waves that have been detected at comets, with an emphasis on the new results from the Rosetta mission to Comet 67P.

4.1. Pickup-Induced Waves

If a cometary neutral is ionized in the solar wind, the resulting ion is moving at a velocity of the negative of the solar wind velocity ($-v_{sw}$) in the solar wind frame of reference. It will therefore be subject to the solar wind convective electric field

$$E_{conv} = -\vec{v}_{sw} \times \vec{B}_{IMF} \quad (15)$$

where $\vec{B}_{IMF}$ is the interplanetary magnetic field. Along with the ion's gyromotion in the magnetic field, this will lead to an E × B-drift. In velocity space, this motion describes a circle with the solar wind ion velocity at its center. If ions are produced over a region greater than the ion gyroradius the cometary ion distribution function will form a full ring distribution; if the ions are produced over a region smaller than the ion gyroradius, the ring distribution is only partial (see, e.g., *Behar,* 2018). If the interplanetary magnetic field is parallel to the solar wind velocity, the convective electric field is zero and the cometary ions are not accelerated. Then, they form a beam distribution in the solar wind frame of reference. Both the beam and ring distribution coexist with the solar wind beam distribution. Therefore, upstream of a comet, two ion distributions can be used to approximate the situation: a ring-beam and a beam-beam distribution. Both these distributions are unstable and give rise to wave activity (e.g., *Coates et al.,* 1993).

These waves were detected at active comets and usually at high cometocentric distances (r_c > 1000 km). In this regime, the large interaction region allows for the ring distribution of the cometary ions to fully develop. This was not the case for most of the Rosetta mission, where the ion gyroradius was larger or on the same order of magnitude as the size of the interaction region. In the high-activity case during the Rosetta mission, the plasma environment was larger than the ion gyroradius, but it was also inhomogeneous at those length scales that prevent the ring distribution from developing. Therefore, a full, classical ring-beam distribution does not develop (*Nicolaou et al.,* 2017). As a result, none of the pickup-induced waves were observed at 67P and most of the results pertaining to ring/ring-beam instabilities stem from earlier works. For more information on those results, the reader is referred to *Ip* (2004).

4.2. Singing Comet Waves

These ultra-low-frequency (ULF) waves were first (and only) detected at Comet 67P at low to medium gas production rates (*Richter et al.,* 2015; *Goetz et al.,* 2020). The singing comet waves are characterized by large-amplitude magnetic field magnitude fluctuations in the frequency range of 10–100 mHz and can be detected ubiquitously in the plasma environment of Comet 67P for gas production rates $Q < 5 \times 10^{26}$ s^{-1}. Figure 12a shows an example of the magnetic field measurements from the two magnetometers of Rosetta and Philae. Their frequency is not correlated with the magnetic field magnitude and it was therefore concluded that it was not due to an ion-cyclotron resonance (for which the frequency correlates with the magnetic field). A new mechanism for the generation of these particular waves was found: an unstable cross-field current in the comet's reference frame. At low gas production rates, the cometary pickup ions have gyroradii much larger than the scale length of the cometary environment (thousands of kilometers vs. hundreds of kilometers) and therefore the ions cannot complete a full ring distribution as was seen at more active comets with much larger cometary environments. Instead, they are accelerated linearly along the electric field. Since all ions are moving in the same direction, this constitutes a current that is parallel to the convective electric field but perpendicular to the magnetic field. Such a configuration is unstable to the ion-Weibel instability and will produce waves. This mechanism was investigated in hybrid simulations (*Koenders et al.,* 2016b) and in an analytical model (*Meier et al.,* 2016). Although the hybrid simulation suggests that the waves should be more ubiquitous in the hemisphere that has a positive convective electric field, this cannot be seen from data, where the waves are detected everywhere, without a preferential location (*Goetz et al.,* 2020). This is not necessarily a contradiction, because the waves could be generated in a region around the nucleus that is larger than the distances covered by the measurements. There are some indications that this is the case, e.g., from two-point measurements that constrain the generation region to between 100 km and more than 800 km in size, which is larger or on the same order of the measurement range [up to 260 km in the interval covered by the study by *Volwerk et al.* (2018)]. But the exact extent of the generation region is as of yet unknown and requires more investigation.

Two point measurements constrained the wavelength to hundreds of kilometers and found that the length scale over which the wave trains are coherent is larger than 50 km (the separation of the two measurement points). This is easily seen in the measurements (Fig. 12a) where both magnetometers show the same wave form with only marginal

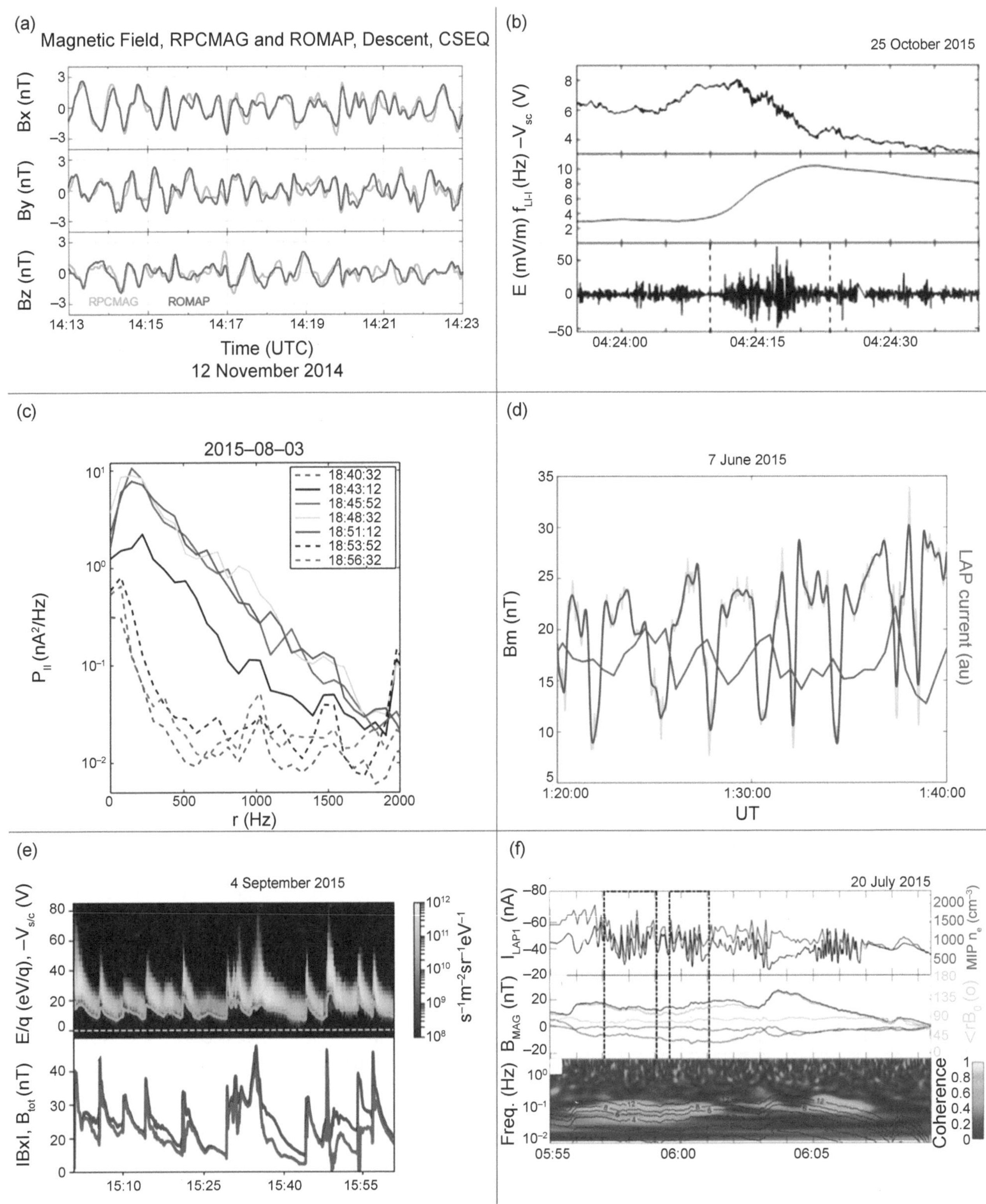

Fig. 12. Examples of wave observations at Comet 67P: **(a)** singing comet waves (from *Richter et al.,* 2016), **(b)** lower hybrid waves (from *Karlsson et al.,* 2017), **(c)** ion acoustic waves (from *Gunnell et al.,* 2017b), **(d)** mirror mode waves (from *Volwerk et al.,* 2016), **(e)** steepened waves (*Stenberg Wieser et al.,* 2017), and **(f)** ion Bernstein waves (from *Odelstad et al.,* 2020). Note that all panels show timeseries except **(c)**, which displays a frequency spectrum for better visibility. For descriptions see text.

deviation. The waves are compressional and, in isolated occasions, also observable in the plasma density (*Breuillard et al.,* 2019). While a case study has found an anti-correlation of the wave frequency with the plasma density, a statistical study covering several months of observations could not confirm this correlation. Therefore, the exact relationship between the plasma density and the magnetic field remains an open question.

4.3. Lower Hybrid Waves

While singing comet waves are mostly detected in the magnetic field observations, electric field measurements also exhibit wave activity. Most prominent among these signatures are waves in the lower hybrid frequency range (a couple of hertz). This particular type of wave can transfer energy between ions and electrons and is typically found in plasmas where the ions are not magnetized but the electrons are. If there are density gradients present in such a plasma, a lower hybrid drift instability (LHDI) can occur and cause lower hybrid waves (LHW) to grow.

The lower hybrid frequency f_{LH} is defined as

$$f_{LH} = \frac{1}{2\pi}\sqrt{\frac{\omega_{gi}\,\omega_{ge}}{1+\frac{\omega_{ge}^2}{\omega_{pe}^2}}} \approx \frac{1}{2\pi}\sqrt{\omega_{ge}\,\omega_{gi}} \quad (16)$$

where ω_{gi} and ω_{ge} are the ion gyrofrequency and the electron gyrofrequency respectively and ω_{pe} is the electron plasma frequency. At Comet 67P, the approximation is usually satisfied as the electron plasma frequency is typically much larger than the electron gyrofrequency. Values for f_{LH} are in the 1–20-Hz range at 67P. Interaction of electrons with lower hybrid waves have been suggested as a possible heating mechanism for the electrons, but no studies have attempted to verify this.

Observations show that wave packets in the electric field are often observed at plasma density gradients and with frequencies near those associated with the LHW (*Karlsson et al.,* 2017; *André et al.,* 2017). In Fig. 12b, the upper panel shows the spacecraft potential (a proxy for the plasma density) during a plasma density gradient, the middle panel shows the derived lower hybrid frequency, and the lower panel shows the electric field measurements with the LHW activity clearly visible. The amplitudes are largest during the density gradient. Estimates of the LHDI criterion and model results show that these observations are consistent with lower hybrid waves generated by a LHDI and that the growth rate can be quite large so that the wave packets can grow to significant amplitudes within a couple of seconds. The LHW can also influence the plasma as a whole by, e.g., forcing the diamagnetic boundary to oscillate slightly. This in turn can lead to a mode conversion, where LHW that are generated at density gradients outside the diamagnetic cavity can be converted to ion acoustic waves (IAW) that propagate in the unmagnetized plasma of the diamagnetic cavity (*Madsen et al.,* 2018).

It should be noted that collisions can significantly inhibit wave growth as they cool and slow down the electrons. Therefore, dense plasmas such as those encountered at Comet 1P/Halley and at Comet 67P close to perihelion and/or very close to the nucleus are not favorable for LHW growth.

4.4. Ion Acoustic Waves

IAWs are compressional waves in an unmagnetized plasma, or in a plasma where the gyrofrequencies are lower than the wave frequency and the gyroradii are larger than the wavelength. They have been observed at Comet 1P/Halley's foreshock (*Oya et al.,* 1986) and at the artificial comet AMPTE (*Gurnett et al.,* 1985). The presence of IAW was reported at Comet 67P at multiple times during the Rosetta mission, all in the plasma in the innermost coma of 67P. IAWs can be observed in a range of frequencies, from hundreds of hertz up to the kilohertz range.

In order to verify that the observed waves are indeed IAW, calculations of the dispersion relation using the observed ion and electron distribution function as well as the measured plasma density can be used. Notably, all observations were made in the inner coma, where a significant cold ion population exists due to ion-neutral collisions. IAW grow if $T_e \gg T_i$, so that any wave activity should be damped quickly in regions where ion cooling is insignificant (*Gunell et al.,* 2017b). In the case that there are accelerated water ions present, this population constitutes a beam-like part of the ion population, which makes the situation unstable and can lead to IAW growth. In the absence of such a beam, a current driven instability can add to the growth rates of the IAW. Rosetta's close flyby of Comet 67P made it possible to observe the large-scale current that is associated with the magnetic field draping near the nucleus (*Koenders et al.,* 2016a) and the wave signatures in that region. It was found that in the presence of this current, IAW are produced, while outside the region containing the current, the waves are propagating away from the current and eventually damped (*Gunell et al.,* 2021).

IAW were also detected inside the diamagnetic cavity, close to the DCB, but not outside of it. Figure 12c shows the power spectral density of the current (a proxy for the density) for short intervals inside (solid lines), outside (dashed lines), and in the boundary (black line) of the diamagnetic cavity. Clearly, the power spectral density is about 2 orders of magnitude higher inside the diamagnetic cavity than it is outside. Here again, the dominance of the cold ion population leads to the growth of the waves. However, an additional current was speculated to be in place. Considering that the diamagnetic cavity boundary is wavy, there might be a current closing through the protruding parts of the diamagnetic cavity that drives the wave generation (*Gunell et al.,* 2017a). Further studies and observations, ideally by multiple spacecraft, are necessary to confirm this generation mechanism. Through a combination of data analysis and modeling of dispersion relations, the wave observations can be used to constrain the plasma parameters in the generation region of the waves.

4.5. Mirror Modes and Magnetic Holes

The pickup process at the comet leads to the generation of a ring or ring-beam distribution of the heavy ions. This distribution is unstable and can cause the generation of mirror-modes in a high-β plasma. Mirror modes are compressional, pressure equilibrium structures. They have large amplitudes and do not propagate in the plasma but rather are convected with the plasma flow. Mirror modes have been observed at Comet 1P/Halley and Comet 21P/Giacobini-Zinner. *Volwerk et al.* (2014) found at 1P/Halley that changes in the dynamic pressure of the solar wind influenced the generation of mirror-mode waves in the cometosheath, as well as outgassing rate changes. Increased solar wind dynamic pressure compresses the magnetosheath and inhibits the growth of mirror modes, but increased outgassing will enhance ion pickup and thereby assist the growth of mirror modes. *Schmid et al.* (2014) showed evidence for the Bohm-type diffusion of mirror modes as they move from the source region further into the magnetosheath. The mirror modes grow in size over time as the high-frequency parts of the structures diffuse faster than the low frequencies.

Mirror modes have been observed in the pileup region at Comet 67P (*Volwerk et al.,* 2016) with timescales of 100 s to 150 s, which corresponds to sizes of tens of water ion gyroradii. Figure 12d shows an example of the magnetic field measurements during a mirror mode wave train; the LAngmuir Probe (LAP) current (a density proxy) is added to show that density and magnetic field are out of phase. Outside the pileup region, the timescale of the mirror modes is smaller (10 s) and the scale size is just a few water ion gyroradii. These mirror mode signatures are more asymmetric with either the decrease or increase of the field being steeper than the other side. The presence of mirror modes indicates that there is a full ring/ring-beam distribution present in the coma, i.e., that the water ions have had enough time to go through an entire gyration before they reach the spacecraft in the inner coma. The larger mirror mode structures are thought to be caused by the diffusion of smaller-scale mirror modes as they are convected through the coma, whereas the asymmetry of the structure could be caused by the interaction of different mirror modes.

Magnetic holes are thought to be a further development stage of these mirror modes (*Winterhalter et al.,* 2000). They are omnipresent in the solar wind (*Volwerk et al.,* 2020, 2021) and therefore should impact the pickup and pileup processes in the coma. Magnetic holes were detected at 67P, inside and outside the solar wind ion cavity (*Plaschke et al.,* 2018), which indicates that the magnetic field structures are moving into the coma along with the electron fluid, while the solar wind ions are substituted by accelerated cometary ions.

4.6. Steepened Magnetosonic Waves

Steepened magnetic field structures were first detected near the diamagnetic cavity of Comet 67P (*Goetz et al.,* 2016a). The diamagnetic cavity entry and exit is also asymmetric, with the former being usually longer than the latter.

High time-resolution observations of the low-energy ion environment in the inner coma showed periodic enhancements in the ion energy with a sharp increase followed by a longer relaxation time, as illustrated in the upper panel of Fig. 12e. Some of this observed increase was due to the spacecraft potential increase, but taking the spacecraft potential into account still allows identification of the asymmetric structures in the measured ion energy and flux (*Stenberg Wieser et al.,* 2017). The occurrence rate of these structures was highest near the diamagnetic cavity.

The magnetic field observations (see lower panel of Fig. 12e) cover a larger time span of the mission time, and therefore a larger study could be performed. Using machine learning, more than 70,000 individual steepened wave structures were detected in the magnetic field (*Ostaszewski et al.,* 2020a). More steepened waves occur when there is more mass-loading of the plasma, which means that the peak in wave activity is around perihelion. During the dayside excursion, the only time that Rosetta left the innermost coma, the number of wave detections decreased. There is no evidence that the occurrence of steepened waves is correlated to the solar wind parameters.

At high activity levels, the waves are steeper but have lower amplitudes, indicating wave evolution based on the interaction region properties. Using a one-dimensional MHD model it is possible to model the steepening of a wave packet in a fluid with non-negligible viscosity and resistivity. It shows that the plasma environment is large enough for the wave packet to steepen to the values of skewness observed in the magnetic field observations. Comparison of the model parameters and the measured wave properties allows inferring the viscosity of the plasma (*Ostaszewski et al.,* 2020b).

Some of these structures can also be detected inside the diamagnetic cavity (*Masunaga et al.,* 2019; *Hajra et al.,* 2017). There, the magnetic field of the structure remains close to zero, but the density and ion flux are similar to the steepened waves upstream of the boundary. This indicates that while the magnetic field remains zero, the diamagnetic cavity boundary is permeable to the heavy ions observed in the inner coma. The exact mechanism of the transmission of those wave packets through the boundary is not yet clear and requires further modeling and analysis.

4.7. Ion Bernstein Waves

A closer inspection of the steepened wave plasma density observations reveals a substructure of wave activity in the descending, longer part of the steepened wave. A corresponding signature in the magnetic field was found to be of lower amplitude and phase shifted by 90°. These were tentatively attributed to ion Bernstein waves, which is an electrostatic wave mode that can be excited by a ring/ring-beam instability (*Odelstad et al.,* 2020). In Fig. 12f the current (plasma density), magnetic field, and coherence of these two parameters is shown. There are clear signatures in the coherence at around 100 Hz for several minutes at a time.

5. SUMMARY AND OUTLOOK

In this chapter we have shown the richness of the processes that arise when the cometary ion cloud interacts with the charged solar wind. The cometary plasma environment is not only highly variable in time, but also in spatial dimension. Depending on the properties of the nucleus and its distance to the Sun, the extension of the coma can vary by 4 orders of magnitude. It therefore is an ideal laboratory to explore processes at different scale sizes and cross-scale interactions in a multi-ion plasma.

Simply put, the environment is created by the ions and electrons of cometary origin and the interaction with the solar wind distributes energy and momentum in this plasma to achieve a stable state, where the solar wind plasma and the cometary plasma are fully mixed. This fundamental process is often referred to as mass-loading and manifests itself in different ways, depending on the scale sizes of the interaction region and the particles in the plasma. Most importantly, the ion gyroradius and collision length scales determine how the solar wind particles and the cometary particles are interacting with each other and among each other. At a high-activity comet, a more fluid-like behavior is common, while a kinetic approach is preferred at low-activity comets and in the inner coma of any comet.

For all comets except the most active, the large gyroradius means that the solar wind ions are not just slowed down, but also deflected. Eventually the cometary ions are picked up and substitute the solar wind ions in the flow. Often, two distinct cometary ion populations are detected: accelerated pickup ions and slow, newborn ions.

On small scales, the electrons and ions decouple, and electrons are accelerated into the inner coma by an ambipolar field. There they can be trapped by collisions and cooled. Therefore, three electron populations are present: cold, warm, and hot. The interplay of electric fields and collisions changes the electron energy.

Transient solar wind events like ICMEs and CIRs increase not only the solar wind pressures but also the compression factor of the cometosphere and the electron impact ionization rate. This leads to an increase in the cometary ion density that is greater than the changes of the solar wind parameters itself. This also increases the magnetic field to unprecedented values.

The interplanetary magnetic field is piled up and draped around the inner coma, creating an induced magnetosphere. Nested draping creates current sheets, and the small gyroradius can lead to draping in a different direction due to ions being deflected. Thus, while the ions are not directly tied to the magnetic field, they still have an influence on it via the electrons. The plasma tail is structured, with cometary rays and density enhancements being visible from Earth remotely. Magnetic reconnection at current sheets could potentially cause tail disconnection events.

There are multiple boundaries that form depending on the gas production rate. The first is a bow shock that is broad and weak due to the mass-loaded nature of the plasma. Here, small-scale processes, like electron impact ionization and charge exchange, can affect the bow shock standoff distance, demonstrating the importance of cross-scale coupling. While a cometopause seems to exist at most comets, a solar wind ion cavity also appears and is the more obvious boundary. Collisionopauses are broad regions where different collisional processes dominate.

The diamagnetic cavity at comets seems highly unstable and asymmetric, and the boundary is often dominated by surface waves. While a diamagnetic cavity exists at both Comet 1P/Halley and at the lower activity Comet 67P, the mechanism that forms this region seems entirely different.

The free energy that is added to the plasma by the creation of heavy water ions modulates the overall wave activity that is observable in the environment as waves are a way to distribute energy. Often, the existence of certain wave modes allows us to learn more about the plasma in which they were generated. Steepening of waves can be used to diagnose the resistivity and viscosity of the plasma and how waves evolve when they travel through it.

All the phenomena described here can be investigated in their own right, but the coupling between processes of different temporal and spatial scales necessitates that a more rounded approach is taken. While the comet nucleus is quite small, the plasma environment can extend up to millions of kilometers, with the plasma tail spanning multiple astronomical units at times. Therefore, a large parameter space is covered, and a multitude of processes can be observed to have effects on the plasma environment.

In the corresponding chapter of *Comets II* (*Ip,* 2004), high hopes were put on the results from the Rosetta mission. However, in hindsight, it is very difficult to compare Rosetta results with those from previous flyby missions to more-active comets. It turned out that the plasma environment at 67P was very different in terms of collisionality and gyroradius effects. Instead of answering the questions posed in *Comets II*, Rosetta has provided a whole new set of results that expand our knowledge of the plasma environment of low to medium activity comets.

Many open questions remain with regard to this topic and future missions to gather a more complete dataset are necessary. First and foremost among these should be a multi-spacecraft mission that will be able to provide spatial and temporal coverage at the same time and allow us to disentangle the influence of the different contributions (solar wind processes and internal processes) to the plasma. Only then can we take full advantage of this intriguing plasma laboratory that presents itself to us any time a comet is explored.

REFERENCES

Alfvén H. (1957) On the theory of comet tails. *Tellus, 9,* 92–96.
Alho M., Simon Wedlund C., Nilsson H., et al. (2019) Hybrid modelling of cometary plasma environments. II. Remote sensing of a cometary bow shock. *Astron. Astrophys., 630,* A45.

Alho M., Jarvinen R., Simon Wedlund C., et al. (2021) Remote sensing of cometary bow shocks: Modelled asymmetric outgassing and pickup ion observations. *Mon. Not. R. Astron. Soc., 506,* 4735–4749.

Altwegg K., Balsiger H., Geiss J., et al. (1993) The ion population between 1300 km and 230,000 km in the coma of Comet P/Halley. *Astron. Astrophys., 279,* 260–266.

Ananthakrishnan S., Bhandai S. M., and Pramesh R. A. (1975) Occultation of radio source PKS 2025-15 by Comet Kohoutek (1973f). *Astrophys. Space Sci., 37,* 275–282.

André M., Odelstad E., Graham D. B., et al. (2017) Lower hybrid waves at Comet 67P/Churyumov-Gerasimenko. *Mon. Not. R. Astron. Soc., 469,* S29–S38.

Balogh A. and Treumann R. A. (2013) *Physics of Collisionless Shocks: Space Plasma Shock Waves.* Springer, New York. 500 pp.

Balogh A., Bothmer V., Crooker N. U., et al. (1999) The solar origin of corotating interaction regions and their formation in the inner heliosphere. *Space Sci. Rev., 89,* 141–178.

Behar E. (2018) Solar wind dynamics within the atmosphere of Comet 67P/Churyumov-Gerasimenko. Ph.D. thesis, Luleå University of Technology, Luleå. 216 pp.

Behar E., Lindkvist J., Nilsson H., et al. (2016a) Mass-loading of the solar wind at 67P/Churyumov-Gerasimenko: Observations and modelling. *Astron. Astrophys., 596,* A42.

Behar E., Nilsson H., Wieser G. S., et al. (2016b) Mass loading at 67P/Churyumov-Gerasimenko: A case study. *Geophys. Res. Lett., 43,* 1411–1418.

Behar E., Nilsson H., Alho M., et al. (2017) The birth and growth of a solar wind cavity around a comet — Rosetta observations. *Mon. Not. R. Astron. Soc., 469,* S396–S403.

Behar E., Nilsson H., Henri P., et al. (2018) The root of a comet tail — Rosetta ion observations at Comet 67P/Churyumov-Gerasimenko. *Astron. Astrophys., 616,* A21.

Berčič L., Behar E., Nilsson H., et al. (2018) Cometary ion dynamics observed in the close vicinity of Comet 67P/Churyumov-Gerasimenko during the intermediate activity period. *Astron. Astrophys., 613,* A57.

Bergman S., Stenberg Wieser G., Wieser M., et al. (2021) Ion bulk speeds and temperatures in the diamagnetic cavity of Comet 67P from RPC-ICA measurements. *Mon. Not. R. Astron. Soc., 503,* 2733-2745.

Beth A., Altwegg K., Balsiger H. et al. (2016) First *in situ* detection of the cometary ammonium ion NH^+ (protonated ammonia NH_3) in the coma of 67P/C-G near perihelion. *Mon. Not. R. Astron. Soc., 462,* S562–S572.

Bieler A., Altwegg K., Balsiger H., et al. (2016) Mass spectrometric characterization of the Rosetta spacecraft contamination. In *Systems Contamination: Prediction, Control, and Performance 2016* (J. Egges et al., eds.), p. 99520E. SPIE Conf. Ser. 9952, Bellingham, Washington.

Biermann L. (1951) Kometenschweife und solare Korpuskularstrahlung. *Z. Astrophys., 29,* 274–286.

Biermann L., Brosowski B., and Schmidt H. U. (1967) The interactions of the solar wind with a comet. *Solar Phys., 1,* 254–284.

Bobrovnikoff N. T. (1930) Halley's comet in its apparition of 1909–1911. *Lick Obs. Publ., 17,* 304–482.

Bodewits D., Juhasz Z., Hoekstra R., et al. (2004) Catching some Sun: Probing the solar wind with cometary X-ray and far-ultraviolet emission. *Astrophys. J. Lett., 606,* L81–L84.

Bodewits D., Lara L. M., A'Hearn M. F., et al. (2016) Changes in the physical environment of the inner coma of 67P/Churyumov-Gerasimenko with decreasing heliocentric distance. *Astron. J., 152,* 130.

Boström R. (1974) Ionosphere-magnetosphere coupling. In *Magnetospheric Physics* (B. M. McCormac, ed.), pp. 45–59. Reidel, Dordrecht.

Brandt J. C. (1982) Observations and dynamics of plasma tails. In *Comets* (L. L. Wilkening, ed.), pp. 519–537. Univ. of Arizona, Tucson.

Brandt J. C. and Snow M. (2000) Heliospheric latitude variations of properties of cometary plasma tails: A test of the Ulysses comet watch paradigm. *Icarus, 148,* 52–64.

Brandt J. C., Caputo F. M., Hoeksema J. T., et al. (1999) Disconnection events (DEs) in Halley's Comet 1985–1986: The correlation with crossings of the heliospheric current sheet (HCS). *Icarus, 137,* 69–83.

Bredikhin T. (1879) *Annales de l'obervatioire de Moscou.* Imprimerie F. Neubürger, Moscow, Russia.

Breuillard H., Henri P., Bucciantini L., et al. (2019) The properties of the singing comet waves in the 67P/Churyumov-Gerasimenko plasma environment as observed by the Rosetta mission. *Astron. Astrophys., 630,* A39.

Broiles T. W., Burch J. L., Clark G., et al. (2015) Rosetta observations of solar wind interaction with the Comet 67P/Churyumov-Gerasimenko. *Astron. Astrophys., 583,* A21.

Broiles T. W., Burch J. L., Chae K., et al. (2016) Statistical analysis of suprathermal electron drivers at 67P/Churyumov-Gerasimenko. *Mon. Not. R. Astron. Soc., 462,* S312–S322.

Buffington A., Bisi M. M., Clover J. M., et al. (2008) Analysis of plasma-tail motions for Comets C/2001 Q4 (NEAT) and C/2002 T7 (LINEAR) using observations from SMEI. *Astrophys. J., 677,* 798–807.

Clark G., Broiles T. W., Burch J. L., et al. (2015) Suprathermal electron environment of Comet 67P/Churyumov-Gerasimenko: Observations from the Rosetta Ion and Electron Sensor. *Astron. Astrophys., 583,* A24.

Coates A. (1997) Ionospheres and magnetospheres of comets. *Adv. Space Res., 20,* 255–266.

Coates A. J., Johnstone A. D., Wilken B., et al. (1993) Velocity space diffusion and nongyrotropy of pickup water group ions at Comet Grigg-Skjellerup. *J. Geophys. Res.–Space Phys., 98,* 20985–20994.

Coates A. J., Burch J. L., Goldstein R., et al. (2015) Ion pickup observed at Comet 67P with the Rosetta Plasma Consortium (RPC) particle sensors: Similarities with previous observations and AMPTE releases, and effects of increasing activity. *J. Phys. Conf. Ser., 642,* 012005.

Combi M. R. and Feldman P. D. (1993) Water production rates in Comet P/Halley from IUE observations of HI Lyman-b. *Icarus, 105,* 557–567.

Combi M. R., Harris W. M., and Smyth W. H. (2004) Gas dynamics and kinetics in the cometary coma: Theory and observations. In *Comets II* (M. C. Festou et al., eds.), pp. 523–552. Univ. of Arizona, Tucson.

Combi M. R., Fougere N., Mäkinen J. T. T., et al. (2014) Unusual water production activity of Comet C/2012 S1 (ISON): Outbursts and continuous fragmentation. *Astrophys. J. Lett., 788,* L7.

Cravens T. E. (1987) Theory and observations of cometary ionospheres. *Adv. Space Res., 7,* 147–158.

Cravens T. E. (1989) Galactic cosmic rays and cell-hit frequencies outside the magnetosphere. *Adv. Space Res., 9,* 293–298.

Cravens T. E. (1991) Collisional processes in cometary plasmas. In *Cometary Plasma Processes* (A. D. Johnstone, ed.), pp. 27–35. AGU Geophys. Monogr. 61, American Geophysical Union, Washington, DC.

Deca J., Divin A., Henri P., et al. (2017) Electron and ion dynamics of the solar wind interaction with a weakly outgassing comet. *Phys. Rev. Lett., 118,* 205101.

Deca J., Henri P., Divin A., et al. (2019) Building a weakly outgassing comet from a generalized Ohm's law. *Phys. Rev. Lett., 123,* 055101.

Degroote P., Bodewits D., and Reyniers M. (2008) Folding ion rays in Comet C/2004 Q2 (Machholz) and the connection with the solar wind. *Astron. Astrophys., 477,* L41–L44.

Delva M., Schwingenschuh K., Niedner M. B., et al. (1991) Comet Halley remote plasma tail observations and *in situ* solar wind properties: Vega-1/2 IMF/plasma observations and ground-based optical observations from 1 December 1985 to 1 May 1986. *Planet. Space Sci., 39,* 697–708.

Delva M., Bertucci C., Schwingenschuh K., et al. (2014) Magnetic pileup boundary and field draping at Comet Halley. *Planet. Space Sci., 96,* 125–131.

Delva M., Volwerk M., Jarvinen R., et al. (2017) Asymmetries in the magnetosheath field draping on Venus' nightside. *J. Geophys. Res.–Space Phys., 122,* 10396–10407.

Divin A., Deca J., Eriksson A., et al. (2020) A fully kinetic perspective of electron acceleration around a weakly outgassing comet. *Astrophys. J. Lett., 889,* L33.

Eberhardt P. and Krankowsky D. (1995) The electron temperature in the inner coma of Comet P/Halley. *Astron. Astrophys., 295,* 795–806.

Edberg N. J. T., Eriksson A. I., Odelstad E., et al. (2015) Spatial distribution of low-energy plasma around Comet 67P/CG from Rosetta measurements. *Geophys. Res. Lett., 42,* 4263–4269.

Edberg N. J. T., Alho M., André M., et al. (2016a) CME impact on

Comet 67P/Churyumov-Gerasimenko. *Mon. Not. R. Astron. Soc.*, *462*, S45–S56.

Edberg N. J. T., Eriksson A. I., Odelstad E., et al. (2016b) Solar wind interaction with Comet 67P: Impacts of corotating interaction regions. *J. Geophys. Res.–Space Phys.*, *121*, 949–965.

Edberg N. J. T., Johansson F. L., Eriksson A. I., et al. (2019) Solar flares observed by Rosetta at Comet 67P/Churyumov-Gerasimenko. *Astron. Astrophys.*, *630*, A49.

Eddington A. S. (1910) The envelopes of Comet Morehouse (1098 c). *Mon. Not. Roy. Astron. Soc.*, *70*, 442–458.

Engelhardt I. A. D., Eriksson A. I., Vigren E., et al. (2018a) Cold electrons at Comet 67P/Churyumov-Gerasimenko. *ArXiV e-prints*, arXiv:1806.09833.

Engelhardt I. A. D., Eriksson A. I., Vigren E., et al. (2018b) Cold electrons at Comet 67P/Churyumov-Gerasimenko. *Astron. Astrophys.*, *616*, A51.

Eriksson A. I., Engelhardt I. A. D., André M., et al. (2017) Cold and warm electrons at Comet 67P/Churyumov-Gerasimenko. *Astron. Astrophys.*, *605*, A15.

Ershkovich A. I. and Heller A. B. (1977) Helical waves in type-1 comet tails. *Astrophys. Space Sci.*, *48*, 365–377.

Feldman W. C., Asbridge J. R., Bame S. J., et al. (1975) Solar wind electrons. *J. Geophys. Res.*, *80*, 4181–4196.

Fleshman B. L., Delamere P. A., Bagenal F., et al. (2012) The roles of charge exchange and dissociation in spreading Saturn's neutral clouds. *J. Geophys. Res.–Planets*, *117*, E05007.

Fougere N., Altwegg K., Berthelier J.-J., et al. (2016a) Three-dimensional direct simulation Monte-Carlo modeling of the coma of Comet 67P/Churyumov-Gerasimenko observed by the VIRTIS and ROSINA instruments on board Rosetta. *Astron. Astrophys.*, *588*, A134.

Fougere N., Altwegg K., Berthelier J. J., et al. (2016b) Direct simulation Monte Carlo modelling of the major species in the coma of Comet 67P/Churyumov-Gerasimenko. *Mon. Not. R. Astron. Soc.*, *462*, S156–S169.

Fuselier S. A., Shelley E. G., Goldstein B. E., et al. (1991) Observations of solar wind ion charge exchange in the Comet Halley coma. *Astrophys. J.*, *379*, 734–740.

Galand M., Héritier K. L., Odelstad E., et al. (2016) Ionospheric plasma of Comet 67P probed by Rosetta at 3 au from the Sun. *Mon. Not. R. Astron. Soc.*, *462*, S331–S351.

Galand M., Feldman P. D., Bockelée-Morvan D., et al. (2020) Far ultraviolet aurora identified at Comet 67P/Churyumov-Gerasimenko. *Nature Astron.*, *4*, 1084–1091.

Galeev A. A. and Sagdeev R. Z. (1988) Alfvén waves in a space plasma and its role in the solar wind interaction with comets. *Astrophys. Space Sci.*, *144*, 427–438.

Galeev A. A., Cravens T. E., and Gombosi T. I. (1985) Solar wind stagnation near comets. *Astrophys. J.*, *289*, 807–819.

Gan L. and Cravens T. E. (1990) Electron energetics in the inner coma of Comet Halley. *J. Geophys. Res.–Space Phys.*, *95*, 6285–6303.

Gapper G. R., Hewish A., Purvis A., et al. (1982) Observing interplanetary disturbances from the ground. *Nature*, *296*, 633–636.

Gilet N., Henri P., Wattieaux G., et al. (2020) Observations of a mix of cold and warm electrons by RPC-MIP at 67P/Churyumov-Gerasimenko. *Astron. Astrophys.*, *640*, A110.

Goetz C., Koenders C., Hansen K. C., et al. (2016a) Structure and evolution of the diamagnetic cavity at Comet 67P/Churyumov-Gerasimenko. *Mon. Not. R. Astron. Soc.*, *462*, S459–S467.

Goetz C., Koenders C., Richter I., et al. (2016b) First detection of a diamagnetic cavity at Comet 67P/Churyumov-Gerasimenko. *Astron. Astrophys.*, *588*, A24.

Goetz C., Volwerk M., Richter I., et al. (2017) Evolution of the magnetic field at Comet 67P/Churyumov-Gerasimenko. *Mon. Not. R. Astron. Soc.*, *469*, S268–S275.

Goetz C., Tsurutani B. T., Henri P., et al. (2018) Unusually high magnetic fields in the coma of 67P/Churyumov-Gerasimenko during its high-activity phase. *Astron. Astrophys.*, *630*, A38.

Goetz C., Plaschke F., and Taylor M. G. G. T. (2020) Singing comet waves in a solar wind convective electric field frame. *Geophys. Res. Lett.*, *47*, e2020GL087418.

Goetz C., Behar E., Beth A., et al. (2021a) The plasma environment of Comet 67P/Churyumov-Gerasimenko. *Space Sci. Rev.*, *218*, 65.

Goetz C., Gunell H., Johansson F., et al. (2021b) Warm protons at Comet 67P/Churyumov–Gerasimenko — Implications for the infant bow shock. *Ann. Geophys.*, *39*, 379–396.

Goldstein R., Burch J. L., Llera K., et al. (2019) Electron acceleration at Comet 67P/Churyumov-Gerasimenko. *Astron. Astrophys.*, *630*, A40.

Gombosi T. I. (1987) Charge exchange avalanche at the cometopause. *Geophys. Res. Lett.*, *14*, 1174–1177.

Gombosi T. I. (2015) Physics of cometary magnetospheres. In *Magnetotails in the Solar System* (A. Keiling et al., eds.), pp. 169–188. AGU Geophys. Monogr. 207, American Geophysical Union, Washington, DC.

Götz C., Gunell H., Volwerk M., et al. (2019) Cometary plasma science: A white paper in response to the Voyage 2050 call by the European Space Agency. *ArXiV e-prints*, arXiv:1908.00377.

Gringauz K. and Verigin M. (1991) Permanent and nonstationary plasma phenomena in Comet Halley's head. In *Cometary Plasma Processes* (A. D. Johnstone, ed.), pp. 107–116. AGU Geophys. Monogr. 61, American Geophysical Union, Washington, DC.

Gringauz K. I., Gombosi T. I., Tátrallyay M., et al. (1986) Detection of a new "chemical" boundary at Comet Halley. *Geophys. Res. Lett.*, *13*, 613–616.

Grün E., Agarwal J., Altobelli N., et al. (2016) The 2016 Feb 19 outburst of Comet 67P/CG: An ESA Rosetta multi-instrument study. *Mon. Not. R. Astron. Soc.*, *462*, S220–S234.

Gulyaev R. A. (2015) Type I cometary tails and the solar wind at the epoch of the Maunder minimum. *Astron. Rept.*, *59*, 791–794.

Gunell H., Goetz C., Eriksson A., et al. (2017a) Plasma waves confined to the diamagnetic cavity of Comet 67P/Churyumov-Gerasimenko. *Mon. Not. R. Astron. Soc.*, *469*, S84–S92.

Gunell H., Nilsson H., Hamrin M., et al. (2017b) Ion acoustic waves at Comet 67P/Churyumov-Gerasimenko: Observations and computations. *Astron. Astrophys.*, *600*, A3.

Gunell H., Goetz C., Simon Wedlund C., et al. (2018) The infant bow shock: A new frontier at a weak activity comet. *Astron. Astrophys.*, *619*, L2.

Gunell H., Lindkvist J., Goetz C., et al. (2019) Polarisation of a small-scale cometary plasma environment: Particle-in-cell modelling of Comet 67P/Churyumov-Gerasimenko. *Astron. Astrophys.*, *631*, A174.

Gunell H., Goetz C., Odelstad E., et al. (2021) Ion acoustic waves near a comet nucleus: Rosetta observations at Comet 67P/Churyumov-Gerasimenko. *Ann. Geophys.*, *39*, 53–68.

Gurnett D. A., Anderson R. R., Häusler B., et al. (1985) Plasma waves associated with the AMPTE artificial comet. *Geophys. Res. Lett.*, *12*, 851–854.

Hajra R., Henri P., Vallières X., et al. (2017) Impact of a cometary outburst on its ionosphere: Rosetta Plasma Consortium observations of the outburst exhibited by Comet 67P/Churyumov-Gerasimenko on 19 February 2016. *Astron. Astrophys.*, *607*, A34.

Hajra R., Henri P., Myllys M., et al. (2018) Cometary plasma response to interplanetary corotating interaction regions during 2016 June–September: A quantitative study by the Rosetta Plasma Consortium. *Mon. Not. R. Astron. Soc.*, *480*, 4544–4556.

Hansen K. C., Altwegg K., Berthelier J.-J., et al. (2016) Evolution of water production of 67P/Churyumov-Gerasimenko: An empirical model and a multi-instrument study. *Mon. Not. R. Astron. Soc.*, *462*, S491–S506.

Haser L. (1957) Distribution d'intensité dans la tête d'une comète. *Bull. Acad. R. Belgique*, *43*, 740–750.

Hayakawa H., Fujii Y. I., Murata K., et al. (2021) Three case reports on the cometary plasma tail in the historical documents. *J. Space Weather Space Clim.*, *11*, 21.

Henri P., Vallières X., Hajra R., et al. (2017) Diamagnetic region(s): Structure of the unmagnetized plasma around Comet 67P/CG. *Mon. Not. R. Astron. Soc.*, *469*, S372–S379.

Heritier K. L., Galand, M., Henri, P., et al. (2018) Plasma source and loss at Comet 67P during the Rosetta mission. *Astron. Astrophys.*, *618*, A77.

Hewish A., Scott P. F., and Wills D. (1964) Interplanetary scintillation of small diameter radio sources. *Nature*, *203*, 1214–1217.

Huang Z., Tóth G., Gombosi T. I., et al. (2016) Four-fluid MHD simulations of the plasma and neutral gas environment of Comet 67P/Churyumov-Gerasimenko near perihelion. *J. Geophys. Res.–Space Phys.*, *121*, 4247–4268.

Huang Z., Tóth G., Gombosi T. I., et al. (2018) Hall effect in the coma of 67P/Churyumov-Gerasimenko. *Mon. Not. R. Astron. Soc.*, *475*, 2835–2841.

Huddleston D. E., Coates A. J., Johnstone A. D., et al. (1993) Mass loading and velocity diffusion models for heavy pickup ions at Comet Grigg-Skjellerup. *J. Geophys. Res.–Space Phys.*, *98*, 20995–

21002.
Hyder C. L., Brandt J. C., and Roosen R. G. (1974) Tail structures far from the head of Comet Kohoutek. I. *Icarus, 23,* 601–610.
Iju T., Abe S., Tokumaru M., et al. (2015) Plasma distribution of Comet ISON (C/2021 S1) observed using the radio scintillation method. *Icarus, 252,* 301–310.
Ip W.-H. (1979) Currents in the cometary atmosphere. *Planet. Space Sci., 27,* 121–125.
Ip W.-H. (1989) On charge exchange effect in the vicinity of the cometopause of Comet Halley. *Astrophys. J., 343,* 946–952.
Ip W.-H. (1994) On a thermodynamical origin of the cometary ion rays. *Astrophys. J. Lett., 432,* L143–L145.
Ip W.-H. (2004) Global solar wind interaction and ionospheric dynamics. In *Comets II* (M. C. Festou et al., eds.), pp. 605–629. Univ. of Arizona, Tucson.
Ip W.-H. and Axford W. I. (1982) Theories of physical processes in the cometary comae and ion tails. In *Comets* (L. L. Wilkening, ed.), pp. 588–634. Univ. of Arizona, Tucson.
Ip W.-H. and Mendis D. A. (1975) The cometary magnetic field and its associated electric currents. *Icarus, 26,* 457–461.
Ip W.-H. and Mendis D. A. (1976) The generation of magnetic fields and electric currents in cometary plasma tails. *Icarus, 29,* 147–151.
Israelevich P. and Ershkovich A. (2014) Magnetic tension in the tails of Titan, Venus and Comet Halley. *Planet. Space Sci., 103,* 339–346.
Itikawa Y. and Mason N. (2005) Cross sections for electron collisions with water molecules. *J. Phys. Chem. Ref. Data, 34,* 1–22.
Jia Y. D., Russell C. T., Jian L. K., et al. (2009) Study of the 2007 April 20 CME-comet interaction event with an MHD model. *Astrophys. J. Lett., 696,* L56–L60.
Jockers K. (1985) The ion tail of Comet Kohoutek 1973 XII during 17 days of solar wind gusts. *Astron. Astrophys. Suppl., 62,* 791–838.
Johnson S. A., Thompson M. C., and Hourigan K. (2004) Predicted low frequency structures in the wake of elliptical cylinders. *Eur. J. Mech. B., 23,* 229–239.
Jones G. H. and Brandt J. C. (2004) The interaction of Comet 153P/Ikeya-Zhang with interplanetary coronal mass ejections: Identification of fast ICME signatures. *Geophys. Res. Lett., 31,* L20805.
Jones G. H., Balogh A., and Horbury T. S. (2000) Identification of Comet Hyakutake's extremely long ion tail from magnetic field signatures. *Nature, 404,* 574–576.
Karlsson T., Eriksson A. I., Odelstad E., et al. (2017) Rosetta measurements of lower hybrid frequency range electric field oscillations in the plasma environment of Comet 67P. *Geophys. Res. Lett., 44,* 1641–1651.
Kinoshita D., Fukushima H., Watanabe J.-I., et al. (1996) Ion tail disturbance of Comet C/Hyakutake 1996B2 observed around the closest approach to the Earth. *Publ. Astron. Soc. Japan, 48,* L83–L86.
Kirsch E., McKenna-Lawlor S., Daly P., et al. (1989) Evidence for the field line reconnection process in the particle and magnetic field measurements obtained during the Giotto-Halley encounter. *Ann. Geophys., 7,* 107–113.
Koenders C., Glassmeier K.-H., Richter I., et al. (2013) Revisiting cometary bow shock positions. *Planet. Space Sci., 87,* 85–95.
Koenders C., Glassmeier K.-H., Richter I., et al. (2015) Dynamical features and spatial structures of the plasma interaction region of 67P/Churyumov-Gerasimenko and the solar wind. *Planet. Space Sci., 105,* 101–116.
Koenders C., Goetz C., Richter I., et al. (2016a) Magnetic field pile-up and draping at intermediately active comets: Results from Comet 67P/Churyumov-Gerasimenko at 2.0 AU. *Mon. Not. R. Astron. Soc., 462,* S235–S241.
Koenders C., Perschke C., Goetz C., et al. (2016b) Low-frequency waves at Comet 67P/Churyumov-Gerasimenko: Observations compared to numerical simulations. *Astron. Astrophys., 594,* A66.
Kramer E. A., Fernandez Y. R., Lisse C. M., et al. (2014) A dynamical analysis of the dust tail of Comet C/1995 O1 (Hale-Bopp) at high heliocentric distances. *Icarus, 236,* 136–145.
Kuchar T. A., Buffington A., Arge C. N., et al. (2008) Observations of a comet tail disruption by the passage of a CME. *J. Geophys. Res.–Space Phys., 113,* A04101.
Lavorenti F., Henri P., Califano F., et al. (2021) Electron acceleration driven by the lower-hybrid-drift instability: An extended quasilinear model. *Astron. Astrophys., 652,* A20.
Lee L. C. (1976) Plasma irregularities in the comet's tails. *Astrophys. J., 210,* 254–257.
Li Y.-F., Li Y.-T., Wang W.-M., et al. (2018) Laboratory study on disconnection events in comets. *Sci. Rept., 8,* 463.
Lishawa C. R., Dressler R. A., Gardner J. A., et al. (1990) Cross sections and product kinetic energy analysis of H_2O^+-H_2O collisions at suprathermal energies. *J. Chem. Phys., 93,* 3196–3206.
Luehr H., Kloecker N., and Acuña M. H. (1988) The diamagnetic effect during AMPTE's tail releases: Initial results. *Adv. Space Res., 8,* 11–14.
Lundin R. (2011) Ion acceleration and outflow from Mars and Venus: An overview. *Space Sci. Rev., 162,* 309–334.
Madanian H., Cravens T. E., Rahmati A., et al. (2016) Suprathermal electrons near the nucleus of Comet 67P/Churyumov-Gerasimenko at 3 AU: Model comparisons with Rosetta data. *J. Geophys. Res.–Space Phys., 121,* 5815–5836.
Madanian H., Burch J., Eriksson A., et al. (2020) Electron dynamics near diamagnetic regions of Comet 67P/Churyumov-Gerasimenko. *Planet. Space Sci., 187,* 104924.
Madsen B., Simon Wedlund C., Eriksson A., et al. (2018) Extremely low-frequency waves inside the diamagnetic cavity of Comet 67P/Churyumov-Gerasimenko. *Geophys. Res. Lett., 45,* 3854–3864.
Mandt K. E., Eriksson A., Edberg N. J. T., et al. (2016) RPC observation of the development and evolution of plasma interaction boundaries at 67P/Churyumov-Gerasimenko. *Mon. Not. R. Astron. Soc., 462,* S9–S22.
Mandt K. E., Eriksson A., Beth A., et al. (2019) Influence of collisions on ion dynamics in the inner comae of four comets. *Astron. Astrophys., 630,* A48.
Marshall D., Rezac L., Hartogh P., et al. (2019) Interpretation of heliocentric water production rates of comets. *Astron. Astrophys., 623,* A120.
Masunaga K., Nilsson H., Behar E., et al. (2019) Flow pattern of accelerated cometary ions inside and outside the diamagnetic cavity of Comet 67P/Churyumov-Gerasimenko. *Astron. Astrophys., 630,* A43.
McComas D. J., Gosling J. T., Bame S. J., et al. (1987) The Giacobini-Zinner magnetotail: Tail configuration and current sheet. *J. Geophys. Res.–Space Phys., 92,* 1139–1152.
Meier P., Glassmeier K.-H., and Motschmann U. (2016) Modified ion-Weibel instability as a possible source of wave activity at Comet 67P/Churyumov-Gerasimenko. *Ann. Geophys., 34,* 691–707.
Mendis D. A. (2007) The solar-comet interactions. In *Handbook of the Solar-Terrestrial Environment* (Y. Kamide and A. Chian, eds.), pp. 493–515. Springer, Berlin.
Mendis D., Smith E., Tsurutani B., et al. (1986) Comet-solar wind interaction: Dynamical length scales and models. *Geophys. Res. Lett., 13,* 239–242.
Mendis D., Flammer K., Reme H., et al. (1989) On the global nature of the solar wind interaction with Comet Halley. *Ann. Geophys., 7,* 99–106.
Meyer-Vernet N., Couturier P., Hoang S., et al. (1986) Physical parameters for hot and cold electron populations in Comet Giacobini-Zinner with the ICE radio experiment. *Geophys. Res. Lett, 13,* 179–181.
Möstl C., Amla K., Hall J. R., et al. (2014) Connecting speeds, directions and arrival times of 22 coronal mass ejections from the Sun to 1 AU. *Astrophys. J., 787,* 119.
Mozhenkov R. T. and Vaisberg O. L. (2017) On the classification of comet plasma tails. *Solar System Res., 51,* 258–270.
Myllys M., Henri P., Galand M., et al. (2019) Plasma properties of suprathermal electrons near Comet 67P/Churyumov-Gerasimenko with Rosetta. *Astron. Astrophys., 630,* A42.
Myllys M., Henri P., Vallières X., et al. (2021) Electric field measurements at the plasma frequency around Comet 67P by RPC-MIP on board Rosetta. *Astron. Astrophys., 652,* A73.
Nemeth Z., Burch J., Goetz C., et al. (2016) Charged particle signatures of the diamagnetic cavity of Comet 67P/Churyumov-Gerasimenko. *Mon. Not. R. Astron. Soc., 462,* S415–S421.
Neubauer F. M. (1988) The ionopause transition and boundary layers at Comet Halley from Giotto magnetic field observations. *J. Geophys. Res.–Space Phys., 93,* 7272–7281.
Neubauer F. M., Glassmeier K. H., Pohl M., et al. (1986) First results from the Giotto magnetometer experiment at Comet Halley. *Nature, 321,* 352–355.
Nicolaou G., Behar E., Nilsson H., et al. (2017) Energy-angle dispersion of accelerated heavy ions at 67P/Churyumov-Gerasimenko:

Implication in the mass-loading mechanism. *Mon. Not. R. Astron. Soc., 469,* S339–S345.

Niedner M. B. (1981) Interplanetary gas. XXVII. A catalog of disconnection events in cometary plasma tails. *Astrophys. J. Suppl. Ser., 46,* 141–157.

Niedner M. B. and Brandt J. C. (1978) Interplanetary gas. XXIII. Plasma tail disconnection events in comets: Evidence for magnetic field line reconnection at interplanetary sector boundaries. *Astrophys. J., 223,* 655–670.

Nilsson H., Stenberg Wieser G., Behar E., et al. (2015) Birth of a comet magnetosphere: A spring of water ions. *Science, 347,* aaa0571.

Nilsson H., Stenberg Wieser G., Behar E., et al. (2017) Evolution of the ion environment of Comet 67P during the Rosetta mission as seen by RPC-ICA. *Mon. Not. R. Astron. Soc., 469,* S252–S261.

Nilsson H., Gunell H., Karlsson T., et al. (2018) Size of a plasma cloud matters: The polarisation electric field of a small-scale comet ionosphere. *Astron. Astrophys., 616,* A50.

Nilsson H., Williamson H., Bergman S., et al. (2020) Average cometary ion flow pattern in the vicinity of Comet 67P from moment data. *Mon. Not. R. Astron. Soc., 498,* 5263–5272.

Nisticò G., Valdimirov V., Nakariakov V. M., et al. (2018) Oscillations of cometary tails: A vortex shedding phenomenon? *Astron. Astrophys., 615,* A143.

Odelstad E., Eriksson A. I., Johansson F. L., et al. (2018) Ion velocity and electron temperature inside and around the diamagnetic cavity of Comet 67P. *J. Geophys. Res.–Space Physics, 123,* 5870–5893.

Odelstad E., Eriksson A. I., André M., et al. (2020) Plasma density and magnetic field fluctuations in the ion gyro-frequency range near the diamagnetic cavity of Comet 67P. *J. Geophys. Res.–Space Phys., 125,* e2020JA028592.

Öpik E. J. (1964) The motion of the condensation in the tail of Halley's Comet June 5–8, 1910. *Z. Astrophys., 58,* 192–201.

Ostaszewski K., Heinisch P., Richter I., et al. (2020a) Pattern recognition in time series for space missions: A Rosetta magnetic field case study. *Acta Astronaut., 168,* 123–129.

Ostaszewski K., Glassmeier K.-H., Goetz C., et al. (2020b) Steepening of magnetosonic waves in the inner coma of Comet 67P/Churyumov-Gerasimenko. *Ann. Geophys., 39,* 721–742.

Oya H., Morioka A., Miyake W., et al. (1986) Discovery of cometary kilometric radiations and plasma waves at Comet Halley. *Nature, 321,* 307–310.

Parker E. N. (1957) Sweet's mechanism for merging magnetic fields in conducting fluids. *J. Geophys. Res., 62,* 509–520.

Parker E. N. (1958) Dynamics of the interplanetary gas and magnetic fields. *Astrophys. J., 128,* 664–676.

Perez-de Tejada H. (1989) Viscous flow interpretation of Comet Halley's mystery transition. *J. Geophys. Res.–Space Physics, 94,* 10131–10136.

Petschek H. E. (1964) Magnetic field annihilation. In *The Physics of Solar Flares* (W. N. Hess, ed.), pp. 425–439. NASA SP-50, Washington, DC.

Pierrard V., Lazar M., Poedts S., et al. (2016) The electron temperature and anisotropy in the solar wind: Comparison of the core and halo populations. *Solar Phys., 291,* 2165–2179.

Plaschke F., Karlsson T., Götz C., et al. (2018) First observations of magnetic holes deep within the coma of a comet. *Astron. Astrophys., 618,* A114.

Price O., Jones G. H., Morrill J., et al. (2019) Fine-scale structure in cometary dust tails I: Analysis of striae in Comet C/2006 P1 (McNaught) through temporal mapping. *Icarus, 319,* 540–557.

Raeder J., Neubauer F. M., Ness N. F., et al. (1987) Macroscopic perturbations of the IMF by P/Halley as seen by the Giotto magnetometer. *Astron. Astrophys., 187,* 61–64.

Rahe J. (1968) The structure of tail rays in the coma region of comets. *Z. Astrophys., 68,* 208–213.

Rahe J. and Donn B. (1969) Ionization and ray formation in comets. *Astron. J., 74,* 256–258.

Ramanjooloo Y. (2014) How comets reveal structure of the inner heliosphere. *Astron. Geophys., 55,* 1.32–1.36.

Ramanjooloo Y. and Jones G. H. (2022) Solar wind velocities at Comets C/2011 L4 Pan-STARRS and C/2013 R1 Lovejoy derived using a new image analysis technique. *J. Geophys. Res.–Space Phys., 127,* e2021JA029799.

Rauer H. and Jockers K. (1993) Doppler measurements of the H_2O^+ ion velocity in the plasma tail of Comet Levy 1990c. *Icarus, 102,* 117–133.

Rème H., Mazelle C., D'Uston C., et al. (1994) There is no "cometopause" at Comet Halley. *J. Geophys. Res.–Space Phys., 99,* 2301–2308.

Richter I., Koenders C., Auster H.-U., et al. (2015) Observation of a new type of low-frequency waves at Comet 67P/Churyumov-Gerasimenko. *Ann. Geophys., 33,* 1031–1036.

Richter I., Auster H.-U., Berghofer G., et al. (2016) Two-point observations of low-frequency waves at 67P/Churyumov-Gerasimenko during the descent of PHILAE: Comparison of RPCMAG and ROMAP. *Ann. Geophys., 34,* 609–622.

Riedler W., Schwingenschuh K., Yeroshenko Y. E., et al. (1986) Magnetic field observations in Comet Halley's coma. *Nature, 321,* 288–289.

Rubin M., Fougere N., Altwegg K., et al. (2014) Mass transport around comets and its impact on the seasonal differences in water production rates. *Astrophys. J., 788,* 168.

Saito T. and Saito K. (1986) Effect of the heliospheric neutral sheet to the kinked ion tail of Comet Halley on 13 May. In *ESLAB Symposium on the Exploration of Halley's Comet* (B. Battrick et al., eds.), pp. 135–143. ESA SP-250, Noordwijk, The Netherlands.

Saito T., Yumoto K., Hirao K., et al. (1987) Structure and dynamics of the plasma tail of Comet P/Halley: I. Knot event on December 31, 1985. *Astron. Astrophys., 187,* 209–214.

Sauer K., Bogdanov A., and Baumgärtel K. (1995) The protonopause — An ion composition boundary in the magnetosheath of comets, Venus and Mars. *Adv. Space Res., 16,* 153–158.

Schmid D., Volwerk M., Plaschke F., et al. (2014) Mirror mode structures near Venus and Comet P/Halley. *Ann. Geophys., 32,* 651–657.

Schmidt H. U. and Wegmann R. (1982) Plasma flow and magnetic fields in comets. In *Comets* (L. L. Wilkening, ed.), pp. 538–560. Univ. of Arizona, Tucson.

Schwenn R., Ip W.-H., Rosenbauer H., et al. (1988) Ion temperature and flow profiles in Comet P/Halley's close environment. In *Exploration of Halley's Comet* (M. Grewing et al., eds.), pp. 160–162. Springer, Berlin.

Silverman S. M. and Limor E. (2021) The great comet of 1577: A Palestinian observation. *Hist. Geo. Space Sci., 12,* 111–114.

Simon Wedlund C., Alho M., Gronoff G., et al. (2017) Hybrid modelling of cometary plasma environments: I. Impact of photoionisation, charge exchange, and electron ionisation on bow shock and cometopause at 67P/Churyumov-Gerasimenko. *Astron. Astrophys., 604,* A73.

Simon Wedlund C., Behar E., Nilsson H., et al. (2019a) Solar wind charge exchange in cometary atmospheres III: Results from the Rosetta mission to Comet 67P/Churyumov-Gerasimenko. *Astron. Astrophys., 630,* A37.

Simon Wedlund C., Bodewits D., Alho M., et al. (2019b) Solar wind charge exchange in cometary atmospheres: I. Charge-changing and ionization cross sections for He and H particles in H_2O. *Astron. Astrophys., 630,* A35.

Sishtla C. P., Divin A., Deca J., et al. (2019) Electron trapping in the coma of a weakly outgassing comet. *Phys. Plasmas, 26,* 102904.

Slavin J. A., Golberg B. A., Smith E. J., et al. (1986a) The structure of a cometary type I tail: Ground-based and ICE observations of P/Giacobini-Zinner. *Geophys. Res. Lett., 13,* 1085–1088.

Slavin J. A., Smith E. J., Tsurutani B. T., et al. (1986b) Giacobini-Zinner magnetotail: ICE magnetic field observations. *Geophys. Res. Lett., 13,* 283–286.

Smith E. J., Tsurutani B. T., Slavin J. A., et al. (1986) International cometary explorer encounter with Giacobini-Zinner: Magnetic field observations. *Science, 232,* 382–385.

Snodgrass C. and Jones G. H. (2019) The European Space Agency's Comet Interceptor lies in wait. *Nature Commun., 10,* 5418.

Stenberg Wieser G., Odelstad E., Wieser M., et al. (2017) Investigating short-time-scale variations in cometary ions around Comet 67P. *Mon. Not. R. Astron. Soc., 469,* S522–S534.

Sweet P. A. (1956) The neutral point theory of solar flares. In *Electromagnetic Phenomena in Cosmical Physics* (B. Lehnert, ed.), pp. 123–134. IAU Symp. 6, Cambridge Univ., Cambridge.

Szegö K., Glassmeier K.-H., Bingham R., et al. (2000) Physics of mass loaded plasmas. *Space Sci. Rev., 94,* 429–671.

Tao C., Kataoka R., Fukunishi H., et al. (2005) Magnetic field variations in the jovian magnetotail induced by solar wind dynamic pressure enhancements. *J. Geophys. Res.–Space Phys., 110,* A11208.

Thomsen M. F., Feldman W. C., Wilken B., et al. (1987) *In-situ*

observations of a bi-modal ion distribution in the outer coma of Comet P/Halley. *Astron. Astrophys., 187,* 141–148.
Timar A., Nemeth Z., Szego K., et al. (2019) Estimating the solar wind pressure at Comet 67P from Rosetta magnetic field measurements. *J. Space Weather Space Clim., 9,* A3.
Verigin M. I., Axford W. I., Gringauz K. I., et al. (1987) Acceleration of cometary plasma in the vicinity of Comet Halley associated with an interplanetary magnetic field polarity change. *Geophys. Res. Lett., 14,* 987–990.
Vigren E. and Galand M. (2013) Predictions of ion production rates and ion number densities within the diamagnetic cavity of Comet 67P/Churyumov-Gerasimenko at perihelion. *Astrophys. J., 772,* 33.
Vigren E. and Eriksson A. I. (2019) On the ion-neutral coupling in cometary comae. *Mon. Not. R. Astron. Soc., 482,* 1937–1941.
Volwerk M., Glassmeier K. H., Delva M., et al. (2014) A comparison between VEGA 1, 2 and Giotto flybys of Comet 1P/Halley: Implications for Rosetta. *Ann. Geophys., 32,* 1441–1453.
Volwerk M., Richter I., Tsurutani B., et al. (2016) Mass-loading, pile-up, and mirror-mode waves at Comet 67P/Churyumov-Gerasimenko. *Ann. Geophys., 34,* 1–15.
Volwerk M., Jones G. H., Broiles T., et al. (2017) Current sheets in Comet 67P/Churyumov-Gerasimenko's coma. *J. Geophys. Res.–Space Phys., 122,* 3308–3321.
Volwerk M., Goetz C., Richter I., et al. (2018) A tail like no other: RPC-MAG's view of Rosetta's tail excursion at Comet 67P/Churyumov-Gerasimenko. *Astron. Astrophys., 614,* A10.
Volwerk M., Goetz C., Behar E., et al. (2019) Dynamic field line draping at Comet 67P/Churyumov-Gerasimenko during the Rosetta dayside excursion. *Astron. Astrophys., 630,* A44.
Volwerk M., Goetz C., Plaschke F., et al. (2020) On the magnetic characteristics of magnetic holes in the solar wind between Mercury and Venus. *Ann. Geophys., 38,* 51–60.
Volwerk M., Mautner D., Wedlund C. S., et al. (2021) Statistical study of linear magnetic hole structures near Earth. *Ann. Geophys., 39,* 239–253.
Vourlidas A., Davis C. J., Eyles C. J., et al. (2007) First direct observation of the interaction between a comet and a coronal mass ejection leading to a complete plasma tail disconnection. *Astrophys. J. Lett., 668,* L79–L82.
Wattieaux G., Henri P., Gilet N., et al. (2020) Plasma characterization at Comet 67P between 2 and 4 AU from the Sun with the RPC-MIP instrument. *Astron. Astrophys., 638,* A124.
Wellbrock A., Jones G., Coates A., et al. (2018) Observations of a solar energetic particle event from inside the Comet 67P coma and upstream of the comet. *European Planet. Sci. Congr., 218,* 964.
Williamson H. N., Nilsson H., Stenberg Wieser G., et al. (2020) Momentum and pressure balance of a comet ionosphere. *Geophys. Res. Lett., 47,* e2020GL088666.
Winterhalter D., Smith E. J., Neugebauer M., et al. (2000) The latitudinal distribution of solar wind magnetic holes. *Geophys. Res. Lett., 27,* 1615–1618.
Witasse O., Sánchez-Cano B., Mays M. L., et al. (2017) Interplanetary coronal mass ejection observed at STEREO-A, Mars, Comet 67P/Churyumov-Gerasimenko, Saturn, and New Horizons en route to Pluto: Comparison of its Forbush decreases at 1.4, 3.1, and 9.9 AU. *J. Geophys. Res.–Space Phys.*, 122, 7865–7890.
Wolff R. S., Siscoe G. L., Sibeck D. G., et al. (1985) Cometary rays: Magnetically channeled outflow. *Geophys. Res. Lett., 12,* 749–752.
Womack M., Homich A., Festou M. C., et al. (1997) Maps of HCO^+ emission in C/1995 O1 (Hale-Bopp). *Earth Moon Planets, 77,* 259–264.
Wurm K. (1949) Die natur der kometen. *Mitt. Hamburg. Sternw. Bergedorf, 8,* 57–92.
Wurm K. (1961) Structure and development of the gas tails of comets. *Astron. J., 66,* 362–367.
Wurm K. (1963) The physics of comets. In *The Moon, Meteorites and Comets* (G. P. Kuiper and B. Middlehurst, eds.), pp. 573–617. Univ. of Chicago, Chicago.
Wurz P., Rubin M., Altwegg K., et al. (2015) Solar wind sputtering of dust on the surface of 67P/Churyumov-Gerasimenko. *Astron. Astrophys., 583,* A22.
Yagi M., Koda J., Terai T., et al. (2015) Initial speed of knots in the plasma tail of C/2013 R1(Lovejoy). *Astron. J., 149,* 97.
Yang L., Paulsson J. J. P., Wedlund C. S., et al. (2016) Observations of high-plasma density region in the inner coma of 67P/Churyumov-Gerasimenko during early activity. *Mon. Not. R. Astron. Soc., 462,* S33–S44.
Young D., Crary F., Nordholt J., et al. (2004) Solar wind interactions with Comet 19P/Borrelly. *Icarus, 167,* 80–88.
Zhao X., Liu Y. D., Hu H., et al. (2017) Propagation characteristics of two coronal mass ejections from the Sun far into interplanetary space. *Astrophys. J., 837,* 4.
Zolotova N., Sizonenko Y., Mokhmyanin M., et al. (2018) Indirect solar wind measurements using archival cometary tail observations. *Solar Phys., 293,* 85.

Part 7:

Dust

Engrand C., Lasue J., Wooden D. H., and Zolensky M. E. (2024) Chemical and physical properties of cometary dust. In *Comets III* (K. J. Meech, M. R. Combi, D. Bockelée-Morvan, S. N. Raymond, and M. E. Zolensky, eds.), pp. 577–619. Univ. of Arizona, Tucson, DOI: 10.2458/azu_uapress_9780816553631-ch018.

Chemical and Physical Properties of Cometary Dust

C. Engrand
Laboratoire de Physique des 2 Infinis Irène Joliot-Curie (IJCLab), Université Paris-Saclay/ Centre National de la Recherche Scientifique

J. Lasue
Institut de Recherche en Astrophysique et Planétologie (IRAP), Université de Toulouse, Centre National de la Recherche Scientifique, Centre National d'Études Spatiales (CNES)

D. H. Wooden
NASA Ames Research Center

M. E. Zolensky
NASA Johnson Space Center

Cometary dust particles are the best preserved remnants of the matter present at the onset of the formation of the solar system. Space missions, telescopic observations, and laboratory analyses advanced the knowledge on the properties of cometary dust. The only samples with an ascertained cometary origin were returned by the Stardust space mission from Comet 81P/Wild 2. The "chondritic porous" (here called "chondritic anhydrous") interplanetary dust particles (CA-IDPs), the chondritic porous micrometeorites (CP-MMs), and the ultracarbonaceous Antarctic micrometeorites (UCAMMs) also show strong evidence for a cometary origin. Astronomical observations show that cometary infrared (IR) spectra can generally be modeled using five families of particles: amorphous minerals of olivine and pyroxene compositions, crystalline olivines and pyroxenes, and amorphous carbon. From the analyses of space missions Giotto, Vega 1 and 2, Stardust, and Rosetta and of CA-IDPs, CP-MMs, and UCAMMs, the elemental composition of cometary dust is generally consistent with the chondritic composition (as defined by the composition of the CI, or Ivuna-type, carbonaceous chondrites), with the notable exception of elevated contents in carbon (and possibly nitrogen) compared to CI. As seen in CA-IDPs, CP-MMs, and UCAMMs, the organic matter of cometary dust is mixed with minor amounts of crystalline (10% to ≳25% of the minerals) and amorphous mineral phases. The most abundant crystalline minerals are ferromagnesian silicates (olivine and low-Ca pyroxenes), but high-Ca pyroxenes, refractory minerals, and low-Ni Fe sulfides are also present. The crystalline olivine and low-Ca pyroxene compositions can vary from their Mg-rich end member (forsterite and enstatite) to relatively Fe-rich compositions. Stardust samples from Comet 81P/Wild 2 contain in particular olivines and pyroxenes with minor-element abundances linking them to unequilibrated ordinary chondrites. Refractory minerals as well as secondary minerals like low-Fe, Mn-enriched (LIME) olivines, unusual Fe sulfides or mineral aggregates of specific compositions like Kosmochloric high-Ca pyroxene and FeO-rich olivine (KOOL grains) are also found in CA-IDPs, CP-MMs, and Stardust samples. The presence of carbonates in cometary dust is still debated (both in astronomical observations and in samples analyzed *in situ* or in the laboratory), as well as the presence of hydrated silicates, as proposed after the Deep Impact mission. A phyllosilicate-like phase was, however, observed in a UCAMMs, and a CP-MM shows a mixing texture with a hydrated part. The abundance of pyroxene to olivine (in numbers) in CA-IDPs, CP-MMs, and UCAMMs is larger than in primitive meteorites (e.g., the Px/Ol ratio is usually larger than 1). Glass with embedded metals and sulfides (GEMS) phases are abundant in cometary dust, although not systematically found. Some of the organic matter present in cometary dust particles resembles the insoluble organic matter (IOM) present in primitive meteorites, but amorphous carbon and exotic (e.g., N-rich in UCAMMs) organic phases are also present. The hydrogen isotopic composition in the organic matter of cometary dust particles analyzed in the laboratory is usually rich in deuterium, tracing a formation at very low temperatures, either in the protosolar cloud or in the outer regions of the protoplanetary disk. The presolar dust concentration in cometary dust can reach about 1% (in CA-IDPs collected during dust streams of Comets 26P/Grigg-Skjellerup and 21P/Giacobini-Zinner), which is the most elevated value observed in extraterrestrial samples. The size distribution of cometary dust in comet trails is well represented by a power-law distribution (differential size distribution) with a mean power index N typically ranging from –3 to –4. Polarimetric and light-scattering studies of cometary dust suggest mixtures of porous agglomerates of submicrometer minerals with organic matter, which is compatible with the *in situ* analyses of 67P/Churyumov-Gerasimenko dust particles by the Micro-Imaging Dust Analysis System (MIDAS) on Rosetta and with the studies of Stardust samples, CA-IDPs, CP-MMs, and UCAMMs. Cometary dust particles have low tensile strength and low density, as deduced from observations and analyses during the Rosetta mission. The presence of high-temperature minerals in cometary dust highlights the need for a large-scale transport mechanism in the early protoplanetary disk. Many gaps in the understanding of the formation of cometary dust remain, such as the incorporation of minerals from evolved asteroids, the possibility of aqueous alteration on the comet, the formation processes of organic matter, as well as the mixing process(es) with minerals and ices and the apparent small presolar heritage.

1. COMETARY DUST : FROM SPACE MISSIONS TO GROUNDBASED OBSERVATIONS AND TO THE LABORATORY

Space missions are very powerful in advancing the understanding of comets. However, the cost and timeline of such missions have only permitted the characterization of a limited number of comets so far. Spacecraft flybys of Comet 1P/Halley by Giotto and Vega 1 and 2 with, respectively, the Particulate Impact Analyzer (PIA) and Dust Impact Mass Spectrometer (PUMA) 1 and 2 instruments, Comet 81P/Wild 2 (Stardust), Comet 9P/Tempel 1 (Deep Impact), Comet 103P/Hartley 2 (Deep Impact-Extended), and the long-duration rendezvous with Comet 67P/Churyumov-Gerasimenko (hereafter referred to as 67P/C-G) (Rosetta) provided key advancements in our understanding about the composition and structure of the cometary dust particles, complementing less-costly telescopic observations and laboratory work on cometary materials captured in the stratosphere and from polar regions, and those returned by the Stardust mission to Comet 81P/Wild 2. Groundbased and spaceborne observations of comets encompass a wide range of spatial scales, spectral ranges, and spectral resolutions. A small perihelion event like with Comet C/2006 P1 (McNaught) (Fig. 1) can produce a spectacular release of small to large particles into the coma and dramatic comet tail structures that can be modeled to assess coma particle sizes and modulations in dust production rates.

As the decades advanced from the late 1970s, the instruments and telescopes provided increases in sensitivity and wavelength coverage that allowed multi-epoch studies in the "10-μm window" wavelength and limited studies near 20 μm from groundbased telescopes as well as uninterrupted wavelength coverage from ~5 to 40 μm from the Infrared Space Observatory (ISO) Short Wave Spectrometer (SWS), Stratospheric Observatory for Infrared Astronomy (SOFIA), and Spitzer, with the majority being single-epoch observations. The James Webb Space Telescope (JWST) will provide a greater span of wavelength coverage, which facilitates the simultaneous use of scattered light and thermal emission studies to characterize the dust composition and particle properties in cometary comae. Thus, the decline in the availability in mid- and far-infrared (IR) instrumentation for photometric and spectroscopic studies of comets on groundbased telescopes is complimented by higher sensitivity and broader wavelength coverage of airborne and spacebased telescopes. Multi-epoch studies have thus far revealed that the dust composition of the coma of an individual can vary significantly with heliocentric distance, potentially due to changes in seasonal illumination that changes the "active" areas and/or changes in jet activity. Also, the coma's dust composition can appear to change composition, i.e., increase

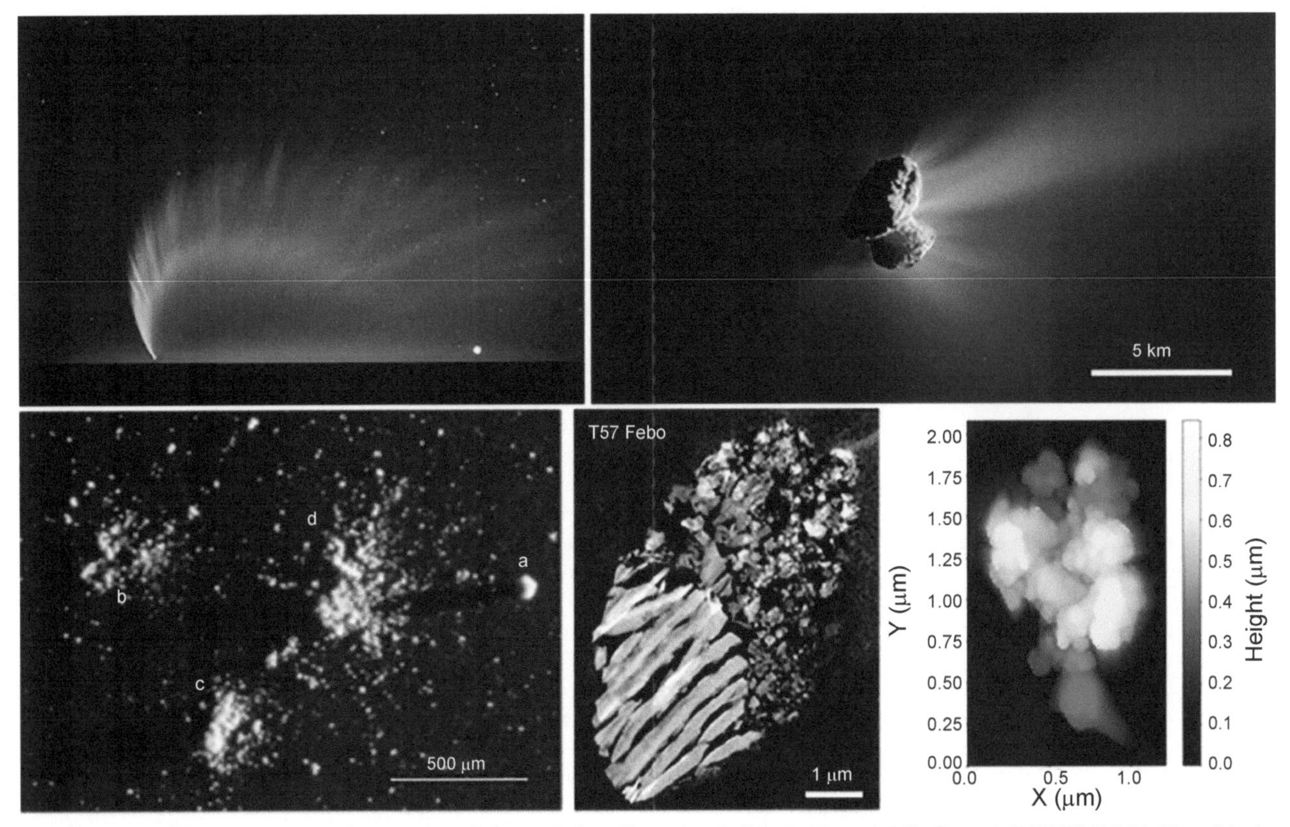

Fig. 1. Cometary dust seen at increasing size resolution (from top left to bottom right): Comet C/2006 P1 McNaught observed from Paranal (commons license), Comet 67P/Churyumov-Gerasimenko (67P/C-G) by Rosetta (commons license), cometary dust particles of 67P/C-G collected by COSIMA (*Langevin et al.,* 2016), section of a Stardust particle from 81P/Wild 2 track 57 (*Matrajt et al.,* 2008), and 67P/C-G dust particles collected by MIDAS on Rosetta (*Mannel et al.,* 2019).

the relative abundance of crystalline silicates because a decrease in heliocentric distance (r_h) causes increases in solar insolation whereby dust components that are less absorbing of sunlight (e.g., more transparent Mg-rich crystalline silicates that are less optically active) may warm sufficiently to gain spectral contrast with respect to more active species (e.g., amorphous carbon and Mg:Fe amorphous silicates).

NASA has been collecting interplanetary dust particles (IDPs) in the stratosphere (*Warren and Zolensky,* 1994), following the pioneering work of Don Brownlee (*Brownlee et al.,* 1977). Several lines of evidence point to a cometary origin for chondritic anhydrous (CA) IDPs (*Bradley and Brownlee,* 1986; *Bradley,* 1994a; *Bradley et al.,* 1999) (Fig. 2), including higher atmospheric entry velocities [as determined by noble gas measurements (*Nier and Schlutter,* 1993)], high particle porosity, anhydrous nature, high bulk carbon content associated with a high abundance of pyroxenes minerals (*Thomas et al.,* 1993), short solar exposure histories, high presolar grain concentrations (*Rietmeijer,* 1998; *Palma et al.,* 2005; *Nguyen et al.,* 2007; *Busemann et al.,* 2009; *Brownlee et al.,* 1995; *Bradley et al.,* 2014). We chose here to name these IDPs "chondritic anhydrous" whereas they are usually quoted as "chondritic porous" (or "chondritic porous anhydrous") in the literature. This choice was made following the observation that many hydrous IDPs are equally porous (*Zolensky et al.,* 1992) so the term "chondritic porous" is somewhat misleading. Moreover, IDPs having a fluffy-like texture when observed as whole particles do not always contain significant porosity when examined in their interior (e.g., by sectioning with ultramicrotomy). In the search for cometary IDPs, timed collections in the stratosphere were performed with the aim of collecting IDPs during dust streams of Comets 26P/Grigg-Skjellerup and 21P/Giacobini-Zinner, but the expected fraction of collected particles arising from those particular sources was only a few percent (*Busemann et al.,* 2009; *Bastien et al.,* 2013). Because of limitations of the collection technique, no particular IDP can be unambiguously identified with a specific small body, so the question of their origin(s) is not yet completely resolved. After 50 years of investigation, links between IDPs and comets remain hazy, and some IDPs might originate from icy asteroids (e.g., *Vernazza et al.,* 2015), which could in turn sample the "asteroid-comet" continuum proposed by *Gounelle* (2011). Nevertheless, a tentative consensus has been formed that the chondritic anhydrous IDPs (CA-IDPs) probably are mainly of cometary origin. Are any of the hydrous chondritic IDPs from comets? The discovery of calcium-aluminum-rich refractory inclusions (CAIs) among 81P/Wild 2 particles could suggest a cometary origin for some refractory IDPs (*Zolensky,* 1987; *McKeegan,* 1987), which have long been ignored. These IDPs are finer grained than typical meteoritic CAIs.

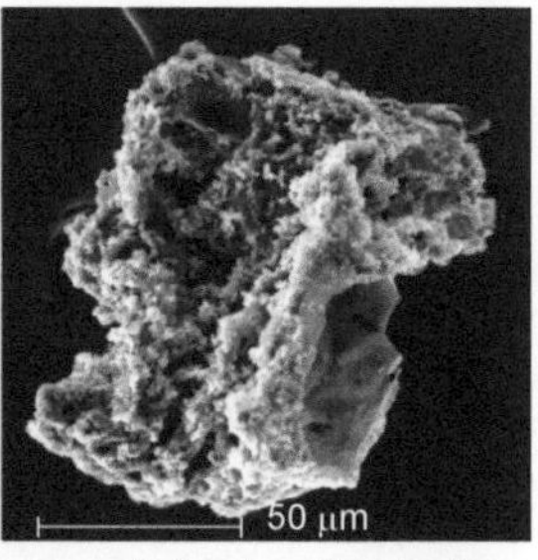

Fig. 2. Secondary electron images of dust collected on Earth with a probable cometary origin: **(a)** chondritic anhydrous interplanetary dust particle (CA-IDP) collected in the stratosphere by NASA; **(b)** ultracarbonaceous Antarctic micrometeorites (UCAMM) from the Concordia collection (*Duprat et al.,* 2007, 2010).

Larger IDPs called micrometeorites (MMs) have also been recovered from polar ice and snow. They were originally found by *Maurette et al.* (1986, 1987) in Greenland ice, and later collected at lower temperatures from Antarctic ice and then snow (e.g., *Maurette et al.,* 1991; *Duprat et al.,* 2007; *Dobrică et al.,* 2009; *Noguchi et al.,* 2015). These larger particles are generally more strongly heated during atmospheric entry than IDPs and may be altered in the terrestrial environment, especially by leaching when collected from ice where they can reside for several tens of thousands of years before collection. The samples collected from snow, however, do not show evidence for extensive aqueous alteration (*Duprat et al.,* 2007). Extraterrestrial dust particles in the size range of MMs (~200 μm) constitute the dominant input of extraterrestrial matter on Earth (*Love and Brownlee,* 1993; *Rojas et al.,* 2021), and they could have played a role in the formation of the terrestrial hydrosphere and the origin of life on Earth (e.g., *Maurette,* 2006). Numerical modeling suggests that ~80% of micrometeorites could originate from comets (*Carrillo-Sánchez et al.,* 2016), but they probably derive from both asteroids and comets. Some MMs have been found to have identical fine-grained components to chondritic-porous IDPs (here called CA-IDPs) and thus these chondritic porous micrometeorites (CP-MMs) sample sources that are most likely cometary (*Noguchi et al.,* 2015, 2017). Ultracarbonaceous Antarctic micrometeorites (UCAMMs) (Fig. 2) also constitute a new family of micrometeorites that was recently discovered in the Concordia and Dome Fuji collections (*Nakamura et al.,* 2005; *Duprat et al.,* 2010; *Yabuta et al.,* 2017). They are dominated by organic matter with a variable (but minor) stony component, show large anomalies of their hydrogen isotopic composition, and most probably originate from comets. They contain an unusual N-rich organic matter that could have formed by galactic cosmic-ray irradiation of N-rich ices in the outer regions of the protoplanetary disk (*Dartois et al.,* 2013, 2018; *Augé et al.,* 2016).

This chapter will describe the chemical and physical properties of cometary dust particles, based on the information recovered from space missions, groundbased telescopic observations, and analyses in the laboratory of CA-IDPs, CP-MMs, and UCAMMs. Since this chapter gathers information from very different collecting and/or analytical methods, the following possible biases should be kept in mind: (1) The *in situ* mass spectrometers at Comet 1P/Halley mostly detected very small particles (nanometers) — possibly subconstituents of larger particles. Rosetta's COmetary Secondary Ion Mass Analyzer (COSIMA) mass spectrometer performed analyses of dust from 67P/C-G at the ~25-μm scale, thus averaging the composition of subcomponents. (2) Samples returned from the Stardust mission were more or less altered by the high-speed collection in the aerogel, depending on the mechanical strength of the initial particles. (3) The identification of cometary dust compositions from astronomical IR spectroscopy requires a careful control on observational and analytical parameters, as presented in section 2.5. (4) Information deduced from CA-IDPs, CP-MMs, and UCAMMs can suffer from small number statistics and from the poor knowledge of their initial parent bodies.

2. CHEMICAL PROPERTIES

2.1. Elementary Composition

The elementary composition of cometary dust particles can be determined from groundbased and spacebased missions, along with dust samples of very probably cometary origin that are collected on Earth: the CA-IDPs (*Bradley et al.,* 2014), the CP-MMs (*Noguchi et al.,* 2015, 2017), and the UCAMMs (*Duprat et al.,* 2010).

Insight into the elemental composition of cometary dust particles was gathered so far for four comets visited by space missions (1P/Halley, 9P/Tempel 1, 103/Hartley 2, 67P/C-G) and from a comet sample return (Stardust mission to 81P/Wild 2). For Comet 1P/Halley, mass spectrometry from Giotto, Vega 1, and Vega 2 of cometary dust impacting at high speeds (~70–75 km s^{-1}) determined two populations: the "rocky" elements of Mg, Fe, O, S (of approximately solar composition) and the carbon, hydrogen, oxygen, and nitrogen (CHON) particles with elemental abundances enhanced over the CI chondritic composition (as defined by the Ivuna-type carbonaceous chondrites) and the Sun (*Jessberger et al.,* 1986; *Jessberger,* 1999). All particles were, in fact, on some level, mixtures of "rocky" and "CHON" materials with about one-quarter being predominantly "rocky," one-quarter predominantly "CHON," and half being mixed with a span of 0.1–10× CHON/rock elemental ratios (*Fomenkova et al.,* 1992; *Lawler and Brownlee,* 1992). The very smallest particles had the greatest C abundances (*Lawler and Brownlee,* 1992). The bulk composition of Halley dust particles was chondritic within a factor of 2, with the exception of carbon, nitrogen, and hydrogen, which were enriched with regard to CI by a factor of 11, 8, and 4, respectively (Fig. 3).

The Stardust mission flyby showed that particle streams in the coma resulted from the disintegration (called "autobrecciation") of larger particles released from the nucleus at slower speeds (*Clark et al.,* 2004). Stardust returned samples revealed the presence of a greater fraction of "hot inner disk materials" than assessed from Halley, including

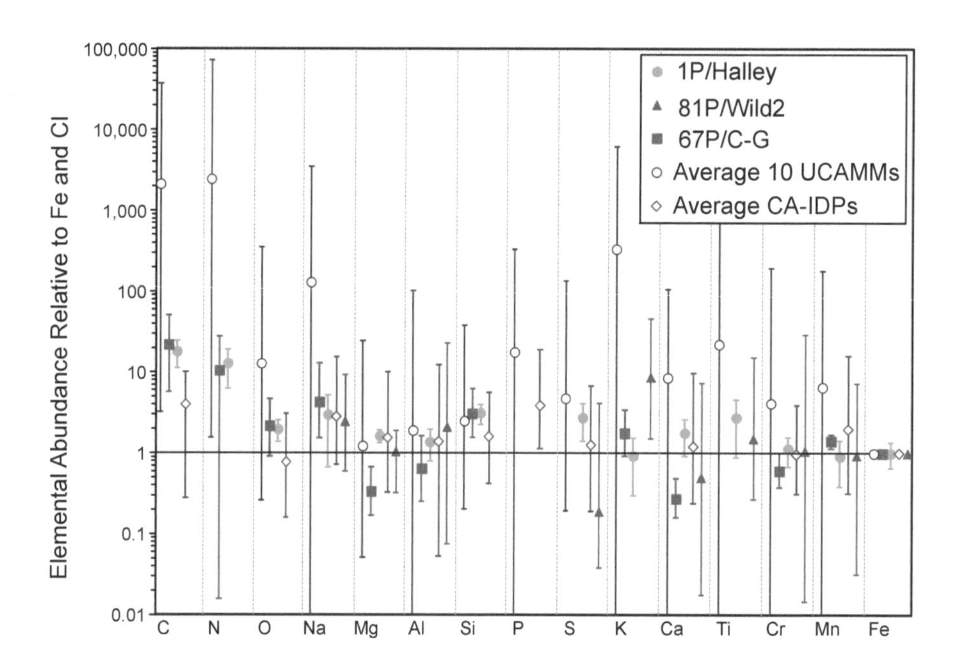

Fig. 3. Average elemental ratios relative to Fe and to CI (*Lodders,* 2010) for dust particles from Comets 1P/Halley (*Jessberger et al.,* 1988), 81P/Wild 2 in aerogel (*Flynn et al.,* 2006; *Ishii et al.,* 2008; *Lanzirotti et al.,* 2008; *Leroux et al.,* 2008; *Stephan,* 2008; *Stephan et al.,* 2008), and 67P/Churyumov-Gerasimenko (67P/C-G) (*Bardyn et al.,* 2017); for 10 ultracarbonaceous Antarctic micrometeorites (UCAMMs) (*Dartois et al.,* 2018, and unpublished data); and for 115 chondritic anhydrous IDPs (CA-IDPs) (*Thomas et al.,* 1993; *Keller et al.,* 2004; *Schramm et al.,* 1989) (not all elements were measured for all CA-IDPs; see text). Error bars represent the variation range of the elemental compositions among cometary dust particles for the CA-IDPs and UCAMMs samples and for Comets 1P/Halley and 81P/Wild 2. In the case of Comet 67P/C-G, the error bars represent uncertainties on the values, which are higher or equal to the variation of composition between particles.

high-temperature CAIs, microchondrules (100-μm-sized) spanning Mg- to Fe-rich olivine (crystals), plagioclase, nepheline, and graphitic carbon (see sections 2.2 and 2.4 for more details) (*Zolensky et al.*, 2006; *Nakamura et al.*, 2008; *De Gregorio et al.*, 2017).

The impact on Comet 9P/Tempel 1, which was created by the Deep Impact mission, released particles into the coma, with coma-gas-accelerated speeds of 200 m s^{-1}, that spectrally appeared to be more similar to the submicrometer-sized silicate-rich and crystal-rich Comet Hale-Bopp. In the hours after impact and from groundbased studies of the inner coma, dust compositions varied between highly silicate- and forsterite-rich to poor relative to (what is fitted as) dark carbonaceous species (*Harker et al.*, 2005, 2007; *Sugita et al.*, 2005). Visible polarization studies revealed the ejection of surface dark carbonaceous particles (*Furusho et al.*, 2007). Spectral studies at lower spatial resolution also revealed smaller and more crystal-rich materials in the coma after impact (*Lisse et al.*, 2005). The fortuitous explosive release of matter into the coma of 17P/Holmes similarly revealed smaller particles and more crystal-rich compositions that were hitherto thought to be associated with Oort cloud comets like Hale-Bopp (*Reach et al.*, 2010).

The flyby over Comet 103P/Hartley 2 showed that the two sources of volatile gas "activity drivers," H_2O and CO_2, produced a different size and composition in the coma, whereby, compared to the H_2O rich mid-region, the CO_2-rich end was ejecting ice chunks and organic gases and probably solid-state organics (*A'Hearn et al.*, 2015; *Feaga et al.*, 2021).

67P/C-G was studied in detail during the rendezvous with the Rosetta mission, which orbited the comet from August 2014 to September 2016. The knowledge base about particle structures and compositions was considerably expanded by the Rosetta investigations. Cometary dust in 67P/C-G coma consists mainly of millimeter-sized and slow-moving [a few meters per second (*Rotundi et al.*, 2015)] hierarchical aggregates (*Mannel et al.*, 2019) that collapsed to various degrees upon collection (*Langevin et al.*, 2016; *Lasue et al.*, 2019) (see also section 3.3). The dust mass analyzer COSIMA (*Kissel et al.*, 2007) collected more than 35,000 particles (*Merouane et al.*, 2016), and about 250 of them were analyzed. The Rosetta Orbiter Spectrometer for Ion and Neutral Analysis (ROSINA) occasionally captured rock-forming elements like Na, Si, K, Ca, and Fe in relative abundances, suggesting a composition enriched in Si compared to CI (*Wurz et al.*, 2015; *Rubin et al.*, 2022). These atomic species in the gas phase could originate from solar wind sputtering of dust particles or directly from nanoparticles present in the coma. Fe and Ni atoms could be associated with organics (see section 2.2). The composition of 67P/C-G dust particles as measured by COSIMA shows some variability (*Bardyn et al.*, 2017; *Sansberro et al.*, 2022) and the average composition of ~30 of these particles was quantified and showed that dust was composed of stony material mixed with high molecular weight solid-state organics (45% by mass) (*Bardyn et al.*, 2017; *Fray et al.*, 2016). The bulk composition of 67P/C-G dust particles is rather chondritic with the exception of a higher content of C and possibly N (Fig. 3). A carbon to silicon atomic ratio C/Si = $5.5^{+1.4}_{-1.2}$ was measured in 67P/C-G dust particles (*Bardyn et al.*, 2017) (Fig. 4). This value is about 1 order of magnitude larger than the CI value (0.76 ± 0.10) and close to the protosolar value [7.19 ± 0.83 (*Lodders*, 2010)]. The H/C ratio is 1.04 ± 0.16 (*Isnard et al.*, 2019), which is higher than in insoluble organic matter (IOM) extracted from the most primitive meteorites, but lower than the value measured in 67P/C-G high molecular weight organic molecules in the gas phase that have an average H/C ratio of 1.56 ± 0.04 (*Hänni et al.*, 2022), which is compatible with that of the soluble organic matter in the Murchison carbonaceous chondrite (*Schmitt-Kopplin et al.*, 2010). The average nitrogen to carbon ratio of 67P/C-G dust particles is N/C = 0.035 ± 0.011 (*Fray et al.*, 2017). This value is in turn compatible with the chondritic value (N/C = ~0.04) (*Alexander et al.*, 2017), but about 1 order of magnitude lower than the protosolar value [N/C = 0.3 ± 0.1 (*Lodders*, 2010)]. The discovery of ammonium salts (in particular ammonium hydrosulfide and fluoride) in 67P/C-G (*Altwegg et al.*, 2020b, 2022; *Poch et al.*, 2020) could account for this missing nitrogen reservoir, as these salts would have sublimated before analysis in COSIMA. Phosphorus and fluorine were detected by COSIMA in the dust particles (*Gardner et al.*, 2020).

The composition of cometary dust in Comet 81P/Wild 2 was measured in samples returned by the Stardust mission (*Brownlee et al.*, 2006; *Hörz et al.*, 2006). Because of the high-speed collection of the samples, the light elements could not be quantified, and volatile elements like S probably

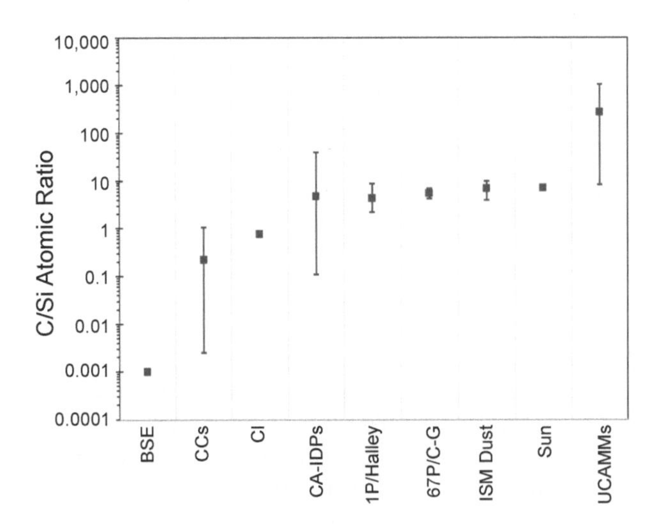

Fig. 4. Atomic C/Si ratios for bulk silicate Earth (BSE) (*Bergin et al.*, 2015), carbonaceous chondrites (CCs) (*Jarosewich*, 1990), the CI value (*Lodders*, 2010), the average of 24 CA-IDPs (*Matrajt et al.*, 2005; *Thomas et al.*, 1993), Comet 1P/Halley (*Jessberger et al.*, 1988), Comet 67P/Churyumov-Gerasimenko (67P/C-G) (*Bardyn et al.*, 2017), ISM dust (e.g., *Dartois et al.*, 2018), the Sun (*Lodders*, 2010) and the average of 10 UCAMMs (*Dartois et al.*, 2018, and unpublished data).

were redistributed around the tracks (*Ishii et al.,* 2008). Other elements show a chondritic composition within a factor of 2 (*Flynn et al.,* 2006; *Ishii et al.,* 2008; *Lanzirotti et al.,* 2008; *Leroux et al.,* 2008; *Stephan,* 2008; *Stephan et al.,* 2008). The bulk composition of cometary dust as discussed here is displayed in Fig. 3, with the abundances normalized to Fe and to CI. This kind of representation allows comparison with reference values, but should be taken with a hint of caution, as apparent enrichment/depletions could depend on the normalizing element (Fe was chosen here). Na and Si seem systematically enriched in cometary dust. The Na enrichment was indeed used as a tracer in COSIMA elementary maps to pinpoint the location of the dust particles and was observed during the entry of Comet C/2013 A1 (Siding Spring) in the martian atmosphere (*Benna et al.,* 2015).

The composition of CA-IDPs was measured for 24 IDPs (*Thomas et al.,* 1993; *Keller et al.,* 2004) for all elements displayed in Fig. 3, except for N, K, and Ti, and for 91 IDPs for major elements (Mg, Al, Si, S, Ca, Fe) (*Schramm et al.,* 1989). These IDPs show a fairly chondritic composition, with an enrichment in C of about 5× the CI value, which seems correlated with a mineralogy dominated by pyroxenes (*Thomas et al.,* 1993). The compositions of 10 UCAMMs were measured by electron microprobe and show large ranges of variation, as seen in Fig. 3. Within this large variation range, the compositions are compatible with CI, except for C and N, which are markedly enriched in UCAMMs with regard to the CI composition.

Figure 4 displays the C/Si atomic ratio in different kinds of solar system material, ordered by increasing values. We can note that data available for cometary dust and CA-IDPs are compatible with that of the Sun and ISM dust. Objects formed in the inner solar system show lower values than the Sun, as noted by *Bergin et al.* (2015). The C/Si atomic ratio of UCAMMs is very high, even higher than that of ISM dust, suggesting a local accumulation process of organics with regard to minerals in the formation regions of UCAMMs (e.g., *Dartois et al.,* 2018).

2.2. Organics

Cometary dust particles are rich in organic matter. These organics are present both as volatile compounds mixed with the host ice phase of the dust particles and as solid organic matter in the dust particles themselves. The organics that are present in the dust particles remain solid when the comet approaches the Sun and are thus quoted as "refractory" organic matter. They can be studied in samples in the laboratory, by astronomical observations (depending on the observations conditions and size of the organics), or by *in situ* analyses during space missions. The volatile organic compounds that have been identified so far in cometary dust particles are associated with ice that sublimates when the comet approaches the Sun and can thus be present in the cometary coma. The Rosetta mission around Comet 67P/C-G allowed a detailed characterization of the volatiles around the comet as a function of heliocentric distance. These volatiles species consist mainly in CH(N)O-bearing molecules, with a great variety of CH-, CHN-, CHS-, CHO_2-, and CHNO-bearing species, both saturated and unsaturated (*Altwegg et al.,* 2017; *Hänni et al.,* 2022), as well as the CN radical and related molecules (*Hänni et al.,* 2020, 2021). Ammonium salts (in particular ammonium hydrosulfide and fluoride) were also found in the coma of 67P/C-G and at the nucleus' surface (*Altwegg et al.,* 2020b, 2022; *Poch et al.,* 2020). Glycine was identified in the gas phase, with a distribution compatible with the sublimation of ices associated with the cometary dust particles rather than a direct sublimation of ices from the nucleus (*Altwegg et al.,* 2016; *Hadraoui et al.,* 2019, 2021). Glycine had also been identified at the surface of Al foil exposed to the coma of 81P/Wild 2 (*Elsila et al.,* 2009).

The "refractory" organic matter was identified as being present in Comet 1P/Halley, but its nature could not be studied by the Giotto and Vega mass spectrometers. Spectral evidence for aromatic organic molecules, i.e., polycyclic aromatic hydrocarbons (PAHs), were suggested from UV spectral analyses of Comet 1P/Halley (*Moreels et al.,* 1994; *Clairemidi et al.,* 2008). The very smallest particles had the greatest C abundances (*Lawler and Brownlee,* 1992). CHON particles had a range of H, O, and N ratios to C (*Fomenkova et al.,* 1992, 1994).

The high-speed collection of 81P/Wild 2 cometary dust did not allow a good preservation of the organics (*Brownlee,* 2014; *Keller et al.,* 2006; *Sandford et al.,* 2010). However, some organic matter, including clumps that were "behind" terminal particles and therefore somewhat protected from the heat generated by impact during collection, revealed a suite of complex organic bonds including mainly alkenes, aromatic C=C, and carboxyl C=O, as well as a variety of textures for the organic matter including organic nanoglobules (Fig. 5) (*Matrajt et al.,* 2012; *De Gregorio et al.,* 2011, 2017). The spectral signature of preserved organic matter in Stardust samples show similarities with that of insoluble organic matter extracted from meteorites (*De Gregorio et al.,* 2011), although a reduced form of carbon was also observed in one Stardust sample (*De Gregorio et al.,* 2017). The concentration of carbon could not be quantified in 81P/Wild 2 samples due to the collection method in aerogel. The low carbon concentration observed in the samples is interpreted as a consequence of the harsh collection of the samples. It could also represent a collection bias of dust from a portion of the coma that was poor in carbon and not representative of the whole comet (*Westphal et al.,* 2017).

The solid organic matter identified in 67P/C-G dust particles also shows similarities with IOM extracted from meteorites (e.g., *Alexander et al.,* 2017), although with a higher atomic H/C ratio (1.04 ± 0.16) (*Fray et al.,* 2016, 2017; *Isnard et al.,* 2019). The higher H/C ratio found in 67P/C-G could suggest less processing and a more primitive origin of the organics present in cometary particles than in IOM. This comparison is, however, made between the bulk organic matter for 67P/C-G and the IOM obtained from meteorites by acid treatments, after removal of soluble organics. Soluble organics represent only a small frac-

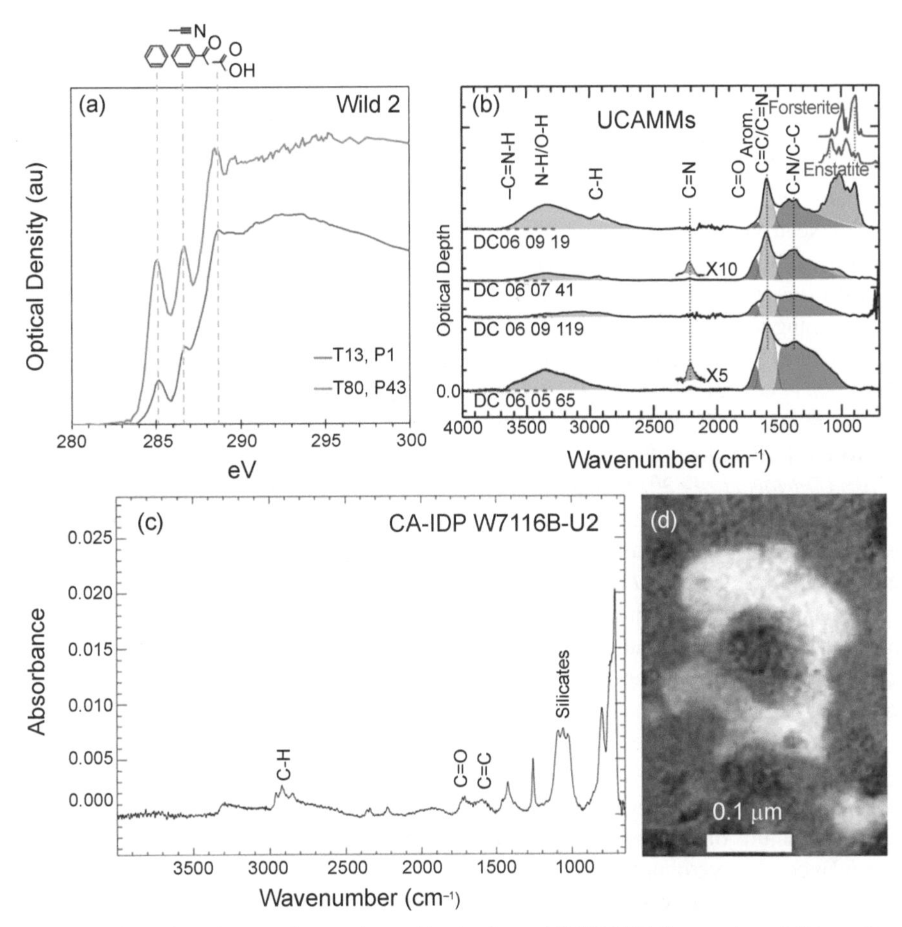

Fig. 5. (a) Spectral absorption signatures of organic matter in Comet 81P/Wild 2 samples at the carbon K-edge in STXM-XANES [adapted from *De Gregorio et al.* (2011)]. (b) μFTIR spectra of four representative UCAMMs (*Dartois et al.*, 2018). (c) μFTIR spectrum of organic matter extracted from a chondritic anhydrous IDP (*Matrajt et al.*, 2005). (d) Energy-filtered image at the carbon edge of a nanoglobule in Comet 81P/Wild 2 samples obtained by transmission electron microscopy (*Matrajt et al.*, 2008).

tion of the carbon in meteorites (~20%), but they could account for some of the difference observed between the H/C ratios of organic matter in 67P/C-G and in IOM. The O/C atomic ratio in 67P/C-G dust is likely higher than in meteoritic IOM (*Bardyn et al.*, 2017). From the COSIMA analyses, the abundance of organic matter in 67P/C-G dust particles was estimated at ~45 wt%, or ~70 vol% (*Bardyn et al.*, 2017; *Levasseur-Regourd et al.*, 2018). The lower abundance of 52 ± 8 vol% of organic matter deduced by *Fulle et al.* (2016b) from density measurement is due to the assumption by these authors of an IDP-like composition that underestimated the carbonaceous content of 67P/C-G. This solid-state organic matter is reminiscent of CHON particles in Halley but the techniques available for analyses better reveal the complexity and details of this organic matter. The ROSINA gas mass spectrometer revealed a huge host of complex molecules, including many S species, through the fortuitous impact with a dust particle that occurred during a close flyby of the nucleus (*Altwegg et al.*, 2017). Glycine, methylamine, and ethylamine, as well as phosphorus, were detected by ROSINA in the gas phase in 67P/C-G (*Altwegg et al.*, 2016). The observations of glycine in the coma can be explained by the presence of this amino acid in sublimating water ice in the dust particles (*Hadraoui et al.*, 2019, 2021). Reflectance spectra of the surface of 67P/C-G also suggest a high abundance of organic matter in the surface material, with a darkening material that could be submicrometer-sized Fe sulfides (*Quirico et al.*, 2016; *Rousseau et al.*, 2018; *Capaccioni et al.*, 2015).

Recent observations showed that Fe and Ni atoms are ubiquitous in cometary atmospheres, even at large heliocentric distances (*Bromley et al.*, 2021; *Hutsemékers et al.*, 2021; *Manfroid et al.*, 2021; *Guzik and Drahus*, 2021). The observed Fe/Ni ratio is about an order of magnitude higher than the solar value. These elements could not result for the sublimation of minerals like silicates or Fe sulfides, due to the low equilibrium temperature of the cometary atmospheres. The correlations observed between the production rates of Fe and Ni and of carbon oxides by *Manfroid et al.* (2021) suggest that these Fe and Ni emissions could be produced by the sublimation of Fe and Ni carbonyls ($Fe(CO)_5$ and $Ni(CO_4)$), which are possible constituents of cometary or interstellar matter.

CA-IDPs, CP-MMs, and UCAMMs are enriched in organic matter compared to primitive meteorites. Organic matter in CA-IDPs was studied by Fourier transform infrared microscopy (μFTIR) (*Flynn et al.*, 2003; *Keller et al.*, 2004; *Matrajt et al.*, 2005; *Muñoz Caro et al.*, 2006; *Merouane et al.*, 2014), Raman microscopy (*Wopenka,* 1988; *Quirico et al.*, 2003, 2005; *Bonal et al.*, 2006; *Muñoz Caro et al.*, 2006; *Sandford et al.*, 2006; *Rotundi et al.*, 2007; *Rotundi and Rietmeijer,* 2008; *Busemann et al.*, 2009), and transmission electron microscopy with electron energy loss spectroscopy (TEM/EELS) (*Flynn et al.*, 2003; *Keller et al.*, 2004). The organic content of CP-MMs was studied by TEM/EELS, Raman microscopy, and scanning transmission X-ray microscopy X-ray near-edge structure (STXM-XANES) absorption microspectroscopy (*Noguchi et al.*, 2015, 2017, 2022; *Yabuta et al.*, 2015). The organic matter of UCAMMs was studied by μFTIR, Raman microscopy, IR nanospectroscopy (AFM-IR) and STXM-XANES (*Dartois et al.*, 2018; *Mathurin et al.*, 2019; *Guérin et al.*, 2020). Carbonaceous materials in CA-IDPs range from hydrocarbon nanoglobules (*Wirick et al.*, 2009) to completely graphitized carbon (*De Gregorio et al.*, 2017), which have been used as thermometers (*Matrajt et al.*, 2013). 100-nm-thick coatings of organics observed on some anhydrous crystals and glass with embedded metals and sulfides (GEMS) are proposed to have facilitated grain aggregation (*Flynn et al.*, 2003). Figure 5 displays the μFTIR signature of organic matter in a CA-IDP after acid treatment (Fig. 5c) and in UCAMMs without any chemical treatment (Fig. 5b). The signature of the organics in CA-IDPs shows a large abundance of polyaromatic organic matter, with aromatic carbon, ketone (C=O), carboxylic groups (COOH), and aliphatic C-H contributions. The organics in CP-MMs show the signature of aromatic/olefinic carbon, aromatic ketone, and carboxylic carbon. Carbonaceous nanoglobules are usually present (*Noguchi et al.*, 2015, 2017, 2022; *Yabuta et al.*, 2015). The signature of organic matter in UCAMMs is unusual, with large amounts of N-bearing species (including nitrile), and a low signature of C=O and aliphatic C-H (*Dartois et al.*, 2013, 2018). STXM-XANES analyses of UCAMMs show that the organics in UCAMMs consist in fact of three distinct organic phases, with different spectroscopic signatures and different amounts of nitrogen (*Engrand et al.*, 2015; *Charon et al.*, 2017; *Guérin et al.*, 2020). The first organic phase of UCAMMs is smooth and N-rich, with N/C atomic ratios up to 0.2. This phase has no equivalent in meteorites and could result from the irradiation by galactic cosmic rays of methane and nitrogen-rich ices at the surface of small bodies in the outer regions of the protoplanetary disk (*Dartois et al.*, 2013, 2018; *Augé et al.*, 2016, 2019; *Rojas et al.*, 2020). The other two organic phases identified in UCAMMs bear similarities with that of chondritic IOM. A carbon-rich clast identified as a cometary xenolith in the LaPaz Icefield 02342 meteorite also shows spectroscopic similarities with meteoritic IOM (*Nittler et al.*, 2019). The peak intensity ratios of CH_2/CH_3 measured by μFTIR in aliphatic C-H in CA-IDPs (*Flynn et al.*, 2003; *Keller et al.*, 2004; *Matrajt et al.*, 2005; *Muñoz Caro et al.*, 2006; *Merouane et al.*, 2014) and in UCAMMs (*Dartois et al.*, 2013, 2018) are higher than the value measured in dust from the diffuse interstellar medium, which is around 1 (*Sandford et al.*, 1991; *Pendleton et al.*, 1994; *Dartois et al.*, 2007; *Godard et al.*, 2012).

The Raman signatures of organic matter in CA-IDPs, CP-MMs, and UCAMMs confirm the polyaromatic nature of the solid organics in these particles and the low thermal metamorphic grade of their organic matter (*Wopenka,* 1988; *Quirico et al.*, 2003, 2005; *Bonal et al.*, 2006; *Muñoz Caro et al.*, 2006; *Busemann et al.*, 2007, 2009; *Rotundi et al.*, 2007; *Rotundi and Rietmeijer,* 2008; *Brunetto et al.*, 2011; *Noguchi et al.*, 2015; *Dobrică et al.*, 2011; *Dartois et al.*, 2013, 2018; *Starkey et al.*, 2013). The potential effect of atmospheric entry heating on the degree of disorder of the organic matter cannot be ruled out, but specific experiments would be needed to study these effects in detail.

Organic nanoglobules seem to be ubiquitous in samples of cometary origin. They are found in 81P/Wild 2 samples (*Matrajt et al.*, 2008; *De Gregorio et al.*, 2010), CA-IDPs (*Matrajt et al.*, 2012), CP-MMs (*Noguchi et al.*, 2015), and UCAMMs (*Charon et al.*, 2017) [and also in chondritic AMMs (*Maurette et al.*, 1995), for which models predict a cometary origin for 80% of them (*Carrillo-Sánchez et al.*, 2016)].

2.3. Isotopes

The Giotto and Vega missions during a flyby around Comet 1P/Halley led to the rough measurement of carbon isotopes in cometary dust, but most of the data on the isotopic compositions of cometary dust particles were gathered from laboratory analyses of returned cometary samples (Stardust mission — 81P/Wild 2 comet), from *in situ* analyses (Rosetta mission — 67P/C-G) or from the analysis of CA-IDPs, CP-MMs, and UCAMMs.

2.3.1. Hydrogen, carbon, nitrogen, and sulfur isotopes. The hydrogen isotopic composition of comets has been measured in the gas phase of comets for an increasingly large number of species (e.g., *Bockelée-Morvan et al.*, 2015; *Altwegg et al.*, 2015; *Müller et al.*, 2022, and references therein), but the *in situ* measurement of D/H ratios in cometary dust was only made possible by the Stardust and Rosetta missions. The D/H ratios of cometary dust particles is displayed in Fig. 6. The D/H ratio of samples from Comet 81P/Wild 2 (Stardust mission) were measured by secondary ion mass spectrometry (SIMS) (*McKeegan et al.*, 2006; *De Gregorio et al.*, 2010, 2011; *Matrajt et al.*, 2008; *Stadermann et al.*, 2008). The bulk D/H isotopic ratios of Stardust samples vary from the terrestrial D/H value [Vienna standard mean ocean water (VSMOW)] of 1.5576×10^{-4} to submicrometer-sized hotspots that can reach values up to $\delta D \sim 2000‰$, which corresponds to 3× VSMOW. The hydrogen isotopic composition was also measured in dust particles from 67P/C-G by COSIMA. The average value measured in the organic matter of 25 cometary particles from 67P/C-G is D/H = $(1.57 \pm 0.54) \times 10^{-3}$ (*Paquette et al.*, 2021), which is about 1 order of magnitude larger than the terrestrial value, and marginally compatible with values measured in organic molecules in cometary comae (*Meier et al.*, 1998; *Müller et

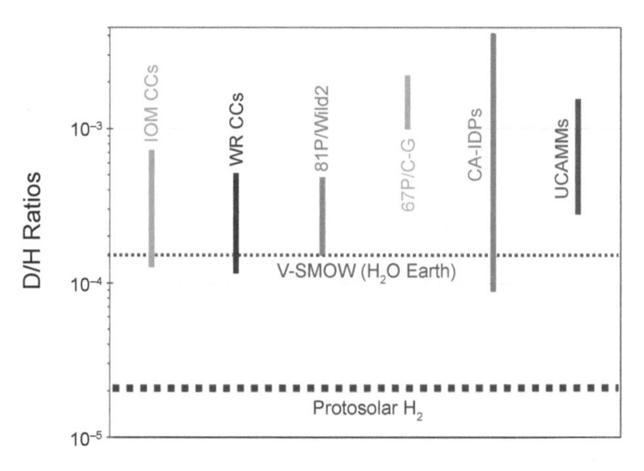

Fig. 6. D/H ratio measured in solid phase in cometary dust particles measured in the Stardust samples (81P/Wild 2), refractory organics by the Rosetta/COSIMA instrument (67P/Churyumov-Gerasimenko — 67P/C-G), chondritic anhydrous IDPs (CA-IDPs), and UCAMMs. The range of composition of D/H ratios measured in insoluble organic matter extracted from carbonaceous chondrites (IOM CCs) and in whole-rock carbonaceous chondrites (WR CCs), as well as the terrestrial value (V- SMOW, in water) and protosolar value (in H_2), are also shown for reference. Data from *Alexander et al.* (2007, 2012), *McKeegan et al.* (2006), *Paquette et al.* (2021), *Zinner et al.* (1983), *McKeegan et al.* (1985), *Messenger* (2000), *Aleon et al.* (2001), *Busemann et al.* (2009), *Duprat et al.* (2010), *Rojas et al.* (2024), and *Geiss and Gloeckler* (1998).

al., 2022). The hydrogen isotopic composition of CA-IDPs varies between D/H ~ 10^{-4} and D/H ~ 4 × 10^{-3} (*Aleon et al.*, 2001; *Busemann et al.*, 2009; *McKeegan et al.*, 1985; *Zinner et al.*, 1983). The bulk D/H value measured for eight UCAMMs vary between ~3 × 10^{-4} and 1.5 × 10^{-3}, with micrometer-sized regions that reach up to 30× the terrestrial value (*Duprat et al.*, 2010; *Rojas et al.*, 2024).

The Giotto and Vega missions to Comet Halley allowed the discovery of isotopically light carbon in the dust particles ($^{12}C/^{13}C$ ~ 5000), providing a possible link to presolar graphite (*Amari et al.*, 1993) or SiC grains, for which a few such light values have been measured (*Hoppe et al.*, 2000; *Lin et al.*, 2002; *Nittler and Alexander*, 2003). The C isotopic composition of Comet Wild 2 dust particles in Stardust samples varies between $\delta^{13}C$ ~–20‰ and ~–50‰ (*McKeegan et al.*, 2006), which are values compatible with that observed in carbonaceous chondrites (*Alexander et al.*, 2007) and CA-IDPs (*Messenger et al.*, 2003). This value is slightly higher than the solar value determined by the Genesis mission at $\delta^{13}C = -105 \pm 20$‰ (*Hashizume et al.*, 2004). The bulk carbon isotopic composition was measured in four UCAMMs and vary from ~–95‰ to ~+35‰. A noticeably low isotopic composition at ~–120‰ is found as a "cold spot" in one UCAMM (*Rojas et al.*, 2024).

The nitrogen isotopic composition of Stardust samples shows moderately elevated values, with hotspots of submicrometric sizes reaching values up to ~500‰ (*McKeegan et al.*, 2006), which are compatible with values measured in CA-IDPs (*Messenger et al.*, 2003; *Aleon et al.*, 2003; *Busemann et al.*, 2009). The bulk nitrogen isotopic composition in four UCAMMs vary from ~–170‰ to ~75‰ with submicrometer-sized regions up to ~500‰ (*Rojas et al.*, 2024) (see Fig. 7). In most cases, there is no correlation between the nitrogen and hydrogen isotopic compositions of cometary dust.

The sulfur isotopic composition measured in Comet 81P/Wild 2 samples is compatible with the solar value, showing an extraterrestrial origin of the impact residue and of the sulfide measured (*Heck et al.*, 2012; *Ogliore et al.*, 2012a). The sulfur isotopic composition of a cosmic symplectite was analyzed in Stardust samples, which showed enrichment in ^{33}S (*Nguyen et al.*, 2017). This composition could result from photochemical irradiation of solar nebular gas. The sulfur isotopic composition (^{34}S and ^{32}S) of 67P/C-G dust was measured during the Rosetta mission by COSIMA

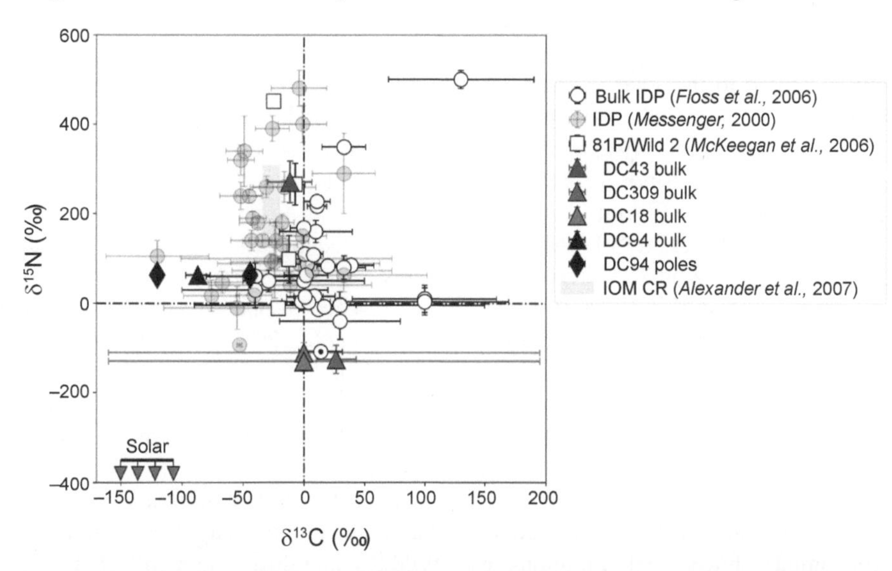

Fig. 7. Nitrogen and carbon isotopic compositions of IDPs, 81P/Wild 2 samples, UCAMMs (DC43, DC309, DC18, DC94), and IOM extracted from CR chondrites. Adapted from *Rojas et al.* (2024).

and is compatible with the reference value (*Paquette et al.,* 2017). Due to mass interferences, the ^{33}S isotope could not be quantified, so a potential ^{33}S excess could not be ruled-out for 67P/C-G dust particles.

2.3.2. Oxygen and silicon isotopes. The oxygen isotopic composition in extraterrestrial matter is used as a taxonomic tool, as most meteorite classes own a given (range of) oxygen isotopic composition(s). On a three-oxygen isotope plot, the oxygen isotopic composition of Stardust samples plot on a slope 1 line and show values that are compatible with carbonaceous chondrite signatures, including ^{16}O enrichments for refractory minerals identified in the 81P/Wild 2 samples (*Nakamura et al.,* 2008; *Nakashima et al.,* 2012a; *Joswiak et al.,* 2014; *Ogliore et al.,* 2015; *Defouilloy et al.,* 2017; *Zhang et al.,* 2021). The oxygen isotopic composition of 81P/Wild 2 samples shows a correlation between $\Delta^{17}O$ and the Mg content of the analyzed minerals, as in CR chondrites (e.g., *Nakashima et al.,* 2012a; *Zhang et al.,* 2021, and references therein).

The $^{18}O/^{16}O$ isotopic ratio could be measured by COSIMA in 67P/C-G dust particles, at $^{18}O/^{16}O = 2.0 \times 10^{-3} \pm 1.2 \times 10^{-4}$ (*Paquette et al.,* 2018). Given the large error bar associated with this value [due to limitations of measuring isotopic compositions with a time-of-flight (TOF)-SIMS method], this value is compatible with the terrestrial VSMOW value, and covers the whole range of values found in meteorites. For reference, the oxygen isotopic composition of H_2O and CO_2 in the 67P/C-G coma are $2.25 \times 10^{-3} \pm 1.77 \times 10^{-4}$ and $2.02 \times 10^{-3} \pm 3.28 \times 10^{-5}$, respectively (*Schroeder et al.,* 2019; *Hässig et al.,* 2017). Other gaseous oxygen-bearing molecules in 67P/C-G can contain heavy oxygen isotopes compared to VSMOW, especially for S-rich molecules (*Altwegg et al.,* 2020a).

The oxygen isotopic composition of CA-IDPs is compatible with that of 81P/Wild 2 samples (*McKeegan,* 1987; *Aléon et al.,* 2009; *Nakashima et al.,* 2012b; *Zhang et al.,* 2021). The silicon isotopic composition of dust at Comet 67P/C-G could be measured by the ROSINA instrument (*Rubin et al.,* 2017) and showed a depletion of heavy silicon isotopes ^{29}Si and ^{30}Si compared to the solar value. Such depletions in heavy isotopes are only found in rare presolar grains identified in meteorites (*Hynes and Gyngard,* 2009).

2.3.3. Magnesium isotopes — aluminum-26. The magnesium isotopic composition of 81P/Wild 2 samples was measured to search for the past presence of ^{26}Al at the time of mineral formation in Comet Wild 2. No resolvable ^{26}Mg excess resulting from the decay of ^{26}Al was found in Wild 2 samples, suggesting either (1) a late formation (a few million years after CAI formation) of minerals in Wild 2, (2) a protoplanetary disk heterogeneous in ^{26}Al, or (3) the formation of Wild 2 minerals before injection of ^{26}Al in the protoplanetary disk (*Matzel et al.,* 2010; *Nakashima et al.,* 2015).

The magnesium isotopic composition measured in olivines from Stardust samples shows small variations in $\delta^{26}Mg$ and $\delta^{25}Mg$ values that are compatible with small mass-dependent fractionation from a chondritic reservoir with respect to the Mg isotopes (*Fukuda et al.,* 2021).

2.3.4. Presolar grains. Presolar grains are found in minute amounts in interplanetary material. They are grains that were present in the molecular cloud that led to the formation of the solar system and survived in the extraterrestrial samples that can be analyzed in the laboratory. At this time, we can only identify the isotopically anomalous grains that were synthesized in previous generations of stars and got incorporated in the protosolar molecular cloud (*Hynes and Gyngard,* 2009; *Stephan et al.,* 2020). *Stricto sensu,* grains that formed in the protosolar molecular cloud before the birth of the Sun are also "presolar," but cannot be identified by isotopic methods, as they carry the solar signature of the initial cloud.

In meteorite samples, the most abundant identified presolar grains are silicates, whereas SiC and graphite were historically the first ones to be identified in the acid residue extracted from meteorites.

As they formed far from the Sun at cold temperatures, comets are expected to have preserved a large abundance of presolar grains. After correction for possible partial destruction during the harsh collection conditions of Wild 2 samples, isotopically anomalous grains remain rare among analyzed Wild 2 materials, occurring at initial abundances of ~700 ppm (*Nguyen et al.,* 2020; *Stadermann et al.,* 2008; *Floss et al.,* 2013). Presolar grains are also found in CA-IDPs, CP-MMs, and UCAMMs in abundances that can reach up to about 1% (*Busemann et al.,* 2009; *Floss et al.,* 2012; *Floss and Haenecour,* 2016).

2.4. Mineralogy

2.4.1. Comets 1P/Halley and 67P/Churyumov-Gerasimenko: Hints at their mineralogy. In Comet 1P/Halley, the "rocky" particles had a wide range of Mg/Fe but with a narrower range of Mg/Si with similarities to Mg-rich silicates (40–60% by number of particles), specifically Mg-rich pyroxenes, Fe(Ni) sulfides, with little Fe metal, and <1% Fe-oxides (*Schulze et al.,* 1997). CAI-like materials were not found in Halley. Few particles could be directly traced to pure mineral grains although the 11.2-μm spectral feature of forsterite was first identified in Halley (*Bregman et al.,* 1987; *Campins and Ryan,* 1989).

There was no instrument on Rosetta that allowed unambiguous identification of minerals in dust from 67P/C-G. The very low reflectance of the nucleus surface suggests the presence of opaque minerals that could be Fe sulfides (*Quirico et al.,* 2016; *Rousseau et al.,* 2018; *Capaccioni et al.,* 2015). The bulk composition and density of 67P/C-G dust particles are also compatible with the presence of silicates, Fe sulfides and carbon (*Bardyn et al.,* 2017; *Fulle et al.,* 2016a).

2.4.2. 81P/Wild 2 mineralogy. To date the only samples that are unambiguously derived from a comet are the 81P/Wild 2 coma dust grains collected by the Stardust spacecraft, returned to Earth in 2006. Well-preserved coma grains from Comet 81P/Wild 2 are dominated by the coarsest components. Fine-grained materials, representing perhaps 90% of the impacting cometary coma grains, were severely altered or vaporized during high-speed (6.1 km s^{-1}) capture in the aerogel capture media (*Brownlee et al.,* 2006). Fine-grained material was only preserved in a minority of cases (e.g., Fig. 8a), but probably sufficiently well to permit elucidation of its general nature (*Ishii et al.,* 2008). It is also possible that the apparent lack of fine-grained amorphous solids could be an artifact of this collection bias. GEMS, frequently abundant in chondritic anhydrous IDPs (*Bradley* 1994a), have not been reliably identified among Stardust materials [although there are unverified reports by *Gainsforth et al.* (2016), possibly also due to destruction during collection and compositional and structural similarities to melted silica aerogel].

As expected, in the Stardust samples the coarse-grained mineral phases are dominated by olivines, pyroxenes, and sulfides. The compositions are distinct from meteorites (*Joswiak et al.,* 2012, 2017; *Frank et al.,* 2014). Olivines exhibit practically the entire range from forsterite to fayalite, with no significant compositional peak. Some terminal olivines are thought to be "microchondrules" by their similarity to type II (Fe >10%) chondrules in primitive chondrites such as CRs and CMs (*Frank et al.,* 2014; *Wooden et al.,* 2017). The minor-element compositions of Stardust olivines link only a subset to low-Fe, Mn-enriched (LIME) olivine (condensates) (*Joswiak et al.,* 2017). One Stardust microchondrule, "Iris," reveals its rapid cooling at high oxygen fugacity and high Na enrichment in the gas phase (*Gainsforth et al.,* 2010). The lack of a pronounced compositional peak at forsterite was very unexpected, as this is a hallmark of chondritic IDPs (*Rietmeijer,* 1998) and the least-equilibrated carbonaceous chondrites (*Frank et al.,* 2014). In terms of the olivine and pyroxene compositions, the closest meteoritic analogs are the unequilibrated ordinary chondrites (*Frank et al.,* 2014). The high abundance of relatively coarse-grained (>30 μm size) (*Frank et al.,* 2014; *Wozniakiewicz et al.,* 2015), well-crystalline ferromagnesian silicates was also unexpected, given laboratory analog experiments and interstellar dust spectroscopic studies, which indicated that silicates should be mostly amorphous (e.g., *Kemper et al.,* 2005).

Refractory minerals found in meteoritic CAIs were identified in Wild 2 samples, which contain olivines, pyroxenes, sulfides, and refractory oxides (e.g., Fig. 8c). Mineral assemblages, mineral chemistries, and measured bulk particle compositions reveal that these grains are similar to refractory materials in chondrites, with mineral chemistries most similar to CAI from CR2 and CH2 chondrites (*Zolensky et al.,* 2006; *Simon et al.,* 2008; *Chi et al.,* 2009) and Al-rich chondrules (*Bridges et al.,* 2012; *Joswiak et al.,* 2014).

Chondrule fragments were also found in Wild 2 (e.g., Fig. 8b). They are similar to chondrules found in carbonaceous chondrites, but with interesting differences. Few type I chondrules (FeO- and volatile-poor) have been found from Wild 2, although these are the most abundant type in meteorites. To date mainly FeO-, MnO-, volatile-rich type II chondrules have been identified from Wild 2 (*Nakamura et al.,* 2008; *Matzel et al.,* 2010; *Joswiak et al.,* 2012; *Ogliore et al.,* 2012b; *Frank et al.,* 2014; *Gainsforth et al.,* 2015). One Al-rich, ^{16}O-rich chondrule fragment has also been identified from Wild 2, as found in carbonaceous chondrites (*Bridges et al.,* 2012).

Minor-element compositions of Wild 2 olivine and pyroxenes, particularly Cr and Mn, suggest that Wild 2 minerals experienced mild secondary thermal metamorphism (*Frank et al.,* 2014), to approximately 400°C. Some Wild 2 olivines and pyroxenes show compositional similarities to those in ordinary chondrites (L/LL), CH chondrites, Rumruti (also known as R) chondrites, and aubrite meteorites (*Frank et al.,* 2019, and in preparation). In addition to diverse nebular components associated with multiple chondrite types, Wild 2 apparently incorporated materials that were liberated from evolved, internally heated asteroids, another hint for the presence of an early large-scale transport mechanism acting from the inner to the outer regions of the protoplanetary disk.

A mineral assemblage found in Wild 2 samples consists of FeO-rich olivines and Na- and Cr-rich clinopyroxenes

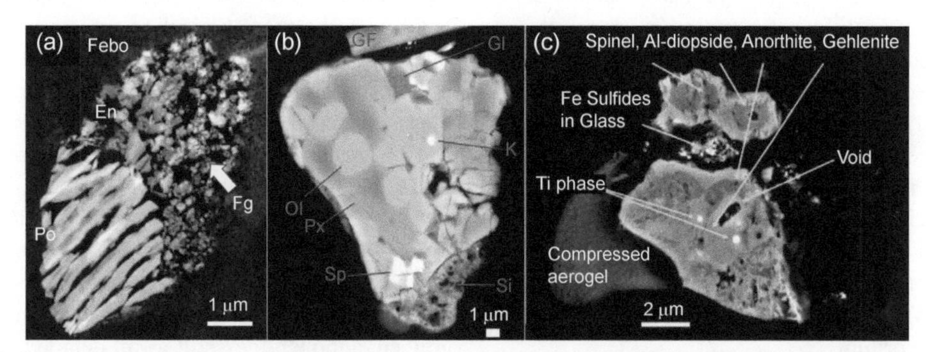

Fig. 8. Mineral diversity observed in 81P/Wild 2 samples brought back by the Stardust spatial mission : **(a)** Scanning transmission electron microscopy dark field image of track 57 terminal grain. Large pyrrhotite (Po) and enstatite (En) crystals are annotated, as well as fine-grained material (Fg). **(b)** Backscattered electron micrograph of a chondrule-like fragment in particle Torajiro, containing olivine (Ol), low-Ca pyroxene (Px), Cr-spinel (Sp), glass (Gl), kamacite (K), and silica aerogel from the collector (Si). A glass fiber (GF) holds the sample (*Nakamura et al.,* 2008). **(c)** Backscattered electron image of a CAI-like particle from track 25 (*Zolensky et al.,* 2006).

(typically augites), sometimes with poorly crystallized albite or albitic glass with spinel (*Joswiak et al.,* 2009). These assemblages have been named "KOOL" (kosmochloric high-Ca pyroxene and FeO-rich olivine) grains and are observed in more than half of all Stardust tracks. KOOL grains are also observed in CA-IDPs and CP-MMs. The textures and mineral assemblages of KOOL grains are suggestive of formation at relatively high temperatures by igneous or metamorphic processes (it is unclear which) and may have formed under relatively high f_{O_2} conditions. KOOL grains have not been observed in chondrites; however, the oxygen isotopic composition of a single Wild 2 KOOL grain is similar to some type II (FeO-rich) chondrule olivines from OC, R, and CR chondrites (*Kita et al.,* 2011; *Isa et al.,* 2011). One type II microchondrule in Wild 2 shows kosmochloric enhancement possibly reinforcing the link between KOOL grains and chondrule-forming processes (*Gainsforth et al.,* 2015). KOOL grains may represent an important precursor material for FeO-rich chondrules.

While no large carbonate grains have been identified among Wild 2 samples, submicrometer carbonate grains have been reported (*Flynn et al.,* 2009), including Mg-Fe-carbonates associated with amorphous silica and iron sulfides (*Mikouchi et al.,* 2007). The observation is interesting because carbonates are typically products of aqueous processes. While Ca carbonate could plausibly be a manufacture contaminant in aerogel, Mg carbonates are unlikely (*Mikouchi et al.,* 2007). However, in principle carbonates also can be formed without the presence of liquid water, in gas-phase reactions in the nebula (*Toppani et al.,* 2005; *Wooden,* 2002; *Wooden et al.,* 2017), so the presence of carbonates is not an unambiguous signature of cometary aqueous alteration.

Sulfides are abundant in Wild 2, at all sizes (*Zolensky et al.,* 2006). These are predominantly pyrrhotite ($Fe_{(1-x)}S$, x = 0 to 0.2), but unusual sulfides are abundant. Some pyrrhotites dominate terminal particles often within assemblages with igneous textures (*Joswiak et al.,* 2012; *Gainsforth et al.,* 2013, 2014). As is generally the case, pyrrhotite often occurs in association with pentlandite $(FeNi)_9S_8$ and Fe-Ni metal (*Joswiak et al.,* 2012). ZnS (probably sphalerite) is unusually abundant in Wild 2 as compared to chondrites. A single report of cubanite ($CuFe_2S_3$) has been interpreted as evidence for aqueous processing (*Berger et al.,* 2011); however, this mineral can form in non-aqueous environments.

The iron oxide magnetite (Fe_3O_4), including a Cr-rich variety, has been identified in a few Wild 2 grains (*Bridges et al.,* 2015). Although magnetite in carbonaceous chondrite meteorites is often ascribed to a secondary origin by aqueous alteration (*Kerridge et al.,* 1979) more detailed observation of Wild 2 magnetite is necessary in order to reliably assess its origin.

It is clear that carbonates, sulfides, and oxides trace a diverse range of formation and processing environments and possibly provide direct evidence for aqueous alteration within Wild 2, although the rarity of these particular phases and the lack of any report of phyllosilicates (*Brownlee and Joswiak,* 2017) limits the overall extent of aqueous alteration.

2.4.3. Chondritic anhydrous interplanetary dust particle mineralogy. Individual IDPs are under 100 μm in diameter and consist of tens to hundreds of thousands of grains, with greatly varying mineralogy and composition, i.e., non-equilibrium phase assemblages. The mineralogy of the anhydrous chondritic IDPs evidences a wide range of protoplanetary disk locations and processes. The most abundant crystalline phases are ferromagnesian silicates, mainly olivine and low-Ca pyroxene with lesser amounts of high-Ca pyroxene, plagioclase, and Fe-Ni-Zn-sulfides (*Rietmeijer,* 1998) (e.g., Figs. 9a,b,c). The (crystalline) olivine and low-Ca pyroxene compositions range from almost pure forsterite and enstatite to relatively high Fe-compositions. While olivine and low-Ca pyroxene in some IDPs is predominantly Mg-rich, a census of anhydrous and hydrous IDPs shows a slight preponderance of Fe-contents ~60% (*Zolensky et al.,* 2008). This is unlike the flat distribution of Fe-contents for terminal olivines ("microchondrules") reported for Wild 2 (*Frank et al.,* 2014). However, additional olivine and pyroxene compositional data for IDPs is required to verify these apparent trends.

LIME olivines are proposed to be high-temperature nebular condensates (*Klock et al.,* 1989). Enstatite "whiskers" in chondritic IDPs, elongated along the [100] crystallographic axis, are consistent with rapid growth from a vapor phase (*Bradley,* 1994c). Most anhydrous chondritic IDPs also contain nanoscale beads of GEMS (*Bradley,* 1994b).

The origins of GEMS is debated, with proposed formation mechanisms including irradiation of crystalline grains (olivine, pyroxene, etc.), formation in the ISM (*Bradley,* 2013), or in the protosolar molecular cloud or outer solar nebula (protoplanetary) disk such that GEMS experienced inheritance of "solar composition" (*Keller and Messenger,* 2011), or a formation by cold processes prior to the aggregation of IDPs (*Ishii et al.,* 2018). In meteorites, GEMS-like phases have been reported in few meteorites, for example, the Paris CM chondrite (*Leroux et al.,* 2015), but this identification is disputed (*Villalon et al.,* 2016). Only in the Ningqiang C3 chondrite is a radiation damage origin demonstrated (*Zolensky et al.,* 2003). GEMS are therefore believed to be more typical of comets than asteroids. It is therefore very unfortunate that GEMS apparently cannot be reliably recognized in 81P/Wild 2 samples because of their similarity to melted silica aerogel (*Ishii et al.,* 2011).

The unanticipated (to say the least) discovery of numerous CAIs among the recovered 81P/Wild 2 grains has refocused attention to refractory IDPs (*Zolensky,* 1987; *McKeegan,* 1987), which had been ignored as they were erroneously assumed to have purely asteroid origins. These refractory IDPs differ from meteoritic analogs principally in being much finer grained, although they still await detailed characterization.

Through seeking IDPs for comparison to Stardust terminal olivine grains or "microchondrules," the giant IDPs became a focus of state-of-the-art studies because they were found to possess a wide range of Mg:Fe as well as minor-element Mn-, Cr-, and Ca-compositions, potentially similar

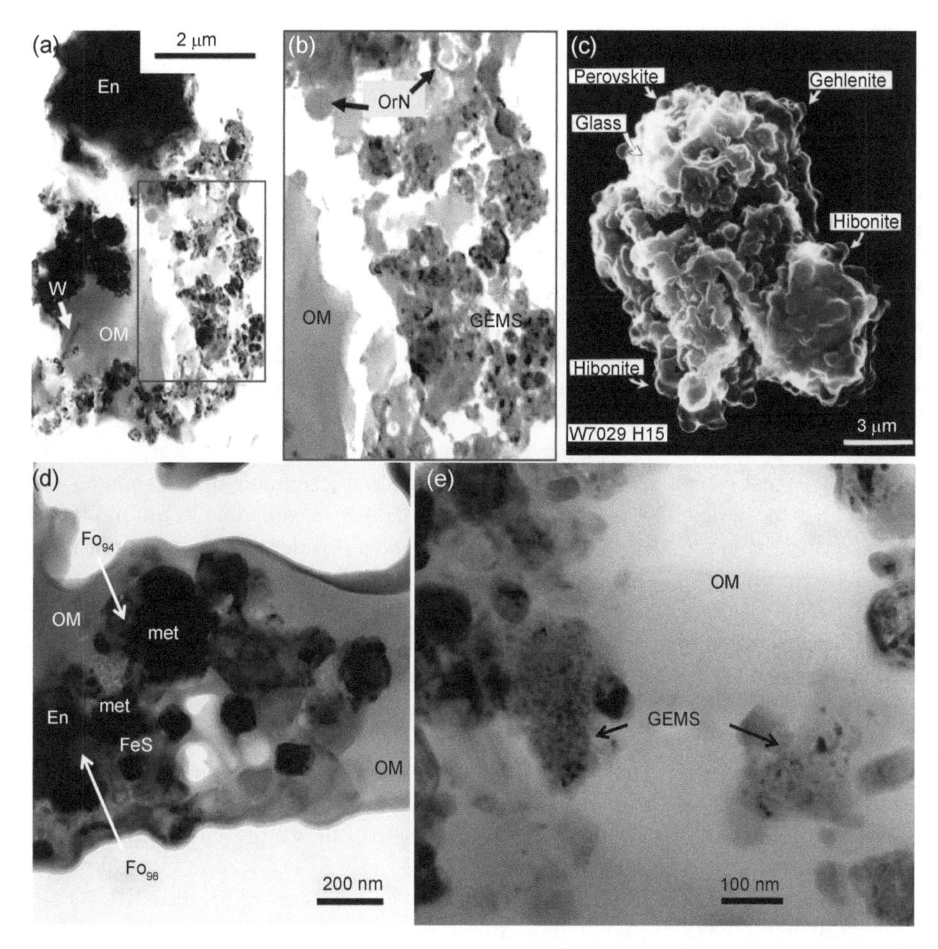

Fig. 9. Bright field transmission electron microscopy (TEM) images [**(a)**,**(b)**,**(d)**,**(e)**] and secondary electron image [**(c)**] illustrating the mineral diversity observed in CA-IDPs [**(a)**,**(b)**,**(c)**] and UCAMMs [**(d)**,**(e)**]. **(a)**,**(b)** Anhydrous chondritic IDP U2153 Cluster particle 1; **(b)** area outlined in **(a)**. Enstatite (En), an enstatite whisker (W), organic material (OM), GEMS, and two organic nanoglobules (OrN) are indicated (TEM image courtesy of K. Nakamura-Messenger). **(c)** Refractory IDP W7029 H15, containing perovskite, hibonite, gehlenite, and a glass. After *Zolensky* (1987). **(d)**,**(e)** UCAMM DC06-05-94. Mg-rich olivines (Fo94 and Fo98), enstatite (En), Fe-Ni metal (met), Fe sulfides (FeS), and organic material (OM) are indicated. (TEM images courtesy of H. Leroux.)

to the Stardust olivines (*Brownlee and Joswiak,* 2017), although additional IDP olivine analyses are still required to demonstrate similarity. As laboratory techniques advanced and focused on giant IDPs, the studies of anhydrous chondritic IDPs with only high-Mg content olivines were set aside. In order to compare the formation conditions of the olivine and pyroxene in these anhydrous chondritic IDPs, similar studies to the giant IDPs, at high spatial resolution and sensitivity, e.g., of elemental compositions of individual grains, are necessary.

2.4.4. Chondritic porous micrometeorite mineralogy. CP-MMs were first reported by *Noguchi et al.* (2015). Their mineralogy is similar to that of CP-IDPS (here referred to as CA-IDPs), and is dominated by GEMS, low-Ca pyroxene (including enstatite whisker/platelet), olivine, and pyrrhotite. These minerals have angular to subrounded shapes and range from 200 nm to 1 μm in size. The olivines and pyroxenes are Fe poor, with compositions ranging from (Mg/Mg + Fe) = 0.7 to 1. Olivines and low-Ca pyroxenes in CP-MMs have compositional peaks at Mg end members (forsterite and enstatite). CP-MMs contain LIME and low-iron chromium-enriched (LICE) ferromagnesian silicates. The most Mn-enriched LIME mineral was a low-Ca pyroxene, with 2.9 wt% MnO, and the most Cr-enriched LICE mineral was also a low-Ca pyroxene with 1.7 wt% Cr_2O_3. Kosmochloric pyroxenes (Ca- and $NaCrSi_2O_6$-rich are occasionally found in CP-MMs. Roedderite with coexisting low-Ca pyroxene and amorphous silicate was also found (*Noguchi et al.,* 2015, 2017). *Noguchi et al.* (2022) reports a unique particle showing both a (anhydrous) CA-IDP-like composition with a hydrated part, suggesting partial hydration of this particle.

2.4.5. Ultracarbonaceous Antarctic micrometeorite mineralogy. The mineral components of UCAMMs consist of isolated minerals or small mineral assemblages embedded in the organic matter (*Dobrică et al.,* 2012; *Charon et al.,* 2017; *Guérin et al.,* 2020; *Yabuta et al.,* 2017). Both crystalline and amorphous phases are present. Crystalline minerals consist of low-Ca Mg-rich pyroxenes (with stoichiometry ranging between En_{60} and En_{97}) and Mg-rich olivines (stoichiometry comprised between Fo_{75} and Fo_{99}) with rare Ca-rich pyroxenes and Fe(Ni) sulfides. Several hypocrystalline-like ("chondrule-like") mineral assemblages were identified in several UCAMMs. Low Ni-Fe metal and Fe sulfides are present in mineral assemblages. In several cases, Fe metal inclusions show a rim of Fe sulfide, suggesting the occurrence of an incomplete sulfidization process. Pentlandite is occasionnally observed. Secondary minerals also include Mn-, Zn-rich sulfide, perryite, as well as small iron oxides, carbonates, and phyllosilicate-like phases (*Dobrică et al.,* 2012; *Guérin et al.,* 2020). Small Na-rich inclusions with stoichiometry close to Na_2S have been observed in the organic matter (*Guérin et al.,* 2020). Glassy phases have been found in several UCAMMs (*Charon et al.,* 2017; *Guérin et al.,* 2020; *Yabuta et al.,* 2017) that resemble GEMS found in primitive IDPs (*Keller and Messenger,* 2011; *Bradley et al.,* 2014). The GEMS-like phases in UCAMMs, however, tend to lack the metal inclusions of GEMS in IDPs, Fe being mostly in the form of Fe sulfides nano-inclusions. As with the Wild 2 grains, the close association of high-temperature crystalline phases with low-temperature carbonaceous matter in UCAMMs supports the hypothesis of a large-scale radial mixing in the early solar nebula (*Brownlee et al.,* 2006).

UCAMMs have an olivine to low-Ca pyroxene ratio of approximately 0.5, similar to that of chondritic anhydrous IDPs (*Zolensky and Barrett,* 1994) and CP-MMs (*Noguchi et al.,* 2022). This ratio is compatible with that of P- and D-type asteroids and comets (*Vernazza et al.,* 2015). Although amorphous minerals like GEMS are ubiquitous in CA-IDPs, CP-MMs, and UCAMMs, their crystalline mineral abundance is higher than the upper limit of a few percent of crystallinity observed in the interstellar medium (*Kemper et al.,* 2004, 2005). *Bradley et al.* (2014) reports that GEMS represent up to 70 vol% of CA-IDPs. The value is not quoted for CP-MMs, but the authors draw a similarity with CA-IDPs. *Dobrică et al.* (2012) report that crystalline materials represent at least 25% of the mineral phases analyzed in UCAMMs.

It is worth noting that in CA-IDPs, CP-MMs, and UCAMMs, olivines and low-Ca pyroxenes have compositional peaks at Mg end members. This is not observed for Wild 2 samples. Either Wild 2 ferromagnesian silicates are not typical of comets, or CA-IDPs, CP-MMs, and UCAMMs derive from parent bodies different from Wild 2-type comets.

2.5. Cometary Dust Compositions from Astronomical Infrared Spectroscopy

2.5.1. Infrared spectroscopy. IR spectroscopic spectral energy distributions (SEDs) (λF_λ vs. λ) of dust thermal emission from cometary comae (~3–40 μm), when fitted with thermal emission models, provide constraints for the composition of the dust particles, their structure (porosity or crystal shape), and their size distribution. The "grain size distribution" (GSD), or equivalently the dust differential size distribution (DSD), has parameters of power law slope N, and either smallest radii limit a_0 or small particle radii a_p where the DSD peaks before rolling off at yet smaller particle radii [often called the Hanner GSD (HGSD) (*Hanner,* 1983)]. The particle porosity is often parameterized by a particle fractal dimension D (see Table 1). The particle porosity and DSD slope (D and N) are co-dependent parameters (*Wooden,* 2002). With broad wavelength coverage, the mid-IR (MIR) (λλ5–13 μm) and far-IR (FIR) (λλ14–40 μm) resonances can be sought and fitted by thermal models to constrain the dust mineralogies. From IR SEDs, the dust compositions are better determined for the more numerous smallest particles (in the DSD) and/or hottest particles because smaller particles produce stronger contrast spectral features (referring to $Q_{\lambda,abs}$) and hotter particles produce more relative flux. The more distinct spectral features arise from submicrometer to micrometer-sized solid particles (P = 0%) or from up to ~20 μm moderately porous particles (P = 85%), which has some implications for comparisons between IR SEDs and cometary samples. For example, the 20-μm and larger Stardust terminal olivine particles would not produce distinct spectral features. Larger extremely porous particles, which are as hot as their small monomers (*Xing and Hanner,* 1997), if present in comae, even if in smaller mass fractions, can produce resonances (*Kolokolova et al.,* 2007) and contribute to the SEDs (*Bockelée-Morvan et al.,* 2017a) (section 2.5.7). Particles in the DSD larger than ~20 μm are present in cometary comae because observed FIR SEDs decline more slowly than a single color temperature assessed for the MIR. As a compliment to this, comae do not solely possess large extremely porous particles because the presence of solely porous aggregate particles as warm as their submicrometer-sized monomers could not explain cometary IR SEDs. Thermal model parameters of composition, which is more sensitive to the properties of the smaller particles, and DSD parameters (P, N, and a_0 or a_p) are constrained by fitting the entire IR SED with a DSD that extends to particle radii out to which the IR SED is sensitive, e.g., millimeter-sized particles for λλ7–40 μm. Above millimeter-sized particles, longer wavelengths such as by Herschel (*Kiss et al.,* 2015) and the Atacama Large Millimeter Array (ALMA) are required, but the mineralogy is constrained by resonances at MIR–FIR wavelengths discussed here.

Identification of dust compositions may be made by comparison of observed spectral features to laboratory absorption spectra but quantifying the relative abundances requires computing the thermal emission models to predict and compare an emission spectrum of an ensemble of particles to the observed IR SED using standard minimization techniques (χ^2-minimization). A key aspect of determining the dust composition, when feasible, is fitting a broad wavelength SED so that multiple spectral resonances (spectral features) can be fitted from a material's vibrational stretch and vibrational bending modes that span, respectively, MIR to FIR wavelengths.

2.5.2. Five primary dust compositions. The compositions of cometary dust as determined by IR spectral analyses has five primary components that suffice to allow

TABLE 1. Table of equations including dust thermally emitted and scattered fluxes, and parameters that quantify observed dust properties.

Dust Equations	Designation	Notes
$P = 1 - f$, given fractional filled volume $f = (a/a_0)^{D-3}$	Porosity, fractal dimension D for $1.7 \lesssim D \leq 3$	*Harker et al.* (2002); *Woodward et al.* (2021); *Lasue et al.* (2019)
Q_{abs}, Q_{sca}	Absorptivity, scattering efficiency	
$C_{abs} = G(a)\, Q_{abs}$, $C_{sca} = G(a)\, Q_{sca}$	Absorption, scattering 578 cross sections	
$Q_{ext} = (Q_{abs} + Q_{sca})$ $F_{emiss}(\lambda) = \frac{1}{1} K \int_{a_{min}}^{a_{max}} G(a) Q_{\lambda,abs}(a) \pi B_\lambda(T_d(a)) n(a) da$	Extinction efficiency thermal flux density	$G(a) = \pi a^2$ for sphere
$\pi B_\lambda(T_d(a) = 2hc^2 \lambda^{-5} \left(\exp^{hc/\lambda k T_d} - 1\right)^{-1}$	Planck function	
$n(a) = \left(1 - \frac{a_0}{a}\right)^M \left(\frac{a}{a_0}\right)^N$	Differential grain size distribution (DSD, GSD)	HGSD: $a_p = \frac{(M+N)}{N}$
$K = \frac{N_{dust}(a_0)}{4\pi(\Delta\ [cm])^2}$ or $K = \frac{N_{dust}(a_p)}{4\pi(\Delta\ [cm])^2}$ for HGSD	Flux scaler, at a_0 (DSD) or at a_p (HGSD, $M \neq 0$)	
$F_{sca}(\lambda, \alpha) = \frac{F_\lambda(T_\odot)}{(r_h\ [au])^2} K \int_{a_{min}}^{a_{max}} G(a) Q_{\lambda,sca}(a) p_{\lambda,sca}(a, \alpha) n(a) da$	Scattered flux	$F_\lambda(T_\odot)$ solar flux at 1 au
θ (°)	Angle from incoming to outgoing ray	
$\alpha = 180° - \theta$	Phase angle between incident sunlight and observer	
$\int_{4\pi} p_{\lambda,sca}\, d\Omega = 1 = \frac{1}{C_{\lambda,sca}(a)} \int_{4\pi} \frac{d(C_{\lambda,sca}(a; \theta, \phi))}{d\Omega} d\Omega$	"Phase function" $p_{\lambda,sca}$, normalized differential, scattering cross section	$C = G(a)Q$
$j_\lambda(a, \alpha') = \frac{p_\lambda(a, \theta = 180° - \alpha')}{p_\lambda(\theta = 180°)} = \frac{p_\lambda(a, 180° - \alpha')}{p_\lambda(\alpha' = 0°)}$	"Phase curve" phase function normalized at backscattering angle $\theta = 180°$	*Hanner et al.* (1981); $\alpha' = 0 \equiv \theta = 180°$
$A = Q_{sca} / Q_{ext}$ $A_p(\alpha') = A_p(\alpha = 0°) \frac{p_\lambda(a, 180° - \alpha')}{p_\lambda(a, 180°)} \equiv A_p(0) j(\alpha')$	Albedo of particle Geometric albedo, ratio of energy backscattered to that of Lambertian surface of equal area	Ref; phase angle; $\alpha' \equiv 180° - \theta$
$A_\lambda(\theta)_{bolo} = \frac{f(\theta)}{(1 + f(\theta))}, f(\theta) = \frac{[\lambda F_{sca}(\lambda, \alpha)] \rvert_{\lambda = \lambda_{max,sca}}}{[\lambda F_{abs}(\lambda)] \rvert_{\lambda = \lambda_{max,abs}}}$	Bolometric albedo, ratio of scattered to sum of scattered and thermal energy at $\lambda_{max,sca}$ and $\lambda_{max,emiss}$	*Woodward et al.* (2015)
$\int_0^{inf} F_{\lambda_{VIS},abs}(Q_{\lambda_{VIS},abs}) d\lambda = \int_0^{inf} F_{\lambda_{IR},emis}(Q_{\lambda_{IR},abs}, T_d) d\lambda$	Radiative equilibrium (absorption = emission)	Radiative equilibrium temperature T_d
$\int_{\lambda_{min}}^{\lambda_{max}} F_{\lambda_{VIS},abs}(Q_{\lambda_{VIS},abs}) d\lambda = \int_{\lambda_{min}}^{\lambda_{max}} \frac{F_\lambda(T_\odot)}{r_h^2} G(a) Q_{\lambda,abs}(a) d\lambda$	Sunlight absorbed (L)	
$\int_{\lambda_{min}}^{\lambda_{max}} F_{\lambda_{IR},emis}(Q_{\lambda_{IR},abs}, T_d) d\lambda = \int_{\lambda_{min}}^{\lambda_{max}} \pi B_\lambda(T_d) G(a) Q_{\lambda,abs}(a) d\lambda$	Thermal emitted (R)	
$g_{col}(\lambda) \equiv \left(\frac{\lambda}{\lambda_{ref}}\right)^{-p_{col}} \propto \int G(a) Q_{\lambda,sca}(a) p_{\lambda,sca}(a, \alpha) n(a) da$	Scattered light color g_{col}	Power p_{col}

TABLE 1. (continued)

Dust Equations	Designation	Notes
$S'_{color} = \frac{2}{(\lambda_{ref} - \lambda)[nm]} \frac{g_{col}(\lambda) - g_{col}(\lambda_{ref})}{g_{col}(\lambda) + g_{col}(\lambda_{ref})}$	Color gradient (slope) (%/100 nm)	
$T_{d,color} \approx 1.1 \times 278\ (K)[r_h/(au)]^{-0.5}$	Dust color temperature, typical behavior	*Hanner et al.* (1997)

the IR SED to be well-fitted by thermal models for most comets: two Mg:Fe amorphous silicates that produce the broad 10-μm and 20-μm features, two Mg-rich crystalline silicates that produce multiple narrow spectral peaks in the range of ~8–33 μm, and a highly absorbing and spectrally featureless and warmer dust component that is ubiquitously observed to dominate the dust's thermal emission in the NIR (~3–7.5 μm). There is a consensus among modelers that this NIR dust emission, sometimes called the NIR dust "continuum" or "pseudo-continuum," is produced by a highly absorbing, carbonaceous dust component. This component is well fitted by the optical properties of amorphous carbon (*Woodward et al.*, 2021; *Bockelée-Morvan et al.*, 2017a; *Harker et al.*, 2023). Aliphatic carbonaceous dust materials are relatively transparent compared to amorphous carbon, and graphitic carbonaceous matter is rare in cometary samples. The potential connections between cometary dust and amorphous carbon, vs. other species, e.g., organics dominated by aromatic bonds, by hydrogenated amorphous carbon (HACs), or by graphite, are discussed in *Wooden et al.* (2017) and *Woodward et al.* (2021).

For dielectric materials like silicates, there are spectral features from which we can deduce the mineralogy in combination with modeled radiative equilibrium temperatures and modeled spectral emission features. These four primary siliceous compositions are Mg:Fe ≃ 50:50 amorphous pyroxene-like $(Mg_{0.5},Fe_{0.5})SiO_3$ and Mg:Fe ≃ 50:50 amorphous olivine-like $(Mg_{0.5},Fe_{0.5})_2SiO_4$, hereafter referred to as Mg:Fe amorphous pyroxene and Mg:Fe amorphous olivine; Mg-olivine (forsterite) $(Mg_x,Fe_{(1-x)})_2SiO_4$ for $1.0 \leq x \leq 0.8$, which produces sharp resonances (spectral "peaks") at or near 11.1–11.2, 19.5, 23.5, 27.5, and 33.5 μm and weaker peaks at 10.5 and 16.5 μm (*Crovisier et al.*, 1997; *Hanner and Zolensky*, 2010; *Wooden et al.*, 2017; *Koike et al.*, 2003, 2010); Mg-orthopyroxene (enstatite) $(Mg_y,Fe_{(1-y)})SiO_3$ for $1.0 \leq y \leq 0.9$ that produces resonances at or near 9.3 and 10.0 μm and with a set of FIR resonances near 20 μm, which is also where Mg:Fe amorphous pyroxene has a broad feature. Depending on the laboratory data and optical constants, the FIR resonances for Mg-pyroxene may be at these sets of wavelengths: 18.5, 19.2, 20.3 μm for ellipsoidal shapes (in our plots) using optical constants from *Jaeger et al.* (1998), which gives peaks measured at 18.2, 20.6, 21.6 μm; also, laboratory spectra of orthoenstatite give peaks at 9.3, 10.7, 19.5, 20.7 μm from *Chihara et al.* (2001). Cometary features from Mg-pyroxene are not yet detected at high spectral contrast in the FIR. For a spectrum with strong Mg-pyroxene one may look at FIR Spitzer IRS spectrum of Herbig Ae/Be star HD179218 (cf. *Juhász et al.*, 2010, Fig. 13). Mg-olivine is a well-established cometary dust component by either a peak at 11.1–11.2 μm or by a shoulder on the broad "10-μm" feature, which makes the feature look "flat-topped." Also, the FIR peaks for Mg-olivine are well separated from the center of the broad Mg:Fe amorphous olivine "20-μm" feature, whose Mg:Fe composition is established by the radiative equilibrium temperatures (*Harker et al.*, 2002, 2004). When detected, the mass fraction of Mg-olivine present in the coma is ≳20%. [Note: Mg:Fe ≃ 50:50 ratios for each of Mg:Fe amorphous olivine and Mg:Fe amorphous pyroxene were determined from radiative equilibrium temperature calculations and SED fitting of Comet Hale-Bopp (*Harker et al.*, 2002, 2004), using optical constants from *Dorschner et al.* (1995), and so Mg:Fe ≃ 50:50 is used by dust modelers (e.g., *Ootsubo et al.*, 2020; *Bockelée-Morvan et al.*, 2017a).]

Crystalline silicate feature wavelengths and feature relative intensities depend on composition and crystal shape. Mg-rich (100–90% Mg) crystalline Mg-olivine (forsterite) is tri-refringent (three optical axes or three sets of indexes of refraction) and has spectral peaks that can shift to somewhat shorter or longer wavelengths depending on crystal shape (*Koike et al.*, 2010; *Lindsay et al.*, 2013). The observed spectral features of forsterite are better fitted by rectangular prisms slightly elongated or flattened along the b-axis (*Lindsay et al.*, 2013) or by ellipsoidal shapes flattened also along the b-axis (*Harker et al.*, 2002, 2004, 2007); note that *Fabian et al.* (2000) shows ellipsoidal shapes and quotes elongation along the b-axis but actually, as computed, these particles are b-axis flattened when the (L_b) parameter used in computing the ellipsoidal particles, as given as by *Bohren and Huffman* (1983), is larger. The optical constants used for forsterite (in figures shown here) are from *Steyer* (1974). See *Juhász et al.* (2009) for a comparison of other optical constants for Mg-olivine. Optical constants of Mg-olivine that are derived from measurements of polished single-crystal samples (*Steyer*, 1974; *Suto et al.*, 2006) are preferred for modeling crystal shapes (*Lindsay et al.*, 2013) rather than optical constants derived from ground samples that inherently have shape-dependencies (*Fabian et al.*, 2001; *Koike et al.*, 2010; *Imai et al.*, 2009). Specific examples of crystal shape are revealed in the absorption spectra of forsterite powders that are prepared by hand-grinding or ball-grinding, where ball-grinding imparts greater sphericity to the particles (*Koike et al.,* 2010; *Lindsay et al.,* 2013; *Imai et al.,* 2009; *Tamanai et al.,* 2006, 2009). A long-standing discrepancy in laboratory data was resolved when it was realized that the preparation of the sample as well as the medium in which the ground mineral sample is embedded (KBr vs. PE) affects the wavelength positions and relative depths of absorption bands and the degree of sensitivity varies for different spectral features as well as with the embedding medium (*Tamanai et al.,* 2009); particles lofted in air also affect the absorption spectra because particles electrostatically cling and agglomerate (*Tamanai et al.,* 2006).

With increasing Fe-content, olivine peaks shift to longer wavelengths (*Koike et al.,* 2003). The observed range of wavelengths for the cometary 11.1–11.2-μm feature restricts Mg-contents from 100–80% for Mg-olivine (see Fig. 12).

Silicates, being dielectrics, emit strongly through their MIR vibrational stretching and FIR vibrational bending modes and possess little absorptivity at wavelengths shorter than ~7.5 μm so, we repeat, the thermal emission at shorter wavelengths is likely from a carbonaceous dust component.

Iron sulfide (FeS) mineral is dark but not as absorbing as amorphous carbon. FeS has yet to be modeled as a major dust component contributing in the NIR for several reasons, including (1) the optical constants are lacking for the full range of wavelengths and/or are contested and are specifically lacking at 3–10 μm, (2) FeS is not yet firmly detected spectroscopically (section 2.5.4), and (3) FeS does not yield the correct scattered light color even if held to 6 vol% of the particle (*Bockelée-Morvan et al.,* 2017a) (see section 2.5.7).

Each of the five dust compositions that dominate SEDs have analogous materials in cometary samples and in anhydrous chondritic IDPs (CA-IDPs). Spectral features are measured in bulk and thin sections of CA-IDPs (*Bradley et al.,* 1992; *Wooden et al.,* 2000; *Matrajt et al.,* 2005), in CP-MMs (*Noguchi et al.,* 2015), and in UCAMMs (*Dartois et al.,* 2018). The amorphous silicates are thought to be akin to the GEMS (*Bradley et al.,* 1992, 1999; *Bradley,* 2013; *Noguchi et al.,* 2015; *Dobrică et al.,* 2012). The Mg-olivine and Mg-pyroxene are spectrally akin to forsterite Fo_{100}–Fo_{80} and ortho-enstatite En_{100}–En_{90}. The amorphous carbon is akin to some phases of disordered carbon and the occasionally quoted amorphous carbon in CA-IDPs (*Muñoz Caro et al.,* 2006; *Wirick et al.,* 2009; *Woodward et al.,* 2021). A fraction of the smallest particles observed in the *in situ* measurements of 1P/Halley were composed solely of carbon (*Lawler and Brownlee,* 1992; *Fomenkova et al.,* 1994). Alternatively, we note that possibly aromatic-bonded carbon (π-bonds or C=C bonds) could produce a significant absorptivity because of its higher UV-VIS cross sections. Organic matter in CA-IDPs and CP-MMs is generally similar to IOM from primitive meteorites, however, with intrinsic characteristics like a higher aliphatic/aromatic ratio in CA-IDPs compared to meteoritic IOM (*Flynn et al.,* 2003; *Keller et al.,* 2004; *Matrajt et al.,* 2005; *Muñoz Caro et al.,* 2006; *Wirick et al.,* 2009; *Matrajt et al.,* 2012; *Noguchi et al.,* 2015). Organic matter in UCAMMs is dominated by polyaromatic matter similar to that in IOM; however, it can show higher abundances of N than in meteoritic IOM (*Dartois et al.,* 2018) (see section 2.2).

Figure 10 shows Comet C/1995 O1 (Hale-Bopp), which is the best example of an IR SED of a coma with a plethora of submicrometer silicate crystals and a higher silicate-to-amorphous carbon ratio (high contrast silicate features relative to the "featureless" emission from a distribution of porous amorphous carbon particles and distribution of larger porous amorphous silicates). Clear and distinctive spectral peaks from (crystalline) Mg-olivine and (crystalline) Mg-pyroxene provide a benchmark for modeling crystalline silicates (shapes and temperatures). Specifically, the "hot crystal" model (*Harker et al.,* 2002, 2004, 2007, 2023; *Woodward et al.,* 2021) increases the radiative equilibrium temperature of the Mg-olivine by a factor of 1.7 over that predicted using the optical constants in order to fit the Hale-Bopp spectra at 2.75 au. Hale-Bopp's Mg-pyroxene is spectrally discernible by its sharp peaks, more so at epochs near perihelion than at r_h =2.75 au, which is attributed to its transparency compared to the other dust compositions (*Wooden et al.,* 1999). The same jets are active at the two epochs at r_h = 1.2 au and 0.93 au (*Hayward et al.,* 2000), and the thermal models fitted to their IR SEDs reveal minor differences in their mineralogy such that Mg-pyroxene has a greater relative abundance compared to Mg-olivine at 1.2 au than at 0.93 au: If the relative strength of the Mg-pyroxene and Mg-olivine features are only attributed to a temperature increase at smaller heliocentric distances then Mg-pyroxene would be expected to be enhanced relative to the Mg-olivine peaks at 0.93 au, but instead by visual comparison the Mg-pyroxene peaks are enhanced at 1.2 au compared to 0.93 au. Also, shown in Fig. 10 is the distinct contribution of the Mg-pyroxene 9.3-μm peak to the short-wavelength shoulders of the broad 10-μm feature from Mg:Fe amorphous pyroxene and Mg:Fe amorphous olivine.

2.5.3. Degeneracies in compositions derived from thermal models. Possible degeneracies in dust compositions that can be overcome by fitting multiple spectral peaks include (1) the 11.1–11.2-μm feature of Mg-olivine and the 11.2-μm PAH emission band; (2) wavelength shifts due to crystal shape (*Koike et al.,* 2010; *Lindsay et al.,* 2013) or increasing Fe-contents (*Koike et al.,* 2003) because the 23.5-μm FIR feature of Mg-olivine is more sensitive to shape [olivine with greater than 40% Fe causes the entire FIR spectrum to change morphology (*Koike et al.,* 2003)]; (3) at $r_h \gtrsim$ 3–3.5 au, the MIR short wavelength shoulder can be fitted with Mg:Fe amorphous pyroxene or by the warm featureless emission from amorphous carbon but fitting the FIR pyroxene band removes this degeneracy; and (4) the Mg-pyroxene 9.3-μm peak and the short wavelength shoulder of the Mg:Fe amorphous pyroxene feature occur at 9.3 μm but sufficient SNR to detect the Mg-pyroxene 10.5-μm peak as well as measuring the far-IR resonances can distinguish between these two compositions. If we contrast Mg-pyroxene with Mg-olivine, Mg-olivine may be assessed even when the 11.1–11.2-μm peak is not clearly discernible against the broad Mg:Fe amorphous olivine or Mg:Fe amorphous pyroxene features

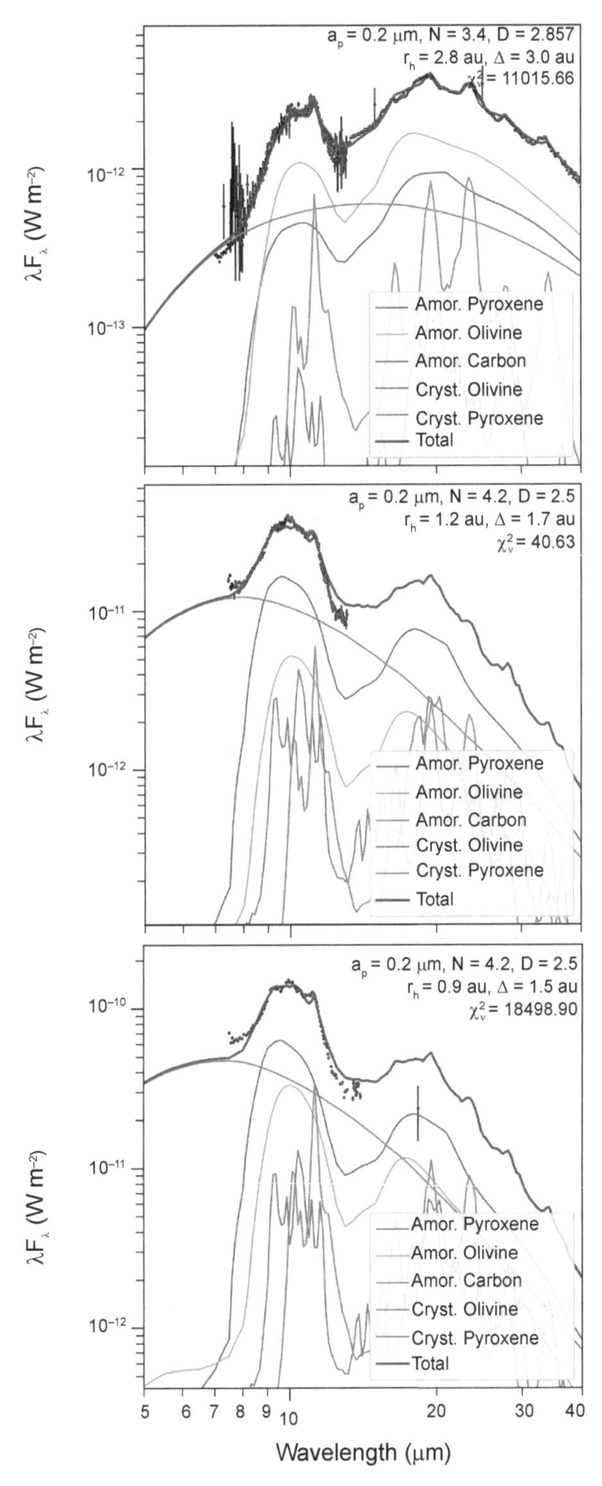

Fig. 10. Comet C/1995 O1 (Hale-Bopp) at three epochs with thermal models (*Woodward et al.*, 2021) fitted with five compositions (top to bottom): 1996-10-11 UT pre-perihelion r_h = 2.75 au from IRTF + HIFOGS and ISO + SWS (*Crovisier et al.*, 1997), IRTF + HIFOGS on 1997-02-14 UT and 1997-04-11 UT at r_h = 1.2 au and r_h = 0.93 au (*Wooden et al.*, 1999; *Harker et al.*, 2002, 2004). Mg-olivine (crystalline) has peaks at 10.0, 11.1–11.2, 16.5, 19.5, 23.5, 27.5, and 33.5 μm. Mg-pyroxene (crystalline) has a triple of peaks at 9.3, 10.5, and 11.0–11.3 μm and has modeled far-IR peaks near 19.6 μm, 18.7 μm, and ~17.9 μm and in λ-vicinity of the FIR Mg:Fe amorphous pyroxene broad feature.

because Mg-olivine's primary peak in the MIR occurs at the longer wavelength side of the broad "10-μm silicate feature" so that if a broad silicate feature appears to not be declining at 11–12.5 μm but instead appears "flat-topped" then Mg-olivine is the candidate.

The FIR peaks from Mg-olivine are at distinctly different wavelengths than the FIR Mg:Fe amorphous olivine feature, thereby allowing for FIR Mg-olivine peaks to not compete with FIR Mg:Fe amorphous olivine in the χ^2-minimization between the model and the data, and the effect is that Mg-olivine FIR peaks are spectrally discernible even when the FIR peaks are weak.

In contrast, the FIR sharp peaks from Mg-pyroxene and the FIR broad feature of Mg:Fe amorphous pyroxene are at nearly the same wavelengths. Moreover, there are some variations in the predicted wavelengths for the Mg-pyroxene FIR peaks partly because there are variations in the optical constants, the peaks depend on shapes of the crystals, and laboratory measurements of feature wavelengths depend on the embedding medium (*Tamanai et al.*, 2009). We have yet to observe a cometary IR SED that has strong Mg-pyroxene peaks in the FIR to confirm the choice of optical constants and crystal shapes in the models. We show for Comet 46P/Wirtanen at r_h = 1.74 au the coincidence of features of Mg-pyroxene and Mg:Fe amorphous pyroxene in the FIR (see Figs. 12b,c and the bottom panel of Fig. 11). In observed cometary IR SEDs, Mg-pyroxene peaks appear more discernible at MIR wavelengths than in the FIR, and this may contribute to Mg-pyroxene having lower relative abundance relative to the other dust components when the MIR and FIR are fitted compared to when only the MIR is fitted. As a demonstration, compare the model fitted to full SED of 46P at 1.7 au (Fig. 12b) with the model fitted only to MIR (Fig. 12c). This aspect of the thermal models seems contrary to the aim of fitting multiple resonances over the fullest wavelength range, and why spectral features of Mg-pyroxene in the FIR are of lower contrast than predicted by thermal models, given the resonances in the MIR, is a puzzle and may depend on their optical properties (see section 2.5.2). When weak Mg-pyroxene features are modeled then there is a significant relative mass fraction ($\gtrsim$30%) of this dust component because its lower absorptivity (Q_{abs}) (*Harker et al.*, 2023) yields cooler temperatures and hence lower fluxes relative to warmer dust components (i.e., a greater relative mass is required to fit the observed features when the particles are cooler) (*Wooden et al.*, 1999). Mg-pyroxene FIR peaks may or may not be fitted by χ^2-minimization of models to the data when the stronger broad feature from Mg:Fe amorphous pyroxene broad feature dominates the flux. These factors need to be considered when contrasting the relatively low number of cometary IR SEDs fitted with Mg-pyroxene compared to the frequent identification of Mg-pyroxene in cometary samples.

Comet 46P at r_h = 1.47 au has a weak silicate feature (Fig. 11b). The fitted thermal model has a DSD that peaks at a_p = 0.6 μm so submicrometer particles are present in the coma, which can produce distinct features when composed

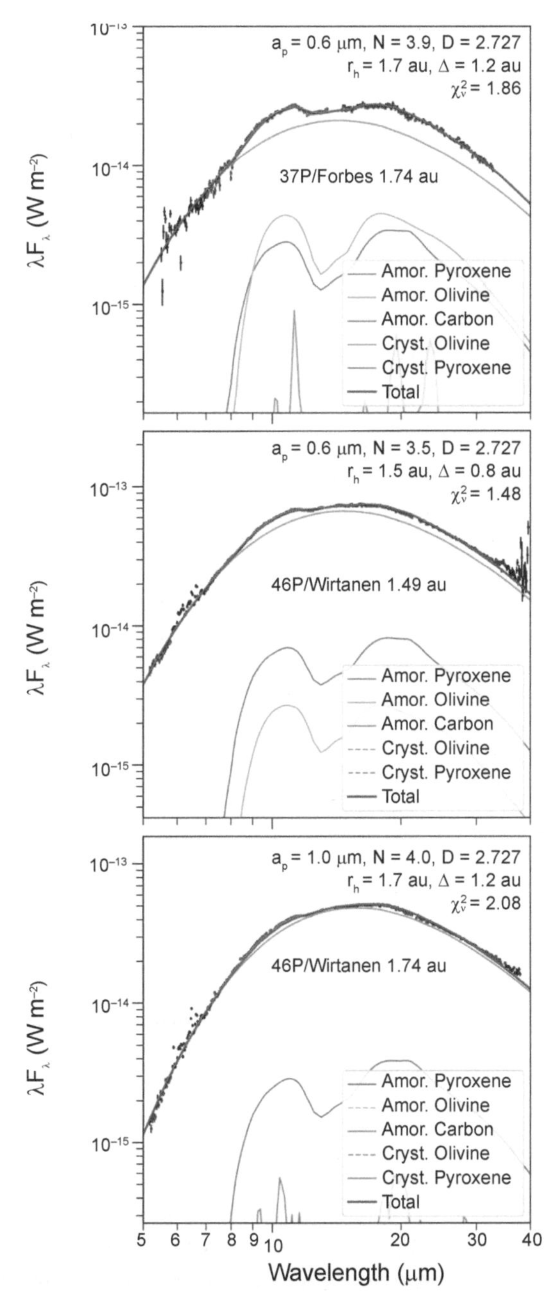

Fig. 11. Spitzer + IRS SEDs of JFCs including 37P/Forbes at r_h = 1.74 au (2005-10-14.63 UT), and two epochs of 46P/Wirtanen showing IR SEDs with decreasing constrast silicate features and decreasing ratios of silicate-to-amorphous carbon in their fitted thermal models (*Harker et al.,* 2022). Amorphous silicates and no crystalline silicates are constrained in the thermal model fit to 46P at r_h = 1.49 au (2008-04-24.64 UT). For Comet 46P/Wirtanen at r_h = 1.74 au (2008-05-24.06 UT), weak features from Mg-pyroxene (crystalline) are evident at 9.3 and 10.5 μm but more difficult to discern in the far-IR and where the Mg-pyroxene resonances overlap in spectral wavelength with the far-IR broad Mg:Fe amorphous pyroxene feature.

of silicates. The weak silicate feature thus is modeled by a high relative mass fraction of amorphous carbon compared to the amorphous silicates whose resonances also are fitted. This epoch of Comet 46P is similar to Comet C/2017 US_{10} (Catalina), which has a high mass fraction of amorphous carbon. The elemental C/Si ratio for Comet C/2017 US_{10} (Catalina) is similar to Comet 67P/C-G, which has organic IOM-like matter, and both of which have C/Si ratios close to the ISM value (*Woodward et al.,* 2021).

2.5.4. Dust compositions not yet firmly detected in infrared spectral energy distributions. As stated above, the five dominant compositions assessed from IR SEDs have analog dust species in cometary IDPs, in Stardust samples, and in *in situ* compositional studies of 1P/Halley and 67P/C-G particles but the contrary is not true: There are dust compositions in cometary samples that are not detected in IR SEDs that include inorganics and organics.

As the Fe-content of olivine increases from 20% to 40%, the MIR peak shifts from 11.2–11.4 μm (*Koike et al.,* 2010), but this wavelength also depends on crystal shape (*Lindsay et al.,* 2013). Detections of peaks in the 16–27-μm FIR region are required to discern Mg:Fe ≈ 60:40 from the peaks we observe and model in comets that span Mg:Fe proportions from 100:0 to 80:20 [see Fig. 12 and *Koike et al.* (2003)]. Due to the enhanced $Q_{abs,UVIS}$ of Fe-olivine, thermal models also will need to demonstrate increased radiative equilibrium temperatures of Fe-olivine in contrast to Mg-olivine.

Fe sulfides are present and abundant in many cometary samples and UCAMMs. FeS may be identified through a very broad 23-μm feature, which is much broader compared to the Mg-olivine 23.5-μm feature, and only is expected for submicrometer FeS grains (*Keller et al.,* 2000). However, different measurements produce different predictions for $Q_{abs,IR}$ that range from strong FIR resonances (*Begemann et al.,* 1994; *Henning and Mutschke,* 1997) to no FIR resonances (*Hofmeister and Speck,* 2003). A single reference suggests a 3.63-μm feature in the NIR (*Tamanai et al.,* 2003), and the dearth of FeS data in the NIR is represented by a dashed line in *Pollack et al.*'s (1994) protoplanetary disk opacities. If FeS has a spectral feature near 3 μm then FeS could not provide the opacity needed to explain the ubiquitous featureless thermal emission (NIR "pseudo-continuum") from warm particles, which currently is ascribed to a highly absorbing carbonaceous component and well-fitted by amorphous carbon.

The MIR spectral features of phyllosilicates overlap with the 10-μm features of anhydrous amorphous silicates, e.g., Mg:Fe amorphous olivine and Mg:Fe amorphous pyroxene, but the 20-μm features from phyllosilicates are significantly different (*Wooden et al.,* 1999). Phyllosilicates including montmorillonite (*Wooden et al.,* 1999), smectite, and serpentine are abundant in hydrated chondritic IDPs (*Bradley et al.,* 1989). At most 5% montmorillonite may be present in the coma of C/1995 O1 (Hale-Bopp) (*Wooden et al.,* 1999). The spectral features of phyllosilicates (smectite, nontronite) are claimed for 9P/Tempel 1 at +45 min post-Deep Impact (*Lisse et al.,* 2007), as well as for Comet C/1995 O1 (Hale-Bopp), but other well-fitted thermal models are without phyllosilicates (*Harker et al.,* 2002, 2004; *Harker and Desch,* 2002; *Min,* 2005) (Fig. 10).

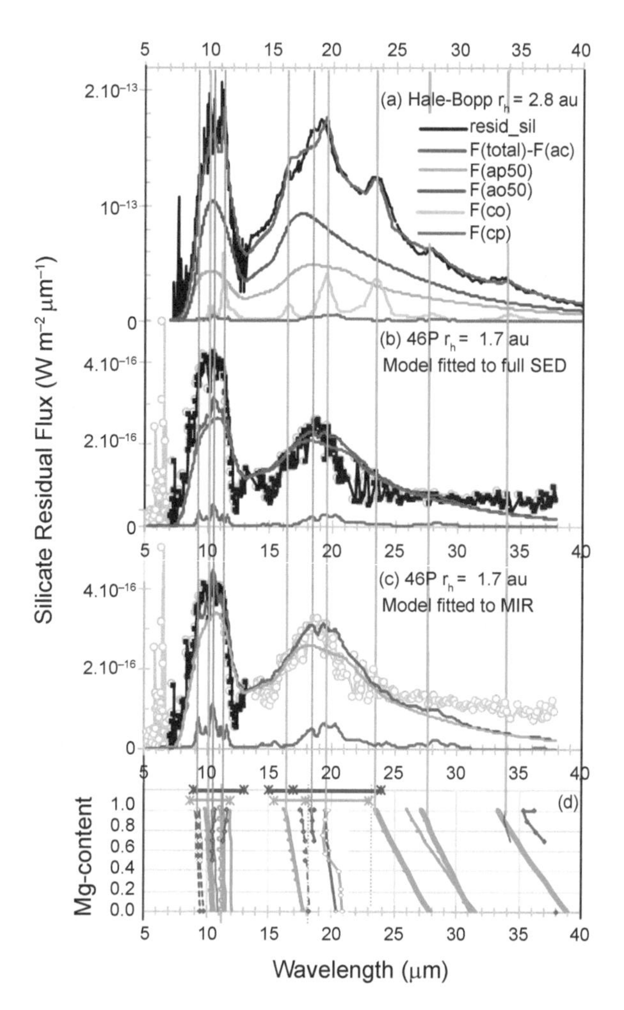

Fig. 12. Silicate residual flux spectra (*Harker et al.,* 2022) shown to **(a)–(c)** highlight the wavelengths of observed features of (crystalline) Mg-olivine (co, in green) and (crystalline) Mg-pyroxene (cp, in magenta) and **(d)** compare with the wavelength dependencies of the laboratory spectra vs. Mg-content = (1−Fe-content). The silicate residual fluxes are the data and model with amorphous carbon (ac) subtracted, i.e., F_{data}–F_{model} (ac) and $F_{model,total}$–F_{model} (ac), respectively, for **(a)** Hale-Bopp at r_h = 2.8 au (1996-10-11 UT) (Fig. 10), **(b)** 46P at r_h = 1.7 au (2008-05-24.06 UT) fitted over the full SED (5–35 µm) black points (as fitted in Fig. 11), and **(c)** 46P at r_h = 1.7 au ((2008-05- 24.06 UT) and fitted only over the MIR 5–13 µm, which shows an increase in Mg-pyroxene relative abundance compared to the model in **(b)**. **(d)** Wavelength dependencies of Mg-olivine (green) and Mg-pyroxene (magenta) from laboratory work by *Chihara et al.* (2002) and *Koike et al.* (2003); thickness of lines relates to strengths of observed peaks. Also are shown are Mg:Fe amorphous olivine (royal blue) and Mg:Fe amorphous pyroxene (cyan). The observed and modeled crystalline feature wavelengths align with Mg = 1.0–0.8.

The low-density amorphous silicates in cometary IDPs, GEMS (*Bradley et al.,* 2022), have limited spectral data (*Bradley et al.,* 1999; *Ishii et al.,* 2018) and are presumed to be spectrally analogous to Mg:Fe amorphous silicates in thermal models, which were derived from rapid cooling of melts (*Dorschner et al.,* 1995). Amorphous silicates are discussed in the context of IR spectra of cometary IDPs (e.g., *Brunetto et al.,* 2011). Experiments have created amorphous silicates, e.g., via particle bombardment of crystals or via the solgel method (*Jäger et al.,* 2003, 2016; *Brucato et al.,* 2004; *Demyk et al.,* 2004; *Wooden et al.,* 2005, 2007), but how the structure of the experimental amorphous silicates or the "amorphization rate" affects the IR signature of the analogs used in thermal models has not been assessed.

Mg-carbonates are relatively rare in cometary samples (section 2.4) and are not definitively detected in cometary IR SEDs. Carbonates (siderite and magnesite) are discussed for Comet 9P/Tempel 1 (*Lisse et al.,* 2007) but the simultaneous occurrence of water vapor emission lines in the overlapping wavelength region, when modeled, yields a marginal detection of the 7-µm carbonate feature and an abundance that is 2–3× lower (*Crovisier and Bockelée-Morvan,* 2008). Water vapor lines must be modeled concurrently with the dust thermal emission, and the 5–8-µm spectral region calls for higher SNR studies such as will be available from the JWST.

Aromatic macromolecules or PAHs are efficient emitters of IR photons and their spectra can be predicted (*Astrochem. org,* 2022). Unlike the free-flying PAH macromolecules, we are asking what the excitation mechanisms are that could produce emission features from aromatic-bonded carbon within organic-rich particles. Can photoprocessing of PAHs, which can alter their size distribution, occur within organic-rich particles (*Clemett et al.,* 2010)? Currently, we rely on computing emission spectra of particles using optical constants that have been determined for a limited number of organic residues (tholin-like materials). Measurements of absorption spectra are more numerous than the suite of available optical constants (n, k) from which we can compute $Q_{abs,UVIS}$ and $Q_{abs,IR}$ for thermal emission models. UCAMMs offer a template for the potential wavelength positions of IR spectral features because UCAMMs have so much organic material, including polyaromatic bonds, that their absorption spectra can be obtained without acid dissolution of the silicates (e.g., *Dartois et al.,* 2018), which is done for meteoritic IOMs (*Alexander et al.,* 2017) and which is known to alter some of the organic fraction in IDPs (e.g., see *Matrajt et al.,* 2005). However, UCAMMs contain a N-rich organic phase that is too N-rich to be akin to meteoritic IOM and 67P/C-G, so UCAMMs cannot serve to predict the presence of organic features in comets such as 67P/C-G.

The PAH spectral features sought are their skeletal vibration bands in the 5–8-µm wavelength region and by their peripheral C-H bonds near 3.28 µm and at longer wavelengths where the skeletal structure affects the feature locations. PAHs are discussed for Comet 9P/Tempel 1 (*Lisse et al.,* 2007). However, thermal models of the same IR SED that simultaneously model the water vapor emission lines and the PAHs in the 5.25–8.5-µm wavelength region produce lower abundances of PAHs by at least 2–3× (*Crovisier and Bockelée-Morvan,* 2008).

Distinct resonances from PAH-like organics contribute to features at ~8.5 µm and ~11.2 µm and possibly aliphatic hydrocarbons at ~9.2 μm (Fig. 13) in the inner coma of Comet 21P/Giacobini-Zinner at r_h = 1.04 au (*Ootsubo et al.*, 2020) as well as Mg-rich amorphous silicates and Mg-silicates. However, the organic bands are not detected in the coma of 21P post-perihelion by lower-spatial-resolution Spitzer + Infrared Spectrograph (IRS) at the larger heliocentric distance of r_h = 2.292 au (*Harker et al.*, 2023) (Fig. 13).

Organics with aliphatic bonds may produce a 3.4-μm emission feature such as seen in CA-IDPs (*Matrajt et al.*, 2005) and in some Stardust particles (*Matrajt et al.*, 2008) and are suggested for Comet 103P/Hartley 2 (*Feaga et al.*, 2021; *Wooden et al.*, 2011) as well as other comets by *Bockelée-Morvan et al.* (1995), who noted a strong correlation with CH_3OH production rates. Lines of sight through the diffuse ISM, e.g, toward the galactic center, reveal the 3.4-μm feature. By comparison of the two primary components of this feature from $-CH_3$ (~2960 cm^{-1}, 3.38 μm) and $-CH_2$ (~2930 cm^{-1}, 3.41 μm), the diffuse ISM has shorter chains and is more processed than in cometary matter (*Matrajt et al.*, 2013).

This solid-state emission feature from organics is in the same spectral region as gaseous emission lines from ethane C_2H_6 and methanol CH_3OH, which are observed at high spectral resolution and for which modeling is a challenge (*Bonev et al.*, 2021, Fig. 5), so a method that combines observing or modeling the emission lines of the gaseous species (*Feaga et al.*, 2021, as mentioned in *Feaga et al.*, 2023) with modeling the solid-state material that typically has multiple resonances from $-CH_2$ and $-CH_3$ bonds is required to further assess the presence of this aliphatic carbon component. Absorption features from aliphatic bonds are common in CA-IDPs (section 2.2). The difference between laboratory studies of IDP absorption spectra and cometary IR spectra is that the 3.4-μm features arising from dust in comae may be in emission. Predicting the emission spectra is at the forefront of thermal modeling developments and we note that optical constants are lacking for carbonaceous materials dominated by aliphatic bonds as opposed to rich in aromatic bonds.

To date, most organics that have measured optical constants are dominated by aromatic bonds (the 3.28-μm PAH feature and features in the 5–9-μm region), and for modeling comets we need organic residues that are rich in aliphatic bonds, if we use CA-IDPs as our guide. The temperatures of the organic-bearing particles that produce the 3.4-μm emission feature and its relative abundance will determine the strength or contrast of the feature relative to the strong NIR "continuum" emission from highly absorbing carbonaceous matter that is modeled by optical constants of amorphous carbon.

2.5.5. Discrete materials or mixed material aggregates? There are compelling reasons to treat cometary particles as aggregates of mixed materials. CA-IDPs show that particles are mixtures, i.e., unequilibrated aggregates of amorphous silicates (GEMS), crystalline silicates, organics of varying compositions (section 2.2), and iron sulfides. Some CA-IDPs are dominated by carbonaceous matter

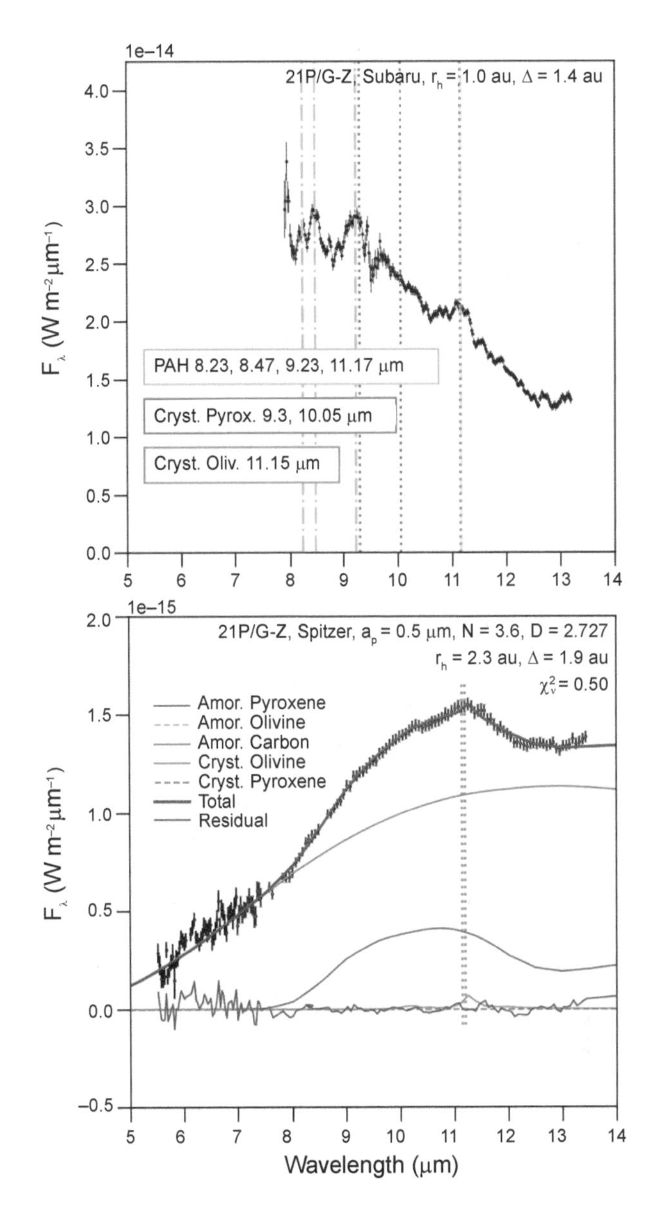

Fig. 13. Comet 21P/Giacobini-Zinner (21P/G-Z) at two different heliocentric distances (r_h) and with higher vs. lower spatial resolution. Upper panel: For 21P/G-Z at r_h = 1.0 au(2005-07-05 UT), Subaru + COMICS spectroscopy of the inner coma reveals emission bands attributed to PAHs (8.23, 9.23, 11.17 μm) that contribute in excess to the thermal model of amorphous carbon, Mg:Fe amorphous olivine, Mg:Fe amorphous pyroxene, Mg-olivine (Crys. Oliv.), and Mg-pyroxene (Crys. Pyrox.) (*Ootsubo et al.*, 2020). Data courtesy of T. Ootsubo. Lower panel (SED): Spitzer + IRS spectroscopy of 21P/G-Z at r_h = 2.3 au, fitted with the five-mineral thermal model providing constraints for amorphous carbon, Mg:Fe amorphous pyroxene, and Mg-olivine (crystalline). 21P is only fitted through 14 μm because of possible calibration issues associated with the data taking sequence (*Harker et al.*, 2022; *Kelley et al.*, 2021). Compared to 2.3 au, PAH bands are present in the coma of 21P at r_h = 1 au as well as hotter particles, which can be seen by the spectral slope change between 1.0 au and 2.3 au.

(*Thomas et al.,* 1993), while others like the giant IDPs are intimate mixtures of mini-chondrules of Fe-olivine with minor Mg-olivine, as well as sulfides and GEMS (*Brownlee and Joswiak,* 2017). Alternatively, there are some observations that strongly motivate our thinking that the carbonaceous materials and the siliceous materials may be discrete components in cometary comae or at least they vary in their relative abundance ratios depending on what part of the nucleus is active. There are several examples of this: (1) Comet C/2001 Q4 (NEAT) revealed a significant drop in the silicate feature contrast in about an hour, which is the jet-crossing time for the observing aperture (*Wooden et al.,* 2004). (2) Comet C/2017 US_{10} (Catalina) had an increase in the amorphous carbon-to-silicate ratio between two epochs separated by about six weeks of time, where the observing epoch refers to the UT date and time and ephemerides of the comet (*Woodward et al.,* 2021). (3),(4) The inner coma studies of 9P/Tempel 1 post-Deep Impact by Gemini Observatory's Michelle spectroscopy show rapidly changing compositions (*Harker et al.,* 2007), and by the Subaru Telescope's Cooled Mid-Infrared Camera and Spectrometer (COMICS) narrow-band imaging show Mg-olivine, then amorphous carbon, and then Mg-olivine in the few hours following the Deep Impact event. (5) The outburst of Comet 17P/Holmes showed crystalline-rich material that then changed composition well after outburst (*Reach et al.,* 2010). (6) Multi-epoch Spitzer IRS spectra of a handful of comets show variable compositions for multi-epochs (*Harker et al.,* 2023) (see also Fig. 11). Complimenting remote sensing results, the *in situ* measurements of cometary particles indicate from 1P/Halley revealed that particles were siliceous-only, carbonaceous-only, and mixed particles (*Schulze et al.,* 1997; *Fomenkova et al.,* 1992; *Lawler and Brownlee,* 1992) and that carbonaceous-only particles dominated within 9000 km of the nucleus (*Fomenkova and Chang,* 1994).

When computing mixed-material aggregates for thermal models to be fitted to cometary IR SEDs, there are a lot of combinations of materials that have to be computed in order to provide the potential suite of relative abundances of discrete materials that will need to be used to fit the IR SEDs. Employing silicate crystals with edges is particularly pertinent to fitting cometary spectra (*Lindsay et al.,* 2013) because spheres notably do not fit the observed spectral features and ellipsoids are adequate (*Harker et al.,* 2002, 2004; *Min,* 2005; *Moreno et al.,* 2003). The work to compute a suite of potential materials in mixed-material aggregates has begun (*Wooden et al.,* 2021) but is as yet insufficient in breath to constrain relative abundances of materials for a sample of cometary IR SEDs. A particle size distribution of an ensemble of discrete materials is currently the state of the art for thermal models.

2.5.6. Thermal emission models and dust equations. Computing the scattered light component and the thermal emission component (typically starting at 3 μm for $r_h <$ 2.5 au) of the flux observed from a cometary coma requires a set of equations (Table 1) and a set of suppositions: Suppose a single DSD, a chosen porosity P, and an ensemble of compositions represent the dust in the coma, and then the models are assessed against the observations using standard minimization (χ^2-minimization) techniques.

The emergent flux is a sum over the particle size distribution of the thermally emitted fluxes per particle of varying compositions and/or particle structures. Per particle, the thermal flux is set by the particle's dust temperature (T_d), which results from radiative equilibrium between absorbed sunlight and emitted thermal radiation. Both the dust temperature and dust flux depend upon the product of the particle's absorptivity ($Q_{abs,IR}$) and its cross sectional area G(a). The particle's wavelength-dependent absorptivity ($Q_{\lambda,abs}$) depends on composition, radius (a), shape, and porosity.

The optical properties are called the absorptivity $Q_{abs,IR}$ and scattering efficiency $Q_{\lambda,sca}$, which are computed from the real and imaginary optical constants, n and k, using various methods. Porous aggregates of mixed compositions, such as amorphous silicates, amorphous carbon, and FeS, can be well modeled by "mixing" optical constants and vacuum such as when Mie theory is combined with effective medium theory (EMT) or Brugemann mixing theory (BM); or by layering such as in distribution of hollow spheres (DHS) (*Min,* 2005); or Rayleigh-Gans-Debye (RGD) theory (*Bockelée-Morvan et al.,* 2017a,b). However, Mg-olivine cannot be well modeled by mixing optical constants with vacuum; the predicted spectral features do not come close to observed spectra or laboratory spectra. This presents the challenge of computing the optical properties of mixed material porous aggregates with crystal monomers with edges and faces, and some progress has been made using the discrete dipole approximation with the Discrete Dipole Scattering (DDSCAT) code (*Moreno et al.,* 2003; *Wooden et al.,* 2021). Scattering efficiencies ($Q_{\lambda,sca}$) have been computed using DDSCAT or T-Matrix because aggregates scatter differently than spheres (*Kimura et al.,* 2016) (see also the chapter in this volume by Kolokolova et al.).

More absorbing particles are warmer and smaller particles are warmer, and the co-dependencies between these dust properties is mitigated by the assessment of the relative fluxes and wavelengths of spectral resonances or spectral "features." Spectral features only arise from particles composed of dielectric materials that are smaller than about 1–3-μm-radii for solid (0% porosity, fractal dimension D = 3) particles and for $\lesssim$10–20-μm-diameter moderately porous particles [~65–85% porosity, D ~ 2.86 to ~2.7 (*Harker et al.,* 2002, 2004)]. For particles of the same composition, the dust temperature depends more weakly on effective radius than on heliocentric distance from the Sun (r_h). When of the same composition, smaller particles are hotter simply because quantum mechanically smaller particles are less efficient emitters of photons at wavelengths larger than their cross sections [G(a)]; historically, this effect has been called *superheat* (*Gehrz and Ney,* 1992). Cometary SEDs do not reveal a single color temperature (Planck function fitted to wavelengths outside resonant features); the color temperature is warmer at shorter NIR wavelengths and cooler at FIR wavelengths. For a color temperature $T_{color}(r_h)$ fitted to ~7.5 µm and ~13 µm, a longstanding relationship vs. heliocentric distance (r_h) is given in Table 1.

For particles of the same effective radius and r_h, there is strong dependence of dust temperature on the dust composition through significant variations in the absorptivity $Q_{abs,UVIS}$ at wavelengths where sunlight is absorbed. To achieve the highest temperatures observed for comae particles (e.g., section 2.5.7), small and highly absorbing particles are required or large extremely porous aggregates that are composed of mainly highly absorbing materials and whose large particle temperatures are equivalent to the temperatures of their submicrometer monomers.

Crystalline materials in the DSD are often limited to radii of less than 1 µm to a few micrometers because predicted resonances of larger crystals do not fit observed spectral features (e.g., see *Lindsay et al.,* 2013). Relative mass fractions are quoted for the ≤1 µm portion of the DSD. The emission spectra of silicate particles or of amorphous carbon particles of increasing larger radii than these 3-µm and 20-µm radii, respectively, for solid and for moderately porous particles have increasingly broader as well as significantly weaker contrast resonances (with respect to wavelengths outside their resonances). Thermal model parameters are co-dependent but are not degenerate when dielectric materials like silicates are present in comae because, in practice, varying the composition, DSD, and particle porosities produces thermal models that are distinguishable when fitted against the observed SEDs with good signal-to-noise ratios using minimization metrics (χ^2-minimization).

There are three albedos of interest for assessing comae particle properties: the particle albedo (A), the geometric albedo, and the bolometric albedo. The geometric albedo is, by definition, assessed at zero degrees phase angle (α = 0° in the observer's frame) (*Hanner et al.,* 1981; *Bockelée-Morvan et al.,* 2017a) (α = 0° translates to θ = 180° or opposition, where θ is the angle between incident sunlight and scattered ray in the particle's frame). The geometric albedo at non-zero phase angles may be extrapolated from comae observations of $A_p(\alpha) = A_p(\alpha = 0^\circ) \times j(\alpha)$ using the "phase curve" $j(\alpha)$ and $A_p(\alpha)$ can be compared to $A_p(\alpha)$ computed for spheres (Mie) and for porous aggregate particles of varying composition and porosity (*Hanner et al.,* 1981; *Kimura et al.,* 2003, 2006, 2016).

Alternatively, to calculate at the observed phase angle $[A_p(\alpha)]$, one can use the thermal model fitting parameters, which include the product $G(a)Q_{\lambda,abs}$ to derive the dust effective area by assuming Q_{abs} = 1

$$\text{dust effective area} = K\int_{a_{min}}^{a_{max}} G(a)\, n(a)\, da$$

Then the geometric albedo is the scattered light flux density divided by the dust effective area G(a) (*Hanner et al.,* 1981; *Tokunaga et al.,* 1986) at the observer's phase angle α. G(a) can be calculated from either the Wein side of the thermal emission in the NIR or by thermal models fitted to a broader-wavelength IR SED. This technique of calculating the geometric albedo, which was popular when the dust emission was studied with narrow-band filter photometry, avoids having to measure or assume knowledge of the dusty coma phase curve $j(\alpha)$.

Finally, the bolometric albedo (*Gehrz and Ney,* 1992) enables an empirical assessment of the particle properties in the comae of many comets at different phase angles (*Woodward et al.,* 2015, 2021). The bolometric albedo $A(\alpha)_{bolo}$ is an approximate measure of the scattered to total incident energy (sunlight) (*Woodward et al.,* 2015), where the total incident energy is assessed by the sum of the thermal (re-emitted) and scattered energies (λF_λ), each measured at the wavelengths of their maximum energy output (Table 1). The opportunity to tie the scattered light to the thermal emission potentially offers additional insights into dust properties and compositions for two reasons: The scattered light may be contributed to by higher albedo (and potentially cooler) dust components such as ice grains or organics, and particle structure affects scattering and thermal in distinct ways (*Tokunaga et al.,* 1986) (sections 2.5.7 and 2.5.8).

2.5.7. Dust composition and differential size distribution from combined scattered and near-infrared thermal: 67P/Churyumov-Gerasimenko outbursts. Rosetta's Visible and Infrared Thermal Imaging Spectrometer (VIRTIS)-H spectra of the quiescent coma and of two short duration outbursts (2015-09-13T13.645 and 2015-09-14T18.828 UT) from Comet 67P/C-G as modeled and presented by *Bockelée-Morvan et al.* (2017a,b) provide an excellent demonstration of how models fitted to dust scattering together with dust thermal emission can predict S'_{color}, $T_{d,color}$, and $A(\theta)_{bolo}$, and thereby provide constraints on dust compositions and DSD parameters.

The scattered light color (S′) (whether the color is "blue," "neutral," or "red" relative to reflected sunlight) provides information about the composition of the dust particles (*Storrs et al.,* 1992; *Zubko et al.,* 2015a; *Hyland et al.,* 2019; *Kulyk et al.,* 2021; *Li et al.,* 2014). The dust color temperature, derived in this case from fitting a scaled Planck function to the 2–5-µm dust continuum measurements, represents either the DSD-weighted temperatures of the warmest particles, i.e., the smallest and/or most highly absorbing ones, or the DSD-weighted temperatures of aggregate mixed particles, or a combination thereof. The bolometric albedo is an approximate measure of the scattered to total incident energy (sunlight) (*Woodward et al.,* 2015) and is assessed at the observers phase angle α = 180°–θ = 108°, 99° for the two dates. The geometric albedo $A_p(\alpha)$ requires the dust effective area to be calculated from dust thermal models fitted to the IR (*Hanner et al.,* 1981, 1985; *Tokunaga et al.,* 1986). When the spectra provide the three metrics of dust scattered light color slope S′, NIR color temperature $T_{d,color}$ from the thermal emission, and either the bolometric albedo $A_\lambda(\theta)_{bolo}$ or the geometric albedo A_p, then these four dust model parameters can be constrained, via χ^2-minimization against the metrics derived from the spectra: two DSD parameters (smallest particle radius a_0 and DSD slope N), the particle porosity P that is parameterized by a fractal dimension D, and the dust composition quantified by q_{frac} where q_{frac}^3 is the volume fraction of inclusions of lesser opaque material within a matrix of more highly opaque material.

First, we summarize the properties of the quiescent coma of 67P/C-G on the two dates prior to the coma outbursts (Fig. 1 in *Bockelée-Morvan et al.,* 2017a). The quiescent coma presents a NIR "red" color slope of S'_{color} = 2.6 ± 0.3%/100 nm, 2.3 ± 0.4%/100 nm for λ_{ref}, λ = 2.0 μm, 2.5 μm on the two respective dates. Compared to the NIR color slope, the visible wavelength scattered light spectra have a steeper color slope 15–18 ± 3%/100 nm with λ_{ref} , λ = 0.45 μm–0.75 μm (*Rinaldi et al.,* 2017), which is typical of cometary comae. The quiescent coma has an estimated bolometric albedo of $A_\lambda(\theta)_{bolo}$ = 0.13 ± 0.2 and a dust color temperature of $T_{d,color} \approx$ 300 K.

A set of models that fit the data include porous spheres of 2-compositions of Mg:Fe amorphous olivine in a matrix of amorphous carbon (q_{frac} = 0.7) and with porosity limited to P ≤ 0.5 [*Mannel et al.* (2016), MIDAS "compact particles") and DSD slope of N ≤ 3. The composition (q_{frac} = 0.7) translates to 66 vol% amorphous carbon and 34 vol% amorphous olivine, with ranges of coupled model parameters that include (a_0, N, q_{frac}, D, P) = (0.3 μm, 2.5, 0.7, 2.5, ≤0.5) and = (0.9 μm, 3.0, 0.7, 2.5, ≤0.5). The model limits P ≤ 0.5, so this limits particles bigger than 0.4 μm to have the porosity that is independent of radii a and this porosity is lower that the porosity equation (Table 1); if D was applied over the full DSD, then for D = 2.5 the porosity values would be, e.g., P(0.4 μm) = 0.5, P(1 μm) = 0.68, P(3 μm) = 0.82.

Alternatively, for a similar porosity prescription and DSD parameters, the data can also be fitted using compositions of pure carbon, or carbon and a lower q_{frac} of more transparent Mg-rich amorphous pyroxene or Mg-olivine (forsterite). However, for pure silicate grains or silicates mixed with 6 vol% FeS [consistent with *Fulle et al.* (2016c)], DSD parameters sufficient to produce the measured $T_{d,color}$ yield NIR neutral colors (as opposed to the observed NIR red colors) and bolometric albedos higher than measured.

Finally, the quiescent coma is also modeled successfully with 25% by number of extremely porous aggregate particles (D = 1.7), which provide less than 1.5% of the particle albedo at 2 μm as well as thermal emission (*Bockelée-Morvan et al.,* 2017a). Extremely porous aggregate particles, often called ballistic cluster cluster aggregates (BCCA), have particle temperatures and spectral resonances similar to their (small and hotter) monomers, i.e., independent of their particle size (*Bockelée-Morvan et al.,* 2017a; *Tazaki et al.,* 2016; *Kimura et al.,* 2016; *Kolokolova et al.,* 2007), and the relative contributions of BCCA to the scattered light albedo is extremely minimal (*Kimura et al.,* 2006). Models for the quiescent comae, which have similar dust compositions as cited for lower porosity particles (D = 2.5, P ≤ 0.4), with 25% BCCA and with steeper size distributions N ≥ 3 (*Bockelée-Morvan et al.,* 2017b), are more commensurate with studies of 67P/C-G's dust by the Grain Impact Analyser and Dust Accumulator (GIADA) (*Lasue et al.,* 2019; *Fulle et al.,* 2015), and the Optical, Spectroscopic, and Infrared Remote Imaging System (OSIRIS) (*Fulle et al.,* 2016c), by MIDAS (*Mannel et al.,* 2016; *Bentley et al.,* 2016a), and by particle topologies (*Langevin et al.,* 2016).

In contrast to the quiescent coma, the outbursts are characterized by sudden increase in dust thermal emission that peaks in a few minutes and then decays toward nominal comae levels in about 30 min, with the outbursting comae having hotter particles $T_{d,color}$ = 550 K, 640 K, bluer NIR scattered light colors (extreme values of S'_{color} = –10%/100 nm with modeled color slope power of p_{col} = 2.30), and higher bolometric albedos $A_\lambda(\theta)_{bolo}$ = 0.6, where all three properties change with time as the outbursts evolve. In the first outburst, all three metrics trend together and follow the light curve. In contrast, during the second outburst $A(\theta)_{bolo}$ remains high (and increasing) as the light curve and $T_{d,color}$ and S'_{color} decay toward quiescent coma values. During outbursts, the visible scattered light colors remain in ranges of 10–15%/100 nm and 6–12%/100 nm for the respective two dates (*Rinaldi et al.,* 2018). The particles are moving relatively fast compared to dust nominal speeds for 67P/C-G, v(a < 10 μm) ≥ 30 m s^{-1}, so subsequent VIRTIS-H spectra sampled different sets of particles.

Particles ejected in the outburst, as modeled, are higher in carbon (smaller q_{frac}) and have smaller minimum particle radii as well as having a steep DSD slope (a_0 = 0.1 μm, 4 ≤ N ≤ 5), which means submicrometer particles dominate. Two of three metrics, S'_{color} and $T_{d,color}$, are fitted but the observed high $A(\theta)_{bolo}$ of ~0.6 is not fitted by submicrometer carbonaceous (dark, i.e., low-particle albedo, and hot) particles. Somewhat larger moderately porous particles (a_0 ~ 0.5 μm) are in the second outburst but the highest values of the bolometric albedo are also not explained. Pure and submicrometer Mg:Fe olivine grains can account for the bolometric albedo but the modeled compact (P = 0) olivine spheres have a color temperature of <530 K, which is too low to match the observed color temperatures (>600 K). If 6 vol% FeS is mixed with Mg:Fe amorphous olivine then the color temperatures increase to 575 K but the S'_{color} cannot be fitted. BCCA may explain the high color temperatures but certainly not the high albedos during outburst. A water ice band cannot be more than ~10% in depth so ice particles are not likely driving the high bolometric albedo. Another possibility is that there could be two distinct compositions of submicrometer grains in the outbursts, one that is cold and bright complimented by one that is dark and hot. *Bockelée-Morvan et al.* (2017a) also propound that some limited lifetime organics may be mixed with the submicrometer Mg:Fe amorphous olivine such that the rapid degeneration of organics may contribute to the enhanced temperatures and enhanced albedos observed during outburst, especially at the onset of the first outburst and the sustained albedo during the second outburst. In summary, a dramatic increase in the numbers of dark (carbonaceous) and smaller particles characterize the outbursts because the "high color temperatures and blue colors imply the presence of Rayleigh-type scatterers in the ejecta, i.e., either very small grains or BCCA type agglomerates" (*Bockelée-Morvan et al.,* 2017a). However, the higher $A(\theta)_{bolo}$ in the outbursts are not explained.

2.5.8. Scattered light and limited lifetime distributed sources. Scattered light observations at UV through NIR wavelengths provides constraints on the structure of dust particles. Polarization (see the chapter by Kolokolova et al.), the scattered light color, and the surface brightness spatial distribution can provide insights into dust composition for those dust components that contribute to scattered light. Here are some examples that do not involve polarization measurements.

Limited-lifetime organic species were discovered following the outburst of C/2000 WM1 (LINEAR) and the Deep Impact impact of 9P/Tempel 1 because they produce wavelength-dependent scattered light "colors" (slope in percent per nanometer) akin to a combination of organics created in the laboratory by irradiating ice mixtures (*Jenniskens et al.,* 1993). [Limited lifetime species associated with the dust, or "extended sources" or "distributed sources," are also known to occur in some comae by the molecular production rates having more extended spatial distributions than the water vapor and these molecules may include CO, formaldehyde (H_2CO), NH_3, and recently discovered C_2H_2 (*Dello Russo et al.,* 2022).]

Another metric is the total scattering cross section (SA), related to the number and intrinsic color of the dust particles

$$SA = \int \left(\Sigma\, A_\lambda(\alpha)\, f \right) d\rho$$

Tozzi et al. (2004, 2015) used this approach to assess limited lifetime dust that contributed the changes in the scattered light for Comets C/2000 WM1 (LINEAR) and 9P/Tempel 1. The significant difference between the two comets is that C/2000 WM1 (LINEAR) has a sublimating component that scatters in the visible and 9P/Tempel 1 does not, so the natures of the sublimating grains are different between the two comets.

For C/2000 WM1 (LINEAR) at r_h = 1.2 au, scale lengths for column densities of limited lifetime organics are assessed for two components with 12250 ± 1625 km and 940 ± 150 km, and adopting a dust velocity $v_{dust} \approx$ 0.2 km s^{-1} implies 1.7 hr and 17 hr coma lifetimes, respectively (*Tozzi et al.,* 2004). The scattered light colors were similar to minimally irradiated and long-exposure irradiated organic residues that may be similar to the UV-irradiated residues created in space [European Retrievable Carrier (EURECA) (*Li and Greenberg,* 1997)]. An impulsive event offers a better opportunity to assess the changes in surface brightness distribution and color, which can then be used to determine lifetimes in the coma.

Prior to the Deep Impact event, 9P/Tempel 1 has a non-sublimating component and sublimating component (with scale length of 6300 km, assuming $v_{dust} \approx$ 0.2 km s^{-1} implies 11 hr at r_h = 1.5 au) and these two sublimating components differ in their NIR colors. The NIR colors are assessed by differencing their brightness measurements between pairs of the 2MASS photometric bands of H, J, and Ks [defined to exclude telluric absorptions and centered at 1.25 μm, 1.65 μm, and 2.17 μm (*Bessell,* 2005)]. Details about post-Deep Impact observations include that the color was neutral from the J to H band but increases by 25% from H to K_s. Three hours later, the scattering efficiency increases by 84% between J and K_s so the reddening increased but the total brightness declined. Also, there was no correlation in the quiescent coma vs. Deep Impact ejecta cloud.

One may wish to compare the "quiescent" coma of Comet 9P/Tempel 1 with the post Deep Impact coma. In the hours following the Deep Impact encounter, polarization images uniquely reveal an expanding front of polarizing particles that are not detected in non-polarized images so these particles are submicrometer-sized (from their ejected velocities) and "dark" carbonaceous grains (*Kadono et al.,* 2007). Spatial imaging via IR photometry as well as scattered light colors and structures in the coma reveal properties about the dust but with more degeneracies in the derived dust composition than with IR spectroscopy thermal modeling, because the IR spectral resonances directly probe the composition of the dust particles, contributing to the thermal emission for those compositions that have resonances.

3. PHYSICAL PROPERTIES

3.1. Sizes and Size Distribution

Cometary dust is ejected from the active cometary nuclei and expands in space following an initially quasi-spherical shell within about 10,000 km from the nucleus, named the coma, and further creating an expanding tail in the direction opposite to the Sun, as illustrated in Fig. 1 (see e.g., *Finson and Probstein* 1968a). The striae visible in the solar direction of the impressive Comet C/2006 P1 McNaught (Fig. 1), named synchrones, represent dust ejected at different times along the orbit of the comet and that disperse in space due to the effect of the β parameter, representing the ratio of the forces of radiation and gravity acting on the dust particles.

$$\beta = \frac{3L_\circ}{16\pi c G M_\circ} \frac{Q_{pr}}{\rho s}$$

where $L_\circ$ and $M_\circ$ are the luminosity and the mass of the Sun, c is the speed of light, G the gravitational constant, and Q_{pr} the radiation pressure efficiency of the dust grain having bulk density ρ and an effective radius s. β therefore depends on the dust grain's composition, shape, structure, and size but is generally proportional to $\frac{1}{\rho s}$, the small grains being easily pushed away by the radiation pressure, while the largest grains, typically 1μm in size and above, will tend to follow the orbit of the cometary nucleus around the Sun (*Burns et al.,* 1979).

The extent of cometary trails is evidence that cometary dust grains present a large range of sizes. Cometary dust size distributions are traditionally estimated by inverting the cometary tails (see e.g., *Finson and Probstein,* 1968a). While cometary dust size distributions are typically inverted bin size by bin size, they are canonically represented by a simpler power-law distribution as it represents well the properties of the observed clouds of dust. This is represented as a DSD with power index, N (also called α). N is related to γ, the power index of the mass distribution of particles by $N = -3\gamma-1$, assuming a constant density for the particles. If $N > -3$, both the mass and brightness depend on the largest ejected grains. Brightness and mass become decoupled if $-4 < N < -3$, in which case the dust mass depends on the largest ejected grains, while the brightness depends on the micrometer-sized grains (*Fulle,* 2004). Typical values for the mean power index, N, range from about −3 to about −4 [see Table 1 in *Fulle* (2004) for a review of groundbased derived DSD]. But one has to recognize that the actual dust size distributions in comets are more complex and can be quite variable with time and space due to outburst and activity (*Fulle,* 1987).

With the advent of space missions to active cometary nuclei, the groundbased observations and models of cometary dust size distributions of the coma are now complemented by direct measurements close to the nucleus or laboratory measurements from returned samples (see Table 2). However, those measurements cannot directly be compared as their relationship is complicated by fragmentation of particles, sublimation of volatiles, differential speed of ejection, and surface inhomogeneous activity (see *Agarwal et al.,* 2007, and references therein).

In situ data on the dust mass distribution was obtained for 1P/Halley by the Vega 1, Vega 2, and Giotto missions in 1986 (*Divine and Newburn,* 1988). *McDonnell et al.* (1987) used the Dust Impact Detection System (DIDSY) onboard Giotto to derive a double size distribution at the nucleus of 1P/Halley with $N = -4.06$ for small particles ($m < 10^{-8}$ kg) and $N = -3.13$ for larger particles ($m > 10^{-8}$ kg). The average DSD prior to close approach was estimated at the nucleus to be $N = -3.49 \pm 0.15$ (*McDonnell et al.,* 1986). *Fulle et al.* (1995) developed a model of 1P/Halley dust emissions to fit the DIDSY fluences and obtained a constant DSD index of $N = -3.5 \pm 0.2$ for grains larger than 20 μm. Later, combined models of optical and impact measurements

TABLE 2. Summary table of cometary dust size distribution indexes N retrieved by space missions and groundbased observations.

Comet	Instrument	DSD Index N	Reference
1P/Halley	DIDSY	-3.49 ± 0.15 (avg)	*McDonnell et al.* (1986)
1P/Halley	DIDSY	-4.06 ($<10^{-8}$ kg)	*McDonnell et al.* (1987)
1P/Halley	DIDSY	-3.13 ($>10^{-8}$ kg)	*McDonnell et al.* (1987)
1P/Halley	DIDSY	-3.5 ± 0.2 (>20 μm)	*Fulle et al.* (1995)
1P/Halley	OPE + DID	-2.6 ± 0.2 ($>10^{-12}$ kg)	*Fulle et al.* (2000)
9P/Tempel 1	IRAS	-3.0 ± 0.45 (pre-impact)	*Lisse et al.* (2005)
9P/Tempel 1	Spitzer	-4.5 ± 0.2 (post-impact)	*Lisse et al.* (2006)
26P/Grigg-Skjellerup	Groundbased	$-4.0 < N < -3.0$	*Fulle* (2004, and references therein)
26P/Grigg-Skjellerup	DIDSY	$-1.81^{+0.39}_{-0.6}$	*McDonnell et al.* (1993)
67P/Churyumov-Gerasimenko	Groundbased	-3.4 ± 0.2 (avg)	*Fulle* (2004, and references therein)
67P/Churyumov-Gerasimenko (far from perihelion)	Groundbased (tail)	−3	*Moreno et al.* (2017)
67P/Churyumov-Gerasimenko (around perihelion)	Groundbased (tail)	$-3.7 < N < -4.3$	*Moreno et al.* (2017)
67P/Churyumov-Gerasimenko	Groundbased (tail)	$-3.6 < N < -4.1$	*Moreno et al.* (2017)
81P/Wild 2	DFMI	−3.25 (closest approach)	*Tuzzolino et al.* (2004); *Green et al.* (2004)
81P/Wild 2	DFMI	$-1.99 < N < -4.39$	*Tuzzolino et al.* (2004); *Green et al.* (2004)
81P/Wild 2	Laboratory analysis	−2.72; −2.89	*Hörz et al.* (2006); *Price et al.* (2010)
103P/Hartley 2	ISOCAM	-3.2 ± 0.1	*Epifani et al.* (2001)
103P/Hartley 2	WISE/NEOWISE	-3.91 ± 0.3	*Bauer et al.* (2011)
103P/Hartley 2	Deep Impact photometry	$-6.6 < N < -4.7$	*Kelley et al.* (2013)
2I/Borisov	Groundbased	-3.7 ± 1.8	*Guzik et al.* (2020)
2I/Borisov	Groundbased	-4.0 ± 0.3	*Cremonese et al.* (2020)

A full table of N values determined from groundbased observations (range and averages) and models is available in *Fulle* (2004).

resulted in a value of N = −2.6 ± 0.2 (*Fulle et al.,* 2000). The interpretation of Halley data did not lead to a general agreement regarding the dust size distribution at the nucleus of the comet; however, it is generally admitted that the coma is dominated by millimeter-sized and larger particles (*Agarwal et al.,* 2007). The Giotto space mission continued its exploration of comets with a close flyby of Comet 26P/Grigg-Skjellerup in 1992. The dust distribution detected in that case corresponds to N = −1.81 for particles with mass larger than m > 10^{-9} kg, indicating for this comet a coma dominated by large particles (*McDonnell et al.,* 1993).

In 2005, the Deep Impact mission collided with Comet 9P/Tempel 1, excavating a crater about 150 m in size and several tens of meters deep, which helped decipher the composition and low strength of the near subsurface layers of cometary nuclei. The impact generated a strong ejection of fresh ice and dust particles that was akin to a cometary activity outburst (*A'Hearn et al.,* 2005). Similarly, the DSD index measured in the coma after the impact [N = −4.5 ± 0.2 (*Lisse et al.,* 2006)] demonstrates a strong increase in the DSD slope as compared to the pre-impact coma value [N = −3.0 ± 0.45 (*Lisse et al.,* 2005)], corresponding to a sharp increase in the number of small particles present in the coma liberated by the impact. Evidence from the changes in activity, gas composition, and dust particles' DSD were used to argue that pristine cometary material was present a few tens of meters below the surface of cometary nuclei (*A'Hearn et al.,* 2005). The Deep Impact mission was then diverted toward Comet 103P/Hartley 2. This very active small comet nucleus ejects very large particles, some of which are made of pure water ice (*A'Hearn et al.,* 2011). Observations from the ground indicated relatively steep DSD indexes, from −3.2 ± 0.1 (*Epifani et al.,* 2001) to −3.91 ± 0.3 (*Bauer et al.,* 2011). The *in situ* dust size distribution obtained by the space probe was even steeper, with a value ranging from −6.6 to −4.7, but applicable to the largest dust particle sizes detected by photometry, from 1 cm to 20 cm in the case of icy particles (*Kelley et al.,* 2013).

In 2006, the Stardust mission delivered samples from Comet 81P/Wild 2 to Earth for laboratory analysis (*Brownlee et al.,* 2006). During the comet flyby, the Dust Flux Monitor Instrument (DFMI) monitored the dust impacts with a high temporal resolution and gave evidence of fragmentation of dust particles within the coma. The size distributions detected by the probe are quite variable with a best-fit value of −3.25 for particle masses lower than 10^{-8} kg at around closest approach, but with values as high as −1.99 and as low as −4.39 depending on the coma region probed (*Tuzzolino et al.,* 2004; *Green et al.,* 2004). However, additional information could be retrieved from the laboratory analyses of the tracks in the aerogel and the craters on the aluminum foils of the return capsule, using calibrations made on Earth by impacting with analog compact particles. The resulting DSD index corresponds to −2.71 (*Hörz et al.,* 2006). Further recalibration on the ground taking into account impacts by aggregates of silica particles slightly revised this value to a lower one of N = −2.89 for particles larger than 10 μm, more compatible with the average value given by DFMI (*Price et al.,* 2010).

Finally, the Rosetta space mission was the only cometary space probe that could survey a comet nucleus over a major part of its orbital trajectory (*Glassmeier et al.,* 2007) for a period spanning 2.5 yr. The groundbased size distribution of 67P/C-G was determined to be N = −3.4 ± 0.2 from an average of previous observations since its discovery (*Fulle,* 2004, and references therein). During the Rosetta mission survey, the dust size distribution showed a strong time evolution, with the optical cross-section dominated by the largest ejected dust far from perihelion with N ≈ −3, while the smallest ejected dust dominates around perihelion with N ≈ −3.9 (*Moreno et al.,* 2017).

The Rosetta space mission also used its many dust analysis instruments [GIADA, COSIMA, OSIRIS, and the Rosetta Lander Imaging System (ROLIS)] to determine the dust size distribution in many complementary size ranges from 1 μm to 1 m. Figure 14 summarizes the results obtained for all these different measurements before the 2015 equinox, around the perihelion time in August 2015 and the size distribution of boulders on the surface of two landing areas studied for Philae. Overall, the DSD index measured by Rosetta is consistent with an average value around −4, with variations due to the timings of measurements with smaller particles typically ejected around perihelion time.

In addition, over the last five years, the solar system was visited by two interstellar objects (1I/Oumuamua and 2I/Borisov). Fortunately, interstellar Comet 2I/Borisov was active enough that its coma and tail could be studied by telescopic observations. In that case, the DSD is also ranging from −3.7 to −4 (see Table 2), which is consistent with most active comets and also "fresh" comets such as C/1995 O1 (Hale-Bopp) or C/1996 B2 (Hyakutake) (*Fulle,* 2004, and references therein).

In summary, cometary dust size distributions usually correspond well to power laws, with typical indexes ranging from −3 to −4. However large variations are detected depending on the activity of the nucleus, the timing of measurements, and the techniques used. The consistency of all measurements made so far are certainly consistent with probing the DSD of primordial building blocks of comets as the DSD index for active comets is consistent over all types of comets, including "fresh" ones, and even interstellar Comet 2I/Borisov.

Further analysis of cometary dust ejected at different times will certainly improve our knowledge of the dust size properties and how they may relate to primordial solar nebula materials.

3.2. Optical and Thermal Properties

Light scattering has historically been the main provider of information on the physical properties of cometary dust through telescopic studies [for more details, see *Kolokolova* (2015) and the chapter in this volume by Kolokolova et

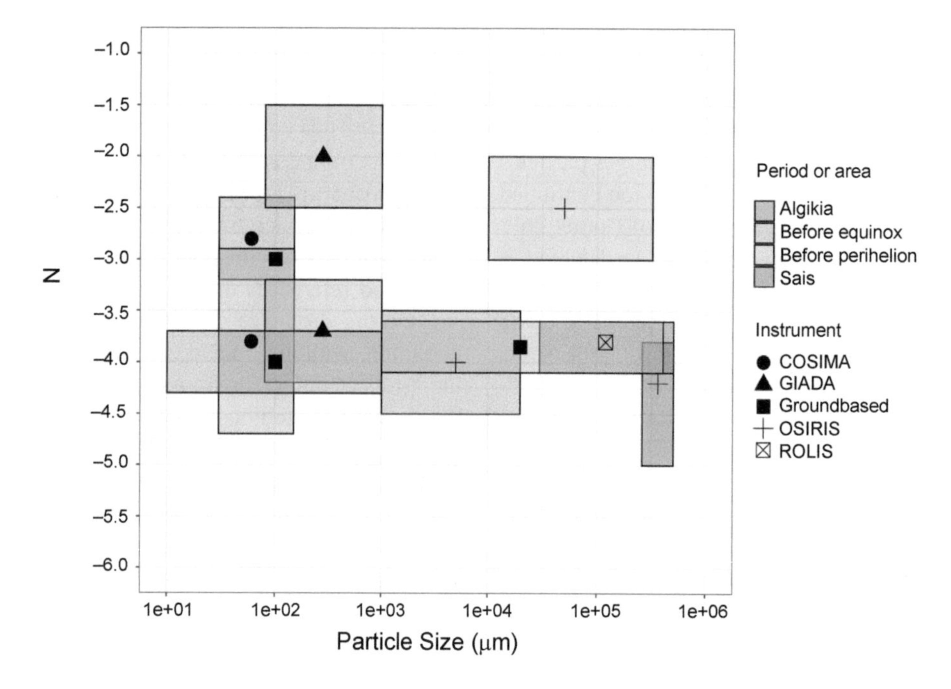

Fig. 14. Power index of the differential size distribution of 67P/C-G. Blue rectangles: power index before the 2015 equinox; green rectangles: power index around the 2015 perihelion. The ranges along the x-axis show the instrument size range sensitivity, the ranges along the y-axis are given by the uncertainty of the power index. [Data from *Merouane et al.* (2017), *Rotundi et al.* (2015), *Fulle et al.* (2016a), *Ott et al.* (2017), *Pajola et al.* (2017), and *Moreno et al.* (2017). Figure adapted from *Levasseur-Regourd et al.* (2018).] Agilkia was the initial landing site of Philae, and Sais was the final landing site of the Rosetta spacecraft.

al.]. The observations made in several domains have been useful in constraining the size, size distribution, structure, and optical indexes of the dust particles. The observations in the visible domain have been used to model the extension of the coma and tail of comets (*Haser et al.,* 2020; *Finson and Probstein,* 1968a,b) and deduce from it the surface activity of the nucleus. In the case of the dust particles in the coma of 67P/C-G, they present a specific scattering phase function with a u-shape and minimum at intermediate phase angles. This is different from the phase function that was usually considered for cometary dust (*Kolokolova et al.,* 2004). The color is consistent with the average of the nucleus surface below 30° of phase angles. There is negligible phase reddening at phase angles <90°, indicating a coma dominated by single scattering (*Bertini et al.,* 2017). Such a phase curve shape may be consistent with crushed primitive meteorites, but even more with analogs developed to simulate the scattering properties of IDPs (*Levasseur-Regourd et al.,* 2019). Photometric studies of single grains with the OSIRIS camera filters from 535 to 882 nm indicate slopes covering the ranges of slopes detected over the reddest to bluest regions of the nucleus (*Frattin et al.,* 2017). Assuming that the majority of dust particles in the zodiacal cloud come from comets, their average geometric albedo toward the Gegenschein has been determined to be 0.06 ± 0.01, similar to the low albedo detected for cometary nuclei (*Ishiguro et al.,* 2013).

The coma and tails of comets are astronomical objects that present some of the largest polarization detected in the solar system [for more details, see *Kolokolova* (2015) and the chapter in this volume by Kolokolova et al.]. Polarimetric observations of comets give complementary information to the scattering properties of the dust particles, in particular on their optical indexes and morphologies. Initial polarimetric observations of Comet Halley combined with Mie light scattering simulations and laboratory work comparisons already indicated that the scattering particles were likely large, rough with a low albedo, and it was shown that material from the Orgueil meteorite was a good scattering analog (*Kikuchi et al.,* 1988; *Mukai et al.,* 1988; *Dollfus,* 1989). It was also recognized that polarimetric properties of dust particles were significantly changed during outbursts (*Dollfus et al.,* 1988) and varied related to jet structures in the coma when the Giotto mission crossed them (*Levasseur-Regourd et al.,* 1999). Since then, improved models of cometary dust particles have been developed that may include a diversity of material mixtures (silicates, organics, ices, etc.) and morphologies, such as hollow spheres, irregular particles, spheroids, and aggregates thereof (see, e.g., *Hanner,* 2003; *Min,* 2005; *Lasue et al.,* 2009; *Kolokolova and Kimura,* 2010; *Zubko et al.,* 2015b, and references therein).

Polarimetric observations of 67P/C-G have been performed during both 2008 and 2015 perihelion passages (*Hadamcik et al.,* 2010, 2016; *Rosenbush et al.,* 2017). Agglomerates of submicrometer-sized grains best fit the higher polarization observed in cometary jets and after fragmentation or disruption events, while a mixture of porous agglomerates of submicrometer-sized Mg-silicates, Fe-silicates, and carbon black grains mixed with compact Mg-silicate grains is generally needed to fit whole comae observations (*Hadamcik et al.,* 2006, 2007). Simple geometric shapes of dust particles are generally poor fits to the observational cometary data (*Kolokolova et al.,* 2004). Numerical simulations strongly suggest that cometary dust is a mixture of (possibly fractal) agglomerates and of compact particles of both non-absorbing silicate-type materials and more absorbing organic-type materials (see, e.g., *Lasue et al.,* 2009; *Kiselev et al.,* 2015). The variety, structure, and size distribution of agglomerates and grains is consistent with the general description of dust particles detected at 67P/C-G by Rosetta (*Güttler et al.,* 2019; *Mannel et al.,* 2019).

Observations in the infrared and thermal wavelength ranges give specific information on the dust particle size distribution and their spatial distribution and dynamics in the cometary coma and tail (*Agarwal et al.,* 2007). In the particular case of 67P/C-G, VIRTIS observations of the dust in the coma from 2 to 5 μm generally show a temperature a few percent above the equilibrium, but it increases three- to fourfold during outbursts. This may be related to the ejection of much smaller dust particles during outbursts (size <100 nm) (*Bockelée-Morvan et al.,* 2017a). Such small particles are not detected by other Rosetta instruments, indicating a collecting bias or a dearth of such particles in the cometary environment.

A general model of light scattering and emission by dust particles consistent with all the observed constraints remains to be elaborated.

3.3. Morphology

The morphology of dust particles corresponds to the spatial arrangement of their constituting components and is often described by parameters such as the porosity, which indicates the ratio of voids to occupied volume, or the fractal dimension, a measure of the self-similarity at different scales of the assemblages. It is critical to study the morphology of cometary dust particles as these properties are essential to better understand the physical properties of the dust, such as strength and thermal and light scattering properties of the dust, but may also be witness to the primitive aggregation processes in the primordial nebula. Initially, polarimetric observations of comets have suggested that the basic morphology of cometary dust particles was best explained by including a combination of porous aggregates and more compact dust particles [see, e.g., *Lasue et al.* (2009), *Kolokolova* (2015), and the chapter by Kolokolova et al.]. As the Stardust samples were analyzed in the laboratory, the impacts on the aluminum foils and the aerogel demonstrated the presence of both compact and porous, easily fragmented dust particles (*Brownlee,* 2014). However, the samples were altered by the speed of collection reaching about 6 m s^{-1}. IDPs collected from the stratosphere also indicate a diversity of morphologies, including very porous fragile aggregates as well as more compact particles; however, the effect of atmospheric traverse, collection by plane, and unknown specific origin of the particles makes it difficult to relate their precise morphologies to a particular solar system body or physics phenomenon (see, e.g., *Brownlee* 2016).

To date, the microscopes onboard the Rosetta space mission have provided the best *in situ* morphological analysis of cometary dust particles collected at low speeds of 1 to 15 m s^{-1} (*Fulle et al.,* 2015) and distances from the nucleus lower than 500 km. The scale ranges accessible to both the atomic force microscope MIDAS and the microscope and mass spectrometer COSIMA were <1 μm to 1 mm with topographic information also available (*Bentley et al.,* 2016b; *Mannel et al.,* 2019; *Hilchenbach et al.,* 2016). These scales also correspond to the smallest [1 μm (*Bentley et al.,* 2016b; *Mannel et al.,* 2019)] and largest [350 μm (*Langevin et al.,* 2016)] particles detected. All particles present textural substructures identified as aggregated monomers, which generally classifies them as "compact aggregates" (*Güttler et al.,* 2019). Additionally, it has been shown that collected particles have a tendency to break up as they impact the plates and produce clusters of fragments with a variety of morphologies ranging from shattered, flattened particles to rubble piles (*Langevin et al.,* 2016). Such a diverse set of morphologies can originate from a single type of aggregates and be related to different incoming velocities and tensile strength of the particles (*Hornung et al.,* 2016; *Ellerbroek et al.,* 2017, 2019; *Lasue et al.,* 2019). In fact, many compact aggregates analyzed by COSIMA fragmented into smaller constituents during analysis due to electrostatic forces (*Hilchenbach et al.,* 2017), demonstrating the relationship between the fragments and their parent compact aggregate particles. Similar fragmentation upon collection have been shown to occur for MIDAS as well (*Bentley et al.,* 2016b). The surface features detected by the two microscopes onboard at their respective scales are reminiscent of those found in chondritic porous IDPs, as illustrated in Fig. 1.

The extension of the morphological analysis to the lowest scales accessible by MIDAS have shown that the smallest aggregate dust particles units are down to 8 nm in size with most of the dust particles being fragile agglomerated dust particles and small micrometer-sized dust particles also formed by those subunits (*Mannel et al.,* 2019). A fragile agglomerate was also determined to have a fractal dimension of about 1.7 (*Mannel et al.,* 2016) consistent with very fluffy dust particles measured by the impact GIADA instrument (*Fulle et al.,* 2015). Even though they do not present a significant fraction of the mass of the cometary nucleus, such fragile particles would not survive most impacts of the early solar system accretion phase and their detection favors an accretion of planetesimals under the gravitational collapse of pebble model (*Fulle and Blum,* 2017; *Blum et al.,* 2017).

In summary, we find that the cometary dust particles present an apparent scale invariance of properties similar to those that would result from a fractal aggregation process, consistent with the one that would be expected to be at work during the early stages of the planetary formation in the early solar system (see, e.g., *Blum* 2018). The results the microscopes obtained on dust collected from 67P/C-G together with many other *in situ* measurements by the Rosetta space mission provide the first view of the hierarchical structure of dust in comets as reviewed in *Güttler et al.* (2019).

3.4. Tensile Strength

Laboratory simulations of macroscopic agglomerates of small silica dust particles (diameters ranging from 0.1 to 10 μm) by ballistic deposition were realized to simulate early aggregation of dust particles similar to the ones forming comets. The tensile strength of such resulting aggregates, the size of which may reach several centimeters, range from 1 to 6 kPa (*Blum and Schräpler,* 2004; *Güttler et al.,* 2009; *Meisner et al.,* 2012). A more realistic set of simulations using silica dust particles and water ice particles under low temperatures (≈150 K) showed that the tensile strength decreases linearly with the particles' diameter, ranging from 4 kPa to 18 kPa, in agreement with previous estimates (*Gundlach et al.,* 2018). Additionally, the experiments demonstrated that under low temperatures, the tensile strength of water ice aggregates was comparable to the data for the silica spheres. This means that at low temperatures water ice presents a specific surface energy similar to the one of silica, which was not expected. Perhaps at temperatures above 150 K, the surface energy of water ice increases steeply, or sintering effects take place. Few direct measurements of the ejected solid material of comets are available; however, the tensile strength of dust particles ejected from 67P/C-G was estimated to be on the order of ≈1 kPa from the study of the fragments' distribution observed with the COSIMA experiment onboard Rosetta (*Hornung et al.,* 2016). Similarly, the meteor showers breakup observed in Earth's atmosphere can give estimates of cometary dust tensile strengths since they are associated with parent comets (*Jenniskens and Jenniskens,* 2006). The derived tensile strengths are again extremely low, and on the same order of magnitude as the other estimates depending on the parent comet: from 40 to 1000 Pa in *Trigo-Rodríguez and Llorca* (2006) and from 0.4 to 150 kPa as presented in Table 2 of *Blum et al.* (2014). A more detailed synthesis of tensile strength values for cometary materials at different scales of the cometary nucleus can be found in Fig. 10 of *Groussin et al.* (2019).

3.5. Density

The density of cometary dust particles is closely related to their morphology and their porosity, defined as 1 minus the volume-filling factor of the particle. As described in section 3.3, two main morphological types of dust particles have been detected from cometary ejection: compact dust particles and very fluffy aggregated dust particles following the classification made by *Güttler et al.* (2019). This was seen first by the Stardust samples brought back to Earth, where high-speed impacts of 81P/Wild 2 particles generated carrot-like aerogel tracks for compact particles (65% of tracks) and bulbous tracks consistent with the disruption of fluffy aggregates (35% of tracks) (*Brownlee,* 2014; *Burchell et al.,* 2008; *Trigo-Rodríguez et al.,* 2008).

These observations are consistent with the particle detections of GIADA for which compact and fluffy dust particles are detected (*Della Corte et al.,* 2015). The strength of the compact particles is consistent with a microporosity ranging from 34% to 85% (*Levasseur-Regourd et al.,* 2018). Dust showers observed by GIADA can only be explained by fractal aggregates with dimensions lower than 2 getting fragmented a few meters from the spacecraft (*Fulle et al.,* 2015); this represents about 30% of the dust detected (*Fulle and Blum,* 2017). MIDAS also detected an extremely porous particle with fractal dimension $D_f = 1.7 \pm 0.1$, which would translate to a porosity around 99% (*Mannel et al.,* 2016, 2019; *Fulle and Blum,* 2017). Following *Güttler et al.* (2019), fluffy aggregated particles are expected to have a microporosity >90%. Studies of the mean free path of light through particles fragments detected by COSIMA have also indicated a microporosity >50% (*Langevin et al.,* 2017).

The GIADA measurements combine both the geometric cross section of the particles and the momentum of impact, which allows determination of the average density of the particles. The value obtained over 271 compact particles detections gives $\rho = 785^{+520}_{-115}$ kg m^{-3}, at 1σ confidence level (*Fulle et al.,* 2017). With a dust microporosity estimated to be 59 ± 8%, this corresponds to a bulk density of compacted and dried dust of 1925^{+2030}_{-560} kg m^{-3}, at 1σ (*Fulle et al.,* 2017). Additionally, a significant fraction of dust particles detected by GIADA has a bulk density larger than 4000 kg m^{-3}. Those are interpreted as single-grain minerals similar to single mineral tracks in Stardust (*Burchell et al.,* 2008).

3.6. Electrical Properties

As dust particles are ejected from a comet, they get exposed to space plasma and UV radiations and become electrically charged, which influences their motion (*Horányi,* 1996). While there were evidence of particle charging in the coma of 1P/Halley from the calculations of particles trajectories (*Ellis and Neff,* 1991), the Rosetta mission provided the first direct evidence of electrically charged nanodust particles in a cometary coma (*Burch et al.,* 2015; *LLera et al.,* 2020). It is typically estimated that a 10^{-19}-kg dust particle will be disrupted by its charging if its tensile strength is less than about 0.5 MPa (*Mendis and Horányi,* 2013).

The dust showers detected by GIADA have been modeled in terms of millimeter-sized fluffy dust aggregates charged by the flux of secondary electrons from the spacecraft decreasing their electric potential by 7 to 15 V (*Fulle et al.,* 2015). The particles are disrupted by their interaction with the electric field of the spacecraft; their deceleration provides the appropriate kinetic energy to explain the Rosetta Plasma Consortium (RPC) Ion and Electron Sensor (IES) charged nanodust detections of 0.2 to 20 keV (*Fulle et al.,* 2016b).

The predicted fractal dimension of such fluffy aggregates is about 2, consistent with the fractal dimensions measured by MIDAS on some particles [$D_f = 1.7 \pm 0.1$ (*Mannel et al.,* 2016, 2019)].

During COSIMA TOF-SIMS measurements, partial charging of the collected dust particles allowed the direct determination of their bulk electrical properties like their specific resistivity ($>1.2 \times 10^{10}$ Ωm) and the real part of their relative electrical permittivity (<1.2) (*Hornung et al.,* 2020). These values are consistent with a dust porosity larger than 80%.

4. FUTURE DIRECTIONS

Analysis of the 81P/Wild 2 coma grains in particular has revealed how similar inorganic comet solids are to some asteroidal materials, at least for Comet Wild 2. The observation of crystalline silicates in Hale-Bopp, and the presence in the coma of Comet Wild 2 of a significant fraction of coarse-grained inorganic mineral grains, including very refractory materials, was largely unexpected, and has profoundly influenced models of early solar system dynamics. The apparent lack of a significant fraction of amorphous and presolar materials in Wild 2 was another surprise, although the collection process in aerogel may have significantly destroyed these materials. The nature of cometary organics was not a major goal of the Stardust mission, and thus great uncertainty remains for this topic. The Rosetta mission contributed to the analysis of cometary organics, confirming that cometary particles can contain a large proportion of organics (as first seen in the CHON grains in Comet Halley), and that this organic matter bears some similarities with the refractory organics present in carbonaceous chondrites. The organic matter present in CA-IDPs and in CP-MMs also bear similarities with that of carbonaceous chondrites, as for two of the three organic phases identified in UCAMMs. The formation of the N-rich organic phase in UCAMMs could have been made by irradiation of N-rich ices in the outer regions of the protoplanetary disk by galactic cosmic rays. The formation and incorporation mechanisms of mineral and organic components in comets are not yet fully understood. Cometary dust (or the ice that initially contained it) also carries soluble organics like amino acids. Cometary dust thus could have contributed to the input of prebiotic matter on the early Earth. The presence of a "continuum" between asteroidal and cometary matter is now considered seriously, asteroidal components being found in cometary material, and cometary activity being observed in asteroids (the so-called "main-belt comets").

Thus, despite great recent progress in our understanding of comets, critical gaps remain concerning the formation and processing of organics, condensation of inorganic volatiles, nature and role of presolar dust in the evolution of early nebular solids, timing and location of condensation and processing of cometary materials, possible role of radiogenic nuclides including ^{26}Al, etc. The comparison between properties of cometary dust deduced by astronomical observations or analyzed by different techniques also sometimes shows discrepancies. In order to address these issues, a cryogenic sample return from a comet nucleus would be a dream, with well-preserved cometary samples available for analysis in terrestrial laboratories (*Bockelée-Morvan et al.,* 2019). For the time being, it however remains just that — a dream.

Acknowledgments. The micrometeorite collections are performed thanks to the logistic support of the French and Italian Polar institutes (IPEV and PNRA). C.E. acknowledges the support in France of ANR (COMETOR ANR-18-CE31-0011), CNRS, IN2P3, LabEx P2IO, DIM-ACAV+ and CNES (Rosetta and MIAMI-H2). J.L. acknowledges the support in France of the Programme National de Planétologie (PNP) of CNRS/INSU, and of CNES for the Rosetta mission. D.H.W. thanks D. E. Harker for sharing thermal models prior to publication in *Harker et al.* (2023) as well as that article's co-authors, C. E. Woodward and M. S. P. Kelley, regarding many subtleties and details discussed here; T. Ootsubo for sharing data on 21P/G-Z; and NASA Ames Space Science and Astrobiology Division for support for time on this chapter. M.E.Z. thanks NASA for supporting astromaterials research and curation, and the Cosmic Dust Program and Stardust Mission. The authors thank the two reviewers, D. Baklouti and B. T. De Gregorio, for their helpful and constructive comments.

REFERENCES

Agarwal J., Müller M., and Grün E. (2007) Dust environment modelling of Comet 67P/Churyumov-Gerasimenko. *Space Sci. Rev., 128,* 79–131.

A'Hearn M. F., Belton M. J. S., Delamere W. A., Kissel J., Klaasen K. P., McFadden L. A., Meech K. J., Melosh H. J., Schultz P. H., Sunshine J. M., Thomas P. C., Veverka J., Yeomans D. K., Baca M. W., Busko I., Crockett C. J., Collins S. M., Desnoyer M., Eberhardy C. A., Ernst C. M., Farnham T. L., Feaga L., Groussin O., Hampton D., Ipatov S. I., Li J.-Y., Lindler D., Lisse C. M., Mastrodemos N., Owen W. M., Richardson J. E., Wellnitz D. D., and White R. L. (2005) Deep Impact: Excavating Comet Tempel 1. *Science, 310,* 258–264.

A'Hearn M. F., Belton M. J. S., Delamere W. A., Feaga L. M., Hampton D., Kissel J., Klaasen K. P., McFadden L. A., Meech K. J., Melosh H. J., Schultz P. H., Sunshine J. M., Thomas P. C., Veverka J., Wellnitz D. D., Yeomans D. K., Besse S., Bodewits D., Bowling T. J., Carcich B. T., Collins S. M., Farnham T. L., Groussin O., Hermalyn B., Kelley M. S., Kelley M. S., Li J.-Y., Lindler D. J., Lisse C. M., McLaughlin S. A., Merlin F., Protopapa S., Richardson J. E., and Williams J. L. (2011) EPOXI at Comet Hartley 2. *Science, 332,* 1396–1400.

A'Hearn M. F., Bodewits D., La Forgia F., Lara L. M., Agarwal J., Bertaux J.-L., Cremonese G., Davidsson B., Fornasier S., Güttler C., Knollenberg J., Lazzarin M., Leyrat C., Lin Z.-Y., Magrin S., Naletto G., Sierks H., Snodgrass C., Thomas N., Tubiana C., and Vincent J.-B. (2015) Rotational variation of outgassing morphology in 67P/Churyumov-Gerasimenko. *IAU General Assembly Abstracts, 29,* 2257858.

Aléon J., Engrand C., Robert F., and Chaussidon M. (2001) Clues to the origin of interplanetary dust particles from the isotopic study of their hydrogen-bearing phases. *Geochim. Cosmochim. Acta, 65,* 4399–4412.

Aléon J., Robert F., Chaussidon M., and Marty B. (2003) Nitrogen isotopic composition of macromolecular organic matter in interplanetary dust particles. *Geochim. Cosmochim. Acta, 67,* 3773–3783.

Aléon J., Engrand C., Leshin L. A., and McKeegan K. D. (2009) Oxygen isotopic composition of chondritic interplanetary dust particles: A genetic link between carbonaceous chondrites and

comets. *Geochim. Cosmochim. Acta, 73,* 4558–4575.
Alexander C. M. O. D., Fogel M., Yabuta H., and Cody G. D. (2007) The origin and evolution of chondrites recorded in the elemental and isotopic compositions of their macromolecular organic matter. *Geochim. Cosmochim. Acta, 71,* 4380–4403.
Alexander C. M. O. D., Bowden R., Fogel M. L., Howard K. T., Herd C. D. K., and Nittler L. R. (2012) The provenances of asteroids, and their contributions to the volatile inventories of the terrestrial planets. *Science, 337,* 721–723.
Alexander C. M. O. D., Cody G. D., De Gregorio B. T., Nittler L. R., and Stroud R. M. (2017) The nature, origin and modification of insoluble organic matter in chondrites, the major source of Earth's C and N. *Chem. Erde, 77,* 227–256.
Altwegg K., Balsiger H., Bar-Nun A., Berthelier J. J., Bieler A., Bochsler P., Briois C., Calmonte U., Combi M., De Keyser J., Eberhardt P., Fiethe B., Fuselier S., Gasc S., Gombosi T. I., Hansen K., Hässig M., Jäckel A., Kopp E., Korth A., LeRoy L., Mall U., Marty B., Mousis O., Neefs E., Owen T., Rème H., Rubin M., Sémon T., Tzou C.-Y., Waite H., and Wurz P. (2015) 67P/Churyumov-Gerasimenko, a Jupiter family comet with a high D/H ratio. *Science, 347,* 1261952.
Altwegg K., Balsiger H., Bar-Nun A., Berthelier J.-J., Bieler A., Bochsler P., Briois C., Calmonte U., Combi M. R., Cottin H., Keyser J. D., Dhooghe F., Fiethe B., Fuselier S. A., Gasc S., Gombosi T. I., Hansen K. C., Haessig M., Jäckel A., Kopp E., Korth A., Roy L. L., Mall U., Marty B., Mousis O., Owen T., Rème H., Rubin M., Sémon T., Tzou C.-Y., Waite J. H., and Wurz P. (2016) Prebiotic chemicals — amino acid and phosphorus — in the coma of Comet 67P/Churyumov-Gerasimenko. *Sci. Adv., 2,* e1600285.
Altwegg K., Balsiger H., Berthelier J., Bieler A., Calmonte U., Fuselier S., Goesmann F., Gasc S., Gombosi T. I., Le Roy L., de Keyser J., Morse A., Rubin M., Schuhmann M., Taylor M. G. G. T., Tzou C.-Y., and Wright I. (2017) Organics in Comet 67P — A first comparative analysis of mass spectra from ROSINA-DFMS, COSAC and Ptolemy. *Mon. Not. R. Astron. Soc., 469,* S130–S141.
Altwegg K., Balsiger H., Combi M., De Keyser J., Drozdovskaya M. N., Fuselier S. A., Gombosi T. I., Hänni N., Rubin M., Schuhmann M., Schroeder I., and Wampfler S. (2020a) Molecule-dependent oxygen isotopic ratios in the coma of Comet 67P/Churyumov-Gerasimenko. *Mon. Not. R. Astron. Soc., 498,* 5855–5862.
Altwegg K., Balsiger H., Hänni N., Rubin M., Schuhmann M., Schroeder I., Sémon T., Wampfler S., Berthelier J.-J., Briois C., Combi M., Gombosi T. I., Cottin H., De Keyser J., Dhooghe F., Fiethe B., and Fuselier S. A. (2020b) Evidence of ammonium salts in Comet 67P as explanation for the nitrogen depletion in cometary comae. *Nature Astron., 4,* 533–540.
Altwegg K., Combi M., Fuselier S. A., Hänni N., De Keyser J., Mahjoub A., Müller D. R., Pestoni B., Rubin M., and Wampfler S. F. (2022) Abundant ammonium hydrosulphide embedded in cometary dust grains. *Mon. Not. R. Astron. Soc., 516,* 3900–3910.
Amari S., Hoppe P., Zinner E., and Lewis R. S. (1993) The isotopic compositions and stellar sources of meteoritic graphite grains. *Nature, 365,* 806–809.
Astrochem.org (2022) The Astrophysics and Astrochemistry Laboratory. Available online at *https://astrochem.org/*.
Augé B., Dartois E., Engrand C., Duprat J., Godard M., Delauche L., Bardin N., Mejía C., Martinez R., Muniz G., Domaracka A., Boduch P., and Rothard H. (2016) Irradiation of nitrogen-rich ices by swift heavy ions — Clues for the formation of ultracarbonaceous micrometeorites. *Astron. Astrophys., 592,* A99.
Augé B., Dartois E., Duprat J., Engrand C., Slodzian G., Wu T. D., Guerquin-Kern J. L., Vermesse H., Agnihotri A. N., Boduch P., and Rothard H. (2019) Hydrogen isotopic anomalies in extraterrestrial organic matter: Role of cosmic ray irradiation and implications for UCAMMs. *Astron. Astrophys., 627,* A122.
Bardyn A., Baklouti D., Cottin H., Fray N., Briois C., Paquette J., Stenzel O., Engrand C., Fischer H., Hornung K., Isnard R., Langevin Y., Lehto H., Le Roy L., Ligier N., Merouane S., Modica P., Orthous-Daunay F.-R., Rynö J., Schulz R., Silén J., Thirkell L., Varmuza K., Zaprudin B., Kissel J., and Hilchenbach M. (2017) Carbon-rich dust in Comet 67P/Churyumov-Gerasimenko measured by COSIMA/Rosetta. *Mon. Not. R. Astron. Soc., 469,* S712–S722.
Bastien R., Broce S., Brown P., Burkett P. J., Campbell-Brown M., Frank D., Gearheart D., Kapitzke M., Moes T., Rodriguez M., Steel D., Williams T., and Zolensky M. E. (2013) The 2012 Draconid storm as observed by the Canadian Meteor Orbit Radar and potentially sampled by ER-2 aircraft. In *44th Lunar and Planetary Science Conference,* Abstract #1622. LPI Contribution No. 1719, Lunar and Planetary Institute, Houston.
Bauer J. M., Walker R. G., Mainzer A. K., Masiero J. R., Grav T., Dailey J. W., McMillan R. S., Lisse C. M., Fernández Y. R., Meech K. J., Pittichova J., Blauvelt E. K., Masci F. J., A'Hearn M. F., Cutri R. M., Scotti J. V., Tholen D. J., DeBaun E., Wilkins A., Hand E., Wright E. L., and the WISE Team (2011) WISE/NEOWISE observations of Comet 103P/Hartley 2. *Astrophys. J., 738,* 171.
Begemann B., Dorschner J., Henning T., Mutschke H., and Thamm E. (1994) A laboratory approach to the interstellar sulfide dust problem. *Astrophys. J. Lett., 423,* L71–L74.
Benna M., Mahaffy P. R., Grebowsky J. M., Plane J. M. C., Yelle R. V., and Jakosky B. M. (2015) Metallic ions in the upper atmosphere of mars from the passage of Comet C/2013 A1 (Siding Spring). *Geophys. Res. Lett., 42,* 4670–4675.
Bentley M. S., Arends H., Butler B., Gavira J., Jeszenszky H., Mannel T., Romstedt J., Schmied R., and Torkar K. (2016a) MIDAS: Lessons learned from the first spaceborne atomic force microscope. *Acta Astronaut., 125,* 11–21.
Bentley M. S., Schmied R., Mannel T., Torkar K., Jeszenszky H., Romstedt J., Levasseur-Regourd A.-C., Weber I., Jessberger E. K., and Ehrenfreund P. (2016b) Aggregate dust particles at Comet 67P/Churyumov-Gerasimenko. *Nature, 537,* 73–75.
Berger E. L., Zega T. J., Keller L. P., and Lauretta D. S. (2011) Evidence for aqueous activity on Comet 81P/Wild 2 from sulfide mineral assemblages in Stardust samples and CI chondrites. *Geochim. Cosmochim. Acta, 75,* 3501–3513.
Bergin E. A., Blake G. A., Ciesla F., Hirschmann M. M., and Li J. (2015) Tracing the ingredients for a habitable Earth from interstellar space through planet formation. *Proc. Natl. Acad. Sci. U.S.A., 112,* 8965–8970.
Bertini I., La Forgia F., Tubiana C., Güttler C., Fulle M., Moreno F., Frattin E., Kovacs G., Pajola M., Sierks H., Barbieri C., Lamy P., Rodrigo R., Koschny D., Rickman H., Keller H. U., Agarwal J., A'Hearn M. F., Barucci M. A., Bertaux J.-L., Bodewits D., Cremonese G., Da Deppo V., Davidsson B., Debei S., De Cecco M., Drolshagen E., Ferrari S., Ferri F., Fornasier S., Gicquel A., Groussin O., Gutierrez P. J., Hasselmann P. H., Hviid S. F., Ip W.-H., Jorda L., Knollenberg J., Kramm J. R., Kührt E., Küppers M., Lara L. M., Lazzarin M., Lin Z.-Y., Moreno J. J. L., Lucchetti A., Marzari F., Massironi M., Mottola S., Naletto G., Oklay N., Ott T., Penasa L., Thomas N., and Vincent J.-B. (2017) The scattering phase function of Comet 67P/Churyumov-Gerasimenko coma as seen from the Rosetta/OSIRIS instrument. *Mon. Not. R. Astron. Soc., 469,* S404–S415.
Bessell M. S. (2005) Standard photometric systems. *Annu. Rev. Astron. Astrophys., 43,* 293–336.
Blum J. (2018) Dust evolution in protoplanetary discs and the formation of planetesimals. *Space Sci. Rev., 214,* 52.
Blum J. and Schräpler R. (2004) Structure and mechanical properties of high-porosity macroscopic agglomerates formed by random ballistic deposition. *Phys. Rev. Lett., 93,* 115503.
Blum J., Gundlach B., Mühle S., and Trigo-Rodriguez J. M. (2014) Comets formed in solar-nebula instabilities! — An experimental and modeling attempt to relate the activity of comets to their formation process. *Icarus, 235,* 156–169.
Blum J., Gundlach B., Krause M., Fulle M., Johansen A., Agarwal J., Von Borstel I., Shi X., Hu X., and Bentley M. S. (2017) Evidence for the formation of Comet 67P/Churyumov-Gerasimenko through gravitational collapse of a bound clump of pebbles. *Mon. Not. R. Astron. Soc., 469,* S755–S773.
Bockelée-Morvan D., Brooke T. Y., and Crovisier J. (1995) On the origin of the 3.2 to 3.6-micron emission features in comets. *Icarus, 116,* 18–39.
Bockelée-Morvan D., Calmonte U., Charnley S., Duprat J., Engrand C., Gicquel A., Hässig M., Jehin E., Kawakita H., Marty B., Milam S., Morse A., Rousselot P., Sheridan S., and Wirström E. (2015) Cometary isotopic measurements. *Space Sci. Rev., 197,* 47–83.
Bockelée-Morvan D., Rinaldi G., Erard S., Leyrat C., Capaccioni F., Drossart P., Filacchione G., Migliorini A., Quirico E., Mottola S., Tozzi G., Arnold G., Biver N., Combes M., Crovisier J., Longobardo A., Blecka M., and Capria M. T. (2017a) Comet 67P outbursts and quiescent coma at 1.3 au from the Sun: Dust properties from Rosetta/VIRTIS-H observations. *Mon. Not. R. Astron. Soc., 469,* S443–S458.

Bockelée-Morvan D., Rinaldi G., Erard S., Leyrat C., Capaccioni F., Drossart P., Filacchione G., Migliorini A., Quirico E., Mottola S., Tozzi G., Arnold G., Biver N., Combes M., Crovisier J., Longobardo A., Blecka M., and Capria M. T. (2017b) Erratum: Comet 67P outbursts and quiescent coma at 1.3 au from the Sun: Dust properties from Rosetta/VIRTIS-H observations. *Mon. Not. R. Astron. Soc., 469,* S842–S843.

Bockelée-Morvan D., Filacchione G., Altwegg K., Bianchi E., Bizzarro M., Blum J., Bonal L., Capaccioni F., Codella C., Choukroun M., Cottin H., Davidsson B., De Sanctis M. C., Drozdovskaya M., Engrand C., Galand M., Güttler C., Henri P., Herique A., Ivanoski S., Kokotanekova R., Levasseur-Regourd A.-C., Miller K. E., Rotundi A., Schönbächler M., Snodgrass C., Thomas N., Tubiana C., Ulamec S., and Vincent J. B. (2019) AMBITION — Comet Nucleus Cryogenic Sample Return: A white paper for ESA Voyage 2050 long-term plan. *ArXiV e-prints,* arXiv:1907.11081.

Bohren C. F. and Huffman D. R. (1983) *Absorption and Scattering of Light by Small Particles.* Wiley, New York. 541 pp.

Bonal L., Quirico E., Montagnac G., and Reynard B. (2006) Interplanetary dust particles: Organic matter studied by Raman spectroscopy and laser induced fluorescence. In *37th Annual Lunar and Planetary Science Conference,* Abstract #2271. LPI Contribution No. 1303, Lunar and Planetary Institute, Houston.

Bonev B. P., Dello Russo N., DiSanti M. A., Martin E. C., Doppmann G., Vervack J., Ronald J., Villanueva G. L., Kawakita H., Gibb E. L., Combi M. R., Roth N. X., Saki M., McKay A. J., Cordiner M. A., Bodewits D., Crovisier J., Biver N., Cochran A. L., Shou Y., Khan Y., and Venkataramani K. (2021) First comet observations with NIRSPEC-2 at Keck: Outgassing sources of parent volatiles and abundances based on alternative taxonomic compositional baselines in 46P/Wirtanen. *Planet. Sci. J., 2,* 45.

Bradley J. (1994a) Mechanisms of grain formation, post-accretional alteration, and likely parent body environments of interplanetary dust particles (IDPs). In *Analysis of Interplanetary Dust* (M. Zolensky et al., eds.), pp. 89–104. AIP Conf. Ser. 310, American Institute of Physics, Melville, New York.

Bradley J. P. (1994b) Chemically anomalous preaccretionally irradiated grains in interplanetary dust from comets. *Science, 265,* 925–929.

Bradley J. P. (1994c) Nanometer-scale mineralogy and petrography of fine-grained aggregates in anhydrous interplanetary dust particles. *Geochim. Cosmochim. Acta, 58,* 2123–2134.

Bradley J. P. (2013) How and where did GEMS form? *Geochim. Cosmochim. Acta, 107,* 336–340.

Bradley J. P. and Brownlee D. E. (1986) Cometary particles — Thin sectioning and electron beam analysis. *Science, 231,* 1542–1544.

Bradley J. P., Germani M. S., and Brownlee D. E. (1989) Automated thin-film analyses of anhydrous interplanetary dust particles in the analytical electron microscope. *Earth Planet. Sci. Lett., 93,* 1–13.

Bradley J. P., Humecki H. J., and Germani M. S. (1992) Combined infrared and analytical electron microscope studies of interplanetary dust particles. *Astrophys. J., 394,* 643–651.

Bradley J. P., Keller L. P., Snow T. P., Hanner M. S., Flynn G. J., Gezo J. C., Clemett S. J., Brownlee D. E., and Bowey J. E. (1999) An infrared spectral match between GEMS and interstellar grains. *Science, 285,* 1716–1718.

Bradley J. P., Holland H. D., and Turekian K. K. (2014) Early solar nebula grains — Interplanetary dust particles. In *Treatise on Geochemistry, Vol. 1, Second Edition: Meteorites and Cosmochemical Processes* (D. Heinrich et al., eds.), pp. 287–308. Elsevier, Oxford.

Bradley J. P., Ishii H. A., Bustillo K., Ciston J., Ogliore R., Stephan T., Brownlee D. E., and Joswiak D. J. (2022) On the provenance of GEMS, a quarter century post discovery. *Geochim. Cosmochim. Acta, 335,* 323–338.

Bregman J. D., Witteborn F. C., Allamandola L. J., Campins H., Wooden D. H., Rank D. M., Cohen M., and Tielens A. G. G. M. (1987) Airborne and groundbased spectrophotometry of Comet P/Halley from 5–13 micrometers. *Astron. Astrophys., 187,* 616–620.

Bridges J. C., Changela H. G., Nayakshin S., Starkey N. A., and Franchi I. A. (2012) Chondrule fragments from Comet Wild 2: Evidence for high temperature processing in the outer solar system. *Earth Planet. Sci. Lett., 341–344,* 186–194.

Bridges J. C., Hicks L. J., MacArthur J. L., Price M. C., Burchell M. J., Franchi I. A., and Gurman S. J. (2015) Magnetite in Stardust terminal grains: Evidence for hydrous alteration in the Wild 2 parent body. *European Planet. Sci. Congr., 2015,* EPSC2015-866.

Bromley S. J., Neff B., Loch S. D., Marler J. P., Országh J., Venkataramani K., and Bodewits D. (2021) Atomic iron and nickel in the coma of C/1996 B2 (Hyakutake): Production rates, emission mechanisms, and possible parents. *Planet. Sci. J., 2,* 228.

Brownlee D. (2014) The Stardust mission: Analyzing samples from the edge of the solar system. *Annu. Rev. Earth Planet. Sci., 42,* 179–205.

Brownlee D. E. (2016) Cosmic dust: Building blocks of planets falling from the sky. *Elements, 12,* 165–170.

Brownlee D. E. and Joswiak D. J. (2017) Diversity of the initial rocky planetary building materials at the edge of the solar system. *Meteorit. Planet. Sci., 52,* 471–478.

Brownlee D. E., Tomandl D. A., and Olszewski E. (1977) Interplanetary dust — A new source of extraterrestrial material for laboratory studies. In *Proceedings of the 8th Lunar and Planetary Science Conference,* pp. 149–160. Pergamon, New York.

Brownlee D., Joswiak D., Schlutter D., Pepin R., Bradley J., and Love S. (1995) Identification of individual cometary IDPs by thermally stepped He release. In *26th Lunar and Planetary Science Conference,* pp. 183–184. LPI Contribution No. 1592, Lunar and Planetary Institute, Houston.

Brownlee D., Tsou P., Aléon J., Alexander C. M. O., Araki T., Bajt S., Baratta G. A., Bastien R., Bland P., Bleuet P., Borg J., Bradley J. P., Brearley A., Brenker F., Brennan S., Bridges J. C., Browning N. D., Brucato J. R., Bullock E., Burchell M. J., Busemann H., Butterworth A., Chaussidon M., Cheuvront A., Chi M., Cintala M. J., Clark B. C., Clemett S. J., Cody G., Colangeli L., Cooper G., Cordier P., Daghlian C., Dai Z., d'Hendecourt L., Djouadi Z., Dominguez G., Duxbury T., Dworkin J. P., Ebel D. S., Economou T. E., Fakra S., Fairey S. A. J., Fallon S., Ferrini G., Ferroir T., Fleckenstein H., Floss C., Flynn G., Franchi I. A., Fries M., Gainsforth Z., Gallien J.-P., Genge M., Gilles M. K., Gillet P., Gilmour J., Glavin D. P., Gounelle M., Grady M. M., Graham G. A., Grant P. G., Green S. F., Grossemy F., Grossman L., Grossman J. N., Guan Y., Hagiya K., Harvey R., Heck P., Herzog G. F., Hoppe P., Hörz F., Huth J., Hutcheon I. D., Ignatyev K., Ishii H., Ito M., Jacob D., Jacobsen C., Jacobsen S., Jones S., Joswiak D., Jurewicz A., Kearsley A. T., Keller L. P., Khodja H., Kilcoyne A. D., Kissel J., Krot A., Langenhorst F., Lanzirotti A., Le L., Leshin L. A., Leitner J., Lemelle L., Leroux H., Liu M.-C., Luening K., Lyon I., MacPherson G., Marcus M. A., Marhas K., Marty B., Matrajt G., McKeegan K., Meibom A., Mennella V., Messenger K., Messenger S., Mikouchi T., Mostefaoui S., Nakamura T., Nakano T., Newville M., Nittler L. R., Ohnishi I., Ohsumi K., Okudaira K., Papanastassiou D. A., Palma R., Palumbo M. E., Pepin R. O., Perkins D., Perronnet M., Pianetta P., Rao W., Rietmeijer F. J. M., Robert F., Rost D., Rotundi A., Ryan R., Sandford S. A., Schwandt C. S., See T. H., Schlutter D., Sheffield-Parker J., Simionovici A., Simon S., Sitnitsky I., Snead C. J., Spencer M. K., Stadermann F. J., Steele A., Stephan T., Stroud R., Susini J., Sutton S. R., Suzuki Y., Taheri M., Taylor S., Teslich N., Tomeoka K., Tomioka N., Toppani A., Trigo-Rodríguez J. M., Troadec D., Tsuchiyama A., Tuzzolino A. J., Tyliszczak T., Uesugi K., Velbel M., Vellenga J., Vicenzi E., Vincze L., Warren J., Weber I., Weisberg M., Westphal A. J., Wirick S., Wooden D., Wopenka B., Wozniakiewicz P., Wright I., Yabuta H., Yano H., Young E. D., Zare R. N., Zega T., Ziegler K., Zimmerman L., Zinner E., and Zolensky M. (2006) Comet 81P/Wild 2 Under a microscope. *Science, 314,* 1711–1716.

Brucato J. R., Strazzulla G., Baratta G., and Colangeli L. (2004) Forsterite amorphisation by ion irradiation: Monitoring by infrared spectroscopy. *Astron. Astrophys., 413,* 395–401.

Brunetto R., Borg J., Dartois E., Rietmeijer F. J. M., Grossemy F., Sandt C., Le Sergeant d'Hendecourt L., Rotundi A., Dumas P., Djouadi Z., and Jamme F. (2011) Mid-IR, far-IR, Raman micro-spectroscopy, and FESEM-EDX study of IDP L2021C5: Clues to its origin. *Icarus, 212,* 896–910.

Burch J. L., Gombosi T. I., Clark G., Mokashi P., and Goldstein R. (2015) Observation of charged nanograins at Comet 67P/Churyumov-Gerasimenko. *Geophys. Res. Lett., 42,* 6575–6581.

Burchell M. J., Fairey S. A. J., Wozniakiewicz P., Brownlee D. E., Hoerz F., Kearsley A. T., See T. H., Tsou P., Westphal A., Green S. F., Trigo-Rodríguez J. M., and Dominguez G. (2008) Characteristics of cometary dust tracks in stardust aerogel and laboratory calibrations. *Meteorit. Planet. Sci., 43,* 23–40.

Burns J. A., Lamy P. L., and Soter S. (1979) Radiation forces on small particles in the solar system. *Icarus, 40,* 1–48.

Busemann H., Alexander C. M. O., and Nittler L. R. (2007)

Characterization of insoluble organic matter in primitive meteorites by microRaman spectroscopy. *Meteorit. Planet. Sci., 42,* 1387–1416.
Busemann H., Nguyen A. N., Cody G. D., Hoppe P., Kilcoyne A. L. D., Stroud R. M., Zega T. J., and Nittler L. R. (2009) Ultra-primitive interplanetary dust particles from the Comet 26P/Grigg-Skjellerup dust stream collection. *Earth Planet. Sci. Lett., 288,* 44–57.
Campins H. and Ryan E. V. (1989) The identification of crystalline olivine in cometary silicates. *Astrophys. J., 341,* 1059–1066.
Capaccioni F., Coradini A., Filacchione G., Erard S., Arnold G., Drossart P., Sanctis M. C. D., Bockelee-Morvan D., Capria M. T., Tosi F., Leyrat C., Schmitt B., Quirico E., Cerroni P., Mennella V., Raponi A., Ciarniello M., McCord T., Moroz L., Palomba E., Ammannito E., Barucci M. A., Bellucci G., Benkhoff J., Bibring J. P., Blanco A., Blecka M., Carlson R., Carsenty U., Colangeli L., Combes M., Combi M., Crovisier J., Encrenaz T., Federico C., Fink U., Fonti S., Ip W.-H., Irwin P., Jaumann R., Kuehrt E., Langevin Y., Magni G., Mottola S., Orofino V., Palumbo P., Piccioni G., Schade U., Taylor F., Tiphene D., Tozzi G. P., Beck P., Biver N., Bonal L., Combe J.-P., Despan D., Flamini E., Fornasier S., Frigeri A., Grassi D., Gudipati M., Longobardo A., Markus K., Merlin F., Orosei R., Rinaldi G., Stephan K., Cartacci M., Cicchetti A., Giuppi S., Hello Y., Henry F., Jacquinod S., Noschese R., Peter G., Politi R., Reess J. M., and Semery A. (2015) The organic-rich surface of Comet 67P/Churyumov-Gerasimenko as seen by VIRTIS/Rosetta. *Science, 347,* aaa0628.
Carrillo-Sánchez J. D., Nesvorný D., Pokorný P., Janches D., and Plane J. M. C. (2016) Sources of cosmic dust in the Earth's atmosphere. *Geophys. Res. Lett., 43,* 11979–11986.
Charon E., Engrand C., Benzerara K., Leroux H., Swaraj S., Belkhou R., Duprat J., Dartois E., Godard M., and Delauche L. (2017) A C-, N-, O-XANES/STXM and TEM study of organic matter and minerals in ultracarbonaceous Antarctic micrometeorites (UCAMMs). In *48th Annual Lunar and Planetary Science Conference,* Abstract #2085. LPI Contribution No. 1964, Lunar and Planetary Institute, Houston.
Chi M., Ishii H. A., Simon S. B., Bradley J. P., Dai Z., Joswiak D., Browning N. D., and Matrajt G. (2009) The origin of refractory minerals in Comet 81P/Wild 2. *Geochim. Cosmochim. Acta, 73,* 7150–7161.
Chihara H., Koike C., and Tsuchiyama A. (2001) Low-temperature optical properties of silicate particles in the far-infrared region. *Publ. Astron. Soc. Japan, 53,* 243–250.
Chihara H., Koike C., Tsuchiyama A., Tachibana S., and Sakamoto D. (2002) Compositional dependence of infrared absorption spectra of crystalline silicates. I. Mg-Fe pyroxenes. *Astron. Astrophys., 391,* 267–273.
Clairemidi J., Moreels G., Mousis O., and Bréchignac P. (2008) Identification of anthracene in Comet 1P/Halley. *Astron. Astrophys., 492,* 245–250.
Clark B. C., Green S. F., Economou T. E., Sandford S. A., Zolensky M. E., McBride N., and Brownlee D. E. (2004) Release and fragmentation of aggregates to produce heterogeneous, lumpy coma streams. *J. Geophys. Res.–Planets, 109,* E12S03.
Clemett S. J., Sandford S. A., Nakamura-Messenger K., Hörz F., and McKay D. S. (2010) Complex aromatic hydrocarbons in Stardust samples collected from Comet 81P/Wild 2. *Meteorit. Planet. Sci., 45,* 701–722.
Cremonese G., Fulle M., Cambianica P., Munaretto G., Capria M. T., Forgia F. L., Lazzarin M., Migliorini A., Boschin W., Milani G., Aletti A., Arlic G., Bacci P., Bacci R., Bryssinck E., Carosati D., Castellano D., Buzzi L., Rubbo S. D., Facchini M., Guido E., Kugel F., Ligustri R., Maestripieri M., Mantero A., Nicolas J., Ochner P., Perrella C., Trabatti R., and Valvasori A. (2020) Dust environment model of the interstellar Comet 2I/Borisov. *Astrophys. J., 893,* L12.
Crovisier J. and Bockelée-Morvan D. (2008) Comment on "Comparison of the composition of the Tempel 1 ejecta to the dust in Comet C/Hale Bopp 1995 O1 and YSO HD 100546" by C. M. Lisse, K. E. Kraemer, J. A. Nuth III, A. Li, D. Joswiak [2007. Icarus 187, 69–86]. *Icarus, 195,* 938–940.
Crovisier J., Leech K., Bockelée-Morvan D., Brooke T. Y., Hanner M. S., Altieri B., Keller H. U., and Lellouch E. (1997) The spectrum of Comet Hale-Bopp (C/1995 O1) observed with the infrared space observatory at 2.9 astronomical units from the Sun. *Science, 275,* 1904–1907.
Dartois E., Geballe T. R., Pino T., Cao A.-T., Jones A., Deboffle D., Guerrini V., Bréchignac P., and d'Hendecourt L. (2007) IRAS 08572+3915: Constraining the aromatic versus aliphatic content of interstellar HACs. *Astron. Astrophys., 463,* 635–640.
Dartois E., Engrand C., Brunetto R., Duprat J., Pino T., Quirico E., Remusat L., Bardin N., Briani G., Mostefaoui S., Morinaud G., Crane B., Szwec N., Delauche L., Jamme F., Sandt C., and Dumas P. (2013) Ultracarbonaceous Antarctic micrometeorites, probing the solar system beyond the nitrogen snow-line. *Icarus, 224,* 243–252.
Dartois E., Engrand C., Duprat J., Godard M., Charon E., Delauche L., Sandt C., and Borondics F. (2018) Dome C ultracarbonaceous Antarctic micrometeorites: Infrared and Raman fingerprints. *Astron. Astrophys., 609,* A65.
Defouilloy C., Nakashima D., Joswiak D. J., Brownlee D. E., Tenner T. J., and Kita N. T. (2017) Origin of crystalline silicates from Comet 81P/Wild 2: Combined study on their oxygen isotopes and mineral chemistry. *Earth Planet. Sci. Lett., 465,* 145–154.
De Gregorio B. T., Stroud R. M., Nittler L. R., Alexander C. M. O., Kilcoyne A. L. D., and Zega T. J. (2010) Isotopic anomalies in organic nanoglobules from Comet 81P/Wild 2: Comparison to Murchison nanoglobules and isotopic anomalies induced in terrestrial organics by electron irradiation. *Geochim. Cosmochim. Acta, 74,* 4454–4470.
De Gregorio B. T., Stroud R. M., Cody G. D., Nittler L. R., David Kilcoyne A. L., and Wirick S. (2011) Correlated microanalysis of cometary organic grains returned by Stardust. *Meteorit. Planet. Sci., 46,* 1376–1396.
De Gregorio B. T., Stroud R. M., Nittler L. R., and Kilcoyne A. L. D. (2017) Evidence for reduced, carbon-rich regions in the solar nebula from an unusual cometary dust particle. *Astrophys. J., 848,* 113.
Della Corte V., Rotundi A., Fulle M., Gruen E., Weissman P., Sordini R., Ferrari M., Ivanovski S., Lucarelli F., and Accolla M. (2015) GIADA: Shining a light on the monitoring of the comet dust production from the nucleus of 67P/Churyumov-Gerasimenko. *Astron. Astrophys., 583,* A13.
Dello Russo N., Vervack R. J., Kawakita H., Bonev B. P., DiSanti M. A., Gibb E. L., McKay A. J., Cochran A. L., Weaver H. A., Biver N., Crovisier J., Bockelée-Morvan D., Kobayashi H., Harris W. M., Roth N. X., Saki M., and Khan Y. (2022) Volatile abundances, extended coma sources, and nucleus ice associations in Comet C/2014 Q2 (Lovejoy). *Planet. Sci. J., 3,* 6.
Demyk K., d'Hendecourt L., Leroux H., Jones A. P., and Borg J. (2004) IR spectroscopic study of olivine, enstatite and diopside irradiated with low energy H^+ and He^+ ions. *Astron. Astrophys., 420,* 233–243.
Divine N. and Newburn R. L. (1988) Modelling Halley before and after the encounters. In *Exploration of Halley's Comet* (M. Grewing et al., eds.), pp. 867–872. Springer, Berlin.
Dobrică E., Engrand C., Duprat J., Gounelle M., Leroux H., Quirico E., and Rouzaud J. N. (2009) Connection between micrometeorites and Wild 2 particles: From Antarctic snow to cometary ices. *Meteorit. Planet. Sci., 44,* 1643–1661.
Dobrică E., Engrand C., Quirico E., Montagnac G., and Duprat J. (2011) Raman characterization of carbonaceous matter in CONCORDIA Antarctic micrometeorites. *Meteorit. Planet. Sci., 46,* 1363–1375.
Dobrică E., Engrand C., Leroux H., Rouzaud J. N., and Duprat J. (2012) Transmission electron microscopy of CONCORDIA ultracarbonaceous Antarctic micrometeorites (UCAMMs): Mineralogical properties. *Geochim. Cosmochim. Acta, 76,* 68–82.
Dollfus A. (1989) Polarimetry of grains in the coma of P/Halley. II — Interpretation. *Astron. Astrophys., 213,* 469–478.
Dollfus A., Bastien P., Le Borgne J.-F., Levasseur-Regourd A.-C., and Mukai T. (1988) Optical polarimetry of P/Halley: Synthesis of the measurements in the continuum. *Astron. Astrophys., 206,* 348–356.
Dorschner J., Begemann B., Henning T., Jaeger C., and Mutschke H. (1995) Steps toward interstellar silicate mineralogy. II. Study of Mg-Fe-silicate glasses of variable composition. *Astron. Astrophys., 300,* 503–520.
Duprat J., Engrand C., Maurette M., Kurat G., Gounelle M., and Hammer C. (2007) Micrometeorites from central Antarctic snow: The CONCORDIA collection. *Adv. Space Res., 39,* 605–611.
Duprat J., Dobrică E., Engrand C., Aléon J., Marrocchi Y., Mostefaoui S., Meibom A., Leroux H., Rouzaud J. N., Gounelle M., and Robert F. (2010) Extreme deuterium excesses in ultracarbonaceous micrometeorites from central Antarctic snow. *Science, 328,* 742–745.
Ellerbroek L. E., Gundlach B., Landeck A., Dominik C., Blum J., Merouane S., Hilchenbach M., Bentley M. S., Mannel T., John H., and van Veen H. A. (2017) The footprint of cometary dust analogues — I. Laboratory experiments of low-velocity impacts and comparison with Rosetta data. *Mon. Not. R. Astron. Soc., 469,*

S204–S216.
Ellerbroek L. E., Gundlach B., Landeck A., Dominik C., Blum J., Merouane S., Hilchenbach M., John H., and van Veen H. A. (2019) The footprint of cometary dust analogues — II. Morphology as a tracer of tensile strength and application to dust collection by the Rosetta spacecraft. *Mon. Not. R. Astron. Soc., 486,* 3755–3765.
Ellis T. A. and Neff J. S. (1991) Numerical simulation of the emission and motion of neutral and charged dust from P/Halley. *Icarus, 91,* 280–296.
Elsila J. E., Glavin D. P., and Dworkin J. P. (2009) Cometary glycine detected in samples returned by Stardust. *Meteorit. Planet. Sci., 44,* 1323–1330.
Engrand C., Benzerara K., Leroux H., Duprat J., Dartois E., Bardin N., and Delauche L. (2015) Carbonaceous phases and mineralogy of ultracarbonaceous Antarctic micrometeorites identified by C- and N-XANES/STXM and TEM. In *46th Lunar and Planetary Science Conference*, Abstract #1902. LPI Contribution No. 1832, Lunar and Planetary Institute, Houston.
Epifani E., Colangeli L., Fulle M., Brucato J. R., Bussoletti E., De Sanctis M. C., Mennella V., Palomba E., Palumbo P., and Rotundi A. (2001) ISOCAM imaging of Comets 103P/Hartley 2 and 2P/Encke. *Icarus, 149,* 339–350.
Fabian D., Jäger C., Henning T., Dorschner J., and Mutschke H. (2000) Steps toward interstellar silicate mineralogy. V. Thermal evolution of amorphous magnesium silicates and silica. *Astron. Astrophys., 364,* 282–292.
Fabian D., Henning T., Jäger C., Mutschke H., Dorschner J., and Wehrhan O. (2001) Steps toward interstellar silicate mineralogy. VI. Dependence of crystalline olivine IR spectra on iron content and particle shape. *Astron. Astrophys., 378,* 228–238.
Feaga L., Sunshine J., Bonev B., Dello Russo N., Di Santi M., and Combi M. (2021) Spatial distribution of organics in the inner coma of 103P/Hartley 2. *AAS/Division for Planetary Sciences Meeting Abstracts, 53,* #208.02.
Feaga L. M., Sunshine J. M., Bonev B. P., and Dello Russo N. (2023) Modeling the near-infrared spectra of 103P/Hartley 2's innermost coma to determine volatile asssociations within its nucleus. In *Asteroids, Comets, Meteors Conference,* Abstract #2476. LPI Contribution No. 2851, Lunar and Planetary Institute, Houston.
Finson M. J. and Probstein R. F. (1968a) A theory of dust comets. I. Model and equations. *Astrophys. J., 154,* 327–352.
Finson M. L. and Probstein R. F. (1968b) A theory of dust comets. II. Results for Comet Arend-Roland. *Astrophys. J., 154,* 353–380.
Floss C. and Haenecour P. (2016) Presolar silicate grains: Abundances, isotopic and elemental compositions, and the effects of secondary processing. *Geochem. J., 50,* 3–25.
Floss C., Noguchi T., and Yada T. (2012) Ultracarbonaceous Antarctic micrometeorites: Origins and relationships to other primitive extraterrestrial materials. In *43rd Lunar and Planetary Science Conference*, Abstract #1217. LPI Contribution No. 1659, Lunar and Planetary Institute, Houston.
Floss C., Stadermann F. J., Kearsley A. T., Burchell M. J., and Ong W. J. (2013) The abundance of presolar grains in Comet 81P/Wild 2. *Astrophys. J., 763,* 140.
Flynn G. J., Keller L. P., Feser M., Wirick S., and Jacobsen C. (2003) The origin of organic matter in the solar system: Evidence from the interplanetary dust particles. *Geochim. Cosmochim. Acta, 67,* 4791–4806.
Flynn G. J., Bleuet P., Borg J., Bradley J. P., Brenker F. E., Brennan S., Bridges J., Brownlee D. E., Bullock E. S., Burghammer M., Clark B. C., Dai Z. R., Daghlian C. P., Djouadi Z., Fakra S., Ferroir T., Floss C., Franchi I. A., Gainsforth Z., Gallien J.-P., Gillet P., Grant P. G., Graham G. A., Green S. F., Grossemy F., Heck P. R., Herzog G. F., Hoppe P., Hörz F., Huth J., Ignatyev K., Ishii H. A., Janssens K., Joswiak D., Kearsley A. T., Khodja H., Lanzirotti A., Leitner J., Lemelle L., Leroux H., Luening K., MacPherson G. J., Marhas K. K., Marcus M. A., Matrajt G., Nakamura T., Nakamura-Messenger K., Nakano T., Newville M., Papanastassiou D. A., Pianetta P., Rao W., Riekel C., Rietmeijer F. J. M., Rost D., Schwandt C. S., See T. H., Sheffield-Parker J., Simionovici A., Sitnitsky I., Snead C. J., Stadermann F. J., Stephan T., Stroud R. M., Susini J., Suzuki Y., Sutton S. R., Taylor S., Teslich N., Troadec D., Tsou P., Tsuchiyama A., Uesugi K., Vekemans B., Vicenzi E. P., Vincze L., Westphal A. J., Wozniakiewicz P., Zinner E., and Zolensky M. E. (2006) Elemental compositions of Comet 81P/ Wild 2 samples collected by Stardust. *Science, 314,* 1731–1735.
Flynn G. J., Durda D. D., Sandel L. E., Kreft J. W., and Strait M. M. (2009) Dust production from the hypervelocity impact disruption of the Murchison hydrous CM2 meteorite: Implications for the disruption of hydrous asteroids and the production of interplanetary dust. *Planet. Space Sci., 57,* 119–126.
Fomenkova M. and Chang S. (1994) Carbon in Comet Halley dust particles. In *Analysis of Interplanetary Dust* (M. Zolensky et al., eds.), pp. 193–202. AIP Conf. Ser. 310, American Institute of Physics, Melville, New York.
Fomenkova M. N., Kerridge J. F., Marti K., and McFadden L. A. (1992) Compositional trends in rock-forming elements of Comet Halley dust. *Science, 258,* 266–269.
Fomenkova M. N., Chang S., and Mukhin L. M. (1994) Carbonaceous components in the Comet Halley dust. *Geochim. Cosmochim. Acta, 58,* 4503–4512.
Frank D. R., Zolensky M. E., and Le L. (2014) Olivine in terminal particles of Stardust aerogel tracks and analogous grains in chondrite matrix. *Geochim. Cosmochim. Acta, 142,* 240–259.
Frank D. R., Huss G. R., Nagashima K., Hellebrand E., Westphal A. J., and Gainsforth Z. (2019) Rumuruti and metamorphosed, ordinary chondrite fragments from Comet 81P/Wild 2. In *50th Annual Lunar and Planetary Science Conference*, Abstract #3281. LPI Contribution No. 2132, Lunar and Planetary Institute, Houston.
Frattin E., Cremonese G., Simioni E., Bertini I., Lazzarin M., Ott T., Drolshagen E., La Forgia F., Sierks H., Barbieri C., Lamy P., Rodrigo R., Koschny D., Rickman H., Keller H. U., Agarwal J., A'Hearn M. F., Barucci M. A., Bertaux J.-L., Da Deppo V., Davidsson B., Debei S., De Cecco M., Deller J., Ferrari S., Ferri F., Fornasier S., Fulle M., Gicquel A., Groussin O., Gutierrez P. J., Güttler C., Hofmann M., Hviid S. F., Ip W.-H., Jorda L., Knollenberg J., Kramm J.-R., Kührt E., Küppers M., Lara L. M., Lopez Moreno J. J., Lucchetti A., Marzari F., Massironi M., Mottola S., Naletto G., Oklay N., Pajola M., Penasa L., Shi X., Thomas N., Tubiana C., and Vincent J.-B. (2017) Post-perihelion photometry of dust grains in the coma of 67P/ Churyumov-Gerasimenko. *Mon. Not. R. Astron. Soc., 469,* S195–S203.
Fray N., Bardyn A., Cottin H., Baklouti D., Briois C., Colangeli L., Engrand C., Fischer H., Glasmachers A., Grün E., Haerendel G., Henkel H., Höfner H., Hornung K., Jessberger E. K., Koch A., Krüger H., Langevin Y., Lehto H., Lehto K., Le Roy L., Merouane S., Modica P., Orthous-Daunay F.-R., Paquette J., Raulin F., Rynö J., Schulz R., Silén J., Siljeström S., Steiger W., Stenzel O., Stephan T., Thirkell L., Thomas R., Torkar K., Varmuza K., Wanczek K.-P., Zaprudin B., Kissel J., and Hilchenbach M. (2016) High-molecular-weight organic matter in the particles of Comet 67P/Churyumov-Gerasimenko. *Nature, 538,* 72–74.
Fray N., Bardyn A., Cottin H., Baklouti D., Briois C., Engrand C., Fischer H., Hornung K., Isnard R., Langevin Y., Lehto H., Le Roy L., Mellado E. M., Merouane S., Modica P., Orthous-Daunay F.-R., Paquette J., Rynö J., Schulz R., Silén J., Siljeström S., Stenzel O., Thirkell L., Varmuza K., Zaprudin B., Kissel J., and Hilchenbach M. (2017) Nitrogen-to-carbon atomic ratio measured by COSIMA in the particles of Comet 67P/Churyumov-Gerasimenko. *Mon. Not. R. Astron. Soc., 469,* S506–S516.
Fukuda K., Brownlee D. E., Joswiak D. J., Tenner T. J., Kimura M., and Kita N. T. (2021) Correlated isotopic and chemical evidence for condensation origins of olivine in Comet 81P/Wild 2 and in AOAs from CV and CO chondrites. *Geochim. Cosmochim. Acta, 293,* 544–574.
Fulle M. (1987) A new approach to the Finson-Probstein method of interpreting cometary dust tails. *Astron. Astrophys., 171,* 327–335.
Fulle M. (2004) Motion of cometary dust. In *Comets II* (M. C. Festou et al., eds.), pp. 565–575. Univ. of Arizona, Tucson.
Fulle M. and Blum J. (2017) Fractal dust constrains the collisional history of comets. *Mon. Not. R. Astron. Soc., 469,* S39–S44.
Fulle M., Colangeli L., Mennella V., Rotundi A., and Bussoletti E. (1995) The sensitivity of the size distribution to the grain dynamics: Simulation of the dust flux measured by Giotto at P/Halley. *Astron. Astrophys., 304,* 622–630.
Fulle M., Levasseur-Regourd A.-C., McBride N., and Hadamcik E. (2000) In situ dust measurements from within the coma of 1P/ Halley: First-order approximation with a dust dynamical model. *Astron. J., 119,* 1968–1977.
Fulle M., Corte V. D., Rotundi A., Weissman P., Juhasz A., Szego K., Sordini R., Ferrari M., Ivanovski S., Lucarelli F., Accolla M., Merouane S., Zakharov V., Epifani E. M., Moreno J. J. L.,

Rodríguez J., Colangeli L., Palumbo P., Grün E., Hilchenbach M., Bussoletti E., Esposito F., Green S. F., Lamy P. L., McDonnell J. A. M., Mennella V., Molina A., Morales R., Moreno F., Ortiz J. L., Palomba E., Rodrigo R., Zarnecki J. C., Cosi M., Giovane F., Gustafson B., Herranz M. L., Jerónimo J. M., Leese M. R., Jiménez A. C. L., and Altobelli N. (2015) Density and charge of pristine fluffy particles from Comet 67P/Churyumov-Gerasimenko. *Astrophys. J. Lett., 802,* L12.

Fulle M., Altobelli N., Buratti B., Choukroun M., Fulchignoni M., Grün E., Taylor M., and Weissman P. (2016a) Unexpected and significant findings in Comet 67P/Churyumov-Gerasimenko: An interdisciplinary view. *Mon. Not. R. Astron. Soc., 462,* S2–S8.

Fulle M., Della Corte V., Rotundi A., Rietmeijer F. J. M., Green S. F., Weissman P., Accolla M., Colangeli L., Ferrari M., Ivanovski S., Lopez-Moreno J. J., Epifani E. M., Morales R., Ortiz J. L., Palomba E., Palumbo P., Rodriguez J., Sordini R., and Zakharov V. (2016b) Comet 67P/Churyumov-Gerasimenko preserved the pebbles that formed planetesimals. *Mon. Not. R. Astron. Soc., 462,* S132–S137.

Fulle M., Marzari F., Corte V. D., Fornasier S., Sierks H., Rotundi A., Barbieri C., Lamy P. L., Rodrigo R., Koschny D., Rickman H., Keller H. U., López-Moreno J. J., Accolla M., Agarwal J., A'Hearn M. F., Altobelli N., Barucci M. A., Bertaux J.-L., Bertini I., Bodewits D., Bussoletti E., Colangeli L., Cosi M., Cremonese G., Crifo J.-F., Deppo V. D., Davidsson B., Debei S., Cecco M. D., Esposito F., Ferrari M., Giovane F., Gustafson B., Green S. F., Groussin O., Grün E., Gutierrez P., Güttler C., Herranz M. L., Hviid S. F., Ip W., Ivanovski S. L., Jerónimo J. M., Jorda L., Knollenberg J., Kramm R., Kührt E., Küppers M., Lara L., Lazzarin M., Leese M. R., López-Jiménez A. C., Lucarelli F., Epifani E. M., McDonnell J. A. M., Mennella V., Molina A., Morales R., Moreno F., Mottola S., Naletto G., Oklay N., Ortiz J. L., Palomba E., Palumbo P., Perrin J.-M., Rietmeijer F. J. M., Rodríguez J., Sordini R., Thomas N., Tubiana C., Vincent J.-B., Weissman P., Wenzel K.-P., Zakharov V., and Zarnecki J. C. (2016c) Evolution of the dust size distribution of Comet 67P/Churyumov-Gerasimenko from 2.2 au to perihelion. *Astrophys. J., 821,* 19.

Fulle M., Della Corte V., Rotundi A., Green S. F., Accolla M., Colangeli L., Ferrari M., Ivanovski S., Sordini R., and Zakharov V. (2017) The dust-to-ices ratio in comets and Kuiper belt objects. *Mon. Not. R. Astron. Soc., 469,* S45–S49.

Furusho R., Ikeda Y., Kinoshita D., Ip W.-H., Kawakita H., Kasuga T., Sato Y., Lin H.-C., Chang M.-S., Lin Z.-Y., and Watanabe J. (2007) Imaging polarimetry of Comet 9P/Tempel before and after the Deep Impact. *Icarus, 190,* 454–458.

Gainsforth Z., Butterworth A., Ogliore R., Westphal A., and Tyliszczak T. (2010) Combined STEM/STXM elemental quantification for cometary particles. *Microsc. Microanal., 16,* 922–923.

Gainsforth Z., McLeod A. S., Butterworth A. L., Dominguez G., Basov D., Keilmann F., Thiemens M., Tyliszczak T., and Westphal A. J. (2013) Caligula, a Stardust sulfide-silicate assemblage viewed through SEM, NanoFTIR, and STXM. In *44th Lunar and Planetary Science Conference*, Abstract #2332. LPI Contribution No. 1719, Lunar and Planetary Institute, Houston.

Gainsforth Z., Brenker F. E., Simionovici A. S., Schmitz S., Burghammer M., Butterworth A. L., Cloetens P., Lemelle L., Tresserras J.-A. S., Schoonjans T., Silversmit G., Solé V. A., Vekemans B., Vincze L., Westphal A. J., Allen C., Anderson D., Ansari A., Bajt S., Bastien R. K., Bassim N., Bechtel H. A., Borg J., Bridges J., Brownlee D. E., Burchell M., Changela H., Davis A. M., Doll R., Floss C., Flynn G., Fougeray P., Frank D., Grün E., Heck P. R., Hillier J. K., Hoppe P., Hudson B., Huth J., Hvide B., Kearsley A., King A. J., Lai B., Leitner J., Leroux H., Leonard A., Lettieri R., Marchant W., Nittler L. R., Ogliore R., Ong W. J., Postberg F., Price M. C., Sandford S. A., Srama R., Stephan T., Sterken V., Stodolna J., Stroud R. M., Sutton S., Trieloff M., Tsou P., Tsuchiyama A., Tyliszczak T., Von Korff J., Zevin D., Zolensky M. E., and >30,000 Stardust@home dusters. (2014) Stardust interstellar preliminary examination VIII: Identification of crystalline material in two interstellar candidates. *Meteorit. Planet. Sci., 49,* 1645–1665.

Gainsforth Z., Butterworth A. L., Stodolna J., Westphal A. J., Huss G. R., Nagashima K., Ogliore R., Brownlee D. E., Joswiak D., Tyliszczak T., and Simionovici A. S. (2015) Constraints on the formation environment of two chondrule-like igneous particles from Comet 81P/Wild 2. *Meteorit. Planet. Sci., 50,* 976–1004.

Gainsforth Z., Butterworth A. L., Jilly-Rehak C. E., Westphal A. J., Brownlee D. E., Joswiak D., Ogliore R. C., Zolensky M. E., Bechtel H. A., Ebel D. S., Huss G. R., Sandford S. A., and White A. J. (2016) Possible GEMS and ultra-fine grained polyphase units in Comet Wild 2. In *47th Lunar and Planetary Science Conference*, Abstract #2366. LPI Contribution No. 1903, Lunar and Planetary Institute, Houston.

Gardner E., Lehto H. J., Lehto K., Fray N., Bardyn A., Lönnberg T., Merouane S., Isnard R., Cottin H., Hilchenbach M., and the COSIMA Team (2020) The detection of solid phosphorus and fluorine in the dust from the coma of Comet 67P/Churyumov-Gerasimenko. *Mon. Not. R. Astron. Soc., 499,* 1870–1873.

Gehrz R. D. and Ney E. P. (1992) 0.7- to 23-μm photometric observations of P/Halley 1986 III and six recent bright comets. *Icarus, 100,* 162–186.

Geiss J. and Gloeckler G. (1998) Abundances of deuterium and helium-3 in the protosolar cloud. *Space Sci. Rev., 84,* 239–250.

Glassmeier K.-H., Boehnhardt H., Koschny D., Kührt E., and Richter I. (2007) The Rosetta mission: Flying towards the origin of the solar system. *Space Sci. Rev., 128,* 1–21.

Godard M., Geballe T. R., Dartois E., and Muñoz Caro G. M. (2012) The deep 3.4 μm interstellar absorption feature toward the IRAS 18511+0146 cluster. *Astron. Astrophys., 537,* A27.

Gounelle M. (2011) The asteroid-comet continuum: In search of lost primitivity. *Elements, 7,* 29–34.

Green S. F., McDonnell J. A. M., McBride N., Colwell M. T. S. H., Tuzzolino A. J., Economou T. E., Tsou P., Clark B. C., and Brownlee D. E. (2004) The dust mass distribution of Comet 81P/Wild 2. *J. Geophys. Res.–Planets, 109,* E12S04.

Groussin O., Attree N., Brouet Y., Ciarletti V., Davidsson B., Filacchione G., Fischer H.-H., Gundlach B., Knapmeyer M., and Knollenberg J. (2019) The thermal, mechanical, structural, and dielectric properties of cometary nuclei after Rosetta. *Space Sci. Rev., 215,* 1–51.

Guérin B., Engrand C., Le Guillou C., Leroux H., Duprat J., Dartois E., Bernard S., Benzerara K., Rojas J., Godard M., Delauche L., and Troadec D. (2020) STEM and STXM-XANES analysis of FIB sections of ultracarbonaceous Antarctic micrometeorites (UCAMMs). In *51st Lunar and Planetary Science Conference*, Abstract #2117. LPI Contribution No. 2326, Lunar and Planetary Institute, Houston.

Gundlach B., Schmidt K. P., Kreuzig C., Bischoff D., Rezaei F., Kothe S., Blum J., Grzesik B., and Stoll E. (2018) The tensile strength of ice and dust aggregates and its dependence on particle properties. *Mon. Not. R. Astron. Soc., 479,* 1273–1277.

Güttler C., Krause M., Geretshauser R. J., Speith R., and Blum J. (2009) The physics of protoplanetesimal dust agglomerates. IV. Toward a dynamical collision model. *Astrophys. J., 701,* 130–141.

Güttler C., Mannel T., Rotundi A., Merouane S., Fulle M., Bockelée-Morvan D., Lasue J., Levasseur-Regourd A.-C., Blum J., and Naletto G. (2019) Synthesis of the morphological description of cometary dust at Comet 67P/Churyumov-Gerasimenko. *Astron. Astrophys., 630,* A24.

Guzik P., Drahus M., Rusek K., Waniak W., Cannizzaro G., and Pastor-Marazuela I. (2020) Initial characterization of interstellar Comet 2I/Borisov. *Nature Astron., 4,* 53–57.

Guzik P. and Drahus M. (2021) Gaseous atomic nickel in the coma of interstellar Comet 2I/Borisov. *Nature, 593,* 375–378.

Hadamcik E., Renard J.-B., Levasseur-Regourd A.-C., and Lasue J. (2006) Light scattering by fluffy particles with the PROGRA^2 experiment: Mixtures of materials. *J. Quant. Spectrosc. Radiat. Transfer, 100,* 143–156.

Hadamcik E., Renard J.-B., Rietmeijer F. J. M., Levasseur-Regourd A.-C., Hill H. G. M., Karner J. M., and Nuth J. A. (2007) Light scattering by fluffy Mg-Fe-SiO and C mixtures as cometary analogs (PROGRA^2 experiment). *Icarus, 190,* 660–671.

Hadamcik E., Sen A. K., Levasseur-Regourd A.-C., Gupta R., and Lasue J. (2010) Polarimetric observations of Comet 67P/Churyumov-Gerasimenko during its 2008–2009 apparition. *Astron. Astrophys., 517,* A86.

Hadamcik E., Levasseur-Regourd A.-C., Hines D. C., Sen A. K., Lasue J., and Renard J.-B. (2016) Properties of dust particles in comets from photometric and polarimetric observations of 67P. *Mon. Not. R. Astron. Soc., 462,* S507–S515.

Hadraoui K., Cottin H., Ivanovski S. L., Zapf P., Altwegg K., Benilan Y., Biver N., Della Corte V., Fray N., Lasue J., Merouane S., Rotundi A., and Zakharov V. (2019) Distributed glycine in Comet 67P/Churyumov-Gerasimenko. *Astron. Astrophys., 630,* A32.

Hadraoui K., Cottin H., Ivanovski S. L., Zapf P., Altwegg K., Benilan Y., Biver N., Della Corte V., Fray N., Lasue J., Merouane S.,

Rotundi A., and Zakharov V. (2021) Distributed glycine in Comet 67P/Churyumov-Gerasimenko (Corrigendum). *Astron. Astrophys., 651,* C2.

Hanner M. S. (1983) The nature of cometary dust from remote sensing. In *Cometary Exploration, Vol. 2* (T. I. Gombosi, ed.), pp. 1–22. Hungarian Academy of Sciences, Budapest.

Hanner M. S. (2003) The scattering properties of cometary dust. *J. Quant. Spectrosc. Radiat. Transfer, 79,* 695–705.

Hanner M. S. and Zolensky M. E. (2010) The mineralogy of cometary dust. In *Astromineralogy* (T. Henning, ed.), pp. 203–232. Lecture Notes in Physics 815, Springer, Berlin.

Hanner M. S., Giese R. H., Weiss K., and Zerull R. (1981) On the definition of albedo and application to irregular particles. *Astron. Astrophys., 104,* 42–46.

Hanner M. S., Knacke R., Sekanina Z., and Tokunaga A. T. (1985) Dark grains in Comet Crommelin. *Astron. Astrophys., 152,* 177–181.

Hanner M. S., Gehrz R. D., Harker D. E., Hayward T. L., Lynch D. K., Mason C. C., Russell R. W., Williams D. M., Wooden D. H., and Woodward C. E. (1997) Thermal emission from the dust coma of Comet Hale-Bopp and the composition of the silicate grains. *Earth Moon Planets, 79,* 247–264.

Hänni N., Altwegg K., Pestoni B., Rubin M., Schroeder I., Schuhmann M., and Wampfler S. (2020) First *in situ* detection of the CN radical in comets and evidence for a distributed source. *Mon. Not. R. Astron. Soc., 498,* 2239–2248.

Hänni N., Altwegg K., Balsiger H., Combi M., Fuselier S. A., De Keyser J., Pestoni B., Rubin M., and Wampfler S. F. (2021) Cyanogen, cyanoacetylene, and acetonitrile in Comet 67P and their relation to the cyano radical. *Astron. Astrophys., 647,* A22.

Hänni N., Altwegg K., Combi M., Fuselier S. A., De Keyser J., Rubin M., and Wampfler S. F. (2022) Identification and characterization of a new ensemble of cometary organic molecules. *Nature Commun., 13,* 3639.

Harker D. E. and Desch S. J. (2002) Annealing of silicate dust by nebular shocks at 10 AU. *Astrophys. J. Lett., 565,* L109–L112.

Harker D. E., Wooden D. H., Woodward C. E., and Lisse C. M. (2002) Grain properties of Comet C/1995 O1 (Hale-Bopp). *Astrophys. J., 580,* 579–597.

Harker D. E., Wooden D. H., Woodward C. E., and Lisse C. M. (2004) Erratum: "Grain Properties of Comet C/1995 O1 (Hale-Bopp)" (ApJ, 580, 579 [2002]). *Astrophys. J., 615,* 1081.

Harker D. E., Woodward C. E., and Wooden D. H. (2005) The dust grains from 9P/Tempel 1 before and after the encounter with Deep Impact. *Science, 310,* 278–280.

Harker D. E., Woodward C. E., Wooden D. H., Fisher R. S., and Trujillo C. A. (2007) Gemini-N mid-IR observations of the dust properties of the ejecta excavated from Comet 9P/Tempel 1 during Deep Impact. *Icarus, 190,* 432–453.

Harker D. E., Wooden D. H., Kelley M. S. P., and Woodward C. E. (2023) Dust properties of comets observed by Spitzer. *Planet. Sci. J., 4,* 242.

Haser L., Oset S., and Bodewits D. (2020) Intensity distribution in the heads of comets. *Planet. Sci. J., 1,* 83.

Hashizume K., Chaussidon M., Marty B., and Terada K. (2004) Protosolar carbon isotopic composition: Implications for the origin of meteoritic organics. *Astrophys. J., 600,* 480–484.

Hässig M., Altwegg K., Balsiger H., Berthelier J. J., Bieler A., Calmonte U., Dhooghe F., Fiethe B., Fuselier S. A., Gasc S., Gombosi T. I., Le Roy L., Luspay-Kuti A., Mandt K., Rubin M., Tzou C.-Y., Wampfler S. F., and Wurz P. (2017) Isotopic composition of CO_2 in the coma of 67P/Churyumov-Gerasimenko measured with ROSINA/DFMS. *Astron. Astrophys., 605,* A50.

Hayward T. L., Hanner M. S., and Sekanina Z. (2000) Thermal infrared imaging and spectroscopy of Comet Hale-Bopp (C/1995 O1). *Astrophys. J., 538,* 428–455.

Heck P. R., Hoppe P., and Huth J. (2012) Sulfur four isotope NanoSIMS analysis of comet-81P/Wild 2 dust in impact craters on aluminum foil C2037N from NASA's Stardust mission. *Meteorit. Planet. Sci., 47,* 649–659.

Henning T. and Mutschke H. (1997) Low-temperature infrared properties of cosmic dust analogues. *Astron. Astrophys., 327,* 743–754.

Hilchenbach M., Kissel J., Langevin Y., Briois C., Von Hoerner H., Koch A., Schulz R., Silén J., Altwegg K., and Colangeli L. (2016) Comet 67P/Churyumov-Gerasimenko: Close-up on dust particle fragments. *Astrophys. J. Lett., 816,* L32.

Hilchenbach M., Fischer H., Langevin Y., Merouane S., Paquette J., Rynö J., Stenzel O., Briois C., Kissel J., Koch A., Schulz R., Silen J., Altobelli N., Baklouti D., Bardyn A., Cottin H., Engrand C., Fray N., Haerendel G., Henkel H., Höfner H., Hornung K., Lehto H., Mellado E. M., Modica P., Le Roy L., Siljeström S., Steiger W., Thirkell L., Thomas R., Torkar K., Varmuza K., Zaprudin B., and the COSIMA Team (2017) Mechanical and electrostatic experiments with dust particles collected in the inner coma of Comet 67P by COSIMA onboard Rosetta. *Philos. Trans. R. Soc. A, 375,* 20160255.

Hofmeister A. M. and Speck A. K. (2003) Absorption and reflection IR spectra of MgO and other diatomic compounds. In *Astrophysics of Dust* (A. N. Witt et al., eds.), p. 148. ASP Conf. Ser. 309, Astronomical Society of the Pacific, San Francisco.

Hoppe P., Strebel R., Eberhardt P., Amari S., and Lewis R. S. (2000) Isotopic properties of silicon carbide X grains from the Murchison meteorite in the size range 0.5–1.5 μm. *Meteorit. Planet. Sci., 35,* 1157–1176.

Horányi M. (1996) Charged dust dynamics in the solar system. *Annu. Rev. Astron. Astrophys., 34,* 383–418.

Hornung K., Merouane S., Hilchenbach M., Langevin Y., Mellado E. M., Della Corte V., Kissel J., Engrand C., Schulz R., Ryno J., Silen J., and the COSIMA Team (2016) A first assessment of the strength of cometary particles collected *in-situ* by the COSIMA instrument onboard Rosetta. *Planet. Space Sci., 133,* 63–75.

Hornung K., Mellado E. M., Paquette J., Fray N., Fischer H., Stenzel O., Baklouti D., Merouane S., Langevin Y., Bardyn A., Engrand C., Cottin H., Thirkell L., Briois C., Modica P., Rynö J., Silen J., Schulz R., Siljeström S., Lehto H., Varmuza K., Koch A., Kissel J., and Hilchenbach M. (2020) Electrical properties of cometary dust particles derived from line shapes of TOF-SIMS spectra measured by the ROSETTA/COSIMA instrument. *Planet. Space Sci., 182,* 104758.

Hörz F., Bastien R., Borg J., Bradley J. P., Bridges J. C., Brownlee D. E., Burchell M. J., Chi M., Cintala M. J., Dai Z. R., Djouadi Z., Dominguez G., Economou T. E., Fairey S. A. J., Floss C., Franchi I. A., Graham G. A., Green S. F., Heck P., Hoppe P., Huth J., Ishii H., Kearsley A. T., Kissel J., Leitner J., Leroux H., Marhas K., Messenger K., Schwandt C. S., See T. H., Snead C., Stadermann F. J., Stephan T., Stroud R., Teslich N., Trigo-Rodríguez J. M., Tuzzolino A. J., Troadec D., Tsou P., Warren J., Westphal A., Wozniakiewicz P., Wright I., and Zinner E. (2006) Impact features on Stardust: Implications for Comet 81P/Wild 2 dust. *Science, 314,* 1716–1719.

Hutsemékers D., Manfroid J., Jehin E., Opitom C., and Moulane Y. (2021) FeI and NiI in cometary atmospheres: Connections between the NiI/FeI abundance ratio and chemical characteristics of Jupiter-family and Oort-cloud comets. *Astron. Astrophys., 652,* L1.

Hyland M. G., Fitzsimmons A., and Snodgrass C. (2019) Near-UV and optical spectroscopy of comets using the ISIS spectrograph on the WHT. *Mon. Not. R. Astron. Soc., 484,* 1347–1358.

Hynes K. and Gyngard F. (2009) The presolar grain database. In *40th Lunar and Planetary Science Conference*, Abstract #1198. LPI Contribution No. 1468, Lunar and Planetary Institute, Houston.

Imai Y., Koike C., Chihara H., Murata K., Aoki T., and Tsuchiyama A. (2009) Shape and lattice distortion effects on infrared absorption spectra of olivine particles. *Astron. Astrophys., 507,* 277–281.

Isa J., Rubin A. E., Marin-Carbonne J., McKeegan K. D., and Wasson J. T. (2011) Oxygen-isotopic compositions of R-chondrite chondrules. In *42nd Annual Lunar and Planetary Science Conference*, Abstract #2623. LPI Contribution No. 1608, Lunar and Planetary Institute, Houston.

Ishiguro M., Yang H., Usui F., Pyo J., Ueno M., Ootsubo T., Kwon S. M., and Mukai T. (2013) High-resolution imaging of the Gegenschein and the geometric albedo of interplanetary dust. *Astrophys. J., 767,* 75.

Ishii H. A., Bradley J. P., Dai Z. R., Chi M., Kearsley A. T., Burchell M. J., Browning N. D., and Molster F. (2008) Comparison of Comet 81P/Wild 2 dust with interplanetary dust from comets. *Science, 319,* 447–450.

Ishii H. A., Wozniakiewicz P. J., Kearsley A. T., Burchell M. J., Bradley J. P., Teslich N., Price M. C., and Cole M. J. (2011) The question of GEMS in Comet 81P/Wild 2: Stardust analogue impacts of fine-grained mineral aggregates. *Meteorit. Planet. Sci. Suppl., 74,* 5213.

Ishii H. A., Bradley J. P., Bechtel H. A., Brownlee D. E., Bustillo K. C., Ciston J., Cuzzi J. N., Floss C., and Joswiak D. J. (2018) Multiple generations of grain aggregation in different environments preceded solar system body formation. *Proc. Natl. Acad. Sci. U.S.A., 115,*

6608–6613.
Isnard R., Bardyn A., Fray N., Briois C., Cottin H., Paquette J., Stenzel O., Alexander C., Baklouti D., Engrand C., Orthous-Daunay F. R., Siljeström S., Varmuza K., and Hilchenbach M. (2019) H/C elemental ratio of the refractory organic matter in cometary particles of 67P/Churyumov-Gerasimenko. *Astron. Astrophys., 630,* A27.
Jaeger C., Molster F. J., Dorschner J., Henning T., Mutschke H., and Waters L. B. F. M. (1998) Steps toward interstellar silicate mineralogy. IV. The crystalline revolution. *Astron. Astrophys., 339,* 904–916.
Jäger C., Dorschner J., Mutschke H., Posch T., and Henning T. (2003) Steps toward interstellar silicate mineralogy*. VII. Spectral properties and crystallization behaviour of magnesium silicates produced by the sol-gel method. *Astron. Astrophys., 408,* 193–204.
Jäger C., Sabri T., Wendler E., and Henning T. (2016) Ion-induced processing of cosmic silicates: A possible formation pathway to GEMS. *Astrophys. J., 831,* 66.
Jarosewich E. (1990) Chemical analyses of meteorites: A compilation of stony and iron meteorite analyses. *Meteoritics, 25,* 323–337.
Jenniskens P. (2006) *Meteor Showers and Their Parent Comets.* Cambridge Univ., Cambridge. 790 pp.
Jenniskens P., Baratta G. A., Kouchi A., Degroot M. S., Greenberg J. M., and Strazzulla G. (1993) Carbon dust formation on interstellar grains. *Astron. Astrophys., 273,* 583–600.
Jessberger E. K. (1999) Rocky cometary particulates: Their elemental, isotopic and mineralogical ingredients. *Space Sci. Rev., 90,* 91–97.
Jessberger E. K., Kissel J., Fechtig H., and Krueger F. R. (1986) On the average chemical composition of cometary dust. In *Proceedings of the Workshop on the Comet Nucleus Sample Return Mission* (O. Melita, ed.), pp. 27–30. ESA SP-249, Noordwijk, The Netherlands.
Jessberger E. K., Christoforidis A., and Kissel J. (1988) Aspects of the major element composition of Halley's dust. *Nature, 332,* 691–695.
Joswiak D. J., Brownlee D. E., Matrajt G., Westphal A. J., and Snead C. J. (2009) Kosmochloric Ca-rich pyroxenes and FeO-rich olivines (KOOL grains) and associated phases in Stardust tracks and chondritic porous interplanetary dust particles: Possible precursors to FeO-rich type II chondrules in ordinary chondrites. *Meteorit. Planet. Sci., 44,* 1561–1588.
Joswiak D. J., Brownlee D. E., Matrajt G., Westphal A. J., Snead C. J., and Gainsforth Z. (2012) Comprehensive examination of large mineral and rock fragments in Stardust tracks: Mineralogy, analogous extraterrestrial materials, and source regions. *Meteorit. Planet. Sci., 47,* 471–524.
Joswiak D. J., Nakashima D., Brownlee D. E., Matrajt G., Ushikubo T., Kita N. T., Messenger S., and Ito M. (2014) Terminal particle from Stardust track 130: Probable Al-rich chondrule fragment from Comet Wild 2. *Geochim. Cosmochim. Acta, 144,* 277–298.
Joswiak D. J., Brownlee D. E., Nguyen A. N., and Messenger S. (2017) Refractory materials in comet samples. *Meteorit. Planet. Sci., 52,* 1612–1648.
Juhász A., Henning T., Bouwman J., Dullemond C. P., Pascucci I., and Apai D. (2009) Do we really know the dust? Systematics and uncertainties of the mid-infrared spectral analysis methods. *Astrophys. J., 695,* 1024–1041.
Juhász A., Bouwman J., Henning T., Acke B., van den Ancker M. E., Meeus G., Dominik C., Min M., Tielens A. G. G. M., and Waters L. B. F. M. (2010) Dust evolution in protoplanetary disks around Herbig Ae/Be stars — The Spitzer view. *Astrophys. J., 721,* 431–455.
Kadono T., Sugita S., Sako S., Ootsubo T., Honda M., Kawakita H., Miyata T., Furusho R., and Watanabe J. (2007) The thickness and formation age of the surface layer on Comet 9P/Tempel 1. *Astrophys. J. Lett., 661,* L89–L92.
Keller L. P. and Messenger S. (2011) On the origins of GEMS grains. *Geochim. Cosmochim. Acta, 75,* 5336–5365.
Keller L. P., Messenger S., and Bradley J. P. (2000) Analysis of a deuterium-rich interplanetary dust particle (IDP) and implications for presolar material in IDPs. *J. Geophys. Res.–Space Phys., 105,* 10397–10402.
Keller L. P., Messenger S., Flynn G. J., Clemett S., Wirick S., and Jacobsen C. (2004) The nature of molecular cloud material in interplanetary dust. *Geochim. Cosmochim. Acta, 68,* 2577–2589.
Keller L. P., Bajt S., Baratta G. A., Borg J., Bradley J. P., Brownlee D. E., Busemann H., Brucato J. R., Burchell M., Colangeli L., d'Hendecourt L., Djouadi Z., Ferrini G., Flynn G., Franchi I. A., Fries M., Grady M. M., Graham G. A., Grossemy F., Kearsley A., Matrajt G., Nakamura-Messenger K., Mennella V., Nittler L., Palumbo M. E., Stadermann F. J., Tsou P., Rotundi A., Sandford S. A., Snead C., Steele A., Wooden D., and Zolensky M. (2006) Infrared spectroscopy of Comet 81P/Wild 2 samples returned by Stardust. *Science, 314,* 1728–1731.
Kelley M. S., Lindler D. J., Bodewits D., A'Hearn M. F., Lisse C. M., Kolokolova L., Kissel J., and Hermalyn B. (2013) A distribution of large particles in the coma of Comet 103P/Hartley 2. *Icarus, 222,* 634–652.
Kelley M. S. P., Harker D. E., Woodward C. E., and Wooden D. H. (2021) *Spitzer Space Telescope Spectroscopy of Comets V1.0.* NASA Planetary Data System, urn:nasa:pds:spitzer:spitzer-spec-comet.
Kemper F., Vriend W. J., and Tielens A. G. G. M. (2004) The absence of crystalline silicates in the diffuse interstellar medium. *Astrophys. J., 609,* 826–837.
Kemper F., Vriend W. J., and Tielens A. G. G. M. (2005) Erratum: "The absence of crystalline silicates in the diffuse interstellar medium" (ApJ, 609, 826 [2004]). *Astrophys. J., 633,* 534.
Kerridge J. F., Mackay A. L., and Boynton W. V. (1979) Magnetite in CI carbonaceous meteorites: Origin by aqueous activity on a planetesimal surface. *Science, 205,* 395–397.
Kikuchi S., Mikami Y., Mukai T., Mukai S., and Hough J. H. (1987) Polarimetry of Comet P/Halley. *Astron. Astrophys., 187,* 689–692.
Kimura H., Kolokolova L., and Mann I. (2003) Optical properties of cometary dust: Constraints from numerical studies on light scattering by aggregate particles. *Astron. Astrophys., 407,* L5–L8.
Kimura H., Kolokolova L., and Mann I. (2006) Light scattering by cometary dust numerically simulated with aggregate particles consisting of identical spheres. *Astron. Astrophys., 449,* 1243–1254.
Kimura H., Kolokolova L., Li A., and Lebreton J. (2016) Light scattering and thermal emission by primitive dust particles in planetary systems. In *Light Scattering Reviews, Vol. 11: Light Scattering and Radiative Transfer* (A. Kokhanovsky, ed.), pp. 363–418. Springer, Berlin.
Kiselev N., Rosenbush V., Levasseur-Regourd A.-C., and Kolokolova L. (2015) Comets. In *Polarimetry of Stars and Planetary Systems* (L. Kolokolova et al., eds.), pp. 379–404. Cambridge Univ., Cambridge.
Kiss C., Müller T. G., Kidger M., Mattisson P., and Marton G. (2015) Comet C/2013 A1 (Siding Spring) as seen with the Herschel Space Observatory. *Astron. Astrophys., 574,* L3.
Kissel J., Altwegg K., Clark B. C., Colangeli L., Cottin H., Czempiel S., Eibl J., Engrand C., Fehringer H. M., Feuerbacher B., Fomenkova M., Glasmachers A., Greenberg J. M., Grün E., Haerendel G., Henkel H., Hilchenbach M., von Hoerner H., Höfner H., Hornung K., Jessberger E. K., Koch A., Krüger H., Langevin Y., Parigger P., Raulin F., Rüdenauer F., Rynö J., Schmid E. R., Schulz R., Silén J., Steiger W., Stephan T., Thirkell L., Thomas R., Torkar K., Utterback N. G., Varmuza K., Wanczek K. P., Werther W., and Zscheeg H. (2007) COSIMA — High resolution time-of-flight secondary ion mass spectrometer for the analysis of cometary dust particles onboard Rosetta. *Space Sci. Rev., 128,* 823–867.
Kita N. T., Huberty J. M., Kozdon R., Beard B. L., and Valley J. W. (2011) High-precision SIMS oxygen, sulfur and iron stable isotope analyses of geological materials: Accuracy, surface topography and crystal orientation. *Surf. Interface Anal., 43,* 427–431.
Klock W., Thomas K. L., McKay D. S., and Palme H. (1989) Unusual olivine and pyroxene composition in interplanetary dust and unequilibrated ordinary chondrites. *Nature, 339,* 126–128.
Koike C., Chihara H., Tsuchiyama A., Suto H., Sogawa H., and Okuda H. (2003) Compositional dependence of infrared absorption spectra of crystalline silicate. II. Natural and synthetic olivines. *Astron. Astrophys., 399,* 1101–1107.
Koike C., Imai Y., Chihara H., Suto H., Murata K., Tsuchiyama A., Tachibana S., and Ohara S. (2010) Effects of forsterite grain shape on infrared spectra. *Astrophys. J., 709,* 983–992.
Kolokolova L., Hanner M. S., Levasseur-Regourd A.-C., and Gustafson B. A. (2004) Physical properties of cometary dust from light scattering and thermal emission. In *Comets II* (M. C. Festou et al., eds.), pp. 577–604. Univ. of Arizona, Tucson.
Kolokolova L., Kimura H., Kiselev N., and Rosenbush V. (2007) Two different evolutionary types of comets proved by polarimetric and infrared properties of their dust. *Astron. Astrophys., 463,* 1189–1196.
Kolokolova L. and Kimura H. (2010) Comet dust as a mixture of aggregates and solid particles: Model consistent with ground-based and space-mission results. *Earth Planets Space, 62,* 17–21.

Kolokolova L., Hough J., and Levasseur-Regourd A.-C., eds. (2015) *Polarimetry of Stars and Planetary Systems*. Cambridge Univ., Cambridge. 487 pp.

Kulyk I., Korsun P., Lukyanyk I., Ivanova O., Afanasiev V., and Lara L. (2021) Optical observations of near isotropic Comet C/2006 OF2 (Broughton) at two different heliocentric distances. *Icarus, 355,* 114156.

Langevin Y., Hilchenbach M., Ligier N., Merouane S., Hornung K., Engrand C., Schulz R., Kissel J., Rynö J., and Eng P. (2016) Typology of dust particles collected by the COSIMA mass spectrometer in the inner coma of 67P/Churyumov Gerasimenko. *Icarus, 271,* 76–97.

Langevin Y., Hilchenbach M., Vincendon M., Merouane S., Hornung K., Ligier N., Engrand C., Schulz R., Kissel J., and Rynö J. (2017) Optical properties of cometary particles collected by the COSIMA mass spectrometer on-board Rosetta during the rendezvous phase around Comet 67P/Churyumov-Gerasimenko. *Mon. Not. R. Astron. Soc., 469,* S535–S549.

Lanzirotti A., Sutton S. R., Flynn G. J., Newville A., and Rao W. (2008) Chemical composition and heterogeneity of Wild 2 cometary particles determined by synchrotron X-ray fluorescence. *Meteorit. Planet. Sci., 43,* 187–213.

Lasue J., Levasseur-Regourd A.-C., Hadamcik E., and Alcouffe G. (2009) Cometary dust properties retrieved from polarization observations: Application to C/1995 O1 Hale-Bopp and 1P/Halley. *Icarus, 199,* 129–144.

Lasue J., Maroger I., Botet R., Garnier P., Merouane S., Mannel T., Levasseur-Regourd A.-C., and Bentley M. S. (2019) Flattened loose particles from numerical simulations compared to particles collected by Rosetta. *Astron. Astrophys., 630,* A28.

Lawler M. E. and Brownlee D. E. (1992) CHON as a component of dust from Comet Halley. *Nature, 359,* 810–812.

Leroux H., Jacob D., Stodolna J., Nakamura-Messenger K., and Zolensky M. E. (2008) Igneous Ca-rich pyroxene in Comet 81P/Wild 2. *Am. Mineral., 93,* 1933–1936.

Leroux H., Cuvillier P., Zanda B., and Hewins R. H. (2015) GEMS-like material in the matrix of the Paris meteorite and the early stages of alteration of CM chondrites. *Geochim. Cosmochim. Acta, 170,* 247–265.

Levasseur-Regourd A.-C., McBride N., Hadamcik E., and Fulle M. (1999) Similarities between *in situ* measurements of local dust light scattering and dust flux impact data within the coma of 1P/Halley. *Astron. Astrophys., 348,* 636–641.

Levasseur-Regourd A.-C., Agarwal J., Cottin H., Engrand C., Flynn G., Fulle M., Gombosi T., Langevin Y., Lasue J., and Mannel T. (2018) Cometary dust. *Space Sci. Rev., 214,* 1–56.

Levasseur-Regourd A.-C., Renard J.-B., Hadamcik E., Lasue J., Bertini I., and Fulle M. (2019) Interpretation through experimental simulations of phase functions revealed by Rosetta in 67P/Churyumov-Gerasimenko dust coma. *Astron. Astrophys., 630,* A20.

Li A. G. and Greenberg J. M. (1997) A unified model of interstellar dust. *Astron. Astrophys., 323,* 566–584.

Li J.-Y., Samarasinha N. H., Kelley M. S. P., Farnham T. L., A'Hearn M. F., Mutchler M. J., Lisse C. M., and Delamere W. A. (2014) Constraining the dust coma properties of Comet C/Siding Spring (2013 A1) at large heliocentric distances. *Astrophys. J. Lett., 797,* L8.

Lin Y., Amari S., and Pravdivtseva O. (2002) Presolar grains from the Qingzhen (EH3) meteorite. *Astrophys. J., 575,* 257–263.

Lindsay S. S., Wooden D. H., Harker D. E., Kelley M. S., Woodward C. E., and Murphy J. R. (2013) Absorption efficiencies of forsterite. I. Discrete dipole approximation explorations in grain shape and size. *Astrophys. J., 766,* 54.

Lisse C. M., A'Hearn M. F., Farnham T. L., Groussin O., Meech K. J., Fink U., and Schleicher D. G. (2005) The coma of Comet 9P/Tempel 1. *Space Sci. Rev., 117,* 161–192.

Lisse C. M., VanCleve J., Adams A. C., A'Hearn M. F., Fernández Y. R., Farnham T. L., Armus L., Grillmair C. J., Ingalls J., Belton M. J. S., Groussin O., McFadden L. A., Meech K. J., Schultz P. H., Clark B. C., Feaga L. M., and Sunshine J. M. (2006) Spitzer spectral observations of the Deep Impact ejecta. *Science, 313,* 635–640.

Lisse C. M., Kraemer K. E., Nuth J. A., Li A., and Joswiak D. (2007) Comparison of the composition of the Tempel 1 ejecta to the dust in Comet C/Hale Bopp 1995 O1 and YSO HD 100546. *Icarus, 191,* 223–240.

LLera K., Burch J. L., Goldstein R., and Goetz C. (2020) Simultaneous observation of negatively and positively charged nanograins at Comet 67P/Churyumov-Gerasimenko. *Geophys. Res. Lett., 47,* e2019GL086147.

Lodders K. (2010) Solar system abundances of the elements. In *Principles and Perspectives in Cosmochemistry* (A. Goswami and B. E. Reddy, eds.), pp. 379–417. Astrophysics and Space Science Proceedings, Springer, Berlin.

Love S. G. and Brownlee D. E. (1993) A direct measurement of the terrestrial mass accretion rate of cosmic dust. *Science, 262,* 550–553.

Manfroid J., Hutsemékers D., and Jehin E. (2021) Iron and nickel atoms in cometary atmospheres even far from the Sun. *Nature, 593,* 372–374.

Mannel T., Bentley M. S., Schmied R., Jeszenszky H., Levasseur-Regourd A.-C., Romstedt J., and Torkar K. (2016) Fractal cometary dust — A window into the early solar system. *Mon. Not. R. Astron. Soc., 462,* S304–S311.

Mannel T., Bentley M. S., Boakes P. D., Jeszenszky H., Ehrenfreund P., Engrand C., Koeberl C., Levasseur-Regourd A.-C., Romstedt J., Schmied R., Torkar K., and Weber I. (2019) Dust of Comet 67P/Churyumov-Gerasimenko collected by Rosetta/MIDAS: Classification and extension to the nanometer scale. *Astron. Astrophys., 630,* A26.

Mathurin J., Dartois E., Pino T., Engrand C., Duprat J., Deniset-Besseau A., Borondics F., Sandt C., and Dazzi A. (2019) Nanometre-scale infrared chemical imaging of organic matter in ultra-carbonaceous Antarctic micrometeorites (UCAMMs). *Astron. Astrophys., 622,* A160.

Matrajt G., Muñoz Caro G. M., Dartois E., d'Hendecourt L., Deboffle D., and Borg J. (2005) FTIR analysis of the organics in IDPs: Comparison with the IR spectra of the diffuse interstellar medium. *Astron. Astrophys., 433,* 979–995.

Matrajt G., Ito M., Wirick S., Messenger S., Brownlee D. E., Joswiak D., Flynn G., Sandford S., Snead C., and Westphal A. (2008) Carbon investigation of two Stardust particles: A TEM, NanoSIMS, and XANES study. *Meteorit. Planet. Sci., 43,* 315–334.

Matrajt G., Messenger S., Brownlee D., and Joswiak D. (2012) Diverse forms of primordial organic matter identified in interplanetary dust particles. *Meteorit. Planet. Sci., 47,* 525–549.

Matrajt G., Messenger S., Joswiak D., and Brownlee D. (2013) Textures and isotopic compositions of carbonaceous materials in A and B-type Stardust tracks: Track 130 (Bidi), track 141 (Coki) and track 80 (Tule). *Geochim. Cosmochim. Acta, 117,* 65–79.

Matzel J. E. P., Ishii H. A., Joswiak D., Hutcheon I. D., Bradley J. P., Brownlee D., Weber P. K., Teslich N., Matrajt G., McKeegan K. D., and MacPherson G. J. (2010) Constraints on the formation age of cometary material from the NASA Stardust mission. *Science, 328,* 483–486.

Maurette M. (2006) *Micrometeorites and the Mysteries of Our Origins*. Springer, Berlin. 330 pp.

Maurette M., Engrand C., Brack A., Kurat G., Leach S., and Perreau M. (1995) Carbonaceous phases in Antarctic micrometeorites and their mineralogical environment: Their contribution to the possible role of micrometeorites as "chondritic chemical reactors" in atmospheres, waters and/or ices. In *26th Lunar and Planetary Science Conference*, pp. 913–914. LPI Contribution No. 1592, Lunar and Planetary Institute, Houston.

McKeegan K. D. (1987) Ion microprobe measurements of H, C, O, Mg and Si isotopic abundances in individual interplanetary dust particles. Ph.D. thesis, Washington University, St. Louis. 201 pp.

McKeegan K. D., Walker R. M., and Zinner E. (1985) Ion microprobe isotopic measurements of individual interplanetary dust particles. *Geochim. Cosmochim. Acta, 49,* 1971–1987.

McKeegan K. D., Aléon J., Bradley J., Brownlee D., Busemann H., Butterworth A., Chaussidon M., Fallon S., Floss C., Gilmour J., Gounelle M., Graham G., Guan Y., Heck P. R., Hoppe P., Hutcheon I. D., Huth J., Ishii H., Ito M., Jacobsen S. B., Kearsley A., Leshin L. A., Liu M.-C., Lyon I., Marhas K., Marty B., Matrajt G., Meibom A., Messenger S., Mostefaoui S., Mukhopadhyay S., Nakamura-Messenger K., Nittler L., Palma R., Pepin R. O., Papanastassiou D. A., Robert F., Schlutter D., Snead C. J., Stadermann F. J., Stroud R., Tsou P., Westphal A., Young E. D., Ziegler K., Zimmermann L., and Zinner E. (2006) Isotopic compositions of cometary matter returned by Stardust. *Science, 314,* 1724–1728.

Meier R., Owen T. C., Jewitt D. C., Matthews H. E., Senay M., Biver N., Bockelée-Morvan D., Crovisier J., and Gautier D. (1998) Deuterium in Comet C/1995 O1 (Hale-Bopp): Detection of DCN.

Science, 279, 1707–1710.
Meisner T., Wurm G., and Teiser J. (2012) Experiments on centimeter-sized dust aggregates and their implications for planetesimal formation. *Astron. Astrophys., 544,* A138.
Mendis D. A. and Horányi M. (2013) Dusty plasma effects in comets: Expectations for Rosetta. *Rev. Geophys., 51,* 53–75.
Merouane S., Djouadi Z., and Le Sergeant d'Hendecourt L. (2014) Relations between aliphatics and silicate components in 12 stratospheric particles deduced from vibrational spectroscopy. *Astrophys. J., 780,* 174.
Merouane S., Stenzel O., Hilchenbach M., Schulz R., Altobelli N., Fischer H., Hornung K., Kissel J., Langevin Y., Mellado E., RynöJ., and Zaprudin B. (2017) Evolution of the physical properties of dust and cometary dust activity from 67P/Churyumov-Gerasimenko measured *in situ* by Rosetta/COSIMA. *Mon. Not. R. Astron. Soc., 469,* S459–S474.
Maurette M., Hammer C., Brownlee D. E., Reeh N., and Thomsen H. H. (1986) Placers of cosmic dust in the blue ice lakes of Greenland. *Science, 233,* 869–872.
Maurette M., Jéhano C., Robin E., and Hammer C. (1987) Characteristics and mass distribution of extraterrestrial dust from the Greenland ice cap. *Nature, 328,* 699–702.
Maurette M., Olinger C., Michel-Levy M. C., Kurat G., Pourchet M., Brandstatter F., and Bourot-Denise M. (1991) A collection of diverse micrometeorites recovered from 100 tonnes of Antarctic blue ice. *Nature, 351,* 44–47.
McDonnell J. A. M., Alexander W. M., Burton W. M., Bussoletti E., Clark D. H., Grard J. L., Gruen E., Hanner M. S., Sekanina Z., and Hughes D. W. (1986) Dust density and mass distribution near Comet Halley from Giotto observations. *Nature, 326,* 338–341.
McDonnell J. A. M., Alexander W. M., Burton W. M., Bussoletti E., Evans G. C., Evans S. T., Firth J. G., Grard R. J. L., Green S. F., and Grun E. (1987) The dust distribution within the inner coma of Comet P/Halley 1982i: Encounter by Giotto's impact detectors. In *Exploration of Halley's Comet* (M. Grewing et al., eds.), pp. 719–741. Springer, Berlin.
McDonnell J. A. M., McBride N., Beard R., Bussoletti E., Colangeli L., Eberhardt P., Firth J. G., Grard R., Green S. F., Greenberg J. M., Grün E., Hughes D. W., Keller H. U., Kissel J., Lindblad B. A., Mandeville J.-C., Perry C. H., Rembor K., Rickman H., Schwehm G. H., Turner R. F., Wallis M. K., and Zarnecki J. C. (1993) Dust particle impacts during the Giotto encounter with Comet Grigg–Skjellerup. *Nature, 362,* 732–734.
Merouane S., Zaprudin B., Stenzel O., Langevin Y., Altobelli N., Della Corte V., Fischer H., Fulle M., Hornung K., Silén J., Ligier N., Rotundi A., Ryno J., Schulz R., Hilchenbach M., Kissel J., and the COSIMA Team (2016) Dust particle flux and size distribution in the coma of 67P/Churyumov-Gerasimenko measured *in situ* by the COSIMA instrument on board Rosetta. *Astron. Astrophys., 596,* A87.
Messenger S. (2000) Identification of molecular-cloud material in interplanetary dust particles. *Nature, 404,* 968–971.
Messenger S., Stadermann F. J., Floss C., Nittler L. R., and Mukhopadhyay S. (2003) Isotopic signatures of presolar materials in interplanetary dust. *Space Sci. Rev., 106,* 155–172.
Mikouchi T., Tachikawa O., Hagiya K., Ohsumi K., Suzuki Y., Uesugi K., Takeuchi A., and Zolensky M. E. (2007) Mineralogy and crystallography of Comet 81P/Wild 2 particles. In *38th Lunar and Planetary Science Conference*, Abstract #1946. LPI Contribution No. 1338, Lunar and Planetary Institute, Houston.
Min M. (2005) Optical properties of circumstellar and cometary grains. Ph.D. thesis, Universiteit van Amsterdam, Amsterdam. 220 pp.
Moreels G., Clairemidi J., Hermine P., Brechignac P., and Rousselot P. (1994) Detection of a polycyclic aromatic molecule in Comet P/Halley. *Astron. Astrophys., 282,* 643–656.
Moreno F., Muñoz O., Vilaplana R., and Molina A. (2003) Irregular particles in Comet C/1995 O1 Hale-Bopp inferred from its mid-infrared spectrum. *Astrophys. J., 595,* 522–530.
Moreno F., Muñoz O., Gutiérrez P. J., Lara L. M., Snodgrass C., Lin Z. Y., Della Corte V., Rotundi A., and Yagi M. (2017) The dust environment of Comet 67P/Churyumov-Gerasimenko: Results from Monte Carlo dust tail modelling applied to a large ground-based observation data set. *Mon. Not. R. Astron. Soc., 469,* S186–S194.
Mukai T., Mukai S., and Kikuchi S. (1988) Complex refractive index of grain material deduced from the visible polarimetry of Comet P/Halley. In *Exploration of Halley's Comet* (M. Grewing et al., eds.), pp. 650–652. Springer, Berlin.
Müller D. R., Altwegg K., Berthelier J. J., Combi M., De Keyser J., Fuselier S. A., Hänni N., Pestoni B., Rubin M., Schroeder I. R. H. G., and Wampfler S. F. (2022) High D/H ratios in water and alkanes in Comet 67P/Churyumov-Gerasimenko measured with Rosetta/ROSINA DFMS. *Astron. Astrophys., 662,* A69.
Muñoz Caro G. M., Matrajt G., Dartois E., Nuevo M., d'Hendecourt L., Deboffle D., Montagnac G., Chauvin N., Boukari C., and Le Du D. (2006) Nature and evolution of the dominant carbonaceous matter in interplanetary dust particles: Effects of irradiation and identification with a type of amorphous carbon. *Astron. Astrophys., 459,* 147–159.
Nakamura T., Noguchi T., Ozono Y., Osawa T., and Nagao K. (2005) Mineralogy of ultracarbonaceous large micrometeorites. *Meteorit. Planet. Sci. Suppl., 40,* 5046.
Nakamura T., Noguchi T., Tsuchiyama A., Ushikubo T., Kita N. T., Valley J. W., Zolensky M. E., Kakazu Y., Sakamoto K., Mashio E., Uesugi K., and Nakano T. (2008) Chondrule-like objects in short-period Comet 81P/Wild 2. *Science, 321,* 1664–1667.
Nakashima D., Ushikubo T., Joswiak D. J., Brownlee D. E., Matrajt G., Weisberg M. K., Zolensky M. E., and Kita N. T. (2012a) Oxygen isotopes in crystalline silicates of Comet Wild 2: A comparison of oxygen isotope systematics between Wild 2 particles and chondritic materials. *Earth Planet. Sci. Lett., 357–358,* 355–365.
Nakashima D., Ushikubo T., Zolensky M. E., and Kita N. T. (2012b) High precision oxygen three-isotope analyses of anhydrous chondritic interplanetary dust particles. *Meteorit. Planet. Sci., 47,* 197–208.
Nakashima D., Ushikubo T., Kita N. T., Weisberg M. K., Zolensky M. E., and Ebel D. S. (2015) Late formation of a Comet Wild 2 crystalline silicate particle, Pyxie, inferred from Al-Mg chronology of plagioclase. *Earth Planet. Sci. Lett., 410,* 54–61.
Nguyen A. N., Stadermann F. J., Zinner E., Stroud R. M., Alexander C. M. O., and Nittler L. R. (2007) Characterization of presolar silicate and oxide grains in primitive carbonaceous chondrites. *Astrophys. J., 656,* 1223–1240.
Nguyen A. N., Berger E. L., Nakamura-Messenger K., Messenger S., and Keller L. P. (2017) Coordinated mineralogical and isotopic analyses of a cosmic symplectite discovered in a Comet 81P/Wild 2 sample. *Meteorit. Planet. Sci., 52,* 2004–2016.
Nguyen A. N., Brownlee D. E., and Joswiak D. J. (2020) High presolar silicate abundance in giant cluster IDP U2-20GCA of probable cometary origin. In *51st Lunar and Planetary Science Conference*, Abstract #2487. LPI Contribution No. 2326, Lunar and Planetary Institute, Houston.
Nier A. O. and Schlutter D. J. (1993) The thermal history of interplanetry dust particles collected in the Earth's stratosphere. *Meteoritics, 28,* 675–681.
Nittler L. R. and Alexander C. M. (2003) Automated isotopic measurements of micron-sized dust: Application to meteoritic presolar silicon carbide. *Geochim. Cosmochim. Acta, 67,* 4961–4980.
Nittler L. R., Stroud R. M., Trigo-Rodríguez J. M., De Gregorio B. T., Alexander C. M. O., Davidson J., Moyano-Cambero C. E., and Tanbakouei S. (2019) A cometary building block in a primitive asteroidal meteorite. *Nature Astron., 3,* 659–666.
Noguchi T., Ohashi N., Tsujimoto S., Mitsunari T., Bradley J. P., Nakamura T., Toh S., Stephan T., Iwata N., and Imae N. (2015) Cometary dust in Antarctic ice and snow: Past and present chondritic porous micrometeorites preserved on the Earth's surface. *Earth Planet. Sci. Lett., 410,* 1–11.
Noguchi T., Yabuta H., Itoh S., Sakamoto N., Mitsunari T., Okubo A., Okazaki R., Nakamura T., Tachibana S., Terada K., Ebihara M., Imae N., Kimura M., and Nagahara H. (2017) Variation of mineralogy and organic material during the early stages of aqueous activity recorded in Antarctic micrometeorites. *Geochim. Cosmochim. Acta, 208,* 119–144.
Noguchi T., Matsumoto R., Yabuta H., Kobayashi H., Miyake A., Naraoka H., Okazaki R., Imae N., Yamaguchi A., Kilcoyne A. L. D., Takeichi Y., and Takahashi Y. (2022) Antarctic micrometeorite composed of CP and CS IDP-like material: A micro-breccia originated from a partially ice-melted comet-like small body. *Meteorit. Planet. Sci., 57,* 2042–2062.
Ogliore R. C., Butterworth A., Gainsforth Z., Huss G. R., Nagashima K., Stodolna J., and Westphal A. J. (2012a) Sulfur isotope measurements of a Stardust fragment. In *43rd Lunar and Planetary Science Conference*, Abstract #1670. LPI Contribution No. 1659, Lunar and Planetary Institute, Houston.
Ogliore R. C., Huss G. R., Nagashima K., Butterworth A. L.,

Gainsforth Z., Stodolna J., Westphal A. J., Joswiak D., and Tyliszczak T. (2012b) Incorporation of a late-forming chondrule into Comet Wild 2. *Astrophys. J. Lett., 745,* L19.

Ogliore R. C., Nagashima K., Huss G. R., Westphal A. J., Gainsforth Z., and Butterworth A. L. (2015) Oxygen isotopic composition of coarse- and fine-grained material from Comet 81P/Wild 2. *Geochim. Cosmochim. Acta, 166,* 74–91.

Ootsubo T., Kawakita H., Shinnaka Y., Watanabe J., and Honda M. (2020) Unidentified infrared emission features in mid-infrared spectrum of Comet 21P/Giacobini-Zinner. *Icarus, 338,* 113450.

Ott T., Drolshagen E., Koschny D., Güttler C., Tubiana C., Frattin E., Agarwal J., Sierks H., Bertini I., and Barbieri C. (2017) Dust mass distribution around Comet 67P/Churyumov-Gerasimenko determined via parallax measurements using Rosetta's OSIRIS cameras. *Mon. Not. R. Astron. Soc., 469,* S276–S284.

Pajola M., Lucchetti A., Fulle M., Mottola S., Hamm M., Da Deppo V., Penasa L., Kovacs G., Massironi M., Shi X., Tubiana C., Güttler C., Oklay N., Vincent J. B., Toth I., Davidsson B., Naletto G., Sierks H., Barbieri C., Lamy P. L., Rodrigo R., Koschny D., Rickman H., Keller H. U., Agarwal J., A'Hearn M. F., Barucci M. A., Bertaux J. L., Bertini I., Cremonese G., Debei S., De Cecco M., Deller J., El Maarry M. R., Fornasier S., Frattin E., Gicquel A., Groussin O., Gutierrez P. J., Höfner S., Hofmann M., Hviid S. F., Ip W.-H., Jorda L., Knollenberg J., Kramm J. R., Kührt E., Küppers M., Lara L. M., Lazzarin M., Moreno J. J. L., Marzari F., Michalik H., Preusker F., Scholten F., and Thomas N. (2017) The pebbles/boulders size distributions on Sais: Rosetta's final landing site on Comet 67P/ Churyumov-Gerasimenko. *Mon. Not. R. Astron. Soc., 469,* S636–S645.

Palma R. L., Pepin R. O., and Schlutter D. (2005) Helium and neon isotopic compositions from IDPs of potentially cometary origin. *Meteorit. Planet. Sci., 40,* A120.

Paquette J. A., Hornung K., Stenzel O. J., Rynö J., Silen J., Kissel J., Hilchenbach M., and the COSIMA Team (2017) The $^{34}S/^{32}S$ isotopic ratio measured in the dust of Comet 67P/Churyumov-Gerasimenko by Rosetta/COSIMA. *Mon. Not. R. Astron. Soc., 469,* S230–S237.

Paquette J. A., Engrand C., Hilchenbach M., Fray N., Stenzel O. J., Silen J., Rynö J., Kissel J., and the COSIMA Team (2018) The oxygen isotopic composition ($^{18}O/^{16}O$) in the dust of Comet 67P/ Churyumov-Gerasimenko measured by COSIMA on-board Rosetta. *Mon. Not. R. Astron. Soc., 477,* 3836–3844.

Paquette J. A., Fray N., Bardyn A., Engrand C., Alexander C. M. O., Siljeström S., Cottin H., Merouane S., Isnard R., Stenzel O. J., Fischer H., Rynö J., Kissel J., and Hilchenbach M. (2021) D/H in the refractory organics of Comet 67P/Churyumov-Gerasimenko measured by Rosetta/COSIMA. *Mon. Not. R. Astron. Soc., 504,* 4940–4951.

Pendleton Y., Sandford S., Allamandola L., Tielens A., and Sellgren K. (1994) Near-infrared absorption spectroscopy of interstellar hydrocarbon grains. *Astrophys. J., 437,* 683–696.

Poch O., Istiqomah I., Quirico E., Beck P., Schmitt B., Theulé P., Faure A., Hily-Blant P., Bonal L., Raponi A., Ciarniello M., Rousseau B., Potin S., Brissaud O., Flandinet L., Filacchione G., Pommerol A., Thomas N., Kappel D., Mennella V., Moroz L., Vinogradoff V., Arnold G., Erard S., Bockelée-Morvan D., Leyrat C., Capaccioni F., De Sanctis M. C., Longobardo A., Mancarella F., Palomba E., and Tosi F. (2020) Ammonium salts are a reservoir of nitrogen on a cometary nucleus and possibly on some asteroids. *Science, 367,* aaw7462.

Pollack J. B., Hollenbach D., Beckwith S., Simonelli D. P., Roush T., and Fong W. (1994) Composition and radiative properties of grains in molecular clouds and accretion disks. *Astrophys. J., 421,* 615–639.

Price M. C., Kearsley A. T., Burchell M. J., Hörz F., Borg J., Bridges J. C., Cole M. J., Floss C., Graham G., Green S. F., Hoppe P., Leroux H., Marhas K. K., Park N., Stroud R., Stadermann F. J., Telisch N., and Wozniakiewicz P. J. (2010) Comet 81P/Wild 2: The size distribution of finer (sub-10 μm) dust collected by the Stardust spacecraft. *Meteorit. Planet. Sci., 45,* 1409–1428.

Quirico E., Raynal P. I., and Bourot-Denise M. (2003) Metamorphic grade of organic matter in six unequilibrated ordinary chondrites. *Meteorit. Planet. Sci., 38,* 795–811.

Quirico E., Borg J., Raynal P.-I., Montagnac G., and d'Hendecourt L. (2005) A micro-Raman survey of 10 IDPs and 6 carbonaceous chondrites. *Planet. Space Sci., 53,* 1443–1448.

Quirico E., Moroz L. V., Schmitt B., Arnold G., Faure M., Beck P., Bonal L., Ciarniello M., Capaccioni F., Filacchione G., Erard S., Leyrat C., Bockelée-Morvan D., Zinzi A., Palomba E., Drossart P., Tosi F., Capria M. T., De Sanctis M. C., Raponi A., Fonti S., Mancarella F., Orofino V., Barucci A., Blecka M. I., Carlson R., Despan D., Faure A., Fornasier S., Gudipati M. S., Longobardo A., Markus K., Mennella V., Merlin F., Piccioni G., Rousseau B., and Taylor F. (2016) Refractory and semi-volatile organics at the surface of Comet 67P/Churyumov-Gerasimenko: Insights from the VIRTIS/ Rosetta imaging spectrometer. *Icarus, 272,* 32–47.

Reach W. T., Vaubaillon J., Lisse C. M., Holloway M., and Rho J. (2010) Explosion of Comet 17P/Holmes as revealed by the Spitzer Space Telescope. *Icarus, 208,* 276–292.

Rietmeijer F. J. M. (1998) Interplanetary dust particles. In *Planetary Materials* (J. J. Papike, ed.), pp. 2-1 to 2-96. Reviews in Mineralogy, Vol. 36, Mineralogical Society of America, Washington.

Rinaldi G., Della Corte V., Fulle M., Capaccioni F., Rotundi A., Ivanovski S. L., Bockelée-Morvan D., Filacchione G., D'Aversa E., Capria M. T., Tozzi G. P., Erard S., Leyrat C., Palomba E., Longobardo A., Ciarniello M., Taylor F., Mottola S., and Salatti M. (2017) Cometary coma dust size distribution from *in situ* IR spectra. *Mon. Not. R. Astron. Soc., 469,* S598–S605.

Rinaldi G., Bockelée-Morvan D., Ciarniello M., Tozzi G. P., Capaccioni F., Ivanovski S. L., Filacchione G., Fink U., Doose L., Taylor F., Kappel D., Erard S., Leyrat C., Raponi A., D'Aversa E., Capria M. T., Longobardo A., Palomba E., Tosi F., Migliorini A., Rotundi A., Della Corte V., and Salatti M. (2018) Summer outbursts in the coma of Comet 67P/Churyumov-Gerasimenko as observed by Rosetta-VIRTIS. *Mon. Not. R. Astron. Soc., 481,* 1235–1250.

Rojas J., Duprat J., Dartois E., Wu T.-D., Engrand C., Augé B., Mathurin J., Guérin B., Guerquin-Kern J.-L., Boduch P., and Rothard H. (2020) Isotopic analyses of ion irradiation-induced organic residues, clues on the formation of organics from UCAMMs. In *51st Lunar and Planetary Science Conference*, Abstract #1630. LPI Contribution No. 2326, Lunar and Planetary Institute, Houston.

Rojas J., Duprat J., Engrand C., Dartois E., Delauche L., Godard M., Gounelle M., Carrillo-Sánchez J., Pokorný P., and Plane J. (2021) The micrometeorite flux at Dome C (Antarctica), monitoring the accretion of extraterrestrial dust on Earth. *Earth Planet. Sci. Lett., 560,* 116794.

Rojas J, Duprat J., Dartois E., Wu T-D., Engrand C., Nittler L. R., Bardin N., Delauche L., Mostefaoui S., Remusat L., Stroud R. M., and Guérin B. (2024) Nitrogen-rich organics from comets probed by ultra-carbonaceous Antarctic micrometeorites. *Nature Astron.,* DOI: 10.1038/s41550-024-02364-y.

Rosenbush V. K., Ivanova O. V., Kiselev N. N., Kolokolova L. O., and Afanasiev V. L. (2017) Spatial variations of brightness, colour and polarization of dust in Comet 67P/Churyumov-Gerasimenko. *Mon. Not. R. Astron. Soc., 469,* S475–S491.

Rotundi A., Ferrini G., Baratta G. A., Palumbo M. E., Palomba E., and Colangeli L. (2007) Combined micro-infrared (IR) and micro-Raman measurements on stratospheric interplanetary dust particles. In *Proceedings of the Workshop on Dust in Planetary Systems* (H. Krueger and A. Graps, eds.), pp. 149–153. ESA SP-643, Noordwijk, The Netherlands.

Rotundi A. and Rietmeijer F. J. M. (2008) Carbon in meteoroids: Wild 2 dust analyses, IDPs and cometary dust analogues. *Earth Moon Planets, 102,* 473–483.

Rotundi A., Sierks H., Corte V. D., Fulle M., Gutierrez P. J., Lara L., Barbieri C., Lamy P. L., Rodrigo R., Koschny D., Rickman H., Keller H. U., López-Moreno J. J., Accolla M., Agarwal J., A'Hearn M. F., Altobelli N., Angrilli F., Barucci M. A., Bertaux J.-L., Bertini I., Bodewits D., Bussoletti E., Colangeli L., Cosi M., Cremonese G., Crifo J.-F., Deppo V. D., Davidsson B., Debei S., Cecco M. D., Esposito F., Ferrari M., Fornasier S., Giovane F., Gustafson B., Green S. F., Groussin O., Grün E., Güttler C., Herranz M. L., Hviid S. F., Ip W., Ivanovski S., Jerónimo J. M., Jorda L., Knollenberg J., Kramm R., Kührt E., Küppers M., Lazzarin M., Leese M. R., López-Jiménez A. C., Lucarelli F., Lowry S. C., Marzari F., Epifani E. M., McDonnell J. A. M., Mennella V., Michalik H., Molina A., Morales R., Moreno F., Mottola S., Naletto G., Oklay N., Ortiz J. L., Palomba E., Palumbo P., Perrin J.-M., Rodríguez J., Sabau L., Snodgrass C., Sordini R., Thomas N., Tubiana C., Vincent J.-B., Weissman P., Wenzel K.-P., Zakharov V., and Zarnecki J. C. (2015) Dust measurements in the coma of Comet 67P/Churyumov-Gerasimenko inbound to the sun. *Science, 347,* aaa3905.

Rousseau B., Érard S., Beck P., Quirico É ., Schmitt B., Brissaud O., Montes-Hernandez G., Capaccioni F., Filacchione G., Bockelée-

Morvan D., Leyrat C., Ciarniello M., Raponi A., Kappel D., Arnold G., Moroz L. V., Palomba E., Tosi F., and the VIRTIS Team (2018) Laboratory simulations of the Vis-NIR spectra of Comet 67P using sub-μm sized cosmochemical analogues. *Icarus, 306,* 306–318.

Rubin M., Altwegg K., Balsiger H., Berthelier J.-J., Bieler A., Calmonte U., Combi M., De Keyser J., Engrand C., Fiethe B., Fuselier S. A., Gasc S., Gombosi T. I., Hansen K. C., Hässig M., Le Roy L., Mezger K., Tzou C.-Y., Wampfler S. F., and Wurz P. (2017) Evidence for depletion of heavy silicon isotopes at Comet 67P/Churyumov-Gerasimenko. *Astron. Astrophys., 601,* A123.

Rubin M., Altwegg K., Berthelier J.-J., Combi M. R., De Keyser J., Dhooghe F., Fuselier S., Gombosi T. I., Hänni N., Müller D., Pestoni B., Wampfler S. F., and Wurz P. (2022) Refractory elements in the gas phase for Comet 67P/Churyumov-Gerasimenko — Possible release of atomic Na, Si, and Fe from nanograins. *Astron. Astrophys., 658,* A87.

Sandford S., Allamandola L., Tielens A., Sellgren K., Tapia M., and Pendleton Y. (1991) The interstellar C-H stretching band near 3.4 microns: Constraints on the composition of organic material in the diffuse interstellar medium. *Astrophys. J., 371,* 607–620.

Sandford S. A., Aléon J., Alexander C. M. O. D., Araki T., Bajt S., Baratta G. A., Borg J., Bradley J. P., Brownlee D. E., Brucato J. R., Burchell M. J., Busemann H., Butterworth A., Clemett S. J., Cody G., Colangeli L., Cooper G., d'Hendecourt L., Djouadi Z., Dworkin J. P., Ferrini G., Fleckenstein H., Flynn G. J., Franchi I. A., Fries M., Gilles M. K., Glavin D. P., Gounelle M., Grossemy F., Jacobsen C., Keller L. P., Kilcoyne A. L. D., Leitner J., Matrajt G., Meibom A., Mennella V., Mostefaoui S., Nittler L. R., Palumbo M. E., Papanastassiou D. A., Robert F., Rotundi A., Snead C. J., Spencer M. K., Stadermann F. J., Steele A., Stephan T., Tsou P., Tyliszczak T., Westphal A. J., Wirick S., Wopenka B., Yabuta H., Zare R. N., and Zolensky M. E. (2006) Organics captured from Comet 81P/Wild 2 by the Stardust spacecraft. *Science, 314,* 1720–1724.

Sandford S. A., Bajt S., Clemett S. J., Cody G. D., Cooper G., Degregorio B. T., de Vera V., Dworkin J. P., Elsila J. E., Flynn G. J., Glavin D. P., Lanzirotti A., Limero T., Martin M. P., Snead C. J., Spencer M. K., Stephan T., Westphal A., Wirick S., Zare R. N., and Zolensky M. E. (2010) Assessment and control of organic and other contaminants associated with the Stardust sample return from Comet 81P/Wild 2. *Meteorit. Planet. Sci., 45,* 406–433.

Sansberro I., Fray N., Cottin H., Baklouti D., Bardyn A., Briois C., Engrand C., Paquette J., Silén J., Stenzel O. J., and Hilchenbach M. (2022) Variability in the elemental composition of the dust particles of 67P/Churyumov-Gerasimenko. *COSPAR Scientific Assembly, 44,* 175.

Schmitt-Kopplin P., Gabelica Z., Gougeon R. D., Fekete A., Kanawati B., Harir M., Gebefuegi I., Eckel G., and Hertkorn N. (2010) High molecular diversity of extraterrestrial organic matter in Murchison meteorite revealed 40 years after its fall. *Proc. Natl. Acad. Sci. U.S.A., 107,* 2763–2768.

Schramm L., Brownlee D., and Wheelock M. (1989) Major element composition of stratospheric micrometeorites. *Meteoritics, 24,* 99–112.

Schroeder I. R. H. G. I, Altwegg K., Balsiger H., Berthelier J.-J., De Keyser J., Fiethe B., Fuselier S. A., Gasc S., Gombosi T. I., Rubin M., Sémon T., Tzou C.-Y., Wampfler S. F., and Wurz P. (2019) $^{16}O/^{18}O$ ratio in water in the coma of Comet 67P/Churyumov-Gerasimenko measured with the Rosetta/ROSINA double-focusing mass spectrometer. *Astron. Astrophys., 630,* A29.

Schulze H., Kissel J., and Jessberger E. K. (1997) Chemistry and mineralogy of Comet Halley's dust. In *From Stardust to Planetesimals* (Y. J. Pendleton, ed.), pp. 397–414. ASP Conf. Ser. 122, Astronomical Society of the Pacific, San Francisco.

Simon S. B., Joswiak D. J., Ishii H. A., Bradley J. P., Chi M., Grossman L., Aléon J., Brownlee D. E., Fallon S., Hutcheon I. D., Matrajt G., and McKeegan K. D. (2008) A refractory inclusion returned by stardust from Comet 81P/Wild 2. *Meteorit. Planet. Sci., 43,* 1861–1877.

Stadermann F. J., Hoppe P., Floss C., Heck P. R., Horz F., Huth J., Kearsley A. T., Leitner J., Marhas K. K., McKeegan K. D., and Stephan T. (2008) Stardust in Stardust — the C, N, and O isotopic compositions of Wild 2 cometary matter in Al foil impacts. *Meteorit. Planet. Sci., 43,* 299–313.

Starkey N. A., Franchi I. A., and Alexander C. M. O. (2013) A Raman spectroscopic study of organic matter in interplanetary dust particles and meteorites using multiple wavelength laser excitation. *Meteorit. Planet. Sci., 48,* 1800–1822.

Stephan T. (2008) Assessing the elemental composition of Comet 81P/Wild 2 by analyzing dust collected by Stardust. *Space Sci. Rev., 138,* 247–258.

Stephan T., Rost D., Vicenzi E. P., Bullock E. S., Macpherson G. J., Westphal A. J., Snead C. J., Flynn G. J., Sandford S. A., and Zolensky M. E. (2008) TOF-SIMS analysis of cometary matter in Stardust aerogel tracks. *Meteorit. Planet. Sci., 43,* 233–246.

Stephan T., Bose M., Boujibar A., Davis A., Dory C., Gyngard F., Hoppe P., Hynes K., Liu N., Nittler L. et al. (2020) The presolar grain database reloaded-silicon carbide. In *51st Lunar and Planetary Science Conference*, Abstract #2140. LPI Contribution No. 2326, Lunar and Planetary Institute, Houston.

Steyer T. R. (1974) Infrared optical properties of some solids of possible interest in astronomy and atmospheric physics. Ph.D. thesis, University of Arizona, Tucson. 168 pp.

Storrs A. D., Cochran A. L., and Barker E. S. (1992) Spectrophotometry of the continuum in 18 comets. *Icarus, 98,* 163–178.

Sugita S., Ootsubo T., Kadono T., Honda M., Sako S., Miyata T., Sakon I., Yamashita T., Kawakita H., Fujiwara H., Fujiyoshi T., Takato N., Fuse T., Watanabe J., Furusho R., Hasegawa S., Kasuga T., Sekiguchi T., Kinoshita D., Meech K. J., Wooden D. H., Ip W.-H., and A'Hearn M. F. (2005) Subaru Telescope observations of Deep Impact. *Science, 310,* 274–278.

Suto H., Sogawa H., Tachibana S., Koike C., Karoji H., Tsuchiyama A., Chihara H., Mizutani K., Akedo J., Ogiso K., Fukui T., and Ohara S. (2006) Low-temperature single crystal reflection spectra of forsterite. *Mon. Not. R. Astron. Soc., 370,* 1599–1606.

Tamanai A., Alexander D. R., Ferguson J. W., and Sedlmayr E. (2003) Optical constants and extinction efficiency of solid materials. In *Stellar Atmosphere Modeling* (I. Hubeny et al., eds.), p. 365. ASP Conf. Ser. 288, Astronomical Society of the Pacific, San Francisco.

Tamanai A., Mutschke H., Blum J., and Neuhäuser R. (2006) Experimental infrared spectroscopic measurement of light extinction for agglomerate dust grains. *J. Quant. Spec. Radiat. Transf., 100,* 373–381.

Tamanai A., Mutschke H., Blum J., Posch T., Koike C., and Ferguson J. W. (2009) Morphological effects on IR band profiles: Experimental spectroscopic analysis with application to observed spectra of oxygen-rich AGB stars. *Astron. Astrophys., 501,* 251–267.

Tazaki R., Tanaka H., Okuzumi S., Kataoka A., and Nomura H. (2016) Light scattering by fractal dust aggregates. I. Angular dependence of scattering. *Astrophys. J., 823,* 70.

Thomas K. L., Blanford G. E., Keller L. P., Klock W., and McKay D. S. (1993) Carbon abundance and silicate mineralogy of anhydrous interplanetary dust particles. *Geochim. Cosmochim. Acta, 57,* 1551–1566.

Tokunaga A. T., Golisch W. F., Griep D. M., Kaminski C. D., and Hanner M. S. (1986) The NASA Infrared Telescope Facility Comet Halley monitoring program. I. Preperihelion results. *Astron. J., 92,* 1183–1190.

Toppani A., Robert F., Libourel G., de Donato P., Barres O., d'Hendecourt L., and Ghanbaja J. (2005) A 'dry' condensation origin for circumstellar carbonates. *Nature, 437,* 1121–1124.

Tozzi G. P., Lara L. M., Kolokolova L., Boehnhardt H., Licandro J., and Schulz R. (2004) Sublimating components in the coma of Comet C/2000 WM1 (LINEAR). *Astron. Astrophys., 424,* 325–330.

Tozzi G.-P., Rinaldi G., Fink U., Doose L., Capaccioni F., Filacchione G., Bockelée-Morvan D., Erard S., Leyrat C., Arnold G., Blecka M., Capria M. T., Ciarniello M., Combi M., Faggi S., Irwin P., Migliorini A., Paolomba E., Piccioni G., and Tosi F. (2015) Study of the characteristics of the grains in the coma background and in the jets in comets 67P/C-G, as observed by VIRTIS-M onboard of the Rosetta mission. *AAS/Division for Planetary Sciences Meeting Abstracts, 47,* #503.07.

Trigo-Rodríguez J. M. and Llorca J. (2006) The strength of cometary meteoroids: Clues to the structure and evolution of comets. *Mon. Not. R. Astron. Soc., 372,* 655–660.

Trigo-Rodríguez J. M., Dominguez G., Burchell M. J., Hoerz F., and Llorca J. (2008) Bulbous tracks arising from hypervelocity capture in aerogel. *Meteorit. Planet. Sci., 43,* 75–86.

Tuzzolino A. J., Economou T. E., Clark B. C., Tsou P., Brownlee D. E., Green S. F., McDonnell J. A. M., McBride N., and Colwell M. T. S. H. (2004) Dust measurements in the coma of Comet 81P/Wild 2 by the dust flux monitor instrument. *Science, 304,* 1776–1780.

Vernazza P., Marsset M., Beck P., Binzel R. P., Birlan M., Brunetto R., Demeo F. E., Djouadi Z., Dumas C., Merouane S., Mousis O., and Zanda B. (2015) Interplanetary dust particles as samples of icy asteroids. *Astrophys. J., 806,* 204.

Villalon K. L., Ishii H. A., Bradley J. P., Stephan T., and Davis A. M. (2016) Resolving the ancestry of GEMS with CHILI. In *47th Lunar and Planetary Science Conference,* Abstract #1796. LPI Contribution No. 1903, Lunar and Planetary Institute, Houston.

Warren J. L. and Zolensky M. E. (1994) Collection and curation of interplanetary dust particles recovered from the stratosphere by NASA. In *Analysis of Interplanetary Dust* (M. Zolensky et al., eds.), pp. 245–253. AIP Conf. Ser. 310, American Institute of Physics, Melville, New York.

Westphal A. J., Bridges J. C., Brownlee D. E., Butterworth A. L., De Gregorio B. T., Dominguez G., Flynn G. J., Gainsforth Z., Ishii H. A., Joswiak D., Nittler L. R., Ogliore R. C., Palma R., Pepin R. O., Stephan T., and Zolensky M. E. (2017) The future of Stardust science. *Meteorit. Planet. Sci., 52,* 1859–1898.

Wirick S., Flynn G. J., Keller L. P., Nakamura-Messenger K., Peltzer C., Jacobsen C., Sandford S. A., and Zolensky M. E. (2009) Organic matter from Comet 81P/Wild 2, IDPs, and carbonaceous meteorites: Similarities and differences. *Meteorit. Planet. Sci., 44,* 1611–1626.

Wooden D. H. (2002) Comet grains: Their IR emission and their relation to ISM grains. *Earth Moon Planets, 89,* 247–287.

Wooden D. H., Harker D. E., Woodward C. E., Butner H. M., Koike C., Witteborn F. C., and McMurtry C. W. (1999) Silicate mineralogy of the dust in the inner coma of Comet C/1995 O1 (Hale-Bopp) pre- and postperihelion. *Astrophys. J., 517,* 1034–1058.

Wooden D. H., Butner H. M., Harker D. E., and Woodward C. E. (2000) Mg-rich silicate crystals in Comet Hale-Bopp: ISM relics or solar nebula condensates? *Icarus, 143,* 126–137.

Wooden D. H., Woodward C. E., and Harker D. E. (2004) Discovery of crystalline silicates in Comet C/2001 Q4 (NEAT). *Astrophys. J. Lett., 612,* L77–L80.

Wooden D. H., Harker D. E., and Brearley A. J. (2005) Thermal processing and radial mixing of dust: Evidence from comets and primitive chondrites. In *Chondrites and the Protoplanetary Disk* (A. N. Krot et al., eds.), p. 774. ASP Conf. Ser. 341, Astronomical Society of the Pacific, San Francisco.

Wooden D., Desch S., Harker D., Gail H.-P., and Keller L. (2007) Comet grains and implications for heating and radial mixing in the protoplanetary disk. In *Protostars and Planets V* (B. Reipurth et al., eds.), pp. 815–833. Univ. of Arizona, Tucson.

Wooden D. H., Woodward C. E., Kelley M. S., Harker D. E., Geballe T. R., and Li A. (2011) Search for organics in 103P/Hartley 2 in the near-IR with GNIRS. *EPSC-DPS Joint Meeting 2011,* 1557.

Wooden D. H., Ishii H. A., and Zolensky M. E. (2017) Cometary dust: The diversity of primitive refractory grains. *Philos. Trans. R. Soc. A, 375,* 20160260.

Wooden D. H., Woodward C. E., Harker D. E., and Kelley M. S. P. (2021) Cometary particles with amorphous carbon: A large carbon reservoir and aggregate particles' optical properties computed with DDSCAT. In *52nd Lunar and Planetary Science Conference,* Abstract #2694. LPI Contribution No. 2548, Lunar and Planetary Institute, Houston.

Woodward C. E., Kelley M. S. P., Harker D. E., Ryan E. L., Wooden D. H., Sitko M. L., Russell R. W., Reach W. T., de Pater I., Kolokolova L., and Gehrz R. D. (2015) SOFIA infrared spectrophotometry of Comet C/2012 K1 (Pan-STARRS). *Astrophys. J., 809,* 181.

Woodward C. E., Wooden D. H., Harker D. E., Kelley M. S. P., Russell R. W., and Kim D. L. (2021) The coma dust of Comet C/2013 US10 (Catalina): A window into carbon in the solar system. *Planet. Sci. J., 2,* 25.

Wopenka B. (1988) Raman observations on individual interplanetary dust particles. *Earth Planet. Sci. Lett., 88,* 221–231.

Wozniakiewicz P. J., Ishii H. A., Kearsley A. T., Bradley J. P., Price M. C., Burchell M. J., Teslich N., and Cole M. J. (2015) The survivability of phyllosilicates and carbonates impacting Stardust Al foils: Facilitating the search for cometary water. *Meteorit. Planet. Sci., 50,* 2003–2023.

Wurz P., Rubin M., Altwegg K., Balsiger H., Berthelier J.-J., Bieler A., Calmonte U., De Keyser J., Fiethe B., Fuselier S. A., Galli A., Gasc S., Gombosi T. I., Jäckel A., Le Roy L., Mall U. A., Rème H., Tenishev V., and Tzou C.-Y. (2015) Solar wind sputtering of dust on the surface of 67P/Churyumov-Gerasimenko. *Astron. Astrophys., 583,* A22.

Xing Z. and Hanner M. S. (1997) Light scattering by aggregate particles. *Astron. Astrophys., 324,* 805–820.

Yabuta H., Noguchi T., Itoh S., Nakamura T., Mitsunari T., Okubo A., Okazaki R., Tachibana T., Terada K., Ebihara M., and Nagahara H. (2015) Variations in organic functional groups between hydrous and anhydrous Antarctic micrometeorites. In *78th Annual Meeting of the Meteoritical Society,* Abstract #5301. LPI Contribution No. 1856, Lunar and Planetary Institute, Houston.

Yabuta H., Noguchi T., Itoh S., Nakamura T., Miyake A., Tsu-jimoto S., Ohashi N., Sakamoto N., Hashiguchi M., Ichi Abe K., Okubo A., Kilcoyne A. D., Tachibana S., Okazaki R., Terada K., Ebihara M., and Nagahara H. (2017) Formation of an ultracarbonaceous Antarctic micrometeorite through minimal aqueous alteration in a small porous icy body. *Geochim. Cosmochim. Acta, 214,* 172–190.

Zhang M., Defouilloy C., Joswiak D. J., Brownlee D. E., Nakashima D., Siron G., Kitajima K., and Kita N. T. (2021) Oxygen isotope systematics of crystalline silicates in a giant cluster IDP: A genetic link to Wild 2 particles and primitive chondrite chondrules. *Earth Planet. Sci. Lett., 564,* 116928.

Zinner E., McKeegan K. D., and Walker R. M. (1983) Laboratory measurements of D/H ratios in interplanetary dust. *Nature, 305,* 119–121.

Zolensky M. E. (1987) Refractory interplanetary dust particles. *Science, 237,* 1466–1468.

Zolensky M. E. and Barrett R. A. (1994) Compositional variations of olivines and pyroxenes in chondritic interplanetary dust particles. *Meteoritics, 29,* 616–620.

Zolensky M. E., Hewins R. H., Mittlefehldt D. W., Lindstrom M. M., Xiao X., and Lipschutz M. E. (1992) Mineralogy, petrology and geochemistry of carbonaceous chondritic clasts in the LEW 85300 polymict eucrite. *Meteoritics, 27,* 596–604.

Zolensky M. E., Nakamura K., Weisberg M. K., Prinz M., Nakamura T., Ohsumi K., Saitow A., Mukai M., and Gounelle M. (2003) A primitive dark inclusion with radiation-damaged silicates in the Ningqiang carbonaceous chondrite. *Meteorit. Planet. Sci., 38,* 305–322.

Zolensky M. E., Zega T. J., Yano H., Wirick S., Westphal A. J., Weisberg M. K., Weber I., Warren J. L., Velbel M. A., Tsuchiyama A., Tsou P., Toppani A., Tomioka N., Tomeoka K., Teslich N., Taheri M., Susini J., Stroud R., Stephan T., Stadermann F. J., Snead C. J., Simon S. B., Simionovici A., See T. H., Robert F., Rietmeijer F. J. M., Rao W., Perronnet M. C., Papanastassiou D. A., Okudaira K., Ohsumi K., Ohnishi I., Nakamura-Messenger K., Nakamura T., Mostefaoui S., Mikouchi T., Meibom A., Matrajt G., Marcus M. A., Leroux H., Lemelle L., Le L., Lanzirotti A., Langenhorst F., Krot A. N., Keller L. P., Kearsley A. T., Joswiak D., Jacob D., Ishii H., Harvey R., Hagiya K., Grossman L., Grossman J. N., Graham G. A., Gounelle M., Gillet P., Genge M. J., Flynn G., Ferroir T., Fallon S., Ebel D. S., Dai Z. R., Cordier P., Clark B., Chi M., Butterworth A. L., Brownlee D. E., Bridges J. C., Brennan S., Brearley A., Bradley J. P., Bleuet P., Bland P. A., and Bastien R. (2006) Mineralogy and petrology of Comet 81P/Wild 2 nucleus samples. *Science, 314,* 1735–1739.

Zolensky M. E., Krot A. N., and Benedix G. (2008) Record of low-temperature alteration in asteroids. *Rev. Mineral. Geochem., 68,* 429–462.

Zubko E., Videen G., Hines D. C., Shkuratov Y., Kaydash V., Muinonen K., Knight M. M., Sitko M. L., Lisse C. M., Mutchler M., Wooden D. H., Li J.-Y., and Kobayashi H. (2015a) Comet C/2012 S1 (ISON) coma composition at ~4 au from HST observations. *Planet. Space Sci., 118,* 138–163.

Zubko E., Shkuratov Y., and Videen G. (2015b) Effect of morphology on light scattering by agglomerates. *J. Quant. Spectrosc. Radiat. Transfer, 150,* 42–54.

Kolokolova L., Kelley M. S. P., Kimura H., and Hoang T. (2024) Interaction of electromagnetic radiation with cometary dust.
In *Comets III* (K. J. Meech, M. R. Combi, D. Bockelée-Morvan, S. N. Raymond, and M. E. Zolensky, eds.), pp. 621–651.
Univ. of Arizona, Tucson, DOI: 10.2458/azu_uapress_9780816553631-ch019.

Interaction of Electromagnetic Radiation with Cometary Dust

Ludmilla Kolokolova and Michael S. P. Kelley
University of Maryland

Hiroshi Kimura
Planetary Exploration Research Center, Chiba Institute of Technology

Thiem Hoang
Korea Astronomy and Space Science Institute

The chapter overviews the recent developments in the remote sensing of cometary dust using visible, near-infrared, and thermal-infrared radiation, as well as interaction of the dust with electromagnetic radiation, which affects the dynamics of dust particles. It considers photometric, polarimetric, and spectral studies of cometary dust, focusing on those observables and correlations between them that allow revealing the composition, size, and structure of the dust particles. The analysis includes the observed brightness and polarization phase curves, color and polarimetric color of the cometary dust, and near- and thermal-infrared spectra. Special attention is paid to the role of gas contamination in the polarimetric and photometric data. A review of modeling efforts to interpret the observational results is also provided, describing the most popular (and some novel) techniques used in the computer modeling of light scattering by dust particles with a focus on modeling the most complex type of cometary particles: fluffy and porous agglomerates. The chapter also considers how properties of the dust particles affect their photoelectric emission and their response to the radiation pressure and radiative torque, including alignment and fragmentation of particles. Results of computer and some laboratory modeling are analyzed for their consistency with the observational and *in situ* data. Also discussed is how the modeling results can be combined with *in situ* data for better characterization of the cometary dust.

1. INTRODUCTION

Despite the great success of space missions studying comets, remote sensing using Earth-based observations (i.e., observations with groundbased and spacebased telescopes) remains the major source of information about comets. It allows studying a variety of comets, enabling further statistical analysis of their properties. Also, Earth-based observations allow monitoring the comets at different orbital positions and apparitions, thus providing information about both short-term and long-term evolution of the cometary environment.

Most of the cometary remote-sensing information is coming from observations of the interaction of cometary material with solar radiation. This chapter focuses on characterization of the dust in cometary coma based on how it scatters, absorbs, and emits radiation. Trying to avoid repetition of the facts and concepts considered in the previous volume, *Comets II* (*Kolokolova et al.,* 2004a, hereafter *K_C2*), this chapter emphasizes the new developments in the field. The observations and their results are overviewed in section 2. The chapter also summarizes the most efficient and broadly used techniques for interpretation of the observations (section 3). It demonstrates how characteristics of the cometary dust and their variations in the coma and along the orbit can be revealed using computational and laboratory modeling. Section 4 considers how the interaction with solar radiation affects motion, spinning, and fragmentation of the dust particles.

The main remote-sensing techniques that provide information about cometary dust are photometry and polarimetry in the visible and near-infrared (NIR), and spectroscopy in the NIR and mid-infrared (MIR) wavelengths. Most comets show similar photometric and polarimetric properties that allowed establishing the following regularities:

In the visible:

1. The color of the cometary dust is mainly red, with the values very similar for long-period and short-period comets and decreasing with increasing wavelength. Statistically

averaged cometary color does not show stable tendencies with phase angle or heliocentric distance (see more in section 2.1).

2. Polarization of cometary dust is characterized by a negative (polarization plane is parallel to the scattering plane) parabolic-shaped branch at phase angles smaller than 20° and a positive (polarization plane is perpendicular to the scattering plane) bell-shaped branch at larger phase angles, reaching the minimum values at about –2% at the phase angle ~10° and the maximum values varying within 10–35% at phase angle ~90°–95°. For details, see section 2.2.
3. In the majority of comets, polarization increases with wavelength, although there are some exceptions.
4. Some comets demonstrate circular polarization. which usually does not exceed 2%.
5. The spectra of cometary dust show no absorption features.

In the NIR:

1. The increase of brightness with wavelength, i.e., red color, is also typical for the NIR; however, it is less red than in the visible
2. The dependence of polarization on phase angle in the NIR looks similar to the one in the visible for comets with high polarization maximum.
3. The majority of comets show a decrease of polarization with wavelength in the NIR.
4. Comets may demonstrate absorption bands of ice at 1.5, 2.2, and 3.1 μm, especially at large heliocentric distances.

In the MIR (thermal emission):

1. The main features in cometary spectra are silicate bands at 10 and 20 μm whose strength varies by comet and can vary with time.
2. The effective continuum temperature (based on the spectral shape) tends to be warmer (up to 30%) than the isothermal temperature of a large blackbody sphere at the same heliocentric distance.

The majority of the above regularities were obtained using aperture observations and were already discussed in *K_C2*. The main advantage of the recent observations is that the aperture observations were replaced by imaging observations with a high spatial resolution. This brought an understanding that some of the listed regularities describe the cometary dust on average, whereas the dust experiences significant changes as the particles move out of the nucleus. Also, the dust was found to show different observational characteristics in cometary jets and other morphological features.

Another recent development in cometary dust studies is a huge progress in computational modeling of light scattering by non-spherical particles, specifically development of powerful computer packages to compute aggregated/agglomerated particles (see section 3). An important role in this development is played by an availability of more powerful computers, especially computer clusters, which allow reaching a significant increase in the efficiency of the computations by using parallelized computer codes, described in section 3.1. Very beneficial for the interpretation of Earth-based observations have been the results of laboratory and *in situ* studies of cometary dust, specifically based on the results of the Stardust and Rosetta missions. Information from those studies, reviewed in the chapters in this volume by Poch et al. and Engrand et al., allows preselecting some characteristics of the dust particles, specifically their size, structure, and composition. This narrows down the parameter pool involved in the modeling, thus increasing the modeling efficiency.

2. OBSERVATIONAL RESULTS

Interaction of cometary dust particles with solar radiation results in radiation being scattered, absorbed, and reemitted by the dust particles. This interaction varies depending on the location in the coma and distance to the Sun, but being averaged, produces the fundamental observational characteristics of the cometary dust: dependence of the brightness and polarization on phase angle (i.e., the Sun-comet-observer angle) and on wavelength.

In general, the measured brightness of the scattered light does not characterize the physical properties of the dust particles as it represents a combined effect of the intrinsic physical properties of the dust particles (size, composition) and their number, i.e., the column density of the dust particles. To cancel the effect of the column density and thus find out the intrinsic properties of the cometary dust, the observational characteristics defined by the ratio of the brightness — color and degree of linear polarization — are used. There are two other fundamental characteristics of the scattered light that are defined by the intrinsic dust properties: the dust albedo and its phase function.

In the light-scattering theory, the particle albedo is usually defined as a single-scattering albedo, i.e., the ratio of the particle scattering efficiency to its extinction efficiency (*van de Hulst,* 1957). However, this characteristic cannot be directly obtained from the observations. In cometary physics, albedo of a dust particle means its reflectivity and is analogous to the geometric albedo in planetary physics, i.e., it is the ratio of the energy scattered at phase angle equal to 0° to that scattered by a white Lambert disk of the same geometric cross section. The detailed description of how to determine the cometary dust albedo from observations as well as some other definitions of the particle albedo can be found in *K_C2*, section 2.1. The geometric albedo of a single particle is defined as $a_0 = \pi S/(Gk^2)$ (*Hanner,* 1981; *K_C2*) where S is the intensity of the light scattered by the particle in the direction of the Sun (zero phase angle) divided by the intensity of the solar light incident to the particle, G is the geometric cross-section of the particle, and k is the wave number related to the radiation wavelength λ as $k = 2\pi/\lambda$. It is clear from the definition that the geometric albedo is an intrinsic particle characteristic defined mainly by the particle composition and probably structure, as the intensity of the scattered light can be reduced by the scattering on the inhomogeneities within the particle. The parameter S depends on size, which in light scattering is characterized through the size parameter x = kr, where r is the particle radius, but for large particles S/G becomes constant, which makes the effect of size negligible for particles of x > 10 (*Mishchenko and Travis,* 1994).

Often the albedo of a particle at other than zero phase angles is considered, i.e., $a(\alpha) = \pi S(\alpha)/(Gk^2)$, where α is phase angle. This is also an intrinsic property of the particle, and $a(\alpha)/a_0$ is identical to the phase function, which demonstrates how the reflectivity of the particle changes with phase angle. Often the phase curve is defined as the normalized intensity curve, i.e., the observed brightness at each phase is divided by the brightness at zero phase angle. In this case the phase curve is identical to $a(\alpha)/a_0$, thus, $a(\alpha)$ is the particle geometric albedo multiplied by the phase function.

The phase curve of the particles strongly depends on their size parameter (*van de Hulst,* 1957; *Bohren and Huffman,* 2008). Small particles with x < 1 scatter light in the Rayleigh regime; their phase function is symmetric, smoothly decreasing from 0° to 90° and then smoothly increasing. For larger particles, the curve becomes more asymmetric with forward scattering increasing with size. Monodisperse compact particles with x between 1 and 20 usually have a phase curve with a complex oscillating structure, which disappears for larger particles where light scattering occurs in the geometric-optics regime. The large particles usually have a strong forward-scattering peak; at smaller phase angles, the curve becomes less steep and at some sizes and composition may show a backscattering peak. The current envision of the cometary phase curve mainly attributes to the composite phase curve compiled by *Schleicher* (2010) based on an empirical function fit to Comet 1P/Halley data from *Schleicher et al.* (1998) and a Henyey-Greenstein model fit to near-Sun comet photometry by *Marcus* (2007). It has a medium-backscattering peak after which the brightness decreases, reaches minimum at about 50°–60° and then increases, becoming steeper as it approaches the forward scattering. Similar phase curves were found for zodiacal dust, Saturn's rings, and protoplanetary and debris disks (Fig. 1) (see also *Hughes et al.,* 2018). A smooth shape with a large forward-scattering and some backscattering peaks can be attributed to a polydisperse distribution of the particles comparable with wavelength, i.e., of size parameter x = 1–5 (*K_C2*).

A rather different shape of the phase curve for Comet 67P (hereafter 67P) with the minimum around 100°, a steeper increase in the backscattering direction, and a rather gradual increase in the direction to the forward scattering (*Bertini et al.,* 2017) was acquired during the Rosetta rendezvous mission. *Moreno et al.* (2018) explained this unusual curve by light scattering on oriented elongated particles. This explanation was supported by a specific position of the camera that was observing the particles affected by the gas flow and radiation torque (see section 4.3). However, later a similar phase curve was observed for some debris disks (*Hughes et al.,* 2018) (see Fig. 1 for examples). The laboratory measurements by *Muñoz et al.* (2020) showed that such curves are typical for large, millimeter-sized, porous particles that can be a reasonable explanation for the case of near-nucleus particles in Comet 67P where the near-nucleus coma is dominated by particles hundreds of micrometers in size (see the chapter in this volume by Engrand et al.).

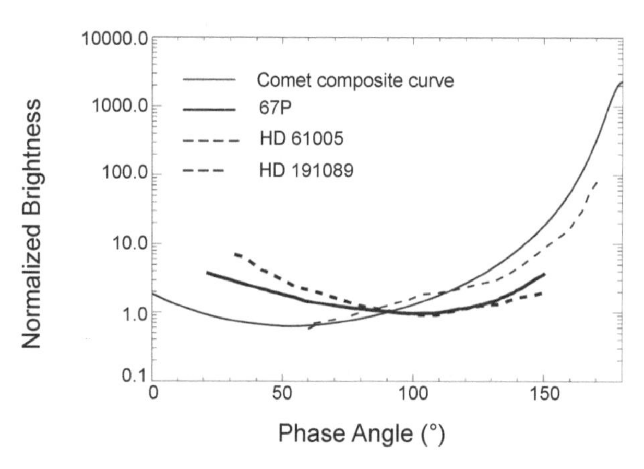

Fig. 1. Phase curves for comets and debris disks. A thin solid line shows the composite dust phase function for comets by *Schleicher* (2010). The thick solid line is the phase curve for Comet 67P measured by the Rosetta OSIRIS camera (*Bertini et al.,* 2017). Dashed lines show the phase curves for debris disks HD 61005 [thin line (*Oloffson et al.,* 2016)] and HD 191089 [thick line (*Ren et al.,* 2019)]. Two types of the phase curves can be seen for both comets and debris disks.

2.1. Color of Cometary Dust

The color of cometary dust is traditionally defined as the difference of the magnitudes in two continuum filters, e.g., $m_c = m_B - m_R$ in the case of B-band and R-band filters. Since the magnitude is –2.5logI where I is the brightness in physical units, the color can be presented as $-2.5 \log(I_B/I_R)$. As it is a ratio of the brightness in two filters, it cancels the effect of dust particle column density and becomes a characteristic of the dust particles per se. It is worth noting that the presented definition of the color contains information on the solar spectrum as it represents the solar light scattered by the comet. Subtraction of the solar color calculated for the same filters, i.e., $m_s = m_{sB} - m_{sR}$, from the observed dust color, gives the intrinsic color of the cometary dust equal to $m_c - m_s$.

The other popular definition of the cometary dust color uses the gradient of the cometary spectrum, i.e., the steepness of the cometary continuum. Presenting the cometary spectrum as $S(\lambda) = F_c(\lambda)/F_{Sun}(\lambda)$, i.e., the comet spectrum $F_c(\lambda)$ divided by the solar spectrum $F_{Sun}(\lambda)$, one can calculate cometary color (*A'Hearn et al.,* 1984) as the normalized gradient of reflectivity:

$$S'(\lambda_1\lambda_2) = \frac{2000}{\lambda_2 - \lambda_1} * \frac{S(\lambda_2) - S(\lambda_1)}{S(\lambda_2) + S(\lambda_1)} \qquad (1)$$

where $S(\lambda_1)$ and $S(\lambda_2)$ correspond to the dust reflectivity at the wavelengths λ_1 and λ_2 under the condition $\lambda_2 > \lambda_1$. $S'(\lambda_1, \lambda_2)$ is expressed in percent per 1000 Å.

Progress in studying cometary colors is characterized by two main developments: (1) collection of many new values of the colors for different comets and at their different positions on the orbit that allow accomplishment of a statistical analysis of the cometary colors; and (2) acquiring images of the brightness in different filters, thus producing maps of cometary colors that reveal changes in the color with the distance from the nucleus and in different cometary features. The color variations in the coma will be considered in section 2.2.1.

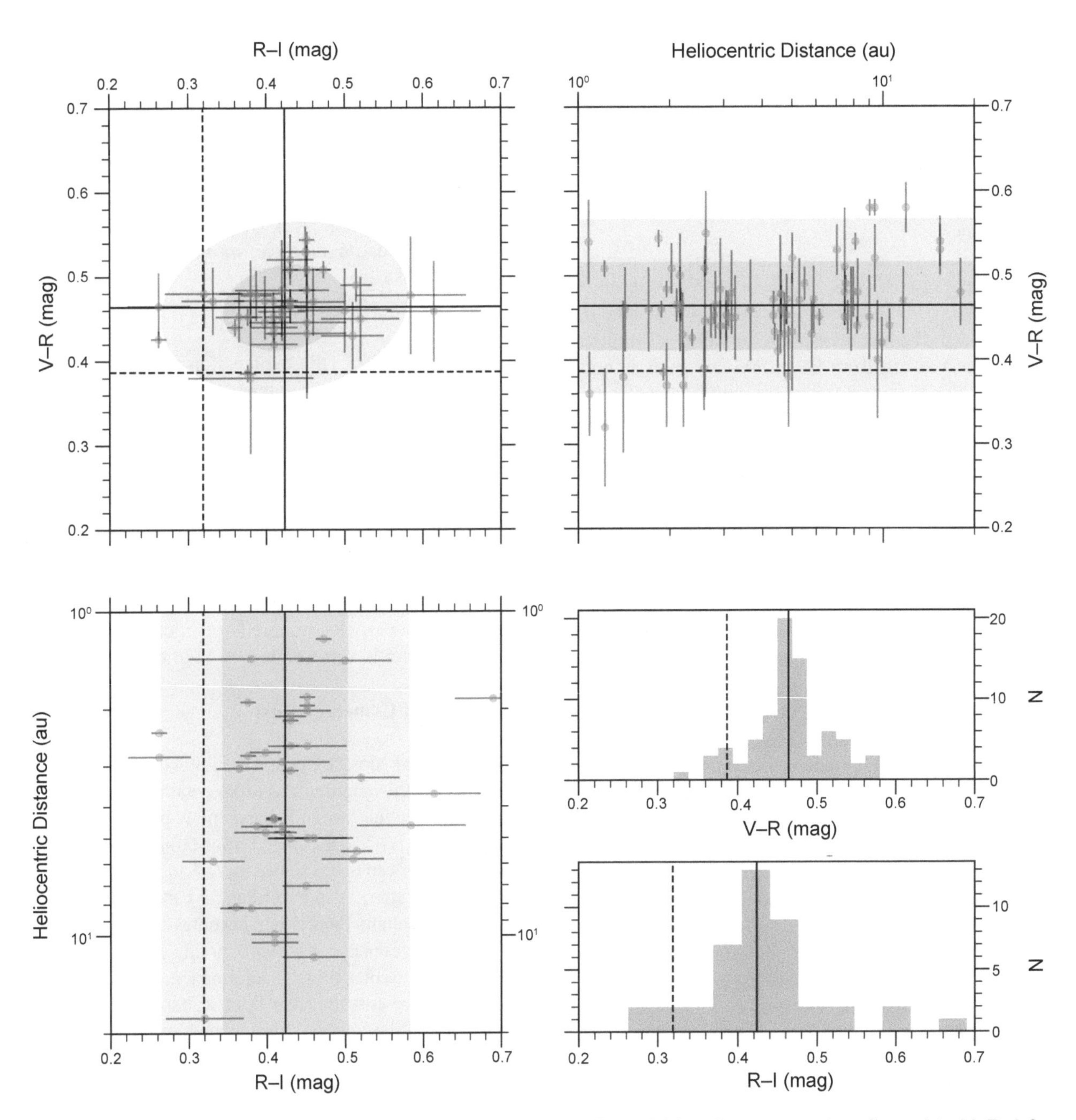

Fig. 2. Broadband coma colors based on photometry in the V, R, and I bandpasses, or transformed to V, R, I from equivalent filters. Solid lines are the mean colors, dashed lines are solar colors. Shaded regions indicate 1 and 2 standard deviations from the mean colors. *Upper left:* Color-color plot and covariance ellipses. *Upper right and lower left:* All V–R and R–I data points vs. heliocentric distance. The axes are configured so that individual points in the V–R vs. R–I plot may be traced to the color-heliocentric distance plots. *Lower right:* Color histograms.

Observations of cometary coma continua span a moderate range of colors, from slightly blue to red. We searched the literature for surveys of broadband optical colors of active comets. Based on the results of *Solontoi et al.* (2012), we added the requirement that any color uncertainties must be better than 0.1 mag. In Fig. 2, we present V–R and R–I colors from *Solontoi et al.* (2012), *Jewitt* (2015), and *Betzler et al.* (2017). A minimum uncertainty of 0.05 mag was adopted for all Beltzer et al. data based on Anderson-Darling tests, which could not distinguish between true variability and a normal distribution for the six comets with time-resolved observations.

Cometary colors are typically measured with broadband filters, which can be affected by the presence of gas emission bands. For example, using spectra of the dusty Comet C/2013 US_{10} (Catalina), *Kwon et al.* (2017) estimated a moderate gas contamination of 8% and 3% in the R- and I-bands (9500 × 1900 km aperture). Observations of gas-rich Comet 252P/LINEAR in the V, r′, BC, and RC filters by *Li et al.* (2017) imply a photometric excess of 25% to 50% in the V and r′ filters due to the presence of gas (2300-km aperture radius, rh = 1.1 au). Therefore, narrowband filter sets designed for cometary comae are better suited for continuum colors; the most recent iteration is the HB filter set (*Farnham et al.,* 2000). But even the narrowband filters may be significantly contaminated. For the HB filter set, C_3 and OH emission is found within the UC filter bandpass (345 ± 40 nm), and C_2 in the GC filter (*Farnham et al.,* 2000; *Rosenbush et al.,* 2002; *Opitom et al.,* 2015).

Gas contamination will be stronger for comets with lower dust-to-gas production rates. *A'Hearn et al.* (1995) presented the ratio of Afρ (the product of albedo, aperture filling factor, and aperture radius, an empirical proxy for dust production rate) and OH production and found that comets span the range $\log_{10}$(Afρ/OH) = –26.5 to –24.9 for units of cm s molecule^{-1} (note that these values are not corrected for dust phase angle effects), thus comets on the low end of the range are said to be gassy and those on the higher end are dusty. Most optical gas emission bands are from molecules produced by photolysis in the coma, which is a function of heliocentric distance. The emissions from these bands have spatial profiles that tend to be shallower than dust spatial profiles (*Combi et al.,* 2004), thus the physical size of the aperture at the distance of the comet also plays an important role.

Altogether, gas contamination is a general problem in photometry of cometary dust, and the cometary astronomer should consider the possibility of gas emission in their data. This statement is especially true when observing unusual comets or comets in unusual circumstances, e.g., see *Bellm et al.* (2019) and *McKay et al.* (2019) for CO^+ emission in g-band and HB UC filter images of Comet C/2016 R2 (PanSTARRS). Of all the optical broadband filters, the redder filters ($\lambda > 600$ nm) tend to be the least affected by gas, but NH_2 and CN bands may still be significant (*Fink,* 2009). Gas contamination can also be estimated or avoided with spectroscopic measurements (e.g., *Jewitt and Meech,* 1986).

The average and standard deviation broadband colors are V–R = 0.46 ± 0.05 mag (N = 66) and R–I = 0.43 ± 0.08 mag (N = 40). Figure 2 shows histograms of the V–R and R–I color sets. The distributions are unimodal and nearly symmetric.

Coma broadband colors may be expected to have a heliocentric distance dependence due to gas contamination as discussed above, the presence of water ice, or the variation of the coma grain size distribution. We compared data taken inside and outside of different heliocentric distance cuts, ranging from 1.5 to 10 au, with the two-sided Kolmogorov-Smirnov test. After accounting for uncertainties, all p-values were >5%, indicating none of the tests found significant differences, consistent with prior results (*Solontoi et al.,* 2012; *Jewitt,* 2015). This dataset may not be useful for realizing these effects if they are much smaller than the observed scatter in the population (~0.08 mag). Moreover, gas contamination may not be apparent given that so few data points are taken near 1 au. Investigating these effects likely requires isolated studies of individual comets with precise color measurements.

In contrast with the groundbased data, spacecraft observations of Comets 67P and 103P/Hartley 2 have shown clear color trends with distance to the nucleus and, for 67P, with distance to the Sun and phase angle. *Filacchione et al.* (2020) show that the optical (0.5 to 0.8 μm) spectroscopic color of the inner (1 to 2.5 km) coma of 67P varied with heliocentric distance, from near 20% to 25%/0.1 μm at 1.2 au, to near 0% to 5%/0.1 μm at 2.8 au, consistent with the presence of icy grains in the coma. This magnitude of a change should be easily observed in groundbased data, but the analysis of *Kwon et al.* (2022) suggests that this variation is likely limited to the innermost coma of this comet. *Bockelée-Morvan et al.* (2019) show a phase angle dependence of Comet 67P's inner coma at 2 to 2.5 μm. The spectral gradient's slope, measured with respect to the continuum flux density at 2 to 2.5 μm, was 0.031%/100 nm/° from 50° to 120° and thus demonstrated a noticeable reddening. They point out that that phase reddening is common on solar system bodies and has also been observed in the zodiacal light, and argue that phase reddening in a cometary coma is likely caused by porous particles much larger than the wavelength of light being observed (i.e., >10 μm in size). Phase reddening for dust particles was indicated in the laboratory measurements by *Escobar-Cerezo et al.* (2017) and was also attributed to the particle structure (roughness) by analogy with the explanation of the phase reddening for rough surfaces (*Schröder et al.,* 2014).

Further examination of the potential effect of water ice on coma optical colors became possible due to the Deep Impact observations of Comet 103P/Hartley 2. Near-infrared observations of this comet showed prominent water ice absorption bands at distances within a few kilometers from the nucleus. *Protopapa et al.* (2014) examined these absorption features in an ice-rich spectrum taken 400 m from the surface and estimated an ice abundance of 5%

by area. *La Forgia et al.* (2017) studied the optical colors of this comet and showed that the ice-enriched regions tend to be bluer than the ice-poor regions in the coma. The azimuthally averaged spectral gradient from the green to red (0.53 to 0.75 μm) ranges from 9% per 0.1 μm near the nucleus to 13% per 0.1 μm 40 km from the nucleus. Given that the water ice is asymmetrically distributed in the coma, the color difference would be even larger if it were not azimuthally averaged. The quoted range corresponds to V–R colors of 0.49 to 0.54 mag, which is fully on the red end of the observed coma colors (Fig. 2), emphasizing that ice is difficult to identify based on absolute colors alone. However, the 0.05-mag change in color is large enough to be apparent in the observed V–R colors of cometary comae. The lack of strong bluing with heliocentric distance in Fig. 2 may suggest that ice is rare in cometary comae, but considerations for competing effects are necessary. The sublimation lifetime of ice compared to the photometric aperture size is of foremost importance. There is some evidence for color gradients in distant comae, reddening by 3% to 8%/0.1 μm with distance from the nucleus, which is suggestive of the presence of water ice grains (*Fitzsimmons et al.,* 1996; *Li et al.,* 2013, 2014).

A changing grain size distribution may also affect coma dust color. As activity increases and overcomes the tensile strengths of the dust agglomerates (*Gundlach et al.,* 2015), small grains could populate the coma, producing a bluer color. Alternatively, grain fragmentation may increase the number of Rayleigh scatterers with increasing nucleus distance.

There are several papers reporting narrowband photometry of cometary comae. We searched the literature for observations with the HB filter set and assembled data on 19 comets from 18 papers (*Bair et al.,* 2018; *Cudnik,* 2005; *Farnham and Schleicher,* 2005; *Ivanova et al.,* 2014; *Knight and Schleicher,* 2013; *Knight et al.,* 2021; *Li et al.,* 2017; *Moulane et al.,* 2018, 2020; *Opitom et al.,* 2015; *Zhang et al.,* 2021; *Schleicher,* 2007, 2008, 2009; *Schleicher and Bair,* 2011; *Schleicher et al.,* 2003). We chose the HB filter set, given its improvements over previous narrowband filter sets, especially with respect to gas contamination mitigation. Figure 3 presents all colors with uncertainties <0.1 mag in color-color plots as a linear spectral slope (equation (1)). It shows that the observations span a moderate range of spectral slopes, from neutral to moderately red (0% to 30%/0.1 μm). However, Comet 96P is a clear outlier with a UC–BC slope of −18%/0.1 μm in accordance with the findings of *Opitom et al.* (2015), that the OH (0–1) band may be a significant contributor to UC filter photometry. A crude correction to the UC reflectance using the relative strength of the OH (0–1) and (0–0) bands (*Schleicher and A'Hearn,* 1988) increased the comet's mean spectral slope from −18 to −4%/0.1 μm. A near-UV spectrum of 96P may provide some clarity on the UC–BC colors of cometary comae, as would further investigations into the magnitude of OH contamination in the UC filter.

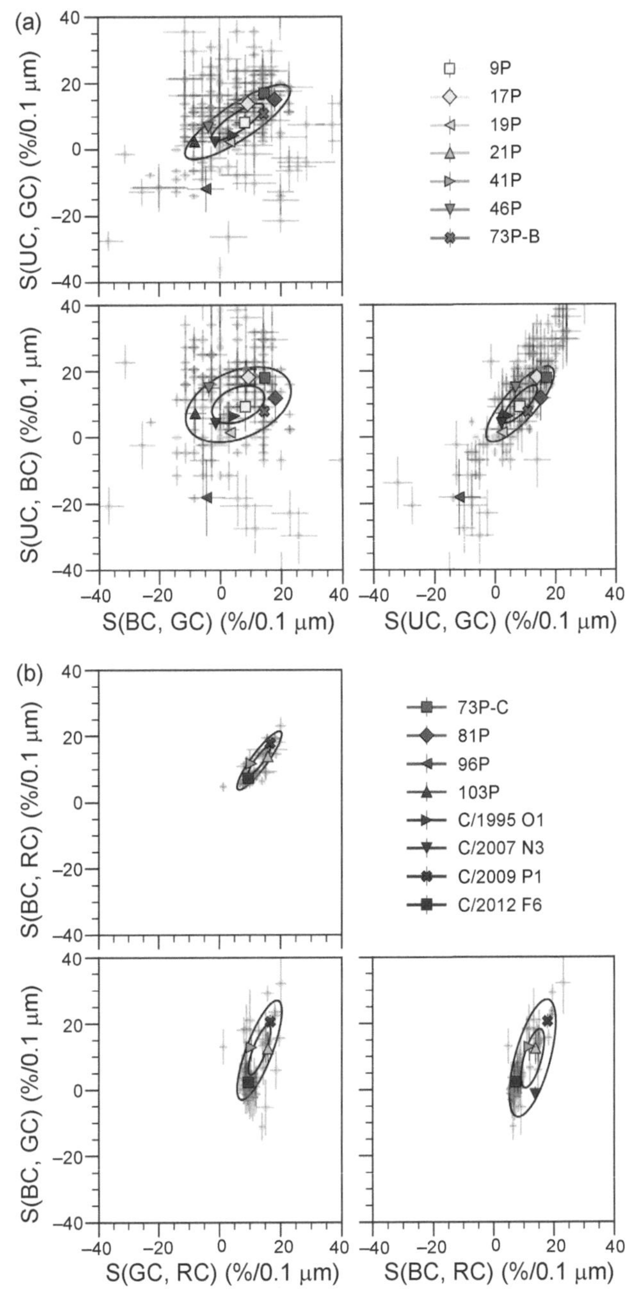

Fig. 3. Narrowband colors of comets based on the HB filter set. Individual measurements with uncertainties better than 10%/0.1 μm are shown as light gray circles. Comet-by-comet averages are also plotted; uncertainties are based on the error on the mean, or the standard deviation of the data, whichever is greater.

2.2. Cometary Polarimetry

In planetary astronomy, the degree of linear polarization (P, hereafter called polarization) is defined as

$$P = (I_{\perp} - I_{\parallel})/(I_{\perp} + I_{\parallel}) \quad (2)$$

where $I_{\perp}$ and $I_{\parallel}$ are the intensity components perpendicular and parallel to the scattering plane (the plane that contains the Sun, comet, and observer). Based on this definition, the polarization with $I_{\perp} > I_{\parallel}$ is called positive, and if $I_{\perp} < I_{\parallel}$ it is called negative. This definition is given here mainly to explain the terms negative and positive polarization and make it clear that polarization is the ratio of intensities and thus depends only on the properties of the dust particles. Currently, with the rise of the Stokes polarimetry (e.g., *Keller et al.,* 2015), polarization is usually defined through the Stokes vector, a mathematical object that is formed by components (I, Q, U, V), where I is the total intensity of the scattered light, Q is the numerator from equation (2), and Q/I is identical to the planetary definition of the linear polarization; U together with Q defines the angle θ that determines the position of the polarization plane (i.e., the plane in which oscillates the linearly polarized part of the electromagnetic wave) as $\tan(2\theta) = Q/U$. Note that a physically correct way to define the linear polarization is through $P = (Q^2 + U^2)^{1/2}/I$ and angle θ; however, in the planetary observations a typical value of θ is close to 0 (negative polarization) or 90° (positive polarization), which justifies the definition given in the beginning of the section. Parameter V defines circular polarization, i.e., the part of the scattered light for which the plane of the electromagnetic wave rotates around the direction of the light propagation. More details on the Stokes vector can be found in *Bohren and Huffman* (2008) or in *K_C2*. The value of the Stokes vector representation is that it allows using matrix algebra to consider light scattering by complex systems. Specifically, the Stokes vector of the radiation source (I_0, Q_0, U_0, V_0) is related to the measured Stokes vector through the 4 × 4 Mueller matrix (often called scattering matrix) as (I, Q, U, V) = $\mathbf{M}$*(I_0, Q_0, U_0, V_0); for the solar radiation the Stokes vector is (I_0, Q_0, U_0, V_0) = (I, 0, 0, 0).

A detailed review of the cometary polarimetric studies can be found in *Kiselev et al.* (2015); below we briefly outline the results reported there, adding the results published after that review.

A comprehensive collection of the results of the cometary polarimetric observations has been recently archived by the NASA Planetary Data System (PDS) as the Database of Cometary Polarimetry (*Kiselev et al.,* 2017). It includes published and unpublished data of 3441 observations for the period 1881–2016. The database includes data for 95 comets with the description of the geometry of observations and characteristics of the polarimetric instrumentation. The data cover the wavelength from 0.26 to 2.32 μm and the ranges of phase angles, heliocentric distances, and geocentric distances 0.0–122.1°, 0.0–7.01 au, and 0.01–6.52 au, respectively. The data from this database and its earlier version (*Kiselev et al.,* 2006) have been already used in several studies where some statistical analysis of the polarimetric data was done; see, e.g., *Mishchenko et al.* (2010), *Borisov et al.* (2015), *Kwon et al.* (2019, 2021), *Rosenbush et al.* (2021). The most interesting observations acquired since the release of the database will be discussed in this section. They have not changed the main characteristics of the dependence of polarization on wavelength and phase angle with the probable exception of the data for interstellar Comet 2I/Borisov (*Bagnulo et al.,* 2021), which showed the values of polarization exceeding the values typical for solar system comets, although similar to those observed for Comet C/1995 O1 Hale-Bopp.

Probably the most significant change in our understanding of the cometary polarization since *K_C2* is reconsideration of the classification of comets in two polarimetric classes: comets with a high polarization maximum (~25%) and those with a low polarization maximum (~10%) (*Chernova et al.,* 1993; *Levasseur-Regourd et al.,* 1996). *Chernova et al.* (1993) related these two classes to different dust/gas ratios (see definition in section 2.1) in the comets; specifically, high-polarization comets were found to have a large dust/gas ratio, and that for the low-polarization comets was at least twice lower. This allowed relating the polarization difference in those two classes to a stronger depolarizing effect of the gas emissions in the low-polarization comets. With many more data acquired after 1996, one can see (Fig. 4) that there is no clear division of the comets into two classes. Although they may be still visible in the blue domain, the comets observed in the red and NIR domain tend to form a single curve. There is a scattering of the data, especially in the blue domain. The main explanation of the scattering relates to the original division of the comets to dusty and gassy types. However, when narrowband filters weakly affected by the gas emissions are used, all comets show similar polarization, typical for the dusty comets.

One more confirmation of the significant influence of gas contamination on the polarimetric data was obtained from observations of cometary polarization with increasing aperture of the observations (*Jockers et al.,* 2005) or using imaging polarimetry (*Kwon et al.,* 2017). These observations demonstrated that for both types of comets the near-nucleus polarization is high, reaching the values 20–25%. However, with increasing aperture or distance from the nucleus, the values of polarization in dusty comets change only slightly, whereas for the gassy comets, a decrease in polarization is significant, bringing the polarization to the values typical for low-polarization comets or even lower, as shown in Fig. 5. *Jockers et al.* (2005) concluded that the low values of polarization of gassy comets resulted from the contamination of the data by gas emissions, which depolarize the light.

Recent observations and detailed analysis of the gas contamination have been reported by *Ivanova et al.* (2017), *Kwon et al.* (2017, 2018), and *Kiselev et al.* (2013, 2020). Spectropolarimetric data for Comet Garradd (*Kiselev et al.,* 2013) showed correlation between the intensity of the gas emissions and polarization. Similarly, *Kwon et al.* (2017) found a strong correlation between the polarization and the intensity of gas emissions in their change with the distance from the nucleus. Obviously, the effect of gas contamination should increase with the distance from the nucleus as the amount of the gas from the distributed gas sources adds to the original gas sublimated from the nucleus.

The dependence of the polarization on the aperture of observations, which can move a comet from the high-polarimetric class to the low-polarimetric class, makes the

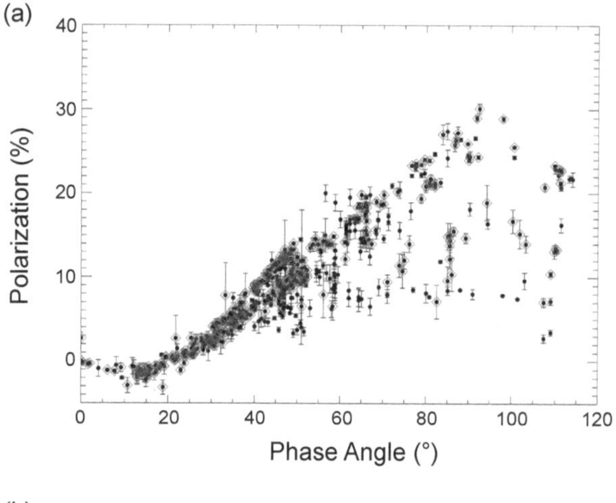

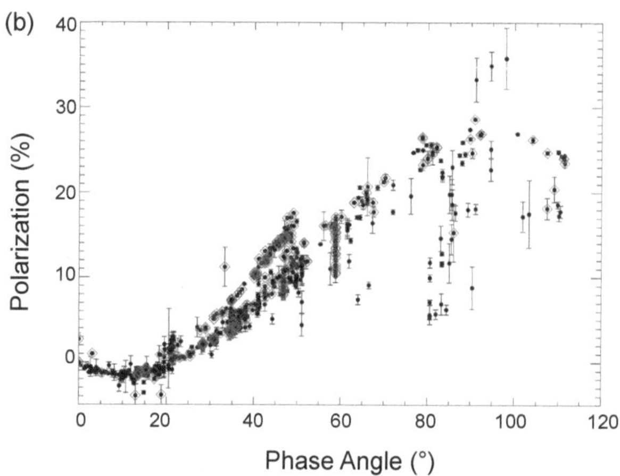

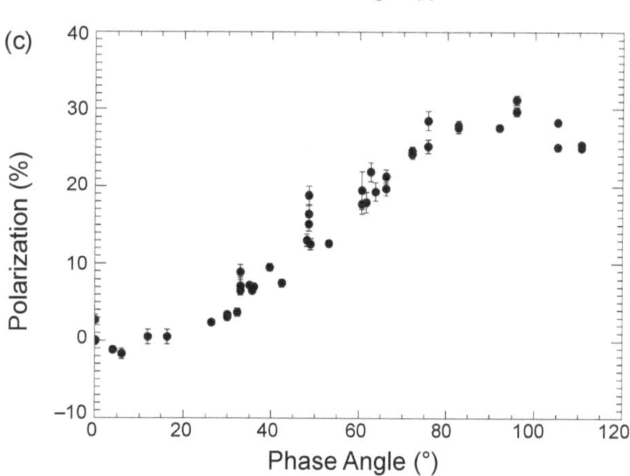

Fig. 4. Compilation of the data from *Kiselev et al.* (2017) for the cometary polarimetric phase curve. **(a)** Data acquired in the blue domain (blue squares are for the BC_IHW filter). **(b)** Data in the red domain (blue squares are for RC_IHW filter). **(c)** Data in the NIR region (JHK filters); the data for Comet West are excluded as they were strongly affected by the thermal emission (*Oishi et al.,* 1978).

division of the comets into two classes questionable, or at least not reflective of the intrinsic properties of the cometary dust. However, a more detailed analysis and especially correlation of the positive polarization of comets with the strength of the silicate infrared feature (see section 2.3.2) allowed the assumption (*Kolokolova et al.,* 2007) that the difference in polarization may reflect a different distribution of the dust particles in the coma; specifically, in the low-polarization comets, the dust is concentrated near the nucleus, thus resulting in a low dust/gas ratio for the overall coma. Such a difference in the dust distribution may result from the different size and porosity of the dust particles, as is shown in section 2.3.2.

An even more detailed picture of the change in polarization throughout the coma came from charge-coupled device (CCD) images with high spatial resolution. Early polarimetric images were reported by *Renard et al.* (1996), then a detailed study of Comet Hale-Bopp with the imaging polarimetry was published by *Jockers et al.* (1997) and *Hadamcik et al.* (1997). A paper by *Hadamcik and Levasseur-Regourd* (2003) presented results of imaging polarimetry for nine comets and showed that an area of lower positive polarization was observed in the near-nucleus area; it was named a "polarimetric halo." A polarimetric halo was observed at small phase angles as an area of more negative polarization in Comet ISON (*Hines et al.,* 2013) and other comets (*Choudhury et al.,* 2015). Since the polarimetric halo has only been observed in some comets, its origin is still not clear. The majority of high-resolution polarimetric images of comets show a rather gradual change in polarization. It appears that for some comets polarization decreases with the distance from the nucleus, probably indicating increasing gas contamination; however, for some other comets polarization increases. Increase in the positive polarization is likely a manifestation of dust particle fragmentation (e.g., *Jones et al.,* 2008) producing an increased abundance of small particles that scatter light in the Rayleigh regime. Both gas emissions and Rayleigh scatterers produce a bell-shaped phase curve that has positive polarization at all phase angles (e.g., *Kiselev et al.,* 2015), thus at low phase angles they can cause a decrease in the magnitude of negative polarization. An increase in the absolute value of the negative polarization with the distance from the nucleus, if observed, has a more complicated origin. Since negative polarization is most likely a result of multiple scattering inside a fluffy/porous particle (e.g., *Muinonen et al.,* 2012), an increase in absolute value of the negative polarization may indicate increasing transparency of the particles due to either evaporation of some dark material (e.g., semivolatile organics) or increasing porosity (*Kimura et al.,* 2006). There are also more complicated cases. For example, Comets 2P/Encke (*Jewitt,* 2004) and 67P (*Rosenbush et al.,* 2017) demonstrated a decrease in polarization that changed to an increase at larger distances from the nucleus, as shown in Fig. 7. A possible explanation of such a behavior is provided in section 2.2.1.

For a detailed discussion of the polarization in the cometary jets and similar features, the reader can check the review by *Kiselev et al.* (2015). Although most often the polarization in the jets is higher than in the ambient coma, suggesting particles of different composition or size in the jets and in the ambient coma, high-resolution images reveal

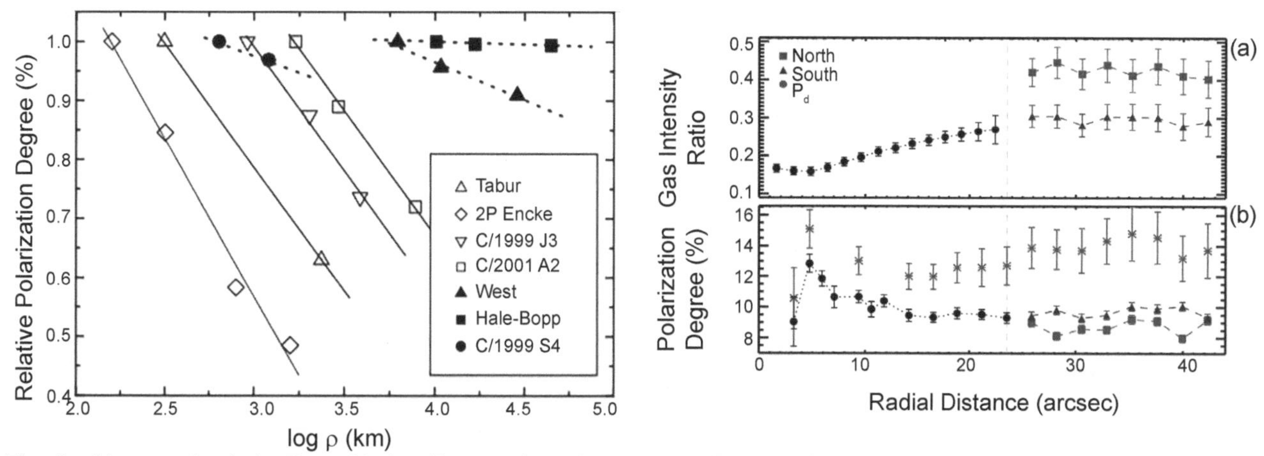

Fig. 5. Change of polarization with the distance from the nucleus. On the left, examples for low-polarization (white symbols) and high-polarization (black symbols) comets (*Kolokolova et al.,* 2007). On the right, the radial profiles of **(a)** the gas intensity ratio, f^g_{Rc}, and **(b)** polarization for Comet C/2013 US_{10} Catalina from *Kwon et al.* (2017); on the left of the gray dashed line at 23″ are aperture polarimetry results, and on the right of this line the sky regions were divided to the north and south wings and were treated separately. The gray asterisks in **(b)** denote polarization corrected with the corresponding values of **(a)**.

that not all jets seen in photometric images are visible in polarization — in this case the particles in the jets and in the ambient coma are similar and the difference between the jets and the coma is in the number density of particles. Examples of both cases were found in Comet C/2011 KP36 (Spacewatch) (*Ivanova et al.,* 2021).

We also want to point a significance of the contribution of the cometary nucleus to the light in the near-nuclear coma. The contribution of the nucleus strongly affected the coma polarization and color up to a distance of 10,000 km from the nucleus for Comet C/2011 KP36 (*Ivanova et al.,* 2021) and up to 3000 km for Comet Encke (*Kiselev et al.,* 2020).

We cannot leave out the polarization measurements performed by the polarimeters onboard the STEREO spacecraft. *Thompson* (2015, 2020) reported that the polarization properties of Comets C/2012 S1 (ISON) and C/2011 W3 (Lovejoy) have a high positive polarization, reaching 60%, and a broad negative branch reaching out to an inversion point between 40° and 50°. Also, dramatic changes in polarization were observed as Comet ISON approached the Sun to distances closer than 10 solar radii, demonstrating that the dust in the Sun-grazing comets undergo significant changes in the vicinity of the Sun. To explain the STEREO data, *Thompson* (2020) suggests fragmentation and sublimation of olivine at those distances. This is in accordance with *Kimura et al.* (2002b) who showed that silicates, such as olivine and pyroxene, sublimate at heliocentric distances below 10 solar radii. The other mechanism can be the dust fragmentation as it is disrupted by the radiative torque (see section 4.3.3).

2.2.1. Polarization of cometary dust in the near-infrared. Adding the polarimetric data in the NIR has resulted in two important conclusions. First, the polarization phase curves in the NIR appeared to be almost identical to those in the visible (see Fig. 4). This indicated that the characteristics of the particles, which affect the polarization, did not change much as the observations move from the submicrometer to micrometer scale, most likely signaling that the dominating size parameter of the particles exceeds unity not only in the visible but also in the NIR, i.e., cometary dust particles are several micrometers or larger in size. The only exception is Comet Hale-Bopp, which showed almost absent negative polarization in the NIR. This behavior can be evidence for particles approaching the Rayleigh regime in the NIR, thus particles <1 μm in size.

The other special feature of the NIR polarization is that the spectral trend in polarization changes from positive to negative, i.e., polarimetric color, defined as $P(\lambda_1)-P(\lambda_2)$ with $\lambda_1 > \lambda_2$, which is usually positive (red) in the visible (see *K_C2*), becomes negative (blue) in the NIR. Explanation of the spectral increase of polarization in the visible is quite straightforward: With increasing wavelength the dust particles or grains (monomers) in agglomerates are approaching the Rayleigh regime of scattering, and that manifests itself in increasing positive polarization. A spectral decrease in polarization requires a different explanation. *Kolokolova and Kimura* (2010a) noted that light scattering on a collection of closely located particles (e.g., monomers in agglomerates) can cause depolarization due to electromagnetic interaction between the monomers (cf. multiple scattering), which is stronger if the number of particles covered by a single wavelength is larger. Approaching the NIR, the wavelength is getting longer and thus it can cover more monomers, decreasing the polarization. This phenomenon was considered in detail in *Kolokolova et al.* (2011), where it was shown that it strongly depends on the porosity: The larger the porosity of the particle, the longer the wavelength at which the electromagnetic interaction becomes significant enough to cause depolarization of light. In a compact agglomerate the longer the wavelength the more monomers it covers, so the interaction between the monomers becomes stronger and the light becomes more depolarized. This results in a decrease of polarization with wavelength (blue polarimetric color). For a porous agglomerate, the number of monomers covered by a single wavelength does not change much as the wavelength increases, i.e., the change in the interaction between the monomers cannot overpower the change in the monomer size parameter, and the polarimetric color stays red. However, as the wavelength reaches some critical value, the number of covered monomers in the porous agglomerate changes significantly and interaction becomes the main factor that defines the polarimetric color, which then becomes blue. Thus, the wavelength at which the decrease in the polarimetric color starts may be determined by the porosity of the agglomerate, and it is smaller for more compact particles. This provides a straightforward explanation of the blue polarimetric color observed in the visible for some comets (*Kiselev et al.,* 2008): Most likely, the dust in those comets is dominated by more compact particles.

2.2.2. Circular polarization. As mentioned above, polarization can also have a circular component. Circular polarization was observed in more than 10 comets and was found at the tenths of percent level (Fig. 6). In the majority of cases, the absolute value of circular polarization (lefthanded polarization is usually considered as negative, and righthanded as positive) increases with the distance from the nucleus (see details in *Kiselev et al.,* 2015). The causes of circular polarization should be within processes or properties of particles that produce a mirror asymmetry of the medium (*van de Hulst,* 1957). Several mechanisms were suggested to explain the origin of circular polarization, among them such an exotic mechanism as homochirality of the molecules in cometary organics. However, T-matrix modeling of light scattering by homochiral molecules (see section 3.1) showed the values of circular polarization several orders lower than the observed values even in the case of the particles consisting entirely of homochiral organics (*Nagdimunov et al.,* 2013; *Sparks et al.,* 2015). More recent studies attribute cometary circular polarization to the alignment of the dust particles under solar magnetic field and radiative torque (see section 4.3).

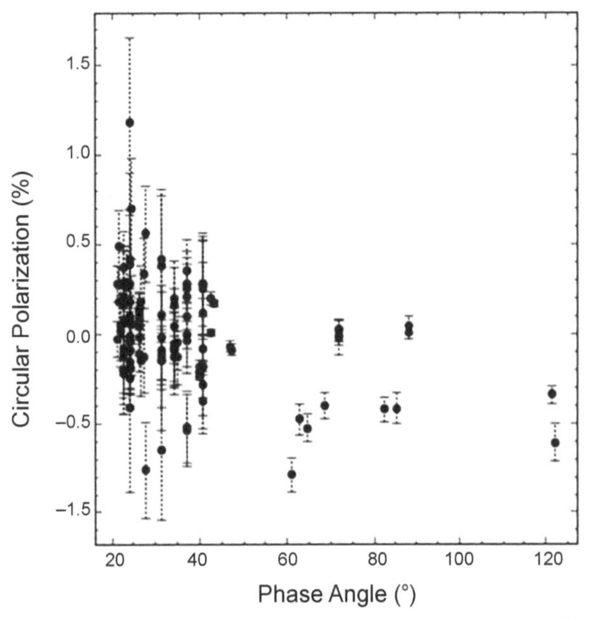

Fig. 6. Circular polarization data from *Kiselev et al.* (2017).

2.2.3. Correlation between color and polarization. An efficient approach in studying properties of cometary dust is combining polarimetric data with observations of the dust color. Color and polarization depend solely on the intrinsic properties of dust particles, specifically their size and composition. Polarization is a more complex characteristic as it also depends on shape and structure of particles; however, if observations show a correlation between color and polarization, this can provide information on the composition or size of the dust particles.

For example, an increase in polarization together with a decrease (bluing) of the color (anticorrelation) was observed for Comet Encke starting with the distances ~3 arcsec (*Jewitt,* 2004) and Comet 67P (*Rosenbush et al.,* 2017) (see Fig. 7). These variations can be explained by a decreasing size of the dust particles (*Kolokolova et al.,* 2002). However, closer to the nucleus, both Encke and 67P showed a decrease in polarization accompanied by a decrease in color (correlation). The currently suggested explanation is that large particles (hundreds of micrometers) dominate close to the nucleus, then they fragment to smaller ones (~10 μm) that are still far from the Rayleigh particles; their color and polarization correlate. Then fragmentation continues, and as particles approach the Rayleigh regime, color and polarization anticorrelate. This scenario is confirmed by the laboratory simulations of light scattering by cometary analog particles (*Hadamcik and Levasseur-Regourd,* 2009). Currently this is only a hypothesis, and more observations are necessary to understand this effect; it is described here mainly to show the remote sensing power of the correlation between color and polarization.

Simultaneous observations of color and polarization and their combined analysis not only allow one to indicate and qualitatively explain changes in the particle properties, but also allow narrowing down the range of the particle parameters at computer modeling, making their results less ambiguous. An example of a comprehensive analysis that involves a simultaneous consideration of the color and polarization variations in the coma together with the changes in the brightness radial profile and Afρ in several filters is presented by *Ivanova et al.* (2019). The unusually deep negative polarization branch (down to −8%) for Comet C/2014 A4 (SONEAR) required a low-absorbing material; this, combined with the red color observed, brought tholin into consideration, which perfectly fit these requirements and was a realistic material for large heliocentric distances. The observed decrease in Afρ with the distance from the nucleus could be caused by decreasing either the albedo or size of particles, but simultaneous decrease in color and negative polarization ruled out particle fading and supported the idea

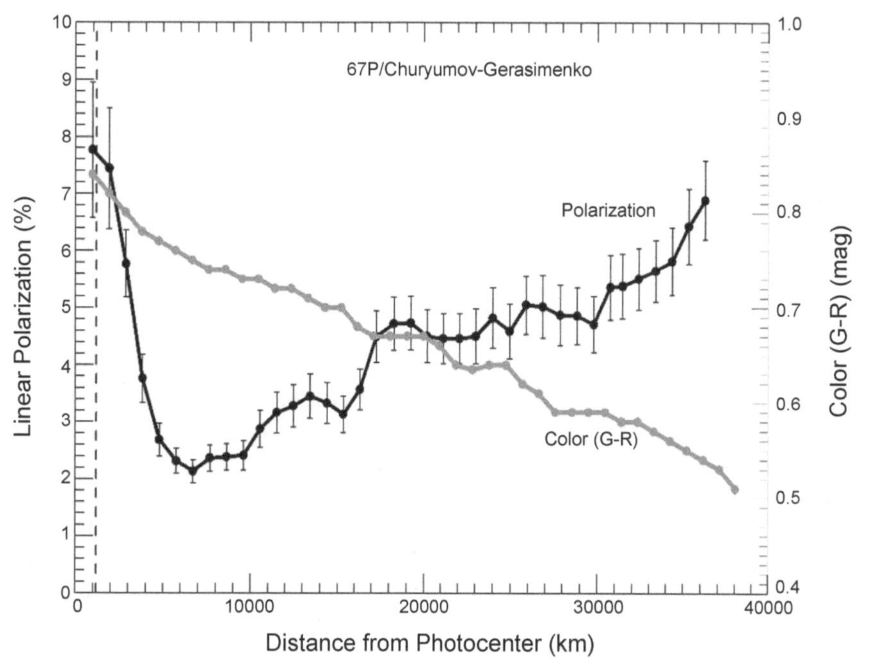

Fig. 7. Change of g-r_{sdss} color and r_{sdss} polarization of Comet 67P with the distance from the nucleus (*Rosenbush et al.,* 2017).

of the dust fragmentation. Thus, the combination of color and polarization provided information about the composition of the dust in Comet C/2014 A4, and the combination of their variations with variations of other observables allowed revealing the evolution of the dust particles in the coma.

2.3. Near-Infrared Scattering and Mid-Infrared Emission

In contrast with optical brightness and color, NIR and MIR (~1–5 μm and ~5–40 μm respectively) wavelengths contain features that are diagnostic of ice and dust composition, although size and porosity also have their effects. Since the publication of *K_C2* and *Hanner and Bradley* (2004), NIR spectroscopy has become more common, and the Spitzer Space Telescope has afforded detailed MIR spectra of several comets.

2.3.1. Near-infrared. The NIR absorption bands due to water ice are located at 1.5, 2.2, and 3.1 μm. The 3-μm band is the strongest feature, but at wavelengths that are difficult to observe from groundbased observatories, due to strong telluric emission and absorption. After suggestive evidence of water ice in the 1980s, the first clear evidence (multiple features at high signal-to-noise ratios) was seen at Comet Hale-Bopp with spectra of the 1.5- and 2.0-μm absorption bands when the comet was at 7 au from the Sun (*Davies et al.,* 1997). Hale-Bopp's 3-μm band was subsequently observed at 2.9 au with the Infrared Space Observatory (ISO), followed by far-infrared emission features at 44 and 65 μm observed at 2.8 au from the Sun (*Lellouch et al.,* 1998). Water ice detections in the NIR became more frequent in the new millennium, with several detections from the ground (*Kawakita et al.,* 2004; *Yang,* 2013; *Yang and Sarid,* 2010; *Yang et al.,* 2009, 2014; *Protopapa et al.,* 2018; *Kareta et al.,* 2021). Even though water appears to be the most abundant volatile in cometary nuclei, there are only a few direct detections of water ice in comae.

The relative shapes and strengths of the water ice absorption bands can be used to retrieve the properties of the ice grains, specifically, their size and chemical phase. Although the techniques to retrieve grain properties are not uniform, most investigations are consistent with micrometer-sized pure water ice grains (*Protopapa et al.,* 2018). The lack of a 1.5-μm feature in the ejecta of the mega-outburst of Comet 17P/Holmes and the subtle presence of the band at Comet C/2011 L4 (PanSTARRS) has been interpreted as indicative of submicrometer grains (*Yang et al.,* 2014). *Mastrapa et al.* (2008, 2009) showed that the NIR spectra have the potential to discriminate between ice in crystalline or amorphous forms. Particular emphasis has been given to a narrow feature at 1.65 μm that has only been observed following a large outburst of Comet P/2010 H2 (Vales) at 3.1 au from the Sun (*Yang and Sarid,* 2010). This feature is strongest in crystalline ice, but only for grain temperatures colder than ~200 K (*Grundy and Schmitt,* 1998). As a consequence of the similarity between warm crystalline ice and amorphous ice at 1.65 μm, but also due to the low crystallization temperature of amorphous ice (~160 K), amorphous water ice has not been definitively observed in a comet (*Kawakita et al.,* 2006, *Protopapa et al.,* 2014). For more on amorphous ice, see the chapter in this volume by Prialnik and Jewitt.

The variation of water ice features with time can also be used to infer coma ice grain properties. *Protopapa et al.* (2018) observed Comet C/2013 US_{10} (Catalina) from 1.3 to 5.8 au. Spectra of the comet at rh ≥ 3.9 au displayed the 1.5-

and 2.0-μm bands, but spectra at ≤2.3 au were featureless in this respect. The variation with heliocentric distance is consistent with the limited lifetimes of water ice in contact with low-albedo material. *Protopapa et al.* (2018) used effective medium approximation (see section 3.1) and an ice sublimation model to conclude that ice grains in the coma of C/2013 US_{10} likely contain a small amount of dust, up to ~1% by volume.

2.3.2. Mid-infrared (thermal) spectra. At longer wavelengths, thermal emission dominates the spectra of comets. The thermal spectral energy distribution, defined by the temperature of the dust, was considered in detail in *K_C2*; here we focus on the studies where the main progress has been achieved: studies of the MIR spectral features.

The MIR silicate features arise from stretching and bending modes in Si–O bonds. The two main silicate types in comets are olivine and pyroxene and the spectral features in cometary comae are consistent with both classes in crystalline and amorphous phases. The "amorphous" materials may not be minerals in the strict sense, but appear to have stoichiometric compositions similar to olivine $[Mg,Fe]_2SiO_4$ and pyroxene $[Mg,Fe]SiO_3$ (see the chapter in this volume by Engrand et al.). The amorphous materials produce broad (≥1 μm) emission features at ~10 and ~20 μm. The shapes of the features vary with Fe-to-Mg ratio, and spectra of Comet Hale-Bopp were in best agreement with model spectra based on amorphous pyroxene and olivine with Fe/Mg = 1.0 (*Harker,* 1999). The crystalline minerals produce a variety of narrow (~0.1–0.5 μm) features throughout the MIR (*Dorschner et al.,* 1995; *Koike et al.,* 2000, 2003). These features are most consistent with Mg-rich crystalline species. A spectrum of Comet 17P/Holmes in Fig. 8 shows prominent peaks near 10.0, 11.2, 16.3, 19.3, 23.5, 28.0, and 33.5 μm arising from Mg-rich olivine. In contrast, crystalline pyroxene features tend to be weaker and not always well separated from Mg-rich olivine features, but evidence for Mg-rich crystalline pyroxene is found at 9.3 μm (*Wooden et al.,* 1997, 1999), ~14.4, ~15.4, and 29.4 μm.

Outside the silicate emission bands, there is a strong pseudo-continuum arising from carbon-bearing materials (see the chapter in this volume by Engrand et al.). The spectral temperature of this pseudo-continuum is up to ~30% warmer than a blackbody sphere at the same heliocentric distance (*Gehrz and Ney,* 1992; *Sitko et al.,* 2004). The continuum temperature is generally measured with a Planck function fit to data near 7.5 to 8.5 and 12 to 13 μm and the best-fit temperature normalized by the temperature of a large blackbody sphere (278 $R_h^{-1/2}$ K, for R_h in astronomical units). The mean flux density of the spectrum in the middle of the silicate band (~10 μm) is normalized by the best-fit Planck function to derive the silicate band strength. *Gehrz and Ney* (1992) and *Sitko et al.* (2004) have shown that strong silicate features apparently correlate with continuum temperature. If the increase in silicate band strength was solely due to an increase in the relative abundance of silicate dust, then an anti-correlation between silicate feature strength and continuum temperature might be expected, as the long-wavelength continuum points (~12.5 μm) have silicate emission, whereas the short-wavelength points (~7.8 μm) have little to no emission, thus cooling the spectral temperature. Therefore, some other coma grain properties (e.g., size distribution or porosity) must be involved (see section 2.3.3).

At shorter wavelengths (3 to 5 μm), *Bockelée-Morvan et al.* (2019) found a correlation between continuum temperature (~0.3 K/° at 1.2 to 1.4 au) and phase angle in the inner coma of Comet 67P and identified thermal gradients

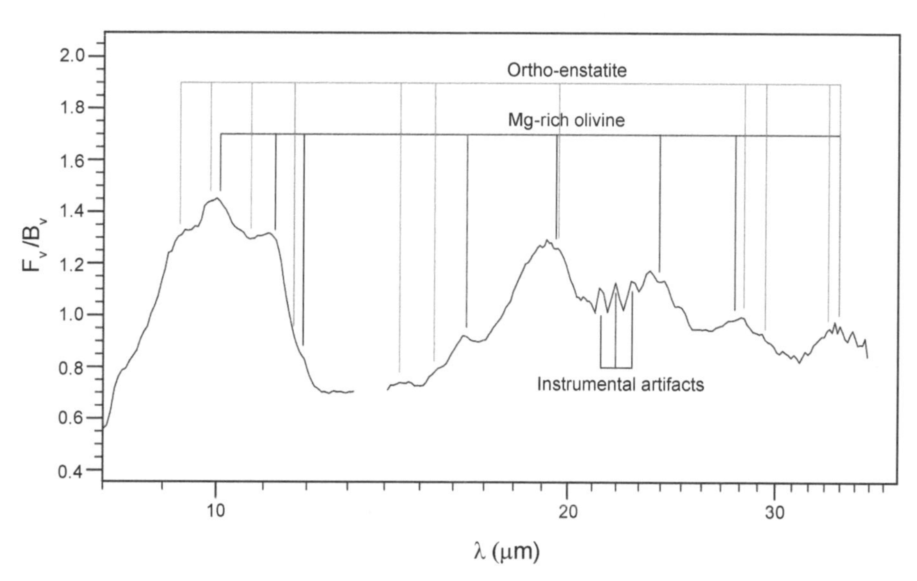

Fig. 8. Spectrum of Comet 17P/Holmes (*Reach et al.,* 2010; *Kelley et al.,* 2021) normalized by a Planck curve fit to the full wavelength range. The expected locations of strong features of Mg-rich olivine and orthoenstatite (Mg-rich pyroxene) from *Koike et al.* (2000, 2003) are marked. Also marked are selected weaker features of orthoenstatite at 14.4, 15.4, and 29.4 that may be present.

supported by large grains (perhaps 100 μm or larger) as the likely cause. This effect has not been identified in ground-based observations but suggests that variations at the 10% level in coma spectral temperature may also be expected in the 10-μm region, depending on observing geometry and grain size distribution.

The presence of several other minerals has been considered based on cometary MIR spectra, and the broad spectral coverage afforded by spacebased or airborne observatories enabled these works. However, results claiming detections are lacking spectrally identifiable features. The challenge to confirming new minerals is identifying individual spectral features, rather than solely relying on the reduced χ-square fitting. The effects of grain shape are also important to consider, especially with anisotropic minerals like Mg-rich olivine (*Lindsay et al.,* 2013). *Lisse et al.* (2007) suggested evidence for carbonates in an ISO spectrum of Hale-Bopp. However, the feature at 7.0 μm is marginal (*Crovisier et al.,* 1997). Carbonates are not seen in other comets, and neither has the 12.7-μm feature been identified (*Woodward et al.,* 2007; *Bockelée-Morvan et al.,* 2009). To date, only the four silicate types (amorphous and crystalline pyroxene and olivine) remain as spectrally identifiable dust materials in MIR observations of comets.

2.3.3. Correlation between polarization and strength of the silicate feature in the thermal infrared. The correlation between the values of maximum polarization and the strength of the 10-μm silicate feature was first considered by Hanner (2003). Lisse et al. (2002) and Sitko et al. (2004) attributed strong silicate features to small particles, whereas weak or absent silicate features were suggested as a signature of large particles. This sounds consistent with the polarimetric data, as small particles can be responsible for higher polarization due to higher contribution of the Rayleigh scattering, whereas the light scattered by large porous particles can experience a strong depolarization. However, this approach cannot explain the change in polarization with the distance from the nucleus (Fig. 5), as this would lead to an unrealistic conclusion that in the gassy comets, the dust is dominated by small particles near the nucleus and by large particles farther from the nucleus.

A more detailed analysis of this correlation was accomplished by *Kolokolova et al.* (2007). They showed that the correlation more likely results from different porosity of the dust particles. Their discrete dipole approximation (DDA) modeling (see section 3.1) of the silicate feature showed that for very fluffy particles, e.g., those presented by the ballistic-cluster-cluster aggregates (BCCA), the silicate feature remains strong no matter what size the particle (defined by the number of submicrometer monomers) is, whereas for less porous ballistic-particle-cluster aggregates (BPCA), the silicate feature becomes weaker with increasing number of monomers (see more on BPCA and BCCA in section 3.2.1). Thus, one can expect that the dust in the comets with a weak silicate feature is dominated by rather large compact particles, while in the comets with a strong silicate feature, the particles can be also large but more porous. At the same time, compact particles tend to be more gravitationally bounded to the nucleus and less accelerated by the gas flow than lighter porous particles of the same size that made them concentrated close to the nucleus. This leads to a decrease in the dust/gas ratio and increase in gas contamination at large apertures. As a result, comets whose dust is dominated by more compact particles (and show a weak silicate feature) should exhibit a decrease in polarization with an increase in aperture and overall lower polarization than comets with the dust dominated by more porous particles. Thus, the size of particles in all comets may be similar, dominated by large particles, but low-polarization and weak-silicate-feature comets are characterized by rather compact particles, whereas for the high-polarization and strong-silicate-feature comets, more porous particles are typical. The conclusion appeared to be consistent with the dynamical characteristics of the comets, specifically, the comets with supposedly more compact particles have smaller perihelia, and thus may be more affected by the solar radiation. The main conclusion of the paper by *Kolokolova et al.* (2007) was that two polarimetric classes of comets may result from a difference in the porosity of the dust particles.

These ideas have been supported by a more careful analysis by *Kwon et al.* (2021), who showed that the compositional differences in comets assumed based on the thermal infrared observations cannot explain the difference in cometary polarization, which is most likely associated with their different porosity.

3. INTERPRETATION TECHNIQUES

3.1. Most Popular Light-Scattering Techniques to Model Cometary Dust

For the cometocentric distances resolved by Earth-based observations, the number density of the cometary dust is low, which allows considering interaction of the dust particles with the solar radiation in the single scattering regime; i.e., one can ignore multiple scattering and other interactions between the particles and use computer models that consider light scattering by individual particles. Of course, one particle cannot represent cometary dust, which is characterized by particles with some size distribution of a variety of shapes and compositions. But combining simulations for different single particles, or even only for those which produce the dominating contribution to the scattered light, allows receiving a rather realistic model of the cometary dust.

Very active comets can have optically thick coma close to the nucleus (e.g., *Rosenbush et al.,* 1997). To simulate these observations, multiple scattering should be accounted for, and radiative transfer should be applied to the modeling of the internal coma of such comets. The ejecta produced by Deep Impact were also optically thick, and radiative transfer was used to model the observed characteristics of the ejecta (*Nagdimunov et al.,* 2014).

The simplest particle used in cometary dust models is a homogeneous sphere. Its light-scattering characteristics (brightness, polarization, and their change with phase angle and wavelength) can be calculated using Mie theory (*Mie,* 1908), which represents an analytical solution of the Maxwell equations for the interaction of electromagnetic waves with a spherical particle. Numerous Mie codes are available online (see *https://scattport.org/index.php/light-scattering-software/mie-type-codes*) and are included in some computer libraries [e.g., interactive data language (IDL) routine Bohren and Huffman Mie (BHMIE) code]. Although it is known that cometary particles are not such spheres, this approach is still regularly used in cometary science, e.g., for interpretation of Rosetta data (e.g., *Fulle et al.,* 2010; *Fink and Rinaldi,* 2015; *Bockelée-Morvan et al.,* 2017), as it provides reasonable results when such properties as scattering and extinction cross-section are of interest (e.g., at computations of spectra) or when fundamental regularities in light scattering by particles are the subject of the study. Notorious resonance structures of Mie results (oscillations in the dependences of the light-scattering characteristics on phase angle or wavelength) can be smoothed out by using a broad size distribution of particles.

A more sophisticated but still a rather simple approach uses non-spherical regular particles, e.g., spheroids, cylinders, or other axisymmetric particles. For modeling such particles, there are direct theoretical solutions of Maxwell equations, reviewed in section 3.3.1 of *K_C2,* as well as solutions that use the T-matrix approach, described in *K_C2,* their section 3.3.2. Considering cometary dust as an ensemble of polydisperse and polyshaped (e.g., spheroids of different axes ratio) particles provides a rather realistic approach to the cometary dust (*Kolokolova et al.,* 2004b), especially if a model counts on the roughness of the particles [see the result for rough spheroids in *Kolokolova et al.* (2015)].

It is well known that cometary particles are not homogeneous, and a typical cometary particle represents an agglomerate of smaller grains (see the chapter in this volume by Engrand et al.). Although a realistic modeling of such particles requires the very sophisticated modeling tools reviewed in section 3.2, quite often such particles can be modeled using effective medium approximations (EMA), also called mixing rules. In EMA, an effective refractive index of a mixture of different materials (including voids) is considered based on the refractive index and volume fraction of each component in the mixture. The most popular mixing rules are those of Maxwell Garnett and Bruggeman [see a review of the various EMA in *Kolokolova and Gustafson* (2001); mixing rules were also considered in detail in section 3.2 of *K_C2*]. *Mishchenko et al.* (2014) showed that EMA can be derived directly from the Maxwell equations, thus representing a rigorous solution for interaction of electromagnetic wave with a media formed by a mixture of materials. *Mishchenko et al.* (2016) also explored the limitations of the EMA, demonstrating that EMA works better for the smaller size parameter of the heterogeneities (inclusions) and for the number of inclusions per unit volume kept within some boundaries. The threshold value of the inclusion size parameter depends on the refractive-index contrast between the host and inclusion materials that often does not exceed several tenths, especially in calculations of the scattering matrix and the absorption cross section. Nevertheless, they showed that even for the materials with strongly contrasting refractive indexes such as hematite and air, there are quite realistic ranges of the sizes of the inclusions and host particle, and volume fraction of inclusions when EMA can be used safely. It is worth noting that one should be careful modeling layered particles using EMA; *Liu et al.* (2014) showed that EMA are reliable only for well-mixed cases down to an inhomogeneity scale, whereas the applicability of the EMA to cases of stratified or weak mixing materials is very limited. Thus, one should be careful using EMA to model porous particles in the visible; however, it is probably safe to use it for mixtures of silicates and organics (see section 3.2.2) or modeling cometary dust in thermal, or even NIR, wavelengths.

We would like to mention one more method that provides reasonable results, as it is an evident simplification of the realistic particles: distribution of hollow spheres (DHS) (*Min et al.,* 2005a). The DHS presents a rigorous solution of the Maxwell equations for coated spheres and considers the particles as a distribution of spherical shells characterized by the fraction, f, occupied by the central vacuum inclusion that changes from zero to some maximum value. Despite the departure of DHS from realistic dust particles, it models realistically looking spectra and even phase curves of polarization of laboratory samples (*Min et al.,* 2005a), allowing derivation of the particle size distribution. DHS was applied to comets only once (*Min et al.,* 2005b), and there are doubts that DHS can reproduce the properties of cometary agglomerates (*Levasseur-Regourd et al.,* 2020; *Tazaki and Tanaka,* 2018). However, this simple technique is extensively used to study other types of cosmic dust, specifically to model the 10-μm silicate feature in debris disks (e.g., *Arriaga et al.,* 2020).

A more comprehensive study of the light scattering by realistic cometary particles can be achieved by solving the Maxwell equations after some adjustments that allow accounting for the shape and structure of complex scatterers. Numerous techniques that make the Maxwell equations solvable for complex particles were reviewed in *K_C2.* Among them, there are two techniques most often used for modeling cometary dust: the T-matrix approach and coupled dipole approximation (more often called discrete dipole approximation, or DDA). The popularity of those techniques is partly related to the fact that T-matrix and DDA codes have been available online for many years; they are regularly updated and are accompanied by a detailed documentation. The main ideas behind those techniques were considered in *K_C2,* so here we focus only on the new developments of these techniques and their application to cometary dust. For a more detailed review of those techniques, as well as other light-scattering approaches to the light scattering by particles, see *Kimura et al.* (2020).

The main idea of the DDA is that a particle is divided into small cells, each of which is considered as a dipole. The DDA technique yields a system of linear equations that describes the fields that excite each dipole: the external field and the fields scattered by all other dipoles. The numerical solution of these equations allows accounting for the contribution of all dipoles in the total scattered field produced by the particle. The main advantage of the DDA approach is its flexibility; it can be applied to particles of arbitrary shape, structure, and composition. See the most popular discrete dipole scattering (DDSCAT) code and its detailed description at *http://ddscat.wikidot.com* and the latest DDSCAT user guide at *https://arxiv.org/abs/1305.6497*. There are two versions of the DDA codes in the DDSCAT package. The newer one utilizes fast Fourier transformation (FFT), which accelerates the computations but significantly increases the required RAM; the code without FFT is slower but requires less computing resources. We also recommend a powerful DDA package, called ADDA (*Yurkin and Hoekstra,* 2011), located at *https://github.com/adda-team/adda*.

To achieve an accurate solution, the number of dipoles should exceed $60|m-1|^3(\Delta/0.1)^{-3}$ (*Draine,* 1988), where m is the particle material refractive index and Δ is the fractional error (the ratio of the error to the quantity being computed). This puts significant requirements on computer resources. Besides, each DDA solution considers only one orientation of the particle, and to calculate a realistic randomly-oriented particle, the computations need to be done for numerous orientations whose number increases with increasing the size of particle and complexity of its structure. Note that some orientationally averaged DDA solutions can be obtained analytically — this and other ways to increase the DDA efficiency are considered in *Kimura et al.* (2016). The DDA technique has been especially successful in modeling thermal infrared spectra; see a review of those modeling efforts in the chapter in this volume by Engrand et al.

Unlike DDA, which provides the solution considering the internal fields of the scatterer, the T-matrix method is based on a solution to the boundary conditions on the particle surfaces and was originally called "extended boundary condition method" (*Waterman,* 1965). The method expands the incident and scattered fields into vector spherical functions with the scattered field expanded outside a sphere circumscribing a non-spherical particle. A significant development of the method was an analytical orientation-averaging procedure (*Mishchenko,* 1991) that made calculations for randomly oriented particles as efficient as for a fixed orientation of the same particle. It has been applied to light scattering by spheroids, cylinders, and Chebyshev particles (*Mishchenko et al.,* 2002); the codes can be downloaded from *https://www.giss.nasa.gov/staff/mmishchenko/tmatrix/*.

In the case of modeling cometary dust, the most popular development of the T-matrix technique is the superposition T-matrix method, which allows computing the light-scattering characteristics of clusters of spheres (i.e., agglomerated particles), calculating the T-matrix for each monomer in the cluster and then calculating the T-matrix for the whole cluster summarizing the scattered external field from other spheres in the cluster. To some extent, it uses the idea of the DDA; however, instead of dipoles, it considers the T-matrixes of individual monomers. The superposition approach allows applying the main T-matrix advantage, namely, analytical averaging over orientations. However, it has been recently showed that in some cases it is more efficient to average over orientations numerically after computations over a set of orientations is done. The main advantage of this type of computation is that it requires less RAM and thus allows considering larger clusters. The most recent version of the superposition T-matrix code, called the multiple sphere T-matrix (MSTM), is available at *https://github.com/dmckwski/MSTM*. It allows considering the clusters formed by spheres of different sizes and compositions, which can be arranged inside or outside of other spheres, thus modeling layered spheres or spheres with spherical inclusions, and/or adjacent to multiple plane boundaries, and/or in two-dimensional periodic lattices. It also allows modeling clusters made of linearly and circularly birefringent and dichroic materials and thus can be applied to crystals and homochiral (e.g., biological) particles (see *Nagdimunov et al.,* 2013).

A significant progress in modeling large particles has been recently achieved with a new development in the T-matrix approach called the fast superposition T-matrix method (FaSTMM), introduced by *Markkanen and Yuffa* (2017). The FaSTMM uses the fast multipole method (FMM) to speed up the STMM solution (which is similar to the MSTM considered above). The FMM forms monomer groups hierarchically and computes electromagnetic interactions between the separate groups in each level of hierarchy. This decreases the costs of computing all the pairwise monomer interactions in a system of N monomers from $O(N^2)$ to $O(N *\log N)$. Note that like the DDA, specifically the DDSCAT code, the FaSTMM does not allow analytical orientational averaging. The FaSTMM code is available at *https://wiki.helsinki.fi/xwiki/bin/view/PSR/Planetary%20System%20Research%20group/*. FaSTMM was successfully used to model Rosetta dust particles (*Markkanen and Agarwal,* 2019); for more information, see section 3.2.1.

One more important development in the light-scattering simulations is the approach to model light scattering by particulate surfaces using the so-called radiative transfer-coherent backscattering (RT-CB) code. Although the approach contains some approximations, it provides quite realistic results on light scattering by particulate surfaces, including such coherent backscattering effects as the opposition spike and negative polarization spike observed at very small phase angles (e.g., *Mishchenko et al.,* 2009). This technique considers a particulate surface as a sparse layer of spheres and is based on the Monte Carlo vector radiative transfer code wherein coherent backscattering is computed by incorporating the reciprocity relation in electromagnetic scattering and keeping the relative phase information of the wave components (*Muinonen et al.,* 2012, 2015). Strong coherent backscattering effects were not observed for cometary surfaces (*Masoumzadeh et al.,* 2019), so the RT-CB code was not used in modeling cometary nuclei. However, a version called radiative transfer with reciprocal transactions (R^2T^2) was successfully combined with FaSTMM (*Muinonen et al.,* 2018; *Markkanen et al.,* 2018a). Unlike RT-CB, R^2T^2 is used for densely packed layers and incorporates the full incoherent extinction, scattering, and absorption properties of a volume element larger than the wavelength. The RT-CB code requires particles to be spherical because it uses the Mueller matrix to compute the scattering interactions. The Mueller matrix does not contain electromagnetic phase information (required by the coherent backscattering CB part) except for spherical particles. R^2T^2 code is basically RT-CB code but the scattering interactions are computed using the T-matrix instead of the Mueller matrix. The FaSTMM solution combined with the R^2T^2 approach was successfully applied to model Rosetta dust particles by *Markkanen et al.* (2018b). For more details and comparison of the different techniques described above, see *Penttilä et al.* (2021).

The other important source of information about light-scattering characteristics of cometary particles is laboratory measurements of cometary dust analogs. Systematic laboratory simulations of light-scattering characteristics of dust particles are performed by four main groups: the Granada-Amsterdam group, specialized in measuring light-scattering characteristics of cosmic analogs; the Paris group, especially known by their low-gravity experiments; and the Bern and Grenoble groups, which perform light-scattering experiments with ice and its mixtures with other materials. A review of the laboratory measurements performed by the Bern and Grenoble groups is presented in the chapter in this volume by Poch et al. The Granada-Amsterdam group has produced a popular database of light-scattering properties of mineral and meteoritic particles (see *https://old-scattering.iaa.csic.es/*). This database provides a lot of photometric and polarimetric phase curves to compare with the astronomical observations described in section 1; they also measured a complete Mueller matrix of the samples that is intensively used to test theoretical tools (see review in *Muñoz and Hovenier,* 2015). The Paris group has collected a large number of the phase curves and wavelength dependences of polarization for particles analogous to the ones expected in comets and other cosmic bodies. This allowed them to find out the main regularities between the polarimetric characteristics of particles and their size and structure (e.g., *Levasseur-Regourd et al.,* 2015), especially in the case of complex particles, which is hard to model theoretically. It is also worth mentioning studies by groups that provide optical constants (complex refractive index) of the materials of astronomical interest. The data for various silicates, oxides, sulfides, and carbonaceous materials produced by the Jena group are presented in *https://www.astro.uni-jena.de/Laboratory/Database/jpdoc/f-dbase.html.* A compilation of several databases of optical constants can be found at the Jena-St. Petersburg Database of Optical Constants (HJPDOC) (*https://www2.mpia-hd.mpg.de/HJPDOC/database.php*), which still contains the largest collection of optical constants of materials of astronomical interest even though it has not been updated since 2008. The MIR refractive indexes for astronomical ices are produced by the Universidade do Vale do Paraíba (UNIVAP), Laboratório de Astroquímica e Astrobiologia (LASA), and can be found at *https://sites.google.com/view/astrow-en/research/databases.* Some collections of optical constants can be found in the NASA PDS, e.g., for water ice by *Mastrapa* (2020).

Fig. 9. Comparison between cometary particles used in numerical simulations and a real one. **(a)** "Agglomerated debris" modeled with DDA (*Zubko,* 2013). **(b)** Agglomerates of layered spheroidal monomers modeled with DDA (*Lasue et al.,* 2009). **(c)** Hierarchically structured cluster of 64 × 16 spheres modeled with MSTM (*Kolokolova et al.,* 2018). **(d)** Real cometary dust particle, imaged by Rosetta atomic force microscope MIDAS (*Mannel et al.,* 2019).

3.2. Applications of Light-Scattering Techniques to Cometary Dust

3.2.1. Size and structure of particles. As was mentioned in the previous section, for interpretation of the remote sensing data on cometary dust, two main techniques, DDA and T-matrix, were used. After a success of modeling the cometary dust as porous aggregates to explain cometary photometric and polarimetric phase curves in Kimura et al. (2003a, 2006), the majority of the interpretation attempts focused on porous and aggregated particles. Figure 9 presents examples of the particles used to model cometary dust with DDA (Figs. 9a,b) and T-matrix (Fig. 9c); Fig. 9d shows an image of a real cometary particle.

The "agglomerated debris" shown in Fig. 9a have been extensively used in the papers by Zubko et al. since the publication of *Zubko et al.* (2005). They are produced by removing cells from a spherical particle, thus creating particles and monomers of irregular shapes, which is the main advantage of the model. However, to achieve a larger size of particles, the model increases the size of the monomers, keeping the larger particles structurally identical to smaller particles; i.e., they have the same number, shape, and position of the monomers, which are scaled to the new size. In reality, larger cometary agglomerates are characterized by a larger number of the monomers of the same size. Thus, Zubko's model does not provide information regarding the size of the monomers and cannot correctly reproduce the electromagnetic interaction between the monomers in realistic particles, which contain hundreds of monomers. As a result, the model attributes the observed differences in light-scattering characteristics for different comets primarily to a difference in the composition of their dust. This makes the model not only too rigid, but also leads to doubtful results.

A more realistic structure for cometary particles is offered by ballistic aggregates, and the following ballistic aggregates are most often used:

- BPCA that have a porosity of about 85% and in the case of large particles approach a fractal dimension of 3;
- BCCA that have porosity larger than 95% and in the case of large particles approach a fractal dimension of 2;
- Ballistic Agglomeration with One Migration (BAM1) and Ballistic Agglomeration with Two Migrations (BAM2) particles are characterized by porosities about 75% and 65% respectively; see *Shen et al.* (2008, 2009) for details.

A detailed description of the particle models and the techniques used for their generation are discussed in *Kimura et al.* (2020).

Closer to the realistic cometary particles is the particle in Fig. 9b where the spheroidal core-mantle monomers are organized in a ballistic aggregate (*Lasue et al.,* 2009). This model considers a size distribution of monomers as well as a size distribution of particles. However, due to a limitation of computer resources, the model was limited to aggregates of 256 monomers, and modeling larger particles required a smaller number of larger monomers.

The particle in Fig. 9c is built of spherical monomers, which is the only opportunity in MSTM modeling. Although these particles are formed by monodisperse monomers, they may better represent a structure of the cometary agglomerates as they are hierarchical BPCAs. To create a hierarchical BPCA, regular BPCA are built and then these clusters are ballistically organized in the clusters of the second order, and then the clusters of each next order are built ballistically from the clusters of the previous order (*Kolokolova et al.,* 2018). Hierarchical structure is confirmed for the dust agglomerates in Comet 67P (*Mannel et al.,* 2019). This model is also consistent with the dust formation modeling in protoplanetary nebulae (*Dominik,* 2009).

Kolokolova and Kimura (2010b) presented cometary dust as a mixture of aggregated and solid non-spherical particles. *Lasue et al.* (2007) used a similar model for modeling interplanetary dust. Lasue et al. used DDA in their modeling, and Kolokolova and Kimura used the MSTM code for aggregates and the T-matrix code for spheroids for compact particles. In both studies the mixture of agglomerated and compact particles allowed the reproduction of brightness and polarization phase curves. The model by *Kolokolova and Kimura* (2010b) also reproduced red color and polarimetric color of the dust and its low albedo. The ratio of compact to fluffy particles appeared to be close to the one found *in situ* for Comet Halley, and the mass ratio of silicate to carbonaceous materials equal to unity in accordance with the elemental abundances found by the Giotto mission in Comet Halley (e.g., *Jessberger et al.,* 1988).

Probably the most comprehensive model of the cometary dust has been presented by *Halder and Ganesh* (2021), in which a mixture of porous, fluffy, and solid (according to the definitions from *Güttler et al.,* 2019) particles was considered. They created high-porosity hierarchical (HA) particles and moderate porosity structures with solids in the core, surrounded by fluffy aggregates called fluffy solids (FS). They also added solids as low-porosity (<10%) particles in the form of agglomerated debris. They studied the mixing combinations: HA and Ssolids; HA and FS; and HA, FS, and solids. Complexity of the model and the efforts of the authors to bring the model close to the properties of cometary dust, as they are known from the Rosetta studies, are very impressive. However, this work also shows a limitation of such complex multi-parameter models: The fit to polarization phase curve for a specific comet (the paper considered Halley, 67P, Hyakutake, Hale-Bopp) could be achieved with different compositions of the particles, depending on their size distribution and the ratio of fluffy to solid particles in the mixture. The authors tried to narrow down the range of the best-fit particle characteristics, adding to the analysis the spectral polarimetric data. They presented a set of the unique particle characteristics that provided the best fit to both angular and spectral polarimetric data. However, a reliability of these unique characteristics is not evident. For example, neither the composition of the Comet 67P dust (50% silicates and 50% carbon) nor the ratio of porous to solid particles (75% solids) are consistent with the Rosetta data (*Bardyn et al.,* 2017; *Güttler et al.,* 2019), and the size distribution (power –2.4) is not consistent either with the Grain Impact Analyser and Dust Accumulator (GIADA) data (*Fulle et al.,* 2016) or with Visible and Infrared Thermal Imaging Spectrometer (VIRTIS) data (*Rinaldi et al.,* 2017), as they derived the power >3 in absolute value for the dates close to the observations of 67P by *Rosenbush et al.* (2017) used for fitting by Halder and Ganesh. One more reason why the obtained best-fit parameters may not reproduce the realistic properties of the dust particles is that the considered particle was <15 μm in size, whereas an abundance of much larger particles was detected by the Rosetta mission and in the tail of Comet 67P (*Fulle et al.,* 2016; *Moreno et al.,* 2017).

The main problem with modeling realistic cometary particles is that their size may reach hundreds of micrometers and even millimeters (*Güttler et al.,* 2019). Unfortunately, the current computer resources limit capabilities of DDA or T-matrix computations, which require enormous RAM and many hours of computation time even for the case of parallelized version of the codes. So far, the largest particles computed with DDA were ~10 μm in size (e.g., *Zubko et al.,* 2020); note that they were agglomerated debris formed just by a dozen of micrometer-sized monomers. The largest particles modeled with the MSTM code are also about 10 μm, consisting of ~2000 submicrometer monomers (*Kolokolova and Mackowski,* 2012). Recent versions of the MSTM code whose efficiency is dramatically increased with the FFT procedure (*Mackowski and Kolokolova,* 2022) allowed computing particles up to 16,000 submicrometer monomers, i.e., reaching particle sizes about 20 μm. However, this is still smaller than the particles studied by the COmetary Secondary Ion Mass Analyzer (COSIMA) and GIADA instruments on the Rosetta spacecraft.

The largest particles, up to 100 μm, were modeled by *Markkanen et al.* (2018b). Their model considered a volume shaped as a Gaussian (see *Muinonen et al.,* 1996) particle, filled with a mixture of submicrometer organic and submicrometer-sized silicate spherical monomers. The monomers were polydisperse, covering the range of radii from 50 to 1000 nm. *Markkanen et al.* (2018b) considered a power-law size distribution of those particles with the power –3 and varied the number of organic and silicate monomers in the mixture as well as the volume fraction of the voids in the particle. The light scattering by such a particle was modeled using the FaSTMM and R^2T^2 techniques described in section 3.1. The best fit to the Rosetta photometric phase curve was found for the dominant particle size between 5 and 100 μm. The total volume of the organic grains was estimated as 0.3 and that of silicate grains estimated as 0.0375, which, being converted to the mass fraction, is reasonably consistent with COSIMA (*Bardyn et al.,* 2017) estimations. As the authors noted, the presence of silicate monomers with organic mantles can easily increase the abundance of the silicates in the dust.

Markkanen et al. (2018b) also modeled the nucleus phase function, presenting the nucleus surface as an ensemble of 1-km-sized Gaussian particles filled with smaller particles. This allowed them to introduce a surface roughness that followed Gaussian statistics, resulting in the corresponding geometric shadowing effect. The results showed that the nucleus could be simulated with the same particles as in the coma but packed densely on a rough surface. Although the main goal of this paper was to model Rosetta's Optical, Spectroscopic, and Infrared Remote Imaging System (OSIRIS) phase curve of the coma and nucleus, photometric, and polarimetric phase curves observed from the Earth are also among the results and are presented by the particles of size <20 μm. This provides strong hope that fitting groundbased observations does not require particles larger than some tens of micrometers. Probably the only questionable feature of this model is that the particle is presented as a volume filled with constituents that are not in direct contact with each other. As was shown in the laboratory measurements with icy particles (see the chapter in this volume by Poch et al.) and in *Kolokolova and Mackowski* (2012), the light-scattering characteristics of the particles assembled in some structures is different from the freely distributed, not connected, particles, probably due to stronger near-field effects.

Thus, all currently used models have some significant limitations that may affect the trustworthiness of the applications to specific comets. However, the developed models can be successfully used to understand the physics of the interaction of the dust particles with radiation and to uncover how the physical properties of the particles affect the observed characteristics. Specifically, one can study dependences of brightness and polarization on the size and composition of monomers and their arrangement in a particle using highly controlled models such as ballistic aggregates. Examples of this type of study can be found in *Kimura et al.* (2006), *Shen et al.* (2009), *Kolokolova and Mackowski* (2012), and *Kolokolova et al.* (2018). Recently *Mackowski and Kolokolova* (2022) have extended such modeling to large BPCAs reaching more than 16,000 monomers. The computations were made using the MSTM code upgraded with FFT capabilities and performed at the NASA High-End Computing Capability (HECC) facility. The results are shown in Fig. 10.

The plots in Fig. 10 allow us to draw some important conclusions. The monomer size (size parameter) strongly affects the polarization and brightness phase curves even if a very narrow range of values is considered. By increasing the number of monomers in the agglomerate, the polarization tends to reach some limit and does not change after the limit is reached. The number of monomers that defines the limit is larger for transparent materials. Thus, it may be plausible to avoid modeling agglomerates of more than 1000 monomers in the case of absorbing materials.

The most interesting aspect is the dependence on the composition. One can see that the polarization maximum is very high for ice, then decreases for silicates, then increases for more absorptive materials like Halley dust (see section 3.2.2) and carbon, but then strongly decreases for even more absorptive materials such as Fe. Thus, the dependence

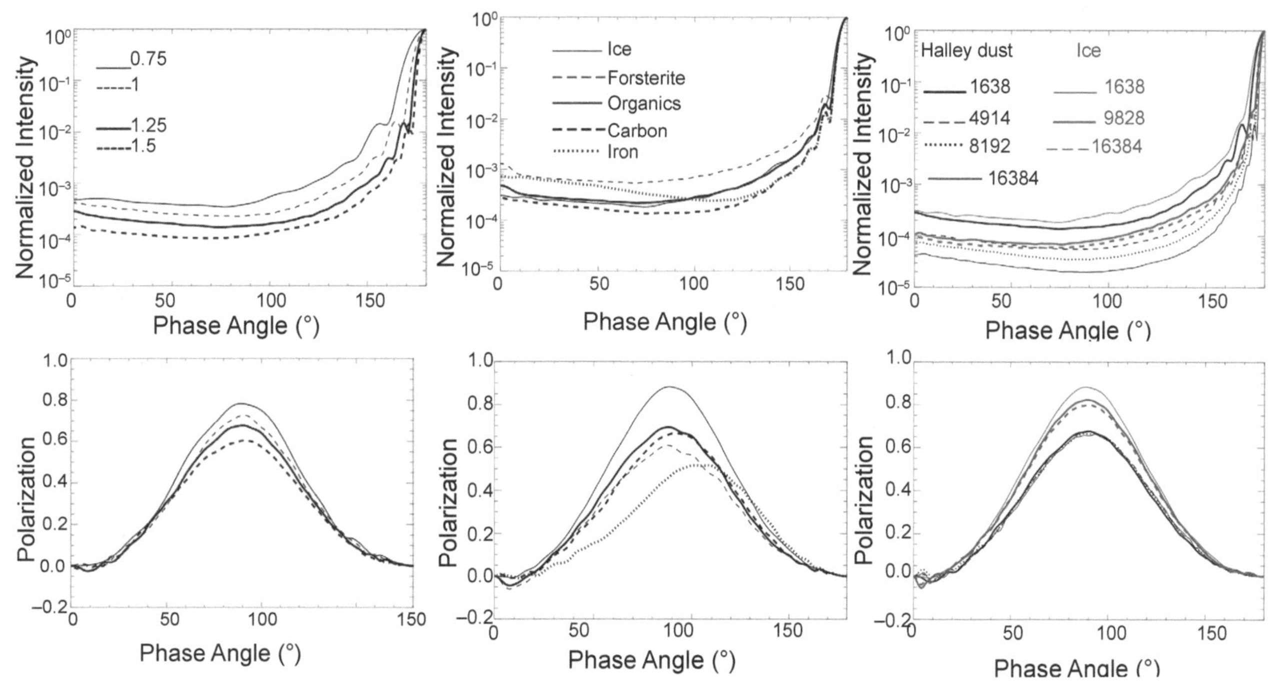

Fig. 10. Modeled phase curves of normalized brightness (top panel) and polarization (bottom panel) for large clusters of particles from *Mackowski and Kolokolova* (2022). *Left column:* Effects of the monomer size parameter for Halley dust (see section 3.2.2). *Middle column:* Results for different materials. *Right column:* Effects of the number of monomers for Halley dust and ice. If not specified, the monomers of radius 0.1 μm and agglomerates of 1638 monomers were used. In all cases the porosity of the agglomerates was ~85% and the wavelength was 0.65 μm.

on the absorption (imaginary part of the refractive index) is not straightforward and can result in an increase or decrease of polarization.

The observed regularities can be explained from the point of view of the polarizability of the particle units. The basic polarizability, α, is polarizability of the molecules in the particle material related to the material refractive index through the Lorentz-Lorenz equation $\alpha \sim (m^2-1)/(m^2+2)$. Thus, in the case of a low imaginary part of the refractive index, polarizability is larger for a larger real part. However, the situation is more complex for agglomerates as stronger polarizability results in stronger interactions between the monomers, causing stronger depolarization of the light. As a result, icy clusters with a low refractive index have weakly interacting monomers, which explains why positive polarization produced by an icy agglomerate is larger than that for silicates. For a larger imaginary part of the refractive index, the polarizability is defined by a complex interplay between the real and imaginary parts of the refractive index and polarization may increase or may decrease (see Fig. 10). More examples can be found in *Kimura et al.* (2006). There, in Fig. 1, for the same imaginary part of the refractive index, polarization of aggregates decreases with an increase in the real part. In Fig. 2 of the same paper, for n = 1.4–1.6, the maximum polarization decreases with an increase in the imaginary part, but the trend is opposite for n = 1.7–2.0.

We also see in the top panel of Fig. 10 a strong dependence on the monomer size. The reason for this can again be in the polarizability, but now in the polarizability of a sphere in the cluster, which is defined by the total number of dipoles (molecules) in the sphere and thus increases for larger monomers consisting of more dipoles. Monomers with larger polarizability interact more strongly, and the effects of electromagnetic interaction produced by them more greatly depolarize the light.

3.2.2. Composition of the cometary dust. Besides the size and structure of the cometary dust particles, an important parameter of the modeling is the particle composition. The chapter in this volume by Engrand et al. considers the composition of cometary dust in detail; this chapter focuses on how the composition of the dust is reflected in the refractive index that is used in light-scattering modeling.

It is now accepted that the main components of cometary dust are carbonaceous materials and silicates. For a long time, it was considered that the most abundant silicates in cometary dust are amorphous silicates, called glass with embedded metal and sulfides (GEMS), and to a lesser extent, anhydrous Mg-rich minerals like forsterite and enstatite (*Hanner and Bradley,* 2004), although *Stenzel et al.* (2017) showed that in the case of Comet 67P the Mg to Fe ratio was close to that of meteorites. The composition of the organics is more obscure. Previously, it was considered that the optical properties of cometary organics can be presented as a mixture of amorphous carbon and so-called cosmic organics (*Li and Greenberg,* 1997). Rosetta findings indicate a material more similar to insoluble organic matter (IOM) although less processed than the IOM found in carbonaceous meteorites (*Paquette et al.,* 2021).

It is still a question how these materials are distributed in the particle. Based on Giotto's dust impact mass spectrometer, PUMA, the IDPs, and Rosetta COSIMA studies, it is very unlikely that cometary particles are pure silicate and pure carbonaceous particles forming two separate populations of the dust particles. Most likely, the silicate grains are embedded in the organic matrix or represent small silicate particles with organic mantles glued together. This allows considering mixtures of small silicate and organic monomers in a single particle (*Markkanen et al.,* 2018b) or a particle composed of an intimate mixture of silicate and metals encased in carbonaceous materials (*Mann et al.,* 2004) as more realistic models. The latter allows applying the EMA (see section 3.1) to calculate the refractive index of the mixture.

Mann et al. (2004) suggested a so called "Halley-like composition" (aka Halley dust) that is an intimate mixture of the materials consistent with the elemental composition of comet dust measured by the mass spectrometer on the Giotto mission (*Jessberger et al.,* 1988). The refractive index based on this composition was successfully used to model cometary light-scattering characteristics in numerous papers. However, it may need to be updated using the *in situ* data for Comet 67P and the elemental composition of the dust consistent with the reported in *Stenzel et al.* (2017), *Bardyn et al.* (2017), and *Paquette et al.* (2021).

To derive the refractive index for the cometary dust material, in both cases of the Halley-like and 67P-like compositions, the silicates are presented by so-called astronomical silicate, an analog material of interstellar silicate with optical properties consistent with those for $MgFeSiO_4$ (e.g., *Laor and Draine,* 1993), which was successfully used to model light scattering by different types of cosmic dust. The organic material was selected based on the results of *Kimura et al.* (2020), who showed that cometary organic matter should be carbonized after the formation of comets, thus acquiring the elastic properties similar to those of hydrogenated amorphous carbon; then a collisional velocity of a few meters per second could result in the easy fragmentation of the dust aggregates, seen in COSIMA's optical microscope (COSISCOPE) images. A lack of the optical constants for carbonized organics forced us to assume that the refractive indexes of the hydrogenated amorphous carbon likely mimic those of carbonized organic matter, because carbonization is characterized by the loss of H, N, and O (*Jenniskens et al.,* 1993). Table 1 presents the optical constants of the materials used to derive the refractive indexes for cometary dust. The ratio of different components in the mixture was chosen to be consistent with the elemental abundances discussed above.

One can see that despite some differences in the composition of those comets, the refractive index and its spectral change are similar, and thus the light-scattering modeling obtained with Halley-like composition is relevant to the dust in 67P and can be recommended for use in light-scattering models.

4. PHYSICAL PROCESSES CAUSED BY INTERACTION OF THE DUST WITH ELECTROMAGNETIC RADIATION

4.1. Radiation Pressure: Physics and Dependence on the Properties of Particles

The interaction of electromagnetic waves with dust particles exerts a force on the particles approximately in the direction of wave propagation, known as radiation pressure (*Burns et al.,* 1979). When dust particles are exposed to solar radiation, the solar radiation pressure pushes the particles outward in the radial direction, i.e., the anti-direction to solar gravity. Because both the solar radiation pressure and the solar gravity are proportional to the inverse square of the distance from the Sun, the ratio β of solar radiation pressure to solar gravity is a non-dimensional quantity useful for studying the relative importance of radiation force to the dynamics of dust particles.

Dust particles in the vicinity of their parent body are also driven away from the surface of the body by radiation pressure due to the reflection of solar radiation and the thermal emission from the body (*Burns et al.,* 1979; *de Moraes,* 1994; *Bach and Ishiguro,* 2021). For the particles moving with respect to the Sun, a relativistic drag force, known as the Poynting-Robertson (P-R) effect, appears in the equation of motion in the direction against particle motion to terms on the order of v/c where v and c denote the velocity of the particle and the speed of light, respectively (*Robertson,* 1937). Owing to the proportionality of the P-R drag to the β ratio, the larger the β ratio of a particle is, the shorter the P-R lifetime of the particle is. As a result, only large particles with small β values can remain near the orbits of their parent bodies, which have been observed as dust trails and meteor showers.

The solar radiation pressure acting on a moving particle differs from that on a particle at rest in the reference frame of the Sun due to the Doppler effect on the radiation pressure cross section, although the effect is typically negligible (*Kimura et al.,* 2002a).

The direction of solar radiation pressure is not exactly radial for non-spherical particles, because scattering and absorption of solar radiation is in general asymmetric around the radial vector (*van de Hulst,* 1957). The radial and non-radial components of radiation pressure on non-spherical particles have been computed by the method of separation of variables for spheroids (*Il'in and Voshchinnikov,* 1998) and the a1-term method for fluffy aggregates (*Kimura et al.,* 2002a). The computations have shown that the non-radial radiation pressure on non-spherical particles is on average a tenth or hundredth of the radial component, although there is a specific orientation of the particles where the non-radial components are comparable to the radial component.

The radial component of solar radiation pressure on fluffy aggregate particles has been computed using the various EMAs, the DDA, the generalized multisphere Mie solu-

TABLE 1. Refractive indexes of the materials and the effective refractive index for the dust in Comets Halley and 67P.

Material name	Wavelength		Volume fraction		Reference
	450 nm	650 nm	67P	Halley	
Amorphous carbon*	1.95 + 0.786i	2.14 + 0.805i	0.5042	0.4379	*Rouleau and Martin* (1991)
Organic refractory**	1.69 + 0.150i	1.71 + 0.149i	0.2521	0.2189	*Li and Greenberg* (1997)
Astronomical silicate	1.69 + 0.0299i	1.68 + 0.0302i	0.2208	0.3176	*Laor and Draine* (1993)
Iron	2.59 + 2.77i	2.90 + 3.02i	0.0228	0.0256	*Johnson and Christy* (1974)
Comet Halley	1.88 + 0.47i	1.98 + 0.48i			*Kimura et al.* (2003a)
Comet 67P	1.901 + 0.526i	2.015 + 0.532i			

* Represents carbonized organics.
** Represents primordial organics.

tion (GMM), and the MSTM [see section 3.1 and review in *Kimura et al.* (2016)]. Figure 11 depicts the β ratios for spheres, BPCAs, and BCCAs consisting of 0.1-μm-radius monomers of amorphous carbon AC1, astronomical silicate, Mg-rich olivine, Mg-rich pyroxene, Mg-rich silicate with Fe inclusions, and silicate-core + organic-mantle, which are a compilation of *Kimura et al.* (2002b, 2003b), *Köhler et al.* (2007), and *Kimura* (2017). The β ratio has a maximum in the submicrometer size range, while the maximum ratio is smaller for fluffy agglomerates than for compact particles. In contrast, values of β for large fluffy agglomerates in the geometrical optics regime are higher than those for compact particles of equal mass, because the geometrical cross section of particles increases with fluffiness. While the maximum β for fluffy agglomerates depends on the size of monomers, the ratios for any pure silicate particles do not exceed unity, irrespective of the particle morphology (*Mukai et al.,* 1992; *Kimura et al.,* 2002b). However, if metallic Fe inclusions are embedded in a Mg-rich silicate matrix, then the maximum values of β could exceed unity (*Altobelli et al.,* 2016; *Kimura,* 2017). The values of β for Mg-rich silicate particles with metallic Fe inclusions are barely distinguishable from those of silicate-core + organic-mantle particles. While solar radiation pressure on particles composed of Fe-rich, Mg-poor silicates is higher than that on particles of Fe-poor, Mg-rich silicates owing to the effect of Fe atoms on the absorption efficiency in the visible wavelength range, solar gravity on the former is also higher than the latter due to the effect of Fe atoms on the density. Accordingly, one should not expect that the ratios of solar radiation pressure to solar gravity for Fe-rich, Mg-poor silicates are significantly different from those for Fe-poor, Mg-rich silicates.

As the particles move away from the sunlit surface, they are gradually decelerated by the solar radiation pressure. Based on Fig. 11, it is reasonable to assume the β ratios of organic-rich and silicate-poor (CHON) particles are lower than those for amorphous carbon (AC1 from *Rouleau and Martin,* 1991), while β for organic-poor and silicate-rich (ROCK) particles exceed those for Mg-rich silicate with Fe inclusions. Thus, the values of β for ROCK particles and CHON particles are expected to be located somewhere between the values for amorphous carbon AC1 and those for Mg-rich silicate with Fe inclusions, although β for ROCK particles should be slightly lower than β for CHON particles. In the results, the spatial variations in the abundance of CHON and ROCK particles may not necessarily be noticeable in observations of cometary comae since β values of CHON particles do not seem to deviate from those of ROCK particles by a factor of 2.

In contrast, observational data may reveal the effect of porosity on the motion of dust particles in the comae, because β for highly porous and non-porous particles could differ by an order of magnitude or more if their size exceeds several micrometers. Therefore, if a variation in the optical properties of dust particles within a coma or among different comets is observed, the variation might be attributed to the difference in the porosities rather than the compositions.

Although β alone cannot constrain the physical properties of the dust particles, it can be used to check the validity of the characteristics of the cometary dust particles derived using other means. For example, an analysis of the temporal evolution in the ejecta of Comet 9P/Tempel 1 created by NASA's Deep Impact mission resulted in $\beta \approx 0.4$, which suggests dust agglomerates of tens of micrometers in size consisting of silicate-core + organic-mantle monomers of 0.1-μm radius (*Kobayashi et al.,* 2013). It turned out that this interpretation is consistent with observational data on the color temperature, silicate feature strength, and degree of linear polarization (*Yamamoto et al.,* 2008). The values of β for dust particles in the tails of sungrazing comets derived from Solar and Heliospheric Observatory (SOHO) observations turned out to be 0.6 at maximum (*Sekanina,* 2000). This was found to be consistent with a predominance of Mg-rich pyroxene and olivine grains or small agglomerates of these grains in the tails of sungrazers remaining after sublimation of organic matter (*Kimura et al.,* 2002b). *Ishiguro et al.* (2016) derived the maximum value of $\beta = 1.6 \pm 0.2$ from the optical observations of Comet 15P/Finlay and concluded that dust particles in the ejecta of multiple outbursts are fluffy agglomerates consisting of silicate-core and organic-mantle grains based on the numerical results of *Kimura et al.* (2003b).

4.2. Photoelectric Emission: Charging of Particles by Radiation

The surface of dust particles is inevitably charged, owing to the interactions of the particles with the UV and plasma. If free electrons are present on the surface of dust particles, then the boundary conditions of particle surface would be modified, which is the case for metallic particles. However, the surface of cometary dust is typically composed of dielectric materials such as ice, organic matter, and silicates. Because the electric conductivity of dielectric materials is extremely small compared to metals, the effect of surface charge on the scattering and absorption of light by cometary dust can be neglected. However, the surface charge might induce an observable effect on the plasma environments around comets through absorption and emission of electrons by the particles, which behave like super-heavy ions in a plasma.

When electromagnetic waves or photons are absorbed by a dust particle, photoelectrons might be emitted from the surface of the particle (*Feuerbacher et al.,* 1972). Photoelectron emission elevates the surface potential of the particle, predominating over other charging processes such as sticking of solar wind plasma and secondary electron emission (e.g., *Wyatt,* 1969; *Whipple,* 1981). The electric current due to photoelectron emission is a function of the absorption cross section and the photoelectric quantum yield, which is the number of photoelectrons per absorbed photon (e.g., *Mukai,* 1981; *Kimura and Mann,* 1998). Simple empirical formulae have been proposed to estimate the photoelectric quantum yield based on experimental data (e.g., *Draine,* 1978). However, the laboratory experiments tend to underestimate the true value of photoelectric quantum yields due to the geometrical limitations on the collection of photoelectrons in the laboratory (*Senshu et al.,* 2015).

Laboratory experiments on photoelectron emission are to a large extent limited to materials specific to the surfaces of satellites and spacecraft, but some measurements with astronomically relevant materials are available. *Baron et al.* (1978) conducted the experiments with a thin film of water ice at 13 and 80 K near the work function. As noted in the pioneering paper on the photoemission by *Einstein* (1905), the work function is a threshold, namely, the minimum energy required to move an electron to infinity (*Kittel,* 1995). *Feuerbacher et al.* (1972) measured the photoelectric quantum yield of glassy carbon, graphite, and lunar fines as a function of wavelength and determined the work function. The photoelectric quantum yields of submicrometer particles for silica, olivine, carbonaceous materials, and lunar samples were provided by *Abbas et al.* (2006, 2007), but the data might have suffered from calibration errors (*Kimura,* 2016).

The physical process of photoelectron emission is well accounted for by a three-step model (*Smith,* 1971): (1) excitation of electrons by absorption of photons at a certain depth below the surface, (2) transportation of the excited electrons to the surface against inelastic collisions, and (3) ejection of the electrons, if their kinetic energies normal to the surface exceed the work function of the surface. A rigorous theory for estimating the photoelectric quantum yield requires the electric field inside the particle to be known. The photoelectric quantum yield has been so far computed only for spheres in the framework of Mie theory or geometrical optics (*Watson,* 1973; *Dwek and Smith,* 1996; *Kimura,* 2016). If dust particles are of submicrometer size or smaller, then the photoelectric quantum yield is in general enhanced due to the surface curvature of the particles and the contribution of the entire volume to absorption of light. On the contrary, the photoelectric quantum yield of very tiny nanometer-sized particles is reduced near the threshold energy of photons, owing to the enhancement of the work function for small particles (e.g., *Wong et al.,* 2003). The application of the DDA to the computation of the photoelectric quantum yield from agglomerates is straightforward, but no one has so far found worth a computation, which is most likely limited to the small particle sizes.

4.3. Radiative Torques: Grain Alignment and Rotational Disruption

The radiation-dust interaction can induce radiative torques (RATs) on a dust grain of non-spherical shape. It is found that RATs can cause the alignment of dust grains and their spinup. Scattering of light by aligned grains induces the circular polarization of scattered light, which may explain circular polarization in comets (see section 2.2.2). For a comprehensive reviews of those effects, see *Lazarian et al.* (2015) and *Andersson et al.* (2015). In cometary comae, RAT can spin up the dust particles to extremely fast rotation so they are disrupted by centrifugal stress, affecting the evolution of dust and ice in cometary coma (see *Hoang,* 2020, for a review).

The RAT induced by the radiation of wavelength λ acting on a grain of radius a is defined as

$$\Gamma_\lambda = \pi a^2 \gamma u_\lambda \left(\frac{\lambda}{2\pi}\right) \mathbf{Q}_\Gamma(\Theta,a,\lambda)$$

where γ is the anisotropy degree of the radiation field, u_λ is the specific energy density, and $\mathbf{Q}_r(\Theta,a,\lambda)$ is the RAT efficiency vector that depends on the angle Θ between the radiation direction and the grain axis of maximum inertia moment (*Draine and Weingartner,* 1996; *Lazarian and Hoang,* 2007). The RAT efficiency vector has three components, Q_{e1}, Q_{e2}, Q_{e3}, with $\mathbf{e}_1$ chosen along the radiation direction **k**.

4.3.1. Grain alignment and spinup by radiative torques. The detailed study of grain alignment by RATs is presented in *Lazarian and Hoang* (2007). The authors found that RATs can induce the alignment along the radiation direction (k-RAT) because the averaging over the fast rotation of the

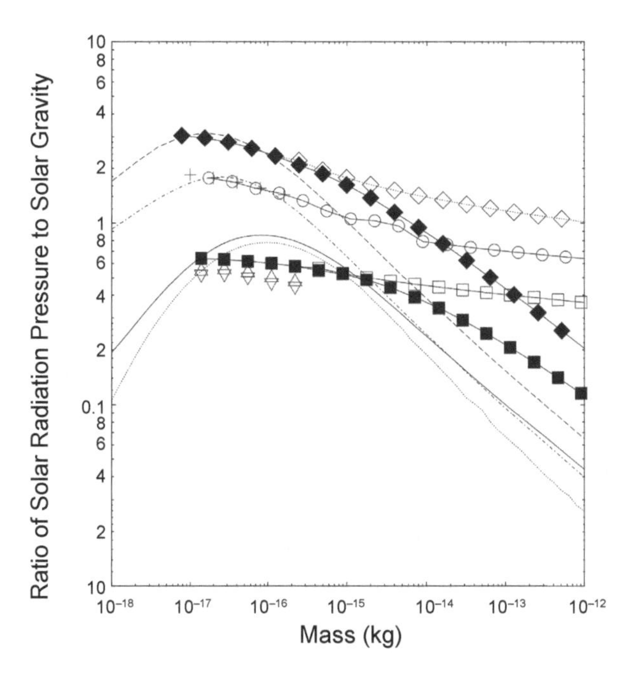

Fig. 11. The values of the parameter β for different dust particles: BPCAs (closed symbols) and BCCAs (open symbols) consisting of 0.1-μm-radius grains (see description of BPCA and BCCA in section 3.2.1). Different symbols represent amorphous carbon AC1 (diamonds), astronomical silicate (squares), Mg-rich olivine (triangles), Mg-rich pyroxene (inverse triangles), amorphous silicate with iron inclusions (circles), and silicate-core + organic-mantle (crosses). Lines show the results for solid spherical particles; the materials can be identified by the overlap of the line with the leftmost symbol.

grain around the axis of major inertia $\mathbf{a}_1$ cancels out the align torque component Q_{e2}, leaving only the component Q_{e1} that spins up the grain. When the grain initially makes an angle Θ with the radiation direction, the aligned component Q_{e2} is not zero and can act to bring the grain back to the stationary point Θ = 0, establishing the k-RAT alignment (Fig. 12a).

If a magnetic field is present, then the axis of grain alignment depends on the precession of the grain angular momentum with the magnetic field and the radiation direction. If an external electric field is also present, under the effect of RATs, grain alignment can occur along one of these three axes, including the radiation direction, magnetic field, and electric field (Fig. 12b).

The magnetic properties control the efficiency of grain alignment in the magnetic field. Paramagnetic relaxation that was first suggested by *Davis and Greenstein* (1951) to align grains with the magnetic field is known as the Davis-Greenstein (D-G) mechanism. The paramagnetic relaxation is based on the dissipation of the rotational energy due to the rotating induced magnetization component perpendicular to the spinning axis, which ultimately leads to both minimum rotational energy with angular momentum and its magnetization vector aligned with the ambient magnetic field only. The D-G mechanism implies that small grains are more efficiently aligned than large grains and that alignment becomes negligible when the grain rotational velocity is comparable to its thermal angular velocity (so-called thermal rotation). Furthermore, numerical simulations of D-G alignment for grains with different magnetic susceptibilities in *Hoang and Lazarian* (2016) demonstrate that, for superparamagnetic grains with embedded Fe inclusions, a maximum alignment degree of the grain axis with the magnetic field can be ~10–20%, although grains have superparamagnetic relaxation because the grain rotation velocity is smaller than the thermal value. This study establishes that efficient grain alignment is only achieved when the grain angular velocity exceeds its thermal angular velocity (i.e., suprathermal rotation), as previously suggested (e.g., *Purcell,* 1979). Such a requirement is satisfied by the spinup effect of RATs. Moreover, grains with enhanced magnetic susceptibility due to Fe inclusions can increase the paramagnetic relaxation rate that acts together with the RAT as well as increases the Larmor precession rate, which leads to B-RAT alignment (Fig. 12b).

4.3.2. Effects of grain alignment on circular polarization in cometary comae. As was shown in section 2.2.2, there are numerous observations of circular polarization of the scattered light in comets, and the most feasible mechanism to describe it is alignment of the dust particles by radiative torque in the presence of the solar magnetic field. This mechanism is considered in this section.

The intensity of circularly polarized radiation resulting from scattering on a dust particle is given by the fourth Stokes parameter V (see section 2.2) and is equal to $V = I_0S_{41}/k^2R^2$ where S_{41} is the 4–1 element of the Mueller matrix, I_0 is the intensity of the incident light, k is the wave number, and R is the distance to the source of light (e.g., heliocentric distance). Using DDA and considering the grains with x < 1 (Rayleigh particles), *Hoang and Lazarian* (2014) showed that

$$V = \frac{I_0k^4}{2R^2} i(\alpha_{\|}\alpha^*_{\perp} - \alpha^*_{\|}\alpha_{\perp})([\mathbf{e}_{inc} \times \mathbf{e}_{sca}].\mathbf{e})(\mathbf{e}_{inc}.\mathbf{e})$$

where $\mathbf{e}_{inc}$ and $\mathbf{e}_{sca}$ are the unit vectors of incident (**k**) and scattering direction, $\alpha_{\|}$ and $\alpha_{\perp}$ are the complex polarizabilities along the grain alignment axis **e** and in the perpendicular direction, respectively. From the above equation one can see that circular polarization can be produced only when the grain has an absorbing material, i.e., its polarizability (or refractive index) contains a non-zero imaginary part. Moreover, circular polarization is not equal to zero only when the grain alignment axis **e** is not parallel to the incident direction $\mathbf{e}_{inc}$. If $\mathbf{e} \parallel \mathbf{e}_{inc}$, the cross-vector $[\mathbf{e}_{inc} \times \mathbf{e}_{sca}]$ becomes perpendicular to **e**, and since $V \propto [\mathbf{e}_{inc} \times \mathbf{e}_{sca}] \cdot \mathbf{e}$, in this case V = 0.

Thus, the k-RAT alignment itself produces zero circular polarization as it aligns the dust particles in the direction where the alignment axis **e** is along the incoming radiation $\mathbf{e}_{inc}$.

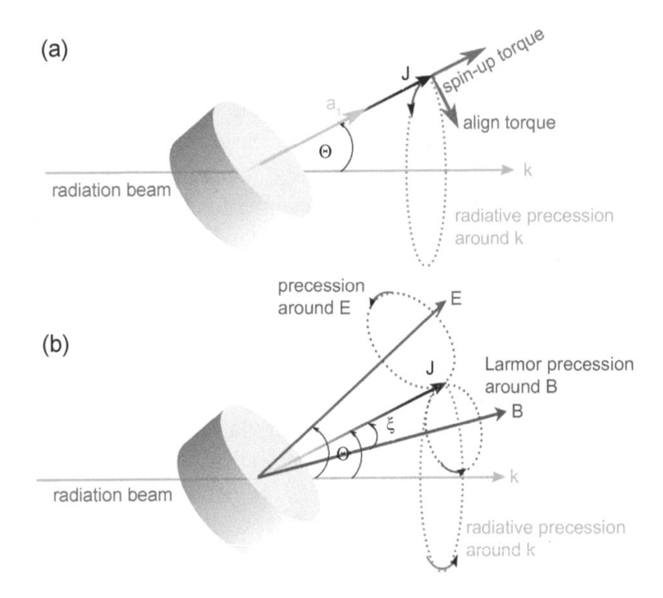

Fig. 12. Illustration of grain alignment and spin-up by radiative torques assuming that the grain shortest axis $\mathbf{a}_1$ is aligned along the angular momentum **J**. **(a)** The grain angular momentum **J** experiences fast radiative precession along the radiation beam **k**. The spinup torque component along the angular momentum acts to spin the grain up, while the align torque component acts to bring **J** in alignment with **k**, resulting in the so-called k-RAT alignment. **(b)** In the presence of the magnetic field (**B**) and electric field (**E**), the grain experiences the precession around **B** (Larmor precession), around the electric field, and the radiation direction, if the grain has a component of the magnetic moment and electric dipole moment parallel to **J**.

However, the presence of a magnetic field induces a magnetic torque on the grain magnetic dipole moment, which causes Larmor precession of the grain angular momentum around the magnetic field (Fig. 12b). If the Larmor precession is faster than the radiative precession, the magnetic field becomes the axis of grain alignment (B-RAT), and the grain short axis **e** deviates from the radiation direction $\mathbf{e}_0$, producing non-zero circular polarization. To have the precession caused by the magnetic field efficient in providing a deviation from the radiative direction, it should be faster than the precession caused by RAT. For solar radiation at the heliocentric distance ~1 au, *Hoang and Lazarian* (2014) showed that the RAT alignment time is $t_{RAT} = 3 \times 10^4$ (a/0.1 μm)$^{1/2}$ s, i.e., for particles of radius a = 0.1 μm, $t_{RAT} = 3 \times 10^4$ s and for particles of radius a = 10 μm, $t_{RAT} = 3 \times 10^5$ s. Thus, although small paramagnetic particles can be efficiently aligned by RAT combined with the magnetic field, alignment of larger particles requires larger values of the magnetic susceptibility. This can be provided by the presence of superparamagnetic, ferromagnetic, and ferrimagnetic inclusions, which increase grain Larmor precession rate and make this mechanism more efficient (*Hoang and Lazarian,* 2016). For a dust particle containing clusters of metallic atoms with $N_{cl} < 10^5$ atoms per cluster, the magnetic susceptibility increases by a factor of N_{cl}, decreasing the timescales of magnetic relaxation and Larmor precession by the same factor. As a result, grain alignment with the magnetic field (B-RAT) is more likely than the k-RAT alignment. *Hoang and Lazarian* (2016) have demonstrated that even a small fraction of metal present in the form of metallic nanoparticles, for example, GEMS present in cometary dust (e.g., *Keller and Messenger,* 2011), can provide the perfect alignment of dust grains that are subject to RATs. The considered mechanism was used to explain circular polarization in comets by *Kolokolova et al.* (2016), who showed that the penetration of the solar magnetic field into cometary coma can provide the alignment of dust particles that causes circular polarization of the light scattered by cometary dust.

4.3.3. Rotational disruption by radiative torques. RATs are very efficient in spinning up a non-spherical grain to suprathermal rotation, i.e., rotation with rates above the thermal rotation rate (*Draine and Weingartner,* 1996; *Hoang and Lazarian,* 2009). *Hoang et al.* (2019) realized that centrifugal stress resulting from such suprathermal rotation can exceed the maximum tensile strength of grain material (S_{max}), resulting in the disruption of the grain into fragments. This new physical mechanism was termed radiative torque disruption (RATD). Since rotational disruption acts to break the loose bonds between the monomers, unlike breaking the strong chemical bonds between atoms in the case of thermal sublimation, RATD can work for the solar radiation field beyond the sublimation zone of refractory dust (*Hoang,* 2020). For the cometary coma, *Hoang* (2021) showed that the disruption size, i.e., the minimum radius starting from which the particles can be disrupted, is $a_{disr} \sim 0.6$ μm for the gas density of $n_H = 10^{10}$cm^{-3} and decreases with increasing cometocentric radius or decreasing gas density. There exists a maximum radius of grains that can still be disrupted by centrifugal stress (*Hoang and Tram,* 2020) of $a_{disr,max} \sim 2.2$ μm for the gas density of $n_H = 10^{10}$ cm^{-3}, assuming $S_{max} = 10^7$ erg cm^{-3}.

Hoang and Tung (2020) showed that RATD is efficient in disrupting large composite grains into smaller ones. Therefore, the RATD mechanism implies the evolution of dust properties, causing a decrease in the abundance of large grains and increase in the abundance of small ones with the cometocentric and heliocentric distances. The efficiency of RATD also increases with decreasing the gas production rate Q_{gas} because the latter determines the rotational damping of grains spun up by RATs. *Herranen* (2020) studied the rotational disruption of agglomerates of radius ~2 μm in sungrazing comets and found that the mechanism could be efficient for those grains. Thus, the size and structure change in the cometary dust under RATD may explain the polarization properties reported in *Thompson* (2020) for Comet ISON.

One more effect of RATs on cometary dust is the rotational desorption of icy mantles on the particles. The current model of ice removal from the dust grains is based on thermal sublimation from icy grains (*Cowan and A'Hearn*, 1979). However, *Hoang and Tung* (2020) found that water ice mantles could be desorbed from the grain core by radiative torques. The rotational desorption can occur at heliocentric distances of $R_{des} \sim 20$ au, much larger than the water sublimation radius, $R_{sub}(H_2O) \sim 3$ au, thus making the RATD mechanism more efficient than sublimation at large heliocentric distances.

5. FINAL REMARKS AND FUTURE WORK

Our understanding of the properties of cometary dust has changed dramatically since 2004 when *Comets II* was published. The chapter in that volume by *Kolokolova et al.* (2004a) ended with the conclusion of a high likelihood of the aggregated structure of the cometary dust particles. Now this is a well-established fact, mainly proven by Stardust and Rosetta mission studies of cometary dust but also supported by computer and laboratory simulations, which have shown that cometary observations, especially polarimetric ones, can be most successfully reproduced if the dust is modeled as a mixture of agglomerated particles of different porosity, from very fluffy to rather compact, which is consistent with the mission-result characteristics of the cometary dust summarized in *Güttler et al.* (2019).

Modeling of agglomerates is in high demand in various studies of cometary dust and nuclei, e.g., in computing radiation pressure, radiative torque, tensile strength, and thermal properties of cometary material. This requires paying due attention to the computer modeling of the realistic cometary dust particles.

One of the main characteristics of a model should be its capability to simulate a comprehensive set of observational data. Polarization phase curve, brightness phase function, and color data in isolation are all ambiguous, and a model that reproduces only one of them has a high probability of being wrong. Thus, modelers always need to check the consistency of their models with all available observational data, angular and spectral, including those in the NIR and thermal infrared as well as with the dynamical properties of the dust particles derived from the morphology of the coma. Also, models should be consistent with ideas about the formation of dust particles in the protosolar nebula and with the cosmic abundances of the elements. Finally, if the images of the coma, including polarimetric maps, are available, the model should be checked for the capability of reproducing the observed trends using physically plausible assumptions on particle fragmentation and sublimation of its material.

We also would like to point out that by using a very complex model with numerous parameters, one can fit almost any observational data. If a model varies the size, structure, and composition of the particles and considers several populations of particles in the dust, a good fit to the observations can be found, but it may not present the realistic characteristics of the cometary dust. To avoid this trap, it is highly recommended to limit the number of model parameters using, where possible, information achieved by cometary missions. The Stardust and Rosetta missions significantly extended our knowledge of cometary dust particles, and the chapter in this volume by Engrand et al. presents a detailed review of those findings. A helpful source of information can be also a comprehensive review of the *in situ* and laboratory data by *Levasseur-Regourd et al.* (2018).

We expect further progress in the computer modeling of the interaction of cometary particles with radiation. The computational tools, specifically such codes as DDA, MSTM, FaSTMM, R^2T^2, etc., are sufficiently powerful to model cometary dust. However, they need more computational resources to be capable of modeling realistically large (hundreds of micrometers) and complex (e.g., hierarchical) particles, which will come as more powerful computers and computer clusters become available.

A better understanding of the interaction of cometary dust and surfaces with radiation will be produced by laboratory modeling, specifically models at low temperatures (see the chapter in this volume by Poch et al.), and by the new theoretical and laboratory developments regarding comet dust formation and evolution (see the chapters in this volume by Aikawa et al. and Simon et al.).

Remote sensing of cometary dust and surfaces remains one of the major resources of information provided by space missions using cameras working in different spectral ranges. For example, the future mission Comet Interceptor (see the chapter in this volume by Snodgrass et al.) has planned several instruments for remote sensing of the cometary environment, among them the Optical Periscopic Imager for Comets (OPIC) and Entire Visible Sky Camera (EnVisS), the latter of which has polarimetric capabilities, along with the Modular Infrared Molecules and Ices Sensor (MIRMIS) for the wavelength range 0.9–25 μm, which, together with the other cameras, may provide significant constraints on dust models. Also, observational information on more comets is expected, allowing statistical analysis of the dust properties and comet classification based on the properties of its dust that may come from the observations with new telescopes such as the James Webb Space Telescope (JWST) as well as groundbased and spacebased surveys (see the chapter in this volume by Bauer et al.). Studies that span a wide range of wavelengths and simultaneously use different techniques (e.g., thermal emission and scattered light) will help to understand the observed comet-to-comet distribution of cometary dust properties.

Finally, important steps forward in remote sensing of comets will be development in two main areas: (1) broadening the wavelength range of observations, which should include more UV observations (sensitive to nanoparticles if present) and microwave radiation capable of providing information about large, millimeter-sized, dust particles; and (2) expanding the observations to other objects related to the solar system comets such as debris disks and interstellar and exosolar comets.

Acknowledgments. L.K. acknowledges helpful discussions with J. Markkanen, N. Kiselev, and O. Shubina. This work was partially supported by NASA Grants No. 80NSSC21K0164 and 80NSSC17K0731 and Grants-in-Aid for Scientific Research (KAKENHI #21H00050) of JSPS.

REFERENCES

Abbas M. M., Tankosic D., Craven P. D., et al. (2006) Photoelectric emission measurements on the analogs of individual cosmic dust grains. *Astrophys. J., 645,* 324.

Abbas M. M., Tankosic D., Craven P. D., et al. (2007) Lunar dust charging by photoelectric emissions. *Planet. Space Sci., 55,* 953–965.

A'Hearn M. F., Dwek E., and Tokunaga A. T. (1984) Infrared photometry of Comet Bowell and other comets. *Astrophys. J., 282,* 803–806.

A'Hearn M. F., Millis R. L., Schleicher D. G., et al. (1995) The ensemble properties of comets: Results from narrowband photometry of 85 comets, 1976–1992. *Icarus, 118,* 223–270.

Altobelli N., Postberg F., Fiege K., et al. (2016) Flux and composition of interstellar dust at Saturn from Cassini's Cosmic Dust Analyzer. *Science, 352,* 312–318.

Andersson B. G., Lazarian A., and Vaillancourt J. E. (2015) Interstellar dust grain alignment. *Annu. Rev. Astron. Astrophys., 53,* 501–539.

Arriaga P., Fitzgerald M. P., Duchêne G., et al. (2020) Multiband polarimetric imaging of HR 4796a with the Gemini planet imager. *Astron. J., 160,* 79.

Bach Y. P. and Ishiguro M. (2021) Thermal radiation pressure as a possible mechanism for losing small particles on asteroids. *ArXiV e-prints*, arXiv:2108.03898.

Bagnulo S., Cellino A., Kolokolova L., et al. (2021) Unusual polarimetric properties for interstellar comet 2I/Borisov. *Nature Commun., 12,* 1–11.

Bair A. N., Schleicher D. G., and Knight M. M. (2018) Coma morphology, numerical modeling, and production rates for comet C/Lulin (2007 N3). *Astron. J., 156,* 159.

Bardyn A., Baklouti D., Cottin H., et al. (2017) Carbon-rich dust in comet 67P measured by COSIMA/Rosetta. *Mon. Not. R. Astron. Soc., 469,* S712–S722.

Baron B., Hoover D., and Williams F. (1978) Vacuum ultraviolet photoelectric emission from amorphous ice. *J. Chem. Phys., 68,* 1997–1999.

Bellm E. C., Kulkarni S. R., Graham M. J., et al. (2019) The Zwicky Transient Facility: System overview, performance, and first results. *Publ. Astron. Soc. Pac., 131,* 18002.

Bertini I., La Forgia F., Tubiana C., et al. (2017) The scattering phase function of comet 67P/Churyumov-Gerasimenko coma as seen from the Rosetta/OSIRIS instrument. *Mon. Not. R. Astron. Soc., 469,* S404–S415.

Betzler A. S., Almeida R. S., Cerqueira W. J., et al. (2017) An analysis of the BVRI colors of 22 active comets. *Adv. Space Res., 60,* 612–625.

Bockelée-Morvan D., Woodward C. E., Kelley M. S., and Wooden D. H. (2009) Water in comets 71P/Clark and C/2004 B1 (LINEAR) with Spitzer. *Astrophys. J., 696,* 1075–1083.

Bockelée-Morvan D., Rinaldi G., Erard S., et al. (2017) Comet 67P outbursts and quiescent coma at 1.3 au from the Sun: Dust properties from Rosetta/VIRTIS-H observations. *Mon. Not. R. Astron. Soc., 469,* S443–S458.

Bockelée-Morvan D., Leyrat C., Erard S., et al. (2019) VIRTIS-H observations of the dust coma of comet 67P: Spectral properties and color temperature variability with phase and elevation. *Astron. Astrophys., 630,* A22.

Bohren C. F. and Huffman D. R. (2008) *Absorption and Scattering of Light by Small Particles*. Wiley, Hoboken. 544 pp.

Borisov G., Bagnulo S., Nikolov P., and Bonev T. (2015) Imaging polarimetry and spectropolarimetry of comet C/2013 R1 (Lovejoy). *Planet. Space Sci., 118,* 187–192.

Burns J., Lamy P. L., and Soter S. (1979) Radiation forces on small particles in the solar system. *Icarus, 40,* 1–48.

Chernova G. P., Kiselev N. N., and Jockers K. (1993) Polarimetric characteristics of dust particles as observed in 13 comets: Comparisons with asteroids. *Icarus, 103,* 144–158.

Choudhury S. R., Hadamcik E., and Sen A. K. (2015) Study of some comets through imaging polarimetry. *Planet. Space Sci., 118,* 193–198.

Combi M. R., Harris W. M., and Smyth W. H. (2004) Gas dynamics and kinetics in the cometary coma: Theory and observations. In *Comets II* (M. C. Festou et al., eds.), pp. 523–552. Univ. of Arizona, Tucson.

Cowan J. J. and A'Hearn M. F. (1979) Vaporization of comet nuclei: Light curves and life times. *Earth Moon Planets, 21,* 155–171.

Crovisier J., Leech K., Bockelée-Morvan D., et al. (1997) The spectrum of Comet Hale-Bopp (C/1995 O1) observed with the Infrared Space Observatory at 2.9 AU from the Sun. *Science, 275,* 1904–1907.

Cudnik B. M. (2005) Observations of the inner coma of C/1995 O1 (Hale Bopp), gas and dust production. *Planet. Space. Sci., 53,* 653–658.

Davies J. K., Roush T. L., Cruikshank D. P., et al. (1997) The detection of water ice in comet Hale-Bopp. *Icarus, 127,* 238–245.

Davis L. Jr. and Greenstein J. L. (1951) The polarization of starlight by aligned dust grains. *Astrophys. J., 114,* 206–240.

de Moraes R. V. (1994) Non-gravitational disturbing forces. *Adv. Space Res., 14,* 45–68.

Dominik C. (2009) Physical processes: Dust coagulation and fragmentation. In *Cosmic Dust — Near and Far* (T. Henning et al., eds.), pp. 494–508. ASP Conf. Ser. 414, Astronomical Society of the Pacific, San Francisco.

Dorschner J., Begemann B., Henning T., et al. (1995) Steps toward interstellar silicate mineralogy. II. Study of Mg-Fe-silicate glasses of variable composition. *Astron. Astrophys., 300,* 503–520.

Draine B. T. (1978) Photoelectric heating of interstellar gas. *Astrophys. J. Suppl. Ser., 36,* 595–619.

Draine B. T. (1988) The discrete-dipole approximation and its application to interstellar graphite grains. *Astrophys. J., 333,* 848–872.

Draine B. T. and Weingartner J. C. (1996) Radiative torques on interstellar grains: I. Superthermal spinup. *ArXiV e-prints,* arXiv:astro-ph/9605046.

Dwek E. and Smith R. K. (1996) Energy deposition and photoelectric emission from the interaction of 10 eV to 1 MeV photons with interstellar dust particles. *Astrophys. J., 459,* 686–700.

Einstein A. (1905) Über einem die erzeugung und verwandlung des lichtes betreffenden heuristischen gesichtspunkt. *Ann. Phys., 322,* 132–148.

Escobar-Cerezo J., Palmer C., Muñoz O., et al. (2017) Scattering properties of large irregular cosmic dust particles at visible wavelengths. *Astrophys. J., 838,* 74–91.

Farnham T. L. and Schleicher D. G. (2005) Physical and compositional studies of Comet 81P/Wild 2 at multiple apparitions. *Icarus, 173,* 533–558.

Farnham T. L., Schleicher D. G., and A'Hearn M. F. (2000) The HB narrowband comet filters: Standard stars and calibrations. *Icarus, 147,* 180–204.

Feuerbacher B., Anderegg M., Fitton B., et al. (1972) Photoemission from lunar surface fines and the lunar photoelectron sheath. *Proc. Lunar Planet. Sci. Conf. 3rd*, pp. 2665–2663.

Filacchione G., Capaccioni F., Ciarniello M., et al. (2020) An orbital water-ice cycle on comet 67P from colour changes. *Nature, 578,* 49–52.

Fink U. (2009) A taxonomic survey of comet composition 1985–2004 using CCD spectroscopy. *Icarus, 201,* 311–334.

Fink U. and Rinaldi G. (2015) Coma dust scattering concepts applied to the Rosetta mission. *Icarus, 257,* 9–22.

Fitzsimmons A. and Cartwright I. M. (1996) Optical spectroscopy of comet C/1995 O1 Hale-Bopp. *Mon. Not. R. Astron. Soc., 278,* L37–L40.

Fulle M., Colangeli L., Agarwal J., et al. (2010) Comet 67P: The GIADA dust environment model of the Rosetta mission target. *Astron. Astrophys., 522,* A63.

Fulle M., Marzari F., Della Corte V., et al. (2016) Evolution of the dust size distribution of comet 67P/Churyumov-Gerasimenko from 2.2 AU to perihelion. *Astrophys. J., 821,* 19.

Gehrz R. D. and Ney E. P. (1992) 0.7- to 23-micron photometric observations of P/Halley 2986 III and six recent bright comets. *Icarus, 100,* 162–186.

Grundy W. M. and Schmitt B. (1998) The temperature-dependent near-infrared absorption spectrum of hexagonal H_2O ice. *J. Geophys. Res.–Planets, 103,* 25809–25822.

Gundlach B., Blum J., Keller H. U., and Skorov Y. V. (2015) What drives the dust activity of comet 67P? *Astron. Astrophys., 583,* A12.

Güttler C., Mannel T., Rotundi A., et al. (2019) Synthesis of the morphological description of cometary dust at comet 67P. *Astron. Astrophys., 630,* A24.

Hadamcik E. and Levasseur-Regourd A.-C. (2003) Imaging polarimetry of cometary dust: Different comets and phase angles. *J. Quant. Spectrosc. Radiat. Transfer, 79,* 661–678.

Hadamcik E. and Levasseur-Regourd A.-C. (2009) Optical properties of dust from Jupiter family comets. *Planet. Space Sci., 57,* 1118–1132.

Hadamcik E., Levassuer-Regourd A. C., and Renard J. B. (1997) CCD polarimetric imaging of Comet Hale-Bopp (C/1995 O1). *Earth Moon Planets, 78,* 365–371.

Halder P. and Ganesh S. (2021) Modelling heterogeneous dust particles: An application to cometary polarization. *Mon. Not. R. Astron. Soc., 501,* 1766–1781.

Hanner M. S. (1981) On the detectability of icy grains in the comae of comets. *Icarus, 47,* 342–350.

Hanner M. S. (2003) The scattering properties of cometary dust. *J. Quant. Spectrosc. Radiat. Transfer, 79,* 695–705.

Hanner M. S. and Bradley J. P. (2004) Composition and mineralogy of cometary dust. In *Comets II* (M. C. Festou et al., eds.), pp. 555–564. Univ. of Arizona, Tucson.

Harker D. E. (1999) Silicate mineralogy of C/1995 O1 (Hale-Bopp) and its implications to the study of pre-main sequence stars and the origins of solar systems. Ph.D. thesis, University of Wyoming, Laramie. 290 pp.

Herranen J. (2020) Rotational disruption of nonspherical cometary dust particles by radiative torques. *Astrophys. J., 893,* 109.

Hines D. C., Videen G., Zubko E., et al. (2013) Hubble Space Telescope pre-perihelion ACS/WFC imaging polarimetry of comet ISON (c/2012 s1) at 3.81 AU. *Astrophys. J. Lett., 780,* L32.

Hoang T. (2021) Variation of dust properties with cosmic time implied by radiative torque disruption. *Astrophys. J., 907,* 37.

Hoang T. and Lazarian A. (2009) Radiative torques alignment in the presence of pinwheel torques. *Astrophys. J., 695,* 1457–1476.

Hoang T. and Lazarian A. (2014) Grain alignment by radiative torques in special conditions and implications. *Mon. Not. R. Astron. Soc., 438,* 680–703.

Hoang T. and Lazarian A. (2016) A unified model of grain alignment: Radiative alignment of interstellar grains with magnetic inclusions. *Astrophys. J., 831,* 159.

Hoang T. and Tram L. N. (2020) Rotational desorption of ice mantles from suprathermally rotating grains around young stellar objects. *Astrophys. J., 891,* 38.

Hoang T. and Tung N.-D. (2020) Evolution of dust and water ice in cometary comae by radiative torques. *Astrophys. J., 901,* 59.

Hoang T., Lee H., Ahn S.-H., et al. (2019) Rotational disruption of dust grains by radiative torques in strong radiation fields. *Nature Astron., 3,* 766–775.

Hughes A. M., Duchêne G., and Matthews B. C. (2018) Debris disks: Structure, composition, and variability. *Annu. Rev. Astron. Astrophys., 56,* 541–591.

Il'in V. and Voshchinnikov N. (1998) Radiation pressure on non-spherical dust grains in envelopes of late-type giants. *Astron. Astrophys. Suppl., 128,* 187–196.

Ishiguro M., Kuroda D., Hanayama H., et al. (2016) 2014–2015 multiple outbursts of 15P/Finlay. *Astron. J., 152,* 169.

Ivanova A. V., Borisenko S. A., and Andreev M. V. (2014) Photometric studies of comet C/2009 P1 (Garradd) before the perihelion. *Solar System Res., 48,* 375–381.

Ivanova O., Rosenbush V., Afanasiev V., and Kiselev N. (2017) Polarimetry, photometry, and spectroscopy of comet C/2009 P1 (Garradd). *Icarus, 284,* 167–182.

Ivanova O., Luk'yanyk I., Kolokolova L., et al. (2019) Photometry, spectroscopy, and polarimetry of distant comet C/2014 A4 (SONEAR). *Astron. Astrophys., 626,* A26.

Ivanova O., Rosenbush V., Luk'yanyk I., et al. (2021) Observations of distant comet C/2011 KP36 (Spacewatch): Photometry, spectroscopy, and polarimetry. *Astron. Astrophys., 651,* A29.

Jenniskens P., Baratta G. A., Kouchi A., et al. (1993) Carbon dust formation on interstellar grains. *Astron. Astrophys., 273,* 583–600.

Jessberger E. K., Christoforidis A., and Kissel J. (1988) Aspects of the major element composition of Halley's dust. *Nature, 332,* 691–695.

Jewitt D. (2004) Looking through the HIPPO: Nucleus and dust in comet 2P/Encke. *Astron. J., 128,* 3061–3069.

Jewitt D. (2015) Color systematics of comets and related bodies. *Astron. J., 150,* 201.

Jewitt D. and Meech K. J. (1986) Cometary grain scattering versus wavelength, or "What color is comet dust?" *Astrophys. J., 310,* 937–952.

Jockers K., Rosenbush V. K., Bonev T., and Credner T. (1997) Images of polarization and colour in the inner coma of comet Hale-Bopp. *Earth Moon Planets, 78,* 373–379.

Jockers K., Kiselev N., Bonev T., et al. (2005) CCD imaging and aperture polarimetry of comet 2P/Encke: Are there two polarimetric classes of comets? *Astron. Astrophys., 441,* 773–782.

Johnson P. B. and Christy R. W. (1974) Optical constants of transition metals: Ti, V, Cr, Mn, Fe, Co, Ni, and Pd. *Phys. Rev. B, 9,* 5056–5070.

Jones T. J., Stark D., Woodward C. E., et al. (2008) Evidence of fragmenting dust particles from near-simultaneous optical and near-infrared photometry and polarimetry of comet 73P/Schwassmann-Wachmann 3. *Astron. J., 135,* 1318–1327.

Kareta T., Harris W. M., and Reddy V. (2021) Probable detection of water ice in the coma of the inbound long-period comet C/2017 K2 (PanSTARRS). *Res. Notes Am. Astron. Soc., 5,* 153.

Kawakita H., Watanabe J.-I., Ootsubo T., et al. (2004) Evidence of icy grains in comet C/2002 T7 (LINEAR) at 3.52 AU. *Astrophys. J. Lett., 601,* L191–L194.

Kawakita H., Ootsubo T., Furusho R., and Watanabe J.-I. (2006) Crystallinity and temperature of icy grains in comet C/2002 T7 (LINEAR). *Adv. Space Res., 38,* 1968–1971.

Keller C. U., Snik F., Harrington D. M., and Packham C. (2015) Instrumentation. In *Polarimetry of Stars and Planetary Systems* (L. Kolokolova et al., eds.), pp. 35–61. Cambridge Univ., Cambridge.

Keller L. P. and Messenger S. (2011) On the origins of GEMS

grains. *Geochim. Cosmochim. Acta, 75,* 5336–5365.

Kelley M. S. P, Harker D. E., Woodward C. E., and Wooden D. H. (2021) *Spitzer Space Telescope Spectroscopy of Comets.* NASA Planetary Data System, urn:nasa:pds:spitzer:spitzer-spec-comet 1.0.

Kimura H. (2016) On the photoelectric quantum yield of small dust particles. *Mon. Not. R. Astron. Soc., 459,* 2751–2761.

Kimura H. (2017) High radiation pressure on interstellar dust computed by light-scattering simulation on fluffy agglomerates of magnesium-silicate grains with metallic-iron inclusions. *Astrophys. J. Lett., 839,* L23.

Kimura H. and Mann I. (1998) The electric charging of interstellar dust in the solar system and consequences for its dynamics. *Astrophys. J., 499,* 454–462.

Kimura H., Okamoto H., and Mukai T. (2002a) Radiation pressure and the Poynting-Robertson effect for fluffy dust particles. *Icarus, 157,* 349–361.

Kimura H., Mann I., Biesecker D. A., and Jessberger E. K. (2002b) Dust grains in the comae and tails of sungrazing comets: Modeling of their mineralogical and morphological properties. *Icarus, 159,* 529–541.

Kimura H., Mann I., and Jessberger E. K. (2003a) Composition, structure, and size distribution of dust in the local interstellar cloud. *Astrophys. J., 583,* 314–321.

Kimura H., Kolokolova L., and Mann I. (2003b) Optical properties of cometary dust: Constraints from numerical studies on light scattering by aggregate particles. *Astron. Astrophys., 407,* L5–L8.

Kimura H., Kolokolova L., and Mann I. (2006) Light scattering by cometary dust numerically simulated with aggregate particles consisting of identical spheres. *Astron. Astrophys., 449,* 1243–1254.

Kimura H., Kolokolova L., Li A., and Lebreton J. (2016) Light scattering and thermal emission by primitive dust particles in planetary systems. In *Light Scattering Reviews, Vol. 11* (A. Kokhanovsky, ed.), pp. 363–418. Springer, Berlin.

Kimura H., Hilchenbach M., Merouane S., et al. (2020) The morphological, elastic, and electric properties of dust aggregates in comets: A close look at COSIMA/Rosetta's data on dust in comet 67P. *Planet. Space Sci., 181,* 104825.

Kiselev N., Velichko S., and Jockers K. (2006) *Database of Comet Polarimetry.* NASA Planetary Data System, EAR-C-COMPIL-5-DB-COMET-POLARIMETRY-V1.0.

Kiselev N., Rosenbush V., Kolokolova L., and Antonyuk K. (2008) The anomalous spectral dependence of polarization in comets. *J. Quant. Spectrosc. Radiat. Transfer, 109,* 1384–1391.

Kiselev N. N., Rosenbush V. K., Afanasiev V. L., et al. (2013) Linear and circular polarization of comet C/2009 P1 (Garradd). *Earth Planets Space, 65,* 1151–1157.

Kiselev N., Rosenbush V., Levasseur-Regourd A.-C., and Kolokolova L. (2015) Comets. In *Polarimetry of Stars and Planetary Systems* (L. Kolokolova et al., eds.), pp. 379–404. Cambridge Univ., Cambridge.

Kiselev N., Shubina E., Velichko S., et al. (2017) *Compilation of Comet Polarimetry from Published and Unpublished Sources.* NASA Planetary Data System, urn:nasa:pds:compil-comet-polarimetry 1.0.

Kiselev N., Rosenbush V., Ivanova O., et al. (2020) Comet 2P/Encke in apparition of 2017: II. Polarization and color. *Icarus, 348,* 113768.

Kittel C. (1995) *Introduction to Solid State Physics, 7th edition.* Wiley, Singapore. 688 pp.

Knight M. M. and Schleicher D. G. (2013) The highly unusual outgassing of comet 103P/Hartley 2 from narrowband photometry and imaging of the coma. *Icarus, 222,* 691–706.

Knight M. M., Schleicher D. G., and Farnham T. L. (2021) Narrowband observations of comet 46P/Wirtanen during its exceptional apparition of 2018/19. II. Photometry, jet morphology, and modeling results. *Planet. Sci. J., 2,* 104.

Kobayashi H., Kimura H., and Yamamoto S. (2013) Dust mantle of comet 9P/Tempel 1: Dynamical constraints on physical properties. *Astron. Astrophys., 550,* A72.

Köhler M., Minato T., Kimura H., and Mann I. (2007) Radiation pressure force acting on cometary aggregates. *Adv. Space Res., 40,* 266–271.

Koike C., Tsuchiyama A., Shibai H., et al. (2000) Absorption spectra of Mg-rich Mg-Fe and Ca pyroxenes in the mid- and far-infrared regions. *Astron. Astrophy., 363,* 1115–1122.

Koike C., Chihara H., Tsuchiyama A., et al. (2003) Compositional dependence of infrared absorption spectra of crystalline silicate. II. Natural and synthetic olivines. *Astron. Astrophys., 399,* 1101–1107.

Kolokolova L. and Gustafson B. Å. S. (2001) Scattering by inhomogeneous particles: Microwave analog experiments and comparison to effective medium theories. *J. Quant. Spectrosc. Radiat. Transfer, 70,* 611–625.

Kolokolova L. and Kimura H. (2010a) Effects of electromagnetic interaction in the polarization of light scattered by cometary and other types of cosmic dust. *Astron. Astrophys., 513,* A40.

Kolokolova L. and Kimura H. (2010b) Comet dust as a mixture of aggregates and solid particles: Model consistent with ground-based and space-mission results. *Earth Planets Space, 62,* 17–21.

Kolokolova L. and Mackowski D. (2012) Polarization of light scattered by large aggregates. *J. Quant. Spectrosc. Radiat. Transfer, 113,* 2567–2572.

Kolokolova L., Gustafson B. S., Jockers K., and Lichtenberg G. (2002) Evolution of cometary grains from studies of comet images. In *Dust in the Solar System and Other Planetary Systems* (S. F. Green et al., eds.), pp. 269–273. Pergamon, Oxford.

Kolokolova L., Hanner M. S., Levasseur-Regourd A.-C., and Gustafson B. Å. S. (2004a) Physical properties of cometary dust from light scattering and thermal emission. In *Comets II* (M. C. Festou et al., eds.), pp. 577–604. Univ. of Arizona, Tucson.

Kolokolova L., Kimura H., and Mann I. (2004b) Characterization of dust particles using photopolarimetric data: Example of cometary dust. In *Photopolarimetry in Remote Sensing* (G. Videen et al., eds.), pp. 431–454. Springer, Dordrecht.

Kolokolova L., Kimura H., Kiselev N., and Rosenbush V. (2007) Two different evolutionary types of comets proved by polarimetric and infrared properties of their dust. *Astron. Astrophys., 463,* 1189–1196.

Kolokolova L., Petrova E., and Kimura H. (2011) Effects of interaction of electromagnetic waves in complex particles. In *Electromagnetic Waves* (V. Zhurbenko, ed.), pp. 173–202. IntechOpen, London.

Kolokolova L., Das H. S., Dubovik O., et al. (2015) Polarization of cosmic dust simulated with the rough spheroid model. *Planet. Space Sci., 116,* 30–38.

Kolokolova L., Koenders C., Goetz C., et al. (2016) Clues to cometary circular polarization from studying the magnetic field in the vicinity of the nucleus of comet 67P/Churyumov-Gerasimenko. *Mon. Not. R. Astron. Soc., 462,* S422–S431.

Kolokolova L., Nagdimunov L., and Mackowski D. (2018) Light scattering by hierarchical aggregates. *J. Quant. Spectrosc. Radiat. Transfer, 204,* 138–143.

Kwon Y. G., Ishiguro M., Kuroda D., et al. (2017) Optical and near-infrared polarimetry of non-periodic comet C/2013 US10 (Catalina). *Astron. J., 154,* 173.

Kwon Y. G., Ishiguro M., Shinnaka Y., et al. (2018) High polarization degree of the continuum of comet 2P/Encke based on spectropolarimetric signals during its 2017 apparition. *Astron. Astrophys., 620,* A161.

Kwon Y. G., Ishiguro M., Kwon J., et al. (2019) Near-infrared polarimetric study of near-Earth object 252P/LINEAR: An implication of scattered light from the evolved dust particles. *Astron. Astrophys., 629,* A121.

Kwon Y. G., Kolokolova L., Agarwal J., and Markkanen J. (2021) An update of the correlation between polarimetric and thermal properties of cometary dust. *Astron. Astrophys., 650,* L7.

Kwon Y. G., Bagnulo S., Markkanen J., et al. (2022) VLT spectropolarimetry of comet 67P: Dust environment around the end of its intense southern summer. *Astron. Astrophys., 657,* A40.

La Forgia F., Bodewits D., A'Hearn M. F., et al. (2017) Near-UV OH prompt emission in the innermost coma of 103P/Hartley 2. *Astron. J., 154,* 185.

Laor A. and Draine B. T. (1993) Spectroscopic constraints on the properties of dust in active galactic nuclei. *Astrophys. J., 402,* 441–468.

Lasue J., Levasseur-Regourd A.-C., Fray N., and Cottin H. (2007) Inferring the interplanetary dust properties-from remote observations and simulations. *Astron. Astrophys., 473,* 641–649.

Lasue J., Levasseur-Regourd A.-C., Hadamcik E., and Alcouffe G. (2009) Cometary dust properties retrieved from polarization observations: Application to C/1995 O1 Hale-Bopp and 1P/Halley. *Icarus, 199,* 129–144.

Lazarian A. and Hoang T. (2007) Radiative torques: Analytical model and basic properties. *Mon. Not. R. Astron. Soc., 378,* 910–946.

Lazarian A., Andersson B.-G., and Hoang T. (2015) Grain alignment: Role of radiative torques and paramagnetic relaxation. In *Polarimetry of Stars and Planetary Systems* (L. Kolokolova et al., eds.), pp. 81–113. Cambridge Univ., Cambridge.

Lellouch E., Crovisier J., Lim T., et al. (1998) Evidence for water ice and estimate of dust production rate in comet Hale-Bopp at 2.9 AU from the Sun. *Astron. Astrophys., 339,* L9–L12.

Levasseur-Regourd A.-C., Hadamcik E., and Renard J. B. (1996) Evidence for two classes of comets from their polarimetric properties at large phase angles. *Astron. Astrophys., 313,* 327–333.

Levasseur-Regourd A.-C., Renard J.-B., Shkuratov Y., and Hadamcik E. (2015) Laboratory studies. In *Polarimetry of Stars and Planetary Systems* (L. Kolokolova et al., eds.), pp. 62–80. Cambridge Univ., Cambridge.

Levasseur-Regourd A.-C., Agarwal J., Cottin H., et al. (2018) Cometary dust. *Space Sci. Rev., 214,* 1–56.

Levasseur-Regourd A.-C., Baruteau C., Lasue J., et al. (2020) Linking studies of tiny meteoroids, zodiacal dust, cometary dust and circumstellar disks. *Planet. Space Sci., 186,* 104896.

Li A. and Greenberg J. M. (1997) A unified model of interstellar dust. *Astron. Astrophys., 323,* 566–584.

Li J.-Y., Kelley M. S. P., Knight M. M., et al. (2013) Characterizing the dust coma of comet C/2012 S1 (ISON) at 4.15 AU from the Sun. *Astrophys. J. Lett., 779,* L3.

Li J.-Y., Samarasinha N. H., Kelley M. S. P., et al. (2014) Constraining the dust coma properties of comet C/Siding Spring (2013 a1) at large heliocentric distances. *Astrophys. J. Lett., 797,* L8.

Li J.-Y., Kelley M. S. P., Samarasinha N. H., et al. (2017) The unusual apparition of comet 252P/2000 G1 (LINEAR) and comparison with comet P/2016 BA14 (PanSTARRS). *Astron. J., 154,* 136.

Lindsay S. S., Wooden D. H., Harker D. E., et al. (2013) Absorption efficiencies of forsterite. I. Discrete dipole approximation explorations in grain shape and size. *Astrophys. J., 766,* 54.

Lisse C. M., A'Hearn M. F., Fernández Y. R., and Peschke S. B. (2002) A search for trends in cometary dust emissions. In *Dust in the Solar System and Other Planetary Systems* (S. F. Green et al., eds.), pp. 259–268. Pergamon, Oxford.

Lisse C. M., Kraemer K. E., Nuth J. A., et al. (2007) Comparison of the composition of the Tempel 1 ejecta to the dust in comet C/Hale Bopp 1995 O1 and YSO HD 100546. *Icarus, 191,* 223–240.

Liu C., Panetta R. L., and Yang P. (2014) Inhomogeneity structure and the applicability of effective medium approximations in calculating light scattering by inhomogeneous particles. *J. Quant. Spectrosc. Radiat. Transfer, 146,* 331–348.

Mackowski D. and Kolokolova L. (2022) Application of the multiple sphere superposition solution to large-scale systems of spheres via an accelerated algorithm. *J. Quant. Spectrosc. Radiat. Transfer, 287,* 108221.

Mann I., Kimura H., and Kolokolova L. (2004) A comprehensive model to describe light scattering properties of cometary dust. *J. Quant. Spectrosc. Radiat. Transfer, 89,* 291–301.

Mannel T., Bentley M. S., Boakes P. D., et al. (2019) Dust of comet 67P collected by Rosetta/MIDAS: Classification and extension to the nanometer scale. *Astron. Astrophys., 630,* A26.

Marcus J. N. (2007) Forward-scattering enhancement of comet brightness. I. Background and model. *Intl. Comet Q., 29,* 39–66.

Markkanen J. and Agarwal J. (2019) Scattering, absorption, and thermal emission by large cometary dust particles: Synoptic numerical solution. *Astron. Astrophys., 631,* A164.

Markkanen J. and Yuffa A. J. (2017) Fast superposition T-matrix solution for clusters with arbitrarily-shaped constituent particles. *J. Quant. Spectrosc. Radiat. Transfer, 189,* 181–188.

Markkanen J., Väisänen T., Penttilä A., and Muinonen K. (2018a) Scattering and absorption in dense discrete random media of irregular particles. *Optics Lett., 43,* 2925–2928.

Markkanen J., Agarwal J., Väisänen T., et al. (2018b) Interpretation of the phase functions measured by the OSIRIS instrument for comet 67P/Churyumov-Gerasimenko. *Astrophys. J. Lett., 868,* L16.

Masoumzadeh N., Kolokolova L., Tubiana C., et al. (2019) Phase-curve analysis of comet 67P at small phase angles. *Astron. Astrophys., 630,* A11.

Mastrapa R. M. E. (2020) *Optical Constants and Lab Spectra of Water Ice V1.0.* NASA Planetary Data System, urn:nasa:pds:gbo.ices.mastrapa.lab-spectra 1.0.

Mastrapa R. M., Bernstein M. P., Sandford S. A., et al. (2008) Optical constants of amorphous and crystalline H_2O-ice in the near infrared from 1.1 to 2.6 μm. *Icarus, 197,* 307–320.

Mastrapa R. M., Sandford S. A., Roush T. L., et al. (2009) Optical constants of amorphous and crystalline H_2O-ice: 2.5–22 μm (4000–455 cm^{-1}) optical constants of H_2O-ice. *Astrophys. J., 701,* 1347–1356.

McKay A. J., DiSanti M. A., Kelley M. S., et al. (2019) The peculiar volatile composition of CO-dominated comet C/2016 R2 (PanSTARRS). *Astron. J., 158,* 128.

Mie G. (1908) Beiträge zur optik trüber medien, speziell kolloidaler metallösungen. *Ann. Phys., 25,* 377–445.

Min M., Hovenier J. W., de Koter A. (2005a) Modeling optical properties of cosmic dust grains using a distribution of hollow spheres. *Astron. Astrophys., 432,* 909–920.

Min M., Hovenier J. W., de Koter A., et al. (2005b) The composition and size distribution of the dust in the coma of Comet Hale-Bopp. *Icarus, 179,* 158–173.

Mishchenko M. I. (1991) Light scattering by randomly oriented axially symmetric particles. *J. Optical Soc. Am., 8,* 871–882.

Mishchenko M. I. and Travis L. D. (1994) Light scattering by polydispersions of randomly oriented spheroids with sizes comparable to wavelengths of observation. *Appl. Opt., 33,* 7206–7225.

Mishchenko M. I., Travis L. D., and Lacis A. A. (2002) *Scattering, Absorption, and Emission of Light by Small Particles.* Cambridge Univ., Cambridge. 462 pp.

Mishchenko M. I., Dlugach J. M., Liu L., et al. (2009) Direct solutions of the Maxwell equations explain opposition phenomena observed for high-albedo solar system objects. *Astrophys. J. Lett., 705,* L118–L122.

Mishchenko M., Rosenbush V., Kiselev N., et al. (2010) Polarimetric remote sensing of solar system objects. *ArXiV e-prints*, arXiv:1010.1171.

Mishchenko M. I., Dlugach Z. M., and Zakharova N. T. (2014) Direct demonstration of the concept of unrestricted effective-medium approximation. *Optics Lett., 39,* 3935–3938.

Mishchenko M. I., Dlugach J. M., and Liu L. (2016) Applicability of the effective-medium approximation to heterogeneous aerosol particles. *J. Quant. Spectrosc. Radiat. Transfer, 178,* 284–294.

Moreno F., Muñoz O., Gutiérrez P. J., et al. (2017) The dust environment of comet 67P/Churyumov-Gerasimenko: Results from Monte Carlo dust tail modelling applied to a large ground-based observation data set. *Mon. Not. R. Astron. Soc., 469,* S186–S194.

Moreno F., Guirado D., Muñoz O., et al. (2018) Models of Rosetta/OSIRIS 67P dust coma phase function. *Astron. J., 156,* 237.

Moulane Y., Jehin E., Opitom C., et al. (2018) Monitoring of the activity and composition of comets 41P/Tuttle-Giacobini-Kresak and 45P/Honda-Mrkos-Pajdusakova. *Astron. Astrophys., 619,* A156.

Moulane Y., Jehin E., Rousselot P., et al. (2020) Photometry and high-resolution spectroscopy of comet 21P/Giacobini-Zinner during its 2018 apparition. *Astron. Astrophys., 640,* A54.

Muinonen K., Nousiainen T., Fast P., et al. (1996) Light scattering by Gaussian random particles: Ray optics approximation. *J. Quant. Spectrosc. Radiat. Transfer, 55,* 577–601.

Muinonen K., Mishchenko M., Dlugach J., et al. (2012) Coherent backscattering verified numerically for a finite volume of spherical particles. *Astrophys. J., 760,* 118.

Muinonen, K., Penttilä, A., and Videen, G. (2015) Multiple scattering of light in particulate planetary media. In *Polarimetry of Stars and Planetary Systems* (L. Kolokolova et al., eds.), pp. 114–129. Cambridge Univ., Cambridge.

Muinonen K., Markkanen J., Väisänen T., et al. (2018) Multiple scattering of light in discrete random media using incoherent interactions. *Optics Lett.,* 43, 683–686.

Mukai T. (1981) On the charge distribution of interplanetary grains. *Astron. Astrophys., 99,* 1–6.

Mukai T., Ishimoto H., Kozasa T., et al. (1992) Radiation pressure forces of fluffy porous grains. *Astron. Astrophys., 262,* 315–320.

Muñoz O. and Hovenier J. W. (2015) Experimental scattering matrices of clouds of randomly oriented particles. In *Polarimetry of Stars and Planetary Systems* (L. Kolokolova et al., eds.), pp. 130–144. Cambridge Univ., Cambridge.

Muñoz O., Moreno F., Gómez-Martín J. C., et al. (2020) Experimental phase function and degree of linear polarization curves of millimeter-sized cosmic dust analogs. *Astrophys. J. Suppl. Ser., 247,* 19.

Nagdimunov L., Kolokolova L., and Mackowski D. (2013) Characterization and remote sensing of biological particles using circular polarization. *J. Quant. Spectrosc. Radiat. Transfer, 131,* 59–65.

Nagdimunov L., Kolokolova L., Wolff M., et al. (2014) Properties of comet 9P/Tempel 1 dust immediately following excavation by deep impact. *Planet. Space Sci., 100,* 73–78.

Oishi M., Kawara K., Kobayashi Y., et al. (1978) Infrared observations of comet West (1975n). I. Observational results. *Publ. Astron. Soc. Japan, 30,* 149–160.

Olofsson J., Samland M., Avenhaus H., et al. (2016) Azimuthal asymmetries in the debris disk around HD 61005: A massive collision of planetesimals? *Astron. Astrophys., 591,* A108.

Opitom C., Jehin E., Manfroid J., et al. (2015) TRAPPIST monitoring of comet C/2012 F6 (Lemmon). *Astron. Astrophys., 574,* A38.

Paquette J. A., Fray N., Bardyn A., et al. (2021) D/H in the refractory organics of comet 67P measured by Rosetta/COSIMA. *Mon. Not. R. Astron. Soc., 504,* 4940–4951.

Penttilä A., Markkanen J., Väisänen T., et al. (2021) How much is enough? The convergence of finite sample scattering properties to those of infinite media. *J. Quant. Spectrosc. Radiat. Transfer, 262,* 107524–107531.

Protopapa S., Sunshine J. M., Feaga L. M., et al. (2014) Water ice and dust in the innermost coma of comet 103P/Hartley 2. *Icarus, 238,* 191–204.

Protopapa S., Kelley M., Yang B., et al. (2018) Icy grains from the nucleus of comet C/2013 US10 (Catalina). *Astrophys. J. Lett., 862,* L16.

Purcell E. M. (1979) Suprathermal rotation of interstellar grains. *Astrophys. J., 231,* 404–416.

Reach W. T., Vaubaillon J., Lisse C. M., et al. (2010) Explosion of comet 17P/Holmes as revealed by the Spitzer Space Telescope. *Icarus, 208,* 276–292.

Ren B., Choquet É., Perrin M. D., et al. (2019) An exo-Kuiper belt with an extended halo around HD 191089 in scattered light. *Astrophys. J., 882,* 64.

Renard J.-B., Hadamcik E., and Levasseur-Regourd A.-C. (1996) Polarimetric CCD imaging of comet 47P/Ashbrook-Jackson and variability of polarization in the inner coma of comets. *Astron. Astrophys., 316,* 263–269.

Rinaldi G., Della Corte V., Fulle M., et al. (2017) Cometary coma dust size distribution from *in situ* IR spectra. *Mon. Not. R. Astron. Soc., 469,* S598–S605.

Robertson H. P. (1937) Dynamical effects of radiation in the solar system. *Mon. Not. R. Astron. Soc., 97,* 423–437.

Rosenbush V. K., Shakhovskoj N. M., and Rosenbush A. E. (1997) Polarimetry of comet Hale-Bopp: Linear and circular polarization, stellar occultation. *Earth Moon Planets, 78,* 381–386.

Rosenbush V., Kiselev N., and Velichko S. (2002) Polarimetric and photometric observations of split comet C/2001 A2 (LINEAR). *Earth Moon Planets, 90,* 423–433.

Rosenbush V. K., Ivanova O. V., Kiselev N. N., et al. (2017) Spatial variations of brightness, colour and polarization of dust in comet 67P/Churyumov-Gerasimenko. *Mon. Not. R. Astron. Soc., 469,* S475–S491.

Rosenbush V., Kiselev N., Husárik M., et al. (2021) Photometry and polarimetry of comet 46P/Wirtanen in the 2018 apparition. *Mon. Not. R. Astron. Soc. 503,* 4297–4308.

Rouleau F. and Martin P. G. (1991) Shape and clustering effects on the optical properties of amorphous carbon. *Astrophys. J., 377,* 526–540.

Schleicher D. G. (2007) Deep Impact's target comet 9P/Tempel 1 at multiple apparitions: Seasonal and secular variations in gas and dust production. *Icarus, 190,* 406–422.

Schleicher D. G. (2008) The extremely anomalous molecular abundances of comet 96p/Machholz 1 from narrowband photometry. *Astron. J., 136,* 2204–2213.

Schleicher D. G. (2009) The long-term decay in production rates following the extreme outburst of comet 17P/Holmes. *Astron. J., 138,* 1062–1071.

Schleicher D. G. (2010) *Composite Dust Phase Function for Comets.* Available online at *https://asteroid.lowell.edu/comet/dustphase/*, accessed September 9, 2024.

Schleicher D. G. and A'Hearn M. (1988) The fluorescence of cometary OH. *Astrophys. J., 331,* 1058–1077.

Schleicher D. G. and Bair A. N. (2011) The composition of the interior of comet 73P/Schwassmann-Wachmann 3: Results from narrowband photometry of multiple components. *Astron. J., 141,* 177.

Schleicher D. G., Millis R. L., and Birch P. V. (1998) Narrowband photometry of Comet P/Halley: Variation with heliocentric distance, season, and solar phase angle. *Icarus, 132,* 397–417.

Schleicher D. G., Woodney L. M., and Millis R. L. (2003) Comet 19P/Borrelly at multiple apparitions: Seasonal variations in gas production and dust morphology. *Icarus, 162,* 415–442.

Schröder S. E., Grynko Y., Pommerol A., et al. (2014) Laboratory observations and simulations of phase reddening. *Icarus, 239,* 201–216.

Sekanina Z. (2000) Solar and Heliospheric Observatory sungrazing

comets with prominent tails: Evidence on dust-production peculiarities. *Astrophys. J. Lett., 545,* L69–L72.
Senshu H., Kimura H., Yamamoto T., et al. (2015) Photoelectric dust levitation around airless bodies revised using realistic photoelectron velocity distributions. *Planet. Space Sci., 116,* 18–29.
Shen Y., Draine B. T., and Johnson E. (2008) Modeling porous dust grains with ballistic aggregates. I. Geometry and optical properties. *Astrophys. J., 689,* 260–275.
Shen Y., Draine B. T., and Johnson E. T. (2009) Modeling porous dust grains with ballistic aggregates. II. Light scattering properties. *Astrophys. J., 696,* 2126–2137.
Sitko M. L., Lynch D. K., Russell R. W., and Hanner M. S. (2004) 3–14 micron spectroscopy of comets C/2002 O4 (Hönig), C/2002 V1 (NEAT), C/2002 X5 (Kudo-Fujikawa), C/2002 Y1 (Juels-Holvorcem), and 69P/Taylor and the relationships among grain temperature, silicate band strength, and structure among comet families. *Astrophys. J., 612,* 576–587.
Smith N. V. (1971) Photoemission properties of metals. *Crit. Rev. Solid State Mater. Sci., 2,* 45–83.
Solontoi M., Ivezić Ž., Jurić M., et al. (2012) Ensemble properties of comets in the Sloan Digital Sky Survey. *Icarus, 218,* 571–584.
Sparks W., Hough J., and Kolokolova L. (2015) Astrobiology. In *Polarimetry of Stars and Planetary Systems* (L. Kolokolova et al., eds.), pp. 462–478. Cambridge Univ., Cambridge.
Stenzel O. J., Hilchenbach M., Merouane S., et al. (2017) Similarities in element content between comet 67P/Churyumov-Gerasimenko coma dust and selected meteorite samples. *Mon. Not. R. Astron. Soc., 469,* S492–S505.
Tazaki R. and Tanaka H. (2018) Light scattering by fractal dust aggregates. II. Opacity and asymmetry parameter. *Astrophys. J., 860,* 79.
Thompson W. T. (2015) Linear polarization measurements of Comet C/2011 W3 (Lovejoy) from STEREO. *Icarus, 261,* 122–132.
Thompson W. T. (2020) Changing linear polarization properties in the dust tail of comet C/2012 S1 (ISON). *Icarus, 338,* 113533.
van de Hulst H. C. (1957) *Light Scattering by Small Particles.* Wiley, New York. 470 pp.
Waterman P. C. (1965) Matrix formulation of electromagnetic scattering. *Proc. IEEE, 53,* 805–812.
Watson W. D. (1973) Photoelectron emission from small spherical particles. *J. Optical Soc. Am., 63,* 164–165.
Whipple E. C. (1981) Potentials of surfaces in space. *Rept. Prog. Phys., 44,* 1197–1250.
Wong K., Vongehr S., and Kresin V. V. (2003) Work functions, ionization potentials, and in between: Scaling relations based on the image-charge model. *Phys. Rev. B, 67,* 35406.
Wooden D. H., Harker D. E., Woodward C. E., et al. (1997) Discovery of Mg-rich pyroxenes in comet C/1995 O1 (Hale-Bopp): Pristine grains revealed at perihelion. *Earth Moon Planets, 78,* 285–291.
Wooden D. H., Harker D. E., Woodward C. E., et al. (1999) Silicate mineralogy of the dust in the inner coma of comet C/1995 01 (Hale-Bopp) pre- and post-perihelion. *Astrophys. J., 517,* 1034–1058.
Woodward C., Kelley M., Bockelée-Morvan D., and Gehrz R. (2007) Water in comet C/2003 K4 (LINEAR) with Spitzer. *Astrophys. J., 671,* 1065–1074.
Wyatt S. P. (1969) The electrostatic charge of interplanetary grains. *Planet. Space Sci., 17,* 155–171.
Yamamoto S., Kimura H., Zubko E., et al. (2008) Comet 9P/Tempel 1: Interpretation with the Deep Impact results. *Astrophys. J. Lett., 673,* L199–L202.
Yang B. (2013) Comets C/2012 S1 (ISON), C/2012 K1 (PanSTARRS) and C/2010 S1 (LINEAR). *Central Bureau Electronic Telegram (CBET), 3622,* 1.
Yang B. and Sarid G. (2010) Crystalline water ice in outburst comet P/2010 H2. *AAS/Division for Planetary Sciences Meeting Abstracts, 42,* 951.
Yang B., Jewitt D., and Bus S. J. (2009) Comet 17P/Holmes in outburst: The near infrared spectrum. *Astron. J., 137,* 4538–4546.
Yang B., Keane J., Meech K., et al. (2014) Multi-wavelength observations of comet C/2011 L4 (PanSTARRS). *Astrophys. J. Lett., 784,* L23.
Yurkin M. A. and Hoekstra A. G. (2011) The discrete-dipole-approximation code ADDA: Capabilities and known limitations. *J. Quant. Spectrosc. Radiat. Transfer, 112,* 2234–2247.
Zhang Q., Ye Q., Vissapragada S., et al. (2021) Preview of comet C/2021 A1 (Leonard) and its encounter with Venus. *Astron. J., 162,* 194.
Zubko E. (2013) Light scattering by cometary dust: Large-particle contribution. *Earth Planets Space, 65,* 139–148.
Zubko E., Petrov D., Shkuratov Y., and Videen G. (2005) Discrete dipole approximation simulations of scattering by particles with hierarchical structure. *Appl. Opt., 44,* 6479–6485.
Zubko E., Videen G., Arnold J. A., et al. (2020) On the small contribution of supermicron dust particles to light scattering by comets. *Astrophys. J., 895,* 110.

Agarwal J., Kim Y., Kelley M. S. P., and Marschall R. (2024) Dust emission and dynamics. In *Comets III* (K. J. Meech, M. R. Combi, D. Bockelée-Morvan, S. N. Raymond, and M. E. Zolensky, eds.), pp. 653–678. Univ. of Arizona, Tucson,
DOI: 10.2458/azu_uapress_9780816553631-ch020.

Dust Emission and Dynamics

Jessica Agarwal
Technische Universität Braunschweig/Max-Planck-Institut für Sonnensystemforschung

Yoonyoung Kim
Technische Universität Braunschweig

Michael S. P. Kelley
University of Maryland

Raphael Marschall
Southwest Research Institute/Centre National de la Recherche Scientifique/ Observatoire de la Côte d'Azur, Laboratoire J.-L. Lagrange

---

When viewed from Earth, most of what we observe of a comet is dust. The influence of solar radiation pressure on the trajectories of dust particles depends on their cross-section to mass ratio. Hence solar radiation pressure acts like a mass spectrometer inside a cometary tail. The appearances of cometary dust tails have long been studied to obtain information on the dust properties, such as characteristic particle size and initial velocity when entering the tail. Over the past two decades, several spacecraft missions to comets have enabled us to study the dust activity of their targets at much greater resolution than is possible with a telescope on Earth or in near-Earth space and added detail to the results obtained by the spacecraft visiting Comet 1P/Halley in 1986. We now know that the dynamics of dust in the inner cometary coma is complex and includes a significant fraction of particles that will eventually fall back to the surface. The filamented structure of the near-surface coma is thought to result from a combination of topographic focusing of the gas flow, inhomogeneous distribution of activity across the surface, and projection effects. It is possible that some larger-than-centimeter debris contains ice when lifted from the surface, which can affect its motion. Open questions remain regarding the microphysics of the process that leads to the detachment and lifting of dust from the surface, the evolution of the dust while traveling away from the nucleus, and the extent to which information on the nucleus activity can be retrieved from remote observations of the outer coma and tail.

1. INTRODUCTION

Dust released from the surface is the most observationally accessible constituent of a comet. When a comet becomes visible to the naked eye, we see primarily sunlight scattered by dust particles in its coma and tail.

Theories explaining the appearance and formation of comet tails date many centuries back. Our concepts to constrain dust properties like size distribution, ejection times, and velocities from the shape of and brightness distribution in a comet tail have their origins in the nineteenth century and primarily exploit the size dependence of solar radiation pressure on dust (section 2.4). These concepts are described in detail in the *Comets II* book chapter by *Fulle* (2004) and mainly yield information on the properties of the dust as it leaves the sphere of influence of the nucleus.

The strongest limit of this approach is set by the finite resolution of the telescope images available. For a ground-based telescope, the typical seeing-limited resolution is on the order of 1″, which corresponds to 725 km at a comet-observer distance of 1 au. Under exceptional circumstances, such as Hubble Space Telescope observations during a close (0.1 au) Earth flyby, a spatial resolution of a few kilometers can be achieved (e.g., *Li et al.,* 2017). However, no contemporary Earth- or near-Earth-based telescope can resolve the nucleus of a comet and the coma immediately above its surface.

The images returned by ESA's Giotto spacecraft from Comet 1P/Halley in 1986 (*Keller et al.,* 1986) were the first to resolve this innermost part of the coma. They showed the comet nucleus as a solid object and the dust as it emerges from its surface. It became clear that cometary nuclei were highly irregular bodies, certainly in shape and potentially in composition, and that the brightness of the innermost dust coma was spatially highly variable. It displayed bright linear features apparently emanating from the surface, embedded in a more diffuse background (section 3.2.3).

Since the publication of *Comets II*, major progress in understanding the motion of dust in the near environment of cometary nuclei was enabled by several space missions. Their returned images confirmed the 1P/Halley results of the detailed fine structure in the coma brightness distribution. Indications were found that the debris itself can be outgassing and subject to physical evolution, and that part of it falls back to the surface. But open questions remain, in particular concerning how activity is distributed across the surface, why it is spatially and temporally variable, and which microphysical processes lead to the ejection of dust. The answers to all these questions are necessary to understand the interior structure and composition of cometary nuclei, and eventually their formation.

ESA's Rosetta mission provided us with a comprehensive, two-year body of data obtained with various complementary techniques, including imaging and spectroscopy of dust and the surface at various wavelengths, and *in situ* analysis of the composition, density, and velocity of both gas and dust. These data represent the best constraints we have to understand how, where, and when dust is released from a cometary surface and its subsequent journey back to the surface or to interplanetary space.

In this chapter, we first review the forces considered relevant for this outward journey of a dust particle (section 2). We outline the still rudimentary knowledge of how dust activity is distributed across the comet surface and what methods can help to address this question (section 3.1). We next describe how the dust particles are accelerated in the gas flow (section 3.2), how solar gravity and radiation pressure take over as the gas dilutes (section 3.3), and finally address the motion of dust in the comet's tail and trail and its transition to the zodiacal cloud (section 3.4). Open questions and potential means to address them are discussed in section 4.

2. FORCES ACTING ON DUST

2.1. Gravity and Tidal Forces

The gravitational force of a massive body acting on a particle of mass m_d is

$$F_g = m_d g \tag{1}$$

where g is the gravitational acceleration of that body. At distance r from the nucleus center of mass, the gravitational acceleration by the nucleus mass, M, is defined as

$$g = \frac{GM}{r^2} \tag{2}$$

where G is the gravitational constant. In the immediate environment of the nucleus, the gravitational field cannot be approximated by that of a point mass as in equation (1). The spatial extent of the nucleus, in combination with its irregular shape and potentially inhomogeneous internal mass distribution, will necessitate also considering higher orders of the gravitational potential for calculations of the dust motion (e.g., *Werner*, 1994).

The solar gravitational force on a particle at heliocentric distance r_h is

$$F_g = \frac{GM_\odot}{r_h^2} m_d \tag{3}$$

where $M_\odot$ is the mass of the Sun. The sphere of gravitational influence (Hill sphere) of the comet nucleus is estimated as

$$R_{Hill} = r_h \left(\frac{1}{3} \frac{M}{M_\odot} \right)^{1/3} \tag{4}$$

Thus, for a typical 2-km-radius Jupiter-family comet with a bulk density ρ_n = 500 kg m^{-3}, perihelion distance of 1 au, and aphelion at 6 au, the Hill radius will range from roughly 200 km at perihelion to 1300 km at aphelion.

The radial velocity component required for an object at distance r from mass M to stop being gravitationally bound to this mass (escape speed) is given by

$$v_{esc} = \sqrt{\frac{2GM}{r}} \tag{5}$$

Typical escapes speeds from the surfaces of kilometer-sized bodies are on the order of 1 m s^{-1}.

For the motion of a dust particle in a frame attached to the nucleus center of mass, the difference in solar gravity between the locations of the particle and of the nucleus (the tidal force) is more relevant than the absolute value of solar gravity. A particle of mass m_d located on the Sun-nucleus line and at a distance r from the nucleus in either direction is subject to a tidal force directed away from the nucleus given by

$$F_{tr} = \frac{2GM_\odot}{r_h^3} m_d r \tag{6}$$

while the tidal force on a particle located above the terminator points toward the nucleus with the magnitude

$$F_{tl} = \frac{GM_\odot}{r_h^3} m_d r \tag{7}$$

2.2. Drag by Surrounding Gas Coma

The main force accelerating dust particles away from the nucleus surface comes from their interaction with the surrounding gas field. As illustrated in Fig. 1, the drag force is strongest within ~10 nucleus radii, R_n. By that distance the gas densities have diluted significantly, making molecule-dust collisions rare and momentum transfer from the gas flow to the dust particles inefficient. The gas dynamics are discussed in the chapter by Marschall et al. in this volume.

Once a dust particle is detached and lofted from the surface, the surrounding gas molecules collide with it and accelerate the particle. The main direction of gas expansion is away from the nucleus surface and therefore so is the net force on the dust particles.

When dust densities are low enough that particles do not exert a back reaction onto the gas flow (e.g., deceleration and/or heating of the gas flow) they can be considered as test particles within the gas flow and thus be treated mathematically separately (*Tenishev et al.,* 2011). This condition is given in most cases but there are exceptions. For example, in the event of a strong outburst, where the dust plume is optically thick, this condition is certainly not satisfied. For such cases, a multi-phase simulation needs to be adopted (e.g., *Shou et al.,* 2017) that also takes into account interparticle collisions.

In the following, we assume the more common case where dust particles do not significantly influence the gas flow. In addition, in most cases, the mean free path of the molecules is much larger than the dust particle size, and therefore free molecular aerodynamics can be applied. In this scenario (*Finson and Probstein,* 1968; *Gombosi et al.,* 1985, 1986; *Gombosi,* 1987; *Sengers et al.,* 2014; *Marschall et al.,* 2016) the drag force, F_D, on a spherical particle is

$$\vec{F}_D = \frac{1}{2} C_D m_g n_g \sigma_d \left|\vec{v}_g - \vec{v}_d\right| \left(\vec{v}_g - \vec{v}_d\right) \tag{8}$$

where σ_d is the geometric cross-section of the dust particle and $\vec{v}_d$ is its velocity. The mass of a gas molecule is m_g, and n_g and $\vec{v}_g$ are their number density and macroscopic velocity, respectively. C_D is called the drag coefficient. For an equilibrium gas flow, and a mean free path of the molecules much larger than the dust size, the drag coefficient (*Bird,* 1994) is defined as

$$C_D = \frac{2\zeta^2 + 1}{\sqrt{\pi}\zeta^3} e^{-\zeta^2} + \frac{4\zeta^4 + 4\zeta^2 - 1}{2\zeta^4} \mathrm{erf}(\zeta) + \frac{2(1-\varepsilon)\sqrt{\pi}}{3\zeta} \sqrt{\frac{T_d}{T_g}} \tag{9}$$

with the gas temperature T_g, the dust particle temperature T_d, the fraction of specular reflection ε, and the molecular speed ratio

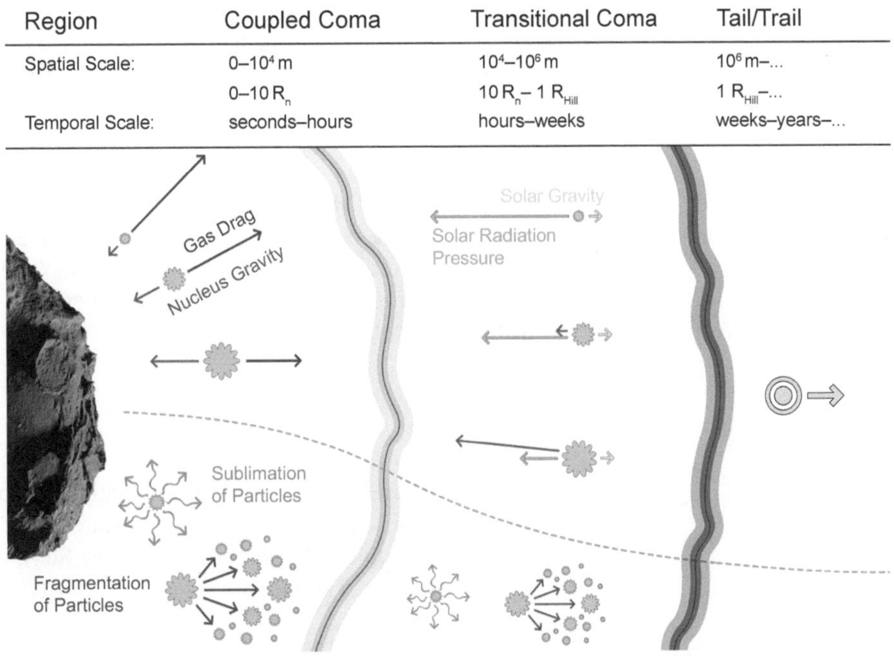

Region	Coupled Coma	Transitional Coma	Tail/Trail
Spatial Scale:	0–10^4 m	10^4–10^6 m	10^6 m–...
	0–10 R_n	10 R_n– 1 R_{Hill}	1 R_{Hill}–...
Temporal Scale:	seconds–hours	hours–weeks	weeks–years–...

Fig. 1. Sketch of the cometary dust environment which we divide into three main dynamical regions: (1) the coupled coma region where the dust dynamics are dominated by local forces connected to the nucleus (gas drag and nucleus gravity) (section 3.2); (2) the transitional coma region where the dust has decoupled from the gas and small particles transition to being dominated by solar forces (gravity and radiation pressure) while large particles are bound in the gravitational field of the nucleus (section 3.3); and (3) the dust tail and trail within which the escaping dust particles are purely governed by solar forces (section 3.4). The boundaries between these regions are complex three-dimensional surfaces. The spatial and temporal scales given are rough estimates for a 67P-like comet at 1 au. R_n is the nucleus radius, and R_{Hill} the Hill radius. This figure is a reproduction of Fig. 4 in *Marschall et al.* (2020c).

$$\zeta = \frac{|\vec{v}_g - \vec{v}_d|}{\sqrt{\frac{2k_b T_g}{m_g}}} \tag{10}$$

where k_b is the Boltzmann constant.

Figure 2 shows the drag coefficient as a function of ζ. For large particles the dust speed is much smaller than the gas speed. At the surface, the gas speed is on the same order as the thermal speed (the denominator in equation (10)) and therefore $\zeta \sim 1$ and $C_D \sim 5$ (Fig. 2). On the other hand, for a typical comet (water-dominated emission well within the snow line) at large cometocentric distances (e.g., at 10 R_n), the temperature of the gas has cooled to a few tens of Kelvin but the gas speed has reached the order of 1 km s^{-1}. In this case slowly moving particles have $\zeta \sim 10$ and thus $C_D \sim 2$ (Fig. 2). Very small, typically submicrometer, particles, may attain a significant fraction of the gas speed already very close to the surface, such that $\zeta < 1$ and $C_D > 5$ (Fig. 2). Because the drag coefficient asymptotically approaches 2 for most particle sizes, a size-independent $C_D = 2$ is often assumed rather than the more complicated equation (10). This choice implies that the acceleration of, in particular, small particles is underestimated.

Equation (9) describes the idealized case of spherical particles in a gas flow when the mean free path of the gas molecules is larger than the dust particle size. Real dust particles are not spherical and, in particular, larger dust grains are porous and fluffy aggregates (*Kolokolova and Kimura,* 2010; *Schulz et al.,* 2015; *Rotundi et al.,* 2015; *Langevin et al.,* 2016; *Bentley et al.,* 2016; *Mannel et al.,* 2016; *Levasseur-Regourd, et al.* 2018). This affects the dynamics of the particles as show by *Skorov et al.* (2016a, 2018). They found that porous aggregates are accelerated to significantly higher speeds than their compact counterparts of the same radius. This behavior can be mimicked in the spherical particle paradigm described above by adjusting the masses and geometric cross-sections to the respective values of the porous particles. In this sense, the spherical particles can be understood as effective particles with the given mass and cross-section.

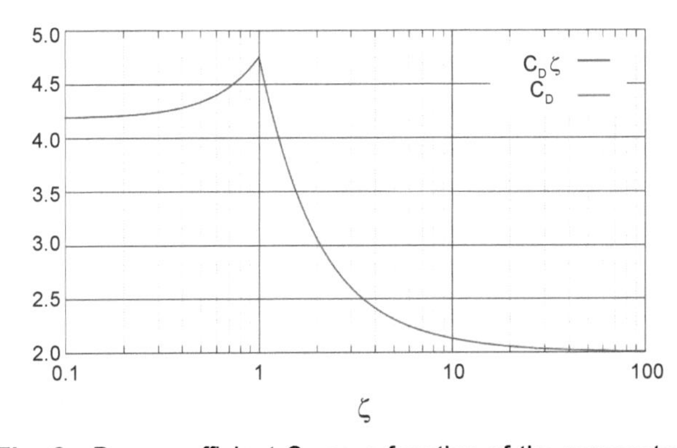

Fig. 2. Drag coefficient C_D as a function of the parameter ζ as given by equations (9) and (10) and assuming $T_d = T_g$ and $\varepsilon = 0$.

Additionally, particles have been observed to rotate in the comae of Comets 103P and 67P (*Hermalyn et al.,* 2013; *Fulle et al.,* 2015b), and the effect of oblate or prolate particle shapes on their dynamics was studied by *Ivanovski et al.* (2017a,b). Not only will non-spherical particles begin to rotate in the gas flow, but they may also accelerate to higher speeds than spherical ones. Unless a particle rotates sufficiently fast, the influence of particle rotation on their dynamics cannot simply be parameterized into a spherical particle paradigm (equation (8)) as described above for porous particles, because spherical particles only experience a force along the direction of the gas flow, while non-spherical, rotating particles also experience a force perpendicular to the flow. This component of the acceleration is not included in equation (8).

2.3. Intrinsic Outgassing

Solid particles in the coma can contain volatile ices. When the grain temperature is sufficiently warm, the ices will sublimate. This loss of vapor accelerates the particle. In the ideal case of an isotropically outgassing spherical particle, the net acceleration is 0. However, coma particles are not spheres, and outgassing may not be uniform, especially from large particles that can sustain a temperature gradient across their surfaces. Rapid rotation of a particle does help distribute the absorbed sunlight across the surface, and smooths out the temperature gradient, but the spin axis may prevent illumination of some portions of the grain. As a result, the acceleration will typically have an anti-solar component. Whether or not a tangential component exists depends on the spin state, shape, porosity, and thermal characteristics of the materials.

The calculation of the acceleration from outgassing is analogous to the non-gravitational acceleration of cometary nuclei (*Marsden et al.,* 1973), except that the microphysics of the particle, e.g., thermal and radiative properties, are very different. Setting these complexities aside, the force from gas sublimation (also called "rocket force"), F_s can be calculated from

$$F_s = \sigma_d m_g Z v_{th} \kappa f_{ice} \tag{11}$$

where Z is the number sublimation rate of the ice per surface area, κ is the degree of asymmetry of outgassing (0 for isotropic, 1 for directly toward the Sun), f_{ice} is the effective fractional surface area of the ice (cf. *Kelley et al.,* 2013), and

$$v_{th} = \frac{\pi}{4}\sqrt{\frac{8k_b T_g}{\pi m_g}} \tag{12}$$

is the mean thermal expansion speed of the gas.

The details hidden within κ may be complex, but equation (11) is still useful for estimating the potential order of magnitude of acceleration by outgassing.

Since the force is generally opposite the Sun, it works similarly to acceleration by radiation pressure, in that the particles are accelerated into the tail direction. For example, *Reach et al.* (2009) examined the acceleration from outgassing water ice at 1 au and found that the resultant force, F_s, can be comparable to or much stronger than that from solar radiation F_r (section 2.4), and estimated $F_s/F_r \sim 100$ for a 1-mm-sized grain with $f_{ice} = 0.0017$. Over small heliocentric distance ranges and when the relevant ice is undergoing free sublimation, the force may be approximated as equivalent to the β-parameter, i.e., a force $\propto r_h^{-2}$ that is addressed with a reduced-gravity solution (section 2.4).

Likely, the outgassing force is observable only for centimeter sized and larger particles, because smaller particles lose their ice content too quickly and hence too close to the surface for the non-gravitational effect on their trajectories to be measured. This conclusion was reached by *Markkanen and Agarwal* (2020) and *Davidsson et al.* (2021) from thermophysical modeling, and, complementarily, by *Reach et al.* (2009) from studying the motion of particles in the debris trail of Comet 73P/Schwassmann-Wachmann 3.

It is possible that asymmetric outgassing changes the rotation rate of an ice-bearing comet fragment, which can lead to disintegration by centrifugal force on timescales of less than a day for a meter-sized fragment (*Jewitt et al.,* 2016). *Steckloff and Jacobson* (2016) propose that fragment disintegration by sublimation torques can lead to the formation of striae observed in the tails of some enigmatic, bright comets. Alternative explanations of striae formation invoke electromagnetic forces (section 2.6).

2.4. Solar Radiation Pressure

The radiation force is proportional to the solar intensity multiplied by the cross-sectional area of the particle

$$F_r = \frac{Q_{pr}}{c}\left(\frac{L_\odot}{4\pi r_h^2}\right)\sigma_d \tag{13}$$

where $L_\odot$ is the solar luminosity, c is the speed of light, and Q_{pr} is the radiation pressure coefficient averaged over the solar spectrum (*Burns et al.,* 1979). Dynamical models commonly use a simplified treatment of comet dust, assuming $Q_{pr} = 1$. A more detailed treatment dependent on dust mineralogies and structures (compact or fluffy) is discussed in the chapter by Kolokolova et al. in this volume.

Since the solar radiation force opposes the solar gravity and both forces are proportional to $1/r_h^2$, the net force can be considered as reduced solar gravity

$$F_{net} = F_r - F_g = (1-\beta)F_g \tag{14}$$

where the β parameter is the ratio of the radiation force, F_r, to the solar gravitational force, F_g

$$\beta \equiv \frac{F_r}{F_g} = \frac{3L_\odot Q_{pr}}{16\pi G M_\odot c \rho_d a_d} = C_\beta \frac{Q_{pr}}{\rho_d a_d} \tag{15}$$

with $C_\beta = 5.77 \times 10^{-4}$ kg m^{-2} and a_d representing the radius of a spherical particle. More generally, $1/a_d$ can be expressed as $4\sigma_d\,\rho_d/(3\,m_d)$, with ρ_d being the dust bulk density.

Calculations show that silicate particles tend to have $\beta < 1$ regardless of aggregate structure (*Silsbee and Draine,* 2016), while absorbing particles may have $\beta > 1$ (*Kimura et al.,* 2016).

2.5. Poynting-Robertson Drag

Small particles in orbit about the Sun are influenced also by radiation pressure tangential to their motion (*Robertson,* 1937; *Wyatt and Whipple,* 1950). The resulting Poynting-Robertson force is given by

$$F_{PR} = \frac{a_d^2 L_\odot}{4c^2}\sqrt{\frac{GM_\odot}{r_h^5}} \tag{16}$$

The Poynting-Robertson effect causes millimeter-sized dust particles in the zodiacal cloud (section 3.4.6) to spiral into the Sun on timescales $\tau_{PR} \gtrsim 6 \times 10^5$ yr (*Kasuga and Jewitt,* 2019).

2.6. Electromagnetic Forces

A dust particle embedded in the cometary or solar wind plasma and interacting with solar ultraviolet radiation is subject to charging by electron and ion collection and by secondary electron and photoelectron emission. Over time, the dust particle will assume the potential at which the involved currents balance. This equilibrium potential depends on the properties of the plasma environment and the dust particle, such as composition and surface roughness (*Horanyi,* 1996). In interplanetary space, the dominant charging process is photoelectron emission, and typical dust potentials, U, range between 0.5 V and 14 V (*Mukai,* 1981). A canonical value of U = 5 V is often used (e.g., *Sterken et al.,* 2012; *Kramer et al.,* 2014). Solar wind interaction with the plasma tail is discussed in the chapter by Götz et al. in this volume.

For a given surface potential and volume, the shape dependence of a grain's charge can be described by the dimensionless parameter $\kappa_e > 1$ that is minimal for a sphere ($\kappa_e = 1$) and can reach values up to $\kappa_e = 5$ for fractal particles (*Auer et al.,* 2007). The integrated charge of a grain can thus be described as

$$q = 4\pi\varepsilon_0 a_d \kappa_e U \tag{17}$$

where a_d is the radius of a volume-equivalent sphere, and ε_0 is the electric permittivity in vacuum.

In the presence of a magnetic field, $\vec{B}$, a charged particle moving with velocity $\vec{v}$ relative to the field is subject to the

Lorentz force

$$\vec{F}_L = q(\vec{v} \times \vec{B}) \tag{18}$$

Outside the immediate environment of the comet, the relevant field is the interplanetary magnetic field (IMF), and the velocity of the dust particle relative to this field can be approximated by the velocity of the solar wind (v_{SW} = 400–800 km s^{-1} radially outward from the Sun) that carries the magnetic field and is at least an order of magnitude faster than typical heliocentric velocities of comets. Splitting the IMF into a radial (B_r), an azimuthal (B_ϕ), and a normal component ($B_\theta = 0$), equation (18) reduces to (*Kramer et al.,* 2014)

$$F_L = \pm q v_{SW} B_\phi = q v_{SW} B_{\phi,0} \frac{r_0}{r_h} \cos\beta_{hg} \tag{19}$$

where β_{hg} is the heliographic latitude, r_0 = 1 au, and $B_{\phi,0}$ = 3 nT is the azimuthal field strength at 1 au. Hence, the Lorentz force decreases with heliocentric distance as $1/r_h$, less steeply than solar gravity and radiation pressure. For a given type of particle, the relative importance of the Lorentz force will increase with heliocentric distance (Fig. 5). *Kramer et al.* (2014) and *Hui et al.* (2019) reported that including the Lorentz force in a model of the dust motion significantly improved reproducing the orientation of the dust tails of Comets Hale-Bopp and C/2010 U3 (Boattini) at heliocentric distances between 15 and 30 au.

Price et al. (2019) find that changes in the appearance of striae in the tail of Comet C/2006 P1 (McNaught) coincided with the comet crossing the heliospheric current sheet and infer that dust in the striae was charged and hence subject to the Lorentz force. Striae are linear features inside a comet's dust tail of unknown origin. They are only seen in comets with very high production rate, typically dynamically new comets. Alternative models of striae formation favor processes of instantaneous disintegration of, e.g., highly non-spherical grains (*Sekanina and Farrell,* 1980) or of large boulders fragmenting under outgassing-induced torques (*Steckloff and Jacobson,* 2016) (see also section 2.3).

Fulle et al. (2015a) find that near the Rosetta spacecraft, fluffy dust particles of extremely low density may get charged by secondary electrons from the spacecraft and disintegrate, leading to the detection of particle swarms by the onboard Grain Impact Analyser and Dust Accumulator (GIADA) instrument. In the vicinity of Comet 67P, charged water clusters smaller than 100 nm were also detected (*Gombosi et al.,* 2015).

2.7. Electrostatic Lofting

Charged dust lofting and transport have been proposed to explain observations of airless bodies in the solar system, such as the lunar horizon glow (*Rennilson and Criswell,* 1974), the "spokes" in Saturn's rings (*Morfill et al.,* 1983), and dust ponds formed on asteroid Eros (*Colwell et al.,* 2005). There is little published work on the relevance of this effect in the presence of outgassing (such as on an active comet), but *Nordheim et al.* (2015) modeled the electrostatic charging of the nucleus of Comet 67P and showed that charged dust grains with radii <50 nm may be electrostatically ejected from the nucleus in situations of weak activity. The electrostatic force on a particle is

$$F_{ES} = qE \tag{20}$$

where E is the local electric field strength. Details on the particle charging equations are presented in *Zimmerman et al.* (2016) and *Wang et al.* (2016).

We here summarize some aspects of dust charging on non-cometary airless bodies: Dust on the lunar surface is levitated due to electrostatic charge gradients resulting from uneven solar illumination. Because on kilometer-sized asteroids gravity is much lower, dust can be electrostatically ejected from such bodies (*Lee,* 1996). Recent asteroid missions have observed rocky surfaces on asteroids Bennu and Ryugu, indicating a lack of regolith (*Jaumann et al.,* 2019; *Lauretta et al.,* 2019). These observations and the particle ejection events observed on Bennu may be partly caused by electrostatic dust lofting and escape (*Hartzell et al.,* 2022; *Nichols and Scheeres,* 2022).

2.8. Interparticle Cohesion

Intermolecular (e.g., Van der Waals) forces between the surfaces of neighboring particles or grains are held responsible for the internal strength of dust aggregates, agglomerates, chunks, and surfaces. The precise form and magnitude of these forces depends on the structure and composition of the material, which are not well known.

For a lunar-type regolith surface with average grain size a, *Sánchez and Scheeres* (2014) derive a strength of

$$T_{reg} = C_{reg} C_{\#} \phi / a_d \tag{21}$$

where $C_{reg} = 4.5 \times 10^{-3}$ N m^{-1}. $C_{\#}$ is the number of neighboring particles that a given grain touches (the coordination number), and ϕ is the volume-filling factor of the dust layer, i.e., the fraction of the volume that is filled by matter.

For a model surface composed of agglomerates of aggregates of dust grains, and using empirical relationships based on laboratory measurements, *Skorov and Blum* (2012) deduce the following expression for the tensile strength of a dust surface

$$T_{agg} = C_{agg} \phi \left(\frac{a_d}{a_0} \right)^{-2/3} \tag{22}$$

with C_{agg} = 1.6 Pa, and a_0 = 1 mm.

Equations (21) and (22) render values that differ by

2 orders of magnitude at $a_d = 100$ μm, which illustrates the sensitive dependence on model assumptions and the lack of well-constrained parameters.

2.9. Relative Importance of Forces

The relative importance of the various forces discussed in sections 2.1–2.8 depends mainly on the particle size and the distances from the Sun and comet, but also on the dust and gas properties. Figures 3–5 illustrate the key dependencies.

Nucleus gravity is always several orders of magnitude weaker than solar gravity, but since both the nucleus and the dust are subject to the solar gravitational acceleration, the relevant quantity with which to compare the nucleus gravitational force is the tidal force. The nucleus distance where tidal force and nucleus gravitational force balance is given by the Hill radius.

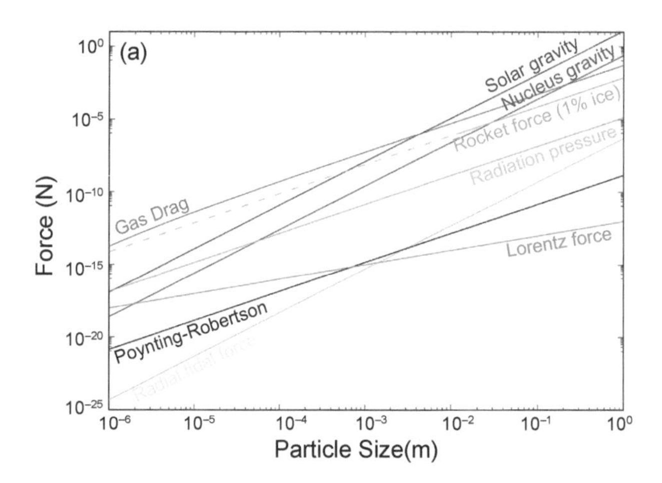

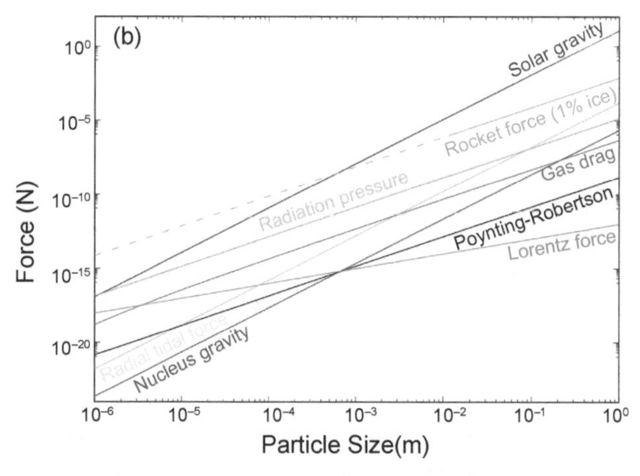

Fig. 3. Forces acting on dust particles (equations (1)–(3), (6), (8), (11), (13), (16)–(19)) as functions of particle size. The parameter values used are listed in Table 1. The heliocentric distance is 1 au in both plots. **(a)** Surface distance of 1 km, displaying the forces well inside the coma. **(b)** Surface distance of 998 km, where dust has essentially decoupled from the nucleus gravity and gas drag. The rocket force is depicted by dashed lines for sizes <1 cm to indicate that ice lifetimes in small particles are too short to influence the dust dynamics on observable timescales.

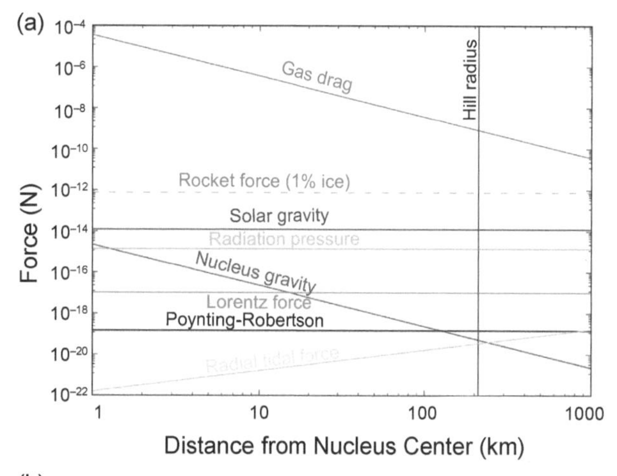

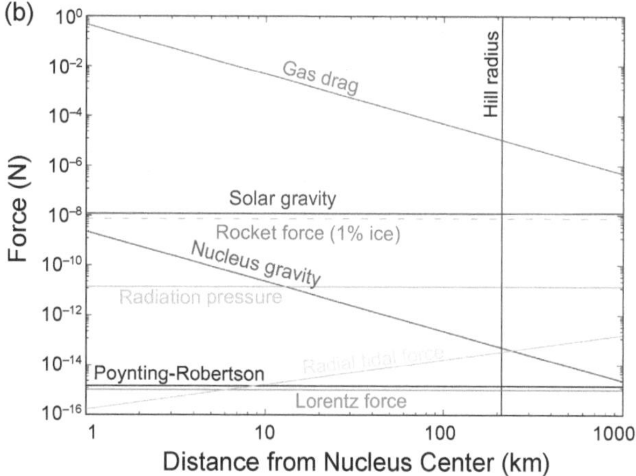

Fig. 4. Same as Fig. 3 for fixed particle sizes [**(a)** 10 μm, **(b)** 1 mm], 1 au from the Sun, and variable nucleocentric distance. The Hill radius (where tidal force and nucleus gravitational force balance) is indicated by a vertical line.

The Poynting-Robertson effect and the Lorentz force are weakest and therefore typically not considered in calculations of cometary dust dynamics. However, far from the Sun, the Lorentz force on small particles can become comparable to radiation pressure.

If present, outgassing-induced "rocket" force tends to be stronger than solar radiation pressure.

Once lifted from the nucleus surface, the initially dominant force on dust particles is gas drag (equation (8)). It decreases roughly quadratically with increasing nucleus distance as the gas dilutes, and at some distance (that depends on particle size and gas production rate), becomes weaker than solar gravity. We refer to this near-surface regime as the acceleration zone (section 3.2) and to the dust velocity at its upper boundary as the terminal velocity. Once the influences of nucleus gravity and gas drag have diminished, the motion of the dust in the outer coma, tail and trail is largely driven by solar gravity and radiation pressure, and by the terminal velocity the dust acquired from the gas drag. In the outer coma (section 3.3), dust that was initially ejected toward the Sun from the sunlit surface where water sublimation is strongest will reverse its direction of motion under the influence

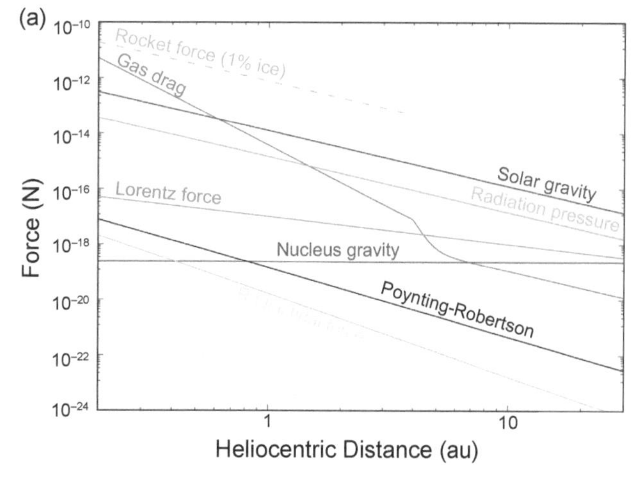

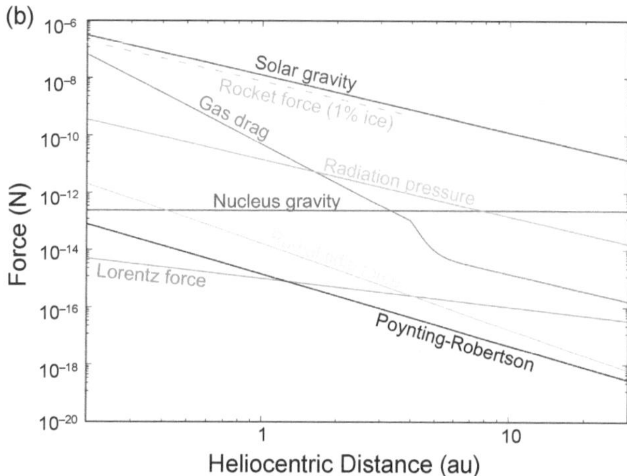

Fig. 5. Same as Fig. 3 for fixed particle sizes [**(a)** 10 μm, **(b)** 1 mm], nucleocentric distance of 100 km (hence outside the zone of acceleration by gas drag) and variable heliocentric distance. The gas drag force drops abruptly near 4 au, where we assume a steep drop of the water production rate. For the same reason, we assume that rocket force from intrinsic outgassing of water ice ceases near 4 au. Rocket force is again represented by a dashed line to indicate that both types of particles are too small to retain ice on dynamically significant timescales if mixed with dust.

of solar radiation pressure, such that eventually almost all dust is driven into the curved tail stretching in the anti-solar direction and along the negative heliocentric velocity vector of the comet, hence outside its orbit (section 3.4).

Generally, if particles contain sublimating ice, the forces on them can implicitly be affected (and become time dependent) by their changing mass, cross section, and temperature. The same applies for fragmentation, spin changes, and changes of temperature.

Often, a maximum liftable grain size is derived from the balance of gravity (equation (1)) and gas pressure at the surface (equation (8)). We here formulate it more generally as minimum liftable cross-section-to-mass ratio

$$\left[\frac{\sigma_d}{m_d}\right]_{max} = \frac{8\pi G}{3C_D}\frac{\rho_N R_N}{m_g v_g Q_g} \tag{23}$$

where $Q_g = n_g v_g$ is the surface gas production rate in molecules per unit time and area. Inserting the values from Table 1 and assuming spherical dust particles, the maximum radius liftable by water vapor alone at 1 au is about 1 m. Since gas production rates can be highly variable and hard to measure on a local scale, the actual maximum liftable grain size for a given situation is difficult to predict.

3. DYNAMICAL REGIMES

3.1. Dust Emission from the Surface

Due to its low optical depth, the surface brightness of dust in the cometary coma is nearly always lower than that of the illuminated surface of the nucleus. With remote sensing methods that measure scattered sunlight or thermal radiation, dust can only be detected against the dark backgrounds of empty space or shadowed surface. It is therefore not straightforward to identify the source regions of dust on the surface, even when resolved images of the dust coma obtained by cameras on spacecraft show a considerable fine structure near the limb (cf. section 3.2.3). Integrated gas production rates indicate that most comets emit only a small fraction of the gas that would be expected from a sublimating surface of pure ice. An exception to this are the so-called hyperactive comets [e.g., 46P/Wirtanen, 103P/Hartley 2 (*A'Hearn et al.* (2011)], in which the global water production is higher than can be explained from pure surface sublimation.

In the following, we describe some of the most common methods used to constrain the distribution of activity across a cometary surface and subsequently outline their findings. A review of local manifestations of cometary activity can be found in *Vincent et al.* (2019).

3.1.1. Methods to locate activity sources. (a) *Inversion/triangulation* — If a bright filament was observed at least twice from different perspectives [separated by ideally 10°–30° of subobserver latitude/longitude (*Vincent et al.,* 2016b)], its three-dimensional orientation and source point on the surface can be identified by triangulation: For each image, the projected central line of the filament and the camera position span a plane in three-dimensional space. For two images, the intersection line of the two planes describes the filament axis in three dimensions, and the intersection point of this axis with the nucleus surface (from a shape model) represents the source location (Fig. 6). If the filament was observed more than twice, all planes should intersect in the same line within the accuracy limits. This technique is called "direct inversion" by *Vincent et al.* (2016b) and relies on the assumptions that (1) the same coma structure can be identified in several images; (2) at least within a certain distance near the surface, the filament can be described by a straight line; and (3) its three-dimensional orientation does not change during the time covered by the observations.

Without making the first two assumptions, source locations can still be identified by "blind inversion" as long as

TABLE 1. Parameters and values used to generate Figs. 3–5.

Quantity	Symbol	Value
Nucleus radius	R_n	2 km
Nucleus density	ρ_n	500 kg m^{-3}
Dust bulk density	ρ_d	500 kg m^{-3}
Drag coefficient	C_D	2
Global water production rate*	$Q_{H_2O}(r_h < 4$ au)	3×10^{28} molecules s^{-1} $(r_h/1\ \text{au})^{-4.5}$
	$Q_{H_2O}(r_h > 4$ au)	4×10^{34} molecules s^{-1} $(r_h/1\ \text{au})^{-15}$
Global CO_2 production rate†	$Q_{CO_2}(r_h)$	4×10^{25} molecules s^{-1} $(r_h/1\ \text{au})^{-2}$
Gas speed in coma	v_g	700 m s^{-1}
Gas speed (from icy chunks)	v_{th}	500 m s^{-1}
Dust speed as a function of radius, a‡	$v_d(a)$	300 m s^{-1} $\sqrt{a/1\ \mu\text{m}}$
Ice fraction in dust	f_{ice}	0.01
Radiation pressure coefficient	Q_{pr}	1
Dust potential	U	5 V
Solar wind speed	v_{SW}	600 km s^{-1}
Azimuthal IMF strength at 1 au	$B_{\Phi,0}$	3 nT
Heliographic latitude	β_{hg}	0°
Volume filling factor	Φ	0.5

* The heliocentric distance dependence of the water production rate is discussed in section 3.3.1. Beyond 4 au we adopt an exponent of –15, which reasonably describes the water production rates between 5 au and 6 au if calculated from the balance of input solar energy, thermal radiation, and sublimation cooling. At larger heliocentric distances, the water production rate drops even more steeply, but becomes negligible anyway compared to more volatile species.

† The CO_2 production rate is here assumed to be proportional to the available solar energy, which scales with r_h^{-2} if radiation cooling is neglected. The true behavior of gas production rates is far more complex (e.g., *Combi et al.,* 2020) and variable between comets.

‡ The size dependence of the dust speed is analogous to equation (26) and Fig. 8.

several images are available. In this approach, the intersection lines of the jet-observer planes with the nucleus surface are calculated for each filament in each image. When combining the surface intersection lines from the different images, lines corresponding to the same filament will intersect in the same point on the surface within the achievable accuracy. Hence the points having multiple intersections are interpreted as the source points of filaments (*Vincent et al.,* 2016b). In the presence of many filaments, the results can also be a source density map rather than a map of individual sources. At Comet 67P, this inversion technique has been used to trace the origins of both diurnally repeating coma structures by, e.g., *Vincent et al.* (2016b), *Shi et al.* (2016), and *Lai et al.* (2019), and of irregular events ("outbursts") by *Vincent et al.* (2016a) and *Shi et al.* (2024). This technique has also been applied to groundbased coma images (e.g., *Farnham et al.,* 2007; *Vincent et al.,* 2010, 2013) (see section 3.2.3).

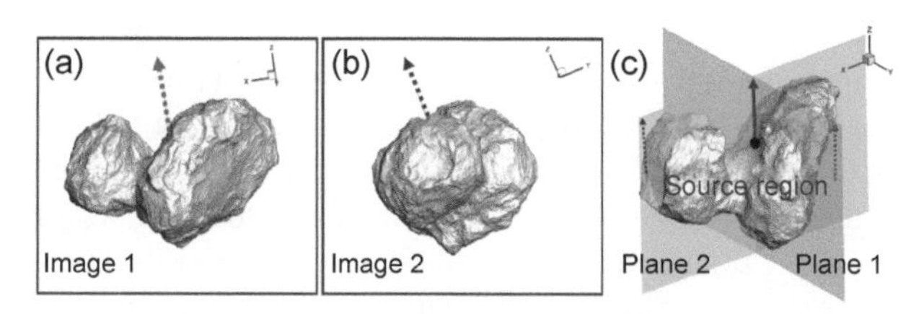

Fig. 6. Example of how the source region of a jet can be inferred from images obtained from at least two different perspectives, shown in **(a)** and **(b)**. Each of these two observations renders a plane that contains the line of sight, the jet, and its source. **(c)** Illustration of how the source region is identified as the point where the intersection line of the two planes crosses the nucleus surface. Credit: Ian Lai, *Astron. Astrophys, 630,* A17, p. 3, 2019, reproduced with permission © ESO.

If the lower part of a filament is brighter than the background nucleus surface, the source points can also be identified directly (e.g., *Agarwal et al.,* 2017; *Fornasier et al.,* 2019b).

(b) *Backtracing of in situ data* — To constrain the distribution of gas or dust sources across the nucleus surface from *in situ* measurements onboard a spacecraft, the measured dust or gas density is often used as a proxy for the activity at the subspacecraft longitude and latitude at the time of observation (e.g., *Hoang et al.,* 2017, 2019; *Della Corte et al.,* 2015, 2016). The underlying assumption is that the activity changes on timescales long compared to the traveling time of the material from the surface to the detector, and that this motion is radial. Since this assumption cannot a priori be taken as justified, in particular for the more slowly moving dust, some authors account for the traveling time by including the near-surface acceleration zone when linking the detection coordinates to the ejection point on the surface (*Longobardo et al.,* 2019, 2020). These approaches help to understand broad regional and diurnal variations of activity but do not generally provide the spatial resolution to, e.g., link coma material to specific landmarks on the surface.

(c) *Forward modeling of the motion of dust embedded in the gas flow* — The source regions of coma dust can be constrained by forward modeling the motion of dust embedded in the gas flow field and iteratively fitting the predicted dust distribution to measurements. Models of the gas dynamics typically either follow a fluid dynamics approach or use the Direct Simulation Monte Carlo (DSMC) method to describe the motion of individual molecules (see the chapter by Marschall et al. in this volume). The comparison of the modeled dust distribution to remote sensing observations requires additional assumptions about the light scattering and/or thermal emission properties of the dust (see the chapter by Kolokolova et al. in this volume), and the projection of the three-dimensional dust distribution onto the image plane by line-of-sight integration. One boundary condition of all coma models is the distribution of gas and dust activity across the cometary surface, which is why they are discussed in the present context. The surface activity distribution needs to be defined such that it enables an optimal reproduction of the data in question. Generally, the obtained solution will not be unique, especially if only the local gas density is used to constrain the models (*Marschall et al.,* 2020a), but fitting the same model to multiple datasets can reduce the degeneracy.

Forward-modeling approaches have been used both to describe temporally and spatially confined phenomena and to understand the global distribution of activity. An example of the former is the study of the influence of topography, local time, and illumination conditions on filament structures emanating from the terminator region in *Shi et al.* (2018). Global models have for example been used to fit data from the Rosetta Orbiter Spectrometer for Ion and Neutral Analysis (ROSINA), from the Optical, Spectroscopic, and Infrared Remote Imaging System (OSIRIS), and from the Visible and Infrared Thermal Imaging Spectrometer (VIRTIS), and to combinations of such datasets (e.g., *Marschall et al.,* 2016, 2017; *Fougere et al.,* 2016a,b; *Zakharov et al.,* 2018a; *Combi et al.,* 2020; *Davidsson et al.,* 2022).

Kramer et al. (2017) and *Läuter et al.* (2019, 2020) instead follow an inverse modeling approach in which the gas emission from each surface element of the shape model is a free parameter and the global distribution of gas in the coma is calculated from the surface production rates and is probed by the measurements (as a function of space and time) of the ROSINA instrument. The vector of the surface emission rates is connected to the vector of measurements through a matrix and is optimized in order to minimize the deviation between model and measurements. The obtained vector of surface production rates describes the geographical distribution of activity.

(d) *Torques and non-gravitational forces* — An additional means of constraining the gas activity distribution across the surface arises from the reaction force that the sublimating gas exerts on the nucleus, similar to a rocket engine. The component of the force crossing the nucleus center of mass leads to an acceleration of the heliocentric orbit, while the component perpendicular to the rotation axis creates a torque that changes the rotation speed and axis orientation. For 67P, a first rough prediction of the change of spin rate during the perihelion passage was made by *Keller et al.* (2015), assuming the local activity to be driven by the illumination conditions and energy balance only. Models aiming to fit simultaneously the rotation state and heliocentric orbit of 67P require a more complex distribution of surface activity (*Kramer et al.,* 2019; *Kramer and Läuter,* 2019; *Attree et al.,* 2019, 2023. 2024), but a single model to fit both constraints has not yet been identified. A highly asymmetric outgassing has been invoked to explain a rapid decrease in rotation rate of Comet 41P/Tuttle-Giacobini-Kresák (*Bodewits et al.,* 2018).

(e) *Surface changes* — At both Comets 9P and 67P, changes on the surface were seen when the same spot was observed multiple times. In particular, the Rosetta mission with its two-year coverage of almost the entire surface offered the possibility to study such changes. Surface changes include cliff collapses, receding scarps, formation and extension of fractures and cavities, displacement of boulders, changes in dust mantle thickness, and the temporary appearance of bright spots (see the chapters by Pajola et al. and Filacchione et al. in this volume). Some of these events, such as cliff collapses and the new exposure of bright surfaces, have been associated with transient dust emission with reasonable certainty (*Pajola et al.,* 2017a; *Agarwal et al.,* 2017). For fractures and cavities, models have been proposed that connect their formation or deepening to the sudden emission of dust (e.g., *Vincent et al.,* 2015; *Skorov et al.,* 2016b), but observational proof of these models has not yet been found. The connection of moving boulders and scarps and of changes in the thickness of a local dust layer to the emission of dust remains likely but also unproven (e.g., *Thomas et al.,* 2013; *El-Maarry et al.,* 2017; *Fornasier et al.,* 2019a).

3.1.2. "Regular" and "irregular" activity. The appearance of the dust coma and the pattern of filament structures changes with time during a rotation of the comet but quite accurately repeats with the diurnal cycle (*Vincent et al.,* 2016a). This, together with the observation that most dust is emitted from the illuminated hemisphere (*Fink et al.,* 2016; *Gerig et al.,* 2020) and that the dust emission follows the subsolar latitude on seasonal timescales (*Vincent et al.,* 2016b; *Della Corte et al.,* 2016; *Lai et al.,* 2019) indicates that direct solar irradiation is the prime driver of the diurnally repeating ("regular") activity.

Insolation is, however, not the only factor determining the strength of local dust and gas activity, because models assuming that a constant fraction of solar energy input is consumed by water ice sublimation, and that the dust production is proportional to the outgassing rate, fail to reproduce the *in situ* measurements of ROSINA, the brightness pattern of coma dust (*Marschall et al.,* 2016), and the non-gravitational forces and torques (*Attree et al.,* 2019;

Kramer et al., 2019). Various patterns of systematic activity enhancements have been proposed, such as enhanced activity from sinkholes (*Vincent et al.,* 2015; *Prialnik and Sierks,* 2017), cliffs (*Vincent et al.,* 2016b; *Marschall et al.,* 2017), fractures (*Höfner et al.,* 2017), and newly illuminated, frost-covered surfaces near receding shadows (*Prialnik et al.,* 2008; *De Sanctis et al.,* 2015; *Fornasier et al.,* 2016; *Shi et al.,* 2018). However, their relative contribution to the global activity has not been firmly established for any of these location types. *Vincent et al.* (2019) point out that generally, activity from close-to-vertical surfaces (having high gravitational slope, i.e., a significant angle between the local surface normal and the negative gravitational acceleration vector) avoids being quenched by a dust mantle, which is an unresolved obstacle to explaining activity from surfaces with low gravitational slopes.

It is further possible that the activity from "pristine" surfaces (cf. the chapter by Pajola et al. in this volume) differs from that originating from terrains that are covered in debris that fell back from the coma. This fallback material did not reach escape speed when accelerated by the gas and reimpacted at locations where the gas pressure was (at least seasonally) sufficiently low (*Thomas et al.,* 2015; *Pajola et al.,* 2017b). Under conditions of higher seasonal irradiation, this material can again be lifted. This effect has been evoked to explain the strong activity from the neck region Hapi on Comet 67P during autumn 2014 that was reported by, e.g., *Lin et al.* (2015), *Pajola et al.* (2019), and *Combi et al.* (2020). At this time, the comet's approach to the Sun near 3 au led to increased sublimation of water ice, and Hapi, located at high northern latitudes, was in local summer. A significant fraction of the dust cover removed from Hapi during this epoch was later resupplied during northern winter in 2015–2016 (*Cambianica et al.,* 2020).

A large-scale trend of enhanced activity from above-average bright and blueish surfaces has been attributed to enrichment of these surfaces with water ice (*Ciarniello et al.,* 2015; *Filacchione et al.,* 2016, 2020; *Fornasier et al.,* 2016). On a smaller scale, a direct link between bright, water-ice rich spots on the surface (e.g., *Pommerol et al.,* 2015; *Barucci et al.,* 2016) and local activity has not been established.

In addition to the diurnally repeating activity, many comets show sudden, short-lived events of increased dust emission that are often called "outbursts." Such irregular activity has been observed on a wide scale of magnitudes, ranging from small, local-scale dust plumes (e.g., *Agarwal et al.,* 2017) to global events easily detectable with Earth-based telescopes (e.g., *Lin et al.,* 2009). The processes triggering such events are not well understood, but a wide range of models have been proposed, and the Deep Impact and Rosetta missions have made it possible to study at which locations small outbursts occur.

Comet 67P has shown irregular activity during the whole comet phase of the Rosetta mission, from April 2014 (*Tubiana et al.,* 2015) to September 2016 (*Altwegg et al.,* 2017). The vast majority of these events were not detected from Earth, with the possible exception of one near-perihelion event (*Boehnhardt et al.,* 2016). It has been suggested that outbursts occurred mainly near morphological boundaries and cliffs (*Vincent et al.,* 2016a; *Fornasier et al.,* 2019b), and some were also observed near pits (*Tenishev et al.,* 2016) and circular features in the Imhotep region (*Knollenberg et al.,* 2016; *Agarwal et al.,* 2017; *Rinaldi et al.,* 2018). One event has been directly linked to the break-off of a cliff face (*Pajola et al.,* 2017a). Temporal concentrations of outburst events have been reported for early morning and local afternoon (*Vincent et al.,* 2016a), but outbursts have also been observed from the deep nightside (*Knollenberg et al.,* 2016; *Pajola et al.,* 2017a; *Rinaldi et al.,* 2019).

The relative contribution of irregular events to the total dust production rate is difficult to estimate because it depends both on a complete knowledge of their frequency and on the amount of material emitted globally and by outburst events. Estimates indicate that individual events contribute no more than a few percent to the global dust production (*Tenishev et al.,* 2016; *Lin et al.,* 2017), such that the major part of the dust would be released by the diurnally repeating activity.

3.1.3. Processes driving the dust lifting. Comets or their precursor planetesimals have resided for several billion years in the cold outer solar system: the Jupiter-family comets in the transneptunian disk at about 30 K, and the long-period and dynamically new comets in the even colder Oort cloud at the limits of the Sun's gravitational influence. When an object from one of these reservoirs enters on a trajectory through the region of the planets, the top layers of the surface get heated and begin to lose their volatile ices, such as initially CO, N_2 (*Delsemme,* 1982; *Läuter et al.,* 2019), then CO_2, and, finally, beginning at roughly 5 au from the Sun, H_2O. Inside 3 au, water ice sublimation likely becomes the dominant driver of activity, but the more-volatile ices like CO_2 keep playing an important role (e.g., *A'Hearn et al.,* 2011; *Combi et al.,* 2020).

The energy input from solar radiation is partially reradiated to space and partially absorbed by the porous surface material. The absorbed energy is conducted and radiated to greater depths, where it can cause phase changes in the ices and the release of gas. This gas percolates through porous material, recondenses on colder surfaces, or eventually escapes through the surface. Typically there is a positive radial temperature gradient: The upper layers are warmer than below. During dusk or in shadowed regions, the temperature gradient may be locally inverted, leading to recondensation of water in the dust mantle. This frost can help to start activity in the morning (*De Sanctis et al.,* 2015; *Fornasier et al.,* 2016; *Shi et al.,* 2018).

The global water production of most comets is much lower than expected from freely sublimating ice surfaces. This observation led *Kuehrt and Keller* (1994) to conclude that the refractory component in the nucleus must be sufficiently abundant to allow a refractory mantle, depleted from volatiles, to form on the surface and quench activity over wide areas. This model implies that the cohesive Van der Waals forces stabilizing the refractory material exceed the gas pressure at the surface by many orders of magnitude, such that the detachment of dust particles from this refractory matrix remains unexplained. *Kuehrt and Keller* (1994) suggest that a heterogeneous surface composition might lead to activity of a small fraction of the surface that would have to be considerably enriched in ice.

Laboratory experiments also demonstrated that insolation of a porous ice-dust mixture leads to sublimation of the ice, and that part of the gas and dust leave the surface (*Kölzer et al.,* 1995). The non-lifted part of the freed dust builds up a porous dust mantle of about 10 Pa failure stress (*Grün et al.,* 1993) that eventually quenches the gas emission. This may be similar on a comet.

Recent estimates of the cohesive strength of cometary material [a few pascals on meter scales (*Attree et al.,* 2018) to kilopascals inside a dust aggregate (*Hornung et al.,* 2016)] and of laboratory analog materials [4–20 kPa (*Gundlach et al.,* 2018)] are indeed larger than the sublimation pressure of water ice in the relevant temperature range (1 Pa at 210 K and 10 kPa at 310 K).

Gundlach et al. (2015) point out that overcoming cohesion becomes easier with the growing size of the particles constituting the cometary surface and predict that the size of ejected particles should grow with heliocentric distance. But their given size ranges do not agree well with observed cometary activity, and the model does not explain the activity of very distant comets such as C/2017 K2, which is thought to have its onset near the orbit of Neptune (*Jewitt et al.,* 2021).

Recent models have tried to overcome the cohesion problem by ascribing a hierarchical structure to the surface, where the material is organized in "pebbles" or aggregates that in turn consist of refractory and optionally H_2O and CO_2 ice particles (which in themselves may have a substructure and be composed of "grains"). On the other hand, the "pebbles" are clumped into "chunks" (*Gundlach et al.,* 2020; *Fulle et al.,* 2020). The ice content of the pebbles would decrease with time and increase with depth. The porosity of the architecture would prevent the formation of an impenetrable dust mantle but increase subsurface pressure, while the small number of contact points between pebbles would facilitate to overcome their cohesion. Alternative or additional processes to counteract interparticle cohesion could be related to thermal fracturing or electrostatic charging (*Jewitt et al.,* 2019a).

Reaching a consolidated understanding and consensus on the processes involved in triggering and maintaining the emission of dust from cometary surfaces remains one of the open topics of the field and is hampered by a lack of constraining data. Laboratory experiments addressing these questions, and the interior structure of comets, are presented in the chapters in this volume by Poch et al. and Guilbert-Lepoutre et al. respectively.

3.2. Acceleration Zone

The dust acceleration region in the general coma extends from the nucleus surface to roughly 10 nucleus radii. At that distance, molecule-dust collisions become rare, and the dust flow decouples from the gas (Fig. 1).

The brightness in the dust coma is often used as a proxy for the density of dust. Indeed, when we have an optically thin coma the reflectance, R, of dust in a given pixel at a certain wavelength, λ, and at a certain scattering angle, Φ, is

$$R(\lambda,\Phi) = \int_{a_{min}}^{a_{max}} n_{col}(a_d)\sigma_d(a_d)q_{eff}(a_d)\frac{p(a_d,\lambda,\Phi)}{4\pi}da_d \quad (24)$$

where the smallest and largest sizes are given by a_{min} and a_{max}, the dust column density along the line of sight is n_{col}, and the geometric dust cross-section is σ_d. The scattering efficiency, q_{eff}, and phase function, p, depend on the material properties of the dust. Equation (24) illustrates that the brightness of the dust cannot, in general, be taken as a proxy for the dust density. There might be a dense part of the coma with particles that have a very low scattering efficiency and thus a low reflectance compared to an area with a smaller number but highly efficiently scattering particles that appear bright.

To understand the structure of the dust coma in the acceleration zone, we will first discuss the radial outflow structure and then go into more detail about how three-dimensional jet-like structures become manifest in the acceleration region.

3.2.1. The extent of the acceleration region. To understand the radial structure of the dust coma, let us first consider a simplified coma where the dust is not accelerated but rather flows out radially from the surface with a constant speed. In this case, the local dust densities decrease with the inverse square of the distance, r. This is due to the fact that mass flux conservation ($n_d v_d A$ = const., where A is the surface area) through closed surfaces around the nucleus is maintained. If one thinks of these surfaces as spherical shells then their surface areas scale with $A \sim r^2$. Because the speed, v_d, is constant in this example, the number density, n_d, therefore needs to scale with $1/r^2$ for the flux to be constant.

The dust brightness measured by a remote observer is proportional to the column, not the local number, density. We can use the above behavior and find that the column density will scale as $n_{col} \sim 1/\rho$, where ρ is now the projected distance to the nucleus in the image plane, rather than r.

In other words, the column density, and by extension the brightness, R, multiplied by ρ is constant for free radial outflow of dust. This is also the basis for the commonly used quantity Afρ (*A'Hearn et al.,* 1984), which — in the case of free radial outflow — is independent of where in the coma it is being measured. The quantity Af stands for the product of albedo and filling factor (optical depth) and is therefore equivalent to R.

Here, being interested in the acceleration region, we want to study deviations from a constant value of the product Rρ. Deviation from a constant Rρ can be caused by a multitude of processes. Sublimation and fragmentation of particles can either increase or decrease Rρ with increasing ρ depending on whether the resulting particles are more- or less-efficient scatterers (Fig. 7). Deviations from a point source nucleus will decrease Rρ, optical depth effects will increase Rρ, and gravitationally bound particles can increase or decrease Rρ depending on the type of orbit they are on (Fig. 7). *Thomas and Keller* (1989) found that the near-nucleus environment of Comet 1P/Halley is dominated by optical depth effects. They observed a behavior similar to that shown in the bottom center panel of Fig. 7. Finally, the acceleration of particles will decrease Rρ (top right panel of Fig. 7) because as the speed of the particles increases, $n_d v_d$ decreases more rapidly than with the $1/r^2$ profile described above for free radial outflow.

We can therefore use Rρ to determine at which point the dust flow transitions from an accelerated to a free radial outflow, corresponding to the outer edge of the acceleration region. As this is an asymptotic process there is no hard boundary. Theoretical calculations (*Zakharov et al.,* 2018b) showed that the dust particles reach 90% of their terminal speed at around 6 nucleus radii. We would therefore stipulate that free radial outflow begins around 10 nucleus radii. This is in very good agreement with observations at and numerical simulations for 67P, where *Gerig et al.* (2018) found the transition to a constant Rρ at around 11 nucleus radii. *Finson and Probstein* (1968) derived an upper limit of 20 nucleus radii for the acceleration zone.

3.2.2. Terminal velocity and transition to the outer coma. As discussed in the previous section, the acceleration region extends to roughly 10 nucleus radii. This is the interface to the part of the coma where the dust motion is controlled by solar radiation pressure and gravity rather than nucleus gravity and gas drag (Fig. 1).

For numerical simulations, these two force regimes require different algorithmic approaches, such that, in practice, the two parts of the coma are most often treated separately. Strictly speaking, there is a transition region (Fig. 1) where nucleus gravity still plays a role (the Hill sphere) and large particles that do not reach escape speed will return to the surface.

In any case, it is possible to define a surface — not necessarily spherical — where the dust has reached terminal speed and decouples from the gas. This surface is the upper boundary of the acceleration region and provides the initial conditions to calculate the dust dynamics thereafter (section 3.3).

In situations where gas drag is the dominant cause of acceleration and where $v_g \gg v_d$, a simplified dependency of the terminal speed, v_{ej}, on particle size, a_d, and global gas

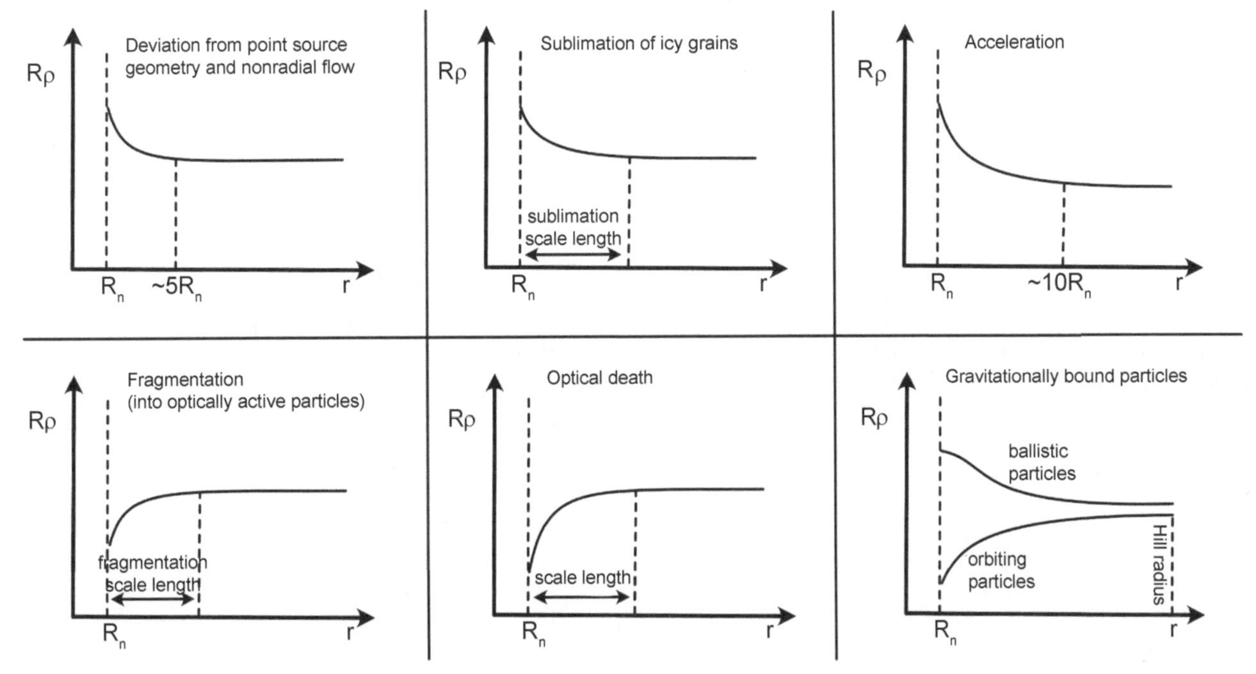

Fig. 7. Schematic of how different processes alter the behavior of the product Rρ of dust coma reflectance, R, and projected distance to nucleus center, ρ, as a function of the distance to the nucleus, r. Deviations from a point source and non-radial flow (top left) will reduce Rρ on a length scale of ~5 R_n, where R_n is the nucleus radius. The sublimation of particles (top center) will also reduce Rρ but the length scale of this effect will depend on the properties of the icy particles (their ice content, size), their ejection speed, and the heliocentric distance. The acceleration of particles (top right) will decrease Rρ and converge to a constant on a length scale of ~10 R_n. Both fragmentation (bottom left) and optical depth effects (bottom center) will increase Rρ. The scale on which these effects act depends on the details of the processes (i.e., the fragmentation rate). Finally, gravitationally bound particles (bottom right) will increase Rρ if they are on bound orbits while decreasing it when on ballistic trajectories. The scale for these gravitational effects is the Hill sphere. The figure was adapted from *Gerig et al.* (2018).

number production rate, Q_g, can be derived analytically. For the given assumptions, equation (8) simplifies to

$$\dot{v}_d \approx \frac{C_D m_g}{2} \frac{\sigma_d}{m_d} n_g v_g^2 \tag{25}$$

where m_d is the dust particle mass. Multiplying with v_d, assuming purely radial motion with nucleus center distance r, describing gas density as $n_g(r) = Q_g/[4\pi r^2 v_g(r)]$, and integrating from the surface ($r = R_n$, $v_d = 0$) to the decoupling distance ($r = r_{max}$, $v_d = v_{ej}$), yields

$$v_{ej}^2 = \frac{\sigma_d}{m_d} \frac{Q_g}{4\pi} m_g \int_{R_n}^{r_{max}} \frac{C_D v_g(r)}{r^2} dr \tag{26}$$

The quantities in the integral depend on the radial distribution of the gas speed and on gas and dust temperatures through C_D. The terminal speed is proportional to the square root of the cross-section-to-mass ratio, σ_d/m_d, and equivalently to $\sqrt{\beta}$ or — for size-independent density — to $a_d^{-1/2}$, and to the square root of the gas production rate, Q_g.

Figure 8a shows the terminal dust speeds as a function of dust size and gas production rate as obtained by numerical simulations. When the dust is much slower than the gas and the dynamics are dominated by gas drag only, then the dust speed scales with $a_d^{-1/2}$ and $Q_g^{1/2}$, consistent with equation (26). Deviations from this behavior are observed at very small and very large sizes. A small dust particle accelerates to almost the gas speed and asymptotically approaches it. The dynamics of large dust particles are significantly influenced by the nucleus gravity and thus their speeds are lower than predicted by equation (26). Above a certain size, the dust does not reach escape speed and will fall back to the surface. Even larger dust cannot be lifted.

Figure 8b shows the phase angle dependency of the dust speed. In the numerical simulations shown there is no nightside activity. Nevertheless, the gas flow and therefore also the dust flow is driven to the nightside. This lateral flow from the dayside to the nightside ensures that even in the absence of nightside activity dust particles reach significant speeds at large phase angles.

We would like to reemphasize that speeds in Fig. 8 are based on spherical dust particles with the same bulk density as the nucleus of 67P. If particles are significantly fluffier or rotating, their speeds can increase beyond the values shown in Fig. 8.

Dust in locally confined gas sources will attain slower speeds than if accelerated by a global gas field due to lateral gas expansion in the plume, for similar gas production rates per unit area (*Jewitt et al.,* 2014).

3.2.3. Jets and other three-dimensional structures in the acceleration region. Reducing the acceleration region to the radial expansion would be overly simplified. Observations from Comet 1P/Halley (e.g., *Keller et al.,* 1987) to Comet 103P/Hartley 2 (e.g., *A'Hearn et al.,* 2011), and Comet 67P (e.g., *Lin et al.,* 2015) have shown intricate dust filament structures.

Whether or not these filaments, also referred to as "jets" (see longer discussion in *Vincent et al.,* 2019), have clear source regions on the surface is a critical question. Three plausible mechanisms can result in the observed filaments:

1. jets with a clearly defined and confined gas and/or dust source on the surface;
2. topographically sculpted filaments that are products of the local topography shaping the dust emission through self-shadowing and/or the underlying convergence of the gas flow;
3. optical illusions, originating from a large area on the surface and appearing as narrow structures only from specific viewing geometries.

The first mechanism encompasses outbursts or exposed icy surfaces with enhanced activity compared to the background. Many outbursts have been spatially and temporally

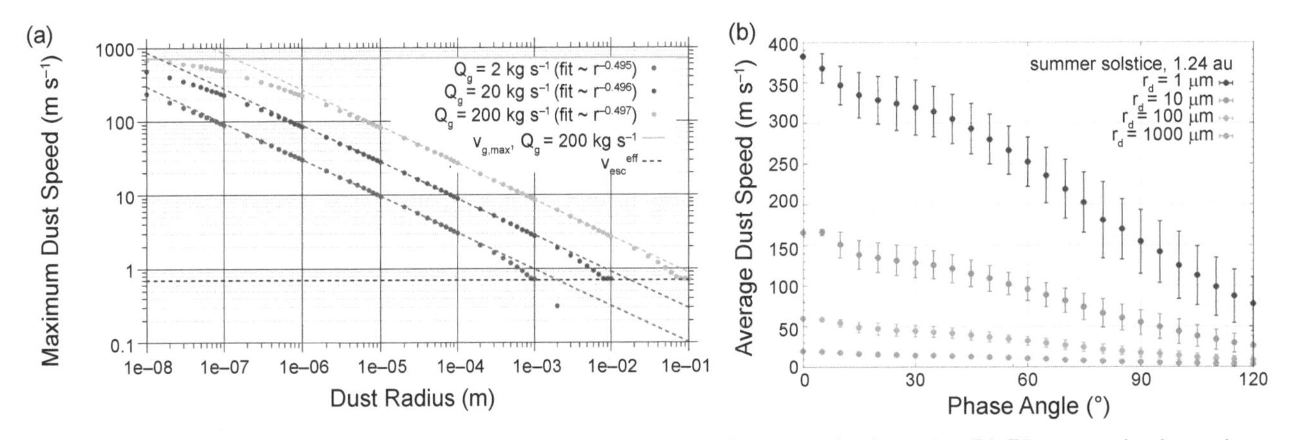

Fig. 8. **(a)** Maximum dust speed as a function of dust radius, and gas production rate. **(b)** Phase angle dependence of the dust speed. Both results shown have been calculated for Comet 67P and figures adaped from *Marschall* (2017). Speeds <2 m s^{-1} for larger-than-centimeter particles have also been reported from 103P/Hartley 2 by *A'Hearn et al.* (2011).

resolved at, e.g., 67P (e.g., *Vincent et al.,* 2016a; *Rinaldi et al.,* 2018) (see also section 3.1.2). They are characterized by a sudden increase of the dust emission, peaking after a few minutes, followed by a smooth decrease of the emission to the pre-outburst level. These events are among the few situations that can with high certainty be characterized as "jets" or "plumes" in the strictly physical sense (see also *Vincent et al.,* 2019).

The second mechanism is related to the irregular topography of the surface. *Crifo et al.* (2002, 2004) have pointed out that dust structures in the coma do not require sources on the surface. Non-spherical nucleus shapes are sufficient to produce such features dynamically. The uneven surface topography collimates the gas and hence the dust flows, resulting in higher-density regions within the coma. More complex nucleus shapes can produce more intricate structures in a total absence of localized sources. For 67P, this was illustrated in *Marschall et al.* (2016, 2017, 2019) and Fig. 9. This work would indicate that the dust filaments observed in the coma of Comet 67P do not have a source area in the traditional sense.

The third mechanism listed above is essentially a mirage. A prime example is the big fan-like structure originating from the northern neck of 67P. *Shi et al.* (2018) demonstrated that this feature likely originates from a sublimating frost front on the morning terminator deep in the neck of 67P. The outflowing dust particles produce a kind of fan that, seen from perpendicular to the plane of the fan, is indistinguishable from the surrounding coma. But with the line of sight inside this plane, the dust column densities in the projected fan are high in contrast to the surrounding coma. The resulting fine filament structure, however, is partly an optical illusion rather than a "jet" in the physical sense.

It appears that optical illusions and topographic sculpting are the rule rather than the exception to explain filament structures in the near-nucleus environment (*Shi et al.,* 2018; *Marschall et al.,* 2019; *Vincent et al.,* 2019). The current state-of-the art modeling suggests that no confined sources of these features are required. Rather, the much simpler assumption of a mostly homogeneous surface, the topography, and the viewing geometry are sufficient to explain the filamentary inner coma environment.

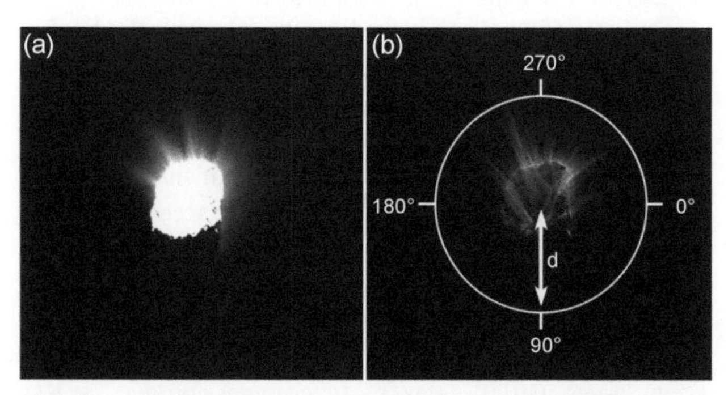

Fig. 9. **(a)** Stretched and cropped OSIRIS-WAC image of the coma of Comet 67P taken on May 5, 2015, 06:28:54 UTC. **(b)** Modeled dust brightness from *Marschall et al.* (2019) but the nucleus is not overexposed as in the actual OSIRIS image.

3.3. Outer Coma

The outer dust coma begins outside the acceleration zone and ends at the tail regime. These distances will vary by grain parameters, but for most active comets with grain radii $\lesssim$1 mm the outer coma spans from nuclear distances of a few nucleus radii to on the order of 10^4 km. The dominant forces acting on the outer coma are nucleus gravity, solar gravity, and radiation pressure. In the absence of grain fragmentation or outgassing, the fine-grained dust coma is typically in a radial outflow. Thus, the grain number density varies with r^{-2}, which produces the canonical ρ^{-1} coma in telescopic observations, where ρ is the projected distance to the nucleus (section 3.2.1). Within the Hill sphere, however, large particles may be bound to and orbiting the nucleus.

3.3.1. Coma size-β-speed relationship. The size of the dust coma depends on the physical properties of the grains (size, mass, and optical properties) and the ejection speeds imparted on the dust by gas pressure in the inner coma, but also from fragmentation processes or intrinsic outgassing, if relevant. In the simple case, solar radiation pressure is the primary non-gravitational force accelerating the dust grains. The acceleration is continuous, so there is no trivial delineation between the coma and tail regimes. One commonly adopted parameter is the apparent turn-back distance, X, which is the distance at which a grain ejected directly at the Sun will reach a speed of 0 in the rest frame of the nucleus. It is derived by integrating the equation of motion of a grain accounting for radiation pressure, ejection speed, and projection onto the sky

$$X = \frac{v_{ej}^2 \sin\theta}{2a_r} = \frac{r_h^2 v_{ej}^2 \sin\theta}{2GM_\odot \beta} = \frac{8\pi a_d \rho_d c r_h^2 v_{ej}^2 \sin\theta}{3Q_{pr}L_\odot} \tag{27}$$

where v_{ej} is the terminal speed of the dust grain after leaving the gas acceleration zone, a_r is the acceleration from radiation pressure [F_r/m_d; equation (13)], θ is the phase (Sun-comet-observer) angle, and β is defined by equation (15). Equation (27) is relevant for heliocentric distances $\gtrsim$4 $R_\odot$ (*Lamy,* 1974). For isotropic expansion, the turn-back distance traces a paraboloid on the sky (*Michel and Nishimura,* 1976).

Based on equation (27), we can estimate an order of magnitude coma upper-limit size. For v_{ej} = 200 m s^{-1}, a_d = 1 μm, ρ_d = 500 kg m^{-3}, and θ = 90° (i.e., no foreshortening due to phase angle), X = 10,000 km at 2 au from the Sun.

However, v_{ej} is a function of gas production rate and dust properties. In section 3.2.2, we showed that when the dynamics are dominated by gas drag, dust terminal speeds tend to scale with $a_d^{-1/2}$ for spherical grains of constant density. Under these assumptions, X is approximately independent of a, but scales with the constant that relates v_{ej}^2 to a_d^{-1}, or alternatively with Q_g.

The dependence of coma size on heliocentric distance is mainly given by the factor $r_h^2 Q_g(r_h)$, where there is no strong consensus regarding the shape of $Q_g(r_h)$. It is often approximated by a power law r_h^k at least in confined intervals of r_h. Calculating Q_g from the balance of sublimation, thermal radiation, and solar irradiation on a perpendicularly illuminated water ice surface gives $Q_g \propto r_h^{-2}$ inside 1 au (where radiation cooling is negligible compared to sublimation cooling). Beyond, the exponent, k, of a locally fitted power law transitions smoothly to k = –4 at 4 au and becomes even steeper beyond. However, beyond 2 au, the sublimation of more volatile species makes a relevant contribution to the total gas production rate, flattening its heliocentric profile. Observations suggest a much steeper than r_h^{-2} profile for water inside 3.5 au: For 67P, *Hansen et al.* (2016) find k = −5.3 before and k = −7.1 after perihelion, while *Marshall et al.* (2017) find k = −3.8 before and k = −4.3 after perihelion. Generally, the exponent k seems to be smaller than –2 in most situations, and hence coma size should rather grow with decreasing heliocentric distance.

The dependency of X on r_h is further — through v_{ej} — affected by the dependency of v_g on r_h. Various approximations exist for $v_g(r_h)$: $v_g \sim r_h^{-0.5}$ (*Tseng et al.*, 2007), $v_g \sim r_h^{-0.4}$ within r_h = 7 au and $v_g \neq v_g$ (r_h) beyond that distance (*Biver et al.*, 2002), while hydrodynamic model calculations by *Müller* (1999) suggest weak dependence of v_g on r_h.

Rather than deriving coma size, equation (27) or related considerations are often applied to the measured coma size in order to estimate coma grain properties from assumptions on velocity and/or acceleration. A few recent comets can serve as examples of the variety of conclusions that can be drawn for this analysis.

Jewitt et al. (2019a) measured a growth in the ρ^{-1} coma size of Comet C/2017 K2 (PanSTARRS) and used it to estimate a timescale of activity. Combining the magnitude of radiation pressure acceleration with this timescale, they estimated from the absence of a detectable tail that the optically dominant dust grains must have $\beta < 0.003$. The paucity of small grains in the coma suggested the influence of particle cohesion, which prevents their release and favors the production of large particles (*Gundlach et al.*, 2015).

Hsieh et al. (2004) estimated ejection speeds ≪45 m s^{-1}, based on the lack of a resolved coma in images of 133P/Elst-Pizarro and an assumption of small, β = 1 particles. *Mueller et al.* (2013) measured the growth of a narrow dust feature in images of Comet 103P/Hartley 2. They found that the source was most likely active for ≈ 22 hr, longer than the long-axis precession of the nucleus, which had consequences on their derived source location for the feature. Finally, *Kelley et al.* (2013) studied point-sources in the coma of Comet 103P/Hartley 2, and with equation (27) concluded that the dynamics of these ≳1-cm-sized particles were not governed by radiation pressure.

3.3.2. Large particles in the coma. Large particles or chunks of nucleus, i.e., centimeter-sized and larger, may be ejected from the nucleus or inner coma with very low speeds and potentially placed into suborbital trajectories. Under the influence of an additional force, the particles can be placed into bound orbits around the nucleus. The force may arise from gas outflow anisotropies in the inner coma, or from outgassing of the large particles themselves.

Evidence for centimeter-sized and larger particles may be found in cometary dust trails, meteor showers associated with comets, and observations of comets at submillimeter to centimeter wavelengths, including radar. See the chapter by Ye and Jenniskens in this volume for a review of cometary meteor showers, and *Harmon et al.* (2004) for a review of radar observations of cometary comae. Dust trails are addressed in section 3.4.6.

Observations of individual particles in the coma, including those in bound orbits, are a more recent phenomenon. *A'Hearn et al.* (2011) presented images from the Deep Impact spacecraft of the inner coma of Comet 103P/Hartley 2 containing thousands of point sources within a few kilometers from the nucleus. Such particles were not reported at 9P/Tempel 1. Depending on the light-scattering properties of the particles, they may be as large as meter-sized. The presence of such large nuclear chunks is a potential solution to the comet's hyperactivity. The large chunks provide additional sublimating surface area, which enhances the water production rate of the comet (*Kelley et al.*, 2013, 2015; *Belton*, 2017).

Point sources were also seen by the Rosetta spacecraft upon its approach to Comet 67P/Churyumov-Gerasimenko. *Rotundi et al.* (2015) estimated their sizes, assuming nucleus-like properties, and found the largest to be meter-sized. The particles are likely in bound orbits, and remnants from the comet's last perihelion passage. Particles seemed to fill the Hill sphere at the time of the observations (radius 318 km at 3.6 au from the Sun).

Outgassing of the large particles seems to be an important dynamical process, at least for those that are freshly ejected from the nucleus. *Kelley et al.* (2013) showed that Comet 103P/Hartley 2 had an asymmetry in its near-nucleus (≲10 km) population of large particles, and concluded that acceleration by outgassing best accounted for their distribution and their speeds as measured by *Hermalyn et al.* (2013). These particles were also likely responsible for Hartley 2's OH-tail observed by *Knight and Schleicher* (2013) and the tailward enhanced rotational temperature seen by *Bonev et al.* (2013).

Agarwal et al. (2016) observed the acceleration of decimeter-sized particles in the vicinity of the nucleus of Comet 67P/Churyumov-Gerasimenko. The acceleration was not strictly in the anti-sunward direction. They concluded that acceleration from outgassing and the ambient coma were the processes most likely to be responsible for the observed motions.

On larger spatial scales, particle outgassing may be less important. *Reach et al.* (2009) argued that on the basis of the width of the trail of Comet 73P/Schwassmann-Wachmann 3, trail particles <10 cm in radius are either ice-free after ejection from the nucleus or are quickly devolatilized.

A fraction of the particles or chunks in the centimeter-to-decimeter size class fall back to the surface where they form smooth layers of fallback material that was observed to cover wide regions of Comet 67P. *Marschall et al.* (2020c) estimate that between 11% and 22% of the debris mass initially lifted off the surface falls back. The fallback material likely still contains substantial amounts of water ice (*Davidsson et al.,* 2021).

3.3.3. Connecting spacecraft and remote observations of comae. If the regular dust structures in the acceleration region do not have local sources in the canonical sense (section 3.2.3), the question arises whether structures observed in the outer coma can be traced back to the surface or not. Generally, from telescope images of outer coma jet features alone, the physical processes causing these features cannot be inferred. But telescopic images have been used to infer the properties of cometary nuclei, including rotational state and number and distribution of active areas (*Vincent et al.,* 2019). These results often rely on the presence of distinct jet-like features in the data.

It has been demonstrated that one can reliably trace large-scale dust coma structures back to a virtual surface that would be the outer edge of the acceleration zone. It seems possible to expand this inversion down to the nucleus surface at the cost of increased spatial uncertainty on the source location, if assuming that the emission is on average perpendicular to the surface and that the emission vector measured at the edge of the acceleration zone is essentially a weighted average of all contributions. From a purely geometrical point of view, the angle between the measured emission vector and the nucleus north pole defines the effective co-latitude of the source on the surface (accounting for large-scale topography). This technique was successfully applied by *Vincent et al.* (2010, 2013) and *Farnham et al.* (2007, 2013) to infer the location of specific sources on the surfaces of Comets 9P, 103P, and 67P from groundbased observations alone. The inverted source locations were confirmed by *in situ* measurements from Deep Impact and Rosetta.

Hence, the connection between dust coma morphology in groundbased observations may be further investigated when *in situ* spacecraft observations are available for comparisons.

A rich set of dust features was observed in the coma of Comet Halley during its 1986 perihelion passage, including shells, arcs, and nearly linear features, which were also observed in previous apparitions (*Larson et al.,* 1987). *Sekanina and Larson* (1986) interpreted the nearly linear features in groundbased images as repeated discrete ejection events, due to their alignment with dust synchrones (cf. section 3.4.2). Images of the comet from the Giotto spacecraft (*Keller et al.,* 1987) show regions of strong activity from small, approximately kilometer-sized regions [nucleus dimensions are (7 × 7 × 15) km^3 (*Merenyl et al.,* 1990)]. The model of *Belton et al.* (1991), with five localized active areas on a nucleus with an excited spin state, successfully combined coma features seen in groundbased data with the inner-coma jets observed by the Giotto and Vega spacecraft.

The next inner coma and nucleus to be imaged by spacecraft was Comet 19P/Borrelly. The comet has a prominent asymmetry due to a jet-like feature in groundbased data (*Farnham and Cochran,* 2002). This feature is not aligned with the sunward or expected dust tail directions, indicating it is produced by directed emission rather than radiation pressure effects. Deep Space 1 images show that this jet-like feature is due to topography (*Soderblom et al.,* 2004). This long bi-lobed nucleus has a rotational axis perpendicular to the long axis of the nucleus, and the telescopically observed jet is due to dust released from its long flat polar region. A similar effect was seen from the flat southern polar region of 9P/Tempel 1 (*Farnham et al.,* 2007).

The inner coma of 81P/Wild 2, imaged by the Stardust spacecraft, presented many jet-like features and filaments (*Sekanina et al.,* 2004). Groundbased images have shown two prominent dust coma asymmetries: (1) a broad fan directed to the north of the orbital plane, mainly active pre-perihelion; and (2) a narrow jet-like feature, directed more sunward and to the south of the orbital plane, mainly active post-perihelion (*Sekanina,* 2003). The Stardust flyby was 98 days after perihelion, and source (1) was not active at the time. Source (2) should have been active, but *Farnham and Schleicher* (2005) could not single out any of the Stardust-observed filaments as candidates for its source. They concluded that source (2) might have been temporarily inactive due to the diurnal rotation of the nucleus or is the result of a combination of several filaments. This highlights one of the potential issues with remotely-observed dust features. As compared to gas, dust coma expansion speeds tend to be low, $\lesssim$100 m s^{-1}, and a broad range of ejection velocities may be imparted on the grains (e.g., Fig. 8). Thus material propagating outward from an active source on a rotating nucleus might trace out an arc or partial spiral for a unimodal speed distribution, but would be less pronounced, blurred, or lost altogether for a broad speed distribution (e.g., *Samarasinha,* 2000).

In addition to the broad southern feature discussed above, enhanced groundbased images of Comet 9P/Tempel 1 show other dust coma features (e.g., *Lara et al.,* 2006). *Vasundhara* (2009) used a spherical nucleus model and groundbased and Deep Impact data to derive four effective dust sources at the nucleus. *Vincent et al.* (2010) used a nucleus shape model based on Deep Impact flyby data to determine source locations for the features, finding six active regions in total with some compatibility with the *Vasundhara* (2009) results.

Deep Impact spacecraft data of Comet 103P/Hartley 2 show a strong active area located on the small end of this bi-lobed nucleus (*A'Hearn et al.,* 2011). Dust coma features in groundbased data were observed, e.g., by *Mueller et al.* (2013) and *Lin et al.* (2013). However, the nucleus is in an excited rotation state, which complicates connecting coma features seen remotely to the inner-coma features and the nucleus seen by Deep Impact. Regardless, *Mueller et al.* (2013) did compare images taken contemporaneously to the Deep Impact spacecraft's closest approach to the comet, and proposed that two secondary features in the coma originated from active areas found along the long axis of the nucleus and near the solar terminator at the time of the Deep Impact flyby.

In contrast to all previous cometary spacecraft missions, the Rosetta mission enabled a much broader comparison to groundbased data, owing to its long (2-yr) residence near the nucleus. Furthermore, the comet's apparent orbit-to-orbit stability in terms of activity (*Snodgrass et al.,* 2017) enables comparisons beyond the 2015/2016 apparition. *Vincent et al.* (2013) studied the morphology of groundbased images of the dust coma in the 2003 and 2009 apparitions and derived a pole orientation and the planetocentric locations of three active areas. Furthermore, *Knight et al.* (2017) found good agreement with their dust coma observations and the predictions of *Vincent et al.* (2013). They indicated one active area was the Hapi region (the "neck" of the bi-lobed nucleus), but concluded the other two were less obvious, with one possibly connected to the southern region, and the other to Imhotep (a flat smooth region on the largest lobe).

3.3.4. Dust size distributions. The size or mass distribution of dust describes the relative abundance of particles of different sizes or masses and is usually approximated by power laws for defined size (mass) intervals, which includes "broken" power laws that have different exponents for different size (mass) ranges. The employed power laws describe either the differential or the cumulative distributions. The conversion between their respective exponents and between size and mass distributions is given by the rules of differential calculus. For such conversions, usually the assumptions of size-independent bulk density and spherical particle shape are made (e.g., *Agarwal et al.,* 2007).

The exponent of the power law determines whether the largest or smallest particles in the concerned interval dominate the mass and scattering cross-section of the particle ensemble. This exponent can not only change with size, but also with a comet's position relative to perihelion (*Fulle et al.,* 2010, 2016; *Merouane et al.,* 2017). Different measurements, being sensitive to different types of particles, can yield different exponents as well (e.g., *Blum et al.,* 2017; *Rinaldi et al.,* 2017).

The size distribution of dust observed in the tail or outer coma will not generally correspond to size distribution of dust lifted from the surface (due to back-falling or orbiting particles), and even less to that of material resting on the surface [e.g., due to particles not liftable; equation (23)] or inside the deep interior. The size distribution of escaping dust may further be affected by fragmentation and sublimation of a potential volatile component.

The dust size distribution in a comet may carry information about its building blocks and formation process, provided that any post-formation changes to that distribution are understood and accounted for.

The unknown fraction of fallback material also complicates attempts to infer the refractory-to-ice ratio in a cometary interior, even when the masses of the escaping dust and gas are known, as is — at least integrated over a whole perihelion passage —the case for Comet 67P (*Choukroun et al.,* 2020).

It is possible that the dust size distribution, and especially the particle size containing most of the light-scattering cross-section, varies between comets. For some prominent, bright, long-period comets, the particles dominating the interaction with light could be micrometer-sized. Assumed indicators of this are the presence of a strong silicate emission feature at wavelengths near 10 μm and a high maximum degree of linear polarization of the scattered light (*Kolokolova et al.,* 2004), although similar characteristics are also expected from aggregates of (sub)micrometer grains. *Fulle* (2004) reports particularly a dominance of micrometer-sized particles in the scattering cross-section of long-period Comets Hyakutake and Hale-Bopp, but also in the active Centaur Chiron. Dust instruments flying by Comet 1P/Halley detected micrometer-sized dust with a size distribution that makes it the optically dominant component of Halley's dust (*McDonnell et al.,* 1989). Also, at Comets 81P/Wild 2 (*Green et al.* 2004) and 9P/Tempel 1 (*Economou et al.* 2013), micrometer-sized dust was detected by *in situ* instruments during flybys. Micrometer-sized particles were also among those returned from Comet Wild 2 to Earth by the Stardust spacecraft (*Brownlee,* 2014).

In other comets, and measured with other methods, the main scattering cross-section seems to rather be in (sub) millimeter-sized grains. One indicator is the absence of a prominent radiation-pressure swept tail in distant Comets C/2014 B1 (*Jewitt et al.,* 2019b), C/2017 K2 (*Jewitt et al.,* 2019a), and the interstellar Comet 2I/Borisov (*Kim et al.,* 2020). *Reach et al.* (2007) investigated thermal infrared emission from the debris trails of some 30, mainly Jupiter-family comets and found that in most of them, the amount of millimeter-sized dust required to explain the brightness of the trails was also sufficient to explain the scattering cross-section in the coma, such that not much scattering should have been contributed by potential additional micrometer-sized particles in the coma.

3.4. Tail, Trail, and Dispersion into the Zodiacal Cloud

3.4.1. Key parameters. The tail regime begins outside the Hill sphere (Fig. 1). The trajectory of a dust particle in the tail region and beyond is determined by the particle size [through the radiation pressure parameter β according to equation (15)] and the ejection terminal velocity v_{ej} (*Finson and Probstein,* 1968).

Most dynamical models use a simplified treatment of comet dust assuming $Q_{pr} = 1$. A more detailed treatment dependent on dust mineralogies and structures (compact or fluffy) is discussed in the chapter by Kolokolova et al. in this volume.

The generally applicable form for particle ejection by gas drag is given by *Whipple* (1950) (see also section 3.2.2)

$$v_{ej} \propto \beta^{1/2} \tag{28}$$

After a cometary dust particle has been ejected from a nucleus and left the dust-gas coupling region (Fig. 1), its motion is mainly controlled by solar gravity and radiation pressure. The size distribution of cometary dust has been inferred from both remote sensing images and *in situ* data. In general, it is assumed that the distribution of particle radii can be approximated by a power law so that the number of particles with radii ranging from a_d to $(a_d + da_d)$ is $n(a)\,da = \Gamma a^{-\alpha} da$. Table 2 provides a summary of α values reported in recent literature. The mean α values in Table 2 is $\alpha = 3.7 \pm 0.2$ and the median is $\alpha \sim 3.4$. Many comets in the past were characterized by $\alpha = 3.5$, typical of particles in collisional equilibrium (*Dohnanyi,* 1969).

3.4.2. General landscape of a dust tail. Assuming that $v_{ej} = 0$, the loci of particles of a given β and different ejection times are defined as a syndyne curve(*Finson and Probstein,* 1968), and the loci of particles with different β but the same ejection time are defined as a synchrone curve. A specific example of a synchrone-syndyne network is shown in Fig. 10. An online tool is available for generating synchrone-syndyne diagrams (*Vincent,* 2014) (see *https://comet.toolbox.com/FP.html*). This two-dimensional model is used for simple analysis of comet tail morphology and has been employed to determine the β range and ejection times of dust, including by sporadic emission events. In reality, v_{ej} is not zero. This leads to an expansion of the syndynamic tube whose width is given by the dust ejection velocity. *Fulle* (2004) pointed out that syndyne analyses tend to yield misleading β values, and thus a three-dimensional dynamical model is needed to consider non-zero ejection velocities.

Cometary dust tails, especially when dust sizes are small, can strongly interact with the solar wind, including the disruption or disconnection of the tail (e.g., *Jones et al.,* 2018) (see also section 2.6).

3.4.3. Tail dynamical modeling. Tail dynamical modeling is a useful technique to explore the properties and ejection processes of cometary dust, which are frequently underconstrained by observations. Several authors have developed various three-dimensional dynamical models to match the observed images (e.g., *Fulle,* 1989; *Lisse et al.,* 1998; *Reach et al.,* 2000; *Ishiguro et al.,* 2007; *Moreno,* 2009). The description of the underlying theory in *Fulle* (2004) is still valid. We present here a simplified model as a guide. For simplicity, we do not consider the heliocentric distance dependence of the ejection velocity and the dust production rate. The ejection terminal velocity of dust particles is given by

$$v_{ej} = V_0 \beta^{1/2} \tag{29}$$

where V_0 is the mean ejection speed of particles with $\beta = 1$. Particles are assumed to be released from the sunlit hemisphere of the nucleus. Assuming that the particle size follows a simple power law with an index α, the dust production rate for a given size and time can be expressed as

$$N(a_{\mu m}, t)\, da = N_0 a_{\mu m}^{-\alpha} da \tag{30}$$

in the size range of $a_{min} \le a_{\mu m} \le a_{max}$ (where $a_{\mu m}$ represents the particle radius in micrometers). The mean dust production rate of 1-µm particles, N_0, can be determined by comparison with the calibrated images. Model images are generated using Monte Carlo simulations by either solving Kepler's equations or using N-body integrators that include solar gravity and radiation pressure, substituting the gravitational constant G by $G(1-\beta)$ (*Chambers,* 1999; *Ye and Hui,* 2014). The cross-sectional area of dust particles in the charge-coupled-device (CCD) pixel coordinate system by integrating over time and particle size is given by

$$C(x,y) = \int_{t_0}^{t_1} \int_{a_{min}}^{a_{max}} N_{cal}(a_{\mu m}, t, x, y) \pi a_{\mu m}^2 da_{\mu m} dt \tag{31}$$

TABLE 2. Size distribution indexes.*

Comet	Method†	Radii (µm)	Index, α	Reference
1P/Halley	*In situ*	>20	3.5 ± 0.2	*Fulle et al.* (1995)
2P/Encke	Optical	>1	3.2 to 3.6	*Sarugaku et al.* (2015)
22P/Kopff	Optical	>1	3.1	*Moreno et al.* (2012)
26P/Grigg-Skjellerup	Optical	>60	3.3	*Fulle et al.* (1993)
67P/Churyumov-Gerasimenko (coma)	*In situ*	>0.01	$3.7^{+0.7}_{-0.1}$	*Marschall et al.* (2020b)
67P/Churyumov-Gerasimenko (trail)	Optical	>100	4.1	*Agarwal et al.* (2010)
81P/Wild 2	Optical	>1	3.45 ± 0.1	*Pozuelos et al.* (2014)
103P/Hartley 2	*In situ*	>10^4	4.7 to 6.6	*Kelley et al.* (2013)
103P/Hartley 2	Optical	>1	3.35 ± 0.1	*Pozuelos et al.* (2014)
209P/LINEAR	Optical	>1	3.25 ± 0.1	*Ishiguro et al.* (2015)

*Differential power-law size distribution index, α. Adapted from *Jewitt et al.* (2021).
†*In situ*: direct measurements from spacecraft in the coma. Optical: determination by fitting tail isophotes in remote sensing data.

Fig. 10. Synchrone (dashed lines)–syndyne (solid lines) network for Comet C/2011 L4 (PanSTARRS) on March 21, 2013. The short near-vertical lines are trailed background stars. Credit: L. Comolli; Model overlay: M. Fulle.

where $N_{cal}(a_{\mu m}, t, x, y)$ is the number of particles projected within a pixel of the CCD image. The modeled image is convolved with a Gaussian function whose full width at half maximum (FWHM) equals the FWHM of the seeing disk or pointspread function of the telescope image as applicable, and the resulting image is compared with the observed image to find a range of possible parameters and infer the underlying ejection process.

3.4.4. Comparison between different types of comets. There has been extensive modeling of dust emissions from different types of comets, such as Jupiter-family comets, long-period comets, main-belt comets, and interstellar objects. As in the outer coma regime (section 3.3), the main scattering cross-section in the tail regime seems to be in submillimeter and larger particles, which were the result of continuous emission that occurred several months to years prior to the observations. Below is a brief summary of the comparison of size, velocity, and dust production rates.

3.4.4.1. Main-belt comets: Main-belt comets are objects that show recurrent mass loss near perihelion yet orbit in the main asteroid belt and may be tracers of ice in the inner solar system (see the chapter by Jewitt and Hsieh in this volume). Their emission characteristics appear as continuous emission of (sub)millimeter-sized grains at <1 m s^{-1} speeds (similar to the nuclear escape speed) over weeks to months, with low dust production rates at <1 kg s^{-1}. Similar model parameters were found for several main-belt comets, indicating that these objects share similar properties of the ejected dust (*Hsieh et al.,* 2009; *Moreno et al.,* 2011; *Jewitt et al.,* 2019c; *Kim et al.,* 2022b,a).

3.4.4.2. Jupiter-family comets: Jupiter-family comets typically emit submillimeter or larger particles at speeds approximately several tens of meters per second [slightly lower than the classical *Whipple* (1950) model but still higher than that of main-belt comets] and dust production rates higher than a few hundred kilograms per second near perihelion, although there are variations between individual objects (*Ishiguro et al.,* 2007; *Kelley et al.,* 2008; *Moreno,* 2009; *Agarwal et al.,* 2010).

3.4.4.3. Long-period comets: Long-period comets show a more diverse distribution of dust parameters than short-period comets. The scattering cross-sectional areas of several long-period cometary comae are dominated by micrometer-sized particles (*Fulle,* 2004; *Lisse et al.,* 1998). However, recently observed distant comets show the absence of small particles, and their comae and tails are composed only of particles larger than a millimeter (*Jewitt et al.,* 2019a,b). Dust speed and dust production rates as function of particle size also show large variance.

3.4.4.4. Interstellar comets: An accurate determination of the orbit of 1I/'Oumuamua revealed the existence of non-gravitational acceleration, for which the most straightforward explanation would be comet-like outgassing (*Micheli et al.,* 2018). However, the outgassing required to supply the non-gravitational acceleration was predicted to be accompanied by a visible dust or gas coma, while 'Oumuamua was always observed as point-like. The morphology of 2I/Borisov is best reproduced by dust dynamical models if the coma is dominated by submillimeter and larger particles, emitted at $\lesssim 9$ m s^{-1} speeds, with total dust production rates estimated from imaging data 35 kg s^{-1} (*Kim et al.,* 2020; *Cremonese et al.,* 2020).

3.4.5. Neckline. The neckline is a substructure detected in dust tails on rare occasions and caused by dynamical effects (*Kimura and Liu,* 1977). A neckline consists of large particles emitted at a true anomaly of 180° before the observation (cf. Fig. 11) that were emitted with a non-zero velocity component perpendicular to the parent body's orbital plane. Their large sizes imply low β and low ejection speeds. Hence their orbits will overall be similar to that of the parent comet but inclined with respect to it due to the non-zero perpendicular velocity component. After initially dispersing in the perpendicular direction, particles ejected at a given time will reassemble in the orbital plane of the comet after 180° of true anomaly and be observable as a thin, bright line of dust. Using neckline photometry and Monte Carlo models, it is possible to determine if there were significant dust emissions at any given time. *Fulle et al.* (2004) identified Comet 67P/Churyumov-Gerasimenko as having a neckline structure and concluded that the comet has significant dust production at 3.6 au pre-perihelion. Taking advantage of the neckline effect, *Ishiguro et al.* (2016) succeeded in detecting the debris cloud ejected from the 2007 outburst of Comet 17P/Holmes.

3.4.6. Trails and dispersion into the zodiacal cloud. Comet debris trails consist of large particles that weakly interact with solar radiation pressure. They were first observed by the Infrared Astronomical Satellite (IRAS) (*Sykes et al.,* 1986), and subsequently in both visible and infrared light (*Ishiguro et al.,* 2002; *Reach et al.,* 2007; *Sykes et al.,* 2004; *Arendt,* 2014). Trails that intersect the Earth's orbit are observed as meteor showers. As of January 2022, there are 44 known comets with remotely observed debris trails whose details are available at the website (*https://www.astro.umd.edu/~msk/science/trails*) and 24 known comets with dust trails implied by meteor showers (see the chapter in this volume by Ye and Jenniskens). Recent models of cometary meteoroid streams show that comet trails can also be observed by *in situ* dust detectors (e.g., *Krüger et al.,* 2020).

Comet debris trails contribute a significant input to the interplanetary dust particle (IDP) cloud complex (*Sykes et al.,* 2004; *Reach et al.,* 2007). *Dikarev et al.* (2004) and *Soja et al.* (2019) quantified that the interplanetary dust cloud at 1 au is sustained mainly by Jupiter-family comets (~90%), with additional contributions by asteroids (~10%) and Halley-type comets (<1%). *Nesvorný et al.* (2010) connected the vertical brightness profile of the observed mid-infrared zodiacal light with that of a numerical model, suggesting that 90% of the zodiacal cloud emission came from comets. *Yang and Ishiguro* (2015) reached similar conclusions by comparing the observed optical properties of zodiacal light (i.e., albedo and optical spectral gradients) with those of other types of small bodies in the solar system. In contrast, a non-negligible fraction of asteroid particles in the IDP cloud is proposed by *Ipatov et al.* (2008) and *Kawara et al.* (2017).

4. FUTURE PERSPECTIVES

In this chapter, we have attempted to describe our current understanding of how dust is released from a cometary surface and transported to interplanetary space. The vast amount of data returned by spacecraft missions and modern telescopes over the past 20 years, since the publication of the *Comets II* book, has shed light on the complexity of this process and left us with a number of open questions that we outline in the following.

One question relates to how activity works at the surface level, i.e., how dust and ice are mixed in the cometary nucleus, and which processes lead to the lifting of dust particles, overcoming cohesive forces (section 3.1.3). We need a theoretical description of this process that is not in conflict with the observation that activity exists also and in particular far from the Sun. In our view, current theoretical efforts are limited by the quality of data that we have to constrain them. To characterize the surface composition, texture, and structure, highly resolved remote sensing data would be needed, especially at mid-infrared wavelengths, where the maximum of thermal emission in the inner solar system occurs, and in polarized light. *In situ* analyses of the surface from a landed laboratory would also greatly help to understand the physical and chemical properties of the surface, and finally, experiments with analog materials in an Earth- or space-station-based laboratory can provide good constraints as well (see the chapter by Poch et al. in this volume).

The second question addresses the extent to which results obtained from spacecraft missions can be generalized to the wider comet population. On the one hand, this can be addressed by comparing results from the different missions we have had until now, searching for similarities, differences, and repeating patterns. Since space missions are costly, however, the major bridge to the comet population in general will be achieved through telescope observations. To link telescope observations and space missions, we need to understand which properties of the early, near nucleus dust dynamics are still reflected in the outer coma and tail and hence accessible to telescopes. Indications are that much information on the details of the activity distribution is lost in the outer coma (*Crifo and Rodionov,* 1997; *Fulle,* 2004), but remote telescope observations do reveal brightness variations in the outer coma that have not yet been linked to features in the inner coma (*Knight et al.,* 2017). A possibility to establish and investigate this connection would be through dedicated modeling of the interface between inner coma and tail for those comets that have been visited by spacecraft. Modeling-wise, these two regions are typi-

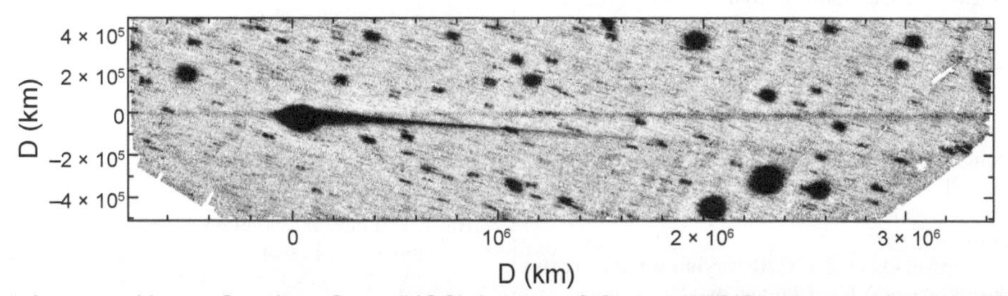

Fig. 11. Subaru telescope Hyper Suprime-Cam (HSC) image of Comet 67P/Churyumov-Gerasimenko (UT March 8, 2016). The trail (parallel to the horizontal axis) and the neckline structure are projected to different sky position angles. After *Moreno et al.* (2017).

cally treated separately according to the prevailing forces (gas drag vs. radiation pressure), but see section 3.3.3 for examples of connecting spacecraft and telescope observations of cometary comae.

A third complex of open questions concerns how dust evolves in its physical properties while it travels away from the nucleus. Potentially relevant processes include outgassing of embedded ice (affecting both dynamics and physical properties) and fragmentation, possibly induced by outgassing and/or fast rotation, and leading to a change in the dust size distribution. Data do not yet give clear evidence for or against any of these processes. High-resolution groundbased observations using complementary techniques such as visible light and thermal infrared spectroscopy and polarimetry could provide stronger constraints on the dust evolution, at least in the outer coma.

Acknowledgments. We thank V. Zakharov, D. Jewitt, X. Shi, F. Keiser, M. Pfeifer, and T. Kasuga, as well as the referees, E. Grün and J.-B. Vincent, for their comments that significantly helped us to improve this manuscript. J.A. and Y.K. acknowledge funding by the Volkswagen Foundation. J.A.'s contribution was made in the framework of project CAstRA funded by the European Union's Horizon 2020 research and innovation program under grant agreement No. 757390. M.S.P.K. acknowledges support from NASA Grant 80NSSC20K0673.

REFERENCES

Agarwal J., Müller M., and Grün E. (2007) Dust environment modelling of comet 67P/Churyumov-Gerasimenko. *Space Sci. Rev., 128,* 79–131.

Agarwal J., Müller M., Reach W. T., et al. (2010) The dust trail of Comet 67P/Churyumov-Gerasimenko between 2004 and 2006. *Icarus, 207,* 992–1012.

Agarwal J., A'Hearn M. F., Vincent J.-B., et al. (2016) Acceleration of individual, decimetre-sized aggregates in the lower coma of comet 67P/Churyumov-Gerasimenko. *Mon. Not. R. Astron. Soc., 462,* S78–S88.

Agarwal J., Della Corte V., Feldman P. D., et al. (2017) Evidence of sub-surface energy storage in comet 67P from the outburst of 2016 July 03. *Mon. Not. R. Astron. Soc., 469,* S606–S625.

A'Hearn M. F., Schleicher D. G., Millis R. L., et al. (1984) Comet Bowell 1980b. *Astron. J., 89,* 579–591.

A'Hearn M. F., Belton M. J. S., Delamere W. A., et al. (2011) EPOXI at comet Hartley 2. *Science, 332,* 1396–1400.

Altwegg K., Balsiger H., Berthelier J. J., et al. (2017) Organics in comet 67P — A first comparative analysis of mass spectra from ROSINA-DFMS, COSAC and Ptolemy. *Mon. Not. R. Astron. Soc., 469,* S130–S141.

Arendt R. G. (2014) DIRBE comet trails. *Astron. J., 148,* 135.

Attree N., Groussin O., Jorda L., et al. (2018) Tensile strength of 67P/Churyumov-Gerasimenko nucleus material from overhangs. *Astron. Astrophys., 611,* A33.

Attree N., Jorda L., Groussin O., et al. (2019) Constraining models of activity on comet 67P/Churyumov-Gerasimenko with Rosetta trajectory, rotation, and water production measurements. *Astron. Astrophys., 630,* A18.

Attree N., Jorda L., Groussin O., et al. (2023) On the activity distribution of comet 67P/Churyumov-Gerasimenko from combined measurements of non-gravitational forces and torques. *Astron. Astrophys., 670,* A170.

Attree N., Gutiérrez P., Groussin O., et al. (2024) Varying water activity and momentum transfer on comet 67P/Churyumov-Gerasimenko from its non-gravitational forces and torques. *Astron. Astrophys., 690,* A82.

Auer S., Kempf S., and Gruen E. (2007) Computed electric charges of grains with highly irregular shapes. In *Workshop on Dust in Planetary Systems* (H. Krueger and A. Graps, eds.), pp. 177–180. ESA SP-643, Noordwijk, The Netherlands.

Barucci M. A., Filacchione G., Fornasier S., et al. (2016) Detection of exposed H_2O ice on the nucleus of comet 67P/Churyumov-Gerasimenko as observed by Rosetta OSIRIS and VIRTIS instruments. *Astron. Astrophys., 595,* A102.

Belton M. J. S. (2017) Hyperactivity in 103P/Hartley 2: Chunks from the sub-surface in Type IIa jet regions. *Icarus, 285,* 58–67.

Belton M. J. S., Julian W. H., Anderson A. J., et al. (1991) The spin state and homogeneity of comet Halley's nucleus. *Icarus, 93,* 183–193.

Bentley M. S., Schmied R., Mannel T., et al. (2016) Aggregate dust particles at comet 67P/Churyumov-Gerasimenko. *Nature, 537,* 73–75.

Bird G. A. (1994) *Molecular Gas Dynamics and the Direct Simulation of Gas Flows.* Clarendon, Oxford. 484 pp.

Biver N., Bockelée-Morvan D., Colom P., et al. (2002) The 1995–2002 long-term monitoring of comet C/1995 O1 (Hale-Bopp) at radio wavelength. *Earth Moon Planets, 90,* 5–14.

Blum J., Gundlach B., Krause M., et al. (2017) Evidence for the formation of comet 67P/Churyumov-Gerasimenko through gravitational collapse of a bound clump of pebbles. *Mon. Not. R. Astron. Soc., 469,* S755–S773.

Bodewits D., Farnham T. L., Kelley M. S. P., et al. (2018) A rapid decrease in the rotation rate of comet 41P/Tuttle-Giacobini-Kresák. *Nature, 553,* 186–188.

Boehnhardt H., Riffeser A., Kluge M., et al. (2016) Mt. Wendelstein imaging of the post-perihelion dust coma of 67P/Churyumov-Gerasimenko in 2015/2016. *Mon. Not. R. Astron. Soc., 462,* S376–S393.

Bonev B. P., Villanueva G. L., Paganini L., et al. (2013) Evidence for two modes of water release in comet 103P/Hartley 2: Distributions of column density, rotational temperature, and ortho-para ratio. *Icarus, 222,* 740–751.

Brownlee D. (2014) The Stardust mission: Analyzing samples from the edge of the solar system. *Annu. Rev. Earth Planet. Sci., 42,* 179–205.

Burns J. A., Lamy P. L., and Soter S. (1979) Radiation forces on small particles in the solar system. *Icarus, 40,* 1–48.

Cambianica P., Fulle M., Cremonese G., et al. (2020) Time evolution of dust deposits in the Hapi region of comet 67P/Churyumov-Gerasimenko. *Astron. Astrophys., 636,* A91.

Chambers J. E. (1999) A hybrid symplectic integrator that permits close encounters between massive bodies. *Mon. Not. R. Astron. Soc., 304,* 793–799.

Choukroun M., Altwegg K., Kührt E., et al. (2020) Dust-to-gas and refractory-to-ice mass ratios of comet 67P/Churyumov-Gerasimenko from Rosetta observations. *Space Sci. Rev., 216,* 44.

Ciarniello M., Capaccioni F., Filacchione G., et al. (2015) Photometric properties of comet 67P/Churyumov-Gerasimenko from VIRTIS-M onboard Rosetta. *Astron. Astrophys., 583,* A31.

Colwell J. E., Gulbis A. A. S., Horányi M., et al. (2005) Dust transport in photoelectron layers and the formation of dust ponds on Eros. *Icarus, 175,* 159–169.

Combi M., Shou Y., Fougere N., et al. (2020) The surface distributions of the production of the major volatile species, H_2O, CO_2 CO and O_2 from the nucleus of comet 67P/Churyumov-Gerasimenko throughout the Rosetta mission as measured by the ROSINA double focusing mass spectrometer. *Icarus, 335,* 113421.

Cremonese G., Fulle M., Cambianica P., et al. (2020) Dust environment model of the interstellar comet 2I/Borisov. *Astrophys. J. Lett., 893,* L12.

Crifo J. F. and Rodionov A. V. (1997) The dependence of the circumnuclear coma structure on the properties of the nucleus. *Icarus, 129,* 72–93.

Crifo J. F., Lukianov G. A., Rodionov A. V., et al. (2002) Comparison between Navier-Stokes and direct Monte-Carlo simulations of the circumnuclear coma. I. Homogeneous, spherical source. *Icarus, 156,* 249–268.

Crifo J. F., Fulle M., Kömle N. I., et al. (2004) Nucleus-coma structural relationships: Lessons from physical models. In *Comets II* (M. C. Festou et al., eds.), pp. 471–503. Univ. of Arizona, Tucson.

Davidsson B. J. R., Birch S., Blake G. A., et al. (2021) Airfall on comet 67P/Churyumov-Gerasimenko. *Icarus, 354,* 114004.

Davidsson B. J. R., Samarasinha N. H., Farnocchia D., et al. (2022) Modelling the water and carbon dioxide production rates of comet 67P/Churyumov-Gerasimenko. *Mon. Not. R. Astron. Soc., 509,* 3065–3085.

Della Corte V., Rotundi A., Fulle M., et al. (2015) GIADA: Shining a light on the monitoring of the comet dust production from the nucleus of 67P/Churyumov-Gerasimenko. *Astron. Astrophys., 583,* A13.

Della Corte V., Rotundi A., Fulle M., et al. (2016) 67P/C-G inner coma dust properties from 2.2 au inbound to 2.0 au outbound to the Sun. *Mon. Not. R. Astron. Soc., 462,* S210–S219.

Delsemme A. H. (1982) Chemical composition of cometary nuclei. In *Comets* (L. Wilkening, ed.), pp. 85–130. Univ. of Arizona, Tucson.

De Sanctis M. C., Capaccioni F., Ciarniello M., et al. (2015) The diurnal cycle of water ice on comet 67P/Churyumov-Gerasimenko. *Nature, 525,* 500–503.

Dikarev V., Grün E., Baggaley J., et al. (2004) Modeling the sporadic meteoroid background cloud. *Earth Moon Planets, 95,* 109–122.

Dohnanyi J. S. (1969) Collisional model of asteroids and their debris. *J. Geophys. Res., 74,* 2531–2554.

Economou T. E., Green S. F., Brownlee D. E., et al. (2013) Dust Flux Monitor Instrument measurements during Stardust-NExT flyby of comet 9P/Tempel 1. *Icarus, 222,* 526–539.

El-Maarry M. R., Groussin O., Thomas N., et al. (2017) Surface changes on comet 67P/Churyumov-Gerasimenko suggest a more active past. *Science, 355,* 1392–1395.

Farnham T. L. and Cochran A. L. (2002) A McDonald Observatory study of comet 19P/Borrelly: Placing the Deep Space 1 observations into a broader context. *Icarus, 160,* 398–418.

Farnham T. L. and Schleicher D. G. (2005) Physical and compositional studies of comet 81P/Wild 2 at multiple apparitions. *Icarus, 173,* 533–558.

Farnham T. L., Wellnitz D. D., Hampton D. L., et al. (2007) Dust coma morphology in the Deep Impact images of comet 9P/Tempel 1. *Icarus, 187,* 26–40.

Farnham T. L., Bodewits D., Li J. Y., et al. (2013) Connections between the jet activity and surface features on comet 9P/Tempel 1. *Icarus, 222,* 540–549.

Filacchione G., Capaccioni F., Ciarniello M., et al. (2016) The global surface composition of 67P/CG nucleus by Rosetta/VIRTIS. (I) Prelanding mission phase. *Icarus, 274,* 334–349.

Filacchione G., Capaccioni F., Ciarniello M., et al. (2020) An orbital water-ice cycle on comet 67P from colour changes. *Nature, 578,* 49–52.

Fink U., Doose L., Rinaldi G., et al. (2016) Investigation into the disparate origin of CO_2 and H_2O outgassing for comet 67/P. *Icarus, 277,* 78–97.

Finson M. J. and Probstein R. F. (1968) A theory of dust comets. I. *Model and equations. Astrophys. J., 154,* 327–352.

Fornasier S., Mottola S., Keller H. U., et al. (2016) Rosetta's comet 67P/Churyumov-Gerasimenko sheds its dusty mantle to reveal its icy nature. *Science, 354,* 1566–1570.

Fornasier S., Feller C., Hasselmann P. H., et al. (2019a) Surface evolution of the Anhur region on comet 67P from high-resolution OSIRIS images. *Astron. Astrophys., 630,* A13.

Fornasier S., Hoang V. H., Hasselmann P. H., et al. (2019b) Linking surface morphology, composition, and activity on the nucleus of 67P/Churyumov-Gerasimenko. *Astron. Astrophys., 630,* A7.

Fougere N., Altwegg K., Berthelier J. J., et al. (2016a) Direct simulation Monte Carlo modelling of the major species in the coma of comet 67P/Churyumov-Gerasimenko. *Mon. Not. R. Astron. Soc., 462,* S156–S169.

Fougere N., Altwegg K., Berthelier J. J., et al. (2016b) Three-dimensional direct simulation Monte-Carlo modeling of the coma of comet 67P/Churyumov-Gerasimenko observed by the VIRTIS and ROSINA instruments on board Rosetta. *Astron. Astrophys., 588,* A134.

Fulle M. (1989) Evaluation of cometary dust parameters from numerical simulations — Comparison with an analytical approach and the role of anisotropic emissions. *Astron. Astrophys., 217,* 283–297.

Fulle M. (2004) Motion of cometary dust. In *Comets II* (M. C. Festou et al., eds.), pp. 565–575. Univ. of Arizona, Tucson.

Fulle M., Mennella V., Rotundi A., et al. (1993) The dust environment of comet P/Grigg-Skjellerup as evidenced from ground-based observations. *Astron. Astrophys., 276,* 582–588.

Fulle M., Colangeli L., Mennella V., et al. (1995) The sensitivity of the size distribution to the grain dynamics: Simulation of the dust flux measured by GIOTTO at P/Halley. *Astron. Astrophys., 304,* 622–630.

Fulle M., Barbieri C., Cremonese G., et al. (2004) The dust environment of comet 67P/Churyumov-Gerasimenko. *Astron. Astrophys., 422,* 357–368.

Fulle M., Colangeli L., Agarwal J., et al. (2010) Comet 67P/Churyumov-Gerasimenko: The GIADA dust environment model of the Rosetta mission target. *Astron. Astrophys., 522,* A63.

Fulle M., Della Corte V., Rotundi A., et al. (2015a) Density and charge of pristine fluffy particles from comet 67P/Churyumov-Gerasimenko. *Astrophys. J. Lett., 802,* L12.

Fulle M., Ivanovski S. L., Bertini I., et al. (2015b) Rotating dust particles in the coma of comet 67P/Churyumov-Gerasimenko. *Astron. Astrophys., 583,* A14.

Fulle M., Marzari F., Della Corte V., et al. (2016) Evolution of the dust size distribution of comet 67P/Churyumov-Gerasimenko from 2.2 au to perihelion. *Astrophys. J., 821,* 19.

Fulle M., Blum J., Rotundi A., et al. (2020) How comets work: Nucleus erosion versus dehydration. *Mon. Not. R. Astron. Soc., 493,* 4039–4044.

Gerig S. B., Marschall R., Thomas N., et al. (2018) On deviations from free-radial outflow in the inner coma of comet 67P/Churyumov-Gerasimenko. *Icarus, 311,* 1–22.

Gerig S. B., Pinzón-Rodríguez O., Marschall R., et al. (2020) Dayside-to-nightside dust coma brightness asymmetry and its implications for nightside activity at comet 67P/Churyumov-Gerasimenko. *Icarus, 351,* 113968.

Gombosi T. I. (1987) Dusty cometary atmospheres. *Adv. Space Res., 7,* 137–145.

Gombosi T. I., Cravens T. E., and Nagy A. F. (1985) Time-dependent dusty gasdynamical flow near cometary nuclei. *Astrophys. J., 293,* 328–341.

Gombosi T. I., Nagy A. F., and Cravens T. E. (1986) Dust and neutral gas modeling of the inner atmospheres of comets. *Rev. Geophys., 24,* 667–700.

Gombosi T. I., Burch J. L., and Horányi M. (2015) Negatively charged nano-grains at 67P/Churyumov-Gerasimenko. *Astron. Astrophys., 583,* A23.

Green S. F., McDonnell J. A. M., McBride N., et al. (2004) The dust mass distribution of comet 81P/Wild 2. *J. Geophys. Res.–Planets, 109,* E12S04.

Grün E., Gebhard J., Bar-Nun A., et al. (1993) Development of a dust mantle on the surface of an insolated ice-dust mixture: Results from the KOSI-9 experiment. *J. Geophys. Res.–Planets, 98,* 15091–15104.

Gundlach B., Blum J., Keller H. U., et al. (2015) What drives the dust activity of comet 67P/Churyumov-Gerasimenko? *Astron. Astrophys., 583,* A12.

Gundlach B., Schmidt K. P., Kreuzig C., et al. (2018) The tensile strength of ice and dust aggregates and its dependence on particle properties. *Mon. Not. R. Astron. Soc., 479,* 1273–1277.

Gundlach B., Fulle M., and Blum J. (2020) On the activity of comets: Understanding the gas and dust emission from comet 67/Churyumov-Gerasimenko's south-pole region during perihelion. *Mon. Not. R. Astron. Soc., 493,* 3690–3715.

Hansen K. C., Altwegg K., Berthelier J. J., et al. (2016) Evolution of water production of 67P/Churyumov-Gerasimenko: An empirical model and a multi-instrument study. *Mon. Not. R. Astron. Soc., 462,* S491–S506.

Harmon J. K., Nolan M. C., Ostro S. J., et al. (2004) Radar studies of comet nuclei and grain comae. In *Comets II* (M. C. Festou et al., eds.), pp. 265–279. Univ. of Arizona, Tucson.

Hartzell C., Zimmerman M., and Hergenrother C. (2022) An evaluation of electrostatic lofting and subsequent particle motion on Bennu. *Planet. Sci. J., 3,* 85.

Hermalyn B., Farnham T. L., Collins S. M., et al. (2013) The detection, localization, and dynamics of large icy particles surrounding comet 103P/Hartley 2. *Icarus, 222,* 625–633.

Hoang M., Altwegg K., Balsiger H., et al. (2017) The heterogeneous coma of comet 67P/Churyumov-Gerasimenko as seen by ROSINA: H_2O, CO_2, and CO from September 2014 to February 2016. *Astron. Astrophys., 600,* A77.

Hoang M., Garnier P., Gourlaouen H., et al. (2019) Two years with comet 67P/Churyumov-Gerasimenko: H_2O, CO_2, and CO as seen by the ROSINA/RTOF instrument of Rosetta. *Astron. Astrophys., 630,* A33.

Höfner S., Vincent J. B., Blum J., et al. (2017) Thermophysics of fractures on comet 67P/Churyumov-Gerasimenko. *Astron. Astrophys., 608,* A121.

Horányi M. (1996) Charged dust dynamics in the solar system. *Annu. Rev. Astron. Astrophys., 34,* 383–418.
Hornung K., Merouane S., Hilchenbach M., et al. (2016) A first assessment of the strength of cometary particles collected *in-situ* by the COSIMA instrument onboard ROSETTA. *Planet. Space Sci., 133,* 63–75.
Hsieh H. H., Jewitt D. C., and Fernández Y. R. (2004) The strange case of 133P/Elst-Pizarro: A comet among the asteroids. *Astron. J., 127,* 2997–3017.
Hsieh H. H., Jewitt D., and Ishiguro M. (2009) Physical properties of main-belt comet P/2005 U1 (Read). *Astron. J., 137,* 157–168.
Hui M.-T., Farnocchia D., and Micheli M. (2019) C/2010 U3 (Boattini): A bizarre comet active at record heliocentric distance. *Astron. J., 157,* 162.
Ipatov S. I., Kutyrev A. S., Madsen G. J., et al. (2008) Dynamical zodiacal cloud models constrained by high resolution spectroscopy of the zodiacal light. *Icarus, 194,* 769–788.
Ishiguro M., Watanabe J., Usui F., et al. (2002) First detection of an optical dust trail along the orbit of 22P/Kopff. *Astrophys. J. Lett., 572,* L117–L120.
Ishiguro M., Sarugaku Y., Ueno M., et al. (2007) Dark red debris from three short-period comets: 2P/Encke, 22P/Kopff, and 65P/Gunn. *Icarus, 189,* 169–183.
Ishiguro M., Kuroda D., Hanayama H., et al. (2015) Dust from comet 209P/LINEAR during its 2014 return: Parent body of a new meteor shower, the May Camelopardalids. *Astrophys. J. Lett., 798,* L34.
Ishiguro M., Sarugaku Y., Kuroda D., et al. (2016) Detection of remnant dust cloud associated with the 2007 outburst of 17P/Holmes. *Astrophys. J., 817,* 77.
Ivanovski S. L., Della Corte V., Rotundi A., et al. (2017a) Dynamics of non-spherical dust in the coma of 67P/Churyumov-Gerasimenko constrained by GIADA and ROSINA data. *Mon. Not. R. Astron. Soc., 469,* S774–S786.
Ivanovski S. L., Zakharov V. V., Della Corte V., et al. (2017b) Dynamics of aspherical dust grains in a cometary atmosphere: I. Axially symmetric grains in a spherically symmetric atmosphere. *Icarus, 282,* 333–350.
Jaumann R., Schmitz N., Ho T. M., et al. (2019) Images from the surface of asteroid Ryugu show rocks similar to carbonaceous chondrite meteorites. *Science, 365,* 817–820.
Jewitt D., Ishiguro M., Weaver H., et al. (2014) Hubble Space Telescope investigation of main-belt comet 133P/Elst-Pizarro. *Astron. J., 147,* 117.
Jewitt D., Mutchler M., Weaver H., et al. (2016) Fragmentation kinematics in comet 332P/Ikeya-Murakami. *Astrophys. J. Lett., 829,* L8.
Jewitt D., Agarwal J., Hui M.-T., et al. (2019a) Distant comet C/2017 K2 and the cohesion bottleneck. *Astron. J., 157,* 65.
Jewitt D., Kim Y., Luu J., et al. (2019b) The discus comet: C/2014 *B1 (Schwartz). Astron. J., 157,* 103.
Jewitt D., Kim Y., Rajagopal J., et al. (2019c) Active asteroid P/2017 S5 (ATLAS). *Astron. J., 157,* 54.
Jewitt D., Kim Y., Mutchler M., et al. (2021) Cometary activity begins at Kuiper belt distances: Evidence from C/2017 K2. *Astron. J., 161,* 188.
Jones G. H., Knight M. M., Battams K., et al. (2018) The science of sungrazers, sunskirters, and other near-Sun comets. *Space Sci. Rev., 214,* 20.
Kasuga T. and Jewitt D. (2019) Asteroid-meteoroid complexes. In *Meteoroids: Sources of Meteors on Earth and Beyond* (G. O. Ryabova et al., eds.), pp. 187–209. Cambridge Univ., Cambridge.
Kawara K., Matsuoka Y., Sano K., et al. (2017) Ultraviolet to optical diffuse sky emission as seen by the Hubble Space Telescope Faint Object Spectrograph. *Publ. Astron. Soc. Japan, 69,* 31.
Keller H. U., Arpigny C., Barbieri C., et al. (1986) First Halley Multicolour Camera imaging results from Giotto. *Nature, 321,* 320–326.
Keller H. U., Delamere W. A., Reitsema H. J., et al. (1987) Comet P/Halley's nucleus and its activity. *Astron. Astrophys., 187,* 807–823.
Keller H. U., Mottola S., Skorov Y., et al. (2015) The changing rotation period of comet 67P/Churyumov-Gerasimenko controlled by its activity. *Astron. Astrophys., 579,* L5.
Kelley M. S., Reach W. T., and Lien D. J. (2008) The dust trail of Comet 67P/Churyumov Gerasimenko. *Icarus, 193,* 572–587.
Kelley M. S., Lindler D. J., Bodewits D., et al. (2013) A distribution of large particles in the coma of comet 103P/Hartley 2. *Icarus, 222,* 634–652.
Kelley M. S. P., Lindler D. J., Bodewits D., et al. (2015) Erratum to "A distribution of large particles in the coma of comet 103P/Hartley 2" [Icarus 222 (2013) 634–652]. *Icarus, 262,* 187–189.
Kim Y., Jewitt D., Mutchler M., et al. (2020) Coma anisotropy and the rotation pole of interstellar comet 2I/Borisov. *Astrophys. J. Lett., 895,* L34.
Kim Y., Agarwal J., Jewitt D., et al. (2022a) Sublimation origin of active asteroid P/2018 P3. *Astron. Astrophys., 666,* A163.
Kim Y., Jewitt D., Agarwal J., et al. (2022b) Hubble Space Telescope observations of active asteroid P/2020 O1 (Lemmon-PANSTARRS). *Astrophys. J. Lett., 933,* L15.
Kimura H. and Liu C. (1977) On the structure of cometary dust tails. *Chinese Astron., 1,* 235–264.
Kimura H., Kolokolova L., Li A., et al. (2016) Light scattering and thermal emission by primitive dust particles in planetary systems. *ArXiV e-prints,* arXiv:1603.03123.
Knight M. M. and Schleicher D. G. (2013) The highly unusual outgassing of comet 103P/Hartley 2 from narrowband photometry and imaging of the coma. *Icarus, 222,* 691–706.
Knight M. M., Snodgrass C., Vincent J.-B., et al. (2017) Gemini and Lowell observations of 67P/Churyumov-Gerasimenko during the Rosetta mission. *Mon. Not. R. Astron. Soc., 469,* S661–S674.
Knollenberg J., Lin Z. Y., Hviid S. F., et al. (2016) A mini outburst from the nightside of comet 67P/Churyumov-Gerasimenko observed by the OSIRIS camera on Rosetta. *Astron. Astrophys., 596,* A89.
Kolokolova L. and Kimura H. (2010) Comet dust as a mixture of aggregates and solid particles: Model consistent with ground-based and space-mission results. *Earth Planets Space, 62,* 17–21.
Kolokolova L., Hanner M. S., Levasseur-Regourd A.-C., et al. (2004) Physical properties of cometary dust from light scattering and thermal emission. In *Comets II* (M. C. Festou et al., eds.), pp. 577–604. Univ. of Arizona, Tucson.
Kölzer G., Grün E., Kochan H., et al. (1995) Dust particle emission dynamics from insolated ice/dust mixtures: Results from the KOSI 5 experiment. *Planet. Space Sci., 43,* 391–395, 397–407.
Kramer E. A., Fernandez Y. R., Lisse C. M., et al. (2014) A dynamical analysis of the dust tail of comet C/1995 O1 (Hale-Bopp) at high heliocentric distances. *Icarus, 236,* 136–145.
Kramer T. and Läuter M. (2019) Outgassing-induced acceleration of comet 67P/Churyumov-Gerasimenko. *Astron. Astrophys., 630,* A4.
Kramer T., Läuter M., Rubin M., et al. (2017) Seasonal changes of the volatile density in the coma and on the surface of comet 67P/Churyumov-Gerasimenko. *Mon. Not. R. Astron. Soc., 469,* S20–S28.
Kramer T., Läuter M., Hviid S., et al. (2019*)* Comet 67P/Churyumov-Gerasimenko rotation changes derived from sublimation-induced torques. *Astron. Astrophys., 630,* A3.
Krüger H., Strub P., Sommer M., et al. (2020) Helios spacecraft data revisited: Detection of cometary meteoroid trails by following *in situ* dust impacts. *Astron. Astrophys., 643,* A96.
Kuehrt E. and Keller H. U. (1994) The formation of cometary surface crusts. *Icarus, 109,* 121–132.
Lai I. L., Ip W.-H., Lee J. C., et al. (2019) Seasonal variations in source regions of the dust jets on comet 67P/Churyumov-Gerasimenko. *Astron. Astrophys., 630,* A17.
Lamy P. L. (1974) Interaction of interplanetary dust grains with the solar radiation field. *Astron. Astrophys., 35,* 197–207.
Langevin Y., Hilchenbach M., Ligier N., et al. (2016) Typology of dust particles collected by the COSIMA mass spectrometer in the inner coma of 67P/Churyumov Gerasimenko. *Icarus, 271,* 76–97.
Lara L. M., Boehnhardt H., Gredel R., et al. (2006) Pre-impact monitoring of comet 9P/Tempel 1, the Deep Impact target. *Astron. Astrophys., 445,* 1151–1157.
Larson S., Sekanina Z., Levy D., et al. (1987) Comet P/Halley near nucleus phenomena in 1986. *Astron. Astrophys., 187,* 639–644.
Lauretta D. S., Dellagiustina D. N., Bennett C. A., et al. (2019) The unexpected surface of asteroid (101955) Bennu. *Nature, 568,* 55–60.
Läuter M., Kramer T., Rubin M., et al. (2019) Surface localization of gas sources on comet 67P/Churyumov-Gerasimenko based on DFMS/COPS data. *Mon. Not. R. Astron. Soc., 483,* 852–861.
Läuter M., Kramer T., Rubin M., et al. (2020) The gas production of 14 species from comet 67P/Churyumov-Gerasimenko based on DFMS/COPS data from 2014 to 2016. *Mon. Not. R. Astron. Soc., 498,* 3995–4004.
Lee P. (1996) Dust levitation on asteroids. *Icarus, 124,* 181–194.

Levasseur-Regourd A.-C., Agarwal J., Cottin H., et al. (2018) Cometary dust. *Space Sci. Rev., 214,* 64.

Li J.-Y., Kelley M. S. P., Samarasinha N. H., et al. (2017) The unusual apparition of comet 252P/2000 G1 (LINEAR) and comparison with comet P/2016 BA14 (PanSTARRS). *Astron. J., 154,* 136.

Lin Z.-Y., Lin C.-S., Ip W.-H., et al. (2009) The outburst of comet 17P/Holmes. *Astron. J., 138,* 625–632.

Lin Z.-Y., Lara L. M., and Ip W.-H. (2013) Long-term monitoring of comet 103P/Hartley 2. *Astron. J., 146,* 4.

Lin Z.-Y., Ip W.-H., Lai I. L., et al. (2015) Morphology and dynamics of the jets of comet 67P/Churyumov-Gerasimenko: Early-phase development. *Astron. Astrophys., 583,* A11.

Lin Z.-Y., Knollenberg J., Vincent J.-B., et al. (2017) Investigating the physical properties of outbursts on comet 67P/Churyumov-Gerasimenko. *Mon. Not. R. Astron. Soc., 469,* S731–S740.

Lisse C. M., A'Hearn M. F., Hauser M. G., et al. (1998) Infrared observations of comets by COBE. *Astrophys. J., 496,* 971–991.

Longobardo A., Della Corte V., Ivanovski S., et al. (2019) 67P/Churyumov-Gerasimenko active areas before perihelion identified by GIADA and VIRTIS data fusion. *Mon. Not. R. Astron. Soc., 483,* 2165–2176.

Longobardo A., Della Corte V., Rotundi A., et al. (2020) 67P/Churyumov-Gerasimenko's dust activity from pre- to post-perihelion as detected by Rosetta/GIADA. *Mon. Not. R. Astron. Soc., 496,* 125–137.

Mannel T., Bentley M. S., Schmied R., et al. (2016) Fractal cometary dust — A window into the early solar system. *Mon. Not. R. Astron. Soc., 462,* S304–S311.

Markkanen J. and Agarwal J. (2020) Thermophysical model for icy cometary dust particles. *Astron. Astrophys., 643,* A16.

Marschall R. (2017) Inner gas and dust comae of comets: Building a 3D simulation pipeline to understand multi-instrument results from the Rosetta mission to comet 67P/Churyumov-Gerasimenko. Ph.D. thesis, University of Bern, Bern. 256 pp.

Marschall R., Su C. C., Liao Y., et al. (2016) Modelling observations of the inner gas and dust coma of comet 67P/Churyumov-Gerasimenko using ROSINA/COPS and OSIRIS data: First results. *Astron. Astrophys., 589,* A90.

Marschall R., Mottola S., Su C. C., et al. (2017) Cliffs versus plains: Can ROSINA/COPS and OSIRIS data of comet 67P/Churyumov-Gerasimenko in autumn 2014 constrain inhomogeneous outgassing? *Astron. Astrophys., 605,* A112.

Marschall R., Rezac L., Kappel D., et al. (2019) A comparison of multiple Rosetta data sets and 3D model calculations of 67P/Churyumov-Gerasimenko coma around equinox (May 2015). *Icarus, 328,* 104–126.

Marschall R., Liao Y., Thomas N., et al. (2020a) Limitations in the determination of surface emission distributions on comets through modelling of observational data — A case study based on Rosetta observations. *Icarus, 346,* 113742.

Marschall R., Markkanen J., Gerig S.-B., et al. (2020b) The dust-to-gas ratio, size distribution, and dust fall-back fraction of comet 67P/Churyumov-Gerasimenko: Inferences from linking the optical and dynamical properties of the inner comae. *Front. Phys., 8,* 227.

Marschall R., Skorov Y., Zakharov V., et al. (2020c) Cometary comae-surface links. *Space Sci. Rev., 216,* 130.

Marsden B. G., Sekanina Z., and Yeomans D. K. (1973) Comets and nongravitational forces. V. *Astron. J., 78,* 211–225.

Marshall D. W., Hartogh P., Rezac L., et al. (2017) Spatially resolved evolution of the local H_2O production rates of comet 67P/Churyumov-Gerasimenko from the MIRO instrument on Rosetta. *Astron. Astrophys., 603,* A87.

McDonnell J. A. M., Green S. F., Grün E., et al. (1989) *In situ* exploration of the dusty coma of comet P/Halley at Giotto's encounter — Flux rates and time profiles from 10^{-19} *kg to* 10^{-5} *kg. Adv. Space Res., 9,* 277–280.

Merenyl E., Foldy L., Szego K., et al. (1990) The landscape of comet Halley. *Icarus, 86,* 9–20.

Merouane S., Stenzel O., Hilchenbach M., et al. (2017) Evolution of the physical properties of dust and cometary dust activity from 67P/Churyumov-Gerasimenko measured *in situ* by Rosetta/COSIMA. *Mon. Not. R. Astron. Soc., 469,* S459–S474.

Michel K. W. and Nishimura T. (1976) The dust coma of comets. In *Interplanetary Dust and Zodiacal Light* (H. Elsässer and F. Fechting, eds.), pp. 328–333. Springer, Heidelberg.

Micheli M., Farnocchia D., Meech K. J., et al. (2018) Non-gravitational acceleration in the trajectory of 1I/2017 U1 ('Oumuamua). *Nature, 559,* 223–226.

Moreno F. (2009) The dust environment of comet 29P/Schwassmann-Wachmann 1 from dust tail modeling of 2004 near-perihelion observations. *Astrophys. J. Suppl. Ser., 183,* 33–45.

Moreno F., Lara L. M., Licandro J., et al. (2011) The dust environment of main-belt comet P/2010 R2 (La Sagra). *Astrophys. J. Lett., 738,* L16.

Moreno F., Pozuelos F., Aceituno F., et al. (2012) Comet 22P/Kopff: Dust environment and grain ejection anisotropy from visible and infrared observations. *Astrophys. J., 752,* 136.

Moreno F., Muñoz O., Gutiérrez P. J., et al. (2017) The dust environment of comet 67P/Churyumov-Gerasimenko: Results from Monte Carlo dust tail modelling applied to a large ground-based observation data set. *Mon. Not. R. Astron. Soc., 469,* S186–S194.

Morfill G. E., Grün E., Goertz C. K., et al. (1983) On the evolution of Saturn's "spokes": Theory. *Icarus, 53,* 230–235.

Mueller B. E. A., Samarasinha N. H., Farnham T. L., et al. (2013) Analysis of the sunward continuum features of comet 103P/Hartley 2 from ground-based images. *Icarus, 222,* 799–807.

Mukai T. (1981) On the charge distribution of interplanetary grains. *Astron. Astrophys., 99,* 1–6.

Müller M. (1999) A model of the inner coma of comets with applications to the comets P/Wirtanen and P/Wild 2. Ph.D. thesis, Universität Heidelberg, Heidelberg.

Nesvorný D., Jenniskens P., Levison H. F., et al. (2010) Cometary origin of the zodiacal cloud and carbonaceous micrometeorites: Implications for hot debris disks. *Astrophys. J., 713,* 816–836.

Nichols K. D. and Scheeres D. J. (2022) Electrostatic lofting conditions for supercharged dust. *Astrophys. J., 931,* 122.

Nordheim T. A., Jones G. H., Halekas J. S., et al. (2015) Surface charging and electrostatic dust acceleration at the nucleus of comet 67P during periods of low activity. *Planet. Space Sci., 119,* 24–35.

Pajola M., Höfner S., Vincent J.-B., et al. (2017a) The pristine interior of comet 67P revealed by the combined Aswan outburst and cliff collapse. *Nature Astron., 1,* 0092.

Pajola M., Lucchetti A., Fulle M., et al. (2017b) The pebbles/boulders size distributions on Sais: Rosetta's final landing site on comet 67P/Churyumov-Gerasimenko. *Mon. Not. R. Astron. Soc., 469,* S636–S645.

Pajola M., Lee J. C., Oklay N., et al. (2019) Multidisciplinary analysis of the Hapi region located on comet 67P/Churyumov-Gerasimenko. *Mon. Not. R. Astron. Soc., 485,* 2139–2154.

Pommerol A., Thomas N., El-Maarry M. R., et al. (2015) OSIRIS observations of meter-sized exposures of H_2O ice at the surface of 67P/Churyumov-Gerasimenko and interpretation using laboratory experiments. *Astron. Astrophys., 583,* A25.

Pozuelos F. J., Moreno F., Aceituno F., et al. (2014) Dust environment and dynamical history of a sample of short-period comets. II. 81P/Wild 2 and 103P/Hartley 2. *Astron. Astrophys., 571,* A64.

Prialnik D. and Sierks H. (2017) A mechanism for comet surface collapse as observed by Rosetta on 67P/Churyumov-Gerasimenko. *Mon. Not. R. Astron. Soc., 469,* S217–S221.

Prialnik D., A'Hearn M. F., and Meech K. J. (2008) A mechanism for short-lived cometary outbursts at sunrise as observed by Deep Impact on 9P/Tempel 1. *Mon. Not. R. Astron. Soc., 388,* L20–L23.

Price O., Jones G. H., Morrill J., et al. (2019) Fine-scale structure in cometary dust tails I: Analysis of striae in comet C/2006 P1 (McNaught) through temporal mapping. *Icarus, 319,* 540–557.

Reach W. T., Sykes M. V., Lien D., et al. (2000) The formation of Encke meteoroids and dust trail. *Icarus, 148,* 80–94.

Reach W. T., Kelley M. S., and Sykes M. V. (2007) A survey of debris trails from short-period comets. *Icarus, 191,* 298–322.

Reach W. T., Vaubaillon J., Kelley M. S., et al. (2009) Distribution and properties of fragments and debris from the split comet 73P/Schwassmann-Wachmann 3 as revealed by Spitzer Space Telescope. *Icarus, 203,* 571–588.

Rennilson J. J. and Criswell D. R. (1974) Surveyor observations of lunar horizon-glow. *Moon, 10,* 121–142.

Rinaldi G., Della Corte V., Fulle M., et al. (2017) Cometary coma dust size distribution from *in situ* IR spectra. *Mon. Not. R. Astron. Soc., 469,* S598–S605.

Rinaldi G., Bockelée-Morvan D., Ciarniello M., et al. (2018) Summer outbursts in the coma of comet 67P/Churyumov-Gerasimenko as observed by Rosetta-VIRTIS. *Mon. Not. R. Astron. Soc., 481,*

1235–1250.
Rinaldi G., Formisano M., Kappel D., et al. (2019) Analysis of night-side dust activity on comet 67P observed by VIRTIS-M: A new method to constrain the thermal inertia on the surface. *Astron. Astrophys., 630,* A21.
Robertson H. P. (1937) Dynamical effects of radiation in the solar system. *Mon. Not. R. Astron. Soc., 97,* 423–437.
Rotundi A., Sierks H., Della Corte V., et al. (2015) Dust measurements in the coma of comet 67P/Churyumov-Gerasimenko inbound to the Sun. *Science, 347,* aaa3905.
Samarasinha N. H. (2000) The coma morphology due to an extended active region and the implications for the spin state of comet Hale-Bopp. *Astrophys. J. Lett., 529,* L107–L110.
Sánchez P. and Scheeres D. J. (2014) The strength of regolith and rubble pile asteroids. *Meteorit. Planet. Sci., 49,* 788–811.
Sarugaku Y., Ishiguro M., Ueno M., et al. (2015) Infrared and optical imagings of the comet 2P/Encke dust cloud in the 2003 return. *Astrophys. J., 804,* 127.
Schulz R., Hilchenbach M., Langevin Y., et al. (2015) Comet 67P/Churyumov-Gerasimenko sheds dust coat accumulated over the past four years. *Nature, 518,* 216–218.
Sekanina Z. (2003) A model for comet 81P/Wild 2. *J. Geophys. Res.–Planets, 108,* 8112.
Sekanina Z. and Farrell J. A. (1980) The striated dust tail of comet West 1976 VI as a particle fragmentation phenomenon. *Astron. J., 85,* 1538–1554.
Sekanina Z. and Larson S. M. (1986) Dust jets in comet Halley observed by Giotto and from the ground. *Nature, 321,* 357–361.
Sekanina Z., Brownlee D. E., Economou T. E., et al. (2004) Modeling the nucleus and jets of comet 81P/Wild 2 based on the Stardust encounter data. *Science, 304,* 1769–1774.
Sengers J. V., Lin Wang Y.-Y., Kamgar-Parsi B., et al. (2014) Kinetic theory of drag on objects in nearly free molecular flow. *Phys. A: Stat. Mech. Appl., 413,* 409–425.
Shi X., Hu X., Sierks H., et al. (2016) Sunset jets observed on comet 67P/Churyumov-Gerasimenko sustained by subsurface thermal lag. *Astron. Astrophys., 586,* A7.
Shi X., Hu X., Mottola S., et al. (2018) Coma morphology of comet 67P controlled by insolation over irregular nucleus. *Nature Astron., 2,* 562–567.
Shi X., Hu X., Agarwal J., et al. (2024) Diurnal ejection of boulder clusters on Comet 67P lasting beyond 3 au. *Astrophys. J. Lett., 961,* L16.
Shou Y., Combi M., Toth G., et al. (2017) A new 3D multi-fluid dust model: A study of the effects of activity and nucleus rotation on dust grain behavior at comet 67P/Churyumov-Gerasimenko. *Astrophys. J., 850,* 72.
Silsbee K. and Draine B. T. (2016) Radiation pressure on fluffy submicron-sized grains. *Astrophys. J., 818,* 133.
Skorov Y. and Blum J. (2012) Dust release and tensile strength of the non-volatile layer of cometary nuclei. *Icarus, 221,* 1–11.
Skorov Y., Reshetnyk V., Lacerda P., et al. (2016a) Acceleration of cometary dust near the nucleus: Application to 67P/Churyumov-Gerasimenko. *Mon. Not. R. Astron. Soc., 461,* 3410–3420.
Skorov Y. V., Rezac L., Hartogh P., et al. (2016b) A model of short-lived outbursts on the 67P from fractured terrains. *Astron. Astrophys., 593,* A76.
Skorov Y., Reshetnyk V., Rezac L., et al. (2018) Dynamical properties and acceleration of hierarchical dust in the vicinity of comet 67P/Churyumov-Gerasimenko. *Mon. Not. R. Astron. Soc., 477,* 4896–4907.
Snodgrass C., A'Hearn M. F., Aceituno F., et al. (2017) The 67P/Churyumov-Gerasimenko observation campaign in support of the Rosetta mission. *Philos. Trans. R. Soc. A, 375,* 20160249.
Soderblom L. A., Boice D. C., Britt D. T., et al. (2004) Imaging Borrelly. *Icarus, 167,* 4–15.
Soja R. H., Grün E., Strub P., et al. (2019) IMEM2: A meteoroid environment model for the inner solar system. *Astron. Astrophys., 628,* A109.
Steckloff J. K. and Jacobson S. A. (2016) The formation of striae within cometary dust tails by a sublimation-driven YORP-like effect. *Icarus, 264,* 160–171.
Sterken V. J., Altobelli N., Kempf S., et al. (2012) The flow of interstellar dust into the solar system. *Astron. Astrophys., 538,* A102.
Sykes M. V., Lebofsky L. A., Hunten D. M., et al. (1986) The discovery of dust trails in the orbits of periodic comets. *Science, 232,* 1115–1117.
Sykes M. V., Grün E., Reach W. T., et al. (2004) The interplanetary dust complex and comets. In *Comets II* (M. C. Festou et al., eds.), pp. 677–693. Univ. of Arizona, Tucson.
Tenishev V., Combi M. R., and Rubin M. (2011) Numerical simulation of dust in a cometary coma: Application to comet 67P/Churyumov-Gerasimenko. *Astron. Astrophys., 732,* 104.
Tenishev V., Fougere N., Borovikov D., et al. (2016) Analysis of the dust jet imaged by Rosetta VIRTIS-M in the coma of comet 67P/Churyumov-Gerasimenko on 2015 April 12. *Mon. Not. R. Astron. Soc., 462,* S370–S375.
Thomas N. and Keller H. U. (1989) The colour of comet P/Halley's nucleus and dust. *Astron. Astrophys., 213,* 487–494.
Thomas N., Davidsson B., El-Maarry M. R., et al. (2015) Redistribution of particles across the nucleus of comet 67P/Churyumov-Gerasimenko. *Astron. Astrophys., 583,* A17.
Thomas P., A'Hearn M., Belton M. J. S., et al. (2013) The nucleus of Comet 9P/Tempel 1: Shape and geology from two flybys. *Icarus, 222,* 453–466.
Tseng W. L., Bockelée-Morvan D., Crovisier J., et al. (2007) Cometary water expansion velocity from OH line shapes. *Astron. Astrophys., 467,* 729–735.
Tubiana C., Snodgrass C., Bertini I., et al. (2015) 67P/Churyumov-Gerasimenko: Activity between March and June 2014 as observed from Rosetta/OSIRIS. *Astron. Astrophys., 573,* A62.
Vasundhara R. (2009) Investigations of the pre-Deep Impact morphology of the dust coma of comet Tempel-1. *Icarus, 204,* 194–208.
Vincent J. (2014) Comet-toolbox: Numerical simulations of cometary dust tails in your browser. In *Asteroids, Comets, Meteors 2014* (K. Muinonen et al., eds.), p. 586. Univ. of Helsinki, Helsinki.
Vincent J.-B., Böhnhardt H., and Lara L. M. (2010) A numerical model of cometary dust coma structures: Application to comet 9P/Tempel 1. *Astron. Astrophys., 512,* A60.
Vincent J.-B., Lara L. M., Tozzi G. P., et al. (2013) Spin and activity of comet 67P/Churyumov-Gerasimenko. *Astron. Astrophys., 549,* A121.
Vincent J.-B., Bodewits D., Besse S., et al. (2015) Large heterogeneities in comet 67P as revealed by active pits from sinkhole collapse. *Nature, 523,* 63–66.
Vincent J.-B., A'Hearn M. F., Lin Z. Y., et al. (2016a) Summer fireworks on comet 67P. *Mon. Not. R. Astron. Soc., 462,* S184–S194.
Vincent J.-B., Oklay N., Pajola M., et al. (2016b) Are fractured cliffs the source of cometary dust jets? Insights from OSIRIS/Rosetta at 67P/Churyumov-Gerasimenko. *Astron. Astrophys., 587,* A14.
Vincent J.-B., Farnham T., Kührt E., et al. (2019) Local manifestations of cometary activity. *Space Sci. Rev., 215,* 30.
Wang X., Schwan J., Hsu H. W., et al. (2016) Dust charging and transport on airless planetary bodies. *Geophys. Res. Lett., 43,* 6103–6110.
Werner R. A. (1994) The gravitational potential of a homogeneous polyhedron or don't cut corners. *Cel. Mech. Dyn. Astron., 59,* 253–278.
Whipple F. L. (1950) A comet model. I. The acceleration of comet *Encke. Astrophys. J., 111,* 375–394.
Wyatt S. P. and Whipple F. L. (1950) The Poynting-Robertson effect on meteor orbits. *Astrophys. J., 111,* 134–141.
Yang H. and Ishiguro M. (2015) Origin of interplanetary dust through optical properties of zodiacal light. *Astrophys. J., 813,* 87.
Ye Q.-Z. and Hui M.-T. (2014) An early look of comet C/2013 *A1 (Siding Spring): Breathtaker or nightmare? Astrophys. J., 787,* 115.
Zakharov V. V., Crifo J. F., Rodionov A. V., et al. (2018a) The near-nucleus gas coma of comet 67P/Churyumov-Gerasimenko prior to the descent of the surface lander PHILAE. *Astron. Astrophys., 618,* A71.
Zakharov V. V., Ivanovski S. L., Crifo J. F., et al. (2018b) Asymptotics for spherical particle motion in a spherically expanding flow. *Icarus, 312,* 121–127.
Zimmerman M. I., Farrell W. M., Hartzell C. M., et al. (2016) Grain-scale supercharging and breakdown on airless regoliths. *J. Geophys. Res.–Planets, 121,* 2150–2165.

Part 8:

Interrelations

Marty B., Bermingham K. R., Nittler L. R., and Raymond S. N. (2024) Clues for solar system formation from meteorites and their parent bodies. In *Comets III* (K. J. Meech, M. R. Combi, D. Bockelée-Morvan, S. N. Raymond, and M. E. Zolensky, eds.), pp. 681–729. Univ. of Arizona, Tucson, DOI: 10.2458/azu_uapress_9780816553631-ch021.

Clues for Solar System Formation from Meteorites and Their Parent Bodies

Bernard Marty
Université de Lorraine, Centre National de la Recherche Scientifique, Centre de Recherches Pétrographiques et Géochimiques

Katherine R. Bermingham
Rutgers University

Larry R. Nittler
Carnegie Institution of Washington (now at Arizona State University)

Sean N. Raymond
Laboratoire d'Astrophysique de Bordeaux,. Université de Bordeaux, Centre National de la Recherche Scientifique

Understanding the origin of comets requires knowledge of how the solar system formed from a cloud of dust and gas 4.567 G.y. ago. Here, a review is presented of how the remnants of this formation process — meteorites and to a lesser extent comets — shed light on solar system evolution. The planets formed by a process of collisional agglomeration during the first hundred million years of solar system history. The vast majority of the original population of planetary building blocks (~100-km-scale planetesimals) was either incorporated into the planets or removed from the system, via dynamical ejection or through a collision with the Sun. Only a small fraction of the original rocky planetesimals survive to this day in the form of asteroids (which represent a total of ~0.05% of Earth's mass) and comets. Meteorites are fragments of asteroids that have fallen to Earth, thereby providing scientists with samples of solar-system-scale processes for laboratory-based analysis. Meteorite datasets complement cometary datasets, which are predominantly obtained via remote observation as there are few cometary samples currently available for laboratory-based measurements. This chapter discusses how analysis of the mineralogical, elemental, and isotopic characteristics of meteorites provides insight into (1) the origin of matter that formed planets; (2) the pressure, temperature, and chemical conditions that prevailed during planet formation; and (3) a precise chronological framework of planetary accretion. Also examined is the use of stable isotope variations and nucleosynthetic isotope anomalies as constraints on the dynamics of the disk and planet formation, and how these data are integrated into new models of solar system formation. It concludes with a discussion of Earth's accretion and its source of volatile elements, including water and organic species.

1. ORIGIN AND DIRECT OBSERVATIONS OF SMALL BODIES

Cosmochemistry — the science of meteorites — has opened a window into the origin of the solar system and the timeframe of its evolution. Outstanding advancements via laboratory-based analytical means now enable the exploration of the composition of primitive material down to the atomic scale. Coupled with astronomical observations and numerical modeling, cosmochemistry has provided a comprehensive framework for not only the inner solar system but also for the comet formation region for which data are scarce and mostly obtained remotely. Besides remote observation, the laboratory analysis of cometary material was restricted to that of a few grains recovered by the Stardust mission and to some dusty grains inferred to be of cometary origin (i.e., some of the so-called interplanetary dust particles). In contrast, meteorites are abundant and diverse in composition. Aside from waiting for a future comet sample return mission, meteorites are the best material available to explore the diversity of solar system material and its stellar precursors, the timing of disk formation and evolution, and the exchange of material between the outer and the inner regions of the disk.

This chapter provides an overview of cosmochemistry with emphasis on the composition and evolution of outer solar system bodies, the precursors of today's comets. We also discuss how cosmochemical data are integrated into models of solar system formation and evolution and provide constraints on the origin of volatile elements on inner planets and their potential relationship with comets. Our chapter complements others in this volume: the chapter on the Sun's birth environment by Bergin *et al.*, the chapter on the physical and chemical properties of protoplanetary disks by Aikawa *et al.*, the chapter on on planetesimal (and cometesimal) formation by Simon *et al.*, the chapter on how cometary reservoirs were sculpted by the dynamics of the early solar system by Kaib and Volk, the chapter on the asteroid-comet continuum by Jewitt and Hsieh, and the chapter on dust by Engrand et al.

1.1 Origin of Small Bodies

The solar system formed from a portion of a molecular cloud (*Adams,* 2010) that was composed of ~99% gas and ~1% dust. The fragment spun into a protoplanetary disk surrounding a nascent star, the proto-Sun (*Turner et al.,* 2014). Within this disk, grains aggregated into increasingly large planetary bodies (*Hayashi,* 1981) (Fig. 1). The demographics of planet-forming disks in stellar clusters of different ages suggest that gaseous disks typically last for a few million years before evaporating (*Haisch et al.,* 2001; *Mamajek et al.,* 2009). While the detailed physical structure and evolutionary pathway of the disk remain debated, models agree on the central steps in the process. The disk around the nascent Sun underwent an episode of viscous spreading to larger orbital radii, reaching a radial extent of at least ~50 au, judging from the outermost planetesimals thought to have formed locally in the cold classical Kuiper belt. From that point onward, the gaseous disk evolved in a relatively quiescent manner. Current thinking is that the disk slowly dissipated by draining onto the central star, while a small portion of the outer disk expanded to conserve angular momentum. The mechanisms of angular momentum transfer remain a subject of active research (e.g., *Armitage,* 2011; *Turner et al.,* 2014). The final phases of evolution were likely dictated by photoevaporation driven by energetic photons (X-rays and UV) from the Sun, with a potential contribution from nearby massive stars (*Ercolano and Pascucci,* 2017). The inner parts of the disk were evaporated first, and the inner edge of the disk swept outward in a final swan song. Here, the inner disk refers to the region inboard of where the gas giants formed, and the outer disk refers to the region outboard of the gas giants. Due to the establishment of a thermal gradient toward the Sun, water ice could not be stable close to the central star, defining a so-called snow line beyond which ice survived. Likewise, organic matter would tend to disintegrate close to the Sun, defining a limit labeled the tar line. Each of the major volatile species (e.g., CO_2) admits a line/distance from the Sun marking a phase change from solid to gaseous.

Material that did not accrete to the Sun was either ejected into interstellar space or formed planetesimals. Some of these planetesimals grew into planets whereas others remained

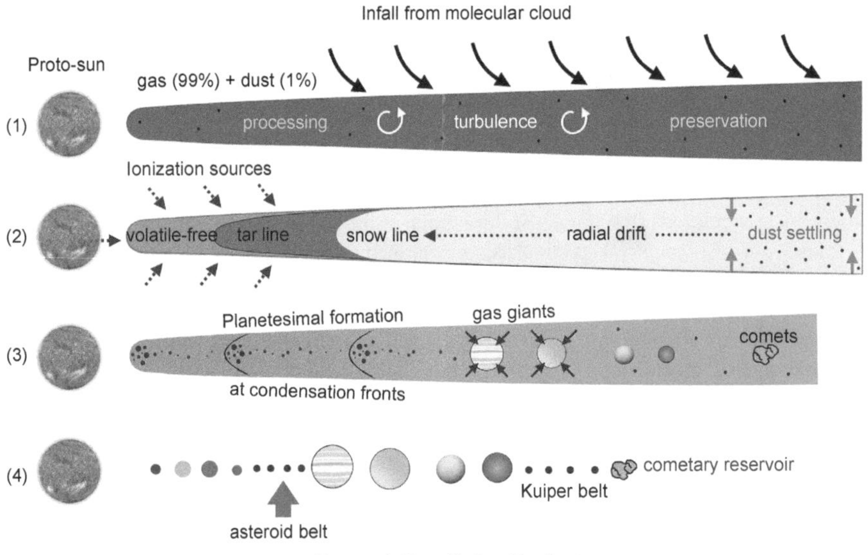

Fig. 1. Schematic evolution of the solar system. (1) A protoplanetary disk formed around a central protostar from the collapse of a molecular cloud core made of gas and dust ~4.56 Ga. (2) The protoplanetary disk was chemically and thermodynamically zoned, with an inner region close to the Sun where volatiles could not condense, a tar line where refractory organic material could be stable, and a snow line beyond which water could condense. This temperature gradient determined the composition of volatiles in small planetary bodies. Those bodies that accreted between the tar line and the snow line did not accrete substantive water ice, whereas those that formed beyond the snow line accreted water ice. (3) Small planetary bodies evolved to yield planetesimals and planetary embryos. Gas giants formed from the accretion of planetesimals and from capture of the protosolar nebula gas. In the inner part of the solar system (the region inboard of where the gas giants formed), planets grew essentially dry as thermodynamic conditions and/or timing (i.e., dissipation of the gas) did not permit gravitational trapping of nebular gas. (4) Meteorites originate from the main asteroid belt located between Mars and Jupiter. Adapted from *Broadley et al.* (2022).

as small bodies (asteroids, Trojans, comets), and these thus preserve a record of the epoch of solar system formation. The surviving planetesimals in the asteroid belt are the primary source of Earth's meteorites.

From the analysis of meteorites, it was realized that small bodies can be divided into two groups: "primitive" or "undifferentiated" bodies and "differentiated" bodies (*Weisberg et al.,* 2006) (Fig. 2). Both groups variably comprised the building blocks of the planets. Primitive bodies, samples of which are also known as chondrites, are those that did not undergo wholescale melting (i.e., differentiation) and thus preserve a mixture of early formed solar system solids. These meteorites provide a unique opportunity to document the first stages of solar system evolution through the analysis of their components. Comets, for which few samples exist, are examples of primitive bodies that formed in the outer bounds of the solar system. Differentiated bodies, samples of which are known as achondrites, are small bodies that underwent wholescale melting at high temperature, which resulted in segregation of the body into a silicate-rich crust and mantle and a metal-rich core. Although these planetary bodies have lost their component's original petrologic characteristics, they provide insight into the earliest stages of planet differentiation.

The highest taxonomic division in meteorite classification defines material as either "carbonaceous chondrite" (CC) or "non-carbonaceous" (NC) (*Warren,* 2011) (Fig. 2). Advancing earlier work by *Trinquier et al.* (2007, 2009), *Warren* (2011) compared the isotopic composition of bulk meteorites and found that all meteorites consistently fell into one of two distinct groups, NC or CC (see section 2.6.3.2). This bimodal grouping has become known as the "NC-CC isotopic dichotomy." The NC-CC notation describes the isotopic or "genetic" character of a parent body regardless of how the meteorite is classified by its mineralogy and petrogenesis. With the addition of several new chondrites and achondrites, it is now realized that meteorites, almost without exception, can be classified as either CC or NC.

From chronological studies of chondrites and achondrites, small bodies were accreted within a few (≤5) million years of the birth of the solar system, which is marked at time zero or t_0. This attests to the rapidity at which a stellar system with its cortege of planets can form (*Morbidelli et al.,* 2012), where such information can only be gained from laboratory-based meteorite analysis. Other primitive materials include comets and outer solar system objects. The lack of ^{26}Al (a useful short-lived chronometer that is described in section 2.6.1) in cometary grains returned by the Stardust mission, and the possible genetic link between CR chondrites and outer solar system objects, suggest a protracted formation of related bodies, but such evidence is extremely weak. Dating cometary matter, however, is of outmost importance and will be a key priority of future cometary missions.

1.2. Knowledge of Small Bodies from Observations

1.2.1. Telescopic and spacecraft observations. Telescopic and spacecraft observations of the composition and structure of small bodies provide the most informative approach to determining the diversity of bodies that formed

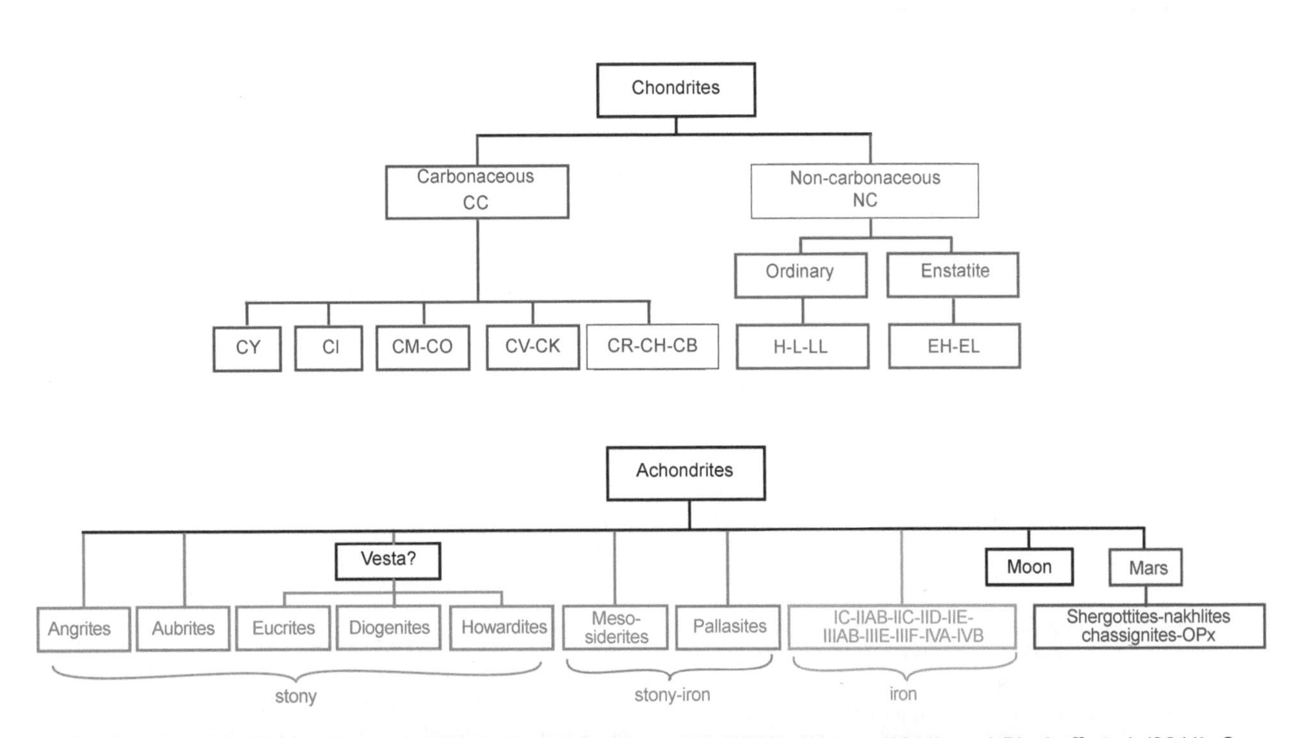

Fig. 2. Classification of meteorites, simplified after *Weissberg et al.* (2006), *Warren* (2011), and *Bischoff et al.* (2011). Some of the meteorite groups (e.g., R-chondrites) are not represented. The classification between carbonaceous chondrites (CC) and non-carbonaceous chondrites (NC) is given for primitive meteorites only (see section 2.3.). Almost all the achondrites can be genetically linked to NC or CC (see *Bermingham et al.,* 2018a; *Spitzer et al.,* 2021; and references therein).

from the protoplanetary disk. The 1801 discovery of 1 Ceres by Giuseppe Piazzi ushered in a string of telescopic discoveries of other small planetary bodies orbiting beyond Mars, which were named "asteroids" by William Herschel in 1802. More than a million asteroids have now been identified, many through large surveys designed to track potential impact hazards. Most orbit the Sun between Mars and Jupiter in the main asteroid belt, but a small fraction have been gravitationally perturbed into orbits that come much closer to the Sun. Those whose perihelion distance is less than 1.3 au are termed "near-Earth" objects. In addition, some asteroids, called "Trojans," share Jupiter's orbit, either leading or trailing the largest planet by 60°.

Telescopic and spacecraft observations of asteroids have revealed a wide diversity of colors, albedos, and spectroscopic features reflecting the variety of parent body compositions in the solar system. The classification scheme derived is based on distinct spectroscopic types of parent body. Notably, the different asteroid types are not uniformly distributed in the asteroid belt (Fig. 3). For example, E- and S-types dominate the inner asteroid belt, while C- and D-types dominate further out. S-types show spectral features of silicate minerals olivine and pyroxene and have been long suggested to be related to ordinary chondrites, the most common type of meteorite found on Earth (Fig. 2). This link has been proven by the Hayabusa mission that sampled a S-type asteroid and returned to Earth ordinary chondrite — type material. In contrast, C- and D-type asteroids are darker, spectrally flat, and thought to be related to carbonaceous chondrite meteorites. Such a compositional gradient likely reflects initial gradients in the protoplanetary disk, but there is clear evidence for substantial mixing of different compositional types throughout the asteroid belt (DeMeo and Carry, 2014).

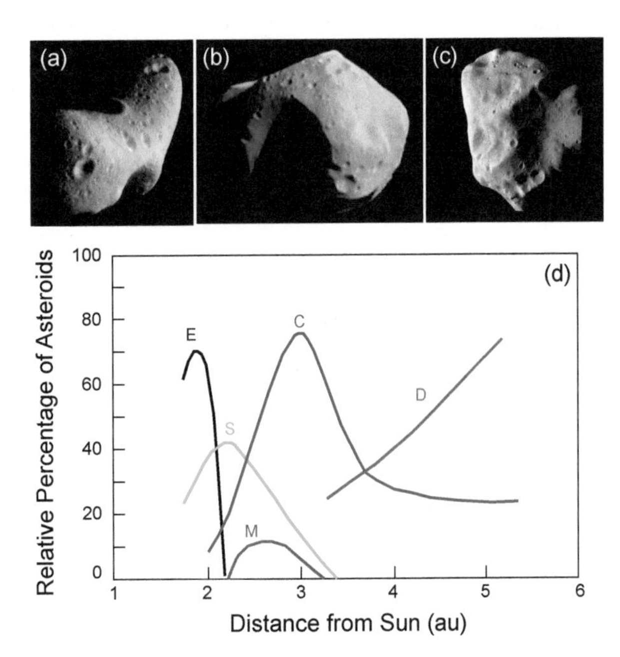

Fig. 3. Asteroid classification and distribution. **(a)** S-type Eros (image from NEAR-Shoemaker, NASA), **(b)** C-type Mathilde (NEAR-Shoemaker, NASA), **(c)** M-type Lutetia (Rosetta, ESA); **(d)** schematic representation of distribution of major asteroid types (after *Gradie and Tedesco,* 1982). The symbols refer to different asteroid spectral types; C-types dominate the mass budget of the belt (*DeMeo and Carry,* 2013).

1.2.2. Sample return missions. Several space missions have returned rocks from planetary bodies, yielding exceptional insight into their nature and their evolution (see the chapter by Snodgrass et al. in this volume). The return of samples from the Moon during the Luna (Russia) and Apollo (USA) missions, although not the main objective of these missions — which was geopolitical — proved the value of sample-based laboratory analysis. Three decades later, NASA's Genesis mission sampled solar wind ions implanted into pure target materials over a period of 27 months and returned its precious cargo in 2004 (*Burnett and the Genesis Science Team,* 2011). Analysis of these samples resolved two of the most pressing problems in cosmochemistry: the triple isotope O (^{16}O, ^{17}O, ^{18}O) composition of the Sun (*McKeegan et al.,* 2011) and the N isotope variability in the solar system (*Marty et al.,* 2011).

The era of small planetary body sampling came with NASA's Stardust mission (Fig. 4), which sampled grains from the coma of Comet 81P/Wild 2 and returned them to Earth in January 2006. The analysis of the cometary samples revealed the presence of high-temperature phases (metal, chondrules, CAIs, olivine, pyroxene) such as those found in primitive meteorites. For the first time, this demonstrated the exchange of matter at the protoplanetary disk scale between the inner and outer regions of the solar system (*Brownlee et al.,* 2006).

The first asteroidal sample return came with the Hayabusa mission from the Japan Aerospace Exploration Agency (JAXA). This technologically complex mission was only half-successful due to engine and sampling problems; however, a few micrograms of dust from S-class asteroid 25143 Itokawa was returned to Earth. A total of 1543 particles, ranging in size from 3 to 40 μm, have been analyzed so far. The sample analysis confirmed that the chemical composition and structure of the particles were similar to those of the most common meteorites on Earth, ordinary chondrites (see section 2.3), which originate from S-type asteroids such as Itokawa (Fig. 4) (*Kawaguchi et al.,* 2008).

The overall success of the Hayabusa mission paved the way for Hayabusa2, which used a similar spacecraft architecture, propulsion, and navigation systems, but with an improved sampling design (*Watanabe et al.,* 2017). Hayabusa2 was a huge technological success and a sophisticated sampling mission that permitted two different sampling sites, of which one was excavated before sampling with an explosive charge. Hayabusa2's target was 162173 Ryugu, a C-type (carbonaceous) asteroid. Unlike S-class Itokawa visited by the first Hayabusa probe, C-type asteroids are thought to be related to carbonaceous chondrites and likely to contain organic materials and volatiles and are therefore a target of choice for investigating the relationship between terrestrial atmosphere and oceans and primitive material.

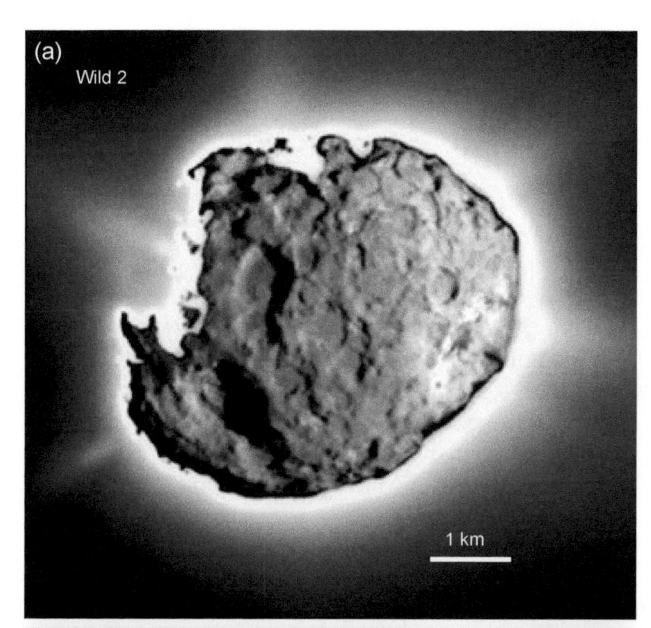

Fig. 4. Small body targets of sample return missions. **(a)** Comet Wild 2, from which coma grains were sampled by the NASA Stardust spacecraft and returned to Earth in 2006. **(b)** Asteroid targets of sample return missions. Thousands of microsized grains were sampled from S-class Itokawa by the Japanese Space Agency (JAXA) Hayabusa spacecraft and returned in 2010. About 5 g of dust and pebbles was sampled on Ryugu, a C-type asteroid, by the JAXA Hayabusa2 spacecraft. The sample return capsule landed in December 2020. The NASA OSIRIS-REx mission sampled about 121.6 g of material from asteroid Bennu, another C-type asteroid, and the sample canister landed in September 2023. Credit: NASA and JAXA.

Ryugu is diamond-shaped with a diameter of about 875 m and a rotation period of 7.63 hr (Fig. 4). Its albedo of 0.047 is very low, possibly indicating a high abundance of organic matter (*Watanabe et al.,* 2019). The mission was launched on December 3, 2014, and the sample return capsule landed in an Australian desert on December 6, 2020. It returned 5.4 g of asteroidal sample, 50× more than the expected mass of 100 mg. Hundreds of scientists across the world are currently proceeding to elemental/isotopic analysis with the latest instrumentation and methods.

The Origins-Spectral Interpretation-Resource Identification-Security-Regolith Explorer (OSIRIS-REx) is a NASA mission aimed at investigating the B-type carbonaceous asteroid Bennu and returning a sample to Earth (*Lauretta et al.,* 2017). The mission launched September 8, 2016, from Cape Canaveral Air Force Station, and the spacecraft reached Bennu in 2018 (*Lauretta et al.,* 2015, 2019) and collected 121.6 g, a success with respect to the planned nominal mass of 60 g. The sample return capsule landed on September 24, 2023.

2. METEORITES

2.1. A Little Bit of History

Although meteorites have been known from the pre-Roman epoch, it took considerable time to recognize their extraterrestrial nature. Iron meteorites were often used for forging arms and decorations since the antiquity, and the origin of these bolides, for which observed falls were often associated with formidable events (lightnings, detonations), was for long thought to be supernatural. The first documented fall of a meteorite in Europe took place near the locality of Ensisheim in Alsace (France) on November 7, 1492. The detonation was heard for several tens of kilometers around and witnesses reported the fall of several stones in the nearby fields. Maximilien of Habsburg, under whose authority Ensisheim was placed, interpreted this extraordinary event as a favorable omen and went to war against Charles VII of France, who had taken his wife away from him a few years before. In 1794, Ernst Florens Friedich Chladni (also known for pioneering work on acoustical physics) proposed that (1) masses of iron and stone fall from the sky, (2) they create balls of fire, and (3) they come from space (*Chladni,* 1794). On April 26, 1803, a rain of stones fell over L'Aigle, a small town in Normandy, France, with remarkable environmental effects observed by numerous witnesses. The French Academy of Science sent Jean-Baptiste Biot, a young promising scientist (also known for his works on electromagnetism that led to the Biot and Savart law), to investigate the cause of this extraordinary event. Through a careful field investigation including collection of numerous pieces and the interview of peasants, Biot could demonstrate not only the extraterrestrial nature of the stones but also could evaluate their trajectories, a method which is now widely used to infer meteoritic origins in space and initial velocities. This study definitively established that meteorites are natural objects within the order of things (*Gounelle,* 2006).

2.2. Falls and Finds

Meteorites are rocks and metal alloys originating from planetary bodies that survive their atmospheric entry and reach the surface of Earth. They provide the most direct access to the chemical and isotopic composition of the nascent solar system. Fragments of planetary surfaces, either large bodies like Mars or the Moon or smaller ones like asteroids or comets, are ejected through gravitational perturbations or impact events. When the meteor enters the atmosphere, heating, mechanical disruption, and chemical interactions with the atmospheric gases cause it to heat up and radiate energy. Consequently, meteor entries are often accompanied by strong visual effects, such as fireballs, and by detonations. Fresh samples can be recovered after impact if the landing site is found by witnesses. Recovered meteorites make only a few percent of the pre-atmospheric masses, the rest being destroyed during passage through the atmosphere (*Zolensky et al.,* 2006). Meteorite chunks obtained immediately after impact are called "falls"; however, most meteorites are found after some time has passed by passers-by or during dedicated meteorite hunts in hot or cold deserts (e.g., deserts located in North Africa, South America, Australia, Antarctica). Estimates for the modern flux of meteorites are variable, on the order of a few thousands to a few tens of thousands of tons per year, depending on the type of objects considered (*Zolensky et al.,* 2006) (see the chapter by Ye and Jenniskens in this volume for more insight into delivery of small bodies).

2.3. Meteorite Classification

2.3.1. Chondrites and achondrites. Chondrites are primitive or undifferentiated stony meteorites that have not been modified by differentiation of their parent body. They are formed from an assemblage of components, including refractory inclusions and chondrules (which are discussed in more depth in section 2.5), that existed before chondrite parent body accretion. As such, chondrites preserve some of the earliest solar-system-formed materials. Carbonaceous chondrites, a well-studied chondrite group, preserve stardust or circumstellar condensates, which are micrometer-sized mineral grains that condensed in stellar outflows or ejecta before the solar system formed.

Achondrites originate from differentiated parent bodies that have undergone wholescale melting to form a core, mantle, and crust. These materials do not preserve components such as chondrules (hence their name), CAIs, or presolar grains, although these components may have originally been present in the parent body prior to differentiation. Silicate achondrites are from the crusts and/or mantles of differentiated asteroids. For example, howardites, eucrites, and diogenites (HEDs) are considered to have been ejected from asteroid 4 Vesta (*Binzel and Xu,* 1993; *McSween et al.,* 2010). Iron meteorites are made mostly of metal (Fe and Ni alloy), and many of these sample cores of differentiated asteroids. Stony irons are meteorites that are made of both silicate and metal and include mesosiderites, which are breccia resulting from impacts and pallasites that are made of large olivine crystals encapsulated into metal and thought to come from core-mantle interfaces or impacts (*Yang et al.,* 2007, 2010; *Scott,* 1977). Primitive achondrites are called primitive because they are achondrites that have retained much of their original chondritic properties, including relic chondrules and chemical compositions close to the composition of chondrites. Their petrology indicates have clearly experienced melting but not substantial separation of silicate and iron melts leading to differentiation. A lunar origin for a specific meteorite clan has been attributed based on their close chemical and petrographic ties with samples returned by the Apollo missions. "Martian" meteorites have been identified as originating from the Red Planet thanks to their relatively young ages and remnants of martian atmospheric gases trapped in glassy veins. The different types of meteorites are described in Fig. 2 and some examples are presented in Fig. 5.

The formation of chondrite parent bodies was protracted by a few million years, typically 1 to 5 m.y. relative to CAIs (*MacPherson,* 2014; *Villeneuve et al.,* 2009). Surprisingly, chondrites were not the first planetary bodies to accrete. They were preceded by the formation of small differentiated planetesimals, some of which formed within the first 1 Ma of solar system history (e.g., *Kruijer et al.,* 2014, 2017). These early formed planetary bodies were destroyed by collisions and only remnants of their cores remain as iron meteorites to testify to their occurrence.

Although meteorites are the easiest to recover and provide the most mass for laboratory study, these cosmochemical samples make up only a small fraction of the flux of extraterrestrial materials on Earth. The peak of the meteorite mass flux to Earth corresponds to grains of a few hundred micrometers in diameter. Particles of this size are usually highly heated during atmospheric entry (producing the commonly observed meteors or "shooting stars") to the point of vaporization or extreme melting. Those that survive to Earth's surface are termed micrometeorites (Fig. 5e). Melted micrometeorites were first found in deep sea sediments in the 1870s and termed "cosmic spherules." In recent decades, melted and unmelted micrometeorites have largely been recovered by melting and filtering snow and ice in Antarctica (*Duprat et al.,* 2010) (Fig. 5e), although some can be found even in urban environments. Many primitive micrometeorites are chemically and mineralogically similar to larger meteorites, especially carbonaceous chondrites, and are thus likely asteroidal in origin, although some are extremely C-rich and may have originated from comets (*Duprat et al.,* 2010). Even smaller grains (<100 μm) are collected in Earth's stratosphere by special collectors on aircraft (*Brownlee,* 1985) or recovered from Antarctica (*Noguchi et al.,* 2015; *Taylor et al.,* 2020), and are known as interplanetary dust particles. These also are considered to originate in both asteroids and comets and are discussed elsewhere (see the chapter in this volume by Engrand et al.).

2.3.2. Constituents of chondrites. Primarily, chondrites consist of three main components: chondrules, refractory inclusions, and matrix (Fig. 6). Chondrules (from the ancient Greek word "chondros," for grain) are spherical objects that dominate the volume of ordinary and enstatite chondrites and are a major fraction of most carbonaceous chondrite groups (with the notable exception of CIs). Typically, they range in size from 0.1 mm to a few millimeters, with different chondrite groups showing distinct characteristic sizes. They are mineralogically and texturally diverse, but in general are composed of crystalline olivine and pyroxene minerals

Fig. 5. Types of meteoritic material from asteroids. **(a)** Ordinary chondrite [Ochansk, *https://en.wikipedia.org/wiki/Ordinary_chondrite#/media/File:Ordinary_chondrite_(Ochansk_Meteorite).jpg*]. **(b)** Carbonaceous chondrite (Allende, *https://en.wikipedia.org/wiki/Allende_meteorite#/media/File:Allende_Matteo_Chinellato.jpg*). **(c)** Iron meteorite (Sikhote-Alin) (photo by H. Raab, licensed under CC BY-SA 3.0). **(d)** Stony iron meteorite (3-kg piece of Brahin pallasite) (photo by S. Jurvetson, licensed under CC BY 2.0). **(e)** Antarctic micrometeorite (courtesy of J. Duprat). **(f)** Hydrated interplanetary dust particle (IDP) (courtesy S. Messenger).

surrounded by Al-rich silicate material that is often glassy, and lower and variable amounts of Fe sulfides, FeNi metal, and/or other minerals. Their spherical shape and textures imply that they formed by rapid (hours) solidification of tiny molten droplets in space. How and where chondrules formed is still mysterious and highly debated. Some models advocate a nebular origin like incomplete fusion of solid precursors (*Tenner et al.,* 2017) or recycling of early condensates (*Marrocchi et al.,* 2018), whereas others call for planetary processes such as collisions between protoplanets (*Libourel and Krot,* 2007) or impact splash (*Lichtenberg et al.,* 2018; *Sanders and Scott,* 2012). Their ubiquity in undifferentiated asteroids suggest they played an important role in planet formation.

Refractory inclusions (RIs) are submillimeter- to centimeter-sized irregular objects consisting of minerals that form at higher temperatures than chondrule constituents. The most common type of refractory inclusions is made of calcium-aluminum-rich minerals (CAIs), amoeboid olivine aggregates (consisting of Mg-rich olivine mineral forsterite — Mg_2SiO_4, the most refractory ferromagnesian silicate — Al-diopside, anorthite, and spinel), and refractory metal nuggets (RNM) (alloys of highly siderophile elements). Calcium-aluminum-rich inclusions are considered to be the first solids to condense from the cooling of an initially hot nebular gas of solar composition (*Grossman and Larimer,* 1974; *MacPherson,* 2014), and their chronometric age of CAIs of 4567.30 ± 0.16 Ma constitutes the anchor point (t_0) of solar system chronology (*Connelly et al.,* 2012). Ages of events in the forming solar system are either expressed in absolute ages from the present or with reference to t_0 taken as the formation of CAIs. The proto-Sun and its parent molecular cloud core were obviously older than CAIs but, in the absence of samples of their formation, events having occurred before the condensation of CAIs rely on analogies with stellar proxies and numerical astrophysical models.

The matrix surrounding CAIs and chondrules is a very fine-grained (submicrometer to a few micrometers) mixture of various phases, including tiny silicates, sulfides, metal grains (including fragments of refractory inclusions), presolar grains, and carbonaceous material. The amount of matrix varies from a few percent in ordinary chondrites to nearly 100% in CI chondrites. Whether the matrixes of different chondrite groups represent a common material that was added to the refractory inclusions or each group has a distinct matrix is debated. Most carbonaceous chondrites also contain hydrated minerals, which attest for liquid water once present in their parent bodies. Notably, both CC and NC stony meteorites contain organics in their matrix, which are more abundant in the CCs than in the NCs. Chondrites

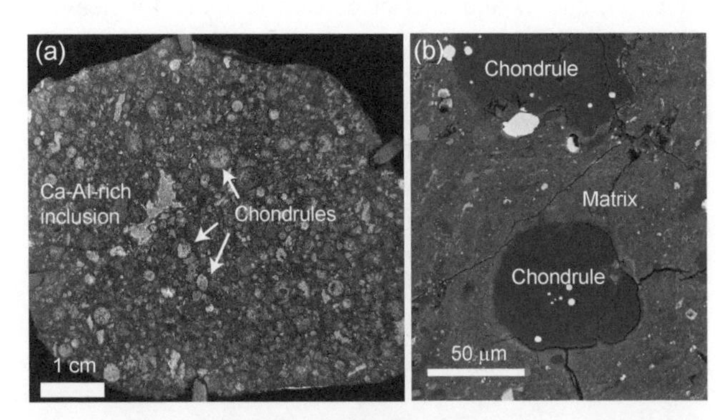

Fig. 6. **(a)** Allende CV chondrite (CC by 2.0, Wikipedia). **(b)** Close up of Asuka 12169 CM chondrite (*Nittler et al.,* 2021).

CB and CH are exceptions since they do not contain interchondrule matrix. This implies that NCs formed in a region between a snow line, beyond which water could condense, and a tar line, within which organics could survive (Fig. 1).

Presolar grains are preserved in the matrix of some carbonaceous chondrites. Presolar grains are microscopic solids that condensed in the outflows or ejecta of stars. These grains survived their passage through the interstellar medium and the protosolar parental molecular cloud and the early stages of solar system formation before they were accreted into parent bodies (Fig. 8). Although only some rare meteorites preserve intact grains, presolar grains were inherited by all celestial bodies in varying proportions. Interstellar grains are also found in chondrites; however, these formed in interstellar space and have no direct link to specific stellar events. The term "presolar grains" includes both circumstellar condensates and interstellar grains; however, much of the literature uses "presolar grain," "circumstellar condensate," and "stardust" interchangeably, without reference to interstellar grains. The existence of presolar material surviving in chondrites was first indicated by isotopically unusual noble gas compositions. This led to a long process of trial-and-error chemical dissolution experiments, spearheaded by Ed Anders at the University of Chicago, that culminated in the discovery of bona fide presolar grains of nanodiamond, graphite, and SiC (see *Anders and Zinner,* 1993, for a history of the discovery of presolar grains). Later technological advances such as the development of the NanoSIMS ion microprobe led to the discovery of many additional presolar phases including oxides and acid-soluble silicates [which can only be identified by time-consuming brute-force secondary ion mass spectrometer (SIMS) analyses]. The defining feature of presolar grains is their highly unusual isotopic compositions compared to everything that formed in the solar system. As discussed below, stable isotope variations for materials that formed in the solar system vary over relatively narrow ranges (tens of percent for O isotopes, factor of a few for N isotopes, parts per 10^6 to 10^4 for many heavy elements), reflecting the large degree of homogenization that took place during the early stages of planet formation. In contrast, presolar grains have isotopic ratios that range over many orders of magnitude for C, N, and O and tens of percent or more for heavier elements, reflecting almost-pure nucleosynthesis signatures. Because presolar grains are essentially pristine products of the stellar environments where nucleosynthesis occurs, they have proven invaluable for probing detailed processes of stellar evolution, nucleosynthesis, stellar dust production, interstellar dust processing, and the initial starting materials of the solar system (*Nittler and Ciesla,* 2016; *Zinner,* 2003; see the chapter by Bergin et al. in this volume).

2.4. Chemical Analysis of Meteorites

The elemental and isotopic compositions of planetary materials reflect both the mixture of their composite building blocks and the products of parent-body processes (e.g., thermal and aqueous alteration processes). Chemical characterization of meteorites is thus of utmost importance. Telescopic or spacecraft-based reflectance spectroscopy is used to obtain some mineralogical and element abundance information about the exposed surfaces of asteroids. Not all minerals, however, exhibit spectral features, and many processes complicate the interpretation of remotely detected spectra. For example, the interaction of airless planetary surfaces with the space environment (space weathering) can induce very strong spectral modifications. Based on the results of the Near Earth Asteroid Rendezvous (NEAR) and Hayabusa missions, S-type asteroids are related to ordinary chondrite meteorites; however, spectra of the former are much more steeply sloped toward longer wavelengths ("reddened") than laboratory chondrite spectra due to space weathering. Remote X-ray and gamma-ray spectroscopy can be also used to directly determine the elemental composition of planetary surfaces from orbiting spacecraft, albeit with limitations in both sensitivity and spatial resolution. Only one small body has been successfully characterized in this way: 433 Eros by the NEAR-Shoemaker mission, from which X-ray and gamma-ray data provided additional evidence that this S-type asteroid is a space-weathered ordinary chondrite (*Nittler et al.,* 2001; *Peplowski et al.,* 2015). Consequently, by far most of the detailed geochemical information we have on small bodies in the solar system comes from laboratory analysis of meteoritic and returned samples.

The goal of laboratory-based sample analysis is to obtain chemical, isotopic, and textural information about a sample to determine its origin. Planetary scientists attempt to derive the composition of celestial bodies and the origin of the solar system from the small fractions of the cosmos available to us in returned samples, whether they have been delivered to Earth naturally or via spacecraft. Although this approach is limited by small samples (grams of material or less), this is somewhat mitigated by the collection of exacting compositional data obtained and interpretation of these data in the context of geochemical and cosmochemical principles that describe element behavior during planet evolution.

The past 30 years has seen a revolution in instrumentation and associated analytical accuracy and precision that is available to planetary scientists. Given the variety of cosmochemical samples available for study, flexibility in instrumentation is required. Samples come in the form of either bulk specimens of a whole rock or its individual components (e.g., chondrules, CAIs). Mineral fractions or chemical leachates of either type of sample can be studied. Analyses of individual chondrules or refractory inclusions provide information about the physical and chemical conditions in the protoplanetary disk before they were accreted to their parent body. Analyses of whole rock samples provide information about the bulk mixture of phases comprising a sample. This informs us about the bulk composition of the parent body and thus the region of the disk that it sampled. The composition of differentiated asteroids cores can be constrained by iron meteorites, whereas silicate achondrites tell us about the silicate portion. Note, the composition of undifferentiated asteroids can typically be constrained by a sample from any part of the parent body.

2.4.1. Non-destructive analytical techniques. Compositional and textural information can be obtained from meteorites at the bulk sample scale to the grain scale using non-destructive techniques that permit analysis with little or no damage to the samples. At the bulk-rock scale, simple visual examination of hand samples helps determine the texture and componentry, thus informing one where to sample further. At the grain scale, the primary instruments for non-destructive characterization are electron microscopes. This includes the scanning electron microscope (SEM), often referred to as an electron microprobe analyzer when optimized for precise elemental abundance measurements. In an SEM, a finely focused beam of electrons (from a few kiloelectron volts to a few tens of kiloelectron volt energy) is scanned over a sample. Various signals are detected, including secondary and backscattered electrons, which reflect topography and average atomic number, X-rays emitted by the atoms present, and cathodoluminescent light emitted by some minerals. X-ray spectrometers allow the quantitative determination and mapping of elemental abundances on scales from 0.5 to hundreds of micrometers. Together with optical microscopy, SEM data are crucial for overall characterization of the mineralogy, textures, and major-element chemistry of planetary samples.

Other widely used non-destructive techniques include laser Raman spectroscopy, which uses incoherent scattering of light off molecular vibrations to characterize certain minerals and phases, synchrotron-based X-ray micro-spectroscopies, and transmission electron microscopy (TEM), which uses high-energy electrons to probe the atomic structure and composition of materials at nanometer to micrometer scales. Note that even if a technique is non-destructive, some sample destruction is often necessary to prepare it for analysis, e.g., through polishing or preparation of thin sections for bulk rocks or slicing of dust particles with a diamond knife attached to an ultramicrotome.

2.4.2. Destructive analytical techniques. *2.4.2.1. Sample preparation:* Destructive analytical techniques involve the partial or complete destruction of a sample to obtain elemental and/or isotopic data. These techniques require careful preparation of the sample prior to introduction into the instrument. To limit contamination of the sample during sample preparation, this work is often performed in a clean laboratory, which is designed to control airborne contaminants by reducing their concentration in the air. A typical city environment contains 35,000,000 particles per cubic meter (0.5 μm and larger in diameter), whereas clean lab operations may require environments of 1000 particles per cubic meter. Some laboratories that specialize in siderophile (iron-loving) element work demand further design restrictions by requiring the laboratory to also be metal-free.

Preparation of a bulk rock sample requires isolation of a rock chip from the main meteorite mass, which is typically housed at a museum or university. If the sample is friable (e.g., carbonaceous chondrites), a section can be broken off by hand. Most other samples, however, require cutting into the main mass using a saw blade and subsequent crushing into a powder. Alternatively, meteorite components can be isolated from a sample by gently crushing it and physically picking out phases of interest using tweezers and a binocular microscope for magnification of the worksite. Microdrilling or microcoring isolates meteorite components by drilling around the desired phase until it can be physically removed. Meteorite components can also be separated using freeze-thaw disaggregation, where the sample is cycled multiple times through immersion in ultrapure water, freezing, and thawing until it naturally fragments and components can be physically removed.

If individual minerals of a bulk rock or meteorite component are of interest, they can be separated via gravity using heavy liquids to isolate minerals of different densities. Alternatively, if minerals have different shapes, they can be separated based on their shape. At the microscopic level, a focused ion beam (FIB) system can be used to extract specific site-selective materials for analysis. This is used frequently for TEM and/or synchrotron methods where very thin samples ($\ll$1 mm) are required. In FIB, a finely focused beam of ions (usually Ga or Xe) is used to sputter material with nanometer-scale precision.

In some cases, extracting microscopic phases that are distributed throughout a sample requires a brute force approach. Some types of presolar grains are isolated by sequentially dissolving or "digesting" the bulk meteorite sample in concentrated mineral acids (e.g., hydrofluoric acid, hydrochloric acid, nitric acid, and perchloric acid) to remove the main mass until a residue remains (methods pioneered by Ed Anders) (e.g., *Lewis et al.,* 1987). This residue contains chemically resistant presolar grains (e.g., nanodiamond, nanospinel, corundum, SiC, graphite) that can be subsequently analyzed as a slurry or individual phases (Fig. 7). Similarly, refractory organic matter that dominates the C budget in primitive meteorites can be chemically purified through dissolution of inorganic phases (*Alexander et al.,* 2017).

Some instruments require the sample to be liquified for introduction into the instrument [e.g., thermal ionization mass spectrometry (TIMS) and inductively coupled plasma mass spectrometry (ICP-MS)]. Liquification is achieved by digesting the rock sample in concentrated mineral acids, the type of which is dependent on the type of sample being dissolved (Fig. 8). Dissolving metal-rich samples can be done using concentrated hydrochloric acid or a mixture of concentrated hydrochloric acid and nitric acid ("aqua regia"). Silicate rock samples are typically dissolved using a combination of concentration hydrofluoric acid and nitric acid. Hydrofluoric acid or cesium fluoride (*Brooks,* 1960) breaks silicon-oxygen bonds, resulting in dissolution of the silicate phases.

Some instruments require elements to be pre-concentrated or "purified" from the sample prior to introduction into the instrument. This may be because the element is low in concentration such that its signal would fall below the detection limit of the instrument or there are interferences

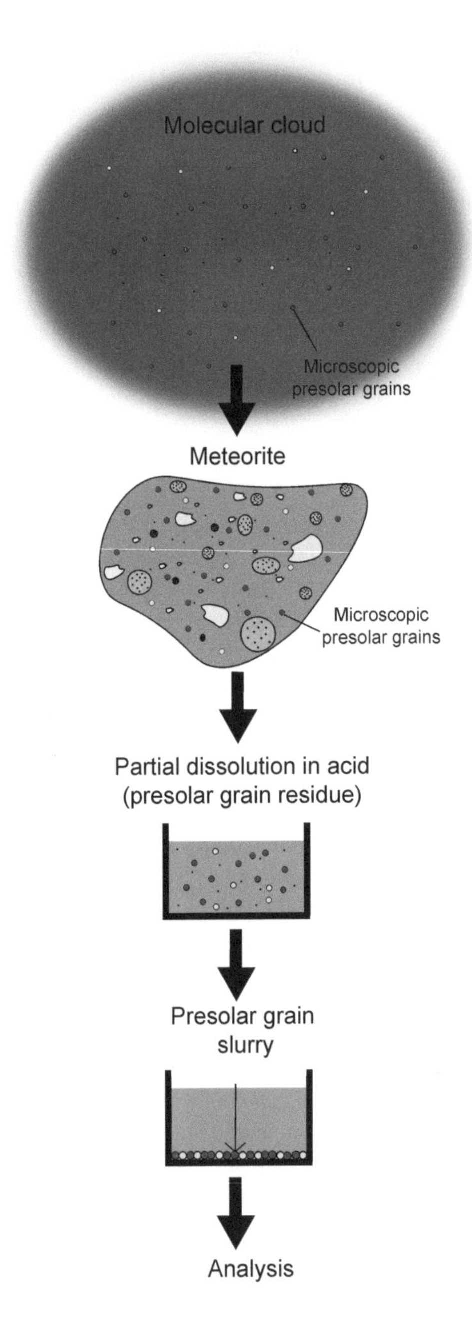

Fig. 7. Schematic depicting the history of presolar grains. Presolar grains produced in stellar events that proceeded the formation of the solar system accumulate in the parent molecular cloud. The protoplanetary disk accretes from a fragment of the molecular cloud. Meteorites that sample the protoplanetary disk thus also sample the mixture of presolar grains that were in the molecular cloud. Presolar grains are isolated from chondrites by dissolving all other phases in the meteorite to leave a slurry of presolar grains. These gains are then analyzed as a mixture of grains or individual grains using high-resolution mass spectrometry. From *Nittler* (1996).

on the element of interest caused by accompanying elements in the bulk sample solution (*Potts,* 1992). A commonly used method for isolating elements from a dissolved sample is ion exchange chromatography (Fig. 8). The principle of ion exchange chromatography is to separate ions via reversable exchange with a solid resin based on the ions' differing affinities to the resin (*Potts,* 1992). Briefly, the sample is dissolved in acid prior to introduction to the resin bed, which is packed into a Teflon or glass column. An ion's affinity to the resin changes by modifying the acidic environment of the column. Once the sample is loaded onto the resin, elements are eluted from the resin sequentially by washing the column with acids that were selected based on calibration tests run prior to the column chemistry. Multiple elements of interest can be isolated from a single elution scheme, each of which are sufficiently pure to analyze using mass spectrometry. The efficiency of ion exchange chromatography makes it one of the most commonly used methods of element purification of meteorite samples.

2.4.2.2. Bulk rock elemental chemistry X-ray fluorescence spectrometry (XRF): Major-element (Na, Ma, Al, Si, P, K, Ca, Ti, Mn, Fe) and trace-element (Ba, Ce, Co, Cr, Cu, Ga, La, Nb, Ni, Rb, Sc, Sr, Rh, U, V, Y, Zr, Zn) compositions of bulk samples can be determined using X-ray fluorescence spectrometry (XRF). This analytical technique was commonly used in silicate rock analysis in the late 1960s following development in the commercial industry (*Norrish and Hutton,* 1969; *Leake et al.,* 1969; for a review, see *Potts,* 1992). Briefly, X-ray fluorescence spectrometry requires a sample to be powdered and fused into glass disks or compressed into pellets. The sample must be thoroughly homogenized to obtain representative bulk rock elemental measurements. Without this, the nugget effect can occur, where there is an uneven sampling of phases that are concentrated in some elements leading to artificially high abundances of those elements in the bulk rock data. After sample preparation, the pellet is exposed to X-ray radiation, which causes ionization of discrete orbital electrons. After exposure, these electrons de-excite to the ground state and emit characteristic fluorescence X-rays particular to the element present. Measurements of the unknown samples are compared to those of previously well-characterized rock standards with similar matrixes to determine the absolute abundances of elements in the sample. Accuracy of this method, therefore, is reliant on a close fit between the standard, the sample matrix, and the composition.

2.4.2.3. Mass spectrometry: Mass spectrometry is a commonly used analytical method that generates isotopic data via the separation of a sample's atoms or molecules based on their mass/charge ratios. Modern mass spectrometry is based on designs by Alfred Otto Carl Nier, who is widely seen as the "father of modern mass spectrometry" (*De Laeter and Kurz,* 2006). Many of Nier's determinations of the isotopic composition of minerals are still in use today. Mass spectrometers measure the isotopic composition of solids, liquids, and gases. This section focuses on four of the most commonly used mass spectrometry techniques in meteoritics: ICP-MS, TIMS, SIMS, and noble-gas mass spectrometry (NGMS). For all these methods measurements of the unknown samples are compared to those of known rock or gas standards to determine the absolute abundances

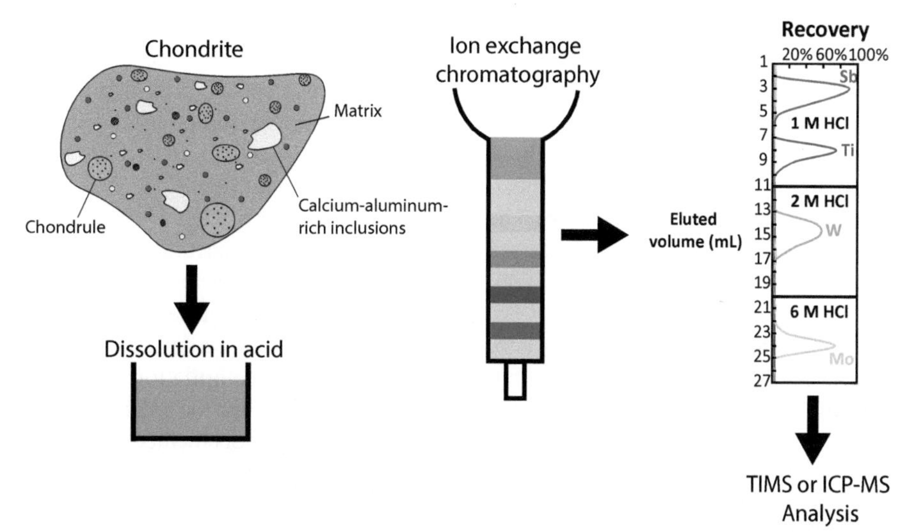

Fig. 8. Schematic depicting complete dissolution (including presolar grains) of a meteorite in concentrated acid. To isolate elements of interest, the resultant solution is passed through a resin bed housed in a column. A sequence of acids is then washed through the column to elute elements of interest that wash off the column in bands. Solutions can then be used for mass spectrometric analysis.

and isotopic ratios of elements in the unknown. Accuracy of mass spectrometric methods (e.g., how close a measurement is to the true or accepted value) is reliant on a close fit between the standard and the sample matrix.

State-of-the-art TIMS and ICP-MS can generate very precise isotopic ratios (±5 ppm, where ppm refers to deviations in atom/atom of isotopic ratios in parts per million). These instruments require a minimum sample mass, which typically requires element purification by ion exchange chromatography prior to sample introduction into the instrument. This limits isobaric interferences, increases ionization, and thus improves sensitivity of the element being analyzed. In TIMS, a purified sample is loaded onto a thin metal filament that is electrically heated under vacuum until the sample ionizes. A magnetic sector mass analyzer is then used to separate the ions based on their mass/charge ratio, and the resultant ion beams are directed into a detector that converts them into voltages. Inductively coupled plasma-MS couples a radiofrequency inductively coupled Ar plasma with a mass spectrometer. Samples are introduced to the plasma via solution, or laser ablation of a solid sample. Inductively coupled plasma-MS is used to measure the isotopic composition of purified solutions, whole rock solutions, or a wide variety of solid materials.

Secondary ion MS is a highly sensitive *in situ* technique that allows trace-element and isotopic measurements of specific phases on submicrometer to tens of micrometer scales, albeit with generally lower precision (±0.1%, depending on element) than obtained by TIMS or ICP-MS. In SIMS a primary beam (generally Cs^+ or O^-) is focused onto a sample and sputtered atoms are ionized and weighed with a mass spectrometer. Secondary ion MS instruments are often referred to as ion probes (or microprobes) and a distinction is often made between conventional instruments [e.g., the Sensitive High-Resolution Ion Microprobe (SHRIMP) or Cameca ims-1280], which are optimized for high precision but limited to spatial resolutions >~1 mm, and the Cameca NanoSIMS, optimized for spatial resolution (~0.1 mm).

Noble-gas mass spectrometry differs in the sense that noble gases are extracted from the sample by heating or crushing, so that the matrix effect does not plays any role. Noble gases are also named rare gases as their abundances are generally very low in rock samples, on the order of 10^{-10}–10^{-18} mole/g. Their analysis therefore requires a different technique where these elements are extracted from rocks and mineral by heating with furnaces or lasers or by crushing at room temperature. Gases are then purified under static vacuum (pumps isolated) and analyzed by static mass spectrometry, i.e., without pumping during analysis. These conditions require drastic treatment of the ultra-high vacuum lines and analyzers including baking, long-term pumping before analysis, and the exclusion of any organic material.

2.5. Using Small Bodies to Constrain the Composition of Solar System Evolution

2.5.1. Approach to defining the solar system composition. An important part of understanding the processes and events that define solar system evolution is to determine the composition of the molecular cloud from which the Sun and planets formed and compare it to the composition of materials we have access to today. Most materials planetary scientists have access to have evolved from their original state. To derive the initial solar system composition, planetary scientists search for objects that either have not undergone significant chemical modification, such as differentiation or aqueous or thermal processing, or sample very large portions of the solar system, thereby providing an average composition. This limits the selection of samples to a few candidates: the Sun, primitive meteorites, and comets.

The Sun contains ~99.8% of the mass of the solar system. Spectroscopic analysis of the solar photosphere (~100-km-deep surface layer of the Sun) is therefore a logical starting point to define the original elemental composition of the solar system. Although fusion reactions are ongoing since the Sun formed, their products either remain deep in the star or their surface expressions can be modeled, and those compositions subtracted from the present-day bulk composition of the Sun to derive an original stellar composition. The isotopic composition of the original solar system is inferred from chondrites because, for the most part, isotopes cannot be measured telescopically.

2.5.2. Goldschmidt's rules and cosmic volatility. Victor Goldschmidt, colloquially known as the "Father of Geochemistry," determined the elemental composition of the solar system by compiling vast quantities of chemical data on terrestrial and extraterrestrial rocks (*Goldschmidt,* 1937). He distilled relationships between elements and isotopes to define a set of principles that describe element behavior under a range of geologic conditions. Elements that tend to form silicates or oxides are called lithophile elements (rock-loving), elements that concentrate in metallic iron or metal alloys are called siderophile elements (iron-loving), elements that react with sulfur are called chalcophile elements (sulfur-loving), and elements that tend to form gaseous species and reside in the atmosphere of planetary bodies are called atmophile (gas-loving). By recognizing these affinities, it became possible to predict where elements would partition during planet formation and evolution. From Goldschmidt's data compilation, he derived a table of "cosmic abundances" (*Goldschmidt,* 1937). The elemental and isotopic composition of the solar system has since been derived for all the elements and stable isotopes of the solar system using Goldschmidt's approach (e.g., *Suess and Urey,* 1956; *Anders and Grevesse,* 1989; *Lodders,* 2003, 2021). Isotopic compositions for elements that are in gas form (e.g., volatile elements) are derived from giant planet atmospheres or from the analysis of the solar wind either *in situ* by spacecraft instruments or implanted in planetary surfaces or directly collected (e.g., the Genesis mission; see section 2.2.). The isotopic compositions reported in these compilations are average compositions and do not account for the small (parts per million) isotopic variations that are documented in terrestrial, lunar, martian, and meteorite samples.

In addition to Goldschmidt's rules, cosmochemistry requires an additional set of chemical principles to predict how elements partition during cooling of the protoplanetary disk. This is known as the cosmic volatility of an element. This involves the condensation of gaseous species to solids (the liquid state is rarely reached due to the very low pressures of space). Based on condensation experiments (*Grossman and Larimer,* 1974; *Davis and Richter,* 2007), the solids that formed via equilibrium condensation from a solar composition gas and their associated temperatures were derived (Fig. 9), and most elements condense into solid solution with major-element condensates. The 50%

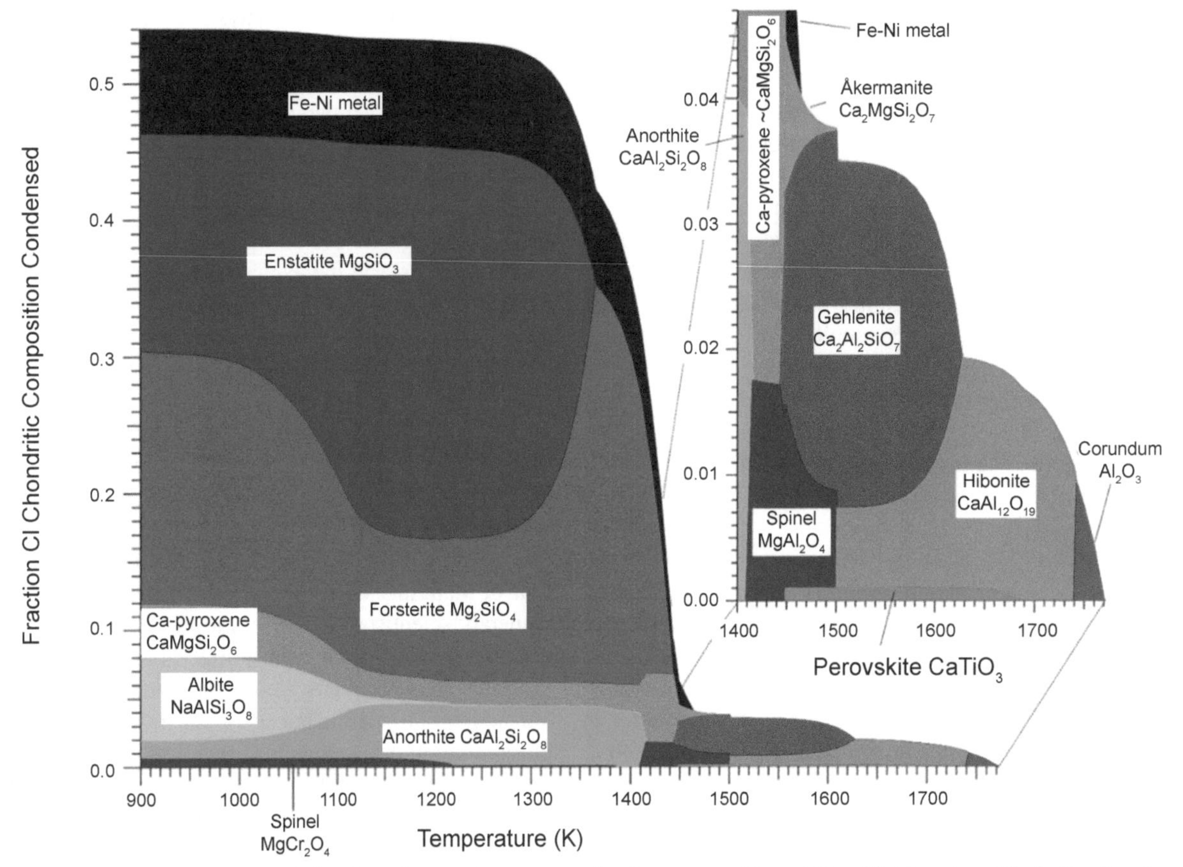

Fig. 9. Condensation of major rock-forming phases from a gas of solar composition at P = 10^{-3} bar (modified after *Davis and Richter,* 2007). The first phases to form upon cooling are Al-Ca-rich, which are effectively observed in CAIs found in chondrites.

condensation temperature of elements are commonly used to describe the temperature where 50% of the element is condensed at a given pressure (e.g., P = 10^{-3}–10^{-4} bar). The cosmic volatility of an element is expressed in relative terms. Highly volatile elements have 50% condensation temperatures below 371 K, volatile elements below 665 K, moderately volatiles between 1335 and 665 K, and refractories above 1335 K [for a gas of solar composition at a total pressure of 10^{-4} bar (*Lodders,* 2003)]. The concept of condensation sequence works very well to explain the composition of refractory inclusions in chondrites as CAIs are predicted to be the first solids to form during cooling of a nebular gas in the canonical conditions described above. Their chemical and mineralogical compositions are in full agreement with the predicted mineral phases to form from a thermodynamical standpoint (Fig. 9). As expected, CAIs are the oldest objects that formed in the solar system (*Connelly et al.,* 2012). This excellent match justifies the concept of the condensation sequence and permits its application over the range of pressure and temperature that prevailed during the formation of planetesimals.

2.5.3. The solar system composition. It has been long recognized that one group of chondrites, Ivuna-type or CI carbonaceous chondrites, have relative elemental abundances that correlate very strongly with those of the solar photosphere (Fig. 10). It is noteworthy that the Hayabusa2 mission has returned grains from C-type asteroid Ryugu, whose composition matches very well that of CIs (*Yokoyama et al.,* 2022; *Tachibana et al.,* 2022). The exceptions are the most-volatile elements, which do not condense easily into rocks (e.g., H, He, N, C, O, Ne, Kr, Xe), and Li, which has been destroyed in the Sun by nuclear reactions. Consequently, CI chondrites are considered to be the meteorite samples that best define the original solar system's elemental composition (*Wasson and Kallemeyn,* 1988). This is despite the fact that these meteorites preserve signatures of extensive aqueous alteration.

On a bulk scale, CI chondrites share the same elemental composition as the bulk solar system (Fig. 10). All the other groups of primitive meteorites (and hence their parent bodies), however, show elemental compositions that are fractionated to some extent (Fig. 11).

The bulk compositional diversity of small bodies indicated by both observations of asteroids and primitive meteorites indicates that it is highly likely that the rocky planets that accreted from smaller bodies also do not have bulk solar composition. The estimated bulk composition of Earth has a chondritic refractory element composition; however, there are significant depletions in volatile elements according to their volatility (*McDonough and Sun,* 1995; *Albarède,* 2009; *Mezger et al.,* 2021). It remains unclear to what extent Earth's depleted volatile budget is a consequence of its accretion from differentiated planetesimals and/or if Earth accreted from chondrites but the process of process of differentiation caused a depletion in volatile elements. Although the chondrite analogy holds for refractory elements, it cannot explain the bulk chemistry of Earth (Fig. 12). Further information can be gained from the isotopic signatures of small bodies compared to Earth's, which is the topic of section 4.

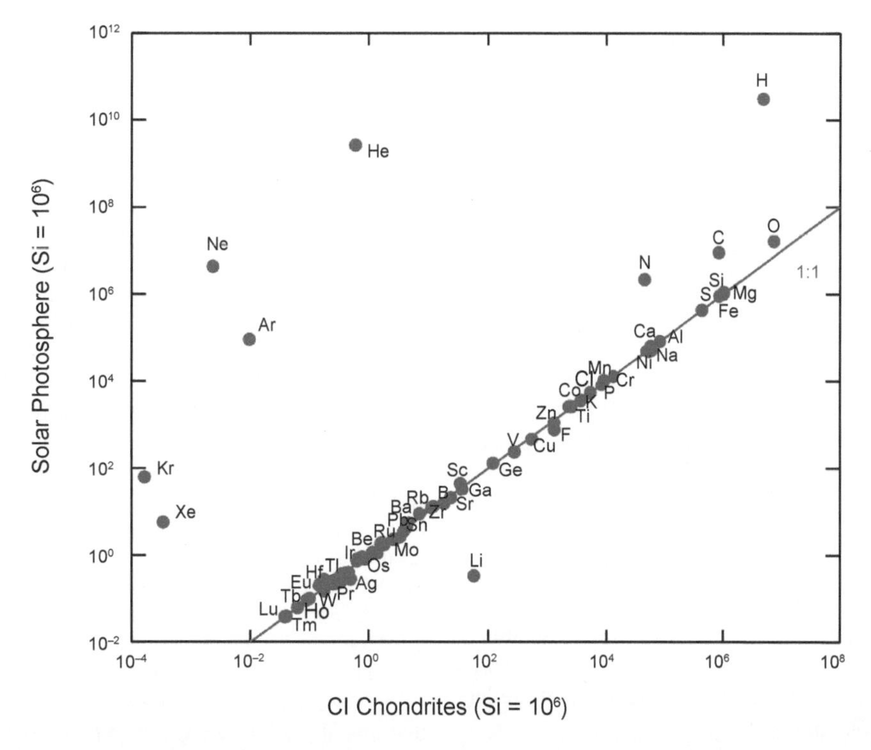

Fig. 10. Abundances of elements as determined spectroscopically in the solar system photosphere are plotted against bulk abundances measured in CI carbonaceous chondrites (both normalized to 10^6 atoms of silicon). Data from *Lodders* (2021).

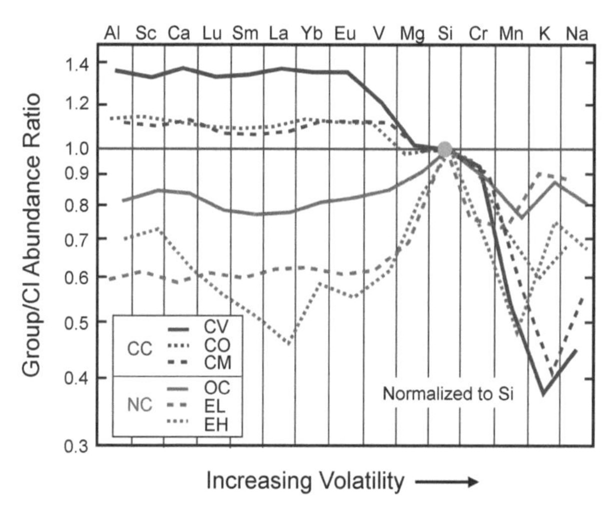

Fig. 11. Si-normalized group/CI abundance ratios for lithophile elements in some of the chondrite groups. The elements are arranged from left to right in order of increasing volatility. CI chondrites are plotted on a horizontal line at unity. All groups, except EL, have a flat refractory lithophile element pattern, implying that these elements were in the same nebular component. In contrast, group differ on the relative abundances of non-refractory elements, as a result of different condensation environment and/or parent body processing. After *Wasson and Kallemeyn* (1988).

2.5.4. Deriving the stellar building blocks of the solar system. The estimated elemental composition of the solar system is presented in Fig. 13 and shows the relative abundance of an element compared to its atomic number (Z). It is characterized by a sawtooth pattern to the curve that gradually decreases in abundance with increasing atomic number, and peaks at the iron group nuclei and double peaks at A = 80 and 90, 130 and 138, and 194 and 208. The sawtooth pattern of the solar elemental abundance curve reflects the structure of atomic nuclei: Nuclei with an even number of protons and neutrons are more stable than those with an odd number (*Oddo,* 1914). This stability translates into higher abundances during the nuclear processes that synthesize elements within stars. When plotting the sequential elemental abundance of a sample, the sawtooth pattern observed obscures geochemical or cosmochemical trends in the elements, which makes it difficult to derive the significance of elemental abundances. The sawtooth pattern is removed if the composition is divided by or "normalized to" a solar system composition (e.g., CI).

The significance of most of the solar elemental abundance curve's structure was realized in the 1950s and summarized in two seminal papers that laid out the basic principles of how the elements are synthesized in the Uuniverse (*Burbidge et al.,* 1957; *Cameron,* 1957, 1973). Most of the basic ideas suggested in those papers have withstood the test of time, although the scientific community has a progressed to a far more detailed understanding of stellar evolution and nucleosynthesis.

Hydrogen (including deuterium), He, and a small amount of Li were synthesized shortly after the Big Bang as the universe rapidly cooled, but the rest of the elements were made later, mostly in stars. Stars spend most of their lives powered by the exothermic fusion of H into He (H-burning) and subsequent burning of He in their cores to form ^{12}C and ^{16}O (He-burning). The low abundances of Li, Be, and B reflect the fact that these are bypassed during He-burning and in fact these elements are made by spallation, essentially fragmentation of heavier elements in interstellar space when they are bombarded by high-energy cosmic rays. Low-mass stars like the Sun do not reach temperatures high enough for fusion reactions to occur beyond He-burning and end their lives as a cooling white dwarf made of C and O. Thus, most of the elements between C and Fe are made in more massive stars (>~10 $M_{\odot}$, where $M_{\odot}$ = solar mass) as their cores burn ever heavier fuels. However, ^{56}Fe has the highest binding energy of any nucleus and this means that exothermic fusion reactions cannot produce energy once the core has burned to iron. At this point, a massive star's core

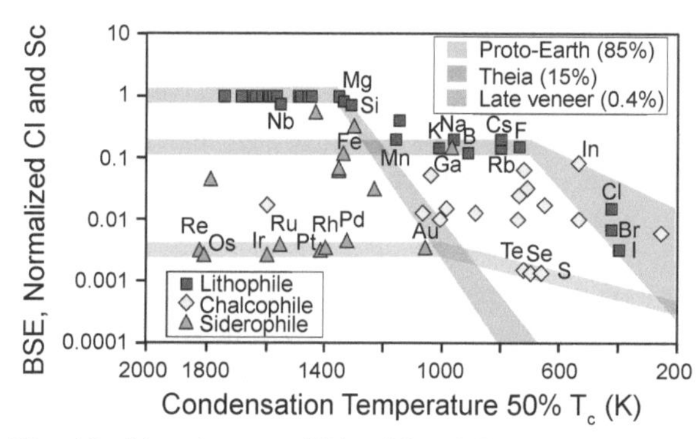

Fig. 12. Abundances of lithophiles (silicate-loving), chalcophile (sulfur-loving), and siderophile (iron-loving) elements of the bulk silicate Earth (BSE) (i.e., bulk Earth excluding core) normalized to CI and Sc (after *Mezger et al.,* 2021). Refractory elements are clearly chondritic in Earth, but non-refractory elements tend to be depleted with increasing volatility (expressed as temperature at which 50% of a given element is condensed in canonical conditions (solar system composition, P = 10^{-3} bar). Note that volatile-element depletions are much stronger than for chondrites. The complicated abundance pattern of the BSE could reflect the contribution of several sources, events, and processes during Earth's accretion (*Albarède,* 2009; *Mezger et al.,* 2021). In the model illustrated here, *Mezger et al.* (2021) propose a three-step process with a dry proto-Earth, to which the Moon-forming event involving collision with an embryo labeled Theia contributes 15% of Earth's mass, followed by the addition of 0.4% chondritic material after this event. The last event, labeled the late accretion, accounts for the non-negligible abundance of siderophile elements in the BSE after core formation since the core should have left the mantle barren of siderophile elements, which is not observed (e.g., *Wang and Becker,* 2013, and references therein). This idea that the Moon-forming impactor brought Earth's volatiles is, however, difficult to envision from a dynamical point of view. Other models propose delivery of volatiles throughout Earth's building and/or during the last stages of accretion (see section 4).

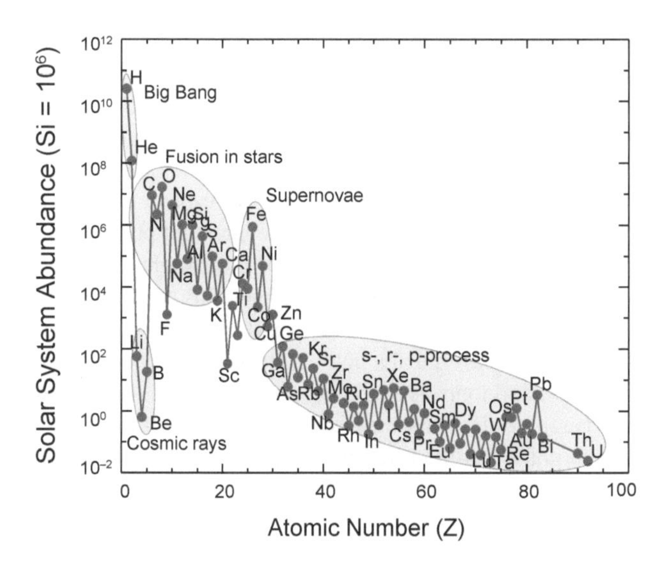

Fig. 13. The solar elemental abundance pattern. Data from *Lodders* (2021). Primary nucleosynthesis processes are indicated.

collapses under its own weight to form a neutron star or black hole and a reverse shock ejects the star's envelope, an explosion known as a Type II supernova. The peak around Fe in the solar abundance pattern reflects the high stability of nuclei in this range of atomic number; these elements are produced in both Type II and Type Ia supernovae, the latter being thermonuclear explosions of white dwarfs due to mass transfer from a binary companion. The slow (*s*-process) or rapid (*r*-process) progressive addition of neutrons are the dominant processes responsible for creation of most elements heavier than Fe. The former takes place in asymptotic giant branch (AGB) stars, the end stage of evolution of stars $<\sim 8\ M_{\odot}$, whereas the *r*-process occurs during mergers of neutron stars and possibly in Type II supernovae. Peaks in the heavy element distribution (e.g., at A = 50, 82, 126) reflect that these nuclei, with "magic" numbers of neutrons and protons, are more stable. A few proton-rich heavy elements are produced by the *p*-process, which is most likely due to photodisintegration of heavier nuclei in supernovae (*Burbidge et al.,* 1957; *Arnould and Goriely,* 2003).

The processes laid out by *Burbidge et al.* (1957) and *Cameron* (1957) were based on deconvolution of the bulk solar abundance pattern and basic principles of nuclear physics. Although it was recognized that the atoms that formed the solar system originated in diverse astrophysical environments, it was long thought that high temperatures in the inner protoplanetary disk would reprocess and homogenize any distinct presolar nucleosynthetic signatures. It took the discovery of isotopic anomalies in Xe and Ne in primitive meteorites (*Reynolds and Turner,* 1964; *Black and Pepin,* 1969) to prove both that the nuclear processes envisioned by the pioneers of nucleosynthesis indeed take place in nature and that preserved presolar grains carrying nucleosynthetic signatures survived solar system formation (see section 2.7.1 as well as the chapter by Bergin et al. in this volume). Isotopic variations or anomalies are defined as excesses or depletions of a given isotope in a sample, and they are expressed as positive or negative deviations of their isotope ratios in parts per 10^3 (permil, ‰-units), 10^4 (epsilon, ε-units) or 10^6 (mu, μ-units, or parts per million) compared to terrestrial reference standards (defined as zero, or "normal").

2.6. ISOTOPIC COMPOSITIONS OF METEORITES

Isotopic anomalies in meteorites provide invaluable information on the origin, processing, and transfer of matter in the solar system, as well as ages of events during its whole history. Several of the most important isotope systems used in cosmochemistry are discussed here. These are divided into (1) isotopes produced or destroyed by nuclear reactions and/or radioactive decay in the nascent solar system; (2) stable isotope variations, which can depend on isotopic masses and provide information on processes and mixings between solar system objects and reservoirs; and (3) signatures of presolar heritage, which record the nature of nucleosynthetic sources that contributed to the solar system (so-called nucleosynthetic isotope anomalies).

2.6.1. Radiometric isotopic data. *2.6.1.1. Principle of dating:* Radioactive isotopes, or radionuclides, are those that undergo spontaneous disintegration to form radiogenic isotopes. Stable isotopes are those that are not destroyed by radioactive decay. Radionuclides are produced during nucleosynthesis, from interactions of energetic particles (e.g., cosmic rays) with dust or gas, and by naturally occurring radioactive chains. Conventionally, the radionuclide is termed the "parent," which decays into the "daughter" isotope. Because the rate of radioactive decay is constant and independent of essentially all environmental factors, radionuclides serve as useful clocks to date a wide variety of processes that involve chemical fractionation of parent and daughter elements. Radioactive decay is fundamental to many aspects of Earth and planetary science, but its relevance here is that it is used to determine absolute and relative ages for processes that occurred in the first few million years of solar system history (including condensation in the protoplanetary disk) and the accretion, alteration, and/or differentiation of meteorite parent bodies. There is a distinction between long-lived radionuclides — those whose lifetimes are longer than the age of the solar system — and short-lived (or extinct) ones — isotopes that were present when the solar system formed but that have completely decayed away by today. While long-lived systems can often provide absolute ages, short-lived ones can only provide relative ages, albeit in some cases with very high time resolution. Notably, the existence of short-lived radioactive nuclides requires that their nucleosynthesis occurred shortly before the formation of the protoplanetary disk (see the chapter by Bergin et al. in this volume).

The basic concepts of radiochronology are illustrated in Fig. 14 for a long-lived (Rb-Sr) and a short-lived (Al-Mg) system. If a melt solidifies with a certain amount of

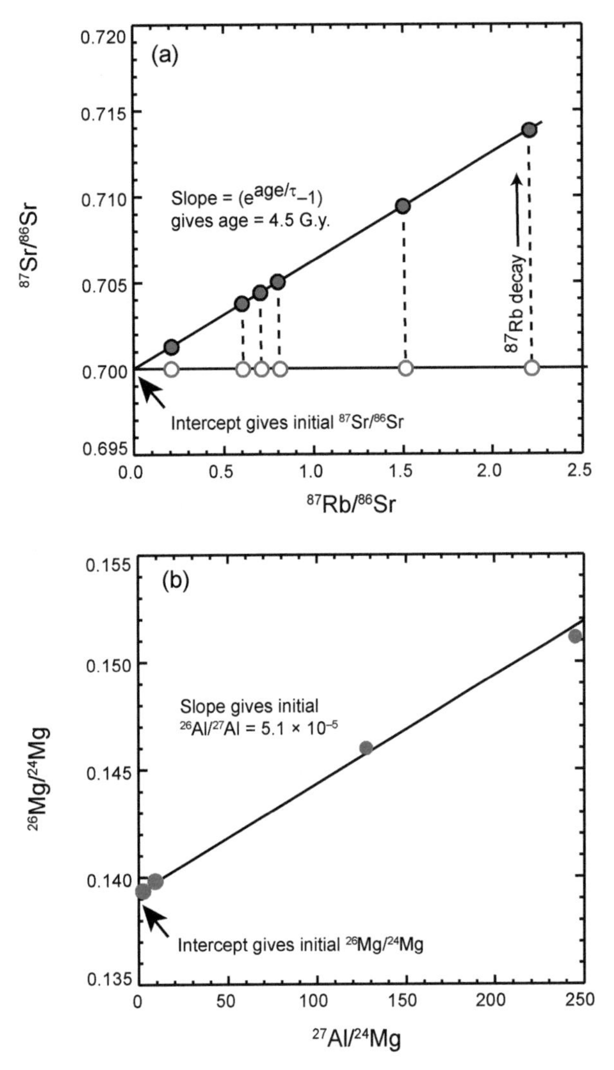

Fig. 14. Examples of radiometric dating systems. **(a)** In the Rb-Sr system, minerals in a sample with higher Rb/Sr ratios will produce higher amounts of daughter isotope ^{87}Sr from ^{87}Rb decay; the slope of the resulting isochron gives the absolute age of the sample. Data are schematic. **(b)** In the Al-Mg system, all initial ^{26}Al has decayed to ^{26}Mg. The slope of the isochron gives the initial ^{26}Al/^{27}Al ratio at the time the sample crystallized. Data are from the original CAI measurements of *Lee et al.* (1977).

radioactive ^{87}Rb [mean-life τ = 71 G.y., where after time t, a radioactive isotope has decreased to $e^{(-\tau/t)}$ of its original abundance] into different minerals with varying Rb/Sr ratios, at some later time, the phase with the highest Rb/Sr would have decayed further into ^{87}Sr and thus there is an increase in ^{87}Sr/^{86}Sr ratio. The different points lie on a line (an "isochron") whose slope can be used to determine the absolute age of the rock and the y-intercept the initial ^{87}Sr/^{86}Sr ratio of the material. Note that the most precise absolute ages of early solar system objects come from U-Pb or Pb-Pb dating, which takes advantage of the decay schemes of both ^{235}U (τ ~ 1 G.y.) and ^{238}U (τ ~ 6.4 G.y.) into ^{207}Pb and ^{206}Pb, respectively. The mathematics is slightly more complicated but fundamentally based on the same concept as Fig. 14a.

For a system containing a short-lived isotope (e.g., ^{26}Al, τ = 1 m.y.), phases that concentrate a higher parent/daughter element ratio end up with a higher abundance of the daughter isotope. In this case, however, one plots a stable isotope of the parent element on the x-axis and the isochron slope gives the ratio of radioactive to stable isotope that the system had at the time of crystallization. The intercept provides the initial stable isotopic composition of the daughter element before radioactive decay. Short-lived systems can be directly tied ("anchored") to a long-lived one if the same sample can be precisely analyzed for both. Note that what is dated in both short and long-lived systems is the last time the parent and daughter elements were set in their current proportions in each phase, a time referred to as "isotopic closure." This often refers to the time that a rock crystallized into distinct minerals but could also refer to a chemical fractionation event in the protosolar disk, or a heating event that induced intergrain diffusion, or loss from a previously formed solid.

Although there are many long-lived and short-lived radiochronological systems used in early solar system studies, the key results for four of the most important are summarized here: U-Pb, Al-Mg, Hf-W, and Mn-Cr. The first is used to provide high-precision absolute and relative ages of meteorites and their components, the second to provide relative ages of meteoritic objects with sub-million-year precision, the third to provide timing of planetary differentiation and core formation, and the fourth to provide information on the timing of hydrothermal alteration of meteorite parent bodies.

2.6.1.2. Chronology of the early solar system: All isotope chronometers are consistent with a solar system age (t_0) of ~4.57 Ga, the best estimate being 4567.30 ± 0.16 Ma for CAIs using the U-Pb system (*Connelly et al.,* 2012). The combined use of the U-Pb system with the ^{26}Al-^{26}Mg extinct radioactivity permits high-resolution chronology (better than 1 m.y.) of events anchored to just after CAI formation. The short-lived ^{182}Hf-^{182}W system (τ = 12.8 m.y.) is particularly important for dating the time of early planetesimal differentiation. This is because W is highly siderophile while Hf is lithophile, so planetary differentiation leads to fractionation of the W/Hf ratio. The ^{53}Mn-^{53}Cr system (τ = 5.3 m.y.) has been found to be useful for dating the time of hydrothermal alteration on carbonaceous chondrite parent bodies since carbonate grains formed during the alteration can have extremely high Mn/Cr ratios and thus measurable excesses of daughter ^{53}Cr. The Mn-Cr system has been also used for dating aqueous alteration of carbonate-free chondrites: H, L, LL, CV, and CO, by measuring hydrothermally formed fayalite and kirschteinite (*MacPherson et al.,* 2017; *Doyle et al.,* 2015).

The main episodes of planetary formation are summarized in Fig. 15, based mainly on the isotope systems described above. Instead of a succession of events, it appears that planetary accretion and differentiation and the assemblage of primitive bodies took place simultaneously. Some differentiated planetesimals formed very early, <1 m.y. after CAIs, and primitive meteorites were assembled within a few million years, implying that their

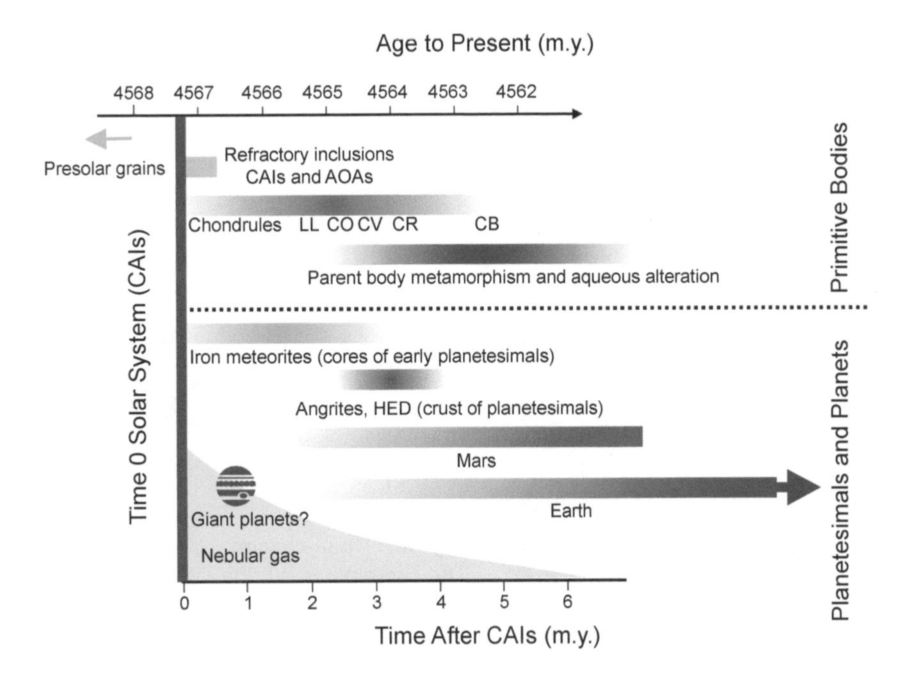

Fig. 15. Chronology of the early solar system (adapted from *Wang and Korotev,* 2019). The solar system formed from the collapse of a molecular cloud core that homogenized stellar dust and gas. Witnesses of past stellar nucleosynthesis can be found in primitive meteorites as refractory presolar grains that survived high enthalpy events. The chronology starts with the condensation of CAIs, the first solids to form during the cooling of the nebular gas (*Bouvier and Wadhwa,* 2010; *Connelly et al.,* 2012, 2017; *Amelin,* 2002; *Kita and Ushikubo,* 2012; *Bollard et al.,* 2019). Iron meteorites that are the remnants of disrupted early planetesimals have Hf-W ages (another extinct radioactivity system where ^{182}Hf decays to ^{182}W with a mean life of 13 m.y.) indistinguishable from the range of CAI ages and extending to a few million years. Chondrules and by extension chondrites formation intervals extending to ~5 m.y. (*Krot et al.,* 2005; *Amelin and Krot,* 2007; *Kita and Ushikubo,* 2012; *Villeneuve et al.,* 2009; *Connelly et al.,* 2017), implying that chondrule formation was contemporary with the occurrence of early formed planetesimals. Angrites (*Zhu et al.,* 2019) and HED [differentiated meteorites thought to originate from asteroid Vesta (*Mittlefehldt,* 2015)] represent remnants of planetary crusts. Mars, a planetary embryo, was mostly accreted within a few million years after CAIs (*Dauphas and Pourmand,* 2011). It took much longer to achieve the accretion of the Earth [≥60 m.y. after CAIs according to *Touboul et al.* (2015)], although recent models propose early accretion contemporary with that of Mars (*Johansen et al.,* 2021). An early accretion of the proto-Earth is also consistent with the occurrence of solar-like neon in the terrestrial mantle from capture of a nebular-like atmosphere (*Yokochi and Marty,* 2004). The age of Jupiter is not determined directly, but models of giant planet formation (*Guillot and Hueso,* 2006) and the purported need to segment the protoplanetary disk early to isolate the NC and CC reservoirs have been interpreted to indicate that Jupiter formed within 1–2 m.y. after CAIs (*Kruijer et al.,* 2017). The solar system nebula lifetime could have lasted a few million years before dissipation (*Podosek and Cassen,* 1994), supplying solar gas to the giant planets and possibly to the terrestrial mantle.

formation was contemporary to that of planetesimals. Planetary embryos leading to the accretion of Mars and proto-Earth were also formed within a few million years after CAIs, when the nebular gas was still present in the protoplanetary disk. The early solar system was a very energetic environment where planetary bodies accreted and were destroyed by collision, and only rare remnants lasted until the present in the form of small bodies providing meteorites to Earth.

2.6.2. Stable isotopes. Hydrogen, He, carbon, N, O, and Ne were the most abundant elements in the molecular cloud from which the solar system formed. Excepting O, however, these elements are drastically depleted in meteorites and terrestrial planets because of their volatility. Furthermore, H, C, N, and O show substantial, and for two of them (H and N) extreme, variations in their stable isotope compositions. A remarkable feature of H, C, N, and O is that inner planet bodies and reservoirs present isotopic compositions strikingly different from those of the protosolar nebula. Synthesizing these observations, it appears that some major reservoirs of the protosolar nebula (e.g., H and N, O) that existed as gaseous molecules (H_2, CO, N_2) were rapidly enriched in their rare and heavy isotopes before or during the collapse of the nebula and the formation and early evolution of the protoplanetary disk. Understanding these processes at work will allow insight into the formation of a stellar system and conditions permitting the rise of habitable planets.

Cosmochemists and geochemists commonly use the delta notation (δ), which represents isotopic deviations of an isotopic ratio (R) relative to a terrestrial standard (R_{std}) in parts per million, in linear form expressed as

$$\delta^i X = \left[\left(\frac{(R)_{sample}}{(R)_{standard}} \right) - 1 \right] \times 1000$$

By convention, the rare and heavy isotopes [^{2}H(D), ^{13}C, ^{15}N, ^{17,18}O] at the numerator are usually normalized to a corresponding abundant isotope (^{1}H, ^{12}C, ^{14}N, and ^{16}O, respectively). Positive δ values indicate that the sample is enriched in the heavy isotope. In nature, most processes follow a mass-dependent isotope fractionation (MDF) law, where the extent of variation is proportional to the relative mass difference between the two considered isotopes. This means that isotopic fractionation (the change of isotopic proportion due to physical or chemical processing) is larger for H, which has a relative mass difference close to 100% between atomic ^{1}H and the heavy hydrogen deuterium (symbol D, equivalent to ^{2}H), than for N (relative mass difference between ^{14}N and ^{15}N: [15–14]/14 = 7.2%). There are, however, other types of isotope variations that do not depend on isotopic masses, which are called mass-independent fractionation (MIF). To unambiguously detect MIF effects, one needs elements having more than two stable isotopes, for example, as discussed below for the case of O, which has three stable isotopes: ^{16}O, ^{17}O and ^{18}O, with ^{16}O being the most abundant one.

2.6.2.1. Oxygen isotopes, the Rosetta stone of the solar system: Oxygen is by far the most abundant element in rock-forming minerals of the solar system [~47% (*Lodders,* 2003)], which makes it a tracer of the origin and evolution of not only volatile-bearing phases in the protosolar nebula, but also nearly any solid material. The University of Chicago is where the geochemistry of stable isotopes was developed before World War II under the guidance of Harold Urey. There, Robert Clayton achieved outstanding advances in this field. With Toshiko Mayeda, Clayton recognized that the isotopic composition of O in meteorites does not always follow a MDF (*Clayton et al.,* 1973). On Earth, most O isotope variations are of the MDF type, i.e., are proportional to the relative mass difference between two isotopes). In a δ^{17}O/^{16}O vs. δ^{18}O/^{16}O diagram (Fig. 16), terrestrial data define a fractionation trend ["terrestrial fractionation line" (TFL)] with a slope close to 1/2 (~0.52), because ^{18}O/^{16}O, which has a relative mass difference ΔM/M of 2/16, varies twice as much as ^{17}O/^{16}O for which the relative mass difference is 1/16. Primitive and differentiated meteorites from a given family originating from a given parent body follow such a trend. In addition to the mass-dependent isotope effect, however, different families form parallel trends to the TFL, implying that there is an additional MIF effect. *Clayton et al.* (1973) analyzed CAIs as the mineral phases forming CAIs are those predicted to be the first to condense in a cooling solar nebula gas. Oxygen isotopic ratios in CAIs define a correlation with a slope close to 1 (Fig. 16). This trend implies that these objects were subject to isotope variations of the same extent for both ^{17}O/^{16}O and ^{18}O/^{16}O. Subsequent O isotopic data from bulk meteorites, lunar samples, and

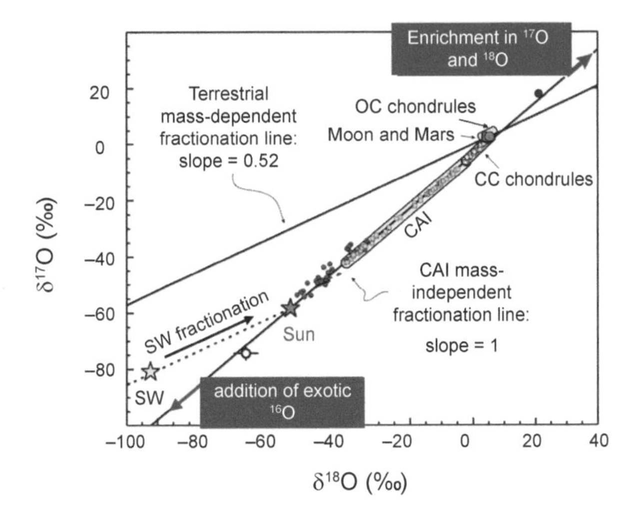

Fig. 16. Three-oxygen isotope diagram of the solar system. The coordinates are the ^{18}O/^{16}O and ^{17}O/^{16}O ratios, in parts per mil relative to terrestrial O (ocean water). Whereas most physical and chemical processes produce mass-dependent fractionations (MDF), here depicted by the slope = 0.52 terrestrial line, the Sun, CAIs, chondrules, and bulk meteorites define a distinct trend of slope ~1, reflecting mass-independent fractionation (MIF) processes. Modified after *McKeegan et al.* (2011).

martian samples defined O isotopic variations that required that small bodies accreted regionally with variable mass and independently fractionated O. This discovery permitted genetic relationships among meteorite family members to be established, leading to the O isotopic compositions of extraterrestrial material becoming one of the most powerful tracers in cosmochemistry.

In the O isotope diagram (Fig. 16), variations could be explained by two different possibilities. They could result from the contribution of an exotic ^{16}O-rich nucleosynthetic component, possibly from the explosion of a nearby supernova, to a "normal" (Earth-like) component, but this possibility was not consistent with the lack of correlation of these O isotopic compositions with those of presolar components (*Fahey et al.,* 1987; *Nittler et al.,* 2008). The alternative view was that the protosolar nebula *was itself* rich in ^{16}O, and that a process was able to conjointly enrich ^{17}O and ^{18}O relative to ^{16}O, either by addition of nucleosynthetic products or by MIF (*Thiemens,* 1999; *Clayton,* 2002; *Yurimoto and Kuramoto,* 2004; *Lyons and Young,* 2005). The best way to decide between these far-reaching possibilities was to determine where the Sun would plot in the O isotope diagram.

McKeegan et al. (2011) succeeded in measuring the three O isotopes of the solar wind (SW; yellow star in Fig. 16) in the returned Genesis solar-wind sample. To obtain the solar value (and therefore that of the proto-solar nebula), the measured SW O isotope composition had to be corrected for isotope fractionation during solar wind acceleration, according to the so-called Coulomb drag model (*Bodmer and Bochsler,* 1998). The resulting composition (red star) falls perfectly as an end member of the MIF line (*McKeegan et al.,* 2011). This composition is fully consistent with data obtained for CAIs, consistent with the latter being the results of early condensation of the nebular gas. Thus, all solar system objects and reservoirs except CAIs (and some other rare meteoritic mineral phases) were enriched to a comparable extent in ^{17}O and ^{18}O relative to ^{16}O, implying that solar system O was not contributed by late-stage injection of nucleosynthetic ^{16}O, but that processes able to enrich solids in the rare, heavy isotopes of O existed. This is further supported by the observation that ^{16}O-rich presolar dust grains found in meteorites are extremely rare, compared to those enriched in ^{17}O and/or ^{18}O (*Zinner,* 2003; *Nittler and Ciesla,* 2016).

The observation of O MIF has led to a new line of interpretation invoking photon-gas interactions and substantiated by laboratory experiments (*Chakraborty et al.,* 2008). One of the possibilities is shelf-shielding during photodissociation of CO, as observed in giant molecular clouds (*Bally and Langer,* 1982). Because ^{16}O is much more abundant than ^{17}O and ^{18}O, UV photons able to dissociate $C^{16}O$ become rapidly depleted by absorption compared to those able to dissociate $C^{17}O$ and $C^{18}O$. Consequently, photodissociation of molecules hosting ^{17}O and ^{18}O takes place deeper in the cloud than for ^{16}O. Self-shielding effects could have taken place in the parent molecular cloud illuminated by nearby stars (*Yurimoto and Kuramoto,* 2004) or at the disk surface of the protosolar nebula with photons from the nascent Sun (*Lyons and Young,* 2005). Photodissociation of gaseous CO could represent a significant source of atomic C and O that could react with H_2 to form hydrocarbons (C_xH_y molecules) and H_2O molecules, respectively. Icy grains formed in this way could then be transported toward the center of the disk where their O isotopes could exchange with those of silicates that eventually got incorporated into primitive meteorites. Hence, CO photodissociation may not only produce the large range of mass-independent O isotope fractionation measured across the solar system, but also provide an effective pathway for the formation of solid phases including ice and organics, which represent the main inventory of volatiles in early forming planetesimals.

The possibility of an irradiation origin for O MIF has been substantiated by key measurements in meteorites. *Sakamoto et al.* (2007) used SIMS to analyze an unusual iron oxide/sulfide phase (now known as cosmic symplectite) in the highly primitive ungrouped carbonaceous chondrite Acfer 094 and found similar $\delta^{17}O$ and $\delta^{18}O$ values as high as +180‰ (Fig. 17). According to these authors, the cosmic symplectite was likely produced by reactions between primordial water and metallic Fe-Ni or iron sulfides and argued that their observation demonstrates the role of water ice as a carrier of the O isotope signature produced by CO self-shielding. More recently, *Nittler et al.* (2019) reported the discovery of an unusual extremely C-rich clast in an Antarctic CR chondrite, Lapaz Icefield 02432. This highly porous clast contained Na-sulfate minerals with a ^{16}O-poor isotope composition similar to that of cosmic symplectite.

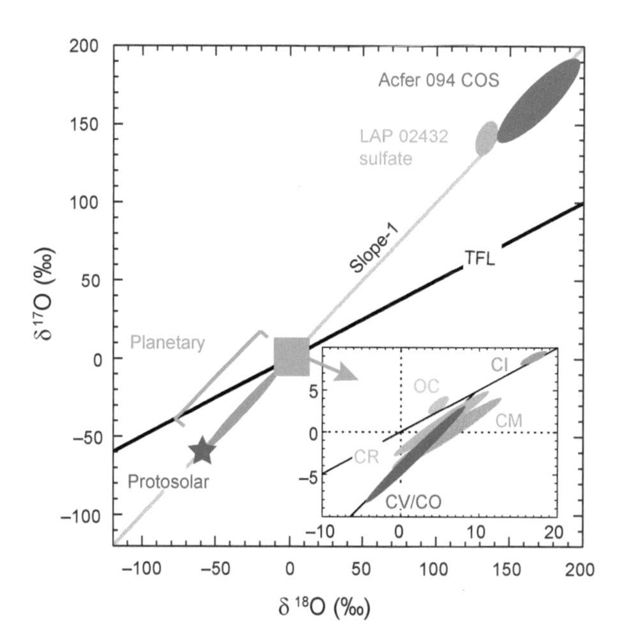

Fig. 17. Large-scale variations of oxygen isotopes in the solar system and relationships between planetary bodies. Protosolar system is bulk Sun inferred from Genesis solar-wind measurements (*McKeegan et al.,* 2011) and "Planetary" refers to bulk chondrites, Earth, Moon, Mars, asteroids, and CAIs. OC, CI, CM, CR, and CV/CO refer to different groups of chondritic meteorites (Fig. 3). Acfer 094 COS is cosmic symplectite (*Sakamoto et al.,* 2007).

These authors proposed that the clast was a cometary xenolith that accreted water ice, and that subsequently reacted with other materials to form the sulfate, providing further evidence for a ^{16}O-poor ice reservoir.

Whatever the origin of O isotope variations, O MIF constitutes an exceptional tool to study genetic relationships between different classes of meteorites and planets. Indeed, meteorites originating from common parent bodies possess a characteristic MIF signature that allows one to identify genetic relationships and define classes of meteorites (see inset in Fig. 17). The diversity of MIF-O signatures and the grouping of values with meteorite classes strongly suggest that small bodies were imprinted very early with specific O isotope signatures and that different classes of meteorites originate from common parent bodies. Enstatite chondrites, the Moon, and Earth have similar MIF-O signatures, suggesting they originated from a common pool of matter. This similarity constitutes an important constraint on the origin of the Moon-Earth system.

2.6.2.2. Hydrogen isotope variations: Hydrogen shows the largest stable isotope variations among the light elements (Fig. 18). From the comparison of H isotopes in the atmosphere of Jupiter and Saturn (*Lellouch et al.,* 2001) and from that of the solar H/He ratio and the $^3He/^4He$ ratio of the solar wind (where solar ^3He has been mainly produced by deuterium burning), it was found that Earth's oceans as well as meteorites are enriched in D by a factor of about 6 with respect to the protosolar nebula (*Geiss and Reeves,* 1972). Furthermore, the D/H ratios appear to increase with

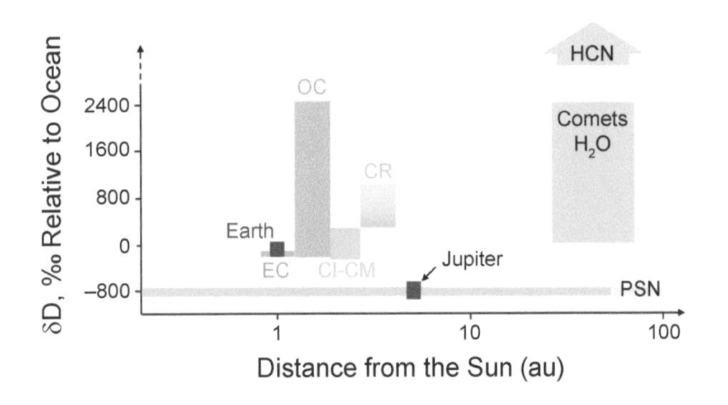

Fig. 18. Hydrogen isotopic variations as a function of present-day radial distance from the Sun (au). Meteorite and planetary data are compiled in *https://doi.org/10.24396/ORDAR-66*. δD is the deviation from the isotopic composition of the standard, which is the D/H of ocean water hydrogen. Jupiter's atmosphere has a δD (*Lellouch et al.,* 2001) similar to that of the protosolar nebula (PSN) (*Geiss and Reeves,* 1972), indicating that hydrogen in the jovian atmosphere originated from gravitational capture of the nebular gas. Cometary data are from *Bockelée-Morvan et al.* (2015), *Altwegg et al.* (2015), *Lis et al.* (2019), and references therein. The different chondrite classes represented here (enstatite chondrites, EC; ordinary chondrites, OC; and CI-CM-, CB-, CR-type carbonaceous chondrites) represent distinct meteorite parent bodies that accreted at different times and locations in the PSN (*Desch et al.,* 2018). Modified after *Broadley et al.* (2022).

radial distance from the Sun, with the highest bulk values being found in cometary ice (*Bockelée-Morvan et al.,* 2015; *Altwegg et al.,* 2015). SIMS measurements have revealed much larger D/H variations at the microscale in meteorites (*Busemann,* 2006) and especially cometary interplanetary dust particles (*Messenger,* 2000) and ultracarbonaceous Antarctic micrometeorites (*Duprat et al.,* 2010). These micrometer-scale variations are largely observed in refractory organic matter (*Alexander et al.,* 2017) but in primitive ordinary chondrites are also seen in hydrated silicates (*Deloule and Robert,* 1995). Ion-molecule exchange reactions at low temperature in dense molecular clouds, which favor the heavy isotopes with the lowest zero-point energy in the reactants (*Aikawa et al.,* 2018), could explain some of the D enrichments found in molecular cloud ices (*Cleeves et al.,* 2014) and meteoritic organic matter. Thus, H isotope variations could represent pristine fingerprints of molecular cloud chemistry, with the outer solar system having better preserved unprocessed (D-rich) material than inner regions. This possibility requires H now found in planetary bodies to have been isolated from the protosolar nebula gas, probably in the form of icy dust that did not equilibrate with nebular H_2 or trapped in refractory organic matter, as suggested by the chondrite data. With respect to the latter, some of the meteoritic classes show large enrichments in D, notably ordinary chondrites (OC in Fig. 18), which are detected using ion probe techniques at the microscale, interbedded within "normal" material (*Deloule and Robert,* 1995). CR and some of the CM chondrites also present elevated D/H ratios (*Alexander et al.,* 2017) whose origin(s) and timing of delivery are actively debated (*McCubbin and Barnes,* 2019).

2.6.2.3. Nitrogen isotopic variations: After H, the two isotopes of N (^{14}N and ^{15}N) present the largest variations among major solar system objects and reservoirs. The $^{15}N/^{14}N$ ratio varies by up to a factor of 6 while other volatile elements like C present modest variations of a few tens of permil. The N isotope variability was posited with the return of the Apollo samples in the 1970s, when N isotope analysis of lunar soils revealed variations of several hundreds of permil, which were at that time attributed to secular change in solar wind composition (*Kerridge,* 1993). A different explanation emerged with the determination of the solar $^{15}N/^{14}N$ ratio, which had originally been assumed to be similar to that of Earth. The analysis of Jupiter's atmosphere by spectroscopy (*Fouchet et al.,* 2000) and mass spectrometry onboard the NASA Galileo probe (*Owen et al.,* 2001) revealed that jovian N is depleted in ^{15}N by about 400‰. The analysis of N in the solar wind-irradiated grains of lunar soils suggested that the Sun is poor in ^{15}N, with an upper limit for $\delta^{15}N$ of –250‰ (*Hashizume et al.,* 2000). Furthermore, an osbornite (TiN) inclusion in a CAI showed a ^{15}N depletion of –360‰ (*Meibom et al.,* 2007). In contrast, enrichments in ^{15}N are common in the solar system, with $\delta^{15}N$ in some meteorites of the CR-CB clan up to +1500‰ (*Grady and Pillinger,* 1990) and N-bearing molecules in comets being systematically rich in ^{15}N by about 1000‰ (*Bockelée-Morvan et al.,* 2015). Nitrogen in chondrites is mainly hosted by organic matter, which could constitute its main carrier from the site of isotope fractionation to the inner solar system where it was incorporated into primitive meteorites.

The NASA Genesis mission confirmed that the Sun is depleted in ^{15}N, relative to planetary materials. The solar wind N was analyzed at the Centre de Recherches Pétrographiques et Géochimiques (CRPG) in Nancy (France) by static mass spectrometry (*Marty et al.,* 2010) and by SIMS (*Marty et al.,* 2011). The results confirmed that solar wind N is ^{15}N-poor, but the precision improved by 1 order of magnitude ($\delta^{15}N_{SW}$ = –407 ± 7‰; 95% confidence interval). After correction for isotope fractionation related to due to solar wind generation, (*Marty et al.,* 2011) proposed that the protosolar nebula, now represented by the Sun, had a composition of $\delta^{15}N_{SW}$ = –383 ± 8‰. This required that the Sun, which contains more than 99.8% of present-day solar system matter, captured the largest share of dust and gas present in the protosolar nebula. By contrast, all solar system objects except the atmospheres of giant planets are rich in ^{15}N by more than 60% (Fig. 19). With the new SW N data, N isotope variations among solar system objects defined an evolution trend with ^{15}N excess relative to the solar composition increasing with the distance from the Sun.

The cause of these remarkable N isotopic variations is debated, and several models have been put forward in

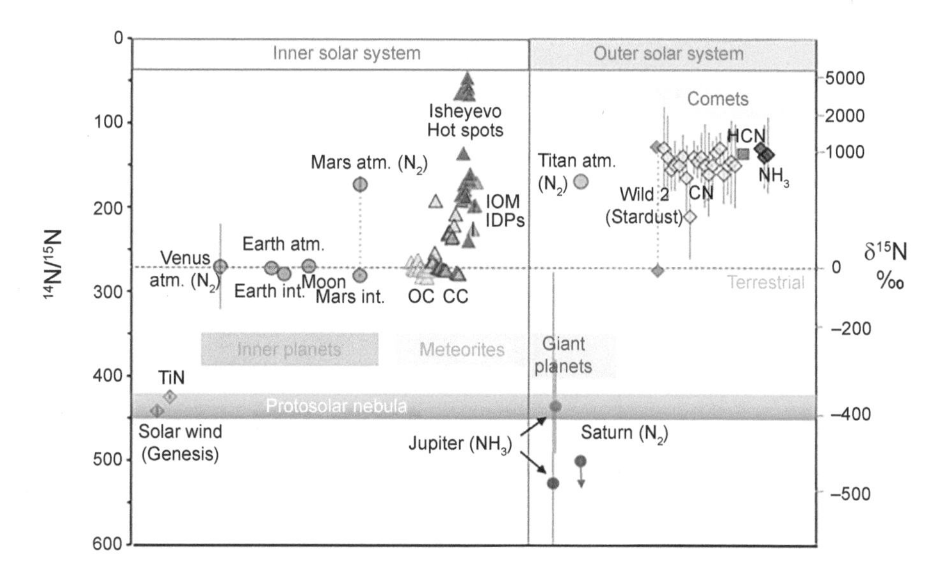

Fig. 19. Nitrogen isotope variation among solar system objects and reservoirs (adapted from *Füri and Marty,* 2015, and references therein). The lefthand axis gives the values as $^{14}N/^{15}N$, a notation used by astronomers, and the righthand axis is the $\delta^{15}N$ notation used by geochemists and cosmochemists. Jupiter atmosphere, a TiN inclusion in a CAI, and solar wind analyzed by Genesis define the composition of the protosolar nebula (horizontal blue bar). Other objects and reservoirs are ^{15}N-rich. Note that some meteorites are as rich as comets in ^{15}N, suggesting a common occurrence of ^{15}N-rich material. The martian atmosphere is also ^{15}N-rich, presumably as a result of isotope fractionation during atmospheric escape.

the community. Isotopic fractionation during ion-molecule exchange could have taken place in the outer parts of the disk or in the molecular cloud before disk collapse, where temperatures of about 10 K would have been required to yield such large isotopic excursions (*Rodgers and Charnley,* 2008). Another possibility requires illumination of the gas by UV photons, capable of triggering photodissociation of N_2 (presumably the main phase of N in the nebula) and photochemistry to form organic molecules. Self-shielding can occur when stellar photons penetrate a cloud of gas and become progressively absorbed by photoreactions, as observed in giant molecular clouds (*Bally and Langer,* 1982). Because ^{14}N is more abundant compared to the rarer ^{15}N ($^{15}N/^{14}N = 0.00365$), UV photons able to dissociate $^{14}N^{14}N$ become rapidly depleted by absorption compared to those able to dissociate $^{15}N^{14}N$ and $^{15}N^{15}N$. Consequently, photodissociation of N_2 hosting ^{15}N takes place deeper in the cloud than for ^{14}N. Self-shielding effects could have taken place in the parent molecular cloud illuminated by nearby stars and/or at the disk surface of the protosolar nebula with photons from the nascent Sun. Isotopically fractionated N would then be isolated from the gas in a solid form (organics or ice grains), hampering isotope reequilibration with nebular N_2. An alternative possibility has been advanced by *Chakraborty et al.* (2014), while experimentally studying photon-induced chemistry in a gas phase made of H_2 and N_2, illuminated by photons below 110 nm from a synchrotron light source. They found enrichments up to 1200% in produced NH_3, observed independently of the experiment temperature, that they attributed to preferential exchange of ^{15}N-bearing molecules in their excited states. Because N has only two stable isotopes, it is not possible to decipher if MIF, as a result of self-shielding, or MDF due to low-temperature isotope fractionation is responsible for N isotope variations.

2.6.2.4. Carbon stable isotope variations: A limited number of data suggests that C may also present relatively large isotope variations (Fig. 20). The most commonly used reference standard for calculating $\delta^{13}C$ values is the $^{13}C/^{12}C$ of terrestrial carbonate (Pee Dee Belemnite). The protosolar nebula value was estimated from measurements of solar-wind-irradiated lunar soil grains (*Hashizume et al.,* 2004) and spectroscopic analyses of photospheric CO (*Lyons et al.,* 2018). *Lyons et al.* (2018) inferred that C in the solar system is an isotopic mixture of C-rich grains enriched in ^{13}C and CO depleted in ^{13}C. Although $\delta^{13}C$ values have been measured for comets [e.g., 65 ± 51‰ for CO_2 in the coma of Comet 67P/Churyumov-Gerasimenko (67P/C-G) (*Hässig et al.,* 2017)], the large associated uncertainties do not allow clear comparison to other planetary objects. Carbon isotopic effects may arise from self-shielding of CO molecules in the molecular cloud (*Clayton,* 2002). The fact that bulk C isotope variations are about 1 order of magnitude lower than those of N could be due to faster back reactions than for the photodissociation of N_2. Some organic particles in chondrites exhibit large C isotope variations (with ^{13}C enrichments and depletions of up to hundreds of permil, again likely due to low temperature reactions) but these are much rarer than seen for H and N anomalies (*Floss and Stadermann,* 2009; *Nittler et al.,* 2018).

2.6.2.5. Noble gases: Solar vs. planetary: There are five stable noble gases: He (atomic mass 4.00 u), Ne (20.18 u), Ar (39.95 u), Kr (83.80 u) and Xe (131.29 u), all of which are excellent tracers of the origin(s) and evolution of small

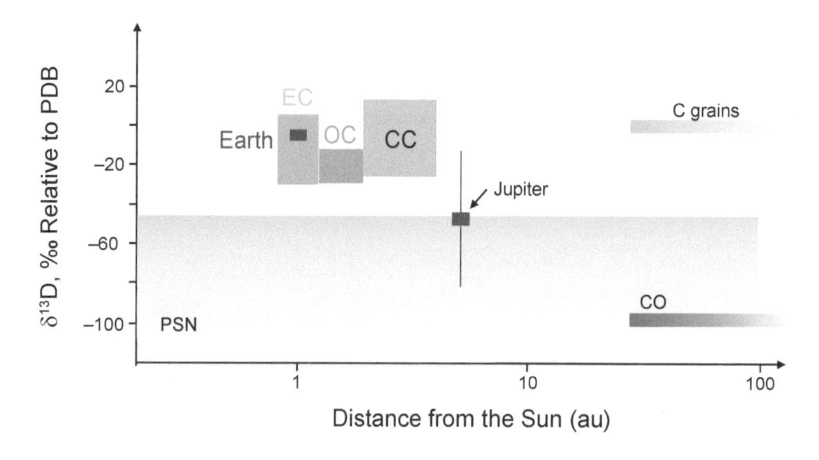

Fig. 20. Carbon isotope variations in the solar system (see text for references). Modified after *Broadley et al.* (2022).

bodies. They are chemically inert in natural environments and their isotopes and abundances can only vary through physical processes such as phase changes, isotopic fractionation, and nuclear reactions. The stable noble gases cover the whole spectrum of masses and their behavior in natural systems, such as solubility in minerals and magmas, is mainly governed by their contrasting atomic sizes. Some of their isotopes are primordial (inherited from the stellar environment), other are produced by nuclear reactions such as spallation, neutron activation, radioactive decay, and fission. Given these characteristics, they have the potential to act as geochronometers as well as excellent tracers of the origin and history of their host phases (*Ozima and Podosek,* 2002; *Ott,* 2014). In this section, we show that noble gases trapped in small bodies are non-solar, i.e., their abundance and their isotopic compositions cannot be related directly to the protosolar nebula gas.

In the protosolar nebula, light noble gases were among the most abundant elements (Fig. 13) (*Lodders,* 2003). The analysis of gases sublimated from the coma of 67P/C-G has shown that they are also abundant in cometary ice, as these elements were cryotrapped in forming icy dust depending on the environmental (cold) temperature (*Balsiger et al.,* 2015; *Marty et al.,* 2017; *Rubin et al.,* 2018). By contrast, noble gases are "rare" on Earth and in meteorites, and for this reason they are also referred to as the rare gases (*Ozima and Podosek,* 2002). Their scarcity in silicates and metal is the result of their chemical inertness, which prevented them from being efficiently trapped in solids at temperatures typical of the inner part of the disk. Noble gases are measured by static mass spectrometry, which permits the quantification of isotopes in amounts as low as a few million atoms. One of the main geochronological tools is the K-Ar dating system, where ^{40}K, a rare isotope of K (0.01167%), decays to ^{40}Ar ($T_{1/2}$ = 1.25 G.y.). Because ^{40}Ar was barely produced by stellar nucleosynthesis, almost all ^{40}Ar found in planetary bodies including meteorites has been produced by the decay of ^{40}K and the K-Ar dating method and its elaborated so-called Ar-Ar method are among the most sensitive geochronological tools to establish the timeframe of solar system formation and evolution (*Lee,* 2015).

The so-called extinct radioactivities were first discovered thanks to the study of noble gases, in particular Xe, in meteorites. Nucleosynthetic models predicted that some of the isotopes synthesized in stars, having half-lives shorter than the age of the solar system, could have been alive when the solar system started to condense. Iodine-129 is a radioisotope that decays to ^{129}Xe with a half-life of a few million years ($t_{1/2}$ = 15.7 m.y.) and should have been produced in massive stars together with the stable isotope of I, ^{127}I. A monoisotopic excess of ^{129}Xe (with respect to the "normal" Xe isotope composition) was found in the Richardton meteorite and correctly attributed to the *in situ* decay of ^{129}I (*Reynolds,* 1960). This was the first experimental evidence that extinct radioactivities were alive when the solar system formed. The ^{129}I-^{129}Xe dating method became one of the most important tools to date the first tens of million years of planetary evolution including the terrestrial atmosphere (*Wetherill,* 1980). Noble gases were subsequently used as tracers for the occurrence of presolar material.

Other examples of isotopically unusual noble gases were found soon thereafter in primitive meteorites and interpreted as signs of surviving refractory presolar grains bearing stellar nucleosynthesis signatures (*Reynolds and Turner,* 1964; *Black and Pepin,* 1969) In order to identify the carriers of exotic noble gases, meteorite samples were subjected to sequential chemical attacks and after each step, noble gases were analyzed (*Lewis et al.,* 1987, 1994; *Amari et al.,* 1990; *Huss et al.,* 2003). This approach eventually permitted the identification of acid-resistant presolar grains including nanodiamonds, graphite, silicon carbide, etc., that were subsequently isolated and analyzed for stable isotopes with ion probes (see section 2.7.1) (*Zinner,* 1998).

The remarkable noble gas isotope variations seen in presolar material do not reflect their composition in bulk meteorites, which is instead dominated by a component common to most meteoritic classes. As for the stable isotopes of H, C, N, and O, noble gases trapped in meteorites and inner planets differ from the solar composition. Not only are their isotopic ratios non-solar, but also their relative abundances are fractionated relative to the solar composition, as approximated by the noble gas composition of the solar wind (see *Ott,* 2014, and references therein).

In primitive meteorites as well as in inner planet atmospheres (Mars and Earth), noble gases are depleted by several orders of magnitude with light noble gases being the most depleted (Fig. 21). They are ubiquitous in primitive meteorites, being hosted by a common phase labeled Phase Q and linked to refractory organic matter. Phase Q noble gases form a rather homogenous reservoir of noble gases having comparable elemental abundance patterns and isotopic ratios that dominate the inventory of Ar, Kr, and Xe in chondrites (*Busemann et al.,* 2000). In addition to Phase Q, several exotic components hosted by presolar phases (nanodiamond, silicon carbide, etc.) are also present in chondrites as discussed above. The origin of Phase Q has been debated for a long time. Ionization experiments are able to reproduce some of the characteristics of Q noble gases (elemental and isotopic fractionation), suggesting that ionization was the main process (apart for cryotrapping) for incorporating efficiently noble gases in solids (*Bernatowicz and Hagee,* 1987; *Matsuda and Maekawa,* 1992). Ionization experiments involving organics also reproduced some of the features of meteoritic organic matter in addition to fractionating the isotopes of noble gases (*Kuga et al.,* 2015, 2017; *Marrocchi et al.,* 2005). Therefore ionization of a reduced gas rich in H_2, CO, N_2, and noble gases has the potential of not only synthesizing organic matter reminiscent of that trapped in primitive meteorites, but also of trapping efficiently noble gases with the required elemental and isotopic fractionation. However, so far these experiments did not reproduce the thermal resistance of meteoritic organics. Q noble gases are often extracted at temperatures >1300 K whereas in experimental runs they often degas below 800 K, and additional processing appears necessary. All these data converge with an important role of gas-photon interactions as the main driver of stable isotope and noble gas fractionation.

The specific isotopic compositions of noble gases in solar system objects and reservoirs permit the investigation of processes of planetary formation and evolution. An example is given in Fig. 22, which represents a comparison of the Xe isotopic compositions between solar and planetary components. Here, meteoritic and Earth's atmospheric Xe present excesses in radiogenic ^{129}Xe (an extinct radioactivity permitting fine chronology of the first hundreds of millions of years) as well as MDF plus nucleosynthetic anomalies. In particular, the terrestrial atmosphere has high MDF [which has been regarded as preferential escape of atmospheric Xe over geological periods of time (*Pujol et al.,* 2011)] as well as depletions in ^{134}Xe, ^{136}Xe, a likely nucleosynthetic anomaly. Such isotope variations allow one to investigate genetic relationships between primitive and evolved materials. For instance, the contribution of cometary matter to Earth's atmosphere was identified for the first time thanks to the specific isotopic composition of Xe measured in 67P/C-G (*Marty et al.,* 2017). Likewise, noble gases trapped in SNC meteorites were the smoking gun for a martian origin of this class of meteorites (*Becker and Pepin,* 1984; *Trei-*

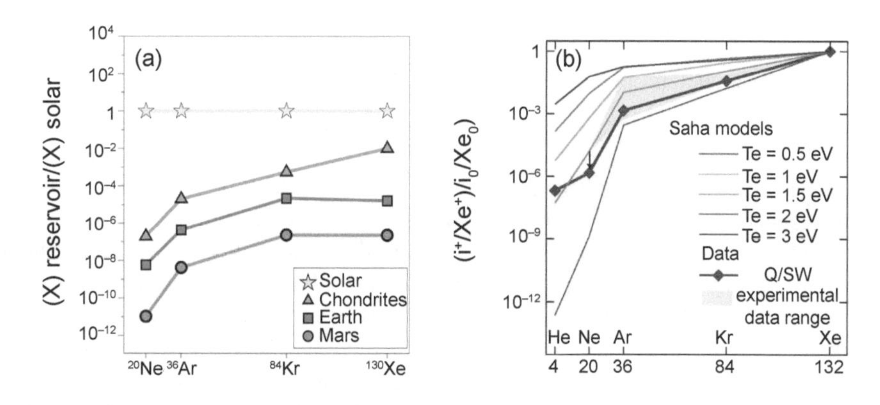

Fig. 21. Noble gas elemental fractionation of planetary bodies with respect to solar composition. **(a)** The solar system composition is approximated by that of solar system wind (*Meshik et al.,* 2014; *Vogel et al.,* 2011) (yellow stars). Chondrites (*Pepin,* 1991), Earth's atmosphere (*Ozima and Podosek,* 2002), and Mars' atmosphere (*Swindle,* 2002) are depleted relative to solar, with neon (and helium, not represented) being the most depleted elements. The depletion pattern of chondrites and inner planets, sometimes labeled misleadingly "planetary," was attributed to the poor retention of chemically inert noble gases into solids at temperatures typical of the inner solar system. The somewhat less-pronounced depletion of the heavy noble gases (e.g., Xe) compared to the light ones (e.g., Ne) was attributed to the better "sticking" efficiency of the former due to van der Walls bonding. Note that the atmospheres of Earth and Mars (and Venus, but venusian data are imprecise) are depleted in xenon relative to krypton and chondritic, which defined the "mission Xe" problem. It was recently shown that xenon was lost from the terrestrial atmosphere by ionization-related escape to space over geological periods of time (*Avice et al.,* 2018) and is therefore not a primordial feature. **(b)** The noble gas fractionation pattern of chondrites, normalized to xenon and solar (thick red line), is compared to fractional ionization in a Saha-type plasma with electron energies of a few electron volts (*Kuga et al.,* 2017).

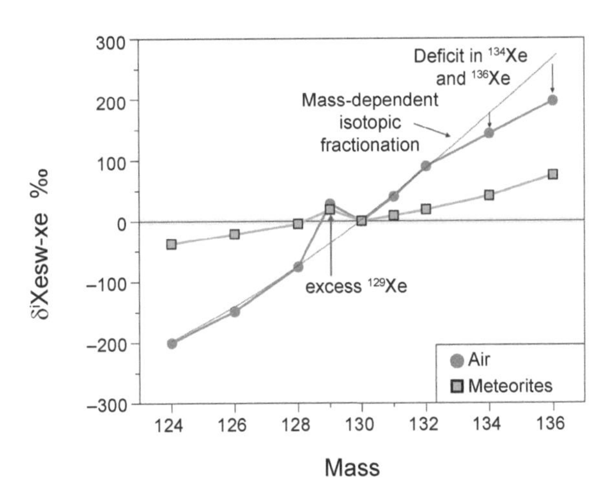

Fig. 22. Xenon isotope diagram for solar system reservoirs (data from *Ott,* 2014). The nine Xe isotopes are normalized to one of them, here ^{130}Xe, and to a specific reference composition, here solar wind (SW) Xe, assumed to represent the composition of the solar nebula from which planetary bodies formed. The isotopic deviations from the reference composition (solar) are given in the classical δ notation: $\delta^i Xe = [(^iXe/^{130}Xe)_{sample}/[(^iXe/^{130}Xe)_{SW}-1] \times 1000$ (in parts per mil). In this format, a solar-like isotope composition yields a horizontal pattern similar to SW ($\delta^i Xe = 0‰$). Earth's atmospheric xenon differs from SW by three important features: (1) it contains a monotopic excess of ^{129}Xe from the decay of ^{129}I (extinct radioactivity $T_{1/2}$ = 15.7 m.y.), (2) it is isotopically fractionated by 35‰ per atomic mass unit relative to SW, and (3) it is depleted in heavy Xe isotopes: ^{134}Xe and ^{136}Xe, relative to a line that passes through the other isotopes, and that characterizes mass-dependent isotopic fractionation. The MDF-Xe has been attributed to atmospheric escape (*Pujol et al.,* 2011) and the deficit in ^{134}Xe and ^{136}Xe is a nucleosynthetic feature inherited from comets (*Marty et al.,* 2017).

man et al., 2000). Some of the noble gas isotopic structures were inherited from previous generations of stars, and noble gases were among the first isotopic systems used to identify presolar material preserved in meteorites. For example, the specific Ne isotopic composition of the terrestrial atmosphere is probably a remnant of the contribution of exotic nanodiamonds trapped in carbonaceous chondrites (*Marty,* 2012; *Mukhopadhyay and Parai,* 2019).

As for stable isotopes described above, noble gases are isotopically non-solar in primitive meteorites and, as for N and partly H, they are associated with organics. Their elemental and isotopic fractionation might have resulted from trapping in organics during interactions between stellar photons, from the proto-Sun or from nearby stars, and the nebular gas rich in H_2 and CO (*Kuga et al.,* 2015; 2017). Such an origin implies large scale transport in the protoplanetary disk from the regions of organic production to those of primitive meteorite building up.

2.6.3. Nucleosynthetic isotopic anomalies and the non-carbonaceous–carbonaceous chondrite isotopic dichotomy. *2.6.3.1. Nucleosynthetic isotopic anomalies:* The so-called nucleosynthetic isotope anomalies are a special class of mass-independent isotope variations found in meteorites and on the planetary scale. Isotopes associated with a growing number of elements (e.g., He, Ne, Kr, Xe, Ca, Ti, Cr, Fe, Ni, Sr, Zr, Mo, Ru, Pd, Ba, Nd, Sm, Er, Yb, Hf, W, and Pt) have been found to show nucleosynthetic isotope anomalies at the bulk meteorite scale (for reviews, see *Ott,* 2014; *Qin and Carlson,* 2016; *Valdes et al.,* 2021; *Bermingham and Kruijer,* 2022). These subtle isotopic anomalies arise from the imperfect homogenization of presolar grains throughout the protoplanetary disk, and they are expressed as positive or negative deviations of their isotope ratios in parts per 10^4 (epsilon, ε-units) or 10^6 (mu, μ-units; or part per million) from terrestrial reference standards.

The nucleosynthetic isotope composition of a sample is used to constrain the bulk (or average) composition of the parent body, and thus the nebular region from which it accreted (e.g., *Walker et al.,* 2015). As discussed above in section 2.6.2.5., noble gas isotopic variations led to the identification of many specific presolar grain types. As nucleosynthetic isotope anomalies are similarly caused by the heterogeneous distribution of presolar grains, these isotope effects were anticipated to be on the same scale as those recorded by noble gases. The bulk effects for noble gases, however, are much larger due to the rarity of these elements in most solar system solids. By contrast, many of the refractory elements that display nucleosynthetic isotope anomalies are much more abundant such that the nucleosynthetic signatures of individual presolar grains are highly diluted (Fig. 23). Nucleosynthetic isotope anomalies are also found in individual meteorite components including CAIs and refractory hibonite inclusions, and in successive acid leachates of different meteorites. These anomalies are, however, typically highly variable and more anomalous than bulk meteorite samples [e.g., Ti isotopes (*Davis et al.,* 2018), Mo isotopes (*Brennecka et al.,* 2020), Ca isotopes (*Valdes et al.,* 2021)], which is expected because the bulk samples are mixtures over large numbers of individual components.

Not all elements display nucleosynthetic isotope variations on the bulk sample scale. For example, nucleosynthetic Os isotope variations are not detected in most bulk meteorites. Ureilites are an exception; however, it has been suggested that Os was mobilized as a result of parent body thermal-based processes rather than this being a bulk characteristic of the parent body (*Goderis et al.,* 2015). Osmium isotopic variations are also documented in acid residues of some chondrites and the extent of an Os isotopic anomaly in an acid residue can correlate with the degree of aqueous alteration of the host chondrite (*Yokoyama et al.,* 2011, 2007). These data indicate that Os isotope anomalies can also be caused by selective destruction/modification of presolar grains carrying Os during progressive aqueous alteration on parent bodies. The generally non-anomalous Os isotopic composition of most bulk samples measured to date (except ureilites), however, demands that the isotopically anomalous Os released from presolar phases during

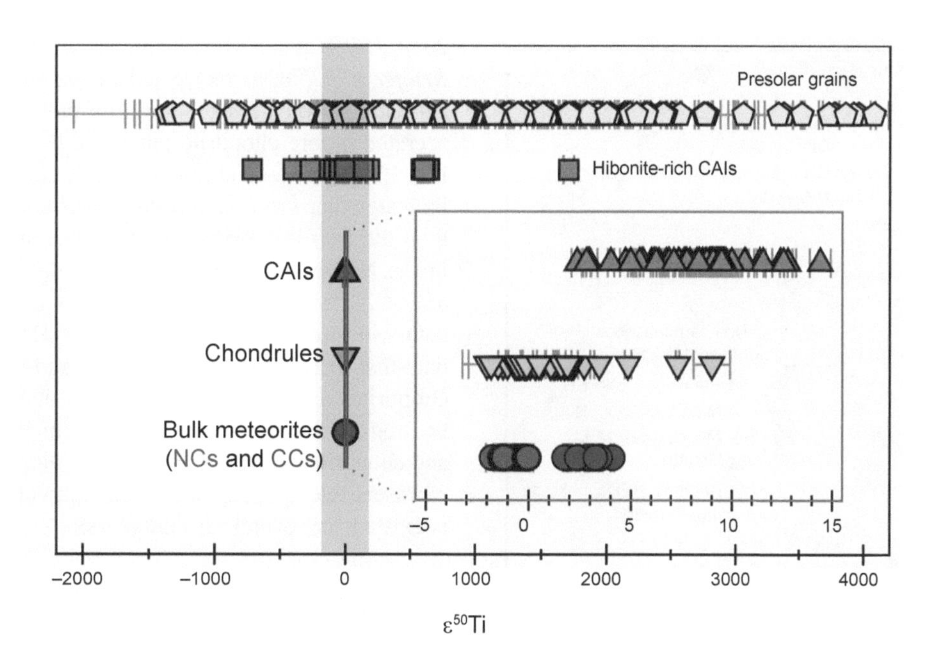

Fig. 23. ε^{50}Ti data for individual presolar grains (PSG; yellow), individual CAIs, bulk terrestrial samples, and NC-CC meteorite samples. The inset figure plots a zoomed-in display of the whole-rock and CAI data. The gray field corresponds to the range of terrestrial sample compositions and analytical precision reached in the respective studies. (*Torrano et al.,* 2019; *Alexander and Nittler,* 1999; *Amari et al.,* 2001; *Davis et al.,* 2018; *Gerber et al.,* 2017; *Gyngard et al.,* 2018; *Hoppe et al.,* 1994; *Huss and Smith,* 2007; *Ireland et al.,* 1991; *Render et al.,* 2019; *Trinquier et al.,* 2009; *Zhang et al.,* 2012; *Zinner et al.,* 2007). Adapted from *Bermingham et al.* (2020).

processing was reincorporated into a phase(s) that was not lost from the part of the parent body where most meteorites were sampled (*Yokoyama et al.,* 2011).

The existence of subtle nucleosynthetic isotope heterogeneity indicates that the protoplanetary disk became reasonably, but not perfectly, homogenized by the time meteorite parent bodies accreted. The cause of the heterogeneous distribution of presolar grains in the disk remains debated. It may be a consequence of imperfect mixing of these grains throughout the protoplanetary disk (e.g., *Clayton,* 1982), their selective destruction via thermal chemical processing in the protoplanetary disk (e.g., *Trinquier et al.,* 2009), and/or size sorting processes in the risk (for a review, see *Dauphas and Schauble,* 2016). Regardless of the cause of the heterogeneous distribution of presolar grains in the disk, the unique nucleosynthetic isotope composition each parent body possesses can be harnessed to trace communication between regions of the solar system during planetary accretion.

2.6.3.2. The non-carbonaceous–carbonaceous chondrite isotope dichotomy: The NC-CC isotope dichotomy has become one of the most utilized and researched applications of nucleosynthetic isotope anomalies in cosmochemistry. Early indication of the NC-CC isotope dichotomy was reported by *Trinquier et al.* (2007, 2009). Subsequently, *Warren* (2011) observed that nucleosynthetic isotope composition of meteorites consistently fell into one of two compositional groups (Fig. 24). The first group comprises primarily carbonaceous chondrites (CO, CK, CV, CM, CB, CR, CI, Tagish Lake), an ungrouped meteorite (Northwest Africa 011), and Eagle Station group pallasites. The second group comprises meteorites and other bodies that are not carbonaceous chondrites (e.g., Earth, Moon, enstatite chondrites, ordinary chondrites, angrites, HEDs, main-group pallasites, mesosiderites, and ureilites). *Warren* (2011) labeled the first group as CC and the second group as NC. The isotopic bimodality has since been extended to siderophile elements (e.g., Mo, Ru, W) in iron meteorites (e.g., *Budde et al.,* 2019; *Kruijer et al.,* 2020; *Poole et al.,* 2017; *Worsham et al.,* 2017; *Bermingham et al.,* 2018a) and several lithophile elements (e.g., Ca, Ti, Cr, Ni, Sr, Zr, Mo, Ru, Ba, Nd, Sm, Hf, W) (for reviews, see *Scott et al.,* 2018; *Burkhardt et al.,* 2019; *Bermingham et al.,* 2018b). Using "carbonaceous chondrite" to describe iron meteorites can be confusing because iron meteorites are not carbonaceous chondrites. The NC-CC terminology, however, describes two isotopic or "genetic" reservoirs in the protoplanetary disk that are sampled by meteorites. It is now considered the highest taxonomic division in meteorite/planetary classification. To date almost all meteorites fall into either the NC or CC group. Thus far, there is only one exception to this rule, the ungrouped iron meteorite Nedagolla, for which the Ni-Mo-Ru isotopic composition of this sample are intermediate between NC and CC (*Spitzer et al.,* 2022). Its ^{182}Hf-^{182}W-based chronology indicates that the mixed NC-CC composition was established >7 m.y. after CAI formation, likely from collisional mixing of preexisting NC-CC bodies.

Using ^{182}Hf-^{182}W-derived core-segregation ages from iron meteorites, *Kruijer et al.* (2017) proposed that the NC and CC groups possess distinct accretion ages, NC <0.4 m.y. after CAI formation and CC 0.9 + 0.4/−0.2 m.y. after CAI

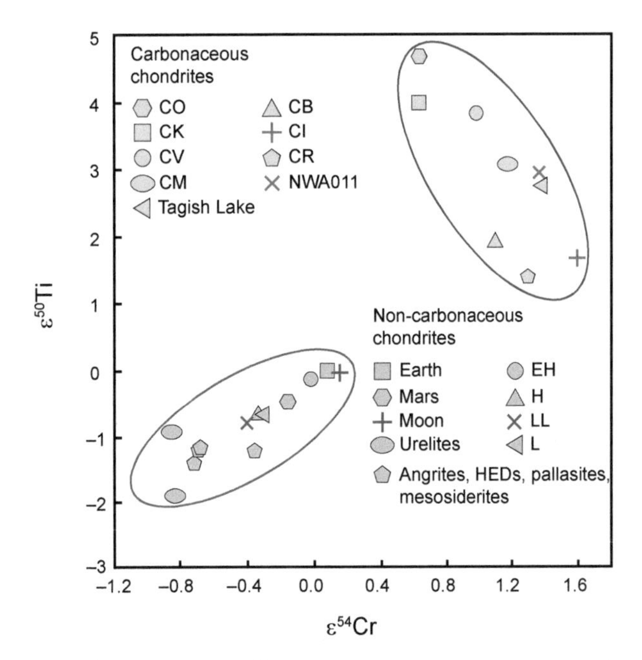

Fig. 24. ε^{54}Cr vs. ε^{50}Ti in bulk noncarbonaceous (NC, red) and carbonaceous (CC, blue) meteorites (after *Warren,* 2011). Non-carbonaceous meteorites include HEDs, ordinary chondrites, angrites, enstatite chondrites, mesosiderites, pallasites, lunar meteorites, martian meteorites, and ureilites. CCs include CI, CO3, CM2, CV3, CR2, CH3, Tagish Lake (C2-ung), CK, and CBa.

formation. Undifferentiated chondrites have not undergone metal-silicate fractionation and thus their accretion ages must be inferred by dating meteorite components that formed before accretion into parent body [e.g., chondrules (*Krot et al.,* 2018; *Alexander et al.,* 2008)]. Robust chronology comes from convergence of ages determined using several chronometers (for a review of this approach, see *Bermingham and Kruijer,* 2022). Accretion ages for differentiated achondrites (e.g., angrites and eucrites) likely underwent multiple stages of Hf-W fractionation. Accordingly, their accretion age must be inferred using Hf-W isotope evolution of mantle sources of these samples (*Kleine et al.,* 2012; *Kruijer et al.,* 2012; *Touboul et al.,* 2015), thus these ages can be less certain than core-segregation ages defined by the ^{182}Hf-^{182}W system. Accretion ages for differentiated achondrites constrain core formation to within the first 1–2 m.y. of solar system formation on the angrite and eucrite parent bodies (*Baker et al.,* 2005; *Kleine et al.,* 2012; *Touboul et al.,* 2015). The recent discovery of an andesitic meteorite, presumably from the crust of a protoplanet and having a ^{26}Al age of 2.2 Ma after CAIs, attests for melting and differentiation of the parent body within ~1 Ma after CAIs (*Barrat et al.,* 2021).

A more recent compilation of age data for multiple bodies, including new data for meteorites that had not previously been studied, showed that the accretion timescales of NC and CC meteorite groups overlap in their uncertainties, meaning the accretion ages of some NC and CC irons can be indistinguishable (*Scott et al.,* 2018; *Hellmann et al.,* 2019; *Hilton et al.,* 2019; *Kruijer et al.,* 2020) (e.g., Fig. 25). *Kruijer et al.* (2020) interpreted a compilation of chronological data to indicate that some iron meteorite parent bodies accreted before chondrite parent bodies (Fig. 25; see also Fig. 15 for a general chronological framework). This dataset indicates that some chondrite parent bodies accreted later than the iron meteorite parent bodies, starting at ~2 m.y. in the NC reservoir and continuing until at least ~4 m.y. after CAI formation in the CC reservoir. Combining these data with nucleosynthetic ^{50}Ti, ^{54}Cr, ^{58}Ni, and Mo isotope data that discriminate between NC and CC, CC irons and chondrites appear to be enriched in nuclides produced in neutron-rich stellar environments compared to the NC irons and chondrites. It was concluded that NC and CC reservoirs represent two spatially separated reservoirs of the disk that coexisted for several million years.

3. FROM STARS TO PLANETS

3.1. A Summary of the Astrophysical Context Based on Studies of Small Bodies

The astrophysical context of the solar system can be constrained using the elemental and isotopic compositions of meteorites. Volatiles, including H, C, N, O, and noble gases trapped in chondrites and comets, have elemental and isotopic compositions that are markedly different from those of the protosolar nebula, precluding a direct genetic relationship between the building blocks that make up the terrestrial planets and the gaseous solar nebula. In addition to having a non-solar volatile composition, these materials show large variations in volatile and isotopic

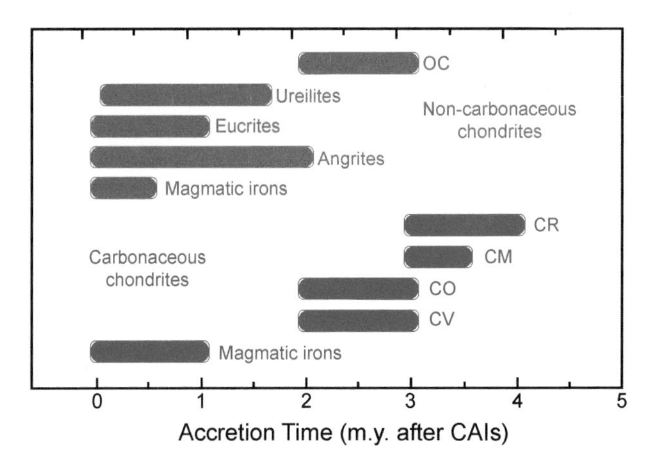

Fig. 25. Accretion timescales in million years after CAI formation of CC (red) and NC (blue) meteorite parent bodies, where the horizontal bars reflect the uncertainty of the accretion age estimates. Accretion timescales for chondrites are based on the chronology of alteration products and chondrule-derived Al-Mg, Hf-W, Pb-Pb, and/or U-Pb ages, which are integrated using thermal modeling. Accretion ages for iron meteorites are model ages derived for differentiated and thermal modeling for internal heating by ^{26}Al of small parent bodies. Figure based on *Kruijer et al.* (2020).

compositions (Figs. 17–22). These isotopic fingerprints were likely inherited from processes having taken place in the interstellar medium or in the molecular cloud. Note, however, that interactions between stellar photon and the nebular gas rich in H_2, CO, and noble gases have the potential to trigger the elemental and isotopic fractionation able to account for some of the observed chemical signatures. It has been proposed that such processes took place when the molecular cloud was illuminated by nearby stars (*Yurimoto* and *Kuramoto,* 2004), knowing that the solar system was probably born in a cluster of several hundreds to thousands of stars (*Adams,* 2010). Another potential site of irradiation is the surface of the disk illuminated by the proto-Sun (*Lyons and Young,* 2005). These possibilities are illustrated in Fig. 26. Icy and organic dusty grains would then be transported along the forming disk by molecular cloud infall, turbulence, and disk spreading.

The molecular cloud also hosted presolar grains and their preservation in meteorites indicates that not all of them were destroyed and homogenized with the gas during solar system formation. When and how the isotopic bimodality defined by nucleosynthetic isotope anomalies (NC and CC isotopic dichotomy) (Fig. 24) was established is debated. It could indicate a heterogeneous distribution of presolar grains in the molecular cloud itself such that the composition of infalling material changed with time. Evidence for early formed refractory grains with large Ti isotope anomalies but no ^{26}Al (e.g., *Kööp et al.,* 2016) may provide evidence for this as it implies late addition of ^{26}Al to the protosolar nebula, possibly contained in presolar grains or direct input from a nearby supernova. Alternatively, it could reflect differential processing of different types of presolar grains (e.g., by evaporation) in different parts of the disk. The early formation of Jupiter might have kept a barrier against transport keeping the segmentation alive for several millions of years. If heterogeneous infall occurred, it was present when the first planetesimals formed (<1 m.y. after CAI formation), as the NC-CC bimodality is recorded in early-formed iron meteorites. The infall is required to have lasted several millions of years, as indicated by the formation of NC and CC bodies (Fig. 25). How this dichotomy could have been maintained for such periods of time is fueling multiple models of solar system evolution. Below, some of these models are developed in the context of dynamical models of planetary formation.

3.2. Formation of Planets

3.2.1. General overview. In this section, models of planetary formation in the light of data obtained from the study of small bodies are reviewed. The main processes of planetary growth as understood today are summarized in Fig. 27 and the following sections. A variety of processes are at play in the growth of micrometer-sized dust grains to millimeter- to centimeter-sized "pebbles" that drift rapidly within the disk, up to 100-km-scale "planetesimals," which are generally considered the macroscopic building blocks of the planets (for a review, see *Johansen et al.,* 2014). Planetesimals may grow by mutual collisions and by accreting drifting pebbles into planetary embryos or cores. The gas giants are constrained to have formed very quickly, within the few-million-year lifetime of the gaseous disk. In contrast,

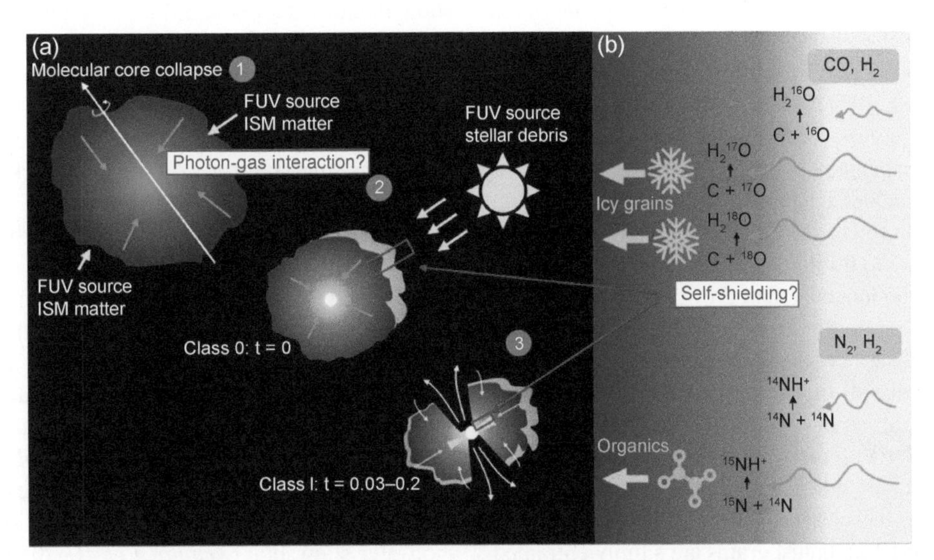

Fig. 26. Possible origins of isotopic signatures of H, C, N, O, and noble gases observed among solar system bodies (adapted from *Lee et al.,* 2008). **(a)** (1) Heritage from the interstellar medium (ISM). (2) Reactions between stellar photons from nearby stars and the nebular gas. Photodissociation of H_2, CO, and N_2 results in the formation of (1) water ice carrying MIF oxygen and (2) organics hosting excesses of ^{15}N and possibly of D. Noble gases can also be fractionated and trapped in organics at this stage. Synthesized ice and organic dusty grains are then transported to the planet-forming regions where they exchange their isotopic signatures with forming solids. (3) Photochemistry could have also taken place at the disk surface illuminated by protosolar photons. **(b)** Possible reactions between stellar photons and nebular gas (made of H_2, CO, N_2) leading to the formation of icy grains and organic dust, both hosting isotopic fractionation.

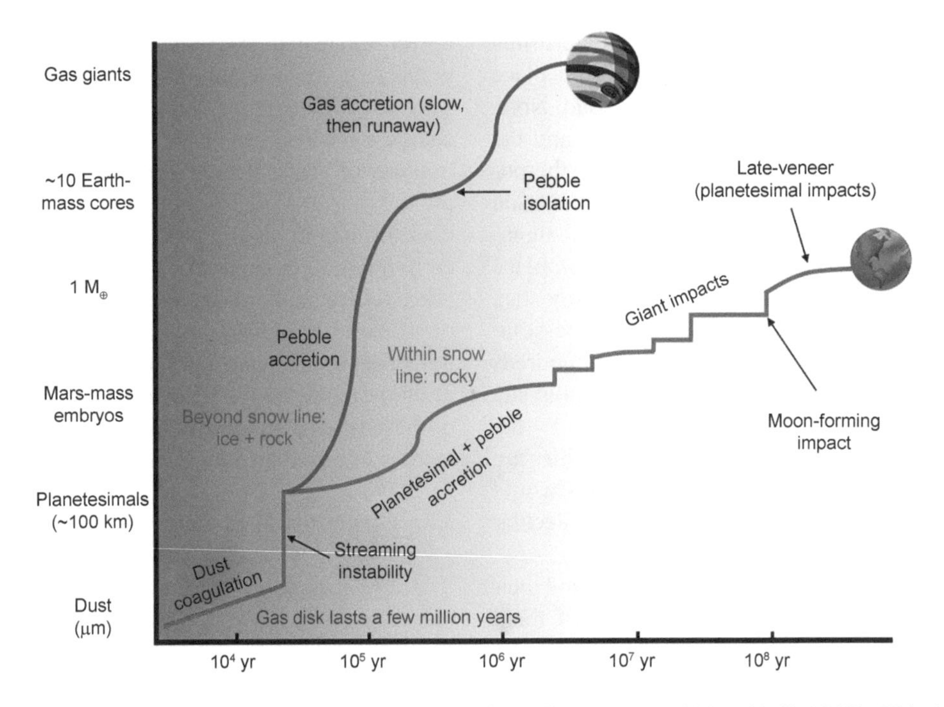

Fig. 27. General framework of planetary formation (adapted from *Raymond and Morbidelli,* 2022). This image illustrates plausible growth pathways of Earth and Jupiter, highlighting the main processes responsible. Current thinking is that the primary branching point between the two growth modes comes from the fact that Jupiter's core accreted beyond the water snow line, where the efficiency of pebble accretion was drastically higher (*Lambrechts et al.,* 2014; *Morbidelli et al.,* 2015).

cosmochemical measurements indicate that Earth probably took ~100 m.y. to complete its formation (e.g., *Kleine et al.,* 2009), although Mars' growth was complete much faster (*Dauphas and Pourmand,* 2011).

3.2.2. Processes of accretion. Although only a minor (~1%) constituent of the protoplanetary disk mass, the dust is of vital importance to us as it represents the solid building blocks of the planets. The first phases of growth involve coagulation of dust by low-velocity collisions (e.g., *Dominik and Tielens,* 1997). As dust grains grow and collide there is a rich variety of outcomes that have been studied in laboratory experiments and using numerical simulations (*Blum and Wurm,* 2008; *Güttler et al.,* 2010; *Zsom et al.,* 2010). Collisional growth is efficient at small sizes but is stifled as dust grows to millimeter to centimer sizes, simply because collisions no longer lead to growth (e.g., *Birnstiel et al.,* 2016). In addition, it is at this size scale that interactions with the gaseous disk increase in importance.

The force balance in the disk is such that, at a given orbital radius, gas orbits the Sun at a speed that is slower than the local Keplerian (purely gravitational) velocity. This is because the force of gravity is slightly counteracted by thermal pressure, as gas closer to the star is hotter and therefore pushes outward. Small dust grains are simply entrained with the gas, but as they grow the pressure force weakens. At a critical size, dust grains start to decouple from the gas' motion. Because the gas is sub-Keplerian, large dust grains feel a headwind. This headwind acts to extract orbital energy and cause the dust to spiral inward (*Weidenschilling,* 1977; *Birnstiel et al.,* 2012). Drifting dust grains are commonly referred to in the literature as "pebbles," although it is noted here that the sizes of what is considered a pebble depends on its aerodynamic properties and thus on the local gas density in the disk (*Johansen and Lambrechts,* 2017; *Ormel,* 2017).

Pebble drift was once considered a catastrophe because all macroscopic solids should simply spiral into the Sun on short timescales (*Weidenschilling,* 1977). Yet newer models show that drifting pebbles are likely the lifeblood of planet formation. The first step is the formation of planetesimals, which can form directly from pebbles by gas-particle instabilities such as the streaming instability (*Johansen et al.,* 2014). When solids are locally concentrated relative to the gas, they collectively create a back-reaction on the gas, accelerating its orbital speed in that location. This removes the headwind and stabilizes the orbits of pebbles in a ring. Other pebbles that drift inward are then trapped in the ring and can enhance this feedback on the gas, increasing the local pebble-to-gas ratio. Once this ratio exceeds a critical value, the streaming instability naturally acts to clump pebbles directly into 100-km-scale planetesimals that follow a characteristic mass distribution (*Carrera et al.,* 2015; *Simon et al.,* 2016; *Yang et al.,* 2017; *Abod et al.,* 2019). Simulations show that planetesimals are likely to form first just outside strong condensation fronts such as the water snow line (*Armitage et al.,* 2016; *Schoonenberg and Ormel,* 2017; *Drążkowska and Alibert,* 2017). While this remains a vigorous area of study, several recent studies have proposed that planetesimals may commonly form in relatively narrow rings rather than broad disks (*Drążkowska et al.,* 2016; *Charnoz et al.,* 2019; *Morbidelli et al.,* 2022; *Izidoro et al.,* 2022). The solar system appears to require at least three such rings to explain the orbital structure of the planets and small bodies. *Izidoro et al.* (2022) suggested that such rings could have been linked with pressure bumps associated with the condensation fronts of silicates, water, and carbon monoxide (corresponding to the inner–outer rings, respectively).

Once large planetesimals have formed, they continue to grow both by colliding with other planetesimals and by accreting pebbles that continue to drift through the disk. Planetesimal accretion has been studied for decades (e.g., *Safronov and Zvjagina,* 1969; *Greenberg et al.,* 1978; *Wetherill and Stewart,* 1989; *Ida and Makino,* 1992), and is generally thought to be the dominant growth mode of the terrestrial planets (*Chambers,* 2016; *Izidoro et al.,* 2021). Starting from a broad disk of planetesimals assumed to have formed early, simulations show that roughly Mars-mass planetary embryos naturally form in the terrestrial region on a million-year timescale, with closer-in embryos forming significantly earlier (*Kokubo and Ida,* 1998; *Kokubo,* 2000; *Leinhardt and Richardson,* 2005; *Walsh and Levison,* 2019). The final phases of growth then involve giant impacts between planetary embryos as well as with remnant planetesimals (e.g., *Chambers,* 2004; *Morbidelli et al.,* 2012; *Raymond et al.,* 2014) (see also the discussion below). This timeline is consistent with the cosmochemically-derived rapid formation of Mars (*Dauphas and Pourmand,* 2011), implying that it can be interpreted as a stranded planetary embryo.

3.2.3. Giant planet formation and migration (during gas disk phase), and instability. Planetesimal accretion is too inefficient to explain the rapid growth of the gas giants, whose several Earth-mass cores must have grown within the few-million-year lifetime of the gaseous disk (*Thommes et al.,* 2003; *Levison et al.,* 2010). Pebble accretion is far more efficient beyond the snow line. As long as there is a sufficient reservoir in millimeter- to centimeter-sized bodies, pebble accretion can grow giant planet cores to several Earth masses in a fraction of the gaseous disk lifetime (*Ormel and Klahr,* 2010; *Lambrechts and Johansen,* 2012; *Levison et al.,* 2015). Pebble accretion is self-limiting: At a critical mass, a growing core creates a pressure bump exterior to its orbit that traps pebbles and halts further growth (*Morbidelli and Nesvorný,* 2012; *Lambrechts et al.,* 2014; *Bitsch et al.,* 2018). The "pebble isolation" mass depends on the properties of the gaseous disk, but for characteristic values it is roughly 20 Earth masses in the Jupiter-Saturn region.

The cores of the gas giants must have undergone orbital migration for the simple reason that they must have grown massive while the gaseous disk was still dense. Decades of analytical and numerical modeling have shown that planet-disk interactions invariably lead to angular momentum exchange between a planet's orbit and the gaseous disk, causing the planet's orbit to shrink or sometimes grow (see reviews by *Kley and Nelson,* 2012; *Baruteau et al.,* 2014). The direction and speed of migration of the gas giants' cores depends on the detailed structure of the disk. In simple viscous-disk models, migration is directed inward in the outer part of the disk but outward closer in, creating zones of convergent migration at a few astronomical units (*Lyra et al.,* 2010; *Bitsch et al.,* 2014, 2015a). Jupiter and Saturn's cores may have converged at such a location, allowing them to accrete gas from the disk at roughly constant orbital radii. Once Jupiter's gaseous envelope reached a critical mass it underwent runaway gas accretion, rapidly growing to hundreds of Earth masses in just $\sim 10^5$ yr (*Pollack et al.,* 1996; *Ida and Lin,* 2004; *Lissauer et al.,* 2009). It gravitationally carved an azimuthal gap in the gaseous disk, thus transitioning to a different mode of migration directly linked to the disk's viscous evolution (*Lin and Papaloizou,* 1986; *Ward,* 1997). Saturn likely underwent the same rapid accretion at a later time, when the gaseous disk was less dense, and so ended up at a lower mass. This view is supported by the different isotopic signatures in H and N between Saturn and Jupiter (e.g., *Mousis et al.,* 2018, and references therein).

The evolution of the gas giants' orbital radii is uncertain. One school of thought proposes that, given the propensity for inward migration, they must have originated much farther from the Sun — at ~15–25 au — to end up at their current radii (*Bitsch et al.,* 2015b). However, it is also possible that they underwent limited migration. Hydrodynamical simulations find that, on its own, each planet would migrate inward (e.g., *D'Angelo et al.,* 2005). Yet both planets together can share a common gap within the disk and enter a different migration. Depending on the disk properties the gas giants can migrate slowly inward, remain on roughly stationary orbits, or even migrate in an outward mode (*Masset and Snellgrove,* 2001; *Morbidelli and Crida,* 2007; *Zhang and Zhou,* 2010; *Pierens et al.,* 2014). The possibility of outward migration is the cornerstone of the Grand Tack model (*Walsh et al.,* 2011), in which Jupiter grew first and migrated inward on its own, and then migrated back outward along with Saturn.

At the end of the gas-disk phase, the gas giants were likely in a resonant configuration regardless of their exact migration history. When two planets migrate toward each other, they invariably become trapped in orbital resonances in which the planets' orbital periods form the ratio of small integers — for instance, in 3:2 resonance the outer planet completes two orbits in the time that the inner planet completes three. The most probable resonances for Jupiter and Saturn were the 3:2 or 2:1 (*Pierens and Nelson,* 2008). The ice giants were likely also trapped in a chain of orbital resonances extending out to 10–15 au (*Morbidelli and Crida,* 2007; *Izidoro et al.,* 2015). This is a far cry from their current orbital locations, with Jupiter and Saturn just inside the 5:2 orbital resonance. The giant planets likely reached their present-day orbits as a result of a dynamical instability (*Tsiganis et al.,* 2005). This dynamical instability, which is colloquially known as the "Nice model," was originally proposed as a delayed event to explain the terminal lunar cataclysm (*Tera et al.,* 1974), a perceived spike in the bombardment rate on the Moon roughly 500–700 m.y. after the planets formed (*Gomes et al.,* 2005). New analyses have called into question the existence of a terminal lunar cataclysm (e.g., *Boehnke and Harrison,* 2016), and several studies propose that the giant planets' instability must have taken place much earlier, anytime within ~100 m.y. of the start of planet formation (*Zellner,* 2017; *Morbidelli et al.,* 2018; *Mojzsis et al.,* 2019). If the instability took place early enough, it may even have played an important role in sculpting the inner solar system (*Clement et al.,* 2019, 2021a) (see discussion in section 4.2). This may very well have been the case: *Liu et al.* (2022) proposed that the instability was triggered by the inside-out dispersal of the Sun's gaseous disk, which would nail down the timing to within 5–10 m.y. of CAIs (e.g., *Hunt et al.,* 2022).

The dynamical instability involved close gravitational encounters between the ice giants, gas giants, and a massive outer disk of planetesimals. The instability spread out the giant planets' orbits, massively depleted the outer planetesimal disk, and likely resulted in the ejection of one or two additional ice giants into interstellar space (*Nesvorný,* 2011; *Batygin and Brown,* 2016). The instability can explain the orbital structure (and in some cases, the very existence) of a number of small body populations in the solar system including Jupiter's Trojan asteroids (*Morbidelli et al.,* 2005; *Nesvorný et al.,* 2013), the giant planets' irregular satellites (*Nesvorný et al.,* 2007), and the populations of the asteroid (*Roig and Nesvorný,* 2015; *Deienno et al.,* 2018) and Kuiper belts (*Levison et al.,* 2008; *Nesvorný,* 2015).

The evidence for the giant planet instability is remarkably strong [although admittedly circumstantial (see review by *Nesvorný,* 2018)], yet the exact evolution of the giant planets' orbits is hard to pin down. Dynamical instabilities are inherently stochastic such that tiny changes in the planets' positions during individual gravitational encounters substantially affect the outcome. The present-day giant planets' orbits are consistent with a broad range of initial conditions and evolutionary paths during the instability (*Nesvorný and Morbidelli,* 2012; *Clement et al.,* 2021a,b). Connecting the exact evolutionary pathway of the giant planets with the orbital structures of small body populations remains another active area of study (e.g., *Clement et al.,* 2020; *Nesvorný,* 2021).

The gas giants' growth likely affected the compositions and orbital distribution of small bodies in several ways. First, once the giant planets' cores reached the pebble isolation mass they blocked the flux of pebbles into the inner solar system and providing a barrier between the reservoirs of pebbles (e.g., *Budde et al.,* 2016; *Kruijer et al.,* 2017). Second, the gas giants' rapid gas accretion represented a drastic change in the gravitational environment for small bodies. For objects already present in the asteroid belt, Jupiter's runaway gas accretion greatly increased the collision velocities and may have induced collisional erosion (*Turrini et al.,* 2012). The gas giants' growth also destabilized the orbits of any planetesimals in a broad region between roughly 4 and 10 au and scattered them in all directions. Under the action of aerodynamic gas drag, a fraction of scattered objects were implanted into the asteroid belt on stable orbits, with a distribution that matches that of C-type asteroids (*Raymond and Izidoro,* 2017a). Third, the gas giants' migration may have shepherded planetesimals across the solar system, inducing large-scale radial mixing (*Walsh et al.,* 2011; *Raymond and Izidoro,* 2017b; *Pirani et al.,* 2019). Finally, while the instability cleared out the majority of the outer solar system's planetesimals, it also sculpted the present-day comet reservoirs, both by scattering them onto very wide orbits in the Oort cloud and by trapping them in select locations within the Kuiper belt and scattered disk (see the chapter by Kaib and Volk in this volume). In the process, a fraction of planetesimals was implanted in the inner solar system — notably in the outer asteroid belt and Jupiter's Trojan swarm (*Levison et al.,* 2009; *Nesvorný et al.,* 2013; *Ribeiro de Sousa et al.,* 2022).

3.2.4. How many pathways could match the sample-based constraints? There is currently a multitude of plausible terrestrial planet formation models. These models fall in two broad categories depending on whether the planets' growth is dominated by planetesimal or pebble accretion. The planetesimal accretion models include the Grand Tack, early instability, and low-mass asteroid belt models (Fig. 28) (see *Raymond et al.,* 2020, for a detailed discussion). Each of these models was created in part to solve the so-called "small Mars problem," so named because simulations of the classical model tend to form Mars analogs that are roughly an order of magnitude more massive that the real Mars (*Wetherill,* 1991; *Chambers,* 2001; *Raymond et al.,* 2009; *Fischer and Ciesla,* 2014). Successful formation models thus include mechanisms to deplete the planetesimal reservoir between roughly Earth's and Jupiter's present-day orbits.

The low-mass asteroid belt model proposes that planetesimals never formed efficiently in the present-day belt. As such, the depletion in Mars' feeding zone was primordial. Models that include pebble drift and a prescription for planetesimal formation within evolving gaseous disks do indeed find that rings of planetesimals may form (*Drążkowska et al.,* 2016; *Charnoz et al.,* 2019; *Izidoro et al.,* 2022). If the terrestrial planets formed predominantly from a ring of planetesimals between ~0.7 and 1 au, their masses and orbits would naturally be reproduced (*Hansen,* 2009; *Kaib and Cowan,* 2015; *Raymond and Izidoro,* 2017b). The outer asteroid belt would have been populated in large part with planetesimals scattered inward during the giant planets' growth, and some of those planetesimals would have delivered water to the terrestrial planets (*Raymond and Izidoro,* 2017a).

The Grand Tack model (*Walsh et al.,* 2011, 2012; *O'Brien et al.,* 2014; *Brasser et al.,* 2016) invokes Jupiter's migration to deplete Mars' feeding zone. In this model, Jupiter is assumed to have formed at a few astronomical units and migrated inward, then turned around (or "tacked") after Saturn migrated inward and became trapped in resonance with Jupiter, triggering the outward migration mechanism discussed in section 2.4.2 (*Masset and Snellgrove,* 2001). If this "tack" happened when Jupiter was at 1.5–2 au then the inner disk of planetesimals and planetary embryos would have been truncated into a narrow ring similar to the one invoked in the low-mass asteroid belt model (*Walsh et al.,* 2011; *Brasser et al.,* 2016). The Grand Tack model can match the orbital structure of the terrestrial planets and asteroid belt and appears consistent with cosmochemical constraints (see *Raymond and Morbidelli,* 2014).

The early instability model invokes an early giant instability (*Clement et al.,* 2018, 2019, 2021a; *Nesvorný et al.,* 2021). The instability would have excited the orbits of planetesimals exterior to roughly 1 au, depleting the asteroid belt and stunting Mars' growth. While the detailed evolutionary pathway of the giant planet instability is hard to pin down (see section 2.4.2), simulations that best match the orbits of the giant planets are the same ones that best match the orbits of the terrestrial planets (*Clement et al.,* 2021b). The early instability model can also match the terrestrial planets' and asteroids' orbits.

Two pebble-driven models appear capable of matching the terrestrial planets' orbits. *Johansen et al.* (2021) showed that the terrestrial planets were consistent with the combined effects of pebble accretion and inward migration if planetesimals only formed at a specific location a little exterior to Mars' orbit (at ~1.6 au). Under such an assumption there is a correlation between the distance from the starting location and the planet mass, which can be matched assuming that this process ended up forming Earth in two parts: the proto-Earth and the Moon-forming body (often called Theia), each of which was less massive than Venus.

Another pebble-driven model invokes convergent orbital migration toward ~1 au (*Brož et al.,* 2021). In that model the most massive planets (Earth and Venus) would have grown closest to the location of convergent migration, whereas the smaller planets (Mercury and Mars) would have been at the outskirts.

The distribution of present-day comet orbits is largely independent of exactly how the terrestrial planets formed. This is because the Oort cloud, Kuiper belt, and scattered disk were all mainly populated during the giant planet instability (*Brasser and Morbidelli,* 2013; *Nesvorný,* 2018; see the chapter by Kaib and Volk in this volume). Current thinking is that the giant planet instability took place, regardless of how the terrestrial planets formed — although, as discussed above, if the instability took place early then it must have affected the rocky planets' growth (*Clement et al.,* 2018; *Liu et al.,* 2022).

3.2.5. Disk evolution informed by the non-carbonaceous–carbonaceous chondrite dichotomy. How the NC-CC isotopic dichotomy was established is debated. The consistency of meteorites falling into the NC or CC group suggests that the NC-CC isotope dichotomy is a pervasive feature of the protoplanetary disk. There are two primary unknowns: (1) What caused the formation of two distinct isotope reservoirs that appear to have been isolated from each other in the protoplanetary disk? (2) What caused the compositional difference between the NC and CC reservoirs?

Regarding unknown (1), *Warren* (2011) proposed that because the NC and CC groups did not overlap in isotopic composition, the dichotomy reflected the sampling of a temporal or spatial compositional dichotomy in the disk. It was speculated that the CC group represents material sourced from the outer solar system and the NC group from the inner solar system. This is because at that time, members of the CC group were understood to be typically more

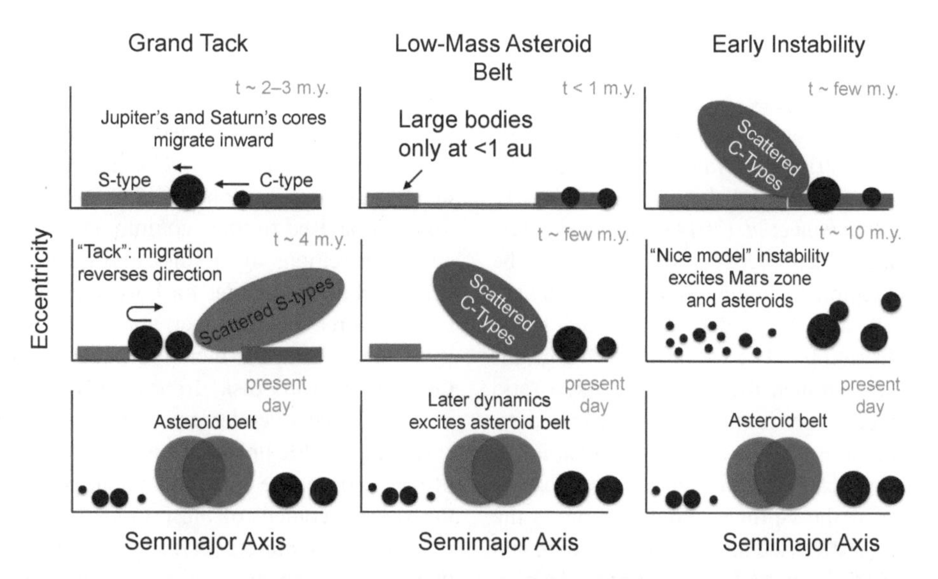

Fig. 28. Cartoon of different global models that can match the orbital architecture of the solar system (adapted from *Raymond et al.,* 2020). They are discussed in section 3.2.4. In context, comets are thought to have originated from ice-rich planetesimals that accreted beyond the giant planets' orbits (*Izidoro et al.,* 2022; see the chapter by Simon et al. in this volume). Comets were scattered by the giant planets into their present-day reservoirs (the Kuiper belt and Oort cloud) during the giant planets' growth and (perhaps) migration (see discussion in section 3.2.6), and most strongly during the giant planet instability (*Nesvorný,* 2018; see the chapter by Kaib and Volk in this volume). Red and blue objects may correspond to the parent bodies of NC and CC meteorites.

volatile rich than NC bodies (e.g., *Wood,* 2005; *Gounelle et al.,* 2006). The author proposed that the two regions may have been physical isolated from each other by the formation of Jupiter; thus, NC meteorites would be sourced from regions inward of Jupiter and CC meteorites from regions outward of Jupiter.

Combining Mo and W nucleosynthetic isotope data with ^{182}Hf-^{182}W-model-derived accretion ages, *Kruijer et al.* (2017) concluded that NC bodies (<0.4 m.y. after CAI formation) formed before CC bodies (0.9 + 0.4/−0.2 m.y. after CAI formation) and inferred an early age for Jupiter (*Kruijer et al.,* 2017). The strict age difference between NC and CC parent bodies is no longer supported following the addition of new data [e.g., Fig. 24 (*Scott et al.,* 2018; *Hilton et al.,* 2019)], as some NC parent bodies are as old as those of CCs (Fig. 25). The cause of segregation has also been revisited. A pressure maximum or a dust trap in the disk near the location at which Jupiter later formed, rather than the early formation of proto-Jupiter, has been proposed as the event that separated the NC and CC reservoirs (*Brasser and Mojzsis,* 2020). *Izidoro et al.* (2021) found that such a barrier was consistent with the structure of the inner solar system, with the terrestrial planets forming from planetesimals rather than pebbles. In contrast, *Johansen et al.* (2021) invoked planetesimal formation at very specific, disjoint orbital radii to explain the dichotomy without invoking a barrier between the inner and outer solar system. *Lichtenberg et al.* (2021) explained the origin of NC and CC planetesimal reservoirs as coming from independent bursts of planet formation separated in both time and orbital radius.

Regarding unknown (2), the compositional difference between the NC and CC reservoirs was originally proposed to result from the addition of *r*-process-rich material to the CC reservoir that did not infiltrate the coexisting, yet spatially separated, NC reservoir (*Kruijer et al.,* 2017). For example, in Mo isotope compositional space (e.g., μ^{94}Mo vs. μ^{95}Mo), the NC and CC groups lie on two close-to parallel lines (e.g., *Budde et al.,* 2016, 2019; *Worsham et al.,* 2017; *Poole et al.,* 2017; *Kruijer et al.,* 2017; *Yokoyama et al.,* 2019). *Budde et al.* (2016), *Worsham et al.* (2017), and *Poole et al.* (2017) proposed that the offset between the lines could be explained by the CC region containing more *r*-process and/or *p*-process material than the NC region. *Stephan and Davis* (2021) contrasted Mo isotopic data for presolar SiC grains and meteorites and determined that there is a fixed ratio between *p*- and *r*-process contributions in all sample data. This was interpreted to indicate that the NC-CC dichotomy, defined by Mo isotopes, can be explained by variations in the isotopic makeup of the *s*-process contribution to the samples. *Desch et al.* (2018) used an extensive isotopic and age data compilation of NC and CC meteorites to develop a numerical disk evolution model that constrains why CC have an overabundance of CAIs relative to NC meteorites. They determined that inside Jupiter's orbit, CAIs could be depleted by aerodynamic drag, which could result in an offset in bulk sample isotopic composition, potentially identifying why NC and CC meteorites had different isotopic compositions. *Alexander* (2019a,b) reported a comprehensive study of the elemental and isotopic characteristics of CC and NC meteorites and identified that different mixtures of the same four meteorite components could reproduce most bulk compositions of CC meteorites, but NC meteorites require different mixtures. *Burkhardt et al.* (2019) used an isotope database that included bulk meteorites, meteorite acid leachates, and presolar grain data and concluded that by mixing "CAI-like" material into an NC-like composition, the chemical and isotopic composition of the CC reservoir could be produced. This conclusion is similar to early reports (e.g., *Gerber et al.,* 2017) that concluded the Ti isotopic compositions of chondrules and CAIs can be attributed to the addition of isotopically heterogeneous CAI-like material to enstatite and ordinary chondrite-like chondrule precursors.

Deriving a unifying theory to explain the NC-CC dichotomy remains elusive. Attempts include those of *Burkhardt et al.* (2019), *Nanne et al.* (2019), and *Spitzer et al.* (2020) who proposed variations of so-called *infall* models. Authors proposed that the change in isotope composition between NC and CC reflects a change in the composition of infalling material from the parental molecular cloud into the disk, coupled with variable mixing and subsequent isolation (by proto-Jupiter) of reservoirs within the disk (Fig. 29). Although infall models are currently the most frequently used models to address questions regarding the origin of the NC-CC isotope dichotomy, their feasibility depends on central assumptions that are yet to be verified (see *Bermingham et al.,* 2020, for further discussion).

Despite the lack of certainty about the cause of the NC-CC isotopic dichotomy, the NC-CC character of parent bodies is used as a tool to cosmolocate material in the protoplanetary disk to recreate the physico-chemical evolution of the protoplanetary disk (e.g., *Kruijer et al.,* (2017, 2020; *Desch et al.,* 2018; *Nanne et al.,* 2019; *Burkhardt et al.,* 2019). This approach contends that NC and CC isotope signatures in meteorites reflect the unique regional characteristics of the inner and outer disk, respectively. Consequently, it is purported that the NC-CC isotope compositions and the timing of accretion can be used to trace communication between the inner and outer regions of the disk and potentially the source of water and other volatiles for Earth and the terrestrial planets. Questions remain, however, about the significance and utility of the NC and CC dichotomy to cosmolocate material in the protoplanetary disk. To use the NC-CC dichotomy as an indicator of volatile-element distribution throughout the solar system, a reliable link between the NC-CC isotope signature of a meteorite, the volatile content of its parent body, and the volatile content of the protoplanetary disk from which it accreted is required (*Bermingham et al.,* 2020). Although there may be an increase in volatile-element abundances when comparing some stony NC and CC meteorites and their inferred accretion locations within the disk, this is not necessarily the general rule. *Bermingham et al.* (2020) contrasted the nucleosynthetic isotope compositions of whole rock meteorite samples with major volatile-element compositions (i.e., C, N, and O), and found that CC parent bodies are not

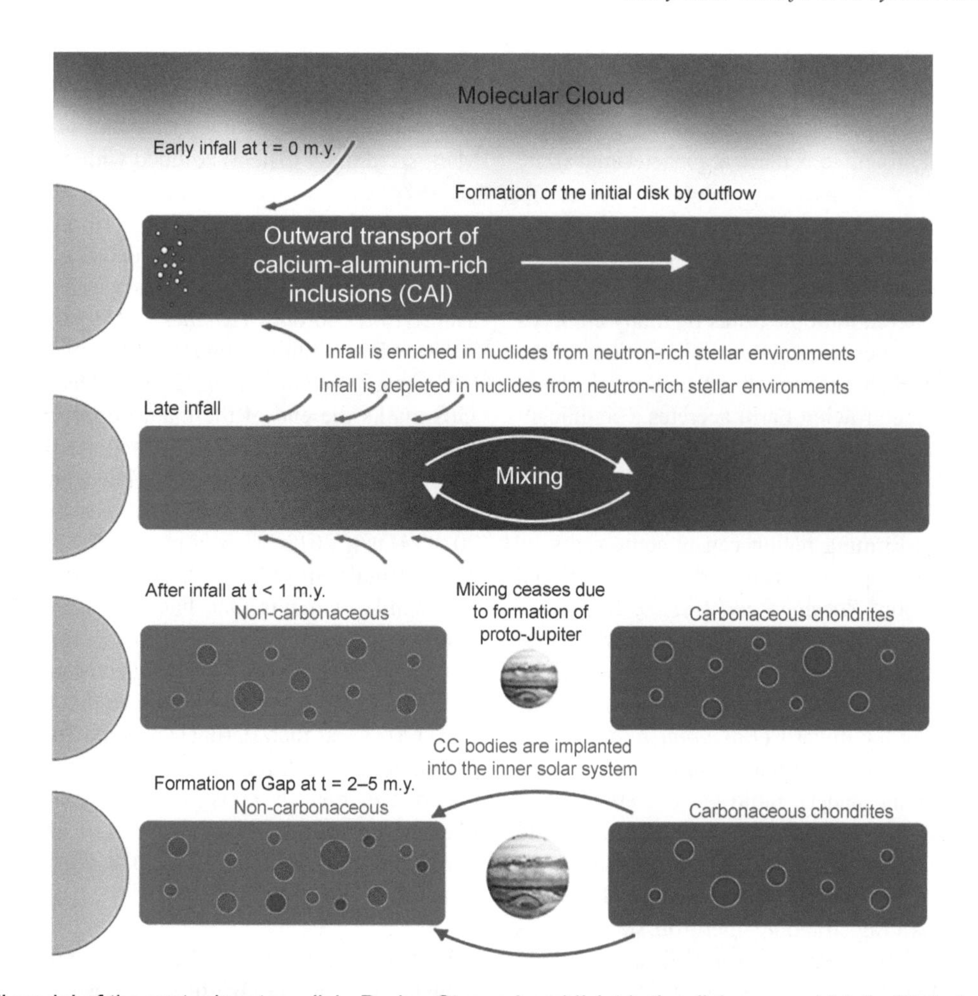

Fig. 29. Infall model of the protoplanetary disk. During Stages I and II (~t_0), the disk grew, which facilitated rapid transport and radial mixing of nebular dust and infalling material from the parental molecular cloud. The growth of proto-Jupiter to the "pebble isolation mass" (see Fig. 31; Stage III, t < 1 m.y.) may have segregated inner and outer protoplanetary disk material, preserving any pre-existing isotope difference between the domains that sourced NC and CC materials. Stage IV (t ~ 2–5 m.y.) saw Jupiter's migration, which facilitated the inward-scattering of planetesimals and their implantation into the main asteroid belt region (*Raymond and Izidoro,* 2017b). Figure modified after *Nanne et al.* (2019), *Kruijer et al.* (2020), and *Bermingham and Kruijer* (2022).

always "volatile-rich" relative to those that formed in the NC. Based on these data, it becomes challenging to assume that all CC meteorites accreted outward and NC meteorites accreted inward of Jupiter. Further research is required to determine how faithfully the volatile-element composition of the disk is preserved in bulk meteorite element abundances and isotopic compositions, and thus if bulk meteorite volatile compositions can be used to assess if the NC-CC dichotomy extends to volatile elements.

3.2.6. Radial mixing of small bodies driven by planet formation and dynamics. Radial mixing of solids likely took place at several different phases of planet formation. If the NC-CC isotopic dichotomy reflects inner vs. outer solar system materials, this dichotomy would not exist if that mixing had been complete, nor would the large differences in compositions between the planets, different types of asteroids, comets, and other small body populations. Large-scale radial (and perhaps temporal) gradients in composition were imprinted onto the constituents of the planets during planet formation.

The emerging paradigm of pebble drift invokes large-scale radial mixing from the outer parts of the solar system inward. The source of inward-drifting pebbles, or "pebble production front," is thought to expand outward in time because the timescale for dust to coagulate to pebble sizes correlates strongly with distance (*Lambrechts and Johansen,* 2012; *Lambrechts et al.,* 2019; *Ida et al.,* 2016). The radial source region of pebbles at a given location is thus continually expanding outward in time. Given their small sizes, inward-drifting pebbles are generally thought to rapidly lose their ices and to "forget" their formation location (e.g., *Morbidelli et al.,* 2015). The bulk compositions of pebbles accreted by a growing planet are thus likely to be dictated by the local temperature. Pebbles' isotopic compositions, however, are unlikely to be altered by their inward drift in the disk. This means that isotopic analyses may be the most promising tool to constrain the role of pebble accretion of different solar system bodies. Yet piecing together the full picture of Earth's growth and whether it was built mostly from planetesimals or pebbles will require isotopic studies of a number of different elements, including simultaneous consideration of constraints from lithophiles, siderophiles, and volatiles.

Additional radial mixing took place at planetesimal sizes, driven by dynamical interactions. Once Mars-mass planetary embryos formed, they excited the orbits of nearby planetesimals (e.g., *Kokubo and Ida,* 1998). Most planetesimals are accreted by an embryo close to their starting location, but a fraction is scattered multiple times by many embryos and can traverse the inner solar system. Simulations in the framework of the "classical model" of terrestrial planet formation find that the growing Earth accretes a significant budget of planetesimals from the outer asteroid belt (*Morbidelli et al.,* 2000; *Raymond et al.,* 2007, 2009). Scattering can also transport material outward, and planetesimals from the terrestrial planet-forming region can in some cases be scattered outward and trapped on stable orbits in the asteroid belt (*Bottke et al.,* 2006; *Raymond and Izidoro,* 2017a).

The giant planets were likely the dominant driver of radial mixing of planetesimals. As the giant planets' cores migrated (see section 3.2.3), they shepherded planetesimals along with them, either inward (*Raymond and Izidoro,* 2017b; *Pirani et al.,* 2019) or possibly outward (*Raymond et al.,* 2016). The giant planets' rapid gas accretion destabilized the orbits of planetesimals in a wide belt, leading to gravitational scattering in all directions. The orbits of some scattered planetesimals were recircularized under the action of aerodynamic gas drag, trapping them on stable orbits in the outer asteroid belt (*Raymond and Izidoro,* 2017b; *Ronnet et al.,* 2018). A much less substantial fraction of planetesimals could have been stabilized on orbits exterior to the growing gas giants' (*Raymond and Izidoro,* 2017b). The orbital migration of Jupiter and Saturn (as well as the ice giants') certainly caused widespread orbital mixing of planetesimals. In the context of the Grand Tack model, Jupiter's inward migration would have scattered out (or pushed inward) all homegrown asteroidal planetesimals [assumed to be associated with present-day S-type asteroids (see Walsh *et al.,* 2011)]. The belt would have been repopulated during Jupiter's and Saturn's subsequent outward migration from a combination of previously scattered S-types and planetesimals originating past the giant planets' orbits (assumed to be associated with present-day C-types). Finally, a side effect of the giant planet instability was to implant a population of asteroids into the outer parts of the main belt, as well as in Jupiter's co-orbital region [the D-types and Jupiter's Trojans (*Morbidelli et al.,* 2005; *Levison et al.,* 2009; *Nesvorný et al.,* 2013; *Vokrouhlický et al.,* 2016)].

4. EARTH'S ACCRETION

4.1. Stages of Earth's Accretion

Contrary to giant planet formation, it took several tens of millions of years for Earth to reach its actual size (Fig. 30). Earth's accretion of Moon-to-Mars-sized planetary embryos can be broadly divided into two stages. The "main-stage accretion" occurred before core formation ceased. The Moon-forming event came at the end of this stage when a Mars-sized body (Theia) collided with proto-Earth (*Benz et al.,* 1986; *Canup and Asphaug,* 2001). Material from Theia merged with the proto-Earth's core and mantle and the Moon formed from the resultant debris cloud (Fig. 30). The timing of the final giant impact on Earth can be constrained using Hf/W isotopes (*Kleine,* 2005; *Fischer and Nimmo,* 2018). Most estimates find a timing of 40–100 m.y. after CAIs (e.g., *Touboul et al.,* 2007; *Thiemens et al.,* 2019). This marks the end of the giant impact phase in the inner solar system. Late accretion (sometimes referred to as "late veneer") followed, the stage during which ~0.5 to 2 wt% of Earth's mass was accreted (*Walker,* 2009; *Bottke et al.,* 2010; *Marty,* 2012; *Marchi et al.,* 2018). Late accretion of planetesimals enriched Earth's crust and mantle with highly siderophile elements that, had they been accreted earlier, would have been sequestered in the core (*Walker,* 2009). Meteorite measurements indicate that Mars accreted ~9× less than Earth during late accretion, and that the Moon accreted 200–1200× less than Earth (*Day et al.,* 2007; *Walker,* 2009). Reconciling the differences in highly siderophile element abundances directly constrains the dynamics of terrestrial planet formation and the properties of its leftovers (*Bottke et al.,* 2010; *Schlichting et al.,* 2012; *Raymond et al.,* 2013; *Morbidelli et al.,* 2018).

4.2. Major-Element (N, C, N) and Noble Gas Constraints on the Origin of Terrestrial Volatiles

4.2.1. Major elements (N, C, N). The causes of D and ^{15}N isotope enrichments among solar system objects and reservoirs (Figs. 18 and 19) are not fully understood, but their variations can be used as tracers of provenance and of exchange between objects and reservoirs. The N and H isotope covariations with heliocentric distance constrain the origin of terrestrial water and the composition of associated volatile elements (*Marty,* 2012; Alexander *et al.,* 2012). Earth, the Moon, and the interior of Mars have D/H and $^{15}N/^{14}N$ compositions within the range of values defined by primitive meteorites and different from either those of the protosolar nebula or of comets (Fig. 31). This similarity indicates a mainly inner solar system — rather than cometary or nebular — origin for the terrestrial atmosphere and oceans. It does not exclude minor contributions from either nebular gas [e.g., D-poor H in the mantle (*Hallis et al.,* 2015; *Olson and Sharp,* 2019); solar-like Ne in the terrestrial mantle] or comets (estimated from the Xe and Kr isotopic compositions of the terrestrial atmosphere; see section 4.4.).

4.2.2. Noble gases. *4.2.2.1. Earth's mantle inventory:* The composition of terrestrial noble gases permits further identification of cosmochemical ancestors of terrestrial volatiles. In particular, the Ne isotope composition of mantle volatiles trapped in oceanic basalts and mineral assemblages of mantle derivation points to the presence of a solar Ne component at depth. Neon has three isotopes, ^{20}Ne, ^{21}Ne,

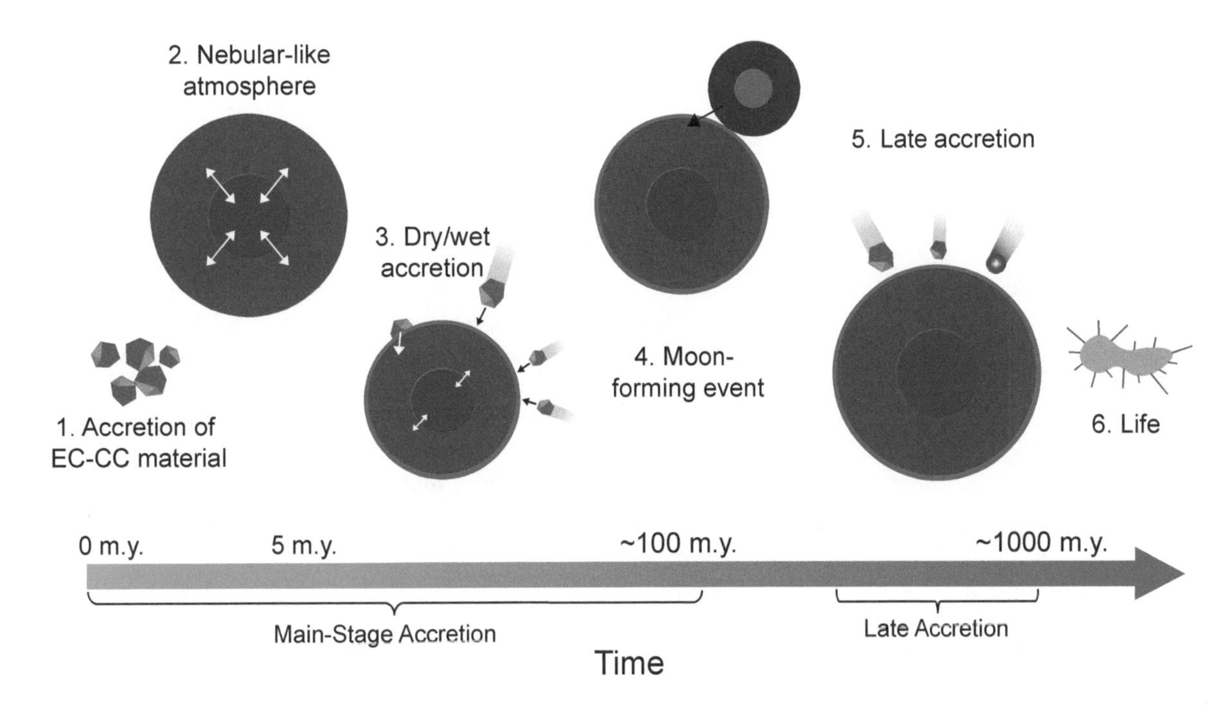

Fig. 30. A schematic depicting Earth's accretion history. (1),(2) Early accreted material formed a proto-Earth with a nebular-like atmosphere. (3) Accretion of planetesimals occurred during ongoing core formation. (4) The Moon-forming impact was the final major addition of mass to proto-Earth (10–20 wt% of Earth's current mass), after which final core segregation occurred. (5) A small fraction (0.5–2 wt% of Earth's current mass) was added to the mantle after final core segregation during late accretion. (6) At some point following this, life began to emerge.

and ^{22}Ne, and, in two-component mixing, ^{20}Ne/^{22}Ne vs. ^{21}Ne/^{22}Ne variations define straight lines linking the component end members (Fig. 32). One of them is atmospheric Ne ("Air" symbol in Fig. 32; ^{20}Ne/^{22}Ne = 9.80). Data for mantle-derived samples define different straight lines having variable slopes and pointing to components rich in ^{21}Ne and ^{20}Ne. "Nucleogenic" ^{21}Ne has been produced in the mantle by neutron and alpha reactions on O and Mg isotopes over eons and is not a primordial component. Neon-22 is also produced by nuclear reaction, but to a lesser extent than ^{21}Ne. Thus, deviations toward increasing ^{21}Ne/^{22}Ne ratios are a function of mantle residence time and not a primordial feature. In contrast, there is no known nuclear process quantitatively producing ^{20}Ne in Earth, and the data define a mantle end-member composition with ^{20}Ne/^{22}Ne up to ~13.0 that is associated with mantle plumes (*Yokochi and Marty,* 2004; *Williams and Mukhopadhyay,* 2019), signifying a primordial composition. This value is notably higher than the range observed in chondrites [8.2 to 12.7 (*Ott,* 2014)], suggesting trapping in the mantle of a solar gas component [solar ^{20}Ne/^{22}Ne = 13.36 ± 0.18 (*Heber et al.,* 2012)]. A way to incorporate solar Ne in the mantle would be dissolution of nebular gas gravitationally trapped by the proto-Earth and dissolved in molten silicates during magma ocean episodes (*Sasaki,* 1990; *Yokochi and Marty,* 2004; *Marty and Yokochi,* 2006; *Williams and Mukhopadhyay,* 2019). This possibility implies that the proto-Earth might have been massive enough to capture a small nebular, Jupiter-like atmosphere that was subsequently blown away by the active proto-Sun. Notably, Xe in the martian mantle appears to have a solar isotopic composition and could have been trapped from a nebular-like atmosphere before gas dissipation, a possibility consistent with both the lifetime of the gas [a few million years (*Haisch et al.,* 2001; *Mamajek et al.,* 2009)] and the timeframe for Mars accretion (*Dauphas and Pourmand,* 2011). However, *Péron and Mukhopadhyay* (2022) recently

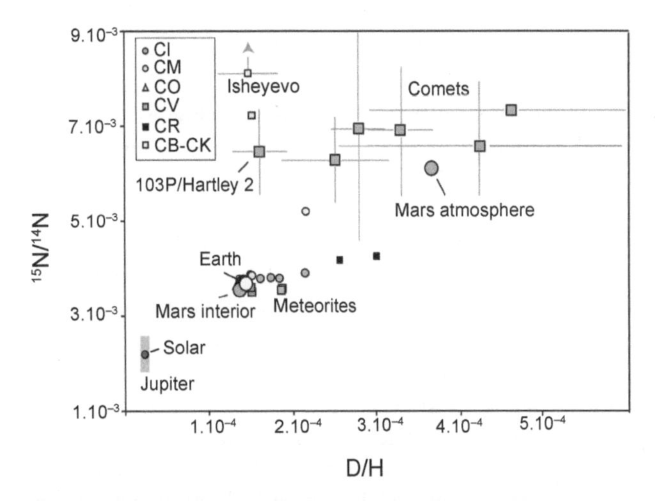

Fig. 31. Nitrogen and hydrogen isotope covariations among solar system objects and reservoirs (adapted from Marty, 2012). Data define three reservoirs: (1) protosolar nebula (solar system and Jupiter), (2) inner planets and chondrites, and (3) comets. This distribution was taken as evidence for a chondritic origin of volatile elements on Earth.

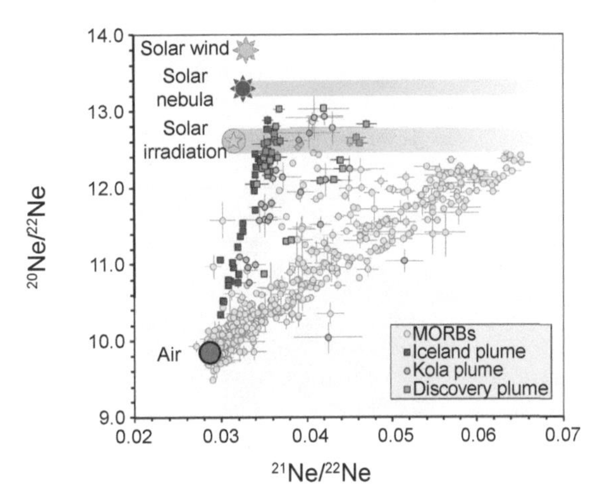

Fig. 32. Neon isotope data of mid-ocean ridge basalts (MORBs) and mantle plume samples. Data source: MORB data from compilation by M. Moreira (personal communication), Icelandic subglacial glass data from *Mukhopadhyay* (2012), Kola mantle plume carbonatite data from *Yokochi and Marty* (2004), and submarine glasses from the Discovery plume area data from *Williams and Mukhopadhyay* (2019). The solar wind value was measured on the Moon and more recently in solar system wind ions collected by the Genesis mission (*Heber et al.,* 2012, and references therein). The solar system nebula value is inferred from the composition of the Sun, corrected for isotope fractionation during solar system wind generation (*Heber et al.,* 2012). The solar irradiation value is the Ne-B component, representing the end-member value found in lunar soils and solar gas-rich meteorites (*Black and Pepin,* 1969; *Wieler et al.,* 1989). Ne-B is isotopically fractionated relative to solar system wind by surface processing on the Moon and on asteroids. In this format, a two-component mixing yields straight lines. In both cases the arrays defined by sample data result from mixing between the composition of atmospheric neon contaminating the samples and a mantle end-member composition. The slopes of correlations relate to the degassing state of the respective mantle reservoirs.

argued that Kr in the martian mantle is chondritic and not solar, casting doubt on a solar origin for Xe trapped in the martian mantle.

Contrary to mantle plumes, the mantle end member seen in mid ocean ridge basalts (MORBs), presumably sampling the shallower convective mantle, appears to have a lower $^{20}Ne/^{22}Ne$ end-member ratio of 12.50–12.75 (*Moreira,* 2013; *Péron et al.,* 2016) than mantle plume Ne. The MORB Ne signature appears comparable to that of the so-called Ne-B component (12.5 ± 0.2) (*Black and Pepin,* 1969). This component, found in planetary regoliths including lunar soils, is thought to be original solar wind Ne that was implanted in regolithic grains and isotopically fractionated during regolithic processing (*Moreira,* 2013; *Trieloff and Kunz,* 2005). *Moreira* (2013) and *Moreira and Charnoz* (2016) argued that the carrier of mantle Ne was dust irradiated by the nascent Sun as turbulence in the disk transported grains in regions that were unshielded from solar irradiation. A dual origin for mantle Ne cannot be discarded: dissolution from a primordial atmosphere in what is now the deep mantle, and contribution of irradiated dust at a later stage of accretion. Resolving the origin of mantle Ne will have far-reaching implications for understanding the evolution of the proto-Earth.

Contrary to Ne (and He), the Kr and Xe isotopic signature of the mantle is dominated by a chondritic component (*Holland et al.,* 2009; *Broadley et al.,* 2020), in line with H and N isotopes. This dichotomy could be due to the lower concentrations of Kr and Xe in the protosolar nebula compared to He and Ne, which, coupled with their lower solubility in basaltic melt, would result in Kr and Xe being less efficiently degassed than He and Ne (*Olson and Sharp,* 2019).

Recently, *Piani et al.* (2020) reported the analysis of H and its D/H ratio in a suite of enstatite chondrites (EC) and argued that EC could constitute an important source of mantle water. This possibility is consistent with the isotopic composition of several key elements like O, Ti, and others, which call for a EC-like source for Earth. However, it also raises a mass balance problem: CC (CI and CM) (5~10% H_2O by mass) are 1 order of magnitude richer in equivalent water than EC (~0.5% H_2O by mass), but only a factor of ~3 richer in trapped noble gases. If, therefore, mantle water was from an EC-like source and assuming closed system condition, the mantle should contain 1 order of magnitude more EC-like noble gases than CC-like noble gases. This possibility cannot be excluded and could be tested with heavy noble gas composition (since light noble gases, e.g., Ne, are dominated by a superimposed solar component). The mantle, however, is drastically depleted in noble gases, by orders of magnitude compared to both EC and CC contents, and has evidently lost its noble gas cargo presumably during accretion and partial melting. One possibility to retain EC-like H while degassing noble gases could be a low oxygen fugacity, consistent with the reduced character of EC-like material. The situation might have evolved with the subsequent delivery of oxidized material to the proto-Earth (*Javoy et al.,* 2010; *Rubie et al.,* 2015), which might have also supplied CC-like noble gases.

4.2.2.2. Earth's surface inventory: The noble gas and stable isotope signatures of the atmosphere and the oceans are markedly different from those of the mantle. This difference is best illustrated by the variations of the $^{20}Ne/^{22}Ne$ ratio as a function of the elemental $^{36}Ar/^{22}Ne$ ratio in different classes of chondrites compared to the atmosphere (*Marty,* 2012; *Williams and Mukhopadhyay,* 2019; *Marty,* 2020, 2022). In this format (Fig. 33), two-component mixing results in a straight line, joining the component end members. Enstatite chondrite data cannot account for mixing between the atmosphere and the mantle (Fig. 33a). Noble gases in ECs often present enrichments in ^{36}Ar compared to other chondritic classes, a pattern labeled "subsolar" (*Crabb and Anders,* 1981). Argon-36 enrichments result in a large spread of data in the figure, which means they are collectively unable to account for the atmospheric composition atm in the figures). Some of the ECs are also rich in solar

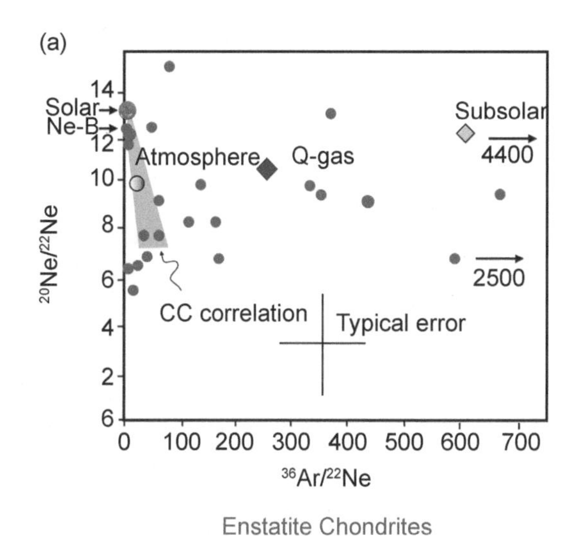

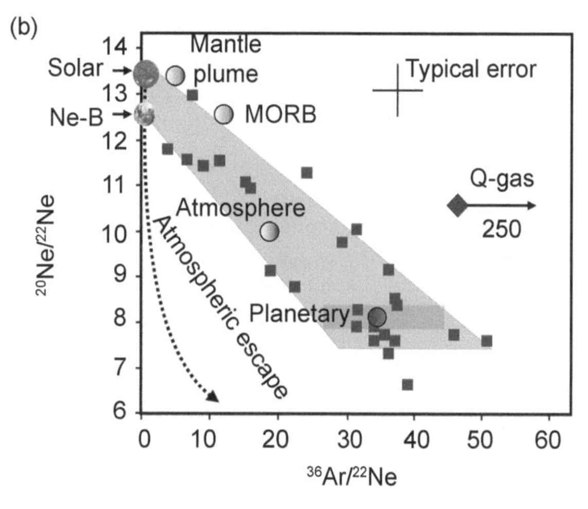

Fig. 33. ^{20}Ne/^{22}Ne ratio as a function of the elemental ^{36}Ar/^{22}Ne ratio in different classes of chondrites compared to terrestrial values. **(a)** Compilation of enstatite chondrite (EC) data. **(b)** Compilation of data for carbonaceous chondrites (CI and CM). The diagrams are based on a compilation of noble gas data in enstatite (NC-type) and carbonaceous (CC-type, here CM and CI) chondrites (for relevant literature, visit *https://zenodo.org/record/3984898#.XzZU7i3M2iA*). Data were corrected for contribution of cosmic-ray-produced Ne and Ar isotopes using the ^{21}Ne/^{22}Ne ratios. Adapted from *Marty* (2022).

Ne (Ne-B) and/or contain the so-called Ne-A (*Black and Pepin,* 1969), also labeled "planetary" component (*Mazor et al.,* 1970), a presolar Ne component found in nanodiamonds and SiC grains. "Q-gas" is the Phase Q component ubiquitously found in chondrites. Error bars are generally large due to correction for cosmic-ray effects and relevant error propagation.

In contrast, CC data define a mixing trend starting at the solar-like mantle Ne component that encompasses nicely the atmospheric composition (Fig. 33b). Data can be well explained by mixing between a solar-like component and Ne-A (or planetary Ne) with low ^{20}Ne/^{22}Ne ratio (*Mazor et al.,* 1970). The composition of the latter is dominated by presolar ^{22}Ne trapped in nanodiamonds. Remarkably, atmospheric and mantle noble gases are intermediate between solar and planetary, which strongly suggests a CC-like origin for terrestrial volatiles (*Marty,* 2012, 2022).

Another important implication of this diagram is that atmospheric escape to space (outlined by the dotted curve in Fig. 33b) cannot account for the composition of atmospheric noble gases, as otherwise often advocated (*Pepin,* 1991).

In summary, light noble gases (Ne and Ar) point to a CC-like origin for Earth's surface inventory. In addition to a limited solar-like Ne (and He, possibly H) component, mantle noble gases (Kr and Xe) are also dominated by a chondritic component, but not necessarily of the CC type. Stable isotope ratios of H and N are also consistent with the occurrence of an EC-like component: H might have been already present in NC-like material (*Piani et al.,* 2020) as could have been C and N hosted by refractory organics. Hence terrestrial volatiles might have been sourced by different types of primitive material both within and outside the snow line.

Deciphering the fractional contributions of cosmochemical sources is difficult and depends on which element or isotope system is considered (see also section 4.4.2.). The bulk Earth (mantle plus surface) H, C, N, noble gas, and halogen inventory is roughly equivalent to the contribution of ~2 ± 1% CC-type material to a dry proto-Earth (*Marty,* 2012), although estimates are variable depending on the considered volatiles (*Hirschmann and Dasgupta,* 2009; *Hirschmann,* 2018). Mass balance involving H and N isotopes suggest a NC/CC mix of about 75–95%/25–5% (*Piani et al.,* 2020; *Marty,* 2020). Establishing the terrestrial inventory of volatiles also requires accounting for the core and its potential to store H (*Wu et al.,* 2018), C (*Fischer et al.,* 2020), N (*Roskosz et al.,* 2013; *Dalou et al.,* 2017), and possibly noble gases (*Bouhifd et al.,* 2013). Accurately anticipating the repartition of volatile elements between the major terrestrial reservoirs requires knowledge of the mode of accretion, the evolution of planetary atmospheres, and the mode of core formation (e.g., *Grewal et al.,* 2019). This is presently an active area of research involving geochemistry, high pressure mineralogy, and planetary dynamics. Given these uncertainties, isotopic tracers are often regarded as being more compelling than elemental abundance ratios to address the origin of terrestrial planetary building blocks.

4.3. Origin of Earth's Volatiles (Oceans, Atmosphere) Inferred from Numerical Modeling

Several numerical models have been developed to explain the origin of Earth's water (see *Meech and Raymond,* 2020). Some models invoke a local source of water at Earth's orbital distance, either via adsorption of H onto silicate grains (e.g., *Sharp,* 2017) or by oxidation of a primordial H-rich atmosphere (*Genda and Ikoma,* 2008). The possibility of Earth having trapped nebular-like water has been substantiated by the claim that the terrestrial mantle contains D-poor H (*Hallis et al.,* 2015). In addition, the recent discovery of abundant H in enstatite chondrite meteorites with Earth-like D/H ratios (*Piani et al.,* 2020) begs the question of whether Earth's water was locally sourced. Many local models, however, struggle to match chemical and isotopic constraints (see discussion in *Meech and Raymond,* 2020), and enstatite chondrites cannot account for Earth's full volatile budget (Fig. 33). Indeed, stable isotopes of volatile elements point to the water delivery by chondritic-like material (*Alexander et al.,* 2012; *Marty,* 2012) (see Fig. 10). As such, attention is focused here on models in which water is "delivered" from more distant parts of the planet-forming disk.

Water delivery is efficient in the context of the classical model, but the model has fallen out of favor. The classical model assumes that the terrestrial planets grew from a broad disk of planetesimals and embryos with minimal intrusion from the giant planets (discussed in *Morbidelli et al.,* 2012; *Clement et al.,* 2020). Radial mixing of planetesimals is driven by scattering by planetary embryos, assumed to have formed throughout the inner solar system (see section 3.2). Simulations find that water is efficiently delivered by this mechanism and the growing Earth typically receives an order of magnitude more water than the present-day budget (*Morbidelli et al.,* 2000; *Raymond et al.,* 2009; *Raymond and Izidoro,* 2017a; *Izidoro et al.,* 2013). The framework of the classical model, however, has a fatal flaw in that it systematically fails to match the terrestrial planets' coupled mass and orbital distribution [see discussion in *Clement et al.* (2020) and the discussion of the "small Mars" problem in section 3.2.4].

In the context of newer terrestrial planet formation models that can match the orbital structure of the terrestrial planets and asteroids (see *Raymond et al.,* 2020), water delivery is driven by the giant planets. When Jupiter and Saturn underwent rapid gas accretion, they scattered nearby planetesimals in all directions; many were implanted into the outer asteroid belt as C-types (see section 3.2). Some planetesimals were scattered past the asteroid belt toward the growing terrestrial planets, delivering water (*Raymond and Izidoro,* 2017a). In the context of the Grand Tack model, Jupiter's outward migration scattered planetesimals inward — again, some were trapped in the main asteroid belt as C-types and others were scattered past the belt to the terrestrial planet-forming region (*Walsh et al.,* 2011; *O'Brien et al.,* 2014). Each of these scenarios can deliver a few times Earth's current water budget, and each predicts that the water delivered should have the same chemical makeup as that of C-type asteroids. Indeed, the D/H ratios and the $^{15}N/^{14}N$ ratios of Earth's water are a good match to that of CC meteorites (e.g., *Marty and Yokochi,* 2006; *Alexander et al.,* 2012; *Marty,* 2012).

Water could in principle have been delivered to the growing Earth in the context of pebble accretion. As the gaseous protoplanetary disk cooled, the snow line swept inward. In most current disk models, the snow line is located interior to 1 au in the late phases of the disk's lifetime, and in some models for the majority of the disk's life (e.g., *Sasselov and Lecar,* 2000; *Lecar et al.,* 2006; *Martin and Livio,* 2012; *Bitsch et al.,* 2015b; *Savvidou et al.,* 2020). If the terrestrial planets grew mostly by accreting inward-drifting pebbles, then that accretion would include any ices contained in the pebbles. Models find that this mechanism can indeed operate in a hypothetical evolving disk (*Sato et al.,* 2016; *Ida et al.,* 2019), and it has been invoked in recent pebble-driven terrestrial planet formation models (*Johansen et al.,* 2021; *Brož et al.,* 2021). This is a delicate balancing act, however, given that most carbonaceous chondrite meteorites contain 10–100× more water by mass than Earth (e.g., *Marty,* 2012; *Alexander et al.,* 2012, 2018). In addition, the total flux in pebbles must remain low enough to avoid having the terrestrial planets transition to a super-Earth growth mode (*Lambrechts et al.,* 2019). To match the terrestrial planets' masses and orbits and simultaneously deliver the appropriate amount of water requires a disk with a fine-tuned evolution of the snow line and pebble flux.

4.4. Late Losses and Deliveries: Chondritic and Cometary Contributions to Earth

4.4.1. Evidence for volatile loss during Earth's accretion. The main accretion of Earth ended with the Moon-forming impact (Giant Impact) 40–100 m.y. after CAIs (e.g., *Lock et al.,* 2020, and references therein), after which our planet evolved toward a habitable world. The Giant Impact was an extremely energetic event that likely resulted in the partial, or complete, melting of the proto-Earth. To what extent existing volatiles survived this event is a matter of debate (*Porcelli et al.,* 2001; *Genda and Abe,* 2005). A dramatic loss of terrestrial volatiles during the Moon-forming event is supported by the observation that our planet contains at most a few percent of ^{129}Xe produced from the decay of short-lived ^{129}I ($t_{1/2}$ = 15.6 m.y.) (*Pepin,* 1991; *Allègre et al.,* 1995; *Avice and Marty,* 2014), indicating that >90% of terrestrial ^{129}Xe was lost before or during the Moon-forming event. The noble gas content of bulk Earth, mostly concentrated in the atmosphere, corresponds to about ~2 ± 1% CC (*Marty,* 2012), attesting independently for efficient volatile loss. On the other hand, the delivery of chondritic material to Earth after the Moon-forming event (the late accretion) may amount for less than 1% of Earth's mass, so that some of the noble gases (and presumably other volatiles) survived the Moon-forming event (*Marty and Yokochi,* 2006).

4.4.2. Tracing volatile accretion using nucleosynthetic isotope anomalies. Deciphering the cosmochemical character and heliocentric source location of Earth's building blocks is complicated by mantle convection, which unquestionably mixes these components in the mantle over time. Studies have shown, however, that the nucleosynthetic isotope composition of Earth's building blocks can be constrained by examining the isotopic composition of elements with differing affinities for metal in mantle-derived materials

(*Dauphas,* 2017). The lithophile (rock-loving) elemental composition of the mantle (e.g., O, Ca, Ti, and Nd) reflects the average composition of materials accreted by Earth throughout its history (*Walker et al.,* 2015; *Dauphas,* 2017). The siderophile (iron-loving) elemental composition of the mantle, however, is biased toward material accreted during late stages of growth (e.g., final core segregation onward). This is because during core formation the siderophile elements partitioned predominately into the core during core segregation. Their accumulation in the mantle thus partly reflects addition after final core formation. Accordingly, siderophile-element isotopic compositions of the mantle can be used to trace the identify of building blocks toward the end of Earth's accretion, i.e., late-stage accretion (e.g., *Dauphas et al.,* 2004; *Walker et al.,* 2015; *Dauphas,* 2017; *Fischer-Gödde and Kleine,* 2017; *Bermingham et al.,* 2018a; *Budde et al.,* 2019; *Hopp et al.,* 2020; *Fischer-Gödde et al.,* 2020). This approach contrasts the nucleosynthetic isotope compositions of meteorites with mantle-derived geochemical data that track different stages of Earth late-stage accretion history (e.g., *Dauphas,* 2017).

Siderophile elements Mo and Ru are well-suited to tracing late-stage accretion due to four primary reasons: (1) Most of the Mo and Ru mantle budgets preserve the chemical signatures of different stages of accretion. The Mo bulk silicate Earth (BSE) budget was likely established during the final 10–20 wt% of Earth's accretion (i.e., during core formation), while the Ru BSE budget was likely established after core formation (i.e., during late accretion) (Fig. 34) (cf. *Dauphas et al.,* 2004; *Dauphas,* 2017). (2) Nucleosynthetic isotope variations in Mo and Ru discriminate between meteorites and the NC-CC character of a parent body (e.g., *Budde et al.,* 2016; *Worsham et al.,* 2017; *Poole et al.,* 2017; *Bermingham et al.,* 2018a). (3) The Mo and Ru nucleosynthetic isotope compositions of many meteorites correlate ("Mo-Ru cosmic correlation"), indicating that the presolar carriers responsible for Mo and Ru isotopic anomalies were the same (*Dauphas et al.,* 2004; *Bermingham et al.,* 2018a; *Worsham et al.,* 2019; *Hilton et al.,* 2019; *Hopp et al.,* 2020; *Tornabene et al.,* 2020). (4) Recent development of high-precision isotopic methods for Ru and Mo (e.g., *Fischer-Gödde et al.,* 2015; *Nagai and Yokoyama,* 2016; *Worsham et al.,* 2016) permit resolution of small (~10–30 ppm) variations in Mo and Ru isotopes in terrestrial materials. With these state-of-the-art analytical tools at hand, the search for remnants of Earth's building blocks has begun.

Few terrestrial samples have been analyzed for Mo and Ru isotopic compositions (*Fischer-Gödde et al.,* 2015, 2020; *Bermingham et al.,* 2018a; *Bermingham and Walker,* 2017; *Budde et al.,* 2019). *Dauphas et al.* (2004) recognized that in a Mo vs. Ru isotopic space, the BSE and meteorites define a common correlation. Subsequent studies have extended the meteorite and terrestrial datasets, and they largely confirm this conclusion (e.g., *Dauphas,* 2017; *Bermingham and Walker,* 2017; *Bermingham et al.,* 2018a). Most terrestrial Ru isotopic data suggest that the BSE is most similar to IAB irons or enstatite chondrites (*Fischer-Gödde et al.,* 2015; *Bermingham and Walker,* 2017; *Bermingham et al.,* 2018a). These data also indicate that late accretion did not provide a significant source of volatiles to Earth, in contrast to the discussion above based on other elements. *Fischer-Gödde and Kleine* (2017) analyzed several chondrites and concluded that all chondrites, including carbonaceous chondrites, have Ru isotopic compositions distinct from that of Earth's mantle. Their data appeared to refute an outer solar system origin for late accretion and indicated that late accretion was not the primary source of volatiles and water on Earth.

Recently, *Fischer-Gödde et al.* (2020) reported the first finding of Ru isotope anomalies in the mantle, in the form of a 22 ppm excess in $^{100}Ru/^{101}Ru$ of Eoarchean ultramafic rocks from southwest Greenland (3.8–3.0 Ga). These authors concluded that these materials sampled a part of the mantle containing a substantial fraction of Ru that was accreted before late accretion and that the mantle beneath southwest Greenland did not fully equilibrate with late accreted material. This interpretation requires that an *s*-process-enriched

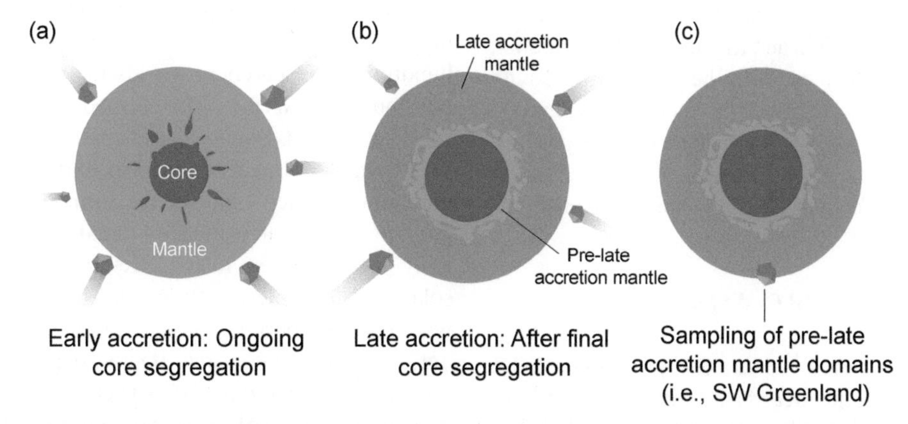

Fig. 34. Possible preservation of pre-late-accretion material in the mantle indicated by nucleosynthetic isotope compositions of terrestrial materials. **(a)** Primary accretion saw most siderophile elements segregate to the core. **(b)** After final core formation, ~0.5–2% of the total percentage weight of Earth's mass accreted from meteorites during late accretion. **(c)** *Fischer-Gödde et al.* (2020) report anomalous isotopic compositions in ancient rocks from southwest Greenland that they conclude is due to the presence of pre-late-accretion mantle material in the rocks. Figure based on *Bermingham* (2020).

reservoir contributed to Earth's growth (Fig. 34), which is contrary to all bulk sample meteoritic Ru isotopic data reported thus far. The excess in $^{100}Ru/^{101}Ru$ Archaean materials compared to modern mantle was reconciled by the authors if late accretion contained substantial amounts of CC-like materials with their characteristic $^{100}Ru/^{101}Ru$ deficits. This finding thus relaxed previous constraints on the composition of late accretion and brings them in line with previous studies that propose a volatile-rich material late-accretion composition (*Marty,* 2012). More Mo and Ru isotopic data are required, however, to ascertain if the southwest Greenland sample suite is representative of the pre-late accretion mantle, if nucleosynthetic fingerprints are observed in the isotopic compositions of other elements in the mantle, and what the identity is of the thus far undetected *s*-process-enriched meteorite group (*Bermingham,* 2020).

4.4.3. Late cometary contribution to Earth? Comets, which are rich in water ice (10–50%) and organic materials, have long been proposed as potential suppliers of water and biologically relevant molecules to early Earth (*Bar-Nun and Owen,* 1998). The D/H value of cometary water is generally higher than the ocean value (*Bockelée-Morvan et al.,* 2015; *Altwegg et al.,* 2015; *Lis et al.,* 2019), and remote measurements of N isotopes in cometary coma have shown that at least some molecules in comets are ^{15}N-rich as well (e.g., *Füri and Marty,* 2015; *Bockelée-Morvan et al.,* 2015). From mass balance, H and N isotopes of the atmosphere and the oceans, together with Ar measurements in 67P/C-G (*Balsiger et al.,* 2015), constrain the cometary water contribution to Earth to 1% or less (*Marty et al.,* 2016). One important finding from the Rosetta mission, however, was that the Xe isotope composition of 67P/C-G exhibits a drastic depletion in the two heaviest isotopes ($^{134,136}Xe$), a feature that had been recognized in the terrestrial atmosphere (*Pepin,* 1991) but had remained unexplained. Based on a binary mixing model between cometary and chondritic components, the cometary contribution to atmospheric Kr and Xe was estimated to be 20 ± 5% (*Marty et al.,* 2017; *Rubin et al.,* 2018). The relative abundances of noble gases fit well such a mixing trend between cometary and chondritic materials (Fig. 35).

Comets are noble-gas-rich, and as such their contribution to the terrestrial inventory can only be distinguished with these elements. For other volatiles, the relative contribution of comets to the water and N to the surface inventory is low, less than 1% of the oceans and the atmosphere, whatever the composition of the chondritic end member is considered to be (e.g., EC or CC; Fig. 36).

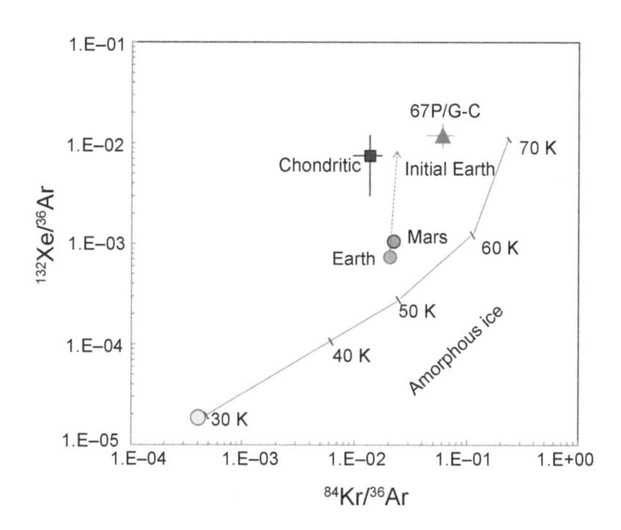

Fig. 35. Noble gas elemental ratios of solar system reservoirs. The solar composition is from *Lodders* (2003). The amorphous ice data as a function of temperature are from extrapolations of *Dauphas* (2003) done using data from *Bar-Nun and Owen* (1998). 67P/C-G data are from *Rubin et al.* (2018). Average chondrite data are from compilation in *https://doi.org/10.24396/ORDAR-66*. The modern Earth's atmosphere (data from *Ozima and Podosek,* 2002) data point is off any mixing relation between solar system, chondritic, and cometary. With respect to a mixing between chondritic and cometary (double arrow line), the initial atmosphere of the Earth, corrected for loss of xenon to space (*Bekaert et al.,* 2020), fits well with chondritic-cometary mixing. This possibility is substantiated by Xe isotopes measured in 67P/C-G indicating the contribution of ~20 cometary Xe to ~80% chondritic Xe to form the initial atmosphere of our planet (*Marty et al.,* 2017).

5. FUTURE WORK

Determining the terrestrial mantle nucleosynthetic isotopic composition and identifying fingerprints of our building blocks will provide the most robust sample-based constraints on our planet's cosmochemical origins. With the extensive nucleosynthetic isotope database from meteorites and high-precision analytical techniques sufficiently developed, the community is now able to undertake these investigations. The exploration of the solar system is tending toward targeting more and more sample return missions, which will greatly aid in corroborating remote observations with sample-derived constraints. Samples already returned from a primitive CC-type asteroid, Ryugu, by the Haybusa2 mission from JAXA are presently being analyzed by several hundreds of researchers worldwide and the OSIRIS-REx mission from NASA also returned material from another CC-like asteroid, Bennu. These missions document the most pristine materials yet returned to Earth. These studies will undoubtedly shed light on the formation of solar system building blocks, as well as on radial mixing that took place early in solar system history. Returning samples from outer solar system bodies, such as comets and D-type asteroids, would address burning questions such as the origin (solar or not?) of outer solar system bodies, degree of presolar grain homogenization and the NC-CC character of the outer solar system, dynamics of accretion, and potential contributions to the inner solar system.

Planet-formation numerical models are in a state of rapid diversification. New global models are being assembled at a rapid rate by connecting various physical mechanisms — for

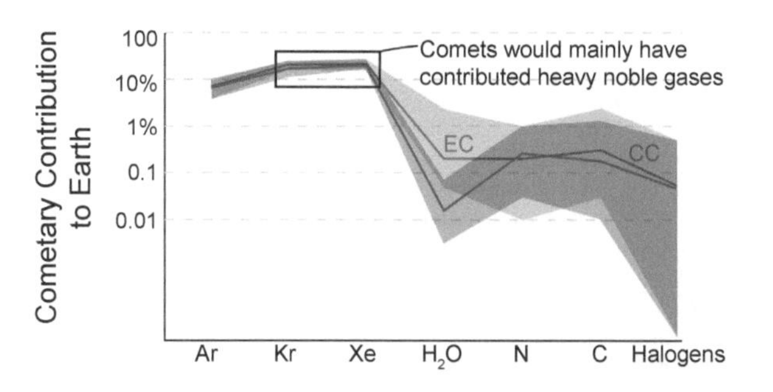

Fig. 36. Contribution of comets to the inventory of terrestrial volatiles (adapted from *Bekaert et al.,* 2020). The figure is based on a two-component mixing between cometary and chondritic (either EC or CC) compared to bulk Earth inventory. Comets only contributed significantly to noble gases. The colored areas correspond to 95% CI.

instance, connecting disk evolution models with streaming instability constraints (e.g., *Drążkowska and Alibert,* 2017), combining pebble accretion with embryo migration (*Brož et al.,* 2021), and connecting the giant planets' instability with a full dynamical picture of the orbital evolution and growth of the terrestrial planets and asteroids (*Clement et al.,* 2018). Evaluating, comparing, and refuting planet-formation models requires a combination of cosmochemistry, disk studies (including modeling and observations), and dynamical simulations. Differentiating between planetesimal- and pebble-driven models requires an understanding of the relative contribution of CC-like material to the growing terrestrial planets, which itself must be based on isotopic studies of different elements in a range of samples. Determining where and when planetesimals formed is of paramount importance and will itself require confronting detailed disk models (combined with dust coagulation, drift, and streaming instability models) with cosmochemical and astronomical constraints. The Grand Tack model is based on a migration mechanism whose robustness is unclear. New complex hydrodynamical modeling can determine whether it is plausible. Finally, the early instability model can best be constrained using a cosmochemical marker of the timing of the instability itself. Indirect markers have already been used [e.g., impact-reset ages like those used in *Mojzsis et al.* (2019)] and perhaps others will be able to pin down the timing of the instability more clearly in the future.

Acknowledgments. This study was supported by the European Research Council (ERC) under the European Union's Horizon 2020 research and innovation program (PHOTONIS Advanced Grant #695618 to B.M.). K.R.B. received support from NASA (80NSSC20K0997), the National Science Foundation (EAR-2051577), and the Department of Earth and Planetary Science, Rutgers University. H. Tornabene (Rutgers University) is thanked for assistance in creating figures. S.N.R. thanks the CNRS's PNP and MITI/80PRIME programs for support. L.R.N thanks the Carnegie Institution for support. We are grateful to M. Zolensky and the reviewers for helpful suggestions and corrections.

REFERENCES

Abod C. P., Simon J. B., Li R., Armitage P. J., Youdin A. N., and Kretke K. A. (2019) The mass and size distribution of planetesimals formed by the streaming instability. II. The effect of the radial gas pressure gradient. *Astrophys. J., 883,* 192.

Adams F. C. (2010) The birth environment of the solar system. *Annu. Rev. Astron. Astrophys., 48,* 47–85.

Aikawa Y., Furuya K., Hincelin U., and Herbst E. (2018) Multiple paths of deuterium fractionation in protoplanetary disks. *Astrophys. J., 855,* 119.

Albarède F. (2009) Volatile accretion history of the terrestrial planets and dynamic implications. *Nature, 461,* 1227–1233.

Alexander C. M. O. (2019a) Quantitative models for the elemental and isotopic fractionations in chondrites: The carbonaceous chondrites. *Geochim. Cosmochim. Acta, 254,* 277–309.

Alexander C. M. O. (2019b) Quantitative models for the elemental and isotopic fractionations in the chondrites: The non-carbonaceous chondrites. *Geochim. Cosmochim. Acta, 254,* 246–276.

Alexander C. M. O. and Nittler L. R. (1999) The galactic evolution of Si, Ti, and O isotopic ratios. *Astrophys. J., 519,* 222–235.

Alexander C. M. O., Grossman J. N., Ebel D. S., and Ciesla F. J. (2008) The formation conditions of chondrules and chondrites. *Science, 320,* 1617–1619.

Alexander C. M. O., Bowden R., Fogel M. L., Howard K. T., Herd C. D. K., and Nittler L. R. (2012) The provenances of asteroids, and their contributions to the volatile inventories of the terrestrial planets. *Science, 337,* 721–723.

Alexander C. M. O., Cody G. D., De Gregorio B. T., Nittler L. R., and Stroud R. M. (2017) The nature, origin and modification of insoluble organic matter in chondrites, the major source of Earth's C and N. *Chem. Erde–Geochem., 77,* 227–256.

Alexander C. M. O., McKeegan K. D., and Altwegg K. (2018) Water reservoirs in small planetary bodies: Meteorites, asteroids, and comets. *Space Sci. Rev., 214,* 36.

Allègre C. J., Manhès G., and Göpel C. (1995) The age of the Earth. *Geochim. Cosmochim. Acta, 59,* 1445–1456.

Altwegg K. et al. (2015) 67P/Churyumov-Gerasimenko, a Jupiter family comet with a high D/H ratio. *Science, 347,* 6220.

Amari S., Anders E., Virag A., and Zinner E. (1990) Interstellar graphite in meteorites. *Nature, 345,* 238–240.

Amari S., Nittler L. R., Zinner E., Lodders K., and Lewis R. S. (2001) Presolar SiC grains of type A and B: Their isotopic compositions and stellar origins. *Astrophys. J., 559,* 463–483.

Amelin Y. (2002) Lead isotopic ages of chondrules and calcium-aluminum-rich inclusions. *Science, 297,* 1678–1683.

Amelin Y. and Krot A. (2007) Pb isotopic age of the Allende chondrules. *Meteorit. Planet. Sci., 42,* 1321–1335.

Anders E. and Grevesse N. (1989) Abundances of the elements: Meteoritic and solar. *Geochim. Cosmochim. Acta, 53,* 197–214.

Anders E. and Zinner E. (1993) Interstellar grains in primitive meteorites: Diamond, silicon carbide, and graphite. *Meteoritics, 28,* 490–514.

Armitage P. J. (2011) Dynamics of protoplanetary disks. *Annu. Rev. Astron. Astrophys., 49,* 195–236.

Armitage P. J., Eisner J. A., and Simon J. B. (2016) Prompt planetesimal formation beyond the snow line. *Astrophys. J., 828,* L2.

Arnould M. and Goriely S. (2003) The p-process of stellar nucleosynthesis: Astrophysics and nuclear physics status. *Phys. Rept., 384,* 1–84.

Avice G. and Marty B. (2014) The iodine-plutonium-xenon age of the Moon-Earth system revisited. *Philos. Trans. R. Soc. A, 372,* 20130260.

Avice G., Marty B., Burgess R., Hofmann A., Philippot P., Zahnle K., and Zakharov D. (2018) Evolution of atmospheric xenon and other noble gases inferred from Archean to Paleoproterozoic rocks. *Geochim. Cosmochim. Acta, 232,* 82–100.

Baker J., Bizzarro M., Wittig N., Connelly J., and Haack H. (2005) Early planetesimal melting from an age of 4.5662 Gyr for differentiated meteorites. *Nature, 436,* 1127–1131.

Bally J. and Langer W. D. (1982) Isotope-selective photodestruction of carbon monoxide. *Astrophys. J., 255,* 143–148.

Balsiger H. et al. (2015) Detection of argon in the coma of comet 67P/

Churyumov-Gerasimenko. *Sci. Adv., 1,* 8.
Bar-Nun A. and Owen T. (1998) Trapping of gases in water ice and consequences to comets and the atmospheres of the inner planets. In *Solar System Ices* (B. Schmitt et al., eds.), pp. 353–366. Astrophys. Space Sci. Library, Vol. 227, Kluwer, Dordrecht.
Barrat J.-A., Chaussidon M., Yamaguchi A., Beck P., Villeneuve J., Byrne D. J., Broadley M. W., and Marty B. (2021) A *4,*565-My-old andesite from an extinct chondritic protoplanet. *Proc. Natl. Acad. Sci. U.S.A., 118,* e2026129118.
Baruteau C., Crida A., Paardekooper S.-J., Masset F., Guilet J., Bitsch B., Nelson R., Kley W., and Papaloizou J. (2014) Planet-disk interactions and early evolution of planetary systems. In *Protostars and Planets VI* (H. Beuther et al., eds.), pp. 667–690. Univ. of Arizona, Tucson.
Batygin K. and Brown M. E. (2016) Evidence for a distant giant planet in the solar system. *Astron. J., 151,* 22.
Becker R. H. and Pepin R. O. (1984) The case for a martian origin of the shergottites: Nitrogen and noble gases in EETA 79001. *Earth Planet. Sci. Lett., 69,* 225–242.
Bekaert D. V., Broadley M. W., and Marty B. (2020) The origin and fate of volatile elements on Earth revisited in light of noble gas data obtained from comet 67P/Churyumov-Gerasimenko. *Sci. Rep., 10,* 5796.
Benz W., Slattery W. L., and Cameron A. G. W. (1986) The origin of the moon and the single-impact hypothesis I. *Icarus, 66,* 515–535.
Bermingham K. R. (2020) Ancient rock bears isotopic fingerprints of Earth's origins. *Nature, 579,* 195–196.
Bermingham K. R. and Kruijer T. S. (2022) Isotopic constraints on the formation of the main belt. In *Vesta and Ceres: Insights from the Dawn Mission for the Origin of the Solar System* (S. Marchi et al., eds.), pp. 212–226. Cambridge Univ., Cambridge.
Bermingham K. R. and Walker R. J. (2017) The ruthenium isotopic composition of the oceanic mantle. *Earth Planet. Sci. Lett., 474,* 466–473.
Bermingham K. R., Worsham E. A., and Walker R. J. (2018a) New insights into Mo and Ru isotope variation in the nebula and terrestrial planet accretionary genetics. *Earth Planet. Sci. Lett., 487,* 221–229.
Bermingham K. R., Gussone N., Mezger K., and Krause J. (2018b) Origins of mass-dependent and mass-independent Ca isotope variations in meteoritic components and meteorites. *Geochim. Cosmochim. Acta, 226,* 206–223.
Bermingham K. R., Füri E., Lodders K., and Marty B. (2020) The NC-CC isotope dichotomy: Implications for the chemical and isotopic evolution of the early solar system. *Space Sci. Rev., 216,* 133.
Bernatowicz T. J. and Hagee B. E. (1987) Isotopic fractionation of Kr and Xe implanted in solids at very low energies. *Geochim. Cosmochim. Acta, 51,* 1599–1611.
Binzel R. P. and Xu S. (1993) Chips off of asteroid 4 Vesta: Evidence for the parent body of basaltic achondrite meteorites. *Science, 260,* 186–191.
Birnstiel T., Klahr H., and Ercolano B. (2012) A simple model for the evolution of the dust population in protoplanetary disks. *Astron. Astrophys., 539,* A148.
Birnstiel T., Fang M., and Johansen A. (2016) Dust evolution and the formation of planetesimals. *Space Sci. Rev., 205,* 41–75.
Bischoff A., Vogel N., and Roszjar J. (2011) The Rumuruti chondrite group. *Chem. Erde–Geochem., 71,* 101–133.
Bitsch B., Morbidelli A., Lega E., and Crida A. (2014) Stellar irradiated discs and implications on migration of embedded planets. *Astron. Astrophys., 564,* A135.
Bitsch B., Johansen A., Lambrechts M., and Morbidelli A. (2015a) The structure of protoplanetary discs around evolving young stars. *Astron. Astrophys., 575,* A28.
Bitsch B., Lambrechts M., and Johansen A. (2015b) The growth of planets by pebble accretion in evolving protoplanetary discs. *Astron. Astrophys., 582,* A112.
Bitsch B., Morbidelli A., Johansen A., Lega E., Lambrechts M., and Crida A. (2018) Pebble-isolation mass: Scaling law and implications for the formation of super-Earths and gas giants. *Astron. Astrophys., 612,* A30.
Black D. C. and Pepin R. O. (1969) Trapped neon in meteorites — II. *Earth Planet. Sci. Lett., 6,* 395–405.
Blum J. and Wurm G. (2008) The growth mechanisms of macroscopic bodies in protoplanetary disks. *Annu. Rev. Astron. Astrophys., 46,* 21–56.
Bockelée-Morvan D. et al. (2015) Cometary isotopic measurements. *Space Sci. Rev., 197,* 47–83.
Bodmer R. and Bochsler P. (1998) The helium isotopic ratio in the solar wind and ion fractionation in the corona by inefficient Coulomb drag. *Astron. Astrophys., 337,* 921–927.
Boehnke P. and Harrison T. M. (2016) Illusory Late Heavy Bombardments. *Proc. Natl. Acad. Sci. U.S.A., 113,* 10802–10806.
Bollard J. et al. (2019) Combined U-corrected Pb-Pb dating and ^{26}Al-^{26}Mg systematics of individual chondrules — Evidence for a reduced initial abundance of ^{26}Al amongst inner solar system chondrules. *Geochim. Cosmochim. Acta, 260,* 62–83.
Bottke W. F., Vokrouhlický D., Rubincam D. P., and Nesvorný D. (2006) The Yarkovsky and Yorp effects: Implications for asteroid dynamics. *Annu. Rev. Earth Planet. Sci., 34,* 157–191.
Bottke W. F., Walker R. J., Day J. M. D., Nesvorný D., and Elkins-Tanton L. (2010) Stochastic late accretion to Earth, the Moon, and Mars. *Science, 330,* 1527–1530.
Bouhifd M. A., Jephcoat A. P., Heber V. S., and Kelley S. P. (2013) Helium in Earth's early core. *Nature Geosci., 6,* 982–986.
Bouvier A. and Wadhwa M. (2010) The age of the solar system redefined by the oldest Pb-Pb age of a meteoritic inclusion. *Nature Geosci., 3,* 637–641.
Brasser R. and Mojzsis S. J. (2020) The partitioning of the inner and outer solar system by a structured protoplanetary disk. *Nature Astron., 4,* 492–499.
Brasser R. and Morbidelli A. (2013) Oort cloud and scattered disc formation during a late dynamical instability in the solar system. *Icarus, 225,* 40–49.
Brasser R., Matsumura S., Ida S., Mojzsis S. J., and Werner S. C. (2016) Analysis of terrestrial planet formation by the Grand Tack model: System architecture and tack location. *Astrophys. J., 821,* 75.
Brennecka G. A., Burkhardt C., Budde G., Kruijer T. S., Nimmo F., and Kleine T. (2020) Astronomical context of solar system formation from molybdenum isotopes in meteorite inclusions. *Science, 370,* 837–840.
Broadley M. W., Barry P. H., Bekaert D. V., Byrne D. J., Caracausi A., Ballentine C. J., and Marty B. (2020) Identification of chondritic krypton and xenon in Yellowstone gases and the timing of terrestrial volatile accretion. *Proc. Natl. Acad. Sci. U.S.A., 117,* 13997–14004.
Broadley M. W., Bekaert D. V., Piani L., Füri E., and Marty B. (2022) Origin of life-forming volatile elements in the inner solar system. *Nature, 611,* 245–255.
Brooks R. R. (1960) Dissolution of silicate rocks. *Nature, 185,* 837–838.
Brownlee D. E. (1985) Cosmic dust: Collection and research. *Annu. Rev. Earth Planet. Sci., 13,* 147–173.
Brownlee D. E. et al. (2006) Comet 81P/Wild 2 under a microscope. *Science, 314,* 1711–1716.
Brož M., Chrenko O., Nesvorný D., and Dauphas N. (2021) Early terrestrial planet formation by torque-driven convergent migration of planetary embryos. *Nature Astron., 5,* 898–902.
Budde G., Kleine T., Kruijer T. S., Burkhardt C., and Metzler K. (2016) Tungsten isotopic constraints on the age and origin of chondrules. *Proc. Natl. Acad. Sci. U.S.A., 113,* 2886–2891.
Budde G., Burkhardt C., and Kleine T. (2019) Molybdenum isotopic evidence for the late accretion of outer solar system material to Earth. *Nature Astron., 3,* 736–741.
Burbidge E. M., Burbidge G. R., Fowler W. A., and Hoyle F. (1957) Synthesis of the elements in stars. *Rev. Mod. Phys., 29,* 547–650.
Burkhardt C., Dauphas N., Hans U., Bourdon B., and Kleine T. (2019) Elemental and isotopic variability in solar system materials by mixing and processing of primordial disk reservoirs. *Geochim. Cosmochim. Acta, 261,* 145–170.
Burnett D. S. and the Genesis Science Team (2011) Solar composition from the Genesis Discovery Mission. *Proc. Natl. Acad. Sci. U.S.A., 108,* 19147–19151.
Busemann H. (2006) Interstellar chemistry recorded in organic matter from primitive meteorites. *Science, 312,* 727–730.
Busemann H., Baur H., and Wieler R. (2000) Primordial noble gases in "phase Q" in carbonaceous and ordinary chondrites studied by closed system etching. *Meteorit. Planet. Sci., 35,* 949–973.
Cameron A. G. W. (1957) *Stellar Evolution, Nuclear Astrophysics, and Nucleogenesis, 2nd edition.* Lecture Series at Purdue University, 167 pp., available online at *https://www.osti.gov/servlets/purl/4318886.*
Cameron A. G. W. (1973) Abundances of the elements in the solar system. *Space Sci. Rev., 15,* 121–146.

Canup R. M. and Asphaug E. (2001) Origin of the Moon in a giant impact near the end of the Earth's formation. *Nature, 412,* 708–712.
Carrera D., Johansen A., and Davies M. B. (2015) How to form planetesimals from mm-sized chondrules and chondrule aggregates. *Astron. Astrophys., 579,* A43.
Chakraborty S., Ahmed M., Jackson T. L., and Thiemens M. H. (2008) Experimental test of self-shielding in vacuum ultraviolet photodissociation of CO. *Science, 321,* 1328–1331.
Chakraborty S., Muskatel B. H., Jackson T. L., Ahmed M., Levine R. D., and Thiemens M. H. (2014) Massive isotopic effect in vacuum UV photodissociation of N_2 and implications for meteorite data. *Proc. Natl. Acad. Sci. U.S.A., 111,* 14704–14709.
Chambers J. E. (2001) Making more terrestrial planets. *Icarus, 152,* 205–224.
Chambers J. E. (2004) Planetary accretion in the inner solar system. *Earth Planet. Sci. Lett., 223,* 241–252.
Chambers J. E. (2016) Pebble accretion and the diversity of planetary systems. *Astrophys. J., 825,* 63.
Charnoz S. et al. (2019) Planetesimal formation in an evolving protoplanetary disk with a dead zone. *Astron. Astrophys., 627,* A50.
Chladni E. F. F. (1794) *Über den Ursprung der von Pallasgefundenen und Anderer ihr Ähnlichen Eisenmassen and Über Einige Damit in Verbindung Stehende Naturerscheinungen.* Johann Friedrich Hartknoch, Riga. 63 pp.
Clayton D. D. (1982) Cosmic chemical memory: A new astronomy. *Q. J. R. Astron. Soc., 23,* 174–212.
Clayton R. N. (2002) Self-shielding in the solar nebula. *Nature, 415,* 860–861.
Clayton R. N., Grossman L., and Mayeda T. K. (1973) A component of primitive nuclear composition in carbonaceous meteorites. *Science, 182,* 485–488.
Cleeves L. I., Bergin E. A., Alexander C. M. O., Du F., Graninger D., Oberg K. I., and Harries T. J. (2014) The ancient heritage of water ice in the solar system. *Science, 345,* 1590–1593.
Clement M. S., Kaib N. A., Raymond S. N., and Walsh K. J. (2018) Mars' growth stunted by an early giant planet instability. *Icarus, 311,* 340–356.
Clement M. S., Kaib N. A., Raymond S. N., Chambers J. E., and Walsh K. J. (2019) The early instability scenario: Terrestrial planet formation during the giant planet instability, and the effect of collisional fragmentation. *Icarus, 321,* 778–790.
Clement M. S., Morbidelli A., Raymond S. N. and Kaib N. A. (2020) A record of the final phase of giant planet migration fossilized in the asteroid belt's orbital structure. *Mon. Not. R. Astron. Soc. Lett., 492,* L56–L60.
Clement M. S., Kaib N. A., Raymond S. N., and Chambers J. E. (2021a) The early instability scenario: Mars' mass explained by Jupiter's orbit. *Icarus, 367,* 114585.
Clement M. S., Raymond S. N., Kaib N. A., Deienno R., Chambers J. E., and Izidoro A. (2021b) Born eccentric: Constraints on Jupiter and Saturn's pre-instability orbits. *Icarus, 355,* 114122.
Connelly J. N., Bizzarro M., Krot A. N., Nordlund Å., Wielandt D., and Ivanova M. A. (2012) The absolute chronology and thermal processing of solids in the solar protoplanetary disk. *Science, 338,* 651–655.
Connelly J. N., Bollard J., and Bizzarro M. (2017) Pb-Pb chronometry and the early solar system. *Geochim. Cosmochim. Acta, 201,* 345–363.
Crabb J. and Anders E. (1981) Noble gases in E-chondrites. *Geochim. Cosmochim. Acta, 45,* 2443–2464.
Dalou C., Hirschmann M. M., von der Handt A., Mosenfelder J., and Armstrong L. S. (2017) Nitrogen and carbon fractionation during core-mantle differentiation at shallow depth. *Earth Planet. Sci. Lett., 458,* 141–151.
D'Angelo G., Bate M. R., and Lubow S. H. (2005) The dependence of protoplanet migration rates on co-orbital torques. *Mon. Not. R. Astron. Soc., 358,* 316–332.
Dauphas N. (2003) The dual origin of the terrestrial atmosphere. *Icarus, 165,* 326–339.
Dauphas N. (2017) The isotopic nature of the Earth's accreting material through time. *Nature, 541,* 521–524.
Dauphas N. and Pourmand A. (2011) Hf-W-Th evidence for rapid growth of Mars and its status as a planetary embryo. *Nature, 473,* 489–492.
Dauphas N. and Schauble E. A. (2016) Mass fractionation laws, mass-independent effects, and isotopic anomalies. *Annu. Rev. Earth Planet. Sci., 44,* 709–783.
Dauphas N., Davis A. M., Marty B., and Reisberg L. (2004) The cosmic molybdenum-ruthenium isotope correlation. *Earth Planet. Sci. Lett., 226,* 465–475.
Davis A. M. and Richter F. M. (2007) Condensation and evaporation of solar system materials. In *Treatise on Geochemistry, Vol. 1* (H. D. Holland and K. K. Turekian, eds.), pp. 1–31. Elsevier, Oxford.
Davis A. M., Zhang J., Greber N. D., Hu J., Tissot F. L. H., and Dauphas N. (2018) Titanium isotopes and rare earth patterns in CAIs: Evidence for thermal processing and gas-dust decoupling in the protoplanetary disk. *Geochim. Cosmochim. Acta, 221,* 275–295.
Day J. M. D., Pearson D. G., and Taylor L. A. (2007) Highly siderophile element constraints on accretion and differentiation of the Earth-Moon system. *Science, 315,* 217–219.
Deienno R., Izidoro A., Morbidelli A., Gomes R. S., Nesvorný D., and Raymond S. N. (2018) Excitation of a primordial cold asteroid belt as an outcome of planetary instability. *Astrophys. J., 864,* 50.
De Laeter J. and Kurz M. D. (2006) Alfred Nier and the sector field mass spectrometer. *J. Mass Spectrom., 41,* 847–854.
Deloule E. and Robert F. (1995) Interstellar water in meteorites? *Geochim. Cosmochim. Acta, 59,* 4695–4706.
DeMeo F. E. and Carry B. (2013) The taxonomic distribution of asteroids from multi-filter all-sky photometric surveys. *Icarus, 226,* 723–741.
DeMeo F. E. and Carry B. (2014) Solar system evolution from compositional mapping of the asteroid belt. *Nature, 505,* 629–634.
Desch S. J., Kalyaan A., and Alexander C. M. O. (2018) The effect of Jupiter's formation on the distribution of refractory elements and inclusions in meteorites. *Astrophys. J. Suppl. Ser., 238,* 11.
Dominik C. and Tielens A. G. G. M. (1997) The physics of dust coagulation and the structure of dust aggregates in space. *Astrophys. J., 480,* 647–673.
Doyle P. M., Jogo K., Nagashima K., Krot A. N., Wakita S., Ciesla F. J., and Hutcheon I. D. (2015) Early aqueous activity on the ordinary and carbonaceous chondrite parent bodies recorded by fayalite. *Nature Commun., 6,* 7444.
Drążkowska J. and Alibert Y. (2017) Planetesimal formation starts at the snow line. *Astron. Astrophys., 608,* A92.
Drążkowska J., Alibert Y., and Moore B. (2016) Close-in planetesimal formation by pile-up of drifting pebbles. *Astron. Astrophys., 594,* A105.
Duprat J. et al. (2010) Extreme deuterium excesses in ultracarbonaceous micrometeorites from central Antarctic snow. *Science, 328,* 742–745.
Ercolano B. and Pascucci I. (2017) The dispersal of planet-forming discs: Theory confronts observations. *R. Soc. Open Sci., 4,* 170114.
Fahey A. J., Goswami J. N., McKeegan K. D., and Zinner E. K. (1987) ^{16}O excesses in Murchison and Murray hibonites: A case against a late supernova injection origin of isotopic anomalies in O, Mg, Ca, and Ti. *Astrophys. J. Lett., 323,* L91.
Fischer R. A. and Ciesla F. J. (2014) Dynamics of the terrestrial planets from a large number of N-body simulations. *Earth Planet. Sci. Lett., 392,* 28–38.
Fischer R. A. and Nimmo F. (2018) Effects of core formation on the Hf-W isotopic composition of the Earth and dating of the Moon-forming impact. *Earth Planet. Sci. Lett., 499,* 257–265.
Fischer R. A., Cottrell E., Hauri E., Lee K. K. M., and Le Voyer M. (2020) The carbon content of Earth and its core. *Proc. Natl. Acad. Sci. U.S.A., 117,* 8743–8749.
Fischer-Gödde M. and Kleine T. (2017) Ruthenium isotopic evidence for an inner solar system origin of the late veneer. *Nature, 541,* 525–527.
Fischer-Gödde M., Burkhardt C., Kruijer T. S., and Kleine T. (2015) Ru isotope heterogeneity in the solar protoplanetary disk. *Geochim. Cosmochim. Acta, 168,* 151–171.
Fischer-Gödde M., Elfers B.-M., Münker C., Szilas K., Maier W. D., Messling N., Morishita T., Van Kranendonk M., and Smithies H. (2020) Ruthenium isotope vestige of Earth's pre-late-veneer mantle preserved in Archaean rocks. *Nature, 579,* 240–244.
Floss C. and Stadermann F. J. (2009) High abundances of circumstellar and interstellar C-anomalous phases in the primitive CR3 chondrites QUE 99177 and MET 00426. *Astrophys. J., 697,* 1242–1255.
Fouchet T., Lellouch E., Bézardt B., Encrenaz T., Drossart P., Feuchtgruber H., and de Graauw T. (2000) ISO-SWS observations of Jupiter: Measurement of the ammonia tropospheric profile and of the $^{15}N/^{14}N$ isotopic ratio. *Icarus, 143,* 223–243.
Füri E. and Marty B. (2015) Nitrogen isotope variations in the solar

system. *Nature Geosci., 8,* 515–522.
Geiss J. and Reeves H. (1972) Cosmic and solar system abundances of deuterium and helium-3. *Astron. Astrophys., 18,* 126–132.
Genda H. and Abe Y. (2005) Enhanced atmospheric loss on protoplanets at the giant impact phase in the presence of oceans. *Nature, 433,* 842–844.
Genda H. and Ikoma M. (2008) Origin of the ocean on the Earth: Early evolution of water D/H in a hydrogen-rich atmosphere. *Icarus, 194,* 42–52.
Gerber S., Burkhardt C., Budde G., Metzler K., and Kleine T. (2017) Mixing and transport of dust in the early solar nebula as inferred from titanium isotope variations among chondrules. *Astrophys. J. Lett., 841,* L17.
Goderis S., Brandon A. D., Mayer B., and Humayun M. (2015) s-process Os isotope enrichment in ureilites by planetary processing. *Earth Planet. Sci. Lett., 431,* 110–118.
Goldschmidt V. M. (1937) The principles of distribution of chemical elements in minerals and rocks: The seventh Hugo Müller Lecture, delivered before the Chemical Society on March 17th, 1937. *J. Chem. Soc.,* 655–673.
Gomes R., Levison H. F., Tsiganis K., and Morbidelli A. (2005) Origin of the cataclysmic Late Heavy Bombardment period of the terrestrial planets. *Nature, 435,* 466–469.
Gounelle M. (2006) The meteorite fall at l'Aigle and the Biot report: Exploring the cradle of meteorites. *Spec. Publ. Geol. Soc. London, 256,* 73–89.
Gounelle M., Spurný P., and Bland P. A. (2006) The orbit and atmospheric trajectory of the Orgueil meteorite from historical records. *Meteorit. Planet. Sci., 41,* 135–150.
Gradie J. and Tedesco E. (1982) Compositional structure of the asteroid belt. *Science, 216,* 1405–1407.
Grady M. M. and Pillinger C. T. (1990) ALH 85085: Nitrogen isotope analysis of a highly unusual primitive chondrite. *Earth Planet. Sci. Lett., 97,* 29–40.
Greenberg R., Wacker J. F., Hartmann W. K., and Chapman C. R. (1978) Planetesimals to planets: Numerical simulation of collisional evolution. *Icarus, 35,* 1–26.
Grewal D. S., Dasgupta R., Sun C., Tsuno K., and Costin G. (2019) Delivery of carbon, nitrogen, and sulfur to the silicate earth by a giant impact. *Sci. Adv., 5,* 1–13.
Grossman L. and Larimer J. W. (1974) Early chemical history of the solar system. *Rev. Geophys., 12,* 71–101.
Guillot T. and Hueso R. (2006) The composition of Jupiter: Sign of a (relatively) late formation in a chemically evolved protosolar disc. *Mon. Not. R. Astron. Soc. Lett., 367,* L47–L51.
Güttler C., Blum J., Zsom A., Ormel C. W., and Dullemond C. P. (2010) The outcome of protoplanetary dust growth: Pebbles, boulders, or planetesimals? *Astron. Astrophys., 513,* A56.
Gyngard F., Amari S., Zinner E., and Marhas K. K. (2018) Correlated silicon and titanium isotopic compositions of presolar SiC grains from the Murchison CM2 chondrite. *Geochim. Cosmochim. Acta, 221,* 145–161.
Haisch K. A. J., Lada E. A., and Lada C. J. (2001) Disk frequencies and lifetimes in young clusters. *Astrophys. J. Lett., 553,* L153–L156.
Hallis L. J., Huss G. R., Nagashima K., Taylor G. J., Halldórsson S. A., Hilton D. R., Mottl M. J., and Meech K. J. (2015) Evidence for primordial water in Earth's deep mantle. *Science, 350,* 795–797.
Hansen B. M. S. (2009) Formation of the terrestrial planets from a narrow annulus. *Astrophys. J., 703,* 1131–1140.
Hashizume K., Chaussidon M., Marty B., and Robert F. (2000) Solar wind record on the Moon: Deciphering presolar from planetary nitrogen. *Science, 290,* 1142–1145.
Hashizume K., Chaussidon M., Marty B., and Terada K. (2004) Protosolar carbon isotopic composition: Implications for the origin of meteoritic organics. *Astrophys. J., 600,* 480–484.
Hässig M. et al. (2017) Isotopic composition of CO_2 in the coma of 67P/Churyumov-Gerasimenko measured with ROSINA/DFMS. *Astron. Astrophys., 605,* A50.
Hayashi C. (1981) Structure of the solar nebula, growth and decay of magnetic fields and effects of magnetic and turbulent viscosities on the nebula. *Progr. Theor. Phys. Suppl., 70,* 35–53.
Heber V. S., Baur H., Bochsler P., McKeegan K. D., Neugebauer M., Reisenfeld D. B., Wieler R., and Wiens R. C. (2012) Isotopic mass fractionation of solar wind: Evidence from fast and slow solar wind collected by the Genesis mission. *Astrophys. J., 759,* 121.
Hellmann J. L., Kruijer T. S., Van Orman J. A., Metzler K., and Kleine T. (2019) Hf-W chronology of ordinary chondrites. *Geochim. Cosmochim. Acta, 258,* 290–309.
Hilton C. D., Bermingham K. R., Walker R. J., and McCoy T. J. (2019) Genetics, crystallization sequence, and age of the South Byron Trio iron meteorites: New insights to carbonaceous chondrite (CC) type parent bodies. *Geochim. Cosmochim. Acta, 251,* 217–228.
Hirschmann M. M. (2018) Comparative deep Earth volatile cycles: The case for C recycling from exosphere/mantle fractionation of major (H_2O, C, N) volatiles and from H_2O/Ce, CO_2/Ba, and CO_2/Nb exosphere ratios. *Earth Planet. Sci. Lett., 502,* 262–273.
Hirschmann M. M. and Dasgupta R. (2009) The H/C ratios of Earth's near-surface and deep reservoirs, and consequences for deep Earth volatile cycles. *Chem. Geol., 262,* 4–16.
Holland G., Cassidy M., and Ballentine C. J. (2009) Meteorite Kr in Earth's mantle suggests a late accretionary source for the atmosphere. *Science, 326,* 1522–1525.
Hopp T., Budde G., and Kleine T. (2020) Heterogeneous accretion of Earth inferred from Mo-Ru isotope systematics. *Earth Planet. Sci. Lett., 534,* 116065.
Hoppe P., Amari S., Zinner E., Ireland T., and Lewis R. S. (1994) Carbon, nitrogen, magnesium, silicon, and titanium isotopic compositions of single interstellar silicon carbide grains from the Murchison carbonaceous chondrite. *Astrophys. J., 430,* 870–890.
Hunt A. C., Theis K. J., Rehkämper M., Benedix G. K., Andreasen R., and Schönbächler M. (2022) The dissipation of the solar nebula constrained by impacts and core cooling in planetesimals. *Nature Astron., 6,* 812–818.
Huss G. R. and Smith J. B. (2007) Titanium isotopic compositions of well-characterized silicon carbide grains from Orgueil (CI): Implications for s-process nucleosynthesis. *Meteorit. Planet. Sci., 42,* 1055–1075.
Huss G. R., Meshik A. P., Smith J. B., and Hohenberg C. M. (2003) Presolar diamond, silicon carbide, and graphite in carbonaceous chondrites: Implications for thermal processing in the solar nebula. *Geochim. Cosmochim. Acta, 67,* 4823–4848.
Ida S. and Makino J. (1992) N-body simulation of gravitational interaction between planetesimals and a protoplanet. *Icarus, 96,* 107–120.
Ida S. and Lin D. N. C. (2004) Toward a deterministic model of planetary formation. I. A desert in the mass and semimajor axis distributions of extrasolar planets. *Astrophys. J., 604,* 388–413.
Ida S., Guillot T., and Morbidelli A. (2016) The radial dependence of pebble accretion rates: A source of diversity in planetary systems. *Astron. Astrophys., 591,* A72.
Ida S., Yamamura T., and Okuzumi S. (2019) Water delivery by pebble accretion to rocky planets in habitable zones in evolving disks. *Astron. Astrophys., 624,* A28.
Ireland T. R., Zinner E. K., and Amari S. (1991) Isotopically anomalous Ti in presolar SiC from the Murchison meteorite. *Astrophys. J. Lett., 376,* L53–L56.
Izidoro A., de Souza Torres K., Winter O. C., and Haghighipour N. (2013) A compound model for the origin of Earth's water. *Astrophys. J., 767,* 54.
Izidoro A., Morbidelli A., Raymond S. N., Hersant F., and Pierens A. (2015) Accretion of Uranus and Neptune from inward-migrating planetary embryos blocked by Jupiter and Saturn. *Astron. Astrophys., 582,* A99.
Izidoro A., Bitsch B., and Dasgupta R. (2021) The effect of a strong pressure bump in the Sun's natal disk: Terrestrial planet formation via planetesimal accretion rather than pebble accretion. *Astrophys. J., 915,* 62.
Izidoro A., Dasgupta R., Raymond S. N., Deienno R., Bitsch B., and Isella A. (2022) Planetesimal rings as the cause of the solar system's planetary architecture. *Nature Astron., 6,* 357–366.
Javoy M. et al. (2010) The chemical composition of the Earth: Enstatite chondrite models. *Earth Planet. Sci. Lett., 293,* 259–268.
Johansen A. and Lambrechts M. (2017) Forming planets via pebble accretion. *Annu. Rev. Earth Planet. Sci., 45,* 359–387.
Johansen A., Blum J., Tanaka H., Ormel C., Bizzarro M., and Rickman H. (2014) The multifaceted planetesimal formation process. In *Protostars and Planets VI* (H. Beuther et al., eds.), pp. 547–570. Univ. of Arizona, Tucson.
Johansen A., Ronnet T., Bizzarro M., Schiller M., Lambrechts M., Nordlund Å., and Lammer H. (2021) A pebble accretion model for the formation of the terrestrial planets in the solar system. *Sci. Adv., 7,* eabc0444.

Kaib N. A. and Cowan N. B. (2015) The feeding zones of terrestrial planets and insights into Moon formation. *Icarus, 252,* 161–174.
Kawaguchi J., Fujiwara A., and Uesugi T. (2008) Hayabusa — Its technology and science accomplishment summary and Hayabusa-2. *Acta Astronaut., 62,* 639–647.
Kerridge J. F. (1993) Long-term compositional variation in solar corpuscular radiation: Evidence from nitrogen isotopes in the lunar regolith. *Rev. Geophys., 31,* 423–437.
Kita N. T. and Ushikubo T. (2012) Evolution of protoplanetary disk inferred from ^{26}Al chronology of individual chondrules. *Meteorit. Planet. Sci., 47,* 1108–1119.
Kleine T. (2005) Hf-W chronometry of lunar metals and the age and early differentiation of the Moon. *Science, 310,* 1671–1674.
Kleine T., Touboul M., Bourdon B., Nimmo F., Mezger K., Palme H., Jacobsen S. B., Yin Q.-Z., and Halliday A. N. (2009) Hf-W chronology of the accretion and early evolution of asteroids and terrestrial planets. *Geochim. Cosmochim. Acta, 73,* 5150–5188.
Kleine T., Hans U., Irving A. J., and Bourdon B. (2012) Chronology of the angrite parent body and implications for core formation in protoplanets. *Geochim. Cosmochim. Acta, 84,* 186–203.
Kley W. and Nelson R. P. (2012) Planet-disk interaction and orbital evolution. *Annu. Rev. Astron. Astrophys., 50,* 211–249.
Kokubo E. (2000) Formation of protoplanets from planetesimals in the solar nebula. *Icarus, 143,* 15–27.
Kokubo E. and Ida S. (1998) Oligarchic growth of protoplanets. *Icarus, 131,* 171–178.
Kööp L., Davis A. M., Nakashima D., Park C., Krot A. N., Nagashima K., Tenner T. J., Heck P. R., and Kita N. T. (2016) A link between oxygen, calcium and titanium isotopes in ^{26}Al-poor hibonite-rich CAIs from Murchison and implications for the heterogeneity of dust reservoirs in the solar nebula. *Geochim. Cosmochim. Acta, 189,* 70–95.
Krot A. N., Amelin Y., Cassen P., and Meibom A. (2005) Young chondrules in CB chondrites from a giant impact in the early solar system. *Nature, 436,* 989–992.
Krot A. N., Nagashima K., Libourel G., and Miller K. E. (2018) Multiple mechanisms of transient heating events in the protoplanetary disk: Evidence from precursors of chondrules and igneous Ca, Al-rich inclusions. In *Chondrules: Records of Protoplanetary Disk Processes* (A. N. Krot et al., eds.), pp. 11–56. Cambridge Univ., Cambridge.
Kruijer T. S., Sprung P., Kleine T., Leya I., Burkhardt C., and Wieler R. (2012) Hf-W chronometry of core formation in planetesimals inferred from weakly irradiated iron meteorites. *Geochim. Cosmochim. Acta, 99,* 287–304.
Kruijer T. S., Touboul M., Fischer-Gödde M., Bermingham K. R., Walker R. J., and Kleine T. (2014) Protracted core formation and rapid accretion of protoplanets. *Science, 344,* 1150–1154.
Kruijer T. S., Burkhardt C., Budde G., and Kleine T. (2017) Age of Jupiter inferred from the distinct genetics and formation times of meteorites. *Proc. Natl. Acad. Sci. U.S.A., 114,* 6712–6716.
Kruijer T. S., Kleine T., and Borg L. E. (2020) The great isotopic dichotomy of the early solar system. *Nature Astron., 4,* 32–40.
Kuga M., Marty B., Marrocchi Y., and Tissandier L. (2015) Synthesis of refractory organic matter in the ionized gas phase of the solar nebula. *Proc. Natl. Acad. Sci. U.S.A., 112,* 7129–7134.
Kuga M., Cernogora G., Marrocchi Y., Tissandier L., and Marty B. (2017) Processes of noble gas elemental and isotopic fractionations in plasma-produced organic solids: Cosmochemical implications. *Geochim. Cosmochim. Acta, 217,* 219–230.
Lambrechts M. and Johansen A. (2012) Rapid growth of gas-giant cores by pebble accretion. *Astron. Astrophys., 544,* A32.
Lambrechts M., Johansen A., and Morbidelli A. (2014) Separating gas-giant and ice-giant planets by halting pebble accretion. *Astron. Astrophys., 572,* A35.
Lambrechts M., Morbidelli A., Jacobson S. A., Johansen A., Bitsch B., Izidoro A., and Raymond S. N. (2019) Formation of planetary systems by pebble accretion and migration. *Astron. Astrophys., 627,* A83.
Lauretta D. S. et al. (2015) The OSIRIS-REx target asteroid (101955) Bennu: Constraints on its physical, geological, and dynamical nature from astronomical observations. *Meteorit. Planet. Sci., 50,* 834–849.
Lauretta D. S. et al. (2017) OSIRIS-REx: Sample return from asteroid (101955) Bennu. *Space Sci. Rev., 212,* 925–984.
Lauretta D. S. et al. (2019) The unexpected surface of asteroid (101955) Bennu. *Nature, 568,* 55–60.
Leake B. E. et al. (1969) The chemical analysis of rock powders by automatic X-ray fluorescence. *Chem. Geol., 5,* 7–86.
Lecar M., Podolak M., Sasselov D., and Chiang E. (2006) On the location of the snow line in a protoplanetary disk. *Astrophys. J., 640,* 1115–1118.
Lee J.-E., Bergin E. A., and Lyons J. R. (2008) Oxygen isotope anomalies of the Sun and the original environment of the solar system. *Meteorit. Planet. Sci., 43,* 1351–1362.
Lee J. K. W. (2015) Ar-Ar and K-Ar dating. In *Encyclopedia of Scientific Dating Methods* (W. J. Rink and J. W. Thompson, eds.), pp 58–73. Springer, Dordrecht.
Lee T., Papanastassiou D. A., and Wasserburg G. J. (1977) Aluminum-26 in the early solar system: Fossil or fuel? *Astrophys. J. Lett., 211,* L107–L110.
Leinhardt Z. M. and Richardson D. C. (2005) Planetesimals to protoplanets. I. Effect of fragmentation on terrestrial planet formation. *Astrophys. J., 625,* 427–440.
Lellouch E., Bézard B., Fouchet T., Feuchtgruber H., Encrenaz T., and de Graauw T. (2001) The deuterium abundance in Jupiter and Saturn from ISO-SWS observations. *Astron. Astrophys., 370,* 610–622.
Levison H. F., Morbidelli A., VanLaerhoven C., Gomes R., and Tsiganis K. (2008) Origin of the structure of the Kuiper belt during a dynamical instability in the orbits of Uranus and Neptune. *Icarus, 196,* 258–273.
Levison H. F., Bottke W. F., Gounelle M., Morbidelli A., Nesvorný D., and Tsiganis K. (2009) Contamination of the asteroid belt by primordial trans-Neptunian objects. *Nature, 460,* 364–366.
Levison H. F., Thommes E., and Duncan M. J. (2010) Modeling the formation of giant planet cores. I. Evaluating key processes. *Astron. J., 139,* 1297–1314.
Levison H. F., Kretke K. A., and Duncan M. J. (2015) Growing the gas-giant planets by the gradual accumulation of pebbles. *Nature, 524,* 322–324.
Lewis R. S., Ming T., Wacker J. F., Anders E., and Steel E. (1987) Interstellar diamonds in meteorites. *Nature, 326,* 160–162.
Lewis R. S., Amari S., and Anders E. (1994) Interstellar grains in meteorites: II. SiC and its noble gases. *Geochim. Cosmochim. Acta, 58,* 471–494.
Libourel G. and Krot A. N. (2007) Evidence for the presence of planetesimal material among the precursors of magnesian chondrules of nebular origin. *Earth Planet. Sci. Lett., 254,* 1–8.
Lichtenberg T., Golabek G. J., Dullemond C. P., Schönbächler M., Gerya T. V., and Meyer M. R. (2018) Impact splash chondrule formation during planetesimal recycling. *Icarus, 302,* 27–43.
Lichtenberg T., Drążkowska J., Schönbächler M., Golabek G. J., and Hands T. O. (2021) Bifurcation of planetary building blocks during solar system formation. *Science, 371,* 365–370.
Lin D. N. C. and Papaloizou J. (1986) On the tidal interaction between protoplanets and the protoplanetary disk. III. Orbital migration of protoplanets. *Astrophys. J., 309,* 846–857.
Lis D. C., Bockelée-Morvan D., Güsten R., Biver N., Stutzki J., Delorme Y., Durán C., Wiesemeyer H., and Okada Y. (2019) Terrestrial deuterium-to-hydrogen ratio in water in hyperactive comets. *Astron. Astrophys., 625,* L5.
Lissauer J. J., Hubickyj O., D'Angelo G., and Bodenheimer P. (2009) Models of Jupiter's growth incorporating thermal and hydrodynamic constraints. *Icarus, 199,* 338–350.
Liu B., Raymond S. N., and Jacobson S. A. (2022) Early solar system instability triggered by dispersal of the gaseous disk. *Nature, 604,* 643–646.
Lock S. J., Bermingham K. R., Parai R., and Boyet M. (2020) Geochemical constraints on the origin of the Moon and preservation of ancient terrestrial heterogeneities. *Space Sci. Rev., 216,* 109.
Lodders K. (2003) Solar system abundances and condensation temperatures of the elements. *Astrophys. J., 591,* 1220–1247.
Lodders K. (2021) Relative atomic solar system abundances, mass fractions, and atomic masses of the elements and their isotopes, composition of the solar photosphere, and compositions of the major chondritic meteorite groups. *Space Sci. Rev., 217,* 44.
Lyons J. R. and Young E. D. (2005) CO self-shielding as the origin of oxygen isotope anomalies in the early solar nebula. *Nature, 435,* 317–320.
Lyons J. R., Gharib-Nezhad E., and Ayres T. R. (2018) A light carbon isotope composition for the Sun. *Nature Commun., 9,* 908.
Lyra W., Paardekooper S.-J., and Mac Low M.-M. (2010) Orbital

migration of low mass planets in evolutionary radiative models: Avoiding catastrophic infall. *Astrophys. J. Lett., 715,* L68–L73.
MacPherson G. J. (2007) Calcium-aluminum-rich inclusions in chondritic meteorites. In *Treatise of Geochemistry, Vol. 1: Meteorites, Comets, and Planets and Cosmochemical Processes* (A. M. Davis, ed.), pp. 1–47. Elsevier-Pergamon, Oxford.
MacPherson G. J., Nagashima K., Krot A. N., Doyle P. M., and Ivanova M. A. (2017) ^{53}Mn-^{53}Cr chronology of Ca-Fe silicates in CV3 chondrites. *Geochim. Cosmochim. Acta, 201,* 260–274.
Mamajek E. E. (2009) Initial conditions of planet formation: Lifetimes of primordial disks. In *Exoplanets and Disks: Their Formation and Diversity* (T. Usuda et al., eds.), pp. 3–10. AIP Conf. Ser. 1158, American Institute of Physics, Melville, New York.
Marchi S., Canup R. M. and Walker R. J. (2018) Heterogeneous delivery of silicate and metal to the Earth by large planetesimals. *Nature Geosci., 11,* 77–81.
Marrocchi Y., Derenne S., Marty B., and Robert F. (2005) Interlayer trapping of noble gases in insoluble organic matter of primitive meteorites. *Earth Planet. Sci. Lett., 236,* 569–578.
Marrocchi Y., Villeneuve J., Batanova V., Piani L., and Jacquet E. (2018) Oxygen isotopic diversity of chondrule precursors and the nebular origin of chondrules. *Earth Planet. Sci. Lett., 496,* 132–141.
Martin R. G. and Livio M. (2012) On the evolution of the snow line in protoplanetary discs. *Mon. Not. R. Astron. Soc. Lett., 425,* L6–L9.
Marty B. (2012) The origins and concentrations of water, carbon, nitrogen and noble gases on Earth. *Earth Planet. Sci. Lett., 313–314,* 56–66.
Marty B. et al. (2017) Xenon isotopes in 67P/Churyumov-Gerasimenko show that comets contributed to Earth's atmosphere. *Science, 356,* 1069–1072.
Marty B. (2020) Origins and early evolution of the atmosphere and the oceans. *Geochem. Perspect., 9,* 135–313.
Marty B. (2022) Meteoritic noble gas constraints on the origin of terrestrial volatiles. *Icarus, 381,* 115020.
Marty B. and Yokochi R. (2006) Water in the early Earth. *Rev. Mineral. Geochem., 62,* 421–450.
Marty B., Zimmermann L., Burnard P. G., Wieler R., Heber V. S., Burnett D. L., Wiens R. C., and Bochsler P. (2010) Nitrogen isotopes in the recent solar wind from the analysis of Genesis targets: Evidence for large scale isotope heterogeneity in the early solar system. *Geochim. Cosmochim. Acta, 74,* 340–355.
Marty B., Chaussidon M., Wiens R. C., Jurewicz A. J. G., and Burnett D. S. (2011) A ^{15}N-poor isotopic composition for the solar system as shown by Genesis solar wind samples. *Science, 332,* 1533–1536.
Marty B., Avice G., Sano Y., Altwegg K., Balsiger H., Hässig M., Morbidelli A., Mousis O., and Rubin M. (2016) Origins of volatile elements (H, C, N, noble gases) on Earth and Mars in light of recent results from the Rosetta cometary mission. *Earth Planet. Sci. Lett., 441,* 91–102.
Masset F. and Snellgrove M. (2001) Reversing type II migration: Resonance trapping of a lighter giant protoplanet. *Mon. Not. R. Astron. Soc. Lett., 320,* L55–L59.
Matsuda J. and Maekawa T. (1992) Noble gas implantation in glow discharge: Comparison between diamond and graphite. *Geochem. J., 26,* 251–259.
Mazor E., Heymann D., and Anders E. (1970) Noble gases in carbonaceous chondrites. *Geochim. Cosmochim. Acta, 34,* 781–824.
McCubbin F. M. and Barnes J. J. (2019) Origin and abundances of H_2O in the terrestrial planets, Moon, and asteroids. *Earth Planet. Sci. Lett., 526,* 115771.
McDonough W. F. and Sun S. (1995) The composition of the Earth. *Chem. Geol., 120,* 223–253.
McKeegan K. D. et al. (2011) The oxygen isotopic composition of the Sun inferred from captured solar wind. *Science, 332,* 1528–1532.
McSween H. Y., Mittlefehldt D. W., Beck A. W., Mayne R. G., and McCoy T. J. (2010) HED meteorites and their relationship to the geology of Vesta and the Dawn mission. In *The Dawn Mission to Minor Planets 4 Vesta and 1 Ceres* (C. Russell and C. Raymond, eds.), pp. 141–174. Springer, New York.
Meech K. and Raymond S. N. (2020) Origin of Earth's water: Sources and constraints. In *Planetary Astrobiology* (V. Meadows et al., eds.), pp. 325–353. Univ. of Arizona, Tucson.
Meibom A., Krot A. N., Robert F., Mostefaoui S., Russell S. S., Petaev M. I., and Gounelle M. (2007) Nitrogen and carbon isotopic composition of the Sun inferred from a high-temperature solar nebular condensate. *Astrophys. J. Lett., 656,* L33–L36.
Meshik A., Hohenberg C., Pravdivtseva O., and Burnett D. (2014) Heavy noble gases in solar wind delivered by Genesis mission. *Geochim. Cosmochim. Acta, 127,* 326–347.
Messenger S. (2000) Identification of molecular-cloud material in interplanetary dust particles. *Nature, 404,* 968–971.
Mezger K., Maltese A., and Vollstaedt H. (2021) Accretion and differentiation of early planetary bodies as recorded in the composition of the silicate Earth. *Icarus, 365,* 114497.
Mittlefehldt D. W. (2015) Asteroid (4) Vesta: I. The howardite-eucrite-diogenite (HED) clan of meteorites. *Chem. Erde–Geochem., 75,* 155–183.
Mojzsis S. J., Brasser R., Kelly N. M., Abramov O., and Werner S. C. (2019) Onset of giant planet migration before 4480 million years ago. *Astrophys. J., 881,* 44.
Morbidelli A. and Crida A. (2007) The dynamics of Jupiter and Saturn in the gaseous protoplanetary disk. *Icarus, 191,* 158–171.
Morbidelli A. and Nesvorný D. (2012) Dynamics of pebbles in the vicinity of a growing planetary embryo: Hydro-dynamical simulations. *Astron. Astrophys., 546,* A18.
Morbidelli A., Chambers J., Lunine J. I., Petit J. M., Robert F., Valsecchi G. B., and Cyr K. E. (2000) Source regions and timescales for the delivery of water to the Earth. *Meteorit. Planet. Sci., 35,* 1309–1320.
Morbidelli A., Levison H. F., Tsiganis K., and Gomes R. (2005) Chaotic capture of Jupiter's Trojan asteroids in the early solar system. *Nature, 435,* 462–465.
Morbidelli A., Lunine J. I., O'Brien D. P., Raymond S. N., and Walsh K. J. (2012) Building terrestrial planets. *Annu. Rev. Earth Planet. Sci., 40,* 251–275.
Morbidelli A., Lambrechts M., Jacobson S., and Bitsch B. (2015) The great dichotomy of the solar system: Small terrestrial embryos and massive giant planet cores. *Icarus, 258,* 418–429.
Morbidelli A., Nesvorný D., Laurenz V., Marchi S., Rubie D. C., Elkins-Tanton L., Wieczorek M., and Jacobson S. (2018) The timeline of the lunar bombardment: Revisited. *Icarus, 305,* 262–276.
Morbidelli A. et al. (2022) Contemporary formation of early solar system planetesimals at two distinct radial locations. *Nature Astron., 6,* 72–79.
Moreira M. (2013) Noble gas constraints on the origin and evolution of Earth's volatiles. *Geochem. Perspect., 2,* 229–403.
Moreira M. and Charnoz S. (2016) The origin of the neon isotopes in chondrites and on Earth. *Earth Planet. Sci. Lett., 433,* 249–256.
Mousis O. et al. (2018) Scientific rationale for Uranus and Neptune *in situ* explorations. *Planet. Space Sci., 155,* 12–40.
Mukhopadhyay S. (2012) Early differentiation and volatile accretion recorded in deep-mantle neon and xenon. *Nature, 486,* 101–104.
Mukhopadhyay S. and Parai R. (2019) Noble gases: A record of Earth's evolution and mantle dynamics. *Annu. Rev. Earth Planet. Sci., 47,* 389–419.
Nagai Y. and Yokoyama T. (2016) Molybdenum isotopic analysis by negative thermal ionization mass spectrometry (N-TIMS): Effects on oxygen isotopic composition. *J. Anal. Atom. Spectrom., 31,* 948–960.
Nanne J. A. M., Nimmo F., Cuzzi J. N., and Kleine T. (2019) Origin of the non-carbonaceous–carbonaceous meteorite dichotomy. *Earth Planet. Sci. Lett., 511,* 44–54.
Nesvorný D. (2011) Young solar system's fifth giant planet? *Astrophys. J. Lett., 742,* L22.
Nesvorný D. (2015) Evidence for slow migration of Neptune from the inclination distribution of Kuiper belt objects. *Astrophys. J., 150,* 73.
Nesvorný D. (2018) Dynamical evolution of the early solar system. *Annu. Rev. Astron. Astrophys., 56,* 137–174.
Nesvorný D. (2021) Eccentric early migration of Neptune. *Astrophys. J. Lett., 908,* L47.
Nesvorný D. and Morbidelli A. (2012) Statistical study of the early solar system's instability with four, five, and six giant planets. *Astrophys. J., 144,* 117.
Nesvorný D., Vokrouhlický D., and Morbidelli A. (2007) Capture of irregular satellites during planetary encounters. *Astrophys. J., 133,* 1962–1976.
Nesvorný D., Vokrouhlický D., and Morbidelli A. (2013) Capture of the Trojans by jumping Jupiter. *Astrophys. J., 768,* 45.
Nesvorný D., Roig F. V., and Deienno R. (2021) The role of early giant-planet instability in terrestrial planet formation. *Astrophys. J., 161,* 50.
Nittler L. R. (1996) Quantitative isotopic ratio ion imaging and its application to studies of preserved stardust in meteorites. Ph.D. thesis, Washington University in St. Louis, St. Louis. 210 pp.

Nittler L. R. and Ciesla F. (2016) Astrophysics with extraterrestrial materials. *Annu. Rev. Astron. Astrophys., 54,* 53–93.

Nittler L. R. et al. (2001) X-ray fluorescence measurements of the surface elemental composition of asteroid 433 Eros. *Meteorit. Planet. Sci., 36,* 1673–1695.

Nittler L. R., Alexander C. M. O., Gallino R., Hoppe P., Nguyen A. N., Stadermann F. J., and Zinner E. K. (2008) Aluminum-, calcium- and titanium-rich oxide stardust in ordinary chondrite meteorites. *Astrophys. J., 682,* 1450–1478.

Nittler L. R., Alexander C. M. O., Davidson J., Riebe M. E. I., Stroud R. M., and Wang J. (2018) High abundances of presolar grains and ^{15}N-rich organic matter in CO3.0 chondrite Dominion Range 08006. *Geochim. Cosmochim. Acta, 226,* 107–131.

Nittler L. R., Stroud R. M., Trigo-Rodríguez J. M., De Gregorio B. T., Alexander C. M. O., Davidson J., Moyano-Cambero C. E., and Tanbakouei S. (2019) A cometary building block in a primitive asteroidal meteorite. *Nature Astron., 3,* 659–666.

Nittler L. R., Alexander C. M. O., Patzer A., and Verdier-Paoletti M. J. (2021) Presolar stardust in highly pristine CM chondrites Asuka 12169 and Asuka 12236. *Meteorit. Planet. Sci., 56,* 260–276.

Noguchi T. et al. (2015) Cometary dust in Antarctic ice and snow: Past and present chondritic porous micrometeorites preserved on the Earth's surface. *Earth Planet. Sci. Lett., 410,* 1–11.

Norrish K. and Hutton J. (1969) An accurate X-ray spectrographic method for the analysis of a wide range of geological samples. *Geochim. Cosmochim. Acta, 33,* 431–453.

O'Brien D. P., Walsh K. J., Morbidelli A., Raymond S. N., and Mandell A. M. (2014) Water delivery and giant impacts in the 'Grand Tack' scenario. *Icarus, 239,* 74–84.

Oddo G. (1914) Die molekularstruktur der radioaktiven atome. *Z. Anorg. Chem., 87,* 253–258.

Olson P. L. and Sharp Z. D. (2019) Nebular atmosphere to magma ocean: A model for volatile capture during Earth accretion. *Phys. Earth Planet. Inter., 294,* 106294.

Ormel C. W. (2017) The emerging paradigm of pebble accretion. In *Formation, Evolution, and Dynamics of Young Solar Systems* (M. Pessah and O. Gressel, eds.), pp. 197–228. Astrophys. Space Sci. Library, Vol. 445, Kluwer, Dordrecht.

Ormel C. W. and Klahr H. H. (2010) The effect of gas drag on the growth of protoplanets. *Astron. Astrophys., 520,* A43.

Ott U. (2014) Planetary and pre-solar noble gases in meteorites. *Chem. Erde–Geochem., 74,* 519–544.

Owen T., Mahaffy P. R., Niemann H. B., Atreya S., and Wong M. (2001) Protosolar nitrogen. *Astrophys. J. Lett., 553,* L77–L79.

Ozima M. and Podosek F. A. (2001) *Noble Gas Geochemistry.* Cambridge Univ., Cambridge. 286 pp.

Pepin R. O. (1991) On the origin and early evolution of terrestrial planet atmospheres and meteoritic volatiles. *Icarus, 92,* 2–79.

Peplowski P. N., Bazell D., Evans L. G., Goldsten J. O., Lawrence D. J., and Nittler L. R. (2015) Hydrogen and major element concentrations on 433 Eros: Evidence for an L- or LL-chondrite-like surface composition. *Meteorit. Planet. Sci., 50,* 353–367.

Péron S. and Mukhopadhyay S. (2022) Krypton in the Chassigny meteorite shows Mars accreted chondritic volatiles before nebular gases. *Science, 377,* 320–324.

Péron S., Moreira M., Colin A., Arbaret L., Putlitz B., and Kurz M. D. (2016) Neon isotopic composition of the mantle constrained by single vesicle analyses. *Earth Planet. Sci. Lett., 449,* 145–154.

Piani L., Marrocchi Y., Rigaudier T., Vacher L. G., Thomassin D., and Marty B. (2020) Earth's water may have been inherited from material similar to enstatite chondrite meteorites. *Science, 369,* 1110–1113.

Pierens A. and Nelson R. P. (2008) On the formation and migration of giant planets in circumbinary discs. *Astron. Astrophys., 483,* 633–642.

Pierens A., Raymond S. N., Nesvorný D. and Morbidelli A. (2014) Outward migration of Jupiter and Saturn in 3:2 or 2:1 resonance in radiative disks: Implications for the Grand Tack and Nice models. *Astrophys. J. Lett., 795,* L11.

Pirani S., Johansen A., Bitsch B., Mustill A. J., and Turrini D. (2019) Consequences of planetary migration on the minor bodies of the early solar system. *Astron. Astrophys., 623,* A169.

Podosek F. A. and Cassen P. (1994) Theoretical, observational, and isotopic estimates of the lifetime of the solar nebula. *Meteoritics, 29,* 6–25.

Pollack J. B., Hubickyj O., Bodenheimer P., Lissauer J. J., Podolak M., and Greenzweig Y. (1996) Formation of the giant planets by concurrent accretion of solids and gas. *Icarus, 124,* 62–85.

Poole G. M., Rehkämper M., Coles B. J., Goldberg T., and Smith C. L. (2017) Nucleosynthetic molybdenum isotope anomalies in iron meteorites — New evidence for thermal processing of solar nebula material. *Earth Planet. Sci. Lett., 473,* 215–226.

Porcelli D., Woolum D., and Cassen P. (2001) Deep Earth rare gases: Initial inventories, capture from the solar nebula, and losses during Moon formation. *Earth Planet. Sci. Lett., 193,* 237–251.

Potts P. J. (1992) *A Handbook of Silicate Rock Analysis.* Springer, New York. 622 pp.

Pujol M., Marty B., and Burgess R. (2011) Chondritic-like xenon trapped in Archean rocks: A possible signature of the ancient atmosphere. *Earth Planet. Sci. Lett., 308,* 298–306.

Qin L. and Carlson R. W. (2016) Nucleosynthetic isotope anomalies and their cosmochemical significance. *Geochem. J., 50,* 43–65.

Raymond S. N. and Morbidelli A. (2014) The Grand Tack model: A critical review. In *Complex Planetary Systems* (Z. Knežević and A. Lemaire, eds.), pp. 194–203. Proc. Intl. Astron. Union, Vol. 9, Cambridge Univ., Cambridge.

Raymond S. N. and Izidoro A. (2017a) Origin of water in the inner solar system: Planetesimals scattered inward during Jupiter and Saturn's rapid gas accretion. *Icarus, 297,* 134–148.

Raymond S. N. and Izidoro A. (2017b) The empty primordial asteroid belt. *Sci. Adv., 3,* e1701138.

Raymond S. N. and Morbidelli A. (2022) Planet formation: Key mechanisms and global models. In *Demographics of Exoplanetary Systems: Lecture Notes of the 3rd Advanced School on Exoplanetary Science* (K. Biazzo et al., eds.), pp. 3–82. Springer, Cham.

Raymond S. N., Quinn T., and Lunine J. I. (2007) High-resolution simulations of the final assembly of Earth-like planets. 2. Water delivery and planetary habitability. *Astrobiology, 7,* 66–84.

Raymond S. N., O'Brien D. P., Morbidelli A., and Kaib N. A. (2009) Building the terrestrial planets: Constrained accretion in the inner solar system. *Icarus, 203,* 644–662.

Raymond S. N., Schlichting H. E., Hersant F., and Selsis F. (2013) Dynamical and collisional constraints on a stochastic late veneer on the terrestrial planets. *Icarus, 226,* 671–681.

Raymond S. N., Kokubo E., Morbidelli A., Morishima R., and Walsh K. J. (2014) Terrestrial planet formation at home and abroad. In *Protostars and Planets VI* (H. Beuther et al., eds.), pp. 595–618. Univ. of Arizona, Tucson.

Raymond S. N., Izidoro A., Bitsch B., and Jacobson S. A. (2016) Did Jupiter's core form in the innermost parts of the Sun's protoplanetary disc? *Mon. Not. R. Astron. Soc., 458,* 2962–2972.

Raymond S. N., Izidoro A., and Morbidelli A. (2020) Solar system formation in the context of extra-solar planets. In *Planetary Astrobiology* (V. Meadows et al., eds.), pp. 287–324. Univ. of Arizona, Tucson.

Render J., Ebert S., Burkhardt C., Kleine T., and Brennecka G. A. (2019) Titanium isotopic evidence for a shared genetic heritage of refractory inclusions from different carbonaceous chondrites. *Geochim. Cosmochim. Acta, 254,* 40–53.

Reynolds J. H. (1960) Determination of the age of the elements. *Phys. Rev. Lett., 4,* 8–10.

Reynolds J. H. and Turner G. (1964) Rare gases in the chondrite Renazzo. *J. Geophys. Res., 69,* 3263–3281.

Ribeiro de Sousa R., Morbidelli A., Gomes R., Neto E. V., Izidoro A., and Alves A. A. (2022) Dynamical origin of the dwarf planet Ceres. *Icarus, 379,* 114933.

Rodgers S. D. and Charnley S. B. (2008) Nitrogen superfractionation in dense cloud cores. *Mon. Not. R. Astron. Soc. Lett., 385,* L48–L52.

Roig F. and Nesvorný D. (2015) The evolution of asteroids in the jumping-Jupiter migration model. *Astron. J., 150,* 186.

Ronnet T., Mousis O., Vernazza P., Lunine J. I., and Crida A. (2018) Saturn's formation and early evolution at the origin of Jupiter's massive moons. *Astron. J., 155,* 224.

Roskosz M., Bouhifd M. A., Jephcoat A. P., Marty B., and Mysen B. O. (2013) Nitrogen solubility in molten metal and silicate at high pressure and temperature. *Geochim. Cosmochim. Acta, 121,* 15–28.

Rubie D. C., Jacobson S. A., Morbidelli A., O'Brien D. P., Young E. D., de Vries J., Nimmo F., Palme H., and Frost D. J. (2015) Accretion and differentiation of the terrestrial planets with implications for the compositions of early-formed solar system bodies and accretion of water. *Icarus, 248,* 89–108.

Rubin M. et al. (2018) Krypton isotopes and noble gas abundances in the coma of comet 67P/Churyumov-Gerasimenko. *Sci. Adv., 4,* eaar6297.

Safronov V. S. and Zvjagina E. V. (1969) Relative sizes of the largest bodies during the accumulation of planets. *Icarus, 10,* 109–115.

Sakamoto N., Seto Y., Itoh S., Kuramoto K., Fujino K., Nagashima K., Krot A. N., and Yurimoto H. (2007) Remnants of the early solar system water enriched in heavy oxygen isotopes. *Science, 317,* 231–233.

Sanders I. S. and Scott E. R. D. (2012) The origin of chondrules and chondrites: Debris from low-velocity impacts between molten planetesimals? *Meteorit. Planet. Sci., 47,* 2170–2192.

Sasaki S. (1990) The primary solar-type atmosphere surrounding the accreting Earth: H_2O-induced high surface temperature. In *Origin of the Earth* (H. E. Newson and J. H. Jones, eds.), pp. 195–209. Oxford Univ., Oxford.

Sasselov D. D. and Lecar M. (2000) On the snow line in dusty protoplanetary disks. *Astrophys. J., 528,* 995–998.

Sato T., Okuzumi S., and Ida S. (2016) On the water delivery to terrestrial embryos by ice pebble accretion. *Astron. Astrophys., 589,* A15.

Savvidou S., Bitsch B., and Lambrechts M. (2020) Influence of grain growth on the thermal structure of protoplanetary discs. *Astron. Astrophys., 640,* A63.

Schlichting H. E., Warren P. H., and Yin Q.-Z. (2012) The late stages of terrestrial planet formation: Dynamical friction and the late veneer. *Astrophys. J., 752,* 8.

Schoonenberg D. and Ormel C. W. (2017) Planetesimal formation near the snowline: In or out? *Astron. Astrophys., 602,* A21.

Scott E. R. D. (1977) Formation of olivine-metal textures in pallasite meteorites. *Geochim. Cosmochim. Acta, 41,* 693–710.

Scott E. R. D., Krot A. N., and Sanders I. S. (2018) Isotopic dichotomy among meteorites and its bearing on the protoplanetary disk. *Astrophys. J., 854,* 164.

Sharp Z. D. (2017) Nebular ingassing as a source of volatiles to the terrestrial planets. *Chem. Geol., 448,* 137–150.

Simon J. B., Armitage P. J., Li R., and Youdin A. N. (2016) The mass and size distribution of planetesimals formed by the streaming instability. I. The role of self-gravity. *Astrophys. J., 822,* 55.

Spitzer F., Burkhardt C., Budde G., Kruijer T. S., Morbidelli A., and Kleine T. (2020) Isotopic evolution of the inner solar system inferred from molybdenum isotopes in meteorites. *Astrophys. J. Lett., 898,* L2.

Spitzer F., Burkhardt C., Nimmo F., and Kleine T. (2021) Nucleosynthetic Pt isotope anomalies and the Hf-W chronology of core formation in inner and outer solar system planetesimals. *Earth Planet. Sci. Lett., 576,* 117211.

Spitzer F., Burkhardt C., Pape J., and Kleine T. (2022) Collisional mixing between inner and outer solar system planetesimals inferred from the Nedagolla iron meteorite. *Meteorit. Planet. Sci., 57,* 261–276.

Stephan T. and Davis A. M. (2021) Molybdenum isotope dichotomy in meteorites caused by s-process variability. *Astrophys. J., 909,* 8.

Suess H. E. and Urey H. C. (1956) Abundances of the elements. *Rev. Mod. Phys., 28,* 53–74.

Swindle T. D. (2002) Martian noble gases. *Rev. Mineral. Geochem., 47,* 171–190.

Tachibana S. et al. (2022) Pebbles and sand on asteroid (162173) Ryugu: In situ observation and particles returned to Earth. *Science, 375,* 1011–1016.

Taylor S. et al. (2020) Sampling interplanetary dust from Antarctic air. *Meteorit. Planet. Sci., 55,* 1128–1145.

Tenner T. J., Kimura M., and Kita N. T. (2017) Oxygen isotope characteristics of chondrules from the Yamato-82094 ungrouped carbonaceous chondrite: Further evidence for common O-isotope environments sampled among carbonaceous chondrites. *Meteorit. Planet. Sci., 52,* 268–294.

Tera F., Papanastassiou D. A., and Wasserburg G. J. (1974) Isotopic evidence for a terminal lunar cataclysm. *Earth Planet. Sci. Lett., 22,* 1–21.

Thiemens M. H. (1999) Mass-independent isotope effects in planetary atmospheres and the early solar system. *Science, 283,* 341–345.

Thiemens M. M., Sprung P., Fonseca R. O. C., Leitzke F. P., and Münker C. (2019) Early Moon formation inferred from hafnium-tungsten systematics. *Nature Geosci., 12,* 696–700.

Thommes E. W., Duncan M. J., and Levison H. F. (2003) Oligarchic growth of giant planets. *Icarus, 161,* 431–455.

Torrano Z. A., Brennecka G. A., Williams C. D., Romaniello S. J., Rai V. K., Hines R. R., and Wadhwa M. (2019) Titanium isotope signatures of calcium-aluminum-rich inclusions from CV and CK chondrites: Implications for early solar system reservoirs and mixing. *Geochim. Cosmochim. Acta, 263,* 13–30.

Touboul M., Kleine T., Bourdon B., Palme H., and Wieler R. (2007) Late formation and prolonged differentiation of the Moon inferred from W isotopes in lunar metals. *Nature, 450,* 1206–1209.

Touboul M., Puchtel I. S., and Walker R. J. (2015) Tungsten isotopic evidence for disproportional late accretion to the Earth and Moon. *Nature, 520,* 530–533.

Treiman A. H., Gleason J. D., and Bogard D. D. (2000) The SNC meteorites are from Mars. *Planet. Space Sci., 48,* 1213–1230.

Trieloff M. and Kunz J. (2005) Isotope systematics of noble gases in the Earth's mantle: Possible sources of primordial isotopes and implications for mantle structure. *Phys. Earth Planet. Inter., 148,* 13–38.

Trinquier A., Birck J. L., and Allegre C. J. (2007) Widespread ^{54}Cr heterogeneity in the inner solar system. *Astrophys. J., 655,* 1179–1185.

Trinquier A., Elliott T., Ulfbeck D., Coath C., Krot A. N., and Bizzarro M. (2009) Origin of nucleosynthetic isotope heterogeneity in the solar protoplanetary disk. *Science, 324,* 374–376.

Tsiganis K., Gomes R., Morbidelli A., and Levison H. F. (2005) Origin of the orbital architecture of the giant planets of the solar system. *Nature, 435,* 459–461.

Turner N. J., Fromang S., Gammie C., Klahr H., Lesur G., Wardle M., and Bai X.-N. (2014) Transport and accretion in planet-forming disks. In *Protostars and Planets VI* (H. Beuther et al., eds.), pp. 411–432. Univ. of Arizona, Tucson.

Turrini D., Coradini A., and Magni G. (2012) Jovian early bombardment: Planetesimal erosion in the inner asteroid belt. *Astrophys. J., 750,* 8.

Valdes M. C., Bermingham K. R., Huang S., and Simon J. I. (2021) Calcium isotope cosmochemistry. *Chem. Geol., 581,* 120396.

Villeneuve J., Chaussidon M., and Libourel G. (2009) Homogeneous distribution of ^{26}Al in the solar system from the Mg isotopic composition of chondrules. *Science, 325,* 985–988.

Vogel N., Heber V. S., Baur H., Burnett D. S., and Wieler R. (2011) Argon, krypton, and xenon in the bulk solar wind as collected by the Genesis mission. *Geochim. Cosmochim. Acta, 75,* 3057–3071.

Vokrouhlický D., Bottke W. F., and Nesvorný D. (2016) Capture of trans-neptunian planetesimals in the main asteroid belt. *Astron. J., 152,* 39.

Walker R. J. (2009) Highly siderophile elements in the Earth, Moon and Mars: Update and implications for planetary accretion and differentiation. *Chem. Erde–Geochem., 69,* 101–125.

Walker R. J., Bermingham K., Liu J., Puchtel I. S., Touboul M., and Worsham E. A. (2015) In search of late-stage planetary building blocks. *Chem. Geol., 411,* 125–142.

Walsh K. J. and Levison H. F. (2019) Planetesimals to terrestrial planets: Collisional evolution amidst a dissipating gas disk. *Icarus, 329,* 88–100.

Walsh K. J., Morbidelli A., Raymond S. N., O'Brien D. P., and Mandell A. M. (2011) A low mass for Mars from Jupiter's early gas-driven migration. *Nature, 475,* 206–209.

Walsh K. J., Morbidelli A., Raymond S. N., O'Brien D. P., and Mandell A. M. (2012) Populating the asteroid belt from two parent source regions due to the migration of giant planets — "The Grand Tack". *Meteorit. Planet. Sci., 47,* 1941–1947.

Wang K. and Korotev R. (2019) Meteorites. In *Oxford Research Encyclopedia of Planetary Science,* Oxford Univ., Oxford. DOI: 10.1093/acrefore/9780190647926.013.16.

Wang Z. and Becker H. (2013) Ratios of S, Se and Te in the silicate Earth require a volatile-rich late veneer. *Nature, 499,* 328–331.

Ward W. R. (1997) Survival of planetary systems. *Astrophys. J. Lett., 482,* L211–L214.

Warren P. H. (2011) Stable-isotopic anomalies and the accretionary assemblage of the Earth and Mars: A subordinate role for carbonaceous chondrites. *Earth Planet. Sci. Lett., 311,* 93–100.

Wasson J. T. and Kallemeyn G. W. (1988) Compositions of chondrites. *Philos. Trans. R. Soc. A, 325,* 535–544.

Watanabe S. et al. (2019) Hayabusa2 arrives at the carbonaceous asteroid 162173 Ryugu — A spinning top-shaped rubble pile. *Science, 364,* 268–272.

Watanabe S., Tsuda Y., Yoshikawa M., Tanaka S., Saiki T., and

Nakazawa S. (2017) Hayabusa2 mission overview. *Space Sci. Rev., 208,* 3–16.

Weidenschilling S. J. (1977) Aerodynamics of solid bodies in the solar nebula. *Mon. Not. R. Astron. Soc.,180,* 57–70.

Weisberg M. K., McCoy T. J., and Krot A. N. (2006) Systematics and evaluation of meteorite classification. In *Meteorites and the Early Solar System II* (D. S. Lauretta and H. Y. McSween, eds.), pp. 19–52. Univ. of Arizona, Tucson.

Wetherill G. W. (1980) Formation of the terrestrial planets. *Annu. Rev. Astron. Astrophys., 18,* 77–113.

Wetherhill G. W. (1991) Occurrence of Earth-like bodies in planetary systems. *Science, 253,* 535–538.

Wetherill G. W. and Stewart G. R. (1989) Accumulation of a swarm of small planetesimals. *Icarus, 77,* 330–357.

Wieler R., Baur H., Pedroni A., Signer P., and Pellas P. (1989) Exposure history of the regolithic chondrite Fayetteville: I. Solar-gas-rich matrix. *Geochim. Cosmochim. Acta, 53,* 1441–1448.

Williams C. D. and Mukhopadhyay S. (2019) Capture of nebular gases during Earth's accretion is preserved in deep-mantle neon. *Nature, 565,* 78–81.

Wood J. A. (2005) The chondrite types and their origins. In *Chondrites and the Protoplanetary Disk* (A. N. Krot et al., eds.), pp. 953–971. ASP Conf. Ser. 341, Astronomical Society of the Pacific, San Francisco.

Worsham E. A., Walker R. J., and Bermingham K. R. (2016) High-precision molybdenum isotope analysis by negative thermal ionization mass spectrometry. *Intl. J. Mass Spectrom., 407,* 51–61.

Worsham E. A., Bermingham K. R., and Walker R. J. (2017) Characterizing cosmochemical materials with genetic affinities to the Earth: Genetic and chronological diversity within the IAB iron meteorite complex. *Earth Planet. Sci. Lett., 467,* 157–166.

Worsham E. A., Burkhardt C., Budde G., Fischer-Gödde M., Kruijer T. S., and Kleine T. (2019) Distinct evolution of the carbonaceous and non-carbonaceous reservoirs: Insights from Ru, Mo, and W isotopes. *Earth Planet. Sci. Lett., 521,* 103–112.

Wu J., Desch S. J., Schaefer L., Elkins-Tanton L. T., Pahlevan K., and Buseck P. R. (2018) Origin of Earth's water: Chondritic inheritance plus nebular ingassing and storage of hydrogen in the core. *J. Geophys. Res.–Planets, 123,* 2691–2712.

Yang C.-C., Johansen A., and Carrera D. (2017) Concentrating small particles in protoplanetary disks through the streaming instability. *Astron. Astrophys., 606,* A80.

Yang J., Goldstein J. I., and Scott E. R. D. (2007) Iron meteorite evidence for early formation and catastrophic disruption of protoplanets. *Nature, 446,* 888–891.

Yang J., Goldstein J. I., and Scott E. R. D. (2010) Main-group pallasites: Thermal history, relationship to IIIAB irons, and origin. *Geochim. Cosmochim. Acta, 74,* 4471–4492.

Yokochi R. and Marty B. (2004) A determination of the neon isotopic composition of the deep mantle. *Earth Planet. Sci. Lett., 225,* 77–88.

Yokoyama T., Rai V. K., Alexander C. M. O., Lewis R. S., Carlson R. W., Shirey S. B., Thiemens M. H., and Walker R. J. (2007) Osmium isotope evidence for uniform distribution of s- and r-process components in the early solar system. *Earth Planet. Sci. Lett., 529,* 567–580.

Yokoyama T., Alexander C. M. O., and Walker R. J. (2011) Assessment of nebular versus parent body processes on presolar components present in chondrites: Evidence from osmium isotopes. *Earth Planet. Sci. Lett., 305,* 115–123.

Yokoyama T., Nagai Y., Fukai R., and Hirata T. (2019) Origin and evolution of distinct molybdenum isotopic variabilities within carbonaceous and noncarbonaceous reservoirs. *Astrophys. J., 883,* 62.

Yokoyama T. et al. (2022) Samples returned from the sateroid Ryugu are similar to Ivuna-type carbonaceous meteorites. *Science, 379,* eabn7850.

Yurimoto H. and Kuramoto K. (2004) Molecular cloud origin for the oxygen isotope heterogeneity in the solar system. *Science, 305,* 1763–1766.

Zellner N. E. B. (2017) Cataclysm no more: New views on the timing and delivery of lunar impactors. *Origins Life Evol. Biospheres, 47,* 261–280.

Zhang H. and Zhou J.-L. (2010) On the orbital evolution of a giant planet pair embedded in a gaseous disk. I. Jupiter-Saturn configuration. *Astrophys. J., 714,* 532–548.

Zhang J., Dauphas N., Davis A. M., Leya I., and Fedkin A. (2012) The proto-Earth as a significant source of lunar material. *Nature Geosci., 5,* 251–255.

Zhu K., Moynier F., Wielandt D., Larsen K. K., Barrat J.-A., and Bizzarro M. (2019) Timing and origin of the angrite parent body inferred from Cr isotopes. *Astrophys. J. Lett., 877,* L13.

Zinner E. (1998) Stellar nucleosynthesis and the isotopic composition of presolar grains from primitive meteorites. *Annu. Rev. Earth Planet. Sci., 26,* 147–188.

Zinner E. (2003) Presolar grains. In *Treatise on Geochemistry, Vol. 1: Meteorites and Cosmochemical Processes* (A. M. Davis, ed.), pp. 181–213. Elsevier-Pergamon, Oxford.

Zinner E. et al. (2007) NanoSIMS isotopic analysis of small presolar grains: Search for Si_3N_4 grains from AGB stars and Al and Ti isotopic compositions of rare presolar SiC grains. *Geochim. Cosmochim. Acta, 71,* 4786–4813.

Zolensky M. E., Bland P. A., Brown P., and Halliday I. (2006) Flux of extraterrestrial materials. In *Meteorites and the Early Solar System II* (D. S. Lauretta and H. Y. McSween, eds.), pp. 869–888. Univ. of Arizona, Tucson.

Zsom A., Ormel C. W., Güttler C., Blum J., and Dullemond C. P. (2010) The outcome of protoplanetary dust growth: Pebbles, boulders, or planetesimals? *Astron. Astrophys., 513,* A57.

Fitzsimmons A., Meech K., Matrà L., and Pfalzner S. (2024) Interstellar objects and exocomets. In *Comets III* (K. J. Meech, M. R. Combi, D. Bockelée-Morvan, S. N. Raymond, and M. E. Zolensky, eds.), pp. 731–766. Univ. of Arizona, Tucson, DOI: 10.2458/azu_uapress_9780816553631-ch022.

Interstellar Objects and Exocomets

Alan Fitzsimmons
Astrophysics Research Centre, Queen's University Belfast

Karen Meech
Institute for Astronomy, University of Hawai'i

Luca Matrà
School of Physics, Trinity College Dublin, The University of Dublin

Susanne Pfalzner
Forschungszentrum Jülich

In this chapter we review our knowledge of our galaxy's cometary population outside our Oort cloud — exocomets and interstellar objects (ISOs). We start with a brief overview of planetary system formation, viewed as a general process around stars. We then take a more detailed look at the creation and structure of exocometary belts, as revealed by the unprecedented combination of theoretical and observational advances in recent years. The existence and characteristics of individual exocomets orbiting other stars is summarized, before looking at the mechanisms by which they may be ejected into interstellar space. The discovery of the first two ISOs is then described, along with the surprising differences in their observed characteristics. We end by looking ahead to what advances may take place in the next decade.

1. INTRODUCTION

At the time of publication of *Comets II* (*Festou et al.,* 2004), the presence of cometary bodies around other stars had been well established. Since that time there has been a significant increase in theoretical investigations of cometary formation in protoplanetary disks (see section 2.3 and the chapter by Simon et al. in this volume). Resolved imaging and spectroscopy of circumstellar disks has become almost commonplace, giving exquisite insights into the birthplaces of comets (see section 2). Increasingly common detections of gas in extrasolar Kuiper belts have opened a new approach to study exocometary gas and composition, combined with a new era of dust detection at infrared (IR) and submillimeter wavelengths. At the same time, observational studies have resulted in the discovery of new exocomet host systems, the detection of dust continuum transits for the first time, and further understanding of extant systems like β Pictoris (hereafter β Pic; see Fig. 1 and section 3.2). White dwarf atmospheric modeling has revealed a variety of compositions of extrasolar small bodies, including potentially volatile-rich exocomets.

Interstellar objects (ISOs) are planetesimals unbound to stars, and generally thought to originate by ejection from their planetary systems by various physical mechanisms. These could be both asteroids that are generally considered to be inert rocky bodies, or comets that contain a large amount of ices, which can lead to outgassing and mass loss, the most common method of telling the difference between them (however, see the chapter by Jewitt and Hsieh in this volume on the blurred boundaries between these objects).

Fig. 1. Artist's impression of the β Pic system, illustrating the extensive families of exocomets observed within this system. Credit: ESO/L. Calçada.

Assuming the same physics and dynamics occurs in other planetary systems as in our own, then the observable ISO population would primarily consist of cometary bodies of similar sizes to those orbiting our Sun, although see the discussion in section 6.4.

In contrast to exocomets, the existence of ISOs was merely hypothesised in *Comets II*, although their discovery was widely anticipated. Detection of ISOs passing through the solar system would in principle allow remote (and eventually *in situ*) sampling of bodies from other planetary systems. The surety that such a discovery would eventually occur grew with increasingly refined models of planetary system evolution and a growing understanding of how cometary bodies are lost to interstellar space and potential evolutionary processes, outlined in section 4.

Increasing observational capabilities of wide-field surveys led to the first discovery of an ISO, 1I/ʻOumuamua, in 2017 (described in section 6). That said, the second ISO discovery, 2I/Borisov, was by an individual effort with a smaller telescope (see section 6.2). 2I/Borisov was clearly an active comet from first detection, matching the expectation that most ISOs would be exocometary in nature. But 1I/ʻOumuamua was very different, with initial observations failing to detect any sign of activity, even though it was observed at a heliocentric distance of 1.2–2.8 au. Hence, it is already clear that ISO properties span a wider range than previously thought, to the extent that it is difficult to predict what will be found in the next decade.

In this chapter we will discuss the formation of and the observational evidence for exocomets, their ejection mechanisms into space to become ISOs, and the effects that this has on their physical properties, including processing in the interstellar medium. We then describe the discovery and characterization of the first two ISOs observed passing through the solar system and the next decade of ISO science.

2. PLANETARY SYSTEM EVOLUTION

2.1. Star and Primordial Disk Formation

ISOs travel through the interstellar medium (ISM), which is the primary galactic repository from which stars, including their surrounding planetary system, form. Since the advent of 1I/ʻOumuamua, it has become clear that the ISM contains, apart from gas and dust, a third component — interstellar objects. These are much smaller than the free-floating planets that have been detected via microlensing (*Mröz et al.,* 2017) and directly imaged (*Miret-Roig et al.,* 2022), but are much more numerous by many orders or magnitude. Thus, ISOs are present during the entire process described in the following text.

The star and planet formation process starts in the ISM when molecular clouds form through turbulent compression and global instabilities. Stars begin to form when denser parts of the clouds become unstable and start to collapse gravitationally. Once a molecular cloud begins to collapse, its central density increases considerably, eventually leading to star formation, with often an entire star cluster emerging (*Bate et al.,* 2003). A large fraction of the formed stars are not single but binaries. The binary fraction is 30–50% for solar-type stars in the local neighbourhood (*Raghavan et al.,* 2010) and up to ≈70% in young clusters (e.g., *Raghavan et al.,* 2010; *Duchêne and Kraus,* 2013). This fact might be important because some of the suggested ISO ejection mechanisms rely on binaries being present (see section 4.1). In this phase, a nascent individual protostar grows in mass via accretion from the infalling envelope until the available gas reservoir is exhausted or stellar feedback effects become important and remove the parental cocoon (for details, see *Klessen and Glover,* 2016, and the chapter by Bergin et al. in this volume).

As a natural consequence of angular momentum conservation, a disk develops during the formation of the protostar. Any initial nudge that imparts some core rotation means that material collapsing from its outer regions (with higher angular momentum) is channeled onto a disk, rather than the protostar itself (*Terebey et al.,* 1984). These young stellar objects (YSOs) consist of several components: the accreting protostar, the cloud (core) it is forming from, a disk surrounding it, and outflows of material that fail to be accreted either to the star or the disk. Observationally, YSOs are classified based on their continuum dust spectrum, thus the developmental stage of its disk. Class I objects are still deeply embedded within the cloud and mainly emit at millimeter and submillimeter wavelengths, leading to a characteristic positive IR spectral index. Class II objects reveal sources with near-infrared (NIR) and mid-infrared (MIR) excess, and Class III objects show only slight NIR excess. This classification scheme mainly reflects the development of the disks surrounding the stars from disk formation up to disk dispersal after several million years.

2.2. Disk Properties and Their Development

Flattened disks of cold dust and gas rotate around almost all low-mass stars shortly after their birth (*Williams and Cieza,* 2011). These are the sites where planetesimals form. ISOs are generally believed to be ejected planetesimals. Therefore, we review the planetesimal formation process in order to understand better the properties of ISOs.

Observations mainly at IR to millimeter wavelengths determine the frequency of disks and their mass, size, structure, and composition as a function of age. The ages of individual young stars can only be determined with a relatively large error margin. Therefore, the temporal development of disks is usually determined in young stellar clusters where many relatively coeval stars are located in close proximity. The mean age of the cluster stars is taken to test the temporal evolution of the disks in this environment.

Disk masses seem to decrease rapidly with system age (*Ansdell et al.,* 2016). In the extremely early phases, the disk mass can be comparable to the mass of the protostar, but after just 1 million years (m.y.), it is often less than 0.1%

of the star's mass. Some material is lost through outflows; some matter accretes onto the star or the star's radiation photoevaporates it. Significantly in this context, some material also condenses into centimeter- or larger-sized bodies, including planetesimals. It also seems that disk dust masses correlate with stellar mass ($R_{disk} \propto M_*^{0.9}$), but this correlation steepens with time, with a more substantial drop for low-mass stars (*Pascucci et al.,* 2016; *Ansdell et al.,* 2016). Observations show that a surprisingly large fraction of disk dust masses appears low compared to the solid content of observed exoplanets (*Najita and Kenyon,* 2014). This has been interpreted as a sign of early planet formation (*Manara et al.,* 2018; *Tychoniec et al.,* 2020). However, recent work suggests that detection biases might play a role here and that disks contain similar amounts of solids as found in exoplanetary systems (*Mulders et al.,* 2021).

Observations suggest a shallower density profile in the inner disk ($\Sigma \propto r^{-1}$) and steeper at larger distances ($\Sigma \propto r^{-3}$) (*Andrews,* 2020). However, the classical assumption of a smooth gas disk is not always appropriate. High-resolution observations with the Atacama Large Millimeter Array (ALMA) provide evidence for rings, gaps, and spiral structures in at least some of the most massive disks. Pronounced gap structures are often interpreted as being carved by protoplanets, implying that planetesimal formation starts early in these environments. The current sample of very high-resolution measurements in the millimeter continuum or scattered light is biased in favor of larger, brighter disks that preferentially orbit more massive host stars (*Andrews et al.,* 2018; *Garufi et al.,* 2018). Therefore, any conclusions about the prevalence of substructure have to be carefully assessed.

The typical disk sizes in dust measured by millimeter continuum observations lie in the range of 10–500 au. Gas disk sizes measured from the CO line emission are often larger than the dust disk size of the same disk. The disk dust radius appears to be correlated with the disk dust mass (*Andrews,* 2020; *Sanchis et al.,* 2021).

The disk lifetime is not only of prime importance for planet formation theory but equally for determining the timescale on which planetesimals form. The mean disk lifetime is mostly determined from the observed disk frequency of clusters of different mean ages. It is found that the disk fraction decreases rapidly with cluster age with <10% of cluster stars retaining their disks for longer than 2–6 m.y. (*Haisch et al.,* 2001; *Mamajek,* 2009; *Richert et al.,* 2018) in solar neighborhood clusters. However, recently several disks have been found with ages >10 m.y., some as old as 40 m.y., that still have the potential of forming planets (*Flaherty et al.,* 2019). Besides, the disk lifetime seems to be much shorter in dense clusters and longer in co-moving groups. The method of using disk fraction in clusters for determining disk lifetimes has been demonstrated to suffer from several biases (*Pfalzner et al.,* 2022). Looking at the transition phase from Class II to Class III objects, *Michel et al.* (2021) found a higher typical disk lifetime of ≈8 m.y. A more complex picture is emerging, where disk lifetimes can range from <1 m.y. to a few tens of millions of years, and the mean age of the field star population lies in the range 1–10 m.y., but this is still not definitely determined.

2.3. Dust and Planetesimal Growth

For planets that reside close to their star, dust accretion is the standard formation scenario. Here planet formation takes place over a variety of size scales (see Fig. 2). Initially, the millimeter-sized dust particles largely follow the movement of the gas. However, Brownian motion means that collisions between the dust particles take place nevertheless. The low relative velocity of the dust particles means that they stick upon contact so that first larger porous fractal aggregates form. These compact in subsequent collisions and settle toward the midplane (*Dubrulle et al.,* 1995), where the growth sequence continues. Once particles reach a size where they start to decouple from the gas aerodynamically, these solids start to migrate radially inward. Further collisions lead to the formation of boulders, and eventually planetesimals emerge. The largest planetesimals accumulate nearby smaller ones, grow further, and become terrestrial planets or the cores of gas giants. However, it is the planetesimals that are of particular interest in this chapter. Today's comets and interstellar objects are thought to be largely unchanged ancient planetesimals (for details, see, e.g., *Blum and Wurm,* 2008; *Morbidelli and Raymond,* 2016; *Armitage,* 2018; and the chapter by Simon et al. in this volume).

Fig. 2. Size scales during dust growth. **(a)** Chondritic interplanetary dust particle from *Jessberger et al.* (2001) (CC License Attribution 2.5); **(b)** dust aggregate simulation from the chapter by Simon et al. in this volume; **(c)** 4.5-G.y.-old Allende meteorite with chondrules (CC Generic License 2.0); **(d)** images of boulders seen on by Rosetta Comet 67P on May 16, 2016, from a distance of 8.9 km (ESA/Rosetta/NAVCAM; CC BY-SA IGO 3.0); **(e)** nucleus of 67P seen by Rosetta on July 7, 2015, from a distance of 154 km with a resolution of 13.1 m/pixel (ESA/Rosetta/NAVCAM; CC BY-SA IGO 3.0).

This model has two potential problems: growth barriers during the accretion process and a relatively long formation timescale. The various growth barriers are the bouncing, fragmentation, and drift barriers. The times expected for the different growth phases are 10^2–10^4 yr to form millimeter- to centimeter-sized pebbles, 10^4–10^6 yr until the planetesimal stage is reached, 10^6–10^7 yr to form terrestrial-type planets, and an additional 10^5 yr for the gas giants to accumulate their gas (*Pollack et al.,* 1996). The cumulative time seems at odds with observations of the disk frequency in young clusters, which indicate the median protoplanetary disk lifetime to be merely 1–3 m.y. for both dust and gas (*Haisch et al.,* 2001).

However, over the last decade, mechanisms have been devised that might solve the timescale problem while at the same time overcoming the growth barriers (*Armitage,* 2018). These mechanisms are the streaming instability (*Youdin and Goodman,* 2005) and pebble accretion (*Lambrechts and Johansen,* 2012).

In the outer parts of the disks, planets also might form via a second process. In the initial phases, these disks are still very massive and can become gravitationally unstable. Simulations show that such disk instabilities lead to spiral arm formation, which can fragment directly to form large protoplanets (*Boss,* 1997). However, the material near the star stays too hot to go unstable, so this process is expected to generate planets typically only at large distances (≥100 au) from the star. Planetesimals would be produced as a by-product of planet formation. Details of the growth process, including relevant references, are described in the chapter on planetesimal formation by Simon et al. in this volume.

2.4. Outer Areas of Young Disks

In section 2.2, we saw that the gas and dust disk differ quite often in size. The difference in size is a result of the grain growth in the disk. Initially, the pressure gradient of a disk exerts an additional force that causes gas to orbit at a slightly subkeplerian speed. As long as the dust is coupled to the disk, it just follows the gas movement. However, as dust grains grow to about millimeter sizes, the orbiting grains experience a frictional force, drifting inward. At the same time, the disks' viscosity means that the gas spreads out to conserve angular momentum and enable gas close in to accrete onto the star. Observations at (sub)millimeter wavelengths typically trace the large dust grains; thus disks appear more extended in gas than dust. The combined effects of growth and vertical and radial migration of dust particles mean that the disks' mean particle size should decrease with distance to the midplane and the central star. Alternatively, in terms of timescales, dust growth should proceed slower off the midplane and in the outer disk areas (*Dullemond and Dominik,* 2004; *Birnstiel and Andrews,* 2014). The question is, where do the planetesimals that are ejected and become ISO primarily form?

In the outer disk areas, at particular distances from the central protostar, the temperature is so low that volatile compounds such as water, ammonia, methane, carbon dioxide, and carbon monoxide can condense into solid ice grains. These distances are referred to as the snow lines. These snow lines are essential in the context of dust growth. Ices can change the effective particle strengths and thereby affect collision outcomes. *Pinilla et al.* (2017) argue that the critical velocity for fragmentation increases at the (CO or) CO_2 and NH_3 snow lines and decreases at the H_2O snow line.

However, direct imaging has observed exoplanets at large distances (greater than several tens of astronomical units) from their parent stars. Either these planets have formed there or migrated outward to these orbits. Among these directly imaged planet hosts, there are several that are thought to be relatively young [<20 m.y.; for example, PDS 70 (*Mesa et al.,* 2019)], which illustrates that even in the outer disks, planetesimals must be able to grow relatively fast. If protoplanets can grow on such a timescale, even in the outer disk, planetesimals and the resulting ISOs should form on even shorter timescales.

3. EXOCOMETS

Exocomets are small bodies orbiting other stars that exhibit signs of activity, through the release of dust and/or gas (*Strøm et al.,* 2020). As such, their definition is broader compared to the solar system population, which is much more refined dynamically and compositionally. Evidence for exocomets dates back to the 1980s, when exocomets were inferred from variable absorption features in the spectrum of β Pic (*Ferlet et al.,* 1987), and when the dust from an exocometary belt was first discovered around Vega (*Aumann et al.,* 1984). In general, exocometary release of gas and dust is detected in both the inner (few stellar radii to few astronomical units) and the outer (~10 au to hundreds of astronomical units) regions of extrasolar planetary systems (as depicted in Fig. 3), and we will distinguish between these two populations in the following sections.

3.1. Exocometary Belts: Extrasolar Cometary Reservoirs

3.1.1. Detection and basic observational properties. Exocometary belts (also known as *debris disks*) are extrasolar analogs of our Kuiper belt and are the reservoirs of icy exocomets in planetary systems. These belts are first detected through the dust produced by exocomets as they collide and grind down to release small, observable dust. Thermal emission as well as scattered starlight by this dust can be detected and used to study these belts. The majority of detectable belts lie at distances of tens of astronomical units from their host star and are cold with typical temperatures of tens of Kelvins (*Chen et al.,* 2014). They are therefore first detected by MIR to far-infrared (FIR) surveys, most sensitive to the peak of the dust's thermal emission, with space telescopes such as the Infrared Astronomical Satellite (IRAS), Spitzer, Wide-field Infrared Survey Explorer (WISE), and Herschel (e.g., *Su et al.,* 2006; *Sibthorpe et al.,*

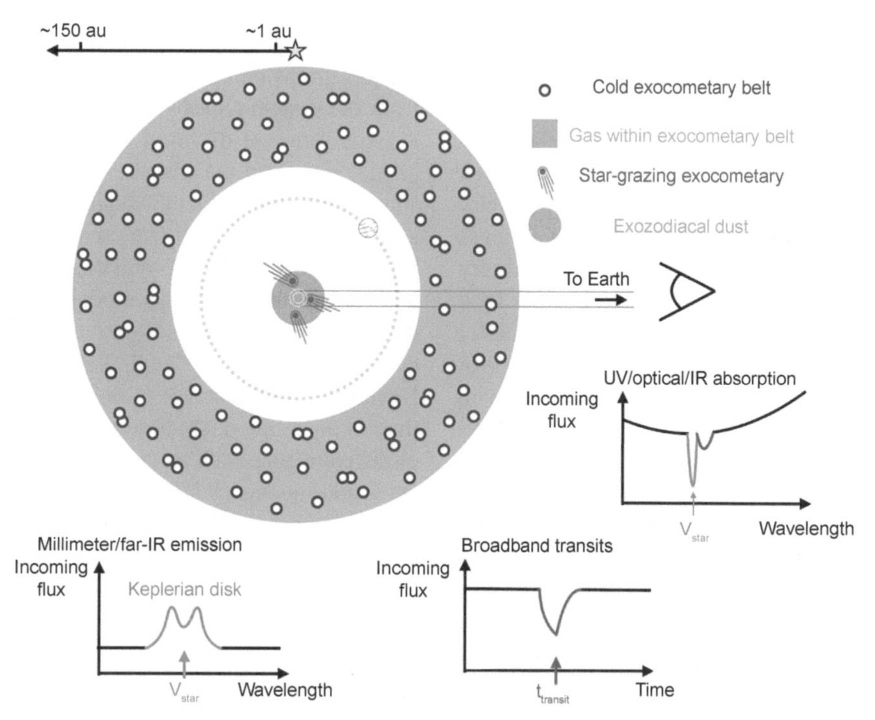

Fig. 3. Top-down view of a young, ~10–100-m.y.-old planetary system after dissipation of the protoplanetary disk. The system has largely formed outer gas and ice giant planet(s) (at few to tens of astronomical units, like the Neptune analog with its orbit depicted as the dashed line), has an exocometary belt at tens of astronomical units producing cold gas and dust, and star-grazing exocomets in the inner region, producing gas and potentially warm exozodiacal dust as they approach and recede from the star. Exocomets can be probed in a variety of ways. In the outer regions (blue/cyan symbols), cold exocomets (extrasolar KBO analogs) can be probed in the belts they formed in, through cold dust and gas emission and cold absorption against the star for edge-on systems. In the inner regions (black symbols), exocomets are probed through warm exozodiacal dust emission, as well as blue- and black-shifted absorption (see also Fig. 5) and asymmetric "shark-fin" transits for edge-on systems.

2018). These surveys show belts to be ubiquitous; they are detected around ~20% of nearby, several-gigayear-old field stars, even though the surveys' sensitivity at best enabled detection of belts a few times more massive than the Kuiper belt (*Eiroa et al.,* 2013). This makes this number very much a lower limit; studies of younger, less evolved belts at a few tens of millions of years indeed show occurrences as high as ~75% (*Pawellek et al.,* 2021).

The typical observable (i.e., not the total) mass of belts, only considering solids of sizes up to approximately centimeters in diameter, range between ~0.1% and 25% of an Earth mass, or about a tenth to 20 times the mass of the Moon (*Holland et al.,* 2017). This is but a small fraction of the belts' total mass, which is dominated by the largest, unobservable bodies. The total mass of the Kuiper belt is estimated from observations of its largest bodies and is about 6% of an Earth mass (*Di Ruscio et al.,* 2020). The Kuiper belt's dust mass is, however, predicted to be much lower than that of exocometary belts, to the point that a true Kuiper belt analog would not be detectable around nearby stars (*Vitense et al.,* 2012). In summary, detectable exocometary belts have dust masses typically ~10–10^5 times higher than inferred for the present-day Kuiper belt, where this broad range represents the spread of belt dust masses observed.

3.1.2. Birth of exocometary belts. Exocometary belts are born at tens to hundreds of astronomical units from the central star within protoplanetary disks, but they cannot be identified until after the protoplanetary disk is dispersed. During dispersal, primordial gas and dust are removed, leaving behind second-generation dust (and gas), produced destructively in collisions of larger bodies. The dispersal of the protoplanetary disk takes place rapidly once the star is a few to ~10 m.y. old (*Ercolano and Pascucci,* 2017) (see also section 2.2). The emergence of an exocometary belt from this dispersal is evidenced by the presence of an IR excess, but with a decrease of ~2 orders of magnitude in dust mass (*Holland et al.,* 2017) compared to protoplanetary levels.

In practice, the largest exocomets within belts must form during the protoplanetary phase to access the gas needed for the planetesimal formation process (see the chapter by Simon et al. in this volume), but observationally validating this is difficult. An interesting proposal is that exocometary belts may originate in (one or more of) the large rings observed in structured protoplanetary disks (*Michel et al.,* 2021), which has been shown could potentially explain the population of observed exocometary belts (*Najita et al.,* 2022). Another possibility is that the photoevaporative dispersal of the gas-rich protoplanetary disk induces high dust-to-gas ratios (*Throop and Bally,* 2005) that may trigger the streaming instability and produce massive belts of planetesimals/(exo)comets (*Carrera et al.,* 2017). However, such massive, large belts are not observed, and other models of dust evolution in a photoevaporating disk indicate that the grains efficiently drift inward faster than they can pile up to trigger planetesimal formation (*Sellek et al.,* 2020).

One difficulty in the protoplanetary disk-exocometary belt observational comparison is that most of the protoplanetary disks in nearby star-forming regions surround low-mass stars, whereas the majority of detectable exocometary belts orbit around A–F stars, likely due to current lack of sensitivity to belts around later-type stars, particularly M dwarfs (*Luppe et al.,* 2020). In general, the presence of evolved Class III disks in young star-forming regions, with masses consistent with young exocometary belts, suggests that exocomets can form early on, within the first ~2 m.y. (*Lovell et al.,* 2021).

A key component to the transition from a protoplanetary disk to an exocometary belt is the presence of gas. This produces drag, slowing collisions and favoring dust growth in protoplanetary disks; on the other hand, gas dispersal enables the more collisionally destructive, dust-producing environment of exocometary belts (*Wyatt et al.,* 2015). Unfortunately, distinguishing primordial from second-generation gas is even harder, due to the surprisingly low CO abundances observed in many protoplanetary disks (*Krijt et al.,* 2020), combined with the presence of gas now discovered in a growing number of exocometary belts (see section 3.1.5) and the difficulty in measuring total gas masses in both evolutionary phases.

3.1.3. Evolution of exocometary belts. The physics of observable (massive) exocometary belts is driven by collisions, causing the grind-down of large exocomets [analogs to the solar system's observable Kuiper belt objects (KBOs)] into small, observable grains. This produces a collisional cascade, which rapidly reaches a quasi-steady state size distribution where the number of solid particles scales as $n(a) \propto a^q$, where $q = -3.5$ for a cascade extending to infinitely large and small bodies (*Dohnanyi,* 1969). More detailed cascade simulations show that this q value can vary between –3 and –4 (*Gáspár et al.,* 2012; *Pan and Schlichting,* 2012). Observations of millimeter to centimeter wave spectral slopes imply values varying between ~–(2.9–3.8) for observable grains in the millimeter to centimeter size regime (*Norfolk et al.,* 2021), although interpretation of measurements in terms of size distributions is complex (*Löhne,* 2020). These slopes imply that what extrasolar observations are sensitive to (the solids' cross-sectional area) is dominated by small, up to approximately centimeter-sized grains, but the total mass is dominated by the largest, unobservable bodies. This is the opposite of observations of the Kuiper belt, where we are most sensitive to the largest bodies.

At the bottom end of the cascade, grains are sensitive to radiation forces, and are typically removed by radiation pressure from the central star (*Burns et al.,* 1979), or by stellar winds, which are particularly important for young late-type stars (*Strubbe and Chiang,* 2006; *Augereau and Beust,* 2006). For a radiation-pressure-dominated environment, the smallest, *blow-out* grain size in the cascade is the one where the radiation pressure force dominates over gravity, pushing smaller grains on unbound orbits and rapidly ejecting them from the system (*Backman and Paresce,* 1993).

This removal of small grains causes a net mass loss of material from the cascade as a function of time. In the simplest model of a collisional cascade, this would cause a belt's mass to initially stay constant — until the largest bodies begin colliding — and then decrease as t^{-1} (*Dominik and Decin,* 2003), although models with more realistic treatment of cascade processes predict somewhat different time decays (*Löhne et al.,* 2008; *Kenyon and Bromley,* 2017). Nevertheless, a decrease in dust emission as a function of time is unambiguously observed in surveys of debris disks of different ages and can largely be explained by collisional evolution models (*Sibthorpe et al.,* 2018). Additional mass removal effects could come into play that may speed up this mass loss, as hinted at by recent observations (*Pawellek et al.,* 2021). Notably, faster mass removal could be caused by gravitational interaction between planets and exocometary belts, potentially ejecting small bodies from the planetary system and contributing to the population of interstellar objects (see section 4). Such interaction could be akin to the late dynamical instability and/or to the outward migration of Neptune inferred to have depleted the vast majority of material in our solar system's Kuiper belt (*Malhotra,* 1993; *Gomes et al.,* 2005).

3.1.4. Resolved dust imaging: Structure and (exo) planet-belt interaction. Resolved imaging is ultimately needed to study the location and structure of exocometary belts. Given the star's luminosity, the observed temperature from the dust's unresolved spectrum can be used to calculate its location/radius (distance from the star), under the assumption that the dust behaves as a blackbody. In reality, small dust dominating the spectrum's emission is an inefficient emitter, and thus, such temperature-derived *blackbody radii* typically underestimate the true belt locations by a factor of a few (*Booth et al.,* 2013).

Belts are detected around stars from as far as nearby star-forming regions (~150 pc) to our nearest neighbors (a few parsecs away), with typical angular sizes of a few tenths to a few tens of arcseconds on-sky, so easily resolvable with the latest generation of high-resolution facilities from optical to millimeter wavelengths. Given their low masses and large on-sky extents, the difficulty in imaging exocometary belts arises often from their low surface brightness, requiring deep integrations. Nevertheless, several famous belts are bright enough for structural characterization at high resolution.

The first resolved exocometary belt was that around β Pic (*Smith and Terrile,* 1984). Almost four decades later we now have a considerable inventory of images for a few tens of belts (e.g., Fig. 4). These have been obtained across the wavelength spectrum, from space and groundbased optical/NIR scattered light images [milliarcsec resolution (e.g., *Apai et al.,* 2015; *Wahhaj et al.,* 2016; *Esposito et al.,* 2020)], to spacebased IR [few arcsecond resolution (e.g., *Booth et*

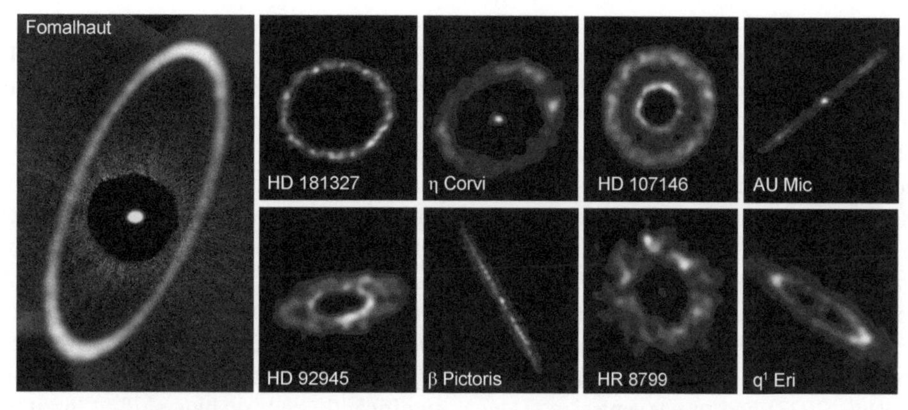

Fig. 4. Debris disks and exocometary belts. Fomalhaut credit: ALMA (ESO/NAOJ/NRAO), M. MacGregor; NASA/ESA Hubble, P. Kalas, B. Saxton (NRAO/AUI/NSF); other images are adapted from *Marino* (2022).

al., 2013)] and millimeter-wave groundbased interferometric images [subarcsecond resolution (e.g., *Marino et al.,* 2018b; *Sepulveda et al.,* 2019)].

The most basic measurable quantity of a resolved belt is its location (radius). Exocometary belts resolved across the wavelength range show typical radii of a few tens to hundreds of astronomical units, with a shallow trend indicating that more luminous stars tend to host larger belts (*Matrà et al.,* 2018a; *Marshall et al.,* 2021). While a combination of observational bias and radius-dependent collisional evolution can explain this positive trend, the low scatter suggests that belts (including our own Kuiper belt) may form at preferential locations in protoplanetary disks. If confirmed with a larger sample of belts, this would imply a potential connection to either temperature-dependent exocomet formation processes (e.g., the CO ice line), or to the timescale of planet formation processes preventing exocomets from growing further into planets at these distances.

While the classic picture of an exocometary belts is that of a narrow ring, recent imaging at IR and millimeter wavelengths shows that broad belts (with width/radius, $\Delta R/R$, approaching $\sim$1) are at least as common as narrow rings ($\Delta R/R \ll 1$; see Fig. 4). Some of these belts may be born narrow but others, like the famous HR 8799 belt, may have evolved to host a scattered disk of high-eccentricity particles akin to our Kuiper belt's, through gravitational interaction with interior planets (*Geiler et al.,* 2019).

Beyond a simple radius and width, exocometary belts observed at sufficiently high resolution show structural features that have long been linked to the gravitational action of planets (or dwarf planets) interior, exterior, or within these belts (e.g., *Pearce et al.,* 2022). (Sub)structures can be broadly separated into radial, vertical, and azimuthal features.

In the radial direction, a belt's inner/outer edge and its sharpness can be linked to dynamical truncation by planets just interior or exterior to it (*Chiang et al.,* 2009). This is particularly true for inner edges with surface density profiles steeper than the r^{α} with $\alpha \sim 2$–2.3 expected from collisional evolution of an undisturbed disk (*Schüppler et al.,* 2016; *Matrà et al.,* 2020).

In the absence of truncating planets, the sharpness of belt edges may also be used to probe the eccentricity dispersion, and hence the level of dynamical excitation of exocomets within the belt (*Marino,* 2021; *Rafikov,* 2023). Dynamical excitation of exocometary material within belts can also be probed by the belts' observed vertical structure in edge-on systems (*Matrà et al.,* 2019b; *Daley et al.,* 2019). Like the Kuiper belt, this has the imprint of the system's dynamical history; for example, the vertical structure of the β Pic belt supports the presence of a double population of exocomet inclinations (*Matrà et al.,* 2019b), akin to the hot and cold populations of KBOs in the solar system (*Brown,* 2001).

While the radial distribution of some wide exocometary belts appears to be smooth (*Matrà et al.,* 2020; *Faramaz et al.,* 2021), at least three systems so far (e.g., *Marino et al.,* 2018b) show the presence of rings separated by a gap, reminiscent of the rings observed in younger protoplanetary disks (*Andrews et al.,* 2018). In the simplest scenario, a single stationary planet at the gap location (tens of astronomical units) can clear its chaotic zone and carve a gap of width and depth related to the planet's mass and the age of the system (*Morrison and Malhotra,* 2015). However, these gaps may be too wide and require planets to migrate (*Friebe et al.,* 2022), or different scenarios such as secular effects due to an eccentric planet interior to both rings (*Pearce and Wyatt,* 2015).

A few narrow belts, most prominently the famous Fomalhaut (*Kalas et al.,* 2005; *MacGregor et al.,* 2017) and HD 202628 (*Krist et al.,* 2012; *Faramaz et al.,* 2019) rings are observed to be significantly eccentric ($e \sim 0.12$). Secular perturbation by eccentric planets interior to the belts can produce and maintain eccentric belts (e.g., *Wyatt et al.,* 1999), although the narrower than predicted width of these rings may point to alternative scenarios, such as belts being born eccentric in the protoplanetary phase of evolution (*Kennedy,* 2020).

Azimuthal asymmetries in the brightness of exocometary belts are often detected in scattered light observations, particularly with the Hubble Space Telescope (HST) (*Schneider et al.,* 2014), although these may be simply produced by a combination of disk eccentricity, grain scattering phase functions, and viewing geometry (*Lee and Chiang,* 2016) rather than true azimuthal asymmetries in their density distribution. Similarly, eccentric disks observed in thermal emission may produce brightness asymmetries due to pericenter and/or apocenter glow (e.g., *MacGregor et al.,* 2022) that are expected to vary with wavelength (*Pan et al.,* 2016; *Lynch and Lovell,* 2022). The strongest evidence for azimuthal asymmetry arises from the edge-on β Pic belt, which displays a strong clump in dust emission at MIR wavelengths (*Telesco et al.,* 2005; *Han et al.,* 2023) and in gas emission (*Dent et al.,* 2014; *Cataldi et al.,* 2018), but not as prominently in scattered light or ALMA dust observations (*Apai et al.,* 2015; *Matrà et al.,* 2019b). Such clumps may be produced by resonant trapping by a migrating planet at the inner edge of the belt (*Wyatt,* 2003), akin to Neptune's migration into the Kuiper belt in the young solar system (*Malhotra,* 1993), but also by giant impacts between planet-sized bodies (*Jackson et al.,* 2014).

With the clear dynamical link between exocometary belt structure and the presence of exoplanets interacting with them, it is natural to ask whether the very existence of a belt is linked to the presence (or not) of exoplanets in the same system. For example, if the brightest debris disks form from the most massive protoplanetary disks that formed exoplanets most efficiently, one may expect a positive correlation between the presence of a belt and planets (*Wyatt et al.,* 2007). On the other hand, the presence of outer giant planets may efficiently eject and deplete exocometary belts (producing ISOs), while enabling terrestrial planet formation to take place interior to them without significant disruption; in this case, one would expect an anticorrelation with outer giant planets and a correlation with low-mass planets (*Raymond et al.,* 2011, 2012). Evidence of trends continues to be debated in the observational literature: A correlation between low-mass planets and belts (through their IR excess) had initially been found (*Wyatt et al.,* 2012; *Marshall et al.,* 2014), but a study with an expanded sample found no correlations with either low- or high-mass planets (*Moro-Martín et al.,* 2015). Similarly, no correlation was found between the presence of radial-velocity planets and the properties of exocometary belts (*Yelverton et al.,* 2020). It is notable, however, that a significant number of long-period, young directly imaged planets are found in systems with bright debris disks, with some evidence of a correlation (*Meshkat et al.,* 2017).

In summary, the structure of exocometary belts at tens of astronomical units probes the orbits of the icy bodies within it, and, as is the case in our solar system's Kuiper belt, can reveal dynamical interactions with mostly unseen outer planets. As proven by dynamical simulations successfully reproducing these structures, these gravitational interactions can remove exocomets in two ways. They can either eject exocomets from outer belts, potentially producing Oort clouds analogs, or directly turning them into ISOs (section 4.1.3), but also scatter them inward, which could eventually produce the transiting exocomets observed close to the central stars (section 3.2) and deliver volatile-rich material to planets in the habitable zones of their host stars.

3.1.5. Evidence for gas within outer exocometary belts. Since their discovery almost four decades ago, exocometary belts have been considered to be gas free, to the point that this was long considered a defining difference between younger protoplanetary and more evolved debris disks (*Zuckerman et al.,* 1995). However, evidence for circumstellar gas in these systems was discovered as early as 1975 around β Pic (*Slettebak,* 1975), in the form of narrow Ca ii H and K absorption at the stellar velocity (i.e., not significantly red- or blue-shifted, as is instead treated in section 3.2) that was later spatially resolved to extend out to hundreds of astronomical units from the star, where the dusty belt lies (*Brandeker et al.,* 2004). For a long time, evidence for cold gas in these systems remained limited to the β Pic and 49 Ceti belts. Their near edge-on geometry allowed detection of gas in both emission and absorption against the central star (*Zuckerman et al.,* 1995; *Dent et al.,* 2005; *Roberge et al.,* 2000, 2014).

The evidence for exocometary gas has rapidly strengthened in the past decade. The sensitivity advance to cold gas emission brought by Herschel and especially ALMA has led to detections of atomic (C I, O I, and/or C II) and/or molecular (CO) emission within over 20 exocometary belts, largely around ~10–40-m.y.-old stars (*Moór et al.,* 2017), but also as old as 440 m.y. [Fomalhaut (*Matrà et al.,* 2017b)]. Early ALMA CO gas detections highlighted a dichotomy between belts with very high CO masses, approaching those of protoplanetary disks [10^{-2}–10^{-1} $M_\oplus$ (e.g., *Kóspál et al.,* 2013)], and more tenuous CO gas disks [10^{-7}–10^{-5} $M_\oplus$ (e.g., *Matrà et al.,* 2017a)].

The origin of the gas remains an open question for high-mass CO systems. In a *primordial* scenario, the gas is a remnant of the protoplanetary disk (contrary to the dust, which is second-generation, leading to the term *hybrid* disk); H_2 dominates the gas mass, allowing CO to remain shielded from stellar and interstellar photodissociation over the lifetime of the system (*Kóspál et al.,* 2013). In a *secondary* scenario, CO gas is continuously produced by exocomets within the belt, just like the dust (*Zuckerman and Song,* 2012; *Matrà et al.,* 2015; *Marino et al.,* 2016). However, the production rate is unlikely to be sufficient to maintain the high CO masses observed, so a shielding agent other than H_2 (which is only produced in trace amounts in solar system comets) is needed. Shielding CO photodissociation (producing C and O) by C photoionization is a promising scenario to produce what would be second-generation, shielded high-mass CO disks (*Kral et al.,* 2019).

For low-mass CO belts, like Fomalhaut, HD 181327 and β Pic, the secondary scenario is heavily favored, as the CO photodissociation timescale, even when accounting for the potential of unseen H_2 with primordial CO/H_2 ratios, is much shorter than the age of the system (e.g., *Marino et al.,* 2016). This simple argument implies that CO must have been recently (and likely continuously) produced, and must be of secondary, exocometary origin.

3.1.6. Gas release processes, composition, and evolution. The gas release process of exocomets within belts from their host stars is likely different from solar system comets. In the very cold (approximately tens of Kelvins) environments at tens of astronomical units, the heating and sublimation-driven process producing gas in single solar system comets and ISOs (as they approach the Sun) is unlikely to be efficient; while activity suggesting CO outgassing takes place out to Kuiper belt distances in at least some long-period comets [e.g., C/2017 K2 (Pan-STARRS) (*Meech et al.,* 2017a)], no activity is detected from KBOs on stable orbits (*Jewitt et al.,* 2008) including Arrokoth [(486958) 2014 MU69 (*Lisse et al.,* 2021)]. Care should be taken in associating locations with temperatures in the Kuiper belt vs. exocometary belts. As exocometary belts are most commonly observed around more luminous, A–F-type stars, they will have warmer temperatures at the same radial location. This is, however, mitigated by the temperature dependence with stellar luminosity being shallow for blackbody-like grains heated by starlight ($T \propto L^{0.25}$), combined with the fact that belts tend to be located at larger radii around more luminous stars (section 3.1.4).

One possibility is that we are observing a vigorous sublimation period for exocomets immediately following the dispersal of the protoplanetary disk (*Steckloff et al.,* 2021; *Kral et al.,* 2021). In this period, hypervolatiles like CO, N_2, and CH_4 sublimate at high rates due to the heating by newly visible starlight, previously shielded by large amounts of protoplanetary dust. In this scenario, Arrokoth-sized objects (approximately tens of kilometers in diameter) at ~45 au from the central star take 10–100 m.y. for the heat to reach their interior, which could explain the higher detection rate around young exocometary belts in that age range. However, in exocometary belts a large range of sizes would have to be considered, and it remains unclear whether the production rates could be sufficiently high to explain the observed CO masses.

Other physical mechanisms have been proposed to release gas in the collisionally active environment of young exocometary belts. While typical collision velocities are unlikely to result in sufficient heating, collisional vaporization of very small grains accelerated by radiation pressure could lead to ice release into the gas (*Czechowski and Mann,* 2007). Additionally, ultraviolet (UV) photodesorption could very effectively remove volatiles such as H_2O from the surface of icy grains/objects (*Grigorieva et al.,* 2007), given that these are continuously produced/resurfaced through the collisional cascade.

Once released, in the absence of shielding, molecular gas is expected to be rapidly destroyed by the stellar and interstellar UV fields. CO (with N_2) is the most photoresistant molecule, which may explain why it is the only one detected so far (*Matrà et al.,* 2018b). The CO ice mass fraction of exocomets can then be estimated from the ALMA mass measurements and steady-state production/photodestruction arguments, indicating an exocometary CO content within an order of magnitude of solar system comets (*Matrà et al.,* 2017b, 2019a), showing that accessing exocometary compositions similar to solar system comets is now possible. Unfortunately, the short photodissociation timescales and/or weak transitions of other molecules make other molecular detections currently challenging (*Matrà et al.,* 2018b; *Klusmeyer et al.,* 2021).

On the other hand, atomic gas is abundant and detected both in emission and absorption for edge-on belts. The detection of atomic N (*Wilson et al.,* 2019) and S, as well as C and O (see *Roberge et al.,* 2006, and references therein) in the β Pic belt suggest that molecules other than CO are being released, arguing against the gas release being limited to hypervolatile molecules. Atomic gas accumulating over time at the belt location is expected to produce a disk that viscously spreads over time, at a rate determined by the poorly-constrained gas viscosity, and eventually form an accretion disk (*Kral et al.,* 2016). This model can largely explain the population of CO masses observed (*Marino et al.,* 2020), including high-CO mass disks through C shielding. However, resolved observations of C i gas by ALMA show inner holes inconsistent with accretion disk profiles, suggesting a more complex evolution (*Cataldi et al.,* 2018, 2020).

3.1.7. Detectability of exo-Oort clouds. Extrasolar Oort cloud analogs may also exist, and exocomets in those stars would be heated by absorption of radiation from their central star, nearby stars and the cosmic microwave background (CMB). The challenge for detecting extrasolar Oort clouds is their low surface density, low temperature, and large sky areas, which is why long-wavelength surveys around nearby stars are best suited for detection. An additional factor is whether dust might be produced in Oort clouds, and if so, how long it might survive. *Stern et al.* (1991) used IRAS to search for excess FIR emission around 17 nearby stars but did not achieve any detections. More recently, *Baxter et al.* (2018) used Planck survey data to average the submillimeter emission around selected stars within 300 pc, and also performed a directed search around Fomalhaut and Vega, but again did not detect any signal. Intriguingly, when stacking 43 hot stars within 40–80 pc, they detected a statistically significant excess flux at radial distances of 10^4–10^5 au. Searching around early-type stars is a key factor, as this will increase stellar heating of the exo-Oort body surfaces and aid detection. But they cautioned that this could be a false positive due to the fine structure of galactic dust on scales of the Planck beam width. Alternatively, it could be a real signal caused by weak nebular emission or dust grains ejected from debris disks near the stars. Future CMB missions such as the Lite (Light) satellite for the studies of B-mode polarization and Inflation from cosmic background Radiation Detection (LiteBIRD) (*Hazumi et al.,* 2020) may approach the sensitivity required for unambiguous detection. However, the 0.5° mission resolution would restrict it to searches around the nearest stars within 10 pc, for which 10^4 au spans ~0.3°. Unresolved emission from more distant stars would be ambiguous as it could also be caused by thermal emission from dust closer to the star (see section 3.1.4).

3.2. Exocomets in Inner Planetary Systems

3.2.1. The β Pictoris system. The star β Pic is a nearby (d = 19.3 pc) A6V star that was the first to have a circumstellar debris disk optically imaged (*Smith and Terrile,* 1984). Soon after, transient, red-shifted absorption features (see Fig. 5) in the UV spectrum were detected in the wings of the stellar Al III, Mg II, and Fe II lines (*Lagrange-Henri et al.,* 1988), and in numerous high-resolution optical spectra that include the strong Ca II H and K lines (*Ferlet et al.,* 1987). These were interpreted as originating from planetesimals undergoing sublimation when within a few stellar radii from β Pic, with most features being redshifted from motion toward the star. They were subsequently termed "falling evaporating bodies" (FEBs). In modern parlance these are exocomets, but care must be taken. The observation of metal lines may imply sublimation of refractory elements, and hence not necessarily a volatile-rich body as is common in our solar system. Subsequent studies of these exocomets have tended to use the strong Ca II H and K lines in high-resolution optical spectra, although Na I D-line absorption is also seen. Unfortunately, only metal lines in these exocomets have been measured and the lack of detections of the volatile species seen in solar system comets has prevented a direct comparison. While studies of the gas in the outer circumstellar disk show it is likely to be released by exocomets in those regions as discussed in section 3.1.6, there is as yet no direct link to the exocomets seen close to the star.

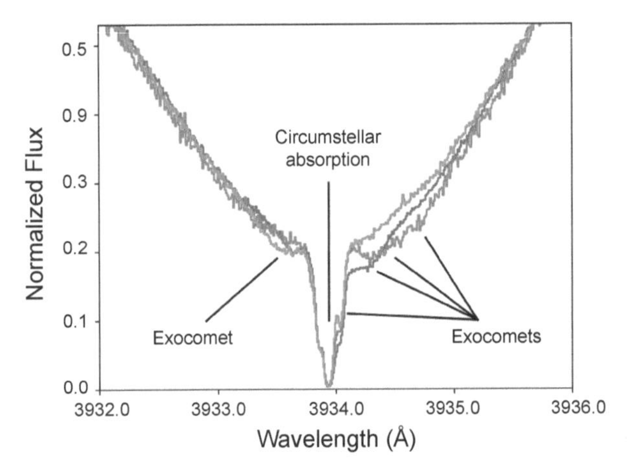

Fig. 5. Spectra of the core of the Ca II K stellar line in β Pic at three separate epochs over 4 days. The stable central narrow absorption is due to gas in the circumstellar disk, marking the stellar radial velocity. Transient Ca II absorption features caused by individual exocomets can be seen both blue-shifted and red-shifted relative to the star (see section 3.1 and Fig. 3). Data from the ESO 3.6-m+HARPS public archive (see acknowledgments).

Akin to solar system comets, these exocomets are clearly undergoing significant mass loss, which argues for a source region in the system. *Beust and Morbidelli* (2000) and *Thébault and Beust* (2001) had previously shown that a Jupiter-sized planet within 20 au could sustain such a population by strong orbital evolution via the 4:1 and/or 3:1 mean-motion resonances. This would place the source region at 4–5 au from β Pic, within the imaged debris disk. The discovery by direct imaging of the jovian exoplanet β Pic b by *Lagrange et al.* (2009) orbiting at a = 9.9 au gave credence to this hypothesis (see section 3.1.4 above). The more recent discovery of the closer massive planet β Pic c at a = 2.7 au (*Lagrange et al.,* 2019) will result in more complex dynamical evolution, but clearly enhances the possibility of scattering exocomets existing in the observed disk into star-approaching orbits.

Although exocomets typically transit the star over a couple of hours, *Kennedy* (2018) was able to measure radial acceleration by individual bodies and constrain their orbital parameters via Markov chain Monte Carlo (MCMC) methods. A larger analysis was made by *Kiefer et al.* (2014) of 252 individual exocomets seen in ~6000 spectra obtained in 2003–2011. The velocities and absorption strengths display a bimodal distribution, implying at least two dynamical populations. Exocomets creating shallow absorption lines tend to have small total absorption and higher velocities, with periastron distances ≃0.08 au (9 stellar radii) over a wide range of longitudes. The less numerous exocomets that produce shallow absorption also exhibit smaller radial velocities due to larger periastron distances of ≃0.16 au (18 stellar radii). A small range of longitudes of periastron imply these less-active objects all share similar orbits and may result from the breakup of a single progenitor similar to the sungrazing Kreutz family (*Jones et al.,* 2018, and references therein).

Exocomets orbiting β Pic have now also been detected via broadband optical light curves using the Transiting Exoplanet Survey Satellite (TESS) mission (*Zieba et al.,* 2019). The transit signals from individual exocomets show a "shark fin" appearance caused by the asymmetric morphology of dust comae and tails, as predicted by *Lecavelier des Etangs et al.* (1999). Analyzing all TESS light curves at that time, *Lecavelier des Etangs et al.* (2022) identified 30 individual transits. Assuming that each exocomet on average has the same fractional surface area sublimating, the differential size distribution has a power-law exponent of -3.6 ± 0.8, similar to that seen in solar system comets.

3.2.2. Other systems exhibiting exocomet signatures. As explored in section 2, the decreasing mass of circumstellar material as a young star begins its main-sequence lifetime implies that exocomets may also be more abundant at early times. Indeed, transient spectroscopic absorption features similar to β Pic are seen around many Herbig Ae/Be stars (*Grady et al.,* 1996). With β Pic as the archetype main-sequence host star, exocomets have been confirmed via similar spectroscopic or photometric transient features around three other stars: 49 Cet, HD 172555, and KIC 3542116 (*Strøm et al.,* 2020, and references therein). In addition, there are likely exocomet systems such as c Aql (HD 183324), which exhibits Ca II variability on timescales of minutes (*Montgomery and Welsh,* 2017; *Iglesias et al.,* 2018). In total, spectroscopic and photometric signatures indicating probable exocomet activity have been reported for another 30 or more stars (*Strøm et al.,* 2020).

The common factor in all these stars is that they are early-type A–F stars with bright apparent magnitudes. The brightness allows high signal-to-noise photometry from spacebased missions such as Kepler and TESS to reveal weak transit signals from stars with less stellar activity than later K/M-type stars, while B/A stars also exhibit relatively clean optical spectra with fewer stellar absorption lines, facilitating spectroscopic detection. In terms of total numbers, a directed spectroscopic survey of 117 B–G stars by *Rebollido et al.* (2020) found ~15% exhibited transient Ca II and Na I absorption. In summary, there is strong evidence that exocomets are commonly associated with early-type B/A stars. The evidence for later-type stars is less extensive, given that a search of 200,000 Kepler light curves by *Kennedy et al.* (2019) only found five potential photometric transits occurring for three stars.

3.2.3. Warm exozodiacal dust. Aside from the more distant cometary belts described in section 3.1, approximately 20% of A–K stars exhibit thermal signatures of warm (T ~ 300 K) dust within a few astronomical units (see *Kral et al.,* 2018). The combination of the Poynting-Robertson effect on large dust grains and radiation pressure on small dust grains implies very short lifetimes without continuous resupply. Potential sources of this dust are either asteroidal collisions or dust ejection from comets. This may especially be the case for systems with colder outer dust disks, where the source of the dust is expected to be ice-rich bodies. Given that the generation of the solar system's zodiacal cloud is dominated by short-period comets (*Rigley and Wyatt,* 2022), it is plausible that this may be also the case for some exosystems (*Marboeuf et al.,* 2016). A survey of exozodiacal dust around 38 stars showed a significant correlation with the existence of cold dust disks but no correlations with stellar age, as would be expected by analogy with the solar system (*Ertel et al.,* 2020).

An exemplar star where this appears to be the case is the main-sequence F2 star η Corvi. This star possesses a cold outer belt centered at ~165 au, plus a warm inner dust ring whose inner boundary is as close as 2–3 au (*Duchêne et al.,* 2014, and references therein). NIR and MIR spectra were modeled by *Lisse et al.* (2012) to reveal large quantities of amorphous carbon and water-ice-rich dust in the inner ring, as well as silicates. They interpret this as the recent break up of a large centaur-like icy body due to a collision with an even larger silicate body, resulting in most of the current observed warm dust. Inward perturbation of icy bodies from the outer cold belt by undetected planets as simulated by *Bonsor et al.* (2014) is supported by the large gap between the inner and outer disks plus the subsequent detection of CO at ~20 au, which could be due to inwardly evolving exocomets undergoing CO sublimation (*Marino et al.,* 2017).

3.3. Late Stages of Stellar Evolution

3.3.1. Exocomets during the late stages of stellar evolution. Once a star evolves off the main sequence, the large changes in luminosity in the red giant/asymptotic giant branch (RG/AGB) stages, together with the mass loss at the end of the AGB phase, will result in significant physical and dynamical changes to any orbiting cometary bodies (*Veras,* 2016). For example, during the Sun's RG phase, the water sublimation boundary will move from ~3 au to ~230 au, leading to cometary activity throughout the transneptunian region (*Stern et al.,* 1990; *Stern,* 2003).

Melnick et al. (2001) used the Submillimeter Wave Astronomy Satellite (SWAS) to discover a large amount of water vapor around the carbon-rich AGB star IRC+10216. This was initially interpreted as originating by ongoing sublimation of exocomets in that star's Kuiper belt. The inferred mass of cometary bodies was ~10 $M_\oplus$ (*Saavik Ford and Neufeld,* 2001). However, a later survey of eight AGB stars with Herschel (*Neufeld et al.,* 2011) showed a cometary source was unlikely, as the water line profiles matched that of the circumstellar CO, with widths as expected from the circumstellar expansion velocity. If the water originated from Kuiper belts, then at least some would be viewed near face-on, and the line widths of cometary-produced water would be much narrower. Instead, water formation appears to be either by shock-induced dissociation of CO, where the oxygen subsequently combines with hydrogen to produce H_2O, or due to UV photodissociation of CO and SiO to again produce oxygen and water vapor (*Lombaert et al.,* 2016). Therefore, there currently exist no strong indicators of exocomets around stars at the RG/AGB stage. However, we know they are present due to observations of stars at the very end of their stellar evolution, white dwarfs.

3.3.2. Evidence for small bodies around white dwarfs. There is now undisputed evidence that planetary systems can (in some form) survive post-main-sequence evolution. The origin of heavy-element absorption lines in white dwarf (WD) photospheres is now accepted as being caused by pollution from infalling planetary material. The paradigm is that small bodies near the WD undergo tidal disruption to form a debris ring, where subsequent collisions create dust that is slowly removed via the Pointing-Robertson effect to "pollute" the stellar photosphere. Strong evidence for this is found in the association between WD IR excesses due to circumstellar dust and atmospheric pollutants (see *Farihi,* 2016, and references therein).

The relatively short timescales for this material to sink below the observed photosphere means this process is currently occurring where observed. The simplicity of WD photospheres results in high-precision elemental abundances being derived via model atmosphere fitting. There is no doubt that almost all small bodies providing this material were volatile-poor (*Gänsicke et al.,* 2012; *Xu et al.,* 2019). However, two WDs have been identified as being polluted by water-rich bodies (*Farihi et al.,* 2013; *Raddi et al.,* 2015). This dominance of silicate-rich small bodies may not reflect the original small-body population that survives into the WD phase. A theoretical study of water retention by *Malamud and Perets* (2016) found that larger water-rich bodies should survive the RGB and AGB stage of stellar evolution. They propose that the observed deficit is due to water being lost prior to tidal disruption or during the disk formation stage. If the WD is in a wide binary system, then dynamical evolution of exocomets from a Kuiper belt via Kozai-Lidov resonances can result in WD pollution over a wide range of timescales of 10^8–10^{10} yr (*Stephan et al.,* 2017). Given that 25–50% of all WDs exhibit pollution by small bodies, exocomets may therefore be relatively common around WDs.

4. DYNAMICAL EJECTION AND EVOLUTION OF COMETARY BODIES

The discovery of 1I/'Oumuamua and 2I/Borisov gives evidence to the hypothesis that over a star's lifetime, part of its remnant small-body population becomes unbound (e.g., *Fernandez and Ip,* 1984; *Duncan et al.,* 1987; *Wyat,t* 2008). Thus, ISO production seems to be a natural consequence of the existence of reservoirs of small bodies in planetary systems. These small bodies can become unbound from their parent system in several ways (see Fig. 6). In the following, we will describe the various ISO production processes in chronological order of their occurrence during a star's lifetime. We give estimates on the order of magnitude level of the total mass of ISOs released by the individual processes. However, these values are highly sensitive to the assumed (debris) disk masses.

4.1. Individual Ejection Mechanisms

4.1.1. Interstellar object formation during disk dispersal. In section 2.2 we saw that disk masses decrease rapidly within the first 1–3 m.y. While initially this mass loss will be in the form of small dust particles, later on, larger particles and eventually planetesimals will contribute to this mass loss. So far, the matter that becomes unbound during this early stage has received little attention. The main challenge is the significant uncertainties concerning the timescale on which planetesimals form. Therefore, it is unclear how many planetesimals are present in disks at this stage. Very young disks have masses comparable to the mass of their host star ($m_d \approx M_s$). Considerable amounts of ISOs could, in principle, already be produced during this stage. The disk masses decrease to ($m_d \approx 0.01$ M_s) within just a few ten thousands of years; if any planetesimals are already present during that phase, they could become released due to the lower potential caused by the gas loss.

In recent years, several old disks (>10 m.y.) have been discovered still containing considerable amounts of gas. These disks likely also contain a sizable planetesimal population. When they eventually disperse their gas, the potential energy change due to the gas loss might be enough that a fraction of the planetesimals becomes unbound from the outer disk regions. Most proposed disk dispersal mechanisms are relatively gentle. Thus, the ISOs would leave their parent star at relatively low velocities ($\ll 0.5$ m s^{-1}). The only process that would potentially lead to asteroidal ISOs ejected at higher velocity are those involving disk jets. However, as jets only exist in the very early phases of disk development (<1 m.y.), it is unlikely that planetesimals had already formed.

4.1.2. Cluster-environment-induced planetesimal release. Planetesimals can also be released during another star's close flybys (*Hands et al.,* 2019). Here mostly planetesimals of the outer disk become unbound so that the ISOs produced by this process are usually icy objects. Close flybys happen most frequently during the first 10 m.y. of a star's life when the star is still part of a cluster (e.g., *Adams et al.,* 2006). However, debris disks can also be affected, as their longer lifetimes can sometimes outbalance the lower frequency of such events during later stages.

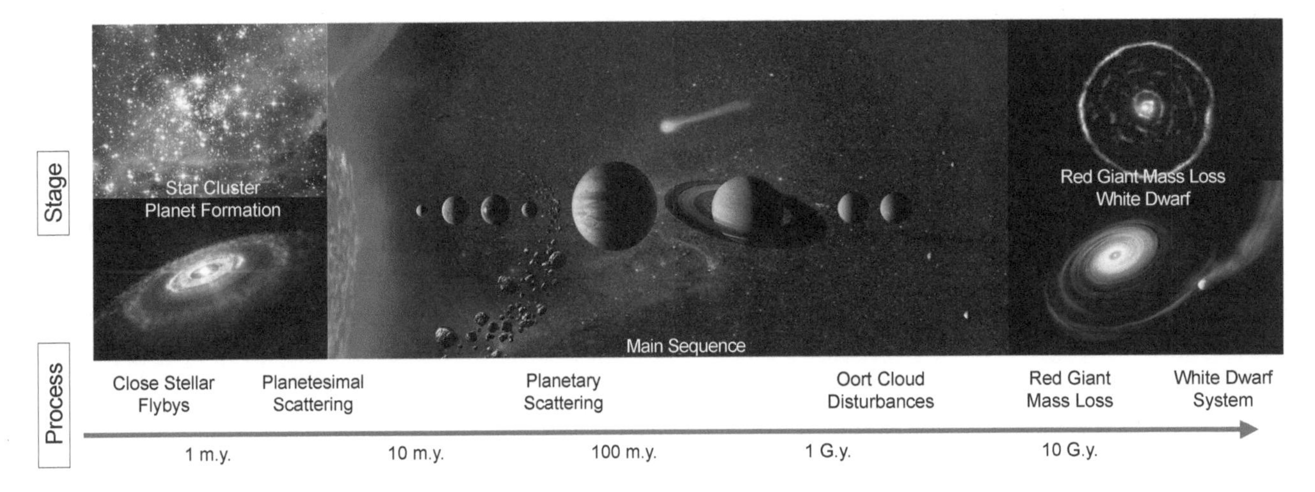

Fig. 6. Schematic picture of various planetesimal ejection mechanisms during the different stages of stellar evolution. Credits left to right, top to bottom; star cluster NGC 346 (NASA/STScI), planet formation (ESO/L. Calçada, solar system (NASA), red giant R Sculptoris (ALMA/ESO/NAOJ/NRAO/M. Maercker et al.), white dwarf planetary evolution (ESO/M. Kornmesser).

The frequency of close flybys varies enormously between different types of stellar groups. In long-lived compact open clusters, stellar flybys are much more common than in the short-lived, more extended clusters typical for the current solar neighborhood. Based on typical disk truncation radii in diverse cluster environments, *Pfalzner et al.* (2021a) find that clusters like the Orion nebula cluster are likely to generate the equivalent of 0.85 $M_\oplus$ of ISOs per star. In contrast, compact clusters like NGC 3603 can produce up to 50 $M_\oplus$ of ISOs per star. Our solar system probably created the equivalent of 2–3 $M_\oplus$ of ISOs by this process. In clusters, ISOs leave their parent system typically with velocities in the range of 0.5–2 km s^{-1}.

External photoevaporation and viscous spreading can, under certain circumstances, strongly influence the gas component of disks (*Adams,* 2010; *Concha-Ramírez et al.,* 2019). However, the decoupling of the gas and planetesimals means that these two processes release only a small number of ISOs. The velocity of these few ISOs should be much smaller than those released due to close flybys.

4.1.3. Early scattering by planets. During the phase when a planet is still accreting, planetesimals can receive gravitational kicks from planets. Such a perturbation can lead to two outcomes: scattering and collisions. The ejection to collision rates (assuming constant density for the planet) are given as $R_e/R_c = (M_p^{4/3} a^2)(M_*^2)$, where M_p and M_* are masses of the star and the planet, and a is the planet's semimajor axis (*Ćuk,* 2018). At large distances, collisions become less likely while scattering and therefore ejection becomes more efficient. The reason is that the Hill radius, which controls scattering, expands linearly with the orbital radius, whereas the physical size is fixed.

A planet can eject a planetesimal if the Safranov number $\frac{1}{2}v_{esc}/v_{orb} \gg 1$, where v_{esc} is the escape velocity and v_{orb} the orbital velocity of the planetesimal (*Raymond et al.,* 2018b). While the other giant planets all have similar Safronov numbers, it is actually the structure of the system itself that makes Jupiter responsible for the vast majority of planetesimals that have been ejected from the solar system (*Dones et al.,* 2015).

Most investigations of this process assumed solar-system-like environments (*Raymond et al.,* 2020). The early solar system likely had a planetesimal disk with a mass of ≈5–65 $M_\oplus$ (*Deienno et al.,* 2017; *Liu et al.,* 2022). The estimates of how many planetesimals have been ejected range from 0.1 $M_\oplus$ (*Raymond et al.,* 2018b) to 20 $M_\oplus$ (*Trilling et al.,* 2017) and even 40 $M_\oplus$ (*Do et al.,* 2018) per star. The differences in estimated total masses are primarily due to the unknown nature of the size distribution of ejected planetesimals. The ejected planetesimals from these regions would leave the systems with a typical velocity of 4–8 km s^{-1} (*Adams and Spergel,* 2005), significantly higher than for the above mechanisms. The amount of planetesimals ejected by scattering depends on the structure of the planetary system. Systems of densely packed giant planets that orbit close to a common plane are most efficient in ejecting planetesimals.

In our solar system, the ejected planetesimals would have been primarily icy because ejection is much more efficient outside the gas giants than closer in. Planetesimal ejection from the asteroid belt will be much less common. It is mainly caused by gravitational perturbations by Jupiter and Saturn and the mutual perturbations amongst the largest asteroids. It is currently an open question whether the early asteroid belt was high-mass and depleted (e.g., *Petit et al.,* 2001) or whether it was born low-mass and later on populated by dynamical processes (*Raymond and Izidoro,* 2017). If it was initially of high mass, ejections have depleted the asteroid belt by a factor of ~100 in number, the original asteroid belt being ≈10–20 times more massive than today (*O'Brien and Greenberg,* 2003). In this case, the total mass of ejected asteroids would amount to ≈0.004–0.008 $M_\oplus$.

However, the system's giant planets eject ISOs not only while still accreting gas (*Raymond and Izidoro,* 2017; *Portegies Zwart et al.,* 2018). In compact planet configurations, interactions between the planets themselves or the planets and the planetesimal disk can trigger instabilities. The planets move positions throughout such a phase of planet-planet scattering. During such an event, the bulk of planetesimal disk would be ejected (*Raymond et al.,* 2010). The structure of today's Kuiper belt and the Oort cloud can be explained by this mechanism.

Many exoplanetary systems are in compact configurations that might become unstable. Moreover, the solar system's giant planets were likely originally on more compact, resonant orbits (*Morbidelli et al.,* 2007; *Pierens and Nelson,* 2008). Originally it was proposed that such an instability happened in the solar system well after planet formation finished (500–700 m.y.); however, a consensus is emerging that the instability happened early, no later than 100 m.y. after the formation of the solar system (*Zellner,* 2017; *Morbidelli et al.,* 2018).

4.1.4. Drifting from Oort clouds. Even though the solar system's Oort cloud formation still has many unknowns, we can also expect exo-Oort clouds to exist around other stars. Oort cloud bodies are icy exocomets and possibly asteroidal objects (*Weissman,* 1999; *Meech et al.,* 2016) and are only weakly bound to their parent system. Being situated at a significant distance from their parent star, they are subject to external influences, like galactic tides and stellar flybys. Consequently, some planetesimals will always drift gently away from the star's Oort cloud over the entire main-sequence lifetime of the star (*Moro-Martín,* 2019b; *Portegies Zwart,* 2021). Thus their escape velocity is on average very low, probably <0.1 m s^{-1}. *Hanse et al.* (2018) estimates the Sun will lose 25–65% of its Oort cloud mass mainly due to stellar encounters. Such former Oort cloud members will contribute, but amount to <10% of the total ISO population (*Portegies Zwart,* 2021).

4.1.5. White dwarf phase. Finally, there is an increased planetesimal ejection rate toward the end of a star's giant branch phase (*Veras,* 2016; *Veras et al.,* 2020). During that phase, expansion of the stellar envelope can directly engulf closely orbiting planets and tidally draw into the envelope planets, which reside beyond the maximum extent of the stellar envelope. While the planetesimal disks are mainly unaffected by engulfment, stellar mass loss plays a role. During the post-main,sequence evolution, all objects, planets, and planetesimals alike move outward due to orbital expansion caused by mass loss (*Villaver et al.,* 2014). In the solar system, planetesimals within $\approx 10^3$ au would all double their semimajor axes (*Veras et al.,* 2020) and therefore be less strongly bound to their host star. The ISOs produced during the giant branch phase gently drift away from their parent star at very low relative velocities of 0.1–0.2 km s^{-1} (*Pfalzner et al.,* 2021a). In addition, mass loss also destabilizes the planetary system, resulting in the gravitational scattering of planetoids by massive perturbers. *Rafikov* (2018) suggests tidal disruption events of (initially bound) planetary objects by WDs as an additional ISO production mechanism. Some of these objects are scattered toward the WD on almost radial orbits and get tidally shredded apart. Some of the fragments that are produced are ejected into interstellar space.

The highest mass WD progenitors would yield the greatest giant branch excitation and are most efficient in the ejection of planetesimals (*Veras et al.,* 2020). This study points out that most stars in the Milky Way are less massive than the Sun and concludes that the production of ISOs from within 40–1000 au of the evolved star is insignificant when compared to the number created during early stellar and planetary evolution.

4.1.6. Binary scenarios. In section 2.1, we noted that binary and multiple stars are very common in the galaxy. Moreover, binaries are also more efficient scatterers, providing more substantial dynamical perturbations than even a gas giant planet (*Raymond and Izidoro,* 2017; *Ćuk,* 2018; *Jackson et al.,* 2018). All the processes discussed before for single stars could also take place in binary systems. So far, more or less exclusively scattering processes have been considered in detail. Close binary stars have well-defined dynamical stability limits, and any objects entering within a critical orbital radius are destabilized. *Jackson et al.* (2018) assume that planetesimals form beyond this stability limit and drift inward across the limit due to aerodynamic gas drag. Their simulations show that all planetesimals that are drifting inside the stability limit are ejected. Especially for close binaries, the snow line is often exterior to the stability limit such that a considerable portion of their ejected planetesimals may be refractory. However, this mechanism is only efficient if a significant fraction of the solid disk mass drifts interior to the binary stability limit. For the aerodynamic forces to work efficiently the planetesimals have to be relatively small ($r < 1$ km). Only then can they enter the dynamically unstable zone close to the binary, rather than piling up at the pressure maximum in the gaseous circumbinary disk. It has been proposed that 1I/'Oumuamua could be a fragment of a planet that formed in a binary system and was tidally disrupted after passing too close to one of the stars (*Ćuk,* 2018). However, tidal disruption is a rare event. Nevertheless, tight binary systems can eject an amount of rocky material comparable to the predominantly icy material thrown out by single- and wide binary-star systems. It is estimated that the mass of ISOs ejected from binary systems is at least the same as by planet scattering (*Jackson et al.,* 2018) and the ISOs leave their parent binary system with $\approx$6.2 km s^{-1} (*Adams and Spergel,* 2005).

4.1.7. Relative importance of the different interstellar object formation mechanisms. All mechanisms mentioned in this section produce ISOs. The question is which of them is/are the dominant ISO production mechanism(s). All the processes discussed above produce on the order of 0.1–30 $M_\oplus$ per star. However, given the strong dependence on the assumed mass of the planetesimal reservoir, these numbers fail to give a clear winner. Thus, different criteria have to be employed. The various models make different predictions concerning the properties of the ISOs. As soon as a large enough sample of ISOs is known, these model predictions can be used to identify the dominant ISO production mechanism. Differences are mainly in terms of the composition and the velocity of the ISOs.

Most mechanisms lead predominantly to icy ISOs that formed outside the snow line in their parent systems. However, 1I/'Oumuamua did not show the cometary coma one would expect from an icy ISO. Therefore, many mechanisms have been tested concerning their ability to produce refractory ISOs. In general, scattering processes tend to produce more asteroidal ISOs than stellar flybys, drifting from the Oort cloud and the processes at the end of the main sequence. However, even scattering processes produce mainly icy ISOs.

There are significant differences in the ejection velocity of the various processes (*Pfalzner et al.,* 2021a). The slowest ISOs are those that drift away from the Oort cloud or are shed during the stellar post-main-sequence phase ($\approx$0.1–0.2 km s^{-1}). Stellar flybys lead to ejection velocities in the range of 0.5–2 km s^{-1}; even higher velocities are achieved by planet scattering ($\approx$4–8 km s^{-1}), only exceeded by processes where two or more stars kick out the planetesimals together.

Most processes discussed have a connection between velocity and composition. Due to their closeness to the star, the refractory planetesimals require a higher velocity to become unbound than the icy planetesimals (unless, of course the rocky planetesimals resided in the Oort cloud of the host star). As soon as a statistically significant sample of ISOs is discovered, the combined information of their observed velocities and composition might help constrain the dominant production process. However, the observed ISO velocity is only indirectly linked to the ejection velocity, and other effects such as dynamical heating for older ISOs may make such studies difficult. The complete ISO velocity distribution contains multiple components reflecting various ejection speeds and the parent system's different ages. In future, disentangling the different components will be one of the significant challenges.

4.2 Resulting Interstellar Object Population

Most of the above described planetesimal ejection mechanisms produce considerably more icy ISOs than rocky ones. The ejection of volatile-free asteroids is possible, but they are thought to be a relatively small fraction of planetesimals that were ejected. However, this impression might be partly caused by our heliocentric view. In the solar system, giant planets all orbit beyond the "snow line." Therefore, the majority of bodies within <2.5 au on unstable orbits end up colliding with the Sun (*Gladman et al.,* 1997; *Minton and Malhotra,* 2010). In the solar system there is indeed a much greater supply of icy planetesimals, however, this may not apply elsewhere (*Cúk,* 2018). If we look at the known exoplanet systems, many of them have relatively massive planets much further in than in our solar system. The ejection of rocky planetesimals would be more efficient where there are close-in massive planets around high-mass stars. However, even then it appears unlikely that the rocky planetesimals could dominate the galactic population of scattered ISOs. The reason is that the large reservoir of volatile planetesimals beyond the snow line is always much larger than that of the rocky planetesimals in the inner systems.

Naturally, the size distribution of ISOs is connected to the size distributions of the reservoir of planetesimals the ISOs originate from. Unfortunately, there is still a large uncertainty of the size distribution of planetesimals. Often a mass function of the form $N_{ISOs} \propto m^{-p}$ is assumed. This means that the number of ejected particles of a certain mass N (m) is then given by

$$N(m) = \frac{2-p}{p-1}\frac{M_{ISO}}{m^{p-1}m_{up}^{2-p}} \tag{1}$$

(*Adams and Spergel,* 2005), where M_{ISO} is the total mass of ISOs and m_{up} is the largest mass possible. Sometimes p = 5/3 is chosen (*Adams and Spergel,* 2005). However, starting from a Dohnanyi distribution and from a distribution obtained by simulations of the streaming instability (p = 1.6), *Raymond et al.* (2018b) argue that the underlying masses are off by 2–4 orders of magnitude. Observational constraints of the small-body population of the solar system can also provide some constraints on the initial size distribution of ISOs. Beyond Neptune, the small-body population is thought to be the least collisionally evolved (*Abedin et al.,* 2022). For these planetesimals a turnover in the size-number distribution is observed around a radius of 50–70 km (*Bernstein et al.,* 2004; *Fraser et al.,* 2014), which likely is of primordial origin (*Vitense et al.,* 2012). Therefore, one can expect that size distribution of the ISOs before expulsion was similar.

The size distribution of the planetesimals before expulsion sets the stage for the ISO size distribution. However, both distributions are not necessarily identical. First, there might be selection effects concerning the planetesimals' mass (and therefore size) in some of the ISO production mechanisms. For example, in three-body scattering processes, ejection is more likely the lower the mass of the body. Second, the ejection processes themselves can alter ISOs, such that the properties of the ISOs differ from those of the planetesimals in the parent system. While ISOs that are relatively gently released from their parent system during the Oort cloud phase will remain unaltered, ISOs ejected in a more violent fashion might be disrupted during ejection (*Raymond et al.,* 2020). For example, planetesimals gravitationally scattered in very close encounters with a giant planet can be subjected to tidal disruption (e.g., *Asphaug and Benz,* 1996). In particular, scattering events that include a close passage to the parent star can lead to the decline or loss of cometary activity (e.g., *Levison and Duncan,* 1997), and sublimation-driven activity may also flatten the shapes of future ISOs (*Seligman and Laughlin,* 2020; *Zhao et al.,* 2021). Third, ISOs might be affected to different degrees by their journey through interstellar space (see section 4.4).

By now, it should have become clear that planet formation and the production of ISOs are intrinsically linked processes. Ever since planets formed, a population of planetesimals were produced at the same time. A considerable portion of these planetesimals are released sooner or later from their parent star(s) and became ISOs. This means that the interstellar medium is steadily enriched with newly released ISOs (*Pfalzner and Bannister,* 2019). Just like the galaxy's stellar metallicity increases over time, so does the density of interstellar objects.

The ISO age distribution should be directly linked to the age distribution of the planets and stars. The stellar age distributions within the Milky Way are shown in Fig. 7. It demonstrates that star formation peaked at the early ages of the galaxy. However, it is unclear whether the first stars produced ISOs at the same rate as today. It is known that planet formation is much more efficient around high-metallicity stars. ISO formation is linked to planet formation. Suppose we take the age distribution of the planet host stars (Fig. 7b) as a proxy for the planetesimal production distribution; this differs considerably from the general stellar age distribution in the Milky Way. The age distribution of planet host stars shows a clear maximum at approximately 4 G.y. The age distribution of ISOs might shift even more to younger ages, as some portion of the ISOs may dissolve during their interstellar journey since the frequency of volatile loss and extinction is far higher for ejected planetesimals than for surviving ones (*Raymond et al.,* 2020).

4.3. Interstellar Object Timescales

A planetesimal that has been ejected from its host star and has become an ISO does not necessarily remain an ISO forever. ISOs may be recaptured. Recapture is most likely if the planetesimal has been ejected from a host star that is a member of a star cluster. This situation is most likely for young host stars because most stars are born in a star cluster environment (*Lada and Lada,* 2003). However, some stars remain for hundreds of millions of years in open clusters

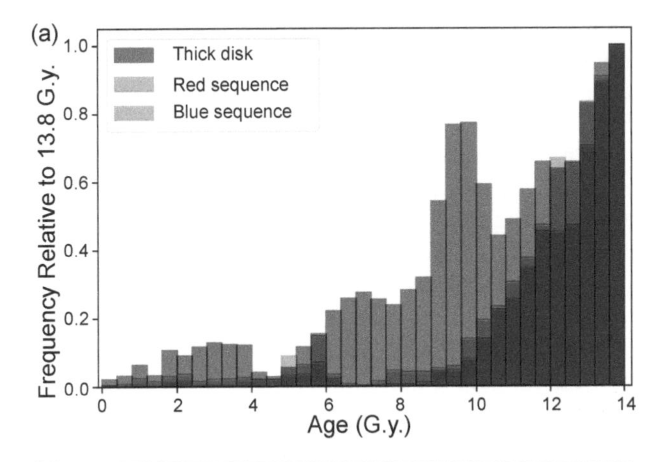

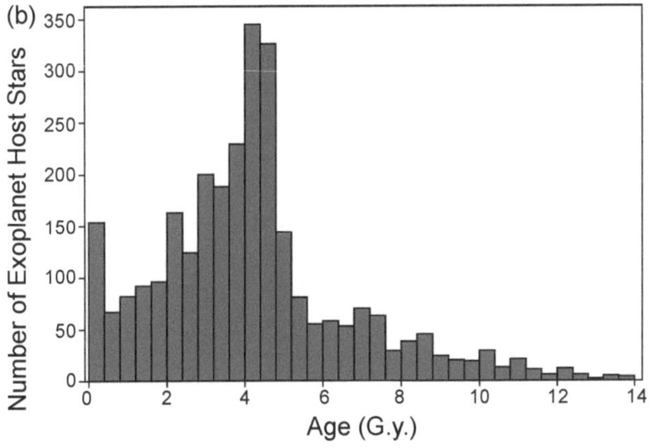

Fig. 7. (a) Stellar age distributions in the Milky Way, adapted from *Gallart et al.* (2019). **(b)** Age distribution of planet hosting stars as listed in the public database at *exoplanets.eu*. The observational bias toward surveying Sun-like stars is clearly evident.

and for several billions of years in globular clusters. In all these situations, recapture is a real option.

In such an environment, the ISO might be very quickly (<1 m.y.) recaptured by a different member of the same star cluster and lose its status as an ISO (*Hands et al.,* 2019). The likelihood of the ISO being recaptured rises with the cluster mass as the escape speed from the cluster increases with cluster mass. Equally, ISOs escaping from stars in the cluster center are more likely to be recaptured. For ISOs that are not in a star cluster environment, the likelihood of recapture is much lower. In any case, the velocity of the ejected ISO is the key parameter for being recaptured. The lower the ISO's velocity, the higher the probability that it is recaptured. This correlation is valid in every capture environment, whether it is another star, a disk surrounding a young star, or a molecular cloud (*Grishin et al.,* 2019; *Pfalzner et al.,* 2021a; *Moro-Martín and Norman,* 2022).

4.4. Physical Processing in the Interstellar Medium

ISOs are often viewed as traveling through interstellar space in calm cryogenic conditions. Therefore, ISOs are expected to be pristine samples of the planetesimals of other planets. As planetesimals are leftovers from the planet-formation process, ISOs should therefore give us direct information about planet formation elsewhere. Generally, an ISO's journey through interstellar space might be not as uneventful as often imagined. The environmental effect on ISOs can be expected to be similar to that of Oort cloud comets. Just like the comets residing in the Oort cloud (*Stern,* 2003), a variety of thermal, collisional, radiation, and ISM processes might affect ISOs during their long journey (*Stern,* 1990). The ISOs' surface is subject to these environmental influences, which can potentially lead to the erosion or even complete destruction of an ISO. How destructive these processes are depends on the composition of the ISO, with a rocky ISO being much less affected than, for example, a hydrogen-rich ISO.

During their journey, ISOs experience cosmic microwave radiation and radiation from the stars. The temperature increase by the cosmic microwave radiation on the surface of the ISO is probably negligible. In contrast, the effect of stellar radiation depends strongly on the ISO's individual journey. Any ISOs passing close to a star can be expected to be modified considerably by such an encounter. The prime effect is the reduction or complete loss of volatiles. Such an event is expected to be rare. However, parsec-range and closer encounters with highly luminous O and supergiant stars can heat ISO surfaces to temperatures capable of removing the most volatile ices, such as neon or oxygen (*Stern,* 2003).

ISOs are also subjected to the effects of cosmic rays and stellar energetic particles. Both have a broad spectrum of energies and interact with the ISO surface and subsurface. Cosmic rays are the primary source of space weathering for the comets in the Oort cloud and therefore likely also for ISOs. While low-energy particles interact only with the ISO surface, the most energetic ones deposit a significant amount of energy down to tens of meters. These processes can modify the isotopic ratios in cometary ices and create secondary compounds through radiolysis (*Gronoff et al.,* 2020). The penetration depth of cosmic rays is relatively small in comparison to the ISOs of size ≥100 m that are likely to be detected by telescopes (see section 5). Even for the extreme case of pure hydrogen ISOs, it would be <10 m (*Hoang and Loeb,* 2020); for rocky material, considerably shallower. Thus, the immediate effect of cosmic rays on ISOs is small, but over time they can erode an ISO gradually from the surface or affect the composition of the upper layers (see section 7.3.2).

The ISM contains large amounts of gas and dust. The effect of this gas and dust on ISOs depends on the gas and dust densities and the relative velocity of the individual particles. Gas and dust densities are relatively low in the interstellar medium itself ($n_H \approx 10$ cm^{-3}) and are considerably higher in molecular clouds ($n_H \approx 10^4$ cm^{-3}) and even higher when passing through protoplanetary disks of newly forming stars ($n_H \approx 10^{5-7}$ cm^{-3}).

Collisions of ISOs with the ambient gas at high speeds can heat the frontal area, possibly resulting in transient evaporation. Such a situation occurs when the ISO passes through a molecular cloud. Each particle directly impacting on the surface of an ISO delivers an energy $E_p = m_p v_p^2/2$. Therefore, ISOs are most impacted by high-velocity particles. At least for Oort cloud comets, simulations show that dust particles could significantly erode the cometary surface in the range of several meters (*Stern,* 1990; *Mumma et al.,* 1993), preferentially removing the submicrometer grains. However, erosion by ISM grains is a complex process depending on the ISM grains' composition and structure.

While moving through a relatively massive molecular cloud forming a large star cluster, there is a non-negligible probability that the most massive stars explode as a supernova. Therefore, an ISO might be subjected to the explosion products from such a supernova. Despite supernovae explosions being brief (≈0.1 yr), they are extremely luminous $L = 10^9 L_{Sun}$ and therefore can heat ISOs from considerable distances. Their intense but shorter thermal pulses could propagate 0.1–2 m into ISO surfaces (*Stern,* 2003).

These processes complicate the interpretation of ISO observations; the surface layer of ISOs might be modified over time. In summary, all these processes erode the surface of ISOs to different degrees. However, as the distinctive properties of the ISOs in terms of their size and composition are not well constrained, it remains unclear whether their surface is affected or whether they decrease in size with age and how many small ISOs are destroyed.

5. INTERSTELLAR OBJECT SEARCHES AND DISCOVERY

5.1. Dynamically Recognizing Interstellar Objects

Upon discovery, ISOs will appear as either comets or asteroids depending on their level of activity. Hence the single most important discriminant is showing the orbit is hyperbolic, with $e > 1$ and $a < 0$ at the 3σ level or higher. Dynamically, this is equivalent of saying that their velocity at an infinite distance $v_\infty > 0$ (also known as the hyperbolic excess velocity). This requires accurate astrometric measurements of the object over a period of time, and hence sensitive wide-field surveys that allow early detection. For ISOs close by Earth it may take only a few days of observations before the orbital arc is large enough to show this unambiguously; more distant ISOs may take weeks (see section 6).

It should be noted that many comets are discovered each year on apparently hyperbolic orbits. These are, in reality, dynamically new comets from the Oort cloud but are listed as hyperbolic due to a number of reasons. First, an orbit that is near-parabolic in the solar system barycentric reference frame may appear to be hyperbolic in the heliocentric reference frame, in which most orbital elements are published. For example, the dynamically new comet C/2021 A9 (PANSTARRS) has an osculating heliocentric orbital eccentricity of e = 1.004. Transforming to the correct barycentric frame gives e = 1.0004, much closer to a parabolic orbit. Second, an originally weakly bound object may become unbound due to gravitational perturbations or non-gravitational forces due to outgassing (see *Królikowska and Dybczyński,* 2010).

To illustrate this, using data from the JPL Small Bodies Database (*https://ssd.jpl.nasa.gov/tools/sbdb query.html*), 17 long-period comets (LPCs) were discovered in 2020–2021 with osculating orbital eccentricity $e \geq 1.0$. Yet all eccentricities were so close to unity that either these comets have $e \leq 1.0$ within the orbital uncertainties and/or within a barycentric reference frame, or they would have possessed an originally parabolic/elliptical orbit before entering the planetary system. Understanding these factors is crucial for confirming the status of suspected ISOs with $e \simeq 1$.

5.2. Interstellar Object Studies Pre-1I/'Oumuamua

There were many papers that made predictions about the space density of ISOs based on the non-detection in surveys. Some very early work was reported by *Safronov* (1972), *McCrea* (1975), *Sekanina* (1976), *Duncan et al.* (1987), *Valtonen and Zheng* (1990), and *Zheng et al.* (1990). The advent of wide-field CCD-based sky surveys in the 1990s allowed quantitative upper limits to be assessed. *Francis* (2005) derived a 95% upper confidence limit of 4.5×10^{-4} au^{-3} using the Lincoln Near Earth Asteroid Research (LINEAR) survey. Just before the discovery of 1I/'Oumuamua, *Engelhardt et al.* (2017) numerically integrated ISOs and linked them to the non-detection by the combined major sky surveys operational at that time: the Catalina Sky Survey, the Mt. Lemmon Survey, and the Panoramic Survey Telescope and Rapid Response System (Pan-STARRS) project. Although they assumed isotropic approach trajectories that disregarded the local standard of rest distribution of stellar velocities, they presciently derived upper limits for both inactive and active (cometary) ISOs. They reported 90% confidence upper limits for 1-km-diameter ISOs of $\leq 2.4 \times 10^{-2}$ au^{-3} and $\leq 1.4 \times 10^{-4}$ au^{-3} respectively.

The important thing to get across here is, why was the first ISO discovered now? To discover these fast-moving objects, you need to survey the whole sky quickly to very faint limiting magnitudes, and we have only had surveys capable of doing this in the last couple of decades. *Heinze et al.* (2021) shows that current near Earth object (NEO) surveys are poorer than expected at finding fast moving objects. However, it is expected that future surveys will do much better (see section 7.3.1).

6. OBSERVED CHARACTERISTICS OF INTERSTELLAR OBJECTS

6.1. 1I/'Oumuamua

The most important questions about the first ISO are: (1) Where did it come from? (2) What is it made of? As discussed below, we still don't know the answers to these questions. 1I/‘Oumuamua was easily observable from the ground for a little over a week, but as it moved away from Earth it faded quickly, and large telescopes were able to observe for only about 1 month. The last HST observations were made in January 2018. It is remarkable that we know as much about this object as we do, because all the large telescope time had to be secured through Director requests. In total, approximately 100 hrs on 2.5–10-m groundbased telescopes were devoted to characterizing this exceptional object. To date, over 200 refereed papers have been written on both interstellar objects, and nearly 450 papers including non-refereed material. It is not practical to cite everything, so key papers have been highlighted and we have included some reviews of the field.

6.1.1. Discovery of 1I/‘Oumuamua. On October 19, 2017, the Pan-STARRS survey found an object, designated P10Ee5V, moving quickly with respect to the stars (see Fig. 8a) during the normal NEO survey. R. Weryk then found pre-discovery images in Pan-STARRS data from October 18. Follow-up data obtained by the 1-m European Space Agency (ESA) ground station on October 20 was rejected by the Minor Planet Center because the data implied a large eccentricity, and P10Ee5V was classified as an Earth-orbit crossing asteroid. Data from October 20 obtained by the Catalina Sky Survey suggested that the object should be classified as a short-period comet. However, observations obtained from the Canada-France-Hawaii Telescope (CFHT) on October 22 showed that the orbit was hyperbolic with an eccentricity of 1.188. On October 24 the object was designated C/2017 U1 [Minor Planet Electronic Circular (MPEC) 2017-U181]. This was corrected on October 26 after deep images from the CFHT from October 22 showed no coma. The object was named A/2017 U1 (MPEC 2017-U183).

Within a week of discovery, a request was made to the Hawai‘ian cultural group that advised the Maunakea observatories management group to propose a name for the new object. They proposed the name ‘Oumuamua, meaning “a messenger from afar arriving first” or a scout or messenger sent from our distant past to reach out to us or build connections with us. In an extraordinarily fast effort, the International Astronomical Union (IAU) approved the name on November 6, 2017, and the new name became 1I/‘Oumuamua. This became the foundation of a project called A Hua He Inoa, which blends traditional indigenous practices into the official naming of astronomical discoveries (*Kimura et al.,* 2019; *Witze,* 2019).

6.1.2. Nuclear characteristics. One of the most straightforward measurements to make is brightness, and from this get an estimate of the object’s radius for an assumed albedo

$$pr_N^2 = 2.235 \times 10^{22} r_h^2 \Delta^2 10^{[0.4(m_\odot - m)]} 10^{0.4(\beta\alpha)} \tag{2}$$

where r_h and Δ are the helio- and geocentric distances (au) and $m_\odot$ and m are the apparent magnitudes of the Sun and comet (*Russell,* 1916). It was apparent from the earliest imaging observations of 1I/‘Oumuamua that it had a very large rotational light curve range so various estimates of the size were reported, even for the same assumed albedo. Combining several nights’ worth of observations from several observatories yields $H_V = 22.4$, which gives an average radius of 0.11 km assuming a typical cometary albedo of 0.04 (*Meech et al.,* 2017b). Spitzer observations of 1I/‘Oumuamua in principle could have provided measurement of both the nucleus size and albedo; however, because of strict solar avoidance angles, these observations could not

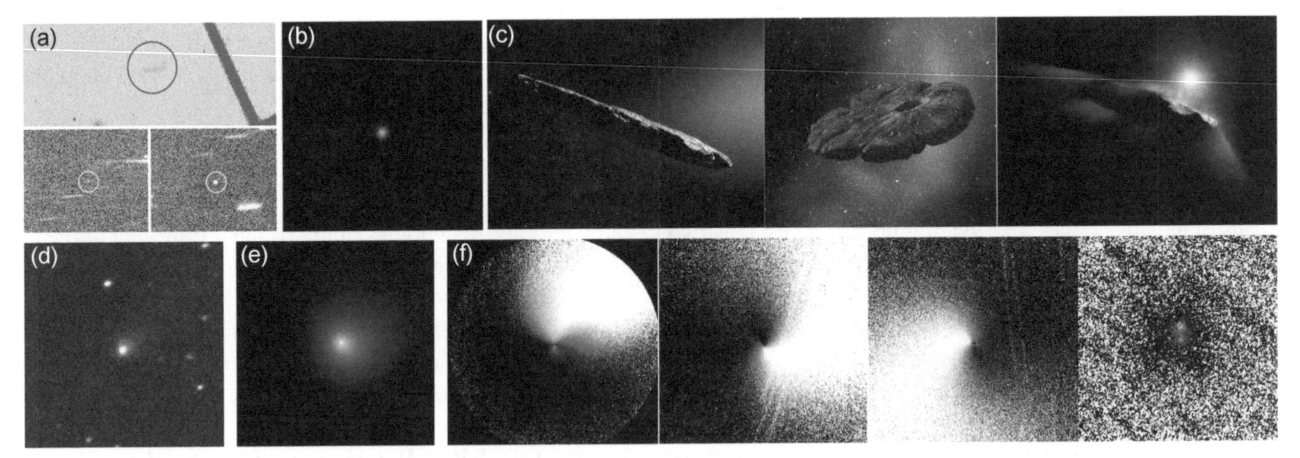

Fig. 8. Two ISOs passing through the inner solar system show the dramatic difference in their characteristics. Top row: 1I/‘Oumuamua: **(a)** Pan-STARRS discovery image (top) taken on October 19, 2017, with follow-up images from the CFHT near min/max brightness on October 27; **(b)** Gemini 8-m color composite image made from 192 images (1.6 hr) showing no hint of a dust coma (Gemini Observatory/AURA/NSF); **(c)** artist’s depiction of two possible nucleus shapes based on the large light curve range (ESO, M. Kornmesser; William Hartmann) and artist’s view of the ISO after the discovery of non-gravitational acceleration (ESA/Hubble, NASA, ESO, M. Kornmesser). Bottom row: 2I/Borisov: **(d)** CFHT image taken 10 days post-discovery (September 10, 2019); **(e)** HST image on October 12, 2019 (NASA/ESA); **(f)** radially normalized ratio HST images median averaged over an orbit from October 12, 2019, December 24 (00:31 UT), and December 24 (02:21 UT) showing outgassing jets, and July 6, 2020, showing the split nucleus (NASA/ESA) (processed by H. Boehnhardt).

be made until late November 2017 and only upper limits on the flux were obtained (*Trilling et al.,* 2018). Thermal models were used to estimate the corresponding radius and albedo, with a preference for an albedo of 0.1 and a radius consistent with the previous estimate.

The rotational light curve shown in Fig. 9 has a brightness range of $\Delta m \sim 3$ mag, implying an axis ratio for an oblate sphere of $\frac{a}{b} = 10^{0.4(\Delta m)} = 15$. However, this does not take into account the phase angle (which can make the object look more elongated than it is; see Fig. 8c). It also does not take into account that the rotation pole position is completely unknown; if the pole was more closely aligned to the ecliptic, then the object could appear even more elongated than the light curve suggests. The consensus is that the ratio is likely a/b $\gtrsim$ 6 (*'Oumuamua ISSI Team et al.,* 2019; *Mashchenko,* 2019). Solar system objects typically do not have axis ratios this large, and the cause of the unusual shape of 1I/'Oumuamua remains an enduring mystery (see section 7.3.2).

Attempts to find a rotation period from the data produced a variety of periods, near 8 hrs, all differing depending on the length of the dataset. The most comprehensive model concluded that 1I/'Oumuamua is in a complex rotation state, rotating around its shortest axis with a period of 8.67 ± 0.34 hr with a period of rotation around the long axis of 54.48 hr (*Belton et al.,* 2018). The damping timescale for a body this small is long enough that an excited rotation state can be preserved from the time of ejection from the host star. The shape of 1I/'Oumuamua as interpreted from the rotational light curve can change significantly, depending on the rotation state (see Fig. 8c).

McNeill et al. (2018) used the shape and rotation of 1I/'Oumuamua to place some constraints on the strength and density of the ISO, finding it was more likely to have a density typical of asteroids.

6.1.3. Constraints on composition. Several groups obtained spectra for 1I/'Oumuamua, both from the ground and from Spitzer, but there was no evidence of outgassing (see *'Oumuamua ISSI Team et al.,* 2019, for a summary). Many groups used either spectra or filter photometry to estimate the spectral reflectivity of the surface. The spectral reflectivity, S, was found to increase with wavelength and ranged between S = (7–23) ± 3%/1000 Å (*Jewitt and Seligman,* 2022). This measurement was challenging because the rotational signature had to be removed. Regardless of the value, the surface of 1I/'Oumuamua was red, typical of organic-rich comet surfaces, but also consistent with iron-rich minerals and space-weathered surfaces.

Deep stacks of images from many nights of data from groundbased and spacebased optical images showed no dust at all to a limit of there being $\lesssim 10^{-3}$ kg s^{-1} of micrometer-sized grains being produced (*Meech et al.,* 2017b; *Jewitt et al.,* 2017). In contrast, the dust production upper limit from Spitzer observations for 10-μm grains was <9kg s^{-1} (*Trilling et al.,* 2018).

6.1.4. Trajectory and potential origins. One of the primary goals for the HST time awarded for 1I/'Oumuamua was to extend the astrometric arc length to be able to determine its orbit with sufficient precision to trace its path backward and determine its home star system. The chances of being able to trace the trajectory backward to find its parent star were low, because gravitational scattering from stellar encounters would limit the search to the past few tens of millions of years (*Zhang,* 2018).

As 1I/'Oumuamua faded, astrometry was obtained with the Very Large Telescope (VLT) in Chile and with HST, with the last observations from HST being obtained on January 2, 2018, when the ISO was fainter than V ~ 27 at 2.9 au. An analysis of the combined groundbased and HST observations showed that the orbit could not be well fit with a gravity-only solution. By adding a radial non-gravitational acceleration term directed radially away from the Sun with a r_h^{-1} dependence, the orbit was well fit (*Micheli et al.,* 2018). The authors ruled out a number of other hypotheses, e.g., the Yarkovsky effect, frictional drag forces, impulsive velocity changes, photocenter offset from a binary or non-uniform albedo, interaction with the solar wind, or radiation pressure. Most were rejected because the acceleration produced by these mechanisms was either too small, in the wrong direction, wouldn't match the continuous nature, or implied a non-physical bulk density for 1I/'Oumuamua. By assuming that the same non-gravitational acceleration was operating on the pre-perihelion trajectory, an attempt was made to find the originating star system for 1I/'Oumuamua. *Dybczyński and Królikowska* (2018) attempted a search for the parent star taking advantage of the Gaia satellite data release 1 (DR1) of stellar positions but did not find an obvious parent star for 1I/'Oumuamua.

The Gaia DR2 data release provided the necessary astrometric position and *velocity* information that was needed to be able to try to trace the path of 1I/'Oumuamua back to its home star using its Keplerian orbit (*Bailer-Jones et*

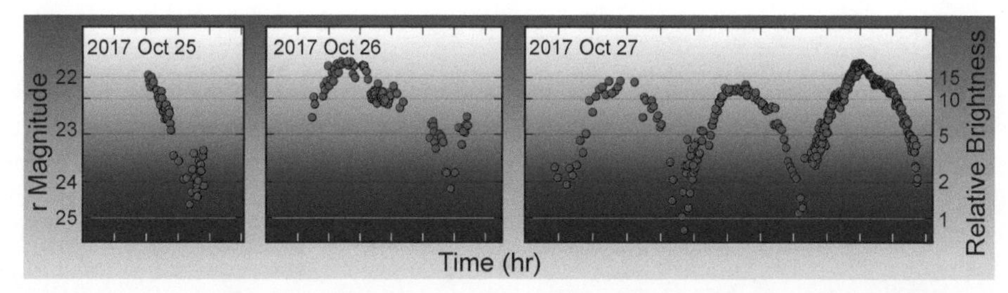

Fig. 9. Rotational light curve of 1I/'Oumuamua compiled from data taken from various observatories in October 2017.

al., 2018). Plausible home stars would be those that had been within 0.5 pc of 1I/'Oumuamua's trajectory (the approximate size of the Sun's Oort cloud), and that had low encounter (ejection) velocities <10 km s^{-1} (see section 4.1). An initial selection of stars that passed within 10 pc of the trajectory was selected and the candidates and the path of 1I/'Oumuamua were integrated through a smooth galactic potential. This resulted in four potential "home" systems with encounter distances between 0.6 and 1.6 pc and encounter velocities between 10.7 and 24.7 km s^{-1}, all with ejection times less than 6.3 m.y. ago. Unfortunately, all these systems had ejection speeds much higher than expected for ISOs.

6.1.5. Cause of the non-gravitational acceleration. In order to explain the non-gravitational acceleration of 1I/'Oumuamua in the absence of any apparent activity (dust coma or gas), *Micheli et al.* (2018) estimated the mass loss needed to accelerate the nucleus to the observed value for a range of comet and asteroid densities. They found mass loss requirements between 0.7 and 140 kg s^{-1} but adopted 10 kg s^{-1} as the best estimate. A thermal model matching the acceleration would require sublimation of water plus a more-volatile species (such as CO_2 or CO). With a high CO/H_2O ratio the model could produce enough mass loss to within a factor of 2–3 of what was needed to accelerate 1I/'Oumuamua. *Trilling et al.* (2018) found 3σ upper limits for the dust production of 9 kg s^{-1} (for 10-μm grains), $Q(CO_2) = 9 \times 10^{22}$ mol s^{-1} and CO 1.1×10^{24} mol s^{-1} (after correction). None of the optical spectra were sensitive enough to detect gas with such a low CO or H_2O production rate. Having consistency within a factor of a few between the model and observations suggests that volatile outgassing is a plausible scenario for the acceleration of 1I/'Oumuamua. *Stern* (1990) postulated that any comet passing through a molecular cloud would have the small grains eroded from the surface, and this could explain the lack of dust in the presence of outgassing.

Other more exotic volatile-sublimation scenarios to explain the non-gravitational acceleration are discussed in section 7.3.2. Finally, *Seligman et al.* (2021) investigated the spin dynamics of 1I/'Oumuamua with outgassing consistent with the non-gravitational acceleration and found that this need not cause a spinup of the nucleus. They were able to reproduce the observed light curve with a model of CO outgassing but not with water-ice jets.

6.2. 2I/Borisov

6.2.1. Discovery of 2I/Borisov. The second ISO was discovered on August 30, 2019, by G. Borisov, an engineer at the Crimean Astronomical Station and amateur astronomer, using a 0.65-m f/1.5 astrograph (*Borisov and Shustov,* 2021). Borisov reported it as a potential new comet with a compact 7-arcsec diameter coma and detected a short 15-arcsec tail two days later. It was placed on the Minor Planet Center Potential Comet Confirmation Page as object gb00234, and by September 11, 2019, sufficient additional astrometry had been reported to give a cometary designation of C/2019 Q4 (Borisov) and an initial orbit with an eccentricity e = 3.08 (MPEC 2019-R106). On September 24, 2019, the permanent designation of 2I/Borisov was announced (MPEC 2019-S72). As further observations were reported, by mid-October the derived orbit had stabilized with e = 3.354, q = 2.006 au, and i = 44.05°, giving discovery circumstances of $r_h = 2.99$ au and $\Delta = 3.72$ au. This allowed *Ye et al.* (2020) to identify precovery images in archival survey data dating back to December 2018 when it was at a heliocentric distance of 7.8 au.

6.2.2. Trajectory and potential origins. The significantly longer observation arc for 2I/Borisov allowed the measurement of standard cometary non-gravitational forces, implying that searches for progenitor systems might suffer less uncertainty than for 1I/'Oumuamua. However, *Hallatt and Wiegert* (2020) did not identify any systems with the golden combination of small encounter distance and low encounter velocity. *Bailer-Jones et al.* (2020) investigated a range of non-gravitational force models and found that 2I/Borisov was ~0.053–0.091 pc from the M0V star Ross 573 910 k.y. ago. They also demonstrated that such a close encounter is unlikely over this short time period. However, the predicted encounter velocity was 22.6 km s^{-1}, significantly higher than expected for ejection of ISOs, even assuming giant planet interactions (*Bailer-Jones et al.,* 2018).

Hence like 1I/'Oumuamua, the home system of 2I/Borisov remains unidentified. It is likely that unless the ejection was from a nearby star very recently, tracing the path to the home system is not possible. As soon as an ISO passes through a molecular cloud, backtracking is no longer possible.

6.2.3. Nuclear characteristics. Due to its activity, there were no direct detections of the nucleus of 2I/Borisov. In their archival analysis, *Ye et al.* (2020) did not detect 2I/Borisov in deep co-added images from November 2018. This gave a strong upper limit to the nucleus radius of $r \leq 7$ km assuming a geometric albedo of 0.04. Rigorous imaging constraints came from HST observations by *Jewitt et al.* (2020a), who measured a strong upper limit of $H_V \geq 16.60 \pm 0.04$ assuming a nominal cometary phase function of 0.04 mag/°. Fitting the coma surface brightness distribution showed no significant excess from a central nuclear point spread function, giving an improved upper limit to the nucleus radius of $r \leq 0.5$ km, or $H_V \geq 19.0$. *Hui et al.* (2020) calculated the non-gravitational parameters for a range of heliocentric distance laws, and combined them with outgassing measurements (see below) and an assumed nuclear density of $\rho = 0.5$ g cm^{-3} to derive $r \leq 0.4$ km. *Xing et al.* (2020) used measured water production rates to derive a minimum radius of $r \geq 0.4$ km assuming the entire sunward surface of the nucleus was active.

Several authors used apparent jet structures visible from HST imaging of the inner coma to attempt to constrain the spin axis. *Kim et al.* (2020) found it difficult to explain the overall dust coma anisotropy in terms of a single isolated emission source and preferred a model of general mass loss from the afternoon nuclear surface, giving a pole orientation of (α, δ) = (205°, +52°). From the same data, *Manzini et al.* (2020) interpreted the appearance of two jet/fan structures as implying localized emission near the rotation equator and derived a spin pole direction of (260°, –35°). Combining the *Jewitt et al.* (2020a) data with additional HST imaging, *Bolin and Lisse* (2020) interpreted the positional evolution of the jets as implying a near-polar source with a direction (322°, +37°). Although all three studies reported uncertainties of at least ±10°, the disparity between these results led *Bolin and Lisse* (2020) to conclude that it was not possible to sufficiently constrain the pole orientation with the extant data. The latter study also reported a tentative detection of a 5.3-hr periodicity in the light curve, but the low amplitude of ~0.05 mag implies a low significance. The lack of precise determinations of these nuclear properties are all unremarkable when compared to remote studies of solar system comets (see the chapter by Knight et al. in this volume).

The similarity of 2I/Borisov to solar system comets extended to the detection by Drahus et al. [The Astronomer's Telegram (ATel) 13549] of two outbursts on March 4 and 8, 2020, brightening the comet by ~0.7 mag. A sequence of HST imaging from March 23 to April 20 to investigate possible nuclear fragmentation was reported by *Jewitt et al.* (2020b). These images showed a discrete nuclear secondary on March 30, 26 days after the first outburst, that was not visible in images 4 days later. They posited this transient nature by the secondary consisting of one or more boulders ejected from the nucleus during the outburst, disrupting weeks later near the time of their observations due to rotational spinup.

The measured properties of 1I/ʻOumuamua and 2I/Borisov are summarized in Table 1.

6.2.4. Composition and evolution. The first characterizations of the coma of 2I/Borisov were obtained even before it was given a provisional designation, with early optical photometry and spectra of the coma giving a red reflectance spectrum typical of normal comets (*Jewitt and Luu,* 2019; *de León et al.,* 2019; *Guzik et al.,* 2020) and similar to

TABLE 1. A summary of measured properties of interstellar objects.

Quantity		1I/ʻOumuamua	Ref†	2I/Borisov	Ref†
Dynamical Properties					
Perihelion date	T_p	2017 September 09.51	[1]	2019 December 08.55	[11]
Perihelion distance	q (au)	0.256	[1]	2.007	[11]
Eccentricity	e	1.201	[1]	3.356	[11]
Earth close approach	Δ (au)	0.162	[1]	1.937	[11]
Incoming velocity*	v_∞ (km s^{-1})	26.420 ± 0.002	[2]	32.288	[11]
Non-grav acceleration	$A_1 \times 10^8$ (au d^{-2})	27.90	[1]	7.09	[11]
Non-grav acceleration	$A_2 \times 10^8$ (au d^{-2})	1.44	[1]	–1.44	[11]
Non-grav acceleration	$A_3 \times 10^8$ (au d^{-2})	1.57	[1]	0.065	[11]
Physical Properties					
Absolute magnitude	H_V	22.4 ± 0.04	[3]	>19.1	[12]
Albedo	p_V	0.01–0.2	[4]	—	
Radius (for p_V = 0.04)	r_N (m)	110	[3]	<500	[12]
Rotation state		complex, long-axis mode	[5]	—	
Rotation period	P (hr)	8.67 ± 0.34 hr	[5]	—	
Axis ratio	a:b	>6:1	[6]	—	
Spectral slope	S_V (% per 100 nm)	(7–23) ± 3	[6,7]	12 ± 1	[8]
H_2O production	$Q(H_2O)$ (molec s^{-1})	<6.1 × 10^{25} @ 0.38 au (obs)	[9]	1.1 × 10^{27} @ 2.0 au (obs)	[13]
H_2O production	$Q(H_2O)$ (molec s^{-1})	4.9 × 10^{25} @ 1.4 au (model)	[10]	—	
Hypervolatile (CO?)	Q(X) (molec s^{-1})	4.5 × 10^{25} @ 1.4 au (model)	[10]	—	
CO_2 production	$Q(CO_2)$ (molec s^{-1})	<9 × 10^{22} @ 2.0 au (obs)	[4]	—	
CO production	Q(CO) (molec s^{-1})	<1 × 10^{24} @ 2.0 au (obs)	[4]	4.4 × 10^{26} @ 2.0 au (obs)	[14]
Dust production	Q(dust) (kg s^{-1})	<10^{-3} @ 1.4 au (obs)	[3,7]	2–50 @ 2.6 au (obs)	[15,16]

*v_∞ is the velocity an object on a hyperbolic orbit has infinitely far from the central body (here, the Sun).

†References: [1] Osculating orbit: JPL Horizons orbital solution #16; [2] speed relative to the Sun while in interstellar space (*Bailer-Jones et al.,* 2018); [3] *Meech et al.* (2017b); [4] the CO production rate is corrected from the number reported by *Trilling et al.* (2018); [5] *Belton et al.* (2018); [6] *ʻOumuamua ISSI Team et al.* (2019); [7] *Jewitt et al.* (2017); [8] *Jewitt and Seligman* (2022); [9] *Hui and Knight* (2019); [10] *Micheli et al.* (2018); [11] JPL Horizons orbital solution #53; [12] *Jewitt et al.* (2020a); [13] *Xing et al.* (2020); [14] *Cordiner et al.* (2020); [15] *Jewitt and Luu* (2019); [16] *de León et al.* (2019).

For production rates of other species for 2I/Borisov, see the summary in *Jewitt and Seligman* (2022).

1I/'Oumuamua (*Fitzsimmons et al.,* 2018). Gas emission via the bright CN(0–0) emission band at 388 nm was first detected on September 20–24, 2019, by *Fitzsimmons et al.* (2019) and *de León et al.* (2019). Spectroscopy and narrow-band photometry over the following month was reported by *Opitom et al.* (2019). As with the earlier spectroscopy, they did not detect the C_2(0–0) emission band at 517 nm and concluded the comet was similar to carbon-chain-depleted solar system comets. However, weak detection of C_2(0–0) was reported by several observers from November (i.e., *Lin et al.,* 2020), and by perihelion on December 8, 2019, the C_2/CN ratio was formally consistent with both depleted and normal carbon-chain cometary abundances (see Fig. 4 of *Aravind et al.,* 2021). These spectra and those presented by *Opitom et al.* (2021) also showed other normal cometary molecular emission features due to species such as NH, NH_2, and CH. Hence the optical species showed a significant evolution during this time as the heliocentric distance decreased from 2.7 au to 2.0 au.

The first constraint on the sublimation rates of primary ice species came from detection of the [O I] 6300 Å line by *McKay et al.* (2020). Assuming this came solely from sublimation and subsequent dissociation of water, they derived a production rate of $Q(H_2O) = (6.3 \pm 1.5) \times 10^{26}$ mol s^{-1} on October 29 at $r_h = 2.38$ au. *Xing et al.* (2020) used the UltraViolet and Optical Telescope (UVOT) onboard the NASA Swift satellite to observe the OH(0–0) and (1–1) emission between 280 and 320 nm, again resulting from dissociation of water molecules. The derived water production rates showed a peak of $Q(H_2O) = (1.1 \pm 0.1) \times 10^{27}$ mol s^{-1} near perihelion, but with higher production rates pre-perihelion than post-perihelion.

Given the significant activity at $r_h \geq 3$ au, it was clear that other volatile species such as CO or CO_2 should be present in 2I. *Bodewits et al.* (2020) used HST to measure the CO fluorescence bands at 140–170 nm and corresponding production rates during and after perihelion. *Cordiner et al.* (2020) used ALMA to measure CO (J = 3–2) emission along with HCN (J = 4–3). Both these teams showed that the CO abundance was exceptionally high in 2I/Borisov, being on par with the H_2O abundance at $Q(CO)/Q(H_2O) \simeq 0.35$–1.55. Only rare CO-rich comets such as C/2016 R2 had previously exhibited such CO-rich compositions (see *McKay et al.,* 2019, and the chapter by Biver et al. in this volume).

Bodewits et al. (2020) found that the CO production rate stayed relatively constant around perihelion, while that of H_2O fell soon after, implying a significant change in the near-surface abundance ratios of these ices. This may have been due to seasonal effects of nucleus insolation, coupled with a non-heterogenous composition or erosion of the surface, similar to that seen by Rosetta in some species at Comet 67P/Churyumov-Gerasimenko (*Läuter et al.,* 2020), although for that comet the CO/H_2O ratio was relatively constant within 2.5 au. Similarly, as mentioned above, the abundance ratios of the primary optical species C_2/CN also changed significantly as the comet moved toward perihelion. Finally, shortly after the discovery of significant neutral metal line emission in optical spectra of non-sungrazing comets (*Hutsemékers et al.,* 2021), these were also found in spectra of 2I/Borisov (*Guzik and Drahus,* 2021; *Opitom et al.,* 2021). The abundance ratio of log[Q(NiI)/Q(FeI)] = 0.21 ± 0.1 was within the range measured for solar system comets, as was Q(NiI + FeI)/Q(CO).

Taken together, although several authors concluded that the optically active species were consistent with normal solar system comet abundances near perihelion, these studies point to a significantly evolving coma composition as 2I/Borisov passed the Sun. It is unclear to what degree this evolution was caused by its interstellar nature, although the CO-rich composition is clearly unusual.

6.3. Interstellar Object Number Densities and Size Distribution

With the discovery of 1I/'Oumuamua and 2I/Borisov, astronomers could finally estimate the true space density of ISOs in the solar neighborhood, albeit with large uncertainties given N = 2. A complication also arises given their different apparent nature during their encounters, as to whether they may come from different populations of ISOs. Inert ISOs like 1I/'Oumuamua appear to have a local space density of ~0.1 to 0.3 au^{-3} at H = 22 (*Meech et al.,* 2017b; *Do et al.,* 2018). This approximately agrees with pre-discovery upper limits for inert ISOs (*Engelhardt et al.,* 2017) when scaled to this absolute magnitude. Some authors reported a possible tension between the inferred mass density contained in ISOs and the galactic stellar density, with 1I/'Oumuamua implying an exceptional amount of mass loss from planetary systems (e.g., *Rafikov,* 2018). However, *'Oumuamua ISSI Team et al.* (2019) demonstrated that the space density of ~10^{15} pc^{-3} ≡ 0.1 au^{-3} was compatible with possible size distributions.

The space density of ISOs exhibiting normal cometary activity like 2I/Borisov is less well constrained, even though most earlier upper limits *assumed* cometary activity. *Engelhardt et al.* (2017) found a pre-2I/Borisov 90% upper limit of $\leq 1.4 \times 10^{-4}$ au^{-3} for a nuclear absolute magnitude $H \leq 19$, assuming the entire sunward surface was active. *Eubanks et al.* (2021) used the statistics of LPC discoveries, together with assuming that ISOs share the same velocity distribution as stars within 100 pc, to derive a lower density of 7×10^{-5} au^{-3}. This is only a factor of 2 less than *Engelhardt et al.* (2017) and is plausible given the extra survey time since that study.

If 1I/'Oumuamua and 2I/Borisov come from the same ISO population, then Fig. 10 provides constraints on the luminosity distribution. We use absolute magnitudes to avoid any assumptions on albedo, although it is important to remember these have been derived using assumed phase laws. If they share the same albedo then this also constrains the size distribution. For the interstellar space density from 1I/'Oumuamua we take the space density and confidence limits from *Levine et al.* (2021), and for 2I/Borisov we take

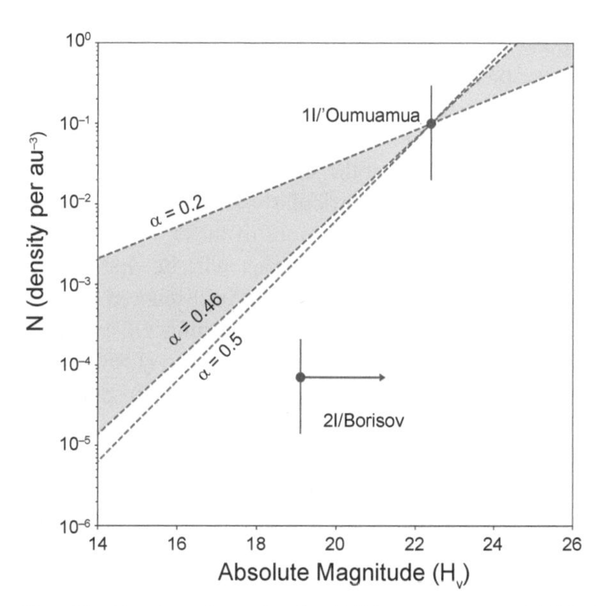

Fig. 10. Number densities for 1I/'Oumuamua and 2I/Borisov as a function of their absolute magnitudes. Also shown are differential brightness power-laws with exponents α = 0.5 for collisional equilibrium, and the range α = 0.46 to 0.2 measured for solar system comets. All are anchored to the absolute magnitude of 1I/'Oumuamua.

the space density of *Eubanks et al.* (2021) together with Poisson uncertainty ranges calculated for a single detection at both magnitudes.

We assume a single power-law relationship for the number density against size, as with only two points a more complex relationship is unwarranted. An obvious size distribution to test is the theoretical differential size distribution for a population in collisional equilibrium, where the number of objects at a radius R is given by $dN/dR \propto R^{-q}$ with q = 3.5 (*Dohnanyi,* 1969). A second size distribution would be that measured for solar system comets, normally given as a cumulative size distribution $N(>R) \propto R^{1-q}$. Several investigators have found $q \simeq 2.9$ (*Meech et al.,* 2004; *Snodgrass et al.,* 2011; *Fernández et al.,* 2013). A more recent analysis of Near-Earth Object Wide-field Infrared Survey Explorer (NEOWISE) comet observations by *Bauer et al.* (2017) found a debiased value of q = 3.3 for JFCs similar to the earlier studies, but a shallower debiased q = 2.0 for LPCs. Note that in comparison with the mass distribution $N(m) \propto m^{-p}$ used in section 4.2, the power-law exponents are related by q = (3p–2).

If we transform these size distributions to differential absolute magnitude distributions $N(H) \propto 10^{\alpha H}$, the exponents of the two forms are related by $\alpha = (q-1)/5$. The above size distributions imply a theoretical magnitude distribution with α = 0.5, and measured values of $\alpha = 0.2 \rightarrow 0.46$. It is clear from Fig. 10 that 1I/'Oumuamua and 2I/Borisov do not match these distributions.

There are a number of possible explanations. First, they could be from different populations of ISOs, each of which has a more standard size distribution but with significantly different space densities. If they instead reflect the true size distribution of ISOs, this implies the size distribution is steeper than expected. As explained in section 4.2, ejection processes can produce an increase in the relative number of smaller bodies, consistent with Fig. 10. An ejection-produced steepening of the size distribution would also explain the mismatch with the observed LPC population that originates from our Oort cloud, where objects have undergone similar long-term exposure to the ISM but not ejection. Finally, there could be a strong variation in albedo between bodies, possibly due to their individual histories in the ISM or, with 1I/'Oumuamua, by its close perihelion passage pre-discovery. Identifying which of these explanations is most likely will require further ISO detections for a much better measurement of the magnitude/size distribution.

6.4. CNEOS 2014-01-08

While 1I/'Oumuamua and 2I/Borisov are *macroscopic* ISOs, the existence of interstellar dust particles — *microscopic* ISOs — has been known for some time from spacecraft detection (see the review by *Sterken et al.,* 2019). The Advanced Meteor Orbit Radar multi-station complex has provided extensive data on interplanetary dust characteristics. The dynamical information that is collected allows identification of discrete source regions. One of the main sources appears to be the β Pic disk (*Baggaley,* 2000). During its cruise phase, the Stardust mission captured particles from the oncoming interstellar dust stream (*Westphal et al.,* 2014).

In between dust and ISO sizes should lie objects ~1 cm to 10 m in size. Fireballs produced by objects at these sizes are regularly detected via U.S. Department of Defense (DoD) satellites, which measure both the luminosity of the fireball and (above a luminosity limit) the entry velocity vector. A search through the publicly accessible fireball catalog led *Siraj and Loeb* (2022) to identify a potential interstellar impactor, CNEOS 2014-01-08. This object entered Earth's atmosphere over the western Pacific with a reported entry velocity of 44.8 km s^{-1}. Although formal measurement uncertainties are not published, DoD personnel communicated that "the velocity estimate reported to NASA is sufficiently accurate to indicate an interstellar trajectory." When integrated backward, the reported velocity vector implied an original orbit with e = 2.4 ± 0.3, hyperbolic at the 3σ level.

While the identification of an interstellar origin appears sound based on the available satellite data, there is a tension with the fireball and meteor data regularly obtained by groundbased optical and radar meteor surveys. *Musci et al.* (2012) identified 2 hyperbolic orbits out of 1739 meteors observed with the Canadian Automated Meteor Observatory, but careful inspection ruled them out as measurement error. Similarly, in an analysis of 824 fireballs detected by the European Fireball Network, *Borovička et al.* (2022) identified two objects that had orbital eccentricities e > 1 at the 3σ level, but the absolute values were near unity and they concluded there was no strong evidence for a hyperbolic nature. In an analysis of nearly 4000 meteors observed by the Fireball Recovery and InterPlanetary Network (FRIPON) project, *Colas et al.* (2020) report an upper limit of interstellar meteors of 0.1% but suspect the true value to be much lower. Collating several meteor orbit catalogs derived from photographic, video, and radar systems, *Hajdukova et al.* (2020) concluded that there was no convincing evidence of any interstellar meteors being detected in 25 years of data covering many thousands of objects. That said, a careful analysis of 4 × 10^5 meteors detected by the Canadian Meteor Orbit Radar led *Froncisz et al.* (2020) to identify five possible interstellar meteors. We note that this possible interstellar fraction of ~1 in 80,000 meteors is significantly lower than the ~1 in 900 implied for the current Center for Near Earth Object Studies (CNEOS) fireball database.

It is clear from these latter studies that measurement uncertainties are extremely important in interstellar meteor identification, due to the short timespan over which the observations are obtained. Until such a time that a quantitative description of the uncertainties on the DoD satellite measurements is forthcoming (and indeed their actual values), some uncertainty unfortunately still remains concerning an interstellar origin for CNEOS 2014-01-08. It therefore follows that the pursuit of interstellar meteor detections with quantifiable uncertainties is a highly worthwhile endeavor.

6.5. Interstellar Object Impact Hazard

As reported in the Planetary Decadal survey (*National Academies of Sciences, Engineering, and Medicine* 2022), as of 2021 about 95% of the NEOs greater than 1 km in diameter have been found. These objects are capable of causing global effects upon impact. Objects that are larger than 140 m in diameter can cause regional destruction and to date, only about 33% of these have been discovered. Estimates show that there might be more than 10^5 objects ≥50 m in diameter that could cause destruction on urban scales, and less than 2% of these have been found. In 2005 Congress directed NASA to find 90% of all NEOs larger than 140 m, and to date this directive has not been met. Once it is met, the threat from impacts from near-Earth asteroids will be minimized, but this will not address the risk from objects on long-period Oort cloud comet (OCC) trajectories. It is likely that OCCs are responsible for most of the very large impacts on Earth (*Jeffers et al.,* 2001).

The typical NEO encounter velocity peaks around 15 km s^{-1}, with a range up to 40 km s^{-1} (*Heinze et al.,* 2021); however, the encounter velocities for OCCs will typically be around 54 km s^{-1} ranging from 16 up to 72 km s^{-1} (*Jeffers et al.,* 2001). ISO trajectories will be very similar to those of OCCs, but the encounter velocities can be even higher. At the time of close approach to Earth on October 14, 2017, the relative velocity of 1I/'Oumuamua with respect to Earth was 60.3 km s^{-1}. It passed within 63 Earth-Moon distances and we did not and could not have discovered it until after it had passed. Objects like 1I/'Oumuamua that are on OCC-like orbits but that do not have any detectable activity, have small nuclei, and have low albedo [such as the Manx comets (see *Meech et al.,* 2016)] are particularly dangerous because they will be harder to detect. *Heinze et al.* (2021) also noted that there is a larger bias than expected against finding small fast-moving objects because of the streak length on the detector, so inactive ISOs will be particularly hard to find, and we are more likely to be missing more of these.

These OCC and ISO trajectories will be distributed across a wide range of inclinations. As discussed in section 7.3.1, the powerful new Vera C. Rubin Observatory (Rubin) Legacy Survey of Space and Time (LSST) will begin in mid-2025. There is a concerted effort to optimize the survey effort to benefit solar system science, including detection of OCCs, NEOs, and potentially hazardous objects (*Schwamb et al.,* 2023). 1I/'Oumuamua was discovered at 1.22 au moving at a rate of 6.2° per day, i.e., typical of a faint nearby NEO, which the LSST will be optimized to detect.

7. THE NEXT DECADE OF EXOCOMET AND ISO STUDIES — WHAT DON'T WE KNOW?

7.1. Exocomet Systems

*7.1.1. **Probing outer exocomet populations.*** Current and under-construction facilities should lead to significant advances in the field of exocomet science. An important area of study will be to understand the dynamical evolution of exocomets in planetary systems, which requires both high angular resolution and sensitivity at dust-emission wavelengths. NIR to MIR dust imaging will take place with the 6.5-m James Webb Space Telescope (JWST); the 39-m European Southern Observatory (ESO) Extremely Large Telescope (ELT) with its Multi-AO Imaging Camera for Deep Observations (MICADO) and Mid-infrared ELT Imager and Spectrograph (METIS) (*Brandl et al.,* 2010; *Davies et al.,* 2018); and large, deep ALMA surveys such as the ongoing ALMA survey to Resolve exo-Kuiper belt Substructures (ARKS) (Marino et al., in preparation). These will deliver unprecedentedly detailed images of dust emission from exocometary belts, revealing planet-belt interaction and allowing us to infer the dynamical fate of exocomets within belts, be it ejection, inward scattering, or survival within the belt.

Gas observations offer a unique window into exocometary compositions in young planetary systems just after formation. NIR to MIR spectroscopy with JWST combined with groundbased NIR high-resolution spectrographs [like the newly-installed CRyogenic high-resolution InfraRed Echelle Spectrograph (CRIRES)+ instrumentation on the VLT and later the High Angular Resolution Monolithic Optical and Near-infrared Integral field spectrograph (HARMONI) on the ELT] will be used to probe ro-vibrational transitions of new molecules undetectable at millimeter wavelengths, such as CO_2, CH_4, and of course H_2O and its photodissociation product OH. This will constrain exocometary molecular gas compositions within outer belts, with the potential to confirm the gas origin, determine the physical release mechanism, and ultimately determine the ice composition of young exocomets, a missing evolutionary link between outer protoplanetary disks and solar system cometary/KBO compositions.

In addition, ALMA CO and C I line imaging will constrain the radial and vertical distribution and hence the evolution of gas within exocometary belts; this will be combined with detailed kinematic analysis, thanks to ALMA's maximum spectral resolution, to reveal the environmental conditions (temperature, bulk density) of the gas and evaluate the dynamical influence of planets on the evolution of exocometary gas.

7.1.2. Probing the inward transport of exocomets. The highest priority for the future of exocomets in the inner regions should be to expand the number of detections and detected systems — currently dominated by β Pic — and study their compositions and dynamics as they enter the innermost regions of a planetary system. This will enable us to probe their formation location and the mechanism leading to their inward transport, with the potential of linking their origin to detected exoplanets and imaged populations of exocomets in outer belts.

JWST is especially sensitive to warm dust emission in the interior regions of nearby systems, either through its direct imaging or aperture-masking modes, and this can be used to set limits on inward scattering and transport of icy exocomets from the outer belt toward the terrestrial region (*Marino et al.,* 2018a). The Roman Space Telescope is a 2.4-m wide-field optical and NIR facility, currently planned to launch in 2027. Its coronagraph instrument is also expected to expand the number of systems with exozodiacal light detections in the terrestrial planet region (*Douglas et al.,* 2022). Imaging of exozodiacal dust may for the first time also be possible around the nearest stars with the ELT (*Roberge et al.,* 2012).

The identification and confirmation of more β Pic-like systems requires high-resolution echelle spectrographs with large-aperture telescopes to provide high signal-to-noise per wavelength bin. Observations with new high-resolution spectrographs on the ELT such as the ArmazoNes high Dispersion Echelle Spectrograph (ANDES) (*Maiolino et al.,* 2013) and METIS (*Brandl et al.,* 2010) will exploit the ELT's large collecting area with their high spectral (R ~ 100,000) resolution and wide wavelength coverage. This will allow more sensitive searches for absorption in edge-on systems, both from narrow, stable gas at the stellar velocity and variable red- and blue-shifted features from exocometary gas. Monitoring observations looking for high-velocity exocomet features should aim to cover volatiles as well as metallic atomic species across a broad wavelength range, as one of the highest priorities is to understand the nature and volatile content of star-grazing exocomets.

Finally, continued exploration of the Transiting Exoplanet Survey Satellite (TESS) and CHaracterising ExOPlanets Satellite (CHEOPS) stellar light curves to look for asymmetric exocomet transits are likely to lead to new discoveries. These will be further supported by the PLAnetary Transits and Oscillations of stars (PLATO) exoplanet mission (*Rauer et al.,* 2014), with its factor of ~5 increased photometric accuracy over TESS, currently scheduled for launch in 2026. This will contribute to understanding the occurrence rate of star-grazing exocomets around large samples of stars, pushing toward smaller transits.

7.2. Galactic Evolution of Interstellar Objects

7.2.1. Galactic effects on interstellar objects. In section 4.2 we saw that ISOs experience various influences from the environment in the form of thermal heating and exposure to radiation and particles of various kinds. The question is whether ISOs also affect the environments they are passing through. The discovery of 1I/'Oumuamua taught us that the ISM contains not only gas and dust but that ISOs are a natural third component of the ISM.

ISOs also pass through molecular clouds (see section 4.3), with young ISOs more likely to make such an experience than older ones (*Pfalzner et al.,* 2020). Such a cloud passage might alter not only the ISO surface and structure but also the direction of their path, while at the same time decelerating or accelerating them. Depending on the size of the actual molecular cloud 10^{19}–10^{20} ISOs reside in them at any time. Being so small in mass, their individual effect is negligible, but it could be different if co-moving streams of ISOs pass through molecular clouds.

However, although every ISO travels through the ISM, certainly not every ISO passes through a molecular cloud and even a protoplanetary disk. However, the probability of ISOs passing through molecular clouds is surprisingly high. In the solar neighborhood, an ISO spends ≈0.1–0.2% of its journey passing through molecular clouds (*Pfalzner et al.,* 2020). This value increases for young ISOs close to the galactic center. In comparison, passing through an existing protoplanetary disk is a much rarer event due to its much smaller cross section. Nevertheless, *Grishin et al.* (2019) estimate that at least 10^4 ISOs larger than 1 km cross any protoplanetary disk around field stars, with the number increasing to 10^5 for star-cluster environments.

Under certain conditions, molecular clouds collapse and form stars. It seems that ISOs take part in this process to some extent. This would mean that ISO could become concentrated in collapsing molecular clouds (*Pfalzner et al.,* 2021b). However, these simulations indicate that there is a competitive capture process at work that favors the capture of ISOs by massive star clusters.

In the current planet-formation scenario, there exist some difficulties in proceeding from meter-sized to kilometer-sized objects (see section 2.3). It has been suggested that ISOs might easily overcome the meter-sized growth barrier by acting as seeds to catalyze planet formation. Two different scenarios have been suggested. *Grishin et al.* (2019) theorize that ISOs could be captured from the local star-formation region into the disks surrounding young stars. Alternatively, *Pfalzner and Bannister* (2019) propose the ISOs in the ISM taking part in molecular cloud collapse would become a natural component of the forming disk, without the need for additional capture. They give a first conservative estimate on the order of 10^{11} ISOs typically being incorporated into forming star-disk systems. *Moro-Martín and Norman* (2022) found a similar total number when considering meter-scale ISOs and larger, but also noted the number in the disk could increase by 2–3 orders of magnitude in cluster environments. If ISOs take part in cloud collapse leading to star formation, this would also mean that ISOs are incorporated into forming stars. The question is, how many ISOs end up in stars?

When ISOs pass through molecular clouds and are captured in planet-forming disks, they can be affected by these environments. Possible mechanisms are devolatilization and erosion. Both will lead to changes in the ISO size distribution. Whether the ISOs are altered depends primarily on the gas/dust density, which increases from an ISM environment to molecular clouds and protoplanetary disks.

Finally, *Do et al.* (2018) point out that exo-Oort clouds around stars that produce supernovae will be irradiated and lose surface volatiles, but the exocomets there will survive. As they then drift away from the supernova remnant they will form a natural population of devolatilized ISOs, possibly somewhat similar to 1I/'Oumuamua. However, supernovae are rare events, and such ISOs should not form a major component of the population.

7.2.2. Interstellar object effects on the galaxy. Over the last four years, it has become apparent that ISOs are a natural component of molecular clouds. The question is whether the presence of the huge number of low-mass ISOs does "contaminate" star-forming regions. This depends on the mass of the ISOs present in molecular clouds, but also on the different chemistry between star-forming systems and on whether ISOs are evenly spread or come in streams from their parent systems.

So far, this question has been studied more from the perspective of individuals already forming planetary systems. This question of whether material can be spread from one planetary system to another is of long-standing interest (*Valtonen and Innanen,* 1982; *Melosh,* 2003; *Adams and Spergel,* 2005; *Brasser et al.,* 2006). More recently, estimates of the likelihood of ISO capture events have been performed for the early phase in young star-forming clusters and the local galactic neighborhood (*Hands et al.,* 2019; *Portegies Zwart,* 2021; *Napier et al.,* 2021). In young clusters the stellar density is much higher than in the field, and therefore the ISO capture rate per million years is naturally much higher. However, whether the capture probability is in general higher during the short cluster phase than over the entire lifetime of a star depends strongly on the assumed cluster properties and the stellar mass. The latter determines the capture cross section and the lifetime of a star. Besides, Gaia data strongly indicate that clusters seem not completely dissolved within 10 m.y. but move together in an unbound state for a considerable length of time. Thus, assuming field densities for the stars and the ISOs underestimates the capture rates in the first few hundred million years. This requires further investigation in the future.

On average, every star will contribute ISOs to the galactic population, and this material might be injected into the inner regions of star systems. *Seligman et al.* (2022a) suggest that this material can impact both the host stars and their planets, enriching their atmospheres, and that understanding the composition of the ISO population will have implications for post-formation exoplanet atmospheric composition.

Currently there exists no definitive evidence that any gravitationally bound objects in the solar system are of extrinsic origin (*Morbidelli et al.,* 2020). However, the discovery of 1I/'Oumuamua and 2I/Borisov sparked wide-ranging speculation regarding the possibility that our solar system is being more broadly contaminated by minor bodies of extrasolar origin (*Siraj and Loeb,* 2019; *Namouni and Morais,* 2020; *Hands and Dehnen,* 2020). There exists a wide range in the estimated masses of planetesimals accreted from other stars while the Sun lived within its birth cluster from 10^{-5} $M_\oplus$ to a third of the Oort cloud population. *Napier et al.* (2021) finds that about 10^6 more ISOs were captured during the cluster phase than accreted subsequently from the field, while the current steady state factor is 10^4 and the total mass of surviving captured ISOs is $\sim 10^{-9}$ $M_\oplus$. To put this into perspective, the mass of the Oort cloud is estimated to range from 0.75 ± 0.25 $M_\oplus$ (*Brasser,* 2008) to 1.9 $M_\oplus$ (*Weissman,* 1996). With a typical comet mass of a few times 10^{12}–10^{14} kg (*Sosa and Fernández,* 2011) the Oort cloud contains approximately 10^{10}–10^{12} comet-sized objects. Thus, the captured ISO population is probably extremely small in comparison to the primordial comet population in the Oort cloud.

Current capture of ISOs into the solar system has also been explored. *Dehnen et al.* (2022) suggest that planets capture ~2 ISOs every 1000 years, which would result in 8 ISOs captured within 5 au of the Sun at any time. A perennial candidate for a captured ISO is Comet 96P/Machholz due to its highly anomalous optical spectrum, which is dominated by NH_2 with strong depletions of carbon-based molecules (*Langland-Shula and Smith,* 2007; *Schleicher,* 2008). Its orbital elements of $q = 0.124$, $i = 58°$, and $e = 0.959$ mark it as one of the lowest-perihelia comets known, although its high inclination is due to strong Kozai-Lidov oscillations (*McIntosh,* 1990). Its nucleus was studied by *Eisner et al.* (2019) to compare with 1I/'Oumuamua to search for possible effects of small perihelion passages. The strongest evidence for it being a captured ISO remains its peculiar coma composition, but this is not a definitive marker.

7.2.3. Intergalactic objects. The existence of ISOs in the Milky Way makes it likely that other galaxies are also populated with large numbers of ISOs. One can also speculate about the number of objects that have been ejected from their parent galaxy and become intergalactic objects. These would clearly be difficult to create, given the huge velocities required; at the Sun's position the escape velocity from the Milky Way is ~500 km s^{-1} (*Koppelman and Helmi,*

2021). While some tens of hypervelocity stars escaping the Milky Way have now been identified (*Li et al.,* 2021), most are thought to have attained their velocities via encounters with massive black holes, whose dynamical and radiation environments are hardly conducive for ISO survival.

However, we note that while escape to intergalactic space is highly unlikely for an ISO, we can let the galaxy carry it to a new home. Galactic mergers have taken place throughout the history of the universe, with five minor mergers identified for the Milky Way (*Kruijssen et al.,* 2020). An unknown fraction of the ISOs in those galaxies would have merged with the galactic population, and of course subsumed stellar systems would have ejected their own ISOs through the usual methods described in section 4. Hence given enough time as ISO detections increase, there is a finite but as-yet-unquantified probability of detecting an ISO carried here by its own galaxy.

7.3. The Range of Interstellar Object Properties

7.3.1. Future detections and surveys. The trajectory of 1I/'Oumuamua brought it in from above the plane of the solar system from the direction of the constellation Lyra (Fig. 11). It came to perihelion inside the orbit of Mercury at q = 0.255 au on September 9, 2017, and was discovered post-perihelion at r = 1.22 au after it had made its close approach to Earth on October 14, 2017. This close approach was at 0.162 au, passing within 63 lunar distances. Images from the Catalina Sky Survey detected it on October 14, but it was noted only after it had been discovered.

1I/'Oumuamua passed through the Solar and Heliospheric Observatory (SOHO) and Solar Terrestrial Relations Observatory (STEREO) fields of view in early September 2017, near perihelion (*Hui and Knight,* 2019). Because of the large phase angle and the extreme forward scattering, this would have enhanced the brightness of any dust around the nucleus by ~2 orders of magnitude but neither satellite detected it. It could not have been discovered any earlier than it was because 1I/'Oumuamua was just coming out of solar conjunction as seen from the ground and was faint (mag ~ 23.7 at 0.79 au on October 2). Prior to moving into solar conjunction in early August it would have been fainter than mag ~ 25 — both times too faint to be discovered by any of the existing surveys. 1I/'Oumuamua was inside the orbit of Jupiter for less than 1.3 yrs. The second ISO was discovered by an amateur searching close to the horizon in twilight — something that large telescope surveys cannot do (see the chapter by Bauer et al. in this volume).

The Rubin observatory in Chile has an 8.3-m survey telescope with a camera that covers a field of view of 9.6 deg^2. It will begin regular survey operations in mid-2025 that will provide a much greater depth than all previous surveys, reaching mag ~ 24.7 and scanning the accessible sky every three days (*Jones et al.,* 2009). *Hoover et al.* (2022) have estimated the number of discoveries that might be made by the LSST by generating a synthetic population of ISOs, assuming no cometary activity. They predict that the survey will find on the order of 1 ISO per year, but the number can be larger than 1 when cometary activity is considered. Ironically, however, even if the LSST survey had been operational in 2017, it would have been unlikely to discover

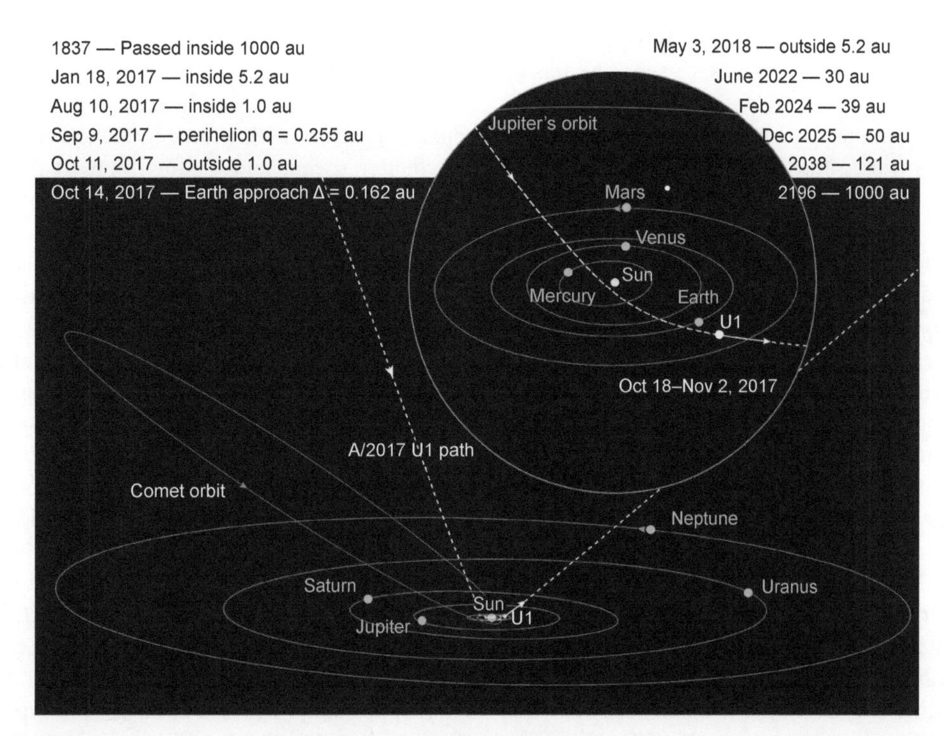

Fig. 11. Path of 1I/'Oumuamua as it entered the solar system, illustrating why it was not discovered sooner.

1I/'Oumuamua because the ISO would have had an *average* brightness brighter than mag ~ 24.7 for less than a week before moving into solar conjunction in early August 2017.

7.3.2. ***Probing origins and evolution.*** The standard mechanisms of ISO creation via ejection of exocomets from their home systems are described in detail in section 4. 1I/'Oumuamua's discovery resulted in the proposal of additional possible creation routes. Many of the formation mechanisms have been suggested in an attempt to explain the origin of the unusual shape for 1I/'Oumuamua. These include fluidization to a Jacobi ellipsoid during the red giant phase of a star (*Katz,* 2018), interstellar ablation (*Vavilov and Medvedev,* 2019), collisional elongation (*Sugiura et al.,* 2019), formation as an icy fractal aggregate (*Moro-Martín,* 2019a) and a tidally disrupted planetesimal that passed close to growing giant planets from which the volatiles were removed during a close stellar passage (*Raymond et al.,* 2018a; *Zhang and Lin,* 2020).

Two 1I/'Oumuamua origin scenarios suggested a volatile composition based on homonuclear diatomic molecules with no dipole moments and no vibrational spectral lines, which would explain why no gas was detected. In an attempt to both explain the shape and acceleration of 1I/'Oumuamua, *Seligman and Laughlin* (2020) suggested that it was composed of molecular hydrogen ice. In their model, mass wasting from sublimation far out in the solar system could explain both the shape and the acceleration. They proposed that the ISO formed in a cold dense molecular cloud core. The freezing point of H_2 is around 14 K, and the lowest-temperature cloud cores are around 10 K. H_2 has no dipole moment and can only be detected in the far-UV or in the IR through its rotational lines from space. An alternate suggestion was made by *Desch and Jackson* (2021) who proposed that 1I/'Oumuamua was a collisional fragment from an exo-Kuiper belt. Many large KBOs in our solar system exhibit N_2 ice on their surfaces. Gaseous N_2 has no bending mode vibration spectral lines and is IR inactive. Its spectral features are in the UV where no observations were taken. This makes it an attractive volatile to explain how 1I/'Oumuamua could have unobserved outgassing. *Levine et al.* (2021) suggests that neither scenario is plausible because large H_2-ice bodies are not likely to form in cloud cores and that the impacts in a Kuiper belt are not sufficient to generate large N_2 fragments.

In order to reconcile some of the observational and model inconsistencies, *Bergner and Seligman* (2023) propose that the acceleration of 1I/'Oumuamua is caused by release of molecular hydrogen that formed by cosmic-ray processing of water ice. These authors argue that laboratory experiments show that H_2 is efficiently produced in water ice during radiation by high-energy particles. They propose that the H_2 gas is released over a wide range of temperatures during annealing of amorphous water ice and that there is sufficient gas released to account for the observed non-gravitational acceleration.

Finally, regarding origin location within a disk, *Seligman et al.* (2022b) propose that measuring the C/O ratio of an ISO can be a tracer of whether the ISO formed inside or outside the ice line in its home star system.

7.4. *In Situ* Observations: Space Missions

1I'/Oumuamua was accessible to groundbased telescopes for less than a month, a little longer using space facilities. After this brief period of observation, it was found that the characteristics were quite different from what was expected from the first ISO, and this one discovery has energized a new interdisciplinary awareness in the study of planet formation. However, a more detailed investigation of ISOs with an *in situ* mission presents unique challenges: The orbits may have high inclinations and the encounter speeds are typically high (tens of kilometers per second), which means a short encounter. There are increased risks with high-velocity flybys if the ISO ejects large dust particles. The Giotto spacecraft was destabilized with a centimeter-sized impactor at a relative velocity of 68 km s^{-1} (*Bird et al.,* 1988).

NASA's competed mission calls are not compatible with missions that are responsive to new discoveries, i.e., missions that do not have a target at the time they are proposed. This is relevant for ISOs as well as OCCs, Manx comets, and hazardous NEOs. Two approaches have been suggested to explore these targets: spacecraft in storage, ready to launch following target discovery and spacecraft in a standby orbit, as is being done by ESA's Comet Interceptor mission (see the chapter by Snodgrass et al. in this volume, as well as *Snodgrass and Jones,* 2019; *Sánchez et al.,* 2021). Typically, the target will be unknown at the time of spacecraft development, and this has an effect on the definition of basic spacecraft capability (e.g., Δv) and on the payload. Launch following the discovery of an ISO offers greater flexibility in terms of target access but requires a very fast turnaround of a launch vehicle. A spacecraft in a standby orbit is more responsive to a target but has a more limited target accessibility.

Using known OCC orbits, an assessment of the phase space of targets with available launch energies, maximum encounter speed, and — for times of flight of less than 10 years — launching while the comet is more than 0.5 years from coming to perihelion shows that for low launch energies, C_3, there are almost no targets. (In astrodynamics, the characteristic energy C_3 km^2 s^{-2} is the measure of the excess specific energy over that required to just barely escape a massive body; $C_3 = v_\infty^2$.) There are also almost no targets unless the encounter speeds are greater than 10 km s^{-1}. This suggests that at present only fast flyby missions are an option for ISOs. For a comprehensive mission, this likely requires significant advancements in payload capabilities (e.g., small deployable satellites and autonomous navigation) (*Donitz et al.,* 2021, 2023).

There were several concepts developed for missions that could reach 1I/'Oumuamua. *Seligman and Laughlin* (2018) estimated that for launch-on-detection scenarios, the wait time would be on the order of 10 years for a favorable mission opportunity. *Hibberd et al.* (2020) explored a range of flyby trajectories that could launch in the early 2030s and deliver a spacecraft to 1I/'Oumuamua with relative velocities between 15 and 20 km s^{-1}, arriving between 2048 and 2052. Finally, *Miller et al.* (2022) proposed high-performance solar-sail scenarios that would allow for a rendezvous with an ISO by maintaining a high-potential-energy position until the detection of an ISO and then matching velocity through a controlled fall toward the Sun.

The LSST survey will increase the number of discoveries by going much fainter and therefore to larger discovery distances. As shown by *Engelhardt et al.* (2017), ISOs are more common at larger perihelion distances. The fainter limiting survey magnitude would enable detection of a 1-km inactive nucleus with 4% albedo out to 5.5 au, active ISOs much further. Although most detected ISOs may be too distant to easily reach with spacecraft, 1I/'Oumuamua and 2I/Borisov demonstrate ISOs exist that come within 2 au, and detection of an incoming ISO at large distances may provide enough warning time for either the storage or standby concepts discussed above.

8. SUMMARY

From the multitude of studies described in this chapter, it is clear that exocomets should be and are common in nearby stellar systems. Together with the physical mechanisms by which they can be lost to interstellar space, the local stellar neighborhood should be rich in ISOs, a prediction finally confirmed through the discovery of 1I/'Oumuamua and 2I/Borisov. So, the question remains, how is 1I/'Oumuamua not visibly active while 2I/Borisov is strongly active; is this due to origin, age, or the unseen 1I/'Oumuamua perihelion passage? From models of our solar system (*Shannon et al.,* 2015), some fraction of ISOs will not be cometary but more like C- or S-type asteroids. Based on the observation of Manx comets (*Meech et al.,* 2016), we should be able to distinguish asteroidal objects from inert comets. Is there a continuum of properties for ISOs, or have we already found two separate populations arriving from different origin mechanisms? This sample of two very different ISOs makes it difficult to predict what will 3I be like. Will it be like planetesimals from our solar system or something different (see Fig. 12)? Given the increasing sensitivity of sky surveys, the current and next generation of optical, IR, and submillimeter facilities, plus the theoretical advances in cometary/exocomet structure and evolution, these open questions have a chance of being answered in *Comets IV.*

Acknowledgments. We thank the reviewers, M. Knight, S. Raymond, and an anonymous referee, for the very thorough and helpful reviews on a very short timescale! K.J.M. acknowledges support through NASA Grant 80-NSSC18K0853 and acknowledges support for HST programs GO/DD-15405, -15447, -16043, -16088, and -16915 provided by NASA through a grant from the Space Telescope Science Institute, which is operated by the Association of Universities for Research in Astronomy under NASA contract NAS 5-26555. A.F. acknowledges support from UK STFC award ST/X000923/1. Figure 5 is based on data obtained from the ESO Science Archive Facility, *https://doi.org/10.18727/archive/33.*

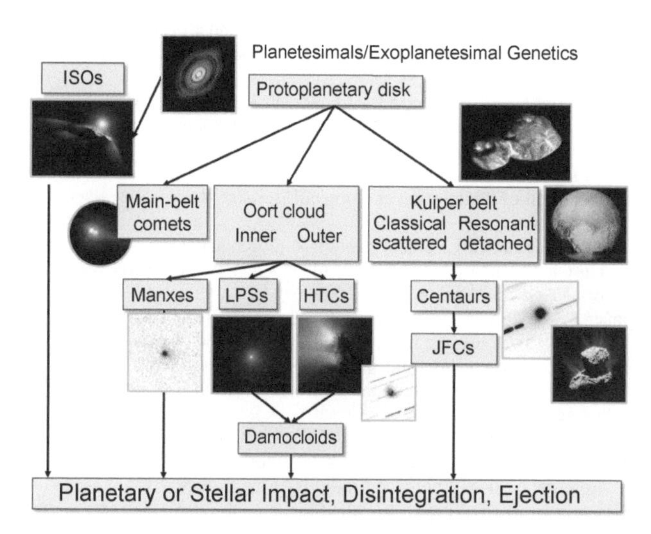

Fig. 12. Genetic relationships of early solar system planetesimals both in our solar system and from exoplanetary systems. We are just at the beginning of exploring the ISO population. Objects on OCC orbits, including ISOs, represent the largest reservoir of objects that has not been explored with an *in situ* mission. LPC = long-period comet, HTC = Halley-type comet, JFC = Jupiter-family comet.

REFERENCES

Abedin A. Y., Kavelaars J. J., Petit J.-M., et al. (2022) OSSOS. XXVI. On the lack of catastrophic collisions in the present Kuiper belt. *Astron. J., 164,* 261.

Adams F. C. (2010) The birth environment of the solar system. *Annu. Rev. Astron. Astrophys., 48,* 47–85.

Adams F. C. and Spergel D. N. (2005) Lithopanspermia in star-forming clusters. *Astrobiology, 5,* 497–514.

Adams F. C., Proszkow E. M., Fatuzzo M., et al. (2006) Early evolution of stellar groups and clusters: Environmental effects on forming planetary systems. *Astrophys. J., 641,* 504–525.

Andrews S. M. (2020) Observations of protoplanetary disk structures. *Annu. Rev. Astron. Astrophys., 58,* 483–528.

Andrews S. M., Terrell M., Tripathi A., et al. (2018) Scaling relations associated with millimeter continuum sizes in protoplanetary disks. *Astrophys. J., 865,* 157.

Ansdell M., Williams J. P., van der Marel N., et al. (2016) ALMA survey of Lupus protoplanetary disks. I. Dust and gas masses. *Astrophys. J., 828,* 46.

Apai D., Schneider G., Grady C. A., et al. (2015) The inner disk structure, disk-planet interactions, and temporal evolution in the β Pictoris system: A two-epoch HST/STIS coronagraphic study. *Astrophys. J., 800,* 136.

Aravind K., Ganesh S., Venkataramani K., et al. (2021) Activity of the first interstellar Comet 2I/Borisov around perihelion: Results from Indian observatories. *Mon. Not. R. Astron. Soc., 502,* 3491–3499.

Armitage P. J. (2018) A brief overview of planet formation. In *Handbook of Exoplanets* (H. J. Deeg and J. A. Belmonte, eds.), pp. 2185–2203. Springer, Cham.

Asphaug E. and Benz W. (1996) Size, density, and structure of Comet Shoemaker-Levy 9 inferred from the physics of tidal breakup. *Icarus, 121,* 225–248.

Augereau J. C. and Beust H. (2006) On the AU Microscopii debris disk: Density profiles, grain properties, and dust dynamics. *Astron. Astrophys., 455,* 987–999.

Aumann H. H., Gillett F. C., Beichman C. A., et al. (1984) Discovery of a shell around Alpha Lyrae. *Astrophys. J. Lett., 278,* L23–L27.

Backman D. E. and Paresce F. (1993) Main-sequence stars with circumstellar solid material: The Vega phenomenon. In *Protostars and Planets III* (E. H. Levy and J. I. Lunine, eds.), pp. 1253–1304. Univ. of Arizona, Tucson.

Baggaley W. J. (2000) Advanced Meteor Orbit Radar observations of interstellar meteoroids. *J. Geophys. Res.–Space Phys., 105,* 10353–10362.

Bailer-Jones C. A. L., Farnocchia D., Meech K. J., et al. (2018) Plausible home stars of the interstellar object 'Oumuamua found in Gaia DR2. *Astron. J., 156,* 205.

Bailer-Jones C. A. L., Farnocchia D., Ye Q., et al. (2020) A search for the origin of the interstellar Comet 2I/Borisov. *Astron. Astrophys., 634,* A14.

Bate M. R., Bonnell I. A., and Bromm V. (2003) The formation of a star cluster: Predicting the properties of stars and brown dwarfs. *Mon. Not. R. Astron. Soc., 339,* 577–599.

Bauer J. M., Grav T., Fernández Y. R., et al. (2017) Debiasing the NEOWISE cryogenic mission comet populations. *Astron. J., 154,* 53.

Baxter E. J., Blake C. H., and Jain B. (2018) Probing Oort clouds around Milky Way stars with CMB surveys. *Astron. J., 156,* 243.

Belton M. J. S., Hainaut O. R., Meech K. J., et al. (2018) The excited spin state of 1I/2017 U1 'Oumuamua. *Astrophys. J. Lett., 856,* L21.

Bergner J. B. and Seligman D. Z. (2023) Acceleration of 1I'/Oumuamua from radiolytically produced H_2 in H_2O ice. *Nature, 615,* 610–613.

Bernstein G. M., Trilling D. E., Allen R. L., et al. (2004) The size distribution of trans-neptunian bodies. *Astron. J., 128,* 1364–1390.

Beust H. and Morbidelli A. (2000) Falling evaporating bodies as a clue to outline the structure of the β Pictoris young planetary system. *Icarus, 143,* 170–188.

Bird M. K., Pätzold M., Volland H., et al. (1988) Giotto spacecraft dynamics during the encounter with Comet Halley. *ESA J., 12,* 149–169.

Birnstiel T. and Andrews S. M. (2014) On the outer edges of protoplanetary dust disks. *Astrophys. J., 780,* 153.

Blum J. and Wurm G. (2008) The growth mechanisms of macroscopic bodies in protoplanetary disks. *Annu. Rev. Astron. Astrophys., 46,* 21–56.

Bodewits D., Noonan J. W., Feldman P. D., et al. (2020) The carbon monoxide-rich interstellar Comet 2I/Borisov. *Nature Astron., 4,* 867–871.

Bolin B. T. and Lisse C. M. (2020) Constraints on the spin-pole orientation, jet morphology, and rotation of interstellar Comet 2I/ Borisov with deep HST imaging. *Mon. Not. R. Astron. Soc., 497,* 4031–4041.

Bonsor A., Raymond S. N., Augereau J.-C., et al. (2014) Planetesimal-driven migration as an explanation for observations of high levels of warm, exozodiacal dust. *Mon. Not. R. Astron. Soc., 441,* 2380–2391.

Booth M., Kennedy G., Sibthorpe B., et al. (2013) Resolved debris discs around A stars in the Herschel DEBRIS survey. *Mon. Not. R. Astron. Soc., 428,* 1263–1280.

Borisov G. V. and Shustov B. M. (2021) Discovery of the first interstellar comet and the spatial density of interstellar objects in the solar neighborhood. *Solar System Res., 55,* 124–131.

Borovička J., Spurný P., and Shrbený L. (2022) Data on 824 fireballs observed by the digital cameras of the European Fireball Network in 2017–2018: II. Analysis of orbital and physical properties of centimeter-sized meteoroids. *Astron. Astrophys., 667,* A158.

Boss A. P. (1997) Giant planet formation by gravitational instability. *Science, 276,* 1836–1839.

Brandeker A., Liseau R., Olofsson G., et al. (2004) The spatial structure of the β Pictoris gas disk. *Astron. Astrophys., 413,* 681–691.

Brandl B. R., Lenzen R., Pantin E., et al. (2010) Instrument concept and science case for the mid-IR E-ELT imager and spectrograph METIS. In *Ground-Based and Airborne Instrumentation for Astronomy III* (I. S. McLean et al., eds.), pp. 969–984. SPIE Conf. Ser. 7735, Bellingham, Washington.

Brasser R. (2008) A two-stage formation process for the Oort comet cloud and its implications. *Astron. Astrophys., 492,* 251–255.

Brasser R., Duncan M. J., and Levison H. F. (2006) Embedded star clusters and the formation of the Oort cloud. *Icarus, 184,* 59–82.

Brown M. E. (2001) The inclination distribution of the Kuiper belt. *Astron. J., 121,* 2804–2814.

Burns J. A., Lamy P. L., and Soter S. (1979) Radiation forces on small particles in the solar system. *Icarus, 40,* 1–48.

Carrera D., Gorti U., Johansen A., et al. (2017) Planetesimal formation by the streaming instability in a photoevaporating disk. *Astrophys. J., 839,* 16.

Cataldi G., Brandeker A., Wu Y., et al. (2018) ALMA resolves C I emission from the β Pictoris debris disk. *Astrophys. J., 861,* 72.

Cataldi G., Wu Y., Brandeker A., et al. (2020) The surprisingly low carbon mass in the debris disk around HD 32297. *Astrophys. J., 892,* 99.

Chen C. H., Mittal T., Kuchner M., et al. (2014) The Spitzer infrared spectrograph debris disk catalog. I. Continuum analysis of unresolved targets. *Astrophys. J. Suppl. Ser., 211,* 25.

Chiang E., Kite E., Kalas P., et al. (2009) Fomalhaut's debris disk and planet: Constraining the mass of Fomalhaut b from disk morphology. *Astrophys. J., 693,* 734–749.

Colas F., Zanda B., Bouley S., et al. (2020) FRIPON: A worldwide network to track incoming meteoroids. *Astron. Astrophys., 644,* A53.

Concha-Ramírez F., Wilhelm M. J. C., Portegies Zwart S., et al. (2019) External photoevaporation of circumstellar discs constrains the time-scale for planet formation. *Mon. Not. R. Astron. Soc., 490,* 5678–5690.

Cordiner M. A., Milam S. N., Biver N., et al. (2020) Unusually high CO abundance of the first active interstellar comet. *Nature Astron., 4,* 861–866.

Ćuk M. (2018) 1I/'Oumuamua as a tidal disruption fragment from a binary star system. *Astrophys. J. Lett., 852,* L15.

Czechowski A. and Mann I. (2007) Collisional vaporization of dust and production of gas in the β Pictoris dust disk. *Astrophys. J., 660,* 1541–1555.

Daley C., Hughes A. M., Carter E. S., et al. (2019) The mass of stirring bodies in the AU Mic debris disk inferred from resolved vertical structure. *Astrophys. J., 875,* 87.

Davies R., Alves J., Clénet Y., et al. (2018) The MICADO first light imager for the ELT: Overview, operation, simulation. In *Ground-Based and Airborne Instrumentation for Astronomy VII* (C. J. Evans et al., eds.), pp. 107021S-1–107021S-12. SPIE Conf. Ser. 10702, Bellingham, Washington.

Dehnen W., Hands T. O., and Schönrich R. (2022) Capture of interstellar objects — II. By the solar system. *Mon. Not. R. Astron. Soc., 512,* 4078–4085.

Deienno R., Morbidelli A., Gomes R. S., et al. (2017) Constraining the giant planets' initial configuration from their evolution: Implications for the timing of the planetary instability. *Astron. J., 153,* 153.

de León J., Licandro J., Serra-Ricart M., et al. (2019) Interstellar visitors: A physical characterization of Comet C/2019 Q4 (Borisov) with OSIRIS at the 10.4 m GTC. *Res. Notes Am. Astron. Soc., 3,* 131.

Dent W. R. F., Greaves J. S., and Coulson I. M. (2005) CO emission from discs around isolated HAeBe and Vega-excess stars. *Mon. Not. R. Astron. Soc., 359,* 663–676.

Dent W. R. F., Wyatt M. C., Roberge A., et al. (2014) Molecular gas clumps from the destruction of icy bodies in the β Pictoris debris disk. *Science, 343,* 1490–1492.

Desch S. J. and Jackson A. P. (2021) 1I/'Oumuamua as an N_2 ice fragment of an exo-Pluto surface II: Generation of N_2 ice fragments and the origin of 'Oumuamua. *J. Geophys. Res.–Planets, 126,* e2020JE006807.

Di Ruscio A., Fienga A., Durante D., et al. (2020) Analysis of Cassini radio tracking data for the construction of INPOP19a: A new estimate of the Kuiper belt mass. *Astron. Astrophys., 640,* A7.

Do A., Tucker M. A., and Tonry J. (2018) Interstellar interlopers: Number density and origin of 'Oumuamua-like objects. *Astrophys. J. Lett., 855,* L10.

Dohnanyi J. S. (1969) Collisional model of asteroids and their debris. *J. Geophys. Res., 74,* 2531–2554.

Dominik C. and Decin G. (2003) Age dependence of the Vega phenomenon: Theory. *Astrophys. J., 598,* 626–635.

Dones L., Brasser R., Kaib N., et al. (2015) Origin and evolution of the cometary reservoirs. *Space Sci. Rev., 197,* 191–269.

Donitz B., Meech K. J., Castillo-Rogez J., et al. (2021) New Frontiers mission concept study to explore Oort cloud comets. *Bull. Am. Astron. Soc., 53,* 344.

Donitz B. P. S., Mages D., Tsukamoto H., et al. (2023) Interstellar object accessibility and mission design. In *2023 IEEE Aerospace Conference Proc.,* Big Sky, Montana, pp. 1–9.

Douglas E. S., Debes J., Mennesson B., et al. (2022) Sensitivity of the Roman Coronagraph instrument to exozodiacal dust. *Publ. Astron. Soc. Pac., 134,* 024402.

Dubrulle B., Morfill G., and Sterzik M. (1995) The dust subdisk in the protoplanetary nebula. *Icarus, 114,* 237–246.

Duchêne G. and Kraus A. (2013) Stellar multiplicity. *Annu. Rev. Astron. Astrophys., 51,* 269–310.

Duchêne G., Arriaga P., Wyatt M., et al. (2014) Spatially resolved

imaging of the two-component η Crv debris disk with Herschel. *Astrophys. J., 784,* 148.
Dullemond C. P. and Dominik C. (2004) Flaring vs. self-shadowed disks: The SEDs of Herbig Ae/Be stars. *Astron. Astrophys., 417,* 159–168.
Duncan M., Quinn T., and Tremaine S. (1987) The formation and extent of the solar system comet cloud. *Astron. J., 94,* 1330–1338.
Dybczyński P. A. and Królikowska M. (2018) Investigating the dynamical history of the interstellar object 'Oumuamua. *Astron. Astrophys., 610,* L11.
Eiroa C., Marshall J. P., Mora A., et al. (2013) DUst around NEarby Stars: The survey observational results. *Astron. Astrophys., 555,* A11.
Eisner N. L., Knight M. M., Snodgrass C., et al. (2019) Properties of the bare nucleus of Comet 96P/Machholz 1. *Astron. J., 157,* 186.
Engelhardt T., Jedicke R., Vereš P., et al. (2017) An observational upper limit on the interstellar number density of asteroids and comets. *Astron. J., 153,* 133.
Ercolano B. and Pascucci I. (2017) The dispersal of planet-forming discs: Theory confronts observations. *R. Soc. Open Sci., 4,* 170114.
Ertel S., Defrère D., Hinz P., et al. (2020) The HOSTS Survey for exozodiacal dust: Observational results from the complete survey. *Astron. J., 159,* 177.
Esposito T. M., Kalas P., Fitzgerald M. P., et al. (2020) Debris disk results from the Gemini Planet Imager Exoplanet Survey's polarimetric imaging campaign. *Astron. J., 160,* 24.
Eubanks T. M., Hein A. M., Lingam M., et al. (2021) Interstellar objects in the solar system: 1. Isotropic kinematics from the Gaia Early Data Release 3. *ArXiV e-prints,* arXiv:2103.03289.
Faramaz V., Krist J., Stapelfeldt K. R., et al. (2019) From scattered-light to millimeter emission: A comprehensive view of the gigayear-old system of HD 202628 and its eccentric debris ring. *Astron. J., 158,* 162.
Faramaz V., Marino S., Booth M., et al. (2021) A detailed characterization of HR 8799's debris disk with ALMA in Band 7. *Astron. J., 161,* 271.
Farihi J. (2016) Circumstellar debris and pollution at white dwarf stars. *New Astron. Rev., 71,* 9–34.
Farihi J., Gänsicke B. T., and Koester D. (2013) Evidence for water in the rocky debris of a disrupted extrasolar minor planet. *Science, 342,* 218–220.
Ferlet R., Hobbs L. M., and Vidal-Madjar A. (1987) The beta Pictoris circumstellar disk: V. Time variations of the Ca II-K line. *Astron. Astrophys., 185,* 267–270.
Fernandez J. A. and Ip W.-H. (1984) Some dynamical aspects of the accretion of Uranus and Neptune: The exchange of orbital angular momentum with planetesimals. *Icarus, 58,* 109–120.
Fernández Y. R., Kelley M. S., Lamy P. L., et al. (2013) Thermal properties, sizes, and size distribution of Jupiter-family cometary nuclei. *Icarus, 226,* 1138–1170.
Festou M. C., Keller H. U., and Weaver H. A., eds. (2004) *Comets II.* Univ. of Arizona, Tucson. 745 pp.
Fitzsimmons A., Snodgrass C., Rozitis B., et al. (2018) Spectroscopy and thermal modelling of the first interstellar object 1I/2017 U1 'Oumuamua. *Nature Astron., 2,* 133–137.
Fitzsimmons A., Hainaut O., Meech K. J., et al. (2019) Detection of CN gas in interstellar object 2I/Borisov. *Astrophys. J. Lett., 885,* L9.
Flaherty K., Hughes A. M., Mamajek E. E., et al. (2019) The planet formation potential around a 45 Myr old accreting M dwarf. *Astrophys. J., 872,* 92.
Francis P. J. (2005) The demographics of long-period comets. *Astrophys. J., 635,* 1348–1361.
Fraser W. C., Brown M. E., Morbidelli A., et al. (2014) The absolute magnitude distribution of Kuiper belt objects. *Astrophys. J., 782,* 100.
Friebe M. F., Pearce T. D., and Löhne T. (2022) Gap carving by a migrating planet embedded in a massive debris disc. *Mon. Not. R. Astron. Soc., 512,* 4441–4454.
Froncisz M., Brown P., and Weryk R. J. (2020) Possible interstellar meteoroids detected by the Canadian Meteor Orbit Radar. *Planet. Space Sci., 190,* 104980.
Gallart C., Bernard E. J., Brook C. B., et al. (2019) Uncovering the birth of the Milky Way through accurate stellar ages with Gaia. *Nature Astron., 3,* 932–939.
Gänsicke B. T., Koester D., Farihi J., et al. (2012) The chemical diversity of exo-terrestrial planetary debris around white dwarfs. *Mon. Not. R. Astron. Soc., 424,* 333–347.
Garufi A., Benisty M., Pinilla P., et al. (2018) Evolution of protoplanetary disks from their taxonomy in scattered light: Spirals, rings, cavities, and shadows. *Astron. Astrophys., 620,* A94.
Gáspár A., Psaltis D., Rieke G. H., et al. (2012) Modeling collisional cascades in debris disks: Steep dust-size distributions. *Astrophys. J., 754,* 74.
Geiler F., Krivov A. V., Booth M., et al. (2019) The scattered disc of HR 8799. *Mon. Not. R. Astron. Soc., 483,* 332–341.
Gladman B. J., Migliorini F., Morbidelli A., et al. (1997) Dynamical lifetimes of objects injected into asteroid belt resonances. *Science, 277,* 197–201.
Gomes R., Levison H. F., Tsiganis K., et al. (2005) Origin of the cataclysmic Late Heavy Bombardment period of the terrestrial planets. *Nature, 435,* 466–469.
Grady C. A., Perez M. R., Talavera A., et al. (1996) The β Pictoris phenomenon among Herbig Ae/Be stars: UV and optical high dispersion spectra. *Astron. Astrophys. Suppl. Ser., 120,* 157–177.
Grigorieva A., Thébault P., Artymowicz P., et al. (2007) Survival of icy grains in debris discs: The role of photosputtering. *Astron. Astrophys., 475,* 755–764.
Grishin E., Perets H. B., and Avni Y. (2019) Planet seeding through gas-assisted capture of interstellar objects. *Mon. Not. R. Astron. Soc., 487,* 3324–3332.
Gronoff G., Maggiolo R., Cessateur G., et al. (2020) The effect of cosmic rays on cometary nuclei. I. Dose deposition. *Astrophys. J., 890,* 89.
Guzik P. and Drahus M. (2021) Gaseous atomic nickel in the coma of interstellar Comet 2I/Borisov. *Nature, 593,* 375–378.
Guzik P., Drahus M., Rusek K., et al. (2020) Initial characterization of interstellar Comet 2I/Borisov. *Nature Astron., 4,* 53–57.
Haisch Karl E. J., Lada E. A., and Lada C. J. (2001) Disk frequencies and lifetimes in young clusters. *Astrophys. J. Lett., 553,* L153–L156.
Hajdukova M., Sterken V., Wiegert P., et al. (2020) The challenge of identifying interstellar meteors. *Planet. Space Sci., 192,* 105060.
Hallatt T. and Wiegert P. (2020) The dynamics of interstellar asteroids and comets within the galaxy: An assessment of local candidate source regions for 1I/'Oumuamua and 2I/Borisov. *Astron. J., 159,* 147.
Han Y., Wyatt M. C., and Dent W. R. F. (2023) Has the dust clump in the debris disc of Beta Pictoris moved? *Mon. Not. R. Astron. Soc., 519,* 3257–3270.
Hands T. O. and Dehnen W. (2020) Capture of interstellar objects: A source of long-period comets. *Mon. Not. R. Astron. Soc., 493,* L59–L64.
Hands T. O., Dehnen W., Gration A., et al. (2019) The fate of planetesimal discs in young open clusters: Implications for 1I/'Oumuamua, the Kuiper belt, the Oort cloud, and more. *Mon. Not. R. Astron. Soc., 490,* 21–36.
Hanse J., Jílková L., Portegies Zwart S. F., et al. (2018) Capture of exocomets and the erosion of the Oort cloud due to stellar encounters in the galaxy. *Mon. Not. R. Astron. Soc., 473,* 5432–5445.
Hazumi M., Ade P. A. R., Adler A., et al. (2020) LiteBIRD satellite: JAXA's new strategic L-class mission for all-sky surveys of cosmic microwave background polarization. In *Space Telescopes and Instrumentation 2020: Optical, Infrared, and Millimeter Wave* (M. Lystrup and M. D. Perrin, eds.), p. 114432F. SPIE Conf. Ser. 11443, Bellingham, Washington.
Heinze A. N., Denneau L., Tonry J. L., et al. (2021) NEO population, velocity bias, and impact risk from an ATLAS analysis. *Planet. Sci. J., 2,* 12.
Hibberd A., Hein A. M., and Eubanks T. M. (2020) Project Lyra: Catching 1I/'Oumuamua — Mission opportunities after 2024. *Acta Astronaut., 170,* 136–144.
Hoang T. and Loeb A. (2020) Destruction of molecular hydrogen ice and implications for 1I/2017 U1 ('Oumuamua). *Astrophys. J. Lett., 899,* L23.
Holland W. S., Matthews B. C., Kennedy G. M., et al. (2017) SONS: The JCMT legacy survey of debris discs in the submillimetre. *Mon. Not. R. Astron. Soc., 470,* 3606–3663.
Hoover D. J., Seligman D. Z., and Payne M. J. (2022) The population of interstellar objects detectable with the LSST and accessible for in situ rendezvous with various mission designs. *Planet. Sci. J., 3,* 71.
Hui M.-T. and Knight M. M. (2019) New insights into interstellar object 1I/2017 U1 ('Oumuamua) from SOHO/STEREO non-detections. *Astron. J., 158,* 256.
Hui M.-T., Ye Q.-Z., Föhring D., et al. (2020) Physical characterization

of interstellar Comet 2I/2019 Q4 (Borisov). *Astron. J., 160,* 92.

Hutsemékers D., Manfroid J., Jehin E., et al. (2021) FeI and NiI in cometary atmospheres: Connections between the NiI/FeI abundance ratio and chemical characteristics of Jupiter-family and Oort-cloud comets. *Astron. Astrophys., 652,* L1.

Iglesias D., Bayo A., Olofsson J., et al. (2018) Debris discs with multiple absorption features in metallic lines: Circumstellar or interstellar origin? *Mon. Not. R. Astron. Soc., 480,* 488–520.

Jackson A. P., Wyatt M. C., Bonsor A., et al. (2014) Debris from giant impacts between planetary embryos at large orbital radii. *Mon. Not. R. Astron. Soc., 440,* 3757–3777.

Jackson A. P., Tamayo D., Hammond N., et al. (2018) Ejection of rocky and icy material from binary star systems: Implications for the origin and composition of 1I/'Oumuamua. *Mon. Not. R. Astron. Soc., 478,* L49–L53.

Jeffers S. V., Manley S. P., Bailey M. E., et al. (2001) Near-Earth object velocity distributions and consequences for the Chicxulub impactor. *Mon. Not. R. Astron. Soc., 327,* 126–132.

Jessberger E. K., Stephan T., Rost D., et al. (2001) Properties of interplanetary dust: Information from collected samples. In *Interplanetary Dust, Volume 1* (E. Grün et al., eds.), pp. 253–294. Springer Berlin, Heidelberg.

Jewitt D. and Luu J. (2019) Initial characterization of interstellar Comet 2I/2019 Q4 (Borisov). *Astrophys. J. Lett., 886,* L29.

Jewitt D. and Seligman D. Z. (2022) The interstellar interlopers. *ArXiV e-prints,* arXiv:2209.08182.

Jewitt D., Garland C. A., and Aussel H. (2008) Deep search for carbon monoxide in cometary precursors using millimeter wave spectroscopy. *Astron. J., 135,* 400–407.

Jewitt D., Luu J., Rajagopal J., et al. (2017) Interstellar interloper 1I/2017 U1: Observations from the NOT and WIYN telescopes. *Astrophys. J. Lett., 850,* L36.

Jewitt D., Hui M.-T., Kim Y., et al. (2020a) The nucleus of interstellar Comet 2I/Borisov. *Astrophys. J. Lett., 888,* L23.

Jewitt D., Kim Y., Mutchler M., et al. (2020b) Outburst and splitting of interstellar Comet 2I/Borisov. *Astrophys. J. Lett., 896,* L39.

Jones G. H., Knight M. M., Battams K., et al. (2018) The science of sungrazers, sunskirters, and other near-Sun comets. *Space Sci. Rev., 214,* 20.

Jones R. L., Chesley S. R., Connolly A. J., et al. (2009) Solar system science with LSST. *Earth Moon Planets, 105,* 101–105.

Kalas P., Graham J. R., and Clampin M. (2005) A planetary system as the origin of structure in Fomalhaut's dust belt. *Nature, 435,* 1067–1070.

Katz J. I. (2018) Why is interstellar object 1I/2017 U1 ('Oumuamua) rocky, tumbling and possibly very prolate? *Mon. Not. R. Astron. Soc., 478,* L95–L98.

Kennedy G. M. (2018) Exocomet orbit fitting: Accelerating coma absorption during transits of β Pictoris. *Mon. Not. R. Astron. Soc., 479,* 1997–2006.

Kennedy G. M. (2020) The unexpected narrowness of eccentric debris rings: A sign of eccentricity during the protoplanetary disc phase. *R. Soc. Open Sci., 7,* 200063.

Kennedy G. M., Hope G., Hodgkin S. T., et al. (2019) An automated search for transiting exocomets. *Mon. Not. R. Astron. Soc., 482,* 5587–5596.

Kenyon S. J. and Bromley B. C. (2017) Variations on debris disks. IV. An improved analytical model for collisional cascades. *Astrophys. J., 839,* 38.

Kiefer F., Lecavelier des Etangs A., Boissier J., et al. (2014) Two families of exocomets in the β Pictoris system. *Nature, 514,* 462–464.

Kim Y., Jewitt D., Mutchler M., et al. (2020) Coma anisotropy and the rotation pole of interstellar Comet 2I/Borisov. *Astrophys. J. Lett., 895,* L34.

Kimura K., Aton K., Baybayan K., et al. (2019) A Hua He Inoa: Hawaiian culture-based celestial naming. *Bull. Am. Astron. Soc., 51,* 135.

Klessen R. S. and Glover S. C. O. (2016) Physical processes in the interstellar medium. In *Star Formation in Galaxy Evolution: Connecting Numerical Models to Reality* (N. Y. Gnedin et al., eds.), pp. 85–249. Saas-Fee Advanced Course, Vol. 43, Springer Berlin, Heidelberg.

Klusmeyer J., Hughes A. M., Matrà L., et al. (2021) A deep search for five molecules in the 49 Ceti debris disk. *Astrophys. J., 921,* 56.

Koppelman H. H. and Helmi A. (2021) Determination of the escape velocity of the Milky Way using a halo sample selected based on proper motion. *Astron. Astrophys., 649,* A136.

Kóspál Á., Moór A., Juhász A., et al. (2013) ALMA observations of the molecular gas in the debris disk of the 30 Myr old star HD 21997. *Astrophys. J., 776,* 77.

Kral Q., Wyatt M., Carswell R. F., et al. (2016) A self-consistent model for the evolution of the gas produced in the debris disc of β Pictoris. *Mon. Not. R. Astron. Soc., 461,* 845–858.

Kral Q., Clarke C., and Wyatt M. C. (2018) Circumstellar discs: What will be next? In *Handbook of Exoplanets* (H. J. Deeg and J. A. Belmonte, eds.), pp. 3321–3352. Springer, Cham.

Kral Q., Marino S., Wyatt M. C., et al. (2019) Imaging [CI] around HD 131835: Reinterpreting young debris discs with protoplanetary disc levels of CO gas as shielded secondary discs. *Mon. Not. R. Astron. Soc., 489,* 3670–3691.

Kral Q., Pringle J. E., Guilbert-Lepoutre A., et al. (2021) A molecular wind blows out of the Kuiper belt. *Astron. Astrophys., 653,* L11.

Krijt S., Bosman A. D., Zhang K., et al. (2020) CO depletion in protoplanetary disks: A unified picture combining physical sequestration and chemical processing. *Astrophys. J., 899,* 134.

Krist J. E., Stapelfeldt K. R., Bryden G., et al. (2012) Hubble Space Telescope observations of the HD 202628 debris disk. *Astron. J., 144,* 45.

Królikowska M. and Dybczyński P. A. (2010) Where do long-period comets come from? 26 comets from the non-gravitational Oort spike. *Mon. Not. R. Astron. Soc., 404,* 1886–1902.

Kruijssen J. M. D., Pfeffer J. L., Chevance M., et al. (2020) Kraken reveals itself — The merger history of the Milky Way reconstructed with the E-MOSAICS simulations. *Mon. Not. R. Astron. Soc., 498,* 2472–2491.

Lada C. J. and Lada E. A. (2003) Embedded clusters in molecular clouds. *Annu. Rev. Astron. Astrophys., 41,* 57–115.

Lagrange A. M., Gratadour D., Chauvin G., et al. (2009) A probable giant planet imaged in the β Pictoris disk: VLT/NaCo deep L'-band imaging. *Astron. Astrophys., 493,* L21–L25.

Lagrange A. M., Meunier N., Rubini P., et al. (2019) Evidence for an additional planet in the β Pictoris system. *Nature Astron., 3,* 1135–1142.

Lagrange-Henri A. M., Vidal-Madjar A., and Ferlet R. (1988) The beta Pictoris circumstellar disk: VI. Evidence for material falling on to the star. *Astron. Astrophys., 190,* 275–282.

Lambrechts M. and Johansen A. (2012) Rapid growth of gas-giant cores by pebble accretion. *Astron. Astrophys., 544,* A32.

Langland-Shula L. E. and Smith G. H. (2007) The unusual spectrum of Comet 96P/Machholz. *Astrophys. J. Lett., 664,* L119–L122.

Läuter M., Kramer T., Rubin M., et al. (2020) The gas production of 14 species from Comet 67P/Churyumov-Gerasimenko based on DFMS/COPS data from 2014 to 2016. *Mon. Not. R. Astron. Soc., 498,* 3995–4004.

Lecavelier des Etangs A., Vidal-Madjar A., and Ferlet R. (1999) Photometric stellar variation due to extra-solar comets. *Astron. Astrophys., 343,* 916–922.

Lecavelier des Etangs A., Cros L., Hébrard G., et al. (2022) Exocomets size distribution in the β Pictoris planetary system. *Sci. Rept., 12,* 5855.

Lee E. J. and Chiang E. (2016) A primer on unifying debris disk morphologies. *Astrophys. J., 827,* 125.

Levine W. G., Cabot S. H. C., Seligman D., et al. (2021) Constraints on the occurrence of 'Oumuamua-like objects. *Astrophys. J., 922,* 39.

Levison H. F. and Duncan M. J. (1997) From the Kuiper belt to Jupiter-family comets: The spatial distribution of ecliptic comets. *Icarus, 127,* 13–32.

Li Y.-B., Luo A. L., Lu Y.-J., et al. (2021) 591 high-velocity stars in the galactic halo selected from LAMOST DR7 and Gaia DR2. *Astrophys. J. Suppl. Ser., 252,* 3.

Lin H. W., Lee C.-H., Gerdes D. W., et al. (2020) Detection of diatomic carbon in 2I/Borisov. *Astrophys. J. Lett., 889,* L30.

Lisse C. M., Wyatt M. C., Chen C. H., et al. (2012) Spitzer evidence for a late-heavy bombardment and the formation of ureilites in η Corvi at ~1 Gyr. *Astrophys. J., 747,* 93.

Lisse C. M., Young L. A., Cruikshank D. P., et al. (2021) On the origin and thermal stability of Arrokoth's and Pluto's ices. *Icarus, 356,* 114072.

Liu B., Johansen A., Lambrechts M., et al. (2022) Natural separation of two primordial planetary reservoirs in an expanding solar protoplanetary disk. *Sci. Adv., 8,* eabm3045.

Löhne T. (2020) Relating grain size distributions in circumstellar discs to the spectral index at millimetre wavelengths. *Astron. Astrophys., 641,* A75.

Löhne T., Krivov A. V., and Rodmann J. (2008) Long-term collisional evolution of debris disks. *Astrophys. J., 673,* 1123–1137.

Lombaert R., Decin L., Royer P., et al. (2016) Constraints on the H_2O formation mechanism in the wind of carbon-rich AGB stars. *Astron. Astrophys., 588,* A124.

Lovell J. B., Wyatt M. C., Ansdell M., et al. (2021) ALMA survey of Lupus class III stars: Early planetesimal belt formation and rapid disc dispersal. *Mon. Not. R. Astron. Soc., 500,* 4878–4900.

Luppe P., Krivov A. V., Booth M., et al. (2020) Observability of dusty debris discs around M-stars. *Mon. Not. R. Astron. Soc., 499,* 3932–3942.

Lynch E. M. and Lovell J. B. (2022) Eccentric debris disc morphologies — I. Exploring the origin of apocentre and pericentre glows in face-on debris discs. *Mon. Not. R. Astron. Soc., 510,* 2538–2551.

MacGregor M. A., Matrà L., Kalas P., et al. (2017) A complete ALMA map of the Fomalhaut debris disk. *Astrophys. J., 842,* 8.

MacGregor M. A., Hurt S. A., Stark C. C., et al. (2022) ALMA images the eccentric HD 53143 debris disk. *Astrophys. J. Lett., 933,* L1.

Maiolino R., Haehnelt M., Murphy M. T., et al. (2013) A community science case for E-ELT HIRES. *ArXiV e-prints,* arXiv:1310.3163.

Malamud U. and Perets H. B. (2016) Post-main sequence evolution of icy minor planets: Implications for water retention and white dwarf pollution. *Astrophys. J., 832,* 160.

Malhotra R. (1993) The origin of Pluto's peculiar orbit. *Nature, 365,* 819–821.

Mamajek E. E. (2009) Initial conditions of planet formation: Lifetimes of primordial disks. In *Exoplanets and Disks: Their Formation and Diversity* (T. Usuda et al., eds.), pp. 3–10. AIP Conf. Ser. 1158, American Institute of Physics, Melville, New York.

Manara C. F., Morbidelli A., and Guillot T. (2018) Why do protoplanetary disks appear not massive enough to form the known exoplanet population? *Astron. Astrophys., 618,* L3.

Manzini F., Oldani V., Ochner P., et al. (2020) Interstellar Comet 2I/Borisov exhibits a structure similar to native solar system comets. *Mon. Not. R. Astron. Soc., 495,* L92–L96.

Marboeuf U., Bonsor A., and Augereau J. C. (2016) Extrasolar comets: The origin of dust in exozodiacal disks? *Planet. Space Sci., 133,* 47–62.

Marino S. (2021) Constraining planetesimal stirring: How sharp are debris disc edges? *Mon. Not. R. Astron. Soc., 503,* 5100–5114.

Marino S. (2023) Planetesimal/debris discs. In *Planetary Systems Now* (L. M. Lara and D. Jewitt, eds.), pp. 381–408. World Scientific (Europe).

Marino S., Matrà L., Stark C., et al. (2016) Exocometary gas in the HD 181327 debris ring. *Mon. Not. R. Astron. Soc., 460,* 2933–2944.

Marino S., Wyatt M. C., Panić O., et al. (2017) ALMA observations of the η Corvi debris disc: Inward scattering of CO-rich exocomets by a chain of 3–30 $M_\oplus$ planets? *Mon. Not. R. Astron. Soc., 465,* 2595–2615.

Marino S., Bonsor A., Wyatt M. C., et al. (2018a) Scattering of exocomets by a planet chain: Exozodi levels and the delivery of cometary material to inner planets. *Mon. Not. R. Astron. Soc., 479,* 1651–1671.

Marino S., Carpenter J., Wyatt M. C., et al. (2018b) A gap in the planetesimal disc around HD 107146 and asymmetric warm dust emission revealed by ALMA. *Mon. Not. R. Astron. Soc., 479,* 5423–5439.

Marino S., Flock M., Henning T., et al. (2020) Population synthesis of exocometary gas around A stars. *Mon. Not. R. Astron. Soc., 492,* 4409–4429.

Marshall J. P., Moro-Martín A., Eiroa C., et al. (2014) Correlations between the stellar, planetary, and debris components of exoplanet systems observed by Herschel. *Astron. Astrophys., 565,* A15.

Marshall J. P., Wang L., Kennedy G. M., et al. (2021) A search for trends in spatially resolved debris discs at far-infrared wavelengths. *Mon. Not. R. Astron. Soc., 501,* 6168–6180.

Mashchenko S. (2019) Modelling the light curve of 'Oumuamua: Evidence for torque and disc-like shape. *Mon. Not. R. Astron. Soc., 489,* 3003–3021.

Matrà L., Panić O., Wyatt M. C., et al. (2015) CO mass upper limits in the Fomalhaut ring — The importance of NLTE excitation in debris discs and future prospects with ALMA. *Mon. Not. R. Astron. Soc., 447,* 3936–3947.

Matrà L., Dent W. R. F., Wyatt M. C., et al. (2017a) Exocometary gas structure, origin and physical properties around β Pictoris through ALMA CO multitransition observations. *Mon. Not. R. Astron. Soc., 464,* 1415–1433.

Matrà L., MacGregor M. A., Kalas P., et al. (2017b) Detection of exocometary CO within the 440 Myr old Fomalhaut belt: A similar $CO+CO_2$ ice abundance in exocomets and solar system comets. *Astrophys. J., 842,* 9.

Matrà L., Marino S., Kennedy G. M., et al. (2018a) An empirical planetesimal belt radius-stellar luminosity relation. *Astrophys. J., 859,* 72.

Matrà L., Wilner D. J., Öberg K. I., et al. (2018b) Molecular reconnaissance of the β Pictoris gas disk with the SMA: A low $HCN/(CO+CO_2)$ outgassing ratio and predictions for future surveys. *Astrophys. J., 853,* 147.

Matrà L., Öberg K. I., Wilner D. J., et al. (2019a) On the ubiquity and stellar luminosity dependence of exocometary CO gas: Detection around M dwarf TWA 7. *Astron. J., 157,* 117.

Matrà L., Wyatt M. C., Wilner D. J., et al. (2019b) Kuiper belt-like hot and cold populations of planetesimal inclinations in the β Pictoris belt revealed by ALMA. *Astron. J., 157,* 135.

Matrà L., Dent W. R. F., Wilner D. J., et al. (2020) Dust populations in the iconic Vega planetary system resolved by ALMA. *Astrophys. J., 898,* 146.

McCrea W. H. (1975) Solar system as space probe. *The Observatory, 95,* 239–255.

McIntosh B. A. (1990) Comet P/Machholz and the Quadrantid meteor stream. *Icarus, 86,* 299–304.

McKay A. J., DiSanti M. A., Kelley M. S. P., et al. (2019) The peculiar volatile composition of CO-dominated Comet C/2016 R2 (PanSTARRS). *Astron. J., 158,* 128.

McKay A. J., Cochran A. L., Dello Russo N., et al. (2020) Detection of a water tracer in interstellar Comet 2I/Borisov. *Astrophys. J. Lett., 889,* L10.

McNeill A., Trilling D. E., and Mommert M. (2018) Constraints on the density and internal strength of 1I/'Oumuamua. *Astrophys. J. Lett., 857,* L1.

Meech K. J., Hainaut O. R., and Marsden B. G. (2004) Comet nucleus size distributions from HST and Keck telescopes. *Icarus, 170,* 463–491.

Meech K. J., Yang B., Kleyna J., et al. (2016) Inner solar system material discovered in the Oort cloud. *Sci. Adv., 2,* e1600038.

Meech K. J., Kleyna J. T., Hainaut O., et al. (2017a) CO-driven activity in Comet C/2017 K2 (PANSTARRS). *Astrophys. J. Lett., 849,* L8.

Meech K. J., Weryk R., Micheli M., et al. (2017b) A brief visit from a red and extremely elongated interstellar asteroid. *Nature, 552,* 378–381.

Melnick G. J., Neufeld D. A., Ford K. E. S., et al. (2001) Discovery of water vapour around IRC+10216 as evidence for comets orbiting another star. *Nature, 412,* 160–163.

Melosh H. J. (2003) Exchange of meteorites (and life?) between stellar systems. *Astrobiology, 3,* 207–215.

Mesa D., Keppler M., Cantalloube F., et al. (2019) VLT/SPHERE exploration of the young multiplanetary system PDS70. *Astron. Astrophys., 632,* A25.

Meshkat T., Mawet D., Bryan M. L., et al. (2017) A direct imaging survey of Spitzer-detected debris disks: Occurrence of giant planets in dusty systems. *Astron. J., 154,* 245.

Michel A., van der Marel N., and Matthews B. (2021) Bridging the gap between protoplanetary and debris disks: Evidence for slow disk dissipation. *ArXiV e-prints,* arXiv:2104.05894.

Micheli M., Farnocchia D., Meech K. J., et al. (2018) Non-gravitational acceleration in the trajectory of 1I/2017 U1 ('Oumuamua). *Nature, 559,* 223–226.

Miller D., Duvigneaud F., Menken W., et al. (2022) High-performance solar sails for interstellar object rendezvous. *Acta Astronaut., 200,* 242–252.

Minton D. A. and Malhotra R. (2010) Dynamical erosion of the asteroid belt and implications for large impacts in the inner solar system. *Icarus, 207,* 744–757.

Miret-Roig N., Bouy H., Raymond S. N., et al. (2022) A rich population of free-floating planets in the Upper Scorpius young stellar association. *Nature Astron., 6,* 89–97.

Montgomery S. L. and Welsh B. Y. (2017) Unusually high circumstellar absorption variability around the δ Scuti/λ Boötis star HD 183324.

Mon. Not. R. Astron. Soc., 468, L55–L58.
Moór A., Curé M., Kóspál Á., et al. (2017) Molecular gas in debris disks around young A-type stars. *Astrophys. J., 849,* 123.
Morbidelli A. and Raymond S. N. (2016) Challenges in planet formation. *J. Geophys. Res.–Planets, 121,* 1962–1980.
Morbidelli A., Tsiganis K., Crida A., et al. (2007) Dynamics of the giant planets of the solar system in the gaseous protoplanetary disk and their relationship to the current orbital architecture. *Astron. J., 134,* 1790–1798.
Morbidelli A., Nesvorny D., Laurenz V., et al. (2018) The timeline of the lunar bombardment: Revisited. *Icarus, 305,* 262–276.
Morbidelli A., Batygin K., Brasser R., et al. (2020) No evidence for interstellar planetesimals trapped in the solar system. *Mon. Not. R. Astron. Soc., 497,* L46–L49.
Moro-Martín A. (2019a) Could 1I/'Oumuamua be an icy fractal aggregate? *Astrophys. J. Lett., 872,* L32.
Moro-Martín A. (2019b) Origin of 1I/'Oumuamua. II. An ejected exo-Oort cloud object? *Astron. J., 157,* 86.
Moro-Martín A. and Norman C. (2022) Interstellar planetesimals: Potential seeds for planet formation? *Astrophys. J., 924,* 96.
Moro-Martín A., Marshall J. P., Kennedy G., et al. (2015) Does the presence of planets affect the frequency and properties of extrasolar Kuiper belts? Results from the Herschel DEBRIS and DUNES Surveys. *Astrophys. J., 801,* 143.
Morrison S. and Malhotra R. (2015) Planetary chaotic zone clearing: Destinations and timescales. *Astrophys. J., 799,* 41.
Mróz P., Udalski A., Skowron J., et al. (2017) No large population of unbound or wide-orbit Jupiter-mass planets. *Nature, 548,* 183–186.
Mulders G. D., Pascucci I., Ciesla F. J., et al. (2021) The mass budgets and spatial scales of exoplanet systems and protoplanetary disks. *Astrophys J., 920,* 66.
Mumma M. J., Weissman P. R., and Stern S. A. (1993) Comets and the origin of the solar system: Reading the Rosetta Stone. In *Protostars and Planets III* (E. H. Levy and J. I. Lunine, eds.), pp. 1177–1252. Univ. of Arizona, Tucson.
Musci R., Weryk R. J., Brown P., et al. (2012) An optical survey for millimeter-sized interstellar meteoroids. *Astrophys. J., 745,* 161.
Najita J. R. and Kenyon S. J. (2014) The mass budget of planet-forming discs: Isolating the epoch of planetesimal formation. *Mon. Not. R. Astron. Soc., 445,* 3315–3329.
Najita J. R., Kenyon S. J., and Bromley B. C. (2022) From pebbles and planetesimals to planets and dust: The protoplanetary disk-debris disk connection. *Astrophys. J., 925,* 45.
Namouni F. and Morais M. H. M. (2020) An interstellar origin for high-inclination Centaurs. *Mon. Not. R. Astron. Soc., 494,* 2191–2199.
Napier K. J., Adams F. C., and Batygin K. (2021) On the fate of interstellar objects captured by our solar system. *Planet. Sci. J., 2,* 217.
National Academies of Sciences, Engineering, and Medicine (2022) *Origins, Worlds, and Life: A Decadal Strategy for Planetary Science and Astrobiology 2023–2032.* National Academies Press, Washington, DC. 800 pp.
Neufeld D. A., González-Alfonso E., Melnick G., et al. (2011) The widespread occurrence of water vapor in the circumstellar envelopes of carbon-rich asymptotic giant branch stars: First results from a survey with Herschel/HIFI. *Astrophys. J. Lett., 727,* L29.
Norfolk B. J., Maddison S. T., Marshall J. P., et al. (2021) Four new planetesimals around typical and pre-main-sequence stars (PLATYPUS) debris discs at 8.8 mm. *Mon. Not. R. Astron. Soc., 507,* 3139–3147.
O'Brien D. P. and Greenberg R. (2003) Steady-state size distributions for collisional populations: Analytical solution with size-dependent strength. *Icarus, 164,* 334–345.
Opitom C., Fitzsimmons A., Jehin E., et al. (2019) 2I/Borisov: A C_2-depleted interstellar comet. *Astron. Astrophys., 631,* L8.
Opitom C., Jehin E., Hutsemékers D., et al. (2021) The similarity of the interstellar Comet 2I/Borisov to solar system comets from high-resolution optical spectroscopy. *Astron. Astrophys., 650,* L19.
'Oumuamua ISSI Team (2019) The natural history of 'Oumuamua. *Nature Astron., 3,* 594–602.
Pan M. and Schlichting H. E. (2012) Self-consistent size and velocity distributions of collisional cascades. *Astrophys. J., 747,* 113.
Pan M., Nesvold E. R., and Kuchner M. J. (2016) Apocenter glow in eccentric debris disks: Implications for Fomalhaut and ε Eridani. *Astrophys. J., 832,* 81.
Pascucci I., Testi L., Herczeg G. J., et al. (2016) A steeper than linear disk mass-stellar mass scaling relation. *Astrophys. J., 831,* 125.
Pawellek N., Wyatt M., Matrà L., et al. (2021) A ~75 per cent occurrence rate of debris discs around F stars in the β Pic moving group. *Mon. Not. R. Astron. Soc., 502,* 5390–5416.
Pearce T. D. and Wyatt M. C. (2015) Double-ringed debris discs could be the work of eccentric planets: Explaining the strange morphology of HD 107146. *Mon. Not. R. Astron. Soc., 453,* 3329–3340.
Pearce T. D., Launhardt R., Ostermann R., et al. (2022) Planet populations inferred from debris discs: Insights from 178 debris systems in the ISPY, LEECH, and LIStEN planet-hunting surveys. *Astron. Astrophys., 659,* A135.
Petit J.-M., Morbidelli A., and Chambers J. (2001) The primordial excitation and clearing of the asteroid belt. *Icarus, 153,* 338–347.
Pfalzner S. and Bannister M. T. (2019) A hypothesis for the rapid formation of planets. *Astrophys. J. Lett., 874,* L34.
Pfalzner S., Davies M. B., Kokaia G., et al. (2020) Oumuamuas passing through molecular clouds. *Astrophys. J., 903,* 114.
Pfalzner S., Aizpuru Vargas L., Bhandare A., et al. (2021a) Significant interstellar object production by close stellar flybys. *Astron. Astrophys., 651,* A38.
Pfalzner S., Paterson D., Bannister M. T., et al. (2021b) Interstellar objects follow the collapse of molecular clouds. *Astrophys. J., 921,* 168.
Pfalzner S., Dehghani S., and Michel A. (2022) Most planets might have more than 5 Myr of time to form. *Astrophys. J. Lett., 939,* L10.
Pierens A. and Nelson R. P. (2008) Constraints on resonant-trapping for two planets embedded in a protoplanetary disc. *Astron. Astrophys., 482,* 333–340.
Pinilla P., Pohl A., Stammler S. M., et al. (2017) Dust density distribution and imaging analysis of different ice lines in protoplanetary disks. *Astrophys. J., 845,* 68.
Pollack J. B., Hubickyj O., Bodenheimer P., et al. (1996) Formation of the giant planets by concurrent accretion of solids and gas. *Icarus, 124,* 62–85.
Portegies Zwart S. (2021) Oort cloud ecology: I. Extra-solar Oort clouds and the origin of asteroidal interlopers. *Astron. Astrophys., 647,* A136.
Portegies Zwart S., Torres S., Pelupessy I., et al. (2018) The origin of interstellar asteroidal objects like 1I/2017 U1 'Oumuamua. *Mon. Not. R. Astron. Soc., 479,* L17–L22.
Raddi R., Gänsicke B. T., Koester D., et al. (2015) Likely detection of water-rich asteroid debris in a metal-polluted white dwarf. *Mon. Not. R. Astron. Soc., 450,* 2083–2093.
Rafikov R. R. (2018) 1I/2017 'Oumuamua-like interstellar asteroids as possible messengers from dead stars. *Astrophys. J., 861,* 35.
Rafikov R. R. (2023) Radial profiles of surface density in debris discs. *Mon. Not. R. Astron. Soc., 519,* 5607–5622.
Raghavan D., McAlister H. A., Henry T. J., et al. (2010) A survey of stellar families: Multiplicity of solar-type stars. *Astrophys. J. Suppl. Ser., 190,* 1–42.
Rauer H., Catala C., Aerts C., et al. (2014) The PLATO 2.0 mission. *Exp. Astron., 38,* 249–330.
Raymond S. N. and Izidoro A. (2017) Origin of water in the inner solar system: Planetesimals scattered inward during Jupiter and Saturn's rapid gas accretion. *Icarus, 297,* 134–148.
Raymond S. N., Armitage P. J., and Gorelick N. (2010) Planet-planet scattering in planetesimal disks. II. Predictions for outer extrasolar planetary systems. *Astrophys. J., 711,* 772–795.
Raymond S. N., Armitage P. J., Moro-Martín A., et al. (2011) Debris disks as signposts of terrestrial planet formation. *Astron. Astrophys., 530,* A62.
Raymond S. N., Armitage P. J., Moro-Martín A., et al. (2012) Debris disks as signposts of terrestrial planet formation: II. Dependence of exoplanet architectures on giant planet and disk properties. *Astron. Astrophys., 541,* A11.
Raymond S. N., Armitage P. J., and Veras D. (2018a) Interstellar object 'Oumuamua as an extinct fragment of an ejected cometary planetesimal. *Astrophys. J. Lett., 856,* L7.
Raymond S. N., Armitage P. J., Veras D., et al. (2018b) Implications of the interstellar object 1I/'Oumuamua for planetary dynamics and planetesimal formation. *Mon. Not. R. Astron. Soc., 476,* 3031–3038.
Raymond S. N., Kaib N. A., Armitage P. J., et al. (2020) Survivor bias: Divergent fates of the solar system's ejected versus persisting planetesimals. *Astrophys. J. Lett., 904,* L4.
Rebollido I., Eiroa C., Montesinos B., et al. (2020) Exocomets: A spectroscopic survey. *Astron. Astrophys., 639,* A11.

Richert A. J. W., Getman K. V., Feigelson E. D., et al. (2018) Circumstellar disc lifetimes in numerous galactic young stellar clusters. *Mon. Not. R. Astron. Soc., 477,* 5191–5206.
Rigley J. K. and Wyatt M. C. (2022) Comet fragmentation as a source of the zodiacal cloud. *Mon. Not. R. Astron. Soc., 510,* 834–857.
Roberge A., Feldman P. D., Lagrange A. M., et al. (2000) High-resolution Hubble Space Telescope STIS spectra of C I and CO in the β Pictoris circumstellar disk. *Astrophys. J., 538,* 904–910.
Roberge A., Feldman P. D., Weinberger A. J., et al. (2006) Stabilization of the disk around β Pictoris by extremely carbon-rich gas. *Nature, 441,* 724–726.
Roberge A., Chen C. H., Millan-Gabet R., et al. (2012) The exozodiacal dust problem for direct observations of exo-Earths. *Publ. Astron. Soc. Pac., 124,* 799–808.
Roberge A., Welsh B. Y., Kamp I., et al. (2014) Volatile-rich circumstellar gas in the unusual 49 Ceti debris disk. *Astrophys. J. Lett., 796,* L11.
Russell H. N. (1916) On the albedo of the planets and their satellites. *Astrophys. J., 43,* 173–196.
Saavik Ford K. E. and Neufeld D. A. (2001) Water vapor in carbon-rich asymptotic giant branch stars from the vaporization of icy orbiting bodies. *Astrophys. J. Lett., 557,* L113–L116.
Safronov V. S. (1972) Ejection of bodies from the solar system in the course of the accumulation of the giant planets and the formation of the cometary cloud. In *The Motion, Evolution of Orbits, and Origin of Comets* (G. A. Chebotarev et al., eds.), pp. 329–334. IAU Symp. 45, Springer, Dordrecht.
Sánchez J. P., Morante D., Hermosin P., et al. (2021) ESA F-Class Comet Interceptor: Trajectory design to intercept a yet-to-be-discovered comet. *Acta Astronaut., 188,* 265–277.
Sanchis E., Testi L., Natta A., et al. (2021) Measuring the ratio of the gas and dust emission radii of protoplanetary disks in the Lupus star-forming region. *ArXiV e-prints,* arXiv:2101.11307.
Schleicher D. G. (2008) The extremely anomalous molecular abundances of Comet 96P/Machholz 1 from narrowband photometry. *Astron. J., 136,* 2204–2213.
Schneider G., Grady C. A., Hines D. C., et al. (2014) Probing for exoplanets hiding in dusty debris disks: Disk imaging, characterization, and exploration with HST/STIS multi-roll coronagraphy. *Astron. J., 148,* 59.
Schüppler C., Krivov A. V., Löhne T., et al. (2016) Origin and evolution of two-component debris discs and an application to the q^1 Eridani system. *Mon. Not. R. Astron. Soc., 461,* 2146–2154.
Schwamb M. E., Jones R. L., Yoachim P., et al. (2023) Tuning the Legacy Survey of Space and Time (LSST) observing strategy for solar system science. *ArXiV e-prints,* arXiv:2303.02355.
Sekanina Z. (1976) A probability of encounter with interstellar comets and the likelihood of their existence. *Icarus, 27,* 123–133.
Seligman D. and Laughlin G. (2020) Evidence that 1I/2017 U1 ('Oumuamua) was composed of molecular hydrogen ice. *Astrophys. J. Lett., 896,* L8.
Seligman D. and Laughlin G. (2018) The feasibility and benefits of in situ exploration of 'Oumuamua-like objects. *Astron. J., 155,* 217.
Seligman D. Z., Levine W. G., Cabot S. H. C., et al. (2021) On the spin dynamics of elongated minor bodies with applications to a possible solar system analogue composition for 'Oumuamua. *Astrophys. J., 920,* 28.
Seligman D. Z., Becker J., Adams F. C., et al. (2022a) Inferring late-stage enrichment of exoplanet atmospheres from observed interstellar comets. *Astrophys. J. Lett., 933,* L7.
Seligman D. Z., Rogers L. A., Cabot S. H. C., et al. (2022b) The volatile carbon-to-oxygen ratio as a tracer for the formation locations of interstellar comets. *Planet. Sci. J., 3,* 150.
Sellek A. D., Booth R. A., and Clarke C. J. (2020) The evolution of dust in discs influenced by external photoevaporation. *Mon. Not. R. Astron. Soc., 492,* 1279–1294.
Sepulveda A. G., Matrà L., Kennedy G. M., et al. (2019) The REASONS survey: Resolved millimeter observations of a large debris disk around the nearby F star HD 170773. *Astrophys. J., 881,* 84.
Shannon A., Jackson A. P., Veras D., et al. (2015) Eight billion asteroids in the Oort cloud. *Mon. Not. R. Astron. Soc., 446,* 2059–2064.
Sibthorpe B., Kennedy G. M., Wyatt M. C., et al. (2018) Analysis of the Herschel DEBRIS Sun-like star sample. *Mon. Not. R. Astron. Soc., 475,* 3046–3064.
Siraj A. and Loeb A. (2022) A meteor of apparent interstellar origin in the CNEOS fireball catalog. *Astrophys. J., 939,* 53.
Siraj A. and Loeb A. (2019) Identifying interstellar objects trapped in the solar system through their orbital parameters. *Astrophys. J. Lett., 872,* L10.
Slettebak A. (1975) Some interesting bright southern stars of early type. *Astrophys. J., 197,* 137–138.
Smith B. A. and Terrile R. J. (1984) A circumstellar disk around β Pictoris. *Science, 226,* 1421–1424.
Snodgrass C. and Jones G. H. (2019) The European Space Agency's Comet Interceptor lies in wait. *Nature Commun., 10,* 5418.
Snodgrass C., Fitzsimmons A., Lowry S. C., et al. (2011) The size distribution of Jupiter family comet nuclei. *Mon. Not. R. Astron. Soc., 414,* 458–469.
Sosa A. and Fernández J. A. (2011) Masses of long-period comets derived from non-gravitational effects — Analysis of the computed results and the consistency and reliability of the non-gravitational parameters. *Mon. Not. R. Astron. Soc., 416,* 767–782.
Steckloff J. K., Lisse C. M., Safrit T. K., et al. (2021) The sublimative evolution of (486958) Arrokoth. *Icarus, 356,* 113998.
Stephan A. P., Naoz S., and Zuckerman B. (2017) Throwing icebergs at white dwarfs. *Astrophys. J. Lett., 844,* L16.
Sterken V. J., Westphal A. J., Altobelli N., et al. (2019) Interstellar dust in the solar system. *Space Sci. Rev., 215,* 43.
Stern S. A. (1990) ISM-induced erosion and gas-dynamical drag in the Oort cloud. *Icarus, 84,* 447–466.
Stern S. A. (2003) The evolution of comets in the Oort cloud and Kuiper belt. *Nature, 424,* 639–642.
Stern S. A., Shull J. M., and Brandt J. C. (1990) Evolution and detectability of comet clouds during post-main-sequence stellar evolution. *Nature, 345,* 305–308.
Stern S. A., Stocke J., and Weissman P. R. (1991) An IRAS search for extra-solar Oort clouds. *Icarus, 91,* 65–75.
Strøm P. A., Bodewits D., Knight M. M., et al. (2020) Exocomets from a solar system perspective. *Publ. Astron. Soc. Pac., 132,* 101001.
Strubbe L. E. and Chiang E. I. (2006) Dust dynamics, surface brightness profiles, and thermal spectra of debris disks: The case of AU Microscopii. *Astrophys. J., 648,* 652–665.
Su K. Y. L., Rieke G. H., Stansberry J. A., et al. (2006) Debris disk evolution around A stars. *Astrophys. J., 653,* 675–689.
Sugiura K., Kobayashi H., and Inutsuka S. (2019) Collisional elongation: Possible origin of extremely elongated shape of 1I/'Oumuamua. *Icarus, 328,* 14–22.
Telesco C. M., Fisher R. S., Wyatt M. C., et al. (2005) Mid-infrared images of β Pictoris and the possible role of planetesimal collisions in the central disk. *Nature, 433,* 133–136.
Terebey S., Shu F. H., and Cassen P. (1984) The collapse of the cores of slowly rotating isothermal clouds. *Astrophys. J., 286,* 529–551.
Thébault P. and Beust H. (2001) Falling evaporating bodies in the β Pictoris system: Resonance refilling and long term duration of the phenomenon. *Astron. Astrophys., 376,* 621–640.
Throop H. B. and Bally J. (2005) Can photoevaporation trigger planetesimal formation? *Astrophys. J. Lett., 623,* L149–L152.
Trilling D. E., Robinson T., Roegge A., et al. (2017) Implications for planetary system formation from interstellar object 1I/2017 U1 ('Oumuamua). *Astrophys. J. Lett., 850,* L38.
Trilling D. E., Mommert M., Hora J. L., et al. (2018) Spitzer observations of interstellar object 1I/'Oumuamua. *Astron. J., 156,* 261.
Tychoniec Ł., Manara C. F., Rosotti G. P., et al. (2020) Dust masses of young disks: Constraining the initial solid reservoir for planet formation. *Astron. Astrophys., 640,* A19.
Valtonen M. J. and Innanen K. A. (1982) The capture of interstellar comets. *Astrophys. J., 255,* 307–315.
Valtonen M. J. and Zheng J. Q. (1990) On the origin or Oort's cloud of comets in a star cluster. In *Nordic-Baltic Astronomy Meeting* (C. I. Lagerkvist et al., eds.), pp. 353–360. Uppsala Univ., Uppsala, Sweden.
Vavilov D. E. and Medvedev Y. D. (2019) Dust bombardment can explain the extremely elongated shape of 1I/'Oumuamua and the lack of interstellar objects. *Mon. Not. R. Astron. Soc., 484,* L75–L78.
Veras D. (2016) Post-main-sequence planetary system evolution. *R. Soc. Open Sci., 3,* 150571.
Veras D., Reichert K., Flammini Dotti F., et al. (2020) Linking the formation and fate of exo-Kuiper belts within solar system analogues. *Mon. Not. R. Astron. Soc., 493,* 5062–5078.
Villaver E., Livio M., Mustill A. J., et al. (2014) Hot Jupiters and cool

stars. *Astrophys. J., 794,* 3.
Vitense C., Krivov A. V., Kobayashi H., et al. (2012) An improved model of the Edgeworth-Kuiper debris disk. *Astron. Astrophys., 540,* A30.
Wahhaj Z., Milli J., Kennedy G., et al. (2016) The SHARDDS survey: First resolved image of the HD 114082 debris disk in the Lower Centaurus Crux with SPHERE. *Astron. Astrophys., 596,* L4.
Weissman P. R. (1999) Diversity of comets: Formation zones and dynamical paths. *Space Sci. Rev., 90,* 301–311.
Weissman P. R. (1996) The Oort cloud. In *Completing the Inventory of the Solar System* (T. Rettig and J. M. Hahn, eds.), pp. 265–288. ASP Conf. Ser. 107, Astronomical Society of the Pacific, San Francisco.
Westphal A. J., Stroud R. M., Bechtel H. A., et al. (2014) Evidence for interstellar origin of seven dust particles collected by the Stardust spacecraft. *Science, 345,* 786–791.
Williams J. P. and Cieza L. A. (2011) Protoplanetary disks and their evolution. *Annu. Rev. Astron. Astrophys., 49,* 67–117.
Wilson P. A., Kerr R., Lecavelier des Etangs A., et al. (2019) Detection of nitrogen gas in the β Pictoris circumstellar disc. *Astron. Astrophys., 621,* A121.
Witze A. (2019) Hawaiian-language experts make their mark on the solar system. *Nature, 565,* 278–279.
Wyatt M. C. (2008) Evolution of debris disks. *Annu. Rev. Astron. Astrophys., 46,* 339–383.
Wyatt M. C. (2003) Resonant trapping of planetesimals by planet migration: Debris disk clumps and Vega's similarity to the solar system. *Astrophys. J., 598,* 1321–1340.
Wyatt M. C., Dermott S. F., Telesco C. M., et al. (1999) How observations of circumstellar disk asymmetries can reveal hidden planets: Pericenter glow and its application to the HR 4796 disk. *Astrophys. J., 527,* 918–944.
Wyatt M. C., Clarke C. J., and Greaves J. S. (2007) Origin of the metallicity dependence of exoplanet host stars in the protoplanetary disc mass distribution. *Mon. Not. R. Astron. Soc., 380,* 1737–1743.
Wyatt M. C., Kennedy G., Sibthorpe B., et al. (2012) Herschel imaging of 61 Vir: Implications for the prevalence of debris in low-mass planetary systems. *Mon. Not. R. Astron. Soc., 424,* 1206–1223.
Wyatt M. C., Panić O., Kennedy G. M., et al. (2015) Five steps in the evolution from protoplanetary to debris disk. *Astrophys. Space Sci., 357,* 103.
Xing Z., Bodewits D., Noonan J., et al. (2020) Water production rates and activity of interstellar Comet 2I/Borisov. *Astrophys. J. Lett., 893,* L48.
Xu S., Dufour P., Klein B., et al. (2019) Compositions of planetary debris around dusty white dwarfs. *Astron. J., 158,* 242.
Ye Q., Kelley M. S. P., Bolin B. T., et al. (2020) Pre-discovery activity of new interstellar Comet 2I/Borisov beyond 5 au. *Astron. J., 159,* 77.
Yelverton B., Kennedy G. M., and Su K. Y. L. (2020) No significant correlation between radial velocity planet presence and debris disc properties. *Mon. Not. R. Astron. Soc., 495,* 1943–1957.
Youdin A. N. and Goodman J. (2005) Streaming instabilities in protoplanetary disks. *Astrophys. J., 620,* 459–469.
Zellner N. E. B. (2017) Cataclysm no more: New views on the timing and delivery of lunar impactors. *Origins Life Evol. Biosphere, 47,* 261–280.
Zhang Q. (2018) Prospects for backtracing 1I/'Oumuamua and future interstellar objects. *Astrophys. J. Lett., 852,* L13.
Zhang Y. and Lin D. N. C. (2020) Tidal fragmentation as the origin of 1I/2017 U1 ('Oumuamua). *Nature Astron., 4,* 852–860.
Zhao Y., Rezac L., Skorov Y., et al. (2021) Sublimation as an effective mechanism for flattened lobes of (486958) Arrokoth. *Nature Astron., 5,* 139–144.
Zheng J.-Q., Valtonen M. J., and Valtaoja L. (1990) Capture of comets during the evolution of a star cluster and the origin of the Oort cloud. *Cel. Mech. Dyn. Astron., 49,* 265–272.
Zieba S., Zwintz K., Kenworthy M. A., et al. (2019) Transiting exocomets detected in broadband light by TESS in the β Pictoris system. *Astron. Astrophys., 625,* L13.
Zuckerman B. and Song I. (2012) A 40 Myr old gaseous circumstellar disk at 49 Ceti: Massive CO-rich comet clouds at young A-type stars. *Astrophys. J., 758,* 77.
Zuckerman B., Forveille T., and Kastner J. H. (1995) Inhibition of giant-planet formation by rapid gas depletion around young stars. *Nature, 373,* 494–496.

Jewitt D. and Hsieh H. H. (2024) The asteroid-comet continuum. In *Comets III* (K. J. Meech, M. R. Combi, D. Bockelée-Morvan, S. N. Raymond, and M. E. Zolensky, eds.), pp. 767–798. Univ. of Arizona, Tucson, DOI: 10.2458/azu_uapress_9780816553631-ch023.

The Asteroid-Comet Continuum

David Jewitt
University of California at Los Angeles

Henry H. Hsieh
Planetary Science Institute

The practical distinctions between asteroids and comets, viewed as products of accretion on either side of the snow line, are less clear-cut than previously understood. In this chapter, we discuss the numerous solar system populations that have physical and dynamical properties that conflict with any simple diagnosis of their nature and origin. Studies of these so-called "continuum" or "transition objects," which include many of the most intriguing bodies in the solar system, have implications for a broad range of scientific topics from the demise of comets and the activation of asteroids to the production of interplanetary debris and the origin of the terrestrial planet volatiles. We present an overview of the current state of knowledge concerning the asteroid-comet continuum and discuss the numerous physical processes behind the activity shown by small bodies in the solar system.

1. INTRODUCTION

Traditionally, small solar system bodies have been classified as either asteroids (complex assemblages of refractory minerals) or comets (ice-rich bodies containing a mix of volatile and refractory solids). Observationally, asteroids and comets are differentiated by the absence or presence of detectable mass loss. Comets, in essence, appear fuzzy while asteroids do not. Cometary activity typically takes the form of comae (unbound ejected material surrounding the central body) and tails.

Meanwhile, the classical dynamical distinction between comets and asteroids is based on the Tisserand parameter (with respect to Jupiter)

$$T_J = \frac{a_J}{a} + 2\left[(1-e^2)\frac{a}{a_J}\right]^{1/2} \cos(i) \quad (1)$$

where a, e, and i are the orbital semimajor axis, eccentricity, and inclination, respectively, and a_J = 5.2 au is the semimajor axis of Jupiter. T_J parameterizes the relative velocity between an object and Jupiter at their closest approach and is approximately conserved in the circular restricted three-body dynamics approximation (*Tisserand,* 1896; *Murray and Dermott,* 2000), even in the event of close encounters (*Carusi et al.,* 1995). Jupiter itself has T_J = 3. Asteroids typically have $T_J > 3$ and comets have $T_J < 3$ (*Vaghi,* 1973).

To make sense of the alphabet soup used to describe small solar system populations, we show in Fig. 1 a simple classification scheme based on T_J and whether or not observable mass loss exists. Within the comet population, long-period comets (LPCs) have $T_J < 2$ and orbital periods $P_{orb} \geq 200$ yr, Halley-type comets (HTCs) have $T_J < 2$ and $P_{orb} < 200$ yr, and Jupiter-family comets (JFCs) have $2 \leq T_J \leq 3$ (*Levison,* 1996). The different orbital properties of

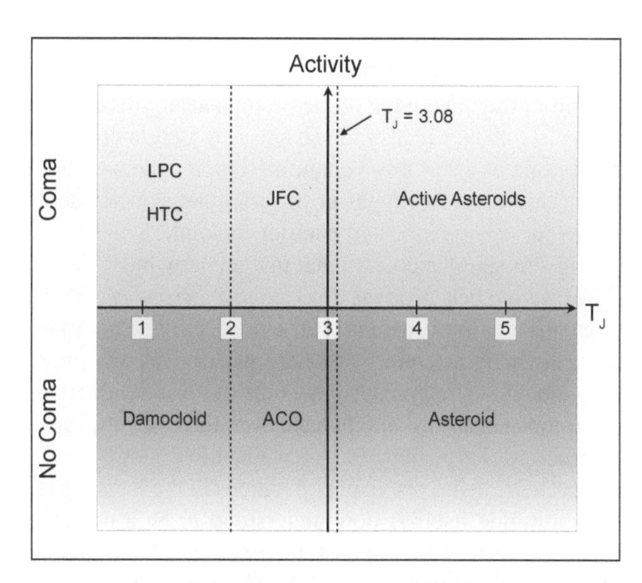

Fig. 1. Classification diagram for small bodies discussed in this chapter, showing the Tisserand parameter with respect to Jupiter (T_J) vs. whether or not a coma has ever been detected. Acronyms are explained in section 1. Centaurs (not shown) can appear anywhere in this plot. The vertical dashed line marked T_J = 3.08 denotes the nominal lower bound for active asteroids.

these populations reflect their origins, with LPCs supplied from the Oort cloud, JFCs from the Kuiper belt, and HTCs from a thus far ambiguous source, with likely contributions from both the Oort cloud and the scattered disk component of the Kuiper belt (*Dones et al.,* 2015; see also the chapter by Kaib and Volk in this volume). Damocloids and asteroids on cometary orbits (ACOs) are inactive bodies with $T_J < 2$ and $2 \le T_J \le 3$, respectively, paralleling the distinction between LPCs and JFCs.

In the classical view of the solar system described above, asteroids and comets formed at different temperatures in locations interior and exterior to the snow line, respectively, and have been preserved more or less in their formation locations for the past 4.5 G.y. In principle, these different formation conditions and source regions should produce objects with distinct physical and dynamical properties. As such, small solar system objects should be uniquely classifiable into one of these categories or the other.

Recently, however, a more nuanced picture has emerged, with classical asteroids and comets now understood as simply end members of observational, physical, and dynamical continua. A key practical problem with the observational classification of objects based only on observations of activity is that it is dependent on instrumental sensitivity (section 2). Furthermore, we now know that many other processes besides ice sublimation can lead to visible mass loss (section 3), meaning that activity is possible for a much broader range of objects beyond just those containing ice.

We also now recognize that asteroids and comets did not originate in wholly distinct regions of the solar system, as was formerly believed (section 4). Comets may be less volatile-rich than the half-rock, half-ice mixture envisioned by *Whipple* (1950), and they contain a curious combination of the most volatile ices from the outermost regions of the protoplanetary disk and high-temperature, crystalline silicates from the inner edge of the protoplanetary disk (*Westphal et al.,* 2009; see also the chapters by Bergin et al. and Filacchione et al. in this volume). Isotopic evidence from meteorites shows two types of material that accreted separately (perhaps interior and exterior to Jupiter's orbit, and perhaps not simultaneously), but that are now intermingled in the asteroid belt (*Warren,* 2011; *Lichtenberg et al.,* 2021). As we discuss later, there is clear evidence for near-surface ice in the main asteroid belt where its survival was previously thought to be implausible. Some of this mixing across the protoplanetary disk might have been forced by the radial migration of the planets, the latter identified first from the unexpectedly dense resonant populations in the Kuiper belt. The snow line itself is now understood to be a dynamic entity that moved in response to time-dependent heating sources in the protoplanetary disk (*Harsono et al.,* 2015).

Finally, dynamically, T_J is an imperfect discriminant due to its neglect of planetary perturbers other than the Sun and Jupiter and lack of consideration of nongravitational perturbations due to asymmetric outgassing. We now know that not only is the dynamical boundary between asteroids and comets better characterized as a range of T_J values rather than a sharp threshold at a single T_J value (i.e., $T_J = 3$), the boundary is also porous, with objects capable of passing through to the other side in both directions, so obscuring their true dynamical origins (section 5).

In this chapter, we review the current state of knowledge concerning objects that sit in the ambiguous space between classical rocky asteroids and classical icy comets. We will consider objects that originate as either classical comets or classical asteroids and evolve observationally, physically, or dynamically into the other type of object, as well as objects for which having characteristics of both asteroids and comets is simply part of their intrinsic nature. While these objects have often been termed "transition objects" in earlier literature, the title of this chapter reflects the fact that not all of these objects are actually in the midst of transitioning from one type of small body to another. Instead, there is simply a broad diversity of objects that span the range of observational, physical, and dynamical properties commonly attributed to either asteroids or comets in a continuous, rather than discrete, fashion (e.g., *Hsieh,* 2017).

In addition to the topics above, we will discuss key examples of objects that have properties of both classical asteroids and comets: active asteroids (section 6), inactive comets (section 7), and Centaurs (section 8). While these do not comprise an exhaustive list, they provide instructive examples relevant to the entire asteroid-comet continuum for small solar system bodies. We will then conclude with a discussion of future research prospects (section 9).

2. ACTIVITY DETECTION

2.1. Overview

One factor in the growing recognition of asteroid-comet continuum objects is a steady improvement in the ability of astronomers to detect weak activity, enabled by the use of larger, more sensitive telescopes, and by increasing numbers of wide-field, time-resolved surveys. A second factor is the recognition that weak activity exists to be detected, which adds an equally important, albeit psychological, dimension to the growing rate with which these objects are perceived. The most direct method of activity detection is resolved imaging, which typically reveals the presence of comet-like features like comae or tails. Other methods, including photometric analysis, spectroscopic detection of gas, detection of debris streams associated with small solar system bodies, and detection of non-gravitational perturbations, can also reveal the presence of activity.

2.2. Resolved Imaging

Resolved imaging of mass loss is the gold standard and most common means by which objects are determined to be active. The optical scattering efficiency (measured as cross-section per unit mass) of dust is much larger than that of resonance fluorescence from common molecules. Therefore, direct imaging typically reveals only ejected dust, whether or not gas is present. Even in classical comets, for which we are sure that gas drag drives mass loss, gas is commonly undetectable against the continuum of sunlight scattered from dust, particularly in observations taken at heliocentric distances $r_H \gtrsim 2$ au.

For weakly active objects, direct imaging of mass loss requires sensitivity to faint near-nucleus surface brightness features and is thus a function of telescope size, angular resolution, sky brightness, and more. Most active asteroids and Centaurs have been discovered in data from wide-field surveys, for which activity detection is typically not prioritized. Unfortunately, the surface brightness detection limits of most of those surveys are not well documented, making it difficult to use their data to quantitatively characterize populations of active bodies. Two exceptions are the wide but shallow survey of *Waszczak et al.* (2013) and the deep but narrow-field "Hawaii Trails Project," where the latter survey resulted in the discovery of a fan-like dust tail associated with 176P/LINEAR, then known as asteroid (118401) 1999 RE_{70}, with an R-band surface brightness of $\Sigma_R = 25.3$ mag arcsec^{-2} (*Hsieh,* 2009). The relatively shallow depths of most wide field surveys, though, mean that low-level activity is not well constrained for the vast majority of asteroids. Many asteroids that are currently considered inactive could in fact be weakly active but have thus far escaped detection (*Sonnett et al.,* 2011).

This view is anecdotally supported by the discovery of activity in (3200) Phaethon, (3552) Don Quixote, and (101955) Bennu. Phaethon (section 6.3.5) and Don Quixote (*Mommert et al.,* 2014) were only found to be active after intensive targeted observational efforts motivated by independent indicators that the objects could be active (i.e., Phaethon's association with the Geminid meteor stream and Don Quixote's dynamically cometary orbit with $T_J = 2.3$), while activity was observed unexpectedly *in situ* for Bennu by the visiting Origins Spectral Interpretation Resource Identification and Security-Regolith Explorer (OSIRIS-REx) spacecraft (Fig. 2; section 6.3.6).

Identification of active objects is also complicated by the fact that activity is usually transient. Active icy objects can become dormant when their surface volatiles are depleted or buried by a non-volatile crust (*Jewitt,* 1996). Additionally, even modest orbital eccentricities can strongly affect water ice sublimation rates (because sublimation is an exponential function of temperature and hence of heliocentric distance) and therefore dust production rates. Thus, even very deep observations of main-belt objects away from perihelion do not exclude the possibility of sublimation-driven activity when closer to the Sun (e.g., *Hsieh et al.,* 2015a). Meanwhile, visible mass loss caused by other processes, e.g., impacts and rotational destabilization, can be very short-lived and can occur at any time (not just near perihelion), meaning that it is almost always discovered by chance in all-sky surveys (e.g., *Birtwhistle et al.,* 2010; *Larson,* 2010; *Smith et al.,* 2019). As such, the vast majority of such events have undoubtedly gone undetected due to the fact that we are, and will continue to be for the foreseeable future, unable to observe all small solar system bodies at all times.

Fig. 2. Reverse polarity image of 0.5 km diameter asteroid (101955) Bennu from OSIRIS-REx showing a swarm of centimeter-sized particles in the lower left of the image. Credit: NASA/Univ. of Arizona.

2.3. Photometric Detection

Very sensitive searches for activity can be conducted by comparing measurements of the apparent magnitude, $m_\lambda(r_H, \Delta, \alpha)$, of an object in a filter with effective wavelength, λ, with predictions based on the absolute magnitude in the same filter, H_λ. The absolute magnitude is defined by

$$H_\lambda = m_\lambda(r_H,\Delta,\alpha) - 5\log_{10}(r_H\Delta) + 2.5\log_{10}(\Phi(\alpha)) \quad (2)$$

where r_H and Δ are the object's instantaneous heliocentric and geocentric distances, respectively, in astronomical units, and $0 < \Phi(\alpha) < 1$ is the phase function. (Note that the phase function is equal to the ratio of the scattered light as phase angle α to that at $\alpha = 0°$. It is affected both by the apparent illuminated fraction of the surface and by microscopic scattering effects in the regolith.) Photometric measurements that are brighter than would be expected from an object's absolute magnitude and rotational light curve amplitude would then imply the presence of unresolved ejected dust within the seeing disk of that object.

Anomalous brightening was first used to identify the activity of (2060) Chiron (*Hartmann et al.,* 1990), with resolved coma directly imaged soon afterwards (*Luu and Jewitt,* 1990; *Meech and Belton,* 1990). Since then, photometry has been used to search for and identify activity in other objects (e.g., *Cikota et al.,* 2014; *Hsieh and Sheppard,* 2015). Unfortunately, absolute magnitudes for most asteroids are computed from imprecise photometric data collected by the International Astronomical Union's (IAU) Minor Planet Center, limiting the accuracy with which their brightnesses can be predicted at any given time. An additional limitation is imposed by the unknown rotational variability of many objects. As a result, while photometric analysis provides very clear detections of activity for bright comets already showing resolved comae and tails (e.g., Fig. 7 of *Ferrín and Orofino,* 2021), results are ambiguous for objects with poorly characterized nucleus properties, as is true of most asteroids. We expect that photometric activity detection will become much more effective when data from the Vera C. Rubin Observatory (see section 9) become available. Those data will enable both the computation of precise absolute magnitudes for large numbers of small solar system bodies and precision photometry in individual observations for activity detection.

2.4. Spectroscopic Detection

Spectroscopy is a powerful tool for studying active objects, as it is essentially the only means for acquiring unequivocal evidence of volatile sublimation and can also provide additional detail about the types of volatiles present. In practice, however, spectroscopy has so far played only a small role in the study of asteroid-comet continuum objects.

At optical wavelengths, two major limitations of spectroscopy are lack of sensitivity to the dominant volatile (water) and the strong heliocentric dependence of resonance fluorescence emission. Water has no observable bands in the optical, forcing the use of trace species (e.g., CN, which has an emission band around 3883 Å) as proxies. For example, one hour of spectroscopic observations targeting CN with the Keck telescope reaches production rates $Q_{CN} \sim 10^{22}$ s^{-1} at r_H = 1 au, falling to $Q_{CN} \sim 10^{24}$ s at r_H = 3 au (Jewitt, 2012). The water production rate in JFCs is 400 times that of CN (*A'Hearn et al.,* 1995), implying spectroscopic detection limits of $Q_{H_2O} \sim 4 \times 10^{24}$ s^{-1} (~0.1 kg s^{-1}) at r_H = 1 au but only $Q_{H_2O} \sim 4 \times 10^{26}$ s^{-1} (~10 kg s^{-1}) at r_H = 3 au. This is why JFCs and LPCs that are visibly active at ≥3 au and active Centaurs at much largely distances commonly show no optical evidence for gas. Worse, there is no fundamental reason why the ratio of CN to water in classical comets should be applicable to other objects like active asteroids, and in fact, reason to believe that it may be far lower (e.g., *Prialnik and Rosenberg,* 2009), meaning that water production rate limits inferred from CN observations are of dubious value.

Spectroscopic observations at non-optical wavelengths are better suited to studying volatile sublimation (see the chapter by Bodewits et al. in this volume) but have not been used extensively for the subjects of this chapter. One exception is the successful detection of outgassing by (1) Ceres in the form of multiple detections of the 1_{10}–1_{01} ground-state transition line of ortho-water at 556.939 GHz using the Herschel Space Observatory (*Küppers et al.,* 2014). Detections of CO rotational transitions at submillimeter and millimeter wavelengths have also been secured in comets and Centaurs at large heliocentric distances (see the chapter by Fraser et al. in this volume).

2.5. Debris Stream Detection

The association between meteoroid streams and active comets has been long recognized (e.g., *Whipple,* 1951). There is also a growing number of identified associations between meteoroid streams and asteroids (see *Jenniskens,* 2008, 2015). The latter associations have been interpreted as indicating unseen mass loss from those otherwise apparently inert bodies. Famously, the association of the Geminid meteoroid stream with Phaethon (*Gustafson,* 1989; *Williams and Wu,* 1993) led to the first suggestions that Phaethon was losing mass (see section 6.3.5). Other suspected asteroidal meteor stream parents (e.g., *Ryabova,* 2002; *Babadzhanov,* 2003; *Babadzhanov et al.,* 2015) could also be experiencing unseen mass loss (see the chapter by Ye and Jenniskens in this volume). Divergence between the orbital elements of debris streams and their parents results from gravitational perturbations by the planets and, potentially, from unknown non-gravitational effects (e.g., *Kaňúchová and Neslušan,* 2007).

Finally, several asteroids, notably the near-Earth asteroids (NEAs) 2201 Oljato and 138175 (2000 EE104), are associated with distinct and repeated disturbances in the magnetic field carried by the solar wind, called interplanetary field enhancements (IFEs) (section 3.8), tentatively interpreted as drag from charged nanodust particles left behind along the orbits of these bodies.

2.6. Non-Gravitational Acceleration

Anisotropic mass loss from a small body creates a recoil force that can measurably perturb the orbit relative to purely gravitational motion, where non-gravitational accelerations are well-known to exist for comets (*Whipple,* 1950). Sublimation-driven mass loss from a spherical nucleus of radius r_n and density ρ at rate M leads to an acceleration

$$\alpha_{ng} = \frac{3k_R V_g \dot{M}}{4\pi\rho r_n^3} \tag{3}$$

in which V_g is the gas velocity and $k_R \sim 0.5$ is a dimensionless constant to account for the fact that the flow is neither perfectly isotropic (k_R = 0) nor perfectly collimated (k_R = 1). For illustration, for plausible values V_g = 500 m s^{-1}, ρ = 500 kg m^{-3}, and $r_n = 10^3$ m, we find $\alpha_{ng} = 10^{-10}$ Mm s^{-2}. Expressed as a fraction of the local gravitational attraction to the Sun at 3 au ($g_\odot = 7 \times 10^{-4}$ m s^{-2}), this is $\alpha_{ng}/g_\odot \sim 1.7 \times 10^{-7}$ M, which should be detectable for the best astrometrically observed asteroids for production rates of M ~ 1 to 10 kg s^{-1} (*Hui and Jewitt,* 2017).

This method is limited in its application mainly by the reliability of astrometry, particularly when plate-era data are used to obtain long astrometric arcs.

3. ACTIVITY MECHANISMS

3.1. Overview

A variety of mechanisms in addition to sublimation can lead to observable mass loss, expanding the range of objects that we must consider to have the potential for activity. We briefly discuss the major mass loss mechanisms, which may operate alone or in conjunction.

3.2. Ice Sublimation

There is now substantial evidence that water ice is currently present (and possibly widespread) in objects with dynamically stable orbits in the main asteroid belt, and not just in objects from the outer solar system. This revelation raises the possibility of being able to conduct present-day studies of ice possibly from a part of the protosolar disk not sampled by the classical comet population, and also presents new opportunities for studies of the thermal and dynamical evolution of small bodies both in the early and present-day solar system (e.g., *Schörghofer,* 2008; *DeMeo and Carry,* 2014; *Hsieh and Haghighipour,* 2016).

Exposed, dirty (low-albedo) water ice at asteroid belt distances is quickly lost by sublimation. However, *Fanale and Salvail* (1989) noted that water ice can persist at relatively shallow depths (e.g., a few centimeters to meters) when protected by a non-volatile surface layer. Its persistence is more likely at high latitudes on objects with larger semimajor axes, slow rotation rates, low thermal conductivity, and low obliquity. *Schörghofer* (2008) confirmed these findings, introducing the concept of a "buried snowline," i.e., the distance from the Sun beyond which subsurface ice can persist over the age of the solar system. That said, near-surface asteroid ice is likely to be dominated by water ice, given the greater thermal instability of other volatiles in the asteroid belt (*Prialnik and Rosenberg,* 2009). *Schörghofer et al.* (2020) combined models of the thermal and dynamical evolution of Themis asteroid family members, finding that most of their subsurface ice would be retained even during transitions to near-Earth orbits, assuming stable spin-pole orientations (Fig. 4). This raises the intriguing possibility that asteroid ice could one day be sampled by a near-Earth mission and reminds us that the outer asteroid belt could be a source of at least some terrestrial planet volatiles.

Buried ice can be exposed and activated by triggering events like impacts (e.g., *Haghighipour et al.,* 2016), rotationally-induced landslides (e.g., *Scheeres,* 2015; *Steckloff et al.,* 2016), and inward drift of the perihelion distance (e.g., *Rickman et al.,* 1990; *Fernández et al.,* 2018). This is only possible, however, if the ice depth is shallow, because deep disturbance of the surface is unlikely. Shallow ice could exist on fragments from the catastrophic disruption of larger parent bodies, as in those that form asteroid families (see *Nesvorný et al.,* 2008, 2015; *Novaković et al.,* 2012). Ice-containing fragments would be more easily triggered by, for example, small impacts excavating just a few meters of the surface (*Hsieh et al.,* 2018c).

Ice sublimation rates are very sensitive to temperature, such that even modest temperature differences between perihelion and aphelion give rise to substantial variations in activity strength and therefore detectability. Neglecting a small term due to conduction, the energy balance equation at the surface of a sublimating gray body may be written

$$\frac{F_\odot}{r_H^2}(1-A) = \chi\left[\varepsilon\sigma T^4 + Lf_D \dot{m}_w(T)\right] \tag{4}$$

where T is the equilibrium surface temperature of the body, $F_\odot = 1360$ W m^{-2} is the solar constant, heliocentric distance r_H is in astronomical units, A = 0.05 is the assumed Bond albedo, the distribution of solar heating is parameterized by χ, σ is the Stefan Boltzmann constant, $\varepsilon = 0.9$ is the assumed effective infrared emissivity, L = 2.83 MJ kg^{-1} is the (nearly temperature-independent) latent heat of sublimation of water ice, f_D represents the reduction in sublimation efficiency caused by a growing rubble mantle, where $f_D = 1$ in the absence of a mantle, and $\dot{m}_w$ is the water mass loss rate due to sublimation. In this equation, $\chi = 1$ applies to a flat slab facing the Sun at all times, known as the subsolar approximation, and produces the maximum attainable temperature, while $\chi = 4$ applies to an isothermal surface (e.g., for a fast-rotating body or low thermal inertia), and corresponds to the minimum expected temperature.

The sublimation rate of ice into a vacuum is given by

$$\dot{m}_w = P_v(T)\left(\frac{\mu m_H}{2\pi kT}\right)^{1/2} \tag{5}$$

where $m_H = 1.67 \times 10^{-27}$ kg is the mass of the hydrogen atom, $\mu = 18$ is the molecular weight of water, and k is the Boltzmann constant. The corresponding ice recession rate, $\dot{\ell}_i$, is given by $\dot{\ell}i = \dot{m}_w/\rho$, for an object with a bulk density ρ. The temperature-dependent sublimation pressure, P(T), is obtained from the Clausius-Clapeyron relation, or from experimental data. Solving equations (4) and (5) iteratively gives the equilibrium temperature and the sublimation rate of a sublimating gray-body at a given heliocentric distance.

Expected water sublimation rates computed using the equations above, assuming $f_D = 1$, for the extreme subsolar and isothermal approximations are plotted in Fig. 3. We see from the figure that even objects whose orbits keep them entirely confined to the main asteroid belt have expected water sublimation rates that can vary by several orders of magnitude from aphelion to perihelion. For main-belt objects that are only weakly active near perihelion, despite their modest eccentricities, it is therefore entirely plausible for sublimation in these objects to be detectably strong near perihelion and undetectably weak near aphelion, just like classical comets with much larger perihelion-to-aphelion heliocentric distance excursions.

Observationally, the expected characteristics of sublimation-driven activity include long-duration mass loss and recurrent activity (e.g., *Hsieh et al.,* 2012a). Unlike mass loss produced by impulsive events like impacts (section 3.3), sublimation-driven activity is expected to persist as long as an object is warm enough for sublimation to occur and volatile material remains. Determining the duration of emission events often requires the use of numerical dust models, however, which are typically underconstrained by observations. Recurrent activity correlated with perihelion strongly suggests thermal modulation of the mass-loss process, which is a natural characteristic of volatile sublimation. Other mechanisms require more contrived conditions to produce similar behavior (e.g., *Hsieh et al.,* 2004), leading us to consider recurrent mass loss near perihelion to be the most dependable indicator of sublimation-driven activity currently available using remote observations (Fig. 4).

3.3. Impacts

Images of asteroids from spacecraft reveal cratered landscapes that record a violent collisional past. Observations of the consequences of impacts can place important constraints on the material properties and internal structure of the bodies that experienced those impacts, motivating decades of crater studies (*Marchi et al.,* 2015), theoretical modeling (*Jutzi et al.,* 2015), and impact experiments in laboratory settings (e.g., *Housen et al.,* 2018), while impacts can also excavate subsurface material, which can then be directly studied. Excitingly, impact experiments on real-world solar system bodies (e.g., *A'Hearn et al.,* 2005; *Schultz et al.,* 2010; *Saiki et al.,* 2017) as well as studies of natural impact events occurring in real time (see section 6.3.2) have also recently become possible. The velocity dispersion in the asteroid belt is ~5 km s^{-1}, resulting in intermittent asteroid-asteroid collisions that are highly erosive or, if sufficiently energetic, destructive. Collisions also occur at the smallest scales, leading to steady erosion by micrometeoroids. Although small and large impacts are part of a continuum, it is useful to describe these extremes separately.

3.3.1. Micrometeoroid erosion. Micrometeoroid impacts set an absolute lower limit to the rate of steady mass loss from a body. Few impacting particles can penetrate deeply because of the steep size and energy distributions of projectiles, and so the erosion is fastest at the surface. On the Moon, where the gravity is non-negligible ($g_☾$ = 1.6 m s^{-2}), micrometeoroid impact-produced fragments are recaptured, forming a regolith that is overturned or "gardened" by continued impacts. Measurements of the lunar gardening rate show that the top 1 m of the regolith is overturned in about 10^9 yr while the top 1 cm is mixed in only 10^6 yr (*Arvidson et al.,* 1975; *Heiken et al.,* 1991; *Horz and Cintala,* 1997). To

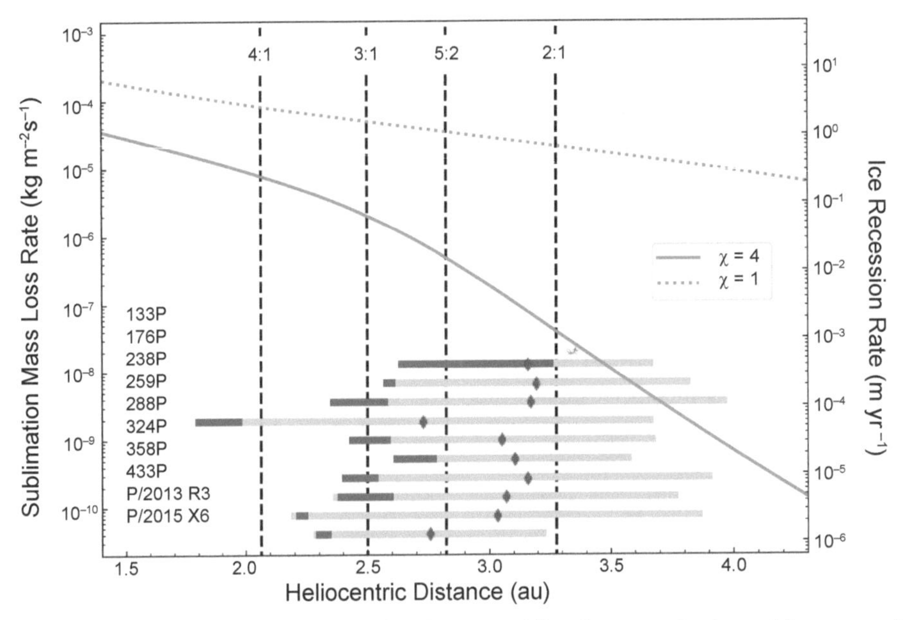

Fig. 3. Plot of mass loss rates due to water sublimation from a sublimating gray body and ice recession rates per year as functions of heliocentric distance over the range of the main asteroid belt using the isothermal approximation (χ = 4; solid curved green line) and subsolar approximation (χ = 1; dotted green line). The positions of the 4:1, 3:1, 5:2, and 2:1 mean-motion resonances with Jupiter are marked with vertical dashed black lines. Also plotted are horizontal light gray bars showing the full heliocentric ranges of a selection of main-belt comet (MBC) orbits (see section 6.2) with overlaid blue bars showing the heliocentric range over which activity was detected for those MBCs, and red diamonds indicating each MBC's semimajor axis distance. After *Hsieh et al.* (2015a).

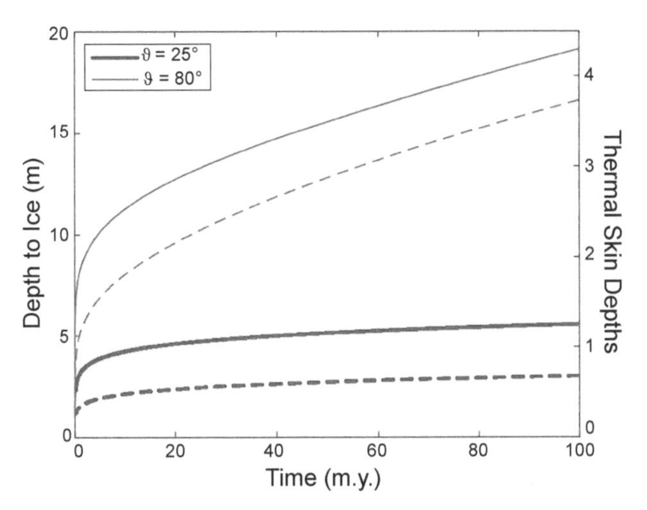

Fig. 4. Plot of depth to ice in a thermal model as a function of time for a main-belt asteroid in a stationary orbit with a = 3 au and e = 0 for obliquities of ϑ = 25° (lower set of curves) and ϑ = 80° (upper set of curves). The assumed thermal diffusivity is $\kappa = 3.8 \times 10^{-7}$ m² s⁻¹. Within the first thermal skin depth, ice retreat is rapid, but total retreat is small for ϑ = 25°, even after 100 m.y. Solid lines mark ice evolution using uniform thermal properties for all regolith, while dashed lines mark ice evolution using different thermal properties for ice-free and ice-rich regolith. From *Schörghofer et al.* (2020).

set a crude upper limit to asteroid erosion, we assume that the material that is gardened on the Moon would instead be launched above escape speed and lost from a kilometer-sized (low-gravity) asteroid, and that the impact rates on the Moon and in the asteroid belt are comparable. We take the higher lunar value, $dr/dt = 10^{-8}$ m yr^{-1}, as a representative rate of loss. With $\rho = 10^3$ kg m^{-3}, the micrometeorite erosion mass flux is $\mathcal{L} = \rho dr/dt \sim 10^{-5}$ kg m^{-2} yr^{-1}.

Sandblasting by micrometeorites, at a rate given by $\dot{M}_\mu = 4\pi r_n^2 \mathcal{L}$ for a spherical body of radius r_n, is evidently very slow. Expressing r_n in kilometers, we obtain

$$\dot{M}_\mu = 4 \times 10^{-6} r_n^2 \text{ kg s}^{-1} \quad (6)$$

A r_n = 1 m boulder would have a ~3×10^7 yr lifetime against micrometeoroid erosion, while a r_n = 1 km body would lose ~4×10^{-6} kg s^{-1}, and could do so indefinitely. Other than for 0.25-km-radius Bennu, whose ~10^{-7} kg s^{-1} mass loss rate (*Hergenrother et al.,* 2020) is ~0.4 $\dot{M}_\mu$, all remote detections of asteroid, comet, and Centaur activity have thus far exceeded $\dot{M}_\mu$ by 3 to 10 orders of magnitude.

3.3.2. Macroscopic impacts. Most of the mass and available impact energy in the asteroid population is carried by the largest projectiles, but large bodies and impacts are rare and their effects are transient. Nevertheless, individual large impactors are easily capable of generating observable quantities of dust. Several impact events have been detected in the main asteroid belt within the last decade. The best example of impact destruction of an asteroid is provided by the 100-m-scale 354P/LINEAR (section 6.3.2) (*Jewitt et al.,* 2010; *Snodgrass et al.,* 2010; *Kim et al.,* 2017a). P/2016 G1 may also be a collisionally disrupted asteroid of similar scale (*Moreno et al.,* 2016b). Similar collisions must also occur in the Kuiper belt but have yet to be observed.

Figure 5 shows the timescale for the impact destruction of a given main-belt asteroid as a function of its diameter, D, where we see that a D = 1 km asteroid has a collisional lifetime ~0.4 G.y., decreasing to ~60 m.y. for D = 0.1 km. Asteroids larger than D ~ 10 km, on average, should survive against collisions over the 4.5-G.y. age of the solar system. The likelihood of witnessing a destructive collision on any *specific* asteroid in a human lifetime is therefore negligibly small. However, the likelihood of observing a collisional destruction *somewhere* in the asteroid belt, when integrated over the asteroid size distribution, is much higher (Fig. 6). Specifically, τ_{CD} (yr), the interval between successive destructions averaged over the whole main belt is well represented by

$$\log_{10}\tau_{CD} = 2.62 + 2.88 \log_{10} D \quad (7)$$

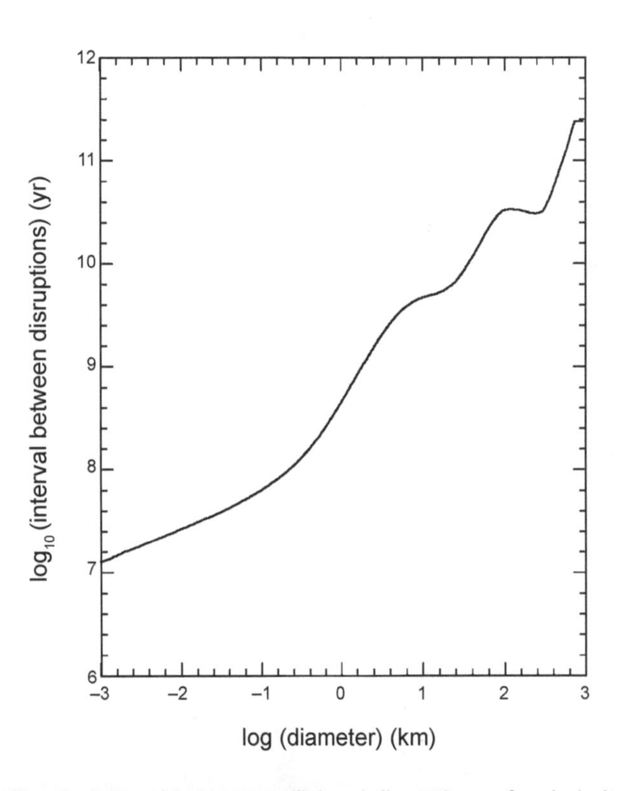

Fig. 5. Interval between collisional disruptions of main-belt asteroids as a function of their diameter in the range 1 m to 1000 km. Adapted from *Bottke et al.* (2005).

for D in kilometers (see Fig. 6). Equation (7) gives $\tau_{CD} \sim 400$ yr for D = 1 km, falling to $\tau_{CD} \sim 1$ yr for D = 0.1 km and $\tau_{CD} \sim 30$ s (!) for D = 1 m. Independent modeling of survey data suggests that up to 10 catastrophic disruption events brighter than V = 18.5 mag may be discoverable each year by current and future asteroid surveys (*Denneau et al.,* 2015).

Compared to catastrophic collisions, collisions resulting in less than complete destruction, so-called "cratering collisions," should be more common. The minimum size of cratering impact that gives rise to an observable signature is poorly modeled, being a function of many uncertain parameters (e.g., *McLoughlin et al.,* 2015). For example, the mass and size distribution of ejecta determine the peak brightness and persistence time (because particles are cleared by solar radiation pressure, which acts in inverse proportion to the particle size). Given better collision rate determinations, we should be able to map the spatial distribution of small (meter-scale) impactors through the asteroid belt.

A cratering impact between a 20-m to 40-m-scale projectile and the 113-km-diameter asteroid (596) Scheila (see section 6.3.2) was a 1 in ~10^4-yr event for this object but, given that there are on the order of 200 asteroids of similar or larger size, should occur once every ~50 yr somewhere in the asteroid belt (*Jewitt,* 2012). As monitoring of the sky improves, we should soon be able to detect cratering collisions throughout the asteroid belt in great abundance (see section 9).

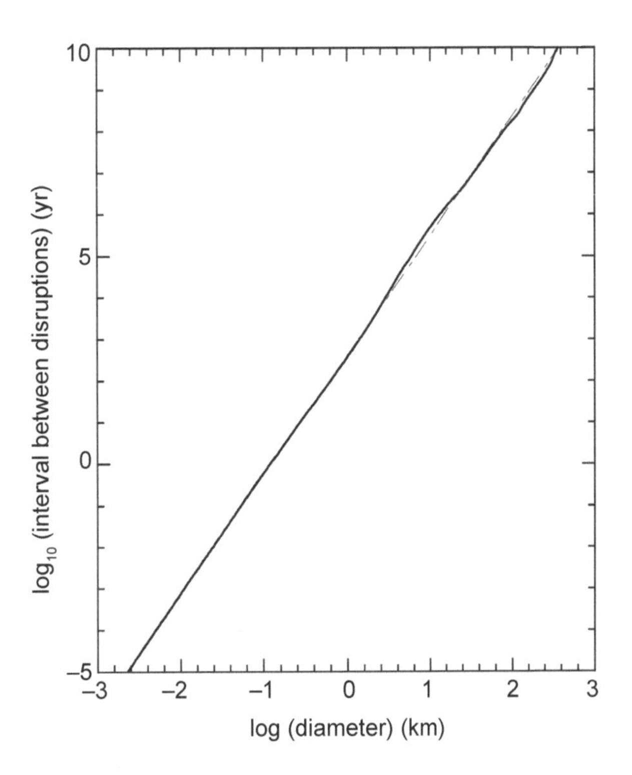

Fig. 6. Interval between collisional disruptions as a function of asteroid diameter, integrated over the asteroid size distribution. Adapted from *Bottke et al.* (2005).

3.4. Rotational Destabilization

Rotational instability causes mass loss when the centrifugal forces exceed the sum of gravitational and cohesive forces acting toward the center of mass. Like impact disruptions (section 3.3), rotational disruptions can place important constraints on a body's interior properties (e.g., *Hirabayashi et al.,* 2015). For a strengthless a × b × c ellipsoid, with a = b ≤ c, bulk density ρ, and rotation about a minor (minimum energy, maximum moment of inertia) axis, the critical, size-independent rotation period is

$$P_c = \left(\frac{3\pi}{G\rho}\right)^{1/2} \left(\frac{c}{a}\right) \tag{8}$$

where G = 6.67 × 10^{-11} N kg^{-2} m^2 is the gravitational constant. For a sphere, c/a = 1, with ρ = 10^3 kg m^{-3}, we find P_c = 3.3 hr. This rises to P_c = 4.0 hr for a body with c/a = 1.2 [the modal axis ratio of small asteroids from *Szabó and Kiss* (2008)], with the instability occurring at the tips of the ellipsoid. Several active asteroids rotate with periods comparable to P_c (cf. Table 1), raising the possibility that rotational instability plays a role in the ejection of material from those objects.

A complicating factor in the application of equation (8) is that even very modest values of material cohesion (used here to mean any combination of tensile strength and shear strength) can hold together a rotating asteroid. To see this, we use the rotational energy density in a uniform sphere as an estimate of the average rotational stress, S_0, obtaining

$$S_0 \sim \rho V_{eq}^2 \tag{9}$$

where $V_{eq} = 2\pi r/P_c$ is the equatorial velocity. With the above parameters and r = 1 km, we find $S_0 \sim 280$ N m^{-2}, which is easily overcome by material cohesion. Most asteroids and comets likely possess a weak, rubble-pile structure in which gravity, rotation, and van der Waals forces all play a role. The magnitude of the latter is ~S_0 (*Scheeres et al.,* 2010; *Sánchez and Scheeres,* 2014; *Sánchez et al.,* 2021), meaning that cohesion can nevertheless substantially control a body's mass loss and shape. Bodies with a weakly cohesive regolith are subject to mass movement and, eventually, mass shedding as particulate material escapes from the equator. Evidence for this process is compelling in "top-shaped" asteroids (Fig. 7), whose shapes are formed by skirts of material that has migrated from higher latitudes toward the equator (*Harris et al.,* 2009).

The spin of an inert body can change because of radiative torques [e.g., the Yarkovsky-O'Keefe-Radzievskii-Paddack (YORP) effect (*Bottke et al.,* 2006)], torques from outgassing if volatiles are present (*Jewitt,* 2021), and impacts (*Marzari*

TABLE 1. Currently known active asteroids.

Object	a*	e†	i‡	T_J§	H_V¶	r_n**	P_{rot}††	Mech.‡‡	Ref.§§
(1) Ceres	2.766	0.078	10.588	3.310	3.53	469.7	9.07	S	[1]
(493) Griseldis	3.116	0.176	15.179	3.140	10.97	20.78	51.94	I	[2]
(596) Scheila	2.929	0.163	14.657	3.209	8.93	79.86	15.85	I	[3]
(2201) Oljato	2.174	0.713	2.522	3.299	15.25	0.90	>26.	—	[4]
(3200) Phaethon	1.271	0.890	22.257	4.510	14.32	3.13	3.60	RTP	[5]
(6478) Gault	2.305	0.193	22.812	3.461	14.81	2.8	2.49	R(I)	[6]
(62412) 2000 SY$_{178}$	3.159	0.079	4.738	3.195	13.74	5.19	3.33	R(SI)	[7]
(101955) Bennu	1.126	0.204	6.035	5.525	20.21	0.24	4.29	(ITP)	[8]
107P/(4015) Wilson-Harrington	2.625	0.632	2.799	3.082	16.02	3.46	7.15	S	[9]
133P/(7968) Elst-Pizarro	3.165	0.157	1.389	3.184	15.84	1.9	3.47	SR(E)	[10]
176P/(118401) LINEAR	3.194	0.193	0.235	3.167	15.45	2.0	22.23	S(IE)	[11]
233P/La Sagra (P/2005 JR$_{71}$)	3.033	0.411	11.279	3.081	~16.6	~1.5	—	—	[12]
238P/Read (P/2005 U1)	3.162	0.253	1.266	3.153	19.40	0.4	—	S(IR)	[13]
259P/Garradd (P/2008 R1)	2.727	0.342	15.899	3.217	20.06	0.30	—	S(IR)	[14]
288P/(300163) 2006 VW$_{139}$	3.051	0.201	3.239	3.203	~17.8, ~18.2	~0.9, ~0.6	—	S	[15]
311P/PANSTARRS (P/2013 P5)	2.189	0.116	4.968	3.660	19.14	0.2	>5.4	R	[16]
313P/Gibbs (P/2014 S4)	3.154	0.242	10.967	3.133	17.1	1.0	—	S(R)	[17]
324P/La Sagra (P/2010 R2)	3.098	0.154	21.400	3.099	18.8	0.55	—	S(R)	[18]
331P/Gibbs (P/2012 F5)	3.005	0.042	9.739	3.228	17.33	1.77	3.24	R	[19]
354P/LINEAR (P/2010 A2)	2.290	0.125	5.256	3.583	—	0.06	11.36	I	[20]
358P/PANSTARRS (P/2012 T1)	3.155	0.236	11.058	3.134	19.9	0.32	—	S(R)	[21]
426P/PANSTARRS (P/2019 A7)	3.188	0.161	17.773	3.103	~17.1	~1.2	—	—	[22]
427P/ATLAS (P/2017 S5)	3.171	0.313	11.849	3.092	18.91	0.45	1.4	SR	[23]
432P/PANSTARRS (P/2021 N4)	3.045	0.244	10.067	3.170	>18.2	<0.7	—	—	[24]
433P/(248370) 2005 QN$_{173}$	3.067	0.226	0.067	3.192	16.32	1.6	—	S(RI)	[25]
P/2013 R3 (Catalina-PANSTARRS)	3.033	0.273	0.899	3.184	—	~0.2 (× 4)	—	SR	[26]
P/2015 X6 (PANSTARRS)	2.755	0.170	4.558	3.318	>18.2	<0.7	—	S(IR)	[27]
P/2016 G1 (PANSTARRS)	2.583	0.210	10.968	3.367	—	<0.4	—	I	[28]
P/2016 J1-A/B (PANSTARRS)	3.172	0.228	14.330	3.113	—	<0.4,<0.9	—	S(IR)	[29]
P/2017 S9 (PANSTARRS)	3.156	0.305	14.138	3.087	>17.8	<0.8	—	—	[30]
P/2018 P3 (PANSTARRS)	3.007	0.416	8.909	3.096	>18.6	<0.6	—	S(IR)	[31]
P/2019 A3 (PANSTARRS)	3.147	0.265	15.367	3.099	>19.3	<0.4	—	—	[32]

TABLE 1. (continued.)

Object	a*	e†	i‡	T_J§	H_V¶	r_n**	P_{rot}††	Mech.‡‡	Ref.§§
P/2019 A4 (PANSTARRS)	2.614	0.090	13.319	3.365	—	0.17	—	(IR)	[33]
P/2020 O1 (Lemmon-PANSTARRS)	2.647	0.120	5.223	3.376	>17.7	<0.9	—	—	[34]
P/2021 A5 (PANSTARRS)	3.047	0.140	18.188	3.147	—	0.15	—	S(IR)	[35]
P/2021 L4 (PANSTARRS)	3.165	0.119	16.963	3.125	>15.8	<2.2	—	—	[36]
P/2021 R8 (Sheppard)	3.019	0.294	2.203	3.179	—	—	—	—	[37]

* Semimajor axis, in au.
† Eccentricity.
‡ Inclination, in degrees.
§ Tisserand parameter with respect to Jupiter.
¶ Measured V-band absolute magnitude, or estimated upper limit based on apparent brightness when active, if available.
** Effective nucleus radius (or radii), in kilometers, as estimated from absolute magnitude or via other means as described in references.
†† Rotation period, in hours, if available.
‡‡ Mechanisms likely contributing to the observed activity (S: sublimation; I: impact; R: rotational destabilization; E: electrostatics; T: thermal fatigue/fracturing; P: phyllosilicate dehydration); mechanisms enclosed in parentheses are not directly indicated but are not excluded by available data.
§§ References: [1] *Küppers et al.* (2014); *Park et al.* (2016); [2] *Tholen et al.* (2015); [3] *Ishiguro et al.* (2011a); [4] *Russell et al.* (1984); *Tedesco et al.* (2004); *Warner et al.* (2019); [5] *Jewitt and Li* (2010); *Ansdell et al.* (2014); [6] *Devogèle et al.* (2021); [7] *Sheppard and Trujillo* (2015); [8] *Lauretta et al.* (2019); [9] *Fernández et al.* (1997); *Licandro et al.* (2009); *Urakawa et al.* (2011); [10] *Hsieh et al.* (2004, 2009a, 2010); [11] *Hsieh et al.* (2009a, 2011a); [12] *Mainzer et al.* (2010); [13] *Hsieh et al.* (2011b); [14] *MacLennan and Hsieh* (2012); *Hsieh et al.* (2021b); [15] *Agarwal et al.* (2020); [16] *Jewitt et al.* (2018b); [17] *Hsieh et al.* (2015b); [18] *Hsieh* (2014b); [19] *Drahus et al.* (2015); [20] *Jewitt et al.* (2010); *Snodgrass et al.* (2010); *Kim et al.* (2017a,b); [21] *Hsieh et al.* (2013, 2018b); [22] *Rudenko et al.* (2021); [23] *Jewitt et al.* (2019c); [24] *Wainscoat et al.* (2021); [25] *Hsieh et al.* (2021a); [26] *Jewitt et al.* (2014a); [27] *Moreno et al.* (2016a); [28] *Moreno et al.* (2016b); *Hainaut et al.* (2019); [29] *Moreno et al.* (2017); *Hui et al.* (2017); [30] *Weryk et al.* (2017); [31] *Weryk et al.* (2018); *Kim et al.* (2019); [32] *Weryk et al.* (2019); [33] *Moreno et al.* (2021); [34] *Weryk et al.* (2020); [35] *Moreno et al.* (2021); [36] *Wainscoat et al.* (2021) [37] *Tholen et al.* (2021).

et al., 2011). Eight asteroids in the published literature display evidence for spinup, namely (1620) Geographos [which interestingly is also a suspected meteor stream parent (*Ryabova,* 2002); also see section 2.5], (1862) Apollo, (3103) Eger, (25143) Itokawa, (54509) YORP, (161989) Cacus, (1685) Toro, and (10115) 1992 SK (*Rozitis and Green,* 2013; *Lowry et al.,* 2014; *Durech et al.,* 2018, 2022). Figure 8 shows characteristic spinup timescales for these asteroids computed using $\tau = \omega/\dot{\omega}$, where ω is the measured angular frequency, as a function of diameter. The 10-m-scale asteroid 2012 TC_4 shows evidence for multiple periods and spinup (*Lee et al.,* 2021) but is not plotted here. Uncertainties for τ_Y are dependent on uncertainties for the albedo, size, shape, density, and thermal properties of the measured asteroids. For illustration, we show error bars indicating ±50% uncertainties.

The timescale for spinup by YORP torque, τ_Y, is expected to vary as $\tau_Y = \Gamma a^2_{au} D^2$, where Γ is a constant, a_{au} is the semimajor axis in astronomical units, and D is the effective diameter. The a^2_{au} dependence results from the fading of sunlight described by the inverse square law. The D^2 dependence arises because the timescale is given by the ratio of the moment of inertia (which varies as D^5) to the torque (given by the product of the surface area and the moment arm for the torque, together $\propto D^3$). The constant, Γ, must be determined empirically, since it is dependent on many unknown details of each asteroid. Under the assumption that the spinup of objects in Fig. 8 is due to YORP, we compute a least-squares fit for the YORP timescale given by

Fig. 7. Three examples of rotationally-shaped asteroids with equatorial skirts. (101955) Bennu, (162173) Ryugu and (66391) Moshup have diameters 500 m, 900 m, and 1300 m, respectively. Bennu and Ryugu were imaged by spacecraft, while the Moshup image was derived from radar observations. Credits: NASA/Goddard/Univ. of Arizona (Bennu); JAXA, Univ. of Tokyo, and collaborators (Ryugu); *Ostro et al.* (2006) (Moshup).

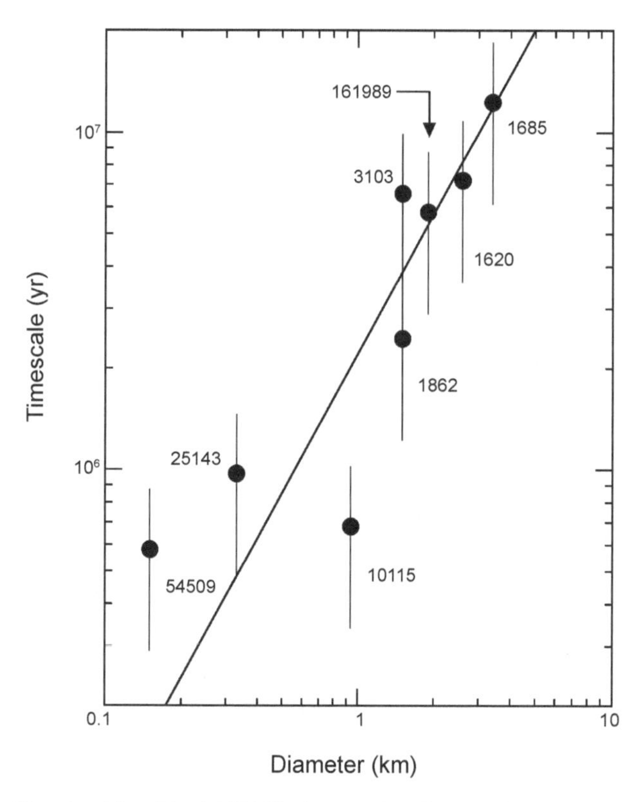

Fig. 8. Empirical YORP timescales for small asteroids (marked by their numerical designations) as a function of diameter, using published measurements of spin-period changes. The plotted objects have semimajor axes a = 1.0 to 1.5 au, and the plotted line is given by equation (10).

$$\tau_Y = (1.4 \pm 0.3) \times 10^6 a_{au}^2 D^{1.4 \pm 0.3} \quad (10)$$

which is shown in the figure as a solid line. The best-fit index, g = 1.4 ± 0.3, is consistent with the expected value of g = 2 to within 2σ. (The small difference, if real, could result from observational selection against the detection of small asteroids with short spinup timescales. Such objects are more likely to have been spun up to destruction and will not survive to be measured.) If we instead force a D^2 dependence, a fit to the data gives $\tau_Y = (0.7 \pm 0.1) \times 10^6 a_{au}^2 D^2$. Both fits (as well as Fig. 8) indicate that a 1-km-diameter object located at $a_{au} = 3$ au should have a characteristic spinup time from $\tau_Y \sim 6$ to 13 m.y., with a considerable scatter due to intrinsic differences (in shape, roughness, thermal diffusivity, spin vector magnitude, and direction) between the asteroids. We take $\tau_Y = 10$ m.y. at $a_{au} = 3$ au as a middle value. This is much shorter than the ~0.4-G.y. collisional lifetime of a 1-km-diameter main-belt asteroid (*Bottke et al.,* 2005), showing the potential importance of rotational breakup in weak asteroids. If sublimating volatiles are present, equally rapid spinup can be achieved by tiny (e.g., 10^{-3} kg s^{-1}) mass loss rates, provided they are sustained over long times (*Jewitt,* 2021).

A major unknown in Fig. 8 and equation (10) concerns the long-term stability of the applied torque. Models show that even very small changes in surface topography can result in significant changes in the magnitude and even direction of the YORP torque (e.g., *Statler,* 2009). Topographical changes can result from YORP itself (leading to feedback in which the change in shape caused by the equatorward movement of material can alter the body shape and YORP torque that caused the movement), or from cratering by impacts. As a result, instead of driving inexorably toward breakup, asteroids more likely migrate erratically toward it. For example, the "top-shaped" objects in Fig. 7 are not necessarily all rotating near break-up; the periods of Bennu, Ryugu, and (66391) Moshup are 4.3, 7.6, and 2.8 hr, respectively. This scatter reflects the temporal instability of the YORP torque and, perhaps, the past loss of angular momentum and subsequent despinning by mass redistribution or shedding in these objects.

The stochastic effects of impacts can rival torques due to YORP for bodies in the size range of interest here (*Wiegert,* 2015). In impact spinup, the dominant effects are from the small number of the largest projectiles (*Holsapple and Housen,* 2019). In the limiting case, an asteroid can be impact-disrupted, forming an unstable disk from which the asteroid reaccretes mass and angular momentum. Reassembly after impact is still capable of producing top-shaped bodies like Bennu and Ryugu (*Michel et al.,* 2020).

3.5. Radiation Pressure Sweeping

Solar radiation can facilitate mass loss in two ways: via direct radiation pressure and thermal radiation pressure. In the first case, a dust grain detached from the surface of a spherical asteroid of mass M_n and radius r_n feels both a gravitational attraction, $g = GM_n/r_n^2$, towards the asteroid and an acceleration away from the Sun, α, caused by solar radiation pressure. We write $\alpha = \beta g_\odot$, where $g_\odot = GM_\odot/r_H^2$ is the gravitational acceleration to the Sun, whose mass is $M_\odot = 2 \times 10^{30}$ kg, and β is the dimensionless radiation pressure efficiency of a particle of radius a. For dielectric spheres, $\beta \sim a_{\mu m}^{-1}$, where $a_{\mu m}$ is the particle radius in units of micrometers. Whereas g is independent of r_H, α varies as r_H^{-2}, meaning that small asteroids close to the Sun are most susceptible to dust loss from radiation pressure sweeping. For a spherical asteroid of density ρ, α/g > 1 is satisfied when

$$a_{\mu m} < \frac{3M_\odot}{4\pi\rho r_n r_H^2} \quad (11)$$

(*Jewitt,* 2012). Equivalently

$$a_{\mu m} \lesssim 10\left(\frac{1\ \text{km}}{r_n}\right)\left(\frac{1\ \text{au}}{r_H}\right)^2 \quad (12)$$

At 3 au, a kilometer-sized asteroid can only lose submicrometer grains but on Phaethon (r_n = 3 km) at perihelion (r_H = 0.14 au), for example, particles up to $a_{\mu m} \sim 170\ \mu m$ can be swept away. This is only possible, however, for particles that are both detached from the asteroid by another process (i.e., not subject to cohesive forces binding them to the surface) and near the illuminated limb (otherwise the radiation pressure accelerates the particles downward into the surface). Meanwhile, sunlight absorbed and thermally reradiated has the same effect as direct radiation pressure, but with the advantage that thermal radiation is always directed outward and so dust can be expelled from the entire dayside of the asteroid (*Bach and Ishiguro,* 2021).

Neither direct nor thermal radiation pressure can detach small dust particles held to an asteroid surface by contact forces. Once these are broken, however, by sublimation drag, thermal fracture, or other processes, radiation pressure acceleration provides a mechanism to remove that dust. As such, while there are currently no known active objects for which radiation pressure sweeping is considered a dominant driver of activity, it is possible that either direct or thermal radiation pressure may serve to enhance the dust ejection efficacy of other mechanisms.

3.6. Electrostatic Ejection

Solid surfaces exposed to solar UV photons become photoionized, with the loss of electrons resulting in a net positive sunlit surface charge (e.g., *Criswell,* 1973). Meanwhile, photoelectrons reimpacting the surface concentrate in shadowed regions. As a result, spatial gradients in illumination of a surface, for example at the terminator or between the sunlit side of a particle and its shadow, produce local electric fields. These fields are capable of moving regolith dust particles, provided that interparticle cohesive forces can be overcome. This has been known since the Lunar Surveyor missions of the mid-1960s, when unexpected horizon glow was detected after sunset and interpreted as forward scattering from ~10-μm-sized regolith particles at heights of h $\lesssim$ 1 m above the lunar surface (*Criswell and De,* 1977). To reach h requires an ejection speed $V = (2g_{☾}h)^{1/2} \sim 1.8\ m\ s^{-1}$, where $g_{☾} = 1.6\ m\ s^{-2}$ is lunar surface gravity. In turn, V is comparable to the gravitational escape speed from a kilometer-sized body, meaning that electrostatic ejection is a possible mechanism behind activity in asteroids (*Jewitt,* 2012). Spacecraft-imaged dust "ponds" settled in local gravitational potential minima on the 17-km-diameter asteroid (433) Eros provide additional evidence for the mobility of surface dust, likely influenced by electrostatic forces (*Colwell et al.,* 2005).

The physics of electrostatic launch is complicated and still under investigation, particularly concerning the question of how electrostatic repulsion can overcome attractive van der Waals contact forces that bind small particles together. One recent finding is that dust ejection can occur because of the development of small-scale, but extremely large (10^5–10^6 V m^{-1}), electric fields in cavities between grains in a porous regolith. This is the so-called "patched charge" model (*Wang et al.,* 2016; *Hood et al.,* 2022). Experiments show that particles up to 10–20 μm can be ejected at meters-per-second speeds.

Unlike on the Moon, where electrostatically launched particles are pulled back by gravity, dust is easily lost from small asteroids so that surface charging constitutes a net loss of surface fines. *Criswell* (1973) estimated the overturning or "churn rate" of lunar dust due to electrostatic levitation to be C = 0.1 kg m^{-2} yr^{-1}. We estimate an upper limit to electrostatic dust losses by assuming that, on a small asteroid, all this churned material will be lost, at a rate $\dot{M} \sim 4\pi r^2$ C. Substituting r_n = 1 km gives $\dot{M} \sim 0.04$ kg s^{-1}. This value is small compared to the ~1-kg s^{-1} rates inferred in the active asteroids, but large enough to be telescopically detectable under ideal circumstances. However, this estimate is an upper limit both because not all ejected particles will be lost, and because electrostatic churning is spatially localized near shadow boundaries, not global, on the Moon.

The fundamental problem for electrostatic ejection from asteroids, however, is that $C \gg \mathcal{L}$, where $\mathcal{L}$ is the rate of production of fresh particles by micrometeoroid impact (section 3.3.1). With no adequate source of replenishment, detachable surface dust on a small asteroid would be quickly depleted. An additional concern is that electrostatic charging is a very general process, dependent only on the presence of ionizing radiation and dust. If electrostatic losses are important, why would only some kilometer-sized asteroids be active rather than all of them? For these reasons, electrostatic loss seems unlikely to be a major contributor to activity on most solar system bodies.

3.7. Thermal Fracture

Thermal fatigue, defined as weakening of a material through the application of a cyclic temperature-related stress, has been suggested as a driver of the particle loss from Phaethon and Bennu (sections 6.3.5 and 6.3.6). The fractional expansion resulting from a temperature change ΔT is $\alpha\Delta T$, where an expansivity of $\alpha \sim 10^{-5}\ K^{-1}$ is typical (*Konietzky and Wang,* 1992). The resulting strain is $S \sim \alpha Y \Delta T$, where Y is Young's modulus and a multiplier on the order of unity, called Poisson's ratio, is ignored. Y values for rock are very large (e.g., 10^{10} N m^{-2} to 10^{11} N m^{-2}), meaning that huge stresses of $S \sim 10^5\ \Delta T$ to $10^6\ \Delta T$ (N m^{-2}) can result from modest temperature excursions. For example, diurnal temperature variations $\Delta T \sim 10^2$ K, common on airless bodies, can generate estimated stresses of $S \sim 10^7$–10^8 N m^{-2}. Since rocks are especially weak in tension (e.g., the tensile strength of basalt is only ~4×10^6 N m^{-2}), thermal fracture is expected to be an important erosive process in space, even at asteroid belt distances and temperatures (*Molaro et al.,* 2015).

In a homogeneous material, the temperature gradients and resulting thermal stresses are largest on length scales comparable to the thermal skin depth, $l_D \sim (\kappa P)^{1/2}$, where κ is the thermal diffusivity and P is the timescale for the insolation. By substitution, the skin depth for a solid rock asteroid ($\kappa = 10^{-6}$ m^2 s^{-1}) with a rotational period of P = 5 hr is $l_D \sim 10$ cm. Boulders of size $f \gg f_D$ can be progressively fractured into smaller rocks but particles with $f \ll f_D$ will be nearly isothermal and thus less susceptible to cracking. As a result, the breakdown of rocks by thermal fracture should progressively modify the size distribution of particles in an asteroid regolith relative to the power-law distribution produced by impact fragmentation. A separate effect arises because many rocks are built from millimeter-sized mineral grains having different compositions and expansivities. Differential expansion between the grains naturally causes weakening and disintegration down to millimeter scales. On Earth, the effects of thermal fracture are amplified by the intervention of liquid water followed by freeze-thaw cracking (*Eppes et al.,* 2016). On asteroids, where there is no liquid water, thermal fracture must be less effective. Still, boulders on asteroids, for example, are often surrounded by skirts of debris likely produced by thermal cracking from the day-night temperature cycle.

Separately, the thermal unbinding of water from hydrated minerals (e.g., hydrous phyllosilicates like serpentine, brucite, muscovite, and talc) causes shrinkage and is another process capable of cracking rocks and producing dust (as in dusty, dry mud lake beds on Earth). The activation energies for dehydration of some minerals correspond to temperatures of $T \sim 10^3$ K (*Bose and Ganguly,* 1994), as reached by (3200) Phaethon and other small perihelion objects (Fig. 9). While Phaethon shows no evidence for hydrated minerals, it is not clear if this is because they have already been lost or were never present (*Takir et al.,* 2020). Already suspected in Phaethon, thermal destruction may play an important role in the reported depletion of low-perihelion asteroids (*Granvik et al.,* 2016).

In order to produce measurable activity as seen in the active asteroids, however, particles produced by thermal fracture must achieve escape velocity. Order of magnitude considerations show that meter-per-second speeds can be generated in 10-μm-sized fragments (*Jewitt,* 2012), sufficient to escape from the gravitational fields of 1-km asteroids, although larger particles will be too slow to escape. In such cases, additional forces, notably radiation pressure sweeping, might combine with thermal fracture to accelerate fresh fragments above the escape speed. Without a removal mechanism, the surface layers of an asteroid would eventually clog with small particles that could not be further fractured.

3.8. Interplanetary Field Enhancements

Several NEAs [e.g., (2201) Oljato and (138175) 2000 EE_{104}] are associated with distinct and repeated disturbances in the magnetic field carried by the solar wind, called IFEs. These disturbances, detected by spacecraft crossing the orbits of the asteroids, can last from minutes to hours and have profiles distinct from magnetic structures emanating from active regions in the Sun (*Russell et al.,* 1984; *Lai et al.,* 2017).

The suggested mechanism underlying IFEs is the impact ejection of nanodust, which quickly becomes photoelectrically charged and then loads the passing solar wind, causing a magnetic field disturbance (*Russell et al.,* 1984; *Lai et al.,* 2017). While nanodust itself is not optically detectable because the scattering efficiency of nanoparticles is negligible, impacts should eject particles with a range of sizes, some of which should be optically detectable. A targeted search for such particles at (138175) 2000 EE_{104} detected none, however, setting limits to the optical depth $\tau \lesssim 2 \times 10^{-9}$ (*Jewitt,* 2020), comparable to faint JFC trails (*Ishiguro et al.,* 2009).

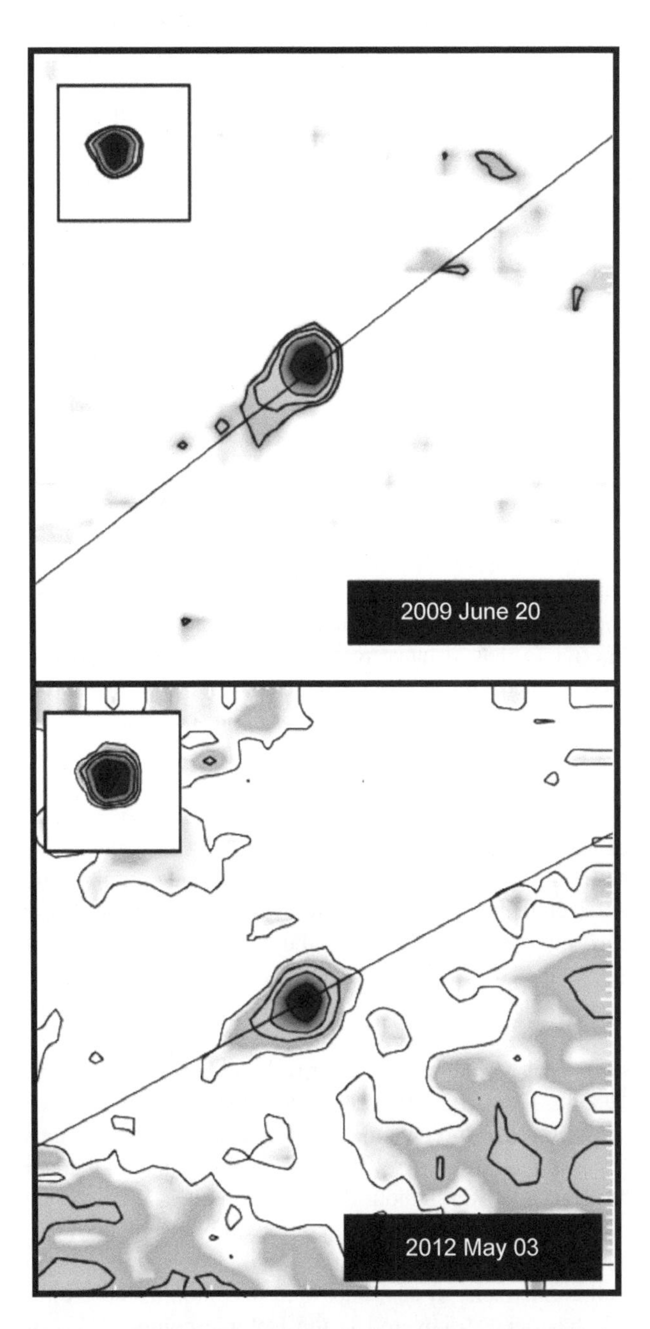

Fig. 9. Contoured STEREO images of (3200) Phaethon against the bright background of the solar corona at perihelion in 2009 and 2012 from *Jewitt et al.* (2013b) showing its faint tail. Inset boxes showing contoured point sources are ~8′ square, corresponding to 350,000 km at Phaethon. Straight lines show the projected Sun-Phaethon line.

Furthermore, the production of sufficient masses of nanodust by impact is problematic (*Jewitt,* 2020). *Lai et al.* (2017) estimated that a mass of $M_0 = 10^5$–10^6 kg of nanodust would be needed to produce a single IFE. Given a range of normal (top-heavy) ejecta power-law size distributions in which most of the mass is contained in the largest particles, nanodust constitutes only $f \sim 10^{-3.8}$–$10^{-5.5}$ of the total mass. Therefore, to supply M_0 would require implausible source masses of $M = M_0/f \sim 10^{8.8}$ kg to $10^{11.5}$ kg, rivaling the $\sim 10^{11}$ kg mass of (138175) itself. Moreover, the planet-crossing asteroid population is comparatively rarefied, and the collision rate for such objects is likely too small to supply IFEs at the measured rate.

While no other explanations have been suggested to date, IFEs appear to be real. If they are caused by charged dust loading of the solar wind, then a solution for the mass problem and a non-impact mechanism for the production of this dust remain to be found. We know from Comet 1P/Halley that nanodust, perhaps produced by the spontaneous fragmentation of larger particles dragged out of the nucleus by gas, can be abundant in comets (*Mann,* 2017). Indeed, IFE-like magnetic field disturbances have been detected along the orbit of Comet 122P/de Vico (*Jones et al.,* 2003).

4. DISTRIBUTION OF VOLATILE MATERIAL

In the classical view, icy objects formed beyond the water snow line, which is the heliocentric distance beyond which temperatures are low enough for ice grains to form and become incorporated into growing planetesimals. Water interior to that distance remained in vapor form and was incorporated into forming planetesimals at a far lower rate (e.g., *Encrenaz,* 2008). Icy bodies are preserved in the Kuiper belt and Oort cloud reservoirs, from which all ice-rich comets were initially believed to originate (e.g., *Duncan,* 2008; *Dones et al.,* 2015). Meanwhile, the inactive nature of asteroids in the inner solar system was historically interpreted to indicate a lack of ice, either because they did not accrete much icy material to begin with, or because any such material has been lost in the heat of the Sun over the age of the solar system.

Initial interpretations of the distribution of asteroid taxonomic types in the asteroid belt placed the water snow line in our solar system at ~2.5 au (e.g., *Gradie and Tedesco,* 1982; *Jones et al.,* 1990), but the migratory nature of the snow line due to changing protoplanetary disk conditions (e.g., *Martin and Livio,* 2012) means that icy material could have accreted throughout the asteroid belt. Moreover, icy objects from the outer solar system, well beyond any plausible location of the snow line, could have been emplaced in the inner solar system after their accretion. Evidence for this has been discovered in the isotopic compositions of meteorites, which fall into two groups identified with accretion in spatially (and, perhaps, temporally) distinct hot and cold reservoirs (*Warren,* 2011). Other observational studies have revealed the otherwise unexpected presence of objects in the main asteroid belt with D-type taxonomic classifications more commonly associated with outer solar system objects (e.g., *DeMeo et al.,* 2014; *Hasegawa et al.,* 2021). Numerical models also suggest that small bodies from the outer solar system could have been scattered inward as a consequence of giant planet migration early in the solar system's history (*Levison et al.,* 2009; *Walsh et al.,* 2011), or perhaps even as a consequence of giant planet formation itself (*Raymond and Izidoro,* 2017).

In addition to finding that some inner solar system objects may contain more ice than typically assumed, recent work has also demonstrated that some outer solar system objects may also contain less ice than typically assumed. Dynamical models of early solar system evolution, both including and excluding planetary migration, predict that some objects originally formed in the terrestrial planet region could have been ejected from the inner solar system, resulting in a small fraction of planetesimals that are predominantly rocky currently residing in the Oort cloud (*Weissman and Levison,* 1997; *Walsh et al.,* 2011; *Izidoro et al.,* 2013; *Shannon et al.,* 2015). When perturbed into the inner solar system, these objects would then be identifiable as possessing orbits characteristic of LPCs, while also being nearly or completely inert (e.g., *Weissman and Levison,* 1997; *Meech et al.,* 2016; *Piro et al.,* 2021).

5. DYNAMICS

As mentioned in section 1, the canonical $T_J = 3$ boundary between asteroids and comets is strictly valid only under the assumptions of the circular restricted three-body approximation (in which the only gravitational perturbers are Jupiter and the Sun, and Jupiter's orbit is exactly circular). A more nuanced understanding of T_J and small body dynamics in general, however, is necessary for properly interpreting the implications of small-body studies that rely on assumptions about the origin of the small bodies being studied based on this parameter.

The real solar system contains other massive bodies of course, and non-gravitational perturbations from asymmetric outgassing from active bodies or radiative effects can also alter an object's motion in ways that are not accounted for in the circular restricted three-body approximation. As a result, real asteroids and comets are not separated by a sharp dynamical boundary at $T_J = 3$, but instead by a less-distinct boundary zone spanning a range of values from $T_J \sim 3.05$ to $T_J \sim 3.10$, with numerical integrations showing that dynamical lifetime distributions of comets with $3.00 < T_J < 3.05$ and comets with $T_J < 3$ are effectively indistinguishable, while comets with $3.05 < T_J < 3.10$ have a mix of short dynamical lifetimes ($t < 2$ m.y.), characteristic of objects with $T_J < 3$, and longer dynamical lifetimes ($t > 200$ m.y.) characteristic of objects with $T_J > 3.10$ (Fig. 10) (*Hsieh and Haghighipour,* 2016). These results are consistent with the comet-asteroid thresholds of $T_J = 3.05$ and $T_J = 3.08$ used by *Tancredi* (2014) and *Jewitt et al.* (2015), respectively.

Dynamical classification systems based on objects' current orbital parameters can of course be made more complex to try to better capture the dynamical nuances of "true" asteroids and comets (e.g., *Tancredi,* 2014), but even these can fail since the idea that an impenetrable boundary in T_J space exists at all between asteroids and comets is fundamentally flawed. Numerical integrations have shown that small bodies can evolve from one side of the T_J boundary to the other, effectively changing their dynamical identities as asteroids or comets as determined by T_J. In a dynamical study of 58 JFCs on near-Earth orbits, *Fernández and Sosa* (2015) found several comets that avoided very close encounters with Jupiter and remained dynamically stable on much longer timescales than other JFCs in the study and hypothesized that they could have originated in the main asteroid belt. Most of these objects also exhibit relatively weak activity for their sizes compared to other JFCs, and some show other asteroid-like physical properties [e.g., 249P/LINEAR (*Fernández et al.,* 2017)], further supporting possible main-belt origins for those objects.

Numerical integrations performed by *Hsieh et al.* (2020) subsequently revealed a pathway by which transitions inferred by *Fernández and Sosa* (2015) could occur on timescales of <100 m.y. In that pathway (see Fig. 11), the 2:1 mean-motion resonance (MMR) with Jupiter can excite the eccentricities of main-belt asteroids (in that work, members of the Themis asteroid family) close to the resonance, lowering their perihelion distances and increasing their aphelion distances. These orbital changes increase the potential for significant interactions with planets other than Jupiter, invalidating the approximation from which the theoretical $T_J = 3$ asteroid-comet boundary is derived, ultimately allowing a small but non-negligible number of objects to evolve from high-T_J orbits to low-T_J orbits.

Statistical evidence (section 7) indicates that main-belt contamination dominates the population of bodies with $2.8 \leq T_J \leq 3.0$. A spectroscopic survey by *Licandro et al.* (2008) found that 7 of 41 observed objects with $T_J < 3$ showed absorption bands suggestive of S-type asteroids [known for their distinctly un-comet-like silicate-containing compositions; see *DeMeo et al.* (2015) for a detailed discussion about asteroid taxonomic classification]. All seven had $T_J > 2.8$. *Geem et al.* (2022) proposed that asteroids and comets might be distinguished by their polarization vs. phase angle relations and gave three examples. One of these, (331471) 1984 QY_1, had a small polarization amplitude deemed more characteristic of an S-type asteroid than of a comet. Future measurements might be able to establish a polarization vs. T_J relation that would help distinguish displaced main-belt asteroids from JFCs.

Meanwhile, dynamical evolution of small bodies in the opposite direction, from cometary orbits to asteroidal orbits, also appears to be possible. *Hsieh and Haghighipour* (2016) found that a small number of synthetic test particles with JFC-like initial orbital elements reached fully main-belt-like orbits in 2-m.y.-long integrations, suggesting that JFCs could occasionally become interlopers in the main asteroid belt. Such an origin has been suggested for P/2021 A5 (PANSTARRS) (*Moreno et al.,* 2021). That said, the JFC interlopers found by *Hsieh and Haghighipour* (2016) are dynamically distinct from the active asteroids in having higher mean eccentricities and inclinations. Efforts to use active asteroids to study primordial volatile material in the asteroid belt (see section 6.2) might be able to mitigate the impact of such interlopers by focusing only on the low-inclination, low-eccentricity portion of the main belt population.

Even if interactions with all other gravitating solar system bodies are taken into account, non-gravitational accelerations from asymmetric outgassing can cause further deviations from gravity-only dynamical evolution. For example, 2P/Encke is a highly active comet with $T_J = 3.025$ whose unusual orbit *Fernández et al.* (2002) found could be produced

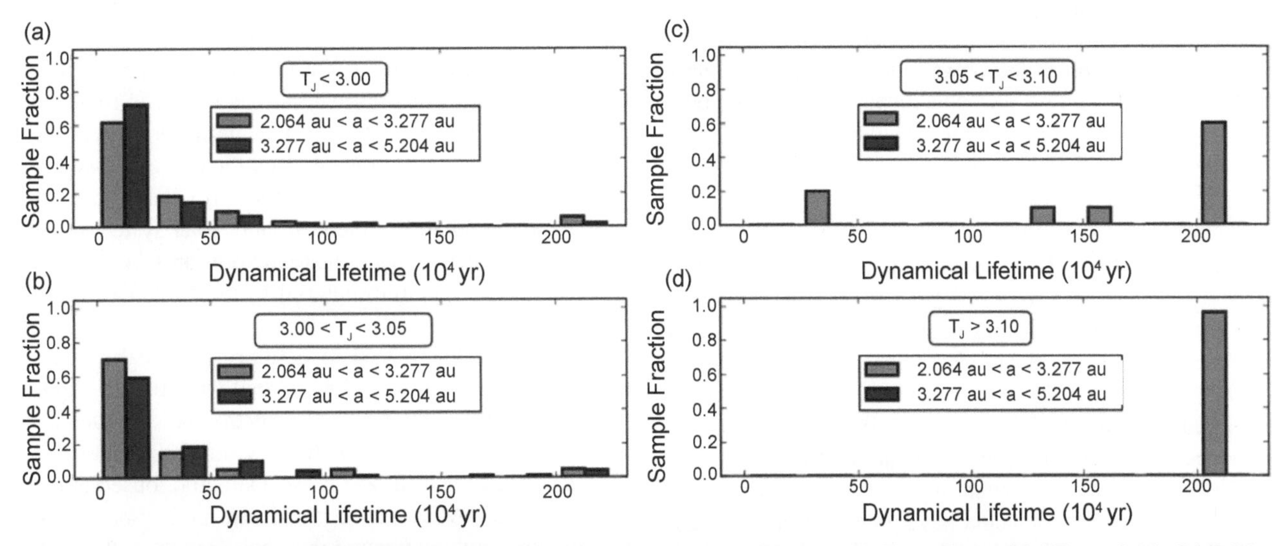

Fig. 10. Histograms of dynamical lifetimes for known comets with $a \leq 5.204$ au with $T_{J,s}$ values of **(a)** $T_{J,s} < 3.00$, **(b)** $3.00 < T_{J,s} < 3.05$, **(c)** $3.05 < T_{J,s} < 3.10$, and **(d)** $T_{J,s} > 3.10$, where light and dark gray bars indicate the fraction of comets with 2.064 au < a < 3.277 au and a > 3.277 au, respectively, that are lost due to ejection or planetary/solar impact within a particular time interval. From *Hsieh and Haghighipour* (2016).

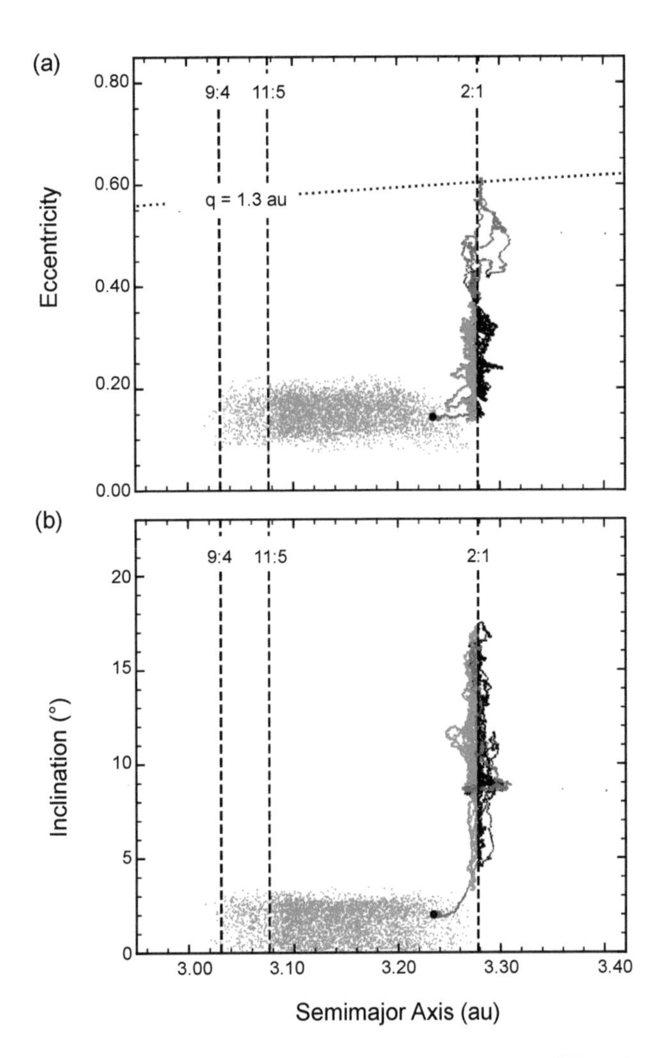

Fig. 11. Plots of the forward dynamical evolution of Themis family asteroid (12360) Unilandes in **(a)** semimajor axis vs. eccentricity space, and **(b)** semimajor axis vs. inclination space. Large black circles in each panel mark the current orbital elements of the asteroid, gray dots mark the current orbital elements of Themis-family asteroids, and other colored dots mark intermediate orbital elements in the forward integrations of the object. Of the latter, blue dots mark intermediate orbital elements meeting criteria for being classified main-belt-like, and red dots mark orbital elements that meet criteria (including $T_J < 3$) for being classified as JFC-like. From *Hsieh et al.* (2020).

from a typical JFC orbit by non-gravitational accelerations, but only if they were sustained for 10^5 yr. This physically implausible conclusion (2P does not contain enough ice to remain active that long) was also reached by *Levison et al.* (2006). An astrometric study of 18 active asteroids demonstrated that only two (313P and 324P) exhibited strong evidence of non-gravitational acceleration (*Hui and Jewitt,* 2017). Nevertheless, the possibility that stronger asymmetric mass loss in the past could have produced their current asteroid-like orbits cannot be ruled out.

In summary, as with other binary criteria for distinguishing comets and asteroids (see sections 2, 3, and 4), dynamical classification is also not always clear-cut. Orbit evolution is an intrinsically chaotic, and therefore non-deterministic, process. Backward and forward numerical integrations can indicate what range of past and future dynamical evolution is possible (with dynamical clones typically used for this purpose) but cannot track an object's exact dynamical origin or fate beyond a certain length of time. Conclusions about a specific object's origin are unavoidably statistical in nature.

6. ACTIVE ASTEROIDS

6.1. Overview

Active asteroids have dynamical properties characteristic of asteroids and the physical appearances of comets. They are one of the more recent groups of objects to be recognized as blurring the lines between asteroids and comets, although many individual objects in this category were known long before the term "active asteroid" was first used. A wide range of potential mechanisms for the observed comet-like mass loss of active asteroids has been identified, including sublimation (section 3.2), impacts (section 3.3), rotational instability (section 3.4), thermal fracture (section 3.7), and combinations of these processes, making these objects valuable for the opportunities they provide to study these processes in the real world. Reviews addressing different aspects of activity in asteroids include *Jewitt* (2004, 2012), *Jewitt et al.* (2015), and *Kasuga and Jewitt* (2019). Table 1 lists the active asteroids known as of January 1, 2022, while their distribution in semimajor axis vs. eccentricity space is shown in Fig. 12.

6.2. Main-Belt Comets

The main-belt comets (MBCs) are a noteworthy subset of the active asteroids, comprising objects for which sublimation appears to plays a role in producing the observed mass loss (*Hsieh and Jewitt,* 2006). The nature of their activity and largely stable orbits in the main asteroid belt indicate that near-surface volatile material has managed to remain preserved in the inner solar system until the present day. They have attracted significant scientific interest since their discovery due to the opportunities they provide to place constraints on the volatile content of inner solar system bodies and their possible connections to the primordial delivery of water and other volatile material to the early Earth (e.g., *Hsieh,* 2014a).

The MBCs discovered to date are too weakly active (production rates $\lesssim$1 kg s^{-1} compared with 10^2–10^3 kg s^{-1} for typical comets) for gas to be spectroscopically detected given the sensitivity limits of current groundbased telescopes (*Jewitt,* 2012). Instead, the most distinctive observational property of the MBCs is repeated near-perihelion activity in different orbits, indicating that their activity is thermally modulated (see section 3.2). Of the many processes considered in section 3, only temperature-dependent sublimation can easily account for repeated near-perihelion activity.

Independent evidence for outgassing comes from the detection of non-gravitational acceleration (caused by recoil from the anisotropic mass loss) in 313P/Gibbs and 324P/La Sagra (*Hui and Jewitt,* 2017). Only gas carries enough momentum to explain the observed non-gravitational accelerations. Evidence for sublimation as the driver of activity in MBCs is thus indirect but compelling. Reviews specifically focusing on MBCs include *Bertini* (2011) and *Snodgrass et al.* (2017). MBCs identified through the repetition of their activity in more than one orbit or through other means (e.g., dust modeling indicating prolonged dust emission events characteristic of sublimation-driven activity) are shown as green symbols in Fig. 12.

The "duty cycle," f_{dc}, defined as the fraction of the total time during which an MBC is active, is an important quantity because it is related to the abundance of ice in the asteroid belt. Values of $f_{dc} \ll 1$ are automatically required in order for the MBCs to have survived against devolatilization (*Hsieh and Jewitt,* 2006) or destruction by spinup (*Jewitt et al.,* 2019c). The most stringent estimates, $f_{dc} \lesssim 3 \times 10^{-5}$, are obtained from surveys conducted in search of asteroid activity (*Gilbert and Wiegert,* 2010; *Waszczak et al.,* 2013). Such small values identify the MBCs as the tip of the proverbial iceberg; each detected MBC represents $f_{dc}^{-1} \gtrsim 30{,}000$ similarly icy asteroids that were not detected as such because they exist in the dormant state. It is entirely possible, for example, that *all* outer-belt asteroids contain ice and that the MBCs are notable simply because we observe them in an unusual outgassing state. Additional characteristics [e.g., fast rotation, small nucleus sizes and therefore low escape velocities, moderate orbital eccentricities, small obliquities, membership in young asteroid families, and so on (*Hsieh et al.,* 2004, 2018c; *Hsieh,* 2009; *Kim et al.,* 2018; *Schörghofer,* 2016)] may aggregate to help produce observable activity.

6.3. Active Asteroid Examples

6.3.1. Sublimation: 133P/(7968) Elst-Pizarro. Comet 133P/(7968) Elst-Pizarro, originally discovered as asteroid 1979 OW_7, was first found to exhibit comet-like activity on August 7, 1996, shortly after passing perihelion on April 18, 1996 (*Elst et al.,* 1996). At the time, 133P was the first and only main-belt asteroid observed to exhibit comet-like activity, leading to uncertainty about whether the observed mass loss, later determined to have occurred over the course of several weeks or months (*Boehnhardt et al.,* 1996), was due to sublimation of volatile material similar to other comets or to a series of impacts (*Toth,* 2000). Follow-up observations in 2002 revealed 133P to be active (*Hsieh et al.,* 2004), again near perihelion, heavily favoring the hypothesis that its activity was due to the sublimation of exposed volatile material instead of a highly improbable succession of impacts. Since then, 133P has continued to exhibit activity at each of its subsequent perihelion passages (e.g., *Hsieh et al.,* 2010; *Jewitt et al.,* 2014b), firmly establishing the recurrent and likely sublimation-driven nature of its activity.

Calculations described in section 3.2 show that sublimation rate differences between perihelion and aphelion for main-belt asteroids can easily explain the preferential appearance of detectable activity only near perihelion, while obliquity-related and surface-shadowing effects may also play a role. To date, seven other active asteroids (Fig. 13)

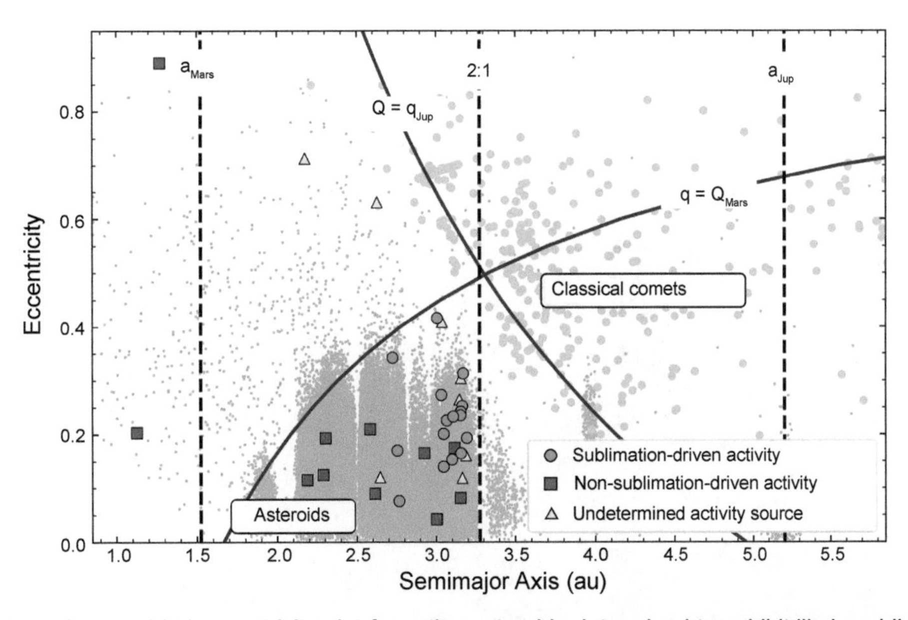

Fig. 12. Semimajor axis vs. orbital eccentricity plot for active asteroids determined to exhibit likely sublimation-driven activity (open green circles), activity driven by processes other than sublimation (open red squares), and activity of currently undetermined origin (open yellow triangles), where classical comets are marked as filled light blue circles and asteroids as small gray dots. Curved arcs show the loci of points having aphelion and perihelion distances equal to the aphelion of Mars and the perihelion of Jupiter, respectively, as marked. Objects above these lines cross the orbits of those planets. Vertical dashed lines mark the orbital radii of Mars and Jupiter and the location of the 2:1 mean-motion resonance with Jupiter.

have been found to exhibit recurrent activity near perihelion (Fig. 3) (see *Hsieh et al.,* 2021a), indicating that they are likely MBCs exhibiting sublimation-driven mass loss, and corroborating the hypothesis that their activity is primarily modulated by temperature and not seasonal effects.

Interestingly, 133P's relatively fast rotation rate of P_{rot} = 3.471 hr (*Hsieh et al.,* 2004) suggests that rotational instability could assist in the ejection of dust particles initially lofted by sublimation. Combined with the hypothesis that 133P's current activity may have required an initial collisional trigger to excavate a subsurface ice reservoir (see section 3.2), 133P presents an excellent illustrative case of an active object for which multiple mechanisms may be responsible for observed activity.

6.3.2. Impact: 354P/LINEAR and (596) Scheila. Comet 354P/LINEAR (formerly P/2010 A2) and (596) Scheila were the first objects discovered immediately following impact events (Fig. 14). The largest component in 354P's debris cloud is ~100 m in scale and is accompanied by a co-moving cluster of smaller bodies embedded in a particle trail (*Jewitt et al.,* 2010, 2013b; *Snodgrass et al.,* 2010). The object was disrupted by a collision nine months before it was discovered, with the delay in the discovery in part due to having an initially small solar elongation (*Jewitt et al.,* 2011a). While rotational destabilization was also initially considered as a possible alternate explanation for 354P's activity (*Jewitt et al.,* 2010), the main nucleus was later found to have a double-peaked rotational period of P_{rot} = 11.36 ± 0.02 hr, ruling out fast rotation as a potential cause of the observed activity (*Kim et al.,* 2017a).

Meanwhile, the 113-km-diameter asteroid (596) Scheila ejected ~10^8 kg of dust due to an apparent cratering impact in late 2010 (*Jewitt et al.,* 2011b), with that dust clearing in a few weeks. The impulsive nature of the brightening and the monotonic fading thereafter are consistent with the impact of a 20-m to 40-m-scale projectile. The unusual dust cloud morphology exhibited by Scheila can be reproduced by an impact ejecta cone and downrange plume produced by an impact at a ~45° angle to the object's surface (*Ishiguro et al.,* 2011a). *Ishiguro et al.* (2011b) and *Bodewits et al.* (2014) also reported post-impact changes in the optical lightcurve while *Hasegawa et al.* (2022) reported a reddening in the 1.0- to 2.5-μm wavelength region, but, curiously, no change from 0.4 to 1.0 μm or from 2.5 to 4.0 μm. If real, the observed light curve modifications may have resulted from fallback of material ejected at speeds less than Scheila's ~60 m s^{-1} escape velocity, while the reported near-infrared color change might suggest burial of a surface previously de-reddened by space weathering.

The discoveries of these two objects, both in 2010, represent a key milestone in our understanding of the comet-asteroid continuum, as they definitively demonstrated for the first time that impact events could be detected in real time, and also that observations of comet-like activity do not always mean that sublimation of volatile material is taking place. This realization helped to pave the way for other non-sublimation-related interpretations of future activity detections (sections 3, 6.3.3, and 6.3.4).

6.3.3. Rotational breakup: P/2013 R3 (Catalina-PANSTARRS). Perhaps the most dramatic example of rotational destabilization found to date is P/2013 R3 (Catalina-PANSTARRS), which was observed in 2013 in the midst of disintegrating into a collection of ~10^2-m-scale bodies with a velocity dispersion of $\Delta V \sim 0.3$ m s^{-1} (Fig. 15) (*Jewitt et al.,* 2014a). The parent body size is estimated at a few $\times 10^2$ m. Bodies this small are susceptible to spinup by radiation torques on timescales $\lesssim$1 m.y., leading to the suspicion that P/2013 R3 was a rotationally disrupted asteroid. Sublimating water ice, if present, could also spin up the body on a very short timescale (*Jewitt,* 2021), and in fact was inferred to be present by the dust ejection behavior of individual components of the object as it disintegrated (*Jewitt et al.,* 2014a). Modeling of the breakup process by

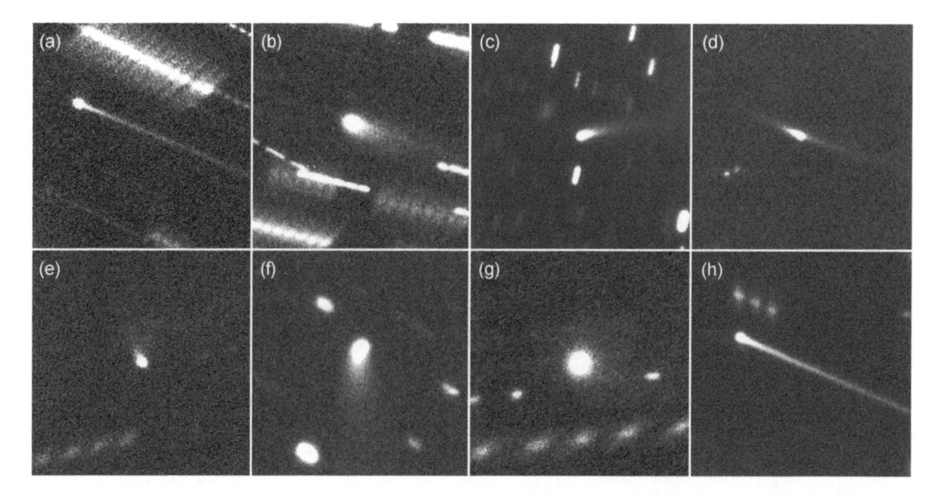

Fig. 13. Gallery of 1′ × 1′ images of MBCs confirmed to exhibit recurrent activity: **(a)** 133P/Elst-Pizarro, **(b)** 238P/Read, **(c)** 259P/Garradd, **(d)** 288P/(300163) 2006 VW_{139}, **(e)** 313P/Gibbs, **(f)** 324P/La Sagra, **(g)** 358P/PANSTARRS, and **(h)** 433P/(248370) 2005 QN_{173}. Images adapted from data published in *Hsieh et al.* (2004, 2009b, 2012b,c, 2013, 2015b, 2021a,b).

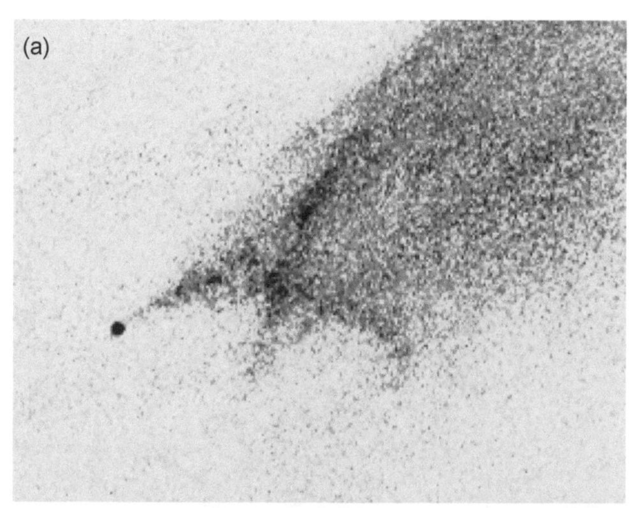

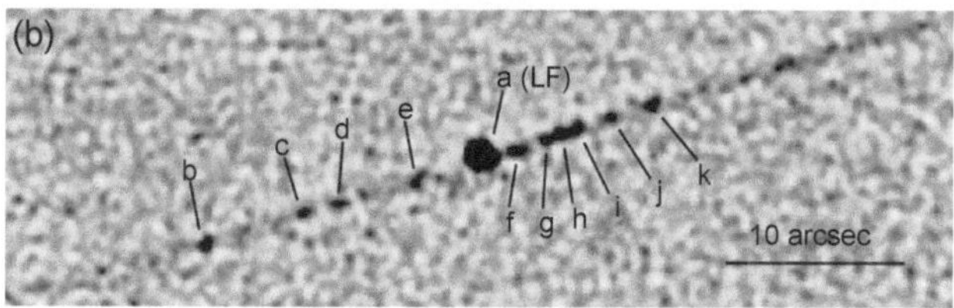

Fig. 14. (a) Head structure of P/2010 A2 on UT 2010 January 25, from *Jewitt et al.* (2010). **(b)** Fragments in the particle trail on UT 2017 January 27 and 28, from *Kim et al.* (2017a).

Hirabayashi et al. (2014) suggested that the parent body had a rotational period of 0.48–1.9 hr, implying that it was likely spinning well beyond the critical disruption limit for a rubble pile prior to its breakup, suggesting that the object had some degree of cohesive strength on the order of 40–210 Pa, comparable to the cohesion expected of a sand pile. Similar to 133P (section 6.3.1), P/2013 R3 presents an interesting case where multiple activity mechanisms are operating, with rotational instability likely being the primary cause of the disintegration of the parent body, but with continued decay and dust release then driven by sublimation of interior ices exposed by the disruption event.

6.3.4. Rotational instability: (6478) Gault, 331P/Gibbs, and 311P/PANSTARRS. Non-catastrophic mass loss events observed for main-belt asteroid (6478) Gault and Comets 311P/PANSTARRS and 331P/Gibbs were all believed to be produced by rotational destabilization yet were simultaneously remarkably diverse compared to one another (as well as compared to the catastrophic rotational disruption of P/2013 R3; section 6.3.3). The diversity in active behavior displayed by these three objects demonstrates the complexity of rotational destabilization as an activity mechanism, with potentially very different outcomes depending on the circumstances of each object and event.

Comet 311P (formerly P/2013 P5) displayed a distinctive set of discrete dust tails (Fig. 16), formed by submillimeter-sized debris ejected in nine bursts over nine months, with no evidence of periodicity in the mass-loss events and no evidence for macroscopic fragments in the ejecta (*Jewitt et al.,* 2013a). The directions of the tails are determined by the ejection epoch and the subsequent action of radiation pressure, creating a size gradient along each tail. 311P's activity likely resulted from the loss of loose surface material that avalanched to the equator in response to rapid rotation. A reliable period has not been determined for 311P. Instead, photometry of its nucleus shows complex structure, suggestive of an eclipsing, possibly contact, binary (*Jewitt et al.,* 2018b).

Meanwhile, when discovered to be active in 2019, Gault exhibited two bright tails (Fig. 17) (e.g., *Kleyna et al.,* 2019), while a third, fainter tail emerged in later follow-up imaging (*Jewitt et al.,* 2019b). Intriguingly, searches of archival data revealed multiple additional past mass-loss events over six years (*Chandler et al.,* 2019). Like for 311P, the episodic nature of Gault's dust ejection events, as well as a lack of evidence that other activity mechanisms such as sublimation could be driving the observed activity, indicated that its activity was likely to be driven by rotational destabilization. This hypothesis was later supported by observations of the nucleus after activity had subsided, revealing an extremely short rotational period of P_{rot} = 2.5 hr with a light-curve amplitude of 0.06 mag (*Devogèle et al.,* 2021; *Luu et al.,* 2021). Together, these observations suggest that Gault is a rapidly spinning asteroid held continuously at the brink of rotational disruption, perhaps by the YORP effect, that periodically undergoes mass-loss events that temporarily restabilize it. Unlike the eight MBCs showing confirmed recurrent activity near perihelion (section 6.2), activity in Gault is unrelated to its heliocentric distance, evidently because sublimation does not play a role in its activity.

331P/Gibbs ejected only a small fraction of the parent body mass ($\Delta M/M \sim 0.01$), visible as a set of fragments embedded in a long debris trail (Fig. 18) (*Jewitt et al.,* 2021). The 0.8-km radius (assuming an albedo of 0.05) primary body again shows clear evidence of rapid rotation, with period P_{rot} = 3.26 hr. At this period, equation (8) gives ρ = 1600 kg m^{-3} as the minimum density required for a cohesionless spherical body to resist mass shedding at its equator. The 19 measurable fragments ejected from 331P travel away from the source at incredibly low speeds, ~0.1 m s^{-1}. The fragments show no sign of further fading or disintegration over a three-year window of observation, consistent with having wholly refractory compositions. The largest, the 100-m-scale 331P-A (visible as the second brightest object in Fig. 18), shows a large-amplitude rotational light curve with morphological characteristics consistent with a contact binary structure. They are embedded in a debris trail, also characterized by a small velocity dispersion. Despite its designation, "comet" 331P is probably a non-volatile, rotationally unstable asteroid.

6.3.5. Thermal breakdown: (3200) Phaethon. The Geminid meteoroid stream consists of millimeter-sized and larger debris with a total mass $M_G \sim (2–7) \times 10^{13}$ kg (*Blaauw,* 2017). The stream is $\tau \sim 10^3$ yr old, as judged by the spread of orbital elements of the Geminid meteors (*Gustafson,* 1989; *Williams and Wu,* 1993) and, to be sup-

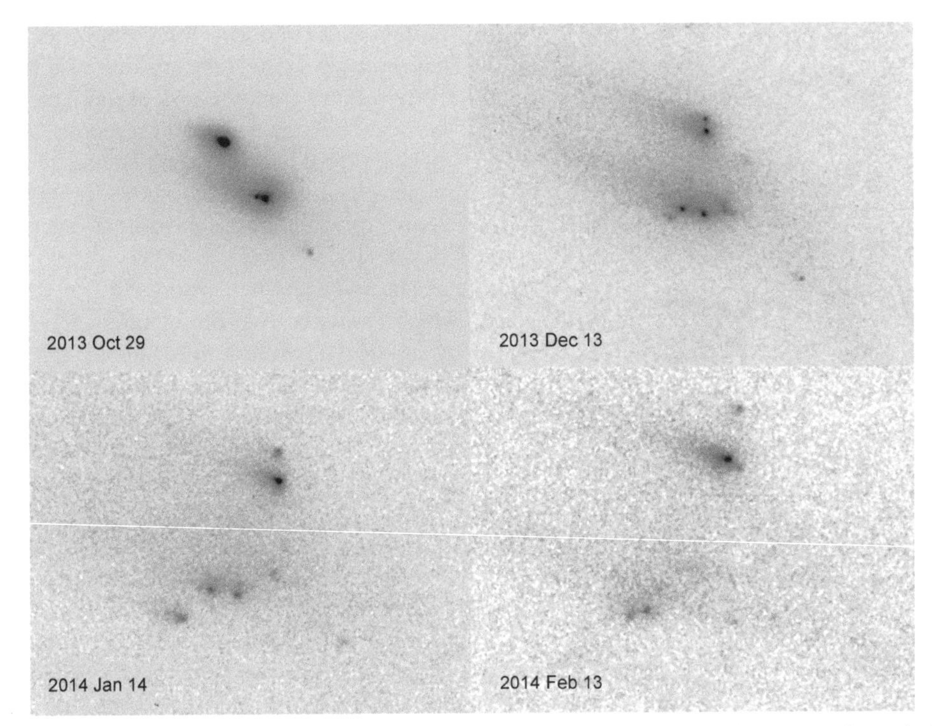

Fig. 15. Fragments of P/2013 R3 on four dates as marked. Each panel shows a region approximately 20,000 km wide. The anti-solar and negative projected orbit vectors are at position angles 67° and 246°, respectively. North to the top, east to the left. From *Jewitt et al.* (2017).

plied in steady state, would require production at the rate $M_G/\tau \sim 700$–2000 kg s^{-1}. Such large rates would be comparable to those of bright comets, but no such bright comet source of the Geminids exists. Instead, the Geminids appear to be supplied by the 5–6-km-diameter B-type asteroid (3200) Phaethon.

Curiously, dark sky observations have consistently shown no visible evidence for ongoing mass loss from Phaethon at the $\sim 10^{-2}$ kg s^{-1} level (*Hsieh and Jewitt,* 2005), or even from the vantage point of a near-Earth (0.07 au) flyby in 2017, where limits $\leq 10^{-3}$ kg s^{-1} were set (*Jewitt et al.,* 2018a, 2019a). However, observations near perihelion at $r_H = 0.14$ au (Fig. 9) show photometric behavior and a resolved tail consistent with the release of micrometer-sized particles at about 3 kg s^{-1} (*Jewitt and Li,* 2010; *Li and Jewitt,* 2013; *Jewitt et al.,* 2013b; *Hui and Li,* 2017). This rate is still 2 to 3 orders of magnitude too small to supply the

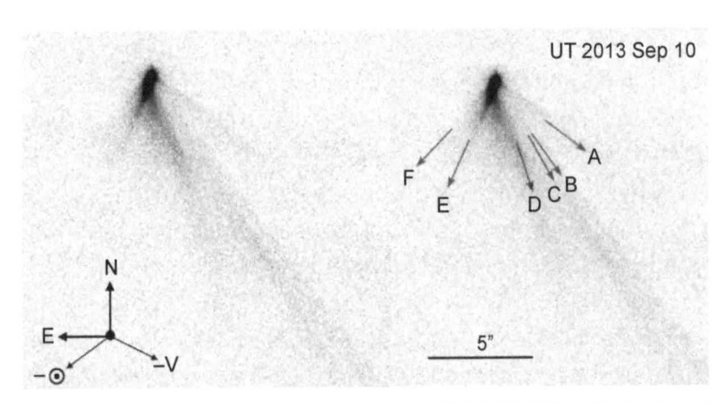

Fig. 16. Multi-tailed active asteroid 311P. The first six of nine tails are labeled. From *Jewitt et al.* (2013a).

Geminids in steady state. Furthermore, the micrometer-sized particles detected near perihelion are so strongly affected by solar radiation pressure that they cannot be confined to the Geminid stream. Either the particle size distribution is shallow, such that unseen large particles dominate the ejected mass, or — more likely — the source of the Geminids is episodic or even catastrophic in nature, and unrelated to the activity at perihelion.

Several mechanisms have been suggested to drive Phaethon's observed repeated perihelion activity. The subsolar temperature on Phaethon is ~1000 K, suggesting that stresses induced by thermal expansion and desiccation shrinkage could eject fragments (*Jewitt and Li,* 2010; *Jewitt,* 2012). Sublimation of mineralogically bound sodium could also eject small dust particles through gas drag (*Masiero et al.,* 2021), provided that there is enough mineralogically bound sodium and that a mechanism exists to replenish quickly-depleted surface materials with sodium-rich minerals. Phaethon's $P_{rot} = 3.6$-hr rotation period (*Ansdell et al.,* 2014) suggests that centrifugal effects might also contribute to mass loss (*Nakano and Hirabayashi,* 2020), as does its likely top-shape (*Hanuš et al.,* 2016; *Taylor et al.,* 2019).

Phaethon's dynamical lifetime is $\lesssim 10^8$ yr, implying delivery from elsewhere in the solar system (*de León et al.,* 2010). A cometary origin is unlikely, given that the orbit is strongly decoupled from Jupiter, and has the very large Tisserand parameter $T_J = 4.5$. However, the specific source region, presumably in the main asteroid belt, is unknown. Suggested sources include the large B-type asteroid 2 Pallas (*de León et al.,* 2010) and, more likely, a collisional family in the inner asteroid belt (*MacLennan et al.,* 2021).

Fig. 17. Gault on UT 2019 February 5, showing two of the three tails ejected in 2019. From *Kleyna et al.* (2019).

The kilometer-sized asteroid 2005 UD appears dynamically related to Phaethon (*Ohtsuka et al.,* 2006) and is coincidentally also a B-type object (*Jewitt and Hsieh,* 2006) [although recent observations show dissimilarity in the near-infrared (*Kareta et al.,* 2021)]. A dynamical relation to another kilometer-sized asteroid, 1999 YC, has also been proposed but is less certain, where 1999 YC is a neutral C-type spectrally distinct from the blue B-types Phaethon and 2005 UD (*Kasuga and Jewitt,* 2008).

Phaethon, 2005 UD, and the Geminid stream are collectively known as the "Phaethon-Geminid Complex" (*Ohtsuka et al.,* 2006). Unfortunately, the genetic relationships between these bodies remain completely unknown, as does the relevance of the current mass loss from Phaethon at perihelion (*Kasuga and Jewitt,* 2019). We hope for clarification of the nature of Phaethon's surface by the upcoming DESTINY+ flyby mission (see section 9), although the 33-km s^{-1} encounter velocity and planned 500-km distance of closest approach will limit the data obtained.

6.3.3. Unknown: (101955) Bennu. Repeated instances of particle ejection from Bennu were observed from the OSIRIS-REx spacecraft (Fig. 2) (*Lauretta et al.,* 2019). Particles ejected from Bennu, estimated to be centimeter-sized, were emitted erratically, with one large burst containing up to 200 particles. Particle speeds were estimated to range from 0.1 m s^{-1} to 3 m s^{-1}, with ~70% of the particles being too slow to escape and falling back to the surface. The peak observed ejection velocity, 3 m s^{-1}, exceeded the 0.2-m s^{-1} escape velocity from Bennu by an order of magnitude. Dissimilar size distributions were measured in three outburst events, with differential power laws indexes varying from –3.3 to –1.2 (*Hergenrother et al.,* 2020). The total ejected mass, however, was very small. *Hergenrother et al.* (2020) estimated a mass loss of ~10 kg per orbit, corresponding to a mean mass loss rate of $\dot{M} \sim 10^{-7}$ kg s^{-1} over the 1.2-yr orbit period. This is ~10^6–$10^7\times$ smaller than the sensitivity limits achieved through groundbased observations of other active asteroids.

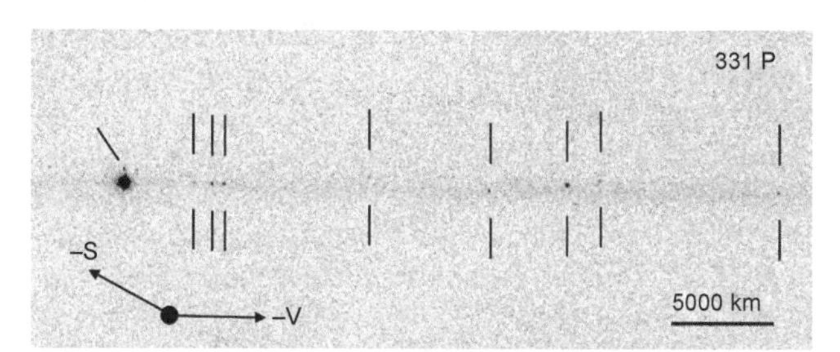

Fig. 18. Some of the 19 fragments of 331P on UT 2013 December 25. The fragments, many of which are very faint and only confirmed by their reappearance in images from different dates, are marked by solid black lines. The 1.6 ± 0.2-km-diameter primary is the bright object to the left. Suspected contact binary 331P-A is third from the right. From *Jewitt et al.* (2021).

The chance, *in situ* discovery of Bennu's activity raises the possibility that many asteroids lose mass at rates that are individually beneath the groundbased detection threshold. To consider the possible magnitude of low-level dust production across the asteroid belt, we scale the surface area of Bennu (~1 km^2) to the total surface area of all asteroids having escape velocities less than the peak ejection velocities of Bennu ejecta. The latter corresponds to a radius of $r_n < 3$ km. This total area, calculated by integrating over the asteroid size distribution, is ~10^8 km^2. Therefore, if all $r_n \leq 3$ km asteroids eject dust at the same rate as Bennu, the total mass loss would be $10^8(\dot{M}) \sim 10$ kg s^{-1}.

While this is a very crude calculation, it serves to show that Bennu-like production of dust in the asteroid belt is, at best, a very minor contributor to the 10^3–10^4-kg s^{-1} production rate needed to sustain the zodiacal cloud (*Nesvorný et al.,* 2011). We also note that the Hayabusa2 spacecraft (*Tsuda et al.,* 2019) did not detect any Bennu-like loss of particles from (162173) Ryugu during its encounter with the asteroid in 2018–2019, despite the fact that Ryugu is also taxonomically classified as a primitive asteroid (*Watanabe et al.,* 2019) and was also observed *in situ* by a visiting spacecraft. We do not know if this reflects an intrinsic difference between the two asteroids, or merely a difference in the sensitivity to small particles between the two spacecraft.

The mechanism driving Bennu's mass loss is currently undetermined, despite the availability of abundant spacecraft data. The primary constraints are the sizes, numbers, and speeds of the ejected fragments, as well as an apparent preference for launch in the local afternoon. Of the mechanisms already proposed for the active asteroids, electrostatic lofting, micrometeorite impact (*Bottke et al.,* 2020), thermal fracture, and dehydration cracking (*Molaro et al.,* 2020) have all been considered in the case of Bennu.

7. INACTIVE COMETS

The average dynamical lifetimes of short-period comets exceed their volatile lifetimes. Therefore, the fate of old comets, if they do not disintegrate completely, is to leave behind inert, asteroid-like bodies traveling in comet-like orbits. These are the ACOs and Damocloids mentioned in section 1. Proper identification of such objects is essential for setting constraints on the physical lifetimes of comets (e.g., *Jewitt,* 2004) and models of cometary source regions (e.g., *Wang and Brasser,* 2014; *Brasser and Wang,* 2015). ACOs and Damocloids are presumed to be comets for which activity has weakened below detectable limits or completely ceased due to the depletion of surface volatiles or mantling (e.g., *Podolak and Herman,* 1985; *Prialnik and Bar-Nun,* 1988; *Jewitt,* 1996), although other processes like hydrostatic compression have also been proposed to contribute to activity quenching (*Gundlach and Blum,* 2016).

Table 2 lists the numbers of asteroids, N_a, and comets, N_c, compiled from the JPL Horizons orbital element database (*https://ssd.jpl.nasa.gov/horizons/*), together with their ratio, $\mathcal{R} = N_a/N_s$, all as functions of the binned Tisserand parameter, T_J. The listed uncertainty on $\mathcal{R}$ has been calculated assuming Poisson counting statistics. In this table, asteroids and comets are distinguished only by whether or not visible activity has ever been reported. Objects with large perihelia, even those known from detailed observations to be active comets, are too cold and weakly active for outgassing to be easily measured. Therefore, in order to make the empirical asteroid/comet distinction meaningful, we only list objects with perihelia $q \le 2.0$ au, where activity is relatively easy to detect. We also exclude from consideration the sungrazing comets, which generally have poorly determined orbits. The table provides a first-order estimate of the asteroid/comet number ratio as a function of a simple dynamical parameter.

As expected, Table 2 shows that almost all objects with $T_J > 3$ are asteroids ($\mathcal{R} \gg 1$), while most of those with $T_J < 2$ are comets ($\mathcal{R} < 1$). In between, $\mathcal{R}$ is roughly constant across the range assigned to the JFCs (namely $2 \le T_J \le 3$) except for a higher value ($\mathcal{R} = 31$) in the range $T_J > 2.8$–3.0. This is presumably because of contamination of the $2.8 \le T_J \le 3.0$ sample by scattered main-belt asteroids (see section 5). An independent study of objects with $q < 1.3$ au also finds that $T_J \sim 2.8$ marks a change in the spectral characteristics, with featureless C- and X-type spectra more common at $T_J < 2.8$ than for larger T_J (*Simion et al.,* 2021). Across the range $2.0 \le T_J \le 2.8$, the asteroid to comet ratio settles to a steady mean value $R = 6.2 \pm 0.8$, meaning that, in this T_J range, asteroids outnumber comets by ~6:1.

In the absence of main-belt contamination, $\mathcal{R}$ would give a measure of the ratio of the dynamical to the active cometary lifetimes. The measured $\mathcal{R} \sim 6$ would result if comets are active for only approximately one-sixth of their dynamical lifetimes. However, this is a lower limit because main-belt contamination is presumably non-zero. Furthermore, $\mathcal{R}$ provides no clue as to whether comets simply stop outgassing (because they exhaust near-surface volatiles), or cycle between active and inactive states [perhaps because of cyclic mantle build-up and blow-off correlated with changes in the perihelion distance and peak surface temperature (*Rickman et al.* 1990)].

TABLE 2. Empirical asteroid: Comet ratio vs. T_J*.

T_J	N_a†	N_c‡	$\mathcal{R}$§
<1	28	372	0.08 ± 0.01
2	453	108	0.24 ± 0.02
2.0–2.2	55	7	7.9 ± 3.2
2.2–2.4	101	23	4.4 ± 1.0
2.4–2.6	202	38	5.3 ± 0.9
2.6–2.8	553	75	7.4 ± 0.9
2.8–3.0	2727	89	30.6 ± 3.3
>3	315331	12	2.6×10^4

* Perihelion $q < 2$ au only.
† Number of asteroids.
‡ Number of comets.
§ Ratio $\mathcal{R} = N_a/N_c$.

Dynamical simulations provide an independent perspective. The mean dynamical lifetime of JFCs is $\tau_d \sim 4 \times 10^5$ yr (*Levison and Duncan,* 1994). Repeated scattering by the terrestrial planets causes the mean inclination of the JFCs to increase with time. In order to reproduce the modest inclinations of the JFCs (the mean and median values are 14.3° and 12.2°, respectively), these authors were forced to assume that the JFC physical lifetime is short compared to τ_d. Specifically, they inferred that the ratio of inactive to active JFCs should be in the range of 5 to 20, barely consistent with the measured $\mathcal{R} \sim 6$ for $2.0 \le T_J \le 2.8$.

Another distinctive feature of JFC nuclei that might be helpful in the statistical identification of such objects when dormant is their rotation period distribution. The rotation periods of comets are highly biased toward long periods (median 15 hr) compared to the period distribution in small NEAs (median ~6 hr) (Fig. 19) (see also *Binzel et al.,* 1992; *Jewitt,* 2021). This difference, which likely reflects the lower bulk density of comets, where the critical period for rotational instability scales as $\rho^{-1/2}$ (equation (8)), would suggest that few ACOs are dead comets. Another pointer is the truncated size distribution and distinct lack of subkilometer cometary nuclei compared to the high abundance of subkilometer asteroids.

Table 2 shows that Damocloids [named after the prototype object (5335) Damocles (*Jewitt,* 2005)] are comparatively rare compared to comets with $T_J < 2$, accounting for only ~20% of the total ($\mathcal{R} = 0.2$) having $q < 2$ au. The ratio becomes even more extreme for smaller values of T_J. The Damocloid population grows larger when the perihelion distance constraint in the table is relaxed, but this might just be because activity is harder to detect when $q \ge 2$ au. The table shows that, whereas apparent asteroids outnumber apparent comets in the range $2 \le T_J \le 3$, the opposite is true for $T_J < 2$. Currently 325 Damocloids are known, almost all of them reddish D-types (*Licandro et al.,* 2018).

The orbital elements of Damocloids are compared with those of JFCs and LPCs in Fig. 20. In all three panels, the Damocloids are intermediate between the JFC and LPC populations. While the inclination distributions of Damocloids and LPCs are similar, albeit with a preponderance of prograde orbits in the former (median inclinations of i = 73° for Damocloids vs. i = 90° for the LPCs), the semimajor axis and eccentricity distributions are distinctly different. Therefore, we can reject the possibility that the Damocloids are exclusively dead or dormant LPC nuclei, and additional sources must be considered. *Wang et al.* (2012) suggest a combination of sources both in the Oort cloud and in the

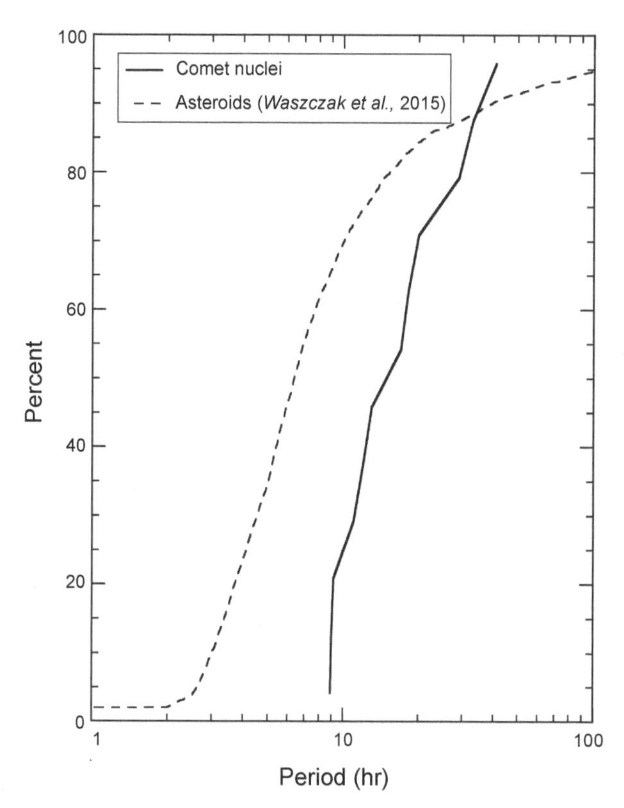

Fig. 19. Comparison of the cumulative distributions of the rotation periods of JFCs (solid line) and asteroids (dashed line) in the 1- to 10-km size range. The much longer mean period of the JFCs is evident. From *Jewitt* (2021).

scattered disk component of the Kuiper belt. Numerical simulations show that transfer to and from the scattered disk population in the Kuiper belt can occur.

Several objects initially classified as Damocloids [e.g., C/2002 CE_{10} (LINEAR) and C/2002 VQ_{94} (LINEAR)] have subsequently been found to show activity. In some cases, the detection of coma simply reflects the acquisition of more sensitive measurements, but in others, it appears that coma is intermittently present. These objects provide firm evidence that Damocloids and comets are compositionally as well as dynamically related.

An observational connection between dynamical and physical properties was established by *Fernández et al.* (2001) and confirmed by the addition of new measurements by *Fernández et al.* (2005) and *Kim et al.* (2014) (see Fig. 21). Specifically, geometric albedos, p_V, are observed to vary systematically with T_J, such that bodies with $T_J < 2$ have comet-like values $p_V = 0.05 \pm 0.02$, regardless of whether or not they show evidence for cometary activity, whereas bodies with $2 \leq T_J \leq 3$ show a wide range of albedos, particularly for $T_J \gtrsim 2.5$. *Kim et al.* (2014) emphasized the existence of high-albedo ($p_V > 0.1$) objects in this population, similar to what is seen in the bona-fide asteroids ($T_J \geq 3$). The albedo trends in Fig. 21 support the conclusions from Table 2 to the effect that Damocloids are mostly inactive comets while the ACO population consists of a mixture of (dark) inactive comets and (on average, brighter) bona-fide asteroids, although is dominated by the latter. Models indicate that objects can be scattered out of the protoplanetary disk by the giant planets from a wide range of initial orbits, even those interior to Jupiter (*Hahn*

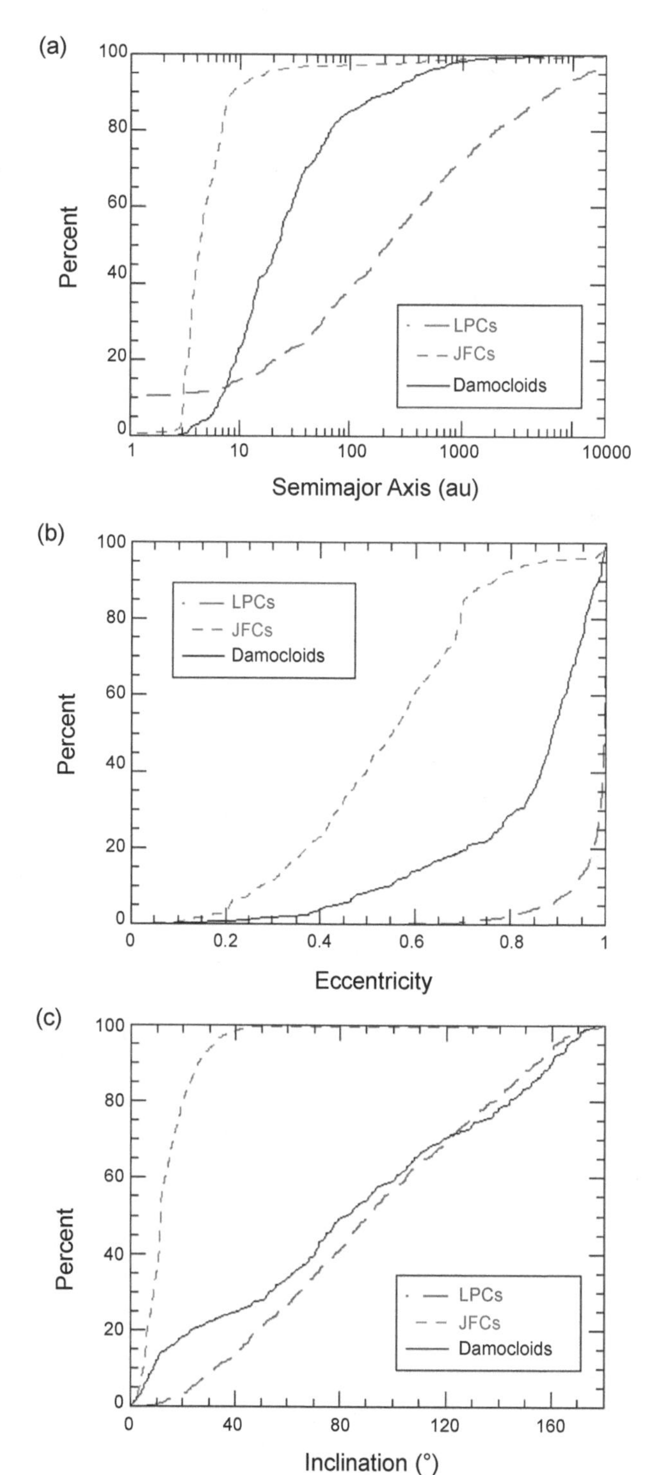

Fig. 20. Comparison of the cumulative distributions of the orbital elements of Damocloids (solid black lines), JFCs (dashed blue lines), and LPCs (long-dashed red lines); **(a)** semimajor axis, **(b)** eccentricity, and **(c)** inclination. Compiled from JPL Horizons.

and Malhotra, 1999). The systematically low (i.e., comet-like) albedos of Damocloids (Fig. 21) indicate that such asteroidal bodies are uncommon.

LPCs showing weak evidence for activity are sometimes referred to as "Manx" objects, an allusion to the short-tailed mutant cats of the same name. It is not clear that the Manx objects are more than semantically different from weakly active LPCs; they could simply be low-activity LPCs in which surface volatiles have been depleted in previous orbits. However, one object, C/2014 S3, shows a spectral continuum downturn near 1 μm similar to that seen in S-type asteroids (*Meech et al.,* 2016). This might suggest the presence of more highly metamorphosed solids like those found in the S-type asteroids, with a possible origin nearer the Sun at higher temperatures than typical for the nuclei of LPCs. Unfortunately, it is not known whether the albedo of C/2014 S3 is like that of an S-type asteroid. Emplacement of objects formed inside the snow line into the Oort cloud is inefficient because Jupiter scatters objects most efficiently to the interstellar medium, not the Oort cloud (*Hahn and Malhotra,* 1999). *Shannon et al.* (2015) estimated that ~8 × 10^9 asteroids could be so emplaced, which would comprise only ~1% of the Oort cloud comet population (~10^{12} total objects). More 1-μm continuum spectra of long-period objects are needed to test any possible association with S-type or other inner-disk objects.

8. CENTAURS

Centaurs are generally considered to be an intermediate population of small bodies that are in dynamical transition from the transneptunian population to the JFCs (*Dones et al.,* 2015; see also the chapter by Fraser et al. in this volume) and thus are scientifically interesting as JFC precursors. Several working definitions of this population exist. Here, we refer to objects with both perihelia and semimajor axes between the orbits of Jupiter (5.2 au) and Neptune (30 au) that are not in MMR with the planets (e.g., excluding Jupiter and Neptune Trojans). The defining object in this class is 2060 Chiron (a = 13.6 au, q = 8.5 au), discovered in 1977 (*Kowal et al.,* 1979), although Comet 29P/Schwassmann-Wachmann 1, which has only recently come to be regarded as a Centaur, was actually discovered five decades before Chiron. About 430 Centaurs are known as of late 2021.

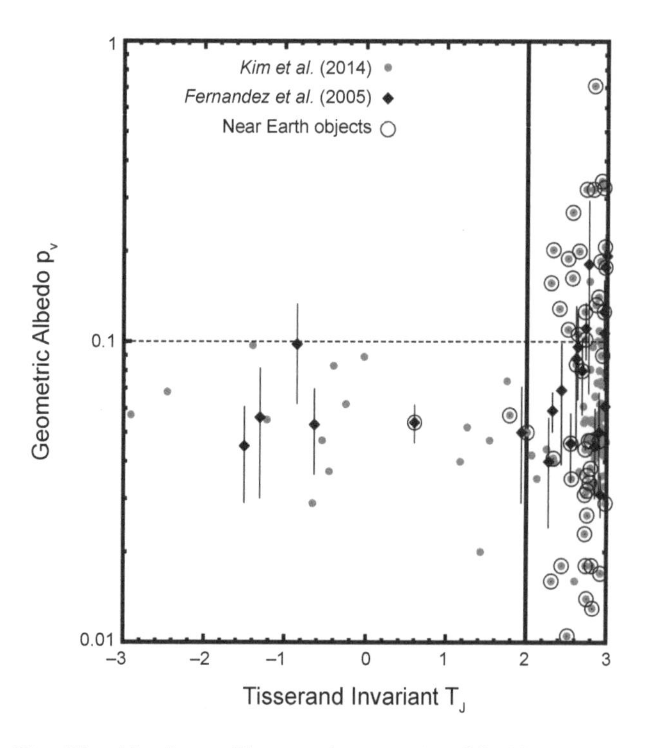

Fig. 21. Albedo vs. Tisserand parameter. All objects with T_J < 2 have low (comet-like) albedos. From *Kim et al.* (2014).

The Centaurs interact strongly with the giant planets and are consequently dynamically short-lived compared to the age of the solar system (*Horner et al.,* 2004; *Volk and Malhotra,* 2008). Centaur dynamical half-lives, $\tau_{1/2}$, are spread over a wide range, becoming shorter as the perihelion distance decreases, which is a reflection of the large mass and gravitational influence of Jupiter. *Horner et al.* (2004) bracketed the range of half-lives (expressed in million years) as

$$0.39e^{0.135q} \lesssim \tau_{1/2} \lesssim 0.06e^{0.275Q} \tag{13}$$

where q and Q are the perihelion and aphelion distances. Substituting q = 5.2 au and Q = 30 au, for example, gives $0.8 \lesssim \tau_{1/2} \lesssim 230$ m.y. Their source, at first unknown, is now firmly recognized as the Kuiper belt, specifically with the scattered disk component of the belt from which objects are destabilized by interactions with Neptune at perihelion (*Volk and Malhotra,* 2008). The fate of the Centaurs, other than a few that collide with a giant planet, is either to be ejected from the solar system to interstellar space, or injected into the terrestrial planet region where they sublimate and are relabeled as JFCs (*Levison and Duncan,* 1994); *Sarid et al.,* 2019). Unfortunately, the current orbit of any particular Centaur provides only a weak guide to its dynamical past because of the role of chaos induced by gas giant planet scattering. While there is a statistical flow of Centaurs from their source beyond 30 au inward, the instantaneous orbit of an individual Centaur may be diffusing inward or outward at any given time.

About 31 of the known Centaurs have a cometary designation, corresponding to about 7% of the total. This is a lower limit to the true incidence of activity because most such objects have not been studied in detail sufficient to reveal activity even if it is present. Furthermore, most Centaur activity is transient, so that observations over a long period can only increase the estimated active fraction. Chiron itself shows intermittent mass loss, unconnected to its heliocentric distance.

Two properties distinguish active Centaurs from others (*Jewitt,* 2009, 2015). First, their perihelia are preferentially small, typically q ≲ 8–10 au (Fig. 22a). While observational selection does favor the detection of activity at small q, a sensitive, high-resolution search for activity in 53 Centaurs with perihelia q > 15 au found none (*Li et al.,* 2020). Second, while Centaurs as a whole show a bimodal distribution of optical colors (*Peixinho et al.,* 2012), this bimodality vanishes for objects with q ≲ 8–10 au (cf. Fig. 23). The loss of bimodality occurs because the Centaurs lose ultrared matter, defined to have color index B–R > 1.6, and thought to consist of highly irradiated organics. [Note that the color index B–R is defined as 2.5× the logarithm of the ratio of the flux density measured in the R filter (center wavelength ~6500 Å) to that in the B filter (4500 Å). Larger values indicate redder colors.] The onset of activity and the disappearance of the ultrared matter appear to be correlated, in the sense that bodies that are active are rarely very red. Active centaur (523676) 2013 UL_{10} with B–R = 1.8 ± 0.1 is a possible exception (*Mazzotta Epifani et al.,* 2018), although the colors are contradicted by *Tegler et al.* (2016), who measured B–R = 1.62 ± 0.03. This could mean that the ultrared matter, whose precise composition is unknown, is thermodynamically unstable at temperatures ≳124 K. Alternatively, a thin layer of ultrared matter could be ejected or simply buried by fallback debris excavated from beneath, as soon as the activity begins (*Jewitt,* 2002). The smaller average perihelia of the active Centaurs correlate as expected from equation (13), with shorter mean dynamical lifetimes (*Melita and Licandro,* 2012; *Fernández et al.,* 2018).

What triggers Centaur activity near 8–10 au, where the isothermal blackbody temperature of a sphere is about 88–98 K? This is far too cold for water ice to sublimate. While water is not volatile enough, exposed supervolatile ices (e.g., CO) are too volatile, in the sense that they would easily drive activity at much larger distances than observed. For example, CO ice would sublimate strongly even at Kuiper belt distances and temperatures, where no activity is observed, and even CO_2 ice is in the strong-sublimation regime at distances ~10 au. Burial of supervolatile ices beneath a refractory mantle could, in principle, delay the onset of sublimation from an incoming body but burial would give no clear reason for a distinct turn-on distance at 8–10 au (cf. Fig. 22). Moreover, thermal evolution models suggest that *pure* CO cannot survive long term in comets (*Davidsson,* 2021). Explanations attempting to match activity to the sublimation of less-volatile, less-abundant materials (such as H_2S) struggle to provide a convincing match to the 8–10 au critical distance (*Poston et al.,* 2018).

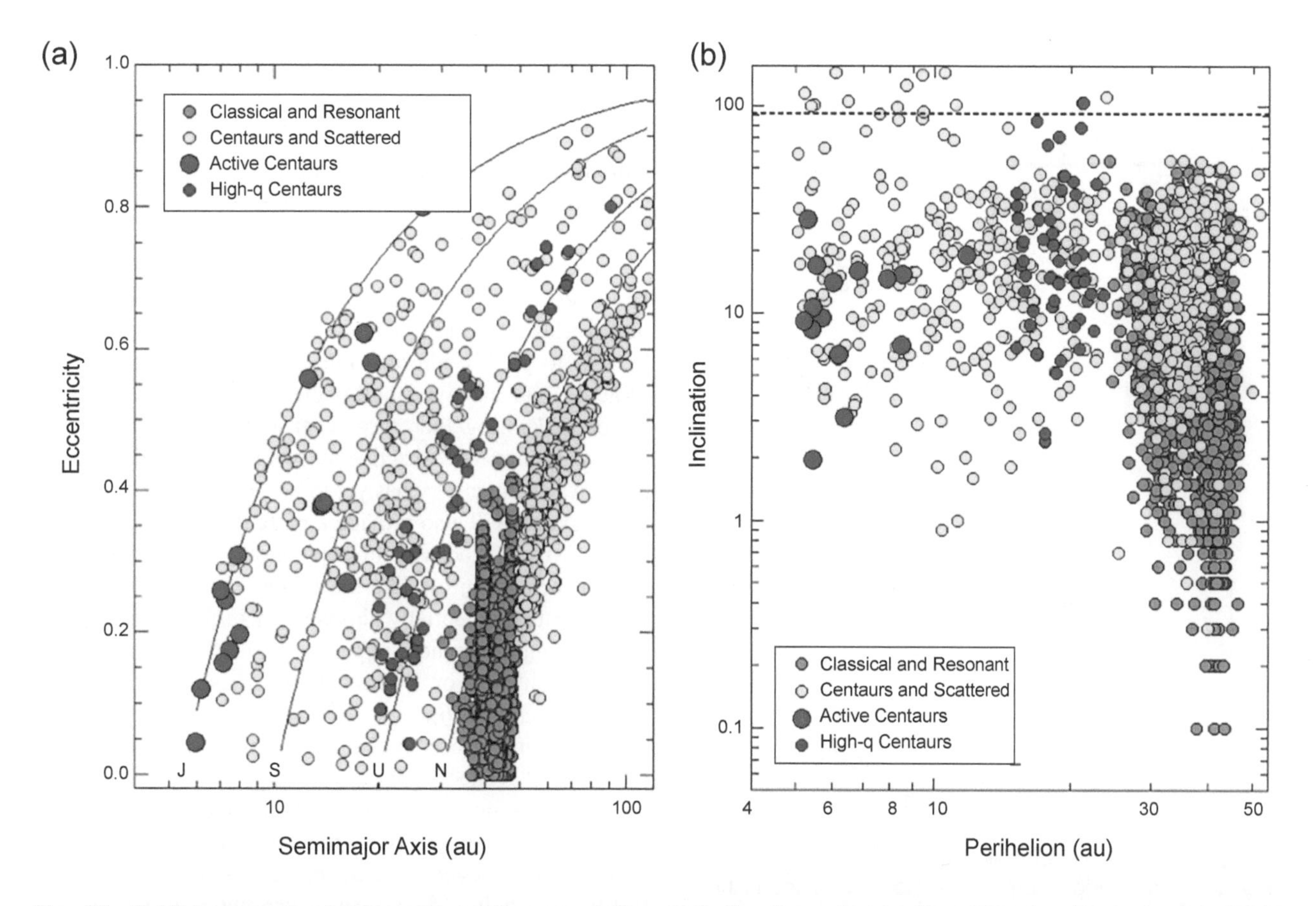

Fig. 22. Orbital elements of outer solar system populations including the active (red) and inactive Centaurs. Semimajor axes of the major planets are marked J, S, U, and N. **(a)** Diagonal arcs show the loci of orbits having fixed perihelia, q = a(1–e), equal to the semimajor axes of the giant planets. **(b)** The dotted horizontal line is at i = 90°; objects above it are retrograde. The active Centaurs cluster within q ≲ 10 au. Adapted from *Li et al.* (2020).

More promisingly, the crystallization timescale for amorphous water ice decreases exponentially with temperature, and first becomes less than the orbital period at ~10 au, suggesting that crystallization might drive Centaur activity (*Jewitt,* 2009; *Guilbert-Lepoutre,* 2012). Amorphous ice is the natural form of ice formed at low temperatures and gas pressures, so its incorporation into the icy bodies of the outer solar system would not be surprising. Crystallization also fits the observed bursty nature of Centaur activity, corresponding to short-lived exothermic runaways in which trapped gases are expelled (see the chapter by Prialnik and Jewitt in this volume). In addition, Centaur activity may be correlated with recent inward drift of the perihelion distance (*Fernández et al.,* 2018), consistent with the crystallization hypothesis. Amorphous ice can trap CO, which has been reported in 29P/Schwassmann-Wachmann 1, (60558) Echeclus, and 2060 Chiron (see review by *Womack et al.,* 2017) but has not been detected in a majority of the Centaurs in which it has been sought (*Bockelée-Morvan et al.,* 2001; *Jewitt et al.,* 2008; *Drahus et al.,* 2017). Absence of evidence is not necessarily evidence of absence, however, and the non-detections could simply reflect the difficulty in measuring CO rotational transitions in bodies far from the Earth. The onset of Centaur activity (Fig. 22) and the collapse of the bimodal color distribution (Fig. 23) are certainly consistent with an origin in the crystallization of amorphous ice, but it must be said that they do not prove that this is the responsible mechanism. Establishing proof of the existence of amorphous ice in JFCs and Centaurs would be of great scientific value. It would imply that ice in the precursor Kuiper belt objects has also survived in the amorphous state, significantly limiting the post-accretion thermal processing of these bodies (*Davidsson,* 2021; see also the chapter by Prialnik and Jewitt in this volume).

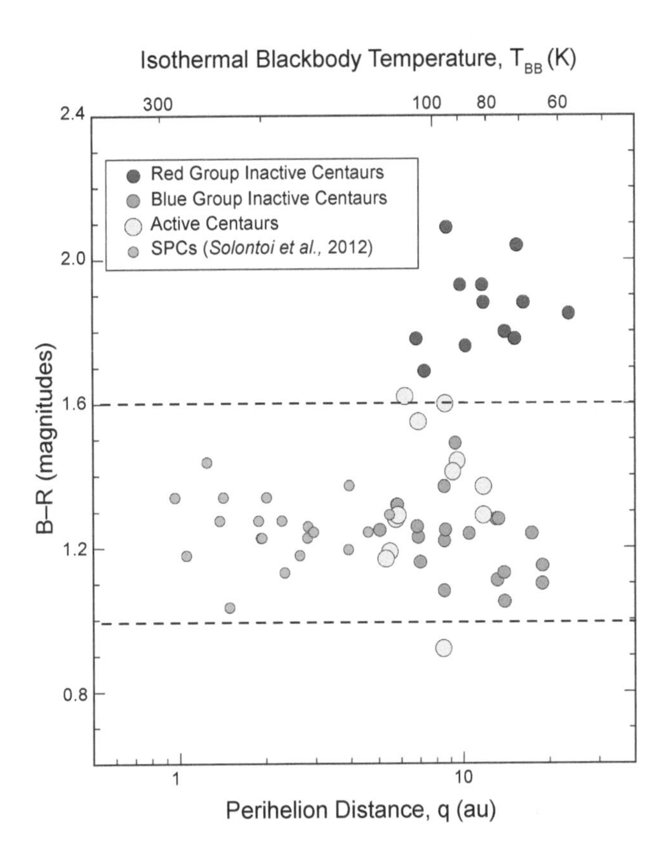

Fig. 23. B–R optical color as a function of perihelion distance for short-period comets (SPCs; synonymous here with JFCs) and active and inactive Centaurs, as marked. Horizontal dashed lines at B–R = 0.99 and B–R = 1.6 mark the colors of the Sun and the nominal beginning of the ultra-red objects, respectively. Adapted from *Jewitt* (2015).

9. FUTURE PROSPECTS AND CONCLUSIONS

Looking ahead, Rubin Observatory's Legacy Survey of Space and Time (LSST) should generate a massive temporal survey of the sky over 10 yrs and is expected to increase the numbers of known small solar system bodies in different populations by an order of magnitude or more (*Jones et al.,* 2009; *Ivezić et al.,* 2019). Its use of a 8.4-m-aperture telescope means that Rubin Observatory should achieve a sensitivity to faint activity that is unprecedented among wide-field surveys. Meanwhile, relevant upcoming NASA spacebased observatories include the Near-Earth Object Surveyor (NEO Surveyor) (*Mainzer et al.,* 2021), which aims to identify near-Earth objects that might present impact threats to Earth, and the Nancy Grace Roman Space Telescope (*Akeson et al.,* 2019), which will conduct a wide-field survey that is nominally focused on dark energy and exoplanets, but that is expected to observe many solar system objects as well (although mostly at high ecliptic latitudes). Observing in the infrared, both spacecraft will be sensitive to thermal emission from long-lasting large dust grains ejected by active small bodies, potentially widening the observability window for detection of active events (e.g., *Holler et al.,* 2018). In particular, NEO Surveyor's focus on discovering near-Earth objects (NEOs) means that it should be able to greatly improve our understanding of the ratio of active and inactive small bodies in the terrestrial planet region. Taken together, these ever more sensitive searches for activity may reveal active objects in small-body populations in which activity has yet to be found (e.g., the Jupiter Trojans) and will also provide a more thorough accounting of the extent of activity in populations already known to contain active objects like the NEOs, main-belt asteroids, and Centaurs, potentially allowing for further classification of active objects in those populations into smaller subgroups based on physical or dynamical properties (e.g., *Hsieh and Haghighipour,* 2016).

Continuing efforts to characterize individual active objects will also shed light on their physical nature, provide needed constraints for thermal models, and help to identify physically plausible activity mechanisms for each object. As most active events are of limited duration, sometimes lasting just a few days, useful characterization of new active objects discovered by LSST, NEO Surveyor, or the Roman Space Telescope will require rapid-response follow-up deep imaging and spectroscopy (e.g., *Najita et al.,* 2016; *Street et al.,* 2018). Long-term monitoring of known active objects (e.g., *Hsieh et al.,* 2018a; *Wierzchos and Womack,* 2020) is also needed to provide insights into both short-term- and long-term-activity evolution, while detailed characterization (e.g., measurements of phase functions, colors, and rotation states) of the nuclei of active objects during periods of inactivity (e.g., see also the chapter in this volume by Knight et al.) provide information essential for thermal modeling studies (e.g. *Schörghofer et al.,* 2020) and also constraining the properties of the population of other potentially active objects. Meanwhile, the recently launched James Webb Space Telescope and the upcoming generation of extremely large (~30-m-class) groundbased telescopes should enable searches for gas with unprecedented sensitivity (*Crampton and Simard,* 2009; *Kelley et al.,* 2016; *Wong et al.,* 2021). Continued targeted and general dynamical studies investigating the connections between small-body populations (see section 5) will also be useful for identifying potential opportunities for close-proximity studies of active objects originally from the main asteroid belt (e.g., *Fernández et al.,* 2017), much as classical comets give us the opportunity to study the composition of objects displaced from the outer solar system.

We can also look forward to a number of *in situ* space missions currently being planned or proposed (see also the chapter by Snodgrass et al. in this volume). The DESTINY+ spacecraft being developed by the Japan Aerospace Exploration Agency (JAXA) will visit Phaethon and has an expected launch date in 2024 (*Arai et al.,* 2021). It will reach a close approach distance of 500 km from the object, briefly permitting imaging of the surface at ≤5 m pixel^{-1}, albeit at a flyby speed of 36 km s^{-1}. An impact ionization time-of-flight mass spectrometer will provide elemental compositions of any dust particles impacting the spacecraft, with a mass resolution of $M/\Delta M$ ~ 100–150 (*Krüger et al.,* 2019). It should be noted, however, that the total dust mass that will be intercepted is modest: less than that of a single 30-μm particle (*Jewitt et al.,* 2018a). Meanwhile, the Chinese National Space Agency's ZhengHe mission aims to visit active asteroid 311P, with an expected launch in 2025. Missions targeting MBCs have been proposed to both NASA and the European Space Agency (ESA) (e.g., *Meech and Castillo-Rogez,* 2015; *Jones et al.,* 2018; *Snodgrass et al.,* 2018), so far without success. *In situ* studies are particularly needed to resolve major issues about the nature of MBC outgassing, because they provide the only means for investigating outgassed material with a mass spectrometer.

In terms of other interesting potential mission concepts, active asteroid 331P ejected ~1% of its mass as recently as 2011, producing a debris trail and a chain of 19 or more fragments that grows in length at ≲10 cm s^{-1} (*Jewitt et al.,* 2021). A single ion-driven spacecraft sent to this object could determine the (presumably) excited rotation state of the kilometer-sized primary body, investigate the geology of the detachment scar at high resolution in order to assess the mode of failure, and then travel the length of the debris trail, visiting the ~19 fragments one by one. The largest fragment, 331P-A, has an effective diameter of ~200 m and shows a light curve consistent with a contact binary, presenting a particularly attractive target for *in situ* investigation. Of course, we have already had one unexpected mission to an active asteroid, OSIRIS-REx's visit to Bennu. Because Bennu's activity was unexpected, only limited relevant observations were possible and the study of the ejected material has been inconclusive. Nevertheless, the discovery of particles too large and sparse to be detected from Earth suggests that many other asteroids may be weakly active when examined closely, and that this possibility should be taken into account for future small body missions in general.

Our understanding of the asteroid-comet continuum has evolved substantially since the publication of *Comets II* (*Jewitt,* 2004). We now recognize that many small bodies possess diverse combinations of physical and dynamical properties as well as evolutionary histories that cannot be cleanly associated with either asteroids or comets. Studies of extant volatile material from the inner portions of the protosolar disk are now possible, as are direct observations of asteroid collisions and rotational disruptions. New observations raise fresh questions about the incredible range of processes that can produce mass loss from small bodies and about the extent to which small bodies have been shuffled radially in the solar system since its formation. We anticipate even more exciting developments in the coming years.

Acknowledgments. We thank M. Ishiguro and F. Moreno for their helpful reviews of this chapter. H.H.H. acknowledges support from NASA grants 80NSSC17K0723, 80NSSC18K0193, and 80NSSC19K0869.

REFERENCES

Agarwal J., Kim Y., Jewitt D. et al. (2020) Component properties and mutual orbit of binary main-belt Comet 288P/(300163) 2006 VW$_{139}$. *Astron. Astrophys., 643,* A152.

A'Hearn M. F., Millis R. C., Schleicher D. O. et al. (1995) The ensemble properties of comets: Results from narrowband photometry of 85 comets, 1976–1992. *Icarus, 118,* 223–270.

A'Hearn M. F., Belton M. J. S., Delamere A. et al. (2005) Deep Impact: A large-scale active experiment on a cometary nucleus. *Space Sci. Rev., 117,* 1–21.

Akeson R., Armus L., Bachelet E. et al. (2019) The Wide Field Infrared Survey Telescope: 100 Hubbles for the 2020s. *ArXiV e-prints,* arXiv:1902.05569.

Ansdell M., Meech K. J., Hainaut O. et al. (2014) Refined rotational period, pole solution, and shape model for (3200) Phaethon. *Astrophys. J., 793,* 50.

Arai T., Yoshida F., Kobayashi M. et al. (2021) Current status of DESTINY$^+$ and updated understanding of its target asteroid (3200) Phaethon. In *52nd Lunar and Planetary Science Conference,* Abstract #1896. LPI Contribution No. 2548, Lunar and Planetary Institute, Houston.

Arvidson R., Drozd R. J., Hohenberg C. M. et al. (1975) Horizontal transport of the regolith, modification of features, and erosion rates on the lunar surface. *Moon, 13,* 67–79.

Babadzhanov P. B. (2003) Meteor showers associated with the near-Earth asteroid (2101) Adonis. *Astron. Astrophys., 397,* 319–323.

Babadzhanov P. B., Kokhirova G. I., and Obrubov Y. V. (2015) Extinct comets and asteroid-meteoroid complexes. *Solar System Res., 49,* 165–172.

Bach Y. P. and Ishiguro M. (2021) Thermal radiation pressure as a possible mechanism for losing small particles on asteroids. *Astron.*

Astrophys., 654, A113.
Bertini I. (2011) Main belt comets: A new class of small bodies in the solar system. Planet. Space Sci., 59, 365–377.
Binzel R. P., Xu S., Bus S. J. et al. (1992) Origins for the near-Earth asteroids. Science, 257, 779–782.
Birtwhistle P., Ryan W. H., Sato H. et al. (2010) Comet P/2010 A2 (LINEAR). IAU Circular 9105.
Blaauw R. C. (2017) The mass index and mass of the Geminid meteoroid stream as determined with radar, optical and lunar impact data. Planet. Space Sci., 143, 83–88.
Bockelée-Morvan D., Lellouch E., Biver N. et al. (2001) Search for CO gas in Pluto, Centaurs and Kuiper belt objects at radio wavelengths. Astron. Astrophys., 377, 343–353.
Bodewits D., Vincent J.-B., and Kelley M. S. P. (2014) Scheila's scar: Direct evidence of impact surface alteration on a primitive asteroid. Icarus, 229, 190–195.
Boehnhardt H., Schulz R., Tozzi G. P. et al. (1996) Comet P/1996 N2 (Elst-Pizarro). IAU Circular 6495.
Bose K. and Ganguly J. (1994) Thermogravimetric study of the dehydration kinetics of talc. Am. Mineral., 79, 692–699.
Bottke W. F., Durda D. D., Nesvorný D. et al. (2005) Linking the collisional history of the main asteroid belt to its dynamical excitation and depletion. Icarus, 179, 63–94.
Bottke Jr. W. F., Vokrouhlický D., Rubincam D. P. et al. (2006) The Yarkovsky and YORP effects: Implications for asteroid dynamics. Annu. Rev. Earth Planet. Sci., 34, 157–191.
Bottke W. F., Moorhead A. V., Connolly H. C. et al. (2020) Meteoroid impacts as a source of Bennu's particle ejection events. J. Geophys. Res.–Planets, 125, e06282.
Brasser R. and Wang J.-H. (2015) An updated estimate of the number of Jupiter-family comets using a simple fading law. Astron. Astrophys., 573, A102.
Carusi A., Kresák L., and Valsecchi G. B. (1995) Conservation of the Tisserand parameter at close encounters of interplanetary objects with Jupiter. Earth Moon Planets, 68, 71–94.
Chandler C. O., Kueny J., Gustafsson A. et al. (2019) Six years of sustained activity in (6478) Gault. Astrophys. J. Lett., 877, L12.
Cikota S., Ortiz J. L., Cikota A. et al. (2014) A photometric search for active main belt asteroids. Astron. Astrophys., 562, A94.
Colwell J. E., Gulbis A. A. S., Horányi M. et al. (2005) Dust transport in photoelectron layers and the formation of dust ponds on Eros. Icarus, 175, 159–169.
Crampton D. and Simard L. (2009) Solar system capabilities of the Thirty Meter Telescope. Earth Moon Planets, 105, 65–72.
Criswell D. R. (1973) Horizon-glow and the motion of lunar dust. In Photon and Particle Interactions with Surfaces in Space (R. J. L. Grard, ed.), pp. 545–556. Astrophys. Space Sci. Library, Vol. 37, Kluwer, Dordrecht.
Criswell D. R. and De B. R. (1977) Intense localized photoelectric charging in the lunar sunset terminator region. 2. Supercharging at the progression of sunset. J. Geophys. Res.–Space Phys., 82, 1005–1007.
Davidsson B. J. R. (2021) Thermophysical evolution of planetesimals in the primordial disc. Mon. Not. R. Astron. Soc., 505, 5654–5685.
de León J., Campins H., Tsiganis K. et al. (2010) Origin of the near-Earth asteroid Phaethon and the Geminids meteor shower. Astron. Astrophys., 513, A26.
DeMeo F. E. and Carry B. (2014) Solar system evolution from compositional mapping of the asteroid belt. Nature, 505, 629–634.
DeMeo F. E., Binzel R. P., Carry B. et al. (2014) Unexpected D-type interlopers in the inner main belt. Icarus, 229, 392–399.
DeMeo F. E., Alexander C. M. O., Walsh K. J. et al. (2015) The compositional structure of the asteroid belt. In Asteroids IV (P. Michel et al., eds.), pp. 13–41. Univ. of Arizona, Tucson.
Denneau L., Jedicke R., Fitzsimmons A. et al. (2015) Observational constraints on the catastrophic disruption rate of small main belt asteroids. Icarus, 245, 1–15.
Devogèle M., Ferrais M., Jehin E. et al. (2021) (6478) Gault: Physical characterization of an active main-belt asteroid. Mon. Not. R. Astron. Soc., 505, 245–258.
Dones L., Brasser R., Kaib N. et al. (2015) Origin and evolution of the cometary reservoirs. Space Sci. Rev., 197, 191–269.
Drahus M., Waniak W., Tendulkar S. et al. (2015) Fast rotation and trailing fragments of the active asteroid P/2012 F5 (Gibbs). Astrophys. J. Lett., 802, L8.
Drahus M., Yang B., Lis D. C. et al. (2017) New limits to CO outgassing in Centaurs. Mon. Not. R. Astron. Soc., 468, 2897–2909.
Duncan M. J. (2008) Dynamical origin of comets and their reservoirs. Space Sci. Rev., 138, 109–126.
Ďurech J., Vokrouhlický D., Pravec P. et al. (2018) YORP and Yarkovsky effects in asteroids (1685) Toro, (2100) Ra-Shalom, (3103) Eger, and (161989) Cacus. Astron. Astrophys., 609, A86.
Ďurech J., Vokrouhlicky D., Pravec P. et al. (2022) Rotation acceleration of asteroids (10115) 1992 SK, (1685) Toro, and (1620) Geographos due to the YORP effect. Astron. Astrophys., 657, A5.
Elst E. W., Pizarro O., Pollas C. et al. (1996) Comet P/1996 N2 (Elst-Pizarro). IAU Circular 6456.
Encrenaz T. (2008) Water in the solar system. Annu. Rev. Astron. Astrophys., 46, 57–87.
Eppes M. C., Magi B., Hallet B. et al. (2016) Deciphering the role of solar-induced thermal stresses in rock weathering. Geol. Soc. Am. Bull., 128, 1315–1338.
Fanale F. P. and Salvail J. R. (1989) The water regime of asteroid (1) Ceres. Icarus, 82, 97–110.
Fernández J. A. and Sosa A. (2015) Jupiter family comets in near-Earth orbits: Are some of them interlopers from the asteroid belt? Planet. Space Sci., 118, 14–24.
Fernández J. A., Gallardo T., and Brunini A. (2002) Are there many inactive Jupiter-family comets among the near-Earth asteroid population? Icarus, 159, 358–368.
Fernández J. A., Licandro J., Moreno F. et al. (2017) Physical and dynamical properties of the anomalous Comet 249P/LINEAR. Icarus, 295, 34–45.
Fernández J. A., Helal M., and Gallardo T. (2018) Dynamical evolution and end states of active and inactive Centaurs. Planet. Space Sci., 158, 6–15.
Fernández Y. R., McFadden L. A., Lisse C. M. et al. (1997) Analysis of POSS images of comet-asteroid transition object 107P/1949 W1 (Wilson-Harrington). Icarus, 128, 114–126.
Fernández Y. R., Jewitt D. C., and Sheppard S. S. (2001) Low albedos among extinct comet candidates. Astrophys. J. Lett., 553, L197–L200.
Fernández Y. R., Jewitt D. C., and Sheppard S. S. (2005) Albedos of asteroids in comet-like orbits. Astron. J., 130, 308–318.
Ferrín I. and Orofino V. (2021) Taurid complex smoking gun: Detection of cometary activity. Planet. Space Sci., 207, 105306.
Geem J., Ishiguro M., Bach Y. P. et al. (2022) A polarimetric study of asteroids in comet-like orbits. Astron. Astrophys., 658, A158.
Gilbert A. M. and Wiegert P. A. (2010) Updated results of a search for main-belt comets using the Canada-France-Hawaii Telescope Legacy Survey. Icarus, 210, 998–999.
Gradie J. and Tedesco E. (1982) Compositional structure of the asteroid belt. Science, 216, 1405–1407.
Granvik M., Morbidelli A., Jedicke R. et al. (2016) Supercatastrophic disruption of asteroids at small perihelion distances. Nature, 530, 303–306.
Guilbert-Lepoutre A. (2012) Survival of amorphous water ice on Centaurs. Astron. J., 144, 97.
Gundlach B. and Blum J. (2016) Why are Jupiter-family comets active and asteroids in cometary-like orbits inactive? How hydrostatic compression leads to inactivity. Astron. Astrophys., 589, A111.
Gustafson B. A. S. (1989) Geminid meteoroids traced to cometary activity on Phaethon. Astron. Astrophys., 225, 533–540.
Haghighipour N., Maindl T. I., Schäfer C. et al. (2016) Triggering sublimation-driven activity of main belt comets. Astrophys. J., 830, 22.
Hahn J. M. and Malhotra R. (1999) Orbital evolution of planets embedded in a planetesimal disk. Astron. J., 117, 3041–3053.
Hainaut O. R., Kleyna J. T., Meech K. J. et al. (2019) Disintegration of active asteroid P/2016 G1 (Pan-STARRS). Astron. Astrophys., 628, A48.
Hanuš J., Delbó M., Vokrouhlický D. et al. (2016) Near-Earth asteroid (3200) Phaethon: Characterization of its orbit, spin state, and thermophysical parameters. Astron. Astrophys., 592, A34.
Harris A. W., Fahnestock E. G., and Pravec P. (2009) On the shapes and spins of "rubble pile" asteroids. Icarus, 199, 310–318.
Harsono D., Bruderer S., and van Dishoeck E. F. (2015) Volatile snowlines in embedded disks around low-mass protostars. Astron. Astrophys., 582, A41.
Hartmann W. K., Tholen D. J., Meech K. J. et al. (1990) 2060 Chiron: Colorimetry and cometary behavior. Icarus, 83, 1–15.
Hasegawa S., Marsset M., DeMeo F. E. et al. (2021) Discovery of two

TNO-like bodies in the asteroid belt. *Astrophys. J. Lett., 916,* L6.
Hasegawa S., Marsset M., DeMeo F. E. et al. (2022) The appearance of a "fresh" surface on 596 Scheila as a consequence of the 2010 impact event. *Astrophys. J. Lett., 924,* L9.
Heiken G. H., Vaniman D. T., and French B. M., eds. (1991) *Lunar Sourcebook: A User's Guide to the Moon.* Cambridge Univ., Cambridge. 736 pp.
Hergenrother C. W., Maleszewski C., Li J. Y. et al. (2020) Photometry of particles ejected from active asteroid (101955) Bennu. *J. Geophys. Res.–Planets, 125,* e06381.
Hirabayashi M., Scheeres D. J., Sánchez D. P. et al. (2014) Constraints on the physical properties of main belt Comet P/2013 R3 from its breakup event. *Astrophys. J. Lett., 789,* L12.
Hirabayashi M., Sánchez D. P., and Scheeres D. J. (2015) Internal structure of asteroids having surface shedding due to rotational instability. *Astrophys. J., 808,* 63.
Holler B. J., Milam S. N., Bauer J. M. et al. (2018) Solar system science with the Wide-Field Infrared Survey Telescope. *J. Astronomical Telesc. Instrum. Syst., 4,* 034003.
Holsapple K. A. and Housen K. R. (2019) The catastrophic disruptions of asteroids: History, features, new constraints and interpretations. *Planet. Space Sci., 179,* 104724.
Hood N., Carroll A., Wang X. et al. (2022) Laboratory measurements of size distribution of electrostatically lofted dust. *Icarus, 371,* 114684.
Horner J., Evans N. W., and Bailey M. E. (2004) Simulations of the population of Centaurs — I. The bulk statistics. *Mon. Not. R. Astron. Soc., 354,* 798–810.
Horz F. and Cintala M. (1997) Impact experiments related to the evolution of planetary regoliths. *Meteorit. Planet. Sci., 32,* 179–209.
Housen K. R., Sweet W. J., and Holsapple K. A. (2018) Impacts into porous asteroids. *Icarus, 300,* 72–96.
Hsieh H. H. (2009) The Hawaii Trails Project: Comet-hunting in the main asteroid belt. *Astron. Astrophys., 505,* 1297–1310.
Hsieh H. H. (2014a) Main-belt comets as tracers of ice in the inner solar system. In *Formation, Detection, and Characterization of Extrasolar Habitable Planets* (N. Haghighipour, ed.), pp. 212–218. IAU Symp. 293, Cambridge Univ., Cambridge.
Hsieh H. H. (2014b) The nucleus of main-belt Comet P/2010 R2 (La Sagra). *Icarus, 243,* 16–26.
Hsieh H. H. (2017) Asteroid-comet continuum objects in the solar system. *Philos. Trans. R. Soc. A, 375,* 20160259.
Hsieh H. H. and Jewitt D. (2005) Search for activity in 3200 Phaethon. *Astrophys. J., 624,* 1093–1096.
Hsieh H. H. and Jewitt D. (2006) A population of comets in the main asteroid belt. *Science, 312,* 561–563.
Hsieh H. H. and Sheppard S. S. (2015) The reactivation of main-belt Comet 324P/La Sagra (P/2010 R2). *Mon. Not. R. Astron. Soc., 454,* L81–L85.
Hsieh H. H. and Haghighipour N. (2016) Potential Jupiter-family comet contamination of the main asteroid belt. *Icarus, 277,* 19–38.
Hsieh H. H., Jewitt D. C., and Fernández Y. R. (2004) The strange case of 133P/Elst-Pizarro: A comet among the asteroids. *Astron. J., 127,* 2997–3017.
Hsieh H. H., Jewitt D., and Fernández Y. R. (2009a) Albedos of main-belt comets 133P/Elst-Pizarro and 176P/LINEAR. *Astrophys. J. Lett., 694,* L111–L114.
Hsieh H. H., Jewitt D., and Ishiguro M. (2009b) Physical properties of main-belt Comet P/2005 U1 (Read). *Astron. J., 137,* 157–168.
Hsieh H. H., Jewitt D., Lacerda P. et al. (2010) The return of activity in main-belt Comet 133P/Elst-Pizarro. *Mon. Not. R. Astron. Soc., 403,* 363–377.
Hsieh H. H., Ishiguro M., Lacerda P. et al. (2011a) Physical properties of main-belt Comet 176P/LINEAR. *Astron. J., 142,* 29.
Hsieh H. H., Meech K. J., and Pittichová J. (2011b) Main-belt Comet 238P/Read revisited. *Astrophys. J. Lett., 736,* L18.
Hsieh H. H., Yang B., and Haghighipour N. (2012a) Optical and dynamical characterization of comet-like main-belt asteroid (596) Scheila. *Astrophys. J., 744,* 9.
Hsieh H. H., Yang B., Haghighipour N. et al. (2012b) Discovery of main-belt Comet P/2006 VW139 by Pan-STARRS1. *Astrophys. J. Lett., 748,* L15.
Hsieh H. H., Yang B., Haghighipour N. et al. (2012c) Observational and dynamical characterization of main-belt Comet P/2010 R2 (La Sagra). *Astron. J., 143,* 104.
Hsieh H. H., Kaluna H. M., Novaković B. et al. (2013) Main-belt Comet P/2012 T1 (Pan-STARRS). *Astrophys. J. Lett., 771,* L1.
Hsieh H. H., Denneau L., Wainscoat R. J. et al. (2015a) The main-belt comets: The Pan-STARRS1 perspective. *Icarus, 248,* 289–312.
Hsieh H. H., Hainaut O., Novaković B. et al. (2015b) Sublimation-driven activity in main-belt Comet 313P/Gibbs. *Astrophys. J. Lett., 800,* L16.
Hsieh H. H., Ishiguro M., Kim Y. et al. (2018a) The 2016 reactivations of the main-belt comets 238P/Read and 288P/(300163) 2006 VW139. *Astron. J., 156,* 223.
Hsieh H. H., Ishiguro M., Knight M. M. et al. (2018b) The reactivation and nucleus characterization of main-belt Comet 358P/Pan-STARRS (P/2012 T1). *Astron. J., 156,* 39.
Hsieh H. H., Novaković B., Kim Y. et al. (2018c) Asteroid family associations of active asteroids. *Astron. J., 155,* 96.
Hsieh H. H., Novaković B., Walsh K. J. et al. (2020) Potential Themis-family asteroid contribution to the Jupiter-family comet population. *Astron. J., 159,* 179.
Hsieh H. H., Chandler C. O., Denneau L. et al. (2021a) Physical characterization of main-belt Comet (248370) 2005 QN173. *Astrophys. J. Lett., 922,* L9.
Hsieh H. H., Ishiguro M., Knight M. M. et al. (2021b) The reactivation of main-belt Comet 259P/Garradd (P/2008 R1). *Planet. Sci. J., 2,* 62.
Hui M.-T. and Jewitt D. (2017) Non-gravitational acceleration of the active asteroids. *Astron. J., 153,* 80.
Hui M.-T. and Li J. (2017) Resurrection of (3200) Phaethon in 2016. *Astron. J., 153,* 23.
Hui M.-T., Jewitt D., and Du X. (2017) Split active asteroid P/2016 J1 (Pan-STARRS). *Astron. J., 153,* 141.
Ishiguro M., Sarugaku Y., Nishihara S. et al. (2009) Report on the Kiso cometary dust trail survey. *Adv. Space Res., 43,* 875–879.
Ishiguro M., Hanayama H., Hasegawa S. et al. (2011a) Interpretation of (596) Scheila's triple dust tails. *Astrophys. J. Lett., 741,* L24.
Ishiguro M., Hanayama H., Hasegawa S. et al. (2011b) Observational evidence for an impact on the main-belt asteroid (596) Scheila. *Astrophys. J. Lett., 740,* L11.
Ivezić Ž., Kahn S. M., Tyson J. A. et al. (2019) LSST: From science drivers to reference design and anticipated data products. *Astrophys. J., 873,* 111.
Izidoro A., de Souza Torres K., Winter O. C. et al. (2013) A compound model for the origin of Earth's water. *Astrophys. J., 767,* 54.
Jenniskens P. (2008) *Meteor Showers and their Parent Comets.* Cambridge Univ., Cambridge. 804 pp.
Jenniskens P. (2015) Meteoroid streams and the zodiacal cloud. In *Asteroids IV* (P. Michel et al., eds.), pp. 281–295. Univ. of Arizona, Tucson.
Jewitt D. (1996) From comets to asteroids: When hairy stars go bald. *Earth Moon Planets, 72,* 185–201.
Jewitt D. (2002) From Kuiper belt object to cometary nucleus: The missing ultrared matter. *Astron. J., 123,* 1039–1049.
Jewitt D. (2004) From cradle to grave: The rise and demise of the comets. In *Comets II* (M. C. Festou et al., eds.), pp. 659–676. Univ. of Arizona, Tucson.
Jewitt D. (2005) A first look at the Damocloids. *Astron. J., 129,* 530–538.
Jewitt D. (2009) The active Centaurs. *Astron. J., 137,* 4296–4312.
Jewitt D. (2012) The active asteroids. *Astron. J., 143,* 66.
Jewitt D. (2015) Color systematics of comets and related bodies. *Astron. J., 150,* 201.
Jewitt D. (2020) 138175 (2000 EE104) and the source of interplanetary field enhancements. *Planet. Sci. J., 1,* 33.
Jewitt D. (2021) Systematics and consequences of comet nucleus outgassing torques. *Astron. J., 161,* 261.
Jewitt D. and Hsieh H. (2006) Physical observations of 2005 UD: A mini-Phaethon. *Astron. J., 132,* 1624–1629.
Jewitt D. and Li J. (2010) Activity in Geminid parent (3200) Phaethon. *Astron. J., 140,* 1519–1527.
Jewitt D., Garland C. A., and Aussel H. (2008) Deep search for carbon monoxide in cometary precursors using millimeter wave spectroscopy. *Astron. J., 135,* 400–407.
Jewitt D., Weaver H., Agarwal J. et al. (2010) A recent disruption of the main-belt asteroid P/2010 A2. *Nature, 467,* 817–819.
Jewitt D., Stuart J. S., and Li J. (2011a) Pre-discovery observations of disrupting asteroid P/2010 A2. *Astron. J., 142,* 28.
Jewitt D., Weaver H., Mutchler M. et al. (2011b) Hubble Space Telescope observations of main-belt Comet (596) Scheila. *Astrophys. J. Lett., 733,* L4.
Jewitt D., Agarwal J., Weaver H. et al. (2013a) The extraordinary multi-

tailed main-belt Comet P/2013 P5. *Astrophys. J. Lett., 778,* L21.
Jewitt D., Li J., and Agarwal J. (2013b) The dust tail of asteroid (3200) Phaethon. *Astrophys. J. Lett., 771,* L36.
Jewitt D., Agarwal J., Li J. et al. (2014a) Disintegrating asteroid P/2013 R3. *Astrophys. J. Lett., 784,* L8.
Jewitt D., Ishiguro M., Weaver H. et al. (2014b) Hubble Space Telescope investigation of main-belt Comet 133P/Elst-Pizarro. *Astron. J., 147,* 117.
Jewitt D., Hsieh H., and Agarwal J. (2015) The active asteroids. In *Asteroids IV* (P. Michel et al., eds.), pp. 221–241. Univ. of Arizona, Tucson.
Jewitt D., Agarwal J., Li J. et al. (2017) Anatomy of an asteroid breakup: The case of P/2013 R3. *Astron. J., 153,* 223.
Jewitt D., Mutchler M., Agarwal J. et al. (2018a) Hubble Space Telescope observations of 3200 Phaethon at closest approach. *Astron. J., 156,* 238.
Jewitt D., Weaver H., Mutchler M. et al. (2018b) The nucleus of active asteroid 311P/(2013 P5) Pan-STARRS. *Astron. J., 155,* 231.
Jewitt D., Asmus D., Yang B. et al. (2019a) High-resolution thermal infrared imaging of 3200 Phaethon. *Astron. J., 157,* 193.
Jewitt D., Kim Y., Luu J. et al. (2019b) Episodically active asteroid 6478 Gault. *Astrophys. J. Lett., 876,* L19.
Jewitt D., Kim Y., Rajagopal J. et al. (2019c) Active asteroid P/2017 S5 (ATLAS). *Astron. J., 157,* 54.
Jewitt D., Li J., and Kim Y. (2021) Fragmenting active asteroid 331P/Gibbs. *Astron. J., 162,* 268.
Jones G. H., Balogh A., Russell C. T. et al. (2003) Possible distortion of the interplanetary magnetic field by the dust trail of Comet 122P/de Vico. *Astrophys. J. Lett., 597,* L61–L64.
Jones G. H., Agarwal J., Bowles N. et al. (2018) The proposed Caroline ESA M3 mission to a main belt comet. *Adv. Space Res., 62,* 1921–1946.
Jones R. L., Chesley S. R., Connolly A. J. et al. (2009) Solar system science with LSST. *Earth Moon Planets, 105,* 101–105.
Jones T. D., Lebofsky L. A., Lewis J. S. et al. (1990) The composition and origin of the C, P, and D asteroids: Water as a tracer of thermal evolution in the outer belt. *Icarus, 88,* 172–192.
Jutzi M., Holsapple K., Wünneman K. et al. (2015) Modeling asteroid collisions and impact processes. In *Asteroids IV* (P. Michel et al., eds.), pp. 711–732. Univ. of Arizona, Tucson.
Kaňuchová Z. and Neslušan L. (2007) The parent bodies of the Quadrantid meteoroid stream. *Astron. Astrophys., 470,* 1123–1136.
Kareta T., Reddy V., Pearson N. et al. (2021) Investigating the relationship between (3200) Phaethon and (155140) 2005 UD through telescopic and laboratory studies. *Planet. Sci. J., 2,* 190.
Kasuga T. and Jewitt D. (2008) Observations of 1999 YC and the breakup of the Geminid stream parent. *Astron. J., 136,* 881–889.
Kasuga T. and Jewitt D. (2019) Asteroid-meteoroid complexes. In *Meteoroids: Sources of Meteors on Earth and Beyond* (G. O. Ryabova et al., eds.), pp. 187–209. Cambridge Univ., Cambridge.
Kelley M. S. P., Woodward C. E., Bodewits D. et al. (2016) Cometary science with the James Webb Space Telescope. *Publ. Astron. Soc. Pac., 128,* 018009.
Kim Y., Ishiguro M., and Usui F. (2014) Physical properties of asteroids in comet-like orbits in infrared asteroid survey catalogs. *Astrophys. J., 789,* 151.
Kim Y., Ishiguro M., and Lee M. G. (2017a) New observational evidence of active asteroid P/2010 A2: Slow rotation of the largest fragment. *Astrophys. J. Lett., 842,* L23.
Kim Y., Ishiguro M., Michikami T. et al. (2017b) Anisotropic ejection from active asteroid P/2010 A2: An implication of impact shattering on an asteroid. *Astron. J., 153,* 228.
Kim Y., JeongAhn Y., and Hsieh H. H. (2018) Orbital alignment of main-belt comets. *Astron. J., 155,* 142.
Kim Y., Agarwal J., Jewitt D. et al. (2019) Hubble Space Telescope investigation of active asteroid P/2018 P3 (PANSTARRS). *EPSC-DPS Joint Meeting 2019,* 986.
Kleyna J. T., Hainaut O. R., and Meech K. (2019) The sporadic activity of (6478) Gault: A YORP-driven event? *Astrophys. J. Lett., 874,* L20.
Konietzky H. and Wang F. (1992) Thermal expansion of rocks. In *Developments in Petroleum Science, Vol. 37: Thermal Properties and Temperature-Related Behavior of Rock/Fluid Systems* (W. H. Somerton, ed.), pp. 29–38. Elsevier, Amsterdam.
Kowal C. T., Liller W., and Marsden B. G. (1979) The discovery and orbit of (2060) Chiron. In *Dynamics of the Solar System* (R. L. Duncombe, ed.), pp. 245 250. Reidel, Dordrecht.
Krüger H., Strub P., Srama R. et al. (2019) Modelling DESTINY+ interplanetary and interstellar dust measurements en route to the active asteroid (3200) Phaethon. *Planet. Space Sci., 172,* 22–42.
Küppers M., O'Rourke L., Bockelée-Morvan D. et al. (2014) Localized sources of water vapour on the dwarf planet (1) Ceres. *Nature, 505,* 525–527.
Lai H. R., Russell C. T., Wei H. Y. et al. (2017) Possible potentially threatening co-orbiting material of asteroid 2000EE104 identified through interplanetary magnetic field disturbances. *Meteorit. Planet. Sci., 52,* 1125–1132.
Larson S. M. (2010) (596) Scheila. *IAU Circular 9188.*
Lauretta D. S., Hergenrother C. W., Chesley S. R. et al. (2019) Episodes of particle ejection from the surface of the active asteroid (101955) Bennu. *Science, 366,* eaay3544.
Lee H.-J., Ďurech J., Vokrouhlický D. et al. (2021) Spin change of asteroid 2012 TC4 probably by radiation torques. *Astron. J., 161,* 112.
Levison H. F. (1996) Comet taxonomy. In *Completing the Inventory of the Solar System* (T. Rettig and J. M. Hahn, eds.), pp. 173–191. ASP Conf. Ser. 107, Astronomical Society of the Pacific, San Francisco.
Levison H. F. and Duncan M. J. (1994) The long-term dynamical behavior of short-period comets. *Icarus, 108,* 18–36.
Levison H. F., Terrell D., Wiegert P. A. et al. (2006) On the origin of the unusual orbit of Comet 2P/Encke. *Icarus, 182,* 161–168.
Levison H. F., Bottke W. F., Gounelle M. et al. (2009) Contamination of the asteroid belt by primordial trans-neptunian objects. *Nature, 460,* 364–366.
Li J. and Jewitt D. (2013) Recurrent perihelion activity in (3200) Phaethon. *Astron. J., 145,* 154.
Li J., Jewitt D., Mutchler M. et al. (2020) Hubble Space Telescope search for activity in high-perihelion objects. *Astron. J., 159,* 209.
Licandro J., Alvarez-Candal A., de León J. et al. (2008) Spectral properties of asteroids in cometary orbits. *Astron. Astrophys., 481,* 861–877.
Licandro J., Campins H., Kelley M. et al. (2009) Spitzer observations of the asteroid-comet transition object and potential spacecraft target 107P (4015) Wilson-Harrington. *Astron. Astrophys., 507,* 1667–1670.
Licandro J., Popescu M., de León J. et al. (2018) The visible and near-infrared spectra of asteroids in cometary orbits. *Astron. Astrophys., 618,* A170.
Lichtenberg T., Drążkowska J., Schönbächler M. et al. (2021) Bifurcation of planetary building blocks during solar system formation. *Science, 371,* 365–370.
Lowry S. C., Weissman P. R., Duddy S. R. et al. (2014) The internal structure of asteroid (25143) Itokawa as revealed by detection of YORP spin-up. *Astron. Astrophys., 562,* A48.
Luu J. X. and Jewitt D. C. (1990) Cometary activity in 2060 Chiron. *Astron. J., 100,* 913–932.
Luu J. X., Jewitt D. C., Mutchler M. et al. (2021) Rotational mass shedding from asteroid (6478) Gault. *Astrophys. J. Lett., 910,* L27.
MacLennan E. M. and Hsieh H. H. (2012) The nucleus of main-belt comet 259P/Garradd. *Astrophys. J. Lett., 758,* L3.
MacLennan E., Toliou A., and Granvik M. (2021) Dynamical evolution and thermal history of asteroids (3200) Phaethon and (155140) 2005 UD. *Icarus, 366,* 114535.
Mainzer A., Read M. T., Scottii J. V. et al. (2010) Comet P/2009 WJ50 (La Sagra). *IAU Circular 9117.*
Mainzer A., Abell P., Bauer J. et al. (2021) Near-Earth Object Surveyor mission: Data products and survey plan. *AAS/Division for Planetary Sciences Meeting Abstracts, 53,* #306.16.
Mann I. (2017) Comets as a possible source of nanodust in the solar system cloud and in planetary debris discs. *Philos. Trans. R. Soc. A, 375,* 20160254.
Marchi S., Chapman C. R., Barnouin O. S. et al. (2015) Cratering on asteroids. In *Asteroids IV* (P. Michel et al., eds.), pp. 725–744. Univ. of Arizona, Tucson.
Martin R. G. and Livio M. (2012) On the evolution of the snow line in protoplanetary discs. *Mon. Not. R. Astron. Soc., 425,* L6–L9.
Marzari F., Rossi A., and Scheeres D. J. (2011) Combined effect of YORP and collisions on the rotation rate of small main belt asteroids. *Icarus, 214,* 622–631.
Masiero J. R., Davidsson B. J. R., Liu Y. et al. (2021) Volatility of sodium in carbonaceous chondrites at temperatures consistent with low-perihelion asteroids. *Planet. Sci. J., 2,* 165.
Mazzotta Epifani E., Dotto E., Ieva S. et al. (2018) 523676

(2013 UL10): The first active red Centaur. *Astron. Astrophys., 620,* A93.
McLoughlin E., Fitzsimmons A., and McLoughlin A. (2015) Modelling the brightness increase signature due to asteroid collisions. *Icarus, 256,* 37–48.
Meech K. J. and Belton M. J. S. (1990) The atmosphere of 2060 Chiron. *Astron. J., 100,* 1323–1338.
Meech K. J. and Castillo-Rogez J. C. (2015) Proteus — A mission to investigate the origins of Earth's water. *IAU General Assembly Abstracts, 29,* 2257859.
Meech K. J., Yang B., Kleyna J. et al. (2016) Inner solar system material discovered in the Oort cloud. *Sci. Adv., 2,* e1600038.
Melita M. D. and Licandro J. (2012) Links between the dynamical evolution and the surface color of the Centaurs. *Astron. Astrophys., 539,* A144.
Michel P., Ballouz R. L., Barnouin O. S. et al. (2020) Collisional formation of top-shaped asteroids and implications for the origins of Ryugu and Bennu. *Nature Commun., 11,* 2655.
Molaro J. L., Byrne S., and Langer S. A. (2015) Grain-scale thermoelastic stresses and spatiotemporal temperature gradients on airless bodies, implications for rock breakdown. *J. Geophys. Res.–Planets, 120,* 255–277.
Molaro J. L., Hergenrother C. W., Chesley S. R. et al. (2020) Thermal fatigue as a driving mechanism for activity on asteroid Bennu. *J. Geophys. Res.–Planets, 125,* e06325.
Mommert M., Hora J. L., Harris A. W. et al. (2014) The discovery of cometary activity in near-Earth asteroid (3552) Don Quixote. *Astrophys. J., 781,* 25.
Moreno F., Licandro J., Cabrera-Lavers A. et al. (2016a) Dust loss from activated asteroid P/2015 X6. *Astrophys. J., 826,* 137.
Moreno F., Licandro J., Cabrera-Lavers A. et al. (2016b) Early evolution of disrupted asteroid P/2016 G1 (Pan-STARRS). *Astrophys. J. Lett., 826,* L22.
Moreno F., Pozuelos F. J., Novaković B. et al. (2017) The splitting of double-component active asteroid P/2016 J1 (Pan-STARRS). *Astrophys. J. Lett., 837,* L3.
Moreno F., Licandro J., Cabrera-Lavers A. et al. (2021) Dust environment of active asteroids P/2019 A4 (Pan-STARRS) and P/2021 A5 (Pan-STARRS). *Mon. Not. R. Astron. Soc., 506,* 1733–1740.
Murray C. D. and Dermott S. F. (2000) *Solar System Dynamics.* Cambridge Univ., Cambridge. 592 pp.
Najita J., Willman B., Finkbeiner D. P. et al. (2016) Maximizing science in the era of LSST: A community-based study of needed U.S. capabilities. *ArXiV e-prints,* arXiv:1610.01661.
Nakano R. and Hirabayashi M. (2020) Mass-shedding activities of asteroid (3200) Phaethon enhanced by its rotation. *Astrophys. J. Lett., 892,* L22.
Nesvorný D., Bottke W. F., Vokrouhlický D. et al. (2008) Origin of the near-ecliptic circumsolar dust band. *Astrophys. J. Lett., 679,* L143–L146.
Nesvorný D., Janches D., Vokrouhlický D. et al. (2011) Dynamical model for the zodiacal cloud and sporadic meteors. *Astrophys. J., 743,* 129.
Nesvorný D., Brož M., and Carruba V. (2015) Identification and dynamical properties of asteroid families. In *Asteroids IV* (P. Michel et al., eds.), pp. 297–321. Univ. of Arizona, Tucson.
Novaković B., Hsieh H. H., and Cellino A. (2012) P/2006 VW139: A main-belt comet born in an asteroid collision? *Mon. Not. R. Astron. Soc., 424,* 1432–1441.
Ohtsuka K., Sekiguchi T., Kinoshita D. et al. (2006) Apollo asteroid 2005 UD: Split nucleus of (3200) Phaethon? *Astron. Astrophys., 450,* L25–L28.
Ostro S. J., Margot J.-L., Benner L. A. M. et al. (2006) Radar imaging of binary near-Earth asteroid (66391) 1999 KW4. *Science, 314,* 1276–1280.
Park R. S., Konopliv A. S., Bills B. G. et al. (2016) A partially differentiated interior for (1) Ceres deduced from its gravity field and shape. *Nature, 537,* 515–517.
Peixinho N., Delsanti A., Guilbert-Lepoutre A. et al. (2012) The bimodal colors of Centaurs and small Kuiper belt objects. *Astron. Astrophys., 546,* A86.
Piro C., Meech K. J., Bufanda E. et al. (2021) Characterizing the Manx candidate A/2018 V3. *Planet. Sci. J., 2,* 33.
Podolak M. and Herman G. (1985) Numerical simulations of comet nuclei II. The effect of the dust mantle. *Icarus, 61,* 267–277.
Poston M. J., Mahjoub A., Ehlmann B. L. et al. (2018) Visible near-infrared spectral evolution of irradiated mixed ices and application to Kuiper belt objects and Jupiter Trojans. *Astrophys. J., 856,* 124.
Prialnik D. and Bar-Nun A. (1988) The formation of a permanent dust mantle and its effect on cometary activity. *Icarus, 74,* 272–283.
Prialnik D. and Rosenberg E. D. (2009) Can ice survive in main-belt comets? Long-term evolution models of Comet 133P/Elst-Pizarro. *Mon. Not. R. Astron. Soc., 399,* L79–L83.
Raymond S. N. and Izidoro A. (2017) The empty primordial asteroid belt. *Sci. Adv., 3,* e1701138.
Rickman H., Fernandez J. A., and Gustafson B. A. S. (1990) Formation of stable dust mantles on short-period comet nuclei. *Astron. Astrophys., 237,* 524–535.
Rozitis B. and Green S. F. (2013) The strength and detectability of the YORP effect in near-Earth asteroids: A statistical approach. *Mon. Not. R. Astron. Soc., 430,* 1376–1389.
Rudenko M., Micheli M., Hsieh H. H. et al. (2021) Comet P/2021 K4 = P/2019 A7 (Pan-STARRS). *Central Bureau Electronic Telegram (CBET) 5010.*
Russell C. T., Aroian R., Arghavani M. et al. (1984) Interplanetary magnetic field enhancements and their association with the asteroid 2201 Oljato. *Science, 226,* 43–45.
Ryabova G. O. (2002) Asteroid 1620 Geographos: II. Associated meteor streams. *Solar Sys. Res., 36,* 234–247.
Saiki T., Imamura H., Arakawa M. et al. (2017) The Small Carry-on Impactor (SCI) and the Hayabusa2 impact experiment. *Space Sci. Rev., 208,* 165–186.
Sánchez P. and Scheeres D. J. (2014) The strength of regolith and rubble pile asteroids. *Meteorit. Planet. Sci., 49,* 788–811.
Sánchez P., Durda D. D., Devaud G. et al. (2021) Laboratory experiments with self-cohesive powders: Application to the morphology of regolith on small asteroids. *Planet. Space Sci., 207,* 105321.
Sarid G., Volk K., Steckloff J. K. et al. (2019) 29P/Schwassmann-Wachmann 1, a Centaur in the gateway to the Jupiter-family comets. *Astrophys. J. Lett., 883,* L25.
Scheeres D. J. (2015) Landslides and mass shedding on spinning spheroidal asteroids. *Icarus, 247,* 1–17.
Scheeres D. J., Hartzell C. M., Sánchez P. et al. (2010) Scaling forces to asteroid surfaces: The role of cohesion. *Icarus, 210,* 968–984.
Schörghofer N. (2008) The lifetime of ice on main belt asteroids. *Astrophys. J., 682,* 697–705.
Schörghofer N. (2016) Predictions of depth-to-ice on asteroids based on an asynchronous model of temperature, impact stirring, and ice loss. *Icarus, 276,* 88–95.
Schörghofer N., Hsieh H. H., Novaković B. et al. (2020) Preservation of polar ice on near-Earth asteroids originating in the outer main belt: A model study with dynamical trajectories. *Icarus, 348,* 113865.
Schultz P. H., Hermalyn B., Colaprete A. et al. (2010) The LCROSS cratering experiment. *Science, 330,* 468–472.
Shannon A., Jackson A. P., Veras D. et al. (2015) Eight billion asteroids in the Oort cloud. *Mon. Not. R. Astron. Soc., 446,* 2059–2064.
Sheppard S. S. and Trujillo C. (2015) Discovery and characteristics of the rapidly rotating active asteroid (62412) 2000 SY178 in the main belt. *Astron. J., 149,* 44.
Simion N. G., Popescu M., Licandro J. et al. (2021) Spectral properties of near-Earth objects with low-jovian Tisserand invariant. *Mon. Not. R. Astron. Soc., 508,* 1128–1147.
Smith K. W., Denneau L., Vincent J.-B. et al. (2019) (6478) Gault. *Central Bureau Electronic Telegram (CBET) 4594.*
Snodgrass C., Tubiana C., Vincent J.-B. et al. (2010) A collision in 2009 as the origin of the debris trail of asteroid P/2010A2. *Nature, 467,* 814–816.
Snodgrass C., Agarwal J., Combi M. et al. (2017) The main belt comets and ice in the solar system. *Astron. Astrophys. Rev., 25,* 5.
Snodgrass C., Jones G. H., Boehnhardt H. et al. (2018) The Castalia mission to main belt Comet 133P/Elst-Pizarro. *Adv. Space Res., 62,* 1947–1976.
Sonnett S., Kleyna J., Jedicke R. et al. (2011) Limits on the size and orbit distribution of main belt comets. *Icarus, 215,* 534–546.
Statler T. S. (2009) Extreme sensitivity of the YORP effect to small-scale topography. *Icarus, 202,* 502–513.
Steckloff J. K., Graves K., Hirabayashi M. et al. (2016) Rotationally induced surface slope-instabilities and the activation of CO_2 activity on Comet 103P/Hartley 2. *Icarus, 272,* 60–69.
Street R. A., Bowman M., Saunders E. S. et al. (2018) General-purpose

software for managing astronomical observing programs in the LSST era. In *Software and Cyberinfrastructure for Astronomy V* (J. C. Guzman and J. Ibsen, eds.), p. 1070711. SPIE Conf. Ser. 10707, Bellingham, Washington.

Szabó G. M. and Kiss L. L. (2008) The shape distribution of asteroid families: Evidence for evolution driven by small impacts. *Icarus, 196,* 135–143.

Takir D., Kareta T., Emery J. P. et al. (2020) Near-infrared observations of active asteroid (3200) Phaethon reveal no evidence for hydration. *Nature Commun., 11,* 2050.

Tancredi G. (2014) A criterion to classify asteroids and comets based on the orbital parameters. *Icarus, 234,* 66–80.

Taylor P. A., Rivera-Valentín E. G., Benner L. A. M. et al. (2019) Arecibo radar observations of near-Earth asteroid (3200) Phaethon during the 2017 apparition. *Planet. Space Sci., 167,* 1–8.

Tedesco E. F., Noah P. V., Noah M. et al. (2004) *IRAS Minor Planet Survey V6.0.* NASA Planetary Data System, IRAS-A-FPA-3-RDR-IMPS-V6.0.

Tegler S. C., Romanishin W., Consolmagno G. J. et al. (2016) Two color populations of Kuiper belt and Centaur objects and the smaller orbital inclinations of red Centaur objects. *Astron. J., 152,* 210.

Tholen D. J., Sheppard S. S., and Trujillo C. A. (2015) Evidence for an impact event on (493) Griseldis. *AAS/Division for Planetary Sciences Meeting Abstracts, 47,* #414.03.

Tholen D., Sheppard S. S., Weryk R. et al. (2021) Comet P/2021 R8 (Sheppard). *Central Bureau Electronic Telegram (CBET) 5079.*

Tisserand F. (1896) *Traité de Mécanique Céleste, Tome IV.* Gauthier-Villars et Fils, Paris. 548 pp.

Toth I. (2000) Impact-generated activity period of the asteroid 7968 Elst-Pizarro in 1996: Identification of the asteroid 427 Galene as the most probable parent body of the impactors. *Astron. Astrophys., 360,* 375–380.

Tsuda Y., Yoshikawa M., Saiki T. et al. (2019) Hayabusa2 — Sample return and kinetic impact mission to near-Earth asteroid Ryugu. *Acta Astronaut., 156,* 387–393.

Urakawa S., Okumura S., Nishiyama K. et al. (2011) Photometric observations of 107P/Wilson-Harrington. *Icarus, 215,* 17–26.

Vaghi S. (1973) Orbital evolution of comets and dynamical characteristics of Jupiter's family. *Astron. Astrophys., 29,* 85–91.

Volk K. and Malhotra R. (2008) The scattered disk as the source of the Jupiter family comets. *Astrophys. J., 687,* 714–725.

Wainscoat R., Weryk R., Cunningham C. et al. (2021) Comet P/2021 L4 (Pan-STARRS). *Central Bureau Electronic Telegram (CBET) 4986.*

Walsh K. J., Morbidelli A., Raymond S. N. et al. (2011) A low mass for Mars from Jupiter's early gas-driven migration. *Nature, 475,* 206–209.

Wang J. H. and Brasser R. (2014) An Oort cloud origin of the Halley-type comets. *Astron. Astrophys., 563,* A122.

Wang S., Zhao H.-B., Ji J.-H. et al. (2012) Possible origin of the Damocloids: The scattered disk or a new region? *Res. Astron. Astrophys., 12,* 1576–1584.

Wang X., Schwan J., Hsu H. W. et al. (2016) Dust charging and transport on airless planetary bodies. *Geophys. Res. Lett., 43,* 6103–6110.

Warner B., Pravec P., and Harris A. P. (2019) *Asteroid Lightcurve Database (LCDB) Bundle V3.0.* NASA Planetary Data System, urn:nasa:pds:ast-lightcurve-database::3.0.

Warren P. H. (2011) Stable-isotopic anomalies and the accretionary assemblage of the Earth and Mars: A subordinate role for carbonaceous chondrites. *Earth Planet. Sci. Lett., 311,* 93–100.

Waszczak A., Ofek E. O., Aharonson O. et al. (2013) Main-belt comets in the Palomar Transient Factory survey — I. The search for extendedness. *Mon. Not. R. Astron. Soc., 433,* 3115–3132.

Watanabe S., Hirabayashi M., Hirata N. et al. (2019) Hayabusa2 arrives at the carbonaceous asteroid 162173 Ryugu — A spinning top-shaped rubble pile. *Science, 364,* 268–272.

Weissman P. R. and Levison H. F. (1997) Origin and evolution of the unusual object 1996 PW: Asteroids from the Oort cloud? *Astrophys. J. Lett., 488,* L133–L136.

Weryk R., Wainscoat R. J., Wipper C. et al. (2017) Comet P/2017 S9 (Pan-STARRS). *Central Bureau Electronic Telegram (CBET) 4448.*

Weryk R., Wainscoat R., Ramanjooloo Y. et al. (2018) Comet P/2018 P3 (Pan-STARRS). *Central Bureau Electronic Telegram (CBET) 4548.*

Weryk R., Wainscoat R., Woodworth D. et al. (2019) Comet P/2019 A3 (Pan-STARRS). *Central Bureau Electronic Telegram (CBET) 4598.*

Weryk R., Wierzchos K. W., Wainscoat R. et al. (2020) Comet P/2020 O1 (LEMMON-Pan-STARRS). *Central Bureau Electronic Telegram (CBET) 4820.*

Westphal A. J., Fakra S. C., Gainsforth Z. et al. (2009) Mixing fraction of inner solar system material in Comet 81P/Wild2. *Astrophys. J., 694,* 18–28.

Whipple F. L. (1950) A comet model. I. The acceleration of Comet Encke. *Astrophys. J., 111,* 375–394.

Whipple F. L. (1951) A comet model. II. Physical relations for comets and meteors. *Astrophys. J., 113,* 464–474.

Wiegert P. A. (2015) Meteoroid impacts onto asteroids: A competitor for Yarkovsky and YORP. *Icarus, 252,* 22–31.

Wierzchos K. and Womack M. (2020) CO gas and dust outbursts from Centaur 29P/Schwassmann-Wachmann. *Astron. J., 159,* 136.

Williams I. P. and Wu Z. (1993) The Geminid meteor stream and asteroid 3200 Phaethon. *Mon. Not. R. Astron. Soc., 262,* 231–248.

Womack M., Sarid G., and Wierzchos K. (2017) CO in distantly active comets. *Publ. Astron. Soc. Pac., 129,* 031001.

Wong M. H., Meech K. J., Dickinson M. et al. (2021) Transformative planetary science with the U.S. ELT Program. *Bull. Am. Astron. Soc., 53,* 490.

Ye Q. and Jenniskens P. (2024) Comets and meteor showers. In *Comets III* (K. J. Meech, M. R. Combi, D. Bockelée-Morvan, S. N. Raymond, and M. E. Zolensky, eds.), pp. 799–821. Univ. of Arizona, Tucson, DOI: 10.2458/azu_uapress_9780816553631-ch024.

Comets and Meteor Showers

Quanzhi Ye
University of Maryland/Boston University

Peter Jenniskens
SETI Institute/NASA Ames Research Center

Earth occasionally crosses the debris streams produced by comets and other active bodies in our solar system. These manifest meteor showers that provide an opportunity to explore these bodies without a need to visit them *in situ*. Observations of meteor showers provide unique insights into the physical and dynamical properties of their parent bodies, as well as into the compositions and the structure of near-surface dust. In this chapter, we discuss the development and current state of affairs of meteor science, with a focus on its role as a tool to study comets and review the established parent body-meteor shower linkages.

1. INTRODUCTION

Comets account for most of the mass of the interplanetary dust cloud and most meteoroids colliding with Earth (*Nesvorný et al.,* 2010; *Jenniskens,* 2015). They actively release gas into interplanetary space, and with it solid particles described alternatively as interplanetary dust, meteoroids, or cometary fragments if they are larger than 1 m in diameter. Lighter materials, such as gas and submicrometer-sized dust (also known as β-meteoroids), are carried by the solar wind directly outward from the Sun into interstellar space. Most heavier materials, such as millimeter-size or larger meteoroids, do not escape the gravity of the Sun and planets, but are still influenced by the radiation forces. A small fraction ends up colliding with Earth's atmosphere. The phenomenon created by the entry of a meteoroid into Earth's atmosphere is known as a *meteor*. What survives to the ground is called a *meteorite*.

Historically, the terms "(cometary/interplanetary) dust" and "meteoroid" have been ambiguous, and they are often used interchangeably. The comet community tends to use "dust," while the meteor community prefers "meteoroid." A clarification issued by the International Astronomical Union (IAU) (*Koschny and Borovička,* 2017) defined a meteoroid to be a solid natural object of a size roughly between ~30 μm and 1 m, with interplanetary dust being the solid matter smaller than meteoroids but acknowledged that the size distribution is continuous (Fig. 1). In this chapter, we use these two terms interchangeably.

After one revolution around the Sun, some meteoroids return early, others later, creating a stream. Those streams are sometimes seen in infrared emission as cometary dust trails (see the chapter in this volume by Agarwal et al.). Earth passing through a stream results in a large number of meteoroids entering the atmosphere from the same direction, forming a *meteor shower*. Due to perspective, observers see shower meteors radiate from a point on the celestial sphere (the *radiant*). The naming of meteor showers is governed by the IAU's Working Group on Meteor Shower Nomenclature. A meteor shower is usually named after the nearest bright star to the radiant at the time of peak activity. The month of the year is sometimes added to distinguish showers in the same constellation, and the word "daytime" is added to showers that are mostly active during daytime. A complete list of meteor showers is maintained by the IAU Meteor Data Center (MDC) (*Jopek and Kaňuchová,* 2017). As of the writing, there are 112 "established" showers and 830 "working list" showers. Showers are assigned a unique name, a three-letter code, and a number, which are all unique to each shower. For instance, the Leonids associated with Comet 55P/Tempel-Tuttle have the code LEO and the number #13, distinguishing it from the other 32 showers with radiants in the constellation of Leo throughout the year.

Many physical parameters are derived from meteor and meteor shower research, including meteoroid orbits, measures of meteoroid deceleration, luminosity, electron density, fragmentation, and elemental abundances, as well as stream dispersion, times of encounters, and flux density. An important observable is the zenith hourly rate (ZHR), defined as the number of meteors seen by a single human observer with standard perception under perfect visual conditions (limiting stellar magnitude of +6.5 and shower radiant at zenith). ZHR is a proxy of the meteoroid stream's flux density, dependent on the magnitude distribution index of meteors (called "population index" in some meteor science literature, often represented by r or χ), which is related to the meteoroid size and mass distribution index, assuming meteoroid density and luminous efficiency are constant in a stream. The measured quantity of luminosity is most closely related to the kinetic

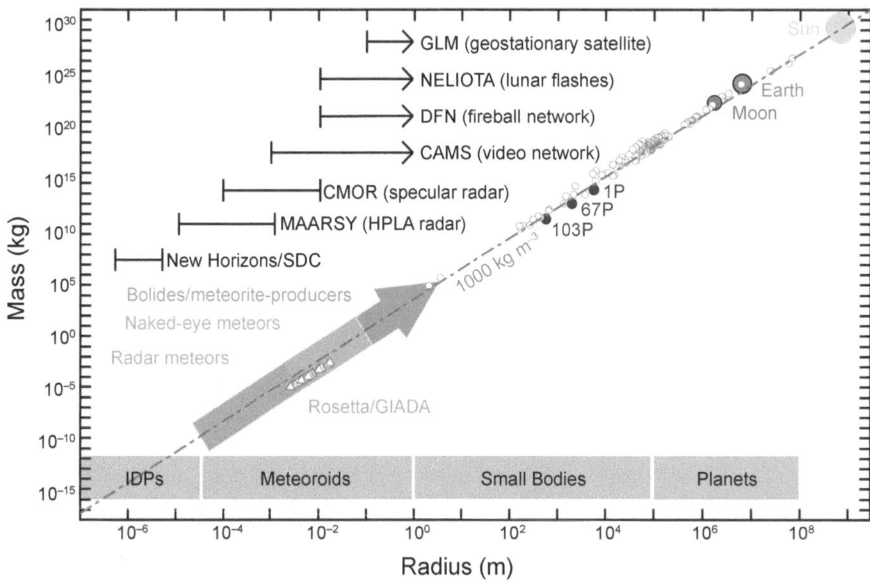

Fig. 1. A diagram of mass and sizes of various bodies in the solar system, labeled with the 2017 IAU definition of meteoroid and interplanetary dust particle (IDP). Also shown are the size regimes of radar, naked-eye (photographic/video) techniques, bolides/meteorite-producers, and examples of several observation programs: the Geostationary Lighting Mapper (GLM) as an example of orbit-based meteor surveillance, Near-Earth object Lunar Impacts and Optical TrAnsients (NELIOTA) for lunar flashes observations, the Desert Fireball Network (DFN) for fireball photography, the Cameras for All-Sky Meteor Surveillance (CAMS) for low-light video camera surveillance, the Canadian Meteor Orbital Radar (CMOR) for meteor orbit radars, the Middle Atmosphere Alomar Radar System (MAARSY) for High Power Large Aperture (HPLA) radar, and the Student Dust Counter (SDC) onboard New Horizons for spacebased dust detector. For reviews of video and radar meteor programs, see *Koten et al.* (2019) and *Kero et al.* (2019).

energy of the meteoroid, not its mass or size (*Koschack and Rendtel,* 1990). The strongest annual meteor showers such as the Quadrantids, Perseids, and Geminids have a ZHR around 100 (a volume density of millimeter-sized meteoroids around 10^{-12}–10^{-11} m^{-3}), while the strongest meteor outbursts in the past two centuries had ZHR over ~10,000 (10^{-10}–10^{-9} m^{-3}). Meteor activity with ZHR over 1000 is called a *meteor storm.*

The history of meteor science has been discussed in great detail in, e.g., *Jenniskens* (2006) (section 1). Like comets, records of meteors go back to ancient times. The earliest confirmed sighting of a meteor shower is recorded in 687 BC in the Chinese chronicle *Zuo Zhuan,* when meteor showers were recognized as periods of higher meteor rates. The radiant phenomenon was widely recognized during the 1833 Leonids meteor storm, providing information on meteoroid orbits. The 1865 return of Comet 55P/Tempel-Tuttle and the 1866 Leonid storm subsequently led to the recognition that meteor showers and comets are linked. Several other major meteor showers were also quickly associated with comets, especially Comet 109P/Swift-Tuttle to the Perseids, Comet C/1861 G1 (Thatcher) with the Lyrids, and 3D/Biela to the Andromedids.

Since then, numerous efforts have been made to map meteor showers and to identify their parent bodies. Looking for radiants, observers in the nineteenth and early twentieth century mostly relied on their naked eyes, a sky chart, and a pencil to record meteor tracks. Photographic and radar techniques were introduced in the 1890s and 1920s, respectively. By the early 2000s, the number of meteoroid orbits was still relatively small. In 2002, the Canadian Meteor Orbit Radar (CMOR) started regular meteor observations in the northern hemisphere in radio wavelength, an effort that was later expanded to the southern hemisphere by the Southern Argentina Agile Meteor Radar (SAAMER). In 2007, early low-light video meteor triangulations were scaled up, first by the SonotaCo network in Japan, later expanded greatly by the European video MeteOr Network Database (EDMOND) and Cameras for All-Sky Meteor Surveillance (CAMS) video surveys. These techniques have ushered in a revolution in our understanding of meteor showers and an increasing number of associations with the parent bodies found by the near-Earth object (NEO) surveys. Table 1 lists the well-established linkages between comets or active asteroids and meteoroid streams known at the time of this writing.

Meteor science connects many disciplines: the production, evolution, and fate of meteoroids are related to astrochemistry and celestial mechanics; ablation of meteoroids is related to aeronomy and atmospheric science; meteorites and extraterrestrial impacts are related to planetary geology; the mineral water and organic matter embedded in meteoroids are related to astrobiology; and the meteoroid impact hazard to space assets is related to aerospace engineering. Of course, as meteoroids are direct products of cometary activity, meteor science is most closely related to comet science and much effort in the past has been in this field.

In this chapter, we review the development and current state of affairs of the science of meteors in the context of the research on comets. We discuss the process that brings

TABLE 1. Established* linkage between comet/asteroids and confirmed meteoroid streams.

Object	Type†	Meteor Shower and IAU ID Number	ZHR‡	Mechanism§
1P/Halley	HTC	η-Aquariids (#31), Orionids (#8)	60, 35	S
2P/Encke¶	ETC	Daytime β-Taurids (#173), N/S Taurids (#17/#2)	10, 5	S, BU?
3D/Biela	JFC	Andromedids (#18)	<2**	BU
7P/Pons-Winnecke	JFC	June Bootids (#170)	<2**	S
8P/Tuttle	JFC	Ursids (#15)	10**	S
15P/Finlay	JFC	Arids (#1130)	<2**	S
21P/Giacobini-Zinner	JFC	October Draconids (#9)	<2**	S
26P/Grigg-Skjellerup	JFC	π-Puppids (#137)	<2**	S
55P/Tempel-Tuttle	HTC	Leonids (#13)	15**	S
73P/Schwassmann-Wachmann 3	JFC	τ-Herculids (#61)	<2**	S? BU?
96P/Machholz 1††	JFC	Daytime Arietids (#171), N/S δ-Aquariids (#26/#5)	60, 30	S + BU?
109P/Swift-Tuttle	HTC	Perseids (#7)	120	S
169P/NEAT and 2017 MB_1	JFC	α-Capricornids (#1)	5	BU?
209P/LINEAR	JFC	Camelopardalids (#451)	<2**	S
289P/Blanpain	JFC	Phoenicids (#254)	<2**	S? + BU?
300P/Catalina	JFC	June ε-Ophiuchids (#459)	<2**	S
C/1861 G1 (Thatcher)	LPC	April Lyrids (#6)	20**	S
C/1911 N1 (Kiess)	LPC	Aurigids (#206)	5**	S
C/1917 F1 (Mellish)	HTC‡‡	December Monocerotids (#19)	3	S
C/1979 Y1 (Bradfield)	LPC	July Pegasids (#175)	<2	S
C/1983 H1 (IRAS-Araki-Alcock)	LPC	η-Lyrids (#145)	<2	S?
(3200) Phaethon	NEA	Geminids (#4)	180	TBU?
(155140) 2005 UD	NEA	Daytime Sextantids (#221)	5	TBU?
(196256) 2003 EH_1	NEA	Quadrantids (#10)	130	BU?

* Only associations that have been investigated/suggested by multiple studies from different research groups have been listed.
† NEA — near-Earth asteroid; ETC — Encke-type comet; HTC — Halley-type comet; JFC — Jupiter-family comet; LPC — Long-period comet.
‡ Zenith hourly rate. See main text for details.
§ S — sublimation; BU — breakup; TBU — thermal breakup.
¶ Also known as the Taurid Complex. May include a couple of NEAs and other meteor showers (*Porubčan et al.,* 2006).
** Known to exhibit occasional outburst.
†† Includes the Marsden comet group (see section 4).
‡‡ This comet has a orbital period of 143 yr, but is designated as a C/ comet.

cometary dust to Earth (section 2) and how we can use this process to study comets (section 3). The known parent-shower associations are reviewed in section 4. Interrelations with comet science, astrobiology, aerospace engineering, and planetary defense are discussed in section 5. The chapter is complete with a list of topics into which we hope to see significant progress between now and *Comets IV* (section 6).

2. FROM COMET TO EARTH

2.1. Formation and Evolution of Meteoroid Streams

Meteoroid ejections are most commonly driven by gas drag from the sublimation of water ice and other volatiles, but can also occur due to breakup, impacts, radiation sweeping, and electrostatic gardening, among others (cf. *Jewitt et al.,* 2015). The activity mechanism dictates the ejection place, velocity distribution, and size distribution of the meteoroids. For sublimation-driven ejection, the ejection speed grossly follows $\propto a_{dust}^{-1/2}$ with a_{dust} the diameter of the dust (*Whipple,* 1950). For impulsive ejections such as rotational disruption or impact-caused ejection, observations show that the ejection speed is largely independent of dust size (*Jewitt et al.,* 2015, Fig. 18). The difference in ejection speed gives the resulting cometary dust trails different looks, allowing activity mechanism to be constrained observationally. This is discussed in detail in the chapter by Agarwal et al. in this volume. Here, we focus on the delivery from comet to Earth and the manifestation of meteor showers at Earth.

Sublimation activity can, in most cases, explain the observed streams (Table 1), but some streams are associated with parents that have either disrupted or have a history of disruption. It is likely that disruption is the dominant mass-loss mechanism. Streams with identified parents only make up 20% of the confirmed streams. The majority of the confirmed streams do not have known corresponding parents, and most of them — especially the short-period ones — are likely the end-product of catastrophic disruptions, given that most sizeable NEOs have been discovered (*Jedicke et al.,* 2015). Taking a typical dispersion timescale of 10^3 yr for JFC streams (see section 2.2) and the number of breakup-driven streams N = 6 from Table 1, we derive a breakup frequency of 6×10^{-3} yr^{-1} for Earth-crossing JFCs. This is in line with the number derived by *Chen and Jewitt* (1994) from comet data.

The early manifestation of meteoroid streams is dictated by the fact that meteoroids escape the comet's weak gravity field at a small relative speed with respect to the comet, typically on the order of a few 1–10 m s^{-1}, which results in a small difference between the orbit of the meteoroid and its parent body. This difference, albeit small, leads to an appreciable difference in the orbital period of the meteoroids. What was ejected as a spherical cloud returns as a trail after even a single orbit. Dynamical processes, such as the gravitational perturbation from major planets, gradually amplified this difference, causing the ejecta to spread along and away from the parent orbit (Fig. 2). This enables meteoroids to arrive at Earth even when the orbit of the parent is not strictly intercepted by Earth. In general, comets with minimum orbit intersection distance (MOID) within ~0.1 au can produce observable meteor showers (*Jenniskens et al.,* 2021b). It usually takes ~10–20 orbits for meteoroids ejected at a certain epoch to spread to the entire orbit (*Ye et al.,* 2016a).

A meteor shower can be detected annually when the material spreads out to the entire orbit, since Earth always sweeps some meteoroids every time it passes the intersection with the stream orbit. Alternatively, meteor "outbursts" are caused by younger meteoroid trails that reside in a small but growing arc of the orbit. Whether or not that trail meets Earth depends on whether the trail is steered in Earth's path and whether that arc is near Earth. Occasionally, meteoroids can be trapped in resonances if their parents are in the proximity of one of the resonance points, leading to a long-lived meteoroid "filament" or "trailet." These trailets can last a few times longer than typical trails, providing a way to probe older materials ejected by the parent. A good example is the Ursid meteor shower, which originated from 8P/Tuttle, of which the 7:6 resonance with Jupiter has produced trailets lasting ~50 orbits (section 4.2.2).

2.2. Demise of Meteoroid Streams and Contribution to the Zodiacal Cloud

All meteoroid streams will eventually disperse into the interplanetary dust complex, known as the "sporadic (meteor) background." The coherence of a stream can be accessed using the dissimilarity criterion, D, which measures the (dis) similarities between a set of meteoroid orbits based on the difference in their orbital elements (see *Williams et al.,* 2019, section 9.2.1, for a review). A threshold of 0.1–0.2 is typically used to separate streams from non-streams, although difficulty can and has arisen to distinguish weak showers from the noise, especially when the statistics are low. The decoherence timescale varies greatly from stream to stream, and can take as little as ~200 orbits if the stream is heavily perturbed (*Ye et al.,* 2016a).

Earth's orbital motion gives rise to several very broad apparent radiant regions in the sporadic background: apex sources (meteors arriving from Earth's direction of motion), helion/antihelion (from the direction of the Sun and the opposite direction), and toroidal (particles on prograde orbits with high inclinations). These have been associated with dispersed meteoroids from mostly long-period comets (LPCs), Jupiter-family comets (JFCs), and

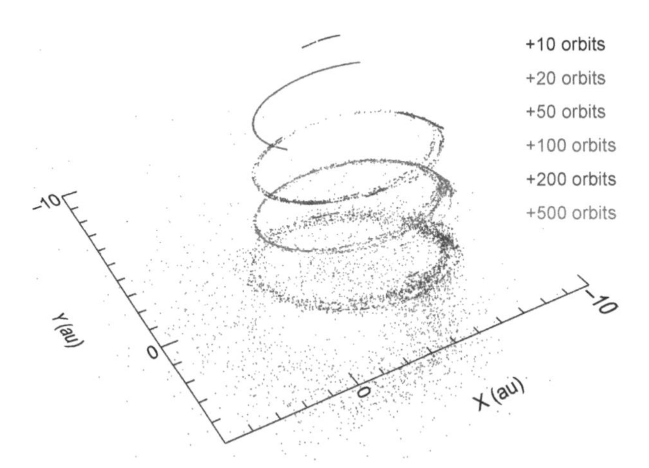

Fig. 2. Simulation of millimeter-sized meteoroids ejected by 21P/Giacobini-Zinner during its 1966 perihelion after (from top to bottom) 10, 20, 50, 100, 200, and 500 orbits, showing the evolution of a cometary ejecta from a meteoroid trail to stream, and eventually blending into the interplanetary dust background.

Halley-type comets (HTCs) respectively (*Wiegert et al.,* 2009; *Nesvorný et al.,* 2010).

As meteoroid streams lose their coherence, meteoroids also degrade through collision with other meteoroids from the sporadic background. The timescale of this process is uncertain. By combining the then-available meteor, lunar crater, and satellite measurements, *Grun et al.* (1985) found that the collisional lifetime of millimeter-sized meteoroids at 1 au is as short as 10^4 yr, although recent results from video and radar meteor surveys with much larger statistics suggested 10^5–10^6 yr, appropriated to meteoroids from submillimeter to centimeter sizes (*Wiegert et al.,* 2009; *Jenniskens et al.,* 2016b). Degradation through other erosion effects, such as thermal fatigue, exposure to solar wind particles, and cosmic rays, are possible but are similarly little understood. The timescale that these erosion processes are on is broadly comparable to the decoherence timescale of JFC streams, but shorter than that of HTC/LPC streams.

Ongoing erosion processes sometimes result in meteor clusters. Only a handful of plausible cases have been reported, mostly from HTCs (*Kinoshita et al.,* 1999; *Watanabe et al.,* 2003) and one from an unknown LPC (*Koten et al.,* 2017). Dynamical modeling showed that these clusters may be produced a few days before entering Earth's atmosphere at a separation speed on the order of 0.1 m s^{-1}. The lower tensile strengths of HTCs/LPCs dust likely contribute to the fact that all observed clusters are exclusively from these comets. A recent search in the short-period Geminid shower produces no statistically significant detection (*Koten et al.,* 2021).

Breakups caused by collision and other erosive processes increase the number of small meteoroids in the sporadic background. About 98% of small 100-μm-class meteoroids are sporadics, but only about 60% of centimeter-sized meteoroids are not associated with showers (*Jenniskens,* 2006, section 26.3). Collisional processes also steepen the size distribution of the stream, an effect that has been observed in both radar and optical observations (*Blaauw et al.,* 2011; *Jenniskens et al.,* 2016b). Eventually, meteoroids are reduced into grains that are small enough to be blown out of the solar system.

3. PROBING COMETS USING METEOR OBSERVATIONS

3.1. Observational Techniques

Meteors and meteoroids can be probed using a wide range of techniques from remote sensing and *in situ* exploration in space or on the ground, but most meteors are studied using remote-sensing in optical or radio wavelengths. Optical observations detect meteors using their thermal and ionization radiation, while radar observations detect meteors using the reflectivity of their plasma trail to radio waves. Given the unpredictable nature of meteors, meteor observations are done in the form of "blind surveys," and virtually all meteor detections are serendipitous in nature. Most meteor surveys typically aim to collect the trajectory and/or light curve (or time-amplitude series for radio observations) of a large sample of meteors. A small number of surveys are specially designed for specific purposes such as meteor spectroscopy (which can broadly distinguish chondrite-like vs. non-chondrite-like meteoroids) or high spatial/temporal resolution morphology. Detail reviews of these techniques are given by *Koten et al.* (2019) and *Kero et al.* (2019).

Since groundbased remote sensing techniques detect meteors using the radiation they release/can reflect, there is an observational bias favoring fast energetic meteors, namely the ones originated from HTCs/LPCs. The luminous power is proportional to the kinetic energy of the meteoroid, i.e., $\propto V^2$, hence meteoroids from HTC/LPC sources (with a mean arriving speed of ~60 km s^{-1}) are around 10× "brighter" than JFC ones (~20 km s^{-1}) with the same sizes.

Most optical and radar observations are conducted from the ground, but advances in technology in recent years have enabled observations from airborne or even spaceborne platforms (e.g., *Jenniskens and Butow,* 1999; *Vaubaillon et al.,* 2013; *Jenniskens,* 2018). Compared to groundbased observations, air- and spaceborne observations have access to windows at ultraviolet and infrared wavelengths and are not limited by weather, and thus are able to explore new regimes with higher efficiency. Progress is also made in the observation of extraterrestrial meteors in the form of impact flashes and meteoric debris layers, which provide valuable information on the distribution of meteoroids beyond Earth's orbit. Examples include impact flashes on the Moon and Jupiter, possible meteor detection by Mars rovers, and the unexpected meteor storm on Mars brought by LPC C/2013 A1 (Siding Spring). Detailed reviews on these topics are given by *Hueso et al.* (2013), *Christou et al.* (2019) and *Madiedo et al.* (2019).

Finally, the observation of newly-ejected meteoroids is an important part of any comet mission, which naturally provides knowledge about the birth of a meteoroid stream. This is discussed in detail in the chapter by Snodgrass et al. in this volume, and we do not repeat here.

3.2. Modeling of Meteoroid Streams

Because much of a meteoroid's orbital evolution is determined by the known forces of gravity and radiation pressure, it is possible to predict when meteoroid streams are steered into Earth's path, or when dense sections of the stream are encountered. An important application of meteoroid stream modeling is to predict meteor outbursts and storms, which can guide dedicated observations to study the outburst itself as well as the parent body. *Jenniskens* (2006, sections 12 and 15) reviewed the early attempts to predict meteor outbursts. It was not until the Leonid storms of the late 1990s that some meteor outbursts could be reliably predicted. The main difficulty was the lack of computing power to calculate the trajectory of a large number of particles with different orbits. This was solved by a simplified method, developed by

Kondrateva and Reznikov (1985) and *McNaught and Asher* (1999), that significantly reduced the calculation time and successfully predicted the 1999 Leonid storm by dynamical modeling of meteoroid streams.

The subsequent advance of computing technology at the turn of the century allowed a large number (up to many millions) of particles to be simulated. *Vaubaillon et al.* (2005a,b) incorporated a hydrodynamical model for meteoroid ejection and utilized realistic dust production rate derived from comet observation (e.g., measurement of the $Af\rho$ quantity), and successfully demonstrated their model during the 2002–2006 Leonids. Those methods have been further developed to study other showers (see *Egal,* 2020, for a review). By carefully accounting for the dynamical and physical evolution of the parent comet, the model can generally achieve minute-level accuracy in predicting the timing of the meteor outbursts. It is worth noting that this approach shares many similarities with the technique used by the comet community to simulate cometary tails, except that it works over a much longer timescale (many years) and addresses a different observable (encounter with Earth, rather than the on-sky distribution of dust).

It is possible to combine the two approaches to achieve a higher degree of realism by replacing the hydrodynamical ejection model with an ejection model derived from comet data. This can be important for certain comets (e.g., low-activity comets), which requires additional tuning to the traditional sublimation theory to accurately describe. In another application, this exercise also allows us to use meteor observations to explore ejection properties of the comet. An example is the observations of low-activity Comet 209P/LINEAR and its associated Camelopardalids meteor shower (Fig. 3; see also section 4.3.1).

Predicting the strength of meteor outbursts is more difficult, since it requires knowledge of the historic activity of the parent comet and the ejection conditions. These questions cannot be easily tackled without a full simulation of the entire meteoroid stream, hence early attempts focused on utilizing the meteoroid flux measurements made during historic encounters of the stream in question to project the flux of future encounters (e.g., *Jenniskens et al.,* 1998; *McNaught and Asher,* 1999). This requires observations of past encounters of the stream, which limits its applicability to a few well-observed streams.

The initial predictions of the Leonid storms using this approach were only accurate to about an order of magnitude. The "full" numerical model, on the other hand, simulates the entire meteoroid stream and therefore can directly translate the dust production of the parent to the fraction of simulated particles arriving at Earth. In return, the measured flux of the meteor outburst can be used to constrain the historic level of activity of the parent and to tune predictions of future encounters. The accuracy of this quantity is usually within a factor of a few from the measured value, depending on the knowledge of the often-poorly-constrained size distribution of the stream, which varies over size range, time, and from comet to comet (*Fulle et al.,* 2016).

3.3. Linking Meteoroid Streams To Their Parents

The paring of meteoroid streams and parent bodies is started by examining the similarities of their orbits, e.g., by using the D-criterion discussed in section 2.2. This is not always straightforward, as planetary dynamics constantly perturb the orbits of both the stream and the parent, gradually erasing/altering their dynamical memories. Pairs in stable dynamical "sweet spots" can be traced over a longer timescale (sometimes $\gtrsim 10^3$ orbits for JFCs), while pairs with frequent close encounters with Jupiter can decouple in only a few orbits.

Thanks to the dramatic increase in the orbital data of small bodies and meteoroid streams, many possible parent-stream linkages have been identified in the past two decades (see *Vaubaillon et al.,* 2019, Table 7.1 for a summary). However, this also brings the challenge of distinguishing genuine linkages from chance alignments, which is further

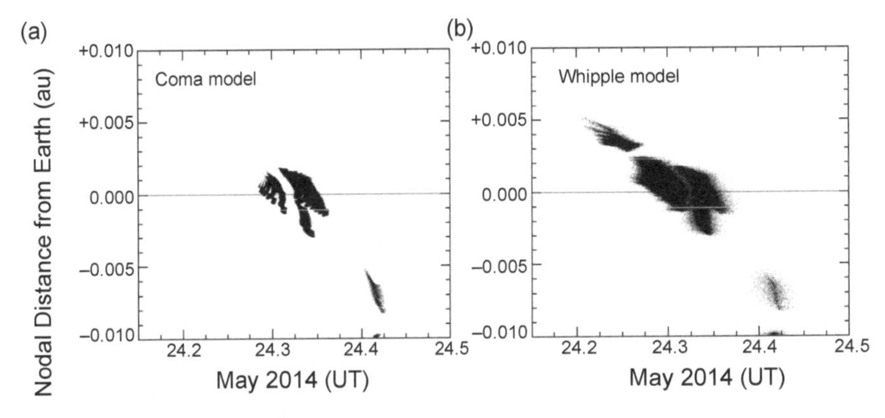

Fig. 3. Footprints of meteoroid trails from low-activity Comet 209P/LINEAR under two different ejection models: **(a)** one derived from fitting the observed coma vs. **(b)** the traditional *Whipple* (1950) sublimation model. Meteor observation can provide constraints to the ejection properties of the meteoroids and provide insights into the activity of the parent comet. In this case, a low terminal speed of the meteoroids [**(a)**] indicates a faster gas-dust decoupling, which implies a low-volatile content on the nucleus.

complicated by the difficulty of obtaining the precise orbit of meteoroid streams and (sometimes) the parents as well as the fragmentary history of some objects. Uncertainties and confusion can also arise when fragmentation results in multiple parent candidates for a given shower. In those cases, it is not always clear if the fragmentation itself, or later activity from the remaining bodies is responsible for meteor shower activity.

The statistical significance of a linkage can be tested using NEO population models. *Wiegert and Brown* (2004b) presented a statistical method that used the *Bottke et al.* (2002) debiased NEO population model to calculate the probability of chance alignment of a small body and a meteoroid stream. *Ye et al.* (2016a) applied this method to all proposed parent-stream linkages and found that only one-fourth of them are statistically significant.

Another challenge is to confirm cometary parents from the more numerous asteroidal candidates. For instance, the probability of 169P/NEAT being a chance alignment parent to the α-Capricornids is either 1 in 3 or 1 in 30 depending on the population being examined — NEOs or near-Earth comets — while the probability for asteroid 2017 MB_1 is 1 in 250 (*Ye,* 2018). A higher significance provides a stronger argument on the ground of statistics but does not exclusively prove the linkage to be real, hence a proposed linkage needs to be critically examined together with other evidence (e.g., stream modeling).

4. KNOWN LINKAGES

Here we review the established parent-stream linkages, with a focus on the knowledge of the parent body enabled by meteor data. We note that some of the cases have been recently reviewed by *Kasuga and Jewitt* (2019), thus for these cases, we only repeat the fundamental findings.

4.1. Complexes

A complex is a group of dynamically associated parent bodies and meteoroid streams that involve multiple parent bodies and/or multiple streams. As of this writing, four complexes have been established.

4.1.1. (3200) Phaethon (155140) 2005 UD, and the Phaethon-Geminid Complex. The Phaethon-Geminid Complex, proposed by *Ohtsuka et al.* (2006), includes asteroid-stream pairs (3200) Phaethon — Geminids (#4; Fig. 4) (155140) 2005 UD — Daytime Sextantids (#221), and possibly asteroid (225416) 1999 YC (*Jenniskens,* 2006, section 22; *Kasuga and Jewitt,* 2008; *Kasuga,* 2009). These bodies and streams share similar orbital characteristics, being short, relatively highly inclined, and having eccentric orbits, with low perihelion that can be explained by an evolutionary sequence by rotating the nodal line (a line defined by the intersection of the asteroid's or stream's orbital plane with the ecliptic plane). They are thought to have originated from a significant fragmentation event (*Ohtsuka et al.,* 2006; *Kasuga,* 2009). A connection to

Fig. 4. Phaethon and various stages of the Geminids: **(a)** image by the Lowell Discovery Telescope showing an inactive Phaethon (*Ye et al.,* 2021); **(b)** the Geminid meteoroid trail in space imaged by Parker Solar Probe (*Battams et al.,* 2020), marked by arrows; **(c)** composite image by Yangwang-1 satellite showing atmospheric entry of the Geminids meteoroids; and **(d)** composite image of the Geminid meteors from the ground, showing the shower radiant (courtesy of S. Yu).

(2) Pallas has been proposed (*de León et al.,* 2010) but is debated (*MacLennan et al.,* 2021).

Radar data show that Phaethon is a spinning-top rubble pile prone to mass loss by spinup (*Taylor et al.,* 2019). Observations of Phaethon by the Solar Terrestrial Relations Observatory (STEREO) revealed a ~2× brightening a few days around perihelion accompanied by a tail, likely due to thermal fracturing or desiccation of surface material (e.g., *Jewitt and Li,* 2010; *Li and Jewitt,* 2013). While Phaethon's linkage to the Geminids is well-established, the mass loss rate for this recurring activity is too small to explain the formation of the Geminids. Recent telescopic observations showed different spectral characteristics between Phaethon and 2005 UD that appears to suggest an independent origin (*Kareta et al.,* 2021) but could also reflect a different history of surface weathering.

The Geminids are the strongest annual meteor shower with a peak around December 13/14 every year. Its current ZHR is about 180 and has been increasing at a rate of +20 per decade (*Koten et al.,* 2019, Fig. 4.8). The ZHR is expected to reach a plateau of 190 in the next few decades as the core of the stream moves into Earth's orbit (*Jones and Hawkes,* 1986; *Jenniskens,* 2006, section 22.8), consistent with the recent upward trend. The formation of the Geminids may have occurred as recently as ~1 k.y. ago and coincided with the epoch that Phaethon reached a minimum q in its cyclic orbital variation (*Williams and Wu,* 1993; *MacLennan et al.,* 2021). Hence, its disruption is likely related to the thermal destruction of near-Sun asteroids (*Granvik et al.,* 2016), but the details remain unclear.

Understanding the dynamics of the Geminid stream has proven to be challenging, mainly due to the difficulty in reconciling the traditional dust ejection theories with the observed timing and duration of the shower. Specifically, various models consistently predict a maximum that is ~1 day later than the observation (cf. *Ryabova,* 2016). A possible explanation, opened up by *Lebedinets* (1985) and further discussed by *Ryabova* (2016) and *Kareta et al.* (2021), is that the orbit of Phaethon has changed significantly during the formation of the Geminid stream. In other words, meteoroid stream modeling that is based on Phaethon's modern-day orbit may not correctly capture the true evolution of the stream. It then becomes a curious question how much of our understanding of the dynamical history of the Geminids, Phaethon, and even PGC itself is still valid.

The Geminid meteoroids are exceptional in that they penetrate relatively deep in Earth's atmosphere and rarely show flares (*Jenniskens,* 2006, section 22.4). Spectroscopy of Geminid meteors revealed a large variation in Na content, with the Na D-line varying from undetectable to strong (*Borovička et al.,* 2005; *Kasuga et al.,* 2005a). This is likely due to the thermal desorption of some Na-bearing minerals, of which the mineralogical identity is unclear. *Kasuga et al.* (2006) showed that the q of the Geminid stream is beyond the distance needed to melt alkaline silicates on meteoroids. Because of the young age of the Geminid stream and the variety of Na contents, the Na desorption likely took place when minerals where heated on the surface of Phaethon.

The Daytime Sextantids is a minor daytime shower with ZHR = 5 and a peak around September 27 every year. Stream modeling showed that the relative position of the nodal line to 2005 UD demands that the Daytime Sextantids are at least $>10^4$ yr old, which is significantly older than the Geminids (*Ohtsuka et al.,* 1997; *Jakubík and Neslušan,* 2015). This suggests that the two showers were formed in separate disruption events and have different parents, instead of both showers originating from the hypothetical breakup that produced Phaethon and 2005 UD (*Jenniskens,* 2006, section 22.1). Running a statistical significance test (*Ye,* 2018) on the phase-synchronized orbits of Phaethon and 2005 UD (see *Ohtsuka et al.,* 2006), we find that the likelihood of chance alignment for the Phaethon-2005 UD pair is extremely low — below 1 in 10,000. However, the issue in the dynamical understanding of the Phaethon-Geminids pair, as discussed above, undermines this conclusion.

4.1.2. 2P/Encke and the Taurid Complex. The Taurid Complex includes Comet 2P/Encke, possibly a dozen NEOs, and four established annual showers: the Northern Taurids (#17), the Southern Taurids (#2), the Daytime β-Taurids (#173), and the Daytime ζ-Perseids (#172). The daytime showers are active from May to July, while the nighttime showers are active from September to December. 2P/Encke is known for its bright dust trail in the thermal infrared (*Kelley et al.,* 2006). The comet currently has a significant mass-loss rate of $\sim 10^{10}$ kg per orbit. Considering that the total mass of the Taurid Complex (10^{13} kg) and a dynamical age of $\sim 10^4$ yr (*Jenniskens,* 2006, section 25.1), this seems to imply that sublimation activity from 2P/Encke alone is sufficient to explain the formation of the Taurid Complex if that activity was as strong as today.

The wide dispersion of the streams and decoupling from Jupiter implies a significant age, and early studies have linked a number of NEOs to the complex. By adding up the masses of the meteoroid complex and these NEOs, *Asher et al.* (1993) suggested that the complex might have been formed by a disrupted 40-km-class comet about 10^4 yr ago. We since understand that most are O- and S-class asteroids and many of these associations are likely chance alignments, contributed in large part by the low inclination of these objects (*Jenniskens,* 2006, section 25.2; *Egal et al.,* 2021).

The northern and southern branches (Northern Taurids and Southern Taurids) are not mirror images in terms of activity and don't represent a full rotation of the nodal line.

The four streams may in fact be a sequence of individual showers that were formed more recently than the time it takes to complete a full rotation of the nodal line. A small number of candidate NEOs have since been proposed associated with the complex, all with a semimajor axis very similar to that of 2P/Encke and the low-albedo members in the complex, with the most notable member being 2004 TG[10] (*Jenniskens,* 2006, section 25.2; *Jenniskens et al.,* 2016b). Each candidate object appears to link to a period of enhanced Taurid rates. However, some of these objects have short orbital arcs, and hence their connection to the Complex remains to be verified. While the Taurids include large meter-sized objects [and some of these were earlier suspected to produce meteorites (cf. *Brown et al.,* 2013)], Encke cometesimals do not appear to be strong enough to survive atmospheric penetration (*Devillepoix et al.,* 2021).

Spectroscopy of Taurid meteors has shown occasional hydrogen lines (carbonaceous matter), a variable Fe content, and low meteoroid strengths, consistent with a cometary origin (*Borovička et al.,* 2007, 2019).

4.1.3. 96P/Machholz 1, (196256) 2003 EH$_1$, and the Machholz Interplanetary Complex. The Machholz Interplanetary Complex is a gigantic system that consists of 96P/Machholz 1; two near-Sun comet groups, the Marsden group and the Kracht group (*Sekanina and Chodas,* 2005); asteroid (196256) 2003 EH$_1$; and at least four established annual showers, the Quadrantids (#10), Daytime Arietids (#171), and Northern (#26) and Southern δ-Aquariids (#5) (*Jenniskens,* 2006, section 23; *Babadzhanov et al.,* 2017). Comet C/1490 Y1 may also be a member of the Complex (*Hasegawa,* 1979). These objects and streams reside in a narrow orbital corridor space and are highly evolved, hence it is challenging to understand the exact relations between these objects and streams (*Wiegert and Brown,* 2005).

96P is known to be an outlier among the solar system comet population, with depleted carbon-chain species and a bluer surface (*Schleicher,* 2008; *Eisner et al.,* 2019). The Marsden and Kracht comets have only been observed by the spacebased Solar and Heliospheric Observatory (SOHO) and are thus poorly understood. Asteroid 2003 EH_1 has not been observed to be active (*Kasuga and Jewitt,* 2015), exhibiting a C-like spectrum but with an absorption band around 1 μm due to unknown reasons (*Kareta et al.,* 2021). All established showers in this Complex produce significant annual meteor activity. The Quadrantids is characterized by a strong (ZHR = 130) but short (only lasts ~0.5 d) maximum, with an annual peak around January 3/4. It is worth noting that observations and dynamical simulations seem to suggest that the Quadrantids are a highly variable shower from year to year due to the perturbation from Jupiter (*Jenniskens,* 2006, section 20.1, Fig. 20.17). Video and radar data revealed a broad and much weaker component of the Quadrantids underlying the main peak (*Jenniskens,* 2006, section 20.1; *Brown et al.,* 2010). The Daytime Arietids is the strongest annual daytime shower, reaching ZHR = 60 around mid-June. The two δ-Aquariids are characterized by a relatively broad peak, reaching ZHR = 30 around late July. The Southern δ-Aquariids also have an underlying broad component (*Jenniskens et al.,* 2016a).

Stream modeling showed that 96P or its progenitor is likely the bulk supplier to all of these streams, especially the broad underlying components, composed of meteoroids ejected from 96P about 12–20 k.y. ago (*Abedin et al.,* 2018). The Marsden and Kracht comets likely have originated from a secondary fragmentation 1–2 k.y. ago (*Abedin et al.,* 2018) and are responsible for some of the Daytime Arietid stream (*Sekanina and Chodas,* 2005), although their contribution may be shadowed by 96P (*Abedin et al.,* 2017). Asteroid (196256) 2003 EH_1 is thought to be the direct parent of the core of the Quadrantid stream. The formation of the Quadrantids core may have occurred as recently as 200–500 yr ago (*Jenniskens,* 2004a; *Wiegert and Brown,* 2004a; *Abedin et al.,* 2015).

Despite having a near-Sun history, meteor spectroscopy showed that the Quadrantid meteors are not as Na-depleted as the Geminids meteors, which have similar history (*Borovička et al.,* 2005; *Kasuga et al.,* 2005a). The underlying cause is unclear (*Kasuga and Jewitt,* 2019), but may be related to the recent formation history during which the perihelion distance was large.

4.1.4. 169P/NEAT and the α-Capricornid Complex. The α-Capricornid Complex consists of JFCs 169P/NEAT, P/2003 T12 (SOHO), asteroid 2017 MB_1, and the α-Capricornid shower (#1). The orbits of 169P and P/2003 T12 are almost identical, and dynamical investigation shows that the spatial distance and relative speed of the pair reached a minimum 2900 yr ago, suggestive of a breakup origin (*Sosa and Fernández,* 2015). Following the approach of *Ye* (2018), we estimate that the likelihood of chance alignment is 1 in 10,000, a remarkably small number. On the other hand, as will be discussed below, there is an appreciable difference between the orbits of the 169P–P/2003 T12 pair and 2017 MB_1, hence it is not clear if all three of these bodies are indeed dynamically related. Additional linkages to working-list showers ξ^2-Capricornids (#623) and ε-Aquariids (#692) have been proposed (*Jenniskens et al.,* 2016a) but are in need of further investigation.

The α-Capricornids reaches a modest ZHR = 5 in late July and is characterized by an abundance in bright meteors. Numerical modeling suggests that the currently observed α-Capricornids were ejected 4500–5000 yr ago by a major disruption event (*Jenniskens and Vaubaillon,* 2010; *Kasuga et al.,* 2010). More recent ejecta have not evolved to Earth's orbit to be observed, hence the comet must have been much more active in the past than the current mass-loss rate of $\sim 10^{-2}$ kg s^{-1} reported by *Kasuga et al.* (2010).

Recently, NEO 2017 MB_1 was found to have an orbit closer to the mean α-Capricornid orbit (*Wiegert et al.,* 2017). The likelihood of chance alignment of the α-Capricornids–2017 MB_1 pair is 1 in 250, a lot lower than that of 169P and P/2003 T12 (1 in 3 and 1 in 5, respectively). It is possible that 2017 MB_1 is one of the macroscopic fragments produced during 169P's major disruption ~5 k.y. ago, similar to the case of 96P and 2003 EH_1 discussed above. Contrary to 169P and P/2003 T12, which have been observed to be active, 2017 MB_1 appears to be inactive. More study is needed to understand the relation between these bodies.

Spectral observations of the α-Capricornid meteors revealed their composition being Mg-rich and Fe-poor compared to typical chondrites (*Madiedo et al.,* 2014a). The same authors also showed a relatively high end height of these meteors, which indicates a low tensile strength compatible with a cometary origin.

4.2. Typical Jupiter-Family Comets

4.2.1. 7P/Pons-Winnecke and the June Bootids. The June Bootids (#170) exhibited significant activity in 1916, 1998, and 2004, with peak ZHR reaching 100, but activity is nearly nonexistent in other years (*Arlt,* 2000; *Jenniskens,* 2004b). Meteor activity in 1998 and 2004 occurred when its parent body, 7P/Pons-Winnecke, had evolved away from Earth (q = 1.26 au in 1998). The erratic behavior of June Bootids is largely due to the rapid orbital evolution of 7P. The comet's frequent close encounter with Jupiter has moved its orbit from q = 0.77 au from 1809 to q = 1.24 au currently, passing Earth's orbit by 0.03 au around 1916. Stream modeling shows that the June Bootids outbursts in 1916, 1998, and 2004 precipitated from meteoroids ejected in 1819–1869 locked in the 2:1 resonance with Jupiter (*Tanigawa and Hashimoto,* 2002; *Jenniskens,* 2004b). No encounter with the June Bootid trails is expected in the next few decades, but 7P will return to the near-Earth space following a close encounter to Jupiter in 2037, hence a resurrection of the June Bootids in the future is possible.

A cometary dust trail was detected in the orbit of 7P by the Infrared Astronomical Satellite (IRAS) (*Sykes and Walker,* 1992). The mass loss rate of 7P can be estimated using the activity of June Bootids. The 1998 June Bootids outburst was dominated by the dust ejected by 7P in 1825 and reached a peak flux of 0.12 ± 0.02 km^{-2} hr^{-1} appropriated to millimeter-sized meteoroids (*Brown and Hocking,*

1998), translating into a mass loss of 6×10^8 kg per orbit, in agreement with the value determined from photometric variation of 7P (*Kresak and Kresakova,* 1987). Hence, it only takes a few orbital revolutions to accumulate a sufficient amount of dust to be detected by IRAS.

4.2.2. 8P/Tuttle and the Ursids. 8P/Tuttle is technically a JFC even though its semimajor axis is considerably larger than most of the JFCs discussed in this chapter (a = 5.7 au vs. a ~ 2–3.5 au). The associated meteoroid stream, the Ursids (#15), is an annual stream known for its "far-type" outbursts — brief heightened meteor activity when 8P is near aphelion. The shower reaches a maximum around December 22 and typically has ZHR = 10, but can exceed ZHR = 100 during outbursts. Stream modeling shows that this is due to the mean-motion resonances (see section 3.2), which caused the meteoroid clumps to slowly drift away from the position of the comet (*Jenniskens et al.,* 2002). Currently observed outburst meteoroids were released by 8P in the ninth through fourteenth centuries (*Jenniskens,* 2006, Table 5a). By comparing the stream model and modern observation of the comet, it can also be inferred that the activity of 8P has not changed appreciably over the past millennium (100 orbits). The outburst Ursids exhibited higher begin/end heights compared to the background Ursids (*Moreno-Ibáñez et al.,* 2017), indicative of higher fragility typically seen in cometary meteoroids. 8P currently has MOID = 0.1 au to Earth, hence more recently released meteoroids will not reach Earth in the near future as long as the stream is not highly perturbed.

4.2.3. 15P/Finlay and the Arids. The Arids (#1130) is an emerging meteor shower that originated from 15P/Finlay. 15P has an Earth MOID of only 0.0097 au, but the associated meteor activity had not been observed until 2021. *Beech et al.* (1999) suggested that the lack of a meteor shower was due to the facts that the planetary perturbation that has driven most meteoroids away from Earth's orbit, and that the comet has not been very active. Recent stream modeling found that ejecta from 1995 to 2014 apparitions of the comet would cross Earth's path in 2021 (*Ye et al.,* 2015), which was confirmed by video networks and radar (*Jenniskens et al.,* 2021a). Remarkably, the detected meteor activity was partially contributed by the two comet outbursts detected during its 2014 return (*Ye et al.,* 2015; *Ishiguro et al.,* 2016). This is the first time that heightened meteor activity can be traced back to telescopically-detected cometary outbursts. Calculation by M. Maslov (unpublished data) suggests more meteor activity is possible in the coming decades, especially in 2047 when Earth directly encounters the 2008 trail.

4.2.4. 21P/Giacobini-Zinner and the October Draconids. Although the activity is nearly nonexistent in most years, the October Draconids (#9; sometimes being referred to as simply the "Draconids") has produced some of the strongest meteor activities ever observed in modern history. Meteor storms have been detected in the optical (1933, 1946, 1998) and by radar (1999, 2012), primarily due to the mass-dependent delivery of meteoroids under the dynamical effects (*Egal et al.,* 2019).

Early efforts to model the October Draconid stream have encountered difficulties in reproducing the timing of the 2005 and 2012 outbursts (*Campbell-Brown et al.,* 2006; *Ye et al.,* 2014), likely due to the orbital uncertainty introduced by 21P's multiple close encounters with Jupiter and/or variations in nongravitational acceleration. *Egal et al.* (2019) showed that this discrepancy can be largely overcome by using an orbital solution appropriated to each apparition rather than the solutions averaged over several apparitions. This highlights the challenge of modeling meteoroid streams from comets that have highly variable activity and/or have frequent close encounters with major planets. However, the same effect can also potentially provide a powerful tool to probe the dynamics of the comet in the past, especially over times when the comet was not observed telescopically.

The October Draconid meteors are known for their unusually high ablation altitude and abundant fragmentation, which can be explained by the extreme fragility of the meteoroids (*Jacchia et al.,* 1950). The begin and end heights of these meteors are typically 10 km higher than other meteors with the same entry speeds (*Ye et al.,* 2016b, Fig. 10). The fragile nature of the meteoroids may be related to the unusual bluing of the dust coma seen in polarimetry (*Kiselev et al.,* 2000). Stereoscopic and spectroscopic observations have revealed physical and chemical heterogeneity of these meteors. The sizes of meteoroid constituents appear to vary appreciably over a factor of 5, but the bulk meteoroids are all very porous with porosities around 90% (*Borovička et al.,* 2007). The chemical abundances of the meteoroids are nearly chondritic. Temporally resolved spectroscopy of a bright October Draconid meteor (with an initial mass of ~5 kg) revealed an early release of Na in the meteoroid ablation, which could indicate the presence of hydrated minerals in the meteoroid (*Madiedo et al.,* 2013).

4.2.5. 26P/Grigg-Skjellerup and the π-Puppids. Similar to June Bootids, the π-Puppids (#137) are showers with minimal activity in normal years but which have exhibited occasional outbursts, notably in 1972, 1977, 1982, and 2003 (*Jenniskens,* 2006, section 19.1). Its parent Comet 26P/Grigg-Skjellerup experiences frequent close encounters with Jupiter, which changed its q from 0.73 au in 1808 to the present 1.12 au. As a result, the π-Puppids stream has been split into many filaments. There are some discrepancies between the stream model and the observations that are difficult to reconcile (*Vaubaillon and Colas,* 2005), possibly due to the limited observation or the poor knowledge of the filaments. 26P will stay beyond Earth's orbit until a close encounter to Jupiter in 2118. Before that time, Earth might still be able to encounter filaments produced by past activity of the comet (*Jenniskens,* 2006, Table 6e).

4.3 Low-Activity Jupiter-Family Comets

4.3.1. 209P/LINEAR and the Camelopardalids. Despite its relatively large size (diameter ~3 km) and stable orbit in the near-Earth space, 209P/LINEAR was not discovered until 2004. Subsequent telescopic observation of the comet revealed a dust-rich comet with an extremely low activity level (*Ye and Wiegert,* 2014; *Schleicher and Knight,* 2016). Modeling of the dust coma showed a dust terminal speed that is 10× lower compared to the Whipple model (*Ye et al.,* 2016b), likely due to fast dust-gas decoupling caused by low ice content in the dust grains. Dynamical investigation shows that the comet is in a semistable orbit, with no close encounter to Jupiter over the past ~10^4 yr (*Fernández and Sosa,* 2015; *Ye et al.,* 2016b).

Meteor activity from 209P, the Camelopardalids (#451), was first detected in 2014 when the comet made a close approach to Earth and a handful of trails generated between 1798 and 1979 were directly crossed by Earth (*Ye and Wiegert,* 2014). Characterization of the optical-sized meteors showed low tensile strengths compatible with a cometary origin, although spectroscopy also revealed a low Fe content indicative of non-chondrite material (*Madiedo et al.,* 2014b). Interestingly, the radar-sized meteoroids are apparently of higher tensile strength (*Ye et al.,* 2016b), implying a difference in meteoroid structure compared to comets such as 21P (although meteoroids are consistently fragile regardless of size). A previously unnoticed minor outburst in 2011 was discovered by *Ye et al.* (2016b). Stream modeling showed that the 2011 outburst was produced by trails generated in 1763 and 1768. The observed meteoroid flux is 100× larger than the number calculated using the current dust production rate of 209P, implying an elevated activity of the comet back in those years. Overall, the Camelopardalids is nearly undetectable in most years, showing that 209P has supplied little dust over the past few thousand years, the dispersion timescale of the meteoroid stream. This evidence supports the idea that the comet is about to become an inert object. The Camelopardalid stream, like its parent, is highly stable, and occasional minor outbursts are expected throughout this century.

4.3.2. 300P/Catalina and the June ε-Ophiuchids. Radar observation shows that 300P ejects large, centimeter-class meteoroids (*Harmon et al.,* 2006). A linkage to the June ε-Ophiuchids (#459) has been proposed (*Rudawska and Jenniskens,* 2014). This very weak shower was initially only detected through a D-criterion clustering search in video data, but a more significant (although still minor) outburst was detected in 2019 (*Matlovič et al.,* 2020). Stereoscopic and spectroscopic observations revealed a chondritic, porous structure of the meteoroids consistent with a cometary origin (*Matlovič et al.,* 2020). The meteors exhibited a relative depletion in Na, a puzzling feature given that 300P has not been in a low-q orbit in the past several thousand years.

4.4. Jupiter-Family Comets — Fragmentation and Breakup

4.4.1. 3D/Biela and the Andromedids. The disruption of 3D/Biela is the first well-studied comet fragmentation event in the telescopic era. The comet was observed to start disintegration during its 1846 apparition and was not seen again after 1852. Instead, meteor storms from the associated Andromedid meteor shower (#18) were seen in 1872 and 1885. Numerical modeling confirmed that these exceptional activities were produced by continuous disintegration of 3D/Biela (*Jenniskens and Vaubaillon,* 2007). However, the same model also suggested that the observed Andromedid stream only accounts for a few percent of the estimated mass of 3D/Biela, prompting the question about the fate of the rest of the mass. One possibility is that one or more macroscopic fragments have survived the disruption and are now inert, but the lack of detection of such objects seems to disagree. Other possibilities include overestimated nucleus size or underestimated stream mass. The current estimate of the nucleus diameter of ~3 km is based on the nongravitational acceleration and brightness derived from the astrometric and photometric data assuming a typical nucleus behavior (*Babadzhanov et al.,* 1991). However, we now know that some small comets can exhibit large nongravitational acceleration and/or high brightness, such as 252P/LINEAR described above. Hence, it is likely that the pre-breakup nucleus of 3D/Biela is smaller, perhaps around 0.5 km in diameter if all the mass has been converted to the meteoroid stream. The Andromedid stream itself remains detectable (*Wiegert et al.,* 2013), albeit mostly at low activity level as the bulk of the stream has been evolved away from Earth's orbit. Planetary perturbation may bring the stream back to Earth's orbit after 2120 (*Babadzhanov et al.,* 1991).

4.4.2. 73P/Schwassmann-Wachmann 3 and the τ-Herculids. 73P/Schwassmann-Wachmann 3 is known for its spectacular fragmentation in 1995 and 2006. The associated τ-Herculid meteor shower (#61) was first detected in 1930, shortly after the discovery of the comet (*Nakamura,* 1930). The shower exhibited elevated activity in its 1930 return but has been quiet in other years. Elevated meteor activity is expected in 2022, 2049, and 2065 due to direct encounters with meteoroids ejected in the twentieth century, including the interesting 1995 apparition. Models are currently divided on whether the 1995 breakup of 73P can produce meteor storms in 2022 and 2049 (*Wiegert and Brown,* 2005; *Horii et al.,* 2008), mainly due to the unknown details of the fragmentation. (Non-)detection of a storm will help refine the details of the 1995 breakup.

Detection of a dust trail from 73P has been reported in Cosmic Background Explorer (COBE) data (*Arendt,* 2014) as well as dedicated observation by Spitzer (*Vaubaillon and Reach,* 2010). Taking ZHR = 2 and a shower duration of 0.5 days of the associated τ-Herculid shower, we obtain a stream mass of 5×10^8 kg. This does not include the mass deposited by the 1995 and 2006 fragmentation of 73P, which may have produced a similar amount of material (*Ishiguro et al.,* 2009).

4.4.3. 289P/Blanpain and the Phoenicids. 289P/Blanpain was originally discovered in 1819 and was subsequently lost (designated as D/1819 W1 for that apparition). More than 100 years years, following a significant outburst of the then-little-known Phoenicid meteor shower (#254) in 1956, *Ridley* (1963) suggested that the meteor shower was related to the lost Comet D/1819 W1 (Blanpain). Finally, asteroid

2003 WY_{25} was discovered in November 2003 and was subsequently linked back to D/1819 W1 (*Jenniskens and Lyytinen,* 2005), receiving the permanent designation as 289P. It remains unclear whether 2003 WY_{25} is D/1819 W1 itself or a fragment of it. The mass of 2003 WY_{25} is only ~one-tenth of that of the Phoenicid stream (*Jenniskens and Lyytinen,* 2005), hence it is possible that either more macroscopic objects remain to be discovered, or one or more catastrophic mass-loss events has occurred in the recent history. The former scenario appears to be unlikely given the high completion rate of NEO surveys.

2003 WY_{25} is still weakly active (*Jewitt,* 2006) and has produced a significant ($\Delta m = -9$ mag) outburst in 2013 while near aphelion (*Ye and Clark,* 2019), one of the largest comet outbursts ever observed. The outburst is remarkable considering 289P's small size [a nucleus radius of 100–160 m (*Jewitt,* 2006)]. The outburst accounted for ~1% of the remaining mass of 289P and is likely driven by runaway sublimation of supervolatiles triggered by rotational spinup. A significant nongravitational motion was detected in 289P's subsequent return in 2020, which may be caused by the 2013 mega-outburst. Minor enhancements of the Phoenicid shower have been observed in 2014 and 2019 (*Fujiwara et al.,* 2017; *Roggemans et al.,* 2020). A small amount of the ejecta from the 2013 outburst will reach Earth in 2036 and 2041.

4.5. Comets Without Showers

4.5.1. 46P/Wirtanen. This comet was recently brought into the near-Earth space due to close encounters with Jupiter in 1972 and 1984. Meteor activity has been predicted but not detected (*Maslov and Muzyko,* 2017). Additionally, no dust trail has been reported despite repeat infrared observations by IRAS, the Infrared Space Observatory (ISO), and Spitzer. A faint optical trail was recently detected in the images taken by the Transiting Exoplanet Survey Satellite (TESS) (*Farnham et al.,* 2019), providing a lower mass limit of 2×10^5 kg. Meteor activity has been predicted but no detection has been reported. Taking a typical detection limit of $\sim 10^{-4}$ km^{-2} hr^{-1} of millimeter-class meteoroids, achievable by routinely-operated meteor surveys nowadays, and a shower duration of 0.5 days, we obtain an upper stream mass limit of 5×10^7 kg. On the other hand, high-cadence observations have revealed at least six minor outbursts of 46P during its 2018 apparition, each producing 10^4–10^6 kg of material (*Kelley et al.,* 2021). If 46P has been active at this level during its past perihelion passages since arriving at its current orbit in 1984, it should have deposited 10^5–10^8 kg of material along its orbit. The absence of a meteor shower appears to argue for a smaller mass loss rate in this range.

4.5.2. 103P/Hartley 2. Similar to 46P, this comet arrived in the near-Earth space after a close encounter with Jupiter in 1971. It has a similar MOID as 46P (both are 0.07 au) but no meteor activity has been detected. Trail mass constrained from the Wide-field Infrared Survey Explorer (WISE) data (*Bauer et al.,* 2011) places it at a similar level as the trail of 2P. A cursory stream simulation suggested a lack of direct encounters to the dust trail until the 2060s (*Jenniskens,* 2006, section 19.8). 103P is known to produce large ejecta of up to 0.2–2 m in size (*Kelley et al.,* 2013).

4.5.3. 107P/Wilson-Harrington. This object is known for its singular episodic activity during its discovery in 1949. A linkage to the unconfirmed September γ-Sagittariids has been proposed by *Jenniskens* (2006, section 9.3), a shower that has only been detected once. Significant nondetection of an associated trail has been reported in optical (*Ishiguro et al.,* 2009) and infrared (*Reach et al.,* 2007). Following the discussion for 46P, we place an upper limit of 5×10^7 kg of any meteoroid stream. 107P, however, is dynamically more stable than 46P as it is quasi-decoupled from Jupiter, hence material from low-level activity (or occasional episodic ejection) has sufficient time to accumulate along its orbit. The lack of meteor activity may indicate a rather infrequent, or even a one-off, ejection.

4.5.4. 249P/LINEAR. This comet is among the most stable cometary objects in the NEO population that may have originated in the asteroid belt (*Fernández and Sosa,* 2015; *Fernández et al.,* 2017). Meteor activity is possible in mid April (from $\alpha = 207°$, $\delta = -19°$ at geocentric speed $v_g = 26.1$ km s^{-1}) and early November (from $\alpha = 217°$, $\delta = -6°$ at $v_g = 26.2$ km s^{-1}) but has not been detected. Further study is needed to understand the visibility of the 249P meteoroid stream and its implication of the past activity of the comet.

4.5.5. 252P/LINEAR. Although originally considered as a low-activity comet, 252P's close approach to Earth in 2016 (at 0.036 au) coincides with a revival in activity, primarily in form of gas emission (*Li et al.,* 2017). Meteoroid delivery to Earth is possible, but no meteor has been detected (*Ye et al.,* 2016a). The comet has a small nucleus with a radius of 300 ± 30 m, but has exhibited strong non-gravitational motion during its 2016 return, possibly due to a strong jet (*Ye et al.,* 2017). The non-gravitational motion may have evolved the comet away from its likely tiny meteoroid stream, but this possibility has not been thoroughly investigated. Future meteor observation may also help understand certain polarimetric features detected in comet data that have been interpreted as signatures of large, compact grains ejected from a desiccated surface (*Kwon et al.,* 2019).

4.5.6. P/2021 HS (PANSTARRS). This comet has one of the smallest steady-state active area ever measured, equivalent to a single circular pit with a radius of 15 m (*Ye et al.,* 2023). With a MOID of 0.04 au, meteor activity is possible but has not been detected, placing an upper limit of 5×10^7 kg of the stream (*Ye et al.,* 2023). The same authors also reported an independently-derived upper limit of $(5–50) \times 10^6$ kg of the stream based on the non-detection of dust trail in the TESS images. Dynamical investigation showed that the comet had a close encounter with Jupiter in 1670, by only 0.05 au, and hence is dynamically unstable as are most JFCs. Its extreme depletion of volatiles may indicate a highly thermally processed nucleus.

4.5.7. (3552) Don Quixote. With a diameter of 18 km, Don Quixote is the largest near-Earth asteroid (NEA) on a comet-like orbit and has long been suspected to be a dormant comet. Recent observations with Spitzer and groundbased optical telescopes have revealed recurring activity (*Mommert et al.,* 2014, 2020). The cause of the activity, however, is inconclusive. Don Quixote's current MOID to Earth is 0.34 au, much larger than typical MOIDs of meteor-producing parents, but some meteoroids may still evolve into Earth-approaching orbits within a few thousand years (*Rudawska and Vaubaillon,* 2015).

4.6. Halley-Type Comets

With the exception of 1P/Halley, most known HTCs have only been observed once or twice since the dawn of modern astronomy and are poorly characterized. Interestingly, three out of the four streams with well-established parents are associated with large and very active comets that have been sighted throughout a significant part of human history. Whether this is due to observational bias or is a reflection of the physical and/or dynamical evolution of the HTC population in general remains to be investigated. Using meteor spectroscopy and by combining the result with *in situ* exploration, it has been recognized that HTC meteoroids are more consistent with the anhydrous, carbon proto-CI material that represents typical cometary dust (*Rietmeijer and Nuth,* 2000; *Borovička,* 2004).

4.6.1. 1P/Halley, the η-Aquariids, and the Orionids. As an active and long-lived comet that periodically visits the inner solar system, 1P/Halley has produced a complex of meteor showers. The most significant activity are the Orionids (#8) in late October at the ascending node and the η-Aquariids (#31) in early May at the descending node. Both showers are among the strongest annual showers: η-Aquariids has a ZHR of 60 and the Orionids' is about 20. These two showers have been identified since ancient times (*Zhuang,* 1977).

The Earth crosses the two branches at different nodal distances. Currently, the ascending node represented by the Orionids lies farther away from Earth's orbit than the descending node represented by the η-Aquariids (0.17 au vs. 0.06 au), therefore the Orionids may represent older and more evolved materials from 1P. Sampling of material at different distances to the node also provides a means by which to explore the structure of a meteoroid stream. The attempt to model the activity profile of both showers led to the realization of *McIntosh and Hajduk*'s (1983) shell model, an incomplete rotation of the nodal line due to liberating about a mean-motion resonance, which correctly describes the structure of meteoroid stream. Compared to the η-Aquariids, the older Orionids seem to have lower tensile strength and have a more symmetric activity profile (*Egal et al.,* 2020a). Both showers show a periodicity of 11–12 yr due to the perturbation of Jupiter, and elevated activity is expected for η-Aquariids in 2023/2024 and 2045/2046 (*Egal et al.,* 2020b). An interesting feature of these streams is the abundance of small, submillimeter-class meteoroids, which differs from many other meteoroid streams. High-power radar observations have confirmed the unusual richness in <100-μm-class meteoroids of the Halleyid stream compared to other streams, including HTC streams (*Kero et al.,* 2011; *Schult et al.,* 2018). This appears to be the intrinsic character of 1P, as a similar abundance of small meteoroids was also detected by the Giotto spacecraft during its visit to 1P in 1986 (*McDonnell et al.,* 1986). Spectra of the Orionid meteors closely resemble those of the Perseids, meteoroids from HTC 109P/Swift-Tuttle (*Halliday,* 1987). High-resolution spectrum (R ~ 10000) of an Orionid meteor has revealed many metal species such as Ti, Cr, Zr, Pd, and W (*Passas et al.,* 2016).

4.6.2. 55P/Tempel-Tuttle and the Leonids. The Leonids (#13) have produced most of the observed meteor storms in modern history, with the most recent ones in 1999–2002. Leonid storms were recorded as far back as 902 AD (*Hasegawa,* 1996), which helps establish the orbital history of the parent comet, 55P/Tempel-Tuttle, beyond its first recorded detection in 1366. The Leonid storm in 1966 may have reached a ZHR of 15,000 (*Jenniskens,* 1995), the strongest meteor outburst ever measured. In normal years, the Leonids are a moderately active shower, reaching ZHR = 15 around November 17 each year. The next cluster of strong Leonid activity may arrive around 2031 when 55P returns to perihelion.

Studies of the Leonids have brought two important landmarks in meteor science: understanding of the nature of meteor showers and their relation to comets (section 1) and successful modeling and prediction of meteor storms (section 3.2). The Leonid Multi-Instrument Aircraft Campaign organized during the 1998–2002 Leonid returns deployed a unique array of new observing techniques and has greatly enhanced our understanding of meteor showers, with mid-infrared and submillimeter meteor observations being conducted for the first time (see the reviews by *Jenniskens and Butow,* 1999; *Jenniskens et al.,* 2000).

A rather unusual detection of a dust trail from 55P was made by *Nakamura et al.* (2000), who detected the optical glow of the 1899 trail using a wide-angle lens when the trail passed very close to Earth in 1998. Different from the typical trails in telescopic data, this detection manifested in the form of a large (a few degrees in radius), diffuse, and largely circular structure in the night sky, which is effectively the cross-section of a meteoroid trail. They derived a radius of 0.01 au and a number density of 1.2×10^{-10} m^{-3}. Assuming a toroidal-like structure of the stream with the uniform spatial distribution of particles of 10 μm in size (compatible with the optical detection of the glow) and a bulk density of 400 kg m^{-3} (*Babadzhanov and Kokhirova,* 2009), we derive a total mass of 3×10^{10} kg for the 1899 trail. Given a typical dispersion timescale of 100 orbits (section 2.2), this agrees well with the total stream mass of 5×10^{12} kg derived by *Jenniskens and Betlem* (2000).

Stereoscopic and spectroscopic observations of Leonid meteors revealed a different behavior compared to the other two well-studied HTC showers, the Orionids and the Perseids. The Leonid meteoroids are more fragile, less abundant in carbonaceous matter, and are slightly more abundant in Na and Mg (*Borovička et al.,* 1999; *Kasuga et al.,* 2005b). The fragile structure makes the Leonids more similar to the October Draconids (a JFC stream) than to the Perseids. Spectroscopy of the rare meteoric afterglow, the long-lasting visible debris train left by bright fireballs, showed many metal lines such as Fe, Mg, Na, Ca, Cr, Mn, K, and possibly Al in the debris (*Borovička and Jenniskens,* 2000).

4.6.3. 109P/Swift-Tuttle and the Perseids. The Perseids (#7) is perhaps the most observed meteor shower in history, with sightings dating back to the first millennium. It is the strongest annual meteor shower with a comet parent, reaching ZHR = 80 around August 12 of each year, but outbursts up to ZHR = 500 have been observed during the return of its parent Comet 109P/Swift-Tuttle in 1989–1995 (*Jenniskens,* 1995).

With an effective diameter of 26 km (*Lamy et al.,* 2004), 109P is the largest known object in the near-Earth comet population. The Perseid meteoroid stream is among the most massive stream crossing Earth's orbit, with a total mass of $\sim 4 \times 10^{13}$ kg (*Jenniskens,* 1994). A dynamical model constructed by *Brown and Jones* (1998) showed a stream age of $\sim 10^5$ yr with a slightly younger core about $(2.5 \pm 1.0) \times 10^4$ yr old. The elevated activity near the perihelion year is primarily due to the young meteoroids ejected in the last few orbits. Stream modeling shows the age of the outburst material is clearly correlated to the distance to the comet: The outbursts within 2–3 years from perihelion were primarily caused by the youngest meteoroids ejected in the last 1–2 orbits, outbursts about 3–4 years were caused by meteoroids ejected about 3–4 orbits ago, and the ones further away from the perihelion were the older materials. Recent studies have also identified a 12-year cycle of the Perseid activity due to the perturbation of Jupiter, which produces weaker outbursts even further away from the perihelion year (*Jenniskens,* 2006, section 17.9). The last outburst in this cycle occurred in 2016 and reached ZHR = 200 (*Rendtel et al.,* 2017). The next outburst is expected in 2028.

Spectroscopic observations showed that the composition of Perseids is more similar to the dust sampled at 1P than to chondrites (*Borovička,* 2004; *Spurný et al.,* 2014). The meteoroids are depleted in certain metal elements like Fe, Cr, and Mn but are enhanced in Si and Na. The abundance in Na may reflect the fact that Na is the interstitial fine-grain material that joins large mineral grains (*Trigo-Rodríguez and Llorca,* 2007).

4.6.4. C/1917 F1 (Mellish) and the December Monocerotids. The December Monocerotids (#19) is a modest annual stream active in mid-December characterized by its relatively small perihelion of q = 0.19 au. A connection to near-Sun Comet C/1917 F1 (Mellish) was first proposed by *Whipple* (1954) and supported by recent meteor data (*Lindblad and Olsson-Steel,* 1990; *Vereš et al.,* 2011). The latter authors also found that a relatively high ejection speed (~ 100 m s^{-1}) is required to form the stream if the ejection occurs at perihelion, a value that is too high to be explained by sublimation activity alone, but this can be easily compensated by the low precision of C/Mellish's currently-available orbit. The comet experienced close encounters with Venus in the past 1 k.y., which further complicates knowledge of its past orbit. Linkages to two minor streams, the Daytime κ-Leonids and April ρ-Cygnids, have also been proposed (*Brown et al.,* 2008; *Neslušan and Hajduková,* 2014). Comet C/Mellish will return around 2060. Recovery and observation of the comet will improve our knowledge of its orbit and physical characteristics, which will help understand its connection to these streams.

4.7. Long-Period Comets

Some LPCs that pass within 0.1 au from Earth's orbit cause occasional outbursts at intervals much shorter than the orbital period of the parents. It is now recognized that many of these outbursts are caused by the meteoroids ejected during the most recent perihelion passage of the parent (*Jenniskens,* 1997; *Lyytinen and Jenniskens,* 2003). Even so, details of when and how the meteoroids are ejected are still unclear. The existence of a few dozens of "orphan" LPC streams (streams without known parents) also needs to be understood.

4.7.1. C/1861 G1 (Thatcher) and the April Lyrids. Record of the April Lyrids (#6) dates back to 1803, when a strong outburst of ZHR = 860 was recorded (*Jenniskens,* 1995). It may have been the meteor shower recorded by the Chinese observers in 687 BC (*Jenniskens,* 2006, p. 11), although without information of the radiant, it is difficult to verify this connection. The annual component of the shower peaks around April 22, reaching ZHR = 20.

The shower appears to exhibit regular outbursts over a ~60-yr interval, most recently in 1982, much shorter than the 415-yr period of C/Thatcher. Proposed explanations included a disrupted fragment in a 60-yr orbit (*Arter and Williams,* 1995) or trapped dust trails in multiple resonances (*Emel'Yanenko,* 2001); however, the ejection speed needed for these scenarios (on the order of several hundred meters per second) is unrealistic based on the observed comet disruptions and sublimation models. Numerical modeling also showed that radiation drag could not bring optical-sized particles from parent orbits to the nearest plausible resonance within typical meteoroid collisional lifetime, or ~50 k.y. (*Kornoš et al.,* 2015). Alternatively, *Jenniskens* (1997) suggested that the outbursts from LPC streams can be explained by the reflex motion of the Sun due to Jupiter (and to a lesser extent, Saturn), a theory that has been verified by the successful prediction of the Aurigids outburst in 2007 (see below). The next April Lyrids outbursts in the cycle are in 2040/2041.

4.7.2. C/1911 N1 (Kiess) and the Aurigids. The Aurigids (#206) typically reaches ZHR = 5 around September 1 every year. Several outbursts have been detected at irregular intervals, in 1935, 1986, 1994, and most recently, 2007. A connection to LPC C/1911 N1 (Kiess) was proposed shortly after the 1935 outburst was observed (*Jenniskens,* 2006, section 13.1), but the fact that the Aurigid outbursts occurred in much shorter intervals compared to the ~2500-yr orbital period of C/Kiess had been a puzzle for many years. *Jenniskens* (1997) introduced the reflex motion theory to explain the outbursts of LPC streams and predicted an Aurigid outburst in 2007 (*Jenniskens and Vaubaillon,* 2007). An outburst of ZHR = 100 occurred as predicted (*Atreya and Christou,* 2009).

The observed height distribution of the Aurigid meteors was in good agreement with the Leonids, which have a similar entry speed, implying that the origin and composition of the Aurigids are likely compatible with that of the Leonids.

4.7.3. C/1979 Y1 (Bradfield) and the July Pegasids. The annual and weakly-active July Pegasids (#175) has been detected in a number of video surveys but none of the radar survey, indicative of the dominance of larger, millimeter-class meteoroids. A connection to C/1979 Y1 (Bradfield) has been proposed by several authors and investigated by *Hajduková et al.* (2019). Numerical simulation has predicted encounters with other filaments, which should produce meteor activity, but this has not been observationally confirmed.

4.7.4. C/1983 H1 (IRAS-Araki-Alcock) and the η-Lyrids. C/IRAS-Araki-Alcock made a close approach to Earth at 0.03 au shortly after its discovery. Prediction of meteor activity was made, but no reliable detection of outbursts has been reported (*Jenniskens,* 2006, section 6.1). A relatively strong annual η-Lyrid stream (#145) has been detected in recent years, first from a few photographic orbits (*Jenniskens,* 2006, section 6.1) and later from many low-light video orbits (*SonotaCo,* 2009; *Jenniskens et al.,* 2016a).

4.7.5. Other proposed linkages. Besides the ones discussed above, there are about a dozen established meteoroid streams with proposed linkages to LPCs, as tabulated in Table 1. Most of these associations involve comets that were observed before modern astrometry became available, and thus it is difficult to critically examine the dynamical history of the comets (and therefore the linkages themselves). On the other hand, progress has been made in studying other LPCs and conducting a targeted search for the streams they may have produced (e.g., *Hajduková and Neslušan,* 2019, 2020, 2021; *Jenniskens et al.,* 2021b). Most of the linkages being proposed involve streams that have not been confirmed. More comet and meteor data are needed to verify these proposed linkages.

4.7.6. C/2013 A1 (Siding Spring) and a meteor storm at Mars. C/Siding Spring passed Mars by a remarkable close distance of 0.0009 au on October 19, 2014, about 15× closer than the close approach of D/1770 (Lexell) to Earth. Despite the initial prediction that the bulk of the dust will miss Mars, significant enhancements of meteoric metal ions were detected by orbiters at Mars during the encounter (cf. *Christou et al.,* 2019, section 5.5.4). This may have pointed to distant activity of this LPC (*Zhang et al.,* 2021, section 3.1), which echoes the recent detection of the distant activity of LPCs like C/2017 K2 (PANSTARRS) (cf. *Jewitt et al.,* 2021).

5. INTERRELATIONS

5.1. Physical and Dynamical Evolution of Comets

Meteoroid streams trace the past evolution and activity of the parent comet, as far back as the coherence timescale of the meteoroid stream (~10^3–10^5 yr, section 2.2). Examples are the asteroidal meteoroid streams, which betray the fact that their parents must have been recently active. The Phaethon-Geminid Complex (section 4.1.1) is an excellent example. Complexes that are simultaneously related to both comets and asteroids, such as the Machholz Interplanetary Complex (section 4.1.3) and the 169P/NEAT Complex (section 4.1.4), may indicate an inhomogeneous compositional structure of the progenitor, which has been suggested from comet observations (see the chapter by Guilbert-Lepoutre et al. in this volume). The low and sometimes lack of meteor activity can be used to probe the recent history of low-activity comets and comet-asteroid transitional objects (see also the chapters by Jewitt and Hsieh and Knight et al. in this volume). Some comet examples were discussed in section 4.3. By combining dynamical modeling, meteor data can provide additional insights into the activity of transitional objects. Investigation of transitional object 107P/(4015) Wilson-Harrington, for instance, has been limited by the singular activity detection in 1949. The absence of meteor activity of this Earth-approaching object (MOID = 0.04 au) may imply a rather infrequent, or even a one-off, ejection. A cued search for meteor activity based on parent orbits revealed a meteor outburst from NEA (139359) 2001 ME_1 (*Ye et al.,* 2016a), which may be similar to the activity of 107P in nature.

Similarly, significant meteor activity can be used to trace mass-loss events, and even the demise, of the parent body. A large number of "orphan" meteoroid streams reveal frequent disruptions in the near-Earth space (*Vaubaillon et al.,* 2006), and the relative abundance of near-Sun meteoroid streams can be explained by thermally disrupted asteroids (*Ye and Granvik,* 2019; *Wiegert et al.,* 2020). The Andromedid meteor storms (section 4.4.1) would have revealed 3D's demise had the comet fragmentation not been observed. Meteor data and modeling of the α-Capricornid and Phoenicid streams revealed previous fragmentations of their parents, 169P and 289P (section 4.1.4 and section 4.4.3). Using meteor activity as a constraint to parent orbit was first demonstrated by *Wiegert et al.* (2013) on the case of 3D. Since meteor activity can often be traced to parent activity well before the telescopic era, this method can provide knowledge about the past orbit of the comet that is otherwise irretrievable. For example, application to the cold case of Comet D/1770 L1 (Lexell), one of the largest near-Earth comets that is now lost, showed that the comet likely still remains in the solar system on an altered orbit (*Ye et al.,* 2018).

Finally, photometric, stereoscopic, and spectroscopic characterization of meteors provide clues to the structure and composition of the materials that make up the parent. Mass photometry of meteors can be used to constrain the dust size distribution of the stream. Comparison with *in situ* data from 1P/Halley and 81P/Wild 2 shows general agreement of the size distribution measured by spacecraft and by meteor observations (*Jenniskens*, 2006, section 18.1; *Trigo-Rodríguez and Blum*, 2021). Meteor spectroscopy shows that most major streams have compositions broadly consistent with chondrites (*Borovička et al.*, 2019). Depletion of Na content is associated with near-Sun streams such as the Southern δ-Aquariids from 96P/Machholz 1 and the Geminids from (3200) Phaethon (*Kasuga et al.*, 2006; *Vojáček et al.*, 2015) due to thermal processing. Case-by-case discussions were given in section 4.

5.2. Dust Trails and Meteoroid Streams

Meteoroid streams and optical/thermal dust trails detected by telescopes are essentially the same structure probed by different techniques. A case-by-case discussion has been given in section 4. Table 2 summarizes the known dust trails originated from NEOs with MOID < 0.2 au and their corresponding meteor showers (or the lack thereof). The lower limit of the total mass of a dust trail can be estimated using the brightness and dimension of the observed trail (*Sykes and Walker,* 1992, section IV). This lower limit tends to be overly conservative as the presumed mass of the typical dust in the trail tends to be underestimated, and the trails often extend beyond the field of view. Compared to the total mass determined from meteor observations, we find that the determining mass of the dust trail is usually too low by 2–4 orders of magnitude. The mass of a meteoroid stream can be estimated using the method described by *Hughes and McBride* (1989). The largest source of uncertainty in this method is the mass influx, which has been better constrained thanks to the considerable progress in meteor surveys in the last few decades.

5.3. Interstellar Objects

Studies of meteors of interstellar origin have been an important research topic in meteor science since the early twentieth century. Despite extensive searches, no definitive detection has been made (see *Hajduková et al.,* 2019, for a review). On the other hand, successful detection of interstellar micrometeoroids has been made with dust detectors flown with a number of deep space missions (e.g., *Westphal et al.,* 2014; *Altobelli et al.,* 2016).

TABLE 2. NEOs with significant (non-)detections of dust trails and/or meteor showers.

Object	Optical?	IR?	Trail Mass	Meteor Shower	Stream Mass
1P/Halley		✓[1]	$> 4 \times 10^7$ kg	η-Aquariids, Orionids	$1-2 \times 10^{12}$ kg[2]
2P/Encke	✓[3]	✓[1,4-7]	$3-7 \times 10^{10}$ kg[1,5]	Taurids complex	1×10^{13} kg[2]
7P/Pons-Winnecke		✓[4]	$> 8 \times 10^8$ kg[4]	June Bootids	6×10^8 kg/orbit
55P/Tempel-Tuttle	✓[8]		3×10^{10} kg/orbit	Leonids	5×10^{12} kg[9]
73P/S-W 3		✓[1,10]	$> 3 \times 10^6$ kg	τ -Herculids	5×10^8 kg
169P/NEAT		✓[1]	$> 3 \times 10^8$ kg	α-Capricornids	9×10^{13} kg[11]
(3200) Phaethon	✓[12]	✓[1]	$> 4 \times 10^{11}$ kg[16]	Geminids	$10^{13}-10^{15}$ kg[13]
46P/Wirtanen	✓[14]	×	$> 2 \times 10^5$ kg	-	$< 5 \times 10^7$ kg
103P/Hartley 2		✓[15]	4×10^{10} kg[15]	-	n/a
107P/W-H	×[16]	×[5]	$\lesssim 10^6$ kg?	-	$< 10^{11}$ kg
P/2021 HS	×[17]		$< (5-50) \times 10^6$ kg[17]	-	$< 5 \times 10^7$ kg[17]

Total trail masses are either quoted from the references given or derived using the measurements reported by the references following the method described by *Sykes and Walker* (1992). ✓ in the optical and/or IR columns indicates detection of a dust trail in either/both wavelengths, while × indicates significant nondetection.

References: [1] *Arendt* (2014); [2] *Jenniskens* (1994); [3] *Ishiguro et al.* (2007); [4] *Sykes and Walker* (1992); [5] *Reach et al.* (2007); [6] *Kelley et al.* (2006); [7] *Reach et al.* (2000); [8] *Nakamura et al.* (2000); [9] *Jenniskens and Betlem* (2000); [10] *Vaubaillon and Reach* (2010); [11] *Jenniskens and Vaubaillon* (2010); [12] *Battams et al.* (2020); [13] *Ryabova* (2017); [14] *Farnham et al.* (2019); [15] *Bauer et al.* (2011); [16] *Ishiguro et al.* (2011); [17] *Ye et al.* (2023).

The discovery of the first two interstellar objects, 1I/'Oumuamua and 2I/Borisov (see the chapter by Fitzsimmons et al. in this volume), the latter of which is an unambiguous comet, raises the possibility that meteoroids ejected from these objects can be detected. On its way in, the meteoroids released can sometimes disperse wide enough to be encountered by Earth. Such detection will allow materials from another planetary system to be sampled at Earth. Comet 2I/Borisov has q = 2.01 au and does not come close to Earth or any other major planets. 1I/'Oumuamua had an Earth MOID of 0.10 au and a minimal flyby distance of 0.16 au, too far for meteoroid delivery driven by conventional water-ice sublimation. Indeed, a targeted search yielded no detection of meteor activity (*Ye et al.,* 2017). The nature of 1I/'Oumuamua is debated, due to the fact that the object has an unusual disk shape and exhibited non-gravitational acceleration with no coma (*Micheli et al.,* 2018), and it can only be agreed that the object is neither a typical asteroid nor a typical comet. A number of hypotheses have been put forward to explain 'Oumuamua's erratic motion, including the sublimation of molecular hydrogen or nitrogen ice (*Hoang and Loeb,* 2020; *Seligman and Laughlin,* 2020; *Desch and Jackson,* 2021). 'Oumuamua's distant flyby only permits the H_2 ice scenario to be tested using meteor observation, and only a loose constraint of <10 kg s^{-1} was derived (*Ye et al.,* 2017), which was several orders of magnitude larger than sustainable considering 'Oumuamua's small mass ($\sim 10^9$ kg). However, future sensitive meteor observations (perhaps through dedicated campaigns if advance warning can be given) could permit more stringent limits to validate these models.

5.4. Sample Return as (Micro-)Meteorites

Meteoroids larger than a few 0.1 m may survive the atmospheric entry and reach the ground as meteorites. Thousands of meteorites have been found on Earth, but only a handful of them (38 at the time of writing) for which orbits of the pre-entry body can be derived have been observed instrumentally, and none have been unambiguously associated with comets. The expected mineral and chemical properties of cometary meteorites are under debate. If analogous to cometary dust, they would likely be nearly chondritic with a high abundance of C, N, and organics (*Campins and Swindle,* 1998). It has long been suspected that CI chondrites may have come from comets (cf. *Lodders and Osborne,* 1999), but this scenario appears increasingly unlikely given that Hayabusa2 samples from (162173) Ryugu, an undisputed asteroid, are consistent with CI chondrites. Recent comet missions such as Rosetta and Stardust have measured cometary dust and revealed some similarities to primitive meteorites (*Stadermann et al.,* 2008; *Hoppe et al.,* 2018), but do not offer substantially more information for the search of cometary meteorites. Meanwhile Philae, designed to land on 67P/Churyumov-Gerasimenko and directly measure surface and subsurface material, only returned limited data. The derived properties are difficult to reconcile with other observations and are hence inconclusive (*Keller and Kührt,* 2020).

Dynamical properties (i.e., orbits) provide an easy but somewhat ambiguous metric to identify possibly cometary meteorites. Based on the orbit derived from unaided witness reports, it has been proposed that the historic Orgueil meteorite may be cometary (*Gounelle et al.,* 2006), but orbits derived this way can be highly uncertain (*Egal et al.,* 2018). The 1908 Tunguska event (*Jenniskens et al.,* 2019) has also been associated with a comet, specifically a fragment of 2P/Encke (cf. *Kelley et al.,* 2009), but that too remains controversial. Analysis of meter-sized impactors showed that 10–15% of them have comet-like orbits (*Brown et al.,* 2016). Compared to asteroids, comets tend to have higher relative velocities with respect to Earth and lower mechanical strength, making any meter-class cometesimals harder to survive atmospheric penetration. Specifically, it has been found that Taurid meteors brighter than –7 mag are in the strength-limited regime and do not penetrate deeper with increasing size (*Devillepoix et al.,* 2021).

At the other end of the spectrum, micrometeoroids (meteoroids smaller than ~200 μm) are too small to ablate during atmospheric entry, and hence can survive to the ground. They are mostly found on ocean floors and in polar sediments where the surrounding environment is largely stable, although fresh micrometeoroids can be collected from an urban environment with careful treatment (*Genge et al.,* 2017; *Suttle et al.,* 2021). Unlike meteorites, the vast majority of micrometeoroids are compositionally compatible with carbonaceous chondrites, indicative of cometary origin (*Engrand and Maurette,* 1998; *Nesvorný et al.,* 2010).

5.5. Astrobiology

Organic matter and water were delivered to Earth during or before the late heavy bombardment about 4 G.y. ago, creating the conditions possible for life. The mass influx of meteoroids is on the same order as that of planetesimals (*Jenniskens et al.,* 2004). Meteors also deliver organics and metallic compounds that can have fertilized oceans and supplied organic compounds. The metallic content in meteoroids can be probed via atomic and ionic emissions of meteors, of which many are accessible in the visible and near-infrared wavelengths. Elements such as Ng, Mg, Fe, Ca, and Si can easily be explored using visible spectroscopy (cf. *Borovička et al.,* 2019; *Koten et al.,* 2019). Organic matter produces CN bands, hydrogen, and OH emissions. Attempts have been made to look for the strong CN emission at 387 nm and the OH (0–0) emission at 308 nm but without clear detection (*Jenniskens and Mandell,* 2004; *Abe et al.,* 2007). *Jenniskens et al.* (2004) derived an upper limit of <0.017 for CN/Fe, more than 2 orders of magnitude lower than if all nitrogen from a cometary dust is converted to CN. Simulation using a thermochemical equilibrium model suggested that most of the carbon atoms from organic matter in plasma excitation temperature around 4000–5000 K quickly bound with atmospheric O and form CO, and few form CN and C_2, making detection even more difficult (*Berezhnoy and Borovička,* 2010). On the other hand, thermochemical equilibrium is not expected during meteoroid entry. Regarding OH, the same model also showed that it is possible to satisfy several constraints derived by *Jenniskens et al.* (2004), and by *Abe et al.* (2007) under certain temperature ranges (see *Berezhnoy and Borovička,* 2010, section 6). However, the presence of the OH (0–0) emission needs further confirmation. It is also not clear how much of the plausible OH feature is produced by the reaction by the ozone-hydrogen reaction rather than the dissociation of meteoric mineral water. The latter scenario, if confirmed, can provide another way to provide hydrated content delivered to Earth from comets and water-rich asteroids.

5.6. Satellite Impact Hazard

Meteoroid impacts can interfere with, and sometimes terminate, the operation of satellites and spacecraft in orbit. Adverse effects from meteoroid impacts include mechanical damage to the hardware and unwanted electronic effects caused by impact-generated plasma, which can result in life-threatening situations for astronauts. Hazards from meteoroid impacts have received great attention in recent years, and various preventive measures are being taken to mitigate the risk of an impact, including the development of meteoroid models, improved spacecraft designs, and operational measures such as changing the orientation of solar panels or performing maneuvers to minimize the surface area exposed to a shower. Cometary streams, especially the ones from HTCs and LPCs, pose a greater threat due to their higher arrival speed at Earth. This is one of the motivations for the study of such meteor showers and outbursts. Detailed reviews of this topic include *Drolshagen and Moorhead* (2019).

5.7. Planetary Defense

Meteoroid streams can also betray undiscovered parents, which are of significance in planetary defense. After several decades of NEO surveying, >95% of kilometer-class NEAs have been cataloged (*Jedicke et al.,* 2015), but kilometer-class LPCs are much less known and now are the dominant remaining impact risk in planetary defense. There are presently several dozens of established HTC/LPC meteoroid streams that have not been identified with any known comets or asteroids, and more await confirmation. The strongest and best-characterized among them is the α-Monocerotids (#246), an annual meteor shower with activity in mid-November that has also produced a handful of intense outbursts (*Jenniskens et al.,* 1997; *Jenniskens and Docters van Leeuwen,* 1997). All LPC showers with such outbursts have parent bodies that oscillate in a similar manner in and out of Earth orbit and thus are a potential impact hazard.

6. FUTURE WORK

Since *Comets II*, our understanding of comets and their meteoroid streams has advanced significantly. This trend is expected to continue given the development on many fronts. On the front of meteor science and observation, more optical and radar meteor networks are entering operation, especially over areas that have had poor coverage in the past. Networks such as CAMS, the Global Fireball Observatory, Fireball Recovery and InterPlanetary Observation Network (FRIPON), and the Global Meteor Network are expanding globally, and new video networks and meteor radars are being set up in historically poorly-covered regions. The accumulation of data allows existing streams to be better characterized and new streams to be verified. Advances in sensor technology pave the way for instruments with higher sensitivity, frame rate, and/or resolution, enabling better constraints on meteoroid properties and orbits.

On the front of traditional telescopic science, the next-generation astrophysical observatories and time-domain surveys, such as the Vera Rubin Observatory/Legacy Survey of Space and Time (LSST), NEO Surveyor, James Webb Space Telescope (JWST), Roman, Euclid, and the Chinese Space Station Telescope (CSST) will enhance our understanding of various small body populations (see the chapter by Bauer et al. in this volume), likely resulting in the detection of more meteoroid stream parents or remnants from past breakup events. LSST alone will increase the number of cataloged small bodies by 5–10× (*Schwamb et al.,* 2018). In addition, LSST and numerous shallower time-domain surveys will continue to monitor episodic ejection events from comets that are responsible for the formation of some streams in the near future. Other multi-bandpass and wavelength surveys (e.g., NEO Surveyor, Roman, Euclid, CSST), as well as multi-usage telescopes (e.g., JWST), will characterize the parents and provide context for meteor observations.

On the *in situ* exploration front, new small-body missions will provide insights into the origin and evolution of the parent bodies as well as the formation of the meteoroid streams. The Rosetta mission has, along with many other things, provided crucial data for understanding the formation of a meteoroid stream at the source, even though 67P/Churyumov-Gerasimenko does not currently produce a meteor shower at Earth. In the foreseeable future, we have at least Comet Interceptor, which will visit a to-be-discovered LPC or interstellar comet, with a handful of backup targets that include many meteor shower parents, as well as DESTINY+ to Phaethon and 2005 UD. Knowledge from meteor observation/stream modeling will complement the data returned from these missions.

Other Earth-based and interplanetary missions can potentially provide unique data for meteor research as well. Weather radars and satellites occasionally assist investigation of bolides [e.g., the Geostationary Lightning Mapper (GLM) (*Jenniskens et al.,* 2018)]. The unexpected flyby of C/2013 A1 (Siding Spring) to Mars allowed the spacecraft at Mars to observe the comet as well as the meteor shower in the Mars atmosphere up close, providing a wealth of data of this dynamically new comet. The event is a warning that more intense meteor showers are possible than we have experienced in recent years. A more distant flyby of C/2021 A1 (Leonard) to Venus at the end of 2021 provided an opportunity to examine the activity of the comet at >30 au (*Zhang et al.,* 2021), made possible by the Akatsuki spacecraft, which is currently orbiting the planet. Mars is sparsely populated with landers and rovers, which can conduct complimentary meteor observations. Other planets and moons with notable atmospheres or exospheres are also ideal places to study meteors, such as Mercury's exosphere and impact flashes on the Moon and Jupiter. These data provide additional points to sample and study dust streams in the solar system in addition to our Earth. With the ever-growing interest to explore our solar system with *in situ* missions, the opportunities to study these "exo" meteors are numerous.

We conclude with some questions for future research:

1. What is the formation mechanism for asteroidal meteoroid streams? How can meteor data be used to distinguish dormant comets from asteroids?
2. What is the underlying cause of the streams with simultaneous linkages to asteroids and comets? Are these linkages coincidental?
3. How can meteor data be used to identify and investigate dormant comets in the NEO population? What is the mechanism that drives episodic ejections of 107P/Wilson-Harrington and (139359) 2001 ME_1?
4. How can meteor data be used to search for hidden LPC impactors? Can we use meteor data to probe the "fading paradox" of the LPCs?
5. What is the cause of a large number of orphan meteoroid streams, especially toward the HTC/LPC regime? Is this due to the observational bias of the telescopic surveys?
6. How can real parent-stream linkages from chance alignments in the big data era be effectively distinguished?

Acknowledgments. We thank M. Fries, J. Vaubaillon, and an anonymous reviewer for their careful review, as well as D. Jewitt, S. X. Han, and M. Knight for their comments, all of which helped improve this chapter. We also thank K. Battams, Z. Wang, and S. Yu for kindly providing their images of the Geminids. P.J. was supported by NASA grant 80NSSC19K0563. Q.Y. was partially supported by STScI grants HST-GO-15357 and HST-GO-15978.

REFERENCES

Abe S., Ebizuka N., Yano H., et al. (2007) Search for OH(A-X) and detection of N_2^+(B-X) in ultraviolet meteor spectrum. *Adv. Space Res., 39,* 538–543.

Abedin A., Spurny P., Wiegert P., et al. (2015) On the age and formation mechanism of the core of the Quadrantid meteoroid stream. *Icarus, 261,* 100–117.

Abedin A., Wiegert P., Pokorný P., et al. (2017) The age and the probable parent body of the daytime Arietid meteor shower. *Icarus, 281,* 417–443.

Abedin A., Wiegert P., Janches D., et al. (2018) Formation and past evolution of the showers of 96P/Machholz complex. *Icarus, 300,* 360–385.

Altobelli N., Postberg F., Fiege K., et al. (2016) Flux and composition of interstellar dust at Saturn from Cassini's Cosmic Dust Analyzer. *Science, 352,* 312–318.

Arendt R. G. (2014) DIRBE comet trails. *Astron. J., 148,* 135.

Arlt R. (2000) The analysis of a weak meteor shower: The June Bootids in 2000. *WGN, J. Intl. Meteor Org., 28,* 98–108.

Arter T. R. and Williams I. P. (1995) The April Lyrids. *Mon. Not. R. Astron. Soc., 277,* 1087–1096.

Asher D. J., Clube S. V. M., and Steel D. I. (1993) Asteroids in the Taurid complex. *Mon. Not. R. Astron. Soc., 264,* 93–105.

Atreya P. and Christou A. A. (2009) The 2007 Aurigid meteor outburst. *Mon. Not. R. Astron. Soc., 393,* 1493–1497.

Babadzhanov P. B. and Kokhirova G. I. (2009) Densities and porosities of meteoroids. *Astron. Astrophys., 495,* 353–358.

Babadzhanov P. B., Wu Z., Williams I. P., et al. (1991) The Leonids, Comet Biela and Biela's associated meteoroid stream. *Mon. Not. R. Astron. Soc., 253,* 69–74.

Babadzhanov P. B., Kokhirova G. I., Williams I. P., et al. (2017) Investigation into the relationship between Comet 96P/Machholz 1 and asteroid 2003 EH1. *Astron. Astrophys., 598,* A94.

Battams K., Knight M. M., Kelley M. S. P., et al. (2020) Parker Solar Probe observations of a dust trail in the orbit of (3200) Phaethon. *Astrophys. J. Suppl. Ser., 246,* 64.

Bauer J. M., Walker R. G., Mainzer A. K., et al. (2011) WISE/NEOWISE observations of Comet 103P/Hartley 2. *Astrophys. J., 738,* 171.

Beech M., Nikolova S., and Jones J. (1999) The 'silent world' of Comet 15P/Finlay. *Mon. Not. R. Astron. Soc., 310,* 168–174.

Berezhnoy A. A. and Borovička J. (2010) Formation of molecules in bright meteors. *Icarus, 210,* 150–157.

Blaauw R. C., Campbell-Brown M. D., and Weryk R. J. (2011) Mass distribution indices of sporadic meteors using radar data. *Mon. Not. R. Astron. Soc., 412,* 2033–2039.

Borovička J. (2004) Elemental abundances in Leonid and Perseid meteoroids. *Earth Moon Planets, 95,* 245–253.

Borovička J. and Jenniskens P. (2000) Time resolved spectroscopy of a Leonid fireball afterglow. *Earth Moon Planets, 82,* 399–428.

Borovicka J., Stork R., and Bocek J. (1999) First results from video spectroscopy of 1998 Leonid meteors. *Meteorit. Planet. Sci., 34,* 987–994.

Borovička J., Koten P., Spurný P., et al. (2005) A survey of meteor spectra and orbits: Evidence for three populations of Na-free meteoroids. *Icarus, 174,* 15–30.

Borovička J., Spurný P., and Koten P. (2007) Atmospheric deceleration and light curves of Draconid meteors and implications for the structure of cometary dust. *Astron. Astrophys., 473,* 661–672.

Borovička J., Macke R. J., Campbell-Brown M. D., et al. (2019) Physical and chemical properties of meteoroids. In *Meteoroids: Sources of Meteors on Earth and Beyond* (G. O. Ryabova et al., eds.), pp. 37–62. Cambridge Univ., Cambridge.

Bottke W. F., Morbidelli A., Jedicke R., et al. (2002) Debiased orbital and absolute magnitude distribution of the near-Earth objects. *Icarus, 156,* 399–433.

Brown P. and Hocking W. K. (1998) June Bootid meteors 1998. *IAU Circular 6966.*

Brown P. and Jones J. (1998) Simulation of the formation and evolution of the Perseid meteoroid stream. *Icarus, 133,* 36–68.

Brown P., Weryk R. J., Wong D. K., et al. (2008) A meteoroid stream survey using the Canadian Meteor Orbit Radar. I. Methodology and radiant catalogue. *Icarus, 195,* 317–339.

Brown P., Weryk R. J., Kohut S., et al. (2010) Development of an all-sky video meteor network in Southern Ontario, Canada: The ASGARD System. *WGN, J. Intl. Meteor Org., 38,* 25–30.

Brown P., Marchenko V., Moser D. E., et al. (2013) Meteorites from meteor showers: A case study of the Taurids. *Meteorit. Planet. Sci., 48,* 270–288.

Brown P., Wiegert P., Clark D., et al. (2016) Orbital and physical characteristics of meter-scale impactors from airburst observations. *Icarus, 266,* 96–111.

Campbell-Brown M., Vaubaillon J., Brown P., et al. (2006) The 2005 Draconid outburst. *Astron. Astrophys., 451,* 339–344.

Campins H. and Swindle T. D. (1998) Expected characteristics of cometary meteorites. *Meteorit. Planet. Sci., 33,* 1201–1211.

Chen J. and Jewitt D. (1994) On the rate at which comets split. *Icarus, 108,* 265–271.

Christou A., Vaubaillon J., Withers P., et al. (2019) Extra-terrestrial meteors. In *Meteoroids: Sources of Meteors on Earth and Beyond* (G. O. Ryabova et al., eds.), pp. 119–135. Cambridge Univ.,

Cambridge.
de León J., Campins H., Tsiganis K., et al. (2010) Origin of the near-Earth asteroid Phaethon and the Geminids meteor shower. *Astron. Astrophys., 513,* A26.
Desch S. J. and Jackson A. P. (2021) 1I/'Oumuamua as an N_2 ice fragment of an exo-Pluto surface II: Generation of N_2 ice fragments and the origin of 'Oumuamua. *J. Geophys. Res.–Planets, 126,* e06807.
Devillepoix H. A. R., Jenniskens P., Bland P. A., et al. (2021) Taurid stream #628: A reservoir of large cometary impactors. *Planet. Sci. J., 2,* 223.
Drolshagen G. and Moorhead A. V. (2019) The meteoroid impact hazard for spacecraft. In *Meteoroids: Sources of Meteors on Earth and Beyond* (G. O. Ryabova et al., eds.), pp. 255–274. Cambridge Univ., Cambridge.
Egal A. (2020) Forecasting meteor showers: A review. *Planet. Space Sci., 185,* 104895.
Egal A., Veljkovic K., Vaubaillon J., et al. (2018) Time perception of a meteorite fall. *WGN, J. Intl. Meteor Org., 46,* 7–23.
Egal A., Wiegert P., Brown P. G., et al. (2019) Meteor shower modeling: Past and future Draconid outbursts. *Icarus, 330,* 123–141.
Egal A., Brown P. G., Rendtel J., et al. (2020a) Activity of the Eta-Aquariid and Orionid meteor showers. *Astron. Astrophys*., 640, A58.
Egal A., Wiegert P., Brown P. G., et al. (2020b) Modeling the past and future activity of the Halleyid meteor showers. *Astron. Astrophys., 642,* A120.
Egal A., Wiegert P., Brown P. G., et al. (2021) A dynamical analysis of the Taurid Complex: Evidence for past orbital convergences. *Mon. Not. R. Astron. Soc., 507,* 2568–2591.
Eisner N. L., Knight M. M., Snodgrass C., et al. (2019) Properties of the bare nucleus of Comet 96P/Machholz 1. *Astron. J., 157,* 186.
Emel'Yanenko V. V. (2001) Resonance structure of meteoroid streams. In *Meteoroids 2001 Conference* (B. Warmbein, ed.), pp. 43–45. ESA SP-495, Noordwijk, The Netherlands.
Engrand C. and Maurette M. (1998) Carbonaceous micrometeorites from Antarctica. *Meteorit. Planet. Sci., 33,* 565–580.
Farnham T. L., Kelley M. S. P., Knight M. M., et al. (2019) First results from TESS observations of Comet 46P/Wirtanen. *Astrophys. J. Lett., 886,* L24.
Fernández J. A. and Sosa A. (2015) Jupiter family comets in near-Earth orbits: Are some of them interlopers from the asteroid belt? *Planet. Space Sci., 118,* 14–24.
Fernández J. A., Licandro J., Moreno F., et al. (2017) Physical and dynamical properties of the anomalous Comet 249P/LINEAR. *Icarus, 295,* 34–45.
Fujiwara Y., Nakamura T., Uehara S., et al. (2017) Optical observations of the Phoenicid meteor shower in 2014 and activity of Comet 289P/Blanpain in the early 20th century. *Publ. Astron. Soc. Japan, 69,* 60.
Fulle M., Marzari F., Della Corte V., et al. (2016) Evolution of the dust size distribution of Comet 67P/Churyumov-Gerasimenko from 2.2 au to perihelion. *Astrophys. J., 821,* 19.
Genge M. J., Larsen J., Van Ginneken M., et al. (2017) An urban collection of modern-day large micrometeorites: Evidence for variations in the extraterrestrial dust flux through the Quaternary. *Geology, 45,* 119–122.
Gounelle M., Spurný P., and Bland P. A. (2006) The orbit and atmospheric trajectory of the Orgueil meteorite from historical records. *Meteorit. Planet. Sci., 41,* 135–150.
Granvik M., Morbidelli A., Jedicke R., et al. (2016) Supercatastrophic disruption of asteroids at small perihelion distances. *Nature, 530,* 303–306.
Grun E., Zook H. A., Fechtig H., et al. (1985) Collisional balance of the meteoritic complex. *Icarus, 62,* 244–272.
Hajduková M. and Neslušan L. (2019) Modeling of the meteoroid stream of Comet C/1975 T2 and λ-Ursae Majorids. *Astron. Astrophys., 627,* A73.
Hajduková M. and Neslušan L. (2020) The χ-Andromedids and January α-Ursae Majorids: A new and a probable shower associated with Comet C/1992 W1 (Ohshita). *Icarus, 351,* 113960.
Hajduková M. and Neslušan L. (2021) Modeling the meteoroid streams of Comet C/1861 G1 (Thatcher), Lyrids. *Planet. Space Sci., 203,* 105246.
Hajduková M., Sterken V., and Wiegert P. (2019) Interstellar meteoroids. In *Meteoroids: Sources of Meteors on Earth and Beyond* (G. O. Ryabova et al., eds.), pp. 235–252. Cambridge Univ., Cambridge.
Halliday I. (1987) The spectra of meteors from Comet P/ Halley. *Astron. Astrophys., 187,* 921–924.
Harmon J. K., Nolan M. C., Margot J. L., et al. (2006) Radar observations of Comet P/2005 JQ5 (Catalina). *Icarus, 184,* 285–288.
Hasegawa I. (1979) Orbits of ancient and medieval comets. *Publ. Astron. Soc. Japan, 31,* 257–270.
Hasegawa I. (1996) Further comments on the identification of meteor showers recorded by the Arabs. *Q. J. R. Astron. Soc., 37,* 75–78.
Hoang T. and Loeb A. (2020) Destruction of molecular hydrogen ice and implications for 1I/2017 U1 ('Oumuamua). *Astrophys. J. Lett., 899,* L23.
Hoppe P., Rubin M., and Altwegg K. (2018) Presolar isotopic signatures in meteorites and comets: New insights from the Rosetta mission to Comet 67P/Churyumov-Gerasimenko. *Space Sci. Rev., 214,* 106.
Horii S., Watanabe J.-I., and Sato M. (2008) Meteor showers originated from 73P/Schwassmann Wachmann. *Earth Moon Planets, 102,* 85–89.
Hueso R., Pérez-Hoyos S., Sánchez-Lavega A., et al. (2013) Impact flux on Jupiter: From superbolides to large-scale collisions. *Astron. Astrophys., 560,* A55.
Hughes D. W. and McBride N. (1989) The mass of meteoroid streams. *Mon. Not. R. Astron. Soc., 240,* 73–79.
Ishiguro M., Sarugaku Y., Ueno M., et al. (2007) Dark red debris from three short-period comets: 2P/Encke, 22P/Kopff, and 65P/Gunn. *Icarus, 189,* 169–183.
Ishiguro M., Usui F., Sarugaku Y., et al. (2009) 2006 fragmentation of Comet 73P/Schwassmann-Wachmann 3B observed with Subaru/Suprime-Cam. *Icarus, 203,* 560–570.
Ishiguro M., Ham J.-B., Tholen D. J., et al. (2011) Search for the comet activity of 107P/(4015) Wilson-Harrington during 2009/2010 apparition. *Astrophys. J., 726,* 101.
Ishiguro M., Kuroda D., Hanayama H., et al. (2016) 2014–2015 multiple outbursts of 15P/Finlay. *Astron. J., 152,* 169.
Jacchia L. G., Kopal Z., and Millman P. M. (1950) A photographic study of the Draconid meteor shower of 1946. *Astrophys. J., 111,* 104–133.
Jakubík M. and Neslušan L. (2015) Meteor complex of asteroid 3200 Phaethon: Its features derived from theory and updated meteor data bases. *Mon. Not. R. Astron. Soc., 453,* 1186–1200.
Jedicke R., Granvik M., Micheli M., et al. (2015) Surveys, astrometric follow-up, and population statistics. In *Asteroids IV* (P. Michel et al., eds.), pp. 795–814. Univ. of Arizona, Tucson.
Jenniskens P. (1994) Meteor stream activity I. The annual streams. *Astron. Astrophys., 287,* 990–1013.
Jenniskens P. (1995) Meteor stream activity. II. Meteor outbursts. *Astron. Astrophys., 295,* 206–235.
Jenniskens P. (1997) Meteor stream activity. IV. Meteor outbursts and the reflex motion of the Sun. *Astron. Astrophys., 317,* 953–961.
Jenniskens P. (2004a) 2003 eh1 is the Quadrantid shower parent comet. *Astron. J., 127,* 3018–3022.
Jenniskens P. (2004b) 2004 June Bootids: Video images and low-resolution spectra of 7P/Pons-Winnecke debris. *WGN, J. Intl. Meteor Org., 32,* 114–116.
Jenniskens P. (2006) *Meteor Showers and their Parent Comets.* Cambridge Univ., Cambridge. 790 pp.
Jenniskens P. (2015) Meteoroid streams and the zodiacal cloud. In *Asteroids IV* (P. Michel et al., eds.), pp. 281–295. Univ. of Arizona, Tucson.
Jenniskens P. (2018) Review of asteroid-family and meteorite-type links. *Proc. Intl. Astron. Union, 14,* 9–12.
Jenniskens P. and Docters van Leeuwen G. (1997) The α-Monocerotids meteor outburst: The cross section of a comet dust trail. *Planet. Space Sci., 45,* 1649–1652.
Jenniskens P. and Butow S. J. (1999) The 1998 Leonid Multi-Instrument Aircraft Campaign — An early review. *Meteorit. Planet. Sci., 34,* 933–943.
Jenniskens P. and Betlem H. (2000) Massive remnant of evolved cometary dust trail detected in the orbit of Halley-type Comet 55P/Tempel-Tuttle. *Astrophys. J., 531,* 1161–1167.
Jenniskens P. and Mandell A. M. (2004) Hydrogen emission in meteors as a potential marker for the exogenous delivery of organics and water. *Astrobiology, 4,* 123–134.
Jenniskens P. and Lyytinen E. (2005) Meteor showers from the debris of broken comets: D/1819 W_1 (Blanpain), 2003 WY_{25}, and the Phoenicids. *Astron. J., 130,* 1286–1290.
Jenniskens P. and Vaubaillon J. (2007) 3D/Biela and the Andromedids: Fragmenting versus sublimating comets. *Astron. J., 134,* 1037–1045.
Jenniskens P. and Vaubaillon J. (2010) Minor planet 2002 EX_{12} (=169P/

NEAT) and the Alpha Capricornid shower. *Astron. J., 139,* 1822–1830.

Jenniskens P., Betlem H., De Lignie M., et al. (1997) The detection of a dust trail in the orbit of an Earth-threatening long-period comet. *Astrophys. J., 479,* 441–447.

Jenniskens P., Crawford C., Butow S. J., et al. (1998) Lorentz shaped comet dust trail cross section from new hybrid visual and video meteor counting technique — Implications for future Leonid storm encounters. *Earth Moon Planets, 82,* 191–208.

Jenniskens P., Butow S. J., and Fonda M. (2000) The 1999 Leonid Multi-Instrument Aircraft Campaign — An early review. *Earth Moon Planets, 82,* 1–26.

Jenniskens P., Lyytinen E., de Lignie M. C., et al. (2002) Dust trails of 8P/Tuttle and the unusual outbursts of the Ursid shower. *Icarus, 159,* 197–209.

Jenniskens P., Schaller E. L., Laux C. O., et al. (2004) Meteors do not break exogenous organic molecules into high yields of diatomics. *Astrobiology, 4,* 67–79.

Jenniskens P., Nénon Q., Albers J., et al. (2016a) The established meteor showers as observed by CAMS. *Icarus, 266,* 331–354.

Jenniskens P., Nénon Q., Gural P. S., et al. (2016b) CAMS newly detected meteor showers and the sporadic background. *Icarus, 266,* 384–409.

Jenniskens P., Albers J., Tillier C. E., et al. (2018) Detection of meteoroid impacts by the Geostationary Lightning Mapper on the GOES-16 satellite. *Meteorit. Planet. Sci., 53,* 2445–2469.

Jenniskens P., Popova O. P., Glazachev D. O., et al. (2019) Tunguska eyewitness accounts, injuries, and casualties. *Icarus, 327,* 4–18.

Jenniskens P., Cooper T., and Lauretta D. (2021a) Arid meteors 2021. *Central Bureau Electronic Telegrams (CBET), 5046,* 1.

Jenniskens P., Lauretta D. S., Towner M. C., et al. (2021b) Meteor showers from known long-period comets. *Icarus, 365,* 114469.

Jewitt D. (2006) Comet D/1819 W1 (Blanpain): Not dead yet. *Astron. J., 131,* 2327–2331.

Jewitt D. and Li J. (2010) Activity in Geminid parent (3200) Phaethon. *Astron. J., 140,* 1519–1527.

Jewitt D., Hsieh H., and Agarwal J. (2015) The active asteroids. In *Asteroids IV* (P. Michel et al., eds.), pp. 221–241. Univ. of Arizona, Tucson.

Jewitt D., Kim Y., Mutchler M., et al. (2021) Cometary activity begins at Kuiper belt distances: Evidence from C/2017 K2. *Astron. J., 161,* 188.

Jones J. and Hawkes R. L. (1986) The structure of the Geminid meteor stream. II — The combined action of the cometary ejection process and gravitational perturbations. *Mon. Not. R. Astron. Soc., 223,* 479–486.

Jopek T. J. and Kaňuchová Z. (2017) IAU Meteor Data Center — The shower database: A status report. *Planet. Space Sci., 143,* 3–6.

Kareta T., Hergenrother C., Reddy V., et al. (2021) Surfaces of (nearly) dormant comets and the recent history of the Quadrantid meteor shower. *Planet. Sci. J., 2,* 31.

Kasuga T. (2009) Thermal evolution of the Phaethon-Geminid stream complex. *Earth Moon Planets, 105,* 321–326.

Kasuga T. and Jewitt D. (2008) Observations of 1999 YC and the breakup of the Geminid stream parent. *Astron. J., 136,* 881–889.

Kasuga T. and Jewitt D. (2015) Physical observations of (196256) 2003 EH1, presumed parent of the Quadrantid meteoroid stream. *Astron. J., 150,* 152.

Kasuga T. and Jewitt D. (2019) Asteroid-meteoroid complexes. In *Meteoroids: Sources of Meteors on Earth and Beyond* (G. O. Ryabova et al., eds.), pp. 187–209. Cambridge Univ., Cambridge.

Kasuga T., Watanabe J., and Ebizuka N. (2005a) A 2004 Geminid meteor spectrum in the visible-ultraviolet region: Extreme Na depletion? *Astron. Astrophys., 438,* L17–L20.

Kasuga T., Yamamoto T., Watanabe J., et al. (2005b) Metallic abundances of the 2002 Leonid meteor deduced from high-definition TV spectra. *Astron. Astrophys., 435,* 341–351.

Kasuga T., Yamamoto T., Kimura H., et al. (2006) Thermal desorption of Na in meteoroids: Dependence on perihelion distance of meteor showers. *Astron. Astrophys., 453,* L17–L20.

Kasuga T., Balam D. D., and Wiegert P. A. (2010) Comet 169P/NEAT(=2002 EX_{12}): The parent body of the α-Capricornid meteoroid stream. *Astron. J., 140, 1806–1813.*

Keller H. U. and Kührt E. (2020) Cometary nuclei — From Giotto to Rosetta. *Space Sci. Rev., 216,* 14.

Kelley M. C., Seyler C. E., and Larsen M. F. (2009) Two-dimensional turbulence, space shuttle plume transport in the thermosphere, and a possible relation to the Great Siberian Impact Event. *Geophys. Res. Lett., 36,* L14103.

Kelley M. S., Woodward C. E., Harker D. E., et al. (2006) A Spitzer study of comets 2P/Encke, 67P/Churyumov-Gerasimenko, and C/2001 HT50 (LINEAR-NEAT). *Astrophys. J., 651,* 1256–1271.

Kelley M. S., Lindler D. J., Bodewits D., et al. (2013) A distribution of large particles in the coma of Comet 103P/Hartley 2. *Icarus, 222,* 634–652.

Kelley M. S. P., Farnham T. L., Li J.-Y., et al. (2021) Six outbursts of Comet 46P/Wirtanen. *Planet. Sci. J., 2,* 131.

Kero J., Szasz C., Nakamura T., et al. (2011) First results from the 2009–2010 MU radar head echo observation programme for sporadic and shower meteors: The Orionids 2009. *Mon. Not. R. Astron. Soc., 416,* 2550–2559.

Kero J., Campbell-Brown M. D., Stober G., et al. (2019) Radar observations of meteors. In *Meteoroids: Sources of Meteors on Earth and Beyond* (G. O. Ryabova et al., eds.), pp. 65–89. Cambridge Univ., Cambridge.

Kinoshita M., Maruyama T., and Sagayama T. (1999) Preliminary activity of Leonid meteor storm observed with a video camera in 1997. *Geophys. Res. Lett., 26,* 41–44.

Kiselev N. N., Jockers K., and Rosenbush V. K. (2000) Organic matter in dust of Comet 21P/Giacobini-Zinner and the Draconid meteoroids. *Earth Moon Planets, 82,* 141–148.

Kondrateva E. D. and Reznikov E. A. (1985) Comet Tempel-Tuttle and the Leonid meteor swarm. *Astron. Vestnik, 19,* 144–151.

Kornoš L., Tóth J., Porubčan V., et al. (2015) On the orbital evolution of the Lyrid meteoroid stream. *Planet. Space Sci., 118,* 48–53.

Koschack R. and Rendtel J. (1990) Determination of spatial number density and mass index from visual meteor observations (I). *WGN, J. Intl. Meteor Org., 18,* 44–58.

Koschny D. and Borovicka J. (2017) Definitions of terms in meteor astronomy. *WGN, J. Intl. Meteor Org., 45,* 91–92.

Koten P., Čapek D., Spurný P., et al. (2017) September epsilon Perseid cluster as a result of orbital fragmentation. *Astron. Astrophys., 600,* A74.

Koten P., Rendtel J., Shrbený L., et al. (2019) Meteors and meteor showers as observed by optical techniques. In *Meteoroids: Sources of Meteors on Earth and Beyond* (G. O. Ryabova et al., eds.), pp. 90–115. Cambridge Univ., Cambridge.

Koten P., Čapek D., Spurný P., et al. (2021) Search for pairs and groups in the 2006 Geminid meteor shower. *Astron. Astrophys., 656,* A98.

Kresak L. and Kresakova M. (1987) The mass loss rates of periodic comets. In *Diversity and Similarity of Comets* (E. J. Rolfe et al., eds.), pp. 739–744. ESA SP-278, Noordwijk, The Netherlands.

Kwon Y. G., Ishiguro M., Kwon J., et al. (2019) Near-infrared polarimetric study of near-Earth object 252P/LINEAR: An implication of scattered light from the evolved dust particles. *Astron. Astrophys., 629,* A121.

Lamy P. L., Toth I., Fernandez Y. R., et al. (2004) The sizes, shapes, albedos, and colors of cometary nuclei. In *Comets II* (M. C. Festou et al., eds.), pp. 223–264. Univ. of Arizona, Tucson.

Lebedinets V. N. (1985) Origin of meteor swarms of the Arietid and Geminid types. *Astron. Vestnik, 19,* 152–158.

Li J. and Jewitt D. (2013) Recurrent perihelion activity in (3200) Phaethon. *Astron. J., 145,* 154.

Li J.-Y., Kelley M. S. P., Samarasinha N. H., et al. (2017) The unusual apparition of Comet 252P/2000 G1 (LINEAR) and comparison with Comet P/2016 BA_{14} (PanSTARRS). *Astron. J., 154,* 136.

Lindblad B. A. and Olsson-Steel D. (1990) The Monocerotid meteor stream and Comet Mellish. *Bull. Astron. Inst. Czech., 41,* 193–200.

Lodders K. and Osborne R. (1999) Perspectives on the comet-asteroid-meteorite link. *Space Sci. Rev., 90,* 289–297.

Lyytinen E. and Jenniskens P. (2003) Meteor outbursts from long-period comet dust trails. *Icarus, 162,* 443–452.

MacLennan E., Toliou A., and Granvik M. (2021) Dynamical evolution and thermal history of asteroids (3200) Phaethon and (155140) 2005 UD. *Icarus, 366,* 114535.

Madiedo J. M., Trigo-Rodríguez J. M., Konovalova N., et al. (2013) The 2011 October Draconids outburst — II. Meteoroid chemical abundances from fireball spectroscopy. *Mon. Not. R. Astron. Soc., 433,* 571–580.

Madiedo J. M., Trigo-Rodríguez J. M., Ortiz J. L., et al. (2014a) Orbit and emission spectroscopy of α-Capricornid fireballs. *Icarus, 239,* 273–280.

Madiedo J. M., Trigo-Rodríguez J. M., Zamorano J., et al. (2014b) Orbits and emission spectra from the 2014 Camelopardalids. *Mon. Not. R. Astron. Soc., 445,* 3309–3314.

Madiedo J. M., Ortiz J. L., Yanagisawa M., et al. (2019) Impact flashes of meteoroids on the Moon. In *Meteoroids: Sources of Meteors on Earth and Beyond* (G. O. Ryabova et al., eds.), pp. 136–158. Cambridge Univ., Cambridge.

Maslov M. P. and Muzyko E. I. (2017) Forecast of the Comet 46P/Wirtanen meteor shower activity in 2017 and 2019. *Earth Moon Planets, 119,* 85–94.

Matlovič P., Kornoš L., Kováčová M., et al. (2020) Characterization of the June epsilon Ophiuchids meteoroid stream and the Comet 300P/Catalina. *Astron. Astrophys., 636,* A122.

McDonnell J. A. M., Alexander W. M., Burton W. M., et al. (1986) Dust density and mass distribution near Comet Halley from Giotto observations. *Nature, 321,* 338–341.

McIntosh B. A. and Hajduk A. (1983) Comet Halley meteor stream: A new model. *Mon. Not. R. Astron. Soc., 205,* 931–943.

McNaught R. H. and Asher D. J. (1999) Leonid dust trails and meteor storms. *WGN, J. Intl. Meteor Org., 27,* 85–102.

Micheli M., Farnocchia D., Meech K. J., et al. (2018) Nongravitational acceleration in the trajectory of 1I/2017 U1 ('Oumuamua). *Nature, 559,* 223–226.

Mommert M., Hora J. L., Harris A. W., et al. (2014) The discovery of cometary activity in near-Earth asteroid (3552) Don Quixote. *Astrophys. J., 781,* 25.

Mommert M., Hora J. L., Trilling D. E., et al. (2020) Recurrent cometary activity in near-Earth object (3552) Don Quixote. *Planet. Sci. J., 1,* 12.

Moreno-Ibáñez M., Trigo-Rodríguez J. M., Madiedo J. M., et al. (2017) Multi-instrumental observations of the 2014 Ursid meteor outburst. *Mon. Not. R. Astron. Soc., 468,* 2206–2213.

Nakamura K. (1930) On the observation of faint meteors, as experienced in the case of those from the orbit of Comet Schwassmann-Wachmann, 1930d. *Mon. Not. R. Astron. Soc., 91,* 204–209.

Nakamura R., Fujii Y., Ishiguro M., et al. (2000) The discovery of a faint glow of scattered sunlight from the dust trail of the Leonid parent Comet 55P/Tempel-Tuttle. *Astrophys. J., 540,* 1172–1176.

Neslušan L. and Hajduková M. (2014) The meteor-shower complex of Comet C/1917 F1 (Mellish). *Astron. Astrophys., 566,* A33.

Nesvorný D., Jenniskens P., Levison H. F., et al. (2010) Cometary origin of the zodiacal cloud and carbonaceous micrometeorites: Implications for hot debris disks. *Astrophys. J., 713,* 816–836.

Ohtsuka K., Shimoda C., Yoshikawa M., et al. (1997) Activity profile of the Sextantid meteor shower. *Earth Moon Planets, 77,* 83–91.

Ohtsuka K., Sekiguchi T., Kinoshita D., et al. (2006) Apollo asteroid 2005 UD: Split nucleus of (3200) Phaethon? *Astron. Astrophys., 450,* L25–L28.

Passas M., Madiedo J. M., and Gordillo-Vázquez F. J. (2016) High resolution spectroscopy of an Orionid meteor from 700 to 800 nm. *Icarus, 266,* 134–141.

Porubčan V., Kornoš L., and Williams I. P. (2006) The Taurid complex meteor showers and asteroids. *Contrib. Astron. Obs. Skalnate Pleso, 36,* 103–117.

Reach W. T., Sykes M. V., Lien D., et al. (2000) The formation of Encke meteoroids and dust trail. *Icarus, 148,* 80–94.

Reach W. T., Kelley M. S., and Sykes M. V. (2007) A survey of debris trails from short-period comets. *Icarus, 191,* 298–322.

Rendtel J., Ogawa H., and Sugimoto H. (2017) Meteor showers 2016: Review of predictions and observations. *WGN, J. Intl. Meteor Org., 45,* 49–55.

Ridley H. B. (1963) The Phoenicid meteor shower of 1956 December 5. *Mon. Not. Astron. Soc. South Africa, 22,* 42–49.

Rietmeijer F. J. M. and Nuth J. A. I (2000) Collected extraterrestrial materials: Constraints on meteor and fireball compositions. *Earth Moon Planets, 82,* 325–350.

Roggemans P., Johannink C., and Martin P. (2020) Phoenicids (PHO#254) activity in 2019. *eMeteorNews, 5,* 4–10.

Rudawska R. and Jenniskens P. (2014) New meteor showers identified in the CAMS and SonotaCo meteoroid orbit surveys. In *Meteoroids 2013* (T. J. Jopek et al., eds.), pp. 217–224. A. M. Univ., *Poznań, Poland.*

Rudawska R. and Vaubaillon J. (2015) Don Quixote — A possible parent body of a meteor shower. *Planet. Space Sci., 118,* 25–27.

Ryabova G. O. (2016) A preliminary numerical model of the Geminid meteoroid stream. *Mon. Not. R. Astron. Soc., 456,* 78–84.

Ryabova G. O. (2017) The mass of the Geminid meteoroid stream. *Planet. Space Sci., 143,* 125–131.

Schleicher D. G. (2008) The extremely anomalous molecular abundances of Comet 96P/Machholz 1 from narrowband photometry. *Astron. J., 136,* 2204–2213.

Schleicher D. G. and Knight M. M. (2016) The extremely low activity Comet 209P/LINEAR during its extraordinary close approach in 2014. *Astron. J., 152,* 89.

Schult C., Brown P., Pokorný P., et al. (2018) A meteoroid stream survey using meteor head echo observations from the Middle Atmosphere ALOMAR Radar System (MAARSY). *Icarus, 309,* 177–186.

Schwamb M. E., Jones R. L., Chesley S. R., et al. (2018) Large Synoptic Survey Telescope solar system science roadmap. *ArXiV e-prints,* arXiv:1802.01783.

Sekanina Z. and Chodas P. W. (2005) Origin of the Marsden and Kracht groups of sunskirting comets. I. Association with Comet 96P/Machholz and its interplanetary complex. *Astrophys. J. Suppl. Ser., 161,* 551–586.

Seligman D. and Laughlin G. (2020) Evidence that 1I/2017 U1 ('Oumuamua) was composed of molecular hydrogen ice. *Astrophys. J. Lett., 896,* L8.

SonotaCo (2009) A meteor shower catalog based on video observations in 2007–2008. *WGN, J. Intl. Meteor Org., 37,* 55–62.

Sosa A. and Fernández J. A. (2015) Comets 169P/NEAT and P/2003 T12 (SOHO): Two possible fragments of a common ancestor? *IAU Gen. Assy. 29,* p. 2255583. Cambridge Univ., Cambridge.

Spurný P., Shrbený L., Borovička J., et al. (2014) Bright Perseid fireball with exceptional beginning height of 170 km observed by different techniques. *Astron. Astrophys., 563,* A64.

Stadermann F. J., Hoppe P., Floss C., et al. (2008) Stardust in Stardust — The C, N, and O isotopic compositions of Wild 2 cometary matter in Al foil impacts. *Meteorit. Planet. Sci., 43,* 299–313.

Suttle M. D., Hasse T., Hecht L., et al. (2021) Evaluating urban micrometeorites as a research resource — A large population collected from a single rooftop. *Meteorit. Planet. Sci., 56,* 1531–1555.

Sykes M. V. and Walker R. G. (1992) Cometary dust trails I. Survey. *Icarus, 95,* 180–210.

Tanigawa T. and Hashimoto T. (2002) The origin of the 1998 June Boötid meteor shower. *Earth Moon Planets, 88,* 27–33.

Taylor P. A., Rivera-Valentín E. G., Benner L. A. M., et al. (2019) Arecibo radar observations of near-Earth asteroid (3200) Phaethon during the 2017 apparition. *Planet. Space Sci., 167,* 1–8.

Trigo-Rodríguez J. M. and Llorca J. (2007) On the sodium overabundance in cometary meteoroids. *Adv. Space Res., 39,* 517–525.

Trigo-Rodríguez J. M. and Blum J. (2021) Learning about comets from the study of mass distributions and fluxes of meteoroid streams. *Mon. Not. R. Astron. Soc., 512,* 2277-2289.

Vaubaillon J. and Colas F. (2005) Demonstration of gaps due to Jupiter in meteoroid streams: What happened with the 2003 Pi-Puppids? *Astron. Astrophys., 431,* 1139–1144.

Vaubaillon J. J. and Reach W. T. (2010) Spitzer Space Telescope observations and the particle size distribution of Comet 73P/Schwassmann-Wachmann 3. *Astron. J., 139,* 1491–1498.

Vaubaillon J., Colas F., and Jorda L. (2005a) A new method to predict meteor showers. I. Description of the model. *Astron. Astrophys., 439,* 751–760.

Vaubaillon J., Colas F., and Jorda L. (2005b) A new method to predict meteor showers. II. Application to the Leonids. *Astron. Astrophys., 439,* 761–770.

Vaubaillon J., Lamy P., and Jorda L. (2006) On the mechanisms leading to orphan meteoroid streams. *Mon. Not. R. Astron. Soc., 370,* 1841–1848.

Vaubaillon J., Koten P., Rudawska R., et al. (2013) Overview of the 2011 Draconids airborne observation campaign. In *Proceedings of the International Meteor Conference, La Palma, Canary Islands, Spain, 20–23 September 2012* (M. Gyssens and P. Roggemans, eds.), pp. 61–64. Intl. Meteor Organization, Hove, Belgium.

Vaubaillon J., Neslušan L., Sekhar A., et al. (2019) From parent body to meteor shower: The dynamics of meteoroid streams. In *Meteoroids: Sources of Meteors on Earth and Beyond* (G. O. Ryabova et al.,

eds.), pp. 161–186. Cambridge Univ., Cambridge.
Vereš P., Kornoš L., and Tóth J. (2011) Meteor showers of Comet C/1917 F1 Mellish. *Mon. Not. R. Astron. Soc., 412,* 511–521.
Vojáček V., Borovička J., Koten P., et al. (2015) Catalogue of representative meteor spectra. *Astron. Astrophys., 580,* A67.
Watanabe J.-I., Tabe I., Hasegawa H., et al. (2003) Meteoroid clusters in Leonids: Evidence of fragmentation in space. *Publ. Astron. Soc. Japan, 55,* L23–L26.
Westphal A. J., Stroud R. M., Bechtel H. A., et al. (2014) Evidence for interstellar origin of seven dust particles collected by the Stardust spacecraft. *Science, 345,* 786–791.
Whipple F. L. (1950) A comet model. I. The acceleration of Comet Encke. *Astrophys. J., 111,* 375–394.
Whipple F. L. (1954) Photographic meteor orbits and their distribution in space. *Astron. J., 59,* 201–217.
Wiegert P. and Brown P. (2004a) The core of the Quadrantid meteoroid stream is two hundred years old. *Earth Moon Planets, 95,* 81–88.
Wiegert P. and Brown P. (2004b) The problem of linking minor meteor showers to their parent bodies: Initial considerations. *Earth Moon Planets, 95,* 19–26.
Wiegert P. and Brown P. (2005) The Quadrantid meteoroid complex. *Icarus, 179,* 139–157.
Wiegert P., Vaubaillon J., and Campbell-Brown M. (2009) A dynamical model of the sporadic meteoroid complex. *Icarus, 201,* 295–310.
Wiegert P. A., Brown P. G., Weryk R. J., et al. (2013) The return of the Andromedids meteor shower. *Astron. J., 145,* 70.
Wiegert P., Clark D., Campbell-Brown M., et al. (2017) Minor planet 2017 MB 1 and the Alpha Capricornids meteor shower. *Central Bureau Electronic Telegrams (CBET), 4415,* 1.
Wiegert P., Brown P., Pokorný P., et al. (2020) Supercatastrophic disruption of asteroids in the context of SOHO comet, fireball, and meteor observations. *Astron. J., 159,* 143.
Williams I. P. and Wu Z. (1993) The Geminid meteor stream and asteroid 3200 Phaethon. *Mon. Not. R. Astron. Soc.*, 262, 231–248.
Williams I. P., Jopek T. J., Rudawska R., et al. (2019) Minor meteor showers and the sporadic background. In *Meteoroids: Sources of Meteors on Earth and Beyond* (G. O. Ryabova et al., eds.), pp. 210–234. Cambridge Univ., Cambridge.
Ye Q. (2018) Meteor showers from active asteroids and dormant comets in near-Earth space: A review. *Planet. Space Sci., 164,* 7–12.
Ye Q. and Wiegert P. A. (2014) Will Comet 209P/LINEAR generate the next meteor storm? *Mon. Not. R. Astron. Soc., 437,* 3283–3287.
Ye Q. and Clark D. L. (2019) Rising from ashes or dying flash? The mega outburst of small Comet 289P/Blanpain in 2013. *Astrophys. J. Lett., 878,* L34.
Ye Q. and Granvik M. (2019) Debris of asteroid disruptions close to the Sun. *Astrophys. J., 873,* 104.
Ye Q., Wiegert P. A., Brown P. G., et al. (2014) The unexpected 2012 Draconid meteor storm. *Mon. Not. R. Astron. Soc., 437,* 3812–3823.
Ye Q., Brown P. G., Bell C., et al. (2015) Bangs and meteors from the quiet Comet 15P/Finlay. *Astrophys. J., 814,* 79.
Ye Q., Brown P. G., and Pokorný P. (2016a) Dormant comets among the near-Earth object population: A meteor-based survey. *Mon. Not. R. Astron. Soc., 462,* 3511–3527.
Ye Q., Hui M.-T., Brown P. G., et al. (2016b) When comets get old: A synthesis of comet and meteor observations of the low activity Comet 209P/LINEAR. *Icarus, 264,* 48–61.
Ye Q., Zhang Q., Kelley M. S. P., et al. (2017) 1I/2017 U1 (‘Oumuamua) is hot: Imaging, spectroscopy, and search of meteor activity. *Astrophys. J. Lett., 851,* L5.
Ye Q., Wiegert P. A., and Hui M.-T. (2018) Finding long lost Lexell’s comet: The fate of the first discovered near-Earth object. *Astron. J., 155,* 163.
Ye Q., Knight M. M., Kelley M. S. P., et al. (2021) A deep search for emission from “Rock Comet” (3200) Phaethon at 1 au. *Planet. Sci. J., 2,* 23.
Ye Q., Kelley M. S. P., Bauer J. M., et al. (2023) Comet P/2021 HS (PANSTARRS) and the challenge of detecting low-activity comets. *Planet. Sci. J., 4,* 47.
Zhang Q., Ye Q., Vissapragada S., et al. (2021) Preview of Comet C/2021 A1 (Leonard) and its encounter with Venus. *Astron. J., 162,* 194.
Zhuang T.-s. (1977) Ancient Chinese records of meteor showers. *Chinese Astron., 1,* 197–220.

Prialnik D. and Jewitt D. (2024) Amorphous ice in comets: Evidence and consequences. In *Comets III* (K. J. Meech, M. R. Combi, D. Bockelée-Morvan, S. N. Raymond, and M. E. Zolensky, eds.), pp. 823–843. Univ. of Arizona, Tucson, DOI: 10.2458/azu_uapress_9780816553631-ch025.

Amorphous Ice in Comets: Evidence and Consequences

Dina Prialnik
Tel Aviv University

David Jewitt
University of California at Los Angeles

Ice naturally forms in the disordered or "amorphous" state when accreted from vapor at temperatures and pressures found in the interstellar medium and in the frigid, low-density outer regions of the Sun's protoplanetary disk. It is therefore the expected form of ice in comets and other primitive bodies that have escaped substantial heating since formation. Despite expectations, however, the observational evidence for amorphous ice in comets remains largely indirect. This is both because the spectral features of amorphous ice are subtle and because the solar system objects for which we possess high-quality data are mostly too close to the Sun and too hot for amorphous ice to survive near the surface, where it can be detected. This chapter reviews the properties of amorphous ice, the evidence for its existence, and its consequences for the behavior of comets.

1. INTRODUCTION

Almost half a century ago, the idea that cometary ice might be amorphous and thus explain the ubiquitous comet outbursts was proposed by *Patashnick* (1974) in a short *Nature* paper: "Observational evidence indicates that comet outbursts require an internal energy source. If at least the surface of a comet nucleus contains a substantial percentage of amorphous ice, then the phase transition of the amorphous ice to a cubic structure provides a release of energy which may be responsible for the outbursts observed in many comets." The idea was pursued a few years later by *Smoluchowski* (1981), who computed temperature profiles through the nucleus and even alluded to the possibility of self-propagation of the crystallization front.

At the time, other volatiles observed in cometary comae were assumed to be included in the nucleus as ices or in the form of clathrates, although it was recognized that the formation of clathrates would require a much higher pressure upon formation than that prevailing in the solar nebula and that impurities could be occluded in limited amounts. The first to consider gas trapping in amorphous water ice were *Bar-Nun et al.* (1985), by developing a dedicated experimental setup. The important result of their study was that gases could be trapped in large amounts and that they were released from the ice upon crystallization, typically at a temperature of ~140 K, too cold for crystalline water ice sublimation, meaning that crystallization in comets may trigger activity at large heliocentric distances, where solar radiation is too weak. This conclusion was soon confirmed by theoretical studies (*Prialnik and Bar-Nun,* 1987). Since then, numerous studies have addressed experimental, observational, and theoretical evidence concerning amorphous ice, as we shall briefly review in this chapter. Related reviews include those by *Mastrapa et al.* (2013) concerning amorphous ice; by *Gudipati et al.* (2015), who focused more generally on laboratory studies of ice; and by Guilbert-Lepoutre et al. (see the chapter in this volume) regarding the structure of cometary nuclei.

1.1. What is Amorphous Ice?

Amorphous water ice is a metastable form of ice, produced when ice is deposited at very low temperatures. The individual molecules have insufficient energy (mobility) to reorient themselves into more energetically favorable positions and a highly disordered solid is formed. Upon warming, the molecules rearrange themselves into lower-energy orientations and somewhat more ordered structures are produced. This process is called annealing, and it ends when a fully ordered crystalline lattice emerges. Macroscopic samples of amorphous and crystalline ice are shown in Fig. 1.

Amorphous ice may include a large fraction of impurities by trapping them in the structureless and porous matrix. When it undergoes crystallization, these impurities are, in part, expelled from the crystal lattice and, in part, retained, more tightly bound. Crystalline ice may occlude at most one foreign molecule per six H_2O molecules (*Whipple and Huebner,* 1976; *Gudipati et al.,* 2015) to form clathrate-hydrates. Trapping processes will be discussed in section 2.

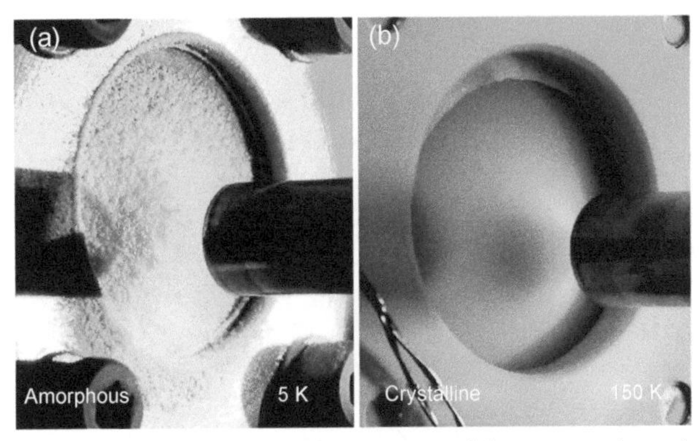

Fig. 1. Water ice samples prepared by vapor deposition onto a 2.5-cm-diameter cold finger at **(a)** 5 K, where amorphous ice is the result and **(b)** 150 K, where the ice is crystalline. Note the granular surface of the amorphous ice indicating its fluffy nature relative to the much smoother surface of the crystalline ice. Figure courtesy of M. Gudipati.

1.2. Is Cometary Ice Amorphous?

There are several reasons why it is reasonable to assume that cometary ice is amorphous, or at least was amorphous when the nucleus formed. First, interstellar ice grains — the building blocks of comet nuclei — are made of amorphous ices (*Öberg et al.,* 2011; *van Dishoeck et al.,* 2013). Secondly, comets eject large amounts of volatiles of various species, and amorphous ice is more capable of trapping larger quantities of impurities than crystalline ice, forming clathrates. Finally, comets are small porous bodies that have spent most of the last 4.5 G.y. in cold storage in the Kuiper belt and Oort cloud. Except perhaps for their surface layers, they have experienced little thermal or structural alteration and hence could have preserved their primordial composition. For example, *A'Hearn et al.* (2008) argue, based on the Deep Impact mission results, that pristine material may be found below a meter-thick surface layer on short-period comets, indicating very low internal temperatures (see also *Herman and Weissman,* 1987). We return to this point in section 4.4. Experiments carried out by *Sandford and Allamandola* (1988) with H_2O and CO mixtures deposited at a low temperature and pressure have shown by infrared spectra that the ice formed is amorphous in structure rather than crystalline or clathrate. Furthermore, *Ciesla* (2014) has shown that even if ice condensed in the solar nebula at sufficiently high temperatures for a crystalline structure to form, dynamical evolution of the icy grains in the outer part of the nebula would cause this ice to be lost and reformed in the amorphous state.

Nevertheless, there is still some controversy in the literature regarding the question of whether cometary ice is amorphous or crystalline, and several studies have argued that the amorphous form of ice may not be a necessary prerequisite (*Marboeuf et al.,* 2012). This point is thoroughly reviewed by *Gudipati et al.* (2015); we will return to it in the conclusions of section 4. Observational evidence for amorphous ice in comets and related objects will be described in section 3. The consequences of the presence of amorphous ice in comet nuclei concerning cometary behavior will be considered in section 4, and prospects for the future in section 5.

2. PROPERTIES OF AMORPHOUS ICE: FROM LABORATORY EXPERIMENTS

2.1. Volatile Trapping and Release

A distinguishing characteristic of amorphous ice is its large surface area per unit mass (cf. Fig. 2), giving it the ability to trap substantial quantities of other gases present upon its formation. This has been experimentally demonstrated by forming amorphous ice at low temperatures and pressures in the presence of other gases, then measuring the progressive expulsion of these gases as the sample temperature is raised (*Bar-Nun et al.,* 1985, 1987, 1988; *Notesco et al.,* 2003). These experiments reveal ice areas per unit mass from ~90 $m^2 g^{-1}$ to ~300 $m^2 g^{-1}$ (*Schmitt and Klinger,* 1987; *Yokochi et al.,* 2012), and even as large as ~400 $m^2 g^{-1}$ (*Mayer and Pletzer,* 1986), and densities of 0.6–0.7 g cm^{-3} (*Bar-Nun et al.,* 1985; *Sandford and Allamandola,* 1988; *Kouchi et al.,* 2016).

Very large trapping efficiencies have been reported, particularly at low temperatures of relevance to the formation of comets. For example, *Bar-Nun and Kleinfeld* (1989) started at $T < 30$ K with 1:1 CO:water ratios in gas, and so much CO was trapped in the amorphous ice formed that in the emitted gas, $CO/H_2O > 1$. Trapping in amorphous ice might be particularly important for the noble gases, which, being chemically unreactive, are difficult to retain otherwise (e.g., *Bar-Nun et al.,* 2007; *Ciesla et al.,* 2018). However, labora-

Fig. 2. The microstructure of amorphous water ice (25 monolayers thickness); the column-like structure at the lower temperatures becomes smoother with increasing temperature, until eventually an entirely smooth structure is obtained at 140–150 K. From *He et al.* (2019).

tory measurements are taken under physical conditions quite different from those prevailing in the interstellar medium and the outer regions of the protoplanetary disk where cometary ice likely accreted. Therefore, some consideration of the nature of the trapping mechanism is appropriate.

Two distinct modes of gas trapping in ice are potentially important. First, as described by *Kouchi et al.* (1992), ice forms in the amorphous state when its hop length across the surface is small compared to the spacing of the ice lattice. Equivalently, molecules are adsorbed on the ice surface at the point of impact and held for a residence time $t_r = \nu^{-1} \exp[E/(kT)]$, where $\nu \sim 10^{12}$ s^{-1} is the vibrational frequency of the molecule in the surface potential well that holds it, E is the binding energy, k is Boltzmann's constant, and T is the temperature. Adsorption thus strongly favors low temperatures. For example, for argon on water ice, $E = 1.4 \times 10^{-20}$ J (*Ciesla et al.,* 2018), giving $t_r = 420$ s at T = 30 K and falling by 7 orders of magnitude to only $t_r = 2 \times 10^{-5}$ s when doubling the temperature to T = 60 K. Secondly, molecules can also be physically trapped by burial under later-arriving monolayers of ice. These two modes of gas trapping have been investigated by *Martin et al.* (2002). Whereas adsorption depends exponentially on the temperature of the sticking surface, burial depends as well on the density-dependent rate of growth of the ice, which determines how long an enveloping monolayer takes to accrete. The efficiency of gas trapping in amorphous ice is therefore a function of both temperature and gas density. The strong dependence on temperature in one set of experiments is illustrated in Fig. 3.

While a detailed balance treatment is needed to properly account for the impact and escape of water molecules at the growing ice surface, we can obtain a rough estimate from t_r, above. If we take, say, $\ell \sim 1$ Å as the order of magnitude scale of a molecule (the van der Waals radius of H_2O is 1.7 Å), then burial is possible if the ice accretes at a rate much larger than ℓ/t_r. For example, argon at 30 K can be buried if the ice accretes at rates $\ell/t_r \gg 10^{-5}$ µm min^{-1}. The laboratory experiments by Bar-Nun and collaborators were conducted at astrophysically relevant temperatures as low as 10 K but at ice growth rates 10^{-5}–10^{-1} µm min^{-1}, orders of magnitude larger than likely to be found in nature (*Cuppen and Herbst,* 2007). *Yokochi et al.* (2012) found that the trapping efficiency depends on the temperature and the partial pressure of the trapped species and suggested that the results of *Bar-Nun and Kleinfeld* (1989) could not be easily interpreted because of experimental concerns. In addition, we should expect strong abundance gradients with respect to the formation distance in the protoplanetary disk, since both temperature and pressure vary radially. High temperature in the inner disk and low pressure in the outer disk suggest the existence of a critical distance at which gas-trapping efficiencies by amorphous ice should be maximized.

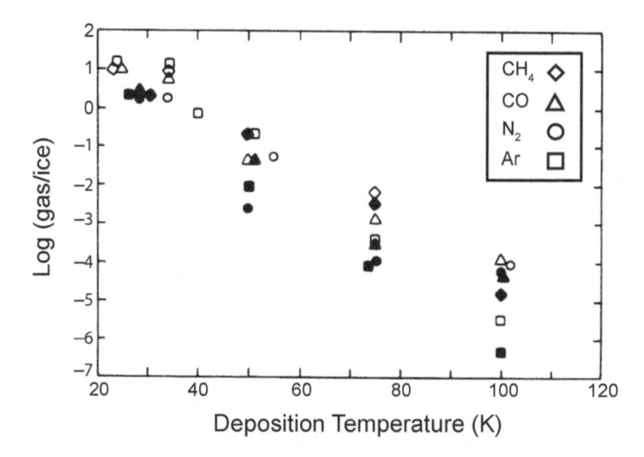

Fig. 3. Total amount of trapped gas vs. deposition temperature for water vapor/gas mixtures: open symbols — deposition of a 1:1 ratio of a single gas species; filled symbols — deposition of gas mixtures, H_2O:CH_4:Ar:CO(or N_2) = 1:0.33:0.33:0.33. Adapted from *Bar-Nun et al.* (1988).

Experimentally, gas release from amorphous ice occurs at a strongly temperature-dependent but non-monotonic rate. The ice transforms into the cubic form at ~140 K and into the hexagonal form at 160 K (*Jenniskens and Blake,* 1994), with each phase transition leading to a pulse of expelled gases. When molecules such as CO, CH_4, N_2, and Ar are trapped in the ice below 30 K, they are released upon warming in several temperature ranges (see Fig. 4) starting at 23 K, where the gas frozen on the surface evaporates, to 44 K, where the monolayer of adsorbed gas evaporates. At ~60 K, gas release ceases altogether, as the remaining gas is locked in an impermeable amorphous ice matrix. It resumes only at 140 K, where the ice becomes temporarily more mobile during its transformation into the cubic form but ceases when the transformation is completed.

The next chance for gas to escape occurs at 160 K, where the cubic ice transforms into a hexagonal crystal. The remaining trapped volatile forms a clathrate (*Sandford and Allamandola,* 1990; *Blake et al.,* 1991; *Marboeuf et al.,* 2012). When water ice undergoes crystallization, it is justified to assume that the rate of clathrate formation is equal to that of crystallization (*Marboeuf et al.,* 2012) because the mobility of the water molecules is high, and the amount of volatile molecules released from the amorphous ice phase is sufficient to form cages in the crystalline phase. Finally, at ~180 K, water vapor and gas evolve simultaneously when the clathrate evaporates. If the gas is trapped in the ice between 40 K and 80 K, the peaks between 140 K and 180 K diminish by orders of magnitude, depending on the formation temperature and on the trapped species.

2.2. Formation of Hydrocarbon Aggregates

Besides volatiles, amorphous ice may also trap surprisingly large hydrocarbon molecules. *Lignell and Gudipati* (2015) used $C_{16}H_{10}$ (four fused carbon rings known as pyrene) to show that upon crystallization, the hydrocarbons form aggregates that remain trapped in the crystalline ice. Subsequent heating leads to emission of the aggregates. Experiments with even more massive C_{60} (Buckminster-

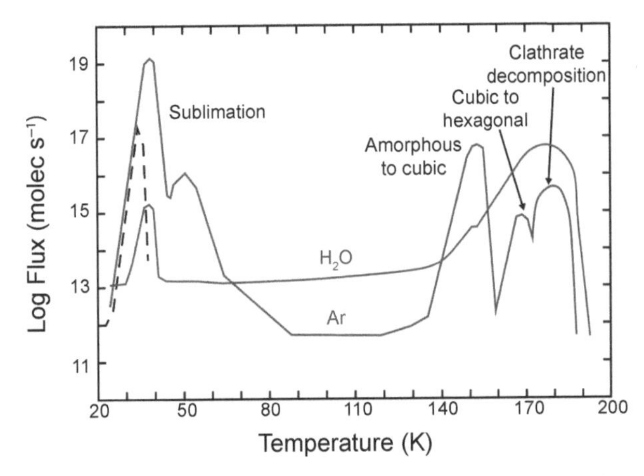

Fig. 4. Fluxes of evolved argon and water vs. temperature in various ranges. Dashed line: sublimation of frozen argon from an ice-free plate, shown for comparison. Adapted from *Bar-Nun et al.* (1987).

fullerene) show similar aggregation upon crystallization (*Halukeerthi et al.,* 2020). In comets, the mobilization and aggregation of hydrocarbons at the phase transition may enhance the formation of an outer crust when the organic molecules become mixed with silicate particles. Chemical processing through micrometeorite impact heating (*Nelson et al.,* 2016) adds to the complexity of this process.

2.3. The Latent Heat of Crystallization

The crystallization of amorphous ice that takes place upon heating, first into a cubic structure and then into a hexagonal lattice, is an irreversible process. The timescale for crystallization, τ_{CR}, is strongly temperature-dependent

$$\tau_{CR}(T) = 9.7 \times 10^{-14} \exp^{5370/T} [s] \tag{1}$$

as was determined experimentally by *Schmitt et al.* (1989).

When pure ice is involved, crystallization is clearly an exothermic process, as the entropy decreases. The first and widely used estimate of the latent heat released, $H_a = 90$ kJ/kg, was obtained experimentally by *Ghormley* (1968) based on temperature measurements (warming curves). A somewhat higher value, 118 kJ/kg, obtained by comparing the mutual binding energies of H_2O molecules in amorphous and crystalline ice, was derived experimentally by *Sandford and Allamandola* (1988, 1990), based on measurements of release rates of occluded gases in amorphous ice samples (sticking coefficients).

The question of whether the process remains exothermic when impurities are occluded in the amorphous ice has been debated for many years, particularly since the experimental results of *Kouchi and Sirono* (2001), who claimed that the process becomes endothermic if even only 3% of occluded CO is released, an extreme conclusion that has been regarded with caution.

The basic question is whether the occluded gas is just trapped in cages of the highly porous amorphous ice structure or rather is bound to water molecules. In the former case, there would be a small amount of energy release when the gas escapes, resulting from the difference in the specific internal energies of the gas and the amorphous ice. For CO, for example, it would amount to ~18 $f_{CO,t}$ kJ/kg, where $f_{CO,t}$ is the fraction of trapped CO. Many models of comet nuclei (e.g., *Prialnik and Bar-Nun,* 1990; *Espinasse et al.,* 1991; *Tancredi et al.,* 1994) have adopted this approach.

There is experimental evidence, however, that the impurities are bound to water molecules. Here we have to distinguish between the fraction of gas that remains trapped in the annealed ice and the fraction of escaping gas. Accordingly, in the case of CO, the energy absorbed in gas release should be proportional to $f_{CO,r}$, the fraction of released CO, and to the binding energy of CO on H_2O, $H_{a,r}$ = 517 kJ/kg (*Sandford and Allamandola,* 1990) or 418 kJ/kg, as obtained by *Manca and Allouche* (2001) from quantum calculations for the absorption of small molecules on ice. At the same time, latent heat should be released due to the increased binding energy of the remaining CO. According to the experimental results of *Sandford and Allamandola* (1988), the difference between the volume binding energies of CO on H_2O in amorphous and crystalline ice yields a latent heat of $H_{a,c} \approx 74$ $f_{CO,c}$ kJ/kg, where $f_{CO,c}$ is the fraction of CO that remains trapped when the ice crystallizes (keeping in mind the large error bars). As a simple, illustrative example, if 10% of CO (by mass) is trapped in amorphous ice, half of which remains trapped when the ice crystallizes, while the other half escapes — in agreement with Fig. 4 — the process will still be highly exothermic. The heat released

$$H = [1-(f_c + f_r)]H_a + f_c H_{a,c} - f_r H_{a,r} \tag{2}$$

will be in this case about 60 kJ/kg. Some studies (e.g., *Davidsson,* 2021; *Enzian et al.,* 1997) adopt the ad hoc assumption that the amount of energy absorbed in gas release is equal to the latent heat of sublimation of the trapped species (for CO, 285 $f_{CO,t}$ kJ/kg), which results in a moderate reduction of the latent heat of pure ice crystallization. *González et al.* (2008) consider the effect of different values of latent heat. The escape of CO (and other species) may be a combination of diffusion through the ice and removal from the ice surface (*Sandford and Allamandola,* 1988), which are energetically different, and may depend on the formation process of the mixture.

In conclusion, reliable quantitative data for the heat released in crystallization of gas-laden amorphous ice are still needed from laboratory experiments. Nevertheless, for the typical amounts of trapped volatiles in amorphous ice that are released upon crystallization, it is safe to assume the process to be exothermic and to play a significant role in cometary outbursts, as we shall describe in section 4.

2.4. Thermal Properties of Amorphous Ice

The thermal conductivity of amorphous ice has been determined both experimentally and from theoretical considerations. The formulae provided by *Klinger* (1980) and *Klinger* (1981) have gained widespread use. *Klinger* (1980) derived a theoretical expression from the classical phonon theory

$$K = \frac{1}{3} v_s \ell_{ph} c\rho \tag{3}$$

where $v_s = 2.5 \times 10^3$ m s^{-1} is the speed of sound in ice and $\ell_{ph} = 5 \times 10^{-10}$ m is the phonon mean free path, ρ is the ice density, and c is the heat capacity, given by an empirical relation c = 7.49T + 90 J kg^{-1} K^{-1}, derived from measurements (*Giauque and Stout,* 1936). Substitution gives K ≈ 0.4 W m^{-1} K^{-1} in the temperature range 70 ≤ T ≤ 135 K. *Kouchi et al.* (1992) measured in their laboratory experiments a much lower conductivity of K = (0.6–4.1) × 10^{-5} W m^{-1} K^{-1} in the range 125 ≤ T ≤ 135 K. *Andersson and Suga* (1994, 2002) found an experimental value for the conductivity of non-porous, low-density, amorphous water ice that is similar to the value derived by Klinger. In a unique experiment with a relatively large sample of fluffy amorphous ice, *Bar-Nun and Laufer* (2003) measured an intermediate value between those of Klinger and Kouchi. This may suggest that differences between experimentally derived thermal conductivity coefficients may be caused by different porosity of the samples.

While additional measurements are clearly needed, the thermal conductivity of amorphous ice is one or more orders of magnitude smaller than that of crystalline ice (cubic or hexagonal). This has significant consequences for the thermal evolution of comet nuclei, since the timescale is inversely proportional to K, and the skin depth is proportional to $\sqrt{K}$.

3. OBSERVATIONAL EVIDENCE

3.1. Stability

The radiation equilibrium temperature of a solar system body depends on the distance from the Sun, as well as optical properties (albedo, emissivity), thermodynamic properties (thermal conductivity, density, and specific heat capacity), and also rotational properties (rotation period, spin axis orientation). For most objects, these parameters are unknown or poorly constrained, and the surface temperature cannot be accurately defined. As a guide, however, we consider T_{BB}, the temperature of a spherical, isothermal body, as a lower limit to the dayside temperature, and T_{SS}, the temperature of a flat plate oriented normal to the Sun, as an upper limit.

Equating the power absorbed from the Sun to the power radiated in equilibrium, we have

$$T_{iso} = \left(\frac{L_\odot (1-A)}{16\pi\varepsilon\sigma r_H^2} \right)^{1/4} \tag{4}$$

and

$$T_{SS} = \sqrt{2} T_{iso} \tag{5}$$

in which $L_\odot = 3.83 \times 10^{26}$ W is the luminosity of the Sun, σ is the Stefan-Boltzmann constant, A and ε are the Bond albedo and emissivity, respectively, and r_H is the heliocentric distance. Substituting A = 0, ε = 1 into equations (4) and (5) and expressing r_H in astronomical units (1 au = 1.5 × 10^{11} m) gives blackbody temperatures

$$T_{BB} = \frac{278}{\sqrt{r_H}} \text{ and } T_{SS} = \frac{393}{\sqrt{r_H}} \tag{6}$$

(in Kelvin). The crystallization time, τ_{CR}, is shown as a function of heliocentric distance for T_{BB} and T_{SS} in Fig. 5. The figure shows (a) a substantial range of crystallization timescales at any heliocentric distance caused by the exponential temperature dependence in equation (1), and (b) that temperatures at Kuiper belt distances (r_H ≥ 30 au) are too low for crystallization to have occurred in the 4.5 × 10^9-yr age of the solar system, no matter which temperature model is used. On this basis, we should reasonably expect that the Kuiper belt objects (KBOs), at least those small enough to have escaped self-heating due to the energy of formation and to radioactive decay, could retain amorphous ice. The nuclei of comets, being derived from the Kuiper belt and the (even colder) Oort cloud reservoirs, are also expected to contain amorphous ice, subject to uncertainties about the temperature-time histories of these bodies.

3.2. Diagnostics

For practical reasons, most spectroscopy of icy bodies has been done at optical and near infrared (1 to 2.5-μm) wavelengths. The vibrational and overtone bands of water and other simple molecules are best accessed in the near-infrared, but differences between amorphous and crystalline ice in this spectral region are, for the most part, subtle. They occur because, while the vibrational and overtone bands due to the OH bond in water molecules are substantially the same, smaller differences exist in the degree of hydrogen bonding between water molecules. Unfortunately, many of the important differences occur in regions of the electro-

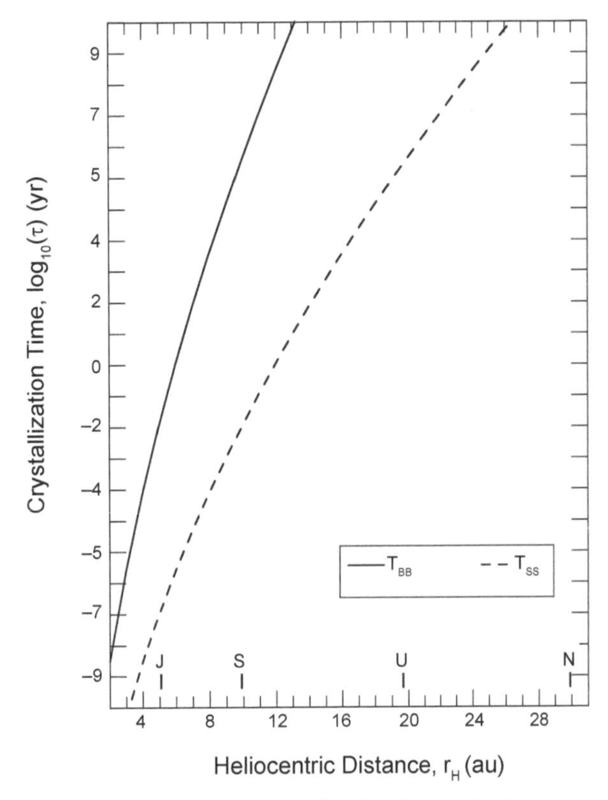

Fig. 5. Logarithm of the crystallization timescale (from equation (1)) as a function of heliocentric distance for high (T_{SS}) and low (T_{BB}) temperature limits given by equation (6). The orbital radii of the giant planets are marked.

magnetic spectrum that are traditionally difficult to access. This is because the bands in water ice largely overlap the corresponding features from water vapor in the atmosphere, rendering groundbased study difficult. The optically thick 3-μm band is strongly affected and can be studied only from the most stable and dry high-altitude sites or, better, from space. A weaker crystalline ice band at 1.65 μm falls in an atmospherically transparent region and is more frequently used as a convenient diagnostic.

Use of this feature is not without problems, however. The band center, width, and depth are all functions of the ice temperature as well as the crystallinity (*Fink and Larson,* 1975; *Grundy and Schmitt,* 1998) and have, indeed, been used as a spectroscopic thermometer (e.g., *Grundy et al.,* 1999). The band is half as deep at 100 K as it is at 20 K (Fig. 6), meaning that without additional information, a weak 1.65-μm band could be due to the presence of amorphous ice or simply a result of higher temperatures. Radiative transfer (scattering and absorption) in granular ice can influence the band parameters depending on the grain size (*Hansen and McCord,* 2004), as can energetic particle irradiation (*Mastrapa and Brown,* 2006). Two of these effects are shown in Fig. 7, where **(a)** shows the spectrum of amorphous ice deposited at 50 K, **(b)** shows the effect of heating the ice to 160 K (to crystallize it) then cooling back down to 50 K, and **(c)** shows the additional effect on the same sample of irradiation by 1-MeV protons up to 16 eV/molecule. The 1.65-μm band is clearly associated with the crystalline phase and diminished, but not erased, by bond destruction due to particle irradiation in spectrum **(c)**. The degree to which the 1.65-μm band is weakened by irradiation varies inversely with the ice temperature.

Figure 8 plots the reflectivity of granular ice in the 3-μm region (*Mastrapa et al.,* 2013). This extremely deep absorption feature shows crystallinity-dependent band shifts and reflectivity differences (e.g., especially near the 3.1-μm Fresnel peak) that will be of practical value once spacebased observations at this wavelength are routine.

3.3. Icy Satellites

The Galilean satellites Europa, Ganymede, and Callisto are natural places to examine the effects of irradiation on ice crystal structure (*Hansen and McCord,* 2004). All three have surfaces rich in water ice and orbit deep within Jupiter's magnetosphere, with orbital radii 9.4 R_J, 15.0 R_J, and 26.4 R_J, respectively, where $R_J = 7.14 \times 10^7$ m is the radius of Jupiter. The jovian magnetosphere rotates with the planet at about 10 hr, which is short compared to the 3.6-, 7.2-, and 21.6-day orbital periods of the satellites. As a result, magnetospheric particles preferentially impact the trailing hemispheres of each satellite, leaving the leading hemispheres relatively unirradiated. Furthermore, because the satellites rotate synchronously, this hemispherically asymmetric bombardment is imprinted as a fixed longitudinal pattern on each satellite.

Figure 9 shows the distribution of amorphous ice on the surface of Europa (*Ligier et al.,* 2016; cf. *Berdis et al.,* 2020). The map shows a clear dichotomy, with the more amorphous ice (less-deep 1.65-μm band) concentrated on the trailing hemisphere (longitudes centered near 90°) where the bombarding magnetospheric particle flux is highest. This observation by itself provides solid evidence for the amorphization of ice by energetic particles in the natural

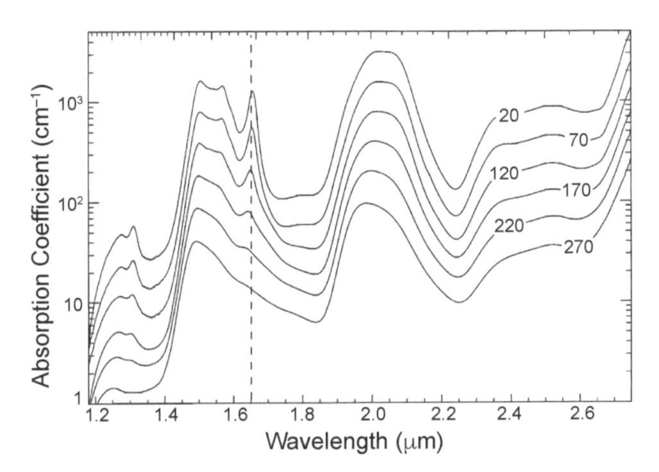

Fig. 6. Temperature dependence of the absorption coefficient in crystalline water ice from 20 K to 270 K, as labeled. Note the strong changes in the 1.65-μm band, marked by a vertical dashed line. The curves are vertically displaced for clarity. Modified from *Grundy and Schmitt* (1998).

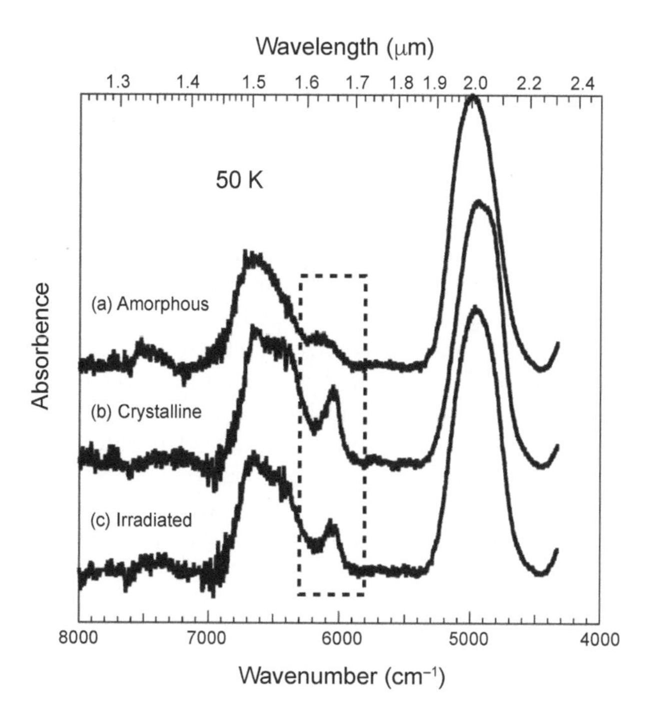

Fig. 7. Three near-infrared spectra showing **(a)** amorphous ice grown at 50 K, **(b)** crystalline ice obtained by heating the sample to 160 K and then cooling to 50 K, and **(c)** the crystallized sample after irradiation by 1-MeV protons to 16 eV per molecule. The dotted box highlights variations in the 1.65-μm band. Modified from *Mastrapa and Brown* (2006).

environment. According to these authors, temperatures on Europa are such that amorphous ice should crystallize on timescales ~10 yr, while the magnetospheric ion flux amorphizes crystalline ice in only ~1 yr.

The magnetospheric particle flux also follows a strong radial gradient with respect to distance from Jupiter, decreasing by a factor of about 300 from Europa to Callisto. It is slightly more complicated to compare ice spectra from the three satellites because not only does the particle flux vary with orbital distance, but the temperatures of the satellites differ because of their different optical properties, principally, the Bond albedo, which controls the radiative temperature balance with sunlight and varies from 0.68 (Europa) to 0.43 (Ganymede) and 0.22 (Callisto). The surface temperature differences are modest (few × 10 K), but because the crystallization rate is exponentially dependent on the temperature, it can have a large effect. Nevertheless, *Hansen and McCord* (2004) concluded that ice on Europa, where the particle fluxes are highest, is on average more amorphous than is ice on Callisto, while Ganymede is an intermediate case. These authors also found evidence for a vertical crystallinity gradient in the top millimeter of the ice. This inference derives from the observation that the 1.65-μm band, which has an absorption length ~1 mm, gives a larger crystalline fraction than the 3.1-μm band, which has an absorption length of only ~10 μm. A vertical crystallinity gradient is expected because of the steep energy spectrum; energetic particles able to penetrate deep into the ice are comparatively rare. For example, the stopping length in ice of 1-MeV electrons is ~10 μm, while their flux is ~10^4× that of the 10-MeV electrons needed to reach ~1-mm depth (*Cooper et al.,* 2001).

The crystalline feature is also present in the spectra of other icy satellites including those of Uranus (*Grundy et al.,* 2006; *Guilbert et al.,* 2009) and Pluto's satellite Charon (*Brown and Calvin,* 2000).

3.4. Kuiper Belt Objects

The KBOs are cold enough for amorphous ice to persist over the age of the solar system. Paradoxically, high signal-to-noise ratio astronomical spectra instead reveal the 1.65-μm absorption band characteristic of crystalline ice (*Jewitt and Luu.* 2004; *Merlin et al.,* 2007; *Trujillo et al.,* 2007; *Barkume et al.,* 2008). Table 1 lists measurements of the "crystallinity factor" (essentially, the fraction of crystalline ice) determined from narrowband photometry of the 1.65-μm feature (*Terai et al.,* 2016). All the objects show substantial crystallinity, regardless of their temperature over the 48 K to 82 K range. The objects with the smallest (90482 Orcus) and the largest (315530 2008 AP129) crystallinity factors both have temperatures so low that amorphous ice, if present, would be indefinitely stable by equation (1).

The detection of the crystalline ice feature is further puzzling because the surfaces of KBOs are exposed to cosmic-ray bombardment, whose effect (just as with particle bombardment of the Galilean satellites) is to destroy the crystalline lattice and convert the ice back to the amorphous form. The timescale for this damage to occur is ~10^6 yr, very short compared to the age of the solar system. At first, this observation was thought to indicate that the surfaces have

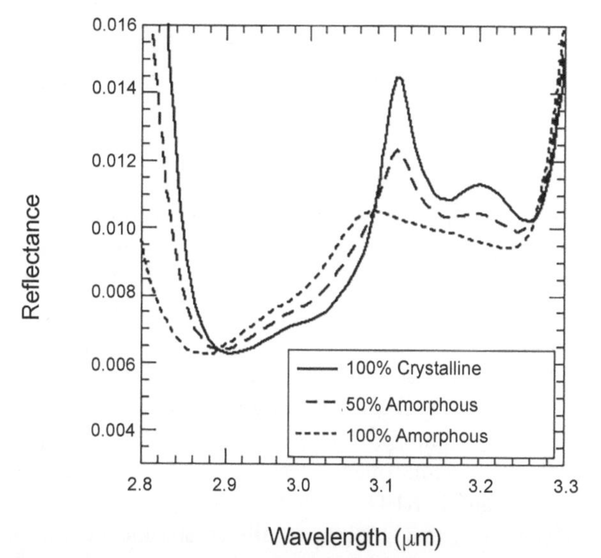

Fig. 8. Three micrometer spectra showing the sensitivity to amorphous vs. crystalline structure in ice at 100 K, with a 20-μm grain size. Modified from *Mastrapa et al.* (2013).

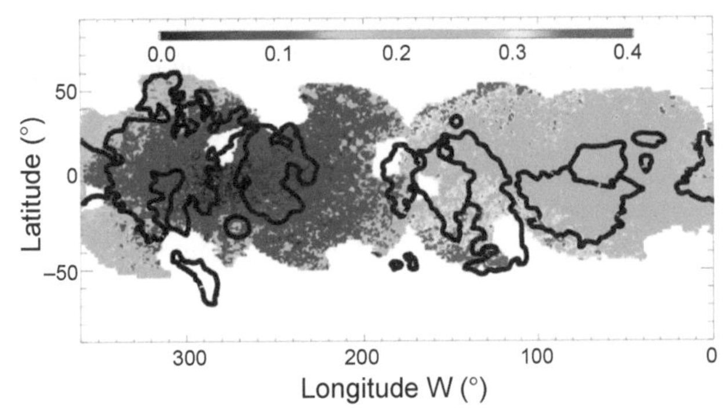

Fig. 9. Map of the water ice crystallinity on Europa, largely deduced from observations in the 1.65-μm band. The amorphous fraction is color coded from black (0) to red (0.4); the ice on the trailing hemisphere (longitudes 0° to 180°) is more amorphous than on the leading hemisphere (180° to 360°). Irregular shapes outlined in black lines show geological surface units, with which the amorphous/crystalline pattern shows no simple correlation. Adapted from *Ligier et al.* (2016).

been recently heated, perhaps through cryovolcanism or, more likely, through micrometeorite bombardment heating (*Porter et al.*, 2010).

One clue to the observational prevalence of crystalline ice comes from spectral models showing that amorphous ice can be difficult to detect if crystalline ice is present. For example, Fig. 10 shows that the addition of 10% amorphous ice to an otherwise crystalline ice model creates only tiny differences in the reflectivity spectrum that require high signal-to-noise ratios to be detected. For this reason, the absence of spectral evidence for amorphous ice is not necessarily evidence of its absence.

Laboratory experiments provide supporting evidence for this conclusion. The degree to which irradiation can scramble the crystalline structure in ice is itself a function of temperature (see Fig. 11). At low temperatures, the damage done to the crystal structure by energetic particles is evidently severe and long-lived, while at higher temperatures, thermal vibrations result in partial resetting of the bonds, offsetting amorphization by the destructive ionizing particle flux (*Zheng et al.*, 2009, cf. Fig. 11). At 50 K, for example, these experiments show that the ratio of the surviving to the original 1.65-μm band area stabilizes at about $\alpha = 0.6$ after ~0.8 G.y. of equivalent cosmic-ray exposure, while only at 10 K is the crystalline band almost obliterated ($\alpha = 0.1$). Although different experimenters (cf. *Mastrapa et al.*, 2013) obtain results for the dependence of α on T that are different in detail from those in *Zheng et al.* (2009), the basic result that the degree of amorphization by cosmic rays depends on temperature is confirmed. Therefore, the prevalence of the 1.65-μm band in KBOs is not evidence that amorphous ice is absent, only that at least partially crystalline ice must be present. Proper interpretation of the 1.65-μm band requires more detailed knowledge of KBO surface properties than we normally possess.

3.5. Centaurs

Centaurs are recently escaped KBOs, defined as having perihelia and semimajor axes between the orbits of Jupiter and Neptune (excluding those trapped in mean-motion resonances with the giant planets, like the Jupiter and Neptune trojans). These objects are precursors to the Jupiter-family comets and are dynamically short-lived because of scattering by the giant planets. In order for their population to remain in steady-state, the Centaurs must be continually replenished. The source region is thought to be the scattered disk component of the Kuiper belt. These are high-eccentricity objects with perihelia within $\lesssim$10 au of the orbit of Neptune. The latter property renders them susceptible to scattering by Neptune, increasing their orbital eccentricities and giving a finite probability of being scattered into Neptune-crossing orbits that ultimately fall under the influence of Uranus and the gas giants. The Centaurs are scattered among the giant planets until they either collide with a planet or, more frequently, are ejected from the planetary system or injected to Jupiter-family-comet orbits. With perihelia beyond Neptune, the scattered KBOs also have isothermal blackbody temperatures $T_{BB} \leq 50$ K, providing indefinite stability for amorphous ice.

Observationally, we should expect that the physical properties of the Centaurs, for example, the sizes and size distribution, the colors, albedos, and rotational period distributions, should be identical to those of the KBOs in the source region from which they are derived. The best-measured of these physical properties is the optical color (Fig 12). For Centaurs with perihelion distances $q \gtrsim 10$ au, the color distributions of Centaurs and scattered KBOs are indeed comparable. In particular, the $q \gtrsim 10$ au Centaurs and KBOs are unique in the solar system in showing a substantial fraction of objects with ultrared colors (color index B–R > 1.6). On the other hand, Centaurs with $q \lesssim 8$–10 au are depleted in ultrared surfaces, indicating a change in the nature of the surface material at isothermal blackbody temperatures $T_{BB} \sim 90$–100 K. Instead, the optical colors of small-q Centaurs are indistinguishable from those of short-period comets with orbits $q < 5$ au,

TABLE 1. Crystallinity factors (CF) and infrared color temperatures (T) from *Terai et al.* (2016); color temperatures are slightly smaller than T_{SS} from equation (6).

Object	r_H (au)	T (K)	CF
(136108) Haumea	51	48	$0.77^{+0.06}_{-0.05}$
(90482) Orcus	48	49	$0.53^{+0.08}_{-0.09}$
(50000) Quaoar	43	52	$0.82^{+0.15}_{-0.15}$
(315530) 2008 AP129	38	56	$1.00^{+0.00}_{-0.47}$
(38628) Huya	29	66	$0.92^{+0.07}_{-0.93}$
(42355) Typhon	19	82	$0.79^{+0.21}_{-0.27}$

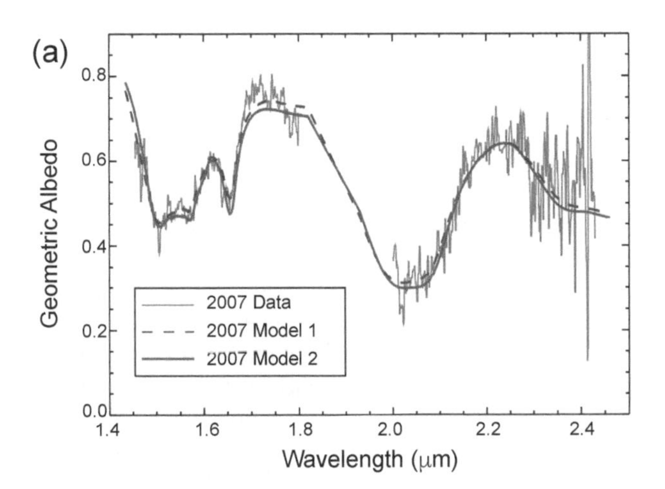

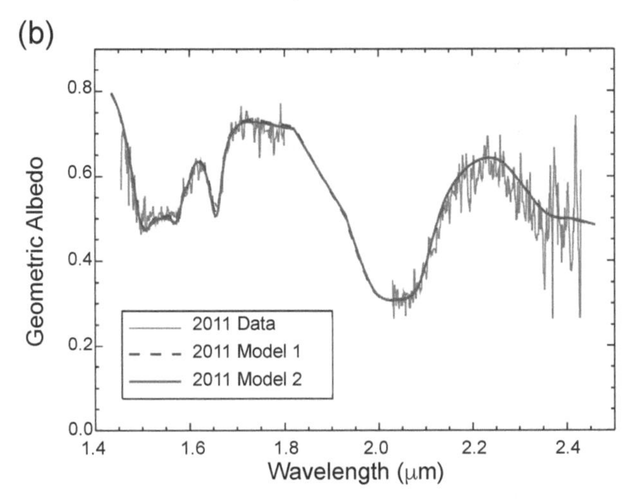

Fig. 10. Near-infrared spectrum of KBO Haumea showing the 1.5-, 1.65-, and 2.0-μm bands of water ice. The dashed blue and solid red lines show models of pure crystalline ice and crystalline ice with a 10% admixture of amorphous ice. Only tiny differences between the models are apparent. From *Gourgeot et al.* (2016).

in the water-sublimation regime. Coincidentally, mass loss from some Centaurs has been detected and is found preferentially in objects with $q \gtrsim 10$ au. The similarity between the perihelion distance at which activity begins and the ultrared matter disappears suggests that activity and color change might have the same origin. A likely mechanism is surface blanketing caused by the deposition of suborbital debris (*Jewitt,* 2015).

Crystallization of exposed amorphous ice occurs on timescales comparable to the orbit period at about 8–15 au (*Guilbert-Lepoutre,* 2012) and therefore offers a plausible, but non-unique, explanation of the dichotomy in the Centaur observations. Water ice is involatile at 10 au and cannot drive the observed mass loss from Centaurs. Exposed supervolatile ices (e.g., CO, N_2, and even CO_2) are too volatile in that they sublimate strongly even beyond 10 au, where no activity is detected in even the most sensitive data (*Li et al.,* 2020). They provide no natural explanation for the observed changes at 10 au. Sublimation could be suppressed and delayed by burial of supervolatiles under an inert mantle, but again, there is no natural explanation for the observed critical distance of 8–10 au. Other species less volatile than CO but more volatile than water have been suggested, including H_2S, but the match to the critical distance is poor, and spectral confirmation of these substances has so far failed (*Wong et al.,* 2019).

3.6. Comets

Most comets become bright only when close to the Sun and therefore remain unobserved at large distances. However, near the Sun, surface cometary ice is too warm to exist in the amorphous state, whether or not it is amorphous inside the cometary nucleus. For this reason, attempts to spectroscopically identify amorphous ice must necessarily be focused on distant, intrinsically faint, and observationally unattractive targets. Few such observations exist.

To see where amorphous ice might be detected, if it is present, consider that the best groundbased resolution on comets is limited to a distance $L \sim 10^3$ km and that small cometary dust grains are well-coupled to the gas flow and so leave the nucleus at a speed comparable to the speed of sound, $V \sim 1$ km s^{-1}. Then, we set $\tau_{CR} = L/V$ and solve for T using equation (1). This gives $T \sim 145$ K, corresponding to the isothermal temperature at $r_H \sim 3.7(1-A)^{1/2}$ au, where A is the Bond albedo. Cometary nuclei have $A \sim 0.01$. Amorphous ice grains with this low albedo in comets with $r_H \sim 3.7$ au would sublimate before traveling a single spatial resolution element in groundbased spectra and so could not be detected. *Zubko et al.* (2017) estimates the

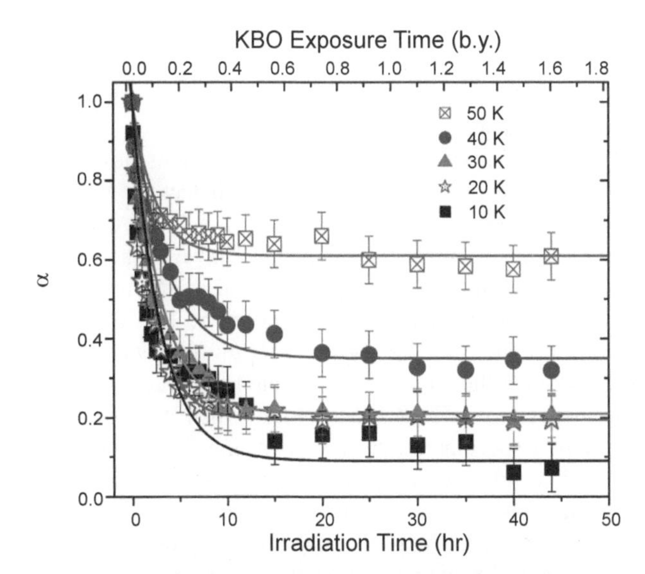

Fig. 11. 1.65-μm band strength variation (plotted as the dimensionless parameter α) as a function of irradiation dose and ice sample temperature. The dose is related to the effective Kuiper belt exposure time on the upper y-axis, scaling from *Cooper et al.* (2003). The 1.65-μm band is most strongly suppressed by a given dose at the lowest temperatures. Figure from *Zheng et al.* (2009).

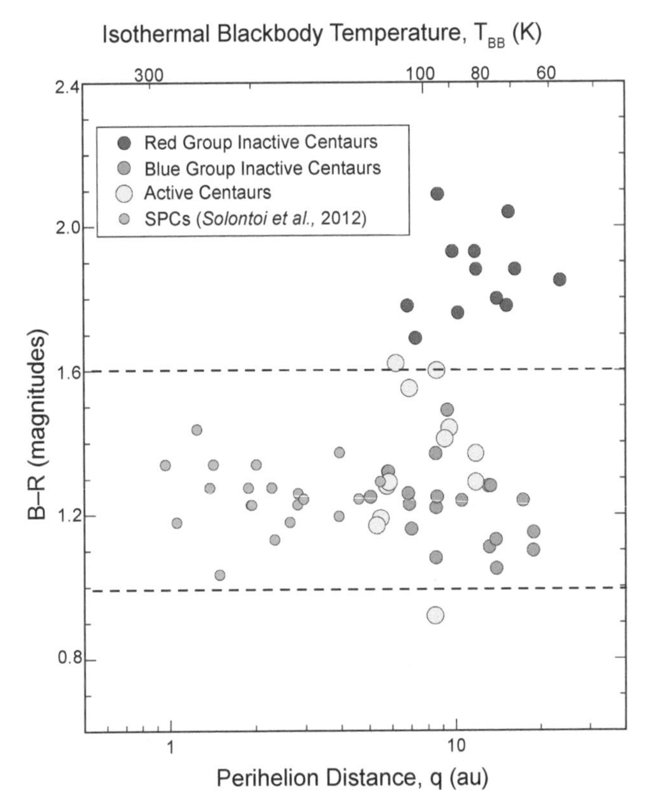

Fig. 12. Optical color vs. perihelion distance for Centaurs and short-period comets, showing a bimodal distribution for q ≳ 8–10 au and a unimodal distribution at smaller q. Dashed horizontal lines show the color of the Sun at B–R = 0.99 and the onset of ultrared colors at B–R = 1.60. Adapted from *Jewitt* (2015).

cometary dust albedo A ~ 0.1–0.2; amorphous ice grains might then be detected down to r_H ~ 3.3 au. Pure ice grains could, in principle, have larger A and survive to smaller distances.

A few relevant spectroscopic observations have been reported. *Davies et al.* (1997) failed to detect the 1.65-μm band in Comet C/Hale-Bopp when at r_H = 7 au and speculated that the ice might be amorphous. However, the 1.5-μm and 2.0-μm bands in their spectra are also very weak, and it is not clear that a significant limit on the 1.65-μm band, or on amorphous ice, can be placed. *Kawakita et al.* (2004) reported that the 1.65-μm crystalline band was missing in the spectrum of C/2002 T7 (LINEAR) when at r_H = 3.52 au. They first speculated that this might mean that the ice is amorphous but later determined that crystalline ice can fit the data equally well, within the observational uncertainties (cf. Fig. 13). The isothermal blackbody temperature at 3.52 au is 148 K, giving a crystallization time τ_{CR} ~ 10 min by equation (1). Outbursting Comet P/2010 H2 (Vales) at r_H = 3.1 au showed a strong 1.65-μm band associated with crystalline ice, which is not surprising given that the spherical blackbody temperature at this distance is T_{BB} = 158 K, and the crystallization time (equation (1)) is only τ_{CR} ~ 55 s (*Yang and Sarid,* 2010).

4. CRYSTALLIZATION AS THE DRIVER OF COMETARY ACTIVITY

4.1. Evolutionary Considerations

Cometary behavior is largely determined by the orbit, and hence activity patterns vary widely among comets; nevertheless, it is possible to assess the effect of crystallization in comets in a general manner, regardless of orbit (*Prialnik,* 1993). A heat wave generated by a crystallization front propagating from the surface of the nucleus inward into a cold isothermal medium of porous ice and generating sublimation causes a sharp temperature gradient to arise between the regions lying in front and behind it, where the temperatures are almost uniform. The uniform low temperature ahead of the front is maintained because the very *low* thermal conductivity of porous amorphous ice prevents heat dissipation. The uniform, relatively high temperature behind it is due to the stabilizing effect of the ice-vapor mixture. The thickness ℓ of the crystallization front may be

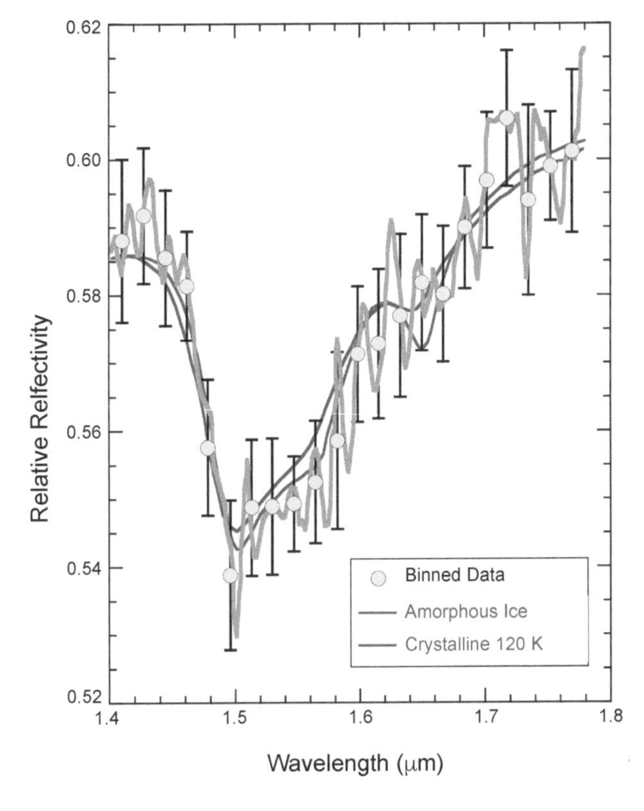

Fig. 13. Near-infrared spectrum of icy grains in the coma of Comet C/2002 T7 at 3.52 au showing the absence of a 1.65-μm band. The raw data are shown as a thick gray line, the yellow circles with error bars show the data binned to resolution 0.016 μm. The spectrum is compared with a model of crystalline ice at 120 K (blue line) and one of amorphous ice at 100 K (red line). Within the noise in the data the model fits are indistinguishable. Adapted from *Kawakita et al.* (2006).

estimated according to the thermal diffusivity $\kappa = K/\rho c$ of the amorphous ice medium into which the front propagates and the time constant τ_{CR} of the reaction, both functions of temperature; defining $\lambda = \tau_{CR}^{-1}$, the velocity of the front $v = \ell/\tau$ is thus obtained

$$\ell(T) = \sqrt{\kappa(T)\tau_{CR}(T)} \text{ and } v(T) = \sqrt{\kappa(T)\lambda(T)} \tag{7}$$

ranging between a few centimeters per year at 100 K and a few kilometers per year around 150 K. Hence, no amorphous ice will be found in any part of a cometary nucleus that has been heated at some point in time to ~140 K. At the surface, this corresponds to the isothermal blackbody temperature at ~4 au.

Given the mass fraction of amorphous ice X_a and the density ρ, the resulting flux of volatiles, for example, CO, released from the amorphous ice may be estimated by

$$J_{CO} = vf_{CO,r}X_a\rho/m_{CO} \tag{8}$$

where m_{CO} is the molecular mass. At T = 140 K, $J_{CO} \sim 10^{19}$ molecules s^{-1} m^{-2} within an order of magnitude, depending on the thermophysical properties assumed and on composition. Although the crystallization rate is a continuous function of temperature, laboratory experiments (*Bar-Nun et al.,* 1987), analytical considerations (*Prialnik,* 1993), and numerical simulations (*Tancredi et al.,* 1994) all show that the process actually occurs within a narrow range of temperatures around 140 K. The reason is that the steep rise of λ with temperature in this range, as shown in Fig. 14, results in depletion of the amorphous ice content before the temperature rises any further. Consequently, one may speak of a crystallization temperature T_c with an uncertainty of less than 10%.

The trapped gas released upon crystallization flows both outward to the surface and inward, and if it encounters sufficiently cold regions, it may refreeze, creating a volatile-ice-enriched layer. Upon subsequent heating, this layer will evaporate independently of crystallization. Hence, the composition of volatile ices mixed with water ice should not necessarily be pristine. The gas flowing outward may be impeded from escaping by the presence of a dense and less permeable outer crust, the result being a high-pressure gas pocket. A model of such a configuration is shown in Fig. 15.

4.2. Spurts of Crystallization and Runaway

As noted in section 2.3, the maximum energy released by the crystallization of pure amorphous water ice is $H_a \sim 10^5$ J kg^{-1}. This is only ~1/25 of the latent heat of sublimation, 2.5×10^6 J kg^{-1}, and for this reason, crystallization is not capable of driving substantial sublimation by itself. However, the warming of the ice by H_a is capable of driving a crystallization runaway, potentially leading to the explosive release of trapped gases, as we now describe.

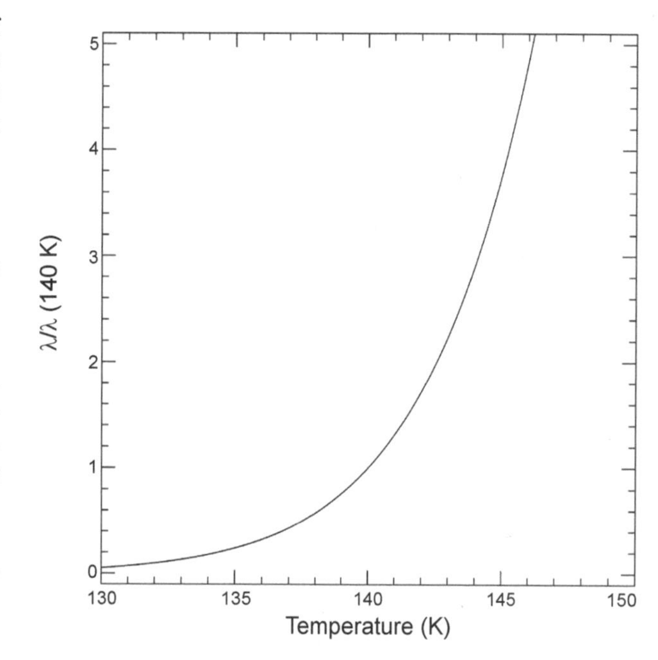

Fig. 14. Rate of crystallization, $\tau_{CR}(T)^{-1}$ as a function of temperature, normalized by the rate at T = 140 K (2.26×10^{-4} s^{-1}).

The progress of crystallization depends on the ambient temperature T_0 of the medium into which it propagates. If a mass element Δm_1 has just crystallized, liberating an amount of heat $H_a\Delta m_1$, of which a fraction η is absorbed by an adjacent mass element Δm_2, raising its temperature up to T_c (the rest being dissipated over a more extended region), then Δm_2, in turn, crystallizes. Thus

$$\eta H_a \Delta m_1 = [u(T_c) - u(T_0)]\Delta m_2 \tag{9}$$

and heat is again released in an amount $H_a\Delta m_2$, of which the absorption of a fraction η causes another mass element to crystallize, and so forth. The total amount of mass Δm that will ultimately crystallize (starting spontaneously with the crystallization of Δm_1) is then given by the sum of a geometric series with the factor $q = \eta H_a/([u(T_c)-u(T_0)]$. If $q \geq 1$, the sum diverges, meaning that crystallization will continue indefinitely. However, if $q < 1$, the sum converges to $\Delta m_1/(1-q)$, meaning that it will stop after a while. A critical initial temperature may be defined, T_{crit}, corresponding to $q = 1$. Accordingly, if $T_0 \geq T_{crit}$, crystallization will proceed continuously in a runaway process; otherwise ($T_0 < T_{crit}$), it will stop and will need to be triggered again. For $\eta = 0.5$, $T_{crit} \approx 100$ K, whereas typically, $T_0 \lesssim 50$ K, as in the Kuiper belt. Therefore, spurts of crystallization should

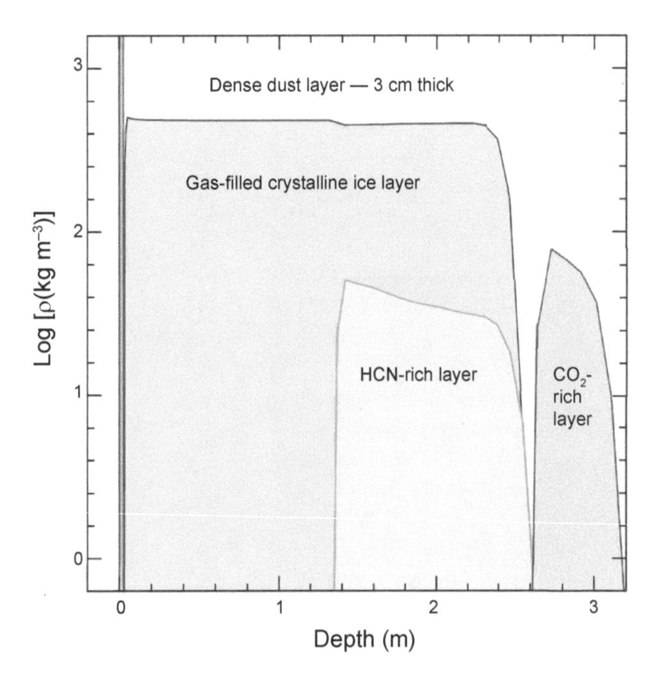

Fig. 15. Densities of volatile ices in the outer layer of a model nucleus following crystallization; the outer dust layer (marked) is depleted of volatiles. Adapted from *Prialnik et al.* (2008).

be more common than runaways.

A burst of crystallization may be initiated by the heat wave propagating inward from the insolated comet surface to the crystalline-amorphous ice boundary, provided that reaching this boundary, it still carries sufficient energy for raising the local temperature to T_c. However, once this has occurred and the boundary has moved deeper into the nucleus, later heat waves originating at the surface will be too weak when reaching the boundary to rekindle crystallization. A quiescent period thus ensues, until the surface recedes (by sublimation) to a sufficiently short distance from the crystalline-amorphous ice boundary for a new spurt of crystallization to take place. Since in the meantime the interior ice temperature has risen to some extent, crystallization advances deeper into the nucleus than at the previous spurt. This, in turn, affects the time span to the next spurt of crystallization. Evolutionary simulations through repeated outbursts triggered by crystallization illustrate this behavior trend (*Prialnik and Bar-Nun,* 1987; *Tancredi et al.,* 1994; *González et al.,* 2008). An example is shown in Fig. 16.

By the time the heat wave reaches the amorphous ice boundary and triggers a new spurt of crystallization, the comet may have moved far away from the Sun. This is how crystallization can explain the distant activity of comets. In new comets, which may have amorphous ice close to the surface, crystallization sets in when the surface temperature approaches T_c, which occurs far beyond the distance where ice sublimation controls cometary activity. New Comet C/2003 A2 (Gleason), for example, was observed to be active at 11.49 au pre-perihelion, and *Meech et al.* (2009) suggested crystallization as the cause. Conversely, inbound comets C/2017 K2 (PANSTARRS) (*Jewitt et al.,* 2017; *Meech et al.,* 2017) and C/2014 UN271 (Bernardinelli-Bernstein) (*Farnham et al.,* 2021) were both observed to be active at 23 au, where temperatures are too low for triggering crystallization.

Perhaps counterintuitively, runaway crystallization is favored by a low conductivity of the amorphous ice, in which case most of the heat released is absorbed near the crystallization front, raising the temperature sufficiently to keep it going. A high conductivity results in spread out and waste of heat. Indeed, *Haruyama et al.* (1993) studied the thermal history of comet nuclei heated by radioactive decay during their residence in the Oort cloud and found that for a very low K, almost all the ice crystallizes, while for a sufficiently large K, the initial amorphous ice is almost completely preserved. The refractory component of the nucleus, presumed to be intimately mixed with the ice, also affects the possibility of runaway crystallization. A large refractory/ice ratio inhibits runaway by effectively reducing η. The likelihood of runaway in any given comet is influenced by many factors.

4.3. Non-Uniform Crystallization

As the comet nucleus surface is not uniformly heated by solar radiation, and since the thermal conductivity is low, we expect the depth of the crystallization front to vary with latitude and be affected by spin axis orientation and surface inhomogeneities. These effects have been studied by means of multi-dimensional evolution codes (e.g., *Guilbert-Lepoutre et al.,* 2016). The dependence on latitude

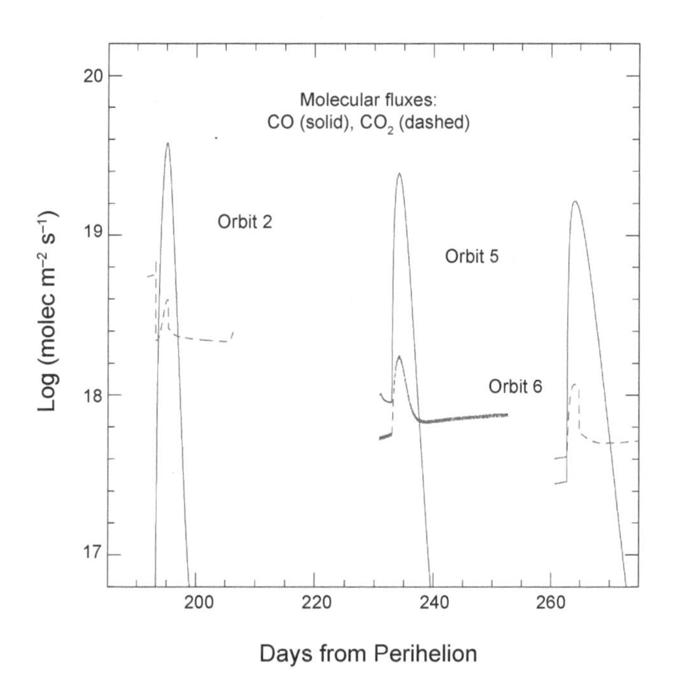

Fig. 16. Examples of post-perihelion outbursts on different orbits for orbital parameters of Comet 67P/Churyumov-Gerasimenko, illustrated by production rates of CO and CO_2 — from evolutionary calculation. Adapted from *Prialnik and Sierks* (2017).

is illustrated in Fig. 17.

The effect of non-uniform surface heating, modeled as a patch of high albedo, was shown by *Guilbert-Lepoutre and Jewitt* (2011) to induce varying temperature gradients in the subsurface layers, as illustrated in Fig. 18. When coupled with rotational and seasonal effects, it is clear that a spatially complex and time-varying subsurface thermal structure can be expected, resulting in local crystallization spurts at various heliocentric distances. The frequent, localized mini-outbursts detected by Rosetta on Comet 67P/Churyumov-Gerasimenko may have resulted from such behavior.

4.4. Crystallization During Early Evolution

Cometary nuclei formed beyond the snowline of the solar nebula, where radiogenic heating was stronger than insolation. This was especially true during the first few million years, when the strongest heat source, ^{26}Al (half-life 0.7 m.y.), was abundant. The length scale over which conduction can transport heat in time τ_c is $\ell \sim (\kappa\tau_c)^{1/2}$, where κ is the thermal diffusivity. We take $\kappa = 10^{-8}$ m^2 s^{-1} as representative. With τ_c = 7 m.y. (i.e., 10 half-lives of ^{26}Al), we find $\ell \sim 1.5$ km. Nuclei larger than 1.5 km should capture the energy released by the decay of ^{26}Al below a 1.5-km-thick outer mantle, and so the core would suffer strong heating comparable to that of the main-belt asteroids, which are known to have formed quickly and to be highly metamorphosed. Setting τ_c = 4.5 G.y., we estimate $\ell \sim 40$ km; bodies larger than ~40 km would capture much of the energy released by the decay of longer-lived radiogenic elements (K, U, Th) and be similarly metamorphosed. These estimates are obviously simplistic (heat can also be transported by radiation, by gas, and if the liquid phase is generated, by convection), but they serve to illustrate the potential importance of radiogenic decay heating in comets.

The extent of nucleus heating and the question of the long-term survival of amorphous ice are difficult to address with confidence because of the lack of knowledge of key nucleus physical parameters. First, the low density [~500 kg m^{-3} (*Jorda et al.,* 2016)] of the nucleus of 67P/Churyumov-Gerasimenko suggests substantial porosity, favoring lower values of the diffusivity, possibly by one or more orders of magnitude. At the same time, a high porosity increases the permeability to gas flow and hence, the efficiency of heat transport by advection. Secondly, the amount of ^{26}Al and the radiogenic power production in a given nucleus are proportional to the dust/ice ratio, $\mathcal{R}$. Modern astronomical data suggest large dust/ice ratios, with $\mathcal{R} > 1$ for 67P/Churyumov-Gerasimenko (*Choukroun et al.,* 2020), $\mathcal{R} \sim 3$ for KBOs (*Fulle et al.,* 2019), and published values reaching $\mathcal{R}$ = 10–30 for 2P/Encke (*Reach et al.,* 2000). These large values indicate that radiogenic heating of cometary ices might have been substantial. Finally, we consider the accretion mechanism itself: Recently, popular models of streaming-instability/pebble accretion (cf. *Davidsson,* 2021) invoke rapid-nucleus-formation scenarios in which short-lived isotopes would

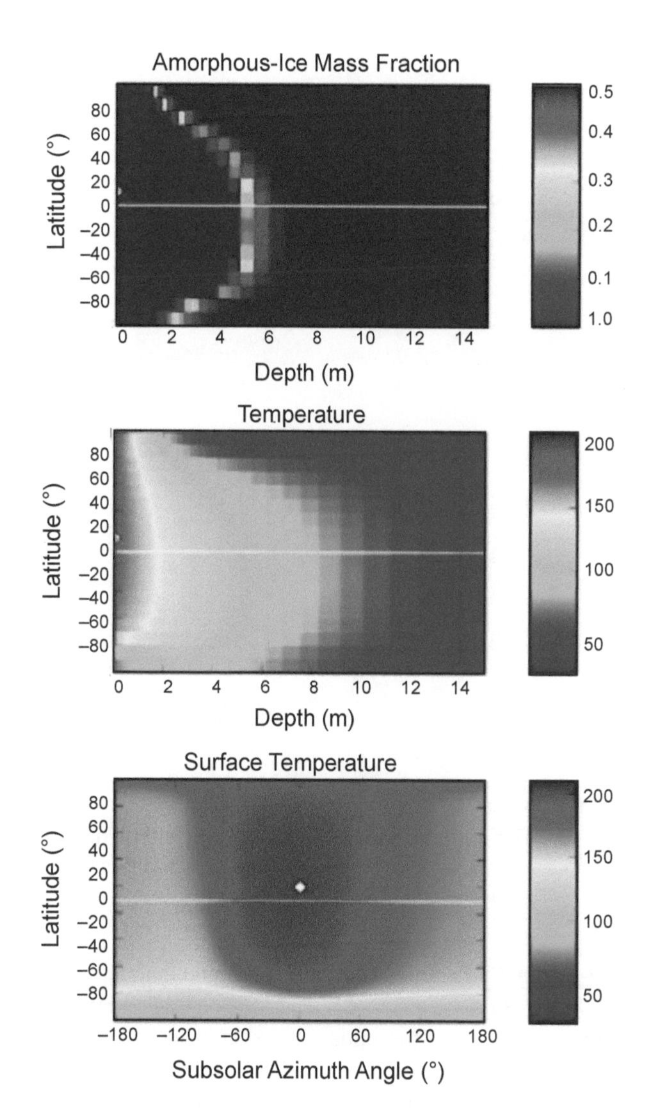

Fig. 17. Advance of the crystallization front in a three-dimensional simulation of a model with a spin period of 12.6 hr, and an axis tilt angle of 15°, illustrating the effect of latitude. Adapted from *Rosenberg and Prialnik* (2007).

be efficiently trapped, leading to high initial abundances of ^{26}Al, but it is still possible that the growth of comets was slow (*Golabek and Jutzi,* 2021), allowing ^{26}Al to decay before its incorporation into the nuclei.

Against the indications of the likely importance of radioactive decay heating stand spectroscopic observations showing that comets preserve supervolatiles like CO and CO_2 in substantial abundance. A plausible resolution of this conflict is that the supervolatiles were trapped in amorphous ice, some of which was destroyed in the epoch of ^{26}Al decay, but part of which survived in the outer, low-temperature regions of radiogenically heated nuclei. This scenario is indeed supported by detailed thermal modeling. The first to have modeled radiogenic heating of comets by ^{26}Al was *Wallis* (1980) using a simple analytic approach; the aim, however, was melting rather than crystallization. *Prialnik and Podolak* (1995) simulated numerically the early evolution of icy bodies in the Kuiper belt region, including

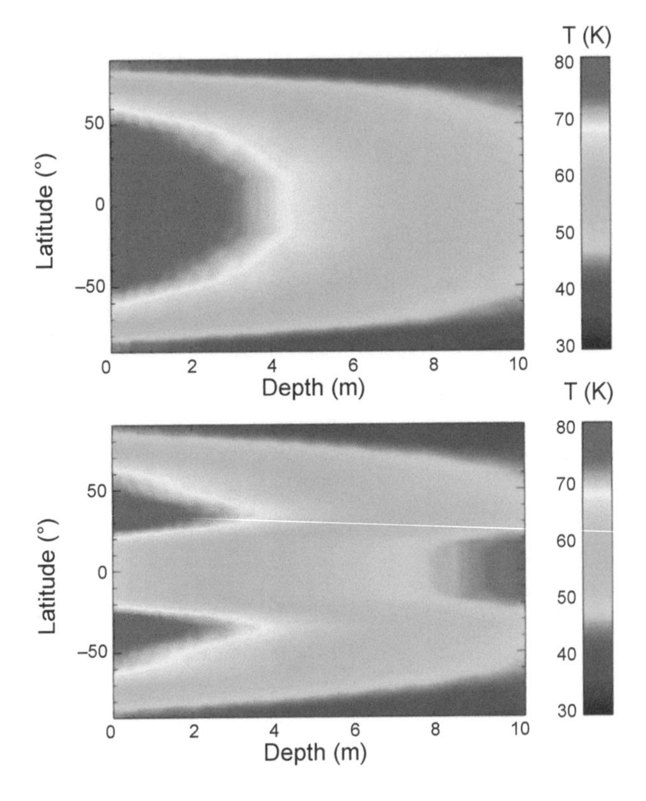

Fig. 18. Subsurface temperature shadows resulting from surface albedo variations on a cometary nucleus. From *Guilbert-Lepoutre and Jewitt* (2011).

long-lived radioisotopes and ^{26}Al, and found that depending on size, physical parameters — such as porosity and thermal conductivity — and the initial ^{26}Al abundance, KBOs may emerge from the long-term evolution in three different configurations: (1) preserving their pristine structure throughout; (2) having a large crystallized core; and (3) having a crystallized core, a layer of frozen gas (originally occluded in the amorphous ice), and an outer layer of unaltered pristine material. *Merk and Prialnik* (2003, 2006) refined the calculations by considering concomitant growing of KBOs by accretion and thermal evolution due to radiogenic heating but neglecting gas flow through the porous nucleus. An example of the results is given in Fig.19. The occurrence of the peak in the extent of the crystalline ice core is due to the competition between the advance of the crystallization front and the accretion front, and its location strongly depends on assumed parameters.

Finally, *Yabushita* (1993) showed that in large objects ($R \geq 40$ km), crystallization in the core may occur by radioactive heating even due to long-lived radioisotopes alone. However, in all cases a sufficiently extended outer layer of amorphous ice would be preserved, to become active when these objects eventually become comets.

4.5. Observed Outbursts Modeled by Crystallization

Crystallization of amorphous ice and release of trapped gases — accompanied by drag of dust particles — has been invoked to explain many observed cometary outbursts, as we describe in this section. Other interpretations are possible and generally involve the buildup of internal pressure due to sublimation of pure supervolatiles within confined spaces in cometary nuclei that maintain significant cohesive strength. Of course, crystallization and gas confinement might occur together, leading to outburst (*Samarasinha,* 2001).

The occurrence of outbursts triggered by crystallization depends on the depth of the amorphous-crystalline ice boundary. A schematic view of such an outburst is shown in Fig. 20. For a new comet, where the boundary is close to the surface, outbursts are expected on the inbound leg of the orbit, at the distance where the surface temperature is close to the crystallization temperature, from equation (6) beyond 8 au. For an old comet, where the boundary lies at a depth below the surface, the heat wave generated by absorption of solar radiation close to perihelion penetrates to deeper layers with a time lag determined by the thermal diffusivity, and crystallization sets in when it reaches the amorphous ice boundary, which may occur at large heliocentric distances post-perihelion (see section 4.2). A simple estimate is shown in Fig. 21. If the boundary is found very deep, a burst of crystallization may occur as late as the next inbound leg of the orbit. Indeed, according to the survey of the distant activity of SPC by *Mazzotta Epifani et al.* (2009), outbursts occur predominantly post-perihelion and only on rare occasions, on the inbound leg of the orbit.

4.5.1. Comet 1P/Halley. In February 1991, when outbound at 14.3 au, the comet was found to have undergone a major outburst that lasted for about three months, with a 300-fold brightness increase and the development of a tail (*West et al.,* 1991), the first comet outburst detected at such a large distance. Crystallization of amorphous ice was

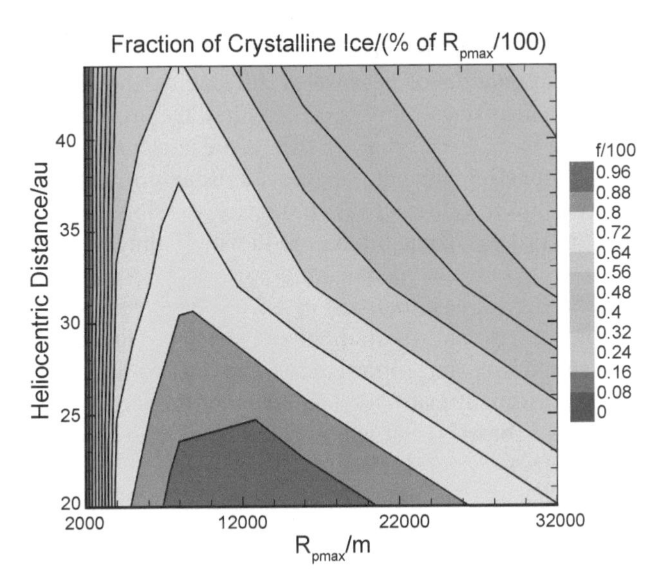

Fig. 19. Fractional size of the crystalline core for ice-rich objects as function of final radius and heliocentric distance. Note that half the total radius means only 12.5% of the volume. Adapted from *Merk and Prialnik* (2006).

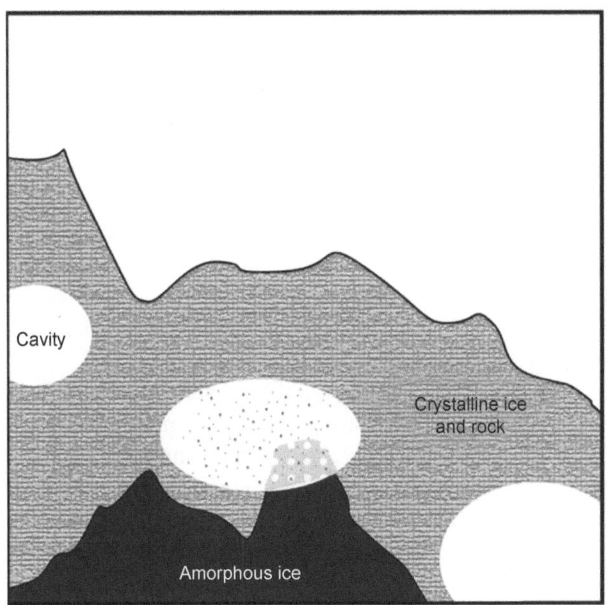

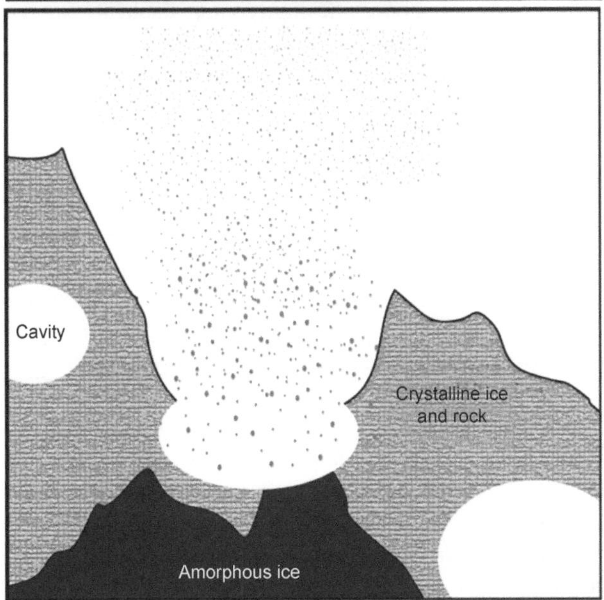

Fig. 20. Schematic diagram showing a cometary outburst produced by pressure build-up in a buried cavity within a nucleus whose surface layers have already crystallized. The yellow patch represents amorphous ice undergoing crystallization and releasing gas into a cavity. Adapted from *Reach et al.* (2010).

immediately invoked as the most plausible mechanism to explain the outburst (*Schmitt et al.,* 1991; *Weissman,* 1991). *Prialnik and Bar-Nun* (1992) showed by numerical simulations that the outburst features — heliocentric distance, characteristic time, and outburst amplitude — are indeed obtained for a model of low density, where the outburst is initiated by crystallization and CO release at a depth of a few tens of meters.

4.5.2. Comet C/Hale-Bopp. This comet was discovered in July 1995, being characterized by an unusually bright dust coma at a distance of about 7 au. In late 1996, *Jewitt et al.* (1996) detected a very large flux of CO molecules that increased rapidly. Such rapid brightening is unlikely to have resulted from surface (or subsurface) sublimation of CO ice in response to insolation. Moreover, CO ice should have been depleted much earlier on the orbit, since at 7 au the surface temperature is already above 100 K, far above the sublimation temperature of CO.

Prialnik (1997) modeled the thermal evolution of Comet Hale-Bopp and showed that a sharp rise in the activity of the nucleus at 7 au pre-perihelion, marked by an increase in the CO flux and in the rate of dust emission by several orders of magnitude, is consistent with runaway crystallization taking place a few meters below the surface, accompanied by the release of the trapped CO. The runaway was eventually quenched, and a period of sustained but variable activity ensued, in agreement with observations. Similar results were obtained by *Enzian et al.* (1998), using a quasi-three-dimensional model of the nucleus. They showed that crystalline ice cannot account for the observed activity. *Capria et al.* (2000b) showed that to account for the observed CO production, one must invoke gas trapped in amorphous ice, a fraction of which remains trapped until the ice sublimates.

4.5.3. Comet 29P/Schwassmann-Wachmann 1. The orbit of 29P is confined to a small range of heliocentric distances around 6 au, just in the temperature range where crystallization occurs on orbital timescales (see Fig. 5). Despite the large distance from the Sun, the comet is always active (*Jewitt,* 1990) with superimposed outbursts by factors up to several hundred (see Fig. 22). Production of CO occurs at about 10^3 kg s^{-1} but, unlike the continuum, has

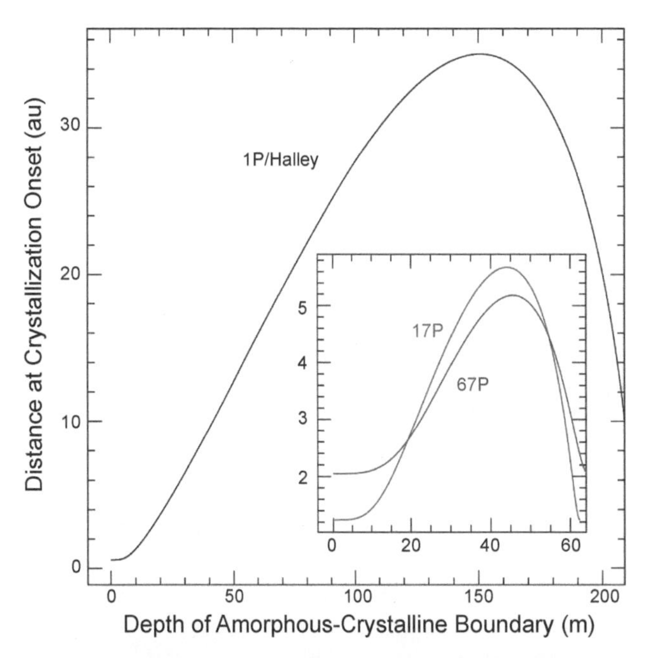

Fig. 21. Illustration of the time lag of crystallization, for orbital parameters of different comets. Large depths may be reached by the heat wave after aphelion, keeping in mind that the wave amplitude decreases with depth.

been observed to vary by only a factor of ~2 since the first detection nearly three decades ago (*Senay and Jewitt,* 1994; *Festou et al.,* 2001; *Gunnarsson et al.,* 2008; *Wierzchos and Womack,* 2020). The nucleus is estimated to be about 32 ± 3 km in radius (*Schambeau et al.,* 2021). The total mass of CO lost since 1994 is $\Delta M_{CO} \sim 10^{12}$ kg, corresponding to a layer ~0.15 m deep over the whole nucleus or, more likely, to deeper excavation or progressive cliff collapse over a smaller portion of the surface.

Cabot et al. (1996) analyzed the activity of the comet and suggested that it shows a maximum after each perihelion passage and that the delay between this peak of activity and perihelion tends to increase from one revolution to the next: less than one year in 1940, more than two years in 1957, and almost five years in 1973. They also noted that the average level of activity increased after the orbit change in the 1980s. They further showed that this erratic activity can be explained by the crystallization of amorphous ice, triggered by the propagation of a heat wave into the nucleus. The modeled activity is correlated with heliocentric distance: The maximum activity is delayed with respect to perihelion, and the delay seems to increase between successive revolutions (see section 4.2). *Enzian et al.* (1997) applied to this comet a 2.5-dimensional model (that considers radial and latitudinal effects, summing over meridians for a spin period) and showed that crystallization of amorphous ice and release of trapped gas are crucial and uniquely successful for explaining the activity features of this comet.

4.5.4. Comet 17P/Holmes. This short-period comet displayed two large post-perihelion outbursts in November 1892 and January 1893. Sixteen uneventful orbits later, an even greater and much more closely observed post-perihelion outburst occurred in October 2007 (*Hsieh et al.,* 2010; *Reach et al.,* 2010; *Stevenson et al.,* 2010; *Li et al.,* 2011). The total brightening was about 19 mag, from likely quiescent magnitude V ~ 17 to a peak near –2 (*Li et al.,* 2011) corresponding to a factor of about 40,000,000 (see Fig. 23). Although the effects were measurable for months, the mass loss rate was sharply peaked with a full-width at half maximum of 0.4 days, centered about 1 day after the start of activity (see Fig. 24) and reaching a maximum mass ejection rate ~3×10^5 kg s^{-1}, with dust ejection speeds ~400 m s^{-1}. Using a quasi-three-dimensional model, *Hillman and Prialnik* (2012) were able to simulate the observed behavior resulting from crystallization of amorphous ice and release of occluded gas species, in particular, the interval between outbursts, the close post-perihelion occurrence, and the production rates of CO, CO_2, and NH_3.

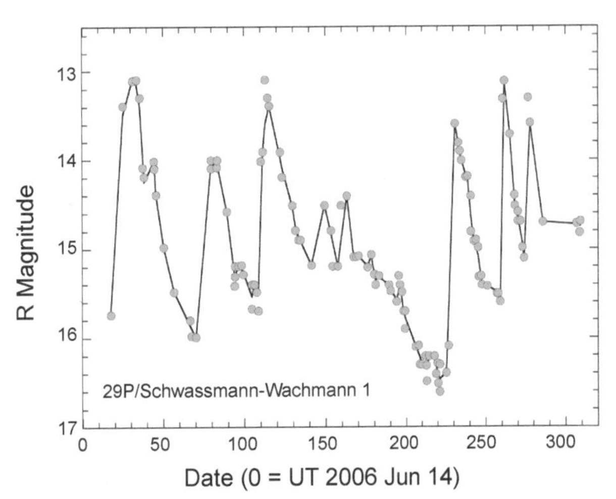

Fig. 22. Approximately one year of photometry of 29P/Schwassmann-Wachmann 1 (circles). The solid line is a running mean added to guide the eye. Adapted from *Trigo-Rodríguez et al.* (2008).

Kossacki and Szutowicz (2010) applied a thermal model to Comet 17P, taking into account amorphous water ice and dust but no other volatiles. They showed that in this case crystallization is unable to account for the observed, powerful outburst. This result, in fact, strengthens the conclusion that supervolatile gases expelled from the ice upon crystallization are crucial to the outburst mechanism, since the heat released is only a few percent of the heat required to sublimate water ice (see section 4.2).

4.5.5. 2060 Chiron. First classified as an asteroid, Chiron is the first object identified as a Centaur, exhibiting cometary activity: developing a coma and undergoing outbursts. Its distance from the Sun is too large for water sublimation to drive this activity. Both *Prialnik et al.* (1995) and *Capria et al.* (2000a) have modeled the activity of this object and showed that the observed characteristics may be accounted for by a composition of porous gas-laden amorphous ice and dust.

4.5.6. Comet 67P/Churyumov-Gerasimenko. *Mousis et al.* (2015) have suggested that pits observed by the Rosetta mission on the surface of Comet 67P may result from disrupting outgassing due to increased gas pressure that could be caused by clathrate destabilization and amorphous ice crystallization. They discarded the possibility of impacts because of inconsistency between the size distributions of pits and impactors as well as the possibility of erosion due to insolation. *Prialnik and Sierks* (2017) explored a possible mechanism that may explain sudden depressions of surface areas on a comet nucleus, as those observed on this comet. Assuming the area is covered by a thin, compact dust layer of low permeability to gas flow compared to deeper, porous layers, gas can accumulate below the surface when a surge of gas release from amorphous ice occurs upon crystallization, as we have shown in Fig. 15. The gas pressure is found to exceed the hydrostatic pressure down to a depth of a few meters. The rapid buildup of pressure may weaken the already fragile, highly porous structure. Eventually, the high-pressure gradient that arises drives the gas out, and the pressure falls well below the hydrostatic pressure. The rapid pressure drop results in collapse. Since the crystallization front lies at some depth below the surface, the location on the orbit when this phenomenon occurs is determined by the thermal lag (see Fig. 21).

4.5.7. Comet 332P/2010 V1 (Ikeya-Murakami). Similarly to Comet 17P/Holmes, the comet was discovered due to

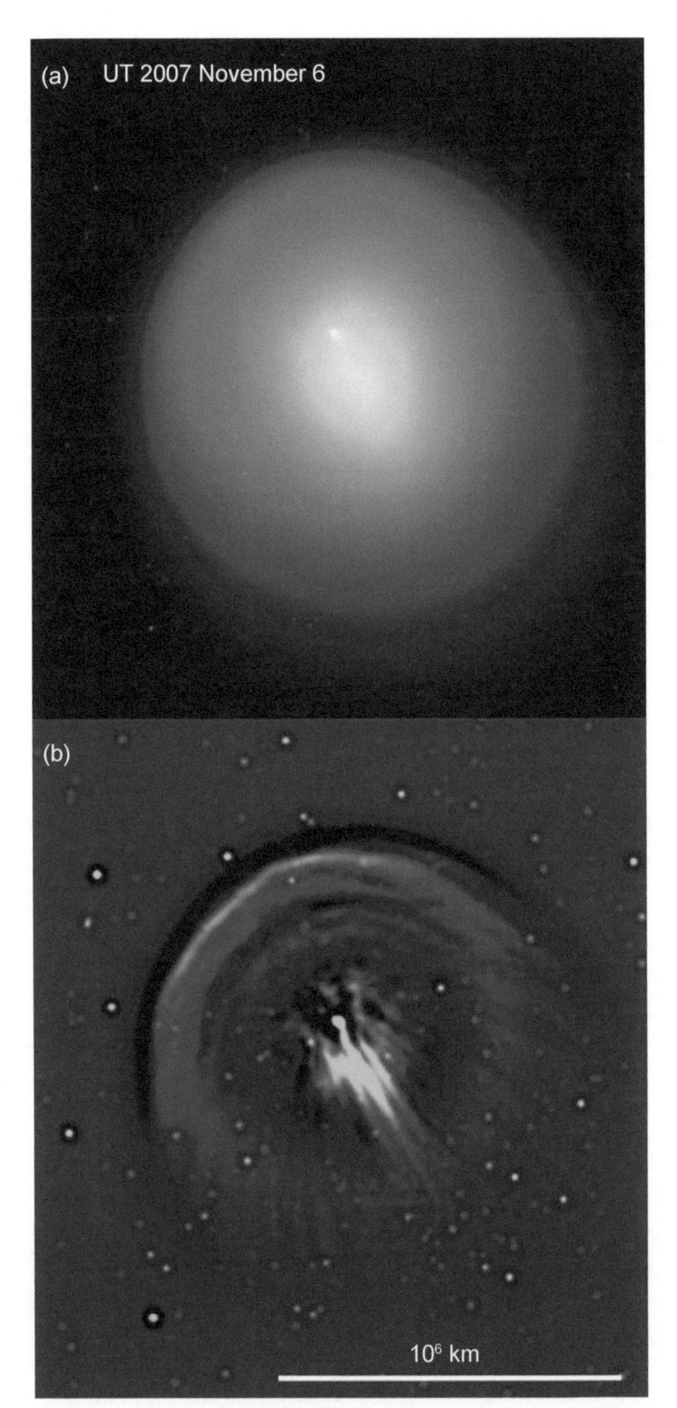

Fig. 23. Outburst of 17P/Holmes on UT 2007 November 6 shown **(a)** as a raw image, revealing the off-center nucleus and the downtail debris cloud, and **(b)** spatially filtered to reveal structure in the coma. About 14 days after the start of the outburst, the ~10^6-km-diameter of the coma indicates an expansion speed ~400 m s^{-1}, a large fraction of the local gas thermal speed. Adapted from *Stevenson et al.* (2010).

a large-scale explosion it underwent between UT 2010 October 31 and November 3 (see Fig. 25). In both cases, the outbursts occurred post-perihelion — 20 days for 332P and 172 days for 17P/Holmes — and the decline times were very similar, 70 and 50 days, respectively. Although the mass and energy were orders of magnitude smaller for 332P than for 17P, the energy per unit mass was similar, ~10^4 J kg^{-1}. The sudden ejection and the derived energy per unit mass of the ejecta were found to be consistent with runaway crystallization of buried amorphous ice as the source of energy to drive the outburst (*Ishiguro et al.,* 2014). The main nucleus of 332P was later determined to be ≤275 m in radius (0.04 albedo assumed) and underwent partial fragmentation in 2015 (*Jewitt et al.,* 2016). The relationship between the outburst in 2010 and the subsequent fragmentation is not known. But stresses induced by crystallization of amorphous ice might lead to fracture in a cometary nucleus (*Prialnik et al.,* 1993), especially in a small one.

4.5.8. Comet P/2010 H2 (Vales). This quasi-Hilda asteroid underwent a spectacular photometric outburst by 7.5 mag (factor of ≳10^3) in 2010, 37 days post-perihelion at a distance of 3.1 au. While the rising phase of the outburst was very steep (brightness doubling time of hours), subsequent fading occurred slowly (fading timescales increasing from weeks to months). The specific energy of the ejecta was estimated at 220 J kg^{-1} (*Jewitt and Kim,* 2020). Low-energy processes known to drive mass loss in active asteroids, including rotational disruption, thermal and desiccation stress cracking, and electrostatic repulsion cannot generate the high particle speeds measured in P/Vales and were discounted. Impact origin was deemed unlikely, given the short dynamical lifetimes of the quasi-Hildas and the low collision probabilities of these objects. *Jewitt and Kim* (2020) conclude that Comet Vales is most likely a temporarily captured comet in which conductive heating of subsurface

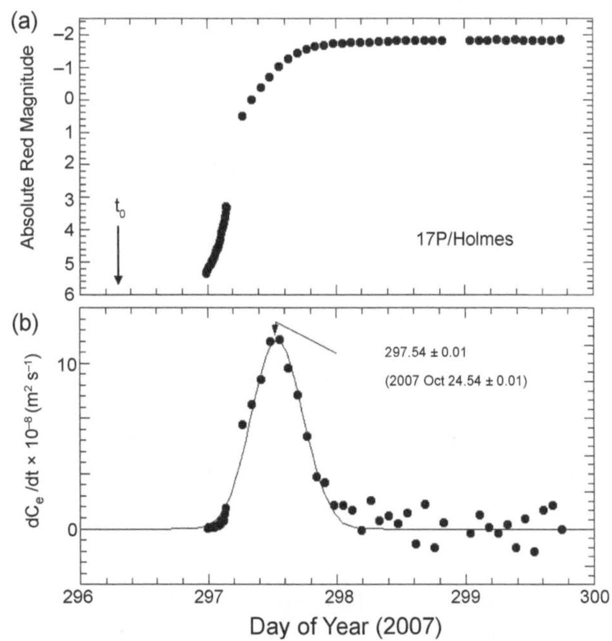

Fig. 24. (a) Integrated photometry of 17P/Holmes showing outburst brightening that began at t_0 = 296.3 (UT 2007 October 23.3). **(b)** Rate of change of dust-scattering cross-section revealing a single release of dust. The peak rate of brightening corresponds to 1000 km^2 s^{-1} and mass loss rate 3 × 10^5 kg s^{-1}. Adapted from *Li et al.* (2011).

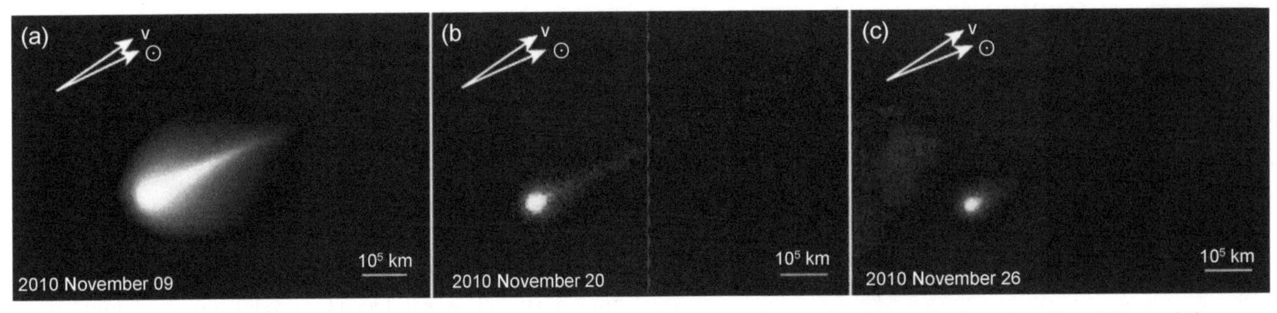

Fig. 25. Outbursting Comet P/2010 V1: time-series RC-band images. The projected antisolar direction (⊙) and the projected negative heliocentric velocity vector (V) are marked. Adapted from *Ishiguro et al.* (2014).

ice has triggered an outburst, probably through exothermic crystallization of amorphous ice. To account for the outburst energy, an ice volume about 0.4 km^2 in areal extent (1% of the nucleus surface) and $\gtrsim$5 m thick, and buried beneath a refractory layer a few meters thick, is inferred.

4.5.9. Comet 9P/Tempel 1. Mini-outbursts observed by the Deep Impact mission on Comet 9P/Tempel 1 (*A'Hearn et al.,* 2005) were found to occur in regions of lowest effective surface gravity; such regions would be the first to yield to internal stresses resulting from gas pressure buildup due to gas release by amorphous ice crystallization (*Belton et al.,* 2008). Furthermore, *Belton and Melosh* (2009) propose that the smooth terrains observed on the nucleus surface are produced by spurts of CO release at the crystallization front at a depth of several tens of meters, which result in fluidization of the dust and ice layer above it and extrude fluidized material onto the surface at low velocity (cometary cryo-volcanism). The inferred CO loss rate of ~4×10^{18} molecules s^{-1} m^{-2} (*Feldman et al.,* 2006; *Belton and Melosh,* 2009) is consistent with gas release upon crystallization.

There are more examples of cometary outbursts explained by the amorphous ice crystallization, but those listed are so different from each other and yet accounted for by the same basic mechanism that they strengthen the hypothesis that links cometary activity to amorphous ice.

The most plausible picture that emerges is the following. Comets formed out of amorphous ice with large amounts of occluded gases in environments that were too hot for extremely volatile species to be wholly frozen as bulk ice but cold enough to preclude clathrate formation. Radiogenic heating, even by powerful short-lived radionuclides, did not affect this composition over most of the comet nucleus. When comets approach the Sun, runaway but transient crystallization of the amorphous ice manifests itself as outbursts in which a fraction of the occluded gas is expelled. The remaining gas is trapped in part in the crystalline ice to form clathrates and in part diffuses toward colder regions of the nucleus, where gases refreeze down the temperature gradient in order of volatility, forming a stratified pattern. Later on, and in subsequent orbits, cometary activity is bound to be complex, as there are now various spatially and compositionally distinct sources of volatiles. The decomposition of newly formed clathrate and the sublimation of newly frozen volatiles are induced by solar energy absorption, but the resulting production rates and the ratios between them, as well as between them and water production, will differ among comets and epochs, depending on size, shape, nucleus spin, and spin axis orientation, and on the dynamical history of the comet. Heat propagating from the surface will trigger further bursts of crystallization of amorphous ice leading to outbursts of various strengths, depending again on the nucleus characteristics. Driven by conducted heat, outbursts may occur at any distance from the Sun, including large distances. As a byproduct, fresh clathrates and ices will form. All this accounts for the enormous complexity and variability of cometary activity, despite the fact that the initial physical and compositional state of the ice is common and simple. The long-standing controversy over whether cometary ice is amorphous or crystalline (e.g., *Luspay-Kuti et al.,* 2016) is thus a false dichotomy.

5. THE FUTURE

Despite hundreds of observations, laboratory studies, results from space missions, and sophisticated numerical simulations, evidence for the role of amorphous ice in comets remains indirect, although this is still the most plausible form of cometary ice. One reason is that, by its very nature, amorphous ice is only long-lived in the more frigid, distant regions of the solar system, precisely where observations are the most difficult to obtain. Another is that, in evolved comets, amorphous ice is expected to be buried below the surface, being inaccessible to direct observation and manifesting itself only indirectly. Moreover, the accessible indicators of amorphous ice, principally the strength of the band at 1.65 μm, provide an imperfect measure of the fraction of amorphous ice because of the confounding influences of ice temperature, irradiation history, grain size, and other variables. More laboratory work is needed to determine the thermophysical properties of amorphous and crystalline ices, especially when loaded with other volatiles.

Observational improvements in characterization of the crystalline state are expected in the era of the James Webb Space Telescope, driven by its greater near-infrared sensitivity and by access to the much stronger 3-μm water ice band (cf. Fig. 8). Ideally, of course, we would like to sample and return cometary ices to Earth while preserving their physical and chemical properties (*Westphal et al.,* 2020;

Vernazza et al., 2021). A cryogenic sampling mission to excavate nucleus material from depth and bring it back to Earth without causing it to crystallize would be one of the most scientifically interesting, but technically challenging, endeavors ever attempted.

Acknowledgments. We thank M. Gudipati for comments on a draft of the manuscript. D.P. gratefully acknowledges support from the Israeli Science Foundation grant 566/17.

REFERENCES

A'Hearn M. F., Belton M. J. S., and Delamere W. A. (2005) Deep Impact: Excavating comet Tempel 1. *Science, 310,* 258–264.
A'Hearn M. F., Belton M. J. S., Collins S. M., Farnham T. L., Feaga L. M., Groussin O., Lisse C. M., Meech K. J., Schultz P. H., and Sunshine J. M. (2008) Deep Impact and sample return. *Earth Planets Space, 60,* 61–66.
Andersson O. and Suga H. (1994) Thermal conductivity of low-density amorphous ice. *Solid State Commun., 91,* 985–988.
Andersson O. and Suga H. (2002) Thermal conductivity of amorphous ices. *Phys. Rev. B, 65,* 140201.
Barkume K. M., Brown M. E., and Schaller E. L. (2008) Near-infrared spectra of Centaurs and Kuiper belt objects. *Astron. J., 135,* 55–67.
Bar-Nun A. and Kleinfeld I. (1989) On the temperature and gas composition in the region of comet formation. *Icarus, 80,* 243–253.
Bar-Nun A. and Laufer D. (2003) First experimental studies of large samples of gas-laden amorphous "cometary" ices. *Icarus, 161,* 157–163.
Bar-Nun A., Herman G., Laufer D., and Rappaport M. L. (1985) Trapping and release of gases by water ice and implications for icy bodies. *Icarus, 63,* 317–332.
Bar-Nun A., Dror J., Kochavi E., and Laufer D. (1987) Amorphous water ice and its ability to trap gases. *Phys. Rev. B, 35,* 2427–2435.
Bar-Nun A., Kleinfeld I., and Kochavi E. (1988) Trapping of gas mixtures by amorphous water ice. *Phys. Rev. B, 38,* 7749–7754.
Bar-Nun A., Notesco G., and Owen T. (2007) Trapping of N_2, CO and Ar in amorphous ice — Application to comets. *Icarus, 190,* 655–659.
Belton M. J. S. and Melosh J. (2009) Fluidization and multiphase transport of particulate cometary material as an explanation of the smooth terrains and repetitive outbursts on 9P/Tempel 1. *Icarus, 200,* 280–291.
Belton M. J. S., Feldman P. D., A'Hearn M. F., and Carcich B. (2008) Cometary cryo-volcanism: Source regions and a model for the UT 2005 June 14 and other mini-outbursts on Comet 9P/Tempel 1. *Icarus, 198,* 189–207.
Berdis J. R., Gudipati M. S., Murphy J. R., and Chanover N. J. (2020) Europa's surface water ice crystallinity: Discrepancy between observations and thermophysical and particle flux modeling. *Icarus, 341,* 113660.
Blake D., Allamandola L., Sandford S., Hudgins D., and Freund F. (1991) Clathrate hydrate formation in amorphous cometary ice analogs in vacuo. *Science, 254,* 548–551.
Brown M. E. and Calvin W. M. (2000) Evidence for crystalline water and ammonia ices on Pluto's satellite Charon. *Science, 287,* 107–109.
Cabot H., Enzian A., Klinger J., and Majolet S. (1996) Complementary studies on the unexpected activity of comet Schwassmann-Wachmann 1. *Planet. Space Sci., 44,* 1015–1020.
Capria M. T., Coradini A., De Sanctis M. C., and Orosei R. (2000a) Chiron activity and thermal evolution. *Astron. J., 119,* 3112–3118.
Capria M. T., Coradini A., De Sanctis M. C., and Orosei R. (2000b) CO emission mechanisms in C/1995 O1 (Hale-Bopp). *Astron. Astrophys., 357,* 359–366.
Choukroun M., Altwegg K., Kührt E., Biver N., Bockelée-Morvan D., Drążkowska J., Hérique A., Hilchenbach M., Marschall R., Pätzold M., Taylor M. G. G. T., and Thomas N. (2020) Dust-to-gas and refractory-to-ice mass ratios of comet 67P/Churyumov-Gerasimenko from Rosetta observations. *Space Sci. Rev., 216,* 44.
Ciesla F. J. (2014) The phases of water ice in the solar nebula. *Astrophys. J. Lett., 784,* L1.
Ciesla F. J., Krijt S., Yokochi R., and Sandford S. (2018) The efficiency of noble gas trapping in astrophysical environments. *Astrophys. J., 867,* 146.
Cooper J. F., Johnson R. E., Mauk B. H., Garrett H. B., and Gehrels N. (2001) Energetic ion and electron irradiation of the icy Galilean satellites. *Icarus, 149,* 133–159.
Cooper J. F., Christian E. R., Richardson J. D., and Wang C. (2003) Proton irradiation of Centaur, Kuiper belt, and Oort cloud objects at plasma to cosmic ray energy. *Earth Moon Planets, 92,* 261–277.
Cuppen H. M. and Herbst E. (2007) Simulation of the formation and morphology of ice mantles on interstellar grains. *Astrophys. J., 668,* 294–309.
Davidsson B. J. R. (2021) Thermophysical evolution of planetesimals in the primordial disc. *Mon. Not. R. Astron. Soc., 505,* 5654–5685.
Davies J. K., Roush T. L., Cruikshank D. P., Bartholomew M. J., Geballe T. R., Owen T., and de Bergh C. (1997) The detection of water ice in comet Hale-Bopp. *Icarus, 127,* 238–245.
Enzian A., Cabot H., and Klinger J. (1997) A 2½ D thermodynamic model of cometary nuclei. I. Application to the activity of comet 29P/Schwassmann-Wachmann 1. *Astron. Astrophys., 319,* 995–1006.
Enzian A., Cabot H., and Klinger J. (1998) Simulation of the water and carbon monoxide production rates of comet Hale-Bopp using a quasi 3-D nucleus model. *Planet. Space Sci., 46,* 851–858.
Espinasse S., Klinger J., Ritz C., and Schmitt B. (1991) Modeling of the thermal behavior and of the chemical differentiation of cometary nuclei. *Icarus, 92,* 350–365.
Farnham T. L., Kelley M. S. P., and Bauer J. M. (2021) Early activity in comet C/2014 UN271 Bernardinelli-Bernstein as observed by TESS. *Planet. Sci. J., 2,* 236.
Feldman P. D., Lupu R. E., McCandliss S. R., Weaver H. A., A'Hearn M. F., Belton M. J. S., and Meech K. J. (2006) Carbon monoxide in comet 9P/Tempel 1 before and after the Deep Impact encounter. *Astrophys. J. Lett., 647,* L61–L64.
Festou M. C., Gunnarsson M., Rickman H., Winnberg A., and Tancredi G. (2001) The activity of comet 29P/Schwassmann-Wachmann 1 monitored through its CO J=2→1 radio line. *Icarus, 150,* 140–150.
Fink U. and Larson H. P. (1975) Temperature dependence of the water-ice spectrum between 1 and 4 microns: Application to Europa, Ganymede and Saturn's rings. *Icarus, 24,* 411–420.
Fulle M., Blum J., Green S. F., Gundlach B., Herique A., Moreno F., Mottola S., Rotundi A., and Snodgrass C. (2019) The refractory-to-ice mass ratio in comets. *Mon. Not. R. Astron. Soc., 482,* 3326–3340.
Ghormley J. A. (1968) Enthalpy changes and heat-capacity changes in the transformations from high-surface-area amorphous ice to stable hexagonal ice. *J. Chem. Phys., 48,* 503–508.
Giauque W. F. and Stout J. W. (1936) The entropy of water and third law of thermodynamics. The heat capacity of ice from 15 to 273 K. *J. Am. Chem. Soc., 58,* 1144–1144.
Golabek G. J. and Jutzi M. (2021) Modification of icy planetesimals by early thermal evolution and collisions: Constraints for formation time and initial size of comets and small KBOs. *Icarus, 363,* 114437.
González M., Gutiérrez P. J., Lara L. M., and Rodrigo R. (2008) Evolution of the crystallization front in cometary models: Effect of the net energy released during crystallization. *Astron. Astrophys., 486,* 331–340.
Gourgeot F., Carry B., Dumas C., Vachier F., Merlin F., Lacerda P., Barucci M. A., and Berthier J. (2016) Near-infrared spatially resolved spectroscopy of (136108) Haumea's multiple system. *Astron. Astrophys., 593,* A19.
Grundy W. M. and Schmitt B. (1998) The temperature-dependent near-infrared absorption spectrum of hexagonal H_2O ice. *J. Geophys. Res.–Planets, 103,* 25809–25822.
Grundy W. M., Buie M. W., Stansberry J. A., Spencer J. R., and Schmitt B. (1999) Near-infrared spectra of icy outer solar system surfaces: Remote determination of H_2O ice temperatures. *Icarus, 142,* 536–549.
Grundy W. M., Young L. A., Spencer J. R., Johnson R. E., Young E. F., and Buie M. W. (2006) Distributions of H_2O and CO_2 ices on Ariel, Umbriel, Titania, and Oberon from IRTF/SpeX observations. *Icarus, 184,* 543–555.
Gudipati M. S., Abou Mrad N., Blum J., Charnley S. B., Chiavassa T., Cordiner M. A., Mousis O., Danger G., Duvernay F., Gundlach B., Hartogh P., Marboeuf U., Simonia I., Simonia T., Theulé P., and Yang R. (2015) Laboratory studies towards understanding comets. *Space Sci. Rev., 197,* 101–150.

Guilbert A., Alvarez-Candal A., Merlin F., Barucci M. A., Dumas C., de Bergh C., and Delsanti A. (2009) ESO-Large Program on TNOs: Near-infrared spectroscopy with SINFONI. *Icarus, 201,* 272–283.

Guilbert-Lepoutre A. (2012) Survival of amorphous water ice on Centaurs. *Astron. J., 144,* 97.

Guilbert-Lepoutre A. and Jewitt D. (2011) Thermal shadows and compositional structure in comet nuclei. *Astrophys. J., 743,* 31.

Guilbert-Lepoutre A., Rosenberg E. D., Prialnik D., and Besse S. (2016) Modelling the evolution of a comet subsurface: Implications for 67P/Churyumov-Gerasimenko. *Mon. Not. R. Astron. Soc., 462,* S146–S155.

Gunnarsson M., Bockelée-Morvan D., Biver N., Crovisier J., and Rickman H. (2008) Mapping the carbon monoxide coma of comet 29P/Schwassmann-Wachmann 1. *Astron. Astrophys., 484,* 537–546.

Halukeerthi S. O., Shephard J. J., Talewar S. K., Evans J. S. O., Rosu-Finsen A., and Salzmann C. G. (2020) Amorphous mixtures of ice and C_{60} fullerene. *J. Phys. Chem. A, 124,* 5015–5022.

Hansen G. B. and McCord T. B. (2004) Amorphous and crystalline ice on the Galilean satellites: A balance between thermal and radiolytic processes. *J. Geophys. Res.–Planets, 109,* E01012.

Haruyama J., Yamamoto T., Mizutani H., and Greenberg J. M. (1993) Thermal history of comets during residence in the Oort cloud: Effect of radiogenic heating in combination with the very low thermal conductivity of amorphous ice. *J. Geophys. Res.–Planets, 98,* 15079–15090.

He J., Clements A. R., Emtiaz S., Toriello F., Garrod R. T., and Vidali G. (2019) The effective surface area of amorphous solid water measured by the infrared absorption of carbon monoxide. *Astrophys. J., 878,* 94.

Herman G. and Weissman P. R. (1987) Numerical simulation of cometary nuclei: III. Internal temperatures of cometary nuclei. *Icarus, 69,* 314–328.

Hillman Y. and Prialnik D. (2012) A quasi 3-D model of an outburst pattern that explains the behavior of Comet 17P/Holmes. *Icarus, 221,* 147–159.

Hsieh H. H., Fitzsimmons A., Joshi Y., Christian D., and Pollacco D. L. (2010) SuperWASP observations of the 2007 outburst of Comet 17P/Holmes. *Mon. Not. R. Astron. Soc., 407,* 1784–1800.

Ishiguro M., Jewitt D., Hanayama H., Usui F., Sekiguchi T., Yanagisawa K., Kuroda D., Yoshida M., Ohta K., Kawai N., Miyaji T., Fukushima H., and Watanabe J. (2014) Outbursting comet P/2010 V1 (Ikeya-Murakami): A miniature comet Holmes. *Astrophys. J., 787,* 55.

Jenniskens P. and Blake D. F. (1994) Structural transitions in amorphous water ice and astrophysical implications. *Science, 265,* 753–756.

Jewitt D. (1990) The persistent coma of comet P/Schwassmann-Wachmann 1. *Astrophys. J., 351,* 277–286.

Jewitt D. (2015) Color systematics of comets and related bodies. *Astron. J., 150,* 201.

Jewitt D. C. and Luu J. (2004) Crystalline water ice on the Kuiper belt object (50000) Quaoar. *Nature, 432,* 731–733.

Jewitt D. and Kim Y. (2020) Outbursting quasi-Hilda asteroid P/2010 H2 (Vales). *Planet. Sci. J., 1,* 77.

Jewitt D., Senay M., and Matthews H. (1996) Observations of carbon monoxide in comet Hale-Bopp. *Science, 271,* 1110–1113.

Jewitt D., Mutchler M., Weaver H., Hui M.-T., Agarwal J., Ishiguro M., Kleyna J., Li J., Meech K., Micheli M., Wainscoat R., and Weryk R. (2016) Fragmentation kinematics in comet 332P/Ikeya-Murakami. *Astrophys. J. Lett., 829,* L8.

Jewitt D., Hui M.-T., Mutchler M., Weaver H., Li J., and Agarwal J. (2017) A comet active beyond the crystallization zone. *Astrophys. J. Lett., 847,* L19.

Jorda L., Gaskell R., Capanna C., Hviid S., Lamy P., Ďurech J., Faury G., Groussin O., Gutiérrez P., Jackman C., Keihm S. J., Keller H. U., Knollenberg J., Kührt E., Marchi S., Mottola S., Palmer E., Schloerb F. P., Sierks H., Vincent J. B., A'Hearn M. F., Barbieri C., Rodrigo R., Koschny D., Rickman H., Barucci M. A., Bertaux J. L., Bertini I., Cremonese G., Da Deppo V., Davidsson B., Debei S., De Cecco M., Fornasier S., Fulle M., Güttler C., Ip W. H., Kramm J. R., Küppers M., Lara L. M., Lazzarin M., Lopez Moreno J. J., Marzari F., Naletto G., Oklay N., Thomas N., Tubiana C., and Wenzel K. P. (2016) The global shape, density and rotation of Comet 67P/Churyumov-Gerasimenko from preperihelion Rosetta/OSIRIS observations. *Icarus, 277,* 257–278.

Kawakita H., Watanabe J., Ootsubo T., Nakamura R., Fuse T., Takato N., Sasaki S., and Sasaki T. (2004) Evidence of icy grains in comet C/2002 T7 (LINEAR) at 3.52 AU. *Astrophys. J. Lett., 601,* L191–L194.

Kawakita H., Ootsubo T., Furusho R., and Watanabe J. (2006) Crystallinity and temperature of icy grains in comet C/2002 T7 (LINEAR). *Adv. Space Res., 38,* 1968–1971.

Klinger J. (1980) Influence of a phase transition of ice on the heat and mass balance of comets. *Science, 209,* 271–272.

Klinger J. (1981) Some consequences of a phase transition of water ice on the heat balance of comet nuclei. *Icarus, 47,* 320–324.

Kossacki K. J. and Szutowicz S. (2010) Crystallization of ice in comet 17P/Holmes: Probably not responsible for the explosive 2007 megaburst. *Icarus, 207,* 320–340.

Kouchi A. and Sirono S. (2001) Crystallization heat of impure amorphous H_2O ice. *Geophys. Res. Lett., 28,* 827–830.

Kouchi A., Greenberg J. M., Yamamoto T., Mukai T., and Xing Z. F. (1992) A new measurement of thermal conductivity of amorphous ice and its implications for the thermal evolution of comets. In *Asteroids, Comets, Meteors 1991* (A. W. Harris and E. Bowell, eds.), pp. 325–328. LPI Contribution No. 1087, Lunar and Planetary Institute, Houston.

Kouchi A., Hama T., Kimura Y., Hidaka H., Escribano R., and Watanabe N. (2016) Matrix sublimation method for the formation of high-density amorphous ice. *Chem. Phys. Lett., 658,* 287–292.

Li J., Jewitt D., Clover J. M., and Jackson B. V. (2011) Outburst of comet 17P/Holmes observed with the solar mass ejection imager. *Astrophys. J., 728,* 31.

Li J., Jewitt D., Mutchler M., Agarwal J., and Weaver H. (2020) Hubble Space Telescope search for activity in high-perihelion objects. *Astron. J., 159,* 209.

Ligier N., Poulet F., Carter J., Brunetto R., and Gourgeot F. (2016) VLT/SINFONI observations of Europa: New insights into the surface composition. *Astron. J., 151,* 163.

Lignell A. and Gudipati M. S. (2015) Mixing of the immiscible: Hydrocarbons in water-ice near the ice crystallization temperature. *J. Phys. Chem. A, 119,* 2607–2613.

Luspay-Kuti A., Mousis O., Hässig M., Fuselier S. A., Lunine J. I., Marty B., Mandt K. E., Wurz P., and Rubin M. (2016) The presence of clathrates in comet 67P/Churyumov-Gerasimenko. *Sci. Adv., 2,* 1501781.

Manca C. and Allouche A. (2001) Quantum study of the adsorption of small molecules on ice: The infrared frequency of the surface hydroxyl group and the vibrational stark effect. *J. Chem. Phys., 114,* 4226–4234.

Marboeuf U., Schmitt B., Petit J. M., Mousis O., and Fray N. (2012) A cometary nucleus model taking into account all phase changes of water ice: Amorphous, crystalline, and clathrate. *Astron. Astrophys., 542,* A82.

Martin C., Manca C., and Roubin P. (2002) Adsorption of small molecules on amorphous ice: Volumetric and FT-IR isotherm co-measurements. Part II. The case of CO. *Surf. Sci., 502-503,* 280–284.

Mastrapa R. M. E. and Brown R. H. (2006) Ion irradiation of crystalline H_2O-ice: Effect on the 1.65-μm band. *Icarus, 183,* 207–214.

Mastrapa R. M. E., Grundy W. M., and Gudipati M. S. (2013) Amorphous and crystalline H_2O-ice. In *The Science of Solar System Ices* (M. S. Gudipati and J. Castillo-Rogez, eds.), pp. 371–408. Astrophys. Space Sci. Library, Vol. 356, Kluwer, Dordrecht.

Mayer E. and Pletzer R. (1986) Astrophysical implications of amorphous ice — A microporous solid. *Nature, 319,* 298–301.

Mazzotta Epifani E., Palumbo P., and Colangeli L. (2009) A survey on the distant activity of short period comets. *Astron. Astrophys., 508,* 1031–1044.

Meech K. J., Pittichová J., Bar-Nun A., Notesco G., Laufer D., Hainaut O. R., Lowry S. C., Yeomans D. K., and Pitts M. (2009) Activity of comets at large heliocentric distances pre-perihelion. *Icarus, 201,* 719–739.

Meech K. J., Kleyna J. T., Hainaut O., Micheli M., Bauer J., Denneau L., Keane J. V., Stephens H., Jedicke R., Wainscoat R., Weryk R., Flewelling H., Schunová-Lilly E., Magnier E., and Chambers K. C. (2017) CO-driven activity in comet C/2017 K2 (PANSTARRS). *Astrophys. J. Lett., 849,* L8.

Merk R. and Prialnik D. (2003) Early thermal and structural evolution of small bodies in the trans-neptunian zone. *Earth Moon Planets, 92,* 359–374.

Merk R. and Prialnik D. (2006) Combined modeling of thermal evolution and accretion of trans-neptunian objects — Occurrence of

high temperatures and liquid water. *Icarus, 183,* 283–295.
Merlin F., Guilbert A., Dumas C., Barucci M. A., de Bergh C., and Vernazza P. (2007) Properties of the icy surface of the TNO 136108 (2003 EL_{61}). *Astron. Astrophys., 466,* 1185–1188.
Mousis O., Guilbert-Lepoutre A., Brugger B., Jorda L., Kargel J. S., Bouquet A., Auger A. T., Lamy P., Vernazza P., Thomas N., and Sierks H. (2015) Pits formation from volatile outgassing on 67P/Churyumov-Gerasimenko. *Astrophys. J. Lett., 814,* L5.
Nelson A. O., Dee R., Gudipati M. S., Horányi M., James D., Kempf S., Munsat T., Sternovsky Z., and Ulibarri Z. (2016) New experimental capability to investigate the hypervelocity micrometeoroid bombardment of cryogenic surfaces. *Rev. Sci. Instrum., 87,* 024502.
Notesco G., Bar-Nun A., and Owen T. (2003) Gas trapping in water ice at very low deposition rates and implications for comets. *Icarus, 162,* 183–189.
Öberg K. I., Boogert A. C. A., Pontoppidan K. M., van den Broek S., van Dishoeck E. F., Bottinelli S., Blake G. A., and Evans N. J. II (2011) The Spitzer ice legacy: Ice evolution from cores to protostars. *Astrophys. J., 740,* 109.
Patashnick H. (1974) Energy source for comet outbursts. *Nature, 250,* 313–314.
Porter S. B., Desch S. J., and Cook J. C. (2010) Micrometeorite impact annealing of ice in the outer solar system. *Icarus, 208,* 492–498.
Prialnik D. (1993) A two-zone steady state crystallization model for comets. *Astrophys. J. Lett., 418,* L49–L52.
Prialnik D. (1997) A model for the distant activity of comet Hale-Bopp. *Astrophys. J. Lett., 478,* L107–L110.
Prialnik D. and Bar-Nun A. (1987) On the evolution and activity of cometary nuclei. *Astrophys. J., 313,* 893–905.
Prialnik D. and Bar-Nun A. (1990) Gas release in comet nuclei. *Astrophys. J., 363,* 274–282.
Prialnik D. and Bar-Nun A. (1992) Crystallization of amorphous ice as the cause of Comet P/Halley's outburst at 14 AU. *Astron. Astrophys., 258,* L9–L12.
Prialnik D. and Podolak M. (1995) Radioactive heating of porous comet nuclei. *Icarus, 117,* 420–430.
Prialnik D. and Sierks H. (2017) A mechanism for comet surface collapse as observed by Rosetta on 67P/Churyumov-Gerasimenko. *Mon. Not. R. Astron. Soc., 469,* S217–S221.
Prialnik D., Egozi U., Ban-Nun A., Podolak M., and Greenzweig Y. (1993) On pore size and fracture in gas-laden comet nuclei. *Icarus, 106,* 499–507.
Prialnik D., Brosch N., and Ianovici D. (1995) Modelling the activity of 2060 Chiron. *Mon. Not. R. Astron. Soc., 276,* 1148–1154.
Prialnik D., Sarid G., Rosenberg E. D., and Merk R. (2008) Thermal and chemical evolution of comet nuclei and Kuiper belt objects. *Space Sci. Rev., 138,* 147–164.
Reach W. T., Sykes M. V., Lien D., and Davies J. K. (2000) The formation of Encke meteoroids and dust trail. *Icarus, 148,* 80–94.
Reach W. T., Vaubaillon J., Lisse C. M., Holloway M., and Rho J. (2010) Explosion of comet 17P/Holmes as revealed by the Spitzer Space Telescope. *Icarus, 208,* 276–292.
Rosenberg E. D. and Prialnik D. (2007) A fully 3-dimensional thermal model of a comet nucleus. *New Astron., 12,* 523–532.
Samarasinha N. H. (2001) A model for the breakup of comet LINEAR (C/1999 S4). *Icarus, 154,* 540–544.
Sandford S. A. and Allamandola L. J. (1988) The condensation and vaporization behavior of H_2O: CO ices and implications for interstellar grains and cometary activity. *Icarus, 76,* 201–224.
Sandford S. A. and Allamandola L. J. (1990) The physical and infrared spectral properties of CO_2 in astrophysical ice analogs. *Astrophys. J., 355,* 357–372.
Schambeau C. A., Fernández Y. R., Samarasinha N. H., Womack M., Bockelée-Morvan D., Lisse C. M., and Woodney L. M. (2021) Characterization of thermal-infrared dust emission and refinements to the nucleus properties of Centaur 29P/Schwassmann-Wachmann 1. *Planet. Sci. J., 2,* 126.
Schmitt B. and Klinger J. (1987) Different trapping mechanisms of gases by water ice and their relevance for comet nuclei. In *Proceedings of the International Symposium on the Diversity and Similarity of Comets* (E. J. Rolfe et al., eds.), pp. 613–619. ESA-SP-278, Noordwijk, The Netherlands.
Schmitt B., Espinasse S., Grim R. J. A., Greenberg J. M., and Klinger J. (1989) Laboratory studies of cometary ice analogues. In *Physics and Mechanics of Cometary Materials* (J. J. Hunt and T. D. Guyenne, eds.), pp. 65–69. ESA-SP-302, Noordwijk, The Netherlands.
Schmitt B., Espinasse S., and Klinger J. (1991) A possible mechanism for outbursts of comet P/Halley at large heliocentric distances. *Meteoritics, 26,* 392.
Senay M. C. and Jewitt D. (1994) Coma formation driven by carbon monoxide release from comet Schwassmann-Wachmann 1. *Nature, 371,* 229–231.
Smoluchowski R. (1981) Amorphous ice and the behavior of cometary nuclei. *Astrophys. J. Lett., 244,* L31–L34.
Stevenson R., Kleyna J., and Jewitt D. (2010) Transient fragments in outbursting comet 17P/Holmes. *Astron. J., 139,* 2230–2240.
Tancredi G., Rickman H., and Greenberg J. M. (1994) Thermochemistry of cometary nuclei: I. The Jupiter family case. *Astron. Astrophys., 286,* 659–682.
Terai T., Itoh Y., Oasa Y., Furusho R., and Watanabe J. (2016) Photometric measurements of H_2O ice crystallinity on trans-neptunian objects. *Astrophys. J., 827,* 65.
Trigo-Rodríguez J. M., García-Melendo E., Davidsson B. J. R., Sánchez A., Rodríguez D., Lacruz J., de Los Reyes J. A., and Pastor S. (2008) Outburst activity in comets: I. Continuous monitoring of comet 29P/Schwassmann-Wachmann 1. *Astron. Astrophys., 485,* 599–606.
Trujillo C. A., Brown M. E., Barkume K. M., Schaller E. L., and Rabinowitz D. L. (2007) The surface of 2003 EL_{61} in the near-infrared. *Astrophys. J., 655,* 1172–1178.
van Dishoeck E. F., Herbst E., and Neufeld D. A. (2013) Interstellar water chemistry: From laboratory to observations. *Chem. Rev., 113,* 9043–9085.
Vernazza P., Beck P., and Ruesch O. (2021) Sample return of primitive matter from the outer solar system. *Experimental Astronomy, 54,* 1051–1075.
Wallis M. K. (1980) Radiogenic melting of primordial comet interiors. *Nature, 284,* 431–433.
Weissman P. (1991) Why did Halley hiccup? *Nature, 353,* 793–794.
West R. M., Hainaut O., and Smette A. (1991) Post-perihelion observations of P/Halley: III. An outburst at r = 14.3 AU. *Astron. Astrophys., 246,* L77–L80.
Westphal A. J., Nittler L. R., Stroud R., Zolensky M. E., Chabot N. L., Dello Russo N., Elsila J. E., Sandford S. A., Glavin D. P., Evans M. E., Nuth J. A., Sunshine J., Vervack J., Ronald J., and Weaver H. A. (2020) Cryogenic cometary sample return. *ArXiV e-prints,* arXiv:2009.00101.
Whipple F. L. and Huebner W. F. (1976) Physical processes in comets. *Annu. Rev. Astron. Astrophys., 14,* 143–172.
Wierzchos K. and Womack M. (2020) CO gas and dust outbursts from Centaur 29P/Schwassmann-Wachmann. *Astron. J., 159,* 136.
Wong I., Brown M. E., Blacksberg J., Ehlmann B. L., and Mahjoub A. (2019) Hubble ultraviolet spectroscopy of Jupiter Trojans. *Astron. J., 157,* 161.
Yabushita S. (1993) Thermal evolution of cometary nuclei by radioactive heating and possible formation of organic chemicals. *Mon. Not. R. Astron. Soc., 260,* 819–825.
Yang B. and Sarid G. (2010) Crystalline water ice in outburst comet P/2010 H2. *AAS/Division for Planetary Sciences Meeting Abstracts, 42,* #5.09.
Yokochi R., Marboeuf U., Quirico E., and Schmitt B. (2012) Pressure dependent trace gas trapping in amorphous water ice at 77 K: Implications for determining conditions of comet formation. *Icarus, 218,* 760–770.
Zheng W., Jewitt D., and Kaiser R. I. (2009) On the state of water ice on Saturn's moon Titan and implications to icy bodies in the outer solar system. *J. Phys. Chem. A, 113,* 11174–11181.
Zubko E., Videen G., Shkuratov Y., and Hines D. C. (2017) On the reflectance of dust in comets. *J. Quant. Spectrosc. Radiat. Transfer, 202,* 104–113.

Index

Page numbers refer to specific pages on which an index term or concept is discussed. "ff" indicates that the term is also discussed on the following pages.

Comets (*continued*)

Comets (*continued*)